CELL BIOLOGY

NOTE TO INSTRUCTORS:

Contact your Elsevier Sales Representative for image banks for *Cell Biology*, 3e, or request these supporting materials at: http://evolve.elsevier.com

THIRD EDITION

CELL BIOLOGY

THOMAS D. POLLARD, MD
Sterling Professor
Department of Molecular, Cellular, and Developmental Biology
Yale University
New Haven, Connecticut

WILLIAM C. EARNSHAW, PhD, FRSE
Professor and Wellcome Trust Principal Research Fellow
Wellcome Trust Centre for Cell Biology, ICB
University of Edinburgh
Scotland, United Kingdom

JENNIFER LIPPINCOTT-SCHWARTZ, PhD
Group Leader
Howard Hughes Medical Institute, Janelia Research Campus
Ashburn, Virginia

GRAHAM T. JOHNSON, MA, PhD, CMI
Director, Animated Cell
Allen Institute for Cell Biology
Seattle, Washington;
QB3 Faculty Fellow
University of California, San Francisco
San Francisco, California

ELSEVIER

ELSEVIER

1600 John F. Kennedy Blvd.
Ste 1800
Philadelphia, PA 19103-2899

CELL BIOLOGY, THIRD EDITION
IE

ISBN: 978-0-323-34126-4
ISBN: 978-0-323-41740-2

Copyright © 2017 by Elsevier, Inc. All rights reserved.

No part of this publication may be reproduced or transmitted in any form or by any means, electronic or mechanical, including photocopying, recording, or any information storage and retrieval system, without permission in writing from the publisher. Details on how to seek permission, further information about the Publisher's permissions policies and our arrangements with organizations such as the Copyright Clearance Center and the Copyright Licensing Agency, can be found at our website: www.elsevier.com/permissions.

This book and the individual contributions contained in it are protected under copyright by the Publisher (other than as may be noted herein).

Notices

Knowledge and best practice in this field are constantly changing. As new research and experience broaden our understanding, changes in research methods, professional practices, or medical treatment may become necessary.

Practitioners and researchers must always rely on their own experience and knowledge in evaluating and using any information, methods, compounds, or experiments described herein. In using such information or methods they should be mindful of their own safety and the safety of others, including parties for whom they have a professional responsibility.

With respect to any drug or pharmaceutical products identified, readers are advised to check the most current information provided (i) on procedures featured or (ii) by the manufacturer of each product to be administered, to verify the recommended dose or formula, the method and duration of administration, and contraindications. It is the responsibility of practitioners, relying on their own experience and knowledge of their patients, to make diagnoses, to determine dosages and the best treatment for each individual patient, and to take all appropriate safety precautions.

To the fullest extent of the law, neither the Publisher nor the authors, contributors, or editors, assume any liability for any injury and/or damage to persons or property as a matter of products liability, negligence or otherwise, or from any use or operation of any methods, products, instructions, or ideas contained in the material herein.

Previous editions copyrighted © 2008, 2004 by Thomas D. Pollard, William C. Earnshaw, Jennifer Lippincott-Schwartz.

Library of Congress Cataloging-in-Publication Data

Names: Pollard, Thomas D. (Thomas Dean), 1942- , author. | Earnshaw, William C., author. | Lippincott-Schwartz, Jennifer, author. | Johnson, Graham T., author.
Title: Cell biology / Thomas D. Pollard, William C. Earnshaw, Jennifer Lippincott-Schwartz, Graham T. Johnson.
Description: Third edition. | Philadelphia, PA : Elsevier, [2017] | Includes bibliographical references and index.
Identifiers: LCCN 2016008034| ISBN 9780323341264 (hardcover : alk. paper) | ISBN 9780323417402 (international edition)
Subjects: | MESH: Cell Physiological Phenomena | Cells
Classification: LCC QH581.2 | NLM QU 375 | DDC 571.6—dc23 LC record available at http://lccn.loc.gov/2016008034

Executive Content Strategist: Elyse O'Grady
Senior Content Development Specialist: Margaret Nelson
Publishing Services Manager: Patricia Tannian
Senior Project Manager: Carrie Stetz
Design Direction: Margaret Reid

Printed in the United States of America

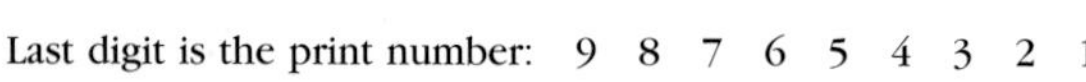

Last digit is the print number: 9 8 7 6 5 4 3 2 1

Working together
to grow libraries in
developing countries

www.elsevier.com • www.bookaid.org

The authors thank their families, who supported this work, and also express gratitude to their mentors, who helped to shape their views of how science should be conducted. Bill is proud to have both his longtime partner and confidante Margarete and his son Charles as advisors on the science for this edition. He would not be surprised if his daughter Irina were added to that panel for our next edition. His contributions are firstly dedicated to them. Bill also would like to thank Jonathan King, Stephen Harrison, Aaron Klug, Tony Crowther, Ron Laskey, and Uli Laemmli, who provided a diverse range of rich environments in which to learn that science at the highest level is an adventure that lasts a lifetime. Graham dedicates the book to his family, Margaret, Paul, and Lara Johnson; the Benhorins; friends Mari, Steve, and Andrew; and his partners Flower and Anna Kuo. He also thanks his mentors at the Scripps Research Institute, Arthur Olson, David Goodsell, Ron Milligan, and Ian Wilson, for developing his career. Jennifer thanks her husband Jonathan for his strong backing and her lab members for their enthusiasm for the project. Tom dedicates the book to his wife Patty, a constant source of support and inspiration for more than five decades, and his children Katie and Dan, who also provided advice on the book. He also thanks Ed Korn and the late Sus Ito for the opportunity to learn biochemistry and microscopy under their guidance, and Ed Taylor and the late Hugh Huxley, who served as role models.

Contributors

Jeffrey L. Corden, PhD
Professor
Department of Molecular Biology and Genetics
Johns Hopkins Medical School
Baltimore, Maryland

David Tollervey, PhD
Professor
Wellcome Trust Centre for Cell Biology
University of Edinburgh
Scotland, United Kingdom

Preface

Our goal is to explain the molecular basis of life at the cellular level. We use evolution and molecular structures to provide the context for understanding the dynamic mechanisms that support life. As research in cell biology advances quickly, the field may appear to grow more complex, but we aim to show that understanding cells actually becomes simpler as new general principles emerge and more precise molecular mechanisms replace vague concepts about biological processes.

For this edition, we revised the entire book, taking the reader to the frontiers of knowledge with exciting new information on every topic. We start with new insights about the evolution of eukaryotes, followed by macromolecules and research methods, including recent breakthroughs in light and electron microscopy. We begin the main part of the book with a section on basic molecular biology before sections on membranes, organelles, membrane traffic, signaling, adhesion and extracellular matrix, and cytoskeleton and cellular motility. As in the first two editions, we conclude with a comprehensive section on the cell cycle, which integrates all of the other topics.

Our coverage of most topics begins with an introduction to the molecular hardware and finishes with an account of how the various molecules function together in physiological systems. This organization allows for a clearer exposition of the general principles of each class of molecules, since they are treated as a group rather than isolated examples for each biological system. This approach allows us to present the operation of complex processes, such as signaling pathways, as an integrated whole, without diversions to introduce the various components as they appear along the pathway. For example, the section on signaling mechanisms begins with chapters on receptors, cytoplasmic signal transduction proteins, and second messengers, so the reader is prepared to appreciate the dynamics of 10 critical signaling systems in the chapter that concludes the section. Teachers of shorter courses may concentrate on a subset of the examples in these systems chapters, or they may use parts of the "hardware" chapters as reference material.

We use molecular structures as one starting point for explaining how each cellular system operates. This edition includes more than 50 of the most important and revealing new molecular structures derived from electron cryomicroscopy and x-ray crystallography. We explain the evolutionary history and molecular diversity of each class of molecules, so the reader learns where the many varieties of each type of molecule came from. Our goal is for readers to understand the big picture rather than just a mass of details. For example, Chapter 16 opens with an original figure showing the evolution of all types of ion channels to provide context for each family of channels in the following text. Given that these molecular systems operate on time scales ranging from milliseconds to hours, we note (where it is relevant) the concentrations of the molecules and the rates of their reactions to help readers appreciate the dynamics of life processes.

We present a wealth of experimental evidence in figures showing micrographs, molecular structures, and graphs that emphasize the results rather than the experimental details. Many of the methods will be new to readers. The chapter on experimental methods introduces how and why scientists use particularly important approaches (such as microscopy, classical genetics, genomics and reverse genetics, and biochemical methods) to identify new molecules, map molecular pathways, or verify physiological functions.

The book emphasizes molecular mechanisms because they reveal the general principles of cellular function. As a further demonstration of this generality, we use a wide range of experimental organisms and specialized cells and tissues of vertebrate animals to illustrate these general principles. We also use medical "experiments of nature" to illustrate physiological functions throughout the book, since connections have now been made between most cellular systems and disease. The chapters on cellular functions integrate material on specialized cells and tissues. Epithelia, for example, are covered under membrane physiology and junctions; excitable membranes of neurons and muscle under membrane physiology; connective tissues under the extracellular matrix; the immune system under connective tissue cells, apoptosis, and signal transduction; muscle under the cytoskeleton and cell motility; and stem cells and cancer under the cell cycle and signal transduction.

The Guide to Figures Featuring Specific Organisms and Specialized Cells that follows the Contents lists figures by organism and cell. The relevant text accompanies these figures. Readers who wish to assemble a unit on cellular and molecular mechanisms in the immune system, for example, will find the relevant material associated with the figures that cover lymphocytes/immune system.

Our Student Consult site provides links to the Protein Data Bank (PDB), so readers can use the PDB accession numbers in the figure legends to review original data, display an animated molecule, or search links to the original literature simply by clicking on the PDB number in the online version of the text.

Throughout, we have attempted to create a view of Cell Biology that is more than just a list of parts and reactions. Our book will be a success if readers finish each section with the feeling that they understand better how some aspect of cellular behavior actually works at a mechanistic level and in our bodies.

Thomas D. Pollard

William C. Earnshaw

Jennifer Lippincott-Schwartz

Graham T. Johnson

Contents

SECTION I
Introduction to Cell Biology

1 Introduction to Cells, 3

2 Evolution of Life on Earth, 15

SECTION II
Chemical and Physical Background

3 Molecules: Structures and Dynamics, 31

4 Biophysical Principles, 53

5 Macromolecular Assembly, 63

6 Research Strategies, 75

SECTION III
Chromatin, Chromosomes, and the Cell Nucleus

7 Chromosome Organization, 107

8 DNA Packaging in Chromatin and Chromosomes, 123

9 Nuclear Structure and Dynamics, 143

SECTION IV
Central Dogma: From Gene to Protein

10 Gene Expression, 165

11 Eukaryotic RNA Processing, 189

12 Protein Synthesis and Folding, 209

SECTION V
Membrane Structure and Function

13 Membrane Structure and Dynamics, 227

14 Membrane Pumps, 241

15 Membrane Carriers, 253

16 Membrane Channels, 261

17 Membrane Physiology, 285

SECTION VI
Cellular Organelles and Membrane Trafficking

18 Posttranslational Targeting of Proteins, 303

19 Mitochondria, Chloroplasts, Peroxisomes, 317

20 Endoplasmic Reticulum, 331

21 Secretory Membrane System and Golgi Apparatus, 351

22 Endocytosis and the Endosomal Membrane System, 377

23 Processing and Degradation of Cellular Components, 393

SECTION VII
Signaling Mechanisms

24 Plasma Membrane Receptors, 411

25 Protein Hardware for Signaling, 425

26 Second Messengers, 443

27 Integration of Signals, 463

SECTION VIII
Cellular Adhesion and the Extracellular Matrix

28 Cells of the Extracellular Matrix and Immune System, 491

29 Extracellular Matrix Molecules, 505

30 Cellular Adhesion, 525

31 Intercellular Junctions, 543

32 Connective Tissues, 555

SECTION IX
Cytoskeleton and Cellular Motility

33 Actin and Actin-Binding Proteins, 575

34 Microtubules and Centrosomes, 593

35 Intermediate Filaments, 613

36 Motor Proteins, 623

37 Intracellular Motility, 639

38 Cellular Motility, 651

39 Muscles, 671

SECTION X
Cell Cycle

40 Introduction to the Cell Cycle, 697

41 G_1 Phase and Regulation of Cell Proliferation, 713

42 S Phase and DNA Replication, 727

43 G_2 Phase, Responses to DNA Damage, and Control of Entry Into Mitosis, 743

44 Mitosis and Cytokinesis, 755

45 Meiosis, 779

46 Programmed Cell Death, 797

Cell SnapShots, 817

Glossary, 823

Index, 851

Acknowledgments

The authors thank their families and colleagues for sharing so much time with "the book." Bill thanks Margarete, Charles, and Irina for sharing their weekends and summer holidays with this all-consuming project. He also thanks the Wellcome Trust for their incomparable support of the research in his laboratory and Melpomeni Platani and the Dundee Imaging Facility for access to the OMX microscope. Graham thanks Thao Do and Andrew Swift for contributions to the illustrations, and colleagues Megan Riel-Mehan, Tom Goddard, Arthur Olson, David Goodsell, Warren DeLeno, Andrej Sali, Tom Ferrin, Sandra Schmid, Rick Horwitz, UCSF, and the Allen Institute for Cell Science for facilitating work on this edition. He has special thanks for Ludovic Autin for programming the embedded Python Molecular Viewer (ePMV), which enabled substantial upgrades of many figures with complex structures. Jennifer thanks her family for sharing time with her part in the book. Tom appreciates four decades of support for his laboratory from the National Institutes of General Medical Sciences.

Many generous individuals generously devoted their time to bring the science up to date by providing suggestions for revising chapters in their areas of expertise. We acknowledge these individuals at the end of each chapter and here as a group: Ueli Aebi, Anna Akhmanova, Julie Ahringer, Hiro Araki, Jiri Bartek, Tobias Baumgart, Wendy Bickmore, Craig Blackstone, Julian Blow, Jonathan Bogan, Juan Bonifacino, Ronald Breaker, Klaudia Brix, Anthony Brown, David Burgess, Cristina Cardoso, Andrew Carter, Bill Catterall, Pietro De Camilli, Iain Cheeseman, Per Paolo D'Avino, Abby Dernburg, Arshad Desai, Julie Donaldson, Charles Earnshaw, Donald Engelman, Job Dekker, Martin Embley, Barbara Ehrlich, Roland Foisner, Nicholas Frankel, Tatsuo Fukagawa, Anton Gartner, Maurizio Gatti, David Gilbert, Gary Gorbsky, Holly Goodson, Jim Haber, Lea Harrington, Scott Hawley, Ron Hay, Margarete Heck, Ramanujan Hegde, Ludger Hengst, Harald Herrmann, Erika Holzbaur, Tim Hunt, Catherine Jackson, Emmanuelle Javaux, Scott Kaufmann, David Julius, Keisuke Kaji, Alexey Khodjakov, Vladimir Larionov, Dan Leahy, Richard Lewis, Kaspar Locker, Kazuhiro Maeshima, Marcos Malumbres, Luis Miguel Martins, Amy MacQueen, Ciaran Morrison, Adele Marston, Satyajit Mayor, Andrew Miranker, Tom Misteli, David Morgan, Peter Moore, Rachel O'Neill, Karen Oegema, Tom Owen-Hughes, Laurence Pelletier, Alberto Pendas, Jonathon Pines, Jordan Raff, Samara Reck-Peterson, Elizabeth Rhoades, Matthew Rodeheffer, Michael Rout, Benoit Roux, John Rubinstein, Julian Sale, Eric Schirmer, John Solaro, Chris Scott, Beth Sullivan, Lee Sweeney, Margaret Titus, Andrew Thorburn, Ashok Venkitaraman, Rebecca Voorhees, Tom Williams, and Yongli Zhang. We thank David Sabatini, Susan Wente, and Yingming Zhao for permission to use their Cell SnapShots and Jason M. McAlexander for help with the final figures.

Special thanks go to our colleagues at Elsevier. Our visionary editor Elyse O'Grady encouraged us to write this third edition and was a champion for the project from beginning to end as it evolved from a simple update of the second edition to an ambitious new book. Margaret Nelson, Content Development Specialist supreme, kept the whole project organized while dealing deftly with thousands of documents. Project Manager Carrie Stetz managed the assembly of the book with skill, patience, and good cheer in the face of many complicated requests for alterations.

Guide to Figures Featuring Specific Organisms and Specialized Cells

Organism/Specialized Cell Type	Figures
PROKARYOTES	
Archaea	1.1, 1.2, 2.1, 2.4, 2.5
Bacteria	1.1, 1.2, 2.1, 2.4, 2.5, 2.7, 5.8, 5.12, 6.11, 7.4, 10.2, 10.5, 10.10, 10.11, 11.16, 12.6, 12.11, 13.9, 14.3, 14.9, 14.10, 15.4, 16.2, 16.3, 16.6, 16.13, 16.14, 18.2, 18.9, 18.10, 19.2, 19.7, 19.9, 20.5, 22.3, 22.10, 22.15, 27.11, 27.12, 27.13, 35.1, 37.12, 38.1, 38.24, 38.25, 42.3, 43.13, 44.27
Viruses	5.10, 5.11, 5.12, 5.13, 22.15, 37.12
PROTOZOA	
Amoeba	2.1, 2.4, 2.8, 22.2, 22.5, 38.1, 38.4, 38.10, 41.7
Ciliates	2.4, 38.1, 38.13
Other protozoa	2.4, 2.7, 36.7, 38.4, 37.10, 38.6, 38.21, 38.23
ALGAE AND PLANTS	
Chloroplasts	18.1, 18.2, 18.6, 19.7, 19.8, 19.9
Green algae	2.8, 37.1, 37.9, 38.13, 38.14, 38.16, 38.18
Plant cell wall	31.4, 32.12, 32.13
Plant (general)	1.2, 2.1, 2.4, 2.7, 2.8, 3.25, 6.6, 31.4, 33.1, 34.2, 36.7, 37.9, 38.1, 40.3, 44.26, 45.8
FUNGI	
Budding yeast	1.2, 2.4, 2.8, 6.15, 6.16, 7.3, 7.4, 7.7, 7.8, 8.22, 34.2, 34.20, 37.11, 42.4, 42.5, 42.15, 43.8
Fission yeast	2.4, 2.8, 6.3, 7.8, 33.1, 40.6, 43.2, 44.23
Other fungi	2.8, 45.6
INVERTEBRATE ANIMALS	
Echinoderms	2.8, 36.13, 40.11, 44.21, 44.22, 44.23
Nematodes	2.8, 36.7, 36.13, 38.11, 45.10, 46.9, 46.10
Insects	2.8, 7.4, 7.8, 7.15, 8.12, 8.13, 9.19, 14.19, 38.5, 38.11, 44.14, 44.12, 44.21, 44.25, 45.2, 45.8, 45.10
VERTEBRATE ANIMALS	
Blood	
Granulocytes	28.1, 28.4, 28.7, 30.13, 38.1
Lymphocytes/immune system	27.8, 28.1, 28.4, 28.9, 28.10, 46.7, 46.9, 46.18
Monocytes/macrophages	28.1, 28.4, 28.7, 32.6, 32.11, 38.3, 46.2, 46.13
Platelets	28.4, 28.5, 30.14, 32.11
Red blood cells	13.8, 13.9, 13.11, 28.4, 32.11
Cancer	34.19, 38.9, 41.2, 41.11, 41.12, 41.15, 42.10
Connective tissue	
Cartilage cells	28.1, 32.2, 32.3, 32.8, 32.9
Extracellular matrix	8.20
Fibroblasts	28.1, 28.2, 29.3, 29.4, 29.15, 32.1, 32.11, 35.1, 35.5, 37.1, 38.1
Mast cells	28.1, 28.8
Bone cells	28.1, 32.4, 32.5, 32.6, 32.7, 32.8, 32.9, 32.10
Fat cells	27.7, 28.1, 28.3
Epithelia	
Epidermal, stratified	29.7, 35.6, 40.1, 41.2, 41.5, 42.10, 46.8
Glands, liver	21.26, 23.6, 34.20, 41.2, 44.2
Intestine	17.2, 31.1, 32.1, 33.1, 33.2, 34.2, 46.19
Kidney	17.3, 29.17, 35.1, 46.6, 46.7
Respiratory system	17.4, 32.2, 34.3, 37.6, 38.17
Vascular	22.6, 29.8, 29.17, 30.13, 30.14, 31.2, 32.11, 46.20
Muscle	
Cardiac muscle	39.1, 39.13, 39.14, 39.18, 39.19, 39.20, 39.21, 39.22
Skeletal muscle	17.9, 29.17, 33.3, 36.3, 36.4, 36.5, 39.1, 39.2, 39.3, 39.4, 39.5, 39.6, 39.7, 39.8, 39.9, 39.10, 39.11, 39.12, 39.13, 39.14, 39.15, 39.16, 39.17
Smooth muscle	29.8, 33.1, 35.8, 39.1, 39.23, 39.24
Nervous system	
Central nervous system neurons	17.9, 17.10, 17.11, 30.8, 34.11, 34.12, 35.9, 37.7, 38.11, 39.12, 23.4
Glial cells	17.7, 17.9, 17.10, 29.17, 37.7
Peripheral nervous system neurons	17.7, 17.9, 26.3, 26.16, 27.1, 27.2, 29.17, 30.15, 33.18, 35.9, 37.1, 37.3, 37.4, 37.5, 38.1, 38.6, 39.12
Synapses	17.9, 17.10, 17.11, 29.17, 39.12
Reproductive system	
Oocytes, eggs	26.15, 34.14, 40.7, 40.8, 40.10, 40.11, 40.12, 45.14
Sperm	38.1, 38.2, 38.14, 38.15, 38.20, 38.22, 45.1, 45.2, 45.4, 45.5, 45.8, 45.11
Other human cells and disease	
Various organs	7.4, 7.6, 7.9, 7.11, 8.20, 9.10, 23.4, 41.2, 42.10

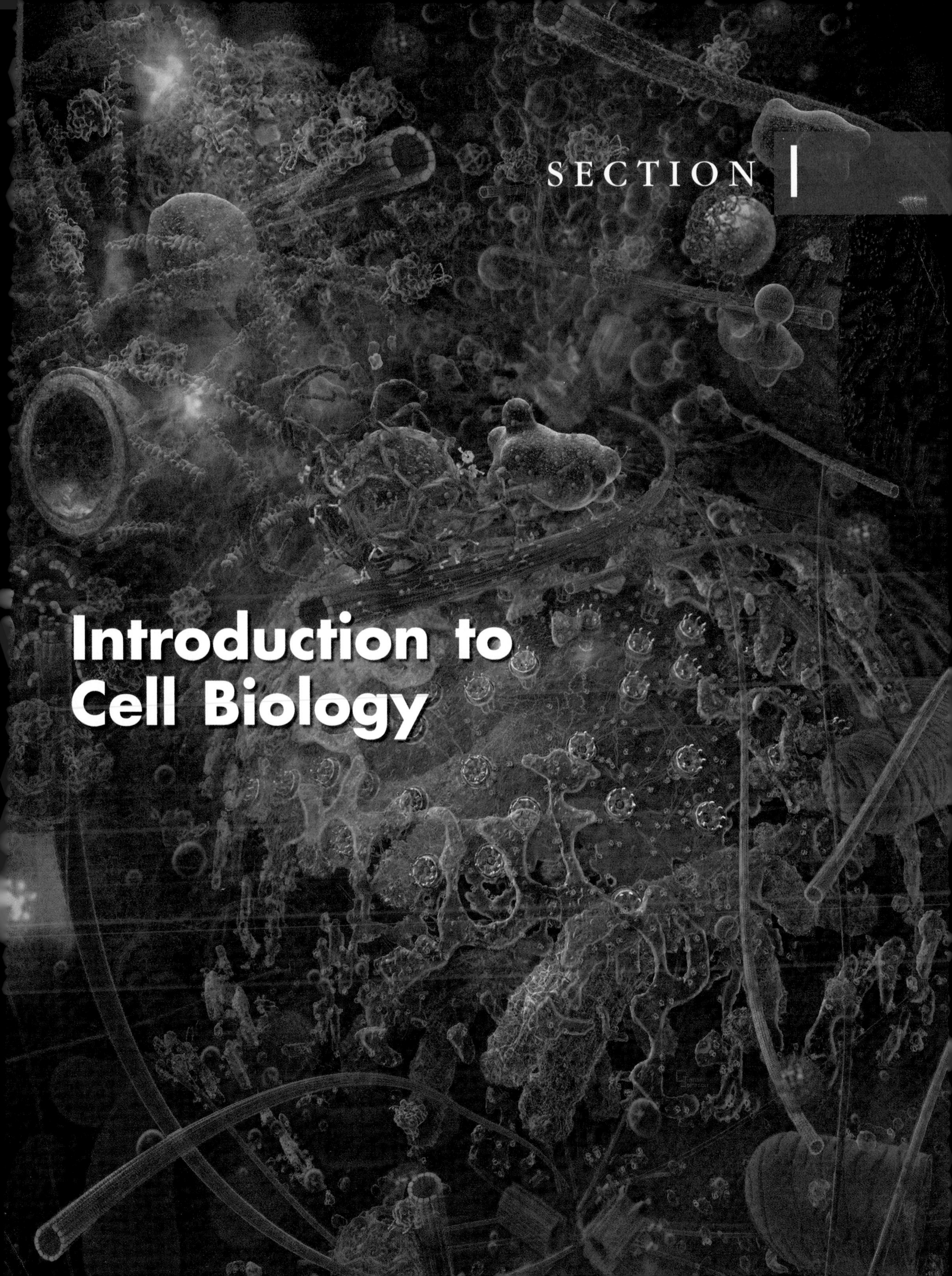

SECTION I

Introduction to Cell Biology

CHAPTER 1

Introduction to Cells

Biology is based on the fundamental laws of nature embodied in chemistry and physics, but the origin and evolution of life on earth were historical events. This makes biology more like astronomy than like chemistry and physics. Neither the organization of the universe nor life as we know it had to evolve as they did. Chance played a central role. Throughout history and continuing today, the genes of all organisms have sustained chemical changes, some of which are inherited by their progeny. Many changes have no obvious effect on the fitness of the organism, but some reduce it and others improve fitness. Over the long term, competition between individuals with random differences in their genes determines which organisms survive in various environments. Surviving variants have a selective advantage over the alternatives, but the process does not necessarily optimize each chemical life process. Thus, students could probably design simpler or more elegant mechanisms for many cellular processes.

Despite obvious differences, all forms of life share many molecular mechanisms, because they all descended from a **common ancestor** that lived 3 to 4 billion years ago (Fig. 1.1). This founding organism no longer exists, but it must have used many biochemical processes similar to those that sustain contemporary cells.

Over several billion years, living organisms diverged from the common ancestor into three great divisions: **Bacteria**, **Archaea**, and **Eucarya** (Fig. 1.1). Archaea and Bacteria were considered to be one kingdom until the 1970s when the sequences of genes for ribosomal RNAs revealed that their ancestors branched from each other early in evolution. The origin of eukaryotes, cells with a nucleus, is still uncertain, but they inherited genes from both Archaea and Bacteria. One possibility is that eukaryotes originated when an Archaea engulfed a Bacterium that subsequently evolved into the mitochondrion. Multicellular eukaryotes (*green, blue,* and *red* in Fig. 1.1) evolved relatively recently, hundreds of millions of years after single-celled eukaryotes appeared. Note that algae and plants branched before fungi, our nearest relatives on the tree of life.

Living things differ in size and complexity and are adapted to environments as extreme as deep-sea hydrothermal vents at temperatures of 113°C or pockets of water at 0°C in frozen Antarctic lakes. Organisms also employ different strategies to extract energy from their environments. Plants, algae, and some Bacteria use photosynthesis to derive energy from sunlight. Some Bacteria and Archaea obtain energy by oxidizing inorganic compounds, such as hydrogen, hydrogen sulfide, or iron. Many organisms in all parts of the tree, including animals, extract energy from organic compounds.

As the molecular mechanisms of life have become clearer, the underlying similarities among organisms are more impressive than their external differences. For example, all living organisms store genetic information in nucleic acids (usually **DNA**) using a common genetic

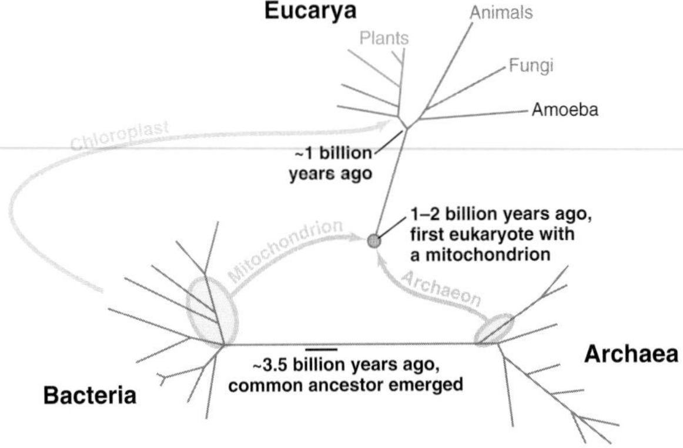

FIGURE 1.1 SIMPLIFIED PHYLOGENETIC TREE. This tree shows the common ancestor of all living things and the three main branches of life Archaea and Bacteria diverged from the common ancestor and both contributed to the origin of Eukaryotes. Note that eukaryotic mitochondria and chloroplasts originated as symbiotic Bacteria.

code, transfer genetic information from **DNA** to **RNA** to **protein,** employ proteins (and some RNAs) to catalyze chemical reactions, synthesize proteins on **ribosomes,** derive energy by breaking down simple sugars and lipids, use adenosine triphosphate **(ATP)** as their energy currency, and separate their cytoplasm from the external environment by means of phospholipid **membranes** containing **pumps, carriers,** and **channels.**

Retention of these common molecular mechanisms in all parts of the **phylogenetic tree** is remarkable, given that the major groups of organisms have been separated for vast amounts of time and subjected to different selective pressures. These ancient biochemical mechanisms could have diverged radically from each other in the branches of the phylogenetic tree, but they worked well enough to be retained during natural selection of all surviving species.

The cell is the only place on earth where the entire range of life-sustaining biochemical reactions can function, so an unbroken lineage stretches from the earliest cells to each living organism. Many interesting creatures were lost to extinction during evolution. The fact that extinction is irreversible, energizes discussions of biodiversity today.

This book focuses on the molecular mechanisms underlying biological functions at the cellular level (Fig. 1.2). The rest of Chapter 1 summarizes the main points of the whole text including the general principles that apply equally to eukaryotes and prokaryotes and special features of eukaryotic cells. Chapter 2 explains what is known of the origins of life and its historic diversification through evolution. Chapter 3 covers the macromolecules that form cells, while Chapters 4 and 5 introduce the chemical and physical principles required to understand how these molecules assemble and function. Chapter 6 introduces laboratory methods for research in cell biology.

Universal Principles of Living Cells

Biologists believe that a limited number of general principles based on common molecular mechanisms can explain even the most complex life processes in terms of chemistry and physics. This section summarizes the numerous features shared by all forms of life.

1. *Genetic information stored in the chemical sequence of DNA is duplicated and passed on to daughter cells (Fig. 1.3).* Long DNA molecules called **chromosomes** store the information required for cellular growth, multiplication, and function. Each DNA molecule is composed of two strands of four different nucleotides (adenine [A], cytosine [C], guanine [G], and thymine [T]) covalently linked in linear polymers. The two strands pair, forming a double helix held together by interactions between complementary pairs of nucleotide bases with one on each strand: A pairs

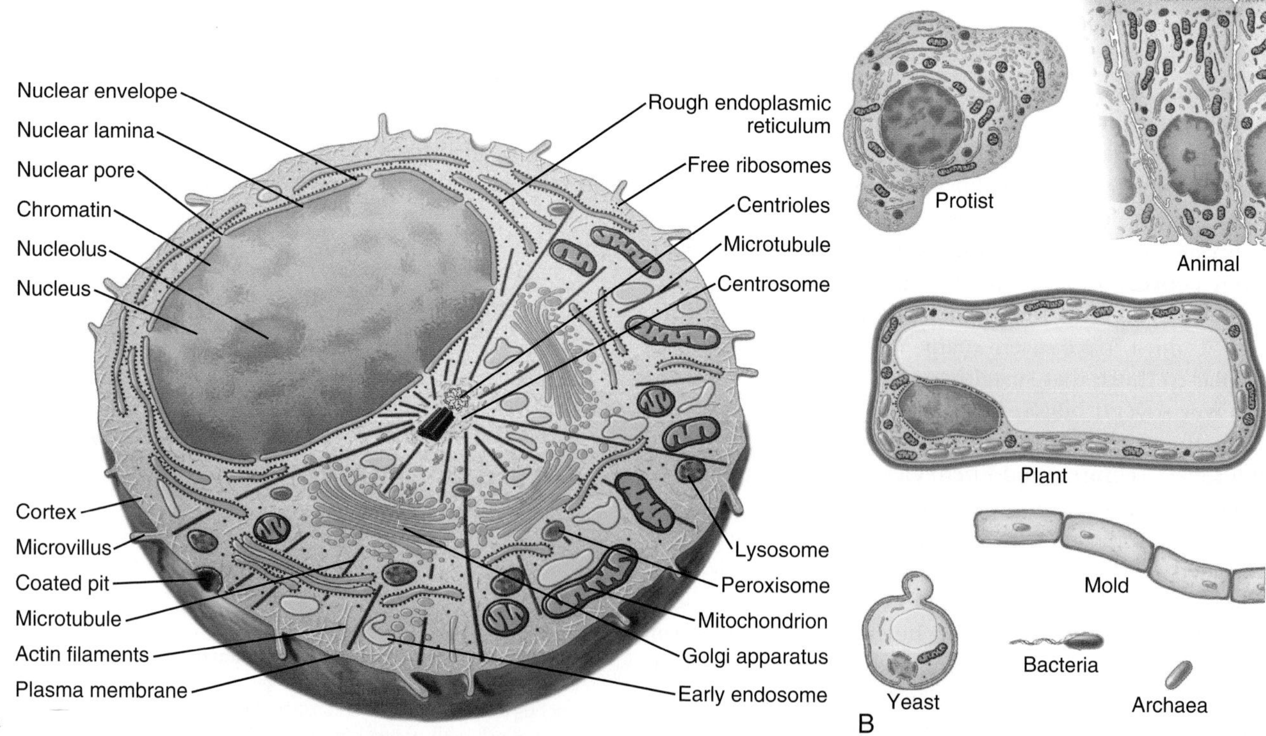

FIGURE 1.2 BASIC CELLULAR ARCHITECTURE. A, Section of a eukaryotic cell showing the internal components. **B,** Comparison of cells from the major branches of the phylogenetic tree.

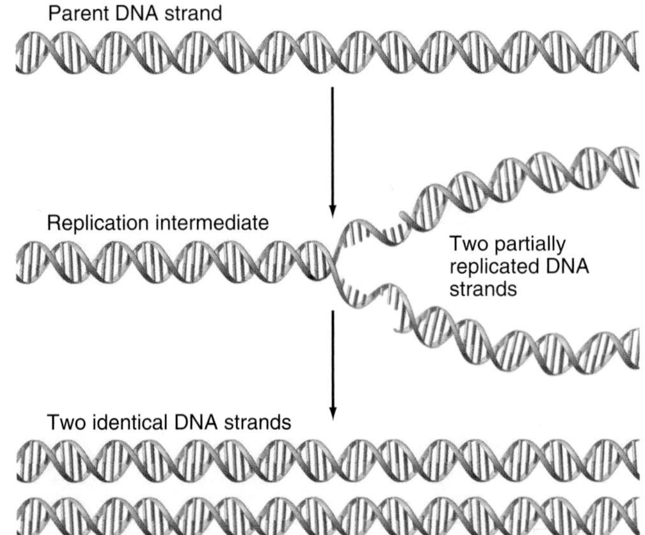

FIGURE 1.3 DNA STRUCTURE AND REPLICATION. Genes stored as the sequence of bases in DNA are replicated enzymatically, forming two identical copies from one double-stranded original.

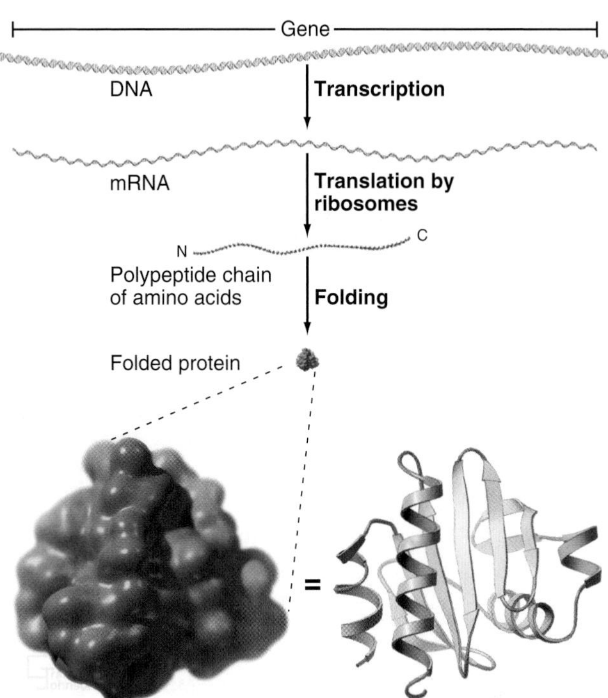

FIGURE 1.4 Genetic information contained in the base sequence of DNA determines the amino acid sequence of a protein and its three-dimensional structure. Enzymes copy (transcribe) the sequence of bases in a gene to make a messenger RNA (mRNA). Ribosomes use the sequence of bases in the mRNA as a template to synthesize (translate) a corresponding linear polymer of amino acids. This polypeptide folds spontaneously to form a three-dimensional protein molecule, in this example the actin-binding protein profilin. (For reference, see Protein Data Bank [www.rcsb.org] file 1ACF.) *Scale drawings of DNA, mRNA, polypeptide, and folded protein:* The folded protein is enlarged at the bottom and rendered in two styles—space-filling surface model *(left)* and a ribbon diagram showing the polypeptide folded into *blue* α-helices and *yellow* β-strands *(right).*

with T and C pairs with G. The two strands separate during enzymatic replication of DNA, each serving as a template for the synthesis of a new complementary strand, thereby producing two identical copies of the DNA. Precise segregation of one newly duplicated double helix to each daughter cell then guarantees the transmission of intact genetic information to the next generation.

2. *Linear chemical sequences stored in DNA code for both the linear sequences and three-dimensional structures of RNAs and proteins (Fig. 1.4).* Enzymes called RNA polymerases copy (transcribe) the information stored in genes into linear sequences of nucleotides of RNA molecules. Many RNAs have structural roles, regulatory functions, or enzymatic activity; for example, ribosomal RNA is by far the most abundant class of RNA in cells. Other genes produce **messenger RNA** (mRNA) molecules that act as templates for protein synthesis, specifying the sequence of amino acids during the synthesis of **polypeptides** by ribosomes. The amino acid sequence of most proteins contains sufficient information to specify how the polypeptide folds into a unique three-dimensional structure with biological activity. Two broad mechanisms control the production and processing of RNA and protein from tens of thousands of genes. **Genetically** encoded control circuits consisting of proteins and RNAs respond to environmental stimuli through signaling pathways. **Epigenetic** controls involve modifications of DNA or associated proteins that affect gene expression. Some epigenetic modifications can be transmitted during cell division and from a parent to an offspring. The basic plan for the cell contained in the genome, together with ongoing regulatory mechanisms (see points 7 and 8 below), works so well that each human develops with few defects from a single fertilized egg into a complicated ensemble of trillions of specialized cells that function harmoniously for decades in an ever-changing environment.

3. *Macromolecular structures assemble from subunits (Fig. 1.5).* Many cellular components form by **self-assembly** of their constituent molecules without the aid of templates or enzymes. The protein, nucleic acid, and lipid molecules themselves contain the information required to assemble complex structures. Diffusion usually brings the molecules together during these assembly processes. Exclusion of water from complementary surfaces ("lock-and-key" packing), as well as electrostatic and hydrogen bonds, provides the energy to hold the subunits together. In some cases, protein chaperones assist with assembly by preventing the aggregation of incorrectly folded intermediates. Important cellular structures assembled in this

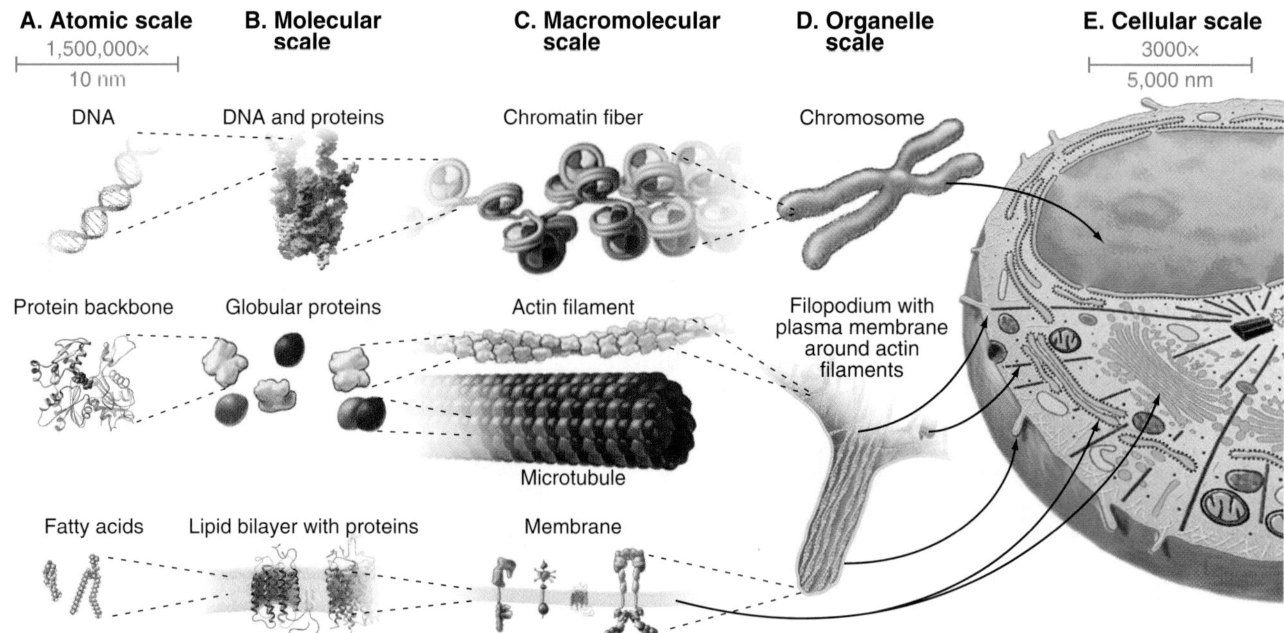

FIGURE 1.5 MACROMOLECULAR ASSEMBLY. Many macromolecular components of cells assemble spontaneously from constituent molecules without the guidance of templates. This figure shows chromosomes assembled from DNA and proteins, a bundle of actin filaments in a filopodium assembled from protein subunits, and the plasma membrane formed from lipids and proteins.

way include **chromatin,** consisting of nuclear DNA packaged by associated proteins; **ribosomes,** assembled from RNA and proteins; cytoskeletal polymers, assembled from protein subunits; and membranes formed from lipids and proteins.

4. *Membranes grow by expansion of preexisting membranes (Fig. 1.6).* Cellular membranes composed of lipids and proteins grow only by expansion of preexisting lipid bilayers rather than forming de novo. Thus membrane-bounded **organelles**, such as **mitochondria** and **endoplasmic reticulum**, multiply by growth and division of preexisting organelles and are inherited maternally from stockpiles stored in the egg. The endoplasmic reticulum (ER) plays a central role in membrane biogenesis as the site of phospholipid synthesis. Through a series of vesicle budding and fusion events, membrane made in the ER provides material for the **Golgi apparatus**, which, in turn, provides lipids and proteins for lysosomes and the plasma membrane.
5. *Signal-receptor interactions target cellular constituents to their correct locations (Fig. 1.6).* Specific recognition signals incorporated into the structures of proteins and nucleic acids route these molecules to their proper cellular compartments. Receptors recognize these signals and guide each molecule to its appropriate compartment. For example, proteins destined for the nucleus contain short amino acid sequences that bind receptors to facilitate their passage through **nuclear pores** into the **nucleus**. Similarly, a peptide signal sequence first targets lysosomal proteins into the lumen of the ER. Subsequently, the Golgi apparatus adds a sugar-phosphate group recognized by receptors that secondarily target these proteins to lysosomes.
6. *Cellular constituents move by diffusion, pumps, and motors (Fig. 1.7).* Most small molecules move through the cytoplasm or membrane channels by diffusion. However, energy provided by ATP hydrolysis or electrochemical gradients is required for molecular pumps to drive molecules across membranes against concentration gradients. Similarly, **motor proteins** use energy from ATP hydrolysis to move organelles and other cargo along microtubules or actin filaments. In a more complicated example, protein molecules destined for mitochondria diffuse from their site of synthesis in the cytoplasm to a mitochondrion (Fig. 1.6), where they bind to a receptor. Energy-requiring reactions then transport the protein into the mitochondrion.
7. *Receptors and signaling mechanisms allow cells to adapt to environmental conditions (Fig. 1.8).* Environmental stimuli modify cellular behavior. Faced with an unpredictable environment, cells must decide which genes to express, which way to move, and whether to proliferate, differentiate into a specialized cell, or die. Some of these choices are programmed genetically or epigenetically, but minute-to-minute decisions generally involve the reception of chemical or physical stimuli from outside the cell and

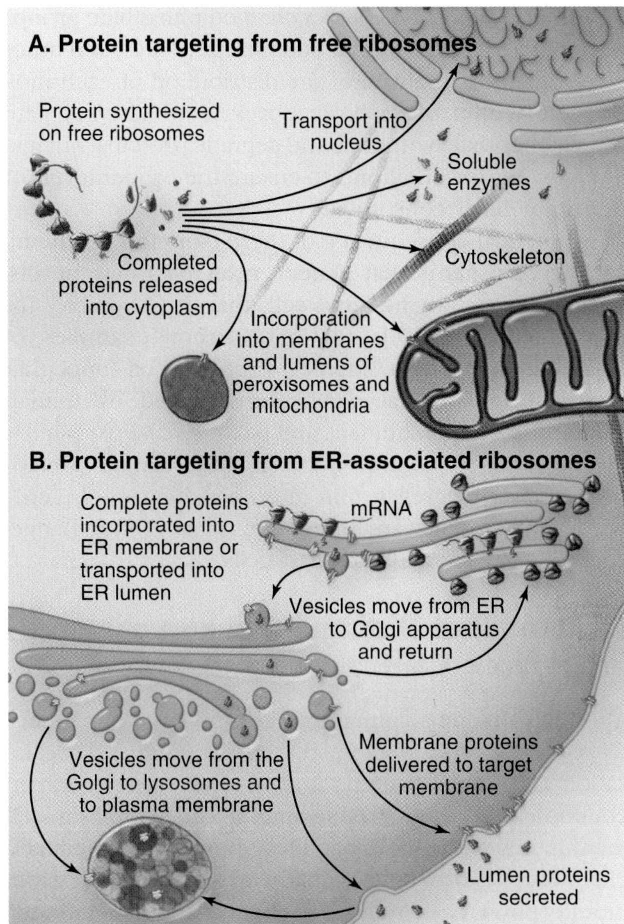

FIGURE 1.6 PROTEIN TARGETING. Signals built into the amino acid sequences of proteins target them to all compartments of the eukaryotic cell. **A,** Proteins synthesized on free ribosomes can be used locally in the cytoplasm or guided by different signals to the nucleus, mitochondria, or peroxisomes. **B,** Other signals target proteins for insertion into the membrane or lumen of the endoplasmic reticulum (ER). From there, a series of vesicular budding and fusion reactions carry the membrane proteins and lumen proteins to the Golgi apparatus, lysosomes, or plasma membrane. mRNA, messenger RNA.

processing of these stimuli to change the behavior of the cell. Cells have an elaborate repertoire of **receptors** for a multitude of stimuli, including nutrients, growth factors, hormones, neurotransmitters, and toxins. Stimulation of receptors activates diverse signal-transducing mechanisms that amplify the message and generate a wide range of cellular responses. These include changes in the electrical potential of the plasma membrane, gene expression, and enzyme activity. Basic **signal transduction** mechanisms are ancient, but receptors and output systems have diversified by gene duplication and divergence during evolution.

8. *Molecular feedback mechanisms control molecular composition, growth, and differentiation (Fig. 1.9).* Living cells are dynamic, constantly fine-tuning their composition in response to external stimuli, nutrient

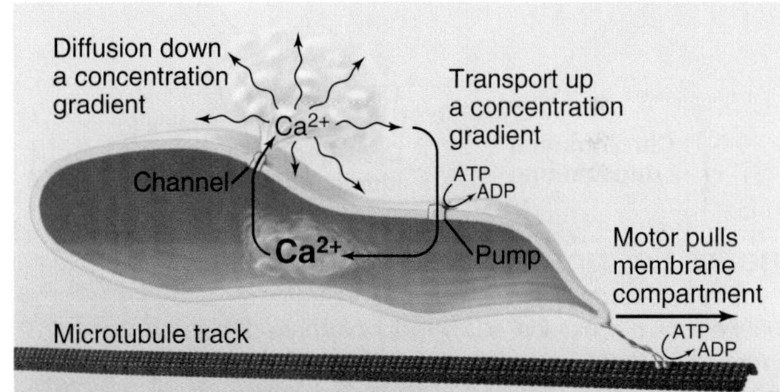

FIGURE 1.7 MOLECULAR MOVEMENTS BY DIFFUSION, PUMPS, AND MOTORS. *Diffusion:* Molecules up to the size of globular proteins diffuse in the cytoplasm. Concentration gradients can provide a direction to diffusion, such as the diffusion of Ca^{2+} from a region of high concentration inside the endoplasmic reticulum through a membrane channel to a region of low concentration in the cytoplasm. *Pumps:* Adenosine triphosphate (ATP)-driven protein pumps transport ions up concentration gradients. *Motors:* ATP-driven motors move organelles and other large cargo along microtubules and actin filaments. ADP, adenosine diphosphate.

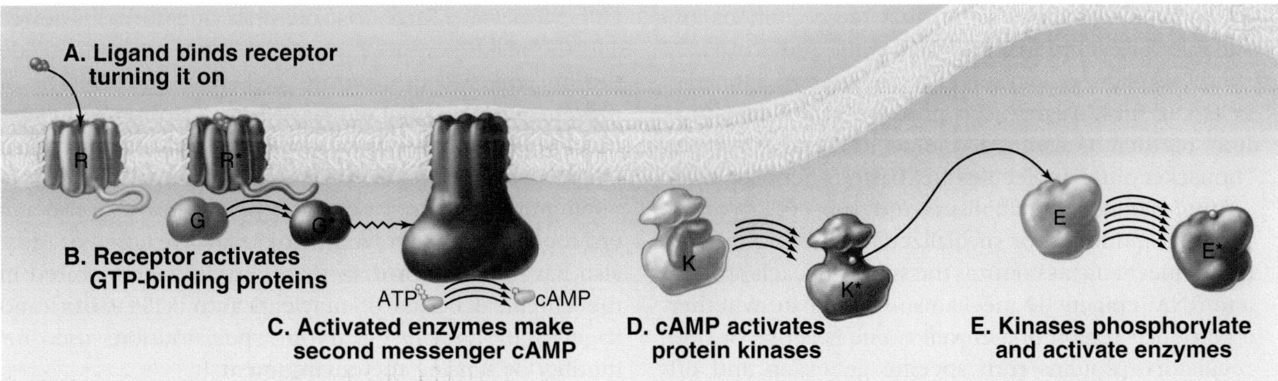

FIGURE 1.8 RECEPTORS AND SIGNALS. Activation of cellular metabolism by an extracellular ligand, such as a hormone. In this example, binding of the hormone **(A)** triggers a series of linked biochemical reactions **(B–E),** leading through a second messenger molecule (cyclic adenosine monophosphate [cAMP]) and a cascade of three activated proteins to regulate a metabolic enzyme. The response to a single ligand is multiplied at steps **B, C,** and **E,** leading to thousands of activated enzymes. GTP, guanosine triphosphate.

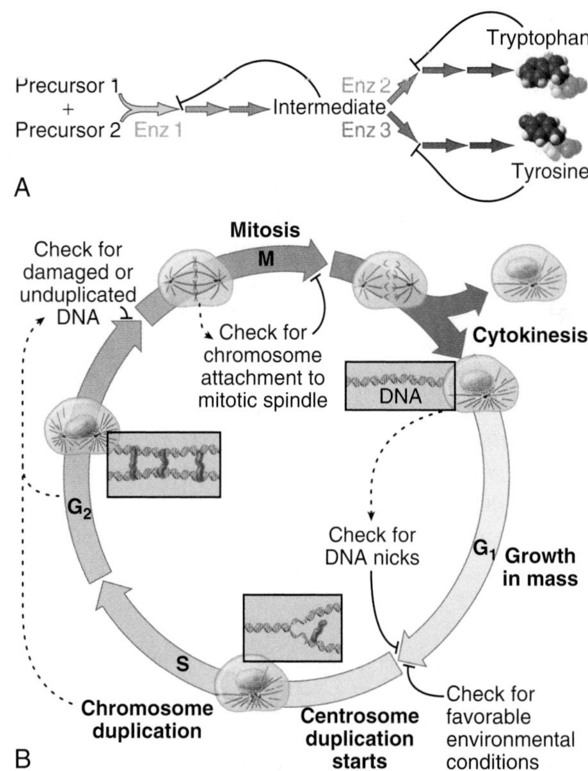

FIGURE 1.9 MOLECULAR FEEDBACK LOOPS. A, Control of the synthesis of aromatic amino acids. An intermediate and the final products of this biochemical pathway inhibit three of nine enzymes (Enz) in a concentration-dependent fashion, automatically turning down the reactions that produced them. This maintains constant levels of the final products, two amino acids essential for protein synthesis. **B,** Control of the cell cycle. The cycle consists of four stages. During the G_1 phase, the cell grows in size. During the S phase, the cell duplicates the DNA of its chromosomes. During the G_2 phase, the cell checks for completion of DNA replication. In the M phase, chromosomes condense and attach to the mitotic spindle, which separates the duplicated pairs in preparation for the division of the cell by cytokinesis. Biochemical feedback loops called checkpoints halt the cycle *(blunt bars)* at several points until the successful completion of key preceding events.

availability, and internal signals. The most dramatic example is the regulation of each step in the cell cycle. Feedback loops assure that the conditions are suitable for each transition such as the onset of DNA synthesis and the decision to begin mitosis. Similarly, cells carefully balance the production and degradation of their constituent molecules. Cells produce "housekeeping" molecules for basic functions, such as intermediary metabolism, and subsets of other proteins and RNAs for specialized functions. A hierarchy of mechanisms controls the supply of each protein and RNA: epigenetic mechanisms designate whether a particular region of a chromosome is active or not; regulatory proteins turn specific genes on and off and modulate the rates of translation of mRNAs into protein; synthesis balanced by the rates of degradation determines the abundance of specific RNAs and proteins; phosphorylation (covalent modification of certain amino acids with a charged phosphate group) regulates protein interactions and activities; and other mechanisms regulate of the distribution of each molecule within the cell. Feedback loops also regulate enzymes that synthesize and degrade proteins, nucleic acids, sugars, and lipids to ensure the proper levels of each cellular constituent.

A practical consequence of these common biochemical mechanisms is that general principles may be discovered by studying any cell that is favorable for experimentation. This text cites many examples of research on bacteria, insects, protozoa, or fungi that revealed fundamental mechanisms shared by human cells. For example, humans and baker's yeast use similar mechanisms to control the cell cycle, guide protein secretion, and segregate chromosomes at mitosis. Indeed, particular proteins are often functionally interchangeable between human and yeast cells.

Features That Distinguish Eukaryotic and Prokaryotic Cells

Although sharing a common origin and basic biochemistry, cells vary considerably in their structure and organization (Fig. 1.2). Bacteria and Archaea have much in common, including chromosomes in the cytoplasm, cell membranes with similar families of pumps, carriers and channels, basic metabolic pathways, gene expression, motility powered by rotary flagella, and lack of membrane-bound organelles. On the other hand, these prokaryotes are wonderfully diverse in terms of morphology and their use of a wide range of energy sources.

Eukaryotes comprise a multitude of unicellular organisms, algae, plants, amoebas, fungi, and animals that differ from prokaryotes in having a compartmentalized cytoplasm with membrane-bounded organelles including a nucleus. The basic features of eukaryotic cells were refined more than 1.5 billion years ago, before the major groups of eukaryotes diverged. The **nuclear envelope** separates the two major compartments: nucleoplasm and cytoplasm. **Chromosomes** carrying the cell's **genes** and the machinery to express those genes reside inside the nucleus. Most eukaryotic cells have **ER** (the site of protein and phospholipid synthesis), a **Golgi apparatus** (adds sugars to membrane proteins, lysosomal proteins, and secretory proteins), **lysosomes** (compartments containing digestive enzymes), and **peroxisomes** (containers for enzymes involved in oxidative reactions). Most also have **mitochondria** that convert energy stored in the chemical bonds of nutrients into ATP. **Cilia** (and flagella) are ancient eukaryotic specializations used for motility or sensing the environment.

Membrane-bounded compartments give eukaryotic cells a number of advantages. Membranes provide a barrier that allows each type of organelle to maintain novel ionic and enzymatic interior environments. Each of these special environments favors a subset of the biochemical reactions required for life as illustrated by the following examples. The nuclear envelope separates the synthesis and processing of RNA in the nucleus from the **translation** of mature mRNAs into proteins in the cytoplasm. Segregation of digestive enzymes in lysosomes prevents them from destroying other cellular components. ATP synthesis depends on the impermeable membrane around mitochondria; energy-releasing reactions produce a proton gradient across the membrane that drives enzymes in the membrane to synthesize ATP.

Overview of Eukaryotic Cellular Organization and Functions

This section previews the major constituents and processes of eukaryotic cells. With this background the reader will be able to appreciate cross-references to chapters later in the book.

Plasma Membrane

The plasma membrane is the interface of the cell with its environment (Fig. 1.2). Owing to the hydrophobic interior of its lipid bilayer, the plasma membrane is impermeable to ions and most water-soluble molecules. Consequently, they cross the membrane only through transmembrane channels, carriers, and pumps (Fig. 1.10). These transmembrane proteins provide the cell with nutrients, control internal ion concentrations, and establish a transmembrane electrical potential. A single amino acid change in one plasma membrane pump and Cl^- channel causes the human disease cystic fibrosis.

Other plasma membrane proteins mediate interactions of cells with their immediate environment. Transmembrane receptors convert the binding of extracellular signaling molecules, such as hormones and growth factors into chemical or electrical signals that influence the activity of the cell. Genetic defects in signaling proteins, which mistakenly turn on signals for growth in the absence of appropriate extracellular stimuli, contribute to human cancers.

Plasma membrane **adhesion proteins** allow cells to bind specifically to each other or to the **extracellular matrix** (Fig. 1.10). These selective interactions allow cells to form multicellular associations, such as epithelia (sheets of cells that separate the interior of the body from the outside world). Similar interactions allow white blood cells to bind bacteria so that they can be ingested and killed. In cells that are subjected to mechanical forces, such as muscle and epithelia, cytoskeletal filaments inside the cell reinforce the plasma membrane adhesion proteins. In skin, defects in these attachments cause blistering diseases.

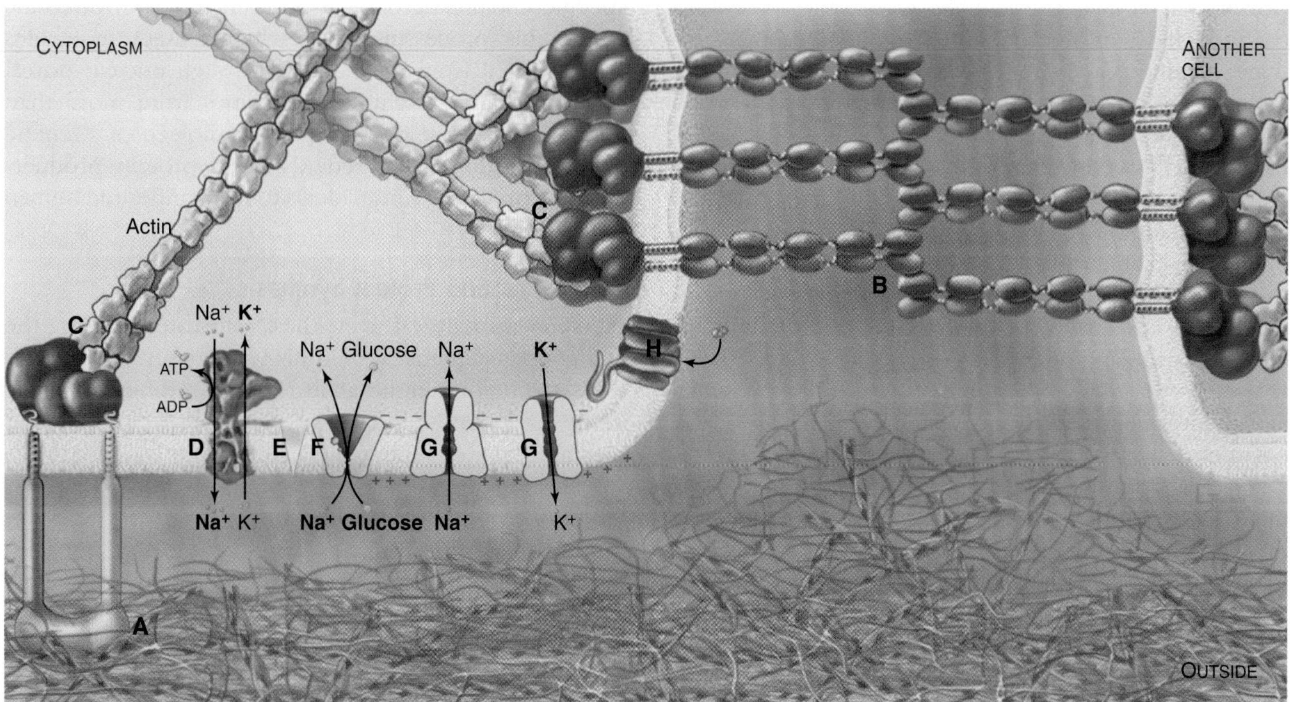

FIGURE 1.10 STRUCTURE AND FUNCTIONS OF AN ANIMAL CELL PLASMA MEMBRANE. The lipid bilayer is a permeability barrier between the cytoplasm and the extracellular environment. Transmembrane adhesion proteins anchor the membrane to the extracellular matrix **(A)** or to like receptors on other cells **(B)** and transmit forces to the cytoskeleton **(C)**. Adenosine triphosphate (ATP)-driven enzymes **(D)** pump Na^+ out of and K^+ into the cell **(E)** to establish concentration gradients across the lipid bilayer. Transmembrane carrier proteins **(F)** use these ion concentration gradients to transport of nutrients into the cell. Selective ion channels **(G)** regulate the electrical potential across the membrane. A large variety of receptors **(H)** bind specific extracellular ligands and send signals across the membrane to the cytoplasm.

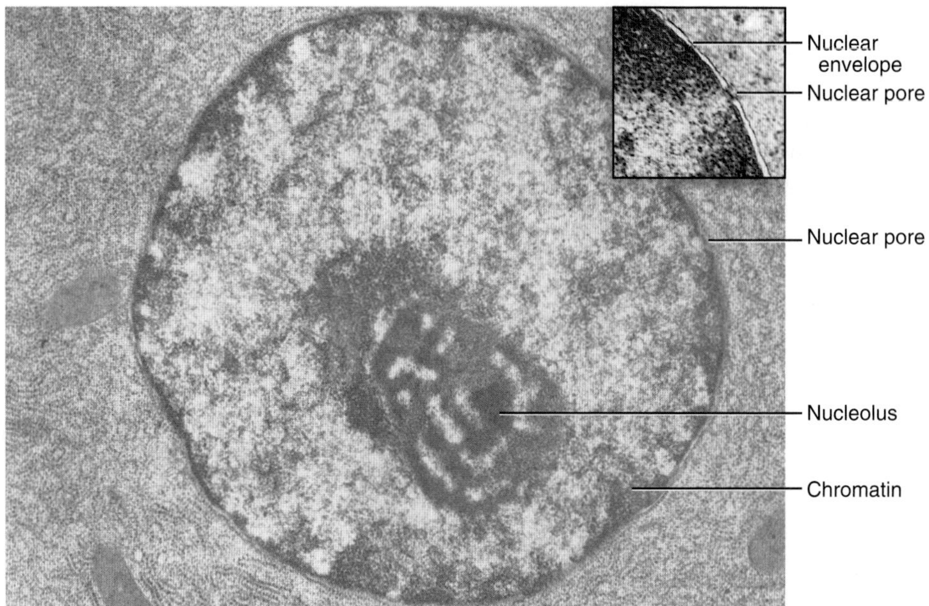

FIGURE 1.11 ELECTRON MICROGRAPH OF A THIN SECTION OF A NUCLEUS. (Courtesy Don Fawcett, Harvard Medical School, Boston, MA.)

Nucleus

The **nuclear envelope** is a double membrane that separates the nucleus from the cytoplasm (Fig. 1.11). All traffic into and out of the nucleus passes through **nuclear pores** that bridge the double membranes. Inbound traffic includes all nuclear proteins and ribosomal proteins destined for the nucleolus. Outbound traffic includes mRNAs and ribosomal subunits.

The nucleus stores genetic information in extraordinarily long DNA molecules called chromosomes. Remarkably, portions of genes encoding proteins and structural RNAs make up only a small fraction (<2%) of the 3 billion nucleotide pairs in human DNA, but more than 50% of the 97 million nucleotide pairs in a nematode worm. Regions of DNA called **telomeres** stabilize the ends of chromosomes, and other DNA sequences organize **centromeres** that direct the distribution of chromosomes to daughter cells when cells divide. Much of the DNA encodes a myriad of RNAs with regulatory activities.

The DNA and its associated proteins are called chromatin (Fig. 1.5). Interactions with histones and other proteins fold each chromosome compactly enough to fit into discrete territories inside the nucleus. During **mitosis,** chromosomes condense and reorganize into separate structural units suitable for sorting into daughter cells (Fig. 1.5).

Regulatory proteins called **transcription factors** turn specific genes on and off in response to genetic, developmental, and environmental signals. Enzymes called polymerases make RNA copies of active genes, a process called **transcription**. mRNAs specify the amino acid sequences of proteins. Other RNAs have structural, regulatory, or catalytic functions. Most newly synthesized RNAs are processed extensively before they are ready for use. Processing involves removal of intervening sequences, alteration of bases, or addition of specific chemical groups at both ends. For cytoplasmic RNAs, this processing occurs before RNA molecules are exported from the nucleus through nuclear pores**.** The **nucleolus** assembles ribosomes from more than 50 different proteins and 3 RNA molecules. Genetic errors resulting in altered RNA and protein products cause or predispose individuals to many inherited human diseases.

Ribosomes and Protein Synthesis

Ribosomes catalyze the synthesis of proteins, using the nucleotide sequences of mRNA molecules to specify the sequence of amino acids (Fig. 1.4). Ribosomes free in the cytoplasm synthesize proteins that are released for routing to various intracellular destinations (Fig. 1.6).

Endoplasmic Reticulum

Ribosomes synthesizing proteins destined for insertion into cellular membranes or for export from the cell associate with the ER, a continuous system of flattened membrane sacks and tubules (Fig. 1.12). Proteins produced on these ribosomes carry **signal sequences** of amino acids that target their ribosomes to receptors on the ER (Fig. 1.6). These regions of the ER are called rough ER owing to the attached ribosomes. As a polypeptide chain grows, its sequence determines whether the protein folds up in the lipid bilayer or translocates across the membrane into the lumen of the ER. Enzymes add sugar

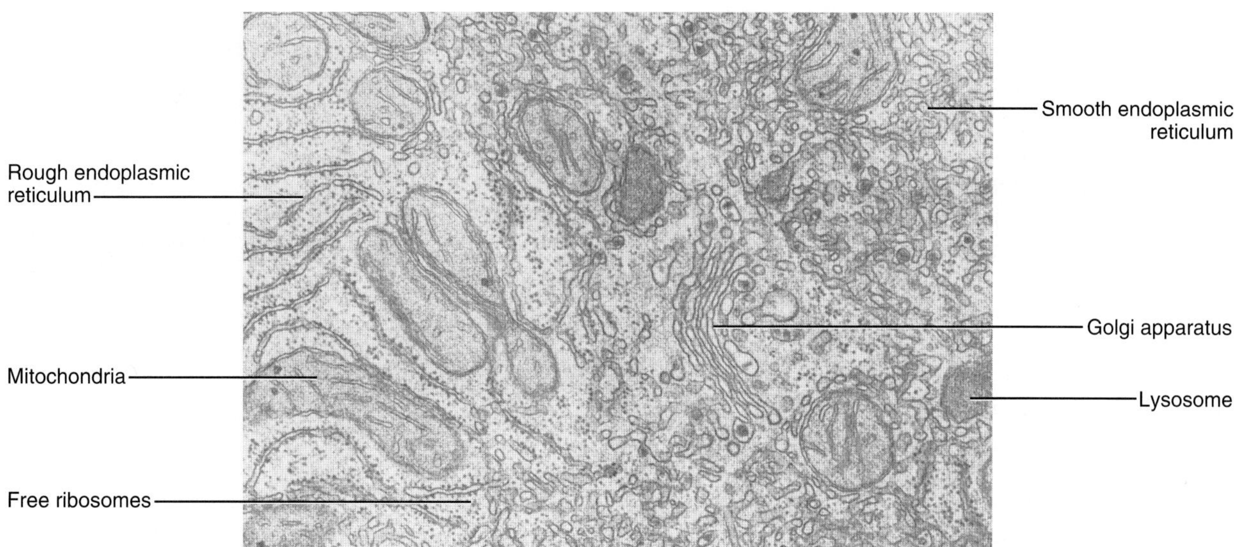

FIGURE 1.12 ELECTRON MICROGRAPH OF A THIN SECTION OF A LIVER CELL SHOWING ORGANELLES. (Courtesy Don Fawcett, Harvard Medical School, Boston, MA.)

polymers to some proteins exposed in the lumen. Some proteins are retained in the ER, but most move on to other parts of the cell.

ER is very dynamic. Motor proteins move along microtubules to pull the ER membranes into a branching network spread throughout the cytoplasm. Continuous bidirectional traffic moves small vesicles between the ER and the Golgi apparatus. These vesicles carry soluble proteins in their lumens, in addition to transporting membrane lipids and proteins. Proteins on the cytoplasmic surface of the membranes catalyze each membrane budding and fusion event. The use of specialized proteins for budding and fusion of membranes at different sites in the cell organizes this membrane traffic and prevents the membrane components from getting mixed up.

The ER also serves as the outer membrane of the nuclear envelope, which can have attached ribosomes. ER enzymes synthesize many cellular lipids and metabolize drugs, while ER pumps and channels regulate the cytoplasmic Ca^{2+} concentration.

Golgi Apparatus

The Golgi apparatus processes the sugar side chains on transmembrane and secreted proteins. It consists of a stack of flattened, membrane-bound sacks with many associated vesicles. The Golgi apparatus is characteristically located in the middle of the cell near the nucleus and the centrosome (Figs. 1.2 and 1.12). Proteins to be processed come in vesicles that detach from the ER and fuse with Golgi apparatus membranes (Fig. 1.6). As proteins pass through the stacked Golgi membranes from one side to the other, enzymes in specific stacks modify the sugar side chains of secretory and membrane proteins.

On the downstream side of the Golgi apparatus, processed proteins segregate into different vesicles destined for lysosomes or the plasma membrane (Fig. 1.6). Many components of the plasma membrane including receptors for extracellular molecules recycle from the plasma membrane to endosomes and back to the cell surface many times before they are degraded. Defects in this process can cause arteriosclerosis.

Lysosomes

An impermeable membrane separates degradative enzymes inside lysosomes from other cellular components (Fig. 1.12). After synthesis by rough ER, lysosomal proteins move through the Golgi apparatus, where enzymes add the modified sugar, phosphorylated mannose (Fig. 1.6). Vesicular transport, guided by phosphomannose receptors, delivers lysosomal proteins to the lumen of lysosomes.

Cells ingest microorganisms and other materials in membrane vesicles derived from the plasma membrane. The contents of these **endosomes** and **phagosomes** are delivered to lysosomes for degradation by lysosomal enzymes. Deficiencies of lysosomal enzymes cause many severe congenital diseases where substrates of the enzyme accumulate in quantities that can impair the function of the brain, liver, or other organs.

Mitochondria

Mitochondrial enzymes use most of the energy released from the breakdown of nutrients to synthesize ATP, the common currency for most energy-requiring reactions in cells (Fig. 1.12). This efficient process uses molecular oxygen to complete the oxidation of fats, proteins, and sugars to carbon dioxide and water. A less-efficient glycolytic system in the cytoplasm extracts energy from the partial breakdown of glucose to make ATP. Mitochondria cluster near sites of ATP utilization, such as membranes engaged in active transport, nerve terminals, and the contractile apparatus of muscle cells.

Mitochondria also respond to toxic stimuli from the environment including drugs used in cancer chemotherapy by activating controlled cell death called **apoptosis.** A toxic cocktail of enzymes degrades proteins and nucleic acids as the cell breaks into membrane-bound fragments. Defects in this form of cellular suicide lead to autoimmune disorders, cancer, and some neurodegenerative diseases.

Mitochondria form in a fundamentally different way from the ER, Golgi apparatus, and lysosomes (Fig. 1.6). Cytoplasmic ribosomes synthesize most mitochondrial proteins. Signal sequences on these mitochondrial proteins bind receptors on the surface of mitochondria. The proteins are then transported into the mitochondrial interior or inserted into the outer or inner mitochondrial membranes.

Mitochondria arose from symbiotic Bacteria (Fig. 1.1) and most of the bacterial genes subsequently moved to the nucleus. However, mitochondrial DNA, ribosomes, and mRNAs still produce a few essential proteins for the organelle. Defects in the maternally inherited mitochondrial genome cause several diseases, including deafness, diabetes, and ocular myopathy.

Peroxisomes

Peroxisomes are membrane-bound organelles containing enzymes that participate in oxidative reactions. Like mitochondria, peroxisomal enzymes oxidize fatty acids, but the energy is not used to synthesize ATP. Peroxisomes are particularly abundant in plants. Peroxisomal proteins are synthesized in the cytoplasm and imported into the organelle using the same strategy as mitochondria but with different targeting sequences and transport machinery (Fig. 1.6). Genetic defects in peroxisomal biogenesis cause several forms of mental retardation.

Cytoskeleton and Motility Apparatus

A cytoplasmic network of three protein polymers—actin filaments, intermediate filaments, and microtubules (Fig. 1.13)—maintains the shape of most cells. Each polymer has distinctive properties and dynamics. Actin filaments and microtubules provide tracks for the ATP-powered motor proteins that produce most cellular movements (Fig. 1.14), including locomotion, muscle contraction, transport of organelles through the cytoplasm, mitosis, and the beating of **cilia** and **flagella.** The proteins are also used for highly specialized motile processes, such as muscle contraction and sperm motility.

Networks of crosslinked actin filaments anchored to the plasma membrane (Fig. 1.10) reinforce the surface of the cell. In many cells, tightly packed bundles of actin filaments support finger-like projections of the plasma membrane (Fig. 1.5). These filopodia or microvilli increase the surface area of the plasma membrane for transporting nutrients and other processes, including sensory transduction in the ear. Genetic defects in a membrane-associated, actin-binding protein called dystrophin cause the most common form of muscular dystrophy.

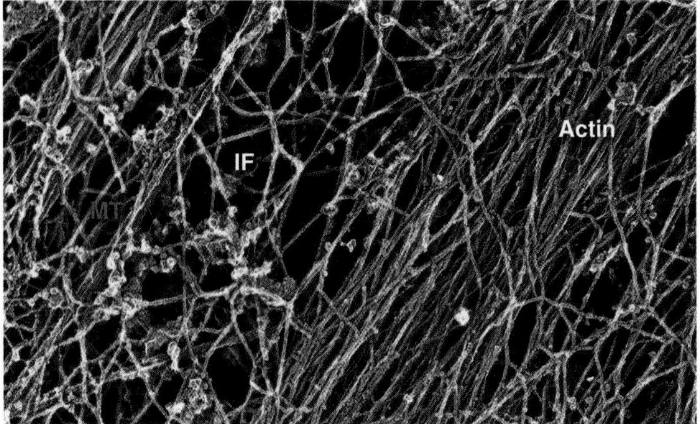

FIGURE 1.13 ELECTRON MICROGRAPH OF THE CYTOPLASMIC MATRIX. A fibroblast cell was prepared by detergent extraction of soluble components, rapid freezing, sublimation of ice, and coating with metal. IF, intermediate filaments; MT, microtubules (shaded red). (Courtesy J. Heuser, Washington University, St. Louis, MO.)

Actin filaments participate in movements in two ways. Assembly of actin filaments produces some movements, such as the protrusion of pseudopods. Other movements result from force generated by **myosin** motor proteins that use the energy from ATP hydrolysis to produce movements along actin filaments. Muscles use a highly organized assembly of actin and myosin filaments to drive forceful, rapid, one-dimensional contractions. Myosin also drives the contraction of the cleavage furrow during cell division. External signals, such as chemotactic molecules, can influence both actin filament organization and the direction of motility. Genetic defects in myosin cause enlargement of the heart and sudden death.

Intermediate filaments are flexible but strong intracellular tendons that reinforce epithelial cells of the skin and other cells subjected to substantial physical stresses. All intermediate filament proteins are related to the keratin molecules found in hair. Intermediate filaments characteristically form bundles that link the plasma membrane to the nucleus. Lamin intermediate filaments reinforce the nuclear envelope. Intermediate filament networks are disassembled during mitosis and cell movements as a result of specific reversible phosphorylation events. Genetic defects in keratin intermediate filaments cause blistering diseases of the skin. Defects in nuclear lamins are associated with some types of muscular dystrophy and premature aging.

Microtubules are rigid cylindrical polymers that resist compression better than actin or intermediate filaments.

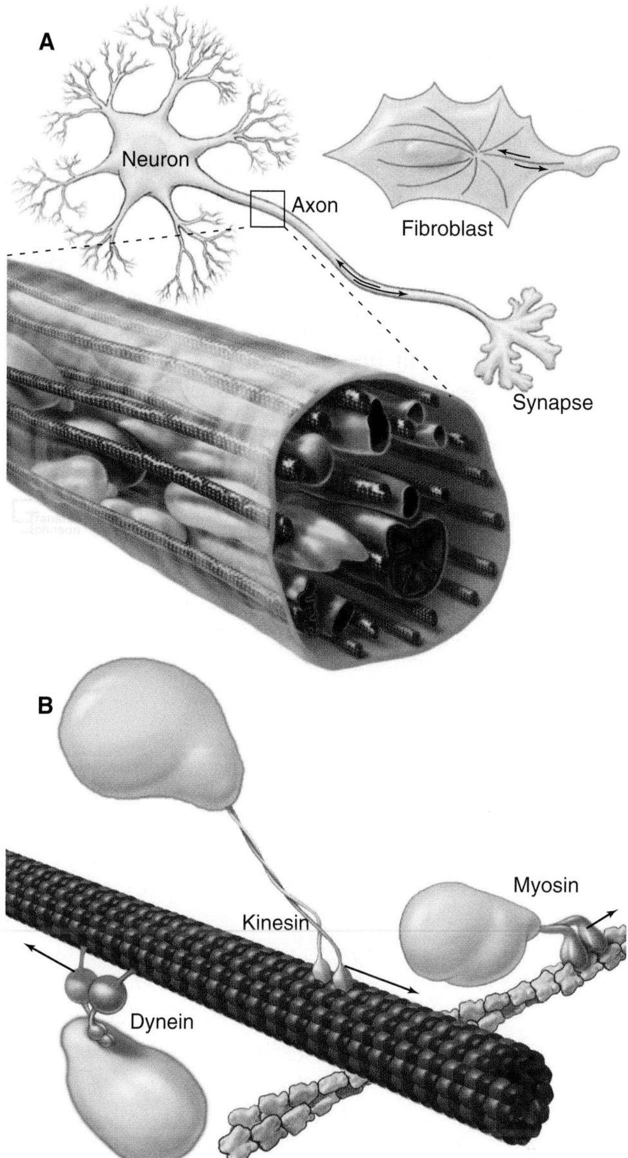

FIGURE 1.14 TRANSPORT OF CYTOPLASMIC PARTICLES ALONG ACTIN FILAMENTS AND MICROTUBULES BY MOTOR PROTEINS. A, Overview of organelle movements in a neuron and fibroblast. **B,** Details of the molecular motors. The microtubule-based motors, dynein and kinesin, move in opposite directions. The actin-based motor, myosin, moves in one direction along actin filaments. (Modified from Atkinson SJ, Doberstein SK, Pollard TD. Moving off the beaten track. *Curr Biol.* 1992;2:326–328.)

The molecular polarity of the microtubule polymer gives the two ends different properties and determines the direction of movement of motor proteins. Most microtubules in cells have the same polarity relative to the organizing centers that initiate their growth (eg, the **centrosome**) (Fig. 1.2). Their rapidly growing ends are oriented toward the periphery of the cell. Individual cytoplasmic microtubules are remarkably dynamic, growing and shrinking on a time scale of minutes.

Microtubules serve as mechanical reinforcing rods for the cytoskeleton and the tracks for two classes of motor proteins that use the energy liberated by ATP hydrolysis to move along the microtubules. **Kinesin** moves its associated cargo (vesicles and RNA-protein particles) along the microtubule network radiating away from the centrosome, whereas **dynein** moves its cargo toward the centrosome. Together, they form a two-way transport system that is particularly well developed in the axons and dendrites of nerve cells. Toxins can impair this transport system and cause nerve malfunctions.

During mitosis, the cell assembles a mitotic apparatus of highly dynamic microtubules and uses microtubule motor proteins to distribute the replicated chromosomes into the daughter cells. The motile apparatus of cilia and flagella is built from a complex array of stable microtubules that bends when dynein slides the microtubules past each other. A genetic absence of dynein immobilizes these appendages, causing male infertility and lung infections.

Microtubules, intermediate filaments, and actin filaments each provide mechanical support for the cell. Interactions of microtubules with intermediate filaments and actin filaments unify the cytoskeleton into a continuous mechanical structure. These polymers also provide a scaffold for some cellular enzyme systems.

Cell Cycle

Cells carefully control their growth and division using an integrated regulatory system consisting of **protein kinases** (enzymes that add phosphate to the side chains of proteins), specific kinase inhibitors, transcription factors, and highly specific protein degradation. When conditions inside and outside a cell are appropriate for cell division (Fig. 1.9B), specific cell cycle kinases are activated to trigger a chain of events leading to DNA replication and cell division. Once DNA replication is complete, activation of cell cycle kinases such as Cdk1 pushes the cell into mitosis, the process that separates chromosomes into two daughter cells. Four controls sequentially activate Cdk1 through a positive feedback loop: (a) synthesis of a regulatory subunit, (b) transport into the nucleus, (c) removal and addition of inhibitory and stimulatory phosphate groups, and (d) repression of phosphatases (enzymes that remove the phosphate groups Cdk1 puts on its protein targets).

Phosphorylation of proteins by Cdk1 leads directly or indirectly to disassembly of the nuclear envelope (in most but not all eukaryotic cells), condensation of mitotic chromosomes, and assembly of the **mitotic spindle** composed of microtubules. Selective proteolysis of regulatory subunits of Cdk1 and key chromosomal proteins then allows the mitotic spindle to separate the previously duplicated identical copies of each chromosome. As cells exit mitosis, the nuclear envelope reassembles on the surface of the chromosomes to reform the daughter nuclei. Then the process of **cytokinesis** cleaves the daughter cells.

A key feature of the cell cycle is a series of built-in quality controls, called **checkpoints** (Fig. 1.9), which ensure that each stage of the cycle is completed successfully before the process continues to the next step. These checkpoints also detect damage to cellular constituents and block cell-cycle progression so that the damage may be repaired. Misregulation of checkpoints and other cell-cycle controls predisposes to cancer. Remarkably, the entire cycle of DNA replication, chromosomal condensation, nuclear envelope breakdown, and reformation, including the modulation of these events by checkpoints, can be carried out in cell-free extracts in a test tube.

Welcome to the Rest of the Book

This overview should prepare the reader to embark on the following chapters, which explain our current understanding of the molecular basis of life at the cellular level. This journey starts with the evolution of the cell and introduction to the molecules of life. The following sections cover membrane structure and function, chromosomes and the nucleus, gene expression and protein synthesis, organelles and membrane traffic, signaling mechanisms, cellular adhesion and the extracellular matrix, cytoskeleton and cellular motility, and the cell cycle. Enjoy the adventure of exploring all of these topics. As you read, appreciate that cell biology is a living field that is constantly growing and identifying new horizons. The book will prepare you to understand these new insights as they unfold in the future.

CHAPTER 2

Evolution of Life on Earth

No one is certain how life began, but the **common ancestor** of all living things populated the earth more than 3 billion years ago, not long (geologically speaking) after the planet formed 4.5 billion years ago (Fig. 2.1). Biochemical features shared by all existing cells suggest that this primitive microscopic cell had about 600 genes encoded in DNA, ribosomes to synthesize proteins from messenger RNA templates, basic metabolic pathways, and a plasma membrane with pumps, carriers, and channels. Over time, mutations in the DNA created progeny that diverged genetically into a myriad of distinctive species, most of which have become extinct. Approximately 1.7 million living species are known to science. Extrapolations predict approximately 9 million eukaryotic species and 10 times more prokaryotic organisms living on the earth today. On the basis of evolutionary histories preserved in their genomes, living organisms are divided into three primary domains: Bacteria, Archaea, and Eucarya.

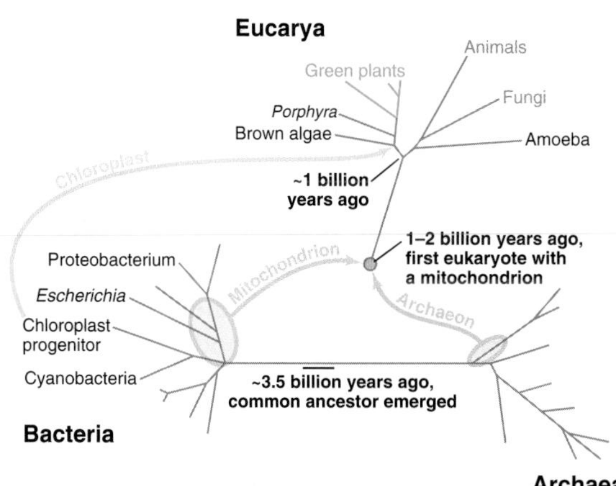

FIGURE 2.1 SIMPLE PHYLOGENETIC TREE WITH THE THREE DOMAINS OF LIFE—BACTERIA, ARCHAEA, AND EUCARYA (EUKARYOTES)—AND A FEW REPRESENTATIVE ORGANISMS. The origin of eukaryotes with a mitochondrion about 2 billion years ago is depicted as a fusion of an α-proteobacterium with an Archaeon. Chloroplasts arose from the fusion of a cyanobacterium with the precursor of algae and plants.

This chapter explains our current understanding of the origin of the first self-replicating cell followed by divergence of its progeny into the two diverse groups of prokaryotes, Bacteria and Archaea. It goes on to consider the origin of Eucarya and their diversification over the past 2 billion years.

Evolution is *the* great unifying principle in biology. Research on evolution is both exciting and challenging because this ultimate detective story involves piecing together fragmentary evidence spread over 3.5 billion years. Data include fossils of ancient organisms and/or chemical traces of their metabolic activities preserved in stone, ancient DNA from historical specimens (going back more than 500,000 years), and especially DNA of living organisms.

Prebiotic Chemistry Leading to an RNA World

Where did the common ancestor come from? A wide range of evidence supports the idea that life began with self-replicating RNA polymers sheltered inside lipid vesicles even before the invention of protein synthesis (Fig. 2.2). This hypothetical early stage of evolution is called the **RNA World.** This attractive postulate solves the chicken-and-egg problem of how to build a system of self-replicating molecules without having to invent either DNA or proteins on their own. RNA has an advantage, because it provides a way to store information in a type of molecule that can also have catalytic activity. Proteins excel in catalysis but do not store self-replicating genetic information. Today, proteins have largely superseded RNAs as cellular catalysts. DNA excels for storing genetic information, since the absence of the 2′ hydroxyl makes it less reactive and therefore more stable than RNA. Readers unfamiliar with the structure of nucleic acids should consult Chapter 3 at this point.

Experts agree that the early steps toward life involved the "prebiotic" synthesis of organic molecules that became the building blocks of macromolecules. To use

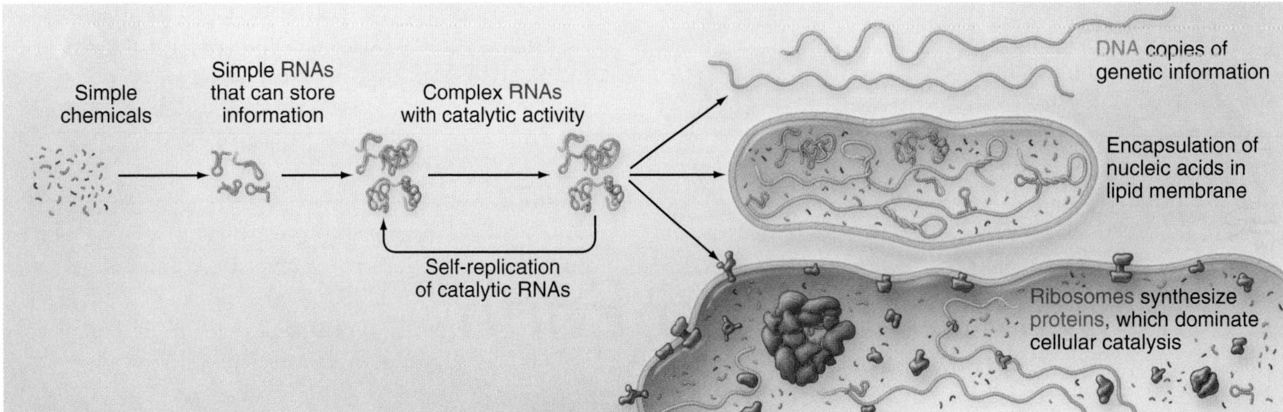

FIGURE 2.2 HYPOTHESES FOR PREBIOTIC EVOLUTION TO LAST COMMON ANCESTOR. Simple chemical reactions are postulated to have given rise to ever more complicated RNA molecules to store genetic information and catalyze chemical reactions, including self-replication, in a prebiotic "RNA world." Eventually, genetic information was stored in more stable DNA molecules, and proteins replaced RNAs as the primary catalysts in primitive cells bounded by a lipid membrane.

RNA as an example, mixtures of chemicals likely to have been present on the early earth can react to form ribose, nucleic acid bases, and ribonucleotides. Minerals can catalyze formation of simple sugars from formaldehyde, and hydrogen cyanide (HCN) and cyanoacetylene or formamide can react to make nucleic acid bases. One problem was the lack of plausible mechanisms to conjugate ribose with a base to make a nucleoside or add phosphate to make a nucleotide without the aid of a preexisting biochemical catalyst. However, new work revealed a pathway to make ribonucleotides directly from cyanamide, cyanoacetylene, glycolaldehyde, glyceraldehyde, and inorganic phosphate. Nucleotides do not polymerize spontaneously into polynucleotides in water, but can do so on the surface of clay called montmorillonite. While attached to clay, single strands of RNA can act as a template for synthesis of a complementary strand to make a double-stranded RNA.

Given a supply of nucleotides, these reactions could have created a heterogeneous pool of small RNAs in special environments such as cracks in rocks heated by hydrothermal vents. These RNAs set in motion the process of natural selection at the molecular level. The idea is that random sequences of RNA were selected for replication on the basis of useful attributes such as the ability to catalyze biochemical reactions. These RNA enzymes are called **ribozymes**.

One can reproduce this process of molecular evolution in the laboratory. Starting with a pool of random initial RNA sequences, multiple rounds of error-prone replication can produce variants that can be tested for a particular biochemical function.

In nature random events would rarely produce useful ribozymes, but once they appeared, natural selection could enrich for RNAs with catalytic activities that sustain a self-replicating system, including synthesis of RNA from a complementary RNA strand. Over millions of years, a ribozyme eventually evolved with the ability to catalyze the formation of peptide bonds and to synthesize proteins. This most complicated of all known ribozymes is the ribosome (see Fig. 12.6) that catalyzes the synthesis of proteins. Proteins eventually supplanted ribozymes as catalysts for most other biochemical reactions. Owing to its greater chemical stability, DNA proved to be superior to RNA for storing the genetic blueprint over time.

Each of these events is improbable, and their combined probability is exceedingly remote, even with a vast number of chemical "experiments" over hundreds of millions of years. Encapsulation of these prebiotic reactions may have enhanced their probability. In addition to catalyzing RNA synthesis, clay minerals can also promote formation of lipid vesicles, which can corral reactants to avoid dilution and loss of valuable constituents. This process might have started with fragile bilayers of fatty acids that were later supplanted by more robust phosphoglyceride bilayers (see Fig. 13.5). In laboratory experiments, RNAs inside lipid vesicles can create osmotic pressure that favors expansion of the bilayer at the expense of vesicles lacking RNAs.

No one knows where these prebiotic events took place. Some steps in prebiotic evolution might have occurred in thermal vents deep in the ocean or in hot springs on volcanic islands where conditions were favorable for some of the reactions. Carbon-containing meteorites have useful molecules, including amino acids. Conditions for prebiotic synthesis were probably favorable beginning approximately 4 billion years ago, but the geologic record has not preserved convincing microscopic fossils or traces of biosynthesis older than 3.5 billion years.

Another mystery is how L-amino acids and D-sugars (see Chapter 3) were selected over their stereoisomers for biological macromolecules. These were pivotal events, since racemic mixtures of L- and D-amino acids are not favorable for biosynthesis. For example, mixtures of nucleotides composed of L- and D-ribose cannot base-pair well enough for template-guided replication of nucleic acids. In the laboratory, particular amino acid stereoisomers (that could have come from meteorites) can bias the synthesis of D-sugars.

Divergent Evolution From the Last Universal Common Ancestor of Life

Shared biochemical features suggest that all current cells are derived from a last universal common ancestor (LUCA) that lived at least 3.5 billion years ago (Fig. 2.1). LUCA could, literally, have been a single cell or colony of cells, but it might have been a larger community of cells sharing a common pool of genes through interchange of their nucleic acids. The situation is obscure, because none of these primitive organisms survived and they left behind few traces. All contemporary organisms have diverged equally far in time from their common ancestor.

Although the features of the LUCA are lost in time, this organism is inferred to have had approximately 600 genes encoded in DNA. It surely had messenger RNAs (mRNAs), transfer RNAs, and ribosomes to synthesize proteins and a plasma membrane with all three families of pumps, as well as carriers and diverse channels, since these are now universal cellular constituents. LUCA probably lived at moderate temperatures and may have used hydrogen as an energy source. The transition from primitive, self-replicating, RNA-only particles to this complicated little cell is, in many ways, even more remarkable than the invention of the RNA World.

During evolution three processes diversify genomes (Fig. 2.3):

- **Gene divergence:** Every gene is subject to random mutations that are inherited by succeeding generations. Some mutations change single base pairs. Other mutations add or delete larger blocks of DNA such as sequences coding a protein domain, an independently folded part of a protein (see Fig. 3.13). These events inevitably produce genetic diversity through divergence of sequences or creation of novel combinations of domains. For example, a typical human genome differs at hundreds of thousands of sites from the the so-called reference genome (see Chapter 7). Many mutations are neutral, but others may confer a reproductive advantage that favors persistence via natural selection. Other mutations are disadvantageous, resulting in disappearance of the lineage. When species diverge, genes with common origins are called **orthologs** (Box 2.1).
- **Gene duplication and divergence:** Rarely, a gene, part of a gene, or even a whole genome is duplicated during replication or cell division. This creates an opportunity for evolution. Some sister genes are eliminated, but others are retained. As these sister genes acquire random point mutations, insertions, or deletions, their structures inevitably diverge, which allows

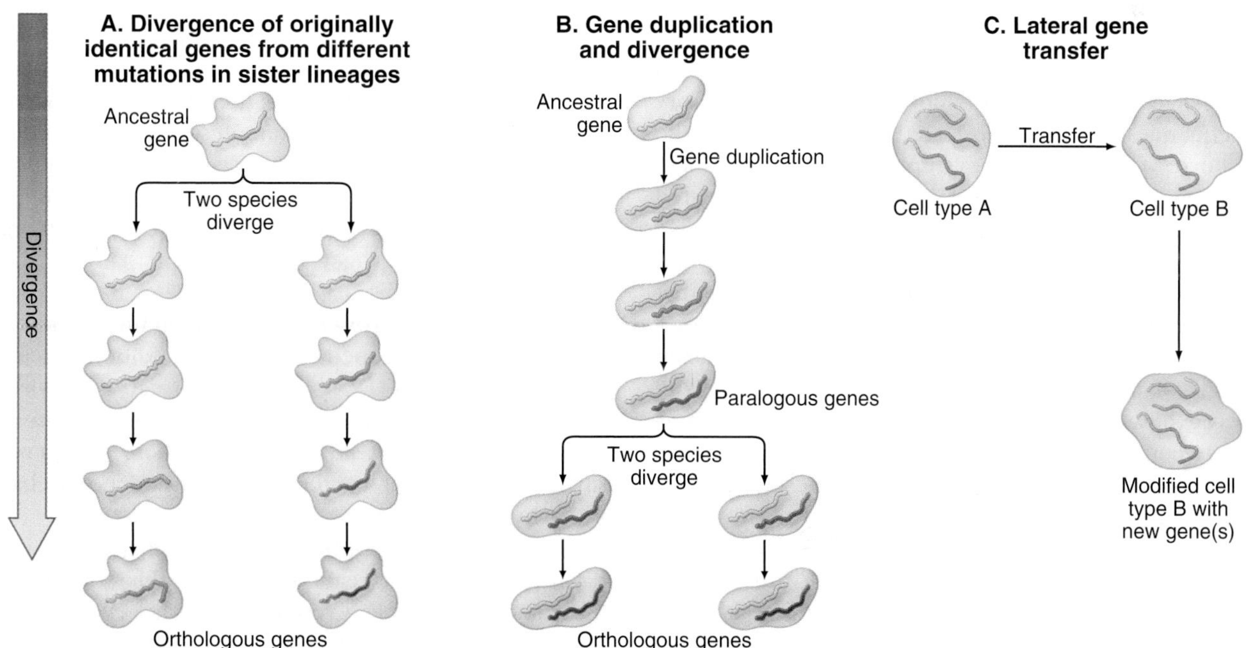

FIGURE 2.3 MECHANISMS OF GENE DIVERSIFICATION. A, Gene divergence from a common origin by random mutations in sister lineages creates orthologous genes. **B,** Gene duplication followed by divergence within and between sister lineages yields both orthologs (separated by speciation) and paralogs (separated by gene duplication). **C,** Lateral transfer moves entire genes from one species to another.

BOX 2.1 Orthologs, Paralogs, and Homologs

Genes with a common ancestor are homologs. The terms *ortholog* and *paralog* describe the relationship of homologous genes in terms of how their most recent common ancestor was separated. If a speciation event separated two genes, then they are orthologs. If a duplication event separated two genes, then they are paralogs. To illustrate this point, let us say that gene A is duplicated within a species, forming paralogous genes A1 and A2. If these genes are separated by a speciation event, so that species 1 has genes sp1A1 and sp1A2 and species 2 has genes sp2A1 and sp2A2, it is proper to say that genes sp1A1 and sp2A1 are orthologs and genes sp1A1 and sp1A2 are paralogs, but genes sp1A1 and sp2A2 are also paralogs because their most recent common ancestor was the gene that duplicated.

for different functions. Some changes may confer a selective advantage; others confer a liability. Multiple rounds of gene duplication and divergence can create huge families of genes encoding related but specialized proteins, such as membrane carrier proteins. Sister genes created by duplication and divergence are called **paralogs.**

- **Lateral transfer:** Another mechanism of genetic diversification involves movement of genes between organisms, immediately providing the host cell with a new biochemical activity. Contemporary bacteria acquire foreign genes in three ways. Pairs of bacteria exchange DNA directly during conjugation. Many bacteria take up naked DNA, as when plasmids move genes for antibiotic resistance between bacteria. Viruses also move DNA between bacteria. Such lateral transfers explain how highly divergent prokaryotes came to share some common genes and regulatory sequences. Laterally transferred genes can change the course of evolution. For example, all the major branching events among Archaea appear to be associated with lateral transfers of genes from Bacteria. Massive lateral transfer occurred twice in eukaryotes when they acquired two different symbiotic bacteria that eventually adapted to form mitochondria and chloroplasts. Lateral transfer continues to this day between pairs of prokaryotes, between pairs of protists, and even between prokaryotes and eukaryotes (such as between pathogenic bacteria and plants).

The genetic innovations created by these processes produce phenotypic changes that are acted on by natural selection. The process depends on tolerance of organisms to change, a feature called "evolvability." After making assumptions about the rates of mutations, one can use differences in gene sequences as a **molecular clock**.

When conditions do not require the product of a gene, the gene can be lost. For example, the simple pathogenic bacteria *Mycoplasma genitalium* has just 470 genes, less than the inferred common ancestor, because it relies on its animal host for most nutrients rather than making them de novo. Similarly, ancient eukaryotes had approximately 200 genes required to assemble an axoneme for a cilium or flagellum (see Fig. 38.13), but most plants and fungi lost them. Vertebrates also lost many genes that had been maintained for more than 2 billion years in earlier forms of life. For instance, humans lack the enzymes to synthesize certain essential amino acids, which must be supplied in our diets.

Evolution of Prokaryotes

Bacteria and Archaea dominate the earth in terms of numbers, variety of species, and range of habitats. They share many features, including a single cytoplasmic compartment with both transcription and translation, basic metabolic enzymes and flagella powered by rotary motors in the plasma membrane. Both divisions of prokaryotes are diverse with respect to size, shape, nutrient sources, and environmental tolerances, so these features cannot be used for classification, which relies instead on analysis of their genomes. For example, sequences of the genes for ribosomal RNAs cleanly identify Bacteria and Archaea (Fig. 2.4). Bacteria are also distinguished by plasma membranes composed of phosphoglycerides (see Fig. 13.2) with F-type adenosine triphosphatases (ATPases) that use proton gradients to synthesize adenosine triphosphate (ATP) or ATP hydrolysis to pump protons (see Fig. 14.5). On the other hand the plasma membranes of Archaea are composed of isoprenyl ether lipids and their V-type ATPases only pump protons (see Fig. 14.5).

Abetted by rapid proliferation and large populations, natural selection allowed prokaryotes to explore many biochemical solutions to life on the earth. Some Bacteria and Archaea (and some eukaryotes too) thrive under inhospitable conditions, such as anoxia and temperatures greater than 100°C as found in deep-sea hydrothermal vents. Other Bacteria and Archaea can use energy sources such as hydrogen, sulfate, or methane that are useless to eukaryotes. Far less than 1% of Bacteria and Archaea have been grown successfully in the laboratory, so many varieties escaped detection by traditional means. Today, sequencing DNA samples from natural environments has revealed vast numbers of new species in the ocean, soil, human intestines, and elsewhere. Only a very small proportion of bacterial species and no Archaea cause human disease.

Chlorophyll-based photosynthesis originated in Bacteria around 3 billion years ago. Surely this was one of the most remarkable events during the evolution of life on the earth, because **photosynthetic reaction centers** (see Fig. 19.8) require not only genes for several transmembrane proteins, but also genes for multiple enzymes, to synthesize chlorophyll and other

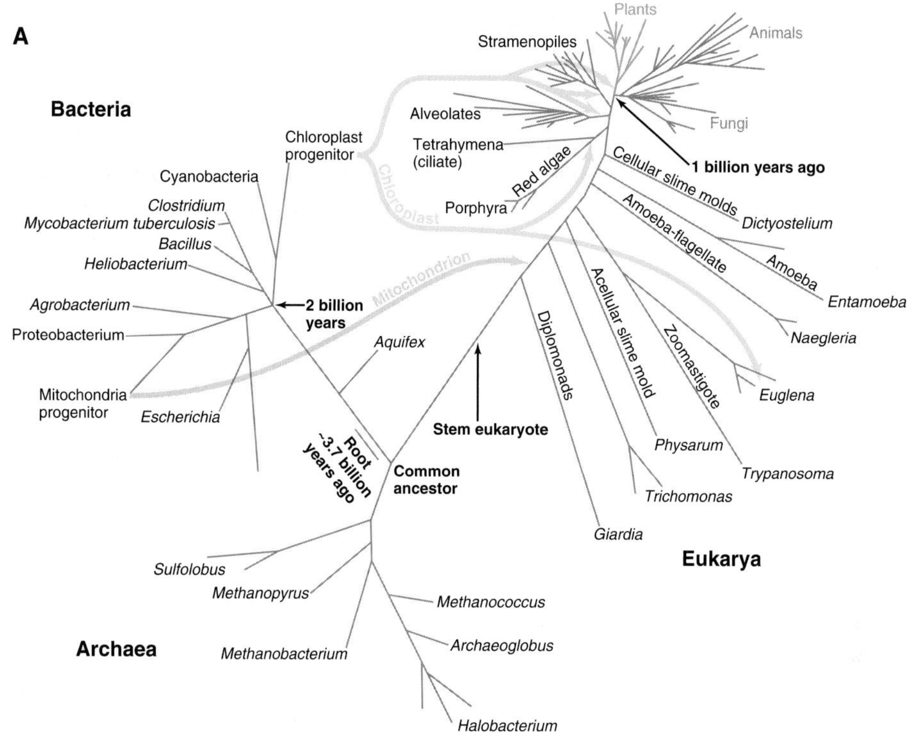

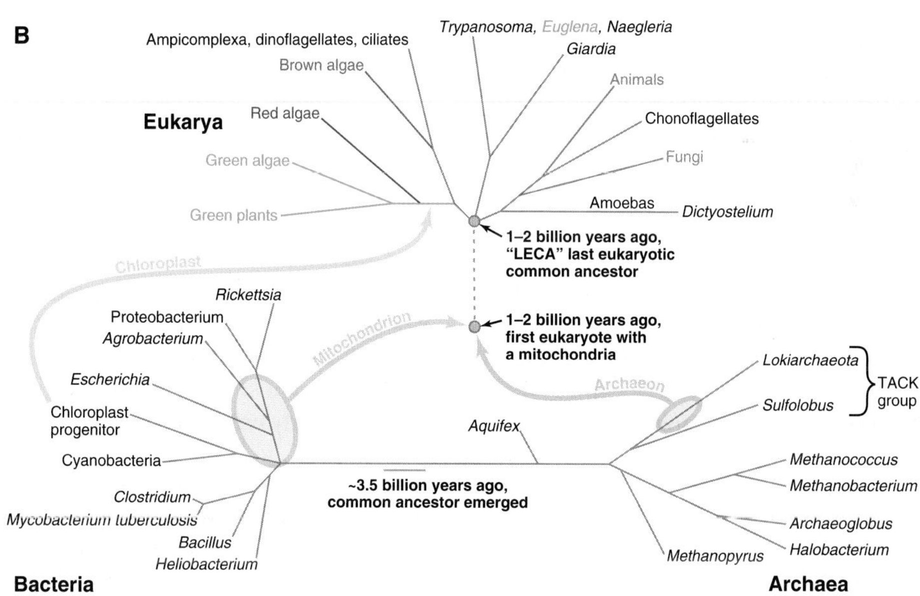

FIGURE 2.4 COMPARISONS OF TREES OF LIFE. A, Universal tree based on comparisons of ribosomal RNA (rRNA) sequences. The rRNA tree has its root deep in the bacterial lineage 3 billion to 4 billion years ago. All current organisms, arrayed at the ends of branches, fall into three domains: Bacteria, Archaea, and Eucarya (eukaryotes). This analysis assumed that the organisms in the three domains diverged from a common ancestor. The lengths of the segments and branches are based solely on differences in RNA sequences. Because the rates of random changes in rRNA genes vary, the lengths of the lines that lead to contemporary organisms are not equal. Complete genome sequences show that genes moved laterally between Bacteria and Archaea and within each of these domains. Multiple bacterial genes moved to Eucarya twice: First, an α-proteobacterium fused with a primitive eukaryote, giving rise to mitochondria that subsequently transferred many of their genes to the eukaryotic nucleus; and second, a cyanobacterium fused with the precursor of algae and plants to give rise to chloroplasts. **B,** Tree based on analysis of full genome sequences and other data showing that eukaryotes formed by fusion of an α-proteobacterium with an Archaeon related to contemporary *Lokiarchaeota*. Chloroplasts arose from the fusion of a cyanobacterium with the eukaryotic precursor of algae and plants. (**A,** Based on a branching pattern from Sogin M, Marine Biological Laboratory, Woods Hole, MA; and Pace N. A molecular view of microbial diversity and the biosphere. *Science.* 1997;276:734–740. **B,** Based on multiple sources, including Adl SM, Simpson AG, Lane CE, et al. The revised classification of eukaryotes. *J Eukaryot Microbiol*. 2012;59:429–493; and Spang A, Saw JH, Jørgensen SL, et al. Complex archaea that bridge the gap between prokaryotes and eukaryotes. *Nature.* 2015;521:173–179.)

complex organic molecules associated with the proteins. Chapter 19 describes the machinery and mechanisms of photosynthesis.

Even more remarkably, photosynthesis was invented twice in different bacteria. A progenitor of green sulfur bacteria and heliobacteria developed photosystem I, while a progenitor of purple bacteria and green filamentous bacteria developed photosystem II. Approximately 3 billion years ago, a momentous lateral transfer event brought the genes for the two photosystems together in **cyanobacteria,** arguably the most important organisms in the history of the earth. Cyanobacteria (formerly misnamed *blue-green algae*) use an enzyme containing manganese to split water into oxygen, electrons, and protons. Sunlight energizes photosystem II and photosystem I to pump the protons out of the cell, creating a proton gradient that is used to synthesize ATP (see Chapters 14 and 19). This form of oxygenic photosynthesis derives energy from sunlight to synthesize the organic compounds that many other forms of life depend on for energy. In addition, beginning approximately 2.4 billion years ago, cyanobacteria produced most of the oxygen in the earth's atmosphere as a by-product of photosynthesis, bioengineering the planet and radically changing the chemical environment for all other organisms as well.

Origin of Eukaryotes

Divergence from the common ancestor explains the evolution of prokaryotes but not the origin of eukaryotes, which inherited genes from both Archaea and Bacteria. The archaeal host cell that gave rise to eukaryotes (Fig. 2.4B) contributed genes for informational processes such as transcription of DNA into RNA and translation of RNA into protein, membrane traffic (Ras family guanosine triphosphatases [GTPases] and ESCRT [endosomal sorting complexes required for transport]-III complex), actin, and ubiquitin-dependent proteolysis. A contemporary archaeon called **Lokiarchaeota** has these genes and is the closest known living relative of the ancient archaeon that became the eukaryote. The original molecular phylogenies based on ribosomal RNA (rRNA) sequences (Fig. 2.4A) did not include Lokiarchaeota, so they missed the direct connection between Archaea and eukaryotes. Those trees accurately represented the relationships among the sampled rRNAs. The long branch originating between Archaea and Bacteria and extending to eukaryotes reflected the extensive divergence of the rRNAs sequences, but not our current understanding of the historical events depicted in Fig. 2.4B.

The bacterial ancestor of mitochondria was an **α-proteobacterium** related to modern-day pathogenic *Rickettsias*. The bacterium established a symbiotic relationship with an ancient archaeal cell, donated genes for many metabolic processes carried out in the cytoplasm and evolved into the mitochondrion. The Bacterium retained its two membranes and contributed molecular machinery for **ATP** synthesis by oxidative phosphorylation (see Fig. 19.5), while the host cell may have supplied organic substrates to fuel ATP synthesis. Together, they had a reliable energy supply for processes such as biosynthesis, regulation of the internal ionic environment, and cellular motility. This massive lateral transfer of genes into the new organism was one of the defining events in the origin of eukaryotes.

This pivotal transfer on the proteobacterial genome to the original eukaryote seems to have occurred just once! The time is uncertain, but may have been as long as 2 billion years ago. The exact mechanism is unknowable and probably irrelevant given its uniqueness (Fig. 2.5). The two prokaryotes may have fused, but more likely an entire bacterium entered into the cytoplasm of its host allowing the two cells to establish a mutually beneficial symbiotic relationship.

All traces of the original eukaryote have disappeared except for the genes donated to its progeny. Thus we do not know if it had a nucleus, organelles, or a cytoskeleton. Microscopic, single-celled eukaryotes called protists have been numerous and heterogeneous throughout evolution, but no existing protist appears to be a good model for the ancestral eukaryote.

The First Billion Years of Eukaryotic Evolution

Ancestral eukaryotes were present on earth more than 2 billion years ago, but current eukaryotes all diverged later from a singular, relatively sophisticated, amoeboid "last eukaryotic common ancestor" (LECA) with most of the specializations that characterize current eukaryotes, including mitochondria, nuclear envelope, linear chromosomes, membrane-bound organelles of the secretory and endocytic pathways, and motile flagella (Fig. 2.4B). The archaeal host brought genes for some of these functions, but early eukaryotes must have tested many different genetic innovations during the long time leading up to LECA. Reconstructing the events between the first eukaryotes and LECA is challenging, because molecular clocks disagree and the fossil record is sparse. The earliest unambiguous eukaryotic fossils are 1.7 billion years old, but LECA could have lived in the range from 2.1 to 0.9 billion years ago. Thereafter LECA swept aside its competitors, since all subsequently diverging species share the full complement of eukaryotic organelles.

Evolution of the Mitochondrion

The mitochondrial progenitor brought along approximately 2000 genes, most of which eventually moved (by a still mysterious process) to the host cell nucleus or were lost. This transfer of mitochondrial genes reduced the size of current mitochondrial genomes variously,

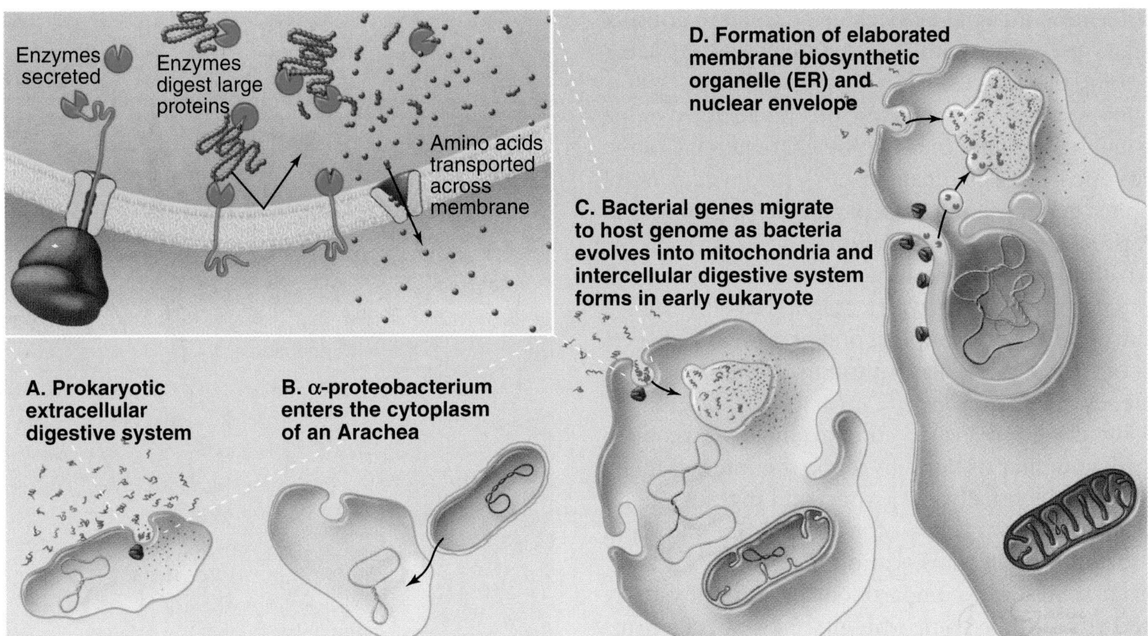

FIGURE 2.5 SPECULATIONS REGARDING THE EVOLUTION OF INTRACELLULAR COMPARTMENTS FROM PROKARYOTES TO PRIMITIVE EUKARYOTES. A–D, Possible stages in the evolution of intracellular compartments. ER, endoplasmic reticulum.

leaving behind between three and 97 protein-coding bacterial genes (see Chapter 19 for more details). Like their bacterial ancestors, **mitochondria** are enclosed by two membranes, with the inner membrane equipped for synthesis of ATP. Mitochondria maintain the capacity to synthesize proteins and a few genes for mitochondrial components. Nuclear genes encode most mitochondrial proteins, which are synthesized in the cytoplasm and imported into the organelle (see Fig. 18.2). The transfer of bacterial genes to the nucleus sealed the dependence of the organelle on its eukaryotic host.

Even though acquisition of mitochondria was an early event in eukaryotic evolution, some eukaryotes, including the anaerobic protozoans *Giardia lamblia* and *Entamoeba histolytica* (both causes of diarrhea), lack fully functional mitochondria. These lineages lost many mitochondrial genes and functions through "reductive evolution" in certain environments that did not favor natural selection for respiration. These reduced organelles have two membranes like mitochondria, but vary considerably in other functions. Such mitochondrial remnants in many organisms synthesize iron–sulfur clusters for cytoplasmic ATP synthesis, while others, called *hydrogenosomes*, make hydrogen.

Evolution of Membrane-Bounded Organelles

Compartmentalization of the cytoplasm into membrane-bounded organelles is one feature of eukaryotes that is generally lacking in prokaryotes. Mitochondria were an early compartment, while chloroplasts resulted from a late endosymbiotic event in algal cells (Fig. 2.7). Endoplasmic reticulum, Golgi apparatus, lysosomes, and endocytic compartments arose by different mechanisms. Compartmentalization allowed ancestral eukaryotes to increase in size, to capture energy more efficiently, and to regulate gene expression in more complex ways.

Prokaryotes that obtain nutrients from a variety of sources appear to have carried out the first evolutionary experiment with compartmentalization (Fig. 2.5A). However, these prokaryotes are compartmentalized only in the sense that they separate digestion outside the cell from biosynthesis inside the cell. They export digestive enzymes (either free or attached to the cell surface) to break down complex organic macromolecules (see Fig. 18.10). They must then import the products of digestion to provide building blocks for new macromolecules. Evolution of the proteins required for targeting and translocation of proteins across membranes was a prokaryotic innovation that set the stage for compartmentalization in eukaryotes.

More sophisticated compartmentalization might have begun when a prokaryote developed the capacity to segregate protein complexes with like functions in the plane of the plasma membrane. Present-day Bacteria segregate their plasma membranes into domains specialized for energy production or protein translocation. Invagination of such domains might have created the endoplasmic reticulum (ER), Golgi apparatus, and lysosomes, as speculated in the following points (Fig. 2.5):

- Invagination of subdomains of the plasma membrane that synthesize membrane lipids and translocate proteins could have generated an intracellular biosynthetic organelle that survives today as the ER.

- Translocation into the ER became coupled to cotranslational protein synthesis, particularly in later-branching eukaryotes.
- The ER was refined to create the nuclear envelope housing the genome, the defining characteristic of the eukaryotic cell. This enabled cells to develop more complex genomes and to separate transcription and RNA processing from translation.
- Internalization of plasma membrane domains with secreted hydrolytic enzymes might have created a primitive lysosome. Coupling of digestion and absorption of macromolecular nutrients would increase efficiency.

This divide-and-specialize strategy might have been employed a number of times to refine the internal membrane system. Eventually, the export and digestive pathways separated from each other and from the lipid synthetic and protein translocation machinery.

As each specialized compartment became physically separated from other compartments, new mechanisms were required to allow traffic between these compartments. The solution was transport vesicles to carry products to the cell surface or vacuole and to import raw materials. Vesicles also segregated digestive enzymes from the surrounding cytoplasm. Once multiple destinations existed, targeting instructions were required to distinguish the routes and destinations.

The outcome of these events (Fig. 2.6) was a vacuolar system consisting of the ER, the center for protein translocation and lipid synthesis; the Golgi complex and secretory pathway, for posttranslational modification and distribution of biosynthetic products to different destinations; and the endosome/lysosome system, for uptake and digestion. Comparative genomics reveals that LECA had a vesicular transport system nearly as complex as humans.

Atmospheric oxygen produced by photosynthetic cyanobacteria allowed eukaryotic cells to synthesize cholesterol (see Fig. 20.15). Cholesterol strengthens membranes without compromising their fluidity, so it may have enabled early eukaryotic cells to increase in size and shed their cell walls. Having shed their cell walls, they could engulf entire prey organisms rather than relying on extracellular digestion. Oxygen also contributed to the precipitation of most of the dissolved iron in the world's oceans, creating ore deposits that are being mined today to extract iron.

The origins of **peroxisomes** are obscure. No nucleic acids or prokaryotic remnants have been detected in peroxisomes, so it seems unlikely that peroxisomes began as prokaryotic symbionts. Peroxisomes arose as centers for oxidative degradation, particularly of products of lysosomal digestion that could not be reutilized for biosynthesis (eg, D-amino acids, uric acid, xanthine). One possibility is that they evolved as a specialization of the ER.

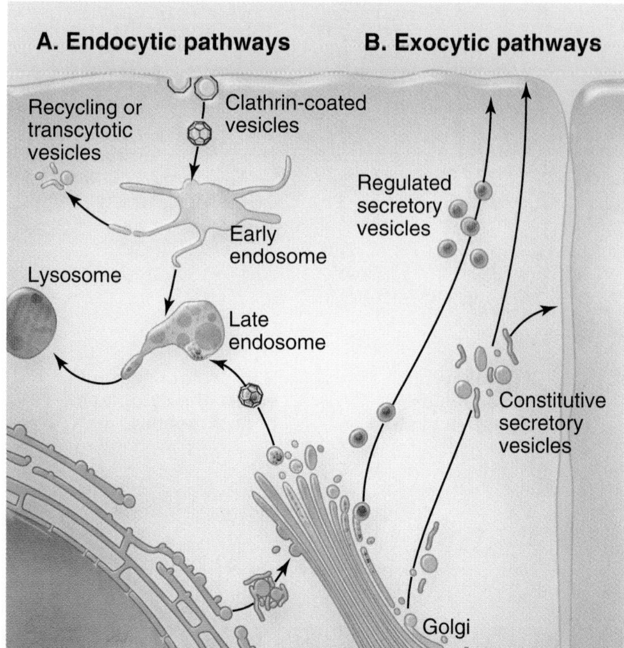

FIGURE 2.6 MEMBRANE-BOUNDED COMPARTMENTS OF EUKARYOTES. A, Pathways for endocytosis and degradation of ingested materials. **B,** Pathways for biosynthesis and distribution of proteins, lipids, and polysaccharides. Membrane and content move through these pathways by controlled budding of vesicles from donor compartments and fusion with specific acceptor compartments. Transport of membranes and content through these two pathways is balanced to establish and maintain the sizes of the compartments.

Given that these internal organelles are found in all branches of eukaryotes, they must all have evolved prior to the diversification of eukaryotes from LECA. One unknown is which organelles appeared before the arrival of the mitochondrion.

Origins and Evolution of Chloroplasts

The acquisition of **plastids,** including chloroplasts, began when a cyanobacterial symbiont brought photosynthesis into an ancient cell that then became an alga (Fig. 2.7). The host cell already had a mitochondrion and depended on external carbon sources for energy. The cyanobacterium provided both photosystem I and photosystem II, allowing energy from sunlight to split water and to drive conversion of CO_2 into organic compounds with O_2 as a by-product (see Fig. 19.8). Symbiosis turned into complete interdependence when most of the genes required to assemble the plastid moved to the nucleus of host cells that continued to rely on the plastid to capture energy from sunlight. This still-mysterious transfer of genes to the nucleus gave the host cell control over the replication of the former symbiont.

Many animal cells and protozoa associate with photosynthetic bacteria or algae, but the original conversion of a bacterial symbiont into a plastid is believed to have been a singular event. The original photosynthetic

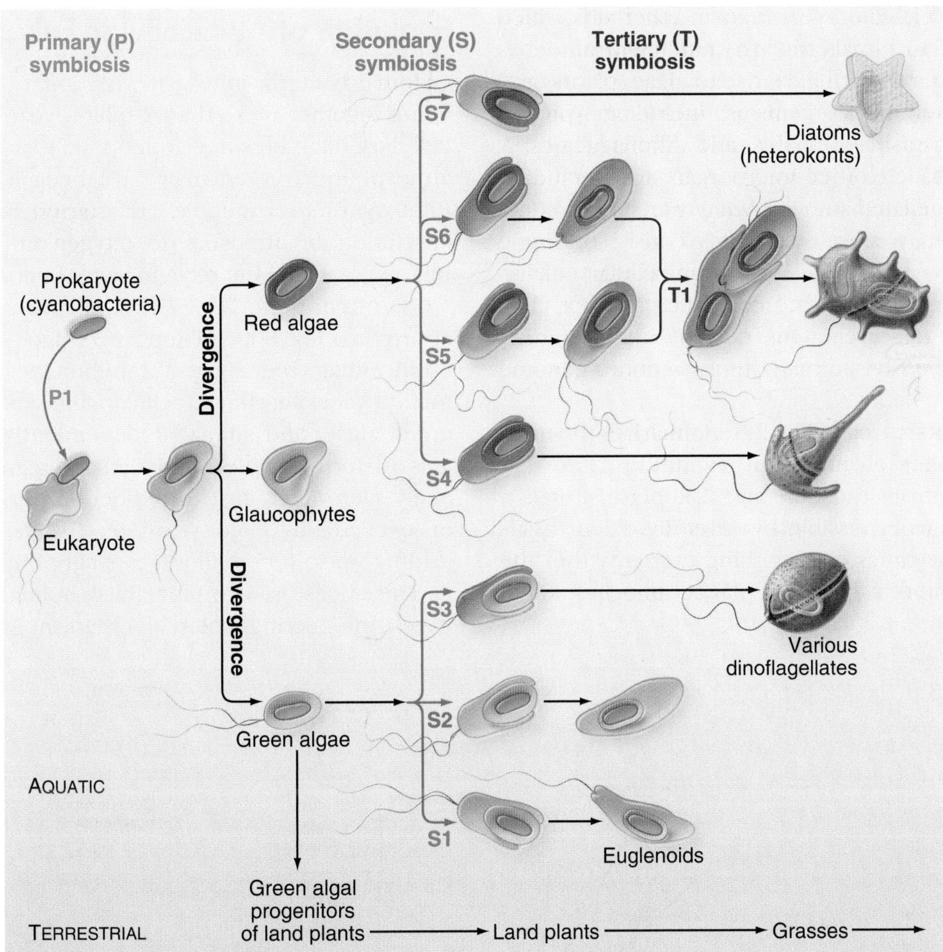

FIGURE 2.7 ACQUISITIONS OF CHLOROPLASTS. This is a timeline from *left* to *right*. The primary event was the ingestion of a cyanobacterium by the eukaryotic cell that gave rise to red algae, glaucophytes, and green algae. Green algae gave rise through divergence to land plants. Diatoms, dinoflagellates, and euglenoids acquired chloroplasts by secondary (S1 through S7) or tertiary (T1) symbiotic events when their precursors ingested an alga with chloroplasts. (Modified from Falkowski PG, Katz ME, Knoll AH, et al. Evolution of modern eukaryotic phytoplankton. *Science.* 2004;305:354–360.)

eukaryote then diverged into four lineages: green algae (such as the experimentally useful model organism *Chlamydomonas* [see Fig. 38.16]), red algae (such as sea weeds and coral symbionts), brown algae (such as kelp), and a minor group of photosynthetic unicellular organisms called glaucophytes (Fig. 2.7). Green algae gave rise through divergence to more than 300,000 species of land plants. We understand those phylogenetic relationships much better than the branching of more than 50,000 species of red, brown, and green algae.

Events following the initial acquisition of chloroplasts were more complicated, since in at least seven instances, other eukaryotes acquired photosynthesis by engulfing an entire green or red alga, followed by massive loss of algal genes. These secondary symbiotic events left behind chloroplasts along with the nuclear genes required for chloroplasts. For example, precursors of *Euglena* (on a different branch than algae and plants) took up a whole green alga, as did one family of dinoflagellates (on another branch). Red algae participated in four secondary and one tertiary symbiotic events, giving rise to photosynthetic diatoms and dinoflagellates. Today, photosynthesis by these marine microbes converts CO_2 into much of the oxygen and organic matter on the earth.

These secondary symbiotic events make phylogenetic relationships of nuclear genes and chloroplast genes discordant in these organisms. The original phylogeny based on rRNA sequences (Fig. 2.4A) assumed incorrectly that these diverse organisms acquired chloroplasts by primary symbiosis by a cyanobacterium. The phylogenetic relationships of dinoflagellates are particularly complex, given that they acquired chloroplasts from three separate sources.

Divergence Eukaryotes From Last Eukaryotic Common Ancestor

Molecular phylogenies indicate that multiple eukaryotic lineages diverged from LECA at an uncertain date between 2.1 and 1 billion years ago and then diversified rapidly (Fig. 2.4B). Animals are on a branch with amoebas and fungi. Another branch gave rise to algae, plants, and a vast number of microorganisms, including Apicomplexa (malaria parasites), ciliates, and dinoflagellates. A third branch yielded other microorganisms, including *Euglena*, the flagellated amoeba *Naegleria*, and trypanosomes. To this day most eukaryotes consist of single cells. Placing some groups of these unicellular eukaryotes on the phylogenetic tree continues to be a challenging field of research. Our current understanding (Fig. 2.4B) will be revised many times as more genome sequences are available.

The phylogenetic tree in Fig. 2.8 summarizes the most recent billion years of eukaryotic evolution. Note that this tree differs from those in Fig. 2.4, because it is a radial timeline made possible by carefully dated fossils that establish the times of branching events within the four major eukaryotic lineages: plants, amoebas, fungi and animals.

Evolution of Multicellular Eukaryotes

Colonial bacteria initiated evolutionary experiments in living together more than 2 billion years ago, but multicellular eukaryotes developed much later. Low levels of atmospheric oxygen may have been a limiting factor. Photosynthetic cyanobacteria started to raise the concentration of atmospheric oxygen approximately 2.2 billion years ago, but oxygen levels fluctuated widely and were often quite low (<1% of present levels) until approximately 800 million years ago. Fossils preserve multicellular red algae 1.2 billion years old. By 750 million years ago fungi, cellular slime molds, brown and green algae, and animals independently evolved strategies to form simple multicellular organisms (Fig. 2.8). Many were likely lost to extinction during two periods of exceedingly cold weather (the "snowball earth") leading up to 635 million years ago, but some survived.

The ancestor of multicellular animals (**metazoans**) was a pioneering colonial organism having much in

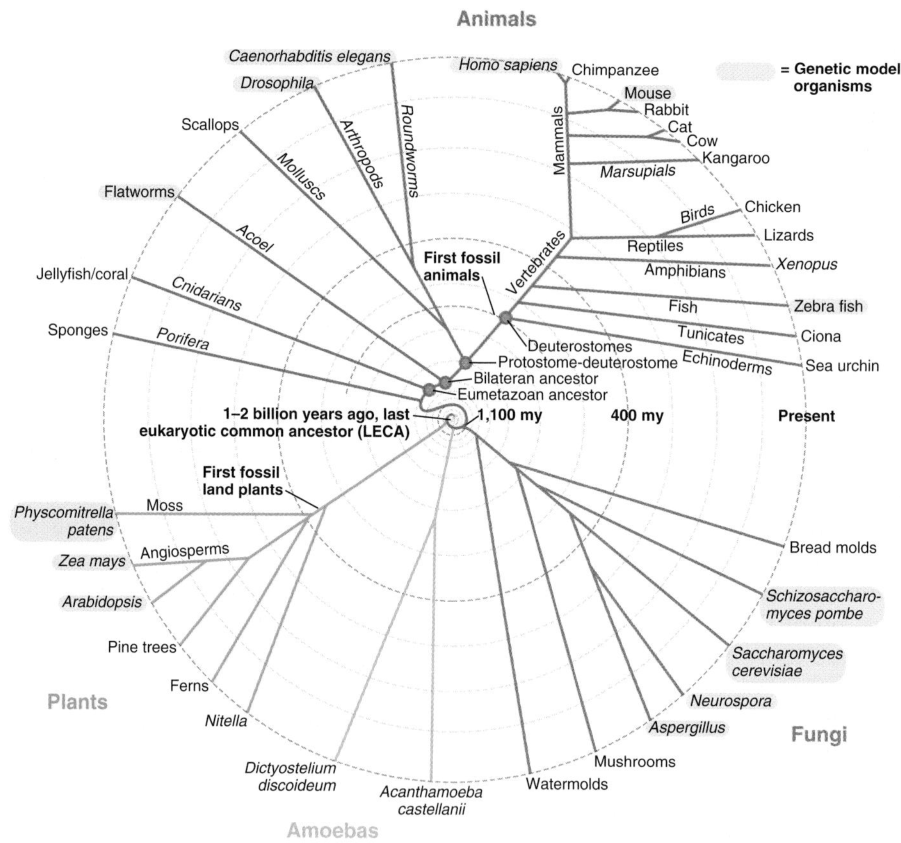

FIGURE 2.8 TIMELINE FOR THE DIVERGENCE OF ANIMALS, PLANTS, AND FUNGI. This tree has a radial timescale originating about 1100 million years (my) ago with the last common ancestor of plants, animals, and fungi. Contemporary organisms and time are at the circumference. Lengths of branches are arbitrary. The order of branching is established by comparisons of gene sequences. The times of the earliest branching events are only estimates because calibration of the molecular clocks is uncertain and the early fossil records are sparse. (Modified from Kuman S, Hedges SB. A molecular timescale for vertebrate evolution. *Nature.* 1998;392:917–920 [for animals]; Green Plant Phylogeny Research Coordination Group at http://ucjeps.berkeley.edu/bryolab/GPphylo [for plants]; and Tree of Life Web Project at http://tolweb.org/tree [for fungi].)

common with contemporary ciliated protozoa called choanoflagellates. Cells of sponges on the earliest surviving branch of animals *(Porifera)* still retain the morphology of choanoflagellates. Earlier stages in the evolution of metazoans are still missing from the fossil record.

Approximately 700 million years ago an organism called the eumetazoan appeared and gave rise to major branches of animals: Cnidarians (jellyfish, sea anemones, Hydra and corals); and all bilaterally symmetrical animals. The genome sequence of a sea anemone showed that the eumetazoan ancestor had a large fraction of the core human genes. In fact many features of the sea anemone genome (including the placement of introns) are more similar to vertebrates than insects. These genes and the properties of contemporary Cnidarians show that the ancient eumetazoan had advanced features including specialized epithelial, nerve, and muscle cells in two layers.

Genome sequences and well-preserved fossils show that the eumetazoan gave rise 600 million years ago to animals that have bilateral symmetry at some time in their lives, three tissue layers (ectoderm, mesoderm, and endoderm), and complex organs. These tiny (180 μm long) animals had a mouth, a gut, a coelomic cavity, and surface specializations that are speculated to be sensory structures. Other 570-million-year-old fossils are similar to contemporary animal embryos. Formation of tissues required plasma membrane proteins for adhesion to the extracellular matrix and to other cells (see Chapter 30). Genes for adhesion proteins—including proteins related to cadherins, integrins, lectins, and immunoglobulin-cellular adhesion molecules (Ig-CAMs)—are found in species that branched before metazoans, so their origins are ancient.

The ancient bilaterans then branched in succession into lineages containing flatworms (including planaria), protostomes (arthropods and nematodes; mollusks, annelid worms, brachiopods, and platyhelminths), and deuterostomes (echinoderms and Chordata, including humans). The common ancestor of chordates had nearly 80% of the classes of human genes but with some important exceptions, such as those for the adaptive immune response (see Chapter 28).

Approximately 540 million years ago, conditions allowed the rapid emergence of macroscopic multicellular animals with skeletons in less than 20 million years. At the time of this "Cambrian explosion," metazoans became abundant in numbers and varieties in the fossil record. Geological factors including large increases in the sea level help to trigger the biological events. It will be interesting to learn how mutations in genes controlling the body plan drove this dramatic appearance of large, complicated animals over a relatively short period.

Looking Back in Time

Viewing contemporary eukaryotic cells, one should be awed by the knowledge that they are mosaics created by historical events that occurred over a vast range of time. Roughly 3.5 billion years ago, the common ancestors of living things already stored genetic information in DNA; transcribed genes into RNA; translated mRNA into protein on ribosomes; carried out basic intermediary metabolism; and were protected by plasma membranes with carriers, pumps, and channels. More than 2.5 billion years ago, bacteria evolved the genes required for oxygenic photosynthesis and donated this capacity to eukaryotes via endosymbiosis more than 1 billion years ago. An α-proteobacterium took up residence in an early eukaryote, giving rise to mitochondria approximately 2 billion years ago. Although some prokaryotes have genes for homologs of all three cytoskeletal proteins, eukaryotes developed the capacity for cellular motility approximately 1.7 billion years ago when they shed their cell walls and evolved genes for molecular motors and many proteins that regulate the cytoskeleton. Multicellular eukaryotes with specialized cells and tissues arose only in the past 1.2 billion years after acquiring plasma membrane receptors used for cellular interactions.

It is also instructive to consider how more complex functions, such as the operation of the human nervous system, have their roots deep in time, beginning with the advent of molecules such as receptors and voltage-sensitive ion channels that originally served their unicellular inventors. At each step along the way, evolution has exploited the available materials for new functions to benefit the multitude of living organisms.

ACKNOWLEDGMENTS

We thank Mike Donoghue, Jim Lake, Leslie Orgel, Daniel Pollard, Katherine Pollard, Mitch Sogin, and Steve Stearns for their suggestions on the second edition, and especially to Martin Embley, Emmanuelle Javaux, and Tom Williams for advice on this third edition.

SELECTED READINGS

Adl SM, Simpson AG, Lane CE, et al. The revised classification of eukaryotes. *J Eukaryot Microbiol*. 2012;59:429-493.

Butterfield NJ. Early evolution of the eukaryota. *Palaeontology*. 2015;58:5-17.

Chen J-Y, Bottjer DJ, Davidson EH, et al. Small bilaterian fossils from 40 to 55 million years before the Cambrian. *Science*. 2004;305: 218-222.

Dawkins R. *The Ancestor's Tale*. New York: Houghton Mifflin; 2004:673.

Deep Green Tree of Life Web Project. Available at <http://tolweb.org/tree/phylogeny.html>.

Embley T, Martin W. Eukaryotic evolution, changes and challenges. *Nature*. 2006;440:623-630.

Eme L, Sharpe SC, Brown MW, Roger AJ. On the age of eukaryotes: evaluating evidence from fossils and molecular clocks. *Cold Spring Harb Perspect Biol*. 2014;6:a016139.

Falkowski PG, Katz ME, Knoll AH, et al. Evolution of modern eukaryotic phytoplankton. *Science*. 2004;305:354-360.

Gerlt JA, Babbitt PC. Divergent evolution of enzymatic function: Mechanistically diverse superfamilies and functionally distinct suprafamilies. *Annu Rev Biochem*. 2001;70:209-246.

Harwood A, Coates JC. A prehistory of cell adhesion. *Curr Opin Cell Biol*. 2004;16:470-476.

Javaux EJ. Early eukaryotes in Precambrian oceans. In: Gargaud MP, Lopez-Garcia P, Martin H, eds. *Origins and Evolution of Life: An Astrobiology Perspective*. Cambridge, UK: Cambridge University Press; 2011:414-449.

Javaux EJ, Marshall CP, Bekker A. Organic-walled microfossils in 3.2-billion-year-old shallow-marine siliciclastic deposits. *Nature*. 2010; 463:934-938.

Joyce GF. Forty years of in vitro evolution. *Angew Chem Int Ed Engl*. 2007;46:6420-6436.

Keeling PJ. The number, speed, and impact of plastid endosymbiosis in eukaryotic evolution. *Annu Rev Plant Biol*. 2013;64:583-607.

Knoll AH. *Life on a Young Planet: The First Three Billion Years of Life on Earth*. Princeton, NJ: Princeton University Press; 2003:277.

Knoll AH. Paleobiological perspectives on early eukaryotic evolution. *Cold Spring Harb Perspect Biol*. 2014;6:a016121.

Koonin EV, Yutin N. The dispersed archaeal eukaryome and the complex archaeal ancestor of eukaryotes. *Cold Spring Harb Perspect Biol*. 2014;6:a016188.

Lyons TW, Reinhard CT, Planavsky NJ. The rise of oxygen in Earth's early ocean and atmosphere. *Nature*. 2014;506:307-314.

Mora C, Tittensor DP, Adl S, et al. How many species are there on earth and in the ocean? *PLoS Biol*. 2011;9(8):e1001127.

Orgel LE. Prebiotic chemistry and the origin of the RNA world. *Crit Rev Biochem Mol Biol*. 2004;39:99-123.

Poole AM, Gribaldo S. Eukaryotic origins: How and when was the mitochondrion acquired? *Cold Spring Harb Perspect Biol*. 2014;6: a015990.

Powner MW, Gerland B, Sutherland JD. Synthesis of activated pyrimidine ribonucleotides in prebiotically plausible conditions. *Nature*. 2009;459:239-242.

Putnam NH, Srivastava M, Hellsten U, et al. Sea anemone genome reveals ancestral eumetazoan gene repertoire and genomic organization. *Science*. 2007;317:86-94.

Rivera MC, Lake JA. The ring of life provides evidence for a genome fusion origin of eukaryotes. *Nature*. 2004;431:152-155.

Schlacht A, Herman EK, Klute MJ, Field MC, Dacks JB. Missing pieces of an ancient puzzle: evolution of the eukaryotic membrane-trafficking system. *Cold Spring Harb Perspect Biol*. 2014;6:a016048.

Schrum JP, Zhu TF, Szostak JW. The origins of cellular life. *Cold Spring Harb Perspect Biol*. 2010;2(9):a002212.

Spang A, Saw JH, Jørgensen SL, et al. Complex archaea that bridge the gap between prokaryotes and eukaryotes. *Nature*. 2015;521:173-179.

True JR, Carroll SB. Gene co-option in physiological and morphological evolution. *Annu Rev Cell Dev Biol*. 2002;18:53-80.

Vogel C, Bashton M, Kerrison ND, et al. Structure, function and evolution of multidomain proteins. *Curr Opin Struct Biol*. 2004;14: 208-216.

Williams TA, Foster PG, Cox CJ, Embley M. An archaeal origin of eukaryotes supports only two primary domains of life. *Nature*. 2013;504:231-236.

Woese CR. A new biology for a new century. *Microbiol Mol Biol Rev*. 2004;68:173-186.

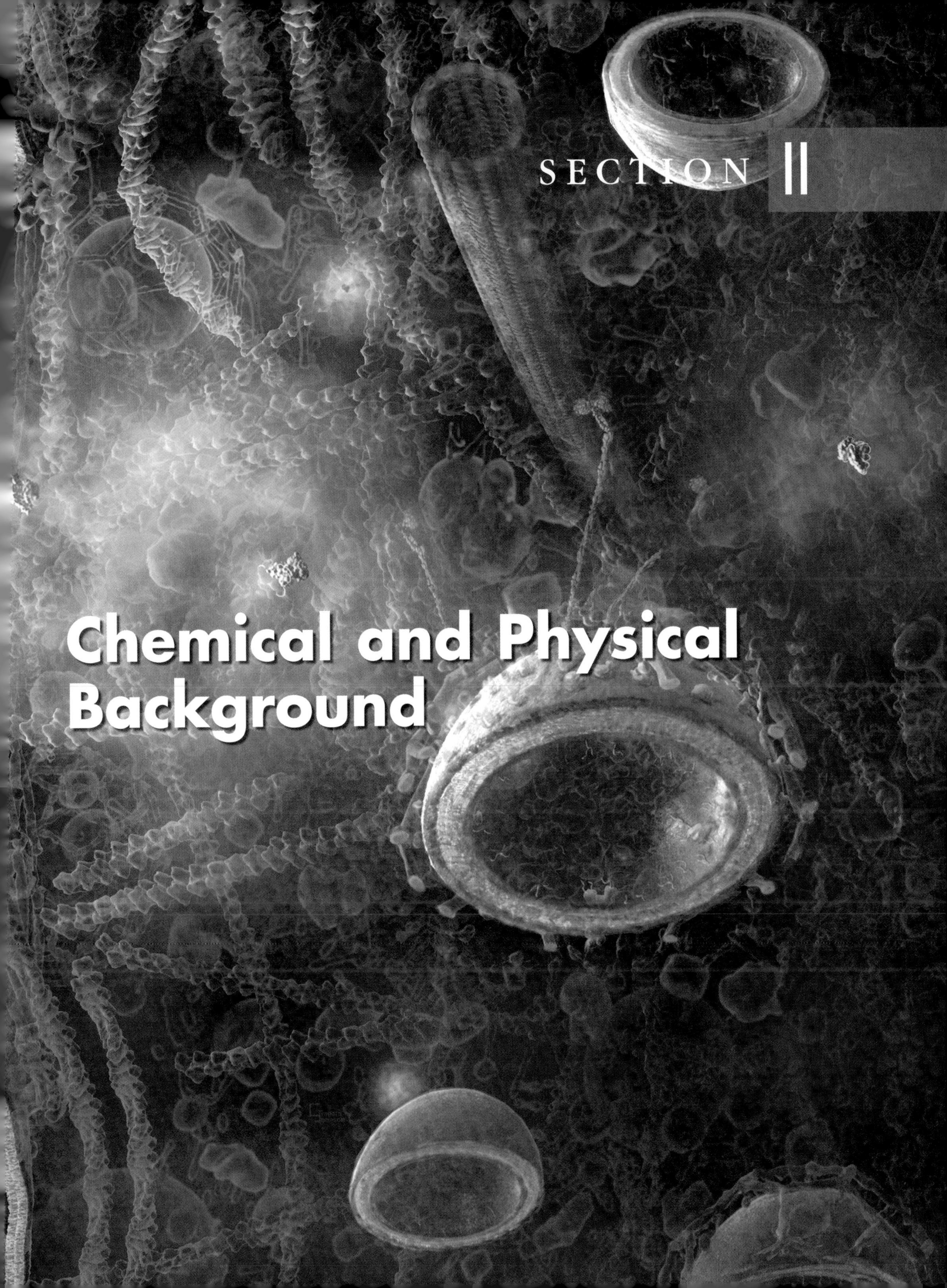

SECTION II

Chemical and Physical Background

SECTION II OVERVIEW

A primary objective of this book is to explain the molecular basis of life at the cellular level. This requires an appreciation of the structures of molecules as well as the basic principles of chemistry and physics that account for molecular interactions. The featured molecules are mostly **proteins,** but **nucleic acids, complex carbohydrates,** and **lipids** are all essential for life.

Chapter 3 explains the design principles of the major biological macromolecules in enough detail that a reader will appreciate the functions of the hundreds of proteins and nucleic acids that are considered in later chapters. Important concepts include the chemical nature of the building blocks of proteins **(amino acids),** nucleic acids **(nucleotides),** and sugar polymers **(monosaccharides);** the chemical bonds that link these units together; and the forces that drive the folding of polypeptides and nucleic acids into three-dimensional structures. Chapter 13 in Section V of the book introduces lipids in the context of the structure and function of biological membranes.

No biological macromolecule operates in isolation in cells, so Chapter 4 explains the physics and chemistry of their interactions. Many readers will never take a physical chemistry course, but they will discover in this chapter that a relatively few general principles can explain the **kinetics** and **thermodynamics** of most molecular interactions that are relevant to cells. For example, just two numbers and the concentrations of the reactants explain the forward and reverse rates of chemical reactions. Just one simple equation relates these two kinetic parameters to the key thermodynamic parameter, the **equilibrium constant**—the tendency of the reaction to go forward or backward. A second simple equation relates the equilibrium constant to the energy of the reactants and products. A third simple equation relates the change in free energy during a reaction to only two underlying parameters, the changes in heat and order in the system. These three equations explain all the chemical reactions that make life possible. The authors hope that Chapter 4 inspires a few readers to try a "P-chem" course to learn more.

Many cellular processes depend on macromolecular catalysts, protein enzymes, or RNA ribozymes. Chapter 4 explains how biochemists analyze enzyme mechanisms,

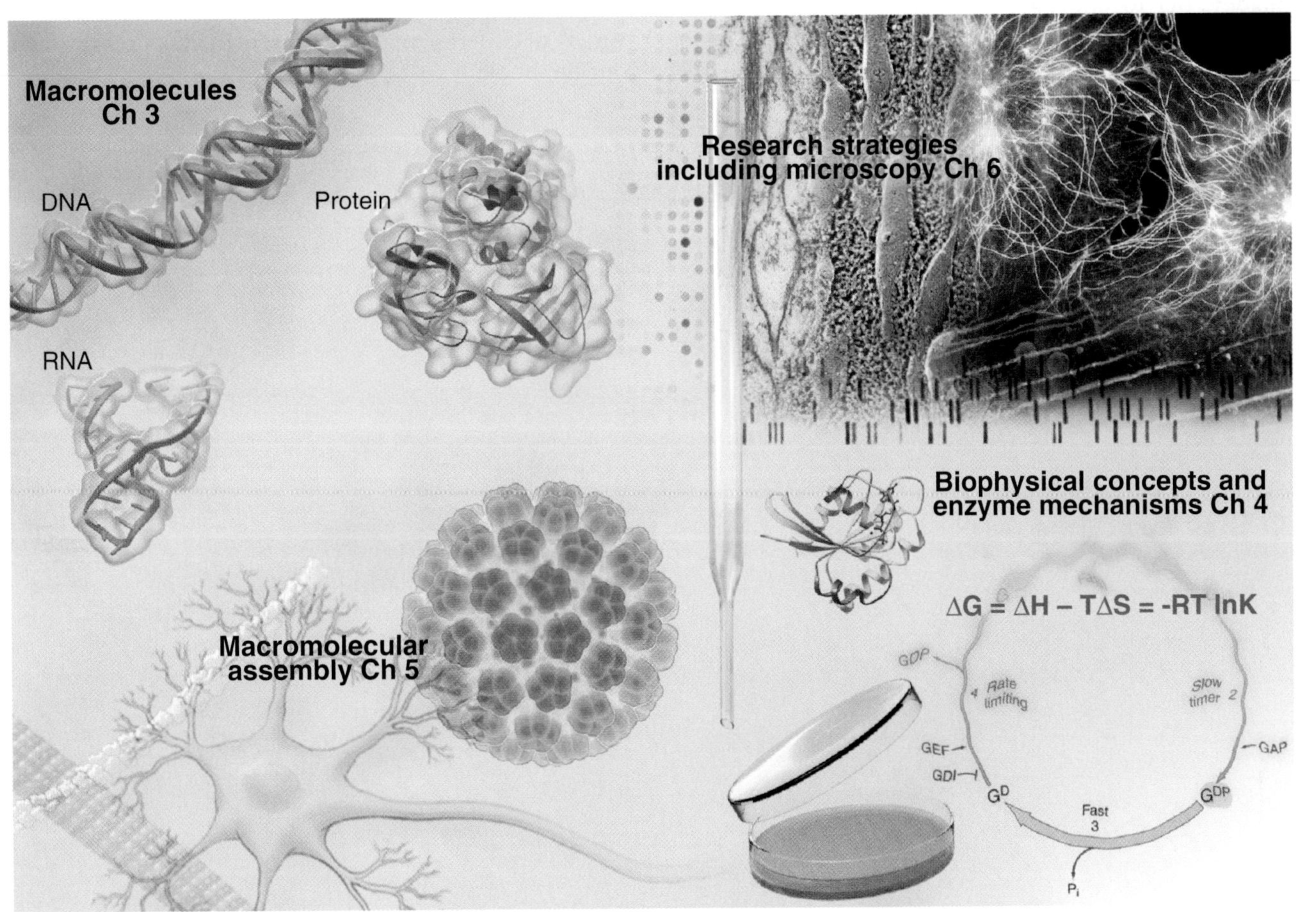

using as the example a protein that binds and hydrolyzes a nucleotide, guanosine triphosphate (GTP). Cells use related guanosine triphosphatases (GTPases) as molecular switches for many processes, including transport of macromolecules into and out of the nucleus (Chapter 9), protein synthesis (Chapter 12), membrane traffic (Chapters 20 to 22), signal transduction (Chapters 25 and 27), regulation of the cytoskeleton (Chapters 33 and 38), and mitosis (Chapter 44).

Macromolecules are polymers that are held together by strong covalent bonds between the building blocks. Templates guide the synthesis of proteins (Chapter 12) and nucleic acids (Chapters 10 and 42), but most macromolecular structures in cells assemble spontaneously from their components without a template. Weak, noncovalent bonds between complementary surfaces hold these macromolecular assemblies together. Chapter 5 explains how simple bimolecular reactions and conformational changes guide the assembly pathways for complexes of multiple proteins and complexes of proteins with nucleic acids. Cells often use adenosine triphosphate (ATP) hydrolysis or changes in protein conformation to control the reversible reactions required to assemble cytoskeletal polymers, signaling machines, coats around membrane vesicles, and chromosomes, among many other examples.

This book is not a manual for experimental cell biology, but to understand the experiments on which modern cell biological understanding is based; readers will want to appreciate the general strategies and the principles behind a few common methods. Chapter 6 explains that the dominant approach in cell biology is a **reductionist** one. Many classical questions in cell biology were defined by the behavior of cells described by early pioneers in the 19th and early 20th centuries. Subsequent microscopic analysis, genetic analysis in "model organisms," and studies of human diseases have further refined these questions in a modern context. Once a cellular process of interest has been identified, biologists use genetics or biochemistry to identify the molecules that are involved. Next, chemical and physical methods are applied to learn enough about each molecule to formulate a hypothesis about mechanisms. In the best-understood situations, these hypotheses are formalized as mathematical models for rigorous comparison with biological observations.

Microscopes are the most frequently used tool in cell biology, so Chapter 6 explains how light and electron microscopes both magnify and produce contrast—the two factors that are required to image cells and molecules. Equally important are the methods that are used to prepare biological specimens for microscopy and to showcase particular molecules for microscopic observation. In particular, fusion of proteins to jellyfish fluorescent proteins has revolutionized the study of protein behavior in living cells. Chapter 6 also explains a number of the basic genetic experiments and methods to manipulate nucleic acids in "molecular cloning" experiments. This background should help readers to understand the variety of experimental data presented in figures throughout the book.

CHAPTER 3

Molecules: Structures and Dynamics

This chapter describes the properties of water, proteins, nucleic acids, and carbohydrates as they pertain to cell biology. Chapter 13 covers lipids in the context of biological membranes.

Water

Water is so familiar that its role in cell biology and its fascinating properties tend to be neglected. Water is the most abundant and important molecule in cells and tissues. Humans are approximately two-thirds water. Water is not only the solvent for most cellular compounds but also a reactant or product in thousands of biochemical reactions catalyzed by enzymes, including the synthesis and degradation of proteins and nucleic acids and the synthesis and hydrolysis of adenosine triphosphate (ATP), to name a few examples. Water is also an important determinant of biological structure, as lipid bilayers, folded proteins, and macromolecular assemblies are all stabilized by the hydrophobic effect derived from the exclusion of water from nonpolar surfaces (see Fig. 4.5). Additionally, water forms hydrogen bonds with polar groups of many cellular constituents, ranging in size from small metabolites to large proteins. It also associates with small inorganic ions.

Physical chemists are still investigating water, one of the most complex liquids. The molecule is roughly tetrahedral in shape (Fig. 3.1A), with two hydrogen bond donors and two hydrogen bond acceptors. The electronegative oxygen withdraws the electrons from the O–H covalent bonds, leaving a partial positive charge on the hydrogens and a partial negative charge on the oxygen. Hydrogen bonds between water molecules are partly electrostatic because of the charge separation (induced dipole) but also have some covalent character, owing to overlap of the electron orbitals. The strength of hydrogen bonds depends on their orientation, being strongest along the lines of tetrahedral orbitals. One can think of the oxygens of two water molecules sharing a hydrogen-bonded hydrogen. Given two hydrogen bond donors and acceptors, water can be fully hydrogen-bonded, as it is in ice (Fig. 3.1C). Crystalline water in ice has a well-defined structure with a complete set of tetragonal hydrogen bonds and a remarkable amount (35%) of unoccupied space (Fig. 3.1D).

Liquid water is very heterogeneous and dynamic, with regions of local order and disorder fluctuating on a picosecond time scale but no well-defined, long-range structure. When ice melts, the volume decreases by only about 10%, so liquid water has considerable empty space too. The heat required to melt ice is a small fraction (15%) of the heat required to convert ice to a gas, in which all the hydrogen bonds are lost. Because the heat of melting reflects the number of bonds broken, liquid water must retain most of the hydrogen bonds that stabilize ice. These hydrogen bonds create a continuous but dynamic, three-dimensional network of water molecules connected at their tetrahedral vertices, allowing water to remain a liquid at a higher temperature than is the case for a molecule of similar size, ammonia.

The properties of water have profound effects on all other molecules in the cell. For example, shells of water organized around ions compete effectively with other ions with which they might interact electrostatically (Fig. 3.1E). These shells of water travel with ions, governing the size of pores that they can penetrate. Similarly, hydrogen bonding with water strongly competes with the hydrogen bonding that occurs between solutes, including macromolecules. By contrast, water does not interact as favorably with nonpolar molecules as it does with itself, so the solubility of nonpolar molecules in water is low, and they tend to aggregate to reduce their

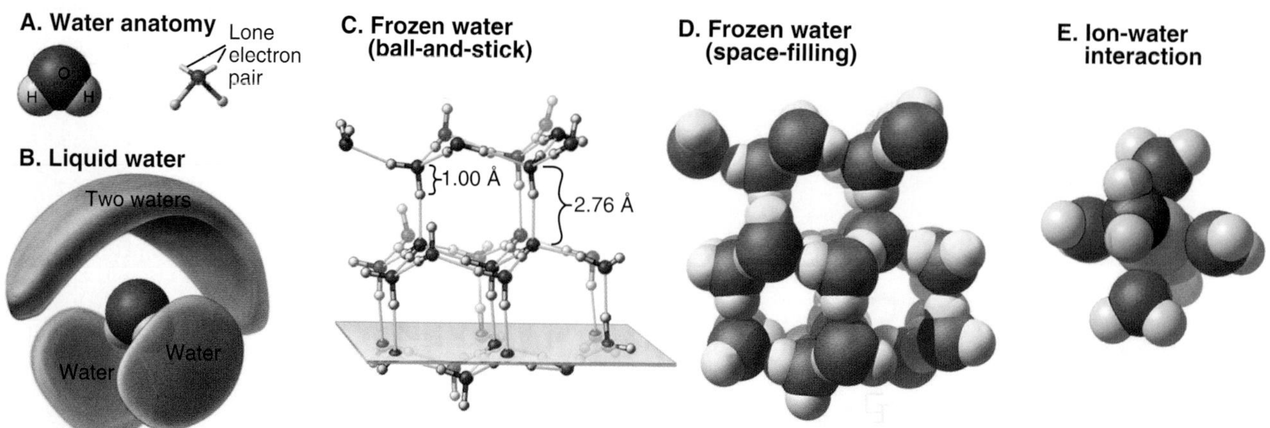

FIGURE 3.1 WATER. A, Space-filling model and orientation of the tetrahedral electron orbitals that define the directions of the hydrogen bonds. **B,** Tetrahedral local order in liquid water revealed by a theoretical calculation of a three-dimensional map of regions around the central water molecule where the local density of oxygen is at least 40% higher than average. Two adjacent water oxygens are centered near the two hydrogen bond donors, and two other waters are positioned in an elongated cap so that their protons can hydrogen-bond with the central water oxygen. **C,** Stick figure of crystallized ice showing the tetrahedral network of hydrogen bonds. **D,** A space-filling model of crystalline ice showing the large amount of unoccupied space. **E,** Shell of water molecules around a potassium ion. Small ions, such as Li^+, Na^+, and F^-, bind water more tightly than do larger ions, such as K^+, Cl^-, and I^-. (**D–E,** From www.nyu.edu/pages/mathmol/library/water, Project MathMol Scientific Visualization Lab, New York University. See "ice.pdb" and "waterbox.pdb.")

surface area in contact with water. Such nonpolar interactions are energetically favorable, because they reduce unfavorable interactions of nonpolar groups with water and increase favorable interactions of water molecules with each other. This is called the hydrophobic effect (see Fig. 4.5). These interactions of water dominate the behavior of solute molecules in an aqueous environment with a water concentration of 55.5 M, where they influence the assembly of proteins, lipids, and nucleic acids into the structures that they assume in the cell. In addition, strategically placed, hydrogen-bonded water molecules can bridge two macromolecules in functional assemblies.

Proteins

Proteins are major components of all cellular systems. This section presents some basic concepts about protein structure that help explain how proteins function in cells. More extensive coverage of this topic is available in biochemistry books and specialized books on protein chemistry.

Proteins consist of one or more linear polymers called **polypeptides** composed of various combinations of 20 different **amino acids** (Figs. 3.2 and 3.3) linked together by **peptide bonds** (Fig. 3.4). When linked in polypeptides, amino acids are referred to as "residues." The sequence of amino acid residues in each type of polypeptide is unique. It is specified by the gene encoding the protein and is read out precisely during protein synthesis (see Fig. 12.9). The polypeptides of proteins with more than one chain are usually synthesized separately. However, in some cases, a single chain is divided into pieces by cleavage after synthesis.

Polypeptides range widely in length. Small peptide hormones, such as oxytocin, consist of as few as nine residues, while the giant structural protein titin (see Fig. 39.7) has more than 25,000 residues. Most cellular proteins fall in the range of 100 to 1000 residues. Without stabilization by disulfide bonds or bound metal ions, approximately 40 residues are required for a polypeptide to adopt a stable three-dimensional structure in water.

The sequence of amino acids in a polypeptide can be determined chemically by removing one amino acid at a time from the amino terminus and identifying the product. This procedure, called **Edman degradation**, can be repeated approximately 50 times before declining yields limit progress. Longer polypeptides can be divided into fragments of fewer than 50 amino acids by chemical or enzymatic cleavage, after which they are purified and sequenced separately. Even easier, one can sequence the gene or a complementary DNA **(cDNA)** copy of the messenger RNA (mRNA) for the protein (Fig. 3.16) and use the genetic code to infer the amino acid sequence. This approach misses posttranslational modifications (Fig. 3.3). Analysis of protein fragments by mass spectrometry identifies posttranslational modifications and can be used to sequence tiny quantities of proteins.

Properties of Amino Acids

Every student of cell biology should know the chemical structures of the amino acids used in proteins (Fig. 3.2). Without these structures in mind, reading the literature and this book is like spelling without knowledge of the alphabet. In addition to their full names, amino acids are frequently designated by three-letter or single-letter abbreviations.

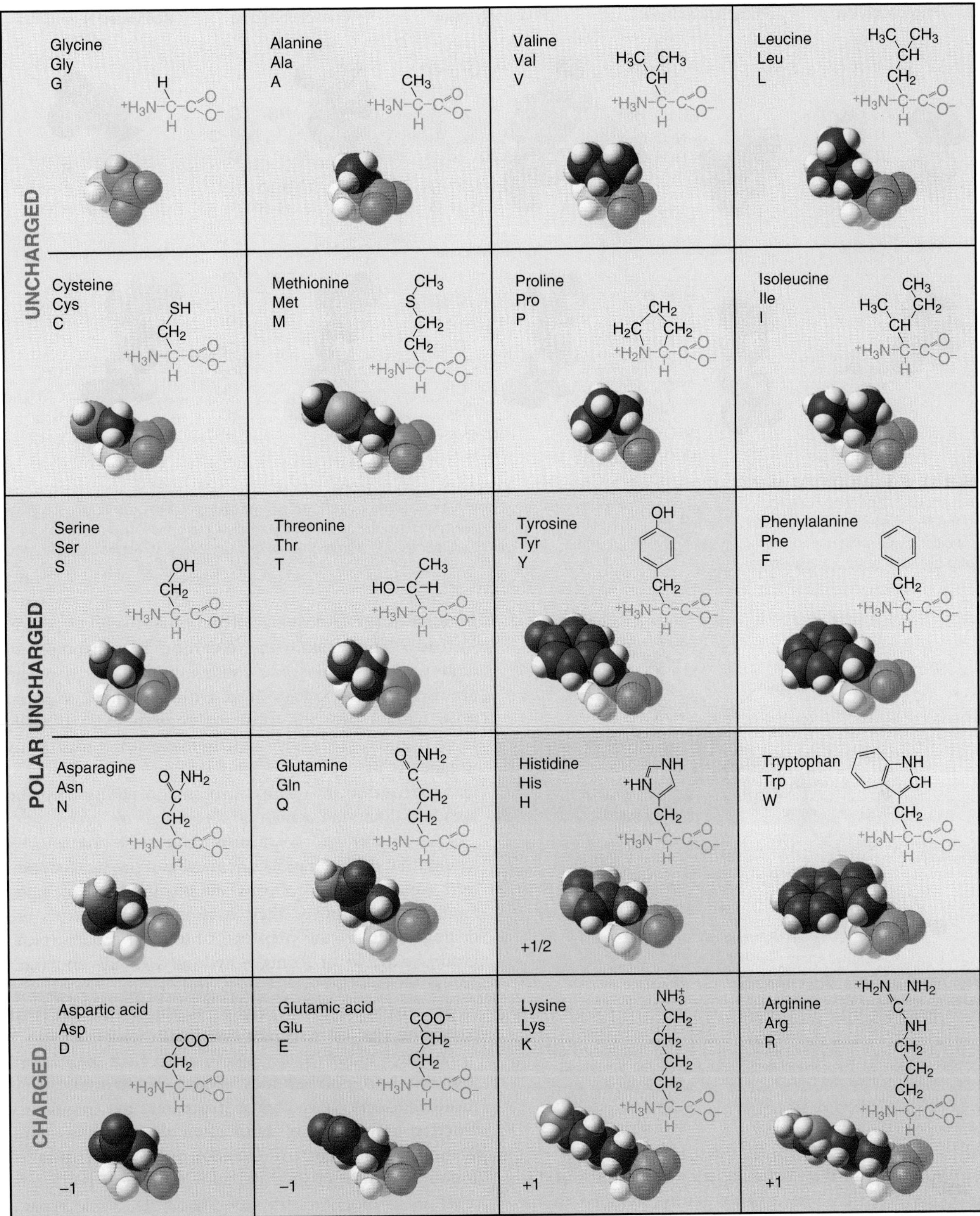

FIGURE 3.2 THE 20 L-AMINO ACIDS SPECIFIED BY THE GENETIC CODE. Shown for each are the full name, the three-letter abbreviation, the single-letter abbreviation, a stick figure of the atoms, and a space-filling model of the atoms in which hydrogen is *white*, carbon is *black*, oxygen is *red*, nitrogen is *blue*, and sulfur is *yellow*. For all, the amino group is protonated and carries a +1 charge, whereas the carboxyl group is ionized and carries a −1 charge. The amino acids are grouped according to the side chains attached to the α-carbon. These side chains fall into three subgroups. *Top,* The aliphatic (G, A, V, L, I, C, M, P) and aromatic (Y, F, W) side chains partition into nonpolar environments, as they interact poorly with water. *Middle,* The uncharged side chains with polar hydrogen bond donors or acceptors (S, T, N, Q, Y) can hydrogen-bond with water. *Bottom,* At neutral pH, the basic amino acids K and R are fully protonated and carry a charge of +1, the acidic amino acids (D, E) are fully ionized and carry a charge of −1, and histidine (pK: ~6.0) carries a partial positive charge. All the charged residues interact favorably with water, although the aliphatic chains of R and K also give them significant nonpolar character.

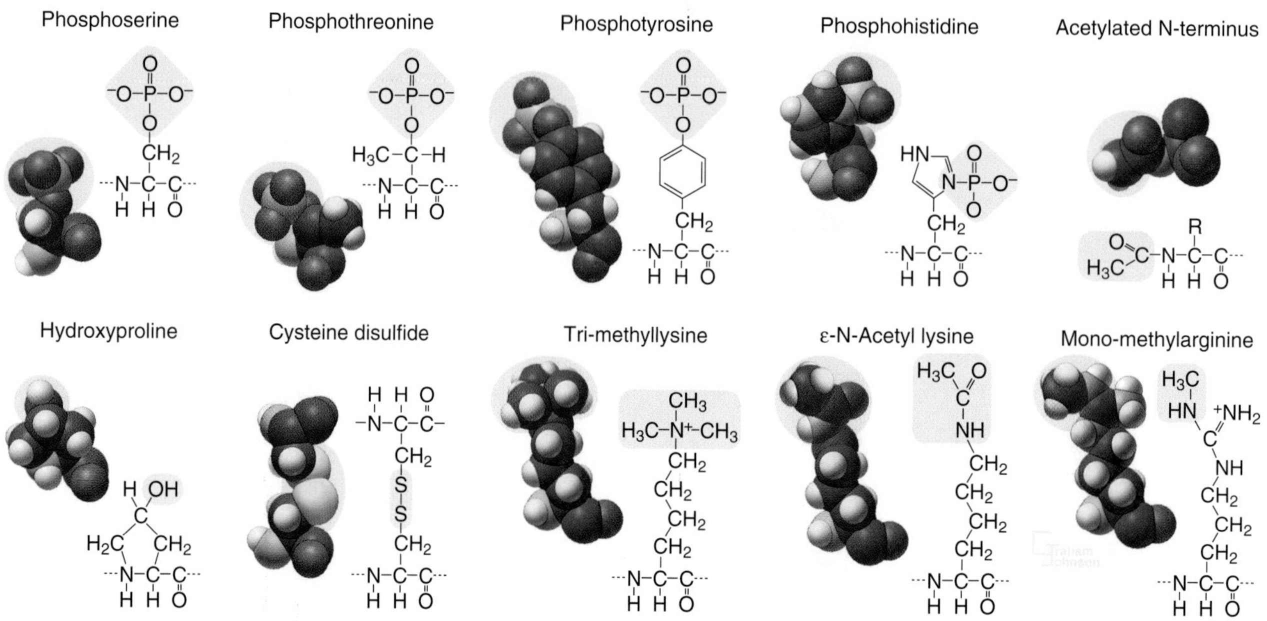

FIGURE 3.3 MODIFIED AMINO ACIDS. Protein kinases add a phosphate group to serine, threonine, tyrosine, histidine, and aspartic acid (not shown). Other enzymes add one or more methyl groups to lysine, arginine, or histidine (not shown); a hydroxyl group to proline; or an acetate to the N-terminus of many proteins. The reducing environment of the cytoplasm minimizes the formation of disulfide bonds, but under oxidizing conditions within the membrane compartments of the secretory pathway (see Chapter 21), intramolecular or intermolecular disulfide (S–S) bonds form between adjacent cysteine residues.

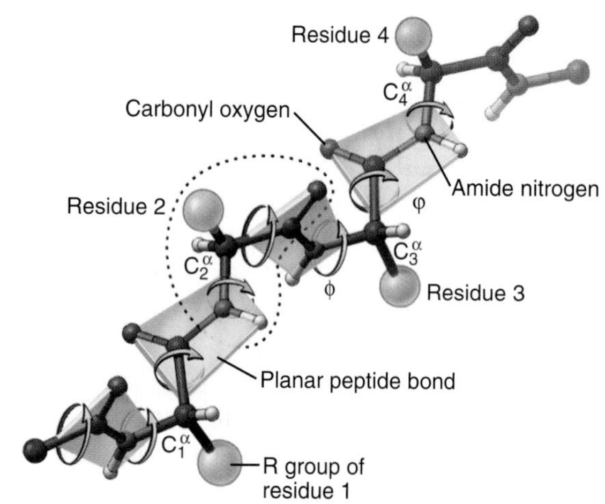

FIGURE 3.4 THE POLYPEPTIDE BACKBONE. This perspective drawing shows four planar peptide bonds, the four participating α-carbons (labeled 1 to 4), the R groups represented by the β-carbons, amide protons, carbonyl oxygens, and the two rotatable backbone bonds (Φ and φ). The *dotted lines* outline one amino acid. (Modified from Creighton TE. *Proteins: Structure and Molecular Principles.* New York: WH Freeman; 1983.)

All but one of the 20 amino acids commonly used in proteins consist of an **amino group,** bonded to the **α-carbon,** bonded to a **carboxyl group.** Proline is a variation on this theme with a cyclic side chain bonded back to the nitrogen to form an imino group. Both the amino group (pK >9) and carboxyl group (pK = ~4) are partially ionized under physiological conditions. Except for glycine, all amino acids have a β-carbon and a proton bonded to the α-carbon. (Glycine has a second proton instead.) This makes the α-carbon an asymmetrical center with two possible configurations. The L-isomers are used almost exclusively in living systems. Compared with natural proteins, proteins constructed artificially from D-amino acids have mirror-image structures. Many organisms incorporate selenocysteine (Se replacing S) and pyrrolizine into a few proteins in addition to the standard 20 amino acids.

Each amino acid has a distinctive **side chain,** or R group, that determines its chemical and physical properties. Amino acids are conveniently grouped in small families according to their R groups. Side chains are distinguished by the presence of ionized groups, polar groups capable of forming hydrogen bonds and their apolar surface areas. Glycine and proline are special cases, owing to their unique effects on the polymer backbone (see later section on protein folding).

Enzymes modify many amino acids after their incorporation into polypeptides. These **posttranslational modifications** have both structural and regulatory functions (Fig. 3.3). This book often refers to these modifications, especially to reversible phosphorylation of amino acid side chains, the most common regulatory reaction in biochemistry (see Fig. 25.1). Some asparagine, serine and threonine residues are conjugated with one or more sugars (Fig. 3.26). Methylated and acetylated lysines are important for chromatin regulation in the nucleus (see Fig. 8.3). In addition to the examples in Fig. 3.3, whole proteins, such as ubiquitin or SUMO (small ubiquitin-like modifier), can be attached through

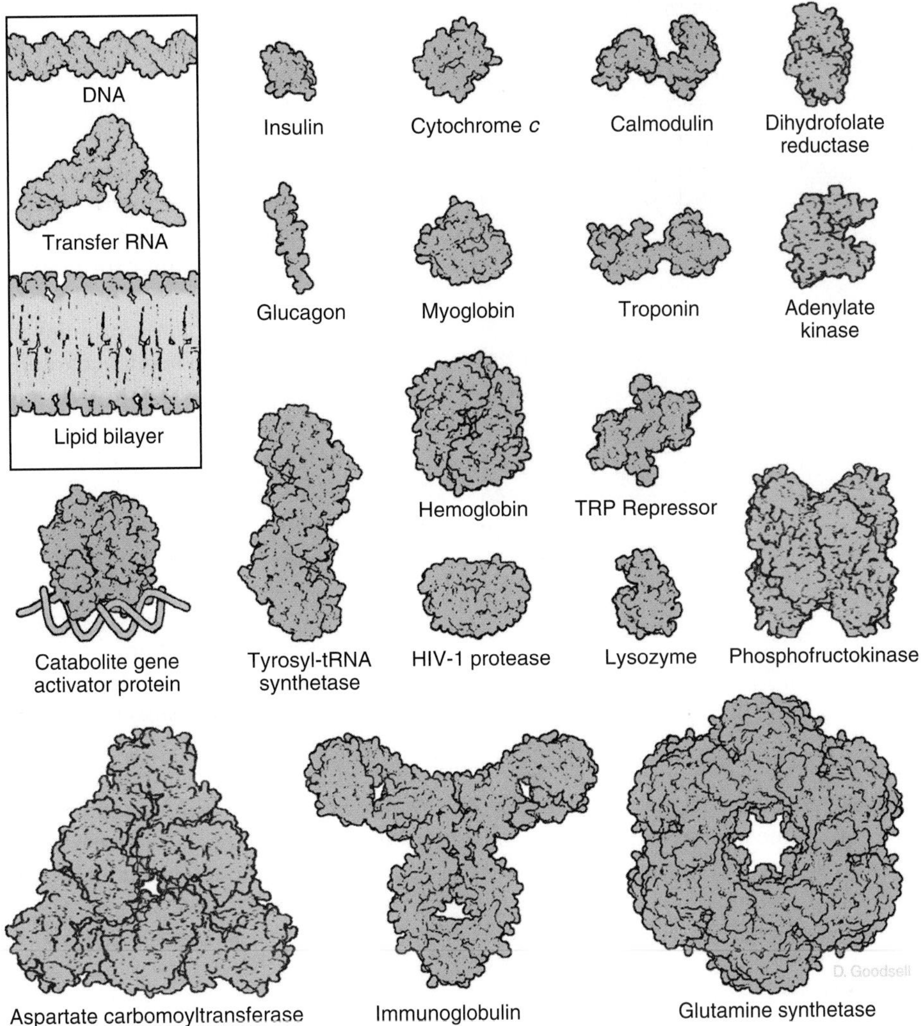

FIGURE 3.5 A GALLERY OF MOLECULES. Space-filling models of proteins compared with a lipid bilayer, transfer RNA (tRNA), and DNA, all on the same scale. TRP, tryptophan. (Modified from Goodsell D, Olsen AJ. Soluble proteins: Size, shape, and function. *Trends Biochem Sci.* 1993;18:65–68.)

isopeptide bonds to lysine ε-amino groups to act as signals for degradation (see Fig. 23.3) or endocytosis (see Fig. 23.5).

This repertoire of amino acids is sufficient to construct millions of different proteins, each with different capacities for interacting with other cellular constituents. This is possible because each protein has a unique three-dimensional structure (Fig. 3.5), each displaying the relatively modest variety of functional groups in a different way on its surface.

Architecture of Proteins

Our knowledge of protein structure is based on x-ray diffraction studies of protein crystals, electron microscopy of single molecules and nuclear magnetic resonance (NMR) spectroscopy studies of small proteins in solution. These methods show the arrangement of the atoms in space, allowing for computer simulations of the atomic motions (molecular dynamics simulations). X-ray diffraction requires three-dimensional crystals of the protein and yields a three-dimensional contour map showing the density of electrons in the molecule (Fig. 3.6). In favorable cases, all the atoms except hydrogens are clearly resolved, along with water molecules occupying fixed positions in and around the protein. NMR requires concentrated solutions of protein and reveals distances between particular protons. Given enough distance constraints, it is possible to calculate the unique protein fold that is consistent with these spacings. Electron microscopy of single molecules can now reveal structures at near atomic resolution (see Figs. 6.7, 14.4, 14.6, 12.7, 16.9, 34.4, 36.10, and 37.2).

Each amino acid residue contributes three atoms to the polypeptide backbone: the nitrogen from the amino group, the α-carbon, and the carbonyl carbon from the carboxyl group. The peptide bond linking the amino acids together is formed by dehydration synthesis (see Fig. 12.9), a common chemical reaction in biological

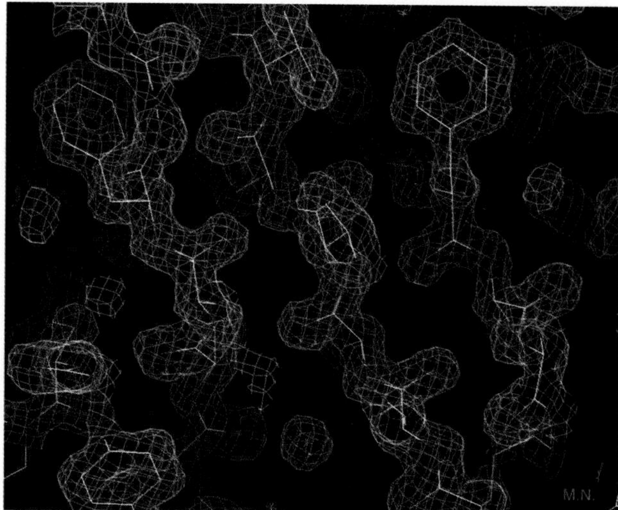

FIGURE 3.6 PROTEIN STRUCTURE DETERMINATION BY X-RAY CRYSTALLOGRAPHY. A small part of an electron density map at 1.5-Å resolution of the cytoplasmic T1 domain of the shaker potassium channel from *Aplysia.* The chicken-wire map shows the electron density. The stick figure shows the superimposed atomic model. (Data from M. Nanao and S. Choe, Salk Institute for Biological Studies, San Diego, CA.)

systems. Water is removed in the form of a hydroxyl from the carboxyl group of one amino acid and a proton from the amino group of the next amino acid in the polymer. Ribosomes (RNA enzymes) catalyze this reaction in cells. Chemical synthesis can achieve the same result in the laboratory. The peptide bond nitrogen has an **(amide) proton,** and the carbon has a double-bonded **(carbonyl) oxygen.** The amide proton is an excellent hydrogen bond donor, whereas the carbonyl oxygen is an excellent hydrogen bond acceptor.

The end of a polypeptide with the free amino group is called the **amino terminus** or **N-terminus.** The numbering of the residues in the polymer starts with the N-terminal amino acid, as the biosynthesis of the polymer begins there on ribosomes. The other end of a polypeptide has a free carboxyl group and is called the **carboxyl terminus** or **C-terminus.**

The peptide bond has some characteristics of a double bond, owing to resonance of the electrons, and is relatively rigid and planar. The bonds on either side of the α-carbon can rotate through 360 degrees, although a relatively narrow range of bond angles is highly favored. Steric hindrance between the β-carbon (on all the amino acids but glycine) and the α-carbon of the adjacent residue favors a *trans* configuration in which the side chains alternate from one side of the polymer to the other (Fig. 3.4). Folded proteins generally use a limited range of rotational angles to avoid steric collisions of atoms along the backbone. Glycine, which lacks a β-carbon, is free to assume a wider range of configurations and is useful for making tight turns in folded proteins.

Folding of Polypeptides

The amino acid sequence of each protein contains all the information required to specify folding into the native structure, just one of a vast number of possible conformations. Although many proteins are flexible enough to undergo conformational changes (Fig. 3.12), polypeptides rarely fold into more than one final stable structure. Exceptions with medical importance are influenza virus hemagglutinin protein and amyloid (see Chapter 12).

Unfolding and refolding proteins in a test tube established that amino acid sequences alone specify the three dimensional structures of proteins. Many, but not all, proteins that are unfolded by harsh treatments (high concentrations of urea or extremes of pH) refold to regain full activity when returned to physiological conditions. Chapter 12 explains how an unfolded polypeptide rapidly samples many conformations through trial and error to select stable intermediates leading to the native structure. Cells use molecular chaperones to guide and control the quality of folding.

The following factors influence protein folding:

1. Hydrophobic side chains pack very tightly in the core of proteins to minimize their exposure to water. Little free space exists inside proteins, so the hydrophobic core resembles a hydrocarbon crystal more than an oil droplet (Fig. 3.7). Accordingly, many of the most conserved residues in families of proteins are found in the interior. Nevertheless, the internal packing is malleable enough to tolerate mutations that change the size of buried side chains, as the neighboring chains can rearrange without changing the overall shape of the protein. Interior charged or polar residues frequently form hydrogen bonds or salt bridges to neutralize their charge.
2. Most charged and polar side chains are exposed on the surface, where they interact favorably with water. Although many hydrophobic residues are inside, roughly half the residues exposed to solvent on the outer surface are also hydrophobic. Amino acid residues on the surface typically appear to play a minor role in protein folding. Experimentally, one can substitute many residues on the surface of a protein with any other residue without changing the stability or three-dimensional structure.
3. The polar amide protons and carbonyl oxygens of the polypeptide backbone maximize their potential to form hydrogen bonds with other backbone atoms, side chain atoms, or water. In the hydrophobic core of proteins, this is achieved by hydrogen bonds with other backbone atoms in two major types of **secondary structures: α-helices** and **β-sheets** (Fig. 3.8).
4. Most elements of secondary structure extend completely across compact domains. Consequently, most loops connecting α-helices and β-strands are on the

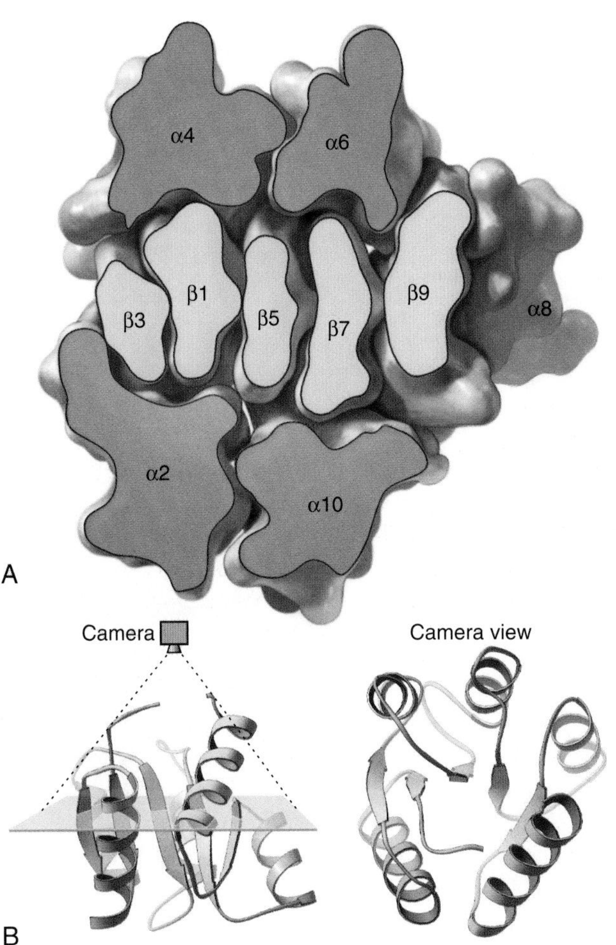

FIGURE 3.7 Space-filling **(A)** and ribbon **(B)** models of a cross-section of the bacterial chemotaxis protein CheY illustrate some of the factors that contribute to protein folding. α Helices pack on both sides of the central, parallel β-sheet. Most of the polar and charged residues are on the surface. The tightly packed interior of largely apolar residues excludes water. The buried backbone amides and carbonyls are fully hydrogen-bonded to other backbone atoms in both the α-helices and β-sheet. (For reference, see Protein Data Bank [PDB; www.rcsb.org] file 2CHF.)

surface of proteins, not in the interior (Fig. 3.9). Exceptions are found in some integral membrane proteins (see Figs. 16.3, 16.13, 16.14, and 16.15), where α-helices can reverse in the interior of the protein.

These factors tend to maximize the stability of folded proteins in one particular "native" conformation, but the native folded state of naturally evolved proteins is relatively unstable. The standard free energy difference (see Chapter 4) between a folded and globally unfolded protein is only about 40 kJ mol^{-1}, much less than that of a single covalent bond! Even the substitution of a single crucial amino acid can destabilize certain proteins, causing a loss of function. Some amino acid substitutions, however, increase the stability of natural proteins, so evolution seems to have selected for marginal stability. In other cases, misfolding results in noncovalent polymerization of a protein into amyloid fibrils associated with serious diseases (see Chapter 12).

Given that protein structures are encoded in their amino acid sequences, a long-range goal has been to predict three-dimensional structures of proteins from sequences alone. The rapid accumulation of genome sequences increased the value of this approach. Although once seen as intractable, advances in computational methods are making structure prediction a reality. Prediction is straightforward if the structure of an ortholog or paralog is available. One builds the amino acid sequence of the unknown protein into the known structure to make a **homology model** that is often accurate enough to make reliable inferences about function. Strategies to predict protein structures from sequence alone include comparisons with sequences of known structures, threading test sequences through structural elements of known proteins and computational searches for folds with the lowest free energy with or without guidance from databases of known protein structures. These methods accurately predict many protein folds, but generally lack fine details provided by x-ray crystallography. The Protein Structure Prediction Center runs competitions to predict structures and lists the most successful publically available methods at http://www.predictioncenter.org/index.cgi?page=links. These prediction methods are also useful for improving the quality of experimental structures when the resolution of the data are limited.

Secondary Structure

Much of the polypeptide backbone of proteins folds into stereotyped elements of secondary structure, especially α-helices and β-sheets (Fig. 3.8). They are shown as spirals and polarized ribbons in "ribbon diagrams" of protein organization used throughout this book. Both α-helices and β-strands are linear, so globular proteins can be thought of as compact bundles of straight or gently curving rods, laced together by surface turns.

α-Helices allow polypeptides to maximize hydrogen bonding of backbone polar groups while using highly favored rotational angles around the α-carbons and tight packing of atoms in the core of the helix (Fig. 3.8). All these features stabilize the α-helix. Viewed with the amino terminus at the bottom, the amide protons all point downward and the carbonyl oxygens all point upward. The side chains project radially around the helix, tilted toward its N-terminus. Given 3.6 residues in each turn of the right-handed helix, the carbonyl oxygen of residue 1 is positioned perfectly to form a linear hydrogen bond with the amide proton of residue 5. This n to $n + 4$ pattern of hydrogen bonds repeats along the whole α-helix.

The orientation of backbone hydrogen bonds in α-helices has two important consequences. First, a

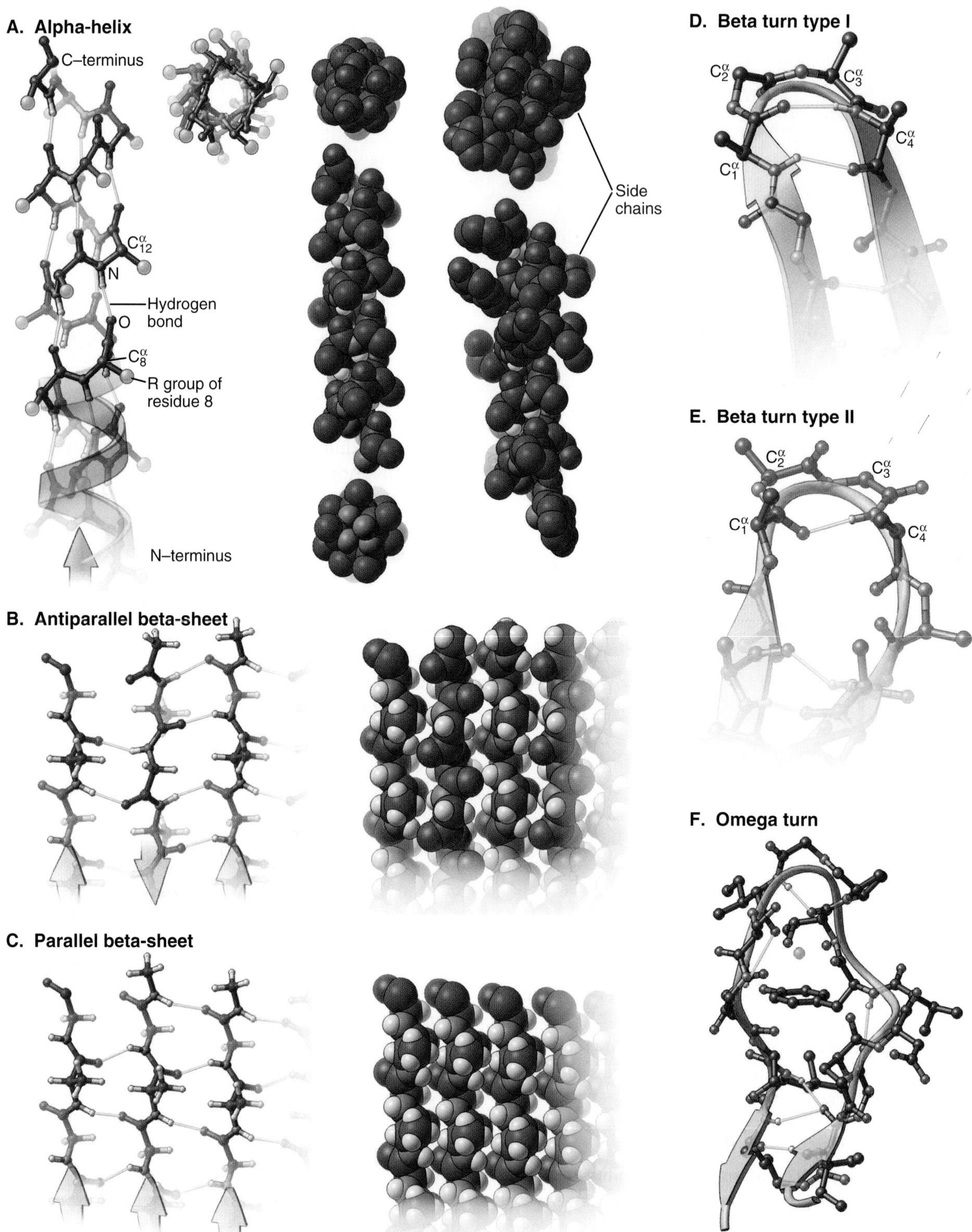

FIGURE 3.8 MODELS OF SECONDARY STRUCTURES AND TURNS OF PROTEINS. A, α-Helix. The stick figure *(left)* shows a right-handed α-helix with the N-terminus at the bottom and side chains R represented by the β-carbon. Hydrogen bonds between backbone atoms are indicated by *blue lines.* In this orientation, the carbonyl oxygens point upward, the amide protons point downward, and the R groups trail toward the N-terminus. Space-filling models *(middle)* show a polyalanine α-helix. The end-on views show how the backbone atoms fill the center of the helix. A space-filling model *(right)* of α-helix 5 from bacterial rhodopsin shows the side chains. Some key dimensions are 0.15 nm rise per residue, 0.55 nm per turn, and diameter of approximately 1.0 nm. (See PDB file 1BAD.) **B,** Stick figure and space-filling models of an antiparallel β-sheet. The *arrows* indicate the polarity of each chain. With the polypeptide extended in this way, the amide protons and carbonyl oxygens lie in the plane of the sheet, where they make hydrogen bonds *(blue lines)* with the neighboring strands. The amino acid side chains alternate pointing upward and downward from the plane of the sheet. Some key dimensions are 0.35 nm rise per residue in a β-strand and 0.45 nm separation between strands. (See PDB file 1SLK.) **C,** Stick figure and space-filling models of a parallel β-sheet. All strands have the same orientation *(arrows).* The orientations of the hydrogen bonds are somewhat less favorable than that in an antiparallel sheet. **D–E,** Stick figures of two types of reverse turns found between strands of antiparallel β-sheets. **F,** Stick figure of an omega loop. (See PDB file 1LNC.)

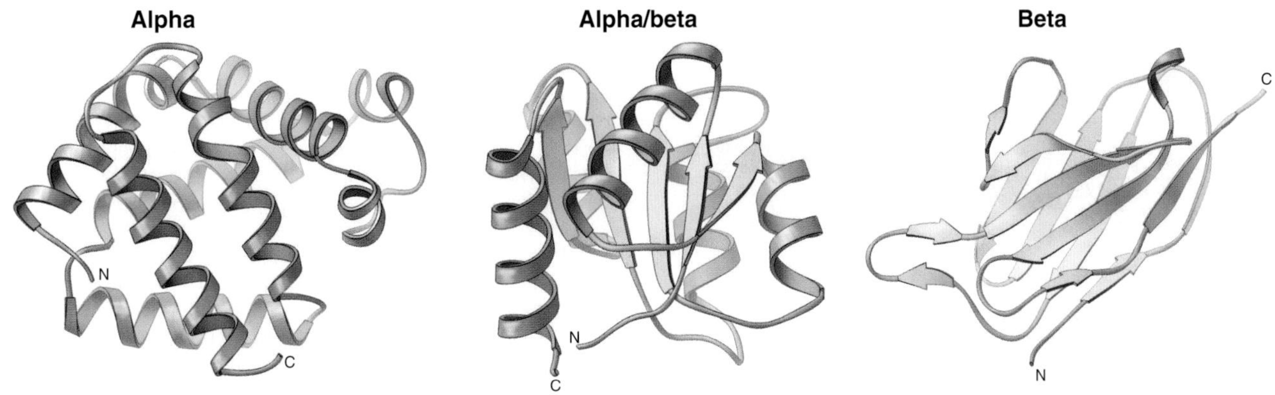

FIGURE 3.9 RIBBON DIAGRAMS OF PROTEIN BACKBONES SHOWING β-STRANDS AS FLATTENED ARROWS, α-HELICES AS COILS, AND OTHER PARTS OF THE POLYPEPTIDE CHAINS AS ROPES. *Left,* The β-subunit of hemoglobin consists entirely of tightly packed α-helices. (See PDB file 1MBA.) *Middle,* CheY is a mixed α/β structure, with a central parallel β-sheet flanked by α-helices. Note the right-handed twist of the sheet (defined by the sheet turning away from the viewer at the upper right) and right-handed pattern of helices (defined by the helices angled toward the upper right corner of the sheet) looping across the β-strands. (Compare the cross section in Fig. 3.7). (See PDB file 2CHF.) *Right,* The immunoglobulin V_L domain consists of a sandwich of 2 antiparallel β-sheets. (See PDB file 2IMM.)

helix has an electrical dipole moment, more negative at the C-terminus. Second, the ends of helices are less stable than the middle, as four potential hydrogen bonds are not completed by backbone interactions at each end. These unmet backbone hydrogen bonds can be completed by interaction with appropriate donors or acceptors on the side chains of the terminal residues. Interactions with serine and asparagine are favored as "caps" at the N-termini of helices, because their side chains can complete the hydrogen bonds of the backbone amide protons. Lysine, histidine, and glutamine are favored hydrogen bonding caps for the C-termini of helices.

All amino acids are found within naturally occurring α-helices. Proline is often found at the beginning of helices and glycine at the end, because they are favored in bends. Both are underrepresented within helices. When present, proline produces bends. Glycine is more common in transmembrane helices, where it contributes to helix–helix packing.

A second strategy used to stabilize the backbone structure of polypeptides is hydrogen bonding of β-strands laterally to form **β-sheets** (Figs. 3.8 and 3.9). In individual β-strands, the peptide chain is extended in a configuration close to *all-trans* with side chains alternating top and bottom and amide protons and carbonyl oxygens alternating right and left. β-Strands can form a complete set of hydrogen bonds, with neighboring strands running in the same or opposite directions in any combination. However, the orientation of hydrogen bond donors and acceptors is more favorable in a β-sheet with antiparallel strands than in sheets with parallel strands. Largely parallel β-sheets are usually extensive and completely buried in proteins. β-Sheets have a natural right-handed twist in the direction along the strands. Antiparallel β-sheets are stable even if the strands are short and extensively distorted by twisting. Antiparallel sheets can wrap around completely to form a **β-barrel** with as few as five strands, but the natural twist of the strands and the need to fill the core of the barrel with hydrophobic residues favors barrels with eight strands.

Up to 25% of the residues in globular proteins are present in bends at the surface (Fig. 3.8D–F). Residues constituting bends are generally hydrophilic. The presence of glycine or proline in a turn allows the backbone to deviate from the usual geometry in tight turns, but the composition of bends is highly variable and not a strong determinant of folding or stability. Turns between linear elements of secondary structure are called **reverse turns,** as they reverse the direction of the polypeptide. Those between β-strands have a few characteristic conformations and are called β-bends.

Many parts of polypeptide chains in proteins do not have a regular structure. At one extreme, small segments of polypeptide, frequently at the N- or C-terminus, are truly disordered in the sense that they are mobile. Many other irregular segments of polypeptide are tightly packed into the protein structure. **Omega loops** are compact structures consisting of 6 to 16 residues, generally on the protein surface, that connect adjacent elements of secondary structure (Fig. 3.8F). They lack regular structure but typically have the side chains packed in the middle of the loop. Some are mobile, but many are rigid. Omega loops form the antigen-binding sites of antibodies. In other proteins, they bind metal ions or participate in the active sites of enzymes.

Packing of Secondary Structure in Proteins

Elements of secondary structure can pack together in almost any way (Fig. 3.9), but a few themes are favored enough to be found in many proteins. For example, two β-sheets tend to pack face to face at an angle of

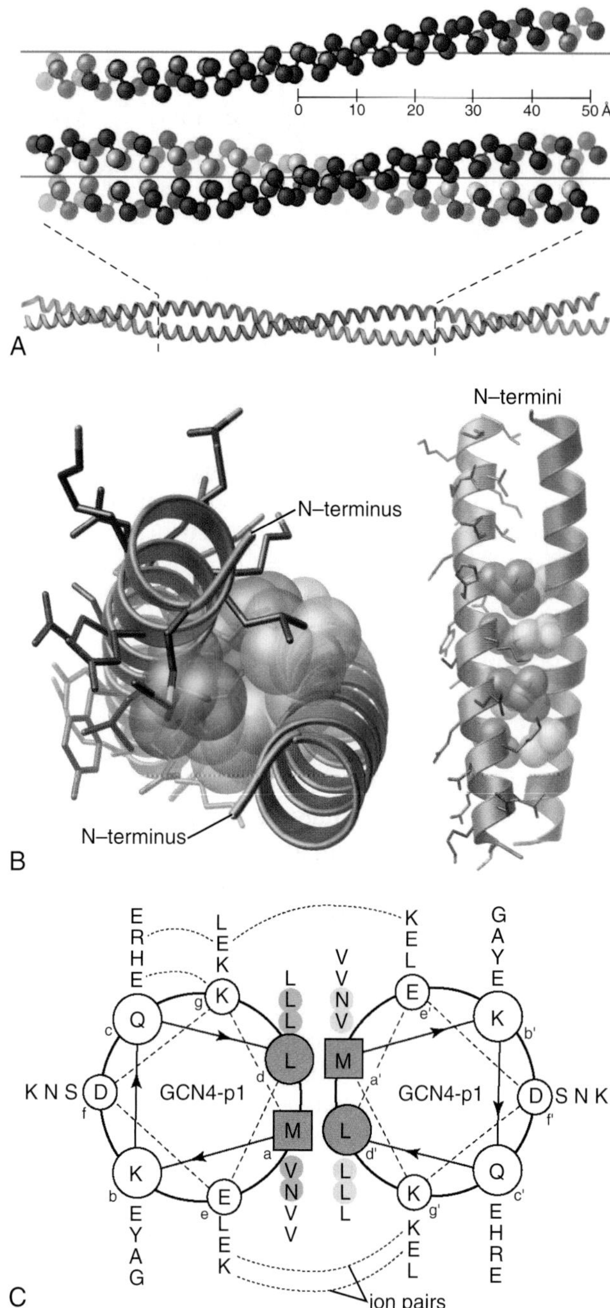

FIGURE 3.10 COILED-COILS. A, Comparison of a single α-helix, represented by spheres centered on the α-carbons, and a two-stranded, left-handed coiled-coil. Two identical α-helices make continuous contact along their lengths by the interaction of the first and fourth residue in every two turns (seven residues) of the helix. (See PDB file 2TMA.) **B,** Atomic structure of the GCN4 coiled-coil, viewed end-on. The coiled-coil holds together two identical peptides of this transcription factor dimer (see Fig. 10.14 for information on its function). Hydrophobic side chains fit together like knobs into holes along the interface between the two helices. (See PDB file 2ZTA.) **C,** Helical wheel representation of the GCN4 coiled-coil. Following the *arrows* around the backbone of the polypeptides, one can read the sequences from the single-letter code, starting with the boxed residues and proceeding to the most distal residue. Note that hydrophobic residues in the first *(a)* and fourth *(d)* positions of each two turns of the helices make hydrophobic contacts that hold the two chains together. Electrostatic interactions *(dashed lines)* between side chains at positions *e* and *g* stabilize the interaction. Other coiled-coils consist of two different polypeptides (see Fig. 10.14), and some are antiparallel (see Fig. 8.18). (**C,** Modified from O'Shea E, Klemm JD, Kim PS, Alber T. X-ray structure of the GCN4 leucine zipper, a two-stranded, parallel coiled-coil. *Science.* 1991;254:539–544.)

approximately 40 degrees with nonpolar residues packed tightly, knobs into holes, in between. α-Helices tend to pack at an angle of approximately 30 degrees across β-sheets, always in a right-handed arrangement. Adjacent α-helices tend to pack together at an angle of either +20 degrees or −50 degrees, owing to packing of side chains from one helix into grooves between side chains on the other helix.

Coiled-coils are a common example of regular superstructure (Fig. 3.10). Two α-helices pair to form a fibrous structure that is widely used to create stable polypeptide dimers in transcription factors (see Fig. 10.14) and structural proteins (see Fig. 39.4). Typically, two identical α-helices wrap around each other in register in a left-handed super helix that is stabilized by hydrophobic interactions of leucines and valines at the interface of the two helices. Intermolecular ionic bonds between the side chains of the two polypeptides also stabilize coiled-coils. Given 3.6 residues per turn, the sequence of a coiled-coil has hydrophobic residues regularly spaced at positions 1 and 4 of a **"heptad repeat."** This pattern allows one to predict the tendency of a polypeptide to form coiled-coils from its amino acid sequence.

β-Sheets can also form extended structures. One called a **β-helix** consists of a continuous polypeptide strand folded into a series of short β-sheets that form a three-sided helix. See Fig. 24.4 for a β-helix in the insulin receptor L2 domain.

Interaction of Proteins with Solvent

The surface of proteins is almost entirely covered with protons (Fig. 3.11). Some protons are potential hydrogen bond donors, but many are inert, being bonded to backbone or side chain aliphatic carbons. Although most of the charged side chains are exposed on the surface, so are many nonpolar side chains. Many water molecules are ordered on the surface of proteins by virtue of hydrogen bonds to polar groups. These water molecules appear in electron density maps of crystalline proteins but exchange rapidly, on a picosecond (10^{-12} second) time scale. Waters in contact with nonpolar atoms on the surface of proteins maximize hydrogen bonding with each other, forming a dynamic layer of water with reduced translational diffusion compared with bulk water. This lowers the entropy of the water by

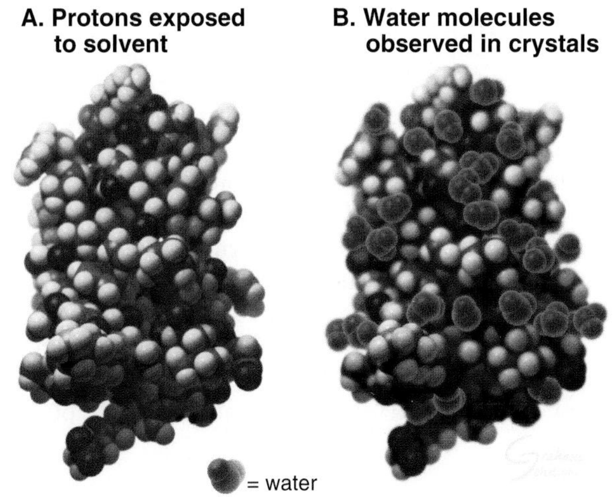

FIGURE 3.11 WATER ASSOCIATED WITH THE SURFACE OF A PROTEIN. A, Protein protons exposed to solvent *(white)* on the surface of a small protein, bovine pancreatic trypsin inhibitor. **B,** Water molecules observed on the surface of the protein in crystal structures. (See PDB file 5BTI.)

increasing its order and provides a thermodynamic impetus to protein folding pathways that minimize the number of hydrophobic atoms displayed on the surface (see Fig. 4.5).

Intrinsically Disordered Regions of Proteins

Although most proteins fold into three-dimensional structures, some regions of proteins, and even whole proteins, can be **intrinsically disordered.** One method predicted disordered segments of 30 or more residues in one-third of eukaryotic protein sequences. Disordered regions were discovered as gaps in electron density maps of crystals of folded proteins, typically loops between elements of secondary structure, between folded domains or at N- and C-termini (see Fig. 8.2). These polypeptides typically have many polar and few large hydrophobic residues, features that allow for accurate prediction of disorder from amino acid sequences. Completely disordered proteins are soluble, but they occupy much larger volumes than folded proteins. NMR and other spectroscopic methods show that they lack stable secondary structure.

Some disordered sequences include "short linear motifs" of less than a dozen amino acids that participate in protein–protein interactions. A few conserved residues within these motifs interact specifically but with low affinity with other proteins, often taking on a specific structure when they bind the partner. Many of these motifs are ligands that bind globular proteins to build signalling pathways (see Fig. 25.10 for SH2, SH3 and other receptors) or link other proteins (see Fig. 9.4 for nuclear localization sequences). Even more frequently these motifs are recognition sites for posttranslational modifications (see Fig. 8.3), which can regulate binding to receptors.

Protein Dynamics

Pictures of proteins tend to give the false impression that they are rigid and static. On the contrary, molecular dynamics simulations show that the atoms of proteins vibrate around their mean positions on a picosecond time scale with amplitudes up to 0.2 nm and velocities of 200 m per second. This motion is a consequence of the kinetic energy of each atom, approximately 2.5 kJ mol^{-1} at 25°C. These motions allow the protein as a whole to explore a variety of subtly different conformations on a fast time scale. Binding to a ligand or a change in conditions may favor one of these alternative conformations.

In addition to relatively small, local variations in structure, many proteins undergo large conformational changes (Fig. 3.12). These changes in structure often reflect a change of activity or physical properties. **Conformational changes** play roles in many biological processes, ranging from opening and closing ion channels (see Fig. 16.5) to cell motility (see Fig. 36.4). Many conformational changes have been observed indirectly by spectroscopy or hydrodynamic methods or directly by crystallography or NMR. For example, when glucose binds the enzyme hexokinase, the two halves of the protein clamp around this substrate by rotating 12 degrees about hinges in two sections of the polypeptide. Guanosine triphosphate (GTP) binding to elongation factor EF-Tu causes a domain to rotate 90 degrees about two glycine residues! Similarly, phosphorylation of glycogen phosphorylase causes a local rearrangement of the N-terminus that transmits a structural change over a distance of more than 2 nm to the active site (see Fig. 27.3). The Ca^{2+} binding regulatory protein calmodulin undergoes a dramatic conformational change when wrapping tightly around a helical peptide of a target protein (also see Chapter 26).

Modular Domains in Proteins

Some proteins consist of a single compact unit (Fig. 3.7), but many others consist of multiple, independently folded, globular regions, or **domains,** connected in a modular fashion in one polypeptide (Fig. 3.13). Most domains consist of 40 to 100 residues, but kinase domains (see Fig. 25.3) and motor domains (see Figs. 36.3 and 36.9) are much larger. Many domains have folds shared by domains in other proteins. The members of such a family of domains are said to be **homologous**, because they evolved from a common ancestor. Through the processes of **gene duplication, transposition,** and **divergent evolution** widely used domains (eg, the immunoglobulin domain) became incorporated into hundreds of different proteins, where they serve unique functions. Homologous domains in different proteins have similar folds but may differ significantly in amino acid sequences. Nevertheless, most related domains can be recognized from characteristic *patterns* of amino acids along their sequences. For example, cysteine

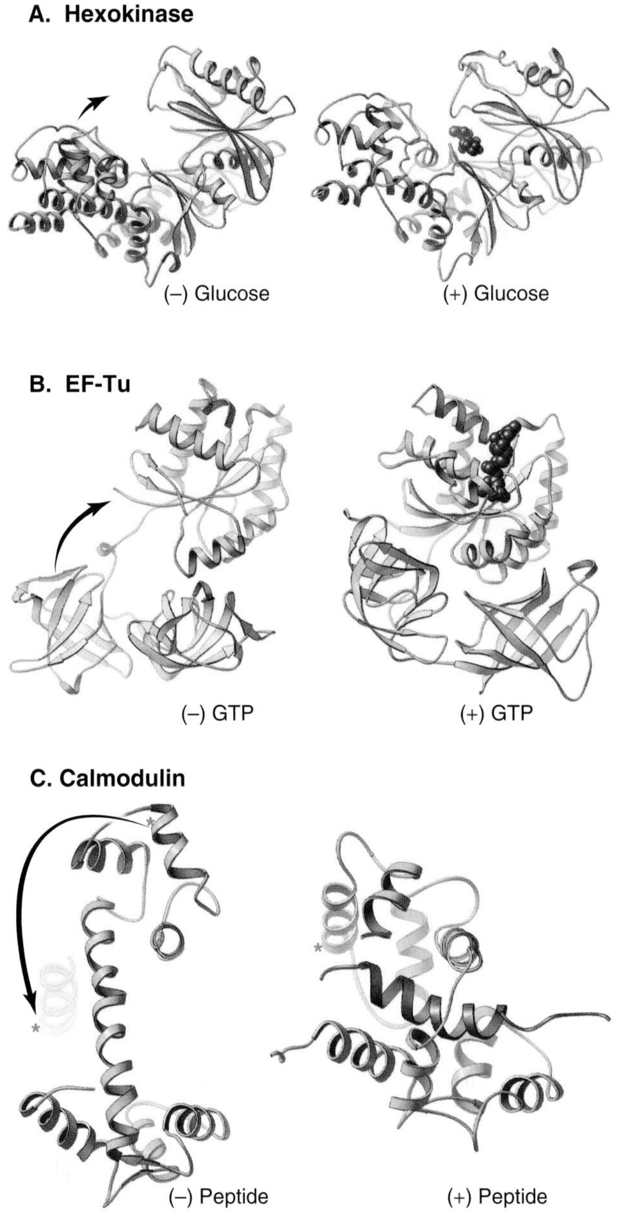

A. Hexokinase

(–) Glucose (+) Glucose

B. EF-Tu

(–) GTP (+) GTP

C. Calmodulin

(–) Peptide (+) Peptide

FIGURE 3.12 CONFORMATIONAL CHANGES OF PROTEINS. **A,** The glycolytic enzyme hexokinase. The two domains of the protein hinge together to surround the substrate, glucose. (See PDB files 2YHX and 1HKG.) **B,** EF-Tu, a cofactor in protein synthesis (see Fig. 12.9), folds more compactly when it binds guanosine triphosphate (GTP). (See PDB files 1EFU and 1EFT.) **C,** Calmodulin (see Chapter 26) binds Ca^{2+} and wraps itself around an α-helix *(red)* in target proteins. Note the large change in position of the helix marked with an *asterisk.* (See PDB files 3CLN and 2BBM.)

residues of immunoglobulin (Ig) G domains are spaced in a pattern required to make intramolecular disulfide bonds (Fig. 3.3).

Rarely, protein domains with related structures arose independently and converged during evolution toward a particularly favorable fold. This is the hypothesis to explain the similar folds of immunoglobulin and fibronectin (FN) III domains, which have unrelated amino acid sequences.

Nucleic Acids

Nucleic acids, polymers of a few simple building blocks called **nucleotides,** store and transfer all genetic information. This is not the limit of their functions. RNA enzymes, **ribozymes,** catalyze some biochemical reactions. Other RNAs are receptors **(riboswitches)** or contribute to the structures and enzyme activities of major cellular components, such as ribosomes (see Fig. 12.7) and spliceosomes (see Fig. 11.15). In addition, nucleotides themselves transfer chemical energy between cellular systems and information in signal transduction pathways. Later chapters elaborate on each of these topics.

Building Blocks of Nucleic Acids

Nucleotides consist of three parts: (1) a **base** built of one or two cyclic rings of carbon and a few nitrogen atoms, (2) a five-carbon sugar, and (3) one or more phosphate groups (Fig. 3.14). **DNA** uses four main bases: the purines **adenine** (A) and **guanosine** (G) and the pyrimidines **cytosine** (C) and **thymine** (T). In **RNA, uracil** (U) is found in place of thymine. Some RNA bases are chemically modified after synthesis of the polymer. The sugar of RNA is **ribose,** which has the aldehyde oxygen of carbon 4 cyclized to carbon 1. The DNA sugar is deoxyribose, which is similar to ribose but lacks the hydroxyl on carbon 2. In both RNA and DNA, carbon 1 of the sugar is conjugated with nitrogen 1 of a pyrimidine base or with nitrogen 9 of a purine base. The hydroxyl of sugar carbon 5 can be esterified to a chain of one or more phosphates, forming **nucleotides** such as adenosine monophosphate **(AMP),** adenosine diphosphate **(ADP),** and ATP.

Covalent Structure of Nucleic Acids

DNA and RNA are polymers of nucleotides joined by **phosphodiester bonds** (Fig. 3.15). The backbone links a chain of five atoms (two oxygens and three carbons) from one phosphorous to the next—a total of six backbone atoms per nucleotide. Unlike the backbone of proteins in which the planar peptide bond greatly limits rotation, all six bonds along a polynucleotide backbone have some freedom to rotate, even that in the sugar ring. This feature gives nucleic acids much greater conformational flexibility than polypeptides, which have only two variable torsional angles per residue. The backbone phosphate group has a single negative charge at neutral pH. The N-C bond linking the base to the sugar is also free to rotate on a picosecond time scale, but rotation away from the backbone is strongly favored. The bases have a strong tendency to stack upon each other, owing to favorable van der Waals interactions (see Chapter 4) between these planar rings.

Each type of nucleic acid has a unique sequence of nucleotides. Laboratory procedures employing the

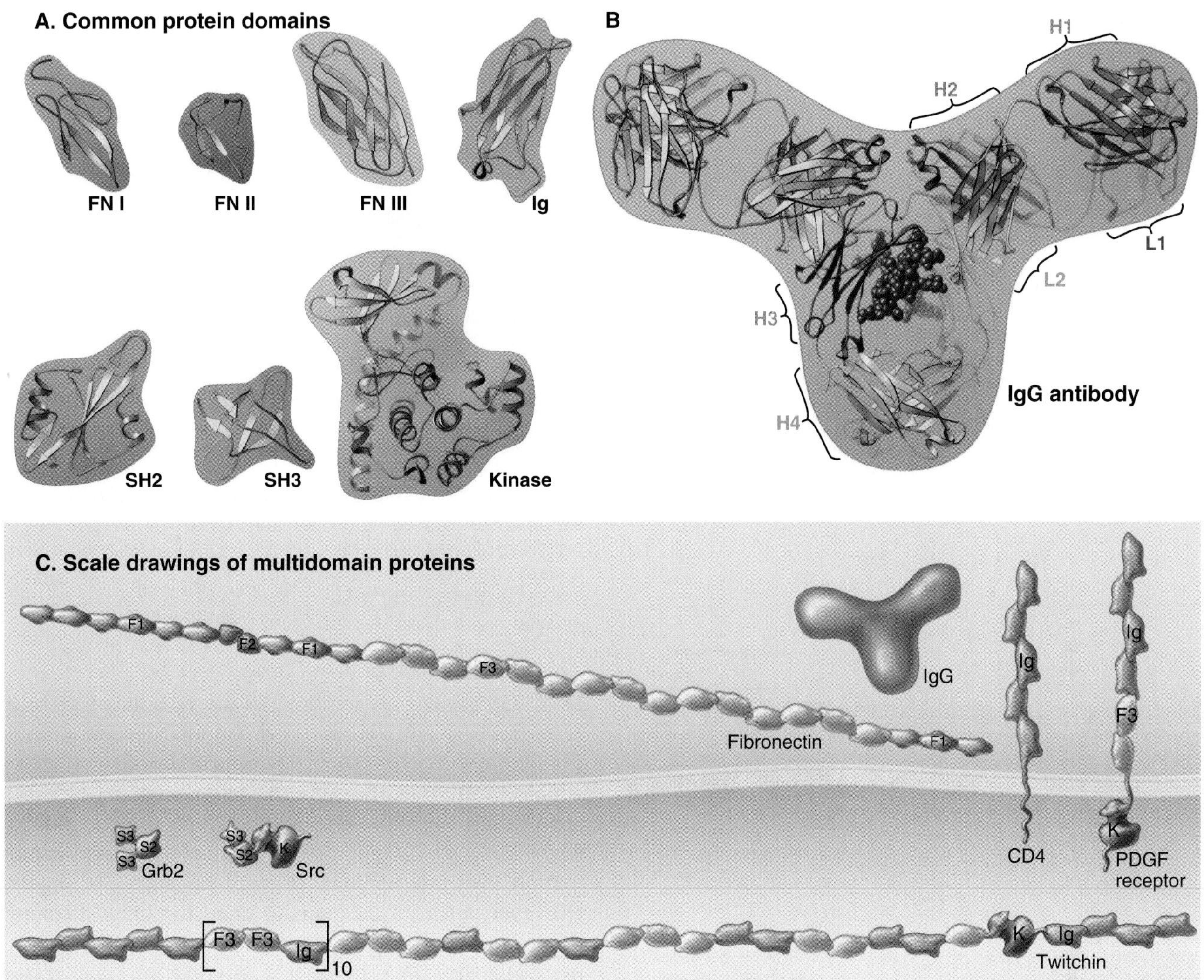

FIGURE 3.13 MODULAR PROTEINS CONSTRUCTED FROM EVOLUTIONARILY HOMOLOGOUS, INDEPENDENTLY FOLDED DOMAINS. A, Examples of protein domains used in many proteins: fibronectin 1 (FN I), fibronectin 2 (FN II), fibronectin 3 (FN III), immunoglobulin (Ig), Src homology 2 (SH2), Src homology 3 (SH3), and kinase. (See PDB files 1PDC, 1FNA, 2IG2, 1HCS, 1PRM, and 1CTP.) **B,** Immunoglobulin G (IgG), a protein composed of 12 Ig domains on four polypeptide chains. Two identical heavy chains (H) consist of four Ig domains, and two identical light chains (L) consist of two Ig domains. The sequences of these six Ig domains differ, but all are folded similarly. The two antigen-binding sites are located at the ends of the two arms of the Y-shaped molecule composed of highly variable loops contributed by domains H1 and L1. (See PDB file 2IG2.) **C,** Examples of proteins constructed from the domains shown in **A**: fibronectin (see Fig. 29.15), CD4 (see Figs. 27.8 and 28.8), platelet-derived growth factor (PDGF) receptor (see Fig. 24.4), Grb2 (see Fig. 27.6), Src (see Fig. 25.3 and Box 27.1), and twitchin (see Chapter 39). Each of the 31 FN III domains in twitchin has a different sequence. F1 is FN I, F2 is FN II, and F3 is FN III.

enzymatic synthesis of DNA allow the sequence to be determined rapidly (Fig. 3.16). Newly synthesized DNA and RNA molecules have a phosphate at the 5′ end and a 3′ hydroxyl at the other end. All DNA and RNA molecules are synthesized biologically in the same direction (see Figs. 10.9 and 42.1) by reaction of a nucleoside triphosphate with the 3′ sugar hydroxyl of the growing strand. Cleavage of the two terminal phosphates from the new subunit provides energy for extension of the polymer in the 5′ to 3′ direction. The 5′ nucleotides of mRNAs are subsequently modified by the addition of a specialized cap structure (see Figs. 11.2 and 12.3).

Secondary Structure of DNA

A few viruses have chromosomes consisting of single-stranded DNA molecules, but most DNA molecules are paired with a complementary strand to form a right-handed **double helix,** as originally proposed by Watson and Crick (Fig. 3.17). Key features of the double helix are two strands running in opposite directions with the sugar-phosphate backbone on the outside and pairs of bases hydrogen-bonded to each other on the inside (Fig. 3.14). Pairs of bases are nearly perpendicular to the long axis of the polymer and stacked 0.34 nm apart from the adjacent bases. This regular structure is called **B-form**

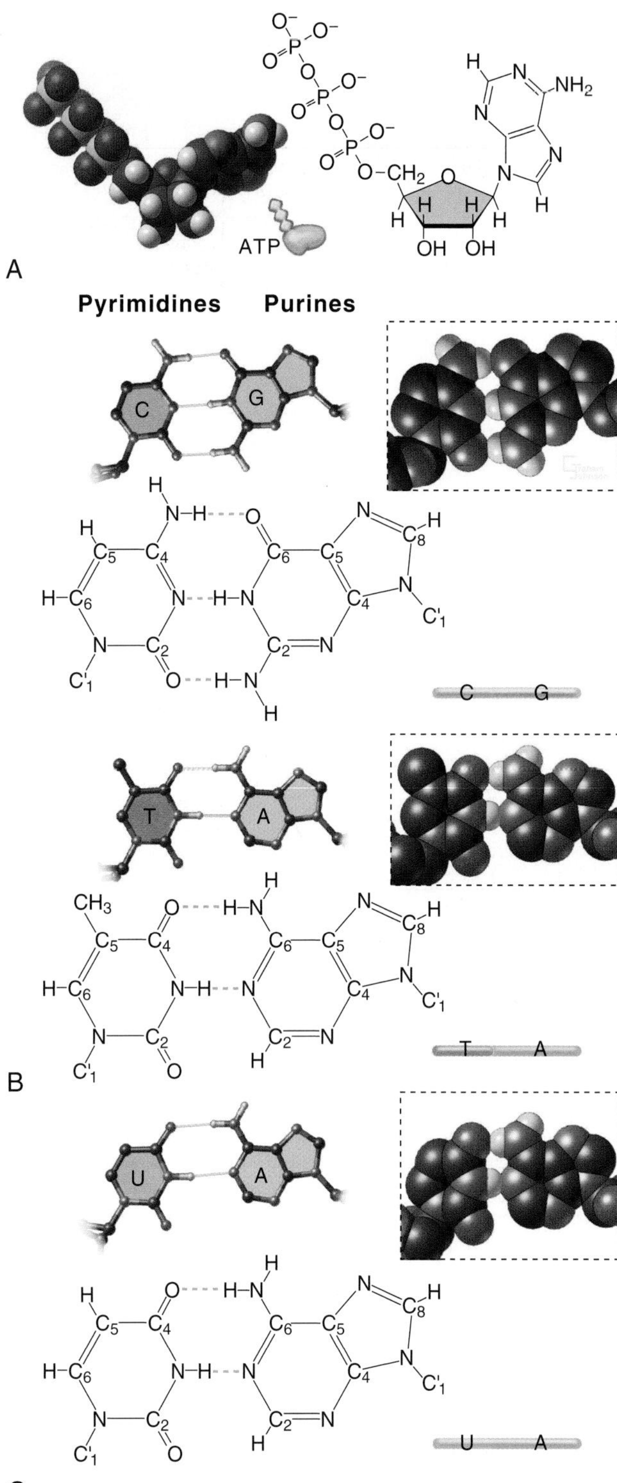

FIGURE 3.14 ADENOSINE TRIPHOSPHATE (ATP) AND NUCLEOTIDE BASES. A, Stick figure and space-filling model of ATP. **B,** Four bases used in DNA. Stick figures show the hydrogen bonds used to form base pairs between thymine (T) and adenine (A) and between cytosine (C) and guanine (G). **C,** Uracil replaces thymine in RNA. C'_1 refers to carbon 1 of ribose and deoxyribose.

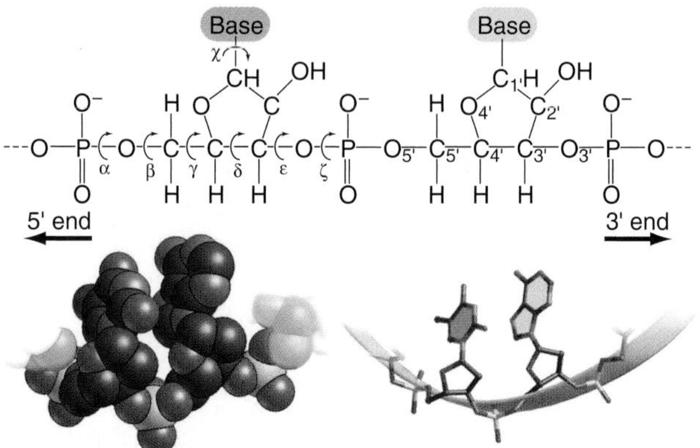

FIGURE 3.15 ROTATIONAL FREEDOM OF THE BACKBONE OF A POLYNUCLEOTIDE, RNA IN THIS CASE. The stick figure of two residues shows that all six of the backbone bonds are rotatable, even the $C_{4'}$—C′ bond that is constrained by the ribose ring. This gives polynucleotides more conformational freedom than polypeptides. Note the phosphodiester bonds between the residues and the definition of the 3′ and 5′ ends. Space-filling and stick figures at the bottom show a uridine (U) and adenine (A) from part of Fig. 3.17. (Modified from Jaeger JA, SantaLucia J, Tinoco I. Determination of RNA structure and thermodynamics. *Annu Rev Biochem.* 1993;62:255–287.)

DNA. On average, in solution, B-form DNA has 10.5 base pairs per turn and a diameter of 1.9 nm, but real DNA is not completely regular. Hydrogen bonds between adenine and thymine and between guanine and cytosine span nearly the same distance between the backbones, so the helix has a regular structure that, to a first approximation, is independent of the sequence of bases. However, a run of As tends to bend the helix. Because the bonds between the bases and the sugars are asymmetrical, the DNA helix is asymmetrical: The major groove on one side of the helix is broader than the other, minor groove. Most cellular DNA is approximately in the B-form conformation, but proteins that regulate gene expression can distort the DNA significantly (see Fig. 10.7).

Under some laboratory conditions, DNA forms stable helical structures that differ from classic B-form DNA. All these variants have the phosphate-sugar backbone on the outside, and most have the usual complementary base pairs on the inside. A-form DNA has 11 base pairs per turn and an average diameter of 2.3 nm. DNA-RNA hybrids and double-stranded RNA also have A-form structure. Z-DNA is the most extreme variant, as it is a left-handed helix with 12 base pairs per turn. The existence of Z-DNA in cells is still in question.

DNA molecules are either linear or circular. Human chromosomes are huge single linear DNA molecules (see Fig. 7.1). Eukaryotic mitochondria and chloroplasts have circular DNA chromosomes (Fig. 3.18) like most bacteria and DNA viruses.

When circular DNAs or linear DNAs with both ends anchored (as in chromosomes; see Chapter 8) are twisted about their long axis, the strain is relieved by

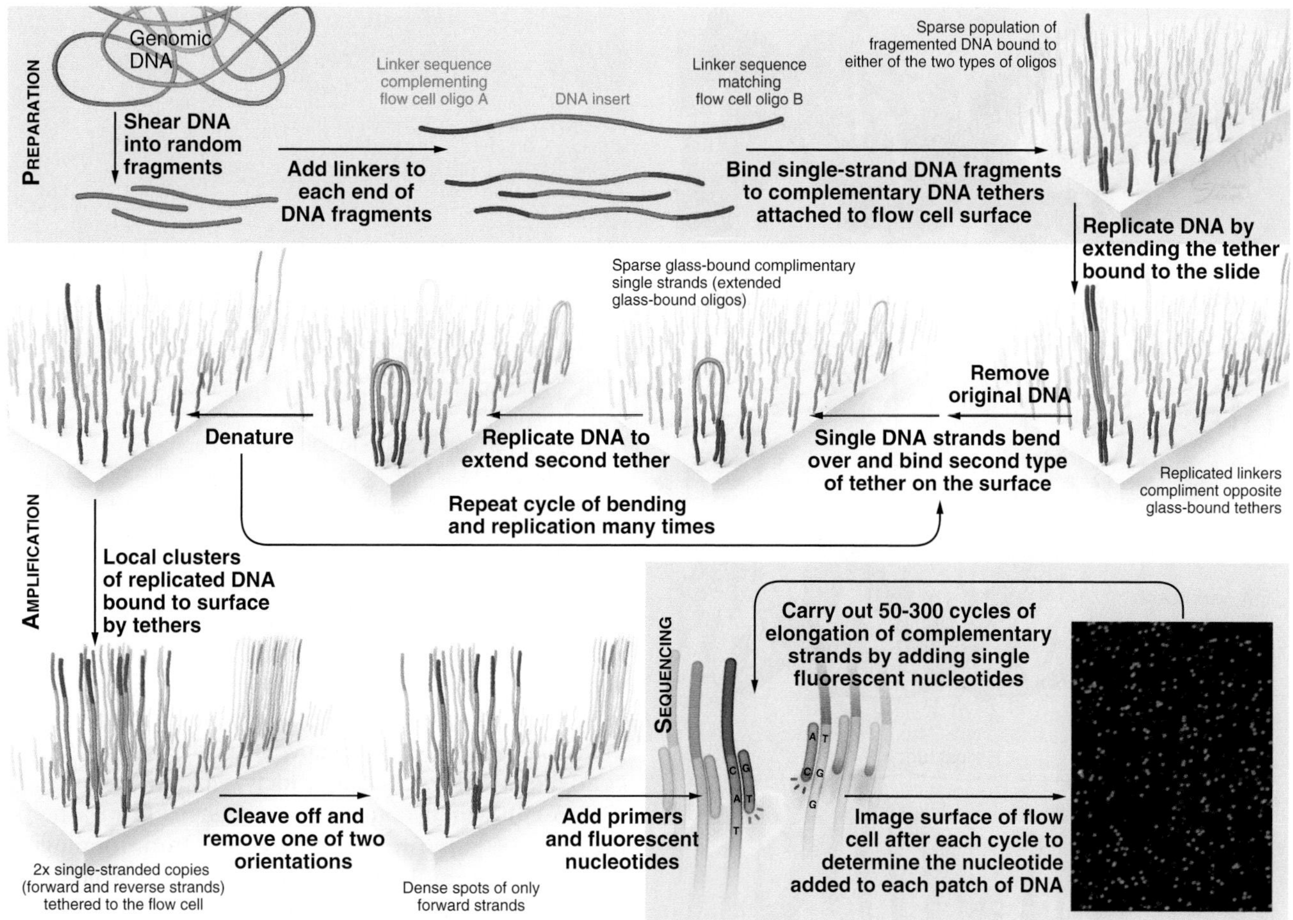

FIGURE 3.16 RAPID PARALLEL DNA SEQUENCING. Random fragments of DNA are physically separated on the surface of a flow cell and then amplified by the polymerase chain reaction. Next, DNA polymerase is used to add one complementary base with a fluorescent dye and a blocking group. After imaging the color of the base added to each molecule, the fluorescent dye and blocking group are removed from each DNA, allowing another round of complementary base addition. (This is the technology used in sequencers from Illumina, Inc.)

the development of long-range bends and twists called **supercoils** or **superhelices** (Fig. 3.18). Supercoiling can be either positive or negative depending on whether the DNA helix is wound more tightly or somewhat unwound. Supercoiling is biologically important, as it can influence the expression of genes. Under some circumstances, supercoiling favors unwinding of the double helix. This can promote access of proteins involved in the regulation of transcription from DNA (see Chapter 10).

The degree of supercoiling is regulated locally by enzymes called **topoisomerases.** Type I topoisomerases nick one strand of the DNA and cause the molecule to unwind by rotation about a backbone bond. Type II topoisomerases cut both strands of the DNA and use an ATP-driven conformational change (called *gating*) to pass a DNA strand through the cut prior to rejoining the ends of the DNA. To avoid free DNA ends during this reaction, cleaved DNA ends are linked covalently to tyrosine residues of the enzyme. This also conserves chemical bond energy, so ATP is not required for religation of the DNA at the end of the reaction.

Secondary and Tertiary Structure of RNAs

RNAs range in size from microRNAs of 20 nucleotides (see Fig. 11.12) to mRNAs with more than 80,000 nucleotides. Because each nucleotide has approximately three times the mass of an amino acid, RNAs with a modest number of nucleotides are bigger than most proteins (see Fig. 1.4). The 16S RNA of the small ribosomal subunit of bacteria consists of 1542 nucleotides with a mass of approximately 460 kD, much larger than any of the 21 proteins with which it interacts (see Fig. 12.6).

Except for the RNA genomes of a few viruses, RNAs generally do not have a complementary strand to pair with each base. Instead they form specific structures by optimizing *intramolecular* base pairing (Figs. 3.19 and 3.20). Comparisons of homologous RNA sequences from a selection of organisms provide much of what

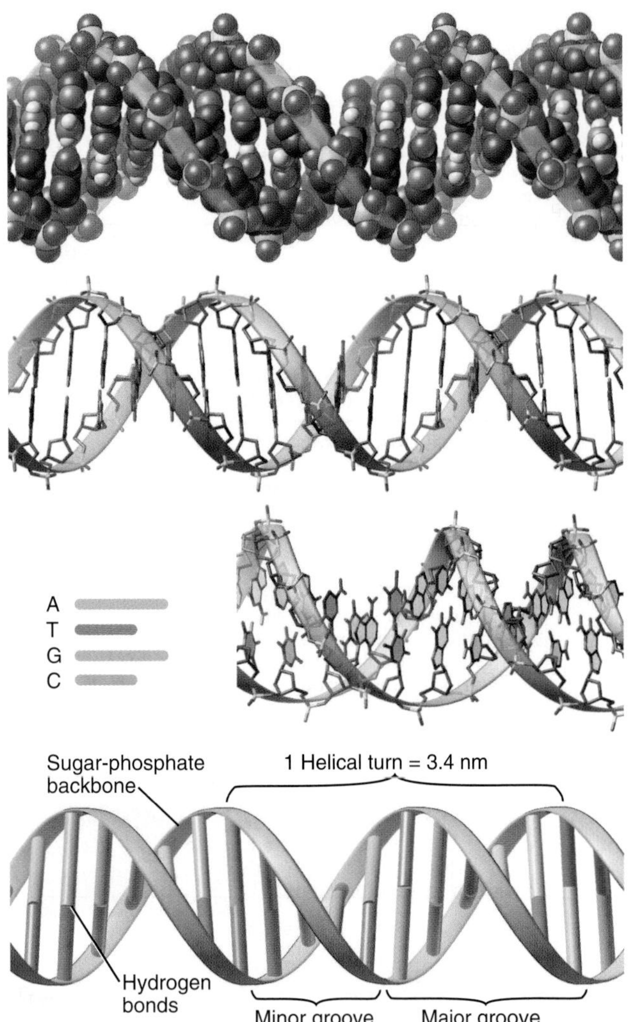

FIGURE 3.17 MODELS OF B-FORM DNA. The molecule consists of two complementary antiparallel strands arranged in a right-handed double helix with the backbone (see Fig. 3.15) on the outside and stacked pairs of hydrogen-bonded bases (see Fig. 3.14) on the inside. *Top,* Space-filling model. *Middle,* Stick figures, with the lower figure rotated slightly to reveal the faces of the bases. *Bottom,* Ribbon representation. (Idealized 24-base pair model built by Robert Tan, University of Alabama, Birmingham.)

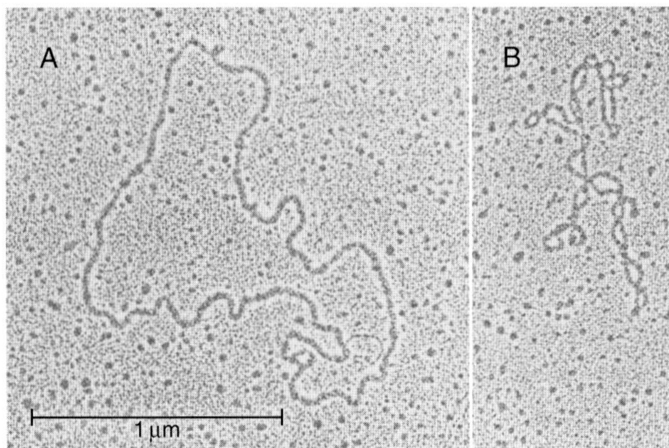

FIGURE 3.18 DNA SUPERCOILING. Electron micrographs of a circular mitochondrial DNA molecule in a relaxed configuration **(A)** and a supercoiled configuration **(B)**. (From David Clayton, Stanford University, Stanford, CA; originally in Stryer L: *Biochemistry,* 4th ed. New York: WH Freeman; 1995.)

is known about this intramolecular base pairing. The approach is to identify pairs of nucleotides that vary together across the phylogenetic tree. For example, if an A and a U at discontinuous positions in one RNA are changed together to C and a G in homologous RNAs, it is inferred that they are hydrogen-bonded together. This **covariance method** works remarkably well, because hundreds to thousands of homologous sequences for the major classes of RNA are available from comparative genomics. Conclusions about base pairing from covariance analysis have been confirmed by experimental mutagenesis of RNAs and direct structure determination.

The simplest RNA secondary structure is an antiparallel double helix stabilized by hydrogen bonding of complementary bases (Figs. 3.19, 3.20, and 3.21). Similarly to DNA, G pairs with C and U pairs with A. Unlike the case in DNA, G also frequently pairs with U in RNA. Helical base pairing occurs between both locally contiguous sequences and widely separated sequences (Fig. 3.19). When contiguous sequences form a helix, the strand is often reversed by a tight turn, forming an antiparallel **stem–loop** structure. These hairpin turns frequently consist of just four bases. A few sequences are highly favored for turns, owing to their compact, stable structures. Bulges due to extra bases or noncomplementary bases frequently interrupt base-paired helices of RNA (Fig. 3.19).

Crystal structures of transfer RNAs **(tRNAs)** (Fig. 3.20) and a hammerhead **ribozyme** (Fig. 3.21) established that RNAs have novel, specific, three-dimensional structures. Crystal structures of ribosomes (see Fig. 12.7) showed that larger RNAs fold into specific structures using similar principles. Crystallization of RNAs is challenging, and NMR provides much less information on RNA than on proteins of the same size, so much is yet to be learned about RNA structures.

As in proteins, many residues in RNAs are in conventional secondary structures, especially stems consisting of base-paired double helices; however, RNA backbones make sharp turns that allow unconventional hydrogen bonds between bases, ribose hydroxyls, and backbone phosphates. Generally, the phosphodiester backbone is on the surface with most of the hydrophobic bases stacked internally. Some bases are hydrogen-bonded together in triplets (Fig. 3.22) rather than in pairs. Clusters of Mg^{2+} ions stabilize regions of tRNA with high densities of negative charge.

Like proteins, RNAs can change conformation. The TAR RNA is a stem–loop structure with a bulge formed by three unpaired nucleotides (Fig. 3.22). TAR is located

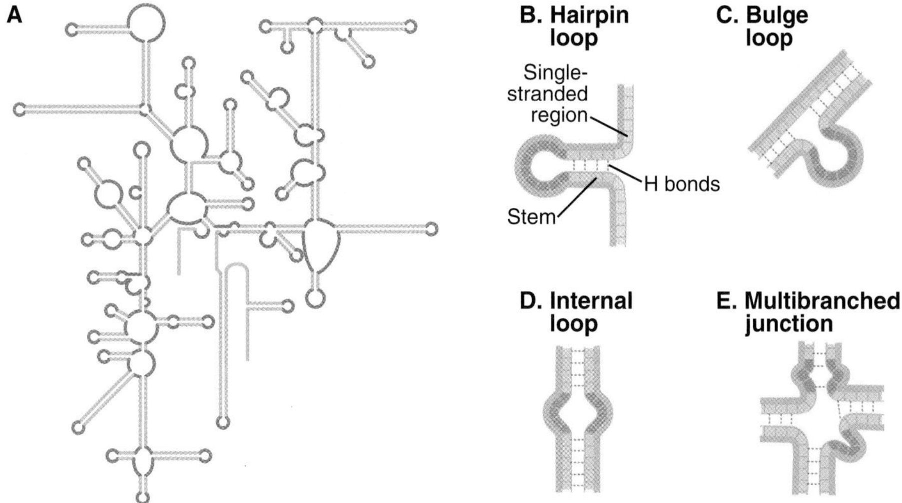

FIGURE 3.19 RNA SECONDARY STRUCTURES. A, Base pairing of *Escherichia coli* 16S ribosomal RNA determined by covariance analysis of nucleotide sequences of many different 16S ribosomal RNAs. The *line* represents the sequence of nucleotides. *Blue* sections are base-paired strands; *pink* sections are bulges and turns; *green* sections are neither base-paired nor turns. **B,** An antiparallel base-paired stem forming a hairpin loop. **C,** A bulge loop. **D,** An internal loop. **E,** A multibranched junction. (**A,** Modified from Huysmans E, DeWachter R. Compilation of small ribosomal subunit RNA sequences. *Nucleic Acids Res.* 1987;14(Suppl):73–118. **B–E,** Modified from Jaeger JA, SantaLucia J, Tinoco I. Determination of RNA structure and thermodynamics. *Annu Rev Biochem.* 1993;62:255–287.)

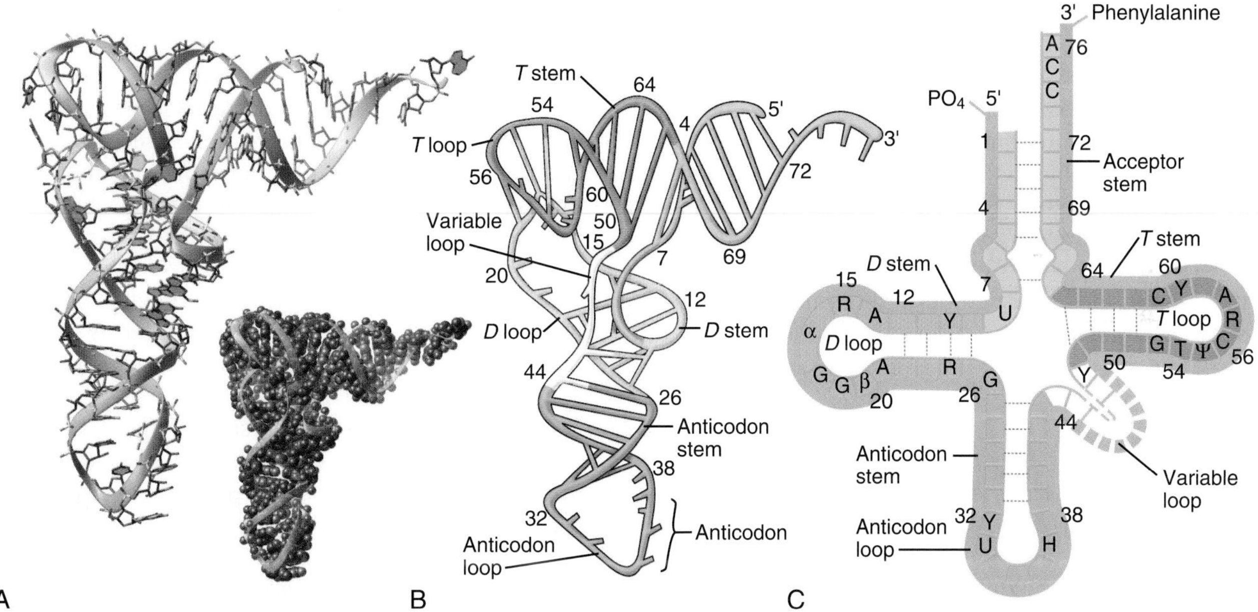

FIGURE 3.20 Atomic structure of phenylalanine transfer RNA (phe-tRNA) determined by x-ray crystallography. **A,** An *orange* ribbon traces the RNA backbone through a stick figure *(left)* and space filling model *(right)*. **B,** Skeleton drawing. **C,** Two-dimensional base-pairing scheme. Note that the base-paired segments are much less regular than is B-form DNA. (For reference, see PDB file 6TNA. **B,** Modified from an original by Alex Rich, Massachusetts Institute of Technology, Cambridge, MA.)

at the 5′ end of all RNA transcripts of HIV, the virus that causes AIDS. Binding of a regulatory protein called TAT changes the conformation of TAR and promotes elongation of the RNA. Binding arginine also changes the conformation of TAR.

Like proteins, RNAs can bind ligands. RNA sequences located in the mRNAs regulate approximately 2% of the genes in the bacterium *Bacillus subtilis*. For example, mRNAs for enzymes used to synthesize purines such as guanine have a guanine-sensitive riboswitch that controls translation (Fig. 3.22C–D). Low concentrations of guanine do not bind the RNA, which assumes a variety of conformations that allow transcription. High concentrations of guanine bind the RNA and favor a conformation that blocks transcription. This negative feedback optimizes the cellular concentration of guanine.

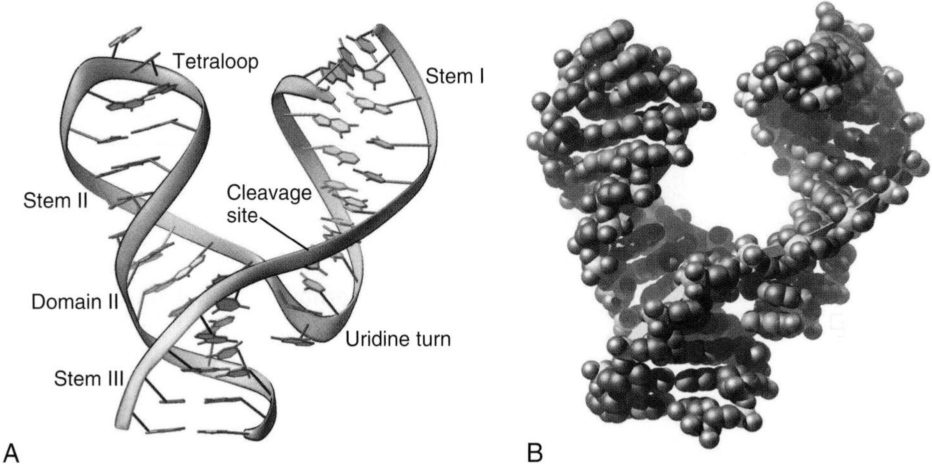

FIGURE 3.21 HAMMERHEAD RIBOZYME, A SELF-CLEAVING RNA SEQUENCE FOUND IN PLANT VIRUS RNAS. A, Ribbon diagram. **B,** Space-filling model. The structure consists of an RNA strand of 34 nucleotides complexed to a DNA strand of 13 nucleotides (in vivo, this is a 13-nucleotide stretch of RNA that would be cleaved by the ribozyme). The RNA forms a central stem–loop structure (stem II) and base pairs with the substrate DNA to form stems I and III. Interactions of the substrate strand with the sharp uridine turn distort the backbone and promote its cleavage. (For reference, see PDB file 1HMH. **A,** Modified from Pley HW, Flaherty KM, McKay DB. Three-dimensional structure of a hammerhead ribozyme. *Nature.* 1994;372:68–74.)

Riboswitches bind other metabolites and even fluoride ions. Binding these ligands regulates the expression of proteins relevant to their physiology.

Like proteins, RNAs catalyze chemical reactions (see Chapter 11 for more details.). Some function entirely on their own, but proteins support some enzymatically active RNAs including RNase P (see Fig. 11.9) and the large ribosomal subunit. Binding of cellular metabolites regulates the activities of some ribozymes. The 14 known classes of naturally occurring ribozymes cleave RNAs and synthesize proteins, while artificial ribozymes developed experimentally can catalyze other reactions, including RNA synthesis.

Carbohydrates

Carbohydrates are a large family of biologically essential molecules made up of one or more sugar molecules that get their name because their chemical composition is often formed from multiples of CH_2O. Sugar polymers differ from proteins and nucleic acids by having branches. Compared with proteins, which are generally compact, hydrophilic sugar polymers tend to spread out in aqueous solutions to maximize hydrogen bonds with water. Carbohydrates may occupy 5 to 10 times the volume of a protein of the same mass. The terms **glycoconjugate** and **complex carbohydrate** are currently preferred for sugar polymers rather than polysaccharide.

Carbohydrates serve four main functions:

1. Covalent bonds of sugar molecules are a primary source of energy for cells.
2. The most abundant structural components on earth are sugar polymers: cellulose forms cell walls of plants; chitin forms exoskeletons of insects; and glycosaminoglycans are space-filling molecules in connective tissues of animals.
3. Sugars form part of the backbone of nucleic acids, and nucleotides participate in many metabolic reactions.
4. Single sugars and groupings of sugars form side chains on lipids (see Fig. 13.3) and proteins (see Figs. 21.26 and 29.13). These modifications provide molecular diversity beyond that inherent in proteins and lipids themselves, changing their physical properties and vastly expanding the potential of these glycoproteins and glycolipids to interact with other cellular components in specific receptor-ligand interactions (see Fig. 30.12). Conversely, other glycoconjugates block inappropriate cellular interactions.

A modest number of simple sugars (Fig. 3.23) form the vast array of different complex carbohydrates found in nature. These sugars consist of three to seven carbons with one aldehyde or ketone group and multiple hydroxyl groups. In water, the common five-carbon **(pentose)** and six-carbon **(hexose)** sugars cyclize by reaction of the aldehyde or ketone group with one of the hydroxyl carbons. Cyclization forms compact structures used in all the glycoconjugates considered in this book. Given several asymmetrical carbons in each sugar, a great many **stereochemical isomers** exist. For example, the hydroxyl on carbon 1 can either be above (β-isomer) or below (α-isomer) the plane of the ring. Proteins (enzymes, lectins, and receptors) that interact with sugars distinguish these stereoisomers.

Sugars are coupled to other molecules by highly specific enzymes, using a modest repertoire of intermolecular bonds (Fig. 3.24). The common *O*-glycosidic (carbon-oxygen-carbon) bond is formed by removal of water from two hydroxyls—the hydroxyl of the carbon

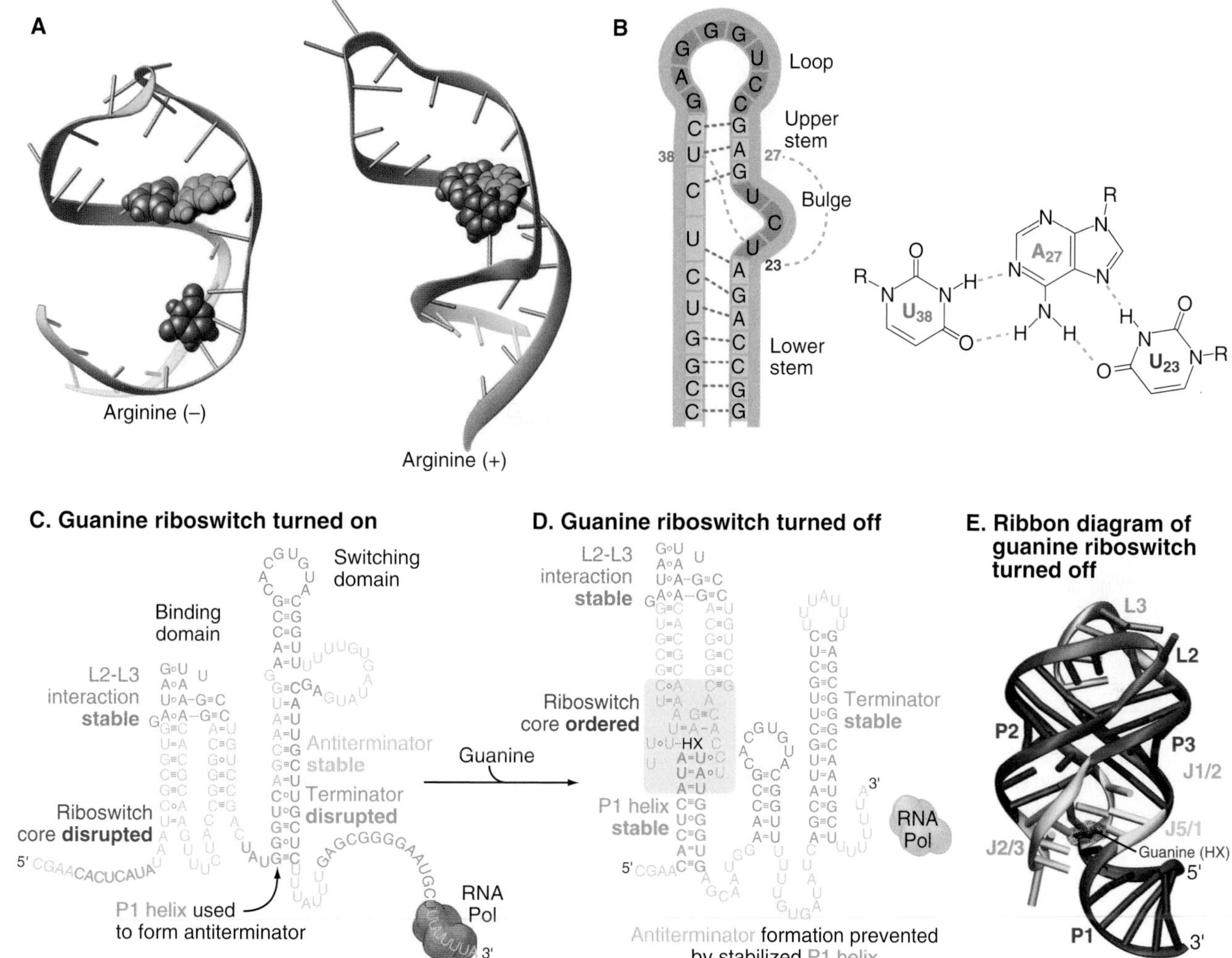

FIGURE 3.22 RNA CONFORMATIONAL CHANGES. A–B, Molecular models of NMR structures of TAR, a stem–loop regulator of HIV messenger RNA (mRNA). Binding of arginine (or a protein called TAT) causes a major conformational change: Two bases twist out of the helix into the solvent *(top).* U23 forms a base triplet with U38 and A27 *(space-filling model),* and the stem straightens. This conformational change promotes transcription of the rest of the mRNA. (**A,** For reference, see PDB files 1ANR and 1AKX.) **C–E,** Guanine-binding riboswitch from *Bacillus subtilis.* **C,** Diagram of the mRNA showing the location of the riboswitch just upstream of the genes for the enzymes required to synthesize guanine. At low guanine concentrations, the RNA is folded in a way that allows transcription of the genes. (See PDB file 4FE5.) **D,** High guanine concentrations (the analog hypoxanthine [HX] is shown here) bind to the riboswitch, causing refolding into a terminator stem loop that prevents transcription of the mRNA. **E,** Ribbon drawing of the crystal structure with bound hypoxanthine. (For reference, see Batey RT, Gilbert SD, Montange RK. Structure of a natural guanine-responsive riboswitch complexed with the metabolite hypoxanthine. *Nature.* 2004;432:411–415 [**C**] and Mandal M, Boese B, Barrick JE, et al. Riboswitches control fundamental biochemical pathways in *B. subtilis* and other bacteria. *Cell.* 2003;113:577–586 [**D**].)

bonded to the ring oxygen of a sugar and a hydroxyl oxygen of another sugar or the amino acids serine and threonine. A similar reaction couples a sugar to an amine, as in the bond between a sugar and a nucleoside base. Sugar phosphates with one or more phosphates esterified to a sugar hydroxyl are components of nucleotides as well as of many intermediates in metabolic pathways.

Glycoconjugates—polymers of one or more types of sugar molecules—are present in massive amounts in nature and are used as both energy stores and structural components (Fig. 3.25). Cellulose (unbranched β-1,4 polyglucose), which forms the cell walls of plants, and chitin (unbranched β-1,4 poly *N*-acetylglucosamine), which forms the exoskeletons of many invertebrates, are the first and second most abundant biological polymers found on the earth. In animals, giant complex carbohydrates are essential components of the extracellular matrix of cartilage and other connective tissues (see Figs. 29.13 and 32.3). Glycogen, a branched α-1,4 polymer of glucose, is the major energy store in animal cells. Starch—polymers of glucose with or without a modest level of branching—performs the same function for plants.

Glycoconjugates differ from proteins and nucleic acids in that they have a broader range of conformations owing to the flexible glycosidic linkages between the sugar subunits. Although extensive intramolecular hydrogen bonds stabilize some sugar polymers and some

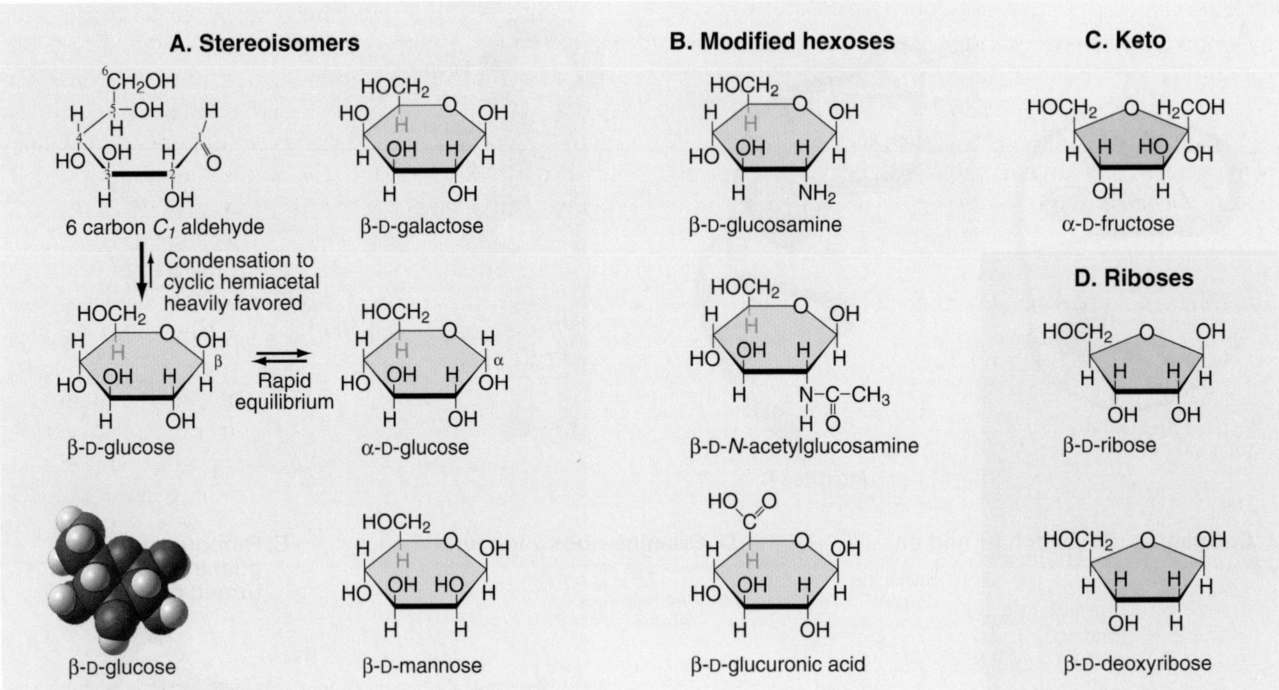

FIGURE 3.23 A–C, Simple sugar molecules. Stick figures and space-filling model of D-glucose showing the highly favored condensation of the carbon 5 hydroxyl with carbon 1 to form a hemiacetal. The resulting hydroxyl group on carbon 1 is in a rapid equilibrium between the α (down) or β (up) configurations. The space-filling model of β-D-glucose illustrates the stereochemistry of the ring; the stick figures are drawn as unrealistic planar rings to simplify comparisons. Stick figures show three stereoisomers of the 6-carbon glucose **(A)**, three modifications of glucose **(B)**, a 6-carbon keto sugar condensed into a 5-membered ring **(C)**, and two 5-carbon riboses **(D)**.

FIGURE 3.24 GLYCOSIDIC BONDS. Stick figures show the formation of *O*- and *N*-glycosidic bonds and a common example of each: the disaccharide sucrose and the nucleoside cytidine. Enzymes catalyze the formation of glycosidic bonds in cells. The chemical name of sucrose [glucose-α(1→2)fructose] illustrates the convention for naming the bonds of glycoconjugates.

glycosidic linkages are relatively rigid, NMR studies revealed that many glycosidic bonds rotate freely, allowing the polymer to change its conformation on a submillisecond time scale. This dynamic behavior limits efforts to determine glycoconjugate structures. They are reluctant to crystallize, and the multitude of conformations does not lend itself to NMR analysis. Structural details are best revealed by x-ray crystallography of a glycoconjugate bound to a protein, such as a lectin or a glycosidase (a degradative enzyme).

Specific enzymes link sugars to proteins in just three different ways (Fig. 3.26). Glycoprotein side chains vary in size from one sugar to polymers of hundreds of sugars. These sugar side chains can exceed the mass of the protein to which they are attached. Chapters 21 and 29 consider glycoprotein biosynthesis.

Compared with the nearly invariant sequences of proteins and nucleic acids, glycoconjugates are heterogeneous because enzymes assemble these sugar polymers without the aid of a genetic template. These glycosyltransferases link high-energy sugar-nucleosides to acceptor sugars. These enzymes are specific for the donor sugar-nucleoside and selective, but not completely specific, for the acceptor sugars. Thus, cells require many different glycosyltransferases to generate the hundreds of types of sugar-sugar bonds found in glycoconjugates. Particular cells consistently produce the same range of specific glycoconjugate structures. This reproducible heterogeneity arises from the repertoire of glycosyltransferases expressed, their localization in specific cellular compartments, and the availability of suitable acceptors. Glycosyltransferases compete with each other for acceptors, yielding a variety of products at many steps in the synthesis of glycoconjugates. For example, the probability of encountering a particular glycosyltransferase depends on the part of the Golgi apparatus (see Fig. 21.14) in which a particular acceptor finds itself.

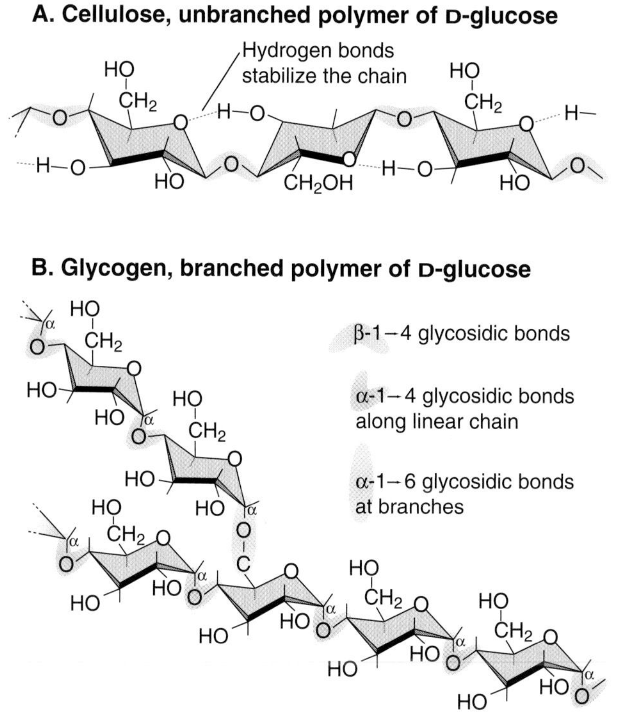

FIGURE 3.25 EXAMPLES OF SIMPLE GLYCOCONJUGATES. Stick figures show the conformations of the sugar rings. **A,** Cellulose, an unbranched homopolymer of glucose used to construct plant cell walls. **B,** Glycogen, a branched homopolymer of glucose used by animal cells to store sugar. Many glycoconjugates consist of several different types of sugar subunits (see Figs. 21.26 and 29.13).

Aqueous Phase of Cytoplasm

The aqueous phase of cells contains a wide variety of solutes, including inorganic ions, building blocks of major organic constituents, intermediates in metabolic pathways, carbohydrate and lipid energy stores, and high concentrations of proteins and RNA. In addition, eukaryotic cells have a dense network of cytoskeletal fibers (Fig. 3.27). Cells control the concentrations of solutes in each cellular compartment, because many (eg, pH, Na^+, K^+, Ca^{2+}, and cyclic AMP) have essential regulatory or functional significance in particular compartments.

The high concentration of macromolecules and the network of cytoskeletal polymers make the cytoplasm a very different environment from the dilute salt solutions that are usually employed in biochemical experiments on cellular constituents. The presence of 300 mg/mL of

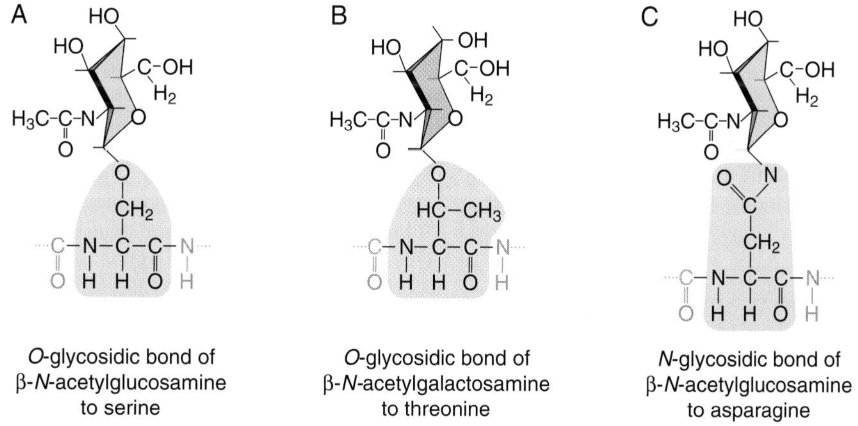

FIGURE 3.26 THREE TYPES OF GLYCOSIDIC BONDS LINK GLYCOCONJUGATES TO PROTEINS. A, An *O*-glycosidic bond links *N*-acetylglucosamine to serine residues of many intracellular proteins. **B,** An *O*-glycosidic bond links *N*-acetylgalactosamine to serine or threonine residues of core proteins, initiating long glycoconjugate polymers called glycosaminoglycans on extracellular proteoglycans (see Fig. 29.13). **C,** An *N*-glycosidic bond links *N*-acetylglucosamine to asparagine residues of secreted and membrane glycoproteins (see Fig. 21.26). A wide variety of glycoconjugates extend the sugar polymer from the *N*-acetylglucosamine. These stick figures illustrate the conformations of the sugar rings.

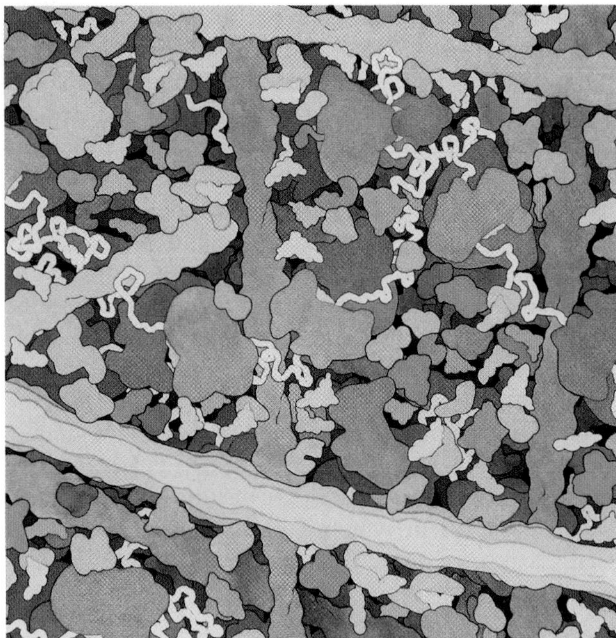

FIGURE 3.27 CROWDED CYTOPLASM. Scale drawing of eukaryotic cell cytoplasm emphasizing the high concentrations of ribosomes *(shades of red)*, proteins *(shades of tan, blue, and green)*, and nucleic acids (*gray*) among cytoskeletal polymers *(shades of blue)*. (From D. Goodsell, Scripps Research Institute, La Jolla, CA.)

protein and RNA causes the cytoplasm to be crowded. The concentration of bulk water in cytoplasm is less than the 55 M in dilute solutions, but the microscopic viscosity of the aqueous phase in live cells is remarkably close to that of pure water. Crowding lowers the diffusion coefficients of the molecules approximately threefold, but it also enhances macromolecular associations by raising the chemical potential of the diffusing molecules through an "excluded volume" effect. Macromolecules take up space in the solvent, so the concentration of each molecule is higher in relation to the available solvent. At cellular concentrations of macromolecules, the chemical potential of a molecule (see Chapter 4) may be one or more orders of magnitude higher than its concentration. (The chemical potential, rather than the concentration, determines the rate of reactions.) Therefore, crowding favors protein–protein, protein–nucleic acid, and other macromolecular assembly reactions that depend on the chemical potential of the reactants. Crowding also changes the rates and equilibria of enzymatic reactions, usually increasing the activity as compared with values in dilute solutions.

ACKNOWLEDGMENTS

We thank Ronald Breaker, Andrew Miranker, and Elizabeth Rhoades for their suggestions on revisions to this chapter.

SELECTED READINGS

Adams PD, Baker D, Brunger AT, et al. Advances, interactions, and future developments in the CNS, Phenix, and Rosetta structural biology software systems. *Annu Rev Biophys*. 2013;42:265-287.

Bryant RG. The dynamics of water-protein interactions. *Annu Rev Biophys Biomol Struct*. 1996;25:29-53.

Chothia C, Hubbard T, Brenner S, et al. Protein folds in the all-beta and all-alpha classes. *Annu Rev Biophys Biomol Struct*. 1997;26:597-627.

Creighton TE. *Proteins: Structure and Molecular Principles*. 2nd ed. New York: WH Freeman; 1993:507.

DNA sequencing. Available at <http://en.wikipedia.org/wiki/DNA_sequencing>.

Doherty EA, Doudna JA. Ribozyme structures and mechanisms. *Annu Rev Biophys Biomol Struct*. 2001;30:457-475.

Dorn M, E Silva MB, Buriol LS, Lamb LC. Three-dimensional protein structure prediction: Methods and computational strategies. *Comput Biol Chem*. 2014;53PB:251-276.

Feizi T, Mulloy B. Carbohydrates and glycoconjugates: Glycomics: The new era of carbohydrate biology. *Curr Opin Struct Biol*. 2003;13:602-604.

Frommer J, Appel B, Müller S. Ribozymes that can be regulated by external stimuli. *Curr Opin Biotechnol*. 2015;31:35-41.

Fürtig B, Nozinovic S, Reining A, Schwalbe H. Multiple conformational states of riboswitches fine-tune gene regulation. *Curr Opin Struct Biol*. 2015;30:112-124.

Komander D, Rape M. The ubiquitin code. *Annu Rev Biochem*. 2012;81:203-229.

Lindorff-Larsen K, Piana S, Dror RO, Shaw DE. How fast-folding proteins fold. *Science*. 2011;334:517-520.

Lupas A. Coiled-coils: New structures and new functions. *Trends Biochem Sci*. 1996;21:375-382.

Moult J, Fidelis K, Kryshtafovych A, et al. Critical assessment of methods of protein structure prediction (CASP)—round x. *Proteins*. 2014;82(suppl 2):1-6.

Murthy VL, Srinivasan R, Draper DE, Rose GD. A complete conformational map for RNA. *J Mol Biol*. 1999;291:313-327.

Oldfield CJ, Dunker AK. Intrinsically disordered proteins and intrinsically disordered protein regions. *Annu Rev Biochem*. 2014;83:553-584.

Onoa B, Tinoco I. RNA folding and unfolding. *Curr Opin Struct Biol*. 2004;14:374-379.

Parak FG. Proteins in action: The physics of structural fluctuations and conformational changes. *Curr Opin Struct Biol*. 2003;13:552-557.

Ponting CP, Russell RR. The natural history of protein domains. *Annu Rev Biophys Biomol Struct*. 2002;31:45-71.

Ramesh A, Winkler WC. Metabolite-binding ribozymes. *Biochim Biophys Acta*. 2014;1839:989-994.

Serganov A, Nudler E. A decade of riboswitches. *Cell*. 2013;152:17-24.

Sosnick TR, Barrick D. The folding of single domain proteins—have we reached a consensus? *Curr Opin Struct Biol*. 2011;21:12-24.

Toor N, Keating KS, Pyle AM. Structural insights into RNA splicing. *Curr Opin Struct Biol*. 2009;19:260-266.

Van Roey K, Uyar B, Weatheritt RJ, et al. Short linear motifs: Ubiquitous and functionally diverse protein interaction modules directing cell regulation. *Chem Rev*. 2014;114:6733-6778.

Vogel C, Bashton M, Kerrison ND, et al. Structure, function and evolution of multi-domain proteins. *Curr Opin Struct Biol*. 2004;14:208-216.

Wolynes PG. Evolution, energy landscapes and the paradoxes of protein folding. *Biochimie*. 2015;119:218-230.

CHAPTER 4

Biophysical Principles

The concepts in this chapter form the basis for understanding all the molecular interactions in chemistry and biology. To illustrate some of these concepts with a practical example, the chapter concludes with a section on the exceptionally important Ras family of enzymes that bind and hydrolyze the nucleotide guanosine triphosphate (GTP). This example provides the background knowledge to understand how guanosine triphosphatases (GTPases) participate in numerous processes covered in later chapters.

Most molecular interactions in cells are driven by diffusion of reactants that simply collide with each other on a random basis. Similarly, dissociation of molecular complexes is a random process that occurs with a probability determined by the strength of the chemical bonds holding the molecules together. Many other reactions occur within molecules or molecular complexes. The aim of biophysical chemistry is to explain life processes in terms of such molecular interactions.

The extent of a chemical reaction is characterized by the **equilibrium constant;** the rates of the reactions are described by **rate constants.** This chapter reviews the physical basis for rate constants and how they are related to the thermodynamic parameter, the equilibrium constant. These simple but powerful principles permit a deeper appreciation of molecular interactions in cells. On the basis of many examples presented in this book, it will become clear to the reader that rate constants are at least as important as equilibrium constants because the rates of reactions govern the dynamics of the cell and many processes are controlled at rate-limiting steps. The chapter includes discussion of the chemical bonds important in biochemistry.

This chapter is adapted in part from Wachsstock. DH, Pollard TD. Transient state kinetics tutorial using KINSIM. *Biophys J.* 1994;67:1260–1273.

First-Order Reactions

First-order reactions have one reactant (R) and produce one or more products (P). The general case is simply

$$R \rightarrow P$$

Some common examples of first-order reactions (Fig. 4.1) include conformational changes, such as a change in shape of protein A to shape A*:

$$A \rightarrow A^*$$

and the dissociation of complexes, such as

$$AB \rightarrow A + B$$

The rate of a first-order reaction is directly proportional to the concentration of the reactant (R, A, or AB in these examples). The rate of a first-order reaction, expressed as a differential equation (rate of change of reactant or product as a function of time *[t]*), is simply the concentration of the reactant times a constant, the rate constant k, with units of s^{-1} (per second):

$$\text{Rate} = -d[R]/dt = d[P]/dt = k[R]$$

The rate of the reaction has units of M s^{-1}, where M is moles per liter and s is seconds (molar per second). As the reactant is depleted, the rate slows proportionally.

A first-order rate constant can be viewed as a **probability** per unit of time. For a conformational change, it is the probability that any A will change to A* in a unit of time. For dissociation of complex AB, the first-order rate constant is determined by the strength of the bonds holding the complex together. This "dissociation rate constant" can be viewed as the probability that the complex will fall apart in a unit of time. The probability of the conformational change of any particular A to A* or of the dissociation of any particular AB is independent of its concentration. *The concentrations of A and AB*

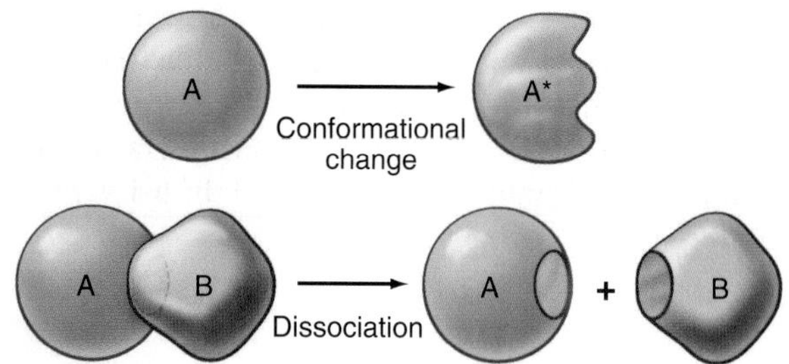

FIGURE 4.1 FIRST-ORDER REACTIONS. In first-order reactions, a single reactant undergoes a change. In these examples, molecule A changes conformation to A* and the bimolecular complex AB dissociates to A and B. The rate constant for a first-order reaction *(arrows)* is a simple probability.

are important only in determining the rate of the reaction observed in a bulk sample (Box 4.1).

To review, the rate of a first-order reaction is simply the product of a constant that is characteristic of the reaction and the concentration of the single reactant. The constant can be calculated from the half-time of a reaction (see Box 4.1).

Second-Order Reactions

Second-order reactions have two reactants (Fig. 4.2). The general case is

$$R_1 + R_2 \rightarrow \text{product}$$

A common example in biology is a bimolecular association reaction, such as

$$A + B \rightarrow AB$$

where A and B are two molecules that bind together. Some examples are binding of substrates to enzymes, binding of ligands to receptors, and binding of proteins to other proteins or nucleic acids.

The rate of a second-order reaction is the product of the concentrations of the two reactants, R_1 and R_2, and the second-order rate constant, k:

$$\text{Reaction rate} = d[P]/dt = k[R_1][R_2]$$

The second-order rate constant, k, has units of $M^{-1}\ s^{-1}$ (per molar per second). The units for the reaction rate are

$$[R_1]\cdot[R_2]\cdot k = M\cdot M\cdot M^{-1}s^{-1} \text{ or } M\ s^{-1}$$

the same as a first-order reaction.

The value of a second-order "association" rate constant, k_+, is determined mainly by the rate at which the molecules collide. This collision rate depends on the rate of diffusion of the molecules (Fig. 4.2), which is determined by the size and shape of the molecule, the viscosity of the medium, and the temperature. These factors are summarized in a parameter called the **diffusion coefficient,** D, with units of $m^2\ s^{-1}$. D is a measure of how fast a molecule moves in a given medium. The rate constant for collisions is described by the Debye-Smoluchowski equation, a relationship that depends only on the diffusion coefficients and the area of interaction between the molecules:

$$k = 4\pi b(D_A + D_B)N_o 10^3$$

where b is the interaction radius of the two particles (in meters), the Ds are the diffusion coefficients of the

BOX 4.1 Relationship of the Half-Time to a First-Order Rate Constant

In thinking about a first-order reaction, it is useful to refer to the half-time of the reaction. The half-time, $t_{1/2}$, is the time for half of the existing reactant to be converted to product. For a first-order reaction, this time depends *only on the rate constant* and therefore is the same regardless of the starting concentration of the reactant. The relationship is derived as follows:

$$d[R]/dt = -k[R]$$

so

$$d[R]/[R] = -kdt$$

Integrating and rearranging, we have

$$\ln[R_t] = \ln[R_o] - kt$$

or

$$[R_t] = [R_o]e^{-kt}$$

where R_o is the initial concentration and R_t is the concentration at time t.

When the concentration at the initial time point R_o is reduced by half,

$$\tfrac{1}{2}[R_o] = [R_o]e^{-kt_{1/2}}$$

or

$$\tfrac{1}{2} = e^{-kt_{1/2}}$$

and

$$2 = e^{kt_{1/2}}$$

Thus,

$$\ln 2 = kt_{1/2}$$

so, rearranging, we have

$$t_{1/2} = 0.693/k$$

and

$$k = 0.693/t_{1/2}$$

Therefore dividing 0.693 by the half-time gives the first-order rate constant and dividing 0.693 by the rate constant gives the half-time. This relationship is independent of the extent of the reaction at the outset of the observations and allows one to estimate the rate constant without knowing absolute concentrations.

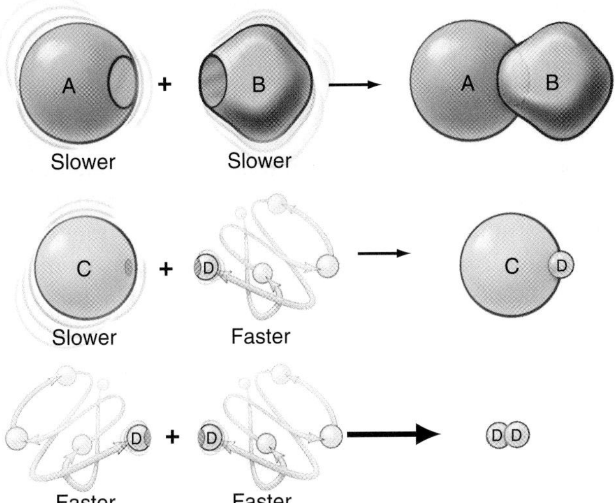

FIGURE 4.2 SECOND-ORDER REACTIONS. In second-order reactions, two molecules must collide with each other. The rate of these collisions is determined by their concentrations and by a collision rate constant *(arrows).* The collision rate constant depends on the sum of the diffusion coefficients of the reactants and the size of their interaction sites. The rate of diffusion in a given medium depends on the size and shape of the molecule. Large molecules, such as proteins, move more slowly than small molecules, such as adenosine triphosphate (ATP). A protein with a diffusion coefficient of 10^{-11} m^2 s^{-1} diffuses about 10 μm in a second in water, whereas a small molecule such as ATP diffuses 100 times faster. The rate constants *(arrows)* are about the same for A + B and C + D because the large diffusion coefficient of D offsets the small size of its interaction site on C. Despite the small interaction size, D + D is faster because both reactants diffuse rapidly.

reactants, and N_o is Avogadro's number. The factor of 10^3 converts the value into units of M^{-1} s^{-1}.

For particles the size of proteins, D is approximately 10^{-11} m^2 s^{-1} and b is approximately 2×10^{-9} m, so the rate constants for collisions of two proteins are in the range of 3×10^8 M^{-1} s^{-1}. For small molecules such as sugars, D is approximately 10^{-9} m^2 s^{-1} and b is approximately 10^{-9} m, so the rate constants for collisions of a protein and a small molecule are approximately 20 times larger than collisions of two proteins, in the range of 7×10^9 M^{-1} s^{-1}. On the other hand, experimentally observed rate constants for the association of proteins are 20 to 1000 times smaller than the collision rate constant, on the order of 10^6 to 10^7 M^{-1} s^{-1}. The difference is attributed to a steric factor that accounts for the fact that macromolecules must be correctly oriented relative to each other to bind together when they collide. Thus the complementary binding sites are aligned correctly only 0.1% to 5% of the times that the molecules collide.

Many binding reactions between two proteins, between enzymes and substrates, and between proteins and larger molecules (eg, DNA) are said to be "diffusion limited" in the sense that the rate constant is determined by diffusion-driven collisions between the reactants. Thus many association rate constants are in the range of 10^6 to 10^7 M^{-1} s^{-1}.

To review, the rate of a second-order reaction is simply the product of a constant that is characteristic of the reaction and the concentrations of the two reactants. In biology, the rates of many bimolecular association reactions are determined by the rates of diffusion-limited collisions between the reactants.

Reversible Reactions

Most reactions are reversible, so the net rate of a reaction is equal to the difference between the forward and reverse reaction rates. The forward and reverse reactions can be any combination of first- or second-order reactions. A reversible conformational change of a protein from A to A* is an example of a pair of simple first-order reactions:

$$A \rightleftarrows A^*$$

The forward reaction rate is k_+A with units of M s^{-1}, and the reverse reaction rate is k_-A^* with the same units. At equilibrium, when the net concentrations of A and A* no longer change,

$$k_+[A] = k_-[A^*]$$

and

$$K_{eq} = k_+/k_- = [A^*]/[A]$$

This equilibrium constant K_{eq} is unitless because the units of concentration and the rate constants cancel out. **Equilibrium constants** are designated by uppercase *K*s.

The same reasoning with respect to the equilibrium constant applies to a simple bimolecular binding reaction:

$$A + B \rightleftarrows AB$$

where A and B are any molecule (eg, enzyme, receptor, substrate, cofactor, or drug). The forward (binding) reaction is a second-order reaction, whereas the reverse (dissociation) reaction is a first-order reaction. The opposing reactions are

$$\text{Rate of association} = k_+[A][B]$$

$$\text{Units: M s}^{-1}$$

$$\text{Rate of dissociation} = k_-[AB]$$

$$\text{Units: M s}^{-1}$$

The overall rate of the reaction is the forward rate minus the reverse rate:

$$\text{Net rate} = \text{association rate} - \text{dissociation rate}$$
$$= k_+[A][B] - k_-[AB]$$

Depending on the values of the rate constants and the concentrations of A, B, and AB, the reaction can go forward, backward, or nowhere.

At equilibrium, the forward and reverse rates are (by definition) the same:

$$k_{+}[A][B] = k_{-}[AB]$$

The equilibrium constant for such a bimolecular reaction can be written in two ways:

Association equilibrium constant:

$$K_a = [AB]/[A][B] = k_{+}/k_{-}$$

$$\text{Units: } M^{-1} = M/M \times M$$

This is the classical equilibrium constant used in chemistry, where the strength of the reaction is proportional to the numerical value. For bimolecular reactions, the units of reciprocal molar are difficult to relate to, so biochemists frequently use the reciprocal relationship:

Dissociation equilibrium constant:

$$K_d = [A][B]/[AB] = k_{-}/k_{+}$$

$$\text{Units: } M = M \times M/M$$

When half of the total A is bound to B, the concentration of free B is simply equal to the dissociation equilibrium constant.

Thermodynamic Considerations

The driving force for chemical reactions is the lowering of the free energy of the system when reactants are converted into products. The larger the reduction in free energy, the more completely reactants will be converted to products at equilibrium. A thorough consideration of thermodynamics is beyond the scope of this text, but an overview of this subject is presented to allow the reader to gain a basic understanding of its power and simplicity.

The change in Gibbs free energy, ΔG, is simply the difference in the chemical potential, μ, of the reactants (R) and products (P):

$$\Delta G = \mu^{P} - \mu^{R}$$

The chemical potential of a particular chemical species depends on its intrinsic properties and its concentration, expressed as the equation

$$\mu = \mu^{0} + RT \ln C$$

where μ^0 is the chemical potential in the standard state (1 M in biochemistry), R is the gas constant (8.3 J mol^{-1} degree^{-1}), T is the absolute temperature in degrees Kelvin, and C is the ratio of the concentration of the chemical species to the standard concentration. Because the standard state is defined as 1 M, the parameter C has the same numerical value as the molar concentration, but is, in fact, unitless. The term $RT \ln C$ adjusts for the concentration. When $C = 1$, $\mu = \mu^0$.

Under standard conditions in which 1 mol of reactant is converted to 1 mol of product, the standard free energy change, ΔG^0, is

$$\Delta G^{0} = \mu^{0P} - \mu^{0R}$$

However, because most reactions do not take place under these standard conditions, the chemical potential must be adjusted for the actual concentrations. This is done by including the concentration term from the definition of the chemical potential. An equation for the free energy change that takes concentrations into account is

$$\Delta G = \mu^{0P} + RT \ln[P] - \mu^{0R} - RT \ln[R]$$

Substituting the definition of ΔG^0, we have

$$\Delta G = \Delta G^{0} + RT \ln[P] - RT \ln[R] = \Delta G^{0} + RT \ln[P]/[R]$$

This relationship tells us that the free energy change for the conversion of reactants to products is simply the free energy change under standard conditions corrected for the actual concentrations of reactant and products.

At equilibrium, the concentrations of reactants and products do not change and the free energy change is zero, so

$$0 = \Delta G^{0} + RT \ln[P_{eq}]/[R_{eq}]$$

or

$$\Delta G^{0} = -RT \ln[P_{eq}]/[R_{eq}]$$

You are already familiar with the fact that the equilibrium constant for a reaction is the ratio of the equilibrium concentrations of products and reactants. Thus that relationship can be substituted in this thermodynamic equation:

$$\Delta G^{0} = -RT \ln K$$

or

$$K = e^{-\Delta G^{0}/RT} = k^{+}/k^{-} = [P_{eq}]/[R_{eq}]$$

This profound relationship shows how the free energy change is related to the equilibrium constant. The change in the standard Gibbs free energy, ΔG^0, specifies the ratio of products and reactants when the reaction reaches equilibrium, *regardless of the rate or path of the reaction.* The free energy change provides no information about whether or not a given reaction will proceed on a time scale relevant to cellular activities. Nevertheless, because the equilibrium constant depends on the ratio of the rate constants, knowledge of the rate constants reveals the equilibrium constant and the free energy change for a reaction. Consider the consequences of various values of ΔG^0:

- If ΔG^0 equals 0, $e^{-\Delta G^{0}/RT}$ equals 1, and at equilibrium, the concentration of products will equal the concentration of reactants (or in the case of a bimolecular reaction, the product of the concentrations of the reactants).

- If ΔG^0 is less than 0, $e^{-\Delta G^0/RT}$ is greater than 1, and at equilibrium, the concentration of products will be greater than the concentration of reactants. Larger, negative free energy changes will drive the reaction farther toward products. Favorable reactions have large negative ΔG^0 values.
- If ΔG^0 is greater than 0, $e^{-\Delta G^0/RT}$ is less than 1, and at equilibrium, the concentrations of reactants will exceed the concentration of products.

It is sometimes said that a reaction with a positive ΔG^0 will not proceed spontaneously. This is not strictly true. Reactants will still be converted to products, although relative to the concentration of reactants, the concentration of products will be small. The size and sign of the free energy change tell nothing about the rate of a reaction. For example, the oxidation of sucrose by oxygen is highly favored with a ΔG^0 of −5693 kJ/mol, but "a flash fire in a sugar bowl is an event rarely, if ever, seen."*

The free energy change is additionally related to two thermodynamic parameters that are important to the subsequent discussion of molecular interactions. The Gibbs-Helmholtz equation is the key relationship:

$$\Delta G = \Delta H - T\Delta S$$

where ΔH is the change in **enthalpy,** an approximation (with a small correction for pressure-volume work) of the bond energies of the molecules. Thus ΔH is the heat given off when a bond is made or the heat taken up when a bond is broken. The change in enthalpy is simply the difference in enthalpy of reactants and products. In biochemical reactions, the enthalpy term principally reflects energies of the strong covalent bonds and of the weaker hydrogen and electrostatic bonds. If no covalent bonds change, as in a binding reaction or a conformational change, ΔH is determined by the difference in the energy of the weak bonds of the products and reactants.

The change in **entropy,** expressed as ΔS, is a measure of the change in the order of the products and reactants. The value of the entropy is a function of the number of microscopic arrangements of the system, including the solvent molecules. Note the minus sign in front of the $T\Delta S$ term. Reactions are favored if the change in entropy is positive, that is, if the products are less-well-ordered than the reactants. Increases in entropy drive reactions by increasing the negative free energy change. For example, the hydrophobic effect, which is discussed later in this chapter, depends on an increase in entropy. Increases in entropy provide the free energy change for many biologic reactions, especially macromolecular folding (see Chapters 3 and 12) and assembly (see Chapter 5).

As emphasized in the case of ΔG, neither the rate of the reaction nor the path between reactants and products is relevant to the difference in enthalpy or entropy of reactants and products. The reader may consult a physical chemistry book for a fuller explanation of these basic principles of thermodynamics.

*Eisenberg D, Crothers D. *Physical Chemistry with Applications to the Life Sciences.* Menlo Park, CA: Benjamin Cummings, 1979.

Linked Reactions

Many important processes in the cell consist of a single reaction, but most of cellular biochemistry involves a series of linked reactions (Fig. 4.3). For example, when two macromolecules bind together, the complex often undergoes some type of internal rearrangement or conformational change, linking a first-order reaction to a second-order reaction.

$$A + B \rightleftarrows AB \qquad AB \rightleftarrows AB^*$$

One of thousands of such examples is GTP binding to a G protein, causing it to undergo a conformational change from the inactive to the active state (Figs. 4.6 and 4.7 later).

Similarly, the basic enzyme reaction considered in most biochemistry books is simply a series of reversible second- and first-order reactions:

$$E + S \rightleftarrows ES \qquad ES \rightleftarrows EP \qquad EP \rightleftarrows E + P$$

where E is enzyme, S is substrate, and P is product. These and more complicated reactions can be described rigorously by a series of rate equations like those explained previously. For example, enzyme reactions nearly always involve one or more additional intermediates between ES and EP, coupled by first-order reactions, in which the molecules undergo conformational changes.

Linking reactions together is the strategy used by cells to carry out unfavorable reactions. All that matters is that the total free energy change for all coupled reactions is negative. An unfavorable reaction is driven forward by a favorable reaction upstream or downstream. For example, a proton gradient across the mitochondrial

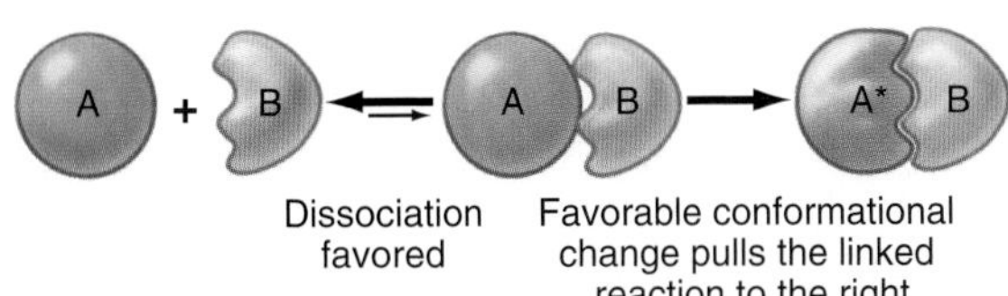

FIGURE 4.3 LINKED REACTIONS. Two molecules, A and B, bind together weakly and then undergo a favorable conformational change. The binding reaction is unfavorable, owing to the high rate of dissociation of AB, but the favorable conformational change pulls the overall reaction far to the right.

membrane is used as an energy source to drive the unfavorable reaction producing adenosine triphosphate (ATP) from adenosine diphosphate (ADP) and inorganic phosphate (see Fig. 14.5). This proton gradient is derived, in turn, from the oxidation of chemical bonds of nutrients. To use a macroscopic analogy, a siphon can initially move a liquid uphill against gravity provided that the outflow is placed below the inflow, so that the overall change in energy is favorable.

An appreciation of linked reactions makes it possible to understand how catalysts, including biochemical catalysts—protein enzymes and ribozymes—influence reactions. They do not alter the free energy change for reactions, but they enhance the rates of reactions by speeding up the forward and reverse rates of unfavorable intermediate reactions along pathways of coupled reactions. Given that the rates of both first- and second-order reactions depend on the concentrations of the reactants, the overall reaction is commonly limited by the concentration of the least-favored, highest-energy intermediate, called a transition state. This might be a strained conformation of substrate in a biochemical pathway. Interaction of this transition state with an enzyme can lower its free energy, increasing its probability (concentration) and thus the rate of the limiting reaction. Acceleration of biochemical reactions by enzymes is impressive. Enhancement of reaction rates by 10 orders of magnitude is common.

Chemical Bonds

Covalent bonds are responsible for the stable architecture of the organic molecules in cells (Fig. 4.4). They are very strong. C—C and C—H bonds have energies of approximately 400 kJ mol^{-1}. Bonds this strong do not dissociate spontaneously at body temperatures and pressures, nor are the reactive intermediates required to form these bonds present in finite concentrations in cells. To overcome this problem, living systems use enzymes, which stabilize high-energy transition states, to catalyze formation and dissolution of covalent bonds. Energy for making strong covalent bonds is obtained indirectly by coupling to energy-yielding reactions. For example, metabolic enzymes convert energy released by breaking covalent bonds of nutrients, such as carbohydrates, lipids, and proteins, into ATP (see Fig. 19.4), which supplies energy required to form new covalent bonds during the synthesis of polypeptides. Metabolic pathways relating the covalent chemistry of the molecules of life are covered in depth in many excellent biochemistry books.

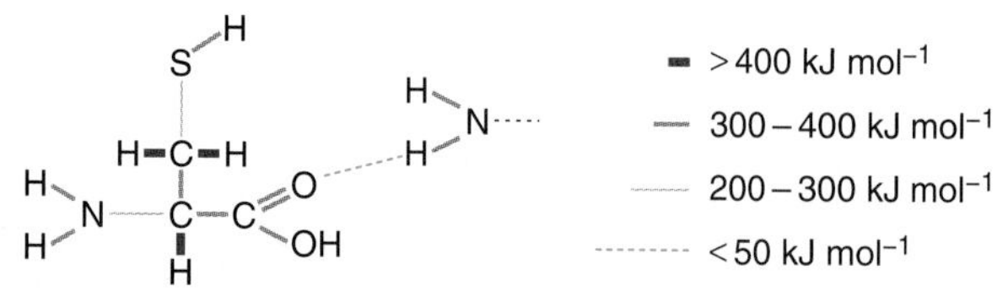

For cell biologists, four types of relatively weak interactions (Fig. 4.5) are as important as covalent bonds because they are responsible for folding macromolecules into their active conformations and for holding molecules together in the structures of the cell. These weak interactions are (a) **hydrogen bonds,** (b) **electrostatic interactions,** (c) the **hydrophobic effect,** and (d) **van der Waals interactions.** None of these interactions is particularly strong on its own. Stable bonding between subunits of many macromolecular structures, between ligands and receptors, and between substrates and enzymes is a result of the additive effect of many weak interactions working in concert.

Hydrogen and Electrostatic Bonds

Hydrogen bonds (Fig. 4.5) occur between a covalently bound donor H atom with a partial positive charge, Δ+ (the result of electron withdrawal by a covalently bonded O or N), and an acceptor atom (usually O or N) with a partial negative charge, Δ–. These bonds are highly directional, with optimal bond energy (12 to 29 kJ mol^{-1}) when the H atom points directly at the acceptor atom. Hydrogen bonds are extremely important in the stabilization of secondary structures of proteins, such as α-helices and β-sheets (see Fig. 3.8), and in the base pairing of DNA and RNA (see Fig. 3.14).

- >400 kJ mol^{-1}
- 300–400 kJ mol^{-1}
- 200–300 kJ mol^{-1}
- <50 kJ mol^{-1}

FIGURE 4.4 COVALENT BONDS. Bond energies for the amino acid cysteine.

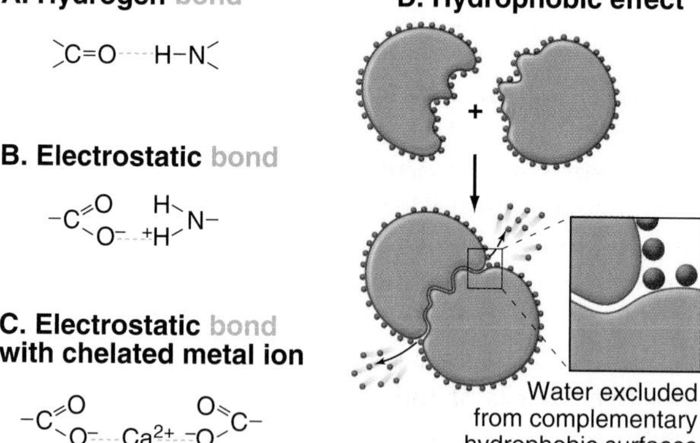

FIGURE 4.5 WEAK INTERACTIONS. A, Hydrogen bond. Opposite partial charges in the oxygen and hydrogen provide the attractive force. **B,** Electrostatic bond. Atoms with opposite charges are attracted to each other. **C,** Ca^{2+} chelated between two negatively charged oxygens. **D,** The hydrophobic effect arises when two complementary, apolar surfaces make contact, excluding water molecules that formerly were associated with the surfaces. The increased disorder of the water increases the entropy and provides the decrease in free energy to drive the association. van der Waals interactions between closely packed atoms on complementary surfaces also stabilize interactions.

Electrostatic (or ionic) bonds occur between charged groups that have either lost or gained a proton (eg, —COO^- and —NH_3^+). Although these bonds are potentially about as strong as an average hydrogen bond (20 kJ mol^{-1}), it has been argued that they contribute little to biological structure. This is because a charged group is usually neutralized by an inorganic counterion (such as Na^+ or Cl^-) that is itself surrounded by a cloud of water molecules. The effect of having the cloud of water molecules is that the counterion does not occupy a single position with respect to the charged group on the macromolecule; consequently, these interactions lack structural specificity.

Electrostatic interactions come into their own in dissociating biological structures, since like charges on potential binding surfaces of macromolecules will repel one another. This allows phosphorylation to control many biological interactions. Enzymatic introduction of a negatively charged phosphate group can disrupt an otherwise stable interaction between two proteins, whereas removal of phosphate allows the interaction (see Chapter 25).

Hydrophobic Effect

Self-assembly and other association reactions that involve the joining together of separate molecules to form more ordered structures might seem unlikely when examined from the point of view of thermodynamics. Nonetheless, many binding reactions are highly favored, and when such processes are monitored in the laboratory, it can be shown that ΔS actually increases.

How can association of molecules lead to increased disorder? The answer is that the entropy of the system—including macromolecules and solvent—increases owing to the loss of order in the *water* surrounding the macromolecules (Fig. 4.5). This increase in the entropy of the water more than offsets the increased order and decreased entropy of the associated macromolecules. Bulk water is a semistructured solvent maintained by a loose network of hydrogen bonds (see Fig. 3.1). Water cannot form hydrogen bonds with nonpolar (hydrophobic) parts of lipids and proteins. Instead, water molecules form "cages" or "clathrates" of extensively H-bonded water molecules near these hydrophobic surfaces. These clathrates are more ordered than is bulk water or water interacting with charged or polar amino acids.

When proteins fold (see Fig. 12.10), macromolecules bind together (see Chapter 5), and phospholipids associate to form bilayers (see Fig. 13.5), hydrophobic groups are buried in pockets or between interfaces that exclude water. The highly ordered water formerly associated with these surfaces disperses into the less-ordered bulk phase, and the entropy of the system increases.

The increase in the disorder of water that results when hydrophobic regions of macromolecules are buried is called the **hydrophobic effect.** Hydrophobic interactions are a major driving force, but they would not confer specificity on an intermolecular interaction except for the fact that the molecular surfaces must be complementary to exclude water. The hydrophobic effect is not a bond per se, but a thermodynamic factor that favors macromolecular interactions.

van der Waals Interactions

van der Waals interactions occur when adjacent atoms come close enough that their outer electron clouds just barely touch. This action induces charge fluctuations that result in a nonspecific, nondirectional attraction. These interactions are highly distance dependent, decreasing in proportion to the sixth power of the separation. The energy of each interaction is only about 4 kJ mol^{-1} (very weak when compared with the average kinetic energy of a molecule in solution, which is approximately 2.5 kJ mol^{-1}) and is significant only when many interactions are combined (as in interactions of complementary surfaces). Under optimal circumstances, van der Waals interactions can achieve bonding energies as high as 40 kJ mol^{-1}.

When two atoms get too close, they strongly repel each other. Consequently, imperfect fits between interacting molecules are energetically very expensive, preventing association if surface groups interfere sterically with each other. As a determinant of specificity of macromolecular interactions, this van der Waals repulsion is even more important than the favorable bonds discussed earlier, because it precludes many nonspecific interactions.

Strategy for Understanding Cellular Functions

One strategy for understanding the mechanism of any molecular process—including binding reactions, self-assembly reactions, and enzyme reactions—is to determine the existence of the various reactants, intermediates, and products along the reaction pathway and then to measure the rate constants for each step. Such an analysis yields additional information about the thermodynamics of each step, as the ratio of the rate constants reveals the equilibrium constant and the free energy change, even for transient intermediates that may be difficult or impossible to analyze separately.

In earlier times, biochemists lacked methods to evaluate the internal reactions along most pathways, but they could measure the overall rate of reactions, such as the steady-state rate of conversion of reactants to products by an enzyme. To analyze these data, they simplified complex mechanisms using relationships such as the Michaelis-Menten equation (described in biochemistry textbooks). Now, abundant supplies of proteins, convenient methods for measuring rapid reaction rates, and computer programs that can be used to analyze complex reaction mechanisms generally make such simplifications unnecessary.

Analysis of an Enzyme Mechanism: The Ras GTPase

This section uses a vitally important family of enzymes called **GTPases** to illustrate how enzymes work. The example is **Ras,** a small GTPase that serves as part of a biochemical pathway linking growth factor receptors in the plasma membrane of animal cells to regulation of the cell cycle. The example shows how to dissect an enzyme reaction by kinetic analysis and how crystal structures can reveal conformational changes related to function. GTPases related to Ras regulate a host of systems, including nuclear transport (see Fig. 9.18), protein synthesis (see Figs. 12.8 and 12.9), vesicular trafficking (see Fig. 21.6), signaling pathways coupled to seven-helix receptors including vision and olfaction (see Fig. 25.9), the actin cytoskeleton (see Figs. 33.13 and 33.19), and assembly of the mitotic spindle (see Fig. 44.8). This section gives the reader the background required to understand the contributions of GTPases to all these processes as they are presented in the following sections of the book.

Having evolved from a common ancestor, Ras and its related GTPases share a homologous core domain that binds a guanine nucleotide and uses a common enzymatic cycle of GTP binding, hydrolysis, and product dissociation to switch the protein on and off (Fig. 4.6). The GTP-binding domain consists of approximately 200 residues folded into a six-stranded β-sheet sandwiched between five α-helices. GTP binds in a shallow groove formed largely by loops at the ends of elements of secondary structure. A network of hydrogen bonds between the protein and guanine base, ribose, triphosphate, and Mg^{2+} anchor the nucleotide. Larger GTPases have a core GTPase domain plus domains required for coupling to seven-helix receptors (see Fig. 25.9) or regulating protein synthesis (see Figs. 12.10 and 25.7).

The bound nucleotide determines the conformation and activity of each GTPase. The GTP-bound conformation is active, as it interacts with and stimulates effector proteins. In the example considered here, the Ras-GTP binds and stimulates a protein kinase, Raf, which relays signals from growth factor receptors to the nucleus (see Fig. 27.6). The guanosine diphosphate (GDP)-bound conformation of Ras is inactive because it does not bind effectors. Thus, GTP hydrolysis and phosphate dissociation switch Ras and related GTPases from the active to the inactive state.

All GTPases use the same enzyme cycle, which involves four simple steps (Fig. 4.6). GTP binding favors the active conformation that binds effector proteins. GTPases remain active until they hydrolyze the bound

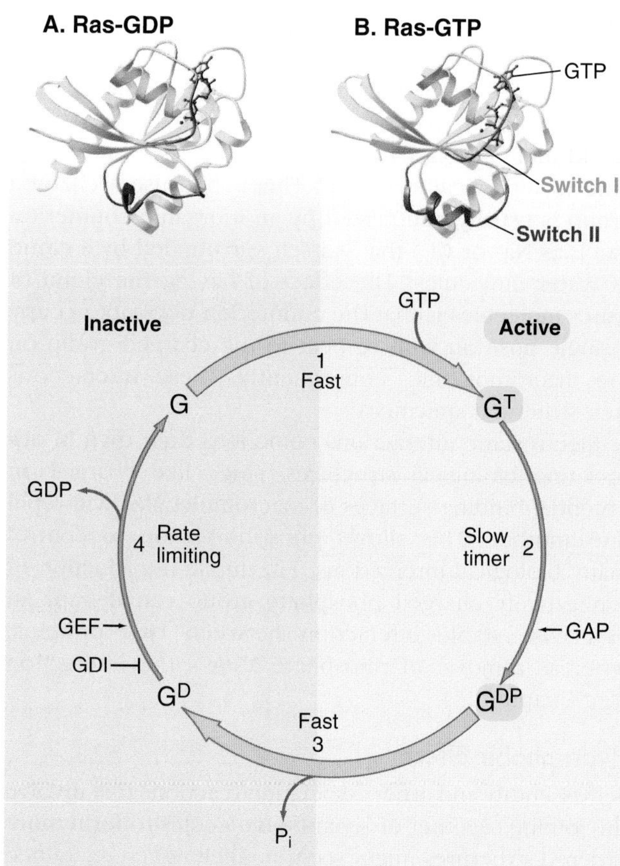

FIGURE 4.6 *Top,* Atomic structures of the small GTPase Ras. GTP hydrolysis and phosphate dissociation change the conformations of the switch loops. (For reference, see Protein Data Bank [www.rcsb.org] files 1Q21 [**A**] and 121P [**B**].) *Bottom,* Generic GTPase cycle. The size of the *arrows* indicates the relative rates of the reactions. GAP, GTPase activating protein; G^D, GTPase with bound GDP; GDI, guanine nucleotide dissociation inhibitor; G^{DP}, GTPase with bound GDP and inorganic phosphate; GEF, guanine nucleotide exchange factor; G^T, GTPase with bound GTP; P_i, phosphate.

GTP. Hydrolysis is intrinsically slow, but binding to effector proteins or regulatory proteins can accelerate this inactivation step. GTPases tend to accumulate in the inactive GDP state, because GDP dissociation is very slow. Specific proteins catalyze dissociation of GDP, making it possible for GTP to rebind and activate the GTPase. Seven-helix receptors activate their associated G-proteins. Guanine nucleotide exchange proteins (GEFs) activate small GTPases.

Figure 4.7 illustrates the experimental strategy used to establish the mechanism of the Ras GTPase cycle.

Step 1: GTP binding. GTP binds rapidly to nucleotide-free Ras in two linked reactions (Fig. 4.7A). The first is rapid but reversible association of GTP with Ras. Second is a slower but highly favorable first-order conformational change, which produces the fluorescence signal in the experiment and accounts for the

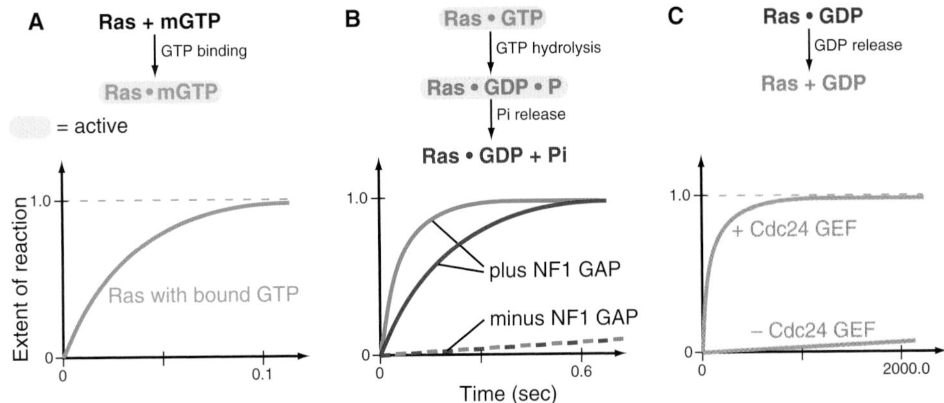

FIGURE 4.7 Kinetic dissection of the Ras GTPase cycle using a series of "single turnover" experiments, in which each enzyme molecule carries out a reaction only once. **A,** GTP binding. Nucleotide-free Ras is mixed rapidly with a fluorescent derivative of GTP (mGTP), and fluorescence is followed on a millisecond time scale. With 100 μM mGTP (approximately 10% of the cellular concentration), binding is fast (half-time less than 5 ms), but the change in fluorescence is slower, approximately 30 s^{-1}, because it depends on a subsequent, slower conformational change. Linking the association reaction to this highly favorable ($K = 10^6$) first-order conformational change accounts for the exceedingly high affinity ($K_d = \sim 10^{-11}$ M) of Ras for GTP. Binding and dissociation of GDP are similar. **B,** GTP hydrolysis and γ-phosphate dissociation. GTP is mixed with Ras, and hydrolysis is followed by collecting samples on a millisecond time scale with a "quench-flow" device, dissociating the products from the enzyme and measuring the fraction of GTP converted to GDP. The Ras-GDP-P intermediate releases γ-phosphate spontaneously in a first-order reaction. A fluorescent phosphate-binding protein is used to measure free phosphate. On this time scale in this figure, Ras alone does not hydrolyze GTP or dissociated phosphate because the hydrolysis rate constant is 5×10^{-5} s^{-1}, corresponding to a half-time of 1400 seconds. The GTPase activating protein (GAP) neurofibromin 1 (NF1) at a concentration of 10 μM increases the rate of hydrolysis to 20 s^{-1} and allows observation of the time course of phosphate dissociation at 8 s^{-1}. **C,** GDP dissociation. Ras with bound fluorescent mGDP is mixed with GTP, which replaces the mGDP as it dissociates. The loss of fluorescence over time gives a rate constant for mGDP dissociation of 0.00002 s^{-1}. The guanine nucleotide exchange factor $Cdc24^{Mn}$ at a concentration of 1 μM increases the rate of mGDP dissociation 500-fold to 0.01 s^{-1}. (Data from Lenzen C, Cool RH, Prinz H, et al. Kinetic analysis by fluorescence of the interaction between Ras and the catalytic domain of the guanine nucleotide exchange factor $Cdc24^{Mn}$. *Biochemistry.* 1998;37:7420–7430; and Phillips RA, Hunter JL, Eccleston JF, Webb MR. Mechanism of Ras GTPase activation by neurofibromin. *Biochemistry*. 2003;42:3956–3965.)

high affinity of Ras for GTP (K_d typically in the range of 10^{-11} M). The conformation change involves three segments of the polypeptide chain called switch I, switch II, and switch III. Folding of these three loops around the γ-phosphate of GTP traps the nucleotide and creates a binding site for the Raf kinase, the downstream effector (see Fig. 27.6).

Step 2: GTP hydrolysis. Hydrolysis is essentially irreversible and slow with a half-time of approximately 4 hours (Fig. 4.7B). Although slow, GTP hydrolysis on the enzyme is many orders of magnitude faster than in solution. Like other enzymes, interactions of the protein with the substrate stabilize the "transition state," a high-energy chemical intermediate between GTP and GDP. In this transition state, the γ-phosphate is partially bonded to both the β-phosphate and an attacking water. Hydrogen bonds between protein backbone amides and oxygens bridging the β- and γ-phosphates and on the γ- and β-phosphates stabilize negative charges that build up on these atoms in the transition state. Hydrolysis is slow in comparison with most enzyme reactions, because none of these hydrogen bonds is particularly strong. Another hydrogen bond from a glutamine side chain helps position a water for nucleophilic attack on the γ-phosphate. The importance of this interaction is illustrated by mutations that replace that glutamine with leucine. This mutation reduces the rate of hydrolysis by orders of magnitude and predisposes to the development of many human cancers by prolonging the active state and thus amplifying growth-promoting signals from growth factor receptors.

Step 3: Dissociation of inorganic phosphate. After hydrolysis, the γ-phosphate dissociates rapidly. This reverses the conformational change of the three switch loops, dismantling the binding site for effector proteins.

Step 4: Dissociation of GDP. On its own, Ras accumulates in the inactive GDP state, because GDP dissociates extremely slowly with a half-time of 10 hours (Fig. 4.7C). GTP cannot bind and activate Ras until GDP dissociates.

Ras and most other small GTPases depend on regulatory proteins to stimulate the two slow steps in the GTPase cycle: GDP dissociation and GTP hydrolysis. For example, when growth factors stimulate their receptors, a series of reactions (see Fig. 27.6) brings a **guanine nucleotide exchange factor (GEF)** to the plasma membrane to activate Ras by accelerating dissociation of GDP. First the GEF binds Ras-GDP and then favors a slow conformational change that distorts a part of Ras that interacts with the β-phosphate. This allows GDP to dissociate on a time scale of seconds to minutes rather than 10 hours (Fig. 4.7C). Once GDP has dissociated, nucleotide-free Ras can bind either GDP or GTP. Binding GTP is more likely in cells, because the cytoplasmic concentration of GTP (approximately 1 mM) is 10 times that of GDP. GTP binding activates Ras, allowing transmission of the signal to the nucleus.

GTPase-activating proteins (GAPs) turn off Ras and related GTPases, by binding Ras-GTP and stimulating GTP hydrolysis, thereby terminating GTPase activation (Fig. 4.7B). Ras GAPs stabilize the transition state, by contributing a positively charged arginine side chain that stabilizes the negative charges on the oxygen bridging the β- and γ-phosphates and on the γ-phosphate. GAPs also help position Gln61 and its attacking water. In the experiment in the figure, a GAP called neurofibromin (NF1) binds Ras with a half-time of 3 ms (not illustrated) and stimulates rapid hydrolysis of GTP at 20 s^{-1}. This is followed by rate-limiting dissociation of γ-phosphate from the Ras-GDP-P intermediate at 8 s^{-1} and rapid dissociation of NF1 from Ras at 50 s^{-1}. NF1 is the product of a human gene that is inactivated in the disease called neurofibromatosis. Lacking the NF1 GAP activity to keep Ras in check, affected individuals develop numerous neural tumors that disfigure the skin and may compromise the function of the nervous system.

ACKNOWLEDGMENT

We thank Martin Webb for his help with GTPase kinetics for the second edition.

SELECTED READINGS

Berg OG, von Hippel PH. Diffusion controlled macromolecular interactions. *Annu Rev Biophys Biophys Chem*. 1985;14:131-160.

Eisenberg D, Crothers D. *Physical Chemistry with Applications to the Life Sciences*. Menlo Park, CA: Benjamin Cummings; 1979.

Garcia HG, Kondev J, Orme N, Theriot JA, Phillips R. Thermodynamics of biological processes. *Methods Enzymol*. 2011;492:27-59.

Garcia-Viloca M, Gao J, Karplus M, Truhlar DG. How enzymes work: analysis by modern rate theory and computer simulations. *Science*. 2004;303:186-194.

Herrmann C. Ras-effector interactions: after one decade. *Curr Opin Struct Biol*. 2003;13:122-129.

Johnson KA. Transient-state kinetic analysis of enzyme reaction pathways. *Enzymes*. 1992;20:1-61.

Lenzen C, Cool RH, Prinz H, et al. Kinetic analysis by fluorescence of the interaction between Ras and the catalytic domain of the guanine nucleotide exchange factor Cdc^{Mn}. *Biochemistry*. 1998;37:7420-7430.

Northrup SH, Erickson HP. Kinetics of protein-protein association explained by Brownian dynamics computer simulation. *Proc Natl Acad Sci USA*. 1992;89:3338-3342.

Phillips RA, Hunter JL, Eccleston JF, Webb MR. Mechanism of Ras GTPase activation by neurofibromin. *Biochemistry*. 2003;42:3956-3965.

Pollard TD. A guide to simple and informative binding assays. *Molec Biol Cell*. 2010;21:4061-4067.

Pollard TD, De La Cruz E. Take advantage of time in your experiments: a guide to simple, informative kinetics assays. *Mol Biol Cell*. 2013;24:1103-1110.

CHAPTER 5

Macromolecular Assembly

The discovery that dissociated parts of viruses can reassemble in a test tube led to the concept of **self-assembly**. Demonstration that purified components of viruses, bacterial flagella, ribosomes, and cytoskeletal filaments assemble in vitro established self-assembly as a central principle in biology. Even large biological structures, such as the mitotic spindle (Fig. 5.1), are constructed from molecules that assemble by defined pathways without external templates. Chromosomes, nuclear pores, transcription initiation complexes, vesicle fusion machinery, and intercellular junctions, assemble by the same strategy. The properties of the constituents determine the assembly mechanism and architecture of the final structure. Weak but highly specific noncovalent interactions hold together the building blocks, which include proteins, nucleic acids, and lipids.

The ability of **subunit molecules** to assemble spontaneously into the complicated structures required for cellular function greatly increases the power of the information stored in the genome. The primary structure of a protein or nucleic acid specifies not only the folding of the individual protein or nucleic acid subunit but also the bonds that it can make in a larger assembly.

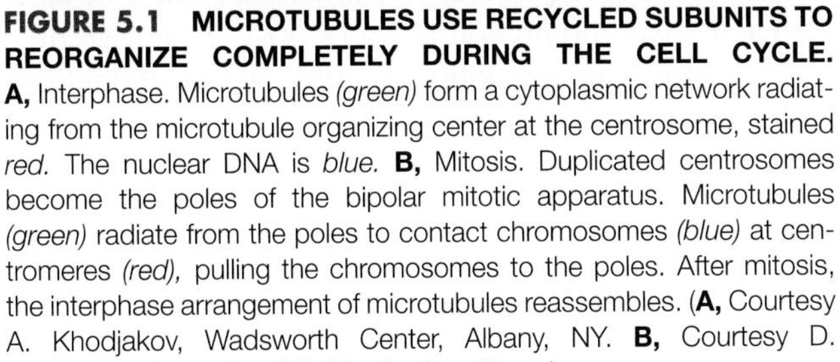

Assembly of macromolecular structures differs fundamentally from the template-specified, enzymatic mechanisms with which cells replicate genes (see Chapter 42), transcribe gene sequences into RNAs, and translate messenger RNA (mRNA) sequences into proteins (see Chapters 10 and 12). Macromolecular assembly does not require templates and rarely involves enzymatic formation or dissolution of covalent bonds between subunits. When enzymatic processing occurs during the assembly of some viruses (see Example 6 below), collagen (see Fig. 29.6), and elastin (see Fig. 29.11), it usually precludes reassembly of the dissociated parts.

After explaining the advantages and general features of self-assembly, this chapter concludes with several model systems illustrating these principles. Subsequent chapters show how these ideas help explain the structure, biogenesis, and function of most cellular components.

A B

FIGURE 5.1 MICROTUBULES USE RECYCLED SUBUNITS TO REORGANIZE COMPLETELY DURING THE CELL CYCLE. **A,** Interphase. Microtubules *(green)* form a cytoplasmic network radiating from the microtubule organizing center at the centrosome, stained *red.* The nuclear DNA is *blue.* **B,** Mitosis. Duplicated centrosomes become the poles of the bipolar mitotic apparatus. Microtubules *(green)* radiate from the poles to contact chromosomes *(blue)* at centromeres *(red),* pulling the chromosomes to the poles. After mitosis, the interphase arrangement of microtubules reassembles. (**A,** Courtesy A. Khodjakov, Wadsworth Center, Albany, NY. **B,** Courtesy D. Cleveland, University of California–San Diego.)

Assembly of Macromolecular Structures From Subunits

Using subunits provides multiple advantages for assembly processes, as originally pointed out by Crane (Box 5.1).

Assembly of large structures from subunits conserves the genome. The assembly of macromolecular structures from identical subunits, like bricks in a wall, obviates the need to specify separate parts. For example, a plant virus, the **tobacco mosaic virus** (TMV; see Example 4 below), consists of 2130 protein subunits of 158 amino acids each and a single-stranded RNA molecule of 6390 nucleotides. Having a separate gene for each viral coat protein would require 1,009,620 nucleotides of RNA, which would be approximately 160-fold longer than the entire viral RNA! The virus conserves its genome by using a single copy of the coat protein gene (474 nucleotides—7.4% of the genome) to make the 2130 identical protein subunits that assemble into the virus coat.

Using small subunits improves the chance of synthesizing error-free building blocks. All biological processes are susceptible to error, and protein synthesis by ribosomes is no exception (see Chapter 12). The error rate of translation is approximately 1 in 3000 amino acid residues. Therefore, the odds that any given amino acid residue is correct are 0.99967. With these odds, the chance that a TMV subunit will be translated correctly is 0.99967^{158}, or 0.949. Thus, approximately 95% of all TMV coat proteins in an infected cell are perfect, providing an ample supply of subunits with which to construct an infectious virus. Of the 5% of subunits with a mistake, some will be functional and others will not, depending on the nature and position of the amino acid substitution. By contrast, the chance of correctly synthesizing the viral coat, if TMV coated its RNA with one huge polypeptide with 336,540 residues, would be only 0.99967^{336540}, or 5.6×10^{-49}.

Construction from subunits provides a mechanism for eliminating faulty components. Given that a significant fraction of all proteins has minor errors, good and bad subunits can be segregated on the basis of their ability to form correct bonds with their neighbors at the time of assembly. Faulty subunits will not bond, are excluded from the final structure, and are usually degraded.

Subunits can be recycled. Many macromolecular structures assemble reversibly, and because they are built of subunits, the subunits can be reused later. For example, the subunits of the mitotic spindle microtubules reassemble into the interphase array of microtubules (Fig. 5.1; see also Chapter 44).

Assembly from subunits provides multiple opportunities for regulation. Simple modifications of subunits can regulate the state of assembly. For example, many intermediate filaments disassemble during mitosis when their subunits are phosphorylated by protein kinases (see Figs. 35.5 and 44.6).

BOX 5.1 Crane's Hypothesis

In 1950, the physicist H.R. Crane predicted in *Scientific Monthly* that macromolecular structures in biology are assembled from multiple subunits and according to the laws of **symmetry.** A symmetric structure is composed of numerous identical **subunits,** all in equivalent environments (ie, making identical contacts with their neighbors). For example, Fig. 5.2A shows a plane hexagonal array, with each subunit making identical contacts with the six surrounding subunits. This is the most efficient way to fill a flat surface with globular subunits.

Crane also predicted that elongated tubular structures are assembled with symmetry. This type of symmetry is known as a **helix.** One way of constructing a helix is to take a plane hexagonal array, cut it along one of its lattice lines, and roll it up into a tube (Fig. 5.2B). The bonds between adjacent subunits are nearly identical in the plane array and the helical tube, except for the fact that each bond is distorted just enough to roll the sheet into a tube. Introduction of fivefold vertices into a hexagonal array allows it to fold up into a closed polygon (Fig. 5.2D–F).

Crane argued further that biological structures could avoid the problem of poisoning by defective subunits if such subunits were recognized and discarded. Crane's thinking about this problem was stimulated by a visit to a factory producing complex parts for vacuum tubes during World War II. When he asked the factory manager how much training the workers needed to assemble such a complex product, he was surprised to learn that the average was only 4 hours. The supervisor explained that they worked on an assembly line where each worker made only one small component (a subunit). If that component was defective, it was simply discarded, so the final product was built only from perfect components. Crane suggested that cells use the same strategy.

Crane's theories led to the hypothesis that cellular structures "build" themselves by self-assembly. Thus, the design of the final structure is somehow incorporated into the shape of the individual subunits. Remarkably, all of Crane's predictions about subunits and assembly turned out to be correct.

Specificity by Multiple Weak Bonds on Complementary Surfaces

Stable macromolecular assemblies require intermolecular interactions stronger than the forces tending to dissociate the subunits. Subunits diffusing independently in an aqueous milieu have a kinetic energy of approximately 2.5 kJ mol^{-1} at 25°C. Interactions in macromolecular assemblies must be strong enough to overcome this thermal energy, which tends to pull them apart.

Specific macromolecular associations are achieved by combining a small repertoire of weak bonds on complex, three-dimensional surfaces. Four weak interactions (see Fig. 4.5)—the hydrophobic effect, hydrogen bonds, electrostatic interactions, and van der Waals interactions—hold together subunits of macromolecular assemblies just like they stabilize folded proteins. Multiple weak interactions suffice, because the free energy changes contributed by all the weak interactions are added together. With a small correction for entropy changes, the overall binding constant for the association of subunits is the product of the equilibrium constants for each weak interaction [$K_A = (K_1)(K_2)(K_3)(\ldots)(K_n)$].

Far from being a liability, multiple weak interactions provide assembly systems with the ability to achieve exquisite specificity that is derived from the "fit" between **complementary surfaces** of interacting molecules (see Examples 4 and 5 below). Complementary surfaces are important for three reasons. First, atoms with the potential to form hydrogen bonds or electrostatic bonds must be placed in a complementary arrangement for

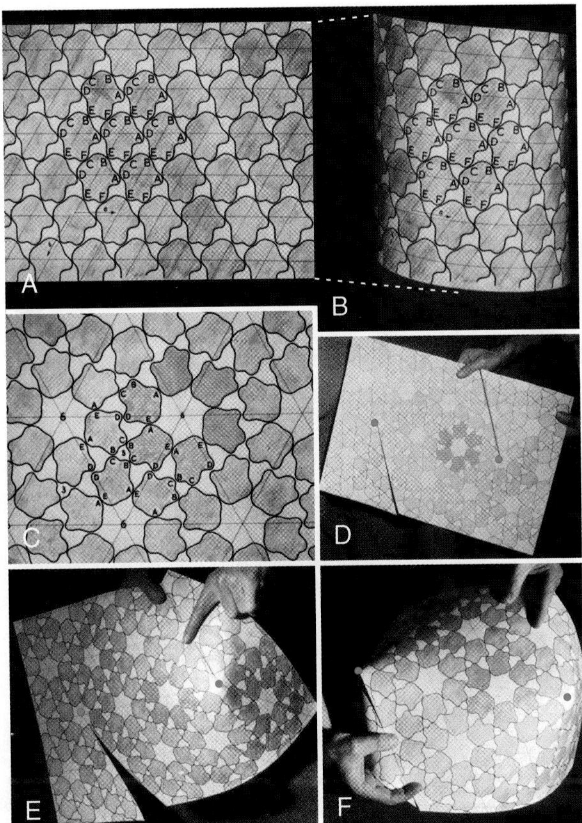

FIGURE 5.2 FOLDING OF PAPER MODELS OF HEXAGONAL ARRAYS OF IDENTICAL PARTICLES INTO A HELIX OR A CLOSED POLYGON. A, A hexagonal array of particles similar to the arrangement of subunits in the tobacco mosaic virus. **B,** The sheet is rolled around onto itself to make a helix similar to the virus. **C,** A hexagonal array of particles with three identical subunits in each triangular unit. The subunits around one sixfold axis are colored pink. **D–F,** The sheet is cut along two lattice lines and folded, creating two fivefold vertices *(green dot)*. Introduction of 12 such fivefold vertices creates an icosahedron. (From Caspar D, Klug A. Physical principles in the construction of regular viruses. *Cold Spring Harb Symp Quant Biol.* 1962;27:1–24.)

the bonds to form. Second, complementary surfaces can exclude water between subunits, as required for the hydrophobic effect. Third and most important, repulsive forces arising from clashes between even a few atoms on imperfectly matching surfaces can preclude interactions between two incorrect bonding partners.

To use a macroscopic analogy, the interactions between subunits of macromolecular assemblies have much more in common with Velcro fasteners than with snaps. Snaps provide an easy way to attach components to one another, and they can attach components whose surfaces touch only at the snaps. A single snap is often enough to hold two items together. By contrast, Velcro fasteners work because many tiny hooks become entrapped in a mesh of fibrous loops. The strength provided by each hook is minuscule, but when hundreds or thousands of hooks work together, bonding is strong. Velcro works best when the two bonding surfaces are smoothed against one another; in the case of rigid objects, a Velcro-like bond is tightest when the surfaces have complementary shapes.

Often short, intrinsically disordered regions of proteins adopt a stable structure when they dock onto the surface of a partner protein. This provides both flexibility and strength to protein assemblies. In some assemblies, flexible polypeptide strands knit subunits together (see Examples 1, 5, and 6 in "Regulation by Accessory Proteins" below). In other cases, assembly is coupled to the folding (or refolding) of the subunit proteins (see Examples 3, 4, and 6 below).

Symmetrical Structures Constructed From Identical Subunits

Studies of relatively simple systems composed of identical subunits, such as actin filaments, bacterial flagella and viruses, provided most of what is known about assembly processes. Their symmetries suit them for analysis by x-ray crystallography and electron microscopy, and their biochemical simplicity facilitates analysis of assembly mechanisms. Subunits in asymmetric assemblies, such as transcription factor complexes (see Fig. 10.7), are likely to behave similarly but are more difficult to study.

The subunits in a symmetrical macromolecular structure make identical bonds with one another. In practice, biological assemblies use only three fundamental types of symmetry. Proteins that assemble on flat surfaces, such as membranes, typically have **plane hexagonal symmetry**; filaments have **helical symmetry**; and closed structures have **polygonal symmetry**.

Subunits Arranged in Hexagonal Arrays in Plane Sheets

The simplest way to pack globular subunits in a plane is to form a hexagonal array with each subunit surrounded by six neighbors. This happens if one puts a layer of marbles in the bottom of a box and then tilts the box. A hexagonal array maximizes contacts between the surfaces of adjacent subunits. Membranes are the only flat surfaces in cells, and a number of membrane proteins crowd together in hexagonal arrays on or within the lipid bilayers. Connexons of gap junctions (Fig. 5.3A), bacteriorhodopsin of purple membranes (see Fig. 13.8B), and porin channels of bacterial membranes (see Fig. 13.8C) all form regular hexagonal arrays in the plane of the lipid bilayer. Clathrin coats form hexagonal nets on the surface of membranes (Fig. 5.3B).

Helical Filaments Produced by Polymerization of Identical Subunits With Like Bonds

Helical arrays of identical subunits form cytoskeletal filaments (see Examples 1 and 2 below), bacterial flagella (see Example 3 below), and some viruses (see Example 4 below). In helices subunits are positioned like steps of

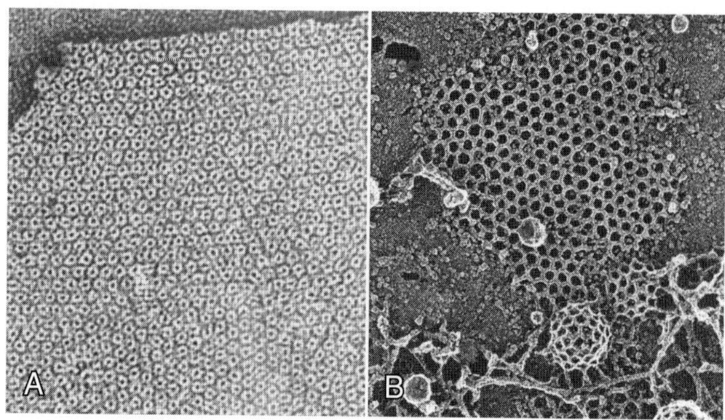

FIGURE 5.3 ELECTRON MICROGRAPHS SHOWING HEXAGONAL NETWORKS OF MEMBRANE PROTEINS. A, Integral membrane protein. Gap junction subunits called connexons span the lipid bilayer. An isolated junction was prepared by negative staining. **B,** Peripheral membrane proteins. Clathrin coats on the surface of a membrane in a hexagonal array. Introduction of fivefold vertices allows this sheet to fold up around a coated vesicle, shown at the bottom of the figure. This is a replica of the inner surface of the plasma membrane. (**A,** Courtesy N.B. Gilula, Scripps Research Institute, La Jolla, CA. **B,** Courtesy J. Heuser, Washington University, St. Louis, MO.)

a spiral staircase. Subunits are spaced a fixed distance apart along the axis and each new subunit is rotated by a fixed angle relative to the previous subunit. Helices can have one or more strands. TMV has one strand of subunits (see Example 4 below), whereas bacterial flagella have 11 strands (see Example 3 below). Helices can be either solid, like actin filaments (see Example 1 below), or hollow, like bacterial flagella (see Example 3 below) and TMV (see Example 4 below).

The asymmetry of protein subunits gives most helical polymers in biology a polarity (see Examples 1, 3, and 4 below). Different bonding properties at the two ends of the polymer can have important consequences for their assembly and functions.

Spherical Assemblies Formed by Regular Polygons of Subunits

Geometric constraints limit the ways that identical subunits can be arranged on a closed spherical surface with equivalent or nearly equivalent contacts between the subunits. By far, the most favored arrangement is based on a net of equilateral triangles. On a plane surface, these triangles will pack hexagonally with sixfold vertices (Fig. 5.2). Since the time of Plato, it has been appreciated that introducing vertices surrounded by three, four, or five triangles will cause such a network of triangles to pucker and, given an appropriate number of puckers, to close up into a complete shell (Fig. 5.4). Four threefold vertices make a tetrahedron, six fourfold vertices make an octahedron, and 12 fivefold vertices make an **icosahedron.** Remarkably, no other ways of arranging triangles will complete a shell. In addition to threefold, fourfold, or fivefold vertices that introduce puckers, a closed polygon can contain additional triangular faces and sixfold vertices to expand the volume. The sixfold vertices can be placed symmetrically with respect to the fivefold vertices to produce a spherical shell or asymmetrically to form an elongated structure (Fig. 5.4G).

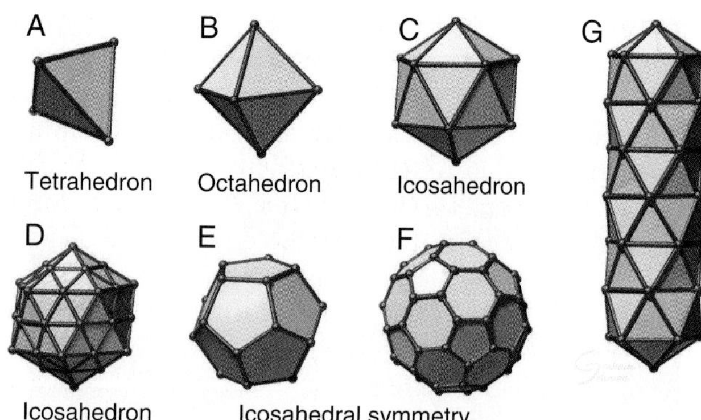

FIGURE 5.4 MODELS OF GEOMETRIC SOLIDS. A, A tetrahedron with four threefold vertices and four triangular faces. **B,** An octahedron with six fourfold vertices and eight triangular faces. **C–H,** Various icosahedral solids with 12 fivefold vertices. Many other arrangements of subunits are possible. **C,** One triangle on each face. **D,** Four triangles on each face. **E,** A dodecahedron with 20 vertices and 12 faces. **F,** An intermediate polyhedron with 60 vertices and 32 faces (12 pentagons and 20 hexagons). **G,** An extended structure made by including rings of hexagons between two icosahedral hemispheres. **H,** R. Buckminster Fuller standing in front of one of his geodesic domes. (From Caspar D, Klug A. Physical principles in the construction of regular viruses. *Cold Spring Harb Symp Quant Biol.* 1962;27:1–24.)

Most closed macromolecular assemblies in biology are polygons with fivefold vertices (see Examples 5 and 6 below). (The cubic iron-carrying protein ferritin is an exception.) An important reason for this is that most structures require some sixfold vertices to provide sufficient internal volume. This favors fivefold vertices for the puckers, as they require much less distortion of the subunits located on the triangular faces of the hexagonal plane sheet than do threefold or fourfold vertices. Furthermore, the distortion in the contacts between the triangles is minimized if the fivefold vertices are in equivalent positions. Closed icosahedral shells can be assembled from any type of asymmetrical subunit given two provisions: (a) The subunit must be able to form bonds with like subunits in a triangular network; and (b) these subunits must be able to accommodate the distortion required to form both fivefold and sixfold vertices. Both fibrous (Fig. 5.3B) and globular subunits (see Examples 5 and 6 below) can fulfill these criteria.

These considerations indicate that subunits in a closed macromolecular assembly must be arranged in rings of five or six. A simple variation has three like protein subunits on each face, but three different protein subunits, or more than three like subunits, can be used on each face to construct icosahedrons. The closest packing is achieved if the protein subunits form pentamers and hexamers, but other arrangements on the 20 faces of an icosahedron are possible (see Example 6 below).

Assembly Pathways

Understanding any assembly mechanism depends on determining the order that the subunits bind together and the rates of these reactions. This section describes some general principles about assembly reactions, but the following examples illustrate that more is generally known about the pathways than the reaction rates.

All self-assembly processes depend on diffusion-driven, random, reversible collisions between the subunits. As is described in Chapter 4, the rate equation for such a second-order bimolecular reaction is

$$\text{Rate} = k_+(\text{A})(\text{B}) - k_-(\text{AB})$$

where k_+ is the association rate constant; k_- is the dissociation rate constant; and *(A)*, *(B)*, and *(AB)* are the concentrations of the reactants and products. In assembly reactions *A* and *B* are the subunit and the structure to which it binds. Elongation of actin filaments (see Example 1 below) illustrates this mechanism.

The association rate is directly proportional to the concentration of subunits and a rate constant (k_+). This rate constant depends on the rates of **diffusion** of the subunits, the size of their complementary surfaces, and the degree of tolerance in orientation permitted for binding. In general, association rate constants are limited by diffusion and are in the range of 10^5 to 10^7 M^{-1} s^{-1} for most protein association reactions.

The rate of dissociation (k_-) determines the stabilities of complexes formed by random collisions. If two macromolecules collide in an orientation that allows a large number of simultaneous weak interactions on complementary surfaces or allows flexible strands to intertwine two subunits, the complex will dissociate slowly. If the surfaces are noncomplementary, few interactions form and the collision complex dissociates rapidly. Collision complexes have a wide spectrum of dissociation rate constants ranging from greater than 1000 s^{-1} for very unstable complexes to less than 0.00001 s^{-1} for very stable complexes. (The former complexes have a half-life of 0.7 ms, whereas the half-life of the latter is 16 hours. See Box 4.1 for an explanation of half-times.) Thus specificity is achieved by rapid dissociation of nonspecific complexes.

The sequence of random collisions, each followed by separation or bonding, can be viewed as a scanning process that allows each molecule to sample a variety of interactions. At cellular concentrations (see Fig. 3.27), macromolecules collide at high rates, but most collisions involve irrelevant molecules or molecules that could bind but collide in the wrong orientation. Given the high frequency of random collisions, it is important that proteins are not intrinsically too "sticky."

Conformational changes following formation of a collision complex between subunits often stabilize interactions. Because the equilibrium constants for all the coupled reactions are multiplied, a favorable conformational change can provide the major change in free energy holding a structure together (see Fig. 4.3). Bacterial flagella provide one clear example (see Example 3 below).

Large structures usually assemble by specific pathways in which new properties emerge at most steps. A new binding site for the next subunit may emerge from a conformational change in a newly incorporated subunit or by juxtaposition of two parts of a binding site on adjacent subunits. Such emergent properties favor addition of subunits in an orderly fashion until the process is completed. The assembly of actin (see Example 1), myosin (see Example 2), tomato bushy stunt virus (see Example 5), and bacteriophage T4 (see Example 6) illustrates control of assembly by emergent properties.

Initiation of assembly is frequently much less favorable than its propagation. Free subunits associating randomly cannot participate in all the stabilizing interactions enjoyed by a subunit joining a preexisting structure. Consequently, assembly of the first few subunits to form a "nucleus" for further growth may be thousands of times less favorable than the steps that follow during the growth of the assembly (see Example 1 below). The chance of dissociation from the assembly is reduced once subunits can engage in the full complement of bonds made possible by conformational changes that stabilize the structure. Cells often solve the **nucleation** problem by constructing specialized structures to nucleate the formation of macromolecular assemblies (see Examples 3 and 6; also see Figs. 33.13 and 34.16).

Regulation at Multiple Steps on Sequential Assembly Pathways

Many assembly reactions proceed spontaneously in vitro, but all seem to be tightly regulated in vivo. For example, at the time of mitosis, cells disassemble their entire microtubule network and reassemble the mitotic spindle with the same subunits (Fig. 5.1). The following are some examples of the mechanisms that cells use to control assembly processes.

Regulation by Subunit Biosynthesis and Degradation

Cells regulate the supply of building blocks for assembly reactions. For example, the concentration of unpolymerized tubulin regulates the stability of tubulin mRNA providing a feedback mechanism that controls the concentration of tubulin subunits available to form microtubules. On the other hand, red blood cells regulate the assembly of their membrane skeleton (see Fig. 13.10) by synthesizing a limiting amount of one subunit of the spectrin heterodimer. Following assembly of the membrane skeleton, proteolysis destroys the leftover unassembled copies of the other subunit.

Regulation of Nucleation

Regulation of a rate-limiting nucleation step is particularly striking in the case of microtubules. Microtubule nucleation from subunits is so unfavorable that most cellular microtubules grow from preformed structures in microtubule organizing centers (see Figs. 5.1 and 34.16). Varying the number, position, and activity of microtubule organizing centers helps cells to produce different microtubule arrays during interphase and mitosis.

Regulation by Changes in Environmental Conditions

Weak bonds between subunits allow cells to regulate assembly processes with relatively mild changes in conditions, such as in pH or ion concentrations. For example, when TMV infects a plant cell, the low concentration of Ca^{2+} in cytoplasm promotes disassembly of the virus because Ca^{2+} links the protein subunits together (see Example 4 below). Uncoating the RNA genome in this way begins a new cycle of replication.

Another example is a requirement for a phospholipase to promote the rapid release of the RNA genome from picornaviruses such as polio and cold viruses. Without the phospholipase, virus particles are engulfed and destroyed by autophagy (see Chapter 23) before they can replicate and propagate an infection.

Regulation by Covalent Modification of Subunits

Phosphorylation of specific serine, threonine, or tyrosine residues (see Fig. 25.1) can regulate interactions of protein subunits in macromolecular assemblies. This strategy is versatile, because the cell cycle and extracellular signals control the activities of the kinases that add phosphate and the enzymes, called protein phosphatases, that reverse the modification. Given the uniform bonding between subunits of symmetrical macromolecular structures, phosphorylation of the same amino acid residue on each subunit can control assembly.

Reversible phosphorylation regulates the assembly of the nuclear lamina, the filamentous network that supports the nuclear envelope (see Fig. 9.8). At the onset of mitosis, a protein kinase adds phosphate groups to the lamina subunits causing the network to fall apart (see Fig. 44.6). Removing these phosphates at the end of mitosis is one step in the reassembly of the nucleus. Other chemical modifications can regulate assembly reactions. Proteolysis is a drastic and irreversible modification used in the assembly of the bacteriophage T4 head (see Example 6 below) and collagen (see Fig. 29.4). Assembly of collagen fibrils is an extreme example, as it requires hydroxylation of prolines and lysines, glycosylation, disulfide bond formation, oxidation of lysines, and chemical crosslinking. Subunits in other assemblies are modified by methylation, acetylation, glycosylation, fatty acylation, tyrosination, polyglutamylation, or linkage to ubiquitin-like proteins.

Regulation by Accessory Proteins

Self-assembly processes were originally thought to require only the components found in the final structure, but many assembly reactions either require or are facilitated by auxiliary factors. **Molecular chaperones** that promote protein folding (see Fig. 12.11) also promote assembly reactions. In fact, bacterial mutations that compromised assembly of bacteriophages led to the discovery of the original chaperonin-60, GroEL (see Fig. 12.14). This class of chaperones also facilitates assembly of oligomeric proteins, such as the chloroplast enzyme RUBISCO. Chaperones may simply prevent aggregation during the folding of subunit proteins prior to their assembly. They may also participate directly in assembly reactions, but this has not been proven.

Bacteriophage T4 depends on accessory proteins coded by the virus to assemble its head. Often, proteolysis destroys these accessory proteins prior to insertion of the viral DNA (see Example 6 below). Bacteriophage P22 uses approximately 250 copies of an accessory **"scaffolding protein"** as a catalyst to guide the initial assembly of its capsid protein into an icosahedral head. Before the DNA is inserted, the scaffolding proteins exit from the interior of the head and recycle to promote the assembly of other viruses.

Accessory molecules specify the size of a few assemblies. The best characterized example is the RNA genome precisely regulating the length of TMV (see Example 4 below).

Numerous proteins regulate assembly of the cytoskeleton, and some are incorporated into the polymer network. Taking actin as an example, different classes of proteins regulate nucleotide exchange, determine the concentration of monomers available for assembly, nucleate and cap the ends of filaments, sever filaments, and crosslink filaments into bundles or random networks (see Fig. 33.10). Proteins with similar activities regulate the assembly of microtubules.

The following examples demonstrate how the general principles govern the assembly of real biological structures.

EXAMPLE 1 ***Actin Filaments: Rate-Limiting Nucleation and the Concept of Critical Concentration***

Actin filaments consist of two strands of subunits wound helically around one another (Fig. 5.5). (The structure can also be described as a single short-pitch helix with all the subunits repeating every 5.5 nm.) Each subunit contacts two subunits laterally and two other subunits longitudinally. Hydrogen bonds, electrostatic bonds, and hydrophobic interactions stabilize contacts between subunits. Subunits all point in the same direction, so the polymer is polar. The appearance of actin filaments with bound myosin (see Fig. 33.8) originally revealed the **polarity** now seen directly at atomic resolution. The filament decorated with myosin looks like a line of arrowheads with a point at one end and a barb at the other.

Actin binds adenosine diphosphate (ADP) or adenosine triphosphate (ATP) in a deep cleft. Irreversible hydrolysis of bound ATP during polymerization complicates the assembly process in a number of important ways (see Fig. 33.9). Here, assembly of ADP-actin, a relatively simple, reversible reaction, illustrates the concepts of nucleation and critical concentration.

Initiation of polymerization by pure actin monomers, also called **nucleation,** is so unfavorable that polymer accumulates only after a lag during which enough filaments accumulate to detect polymerization (Fig. 5.6C). Initiation of each new filament is slow, because small actin oligomers are exceedingly unstable. Actin dimers dissociate on a microsecond time scale, so their concentration is low, making addition of a third subunit rare. Actin trimers are more stable than dimers and serve as the nucleus for filament growth by adding more subunits (Fig. 5.6A). A trimer makes sense as a nucleus because it is the smallest oligomer with a complete set of intermolecular bonds. Unfavorable nucleation minimizes the spontaneous formation of filaments and enables the cell to control this reaction with specific nucleating proteins (see Figs. 33.12 and 33.14).

Elongation of actin filaments is a bimolecular reaction between monomers and a single site on each end of the filament (Fig. 5.6B–D). The growth rate of each filament is directly proportional to the concentration of subunits. If the rate of assembly is graphed as a function of the concentration of actin monomer, the slope is the association rate constant, k_+. The y-intercept is the dissociation rate constant, k_-. The elongation rate is zero where the plot crosses the x-axis. This monomer concentration is called the **critical concentration.** Above this concentration, polymers grow longer. Below this concentration, polymers shrink. Polymers grow until the monomer concentration falls to the critical concentration. At the critical concentration, subunits bind and dissociate at the same rate. The rates of association and dissociation are

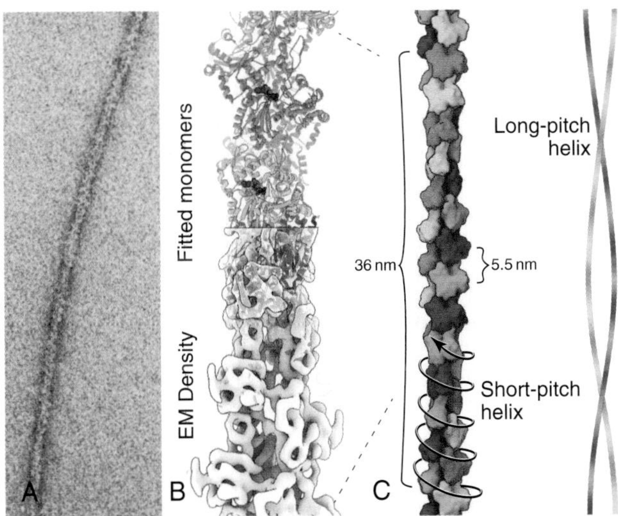

FIGURE 5.5 ACTIN FILAMENT STRUCTURE. A, Electron micrograph of a negatively stained actin filament. **B,** Model of the actin filament. The lower part is a reconstruction from electron micrographs. The upper part is a ribbon diagram showing the subunits using different colors corresponding to **C**. **C,** Model showing two ways to describe the helix: (1) two long-pitch helices (*orange/yellow* and *blue/green*) or (2) a one-start short-pitch helix including all the subunits (*yellow* to *green* to *orange* to *blue*).

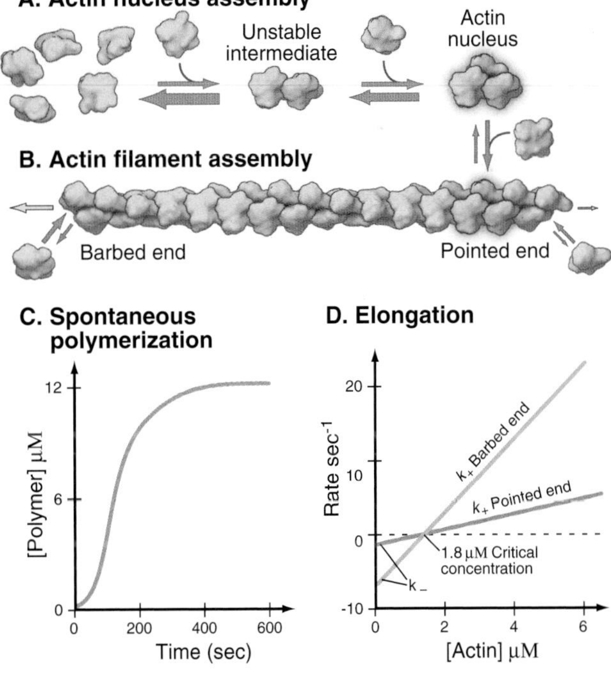

FIGURE 5.6 ACTIN FILAMENT ASSEMBLY. A, Formation of a trimeric nucleus from monomers. **B,** Elongation of the two ends of a filament by association and dissociation of monomers. **C,** Time course of spontaneous polymerization of purified adenosine diphosphate (ADP)-actin under physiological conditions. **D,** Dependence of the rates of elongation at the two ends of actin filaments on the concentration of ADP-actin monomers. (Data from Pollard TD. Rate constants for the reactions of ATP- and ADP-actin with the ends of actin filaments. *J Cell Biol*. 1986;103:2747–2754.)

somewhat different at the two ends of the polar filament. The rapidly growing end is called the barbed end, and the slowly growing end is called the pointed end.

EXAMPLE 2 *Myosin Filaments: New Properties Emerge as the Filaments Grow*

Myosin-II forms bipolar filaments held together by interactions of the α-helical, coiled-coil tails of the molecules (Fig. 5.7). Antiparallel overlap of tails forms a central bare zone flanked by filaments with protruding heads. On either side of the bare zone, parallel interactions extend the filament. The simplest myosin-II minifilaments from nonmuscle cells consist of just eight molecules (Fig. 5.7B). Muscle myosin filaments are much larger but are built on the same plan (Fig. 5.7A). Molecules are staggered at 14.3-nm intervals in these filaments. This arrangement maximizes the ionic bonds between zones of positive and negative charge that alternate along the tail. Hydrophobic interactions are also important; 170 water molecules dissociate from every molecule incorporated into a muscle myosin filament.

Both types of bipolar myosin-II filaments grow from the center, with molecules adding to both ends. Growth of cytoplasmic myosin-II minifilaments is self-limited. Filaments of muscle myosin-II grow longer by adding molecules to the ends of filaments in a diffusion-limited, bimolecular reaction. The reaction is unusual in that the dissociation rate constant increases with the length of the filament, eventually limiting the length of the polymer at the point where the dissociation rate equals the association rate.

EXAMPLE 3 *Bacterial Flagella: Assembly with a Rate-Limiting Folding Reaction*

Bacterial flagella are helical polymers of the protein flagellin (Fig. 5.8). Eleven strands of subunits surround a narrow central channel.

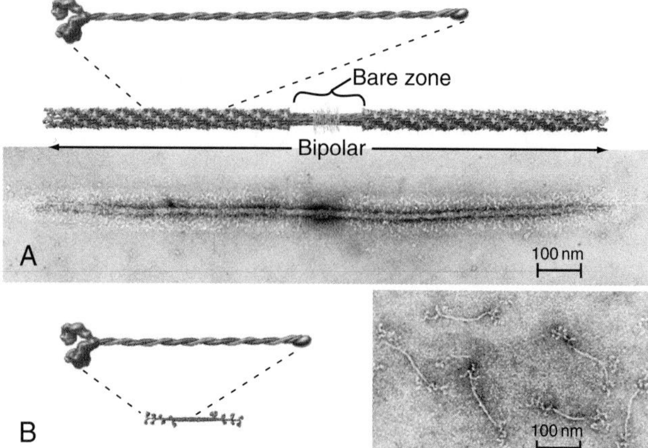

FIGURE 5.7 STRUCTURE OF MYOSIN FILAMENTS. A, Skeletal muscle myosin filament. Drawing and electron micrograph of a negatively stained filament. **B,** *Acanthamoeba* myosin-II minifilament. Drawing and electron micrograph of a negatively stained filament. (**A,** Courtesy J. Trinick, Bristol University, United Kingdom.)

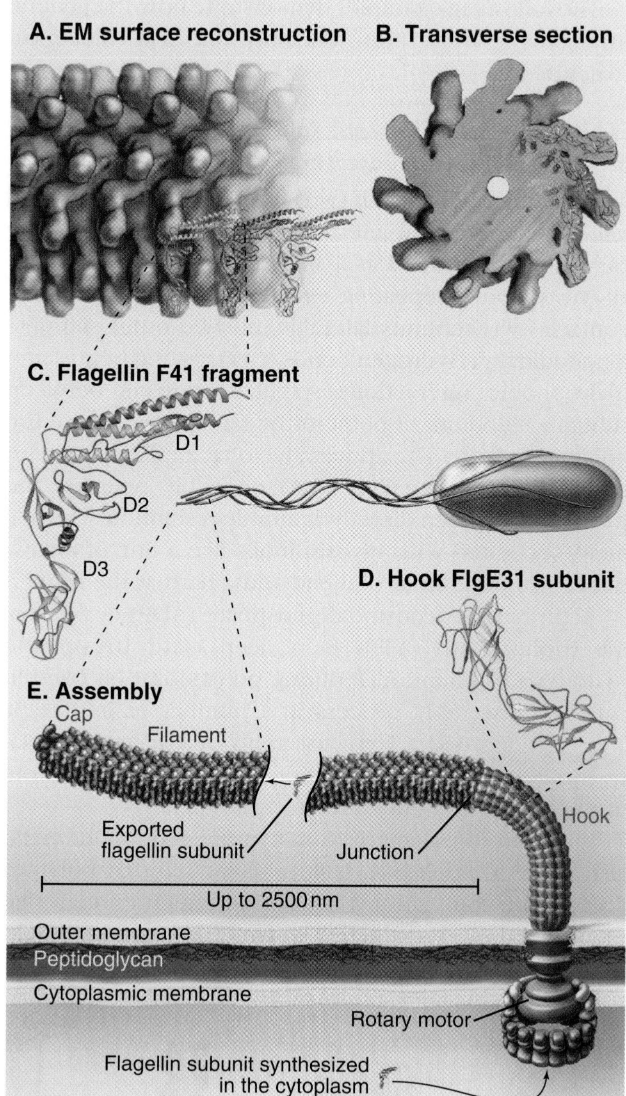

FIGURE 5.8 STRUCTURE OF THE FLAGELLA FROM THE BACTERIUM *Salmonella typhimurium*. A, Surface rendering from reconstructions of electron micrographs with superimposed ribbon diagrams of the structure of the flagellin subunit. **B,** Cross section from image processing of electron micrographs, showing the central channel and superimposed ribbon diagrams of the structure of the flagellin subunit. (For reference, see Protein Data Bank [PDB; www.rcsb.org] file 1IO1.) **C,** Ribbon diagram of part of the flagellin subunit. (See PDB file 1WLG.) **D,** Ribbon diagram of the hook subunit, FlgE31. **E,** Drawing of a flagellar filament attached via the hook segment to the basal body, the rotary motor that turns the flagellum. The cap structure is found at the distal end of the filament. A flagellin subunit in transit through the central channel from its site of synthesis in the cytoplasm to the distal tip is shown in the break in the filament. (**A–B,** From Mimori-Kiyosue Y, Yamashita I, Fujiyoshi Y, et al. Role of the outermost subdomain of *Salmonella* flagellin in the filament structure revealed by electron cryomicroscopy. *J Mol Biol*. 1998;284:521–530. **B,** Data from Samatey FA, Imada K, Nagashima S, et al. Structure of the bacterial flagellar protofilament and implications for a switch for supercoiling. *Nature*. 2001;410:331–337. **C,** Data from Samatey FA, Matsunami H, Imada K, et al. Structure of the bacterial flagellar hook and implication for the molecular universal joint mechanism. *Nature*. 2004;431:1062–1068.)

Isolated flagella elongate by addition of flagellin. As expected for a bimolecular reaction the rate is proportional to the concentration of flagellin monomers at low concentrations (Fig. 5.9A), but unexpectedly, the rate of elongation plateaus at a maximum of approximately three monomers per second at high flagellin concentrations (Fig. 5.9B). This plateau occurs because a rate-limiting step consisting of a relatively slow conformational change is required before the next subunit can bind. The slow step may involve folding of disordered parts of flagellin into α-helices that interact to form the two concentric cylinders inside the flagellum.

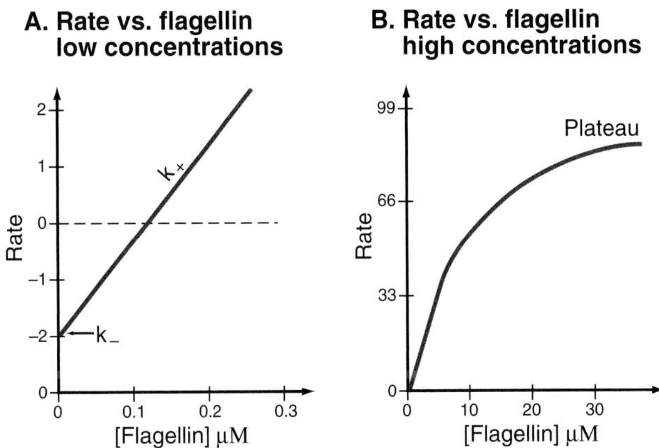

FIGURE 5.9 ELONGATION OF FLAGELLAR FILAMENTS FROM SEEDS (FRAGMENTS OF FLAGELLA) IN VITRO. The plots show the dependence of the elongation rate on subunit concentration. **A,** Low concentrations. **B,** High concentrations. (Modified from Asakura S. A kinetic study of in vitro polymerization of flagellin. *J Mol Biol.* 1968;35:237–239.)

Bacteria use structures called the base plate and hook assembly to initiate flagellar growth and to anchor the flagellum to the rotary motor that turns it (see Fig. 38.25). This overcomes extremely unfavorable nucleation reactions. Amazingly, flagella grow only at the end located farthest from the cell. Flagellin subunits synthesized in the cytoplasm diffuse through the narrow central channel of the flagellum (Fig. 5.9) out to the distal tip, where a cap consisting of an accessory protein prevents their escape before assembly.

EXAMPLE 4 *Tobacco Mosaic Virus: A Helical Polymer Assembled With a Molecular Ruler of RNA*

TMV was the first biological structure recognized to be a helical array of identical subunits, and it was the first helical protein structure to be determined at atomic resolution (Fig. 5.10). Production of infectious TMV from RNA and protein subunits was the first self-assembly reaction reproduced from purified components. At the time, during the 1950s, newspapers proclaimed, "Scientists create life in a test tube!"

The virus is a cylindrical copolymer of one RNA molecule (the viral genome) and 2130 protein subunits. The protein subunits are constructed from a bundle of four α-helices, shaped somewhat like a bowling pin. These subunits pack tightly in the virus and are held together by hydrophobic interactions, hydrogen bonds, and salt bridges. The RNA follows the protein helix in a spiral from one end of the virus to the other, protected in a groove in the protein subunits. Arginine residues lining the groove neutralize the negative charges along the RNA backbone (Fig. 5.10B). Each protein subunit

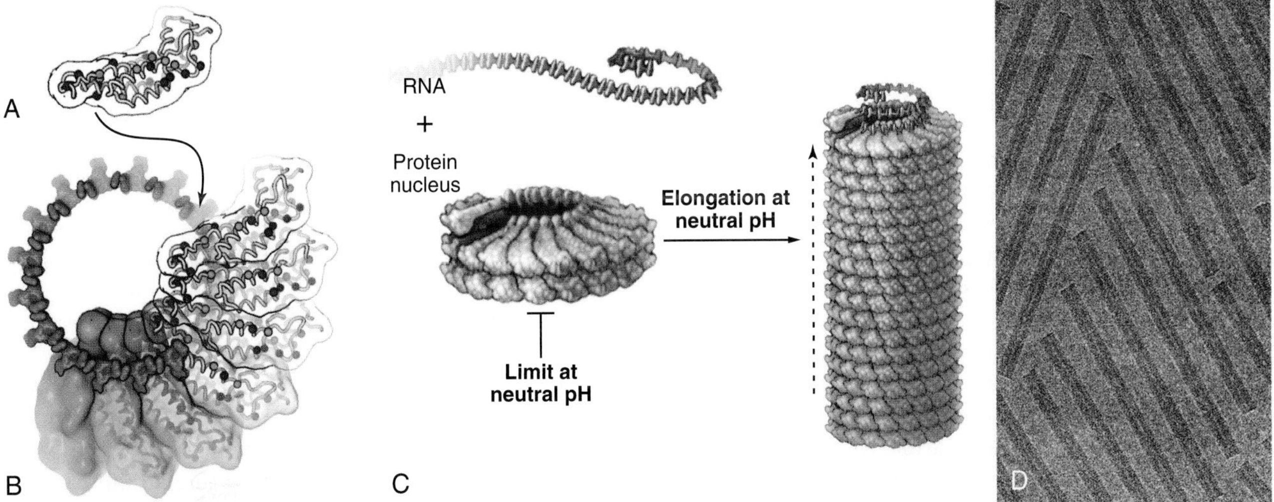

FIGURE 5.10 STRUCTURE AND ASSEMBLY OF TOBACCO MOSAIC VIRUS. A, Atomic structure of the protein subunit with the backbone in *gray,* beta carbons of acidic residues in *red* and beta carbons of basic residues in *blue*. **B,** The helical arrangement of protein subunits and their interactions with the individual nucleotides of RNA in *red*. Note the basic residues in the protein groove that binds the RNA. **C,** The subunit protein forms small oligomers of two plus turns at neutral pH that can elongate in the presence of RNA. **D,** Electron micrograph of tobacco mosaic virus (TMV) frozen in amorphous ice. (For reference, see PDB file 2TMV. **A–C,** Modified from drawings of D. Caspar, Florida State University, Tallahassee, FL. **D,** Courtesy R. Milligan, Scripps Research Institute, La Jolla, CA.)

also makes hydrophobic and electrostatic interactions with three RNA bases.

RNA regulates assembly of the protein subunits in two ways. First, RNA allows the protein to polymerize at a physiological pH. Protein alone forms helical polymers of varying lengths at nonphysiological acidic pH, but at neutral pH it forms only unstable oligomers of 30 to 40 protein subunits, slightly more than two turns of the helix (Fig. 5.10C). RNA promotes folding of disordered loops lining the central channel of these oligomers, acting as a switch to drive propagation of the helix by incorporating additional protein subunits. Second, RNA is the molecular ruler that determines the precise length of the assembled virus. Only after interacting with RNA at the growing end of the polymer can subunits fold into a structure compatible with a stable virus.

EXAMPLE 5 *Tomato Bushy Stunt Virus: Quasi-equivalent Bonding Between Protein Subunits*

The first atomic structure of a spherical virus (tomato bushy stunt virus, TBSV) revealed that the flexibility required to form both fivefold and sixfold icosahedral vertices lies within the protein subunit rather than in the bonds between subunits. The 180 identical subunits associate in pairs in two different ways, distinguished in Fig. 5.11 by the green-blue and red colors. The blue subunit of the green-blue pairs is used exclusively for fivefold vertices. Three red subunits and three green subunits form sixfold vertices. External contacts of both green-blue and red pairs with their neighbors are similar, but the contacts between pairs of red subunits differ from pairs of green-blue subunits. The difference is achieved by changing the position of the amino-terminal portion of the coat protein polypeptide chain. Two subunits in green-blue pairs pack tightly against each other, providing the sharp curvature required at fivefold vertices. In red dimers, the amino-terminal peptide acts as a wedge to pry the inner domains of the subunits apart and flatten the surface, as is appropriate for sixfold vertices. Thus, the flexible arm acts like a switch to determine the local curvature. This subunit flexibility accommodates the 12-degree difference in packing at fivefold and sixfold vertices. Other spherical viruses use a similar strategy to achieve **quasiequivalent** packing of identical subunits.

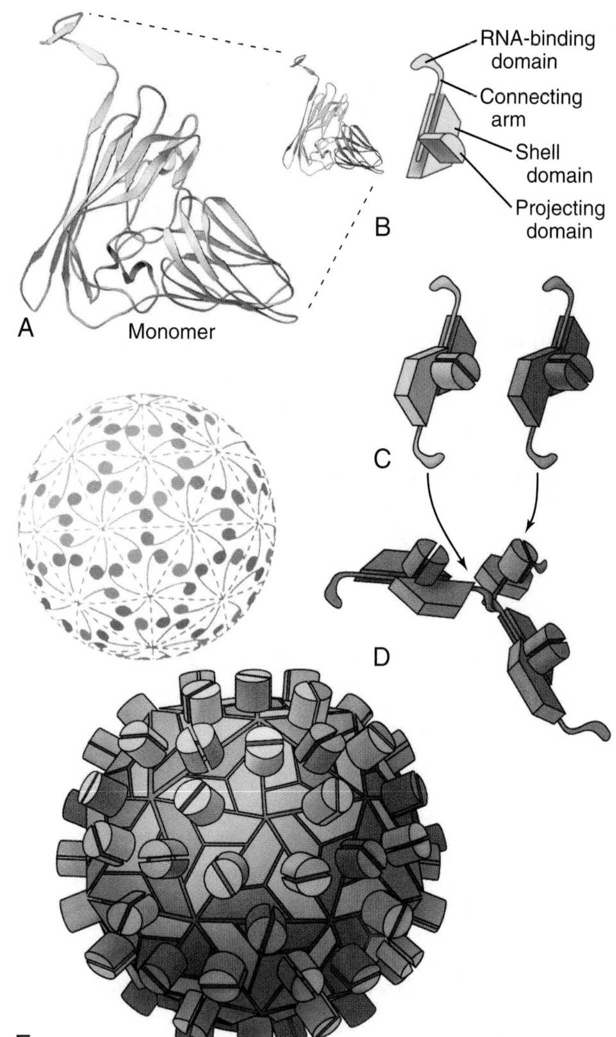

FIGURE 5.11 TOMATO BUSHY STUNT VIRUS STRUCTURE AND ASSEMBLY PATHWAY. A, Ribbon diagram of a coat protein subunit. (See PDB file 2TBV.) **B,** Block diagram of one subunit. **C,** Block diagrams of dimers of coat protein subunits. **D,** Proposed nucleus for a sixfold vertex with three dimers *(red).* Three additional dimers *(green-blue)* are proposed to add to complete a sixfold vertex. Five blue subunits associate to make a fivefold vertex. **E,** Two different surface representations of the viral capsid showing the quasi-equivalent positions occupied by red, blue, and green subunits. (**C–D,** Modified from Olson A, Bricogne G, Harrison S. Structure of tomato bushy stunt virus IV. The virus particle at 2.9 Å resolution. *J Mol Biol.* 1983; 171:61–93.)

TBSV provided the first of many examples of flexible arms that lace subunits together. Amino-terminal extensions of three red subunits intertwine at sixfold vertices. As if holding hands, these arms form a continuous network on the inner surface, reinforcing the coat.

Icosahedral plant viruses like TBSV can assemble from pure protein and RNA, although in cells they assemble in association with intracellular membranes, reactions facilitated by ESCRT, the endosomal sorting complex required for transport (see Fig. 22.17). An attractive hypothesis for self-assembly from dimers of coat protein is that local information built into the growing shell specifies the pathway. Three dimers in the red conformation bind a specific viral RNA sequence, forming a nucleating structure similar to a sixfold vertex. Folding of the arms in this nucleus forces the next three dimers to take the green-blue conformation, since no intermolecular binding sites are available for their arms. The greater curvature of the green-blue dimers dictates that fivefold vertices form at regular positions around the nucleating sixfold vertex. Additional fivefold vertices form appropriately as positions for this more favored association become available around the growing shell. The beauty of this idea is that local information (the availability of intermolecular binding sites for strands) automatically favors the insertion of green-blue or red dimers, as appropriate, to complete the icosahedral shell.

EXAMPLE 6 *Bacteriophage T4: Three Irreversible Assembly Pathways Form a Metastable Structure*

Bacteriophage T4 is a virus of the bacterium *Escherichia coli* (Fig. 5.12). Genetic analysis established that more than 49 distinct gene products contribute to assembly of this virus. Three separate, multicomponent substructures—heads, tails, and tail fibers—assemble along independent pathways and combine to form the virus (Fig. 5.13). Emergence of new properties automatically orders the steps along each pathway, so assembly occurs sequentially even in the presence of reactive pools of all the subunits. A good product is ensured, because defective subassemblies fail to attach and are rejected.

A protein complex nucleates the growth of a preliminary version of the icosahedral head and later attaches one vertex of the head to the tail. A complex of the major head protein with several accessory proteins adds to the growing head. The accessory proteins end up inside the precursor head. After proteolysis cleaves 20% of the peptide from the N-terminus of the major head protein and degrades the accessory proteins, a major conformational change shifts part of the head protein from inside to outside and expands the volume of the head by 16%. Then, an ATP-driven rotary motor inserts the 166,000-base-pair DNA molecule into the head through a hole in a vertex. This motor, one of the strongest in nature, can produce a force of 70 pN, enough to compress the DNA inside the head to a pressure of 60 atmospheres.

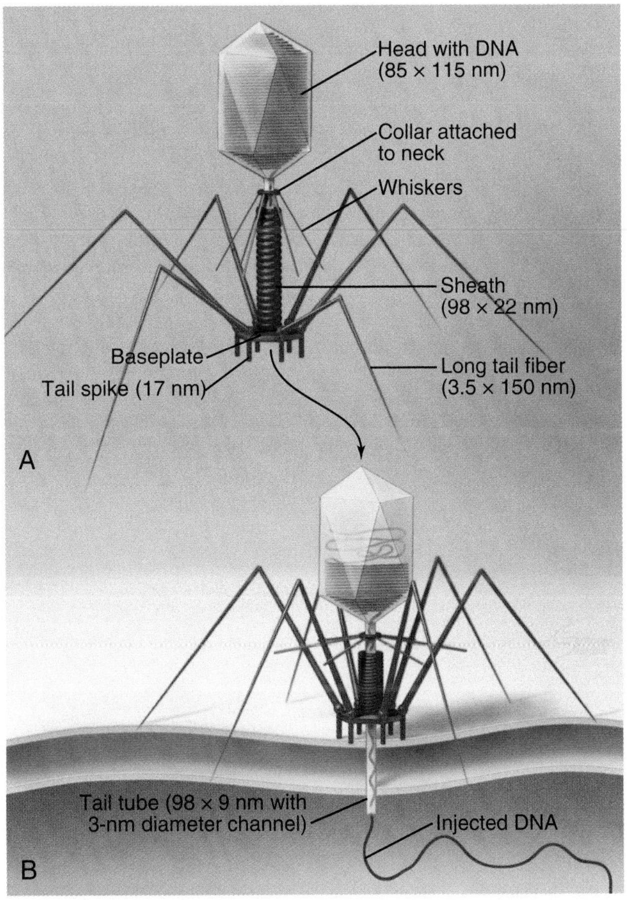

FIGURE 5.12 STRUCTURE OF BACTERIOPHAGE T4. A, Infectious phage particle. **B,** Association with *Escherichia coli* and injection of DNA by contraction of the sheath. (Data from Leiman PG, Chipman PR, Kostyuchenko VA, et al. Three-dimensional rearrangement of proteins in the tail of bacteriophage T4 on infection of its host. *Cell.* 2004;118:419–429.)

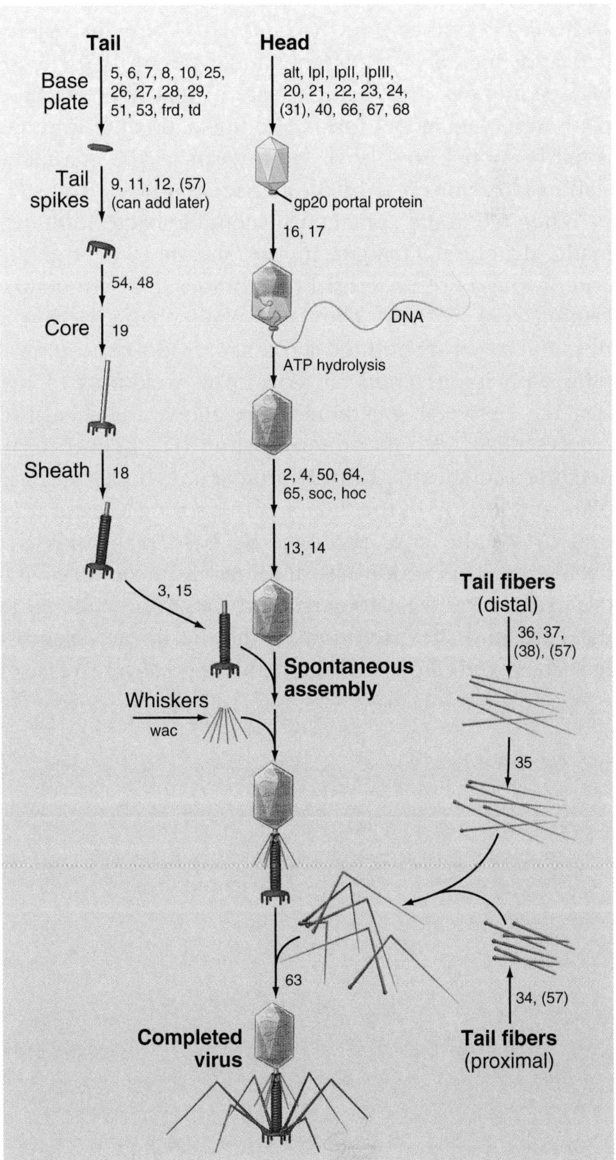

FIGURE 5.13 ASSEMBLY PATHWAY OF BACTERIOPHAGE T4. The numbers refer to genes required at each step. ATP, adenosine triphosphate. (Modified from Wood WB, Edgar RS, King J, et al. Bacteriophage assembly. *Fed Proc.* 1968;27:1160–1166.)

Within the head, the pressurized DNA is restrained in a near-crystalline, metastable state until it is released during infection of the *E. coli* host.

The tail is a double cylinder of a rod-like, helical core and a loosely fitting helical sheath, both attached to a base plate. A complicated pathway involving at least 15 gene products and 13 steps assembles the hexagonal base plate. One of these proteins, acting like a "safety" on a gun, stabilizes its shape. A plug in the middle of the hexagonal base plate nucleates the polymerization of core subunits. Next, the sheath subunits polymerize into a helical lattice that mimics the underlying core. In mutants that lack base plates, sheath subunits assemble inefficiently into a shorter and fatter helix.

The three assembly lines converge, joining heads to tails and then adding the six long, independently assembled tail fibers that give the completed virus its spider-like appearance. Attachment of tail fibers to the base plate somehow removes the "safety" that held the base plate in its hexagonal form. The finished bacteriophage is hardy enough to survive for 20 years at 4°C in a metastable state, poised to infect its bacterial host.

When tail fibers contact a susceptible bacterium, dramatic structural changes in the sheath force the tail core through both bacterial membranes in a syringe-like fashion (Fig. 5.13B). The base plate changes from a hexagon into a six-pointed star that cuts loose the central plug with its attached tail core. The weakness of the contacts between sheath and core allows the sheath to "recrystallize" into its preferred short, fat, helical form. Because the sheath is firmly attached at both the base plate and the top of the tail core, this spring-like contraction drives the core through the base plate into the bacterium. This action also unplugs the head, allowing the pressurized DNA to extrude through the channel in the core into the bacterium. Thus the linear assembly reactions and an adenosine triphosphatase (ATPase) motor produce a machine that does physical work when triggered.

ACKNOWLEDGMENT

We thank Tony Crowther for his suggestions on revisions to this chapter for the second edition.

SELECTED READINGS

Caspar DLD. Virus structure puzzle solved. *Curr Biol.* 1992;2: 169-171.

Caspar DLD, Klug A. Physical principles in the construction of regular viruses. *Cold Spring Harb Symp Quant Biol.* 1962;27:1-24.

Harrison SC. What do viruses look like? *Harvey Lect.* 1991;85: 127-152.

Leiman PG, Chipman PR, Kostyuchenko VA, et al. Three-dimensional rearrangement of proteins in the tail of bacteriophage T4 on infection of its host. *Cell.* 2004;118:419-429.

Liddington RC, Yan Y, Moulai J, et al. Structure of simian virus 40 at 3.8 A resolution. *Nature.* 1991;354:278-284.

Namba K, Stubbs G. Structure of tobacco mosaic virus at 3.6 A resolution: Implications for assembly. *Science.* 1986;231:1401-1406.

Oosawa F, Asakura S. *Thermodynamics of the Polymerization of Protein.* New York: Academic Press; 1975.

Pollard TD, Blanchoin L, Mullins RD. Biophysics of actin filament dynamics in nonmuscle cells. *Annu Rev Biophys Biomol Struct.* 2000;29:545-576.

Rossmann MG, Mesyanzhinov VV, Arisaka F, Leiman PG. The bacteriophage T4 DNA injection machine. *Curr Opin Struct Biol.* 2004;14:171-180.

Simpson AA, Tao Y, Leiman PG, et al. Structure of the bacteriophage phi29 DNA packaging motor. *Nature.* 2000;408:745-750.

Sinard JH, Pollard TD. *Acanthamoeba* myosin-II minifilaments assemble on a millisecond time scale with rate constants greater than those expected for a diffusion limited reaction. *J Biol Chem.* 1990; 265:3654-3660.

Smith DE, Tans SJ, Smith SB, et al. The bacteriophage straight phi29 portal motor can package DNA against a large internal force. *Nature.* 2001;413:748-752.

Wood WB. Genetic control of bacteriophage T4 morphogenesis. *Symp Soc Dev Biol.* 1973;31:29-46.

CHAPTER 6

Research Strategies

Research in cell biology aims to discover how cells work at the molecular level. Powerful tools are available to achieve this goal. To understand how these methods contribute to explaining cellular function, this chapter begins with a brief account of the synthetic approach used in cell biology. This strategy is based on the premise that one can understand a complex cellular process by reducing the system to its constituent parts and characterizing their properties to generate mechanistic hypotheses for testing in live cells. This approach, also called **reductionism,** has dominated cell biology research since the middle of the 20th century and has succeeded time after time. For example, most of what is understood about protein synthesis has come from isolating and characterizing ribosomes, messenger RNAs (mRNAs), transfer RNAs (tRNAs), and accessory factors. In this and many other cases, proof of function has been established by reconstituting a process from isolated parts of the molecular machine, verifying these conclusions with genetic experiments and quantitative measurements in live cells. Most processes are sufficiently complicated that computer simulations of mathematical models are an important part of interpreting the observations.

This reductionist approach involves much more than simply identifying the molecular parts of a cellular machine. Essential tasks include the following (note that after item 1, the order can vary):

1. Defining a biological question
2. Making a complete inventory of the participating molecules
3. Localizing the molecules in cells
4. Measuring the cellular concentrations of the molecules
5. Determining atomic structures of the molecules
6. Identifying molecular partners and pathways in the system
7. Measuring rate and equilibrium constants for the reactions
8. Reconstituting the biological process from purified molecules
9. Testing for physiological function with genetics, drugs, or other approaches
10. Formulating a mathematical model and simulating the behavior of the system

This full agenda is complete for a few biological processes, such as bacterial chemotaxis (see Figs. 27.12 and 27.13). Often, much is known about some aspects of a process, such as a partial list of participating molecules, the localization of these molecules in a cell, or functional tests by removing the genes for one or more molecules from an experimental organism. Less often is enough information available about molecular concentrations and reaction rates to formulate and simulate a dynamical mathematical model of the process to verify that the whole system actually works as anticipated. Thus, much work remains to be done.

Imaging

Microscopy of live and fixed cells often provides initial hypotheses about the mechanisms of cellular processes. Imaging is also a valuable adjunct to genetic analysis and testing mechanisms.

Microscopy is useful for cell biologists, owing to fortunate coincidences within the electromagnetic spectrum. First, the wavelength of visible light (390 to 700 nm) is suitable for imaging cells and their membrane bounded organelles (0.5 μm to tens of micrometers), and the wavelength of electrons (~0.004 nm) is right for imaging macromolecular assemblies (angstroms to nanometers) and larger objects such as cellular organelles. Second, one can focus visible light with glass lenses and electrons with electromagnetic lenses.

Resolution, the ability to discriminate two points, is directly related to the wavelength of the light. The equation is

$$D = 0.61\lambda / N \sin\alpha$$

where D is the resolution, λ is the wavelength of light, N is the refractive index of the medium between objects

(~1 for cells), and *sin* α is the numerical aperture of the lens (up to 1.4 for light microscopes). The limit of resolution with visible light and glass lenses is normally approximately 0.2 μm. Fortunately, various superresolution methods described below allow much higher resolution imaging with visible light. Soft x-rays with a wavelength of approximately 3 nm have the potential to provide high resolution, but are not practical for routine imaging because the lenses are relatively crude. However, analysis of molecular crystals by diffraction of higher energy x-rays (wavelength ~0.1 nm) is a powerful method for determining structures of macromolecules at atomic resolution. The wavelength of electrons accelerated at 100 kiloelectron volt (keV) is small, and with new detectors and image averaging, researchers are now able to achieve resolutions of less than 1 nm, making them preferable to x-rays for visualizing large macromolecular assemblies.

Microscopes have two functions. The first is to enlarge an image of the specimen so that it can be seen with the eye or a camera. Just as important, but less appreciated, microscopes must produce **contrast** so that details of the enlarged image stand out from each other.

Light Microscopy Methods

Six methods are used to produce contrast in light micrographs of biological specimens (Table 6.1 and Fig. 6.1). These are called wide-field methods, as a broad beam of illuminating light is focused on the specimen by a condenser lens.

The classic light microscopic method is **bright field,** whereby the specimen is illuminated with white light. However, most cells absorb very little visible light and thus show little contrast with bright-field illumination (Fig. 6.2A). For this reason, specimens are fixed with crosslinking chemicals and permeabilized before staining with organic dyes that absorb light and create contrast. Three-dimensional tissues are fixed and embedded in paraffin or plastic, before cutting sections with a microtome (a device that cuts a series of thin slices from the surface of a specimen), and staining with a variety of dyes (for examples, see Figs. 28.3, 29.3, 29.8, 31.1, 32.1, 32.2, 32.5, 32.7, 32.9, and 40.1). Alternatively, sections may be taken from frozen tissue and then stained. In either case, the cells are killed by fixation or sectioning prior to observation.

Observations of live cells require other methods to produce contrast. Most of these methods are also useful for fixed cells. **Phase-contrast** microscopy generates contrast by interference between light scattered by the specimen and a slightly delayed reference beam of light. Small variations in either thickness or refractive index (speed of light) can be detected, even within specimens that absorb little or no light (Fig. 6.2B). **Differential interference contrast** (DIC) produces an image that looks as though it is illuminated by an oblique shaft of light (Fig. 6.2C) but is actually a thin optical section of the specimen. Two nearby beams interfere with each other, producing contrast in proportion to the gradient of local differences in the refractive index across the specimen. Thus, an organelle with a high refractive index (slow speed of light) in cytoplasm will appear light on one side (where the refractive index is increasing with respect to the cytoplasm) and dark on the other (where the refractive index is decreasing). Computer processing can greatly enhance contrast and remove optical artifacts from images. For example, computer-enhanced DIC can increase the contrast enough to image single microtubules (see Fig. 34.6).

Dark-field microscopy and **polarization microscopy** have specialized uses in biology. In dark-field microscopy, the specimen is illuminated at an oblique angle so that only light scattered by the specimen is collected by the objective lens. Recall how easy it is to detect tiny dust particles in a beam of light in a dark room. The contrast is so great that a single isolated microtubule stands out brightly from the dark background. However, a dark-field image of something as

TABLE 6.1 Methods for Producing Contrast in Light Microscopy

Type	Principle	Requirements	Live Cells	Fixed Cells
Bright field	Absorption of visible light	Light-absorbing stains on a thin specimen	No	Yes
Fluorescence	Emission of light by fluorescent molecule	Cellular molecules labeled with fluorescent dyes or expression of fluorescent proteins	Yes	Yes
Phase contrast	Variations in thickness and refractive index within specimen	Relatively flat cells	Yes	Yes
Differential interference contrast (DIC)	Gradient of refractive index across the specimen	May be used on thick, unstained specimens	Yes	Yes
Dark field	Scattering of light	Relatively thin, simple specimen	Yes	Yes
Polarization	Differences in refractive index for perpendicular beams of polarized light	Birefringent (highly ordered along a linear axis) elements in specimen	Yes	Yes

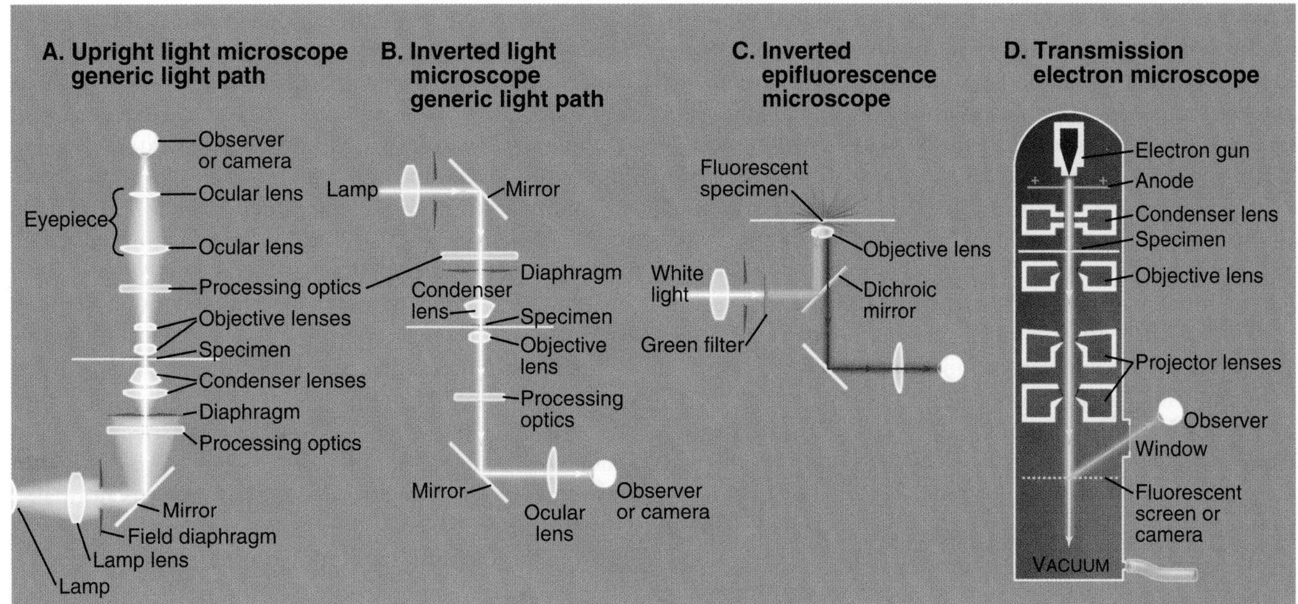

FIGURE 6.1 LIGHT PATHS THROUGH LIGHT AND ELECTRON MICROSCOPES. A, Optical path in an upright light microscope. The condenser lens focuses light on the specimen. Light interacts with the specimen. The objective lens collects and recombines the altered beam. An ocular lens projects the enlarged image onto the eye or a camera. Processing optics produce contrast by phase contrast, differential interference, or polarization. **B,** Optical path in an inverted light microscope. **C,** Epi-illumination for fluorescence microscopy. The objective lens acts as the condenser to focus the exciting, short-wavelength light (*green,* in this example) on the specimen. Fluorescent molecules in the specimen absorb the exciting light and emit longer-wavelength light (*red,* in this example). The same objective lens collects emitted long-wavelength light. A dichroic mirror in the light path reflects the exciting light and transmits emitted light. An additional filter (not shown) blocks any short-wavelength light from reaching the viewer. **D,** Optical path in a transmission electron microscope. Electromagnetic lenses carry out the same functions as glass lenses in a light microscope. The image may be observed visually when the electrons produce visible light from a fluorescent screen or recorded on film or by a digital camera.

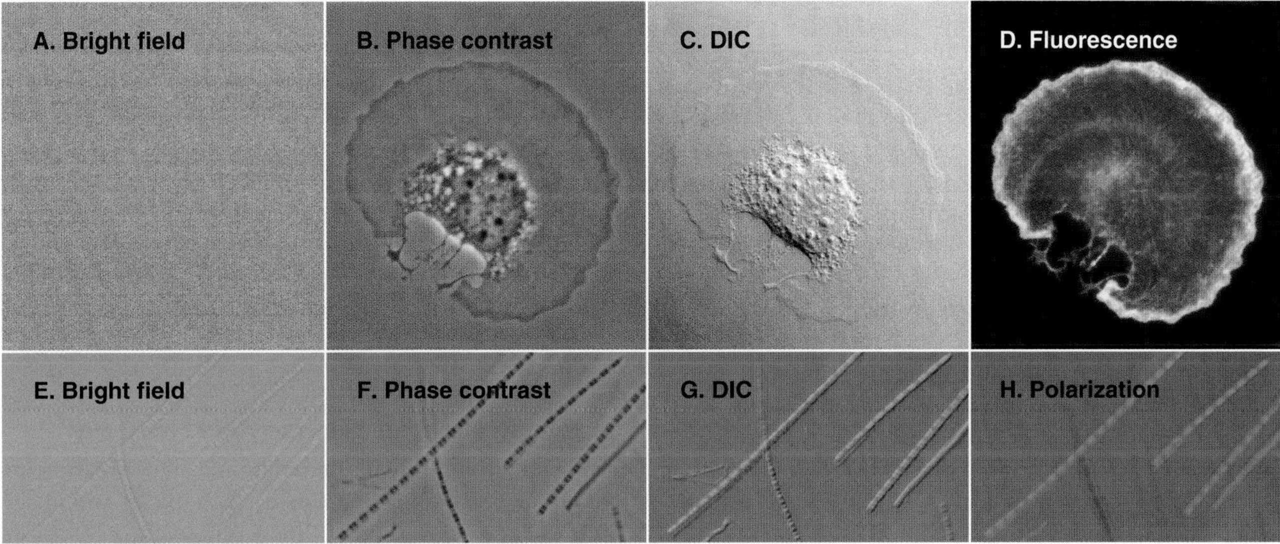

FIGURE 6.2 COMPARISON OF METHODS TO PRODUCE CONTRAST. A–D, Micrographs of a spread mouse 3T3 cell grown in tissue culture on a microscope slide, then fixed and stained with rhodamine-phalloidin, a fluorescent peptide that binds actin filaments. Contrast methods include bright field **(A)**, phase contrast **(B)**, differential interference contrast (DIC) **(C)**, and fluorescence **(D)**. **E–H,** Micrographs of myofibrils isolated from skeletal muscle. Contrast methods include bright field **(E)**, phase contrast **(F)**, differential interference contrast **(G)**, and polarization **(H)**. The A-bands, consisting of parallel thick filaments of myosin (see Fig. 39.3), appear as dark bands with phase contrast and are birefringent (either bright or dark, depending on the orientation) with polarization. (**A–D,** Courtesy R. Mahaffy, Yale University, New Haven, CT.)

complicated as cytoplasm is confusing, owing to multiple overlapping objects that scatter light.

In polarization microscopy, the specimen is placed between two crossed polarizing filters so no light passes through the second polarizer unless the specimen modifies its polarization state. This happens if the polarized light passes more slowly through the specimen when vibrating in one plane than when vibrating in the perpendicular plane (much as a saw cuts wood faster with the grain than across it). The filaments in striated muscle (Fig. 6.2H) and microtubules in a mitotic spindle are among the few biological specimens aligned well enough to be birefringent and produce contrast in a polarization microscope. New innovations are expanding the capabilities of this approach.

Fluorescence Microscopy

Remarkable sensitivity makes **fluorescence microscopy** a powerful tool. Digital cameras can image a single fluorescent molecule. When a fluorescent molecule absorbs a photon, an electron is excited into a higher state. Nanoseconds later, the electron falls back to its ground state and most of the energy is converted into a longer-wavelength (lower-energy) photon. For example, the fluorescent dye rhodamine absorbs green light (shorter wavelength) and emits red light (longer wavelength).

Fluorescent Probes

Fluorescence microscopy requires a fluorescent molecule, either an organic dye or fluorescent protein in the specimen. The historic approach was to target molecules in fixed, permeabilized cells with a protein or nucleic acid labeled with a fluorescent dye. A powerful version of this strategy uses antibodies, proteins produced by the immune system (see Fig. 28.8), to react with specific molecular targets. Antibodies can be tagged with fluorescent dyes and used to localize molecules in cells by fluorescence microscopy (Fig. 6.3E). This is called fluorescent antibody staining or **immunofluorescence**. Another approach is to label an oligonucleotide with a fluorescent dye to probe for nucleic acids with complementary sequences in fixed cells, a process called fluorescence in situ hybridization or **FISH** (see Fig. 8.10). Similarly, one can attach a fluorescent dye to a small molecule that binds tightly to a cellular component, such as phalloidin binding to actin filaments, to localize them in cells (Fig. 6.2D).

The application of fluorescence microscopy to live cells began with labeling a purified lipid, protein, or nucleic acid with a fluorescent dye. When introduced into a live cell by microinjection or other means, the tagged molecule often seeks its natural location (see Figs. 37.6 and 38.9).

The discovery of naturally fluorescent proteins, such as **green fluorescent protein** (GFP) from jellyfish, made it possible to genetically encode fluorescent tags and track individual proteins in live cells. DNA-encoding GFP is joined (usually to one end) to the coding sequence for a protein and introduced into cells where it directs the synthesis of a fusion protein consisting of GFP linked to the protein of interest. Ideally, homologous recombination or genome editing (Fig. 6.16) is used to replace the wild-type gene with the coding sequence for GFP fusion protein in the genome of the test cell. Where this is difficult or impossible, the GFP fusion protein can be produced from exogenous DNA or RNA introduced into the cell. A critical but often neglected aspect of these studies is to demonstrate by genetic or biochemical experiments that the fusion protein functions normally. GFP fluorescence marks the fusion protein wherever it goes in the cell and can be measured to count labeled molecules (Fig. 6.3A–C).

Mutations in GFP can change its fluorescence properties, providing fluorescent proteins with a range of colors. When attached to different protein types, these probes allow two or more protein species to be visualized simultaneously in the same cell (Fig. 6.3D). Other mutations allow UV light to turn on the fluorescence (photoactivation) or change the wavelength of the emitted light (photoswitching). Fluorescent proteins have been engineered into "biosensors" to measure pH or Ca^{2+} concentration or a protein's behavior/interactions.

Inventive imaging techniques make good use of these new optical probes. For example, one can bleach the GFP in part of the cell with strong light and observe over time how GFP from other parts of the cell fills in the dark area (Fig. 6.3F). Such a fluorescence recovery after photobleaching (**FRAP**) experiment reveals how molecules move by diffusion in the cytoplasm or in the plane of cellular membranes. Other applications include fluorescence resonance energy transfer (FRET) to measure the distance between fluorophores and fluorescence correlation spectroscopy (FCS) to measure diffusion coefficients of molecules in a narrow beam of light.

Imaging Methods for Fluorescence Microscopy

The standard method of illumination, called **epifluorescence,** uses the objective lens to both excite and image fluorescence (Fig. 6.1C). Filters and dichroic mirrors that reflect short wavelengths direct the exciting light through the objective to the specimen. Fluorescent molecules in the specimen emit longer-wavelength light in every direction, some of which is collected by the objective. The emitted light passes through the dichroic mirror and a camera records the image. Emission filters remove any exciting light scattered by the specimen. Because the exciting light passes through the entire

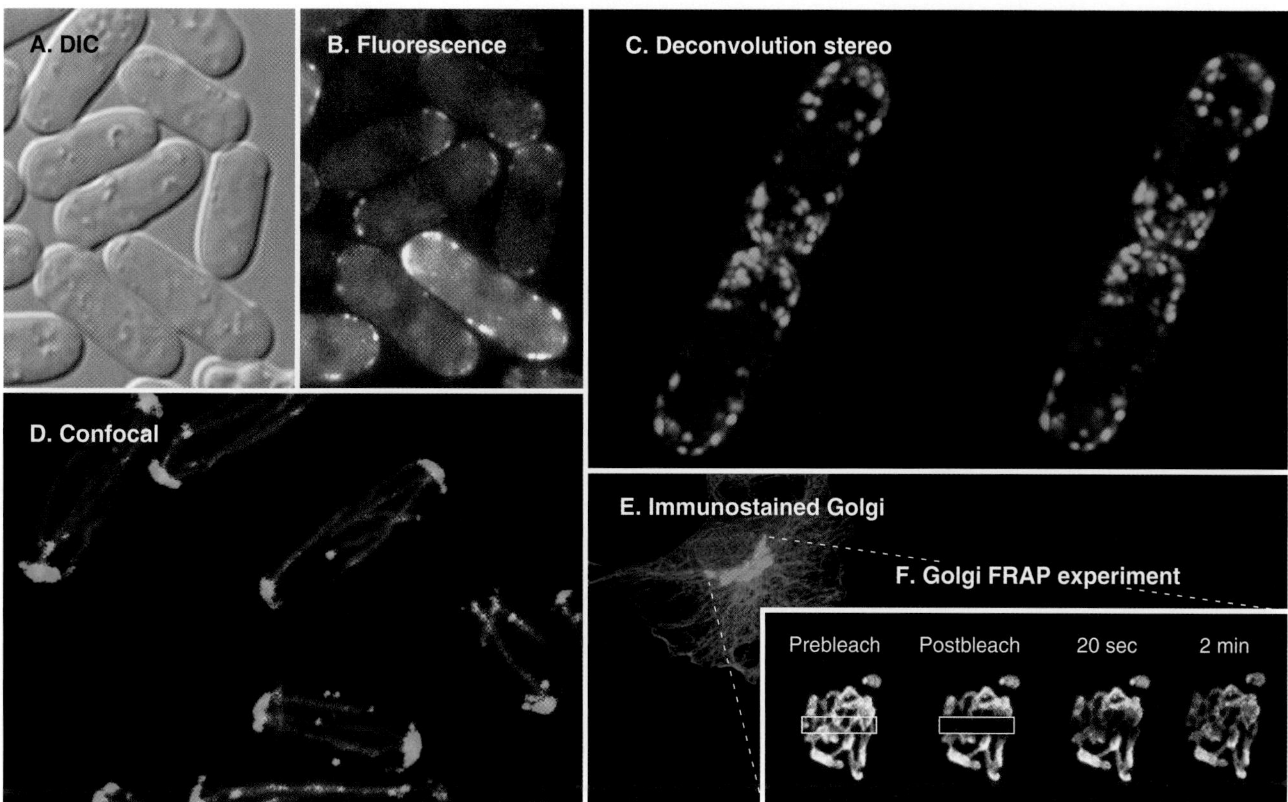

FIGURE 6.3 FLUORESCENCE MICROSCOPY METHODS. A–C, Light micrographs of live fission yeast expressing green fluorescence protein (GFP) fused to myosin-I. **A,** Differential interference contrast (DIC). **B,** Standard wide-field fluorescence of the same cells. **C,** Stereo pair of a three-dimensional reconstruction of a stack of optical sections made by deconvolution of wide-field images. Removal of out-of-focus blur improves the resolution and contrast of small patches enriched in myosin-I. A stereo view is obtained by focusing your left eye on the left image and right eye on the right image. This can be achieved by holding the micrographs close to your eyes and then gradually withdrawing the page about 12 inches. **D,** Scanning confocal fluorescence micrograph of fission yeast cells showing *red* microtubules and *green* Tea 1 protein (a protein involved in determining cell shape). This thin optical section eliminates the blur from fluorescence in other planes of focus. **E–F,** Fluorescence recovery after photobleaching (FRAP). **E,** A fibroblast cell in tissue culture stained with fluorescent antibodies for the Golgi apparatus *(yellow)* and microtubules *(green)* and with the fluorescent dye DAPI (4,6-diamidino-2-phenylindole) for DNA *(blue).* **F,** A series of fluorescence micrographs of a fibroblast cell expressing GFP-galactosyltransferase, which concentrates in the Golgi apparatus. The GFP in a bar-shaped zone is bleached with a strong pulse of light, and the fluorescence is followed over time. After 2 minutes GFP-galactosyltransferase redistributes by lateral diffusion in the membranes to fill in the bleached zone. (**A–C,** From Lee W-L, Bezanilla M, Pollard TD. Fission yeast myosin-I, Myo1p, stimulates actin assembly by Arp2/3 complex and shares functions with WASp. *J Cell Biol* 2000;151:789–800. **D,** Courtesy Hilary Snaith and Kenneth Sawin, University of Edinburgh, United Kingdom. **E–F,** Courtesy J. Lippincott-Schwartz, N. Altan, and K. Hirschberg, National Institutes of Health, Bethesda, MD.)

specimen, out of focus light emitted by molecules above and below the focal plane blurs the image (Fig. 6.3B).

Superimposition and out-of-focus noise can be minimized either computationally or optically. An image processing method called **deconvolution** produces clear fluorescence images of thick specimens by using an iterative computer process to restore light that is blurred out of focus to its proper focal plane. Starting with a stack of blurry images taken with a traditional wide-field microscope at different focal planes all the way through the specimen, this method produces a remarkably detailed three-dimensional image in sharp focus throughout (Fig. 6.3C).

Several optical methods are used to image thin sections of a specimen. In **total internal reflection fluorescence microscopy** (Fig. 6.4G–H), an oblique beam of exciting light is reflected from the interface between the slide and the aqueous specimen, setting off an evanescent wave that penetrates the specimens only approximately 100 nm. This excites molecules only near the surface of the slide and avoids fluorescent molecules deeper in the specimen.

Confocal microscopy produces thin optical sections of fluorescent specimens by illuminating with one or many points of laser light sharply focused in all three directions: *x, y,* and *z* (Fig. 6.4A–D). These points of light are scanned across the specimen in a raster (parallel lines) pattern to excite fluorescent molecules. Light emitted at each consecutive point in the specimen passes through a pinhole in front of the camera to remove out-of-focus light. A photomultiplier detects the light from each raster (Fig. 6.4A). A computer reassembles the

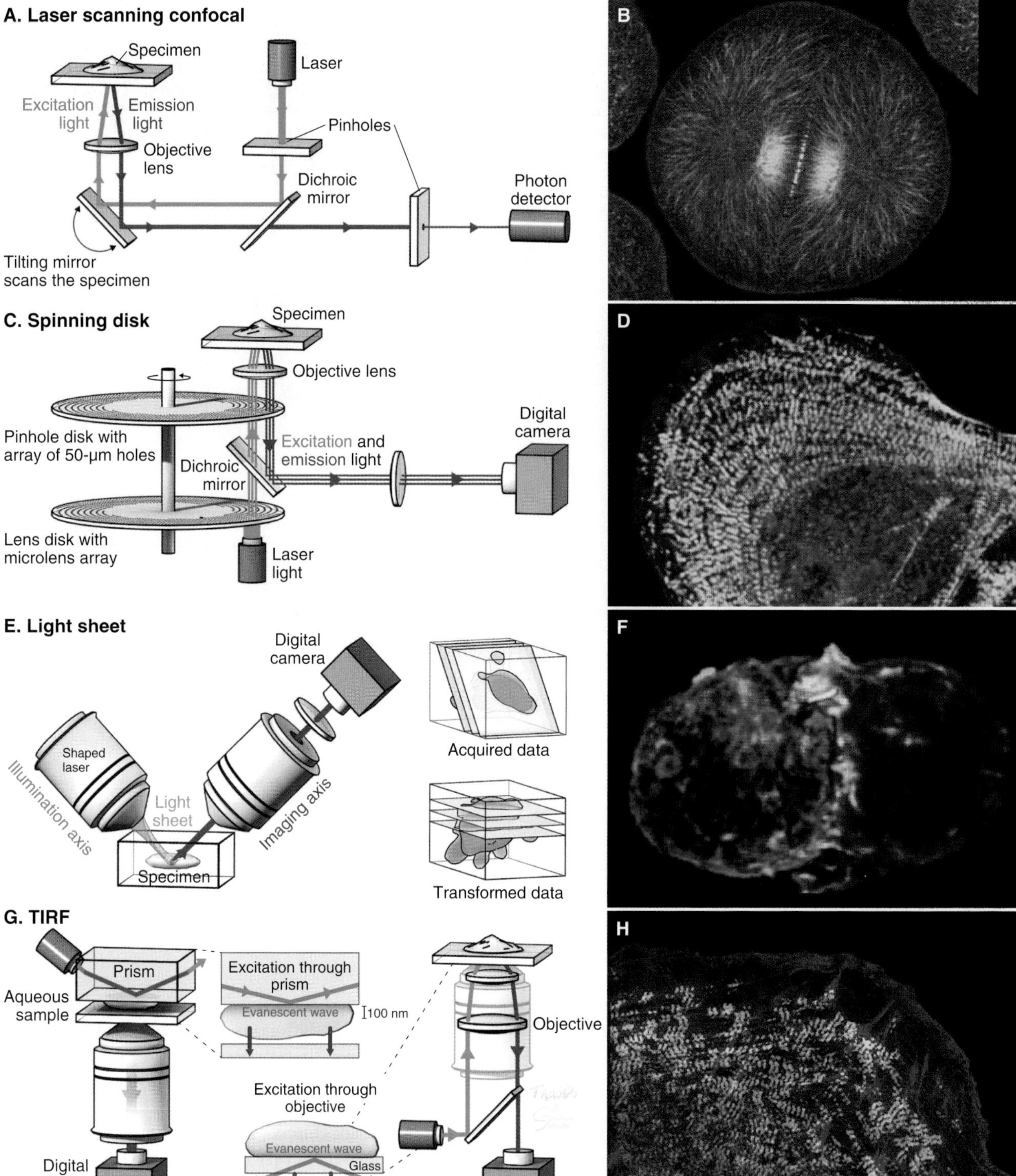
A. Laser scanning confocal
Specimen
Laser
Excitation light
Emission light
Pinholes
Objective lens
Dichroic mirror
Photon detector
Tilting mirror scans the specimen
B
C. Spinning disk
Specimen
Objective lens
Digital camera
Pinhole disk with array of 50-μm holes
Excitation and emission light
Dichroic mirror
Lens disk with microlens array
Laser light
D
E. Light sheet
Digital camera
Acquired data
Shaped laser
Illumination axis
Light sheet
Imaging axis
Specimen
Transformed data
F
G. TIRF
Prism
Excitation through prism
Aqueous sample
Evanescent wave
100 nm
Objective
Excitation through objective
Evanescent wave
Glass
Digital camera
H

FIGURE 6.4 METHODS TO MAKE THIN OPTICAL SECTIONS OF FLUORESCENT SPECIMENS. A, Imaging strategy for a laser scanning confocal fluorescence microscope. **B,** Optical section taken with a laser scanning microscope through a live, dividing starfish embryo expressing 3 × GFP SpEct2 (gold; a guanine nucleotide exchange factor of Rho-GTPases) and 2 × mCh EMTB (cyan; ensconsin microtubule-binding domain to mark microtubules). The field is 180 μM wide. **C,** Imaging strategy for a spinning disk confocal fluorescence microscope. **D,** Image taken with a spinning disk confocal microscope of a fixed U2OS cell stained with rhodamine-phalloidin (red; to mark actin filaments) and Alexa Fluor 488–antibodies to myosin-IIA (green). **E,** Imaging strategy for light sheet microscopy. **F,** Image from a lattice light sheet movie of a cytotoxic T lymphocyte expressing Lifeact-mEmerald attacking a target cell expressing membrane-targeted mTagBFP2. **G,** Total internal reflection fluorescence (TIRF) microscopy with structured illumination (SIM). The exciting laser beam is reflected from the glass–water interface, producing a thin (100-nm) evanescent wave that excites fluorophores in the specimen. The exciting light may be directed through a prism or the microscope objective. **H,** Image taken by TIRF-SIM microscopy of a U2OS cell expressing enhanced green fluorescent protein (EGFP)-nonmuscle myosin-IIA (green) and mAppleFtractin (red, to mark actin filaments). In **D** and **H** the fluorescence from myosin-II appears as pairs of spots marking the two ends of the minifilaments (see Fig. 5.7B). (**A,** Modified from http://malone.bioquant.uni-heidelberg.de/methods/imaging/imaging.html#CLSM. **B,** From Su KC, Bement WM, Petronczki M, von Dassow G. An astral simulacrum of the central spindle accounts for normal, spindle-less, and anucleate cytokinesis in echinoderm embryos. *Mol Biol Cell.* 2014;25:4049–4062. **C,** Courtesy Carl Zeiss Microscopy GmbH. **D,** Courtesy Dylan Burnette, Vanderbilt University, Nashville, TN. **E,** Modified from Rozbicki E, Chuai M, Karjalainen AI, et al. Myosin-II-mediated cell shape changes and cell intercalation contribute to primitive streak formation. *Nat Cell Biol.* 2015;17:397–408. **F,** From Alex Ritter and Jennifer Lippincott-Schwartz [National Institutes of Health, Bethesda, MD], Gillian Griffiths [Cambridge Institute for Medical Research, United Kingdom], and Eric Betzig [Janelia Farm Research Campus, Ashburn, VA]. **G,** Modified from www.nikon.com/products/microscope-solutions/lineup/inverted/wtirf/index.htm. **H,** Courtesy Jordan Beach and John Hammer, National Institutes of Health. For reference, see Beach JR, Shao L, Remmert K, et al. Nonmuscle myosin II isoforms coassemble in living cells. *Curr Biol.* 2014;24: 1160–1166.)

TABLE 6.2 Superresolution Fluorescence Microscopy

Name	Principle	Fluorophores	Resolution (X-Y Plane)	Time to Image
Localization microscopy (FPALM, PALM, STORM)	Wide-field illumination is used to activate a small subset of widely separated photoconvertible fluorescent molecules, their positions are determined precisely, building up a two-dimensional image over many successive cycles	Photoconvertible fluorescent dyes and proteins that turn on and off or change color	20–40 nm	Seconds to minutes
Structured illumination (SIM)	Superimposition of the fluorescence image with an intense scanned and rotated bar pattern improves the resolution	Any photostable fluorophore	~100 nm	Seconds
Stimulated emission depletion (STED)	Two superimposed beams scan the specimen with one suppressing emission from all but a tiny spot	Fluorescent dyes, some fluorescent proteins	~30 nm	30 frames/sec with 62-nm resolution

FPALM, fluorescence photoactivation localization microscopy; PALM, photoactivated localization microscopy; STORM, stochastic optical reconstruction microscopy.

image by assigning the fluorescence intensity measured at each point along the raster lines to the corresponding point in the cell (Fig. 6.4B). Scanning the specimen rapidly with spinning disks of small lenses and corresponding pinholes allows rapid imaging with a digital camera (Fig. 6.4C–D). A series of confocal images taken at different planes of focus can be used for three-dimensional reconstructions.

Light sheet microscopy creates thin optical sections of fluorescent specimens by focusing laser light into a sheet 2–8 μm thick and passing it through an illumination objective to focus onto the sample (Fig. 6.4E–F). A detection objective sits at right angles to the illumination objective to collect emitted fluorescence for the camera. The sample resides between the two objectives on a rotatable stage that allows light sheet illumination of successive planes. Acquisition of three-dimensional images at high imaging speeds is possible, allowing all the cells in thick specimens, including intact embryos, to be imaged at high resolution.

Superresolution Fluorescence Microscopy

Three methods have extended the resolution of fluorescence microscopy well beyond the classic limit of 0.2 μm. Each has strengths and weaknesses for different applications (Table 6.2).

Localization microscopy (independently named FPALM [fluorescence photoactivation localization microscopy], PALM [photoactivated localization microscopy], and STORM [stochastic optical reconstruction microscopy]) depends on the availability of organic dyes and fluorescent proteins that can be switched by light between dark and fluorescent states (photoactivation) or between two fluorescent states with different emission wavelengths (photoconversion). This makes it possible to turn on the fluorescence of just a few widely separated individual

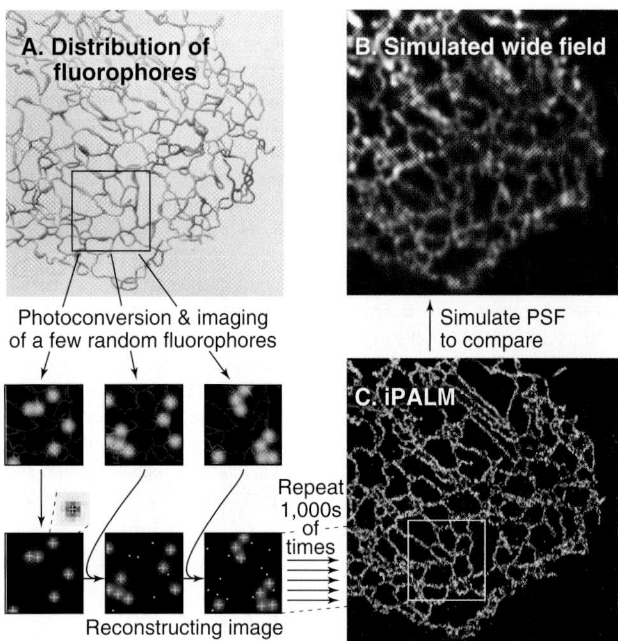

FIGURE 6.5 SUPERRESOLUTION FLUORESCENCE MICROSCOPY. Superresolution fluorescence localization microscopy. **A,** Imaging strategy. A specimen containing thousands of photoconvertible fluorophores is pulsed with UV light to activate a few of the molecules. A wide-field image shows the position of each as a blurred spot *(green)*. The center of each is located precisely *(cursor and white point)*. Each fluorescent molecule is bleached and the process is repeated many times to assemble a high-resolution image. **B** and **C,** Image of endoplasmic reticulum in a cultured cell stained with fluorescent antibodies to reticulon. **B,** Image showing how the localizations would appear in a conventional wide-field fluorescence microscope. **C,** Superresolution localization microscopy. (**B–C,** Courtesy Dylan Burnette, Vanderbilt University, Nashville, TN.)

fluorescent molecules (Fig. 6.5). Although the image of each fluorescent molecule is blurred by the point spread function of the microscope, the center of each point of light can be determined precisely by fitting the distribution with a computer. An image is built up of these point localizations through thousands of cycles of photoactivation/conversion, imaging, and photobleaching. Initially this process took many minutes, but high-speed digital cameras can now collect hundreds of images per second, making the method useful for live cells. The initial reports used epifluorescence and total internal reflection for photoconversion and localization. Light sheet illumination offers a new alternative. The improvement in the resolution from 200 nm to 30 nm offered by these methods reveals many structural details of interest to cell biologists (Fig. 6.5B). Because single molecules are imaged, localization microscopy is also used to estimate the stoichiometry and spatial correlations of molecules within cells, and to track their motions.

Stimulated emission depletion (STED) microscopy improves resolution by reducing the focal spot size with specialized optics. Concentric beams of laser light narrow the fluorescence emission to a central focal spot by depleting fluorescence in surrounding region of the sample. Scanning this focal spot across the sample allows formation of an image with ~70- to 90-nm resolution. STED initially employed fluorescent dyes but can now be performed with fluorescent proteins and in live cells.

Structured illumination microscopy (SIM) increases spatial resolution by illuminating a sample with patterned light and computationally analyzing the interference of the illumination pattern and the sample. It can image live cells faster and with much less light than that required by other superresolution approaches. Because the structured illumination pattern cannot be focused beyond half the wavelength of the excitation light, SIM originally enhanced resolution only by a factor of two. Nonlinear approaches have extended the resolution to better than 100 nm, making SIM a powerful superresolution method for imaging live cells (Fig. 6.4H).

Electron Microscopy

A **transmission electron microscope** (Fig. 6.1D) can resolve points below 0.3 nm, but the practical resolution was historically limited by the methods used to prepare specimens and damage from the electron beam. The initial method to prepare cells and tissues for electron microscopy was to fix with chemicals, dehydrate with organic solvents, embed in plastic, cut with a diamond knife into **thin sections,** and stain the sections with heavy metals (Fig. 6.6A). The resolution of approximately 3 nm was sufficient to bridge the gap between light microscopy and macromolecular structures. Between 1950 and 1970 electron micrographs of thin sections of cells and tissues revealed most of what is known about the organization of their organelles.

Electron microscopy advanced with the introduction of new methods to prepare cells and molecules. The highest resolution of whole cells is attained by directly viewing rapidly frozen specimens embedded in vitreous ice (amorphous water frozen so rapidly that ice crystals do not form) (see Fig. 5.11). This is called **electron cryomicroscopy**, because the stage holding the frozen specimen is cooled to the temperature of liquid nitrogen. Imaging requires relatively thin specimens. Frozen specimens can be cut with a microtome or by ion beam etching into sections thin enough for direct viewing.

Low contrast and superimposition of details limited electron cryomicroscopy until image-processing methods called **tomography** were developed to reconstruct three-dimensional volumes. The specimen is tilted inside the microscope, and micrographs taken from a wide range of angles. The information is merged computationally into a three-dimensional map, for viewing in thin slices from any angle (Fig. 6.6D). Structures as complex as entire cells can be visualized with a

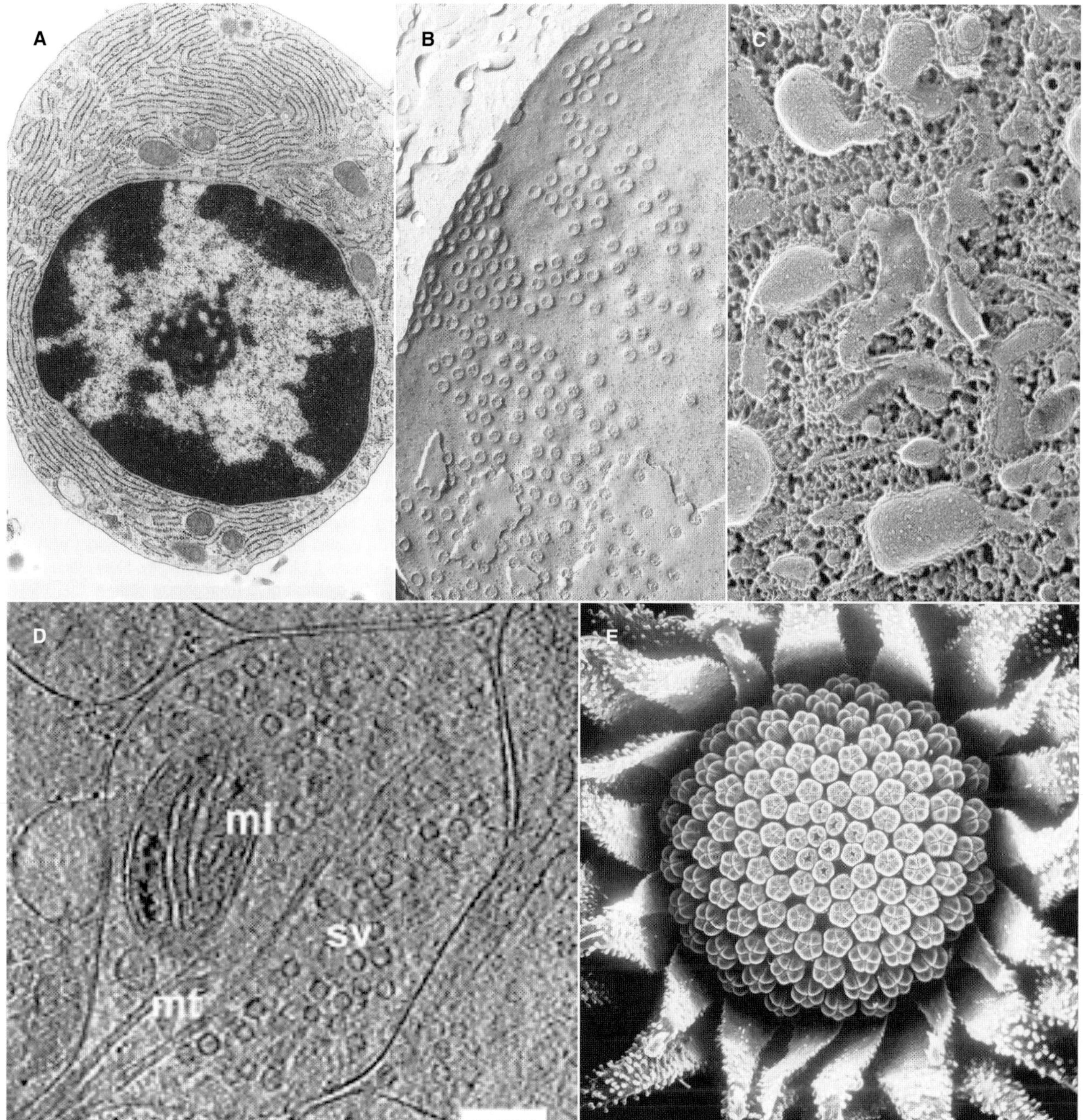

FIGURE 6.6 ELECTRON MICROGRAPHS OF CELLS. A–D, Transmission electron micrographs. **A,** Thin section of a plasma cell, an immune cell specialized to synthesize and secrete antibodies. **B,** Freeze-fracturing. The cleavage plane passed through the cytoplasm and then split apart the two halves of the bilayer of the nuclear envelope. This fractured surface was then shadowed with platinum. The cytoplasm is in the upper left. Nuclear pores are prominent in the nuclear envelope. **C,** A cultured cell prepared by rapid freezing, fracturing, deep etching, and rotary shadowing with platinum. Membranes of the endoplasmic reticulum stand out against the porous cytoplasmic matrix. **D,** Tomographic reconstruction of a thin slice through a presynaptic terminal of a cultured neuron that was rapidly frozen and thinned by focused ion beam milling. The image shows a mitochondrion (mi), microtubules (mt), and synaptic vesicles (sv) inside the plasma membrane. **E,** Scanning electron micrograph of developing flowers of the Western mountain aster. (**A–B,** Courtesy Don W. Fawcett, Harvard Medical School, Boston, MA. **C,** Courtesy John Heuser, Washington University, St. Louis, MO. **D,** From Lučič V, Rigort A, Baumeister W. Cryo-electron tomography: the challenge of doing structural biology in situ. *J Cell Biol.* 2013;202:407–419. **E,** Courtesy J.-L. Bowman, University of California, Davis.)

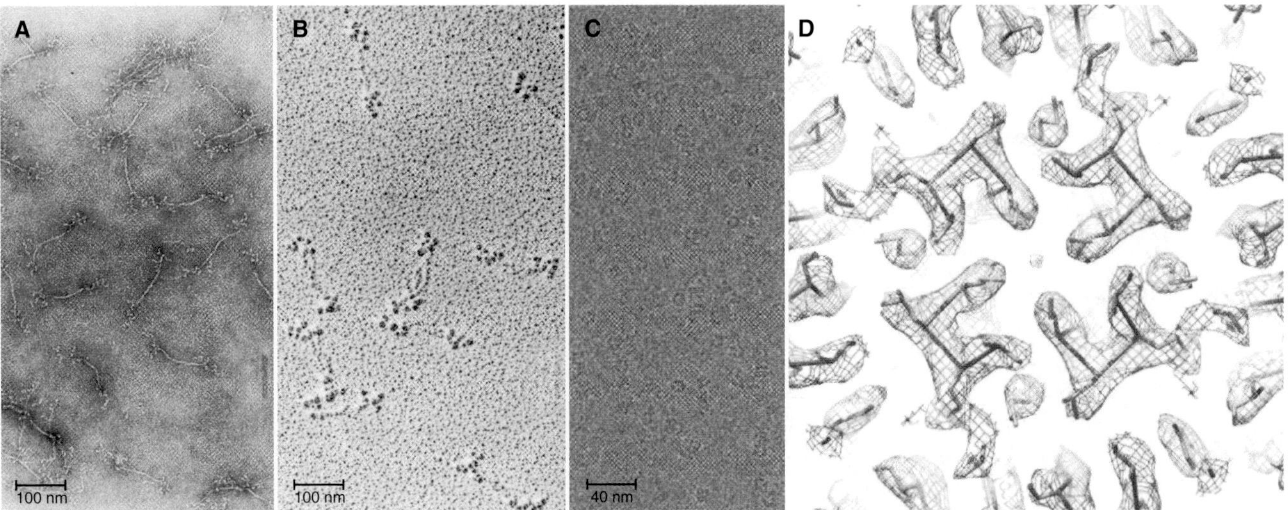

FIGURE 6.7 ELECTRON MICROGRAPHS OF MOLECULES. A–B, Transmission electron micrographs of myosin-II minifilaments. **A,** Filaments on a thin carbon film prepared by negative staining with uranyl acetate. **B,** Filaments on a mica surface prepared by rotary shadowing with platinum. **C,** Low-dose electron micrograph of single, purified TRPV1 (transient receptor potential vanilloid-1) channel proteins in a thin film of ice. The contrast is so low that identifying the molecules is difficult. **D,** Reconstruction the TRPV channel structure at 3.4 Å resolution from 10,000 single molecules. Slice through the three-dimensional reconstruction showing amino acid side chains and the orange atomic model with the central ion pore surrounded by one α-helix from each of the four identical subunits. (**A–B,** Courtesy J. Sinard, Yale University, New Haven, CT. **C–D,** From Liao M, Cao E, Julius D, Cheng Y. Structure of the TRPV1 ion channel determined by electron cryo-microscopy. *Nature.* 2013;504: 107–112.)

resolution of a few nanometers. Tomography can also be applied to sections of plastic-embedded specimens.

The **freeze–fracture** method provides a different view inside cells. A frozen specimen is cleaved by cracking the ice. Surfaces exposed by the fracture are rotary shadowed with a thin coat of platinum for viewing with a transmission electron microscope (Fig. 6.6B). Frequently, the cleavage plane splits apart the two halves of lipid bilayers to reveal proteins embedded in the plane of the membrane. Subliming some frozen water from the fracture surface before shadowing reveals three-dimensional details of the cytoskeleton deeper in the cytoplasm, if soluble molecules are extracted before freezing (Fig. 6.6C; see also Figs. 1.13 and 5.3B).

Electron microscopy is valuable to study macromolecules, macromolecular assemblies, polymers, and two-dimensional crystals. These specimens can be frozen in vitreous ice for direct imaging by electron cryomicroscopy (see Figs. 5.10A, 6.7C, 34.6B, and 36.4A). Alternatively, macromolecules can be rotary shadowed as in freeze fracturing after drying on a smooth surface (Fig. 6.7C) or rapidly freezing and subliming away the ice (see Figs. 30.4 and 34.10). Another method is negative staining, in which specimens are dried from aqueous solutions of heavy metal salts (Fig. 6.7A). A shell of dense stain encases particles on the surface of a thin film of carbon and can preserve structural details at a resolution of approximately 2 nm.

Computer image processing of electron micrographs is used to reconstruct three-dimensional structures of macromolecules and macromolecular assemblies (Fig. 6.7C). Single particles are reconstructed by first classifying images of tens of thousands of randomly oriented particles into categories corresponding to different views. Then, an average three-dimensional structure is calculated computationally from this ensemble (Fig. 6.7D). The exposure to electrons is kept low to avoid radiation damage.

Micrographs of particles with helical symmetry, such as actin filaments (see Fig. 33.7) and bacterial flagella (see Fig. 5.8), are analyzed in two ways: either helical image processing or by dividing the polymer into short segments for analysis by single particle methods. A few proteins form two-dimensional crystals naturally or in the laboratory. Computational analysis of electron micrographs and electron diffraction patterns of such two-dimensional crystals with methods related to those for x-ray diffraction have produced structures of bacteriorhodopsin (see Fig. 13.9), aquaporin water channels (see Fig. 16.15), and tubulin (see Fig. 34.4B) at steadily improving resolutions.

Improvements in freezing technology, image recognition and averaging algorithms, microscopes and, most importantly, cameras that detect electrons directly (analogous to the CCD cameras used to detect fluorescent light) have pushed the resolution in favorable specimens to less than 0.3 nm (Fig. 6.7C–D). It is now possible to visualize amino acid side chains in many large macromolecular structures that were difficult or impossible to study by x-ray diffraction. These recent advances in electron microscopy have revolutionized studies of complex biological structures.

A **scanning electron microscope** (SEM) can be used on thicker specimens, such as whole cells or tissues that have been fixed, dried, and coated with a thin metal film. Here, an electron beam scans a raster pattern over the surface of specimens, and secondary electrons emitted from the surface at each point are collected and used to reconstruct an image (Fig. 6.6E). The resolution of conventional SEM is limited, but nonetheless valuable for studying surface features of cells and their three-dimensional relationships in tissues.

SEMs with high-energy (field emission) guns to produce the electron beam have higher resolution suitable for studying cellular substructures, such as nuclear pores (see Fig. 9.6B). One may use a focused ion beam to etch away the surface of an embedded specimen to expose internal details for imaging of the surface by SEM. Many cycles of etching and imaging can be used to reconstruct the entire specimen in three dimensions.

Choice of Organisms for Biological Research

Given that life on earth arose from a common ancestor (see Fig. 2.1), one can learn about basic cellular processes in any organism by studying the molecules of interest. It is useful to select an organism that specializes in the process, such as skeletal muscle to study contractile proteins (see Chapter 39) or *Chlamydomonas* to study flagella (see Fig. 38.18). Some organisms have the advantage that communities of scientists invested years to develop genetic, molecular genetic, and biochemical methods for experimentation. These valuable experimental tools have attracted investigators to a growing number of "model" organisms (Table 6.3).

Model Organisms

Most model organisms have completely sequenced genomes and facile methods to manipulate the genes, including replacement of a gene with a modified gene by the process of homologous recombination or genome editing (Fig. 6.16). Haploid organisms with one copy of each chromosome after mitotic division are particularly favorable for detecting the effects of changes in genes, called mutations (Box 6.1). It is useful for a haploid organism to have a diploid stage with two copies of each chromosome and a sexual phase, during which meiotic recombination occurs between the chromosomes from the two parents. (See Fig. 45.3 for details on recombination.) This allows one to construct strains with a variety of mutations and facilitates mapping mutations to a particular gene. In addition, diploids carrying a lethal mutation of a gene that is essential for life can be propagated, provided that the mutation is recessive.

Budding yeast and fission yeast meet all these criteria, so they are widely used to study basic cellular functions. Moving between haploid and diploid stages greatly simplifies the process of creating and analyzing recessive mutations. This is important, because most loss-of-function mutations are recessive. Research combining genetic, biochemical, and microscopic analysis have

TABLE 6.3 Model Genetic Organisms

Organism	Genome Size and Ploidy	Number of Genes	Gene Targeting by Homologous Recombination	Genome Editing With Nucleases	Meiotic Recombination	Biochemistry
Gram-negative bacterium, *Escherichia coli*	4.6 Mb, haploid	4288	Yes	Yes	No	Excellent
Cellular slime mold, *Dictyostelium discoideum*	34 Mb, haploid	~12,000	Yes	Yes	No	Excellent
Budding yeast, *Saccharomyces cerevisiae*	12.1 Mb, haploid	~6604	Yes	Yes	Yes	Good
Fission yeast, *Schizosaccharomyces pombe*	14 Mb, haploid	~4900	Yes	Yes	Yes	Good
Mustard weed, *Arabidopsis thaliana*	100 Mb, diploid	~25,706	No	Yes	Yes	Poor
Nematode worm, *Caenorhabditis elegans*	97 Mb, diploid	~18,266	Difficult	Yes	Yes	Poor
Fruit fly, *Drosophila melanogaster*	180 Mb, diploid	~13,338	Difficult	Yes	Yes	Fair
Zebrafish, *Danio rerio*	1400 Mb, diploid	~26,206	Difficult	Yes	Yes	Good
Mouse, *Mus musculus*	3000 Mb, diploid	~25,000	Yes	Yes	Yes	Good
Human, *Homo sapiens*	3000 Mb, diploid	~25,000	Yes, cultured cells	Yes, cultured cells	Yes	Good

BOX 6.1 Key Genetic Terms

Allele. A version of a gene.
Complementation. Providing gene function in *trans* (ie, by another copy of a gene).
Conditional mutation. Mutation giving an altered phenotype only under certain conditions, such as temperature, medium composition, and so on.
Diploid. Genome with two copies of each chromosome, one from each parent.
Dominant mutation. Mutation giving an altered phenotype, even in the presence of a copy of the wild-type gene on another chromosome.
Essential gene. Gene required for viability.
Gene. Nucleotide sequence required to make a protein or RNA product, including the coding sequence, flanking regulatory sequences, and any introns.
Genome. Entire genetic endowment of an organism.
Genotype. Genetic complement, including particular mutations.
Haploid. Genome with single copies of each chromosome.
Mutant. Organism with a mutation of interest.
Mutation. A change in the chemical composition of a gene, including changes in nucleotide sequence, insertion, deletions, and so on.
Pedigree. Family history of a genetic trait.
Phenotype. (From the Greek term for "shining" or "showing.") Appearance of the organism as dictated by its genotype.
Plasmid. Circular DNA molecule that self-replicates in the cytoplasm of a bacterium or nucleus of a eukaryote.
Recessive mutation. Mutation giving an altered phenotype only when no wild-type version is present.
Recombination. Physical exchange of regions of the genome between homologous chromosomes or between a plasmid and a chromosome.
Wild type. The naturally occurring allele of a gene; the phenotype of the naturally occurring organism.

made fundamental contributions to cell biology, but yeast are solitary cells with specialized lifestyles.

Multicellular organisms are required to study the development and function of tissues and organs. Flies, nematode worms, mice, and humans share many ancient, conserved genes that control their cellular and developmental systems, so flies and worms are popular for basic studies of animal development and tissue function. However, vertebrates evolved a substantial number of new gene families (roughly 7% of total genes) and a large number of new proteins by rearranging ancient domains in new ways. Therefore, mice are often used for experiments on specialized vertebrate functions, especially the nervous system. Although not an experimental organism, humans are included on this list, because much can be learned by analysis of human genetic variation and its relationship to disease. Humans are, of course, much more eloquent than the model organisms when it comes to describing their medical problems, many of which have a genetic basis that can be documented by analysis of pedigrees and DNA samples. Furthermore, traits can be studied in billions of humans. The mustard *Arabidopsis thaliana* is the most popular plant for genetics, because its genome is small, reproduction is relatively rapid, and methods for genetic analysis are well developed. The moss *Physcomitrella patens* has experimental advantages for studying basic plant biology. One challenge has been the lack of methods to replace genes by homologous recombination in plants (see later section).

By focusing on a limited number of easy-to-use model organisms, biological research has raced forward beginning during the last quarter of the 20th century. This focus does have liabilities. For one, these organisms represent a very limited range of lifestyles. Thousands of other solutions to survival exist in nature, and they tend to be ignored. At the cellular level, these liabilities are less severe, as most cellular adaptations are ancient and shared by most organisms.

Cell Culture

Regardless of the species to be studied, growing large populations of isolated cells for biochemical analysis and microscopic observation is helpful. This is straightforward for the unicellular organisms such as fungi or bacteria, which can be grown suspended in a nutrient medium. These organisms can also be grown on the surface of gelled agar in a petri dish. When single cells are dispersed widely on an agar surface, each multiplies to form a macroscopic colony, all descendants of a single cell. This family of cells is called a clone.

It is often possible to isolate single live cells from multicellular organisms by dissociating a tissue with proteolytic enzymes under conditions that weaken adhesions between the cells. Many isolated cells grow in sterile media, a method called **tissue culture** or cell culture. Terminally differentiated cells such as muscle or nerve cells do not reenter the cell cycle and grow. Cells that are predisposed to grow in the body, including fibroblasts (see Fig. 28.2) and endothelial cells from blood vessels (see Fig. 30.13), will grow if the nutrient medium is supplemented with growth factors to drive the cell cycle (see Fig. 41.9). This is accomplished by adding fetal calf serum, which contains a rich mixture of growth factors. Some cultured cells grow in suspension, but most prefer to grow on a surface of plastic or glass (Fig. 6.2), often coated with extracellular matrix molecules for adhesion (see Fig. 30.11). This is the origin of the term in vitro, meaning "in glass," used to describe cell culture. Normal cells grow until they cover the artificial surface, when contacts with other cells arrest further growth (see Fig. 41.3). Dissociation and dilution of the cells onto a fresh surface allow growth to resume. Most "primary cells" isolated directly from tissues divide a limited number of times (see Fig. 7.15). Primary cells can become immortal, either through mutations or transformation by a tumor virus that overcomes cell-cycle controls. Such immortal cells are called **cell lines**. Similar changes allow cancer cells to grow indefinitely. HeLa cells, the very first cell line, were derived from a cervical cancer that afflicted the African-American patient Henrietta Lacks. HeLa cells have been growing worldwide in laboratories for more than 60 years.

A variation on cell culture is to grow a whole organ or part of an organ in vitro. The requirements for organ culture are often more stringent than those for growing individual cells, but the method is used routinely for experiments on slices of brain tissue and for studying the development of embryonic organs from stem cells (see Box 41.2).

Inventory: Gene and Protein Discovery

Classical Genetics: Identification of Genes Through Mutations

The strategy in classical genetics is to make random mutations that compromise a particular cellular function and then to find the mutated gene(s). This approach is extremely powerful, especially when little or nothing is known about a process or when the gene product (usually a protein) is present at low concentrations. Genetic analysis of yeast has been spectacularly successful in mapping out complex pathways, including the cell cycle (see Chapters 40 to 44) and secretory pathway (see Chapter 21).

Because one generally does not know the relevant genes in advance, it is important that mutations are introduced randomly into the genome and, ideally, limited to one mutation in each organism tested. A prerequisite for such a genetic screen is a good assay for the biological function of interest. Simplicity and specificity are essential, as interesting mutations may be rare, and much effort may be expended characterizing each mutation. The assay may test the ability to grow under certain conditions, drug resistance, morphologic changes, cell-cycle arrest, or abnormal behavior. Mutations arise spontaneously at low rates, so often a chemical (eg, ethyl methyl sulfonate or nitrosoguanidine) or radiation is used to increase the frequency of damage. Another strategy is to insert an identifiable segment of DNA randomly into the genome to disrupt genes and mark them for subsequent analysis.

Haploid organisms are favorable for detecting mutations, because damage to the single copy of a gene will alter function, so either a loss or a gain of function can be detected with suitable test conditions (ie, the ability to grow under certain conditions), biochemical assay, or morphologic assay. A disadvantage of haploid organisms is that they are not viable following the loss of function of an essential gene. Consequently, one selects for **conditional mutant** alleles that allow a haploid organism to survive mutation of an essential gene under **permissive conditions** (eg, low temperatures) but not under **restrictive conditions** (eg, high temperatures).

One can often identify a mutated gene in a haploid organism by a **complementation** experiment. A population of mutant cells is induced to take up DNA fragments contained in a plasmid library constructed from the wild-type genome or complementary DNAs (cDNAs). **Plasmids** are circular DNA molecules that can be propagated readily in bacteria and, if suitably designed, in eukaryotes as well. Plasmids carrying the wild-type gene will correct loss-of-function mutations, allowing cells to grow normally. Plasmids complementing the mutation are isolated and sequenced. The mutant gene can be sequenced to determine the nature of the damage. This complementation test can also be used to discover genes from other species that correct the mutation in the model organism. For example, genes for human cell-cycle proteins can complement many cell-cycle mutations in yeast (see Chapter 40).

Genetics in obligate diploid organisms is more complicated. Many mutations will appear to have no effect, provided that the corresponding gene on the other chromosome functions normally. These **recessive mutations** produce a phenotype only after crossing two mutant organisms, yielding 25% of offspring with two copies of the mutant gene. (Consult a genetics textbook for details on mendelian segregation.) Other mutations will yield an altered phenotype even when only one of the two genes is affected. These **dominant mutations** can include simple loss of function alleles when two wild-type genes are required to make sufficient product for normal function (called **haploinsufficiency**); production of an altered protein that compromises the formation of a multimeric assembly by normal protein subunits produced by the wild-type gene (called **dominant negative**); and production of an unregulated protein that cannot be controlled by partners in the cell (another type of dominant negative).

If the genome is small, a mutation can be found by sequencing the entire genome, but the classic method for identifying a mutated gene is **genetic mapping.** One observes the frequency of recombination between known genetic markers and the mutation of interest in genetic crosses. This is usually sufficient to map a gene to a broad region of a particular chromosome. If a complete genome sequence is available, the database of sequenced genes in the area highlighted by mapping is examined to look for candidate genes that might carry the mutation. If the mutation was made by inserting a piece of DNA, such as a transposable element, randomly into the genome, one can recover the transposable element together with some of the surrounding chromosome, which is sequenced to identify the disrupted gene.

Once a gene required for the function of interest is identified and sequenced (see Fig. 3.16), the primary structure of the protein (or RNA) is deduced from the coding sequence. Searching for proteins with similar sequences or domains in the same or other species is often informative, particularly if something is known about the function of the corresponding gene product. Protein can often be expressed from a cDNA copy of the mRNA, tested for activity and binding partners, and (when fused to GFP or when used to make an antibody) localized in cells.

Genomics and Reverse Genetics

Complete sequences of the coding regions of most popular experimental organisms are now available. Work is continuing to annotate this data to provide definitive inventories of genes. The task has been aided by sequencing cDNA copies of expressed genes (**expressed sequence tags** [ESTs]), which help document the diversity of products created by transcription and RNA processing (see Chapter 10).

The sequences of proteins with known functions are used to search sequenced genomes for new proteins with related functions. These searches are surprisingly fruitful, as many genes arose by gene duplication and occur as extended families. Once a predicted sequence has been identified, one can check when and where the gene is expressed in the organism, test the consequences of deleting the gene, or test for interactions of the protein with other proteins (see later section). These tests can be done one gene at a time or on a genome-wide scale. For example, investigators created strains of budding yeast lacking each of the 6000 genes and discovered that only 19% are essential for viability. They also tested for interaction between the deletion mutations and other genes, and for interactions of the product of each gene with the products of all other genes. These preliminary screening tests often yield clues about function. Ultimately, however, function is understood only when representatives of each protein family are studied in detail by the biophysical, biochemical, and cellular methods described in the following sections.

Reverse genetics refers to the process of starting with a known gene and selectively disrupting its function by deleting the gene by homologous recombination or genome editing. Depleting the RNA product by RNAi (RNA interference; discussed later in the section titled "Three Options to Test for Physiological Function") is simpler experimentally but often more complicated to interpret due to "off-target" effects.

Biochemical Fractionation

The biochemical approach (to the inventory) is to purify active molecules for analysis of structure and function. This requires a sensitive, quantitative assay to detect the component of interest in crude fractions, an assay to assess purity, and methods to separate the molecule from the rest of the cellular constituents. Assays are as diverse as the processes of life. Enzyme activity is often easy to measure. Many molecules are detected by binding a partner molecule. For example, nucleic acids bind complementary nucleotide sequences and sequence-specific regulatory proteins; receptors bind ligands; antibodies bind their antigens; and many proteins bind partner proteins. More difficult assays reconstitute a cellular process, such as membrane vesicle fusion, nuclear transport, or molecular motility. Devising a sensitive and specific assay requires creativity. A second prerequisite for purification is a simple method for assessing purity. Various types of **gel electrophoresis** often work brilliantly (Box 6.2 and Fig. 6.8).

With a functional assay and a method to assess purity, one sets about purifying the molecule of interest. Highly abundant constituents, such as actin or tubulin, may require purification of only 20- to 100-fold, but many important molecules, such as signaling proteins and transcription factors, constitute less than 0.1% of the cell protein, so extensive purification is required.

One may start with the organism, if it is available in large quantities (tens of grams) and amenable to fractionation. Alternatively, many proteins can be expressed from cDNAs in bacteria, yeast, or virus-infected insect cells. An advantage of this approach is that mutations can be made at will, including substitution of one or more amino acids, deletion of parts of the protein or adding domains that facilitate purification.

First, the cell is disrupted gently to avoid damage to the molecule of interest. This may be accomplished physically by mechanical shearing with various types of **homogenizers** or, where appropriate, chemically, with mild detergents that extract lipids from cellular membranes. Next, the homogenate is **centrifuged** to separate particulate and soluble constituents.

Purification of Organelles

If the molecule of interest is part of an organelle, centrifugation can be used to isolate the organelle. Typically, the first step is to centrifuge the crude cellular homogenate multiple times at a succession of higher speeds (and therefore forces). Large particles such as nuclei pack into a pellet at the bottom of the centrifuge tube at low speeds, whereas high speeds are required to pellet small vesicles. These pellets may be enriched in particular organelles but are never pure.

Two complementary centrifugation methods can improve the purity. The motion of particles in a centrifugal force field depends on their mass and shape, but also on the difference between their density and that of the surrounding medium. Particles do not move when the density of the medium matches their own density. Therefore, organelles with different densities can be separated from each other by centrifuging for many hours in a tube containing a concentration gradient of sucrose (eg, 5% sucrose in buffer at the top of the tube, increasing to

BOX 6.2 Gel Electrophoresis

An electrical field drives molecules in a sample through a gel matrix. **Agarose gels** (Fig. 6.8B) are used commonly for nucleic acids, whereas **polyacrylamide gels** are used for both nucleic acids and proteins (Fig. 6.8C). Most often, the ionic detergent sodium dodecylsulphate (SDS) is employed to dissociate the components of the sample from each other, making their rate of migration through the gel depend on their size. SDS binding unfolds polypeptide chains and gives them a uniform negative charge per unit length. Small molecules move rapidly and separate from slowly moving large molecules, which are more impeded by the matrix. By the time small molecules reach the end of the gel, all the components in the sample are spread out according to size. Buffers containing high concentrations of the nonionic, denaturing agent urea also dissociate and unfold protein molecules. Electrophoresis in urea separates the proteins depending on both their charge *and* size. Negatively charged proteins move toward the positive electrode, whereas positively charged proteins move in the opposite direction. Another approach, called **isoelectric focusing,** uses a buffer containing molecules called Ampholines, which have both positive and negative charges. In an electrical field across a gel, Ampholines set up a pH gradient. Proteins (usually dissociated in urea) migrate to the pH where they have a net charge of zero, their isoelectric point. This is a sensitive approach to detect charge differences in proteins, such as those introduced by phosphorylation. Isoelectric focusing in one gel followed by SDS-gel electrophoresis in a second dimension can resolve hundreds of individual proteins in complex samples (see Fig. 38.14A).

Many methods are available to detect molecules separated by gel electrophoresis. Fluorescent dyes, such as ethidium bromide, bind nucleic acids (Fig. 6.8B). Following blotting of separated nucleic acids from the gel onto nitrocellulose or nylon films, specific sequences can be detected by hybridization with complementary oligonucleotides or longer sequences of cloned DNA (probes) labeled with radioactivity or fluorescent dyes.

Proteins are detected by binding colored dyes or more sensitive metal reduction techniques. Obtaining a single stained band on a heavily loaded SDS gel is the goal of those purifying proteins. Of course, some pure proteins consist of multiple polypeptide chains (Fig. 6.8C); in such cases, multiple bands in characteristic ratios are seen. Specific proteins are often detected with antibodies. To do this, proteins are transferred electrophoretically from the polyacrylamide gel to a sheet of nitrocellulose or nylon before reaction with antibodies. This transfer step is called **blotting.** Antibodies labeled with radioactivity are detected by exposing a sheet of x-ray film. Antibodies are also detected by reaction with a second antibody conjugated to an enzyme that catalyzes a light-emitting reaction **(chemiluminescence)** or by direct conjugation with fluorescent molecules, which exposes a sheet of x-ray film or a digital detector. Some proteins can be detected by reaction with naturally occurring binding partners.

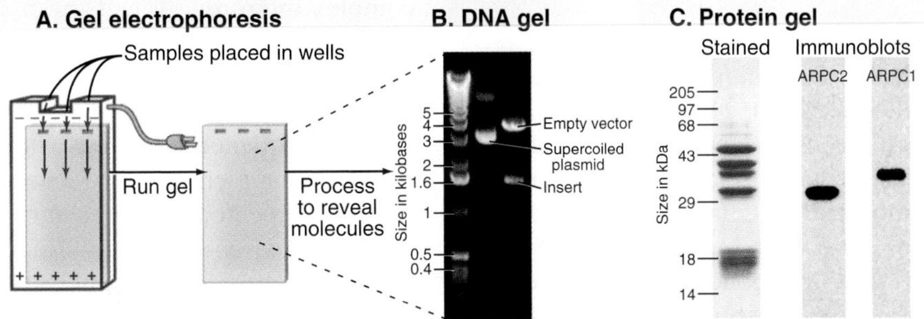

FIGURE 6.8 GEL ELECTROPHORESIS. A, Schematic showing a (generic) gel with three sample wells and an electric field. **B,** Agarose gel electrophoresis of DNA samples stained with ethidium bromide. The lane on the left shows size standards. The middle lane has a bacterial plasmid, a supercoiled (see Fig. 3.18) circular DNA molecule carrying an insert (Fig. 6.11 provides details). The right lane has the same plasmid digested with a restriction enzyme that cleaves the DNA twice, releasing the insert. Although smaller than the circular plasmid, the empty vector runs more slowly on the gel because the linear DNA offers more resistance to movement than the supercoiled circular plasmid. **C,** Polyacrylamide gel electrophoresis of the Arp2/3 complex, an assembly of seven protein subunits involved with actin polymerization (see Fig. 33.12). All three samples are identical. In the left lane, the proteins are stained with the nonspecific protein dye Coomassie blue. The proteins in the other two lanes were transferred to nitrocellulose paper; each reacted with an antibody to one of the subunit proteins (ARPC2 and ARPC1). The position of the bound antibody is determined with a second antibody coupled to an enzyme that produces light and exposes a piece of film black. This method is called chemiluminescence. (**B,** Courtesy V. Sirotkin, Yale University, New Haven, CT. **C,** Courtesy H. Higgs, Dartmouth Medical School, Hanover, NH.)

20% sucrose at the bottom). In such a **sedimentation equilibrium** gradient, particles such as membrane bound organelles move until their density equals that of the gradient, at which point they move no farther, regardless of how long or hard they are spun. Very dense particles containing DNA or RNA can be purified by centrifugation to equilibrium in gradients of dense salts, such as cesium chloride. An alternative is a **sedimentation velocity** experiment, where one centrifuges for a shorter time, so particles separate on the basis of their sedimentation rates rather than coming to equilibrium. Gradients of sucrose or glycerol are used to give the particles more or less constant velocities, with the higher density near the bottom counteracting the higher centrifugal force, which increases with the square of the distance from the center of the rotor (think of a spinning ice skater).

Additional methods are useful for purifying organelles from subcellular fractions obtained by sedimentation velocity and sedimentation equilibrium. For example, antibodies specific for a molecule on the surface of an organelle can be attached to a solid support and used to bind the organelle. Contaminating material can then be washed away.

Purification of Soluble Proteins

Given sufficient starting material most soluble proteins can be purified by **chromatography** (Box 6.3 and Fig. 6.9). The most powerful of these methods is affinity chromatography, where a molecule attached to a bead binds specifically to the soluble macromolecule of interest. Adding a binding site to a recombinant protein facilitates its purification. Popular examples include an **epitope tag,** which is a short peptide that binds a particular antibody; a **His-tag,** which is a short sequence of histidine residues that binds to a metal chelate; **GST** is the enzyme glutathione-*S*-transferase that binds tightly to glutathione; and maltose-binding protein binds to maltose.

From Protein to Gene

Once a protein of interest has been purified, the path to its gene(s) is relatively direct. The pioneering approach was to cut the polypeptide into fragments by proteolytic enzymes. These fragments were isolated by chromatography and their amino acid sequences determined by Edman degradation (see Chapter 3). Given part of the amino acid sequence, the corresponding gene was identified in a genomic database or isolated by using oligonucleotide probes as the assay (see "Isolation of Genes and Complementary DNAs" below).

Mass spectrometry is now the dominant method to identify protein sequences. A purified polypeptide is cleaved at specific sites with a proteolytic enzyme such as trypsin, and the masses of the fragments are measured precisely with a mass spectrometer. If the protein comes from an organism with a sequenced genome, the gene encoding the protein can be identified by matching the experimental masses of the tryptic fragments with masses of all the peptides predicted from the genome sequence. More frequently, certain initial peptide fragments are selected within the mass spectrometer and diverted into a chamber inside the machine where they are bombarded under conditions that randomly break the peptide backbone. The masses of the overlapping fragments reveal the amino acid sequence of the peptide. This approach can be applied to identify thousands of proteins in complex mixtures by digesting the heterogeneous sample with trypsin, fractionating by chromatography, and analyzing the column fractions by mass spectrometry.

Recent important advances in mass spectrometry include the ability to provide accurate quantitation of relative amounts of large numbers of proteins in different samples (eg, two different mutant cell lines) and measuring the mass of large protein complexes. Mass spectrometry can also identify the positions of chemical crosslinks between protein subunits in macromolecular complexes. This method provides information about the organization of protein complexes that cannot be studied by x-ray crystallography.

Isolation of Genes and Complementary DNAs

A variety of methods make isolation of specific nucleic acids relatively routine. Genomic DNA is isolated from whole cells by selective extraction. mRNAs are purified by affinity chromatography, taking advantage of their polyadenylate (poly[A]) tails (see Fig. 11.3), which bind by base pairing to poly(dT) attached to an insoluble matrix (Fig. 6.9A). DNA is easier to work with than RNA (eg, it can be cleaved by restriction endonucleases and cloned), so RNAs are usually converted to a complementary DNA (**cDNA**) by reverse transcriptase, a viral DNA polymerase that uses RNA as a template.

Several options exist to obtain a particular DNA from a complex mixture:

1. The **polymerase chain reaction (PCR)** uses a heat-stable DNA polymerase and two primers (oligonucleotides, each complementary to one of the ends of a DNA sequence of interest) to synthesize a strand of DNA complementary to another DNA strand (Fig. 6.10A). Repeated steps of synthesis and denaturation allow an exponential amplification in the amount of the DNA between the primers. Designing the primers requires knowledge of the sequence of the gene of interest. This may be available from databases or may be guessed from the protein sequence or the sequence of the same gene in a related species or a similar gene in the same species. If the reaction is successful, a single sequence is amplified in quantities sufficient for cloning, sequencing, or large-scale biological production by expression in a bacterium (see later discussion). PCR is so sensitive that DNA sequences from a single cell can be isolated and characterized.
2. A DNA segment of interest can be isolated by **cloning** in a bacterial virus or plasmid (Fig. 6.11A). Such cloning strategies use **"libraries"** of DNA sequences, highly complex mixtures that may include more than 10^6 different cDNAs or genomic DNA fragments. These DNA molecules are transferred into a plasmid, a circular DNA molecule that is capable of replication in a host bacterium, or less often into the genome of

BOX 6.3 Chromatography

Affinity chromatography (Fig. 6.9) is the most selective purification method. A ligand that binds the target molecule is attached covalently to a solid matrix. When a complex mixture of molecules passes through the column, the target molecule binds, whereas most of the other molecules flow through. After the column is washed, the target protein is eluted by competition with free ligand or changing conditions, such as changes in pH or salt concentration. The ligand and target in Fig. 6.9 are both nucleic acids, but they can be any molecules that bind together, including pairs of proteins, drugs and proteins, proteins and nucleic acids, and so on.

Gel filtration separates molecules on the basis of size. Inert beads of agarose, polyacrylamide, or other polymers are manufactured with pores of a particular size. Large molecules are excluded from the pores and elute first from the column in a volume (void volume) equal to the volume of buffer outside the beads in the column. Small molecules, such as salt, penetrate throughout the beads and elute much later in a volume equal to the total volume of the column. Molecules of intermediate size penetrate the beads to an extent that depends on their molecular radius. This parameter, called the **Stokes' radius,** can be measured quantitatively if the column is calibrated with standards of known size. Such molecules elute between the void volume and the total volume.

Ion exchange chromatography uses charged groups attached covalently to inert beads. These charged groups may be positive (eg, the tertiary amine diethylaminoethyl [DEAE]) or negative (eg, carboxylate or phosphate). Ionic interactions retain oppositely charged solutes on the surface of the column particles, provided that the ionic strength of the buffer is low. Typically, a gradient of salt is used to elute bound solutes.

Other types of chromatography media are widely used. Crystals of calcium phosphate, called hydroxyapatite, bind both proteins and nucleic acids, which can be eluted selectively by a gradient of phosphate buffer. Beads with hydrophobic groups, such as aromatic rings, absorb many proteins in concentrated salt solutions. They can be eluted selectively by a declining gradient of salt.

The resolution of all chromatography methods depends on the size of the particles (usually beads) that form the immobile phase in the column. Small particles give better resolution, but also resist flow. Therefore, high-resolution systems require high pressures to maintain good flow rates (eg, high-pressure liquid chromatography [HPLC]).

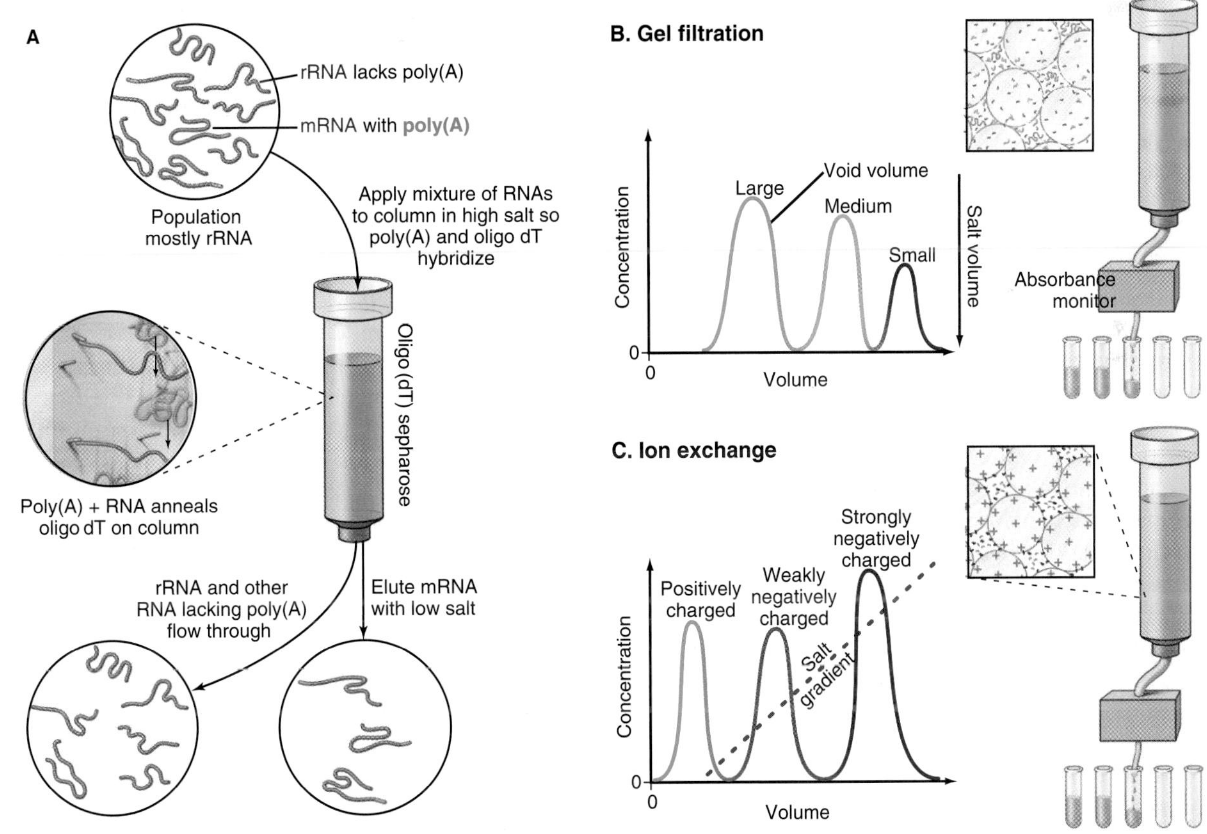

FIGURE 6.9 CHROMATOGRAPHY. A, Affinity chromatography to purify poly(A) mRNAs with poly(dT) attached to beads. A mixture of RNAs is extracted from cells and applied to the column in a buffer containing a high concentration of salt. Only poly(A)$^+$ mRNA binds and is then eluted with buffer containing a low concentration of salt. (mRNA, messenger RNA; rRNA, ribosomal RNA.) **B,** Gel filtration chromatography separates molecules on the basis of size. Large molecules *(blue)* are excluded from the beads and travel through the column in the void volume outside the beads. Smaller molecules *(green)* penetrate the beads depending on their size. Tiny molecules *(red),* such as salt, completely penetrate the beads and elute in a volume (the salt volume) equal to the size of the bed of beads. Material eluting from the column is monitored for absorbance of ultraviolet light (260 nm for nucleic acids, 280 nm for proteins) to measure concentration and then collected in tubes in a fraction collector. **C,** Anion exchange chromatography. The beads in the column have a positively charged group that binds negatively charged molecules. A gradient of salt elutes bound molecules depending on their affinity for the beads. For cation exchange chromatography, the beads carry a negative charge.

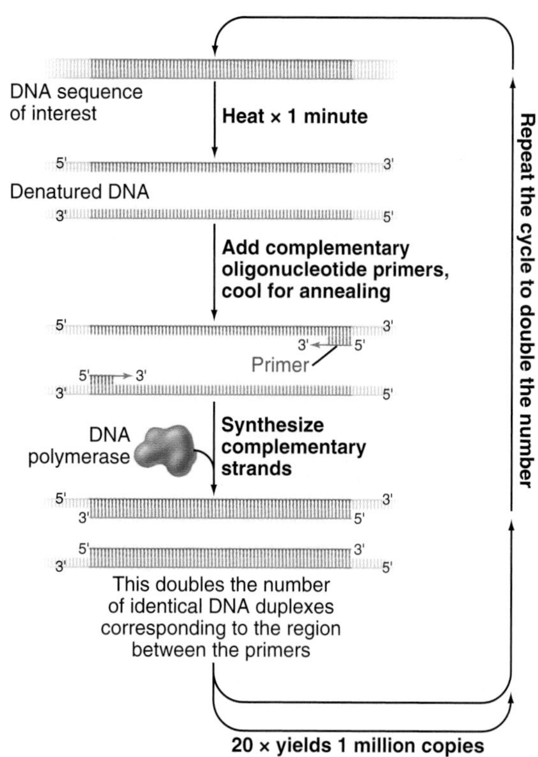

FIGURE 6.10 POLYMERASE CHAIN REACTION. From the top, double-stranded DNA with a sequence of interest is denatured by heating to separate the two strands. An excess of oligonucleotide primers complementary to the ends of the sequence of interest are added and allowed to bind by base pairing. DNA polymerase synthesizes complementary strands, starting from the primers. This cycle is repeated many times to amplify the sequence of interest. Use of a DNA polymerase from a thermophilic bacterium allows many cycles at high temperature without losing activity.

a virus (usually a bacteriophage). The plasmids are introduced into susceptible bacteria, which grow on agar in petri dishes. Conditions are chosen in which only those bacteria carrying a plasmid will grow to form a colony. Some of the cells are picked up from the agar with a nylon membrane, and the DNA they carry is tested for hybridization to a DNA probe complementary to the sequence of interest. This probe may be a chemically synthesized oligonucleotide based on a sequence in a database or may be inferred from the amino acid sequence of the protein of interest. Commonly, the probe is a small piece of cloned DNA generated by PCR or obtained from an EST repository. Colonies that react with the probe are recovered from the petri dish. Initially, these isolates are complex mixtures of cells bearing plasmids. A uniform population (clone) is obtained by successive rounds of dilution, recovery, and replating until all the cells carry the plasmid of interest.

3. DNA synthesis is an increasingly attractive approach to avoid the cloning procedure entirely. Modern technologies permit the synthesis of DNAs greater than 10,000 base pairs long, so if the desired sequence is known, a commercial company can synthesize the DNA fragment in a plasmid backbone. An attractive feature of this approach is that one can make specific or random mutations in the DNA by adjusting the synthesis conditions. This avoids the need to carry out site-directed mutagenesis, as discussed in the following paragraph.

Once a gene or cDNA has been cloned, it is sequenced and used to deduce the sequence of the encoded protein. Of course, analysis of a DNA sequence cannot reveal posttranslational modifications of a protein, such as phosphorylation, glycosylation, or proteolytic processing. Such modifications, which are often critical for function, can be identified only by mass spectrometry analysis of proteins isolated from cells.

Cloned cDNAs are used to express native or modified proteins in bacteria or other cells for biochemical analysis or antibody production. This approach has two advantages. First, the quantity of protein produced is often far greater than that from the natural source. Second, cloned DNA can readily be modified by **site-directed mutagenesis** to make specific amino acid substitutions and other alterations that are useful for studying protein function (Fig. 6.12). The behavior of mutant proteins in cells can provide evidence for the role of a given protein in particular cellular functions. Thus, biochemical, genetic, and molecular cloning approaches may be applied collectively to reveal the function of proteins.

Genome Engineering

Classical random mutagenesis has created many useful mutations that change the activity of genes, but genome editing modifies genes precisely and directly. These methods allow one to remove, insert, or modify DNA sequences at will. All the directed methods depend on creation of single or double strand breaks in DNA. Cells repair double strand breaks by two methods (see Fig. 43.12). **Nonhomologous end joining** brings together any two broken ends, often with the loss of some base pairs. This may be exploited to eliminate the expression or normal function of a gene. Alternatively **homology-directed repair** uses a homologous DNA sequence on the sister chromatid after DNA replication (see Box 43.1) to guide the repair. One can make use of homology-directed repair by providing exogenous DNA with homology to the genome on both sides of a region to be modified. The region between these targeting sequences is used to remove, alter, or add a particular sequence. Traditionally, a plasmid was created containing two substantial regions of the chromosome (usually several thousand base pairs) flanking the region to be engineered. When introduced into cells capable of homologous recombination, the targeting sequences can recombine into the chromosome, thereby replacing the

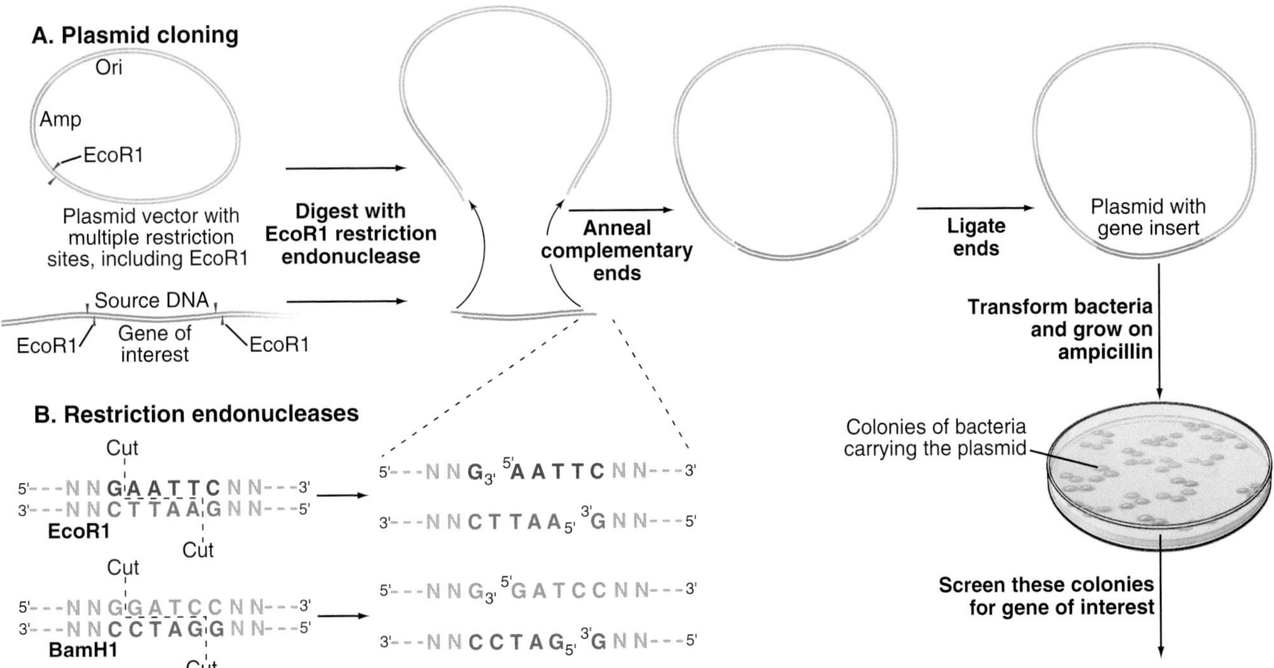

FIGURE 6.11 DNA CLONING. A, Cloning of a segment of DNA into a plasmid vector. The vector is a circular DNA molecule with an origin of replication (Ori) that allows it to replicate in a host bacterium. Most vectors also include one or more genes conferring antibiotic resistance—in this example, resistance to ampicillin (Amp). This enables one to select only those bacteria carrying the plasmid by the ability to grow in the presence of ampicillin. Vectors also contain a sequence of DNA with multiple restriction enzyme digestion sites (see part **B**) for the insertion of foreign DNA molecules. In this example, a single restriction enzyme, EcoR1, is used to cut both the source DNA and the plasmid vector, leaving both with identical single-strand overhangs. The ends of the insert and the cut vector anneal together by base pairing and are then covalently linked together by a ligase enzyme, forming a complete circle of DNA. Plasmids are introduced into bacteria, which are then grown on ampicillin to select those with plasmids. Colonies of bacteria are screened for those containing the desired insert using, for example, DNA probes for sequences specific to the gene of interest. Fig. 6.8B shows gel electrophoresis of a plasmid carrying an insert before and after digestion with a restriction enzyme to liberate the insert from the vector. **B,** Sequence-specific cutting of DNA with restriction enzymes. EcoR1 and BamH1 are two of the hundreds of different restriction enzymes that recognize and cleave specific DNA sequences. Both of these restriction enzymes recognize a palindrome of six symmetrical bases. Note that each enzyme leaves ends with characteristic sequences on both cut ends that are useful for base pairing with DNA having the same cut. Other restriction enzymes recognize and cut from 4 to 10 bases.

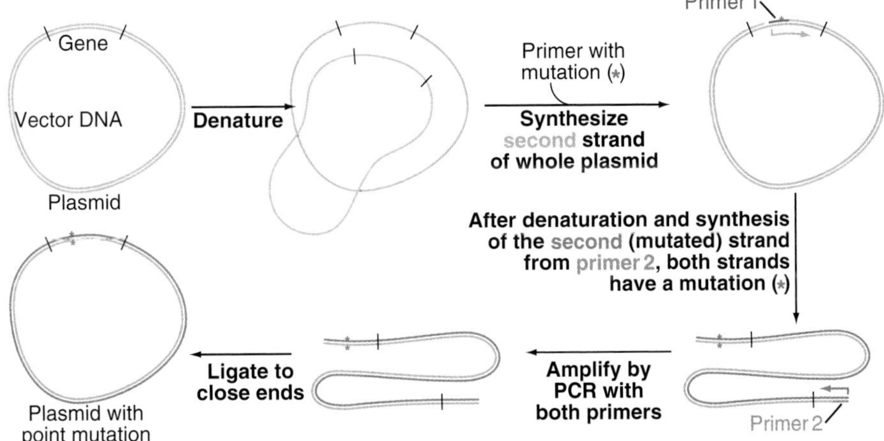

FIGURE 6.12 IN VITRO MUTAGENESIS OF CLONED DNA. This is one of several types of PCR methods used to change one or more nucleotides (the *asterisk* in this example) in a cloned gene using a primer with altered bases. In this particular method, primer 1 has the altered base and is used to duplicate the entire plasmid. Primer 2 is used to synthesize the whole plasmid from the other end. After amplification with both primers, the two ends are ligated together, and the plasmid is produced in quantity by growth in bacteria.

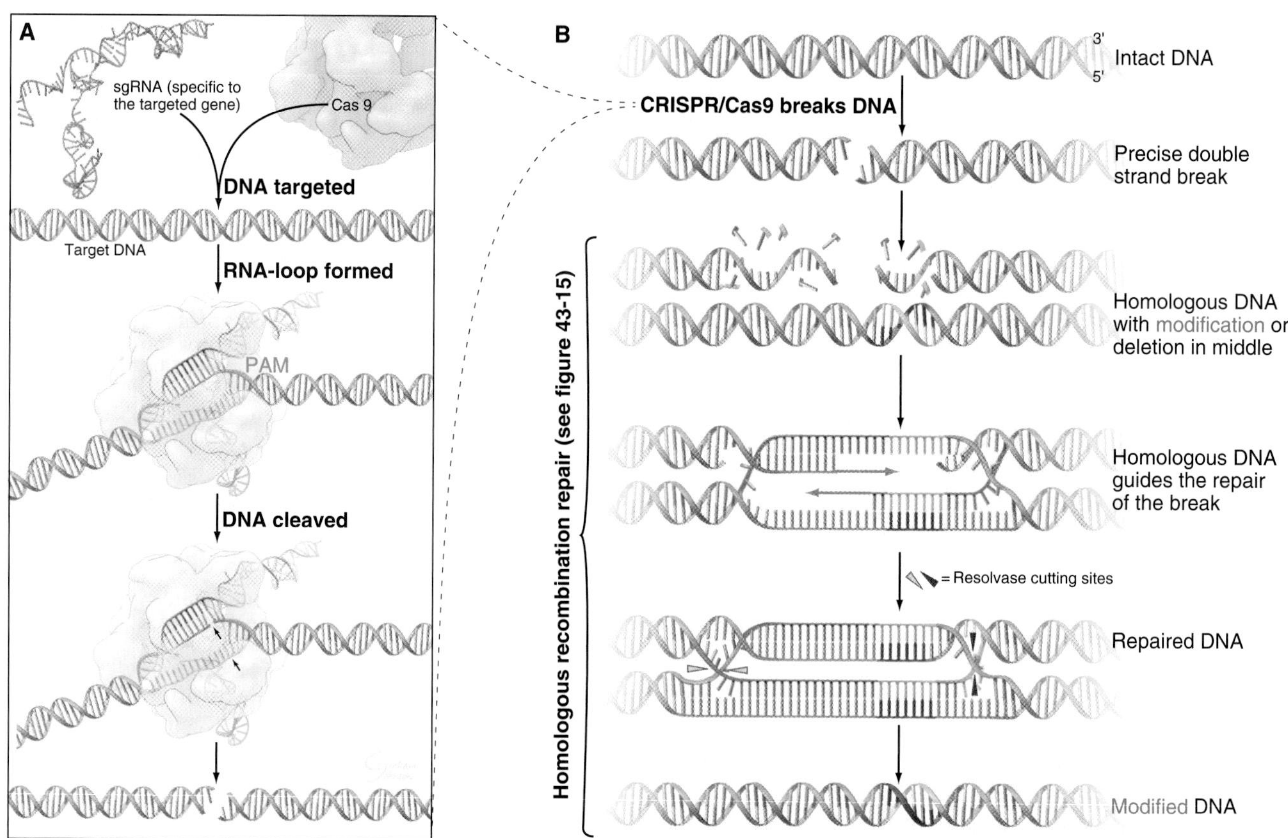

FIGURE 6.13 GENOME MODIFICATION BY HOMOLOGOUS RECOMBINATION AND EDITING. A, Creation of a precise double strand break by the CRISPR/Cas9 system. The guide RNA may consist of two pieces or be joined together by the *blue* loop. The red stem loops target the guide RNA to Cas9. The guide sequence *(green)* binds a complementary DNA sequence and separates the strands, allowing the two nuclease active sites of Cas9 to cleave the two strands of the target DNA. Cellular machinery repairs the double strand break either with a small deletion by nonhomologous end joining or with the insertion of donor DNA as in **B** by homology directed repair. **B,** Homologous recombination to insert a new DNA sequence between a pair of DNA sequences homologous to sequences in the genome. (**A,** For reference, see Doudna JA, Charpentier E. Genome editing. The new frontier of genome engineering with CRISPR-Cas9. *Science.* 2014;346:1077.)

endogenous sequence with the new DNA in between the targeting sequences. The new DNA can change a single base pair, add codons for amino acids, including epitope tags or fluorescent proteins, or even replace the entire coding sequence to create a **null mutation.** When making a null mutation, one places between the targeting sequences a selectable marker, such as a gene encoding resistance to a drug that would normally kill the cells. This selectable marker replaces all or a portion of the coding sequence of the target gene and is used to enrich for cells with the disrupted gene. Resistance to the selectable marker does not ensure that the targeted gene is disrupted, as the exogenous DNA with the selectable marker may integrate elsewhere in the genome. Thus careful analysis must be performed to verify the products of the experiment.

Simply introducing a DNA molecule with targeting and modification sequences into some cells suffices to make precise genome modifications by **homologous recombination** (Fig. 6.13A). The frequency of spontaneous breaks followed by homology-directed repair guided by an exogenous DNA is high enough in yeast, chicken DT40 cells, and mammalian embryonic stem cells to make gene targeting practical. However, until recently the process was inefficient or impossible in most eukaryotes including plants and popular metazoan model organisms (Table 6.3).

New **genome editing** methods allow efficient gene disruption or modification in most organisms (Table 6.3). The essential trick was to find a way to cut a single defined site in huge genomes to create an opportunity for cellular repair machinery to add or remove a few nucleotides and rejoin the cut by nonhomologous end joining or insert modified DNA by homology-directed repair. Identification of a unique site in a whole genome requires recognition of 20 or more bases in the DNA. Initial success was achieved coupling a nuclease to proteins that recognize DNA sequences, either zinc-finger domains from transcription factors (see Fig. 10.14B) or transcription activator-like effector nucleases (TALENs).

A revolutionary new technology for genome editing, the **CRISPR/Cas9 system**, uses a guide RNA to direct the Cas9 nuclease to a specific location in the genome (Fig. 6.13B). The Cas9 protein has two different nuclease domains (see Fig. 11.16), each of which makes a single strand break in the DNA. CRISPR stands for clustered regularly interspaced short palindromic repeats of DNA. Bacteria and Archaea evolved the system for defense against viruses. CRISPR guide RNAs include sequences that bind Cas9 and a sequence complementary to a viral RNA or DNA. In the guide RNAs used for genome editing, the viral sequence is replaced with a sequence of 19 or 20 nucleotides that can base pair with high fidelity with a complementary genomic DNA sequence. This genomic sequence must be followed by a PAM (protospacer adjacent motif) sequence that is specific for each CRISPR system. This PAM sequence can be as simple as "NGG" (where N is any base) as observed in *Streptococcus pyogenes*. The guide RNA binds one strand of DNA, separates the two strands, and presents them to the two Cas9 nuclease sites. Repair by nonhomologous end joining creates local additions or deletions that may inactivate or modify the gene by altering the codon reading frame. Adaptations of the system allow workers to efficiently introduce single insertions, deletions or mutations, or to introduce larger insertions. When homology-directed repair fixes the double strand break, DNA may be removed or exogenous DNA provided by the experimentalist can be inserted to modify the genome sequence or add a desired sequence such as that for GFP. This method works in most cells tested, including all the popular experimental organisms (including adult mice!), greatly expanding the reach of genome engineering for basic research and showing promise for clinical applications. However, the efficiency varies, because homology-directed repair activity is low in many cells and tissues. Alternative strategies use pairs of guide RNAs to make to cuts in a gene or Cas9 with only one active nuclease domain results in a single strand break that the cell repairs by the base excision repair pathway (see Fig. 43.12). These methods are rapidly evolving, and creating revolutionary opportunities to study gene function.

Molecular Structure

Primary Structure

DNA sequences are now determined by automated methods (see Fig. 3.16) and used to deduce the sequence of proteins and structural RNAs. Mass spectrometry is now the method of choice to detect modified amino acids (see Fig. 3.3), which are not revealed by the DNA.

Subunit Composition

Gel electrophoresis of many isolated proteins has revealed that they consist of more than one polypeptide chain (eg, Fig. 6.8C). Their stoichiometry can be determined from their masses and intensities of the stained bands on the gel, but the only way to determine the total number of subunits is to measure the molecular weight of the native protein or protein assembly. A sedimentation velocity experiment in an analytical ultracentrifuge provides both parameters required to measure the molecular weight of a purified macromolecule: the **sedimentation coefficient** (from the velocity of the moving particles); and the diffusion coefficient (from the spreading of the boundary of particles). The diffusion coefficient can also be calculated from the Stokes radius measured by gel filtration (Fig. 6.9B). The combination of gel filtration with **multiangle laser light scattering** also gives accurate molecular weights, but neither the diffusion coefficient nor sedimentation coefficient alone provides enough information to measure a molecular weight. An analytical ultracentrifuge can also measure the molecular weight with a sedimentation equilibrium experiment. A sample of purified material is centrifuged in a physiological salt solution at relatively low speed in a rotor that allows the measurement of the mass concentration from the top to bottom of the sample cell. At equilibrium, the sedimentation of the material toward the bottom of the tube is balanced by diffusion from the region of high concentration at the bottom of the tube. This balance between sedimentation and diffusion uniquely defines the molecular weight of the particle.

Atomic Structure

Three complementary methods, **x-ray crystallography** (see Fig. 3.6), electron microscopy (Fig. 6.7D), and **nuclear magnetic resonance (NMR)** spectroscopy, are used to determine the structure of proteins and nucleic acids at atomic resolution. X-ray crystallography has the highest resolution but not all proteins can be crystallized. Both NMR and electron microscopy avoid the requirement to crystallize the sample. Electron microscopy is particularly useful for large structures such as the phage T4 tail baseplate, where 56,082 amino acid residues were mapped. NMR provides more information about the dynamics of the molecule in solution, but the protein must be soluble at high concentrations, and NMR is difficult for proteins larger than 30 kD.

Partners and Pathways

Most cellular components are parts of assemblies, networks, or pathways, so a major challenge in defining biological function is to place each molecule in its physiological context with all of its molecular partners. The classic example of such an endeavor is the biochemical mapping of major metabolic pathways (see Fig. 19.4 or a biochemistry textbook). Genetics played a prominent role in the discovery of the network of proteins that control the cell cycle (see Fig. 40.2). Currently, signaling, regulation of gene expression, membrane trafficking, and the control of development are pathways of particular interest.

Biochemical Methods

Once a molecule of interest has been purified, finding its cellular partners is often the next step. Antibodies are frequently used to separate a protein and its partners from crude extracts. An antibody specific for the molecule of interest can be attached directly or indirectly to a bead and used to bind the protein of interest along with any associated molecules. This is called **immunoprecipitation.** After a gentle wash, bound proteins are analyzed by gel electrophoresis and identified with antibodies or mass spectrometry. Bound nucleic acids are cloned and sequenced.

For soluble proteins, the molecule of interest can be attached to an insoluble support, such as small beads, and used for **affinity chromatography** (Fig. 6.9A). If purified protein is available, it can be attached to beads by chemical crosslinking. Alternatively, the protein of interest can be expressed as a fusion protein with a protein (eg, GST) or peptide epitope tag that binds to another molecule attached to the beads. The beads with the attached protein are mixed with a crude cellular extract to allow other proteins to bind. Then unbound molecules are washed away in chromatography column or by pelleting in a centrifuge. Varying the concentration of such beads is a simple way to measure the affinity of the probe for its various partners.

A variation of this method called **TAP (tandem affinity purification) tagging** is used to purify stable protein complexes from crude whole-cell lysates. A recombinant protein is tagged with two different peptide epitopes separated by a cleavage site for a highly specific viral protease. Beads with antibodies to the outermost tag are used to capture the doubly tagged protein along with associated proteins from a crude cellular extract. The TEV (tobacco etch virus) protease, which has no natural targets in the cell, cleaves the tagged protein from the immobilized antibody. Then the remaining tag is used for a second round of affinity purification to remove most nonspecifically bound proteins.

Genetics

Given a mutation in a gene of interest, two genetic tests are used to search for partners: (a) identification of a second mutation that ameliorates the effects of the primary mutation (a **suppressor mutation,** Fig. 6.14A–B) and (b) identification of a second mutation that makes the phenotype more severe, often lethal (an **enhancer mutation** [Fig. 6.14C–E]). A specialized class of enhancer

A. Bypass suppression

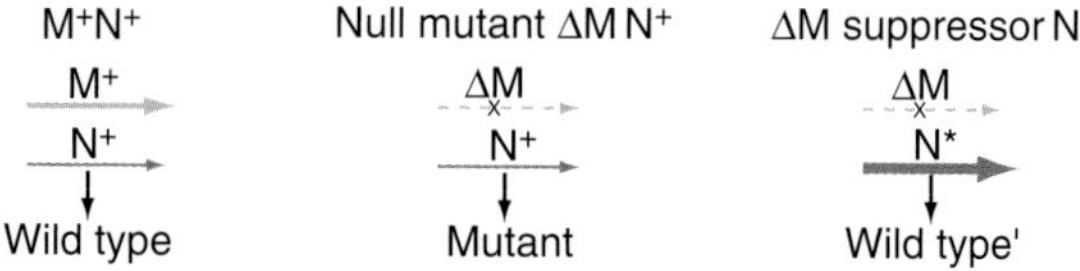

B. Suppression by epistasis

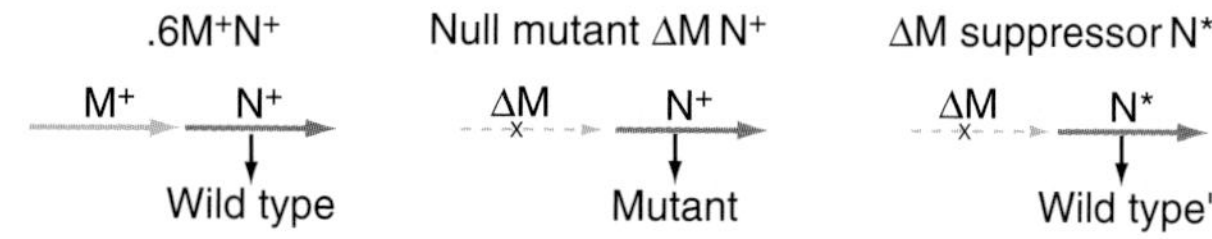

C. Interactional suppression

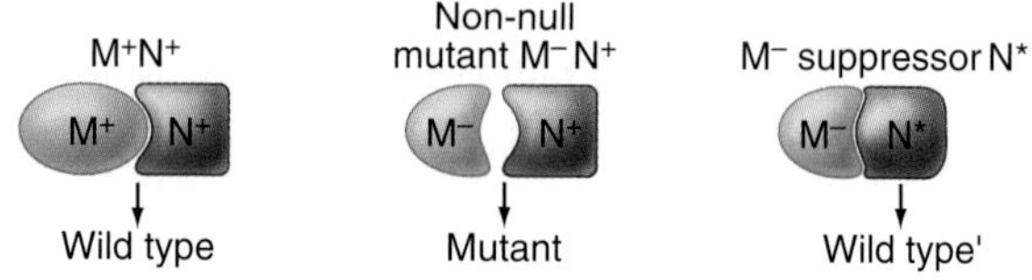

D. Synthetic lethal interaction when mutations in either M or N are viable

Wild type M+N+ — M+, N+ — Viable
Null mutant ΔM or ΔN — ΔM, N+ — Viable
Double null mutant ΔM ΔN — ΔM, ΔN — Lethal

E. Synthetic lethal interaction when null mutations in either M or N are lethal

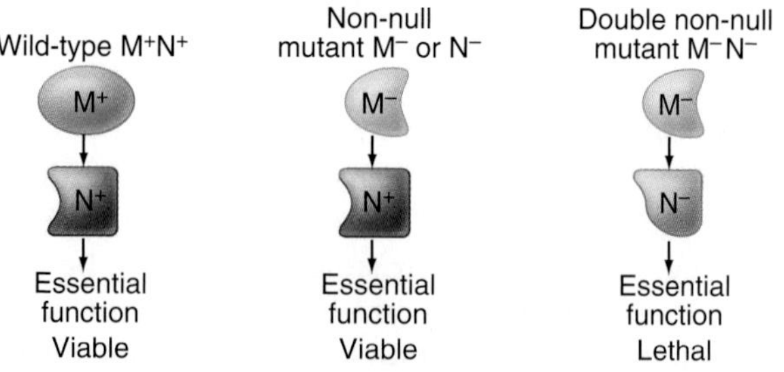

FIGURE 6.14 ANALYSIS OF GENETIC INTERACTIONS BETWEEN TWO GENES, M AND N. The sizes of the *arrows* indicate the level of function of the gene product, usually a protein. The phenotype is indicated for each example. Mutant phenotype means an altered function dependent on gene products M and N. In the diagram, the *plus sign* indicates a wild-type allele, the *asterisk* indicates a suppressor allele, and Δ indicates a null mutation. **A,** Bypass suppression. Gene products M and N operate in parallel, with M making the larger contribution. Loss of M yields a mutant phenotype because N alone does not provide sufficient function. Mutation N* enhances the function of N, allowing it to provide function on its own. **B,** Suppression by epistasis. Products M and N act in series on the same pathway. Loss of M function blocks the pathway. Mutation N* allows N to function without stimulation by product M. **C,** Interactional suppression. Function requires interaction of gene products M and N. Mutation M⁻ interferes with the interaction. Suppressor mutation N* allows product N* to interact with M⁻. **D,** Synthetic lethal interaction when null mutations in either M or N are viable. The products of genes M and N operate in parallel to provide function. N provides sufficient function in the absence of M (ΔM) and vice versa. Loss of both M and N is lethal. **E,** Synthetic lethal interaction when null mutations in either M or N are lethal. Products M and N function in series. N can provide residual function even when M is compromised by mutation M⁻, and vice versa. When both M and N are compromised (M⁻, N⁻), the pathway provides insufficient function for viability. (Modified from Guarente L. Synthetic enhancement in gene interaction: a genetic tool comes of age. *Trends Genet.* 1993;9:362–366.)

mutations, called **synthetic lethal mutations,** is particularly useful in the analysis of genetic pathways in yeast. In this case, mutations in two genes in the same pathway, if present in the same cell, even as heterozygotes (ie, each cell having one good and one mutant copy of each gene), cannot be tolerated, so the cell dies. It is thought that each mutation lowers the level of production of some critical factor just a bit and that the combination of the two effectively means that the output of the pathway is insufficient for survival. These tests can be made with existing collections of mutations by genetically crossing mutant organisms. Alternatively, one can seek new mutations created by a second round of mutagenesis. The results depend on the architecture of the particular pathway. If the products of the genes in question operate in a sequence, analysis of single and double mutants can often reveal their order in the pathway. For essential genes in haploid organisms, a conditional allele of the primary mutation simplifies the experiment. Synthetic interactions (suppression or lethality) may also be discovered by overproduction of wild-type genes on a plasmid. Caution is required in interpreting suppressor and enhancer mutations, given the complexity of cellular systems and the possibility of unanticipated consequences of the mutations.

Another approach to find protein partners is called a **two-hybrid assay** (Fig. 6.15). This assay depends on the observation that some activators of transcription have two modular domains with discrete functions: One domain binds target sites on DNA, and the other recruits the transcriptional apparatus (see Fig. 10.15). The target gene is expressed if both activities are present at the transcription start site, even if the activities are on two different proteins. For the two-hybrid assay, the coding sequence of the protein whose partners are to be identified is fused to the coding sequence of a yeast protein that recognizes a target DNA sequence upstream of a gene that provides the readout of the assay. This so-called bait protein is expressed constitutively in yeast cells. A plasmid library is constructed consisting of cDNA sequences of all possible interaction partners ("prey"), each fused to the coding sequence of an "activator domain" and a nuclear localization sequence. This library of "prey" proteins is introduced into a population of the "bait" yeast strain. The readout gene is expressed in a cell if a "prey" protein binds the "bait" protein and recruits the transcriptional apparatus. Many variations of this assay exist. One produces an enzyme that makes a colored product, so colonies of yeast with interacting proteins can be identified visually. In another version, the target gene encodes a gene essential for production of a particular amino acid, so only cells with a bait–prey interaction will grow on agar plates lacking that amino acid. Putative interactions must subsequently be tested carefully to define specificity, as false-positive results are common. Moreover, some valid interactions are missed owing to false-negative results.

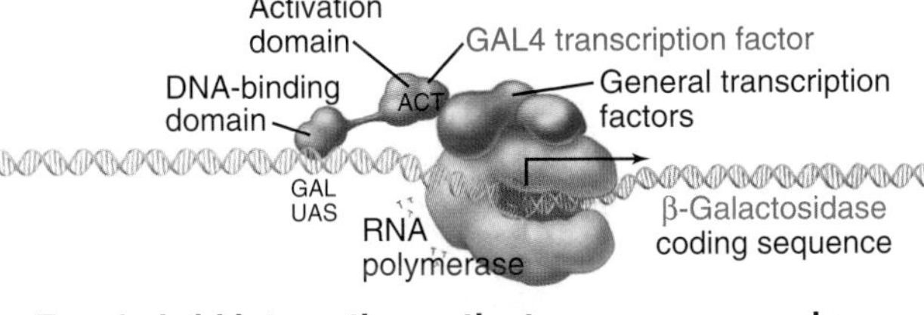

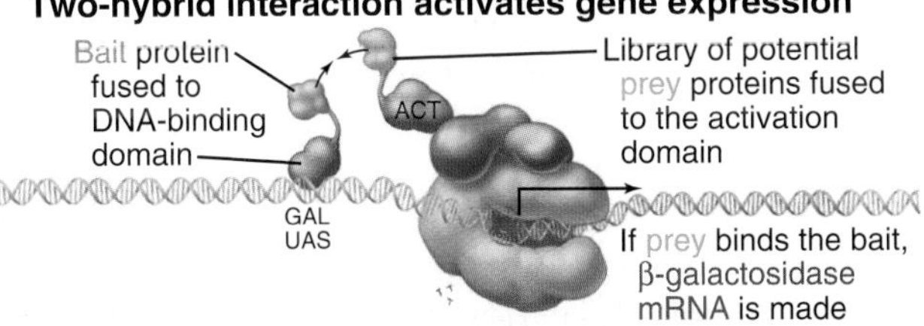

FIGURE 6.15 ONE VERSION OF THE YEAST TWO-HYBRID ASSAY FOR INTERACTING PROTEINS. Interaction between "bait" protein and "prey" protein *(bottom)* brings together the two halves of a transcription factor required to turn on the expression of β-galactosidase. The DNA-binding domain of the GAL4 transcription factor binds a specific DNA sequence: GAL UAS. Generally, a library of random cDNAs or gene fragments is used to express test prey proteins as fusions with the activation domain.

Large-Scale Screening With Microarrays

Microarrays display thousands of tiny spots on a glass slide, each with a particular DNA sequence or protein (Fig. 6.16). This allows many reactions to be monitored in parallel. One type of microarray has cDNAs or oligonucleotides for thousands of genes. Probing such an array with complementary copies of mRNAs from a test sample reveals which genes are expressed. This assay can be used to find partners, because expression of genes contributing proteins to a particular pathway is often coordinated as conditions change. For example, unfolded proteins in the lumen of the endoplasmic reticulum trigger the expression of nearly 300 genes for proteins of the endoplasmic reticulum (see Fig. 20.13). Microarrays of thousands of different proteins can be used to test for interactions. For example, reaction of protein arrays with each yeast protein kinase, one kinase per slide, identified the substrates phosphorylated by each kinase (Fig. 6.16B).

Rates and Affinities

Information about reaction rates is important for two reasons. First, reaction rates are required to account for the dynamic aspects of any biological system. Second, although the methods in the previous section usually provide initial clues about the integration of proteins into pathways, knowledge of reactant concentrations and **rate constants** is the only way to fully understand biochemical pathways. Fortunately, just two types of

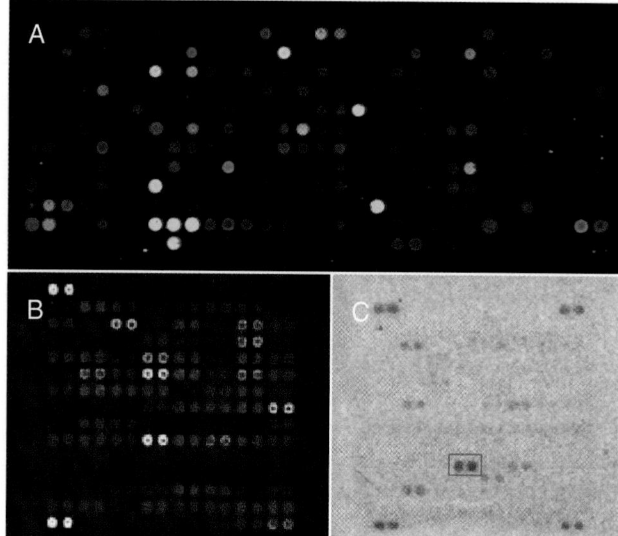

FIGURE 6.16 LARGE-SCALE ANALYSIS OF GENE EXPRESSION AND KINASE ACTIVITY WITH MICROARRAYS. A, Gene expression. PCR was used to make cDNA copies of mRNAs from two parts of the human brain. The cDNAs from cerebral cortex mRNAs were labeled with a *red* fluorescent dye, whereas those from the cerebellum were labeled with a *green* fluorescent dye. A mixture of equal proportions of the two fluorescent cDNA preparations was reacted with 384 different known cDNAs arrayed in tiny spots on a glass slide. The fluorescence-bound cDNAs were imaged with a microscopic fluorescent scanner similar to a confocal microscope. *Yellow* spots bound equal quantities of cDNAs from the two sources. *Red* spots bound more cDNA from the cortex, indicating a higher concentration of those mRNAs. *Green* spots bound more cDNA from the cerebellum, indicating a higher concentration of those mRNAs. **B–C,** Large-scale identification of substrates for a protein kinase. Thousands of different budding yeast proteins tagged with GST– and 6 histidines were overexpressed in yeast and purified by affinity chromatography. Each protein was spotted in duplicate on a glass slide, a small portion of which is shown here. **B,** The amount of bound protein in each spot was detected with a fluorescent antibody to GST (indicated by varying intensity of fluorescence from *dark red* to *white*). **C,** The slide was incubated with a yeast kinase in the presence of ^{33}P-adenosine triphosphate (ATP). Radioactive phosphorylated proteins were detected as pairs of dark spots by autoradiography. One pair is boxed. (**A,** Courtesy C. Barlow and M. Zapala, Salk Institute, La Jolla, CA. **B–C,** Courtesy Geeta Devgan and Michael Snyder, Yale University, New Haven, CT. For reference, see Zhu H, Bilgin M, Bangham R, et al. Global analysis of protein activities using proteome chips. *Science.* 2001;293:2101–2105.)

reactions occur in biology: first-order reactions, such as conformational changes and dissociation of molecular complexes, and second-order reactions between two molecules. Chapter 4 explains the rate constants for such reactions, the relationship of rate constants to the equilibrium constant for a reaction, and the relationship of the equilibrium constant to thermodynamics. Fig. 4.7 illustrates how transient kinetics experiments were used to determine the mechanism of the Ras guanosine triphosphatase (GTPase) (see Fig. 4.6).

Despite their importance, rate constants and the physiological concentrations of the molecules in a pathway are usually the least understood aspects of most biological systems. A common impediment is the lack of an assay with sufficient sensitivity and time resolution to measure reaction rates. Optical methods, such as those using fluorescence, are usually the best and can be devised for most processes.

Reconstitution of Function From Isolated Components

The classic biochemical test of function is **reconstitution** of a biological process from purified components. This involves creating conditions in the test tube in which isolated molecules can perform a complex process normally carried out by a cell. The difficulty of the task depends on the complexity of the function. Successful reconstitution experiments reveal the molecular requirements and mechanisms involved in a process. Examples of successful tests include reconstitution of ion channel function in pure lipid membranes (see Chapter 16), protein synthesis and translocation of proteins into the endoplasmic reticulum (see Fig. 20.9), and motility of bacteria powered by assembly of actin filaments (see Fig. 37.12).

Anatomic Tests of Physiological Function

No biological process can be understood without knowledge of where the components are located in the cell. Often, cellular **localization** of a newly discovered molecule provides the first clue about its function. This accounts for why cell biologists put so much effort into localizing molecules in cells. Cell fractionation, fluorescent antibody staining, and expression of GFP fusion proteins are all valuable approaches, illustrated by numerous examples in this book. For more detailed localization, antibodies can be adsorbed to small gold beads and used to label fixed specimens for electron microscopy (see Fig. 29.7).

GFP fusion proteins are particularly valuable, because of the ease of their construction and expression and because they can be used to monitor both the behavior and dynamics of molecules within living cells. For example, the time course of fluorescence recovery of a photobleached area (Fig. 6.3F) provides information on the mobility of the fusion protein (ie, whether it diffuses freely, is immobilized on a scaffold, or is actively transported) and its interaction properties within the cell (see Figs. 13.12 and 9.4). However, attaching GFP might affect either the localization or function of the protein, so it is important to demonstrate that a GFP fusion protein is fully functional by genetic replacement of the native protein. This is done routinely in yeast but was done rarely for animal proteins until genome editing methods became available.

Proteins and other cellular components, including DNA, RNA, and lipids, can be labeled with fluorescent dyes to study their intracellular localization and dynamics. Fluorescent RNAs and proteins can be microinjected into cells. Fluorescent lipids can be inserted into the outer leaflet of the plasma membrane in living cells; from there, they move to appropriate membranes and then mimic the behavior of their natural lipid counterpart.

Three Options to Test for Physiological Function

Although often obscured by technical jargon, just three methods are available to test for physiological function: (a) reducing the concentration of active protein (or other molecule), (b) increasing the concentration of active molecule, and (c) replacing a native molecule with a molecule with altered biochemical properties. Biochemical, pharmacological, and genetic methods are available for each test, the genetic methods often yielding the cleanest results. Interpreting these experiments is most reliable when robust assays are available to measure quantitatively how the cellular process under investigation functions when the concentration of a native molecule is varied or an altered molecule replaces the native molecule. When done well, these experiments provide valuable constraints for quantitative models of biological systems, as described below.

Reducing the Concentration of a Macromolecule

Disrupting a gene by classical, random genetics or by genome engineering usually provides definitive information about the functions of the gene and its products. This option is available if the molecule is not required for viability. If a protein is essential, there are four options for eliminating its activity. The mRNA can be depleted, the protein can be depleted, the protein can be inhibited, or the protein can be replaced with an altered version that is fully active under a certain set of conditions and completely inactive under other conditions (a **conditional mutant**).

- *Depletion of RNAs:* **RNAi** is widely used to deplete mRNAs from many cells and organisms, including nematodes and cultured cells of flies and humans (see Fig. 11.12 for details). Animals, fungi, and plants naturally use this process to suppress expression of foreign RNAs, such as those introduced by viruses. To suppress a particular RNA in human cells experimentally, one synthesizes a double-stranded RNA, including a sequence of at least 21 nucleotides matching the target cellular RNA such as an mRNA. These may be on separate strands or a single hairpin molecule. When introduced into cells, an enzyme cuts the double-stranded RNA into pieces of approximately 21 nucleotides and one strand is loaded onto a protein complex (see Fig. 11.12). If the bound RNA base pairs with the complementary sequence of a cellular RNA (usually an exact match is required), the protein cleaves the target RNA, initiating its degradation. If depletion of the mRNA is successful, the level of the targeted protein falls 5- to 10-fold as it is degraded naturally over several days. Loss of the protein may produce a cellular phenotype. The simplicity of this approach made RNAi very popular and suitable for scaling up to study thousands of genes. However, false-negative results are common, because some targeted protein usually remains. Furthermore, off-target effects, where an unexpected second mRNA is also targeted, are also common and difficult to detect. One must thus be very cautious in interpreting the results with this method. Nevertheless, many are attempting to use RNAi therapeutically.

  A second option is to put the expression of the protein or RNA under the control of regulatory proteins that are sensitive to the presence of a small molecule, such as a vitamin or hormone. Then, expression of the molecule can be turned on and off at will. This is commonly done for vertebrate cells by using promoters of gene expression engineered so that they can be turned on or off by the antibiotic tetracycline, which alters the ability of a bacterial protein (the tetracycline repressor) to bind particular regulatory sequences added to the DNA. A limitation of this technology is that some proteins are so stable that days are required to reduce their concentrations. During this time, cells may be able to compensate for the loss of the protein of interest.
- *Selective protein degradation:* Proteins can be removed from cells by targeted degradation when linked to a **degron**–an amino acid sequence that under certain conditions triggers rapid proteolysis of the fusion protein by the ubiquitin-proteasome system (see Fig. 23.3). For example, plant cells respond to the hormone auxin by destroying certain transcriptional repressors carrying a degron sequence. Because animal cells do not use auxin, one can add the degron sequence to target proteins (preferably by editing the degron into the chromosomal copy of the gene) and use auxin to trigger their specific destruction, often in less than 1 hour.
- *Inhibitors:* A time-tested strategy is to inhibit a particular protein with a drug, inhibitory peptide, antibody, or inactive partner protein. Drugs as probes for function have a long and distinguished history in biology, but their use is hampered by the difficulty of ruling out side effects, including action on other unknown targets. One wag even asserted, "drugs are only specific for about a year," roughly the time it takes someone to find an unexpected second target. Nevertheless, many drugs are useful because their

onset of their action is rapid and their effects are reversible, so one can follow the process of recovery when they are removed. The use of libraries of small molecules to probe biological processes has been called *chemical genetics.*

If microinjected into cells, antibodies can be very specific, but the effects on their target must be fully characterized, and sufficient antibody must be introduced into the target cell to inactivate the target molecule. Some arginine-rich peptides, such as one from the HIV Tat protein, can be used to carry inhibitory peptides across the plasma membrane into the cytoplasm. Other peptides can guide experimental peptides into various cellular compartments. It is also possible to inactivate pathways by expressing **dominant negative mutants** that can do part, but not all, of the job of a given protein. Dominant negative mutants of protein kinases are particularly effective. The active site is modified to eliminate enzymatic activity, but the modified protein can still bind to its regulatory proteins and substrates. This can interfere with signal transduction pathways very effectively by competing with functional endogenous kinases for regulatory factors and substrates. Dominant negative mutants offer the advantage that they can be expressed in many types of cells. All too often, however, little is known about the concentrations of these dominant negative agents or the full range of their targets.

- *Conditional mutations:* Some mutant proteins are active under limited conditions. For example, one class of **conditional mutations** allows a protein to be active at one temperature and inactive at another, typically a high temperature. Such temperature-sensitive mutations have been invaluable to study essential genes in prokaryotes and fungi, but are used much less often in plants and metazoan organisms. Although one must control for the effects of temperature on other cellular processes, the rapid onset and reversibility of the effects of conditional mutations reduces the chance that the cell adapts with genetic changes.

Increasing the Concentration of a Macromolecule

The concentration of active protein can be increased by **overexpression,** for example, driving the expression of a cDNA from a very active viral promoter. Some expression systems are conditional, being turned on, for example, by an insect hormone that does not activate endogenous genes. Interpreting the consequences of overexpression tends to be less straightforward than other approaches, as specificity of interactions with other cellular components can be lost at high concentrations.

Altering the Activity of a Macromolecule

Genetics or genome editing are used to replace a native protein or RNA with a version with altered biochemical properties (Table 6.2). Examples of altered proteins include an enzyme with altered catalytic function or a protein with altered affinity for a particular cellular partner. Ideally, the altered protein is fully characterized before its coding sequence is used to replace that of the wild-type protein. Amino acid residues to be mutated are often determined based on atomic structures. When these experiments are conducted in vivo, it is important that the cellular concentration of the altered protein is confirmed to be the same as the wild-type protein. On the relatively long time scale of such experiments (months in vertebrates), interpreting the outcome may be compromised by the ability of cells to adapt genetically in unknown ways to the change imposed by the gene substitution.

Mathematical Models of Systems

An inventory of molecular components; their structures, concentrations, molecular partners, and reaction rates; and genetic tests for their contributions to a physiological process will suggest hypotheses for how the system works. However, one does not know if the proposed mechanism works unless simulations based on a mathematical model can match the performance of the cellular system over a range of conditions, including mutation and inhibition of one or more components. Even in the best cases (bacterial metabolic pathways, bacterial chemotaxis, yeast cell cycle, muscle calcium transients, and muscle crossbridges), initial mathematical models fell short of duplicating the physiological process. This meant that some aspect of the process was incompletely understood or that assumptions in the mathematical model were incorrect. Regardless, such failures offer important clues about the shortcomings of current knowledge and point the way toward improvements in underlying assumptions, experimental parameters, or mathematical models. By cycling from theory to simulation to experiment and back to improved theory, investigators converge on the underlying truth.

SELECTED READINGS

Altieri AS, Byrd TA. Automation of NMR structure determination of proteins. *Curr Opin Struct Biol.* 2004;14:547-553.

Bader GD, Heilbut A, Andrews B, et al. Functional genomics and proteomics: Charting a multidimensional map of the yeast cell. *Trends Cell Biol.* 2003;13:344-356.

Brent R, Finley RLJ. Understanding gene and allele function with two-hybrid methods. *Annu Rev Genet.* 1997;31:663-704.

Bruckner A, Polge C, Lentze N, Auerbach D, Schlattner U. Yeast two-hybrid, a powerful tool for systems biology. *Int J Mol Sci.* 2009;10: 2763-2788.

Chen BC, Legant WR, Wang K, et al. Lattice light-sheet microscopy: imaging molecules to embryos at high spatiotemporal resolution. *Science.* 2014;346:349.

Cheng Y. Single-particle cryo-EM at crystallographic resolution. *Cell.* 2015;161:450-457.

Costanzo M, Baryshnikova A, Myers CL, Andrews B, Boone C. Charting the genetic interaction map of a cell. *Curr Opin Biotechnol*. 2011; 22:66-74.

Cox S. Super-resolution imaging in live cells. *Dev Biol*. 2015;401: 175-181.

Danuser G, Waterman-Storer CM. Quantitative fluorescent speckle microscopy of cytoskeleton dynamics. *Annu Rev Biophys Biomol Struct*. 2006;35:361-387.

Doudna JA, Charpentier E. Genome editing. The new frontier of genome engineering with CRISPR-Cas9. *Science*. 2014;346:1077.

Frey TG, Perkins GA, Ellisman MH. Electron tomography of membrane-bound cellular organelles. *Annu Rev Biophys Biomol Struct*. 2006; 35:199-224.

Giepmans BN, Adams SR, Ellisman MH, Tsien RY. The fluorescent toolbox for assessing protein location and function. *Science*. 2006; 312:217-224.

Godin AG, Lounis B, Cognet L. Super-resolution microscopy approaches for live cell imaging. *Biophys J*. 2014;107:1777-1784.

Green MR, Sambrook J. *Molecular Cloning*. 4th ed. Plainview, NY: Cold Spring Harbor Laboratory; 2001.

Guarente L. Synthetic enhancement in gene interaction: A genetic tool come of age. *Trends Genet*. 1993;9:362-366.

Hsu PD, Lander ES, Zhang F. Development and applications of CRISPR-Cas9 for genome engineering. *Cell*. 2014;157:1262-1278.

Huang B, Bates M, Zhuang X. Super-resolution fluorescence microscopy. *Annu Rev Biochem*. 2009;78:993-1016.

Janes KA, Lauffenburger DA. Models of signaling networks-what cell biologists can gain from them and give to them. *J Cell Sci*. 2013; 126:1913-1921.

Larance M, Lamond AI. Multidimensional proteomics for cell biology. *Nat Rev Mol Cell Biol*. 2015;16:269-280.

Lewis NE, Nagarajan H, Palsson BO. Constraining the metabolic genotype-phenotype relationship using a phylogeny of in silico methods. *Nat Rev Microbiol*. 2012;10:291-305.

Li D, Shao L, Chen BC, et al. ADVANCED IMAGING. Extended-resolution structured illumination imaging of endocytic and cytoskeletal dynamics. *Science*. 2015;349:aab3500.

Lučič V, Rigort A, Baumeister W. Cryo-electron tomography: the challenge of doing structural biology in situ. *J Cell Biol*. 2013;202: 407-419.

McIntosh JR, Nicastro D, Mastronarde D. New views of cells in 3D: An introduction to electron tomography. *Trends Cell Biol*. 2005;15: 43-51.

Mogilner A, Wollman R, Marshall WF. Quantitative modeling in cell biology: What good is it? *Dev Cell*. 2006;11:1-9.

Mohr SE, Smith JA, Shamu CE, Neumüller RA, Perrimon N. RNAi screening comes of age: improved techniques and complementary approaches. *Nat Rev Mol Cell Biol*. 2014;15:591-600.

Murphy DB. *Fundamentals of Light Microscopy and Electronic Imaging*. New York: Wiley-Liss; 2001.

Nishimura K, Fukagawa T, Takisawa H, Kakimoto T, Kanemaki M. An auxin-based degron system for the rapid depletion of proteins in nonplant cells. *Nat Methods*. 2009;6:917-922.

Pollard TD. No question about exciting questions in cell biology. *PLoS Biol*. 2013;e1001734.

Pollard TD, De La Cruz E. Take advantage of time in your experiments: a guide to simple, informative kinetics assays. *Mol Biol Cell*. 2013;24:1103-1110.

Pratsch K, Wellhausen R, Seitz H. Advances in the quantification of protein microarrays. *Curr Opin Chem Biol*. 2014;18:16-20.

Ramsey JD, Flynn NH. Cell-penetrating peptides transport therapeutics into cells. *Pharmacol Ther*. 2015;154:78-86.

Sample V, Mehta S, Zhang J. Genetically encoded molecular probes to visualize and perturb signaling dynamics in living biological systems. *J Cell Sci*. 2014;127:1151-1160.

Sioud M. RNA interference: mechanisms, technical challenges, and therapeutic opportunities. *Methods Mol Biol*. 2015;1218:1-15.

Slayter EM. *Optical Methods in Biology*. New York: Wiley-Interscience; 1970.

Slepchenko BM, Schaff JC, Carson JH, Loew LM. Computational cell biology: Spatiotemporal simulation of cellular events. *Annu Rev Biophys Biomol Struct*. 2002;31:423-441.

Westermarck J, Ivaska J, Corthals GL. Identification of protein interactions involved in cellular signaling. *Mol Cell Proteomics*. 2013;12: 1752-1763.

Wijdeven RH, Neefjes J, Ovaa H. How chemistry supports cell biology: the chemical toolbox at your service. *Trends Cell Biol*. 2014;24: 751-760.

Wu RZ, Bailey SN, Sabatini DM. Cell-biological applications of transfected-cell microarrays. *Trends Cell Biol*. 2002;12:485-488.

Internet

Biophysical Society. *Biophysical Techniques*. <http://www.biophysics.org/Education/SelectedTopicsInBiophysics/BiophysicalTechniques/tabid/2313/Default.aspx>

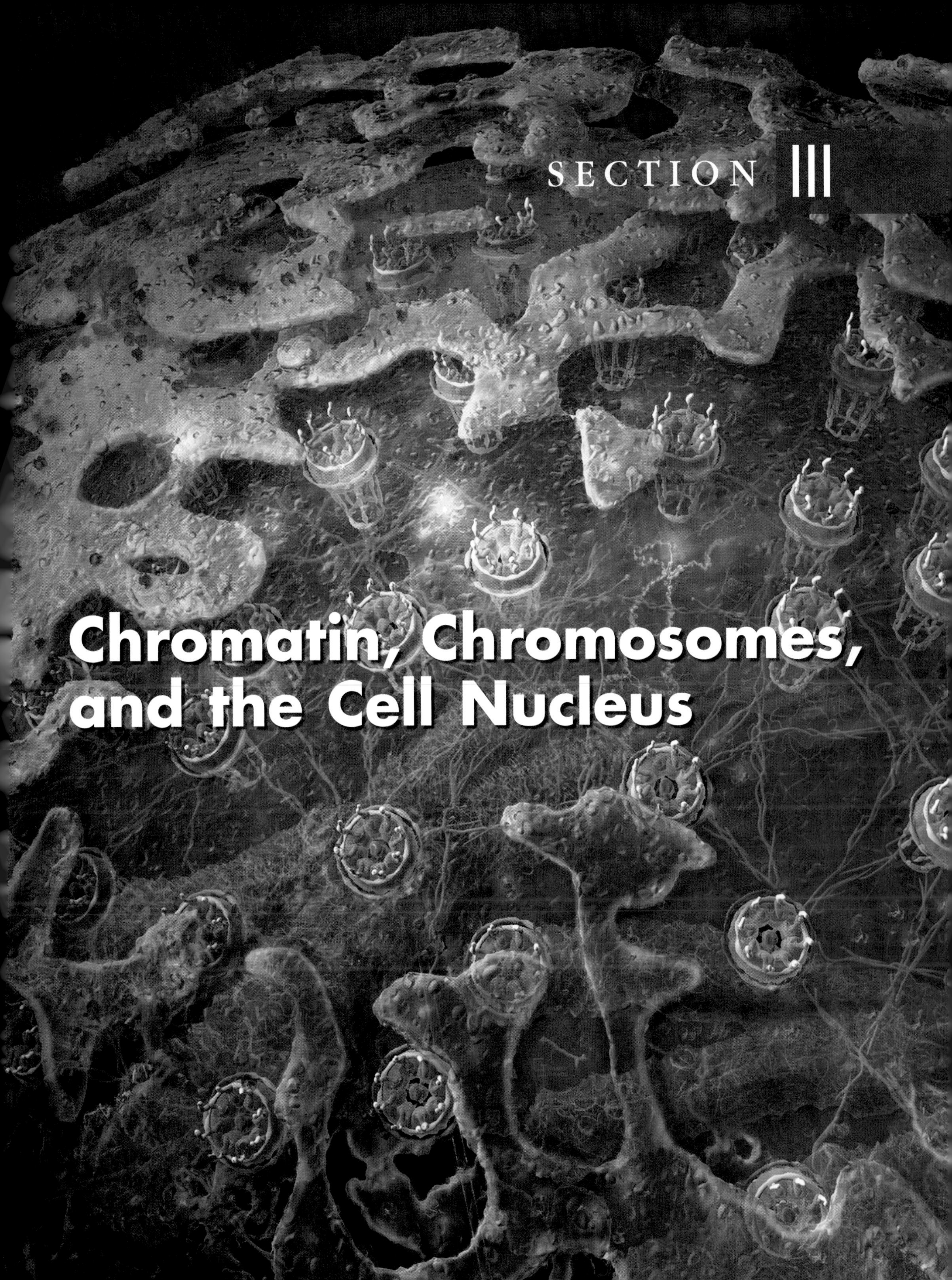

SECTION III

Chromatin, Chromosomes, and the Cell Nucleus

SECTION III OVERVIEW

Every organism is defined by a blueprint consisting of information stored in its **chromosomes.** With the exception of a few viruses, these chromosomes are composed of enormously long circular or linear molecules of DNA. (Those few viruses use RNA instead.) Chromosomes have fascinated biologists ever since it was understood that they contain the genetic information that defines each organism—its **genome.** After Watson and Crick's proposal of a structure for DNA in 1953, it was realized that the DNA is a linear sequence of A, T, G, and C bases that can be thought of as a code to describe the physical attributes for every organism.

This code was originally thought to be impossibly complex and so vast that it could never be completely understood, but recent technological advances have permitted scientists to determine the complete sequences of enormous DNA molecules. Since 1996, the ENSEMBL database (the official website coordinating genome information) has grown to contain the sequences of DNA molecules that make up the genomes of more than 40 plants, 50 animals (from aardvarks to zebrafish), 65 "model organisms" including nematode worms such as *Caenorhabditis elegans* and fruit flies such as *Drosophila melanogaster*, 160 protists, 600 fungi (including many species and strains of budding and fission yeast) and over 30,000 bacteria. The genomes of thousands of humans have been sequenced as well as much of the genome of Neanderthal man. These genome sequences not only reveal much about the biology of living organisms, but also are the most important source of information about the evolution of life on Earth (see Chapter 2).

This does not mean that we understand everything about chromosomes, however. Far from it. We still know very little about how chromosomal DNA molecules are packaged so that they not only fit into cells but also allow access to the library of genetic information that they contain. In prokaryotes, the single chromosome is concentrated in a specialized region of the cytoplasm called the nucleoid. In eukaryotes, the chromosomes are packaged in a specialized membrane-bounded compartment known as the **nucleus.** This difference in organization has important consequences for the regulation of gene expression.

Chapter 7 describes the organization of chromosomal DNA molecules. Every species has a characteristic number of chromosomes that occupy distinct territories within the nucleus and can be visualized as separate entities only during cell division. For example, humans have 46 chromosomes that contain, in total, about 6.2×10^9 base pairs of DNA. Perhaps the most surprising characteristic of this chromosomal DNA is how variable it is from person to person. A "typical" human genome has more than 400,000 differences from the "reference genome" stored in ENSEMBL!

Analysis of the human genome sequence revealed that the genes encoding proteins and RNAs are often surrounded by huge noncoding deserts. In fact, the vast majority of the chromosomal DNA in humans has no coding function, although much of it is transcribed into noncoding RNAs. Some of these noncoding RNAs have regulatory functions (Chapter 11), but the function of other long noncoding RNAs remains enigmatic. Two specific DNA structures are essential for the maintenance of a constant chromosome complement in a given species: **centromeres** and **telomeres.** Centromeres consist of DNA sequences that, together with 90 or more proteins (Chapter 8), direct the segregation of chromosomes during cell division. Telomeres are specialized structures that protect the ends of chromosomes and permit complete replication of the chromosomal DNA.

Given the spacing of 3.4 Å per base pair in B-form DNA, each human cell contains more than 2 m of DNA packaged into a nucleus only 5 to 20×10^{-6} m in diameter! Chapter 8 explains how DNA is extensively folded to fit into the nucleus. The first levels of packaging shorten the DNA about 40-fold by wrapping it around histone proteins to form nucleosomes. Higher levels of packaging of the chromatin fiber are just beginning to be understood using powerful genomics methods such as Hi-C (a method that identifies DNA sequences that are close to one another in the nucleus), which has revealed that the genome is packaged into local domains of 100,000 to 1 million base pairs known as topologically associating domains (TADs).

The complex of DNA with its packaging proteins is called **chromatin.** Nuclei contain two broad classes of chromatin: **heterochromatin,** which is highly condensed throughout the cell cycle and is generally inactive in transcription, and **euchromatin,** which is less condensed and contains actively transcribed genes. Different types of chromatin are defined by complex patterns of posttranslational modifications of the histone proteins. These modifications direct the binding of protein readers that establish chromatin states to promote or repress gene expression or serve other structural roles.

Chapter 9 discusses the structure and physiology of the nucleus. The boundary of the nucleus is a **nuclear envelope** composed of inner and outer nuclear membranes, separated by a perinuclear space that is continuous with the lumen of the endoplasmic reticulum. The inner nuclear membrane is supported by a protein layer called the **nuclear lamina.** Mutations in the lamina and other nuclear envelope proteins cause a wide spectrum of inherited human diseases, with mutations in the lamin genes causing approximately 16 different diseases.

Traffic into and out of the nucleus moves through **nuclear pore complexes** that span the two membrane bilayers of the nuclear envelope. Newly processed RNAs head out to the cytoplasm. So do the ribosomal subunits that will translate them into proteins, some of which then wend their way back into the nucleus. Proteins destined for transport across the nuclear envelope (either alone or associated with RNA molecules) typically contain short stretches of amino acids, called **nuclear localization sequences** or **nuclear export sequences,** that bind to specific adapter and receptor proteins to facilitate transport across the nuclear pore. A small guanosine triphosphatase (GTPase) called **Ran** regulates the directionality of this transport, because it is present primarily in its GTP-bound form in the nucleus and its GDP-bound form in the cytoplasm. Ran-GTP in the nucleus causes imported cargos to dissociate from their transporters and cargos destined for export to bind to their carriers.

The nucleus contains a number of substructures. The most prominent of these is the **nucleolus,** a versatile factory for transcription of ribosomal RNA (rRNA) from a tandem array of genes and processing of rRNA and other noncoding RNAs, as well as ribosome assembly. Nuclei also contain several other specialized regions. These serve a range of functions, including small nuclear ribonucleoprotein (snRNP) and small nucleolar ribonucleoprotein (snoRNP) assembly (in Cajal bodies) and serving as assembly sites for certain transcriptional corepressor complexes (PML and Polycomb group bodies). Other nuclear substructures are sites of DNA damage that are marked for repair (53BP1 nuclear bodies). Studies of these specialized subdomains reveal that compartmentalization of the nucleus contributes to the regulation of nuclear functions.

CHAPTER 7

Chromosome Organization

Chromosomes are enormous DNA molecules that can be propagated stably through countless generations of dividing cells (Fig. 7.1). Genes are the reason for the existence of the chromosomes, but in higher eukaryotes, they make up a relatively small fraction of the chromosomal DNA. Cells package chromosomal DNA with roughly twice its weight of protein. This DNA-protein complex, called chromatin, is discussed in Chapter 8.

FIGURE 7.1 ELECTRON MICROGRAPH OF A CHROMOSOME FROM WHICH MOST PROTEINS WERE EXTRACTED. This allows DNA (*thin lines*) to spread out from the residual scaffold. Enormous amounts of DNA are packaged in each chromosome. This image shows less than 30% of the DNA of this chromosome. (From Paulson JR, Laemmli UK. The structure of histone-depleted chromosomes. *Cell*. 1977;12:817–828.)

In addition to the genes, only three classes of specialized DNA sequences are needed to make a fully functional chromosome: (a) a centromere, (b) two telomeres, and (c) an origin of DNA replication for approximately every 100,000 base pairs (bp). Centromeres regulate the partitioning of chromosomes during mitosis and meiosis. Telomeres protect the ends of the chromosomal DNA molecules and ensure their complete replication. Chapter 42 discusses DNA replication. Chapter 10 considers the structure of genes. Box 7.1 lists a number of key terms presented in this chapter.

Chromosome Morphology and Nomenclature

With few specialized exceptions, chromosomes from somatic cells of higher eukaryotes are visualized directly only during mitosis. Each mitotic chromosome consists of two **sister chromatids** (corresponding to the two copies of the replicated DNA) that are held together at a waist-like constriction called the **centromere.** The portions of the chromosomes that are not in the centromere itself are called chromosome "arms" (Fig. 7.2).

One DNA Molecule Per Chromosome

Most prokaryotic and mitochondrial chromosomes are circular DNA molecules that lack telomeres, but naturally occurring eukaryotic nuclear chromosomes are generally one linear DNA molecule that stretches between the **telomeres** at either end. The clearest proof that each chromosome is composed of a single DNA molecule was obtained for budding and fission yeasts, where intact chromosomal DNA molecules may be visualized by pulsed-field gel electrophoresis as a characteristic series of bands (Fig. 7.3). This technique can display the largest chromosome of fission yeast at 5,579,133 bp, but

BOX 7.1 Key Terms

Centromere: The chromosomal locus that regulates the movements of chromosomes during mitosis and meiosis. The centromere is defined by specific DNA sequences plus proteins that bind to them, although epigenetic factors also play a key role. In higher eukaryotes, the centromere of mitotic chromosomes can be visualized as a constricted region where sister chromatids are held together most closely.

Chromatin: DNA plus the proteins that package it within the cell nucleus.

Chromosome: A DNA molecule with its attendant proteins that moves as an independent unit during mitosis and meiosis. Before DNA replication, each chromosome consists of a single DNA molecule plus proteins and is called a *chromatid.* After replication, each chromosome consists of two identical DNA molecules plus proteins. These are called *sister chromatids.* Chromosomal DNA molecules are usually linear but can be circular in organelles, bacteria, and viruses.

Kinetochore: The centromeric substructure that binds microtubules and directs the movements of chromosomes in mitosis.

Telomere: The specialized structure at either end of the chromosomal DNA molecule that ensures the complete replication of the chromosomal ends and protects the ends within the cell.

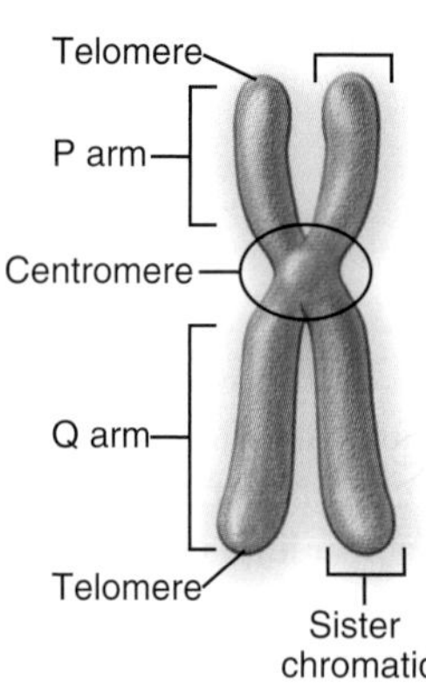

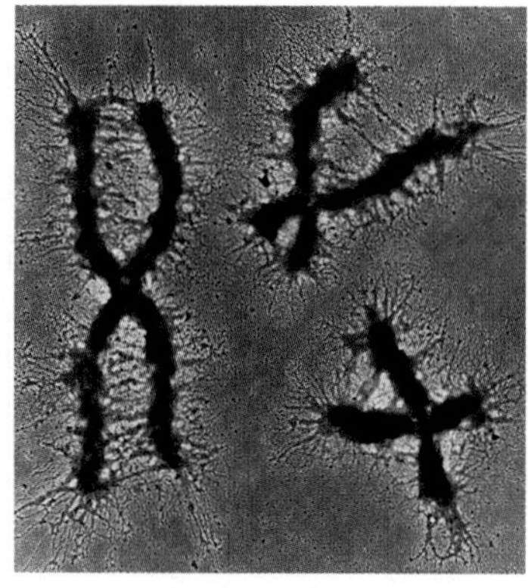

FIGURE 7.2 ANATOMY OF MITOTIC CHROMOSOMES FROM HIGHER EUKARYOTES. *Left,* The principal structural features of chromosomes. *Right,* An electron micrograph of human mitotic chromosomes. *Bottom,* A diagram of the various classes of chromosomes. At mitosis, chromosomes of higher eukaryotes consist of sister chromatids held together at the centromeric region. Chromosomes are classified on the basis of the position of the centromere relative to the arms. In *metacentric* chromosomes, the centromere is located midway along the chromatid. In *submetacentric* chromosomes, the centromere is located asymmetrically so that each chromatid can be divided into short (P) and long (Q) arms. In *acrocentric* chromosomes, the centromere is located near the end of the arms. In *telocentric* chromosomes, the centromere appears to be located very near the end of the chromatid. (Micrograph courtesy William C. Earnshaw.)

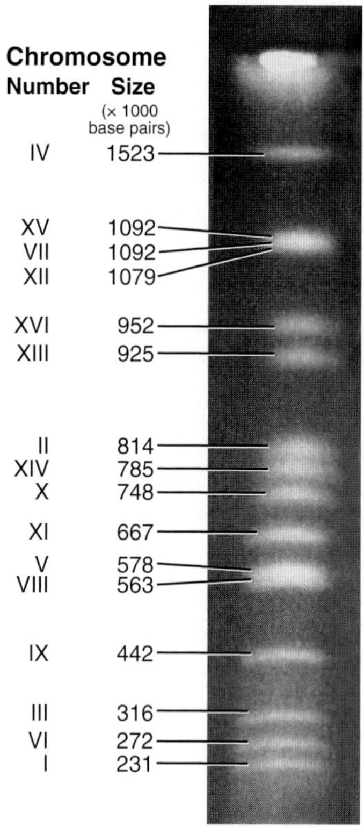

FIGURE 7.3 PULSED-FIELD GEL ELECTROPHORESIS OF BUDDING YEAST CHROMOSOMES. Intact cells embedded in a block of agarose are treated under very gentle conditions with proteases and detergents to free the chromosomal DNA from other cellular constituents. The DNA is then moved under the influence of an electrical field out of the agarose block and directly into an agarose gel. The technique uses a specialized gel apparatus in which the direction and strength of the electrophoretic field is varied periodically. This technique permits the separation of very long DNA molecules (of up to several million base pairs). (Courtesy P. Hieter, University of British Columbia, Vancouver, Canada.)

even the smallest human chromosome, which is about 48 million bp long, is too large to resolve in this way.

Organization of Genes on Chromosomes

The first chromosome to be completely sequenced (in 1977) was that of the bacterial virus *ϕx174* (Table 7.1). Starting in the 1990s much effort has been devoted to determining the complete sequences of the chromosomes of a wide variety of organisms including thousands of microbial species and well over 1000 humans. The complex genomes sequenced thus far range in size from 580,000 bp for *Mycoplasma genitalium,* which causes urinary tract infections in humans, to 3,547,121,844 bp for humans themselves (this figure is arbitrary as explained below). Numbers of protein-coding genes identified range from 480 in *M. genitalium* to 20,296 for humans (Table 7.1). However, because gene-detection algorithms are still being perfected,

TABLE 7.1 DNA Content of Various Genomes

Organism	Haploid Genome Size (bp)	Predicted Number of Protein-Coding Genes
fX174 (bacterial virus)	5386	11
Mycoplasma genitalium (pathogenic bacterium)	580,070	480*
Rickettsia prowazekii (endoparasitic bacterium)	1,111,523	834
Escherichia coli (free-living bacterium)	4,639,221	4288
Bacillus subtilis (free-living bacterium)	4,214,810	4100
Saccharomyces cerevisiae (budding yeast)	12,157,105	6692
Schizosaccharomyces pombe (fission yeast)	13,800,000	4970
Caenorhabditis elegans (nematode worm)	10.3×10^7	20,447
Drosophila melanogaster (fruit fly)	1.4×10^8	13,918
Arabidopsis thaliana (plant)	1.25×10^8	27,000
Anopheles gambiae (malaria mosquito)	2.78×10^8	14,000
Oryza sativa japonica (rice)	4.3×10^8	16,941
Mus musculus (house mouse)	3.4×10^9	22,547
Rattus norvegicus (Brown Norway rat)	3.0×10^9	22,293
Xenopus tropicalis (South African clawed frog)	1.3×10^9	18,442
Homo sapiens (human)	3.5×10^9	20,296
Amoeba dubia (single-celled protozoan)	670×10^9	?

*It appears that only 265 to 350 of these genes are essential for life.
In most higher eukaryotes, the huge tracts of repeated DNA sequences in and around centromeres are poor in genes and beyond the limits of present technology to sequence. Thus, when statistics are given on chromosome sizes in descriptions of genome sequencing projects, these portions are often omitted. Where possible, the genome size figures given here reflect the entire genome (sequenced and unsequenced). Predicted gene numbers constantly change as genome sequences are reanalyzed.

estimates of gene numbers are constantly changing, even for completely sequenced genomes.

As a rule of thumb, bacterial genomes tend to make very efficient use of space, with approximately 90% of the genome being devoted to coding sequences. The remaining 10% is mostly taken up by sequences involved in gene regulation. *Rickettsia prowazekii* is a notable exception with only 76% of the genome devoted to coding sequences. Because this intracellular parasite derives many of its metabolic functions from the host cell, much of its noncoding DNA may be remnants of unneeded genes undergoing various stages of gradual loss from the genome.

The first fully sequenced, eukaryote genome was from budding yeast *Saccharomyces cerevisiae.* The 12 million bp yeast genome is subdivided into 16 chromosomes ranging in size from 230,000 bp to more than 1 million bp (Fig. 7.3). This genome has a dramatic history. Ancestral budding yeast apparently had eight chromosomes but at one point underwent a duplication of the entire genome. This event was followed by numerous small deletions that resulted in the subsequent loss of approximately 90% of the duplicated genes. As a result, the modern budding yeast genome contains approximately 6692 predicted genes, many of which are paralogs (genes produced by duplication that have evolved to take on distinct functions; see Box 2.1). Remarkably, only about 1000 of these genes are indispensable for life. Approximately 5% of yeast genes are segmented, containing regions that appear in mature RNA molecules **(exons)** and regions that are removed by splicing **(introns)** (discussed in detail in Chapter 11). Exons occupy approximately 75% of the budding yeast genome, with the remainder in regulatory regions, repeated DNAs, and introns (Fig. 7.4).

The fission yeast genome yielded some surprises. Fission yeast has substantially fewer genes than budding yeast, but the genes that it does have exhibit greater diversity. Furthermore, 43% of those genes have introns. During the more than 500 million years of evolution since the two yeasts diverged, the fission yeast genome was not duplicated and trimmed down, so it has fewer sister (paralogous) genes and has retained more ancient genes. The biggest difference between the fission and budding yeast chromosomes is the structure of their centromere regions (see later).

Two other important milestones were the complete genome sequences of two "model" organisms that are widely used by cell and developmental biologists: the nematode worm *Caenorhabditis elegans* and the fruit fly *Drosophila melanogaster.* These metazoan sequences revealed many important organizational differences from fungi. Although its genome is eight times larger than that of budding yeast (103 million bp distributed in six chromosomes), the nematode has only about three times more genes. Surprisingly, the fly, with an even larger genome and more complex body plan and life cycle, has about one-third fewer genes than the worm. In fact, only approximately 27% of the *C. elegans* genome and 13% of the *Drosophila* genomic DNA code for proteins.

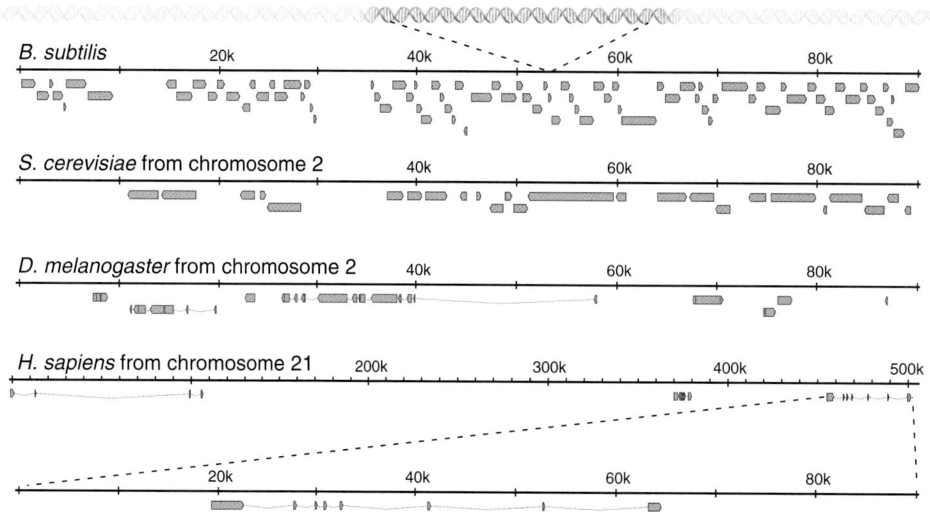

FIGURE 7.4 COMPARISON OF THE DISTRIBUTION OF GENES OVER 90,000 BP OF THE CHROMOSOME OF A TYPICAL BACTERIUM *(Bacillus subtilis)*, THE BUDDING YEAST *(Saccharomyces cerevisiae)*, THE FRUIT FLY *(Drosophila melanogaster)*, AND HUMANS *(Homo sapiens)*. To give a more accurate representation of the distribution of human genes, we also show a stretch of chromosome 21 spanning 500,000 bp. *Arrows* show the direction of transcription. Regions of genes encoding a product are shown as *thick orange bars*. Intervening sequences (introns) are shown as *thin lines*. (Courtesy A. Kerr, University of Edinburgh, United Kingdom.)

Instead, the fly has much more noncoding DNA than the worm.

The "finished" sequence of the human genome published in 2004 (and which still contains a number of unresolved "gaps") revealed an even lower gene density. Humans have far fewer genes than the up to 100,000 that had been predicted (current total 20,296, although this is subject to change) (Table 7.1). Protein-coding regions occupy only approximately 1.5% of the chromosomes, although genes themselves occupy up to approximately 46% of the genome (see next paragraph). Various repeated-sequence elements and pseudogenes occupy approximately 50% of the genome, as is discussed in a later section.

To put this all in perspective, every million base pairs of DNA sequenced yielded 483 genes in *S. cerevisiae,* 197 genes in *C. elegans,* 117 genes in *D. melanogaster,* and only 7 to 9 genes in humans. If the *Escherichia coli* chromosome were the size of chromosome 21, the smallest human chromosome at approximately 48×10^6 bp, it would have nearly 44,000 genes—more than the entire human complement! In fact, chromosome 21 has only 225 genes. As a result of this organization, a common strategy is to sequence the **exome** of an individual (the 1.5% of the genome found in exons) to reveal all changes (mutations) in protein sequences. However this strategy misses many mutations in noncoding regulatory regions that cause disease.

Human genes range in size from a few hundred base pairs to well over 10^6 bp, the average being about 28,000 bp and the longest (encoding dystrophin; see Fig. 39.9) being 2.2×10^6 bp. Most human protein-coding genes have introns separating an average of nine exons averaging only 145 bp each, but the variability is enormous. Genes can have more than 100 exons or only one. The average intron is a bit over 3000 bp long, but the human genome has more than 3000 introns that are greater than 50,000 bp and nine that are greater than 500,000 bp long. In total, approximately 25% of the genome is transcribed as introns. As a result, discovering new genes in genomic DNA sequence is a complex art.

The distribution of protein-coding genes along chromosomes is also highly variable. For example, on chromosome 9, gene density ranges from 3 to 22 genes per 10^6 bp. On chromosome 21, one region of 7×10^6 bp, encompassing nearly 20% of the whole chromosome, has no identified protein-coding genes at all. This region is almost twice the size of the entire *E. coli* chromosome! Approximately 25% of the genome is made up of regions of greater than 5×10^5 bp that are devoid of protein-coding genes and are termed *gene deserts.*

Much of the "noncoding DNA" is transcribed into RNA, so that overall approximately 80% of the genome is transcribed. Some **long noncoding RNAs** (lncRNAs) have roles in chromosome structure or gene regulation, but the functions (if any) of most lncRNAs remain to be established and many may be transcriptional "noise."

Transposable Elements Make up Much of the Human Genome

Eukaryotic genomes contain large amounts of **repetitive DNA** sequences that are present in many copies (thousands, in some cases). By contrast, coding regions of genes (which are typically present in a single copy per haploid genome) are referred to as **unique-sequence DNA.**

Repetitive DNA shows two patterns of distribution in the chromosomes. **Satellite DNAs** are clustered in discrete areas, often at centromeres (see "Pseudogenes" below). Other types of repetitive DNA are dispersed throughout the genome. In humans, most of this dispersed repetitive DNA is composed of **transposable elements**—small, discrete DNA elements that either are now or were formerly capable of moving from place to place within the DNA. There are many types of these elements, but for purposes of simplicity, they can be divided into two overall classes. **Transposons** move via DNA intermediates, and **retrotransposons** move via RNA intermediates. Transposons generally move by a cut-and-paste mechanism in which the starting element cuts itself out of one location within the genome and inserts itself somewhere else. There is currently no evidence for active transposons in humans, but in *Drosophila,* transposition by transposons, such as the **P element,** accounts for at least half of spontaneous mutations.

Even though humans no longer have active transposons, we still use at least two functional vestiges of these elements. One of the ways in which the diversity of the immune system is generated is by cutting and pasting portions of the genes that encode the variable regions of the immunoglobulin chains (see Fig. 28.10). This process involves moving segments of DNA around, and it now appears that the enzymes that accomplish this process were originally encoded by ancient transposons. In addition, CENP-B (centromere protein B; see Fig. 8.20), an abundant protein that binds to the α-satellite DNA repeats in primate centromeres, is closely related to a transposase enzyme encoded by one family of transposons.

Retrotransposons move (transpose) from one place in the DNA to another via production of an RNA intermediate. They then convert this RNA into DNA as it is being inserted at another site in the genome. Thus, on completion of a transposition event, the original retrotransposon remains in its original chromosomal location, and a newly generated element (which may be either full-length or partial) is inserted at a new site in the genome. The copying of RNA into DNA is carried out by a specialized type of DNA polymerase called a **reverse transcriptase.** These enzymes were discovered in tumor viruses with RNA chromosomes, but human cells also have a number of genes encoding reverse transcriptases.

The best-known retrotransposons are **LINES** (long interspersed nuclear elements) and **SINES** (short interspersed nuclear elements). Reverse transcriptases encoded by LINES are responsible for movements of both LINES and SINES. The L1 class of LINES encodes two proteins, one of which has reverse transcriptase activity (Fig. 7.5). All DNA polymerases, including reverse transcriptases, work by elongating a preexisting stretch

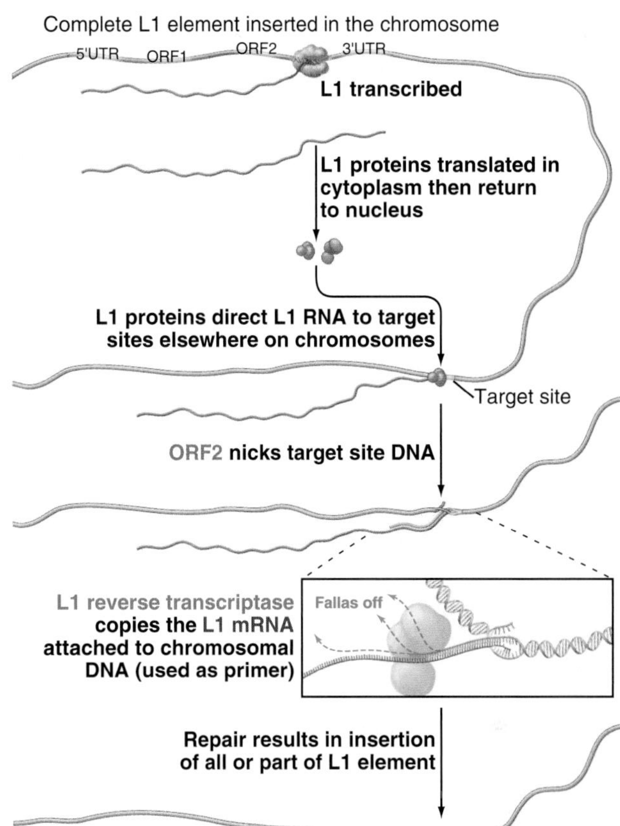

FIGURE 7.5 MECHANISM OF TRANSPOSITION OF AN L1 ELEMENT. The element is transcribed by RNA polymerase II (see Fig. 10.4). Proteins encoded by the element nick the chromosome, promote base pairing of the L1 transcript with the target site, and reverse transcribe the RNA into DNA. The L1 DNA is synthesized as an extension of the chromosome. The mechanism of final closing up of the nicks and gaps is not yet fully understood. mRNA, messenger RNA; ORF, open reading frame; UTR, untranslated region.

of double-stranded nucleic acid (see Chapter 42 for a discussion of the mechanism of DNA synthesis). L1 elements insert themselves into the chromosome by first nicking the chromosomal DNA, then using the newly created end as a primer for synthesis of a new DNA strand (Fig. 7.5). The template for this DNA synthesis by the reverse transcriptase is the LINE RNA, and the newly synthesized DNA is made as a direct extension of the chromosomal DNA molecule. Most LINE insertions are partial copies of the full-length element. Apparently, the reverse transcriptase usually falls off before it completes copying the entire element.

LINES and SINES plus other remnants of transposable elements account for up to 50% of the human genome. LINES, with a consensus sequence of 6 to 8 kb, make up approximately 20% of the genome. (A consensus sequence is the average arrived at by comparing a number of different sequenced DNA clones.) Approximately 80% of human genes have at least one segment of L1 sequence inserted, typically in an intron. The Alu class of SINES, with a consensus sequence of about

300 bp, constitutes approximately 13% of the total DNA—almost a million copies scattered throughout the genome. Alu elements are derived from the 7SL RNA gene, which encodes the RNA component of signal recognition particle (see Fig. 20.5). They are actively transcribed by RNA polymerase III (see Fig. 10.8) but are short and do not have enough coding capacity to encode for complex proteins. They therefore rely on the L1 machinery to move around. It is therefore somewhat paradoxical that SINES and LINES have quite different distributions along the chromosomes. LINES are concentrated in gene-poor regions of the chromosomes with a relatively higher content of A + T base pairs. In contrast, the Alu SINES are concentrated in gene-rich regions with a relatively higher content of G + C base pairs.

Transposition can be harmful, as along the way, genes can be disrupted, deleted, or rearranged. Because of their tendency to insert into gene-rich regions of chromosomes, Alu elements are one of the most potent endogenous human mutagens, with a new Alu insertion occurring once in every 100 births. In contrast, although LINES can cause genome instability when they move, and despite the large fraction of the human genome that is derived from LINES, they cause only 0.07% of spontaneous mutations seen in humans, owing to several mitigating features: (a) Only about 100 L1 elements are active, and these appear to be active in the germline (ie, during production of gametes) and in brain (where they may promote neuronal diversity). (b) LINE elements prefer to move into gene-poor areas of chromosomes. (c) Most LINE sequences are only fragments of the complete element. In contrast, mice apparently have many more active L1 elements (~3000), and L1 transposition causes approximately 2.5% of spontaneous mutations in mice. One of the ancestral roles of the RNA interference (RNAi) machinery (see Fig. 11.13) might have been to suppress the deleterious activity of transposable elements.

The physiological role, if any, of these elements is much debated. One long-favored possibility is that they do nothing advantageous and are analogous to an infection of the DNA that is tolerated as long as it does not disrupt genes that are essential for life. This is called the "**selfish DNA**" hypothesis. This notion has been challenged for Alu sequences, which are efficiently transcribed into RNA. Alu transcripts accumulate under conditions of cellular stress such as viral infection. This is interesting because Alu transcripts can bind very efficiently to a protein kinase called **PKR,** which is induced by interferon as part of the cell's antiviral protection pathways. The best-known function of PKR is phosphorylation of eukaryotic initiation factor 2α-subunit (eIF-2α; see Fig. 12.8). This profoundly inhibits protein synthesis. PKR is generally activated by double-stranded RNAs (dsRNAs), and this is presumably important for its antiviral role, as many viruses have RNA chromosomes. Alu transcripts at low levels activate PKR (ie, suppress protein synthesis), but at higher levels, they inactivate the enzyme (ie, promote protein synthesis). Thus, it has been suggested that Alu transcripts might be natural regulators of protein synthesis under conditions of cellular stress.

LINES can also modulate transcription of genes by influencing the behavior of RNA polymerase as it passes through them. Thus, they might have a role in the control of gene expression. As discussed at the end of this chapter, the structure of telomeres (the ends of chromosomes) is in part maintained by **telomerase**, a specialized form of reverse transcriptase, whose mechanism is closely related to that of the L1 reverse transcriptase.

Pseudogenes

One surprise that emerged from analysis of the eukaryotic genome sequences was the presence of **pseudogenes:** more than 14,000 in humans. Pseudogenes are derived from genes but are no longer functional. They arise in two ways, both involving transposable elements. **Processed pseudogenes,** the more common variety, are created by reverse transcription of mature messenger RNA (mRNA) sequences into DNA, apparently by a LINE reverse transcriptase that inserts the copy back into the genome. Because these sequences come from mature mRNA, they lack introns. They also lack sequences that regulate transcription initiation and termination (see Chapter 10), so they are not expressed. **Unprocessed pseudogenes** are created either by reverse transcription of unspliced precursor mRNAs or by local duplications of the chromosome that can occur as a result of recombination between transposable elements. The duplications can initially create bona fide functional gene copies that may become pseudogenes as they accumulate mutations that render their transcripts nonfunctional. Because pseudogenes are not functional, mutation of their DNA is not selected against during evolution, as are harmful mutations in the coding sequences of genes. Thus, over time, pseudogenes become decreasingly recognizable and eventually are lost from recognition in the sea of noncoding DNA.

Segmental Duplications in the Human Genome

Approximately 5% of the human genome is composed of regions of **segmental duplication** that have formed relatively recently in evolutionary time. Segmental duplications are regions of 1000 or more base pairs with a DNA sequence identity of 90% or greater that are present in more than one copy but are not transposons. They are interesting, because they can have a significant impact on human health. Regions of highly related DNA sequence can base-pair with one another and can

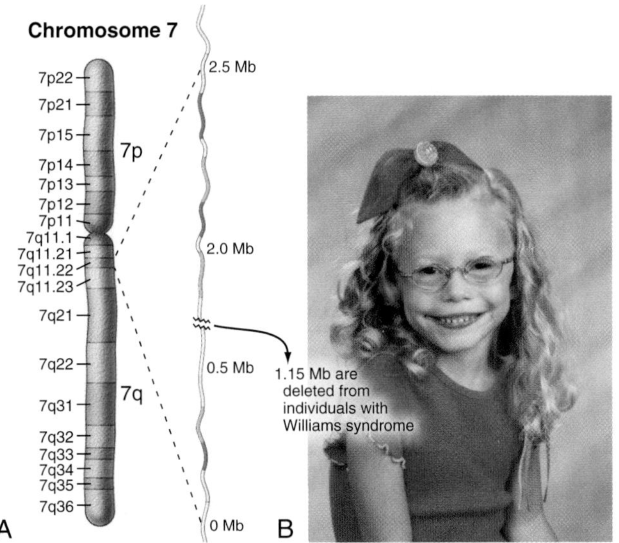

FIGURE 7.6 SEGMENTAL DUPLICATIONS WITHIN A REGION OF HUMAN CHROMOSOME 7 GIVE RISE TO WILLIAMS-BEUREN SYNDROME. A, Inappropriate recombination between duplicated sequences causes the deletion of a region of the chromosome. **B,** Williams-Beuren syndrome is a rare congenital disorder that is characterized by an outgoing personality, a characteristic elfin-like facial appearance, moderate to mild intellectual disability, and a range of physical problems. (Courtesy the Williams Syndrome Association, http://www.williams-syndrome.org/.)

subsequently recombine. Depending on how these regions are distributed on the chromosomes, this recombination can eliminate intervening regions of nonduplicated DNA. If the deleted region contains genes important for human health, then the result can be disease.

One example of this is found on chromosome 7, where deletion of a portion of the long arm is associated with Williams-Beuren syndrome, a complex developmental disorder associated with a highly variable range of symptoms that can include elfin-like facial features, defects in certain mental skills, and a wide range of physical problems (Fig. 7.6). These deletions occur because large (>140,000 bp) segmental duplications of DNA distributed across a region of 2×10^6 bp flank a unique sequence region of approximately 1×10^6 bp. If recombination occurs between the segmental duplications, approximately 1.6×10^6 bp, including the unique sequence DNA, are lost. Because of the highly complex organization of this region and the large size of the duplications, this turned out to be the most difficult region of chromosome 7 to sequence.

The Human Genome: Variations on a Theme

The human "reference genome" sequence does not come from a single person, but is instead an idealized assembly derived from the DNA of a number of people. Constructing an artificial reference genome is necessary, because although we might imagine that there is only one "human genome," data from sequencing many thousands of genomes have shown that there are dramatic variations in DNA content and sequence among individuals. Famously, analysis of some particularly variable regions of repetitive sequences forms the basis for DNA testing in criminology and paternity testing. Given the large number of genomes sequenced to date, it makes sense to talk of a "typical" genome and how this differs from the reference. Prepare to be amazed. A typical genome has 4 to 5×10^6 differences from the reference! The largest number of affected base pairs are in 2100 to 2500 "structural variants" (changes involving >50 bp). These include deletions, more than 120 LINE and more than 900 SINE insertions, and other changes not found in the reference genome. Overall, they encompass 20×10^6 bp and often occur in regions of repeated DNA sequence. Other variations occur in genes, with a typical genome having approximately 165 mutations that truncate proteins, approximately 11,000 mutations that change protein sequences, and a staggering 520,000 mutations in regions thought to be involved in regulating gene expression. Occasionally, these variations are linked to inherited human disease, and genome-wide association studies (GWAS) correlating sequence changes with human disease are a major ongoing focus of these sequencing efforts. At centromere regions of chromosomes, the content of repeated DNA sequences commonly varies by over 10^6 bp between different individuals. Overall, this rather staggering variability leads to the question, "What is a 'normal' human genome?"

The Centromere: Overview

The centromere is at the heart of all chromosomal movements in mitosis and meiosis, as it nucleates on its surface the formation of the button-like **kinetochore** (see Fig. 8.21), the structure that attaches chromosomes to the mitotic spindle (the microtubule-based apparatus upon which chromosomes move; see Chapter 44). In mitotic chromosomes of most higher eukaryotes, the **centromere** forms a waist-like stricture or **primary constriction** where the two sister chromatids are most intimately paired. The centromere is a chromatin structure, and both DNA and proteins are essential to its function.

Variations in Centromere Organization Among Species

In budding yeast, autonomous **CEN (centromere)** sequences specify protein-binding sites required for assembly of the kinetochore; if inserted into circular DNA molecules (plasmids), they render them capable of interacting with the mitotic spindle and segregating during mitosis (Fig. 7.7). In other organisms, including the fission yeast *Schizosaccharomyces pombe,* centromere sequences require an activation event to nucleate kinetochore formation. This event appears to

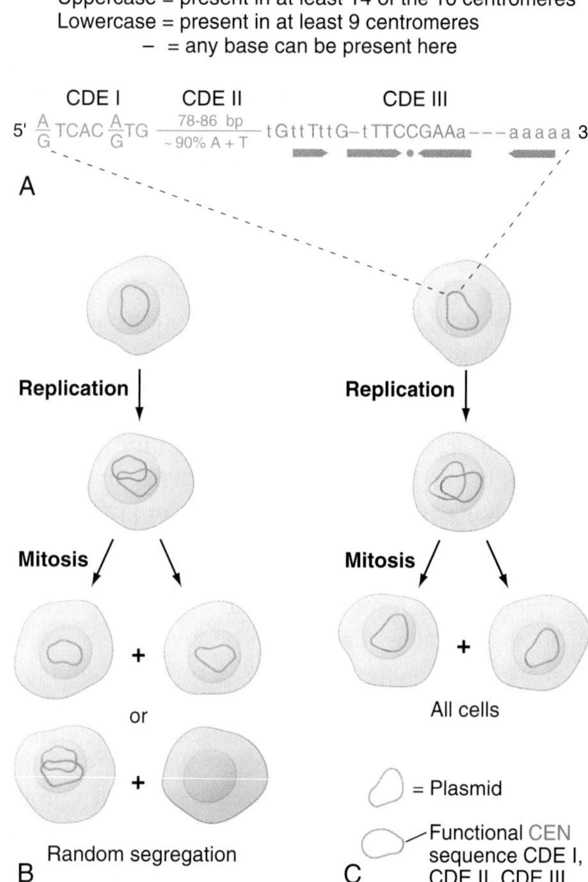

FIGURE 7.7 THE BUDDING YEAST CENTROMERE (CEN) IS SPECIFIED BY A 125-BP SEQUENCE. A, Three conserved DNA elements (CDE I to CDE III). CDE I and CDE III bind proteins in a sequence-specific manner. CDE III has mirror symmetry: a central C *(dot)* is flanked by two regions of complementary DNA sequence *(arrows)*. All that seems to be important about CDE II is its abundance of A and T nucleotides and its overall length. **B–C,** The assay for mitotic stability of a plasmid used to clone CEN DNA from most budding yeast chromosomes. The plasmid carries a gene encoding an enzyme involved in adenine metabolism. When the plasmid is present, colonies are white. If the plasmid is lost, the colonies become red as a result of the accumulation of a metabolic by-product. If the plasmid is capable of replication but lacks a centromere, the colonies will be mostly red, reflecting the inefficient segregation of the plasmid at mitosis **(B).** If the plasmid carries a functional centromere, the colonies will be white, as the plasmid will be successfully transmitted at nearly every division **(C).**

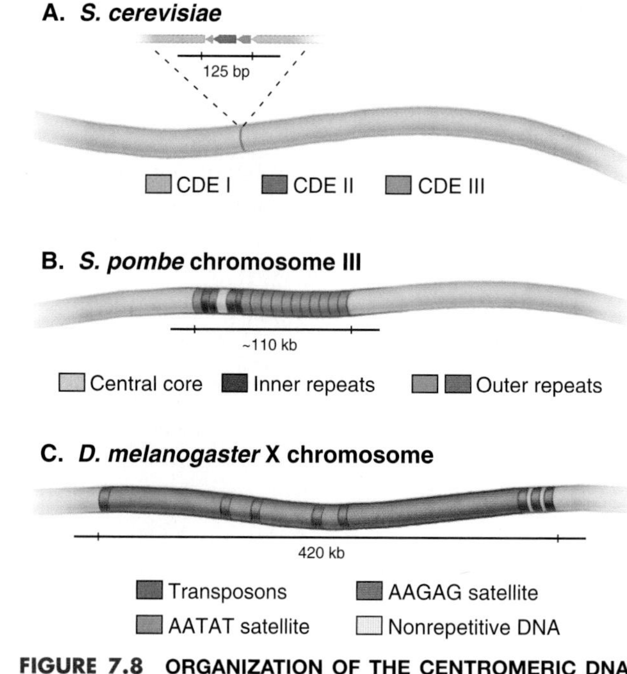

FIGURE 7.8 ORGANIZATION OF THE CENTROMERIC DNAS OF BUDDING YEAST, FISSION YEAST, AND FRUIT FLY. A, The budding yeast *(Saccharomyces cerevisiae)* point centromere is specified by a 125-bp sequence. **B,** The fission yeast *(Schizosaccharomyces pombe)* regional centromeres all contain central core DNA flanked by complex arrays of repeated sequences. Embedded within these repeated sequences are a number of genes encoding transfer RNAs, not shown here. The minimum region required to construct a functional centromere in fission yeast artificial chromosomes is approximately 10 kb in length and includes the central core DNA plus a portion of the flanking repeated DNA. **C,** The fruit fly *(Drosophila melanogaster)* also has a regional centromere encompassing 420 kb. This is rich in satellite DNA and contains a number of transposable elements. The same satellite DNAs and transposons are also found at other, noncentromeric, regions of the chromosomes.

involve epigenetic modification of the DNA and/or chromatin (discussed later).

CEN sequences from all 16 budding yeast chromosomes have a common organization based around three conserved sequence elements (Fig. 7.7). These are designated (in the 5′ to 3′ direction) **CDE I** (centromere DNA element I, 8 bp), **CDE II** (78 to 86 bp), and **CDE III** (25 bp). A 125-bp region spanning CDE I to CDE III is sufficient to direct the efficient segregation of a yeast chromosome, which can reach a size of more than 1 million bp. This type of centromere, in which the kinetochore is assembled as a result of protein recognition of specific DNA sequences, is known as a **point centromere** and to date has been found only in budding yeasts. Kinetochores assembled on point centromeres bind a single microtubule.

Even though the average size of *S. pombe* chromosomes is only fivefold larger than their counterparts in *S. cerevisiae* (4.6 Mb vs. 0.87 Mb), fission yeast centromeres are 300- to 600-fold larger (Fig. 7.8). The smallest *S. pombe* centromere consists of 35,000 bp, whereas the largest spans 110,000 bp. Fission yeast centromeric DNA is much more complex than its budding yeast counterpart, containing a central core of 4 to 7 kb of unique-sequence DNA flanked by complex arrays of repeated sequences. This type of centromere is known as a **regional centromere.** Kinetochores assembled on regional centromeres bind multiple microtubules (two to four in the case of *S. pombe*).

Studies of *S. pombe* centromeres revealed in addition to the primary DNA sequence, an **epigenetic** activation step is required for CEN DNA to function as a centromere. Epigenetic events are inheritable properties of chromosomes that are not directly encoded in the nucleotide sequence. They are typically explained either by enzymatic modification of the DNA (eg, methylation of cytosine) or by modification of proteins that are stably associated with the DNA. Epigenetic mechanisms also play an essential role in the assembly of centromeres in higher eukaryotes, including humans. In both *S. pombe* and metazoans, these epigenetic changes involve the construction of a special chromatin environment at centromeres. What this means in practice is that (except budding yeast), no single DNA sequence can be put into cells and function directly as a centromere. If a piece of *S. pombe* centromeric DNA is introduced into cells, it must undergo a series of packaging events and modifications that turn it into a functional centromere. These events are so rare that when candidate DNA molecules with CEN sequences are introduced into *S. pombe* cells, only about 1 in 10^5 assembles into a functional centromere.

Regional centromeres are typically organized around a core region that nucleates kinetochore formation during mitosis. This core consists of a specialized form of chromatin called centrochromatin containing CENP-A, a specialized form of histone H3 that can replace H3 in nucleosomes (see Fig. 8.21). How the centrochromatin is organized varies dramatically between species (Fig. 7.8). Centrochromatin is typically flanked by **constitutive heterochromatin,** a form of chromatin that generally suppresses gene transcription and remains condensed throughout the cell cycle (see Fig. 8.7). Constitutive heterochromatin is characterized by the presence of special modifications of the histone proteins and other proteins that "read" (bind to) those modifications. (Chapter 8 discusses heterochromatin.) Both the core of the centromere and flanking heterochromatin are usually (but not invariably) comprised of repeated DNA sequences.

The first fully sequenced centromere of a metazoan was that of rice chromosome 8. Sequencing was possible, because the rice centromere contains limited amounts of a centromeric satellite DNA (CentO) dispersed in blocks separated by transposons, retrotransposons, and fragments. All in all, 72% of this centromere is composed of repetitive sequences. The kinetochore, as defined by sequences associated with CENP-A, spans 750 kb and is interspersed with regions of chromatin containing normal histone H3 that is apparently packaged into heterochromatin. Surprisingly, this centromere region contains at least four genes that are actively transcribed.

More recently it was discovered that chickens have three and the horse one chromosome with sequences composed of nonrepetitive DNA and lacking flanking heterochromatin. These centromeres are thought to be evolutionarily new, and may have originated from neocentromeres (see later). It is thought that such evolutionarily new centromeres gradually acquire repetitive DNA sequences, possibly because they provide as-yet unknown advantages over evolutionary time. The rice centromere is not evolutionarily new, having had its present organization for at least the last 10,000 years (since the *indica* and *japonica* cultivars of rice were separated) and appears to be intermediate between a canonical metazoan centromere and a neocentromere.

The centromere organization of the fruit fly *D. melanogaster* shows important similarities and differences to the rice centromere. The centromere of the fly's X chromosome occupies a stretch of roughly 420,000 bp (Fig. 7.8) that is composed mostly of simple-sequence satellite DNAs interspersed with transposable DNA elements. This resembles the situation in plants; however, in *Drosophila,* no sequences were found in this region that are unique to the fly centromeres; all sequences found at centromeres could also be found on the chromosome arms. Thus, it appears that something other than the DNA sequence alone must be responsible for conferring centromere activity on this region of the chromosome.

In addition to point centromeres in budding yeast and regional centromeres found in most metazoans, many plants and insects as well as in the nematode *C. elegans* have a third variant, in which centromere activity is distributed along the whole length of the mitotic chromosomes. These **holocentric** chromosomes have binding sites for about 20 microtubules distributed along the whole poleward-facing surface of the chromosome during mitosis rather than a disk-like kinetochore at a centromeric constriction, as in humans. If a holocentric chromosome is fragmented, every piece can bind microtubules and segregate in mitosis. Perhaps surprisingly, the proteins of the holocentric kinetochore are the same as those found at disk-like regional kinetochores (see Chapter 8). Accordingly CENP-A is found in domains scattered across half of the worm genome that are characterized by low levels of transcription in the germline. One possibility is that in these chromosomes, any chromatin with the right transcriptional profile can serve to nucleate kinetochore assembly—perhaps the requirement for special epigenetic marks has been relaxed.

Vertebrate Centromere DNA

Vertebrate centromeres initially proved extremely difficult to characterize in molecular detail, largely due to their large size and complex, highly repetitious organization. For example, the centromere of chromosome 21 (the smallest human chromosome at ~48 million bp) has been estimated to encompass more than 5 million bp. This entire region is composed of many thousands of copies of short DNA repeat sequences clustered together in head-to-tail arrays known as satellite DNA. Many lines of research have now converged to reveal that this centromere-associated satellite DNA is a preferred site of

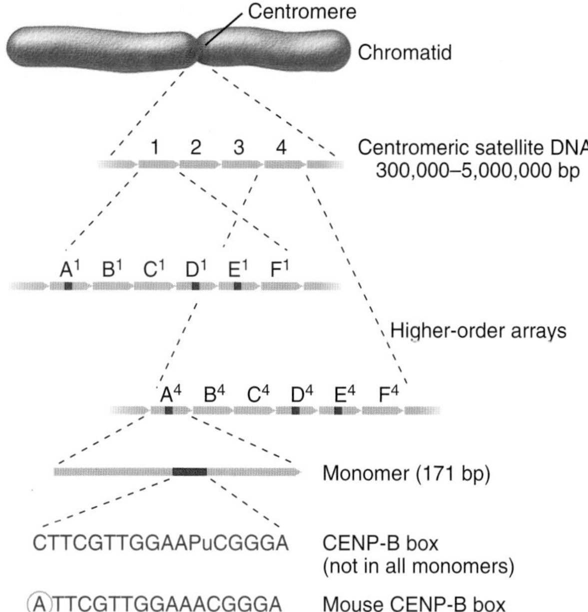

FIGURE 7.9 HIERARCHICAL ORGANIZATION OF α-SATELLITE DNA AT HUMAN CENTROMERES. The numbers (1 to 4) indicate higher-order repeats of α-satellite DNA. These may contain from 2 to 32 monomers (indicated by A^1, B^1, and so on). DNA sequences of adjacent monomers within a repeat (eg, A^1, B^1, C^1) may differ by as much as 40% from one another. DNA sequences of monomers occupying identical positions within the higher-order repeats (A^1, A^4, etc.) are nearly 99% identical to one another. The *red* sequence shown at the bottom represents the binding site for centromeric protein CENP-B.

centromere formation, but that centromeres can (rarely) form elsewhere. The trigger that seems to define any particular region of the chromosome as a centromere involves epigenetic modifications of the DNA and chromatin, at least one of which is the binding of the specialized histone H3 variant CENP-A.

The major human centromeric satellite DNA, **α-satellite,** is a complex family of repeated sequences that constitutes approximately 5% of the genome. Monomers averaging approximately 171 bp long are organized into higher-order repeats (Fig. 7.9). Some of the monomers have a conserved 17-bp sequence (the CENP-B box) that forms the binding site for the centromeric protein CENP-B (mentioned earlier as having its origin in an ancient DNA transposon). The organization of higher-order repeats varies greatly from chromosome to chromosome, and numerous repeat patterns, comprising 2 to 32 monomers, have been described. Each chromosome has one or a few types of higher-order repeats of α-satellite DNA.

The entire centromeric region of certain chromosomes may be composed of α-satellite monomers, apparently with little or no interspersed DNA of other types. The amount of α-satellite DNA at different centromeres varies widely: from as little as 300,000 bp on the Y chromosome to up to 5 million bp on chromosome 7. In addition, the α-satellite DNA content of a given chromosome can vary by more than a million bp between different individuals. Clearly, a wide variation in the local organization of α-satellite DNA is tolerated.

Human chromosomes also contain several other families of satellite DNA. Classical satellites I to IV, which together constitute 2% to 5% of the genome, are composed of divergent repeats of the sequence GGAAT. These satellites occur in blocks more than 20,000 bp long that are immediately adjacent to the centromeres of a number of chromosomes and may be found at lower levels near most centromeres. The so-called pericentromeric region adjacent to the centromere of chromosome 9 apparently contains 7 to 10 million bp of satellite III sequence. The long arm of the Y chromosome also contains huge amounts of satellite III DNA (up to 40% of its total DNA).

If α-satellite DNA arrays longer than about 50,000 bp are introduced into cultured human cells, they occasionally form tiny minichromosomes with functional centromeres. For this to work, the α-satellite DNA arrays must have a highly regular organization, and some of the monomers must contain binding sites for CENP-B. Formation of these mammalian artificial chromosomes is very inefficient, so it is clear that α-satellite DNA arrays cannot automatically function as CEN DNA—some type of epigenetic activation is required.

There is an interesting corollary of this role of epigenetic modifications in assembly of a functional centromere. Suppose a bit of noncentromeric DNA somehow acquired the right set of modifications. Could that now function as a centromere? The answer is yes. The formation of **neocentromeres** on noncentromeric DNA has been seen in *S. pombe*, fruit flies, chickens, and humans and was first described in plants.

Rare individuals have a chromosome fragment that segregates in mitosis, despite loss of the normal centromere. Such chromosomes have acquired a new centromere or neocentromere in a new location on one of the chromosome arms. Remarkably, neocentromeres are composed of the normal DNA that exists at that location on the chromosome arm and yet somehow has acquired centromere function. Neocentromeres are bona fide centromeres; for example, they bind all known centromeric proteins except for CENP-B, which requires specific sequences on α-satellite DNA for binding. Different neocentromeres need have no sequences in common. These observations strongly support the hypothesis that epigenetic markers rather than the exact DNA sequence specify the centromere. The natural occurrence of α-satellite DNA at centromeres may reflect a propensity of α-satellite chromatin to acquire the epigenetic mark, rather than a sequence-specific mechanism as occurs in *S. cerevisiae.*

In one study of a chicken cell line, more than 100 independent new neocentromeres formed after the normal centromere was deleted experimentally (Fig. 7.10). Amazingly, every neocentromere formed on a different DNA sequence with no common underlying sequence features. Regions containing or lacking genes could be incorporated into a neocentromere. The only common feature was a domain of chromatin roughly 40,000 bp long containing CENP-A nucleosomes. This corresponds to the size of the centromere on the starting chromosome.

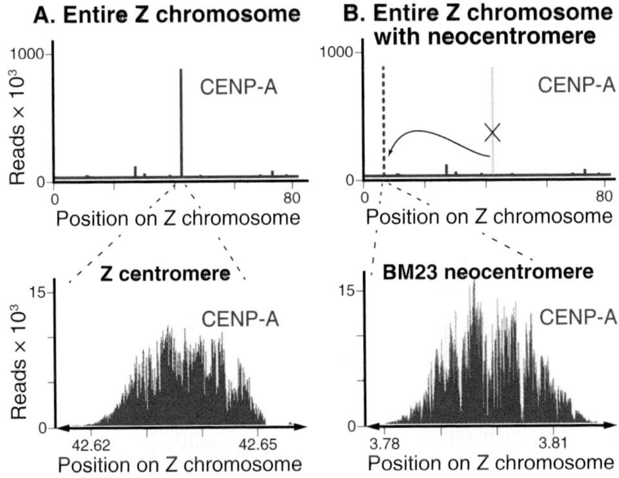

FIGURE 7.10 DISTRIBUTION OF CENTROMERE HISTONE CENP-A AT THE NATURAL CENTROMERE AND AT A NEOCENTROMERE ON THE CHICKEN Z CHROMOSOME. Cells were treated with formaldehyde to crosslink proteins to DNA. Isolated DNA was fragmented into short pieces of a few hundred base pairs and an antibody used to pull down the DNA fragments crosslinked to CENP-A. Thousands of DNA fragments associated with CENP-A were then sequenced. These sequences were mapped along the Z chromosome (the female sex chromosome of the chicken). (Data from Hori T, Shang W-H, Toyoda A, Misu S, et al. Histone H4 Lys 20 mono-methylation of the CENP-A nucleosome is essential for kinetochore assembly. *Dev Cell.* 2014;29:740–749.)

The epigenetic mark that defines an active centromere can be lost as well as gained. Thus, it is possible for a centromere to retain its normal DNA composition and yet lose the ability to assemble a kinetochore. This has been seen most clearly in naturally occurring human dicentric chromosomes. The chromosome shown in Fig. 7.11 arose through a breakage and fusion near the long arm of chromosome 13 and has two centromeres. As shown in the figure, one of these lost the ability to assemble a kinetochore even though it retained its α-satellite.

What is the elusive epigenetic mark and how does it "magically" mark a region of the chromosome as a centromere? At present, all evidence suggests that the epigenetic mark has something to do with low level transcription of the CENP-A-containing DNA *during mitosis*. This is remarkable, because transcription is supposed to be entirely shut off during mitosis, and indeed, it seems that centromeres are the only region of the genome that is transcribed at that time. We do not yet know whether it is the process of transcription that is important or whether the RNA transcripts themselves serve an important role in specifying centromere chromatin.

Once a DNA sequence has acquired the proper epigenetic mark, it can assemble a functional kinetochore that can regulate chromosome behavior in mitosis. This involves the binding and function of 100 or more proteins as discussed in Chapter 8 (see Fig. 8.21).

Ends of the Chromosomes: Why Specialized Telomeres Are Needed

The ends of chromosomal DNA molecules pose at least two problems that cells solve by packaging the chromosome ends into specialized structures called **telomeres.**

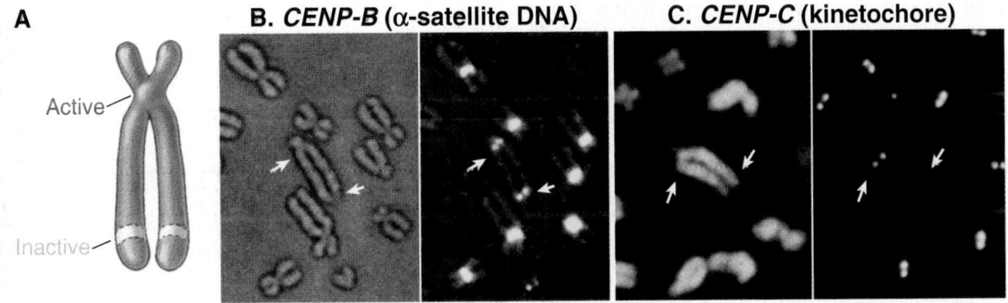

FIGURE 7.11 EPIGENETIC REGULATION OF HUMAN CENTROMERE FUNCTION. An unusual chromosome was discovered during prenatal screening of a fetus that sonography had indicated to be abnormal. This chromosome consisted of two copies of the maternal chromosome 13 linked end to end. It thus contained two centromeres and so was termed *dicentric.* Such dicentric chromosomes are normally unstable during mitosis, as the two centromeres on one chromatid often become attached to opposite spindle poles. This causes the chromosome to be stretched between opposite spindle poles and ultimately break. In the case of this particular dicentric chromosome, one of the centromeres was inactivated (presumably, it lost its epigenetic mark). This chromosome thus behaves perfectly normally in mitosis. When the distribution of centromere proteins at the active and inactive centromeres was compared, it was found that CENP-B was present at both but that CENP-C, a marker for kinetochores, was present only at the active centromere. **A,** Organization of the dicentric chromosome. **B,** Phase-contrast view of chromosomes from the amniocytes *(left).* Phase-contrast view taken with superimposed antibody staining for CENP-B *(right).* **C,** DNA stain of a different chromosome spread *(left).* Staining with antibody specific for CENP-C *(right).* (**B–C,** Courtesy William C. Earnshaw.)

First, it is essential that cells distinguish the ends of a chromosome from breaks in DNA. When cells detect DNA breaks, they stop their progression through the cell cycle and repair the breaks by joining the ends together (see Box 43.1). Telomeres keep normal chromosome ends from inducing cell cycle arrest and from being joined to other DNA ends by the repair machinery. Second, telomeres permit the chromosomal DNA to be replicated out to the very end.

Structure of Telomeric DNA

Telomeres in all eukaryotes tested to date (with the exception of several insect species including *Drosophila*) are composed of many repeats of short DNA sequences. The sequence 5′ TTAGGG 3′ is found at the ends of chromosomes in organisms ranging from human to rattlesnake to the fungus *Neurospora crassa.* In the human, roughly 650 to 2500 copies of this sequence are found at the ends of each chromosome, a total length of approximately 4000 to 15,000 bp (this varies in different tissues). Higher plant telomeres have the sequence TTTAGGG, and other variations of this repeat sequence have been noted in protozoans and yeasts.

The telomeric repeat is organized in a unique orientation with respect to the chromosome end. Thus, the end of every chromosome has one G-rich strand and one complementary C-rich strand. The G-rich strand always makes up the 3′ end of the chromosomal DNA molecule. Thus, the very 3′ end of the chromosome always has the following structure: …(TTAGGG)-OH. Furthermore, the end of the chromosome is not a blunt structure; the G-rich strand ends in a single-stranded overhang 30 to 400 bp long. This single strand of DNA is critical for telomere structure and function. It regulates telomerase activity and also "invades" the double helix of telomeric repeats, base-pairing and causing the ends of chromosomes to form large loops, called T loops that protect chromosome ends (see later discussion). A surprisingly complex balance of enzymatic activities maintains this single strand of DNA. These activities change throughout the cell cycle in dividing cells.

How Telomeres Replicate the Ends of the Chromosomal DNA

Telomeres prevent the erosion of the end of the chromosomal DNA molecule during each round of replication (for a more extensive discussion of DNA replication, see Chapter 42). All DNA replication proceeds with a polarity of 3′ to 5′ on the template DNA (5′ to 3′ in the newly synthesized DNA). Furthermore, all DNA polymerases (but not RNA polymerases) work by elongating a pre-existing stretch of double-stranded nucleic acid. During cellular DNA replication, this is achieved by making a short RNA primer and then elongating the RNA: DNA duplex with DNA polymerase. The primer is subsequently removed, and DNA polymerase fills the gap by elongation from the next upstream DNA end (Fig. 7.12).

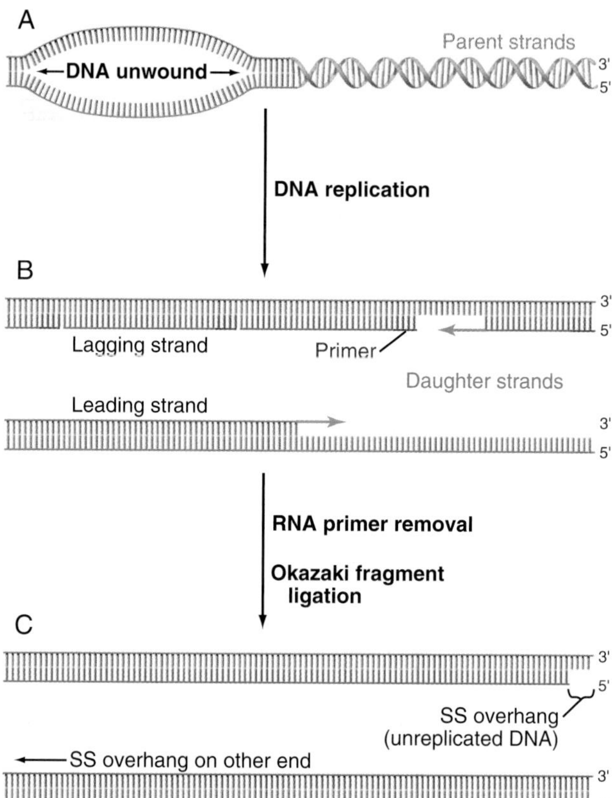

FIGURE 7.12 DNA REPLICATION PROBLEM AT CHROMOSOME ENDS. DNA polymerases cannot initiate the formation of DNA on a template de novo; they can only extend preexisting nucleotide strands (see Chapter 42). In contrast, RNA polymerases can initiate synthesis without a primer. All replicating DNA chains start from a short region of RNA, which is used to "prime" DNA polymerase. **A,** DNA strand separation. **B,** RNA primer synthesis. Replication of DNA starts with the synthesis of an RNA primer *(magenta)* complementary to a short sequence of DNA, which is extended by DNA polymerase. **C,** The RNA primer is degraded and the gap is filled in by DNA polymerase. This being true, how can the DNA underneath the very last RNA primer replicated? SS, single stranded.

If the terminus of the chromosomal DNA is replicated from an RNA primer that sits on the very end of the DNA molecule, it follows that when this primer is removed, there is no upstream DNA on which to put a primer. How, then, is the DNA underneath the last RNA primer replicated? Years of searching for a DNA polymerase that could operate in the opposite direction proved fruitless. The answer that ultimately emerged turned out to be both elegant and unexpected.

Most organisms have an enzyme called **telomerase** that specifically lengthens the 3′ end of the chromosomal DNA. Telomerases contain both protein and RNA subunits. The sequences of the RNA component provided an essential clue to how this enzyme works.

The RNA component of human telomerase contains the sequence AUCCCAAUC, which can base-pair with the TTAGGG telomere repeat at the ends of the chromosome. The enzyme uses its own RNA as a template for the synthesis of DNA, which it "grows" from the end of the chromosome (Fig. 7.13). This hypothesis was confirmed by showing that changing the sequence of the telomerase RNA alters the telomere sequence at the end of the chromosome.

According to this model, the telomerase actually synthesizes DNA using an RNA template. Thus, telomerase is a reverse transcriptase similar to that involved in the movement of the LINE retrotransposons (Fig. 7.5). When L1 family LINE retrotransposons insert themselves into the chromosome, a DNA end created at a nick in the chromosome is used to prime synthesis of a DNA strand using the LINE RNA as template, the newly synthesized DNA being a direct extension of the chromosomal DNA molecule.

Human telomerase consists of hTERT (the telomerase reverse transcriptase) complexed to hTERC, the telomerase RNA, which is 450 nucleotides long. Telomerase RNA varies in size and sequence between species. Active human telomerase can be reconstituted in vitro from purified hTERC and hTERT in the presence of a cell-free lysate from reticulocytes (which appears to provide essential protein-folding factors). In cells, telomerase is associated with auxiliary protein subunits that are involved in telomerase RNA processing and maturation.

Telomerase is subject to tight biological regulation. Active enzyme is detected in only a few normal tissues of adult humans. These include the stem cells of various tissues and male germ cells. In addition, approximately 90% of cancer cells express active telomerase and abnormal expression of telomerase has been linked to cancer. This telomerase is thought to enable the cancer cells to grow indefinitely without undergoing erosion of the ends of the chromosomes.

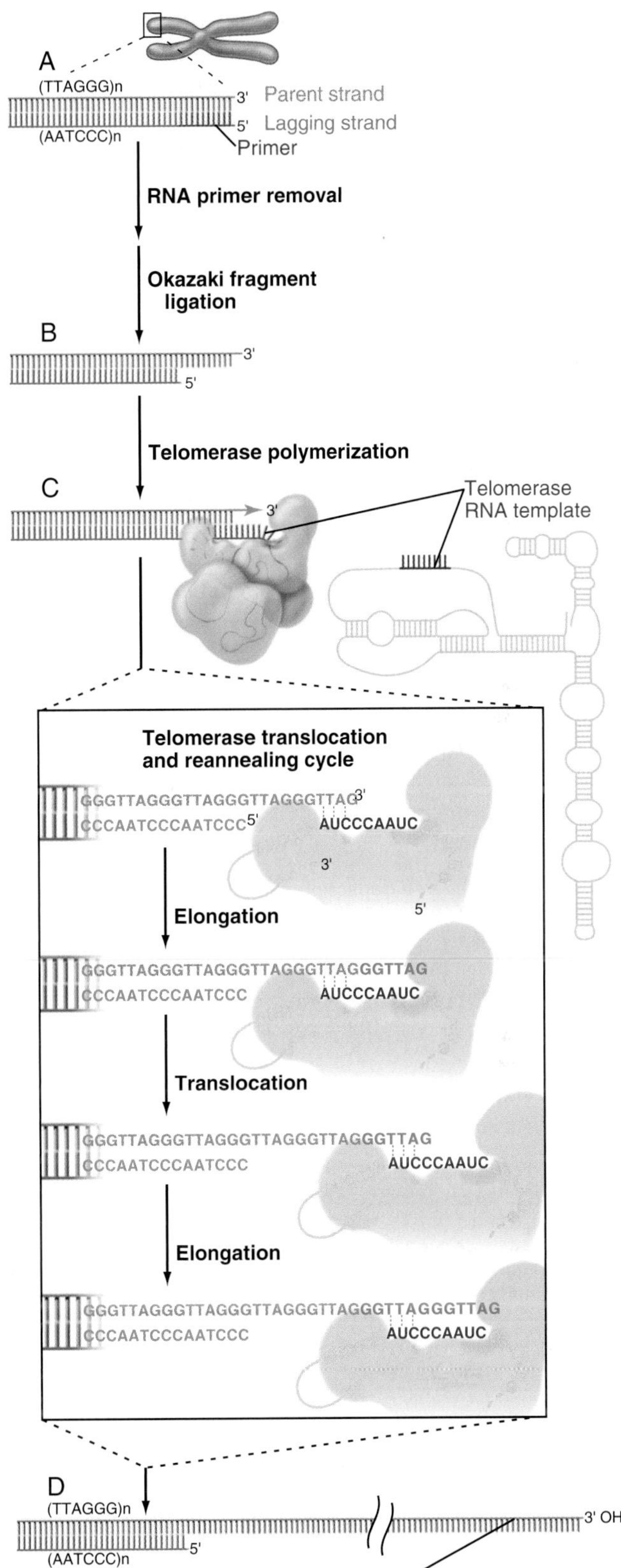

FIGURE 7.13 TELOMERASE PROVIDES A ECHANISM FOR LENGTHENING CHROMOSOMAL ENDS. A–B, Normal mechanisms of DNA replication are unable to replicate the very 3′ end of the chromosomal DNA. **C,** Telomerase solves this problem by providing its own template in the form of an intrinsic RNA subunit. This RNA subunit contains a sequence complementary to that found at the chromosome terminus on the 3′ strand. This sequence is able to base-pair with the DNA at the chromosome terminus and act as a template for DNA synthesis. In this case, the primer is the 3′ end of the chromosomal DNA, and the template is the RNA of the telomerase enzyme. Thus, the process of telomere elongation is a specialized form of reverse transcription (copying RNA into DNA), a process similar to that occurring during transposition of LINES (long interspersed nuclear elements) (Fig. 7.5), and during the life cycle of certain RNA-containing tumor viruses. The telomerase enzyme releases and rebinds its template after each 6 to 7 bp of new DNA has been synthesized. Up to several hundred base pairs may be added to the telomere in this way. **D,** In most cells, the 3′ end of the chromosomal DNA terminates in a single-stranded G-rich strand 30 to 400 nucleotides long that is essential for telomere structure and function.

Paradoxically, hTERC is not tightly regulated. The hTERC RNA is detected in many tissues, most of which lack telomerase activity. By contrast, the expression of hTERT correlates tightly with telomerase activity. Indeed, introduction of a DNA-encoding hTERT into telomerase-negative cells produces telomerase activity. This can have extremely important consequences for the proliferation of the cells (Fig. 7.16).

In cells that lack telomerase, a second pathway can help maintain the telomeric repeats at chromosome ends. This **ALT** (alternative lengthening of telomeres) process involves DNA recombination between telomeres. Cancer cells that lack telomerase expression have an activated ALT pathway.

A third solution to this problem was taken by dipterans such as *D. melanogaster,* in which the ends of the chromosomes are composed of transposable elements. In the fly, a few bp are lost from the end of the chromosome at every round of replication. This erosion of the chromosome ends is remedied by the occasional transposition of specialized transposable elements to the chromosome end. Thus, this appears to be an example of an originally "selfish DNA" that has become recruited for an essential cellular function.

Structural Proteins of the Telomere

Telomeres provide special protected ends for the chromosomal DNA molecule, in part by coating the end of the DNA molecules with protective proteins and by adopting a specialized DNA loop structure. In organisms with relatively short telomeric DNA sequences, those sequences are packaged into a specialized chromatin structure. In mammals, in which the telomeric sequences are much longer, the bulk of the telomeric DNA is packaged into conventional chromatin (see Chapter 8).

A complex of six proteins called **shelterin** associates with telomeres in most organisms that have a telomerase (Fig. 7.14). Two subunits directly bind the TTAGGG duplex while one binds to the single stranded overhang. The other two subunits bridge the DNA binding subunits. *S. cerevisiae* has homologous subunits that bind to the telomeric repeats and the G-strand overhang. They protect the end of the recessed C-rich strand at telomeres, and this strand is rapidly degraded if these proteins are missing, with lethal consequences for the cell. Shelterin appears to both regulate telomerase activity and play an essential role in protecting chromosome ends.

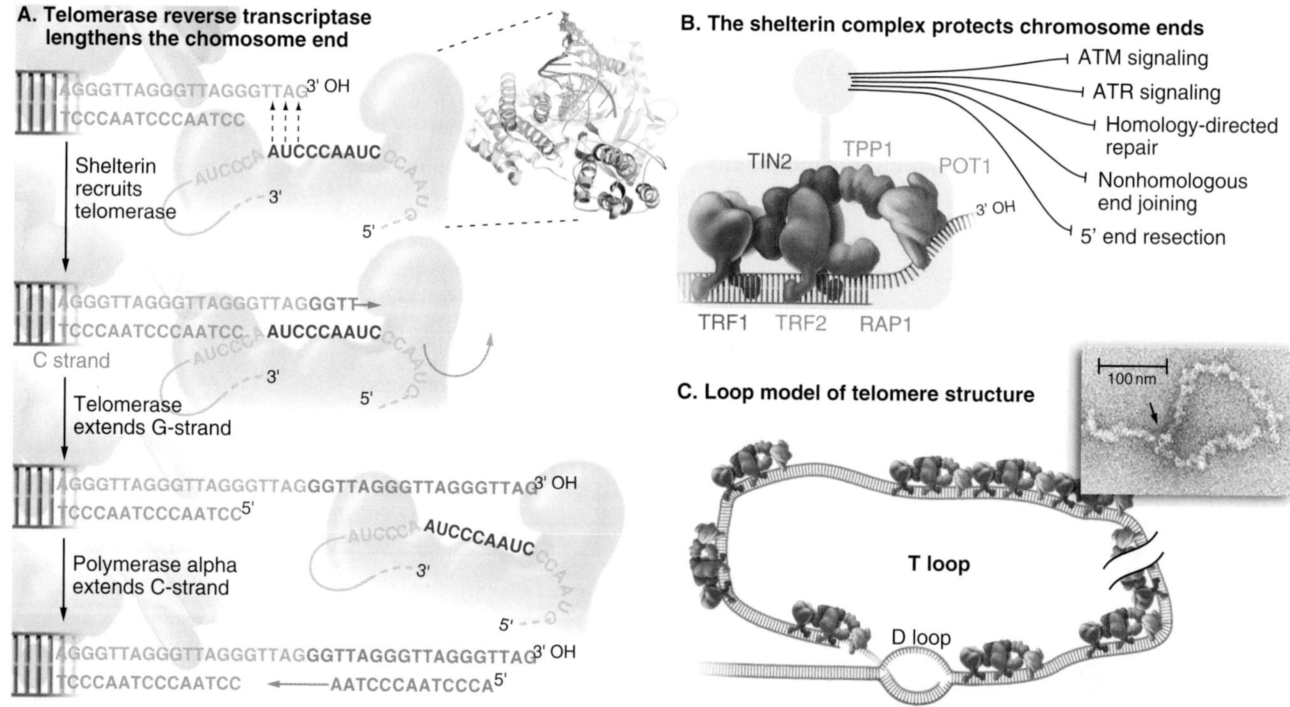

FIGURE 7.14 TELOMERE STRUCTURES. A, Structure of telomerase. **B,** Organization and functions of shelterin, a complex of six subunits. TRF1 and TRF2 dimers bind to the double-stranded $(TTAGGG)_n$ repeats at telomeres. Together they bind TIN2, which in turn binds TPP1, which helps recruit POT1 to the single-stranded DNA at the chromosome end. If shelterin is lost, chromosomes fuse with one another, and many abnormalities are seen. **C,** T-loop model for vertebrate telomeres. Chromosomal ends may form a T-loop structure when a single-stranded G-rich 3′ end of the chromosome "invades" a double-stranded portion of the telomere, base-pairing with one strand and displacing the other strand (D loop). *Inset,* A T loop excised together with its chromatin proteins from a chicken erythrocyte chromosome. (*Inset,* From Nikitina T, Woodcock CL. Closed chromatin loops at the ends of chromosomes. *J Cell Biol.* 2004;166:161–165.)

The Ku70/80 and MRN complexes are additional components of telomeres that are conserved from yeast to human. If mutations inactivate these complexes, telomeres frequently fuse together. This poses a conundrum, because elsewhere on chromosomes, these same proteins recognize DNA ends and participate in the repair of DNA breaks by joining bits of broken DNA together, a pathway known as **nonhomologous end joining (NHEJ)** (see Chapter 43). This is exactly the opposite of their role at telomeres. It thus appears that the breakage repair machinery recognizes chromosome ends, but the shelterin complex somehow changes its function from an end-joining role to an end-blocking protective role.

Loss of shelterin results in a loss of the G-strand overhangs and a dramatic increase in the tendency of chromosomes to fuse end to end. This is because the chromosome ends are now recognized as DNA breaks, and the cell attempts to repair them using several of the DNA repair pathways discussed in Chapter 43. Fig. 7.15 shows fused chromosomes in a *Drosophila* mutant lacking a protein essential for the assembly of the fly equivalent of the shelterin complex at telomeres. In organisms with shelterin, the end protection may occur in part because subunit TRF2 can promote the formation of a special looped configuration of DNA in which the single-stranded G-strand overhang is base-paired with "upstream" TTAGGG DNA (Fig. 7.14C).

Telomeres may also direct chromosome ends to their proper location within the cell. In budding yeast (and many other species), telomeres prefer to cluster together at the nuclear periphery. Mutants in telomere-binding proteins, or in regions of the histones with which they interact, disrupt this clustering in yeast. This results in activation of genes that are normally silenced when located in close proximity to telomeres. Thus, positioning of the telomere within the nucleus may be used to sequester genes into compartments where their transcriptional activity is repressed.

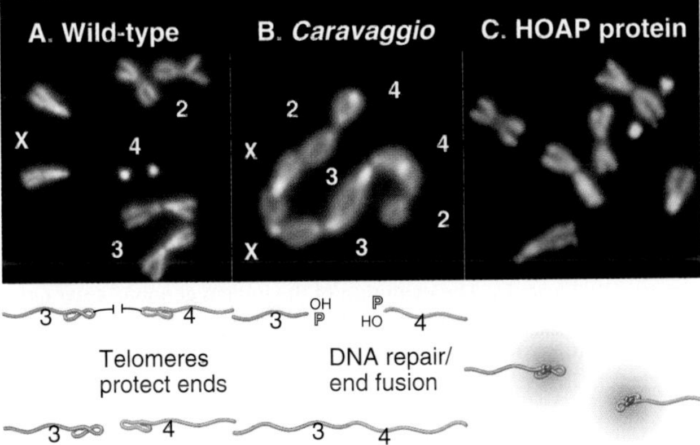

FIGURE 7.15 DISRUPTION OF THE PROTECTIVE COMPLEX AT TELOMERES RESULTS IN CHROMOSOME FUSIONS. A, The chromosomes of a wild-type female *Drosophila melanogaster* seen at mitotic metaphase (see Chapter 44). **B,** The *Caravaggio* mutant is characterized by a "train" of chromosomes generated by telomere-telomere fusions. (Caravaggio is the name of an Italian train.) **C,** The *cav* gene encodes HP1/Orc2-associated protein (HOAP), which specifically localizes at all *Drosophila* telomeres. (**A,** Courtesy Gianni Cenci and Maurizio Gatti, University of Rome, Italy. **B,** From Cenci G, Siriaco G, Raffa GD, et al. *Drosophila* HOAP protein is required for telomere capping. *Nat Cell Biol*. 2003;5:82–84. **C,** Courtesy *Nature Cell Biology*.)

Telomeres, Aging, and Cancer

Although the average length of telomeric repeats in humans is approximately 4000 bp, this length varies. Chromosomes of older individuals have shorter telomeres, and gametes have longer telomeres. This suggested the interesting possibility that chromosomes might lose telomeric sequences during the life of an individual.

The relationship between telomere length and aging can be studied in cultured cells. Normal cells in culture grow for only a limited number of generations (often called the Hayflick limit) before undergoing **senescence** (this involves permanent cessation of growth, enlargement in size, and expression of marker enzymes, such as β-galactosidase). Because normal somatic cells lack telomerase activity, their telomeres shorten and eventually reach a critically short threshold before the cells senesce. In some cases, it is possible to force senescent cells to resume proliferation (eg, by expressing certain viral oncogenes). These "driven" cells continue to divide and their telomeres continue to shorten until a **crisis** point is reached. In crisis, cells suffer chromosomal instability (chromosomal fusions and breaks can occur) and cell death. In populations of human cells in crisis, very rarely (in approximately 1 in 10^6 cases), cells appear that once again grow normally. These cells now express telomerase. These observations with cultured cells led to the suggestion that senescence might occur in cells when the telomeric repeats of one or more chromosomes are reduced to a critical level.

If correct, this model suggests very interesting (and controversial) implications for the regulation of cell life. Suppose that telomerase is active in the germline, so that all gametes have long telomeres. Now, if the enzyme were inactivated in somatic cells, this would effectively provide every cell lineage with a limitation on how many times it could divide before loss of telomeric sequences caused it to become senescent. Provided that the starting telomeres were sufficiently long and that telomerase was expressed in stem cells of tissues like testis and intestine, in which rapid division occurs throughout the life of the individual, this lack of telomerase in most cells would have no deleterious effect on the life span of the organism. In fact, such a mechanism might provide an important advantage by minimizing the chances that a clone of cells would escape from the normal regulation of growth control and become cancerous.

This model has been tested in two ways. First, mice were prepared in which the gene coding for the RNA component of telomerase or the telomerase reverse transcriptase was disrupted. These mice were healthy and fertile for six generations in the complete absence of telomerase but then subsequent generations became sterile as a result of cell death in the male germline. The cell death occurred when the telomeres shortened below a critical threshold. Having telomeres approximately seven times longer than humans might have contributed to their initial survival through several generations. Other studies show that mice age prematurely, when their telomeres shorten below a certain length. Remarkably, this ageing phenotype can be cured over the course of several weeks by activating hTERT in those mice. However, this "cure" can be a two-edged sword, as depending on the genetic makeup of the mice, the activation of hTERT can result in the formation of aggressive tumors! These experiments show that telomerase is not essential for the day-to-day life of the mouse, but clearly it is needed for the long-term survival of the species. In humans, a number of diseases (collectively termed "telomeropathies") are associated with inheritance of mutant alleles of telomere components. These diseases include dyskeratosis congenita (a complex condition affecting the skin and nails that is associated with a complex array of other life-threatening conditions), aplastic anemia (loss of blood cell formation), bone marrow failure and others. These diseases are all associated with failures in cell proliferation.

In a second experiment, the hTERT reverse transcriptase subunit of telomerase was introduced into normal cells growing in culture. This caused an increase in the level of active telomerase with dramatic results. Instead of undergoing senescence, these cells kept dividing in culture, apparently indefinitely (Fig. 7.16). However, unlike cancer cells, which are also immortal, these cells did not acquire the ability to cause tumors. Thus, this experiment showed convincingly that telomeres are part of a mechanism that regulates the proliferative capacity of somatic cells.

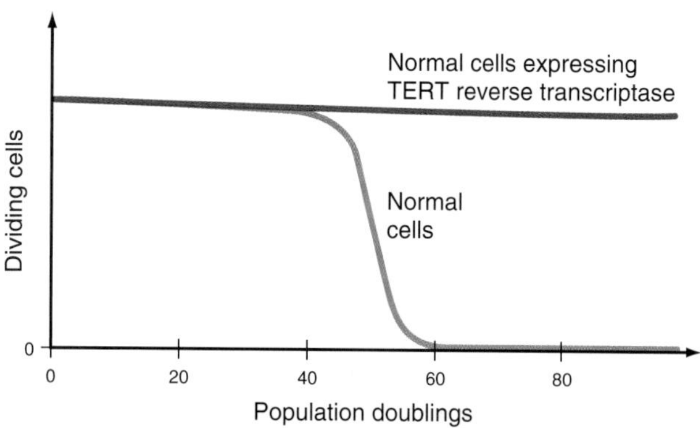

FIGURE 7.16 INTRODUCTION OF hTERT INTO NORMAL CELLS IS SUFFICIENT TO OVERCOME THE SENESCENCE LIMIT AND IMMORTALIZE THE CELLS. Following expression of hTERT, the human reverse transcriptase subunit of telomerase, cells act like normal cells (they are not transformed into cancer cells), but they can grow indefinitely. TERT, telomerase reverse transcriptase.

ACKNOWLEDGMENTS

We thank Beth Sullivan, Rachel O'Neill, Vladimir Larionov, Maurizio Gatti, and Lea Harrington for their advice during revision of this chapter.

SELECTED READINGS

Aitman TJ, Boone C, Churchill GA, et al. The future of model organisms in human disease research. *Nat Rev Genet.* 2011;12:575-582.

Beck CR, Garcia-Perez JL, Badge RM, Moran JV. LINE-1 elements in structural variation and disease. *Annu Rev Genomics Hum Genet.* 2011;12:187-215.

Birchler JA, Gao Z, Sharma A, Presting GG, Han F. Epigenetic aspects of centromere function in plants. *Curr Opin Plant Biol.* 2011;14: 217-222.

Bloom KS. Centromeric heterochromatin: the primordial segregation machine. *Annu Rev Genet.* 2014;48:457-484.

Doolittle RF. Microbial genomes opened up. *Nature.* 1998;392: 339-342.

Fukagawa T, Earnshaw WC. The centromere: chromatin foundation for the kinetochore machinery. *Dev Cell.* 2014;30:496-508.

Gent JI, Dawe RK. RNA as a structural and regulatory component of the centromere. *Annu Rev Genet.* 2012;46:443-453.

Heidenreich B, Rachakonda PS, Hemminki K, Kumar R. TERT promoter mutations in cancer development. *Curr Opin Genet Dev.* 2014;24: 30-37.

Huang CR, Burns KH, Boeke JD. Active transposition in genomes. *Annu Rev Genet.* 2012;46:651-675.

Martínez P, Blasco MA. Replicating through telomeres: a means to an end. *Trends Biochem Sci.* 2015;40:504-515.

Palm W, de Lange T. How shelterin protects mammalian telomeres. *Annu Rev Genet.* 2008;42:301-334.

Schueler MG, Sullivan BA. Structural and functional dynamics of human centromeric chromatin. *Annu Rev Genomics Hum Genet.* 2006;7: 301-313.

Simonti CN, Capra JA. The evolution of the human genome. *Curr Opin Genet Dev.* 2015;35:9-15.

Smit AF. Interspersed repeats and other mementos of transposable elements in mammalian genomes. *Curr Opin Genet Dev.* 1999;9: 657-663.

Stanley SE, Armanios M. The short and long telomere syndromes: paired paradigms for molecular medicine. *Curr Opin Genet Dev.* 2015;33:1-9.

Yan H, Jiang J. Rice as a model for centromere and heterochromatin research. *Chromosome Res.* 2007;15:77-84.

The 1000 Genomes Project Consortium. A global reference for human genetic variation. *Nature.* 2015;526:68-74.

CHAPTER 8

DNA Packaging in Chromatin and Chromosomes

Eukaryotic chromosomal DNA molecules are thousands of times longer than the diameter of the nucleus and therefore must be highly compacted throughout the cell cycle. This compaction is accomplished by combining the DNA with structural proteins to make **chromatin.** Chromatin folding must compact the DNA but still permit access of the transcriptional machinery to regions of the chromosome required for gene expression.

The first level of folding involves coiling DNA around a protein core to yield a **nucleosome.** The string of nucleosomes, known as a 10-nm fiber, shortens DNA approximately sevenfold relative to naked DNA. In some specialized cell types this is further condensed into a 30-nm fiber that shortens the DNA six- to sevenfold more. However, it appears that in most cells the further folding of the 10-nm fiber involves coils and looping, and is remarkably irregular and dynamic.

First Level of Chromosomal DNA Packaging: The Nucleosome

The continuous DNA fiber of each chromosome is packaged into many hundreds of thousands of nucleosomes linked in series. Individual nucleosomes can be isolated following cleavage of DNA between neighboring particles by DNA-cutting enzymes called nucleases. Random digestion of chromatin initially yields a mixture of particles consisting of short chains of nucleosomes containing multiples of approximately 200 base pairs of DNA (Fig. 8.1). Continued nuclease cleavage yields a stable particle with 146 base pairs of DNA (1.75 turns of the DNA wrapped around the protein core). This is called a **nucleosome core particle.**

The nucleosome core particle is disk-shaped, with DNA coiled in a left-handed superhelix around an octamer of **core histones.** This octamer consists of a central tetramer composed of two closely linked **H3:H4** heterodimers, flanked on either side by two **H2A:H2B** heterodimers. High-resolution crystal structures of nucleosome core particles revealed that each core histone has a compact domain of 70 to 100 amino acid residues that adopts a characteristic Z-shaped "histone fold" consisting of a long α-helix flanked by two shorter α-helices (Fig. 8.2).

The amino-terminal approximately 30 amino acid residues of the core histones (referred to as **N-terminal tails**) are important for interactions both inside and outside the nucleosome. They project outward from the cylindrical faces of the nucleosomal core as well as between the adjacent winds of the DNA on the nucleosome surface. Although these N-terminal tails are not ordered either in crystals of nucleosome core particles or in solution, they are among the most highly conserved regions of these very highly conserved proteins. This is because they serve as signaling platforms and mediate packing interactions between nucleosomes. Modifications of the N-terminal tails regulate DNA accessibility within the chromatin fiber to the transcription, replication, and repair machinery.

Chromatin Modifications and Regulation of Chromatin Function

The discovery that the sequence of bases in DNA provides a code to specify the primary structure of proteins triggered a revolution that culminated 50 years later with the near-complete sequencing of the human genome. To fully exploit this coding information, cells must control when to use it. Initial studies of the processes controlling **gene expression** focused on regulation of transcription by proteins that recognize specific DNA sequences at the 5′ end of genes (see Chapter 10), as this is how bacteria regulate gene expression. Eukaryotic gene regulation is much more elaborate.

Human nuclei contain roughly 3.3×10^7 nucleosomes distributed along the DNA. Although more than 70% of the molecular surface of nucleosomal DNA is accessible

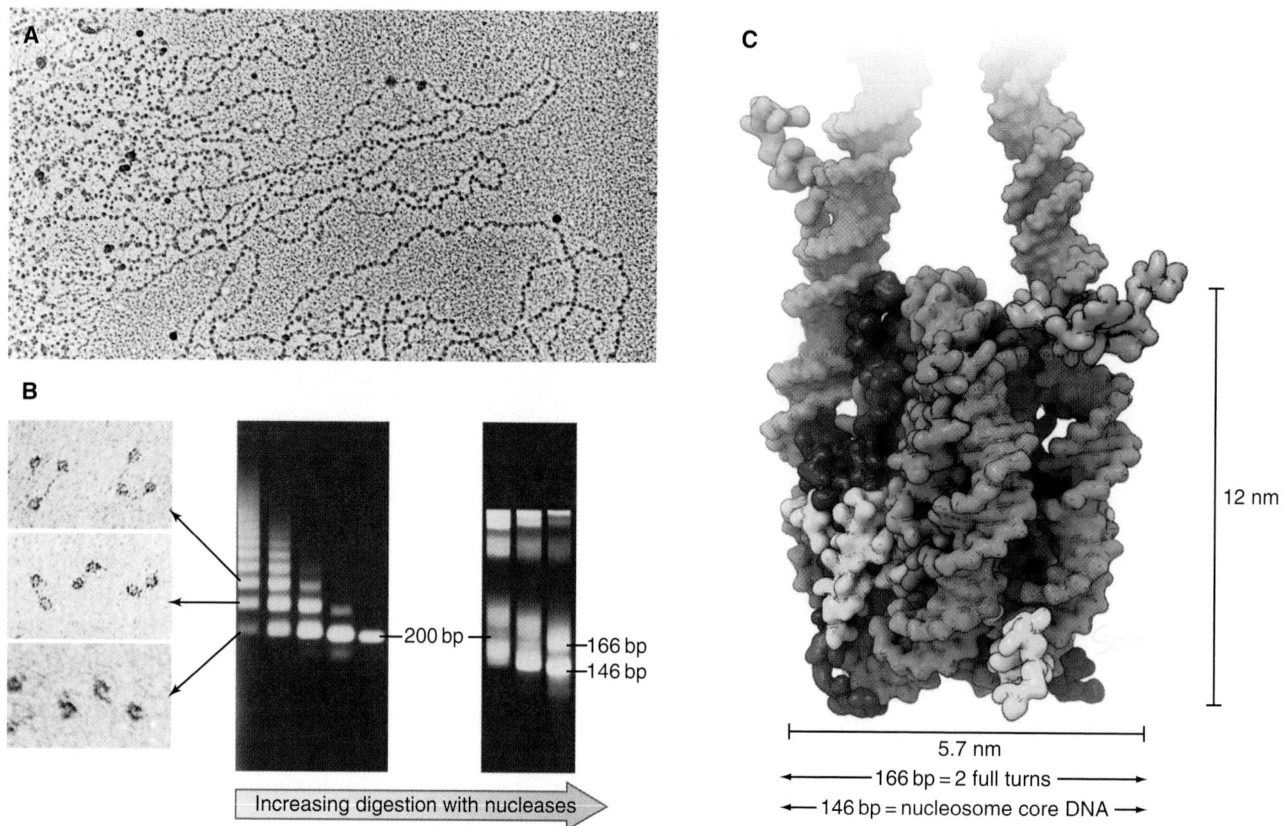

FIGURE 8.1 NUCLEOSOMES. A, Electron micrograph showing chromosomal loops covered in nucleosomes, which look like beads on a string. **B,** Nuclease digestion of chromosomes releases fragments containing varying numbers of nucleosomes *(left)* in which the DNA fragments vary by multiples of 200 base pairs *(center).* More extensive nuclease digestion results in production of the nucleosome core particle, with 146 base pairs of DNA *(right).* **C,** Crystal structure of a nucleosome core particle. The DNA wraps around a compact core of histones. (**A,** Courtesy William C. Earnshaw. **B,** *Left panel,* modified from Woodcock CL, Sweetman HE, Frado LL. Structural repeating units in chromatin. II: Their isolation and partial characterization. *Exp Cell Res*. 1976;97:111–119. **B,** *Center and right panels,* modified from Allan J, Cowling GJ, Harborne N, et al. Regulation of the higher-order structure of chromatin by histones H1 and H5. *J Cell Biol*. 1981;90:279–288. **C,** For reference, see Protein Data Bank [PDB; www.rcsb.org] file 1KX5.)

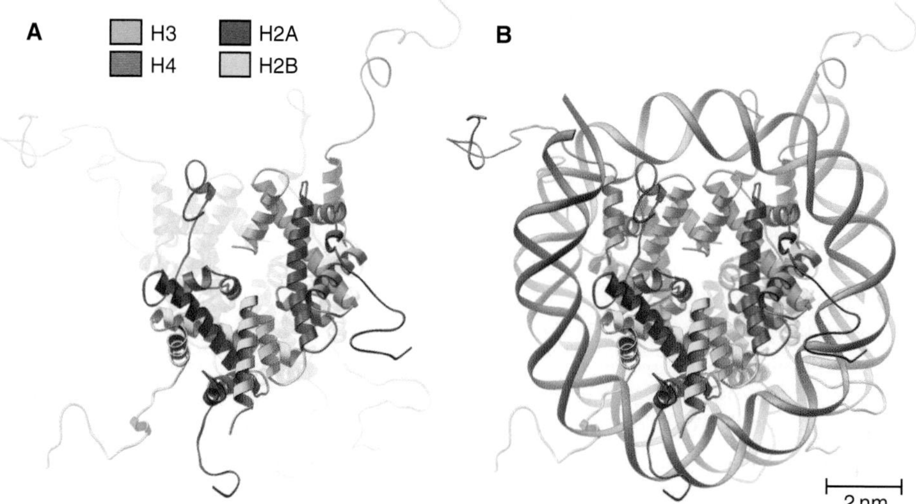

FIGURE 8.2 SECONDARY STRUCTURE OF THE HISTONES WITHIN THE CORE PARTICLE. A, A ribbon diagram shows that each histone protein in the octameric core of the nucleosome has a characteristic Z-shaped α-helical structure (the histone-fold). The flexible N-terminal portions of the histones, which have a critical role in regulating chromatin structure, did not occupy a unique location in the crystal and do not appear in this structure. **B,** The histone octamer surrounded by one of the two turns of DNA. (Modified from PDB file 1KX5 and Luger K, Mäder AW, Richmond RK, et al. Crystal structure of the nucleosome core particle at 2.8 Å resolution. *Nature*. 1997;389:251–260.)

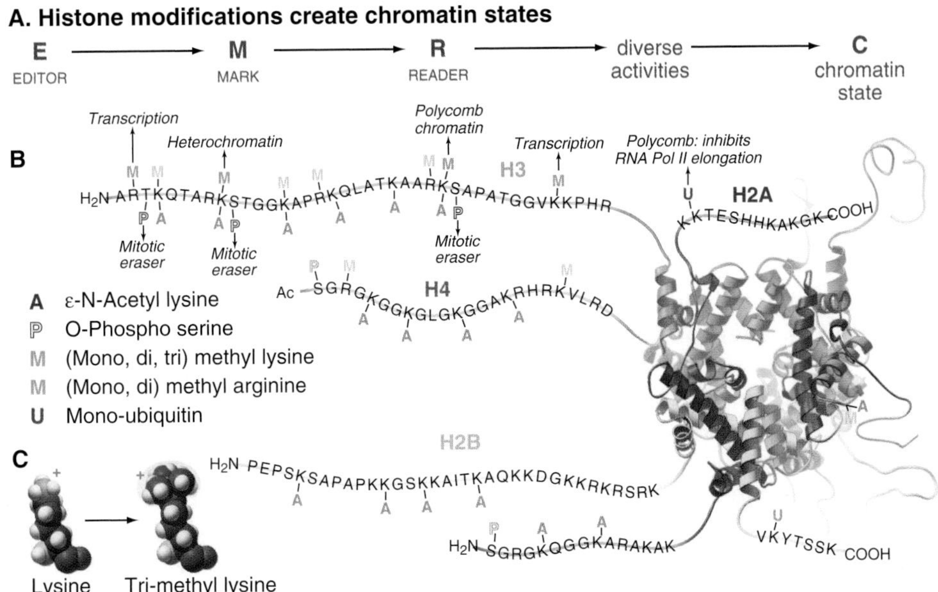

FIGURE 8.3 HISTONE MODIFICATIONS. A, E → M → R → C pathways use posttranslational modifications to create different chromatin states. **B,** Modification of the amino- and carboxyterminal domains of the histones regulates nucleosome assembly, transcription, and mitotic chromosome condensation. Highlighted here are methylations of three lysines, which are associated with transcription, heterochromatin, and facultative heterochromatin respectively. Note that each residue is immediately adjacent to a residue phosphorylated in mitosis, which knocks the READER off the methylation mark. The modifications are described in the figure key. **C,** Structure of tri-methyl lysine. For other structures of modified amino acids see Fig. 3.3. Arginine, R; lysine, K; serine, S. (Modified from PDB file 1KX5 and Khorasanizadeh S. The nucleosome: from genomic organization to genomic regulation. *Cell*. 2004;116:259–272.)

to solvent, most nonhistone proteins involved in gene regulation bind nucleosomal DNA 10- to 10^4-fold less well than naked DNA. Thus, nucleosomes establish a general environment in which DNA replication and gene transcription are repressed unless signals are given to the contrary.

The access of proteins to DNA in chromatin is regulated both by the density and specific localization of nucleosomes, and by specific modifications of the histones. The histones are acted on by enzymes we will call **EDITORS** (Fig. 8.3). EDITORS either place a **MARK** (a posttranslational modification) often, but not exclusively, on the histone N-terminal tail or remove an existing MARK. **READERS** then bind specifically to the MARK and recruit a variety of other activities. In some cases MARKS can act directly by influencing the charge properties of chromatin. The net result is the creation of a specific **CHROMATIN STATE**, of which two examples are: "open for transcription" and "inaccessible to transcription factors."

It has been proposed that the combination of MARKS on histones makes a kind of "code" that specifies the activity of various chromatin regions. This is disputed, however, as depending on context, individual MARKS can recruit different READERS with very different outcomes. Thus, if there is a code, it is far from simple and the significance of many of the histone MARKS remains to be deciphered. It has also been widely proposed that histone MARKS, together with methylation of the DNA itself are the basis of **epigenetic** regulation (see Fig. 7.11): the stable, heritable regulation of chromosomal functions by information that is not encoded in the DNA sequence. DNA methylation can be propagated through many cell divisions, but it is less clear that histone modifications are normally propagated in this way. Thus the role of histone modifications in epigenetic memory should be regarded as a popular hypothesis rather than an accepted fact.

Regulation of Chromatin Structure by the Histone N-Terminal Tails

The N-terminal histone tails provide a molecular "handle" to manipulate DNA accessibility in chromatin. This complex area can only be outlined here. A wide range of MARKS has been identified at many sites in the histone N-terminal tails and elsewhere (Fig. 8.3; see Cell SnapShot 1). These modifications include acetylation, methylation and ubiquitination of lysine residues, phosphorylation of serine and threonine, and poly(ADP) ribosylation. Histones with acetylated lysines are generally associated with "open" chromatin that is permissive for RNA transcription, while histones with methylated lysines can be associated with either "open" or "closed" chromatin states.

Because the histone modifications are read as combinations, individual modifications do not necessarily always have the same consequences. One example of this is the phosphorylation of histone H3 on serine 10 (H3-S10ph). In mitotic cells, this correlates with chromatin compaction, but when combined with acetylation of surrounding amino acid residues, it can also be associated with the activation of gene transcription as nonproliferating cells reenter the cell cycle (see Chapter 41). During mitosis, phosphorylation of threonine 3, serine 10, and serine 28 disrupts the binding of READERS to methylation MARKS on lysines 4, 9, and 28, respectively (Fig. 8.3B). Thus, one MARK can regulate the activity of an adjacent MARK.

Acetylation involves the transfer of acetate groups from acetyl coenzyme A to the ε-amino groups of lysine. This reduces the net positive charge of the N-terminal domain, causing chromatin to adopt an "open" conformation that is more favorable to transcription. The acetylation MARK acts as a binding site for protein READERS, one example of which is an approximately 100-amino-acid sequence motif called a bromodomain. Various bromodomain-containing READERS recruited to chromatin by acetylated histones often further modify histones in other ways that either promote or limit the accessibility of the DNA for transcription into RNA.

Proteins called transcription factors regulate gene expression by binding specific DNA sequences and recruiting the transcriptional machinery (RNA polymerases and associated proteins) to activate gene expression (see Fig. 10.12). Many transcription factors recruit a protein complex, called a **coactivator,** that facilitates loading of the transcriptional apparatus onto the gene. Often, coactivators possess domains that recognize histone MARKS and have EDITOR activities to lay down new MARKS on N-terminal histone tails. For example, the yeast SAGA complex contains over 10 proteins, including READERS that recognize histone methylation and acetylation. It also has an EDITOR activity that removes the protein MARK ubiquitin (see Fig. 23.2) from target proteins plus a histone acetyltransferase EDITOR activity that acetylates lysine-14 and lysine-8 in the N-terminal tails of histone H3 (Fig. 8.4).

Histone acetylation is dynamic. Just as transcriptional coactivators contain histone acetyltransferases that add acetyl groups to nucleosomes and promote gene activation, so *corepressors,* which are recruited in an analogous manner, can contain histone deacetylases that remove acetyl groups from selected lysine residues. Deacetylation tends to repress gene expression and is one strategy used to regulate cell-cycle progression during the G_1 phase of the cell cycle (see Fig. 41.9). Histone acetylation is crucial for life. Yeast cells die if certain key lysines are mutated to arginines, thus preserving their positive charge but preventing them from being acetylated.

In addition to marking nucleosomes by modification of their N-terminal tails, cells also use the energy provided by adenosine triphosphate (ATP) hydrolysis to actively remodel nucleosomes. This involves complex

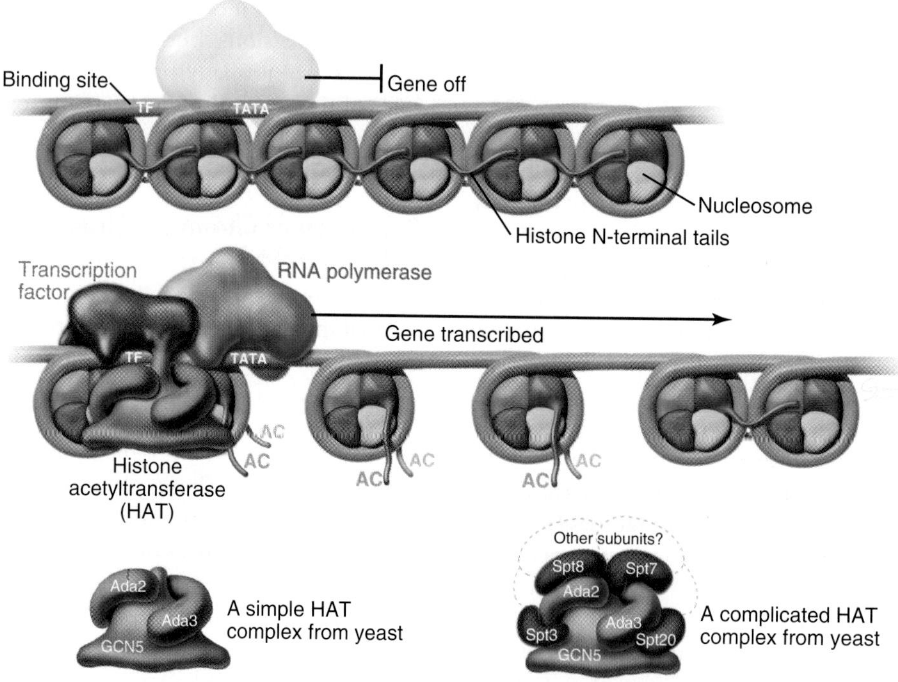

FIGURE 8.4 Transcription factors *(purple)* bind specific DNA sequences and recruit coactivators to the 5′ ends of genes. Many of these coactivators work by acetylating the N-terminal tails or body of the core histones, thereby loosening the chromatin structure and promoting the binding and activation of the RNA polymerase holoenzyme (see Chapter 10). The coactivators vary in composition and complexity from relatively simple histone acetyltransferase complexes *(bottom left)* to the huge and elaborate SAGA complex *(bottom right).* In this side view, only one of the two turns of DNA around the nucleosome is seen. GCN5, Ada2, Ada3, Spt3, Spt7, Spt8, and Spt20 are the names of budding yeast genes whose products are found in these complexes. AC, acetylation; TATA, DNA sequence in the gene promoter [see Chapter 10]).

protein "machines" that include a catalytic subunit that couples ATP hydrolysis to DNA translocation. All eukaryotes possess approximately 20 different classes of these chromatin remodeling enzymes. These different subclasses are capable of directing a range of different changes to nucleosome organization. For example, some enzymes reposition nucleosomes so that they are evenly spaced along DNA. Others remove histones from DNA. Still others direct replacement of core histone proteins with specialized variants.

Histone Deposition During Nucleosome Assembly

During DNA replication, existing nucleosomes are partitioned randomly between daughter DNA strands. Newly assembled nucleosomes then fill the gaps. When not associated with DNA, histones are always bound to protein chaperones. Newly translated H3 and H4, which are acetylated on lysine-9 of H3 and lysine-5 and lysine-12 of H4, associate with a chromatin assembly factor, called CAF1. One of the three subunits of CAF1 is a chaperone called retinoblastoma-associated protein of 48 kD (RbAp48). CAF1 is targeted to sites of DNA replication by interaction with proliferating cell nuclear antigen (PCNA), a doughnut-shaped protein that encircles the DNA and helps DNA polymerase slide along it during replication (see Fig. 42.12). Thus, CAF1 delivers newly synthesized histones to sites on the chromosome where new nucleosomes are required as DNA is synthesized during the S phase of the cell cycle (see Chapter 42). H3 and H4 are deposited first on the new DNA, followed by two H2A:H2B heterodimers to complete the assembly of the nascent nucleosome.

Histone Variants

Approximately 75% of histone H3 in chromatin is deposited during DNA replication by CAF1. The remaining 25% is a special isoform of H3, called H3.3, that is encoded by a different gene and deposited on chromatin by a different mechanism. Histone H3.3 is transcribed throughout the cell cycle and is not coordinated with DNA synthesis. Newly synthesized H3.3 binds to the RbAp48 chaperone, but they then associate with a protein called histone regulator A (HIRA) instead of the two CAF1 subunits. Some H3.3 assembles into nucleosomes at the time of DNA replication, just like the canonical H3. However, H3.3 can also be inserted into chromatin at other times of the cell cycle. For example, the HIRA-RbAp48 complex swaps H3.3/H4 dimers for H3/H4 dimers in chromatin during transcription, when the nucleosomes on the underlying gene are transiently perturbed. Although demethylases can remove the methyl groups from histone H3, replacement of histone H3 methylated on lysine 9 (H3-K9me) with unmethylated H3.3 is an efficient way to convert "closed" chromatin, where transcription is disfavored, into "open" chromatin that is favorable for transcription.

Other specialized histone variants also contribute to the microdiversity of chromatin. For example, the H3 isoform CENP-A is a key component of the kinetochore, the structure that assembles at centromeres to promote chromosome segregation during mitosis (see Fig. 8.21 below).

The largest number of variant forms has been described for H2A. Interaction between the N-terminus of H4 and an acidic patch on the surface of H2A on the adjacent nucleosome has an important role in promoting chromatin fiber compaction. Therefore, altering the local H2A composition, which influences the strength of this interaction, provides another way to vary the accessibility of the DNA for gene expression. One variant, H2AX, which constitutes approximately 15% of the cellular H2A, helps maintain genome integrity. At sites of DNA damage, H2AX is rapidly phosphorylated by protein kinases. This serves as a MARK for the assembly of multiprotein complexes that signal and repair the damage (see Box 43.1).

Linker DNA and the Linker Histone H1

When examined by electron microscopy at low ionic strength, nucleosomal chromatin resembles a string of 10 nm diameter beads with **linker DNA** extended between adjacent nucleosomes (Fig. 8.1). Each nucleosome in chromosomes is typically associated with approximately 200 base pairs of DNA. Subtracting 166 base pairs for two turns around the histone octamer leaves 34 base pairs of linker DNA between adjacent nucleosomes. Linker DNA can vary widely in length in different tissues and cell types.

A fifth histone, **H1** or **linker histone,** binds to linker DNA where the DNA molecule enters and exits the nucleosome (Fig. 8.5). H1 histones have a "winged helix" central domain flanked by unstructured basic domains at both the N- and C-termini (Fig. 8.5). Mammals have at least eight variant forms (called subtypes) of H1 histones ($H1_{a\text{-}e}$, $H1^0$, $H1_t$, and $H1_{oo}$). The amino acid sequences of these variants differ by 40% or more. $H1^0$ is found in cells entering the nondividing G_0 state (see Chapter 41), whereas $H1_t$ and $H1_{oo}$ are found exclusively in developing sperm and oocytes, respectively.

The role of H1 linker histone in chromatin remains enigmatic. The protein was originally assumed to regulate chromatin compaction, yet it is mobile in the nucleus, spending no more than a few minutes at any given location. Deletion of the sole linker histone genes from yeast and *Tetrahymena* (a ciliated protozoan) causes no obvious ill effects, but H1 is essential in mice. Although genes that encode individual H1 isoforms can be deleted in mice, simultaneous deletion of the genes for three isoforms causes embryonic death, apparently the consequence of alterations in chromatin structure that perturb normal patterns of gene expression.

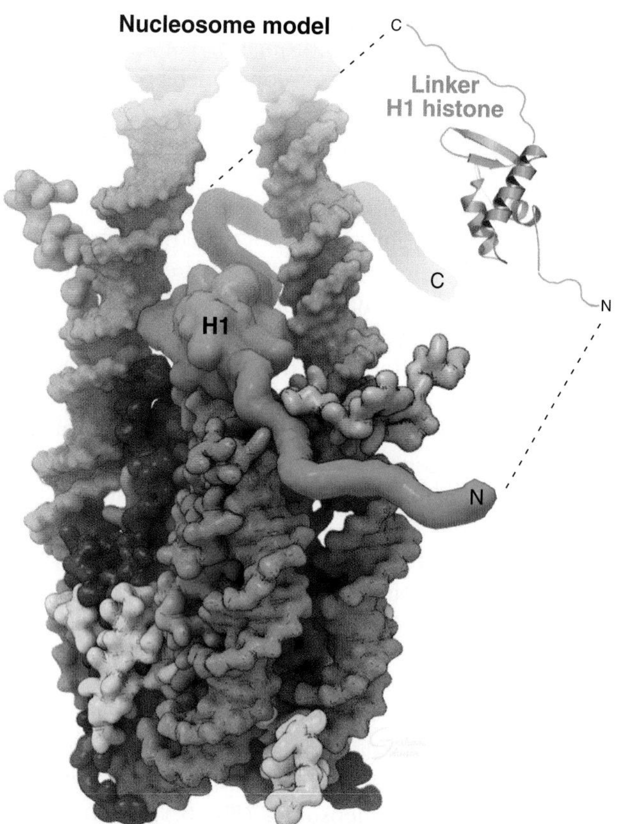

FIGURE 8.5 THE BINDING SITE OF HISTONE H1 ON THE NUCLEOSOME, NEAR THE SITE WHERE DNA STRANDS ENTER AND EXIT THE CORE PARTICLE. *Orange,* DNA; *blue,* H3; *purple,* H4; *red,* H2A; *yellow,* H2B. (For reference, see PDB files 1KX5 and 1HST.)

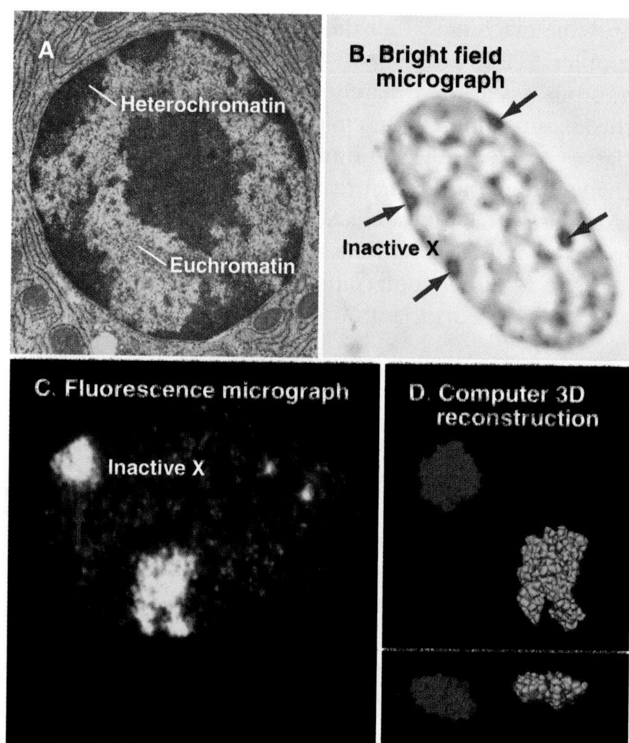

FIGURE 8.6 EUCHROMATIN AND HETEROCHROMATIN. A, Electron micrograph of a thin section of a plasma cell nucleus. Euchromatin is decondensed. Heterochromatin (mostly clumped near the nuclear envelope and central nucleolus) remains condensed. **B,** Light micrograph of a female nucleus with four Barr bodies *(arrows)* (facultative heterochromatin composed of the inactive X chromosome). This woman has a highly abnormal genetic makeup, with five X chromosomes. The X chromosome inactivation system has a built-in counting mechanism that ensures that only one X chromosome remains active. **C–D,** The Barr body is structurally distinct from the active X chromosome. This figure is from a three-dimensional study in which the X chromosome was identified by in situ hybridization "painting" with probes that covered the entire chromosome. **C,** One slice through the three-dimensional data set. **D,** Two different views of the X chromosomes reconstructed in three dimensions. The inactive X chromosome is shown in *red*. Because of X chromosome inactivation, females are mosaic for functions encoded on the X chromosome. Each female embryo has two X chromosomes: X^{pat} and X^{mat} (for paternal and maternal). Following X chromosome inactivation, some cells will express genes from X^{pat} and others will express genes from X^{mat}. The inactivation is permanent; eg, all progeny of a cell with X^{pat} inactivated will also have X^{pat} inactivated. This inactivation occurs randomly in different cells of the embryo. In cats, genes responsible for coat color are encoded on the X chromosome. The patchy color pattern of calico cats reflects the underlying pattern of X chromosome inactivation. All classic calico cats are females. (**A,** From Fawcett DW. *The Cell.* Philadelphia: WB Saunders; 1981. **B,** Courtesy Barbara Hamkalo, University of California, Irvine. **C–D,** From Eils R, Dietzel S, Bertin E, et al. Three-dimensional reconstruction of painted human interphase chromosomes: active and inactive X chromosome territories have similar volumes but differ in shape and surface structure. *J Cell Biol.* 1996;135:1427–1440.)

Functional Compartmentation of Chromatin: Heterochromatin and Euchromatin

Chromatin has traditionally been categorized into two main classes based on structural and functional criteria. **Euchromatin** contains almost all genes, both actively transcribed and quiescent. **Heterochromatin** is transcriptionally repressed and is generally more condensed than euchromatin (Fig. 8.6). Heterochromatin was initially recognized because it stains more darkly with DNA-binding dyes than the remainder of the interphase nucleus.

More recent analyses based on mapping patterns of the modifications of the histone N-terminal tails now suggest that there are at least five classes of chromatin environments in nuclei. These classes, defined somewhat arbitrarily and given colors for names, are green (classic heterochromatin with HP1; described later), yellow (active chromatin rich in H3-K4me3), red (active chromatin rich in histone remodelers), blue (facultative heterochromatin repressed by polycomb proteins), and black (repressed, but not via HP1). Another approach has identified chromatin regions associated with the inner surface of the nuclear envelope. These lamina-associated domains tend to average approximately 10^6 base pairs in size and are mostly transcriptionally inactive.

A typical nucleus has both euchromatin and heterochromatin, the latter often being concentrated near the nuclear envelope and around nucleoli. Much of the nuclear interior is occupied by pale-staining euchromatin rich in actively transcribing genes. Nuclei with low transcriptional activity have relatively more heterochromatin. Classically two types of heterochromatin, constitutive and facultative, have been recognized.

Constitutive heterochromatin is typically associated with repetitive DNA sequences, such as satellite DNAs (see Fig. 7.9), that are packaged into "closed" (green) chromatin in every cell type. In at least some cases, establishment of constitutive heterochromatin involves transcription of those repeated DNA elements to produce double-stranded RNAs that are cleaved into short fragments by the RNA interference (RNAi) machinery (see Fig. 11.13). The resulting short RNAs are thought to target activities that promote heterochromatin formation to their sites of transcription in the chromosome (see Fig. 11.14).

The MARK that best defines constitutive heterochromatin is H3-K9me3 (Fig. 8.7), which is recognized by the READER **heterochromatin protein 1 (HP1)**. The HP1 amino-terminus contains a 50-amino-acid motif called a **chromodomain** (*chro*matin *mo*dification organizer) that binds to H3-K9me3. Surprisingly, most HP1 is highly mobile in nuclei, moving on a time frame of seconds. It also binds and recruits enzymes that tri-methylate histone H3 lysine 9.

HP1 can promote the lateral spreading of heterochromatin along the chromosome by recruiting other proteins that further modify the histone aminoterminal tails (Fig. 8.7). For example, the enzyme that trimethylates H3-K9 itself binds to HP1 and can modify adjacent nucleosomes. As a result, heterochromatin is not a static "closed" chromatin compartment but can "invade" nearby genes along the chromosome. If a chromosomal rearrangement moves an actively transcribed gene close to constitutive heterochromatin, heterochromatin may spread across it and repress transcription (Fig. 8.7). This is called **position effect.**

HP1 and other repressive proteins can also recruit DNA methyltransferases that modify the underlying DNA by adding a methyl group to the 5′ position on cytosine in the dinucleotide CpG (cytosine phosphate guanine). Methylation can recruit READERS that inactivate gene transcription if it occurs near the 5′ promoters of genes (see Chapter 10) in regions with an above average concentration of CpG called **CpG islands**. Among the several binding proteins that recognize DNA containing 5-methyl-cytosine, *me*thyl-cytosine binding *p*rotein (MeCP2) can repress expression of nearby genes by recruiting a histone deacetylase complex that removes acetyl groups from the core histone N-terminal tails (Fig. 8.7). MeCP2 is highly abundant in neurons, and mutations in the protein cause Rett syndrome, an X-linked

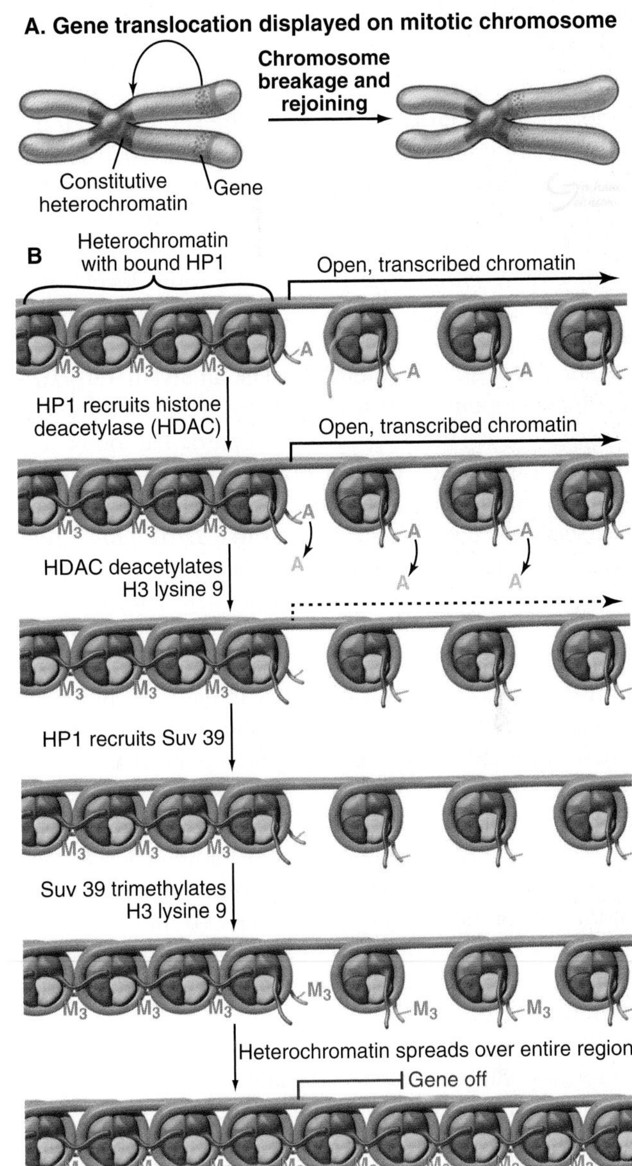

FIGURE 8.7 POSITION EFFECT AND THE SPREADING OF HETEROCHROMATIN. A, If a transcriptionally active gene is moved next to a region of heterochromatin, it may be repressed as the heterochromatin spreads. The relative position of the gene is shown on mitotic chromosomes as it would be determined by in situ hybridization (Fig. 8.10). **B,** Diagrammatic representation of stages in the spreading of heterochromatin and silencing of the active gene: removal of acetyl groups from the histones by a histone deacetylase; addition of two or three methyl groups to lysine 9 of H3; and binding of HP1, which recruits a DNA methyltransferase plus other heterochromatin proteins to create heterochromatin. A, acetylation; Me, methylation.

neurodevelopmental disorder in which female infants develop apparently normally for 6 to 18 months, but then regress, losing language and adopting stereotypical postures and movements. How the MeCP2 defects lead to Rett syndrome is not known.

Facultative heterochromatin consists of sequences that are in heterochromatin in some cell types and in euchromatin in others. X chromosome inactivation is the classic example of facultative heterochromatin in mammals. In females, one X chromosome in each cell (selected at random) is inactivated early in development prior to implantation of the embryo. The inactivated X chromosome forms a discrete patch of heterochromatin at the nuclear periphery known as the **Barr body** (Fig. 8.6). Because most genes carried on the inactivated X chromosome become transcriptionally silent, females with two X chromosomes have the same levels of X chromosome-linked gene expression as males with a single X chromosome.

Polycomb group proteins form facultative heterochromatin by modifying histones. They were identified in *Drosophila* as mutants in which particular body segments "forgot" their identity during development due to reactivation of the expression of several homeodomain transcription factors (see Fig. 10.14). *Drosophila* polycomb chromatin apparently locks genes that have been switched off in a silent epigenetic state that is stable through many generations of cell division.

Two PRCs (*p*olycomb *r*epressive *c*omplexes) regulate transcription. The PRC2 complex initiates silencing by tri-methylating histone H3 on lysine 27 (H3-K27me3). Then chromodomain-containing members of the PRC1 complex bind specifically to H3-K27me3 (note the difference from the HP1 chromodomain, which READS H3-K9me3). PRC1 contains an E3 ubiquitin ligase (see Fig. 23.3) that transfers a single ubiquitin molecule to lysine 119 of histone H2A (H2A-K119ub). PRC1 binding also causes nucleosomes to form dense clumps that are resistant to remodeling and "opening" by ATP-dependent remodeling "machines."

Polycomb group proteins also function in X chromosome inactivation, in stem cell maintenance, and possibly in cancer stem cells. In mammals, the *inactive* X chromosome expresses a large (15 kb) noncoding RNA called XIST that associates with and "coats" the inactive X chromosome. Next, the PRC2 complex associates with the inactive X, transiently recruiting PRC1, which produces H2A-K119ub. This, together with low levels of histone acetylation, enrichment for the H2A variant macroH2A, and high levels of CpG methylation in many CpG islands combines to inhibit transcription of most genes.

Several polycomb group proteins are required for the self-renewal of blood and neural stem cells, and may also perform a similar function in cancer stem cells. In stem cells, polycomb group proteins regulate transcription of factors that control the cyclin-dependent kinases that drive cell-cycle progression (see Chapter 40). They may also participate in the DNA damage response (see Chapter 43).

Imprinting: A Specialized Type of Gene Silencing

The factors that produce heterochromatin are also involved in a very specific type of gene silencing known as **imprinting.** An imprinted gene is stably turned off during formation of the egg or sperm. For example, if the maternal copy of a gene is imprinted, then expression can come only from the corresponding homologous chromosome contributed by the father. Currently, approximately 80 imprinted genes are known.

One well-studied imprinting system involves the genes for insulin-like growth factor-2 (IGF2) and a noncoding RNA H19 in the mouse (Fig. 8.8). The DNA between these genes has an insulator element with binding sites for the *CCTC*-binding *f*actor CTCF. Binding of CTCF to the insulator differs, depending on whether the chromosome is derived from the egg or sperm. On the maternal chromosome, it allows the H19 gene to be expressed but turns off the IGF2 gene by preventing access to a transcriptional enhancer. On chromosomes derived from the sperm, methylation of CpG sequences in the control region stops CTCF from binding. As a result, the paternal copy of IGF2 has access to its enhancer and is expressed, but the H19 gene is not expressed. This simple switch ensures that the offspring expresses only the paternal copy of the IGF2 gene and the maternal copy of H19.

Higher-Order Structure of Chromosomes

Higher Levels of Chromosomal DNA Packaging in Interphase Nuclei

Levels of chromatin structure beyond the nucleosome are poorly understood. This lack of clarity arises in part because dense packing of macromolecules in the nucleus makes it difficult to observe the details of higher-level folding of chromatin fibers directly. For more than 35 years the accepted dogma was that the next level of chromatin compaction beyond the 10-nm fiber was a solenoidal **30-nm fiber**. Recent results, primarily using cryoelectron microscopy, now strongly question the existence of the 30-nm fiber in vivo in most cells.

Visualization of specific DNA loci within fixed interphase nuclei by in situ hybridization (introduced in Fig. 8.10) can be used to estimate the degree of chromatin compaction by comparing the physical distance between two DNA sequences with a known number of base pairs between them. For regions of DNA up to approximately 250,000 base pairs apart, the chromatin fiber is shortened approximately 80- to 100-fold. When sequences are separated by tens of millions of base pairs, the shortening increases by another 20- to 30-fold. This suggests at least two levels of chromatin folding beyond the 10-nm fiber.

The organization of chromatin fibers can be observed by superresolution fluorescence microscopy of living cells after labeling with a fluorescent marker, such as the jellyfish green fluorescent protein (GFP [see Fig. 6.3]) (Fig. 8.9). Individual nucleosomes are locally dynamic, changing their packing and locations as cells traverse the

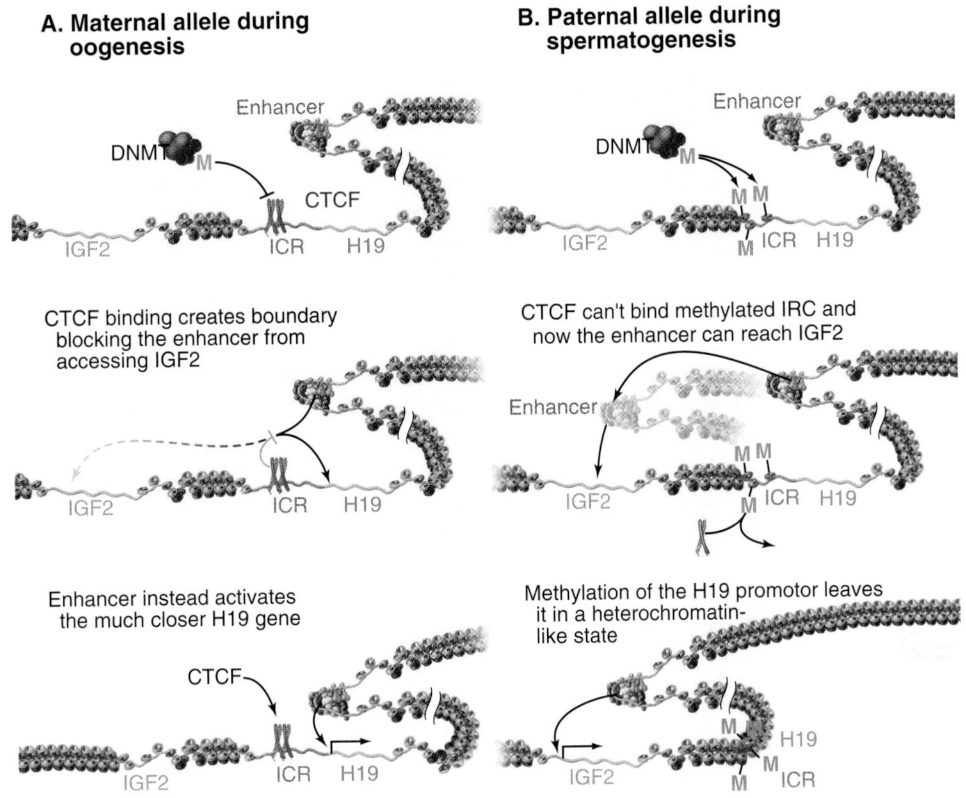

FIGURE 8.8 IMPRINTING OF THE INSULIN-LIKE GROWTH FACTOR-2 AND H19 LOCI. A, During oogenesis, CTCF binding to the imprinting control region (ICR) prevents methylation of the DNA. In the zygote, this methylated chromosome from the mother is bound by CTCF, which acts as an insulator, blocking the IGF2 (insulin-like growth factor-2) gene from gaining access to its enhancer. As a result, the maternal chromosome expresses H19 and not IGF2. **B,** During spermatogenesis, the ICR is methylated. In the zygote, the ICR on the chromosome derived from the father cannot bind CTCF. As a result, the IGF2 gene gains access to its enhancer and is expressed. The H19 gene is off.

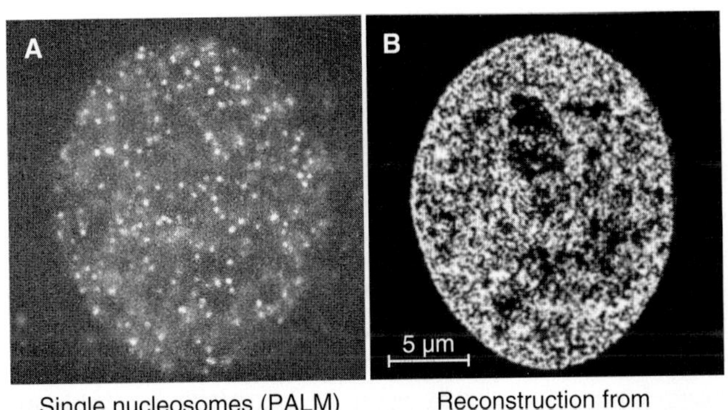

FIGURE 8.9 SUPERRESOLUTION VISUALIZATION OF NUCLEOSOMES IN A LIVING HUMAN CELL. This experiment uses a clone of HeLa (Henrietta Lacks) cells expressing photoactivatable red fluorescent protein (mCherry) linked to histone H2B. This form of red fluorescent protein usually has almost no fluorescence, but some fractions become highly fluorescent by spontaneous activation. **A,** When imaged by photoactivated localization microscopy (PALM) ultraviolet laser microbeam, individual nucleosomes can be observed. **B,** When an image corresponding to approximately 100,000 nucleosomes is reconstructed, the nucleus is seen to be organized into chromatin domains. (Courtesy K. Maeshima, National Institute of Genetics, Japan.)

cell cycle. Some electron microscopy studies observed a fiber, 100 to 300 nm in diameter, which was called a chromonema fiber. In most studies, however, the chromatin appears to be relatively disordered.

Together all these analyses are leading to a view of interphase chromatin as composed of irregular 10-nm chromatin fibers that are organized in dynamic loops. The 30-nm fibers and chromonema filaments may occur only under specialized conditions.

Large-Scale Structural Compartmentation of the Nucleus

Although interphase nuclei lack a high degree of order, a number of general organizational principles are recognized. First, individual chromosomes tend to concentrate within discrete territories and intermingle with one another only to a limited extent. This is seen most clearly in human somatic cell nuclei when individual chromosomes are visualized by a special type of in situ hybridization called chromosome painting (Fig. 8.10). The volume of territories occupied by individual chromosomes correlates with the proportion of actively transcribing genes. In some cases, active genes are

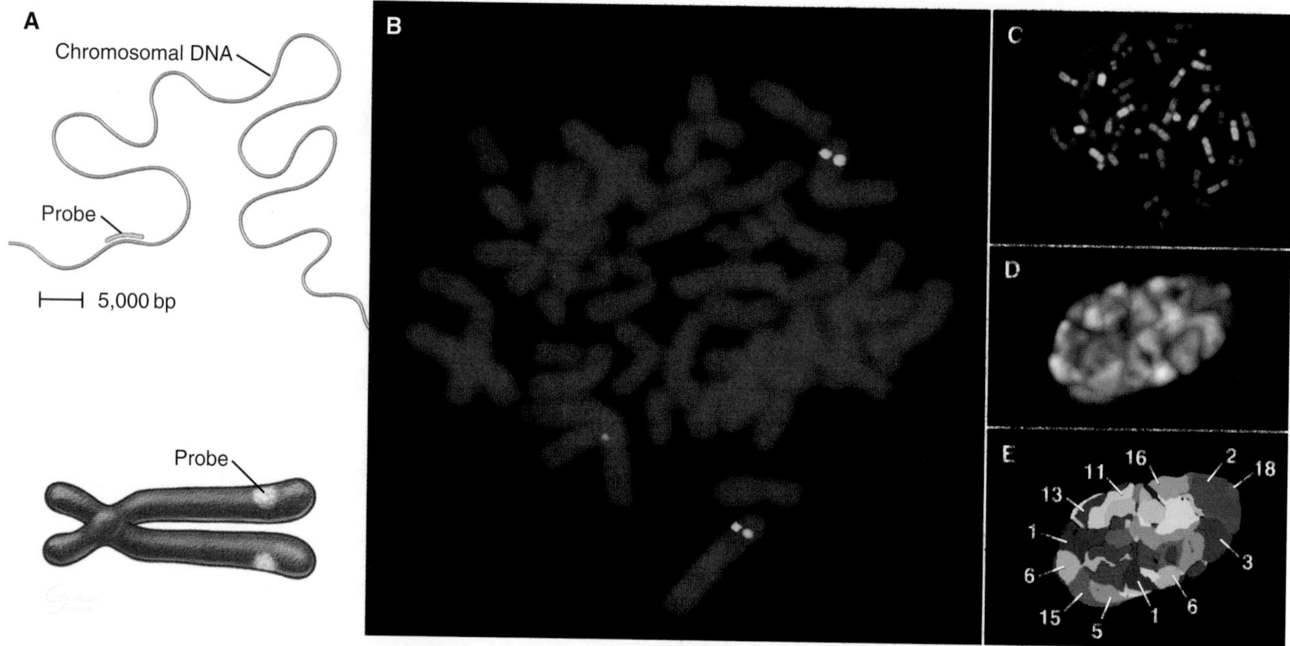

FIGURE 8.10 FLUORESCENCE IN SITU HYBRIDIZATION REVEALS THAT CHROMOSOMES OCCUPY DISCRETE TERRITORIES IN INTERPHASE NUCLEI. A, Chromosomes are spread on a slide as in Fig. 8.15. Following chemical fixation steps to preserve the chromosomal structure, the chromosomal proteins are removed by digestion with proteases and the genomic DNA strands are melted (separated) by heating. Next, a "probe DNA" *(yellow)* is added. This probe DNA is single-stranded so that it can base-pair (hybridize) to its complementary sequences in the chromosome. The probe DNA is chemically labeled with biotin. Next, the sites of hybridization on the chromosomes are detected with fluorescently labeled avidin, a protein from egg white that binds to biotin with extremely high affinity. The sites of avidin-binding appear *yellow,* whereas the remainder of the chromosomal DNA is counterstained with a *red* dye. **B,** The micrograph shows fluorescence in situ hybridization (FISH) analysis using a probe from near the von Hippel–Lindau locus on chromosome 3. **C,** Metaphase chromosome labeled by FISH using chromosome paint probes (probes distributed all along the chromosome, excluding repetitive DNA). In this 24-color FISH image, every chromosome is marked with two or three fluorochromes (true color image). **D,** The same combinatorial probe was used in 24-color FISH on a fibroblast nucleus under conditions preserving the 3D architecture. Every chromosome forms distinct chromosome territory. **E,** Every chromosome territory of the same nuclear optical section as on **B** was identified and false-colored after classification. (**B,** Courtesy Jeanne Lawrence, University of Massachusetts, Worcester. **C–E,** Courtesy I. Solovei, A. Bolzer, and T. Cremer, University of Munich, LMU, Germany.)

located well outside of the territories, as though their activation involved looping out a much larger domain from the remainder of the chromosome. These movements during gene activation may involve relocation from compartments where transcription is relatively infrequent to compartments where transcription is favored (Fig. 8.11C–D).

Silent chromatin tends to be concentrated near the nuclear periphery in a wide range of cell types. This supports the hypothesis that particular chromosomal regions (eg lamina-associated domains near the nuclear lamina; see Chapter 9) might have preferred locations within the nucleus. As a result, chromosomes that are rich in actively transcribed genes tend to be localized toward the nuclear interior, while chromosomes with a lower gene content tend to be found near the nuclear periphery (Fig. 8.11A–B).

These positions of chromosomes are mostly established by where chromosomes are located during the exit from the previous mitosis. Most movements of the chromatin during interphase are of 0.5 μm or less. These movements likely occur within topologically associating domains (TADs) (see the next section), while the chromosomes overall remain relatively stationary.

Special Interphase Chromosomes With Clearly Resolved Loop Structures

Studies of specialized chromosomes from organisms ranging from flies to mammals originally revealed a link between chromatin loops and regulated gene expression. Loops are clearly seen in **lampbrush chromosomes** during meiotic prophase in oocytes of many species (Fig. 8.12A). These loops are sites of intense transcriptional activity as oocytes stockpile huge stores of the components needed for rapid cell divisions during early development of the fertilized egg. The loops are easily seen because the DNA is coated with many RNA transcripts, together with proteins that package and process them.

Similar loops are present in the giant **polytene chromosomes** found in some tissues of *Drosophila* larvae. Each polytene chromosome consists of more than 1000 identical DNA molecules packed side-by-side in precise linear register. Polytene chromosomes have a complex

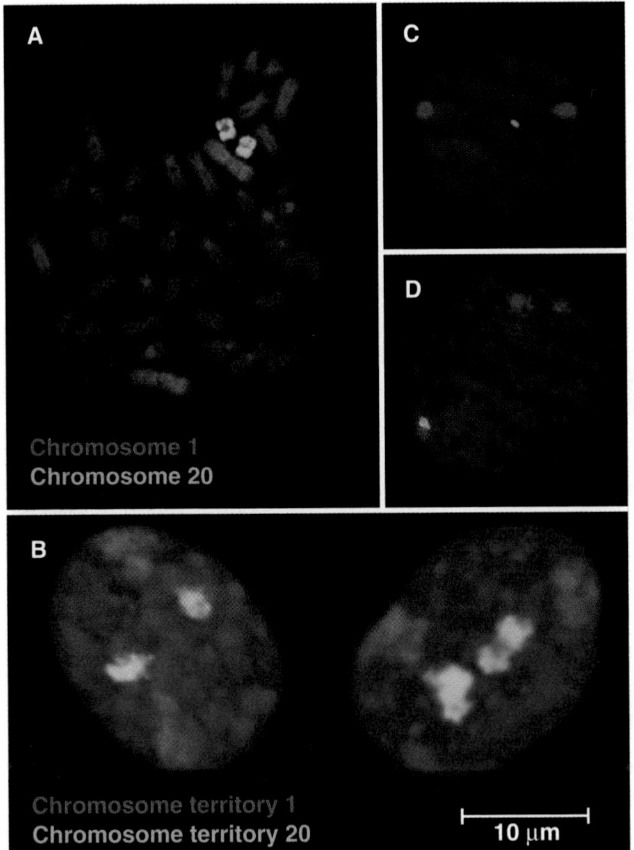

FIGURE 8.11 CHROMOSOME POSITION IN THE NUCLEUS CORRELATES WITH TRANSCRIPTIONAL ACTIVITY. A, Metaphase chromosome spread from a healthy donor with painted chromosomes 1 *(red)* and 20 *(green)*. **B,** The same paint probes were used in fluorescence in situ hybridization (FISH) experiments on three-dimensionally preserved fibroblast nuclei (3D-FISH): they revealed two pairs of chromosome territories. Note the more central positioning of chromosome 20 territories and the more peripheral positioning of chromosome 1 territories. **C–D,** The CD4 gene *(green)* is located in the nucleoplasm in cells where it is expressed **(C)** but is associated with centromeric heterochromatin in cells where it is silent **(D).** (**A–B,** Courtesy I. Solovei, A. Bolzer, and T. Cremer, University of Munich, LMU, Germany. **C–D,** From Lamond AI, Earnshaw WC. Structure and function in the nucleus. *Science*. 1998;280:547–553; and Brown KE, Guest SS, Smale ST, et al. Association of transcriptionally silent genes with Ikaros complexes at centromeric heterochromatin. *Cell*. 1997;91[6]:845–854.)

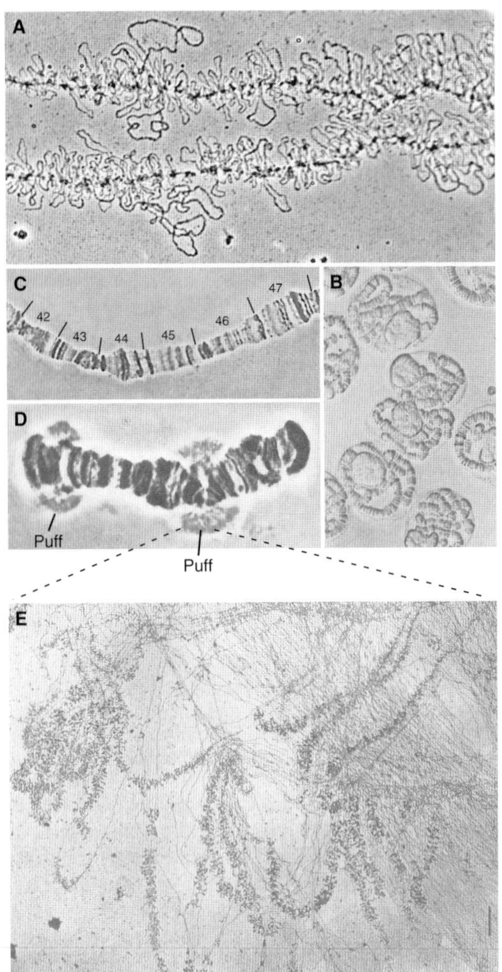

FIGURE 8.12 CHROMATIN LOOPS IN SPECIAL INTERPHASE CHROMOSOMES. A, Phase contrast view of the left end of meiotic lamp brush chromosome 6 from the newt *Notophthalmus viridescens*. **B–D,** Domain organization of polytene chromosomes. Once *Drosophila* larvae reach a certain developmental stage, most cells stop dividing, and larval growth proceeds via an increase in the size of individual cells. To keep the protein synthesis machinery of these huge cells supplied with messenger RNA, DNA replication is uncoupled from cell division so that ultimately, the cells contain many times the normal complement of cellular DNA (ie, they are *polyploid*). In certain tissues, the numerous copies of the chromosomes are maintained in strict alignment with respect to one another, making giant *polytene* chromosomes, the best known of which occur in the salivary gland. **B,** Giant polytene chromosomes are visible within isolated salivary gland nuclei. **C,** A portion of a high-resolution map of the *Drosophila* polytene chromosomes. **D,** Polytene chromosome showing puffs. The *inset box* shows an area analogous to that used in *panel E*. **E,** Electron micrograph of puff showing transcribing DNA loops. These loops are covered with a "fuzz" corresponding to growing RNA chains coated with proteins. (**A,** From Roth MB, Gall JG. Monoclonal antibodies that recognize transcription unit proteins on newt lampbrush chromosomes. *J Cell Biol*. 1987;105:1047–1054. **B,** From Robert M. Isolation and manipulation of salivary gland nuclei and chromosomes. *Methods Cell Biol*. 1975;9:377–390. **C,** Courtesy Margarete Heck, University of Edinburgh, United Kingdom. **D,** From Andersson K, Mahr R, Bjorkroth B, et al. Rapid reformation of the thick chromosome fiber upon completion of RNA synthesis at the Balbiani ring genes in *Chironomus tentans*. *Chromosoma*. 1982;87:33–48. **E,** From Lamb MM, Daneholt B. Characterization of active transcription units in Balbiani rings of *Chironomus tentans*. *Cell*. 1979;17:835–848.)

pattern of thousands of bands (Fig. 8.12B–D). Stress or stimulation of gene expression by hormones causes certain bands to lose their compact shape and puff out laterally. Each puff is composed of hundreds of identical, actively transcribed chromatin loop domains (Fig. 8.12E).

Chromatin Conformation Capture and Topologically Associating Domains

Powerful insights into the organization of chromatin fibers in somatic cell nuclei have followed from the development of a technique called 3C (chromosome

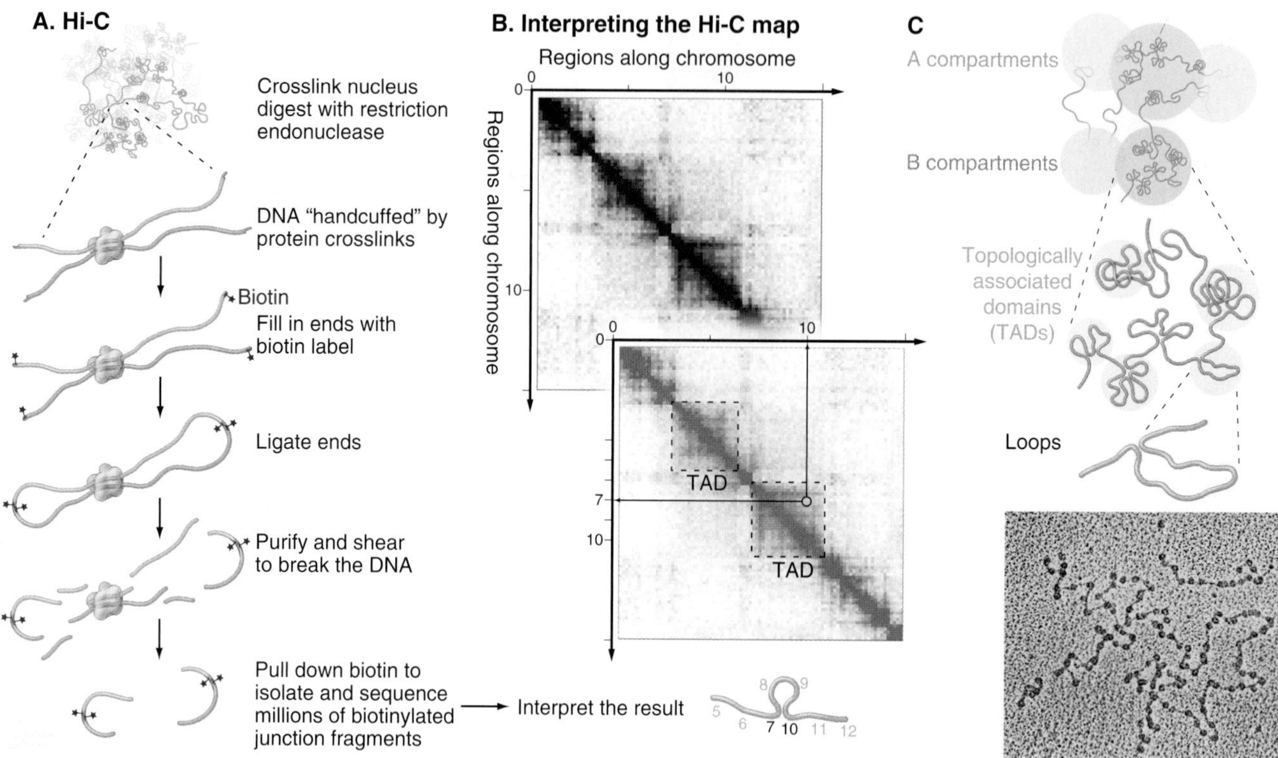

FIGURE 8.13 HI-C REVEALS CHROMATIN FOLDING PATTERNS IN NUCLEI. A, Diagram of important steps during the Hi-C procedure. **B,** Example of a Hi-C map with a diagram of how it is interpreted. The numbers along the axes are arbitrary and are supplied for demonstration purposes only. Each time two sequences are linked together, a red dot is inserted in the matrix. This map contains many millions of those dots. TADS are the square groupings of dark dots which indicate regions that are often linked together. **C,** Hi-C reveals the presence of both very long-range compartments and topologically associating domains (TADs) in chromosomes. (**C,** Modified from Dekker J, Marti-Renom MA, Mirny LA. Exploring the three-dimensional organization of genomes: interpreting chromatin interaction data. *Nat Rev Genet.* 2013;14:390–403. Micrograph from Thoma F, Koller T. Influence of histone H1 on chromatin structure. *Cell.* 1977;12:101–107.)

conformation capture) and its many derivatives, including Hi-C (Fig. 8.13A). In Hi-C, cells are treated with the fixative formaldehyde, which nonspecifically crosslinks proteins to the DNA. The idea is to "handcuff" together adjacent stretches of DNA. After cleaving the DNA with a restriction endonuclease, the ends are labeled with a biotinylated nucleotide, and then ligated together. This links pieces of DNA that were captured together by the crosslinking procedure and were therefore physically close to one another in the nucleus. Importantly, these sequences may be very far apart on the chromosomal DNA or even on different chromosomes! The biotin-containing DNAs are then sequenced by high-speed parallel methods (Fig. 3.16), yielding several hundred million sequence "reads" that generate a map (Fig. 8.13B) with an approximately 100-kb resolution of all regions of chromosomes that are close to other regions of chromosomes.

This analysis reveals two levels of chromatin organization: **TADs** and **compartments** (Fig. 8.13C). A TAD is a region of the chromosome—usually spanning 100,000 to 1,000,000 base pairs—whose DNA sequences are preferentially captured together, presumably, because they form a cluster of loops. The bands seen in polytene chromosomes correspond to TADs. A defining feature is that sequences in two adjacent TADs rarely come in contact with one another even though they may be closer to one another along the DNA strand than two more distant sequences that are found within the same TAD. One current model is that TADs form because the DNA is looped locally by CTCF and the cohesin complex (see later).

Hi-C maps also show longer-range interactions known as **compartments**. They are thought to involve interactions between many TADs, and may correspond to larger domains of euchromatin and heterochromatin.

The protein CTCF (CCCTC binding factor) marks approximately 75% of TAD boundaries at binding sites that define functional elements termed **insulators**. These were originally identified as short DNA sequence elements that frequently separate regions with active and inactive genes. For example, an insulator region containing CTCF and rich in H3 acetylated on K9 often separates an active gene cluster from an adjacent region of heterochromatin, Acetylation of H3 blocks methylation of H3-K9, thereby providing a barrier to the spreading of heterochromatin marked with H3-K9me3. CCTF binding in the insulator can physically block the DNA from being methylated, providing another defense against the spread of heterochromatin. Other TAD boundaries correspond to housekeeping genes undergoing active transcription, or to the presence of other insulators associated with transfer RNA genes.

CTCF can recruit a ring shaped complex called cohesin, which is a key architectural factor in chromosomes (Fig. 8.18). Cohesin was originally identified, because it regulates the pairing between replicated DNA molecules (sister chromatids) when cells divide. However, defects in the cohesin loading machinery cause Cornelia de Lange syndrome, a group of developmental disorders characterized by abnormalities in regulation of gene expression but (surprisingly) no dramatic effects on sister chromatid segregation during mitosis. It later emerged that cohesin is associated with up to half of all actively transcribed genes.

In mammals, 50,000 to 70,000 CTCF binding sites have been mapped, most of which are actually within TADs. One prominent CTCF binding site is found within Alu SINES, the short mobile genetic elements that comprise up to 15% of the human genome (see Chapter 7). Because CTCF and cohesin are thought to function together to bring regulatory elements together with genes, this association with a mobile DNA element has been suggested to be one factor that contributed to humans developing complex patterns of gene regulation.

In terms of function, it seems likely that clustering of loops in TADs may provide a mechanism to coordinate the regulation of gene expression and, possibly, DNA replication. Clusters of loops have been suggested to form **active chromatin hubs** associated with locus control regions, which are responsible for coordinating the expression of groups of genes.

Locus control regions (LCRs) were identified, because they could influence the transcriptional activity of cloned DNA sequences in transgenic animals. When genes are introduced into cultured cells, they normally insert at random into the chromosomes. Expression of such foreign *transgenes* depends on the site of insertion into the host chromosome. Transgenes are usually expressed when they insert into an active chromosomal domain but repressed when they insert into an inactive region. LCRs are DNA sequences that permit transgenes to be expressed no matter where they insert into the chromosomes, suggesting that they create active chromatin hubs independent of the surrounding chromosome.

LCRs typically consist of clusters of multiple short 150 to 300 base pair regions that are rich in binding sites for transcriptional regulators (see Chapter 9). Experiments in which a single LCR drives the expression of a cluster of several genes reveal that the LCR stimulates the expression of only one gene at a time. Thus, LCRs appear to work by physically associating with a gene, forming a loop in the chromatin and establishing an active chromatin hub that turns on its expression. Because cohesin can encircle pairs of DNA strands, it is now thought that this complex may anchor these DNA loops.

Organization of Mitotic Chromosomes

When cells divide, the chromatin is dramatically reorganized, forming mitotic chromosomes that can be segregated efficiently to daughter cells. The formation of mitotic chromosomes involves two steps; compaction of the chromatin roughly threefold and organization of each sister chromatid (the replicated DNA molecule and proteins that package it) into a robust structure that can move as a unit when cells divide.

It is still not known how the chromatin fiber is organized in mitotic chromosomes. Classic hierarchical coiling models suggested that the 30-nm chromatin fiber coils on itself, reaching larger and larger diameters and higher degrees of compaction. The 30-nm fiber is now largely disbelieved, but high-resolution Hi-C data reveal that chromatin fiber coiling is an important feature of mitotic chromosome formation.

Hi-C technology also reveals that TADs disappear from chromosomes as cells enter mitosis and are replaced by a more-or-less uniform distribution of approximately 80,000 to 120,000 base pair loops. A variety of microscopy experiments had previously suggested that chromatin loops containing 15,000 to 100,000 base pairs provide the structural basis for large scale chromatin compaction in mitotic chromosomes. These loops radiate outward from the central chromatid axis and can be seen when metaphase chromosomes are swelled in hypotonic solutions (Fig. 8.14C).

We favor a model proposing that mitotic chromosome formation involves both hierarchical coiling and looping of the chromatin fiber. During this process, key proteins become concentrated along the axial regions of the condensing chromosome arms and stabilize the overall structure (Fig. 8.14A). The mechanism of chromatin folding in mitotic chromosomes remains an area of active investigation and controversy.

Although much less ordered than polytene chromosomes, the arms of typical diploid mitotic chromosomes nonetheless have a more-or-less reproducible substructure. If mammalian chromosomes from the early (prometaphase) stage of mitosis are subjected to a staining procedure called G-banding, up to 2000 discrete bands are observed (Fig. 8.15). Although the structural basis for the bands is not known, the pattern is highly reproducible. Dark G-bands tend to be gene-poor regions relatively enriched for DNA with a low A:T content and rich in long interspersed nuclear elements (see Fig. 7.5). They tend to replicate later in S phase than light G-bands (also called R, or reverse, bands). Cytogeneticists used these highly reproducible banding patterns for many years to identify individual human chromosomes.

The quasi-reproducible higher-order structure of mitotic chromosomes is also seen when specific DNA

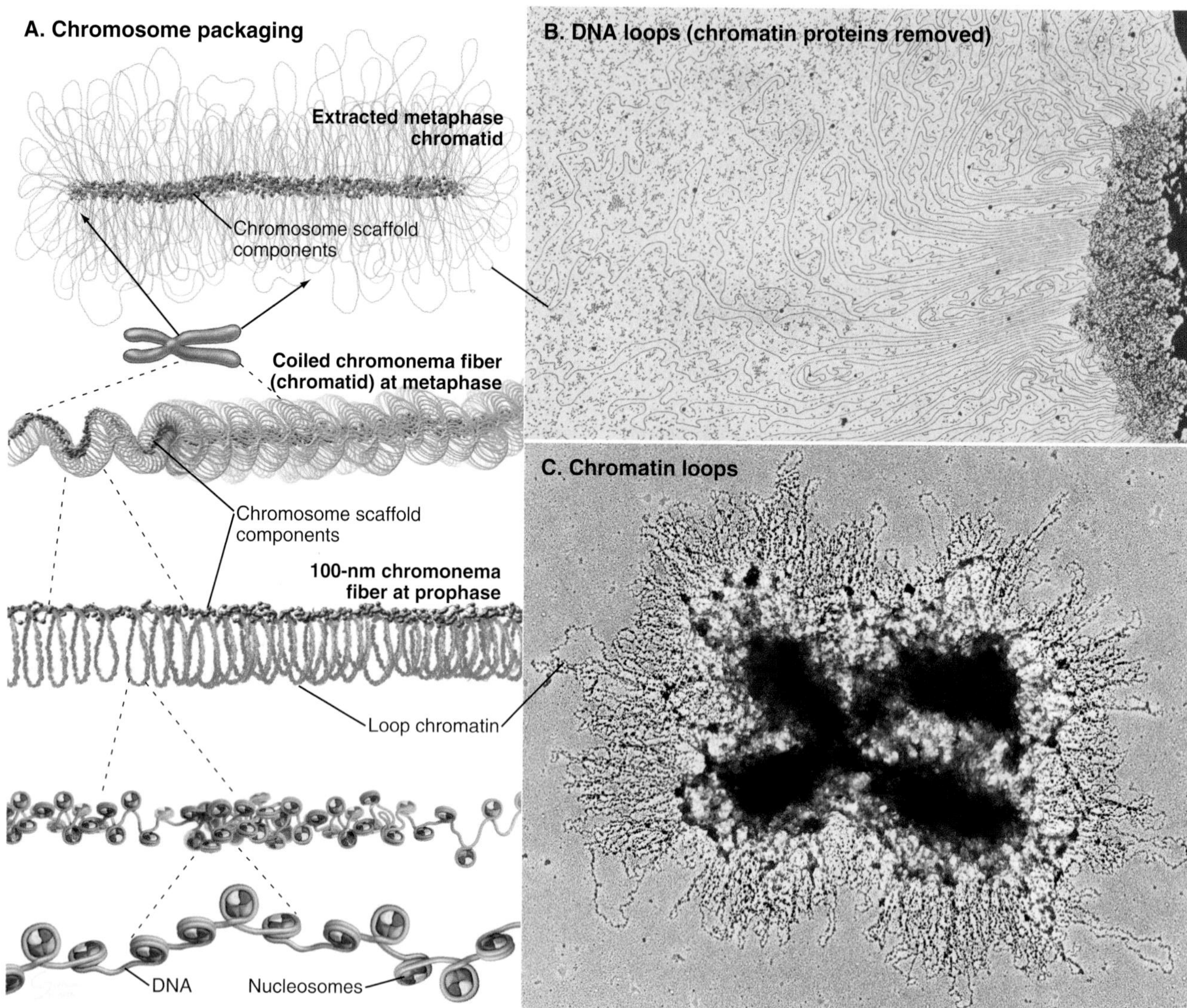

FIGURE 8.14 CURRENT MODEL OF MITOTIC CHROMOSOME STRUCTURE. A, Filament of nucleosomes, chromatin looping, clustering of chromatin loops into coiled fiber. Nonhistone proteins complexes *(blue dots)* bind and end up concentrated along the central axis of the chromatid arm. Crosslinks between these complexes create the chromosome scaffold. When chromosomes are swollen or extracted, the scaffold remains compact, and loops of chromatin or DNA radiate out from it. **B,** DNA loops seen in a human mitotic chromosome from which the histones had been removed. **C,** Human chromosome showing loop domains. (**B,** From Paulson JR, Laemmli UK. The structure of histone-depleted chromosomes. *Cell.* 1977;12:817–828. **C,** Courtesy William C. Earnshaw.)

sequences marked by in situ hybridization appear as pairs of spots on the sister chromatids (Fig. 8.10). The two spots are distributed approximately symmetrically, indicating that the chromatin fiber is folded similarly, though not identically, in both chromatids.

Role of Nonhistone Proteins in Chromosome Architecture

Mitotic chromosomes are composed of roughly equal masses of DNA, histones, and **nonhistone proteins.** Early evidence suggesting that nonhistone proteins might contribute to mitotic chromosome structure came from experiments in which chromosomes were treated with nucleases to digest the DNA and extracted to remove most chromosomal proteins, including essentially all the histones. The surviving remnant of the chromosome contained approximately 5% of the proteins and less than 0.1% of the DNA, but still looked like a chromosome (Fig. 8.16). If the DNA was not digested, loops of DNA protruded from the protein (Fig. 8.14B). The protein remnant was called the **chromosome scaffold** because it looked like a structural backbone for the chromosome. Indeed, chromosome scaffold preparations contain several proteins with essential roles in the structure and maintenance of mitotic chromosomes.

If isolated nuclei are subjected to the procedures used to isolate mitotic chromosome scaffolds, a residual structure is also obtained. This has been termed the **nuclear**

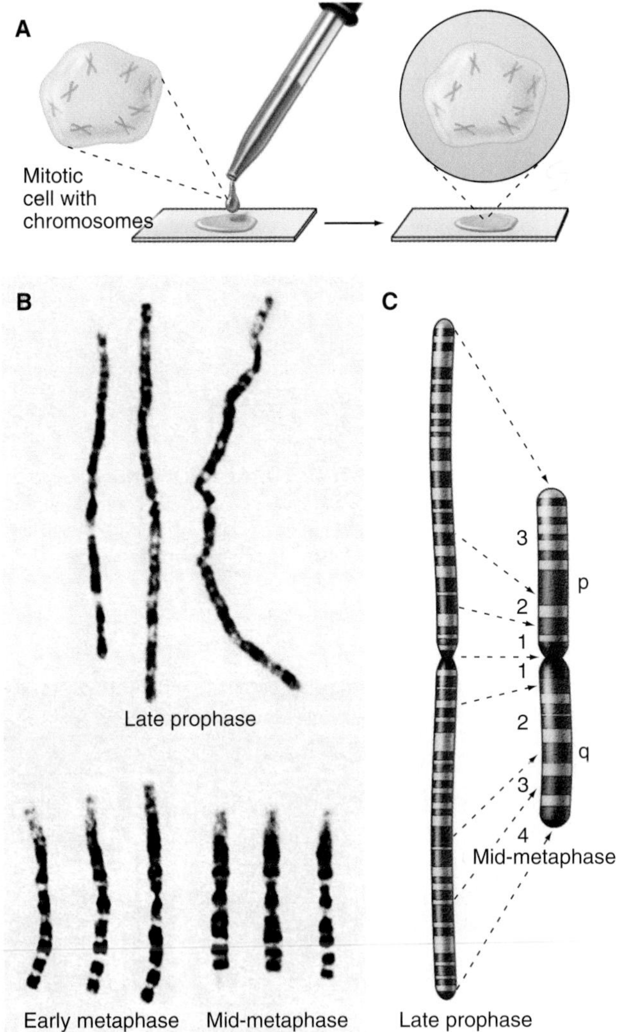

FIGURE 8.15 CHROMOSOME BANDING REVEALS A COMPLEX AND REPRODUCIBLE MULTIDOMAIN SUBSTRUCTURE OF MITOTIC CHROMOSOME ARMS. A, Mitotic cells in a hypotonic medium are dropped onto a slide to spread the chromosomes. In G-banding, chromosomes are given harsh treatments, such as exposure to concentrated sodium hydroxide, proteases, or high temperatures, and then stained with Giemsa dye. The chromosome arms then exhibit a characteristic pattern of light and dark bands. **B,** Photographs of G-banded human chromosome 2 from cells in late prophase, early metaphase, and mid-metaphase. Several examples are shown for each stage, illustrating the reproducibility of the banding patterns. **C,** Diagram summarizing the metaphase and prophase patterns. Because G-banding patterns are reproducible, this technique provides a way to identify individual chromosomes unambiguously. This was a major factor in the development of the field of cytogenetics, which is the study of the correlation between the structure of the chromosomes and genetics. (**B–C,** Modified from Yunis JJ, Sawyer JR, Ball DW. The characterization of high-resolution G-banded chromosomes of man. *Chromosoma*. 1978;67:293–307.)

matrix or nucleoskeleton. Although the existence and function of a nuclear matrix in vivo remains controversial, some components of the mitotic chromosome scaffold (eg, cohesin and condensin; discussed here) have roles in organizing chromosome territories and chromatin loops. For example, cohesin and CTCF are thought to have important roles in organizing the architecture of interphase chromatin into TADs.

Members of the **SMC protein** family have several important roles in chromosome dynamics. The name derives from their roles in the *s*tructural *m*aintenance of *c*hromosomes. SMC proteins are components of multiprotein complexes, such as **condensin** and **cohesin,** that are essential for mitotic chromosome architecture, the regulation of sister chromatid pairing, DNA repair and replication, and the regulation of gene expression.

The two *condensin* complexes are composed of two SMC proteins (SMC2 and SMC4), plus two sets of three auxiliary subunits. Each SMC polypeptide folds back on itself at a hinge region to form a long antiparallel coiled-coil. This brings together two globular domains from either end of the molecule, each with half of an ATP-binding site (Figs. 8.17 and 8.18). ATP binding causes the two globular domains to associate with one another. This association is then reinforced by binding of a strap-like **kleisin** (from the Greek for closure) subunit. The other auxiliary subunits bind to the kleisin and appear to regulate association of the complex with DNA. Condensin I and condensin II are thought to regulate distinct aspects of mitotic chromosome architecture.

Condensin has a complex role in establishing the architecture of mitotic chromosomes. Condensin I regulates the timing of chromosome condensation and has an essential role in changing the genome organization from TADs to a brush-like array of loops as chromosomes form during entry of cells into mitosis. Condensin II apparently drives the compaction of the chromosome loops along the sister chromatid axes.

The cell-cycle kinase Cdk1:cyclin B (see Chapter 40) regulates condensin binding to chromosomes by phosphorylation of an auxiliary subunit. During mitosis condensin is concentrated along the central axis of chromosome arms. When condensin binds to naked DNA in a test tube, it can use the energy of ATP hydrolysis to supercoil the DNA. The cellular role of this activity is unknown, but it may contribute to changing the conformation of chromatin loops. When condensin is depleted, mitotic chromatin condenses (apparently driven by changes in histone modifications), but the resulting chromosomes are fragile and appear disorganized if condensin depletion is rapid and complete.

Cohesin is the second major SMC-containing protein complex of interphase and mitotic chromosomes. Cohesin is a tetramer containing SMC1 and SMC3 plus two auxiliary subunits. Cleavage of the kleisin Scc1, by a protease called separase initiates sister chromatid separation in mitotic anaphase (see Fig. 44.16). Cohesin, like condensin, is a ring-like molecule (Fig. 8.18). How cohesin holds the two sister chromatids together is still debated, although it is generally thought to physically encircle two sister DNA molecules. Cohesin

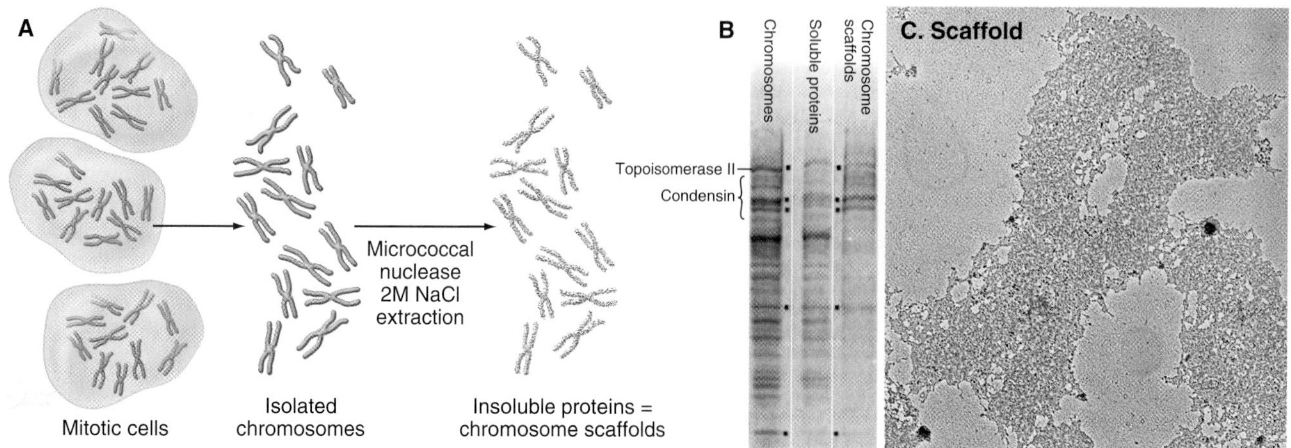

FIGURE 8.16 ISOLATION OF MITOTIC CHROMOSOME SCAFFOLDS REVEALS IMPORTANT STRUCTURAL PROTEINS. A, Diagram of the procedure used to isolate mitotic chromosomes. **B,** Sodium dodecylsulfate polyacrylamide gel showing proteins of isolated chromosomes, proteins extracted during scaffold isolation, and proteins of isolated scaffolds. **C,** Chromosome scaffold centrifuged onto a thin carbon film and rotary-shadowed with Pt : Pd (platinum : palladium). The structure, which is approximately 95% protein, retains the overall shape of the mitotic chromosome. (**B–C,** Courtesy William C. Earnshaw.)

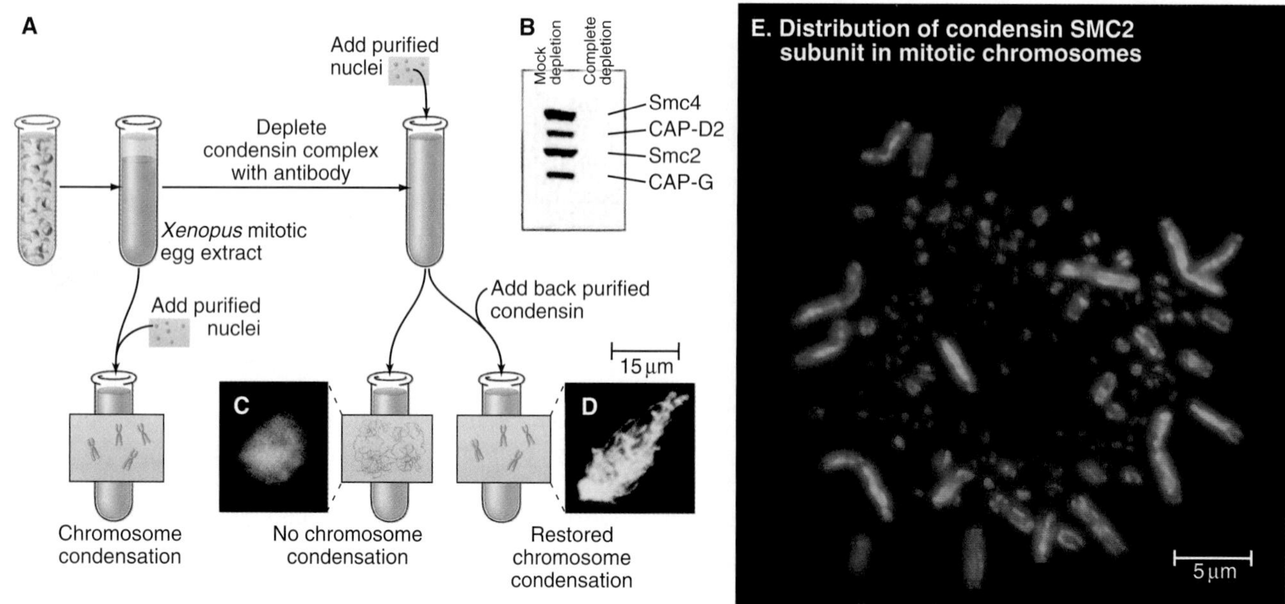

FIGURE 8.17 IDENTIFICATION OF THE CONDENSIN COMPLEX. A, Experimental protocol showing that condensin is required for mitotic chromosome condensation in vitro. **B,** Sodium dodecylsulfate (SDS) polyacrylamide gel reveals the members of the condensin complex and demonstrates that they can be depleted from egg extract using a specific antibody. **C–D,** Chromatin lacking condensin does not form mitotic chromosomes in vitro, and this is restored by adding back condensin. **E,** Immunofluorescence micrograph showing the distribution of condensin subunit SMC2 on mitotic chromosomes of the chicken. The tiny chromosomes, called microchromosomes, are normal bird microchromosomes. (**A–D,** From Hirano T, Kobayashi R, Hirano M. Condensins, chromosome condensation protein complexes containing XCAP-C, XCAP-E and a *Xenopus* homolog of the *Drosophila* Barren protein. *Cell*. 1997;89:511–521. **D,** Micrograph courtesy William C. Earnshaw.)

assembles on chromosomes during DNA replication and is recruited to regions of heterochromatin by HP1. Recent evidence also suggests that cohesin also has an important role in regulating gene expression during interphase, possibly by stabilizing chromatin loops that assemble active chromatin hubs. CTCF can bind cohesin, and the two proteins are found at most boundaries between TADs (though there are many more binding sites for the two proteins within TADs).

DNA topoisomerase IIα, an enzyme that alters DNA topology by passing one double-helix strand through another, is a very abundant component of mitotic chromosomes. In mitosis, topoisomerase IIα is concentrated at centromeres and in axial regions along the

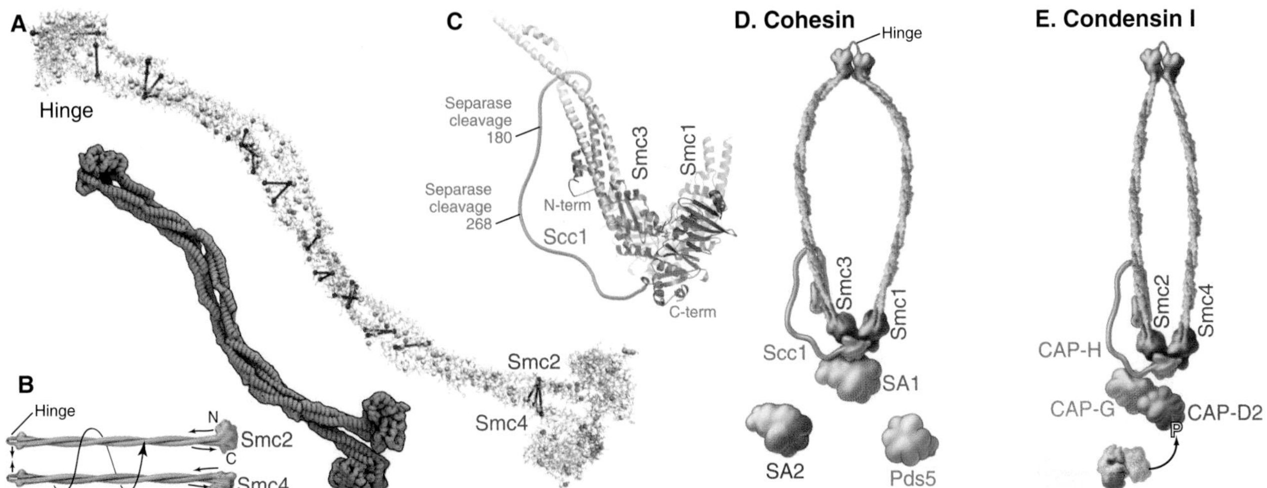

FIGURE 8.18 CONDENSIN AND COHESIN COMPLEXES. A–B, Model of the isolated dimer of SMC2 and SMC4 from condensin. Chemical crosslinks between SMC2 and SMC4 used to constrain the modeling are shown in red. Colored spheres represent lysines involved in crosslinks. **B,** Structure of a portion of the cohesin complex showing the paired heads with a bound fragment of the Scc1 kleisin. **C–D,** Subunit composition and structural organization of the cohesin and condensin complexes. (**B,** From Gligoris TG, Scheinost JC, Bürmann F, et al. Closing the cohesin ring: structure and function of its Smc3-kleisin interface. *Science*. 2014;346:963–967.)

chromosome arms. Topoisomerase IIα is very dynamic in vivo, moving on and off chromosomes in a time frame of seconds. Mitotic chromosomes from cells lacking topoisomerase II are long and thin, and the protein is thought to have a role in untangling the DNA as the loops condense along the chromosome axis during chromosome formation. Topoisomerase II is also required for replicated sister chromatids to separate from one another during mitotic anaphase. Presumably, the enzyme separates tangles and intertwinings of DNA created during DNA replication.

Remarkably, of the more than 4000 proteins found in mitotic chromosomes, only the histones, and fewer than 20 nonhistone proteins are known to have a role in mitotic chromosome formation. This does not count the more than 100 proteins that are required to form the kinetochores, which direct chromosomal movements in mitosis.

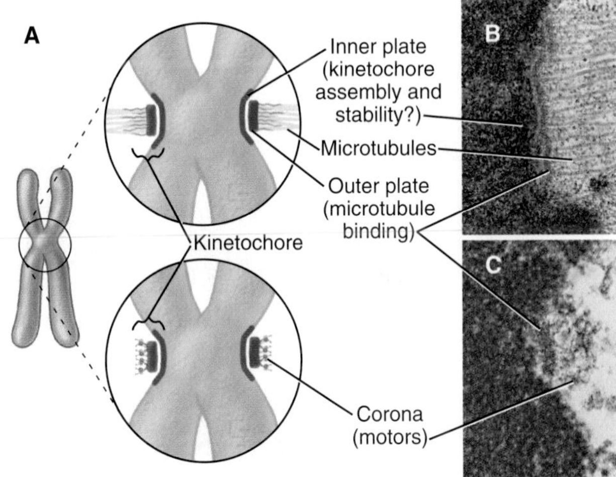

FIGURE 8.19 KINETOCHORE STRUCTURE. A, Diagram of the major layers of the kinetochore. **B,** Thin-section electron micrograph of a kinetochore with attached microtubules. **C,** Thin-section micrograph of an unattached kinetochore. (**B,** Courtesy J.B. Rattner, University of Calgary, Alberta, Canada. **C,** Courtesy Rebecca L. Bernat and William C. Earnshaw.)

The Chromosome's Control Center: The Kinetochore

The centromere is the genetic locus that specifies the site where a kinetochore assembles on the chromosomal DNA molecule. The **kinetochore** is a button-like structure embedded in the surface of the centromeric chromatin of most eukaryotic mitotic chromosomes (Fig. 8.19). When thin sections of centromeres are examined by electron microscopy, the kinetochore often appears to have several layers. The **inner kinetochore** is embedded in the surface of the centromere and is composed of a specialized form of chromatin. The **outer kinetochore** consists of an **outer plate** with a **fibrous corona** on its outer surface. It is constructed from protein complexes that link the chromatin to microtubules of the mitotic spindle.

During interphase, the kinetochore persists as a condensed ball of heterochromatin that resembles other areas of condensed chromatin within the nucleus. The distinct multilayered kinetochore structure forms on the surface of the centromere during an early stage of mitosis called prophase (see Chapter 44), reaching its mature state following nuclear envelope breakdown when the chromosome comes into contact with microtubules at the onset of mitotic prometaphase.

Chapter 7 describes the three types of centromeres known in eukaryotes (see Fig. 7.9). Point centromeres found in budding yeasts assemble kinetochores on defined DNA sequences and do not require epigenetic activation to function. They bind one microtubule. Regional centromeres, found in organisms ranging from fission yeast to humans, are based on preferred DNA sequences but require epigenetic activation to function. They bind two to 20 or more microtubules. In holocentromeres, as found in *Caenorhabditis elegans* and many plants and insects, the microtubules (roughly 20 in *C. elegans*) bind all along the poleward-facing surface of the mitotic chromosome. Given this diversity of centromeres, it is remarkable that the proteins responsible for kinetochore assembly and function are well conserved across evolution.

Mammalian Kinetochore Proteins

The first three specific kinetochore proteins identified in any species were discovered in humans using autoantibodies present in the sera of patients with rheumatic disease (Fig. 8.20). These proteins, designated CENP-A (*cen*tromere *p*rotein), CENP-B, and CENP-C, are conserved from humans to yeasts. They are part of the 16-protein constitutive centromere-associated network, which is composed of proteins that remain bound to the inner kinetochore throughout the cell cycle. The inner kinetochore chromatin is based on specialized nucleosomes with the histone H3 variant **CENP-A** (Fig. 8.21). How CENP-A targets the DNA to assemble kinetochore-specific nucleosomes is unknown, but some factors include a specialized chaperone, histone modifications and specialized RNA transcription, which, remarkably, occurs during mitosis.

CENP-B binds specifically to a 17-base pair sequence (the CENP-B box) in α-satellite centromere DNA (see Fig. 7.9) and is required for efficient kinetochore assembly, but the mechanistic details are unknown. CENP-B probably originated as the enzyme responsible for movement of an ancient transposon.

CENP-C and CENP-T (identified much later) are essential DNA-binding proteins that bridge between the inner chromatin and outer microtubule-binding components

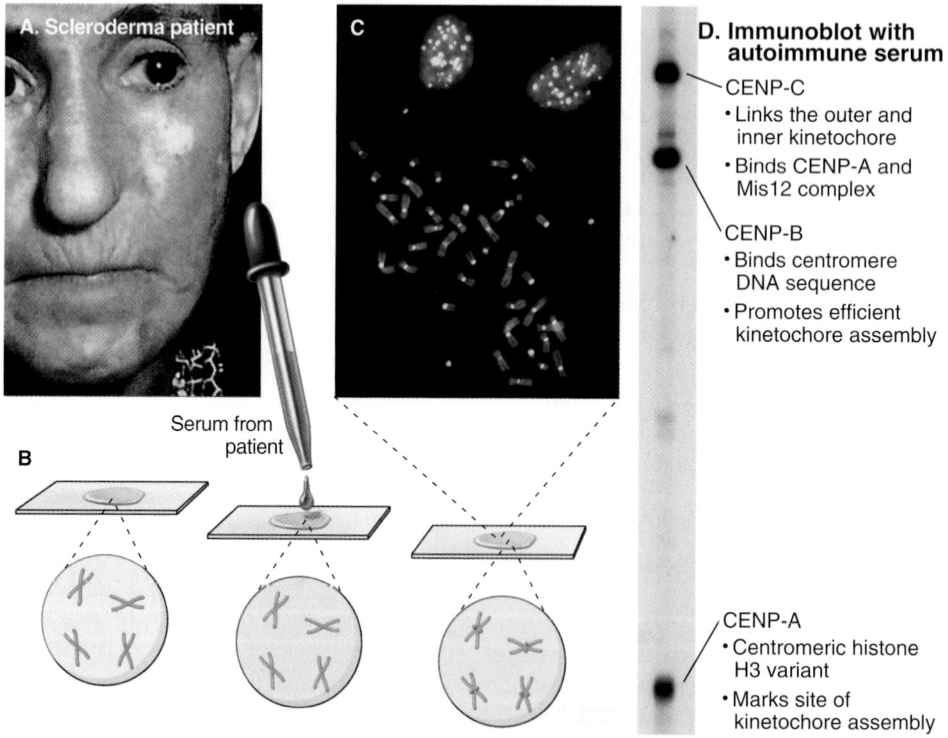

FIGURE 8.20 SOME PATIENTS WITH SCLERODERMA HAVE AUTOANTIBODIES THAT RECOGNIZE CENTROMERIC PROTEINS. Scleroderma ("hard skin") is a serious connective tissue disease associated with excessive deposition of collagen in the skin and walls of blood vessels. Note the "purse string" appearance of the skin surrounding the mouth of this patient **(A)**. When serum from a patient with anticentromere antibodies is added to chromosomes on a slide **(B)** and bound antibodies are detected with a fluorescent probe, the centromeric regions of the chromosomes "light up" **(C)**. Anticentromere antibodies are useful to identify patients who are at risk for serious autoimmune disease. Up to 20% of the population has a mild condition—Raynaud phenomenon (hypersensitivity of the skin to cold)—that is very rarely a precursor to scleroderma. Sensitive assays for anticentromere antibodies revealed that patients with Raynaud phenomenon who also have these autoantibodies have an increased risk of progression to scleroderma. **D,** Centromere proteins (CENPs) detected with anticentromere antibodies from a scleroderma patient on an immunoblot following sodium dodecylsulfate gel electrophoresis of chromosomal proteins. (**A,** From Dana R. Scleroderma. In: Albert DM, ed. *Albert & Jakobiec's Principles & Practice of Ophthalmology*, 3rd ed. Philadelphia: Elsevier; 2008. **C–D,** Courtesy William C. Earnshaw.)

of the kinetochore (Fig. 8.21). It has been suggested that the CENP-C link is primarily involved in enabling chromosome movements, whereas the CENP-T linkage may monitor the status of kinetochore attachment.

The best-characterized component of the outer kinetochore complex is the NDC80 complex—an elongated rod with globular ends linked by central coiled-coils. One end of this complex binds to microtubules. Some copies of the NDC80 complex form a network with six other components that is thought to be the main mechanical link between chromosomes and microtubules. Both CENP-C and CENP-T independently link the inner chromatin to this outer NDC80-associated network. One protein from the NDC80 network also recruits to kinetochores the signaling components of the mitotic checkpoint pathway that regulates progression of the cell through mitosis without errors (see Fig. 44.11).

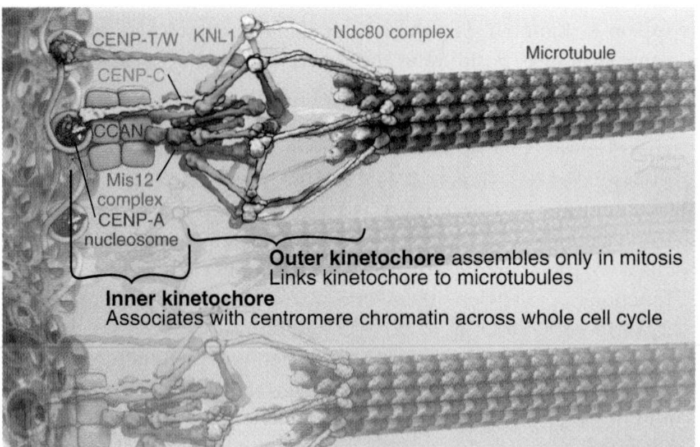

FIGURE 8.21 HYPOTHETICAL MODEL FOR THE ORGANIZATION OF THE VERTEBRATE KINETOCHORE. Protein complexes discussed in the text are indicated.

Centromere Proteins of the Budding Yeast

The best-characterized kinetochores come from budding yeast. Yeast kinetochores have been isolated and subjected to both biochemical and biophysical characterization (Fig. 8.22). More than 65 kinetochore-associated proteins assemble a structure at least the size and complexity of a ribosome.

Specific centromere DNA-binding factors (CBF) recognize the DNA sequences (CDE I and CDE III) that specify the point centromere (see Fig. 7.7) and wrap around a specialized nucleosome containing a centromere-specific histone H3 variant related to CENP-A (Fig. 8.22). A stretch of A:T-rich DNA called CDE II completes one turn around this nucleosome, juxtaposing the flanking CDE I and CDE III elements and their associated proteins.

Several large complexes bind to this nucleosome/CBF platform (Fig. 8.22). Although the CBF proteins are unique to yeast, the other complexes are all conserved from yeast to humans. These include the NDC80 complex (which was first identified by yeast genetics). The NDC80 complex binds to a ring made by the 10-subunit Dam1 complex that encircles the microtubule. This may help the kinetochore hold onto microtubules that are shrinking as the chromosome moves poleward during anaphase. The Dam1 complex is poorly conserved during evolution, although a possible vertebrate counterpart has been identified.

Role of RNA Interference at Fission Yeast Centromeres

The fission yeast *Schizosaccharomyces pombe* has the simplest well-characterized regional centromere. It assembles a kinetochore that binds two to four

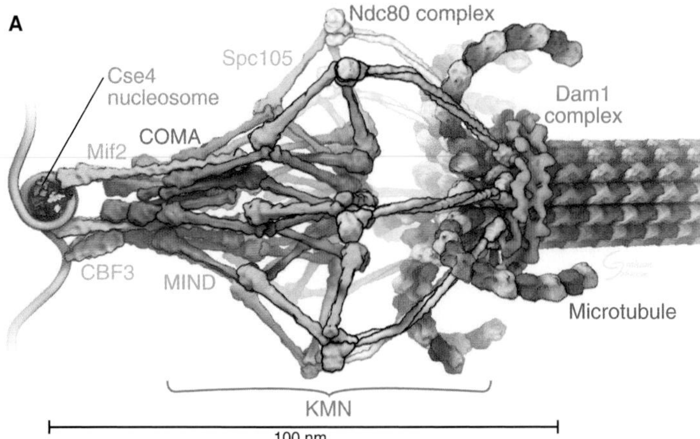

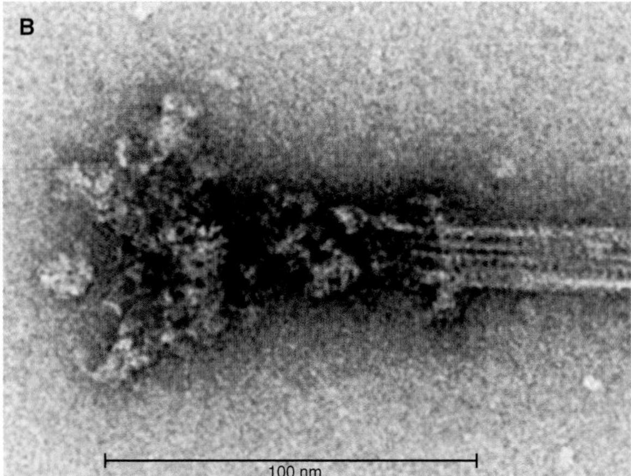

FIGURE 8.22 MODEL FOR THE ORGANIZATION OF THE BUDDING YEAST KINETOCHORE AND MICROGRAPH OF AN ISOLATED KINETOCHORE. A, Hypothetical diagram of budding yeast kinetochore (for discussion of the yeast centromere DNA, see Chapter 7). **B,** Electron micrograph of a budding yeast kinetochore attached to a microtubule. (**B,** From Gonen S, Akiyoshi B, Iadanza MG, et al. The structure of purified kinetochores reveals multiple microtubule-attachment sites. *Nat Struct Mol Biol*. 2012;19:925–929.)

microtubules. Fission yeast have orthologs of all the proteins and protein complexes described here. The fission yeast centromere provides insights into the formation of centromeric heterochromatin. The "silent" repeated DNA in the *S. pombe* centromere is transcribed from both DNA strands, yielding short double-stranded RNAs that are processed by the RNAi machinery. This RNAi response is part of the pathway for assembly of centromeric heterochromatin (see Fig. 11.14). A wide range of *S. pombe* mutants affecting the RNAi machinery all compromise centromere function and mitotic chromosome segregation.

Whether RNAi is also essential for centromere function in metazoans has been more difficult to determine, as genetic analysis is complicated by multiple redundancies in the genes encoding the RNAi machinery. However, careful analysis reveals that centromeric satellite DNAs are indeed transcribed. Remarkably, this transcription occurs during mitosis, and is the only transcription known to occur during that cell-cycle phase. The specialized regulation that enables mitotic centromere transcription is unknown, as is whether the transcripts participate in a functional RNAi pathway like that observed in yeast.

Conclusions

Ironically, just as the sequence of the euchromatic portion of the human genome was completed, a shift in our understanding revealed that essential aspects of the control of gene activity and chromosome structure cannot be revealed by analysis of the DNA sequence alone, as these regulatory processes are "encoded" in transient epigenetic modifications of DNA and histones. Understanding the extraordinarily elaborate epigenetic code has only just begun, so watch this space for further exciting developments.

ACKNOWLEDGMENTS

We thank Julie Ahringer, Wendy Bickmore, Job Dekker, Margarete Heck, Kazuhiro Maeshima, and Tom Owen-Hughes for their suggestions on revisions to this chapter.

SELECTED READINGS

Allshire RC, Ekwall K. Epigenetic regulation of chromatin states in *Schizosaccharomyces pombe*. *Cold Spring Harb Perspect Biol*. 2015;7:a018770.

Bannister AJ, Kouzarides T. Regulation of chromatin by histone modifications. *Cell Res*. 2011;21:381-395.

Belmont AS. Large-scale chromatin organization: the good, the surprising, and the still perplexing. *Curr Opin Cell Biol*. 2014;26:69-78.

Bickmore WA. The spatial organization of the human genome. *Annu Rev Genomics Hum Genet*. 2013;14:67-84.

Biggins S. The composition, functions, and regulation of the budding yeast kinetochore. *Genetics*. 2013;194:817-846.

Chaligné R, Heard E. X-chromosome inactivation in development and cancer. *FEBS Lett*. 2014;588:2514-2522.

de Graaf CA, van Steensel B. Chromatin organization: form to function. *Curr Opin Genet Dev*. 2013;23:185-190.

Fukagawa T, Earnshaw WC. The centromere: chromatin foundation for the kinetochore machinery. *Dev Cell*. 2014;30:496-508.

Gibcus JH, Dekker J. The hierarchy of the 3D genome. *Mol Cell*. 2013;49:773-782.

Huang H, Sabari BR, Garcia BA, et al. SnapShot: histone modifications. *Cell*. 2014;159:458-458.e1.

Hudson DF, Marshall KM, Earnshaw WC. Condensin: architect of mitotic chromosomes. *Chromosome Res*. 2009;17:131-144.

Jeppsson K, Kanno T, Shirahige K, et al. The maintenance of chromosome structure: positioning and functioning of SMC complexes. *Nat Rev Mol Cell Biol*. 2014;15:601-614.

Maze I, Noh KM, Soshnev AA, et al. Every amino acid matters: essential contributions of histone variants to mammalian development and disease. *Nat Rev Genet*. 2014;15:259-271.

Merkenschlager M, Odom DT. CTCF and cohesin: linking gene regulatory elements with their targets. *Cell*. 2013;152:1285-1297.

Narlikar GJ, Sundaramoorthy R, Owen-Hughes T. Mechanisms and functions of ATP-dependent chromatin-remodeling enzymes. *Cell*. 2013;154:490-503.

Pombo A, Dillon N. Three-dimensional genome architecture: players and mechanisms. *Nat Rev Mol Cell Biol*. 2015;16:245-257.

Simon JA, Kingston RE. Occupying chromatin: polycomb mechanisms for getting to genomic targets, stopping transcriptional traffic, and staying put. *Mol Cell*. 2013;49:808-824.

Takeuchi K, Fukagawa T. Molecular architecture of vertebrate kinetochores. *Exp Cell Res*. 2012;318:1367-1374.

Thadani R, Uhlmann F, Heeger S. Condensin, chromatin crossbarring and chromosome condensation. *Curr Biol*. 2012;22:R1012-R1021.

Westhorpe FG, Straight AF. Functions of the centromere and kinetochore in chromosome segregation. *Curr Opin Cell Biol*. 2013;25:334-340.

Zhang T, Cooper S, Brockdorff N. The interplay of histone modifications-writers that read. *EMBO Rep*. 2015;16:1467-1481.

CHAPTER 9

Nuclear Structure and Dynamics

The nucleus houses the chromosomes together with the machinery for DNA replication and RNA transcription and processing (Fig. 9.1). Immature RNAs must be kept apart from the translational apparatus because eukaryotic genes are transcribed into RNAs containing noncoding intervening sequences that are removed by splicing to assemble mature RNA molecules with a continuous open reading frame. Sequestration of immature RNAs is one function of the nuclear envelope, two concentric membrane bilayers that separate the nucleus and cytoplasm. The nuclear envelope also regulates the bidirectional transport of macromolecules in and out of the nucleus, participates in chemical, protein and mechanical signaling pathways, contributes to genome organization, and provides mechanical stability to the nucleus.

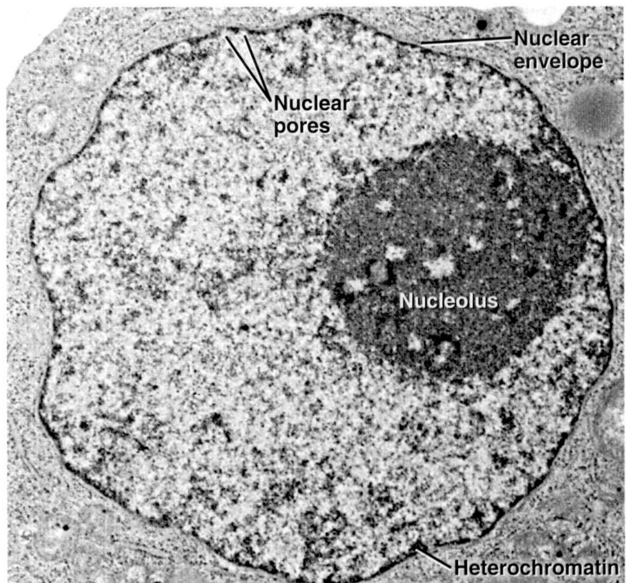

FIGURE 9.1 ELECTRON MICROGRAPH OF A THIN SECTION OF A CANCER CELL NUCLEUS WITH MAJOR FEATURES LABELED. (Courtesy Scott Kaufmann, Mayo Clinic, Rochester, MN.)

This chapter describes what is known about the structure of the nucleus, the nuclear envelope, and the transport of macromolecules into and out of the nucleus, and discusses their links to human diseases. Aspects of nuclear structure and function that are discussed elsewhere include genome and chromosome organization (Chapter 7), chromatin structure (Chapter 8), DNA replication (Chapter 42) and RNA transcription and processing (Chapters 10 and 11).

Overall Organization of the Nucleus

Studies in which entire individual chromosomes are labeled by in situ hybridization (chromosome painting; see Fig. 8.10) reveal that chromosomes tend to occupy discrete regions within the nucleus called **chromosome territories.** The boundaries of adjacent territories, where more actively transcribed regions are generally located, overlap with one another such that approximately 40% of each territory intermingles with adjacent territories. The chromatin of these overlapping regions tends to be less compact than in the rest of the territory, and is referred to as the **interchromosomal domain.** Most RNA transcription and processing are thought to occur within this domain. Although the nucleoplasm is very crowded with chromosomes and ribonucleoproteins (RNPs), proteins can nonetheless diffuse surprisingly rapidly through the nucleus, possibly by moving in the interchromosomal domain. Evidence is accumulating that actin is present in nuclei, presumably in the interchromosomal domain. Although the role of this actin is unknown, an attractive hypothesis is that it forms a scaffold for other processes. Nuclear actin can influence the positioning of nuclear subdomains.

Specialized Subdomains of the Nucleus

Cell nuclei contain numerous discrete subdomains or bodies with distinctive structural organizations and/or biochemical composition (Fig. 9.2 and Table 9.1).

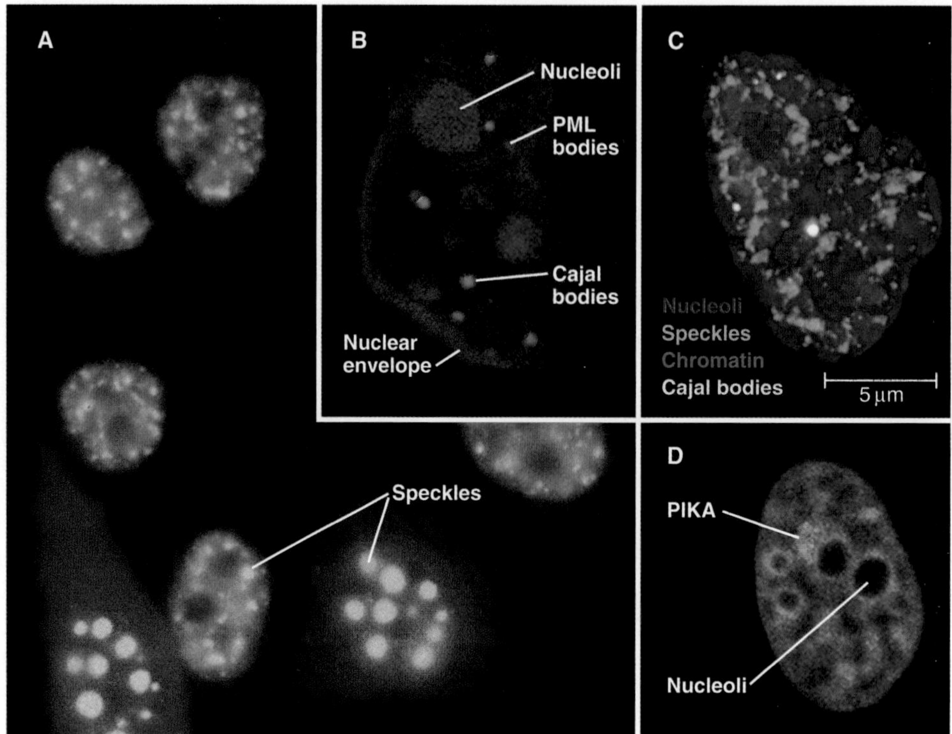

FIGURE 9.2 EXAMPLES OF MAJOR SUBNUCLEAR STRUCTURES. A, Components involved in RNA processing are scattered throughout the nucleus but concentrated in domains called speckles that are rich in interchromatin granules. Inhibition of RNA processing causes splicing components to accumulate in enormous concentrations of interchromatin granule clusters. Several cells were injected with a short oligonucleotide that disrupts the function of the U1 small nuclear ribonucleoprotein (snRNP) in RNA processing (see Fig. 11.15), and were then stained with an antibody recognizing the Sm splicing components *(green).* The injected cells were marked by introducing an inert fluorescent dextran marker into the cytoplasm *(red).* **B,** Nucleus with simultaneous staining of nucleoli *(blue),* PML (promyelocytic leukemia) nuclear bodies *(red),* Cajal bodies *(green),* and the nuclear envelope *(purple).* **C,** Nucleus with simultaneous staining of chromatin *(blue),* nucleoli *(red),* speckles *(green),* and Cajal bodies *(white).* **D,** Nucleus with simultaneous staining of DNA *(blue)* and the polymorphic interphase karyosomal association (PIKA)/53BP1 nuclear body/OPT (Oct1/PTF/transcription) domain *(red).* Nucleoli appear as unstained areas. A number of proteins involved in the sensing and repair of DNA damage concentrate in the PIKA. (**A,** Courtesy David Spector, Cold Spring Harbor Laboratory, Cold Spring Harbor, NY. **B–C,** Courtesy Angus Lamond, University of Dundee, United Kingdom. **D,** Courtesy William S. Saunders and William C. Earnshaw.)

The most prominent of these is the nucleolus, discussed in the next section. Although often referred to as organelles, nuclear subdomains, unlike cytoplasmic organelles, are not membrane-bounded. In fact, many proteins that have been examined by the fluorescence recovery after photobleaching technique (see Fig. 6.3) exchange relatively rapidly between a nuclear body and a nucleoplasmic pool. Therefore, these bodies represent highly dynamic associations of macromolecular complexes. By concentrating particular RNAs and proteins with enzymes involved in their maturation, they accelerate macromolecular assembly and maturation processes. They may also concentrate components involved in gene regulation or repair at particular chromosomal loci.

Structures associated with RNA transcription and processing are found at up to 10,000 sites spread throughout the typical mammalian nucleus as well as in a few more prominent domains. The dispersed sites likely correspond to structures called **perichromatin fibrils**, originally observed by electron microscopy on the surface of regions of condensed chromatin. Perichromatin fibrils contain various splicing factors and RNA-packaging proteins.

When factors involved in RNA processing are detected by fluorescence microscopy, 20 to 50 bright **speckles** are seen against a diffuse background of nucleoplasmic staining (Fig. 9.2). The diffuse staining probably corresponds to splicing factors associated with perichromatin fibrils at dispersed sites. Most speckles correspond to clusters of **interchromatin granules,** particles 20 to 25 nm in diameter distributed throughout the interchromosomal domain. Proteomic analysis reveals that isolated interchromatin granules contain more than 200 stably associated proteins, most involved with various aspects of RNA processing. When tagged with green fluorescent protein, components involved in pre-mRNA (messenger RNA) processing cycle between speckles and sites of transcription in less than 1 minute in live cells.

Metabolic labeling experiments indicate that speckles are not major sites of active transcription, although most transcription sites are associated with the periphery of speckles. Speckles are less prominent in cells that transcribe RNA at high levels, and become strikingly

TABLE 9.1 Major Nuclear Subdomains

Structure	Comments
Nucleolus	The nucleolus (typically 1 to 5 structures of 0.5 to 5 μm diameter in mammalian cell nuclei) is the site of ribosomal RNA (rRNA) transcription and processing, as well as of preribosomal assembly. It is also the site of processing of several other noncoding RNAs, including the RNA component of the signal recognition particle (SRP; see Fig. 20.5). It plays an important role in helping organize the genome during interphase, as well as in regulating the stability of p53, a critical transcription factor that is involved in regulating the cell cycle, particularly when DNA damage occurs.
Speckles	Speckles are concentrations of components involved in RNA processing. They often correspond to clusters of interchromatin granules seen by electron microscopy. They may serve as storage depots of splicing factors, or they may play a more active role in splicing factor modification and/or assembly.
Cajal bodies	Formerly known as coiled bodies. Approximately 0.2 to 1.0 μm in diameter, Cajal bodies have a coiled fibrous substructure. First identified by electron microscopy, up to 10 of these structures are seen in transformed cells. They are usually absent from nontransformed normal cells. They contain the human autoantigen p80-coilin and survival of motor neurons (SMN) protein, which is encoded by the gene mutated in spinal muscular atrophy, a severe, inherited, human, muscular wasting disease. They are involved in small nuclear ribonucleoprotein (snRNP) and small nucleolar ribonucleoprotein (snoRNP) assembly and in maturation of telomerase (which also contains an RNA component).
PML bodies	Also known as PODs and ND10, 10 to 30 of these structures are scattered throughout the nucleus. They are thought to enhance gene repression by serving as assembly sites for certain transcriptional corepresser complexes. They also appear to be targeted during viral infections. Fusion of the marker protein PML to the α-retinoic acid receptor is often found in acute promyelocytic leukemia (hence the name PML), in which the PML bodies appear highly fragmented. Treatments that are effective against PML restore the normal morphology of PML bodies (see text).
Polycomb group bodies	Concentrations of the PRC1 and PRC2 complexes (see Chapter 8) involved in the silencing of facultative heterochromatin. One mechanism for gene inactivation may be translocation into these inactive domains.
53BP1 nuclear bodies	Defined as concentrations of the DNA repair-associated protein 53BP1. Also known as PIKA (polymorphic interphase karyosomal association) and OPT (Oct1/PTF/transcription) domain. These domains may be up to 5 μm in diameter during G_1 phase, but their morphology and number vary across the cell cycle. They appear to correspond to sites of DNA damage during mitosis that arise as a result of incomplete DNA replication during S-phase.

prominent when RNA processing is inhibited (Fig. 9.2). Together these observations suggest that speckles may be dynamic depots where RNA processing factors accumulate then they are not active. They may also have a role in rendering mRNAs competent for export to the cytoplasm.

Cajal bodies (formerly known as coiled bodies) are compact structures approximately 0.3 to 1.0 μm in diameter (Fig. 9.2B) that resemble balls of tangled threads in the electron microscope. Nuclei of rapidly growing transformed cells typically have one to 10 prominent Cajal bodies. These structures are absent from most nontransformed (normal) cells. They contain an 80-kD human autoantigen called p80-coilin and the survival of motor neurons (SMN) protein, which is encoded by a gene mutated in spinal muscular atrophy, a severe inherited human muscular wasting disease. The SMN protein participates in importing immature small nuclear ribonucleoproteins (snRNPs) into the nucleus after their assembly in the cytoplasm. p80-coilin recruits the SMN complex to Cajal bodies, where the snRNPs are further processed to render them functional in RNA splicing reactions (see Chapter 11). Cajal bodies also have a role in the maturation of the RNP enzyme telomerase as well as other functions.

Mammalian nuclei also contain approximately 10 to 30 bodies, varying in size from 0.3 to 1.0 μm, known as **promyelocytic leukemia (PML) bodies** (Fig. 9.2B). PML bodies were initially defined by the presence of a protein called PML, an important regulator of cell growth and genome stability. PML has a RING-finger amino acid sequence motif and is therefore probably an E3 ligase for ubiquitin or ubiquitin-like proteins (see Fig. 23.2). Its targets are unknown. In normal cells, PML bodies apparently have a role in assembling corepressor complexes that modify chromatin to repress transcription (see Chapter 8). Their other functions are not known.

The PML gene was identified by analysis of a chromosome translocation between chromosomes 15 and 17 found in patients with acute promyelocytic leukemia (APL). In many patients, this translocation produces a gene fusion between PML and the retinoic acid receptor alpha (RARα). The fusion protein, PML-RARα, blocks differentiation of hematopoietic precursors and causes APL. In APL cells, PML-RARα is distributed in tiny punctate foci scattered throughout the nucleus. When APL cells are treated with arsenic trioxide, which is clinically effective in treating APL, PML-RARα aggregation causes prominent PML bodies to reform. PML-RARα is ubiquitylated within those bodies and subsequently degraded.

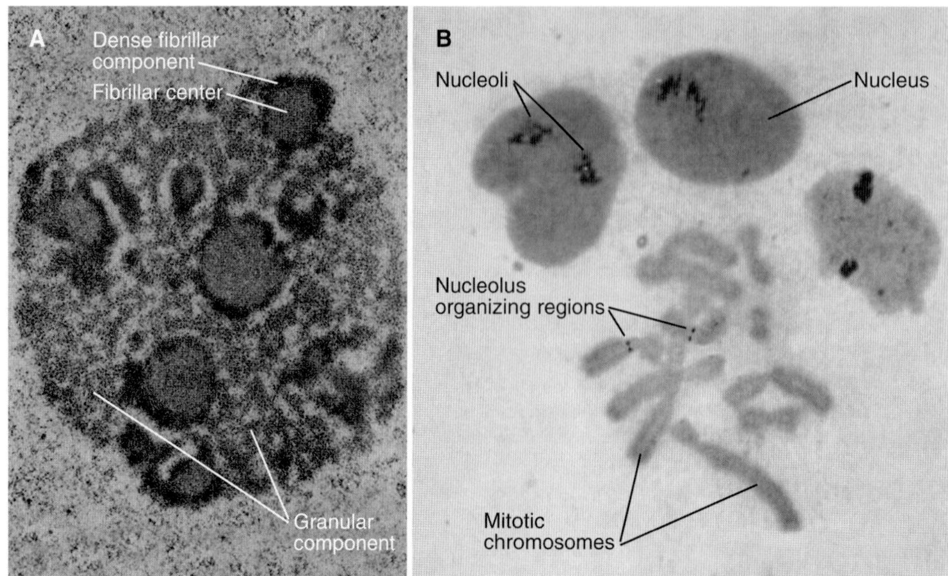

FIGURE 9.3 NUCLEOLUS AND NUCLEOLAR ORGANIZER REGION. A, Electron micrograph of a thin section of a typical nucleolus. The fibrillar centers, dense fibrillar component, and granular component are indicated. **B,** Use of silver staining to visualize the nucleolus in interphase nuclei and the nucleolar organizer regions on mitotic chromosomes of the rat kangaroo. (**A,** From Fawcett DW. *The Cell*. Philadelphia: WB Saunders; 1981. **B,** From Robert-Fortel I, Junéra HR, Géraud G, et al. Three-dimensional organization of the ribosomal genes and Ag-NOR proteins during interphase and mitosis in PtK1 cells studied by confocal microscopy. *Chromosoma*. 1993;102:146–157.)

This allows the hematopoietic precursors to differentiate and cures the cancer.

The Nucleolus: The Most Prominent Nuclear Subdomain

The nucleolus, first described only 5 years after the nucleus, in 1835, is the most conspicuous and best-characterized nuclear subdomain (Figs. 9.1 and 9.3). Most mammalian cells have one to five nucleoli, which are specialized regions 0.5 to 5.0 μm in diameter surrounding transcriptionally active ribosomal RNA (rRNA) gene clusters. Nucleoli are the sites of most steps in ribosome biogenesis, from the transcription and processing of rRNA to the initial assembly of ribosomal subunits. The ribosome is a complex macromolecular machine with four different structural rRNA molecules and approximately 85 proteins that are assembled into two subunits (see Figs. 12.6 and 12.7). Transcription of rRNA by RNA polymerase I comprises nearly half of total cellular RNA synthesis in some cell types. This high level of synthesis is necessary to produce approximately 5 million ribosomes in each cell cycle, more than 30 every second in budding yeast.

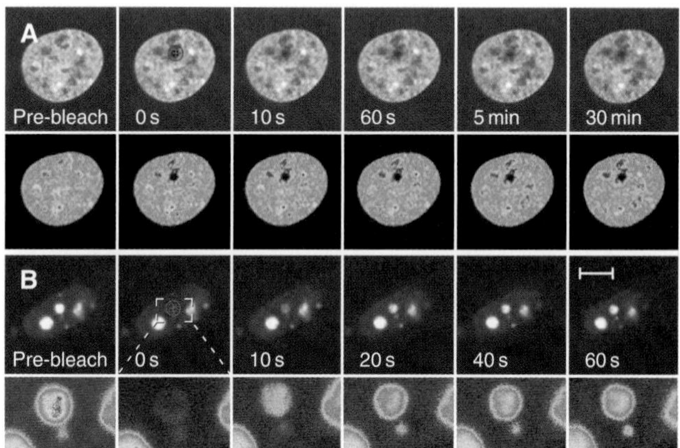

FIGURE 9.4 ANALYSIS OF DYNAMICS OF CHROMATIN AND A MAJOR NUCLEOLAR COMPONENT. A, Fluorescence recovery after photobleaching (FRAP) of H2B-GFP (green fluorescent protein) shows that chromatin is immobile within the cell nucleus. **B,** FRAP of fibrillarin-GFP shows that this major component of nucleoli is highly dynamic. Scale bar: 5 μm. (**A–B,** Courtesy Tom Misteli. **B,** From Phair RD, Misteli T. High mobility of proteins in the mammalian cell nucleus. *Nature*. 2000;404:604–609.)

Nearly 700 proteins associate stably with human nucleoli. Many more may associate transiently, and this composition changes to reflect different metabolic states of the cell (Fig. 9.4). Many nucleolar proteins are involved with either rRNA synthesis and modification or with ribosome subunit assembly. The functions of many other nucleolar proteins remain unknown and may reflect the involvement of nucleoli in other biological processes. Other stable RNAs, including the RNA component of the signal recognition particle (see Fig. 20.5), are also processed in the nucleolus.

Intriguingly, the nucleolus is involved in controlling the stability of the critical cell-cycle regulator protein p53 (see Fig. 41.5). Healthy cells keep p53 levels low by using ubiquitylation in the nucleolus to destabilize it. Under certain types of stress, cells defend themselves by activating p53. They do this by having nucleolar proteins bind and inactivate Mdm2, the key factor that ubiquitylates p53.

Ribosomal Biogenesis in Functionally Distinct Regions of the Nucleolus

Transmission electron micrographs of thin sections show three morphologically distinct regions in the nucleolus (Fig. 9.3). **Fibrillar centers** contain concentrations of rRNA genes, together with RNA polymerase I and its associated transcription factors. Actively transcribed ribosomal genes are found near the border between the fibrillar centers and a **dense fibrillar component** that surrounds them. The **granular component** is the site for many steps in ribosome subunit assembly and is made up of densely packed clusters of ribosomal precursors called preribosomal particles 15 to 20 nm in diameter.

rRNA loci have a modular organization. Genes alternate with spacer regions in large tandemly arranged clusters (see Fig. 11.10). The repeat unit in this array (gene plus spacer) is approximately 40,000 base pairs in humans. Humans have approximately 300 to 400 copies of the ribosomal DNA (rDNA) repeat unit located in clusters on chromosomes 13, 14, 15, 21, and 22. Usually, only a fraction of these genes is actively transcribed. An additional rRNA, 5S RNA, is encoded by distinct genes and transcribed by RNA polymerase III (see Fig. 10.8).

A simple yet efficient mechanism guarantees a balance between the RNA components of the two ribosomal subunits. The major rRNA components are encoded by a single precursor RNA molecule. In humans, this 13,000-base precursor is commonly described by its sedimentation coefficient in sucrose gradients as 45S. Following its transcription, the RNA precursor is processed in a series of cleavages to yield the 18S, 5.8S, and 28S rRNA molecules (see Fig. 11.10). In addition to the cleavages, rRNA processing also involves extensive base and sugar modifications, including approximately 100 2′-*O*-methyl ribose and approximately 90 pseudouridine residues per molecule. The earliest stages of rRNA processing probably occur in the dense fibrillar component of the nucleolus. Later stages take place in the granular component. Ribosomal protein synthesis occurs in the cytoplasm on free ribosomes. The newly synthesized proteins are transported into the nucleus for assembly into ribosomes, predominantly in the granular component.

Disassembly of the Nucleolus During Mitosis

The nucleolus disassembles during each mitotic cycle, starting with the dispersal of the dense fibrillar and granular components during prophase. This disassembly is driven by specific phosphorylation of nucleolar proteins. Ultimately, the fibrillar centers alone remain associated with the mitotic chromosomes, forming what are termed **nucleolus-organizing regions (NORs** [Fig. 9.3B]), which often form a prominent **secondary constriction** of the chromosome. (The primary constriction is the centromere.) Several nucleolar proteins and RNA polymerase I remain bound at NORs as cells enter and exit mitosis but most nucleolar proteins coat the surface of the mitotic chromosomes forming a **perichromosomal layer** or "skin."

Nucleolar reformation begins in mitotic telophase as processing factors and unprocessed pre-RNA remaining from the previous cell cycle associate with NORs (10 in human), which then cluster into one to five foci. Next, a wide variety of nucleolar components assemble into particles termed **prenucleolar bodies** that associate with the NORs in a process requiring transcription of the rRNA genes. Normally, nascent transcripts, rather than ribosomal genes, nucleate assembly of the nucleolus in each cell cycle. If antibodies to RNA polymerase I are microinjected into mitotic cells, rRNA transcription is blocked, and nucleoli do not reform in the next G_1 phase.

Structure of the Nuclear Envelope

The nuclear envelope provides a selective permeability barrier between the nuclear compartment and the cytoplasm and acts as a platform that helps organize the chromosomes in discrete functional domains (Fig. 9.5). The barrier keeps pre-mRNAs in the nucleus until they are fully processed and licensed for export so that only mature mRNAs are delivered to ribosomes in the cytoplasm for translation into protein. It also provides an

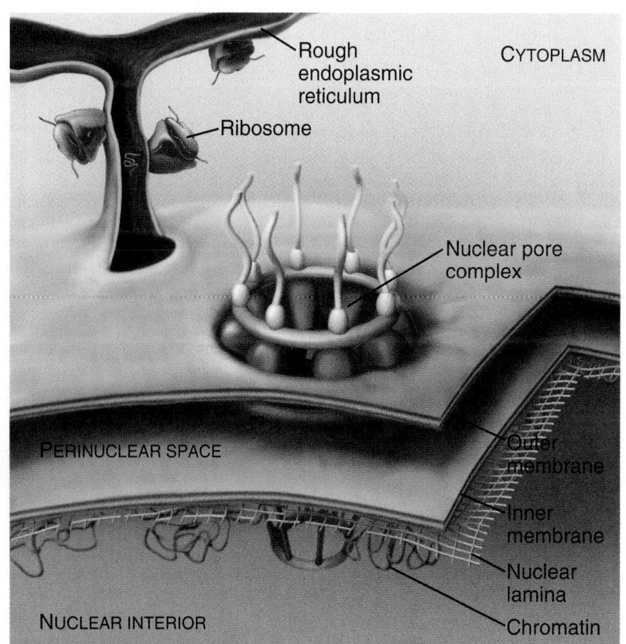

FIGURE 9.5 OVERVIEW OF NUCLEAR ENVELOPE ORGANIZATION.

additional level of genetic protection and control since various chromosomal events, including DNA replication and expression of certain genes, are regulated, at least in part, by changes in the ability of factors to move between the cytoplasm and nucleus.

The nuclear envelope is composed of two concentric lipid bilayers termed the **inner** and **outer nuclear membranes.** The outer nuclear membrane is continuous with the rough endoplasmic reticulum and shares some of its functions, including the presence of ribosomes. It also has unique proteins and functions. For example, it contains proteins that help link the nuclear interior with the cytoskeleton. A fibrous **nuclear lamina** of intermediate filaments supports the inner nuclear membrane in many eukaryotes. These and other inner nuclear membrane proteins mediate interactions of the envelope with chromatin. The inner and outer nuclear membranes are separated by an approximately 50-nm **luminal space** that is continuous with the lumen of the endoplasmic reticulum. **Nuclear pore complexes** and the associated pore membrane bridge both nuclear membranes and provide the primary route for communication between the nucleus and cytoplasm during interphase.

Structure and Assembly of the Nuclear Lamina

The nuclear lamina is a thin protein meshwork composed of type V intermediate filament proteins called **nuclear lamins** (Figs. 9.6 and 9.7). Lamins can be divided into two families. **Lamin A** is encoded by a gene that gives rise to four major polypeptides (including **lamin C**) by alternative splicing (see Fig. 11.6). Members of the **lamin B** family are the products of two distinct genes. The various families of lamin proteins assemble into distinct fibrous networks (Fig. 9.6), exhibit different patterns of gene expression, and appear to have distinct roles in nuclear structure.

The pattern of lamin gene expression depends on the cell type and stage of development. The lamina of embryonic stem cells and early embryos is comprised of B-type lamins. Lamins A and C typically appear later in development as cells begin to differentiate, and their expression varies in different cell types. This variation in lamina composition may contribute to different patterns of gene expression and mechanical stability of the nucleus. Lamins A/C promote nuclear stiffness, whereas nuclei containing only B-type lamins are more elastic.

Like other intermediate filament proteins (see Fig. 35.2), nuclear lamins have a central, rod-like domain that

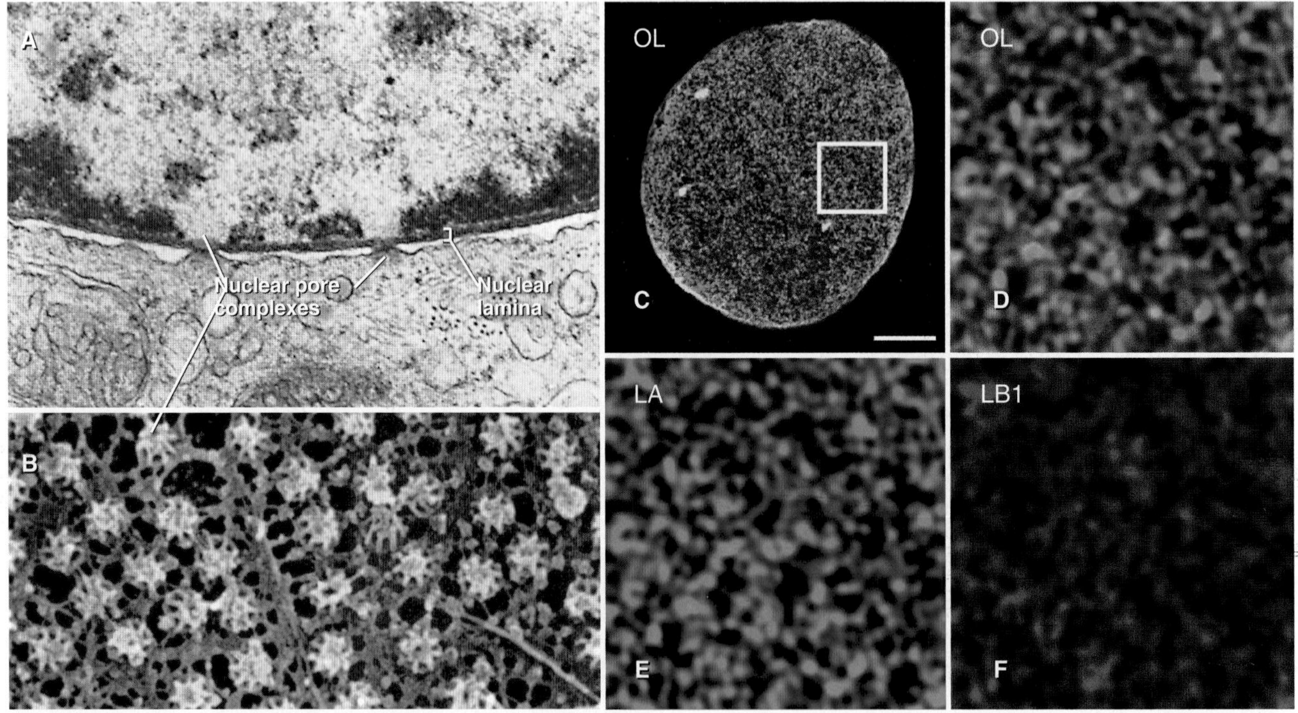

FIGURE 9.6 NUCLEAR LAMINA. A, Thin-section electron micrograph of a nuclear envelope with a prominent nuclear lamina and nuclear pores. **B,** Field emission scanning electron micrograph of the inner surface of an amphibian oocyte nuclear envelope. The nuclear pores are prominent, protruding above the underlying nuclear lamina. **C–F,** Visualization of lamins A and B1 in a HeLa (Henrietta Lacks) cell nucleus by structured illumination superresolution microscopy. Both lamins form short filaments that mostly do not colocalize. OL, overlay. (**A,** For reference, see Fawcett DW. *The Cell.* Philadelphia: WB Saunders; 1981, Fig. 156 [top]. **B,** From Zhang C, Jenkins H, Goldberg MW, et al. Nuclear lamina and nuclear matrix organization in sperm pronuclei assembled in *Xenopus* egg extract. *J Cell Sci.* 1996;109:2275–2286. **C–F,** From Shimi T, Kittisopikul M, Tran J, et al. Structural organization of nuclear lamins A, C, B1, and B2 revealed by superresolution microscopy. *Mol Biol Cell.* 2015;26:4075–4086.)

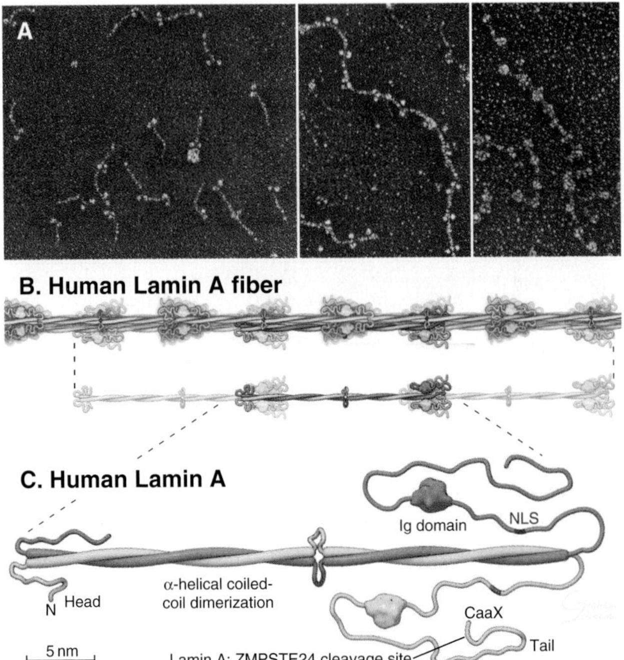

FIGURE 9.7 LAMIN ORGANIZATION AND ASSEMBLY. A, Several stages in the assembly of isolated lamin B dimers into filaments in vitro. The dimers at left have two globular heads at the C-terminal end of a rod that is 52 nm long. **B–C,** Diagram of the structural organization of the nuclear lamins. The sequence CaaX (see text) is a signal for the attachment of a farnesyl group. NLS, nuclear localization sequence. (**A,** From Heitlinger E, Peter M, Haner M, et al. Expression of chicken lamin B2 in *Escherichia coli*: characterization of its structure, assembly, and molecular interactions. *J Cell Biol*. 1991;113: 485–495.)

is largely α-helical (Fig. 9.7). The basic building block of lamin assembly is a dimeric α-helical coiled-coil (see Fig. 3.10) of two identical parallel polypeptides. Lamin dimers self-associate end to end to form protofilaments that associate laterally in a process that is under active investigation.

The coiled-coil is followed by a large C-terminal domain with a central globular fold and containing a nuclear localization sequence (see later section) that promotes the rapid import of newly synthesized lamin precursors into the nucleus. In most lamins, the C-terminus acquires a lipid posttranslational modification that targets them to the nuclear membrane. This involves enzymatic addition of the C15-isoprenoid hydrocarbon tail **farnesyl** (see Figs. 13.10 and 20.15). The farnesyl group is added to a cysteine side chain in an amino acid motif called the CaaX box (Ca_1a_2X, where C is a cysteine located four amino acids from the carboxyl terminus; a_1 is any aliphatic amino acid; a_2 is valine, isoleucine or leucine; and X is usually methionine or serine) at the carboxyl terminus of the protein. This motif was first recognized in the *Ras* proteins (see Fig. 25.7). The aaX residues are removed after addition of the farnesyl group. B-type lamins are not processed further, leaving the farnesylated cysteine at the carboxyl terminus.

In contrast, once it is at the nuclear membrane, prelamin A is processed by a protease called Zmpste24 (zinc metalloprotease similar to yeast Sterile 24) that clips off 15 additional amino acids from the C-terminus including the farnesylated cysteine, thereby loosening its association with the membrane. Possibly as a result of this, some A-type lamins also distribute throughout the nucleoplasm. These intranuclear lamins have been suggested to have roles in cell-cycle regulation (see next section).

The assembled lamina is tethered to the inner nuclear membrane both by the farnesyl group and by interactions with integral membrane proteins (see next section). The surface of the lamina facing the nuclear interior also interacts with the chromosomes. Thus, the lamina and its associated proteins both serve as a structural support for the nuclear envelope and influence chromosome distribution and function within the nucleus (see later).

Proteins of the Inner Nuclear Membrane

Several hundred integral membrane proteins are associated with the inner nuclear membrane, often in in a tissue specific manner. Of these, the lamin B receptor, LAP2 (lamina-associated protein 2), emerin, MAN1, SUN1, and SUN2 have been characterized in detail. Some inner nuclear membrane proteins bind lamins to help anchor the lamina polymer to the membrane and many can interact with chromatin. The lamin B receptor binds heterochromatin protein HP1 (Fig. 9.8) and links the envelope to condensed chromatin. Codisruption of the lamin B receptor and lamin A releases most heterochromatin from the nuclear periphery.

The LEM domain, a 40-amino-acid motif common to several nuclear proteins, including LAP2, emerin, and MAN1, binds to an abundant small protein called barrier-to-autointegration factor (BAF), so named for a separate role facilitating viral genome integration for HIV. BAF binds directly to DNA and to histones and functions in organizing chromatin across the cell cycle.

LAP2 can affect chromatin organization in multiple ways. Some of its several splice variants lack the transmembrane region for inner nuclear membrane association and are soluble. Both soluble and transmembrane forms have the LEM domain and bind BAF, but the transmembrane forms can also bind a histone deacetylase. Interactions of intranuclear lamin A and a soluble splice variant of LAP2 are important for cell-cycle regulation by forming a complex with the tumor suppressor retinoblastoma protein (pRb; see Chapter 41). This, in turn, regulates the transcription factor E2F (see Fig. 41.9), which is important for activating the G_2-to-S transition. Several other inner nuclear membrane proteins also bind transcriptional activators, in some cases sequestering them at the nuclear periphery away from their gene targets.

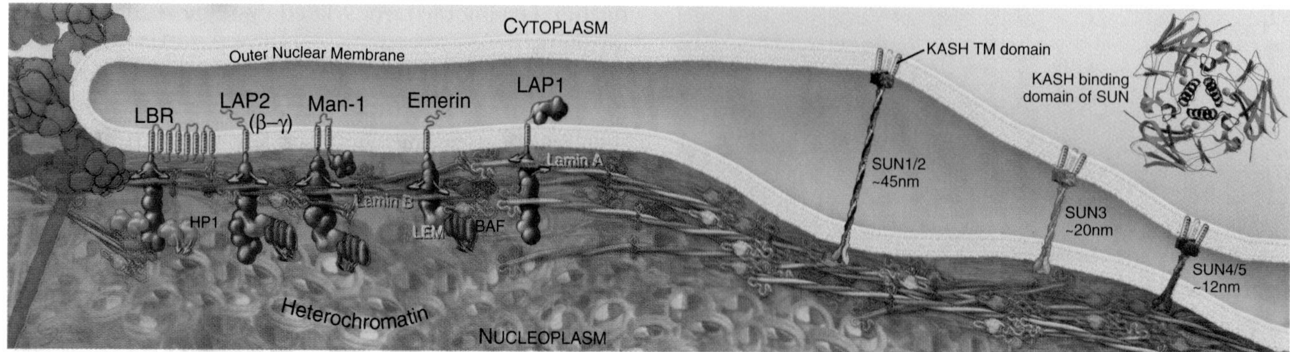

FIGURE 9.8 INTEGRAL PROTEINS OF THE INNER NUCLEAR MEMBRANE. Lamin B receptor (LBR), lamina-associated protein 2 (LAP2), Man-1, and emerin all bind lamin B. LBR associates with chromatin via HP1. The other three associate with chromatin via the barrier-to-autointegration factor (BAF). Emerin and lamina-associated protein 1 (LAP1) also bind to lamin A. The α form of LAP2 is not membrane associated and is not shown here. Three isoforms of SUN proteins link the inner nuclear membrane to KASH domain proteins of the outer nuclear membrane. KASH proteins asssociate with the cytoskeleton. (SUN proteins from Sosa BA, Kutay U, Schwartz TU. Structural insights into LINC complexes. *Curr Opin Struct Biol.* 2013;23[2]:285–291. For reference, see Protein Data Bank [www.rcsb.org] file 4DXT [ribbon diagram of SUN/KASH].)

The SUN proteins bind lamin A, and then connect across the nuclear envelope lumen to huge KASH domain proteins in the outer nuclear membrane that in turn bind to all three major cytoplasmic filament systems, actin filaments, intermediate filaments, and microtubules (see Fig. 9.8 and Chapters 33 to 35). Thus, the lamina is linked to the rest of the cytoskeleton. Deletions or mutations in several lamina proteins, including the SUN proteins, reduce the mechanical stability of the cell and interfere with cell migration. SUN proteins are also important for maintaining the uniform spacing of the nuclear envelope lumen. Their disruption results in uneven separation of the inner and outer nuclear membranes.

These diverse functions in genome organization and regulation, cell cycle regulation, signaling cascades and cell and nuclear mechanical stability could explain the link between mutations in nuclear envelope proteins and human disease (see later).

Role of the Nuclear Envelope in Genome Organization

A high-throughput method revealed that the nuclear lamina has an important role in chromosome organization within nuclei. The method uses a DNA-modifying enzyme fused to a lamin protein, so nearby DNA is modified and can be mapped along the genome. In human cells, approximately 40% of the genome is found in 1000 to 1500 LADs (lamina-associated domains; Fig. 9.9) ranging in size from 10 kb to approximately 10 Mb. Analysis of single cells revealed that approximately 15% of LADs are associated with the lamina in most cells, with the remainder varying from cell to cell. The constitutive LADs have low transcriptional activity. Indeed, heterochromatin-associated histone marks such as H3K9me3 (see Fig. 8.7) promote association of particular chromosomal regions with the lamina. The LADs at a distance from the nuclear periphery are either associated with a similar repressive compartment surrounding nucleoli or are in the nuclear interior. Interestingly, association of the chromosomes with the nuclear envelope is perturbed in some nuclear envelope-associated diseases (see the next section).

Chromatin interactions with nuclear pores can have both positive and negative effects on gene expression. In mammalian cells, the chromatin near pores appears less condensed (less heterochromatic) than most chromatin adjacent to the lamina. The significance of these interactions is still under study.

Disassembly of the nuclear envelope during mitosis in metazoa releases the chromosomes so that they can be segregated to the daughter cells by the cytoplasmic mitotic spindle (see Fig. 44.1). Mitotic segregation of chromosomes to daughter cells takes place within the nucleus in many other eukaryotes, including yeasts.

Nuclear Envelope Defects Lead to Human Diseases

In 1994, a gene mutated in patients with human X-linked Emery-Dreifuss muscular dystrophy was found to encode a protein of the inner nuclear envelope. The gene was named emerin. This link between the nuclear envelope and human disease was the tip of a huge iceberg. Genetic defects in nuclear envelope proteins cause at least 20 disorders, including muscular dystrophies, lipodystrophies, and neuropathies (diseases of striated muscle, fatty tissue, and the nervous system, respectively). The most dramatic of these is Hutchinson-Gilford progeria syndrome (Fig. 9.10). Affected individuals are essentially normal at birth, but they appear to age rapidly and die in their early teens of symptoms (including atherosclerosis and heart failure) that are typically associated with extreme age.

More than 500 mutations scattered across the gene encoding lamin A/C cause at least 15 different diseases,

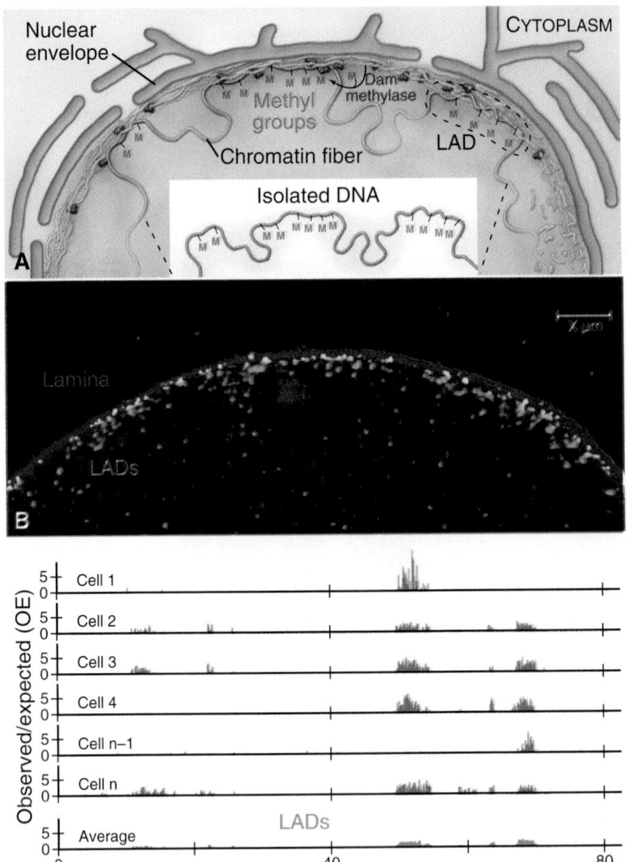

FIGURE 9.9 NUCLEAR LAMINA HELPS ORGANIZE THE CHROMATIN INTO FUNCTIONAL DOMAINS. A, Dam methylase fused to a lamin protein methylates the DNA in chromatin that is closely associated with the nuclear lamina. Isolation of the DNA allows the methylation sites to be mapped. **B,** Constitutive LADs associate with the lamina even after the cell has gone through mitosis and the lamina has been disassembled and reassembled. **C,** Sites of increased methylation on chromosome 17 from six single human cells, plus an average of the entire population. Lamina-associated domains (LADs) are indicated. (**B–C,** Modified from Kind J, Pagie L, de Vries SS, et al. Genome-wide maps of nuclear lamina interactions in single human cells. *Cell*. 2015;163:134–147.)

collectively termed laminopathies, some of which are variants of the diseases mentioned above (Fig. 9.10). At least two laminopathies are also linked to mutations in the Zmpste24 protease. Some of the symptoms of laminopathies can be modeled in mice, where loss of lamin A causes nuclear envelope defects and leads to a type of muscular dystrophy.

The most surprising aspect of the laminopathies is the fact that except for premature aging, the defects linked to each mutation are limited to a few tissues such as striated muscle, even though lamins A/C are ubiquitous in differentiated cells throughout the body. Lamin mutations appear to compromise the stability of the nuclear envelope, so it has been suggested that muscle nuclei might be particularly sensitive to these mutations, owing to mechanical stress during contraction. However, this mechanism cannot account for the link between lamin mutations and lipodystrophy—fat is not a force-generating tissue—neuropathy, or progeria.

An alternative suggestion is that these mutations change interactions between the inner nuclear membrane and chromatin and this alters gene expression patterns. Cells from patients with Hutchinson-Gilford progeria syndrome show signs of aging in culture that are accompanied by dramatic alterations in heterochromatin (see Fig. 8.6), but changes in gene expression are relatively small and vary between patients.

Nuclear Pore Complexes

In a typical growing cell, nearly all traffic between the nucleus and cytoplasm passes through approximately 3000 channels, called **nuclear pore complexes,** that bridge the inner and outer nuclear membranes (Fig. 9.11). Nuclear pore complexes have a scaffold consisting of three stacked rings each with eightfold symmetry. **Cytoplasmic** and **nuclear rings** flank a prominent **spoke ring** that is intimately associated with the pore membrane linking the inner and outer nuclear membranes. The nuclear ring is anchored to the nuclear lamina. A less-prominent fourth **luminal ring** surrounds the pore membrane in the NE lumen. The minimum diameter of the central channel through the pore is approximately 40 nm, and the channel is approximately 50- to 70-nm long.

Eight filaments project outward from both the nuclear and cytoplasmic rings. These are involved in docking of macromolecules to be transported through the pore. The nuclear filaments are linked at their outer ends by a terminal ring, much like the wire that secures the cork on a champagne bottle. This structure is called the **nuclear basket.**

Vertebrate nuclear pore complexes are large structures with a mass of approximately 90 to 120 million Da as assessed by electron cryomicroscopy. Core components identified by mass spectrometry account for approximately 70 million Da of the mass. Yeast nuclear pores are similar in overall structure but about half the mass.

The protein composition of nuclear pore complexes is remarkably conserved. Approximately 30 core proteins, called **nucleoporins** (Fig. 9.12), are present in multiples of eight copies. Mass differences between electron cryomicroscopy and mass spectrometry measurements may be accounted for by transport factors and other auxiliary subunits that do not have a key structural role.

Two multiprotein complexes, the 10-member Y complex (named because of its shape) and the five-member Nup93 complex make up the scaffold of the pore. The cytoplasmic and nuclear rings are assembled

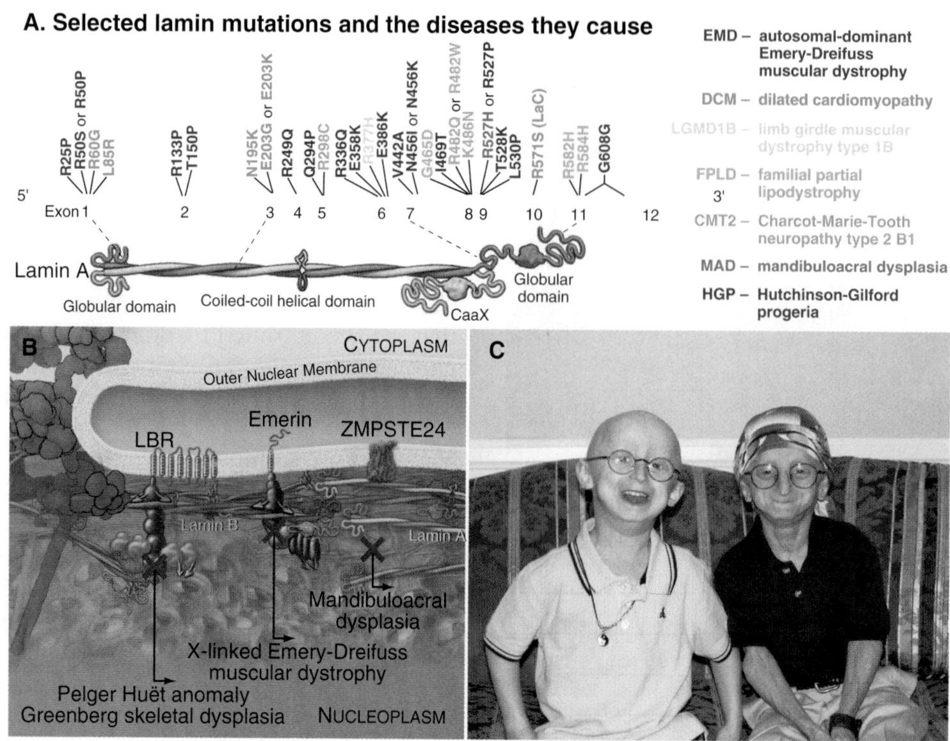

FIGURE 9.10 HUMAN DISEASES ASSOCIATED WITH NUCLEAR ENVELOPE ABNORMALITIES. A, Some of the mutations in the gene encoding lamin A that are associated with human disease. The G608G mutation makes no change in the protein sequence but creates a splice site leading to the loss of 50 amino acid residues from lamin A, leading to an impaired processing of prelamin A and generation of stably farnesylated lamin A. This mutation causes Hutchinson-Gilford progeria. Colored numbers at the top refer to the amino acid that has been changed by each mutation. Mutations are distributed in exons all across the gene as shown. **B,** Mutations in other nuclear envelope proteins cause similar diseases. Three examples are shown. **C,** Two young boys with the premature aging disorder Hutchinson-Gilford progeria. Sam Berns *(left)* with friend John Tacket, Progeria Research Foundation Youth Ambassador. (**A,** Modified from Mounkes L, Kozlov S, Burke B, et al. The laminopathies: nuclear structure meets disease. *Curr Opin Genet Dev*. 2003;13:223–230. **C,** Courtesy the Progeria Research Foundation, Peabody, MA; http://www.progeriaresearch.org.)

from 16 copies each of the Y complex. The Nup93 complex forms the framework of the spoke ring and interacts with four nucleoporins having transmembrane domains that bend and fuse the pore membrane. Several of these proteins share structural features with clathrin-like proteins that coat membrane transport vesicles (see Fig. 21.8), so they may have a common evolutionary origin.

Eleven of the 30 nucleoporins contain repeats of the dipeptide FG (phenylalanine-glycine). Two common versions include XFXFG and GLFG. In all, the pore contains approximately 5000 of these FG repeats in highly flexible intrinsically disordered regions of the proteins. FG nucleoporins are anchored to the pore scaffold with their FG repeat regions projecting into the central pore, where they form the transport barrier (see later).

Three experiments show that nucleoporins are required to transport proteins into the nucleus. First, antibodies to nucleoporins inhibit transport when added to isolated nuclei or when injected into live cells. Second, lectins, such as wheat germ agglutinin (which binds specifically to sugars attached to many nucleoporins), inhibit transport in similar experiments. Third, nuclear pore complexes assembled in *Xenopus* egg extracts (see Box 40.3) in the absence of the highly conserved nucleoporin **p62,** the defining member of the FG-rich p62 complex of three nucleoporins, appear structurally normal but are inactive in transport.

In metazoans, nuclear pore complexes are remarkably stable, with the proteins apparently persisting for the lifetime of the cell. New pore complexes continue to assemble throughout interphase, but they disassemble into soluble subcomplexes during mitosis. During the telophase stage of mitosis, pore complex reassembly begins with binding of the Y complex to chromatin. The Y complex then interacts with transmembrane nucleoporins and the Nup93 complex, which recruits factors that bend and fuse the membranes, forming the pore. If the Nup93 complex is depleted from *Xenopus* egg extracts, nuclear membranes form around added nuclei but are devoid of pores.

Traffic Between Nucleus and Cytoplasm

The nuclear pore complex is a highly efficient conduit that can allow the passage of up to approximately 100 MDa of cargo per second. Traffic leaving the nucleus

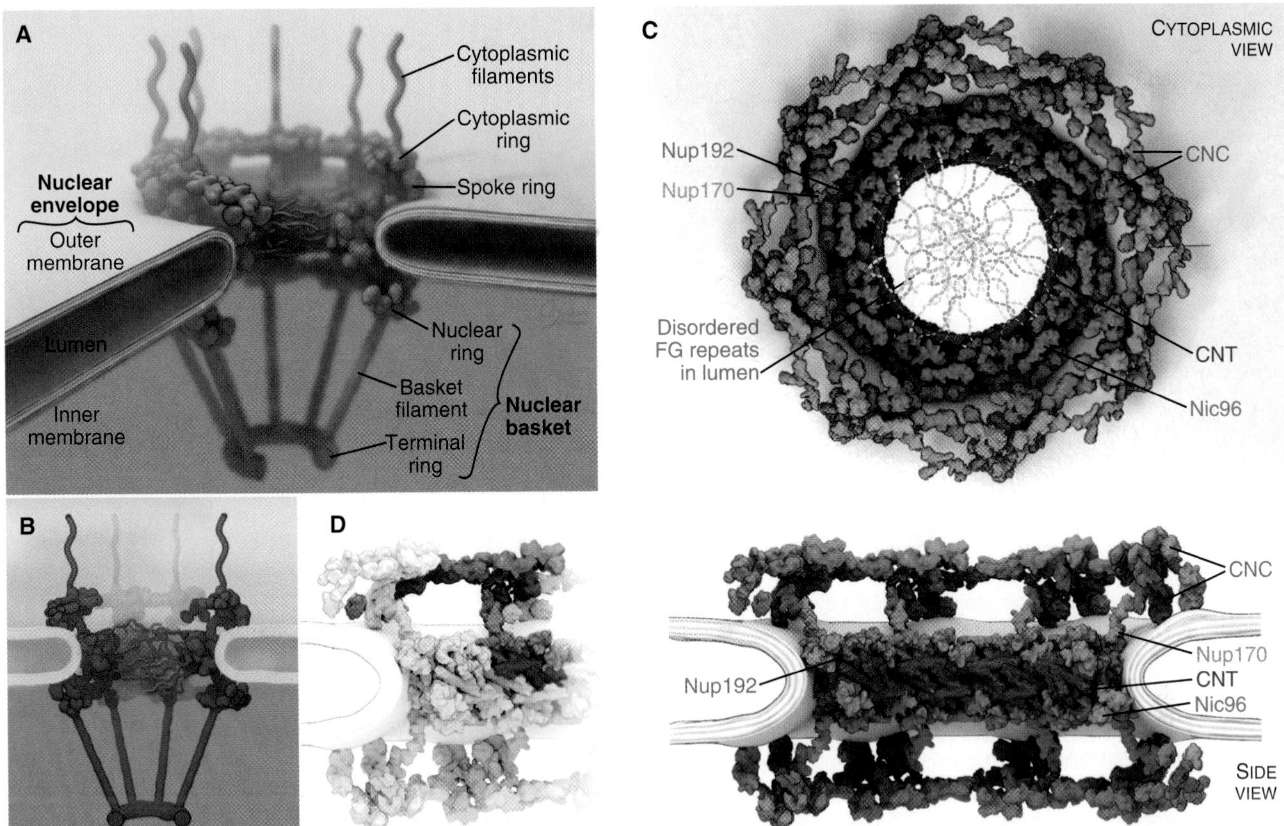

FIGURE 9.11 NUCLEAR PORE COMPLEX. A–B, Three-dimensional and central section views of models of the human nuclear pore complex. **C,** Two views of the molecular organization of the nuclear pore based on three-dimensional reconstructions of cryoelectron micrographs. The pore has eight-fold symmetry in the plane of the nuclear envelope and two-fold symmetry perpendicular to the nuclear envelope. Colored protein subunits are identified with labels; the membrane is gray. Disordered FG repeats *(green dotted lines)* fill the central pore. **D,** Detail of the molecular model illustrating the two-fold symmetry of the protein subunits perpendicular to the nuclear envelope, with colored subunits above and gray subunits below. (**C–D,** For reference, see Protein Data Bank file 5A9Q and von Appen A, Kosinski J, Sparks L, et al. In situ structural analysis of the human nuclear pore complex. *Nature.* 2015;526:140–143.)

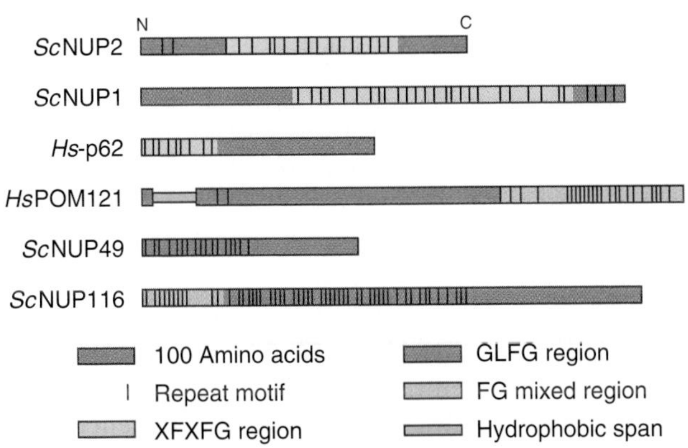

FIGURE 9.12 SEQUENCE ORGANIZATION OF SEVERAL NUCLEOPORINS, THE STRUCTURAL COMPONENTS OF THE NUCLEAR PORES. Nucleoporins contain combinations of repeated sequences as shown. Letters refer to the amino acids (see Fig. 3.2). The hydrophobic FG (phenylalanine-glycine) repeats facilitate nuclear trafficking through the pores by interacting specifically with transport factors carrying cargo.

includes messenger ribonucleoproteins (mRNPs), ribosomal subunits, and transfer RNAs (tRNAs), all of which must be transported to the cytoplasm to function in protein synthesis. Traffic entering the nucleus includes transcription factors, chromatin components, and ribosomal proteins. Other molecules follow more complex routes. Small nuclear RNAs (snRNAs) are exported to the cytoplasm to acquire essential protein components; they are then reimported into the nucleus, where they undergo further maturation steps before functioning in RNA processing. Individual pores can simultaneously transport components in both directions.

Nuclear pores have constitutive peripheral channels through which solutes and small proteins of up to 30 to 40 kD (~5–10 nm) can diffuse passively. However, the pores can also actively transport much larger macromolecular complexes via the central channel. Almost all physiological traffic through the pores, even of small molecules, is a facilitated process that involves specific carrier proteins traversing the central channel. For example, the 28-kD NTF2 dimer (the Ran transporter; see later) traverses the pore approximately 120 times

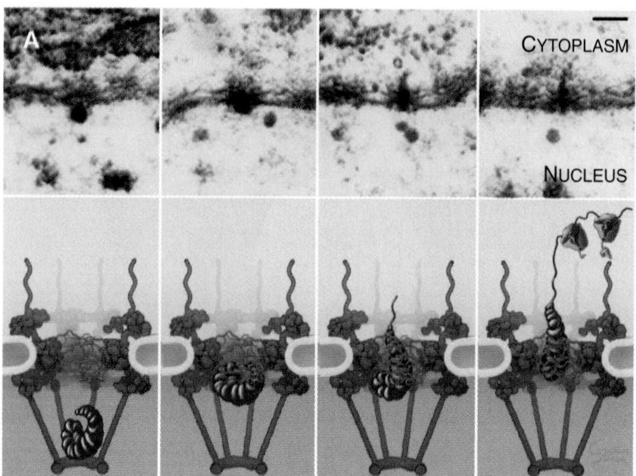

FIGURE 9.13 Electron micrographs *(upper panels)* and an artist's rendition *(lower panels)* show deformation of a large RNP particle as it passes through the nuclear pore complex (cytoplasm *[top]*; nucleus *[bottom]*). This RNA encodes a secreted protein, with a molecular weight of about 1 million Da, from the salivary gland of the fly *Chironomus tentans.* Once in the cytoplasm, the 5′ end of the RNA docks with ribosomes and begins synthesis of its protein even before the passage of the remainder of the RNP through the pore has been completed. (From Mehlin H, Daneholt B, Skoglund U. Translocation of a specific premessenger ribonucleoprotein particle through the nuclear pore studied with electron microscope tomography. *Cell.* 1992;69:605–613.)

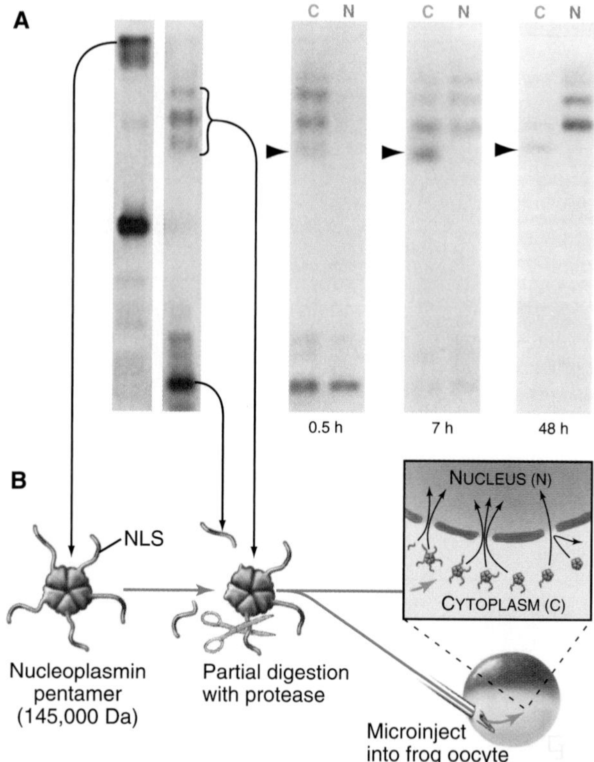

FIGURE 9.14 IDENTIFICATION OF A NUCLEAR LOCALIZATION SEQUENCE ON THE PROTEIN NUCLEOPLASMIN. This 29,000-Da protein exists in vivo as a pentameric complex with a molecular weight of 145,000. The monomer is small enough to diffuse passively through the nuclear pores, but the pentamer is too large to do so. **A,** Gentle cleavage of the pentamer with a protease removes a relatively small peptide from one end of the protein *(left two gel lanes).* When the cleaved pentamers were labeled with radioactivity and injected into the cytoplasm of a *Xenopus* oocyte, it was found that four species were produced that could still migrate into the nucleus and one species was produced that could not *(right three pairs of gel lanes).* **B,** The interpretation of this experiment is that each nucleoplasmin polypeptide contains a "tail" that can be removed by proteolysis and that this tail contains a nuclear localization sequence. Each pentamer can migrate into the nucleus as long as it retains at least one polypeptide with a tail. Tailless pentamers remain stuck in the cytoplasm. (**A,** From Dingwall C, Sharnick SV, Laskey RA. A polypeptide domain that specifies migration of nucleoplasmin in the nucleus. *Cell.* 1982;30:449–458.)

more rapidly than does the 27-kD green fluorescent protein.

The pore gate opens to a maximum of approximately 40 nm, but larger particles can squeeze through, provided that they are deformable. This is well documented for export of a well-studied enormous RNA that associates with roughly 500 packaging proteins to make an RNP particle approximately 50 nm in diameter. The RNP is deformed into a rod-shaped structure as it squeezes through the pore (Fig. 9.13). Rigid particles cannot usually exceed the 30- to 40-nm limit.

Integral proteins of the inner nuclear membrane enter the nucleus by diffusion in the plane of the membrane. The lamin B receptor is highly mobile in the endoplasmic reticulum (ER), its site of synthesis, and rapidly diffuses to the nuclear envelope. There, it transits to the inner membrane through the peripheral channels of the nuclear pore complex. Once in the inner nuclear membrane, it becomes fixed in place, presumably by binding to the lamina and/or chromatin. This mechanism involving lateral diffusion and retention is a common mode of membrane protein translocation into the nucleus, although conventional transport through the pore may also occur (see later).

Proteins that are imported into the nucleus bear a **nuclear localization sequence (NLS),** also called a nuclear localization *signal,* that is recognized by specific carrier proteins called transport receptors (Figs. 9.14 and 9.15). The best-studied NLS is a patch of basic amino acids similar to the sequence PKKKRKV (single-letter amino acid code; see Fig. 3.2), first identified on the simian virus 40 (SV40) large T antigen. A point mutation, yielding PKNKRKV, inactivates this sequence as an NLS. A related type of bipartite NLS features two smaller patches of basic residues separated by a variable spacer (<u>KRP</u>AATKKAGQA<u>KKKK</u> [critical residues are underscored]). These two types of sequences are referred to as basic NLSs. Basic NLSs function autonomously and can direct the migration of a wide range of molecules into the nucleus in vivo. In one example, colloidal gold particles up to 23 nm in diameter coated with nucleoplasmin (a protein with a bipartite basic NLS) are transported through nuclear pores (Fig. 9.16). NLSs vary

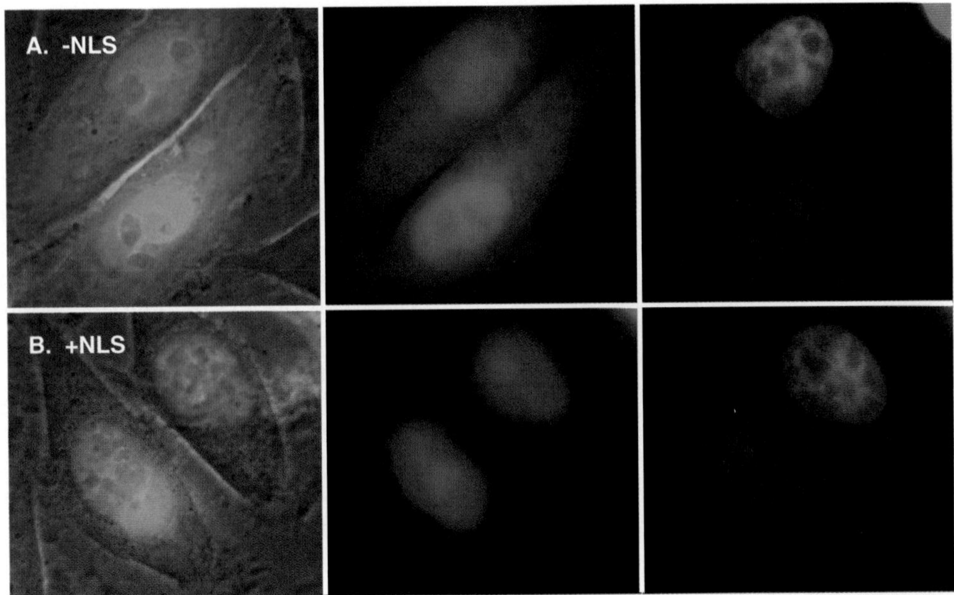

FIGURE 9.15 ICAD (inhibitor of caspase-activated DNase) protein (see Fig. 46.13) was fused to the green fluorescent protein (GFP; *green* here) and expressed in cultured cells. The DNA is *blue*. **A,** A mutant form of ICAD:GFP fusion protein lacking the ICAD nuclear localization sequence (NLS) accumulates randomly throughout the cell. **B,** The intact ICAD:GFP fusion protein with NLS accumulates quantitatively in the nucleus. (Courtesy K. Samejima, University of Edinburgh, United Kingdom.)

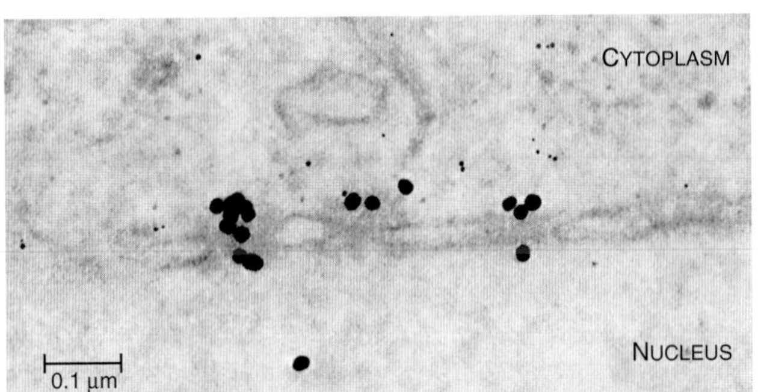

FIGURE 9.16 NUCLEAR LOCALIZATION SEQUENCE OF NUCLEOPLASMIN CAN CAUSE COLLOIDAL GOLD PARTICLES TO BE TRANSPORTED INTO THE CELL NUCLEUS. A thin-section electron micrograph shows gold particles coated with nucleoplasmin crossing the nuclear envelope by passing through the nuclear pore complexes. Much smaller gold particles coated with bovine serum albumin (BSA) remain in the cytoplasm. Both sets of gold particles were microinjected into the cytoplasm of *Xenopus* oocytes, and the cells were processed 1 hour later for electron microscopy. Scale bar: 0.1 μm. (From Dworetzky SI, Lanford RE, Feldherr CM. The effects of variations in the number and sequence of targeting signals on nuclear uptake. *J Cell Biol*. 1988;107:1279–1287.)

in size and sequence, and are recognized by a number of different kinds of transport receptors. For example, an alternative type of NLS rich in glycine promotes nuclear import by a similar mechanism (see later) but using a different transport receptor.

Many proteins exported from the nucleus bear a **nuclear export sequence (NES)** that is recognized by transport receptors related to those used for nuclear import (Fig. 9.17). Like import signals, these signals vary in size and complexity. The HIV I Rev protein provides one example of a leucine-rich sequence (LQLPPLERLTL) that is recognized by the carrier CRM1. Certain RNA sequences or structures may also serve as NESs.

The following is a brief thumbnail of protein import into the nucleus (Fig. 9.18). A protein with an NLS (known as **cargo**) binds to an **import receptor** either by itself or in combination with an **adapter** molecule, forming a complex that then passes through pores into the nucleus. There, the cargo and adapter (if used) are displaced from the import receptor. The adapter then releases its cargo and is transported back to the cytoplasm as the cargo of an export receptor. Import receptors also shuttle back through pores, where they can meet more cargo or cargo/adapter complexes. Molecules exported from the nucleus use a variation of this cycle, being picked up by the transport machinery in the nucleus and discharged in the cytoplasm.

The key to this system is that it is *vectorial:* Nuclear components are transported into the nucleus while components that function in the cytoplasm are transported out. This means that each carrier picks up its cargo on one side of the nuclear envelope and deposits it on the other. This directionality is regulated by a simple yet elegant system involving Ran, a small guanine triphosphatase (GTPase [see Figs. 4.6 and 4.7 for background material on GTPases]), and associated factors.

Components of Nuclear Import and Export

The nuclear import and export system involves many components, but the general principles of its operation are simple. To understand how it works, this section

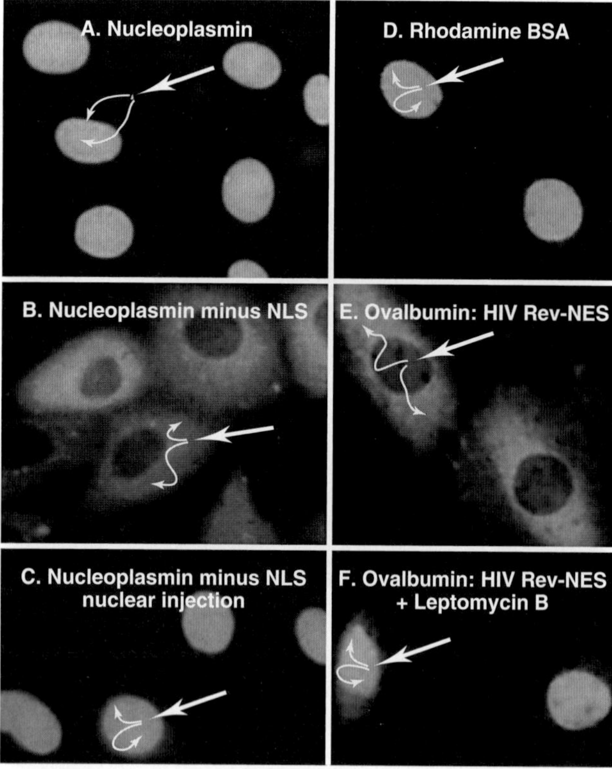

FIGURE 9.17 DEMONSTRATION OF THE EXISTENCE OF SPECIFIC NUCLEAR IMPORT AND EXPORT SIGNALS ON PROTEINS. *Left,* Nuclear import. **A,** Nucleoplasmin microinjected into the cytoplasm rapidly migrates into the nucleus. **B,** Nucleoplasmin lacking its nuclear localization sequence (NLS), when microinjected into the cytoplasm, stays in the cytoplasm. **C,** Nucleoplasmin lacking its NLS microinjected into the nucleus stays in the nucleus. *Right,* Nuclear export. **D,** Fluorescently labeled bovine serum albumin (BSA) microinjected into the nucleus stays in the nucleus. **E,** When ovalbumin conjugated to the nuclear export sequence (NES) of the HIV (the virus that causes AIDS) Rev protein is microinjected into the nucleus, it rapidly migrates into the cytoplasm. **F,** In the presence of leptomycin B (a drug that inhibits the activity of the nuclear export receptor CRM1), ovalbumin conjugated to the NES of HIV Rev protein stays in the nucleus after microinjection. (**A–C,** From Dingwall C, Robbins J, Dilworth SM, et al. The nucleoplasmin nuclear location sequence is larger and more complex than that of SV-40 large T antigen. *J Cell Biol.* 1988;107:841–849, copyright the Rockefeller University Press. **D–F,** From Fukuda M, Asano S, Nakamura T, et al. CRM1 is responsible for intracellular transport mediated by the nuclear export signal. *Nature.* 1997;390:308–311.)

first introduces several of the components (see Cell SnapShot 2) and then describes one transport event in detail.

Adapters

Adapters bind to the NLS or NES sequences on some cargo molecules and also to particular regions on receptors. The best-characterized adapter is **importin α,** which is responsible for recognition of small basic NLS sequences and works together with the transport receptor importin β (see later) in nuclear transport. Importin α consists of a highly flexible N-terminal NLS-like importin β-binding domain followed by 10 repeats of a helical motif (the Armadillo repeat [Fig. 9.18D]) that give the structured portion of the molecule a slug-like shape. The importin β-binding motif can bind either the NLS-binding region on importin β or the NLS-binding domain on importin α itself (the "belly" of the slug). The latter provides an autoinhibitory mechanism that is thought to be important in regulating the release of cargo in the nucleus at the end of an import cycle. Binding to importin β uncovers the NLS binding site on importin α so that it can bind cargo more efficiently.

Other nuclear trafficking pathways use different adapters. For example, two adapters bridge between snRNA and the export receptor CRM1 during snRNA export from the nucleus.

Nuclear Transport Receptors

Except for mRNP export from the nucleus (which uses special transport factors), all nuclear trafficking receptors are related to **importin** β, the import receptor for proteins bearing a basic NLS. At least 20 nuclear transport receptors are known in vertebrates (14 in yeast). These proteins are also called **karyopherins.** Some function in nuclear import, but others function in export. Importin β consists entirely of 19 copies of a helical protein interaction motif called a HEAT repeat, giving the protein the shape of a snail-like superhelix with the potential to interact with a large number of protein ligands. All importin β family members have a binding site for the Ran GTPase (Fig. 9.18D). Importin β binds many NLSs directly but also interacts with other cargoes via the importin α adapter. Nucleoporin FG repeats sandwich between importin β HEAT repeat helices during passage through the pore channel.

Directionality/Recycling Factors

Ran-GTPase and its bound nucleotides inform nuclear trafficking receptors whether they are located in the nucleus or cytoplasm. Ran-GTP (Ran with bound guanosine triphosphate [GTP]) *dissociates* import complexes but is required to *form* export complexes. The system imparts directionality because Ran-GTP is converted to Ran-GDP (guanosine diphosphate) in the cytoplasm and Ran-GDP is converted to Ran-GTP in the nucleus.

Like other small GTPases, Ran has low intrinsic GTPase activity, but interactions with binding proteins (Ran-BP1 or Ran-BP2) and a GTPase-activating protein called **Ran-GAP1** stimulate GTP hydrolysis. Ran-BP1 is anchored in the cytoplasm. Ran-BP2 is a component of the fibers projecting from the nuclear pore into the cytoplasm. This huge (>350 kD) protein can bind up to four Ran molecules as well as Ran-GAP1 and provides a structural scaffold for the conversion of Ran-GTP into Ran-GDP at the surface of the pore. Because Ran-BP1 and Ran-BP2 are both anchored in the cytoplasm, Ran-GTP is efficiently converted to Ran-GDP only in the cytoplasm,

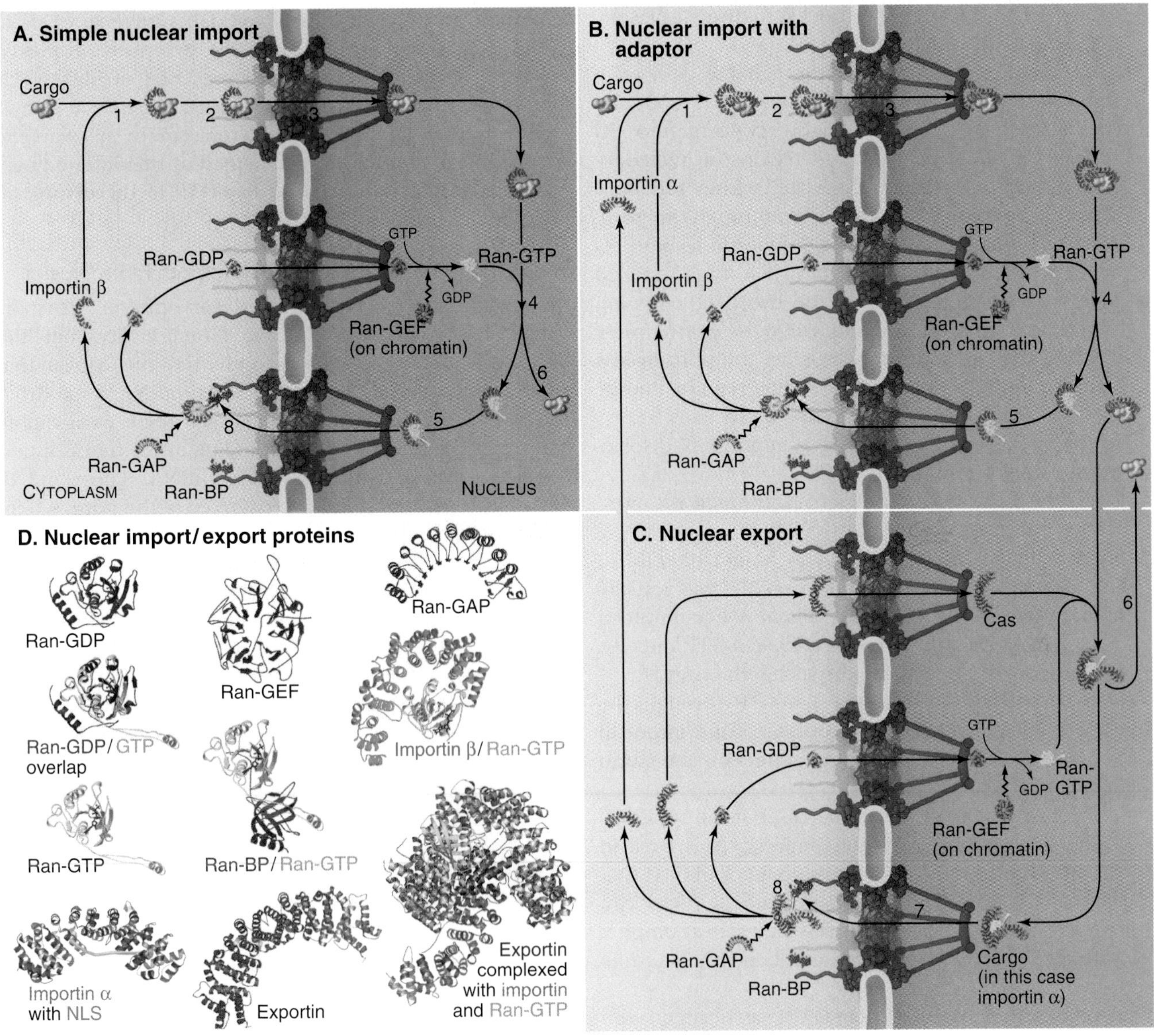

FIGURE 9.18 NUCLEAR TRAFFICKING OF MACROMOLECULES. Nuclear import of a cargo by the import receptor importin β without **(A)** or with **(B)** the use of an adapter protein. **C,** Export of a cargo by the importin β-related export receptor Cas. In this case, the cargo is the import adapter importin α. Directionality is given by Ran. Ran-GTP (guanosine triphosphate) releases import cargoes in the nucleus and is required for formation of the export complex. Numbers refer to the steps described in the text. **D,** Crystal structures of several of the components involved in nuclear transport. (Ribbon models courtesy F. Wittinghofer, MPI Dortmund, Germany.)

yielding a nuclear/cytoplasmic ratio of Ran-GTP of approximately 200:1.

Ran-GDP must reenter the nucleus to be recharged with GTP. Efficient Ran-GDP transport into the nucleus requires nuclear transport factor 2 (NTF2). Back in the nucleus, Ran must release its bound GDP to acquire GTP. GDP dissociation is intrinsically slow but is stimulated by a guanine nucleotide exchange factor (GEF). This protein, called **regulator of chromosome condensation 1 (RCC1),** is tightly associated with chromatin throughout the cell cycle. This allows nuclear import to resume immediately after the nuclear envelope reforms at the end of mitosis. Because Ran is involved in essentially every nuclear trafficking event, the flux of this small protein across the nuclear envelope is enormous—several million molecules per minute in cultured cells.

Description of a Single Import Cycle in Detail

Consider the import into the nucleus of a typical protein (Fig. 9.18):

1. In the cytoplasm, the import complex forms as importin β binds to cargo either directly or complexed with an importin α adapter (the latter is true for cargos containing the very widely studied basic NLS discussed previously).

2. The import complex binds (docks) to the cytoplasmic filaments of the nuclear pore.
3. The complex is transferred through the pore in a process that is still under investigation. A popular model proposes that the highly concentrated FG repeat-containing unstructured regions of nucleoporins associate to form a hydrogel within the pore channel that blocks most diffusion through the pore. Nuclear transport receptors (eg, importin β) bind FG repeats by trapping them between their packed helices. This locally "melts" the hydrogel, allowing the receptor and its bound cargo to drift rapidly through the gel, ultimately crossing the pore in less than 20 ms. This process does not require energy from nucleoside triphosphate hydrolysis.
4. In the nucleus, Ran-GTP binds to importin β, displacing the cargo from it.
5. Importin β/Ran-GTP shuttles back through the pore to the cytoplasm.
6. In the nucleus, if the cargo was bound directly to importin β, it is now free to function. If it was actually a cargo/importin α complex, this now encounters a nuclear export receptor called **CAS.** Ran-GTP and CAS bind tightly to importin α, displacing the cargo.
7. CAS carries importin α and Ran-GTP through the nuclear pores back to the cytoplasm. Thus, importin α functions as an adapter in one direction and cargo in the other.

   The cargo is now in the nucleus, but the system is stalled. The import receptor, importin β, is back in the cytoplasm, but in a complex with Ran-GTP that cannot bind new cargo. The import adapter, importin α, is also in the cytoplasm, but it is locked in a complex with the CAS export receptor and Ran-GTP. The solution to this problem is simple.
8. Ran-BP1, Ran-BP2, and Ran-GAP1 associated with cytoplasmic filaments of the nuclear pore catalyze the hydrolysis of GTP bound to Ran. Ran-GDP dissociates from importin α, readying it for further cycles of nuclear import. In addition, GTP hydrolysis causes the importin α/CAS/Ran-GDP complex to dissociate, allowing CAS to return to the nucleus for further cycles as an export receptor and making importin α available in the cytoplasm to bind more cargo and function as an import adapter. The hydrolysis of GTP on Ran is the only source of chemical energy required to drive the accumulation of proteins in the nucleus against a concentration gradient.

Although there are several names to remember, the nuclear trafficking system is actually quite straightforward, being regulated by the state of the guanine nucleotide bound by Ran. The key point is that the GEF that charges Ran-GDP with GTP is in the nucleus and the Ran-GAPs that promote hydrolysis of GTP bound to Ran are cytoplasmic. Cargo that is meant to be imported into the nucleus is released from its carriers in the presence of high levels of nuclear Ran-GTP. Conversely, cargo that is destined for export to the cytoplasm is picked up by its carriers only in the presence of high levels of nuclear Ran-GTP and is released when the Ran is converted to Ran-GDP in the cytoplasm. In this way, the directionality of transport is defined by the different concentrations of Ran-GDP and Ran-GTP in the cytoplasm and nucleus.

A Distinct Pathway for mRNA Export From Nuclei

Small RNAs are exported by karyopherin transport receptors using Ran-GTP for directionality, but the export of mRNA depends on a different mechanism that includes numerous quality controls. mRNA is exported as very large mRNP complexes that begin to assemble during RNA processing with binding of the transcription export (TREX) complex to the mRNA. These mRNP complexes dock on the inner surface of the pore, where they are subjected to quality control by the exosome (see Fig. 11.8) and other surveillance activities. Incorrectly processed mRNAs are degraded. Correctly processed mRNAs are guided through the nuclear pore by a dimeric transport receptor, Nxf1-Nxt, which is not related to karyopherins, but also interacts with FG repeats. Adenosine triphosphate (ATP), rather than GTP hydrolysis gives directionality to the process. The ATP is used in the cytoplasm by enzymes that change the RNA structure and dissociate Nxf1-Nxt1, thus preventing the RNP from reentering the pore.

Regulation of Transport Across the Nuclear Envelope

Cells regulate nuclear trafficking in several ways. The first of these is to change the number of pores. In rat liver, there are 15 to 20 pores per square micrometer of nuclear envelope (~4000 per nucleus), whereas nuclei of transcriptionally quiescent avian erythrocytes have very few nuclear pore complexes.

Nuclear trafficking is often regulated by phosphorylation near the NLS on the cargo. Phosphorylation adjacent to a basic NLS inhibits nuclear import. This provides a mechanism to regulate the ability of a particular cargo to enter the nucleus in response to cell cycle (see Fig. 43.6) or other signals that can be coupled to specific protein kinase activation.

Traffic across the nuclear envelope is also regulated by masking or unmasking NLSs. A "nuclear" protein whose NLS is covered up is trapped in the cytoplasm. A good example is the regulation of transcription factor nuclear factor κB (NF-κB) by inhibitor of nuclear factor κB (IκB; Fig. 9.19). IκB binds to NF-κB and covers up its NLS. Because IκB also has a nuclear export signal, the NF-κB:IκB complex is entirely cytoplasmic. Following an appropriate signal (see Fig. 10.21C), IκB is degraded. This uncovers the NLS on NF-κB, allowing it to enter the nucleus. This mechanism regulates gene expression

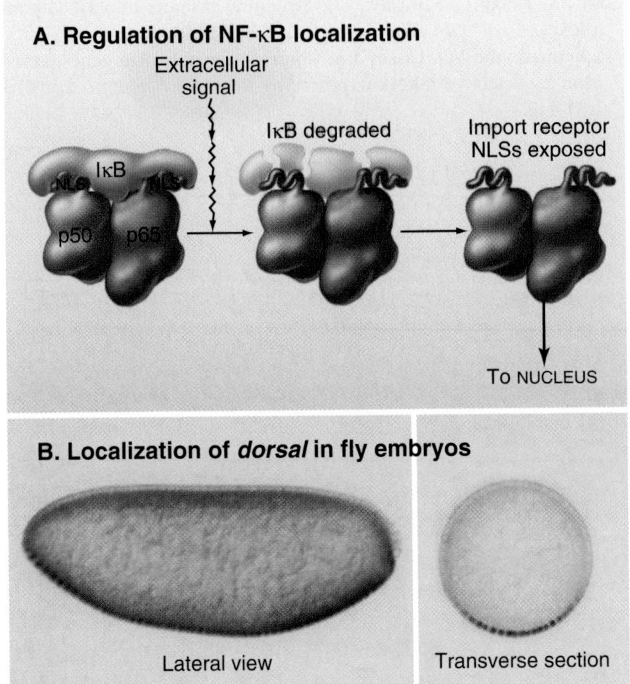

FIGURE 9.19 REGULATION OF NUCLEAR FACTOR κB (NF-κB) LOCALIZATION. A, The transcription factor NF-κB is kept in the cytoplasm as a result of interactions with its inhibitor IκB (inhibitor of nuclear factor κB). IκB holds NF-κB in the cytoplasm in two ways. When it binds NF-κB, it covers up the NF-κB nuclear localization sequence (NLS). Second, IκB contains a nuclear export signal, so that any NF-κB associated with it that happens to enter the nucleus is rapidly exported to the cytoplasm. **B,** Localization of the *dorsal* transcription factor (a relative of NF-κB) in *Drosophila* embryos. These images represent a longitudinal *(left)* and cross-sectional *(right)* view of wild-type embryos. The *dorsal* protein is stained with specific antibody, which appears as dark spots where it has become concentrated in the cell nuclei in the ventral portion of the embryo. (**B,** From Roth S, Stein D, Nusslein-Volhard C. A gradient of nuclear localization of the dorsal protein determines dorsoventral pattern in the *Drosophila* embryo. *Cell.* 1989;59:1189–1202.)

during development (Fig. 9.19B) and activation of immune cells (see Fig. 27.8), among other examples.

Disorders Associated With Defective Nuclear Trafficking

In many instances, protein function appears to be regulated by adjusting its location in the cell, and nuclear transport is one mechanism controlling localization. Thus, a myriad of examples undoubtedly exist in which disruption of transport leads to disease. This area has yet to be explored systematically, but in one interesting example, human sex determination is disrupted by mutations of an NLS on the SRY (sex-determining region Y) transcription factor, a master regulator of sex determination. These NLS mutants apparently disrupt the accumulation of SRY in the nucleus at a critical stage during development, causing individuals with a 46XY karyotype (normal male) to develop as females. Mutations in nuclear pore proteins are also associated with developmental diseases and chromosomal translocations involving pore components and are implicated in a variety of cancers. Nuclear transport defects are also found in numerous human neurodegenerative diseases (eg, Alzheimer disease), but the mechanism is not known.

Other Uses of the Importin/Ran Switch

The ability of Ran-GTP to release substrates bound to importin β provides a highly efficient switch for regulating protein availability. Cells use this system to regulate several supramolecular assembly processes, including assembly of the nuclear envelope, nuclear pore, and mitotic spindle.

In these processes, importin β (and occasionally importin α) acts as a negative regulator of assembly by binding to and sequestering key proteins. In the case of mitotic spindle assembly in large cells such as eggs that lack centrosomes, sequestration of key proteins blocks spindle assembly. In eggs, this block is overcome in the vicinity of chromosomes, which bind high concentrations of the GEF RCC1. Spindle assembly is triggered only after nuclear envelope breakdown, when the chromosomes come in contact with the cytoplasm (see Fig. 44.2). Conversion of Ran-GDP to Ran-GTP near the chromosomes results in Ran-GTP binding to importin β. This releases bound proteins and triggers mitotic spindle formation. Importin β and Ran also appear to regulate nuclear pore assembly in a similar way by sequestering key pore components, including the Nup107-160 complex, until they are released by Ran-GTP.

ACKNOWLEDGMENTS

We thank Roland Foisner, Harald Herrmann, Tom Misteli, Michael Rout, and Eric Schirmer for their advice on revisions to this chapter.

SELECTED READINGS

Amendola M, van Steensel B. Mechanisms and dynamics of nuclear lamina-genome interactions. *Curr Opin Cell Biol.* 2014;28:61-68.

Azuma Y, Dasso M. The role of Ran in nuclear function. *Curr Opin Cell Biol.* 2000;12:302-307.

Dundr M. Nuclear bodies: multifunctional companions of the genome. *Curr Opin Cell Biol.* 2012;24:415-422.

Fernandez-Martinez J, Rout MP. A jumbo problem: mapping the structure and functions of the nuclear pore complex. *Curr Opin Cell Biol.* 2012;24:92-99.

Forbes DJ, Travesa A, Nord MS, et al. Nuclear transport factors: global regulation of mitosis. *Curr Opin Cell Biol.* 2015;35:78-90.

Gruenbaum Y, Foisner R. Lamins: nuclear intermediate filament proteins with fundamental functions in nuclear mechanics and genome regulation. *Annu Rev Biochem.* 2015;84:131-164.

Kabachinski G, Schwartz TU. The nuclear pore complex—structure and function at a glance. *J Cell Sci.* 2015;128:423-429.

Lamond AI, Earnshaw WC. Structure and function in the nucleus. *Science*. 1998;280:547-553.

Pombo A, Dillon N. Three-dimensional genome architecture: players and mechanisms. *Nat Rev Mol Cell Biol*. 2015;16:245-257.

Schmidt HB, Görlich D. Transport selectivity of nuclear pores, phase separation, and membraneless organelles. *Trends Biochem Sci*. 2016;41:46-61.

Sosa BA, Kutay U, Schwartz TU. Structural insights into LINC complexes. *Curr Opin Struct Biol*. 2013;23:285-291.

Wickramasinghe VO, Laskey RA. Control of mammalian gene expression by selective mRNA export. *Nat Rev Mol Cell Biol*. 2015;16:431-442.

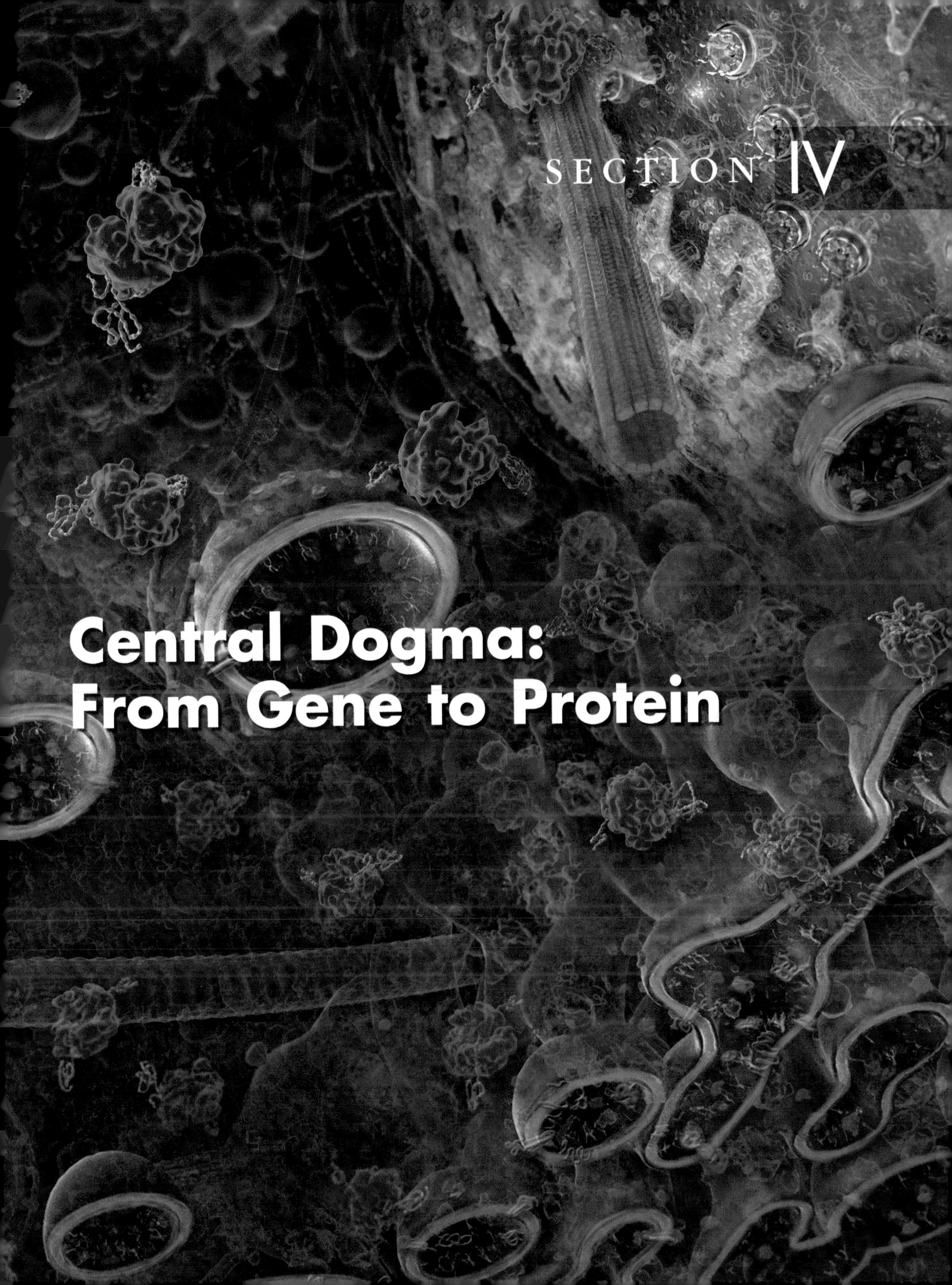

SECTION IV

Central Dogma: From Gene to Protein

SECTION IV OVERVIEW

The hugely important prediction of a structure for DNA not only led Crick and Watson to propose a general strategy for the replication of DNA (discussed in Chapter 42) but also led Francis Crick to propose the central dogma of molecular biology: that DNA is transcribed into RNA and that this RNA is then translated into protein. Chapters 10 to 12 present how this central dogma plays out at the cellular level, with one crucial addition that could not have been foreseen by Crick. This new element is the complex battery of processing events that RNAs undergo before they function as messengers, transfer vehicles, processing machines, or protein synthesizing machines in the ribosome.

Chapter 10 discusses transcription of DNA sequences into RNA, the initial step in recovering the information encoded in the genome. Three eukaryotic cellular RNA polymerases have distinct specialized tasks: **polymerase I** transcribes ribosomal RNAs; **polymerase II** transcribes all messenger RNAs (mRNAs) plus a number of small RNA molecules that are involved in RNA processing; and **polymerase III** transcribes transfer RNAs (tRNAs) and the smallest ribosomal RNAs. These three polymerases evolved from a common ancestor and retain many shared features. However, they have acquired significant differences in the ways they act on their target genes.

Eukaryotic genes contain both upstream (5′) and downstream (3′) regulatory regions that are not transcribed into RNA. Each gene has a **promoter** located just upstream from the site where transcription begins. **Enhancers** are DNA sequences that regulate transcription from a distance. Both promoter and enhancer sequences form binding sites for regulatory proteins that either stimulate or repress transcription. The chromatin organization of the DNA template and its organization within the nucleus also influence the efficiency of transcription.

Fundamental differences in the ways in which eukaryotes and prokaryotes store their genomes have had a profound influence on the structure of genes and the fate of cellular RNAs. In prokaryotes, the DNA occupies a distinct region of cytoplasm that is not bounded by a membrane. This means that transcription of DNA sequences into mRNAs and translation of mRNAs into proteins can be coupled directly, with ribosomes attaching to nascent mRNAs even before they are fully copied from the DNA template. In contrast, eukaryotes house their genomes and the machinery for RNA transcription and processing in a nucleus bounded by a nuclear envelope. Eukaryotic protein-coding RNAs must be transported across the nuclear membrane prior to their translation by ribosomes in cytoplasm. This geographic segregation, in which mRNAs are created in one subcellular compartment and used in another, has allowed the evolution of structurally complex genes whose RNA products are spliced before use.

The initial RNA products of transcription of most eukaryotic genes require extensive modifications by **RNA processing** before they are ready to function. Chapter 11 explains that most protein-coding genes of higher eukaryotes contain protein-coding regions called **exons** separated by noncoding **intron** regions. Consequently, the initial RNA copy of these genes must be processed to remove the introns before the finished mRNA is exported from the nucleus.

The nucleus is the site of many other essential RNA-processing events. These include the addition of 5′ cap structures to mRNAs, polyadenylation of the 3′ end of mRNAs, cleavage of some RNAs into functional smaller pieces, modification of RNA bases, and a host of sometimes bizarre editing events. Both the RNA substrates for these events and many enzymes that carry out the reactions are packaged into ribonucleoprotein particles by

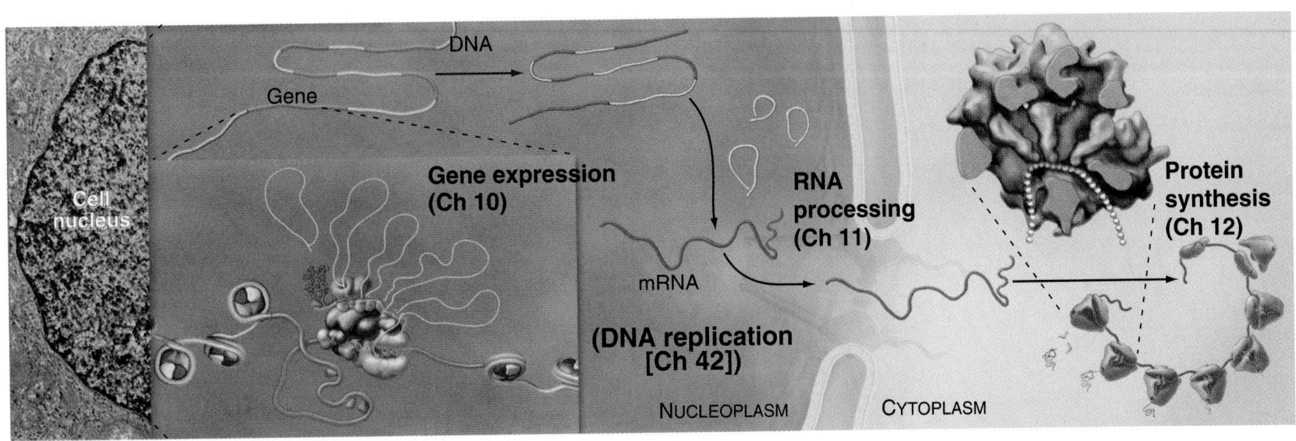

specific proteins, but RNAs themselves carry out a number of enzymatic reactions, including catalysis of peptide bond formation by the ribosome.

Cells also contain enzymes that fragment double-stranded RNAs into small pieces, used by other proteins to direct the silencing of the genes that encoded them. This process of **RNA interference (RNAi)** is critical for defense against RNA viruses and in chromatin regulation. Cell biologists also use RNAi as a technique to study gene function in the laboratory.

Chapter 12 describes how ribosomes translate the sequence of nucleotide triplets in mRNAs into proteins. tRNAs act as adapters, matching specific amino acids with triplet codons in the mRNA. The RNA component of ribosomes catalyzes the transfer of each successive amino acid from its tRNA onto the C-terminus of the growing polypeptide. Every step in the process is carefully regulated to ensure quality control of the finished polypeptide. Initiation factors select the proper AUG codon in the mRNA to begin the polypeptide with a methionine residue (or formylmethionine in the case of bacteria). Elongation factors check that the proper tRNA is matched with each codon before peptide bonds are formed. Although polypeptides grow at 20 residues per second, errors occur at a rate of less than one residue in a thousand. Termination factors bring protein synthesis to a close at the C-terminus of the polypeptide and recycle the ribosomal subunits for another round of translation.

Although some proteins fold spontaneously into their mature form following release from a ribosome, many proteins require a helping hand to reach their properly folded state. Chapter 12 covers four types of chaperones that help proteins fold by different mechanisms. Trigger factor, which is associated with ribosomes, provides a hydrophobic groove for protein folding. Heat shock protein (Hsp) 70 and Hsp90 chaperones bind hydrophobic residues in nascent polypeptides, prevent the unfolded protein from aggregating, and thereby promote folding. Cycles of binding and release are accompanied by hydrolysis of adenosine triphosphate (ATP). Chaperonins related to GroEL provide chambers to protect proteins during folding. ATP hydrolysis releases the protein from this chamber.

CHAPTER 10

Gene Expression*

Each organism, whether it has 600 genes *(Mycoplasma)*, 6000 genes (budding yeast), or 25,000 genes (humans), depends on reliable mechanisms to regulate the expression of these genes (ie, turn them on and off). This is called regulation of gene expression. In simple organisms, such as bacteria and yeast, environmental signals, such as temperature or nutrient levels, control much of gene expression. In multicellular organisms, genetically programmed gene expression controls development starting from a fertilized egg. Within these organisms, cells send each other signals that control gene expression either through direct contact or via secreted molecules, such as growth factors and hormones.

Given the vast numbers of genes, even in simple organisms, regulation of gene expression is complicated. Control is exerted at multiple steps, including production of messenger RNA (mRNA), translation, and protein turnover. This chapter focuses on the first of these regulatory steps: the transcription mechanisms that lead to the production of mRNA and other RNA transcripts.

Proteins called **transcription factors (TFs)** turn genes on or off by binding to DNA regulatory sequences associated with sequences encoding the protein or RNA product of the gene. The paradigm of this level of regulation is the bacterial repressor that controls expression of genes required for lactose metabolism in *Escherichia coli.* In eukaryotes, TFs are numerous, representing approximately 6% of human genes. They are also quite diverse, binding to a wide range of DNA regulatory sites. Fortunately, they fall into a limited number of families with similar structures and binding mechanisms. Three types of eukaryotic DNA-dependent RNA polymerases respond to these regulatory proteins and transcribe DNA sequence into RNA. Regulation of TFs is achieved by variations in a limited number of mechanisms that control their synthesis, transport from the cytoplasm into the nucleus, activity through posttranslational modifications or binding to small molecular ligands.

One key level of regulation is transcription initiation, the first step in production of RNA transcripts. This chapter examines the basic features of both prokaryotic and eukaryotic transcription units and the transcription machinery. Regulatory TFs that control the expression of groups of genes are discussed in the context of how external signals can reprogram patterns of gene expression. Finally, the chapter addresses the mechanisms that couple transcription to the downstream processing of nascent transcripts.

Transcription Cycle

Synthesis of RNA by **RNA polymerases** is a cyclic process that can be broken down into three sets of events: initiation, elongation, and termination (Fig. 10.1). Each of these events consists of multiple steps that can be regulated independently. In the first step of the **initiation** process, RNA polymerase binds to the chromosome near the beginning of the gene, forming a **preinitiation complex** at a sequence termed a **promoter.** This binding must be highly specific to

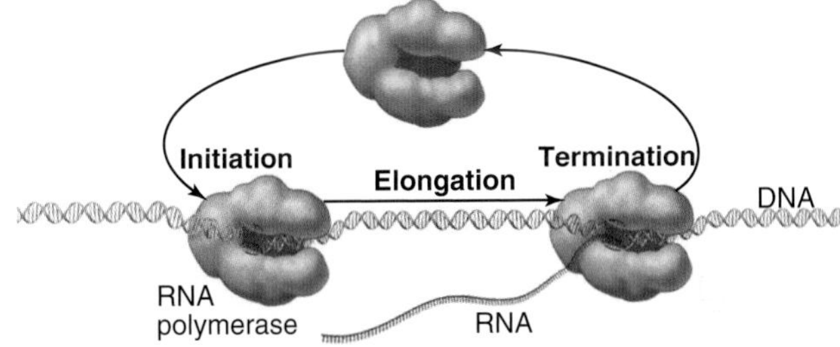

FIGURE 10.1 THE TRANSCRIPTION CYCLE. The transcription reaction consists of three basic steps in which the RNA polymerase initiates transcription at the promoter, elongates the nascent RNA copy of one of the DNA strands, and terminates transcription recognition of the appropriate signals.

*This chapter was written by Jeffry L. Corden.

distinguish promoter from nonpromoter DNA. Next, a conformational change in the polymerase–promoter complex separates the DNA strands. This **open complex** allows RNA polymerase access to single-stranded nucleotide bases that serve as the template to start the transcript. After formation of a phosphodiester bond between the first two complementary ribonucleotides, the polymerase translocates one base and repeats the process of phosphodiester bond formation, resulting in **elongation** of the nascent RNA. The elongation reaction cycle continues at an average rate of approximately 20 to 30 nucleotides per second until the complete gene has been transcribed. However, the rate of elongation is not uniform, as RNA polymerase pauses at certain sequences. The final step in the transcription cycle, **termination,** occurs when the polymerase reaches a signal on DNA that causes an extended pause in elongation. Given enough time, the appropriate sequence context and factors, the nascent transcript dissociates from the elongating RNA polymerase, and the DNA template returns to a base-paired duplex conformation. Ultimately, RNA polymerase dissociates from the template and is free to search for a new promoter.

Regulatory molecules target each of the steps in the transcription cycle. The frequency of initiation from different promoters varies as dictated by the need for the gene product. The initiation reaction is most often regulated, presumably because this prevents synthesis of transcripts that are not needed. Elongation and termination can also be regulated, as can splicing and further processing of mRNAs and noncoding RNAs (ncRNAs) (see Chapter 11). In eukaryotes, the sum of these nuclear regulatory steps, together with cytoplasmic regulation of mRNA stability and translation efficiency, contributes to the wide variation in the abundance of various mRNAs and proteins in particular types of cells.

A. Procaryotic transcription unit

DNA *I* *Z* *Y* *A*

Transcription

mRNA 5' *i* *z* *y* *a* 3'

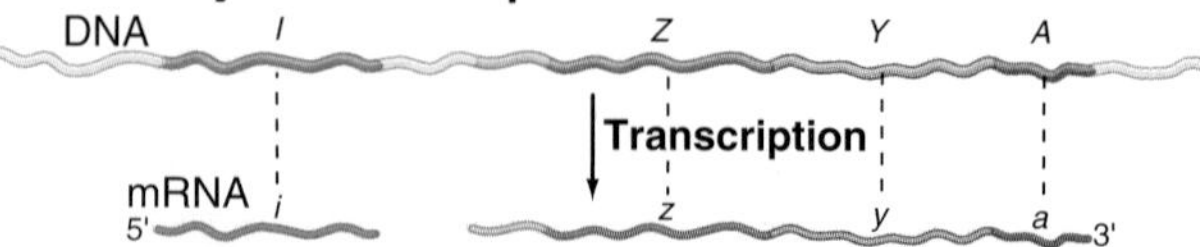

B. Eukaryotic transcription unit

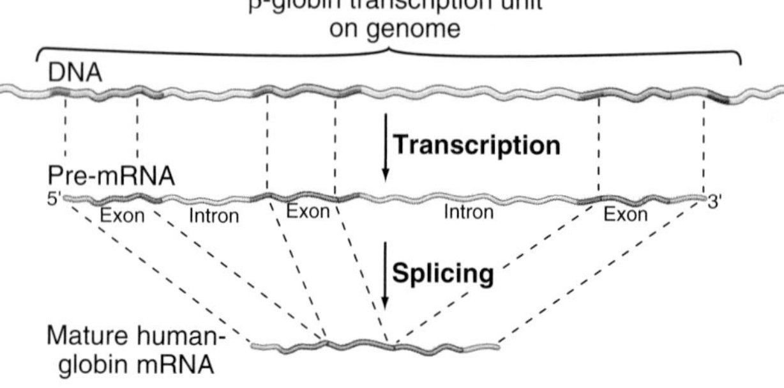

■ Promoter mutations result in lower level of mRNA

■ Nonsense, frameshift, missense mutations yield unstable or inactive protein

■ Splice-site mutations result in aberrantly spliced mRNA

■ 3' processing site mutations result in failure to polyadenylate mRNA

FIGURE 10.2 PROKARYOTIC AND EUKARYOTIC TRANSCRIPTION UNITS. A, The two transcription units required for regulation of lactose metabolism in *Escherichia coli.* The *I* gene encodes the lac repressor, while the *Z, Y,* and *A* genes encode β-galactosidase, lactose permease, and thiogalactoside transacetylase. All three genes are required for the cell to grow on media containing lactose and are coregulated as the *lac* operon. **B,** The nucleotide sequence of one of the two DNA strands is transcribed into a complementary pre-messenger RNA (mRNA) copy. The pre-mRNA is processed by removing introns and splicing together the protein-coding exons *(orange).* The DNA sequences required for expression of a functional β-globin protein are indicated in different colors (see key). Mutations in any of these sequences can lead to decreased β-globin expression.

Transcription Unit

Genetic information in DNA is transcribed in segments corresponding to one or a few genes. Gene-coding and regulatory (*cis*-acting) DNA sequences that direct the initiation of transcription, elongation, and termination are collectively called a **transcription unit.** Prokaryotic transcription units, called **operons,** contain more than one gene, often encoding proteins with related physiological functions (Fig. 10.2A). DNA sequences flanking the operon direct the initiation and termination of transcription.

A simple eukaryotic transcription unit, such as that encoding the human hemoglobin β-chain, also has flanking regulatory sequences, but the region encoding the polypeptide is interrupted by exons (Fig. 10.2B). Mutations that reduce β-globin levels in patients with β-thalassemias can occur either in the coding region, resulting in an unstable or truncated polypeptide, or in the adjacent control regions, leading to low levels of transcription or aberrant processing of the newly synthesized RNA (see Chapter 11). Thus, the transcription unit can be thought of as a linked series of modules, all of which must be functional for the gene to be transcribed at the correct level.

Biogenesis of RNA

Typical cells contain more RNA than genomic DNA. The population of RNA molecules range in size from several tens to several thousand nucleotides. In prokaryotes, translation is initiated on newly synthesized mRNA during transcription. In eukaryotes, RNA is transported from its site of synthesis in the nucleus to the cytoplasm, where most RNA is used to synthesize proteins. Eukaryotic cells have four different types of RNA:

1. **Ribosomal RNA (rRNA** [see Fig. 11.9]) makes up approximately 75% of the total.
2. Small, stable RNAs, such as **transfer RNA (tRNA** [see Fig. 12.4]), **small nuclear RNAs (snRNA** [see

Chapter 11]) involved in splicing, and **5S rRNA,** make up approximately 15% of the total.

3. **mRNA** and its precursor **heterogeneous nuclear RNA (hnRNA)** account for only 10% of the total.
4. **ncRNAs,** including **micro RNAs (miRNAs),** are not abundant but regulate a variety of RNA-based processes.

Transcription of eukaryotic DNA in the nucleus is linked to subsequent steps that process the nascent transcript in preparation for its eventual function (see Chapter 11 for a complete discussion of these steps). Processing of mRNA precursors includes capping and methylation of the 5′ end of the nascent transcript, splicing to remove introns and modifying the 3′ end by cleavage and addition of a stretch of adenosine residues. The finished mRNA is then transported to the cytoplasm, where it serves as the template for protein synthesis.

Eukaryotic ribosomal RNA is encoded in tandemly repeated genes and each gene is transcribed as a long precursor molecule, which is cleaved and modified to give the final 28S, 5.8S, and 18S RNAs (Fig. 10.3). These RNAs are assembled, together with 5S RNA and approximately 80 proteins, into ribosomes in the nucleolus.

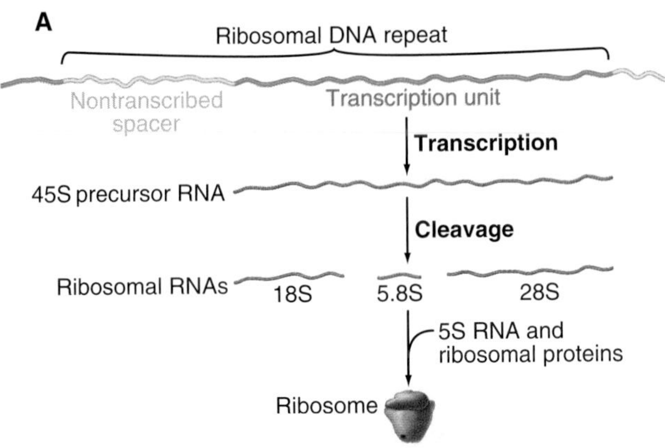

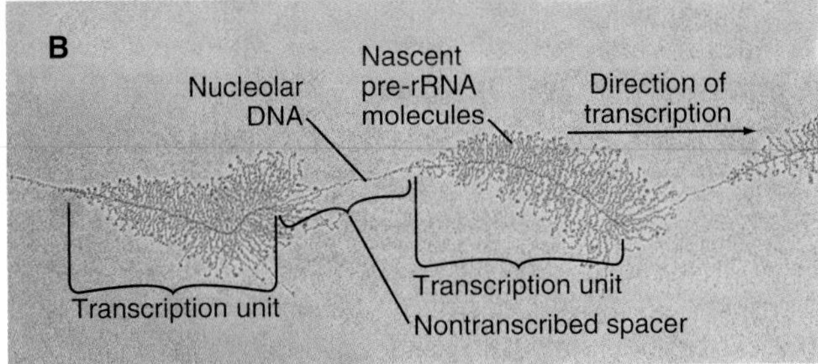

FIGURE 10.3 RIBOSOMAL RNA TRANSCRIPTION UNIT. Ribosomal RNA (rRNA) is transcribed from a set of transcription units arrayed as tandem copies of the same transcription unit. **A,** Map showing the arrangement of sequences in a typical ribosomal DNA repeat. **B,** Electron micrograph showing two active rRNA transcription units. Note that each transcription unit is transcribed by multiple RNA polymerases. As the polymerases traverse the gene, the attached nascent RNA is extended, giving a tree-like appearance. (**B,** Courtesy of Yvonne Osheim, University of Virginia, Charlottesville.)

Transfer RNA is synthesized in the nucleus and transported to the cytoplasm, where it is charged with amino acids prior to participating in protein synthesis (see Fig. 12.5). snRNAs are synthesized and processed in the nucleus. From there, they migrate to the cytoplasm, where they acquire essential proteins, and then return to the nucleus to catalyze RNA splicing reactions (see Fig. 11.11). The postsynthetic processing pathway that a particular transcript follows is dictated, in part, by the transcription machinery that is used to initiate and elongate the transcript and by certain features of the nascent RNA.

RNA Polymerases

Cellular RNA polymerases synthesize a strand of nucleic acid in the 5′ to 3′ direction that is complementary to one of the chromosomal DNA strands. Even though the enzymatic reaction is similar to DNA replication (see Fig. 42.1), there are several important differences. First, RNA polymerases synthesize a strand of ribonucleotides. Second, unlike DNA polymerase, RNA polymerases can initiate transcription without a primer. Finally, unlike replication, the newly transcribed sequences do not remain base-paired with the template but are displaced from the template approximately 10 base pairs (bp) from the growing end of the nascent RNA. All known RNA polymerases share these properties and have similar structures, since they arose from a common ancestor during evolution.

Bacteria have a single RNA polymerase composed of six polypeptides. Two copies of the α subunit and one each of the β, β', and ω subunits form a five-subunit **core enzyme** that synthesizes RNA. The sixth subunit, σ, binds to the core enzyme to form a **holoenzyme** that is able to recognize promoter sequences and initiate transcription.

Most eukaryotes have three different RNA polymerases (some species of plants contain four) with the largest subunits closely related to bacterial β and β' subunits. RNA polymerases I, II, and III each have 10 core subunits, most of which are unique to each enzyme (Fig. 10.4A). RNA polymerases I and III have additional subunits similar to RNA polymerase II general TFs discussed in a following section.

RNA polymerase I concentrates in the nucleolus, where it synthesizes rRNA. Throughout the nucleoplasm RNA polymerase II synthesizes mRNA and several classes of ncRNAs including some snRNAs involved in RNA splicing, and long noncoding RNAs (lncRNAs) and miRNAs implicated in gene regulation. RNA polymerase III synthesizes tRNA, 5S rRNA, and the 7S RNA of the signal recognition particle (see Fig. 21.5). RNA polymerase IV is present only in plants, where it is involved in heterochromatin formation and gene silencing.

The multiple eukaryotic RNA polymerases apparently originated through duplication of primordial genes,

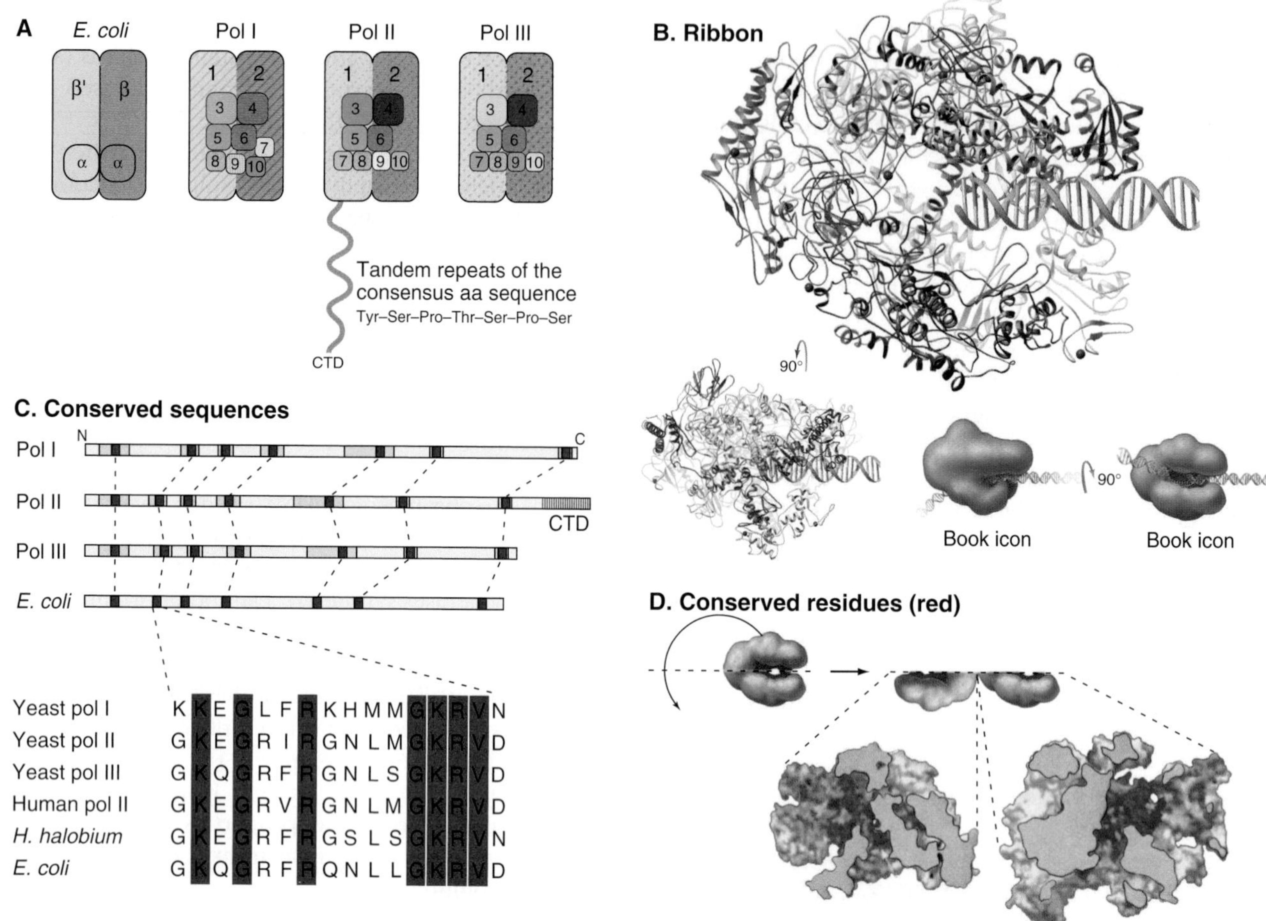

FIGURE 10.4 MULTIPLE RNA POLYMERASES. A, Eukaryotic cells have three different polymerases (Pol) that share three common subunits (numbers 5, 6, and 8) and have a number of other related, but distinct, subunits (indicated by *related colors* and *distinct shading*). **B,** A ribbon diagram of the structure of RNA polymerase II showing the arrangement of different subunits (colored as in part A). Metal ions are indicated as *red balls.* A prominent cleft, large enough to accommodate a DNA template, is formed between the two largest subunits. The model DNA fragment is shown for size comparison only. **C,** Conserved amino acid sequences are dispersed throughout the largest subunits. *Red* indicates sequences that are conserved among both prokaryotes and eukaryotes. *Yellow* represents sequences that are conserved among the three different eukaryotic RNA polymerases. **D,** Conserved residues are located on the inner surface of the RNA polymerase cleft. *E. coli, Escherichia coli; H. halobium, Halobacterium halobium.* (**B,** For reference, see Protein Data Bank [PDB; www.rcsb.org] file 1I50 and Cramer P, Bushnell DA, Kornberg RD. Structural basis of transcription: RNA polymerase II at 2.8 angstrom resolution. *Science.* 2001;292:1863–1876. **D,** From Zhang G, Campbell EA, Minakhin L, et al. Crystal structure of *Thermus aquaticus* core RNA polymerase at 3.3 Å resolution. *Cell.* 1999;98:811–824.)

followed by evolution of specialized functions. RNA polymerase II is the most versatile, because it must transcribe approximately 25,000 different species of human mRNAs and perhaps an equal number of ncRNAs. The relative abundance of individual mRNAs can vary widely, often in response to external signals, from just a few copies to more than 10,000 copies per cell. Thus, RNA polymerase II must recognize thousands of different promoters and transcribe them with widely varying efficiencies. In contrast, RNA polymerase I is specialized to transcribe more than 100,000 copies of rRNA per cell and RNA polymerase III synthesizes several hundred species of highly abundant transcripts.

Specialization has been balanced, however, by the need to retain the structural elements required for RNA synthesis. The subunits of both prokaryotic and eukaryotic enzymes assemble into a roughly spherical structure with a diameter of approximately 150 Å and a cleft 25 Å wide, to accommodate the DNA template (Fig. 10.4B). The two largest subunits form the framework of the structure, with two lobes that clamp down on the template DNA and form the catalytic core (Fig. 10.4C). The most conserved residues are located on the inner surfaces of the enzymes with the site of nucleotide addition on the back wall of the cleft (Fig. 10.4D).

Transcription does not necessarily require such large enzymes. Bacteriophages have evolved structurally distinct, DNA-dependent RNA polymerases that are one-fifth the size of the eukaryotic enzymes yet are able to carry out complete transcription cycles. The complexity of the eukaryotic enzymes is likely attributable to the need for regulation, with additional subunits acting as sites for interaction with regulatory proteins. Domains that differ among the three types of eukaryotic RNA polymerases are likely to interact with factors that are unique to a particular class of polymerase. One example of a class-specific domain is the **carboxyl-terminal domain (CTD)** of the largest subunit of RNA polymerase II, which is composed of tandem repeats of the consensus heptapeptide TyrSerProThrSerProSer. The CTD is highly phosphorylated in vivo, and the timing of CTD phosphorylation suggests that this modification may be involved in the transition between the initiation and elongation steps of transcription. By serving as a scaffold binding numerous auxiliary factors, the CTD also couples transcription with the subsequent processing of the nascent mRNA as is discussed in a later section.

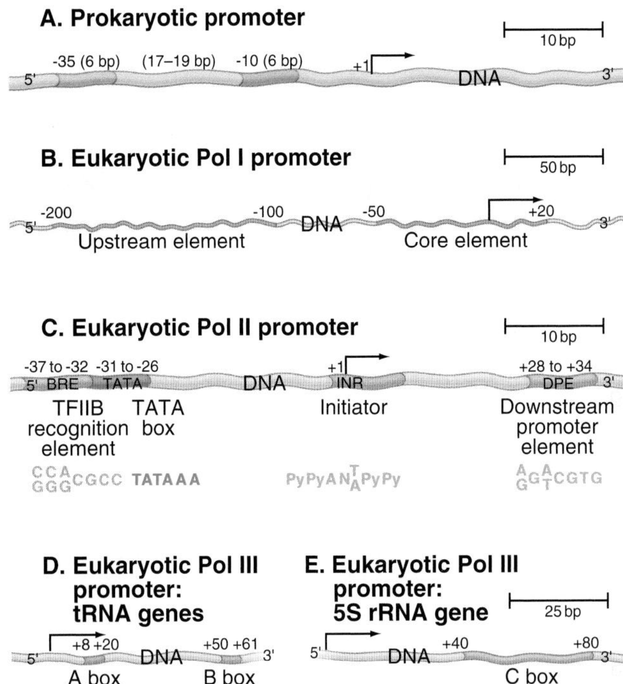

FIGURE 10.5 PROKARYOTIC AND EUKARYOTIC PROMOTERS. The prokaryotic **(A)** and three eukaryotic **(B–E)** RNA polymerases recognize different promoter sequences. Positions of promoter elements are indicated with respect to the start of transcription (+1). For the RNA polymerase II (Pol II) promoter elements, the consensus sequences are shown. Not all polymerase II promoters contain all these elements. Pol, polymerase; rRNA, ribosomal RNA; TF, transcription factor; tRNA, transfer RNA.

RNA Polymerase Promoters

Initiation of transcription requires loading of RNA polymerase onto the chromosome at the promoter of a gene or operon. A promoter can be loosely defined as a DNA sequence where RNA polymerase binds, unwinds the template and initiates transcription. Strong promoters drive the expression of genes whose products are required in abundance. Weaker promoters regulate the expression of rare proteins or RNAs. In multicellular organisms, a promoter may direct expression at a high level in some cells, at an intermediate level in others, and be repressed in yet others.

Promoters in bacteria are recognized by direct interactions of the RNA polymerase σ factor with specific DNA sequences. The most common σ factor in *E. coli* (σ 70) recognizes two conserved six-base sequences located 10 bases (minus 10) and 35 (minus 35) upstream of the transcription start site (Fig. 10.5A). Once initiation has occurred, σ is no longer required and can dissociate from the core enzyme. Bacterial cells have several distinct σ factors, each of which binds the core enzyme and directs RNA polymerase to a subset of promoters that contain different recognition sequences, thereby promoting independently regulated transcription of genes with diverse functions.

Eukaryotic promoter sequences for RNA polymerases I and II are also situated upstream of the transcription start site. RNA polymerase I recognizes a single type of promoter located upstream of each copy of the long tandem array of pre-rRNA coding sequences (Figs. 10.3B and 10.5B). The core element of this promoter overlaps the transcription start site, while an upstream control element located approximately 100 bp from the start site stimulates transcription.

Comparison of the first eukaryotic protein-coding gene sequences revealed a conserved consensus sequence—TATAAAA—called a **TATA box,** located approximately 30 bp upstream of the transcription start site of many genes transcribed by RNA polymerase II (Fig. 10.5C). In addition to the TATA box, a less-conserved promoter element, the **initiator,** is found in the vicinity of the transcription start site of many genes. Some genes transcribed by polymerase II do not contain TATA boxes but may contain strong initiator elements. Together, these two elements account for the basal promoter activity of most protein-coding genes.

Two types of RNA polymerase III promoters have key elements within the transcribed sequences (Fig. 10.5D–E). tRNA genes contain two 11-bp elements, the A box and B box, centered approximately 15 bp from the 5′ and 3′ ends of the coding sequence, respectively. The 5S-rRNA gene contains a single internal element, the C box, located in the center of the coding region. Given the differences in classes of eukaryotic promoters, it is not surprising that each type of polymerase uses different proteins to recognize the promoter sequences.

Transcription Initiation

The loading of RNA polymerase onto double-stranded genomic DNA at a promoter sequence is best understood in prokaryotes and is discussed first before initiation by eukaryotes. Initiation takes place in a series of defined

A. Closed complex (binding) **B. Open complex (melting)** **C. Transcribing complex**

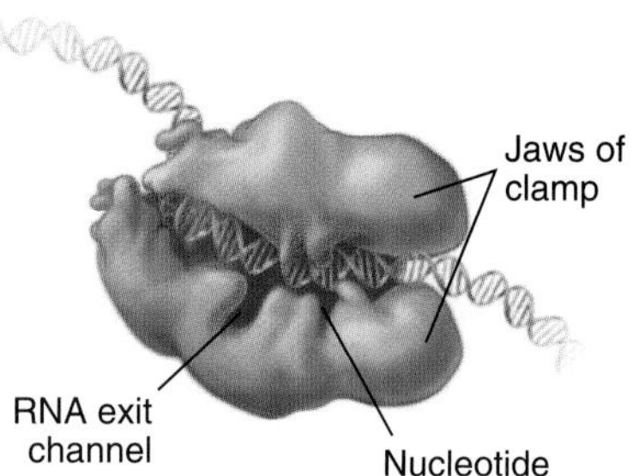

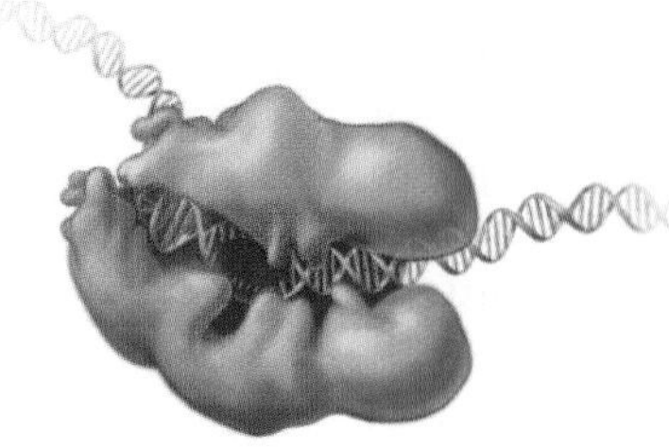

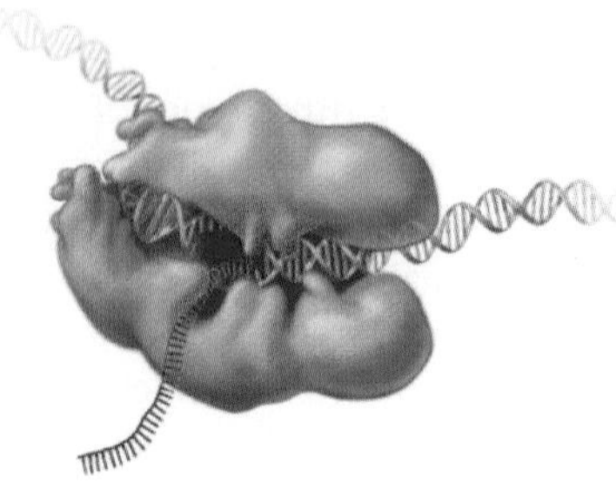

FIGURE 10.6 THREE STEPS IN RNA POLYMERASE INITIATION. A, In the closed complex, the double-stranded promoter DNA is recognized by σ factor domains on the surface of the holoenzyme. Double-stranded DNA then transfers into the active site shown here. **B,** The open complex forms by unwinding DNA surrounding the transcription start site and positioning the single-stranded template in the active site of the polymerase. **C,** The initiation reaction in the context of the transcription cycle.

TABLE 10.1 Summary of Eukaryotic RNA Polymerase II General Transcription Factors

Factor	Number of Subunits	Subunit M (kD)	Functions
TFIIA	3	12, 19, 35	Stabilizes binding of TBP and TFIIB
TFIIB	1	25	Binds TBP, selects start site, and recruits polymerase II
TFIID	12	15–250	Interacts with regulatory factors
(TBP)	1	38	Subunit of TFIID; specifically recognizes the TATA box
TFIIE	2	34, 57	Recruits TFIIH
TFIIF	2	30, 74	Binds polymerase II and TFIIB
TFIIH	9	35–98	Unwinds promoter DNA; phosphorylates CTD (C-terminal domain of RNA polymerase II)
Polymerase II	12	10–220	Catalyzes RNA synthesis
TOTALS	42	~1000	

TBP, TATA box–binding protein.

steps (Fig. 10.6). First, RNA polymerase holoenzyme binds to the double-stranded promoter, forming what is called the **closed complex.** Interactions between the σ factor and bases in the −10 and −35 elements of the promoter determine the specificity and strength of this interaction (Fig. 10.5). The second step in initiation is the formation of an **open complex** by separation of the two strands of DNA around the transcription start site producing a 14 base transcription bubble. This unpairing is accompanied by a conformational change in the polymerase that positions the single-stranded DNA template in the active site and narrows the DNA-binding cleft, effectively closing the polymerase clamp. In the next step, the DNA template in the active site base-pairs with the first two ribonucleotides, and the first phosphodiester bond is catalyzed. The process of single nucleotide addition is repeated until the nascent RNA is eight to nine bases long, at which point addition of bases to the growing RNA chain results in the unpairing of the 5′ RNA base of the RNA-DNA hybrid, and the nascent RNA begins to exit through a channel on the surface of the polymerase. The resulting conformational change in polymerase leads to the release of σ factor and formation of a stable ternary (three-way) complex containing RNA polymerase, the DNA template, and the nascent RNA.

General Eukaryotic Transcription Factors

Eukaryotic RNA polymerases require multiple initiation factors to start transcription. All the RNA polymerases use a **TATA box–binding protein**, but most of the other initiation factors are unique for each class. On the other hand, each RNA polymerase uses the same **general transcription factors (GTFs)** for most promoters. GTFs are remarkably conserved among eukaryotes. The next sections describe transcription initiation by the three forms of eukaryotic RNA polymerase.

RNA Polymerase II Factors

Initiation of transcription by RNA polymerase II in vitro depends on the ordered assembly of more than 20 GTFs at the promoter (Table 10.1). Assembly of this RNA polymerase II **preinitiation complex** begins with binding of **TFIID,** a large protein complex (~700 kD) consisting of TATA box–binding protein (TBP) and **TBP-associated factors** called **TAFIIs** (Fig. 10.7A). TBP alone is sufficient for basal transcription, while

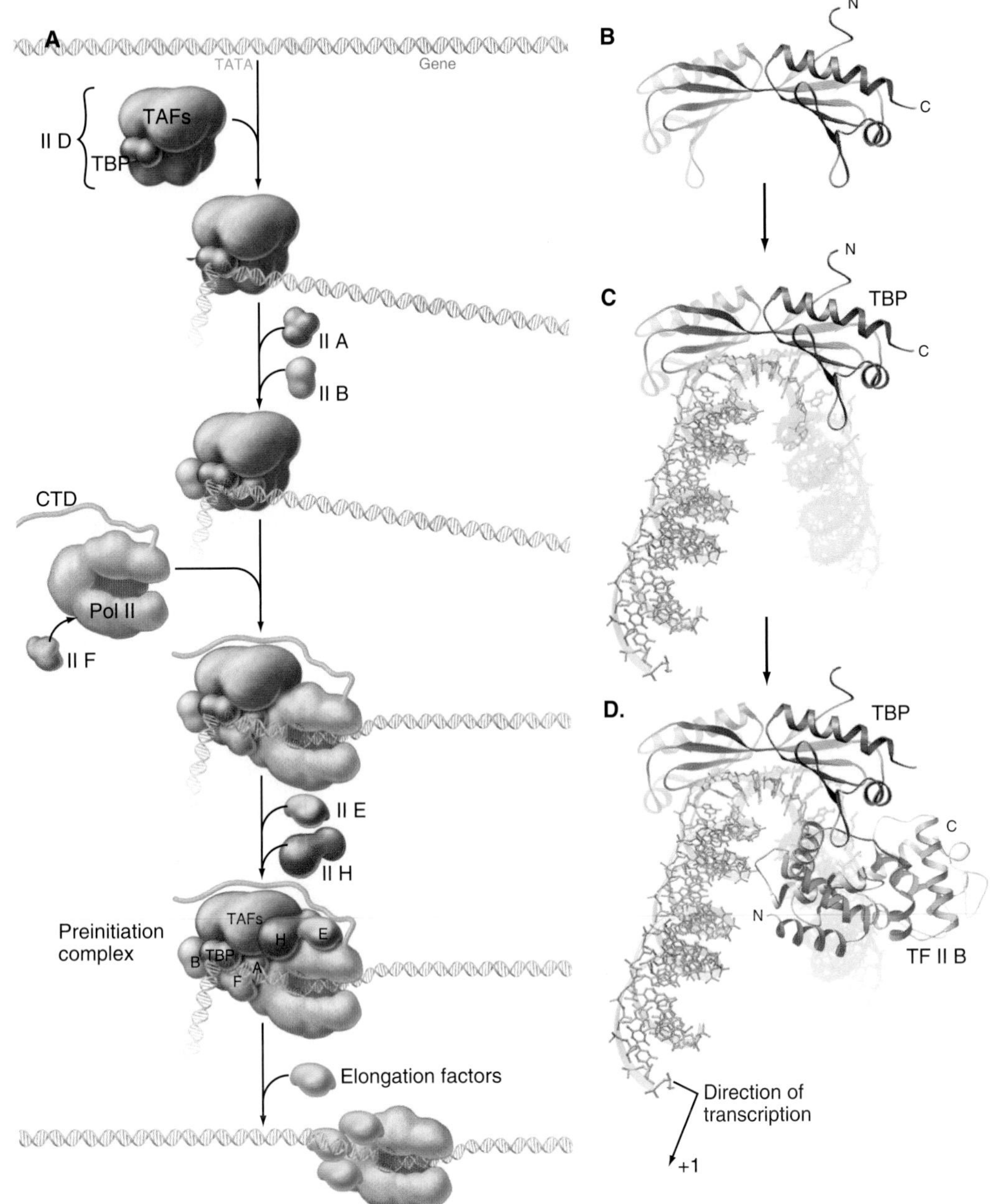

FIGURE 10.7 RNA POLYMERASE II PREINITIATION COMPLEX ON THE ADENOVIRUS-2 MAJOR LATE PROMOTER DNA. A, The sequential assembly of general transcription factors leads to a preinitiation complex with the promoter region in the closed complex. Helicase activities present in transcription factor IIH (TFIIH) use the energy of adenosine triphosphate (ATP) to unwind the promoter, leading to formation of an open complex. **B,** Binding of the TATA box–binding protein (TBP) leads to **C,** a pronounced bend in the DNA. **D,** TFIIB interacts both upstream and downstream of the TATA box and directs RNA polymerase to the transcription start site. (**B–D,** For reference, see PDB file 1VOL. TBP + DNA coordinates courtesy Stephen Burley, Rockefeller University, New York.)

TBP-associated factors (TAFs) apparently serve as targets for further activation of transcription (see subsequent sections). DNA binding by TBP is provided by a highly conserved C-terminal of 180 amino acids, which forms a saddle-shaped monomer with an axis of dyad symmetry (Fig. 10.7B). The underside of the TBP "saddle" binds to the minor groove of the TATA sequence, which is splayed open in the process. A pronounced DNA bend is produced at each end of the TATAAA element by the intercalation of phenylalanine side chains (Fig. 10.7C).

The TFIID-TATA box complex serves as a binding site for additional GTFs and positive and negative regulators. **TFIIA** binding stabilizes the TBP-DNA interaction and prevents the binding of repressors that would otherwise block further initiation complex formation.

The next step in assembly of the initiation complex is adding **TFIIB,** which binds to one side of TBP, making contacts with DNA upstream and downstream of the TATA box (Fig. 10.7D). Mutations in the yeast gene encoding TFIIB alter mRNA start-site selection, indicating that TFIIB establishes the spacing between the TATA box and the transcription start site. TFIIB interacts directly with TBP and RNA polymerase II and is essential for the next steps in initiation complex assembly.

RNA polymerase II joins the preinitiation complex (Fig. 10.7A) associated with **TFIIF.** This factor stabilizes the interaction of RNA polymerase II with TFIIB and TBP. TFIIF also binds to free polymerase and prevents interactions with nonpromoter DNA sites.

TFIIH and its stimulatory factor **TFIIE** are the final general factors to enter the preinitiation complex. Their binding stabilizes contacts between proteins and DNA in the vicinity of the transcription start site. TFIIH contains eight polypeptides, several of which have functions outside of transcription initiation. **Helicases** associated with TFIIH use energy from adenosine triphosphate (ATP) hydrolysis to unwind a short stretch of promoter DNA at the transcription start site. This separation of DNA strands allows RNA polymerase II to recognize the template strand, bind the complementary nucleotides, and synthesize the first few phosphodiester bonds.

TFIIH also contains a protein kinase that phosphorylates the CTD. This is Cdk-activating kinase (CAK), itself a Cdk-cyclin complex that phosphorylates and activates other cyclin-dependent kinases (see Fig. 40.14). In the initiation complex, phosphorylation of the CTD releases it from interactions with GTFs and mediator (see later section) allowing it to leave the promoter and enter the transcription elongation phase. Other TFIIH subunits have been identified as components of the DNA repair machinery. Several genes encoding TFIIH subunits are mutated in xeroderma pigmentosa, a human disease with defects in DNA excision repair. This suggests that TFIIH might link transcription to DNA repair (see Box 43.1).

Initiation by RNA Polymerases I and III

Distinct initiation complexes initiate transcription at RNA polymerase I and III promoters (Fig. 10.8). RNA polymerases I (Pol I) and III (Pol III) contain subunits related to polymerase II (Pol II) GTFs TFIIF and TFIIE. Unique TFIIB-related factors provide additional GTF functions for Pol I and Pol III.

The Pol I upstream binding factor binds to both the upstream control element and part of the core element of the promoter (Fig. 10.8A). A protein complex called SL1 stabilizes this initial complex. SL1 consists of TBP and TAFs specific to RNA Pol I, including one related to TFIIB. A unique factor Rrn3 binds Pol I and modulates rRNA transcription in response to nutrient availability.

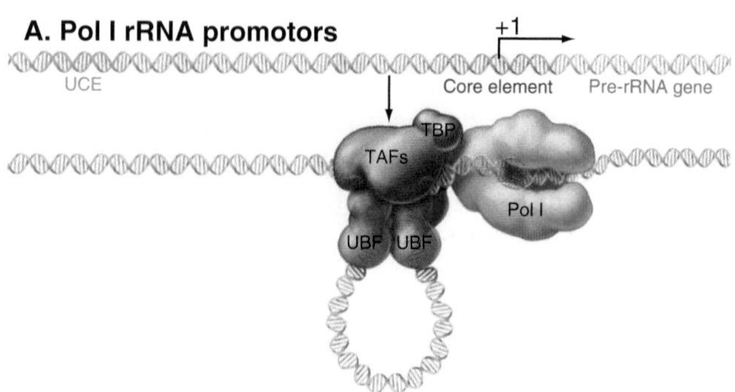

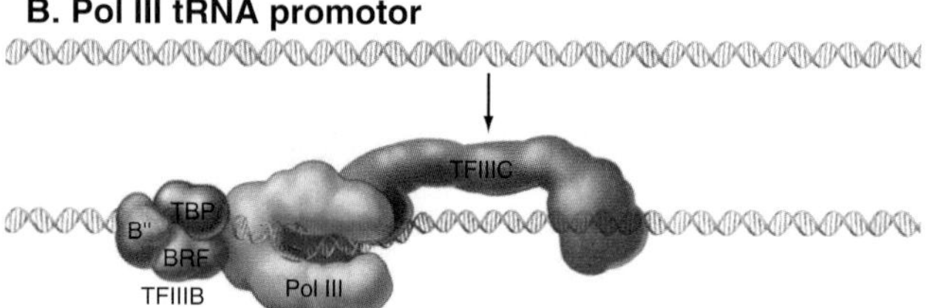

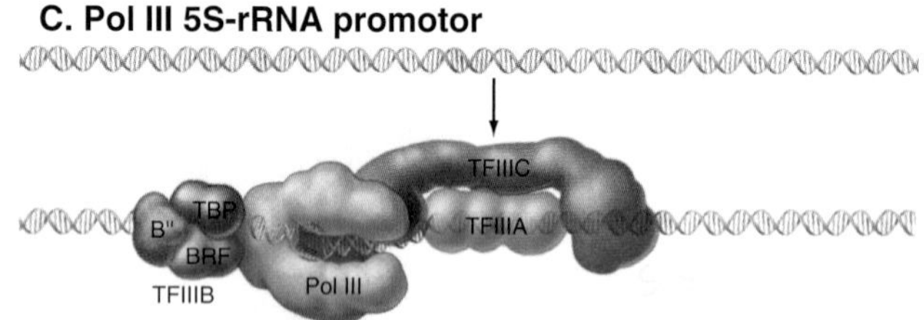

FIGURE 10.8 RNA POLYMERASE I AND III PREINITIATION COMPLEXES. A, Ribosomal RNA promoters assemble a preinitiation complex. (UCE, upstream control element.) This complex consists of an upstream binding factor (UBF) and a multisubunit factor that contains TATA box–binding protein (TBP). Together, these factors recruit RNA polymerase I. **B–C,** Initiation at RNA polymerase III promoters requires recognition of sequences within the transcribed sequences. These sequences differ for transfer RNA (tRNA) and 5S ribosomal genes. **B,** In the case of tRNA genes, only TFIIIC is required for specific binding. **C,** For 5S genes, the internal element is recognized by the specific DNA-binding factor TFIIIA. BRF, TFIIB-related factor.

The assembly of RNA Pol III initiation complexes differs at various promoters. Initiation at tRNA genes begins with TFIIIC binding to the A and B boxes (Fig. 10.8B); TFIIIB then binds upstream of the A box at a sequence determined both by an interaction with TFIIIC and through the DNA-binding capacity of TBP. Once the TFIIIC–TFIIIB complex has assembled, RNA Pol III initiates transcription. Multiple rounds of initiation can occur on the stable transfer DNA (tDNA)–TFIIIC–TFIIIB complex. Transcription of 5S rRNA genes requires an additional factor called TFIIIA that recognizes the C box located near the center of the 5S rRNA coding region. TFIIIC then binds with contacts on each side of TFIIIA, similar to the A and B boxes contacting tRNA genes. Finally, TFIIIB binds through interactions with TFIIIC and DNA, and the resulting preinitiation complex is recognized by RNA Pol III.

Summary of the Eukaryotic Basal Transcription Machinery

Despite the evolutionary divergence of the multiple eukaryotic RNA polymerases and the specialization of each polymerase for a unique set of promoters, the fundamental mechanisms of transcription have been conserved. This conservation is reflected not only in similar sequences of the subunits of the polymerases themselves but also in the presence of TBP and TFIIB homologs among the GTFs used by each class of polymerase. Indeed, Archaea, which have only a single RNA polymerase, contain both TBP and TFIIB suggesting that initiation mechanisms employing GTFs evolved before the duplication of the RNA polymerases.

Why are so many factors required to make a transcript? Part of the complexity might be necessary to generate multiple sites for interaction with regulatory factors that could either activate or repress the assembly or function of the preinitiation complex. A second role for the complex set of factors could be to target polymerases to specific sites in the nucleus. Finally, some factors could help load elongation, splicing, or termination factors onto the RNA polymerases.

Transcription Elongation and Termination

The final stage of initiation leads to elongation and movement of the polymerase away from the promoter. This process of **promoter clearance** is associated with structural changes in the polymerase, which prepare it for efficient RNA synthesis and render it susceptible to the action of factors that regulate elongation. Such regulatory factors, together with structural features of the nascent transcript, influence elongation and can trigger the termination of transcription and the dissociation of the ternary elongation complex containing the DNA template, nascent RNA, and RNA polymerase. The termination reaction typically occurs at the 3′ end of the gene or operon and serves both to recycle RNA polymerase for additional initiation reactions as well as to ensure that adjacent genes are not inadvertently transcribed.

Transcription Elongation Complex

Efficient synthesis of RNA requires balancing two competing demands. First, the elongation complex must be very stable, because premature dissociation from DNA produces defective partial transcripts and requires the polymerase to restart transcription from the promoter. However, the complex must also be bound loosely enough so that the polymerase can easily translocate along the DNA template.

RNA polymerase evolved to meet these needs. The cleft formed at the interface between the two largest subunits is open when the polymerase is in the initiation complex. Once the first few RNA phosphodiester bonds form, the polymerase undergoes a conformational change. Subunits at the outer edge of the cleft close like jaws to encircle the DNA template. In this structure, the front end of the transcription "bubble" (an unpaired segment of the DNA template) is positioned at the back wall of the cleft, close to the catalytic center. The elongation complex is highly efficient and can function continuously for the 17 hours required to transcribe the more than 2 million-bp mammalian dystrophin gene (see Fig. 39.17).

Catalytic Cycle

The DNA-dependent RNA polymerases catalyze synthesis of an RNA polymer from ribonucleoside 5′-triphosphates (ATP, guanosine triphosphate [GTP], cytidine triphosphate [CTP], and uridine triphosphate [UTP]) according to the following reaction:

$$(\mathrm{NMP})_n + \mathrm{NTP} \rightarrow (\mathrm{NMP})_{n+1} + \mathrm{PP_i}$$

where $(\mathrm{NMP})_n$ is the RNA polymer; NTP is ATP, UTP, CTP, or GTP; and $\mathrm{PP_i}$ is pyrophosphate. Polymerase extends the RNA chain in the 5′ to 3′ direction by adding ribonucleotide units to the chain's 3′ OH end. Selection of the incoming nucleoside triphosphate (NTP) is directed by the DNA template and takes place at the transcription bubble (Fig. 10.9). The 3′ hydroxyl group acts as a nucleophile, attacking the α-phosphate of the incoming NTP in a reaction similar to that seen in DNA replication (see Fig. 42.1). The chain elongation reaction proceeds in vivo at a rate of 30 to 100 nucleotides per second and is facilitated by a set of flexible protein modules surrounding the polymerase active site.

Pausing, Arrest, and Termination

Following the addition of each nucleotide, RNA polymerase may add an additional nucleotide, pause, move in reverse, or terminate (Fig. 10.9B). The relative probabilities of these alternative reactions depend on interactions between the transcription complex and the template, the nascent RNA transcript, and regulatory TFs.

RNA polymerase does not elongate at a constant rate. Instead, it synthesizes RNA in short spurts between pauses. A pause of short duration can be caused by low NTP concentrations or alternatively by the transient unpairing of the 3′ end of the nascent transcript and template. Longer pauses are provoked by the presence, in the nascent RNA, of short (~20 bp) self-complementary sequences that can fold to form a stem-loop or hairpin, or the presence of a weak RNA–DNA hybrid. The presence of an unstable RNA–DNA hybrid can arise from the misincorporation of an NTP leading to an unpaired base in the hybrid. In this case, the RNA polymerase can backtrack or slide backward on the template (Fig. 10.9C).

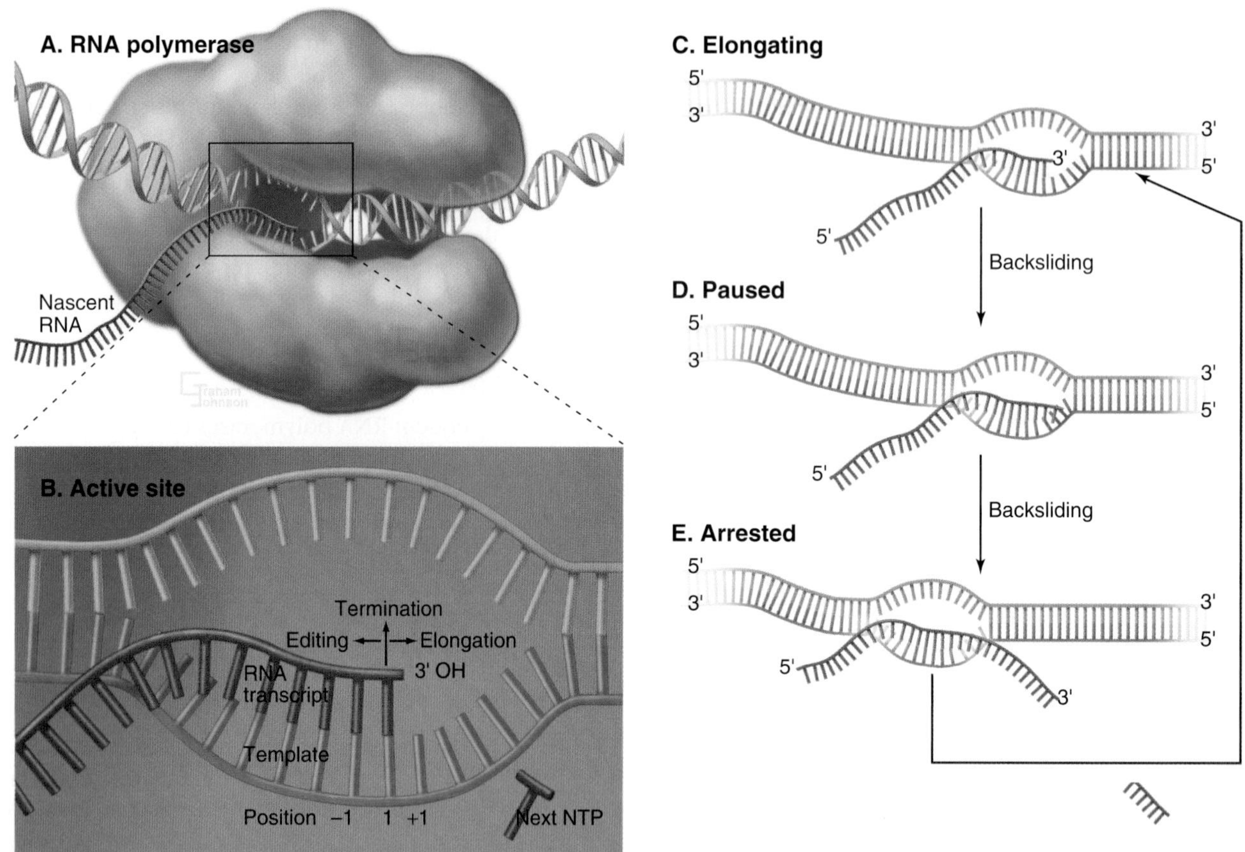

FIGURE 10.9 TRANSCRIPTION ELONGATION. A, Model of the transcription elongation complex consisting of RNA polymerase, template DNA, and nascent RNA transcript. RNA polymerases interact with the template upstream and downstream of the transcription bubble. **B,** The active site of RNA polymerase positions the growing end of the nascent transcript in the appropriate location for the addition of the next nucleoside triphosphate (NTP). After each single nucleotide addition, the polymerase may translocate forward and repeat the nucleotide addition **(C),** slide backward and pause for a variable time **(D),** or slide further backward, causing a transcription arrest that is reversed when the polymerase cleaves the nascent RNA **(E).**

Backward movement of the transcription bubble is accompanied by a zippering movement of the RNA-DNA hybrid in which the nascent RNA in the exit channel rehybridizes with upstream template sequences and the 3′ end of the transcript unpairs from the hybrid and is extruded through the same channel that NTPs use to enter the active site. If the polymerase backtracks more than a few nucleotides the complex becomes arrested and cannot resume elongation without assistance of additional factors. For example, transcription elongation factors can bind in the NTP channel of arrested complexes and activate the RNA polymerase to cleave the backtracked RNA. The new 3′ terminal residue is correctly positioned for incorporation of the next complementary NTP (Fig. 10.9C–E). This editing process increases the fidelity of transcription. Pausing also occurs following transcription of U-rich sequences, and in prokaryotes this is often associated with transcription termination.

Termination

When elongating RNA polymerase reaches the end of a gene or operon, specific sequences in the RNA called terminators trigger the release of the transcript and dissociation of the RNA polymerase. Bacteria have two types of terminators. The first are called **intrinsic** (or **rho-independent**) **terminators,** because they function in the absence of any protein factors (Fig. 10.10A). Intrinsic terminators consist of two sequence elements in the RNA: a stable GC-rich hairpin and a run of about eight consecutive U residues. As the first of these elements is synthesized, it forms a hairpin, causing polymerase to pause with less stable U:A bp (with only two H bonds [see Fig. 3.14]) in the hybrid. The nascent transcript is released from this complex, terminating transcription. The second type of prokaryotic termination requires a protein factor called rho (Fig. 10.10B). Rho is a hexameric helicase that binds cytosine-rich sequences and uses ATP hydrolysis to translocate along the nascent transcript in the 5′ to 3′ direction, essentially chasing the RNA polymerase. When polymerase pauses, rho can catch up and use the energy derived from ATP hydrolysis to pull the RNA out of the transcription elongation complex.

Eukaryotic RNA polymerases evolved distinct mechanisms for termination. RNA Pol III requires no protein

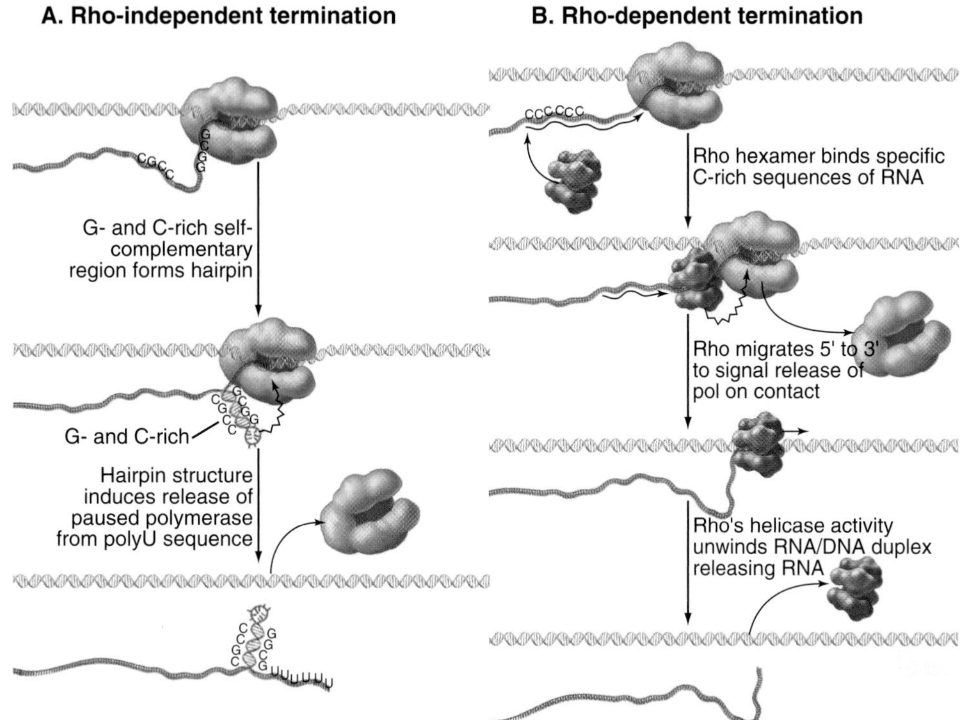

FIGURE 10.10 PROKARYOTIC TRANSCRIPTION TERMINATION. A, Rho-independent termination is directed by sequences in the nascent transcript that operate in the absence of any additional factors. **B,** The bacterial termination factor rho translocates along the nascent RNA and on reaching the RNA polymerase (pol) causes the disassembly of the elongation complex.

factors but terminates efficiently after transcribing four to six consecutive U residues, presumably owing to instability of the RNA-DNA hybrid in the enzyme active site. RNA Pol I terminates in response to a protein factor that blocks further elongation by binding to a DNA sequence downstream of the termination site, leaving an inherently unstable U-rich RNA-DNA hybrid in the active site. The RNA Pol II termination mechanism is more complex, requiring a large multiprotein complex that recognizes the poly(A) addition signal sequence in the nascent transcript (see Fig. 11.3 for pre-mRNA processing). Deletion or mutation of the poly(A) signal results in a failure to terminate messages at the appropriate site. Thus, RNA Pol II termination is coupled to 3′-end processing (see Chapter 11).

Gene-Specific Transcription Regulation

Transcription initiation is the critical first step in determining that each gene is expressed at the appropriate level in each cell. Depending mainly on the sequence of the promoter and other regulatory sequences, expression can be constitutive or influenced by regulatory proteins. This section discusses proteins that regulate transcription of specific genes either positively or negatively. The discussion starts with a prokaryotic example and then covers a variety of eukaryotic regulators. Although the details differ in prokaryotes and eukaryotes, many of the basic principles are the same.

Regulation of Transcription Initiation in Prokaryotes

Prokaryotes typically regulate gene expression in response to environmental cues such as the presence of nutrients in the growth medium (see Fig. 27.11). These signals are transmitted to the appropriate genes through regulatory proteins that bind to specific sequences near the genes they control to either activate or repress transcription. Both of these regulatory mechanisms come into play in regulation of the *E. coli* lactose (lac) operon (Figs. 10.2A and 10.11). The genes expressed from this operon are required for cells to metabolize lactose but are not expressed in the absence of lactose. Genetic studies in the 1960s showed that the gene upstream of the lac operon (*I* in Fig. 10.2A) encodes a repressor (lac repressor) that blocks expression of the lac operon in the absence of lactose (Fig. 10.11). The lac repressor binds to a site called an operator that overlaps the RNA polymerase binding site in the lac promoter. Lactose binding changes the conformation of the repressor, so it dissociates from DNA, allowing RNA polymerase to bind the promoter.

Full expression of the lac operon requires the catabolite activator protein (CAP), another allosteric protein that binds DNA just upstream of the lac promoter. CAP is activated by a conformational change induced when

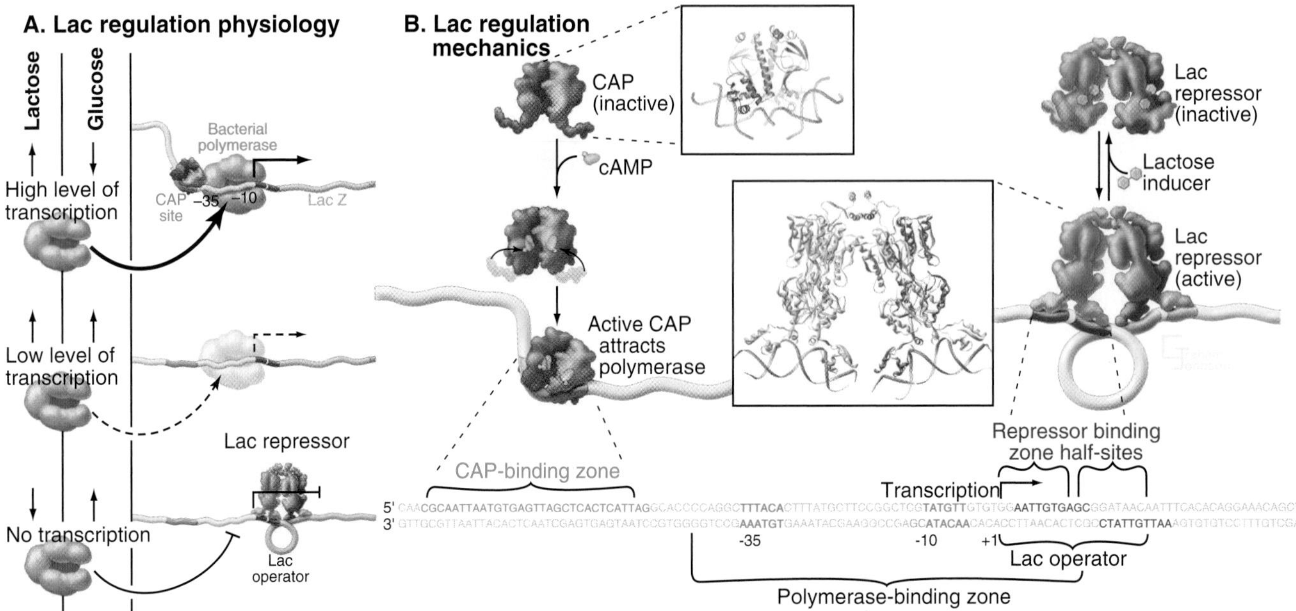

FIGURE 10.11 REGULATION OF THE LACTOSE (LAC) OPERON. A, RNA polymerase *(green)* binding to the lac promoter is regulated by the binding of repressor or activator (catabolite activator protein [CAP]). **B,** Binding sites for CAP and the repressor at the lac operon. The main repressor-binding site overlaps the promoter and blocks access of RNA polymerase. Additional lac repressor-binding sites are located upstream and downstream of the promoter. Lac repressor can form a tetramer and thus bind two operators, forming a loop in the lac operon DNA. Inducer (eg, lactose) binding dramatically alters the conformation of the lac repressor diminishing its affinity for the operator. CAP binds just upstream of the promoter where it can stabilize the bound RNA polymerase.

it binds cyclic adenosine monophosphate (cAMP), which the cell produces when the intracellular glucose concentration is low. Active CAP bound to its site stabilizes the otherwise weak interaction of RNA polymerase with the promoter. The resulting activation allows maximum expression of the lac operon in the presence of lactose and the absence of glucose.

Control of lac gene expression by opposing repressors and activators is an example of regulation at the first step in transcription initiation, binding of RNA polymerase to the promoter. Regulating access of RNA polymerase to promoters is a common form of transcription regulation in both prokaryotes and eukaryotes.

Overview of Eukaryotic Gene-Specific Transcription

While recruitment of RNA polymerase to the promoter remains a key step in eukaryotic transcription regulation, there are additional layers of complexity. First, DNA is bound by histones and packaged in nucleosomes (see Fig. 8.1) that can block binding of TFs and RNA polymerase. Overcoming this generalized repressive effect requires activators that alter chromatin structure allowing the recruitment of RNA polymerase.

Another major difference is that eukaryotic TFs bound tens to hundreds of kilobases away from the promoter

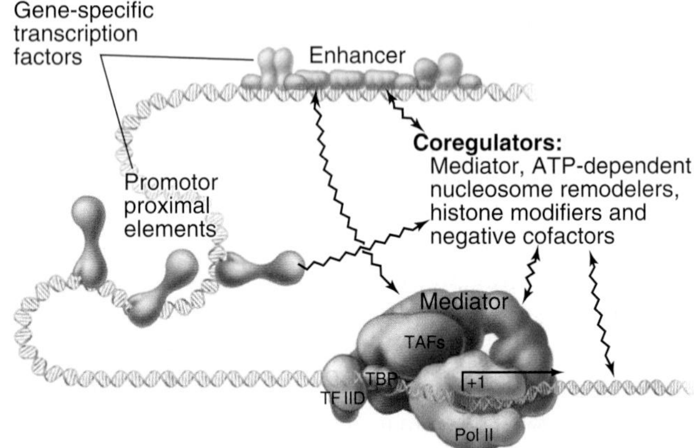

FIGURE 10.12 NETWORK OF INTERACTIONS THAT REGULATE RNA POLYMERASE II. Input comes from transcription factors bound to promoter proximal elements and enhancers and from coregulators that modify chromatin.

can activate transcription. In many cases these gene-specific TFs do not act directly on polymerase but require coregulators that act as a bridge between gene-specific factors, the chromatin template and RNA polymerase with its associated GTFs (Fig. 10.12).

The following sections explain how detailed mechanistic studies of a small set of model genes provided

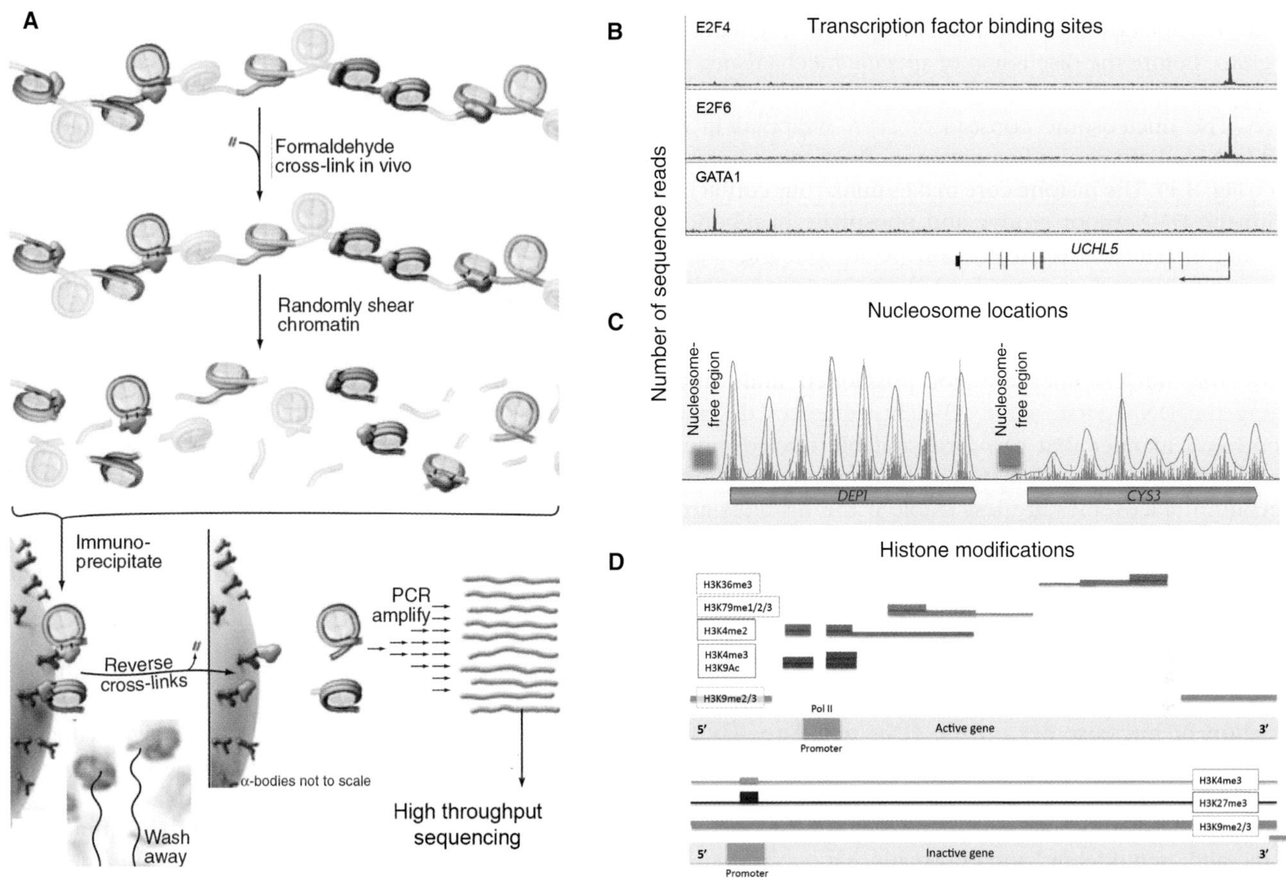

FIGURE 10.13 CHROMATIN IMMUNOPRECIPITATION COUPLED WITH HIGH-THROUGHPUT SEQUENCING (CHIP-SEQ) MAPS PROTEIN BINDING SITES AND HISTONE MODIFICATIONS. A, Experimental protocol. **B,** Frequency of DNA reads of DNA associated with three transcription factors associated with the UCHL5 gene. **C,** Frequency of DNA reads of DNA associated nucleosomes along two budding yeast genes. Nucleosomes are spaced regularly along the DEP1 gene but not along the CYS3 gene. **D,** Histone modifications along an active and an inactive gene. The thickness of the bar represents the frequency of each modification. (**B,** From Farnham P. Insights from genomic profiling of transcription factors. *Nat Rev Gen.* 2009;10:605–616. **C,** From Barth TK, Imhof A. Fast signals and slow marks: the dynamics of histone modifications. *Trends Biochem Sci.* 2010;35:618–626. **D,** Based on data from Jiang C, Pugh F. Nucleosome positioning and gene regulation: advances through genomics. *Nat Rev Gen.* 2010;10:161–172.)

the concepts for our current understanding of how thousands of different proteins combine to regulate tens of thousands of different promoters. Genome-wide studies have refined our understanding of how these regulatory mechanisms function in more global gene regulatory networks. Before addressing specific mechanisms, we consider techniques for mapping regulatory proteins to specific sites in the eukaryotic genome.

Mapping Transcription Components on the Genome

One of the key advances in transcription research has been to map transcription regulators and transcripts on a genome-wide basis. Fig. 10.13 describes one of these approaches: chromatin immunoprecipitation coupled with high-throughput sequencing (ChIP-seq). This approach yields a genome-wide snapshot of the positions of RNA polymerase, TFs, and histones on DNA and the modifications of these components. This comprehensive view of the distributions of transcription components has yielded novel insights about the locations of regulatory sequences and the presence of different combinations of histone modifications. This information will undoubtedly guide future experiments where the regulatory mechanisms are not yet clear.

Chromatin and Transcription

DNA in eukaryotic cells associates with an equal mass of protein to form chromatin (see Chapter 8). Packaging DNA in arrays of nucleosomes compacts the DNA and restricts access of transcription proteins to the DNA template. Understanding how the transcription machinery interacts with nucleosomes is a key to understanding eukaryotic transcription regulation.

Gene activation often involves disruption or displacement of nucleosomes located on specific regulatory regions. Before the discussion of specific mechanisms, it is useful to consider some aspects of nucleosome structure. The nucleosome consists of DNA wrapped in a left-handed helix around an octamer of histone subunits (see Fig. 8.1). The histone core makes numerous contacts with the DNA minor groove and phosphate backbone, leading to tight but relatively nonspecific binding. This aspect of the nucleosome allows for a dynamic association with DNA, because binding of the histone core to DNA is nearly as energetically favorable for all sequences. However, nucleosomes are not positioned uniformly along the DNA. First, some AT-rich sequences do not bend in a manner that can form a stable nucleosome. Such sequences are often found in promoter regions. Second, nucleosomes are less stable if the histones are modified, for example by acetylation or the inclusion of variant histone proteins. The presence of unstable nucleosomes enables the transcription machinery to access key regulatory sequences.

Nucleosome remodeling complexes can either expose or shield regulatory elements by altering the location of nucleosomes on the DNA template. These multiprotein remodeling complexes use energy from ATP hydrolysis to destabilize interactions between histones and DNA thus altering the position of the nucleosome and "remodeling" the chromatin. One example is the SWI/SNF (yeast mating type switching defective/sucrose nonfermenting) complex that is recruited to a specific subset of genes through interactions with transcription activators. The resulting remodeling of nucleosomes in the vicinity of promoters may be required to form a stable preinitiation complex. Genomic mapping of histones (Fig. 10.13) shows that most Pol II promoters are free of nucleosomes.

Histone Modifications and Gene Expression

Specific enzymes modify the histone tails with diverse chemical groups, often on lysine residues. Gene regulatory proteins recruit the modifying enzymes to chromatin generally as part of larger complexes (Table 10.2). Activator proteins generally recruit histone acetyltransferases, while histone deacetylases are part of corepressor complexes.

The hundreds of chromatin regulatory complexes in cells give rise to different chromatin states defined both by their pattern of histone modification and by their transcription (Fig. 10.13). Silent chromatin is not transcribed and has nucleosomes with H3K9me3 or H3K27me3 modifications spanning multiple genes in heterochromatin (see Chapter 8). Active chromatin often contains nucleosomes with H3K4ac or H3K4me, H4K8ac modifications, often in promoter-proximal nucleosomes (Fig. 10.13). In stem cells (see Box 41.2), many promoter-proximal regions contain both activating and repressing marks, so the genes are thought to be poised to be either activated or repressed as downstream signals dictate.

Most chromatin regulators are parts of larger complexes containing protein modules that recognize histone modifications such as **bromodomains** that interact with acetylated tails or **chromodomains** that bind methylated tails. For example, the SAGA histone acetyltransferase complex contains a bromodomain that anchors the complex to chromatin, facilitating further modification of regions that are already acetylated. A subunit of TFIID also contains a bromodomain that can facilitate the binding of TFIID to acetylated nucleosomes associated with active chromatin. Similarly, a number of histone methyltransferases contain chromodomains and are therefore targeted to their substrates by preexisting histone methylation.

The following sections describe examples of how TFs, chromatin regulators, and the general transcription machinery interact to regulate eukaryotic genes.

Gene-Specific Eukaryotic Transcription Factors

Eukaryotic TFs bind specific DNA sequences associated with the genes they regulate. This binding leads to activation or repression of transcription in a spatially and temporally controlled manner. In the simplest cases, the TF interacts directly with RNA Pol II and the GTFs but in more complex cases, the interaction may involve a coactivator or corepressor (see the following section). Current estimates indicate that approximately 6% of

TABLE 10.2 Nucleosome-Modifying Complexes

Name	Subunits	Catalytic Activity	Histone-Interacting Domain	Target Histone(s)
SAGA	15	Histone acetylase	Bromodomain	H3, H2B
NuA4	6	Histone acetylase	Chromodomain	H4
P300	1	Histone acetylase	Bromodomain	H2A, H2B, H3, H4
NuRD	9	Histone deacetylase	Chromodomain	?
SIR2	3	Histone deacetylase	Neither	H4
MLL	7	Histone methylase	Neither	H3 (lysine 4)

the coding capacity of the human genome (more than 1000 genes) is devoted to TFs that recognize specific DNA sequences. The following sections discuss the functional organization of these proteins, how they recognize DNA and how they interact with chromatin and the GTFs.

DNA-Binding Domains

Binding proteins to specific DNA sequences requires recognition of a pattern of bases along the double helix. The richest source of DNA sequence specificity comes from the chemical groups exposed in the major groove. Most specific DNA-binding proteins probe the major groove of the double helix with a small structural element (usually, an α-helix) with a shape complementary to the surface topography of a particular DNA sequence. The correct DNA sequence is recognized through multiple interactions between amino acid side chains in the recognition helix and the chemical groups on the DNA bases in the major groove. Single amino acid changes in the recognition helix can change the DNA sequence that is recognized. Protein-DNA complexes are stabilized by additional contacts between amino acid side chains and deoxyribose rings and phosphate groups or by bending the DNA.

DNA recognition domains of specific TFs typically interact with only 3 to 6 bp of DNA. Given the size and complexity of the typical mammalian genome, a sequence must be approximately 16 bp long to occur by chance only once. How then can TFs recognize specific genes among the vast number of close but nonidentical sequences? Two strategies increase the length of the specific sequence to be recognized. The protein can either use several recognition elements or dimerize with itself or other DNA-binding proteins. Protein dimers can recognize sequences with twofold rotational symmetry.

DNA-binding proteins can be grouped into families based on the structure of the domains used for DNA sequence recognition (Fig. 10.14). These include the **helix-turn-helix (HTH)** proteins, **homeodomains, zinc finger** proteins, **steroid receptors, leucine zipper** proteins, and **helix-loop-helix** proteins. Although these families include most known TFs, there remain other, less-common recognition domains. Within a given family, the recognition domain of each TF has an amino acid sequence that targets the protein to a particular DNA sequence. Conversely, different families of TFs can recognize the same regulatory sequence. The following sections discuss several of the more common eukaryotic DNA-binding domains.

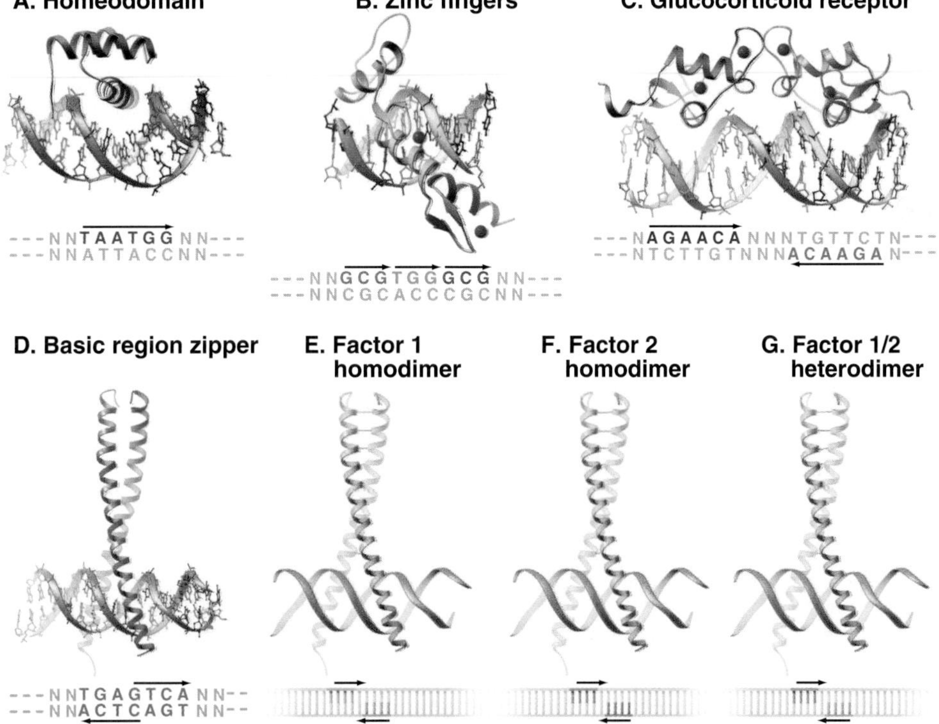

FIGURE 10.14 MOLECULAR STRUCTURES OF TRANSCRIPTION FACTOR DNA-BINDING DOMAINS. Recognition of specific DNA sequences requires interactions between amino acid side chains in the protein and chemical groups on the DNA bases. In each of the examples shown here, an α-helix interacts with specific bases through contacts in the major groove. **A,** The homeodomain α-helix recognizes a specific 6-bp sequence. **B,** A protein with three zinc fingers recognizes three consecutive 3-bp sequences. **C,** The glucocorticoid receptor forms a dimer that recognizes the same 6-bp sequence (a hormone response element) in opposite orientations spaced 3 bp apart. **D,** A leucine zipper factor dimerizes to recognize a pair of 4-bp sites with opposite orientation spaced 1 bp apart.

Homeodomain

This motif of 60 amino acids was discovered in *Drosophila* proteins that regulate development and is found in many eukaryotic TFs, including more than 150 human genes. Recognition is provided by an HTH motif composed of two helices, one of which sits in the major groove of the DNA-binding site contacting a recognition sequence of 6 bp (Fig. 10.14A). The HTH structure is not a stable domain on its own, but functions as part of a larger DNA-binding domain, such as the homeodomain. A flexible arm interacting with the minor groove provides the homeodomain with additional binding affinity.

Zinc Finger Proteins

The zinc finger protein sequence motif (Fig. 10.14B) is found in more than 600 human TFs. Each "finger" consists of 30 residues with conserved pairs of cysteines and histidines that bind a single zinc ion. The tip of the finger sticks into the DNA major groove, where it contacts three bases. Most zinc finger proteins contain multiple fingers, allowing longer sequences to be recognized to increase specificity. A related structure is present in the steroid hormone receptor family, although in this case, four cysteine residues coordinate the zinc ion and the finger is composed of two helices rather than one. Steroid hormone receptors also contain a dimerization domain, allowing recognition of sequences with dyad symmetry (Fig. 10.14C). Artificial zinc fingers can now be designed enabling synthetic proteins to recognize any desired DNA sequence for experimental manipulations.

Leucine Zipper Proteins

Leucine zipper domains are made up of two motifs: a basic region that recognizes a specific DNA sequence and a series of leucines spaced 7 residues apart along an α-helix (leucine zipper) that mediate dimerization. These motifs form a continuous α-helix that can dimerize through formation of a coiled-coil structure involving paired contacts between hydrophobic leucine zipper domains (Fig. 10.14D; also see Fig. 3.10). Dimers of leucine zipper proteins recognize short, inverted, repeat sequences. The zipper family comprises many members, some of which can cross-dimerize and recognize asymmetrical sequences. Another family of factors comprises the helix-loop-helix proteins, which have the same type of basic region but differ in that they have two helical dimerization domains separated by a loop region.

Transcription Factors as Modular Proteins

Binding of a TF to DNA per se does not activate transcription. A separate domain provides this function by interacting directly or indirectly with the basal transcription machinery to elevate the rate of transcription (Fig. 10.15). The best-characterized activation domain is an acidic region derived from the herpesvirus VP16 protein.

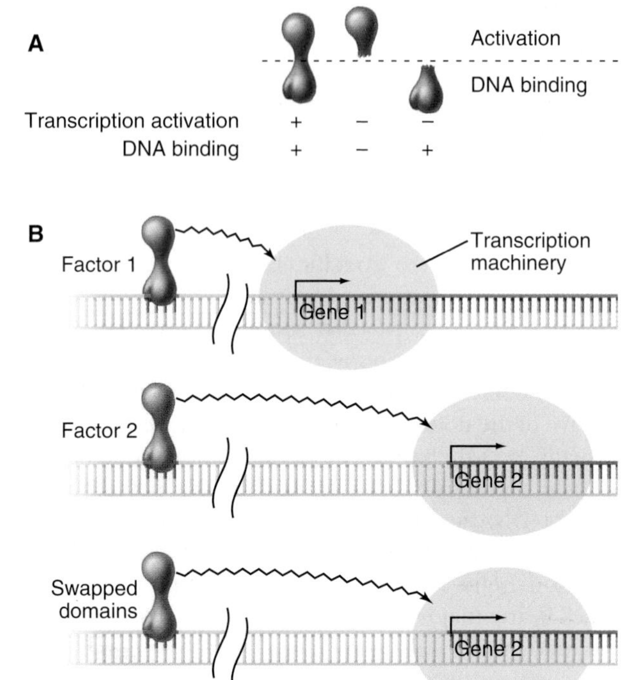

FIGURE 10.15 TRANSCRIPTION FACTORS CONSIST OF DISCRETE, FUNCTIONAL MODULES. A, Domain characterization. Although the entire factor is required for activation, the bottom domain is sufficient for DNA binding. **B,** Domain swapping. The activation domain of one factor (activating gene 1) can be fused to the DNA-binding domain of a heterologous factor (activating gene 2). The resulting chimeric factor will activate only genes containing the recognition site for the DNA-binding domain (gene 2).

Acidic activation domains are generally unstructured segments of polypeptide consisting of multiple acidic residues dispersed among a few key hydrophobic residues. Such domains activate transcription when experimentally grafted to a wide variety of different DNA-binding domains in a number of different cell types. Other types of activator domains have been characterized as being rich in proline or glutamine.

The diverse activation domains use several mechanisms to activate transcription, the most direct being recruitment of the basal transcription machinery. For example, the glutamine-rich activation domain of the SP1 factor (see the next section) interacts with TFIID to recruit GTFs to the promoter. In many cases transcription activators and repressors do not contact the GTFs or Pol II directly but rather act via interactions with coregulator complexes as discussed in a following section.

Transcription Factor Binding to Eukaryotic Promoter Proximal and Enhancer Elements

Experiments analyzing eukaryotic promoter function in living cells revealed numerous DNA regulatory sequence elements in addition to the basal promoter elements recognized by the GTFs. The regulatory sequence elements fall into two broad categories based on their

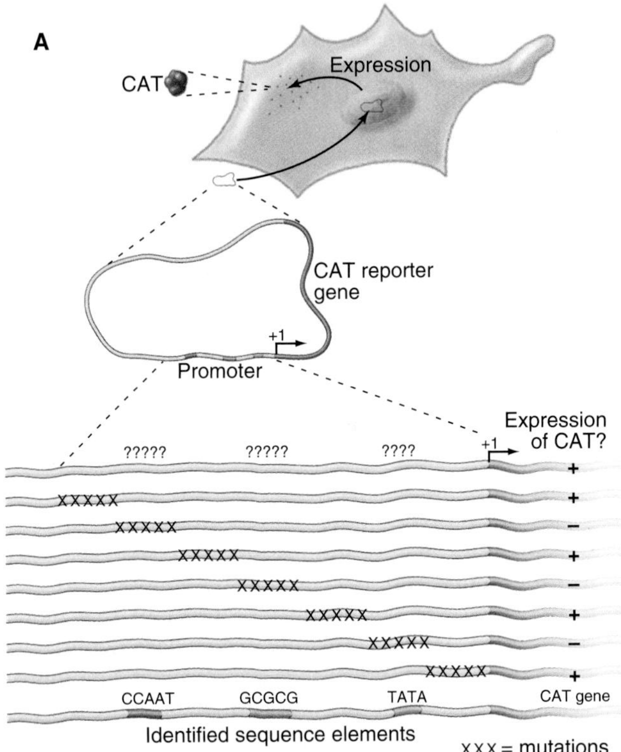

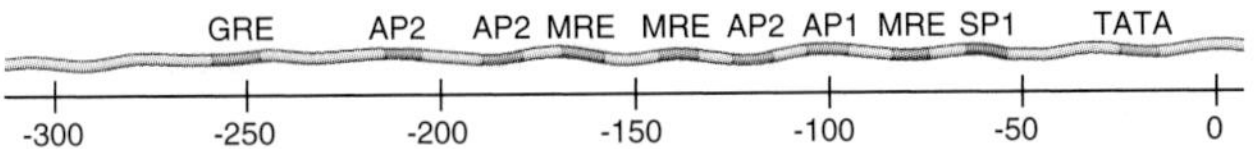

FIGURE 10.16 RNA POLYMERASE II PROMOTER REGULATORY ELEMENTS. A, In vivo assays are used to identify key regulatory sequences. In the example shown, a promoter is placed in front of a gene encoding chloramphenicol acetyltransferase (CAT), and the resulting plasmid is transfected into cultured cells. This bacterial enzyme is easily assayed in eukaryotic cells because there is no corresponding endogenous activity. Targeted clusters of mutations, strategically placed throughout the promoter region, are tested for their effect on expression of the reporter gene. Mutations that reduce expression define important regulatory elements. **B,** The region immediately upstream of the metallothionein gene contains binding sites for several transcription factors. Each element is named for the factor that binds there: GRE (glucocorticoid response element), MRE (metal response element), and AP1, AP2, and SP1 (which bind protein factors with the same names as the DNA elements).

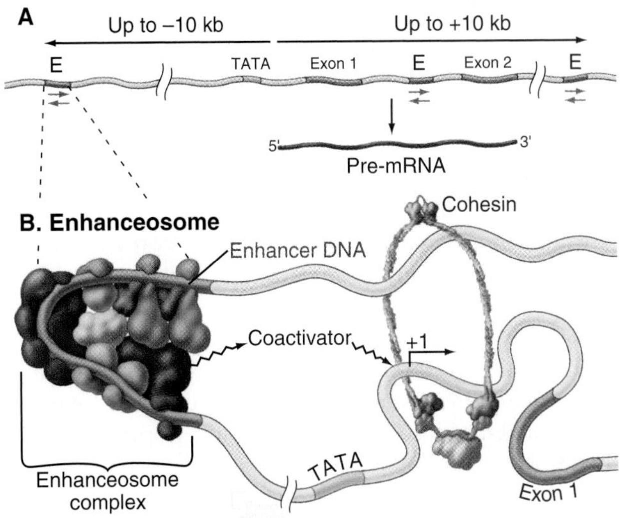

FIGURE 10.17 ENHANCER ELEMENTS. A, These clusters of factor-binding sites can influence expression when located far from the promoter in either the upstream or downstream position. In addition, they work in either orientation with respect to transcription. **B,** Model enhancer showing the tight packing of several different DNA-binding proteins. These complexes fold into structures that have been called enhanceosomes.

distances from the promoter. **Promoter proximal elements** are located within a few hundred base pairs upstream of the transcription start site. **Enhancer** sequences can be located from tens to hundreds of kilobases from the start of transcription.

One example of a promoter proximal element is the CCAAT box in the promoter of the herpes simplex virus thymidine kinase gene. This site was identified by a technique called linker-scanning, in which clustered mutations are introduced at regular intervals in the promoter (Fig. 10.16A). In the case of the thymidine kinase promoter, the CCAAT and TATAAA sequences are required for full transcription. Thymidine kinase expression also requires the sequence GGCGCC, which was subsequently shown to serve as the binding site for SP1, a TF involved in expression of a number of so-called housekeeping genes, whose products are involved in constitutive cellular functions. Other promoter proximal elements are involved in regulated expression, for example, in response to cellular stress or exposure to heavy metals. Most promoters are paired with several different promoter proximal elements. This allows for regulation of transcription levels by varying the relative abundance or activity of the various factors. A good example is the human metallothionein gene, whose product protects cells from the toxic effects of metals (Fig. 10.16B). The location of numerous regulatory elements directly upstream of the TATA box suggests that a variety of different mechanisms regulate this gene.

Enhancers are clusters of regulatory DNA sequences that resemble promoter proximal elements, but are considerably more complicated and have several distinguishing features. First, an enhancer can increase the rate of initiation from a basal promoter even if it is located up to 100 kb away along the chromosome. Second, the enhancer element will work in either orientation relative to the promoter (Fig. 10.17A). Third, enhancers can function with a heterologous promoter. Figure 10.17B shows an example of an enhancer sequence with a number of TFs bound, forming a complex called an enhanceosome. Many genes are associated with multiple enhancers. Each enhancer usually works in a cell type–specific fashion. An example is a sequence in an intron of the immunoglobulin heavy-chain gene that enhances transcription in lymphocytes but not in other cells. This regulation of enhancer function is accomplished by varying the levels of various enhancer-binding TFs in different tissues. In addition, enhancer chromatin structure is characterized by a nucleosome free region that allows TFs to bind, flanked by nucleosomes bearing histone H3K27ac and H4K3me1 modifications. The following sections discuss how enhancers interact with TFs and coregulators to increase transcription from promoters.

Coactivators

Coactivators are complexes of regulatory proteins that do not bind DNA themselves but are recruited by gene-specific TFs. These complexes contain proteins that recruit the GTFs, alter chromatin structure and assist in the early stages of transcription. The most common coactivator is the Mediator, a complex of 26 proteins in human cells that bridges DNA-bound TFs to Pol II (Fig. 10.18A). Different Mediator subunits bind particular TFs and communicate regulatory signals to the initiation complex. One example: an interaction of Mediator with the Pol II CTD helps stabilize the preinitiation complex. Mediator also stimulates the CTD kinase activity of TFIIH thus releasing the CTD from the Mediator (Fig. 10.18A)

Another class of coactivators has histone acetyltransferase activity that modifies histones and other proteins (Fig. 10.18B). One example is p300/CBP. This was initially identified as a protein interacting with a TF that binds cAMP response elements. Many different TFs recruit p300 to chromatin to locations generally assumed to be enhancers. Histone H3K27 is one of the main targets of p300. This same H3 residue is the target of the polycomb repressive complexes (see Chapter 8), suggesting that p300 plays a role in switching between active and repressed chromatin states.

A third class of chromatin coactivators regulates access to DNA by moving, ejecting, or altering the composition of nucleosomes. The SWI/SNF complex is an example of a nucleosome remodeler. When recruited to chromatin by TFs, this complex moves nucleosomes that block regulatory or promoter sequences. Coactivators often work together to activate genes. Initial binding of a TF may recruit a chromatin remodeler that exposes a second TF binding site. This second TF may then recruit mediator, thereby recruiting Pol II and the GTFs.

Corepressors act in opposition to coactivators by repressing transcription. The most common form of repression involves chromatin modifications that block TF access (Fig. 10.18C). Histone deacetylase complexes like Sir2 and NuRD (Table 10.2) remove acetyl modifications leading to chromatin compaction and repression of transcription. Polycomb is another pair of corepressor complexes that methylates H3K27 leading to inhibition of RNA polymerase elongation. Polycomb repressive complexes play critical roles in early embryonic development.

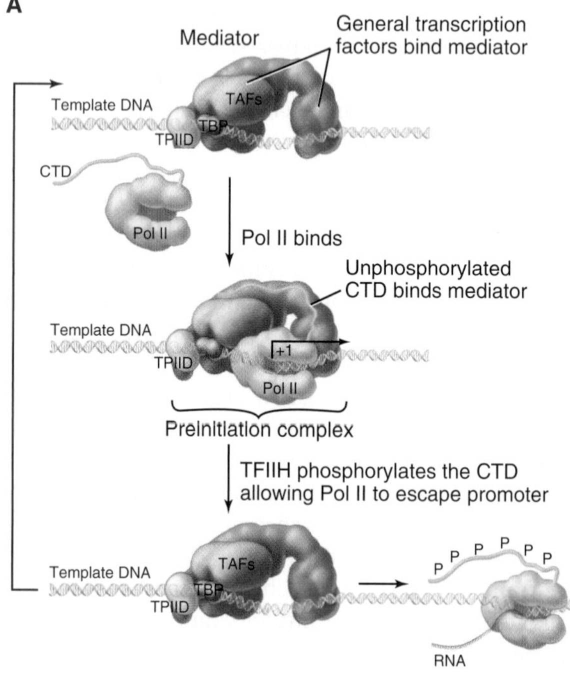

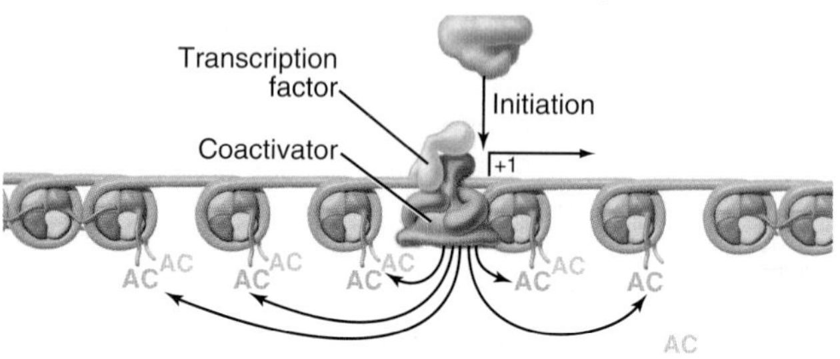

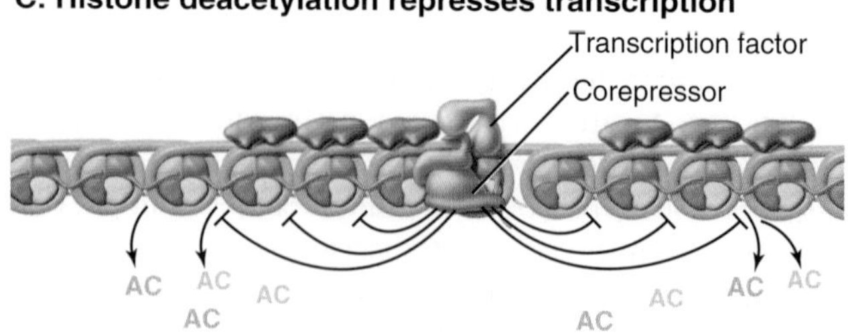

FIGURE 10.18 TRANSCRIPTION ACTIVATION MECHANISMS. A, General transcription factors and mediator form a scaffold for binding RNA polymerase II with unphosphorylated C-terminal domain (CTD) to form a preinitiation complex. Phosphorylation of CTD by TFIIH releases polymerase and starts transcription. **B,** Histone acetylases in a coactivator loosen chromatin in the vicinity of the promoter, allowing assembly of preinitiation complexes. **C,** Recruitment of histone deacetylases in a corepressor represses transcription by compacting the chromatin in the vicinity of the promoter.

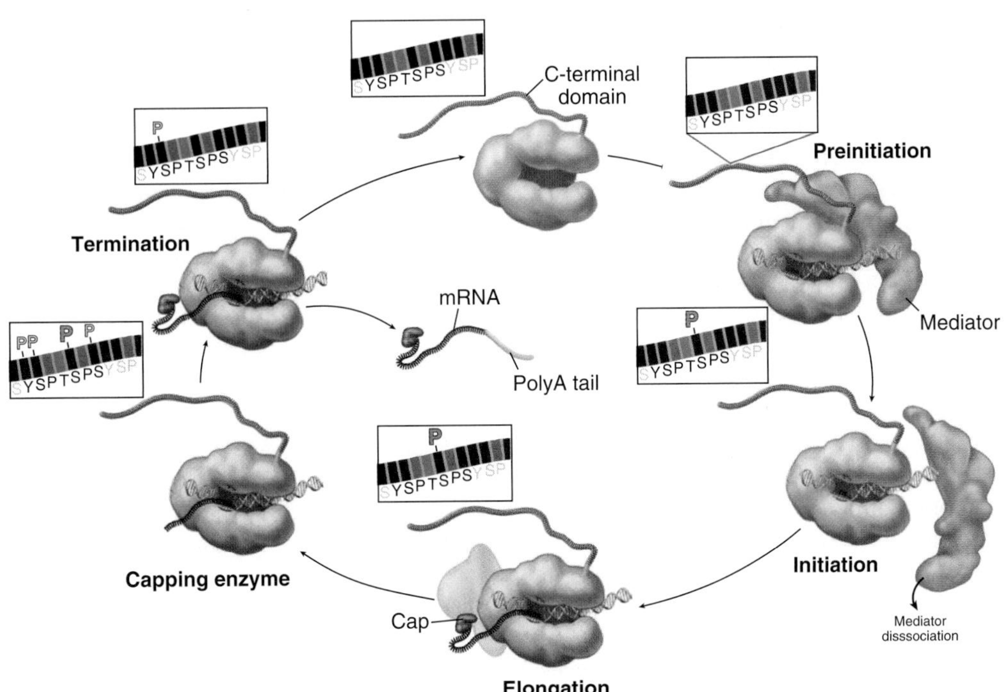

FIGURE 10.19 PHOSPHORYLATION OF THE C-TERMINAL DOMAIN OF RNA POLYMERASE II REGULATES TRANSCRIPTION. This cycle illustrates how phosphorylation influences each step in mRNA transcription. See the text for details.

Long-Range Regulatory Interactions

Most genes are regulated by enhancers located many thousands of bases away from their promoter, so some means of communication between enhancer and promoter is required for gene activation. This is most commonly achieved through direct interaction when the chromatin fiber forms a loop bringing the enhancer and promoter into close contact. Such interactions between enhancers and promoters involve the cohesin complex discovered because it regulates separation of sister chromatids during cell division (see Fig. 8.18). The cohesin complex forms a ring around two DNA strands thus stabilizing the loop (Fig. 10.17). ChIP-seq screening for cohesin and high levels of Mediator has identified several hundred intergenic regions containing multiple enhancers clustered together. These "super enhancers" direct transcription of genes that specify cell fate and when associated with oncogenes lead to tumor pathogenesis.

Post Initiation Regulation of Polymerase II Transcription

After formation of a transcription preinitiation complex several steps lead to promoter clearance, elongation and termination. The CTD of Pol II not only orchestrates promoter proximal events but is also important for coupling transcription to splicing of the nascent transcript and to 3′-end formation (see Chapter 11). The CTD (Fig. 10.4C) is comprised of tandem repeats of the repeat of seven amino acids ($Y_1S_2P_3T_4S_5P_6S_7$). The three serines are phosphorylated at different stages of the transcription cycle (Fig. 10.19). TFIIH kinase CDK7 phosphorylates serine 5 (Ser5) in the preinitiation complex. This releases the Mediator from Pol II and creates a binding site for the capping enzyme that modifies the 5′ end of the message (see Fig. 11.2). At most promoters, inhibitory factors pause the early Pol II elongation complex after synthesizing approximately 30 nucleotides. Phosphorylation of CTD Ser2 by Cdk9 releases the paused polymerase and allows elongation to proceed. Signaling pathways regulate this process at many genes. Ser2 phosphorylation also helps recruit the RNA splicing machinery to the nascent transcript. At the 3′ end of the gene, Ser2 phosphorylation recruits the cleavage and polyadenylation machinery leading to formation of the mature mRNA and termination of elongation.

Combinatorial Control

The complexity of eukaryotic regulatory systems allows for the integration of multiple regulatory signals at individual genes. Such combinatorial control is seen in a limited way in prokaryotes. For example, the *E. coli* lac genes are regulated by both lactose and glucose. Only when glucose is low and lactose is present do the activator (CAP) and repressor (lac repressor) function to maximize lac expression.

Regulation of transcription initiation in eukaryotes is based on similar principles with DNA-binding activators and repressors controlling individual genes. Each eukaryotic gene typically has binding sites for multiple factors. Integration of the individual binding events can take place in several ways. First, there is a degree of synergism to the binding of multiple factors. The enhanceosome is an example where binding of proteins that bend the DNA promotes binding of additional proteins. The key characteristic of the enhanceosome complex is that it stimulates transcription more strongly than the sum of the individual TFs. Synergy can also result from multiple interactions between activators bound to DNA at different upstream sites or different enhancers and targets in coactivators such as the mediator or nucleosome remodeling complexes. Many of the same mechanisms also can occur with repressors.

Combinatorial control also can result from the interplay between factors that alter chromatin structure. For example, modification of histone tails by a histone acetyltransferase tethered to a DNA-bound TF can loosen chromatin at a particular site and create binding sites for additional factors. Subsequent binding of a nucleosome-remodeling complex can render sequences more accessible to the transcriptional machinery.

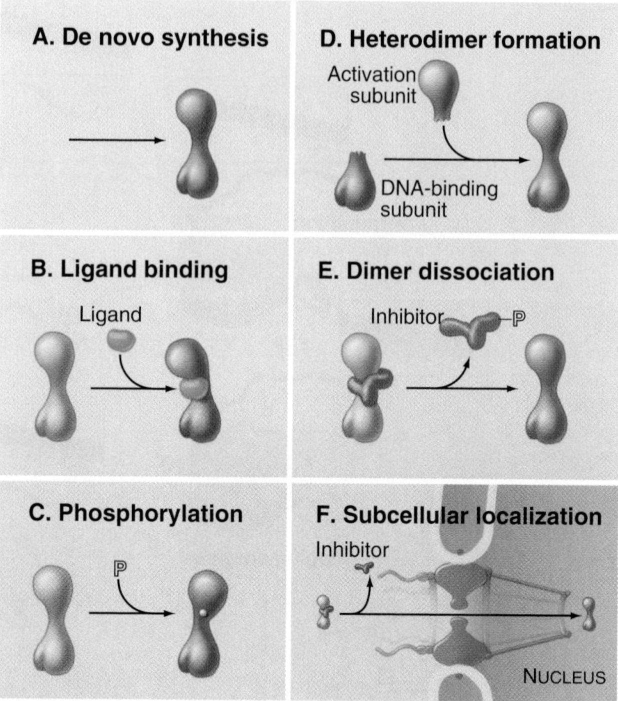

FIGURE 10.20 REGULATION OF TRANSCRIPTION FACTOR ACTIVITY. Many strategies have evolved to regulate transcription factors in response to specific signals. **A,** The availability of a factor may be controlled by expressing it, de novo, only when it is needed. **B,** Factors may be synthesized in an inactive state and depend on a small molecule (ligand) for activity. **C,** Transcription factors that are synthesized in an inactive state can be activated by postsynthetic modification, such as phosphorylation. **D,** Some factors require an appropriate partner for activity. **E,** Constitutively active factors can be held in check by associating with inhibitory subunits. **F,** Active factors can be sequestered in the cytoplasm by blocking their transport to the nucleus.

Modulation of Transcription Factor Activity

Regulation of transcription initiation is fundamentally important in controlling gene expression. In many cases, the availability of factors that bind to specific sites in promoters is the switch that turns a gene on. Various strategies control the binding of specific factors to DNA regulatory elements (Fig. 10.20). One of the most straightforward is de novo synthesis of the specific factor (Fig. 10.20A). This requires an additional level of regulation of transcription and translation of the mRNA that encodes the specific factor. These steps take time, so this regulatory strategy is used more commonly to regulate developmental pathways than situations where rapid responses are required.

Several mechanisms are used for rapid regulation of the activity of existing TFs. One mechanism involves the formation of an active factor from two inactive subunits (Fig. 10.20D). This association can be regulated through synthesis or by modification of preexisting subunits, leading to their association. Binding of small-molecule ligands is another means of controlling TF activity (Fig. 10.20B). In this case, the binding of the ligand induces a conformational change that leads to DNA binding and transcription activation. Interaction of TFs with inhibitory subunits is also used to regulate factor activity (Fig. 10.20E). The DNA binding or activation potential is held in check until the appropriate signal leads to dissociation or destruction of the inhibitory factor. Covalent modification—for example, by phosphorylation—is also used to convert inactive TFs to a functional form (Fig. 10.20C). Finally, the ability of TFs to bind DNA may be regulated by restricting their localization to the cytoplasm (Fig. 10.20F). These regulatory schemes are not mutually exclusive, and many regulatory pathways (see the examples that follow) employ several different levels of regulation.

Transcription Factors and Signal Transduction

One hallmark of eukaryotic gene regulation is the ability of cells to respond to a wide range of external signals. Cells detect the presence of hormones, growth factors, cytokines, cell surface contacts, and many other signals. They transmit this information to the nucleus, where changes in expression of specific genes are executed (see Fig. 27.4 for the three types of signaling pathways to the nucleus). TFs often execute the final step in these signal transduction pathways; the following sections discuss several examples not covered in Chapter 27.

Steroid Hormone Receptors

Regulation of gene expression by steroid hormone receptors involves both ligand-binding and inhibitory subunits. This family of nuclear receptors includes TFs with a common sequence organization consisting of a specific DNA-binding domain, a ligand-binding domain

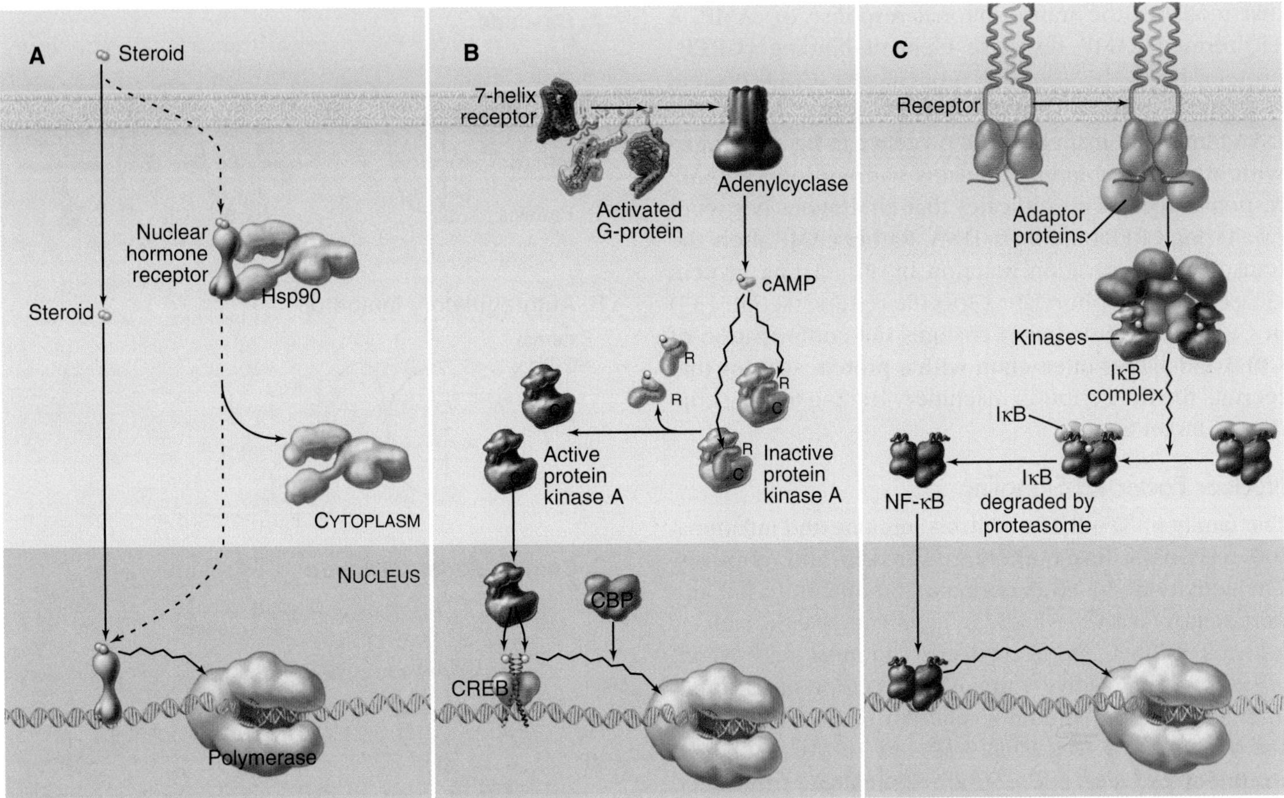

FIGURE 10.21 TRANSCRIPTION FACTORS AS TARGETS OF SIGNAL TRANSDUCTION PATHWAYS. External signals are transmitted by a variety of pathways that eventually impinge on transcription factors. **A,** Steroid hormones diffuse through the cell membrane and bind to the hormone receptor in the cytoplasm (estrogen) or, more commonly, the nucleus. Hormone binding induces a conformational change that renders the receptor competent to activate transcription. **B,** Ligands bound to the extracellular surface of seven-helix receptors initiate a pathway that leads to the activation of protein kinase A, which then moves to the nucleus, where it phosphorylates transcription factor CREB (cyclic adenosine monophosphate [cAMP] response element–binding). (C, catalytic subunit of protein kinase A [PKA]; R, regulatory subunit of PKA that is dissociated from C by binding cAMP [R is shown smaller than actual size].) **C,** In a third strategy, constitutively active transcription factors are kept sequestered in the cytoplasm until a signaling pathway is activated. In this example, the transcription factor nuclear factor κB (NF-κB) is bound to an inhibitor called IκB (inhibitor of nuclear factor κB). Activation of the pathway leads to phosphorylation of IκB, which targets the inhibitory subunit for destruction by the proteasome. The free NF-κB is transported to the nucleus, where it activates the transcription of target genes.

that regulates DNA binding, and one or more transcription activation domains. The ligands that regulate these factors are small, lipid-soluble hormone molecules that diffuse through cell membranes and bind directly to the TF in the cytoplasm (Fig. 10.21A). Steroid hormones, retinoids, thyroid hormone, and vitamin D bind to distinct nuclear receptors, enabling them to recognize sequences in the promoters of a range of target genes. The specific sites of action in promoter DNA, termed **hormone response elements,** are related to either AGAACA or AGGTCA (Fig. 10.14C). Specificity of the response is generated by the spacing and relative orientation of the binding sites. Nuclear receptors can bind as homodimers, although some form heterodimers. In addition to heterodimerizing with other members of the nuclear receptor family, interactions with other types of TFs can link the steroid response to other pathways that signal through cell surface receptors.

Heat shock protein 90 (Hsp90) blocks inactive steroid hormone receptors from interacting with DNA (Fig. 10.21A and see Fig. 17.13). This chaperone keeps the receptor ligand-binding domain in a conformation ready to bind the ligand but unable to enter the nucleus. Hormone binding to the receptor dissociates Hsp90 and frees the receptor's DNA-binding domain. The free ligand–bound receptor moves from the cytoplasm to the nucleus, where it binds its DNA target and activates transcription.

Cyclic Adenosine Monophosphate Signaling

Changes in gene expression often develop in response to the binding of signal molecules to cell surface receptors. Binding of ligand induces a structural change in the receptor that sets off a chain of events leading to changes in transcription. Protein phosphorylation plays an important role in this process.

The adenyl cyclase system controls not only metabolism (see Fig. 27.3) but also gene expression (Fig. 10.21B). Ligand binding to some seven-helix receptors **leads to cAMP** synthesis, which, in turn, activates **protein kinase A** (see Fig. 27.3). The promoters of cAMP-regulated genes contain a conserved DNA sequence element, called a **cAMP response element,** that mediates the transcriptional response to cAMP. A TF, termed cAMP response element-binding **(CREB)** protein, binds this sequence specifically. CREB protein is a leucine zipper TF that binds DNA as a dimer. The DNA-binding domain of CREB protein can be exchanged with other DNA-binding domains without loss of cAMP responsiveness. This indicates that cAMP does not work by altering CREB binding to DNA. Rather cAMP alters the transcription activation function by stimulating protein kinase A to phosphorylate a specific residue (serine 133) in CREB. Phosphorylation changes the conformation of CREB and allows interaction with a protein adaptor that recruits the transcription machinery leading to transcription of target genes.

Nuclear Factor κB Signaling

The family of NF-κB TFs controls immune and inflammatory responses, development, cell growth, and apoptosis. The activity of NF-κB is normally tightly controlled and persistently active NF-κB is associated with cancer, arthritis, asthma, and heart disease. In most cells, NF-κB is held in an inactive form in the cytoplasm through interaction with an inhibitor called **inhibitor of nuclear factor κB (IκB)** (see Figs. 9.19 and 10.21C). When B lymphocytes (see Fig. 28.9) are stimulated to produce antibody, NF-κB binds to an enhancer in the immunoglobulin κ-chain gene and activates transcription. The stimulatory signal leading to NF-κB activity is transmitted through a protein kinase cascade that ultimately phosphorylates I-κB, signaling its destruction by proteolysis. I-κB destruction unmasks the NF-κB nuclear localization signal, leading to its transport to the nucleus, where it activates transcription of immunoglobulin genes.

Transcription Factors in Development

The previous discussion focused on how external signals can lead to changes in gene expression in the nucleus, which, in turn, changes cellular functions. A critical step in this genetic program is the regulation of one TF by another. Such cascades of TF activity are fundamental to gene regulation in development.

Early cell divisions in multicellular organisms create different types of daughter cells that express distinct sets of genes. In this case, two types of information govern the expression of a gene. First, the interaction of the cell with its environment sends signals that are transduced to the nucleus and change the pattern of gene expression. How the nucleus interprets the transduced signals depends on the set of TFs that preexist within it. Thus, in addition to external signals, the history of the cell dictates which genes will respond to incoming signals.

The programs of TF interaction during development are complicated, but the underlying principles of these pathways are well conserved. Many developmentally regulated TFs are autoregulated (Fig. 10.22), allowing

A. Cascade

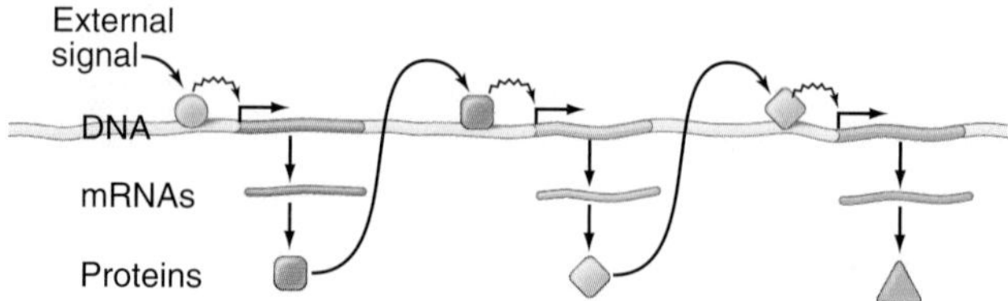

B. Autoregulatory inhibition

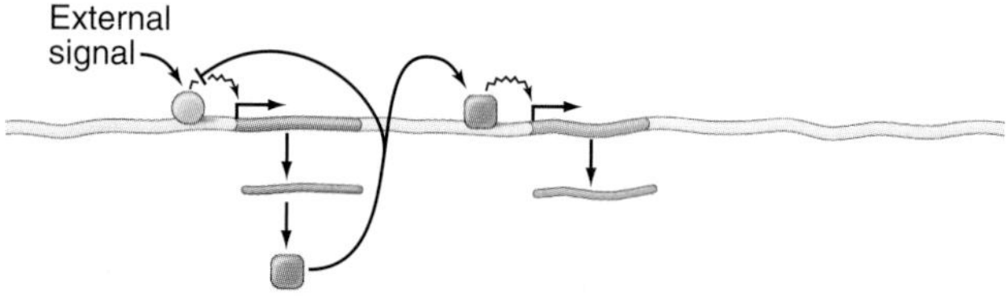

C. Combinatorial activation

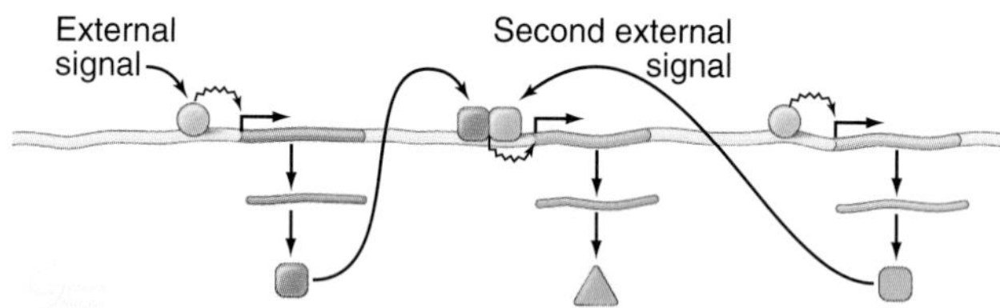

FIGURE 10.22 GENE REGULATORY CIRCUITS. The complex patterns of gene expression observed in multicellular organisms arise from interactions among thousands of transcription activators and repressors, as illustrated by three examples. **A,** Transcription factors activate the expression of other factors leading to a cascade of changes in gene expression following the initial external signal. **B,** Some transcription factors can act as both activators or repressors. In this example, the external signal leads to expression of a transcription factor that goes on to activate the expression of other genes and repress its own expression. **C,** Multiple transcription factors regulate most genes, so activation requires more than one external signal—two in this example.

TFs to activate their own expression. This positive feedback creates a switch that leads to continued expression after the initial stimulus is gone. Other developmentally regulated TFs are, in turn, regulated by several different factors. This allows combinatorial signals to dictate expression (Fig. 10.22C). For example, some TFs activate certain promoters while repressing others. The basis of this contradictory property is thought to be the ability of TFs to cooperate with each other when bound at the same promoter. This cooperation can be either positive or negative. This allows the expression of a target gene to be regulated both by external signals (eg, proximity of an adjacent cell that expresses a signaling molecule) and by the preexistence of a given factor in the cell. In this way, only cells of a given lineage that are located in a certain area of an embryonic segment express the gene. As new TFs involved in development are discovered, the challenge will be to decipher the complicated combinatorial interactions among them.

SELECTED READINGS

Corden JL. RNA polymerase II C-terminal domain: Tethering transcription to transcript and template. *Chem Rev*. 2013;113:8423-8455.

de Laat W, Duboule D. Topology of mammalian developmental enhancers and their regulatory landscapes. *Nature*. 2013;502:499-506.

Dekker J, Mirny L. The 3D genome as moderator of chromosomal communication. *Cell*. 2016;164:1110-1121.

Delest A, Sexton T, Cavalli G. Polycomb: a paradigm for genome organization from one to three dimensions. *Curr Opin Cell Biol*. 2012;24:405-414.

Hnisz D, Abraham BJ, Lee TI, et al. Super-enhancers in the control of cell identity and disease. *Cell*. 2013;155:934-947.

Landick R. The regulatory roles and mechanism of transcriptional pausing. *Biochem Soc Trans*. 2006;34:1062-1066.

Levine M. Transcriptional enhancers in animal development and evolution. *Curr Biol*. 2010;20:R754-R763.

Murakami K, Calero G, Brown CR, et al. Formation and fate of a complete 31-protein RNA polymerase II transcription preinitiation complex. *J Biol Chem*. 2013a;288:6325-6332.

Murakami K, Elmlund H, Kalisman N, et al. Architecture of an RNA polymerase II transcription pre-initiation complex. *Science*. 2013b; 342:1238724.

Ong CT, Corces VG. CTCF: an architectural protein bridging genome topology and function. *Nat Rev Genet*. 2014;15:234-246.

Ruthenburg AJ, Li H, Patel DJ, et al. Multivalent engagement of chromatin modifications by linked binding modules. *Nat Rev Mol Cell Biol*. 2007;8:983-994.

Sainsbury S, Bernecky C, Cramer P. Structural basis of transcription initiation by RNA polymerase II. *Nat Rev Mol Cell Biol*. 2015;16: 129-143.

Spitz F, Furlong EE. Transcription factors: from enhancer binding to developmental control. *Nat Rev Genet*. 2012;13:613-626.

Teves SS, Weber CM, Henikoff S. Transcribing through the nucleosome. *Trends Biochem Sci*. 2014;39:577-586.

Vannini A, Cramer P. Conservation between the RNA polymerase I, II, and III transcription initiation machineries. *Mol Cell*. 2012;45: 439-446.

Yan J, Enge M, Whitington T, et al. Transcription factor binding in human cells occurs in dense clusters formed around cohesin anchor sites. *Cell*. 2013;154:801-813.

Zaret KS, Carroll JS. Pioneer transcription factors: establishing competence for gene expression. *Genes Dev*. 2011;25:2227-2241.

Zhou Q, Li T, Price DH. RNA polymerase II elongation control. *Annu Rev Biochem*. 2012;81:119-143.

CHAPTER 11

Eukaryotic RNA Processing*

In all organisms, the genetic information is encoded in the sequence of the DNA. However, to be used, this information must be copied or transcribed into the related polymer, RNA. Eukaryotes synthesize many different types of RNA, but no RNA is simply transcribed as a finished product. The mature, functional forms of all eukaryotic RNA species are generated by posttranscriptional processing. These processing reactions are the major topic of this chapter.

The major RNAs can be assigned to three major classes: (1) The cytoplasmic messenger RNAs (mRNAs) and their nuclear precursors (pre-mRNAs) carry the information that is used to specify the sequence, and therefore ultimately the structure, of all proteins in the cell. (2) Other RNAs do not encode protein but function directly, playing major roles in various metabolic pathways, including protein synthesis. These include the ribosomal RNAs (rRNAs) and transfer RNAs (tRNAs), which are the key components of the protein synthesis machinery; the small nuclear RNAs (snRNAs), which form the core of the pre-mRNA splicing system; and the small nucleolar RNAs (snoRNAs), which are important factors in ribosome biogenesis. These RNAs are generally much longer-lived than mRNAs and therefore often are referred to as *stable* or *noncoding* RNAs (ncRNAs). (3) The third and most recently identified class of RNA comprises several structurally related groups of very small (21 to 25 nucleotides) RNA species that play important roles in regulating gene expression. Base pairing between endogenous micro-RNAs (miRNAs) and target mRNAs in the cytoplasm represses their translation into protein. The packaging of DNA into a nontranscribed form termed *heterochromatin* (see Fig. 8.7) is promoted by a class of nuclear, small centromeric RNAs. The introduction of small double-stranded RNAs into many cell types and organisms results in cleavage of the target mRNA and consequent silencing of gene expression. This phenomenon is described as RNA interference (RNAi), and the RNAs are referred to as small interfering RNAs (siRNAs).

In addition, a heterogeneous set of longer ncRNAs (lncRNAs) have been implicated in a variety of nuclear events.

*This chapter was written by David Tollervey and includes some text and figures from a chapter in the first edition written by Barbara Sollner-Webb, with contributions from Christine Smith.

Synthesis of Messenger RNAs

Fig. 11.1 shows an overview of mRNA synthesis and degradation.

Messenger RNA Capping and Polyadenylation

Two distinguishing features set mRNA apart from other RNAs: a 5′ cap structure and a 3′ poly(A) tail. These elements help protect the mRNA against degradation and act synergistically to promote translation in the cytoplasm.

The mRNA cap is an unusual structure. It consists of an inverted 7-methylguanosine residue, which is joined onto the body of the mRNA by a 5′-triphosphate–5′ linkage (Fig. 11.2). Cap addition involves three enzymatic activities: (a) a 5′ RNA triphosphatase cleaves the 5′ triphosphate on the nascent transcript to a diphosphate; (b) RNA guanylyltransferase forms a covalent enzyme–guanosine monophosphate (GMP) complex and then caps the RNA by transferring this to the diphosphate; and (c) RNA (guanine-7) methyltransferase covalently alters the guanosine base by methylation, generating m^7G. In addition, the first encoded nucleotides are frequently modified by methylation of the 2′ hydroxyl position on the ribose group, but the functional significance of these internal modifications is currently unclear.

During 3′ end processing, the nascent pre-mRNA is cleaved by an endonuclease, and a tail of adenosine residues is added by poly(A) polymerase. Approximately

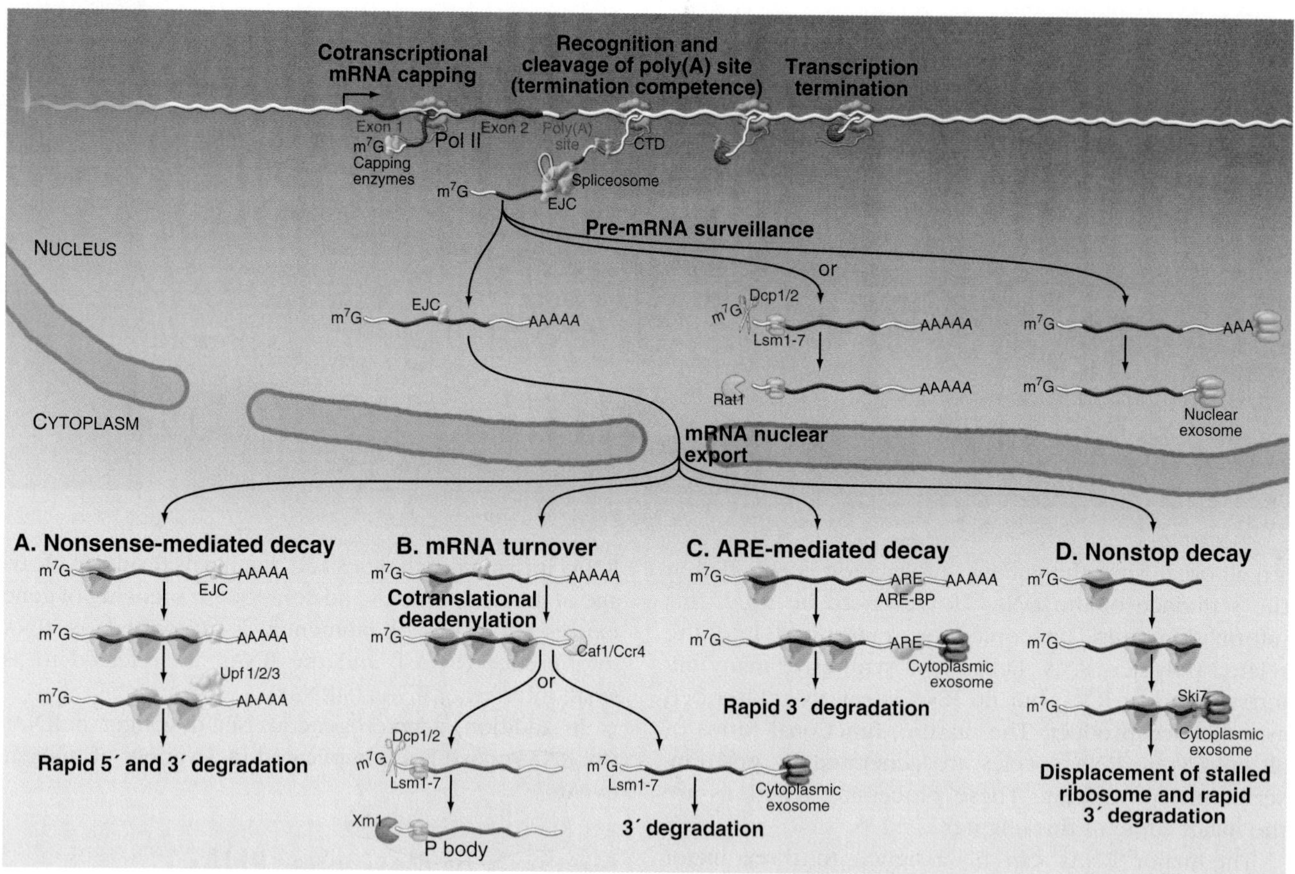

FIGURE 11.1 SYNTHESIS AND DEGRADATION OF EUKARYOTIC MESSENGER RNAS. Nascent messenger RNA (mRNA) transcripts are transcribed by RNA polymerase II. Formation of the 5′ cap structure and cleavage and polyadenylation of the 3′ end of the mRNA both occur cotranscriptionally and involve factors that are recruited by the C-terminal domain (CTD) of the transcribing polymerase (see Fig. 10.4). The termination of transcription requires both the recognition of the site of polyadenylation and the activity of the 5′-exonuclease Rat1, which degrades the nascent RNA transcripts. Rat1 binds to the polymerase CTD via a linker protein. Pre-mRNA splicing can either be cotranscriptional or occur shortly after transcript release, and recruitment of splicing factors is not strongly dependent on the CTD. In human cells, the spliceosome deposits the exon-junction complex (EJC) around 24 nucleotides upstream of the site of splicing. Several steps in nuclear mRNA maturation are subject to surveillance. In yeast, nuclear pre-mRNAs can be either 3′ degraded by the nuclear exosome complex or decapped and 5′ degraded by the exonuclease Rat1. Nuclear decapping requires the Lsm2–8 complex and is probably performed by the Dcp1/2 decapping complex. Once in the cytoplasm, the mRNA is translated into proteins and undergoes degradation. Several different mRNA degradation pathways have been identified. **A,** Nonsense-mediated decay (NMD). If the EJCs all lie within or very close to the open reading frame (ORF), they will be displaced by the translating ribosomes. However, if an EJC lies beyond the end of the ORF, it will remain on the translated mRNA. This is taken as evidence that translation has terminated prematurely and triggers the NMD pathway. Recognition of the EJC requires the Upf1/2/3 surveillance complex, which also interacts with the ribosomes as they terminate translation. In yeast, NMD triggers both rapid decapping and 5′ degradation, without prior deadenylation, and 3′ degradation by the exosome. **B,** General mRNA turnover. During translation, most mRNAs undergo progressive poly(A) tail shortening. Loss of the poly(A) tail leads to rapid degradation. As in the nucleus, cytoplasmic mRNAs can be degraded from either the 5′ or the 3′ end. The 5′ degradation occurs largely in a specialized cytoplasmic region termed the P body in yeast or cytoplasmic foci in human cells. Here, the mRNAs are decapped by the Dcp1/2 heterodimer and then degraded by the cytoplasmic 5′-exonuclease Xrn1. Both activities are strongly stimulated by the cytoplasmic Lsm1–7 complex. Alternatively, deadenylated mRNAs can be 3′ degraded by the cytoplasmic exosome. **C,** ARE-mediated decay. In this pathway, specific A+U rich elements (AREs) are recognized by ARE-binding proteins (ARE-BPs) in the nucleus. These are transported to the cytoplasm in association with the mRNA and recruit the cytoplasmic exosome to rapidly degrade the RNA. **D,** Nonstop decay. If the mRNA lacks a translation termination codon, the first translating ribosome will stall and be trapped at the 3′ end of the RNA. The Ski7 protein, which is associated with the cytoplasmic exosome complex, is believed to release the stalled ribosome and target the RNA for 3′ degradation by the exosome. Note that this legend provides more complex information than that given in the text for interested readers.

200 to 250 A residues are added to mRNAs in human cells, while 70 to 90 are added in yeast. Cleavage and polyadenylation are performed by a large complex containing approximately 20 proteins that recognizes sequences in the mRNA, of which the best defined is a highly conserved AAUAAA motif located upstream of the site of polyadenylation (Fig. 11.3).

Links Between Messenger RNA Processing and Transcription

The processes of cap addition and 3′ cleavage and polyadenylation are both linked to transcription of the mRNA by RNA polymerase II and occur cotranscriptionally on the nascent RNA (Fig. 11.1). The **C-terminal domain**

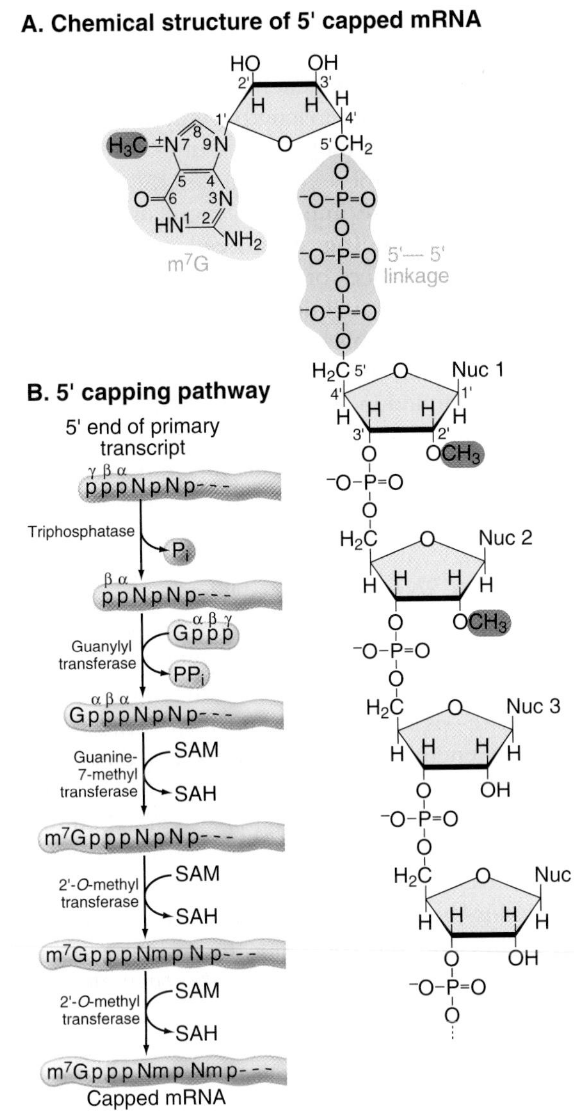

A. Chemical structure of 5' capped mRNA

m^7G — 5'—5' linkage — Nuc 1 — Nuc 2 — Nuc 3 — Nuc 4

B. 5' capping pathway

5' end of primary transcript

pppNpNp---

Triphosphatase → P_i

ppNpNp---

Guanylyl transferase ← Gppp → PP_i

GpppNpNp---

Guanine-7-methyl transferase: SAM → SAH

m^7GpppNpNp---

2'-O-methyl transferase: SAM → SAH

m^7GpppNmpNp---

2'-O-methyl transferase: SAM → SAH

m^7GpppNmpNmp---

Capped mRNA

FIGURE 11.2 MESSENGER RNAS HAVE A DISTINCTIVE 5′ CAP STRUCTURE. A, The 5′ ends of messenger RNAs (mRNAs) are blocked by an inverted guanosine residue that is attached to the body of the mRNA by a 5′–5′ triphosphate linkage. The N7 position of the guanosine is methylated *(red).* The first encoded nucleotide of the mRNA (Nuc 1) is also methylated on the 2′-hydroxyl of the ribose ring. The second nucleotide (Nuc 2) may also be methylated. **B,** Capping of mRNAs is a multistep process.

(CTD) of the largest subunit of RNA polymerase II consists of many copies of a seven-amino-acid repeat (YS_2PTS_5PS), which undergo reversible modification by phosphorylation (see Fig. 10.4). A pronounced change in the CTD phosphorylation pattern coincides with the release of the polymerase from initiation mode into processive elongation mode. Immediately following transcription initiation, the repeats are largely phosphorylated on the serine residue at position 5. This modification is lost, while serine 2 phosphorylation increases, as the polymerase moves along the transcript. Capping of the 5′ end of the mRNA occurs by the time the transcript is approximately 25 to 30 nucleotides long, and the capping enzyme is recruited by the serine 5 phosphorylated CTD. This and other interactions with the polymerase result in strong allosteric activation of capping activity. In contrast, the cleavage and polyadenylation factors involved in 3′ end processing are recruited by interaction with the CTD phosphorylated at serine 2.

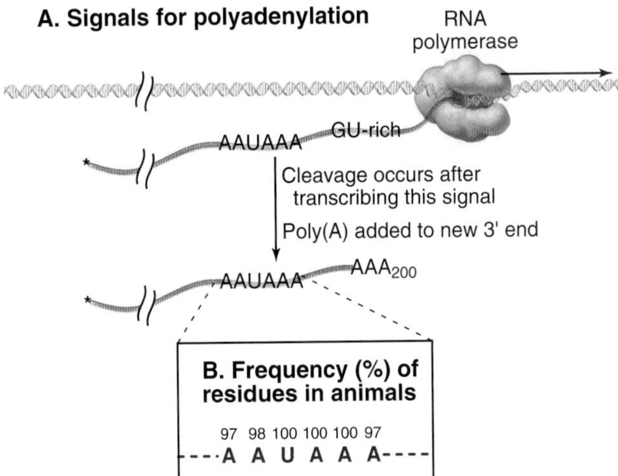

FIGURE 11.3 SIGNALS FOR PRE–MESSENGER RNA POLYADENYLATION. A, Poly(A) tails are added to pre–messenger RNAs (mRNAs) following transcription. After pol II transcribes the protein-coding region of the mRNA, it encounters two sequence elements: AAUAAA and a GU-rich element. These act as signals for the assembly of a large 3′ processing complex that cleaves the nascent pre-mRNA, releasing it from the transcription complex, and adds a tail of up to 200 adenosine residues. **B,** The poly(A) signal is highly conserved in vertebrates.

The major termination pathway for RNA polymerase II on mRNAs is dependent on 3′ processing. Termination requires recognition of the poly(A) site by the cleavage and polyadenylation factors. These are carried along with the transcribing polymerase, and their offloading might make the polymerase competent for termination. Cleavage of the nascent transcript also allows the entry of a **5′ exonuclease**—an enzyme that can degrade RNA from the 5′ end in a 3′ direction. This enzyme, which is called Rat1 in yeast and Xrn2 in humans, then chases after the transcribing polymerase, degrading the newly transcribed RNA strand as it goes. When the exonuclease catches the polymerase, it stimulates termination of transcription. This is referred to as the *Torpedo model* for transcription termination.

Regulated 3′ End Formation on Histone Messenger RNAs

A different 3′ end processing system operates for mRNAs encoding the major, replication-dependent histone proteins. These are highly expressed only during DNA replication, when they must package the newly synthesized DNA. A sequence in the **3′ untranslated region (3′**

UTR) of these mRNAs is recognized by base pairing to a small RNA: the U7 snRNA. In addition, a specific stem-loop structure is recognized by a stem-loop binding protein. Endonuclease cleavage generates the mature 3′ end of the mRNA, which is not polyadenylated but is protected from degradation by the stem-loop binding protein. The efficiency of histone mRNA synthesis is increased during DNA replication at least in part by increased abundance of stem-loop binding protein. Other minor histone variants that are synthesized throughout the cell cycle are polyadenylated like other mRNAs.

Pre–Messenger RNA Splicing

Important experiments in the 1950s and 1960s established that genes are collinear with their protein products. It therefore came as a considerable surprise when, in the late 1970s, it emerged that genes in animals and plants frequently had numerous strikingly large inserts whose sequence was not included in the mature mRNA or the protein product. It turns out that most human pre-mRNAs undergo **splicing** reactions, in which specific regions are cut out and the flanking RNA is covalently rejoined. The regions that will form the mRNA are termed **exons,** and the bits that are cut out (and are normally degraded) are called **introns.** In unicellular eukaryotes, introns are generally a few hundred nucleotides in length or shorter. In metazoans, however, they are often several kilobases in length, and pre-mRNAs can contain many introns. It is therefore remarkable that all the sites can be precisely identified and spliced.

Signals for Splicing

The signals in the pre-mRNA that identify the introns and exons are recognized by a combination of proteins and a group of small RNAs called the **snRNAs**. The snRNAs function in complexes with proteins in small nuclear ribonucleoprotein **(snRNP)** particles. Splicing occurs in a large complex termed the spliceosome, within which the pre-mRNA assembles together with five **snRNAs (U1, U2, U4, U5, and U6)** and approximately 100 different proteins. Particularly important protein-splicing factors are members of a large group of **SR-proteins**—so named because they contain domains rich in serine-arginine dipeptides.

Three conserved sequences within introns play key roles in their accurate recognition by the splicing machinery (Fig. 11.4). These lie immediately adjacent to the **5′ splice site** and **the 3′ splice site** and surround an internal region that will form the **intron branch point** during the splicing reaction. The U1 and U6 snRNAs have sequences that are complementary to the 5′ splice site, whereas U2 is complementary to the branch point region.

Although the spliceosome will finally bring together the sequences at each end of the intron, it is thought that the splicing machinery initially recognizes the *exons* in a reaction termed **exon definition.** This makes sense because mRNA exons are generally quite small—up to a few hundred nucleotides in length—whereas the introns can be many kilobases long.

No sequences in the exons are strictly required for splicing, but there are important stimulatory elements termed **exonic splicing enhancers (ESEs),** which generally bind members of the SR protein family. The ESEs have two major functions: to stimulate the use of the flanking 5′ and 3′ splice sites, promoting exon definition, and to prevent the exon in which they are located from being included in an intron. This latter function is particularly important in ensuring that all introns are spliced out without the splicing machinery skipping from the 5′ end of one intron to the 3′ end of a downstream intron.

Pre–Messenger RNA Splicing Reaction

The splicing reaction proceeds in two steps (Fig. 11.4). In the first, the 5′-3′ phosphate linkage that joins the 5′ exon to the first nucleotide of the intron—at the 5′ splice site—is attacked and broken. This reaction leaves the 5′ end of the intron attached to a downstream adenosine residue via an unusual 5′-2′ phosphate linkage. Because this adenosine remains attached to the flanking nucleotides by conventional 5′ and 3′ phosphodiester bonds, this creates a circular molecule with a tail that includes the 3′ exon. This structure is termed the intron **lariat,** and the adenosine to which the 5′ end of the intron is attached is termed the **branch point,** because it has a branched structure. In the second step of splicing, the free 3′ hydroxyl on the 5′ exon is used to attack and break the linkage between the last nucleotide of the intron and the 3′ exon—at the 3′ splice site. This leaves the 5′ and 3′ exons joined by a conventional 5′-3′ linkage and releases the intron as a lariat. This is linearized by the **debranching enzyme** and is probably rapidly degraded from both ends by exonucleases.

The initial steps in splicing are the recognition of the 5′ splice site by the U1 snRNA and the binding of U2 snRNA to the branch point region, assisted by SR proteins (Fig. 11.5). Base pairing between U2 and the pre-mRNA leaves a single adenosine bulged out of a helix and available for interaction with the 5′ splice site. The U4 and U6 snRNAs then join the spliceosome as a base-paired duplex, within a large complex that also contains the U5 snRNA. The U4 and U6 base pairing is opened, and the liberated U6 sequences displace U1 at the 5′ splice site. They also bind to U2—bringing the 5′ splice site and branch point into close proximity. At this point, the first enzymatic step of splicing occurs. This reaction is believed to be directly catalyzed by the intricate structure of the snRNA/pre-mRNA interactions rather than by the protein components of the spliceosome. The 5′

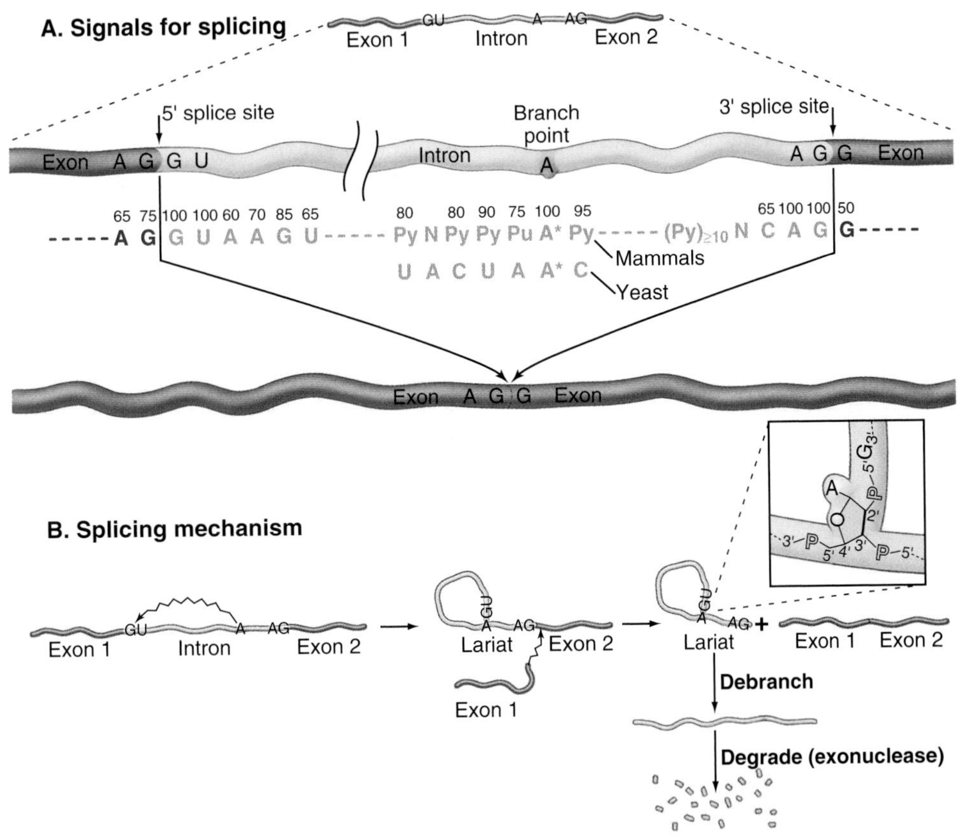

FIGURE 11.4 SIGNALS AND MECHANISM OF PRE–MESSENGER RNA SPLICING. The precursors to most messenger RNAs (mRNAs) in humans and other eukaryotes contain regions (introns) that will not form part of the mature mRNA and do not encode protein products. During pre-mRNA splicing, the introns are removed and flanking regions (exons) are ligated. **A,** Introns contain three conserved sequence elements that are recognized during splicing. These lie at the 5′ and 3′ splice sites and surrounding the branch point adenosine within the intron. Numbers indicate the degree of conservation at each position in mammalian pre-mRNAs. The branch point sequence is much more highly conserved between different pre-mRNAs in yeast. The region between the branch point and the 3′ splice site frequently contains a run of pyrimidine residues, which is referred to as the polypyrimidine tract. **B,** Pre-mRNA splicing involves two catalytic steps. An attack by the branch point adenosine on the 5′ splice site releases the 5′ exon and intron as a circularized molecule (referred to as the intron lariat) joined to the 3′ exon. In the second step, the 3′ end of the 5′ exon attacks the 3′ splice site releasing the joined exons and the free intron lariat. The lariat is subsequently linearized (debranched) and degraded.

splice site is attacked and broken by the ribose 2′ hydroxyl group of the adenosine residue that is bulged out of the U2-intron duplex. The U5 snRNA and its associated proteins are responsible for holding onto the now free 5′ exon and correctly aligning it with the 3′ exon for the second catalytic step of splicing.

Both catalytic steps in splicing are technically termed **transesterification** reactions, because nucleotides are linked by phosphodiester bonds, and the new bond is made at the same time as the old bond is broken. For this reason, the splicing reactions do not, in principle, require any input of energy. However, the assembly and subsequent disassembly of the spliceosome require numerous adenosine triphosphatases (ATPases). Most of these belong to a family of proteins that are generally termed **RNA helicases.** These are believed to use the energy of adenosine triphosphate (ATP) hydrolysis to catalyze structural rearrangements within the assembling and disassembly spliceosome.

AT-AC Introns

The large majority of human mRNA splice sites have a GU dinucleotide at the 5′ splice site and AG at the 3′ splice site (Fig. 11.4). However, a minor group of introns contain different consensus splicing signals and are termed AT-AC (pronounced “attack”) introns because of the identities of the nucleotides located at the 5′ and 3′ splice sites. The splicing of the AT-AC introns involves a distinct set of snRNAs—U11, U12, $U4_{ATAC}$, and $U6_{ATAC}$—which replace U1, U2, U4, and U6, respectively. Only U5 is common to both spliceosomes. However, the underlying splicing mechanism is believed to be the same for both classes of intron.

Alternative Splicing

A surprising finding from the human genomic sequencing project was the relatively low number of predicted protein-coding genes, currently estimated at fewer than

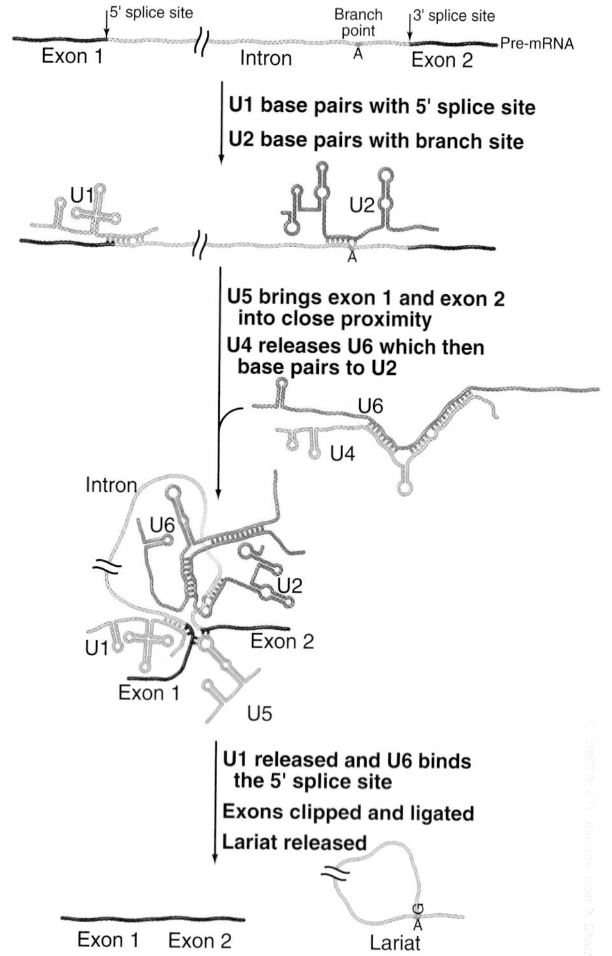

FIGURE 11.5 SMALL NUCLEAR RNAS PLAY KEY ROLES IN PRE–MESSENGER RNA SPLICING. Although shown here as RNAs, the small nuclear RNA (snRNA)s function in large RNA-protein complexes termed snRNPs. Despite this fact, the major steps in both intron recognition and catalysis are believed to be performed by the snRNAs. The 5′ splice site and intron branch point are recognized by base pairing to the U1 and U2 snRNAs, respectively. The U5 snRNA enters the spliceosome in a complex with U4 and U6, which are tightly base-paired. U5 contacts both the 5′ and 3′ exons. U4 releases U6, which base-pairs to U2 and then displaces U1 in binding to the 5′ splice site. Within this very complex RNA structure, the 2′ hydroxyl group on the branch point adenosine, which is bulged out of the duplex between U2 and the pre-mRNA, attacks the phosphate group at the junction between the 5′ exon and the intron. In a transesterification reaction, the phosphate backbone is broken at the 5′ splice site. The 5′ exon is released with a 3′ OH group, and the 5′ phosphate of the intron is transferred onto the 2′ position of the ribose on the branch point adenosine, creating the intron lariat structure. U5 retains the 5′ exon and aligns it for a second transesterification reaction, during which the 3′ hydroxyl on the 5′ exon attacks the 3′ splice site, joining the exons and releasing the intron lariat.

20,000. This result has caused increased interest in the phenomenon of alternative splicing, which allows the production of more than one mRNA, and therefore more than one protein product, from a single gene. Several general forms of alternative splicing are commonly found. Exons can be excluded from the mRNAs, or introns can be included. Some genes have arrays of multiple alternative exons, only one of which is included in each mRNA. In addition, the use of alternative splice sites can generate longer or shorter forms of individual exons (Fig. 11.6).

Current estimates for the proportion of human genes that are subject to alternative splicing range from 30% to 75%. In some cases, this could potentially give rise to a very large number of different protein **isoforms.** Alternatively spliced proteins can have antagonistic functions, such as transcription activation versus transcription repression. For the vast majority of human genes, no information is available on the relative activities of different spliced isoforms. Compounding the difficulty in understanding is the fact that many genes show tissue-specific splicing. Thus, a gene could be transcribed in, say, both the liver and brain but generate products with substantially different functions in each tissue. In addition to generating protein diversity, alternative splicing can generate mRNAs with premature translation termination codons—"nonsense" codons. These are subject to rapid degradation by the nonsense-mediated decay (NMD) surveillance pathway (see later). Switching splicing into a pathway that generates an NMD target is therefore a means of downregulating gene expression.

It is likely that alterations in the activities of many different factors can lead to the preferential use of alternative splice sites. In at least some cases, changes in the abundance of a general splicing factor generates tissue-specific patterns of splicing.

Localization of Pre–Messenger RNA Splicing

The location of the splicing reaction within the nucleus was long a contentious topic. The snRNAs can be detected dispersed in the nucleoplasm but concentrate in small structures referred to as **nuclear speckles** or **interchromatin granules,** as well as in discrete larger structures known as **Cajal bodies** (see Fig. 9.2). It is now widely accepted that most splicing is performed by the dispersed snRNA population and can occur either cotranscriptionally or immediately following transcript release. Consistent with this, there is evidence that the recruitment of some splicing factors is promoted by association with the CTD of the transcribing polymerase. The speckles are likely to represent sites at which splicing factors are stockpiled ready for use. The Cajal bodies, in contrast, represent sites of maturation in which the snRNAs undergo site-specific nucleotide modification and perhaps assembly with specific proteins.

Modification of Messenger RNAs

Site-specific nucleotide modifications take place at many sites on mRNAs, notably including formation of pseudouracil, 5-methylcytosine, N1-methyladenosine,

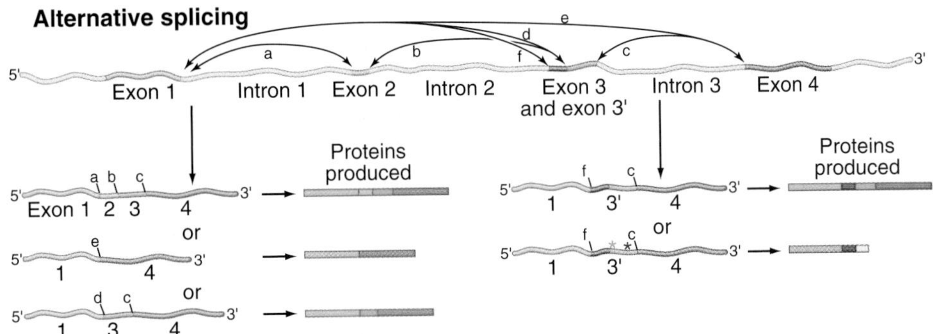

FIGURE 11.6 ALTERNATIVE SPLICING CAN GENERATE MULTIPLE DIFFERENT PROTEINS FROM A SINGLE GENE. Here are some of the possible mRNA and protein products of a gene whose pre-mRNA is subject to alternative splicing. *Left,* Examples show the effects of skipping one or more internal exons, which produces a set of related proteins with different combinations of "modules." *Right,* Examples show the effects of alternative splice sites. In the case shown, the use of alternative 3′ splice sites redefines the 5′ end of the downstream exon. This can lead to the inclusion of additional amino acids in the protein product. Use of an alternative splice site can also cause the exon to be read in a different reading frame *(green asterisk),* changing the amino acid sequence. If the alternative reading frame contains a translation stop codon *(red asterisk),* a truncated protein will be produced, and the mRNA will generally be targeted for rapid degradation by the nonsense-mediated decay (NMD) pathway (Fig. 11.1).

and N6-methyladenosine (m6A). Proteins that interact with m6A have been characterized as WRITERS (methyltransferases that create the modification), READERS (proteins that specifically bind mRNAs with m6A modification and alter the processing, stability, or translation) and ERASERS (proteins that remove the modification by oxidative demethylation). Nuclear binding of m6A READERS can alter pre-mRNA alternative splicing and export. However, most m6A modifications in human mRNAs are close to the 3′ end and cytoplasmic binding of READERS can promote rapid mRNA turnover and translation repression. Methylation and demethylation of m6A are important in human development, particularly during spermatogenesis.

Editing of Messenger RNAs

The term *RNA editing* in humans refers to covalent modifications that are made to individual nucleotides, which alter the base-pairing potential. Because the process of translation involves base pairing between mRNA and tRNAs, editing of the mRNA can have the effect of changing the amino acid that is incorporated and therefore the function of the protein. Like alternative splicing, this increases the diversity of protein products that can be synthesized from the genome.

Slightly confusingly, the term *editing* is also used for quite different mechanisms that insert and delete nucleotides from RNAs in some single-celled eukaryotes. The best-characterized example is in the mitochondria of trypanosomes, which are protozoans that cause major human diseases, including African trypanosomiasis, Chagas disease, and leishmaniasis. Uracil residues are added and, less frequently, deleted from the mitochondrial mRNAs at many sites. These changes are specified by a large number of small guide RNAs. This form of editing is not known to occur in higher eukaryotes.

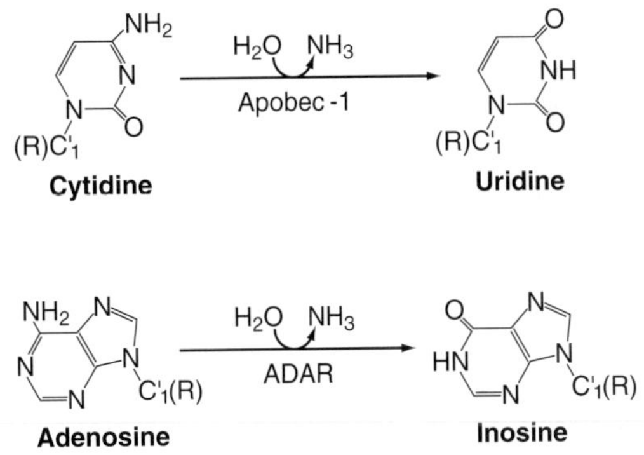

Cytidine → Uridine (H_2O, NH_3; Apobec-1)

Adenosine → Inosine (H_2O, NH_3; ADAR)

FIGURE 11.7 RNA EDITING CHANGES NUCLEOTIDE BASE PAIRING. The coding potential of an mRNA can be altered by deamination. In C-to-U editing, the amino group at position 4 of the cytosine base is replaced with a carbonyl group, creating uracil. In A-to-I editing, replacement of the amino group at position 2 of adenosine creates inosine, which base-pairs with C residues rather than with U. ADAR, adenosine deaminase acting on RNA.

C-to-U Editing

Deamination of cytosine to uracil is performed by an editing complex, sometimes referred to as the editosome, which includes the deaminase Apobec-1 (Fig. 11.7). Only a small number of nuclear-encoded targets have been identified, and in these, editing generates translation termination codons, producing shorter forms of the encoded proteins. The best-characterized example of C-to-U RNA editing involves the mRNA encoding intestinal apolipoprotein B (ApoB), where CAA-to-UAA editing in the loop of a specific stem-loop structure generates a stop codon. The truncated protein, ApoB48, has an important role in lipoprotein metabolism. In other cases editing may generate mRNAs that are targets for NMD (see later), leading to downregulation of protein expression.

A-to-I Editing

The enzyme **ADAR** (adenosine deaminase acting on RNA) can convert adenine residues to inosine by deamination of the base (Fig. 11.7). Inosine acts like guanosine and base-pairs with cytosine rather than uracil, potentially altering the protein encoded by the mRNA. Most of the transcripts edited by ADAR encode receptors of the central nervous system, and RNA editing is required to create the full receptor repertoire. The amino acid substitutions that result from editing of the mRNAs can greatly alter the properties of ion channels, and aberrant editing occurs in various disorders ranging from epilepsy to malignant brain gliomas. ADAR binds as a dimer to imperfect double-stranded RNA duplexes, which are formed between the target site and sequences in a flanking intron. Editing is generally not 100% efficient, so heterogeneous populations of proteins are generated.

In addition to specific editing of individual nucleotides, ADARs can hyperedit long double-stranded RNAs (dsRNAs). In mammals, dsRNAs elicit a strong antiviral response from the innate immune system and hyperediting is important to avoid inappropriate recognition of endogenous dsRNAs.

Cytoplasmic Polyadenylation

The early steps of embryogenesis in metazoans occur before transcription of the genome commences. All mRNAs that are present in early embryos were therefore inherited from the mother. These "maternal messages" are frequently translationally inactive, at least in part because they lack a poly(A) tail. They can be activated for translation by polyadenylation in the cytoplasm. Cytoplasmic polyadenylation events are critical for many developmental decisions in oocytes and embryos. In addition, regulated cytoplasmic polyadenylation at synapses controls local translation in neuronal cells. This involves a family of distinct cytoplasmic polymerases. Their association with substrates and activity are both regulated by specific RNA-binding proteins.

Messenger RNA Degradation and Surveillance

Exosome Complex

The RNA exosome is a multiprotein complex with exonuclease and endonuclease activities. The complex has a barrel-like structure. Substrates are threaded through the lumen of the barrel to reach the active site of the major 3′ exonuclease (DIS3/Rrp44) (Fig. 11.8). In addition, DIS3/Rrp44 harbors an endonuclease activity that is not accessed through the central channel. The nuclear exosome complex is associated with an additional 3′ to 5′ exonuclease (Rrp6 in yeast, PM-Scl100 in humans). In both the nucleus and cytoplasm, the activity of the exosome is dependent on cofactors. Chief among these are two related RNA helicases (proteins that can open RNA and RNA-protein structures using energy derived from ATP hydrolysis); Mtr4 in the nucleus and Ski2 in the cytoplasm. In the nucleus, Mtr4 is a component of the TRAMP (Trf4/5-Air1/2-Mtr4 polyadenylation) and NEXT (nuclear exosome targeting) complexes together with RNA binding proteins, while Ski2 forms the SKI complex, which is required for all known functions of the exosome in mRNA degradation. The nuclear exosome and Mtr4 participate in RNA maturation, notably in the processing of the 5.8S rRNA. However, the major functions of the nuclear exosome are probably in the surveillance and degradation of many different types of defective nuclear RNAs and RNA-protein complexes, and the clearance of numerous classes of

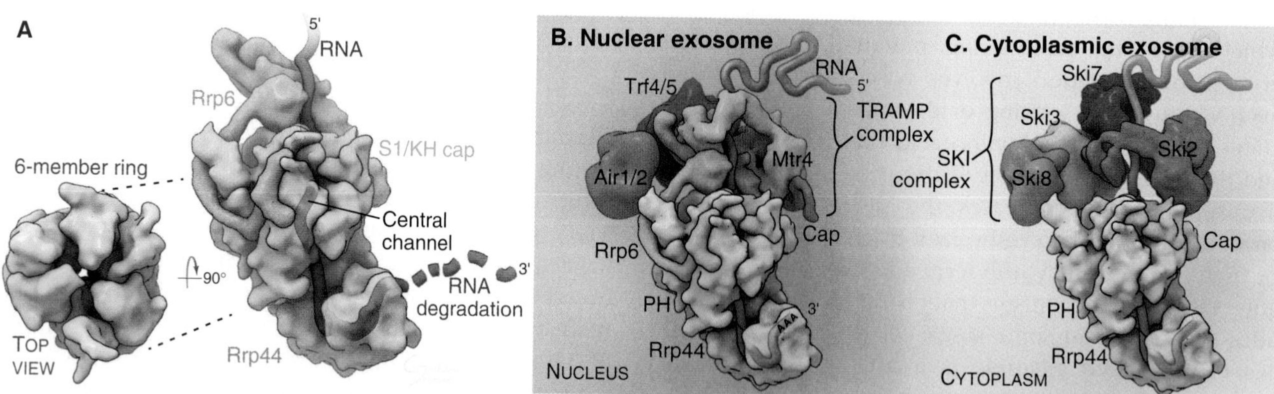

FIGURE 11.8 THE EXOSOME COMPLEX AND COFACTORS. The exosome has a barrel structure with the major active site in the 3′ exonuclease Rrp44/Dis3 located at the base of the central lumen of the complex. An additional 3′ exonuclease Rrp6 is located close to the entrance of the central channel, while an endonuclease active site on Rrp44 is located on the exterior of the structure. The exosome barrel is composed of an RNA-binding cap structure and a core containing six proteins that show sequence similarities to *Escherichia coli* RNase PH, but which have, surprisingly, all lost their catalytic activities in eukaryotes. Substrate RNAs are inserted into the complex by related nuclear and cytoplasmic localized RNA helicases: Mtr4 and the TRAMP (Trf4/5-Air1/2-Mtr4 polyadenylation) complex in the nucleus and Ski2 in the SKI complex in the cytoplasm.

ncRNAs. The cytoplasmic exosome functions, together with the SKI complex, in several different mRNA turnover pathways.

Degradation of Messenger RNA

Most analyses of the regulation of gene expression have concentrated on changes in the levels of mRNA transcription. However, the rate at which mRNAs are degraded is also important, influencing both the total amount of protein synthesized and the timing of protein synthesis following a transcription event. mRNAs are frequently described as having half-lives, but this is generally quite misleading. Degradation is not stochastic, and it is probably better to think of mRNA lifetimes. There are enormous variations in the lifetimes of different human mRNAs—from a very few minutes to many days—that have a large impact on protein expression levels.

Different pathways of mRNA degradation can be classified as (a) the default pathway (ie, when we do not yet know of any specific activator or repressor of degradation), (b) regulated degradation pathways that respond to developmental or other signals, and (c) surveillance pathways that identify and rapidly degrade aberrant mRNAs or pre-mRNAs. A theme emerging from studies of all mRNA decay pathways is that RNA-binding proteins, which associate with the newly transcribed precursor in the nucleus, can be retained when the mRNA is exported to the cytoplasm. These proteins maintain a record of the nuclear history of the RNA that can be "read" by the cytoplasmic degradation machinery, and this plays a key role in determining the cytoplasmic fate of the mRNA.

A key step in the timing of degradation of most mRNA is the slow, stepwise removal of the poly(A) tail by enzymes called **deadenylases.** The intact poly(A) tail is bound by multiple copies of the poly(A)-binding protein (PABP), at a stoichiometry of around one molecule per 10 to 20 A residues. Surprisingly, PABP antagonizes 5′ cap removal, probably via interactions with the translation initiation factor eIF4G, which, in turn, stabilizes the cap-binding protein eIF4E. These interactions effectively circularize the mRNA and strongly stimulate translation initiation (see Fig. 12.8). When the tail becomes too short for the last PABP molecule to bind, these interactions are lost. The cap can then be removed by a **decapping complex,** which cleaves the triphosphate linkage to the body of the mRNA, releasing m^7GDP. Cap removal allows rapid 5′ to 3′ degradation of the mRNA by the 5′ exonuclease Xrn1. In addition, loss of the PABP/poly(A) complex allows 3′-degradation of the mRNA by the cytoplasmic exosome.

A+U Rich Element–Mediated Degradation

The degradation of many mRNA species in human cells is triggered by the presence of sequence motifs referred to as A+U rich elements (AREs) (Fig. 11.1C). These are generally located in the 3′ UTR of the mRNA, where bound proteins will not be displaced by the translating ribosomes. This pathway plays an important regulatory role in gene expression, as it targets for rapid turnover mRNAs that encode proteins such as cytokines, growth factors, oncogenes, and cell-cycle regulators, for which limited and transient expression is important. Computational analyses indicate that up to 8% of human mRNAs carry AREs, and there is evidence that alterations in the activity of this pathway are associated with both developmental decisions and cancer. ARE-binding proteins associate with the nuclear pre-mRNAs and are exported to the cytoplasm, where they can either activate or repress ARE-mediated decay. Some ARE-binding proteins that activate degradation function by directly recruiting the exosome complex to degrade the mRNA from the 3′ end.

Surveillance of Messenger RNAs

Nonsense-Mediated Decay

The surveillance of mRNA integrity is important because defective molecules can encode truncated proteins, which are frequently toxic to the cell. The presence of a premature translation termination signal (or nonsense codon) strongly destabilizes mRNA via the **NMD** pathway (Fig. 11.1A). In human cells, termination codons are identified as being located in a premature position by reference to the sites of pre-mRNA splicing. Normal termination codons are within, or very close to, the 3′ exon, so no former splice sites lie far downstream. If any former splice site is located more than approximately 50 nucleotides downstream of the site of translation termination, the mRNA is targeted for degradation. The sites of former splicing events can be identified in the spliced mRNA product, because the spliceosome deposits a specific protein complex on the mRNA during the splicing reaction (Fig. 11.1). This **exon-junction complex (EJC)** binds to the 5′ exon sequence approximately 24 nucleotides upstream of the splice site. Several of the EJC components remain associated with the mRNA following its export to the cytoplasm. In normal mRNAs, the EJCs will all be displaced by the first translating ribosome, so if one (or more) remains on the mRNA, then translation has terminated too soon and NMD is activated.

The identification of premature termination codons in yeast and *Drosophila* does not rely on cues provided by splice sites but probably involves recognition of other nuclear RNA-binding proteins that are retained on the cytoplasmic mRNAs. In all organisms tested, NMD also requires a **surveillance complex,** which bridges interactions between the terminating ribosome and the "place markers" on the mRNAs.

In yeast and probably in humans, recognition of an mRNA as prematurely terminated activates both 5′ and

3′ degradation. The mRNA can be decapped and 5′-degraded by Xrn1 without prior deadenylation or can be rapidly deadenylated and 3′-degraded by the exosome. In contrast, the degradation of mRNAs targeted by the NMD pathway in *Drosophila* is initiated by an endonucleolytic cleavage.

Nonstop Decay

Some mRNAs lack any translation termination codon because they have been inappropriately polyadenylated, inaccurately spliced, or partially 3′-degraded. Translating ribosomes efficiently stall at the ends of such **nonstop mRNAs.** This inhibits the repeated synthesis of truncated proteins (Fig. 11.1D). The cytoplasmic form of the exosome complex is associated with Ski7p, which is homologous to the guanosine triphosphatases that function in translation. The interaction of Ski7p with the stalled ribosome is believed to both release the ribosome and target the mRNA for rapid degradation.

Nuclear RNA Degradation

Analyses of RNA degradation have focused largely on cytoplasmic mRNA turnover, but most RNA synthesized in a eukaryotic cell is actually degraded within the nucleus. Pre-mRNAs are predominantly composed of intronic sequences, and almost all stable RNAs are synthesized as larger precursors that undergo nuclear maturation. In contrast to the role of poly(A) tails in stabilizing mRNAs in the cytoplasm, there is evidence that short, oligo(A) tails can act as *destabilizing* features during RNA degradation in the nucleus. The TRAMP complex cofactors include nuclear poly(A) polymerases and activate the exosome complex, probably by providing a single-stranded "landing pad," during surveillance and degradation of many defective nuclear RNAs, including pre-mRNAs, pre-tRNAs and pre-rRNAs. In bacteria such as *Escherichia coli,* poly(A) tails are added to RNAs to make them better substrates for degradation. This has led to the proposal that the original function of polyadenylation was in RNA degradation, and this role is maintained in the eukaryotic nucleus. Following the appearance of the nuclear envelope in early eukaryotes, poly(A) tails took on a distinctly different function in promoting mRNA stability and translation in the cytoplasm.

Synthesis of Stable RNAs

Transfer RNA Synthesis

All tRNAs are excised from the interior of larger precursors (pre-tRNAs) (Fig. 11.9). Some pre-tRNAs are polycistronic, with two or more tRNAs excised from the same precursor. In yeast, at least, the genes that encode tRNAs cluster around the surface of the nucleolus, and pre-tRNA processing appears to occur largely within the nucleolus.

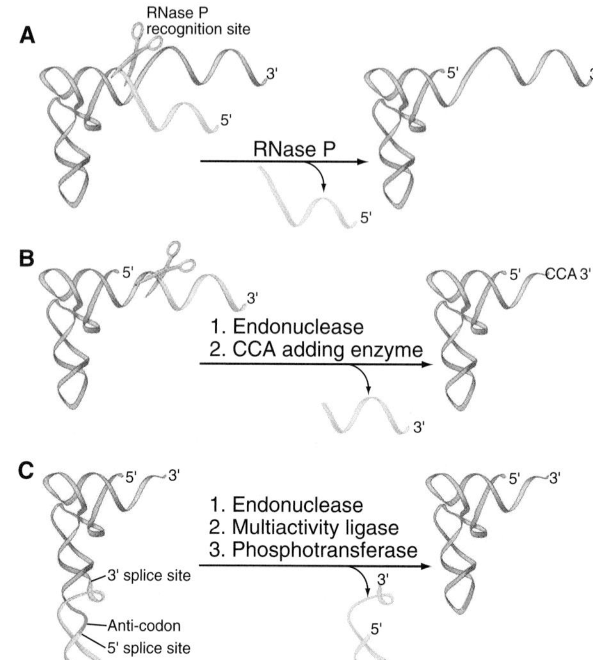

FIGURE 11.9 MATURE TRANSFER RNAS ARE GENERATED BY PROCESSING. A, Transcription by RNA polymerase III generates a pre–transfer RNA (tRNA) that is 5′ and 3′ extended and may also contain an intron. Cleavage by RNase P generates the mature 5′ end. **B,** The 3′ end is cleaved by an unidentified nuclease, and the sequence CCA is added by a specific RNA polymerase. This sequence forms a single stranded 3′ end on all tRNAs. **C,** If an intron is present, it is removed in a splicing reaction that is distinct from pre-mRNA splicing and does not involve small RNA cofactors. The anticodon *(green)* is generally located 1 nucleotide away from the splice site.

The 5′ end of the mature tRNA is generated by cleavage by the ribozyme endonuclease RNase P, which recognizes structural elements that are common to all tRNAs. The 3′ ends of all mature tRNAs have the sequence Cp-Cp-A_{OH}, to which the aminoacyl group is covalently attached. However, this CCA sequence is not encoded by the tRNA gene in eukaryotes, although it is encoded by tRNA genes in many bacteria. Instead, the pre-tRNA is initially trimmed, and the CCA sequence is then added by a specific RNA polymerase that belongs to the same family as the poly(A) polymerases that add tails to mRNAs.

Many pre-tRNAs contain a single, short intron, which is removed by splicing. The enzymology of tRNA splicing is quite different from that of pre-mRNA splicing. The pre-tRNA is cleaved at the 5′ and 3′ splice sites by a tetrameric protein complex containing two endonucleases and two targeting factors. The cleavages leave products with 5′ hydroxyl residues and 2′ to 3′ cyclic phosphate. A separate tRNA ligase then recognizes these termini and rejoins the exons.

In addition, tRNAs are subject to a bewildering array of covalent nucleotide modifications. Almost 100 different modified nucleotides have been identified in tRNAs, ranging from simple methylation to the addition of very elaborate molecules. All are added without breaking the phosphate backbone of the RNA. The structures of all mature tRNAs are very similar, since each must fit exactly into the A, P, and E sites of the translating ribosome (see Fig. 12.7). It is likely that the modifications help the tRNAs fold into precisely the correct shape. They also aid accurate recognition of different tRNAs by the aminoacyl-tRNA synthases, which are responsible for charging each species of tRNA with the correct amino acid.

Ribosome Synthesis

The synthesis of ribosomes is a major activity of any actively growing cell. Three of the four rRNAs—the 18S, 5.8S, and 25S/28S rRNAs—are cotranscribed by RNA polymerase I as a polycistronic transcript. This pre-rRNA is the only RNA synthesized by RNA polymerase I and is transcribed from tandemly repeated arrays of the ribosomal DNA (rDNA). In humans, approximately 300 to 400 rDNA repeats are present in five clusters (on chromosomes 13, 14, 15, 21, and 22). These sites often are referred to as *nucleolar organizer regions,* reflecting the fact that nucleoli assemble at these locations in newly formed interphase nuclei. The pre-rRNAs are very actively transcribed and can be visualized as "Christmas trees" in electron micrographs taken following spreading of the chromatin using low-salt conditions and detergent (Fig. 11.10A). The 5S rRNA is independently transcribed by RNA polymerase III. In most eukaryotes, the 5S rRNA genes are present in separate repeat arrays.

Nucleolus

Most steps in ribosome synthesis take place within a specialized nuclear substructure, the nucleolus (see Fig. 9.3). In micrographs, the nucleolus appears to be a very large and stable structure, but kinetic experiments indicate that it is in fact highly dynamic, with most nucleolar proteins rapidly exchanging with nucleoplasmic pools. A current view of the nucleolus is that its assembly is the consequence of many relatively weak and transient interactions between the nucleolar proteins. The result is a self-assembly process that greatly increases the local concentration of ribosome synthesis factors. This is envisaged to promote efficient preribosome assembly and maturation while allowing the rapid and dynamic changes in preribosome composition involved in this pathway. Similar mechanisms may generate other subnuclear structures such as Cajal bodies.

The key steps in ribosome synthesis are (a) transcription of the pre-rRNA, (b) covalent modification of the mature rRNA regions of the pre-rRNA, (c) processing of the pre-rRNA to the mature rRNAs, and (d) assembly of the rRNAs with the ribosomal proteins (Fig. 11.10D). During ribosome synthesis, the maturing preribosomes move from their site of transcription in the dense fibrillar component of the nucleolus, through the granular component of the nucleolus. They are then released into the nucleoplasm prior to transport through the nuclear pores to the cytoplasm. Here, the final maturation into functional 40S and 60S ribosomal subunits takes place.

Pre–Ribosomal RNA Processing

The posttranscriptional steps in ribosome synthesis are extraordinarily complex, involving approximately 200 proteins and approximately 100 snoRNA species, in addition to the four rRNAs and approximately 80 ribosomal proteins. Ribosome synthesis is best understood in budding yeast, but all available evidence indicates that it is highly conserved throughout eukaryotes. A combination of endonuclease cleavages and exonuclease digestion steps generates the mature rRNAs in a complex, multistep processing pathway. Many pre-rRNA processing enzymes have been identified, although others remain to be found (Fig. 11.10E). The remaining species, 5S rRNA, is independently transcribed and undergoes only 3′ trimming.

Modification of the Pre–Ribosomal RNA

The rRNAs are subject to covalent nucleotide modification at many sites. Modification takes place on the pre-rRNA, either on the nascent transcript or shortly following transcript release from the DNA template. The most common modifications are methylation of the 2′-hydroxyl group on the sugar ring **(2′-*O*-methylation)** and conversion of uracil to **pseudouridine** by base rotation. The sites of these modifications are selected by base pairing with two groups of **snoRNAs.** The **box C/D snoRNAs** direct sites of 2′-*O*-methylation and carry the methyltransferase (called fibrillarin in humans and Nop1 in yeast) (Fig. 11.10B). The **box H/ACA snoRNAs** select sites of pseudouridine formation and carry the pseudouridine synthase (called dyskerin in humans and Cbf5 in yeast [Fig. 11.10C]).

A small number of snoRNAs do not direct RNA modification but are required for pre-rRNA processing. The best characterized is the U3 snoRNA, which binds cotranscriptionally to the 5′-external transcribed spacer (ETS) region of the pre-rRNA. Base pairing between U3 and the pre-rRNA is required for the early processing reactions on the pathway of 18S rRNA synthesis and directs the assembly of a large pre-rRNA processing complex called the small subunit processome. This complex can be visualized as a "terminal knob" in micrographs of spread pre-rRNA transcripts (Fig. 11.10A).

A subset of ribosome synthesis factors interacts with both the rDNA and RNA polymerase I. These interactions might promote both efficient pre-rRNA transcription and recognition of the nascent pre-rRNA. This is reminiscent of the association of mRNA processing factors with RNA polymerase II and suggests that maturation of different

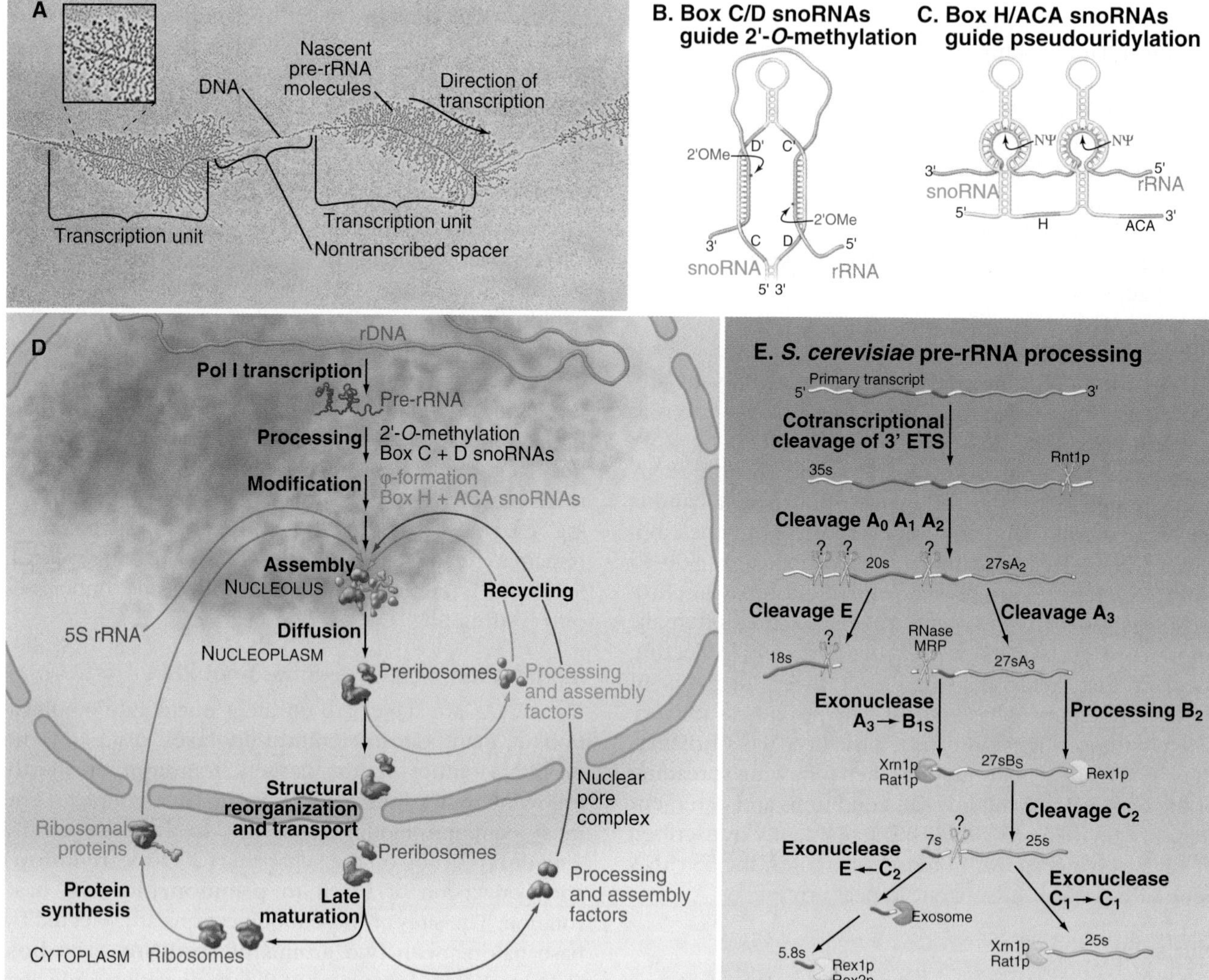

FIGURE 11.10 RIBOSOME SYNTHESIS. A, "Christmas trees" of nascent pre-rRNA transcripts. This electron micrograph shows ribosomal DNA (rDNA) genes in the process of transcription. Note the numerous molecules of RNA polymerase I (Pol I) along the rDNA, each associated with a pre–ribosomal RNA (rRNA) transcript. In the enlarged *inset,* the terminal balls can be seen on the transcripts. These large pre–rRNA-processing complexes (small subunit processomes) assemble around the binding site for the U3 small nucleolar RNA (snoRNA) and are required for the early pre-rRNA processing steps. **B–C,** Roles of the modification guide snoRNAs. The pre-rRNAs undergo extensive covalent modification. Most modification involves methylation of the sugar 2′ hydroxyl group (2′-*O*-methylation) or pseudouridine (Ψ) formation, at sites that are selected by base pairing with a host of small nucleolar ribonucleoprotein (snoRNP) particles. Human cells contain well over 100 different species of snoRNPs, and each pre-rRNA molecule must transiently associate with every snoRNP in order to mature properly. Sites of 2′-*O*-methylation are selected by base pairing with the box C/D class of snoRNAs, which carry the methyltransferase Nop1/fibrillarin. Sites of pseudouridine formation are selected by base pairing with the box H/ACA class of snoRNAs, which carry the pseudouridine synthase Cbf5/dyskerin. **D,** Key steps in eukaryotic ribosome synthesis. Following transcription of the pre-rRNAs, most steps in eukaryotic ribosome synthesis take place within the nucleolus. The preribosomes are then released from association with nucleolar structures and are believed to diffuse to the nuclear pore complex (NPC). Passage through the NPC is preceded by structural rearrangements and the release of processing and assembly factors. Further ribosome synthesis factors are released during late structural rearrangements in the cytoplasm that convert the preribosomal particles to the mature ribosomal subunits. During pre-rRNA transcription and processing, many of the approximately 80 ribosomal proteins assemble onto the mature rRNA regions of the pre-RNA. **E,** The pre-rRNA processing pathway. The pathway is presented for the budding yeast *Saccharomyces cerevisiae,* but extensive conservation is expected throughout eukaryotes. The mature rRNAs are generated by sequential endonuclease cleavage, with some of the mature rRNA termini generated by exonuclease digestion. Scissors with question marks indicate that the endonuclease responsible is unknown.

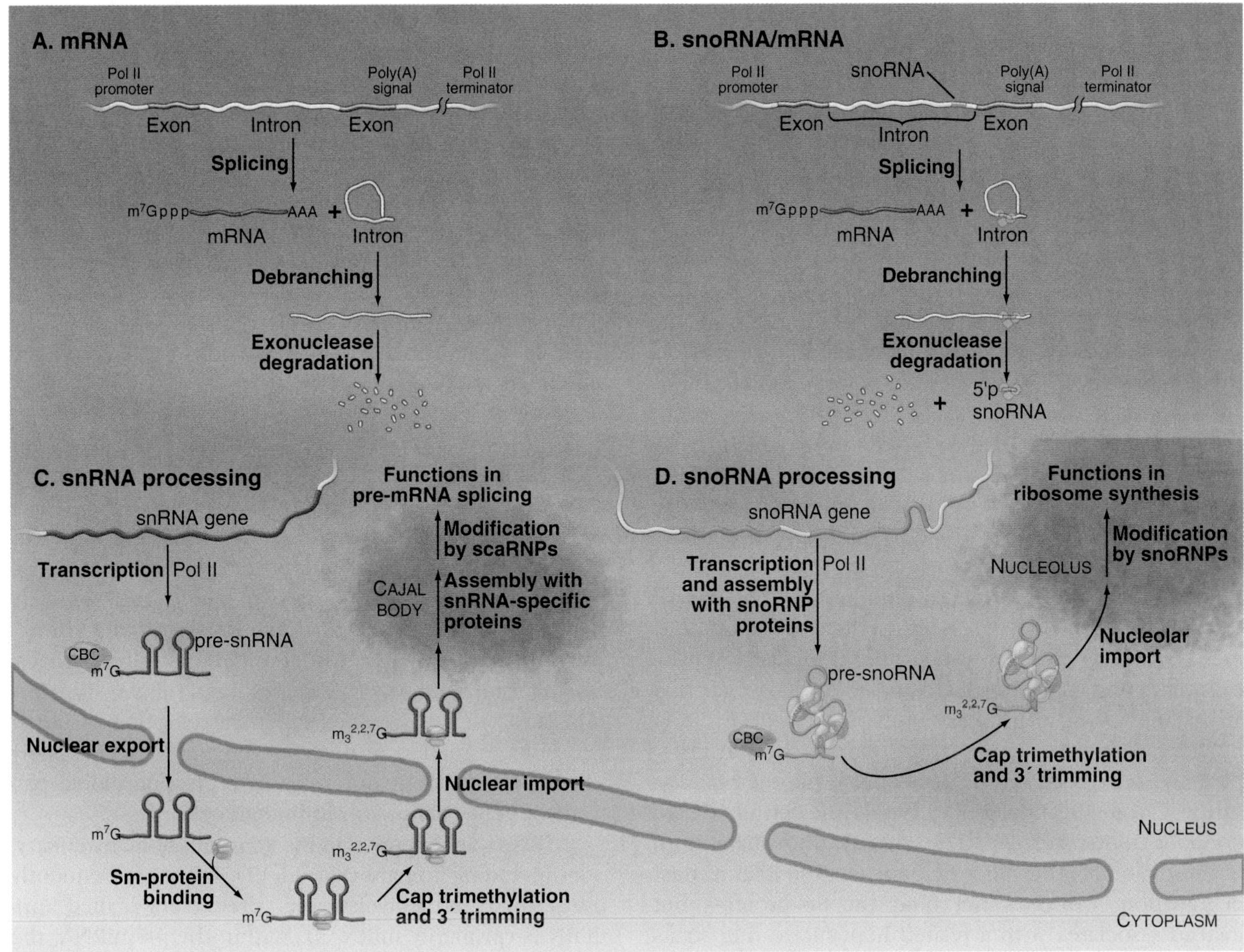

FIGURE 11.11 DIFFERENT PATTERNS OF STABLE RNA SYNTHESIS BY RNA POLYMERASE II. A, Primary transcripts encoding messenger RNAs (mRNAs) generally contain one or more introns, which are removed and degraded to produce the mature mRNA. **B,** In human cells, the small nucleolar RNAs (snoRNAs) that are involved in ribosomal RNA (rRNA) modification are generally synthesized by excision from the introns of highly transcribed protein-coding genes. The small nucleolar ribonucleoproteins (snoRNPs) bind to the snoRNA sequence within the pre-mRNA and protect it from degradation. **C,** The spliceosomal U1, U2, U4, and U5 small nuclear RNAs (snRNAs) are transcribed by RNA polymerase II (Pol II) and, like mRNAs, are capped with 7-methylguanosine and bound by the nuclear cap-binding complex (CBC). The pre-snRNA is exported to the cytoplasm, where it associates with the Sm-protein complex and is 3′ trimmed. The cap is then hypermethylated to 2,2,7-trimethylguanosine, and the RNA-protein complex is reimported into the nucleus. The newly imported snRNPs localize to the Cajal bodies, where the snRNA is covalently modified at sites selected by base pairing to the small Cajal RNAs (scaRNAs), another class of modification guide RNAs. Assembly with specific proteins then generates the mature snRNPs. **D,** Some snoRNAs, including U3, are individually transcribed by RNA polymerase II. Like the snRNAs, they are initially capped by with 7-methylguanosine and bind CBC. Following association with a set of snoRNA-specific proteins, they undergo cap-trimethylation and 3′ trimming. The snoRNPs then localize to the nucleolus, where they themselves undergo snoRNP-dependent modification and then participate in rRNA processing.

classes of RNA and their assembly with specific proteins might be functionally coupled to transcription.

Small Nuclear RNA Maturation

The U1, U2, U4, and U5 snRNAs are encoded by individual genes transcribed by RNA polymerase II (Fig. 11.11C). Like mRNAs, the snRNA precursors undergo cotranscriptional capping with 7-methylguanosine, but they are not polyadenylated. In human cells, the newly synthesized precursors to these snRNAs are then exported to the cytoplasm. Once in the cytoplasm, the snRNAs form complexes with the **Sm-proteins.** This set of seven different, but closely related, proteins assembles into a heptameric ring structure. Sm-proteins are named after the human autoimmune serum that was initially used in their identification. On their own, the Sm-proteins show low substrate specificity in RNA binding. However, in human cells, the assembly of the snRNAs with the Sm-proteins is highly specific and is mediated by a large protein complex. This complex includes the **SMN** protein **(survival of motor neurons),** which is the target of mutations in the relatively common genetic disease spinal muscular atrophy. While in the cytoplasm the snRNAs are further processed; the 3′ end of the RNA is trimmed, and the cap structure undergoes additional methylation to generate 2,2,7-trimethylguanosine. This hypermethylated cap structure is also present on snoRNAs (see later) and might be important to allow resident nuclear RNAs to be distinguished from mRNA precursors.

Once the cap is trimethylated and bound by the Sm-proteins, the snRNAs can be reimported into the nucleus, where they initially localize to discrete subnuclear structures termed **Cajal bodies** (see Fig. 9.2). Within the Cajal bodies, specific nucleotides in the snRNAs are modified by 2′-*O*-methylation and pseudouridine formation. The sites of these modifications are selected by base pairing with a group of resident **small Cajal body RNAs (scaRNAs),** which carry the RNA-modifying enzymes. The scaRNAs closely resemble the snoRNAs except that single scaRNAs can frequently direct both 2′-*O*-methylation and pseudouridine formation.

Maturation of U6 snRNA is quite different from that of the other snRNAs. U6 is transcribed by RNA polymerase III and is not exported to the cytoplasm. Mature U6 retains the 5′ triphosphate and 3′ poly(U) tract that are characteristic of primary transcripts made by RNA polymerase III (see Chapter 10). However, the 5′ triphosphate is methylated on the γ-phosphate (ie, the position furthest from the nucleotide), while the terminal U of the poly(U) tract carries a 2′ to 3′ cyclic phosphate. Both of these modifications may help protect the RNA against degradation. U6 does not bind the Sm-proteins but instead associates with a related heptameric ring structure that is comprised of seven **Lsm proteins** ("like Sm"). Two distinct but related heptameric Lsm complexes are present in the nucleus and cytoplasm. The nuclear Lsm2–8 complex binds to the U6 snRNA and participates in the decapping of mRNA precursors that are destined for degradation in the nucleus (Fig. 11.1). In contrast, the Lsm1–7 complex participates in mRNA decapping and 5′ degradation in the cytoplasm. Nucleotides within the U6 snRNA are also modified at positions that are selected by guide RNAs, but this modification occurs in the nucleolus rather than the Cajal body.

Small Nucleolar RNA Maturation

The snoRNAs are generally transcribed by RNA polymerase II (except in some plants in which polymerase III–transcribed snoRNAs can be found). However, the genes encoding snoRNAs can have a surprising variety of different organizations. In human cells, most snoRNAs are excised from the introns of genes that also encode proteins in their exons (Fig. 11.11B). The introns that encode snoRNAs are released by splicing and then linearized by debranching. The mature snoRNA is then generated by controlled exonuclease digestion. In contrast, most characterized snoRNAs in higher plants and several yeast snoRNAs are processed from polycistronic precursors that encode multiple snoRNA species. Individual pre-snoRNAs are liberated by cleavage of the precursor by the double-strand–specific endonuclease RNase III (Rnt1 in yeast) and then trimmed at both the 5′ and 3′ ends. SnoRNAs can also be processed from single transcripts, and these have many features in common with snRNA transcripts. Like snRNAs, these individually transcribed snoRNAs carry trimethylguanosine cap structures (Fig. 11.11D). However, unlike snRNAs, which have a cytoplasmic phase, the maturation of snoRNAs and assembly of snoRNPs take place entirely within the nucleus, most steps probably occurring in the nucleolus.

Synthesis and Function of Micro-RNAs

The terms **siRNAs** and **miRNAs** are used to describe groups of RNAs that are physically similar but have distinct functions and a variety of different names. All are approximately 22 nucleotides in length and associate with a protein complex called the **RNA-induced silencing complex (RISC).** Under different circumstances, siRNAs can lead to cleavage of target RNAs, repress translation of mRNAs, or inhibit transcription of target genes via formation of heterochromatin. It seems likely that miRNAs play major roles in regulating global patterns of gene expression in human cells.

miRNAs are encoded in the genomes of many eukaryotes, including humans (Fig. 11.12). These are frequently transcribed as polycistronic precursors called pri-miRNAs (primarily miRNAs). Within the pri-miRNA, the precursors to the individual miRNAs (pre-miRNAs) form stem-loop structures. The stems are first cleaved by a nuclear double-strand–specific endonuclease called **Drosha,** releasing the individual pre-miRNAs. These are then exported to the cytoplasm, where cleavage by a second double-strand–specific endonuclease, **Dicer,** releases the miRNA in the form of a duplex with characteristic two-nucleotide 3′ overhangs and 5′ phosphate groups. These duplexes are incorporated into the RISC complex, where one of the strands becomes the functional miRNA.

If the target mRNA sequence is incompletely complementary to the miRNA, its translation is repressed (Fig. 11.12). This is likely to be the normal function of most endogenous miRNAs. It has recently been estimated that 30% or more of human mRNAs are targets of miRNA regulation. miRNAs show tissue-specific patterns of expression and dynamic changes in expression during differentiation. Individual miRNAs can modulate the expression of many different mRNAs. Changes in miRNA expression levels have been correlated with many human developmental transitions and numerous cancers. The effects of individual miRNAs on the expression levels of target RNAs are generally quite small (less than twofold), but they play important roles by reinforcing or

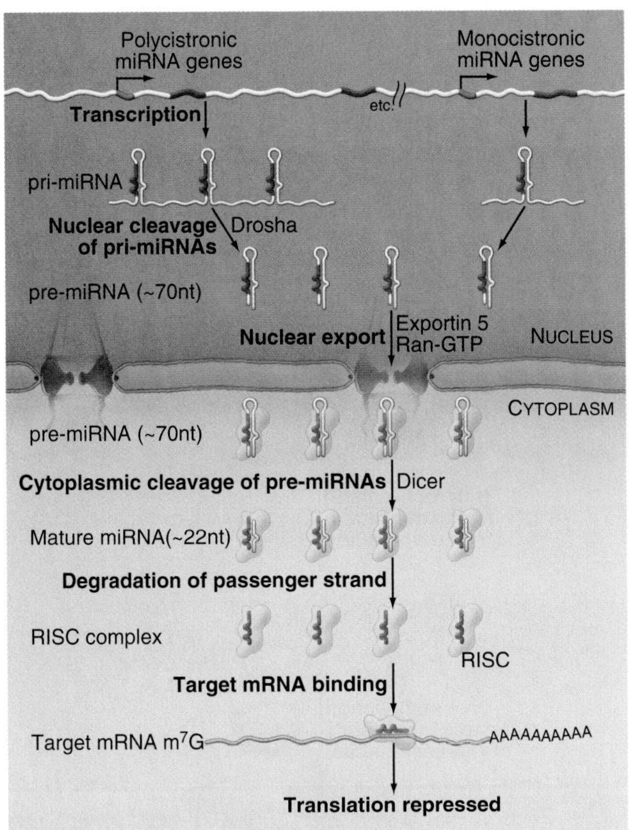

FIGURE 11.12 microRNA MATURATION. The polycistronic micro-RNA (miRNA) precursors (termed primary-miRNAs, or pri-miRNAs) are cleaved by the double-strand-specific endonuclease Drosha within the nucleus. The individual pre-miRNAs are then exported to the cytoplasm by the export factor Exportin 5 in complex with Ran-GTP (see Fig. 9.18). Once in the cytoplasm, the pre-miRNAs are cleaved by the double-strand–specific endonuclease Dicer. One strand of the resulting duplex is then incorporated into the RNA-induced silencing complex (RISC) and becomes the functional miRNA. Imperfect duplexes are formed between the miRNA and target messenger RNAs (mRNAs); this results in the inhibition of the mRNA translation.

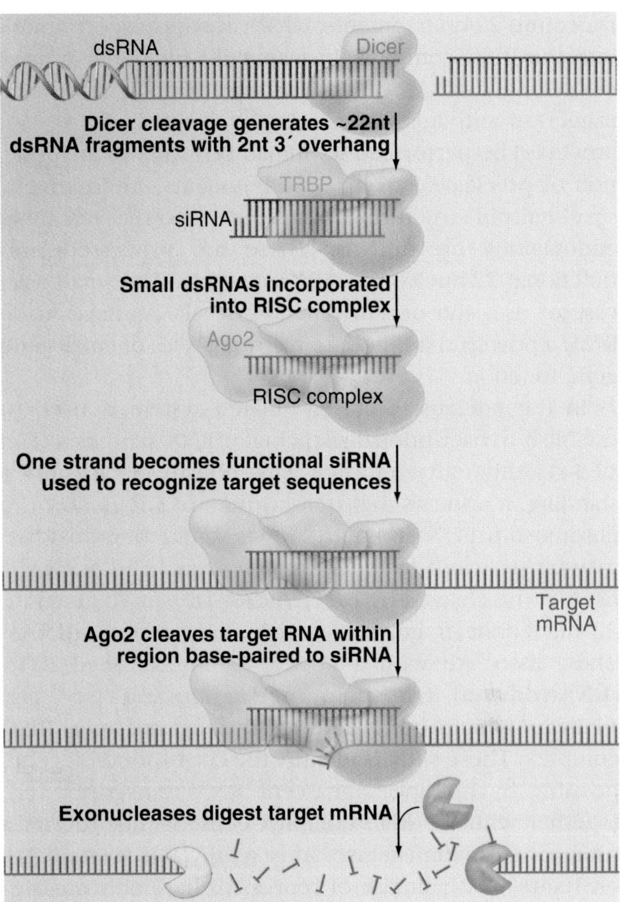

FIGURE 11.13 SMALL INTERFERING RNAS FUNCTION IN MESSENGER RNA CLEAVAGE. In contrast to the endogenous micro-RNAs (miRNAs), exogenously added small interfering RNAs (siRNAs) are designed by experimenters to be perfectly complementary to the target RNA, which is then cleaved by the Ago-2 component of the RNA-induced silencing complex (RISC) complex. In many organisms (including the nematode worm *Caenorhabditis elegans* and insects such as *Drosophila*), long double-stranded RNAs can be used. These are processed to approximately 22-nucleotide duplexes. In human cells, siRNAs are generally introduced as preformed 22-nucleotide duplexes or as stem-loops with structures that resemble endogenous pre-miRNAs. In either case, the siRNAs associate with Dicer, the double-strand RNA-binding protein TRBP, and Argonaut 2 to form the RISC complex. One strand becomes the functional siRNA, while the "passenger" strand is lost from the complex.

suppressing changes in gene expression programs that underlie cell fate decisions.

The synthesis and stability of miRNAs are subject to functionally important regulation. Proteins binding to the pre-miRNAs can enhance of inhibit maturation, imposing tissue-specific expression patterns. In some cases, miRNA or pre-miRNA degradation is strongly stimulated by the activity of terminal uracil transferases (TUTases) that add uracil nucleotides to the 3′ ends of substrate RNAs. This targets them for degradation by a cytoplasmic exonuclease Dis3L2, which shows high specificity for RNAs with 3′ terminal U tracts. This pathway is important, for example, in regulating the abundance of the Let-7 miRNA, which is highly conserved in evolution. Let-7 functions as an oncogene in humans and is frequently overexpressed in cancers.

If a target RNA sequence is perfectly complementary to the miRNA, it is cleaved by a component of the RISC complex, Ago2 (**"Slicer"**). Target RNA cleavage occurs within the miRNA: mRNA duplex at a fixed distance (between nucleotides 10 and 11) from the 5′ end of the miRNA, which is specifically bound and used to precisely position the duplex relative to the catalytic site.

This pathway can be exploited in a technique for the specific inactivation of target mRNAs, termed **RNAi** (Fig. 11.13). RNAi uses exogenously provided RNAs that are generally fully complementary to the target, typically provided as 22-nucleotide RNAs termed **siRNAs.** In many organisms (eg, in *Drosophila* or the nematode *Caenorhabditis elegans*), RNAi can be performed by introducing long dsRNAs. These are cleaved in vivo by

Dicer into 22-bp fragments, which are then incorporated into the RISC complex. In mammals, including human cells, long dsRNAs cannot be used for RNAi, as they trigger an antiviral response and cell death. RNAi can, however, be performed in human cells by the introduction of precleaved 22-bp RNA fragments. Alternatively, small hairpin structures can be expressed that resemble endogenous pre-miRNAs. These are processed into functional 22-nucleotide siRNAs in vivo. The small size, ease of use, and potent function of siRNAs have made RNAi a powerful method for many analyses of eukaryotic gene function.

In the nucleus, a closely related system is used to establish transcriptional silencing of RNA synthesis (Fig. 11.14). Although important gaps remain in our understanding, it appears that transcription of a region of the chromosomal DNA on both strands, generating a dsRNA, may be sufficient to induce its silencing. The dsRNA is likely to be cleaved by Dicer and/or Drosha to generate 22 nucleotide fragments, in this case termed **siRNAs**. These associate with a nuclear complex called **RITS (RNA-induced transcriptional silencing** [see Fig. 11.14]), which is related to the cytoplasmic RISC complex. These siRNAs identify the corresponding gene, possibly by binding to nascent RNA transcripts and, together with the RITS complex components, recruit a protein methyltransferase. This methylates histone H3 on lysine 9, a hallmark of repressive heterochromatin, which in turn recruits other heterochromatin proteins such as HP1 (see Fig. 8.7). The RITS complex includes an RNA-dependent RNA polymerase, and this may be able to generate new siRNAs, allowing the spreading of the heterochromatin into flanking sequences. The tendency of heterochromatin to spread into the flanking euchromatin has long been recognized and gives rise to the phenomenon of position effect variegation (see Fig. 8.7). In some eukaryotes, the methylated histone H3 can also recruit **DNA methyltransferases** that modify cytosine residues to 5′-methylcytosine. This reinforces heterochromatin formation and makes it heritable by daughter cells. This system may be important for the establishment of heterochromatin domains, such as those surrounding the centromeres in higher eukaryotes. It might also function as a defense system against the amplification of transposable elements.

The irony is that it now seems likely that the large-scale organization and transcriptional activity of the genome in many eukaryotes will involve RNAs that long eluded detection because they are so small.

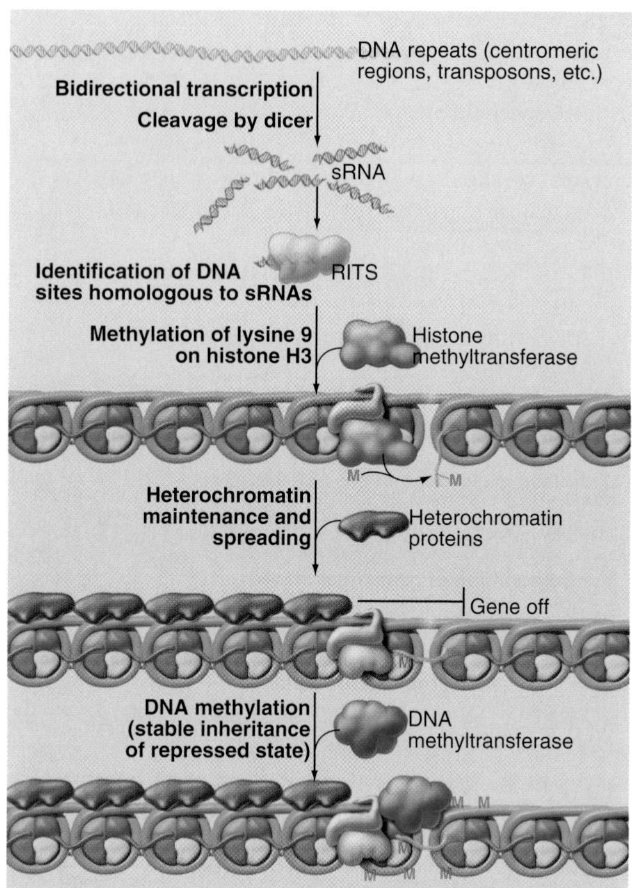

FIGURE 11.14 SMALL HETEROCHROMATIC RNAS FUNCTION IN HETEROCHROMATIN FORMATION. The targets of microRNAs (miRNAs) and small interfering RNAs (siRNAs) are cytoplasmic messenger RNAs (mRNAs). However, siRNAs can also function in the nucleus. Small double-stranded RNAs (dsRNAs) in the nucleus can associate with the RNA-induced transcriptional silencing (RITS) complex. The siRNA-RITS complex then identifies the genomic site of transcription, possibly by recognition of the nascent transcripts. This leads to the establishment of heterochromatin at this location, via the recruitment of protein methyltransferases that methylate lysine 9 on histone H3, a hallmark of repressive heterochromatin (see Fig. 8.7). In some organisms, this is followed by methylation of the DNA, which makes the repressed heterochromatic state more stable and heritable.

Synthesis and Function of Piwi-Interacting RNAs

The **Piwi-interacting RNAs (piRNAs)** form a distinct class of small RNAs, approximately 26 to 31 nucleotides in length. They are named because of their interaction with PIWI proteins that were first identified in *Drosophila* and are related to Argonaut. In invertebrates, piRNAs function in a silencing system in germline cells that blocks expression of transposons, thus protecting the DNA against recombination and mutation. In humans, piRNAs are also strongly expressed in the germline and are necessary for spermatogenesis.

Synthesis and Function of Other Noncoding RNAs

RNA sequencing in human cells revealed that the majority (70%–90%) of the genome is detectably transcribed, even though only less than 2% encodes proteins. This phenomenon is termed pervasive transcription, and it generates a bewildering array of ncRNA species. The definition of "ncRNAs" is somewhat vague, but it has come to mean the collection of transcripts that do not encode proteins, and do not fall neatly into one of the other major RNA classes (tRNA, rRNA, snRNA, etc). Human cells synthesis many thousands of different lncRNAs, defined as being longer than 200 nucleotides in length. These generally resemble mRNAs in being transcribed by RNA polymerase II and carrying cap and poly(A) modifications. The number of different lncRNAs increases with organismal complexity, suggesting that they may play important roles in generating tissue diversity; however, relatively few lncRNAs have been functionally characterized in detail. The best understood human lncRNA is Xist, which plays a key role in silencing one copy of the X chromosome in most female mammals, including humans. This is required to balance gene expression compared to males, which have only a single X chromosome. Xist RNA synthesized from a gene present on the silenced X forms protein complexes that appear to coat the compacted chromosome (see Fig. 8.6). Transcription silencing is achieved, at least in part, by recruitment of the polycomb repressive complex, which modifies specific residues in the histones that package the DNA.

Ribozymes

Some RNAs have catalytic activity in the absence of proteins. Such RNA enzymes are termed **ribozymes,** and they play a number of key roles.

Group I and Group II Self-Splicing Introns

Two classes of introns can catalyze their own excision from precursor RNAs. These ribozymes are referred to as **group I** and **group II self-splicing introns.** Both classes of RNA fold into complex structures that catalyze splicing via two-step transesterification pathways (Fig. 11.15).

The first group I intron was identified in 1981 as a 413-nucleotide fragment that was able excise itself from the pre-rRNA synthesized in the ciliate *Tetrahymena.* This was a major surprise, because at that time all known enzymes were proteins. The demonstration that an RNA could function as an enzyme had a major impact on subsequent RNA research. Group I introns are found in the pre-rRNAs of other unicellular eukaryotes, in the mitochondria and chloroplasts of many lower eukaryotes, and in the mitochondria of higher plants.

Group II introns have been found in mitochondria of plants and fungi and in chloroplasts. The splicing mechanism of group II introns strikingly resembles nuclear pre-mRNA splicing (Fig. 11.15C–D). This led to the proposal that the nuclear pre-mRNA splicing system derived from ancestral group II introns. During early eukaryotic evolution, the catalytic center of the group II intron might have become fragmented and separated into the present spliceosomal snRNAs. This would have converted a system that could work only on its own transcript into a system that could process other RNAs, greatly increasing the potential range of spliced RNAs.

RNase P and RNase MRP

Shortly after the identification of the group I intron in *Tetrahymena,* the RNA component of RNase P was also shown to function as a ribozyme. RNase P is an RNA-protein complex that cleaves pre-tRNAs at the 5′ end of the mature tRNA sequence in all organisms. The bacterial enzyme has one RNA component and one protein, but the RNA can cleave pre-tRNAs in vitro in the absence of the protein. In eukaryotes, RNase P has become more complicated, with one RNA and nine protein components. The eukaryotic RNA has not been shown to be active in the absence of proteins, but it does show structural similarities to the bacterial RNA, and it is assumed to be the catalyst.

Eukaryotes also contain a second RNA-protein enzyme, called RNase MRP, which is closely related to RNase P. The RNA components share common structural features, and the complexes share eight common proteins. RNase MRP cleaves the pre-rRNA between the small and large subunit rRNAs (Fig. 11.10E). Notably, in many bacteria, RNase P can cleave the pre-rRNA at a similar position because of the presence of a tRNA within the pre-rRNA transcript. This suggests that RNase MRP arose in an early eukaryote as a specialized form of RNase P, with a specific function in pre-rRNA processing. By analogy to RNase P, cleavage by RNase MRP is predicted to be RNA catalyzed. RNase MRP also functions in mRNA turnover, at least in yeast, initiating the cell-cycle-regulated degradation of a small number of mRNAs.

Large Subunit Ribosomal RNA

The most important ribozyme is the rRNA component of the large ribosomal subunit, which does not participate in RNA processing but catalyzes peptide bond formation (see Fig. 12.9). During translation elongation, the peptidyltransferase reaction (the reaction by which amino acid residues are attached to each other to form proteins) is catalyzed by the rRNA itself. The peptidyltransferase reaction is energetically favorable, and it is currently thought that the catalytic activity derives primarily from the precise spatial positioning of the A-site and P-site tRNAs by the rRNA. The ribosomal proteins act as chaperones in ribosome assembly and as cofactors to increase the efficiency and accuracy of translation.

RNA-Based Gene Editing

Rapid progress has been made in experimental gene editing techniques, driven by the development of

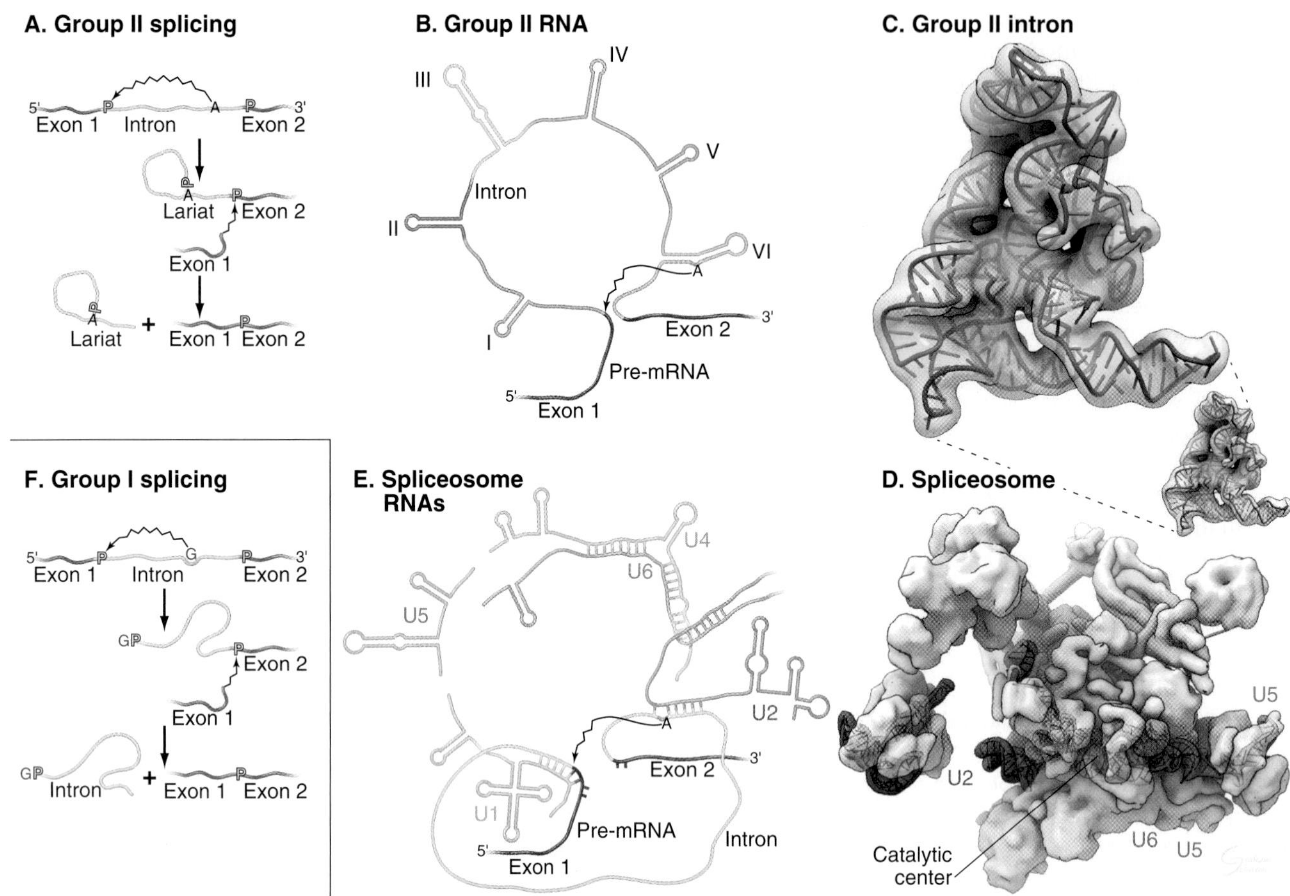

FIGURE 11.15 COMPARISON OF SELF-SPLICING WITH PRE–MESSENGER RNA SPLICING. Group I and group II introns are catalytic RNAs or ribozymes that can excise themselves from precursor RNAs in the absence of proteins. **A,** The removal of group I introns is mechanistically distinct from nuclear pre–messenger RNA (mRNA) splicing and commences with the binding of an exogenous guanosine nucleotide (*red* G) within a pocket created by the intronic RNA structure. This G is used to attack and break the phosphate backbone at the 5′ splice site. Subsequently, the free 3′ end of exon 1 attacks the phosphodiester bond at the 3′ splice site, leading to exon ligation and the release of the linear intron. **B,** In contrast, the mechanism of splicing group II introns is very similar to pre-mRNA splicing. An adenine residue (A) near the 3′ end of the intron attacks the 5′ splice site, leading to the formation of a lariat intermediate. The subsequent attack of the free 3′ end of exon 1 on the phosphodiester bond at the 3′ splice site leads to exon ligation and the release of the intron lariat (compare to Fig. 11.4). **C–D,** Parallels can be drawn between structure and mechanism of group II self-splicing introns and pre-mRNA splicing. This suggested the model that group II introns gave rise to the nuclear pre-mRNA splicing system. The small nuclear RNAs (snRNAs) may be derived from fragments of a group II intron that developed the ability to function in *trans* (ie, on other RNAs) rather than acting only in *cis* on its own sequence. Specifically, domain VI of the group II introns functions like the U2-branch point duplex in activating the branch-point adenosine by bulging it out of a helix. Domain V acts like the U2-U6 duplex in bringing this adenosine to the 5′ splice site. Domain III resembles the U5 snRNA in base pairing to both the 5′ and 3′ exons at the splice sites.

RNA-based, site-specific DNA cleavage systems. These are derived from immunity systems that are present in many bacteria and most *Archaea*. DNA sequencing identified "clustered regularly-interspaced short palindromic repeats" (CRISPR) in which the regions between the repeats are generally derived from the genomes of bacteriophages (viruses that infect bacteria) or plasmids. Conserved proteins encoded adjacent to the CRISPR loci assemble into the Cascade complex. Cascade recognizes and cleaves novel, incoming phage DNA and incorporates small fragments into the CRISPR genomic locus. On subsequent phage infection, Cascade proteins use the CRISPR RNA transcript to recognize and degrade the phage DNA. Numerous variants of the CRISPR/Cascade system exist, probably reflecting evolutionary pressure from the development of antagonistic phage systems.

It was subsequently shown that a single Cascade protein, together with a suitably engineered guide RNA, can perform highly specific, double-stranded DNA cleavage in almost any genome, including humans (Fig. 11.16). This has greatly facilitated genetic manipulation in many systems. At the time of writing, the Cas9 protein from *Streptococcus pyogenes* is predominately used as the RNA-directed endonuclease, but the field is developing rapidly. Future advances in genetic engineering of cells

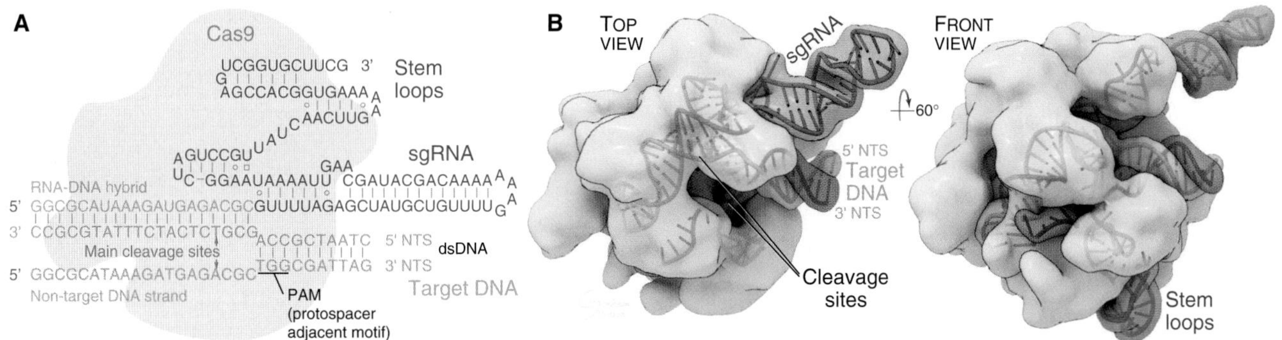

FIGURE 11.16 RNA-GUIDED DNA CLEAVAGE. The Cas9 protein has double-strand DNA (dsDNA) endonuclease activity at sites selected by base pairing with specific guide RNAs. Eukaryotic genome engineering commonly makes use of a complex between the *Streptococcus pyogenes* Cas9 double-strand nuclease and a single-guide RNA (sgRNA) guide. Cas9 can open the DNA duplex allowing potential base-pairing to the sgRNA. If the 20 nt complementary sequence is identified in the DNA together with a conserved element, called the protospacer adjacent motif (PAM), Cas9 can then cleave both DNA strands.

and organisms will revolutionize cell biology and medicine over coming decades.

Conclusions

Eukaryotic cells have a bewildering array of RNA species that perform many different, key functions in gene expression. The mature forms of all these RNAs are generated by RNA processing reactions, so the RNA processing machinery is of considerable importance. Probably for this reason, RNA-processing enzymes and cofactors are generally highly conserved during eukaryotic evolution. For many RNA species, transcription and maturation are closely coupled and can be thought of as an integrated system.

Finally, it is notable that many RNA species and functionally important modifications have only recently been discovered, so there is every reason to think that additional classes of RNA remain to be identified.

SELECTED READINGS

Cech TR, Steitz JA. The noncoding rna revolution—trashing old rules to forge new ones. *Cell*. 2014;157:77-94.

Ebert MS, Sharp PA. Roles for microRNAs in conferring robustness to biological processes. *Cell*. 2012;149:515-524.

Henras AK, Plisson-Chastang C, O'Donohue M-F, et al. An overview of pre-ribosomal RNA processing in eukaryotes. *Wiley Interdiscip Rev RNA*. 2015;6:225-242.

Kilchert C, Wittmann S, Vasiljeva L. The regulation and functions of the nuclear RNA exosome complex. *Nat Rev Mol Cell Biol*. 2016;17:227-239.

Lee M, Kim B, Kim VN. Emerging roles of RNA modification: m6A and U-tail. *Cell*. 2014;158:980-987.

Papasaikas P, Valcárcel J. The spliceosome: the ultimate RNA chaperone and sculptor. *Trends Biochem Sci*. 2016;41:33-45.

CHAPTER 12

Protein Synthesis and Folding*

Whatever their final destination—cytoplasm, membranes, or extracellular space—proteins are synthesized in the cytoplasm of both prokaryotic and eukaryotic cells. The only exceptions are proteins encoded by genes in mitochondria and chloroplasts, which are synthesized in those organelles.

The biochemical synthesis of proteins is called **translation,** as the process translates sequences of nucleotides in a **messenger RNA (mRNA)** into the sequence of amino acids in a polypeptide chain (Fig. 12.1). Translation of mRNA requires the concerted actions of small **transfer RNAs (tRNAs)** linked to amino acids, **ribosomes** (complexes of RNA and protein), and many soluble proteins. Guanosine triphosphate (GTP) binding and hydrolysis regulate several proteins that orchestrate the interactions of these components. Ultimately, RNA molecules in the ribosome catalyze the formation of peptide bonds. Some newly synthesized polypeptides fold spontaneously into their native structure in the cellular environment, but many require assistance from proteins called **chaperones.**

All contemporary organisms share a common translation apparatus, so the mechanism of peptide bond formation must predate the common ancestor approximately 3.5 billion years ago. By the time of the common ancestor, many relatively complicated regulatory features were in place and were inherited across the phylogenetic tree.

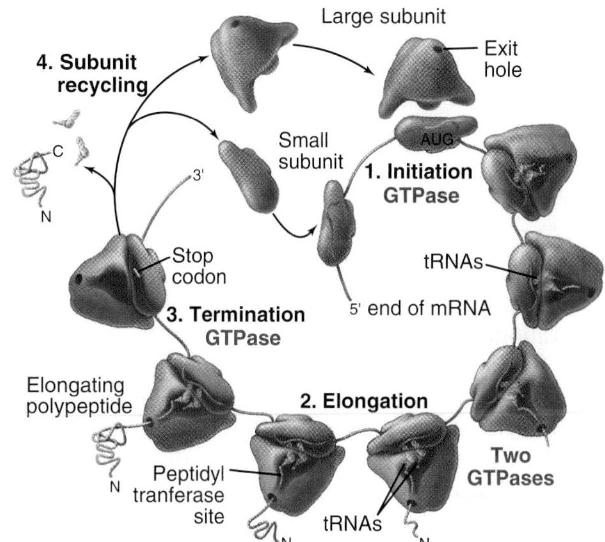

FIGURE 12.1 OVERVIEW OF THE TRANSLATION CYCLE SHOWING SIX RIBOSOMES ON A SINGLE mRNA. *1,* Initiation. Initiator $tRNA^{Met}$, mRNA, and accessory soluble factors assemble on the small subunit, which then joins with a large subunit. Met is the three-letter code for methionine. *2,* Elongation. The polypeptide chain is synthesized, in the order specified by the mRNA, in sequential steps by recruitment of new aa-tRNAs that match the coding sequence of the mRNA, formation of peptide bonds, and dissociation of free tRNA. *3,* Termination. Release factors recognize the stop codon *(yellow)* and terminate translation. The ribosome releases the polypeptide for folding in the cytoplasm. *4,* Subunit recycling. The ribosomal subunits dissociate and are available for another round of translation. aa-tRNA, Aminoacyl-tRNA; AUG, initiation codon; GTPase, guanosine triphosphatase; tRNA, transfer RNA.

Protein Synthetic Machinery

Messenger RNA

mRNAs have three parts: Nucleotides at the 5′ end provide binding sites for proteins that initiate polypeptide synthesis; nucleotides in the middle specify the sequence of amino acids in the polypeptide; and nucleotides at the 3′ end regulate the stability of the mRNA (Fig. 1.1). Within the protein-coding region, successive triplets of three nucleotides, called **codons,** specify the sequence of amino acids. The **genetic code** relating nucleotide triplets to amino acids is, with a few minor exceptions, universal. One to six different triplet codons encode each amino acid (Fig. 12.2). An **initiation codon** (AUG) specifies methionine, which begins

*This chapter was revised using material from the first edition written by William E. Balch, Ann L. Hubbard, J. David Castle, and Pat Shipman.

First Position (5' end)	Second Position: U	C	A	G	Third Position (3' end)
U	UUU Phe	UCU Ser	UAU Tyr	UGU Cys	U
	UUC Phe	UCC Ser	UAC Tyr	UGC Cys	C
	UUA Leu	UCA Ser	● UAA Stop	● UGA Stop	A
	UUG Leu	UCG Ser	● UAG Stop	UGG Trp	G
C	CUU Leu	CCU Pro	CAU His	CGU Arg	U
	CUC Leu	CCC Pro	CAC His	CGC Arg	C
	CUA Leu	CCA Pro	CAA Gln	CGA Arg	A
	CUG Leu	CCG Pro	CAG Gln	CGG Arg	G
A	AUU Ile	ACU Thr	AAU Asn	AGU Ser	U
	AUC Ile	ACC Thr	AAC Asn	AGC Ser	C
	AUA Ile	ACA Thr	AAA Lys	AGA Arg	A
	● AUG Met	ACG Thr	AAG Lys	AGG Arg	G
G	GUU Val	GCU Ala	GAU Asp	GGU Gly	U
	GUC Val	GCC Ala	GAC Asp	GGC Gly	C
	GUA Val	GCA Ala	GAA Glu	GGA Gly	A
	GUG Val	GCG Ala	GAG Glu	GGG Gly	G

● (dark) = Chain-terminating codon
● (light) = Initiation codon

FIGURE 12.2 THE GENETIC CODE. The locations of the nucleotide in first, second, and third positions define the amino acid specified by the code.

all polypeptide chains, but may subsequently be removed. In addition, any one of three **termination codons** (UAA, UGA, UAG) stops peptide synthesis.

Eukaryotic and bacterial mRNAs differ in three ways. First, eukaryotic mRNAs encode one protein, whereas bacterial mRNAs generally encode more than one protein. Second, most eukaryotic (and eukaryotic viral) mRNAs are capped by an inverted 7-methylguanosine residue joined onto the 5′ end of the mRNA by a 5′-triphosphate-5′ linkage (see Fig. 11.2 and Fig. 12.3). This **5′ cap** is stable throughout the life of the mRNA. It provides a binding site for proteins and protects the 5′ end against attack by nucleases. Third, most metazoan mRNAs require processing to remove introns (see Fig. 11.4).

Most eukaryotic mRNAs have a 3′ tail of 50 to 200 adenine residues added posttranscriptionally to the 3′ end (see Fig. 11.3). This **poly(A) tail** binds a protein that promotes export from the nucleus and protects the mRNA from degradation in the cytoplasm. The 3′ poly(A) tails are shorter or absent on bacterial mRNAs. Many single-stranded mRNAs have some double-stranded secondary structure (see Fig. 3.19) that must be disrupted during translation to allow reading of each codon.

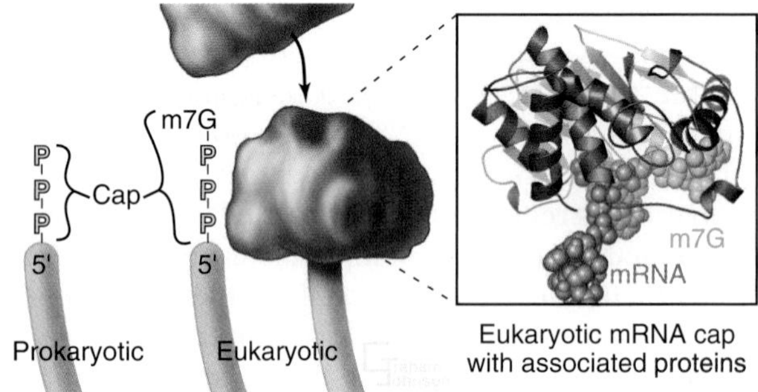

FIGURE 12.3 mRNA CAP STRUCTURES. Prokaryotic mRNAs (messenger RNAs) end with a 5′ triphosphate. The 5′ cap of eukaryotic mRNAs consists of a 7-methylguanosine residue (m7G) linked to the mRNA by three phosphates. The protein eIF-4E binds the cap and protects against degradation by nucleases. (See Protein Data Bank [PDB; www.rcsb.org] file 1EJ1.)

Transfer RNA

tRNAs are adapters that deliver amino acids to the translation machinery by matching mRNA codons with their corresponding amino acids as they are incorporated into a growing polypeptide (Fig. 12.4). One to four different tRNAs are specific for each amino acid, generally reflecting their abundance in proteins. Specialized tRNAs carrying methionine (formylmethionine in bacteria) initiate protein synthesis.

Transfer RNAs consist of ~76 nucleotides that base-pair to form four stems and three intervening loops. These elements of secondary structure fold to form an L-shaped molecule. A "decoding" triplet (the **anticodon**) is at one end of the L (the anticodon arm), and the amino acid acceptor site is at the other end of the L (the acceptor arm).

Enzymes called **aminoacyl-tRNA (aa-tRNA) synthetases** catalyze a two-step reaction that couples a specific amino acid covalently to its cognate tRNA (Fig. 12.5).

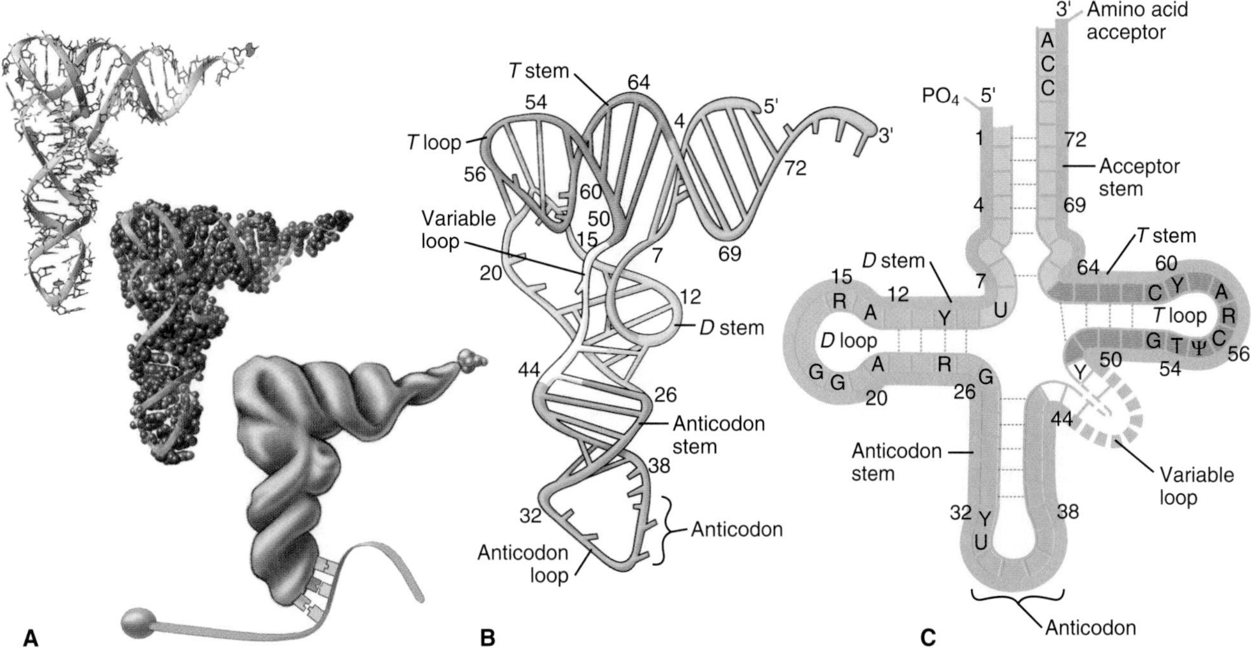

FIGURE 12.4 tRNA STRUCTURE. tRNAs (transfer RNAs) match an amino acid attached at the 3′ end with the mRNA (messenger RNA) triplet coding for that amino acid. **A,** Ribbon model, space-filling model, and textbook icon showing base pairing of the anticodon to an mRNA codon. **B,** Backbone model. **C,** Planar model showing stem loops of a generic tRNA. Single-letter code for the bases: adenine (A), any purine (R), any pyrimidine (Y), cytosine (C), guanine (G), pseudouridine (ψ), thymine (T), and uracil (U). (See PDB file 6TNA.)

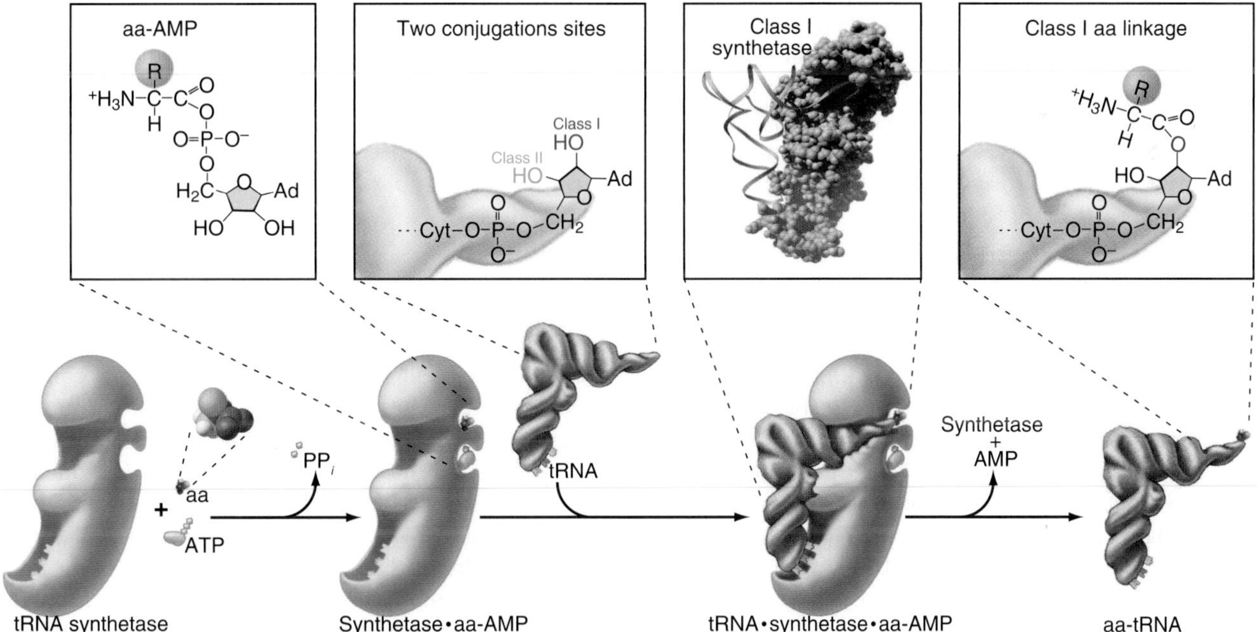

FIGURE 12.5 CHARGING A tRNA WITH ITS CORRECT AMINO ACID. tRNA synthetases (shown schematically and as a space-filling atomic model in *purple*) provide a docking platform for a specific amino acid and its cognate tRNA (shown in *orange* as a schematic model and as a ribbon model bound to a synthetase). The amino acid is first activated by reaction with adenosine triphosphate (ATP). The carboxyl group of the amino acid is coupled to the α-phosphate of adenosine monophosphate (AMP) with the release of pyrophosphate. The synthetase then transfers the amino acid from the aminoacyl-AMP (aa-AMP) to a high-energy ester bond with either the 2′ (illustrated here) or 3′ hydroxyl of the adenine at the 3′ end of the tRNA. aa, Aminoacyl; PP$_i$, inorganic phosphate. (See PDB file 1QTQ.)

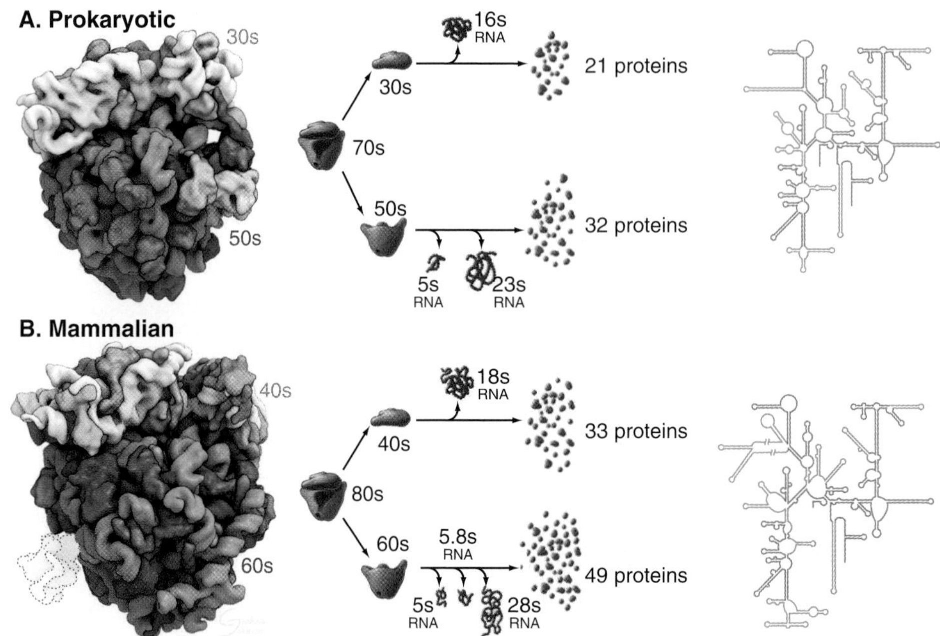

FIGURE 12.6 MOLECULAR COMPONENTS OF RIBOSOMES. RNA is light gray in the small subunits and dark gray in the large subunits. Proteins are colored. **A,** Crystal structure of the 70S ribosome from the thermophilic bacterium *Thermus thermophilus*. **B,** Cryoelectron microscopic structure of the 80S ribosome from pig pancreas actively synthesizing protein. The three columns show space-filling models *(left),* inventories of ribosomal RNAs (rRNAs) and proteins *(middle),* and maps of the secondary structures of prokaryotic 16S rRNA and 18S eukaryotic rRNAs to illustrate their similarities despite divergent sequences. (**A,** See PDB file 4W2F. **B,** Cryo-EM density maps are EMData Bank [EMDB; www.emdatabank.org] files 2644, 2646, 2649, and 2650. Also see PDB file 3J7O.)

In the first step, adenosine triphosphate (ATP) and the amino acid react to form a high-energy aminoacyl (aa) adenosine monophosphate (AMP) intermediate with the release of pyrophosphate. The second step transfers the amino acid to the 3′ adenine of tRNA, forming an aa-tRNA. This reaction is called *charging,* as the high-energy ester bond between the amino acid and the tRNA activates the amino acid, preparing it to form a peptide bond with an amino group in the growing polypeptide chain. Each of the 20 aa-tRNA synthetases couples a particular amino acid to its several corresponding tRNAs. The two classes of aa-tRNA synthases have different evolutionary origins and attach their amino acids to two different hydroxyls of the adenine at the 3′ end of the tRNA (Fig. 12.5).

The fidelity of protein synthesis depends on near-perfect coupling of amino acids to the appropriate tRNAs. Synthetases make this selection by interacting with as many as three areas of their cognate tRNAs: anticodon, 3′ acceptor stem, and the surface between these sites (Fig. 12.5). Some synthetases use proofreading steps to remove incorrectly paired amino acids from tRNAs.

Ribosomes

Ribosomes are giant macromolecular machines that bring together an mRNA and aa-tRNAs to synthesize a polypeptide. Base pairing between mRNA codons and tRNA anticodons ensures that the sequences of the polypeptides synthesized are those prescribed by the sequences of codons in the corresponding mRNAs. After many years of effort, crystal and cryoelectron microscopic (cryo-EM) structures are now available for many kinds of ribosomes (Fig. 12.6). Some surprises emerged from these structures.

Ribosomes consist of a **small subunit** and a **large subunit** that bind together during translation of an mRNA (Fig. 12.7). Each subunit includes one or more **ribosomal RNA (rRNA)** molecules and many distinct proteins (Fig. 12.6). The sizes of these subunits and rRNAs are traditionally given in units of S (Svedburg), the sedimentation coefficient measured in an ultracentrifuge. Although all ribosomes derive from a common ancestor and have similar mechanisms of action, their structures have diverged. Mammalian ribosomes have larger RNAs and more proteins than prokaryotic and mitochondrial ribosomes.

Ribosomal RNAs constitute the structural core of each ribosomal subunit (Fig. 12.7). The 18S rRNA of the small subunit of mammalian ribosomes contains approximately 1900 nucleotides, most of which are folded into base-paired helices. The large subunit of mammalian ribosomes includes three RNAs: a 28S rRNA consisting of approximately 5000 nucleotides, a 5.8S rRNA of 156 nucleotides, and a 5S rRNA of 121 nucleotides. The rRNAs fold into many based-paired helices, as first predicted by comparing the sequences of rRNAs from many different species (Fig. 12.6). These helices and their intervening loops pack to form a compact structure.

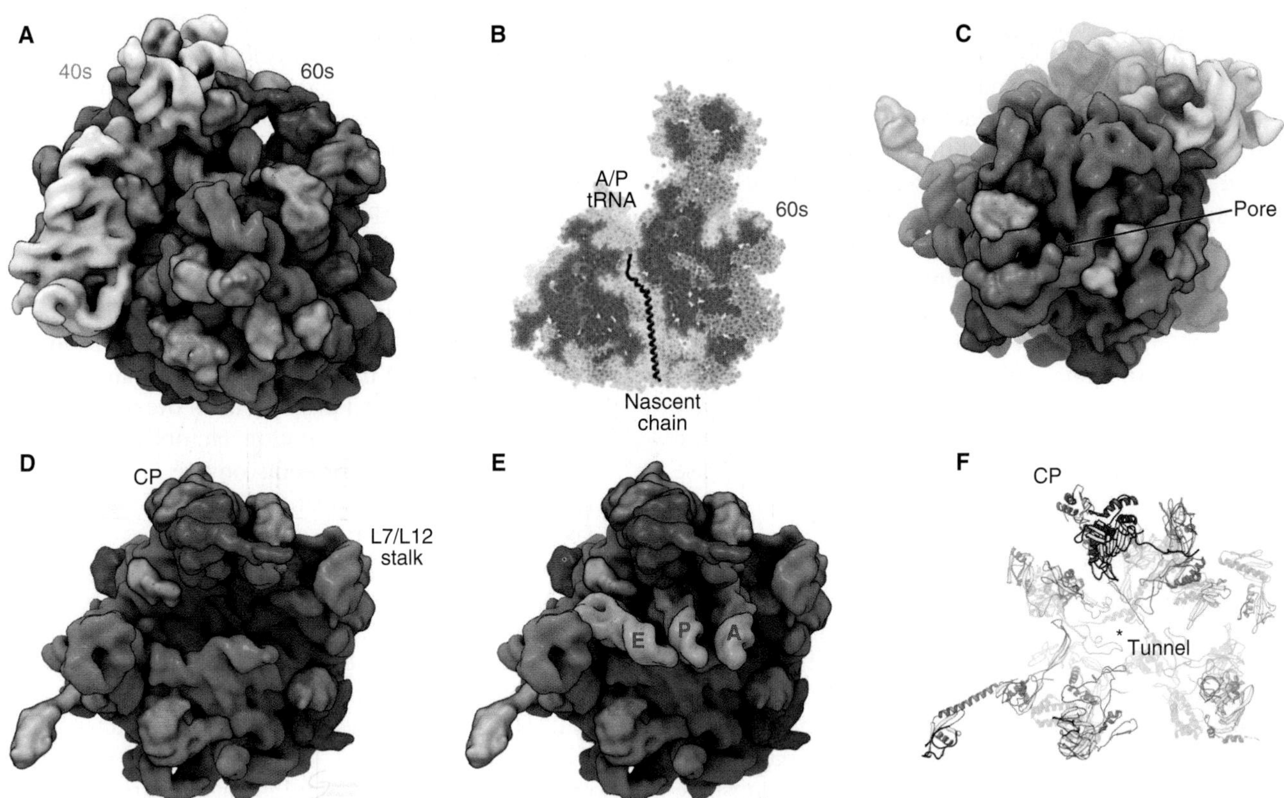

FIGURE 12.7 STRUCTURE OF THE MAMMALIAN RIBOSOME. The structure the pig *Thermus thermophilus* ribosome. RNA is shown in light gray for the small subunit and dark gray for the large subunit, and proteins are shown in a range of colors. **A,** Side view of a space-filling model. **B,** Cutaway side view of the large subunit showing the elongating polypeptide (as a black zigzag) in the exit channel through the core of the large subunit. **C–E,** Space filling models from different points of view. **C,** Bottom view showing the pore of the exit channel. **D,** Crown view showing the active site. **E,** Crown view showing three transfer RNAs (orange) bound in the active site. **F,** Crown view with ribbon diagrams of the proteins minus RNA. (Based on PDB file 4W2F.)

Although prokaryotic rRNAs differ in size and sequence from eukaryotic rRNAs, they fold similarly. Many features of rRNAs have been conserved during evolution, including the surfaces where subunits interact, the sites for binding tRNA, mRNA, and protein cofactors, and the nucleotides involved with peptide bond formation.

Most ribosomal proteins associate with the surface of the rRNA core, although several extend peptide strands into the core (Fig. 12.7). Ribosomal proteins are generally small (10 to 30 kD) and basic. With one exception, ribosomes have just one copy of each protein.

Decoding of mRNAs and polypeptide synthesis take place in the cavity between the subunits. The surfaces of this cavity are generally free of proteins, so (amazingly) rRNAs—not proteins—are largely responsible for mRNA binding, tRNA binding, and catalysis of peptide bond formation. tRNAs move sequentially through three sites shared by the two subunits: the **A site** (aa-tRNA), the **P site** (for peptidyl-tRNA), and the **E site** (for exit). The growing polypeptide chain exits in a tunnel that passes through the RNA core of the large subunit.

The synthesis and assembly of a yeast ribosome requires the participation of all three RNA polymerases, 75 small nucleolar RNAs (snoRNAs), and more than 200 protein factors, in addition to the 80 ribosomal proteins and 4 rRNAs present in the mature ribosomes. Precursor RNAs are cleaved and modified to form the rRNAs (see Fig. 11.10). Assembly factors consisting of snoRNAs and numerous proteins then orchestrate the stepwise assembly of rRNAs and ribosomal proteins into the small and large subunits and guide their export from the nucleus into the cytoplasm.

Although the genes for many ribosomal proteins are essential for viability, mutations in some can cause remarkably specific defects. For example, humans with just one functional gene for ribosomal protein RPSA are missing their spleen, but are otherwise normal. Mutations in genes for other subunits cause anemia and mutations in genes for certain assembly proteins cause liver disease.

Soluble Protein Factors

Many soluble proteins cycle on and off ribosomes during protein synthesis, enhancing the rate and/or the fidelity of the reactions that occur there. The following sections highlight the role(s) of these soluble factors.

Mechanism of Protein Synthesis

Organisms in all three domains of life use homologous components and similar reactions to synthesize proteins, although the details differ as expected after 3 billion years of evolutionary divergence. In all three domains, protein synthesis takes place in four steps: initiation, elongation, termination, and subunit recycling (Fig. 12.1). Conformational changes move ribosomes along the mRNA as the gene sequence is read out. Few errors are made thanks to precise pairing of tRNAs with their amino acids and codons in the mRNA that occur on the ribosome. Guanosine triphosphatase (GTPase) proteins regulate the progress and fidelity each step (see Fig. 4.6 for details on GTPase cycles).

Initiation Phase

The goal of **initiation** is to bring together the initiator tRNA carrying methionine (or *N*-formylmethionine, fMet, in Bacteria) and the AUG initiator codon of the mRNA in the appropriate site on the ribosome (Fig. 12.8). First, the two RNAs form a ternary complex on a small ribosomal subunit, which then associates with a large subunit to form a 70S ribosome in Bacteria and an 80S ribosome in eukaryotes. Eukaryotes use more than 10 soluble protein factors (eukaryotic **initiation factors** [eIFs]) to coordinate the RNA interactions. Fewer protein factors (designated IF) participate in prokaryotes. In eukaryotes, several steps occur in succession:

Step 1. Initiator Met-tRNA and the GTPase eIF-2A (with bound GTP) form a **preinitiation complex** on a small ribosomal subunit.

Step 2. Several protein initiation factors assemble on the 5′ cap of the mRNA. The RNA **helicase** eIF-4A in this complex uses ATP hydrolysis to remove any secondary structure or bound proteins from the 5′ end of the mRNA. These cap recognition factors also interact with poly(A)-binding proteins on the far end of the mRNA, forming a circular complex that can either favor or inhibit initiation of translation.

Step 3. The **cap recognition complex** targets the mRNA to a preinitiation complex. The order of these first three steps is still being investigated. For example, mRNA may bind to the small subunit before the initiation factors and Met-tRNA.

Step 4. The small subunit and two initiation factors form a tunnel on the small subunit through which the mRNA is allowed to slide as the initiator tRNA in

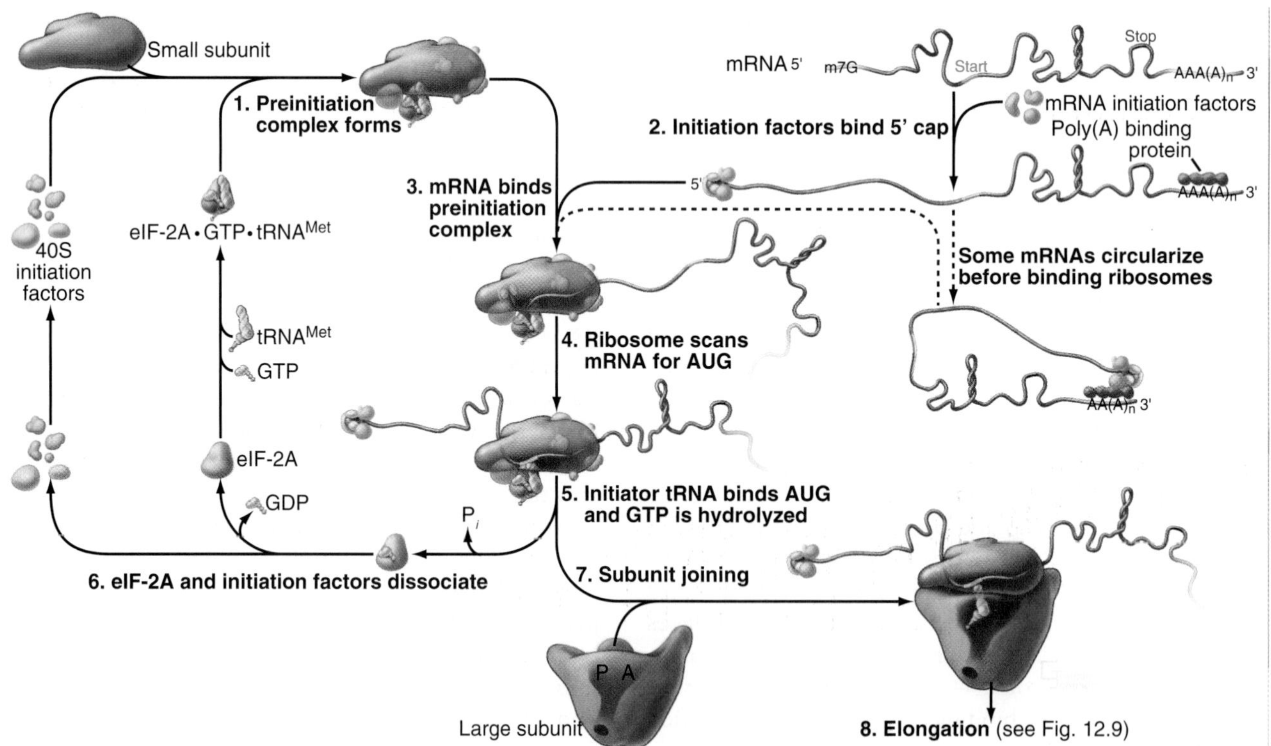

FIGURE 12.8 STEPS IN INITIATION IN EUKARYOTES. *1,* Initiation factors *(green)* assemble with mRNA (messenger RNA), eIF-2A (*purple,* activated with GTP), and $tRNA^{Met}$ on a small ribosomal subunit to form the preinitiation complex. *2,* Other initiation factors *(blue)* bind the 5′ cap of the mRNA. For some mRNAs, these 5′ cap-binding factors interact with poly(A)-binding proteins at the 3′ end of the mRNA. This circularization promotes initiation of some mRNAs and inhibits initiation of other mRNAs. *3,* The preinitiation complex binds an mRNA. *4,* The small subunit scans the mRNA for the AUG start codon *(green)*. *5,* When the initiator tRNA binds the start codon, eIF-2a hydrolyzes its bound GTP (guanosine triphosphate). *6,* Phosphate, GDP (guanosine diphosphate), eIF-2a, and other initiation factors dissociate and recycle for further rounds of initiation. *7,* The small subunit binds a large subunit. *8,* Elongation begins. m7G, 7-Methylguanosine.

the P-site scans for the initiator AUG codon. This movement depends on ATP hydrolysis, but its role is not clear. Eukaryotic mRNAs tend to begin translation at the first AUG codon encountered, but the local sequence of the mRNA may also contribute to the specificity as it does in *Bacteria*.

Step 5. When Met-tRNA base-pairs with the initiator AUG codon, eIF-2A hydrolyzes its bound GTP.

Step 6. eIF-2A and the other initiation factors dissociate from the small subunit for recycling.

Step 7. A large ribosomal subunit binds the small subunit complexed with both the mRNA and Met-tRNA. Another GTPase called eIF-5B hydrolyzes its bound GTP before elongation of the polypeptide begins.

Initiation is the slowest and most highly regulated step in protein synthesis, frequently involving phosphorylation of initiation factors. For example, cells that are subjected to various stresses use phosphorylation of eIF-2A to inhibit translation. Phosphorylation increases the affinity of eIF-2A for its guanine nucleotide-exchange factor (eIF-2B), which competes with the initiator tRNA. In contrast, phosphorylation of eIF-4F favors translation by enhancing the interaction of this initiation factor with the 5′ cap of mRNAs. This mechanism can influence the selective translation of particular mRNAs, since the 5′ caps of mRNAs vary in affinity for eIF-4F.

Elongation Phase

During **elongation,** the ribosome sequentially selects aa-tRNAs from the cellular pool in the order specified by the sequence of codons in the mRNA it is translating (Fig. 12.9). The ribosome catalyzes formation of a peptide bond between the amino group of the amino acid part of each new aa-tRNA and the carboxyl group at the C-terminus of the growing polypeptide chain and then moves on to the next codon. Codon-directed incorporation of amino acids into the polypeptide chain begins once the two ribosomal subunits are joined with an initiator tRNA and mRNA properly in place (Fig. 12.8).

The elongation reactions occur in the cavity between the two ribosomal subunits. mRNA is threaded, codon by codon, along a bent path between the subunits. aa-tRNAs enter on one side of the cavity and bind successively to three sites between the two ribosomal subunits. Interactions with both subunits allows the tRNA to maintain contact with the ribosome as it moves, step by step, from the A site to the P site to the E site prior to dissociation. When the tRNAs are bound in the A and P sites, their anticodons base-pair with mRNA codons. Peptide bonds form at the other end of the tRNAs, which position the amino acid on the tRNA in the A site adjacent to peptidyl chain on the tRNA in the P site of the large subunit. The growing polypeptide exits through a 10-nm-long tunnel in the large subunit.

Two GTPases called **elongation factors** (EF; eEF for eukaryotic elongation factors) bind near the A site and favor movements of the subunits relative to each other that facilitate the movements of the mRNA and tRNAs through the ribosome. Some of the energy from GTP hydrolysis also increases the accuracy, but makes elongation the most expensive phase of translation in terms of energy expenditure.

The following paragraphs summarize the current understanding of the four elongation steps: (1) an aa-tRNA binds to the A site on the ribosome; (2) proofreading ensures that it is the correct aa-tRNA; (3) a peptide bond forms; and (4) translocation advances the mRNA by one codon and moves the peptidyl-tRNA from the A site to the P site on the ribosome. New structures and spectroscopic observations of single ribosomes will continue to reveal more details.

Step 1. aa-tRNA binding. The first GTPase (called eEF1A in eukaryotes and EF-Tu in *Bacteria*; see Fig. 25.7) is charged with GTP by a nucleotide-exchange factor (called eEFX in eukaryotes and EF-Ts in *Bacteria*). This prepares eEF1A to bind an aa-tRNA, which it delivers to an empty A site of a ribosome. Cells contain enough eEF1A-GTP to bind all the aa-tRNAs and protect the labile ester bond anchoring the amino acid.

Step 2. Proofreading. A proofreading mechanism retains aa-tRNAs in the A site if they are correctly base paired with the mRNA codon and allows other aa-tRNAs to dissociate. This "kinetic proofreading mechanism" uses two first-order reactions to discriminate between correct and incorrect aa-tRNAs: hydrolysis of GTP bound to eEF1A; and dissociation of guanosine diphosphate (GDP)-eEF1A from the aa-tRNA and the ribosome. If the aa-tRNA anticodon is base paired with the correct mRNA codon, then the ribosome stimulates GTP hydrolysis, phosphate release, a massive conformational change (see Fig. 25.7), and eEF1A dissociation in a few milliseconds. This allows the aminoacyl end of the aa-tRNA to move into the peptidyl transfer site on the large subunit and form a peptide bond. Those aa-rRNAs with weak, imperfect codon-anticodon pairs dissociate from the A site before eEF1A can hydrolyze GTP and dissociate from the aminoacyl end of the tRNA.

Step 3. Peptidyl transfer. The RNA of the large subunit forms the highly conserved active site that catalyzes the formation of peptide bonds (Fig. 12.9). This reaction eliminates water and transfers the carboxyl group esterified to the peptidyl-tRNA in the P site to the free amino group of the aa-tRNA in the A site. Catalysis of peptide bond formation depends on a combination of precise orientation of the substrates and stabilization of the transition state (just like protein enzymes). The chemistry is similar, but in reverse, to the hydrolysis of peptide bonds by

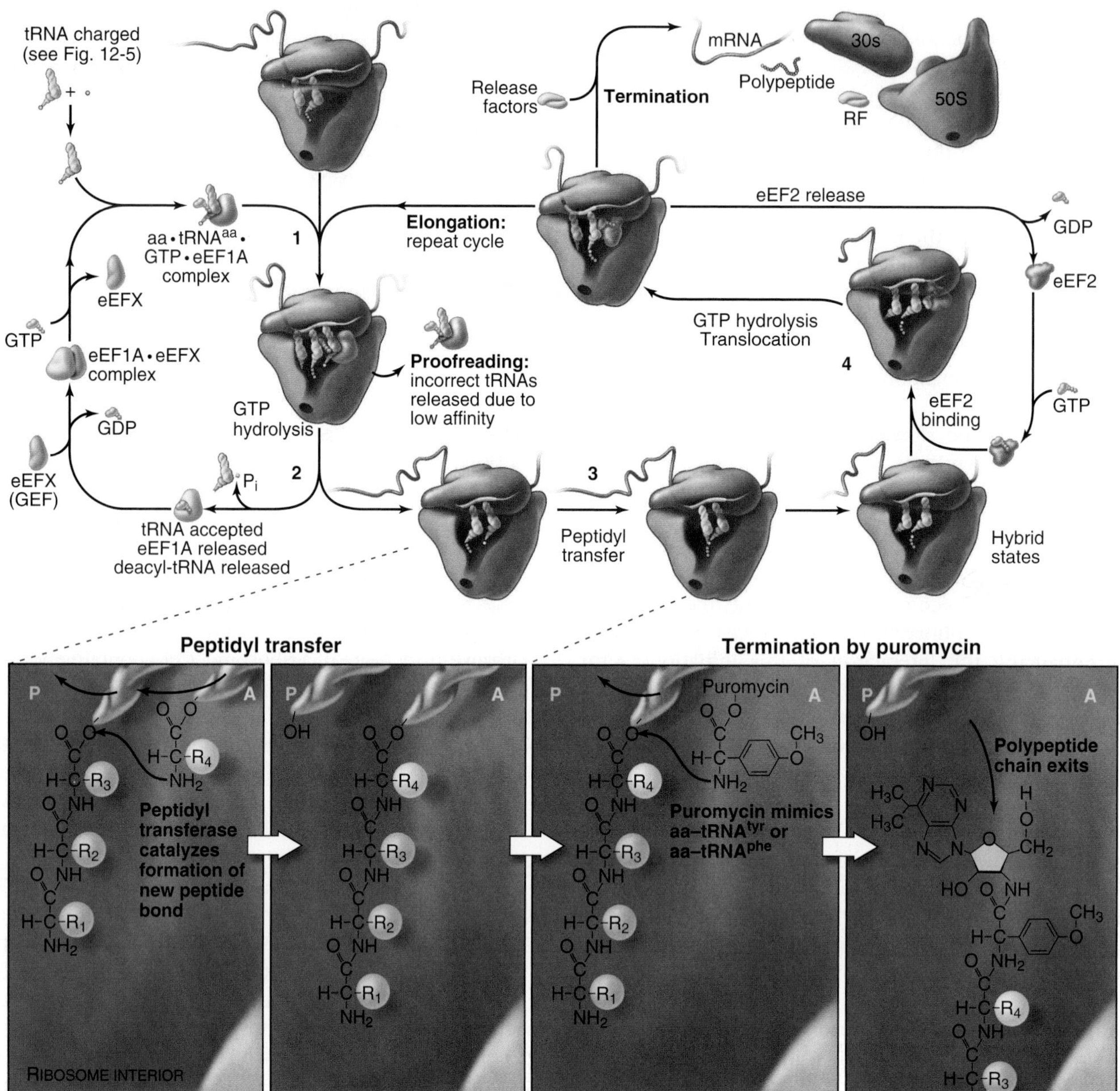

FIGURE 12.9 STEPS IN ELONGATION AND TERMINATION IN EUKARYOTES. Starting in the *upper left,* elongation factor eEF1A (EF-Tu in Bacteria) forms a ternary complex with GTP and each amino acyl-tRNA^{aa} for (1) delivery to the matching the mRNA codon in the A site of the ribosome. This ternary complex dissociates rapidly if the anticodon–codon match is incorrect. (2) If the anticodon–codon match is correct, the ternary complex remains bound to the A site long enough for eEF1A to hydrolyze its bound GTP and dissociate from the tRNA still bound to the A site. (3) The ribosome catalyzes formation of a new peptide bond *(inset).* (4) After eEF2 (EF-G in *Bacteria*) binds the A site, GTP hydrolysis causes a conformational change that facilitates translocation of the tRNAs and mRNA through the ribosome. Release factors (RF, *green*) recognize the stop codon and terminate the polypeptide chain *(blue),* allowing the mRNA and ribosomal subunits to dissociate. The guanine nucleotide-exchange factor eEFX promotes the exchange of GDP for GTP on eEF1A. The enlargements at the bottom show details of peptidyl transfer and the mechanism whereby the antibiotic puromycin terminates translation prematurely by mimicking the terminus of amino acyl-tRNA^{Tyr} or tRNA^{Phe}. It is incorporated on the C-terminus of the polypeptide, which then dissociates from the ribosome, because it lacks an activated carboxyl group.

proteolytic enzymes such as chymotrypsin. After formation of the new peptide bond, the tRNA in the A site has the polypeptide on one end and its anticodon arm still base-paired to its mRNA codon on the small subunit. The antibacterial agent **puromycin** can disrupt elongation by mimicking a tRNA^{Phe} or tRNA^{Tyr} (Fig. 12.10). Puromycin attacks the esterified carboxyl group of a peptidyl-tRNA in the P site, but lacking an appropriate acceptor site for further peptidyl transfer reactions, it terminates elongation. This results in premature release of the polypeptide chain from the ribosome.

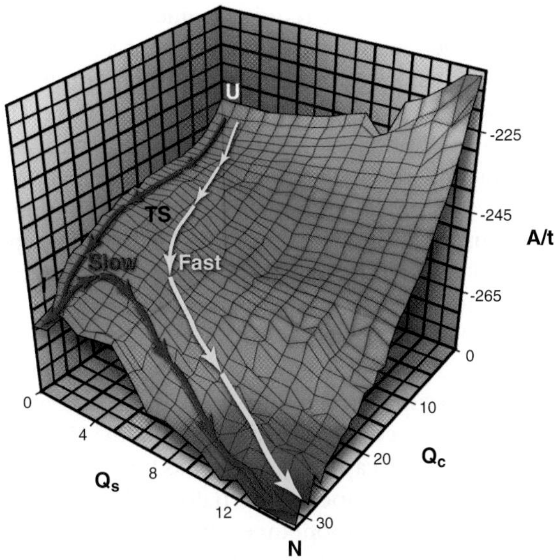

FIGURE 12.10 ENERGY LANDSCAPE IN PROTEIN FOLDING. As a protein matures from the unfolded state (U) through transition states (TS) to the native folded state (N), native-like contacts form, and the free energy of the system decreases. The two paths *(folding trajectories)* illustrate that fast protein folding *(yellow line)* is observed when more native-like contacts are made. When proteins become trapped in partially folded intermediate states, folding is slower *(pink line)* because energy barriers must be overcome. (Modified from Radford SE, Dobson CM. Computer simulations to human disease: emerging themes in protein folding. *Cell.* 1999;97:291–298.)

Step 4. Translocation. The second GTPase elongation factor (eEF2 in eukaryotes and EF-G in *Bacteria*) promotes three linked reactions that complete the elongation cycle. These GTPases have domains similar to domains 1 and 2 of EF-Tu (see Fig. 25.7) plus three domains that mimic the size and shape of a tRNA. Domain 1 binds and hydrolyzes GTP. Domains 3 to 5 target GTP-eEF2 to an empty A site on the ribosome. Binding of GTP-eEF2 to an empty A site favors rotation of the small subunit approximately 6 degrees relative to the large subunit. Hydrolysis of the GTP bound to eEF2 and phosphate dissociation promote the reverse rotation of the small subunit and movement of peptidyl-tRNA from the A site to the P site on the small subunit together with sliding of the mRNA three bases forward on the small subunit. This translocation step produces relatively large forces of approximately 13 pN (piconewtons) with the energy coming from peptide bond formation. At the same time, the deacylated tRNA in the P site is moved to the exit (E) site, where it dissociates later from the ribosome. Finally eEF2 with bound GDP dissociates from the A site, allowing another round of elongation.

Addition of each new amino acid pushes the growing peptide through a 10-nm-long tunnel in the large subunit lined with RNA (Figs. 12.1, 12.6, and 12.7). The tunnel accommodates an extended polypeptide approximately 40 residues long with the N-terminus in the lead. The distal parts of the tunnel are wide enough to pass an α-helix, but most folding of the polypeptide takes place outside the ribosome. Peptides longer than 40 residues protrude from the large subunit.

Cells balance speed and accuracy during translation to achieve an error rate of about 1 in 10^4 incorrect amino acids. As a result of this compromise, ribosomes add about 20 amino acids per second to a polypeptide at 37°C, so synthesis of a protein of average size (300 amino acids) takes only 15 seconds. Greater precision could be achieved by slowing translation, but slower cellular growth might be an evolutionary disadvantage.

Termination Phase

Termination occurs when the ribosome encounters a termination codon (UAA, UAG, or UGA) at the 3′ end of the coding sequence. Assembly of the polypeptide stops because a protein release factor, rather than an aa-tRNA, binds in the A site on the small subunit of the ribosome (Fig. 12.9). These release factors (called eRF1 in eukaryotes and RF1 or RF2 in bacteria) recognize stop codons and induce the ribosome active site to hydrolyze the peptidyl ester between the C-terminal amino acid of the polypeptide chain and the tRNA in the P site. The completed polypeptide chain threads through the ribosome and is released. Then a GTPase uses energy from GTP hydrolysis to promote dissociation of the mRNA and the ribosomal subunits, which are available for **recycling** to initiate translation of another mRNA.

Further Features of Protein Synthesis

Most mRNAs support protein synthesis by multiple ribosomes, forming **polysomes** (Fig. 12.1). Approximately 40 to 50 nucleotides of mRNA are associated with each ribosome. Consequently, once a ribosome has read approximately 60 nucleotides the initiation codon emerges and is available to assemble another ribosome-tRNA complex and start translation. Ribosomes can pack close together on one mRNA with all of newly synthesized polypeptides emerging around the periphery. This multiple occupancy of mRNAs explains why ribosomes are more abundant than mRNAs and how one mRNA molecule can guide the synthesis of several copies of its protein product simultaneously.

This account of protein synthesis may give the impression of a homogeneous population of ribosomes moving steadily on mRNAs, but variation exists at every level. For example, three types of experiments show that ribosomes can pause during translation. Biochemical experiments and observations of single ribosomes showed that certain sequences, such as several consecutive prolines or mRNA secondary structures, can stall translation. Cells have a special elongation factor (called EF-P in bacteria and a/eIF-5A in eukaryotes) that binds stalled ribosomes and promotes peptide bond formation, so the ribosome can move on.

New experiments using high-throughput DNA sequencing have documented pauses and revealed many other features of translation for all the mRNAs in a cell. This method, called "ribosomal profiling" or "ribosome footprints," takes advantage of the fact that a ribosome protects approximately 30 bases of the associated mRNA from digestion by nucleases. Therefore, one can isolate polysomes, digest with a nuclease, and isolate the protected mRNA sequences. After copying into DNA, millions of fragments are sequenced in parallel (see Fig. 3.16) to show precisely to the nucleotide where ribosomes are located on mRNAs. The number of DNA reads is higher if ribosomes stall at certain positions. This broad view also revealed many surprising events that take place during translation, including unconventional start sites, pauses caused by environmental conditions and the association of many small RNAs with ribosomes.

Ribosomes can vary in composition and posttranslational modifications. Single genes encode most mammalian ribosomal proteins, but plants have multiple genes for isoforms that are expressed in different cells. As they mature (see Fig. 11.10), rRNAs are methylated and some uridines are converted to pseudouridine (Fig. 12.4). Ribosomal proteins are modified by acetylation, methylation, phosphorylation, *O*-linked β-D-*N*-acetylglucosamine and ubiquitylation. These differences each have the potential to influence protein synthesis, although few examples have been characterized in detail. A large number of proteins associate with ribosomes and may also influence their activities.

Spontaneous Protein Folding

Termination is the final step in translation, but just the beginning for a new protein. A polypeptide begins to experience its new environment while still being synthesized. When it is approximately 40 residues long, its N-terminus emerges from the protected tunnel of the large ribosomal subunit into cytoplasm, where it must fold into a three-dimensional structure (see Fig. 3.5) and find its correct cellular destination.

The structure of folded proteins and the folding mechanism are both encoded in the amino acid sequence, making folding spontaneous under suitable conditions. For the soluble proteins, these conditions are aqueous solvent at physiological temperature, neutral pH, and moderate ionic strength. Folding of transmembrane proteins in a lipid bilayer is quite different (see Chapter 20). In test tube experiments, small soluble proteins can be denatured with high temperature, extremes of pH, or high concentrations of urea or guanidine. Denatured proteins exist as ensembles of unfolded polymers with little residual secondary structure.

When denatured polypeptides of modest length are transferred to physiological conditions, many fold spontaneously into their native three-dimensional structures on a microsecond to millisecond time scale. (Proteins such as collagen, which require isomerization of prolines, fold much more slowly; see Fig. 29.4.) Starting from many initial denatured states, a polypeptide converges toward a single low-energy native state (Fig. 12.10) driven by energy from numerous noncovalent interactions and the hydrophobic effect (see Fig. 4.5). The number of possible pathways to the native state is so numerous that if they were sampled individually, proteins would never fold. Thus, both theory and experiment indicate that folding involves a subset of the potential pathways, including the formation of an ensemble of loosely folded transition states with elements of secondary structure, certain turns, and hydrophobic contacts found in the core of the native protein. However, the free energy landscape for folding has hills and valleys, so proteins can be trapped in partially folded states.

Many proteins fold spontaneously without assistance during biosynthesis in vivo. Folding begins when the N-terminus of the nascent polypeptide emerges from the ribosome. The vectorial nature of this very slow "cotranslational folding" has both advantages and liabilities. An advantage is that folding before the polypeptide is complete limits the routes to the folded state and might account for why many proteins fold more efficiently during biosynthesis than from the denatured state. On the other hand, vectorial folding precludes interactions of N-terminal sequences with C-terminal sequences until they have emerged from the ribosome. Such interactions are common in folded proteins.

Folding of larger proteins is more complicated, especially in the crowded cytoplasm where partially folded proteins expose hydrophobic segments that are normally buried in the core of native proteins. These exposed core elements can aggregate irreversibly before folding is complete. Thus, many newly synthesized native proteins need assistance to avoid irreversible denaturation, aggregation, or destruction by proteolysis during folding.

Misfolding of mutant proteins contributes to many human diseases. For example, the most common cause of cystic fibrosis is genetic deletion of the codon for a single amino acid in cystic fibrosis transmembrane regulator (CFTR), resulting in failure of the protein to fold properly (see Fig. 17.4). Beyond lacking function, misfolded proteins also poison the assembly of native proteins in blistering skin diseases (see Fig. 35.6), hypertrophic cardiomyopathies (see Table 39.1), and other "dominant negative" conditions. Folding of proteins into nonnative states causes prion and amyloid diseases (Box 12.1).

Chaperone-Assisted Protein Folding

Several families of molecular chaperones (Fig. 12.11) facilitate folding of newly synthesized and denatured proteins. These chaperones do not fold polypeptides by directing the formation of secondary or tertiary

BOX 12.1 Protein Misfolding in Amyloid Diseases

Misfolding of diverse proteins and peptides results in spontaneous assembly of insoluble **amyloid fibrils.** Such pathological misfolding is associated with transmission of HIV, **Alzheimer disease, Parkinson disease**, transmissible spongiform encephalopathies (such as "mad cow disease"), and polyglutamine expansion diseases (such as Huntington disease, in which genetic mutations encode abnormal stretches of the amino acid glutamine). Accumulation of amyloid fibrils in these diseases is associated with slow degeneration of the brain. Pathological misfolding also results in amyloid deposition in other organs such as the endocrine pancreas in Type II diabetes. Some, but not all, amyloids are intrinsically toxic to cells. Some amyloid precursors are more toxic than the fibrils themselves.

The precursor of a given amyloid fibril may be the wild-type protein or a protein modified through mutation, polyglutamine expansion, proteolytic cleavage, or posttranslational modification. In all cases, fibril initiation is unfavorable owing to very slow assembly of the first few molecules, but once formed, fibrils elongate quickly by adding protein subunits. Amyloid fibrils are extremely stable and resistant to proteolysis.

Given that many unrelated proteins and peptides form amyloid, it is remarkable that these twisted fibrils all have similar structures: narrow sheets up to 10 μm long consisting of thousands of short β-strands that run across the width of the fibril. The β-strands can be either parallel or antiparallel, depending on the particular protein or peptide. Some amyloid fibrils consist of multiple layers of β-strands. The structures of the various parent proteins have nothing in common with each other or with amyloid cross-β-sheets, so these are examples of polypeptides with two stable folds.

To form amyloid, the native protein must either be partially unfolded or cleaved into a fragment with a tendency to aggregate. In the common form of dementia called Alzheimer disease, proteolytic enzymes cleave a peptide (Aβ) from a transmembrane protein called β-amyloid precursor protein whose normal role is to participate in signal transduction. Aβ forms toxic oligomers and amyloid fibrils that accumulate in the brain as neurons degenerate. Similarly, proteolytic fragments of an enzyme normally found in human semen form amyloid fibrils that enhance the transmission of HIV by many orders of magnitude. Therapeutic strategies include small molecules that stabilize native proteins or inhibit amyloid polymerization.

"Infectious proteins" called **prions** cause transmissible spongiform encephalopathies, such as "mad cow" disease. Normally, these proteins do no harm, but once misfolded, the protein can act as a seed to induce other copies of the protein to form insoluble amyloid-like assemblies that are toxic to nerve cells. Such misfolding rarely occurs under normal circumstances, but the misfolded seeds can be acquired by ingesting infected tissues.

Other proteins, including the peptide hormone insulin, the actin-binding protein gelsolin, the receptor protein β_2-microglobin and the blood-clotting protein fibrinogen, form amyloid in certain diseases. An inherited point mutation makes the secreted form of gelsolin susceptible to cleavage by a peptide processing protease in the *trans*-Golgi network. Fragments from the protein form extracellular amyloid fibrils in several organs. Exposure to copper during renal dialysis promotes β_2-microglobin to form amyloid fibrils in joints.

Given that amyloid fibrils form spontaneously and are exceptionally stable, it is not surprising that functional amyloids exist in organisms ranging from bacteria to humans. For example, formation of the pigment granules responsible for skin color depends on a proteolytic fragment of a lysosomal membrane protein that forms amyloid fibrils as a scaffold for melanin pigments. Budding yeast has approximately 10 proteins known to either assume their "native" fold or assemble into amyloid fibrils. The native fold of the protein Sup35p serves as a translation termination factor that stops protein synthesis at the stop codon (see Fig. 12.9). Rarely, Sup35p misfolds and assembles into an amyloid fibril. These fibrils sequester all the Sup35p in fibrils, where it is inactive. The faulty translation termination that occurs in its absence has diverse consequences that are inherited like prions from one generation of yeast to the next.

structure. Rather, chaperones inhibit aggregation by binding exposed hydrophobic segments of nonnative polypeptides or providing sequestered environments. They release polypeptides in a folding-competent state for attempts at folding. If folding fails, the cycle of binding and release can be repeated. The following sections cover **trigger factor** (and other chaperones associated with ribosomes), **Hsp70, Hsp90,** and cylindrical **chaperonins.** In addition, specialized chaperones assist with the folding of particular proteins such as tubulin and actin. Mutations in several of these chaperones have been associated with human disease. See Fig. 20.6 for chaperones in the endoplasmic reticulum.

Trigger Factor

Hydrophobic segments of the nascent polypeptide chain must be protected from aggregation until enough of the chain has emerged from the ribosome to participate in folding. Each growing polypeptide first encounters a chaperone bound next to the exit tunnel on the large ribosomal subunit. The chaperone associated with bacterial ribosomes is called trigger factor (Fig. 12.11). A structurally unrelated protein called nascent polypeptide-associated complex has a similar function in *Archaea* and eukaryotes. An extended array of hydrophobic patches on trigger factor binds hydrophobic features on the nascent polypeptide chain. These weak, rapidly reversible interactions prevent folding and protect the unfolded peptide from aggregation. The signal recognition particle binds on the other side of the exit tunnel, positioned so that its methionine-rich groove (see Fig. 20.5) also interacts with the growing polypeptide. Most bacterial polypeptides fold successfully after being released from trigger factor, while most

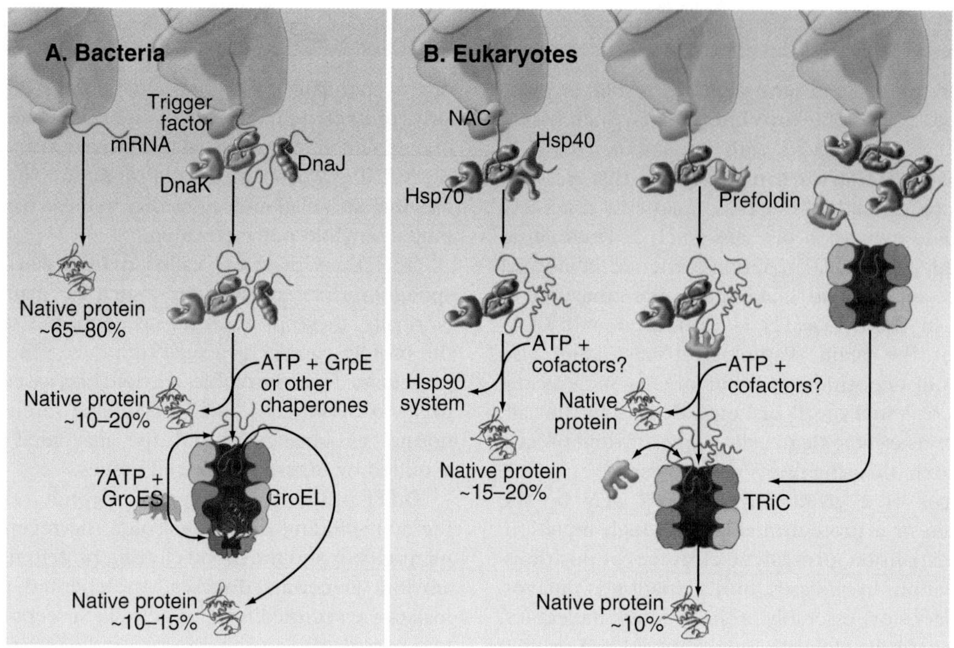

FIGURE 12.11 COMPARISON OF CHAPERONE-ASSISTED FOLDING PATHWAYS. A, Bacteria. **B,** Eukaryotes. The percentages refer to estimates of the fraction of proteins using each pathway. Most proteins fold without the assistance of chaperones. Hsp, heat shock protein; NAC, nascent polypeptide-associated complex. (Modified from Hartl FU, Hayer-Hartl M. Molecular chaperones in the cytosol: From nascent chain to folded protein. *Science.* 2002;295:1852–1858. Copyright 2002 American Association for the Advancement of Science.)

eukaryotic polypeptides require assistance from additional chaperones.

Hsp70 Chaperones

The most widespread chaperones are members of the heat shock protein 70 (Hsp70) family (Fig. 12.12). Their name came from the observation that cells subjected to stresses, such as elevated temperature, increase the synthesis of these proteins to protect against denatured proteins. Hsp70s are present in *Archaea*, *Bacteria* (called DnaK), and most compartments of eukaryotes. The family includes Hsp70 in mitochondria and BiP in endoplasmic reticulum (see Fig. 20.6). Budding yeasts have genes for 14 Hsp70s; vertebrates have more.

Hsp70s enzymes consist of two domains: an N-terminal domain (folded like actin) binds and hydrolyzes ATP. It is connected by flexible hinge to a C-terminal domain that uses a clamp to bind and release a wide range of nascent segments of unfolded polypeptides with approximately eight hydrophobic residues. ATP hydrolysis and phosphate release close the clamp on the hydrophobic polypeptides, while ATP binding opens the clamp and releases the polypeptide. This cycle of peptide binding and release, protects hydrophobic peptides from aggregation during attempts at folding, delivery to mitochondria and chloroplasts, and import into these organelles (see Figs. 18.4 and 18.6).

Hsp70 cooperates with other chaperones. Members of another family of heat shock proteins (Hsp40, called DnaJ in *Bacteria*) deliver unfolded proteins to bacterial Hsp70 (DnaK) and promote their binding by stimulating DnaK to hydrolyze ATP. Another co-chaperone called GrpE promotes exchange of adenosine diphosphate (ADP) for ATP, which opens the clamp and releases the bound peptide. Animal Hsp70s have a mechanism of action similar to that of DnaK except that they have intrinsic nucleotide-exchange activity and do not require a nucleotide-exchange protein such as GrpE.

Remarkably, Hsp70 can cooperate with an AAA adenosine triphosphatase found in bacteria, plants, and fungi to unfold aggregated proteins. Energy from ATP hydrolysis is used to pull a polypeptide from the aggregate through the central channel of the adenosine triphosphatase (ATPase). The polypeptide has a chance to fold once it emerges from the channel.

Hsp90 Chaperones

Hsp90 cooperates with other chaperones to stabilize steroid-hormone receptors such as those for progesterone, glucocorticoids, estrogens, and androgens, before they bind their ligands (Fig. 12.13). The chaperones use cycles of ATP hydrolysis to maintain receptors in an "open" state, ready to bind hydrophobic steroids. Steroid binding completes the folding of the receptors and displaces the Hsp90 complex. Then the receptors move to the nucleus to regulate gene expression (see Fig. 10.21). Hsp90 also interacts with other signaling proteins including protein kinases.

Chaperonins

The chaperonin family of barrel-shaped particles promotes efficient protein folding (Fig. 12.14). They allow

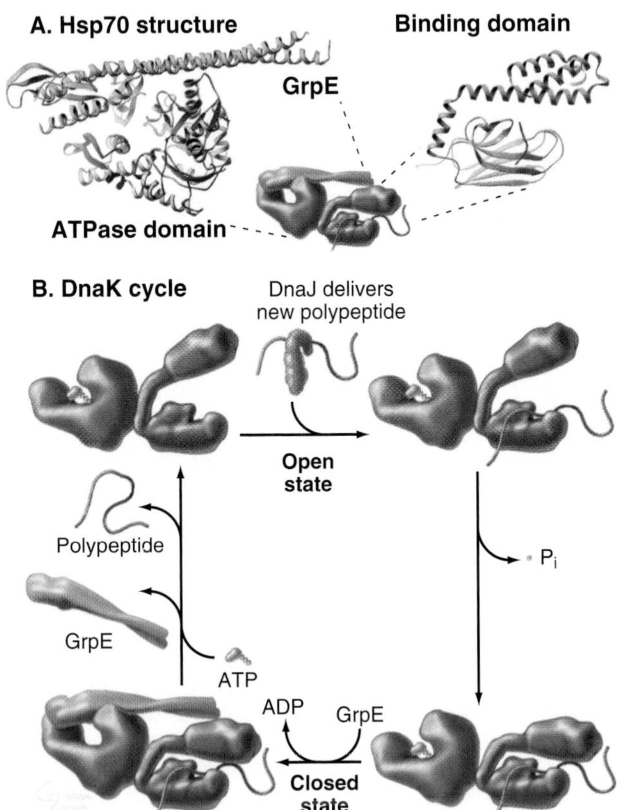

FIGURE 12.12 HEAT SHOCK PROTEIN 70 STRUCTURE AND FUNCTION. A, Ribbon diagrams of the atomic structures of DnaK *(blue)* and GrpE *(green)*. **B,** The heat shock protein (Hsp) 70 folding cycle with bacterial DnaK as the example. DnaJ (Hsp40) delivers an unfolded peptide to the ATP-bound open state of DnaK and promotes ATP hydrolysis. The ADP-bound closed state of DnaK binds the peptide strongly. GrpE promotes dissociation of ADP. Rebinding of ATP dissociates GrpE and the peptide, which is free to attempt folding. Multiple Hsp70 cycles are usually required to complete protein folding. (For reference, see Zhu X, Zhao X, Burkholder WF, et al. Structural analysis of substrate binding by the molecular chaperone DnaK. *Science.* 1996;272:1606–1614; and Harrison CJ, Hayer-Hartl M, Hartl F, et al. Crystal structure of the nucleotide exchange factor GrpE bound to the ATPase domain of the molecular chaperone DnaK. *Science.* 1997;276:431–435.)

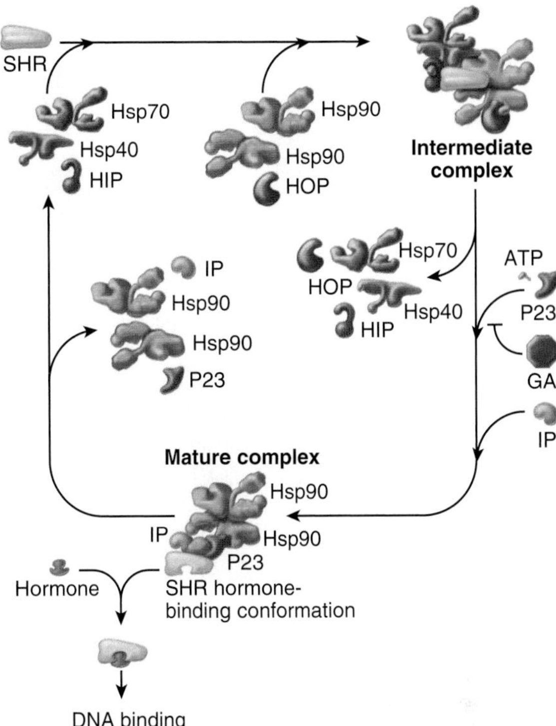

FIGURE 12.13 STABILIZATION OF LIGAND-FREE STEROID HORMONE RECEPTORS BY HSP70, HSP90, AND VARIOUS ACCESSORY FACTORS (HOP, HIP, P23, GA, AND IP). Hormone binding releases the chaperones and allows the receptor-steroid complex to move to the nucleus. SHR, steroid hormone receptor. (For reference, see Buchner J. Hsp90 & Co.—a holding for folding. *Trends Biochem Sci.* 1999;24:136–142.)

nascent and denatured polypeptides to fold or refold while sequestered in a cylindrical cavity protected from the complex environment of the cytoplasm. Although 85% of newly synthesized bacterial proteins fold spontaneously or with the assistance of Hsp70s, the remainder require the more isolated folding environment provided by chaperonins (Fig. 12.11). The mechanism of chaperonins is best understood for *Escherichia coli* **GroEL** and its co-chaperonin **GroES.** They assist with folding of nascent polypeptides, which in bacteria occurs largely after translation is complete.

The GroEL/GroES complex consists of a cylinder with a central cavity composed of GroEL and a cap structure made of GroES. GroEL forms two rings of seven identical subunits. Mitochondrial (Hsp60/Hsp10), chloroplast (Cpn60/Cpn10), and eukaryotic chaperonins (TriC) are similar in design but more elaborate than GroEL/GroES, containing up to eight different gene products. This complexity represents evolutionary diversification for regulation of chaperonin function.

ATP binding and hydrolysis set the tempo for folding cycles. Unfolded polypeptides interact with hydrophobic patches on the inner wall of the GroEL cylinder. Cooperative binding of ATP to each of the subunits in one of the two rings of seven changes their conformation (compare the upper and lower rings in Fig. 12.14B), expanding the internal volume by twofold and favoring binding of a heptameric ring of 10-kD GroES subunits. This closes the top of the cylinder and creates a folding cavity for proteins up to approximately 70 kD. After ATP hydrolysis on the ring surrounding the folding protein and ATP binding to the opposite ring of seven GroEL subunits, the GroES cap releases, and the cage opens. Folded polypeptides escape into the cytoplasm, whereas incompletely folded intermediates can rebind GroEL for another attempt at folding.

ACKNOWLEDGMENT

We thank Peter Moore for his suggestions on revisions to this chapter.

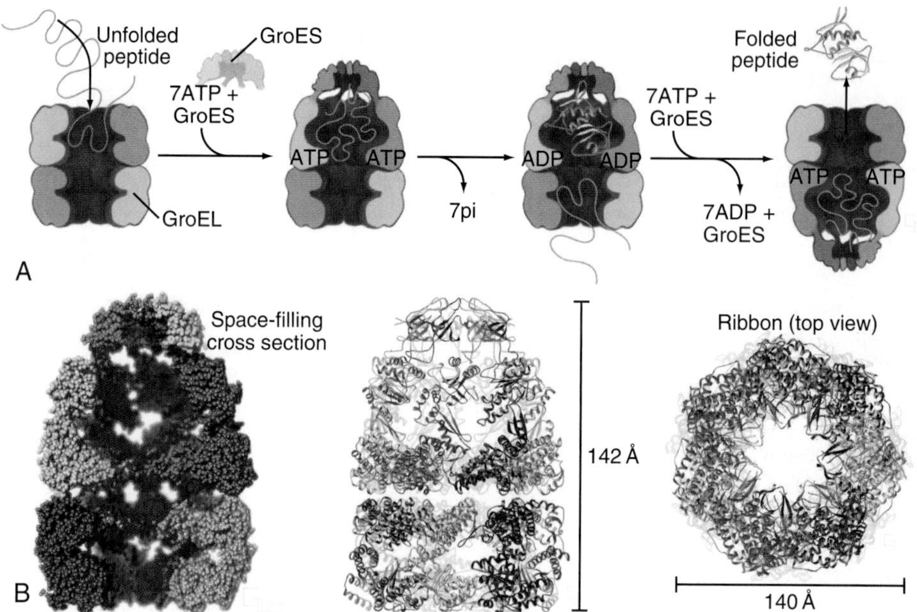

FIGURE 12.14 CHAPERONIN-MEDIATED FOLDING BY GroEL AND GroES. A, One folding cycle. **B,** Crystal structure of GroEL with a GroES cap bound to the upper, adenosine triphosphate (ATP)-bound ring of seven subunits. Unfolded polypeptides bind the rim of an uncapped ring. Cooperative binding of ATP to each of the seven GroEL subunits in one ring changes their conformation, favors GroES binding, and doubles the volume of the central cavity, where the protein folds. Following ATP hydrolysis, binding of ATP and GroES to the lower ring structure dissociates the upper GroES and discharges the folded protein. (**B,** Modified from Xu Z, Horwich AL, Sigler PB: The crystal structure of the asymmetric GroEL-GroES-(ADP)7 chaperonin complex. *Nature.* 1997;388:741–750. See PDB file 1AON.)

SELECTED READINGS

Castellano LM, Shorter J. The surprising role of amylod fibrils in HIV infection. *Biology (Basel)*. 2012;1:58-80.

Chiti F, Dobson CM. Protein misfolding, functional amyloid, and human disease. *Annu Rev Biochem*. 2006;75:333-366.

Daggett V, Fersht AR. Is there a unifying mechanism for protein folding? *Trends Biochem Sci*. 2003;28:18-25.

Dobson CM. Protein folding and misfolding. *Nature*. 2003;426:884-890.

Hayer-Hartl M, Bracher A, Hartl FU. The GroEL-GroES chaperonin machine: A nano-cage for protein folding. *Trends Biochem Sci*. 2016;41:62-76.

Hinnebusch AG. Molecular mechanism of scanning and start codon selection in eukaryotes. *Microbiol Mol Biol Rev*. 2011;75:434-467.

Ibba M, Söll D. Aminoacyl-tRNAs: Setting the limits of the genetic code. *Genes Dev*. 2004;18:731-738.

Ingolia NT. Ribosome profiling: new views of translation, from single codons to genome scale. *Nat Rev Genet*. 2014;15:205-213.

Kim YE, Hipp MS, Bracher A, et al. Molecular chaperone functions in protein folding and proteostasis. *Annu Rev Biochem*. 2013;82:323-355.

Liu T, Kaplan A, Alexander L, et al. Direct measurement of the mechanical work during translocation by the ribosome. *Elife*. 2014;3:e03406.

May BC, Govaerts C, Prusiner SB, Cohen FE. Prions: So many fibers, so little infectivity. *Trends Biochem Sci*. 2004;29:162-165.

Mazumder B, Seshadri V, Fox PL. Translational control by the 3′-UTR: The ends specify the means. *Trends Biochem Sci*. 2003;28:91-98.

Moore PB. How should we think about the ribosome? *Annu Rev Biophys*. 2012;41:1-19.

Mumtaz MA, Couso JP. Ribosomal profiling adds new coding sequences to the proteome. *Biochem Soc Trans*. 2015;43:1271-1276.

Myers JK, Oas TG. Mechanisms of fast protein folding. *Annu Rev Biochem*. 2002;71:783-815.

Ow SY, Dunstan DE. A brief overview of amyloids and Alzheimer's disease. *Protein Sci*. 2014;23:1315-1331.

Pearl LH, Prodromou C. Structure and mechanism of the Hsp90 molecular chaperone machinery. *Annu Rev Biochem*. 2006;75:271-294.

Piper M, Holt C. RNA translation in axons. *Annu Rev Cell Dev Biol*. 2004;20:505-523.

Ramakrishnan V. The ribosome emerges from a black box. *Cell*. 2014;159:979-984.

Ramakrishnan Lab. Ribosome Structure and Function. Movies and Overview Figures of the Ribosome. <http://www.mrc-lmb.cam.ac.uk/ribo/homepage/mov_and_overview.html>.

Rodnina MV. The ribosome as a versatile catalyst: reactions at the peptidyl transferase center. *Curr Opin Struct Biol*. 2013;23:595-602.

Saibil HR. Biochemistry. Machinery to reverse irreversible aggregates. *Science*. 2013;339:1040-1041.

Saio T, Guan X, Rossi P, Economou A, Kalodimos CG. Structural basis for protein antiaggregation activity of the trigger factor chaperone. *Science*. 2014;344:1250494.

Selkoe DJ. Folding proteins in fatal ways. *Nature*. 2003;426:900-904.

Sonenberg N, Dever TE. Eukaryotic translation initiation factors and regulators. *Curr Opin Struct Biol*. 2003;13:56-63.

Voorhees RM, Ramakrishnan V. Structural basis of the translational elongation cycle. *Annu Rev Biochem*. 2013;82:203-236.

Wilkie GS, Dickson KS, Gray NK. Regulation of mRNA translation by 5′- and 3′-UTR-binding factors. *Trends Biochem Sci*. 2003;28:182-188.

Xue S, Barna M. Specialized ribosomes: a new frontier in gene regulation and organismal biology. *Nat Rev Mol Cell Biol*. 2012;13:355-369.

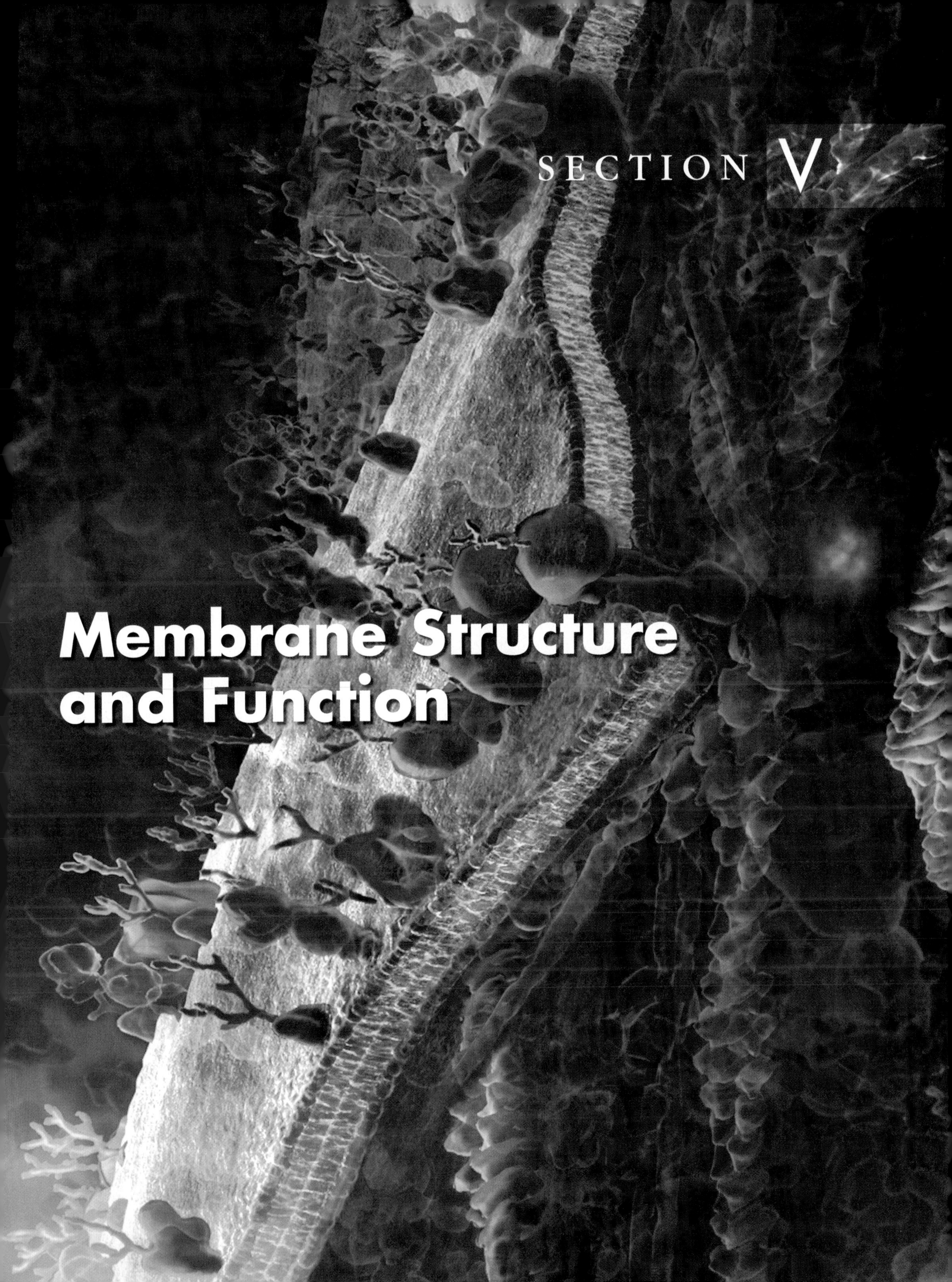

SECTION V

Membrane Structure and Function

SECTION V OVERVIEW

Life, as we know it, depends on a thin membrane that separates each cell from the surrounding world. These membranes, composed of two layers of lipids, are generally impermeable to ions and macromolecules. Proteins embedded in the lipid membrane facilitate the movement of ions, allowing cells to create an internal environment different from that outside. Membranes also subdivide the cytoplasm of eukaryotic cells into compartments called organelles. Chapter 13 introduces the features that are shared by all biological membranes: a **bilayer** of lipids, **integral proteins** that cross the bilayer, and **peripheral proteins** associated with the surfaces.

Membranes are a planar sandwich of two layers of lipids that behave like two-dimensional fluids. Each lipid has a polar group coupled to hydrocarbon tails that are insoluble in water. The hydrocarbon tails are in the middle of the membrane bilayer with polar head groups exposed to water on both surfaces. Despite the rapid, lateral diffusion of lipids in the plane of the membrane, the hydrophobic interior of the bilayer is poorly permeable to ions and macromolecules. This impermeability makes it possible for cellular membranes to form barriers between the external environment, cytoplasm, and organelles. The selectively permeable membrane around each organelle allows the creation of a unique interior space for specialized biochemical reactions that contribute to the life of the cell. Chapters 18 to 23 consider in detail all the organelles, including mitochondria, chloroplasts, peroxisomes, endoplasmic reticulum, Golgi apparatus, lysosomes, and the vesicles of the secretory and endocytic pathways.

Peripheral membrane proteins found on the surfaces of the bilayer often participate in enzyme and signaling reactions. Others form a membrane skeleton on the cytoplasmic surface that reinforces the fragile lipid bilayer and attaches it to cytoskeletal filaments.

Integral membrane proteins that cross lipid bilayers feature prominently in all aspects of cell biology. Some are enzymes that synthesize lipids for biological membranes (Chapter 20). Others serve as adhesion proteins that allow cells to interact with each other or extracellular substrates (see Chapter 30). Because cells need to sense hormones and many other molecules that cannot penetrate a lipid bilayer, they have evolved thousands of protein receptors that span the bilayer (Chapter 24). Hormones or other extracellular signaling molecules bind selectively to receptors exposed on the cell surface. The energy from binding is used to transmit a signal across the membrane and regulate biochemical reactions in the cytoplasm (Chapters 25 to 27).

A large fraction of the energy that is consumed by organs such as our brains is used to create ion gradients across membranes. Several large families of integral membrane proteins control the movement of ions and other solutes across membranes. Chapter 14 introduces three families of **pumps** that use adenosine triphosphate (ATP) hydrolysis as the source of energy to transport ions or solutes up concentration gradients across membranes. For example, pumps in the plasma

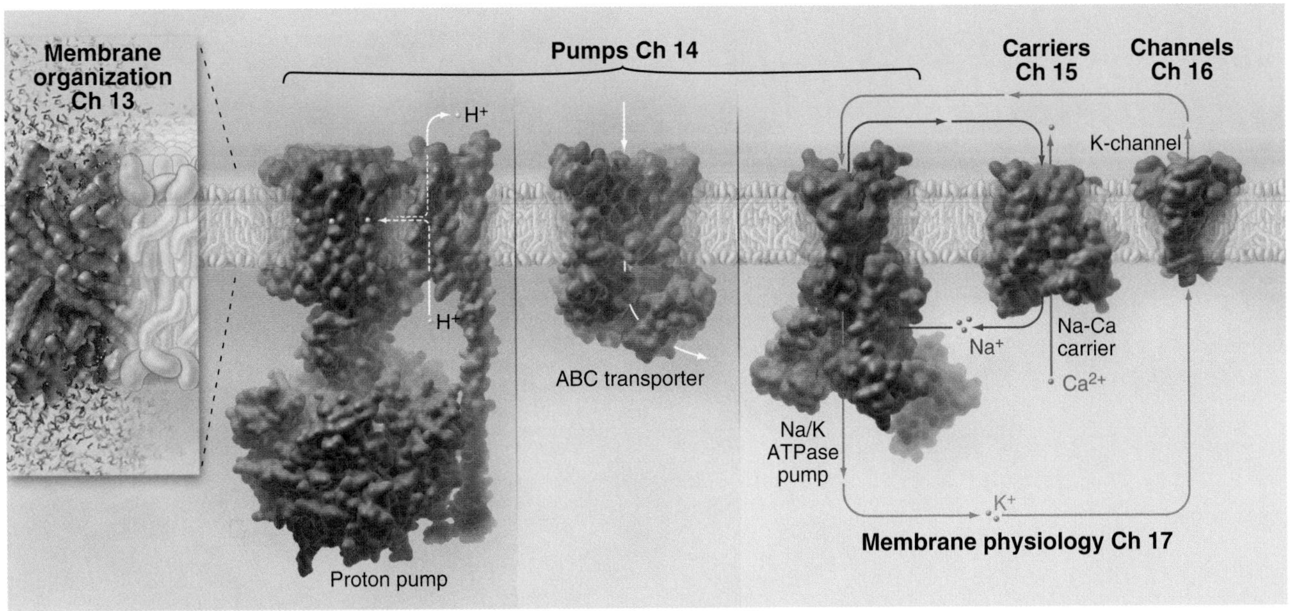

membranes of animal cells use ATP hydrolysis to expel Na^+ and concentrate K^+ in the cytoplasm. Another type of pump creates the acidic environment inside lysosomes. A related pump in mitochondria runs in the opposite direction, taking advantage of a proton gradient across the membrane to synthesize ATP. A third family, called ABC transporters, use ATP hydrolysis to move a wide variety of solutes across plasma membranes.

Carrier proteins (Chapter 15) facilitate the movement of ions and nutrients across membranes, allowing them to move down concentration gradients much faster than they can penetrate the lipid bilayer. Some carriers couple movement of an ion such as Na^+ down its concentration gradient to the movement of a solute such as glucose up a concentration gradient into the cell. Carriers change their shape reversibly, opening and closing "gates" to transport their cargo across the membrane one molecule at a time.

Channels are transmembrane proteins with selective pores that allow ions, water, glycerol, or ammonia to move very rapidly down concentration gradients across membranes (Chapter 16). Taking advantage of ion gradients created by pumps and carriers, cells selectively open ion channels to create electrical potentials across the plasma membrane and some organelle membranes. Many channels open and close their pores in response to local conditions. The electrical potential across the membrane regulates voltage-gated cation channels. Binding of a chemical ligand opens other channels. For instance, nerve cells secrete small organic ions (called neurotransmitters) to stimulate other nerve cells and muscles by binding to an extracellular domain of cation channels. The bound neurotransmitter opens the pore in the channel. In the cytoplasm, other organic ions and Ca^{2+} can also regulate channels. Cyclic nucleotides open plasma membrane channels in cells that respond to light and odors. Inositol triphosphate and Ca^{2+} control channels that release Ca^{2+} from the endoplasmic reticulum. Through these diverse activities channels participate in all aspects of membrane physiology.

All living organisms depend on combinations of pumps, carriers, and channels for many physiological functions (Chapter 17). Cells use ion concentration gradients produced by pumps as a source of potential energy to drive the uptake of nutrients through plasma membrane carriers. Epithelial cells lining our intestines combine different carriers and channels in their plasma membranes to transport sugars, amino acids, and other nutrients from the lumen of the gut into the blood. Many organelles use carriers driven by ion gradients for transport. Most cells use ion channels and transmembrane ion gradients to create an electrical potential across their plasma membranes. Nerve and muscle cells create fast-moving fluctuations in the plasma membrane potential for high-speed communication; operating on a millisecond time scale, voltage-gated ion channels produce waves of membrane depolarization and repolarization called action potentials. Each of our physiological systems depends on this cooperation among pumps, carriers and channels.

Our abilities to perceive our environment, think, and move depend on transmission of electrical impulses between nerve cells and between nerves and muscles at specialized structures called synapses. When an action potential arrives at a synapse, voltage-gated Ca^{2+} channels trigger the secretion of neurotransmitters. In less than a millisecond, the neurotransmitter stimulates ligand-gated cation channels to depolarize the plasma membrane of the receiving cell. Muscle cells respond with an action potential that sets off contraction. Nerve cells in the central nervous system integrate inputs from many synapses before producing an action potential. Pumps and carriers cooperate to reset conditions after each round of synaptic transmission.

CHAPTER 13

Membrane Structure and Dynamics

Membranes composed of lipids and proteins form the barrier between each cell and its environment. Membranes also partition the cytoplasm of eukaryotes into compartments, including the nucleus and membrane-bounded organelles. Each type of membrane is specialized for its various functions, but all biological membranes have much in common: a planar fluid **bilayer** of lipid molecules, **integral membrane proteins** that cross the lipid bilayer, and **peripheral membrane proteins** on both surfaces. This chapter opens with a discussion of the lipid bilayer. It then considers examples of integral and peripheral membrane proteins before concluding with a discussion of the dynamics of both lipids and proteins. The following three chapters introduce three large families of membrane proteins: pumps, carriers, and channels. Chapter 17 explains how pumps, carriers, and channels cooperate in a variety of physiological processes. Chapters 24 and 30 cover plasma membrane receptor proteins.

Development of Ideas About Membrane Structure

Our current understanding of membrane structure began with E. Overton's proposal in 1895 that cellular membranes consist of lipid bilayers (Fig. 13.1A). In the 1920s it was found that the lipids extracted from the plasma membrane of red blood cells spread out in a monolayer on the surface of a tray of water to cover an area sufficient to surround the cell twice. (Actually, offsetting errors—incomplete lipid extraction and an underestimation of the membrane area—led to the correct answer!) X-ray diffraction experiments in the early 1970s established definitively that membrane lipids are arranged in a bilayer.

During the 1930s, cell physiologists realized that a simple lipid bilayer could not explain the mechanical properties of the plasma membrane, so they postulated a surface coating of proteins to reinforce the bilayer

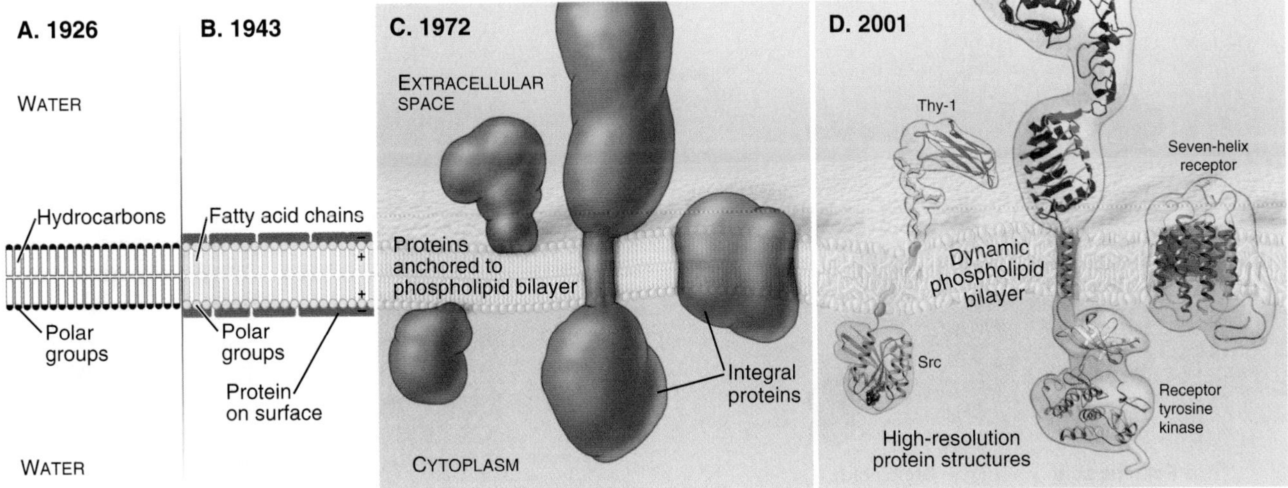

FIGURE 13.1 DEVELOPMENT OF CONCEPTS IN MEMBRANE STRUCTURE. A, Gorder and Grendel model from 1926. **B,** Davson and Danielli model from 1943 reflecting beliefs of the time about the small sizes of proteins. **C,** Singer and Nicholson fluid mosaic model from 1972. **D,** Contemporary model with peripheral and integral membrane proteins. The lipid bilayer shown here and used throughout the book is based on a dynamic computational model (Fig. 13.5). The density of proteins in actual membranes is higher than shown here.

(Fig. 13.1B). Early electron micrographs of thin sections of cells strengthened this view, since all membranes appeared as a pair of dark lines (interpreted as surface proteins and carbohydrates) separated by a lucent area (interpreted as the lipid bilayer). By the early 1970s, two complementary approaches showed that proteins cross the lipid bilayer. First, electron micrographs of membranes that are split in two while frozen (a technique called freeze-fracturing; see Fig. 6.6C) revealed protein particles embedded in the lipid bilayer. Later, chemical labeling showed that many membrane proteins are exposed on both sides of the bilayer. Light microscopy with fluorescent tags demonstrated that membrane lipids and some membrane proteins diffuse in the plane of the membrane. Quantitative spectroscopic studies showed that lateral diffusion of lipids is rapid but that flipping from one side of a bilayer to the other is slow. The fluid mosaic model of membranes (Fig. 13.1C) incorporated this information, showing transmembrane proteins floating in a fluid lipid bilayer. Subsequent work revealed structures of many proteins that span the lipid bilayer, the existence of lipid anchors on some membrane proteins, and a network of cytoplasmic proteins that restricts the motion of many integral membrane proteins (Fig. 13.1D). The density of proteins in actual membranes is higher than illustrated in the figure.

Lipids

Lipids form the framework of biological membranes, anchor soluble proteins to the surfaces of membranes, store energy, and carry information as extracellular hormones and as intracellular second messengers. Lipids are organic molecules generally less than 1000 Da in size that are much more soluble in organic solvents than in water. They consist predominantly of aliphatic or aromatic hydrocarbons.

This chapter explains the structures of the major lipids found in biological membranes and how the hydrophobic effect drives lipids to self-assemble stable bilayers. Membranes also contain hundreds of minor lipid species, some of which may also have important biological functions.

Phosphoglycerides

Phosphoglycerides (also called glycerophospholipids) are the main constituents of membrane bilayers (Fig. 13.2). (These lipids are often called phospholipids, an

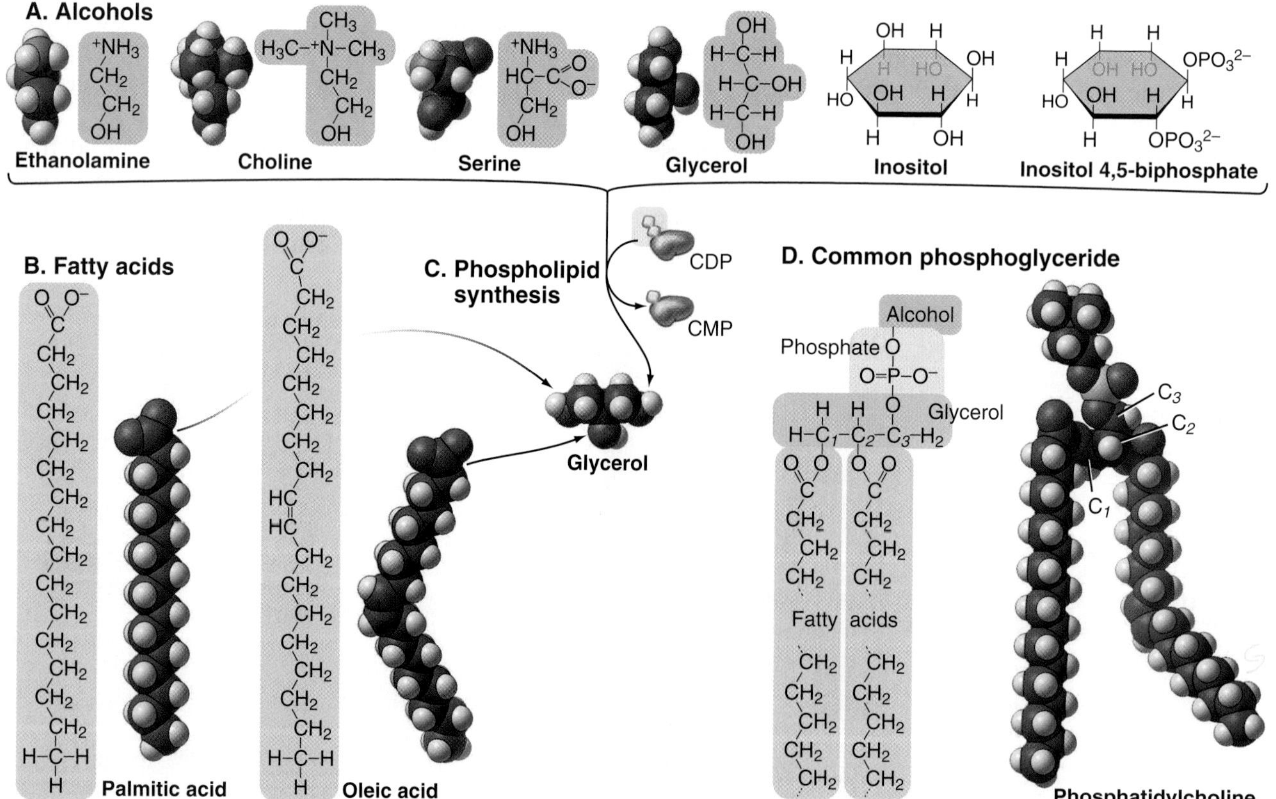

FIGURE 13.2 STRUCTURE AND SYNTHESIS OF PHOSPHOGLYCERIDES. A, Stick figures and space-filling models of the alcohol head groups. **B,** Stick figures and space-filling models of two fatty acids. **C,** An alcohol, glycerol, and two fatty acids combine to make a phosphoglyceride. In some cases cytidine diphosphate (CDP) provides the phosphate linking glycerol to the alcohol. CMP, cytidine monophosphate. **D,** Diagram of a phosphoglyceride and a space-filling model of phosphatidylcholine.

TABLE 13.1 Common Fatty Acids of Membrane Lipids

Name	Carbons	Double Bonds (Positions)
Myristate	14	0
Palmitate	16	0
Palmitoleate	16	1 (Δ9)
Stearate	18	0
Oleate	18	1 (Δ9)
Linoleate	18	2 (Δ9, Δ12)
Linolenate	18	3 (Δ9, Δ12, Δ15)
Arachidonate	20	4 (Δ5, Δ8, Δ11, Δ14)

imprecise term, as other lipids contain phosphate.) Phosphoglycerides have three parts: a three-carbon backbone of glycerol, two long-chain fatty acids esterified (or attached via an ether link in Archaea) to hydroxyl groups on carbons 1 and 2 (C_1 and C_2) of the glycerol, and phosphoric acid esterified to the C_3 hydroxyl group of glycerol. Most also have an alcohol head group esterified to the phosphate. Fatty acids have a carboxyl group at one end of an aliphatic chain of 13 to 19 additional carbons (Table 13.1). More than half of the fatty acids in membranes have one or more double bonds.

Fatty acids and phosphoglycerides are **amphiphilic,** as they have both **hydrophobic** (fears water) and **hydrophilic** (loves water) parts. The aliphatic chains of fatty acids are hydrophobic. The carboxyl groups of fatty acids and the head groups of phosphoglycerides are hydrophilic. The hydrophobic effect (see Fig. 4.5) drives amphiphilic phosphoglycerides to assemble bilayers (see later).

Cells make more than 100 major phosphoglycerides using many different fatty acids and esterifying one of five different alcohols to the phosphate. In general, the fatty acids on C_1 have no or one double bond, whereas the fatty acids on C_2 have two or more double bonds. Each double bond creates a permanent bend in the hydrocarbon chain that contributes to the fluidity of the bilayer. The alcohol head groups give phosphoglycerides their names:

phosphatidic acid [PA] (no head group)
phosphatidylglycerol [PG] (glycerol head group)
phosphatidylethanolamine [PE] (ethanolamine head group)
phosphatidylcholine [PC] (choline head group)
phosphatidylserine [PS] (serine head group)
phosphatidylinositol [PI] (inositol head group)

The various head groups confer distinctive properties to the various phosphoglycerides. All head groups have a negative charge on the phosphate esterified to glycerol. Neutral phosphoglycerides—PE and PC—have a positive charge on their nitrogens, giving them a net charge of zero. PS has extra positive and negative charges, giving it a net negative charge like the other acidic phosphoglycerides (PA, PG, and PI). PI can be modified by esterifying one to five phosphates to the hexane ring hydroxyls. These **polyphosphoinositides** are highly negatively charged.

The complicated metabolism of phosphoglycerides can be simplified as follows: Enzymes can interconvert all phosphoglyceride head groups and remodel fatty acid chains. For example, three successive enzymatic methylation reactions convert PE to PC, whereas another enzyme exchanges serine for ethanolamine, converting PS to PE. Other enzymes exchange fatty acid chains after the initial synthesis of a phosphoglyceride. These enzymes are located on the cytoplasmic surface of the smooth endoplasmic reticulum. Biochemistry texts provide more details of these pathways.

Several minor membrane phospholipids are variations on this general theme. Plasmalogens have a fatty acid linked to carbon 1 of glycerol by an ether bond rather than an ester bond. They serve as sources of arachidonic acid for signaling reactions (see Fig. 26.9). Cardiolipin has two glycerols esterified to the phosphate of PA.

Sphingolipids

Sphingolipids get their name from **sphingosine,** a nitrogen-containing base synthesized from serine and a fatty acid (Fig. 13.3). Sphingosine acts like the structural counterpart of glycerol plus one fatty acid of phosphoglycerides. Sphingosine carbons 1 to 3 have polar substituents. A double bond between C_4 and C_5 begins the hydrocarbon tail. Two variable features distinguish the various sphingolipids: the fatty acid (often lacking double bonds) attached by an amide bond to C_2 and the nature of the polar head groups esterified to the hydroxyl on C_1.

Most sugar-containing lipids of biological membranes are sphingolipids. The head groups of **glycosphingolipids** consist of one or more sugars. Some are neutral; others are negatively charged. All of these head groups lack phosphate. Sugar head groups of some glycosphingolipids serve as receptors for viruses. Alternatively, a phosphate ester can link a base to C_1. These so-called **sphingomyelins** have phosphorylcholine or phosphoethanolamine head groups just like PC and PE. Receptor-activated enzymes remove phosphorylcholine from sphingomyelin to produce the second messenger ceramide (see Fig. 26.11).

Sphingolipids are longer than most phosphoglycerides and much more abundant in the thicker plasma membrane than in membranes inside cells (see Fig. 21.3). The hydrocarbon tails of sphingosine and the fatty acid contribute to the hydrophobic bilayer, and polar head groups are on the surface.

Sterols

Sterols are the third major class of membrane lipids. **Cholesterol** (Fig. 13.4) is the major sterol in animal

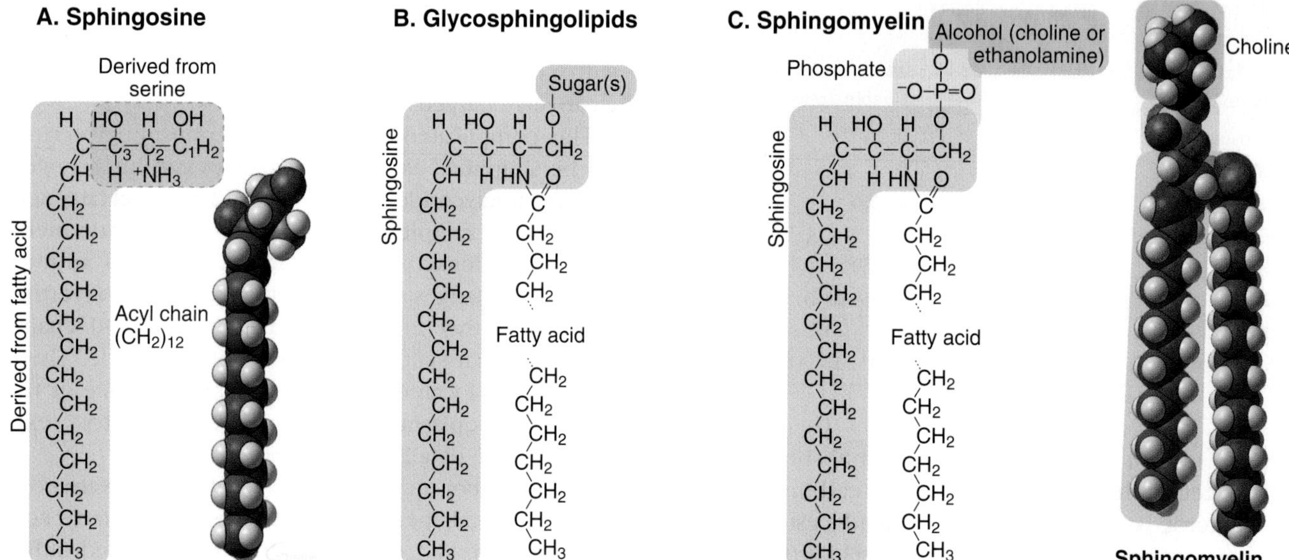

FIGURE 13.3 SPHINGOLIPIDS. A, Stick figure and space-filling model of sphingosine. **B,** Diagram of the parts of a glycosphingolipid. Ceramide has a fatty acid but no sugar. **C,** Stick figure and space-filling model of sphingomyelin.

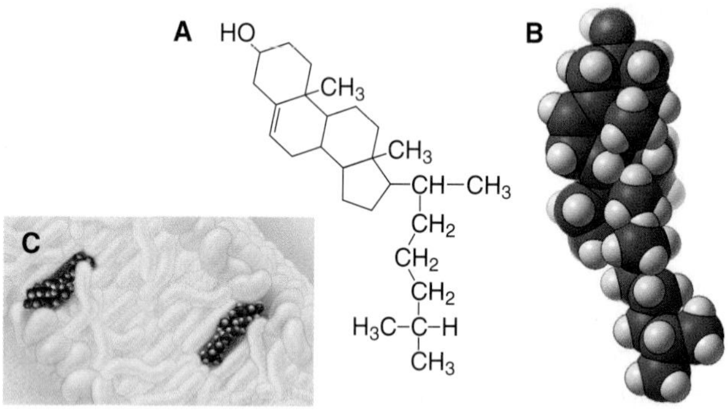

FIGURE 13.4 CHOLESTEROL. A, Stick figure. **B,** Space-filling model. **C,** Disposition of cholesterol in a lipid bilayer with the hydroxyl oriented toward the surface. The rigid sterol nucleus tends to order fluid bilayers in the region between C_1 and C_{10} of the fatty acids but promotes motion of the fatty acyl chains deeper in the bilayer owing to its wedge shape.

plasma membranes, with lower concentrations in internal membranes. Plants, lower eukaryotes, and bacteria have other sterols in their membranes. The rigid four-ring structure of cholesterol is apolar, so it inserts into the core of bilayers with the hydroxyl on C_3 oriented toward the surface.

Cholesterol is vital to metabolism, being situated at the crossroads of several metabolic pathways, including those that synthesize steroid hormones (such as estrogen, testosterone, and cortisol), vitamin D, and bile salts secreted by the liver. Cholesterol itself is synthesized (see Fig. 20.15) from **isopentyl** (5-carbon) building blocks that form 10-carbon **(geranyl),** 15-carbon **(farnesyl),** and 20-carbon **(geranylgeranyl) isoprenoids.** As is described later, these isoprenoids are used as hydrocarbon anchors for many important membrane-associated proteins. Isoprenoids are also precursors of natural rubber and of cofactors present in visual pigments.

Glycolipids

Cells have three types of glycolipids: (a) sphingolipids (the predominant form), (b) glycerol glycolipids with a sugar chain attached to the hydroxyl on C_3 of a diglyceride, and (c) **glycosylphosphatidylinositols (GPIs).** Some GPIs simply have a short carbohydrate chain on the hydroxyl of inositol C_2. Others use a short sugar chain to link C_6 of PI to the C-terminus of a protein (Fig. 13.10C).

Triglycerides

Triglycerides are simply glycerol with fatty acids esterified to all three carbons. Lacking a polar head group, they are not incorporated into membrane bilayers. Instead, triglycerides form oily droplets in the cytoplasm of cells to store fatty acids as reserves of metabolic energy (see Fig. 28.3). Mitochondria oxidize fatty acids and convert the energy in their covalent bonds into adenosine triphosphate (ATP) (see Fig. 19.4).

Physical Structure of the Fluid Membrane Bilayer

Physical Properties of Bilayers of a Single Lipid

In an aqueous environment, amphiphilic lipids spontaneously self-assemble into ordered structures in microseconds. The amphiphilic nature of phosphoglycerides and sphingolipids favors formation of lamellar bilayers, planar structures with fatty acid chains lined up more or less normal to the surface and polar head groups on the

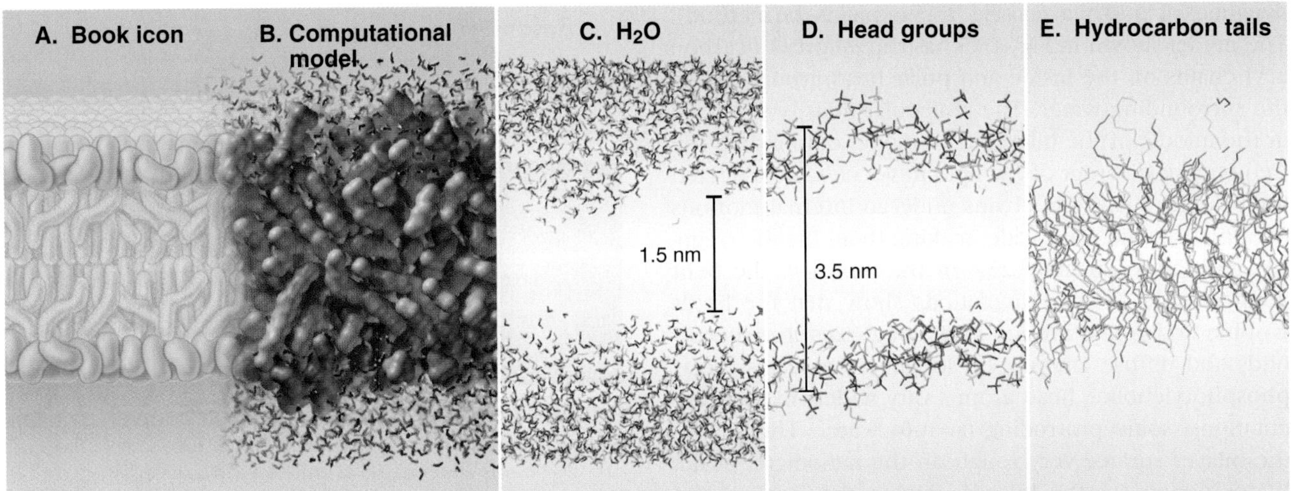

FIGURE 13.5 COMPUTATIONAL MODEL OF A HYDRATED DIMYRISTOYLPHOSPHATIDYLCHOLINE BILAYER. A, Icon of the lipid bilayer used throughout this book, based on the model shown in **B. B,** Space-filling model of all the lipid atoms in the simulation. Stick figures of the water molecules are *red.* The polar regions of phosphatidylcholine (PC) from the carbonyl oxygen to the choline nitrogen are *blue.* Hydrocarbon tails are *yellow.* **C,** Stick figures of the water molecules only. **D,** Stick figures of the polar regions of PC from the carbonyl oxygen to the choline nitrogen only. **E,** Stick figures of the hydrocarbon tails only. This model was calculated from first principles starting with 100 PC molecules (based on an x-ray diffraction structure of PC crystals) in a regular bilayer with 1050 molecules of bulk phase water on each side. Taking into account surface tension and distribution of charge on lipid and water, the computer used simple Newtonian mechanics to simulate the molecular motion of all atoms on a picosecond time scale. After less than 100 picoseconds of simulated time, the liquid phase of the lipids appeared. The model shown here is after 300 picoseconds of simulated time. Such models account for most molecular parameters (electron density, surface roughness, distance between phosphates of the two halves, area per lipid [0.6 nm^2], and depth of water penetration) of similar bilayers obtained by averaging techniques, including nuclear magnetic resonance (NMR), x-ray diffraction, and neutron diffraction. (Courtesy E. Jakobsson, University of Illinois, Urbana. Modified from Chiu S-W, Clark M, Balaji V, et al. Incorporation of surface tension into molecular dynamics simulation of an interface: a fluid phase lipid bilayer membrane. *Biophys J.* 1995;69:1230–1245.)

surfaces exposed to water (Fig. 13.5A). The two halves of the bilayer are called leaflets.

Bilayer formation is favored energetically by the increase in entropy when the hydrophobic acyl chains interact with each other and exclude water from the core of the bilayer. The head groups of PC and PS have about the same cross-sectional areas as the aliphatic tails, so they are approximately cylindrical in shape, appropriate for flat bilayers. The hydrophobic effect is so strong that it drives lipid head groups into close packing, depleting water from the head group layer. The area per lipid molecule of a given type tends to be constant, so bilayers bend in response if molecules are added asymmetrically to one leaflet. The smaller head group makes PE adopt a slightly conical shape, favoring a curved bilayer.

Bilayers of pure lipids are of two physical states depending on the temperature. The **liquid disordered phase** is a flexible, two-dimensional fluid with disordered acyl chains and the lipids diffusing rapidly (Fig. 13.5A). Tight packing of acyl chains in the **gel state** limits lateral diffusion. Low temperatures favor the gel state. Above a critical temperature the gel melts and transitions to the liquid disordered phase.

The transition temperature depends on the saturation and lengths of the acyl chains. Short acyl chains favor the liquid state. Fatty acids with 18 or more carbons are solid at physiological temperatures unless they contain double bonds that create a permanent bend and favor the liquid state by preventing tight packing of fatty acid tails in the middle of the bilayer.

Phosphoglycerides in biological membranes are largely in the liquid phase owing to their compositions. The C14 and C16 fatty acids are saturated, but C18 fatty acids usually have one to three double bonds and C20 fatty acids have four double bonds (Table 13.1). The phosphoglycerides in particular biological membranes vary in both the lengths and saturation of the acyl chains. For example, abundant polyunsaturated acyl chains in synaptic vesicles (see Fig. 17.8) make the bilayer flexible and facilitate membrane traffic.

Biophysical methods, including fluorescence recovery after photobleaching (Fig. 13.12), show that phosphoglycerides diffuse rapidly in the plane of a bilayer with a lateral diffusion coefficient *(D)* of about 1 $\mu m^2\ s^{-1}$. Given that the rate of diffusion is $2(Dt)^{1/2}$ (t = time), a phosphoglyceride moves laterally about 1 µm/s in the plane of the membrane, fast enough to circumnavigate the membrane of a bacterium in a few seconds. Rarely ($\sim10^{-5}\ s^{-1}$, corresponding to a half-time of 20 hours), a neutral phosphoglyceride, such as PC, flips unassisted from one side of a bilayer to the other. Flipping of charged phosphoglycerides is even slower.

A computational method called molecular dynamics simulation is used to study the organization and dynamics of lipid bilayers (Fig. 13.5 explains the method). The model shown in Fig. 13.5 has the (short) 14-carbon acyl chains on the inside and polar head groups facing the surrounding water. The molecular density is lowest in the middle of the bilayer. The model emphasizes the tremendous disorder of the lipid molecules, as expected for a liquid. Fatty acid chains undergo internal motions on a picosecond time scale, making them highly irregular, with approximately 25% of the bonds in the bent configuration. Longer simulations show that the lipids wobble and rotate around their long axes in nanoseconds and diffuse laterally on longer time scales. Polar phosphorylcholine head groups vary widely in their orientations, some protruding far into water. This makes the bilayer surface very rough on the nanometer scale. Water penetrates the bilayer only to the level of the deepest carbonyl oxygens, leaving a dehydrated layer approximately 1.5 nm thick in the middle of the bilayer.

Bilayers of phosphoglycerides have an electrical potential between the hydrocarbon (positive inside) and the aqueous phase, arising from the orientations of the carbonyl groups and the tendency of water molecules near the bilayer to orient with their positive dipole toward the hydrocarbon interior. These factors dominate over an oppositely oriented electrical dipole between the P and N atoms of the head groups. This inside positive potential may contribute to the barrier to the transfer of positively charged ions and polypeptides across membranes.

Despite the disorder and lateral movement of the molecules, bilayers of phosphoglycerides are stable and impermeable to polar or charged compounds, even those as small as Na^+ or Cl^-. This poor electrical conductivity is essential for many biological processes (see Fig. 17.6). Small, uncharged molecules, such as water, ammonia, and glycerol, penetrate the hydrophobic core in small numbers passing only slowly across bilayers and much more rapidly through channels (see Figs. 16.14 and 16.15).

Although bilayers neither stretch nor compress readily, they are very flexible, owing to rapid fluctuations in the arrangement of the lipids. Molecular dynamics simulations accurately reproduce these mechanical properties. Thus, one can draw out a narrow tube of membrane with little resistance by pulling gently on a vesicle composed of a simple bilayer (Fig. 13.6).

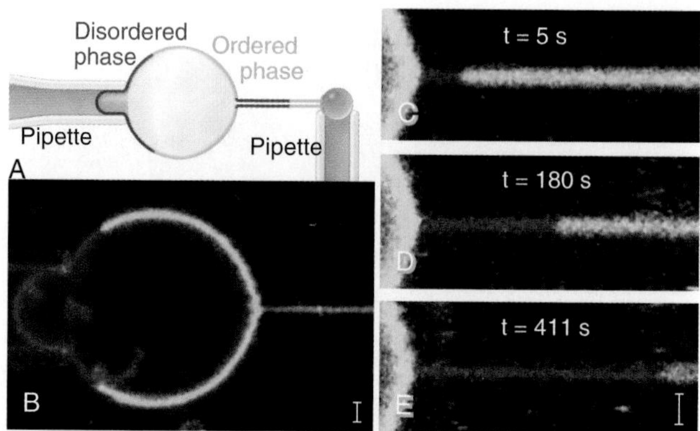

FIGURE 13.6 LIPID SORTING IN DOMAINS DRIVEN BY MEMBRANE CURVATURE. A, Schematic of the experiment. The giant lipid vesicle was composed of 37 mol% 1,2-dipalmitoyl-sn-glycero-3-phosphocholine, 33 mol% cholesterol, 30 mol% 1,2-dioleoyl-sn-glycero-3-phosphocholine and 1 mol% of ganglioside GM_1. This mixture of lipids spontaneously sorts into two domains: a disordered liquid domain marked with PE tagged with a red fluorescent dye; and an ordered liquid domain marked with a protein tagged with a green fluorescent dye that binds GM_1. A suction micropipette on the left holds the vesicle. A second pipette pulled a narrow tube of membrane from the ordered domain. **B,** Immediately after the tube was pulled. **C, D–E,** successive time points showing partitioning of the disordered liquid domain into the tubule. Scale bars are 1 μm.

Physical Properties of Bilayers of Two or More Lipids

All biological membranes consist of mixtures of lipids. Experiments on bilayers reconstituted from purified lipids revealed the physical properties of mixtures of two or more lipids. As expected from first principles, bilayers composed of mixtures of lipids can sort into domains with different compositions. For example, Fig. 13.6 shows a large vesicle formed from cholesterol and 2 forms of PC. The PC with saturated acyl chains segregated into a more **ordered liquid phase** with a high melting temperature distinct from PC with unsaturated acyl chains in a less-ordered liquid phase with a low melting temperature. Cholesterol has opposite effects on liquid and gel phases of phosphoglycerides, favoring the ordered liquid phase above the transition temperature but disrupting the order of the gel state. The presence of cholesterol in a bilayer makes the acyl chains pack more compactly. This allows lateral mobility of the lipids but restricts movement of small molecules across the bilayer.

Sphingolipids are taller than most phosphoglycerides and tend to separate with cholesterol into thicker domains of the bilayer (Fig. 13.7B). These domains are much more abundant in the plasma membrane than in thinner membranes inside cells (see Fig. 21.3).

Structure and Physical Properties of Biological Membranes

Biological membranes vary considerably in lipid composition. In addition to a variety of phosphoglycerides, plasma membranes of animal cells are approximately 35% cholesterol and more than 10% sphingolipids (Fig. 13.7), while internal membranes have lower amounts of these lipids. Prokaryotic membranes have different lipid compositions. Bacterial membranes consist of PE, PG, cardiolipin, and other lipids. Archaeal membranes have a mixture of glycolipids, neutral lipids, and ether-linked lipids, and some include single fatty acids.

Most lipids are distributed asymmetrically between the halves of biological membranes. In animal cell plasma membranes, glycosphingolipids are outside, while PS,

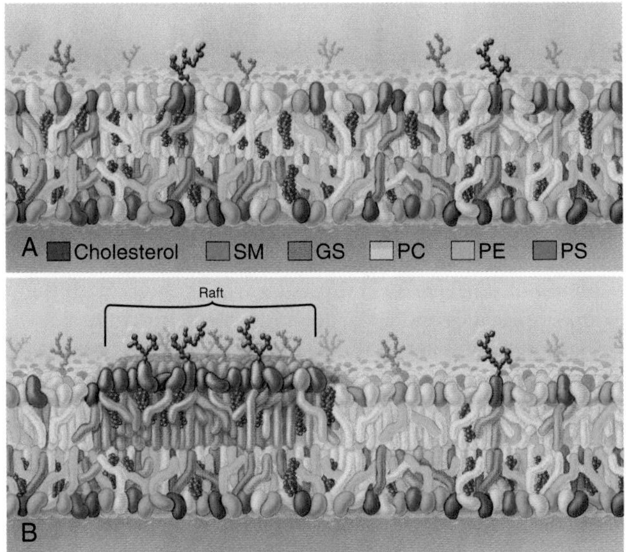

FIGURE 13.7 ASYMMETRICAL DISTRIBUTION OF LIPIDS IN THE PLASMA MEMBRANE OF AN ANIMAL CELL. A, Sphingomyelin (SM) and cholesterol form a small cluster in the external leaflet. GS, glycosphingolipid; PC, phosphatidylcholine; PE, phosphatidylethanolamine; PS, phosphatidylserine. PS is enriched in the inner leaflet. **B,** Lipid raft in the outer leaflet of the plasma membrane enriched in cholesterol and sphingolipids. The lipids in the inner leaflet next to the raft are less well characterized.

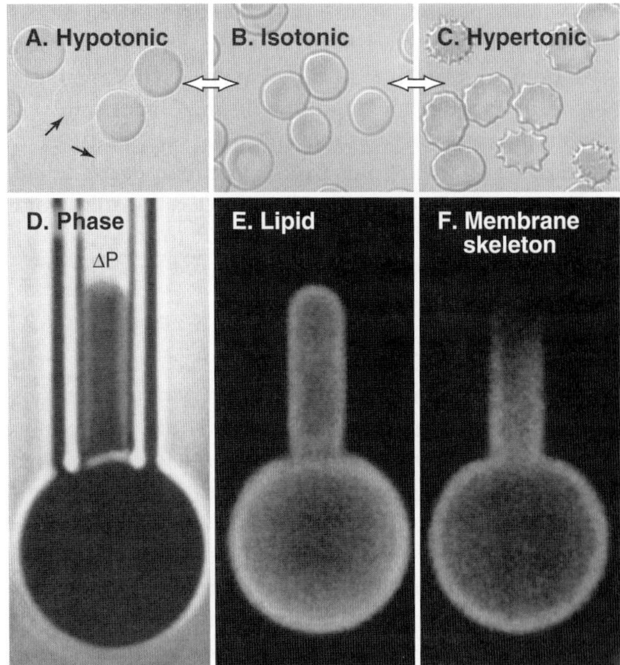

FIGURE 13.8 MEMBRANE DEFORMABILITY ILLUSTRATED BY THE PLASMA MEMBRANE OF HUMAN RED BLOOD CELLS. A–C, Differential interference contrast light micrographs. In an isotonic medium, the cell is a biconcave disk. In a hypotonic medium, water enters the cytoplasm, and the cell rounds up and bursts *(arrows)* if the area of the membrane cannot accommodate the volume. In a hypertonic medium, water leaves the cell and the membrane is thrown into spikes and folds. **D,** Phase-contrast micrograph showing that the plasma membrane is flexible enough to be drawn by suction into a capillary tube. **E,** Fluorescence micrograph showing that membrane lipids, marked with a fluorescent dye, evenly surround the membrane extension. **F,** The elastic membrane skeleton, marked with another fluorescent dye, stretches into the capillary but not to the tip of the extension. (**D–F,** Courtesy N. Mohandas, Lawrence Berkeley Laboratory, Berkeley, CA. For reference, see Discher D, Mohandas N, Evans E. Molecular maps of red cell deformation. *Science.* 1994;266:1032–1035.)

PE, and PI are enriched in the cytoplasmic half of the bilayer (Fig. 13.7). PS asymmetry gives the cytoplasmic surface of the plasma membrane a net negative charge. Less is known about the **lipid asymmetry** of organelle membranes. Transmembrane proteins bind lipids with some specifically, so they also influence the lipid composition of membranes. Cholesterol is distributed more evenly between the two leaflets of membranes because it flips between the two sides of a bilayer on a second time scale. This happens because much less energy is required to bury its single hydroxyl than a polar head group.

Lipid asymmetry is initially established during biosynthesis in the cytoplasmic leaflet of the endoplasmic reticulum (ER) (see Chapter 20). A protein (not yet identified) passively redistributes lipids synthesized on the cytoplasmic side of ER between the halves of the bilayer. Lipid asymmetry is reestablished along the secretory pathway and maintained in the plasma membrane by two families of enzymes that use energy from ATP hydrolysis to move lipid molecules from one side of a bilayer to the other. **Flippases** are P-type adenosine triphosphatase (ATPase) pumps for lipids (see Fig. 14.7). One isoform of the P4 ATPase pumps is found in the Golgi apparatus, while other isoforms are found in secretory vesicles, endosomes, or the plasma membrane to concentrate PS on the cytoplasmic sides of these membranes. A second family called floppases are ABC transporter pumps that move lipids from cytoplasmic leaflet to the extracellular leaflet. The same activity that mixes lipids in the ER also exposes PS on the outer surface of activated platelets (see Fig. 30.14) and on cells marked for phagocytosis during programmed cell death (see Fig. 46.7).

Because they interact favorably, cholesterol and sphingolipids form small domains in the outer leaflet of plasma membranes called **rafts** (Fig. 13.7B). Special invaginations of the plasma membrane called caveolae (see Fig. 22.7) are the best-characterized example of sphingolipid-cholesterol rafts. Some transmembrane proteins, GPI-anchored proteins, and fatty acid–anchored proteins (Fig. 13.10) associate with sphingolipids and cholesterol in artificial bilayers, so rafts are thought to participate in signaling.

Like bilayers of pure phosphoglycerides cellular membranes have limited permeability to ions, high electrical resistance, and the ability to self-seal. Little force is required to deform bilayers into complex shapes. These features are illustrated by the response of a red blood cell plasma membrane to changes in volume (Fig. 13.8). The membrane area is constant, so a reduction in volume throws the membrane into folds, whereas swelling distends it to a spherical shape until it eventually bursts. If osmotic forces rupture a lipid bilayer, it will reseal.

The lipid molecules comprising membranes are not soluble in water, but cytoplasmic lipid-binding proteins can take up specific lipids from a membrane and deliver them to another membrane. This process transfers lipids from their sites of synthesis in the ER to mitochondria as well as between other organelles (see Fig. 20.17).

Membrane Proteins

Proteins are responsible for most membrane functions. The variety of membrane proteins is great, comprising more than one-third of proteins in sequenced genomes. Integral membrane proteins cross the lipid bilayer, and peripheral membrane proteins associate with the inside or outside surfaces of the bilayer. Transmembrane segments of integral membrane proteins interact with hydrocarbon chains of the lipid bilayer and have few hydrophilic residues on these surfaces. Like other soluble proteins, peripheral membrane proteins have hydrophilic residues exposed on their surfaces and a core of hydrophobic residues. Chemical extraction experiments distinguish these two classes of membrane proteins. Alkaline solvents (eg, 0.1 M carbonate at pH 11.3) solubilize most peripheral proteins, leaving behind the lipid bilayer and integral membrane proteins. Detergents, which interact with hydrophobic transmembrane segments, solubilize integral membrane proteins.

Integral Membrane Proteins

Atomic structures of a growing number of integral membrane proteins and primary structures of thousands of others show how proteins associate with lipid bilayers (Fig. 13.9). Many integral membrane proteins have a single peptide segment that fulfills the energetic criteria (Box 13.1) for a membrane-spanning α-helix. Glycophorin from the red blood cell membrane was the first of these proteins to be characterized (Fig. 13.9A). Nuclear magnetic resonance experiments established that the single **transmembrane segment** of glycophorin is an α-helix. This helix interacts more favorably with lipid acyl chains than with water. By analogy with glycophorin, it is generally accepted that single, 25-residue hydrophobic segments of other transmembrane proteins fold into α-helices. In many cases, independent evidence has confirmed that the single segment crosses the bilayer. For example, proteolytic enzymes might cleave the peptide at the predicted membrane interface but cannot access the membrane interior. Potential glycosylation sites might be located outside the cell. Chemical or antibody labeling might identify parts of the protein inside or outside the cells.

Transmembrane segments of integral membrane proteins that cross the bilayer more than once are folded into α-helices or β-strands. Hydrogen bonding of all backbone amides and carbonyls in the secondary structure minimizes the energy required to bury the backbone in the hydrophobic lipid bilayer. For the same reason, most amino acid side chains in contact with fatty acyl chains in the bilayer are hydrophobic. Membrane proteins can bind specific types of lipids that stabilize the protein. Chapter 20 considers how transmembrane proteins fold during their biosynthesis.

Integral membrane proteins with all α-helical transmembrane segments are the most common. Examples are bacteriorhodopsin (Fig. 13.9B; see also Fig. 27.2), pumps (see Figs. 14.3, 14.4, 14.7, and 14.10), carriers (see Fig. 15.4), channels (see Fig. 16.3), cytochrome oxidase (see Fig. 19.5), and photosynthetic reaction centers (see Fig. 19.9). Where these proteins have polar and charged residues in the plane of the bilayer, they generally face away from the lipid toward the interior of the protein, in contrast to the opposite arrangement in water-soluble proteins.

BOX 13.1 Amino Acid Sequences Identify Candidate Transmembrane Segments

Amino acid sequences of integral membrane proteins provide important clues about segments of the polypeptide that cross the lipid bilayer. Each crossing segment must be long enough to span the bilayer with a minimum of charged or polar groups in contact with the lipid (Fig. 13.8). Polar backbone amide and carbonyl atoms are buried in α-helices or β-sheets to avoid contact with the lipid. Aromatic residues frequently project from transmembrane segments into the lipid near the level where acyl chains are bonded to the lipid head groups (red side chains in Fig. 13.8). A helix of 20 to 25 residues or a β-strand of 10 residues is long enough (3.0 to 3.8 nm) to span a lipid bilayer depending on the thickness of the bilayer.

Quantitative analysis of the side chain and backbone **hydropathy** (aversion to water) of the sequence of an integral membrane protein usually identifies one or more hydrophobic sequences long enough to cross a bilayer (see the legend for Fig. 13.8 for details). The approach works best for helices that are inserted directly in the lipid, like the single transmembrane helix of glycophorin A that has mostly apolar side chains. If a protein has multiple transmembrane helices, some may escape detection by hydrophobicity analysis because they form a hydrophilic cavity lined with charged and polar side chains. For example, two of seven transmembrane helices of bacteriorhodopsin contain charged residues facing the interior of the protein, so they are less hydrophobic than the other transmembrane helices. Transmembrane β-strands are more challenging, as only half of the side chains face the membrane lipids. None of the transmembrane strands of porin qualify as transmembrane segments by hydrophobicity criteria. They are short, and many contain polar residues facing the central cavity.

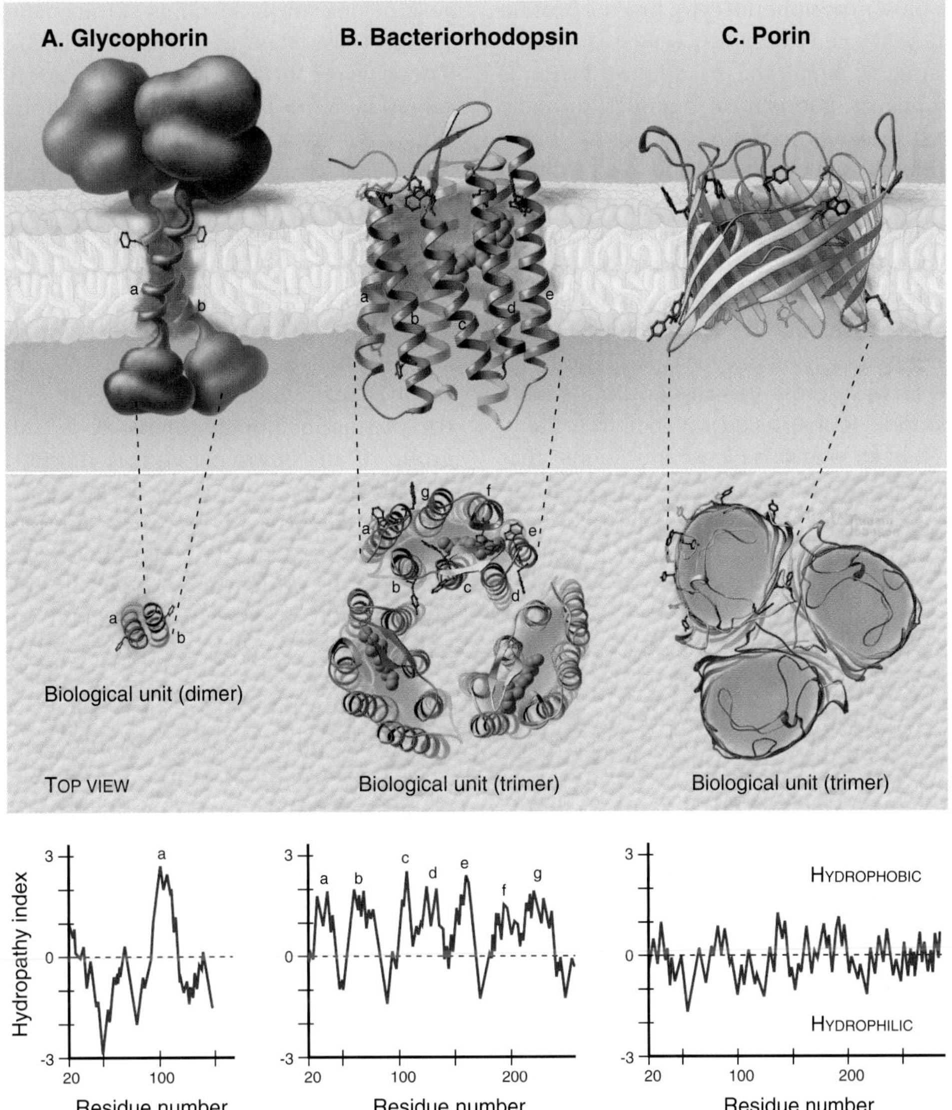

FIGURE 13.9 STRUCTURES OF REPRESENTATIVE INTEGRAL MEMBRANE PROTEINS. *Top,* Views across the lipid bilayer. *Middle,* Views in the plane of the lipid bilayer. *Bottom,* Hydrophobicity analysis. **A,** Glycophorin, a human red blood cell protein, has a single transmembrane α-helix. The extracellular and cytoplasmic domains are artistic conceptions. The transmembrane helices have a strong tendency to form homodimers in the plane of the membrane (see Protein Data Bank [PDB; www.rcsb.org] file 1MSR). **B,** Bacteriorhodopsin, a light-driven proton pump from the plasma membrane of a purple bacterium (see Fig. 14.3), has seven transmembrane helices. The green space-filling structure is retinal, the covalently bound, light-absorbing "chromophore." This structure was first determined by electron microscopy of two-dimensional crystals and extended to higher resolution by x-ray diffraction (see PDB file 1AT9). **C,** Porin, a nonselective channel protein from the outer membrane of a bacterium, is composed largely of transmembrane β-strands. This structure was determined by x-ray crystallography of three-dimensional crystals (see PDB file 1PRN). Hydropathy plots are calculated from the energy required to transfer an amino acid from an organic solvent to water. One sums the transfer free energy for segments of 20 residues. Segments with large, positive (unfavorable) transfer free energies (around 1.5 on this scale) are more soluble in the hydrophobic interior of a membrane bilayer than in water and thus are candidates for membrane-spanning segments.

Many transmembrane proteins consist of multiple subunits that associate in the plane of the bilayer (Fig. 13.9). The transmembrane helix of glycophorin A has a strong tendency to form homodimers in the plane of the membrane. Dimers are favored because complementary surfaces on a pair of helices interact more precisely with each other than with lipids. The positive entropy change associated with dissociation of lipids from interacting protein surfaces (comparable to the hydrophobic effect in water) drives the reaction. Backbone carbonyl oxygens also form unconventional hydrogen bonds with C-α hydrogens that stabilize dimers. Bacteriorhodopsin molecules self-associate in the plane of the membrane to form extended two-dimensional crystals. Many membrane channels form by association of four similar or identical subunits with a pore at their central interface (see Fig. 16.2). Bacterial cytochrome oxidase is an assembly of four different subunits with a total of 22 transmembrane helices (see Fig. 19.5). The purple bacterium photosynthetic reaction center consists of three unique helical subunits plus a peripheral cytochrome protein (see Fig. 19.9).

A minority of integral membrane proteins use β-strands to cross the lipid bilayer. Porins form channels for many substances, up to the size of proteins, to cross the outer membranes of Gram-positive bacteria and their eukaryotic descendents, mitochondria and chloroplasts. Porins consist of an extended β-strand barrel with a hydrophobic exterior surrounding an aqueous pore (Fig. 13.9C). These subunits associate as trimers in the lipid bilayer.

In addition to transmembrane helices or strands, many integral membrane proteins have structural elements that pass partway across the bilayer. Porins have extended polypeptide loops inside the β-barrel. Many channel proteins have short helices and loops that reverse in the middle of the membrane bilayer. These structural elements help form pores specific for potassium (see Fig. 16.3), chloride (see Fig. 16.13), and water (see Fig. 16.15).

Peripheral Membrane Proteins

Six strategies bind peripheral proteins to the surfaces of membranes (Fig. 13.10). One of three different types of hydrophobic acyl chains can anchor a protein to a membrane by inserting into the lipid bilayer. Other proteins bind electrostatically to membrane lipids, and some insert partially into the lipid bilayer. Many peripheral proteins bind directly or indirectly to integral membrane proteins.

Isoprenoid Tails

A 15-carbon isoprenoid (farnesyl) tail (see Fig. 20.15) is added posttranslationally to the side chain of a cysteine residue near the C-terminus of the guanosine triphosphatase (GTPase) Ras (see Fig. 4.6) and many other proteins. The enzyme making this modification recognizes the target cysteine followed by two aliphatic amino acids plus any other amino acid (a CAAX recognition site). Another enzyme cleaves off the AAX residues. This membrane attachment is required for Ras to participate in growth factor signaling (see Fig. 27.6).

Myristoyl Tails

Myristate, a 14-carbon saturated fatty acid, anchors the tyrosine kinase Src (see Box 27.5) and other proteins involved in cellular signaling to the cytoplasmic face of the plasma membrane. Myristate is added to the amino group of an N-terminal glycine during the biosynthesis of these proteins. Insertion of this short, fatty acyl chain into a lipid bilayer is so weak (K_d: ~10^{-4} M) that additional electrostatic interactions between basic side chains of the protein and head groups of acidic phosphoglycerides are required to maintain attachment to the membrane.

Glycosylphosphatidylinositol Tails

A short oligosaccharide-phosphoglyceride tail links a variety of proteins to the outer surface of the plasma membrane. The C-terminus of the protein is attached covalently to the oligosaccharide, and the two fatty acyl chains of PI are in the lipid bilayer. In animal cells, this GPI anchors important plasma membrane proteins, including enzymes (acetylcholine esterase; see Fig. 17.9), adhesion proteins (T-cadherin; see Fig. 30.5), and cell surface antigens (Thy-1). The protozoan parasite

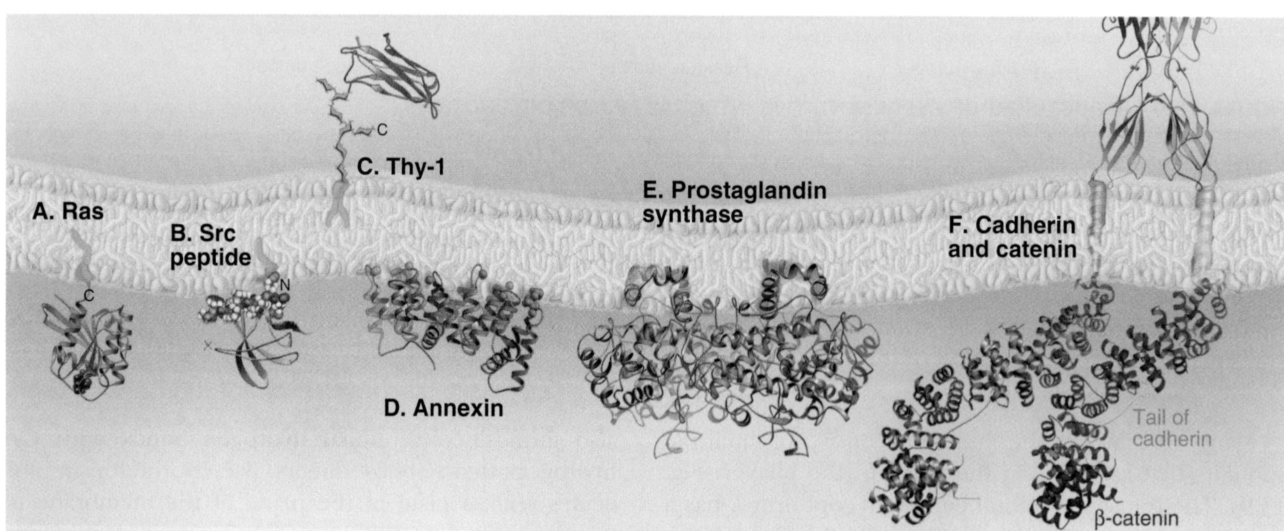

FIGURE 13.10 SIX MODES OF ASSOCIATION OF PERIPHERAL MEMBRANE PROTEINS WITH LIPID BILAYERS. A, A C-terminal isoprenoid tail attaches Ras to the bilayer (see PDB file 121P). **B,** An N-terminal myristoyl tail binds Src weakly to the bilayer. Electrostatic interactions between acidic lipids and basic amino acids stabilize the interaction. **C,** A C-terminal glycosylphosphatidylinositol (GPI) tail anchors Thy-1 (similar to an immunoglobulin variable domain) to the bilayer. **D,** Electrostatic interactions with phospholipids bind annexin to the bilayer (see PDB file 1A8A). **E,** Hydrophobic helices of prostaglandin H_2 synthase partially penetrate the lipid bilayer (see PDB file 1CQE). **F,** The peripheral protein β-catenin (*blue* and *purple*; see PDB file 1CQE) associates with the cytoplasmic portion of the transmembrane adhesion protein cadherin (*red* and *green*; see PDB file 1FF5).

Trypanosoma brucei covers itself with a high concentration of a GPI-anchored protein. If challenged by an antibody response from the host, the parasite sheds the protein by hydrolysis of the lipid anchor and expresses a variant protein to evade the immune system.

Electrostatic Interaction With Phospholipids

As postulated in the 1930s (Fig. 13.1), some soluble cytoplasmic proteins bind the head groups of membrane lipids. Annexins, a family of calcium-binding proteins implicated in membrane fusion reactions, bind tightly to PS (Fig. 13.10D). A second example is the "BAR" domain found in a variety of proteins. Positively charged residues on the concave surface of curved, dimeric BAR domains bind electrostatically to curved membranes or deform flat membranes into tubules (see amphiphysin in Fig. 22.11). Myosin-I motor proteins (see Fig. 36.7) also bind strongly to acidic phosphoglycerides of cellular membranes.

Partial Penetration of the Lipid Bilayer

Hydrophobic α-helices of prostaglandin H_2 synthase (see Figs. 13.10E and 26.9) anchor the enzyme to membranes by partially penetrating the lipid bilayer. Another example is reticulons, proteins that insert into the cytoplasmic leaflet of the ER membrane and promote bending into narrow tubules and sharply curved edges of sheets (see Fig. 20.3).

Association With Integral Proteins

Many peripheral proteins bind cytoplasmic domains of integral membrane proteins. For example, catenins bind transmembrane cell adhesion proteins called cadherins (Fig. 13.10F). These protein–protein interactions provide more specificity and higher affinity than do the interactions of peripheral proteins with membrane lipids. Such protein–protein interactions anchor the cytoskeleton to transmembrane adhesion proteins (see Fig. 31.8) and guide the assembly of coated vesicles during endocytosis (see Fig. 22.9). Protein–protein interactions also provide a way to transmit information across a membrane. Ligand binding to the extracellular domain of a transmembrane receptor can change the conformation of its cytoplasmic domain, promoting interactions with cytoplasmic, signal-transducing proteins (see Chapter 24).

The **membrane skeleton** on the cytoplasmic surface of the plasma membrane of human red blood cells (Fig. 13.11) provided the first insights regarding interaction of peripheral and integral membrane proteins. Two types of integral membrane proteins—an anion carrier called Band 3 and glycophorin—anchor a two-dimensional network of fibrous proteins to the membrane. The main component of this network is a long, flexible, tetrameric, actin-binding protein called **spectrin** (after its discovery in lysed red blood cells, "ghosts"; see Fig. 33.17). A linker protein called ankyrin binds tightly to both Band 3 and spectrin. Approximately 35,000 nodes consisting of a short actin filament and associated proteins interconnect the elastic spectrin network. This membrane skeleton reinforces the bilayer, allowing a cell to recover its shape elastically after it is distorted by squeezing through the narrow lumen of blood capillaries.

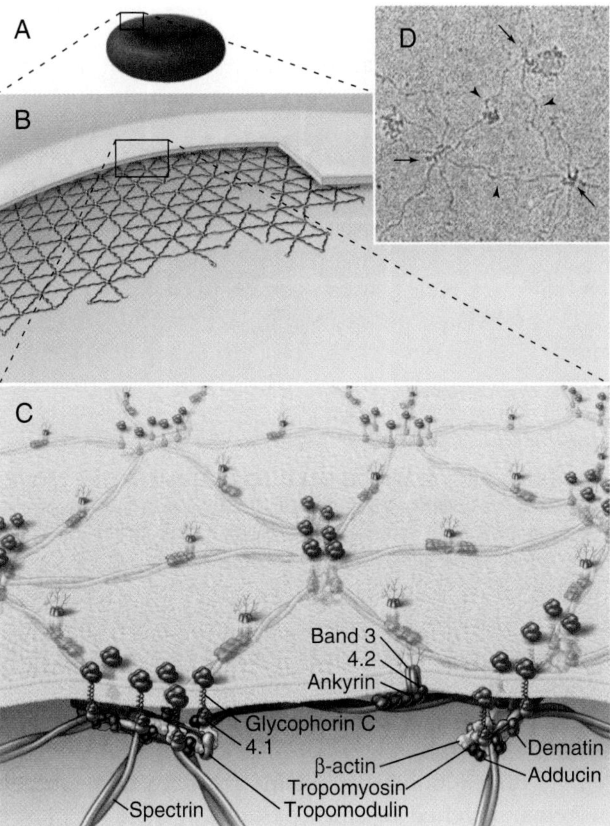

FIGURE 13.11 THE MEMBRANE SKELETON ON THE CYTOPLASMIC SURFACE OF THE RED BLOOD CELL PLASMA MEMBRANE. A, Whole cell. **B,** Cutaway drawing. **C,** Detailed drawing. Nodes consisting of a short actin filament and associated proteins interact with multiple spectrin molecules, which, in turn, bind to two transmembrane proteins: glycophorin and (via ankyrin) Band 3. **D,** An electron micrograph of the actin-spectrin network. (**D,** Courtesy R. Josephs, University of Chicago, IL.)

Membrane Protein Dynamics

Several complementary methods can monitor movements of plasma membrane proteins (Fig. 13.12A). The original approach was to label proteins with a fluorescent dye, either by covalent modification or by attachment of an antibody with a bound fluorescent dye. After a spot of intense light irreversibly bleaches the fluorescent dyes in a small area of the membrane, one observes the fluorescence over time with a microscope. If the test protein is mobile, unbleached proteins from surrounding areas move into the bleached area. The rate and extent of fluorescence recovery after **photobleaching** (FRAP) revealed that a fraction of the population of most membrane proteins diffuses freely in two dimensions in the plane of the membrane, but that a substantial fraction is immobilized because the recovery from photobleaching

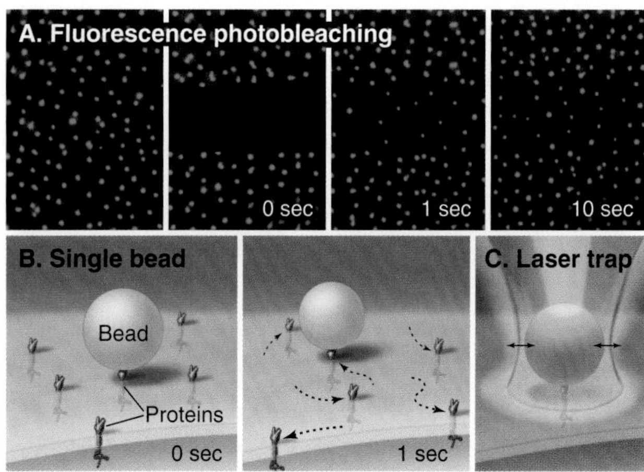

FIGURE 13.12 METHODS USED TO DOCUMENT THE MOVEMENTS OF MEMBRANE PROTEINS. A, Fluorescence recovery after photobleaching. Simulated experimental data with individual molecules are shown as green dots. **B,** Single-particle tracking. **C,** Optical trapping.

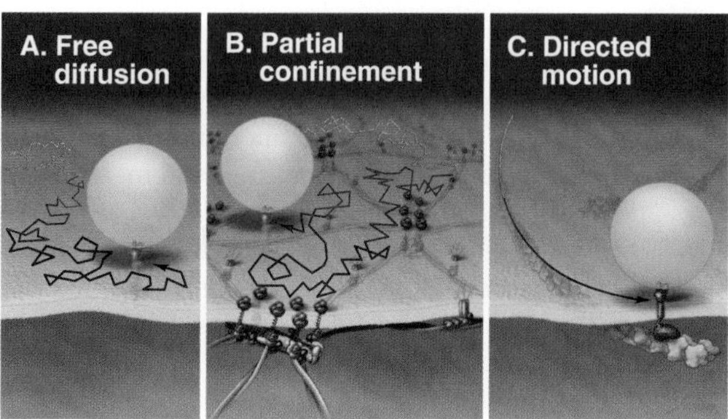

FIGURE 13.13 MOVEMENTS OF PROTEINS IN THE PLANE OF MEMBRANES. A, Free diffusion. **B,** Partial confinement by obstacle clusters, some associated with the membrane skeleton. **C,** Directed movement by a motor on an actin filament. (For reference, see Jacobson K, Sheets ED, Simson R. Revisiting the fluid mosaic model of membranes. *Science*. 1995;268:1441–1442.)

is incomplete. The same photobleaching method is used to study the mobility of fluorescent fusion proteins targeted to any cellular membrane (see Fig. 6.3). The second approach is to label individual membrane proteins with antibodies or lectins (carbohydrate-binding proteins) attached to small particles of gold or plastic beads (Fig. 13.12B). High-contrast light microscopy can follow the motion of a particle attached to a membrane protein. Despite their size, the particles have minimal effects on diffusion of membrane proteins. The third method is an extension of single-particle tracking. Instead of merely watching spontaneous movements, the investigator can grab a particle in an **optical trap** created by focusing an infrared laser beam through the microscope objective (Fig. 13.12C). Manipulation of particles with an optical trap reveals what happens when force is applied to a membrane protein.

Membrane proteins exhibit a wide range of dynamic behaviors (Fig. 13.13). Some molecules diffuse freely. Others diffuse intermittently, alternating with periods of restricted movement. Substantial numbers of membrane proteins are immobilized, presumably by direct or indirect associations with the membrane skeleton or the cytoskeleton, or by forming large arrays through mutual interactions.

The population of a given type of membrane protein (eg, a cell adhesion protein) may exhibit more than one class of dynamic behavior. For example, most proteins with GPI anchors diffuse freely, as is expected from their association with the lipid bilayer, but a fraction of any GPI-anchored protein has restricted mobility. Some transmembrane proteins also diffuse freely, but a fraction may become trapped or immobilized at any time. Diffusing proteins must be free of interactions with the membrane skeleton and with anchored membrane proteins. Cell adhesion proteins (cadherins; see Fig. 30.5) and nutrient receptors (transferrin receptors; see Fig. 22.15) are examples of transmembrane proteins that diffuse intermittently. They alternate between free diffusion and temporary trapping for 3 to 30 seconds in local domains measuring less than 0.5 μm in diameter. In some cases, trapping depends on the cytoplasmic tails of transmembrane proteins, which are thought to interact reversibly with the cytoskeleton or with immobilized membrane proteins. Tugs with an optical trap show that the cages that confine these particles are elastic, as expected for cytoskeletal networks. Extracellular domains of these proteins may also interact with adjacent immobilized proteins. Immobilized proteins do not diffuse freely, and particles attached to them resist displacement by optical traps.

The lipid bilayer can flow past immobilized transmembrane proteins without disrupting the membrane. If the plasma membrane of a red blood cell is sucked into a narrow pipette (Fig. 13.8D), lipids of the fluid membrane bilayer extend uniformly over the protrusion, leaving behind the immobilized membrane proteins and the membrane skeleton.

Some membrane proteins undergo long-distance translational movements in relatively straight lines. Because disruption of cytoplasmic actin filaments by drugs impedes these movements, myosins (see Fig. 36.7) are the most likely motors for these movements. In some instances, members of the integrin family of adhesion proteins (see Fig. 30.9) use this transport system.

Movements of membrane proteins in the plane of the membrane are essential for many cellular functions. Transmembrane receptors concentrate in coated pits before internalization during receptor-mediated endocytosis (see Fig. 22.12). Similarly, transduction of many signals from outside the cell depends on the formation of receptor dimers or trimers (see Figs. 24.5, 24.7, 24.8, 24.9, 24.10, and 46.18). Bound extracellular ligands stabilize collisions between freely diffusing receptor proteins, juxtaposing their cytoplasmic domains and activating downstream signaling mechanisms. Similarly, movements in the plane of the plasma membrane allow clustering of adhesion receptors that enhances binding of cells to their neighbors or to the extracellular matrix (see Figs. 30.6 and 30.11).

ACKNOWLEDGMENTS

We thank Tobias Baumgart and Donald Engelman for their suggestions on this chapter.

SELECTED READINGS

Blaskovic S, Blanc M, van der Goot FG. What does S-palmitoylation do to membrane proteins? *FEBS J*. 2013;280:2766-2774.

Curran AR, Engelman DM. Sequence motifs, polar interactions and conformational changes in helical membrane proteins. *Curr Opin Struct Biol*. 2003;13:412-417.

Engelman DM. Lipid bilayer structure in the membrane of *Mycoplasma laidlawii* [bilayer structure established by x-ray diffraction]. *J Mol Biol*. 1971;58:153-165.

Fleming KG. Energetics of membrane protein folding. *Annu Rev Biophys*. 2014;43:233-255.

Forrest LR. Structural symmetry in membrane proteins. *Annu Rev Biophys*. 2015;44:311-337.

Kiessling LL, Splain RA. Chemical approaches to glycobiology. *Annu Rev Biochem*. 2010;79:619-653.

McNeil PL, Steinhardt RA. Plasma membrane disruption: repair, prevention, adaptation. *Annu Rev Cell Dev Biol*. 2003;19:697-731.

Nagle JF, Tristram-Nagle S. Structure of lipid bilayers. *Biochim Biophys Acta*. 2000;1469:159-195.

Owen DM, Magenau A, Williamson D, Gaus K. The lipid raft hypothesis revisited—new insights on raft composition and function from super-resolution fluorescence microscopy. *Bioessays*. 2012;34: 739-747.

Pandit SA, Scott HL. Multiscale simulations of heterogeneous model membranes. *Biochim Biophys Acta*. 2009;1788:136-148.

Robertson JD. Membrane structure [historical perspective]. *J Cell Biol*. 1981;91:1895-2045.

Sachs JN, Engelman DM. Introduction to the membrane protein reviews: the interplay of structure, dynamics, and environment in membrane protein function. *Annu Rev Biochem*. 2006;35: 707-712.

Shevchenko A, Simons K. Lipidomics: coming to grips with lipid diversity. *Nat Rev Mol Cell Biol*. 2010;11:593-598.

Simons K, Geri MJ. Revitalizing membrane rafts: new tools and insights. *Nat Rev Mol Cell Biol*. 2010;11:688-699.

Stoeckenius W, Engelman DM. Current models for the structure of biological membranes [historical perspective]. *J Cell Biol*. 1969;42: 613-646.

Wang L. Measurements and implications of the membrane dipole potential. *Annu Rev Biochem*. 2012;81:615-635.

Wollam J, Antebi A. Sterol regulation of metabolism, homeostasis and development. *Annu Rev Biochem*. 2011;80:885-916.

Zverina EA, Lamphear CL, Wright EN, Fierke CA. Recent advances in protein prenyltransferases: substrate identification, regulation, and disease interventions. *Curr Opin Chem Biol*. 2012;16:544-552.

CHAPTER 14

Membrane Pumps

Introduction to Membrane Permeability

Lipid bilayers provide a barrier to diffusion of ions and polar molecules larger than about 150 Da, so transmembrane proteins are required for selective passage for ions, and other larger molecules across membranes. The proteins controlling membrane permeability fall into three broad classes—pumps, carriers, and channels—each with distinct properties (Fig. 14.1). These proteins control traffic of small molecules across membranes, an essential of many physiological processes.

- **Pumps** are enzymes that use energy from adenosine triphosphate (ATP), light, or (rarely) other sources to move ions (generally, cations) and other solutes across membranes at relatively modest rates. Pumps are also called **primary active transporters,** because they transduce electromagnetic or chemical energy directly into transmembrane concentration gradients between membrane-bound compartments.
- **Carriers** provide passive pathways for solutes to move across membranes down their concentration gradients from a region of higher concentration to one of lower concentration. Each conformational change in a carrier protein translocates a limited number of small molecules across the membrane. Carriers use ion gradients created by pumps as a source of energy to perform a remarkable variety of work. Translocation of an ion down its concentration gradient can drive another ion or solute up a concentration gradient, so these are called **secondary transporters** (see Chapter 15).
- **Channels** are ion-specific pores that typically open and close transiently in a regulated manner. Open channels are highly specific but **passive transporters**, allowing a flood of an ion or other small solute to pass quickly across the membrane, driven by electrical and concentration gradients. The movement of ions through open channels controls the electrical

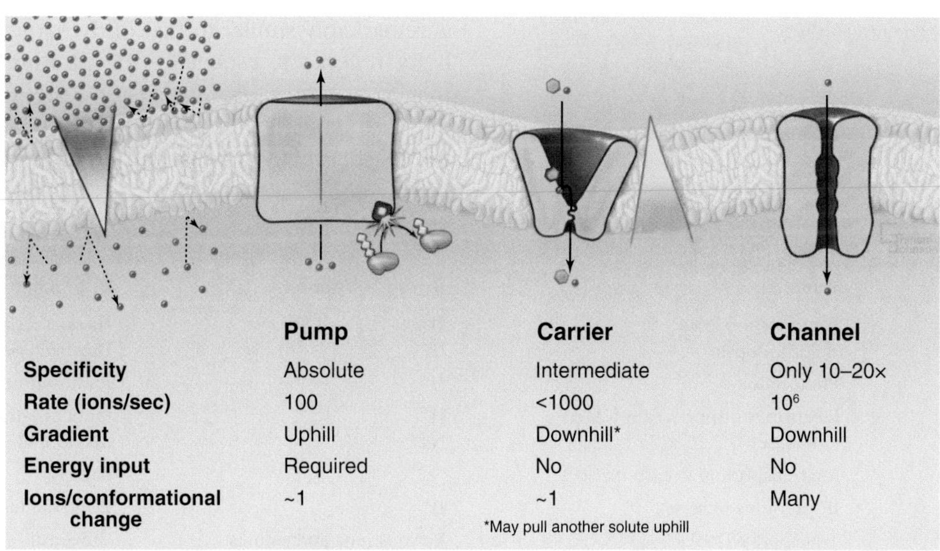

	Pump	Carrier	Channel
Specificity	Absolute	Intermediate	Only 10–20×
Rate (ions/sec)	100	<1000	10^6
Gradient	Uphill	Downhill*	Downhill
Energy input	Required	No	No
Ions/conformational change	~1	~1	Many

*May pull another solute uphill

FIGURE 14.1 PROPERTIES OF THE THREE TYPES OF PROTEINS THAT TRANSPORT IONS AND OTHER SOLUTES ACROSS MEMBRANES. The triangles represent the concentration gradients of Na^+ *(blue)* and glucose *(green)* across the membrane.

potential across membranes (see Fig. 11.17), so that changes in channel activity produce rapid electrical signals in excitable membranes of nerves, muscles, and other cells (see Fig. 12.6).

This chapter starts with pumps, because they create the solute gradients required for transport by carriers (see Chapter 15) and channels (see Chapter 16). Chapter 17 concludes this section with examples of how pumps, carriers, and channels work together to perform a remarkable variety of functions. An important point is that differential expression of a subset of isoforms of these proteins in specific membranes allows differentiated cells to perform a wide range of complex functions.

Diversity of Membrane Pumps

Protein pumps transport ions and other solutes across membranes up concentration gradients. Energy can come from a variety of sources: light, oxidation–reduction (redox) reactions, or, most commonly, hydrolysis of ATP (Table 14.1). Energy is conserved in the form of transmembrane electrical or chemical gradients of the transported ion or solute. The potential energy in these ion gradients drives a variety of energy-requiring processes (Fig. 14.2). Most known biological pumps translocate cations. Although they could just as well move anions, cations were selected during the evolution of early life forms 3 billion years ago.

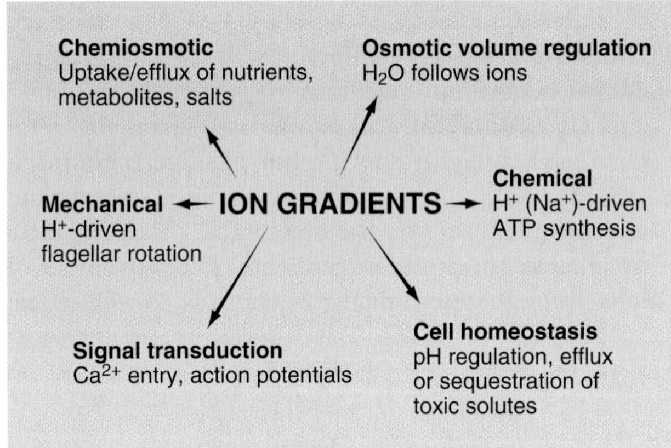

FIGURE 14.2 PROCESSES DRIVEN BY ENERGY STORED IN ION GRADIENTS.

Each family of pumps (Table 14.1) had a single source during evolution, so all members of each family share similar structures and mechanisms. The small number of families is remarkable given the importance of pumps in establishing transmembrane electrochemical gradients. Of course, a host of downstream carriers can take advantage of a limited number of ion concentration gradients to carry out many secondary reactions.

This chapter considers four types of pumps, emphasizing examples in which both high-resolution structures and detailed biochemical analysis of pathways are available. Chapter 19 provides additional details on H^+ translocation by redox-driven cytochrome *c* oxidase and the role of F-type pumps in ATP synthesis by mitochondria and chloroplasts. Microbiology texts provide more information on pumps driven by decarboxylases and pyrophosphatases.

Light-Driven Proton Pumping by Bacteriorhodopsin

Owing to its simplicity, its small size, and the availability of many high-resolution structures (Fig. 14.3), more is known about light-driven transport of protons by **bacteriorhodopsin** than about any other pump. This pump converts light energy into a proton gradient across the plasma membrane of the halophilic (salt-loving) Archaea *Halobacterium halobium*. A proton-driven ATP synthase uses this proton gradient to make ATP (Fig. 14.5). The 26-kD polypeptide is folded into seven α-helices that cross the lipid bilayer. The light-absorbing chromophore retinal (vitamin A aldehyde) is bound covalently to the side chain of lysine 216 (Lys216) via a Schiff base, a C=N—C bond. The design of this seven-helix transporter is remarkably similar to that of the large family of seven-helix receptors, especially the photoreceptor proteins that vertebrates use for vision (see Fig. 27.2). Two-dimensional crystalline arrays of bacteriorhodopsin give the plasma membrane its purple color.

TABLE 14.1 Diversity of Membrane Pumps*

Energy Source	Pump	Driven Substance	Distribution
Light	Bacteriorhodopsin Halorhodopsin	H^+ Cl^-	Halobacteria Halobacteria
Light	Photoredox	H^+	Photosynthetic organisms
Redox potential	Electron transport chain NADH oxidase	H^+ Na^+	Mitochondria, bacteria Alkalophilic bacteria
Decarboxylation	Ion-transporting decarboxylases	Na^+	Bacteria
Pyrophosphate	H^+-pyrophosphatase	H^+	Plant vacuoles, fungi, bacteria
ATP	Transport ATPases	Various ions and solutes	Universal

ATPase, adenosine triphosphatase; NADH, reduced form of nicotinamide adenine dinucleotide; Redox, oxidation–reduction.
*Each class of pumps has a different evolutionary origin and structure.

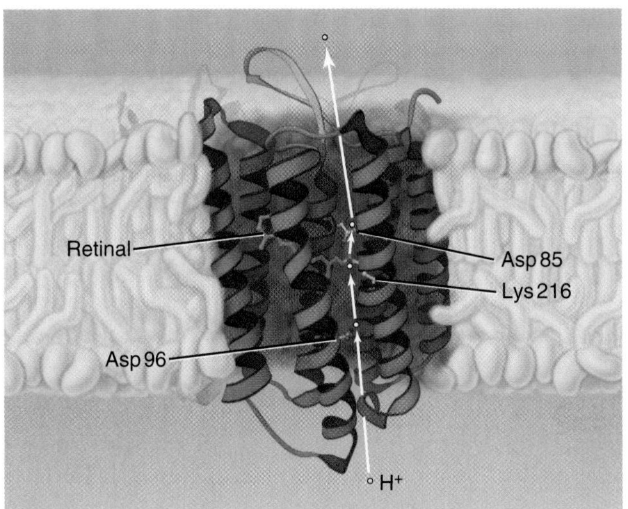

FIGURE 14.3 PROTON PATHWAY ACROSS THE MEMBRANE THROUGH BACTERIORHODOPSIN. Numerous atomic structures, fast spectroscopic measurements of reaction intermediates, and analysis of a wide array of mutations, revealed the pathway for protons through the middle of the bundle of seven α-helices. A cytoplasmic proton binds successively to Asp96, the Schiff base linking retinal to lysine 216 (Lys216), Asp85, and Glu204 before releasing outside the cell. Absorption of light by retinal drives conformational changes in the protein that favor the transfer of the proton across the membrane up its concentration gradient. (For reference, see Protein Data Bank [PDB; www.rcsb.org] file 1C3W and Wickstrand C, Dods R, Royant A, Neutze R. Bacteriorhodopsin: would the real structural intermediates please stand up? *Biochim Biophys Acta*. 2015;1850:536–553.)

The proton pathway across the membrane includes the side chains of aspartate 96 (Asp96), aspartate 85 (Asp85), glutamate 204 (Glu204), the Schiff base, and a water molecule next to the Schiff base, a C=N—C bond. Local environments give the two aspartates remarkably different ionization constant (pK_a) values. Asp96 has a very high pK_a of approximately 10, so it can serve as a proton donor. Asp85 has a low pK_a of approximately 2, so it serves as a proton acceptor. Absorption of a photon induces reversible changes in the conformations of the retinal and the protein that allow for the transfer a single proton from the cytoplasm to the extracellular space.

1. The mechanism starts with retinal in the all-*trans* configuration and protons bound to the Schiff base and Asp96 at the hydrophobic, cytoplasmic end of the proton pathway.
2. Absorption of a photon isomerizes retinal to the 13-*cis* configuration. This changes the conformation of the protein, lowers the pK_a of the Schiff base, and favors transfer of its proton to Asp85.
3. Asp85 transfers the proton to Glu204, which releases the proton outside the cell.
4. A further conformational change of the protein reorients the Schiff base toward Asp96. The pK_a of Asp96 is lower in this conformation, so a proton transfers from Asp96 to the Schiff base.
5. Asp96 is reprotonated from the cytoplasm.
6. The retinal reisomerizes to the all-*trans* configuration in preparation for another cycle.

The net result of this cycle is rapid vectorial transport of a proton from the cytoplasm out of the cell. Steps 4 to 6 are rate limiting, occurring at a rate of about 100 s^{-1}. The other reactions are fast, provided that there is an adequate flux of light. Retinal not only captures energy by absorbing a photon but also acts as a switch that changes both the accessibility and affinities of the proton-binding groups in a sequential fashion.

In addition to bacteriorhodopsin, halobacterial plasma membranes contain two related proteins: halorhodopsin and sensory rhodopsin. Halorhodopsin absorbs light and pumps chloride into the cell. Interestingly, a single amino acid substitution can reverse the direction of pumping. Sensory rhodopsin couples light absorption by its bound retinal to phototaxis (swimming toward light) with a tightly coupled transducer protein. In the absence of this transducer, sensory rhodopsin transports protons out of the cell much like bacterial rhodopsin.

Channelrhodopsins are another member of this family with bound retinal and seven transmembrane helices. Absorbance of a photon transiently opens a cation channel across the membrane as part of chemotaxis of green algae (see Fig. 38.16). Light can control the excitability of cells engineered to express this light-gated channel, a method called **optogenetics**.

Overview of Adenosine Triphosphate–Driven Pumps

Three families of transport adenosine triphosphatases (ATPases) (Table 14.2) are essential for the physiology of all forms of life. The rotary ATPases and P-type ATPases both generate electrical and/or chemical gradients across membranes. ABC (adenosine triphosphate binding cassette) transporters not only produce ion gradients but also transport a much wider range of solutes across membranes. Chemical inhibitors have been useful in characterizing these pumps, and some are also used therapeutically (Table 14.3).

Free energy released by ATP hydrolysis sets the limit on the concentration gradient that these pumps can produce. If transport is electrically neutral (ie, if it does not produce a membrane potential; see Fig. 11.17), the maximum gradient is about 1 million-fold. The electrically neutral, P-type H^+K^+-ATPase of gastric epithelial cells actually produces such an extraordinary gradient when it acidifies the stomach down to a pH of 1.

Rotary ATPase Families

The rotary ATPases originated in the last universal common ancestor of life on earth (see Fig. 1.1) so they share many common features. Over time these genes diverged, giving rise to three families of rotary ATPases:

TABLE 14.2 Adenosine Triphosphate–Driven Transport Adenosine Triphosphatase Pumps

Pump	Subunits	Distribution	Substrate	Function
Rotary ATPases				
A_OA_1	8 or more	Archaea, some bacteria	H^+ (some cases Na^+)	H^+- or Na^+-driven ATP synthesis
F_OF_1	8 or more	Mitochondria, chloroplasts, bacterial	H^+	H^+-driven ATP synthesis or H^+ pumping
V_OV_1	8 or more	Eukaryotic endomembranes	H^+	ATP-driven H^+ pumping
P-Type ATPase Family				
Na^+K^+-ATPase	2	Plasma membrane	3 Na^+ for 2 K^+	Generate Na^+, K^+ gradient
H^+K^+-ATPase	2	Stomach and kidney plasma membranes	1 H^+ for 1 K^+	Gastric and renal H^+ secretion
SERCA Ca-ATPase	1	Sarcoplasmic reticulum, endoplasmic reticulum	2 Ca^{2+} for 2 H^+	Lowers cytoplasmic Ca^{2+}
PMCA Ca-ATPase	1	Plasma membrane	1 Ca^{2+} for 1 H^+	Lowers cytoplasmic Ca^{2+}
H^+-ATPase	1	Plasma membrane in yeast, plants, protozoa	1 H^+	Generate proton gradient
ABC Transporters				
MDR1 P-glycoprotein	1	Plasma membrane	Drugs	Drug secretion
CFTR	1	Respiratory tract and pancreatic epithelial plasma membranes	ATP, Cl^-	Cl^- secretion
TAP1, 2	2	Endoplasmic reticulum	Antigenic peptides	Transport antigenic peptides into ER
MDR2	1	Liver cell apical plasma membrane	Phosphatidylcholine	Phosphoglyceride flippase, bile secretion
STE6	1	Yeast plasma membrane	Mating pheromone peptide	Secretion
HisQMP	4 + pp	Bacteria plasma membrane	Histidine	Histidine uptake
PstSCAB	4 + pp	Bacteria plasma membrane	Phosphate	Phosphate uptake
OppDFBCA	4 + pp	Bacteria plasma membrane	Oligopeptides	Peptide uptake
HlyB	2	*Escherichia coli* plasma	Hemolysin A (107-kD protein)	Hemolysin A uptake

CFTR, cystic fibrosis transmembrane regulator; MDR1, 2, multidrug resistance 1, 2; PMCA, plasma membrane calcium ATPase; pp, periplasmic protein; SERCA, sarcoplasmic-endoplasmic reticulum calcium ATPase; TAP1, 2, transport-associated protein 1, 2.

TABLE 14.3 Tools for Studying Pumps

Agent	Target
Cardiac glycosides* (digitalis)	Na^+K^+-ATPase
Omeprazole*	H^+K^+-ATPase (parietal cell)
Oligomycin	F_OF_1-ATP synthase

*Used clinically as drugs.

A-, F-, and V-types. Contemporary rotary ATPases either use proton gradients to synthesize ATP or run in reverse using ATP hydrolysis to pump protons. This section begins with the common features of rotary ATPases. Concluding sections cover the distribution of the three specialized families among organisms, their structures, and their functions.

Architecture of Rotary Adenosine Triphosphatases

Rotary ATPases consist of two rotary motors (R_1 and R_O), as illustrated by the structure of the proton-driven, A-type ATP synthase from the Archaea *Thermus thermophilus* (Fig. 14.4A–B). The water-soluble, globular R_1 motor catalyzes ATP hydrolysis or synthesis. The R_O (for oligomycin-sensitive factor) motor is embedded in the membrane and conducts protons across the lipid bilayer. Two connections link the R_1 and R_O motors. A centrally located shaft fixed to a ring of transmembrane c-subunits extends through the middle of R_1. Two peripheral stalks anchor R_1 to the a-subunit, an integral membrane protein located in the membrane bilayer next to the ring of c-subunits. The central shaft rotates inside R_1, and the ring of c-subunits rotates relative to the adjacent a-subunit.

Pioneering biochemical studies and a crystal structure of mitochondrial F-type ATP synthase (Fig. 14.4C) first suggested that rotation of the central shaft couples proton fluxes through the transmembrane R_O to ATP synthesis or, alternatively, couples ATP hydrolysis in R_1 to proton pumping by R_O. Light microscopy confirmed this rotation (Fig. 14.4D). Mitochondrial R_1 was attached to a glass coverslip, and a tiny bead or actin filament was attached to the free end of the central shaft. ATP hydrolysis by R_1 drove the rotation of the bead or filament on the central shaft. If the bead is magnetic, a rotating magnetic field can drive the shaft and synthesize ATP from adenosine diphosphate (ADP) and phosphate. The central shaft rotates up to an astounding 130

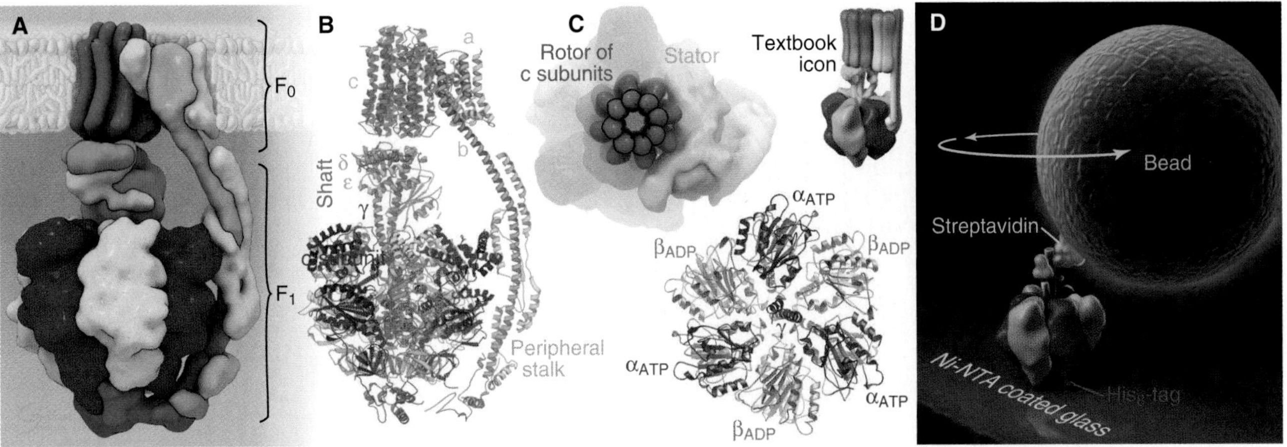

FIGURE 14.4 CRYSTAL STRUCTURE AND MECHANICS OF ROTARY ATP SYNTHASES. A–B, Three-dimensional maps of the *Thermus thermophilus* ATP synthase based on reconstructions of cryoelectron micrographs. Color code: *red,* α-subunits; *yellow,* β-subunits; *blue,* γ-subunits; *green*, a-subunits; *magenta,* c-subunits. (See PDB file 3J0J.) **A,** Side view of a semitransparent gray map with ribbon diagrams based on crystal structures and models of individual subunits. Scale bar is 2.5 nm. **B,** Side view *(top)* and top view *(bottom)* of the a-subunit and ring of c-subunits. **C,** A ribbon diagram of a crystal structure of the bovine mitochondrial ATP synthase viewed from the membrane *(bottom)* side with α-subunits *(red),* β-subunits *(yellow),* and the γ-subunit *(blue).* All three α-subunits have a bound ATP. The β-subunits are empty or bind ATP or adenosine diphosphate (ADP). (See PDB file 1BMF.) **D,** An experiment showing ATP-driven rotation of the γ-subunit relative to the α- and β-subunits. Streptavidin and biotin link the γ-subunit to a bead, which is observed to rotate by light microscopy. (For reference, see Lau WC, Rubinstein JL. Subnanometre-resolution structure of the intact *Thermus thermophilus* H+-driven ATP synthase. *Nature*. 2011;481:214–218; Abrahams JP, Leslie AGW, Lutter R, Walker JE. Structure at 2.8 Å resolution of F_1-ATPase from bovine heart mitochondria. *Nature*. 1994;370:621–628; and Nishizaka T, Oiwa K, Noji H, et al. Chemomechanical coupling in F_1-ATPase revealed by simultaneous observation of nucleotide kinetics and rotation. *Nat Struct Mol Biol*. 2004;11:142–148.)

times per second (8000 rpm) in mitochondria and twice as fast in chloroplasts.

In the simplest case, bacterial R_1 consists of five different types of polypeptides in the ratio $\alpha_3\beta_3\gamma\Delta\varepsilon$. Mitochondrial R_1 has additional subunits. The α- and β-subunits are folded similarly and arranged alternately like six segments of an orange. The γ-subunit is a long, antiparallel, α-helical coiled-coil forming the central shaft. This hydrophobic shaft fits tightly in a hydrophobic sleeve in the middle of the hexamer of α- and β-subunits. To accommodate the asymmetrical shaft, each of the surrounding α- and β-subunits has a slightly different conformation. Each α- and β-subunit has an adenine nucleotide-binding site at the interface with its neighbor. ATP bound stably to α-subunits does not participate in catalysis. Nucleotide-binding sites on β-subunits catalyze ATP synthesis and hydrolysis.

Mechanical rotation of the γ-subunit inside R_1 is tightly coupled to ATP synthesis or hydrolysis (Fig. 14.5). A flux of protons through the R_O complex produces clockwise (when viewed from R_O) rotation inside the αβ hexamer, like the camshaft in a motor. The mechanical force produced by the asymmetric camshaft drives conformational changes in β-subunits that synthesize ATP. When operating in the other direction, ATP hydrolysis drives counterclockwise rotation of the shaft, which can pump protons through R_O.

During ATP hydrolysis or synthesis, catalytic sites on the β-subunits participate cooperatively in a sequence of steps coupled to rotation of the central shaft (Fig. 14.5). The β-subunits are in one of three conformations—open, loose, and tight—designating a range of affinities for adenine nucleotides. At any given time an R_1 molecule has one open β-subunit, one loose β-subunit, and one tight β-subunit. All three subunits pass in lock step through the sequence of three states as the rotor turns.

When hydrolyzing ATP, the rate-limiting step is ATP binding to the open β-subunit. The energy from ATP binding causes a conformational change driving an 80-degree counterclockwise rotation of γ-subunit in less than a millisecond. This favors ATP hydrolysis on the β-subunit that just bound ATP. After approximately 2 milliseconds, another reaction, possibly ADP dissociation from the third β-subunit, rapidly rotates the γ-subunit another 40 degrees, completing a 120-degree step of the motor.

When synthesizing ATP, the β-subunit in the loose conformation binds ADP and inorganic phosphate (P_i). Energy provided by a 120-degree rotation of the γ-subunit converts this site to the tight conformation, which creates an environment in the active site where ADP and P_i spontaneously form ATP. Energy from a subsequent 120-degree rotation of the shaft drives the tight state to the open state, which allows ATP to dissociate.

The membrane-embedded R_O complexes are a second rotary motor consisting of 12 to 18 protein subunits in the ratio ab_2c_{8-15} in bacteria (Fig. 14.4B). The mitochondrial F_O complex has a single "b" subunit with two transmembrane strands. The c-subunits are hairpins of two α-helices with a conserved residue (either aspartic acid or glutamic acid depending on the species). Eight to

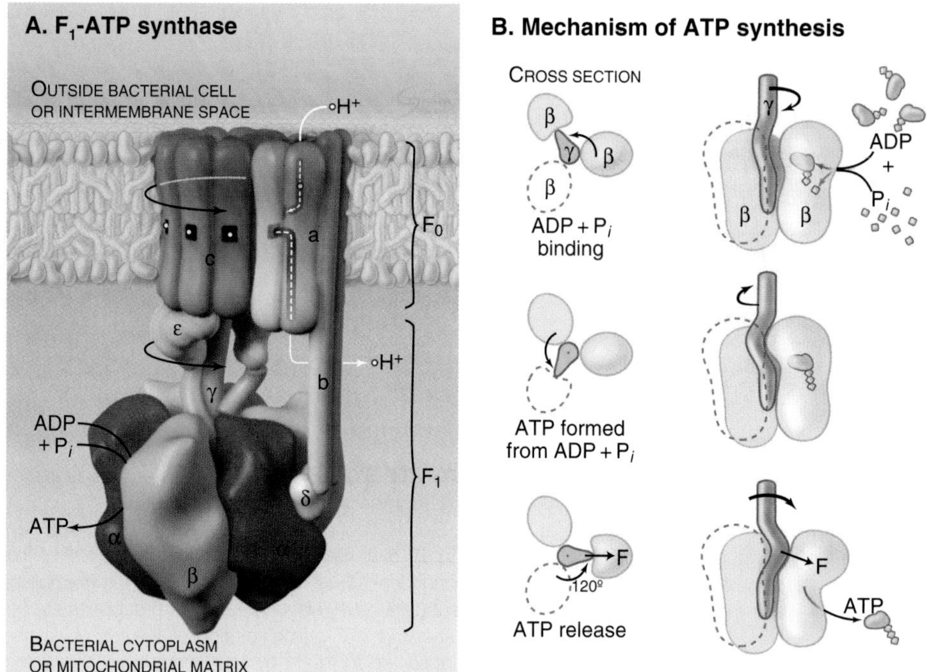

FIGURE 14.5 **A–B,** Model of the mitochondrial F_OF_1-ATP synthase. F_O is the oligomycin-sensitive factor. F_1 is the adenosine triphosphatase (ATPase). Proton transfer across the membrane can drive ATP synthesis, or ATP hydrolysis can pump protons across the membrane. The text explains the reversible ATPase reaction. ADP, adenosine diphosphate; P_i, phosphate. (Data from Elston T, Wang H, Oster G. Energy transduction in ATP synthase. *Nature*. 1998;391:510–513.)

15 c-subunits (depending on the species) form a ring with the acidic residue either exposed to the surrounding lipids or buried at the interface with the a-subunit. The single a-subunit is closely apposed to the ring of c-subunits and linked to the α- and β-subunits by one of three peripheral shafts. The a-subunit consists of multiple α-helices that form two half channels for protons to move across the membrane. Other protein subunits attach the ring of c-subunits to the γ-subunit camshaft that rotates inside the ring of α- and β-subunits.

The R_O motor couples proton translocation to rotation during ATP synthesis as follows. The conserved acidic residues of the c-subunits (Glu63 in *Thermus*, Asp61 in mitochondria) in contact with the interior of the lipid bilayer are protonated, so they are uncharged. As a c-subunit rotates into contact with the a-subunit, the proton on Glu63 escapes through the first half channel into the bacterium or mitochondrion. Thermal motion drives random Brownian motions of the c-ring, but the negatively charged side chain of Glu63 favors rotation that brings it into alignment with the second half channel of the a-subunit. There a proton enters from outside the bacterium or mitochondrion and neutralizes the carboxyl group. The number of protons transported for each ATP synthesized depends on the number of c-subunits, but ranges from three to five.

The reactions in the R_O and R_1 complexes are tightly coupled. For example, reaction of the chemical DCCD (dicyclohexylcarbodiimide) with the conserved acidic residue on a single c-subunit blocks proton conduction and ATP hydrolysis or synthesis. All the transitions are reversible in bacterial ATP synthases. Given a higher concentration of protons outside than inside, protons pass through R_O and drive the synthesis of ATP. Conversely, ATP hydrolysis by R_1 can drive protons out of the cell when required.

Evolutionary Origins and Diversity of Rotary Adenosine Triphosphatases

The original rotary ATPase in the common ancestor of life on earth may have evolved from enzymes called helicases consisting of a hexamer of subunits structurally homologous to the β/B subunits of rotary ATPases. These helicases use energy from ATP hydrolysis to thread DNA or RNA through the channel in the middle of the hexamer. Some helicases associate with channels that could have been the precursor of the R_O. No example of the primordial enzyme still exists and the initial pathways of divergence are unclear, but the rotary ATPases diversified into three distinct families: F-type ATP synthases in many bacteria and eukaryotic mitochondria and chloroplasts; reversible A-type ATPases in other bacteria and archaea; and eukaryotic V-type proton pumps. The eukaryotic V-type pumps likely came from the Archaeal progenitor, but diverged in structure and became specialized to pump protons. Two versions of F-type ATP synthases came to eukaryotes with the symbiotic bacteria that gave rise to mitochondria and chloroplasts.

F-Type Adenosine Triphosphate Synthases

F-type ATPases of mitochondria, chloroplasts, and bacterial plasma membranes (Figs. 14.4C and 14.5) produce most of the world's ATP during aerobic metabolism (see Chapter 19). Their architectures resemble the A- and V-type ATPases, but only a single peripheral stalk connects the R_1 and R_O complexes. Redox-driven and light-driven pumps create proton gradients to drive ATP synthesis by F-type ATPases (see Figs. 19.5 and 19.8). Purified F-type ATPases are freely reversible, and when required by circumstances, many bacteria use their F-type ATPase to produce a proton gradient at the expense of ATP hydrolysis. Eukaryotes have elaborate mechanisms to prevent futile cycles of ATP synthesis and hydrolysis. For example, an inhibitory protein binds the mitochondrial ATP synthase and prevents ATP hydrolysis, if the oxygen supply required to generate the proton gradient is compromised.

V-Type Rotary Adenosine Triphosphatases

V-type ATPases are composed of subunits homologous to other rotary ATPases (Fig. 14.6). The A- and B-subunits are homologs of F-type α- and β-subunits. Three peripheral stalks anchor the channel forming an a-subunit to the hexamer of A- and B-subunits. An ancient gene duplication made the c-subunits of eukaryotic V-type pumps twice as large as those of F-type ATPases and archaeal A-type ATPases. Each of 10 c-subunits has four α-helices with one conserved aspartic acid residue to transfer the proton. Accordingly, a V-type ATPase typically hydrolyzes three ATPs to transport 10 protons.

V-type pumps acidify the interiors of many membrane-bound compartments in eukaryotic cells, including the Golgi apparatus (see Chapter 21), secretory vesicles, clathrin-coated vesicles, endosomes, lysosomes (see Fig. 22.15), and plant vacuoles. The acidic pH promotes ligand dissociation from receptors in endosomes and activates lysosomal hydrolases, as well as many other reactions (see Fig. 22.15). Proton gradients created across these internal membranes also provide the energy source to drive H^+-coupled transport of other solutes by carriers, such as the uptake of neurotransmitters by synaptic vesicles (see Figs. 17.8D and 17.9D). V-type pumps are also present in the plasma membranes of cells specialized to secrete protons, such as osteoclasts (see Fig. 32.6), macrophages, and kidney tubule intercalated cells.

A-Type Rotary Adenosine Triphosphatases

Archaea and some bacteria have A-type ATP synthases (Fig. 14.4A–B) powered by gradients of protons or Na^+ ions. Lipid bilayers are less permeable to Na^+ than H^+,

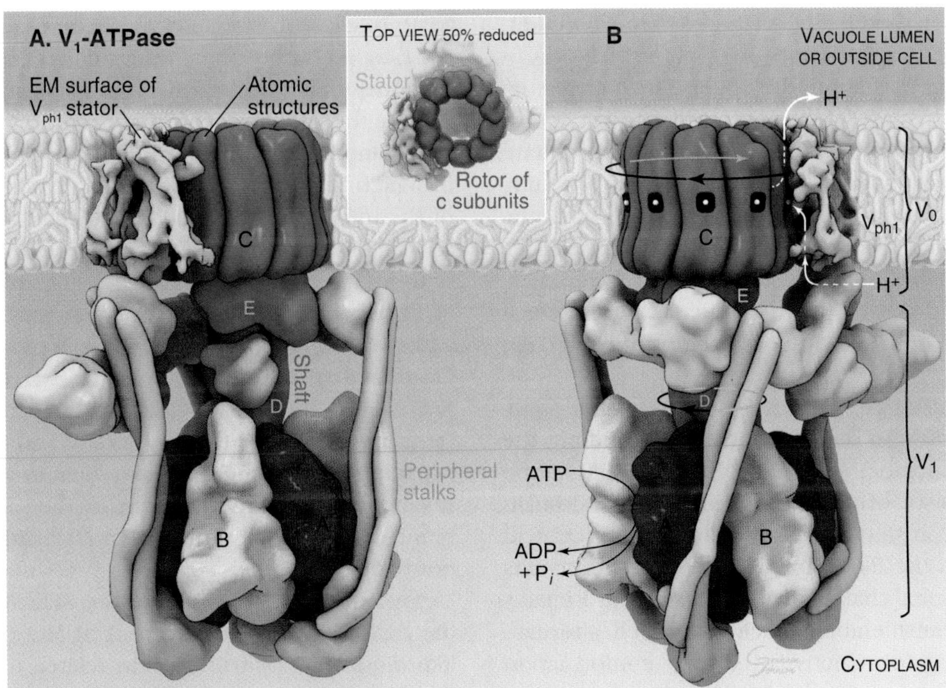

FIGURE 14.6 STRUCTURE OF THE BUDDING YEAST V-TYPE ADENOSINE TRIPHOSPHATASE. A–B, Side views of a space-filling three-dimensional model from two points of view based on reconstructions from cryoelectron micrographs. The stator subunit is shown as a light blue surface view of the electron microscopic density. The subunits are color coded and their names are indicated by letters. **Inset,** Top view of the stator subunit and a ring of 10 C-subunits, each with four transmembrane helices. **B,** Mechanism of the V-type pump. ATP hydrolysis rotates the central shaft and pumps protons across the membrane through the stator and C-subunits. (For reference, see Zhao J, Benlekbir S, Rubinstein JL. Electron cryomicroscopy observation of rotational states in a eukaryotic V-ATPase. *Nature*. 2015;521:241–245.)

which may avoid dissipating energy in some environments. These enzymes have two peripheral stalks.

P-Type Cation Pumps: E_1E_2-Adenosine Triphosphatases

All living organisms depend on P-type ATPases (Table 14.2) to pump cations across membranes. Their name comes from the fact that they use a high-energy covalent **β-aspartyl phosphate intermediate.** They are also called E_1E_2-ATPases from a description of the conformational changes that they undergo during the course of their mechanism.

Eukaryotic P-type ATPases generate primary ion gradients across the plasma membrane that are required for the function of ion channels (see Chapter 16) and most cation-coupled transport mediated by carrier proteins (see Chapter 15). Producing these primary ion gradients is expensive, consuming about one-third of the human body's energy. In animal cells, **Na^+K^+-ATPase** produces the primary gradients of Na^+ and K^+. In plants and fungi, the functional homolog H^+-ATPase generates a proton gradient. Other eukaryotic P-type ATPases acidify the stomach and clear the cytoplasm of the second messenger, Ca^{2+} (see Fig. 26.12). Bacterial P-type ATPases scavenge K^+ and Mg^{2+} from the medium and export Ca^{2+}, Cu^{2+}, and toxic heavy metals. Most P-type ATPases pump cations, but the P4 class (flippases) move lipids from one side of a membrane to the other (see Chapter 13).

The best understood P-type ATPase is the sarco(endo) plasmic reticulum **Ca^{2+}-ATPase** (SERCA1), which pumps Ca^{2+} from the cytoplasm into the endoplasmic reticulum (Fig. 14.7). ATP hydrolysis provides energy to move Ca^{2+} across the membrane up a steep concentration gradient. The mechanism is understood in detail, thanks to extensive biochemical analysis and crystal structures of all the chemical intermediates along the pathway (Fig. 14.8). This thorough analysis was possible because the enzyme is abundant in the sarcoplasmic reticulum of skeletal muscle (see Fig. 39.13), allowing it to be purified in large quantities.

The Ca^{2+}-ATPase has 10 α-helices that cross the membrane bilayer and bind two Ca^{2+} ions side by side in the middle of the membrane. The globular region in the cytoplasm consists of three domains. The N-domain binds ATP and transfers its γ-phosphate to aspartic acid 351 (Asp351) in the P-domain. The A-domain transmits large conformational changes in the cytoplasmic domains to six of the transmembrane helices, which alternate between two conformations. In the E_1 conformation cytoplasmic Ca^{2+} has access to binding sites among the transmembrane helices (Fig. 14.8A). The E_2 conformation opens the Ca^{2+} binding sites to the lumen of the endoplasmic reticulum (Fig. 14.8B). Each step along the pathway—Ca^{2+} binding on the cytoplasmic side, ATP binding, phosphorylation of the enzyme, dissociation of ADP, release of Ca^{2+} into the lumen, and hydrolysis of the β-aspartyl phosphate intermediate—is linked conformational changes in both the cytoplasmic domains and the six transmembrane helices. The transition between the E_1 and E_2 conformations involves an "occluded" state in which the bound Ca^{2+} is not accessible on either side of the membrane. This occluded state allows the pump to transport against a large concentration gradient without a leak.

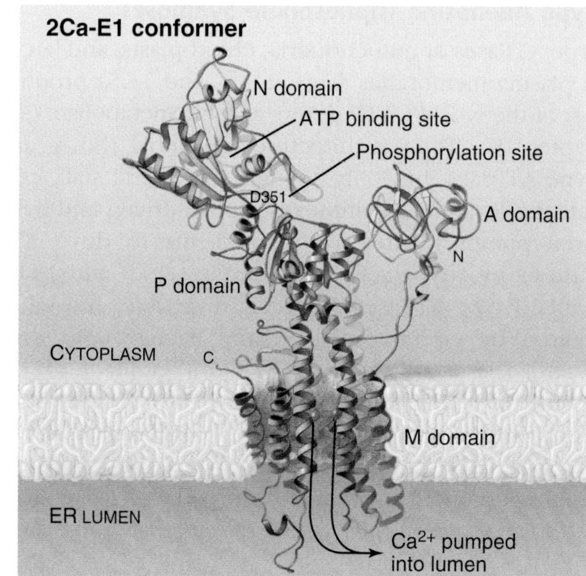

FIGURE 14.7 STRUCTURE OF A P-TYPE ADENOSINE TRIPHOSPHATASE PUMP. Crystal structure of sarcoendoplasmic reticulum calcium–ATPase (SERCA) from skeletal muscle in the 2Ca-E1 conformation with two Ca^{2+} ions bound among four of the 10 transmembrane helices near the middle of the membrane bilayer. In the cytoplasm, the N-domain binds nucleotide (ATP) and transfers the γ-phosphate to Asp351 (D351) in the P-domain. (For reference, see PDB file 1SU4 and Toyoshima C, Nakasako M, Nomura H, Ogawa H. Crystal structure of the calcium pump of sarcoplasmic reticulum at 2.6 Å resolution. *Nature*. 2000;405:647–655.)

Because the cycle transfers two Ca^{2+} into the lumen and two H^+ out, it generates an electrical potential (see Chapter 16). In the steady state, the Ca^{2+} gradient is maintained, but the H^+ gradient dissipates, owing to H^+ permeability across the membrane and to the buffering capacity of the lumen. All reactions in the pathway are reversible, so a large gradient of Ca^{2+} across the membrane can drive the synthesis of ATP, although this does not happen in cells.

Two mechanisms regulate the SERCA2 Ca-ATPase in the heart. Covalent addition of SUMO (small ubiquitin-like modifier), a small protein related to ubiquitin (see Fig 23.2), to two lysines is required for full activity. Phospholamban, a small accessory subunit with one transmembrane helix, inhibits SERCA2 activity. Cyclic adenosine monophosphate (AMP)-protein kinase phosphorylates phospholamban, overcoming this inhibition and stimulating calcium pumping (see Chapter 39).

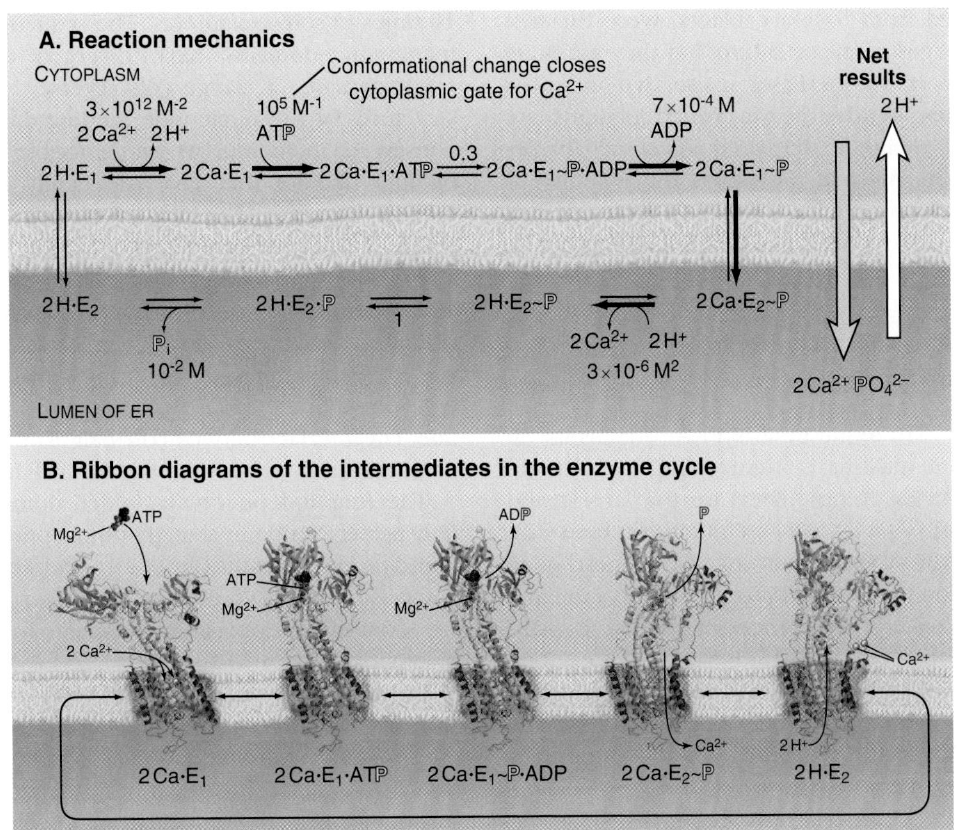

FIGURE 14.8 MECHANISM OF THE SARCOPLASMIC RETICULUM CALCIUM–ATPASE (SERCA). A, Biochemical pathway. The E stands for enzyme, having two conformations: E_1 and E_2. E_1 has the Ca^{2+}-binding site oriented toward the cytoplasm. E_2 has the Ca^{2+}-binding site oriented toward the lumen of the endoplasmic reticulum. Ca^{2+} binds E_1 on the cytoplasmic side. Subsequent binding of ATP and phosphorylation of the enzyme drive the enzyme toward the E_2 state and transport Ca^{2+} up a steep concentration gradient into the lumen of the endoplasmic reticulum. Dephosphorylation of the enzyme favors a return to the E_1 state. **B,** Ribbon diagrams of structures along the biochemical pathway. Conformational changes are coupled to ATP hydrolysis and transport of Ca^{2+}. Starting on the left side, the pump without bound Ca^{2+} or ATP vacillates between the E_2 and E_1 conformations, alternatively exposing the Ca^{2+}-binding sites on the two sides of the membrane. If Ca^{2+} ions are available in the cytoplasm, such as after activation of muscle (see Fig. 39.15), they bind cooperatively to the E_1 conformation with micromolar affinity. Ca^{2+} binding favors Mg-ATP binding to the N-domain and striking conformational changes that rotate the N-domain and close a gate between the bound Ca^{2+} and the cytoplasm when the A-domain pulls on a transmembrane helix. This conformation with the γ-phosphate of ATP close to the side chain of Asp351 allows formation of the phosphoenzyme intermediate. The equilibrium constant for phosphorylation is near unity, so most of the energy from ATP hydrolysis is stored in a high-energy conformation of the protein. Then ADP (adenosine diphosphate) dissociates resulting in a rotation of the A-domain that moves several transmembrane helices to expose the Ca^{2+}-binding sites to the lumen of the endoplasmic reticulum and reduces the affinity for Ca^{2+} by orders of magnitude. Thus Ca^{2+} dissociates into the lumen, completing its uphill transfer from the cytoplasm. Hydrolysis of the phosphorylated intermediate and dissociation of phosphate reverse the conformational changes in both the cytoplasmic and transmembrane domains, completing the cycle. (For reference, see PDB files 1SU4, 1T5S, 1T5T, 1WPG, and 1IWO and Toyoshima C, Nomura H, Tsuda T. Lumenal gating mechanism revealed in calcium pump crystal structures with phosphate analogues. *Nature.* 2004;432:361–368; and Soerensen T, Moeller JV, Nissen P. Phosphoryl transfer and calcium ion occlusion in the calcium pump. *Science.* 2004;304:1672–1675.)

Other P-type ATPases consist of structurally homologous α-subunits with large cytoplasmic domains and a minimum of six transmembrane helices. Eukaryotic P-type ATPases, such as the Na^+K^+-ATPase, have 10 transmembrane helices. These pumps work the same way as the Ca^{2+}-ATPase, but with adaptations to pump other ions. For example, the E_1 conformation of the Na^+K^+-ATPase picks up three Na^+ from the cytoplasm and the P-E_2 conformation releases these Na^+ sequentially outside the cell. Binding of two extracellular K^+ leads to dephosphorylation of the enzyme and results in the occlusion of K^+ in the KE_2 conformation. ATP binding releases this K^+ on the cytoplasmic side and regenerates the E_1 conformation, so the cycle can start over again. An extra 50-kD β-subunit with a single transmembrane helix and a glycosylated extracellular globular domain is required for both transport and intracellular trafficking of the Na^+K^+-ATPase and H^+K^+-ATPase.

P-type ATPases are involved in human diseases. Mutations in the SERCA1 Ca^{2+}-ATPase cause muscle stiffness and cramps. Malfunctions of the SERCA2 contribute to heart failure. Drugs called **cardiac glycosides** can strengthen the heartbeat by inhibiting the cardiac isoform of Na^+K^+-ATPase (see Fig. 39.22 for details). These drugs, originally derived from foxglove plants, were the first used to treat congestive heart failure, but they are quite toxic. Mutations in Cu^{2+}-ATPases cause two inherited diseases: **Menkes syndrome,** in which patients are copper-deficient owing to impaired intestinal absorption, and **Wilson disease,** in which the inability to remove copper from the liver is toxic. Omeprazole and related drugs are used to treat ulcers by inhibiting gastric H^+K^+-ATPase.

Adenosine Triphosphate–Binding Cassette Transporters

ABC transporters are found in all known organisms, so the founding gene must have originated in the common ancestor of all living things. They are the largest and most diverse family of ATP-powered pumps (Table 14.2). For example, ABC transporters are the largest gene family in the colon bacterium *Escherichia coli.* Similarly, the genome of baker's yeast encodes at least 30 ABC transporters, compared with 16 P-type ATPases, one F-type ATPase, and one V-type ATPase. In eukaryotes particular family members are located in the plasma membrane, endoplasmic reticulum, mitochondria and, most likely, other membranes.

Most ABC transporters are specific for one or a few related substrates, but the family as a whole has an enormous range of substrates, including inorganic ions, sugars, amino acids, lipids, complex polysaccharides, peptides, and even proteins. Specialized members of the family act as ion channels (eg, the **cystic fibrosis transmembrane regulator [CFTR]**) or regulate other membrane proteins, such as the sulfonylurea receptor.

ABC transporters have a modular design with two domains that cross the membrane and two cytoplasmic domains that hydrolyze ATP (Figs. 14.9 and 14.10). Each transmembrane domain consists of a bundle of α-helices that spans the bilayer: typically six times but up to 10 times in some examples. The sequences of the transmembrane domains have diverged considerably to accommodate a range of diverse substrates. The structures of the nucleotide binding domains are more conserved, including two sequences in the nucleotide-binding domain that give the family its name (ATP-binding cassette). The Walker A motif (GXXGXGKS/T, where X is any residue) is also called a P loop because it binds the γ-phosphate of ATP in ABC transporters and other ATP-binding proteins. The Walker B motif ($RX_{6-8}F_4D$, where F is any hydrophobic residue) interacts with the Mg^{2+} bound to ATP. Approximately 100 residues typically separate motif B from motif A in the polypeptide. The interface between the pair of nucleotide-binding domains forms two active sites for ATP hydrolysis.

The four independently folded domains of an ABC transporter can be in a single polypeptide or as many as four different subunits (Fig. 14.9). The single polypeptide comprising some vertebrate ABC transporters includes an additional cytoplasmic R-domain for regulation by phosphorylation.

Crystal structures of several ABC transporters and analysis of the enzyme mechanism suggest how ABC transporters might work (Fig. 14.10). The two transmembrane subunits form two chambers for the substrate. In one conformation a chamber lined by hydrophobic side chains is open outside the cell (the periplasmic space in the case of prokaryotes) (Fig. 14.10A). A gate composed of conserved residues blocks access of the substrate to the cytoplasm. In the other conformation, the chamber is open to the cytoplasm (Fig. 14.10B). Each nucleotide-binding cytoplasmic domain partners with a transmembrane subunit and the other nucleotide-binding domain. Two bound ATP molecules link the nucleotide-binding subunits together.

Prokaryotes often use a periplasmic subunit that binds a specific substrate and delivers it to the pump. Because the substrate binding proteins are asymmetric,

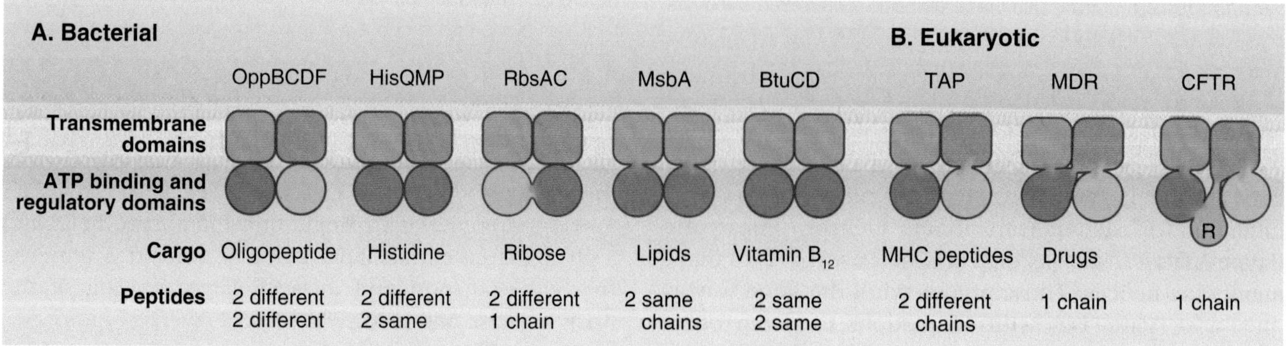

FIGURE 14.9 DOMAIN ARCHITECTURE OF ADENOSINE TRIPHOSPHATE BINDING CASSETTE (ABC) TRANSPORTERS. A, Bacterial transporters. **B,** Eukaryotic transporters. Each transporter has two ATP-binding domains in the cytoplasm *(purple circles)* and two transmembrane domains, each consisting of 6 to 10 α-helices (*blue* or *pink squares*). The cystic fibrosis transmembrane regulator (CFTR) has an additional regulatory (R) domain in the cytoplasm. The four domains required for activity may be four separate polypeptides or may be incorporated in several ways into polypeptides with two or four domains. MHC, major histocompatibility complex.

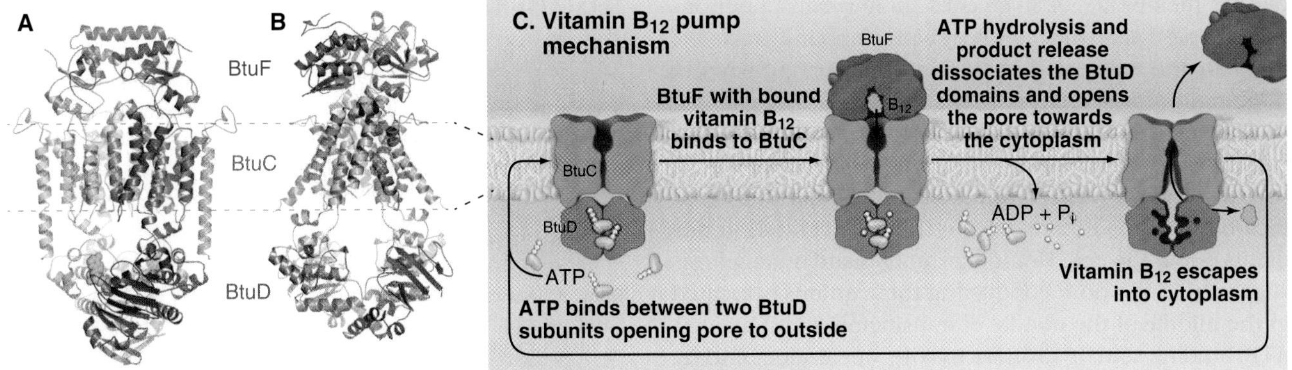

FIGURE 14.10 STRUCTURE OF TWO ADENOSINE TRIPHOSPHATE BINDING CASSETTE (ABC) TRANSPORTERS AND PROPOSED MECHANISM OF TRANSPORT. A, Ribbon diagram of the crystal structure of the *Escherichia coli* BtuCDF vitamin B_{12} transporter with bound adenylyl-imidodiphosphate (AMP-PNP). Two identical BtuC subunits *(blue)* traverse the membrane with a central cavity open to the outside. Two identical BtuD subunits *(red)* bind and hydrolyze ATP. A periplasmic protein BtuF binds and delivers vitamin B_{12} to BtuCD. (See PDB file 4F13.) **B,** Ribbon diagram of the *Archaeoglobus fulgidus* of ModBC molybdate transporter without bound nucleotide and the cavity between the transmembrane domains of MobB open to the cytoplasm. **C,** Proposed mechanism of vitamin B_{12} transport: BtuCD begins on the far right free of both vitamin B_{12} and ATP; two ATP molecules bind between the BtuD subunits, bringing them together and opening the cavity between the transmembrane subunits to the outside; BtuF with bound vitamin B_{12} binds to BtuC and delivers vitaminB_{12} to BtuC; after ATP hydrolysis, adenosine diphosphate (ADP) and inorganic phosphate (P_i) dissociate and separation of the BtuD domains is coupled to conformational changes that open the gate of BtuC transiently, releasing vitamin B_{12} into the cytoplasm. (For reference, see Hollenstein K, Frei DC, Locher KP. Structure of an ABC transporter in complex with its binding protein. *Nature*. 2007;446:213–216; and Korkhov VM, Mireku SA, Locher KP. Structure of AMP-PNP-bound vitamin B12 transporter BtuCD-F. *Nature*. 2012;490:367–372.)

they interact differently with the two transmembrane domains.

ATP binding and hydrolysis drive a cycle of conformational changes that allows a substrate to bind on one side of the membrane and to escape on the other side (Fig. 14.10C). The most complete set of structures is available for the bacterial vitamin B_{12} transporter BtuCD. ATP binding promotes association of the BtuD subunits, changing their orientations and opening a cavity in the transmembrane BtuC subunits facing the periplasm. A BtuF protein carrying vitamin B_{12} binds and unloads the substrate into the open cavity. ATP hydrolysis and release of ADP and P_i cause dissociation of the cytoplasmic domains resulting in a conformational change that transiently opens a pathway for vitamin B_{12} to exit into the cytoplasm. Two ATPs are hydrolyzed for each vitamin transported.

ABC transporters with similar mechanisms include bacterial permeases that pump nutrients into the cell, and **TAP1** and **TAP2** that pump peptide fragments of antigenic proteins into the lumen of the endoplasmic reticulum. Some substrates of ABC transporters are membrane bound. For example, the *E. coli* "flippase" MsbA moves phospholipids from one leaflet of the bilayer to the other, whereas yeast STE6 transports a small, prenylated pheromone peptide out of the cell.

The **multiple drug resistance** proteins (**MDR1**, **MDR2** and related proteins) are ABC transporters that provide a substantial challenge for cancer chemotherapy (Fig. 14.11). In about half of cases in which

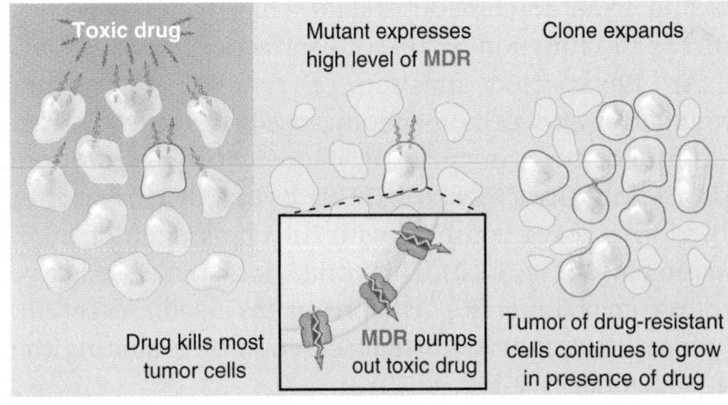

FIGURE 14.11 MULTIPLE DRUG RESISTANCE (MDR) IN CANCER CHEMOTHERAPY. In a population of tumor cells, most are sensitive to killing by a chemotherapeutic drug. However, variants that express high levels of the ABC transporter, MDR, can clear the cytoplasm of the drug. A clone of these variant cells may expand, allowing the tumor to grow in the presence of the drug.

chemotherapy fails to cure a human cancer, the cause is the emergence of clones of tumor cells that overexpress an MDR. Normal cells use a low level of MDR1 to export unknown natural substrates, perhaps a steroid, a phospholipid, or another hydrophobic molecule. MDR can also transport many hydrophobic compounds, including some chemotherapeutic drugs. These drugs enter cells by crossing the plasma membrane and poison vital cellular processes. Cells that overexpress MDR survive by pumping the drug out of the cell.

The multiple drug resistance protein 2 (MDR2) is another unconventional pump located in the apical plasma membrane of liver cells. It normally functions as a flippase that moves phosphatidylcholine from the inner to the outer half of the lipid bilayer, perhaps in preparation for secretion in bile.

The vertebrate cystic fibrosis transmembrane conductance regulator (CFTR [see Fig. 14.9B]) looks like an ABC transporter but acts like a channel. It allows Cl^- and bicarbonate ions to move down their concentration gradients out of the cell. Homology models and mutagenesis suggest that the pore conducting these anions is located in the middle of the bundle of transmembrane α-helices. ATP binding and hydrolysis by the nucleotide-binding domains regulate gating of the channel, but the mechanism is incompletely understood. Mutations in CFTR are responsible for cystic fibrosis (see Fig. 12.4). Of more than 1000 known disease-causing mutations by far the most common is deletion of phenylalanine 508. This position corresponds to a highly conserved hydrophobic residue in the vitamin B_{12} transporter that is important for interaction of BtuD with BtuC. Mutant ΔF508 CFTR folds very slowly and is retained in the endoplasmic reticulum (ER) where it is destroyed, depriving the plasma membrane of chloride channel activity. Depleting Ca^{2+} from the ER by inhibiting the SERCA Ca^{2+}-ATPase can apparently allow some ΔF508 CFTR to escape from calcium-dependent chaperones (see Fig. 20.12) and function on the cell surface.

An unexpected function of an ABC transporter emerged from work on sulfonylurea drugs used to treat forms of diabetes involving inadequate insulin secretion. The sulfonylurea receptor (SUR) is an ABC transporter that forms a 4:4 complex with Kir6 potassium channel subunits (see Fig. 11.2). This complex is an ATP-sensitive potassium channel (K_{ATP}) that regulates insulin secretion. Sulfonylureas activate insulin secretion by inhibiting the K_{ATP} channel (see Fig. 11.2).

ACKNOWLEDGMENTS

We thank Kaspar Locker and John Rubinstein for suggestions on revisions of this chapter.

SELECTED READINGS

Hollenstein K, Dawson RJ, Locher KP. Structure and mechanism of ABC transporter proteins. *Curr Opin Struct Biol.* 2007;17:412-418.

Janas E, Hofacker M, Chen M, et al. The ATP hydrolysis cycle of the nucleotide-binding domain of the mitochondrial ATP-binding cassette transporter Mdl1p. *J Biol Chem.* 2003;278:26862-26869.

Jih KY, Hwang TC. Nonequilibrium gating of CFTR on an equilibrium theme. *Physiology (Bethesda).* 2012;27:351-361.

Junge W, Nelson N. ATP synthase. *Annu Rev Biochem.* 2015;84:631-657.

Kanai R, Ogawa H, Vilsen B, Cornelius F, Toyoshima C. Crystal structure of a Na^+-bound Na^+,K^+-ATPase preceding the E1P state. *Nature.* 2013;502:201-206.

Kho C, Lee A, Jeong D, et al. SUMO1-dependent modulation of SERCA2a in heart failure. *Nature.* 2011;477:601-605.

Kinosita K Jr, Adachi K, Itoh H. Rotation of F_1-ATPase: How an ATP-driven molecular machine may work. *Annu Rev Biophys Biomol Struct.* 2004;33:245-268.

Korkhov VM, Mireku SA, Veprintsev DB, Locher KP. Structure of AMP-PNP-bound BtuCD and mechanism of ATP-powered vitamin B12 transport by BtuCD-F. *Nat Struct Mol Biol.* 2014;21:1097-1099.

Kuhlbrandt W. Biology, structure and mechanism of P-type ATPases. *Nat Rev Mol Cell Biol.* 2004;5:282-295.

Lanyi JK. Bacteriorhodopsin. *Annu Rev Physiol.* 2004;66:665-688.

Lórenz-Fonfría VA, Heberle J. Channelrhodopsin unchained: structure and mechanism of a light-gated cation channel. *Biochim Biophys Acta.* 2014;1837:626-642.

Mulkidjanian AY, Makarova KS, Galperin MY, Koonin EV. Inventing the dynamo machine: the evolution of the F-type and V-type ATPases. *Nat Rev Microbiol.* 2007;5:892-899.

Nichols CG, Singh GK, Grange DK. KATP channels and cardiovascular disease: suddenly a syndrome. *Circ Res.* 2013;112:1059-1072.

Stewart AG, Laming EM, Sobti M, Stock D. Rotary ATPase-dynamic molecular machines. *Curr Opin Struct Biol.* 2014;25:40-48.

Toyoshima C, Iwasawa S, Ogawa H, et al. Crystal structures of the calcium pump and sarcolipin in the Mg^{2+}-bound state. *Nature.* 2013;495:260-264.

Toyoshima C, Nomura H, Tsuda T. Lumenal gating mechanism revealed in calcium pump crystal structures with phosphate analogues. *Nature.* 2004;432:361-368. (See a movie of the structural changes during the ATPase cycle at <https://legacy.wlu.ca/documents/29450/nature06418-s2.mov>.)

Wickstrand C, Dods R, Royant A, Neutze R. Bacteriorhodopsin: Would the real structural intermediates please stand up? *Biochim Biophys Acta.* 2015;1850:536-553.

Winther AM, Bublitz M, Karlsen JL, et al. The sarcolipin-bound calcium pump stabilizes calcium sites exposed to the cytoplasm. *Nature.* 2013;495:265-269.

CHAPTER 15

Membrane Carriers

Carriers are integral membrane proteins that move select chemical substrates across all cellular membranes (Fig. 15.1). Common substrates for carriers are ions and small, soluble organic molecules, but some substrates are lipid soluble. The energy to transport substrates comes from electrochemical gradients across the membrane. Some carriers transport substrates down concentration gradients, but others use transmembrane ion gradients created by pumps to transport across a membrane up a concentration gradient.

Carriers are also known as facilitators, transporters, or porters. This book uses "carrier" because the term unambiguously identifies them, whereas "transporter" is also used to describe pumps, leading to unnecessary confusion.

Basic Carrier Mechanism

Atomic structures of carriers reveal a variety of protein folds, so multiple carrier genes must have arisen during the early days of life on earth. Remarkably, these proteins converged during evolution on a common mechanism.

All carriers are composed of bundles of transmembrane helices that form a binding site for one or more substrates in the middle of the membrane bilayer (Fig. 15.1). Carrier proteins have at least two conformations: one where the substrate has access to the binding site from outside the cell and one where access is available from inside the cell. Both conformations have tight seals formed by hydrophobic residues that keep solutes from crossing the membrane. Some carriers have an additional occluded conformation, where the substrate-binding site is closed on both sides of the membrane. By alternating between the open-out and open-in conformations, the carrier provides a pathway across the membrane.

Carriers work step by step like enzymes, binding substrates on one side of membranes, undergoing a conformational change that reorients this binding site, and releasing substrate on the opposite side of the

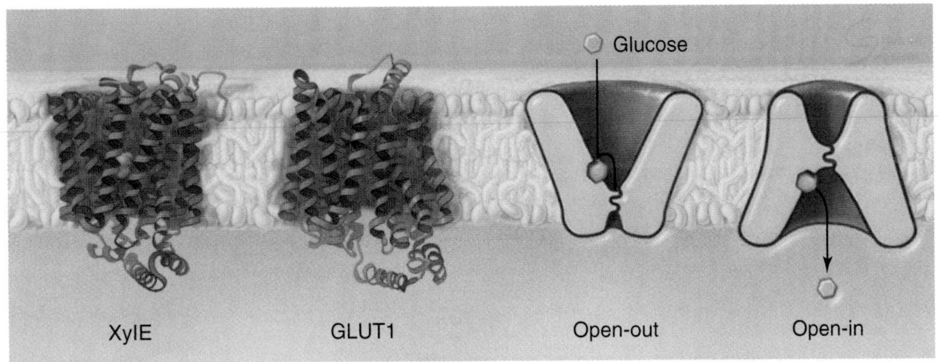

FIGURE 15.1 TRANSPORT BY A CARRIER PROTEIN. Structural basis for transport of glucose *(green)* across the plasma membrane by a carrier protein. Ribbon diagrams show the structures of the glucose carriers XylE with the glucose binding site open outside the cell and glucose transporter 1 (GLUT1) with the glucose binding site open to the cytoplasm. The diagrams show how glucose moves down its concentration gradient by binding to open-out conformation and being released into the cytoplasm by the open-in conformation. The XylE H^+-glucose symporter is shown as a model for the open-out conformation of the GLUT1 uniporter. (For reference, see Protein Data Bank [PDB; www.rcsb.org] files 4GC0 and 4PYP and Deng D, Xu C, Sun P, et al. Crystal structure of the human glucose transporter GLUT1. *Nature.* 2014;510:121–125.)

membrane. In many cases a simple rocking motion, shown diagrammatically in Fig. 15.1, explains this alternating access mechanism. The conformational change that moves substrates across the membrane is rate limiting (on the order of 0.1 to 1000 events per second), whereas channels transfer ions at rates of 10^6 to 10^9 s^{-1} during the brief times that they open (see Chapter 16).

Three Transport Strategies

Transport by carriers depends on energy provided by gradients of solutes across the membrane (Fig. 15.2). All carriers move solutes down concentration gradients. Remarkably, many carriers transport a second substrate up a concentration gradient powered by the transport of another substrate down its electrochemical gradient. All carrier-mediated reactions are reversible, so substrate concentrations on the two sides determine the rates of the second order binding reactions and thus the direction of net transfer across the membrane.

Classic physiological experiments done well before the first carrier structures were determined in the early 2000s led to the discovery of the three types of carriers, characterized their mechanisms, and even correctly predicted some of the molecular details (Box 15.1). Carriers are classified into three broad categories (Fig. 15.2) depending on how transport is coupled to the energy source:

- **Uniporters** transport a single substrate across the membrane and down its electrochemical gradient, hence the prefix "uni-." This reaction is also called **facilitated diffusion**—facilitated in the sense that the carrier provides a low-resistance pathway across a poorly permeable lipid bilayer. Nonelectrolytes, such as glucose, use uniporters simply to move across membranes down a chemical gradient. The substrate binding sites are more fully occupied on the uphill side of the concentration gradient, so the net movement of substrate is downhill. When substrate concentrations are the same on the two sides, exchange continues without net movement because the carrier is equally saturated on both sides of the membrane. Glucose transporter (GLUT) carriers for glucose are an example of uniporters found in mammalian tissues (Fig. 15.1). Movement of charged substrates is influenced by the membrane potential and by pH gradients in the case of weak acids or bases.
- **Antiporters** exchange two substrates in opposite directions across a membrane. Movement of one substrate down its concentration gradient drives the other substrate up its concentration gradient (Fig. 15.2). Antiporters generally exchange like-for-like substrates: cations for cations, anions for anions, sugars for sugars, and so on. Typically, the two substrates compete for binding antiporters, so only one binds at a time. A bound substrate is required for the reversible, conformational change that exposes the substrate-binding site on one or the other side of a membrane. These features make transport of two substrates dependent on each other in an obligate fashion. For example, the adenine nucleotide carrier (ANC) antiporter exchanges adenosine diphosphate (ADP) for adenosine triphosphate (ATP) across the mitochondrial inner membrane (Fig. 15.4A).
- **Symporters** allow two or more substrates to move together in the same direction across a membrane. A concentration gradient of one substrate, often Na^+, provides the energy to transport the other substrate

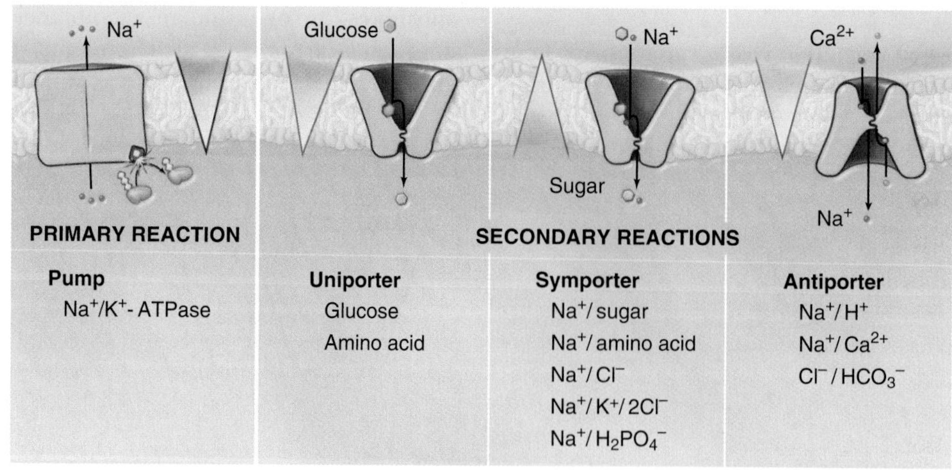

FIGURE 15.2 PRIMARY AND SECONDARY TRANSPORT REACTIONS. An adenosine triphosphate (ATP)-driven pump produces a gradient of an ion, such as Na^+, across a membrane. The triangles represent gradients of Na^+ *(purple)*, glucose *(green)*, sugar *(blue)*, and Ca^{2+} *(light blue)* across the membrane. The Na^+ gradient drives secondary transport reactions mediated by carriers. Uniporters allow an ion or other solute to move across the membrane down its concentration gradient. Symporters and antiporters couple transport of an ion (Na^+ in this example) down its concentration gradient with the transport of a solute (glucose or Ca^{2+} in these examples) up its concentration gradient. Antiporters carry out these reactions in succession, picking up Na^+ outside, reorienting, dissociating Na^+ inside, picking up Ca^{2+} inside, reorienting, and releasing Ca^{2+} outside. ATPase, adenosine triphosphatase.

BOX 15.1 Discovery of Carrier Proteins and Mechanisms

Classic experiments with **GLUT1** in the plasma membrane of red blood cells led to the carrier concept in the 1950s. Human red blood cells are convenient to use, because they express high concentrations of GLUT carriers. The time course of radioactive glucose accumulation (Fig. 15.3A) showed that transport is stereospecific for D-glucose and that net transport stops when the concentrations of D-glucose are equal inside and out. Slow equilibration of L-glucose across the membrane probably represents passive diffusion across the lipid bilayer, as this rate can be predicted from the solubility of glucose in membrane lipids. This experiment showed that something in the membrane accelerates the rate of glucose entry, giving rise to the concept of facilitated diffusion.

The dependence of the initial rate of D-glucose entry on its concentration (Fig. 15.3B) provided evidence for a specific, saturable, carrier molecule in the membrane. Because the rate of D-glucose entry includes both diffusion across the bilayer and movement through a carrier, the L-glucose rate was used to correct for the rate of diffusion. Once this was been done, the rate of facilitated D-glucose entry had a hyperbolic dependence on the concentration of D-glucose (Fig. 15.3C). This concentration dependence is just like a bimolecular binding reaction (see Fig. 4.2) or the rate of a simple enzyme mechanism that depends on the rate of substrate binding to the enzyme. Thus, the substrate concentration at the half-maximal velocity provides an estimate of the affinity of the carrier for the substrate. At high substrate concentrations, the substrate-binding sites on the carrier are saturated, and the rate plateaus at a maximal velocity owing to rate-limiting conformational changes. These enzyme-like properties, along with the ability to stop facilitated transport with protein inhibitors, suggest that carriers are proteins with specific binding sites for their substrates.

Two key experiments (Fig. 15.3D–E) established the symporter concept. The first demonstrated that extracellular Na^+ was required for membranes from intestinal cells with the SGLT1 carrier to accumulate D-glucose against a concentration gradient. This experiment left open the possibility that Na^+ simply activates the carrier in some way without being used directly to drive glucose accumulation. Second, an experiment with LacY demonstrated that sugar and cation move across the membrane together. Not only was a proton gradient required for sugar transport, but also a high concentration of another sugar (a nonmetabolizable lactose analog) in the medium could drive H^+ into the cell along with the sugar. Additional experiments confirmed that the stoichiometry of the reaction was one lactose transported in for every H^+ transported in. (Because this transport reaction is not electrically neutral, the membrane potential is a factor, and another pathway must be available to balance the charge—eg, by carrying K^+ in the opposite direction.) Parallel experiments on plasma membrane vesicles isolated from vertebrate kidney cells showed that Na^+ moving inward down its electrochemical gradient can carry glucose in with it. When the Na^+ concentration is equal on the two sides, the carrier facilitates the movement of glucose across the membrane but not its net accumulation.

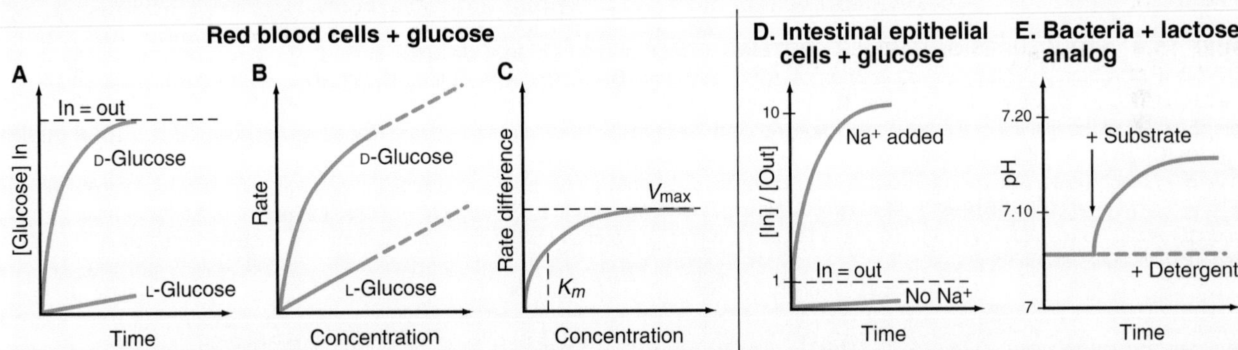

FIGURE 15.3 EXPERIMENTS THAT REVEALED THE EXISTENCE OF MEMBRANE CARRIERS. A-C, Experiments on human red blood cells. **A,** Time course of the uptake of D- and L-glucose. **B,** Rate of uptake of D- and L-glucose as a function of extracellular concentration. Uptake of L-glucose was by diffusion across the lipid bilayer. **C,** Rate of uptake of D-glucose corrected for diffusion as a function of extracellular concentration. The curve is similar to the dependence of an enzyme on substrate concentration, yielding the maximum rate (V_{max}) at high substrate concentration and the apparent affinity of the carrier (K_m) for the substrate at half the maximal rate. **D, E,** Experiments revealing the existence of symporters. **D,** The effect of Na^+ on the uptake of radioactive glucose by apical plasma membrane vesicles isolated from intestinal epithelial cells containing the Na^+/glucose symporter SGLT1. The addition of Na^+ to the external buffer strongly favors glucose uptake against its concentration gradient. **E,** Cotransport of protons and lactose by bacteria expressing LacY H/lactose symporter. Bacteria were suspended in a weakly buffered medium. Lactose added to the medium moved into the cells and down its concentration gradient. Protons accompanied lactose through the symporter, raising the pH of the medium. If detergent made the membrane permeable, the pH did not change.

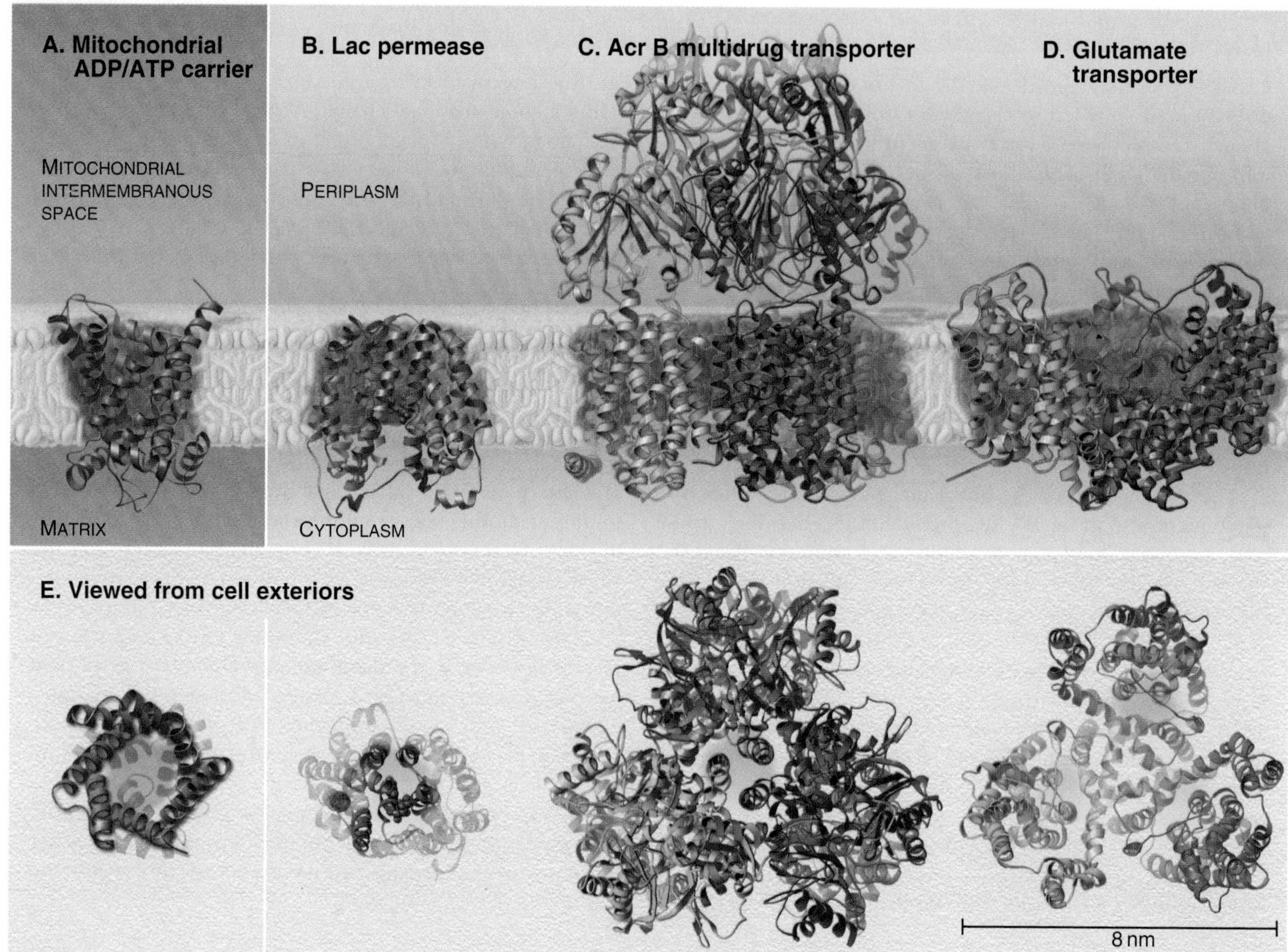

FIGURE 15.4 STRUCTURES OF CARRIER PROTEINS. Ribbon diagrams of four carriers illustrate their diversity. **A,** Bovine mitochondrial adenosine diphosphate (ADP)/adenosine triphosphate (ATP) transporter. **B,** *Escherichia coli* Lac Y. **C,** *E. coli* AcrB multidrug transporter. The three identical subunits are shown in different colors. **D,** *Pyrococcus horikoshii* glutamate transporter. The three identical subunits are shown in different colors. **E,** Views of the exterior of the cell or mitochondrion with the transport pathway shaded tan. (For reference, see PDB files 1OKC [**A**], 1PV7 [**B**], 1OY8 [**C**], and 1XFH [**D**] and Pebay-Peyroula E, Dahout-Gonzalez C, Kahn R, et al. Structure of mitochondrial ADP/ATP carrier in complex with carboxyatractyloside. *Nature*. 2003;426:39–44; Abramson J, Smirnova I, Kasho V, et al. Structure and mechanism of the lactose permease of *E. coli*. *Science*. 2003;301:610–615; Murakami S, Nakashima R, Yamashita E, Yamaguchi A. Crystal structure of bacterial multidrug efflux transporter AcrB. *Nature*. 2002;419:587–593; and Reyes N, Ginter C, Boudker O. Transport mechanism of a bacterial homologue of glutamate transporters. *Nature*. 2009;462:880–885.)

across the membrane (Fig. 15.2). This is also known as **cotransport.** The two substrates bind cooperatively, and the conformational change that reorients the substrate-binding sites is much more favorable for free carrier and carrier with two bound substrates than for carrier with only one bound substrate. These features minimize leaks of one substrate across the membrane. *E. coli* LacY (Fig. 15.4B) and mammalian SGLT (sodium-dependent glucose transporter) Na^+-coupled GLUTs are examples of symporters.

Whether a carrier is a uniporter, antiporter, or symporter depends on the number of substrate-binding sites and the rate and equilibrium constants for the various species to reorient across the membrane. The net rate of transfer depends on the concentrations of substrates.

When a carrier uses an ion gradient to provide the energy to transport a second substrate, it is said to catalyze a secondary reaction. In this sense, pumps catalyze primary transport reactions (Fig. 15.2), using energy from ATP hydrolysis, electron transport, or absorption of light to create ion gradients (see Chapter 14). Coupling an ion gradient created by pumps to drive transport by a carrier is called a **chemiosmotic cycle** (see Fig. 17.1).

A few carriers are more complicated than is indicated by this classification. For example, excitatory neurotransmitter carriers catalyze both antiporter and symporter reactions, with Na^+ and Cl^- going in one direction and a neurotransmitter in the opposite direction (Fig. 15.4D). This example also makes the important general point that the stoichiometry of antiporter and symporter

TABLE 15.1 Examples of Carrier Proteins

Carrier	Type	Distribution	Substrate	Function
Mitochondrial Carriers				
ANC	Antiporter	Mitochondria	ADP/ATP	ATP, ADP exchange
UCP	Uniporter	Brown fat mitochondria	H+	Uncoupling protein, heat production
Sweet				
SemiSWEET	Uniporter	Bacteria	Glucose	Glucose uptake
SWEET	Uniporter	Plants	Glucose	Glucose secretion
Glutamate Carrier Family				
Glutamate carrier	Symporter	Neurons	2 Na^+/glutamate	Glutamate uptake at synapses
GltPh	Symporter	*Pyrococcus horikoshii*	2 Na^+/glutamate	Glutamate uptake
APC (Amino Acid-Polyamine-Organocation)				
Band 3	Antiporter	Red blood cells	HCO_3^-/Cl^-	Acid-base balance
GABA carrier	Symporter	Neurons, glia	Na^+/Cl^-/GABA	GABA uptake at synapses
LeuT	Symporter	*Aquifex aeolicus*	Na^+/Cl^-/aspartate	Aspartate uptake
SGLT1	Symporter	Intestine	Na^+/glucose	Glucose uptake
NHA				
ASBT	Symporter	Intestine	Na^+/bile acid	Bile acid retrieval
Major Facilitator Superfamily (MFS)				
GLUT1	Uniporter	Red blood cells	Glucose	Glucose uptake
GLUT4	Uniporter	Fat, muscle	Glucose	Insulin-responsive glucose uptake
LacY	Symporter	*Escherichia coli*	H^+/lactose	Lactose uptake
NCE	Antiporter	Muscle	$3Na^+/Ca^{2+}$	Ca^{2+} regulation of heart
NHE-1	Antiporter	Kidney, intestine	Na^+/H^+	Acid-base balance
NKCC1	Symporter	Kidney, gut, lung	$Na^+/K^+/2Cl^-$, water	NaCl regulation, fluid secretion
UhpT	Antiporter	*E. coli*	P_i/glucose 6-phosphate	Glucose 6-phosphate uptake
Bacterial Multidrug Carriers				
ArcB	Antiporter	*E. coli*	Drug/H^+	Drug export

ADP, adenosine diphosphate; ATP, adenosine triphosphate; GABA, γ-aminobutyric acid; GLUT, glucose transporter; P_i, inorganic phosphate; SGLT, sodium-dependent glucose transporter.

reactions need not be one-to-one. Table 15.1 provides other examples.

Diversity of Carrier Proteins

Biological experimentation and exploration of genomes have revealed more than a hundred families of carriers, many of which can be grouped into superfamilies. The full extent of carrier diversity in nature is not yet known. Crystal structures of a sample of carriers (Fig. 15.4) illustrate their diversity. These proteins differ in evolutionary origins but converged toward common mechanisms implemented by different structures. In all cases, transport depends on conformational changes, so understanding carrier mechanisms has depended on crystal structures. A number of specific inhibitors (Table 15.2) have been useful in establishing physiological functions of carriers.

Mitochondrial Carrier Family

Members of an extensive family of mitochondrial carrier proteins are the simplest known. The founding members of the family are antiporters that transport ATP and ADP across the membranes of mitochondria, chloroplasts, and other eukaryotic organelles (Fig. 15.4A). Their genes were formed by a threefold duplication and divergence of a sequence that encodes a pair of transmembrane helices connected by a short helix in the mitochondrial matrix. The six transmembrane helices form a cup with a common binding site for ATP and ADP in the middle of the bilayer. Conformational changes are proposed to expose this binding site alternatively on the two sides of the inner mitochondrial membrane without allowing protons to leak across the membrane. Nucleotides are transported down their concentration gradients, ADP

TABLE 15.2 Tools for Studying Carriers

Agent	Target
Furosemide*	$Na^+/K^+/2Cl^-$ symporter
Amiloride*	Na^+/H^+ antiporter
DIDS, SITS	HCO_3^-/Cl^- antiporter
Cytochalasin B	GLUT isoforms
Phloretin	GLUT isoforms
Phlorizin	SGLT isoforms

*Used clinically as a drug.
DIDS, 4,4-diisothiocyanatostilbene-2,2′-disulfonic acid; SITS, 4-acetamido-4′-isothiocyanostilbene-2,2′-disulfonic acid.

moves into mitochondria and ATP moves out. Other family members transport amino acids and ketoacids. Uncoupling protein 1 is a carrier that allows protons to escape from mitochondria in brown fat to generate heat when animals arise from hibernation and mammalian infants are born (see Fig. 28.3).

SWEET/SemiSWEET Sugar Carriers

A second family of tiny carriers also consists of six transmembrane helices, but the bacterial SemiSWEET uniporters are built from two subunits each with three transmembrane helices. A rocking motion opens the central binding site alternatively on the two sides of the membrane similar to other carriers. The homologous SWEET carriers of plants have a similar architecture, but all six transmembrane helices are formed by one polypeptide with an additional helix to link the two halves. These SWEET carriers transport glucose out of plant cells into the phloem.

Bacterial Multidrug Carriers

Multidrug carriers help bacteria in hostile environments by expelling a wide variety of toxic hydrophobic chemicals (Fig. 15.4C). Substrates include bile salts, dyes, detergents, and lipid-soluble antibiotics. When overexpressed, these carriers can make bacteria resistant to antibiotics. The protein is a homotrimer of huge subunits, each with 12 transmembrane helices. Large domains above the membrane help span the periplasmic space. Substrates that are soluble in the membrane diffuse into a hydrophobic binding site in the center of the carrier and are transported out of the cell by an antiporter mechanism that depends on a proton gradient across the plasma membrane.

Excitatory Neurotransmitter Carriers

Glutamate carriers remove the excitatory neurotransmitter glutamate from the synaptic cleft after a nerve impulse (see Fig. 17.10). Homologous carriers transport glutamate and aspartate into bacteria. The protein is a trimer of identical subunits, each composed of eight transmembrane helices that form independent transport pathways (Fig. 15.4D). Two α-helical hairpin loops in each subunit partially cross the bilayer and bind a glutamate. For each glutamate transported across the membrane three Na^+ and one H^+ move in the same direction and one K^+ moves in the opposite direction, so this is an example of symporter with features of an antiporter. Starting with the pathway open outside the cell, Na^+ and glutamate bind cooperatively to the carrier, inducing a conformational change that occludes the binding site. Then the parts of the protein that bind glutamate and Na^+ move as a block approximately 1.8 nm to release these substrates inside the cell, while the parts of the protein at the trimeric interface are stationary. Each of the three subunits in a trimer operates independently.

APC Superfamily of Carriers

Crystal structures revealed that nine different families of carriers have the same fold as the bacterial LeuT sodium-leucine symporter. The proteins have 10 transmembrane helices with the first five folded like the second five, but inverted in the membrane. The founding gene likely formed by duplication of a gene encoding five helices. Subsequent duplication and divergence of this original gene created many paralogs with specificities for a range of substrates, while retaining the 10 helix fold even as the sequences diverged so much that they are no longer recognized as homologous. Some family members have additional transmembrane helices, bringing the total to 11 to 14. Substrates bind deep within the bundle of helices flanked by rings of hydrophobic residues that can, depending on the conformation, block access from either or both sides. Different conformations expose the binding site externally or to the cytoplasm.

The APC superfamily includes both antiporters such as proton/Ca^{2+} exchangers and symporters such as the SGLT1 Na^+/glucose carrier. SGLT uses a gradient of Na^+ established by the Na^+K^+-ATPase (adenosine triphosphatase) pump (see Chapter 14) to move glucose up its concentration gradient into intestinal cells (see Fig. 17.2). Na^+ binds first, promoting glucose binding and closing the external seal. Conformational changes open an inward facing cavity, open the cytoplasmic seal and release Na^+ followed by glucose.

APC sodium symporters in the central nervous system clear synaptic clefts of neurotransmitters, including norepinephrine, dopamine, serotonin, and γ-aminobutyric acid (GABA) (see Fig. 17.8). Dysfunction of these carriers contributes to human depression and other illnesses. Cocaine and widely used drugs including tricyclic antidepressants inhibit these carriers. Antidepressants occupy the substrate binding site and lock the carrier in the open-out conformation.

NhaA Family of Carriers

Crystal structures showed that another group of carriers share a common fold with two related segments of five transmembrane helices, but with a topology different than the APC superfamily. This was surprising, since their sequences lacked similarity. Two of the best characterized examples are the bacterial Na^+/H^+ antiporter and the mammalian Na^+ symporter that transports bile acids from the intestine back into the blood. Although the substrates and their binding sites differ in these two carriers, similar conformational changes expose the binding sites on opposite sides of the membrane.

MFS Carrier Proteins

The **major facilitator superfamily** (MFS) includes many of the best-characterized carriers. MFS carriers consist of single polypeptides that form 12 transmembrane helices with a central substrate-binding pocket

(Figs. 15.1 and 15.4B). A rocking motion around the substrate produces alternate two conformations that expose the substrate-binding site(s) on the opposite side of the membrane.

The sequences and structures of the two halves of each protein are homologous, so it is believed that the original gene was created by duplication of an ancestral gene, which coded for a six-helix protein that formed functional dimers. As the MFS gene family grew during evolution, the ancient gene duplication and fusion had two advantages. First, it allowed the two halves of each gene to diversify separately to increase specificity for a wide variety of substrates. Second, a single polypeptide simplifies assembly of a functional carrier, as two half-sized subunits do not have to find each other. If the two halves of a 12-helix MFS carrier are expressed in the same cell, they can assemble functional carriers, but less efficiently than the intact protein.

The MFS family includes uniporters, symporters, and antiporters. The following examples illustrate their versatility.

Mammalian GLUT uniporters allow glucose to move down its concentration gradient into cells: red blood cells use GLUT1 (Fig. 15.1); and fat and muscle use GLUT4 in response to insulin (see Fig. 27.7). Figure 17.2 shows how epithelial cells use the MFS uniporter GLUT5 and APC superfamily glucose symporter SGLT1 to take up glucose from the intestine after a meal.

The mammalian heart $3Na^+/Ca^{2+}$ antiporter binds either Na^+ or Ca^{2+} and uses the large Na^+ gradient across the plasma membrane to drive the transport of Ca^{2+} out of the cytoplasm up a concentration gradient (see Fig. 39.22). In the outward facing conformation the carrier binds three Na^+. After the conformational change reorients the binding site, the three Na^+ dissociate inside and one Ca^{2+} binds. Reorientation of the binding site carries this Ca^{2+} to the outer surface of the cell, where it dissociates. In addition to the substrate concentrations, the membrane potential must also be taken into account, as this exchange is not electrically neutral, and the potential may affect the binding of one or both substrates to the carrier.

The Na^+/H^+ antiporter allows cells of the mammalian kidney, gut, and most other organs to manipulate their internal pH. Band 3 antiporter of red blood cells exchanges Cl^- for HCO_3^-. Carbon dioxide produced in tissues by oxidative reactions diffuses into red blood cells, where a cytoplasmic enzyme—carbonic anhydrase—transforms carbon dioxide into HCO_3^-. The antiporter provides a way for the HCO_3^- to return to the plasma, where it is carried to the lungs as the bicarbonate anion.

E. coli LacY symporter uses a proton gradient across the plasma membrane to drive accumulation of lactose (Fig. 15.3B). The proton gradient is created by the respiratory chain under aerobic conditions (see Fig. 19.5) or by the F-type ATPase (adenosine triphosphatase) pump under anaerobic conditions (see Fig. 14.5). Protons move down their concentration gradient as lactose moves up its concentration gradient into the cell.

Bacterial GlpT is a glycerol-3-phosphate–phosphate antiporter. GlpT has a pair of conserved arginines near the middle of the bilayer that bind phosphate. When the binding site is open on the periplasmic side of the membrane, GlpT binds glycerol-3-phosphate preferentially, since its affinity is higher than that of phosphate. When the binding site is exposed to the cytoplasm, glycerol-3-phosphate dissociates and is replaced by phosphate, which is present at a higher concentration in the cytoplasm. Another antiporter UhpT (uptake of hexose phosphate transporter) allows *E. coli* to scavenge glucose 6-phosphate from the medium in exchange for inorganic phosphate.

ACKNOWLEDGMENTS

We thank Peter Maloney for material used in the first edition and Michael Caplan for his suggestions on the second edition.

SELECTED READINGS

Deng D, Xu C, Sun P, et al. Crystal structure of the human glucose transporter GLUT1. *Nature*. 2014;510:121-125.

Faham S, Watanabe A, Besserer GM, et al. The crystal structure of a sodium galactose transporter reveals mechanistic insights into Na+/sugar symport. *Science*. 2008;321:810-814.

Guan L, Kaback HR. Lessons from lactose permease. *Annu Rev Biophys Biomol Struct*. 2006;35:67-91.

Madej MG, Kaback HR. Evolutionary mix-and-match with MFS transporters II. *Proc Natl Acad Sci USA*. 2013;110:E4831-E4838.

Murakami S, Nakashima R, Yamashita E, Yamaguchi A. Crystal structure of bacterial multidrug efflux transporter AcrB. *Nature*. 2002;419:587-593.

Nury H, Dahout-Gonzalez C, Trézéguet V, et al. Relations between structure and function of the mitochondrial ADP/ATP carrier. *Annu Rev Biochem*. 2006;75:713-741.

Reyes N, Ginter C, Boudker O. Transport mechanism of a bacterial homologue of glutamate transporters. *Nature*. 2009;462:880-885.

Ruprecht JJ, Hellawell AM, Harding M, et al. Structures of yeast mitochondrial ADP/ATP carriers support a domain-based alternating-access transport mechanism. *Proc Natl Acad Sci USA*. 2014;111:E426-E434.

Shi Y. Common folds and transport mechanisms of secondary active transporters. *Annu Rev Biophys*. 2013;42:51-72.

Traba J, Satrústegui J, del Arco A. Adenine nucleotide transporters in organelles: novel genes and functions. *Cell Mol Life Sci*. 2011;68:1183-1206.

Transporter Classification Database. <http://www.tcdb.org/>.

Yan N. Structural Biology of the Major Facilitator Superfamily Transporters. *Annu Rev Biophys*. 2015;44:257-283.

Yernool D, Boudker O, Jin Y, Gouaux E. Structure of a glutamate transporter homologue from *Pyrococcus horikoshii*. *Nature*. 2004;431:811-818.

Yu EW, McDermott G, Zgurskaya HI, et al. Structural basis of multiple drug-binding capacity of the AcrB multidrug efflux pump. *Science*. 2003;300:976-980.

CHAPTER 16

Membrane Channels

Channels are integral membrane proteins with transmembrane pores that allow particular ions or small molecules to cross a lipid bilayer. Some channels are open constitutively, but most open just part time. Each time a channel opens, thousands to millions of ions diffuse down their electrochemical gradient across the membrane. Carriers and pumps are orders of magnitude slower, since they use rate-limiting conformational changes to transport each ion (see Chapters 14 and 15).

The ability to control diffusion across membranes allows channels to perform three essential functions (Fig. 16.1). First, certain channels cooperate with pumps and carriers to transport water and ions across cell membranes. This is required to regulate cellular volume and for secretion and absorption of fluid, as in salivary glands, kidney, inner ear, and plant guard cells. Second, ion channels regulate the **electrical potential** across membranes. The sign and magnitude of the membrane potential depend on ion gradients created by pumps and carriers and the relative permeabilities of various channels (Appendix 16.3). Open channels allow unpaired ions to diffuse down concentration gradients across a membrane, separating electrical charges and producing a **membrane potential.** Coordinated opening and closing of channels change the membrane potential and produce an electrical signal that spreads rapidly over the surface of a cell. Nerve and muscle cells use these **action potentials** (see Fig. 17.6) for high-speed communication. Third, other channels admit Ca^{2+} from outside the cell or from the endoplasmic reticulum into the cytoplasm, where it triggers a variety of processes (see Fig. 26.12), including secretion (see Fig. 21.19) and muscle contraction (see Fig. 39.16).

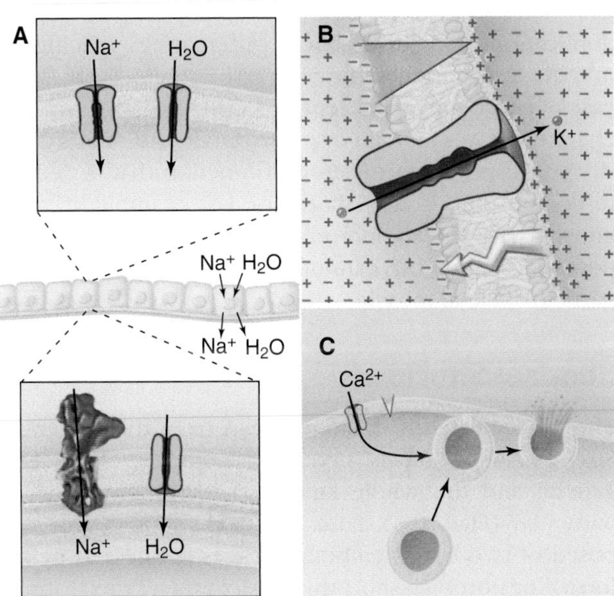

FIGURE 16.1 FUNCTIONS OF MEMBRANE CHANNELS. **A,** Transport of salt and water across an epithelium by water channels in both the apical and basolateral membranes, a Na^+ channel in the apical membrane, and a Na^+ pump in the basolateral membrane. **B,** Regulation of membrane potential. The triangle represents the concentration difference of K^+ across the membrane. The *zigzag arrow* represents the membrane potential, negative inside. **C,** Ca^{2+} signaling in secretion.

Cells control channel activity in two ways. In the long term, each cell type expresses a unique repertoire of channels from among hundreds of channel genes. **Excitable cells,** such as nerve and muscle, express plasma membrane voltage-gated channels to produce action potentials. Epithelial cells express channels for Na^+, Cl^-, K^+, and water to produce the salt and water fluxes required for secretion and reabsorption of fluids in glands and the kidney. In the short term, cells open and shut specific types of channels in response to physiological or environmental stimuli. Some channels respond to changes in membrane potential. Others respond to intracellular or extracellular ligands or to mechanical forces. Still others, such as kidney water channels, are shifted from one membrane compartment to another to mediate physiological functions.

Channels are important in medicine. Ion channels are targets of powerful drugs and toxins, including curare, tetrodotoxin ("voodoo toxin"), paralytic shellfish toxins, cobra toxin, local anesthetics, antiarrhythmic agents, and probably general anesthetics (Appendix 16.1). Defects in ion channel genes cause many inherited disorders, ranging from some cardiac arrhythmias to kidney stones. Antibodies target ion channels in the human autoimmune disorder myasthenia gravis.

This chapter discusses in detail 12 large families of plasma membrane channels. A note at the end of the chapter briefly introduces four additional types of channels. Other chapters discuss cystic fibrosis transmembrane regulator Cl^- channels (see Fig. 17.4), gap junction channels used for communication between adjacent cells (see Fig. 31.7), and intracellular Ca^{2+} release channels that participate in signal transduction (see Figs. 26.13 and 39.15). Understanding channels requires not only information about their structure and activity but also some knowledge of electrical phenomena. Appendixes 16.2 to 16.4 contain essential material about electrophysiology.

Physiologists introduced the concept of channels in the 1950s to explain ion currents during action potentials. Proof that channels are integral membrane proteins followed in the 1970s with isolation of the nicotinic acetylcholine receptor channel and the voltage-gated Na^+ channel. The great diversity of channels was revealed initially by cloning complementary DNAs (cDNAs) using functional assays and homology with known channels. Ultimately, the full repertoire of channels emerged from sequenced genomes.

A new channel protein can be characterized by expressing its cDNA in a test cell and then making electrical recordings of ion currents from the cell or patches of its membrane (Appendix 16.2). If expression of a single-channel protein fails to reproduce the channel activity observed in the cell of origin, auxiliary subunits are probably required. Historically, investigation of channel functions has relied on toxins and drugs that inhibit particular channels more or less specifically (Appendix 16.1). Mutations, including those in human disease, provide definitive tests for physiological functions and have yielded some surprising results.

Channel Diversity and Evolution

Humans have approximately 400 genes encoding channel proteins. The historical channel nomenclature based variously on the ion transported, mode of regulation, physiological role, or drug sensitivity is often ambiguous. Fortunately, knowledge of channel protein structures clarified evolutionary relationships and provided a framework to classify most plasma membrane channels into a few large families (Fig. 16.2).

Channels are integral membrane proteins, usually with two or more α-helices crossing the lipid bilayer. Porins are an exception; they are built from transmembrane β-strands (see Fig. 13.9C). Channels generally consist of two to six subunits, but some are single, large polypeptides. The transmembrane pores for conducting ions or other substrates are often located in the middle of a group of subunits or subunit-like domains, but the pores of chloride, water, and ammonia channels are located within single subunits (Fig. 16.2).

A limited number of genes in early forms of life appear to have given rise to most channel genes. For example, the gene for a simple prokaryotic channel with just two transmembrane segments (confusingly called S5 and S6) was the progenitor of a huge family of channels with 2 to 24 transmembrane segments per subunit (Fig. 16.2). A simple duplication of one of these genes yielded channels with four segments. Even before the emergence of eukaryotes, the addition of four transmembrane segments (S1 to S4) yielded channels with six transmembrane segments. Acquisition of positive charges by S4 provided for voltage sensitivity. Two rounds of gene duplication and divergence produced voltage-gated channels, such as voltage-gated Na^+ channels, consisting of four domains, each with six transmembrane segments. Already in bacteria a gene in this family added a domain that binds glutamate to become a glutamate-gated channel. Water channels originated in prokaryotes by duplication of a gene that encoded three hydrophobic transmembrane segments. Pentameric neurotransmitter channels, ammonia channels and double-barreled Cl^- channels all had bacterial ancestors.

Channels in higher eukaryotes are products of multigene families that arose from multiple rounds of gene duplication and divergence. Alternative splicing of mRNAs also enriches the variety of channels. In some cases, combining different subunit isoforms in one channel creates increased specialization. All of this diversity suggests a sophistication of function that is difficult to demonstrate in the laboratory. For example, it is not known why the sodium channels that produce action potentials in neurons cannot substitute for their counterparts in skeletal muscle.

Channel Structure

The pH-regulated K^+ channel **KcsA** from the bacterium *Streptomyces lividans* serves as a model for channels in general and the whole family of **S5/S6 channels** in particular (Fig. 16.3). Four identical subunits are composed of two transmembrane helices connected by a **P loop** (for pore)—a short third helix and a crucial strand that makes the **selectivity filter.** The transmembrane helices are packed close together on the cytoplasmic side of the bilayer forming a **gate** that can be either open or closed in response to conditions. Local competition between the favorable ion–ligand interactions and unfavorable ligand–ligand interactions yields robust ion selectivity despite thermal fluctuations of the protein

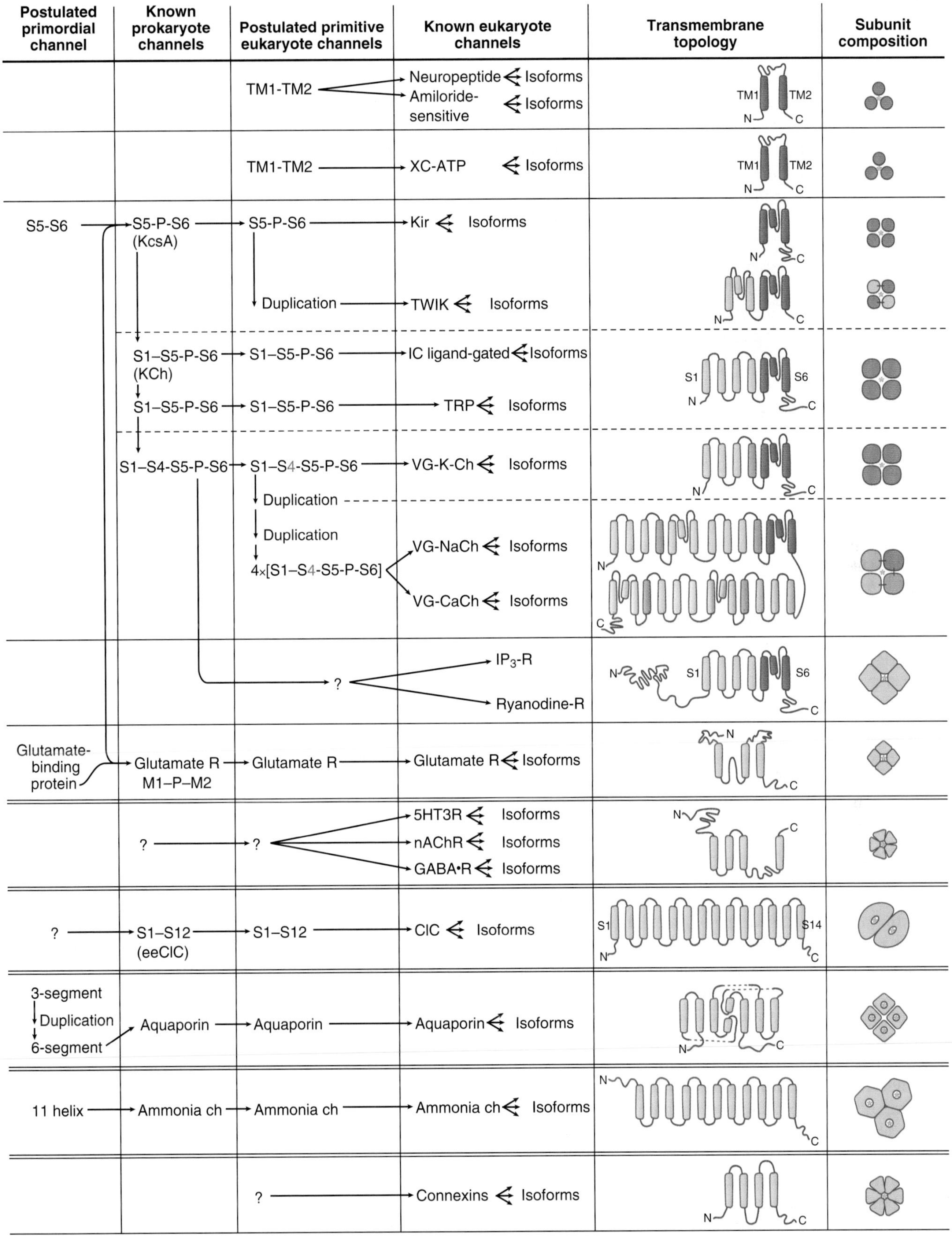

Postulated primordial channel	Known prokaryote channels	Postulated primitive eukaryote channels	Known eukaryote channels	Transmembrane topology	Subunit composition
		TM1-TM2	Neuropeptide ← Isoforms Amiloride-sensitive ← Isoforms	TM1 TM2 N C	
		TM1-TM2	XC-ATP ← Isoforms	TM1 TM2 N C	
S5-S6	S5-P-S6 (KcsA)	S5-P-S6	Kir ← Isoforms	N C	
		Duplication	TWIK ← Isoforms	N C	
	S1–S5-P-S6 (KCh)	S1–S5-P-S6	IC ligand-gated ← Isoforms	S1 S6 N C	
	S1–S5-P-S6	S1–S5-P-S6	TRP ← Isoforms		
	S1–S4-S5-P-S6	S1–S4-S5-P-S6	VG-K-Ch ← Isoforms	N C	
		Duplication Duplication 4×[S1–S4-S5-P-S6]	VG-NaCh ← Isoforms VG-CaCh ← Isoforms	N C	
		?	IP_3-R Ryanodine-R	N S1 S6 C	
Glutamate-binding protein	Glutamate R M1–P–M2	Glutamate R	Glutamate R ← Isoforms	N C	
	?	?	5HT3R ← Isoforms nAChR ← Isoforms GABA•R ← Isoforms	N C	
?	S1–S12 (eeClC)	S1–S12	ClC ← Isoforms	S1 S14 N C	
3-segment Duplication 6-segment	Aquaporin	Aquaporin	Aquaporin ← Isoforms	N C	
11 helix	Ammonia ch	Ammonia ch	Ammonia ch ← Isoforms	N C	
		?	Connexins ← Isoforms	N C	

FIGURE 16.2 CLASSIFICATION OF CHANNEL PROTEINS. This scheme is based on atomic structures and genome sequences. The transmembrane topology has the extracellular side at the top and uses rectangles to indicate helices labeled "S." P loops are shown as a short helix and loop between two transmembrane helices. S4 voltage-sensing helices are *pink;* pore positions are *orange.* The *last column* shows the subunit compositions. The origins of most channel families can be traced back to prokaryotes. Relatively recent gene duplications and divergence have given rise to multiple isoforms of most types of channels in eukaryotes. ClC, chloride channel; GABA, γ-amino butyric acid; 5HT, 5-hydroxytryptamine; IC, intracellular ligand; IP_3, inositol triphosphate; Kir, potassium inward rectifier; nACh, nicotinic acetylcholine; R, receptor; Ryanodine, a chemical that binds calcium-release channels; TRP, transient receptor potential; VG, voltage-gated; XC-ATP, extracellular ATP-gated channel.

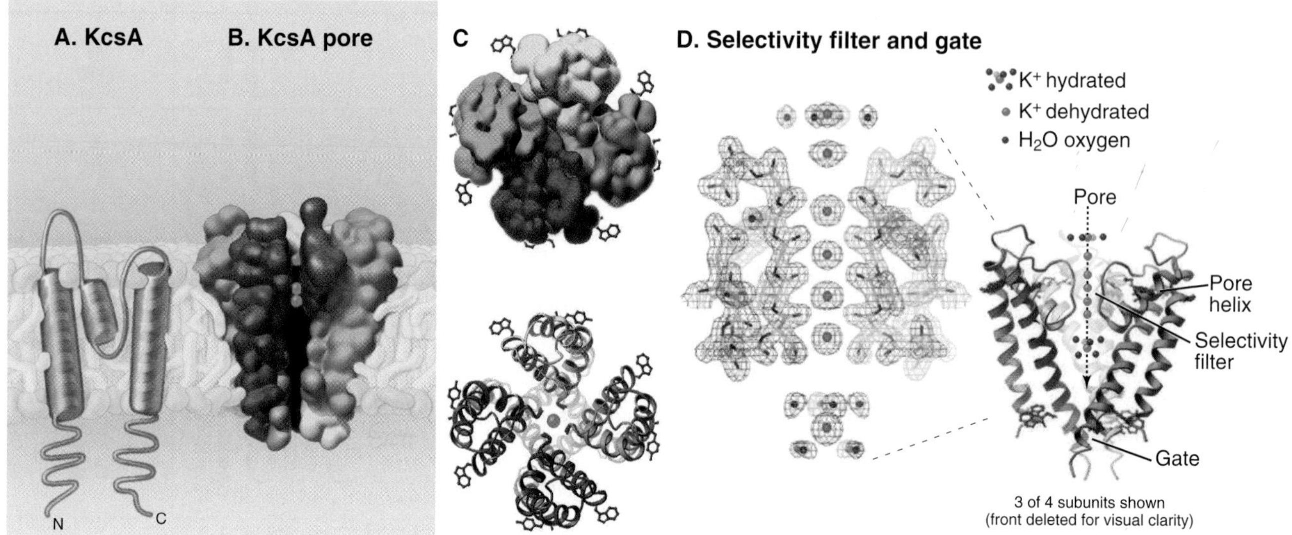

FIGURE 16.3 ATOMIC STRUCTURE OF KCSA, A K$^+$ CHANNEL FROM *STREPTOMYCES LIVIDANS*. A, Transmembrane topology. The short helix and loop between the two transmembrane helices are called a P loop because they form the pore. **B,** Space-filling model with each subunit shaded a different color and with a cutaway view to expose the central pore, which contains three K$^+$ ions *(blue)*. **C,** Views from outside the cell. Aromatic side chains (shown as stick figures) at both ends of the transmembrane helices of each subunit project radially into the lipid. **D,** Selectivity filter and gate. The blue mesh shows the electron density of the region of two subunits flanking the selectivity filter. The stick figure is the molecular model with five *(red)* carbonyl oxygens lining the pore and coordinating with four bound K$^+$ ions *(green)*. *Red* waters surround K$^+$ ions on both sides of the pore. The ribbon model has the front subunit removed to reveal the central pore. Starting on the extracellular side, the 4.5-nm–long pore consists of a negatively charged vestibule; the 1.2-nm–long selectivity filter with binding sites for four dehydrated K$^+$ ion; a central cavity with space for a single hydrated K$^+$; a gate (closed here); and a negatively charged cytoplasmic vestibule. (For reference, see Protein Data Bank [PDB; www.rcsb.org] file 1BL8 and Doyle DA, Morais-Cabral J, Pfuetzner RA, et al. The structure of the potassium channel: molecular basis of K$^+$ conduction and selectivity. *Science*. 1998;280:69–77; and Zhou Y, Morais-Cabral JH, MacKinnon R. Chemistry of ion coordination and hydration revealed by a K$^+$-channel-Fab complex at 2.0 Å resolution. *Nature*. 2001;414:43–48. Also see http://www.sciencemag.org/content/suppl/2014/10/15/346.6207.352.DC1/1254840s1.mov for a molecular dynamics simulation of K$^+$ movements through KcsA.)

structure. On the extracellular side the transmembrane helices splay apart to make room for the **pore helices** and selectivity filter.

The selectivity filter is a centrally located pore where the four identical subunits meet, each subunit contributing a quarter of the wall. Highly conserved residues (glycine-tyrosine-glycine) in an unusual linear conformation line the pore. The backbone carbonyl oxygens (C═O) of four successive residues of each subunit all point toward the pore. The pore is 1.2 nm long and approximately 0.2 nm in diameter—just wide enough to accommodate a dehydrated K$^+$ ion. Physiological experiments show that this passage distinguishes between K$^+$ and Na$^+$ with a fidelity of 1000 : 1 even though Na$^+$ (with a diameter of 0.095 nm) is smaller than K$^+$ (0.133 nm in diameter). K$^+$ fits so perfectly into the pore that the carbonyl oxygens replace its water shell without an energy penalty, whereas the smaller Na$^+$ binds more strongly to its hydration shell than to the pore. Ions that fit poorly in the pore are rejected, as it is energetically unfavorable to shed their hydration shell. Local electrostatic interactions between ions and carbonyl oxygens also contribute to selectivity. Carbonyl oxygens carry the optimal electric dipole to favorably counterbalance the hydration free energy of K$^+$ over that of Na$^+$, thereby giving robust selectivity despite thermal fluctuations of the protein.

The selectivity filter accommodates up to four K$^+$ ions, a local concentration exceeding that inside or outside the cell by more than 10-fold, so it actually concentrates K$^+$. However, this does not impede diffusion through the pore, as electrostatic repulsion between these closely spaced ions forces them apart. Outside the filter, the pore is lined with hydrophobic groups, but a cavity in the middle of this passage accommodates a hydrated K$^+$ in an environment with a negative electrostatic potential that is thought to reduce the electrostatic barrier to the ion as it crosses the membrane.

Open channels vary widely in their ability to discriminate among ions. Highly selective channels, such as KcsA and voltage-gated K$^+$ channels, pass K$^+$ ions without bound water. Less-selective channels, such as the nicotinic acetylcholine receptor, are equally permeable to Na$^+$ and K$^+$, which probably pass through as hydrated ions. Gap junction channels pass most molecules smaller than 800 Da without discrimination (see Fig. 31.7).

The **ion flux** through an open channel (at a fixed membrane potential) is approximately proportionate to the ion concentration on the side from which the ions migrate. The maximum rate of ion flux—10^6 to 10^8 ions/s—is limited by the time required for binding and dissociation at specific sites as an ion traverses the pore. At this high rate, the interactions that discriminate between selected and rejected ions last only 10 to 100 nanoseconds! Ions move in single file through the pore, driven in part by their mutual electrostatic repulsion.

Channel Activity

Electrical recordings of single channels (see Fig. 16.16 for the method) show that the pore across the membrane is either fully open or closed (Fig. 16.4). Open channels, called the **active state,** pass selected ions across the membrane through the filter and a gate at

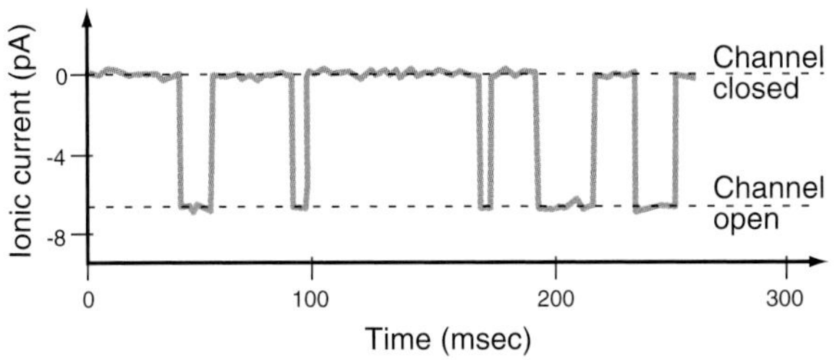

FIGURE 16.4 PATCH RECORDING OF A SINGLE-CATION CHANNEL MOLECULE. This time course shows the current that results when a single channel opens and closes at random. When open, it conducts Na^+ ions at a rate of approximately 36 × 10^6 per second, yielding a current of −6 pA (picoamperes). The transitions between open and closed are so fast that they appear instantaneous on this millisecond time scale.

rates approaching their diffusion in water. In closed channels the gate, filter or both are closed (Fig. 16.5). Switching between conducting and nonconducting states is called **gating.** Transitions between closed and open states are so fast that they appear to be instantaneous on the experimental timescale of single channel recordings (Fig. 16.4). Channels may rapidly flicker open and closed, but generally do not open partway or change their ion selectivity.

A hinge motion of the cytoplasmic ends of the transmembrane helices opens or closes the gate of KcsA and related channels (Fig. 16.5A). Cytoplasmic pH regulates the gate of KcsA. Examples later in this chapter illustrate how energy to open other channels comes from mechanical force, binding of cytoplasmic or extracellular ligands, or changes in the membrane potential.

A process called **inactivation** operates in parallel with gating to stop the flux of ions through channels. Inactivation makes a channel unresponsive to conditions that otherwise favor the active state. Typically channels cycle from closed to open and then inactivate before returning to the closed state (Fig. 16.5C). So-called C-type inactivation results from a subtle change in the conformation of the filter of a channel with an open gate (Fig. 16.5B). In N-type inactivation, flexible cytoplasmic parts of the protein plug the open pore of an otherwise open channel (Fig. 16.5D).

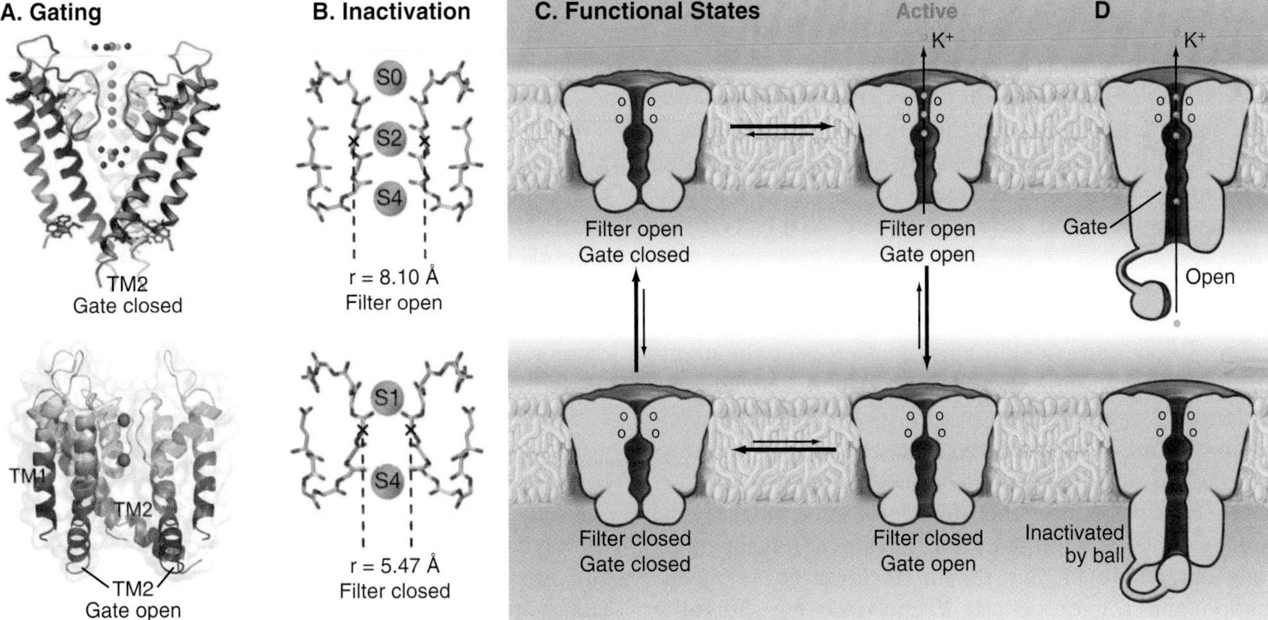

FIGURE 16.5 FUNCTIONAL STATES OF KCSA. A, Gating. Ribbon diagram showing three of the four subunits with the gate closed (see PDB file 1BL8) and open (see PDB file 1K4C). **B,** C-type inactivation. Stick figures show the selectivity filter in the open (see PDB file 1K4C) and closed (see PDB file 1K4D) conformations. **C,** Drawings of four functional states. Closed channels are inactive owing the gate or filter being closed. Both the gate and filter are open in active channels, forming a selective pore for K^+ ions across the bilayer. The *arrows* show the probabilities of transitions between the states; *red arrows* are probabilities in the presence of an external stimulus (acid pH in this case), while *black arrows* are probabilities in the absence of the stimulus. **D,** N-type inactivation of other channels. An inactivation "ball" at the N-terminus of voltage-gated channels blocks the pore of an active channel. (For reference, see Cuello LG, Jogini V, Cortes DM, et al. Structural mechanism of C-type inactivation in K^+ channels. *Nature.* 2010;466:203–208; and Ostmeyer J, Chakrapani S, Pan AC, et al. Recovery from slow inactivation in K^+ channels is controlled by water molecules. *Nature.* 2013;501:121–124.)

Large organic or inorganic ions, such as polyamines and Mg^{2+}, block other open channels simply by occluding the pore. Depending on the channel, blocking ions bind to sites on the outside, the inside, or both sides of the membrane. The membrane potential drives blocking ions into or out of channels. Blocking ions that dissociate on a millisecond time scale cause the current through the channel to flicker on and off multiple times every time the channel opens, while slowly dissociating ions can turns off a channel for a longer time. Local anesthetics such as lidocaine are pharmacological channel blockers.

Transition Between the Closed, Open, and Inactivated States

The steady-state **probability of being open** (P_o) is simply the fraction of the total time that the channel is open. For a given channel, the fraction of time in the open state determines the ion flux. Channels act independently, so the total flux across a membrane depends on the number of channels that are open at a given time.

Local physiological conditions, considered in detail in the following sections, control gating from moment to moment. Cells also use the full range of signaling mechanisms (see Chapters 24 to 26) from phosphorylation to second messengers to guanosine triphosphate (GTP)–binding proteins to influence the probability that particular channels open or close. By modulating the sensitivity of various channels, cells modify the behavior of their membranes and their responses to internal and external conditions. This modulation makes channels in general, and membrane excitability in particular, highly adaptable. Chapter 17 illustrates how channel modulation regulates the heart rate, changes the efficiency of communication between nerve cells, and adapts cells to some stresses.

Opening a channel for a few milliseconds can change the membrane potential but not the cytoplasmic ion composition, because only a few ions crossing the membrane is enough to produce a large change in membrane potential (Appendixes 16.3 and 16.4). This conserves energy because ion gradients created by energy-requiring pumps are not dissipated. Longer openings of tens of milliseconds can alter the ion composition of the cell. For example, voltage-gated Ca^{2+} channels remain open long enough to change the intracellular Ca^{2+} concentration and trigger cellular events (see Fig. 39.16B). In this way, they convert an electrical signal to a chemical signal.

Channels With One Transmembrane Segment

The simplest known channel is found in the membrane envelope of influenza virus. This **M_2 channel** consists of four copies of a small subunit with one transmembrane helix. After an infected cell takes the virus into an endosome (see Chapter 22), the acidic environment opens the channel, allowing protons to enter the virus and to initiate disassembly of the protein shell that surrounds the genome. The antiviral drug amantadine blocks these channels.

Bacteria secrete peptides (gramicidin, alamethicin, and colicins) that are designed to kill other species by forming highly selective and conductive channels. Only 13 amino acids are required for gramicidin A to form β-helical homodimers that function as K^+-selective channels.

Vertebrates also have simple channel proteins of approximately 130 residues with a single transmembrane segment and no sequence homology with other known channels. These **minK** molecules do not form channels on their own but serve as accessory subunits for a conventional P loop, voltage-gated, K^+-channel in the heart and some epithelial cells. Mutations in the minK gene cause defects in excitation of the heart and deafness.

Channels With Two Transmembrane Segments

Evolution produced three unique families of channels with two transmembrane helices. They have the same topologies with both termini in the cytoplasm but different structures, so their evolutionary origins were different.

Mechanosensitive Channels

MscL from bacteria (Fig. 16.6A) is a simple channel of five subunits. One of the two transmembrane helices forms the wall of the pore. A third C-terminal helix extends the pore into the cytoplasm. The pore is lined with polar residues except for a gate at its narrowest constriction, where an isoleucine and a valine reduce the diameter to approximately 0.2 nm. Tension in the plane of the membrane created by osmotic stress is believed to rearrange these helices and open the channel. Open channels have large pores lacking a selectivity filter, so ions, water and even large organic molecules pass through them. This response avoids osmotic lysis of the cell. MscL channels are widespread in prokaryotes and are also found in eukaryotes. Bacteria and most eukaryotes also have functionally similar but structurally unrelated mechanosensitive channels with three transmembrane helices.

Amiloride-Sensitive Channels

A family of trimeric channels is sensitive to inhibition by the diuretic drug **amiloride**. They likely arose in the last common ancestor of eukaryotes. The family includes chordate acid–sensing channels, metazoan epithelial Na^+

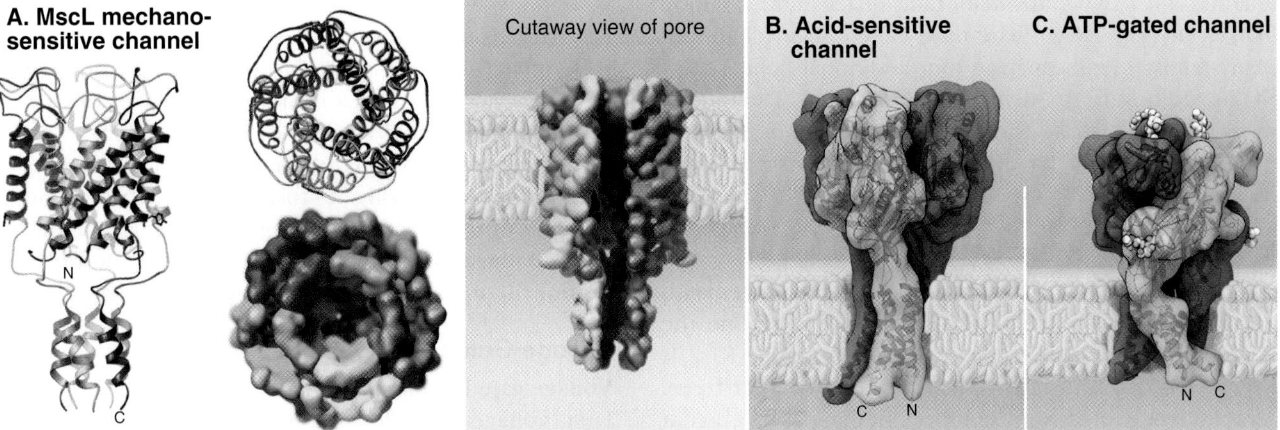

FIGURE 16.6 ATOMIC STRUCTURES OF THREE CHANNELS WITH TWO TRANSMEMBRANE HELICES. A, The mechanosensitive MscL channel from *Mycobacterium tuberculosis*. Ribbon and space-filling models with each subunit shaded a different color, showing both views from the side and from outside the cell. The cutaway view exposes the central pore, which is 8 nm long. **B,** Ribbon model of the trimeric acid-sensitive channel from chicken. **C,** Ribbon model of the trimeric adenosine triphosphate (ATP)-gated $P2X_4$ channel from zebrafish. (**A,** For reference, see PDB file 2OAR and Chang G, Spencer RH, Lee AT, et al. Structure of the MscL homolog from *Mycobacterium tuberculosis:* a gated mechanosensitive ion channel. *Science*. 1998;282:2220–2226. **B,** See PDB file 2QTS and Jasti J, Furukawa H, Gonzales EB, et al. Structure of acid-sensing ion channel 1 at 1.9 A resolution and low pH. *Nature*. 2007;449:316–323. **C,** See PDB file 3I5D and Kawate T, Michel JC, Birdsong WT, et al. Crystal structure of the ATP-gated P2X(4) ion channel in the closed state. *Nature*. 2009;460:592–598.)

channels and metazoan peptide-gated channels. Three identical subunits with two transmembrane helices form a centrally located pore (Fig. 16.6B). This architecture and the lack of a P-loop show that the gene for these channels originated separately from the large family of S5-S5 channels. Judging from homology epithelial Na^+ channels are likely to be heterotrimers of α, β, and γ subunits.

Epithelial Na^+ channels accelerate the rate-limiting step in the transport of Na^+ and water across many types of epithelia (Fig. 16.1A). Typically, Na^+ diffuses down its concentration gradient through these channels into the cytoplasm and is pumped out of the cell into the underlying extracellular space by Na^+/K^+-ATPases (adenosine triphosphatases) in the basolateral plasma membrane (see Chapter 14). Water follows Na^+ through water channels. Renal collecting tubules use this strategy to resorb salt and water. Lung epithelial cells use the same mechanism to clear fluid from air spaces. Mice with knockout mutations in the lung epithelial Na^+ channel gene die at birth with fluid in their lungs.

Amiloride binds and inhibits epithelial Na^+ channels, increasing the output of urine, while the cardiac peptide **atrial natriuretic peptide** inhibits these channels indirectly (see Fig. 24.8). If salt is lost from the body, production of the steroid hormone **aldosterone** increases the plasma membrane content of Na^+ channels and salt resorption by the kidney.

These channels open and close randomly for relatively long periods, between 0.5 and 5 seconds. The pH gates acid-sensing channels involved with pain perception, but neither the membrane potential nor any known natural ligand activate epithelial Na^+ channels.

Mutations in patients with **Liddle syndrome** increase the open time of epithelial Na^+ channels. This results in excess resorption of salt and water by the kidney and severe high blood pressure. Excess secretion of aldosterone by adrenal tumors has similar effects. When the blood supply to the brain is compromised in a stroke, the acidic environment activates one type of epithelial sodium channel that then admits both Na^+ and Ca^{2+}, causing much of the damage in a stroke.

Small peptides gate the amiloride-sensitive Na^+ channels in the nervous systems of invertebrates. Related channels in the human brain are not well characterized.

Adenosine Triphosphate–Gated Channels

Extracellular adenosine triphosphate (ATP) activates a family of homotrimeric cation channels with two transmembrane helices (Fig. 16.6C). Many neurons make synaptic vesicles containing both ATP and neurotransmitters, so they release ATP at synapses. ATP binding to the extracellular domains opens a gate composed of leucine side chains in the middle of the pore. One example is sympathetic nerves that innervate blood vessels and transmit pain perception. This function makes these channels attractive targets for treating pain. These ATP-gated "purinergic" channels are called P2X receptors to distinguish them from P2Y ATP receptors, members of the seven-helix family of receptors coupled to trimeric GTP-binding proteins.

S5-S6 Channel Family

The largest and most diverse family of ion channels arose early during evolution from a gene for a simple channel with two transmembrane helices and a P-loop like KcsA (Fig. 16.2). They are called S5-S6 channels. Many family members have four additional helices (S1–S4) in addition to the defining S5-P-S6 motif. The following sections illustrate the diversity and unique features of these channels having 2, 4, or 6 transmembrane helices.

Inward Rectifier Potassium Channels

Kir channels consist of two transmembrane helices with a P loop in between. The P loop and helix S6 line the K^+-selective pore. Several channels in this family (Kir2.1, Kir2.3, Kir3 family, and Kir4.1) are **inward rectifiers.** A rectifier is an electronic component that passes current preferentially in one direction. Inward rectifier K^+ channels would pass K^+ into the cell if the membrane potential were below E_K (Appendix 16.3), but such membrane potentials are not achieved physiologically. Above the resting potential, open Kir channels pass only a small K^+ current out of the cell. The reason is that impermeant cytoplasmic cations, Mg^{2+}, and polyamines (ornithine metabolites having net positive charges of 2+ to 4+) bind to negatively charged residues on the cytoplasmic end of the S6 segment of open channels and block the passage of K^+. Despite their low permeability, these channels help maintain the resting membrane potential in many cells and to repolarize excitable cells during an action potential.

Divergence from a common ancestor created a number of channels with differing physiological properties:

- Kir1.1 channels provide a pathway for K^+ to leave kidney collecting duct cells for the urine. They are constitutively open and not blocked by cytoplasmic ions.
- Kir2 channels in the heart and brain contribute to maintaining the resting membrane potential by keeping it from being hyperpolarized. They are constitutively active with inward rectification sensitive to membrane potential.
- Kir3 or Kir3.1 and Kir3.4 channels in the pacemaker cells of the heart regulate heart rate under the control of trimeric GTP-binding proteins (see Fig. 39.21).
- *Cytoplasmic* ATP regulates Kir6.2 channels. These channels, also called K_{ATP} channels, have a novel function in the pancreas requiring interaction with a member of the ABC (adenosine triphosphate binding cassette) transporter family, the **sulfonylurea receptor (SUR)** (see Chapter 15). High blood glucose levels raise *intracellular* ATP concentrations, which closes Kir6.2 channels. This pushes the membrane potential toward the threshold for opening a voltage-sensitive Ca^{2+} channel, which triggers insulin secretion. Sulfonylurea drugs used to treat diabetes mellitus act like glucose, which promotes insulin secretion by inhibiting these ATP-sensitive channels.

S5-S6 Channels With Four Transmembrane Helices

K^+ channels with four transmembrane segments and two P loops (Fig. 16.2, TWIK) are abundant in animal genomes, with 15 genes in humans. Two of these subunits form a channel with four domains similar to KcsA. They help establish the resting potential of the plasma membrane by allowing K^+ to leak out of the cell, independent of the membrane potential. Volatile anesthetics activate these leak channels, leading to hyperpolarization of the membrane and reduced excitability.

Voltage-Gated Cation Channels

Voltage-gated channels have two main functions. First, voltage-gated Na^+ and K^+ channels produce action potentials in excitable cells (see Fig. 17.6). Depolarization of the membrane transiently opens these channels in sequence, driving the membrane potential first toward the Na^+ equilibrium potential (Appendix 16.3) and then back toward the K^+ equilibrium potential. Second, voltage-gated Ca^{2+} channels convert electrical signals into chemical signals when they admit Ca^{2+} to the cytoplasm, where it acts as a second messenger (see Figs. 17.9, 17.10, and 26.12) to stimulate secretion, activate protein kinases, trigger muscle contraction, or influence gene expression.

Voltage-gated K^+ channels are tetramers of four identical or mixed subunits with a common domain organization (Fig. 16.2). Voltage-gated Na^+ and Ca^{2+} channels consist of four identical subunits in bacteria, but in animals the four homologous but distinct domains (each with S1 to S6) are linked in a huge polypeptide (Fig. 16.2). Voltage-gated channels may have additional specialized domains and/or subunits, but the four main domains carry out the basic functions.

Hydrophobic segments S5 and S6 are transmembrane helices with a P loop in between (Fig. 16.7). The P loop forms the selectivity filter, since a point mutation in the P loop can change the ion selectivity of the channel, for example from Na^+ to Ca^{2+}. The cytoplasmic ends of the four S6 helices form the gate, like the homologous helices of KcsA (Fig. 16.5A).

Helices S1 to S4 form a separate **voltage-sensing domain** embedded in the lipid bilayer lateral to the pore-forming domain of the adjacent subunit. Helix S4 is a key part of the sensor by virtue of four positively charged residues (most often arginine) on one side. A helix between S4 and S5 connects the gate to the voltage-sensing domain in the bilayer (Fig. 16.7A).

The voltage-sensing domains couple the membrane potential to channel gating. Voltage-sensitive channels are intrinsically open in the absence of a membrane potential. A negative internal membrane potential stabilizes the closed state with the voltage-sensing domain closer to the cytoplasmic side of the bilayer. The transition between closed and open occurs over a narrow voltage range (Fig. 16.8), likely because the four domains

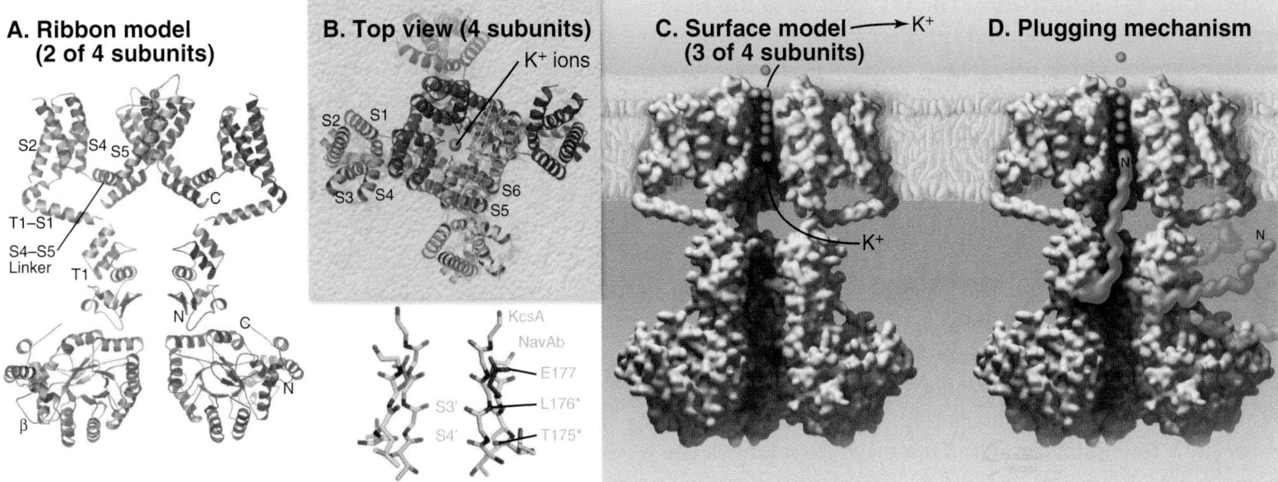

FIGURE 16.7 VOLTAGE-GATED POTASSIUM CHANNEL. Crystal structure of the Kv1.2 K^+ channel with β_2-subunits from rat brain. **A,** Ribbon diagram of two of the four α-subunits and two of the four β-subunits viewed from the side. Each of the four α-subunits contributes an S5 helix, P loop, and S6 helix to form a channel similar to KcsA. Helices S1 through S4 form a separate domain connected to the channel domain by the S4–S5 linker helix. Movements of the S4 helix in response to the membrane potential pull the channel gate open and closed. The N-terminal T1 domains of each α-polypeptide are located in the cytoplasm, interacting with the tetrameric β-subunit. **B,** Ribbon diagram viewed from outside the cell, illustrating the central channel domains and the peripheral voltage-sensing domains. **C,** Side view of a space-filling model of three of the four subunits, showing K^+ ions *(blue),* one above the membrane, four in the selectivity pore, and one in the vestibule. **D,** Space-filling model with an artist's conception of the N-terminal "ball-and-chain" plugging the open gate of an inactivated channel. (For reference, see PDB file 2A79 and Long SB, Campbell EB, MacKinnon R. Crystal structure of a mammalian voltage-dependent Shaker Family K^+ channel. *Science.* 2005;309:897–902.)

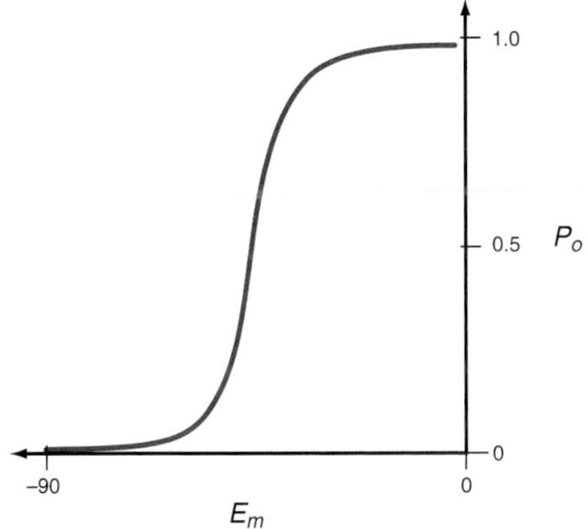

FIGURE 16.8 Graph of the open probability *(P_o)* of a voltage-gated Na^+ channel as a function of membrane potential *(E_m)*. Essentially, all channels are closed at the resting potential of –70 mV, and all are open above a threshold potential of approximately –40 mV.

respond cooperatively. Biophysical measurements show that as the membrane depolarizes, the S4 helix moves approximately 1 nm through the voltage-sensing domain within the bilayer toward the outside of the cell and pulls on S5 and S6 to open the gate. This movement of positive charges on the S4 helix along with the rest of the voltage-sensing domain is detected as a **"gating current."**

These channels inactivate rapidly when an **inactivation peptide** on a flexible chain occludes the open pore and blocks conduction (Figs. 16.5D and 16.7D). Inactivation peptides are located at the N-terminus of K^+ channels (Fig. 16.7D) or in a loop between the domains of Na^+ channels. Amputation of the inactivation peptide eliminates inactivation, but an experimentally provided soluble inactivation peptide can restore inactivation by binding open channels. Less is known about the transition from the inactivated state to the closed state, but a conformational change must occlude the pore before the ball dissociates from the cytoplasmic side of the pore.

Voltage-Gated Potassium Channels

All voltage-gated K^+ channels assemble from four α-subunits. Depolarization of the membrane opens these channels and drives the membrane potential toward the K^+ resting potential.

Animals use two strategies to produce voltage-gated K^+ channels with diverse physiological properties. First, humans express many different K^+ channel proteins from 40 genes, augmented by alternate splicing of messenger RNAs. Some K^+ channels are heterotetramers, a combinatorial strategy with the potential to produce thousands of different tetramers. T1 domains at the N-termini of these subunits (Fig. 16.7A) restrict formation of tetramers from some isoforms. Second, cytoplasmic β-subunits associate with some types of α-subunits (Fig. 16.7A) and modify their physiology. One type of voltage-gated K^+ channel has a Ca^{2+} binding site at the C-terminus. Signaling events that raise cytoplasmic Ca^{2+} make these channels more sensitive to membrane depolarization, reducing the excitability of the membrane.

Mutations in the gene for the cardiac K^+ channel, called *HERG* (human ether-a-go-go-related gene), are the most common cause of an autosomal-dominant disease called **long QT syndrome.** The QT interval is the time between depolarization and repolarization of the heart muscle on electrocardiograms. *HERG* is responsible for repolarizing the membrane during action potentials (see Fig. 39.20). Mutant channels open more slowly in response to depolarization of the membrane. This prolongs the action potential and predisposes to abnormal cardiac rhythms and sudden death. Some *HERG* mutations also cause deafness.

Voltage-Gated Na^+ Channels

When activated by membrane depolarization, voltage-gated Na^+ channels open synchronously over a narrow range of membrane potentials (Fig. 16.8) and then inactivate in 1 to 2 milliseconds. This cycle depolarizes the plasma membrane transiently during action potentials (see Fig. 17.6), so the distribution of these channels effectively defines the excitable regions of nerve cell membranes (see Fig. 17.10A). The selectivity filter favors Na^+ over K^+ by more than 10-fold. The filter consists of two parts: a ring of four negatively charged glutamic acid side chains in bacteria, and a ring of four water molecules. Human Na^+ channels have a ring of two acidic residues, a lysine and an alanine. These rings allow partially hydrated Na^+ ions to diffuse rapidly across the membrane. An open channel rapidly inactivates when a short flexible cytoplasmic segment between domains III and IV binds to and blocks the open pore. The channel remains inactivated until the membrane repolarizes. Then the channel rearranges to the closed state without reopening. Inactivation does not depend steeply on membrane potential, but because it occurs much more frequently when the channel is open, inactivation appears to be voltage dependent. A high frequency of membrane depolarization causes a slower type of inactivation that results from rearrangement of the four subunits (or domains) to change the shape of the filter.

Vertebrates express more than 10 Na^+ channel isoforms with some variations of these common features. A polypeptide insert between domains I and II of isoforms expressed in neurons, cardiac muscle, and neonatal skeletal muscle contains multiple phosphorylation sites that modulate channel activity. In mammals, one or more small β-subunits help target α-subunits to their proper places in the cell or modify channel behavior.

Certain voltage-gated Na^+ channels participate in the perception of pain produced by noxious stimuli such as chemicals, toxins, and extreme temperatures. Remarkably, humans with mutations of the $Na_v1.7$ channel feel no pain and mice with mutations of the $Na_v1.8$ channel are insensitive to scorpion toxin. **Local anesthetics,** such as lidocaine and procaine, prevent the perception of painful stimuli by blocking Na^+ channels in sensory nerves and inhibiting generation of action potentials. $Na_v1.7$ and $Na_v1.8$ are targets for new pain killing drugs. Voltage-gated Na^+ channels of cardiac muscle cells are also sensitive to local anesthetics, which are used for the emergency treatment of potentially fatal disorders of cardiac rhythm. Voltage-gated Na^+ channels are also the targets of puffer fish toxins and paralytic toxins that scorpions, spiders, and snails use to immobilize their prey.

Mutations in the gene for a heart Na^+ channel are another cause of long QT syndrome in humans. Patients have a mixture of normal and defective Na^+ channels. Most of the time, the mutant channels open and close normally, but occasionally, they fail to inactivate, sustaining the inward Na^+ current that depolarizes the membrane. These rare abnormal events in a large population of Na^+ channels delay the repolarization of the membrane, prolong the action potential, and predispose the patient to abnormal cardiac rhythms and sudden death.

Voltage-Gated Calcium Channels

Like voltage-gated Na^+ channels voltage-gated Ca^{2+} channels consist of four homologous domains in vertebrates. These channels favor Ca^{2+} over Na^+ by more than 500-fold, even though both ions are the same size. Remarkably, just three amino acid substitutions in each subunit convert a Na^+ channel into a Ca^{2+} selective channel. A structure of one of these mutant channels shows that the selectivity filter consists of three binding sites for fully hydrated Ca^{2+} ions along the axis of the pore. Higher affinity of these sites for Ca^{2+} than Na^+ explains the selectivity. Like voltage-gated Na^+ channels, Ca^{2+} channels are activated by membrane depolarization, inactivated by a conformational change, and returned to the resting state when the membrane repolarizes. Inactivation is generally slower than that for Na^+ channels.

Voltage-sensitive Ca^{2+}-selective channels have other subunits in addition to the large α_1-subunit with four internally homologous domains. The glycoprotein α_2-subunit is essential for the assembly and normal gating kinetics of α_1. Other, smaller peptide subunits designated β, γ, and Δ, promote delivery of the α_1-subunit to the cell surface and modify the voltage-dependence of the channel.

Ca^{2+} channels have numerous functions. First, Ca^{2+} channels contribute to membrane depolarization during action potentials in some cells. Ca^{2+} currents supplement Na^+ currents in vertebrate heart cells and replace them in heart pacemaker cells (see Fig. 39.20) and some invertebrate neurons. Given the very low Ca^{2+} concentration in the cytoplasm (see Fig. 26.12), open Ca^{2+} channels have a powerful effect on the membrane potential.

Second, Ca^{2+} channels convert electrical signals (membrane depolarization) into chemical signals by raising the local concentration of Ca^{2+}. Action potentials in nerve terminals open plasma membrane calcium channels and incoming Ca^{2+} stimulates the secretion of neurotransmitters (see Figs. 17.9 and 17.10). In cardiac muscle, action potentials cause an influx of Ca^{2+} that triggers the release of additional Ca^{2+} from the endoplasmic reticulum to activate contraction (see Fig. 39.15).

TABLE 16.1 Calcium Channel Classification

Type	Distribution	Functions	Blockers
L-type	Heart; skeletal muscle	Excitation-contraction coupling	Dihydropyridines
N-type	Heart; sympathetic neurons; CNS presynaptic terminals	Neurotransmitter secretion	ω-Conotoxin
P/Q-type	Synapses	Neurotransmitter secretion	
T-type	Neurons	Neuron excitation	Ni^{2+}

Third, voltage-gated Ca^{2+} channels act as voltage sensors in skeletal muscle. Action potentials stimulate voltage-gated Ca^{2+} channels, which use direct physical contact to activate **Ca^{2+} release channels** in the endoplasmic reticulum to release more Ca^{2+} that stimulates contraction (see Fig. 39.15).

To carry out these diverse physiological functions, vertebrate cells express a variety of Ca^{2+} channel proteins with different physiological properties. Traditionally, Ca^{2+} channels were divided into several classes, termed N, T, L, and P/Q, based on their sites of expression, voltage required for activation, open channel currents, inactivation kinetics, and sensitivity to drugs (Table 16.1). For example, only L-type calcium currents are sensitive to dihydropyridines, which are used therapeutically to dilate blood vessels by relaxing smooth muscle. N-type Ca^{2+} currents resist dihydropyridines but are blocked selectively by ω-conotoxin, which prevents neurotransmitter release at some synapses. This classification is still useful, but the discovery and characterization of channel isoforms with novel properties blurred these distinctions.

Transient Receptor Potential Channels

Diverse chemical and physical stimuli activate various transient receptor potential (TRP) channels. These share many features with voltage-gated cation channels, including four subunits, each with six transmembrane helices, a central selectivity filter formed by a P-loop between helices S5 and S6 (Fig. 11.3) and a peripheral domain composed of helices S1 to S4 (Fig. 16.9). It is likely that a common ancestor gave rise to both families which diverged widely. Helix S4 has aromatic residues rather than the arginines characteristic of voltage-gated channels, so TRP channels are generally not voltage-sensitive. The large cytoplasmic domains on both the N- and C-termini of TRP channels are not shared by voltage-gated cation channels.

When active, TRP channels admit modest amounts of both extracellular Na^+ and Ca^{2+} through short selectivity filters that include a ring of four backbone carbonyls. Both the selectivity filter and contacts between the S6 helices contribute to closing the pore of inactive channels. Structures of the heat-sensitive transient receptor potential vanilloid-1 (TRPV1) channel show that binding of a spider toxin to the extracellular side of the pore and capsaicin next to the S5-S6 linker helix each contribute to opening the pore.

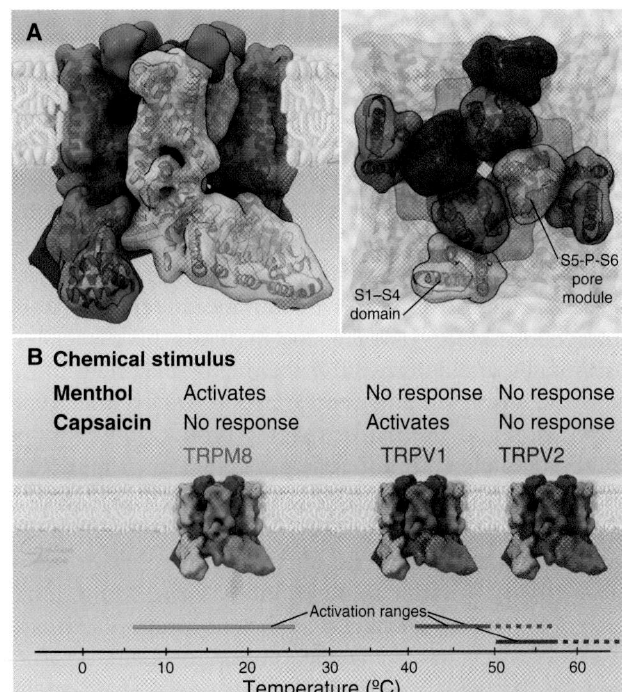

FIGURE 16.9 TRANSIENT RECEPTOR POTENTIAL (TRP) CHANNELS. A, Structure of the heat-activated TRPV1 (transient receptor potential vanilloid-1) channel determined by electron cryomicroscopy. Ribbon diagrams of views from the side and top show the central pore and peripheral domains. **B,** Responses of three TRP channels to chemicals and temperature. Cold temperatures and the cooling chemical menthol activate TRPM8. Hot temperatures and capsaicin, the active ingredient of hot chili peppers, activate TRPV1. (For reference, see Liao M, Cao E, Julius D, et al. Structure of the TRPV1 ion channel determined by electron cryo-microscopy. *Nature*. 2013;504:107–112 [**A**] and McKemy DD, Neuhausser WM, Julius D. Identification of a cold receptor reveals a general role for TRP channels in thermosensation. *Nature*. 2002;416:52–58 [**B**].)

Organisms from most parts of the phylogenetic tree use the TRP family of channels for sensation of diverse stimuli, including chemicals, osmolarity of their environment, and temperature. TRP channels enable humans to sense bitter and sweet tastes, pain, high temperature (and hot spices), and cool stimuli. For example, both high temperatures and chemicals from hot spices activate TRPV1 channels, accounting for the perception of such spices as being "hot" (Fig. 16.9B). The brain cannot discern whether heat or a spice activated the TRPV1 channels in a sensory nerve. Similarly, cold temperatures activate the CMR1 channel that also responds to cooling chemicals such as menthol. Signaling mechanisms downstream from seven-helix receptors and receptors tyrosine kinases (see Chapter 24) activate other TRP channels, in some cases by producing a second messenger (see Chapter 26) such as the membrane lipids phosphatidylinositol 4,5-bisphosphate (PIP_2) and diacylglycerol.

Channels Gated by Intracellular Ligands

Cytoplasmic Ca^{2+}, cyclic nucleotides, or β/γ-subunits of trimeric G-proteins (see Fig. 25.9) activate a group of channels with six transmembrane segments and a P loop (LC ligand-gated [see Fig. 16.2]). Genes for these channels diverged from K^+ channels relatively recently in evolution, about the time when animals diverged from fungi. Their sequences are similar to each other.

Ca^{2+}-activated K^+ channels are first cousins of voltage-gated K^+ channels with the Ca^{2+}-binding protein, **calmodulin** (see Fig. 3.12C), bound constitutively to the cytoplasmic tail following S6. Ca^{2+}, entering the cytoplasm through the plasma membrane or released from intracellular stores (see Fig. 26.12), binds this associated calmodulin and activates the channel by making it more sensitive to membrane depolarization. Differential gene expression and alternative splicing produce a variety of these channels with different physiological properties.

Cyclic nucleotide–gated ion channels have a C-terminal cyclic nucleotide-binding domain homologous with bacterial cyclic nucleotide-binding proteins (Fig. 16.10). Four of these subunits form a functional channel. Binding of cyclic adenosine monophosphate (cAMP) or cyclic guanosine monophosphate (cGMP) (see Fig. 26.1) to the cytoplasmic receptor domain opens a pore for Na^+ and Ca^{2+} and depolarizes the membrane. Changes in cyclic nucleotide concentration provide a sharp on-off switch, as ligand must occupy at least three of the four subunits to open the channel. Ca^{2+} entering the cytoplasm binds to calmodulin associated with the N-terminal cytoplasmic part of the protein. This provides negative feedback to the channel.

Ion channels gated by intracellular cyclic nucleotides are particularly important in sensory systems, including olfaction (see Fig. 27.1) and vision (see Fig. 27.2). Odorant molecules stimulate olfactory sensory neurons by binding seven-helix receptors in the plasma membrane. These receptors work through trimeric G-proteins to increase the cytoplasmic concentration of cAMP. cAMP opens cAMP-gated cation channels, depolarizes the membrane, and activates voltage-gated Na^+ channels to fire an action potential. Signal transduction in photoreceptor cells also uses a cyclic nucleotide-gated channel. Light activates a seven-helix receptor, leading to a decline in cytoplasmic cGMP. This sends a positive signal by closing cGMP-gated channels, hyperpolarizing the photoreceptor plasma membrane, and reducing the secretion of the neurotransmitter glutamate.

A sister group of channels, called *hyperpolarization-activated cyclic nucleotide-gated channels,* regulate the electrical activity of the heart. These "HCN channels" open in response to extremes of membrane potential. Binding of cyclic nucleotides makes these channels more sensitive to the membrane potential.

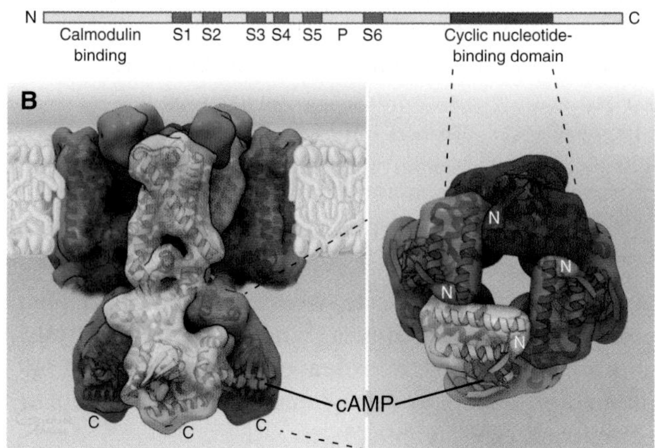

FIGURE 16.10 CYCLIC NUCLEOTIDE–GATED CATION CHANNELS. A, Domain architecture with an N-terminal calmodulin binding site, six predicted transmembrane helices (S1 to S6), a P loop, and a C-terminal cyclic nucleotide–binding domain. **B,** Ribbon diagram of a model of the cyclic adenosine monophosphate (cAMP)-gated channel based on the homologous transmembrane elements of the TRPV1 channel and a structure on the cytoplasmic cyclic nucleotide binding domain that crystalized as a tetramer. (For reference, see PDB file 1Q5O and Zagotta WN, Olivier NB, Black KD, et al. Structural basis for modulation and agonist specificity of HCN pacemaker channels. *Nature.* 2003;425:200–205.)

Ca-Release Channels

The most massive members of the S5-S6 family of channels release Ca^{2+} from the endoplasmic reticulum of eukaryotes (see Fig. 26.13). These huge genes originated in ancient metazoans and diverged into two families called IP_3 receptors (for the ligand inositol 1,4,5-triphosphate) and Ryanodine receptors (a drug) that inactivates them (Fig. 16.2). Huge cytoplasmic domains are linked to channel-forming domains of other similar members of the S5-S6 family (see Fig. 26.13).

Ion Channels Gated by Extracellular Ligands

Channels gated by chemicals mediate communication between nerve terminals and other nerves or muscles. This communication takes place at specializations called synapses, which facilitate chemical transmission (see Figs. 17.9 and 17.10). On the sending side, presynaptic terminals are specialized for exocytosis of chemicals called neurotransmitters, which they package in small vesicles. Neurotransmitters include acetylcholine, serotonin, glutamic acid, glycine, and γ-aminobutyric acid (GABA) (see Fig. 17.8). When an action potential arrives at a nerve terminal, voltage-gated Ca^{2+} channels admit Ca^{2+} to the cytoplasm. This causes synaptic vesicles to fuse with the plasma membrane, releasing neurotransmitters outside the cell. Neurotransmitters diffuse to the postsynaptic membrane in microseconds.

On the receiving side, the neurotransmitter activates ligand-gated ion channels in the postsynaptic membrane. Some ligand-gated channels trigger action potentials in the postsynaptic membrane by admitting cations, which drive the membrane potential toward threshold. Others inhibit action potentials by admitting Cl^-, which hyperpolarizes the postsynaptic membrane.

Stimulation of ligand-gated channels is transient because of inactivating conformational changes called desensitization, and because neurotransmitters are rapidly removed from the synaptic cleft between the cells (see Figs. 17.9 and 17.10). An extracellular enzyme degrades acetylcholine. Carriers (see Chapter 15) remove other neurotransmitters by pumping them back into the presynaptic cell.

Glutamate Receptors

The amino acid glutamate is the main neurotransmitter used by neurons in the brain to excite other neurons. Binding of glutamate to glutamate-gated channels provides the energy to open a cation channel permeable to both Na^+ and K^+, depolarizing the postsynaptic membrane (see Fig. 17.10). This depolarization excites the cell by activating voltage-gated sodium channels to trigger an action potential. Thus these channels are vital to every aspect of brain function and their dysfunction contributes to human diseases including psychiatric disorders. Plant glutamate receptors participate in the response of developing plants to light.

Glutamate receptors are highly divergent members of the S5-S6 family of channels. Their genes apparently originated by fusion of genes for an S5-P-S6 channel and a bacterial amino acid–binding protein (Fig. 16.2).

Eukaryotic glutamate receptor channels are tetramers, usually of two pairs of homologous subunits (Fig. 16.11). Each subunit has an extracellular aminoterminal domain involved with channel assembly, a ligand-binding domain and four hydrophobic segments. M1, M2, and M3 correspond to S5-P-S6 in other family members, although the orientation of the pore in the plasma membrane is inverted compared with KcsA. Like KcsA, the M3 helices cross to form the gate located on the cytoplasmic side of the membrane owing to the inverted orientation of the pore. The available structures have yet to resolve the details of the selectivity filter.

Binding of glutamate (and various drugs) in a cleft in the ligand-binding domain provides the energy to close the domain like a clamshell. This motion produces tension on the transmembrane domain and opens the gate. After the pore opens, spontaneous conformational changes in the extracellular domains rapidly reclose the pore resulting in an inactive "desensitized" state with bound ligand.

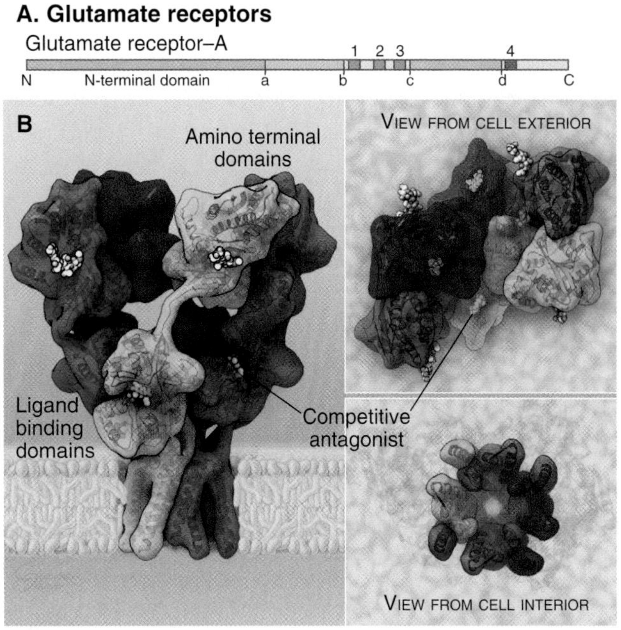

FIGURE 16.11 GLUTAMATE-GATED ION CHANNEL. A, Domain organization of the α-amino-3-hydroxy-5-methylisoxazole-4-propionate (AMPA)–type glutamate receptor with the glutamate-binding domain between a and b and four predicted transmembrane segments, M1 to M4. M2 is a P loop oriented toward the cytoplasm. **B,** Ribbon diagrams of the atomic structure of the inactive AMPA-A channel with each of the four identical subunit polypeptides a different color. Space-filling models show a competitive antagonist bound to the glutamate binding site. The side view shows the three domains. Note that the cytoplasmic domains of the blue and green subunits are oriented vertically, but that the yellow and red subunits swap their amino terminal domains. Top views show that the cytoplasmic domains have twofold symmetry, a dimer of dimers, while the transmembrane domains have fourfold symmetry. (For reference, see PDB file 3KG2 and Sobolevsky AI, Rosconi MP, Gouaux E. X-ray structure, symmetry and mechanism of an AMPA-subtype glutamate receptor. *Nature*. 2008;462:745–756.)

Multiple genes, alternative splicing, and RNA editing (see Fig. 11.7) contribute to the diversity of glutamate receptor subunits, which assemble into homo- and heterotetrameric channels used in different parts of the nervous system. In addition to glutamate receptors, three families of isoforms are sensitive to different pharmacologic agonists: *N*-methyl-D-aspartate **(NMDA),** α-amino-3-hydroxy-5-methyl-4-isoxazole propionate **(AMPA),** or **kainate.** NMDA receptors are obligate heterotetramers involved with learning and memory (see Fig. 17.11). They are more permeable to Ca^{2+} than to Na^+ and K^+ and desensitize slower than the other isoforms. Because excess intracellular Ca^{2+} can be damaging, overstimulation of NMDA receptors by glutamate released from cells during strokes or constitutive activation of NMDA receptors by point mutations can kill nerve cells.

Pentameric Ligand-Gated Ion Channels

Several important neurotransmitters activate members of a family of channels with five subunits (Fig. 16.12). They are also called Cys-loop receptors. A subset of bacteria

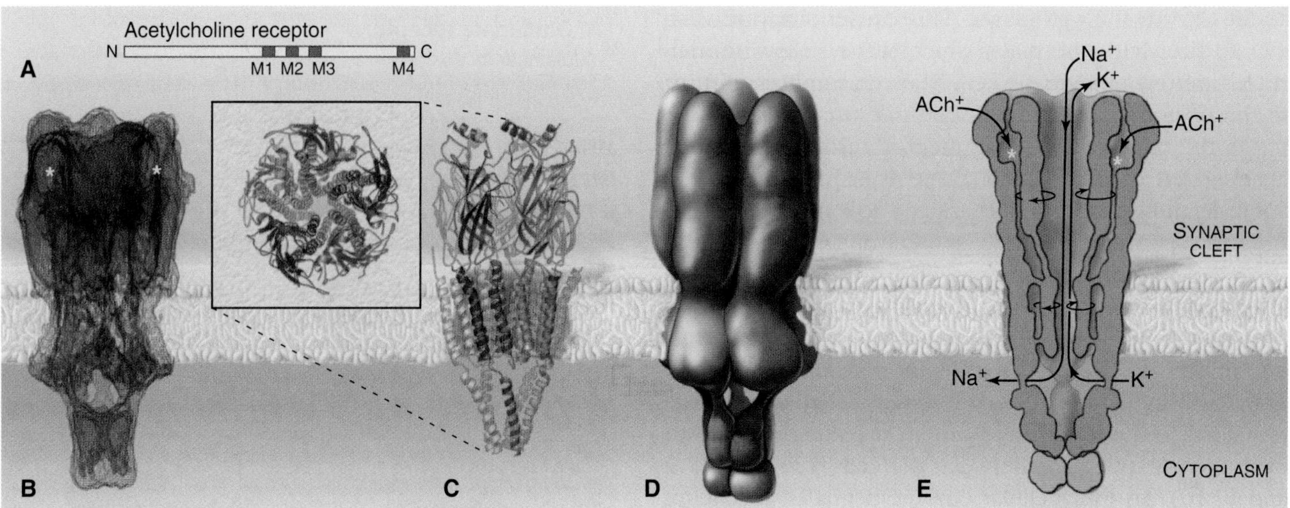

FIGURE 16.12 NICOTINIC ACETYLCHOLINE RECEPTOR. A, Domain organization. All four hydrophobic segments, M1 to M4, form transmembrane helices. **B–C,** Structure of the pentameric nicotinic acetylcholine receptor from the electric organ of the electric ray determined by electron microscopy. **B,** Reconstruction at a resolution of 0.46 nm. **C,** Ribbon diagrams of the nicotinic acetylcholine receptor from a structure at a resolution of 0.40 nm. *Left,* View from the extracellular side, showing five M2 helices lining the central pore. *Right,* Side view of model. The extracellular domains of two *red* α-subunits bind acetylcholine. **D,** Space-filling surface representation. **E,** Diagram showing the acetylcholine (Ach)-binding sites, the proposed conformational changes following activation, and the passages for Na^+ into the cell and K^+ out of the cell. A 43-kD protein called rapsyn *(blue)* binds on the cytoplasmic side. (For reference, see PDB file 1OED and Miyazawa A, Fujiyoshi Y, Unwin N. Structure and gating mechanism of the acetylcholine receptor pore. *Nature*. 2003;423:949–955.)

and one *Archaeon* have genes for structurally related channels, but metazoans are the only eukaryotes with these channels. Perhaps an early metazoan acquired the gene from a bacterium by lateral transfer.

The nicotinic acetylcholine receptor from the plasma membrane of skeletal muscle cells was the first ligand-gated channel to be characterized in detail. This receptor triggers action potentials that stimulate muscle contraction (see Figs. 17.9 and 39.14). It is called the nicotinic acetylcholine receptor because it also binds the tobacco alkaloid **nicotine.** Related nicotinic acetylcholine receptors in the central nervous system are the targets in tobacco addiction.

The muscle nicotinic acetylcholine receptor is a pentamer of four different, but homologous, subunits with the composition $\alpha_2\beta\gamma\varepsilon$ (Fig. 16.12). Each subunit has a large N-terminal extracellular segment, four transmembrane α-helices (M1 to M4), and a large cytoplasmic segment between M3 and M4. M2 α-helices from the five subunits line a central transmembrane pore like staves of a barrel. Hydrophobic side chains line this pore except for a few negative charges that may contribute to cation selectivity. The other three transmembrane α-helices of each subunit separate the M2 helices from the surrounding lipid. The N-terminal segments of each subunit form massive extracellular domains, each folded into similar, highly twisted β-sandwiches. Deep cavities in the α-subunits bind acetylcholine. Similar transmembrane domains form the pores of bacterial pentameric channels, but the extracellular ligand-binding domains are smaller and they lack cytoplasmic domains.

Gating and ion selectivity of acetylcholine receptors differ in concept from the P-loop family of channels. In closed channels, the narrowest part of the closed pore is less than 0.7 nm in diameter, too small for hydrated K^+ and Na^+ ions, and the hydrophobic pore does not provide a passage for unhydrated ions. Acetylcholine binding to the two α-subunits provides energy to change the conformations of the extracellular domains, which rotate the M2 helices and open a channel that is more permeable to K^+ and Na^+ than to Ca^{2+}. The resulting permeability to all three ions causes the membrane potential to collapse toward a reversal potential (see the section in Appendix 16.3 entitled "Net Current through Ion-Selective Channels") around 0 mV. This triggers voltage-gated Na^+ channels to initiate a self-propagating action potential in the muscle plasma membrane with nearly 100% efficiency (see Fig. 17.9).

Muscle cells and some central nervous system neurons express more than two dozen different isoforms of nicotinic acetylcholine receptors. Most are composed of a mixture of subunits but some have five identical subunits.

Many toxins bind nicotinic acetylcholine receptors, blocking communication between motor nerves and skeletal muscle (Appendix 16.1). **α-Bungarotoxin** has been used to characterize the receptor. **Curare** is a powerful muscle relaxant used during surgery because it blocks acetylcholine-binding sites without opening the channel. Local anesthetics, such as procaine, bind within the channel and block ion conductance.

Some people produce autoimmune antibodies to nicotinic acetylcholine receptors, resulting in a disease called **myasthenia gravis.** When antibody binds the receptor, the skeletal muscle cell internalizes the receptor, reducing its response to acetylcholine and causing weakness.

The neurotransmitter serotonin (5-hydroxytryptamine) activates excitatory Na^+/K^+ channels similar in architecture to nicotinic acetylcholine receptors. The pentameric receptors that respond to GABA and glycine are Cl^- channels that hyperpolarize the postsynaptic membrane, so they inhibit excitability. Several isoforms of GABA receptors bind **benzodiazepines,** drugs used to treat depression. They increase the probability that the channel will open. **Strychnine** inhibits glycine receptors, making neural circuits oversensitive to stimulation.

Chloride Channels

ClC Channels

Bacteria, yeast, and animals have genes for members of a large family of ClC chloride channels with a unique evolutionary origin and structure. They are selective for Cl^- and gated by the membrane potential. ClCs control membrane excitability and contribute to volume regulation and epithelial transport.

ClC subunits are triangular transmembrane proteins formed from 18 α-helices (Fig. 16.13). These helices surround a pore that passes through the middle of each subunit, like the pores of ammonia channels (Fig. 16.14), aquaporins (Fig. 16.15), and porins (see Fig. 13.9C). Several helices around the pore extend only part way across the lipid bilayer. Highly conserved residues in the loops between these helices form the selectivity filter for Cl^- in the middle of the membrane bilayer. Two subunits associate tightly in the lipid bilayer, so each channel has two pores that independently conduct Cl^- when active.

ClC0 from skeletal muscle is a well characterized member of the family. Like voltage-gated cation channels, ClC0 channels open when the membrane depolarizes and spontaneously close shortly thereafter. A negatively charged glutamate side chain is believed to block the pore of inactive channels and to swing out of the way in active channels. In an unexpected turn of events, physiological analysis of the bacterial ClC channel used for structural studies revealed it has many features of a carrier that exchanges Cl^- for H^+ rather than behaving like a typical ion channel like other members of this family. This is an example of blurred distinctions among channels, carriers, and pumps.

Mutations of Cl^- channel genes cause several human diseases. Defective skeletal muscle ClC1 channels cause recessive and dominant myotonias. Mutations in kidney ClC5 channels predispose individuals to the formation of kidney stones.

Bestrophin Channels

These channels are another unique family with their own evolutionary origin, distinct structure, and wide distribution in prokaryotes and eukaryotes. They were discovered as the protein mutated in some patients with retinal degeneration. Affected patients have one of more than 120 different mutations spread throughout the protein. Most mutations causing retinal degeneration encode proteins in photoreceptor cells, such as visual pigments, but bestrophin is expressed in supporting cells of the retinal pigment epithelium.

Channels formed by five subunits looks superficially like nicotinic acetylcholine receptors but their fold is different (Fig. 16.13B). The isoform in the human eye is a Cl^- channel activated by intracellular Ca^{2+}, which binds a site near the pore. The selectivity filter is formed by two rings of phenylalanine side chains with the slightly electropositive edges of their aromatic rings in the pore.

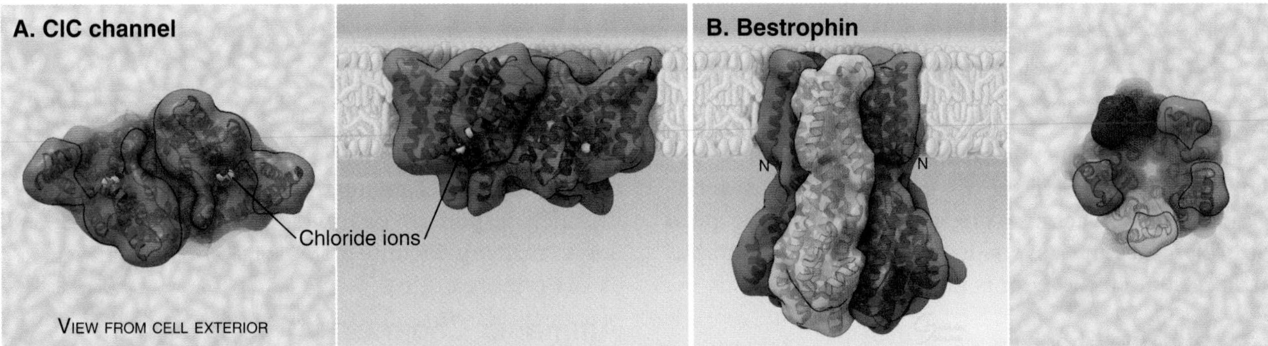

FIGURE 16.13 CHLORIDE CHANNELS. A, Crystal structure of a ClC channel StClC from the bacterium *Salmonella typhimurium.* These ribbon diagrams show one subunit in *red* and the other in *blue.* The *white* sphere shows the position of a Cl^- ion in the selectivity filter. This structure is the model for other chloride channels, but it actually works more like a carrier than a channel. **B,** Crystal structure of bestrophin from the bacterium *Klebsiella pneumoniae*. The ribbon diagrams show each of the five subunits in a different color. (**A,** For reference, see PDB file 1KPK and Dutzler R, Campbell EB, Cadene M, et al. X-ray structure of a ClC chloride channel at 3.0 Å reveals the molecular basis of anion selectivity. *Nature.* 2002;415:287–294. **B,** See PDB file 4WD8 and Yang T, Liu Q, Kloss B, et al. Structure and selectivity in bestrophin ion channels. *Science.* 2014;346:355–359.)

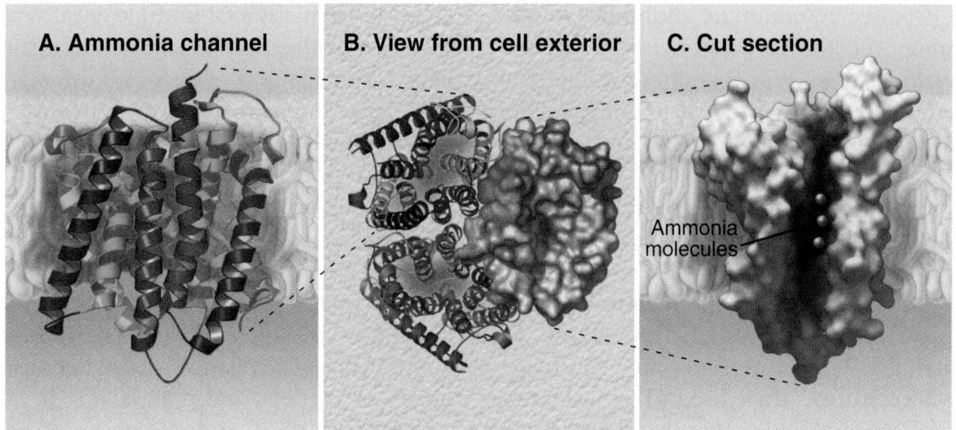

FIGURE 16.14 AMMONIA CHANNELS. A, Ribbon diagram of one subunit of the trimeric AmtB ammonia channel from *Escherichia coli* with the extracellular side at the top. **B,** Ribbon and space-filling diagram of the channel viewed from outside the cell. Each of the three identical subunits has a pore for ammonia to cross the lipid bilayer. **C,** Space-filling cutaway drawing of one subunit exposing the channel for passage of ammonia *(blue)*. In the vestibule facing outside the cell, an ammonium ion gives up a proton to an amino acid side chain before passing through the hydrophobic pore through the core of the protein as uncharged ammonia. At the narrowest point of the pore, hydrogen bonds between the ammonia and two histidines contribute to the specificity. (For reference, see PDB file 1U77 and Khademi S, O'Connell J, Remis J, et al. Mechanism of ammonia transport by Amt/MEP/Rh: structure of AmtB at 1.35 Å. *Science*. 305:1587–1594, 2004.)

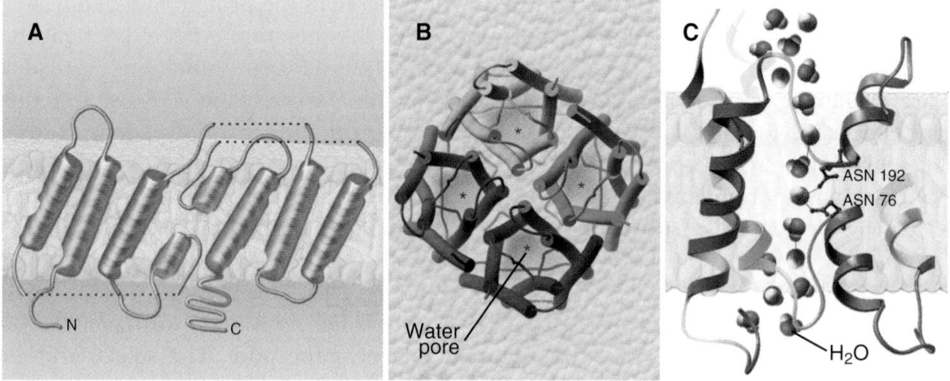

FIGURE 16.15 WATER CHANNELS. A, Membrane topology of aquaporin-1 deduced from the primary structure. The two halves of the polypeptide have similar sequences but are inverted relative to each other. **B,** Structure determined by electron crystallography, showing four identical units, each with a pore *(red asterisk)*. Helices are depicted as cylinders. **C,** Detail of the water pore, with a chain of water molecules crossing the membrane. Two asparagines in the middle of the pore hydrogen bond one water. (Courtesy P. Agre, Johns Hopkins Medical School, Baltimore, MD. For reference, see PDB file 1FQY and Murata K, Mitsuoka K, Hirai T, et al. Structural determinants of water permeation through aquaporin-1. *Nature*. 2000;407:599–605.)

The bacterial channel shown in Fig. 16.13B is a Na^+ channel, but replacing a single isoleucine in the pore with phenylalanine coverts it to a Cl^- channel. Another single residue in the pore is important for gating.

Other Chloride Channels

In addition to the bestrophin family, researchers have characterized three other types of calcium-gated Cl^- channels. TMEM16A channels are dimers of subunits with 10 transmembrane helices, a novel channel structure. Some of the ten members of the TMEM16A family have additional functions including exchange of lipids between the two leaflets of membranes. Volume-regulated anion channels respond to cell swelling by conducting Cl^- and some small organic molecules (Fig. 11.5).

Ammonia Channels

Another ancient family of channel proteins evolved in early prokaryotes to conduct ammonia across the cell membrane. Ammonia can directly penetrate lipid bilayers, but these channels allow low concentrations of ammonia to serve as a source of nitrogen that prokaryotes use to synthesize proteins and nucleic acids.

BOX 16.1 Rh Antigens

Before anything was known about membrane proteins or ammonia transport, immunologists discovered that injection of rhesus monkey red blood cells into rabbits produced antibodies that reacted with most, but not all, human red blood cells. This Rh antigen, now known to be the most common isoform of the human ammonia channel, is clinically relevant because the red blood cells of an "Rh-positive" fetus inheriting this isoform from the father can provoke an immunologic response from the mother if she lacks this isoform and is "Rh negative." During subsequent pregnancies, these maternal antibodies can attack the red blood cells of an Rh-positive fetus, causing serious complications.

Bacteria, Archaea, and eukaryotes still depend on these channels. Humans use these channels to conduct both ammonia and carbon dioxide across the plasma membranes of red blood cells, where they are known as **Rh antigens** (Box 16.1). These channels are also important for transporting ammonia in the kidney and liver.

Ammonia channels consist of three identical subunits, each composed of 11 transmembrane helices (Fig. 16.14). The interfaces between these subunits are tightly sealed, but each subunit has a narrow internal pore that is highly selective for ammonia and methylammonium. Both substrates (ammonium NH_4^+ and methylammonium $CH_3NH_3^+$) are charged in aqueous solution and must leave behind a proton to pass through the pore as uncharged species (NH_3, CH_3NH_2). They pick up a replacement proton on the other side of the membrane as they exit the channel. Selectivity is achieved by the tight fit of the substrates in the hydrophobic pore and by transient formation of a novel hydrogen bond within the pore. With millimolar ammonium on one side of a membrane, these channels conduct hundreds of ammonia molecules per second without leaking water, protons, or other charged species.

Water Channels

Water diffuses relatively slowly across lipid bilayers, so membranes are barriers to water movement unless they contain water channels. Such channels were postulated years ago, but they eluded identification until investigators tested a small hydrophobic protein from red blood cells for water channel activity. Expression of this protein in frog eggs made them permeable to water, so they swelled and burst when placed in hypotonic media. Discovery of this **aquaporin** rapidly led to the characterization of a family of related water channels from many species, including bacteria, fungi, and plants. Related channels, called *aquaglyceroporins*, transport glycerol and ammonia across bacterial membranes. Other family members transport CO_2 in plants.

Four identical aquaporin subunits form a stable tetramer in the plane of the membrane (Fig. 16.15). The two halves of the protein arose by gene duplication, as their structures and sequences are remarkably similar. Each subunit has a narrow pore that is selective for water passing through the middle of a bundle of α-helices. Ten water molecules line up in a pore approximately 0.3 nm in diameter. Hydrogen bonding of waters with a pair of asparagine residues at a narrow point in the pore allows the channel to be selective for water. Water channels have no gates, so they are open constitutively.

Osmotic pressure created by pumps, carriers, and the macromolecular composition of the cytoplasm drives water through aquaporins at rates exceeding 10^9 molecules per second. This explains why red blood cells rapidly swell and shrink passively, depending on the osmolarity of the surrounding fluid (see Fig. 13.8).

Various human tissues express 12 different aquaporin isoforms. Aquaporin-1 is found in red blood cells, renal proximal tubules, blood vessel endothelial cells, and the choroid plexus (which makes spinal fluid in the brain). A few humans carry mutations that inactivate aquaporin-1; remarkably, homozygotes have no symptoms, despite the low water permeability of their red blood cells (and presumably other tissues that depend on this isoform). Aquaporin-2 is required for the epithelial cells lining collecting ducts in the kidney to reabsorb water. **Antidiuretic hormone** (vasopressin) controls the placement of aquaporin-2 in the collecting duct plasma membrane. It activates a seven-helix receptor, causing cytoplasmic vesicles storing aquaporin-2 to fuse with the plasma membrane. This allows water to move from the urine into the hypertonic extracellular space of the renal medulla. Inactivating mutations in both aquaporin-2 genes results in severe water loss, called nephrogenic diabetes insipidus. Reaction of mercuric chloride with sensitive cysteine residues closes the aquaporin water pores, explaining how mercurials, used therapeutically as diuretics in the past, inhibit the reabsorption of water filtered by the kidney.

Plants are particularly dependent on water, so they have evolved a large family of aquaporin genes. These isoforms are located in most cellular membranes. Plants must move water to maintain turgor and expand cells in growing tissues. Water moves continuously from roots through xylem vessels and cells in tissues to exit from stomata in leaves as vapor. These movements across cell and tonoplast membranes depend on aquaporins. Plants also depend on aquaporins to transport CO_2 for photosynthesis.

Porins

Porins are channels with wide, water-filled pores found in the outer membranes of gram-negative bacteria and mitochondria. The subunits are composed of an antiparallel barrel of 16 or 18 β-strands that cross the membrane (see Fig. 13.9C). One to three of the loops connecting the strands extend into the center of the barrel and line the pore. The functional molecule consists of three identical subunits.

Most porins are relatively nonselective pores for small, water-soluble molecules, although some are specific for certain solutes, such as sugars. *E. coli* uses a related protein with 22 transmembrane β-strands to transport iron complexes across the outer membrane. A central "cork" domain occludes the lumen of this β-barrel. Interactions across the periplasmic space with plasma membrane proteins open and close the pore. A variety of viruses use bacterial porins as receptors.

Other Families of Channels

The 12 families of channels covered above have received the most attention, but researchers continue to discover new channels that do not fit into any of these families. TMBIM channels are monomers of just seven transmembrane helices that form a pH-sensitive calcium leak channel through the middle of the tiny protein. Piezo2 forms pressure-sensitive, non-selective cation channels that detect the touch sensation in skin. Piezo genes in a wide range of eukaryotes encode large polypeptides with 24–36 transmembrane segments. Bacteria have a family of fluoride channels consisting of a homodimer of subunits with four transmembrane helices. Both bacteria and eukaryotes have novel channels selective for urea. They are homotrimers of subunits with 12 transmembrane helices. A pore for urea passes through each of the subunits. The human urea channel is vital for the function of the kidney.

ACKNOWLEDGMENTS

We thank Fred Sigworth for advice on earlier editions and Bill Catterall, David Julius, and Benoit Roux for suggestions on revisions to this chapter.

SELECTED READINGS

Biel M. Cyclic nucleotide-regulated cation channels. *J Biol Chem*. 2009;284:9017-9021.

Birnbaum SG, Varga AW, Yuan LL, et al. Structure and function of Kv4-family transient potassium channels. *Physiol Rev*. 2004;84:803-833.

Catterall WA. Ion channel voltage sensors: structure, function, and pathophysiology. *Neuron*. 2010;67:915-928.

Catterall WA, Swanson TM. Structural basis for pharmacology of voltage-gated sodium and calcium channels. *Mol Pharmacol*. 2015;88:141-150.

Dolphin AC. G protein modulation of voltage-gated calcium channels. *Pharmacol Rev*. 2003;55:607-627.

Dutzler R. The structural basis of ClC chloride channel function. *Trends Neurosci*. 2004;27:315-320.

Haswell ES, Phillips R, Rees DC. Mechanosensitive channels: what can they do and how do they do it? *Structure*. 2011;19:1356-1369.

Hilf RJ, Dutzler R. A prokaryotic perspective on pentameric ligand-gated ion channel structure. *Curr Opin Struct Biol*. 2009;19:418-424.

Hille B. *Ion Channels of Excitable Membranes*. 3rd ed. Sunderland, MA: Sinauer Associates; 2001.

Hummer G. Potassium ions line up. Do K^+ ions move in single file through potassium channels? *Science*. 2014;346:303.

Jasti J, Furukawa H, Gonzales EB, Gouaux E. Structure of acid-sensing ion channel 1 at 1.9 Å resolution and low pH. *Nature*. 2007;449:316-323.

Jiang Y, Lee A, Chen J, et al. Crystal structure and mechanism of a calcium-gated potassium channel. *Nature*. 2002;417:515-522.

Jiang Y, Lee A, Chen J, et al. X-ray structure of a voltage-dependent K^+ channel. *Nature*. 2003;423:33-41.

Julius D. TRP channels and pain. *Annu Rev Cell Dev Biol*. 2013;29:355-384.

Kane Dickson V, Pedi L, Long SB. Structure and insights into the function of a Ca^{2+}-activated Cl^- channel. *Nature*. 2014;516:213-218.

Kawate T, Michel JC, Birdsong WT, Gouaux E. Crystal structure of the ATP-gated P2X(4) ion channel in the closed state. *Nature*. 2009;460:592-598.

Khademi S, O'Connell J, Remis J, et al. Mechanism of ammonia transport by Amt/MEP/Rh: Structure of AmtB at 1.35 Å. *Science*. 2004;305:1587-1594.

King LS, Kozono D, Agre P. From structure to disease: The evolving tale of aquaporin biology. *Nat Rev Mol Cell Biol*. 2004;5:687-698.

Kunzelmann K. TMEM16, LRRC8A, bestrophin: chloride channels controlled by Ca(2+) and cell volume. *Trends Biochem Sci*. 2015;40:535-543.

Lee CH, Lü W, Michel JC, et al. NMDA receptor structures reveal subunit arrangement and pore architecture. *Nature*. 2014;511:191-197.

Lewin GR. Natural selection and pain meet at a sodium channel. *Science*. 2013;342:428-429.

Liao M, Cao E, Julius D, Cheng Y. Structure of the TRPV1 ion channel determined by electron cryo-microscopy. *Nature*. 2013;504:107-112.

Lu Z. Mechanism of rectification in inward-rectifier K^+ channels. *Annu Rev Physiol*. 2004;66:103-129.

Matulef K, Zagotta WN. Cyclic nucleotide-gated ion channels. *Annu Rev Cell Dev Biol*. 2003;19:23-44.

Meyerson JR, Kumar J, Chittori S, et al. Structural mechanism of glutamate receptor activation and desensitization. *Nature*. 2014;514:328-334. For an animation of the transition of GluA2 from closed to open see <http://www.nature.com/nature/journal/v514/n7522/fig_tab/nature13603_SV1.html>.

Payandeh J, Scheuer T, Zheng N, Catterall WA. The crystal structure of a voltage-gated sodium channel. *Nature*. 2011;475:353-358.

Sachdeva R, Singh B. Insights into structural mechanisms of gating induced regulation of aquaporins. *Prog Biophys Mol Biol*. 2014;114:69-79.

Sobolevsky AI, Rosconi MP, Gouaux E. X-ray structure, symmetry and mechanism of an AMPA-subtype glutamate receptor. *Nature*. 2009;462:745-756.

Stroud RM, Miercke LJ, O'Connell J, et al. Glycerol facilitator GlpF and the associated aquaporin family of channels. *Curr Opin Struct Biol*. 2003;13:424-431.

Vargas E, Yarov-Yarovoy V, Khalili-Araghi F, et al. An emerging consensus on voltage-dependent gating from computational modeling and molecular dynamics simulations. *J Gen Physiol*. 2012;140:587-594.

Verdoucq L, Rodrigues O, Martinière A, et al. Plant aquaporins on the move: reversible phosphorylation, lateral motion and cycling. *Curr Opin Plant Biol*. 2014;22:101-107.

Yarov-Yarovoy V, DeCaen PG, Westenbroek RE, et al. Structural basis for gating charge movement in the voltage sensor of a sodium channel. *Proc Natl Acad Sci USA*. 2012;109:E93-E102. For a movie of the structural change see <http://www.pnas.org/content/suppl/2011/12/08/1118434109.DCSupplemental/sm01.mov>.

APPENDIX 16.1

Examples of Channel-Blocking Agents

Compound (Chemical Class)	Source	Physiological Effect
Sodium Channel Blockers		
Tetrodotoxin (alkaloid)	Japanese puffer fish	Paralyzes skeletal muscle
Saxitoxin (alkaloid)	Dinoflagellates	Paralyzes skeletal muscle
μ-Conotoxins (peptide)	Maine snails	Paralyzes skeletal muscle
Batrachotoxin (alkaloid)	Arrow poison frogs	Opens Na-channels, paralyzes skeletal muscle
Lidocaine	Chemical synthesis	Reduces cardiac and nerve excitability
Potassium Channel Blockers		
Quaternary amino alkanes	Chemical synthesis	Blocks K-currents, increases nerve excitability
Scorpion toxin	Scorpions	Blocks K-currents, increases nerve excitability
Calcium Channel Blockers		
Dihydropyridines	Chemical synthesis	Reduces excitability of L-type channels of striated muscles
ω-conotoxin (peptide)	Pacific cone snail	Inhibits nervous system N-type channels; blocks synaptic transmission
Nicotinic Acetylcholine Receptor		
α-Bungarotoxin (peptide)	Snake, *Bungarus multicinctus*	Blocks neuromuscular transmission; paralyzes skeletal muscle
α-Cobra toxin	Cobra	Blocks neuromuscular transmission; paralyzes skeletal muscle
Curare	Plant, *Strychnos toxifera*	Blocks neuromuscular transmission; paralyzes skeletal muscle

APPENDIX 16.2

Electrical Recordings in Biology

Analysis of electrical activity across biological membranes requires sensitive methods to detect electrical potential differences and the flow of current on a rapid time scale. Physiologists and clinicians use four general methods, with different sensitivities, to detect electrical activity of single channels (patch electrodes), cell membranes (microelectrodes and fluorescent dyes), and whole tissues (extracellular electrodes).

Single-Channel Recordings With Patch Electrodes

Patch-clamp microelectrodes (Fig. 16.16A) provide the best way to characterize the behavior of individual channels. A small-diameter, fire-polished glass capillary is pressed onto the surface of a cell and suction is used to form a high-resistance seal (10 to −50 gigaohms). The membrane patch is small enough to contain just a few ion channels. The electrode becomes part of an electric circuit that can measure current or voltage across the membrane. The high-resistance seal between micropipette and membrane ensures that more electrical current (composed of ions) flows through a single open channel than leaks in around the side of the electrode. When a channel opens, a sensitive ammeter connected to the micropipette records the direction and magnitude of ion flow through the channel as an electrical current. Patch electrodes give direct information about both current and the time that individual channels spend open or closed.

Variations of the patch-clamp technique provide access to channel properties. Leaving the membrane patch on the cell (cell-attached configuration) reveals properties of the channels in their cellular context. Lifting the membrane patch off the cell (excised-patch configuration) exposes the cytoplasmic surface of the membrane to ions, enzymes, or second messengers that the investigator adds to the bath. Similarly, the investigator can test the effects of potential ligands, drugs, and ions in the micropipette.

Measurement of Membrane Potentials With Intracellular Microelectrodes and Fluorescent Dyes

A glass capillary is drawn to a fine tip (~0.5 μm), filled with a conducting solution (3 *M* KCl), and inserted through the plasma membrane. The tip penetrates the cell with minimal damage, and the membrane seals tightly around it. The microelectrode is connected to a meter to record current and voltage (Fig. 16.16B). Alternatively, the investigator can apply a patch electrode to the cell surface and suck forcefully to breach the membrane, putting the micropipette in continuity

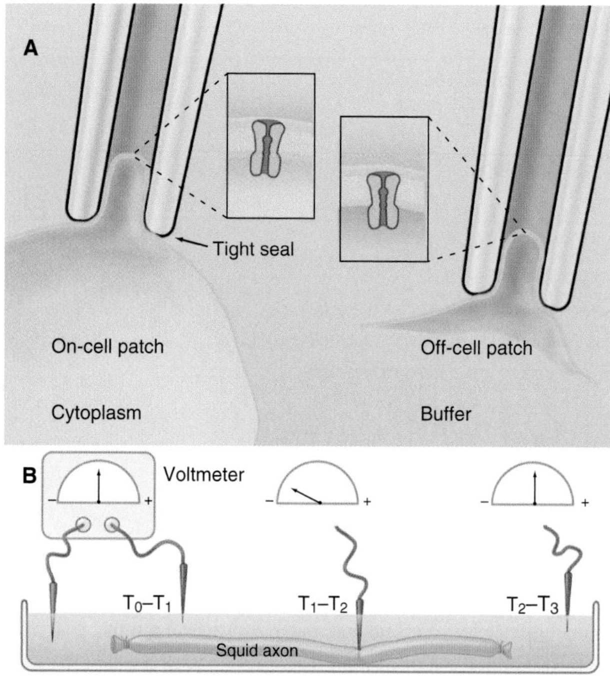

FIGURE 16.16 ELECTROPHYSIOLOGICAL MEASUREMENTS. A, Patch electrode. Fine-tipped glass micropipettes form a tight seal with a small patch of plasma membrane. The salt solution inside the pipette conducts current flowing through an open channel in the patch for recording. Lifting the patch of membrane off the cell exposes the cytoplasmic surface of the membrane to experimental manipulation from the bath. Solutes in the micropipette can stimulate the extracellular face of the membrane. **B,** Measurement of membrane potentials and currents with microelectrodes. A fine-tipped micropipette penetrates the plasma membrane of a cell and is part of a circuit that can record either membrane potential or current flowing across the membrane. *To measure membrane potential,* a voltmeter in the circuit records the voltage inside the cell relative to the bath and follows any changes that occur when ion channels open and close. *To measure current,* an electronic feedback device is placed in the circuit to hold the membrane potential at a constant value. Under these "voltage-clamped" conditions, the feedback device provides current to balance any current that results from opening of membrane channels. The current from the feedback device is a record of current across the membrane channels. The membrane potential is an ensemble property of a large number of individual molecules. Microelectrodes can measure the membrane potential on a submillisecond time scale.

with the cytoplasm for recordings from the rest of the membrane.

Two microelectrodes inserted into a beaker of saline register no potential difference. If one electrode is inserted into a cell, the meter registers a potential difference of −60 to −90 mV inside the cell relative to the bath. This **membrane potential** arises from the combined action of many membrane pumps, carriers, and channels.

Fluorescent dyes can provide an optical signal that is sensitive to membrane potential. This is the only convenient approach for acquiring information about the spatial distribution of potential charges within cells.

Extracellular Electrical Measurements

Synchronous electrical activity of thousands of cells produces small electrical currents outside the cells, which can be recorded with extracellular electrodes or even with electrodes on the surface of the body. Physicians take advantage of this phenomenon to record the ensemble electrical activity of the heart (**electrocardiogram** [ECG]), brain (**electroencephalogram** [EEG]), and muscle (**electromyogram** [EMG]). These recordings reflect the behavior of thousands of cells, so they provide little information about individual events at molecular or cellular levels.

APPENDIX 16.3

Biophysical Basis of Membrane Potentials

The membrane potential arises from separation of charges across an insulating surface (Fig. 16.17). The lipid bilayer provides the insulation required to separate charges. Either pumps or channels can produce unpaired charges. Pumps that transport unpaired ions generate membrane potentials directly. Channels that pass unpaired ions can use ion concentration gradients across membranes to generate membrane potentials. The concentration gradient provides a diffusional force to drive ions through channels. Because channels are ion specific, an excess of charge builds up after very few ions cross a membrane. This excess charge creates a membrane potential and stops the net movement of additional ions across the membrane.

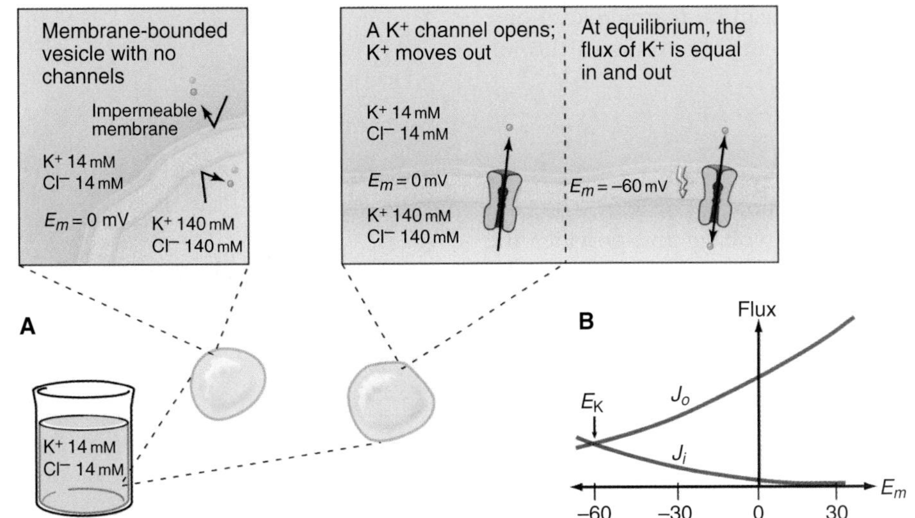

FIGURE 16.17 MEMBRANE POTENTIAL. A, Production of a membrane potential, E_m, by a K^+-selective ion channel and a 10-fold potassium chloride concentration gradient across a membrane. The three panels illustrate the situations without a channel, when a channel first opens, and at equilibrium. **B,** Dependence of K^+ fluxes out of the vesicle (J_o) and into the vesicle (J_i) as a function of membrane potential (E_m). K^+ passes out of the vesicle (driven by the concentration gradient) and into the vesicle (driven by the voltage across the membrane). At a potential of −60 mV, these fluxes are balanced. This is called the resting potential, or E_K. If the membrane potential is more positive than −60 mV, the flux of K^+ out of the vesicle (driven by the concentration gradient) exceeds the flux into the vesicle (driven by the voltage across the membrane). This pushes the potential toward E_K.

This discussion starts with a qualitative description of forces behind membrane potentials and then develops a quantitative account of membrane potentials with single or multiple types of ion channels.

Diffusion Potentials

An impermeable membrane enclosing concentrated potassium chloride is suspended in a bath of more dilute potassium chloride (Fig. 16.17). If the membrane contains a pore that is *selectively permeable* for bidirectional diffusion of K^+, the concentration gradient drives K^+ out of the membrane compartment. Because Cl^- cannot pass through this selective pore or the membrane bilayer, the inside compartment loses positive charge. Charge imbalance creates an electrical field, negative inside, called the **membrane potential.** By convention, extracellular voltage is defined as zero.

Force provided by the membrane potential influences the diffusion of ions through the pore in both directions. The positive potential outside opposes the diffusion of K^+ out of the vesicle and drives K^+ into the vesicle, up its concentration gradient. Net K^+ efflux continues until a charge imbalance builds up a membrane potential large enough to drive K^+ influx at the same rate that the concentration gradient drives K^+ efflux. The electrical potential required to stop net ion movement is called the **equilibrium potential** for K^+, or **Nernst potential,** E_K.

Quantitative Relationships

The quantitative description of membrane potentials by the Nernst equation is *the* central concept of electrophysiology. This relationship between an ion concentration gradient and a balancing membrane potential is derived as follows, using K^+ as an example.

The concentration gradient provides the first force. J_o is the rate (expressed in ions per second) of efflux through the K^+-selective pore. J_i is the rate of influx. The fluxes are proportionate to the concentrations on the side from which the ions come. The ratio of these rates is equal to the ratio of the inside and outside K^+ concentrations, K_i and K_o:

$$\frac{J_o}{J_i} = \frac{K_i}{K_o}$$

A typical cell has a K_i:K_o ratio of approximately 35.

The membrane potential provides a second force. A positive potential gives a positive ion a higher energy, driving it down the electrical gradient. A negative potential has the opposite effect. The difference in electrical energy per mole of ions is equal to zFE, where z is the valence (+1 for K^+), F is the Faraday constant (10^5 coulombs [C]/mol), and E is the potential in volts. This difference in energy enters the equation for the flux ratio as an exponential term (the "Boltzmann factor"), with the electrical energy difference divided by the thermal energy:

$$\frac{J_o}{J_i} = \frac{K_i e^{zFE/RT}}{K_o}$$

where R is the gas constant and T is the absolute temperature.

The K^+ fluxes in and out are equal when

$$\frac{K_i e^{zFE/RT}}{K_o} = 1$$

This famous Nernst equation can be rearranged to give the equilibrium (Nernst) potential in terms of the ion concentrations.

$$E_K = \frac{RT}{zF} \ln K_o / K_i$$

RT is the thermal energy of a mole of particles. The ratio *RT/zF* has the dimensions of voltage and provides the electrical potential that gives a mole of charged particles with valence *z* an electrical energy *(zFE)* equal to the thermal energy *(RT)*. At physiological temperatures, its value is approximately 25 mV for univalent ions where $z = 1$. The ratio of *RT/zF* establishes the range of potentials (tens of millivolts) that occur in cells.

Another form of the Nernst equation is more convenient. Because $\ln(x) = 2.3 \log(x)$ and $2.3\ RT/F = 60$ mV at 30°C, the Nernst equation can be rewritten as

$$E_K = \frac{60 \text{ mV}}{z} \log K_o / K_i$$

Thus, the membrane potential is –60 mV when the K^+ concentration inside is 10 times the concentration outside.

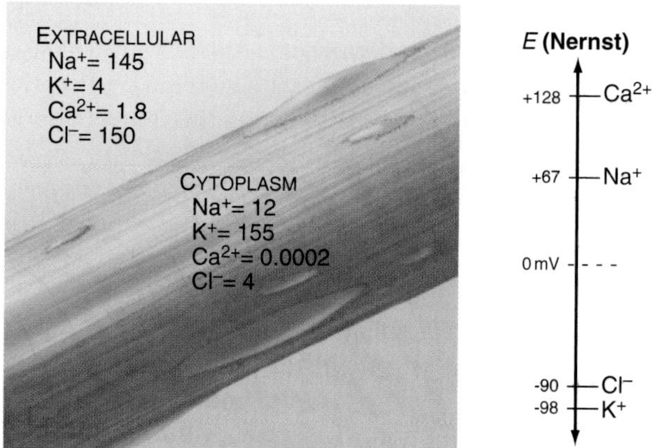

FIGURE 16.18 PHYSIOLOGICAL ION CONCENTRATIONS AND MEMBRANE POTENTIALS. Ion concentrations in the cytoplasm and outside a vertebrate muscle cell. The scale on the right shows the corresponding equilibrium membrane potentials (Nernst potentials) that would result if channels for each one of these ions opened.

Nernst Potential for Various Ions

The Nernst potential can be calculated for each ion known to have a selective channel in cell membranes: Na^+, K^+, Ca^{2+}, and Cl^- (Fig. 16.18). Given physiological gradients of these ions across the plasma membrane, the membrane potential could range from –98 to +128 mV, depending on which channels are open. In resting cells, only K^+ channels are open, so the resting membrane potential is close to E_K. Thus, variation of extracellular K^+ concentration changes the membrane potential. In vertebrates, the normal extracellular K^+ concentration is approximately 4 mM, but it varies from 2 mM to >8 mM in disease states. This fourfold variation in K_o changes the membrane potential by 30 to 37 mV, enough to affect cellular processes that are sensitive to the membrane potential. Other channels open and close selectively in response to extracellular or intracellular ligands, membrane potential, physical forces, or other factors (see text). Selective activation of channels is responsible for action potentials and other behavior of excitable membranes (see Fig. 17.6).

APPENDIX 16.4

Charging and Discharging the Membrane

Opening or closing ion channels influences the membrane potential and the flux of ions across the membrane. This discussion explains how movement of just a few ions allows cells to change their membrane potential without dissipating ion gradients across the membrane. Consequently, flux through a few ion channels rapidly changes the membrane potential during action potentials. The result of opening multiple channels with different ion selectivities and concentration gradients is also explained.

Membrane Capacitance

The membrane potential *(E)* produced by a given net charge inside the cell *(Q)* depends on the physical properties of the membrane, summarized in a constant called **capacitance** *(C)*:

$$E = \frac{Q}{C}$$

Capacitance depends on membrane area, thickness (physical separation between internal and external charges), and dielectric constant. If the capacitance is large, many ions must move to change the membrane potential. For cell membranes, the capacitance is approximately 1 mF/cm^2. One farad is 6×10^{18} charges per volt.

Charge Movement for a Small Cell

The following calculation shows why *ion concentration gradients change little during most electrical events in cells.* This is important to eliminate the requirement for excessive energy to restore ion gradients. A cell that is 18 μm in diameter might have a capacitance of 10^{-11} F, or 6×10^7 charges per volt of membrane potential. Thus, movement of 6 million positive charges out of the cell produces a membrane potential of −0.1 V, or −100 mV. A cell of this size with an internal concentration of 150 mM K^+ contains approximately 2.7×10^{11} K^+, so movement of fewer than one in 40,000 K ions (0.0025%) from inside to outside creates a large membrane potential. This fraction of ions is far less for large cells, owing to their smaller ratio of surface area to volume. Thus, little energy is required for a large change in membrane potential, such as an action potential. When ion channels open, few ions cross the membrane before an opposing electrical field develops and retards further flux.

In Chapters 14 and 15, pumps and carriers were also noted to produce opposing membrane potentials when moving ions across membranes. This can be avoided by opening ion channels that short-circuit the change in membrane potential by providing pathways for counterions to move in the same direction or similar ions to move in the opposite direction across the membrane.

Rate of Charge Movement Through Channels

A current is the rate of movement of charge. The ionic current *(I)* across a membrane is taken as *positive* when charges move *outward.* According to this definition, the equation for conservation of charge in a cell is

$$\frac{dQ}{dt} = -I$$

A positive current reduces the net charge inside the cell, and vice versa. Including the relationship for capacitance ($E = Q/C$), the equation relates the current to the rate of change of membrane potential:

$$\frac{dE}{dt} = \frac{-I}{C}$$

Because channels conduct about 6×10^6 charges per second, a single open channel changes E at a rate of −100 mV/sec on this 18-μm cell.

Because most channels occur at densities of 50–200/μm², an 18-μm cell will have 50,000 to 200,000 channels. If a few channels open together, the membrane potential rapidly approaches the Nernst potential for the selected ion. This explains why most electrical events in cells transpire in a millisecond time frame. Because the rate of current flow through ion channels is not limiting, the time course of electrical events depends on the kinetics of channel opening and closing. This focuses attention on factors that control whether channels are open or closed, also known as **gating.**

Net Current Through Ion-Selective Channels

Another way to describe ionic current across a membrane is

$$I = ze_o(J_o - J_i)$$

where e_o is the elementary charge. The dependence of current on membrane potential for real channels is complicated (Fig. 16.17B), so electrophysiologists approximate this current-voltage relationship of channels by a linear relationship, such as Ohm's law ($E = IR$):

$$I = g(E - E_{ion})$$

where g is **conductance** (inverse of resistance) and E_{ion} is the **reversal potential** of a particular ion channel (the potential at which current reverses from out to in). For perfectly selective pores, the reversal potential for each ion equals its Nernst potential, even in the face of other ionic gradients. The unit used for current is siemens (equivalent to 1 ampere per volt). Most channels have currents in the picosiemens range (10^{-12} S).

For a simple pore, a plot of current versus membrane potential is linear, with no current at E_{ion}; real channels are more complicated. Typical plots of current versus voltage deviate from a straight line. This is called **rectification.** Deviation may be attributable to voltage-dependent conformational changes in the channel protein or to nonpermeant ions blocking the pore.

Each channel contributes independently to the total current, so given n channels on a cell membrane, the total current is

$$I = ng(E - E_{ion})$$

Opening Na^+ and K^+ channels has opposite effects because the ion concentration gradients are reversed. The Nernst potential for Na^+ is approximately +65 mV in a typical cell, given a 10-fold excess of Na^+ outside the cell. Current through an Na^+ channel is negative (ie, inward) at membrane potentials below E_{Na}. Thus, if a Na^+ channel opens on a cell in which E equals 0, the membrane potential rises toward E_{Na}.

Consequence of Multiple Channel Types Opening Simultaneously

More than one type of open channel creates a situation more complicated than the *equilibrium* described by the Nernst potential for a single-ion species (Figs. 16.18 and 16.19). Consider a cell with physiological ion gradients and two channels—one open K^+ channel and one open Na^+ channel—having conductances of g_K and g_{Na}. The total current through these two channels is the sum of the individual currents:

$$I_{total} = g_K(E - E_K) + g_{Na}(E - E_{Na})$$

Note from this relationship that current is zero at the midpoint between E_K and E_{Na}, and the line has twice the slope of a single channel (ie, twice the conductance).

Which channel predominates? The equation for I_{total} can also be written as

$$I_{total} = g_{eff}(E - E_{eff})$$

where the effective conductance g_{eff} and reversal potential E_{eff} are given by

$$g_{eff} = g_K + g_{Na}$$

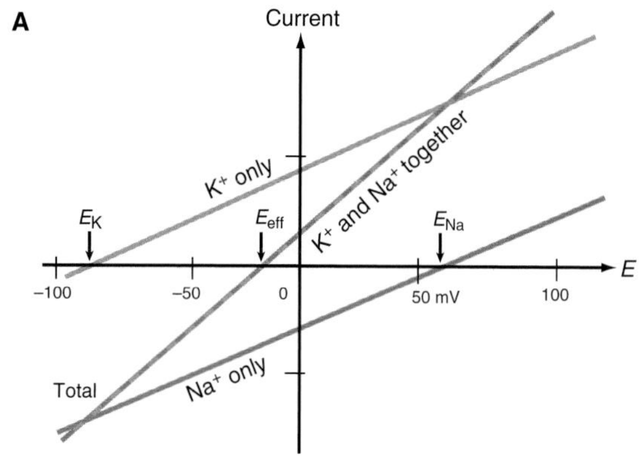

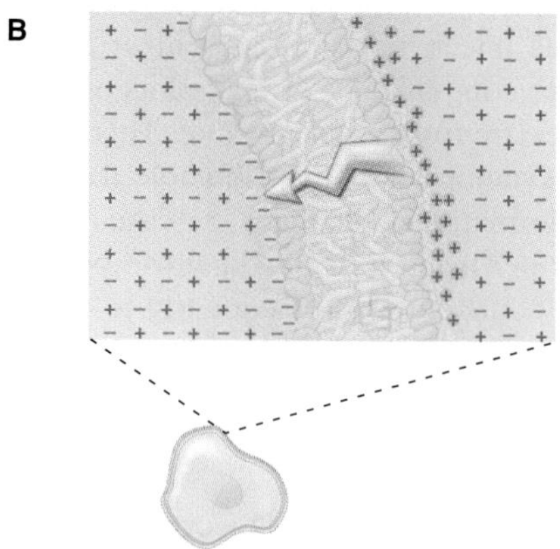

FIGURE 16.19 MEMBRANE POTENTIAL AND CURRENTS ACROSS A MEMBRANE WITH TWO TYPES OF CHANNELS. **A,** Dependence of currents on membrane potential resulting from opening either K^+ channels or Na^+ channels individually or together. In contrast to Fig. 16.17, which shows ion fluxes in each direction, this is a plot of net current. E_K and E_{Na} are the equilibrium potentials (zero current) when only potassium or sodium channels are open. When both types of channels are open, the equilibrium potential (E_{eff}) is midway between the equilibrium potentials of the two types of channels. **B,** Distribution of positive *(red)* and negative *(blue)* ions across the plasma membrane and around a cell having a negative membrane potential. Excess negative charge builds up near the inside of the membrane, with the excess positive charge near the outside.

and

$$E_{eff} = \frac{g_K E_K}{g_K + g_{Na}} + \frac{g_{Na} E_{Na}}{g_K + g_{Na}}$$

The two channels together act like a single channel with an effective conductance equal to the sum of their conductances and a reversal potential that is the weighted average of their reversal potentials, that is, weighted by their relative conductances (Fig. 16.19A).

Goldman, Hodgkin, and Katz formulated another equation for E. It uses permeability (P, in units of cm/s) to describe the membrane potential:

$$E = \frac{RT}{F} \ln \frac{P_{Na}[Na]_o + P_K[K]_o + P_{Cl}[Cl]_o + \cdots}{P_{Na}[Na]_i + P_K[K]_i + P_{Cl}[Cl]_i + \cdots}$$

This equation summarizes the concepts presented here about membrane potentials. Just *two factors* determine the membrane potential: (a) the **concentration gradients** of different ions (eg, the Nernst potentials for each ion) and (b) the relative permeabilities of the membrane to these ions. When all Na^+ and Cl^- channels are closed (P_{Na}, $P_{Cl} = 0$), the equation reduces to the Nernst relationship for K. When all K^+ and Cl^- channels are closed (P_K, $P_{Cl} = 0$), the equation collapses to the Nernst relationship for Na^+.

In nerve cells, the resting membrane is most permeable to K^+ but also slightly permeable to Na^+, so the resting potential is near E_K. Opening more K^+ channels or lowering extracellular K^+ makes the resting potential more negative. Opening more Na^+ channels or raising extracellular Na^+ makes the resting potential more positive.

Charge Redistribution by Electrical Conduction

Most cellular ions have balancing counterions and are distributed randomly in solution, whereas unpaired ions contributing to membrane potentials are confined to boundary layers near the membrane (Fig. 16.19B). Like-charged ions repel one another, so unpaired ions tend to accumulate at boundaries where they can move no farther.

During electrical events, unpaired ions redistribute over membrane surfaces by electrical conduction at rates much faster than diffusion. This works as follows: Ions are always in motion, exchanging places. Introduction of extra ions sets off a chain of movements as neighbors repel each other, resulting in rapid spread of unbalanced charge near the membrane. Diffusion of the entering ions over the plane of the membrane would take much longer than this electrical wave. Thus, electrical signaling is the fastest signaling process in cells.

CHAPTER 17

Membrane Physiology

This chapter describes how pumps, carriers, and channels cooperate in living systems. These three components often work together in circuits or cycles. Pumps establish gradients of ions across membranes (see Chapter 9). Channels regulate membrane permeability to these ions to maintain the electrical potential (see Chapter 11) required for membrane excitability. Carriers use ion gradients as a source of energy to drive transport as well as to do other work (see Chapter 10). Coupling ion fluxes through pumps and carriers to do work is called a **chemiosmotic cycle.**

Selective expression of a repertoire of pumps, carriers, and channels in specific membrane compartments enables cells to build sophisticated machines from a stockpile of standard components. If the pumps, carriers, and channels produced by a cell are known, it is relatively easy to explain complicated physiological processes. The examples in this chapter also show how defects in pumps and channels cause disease and how drugs can alleviate the symptoms.

Chemiosmotic Cycles

A simple chemiosmotic cycle couples a cation transporting pump to solute transport by a carrier across the plasma membrane or an organelle membrane (Fig. 17.1). The driving reaction is called the primary transport step. This involves an input of energy and, in most cases, some chemical reaction. Other steps called secondary transport reactions depend on ion gradients. The transported substrate is the same chemically on both sides of the membrane. Although chemiosmotic cycles are simple in concept, their importance and power should not be underestimated. They operate in every membrane of every cell.

Pumps use energy derived from adenosine triphosphate (ATP) hydrolysis, light absorption, or another chemical reaction (see Table 9.1) to move ions across a membrane. This raises the concentration of a cation (C^+) on one side and depletes it on the other side of a membrane-bounded compartment. An ion gradient is characterized by a chemical term, the concentration gradient, and an electrical gradient (the membrane potential explained in Fig. 11.17). The electrochemical potential across a membrane represents a reservoir of power and a capacity to do work, also known as an ion-motive force. A macroscopic analogy is using a pump to fill an elevated reservoir with fluid.

Carriers and other membrane proteins use the potential energy of ion gradients to drive other processes. This is analogous to using fluid flow out of a reservoir to drive a turbine, which uses the energy for other types of work. Many carriers use energy derived from the downhill passage of one substrate to transport one or more other substances up their concentration gradients across the same membrane barrier. In Fig. 17.1, the carrier links

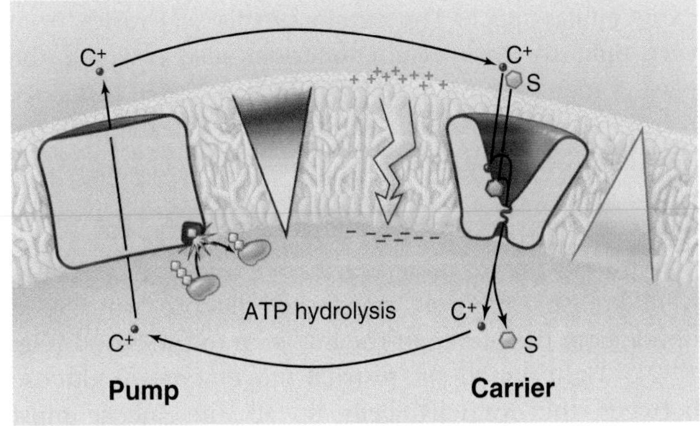

FIGURE 17.1 A MODEL CHEMIOSMOTIC CYCLE IN A MEMBRANE SURROUNDING A CLOSED SPACE. An adenosine triphosphate (ATP)-driven pump transports a cation C^+ out of the compartment. The energy derived from ATP is stored as a concentration gradient of C^+ *(red triangle)* and a membrane potential *(yellow arrow)* across the membrane. The carrier uses the electrochemical gradient of C^+ to drive the transport of both C^+ and a solute up a concentration gradient *(green triangle)* across the membrane.

the transport of solute S to the movement of cation C^+ down its gradient. Recirculation of cations allows a cell to accumulate solute against its concentration gradient. In addition to the osmotic work illustrated in Fig. 17.1, chemiosmotic cycles can do chemical work. During both oxidative and photosynthetic phosphorylation, proton cycles drive ATP synthesis by rotary ATP synthases (see Figs. 9.5, 19.5, and 19.8). Chemiosmotic cycles can also perform mechanical work. An electrochemical gradient of protons across the plasma membrane drives the rotation of bacterial flagella (see Fig. 38.25).

Chemiosmotic cycles using protons dominate the biological world. Most bacterial cycles involve proton pumps, proton-linked carriers, or other proton-linked events. The same is true of early-branching eukaryotes, fungi, and plants. Plasma membranes of plant cells have a powerful proton pump and a collection of proton carriers. Proton chemiosmotic cycles are also characteristic of most eukaryotic organelles, including the Golgi apparatus, endosomes, lysosomes, mitochondria, and chloroplasts. Animal cell plasma membranes are a major exception, because they use predominantly sodium ions for their chemiosmotic cycles.

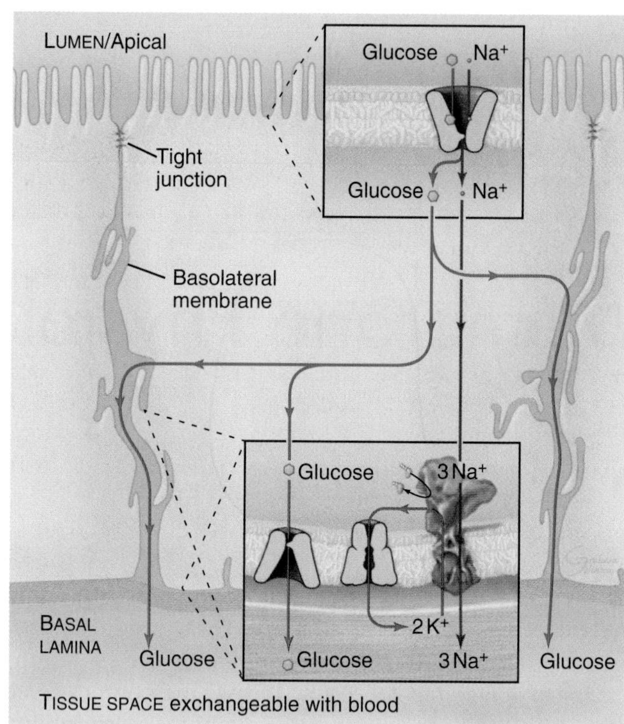

FIGURE 17.2 GLUCOSE TRANSPORT BY THE INTESTINAL EPITHELIUM. Tight junctions seal the epithelium of polarized epithelial cells. Na^+K^+-adenosine triphosphatase (ATPase) pumps (space-filling model) in the basolateral plasma membrane drive Na^+/glucose symporters in the apical plasma membrane *(upper inset)* and glucose uniporters in the basolateral plasma membrane *(left icon in lower inset)* to move glucose from the lumen of the intestine to the blood. Basolateral K^+ channels *(middle icon)* recycle K^+ pumped into the cell.

Epithelial Transport

Net transport across an epithelium depends on tight junctions (see Fig. 31.2) that seal the extracellular space between the cells (Fig. 17.2). These junctions separate two extracellular compartments. The **apical compartment** is the free surface or lumen of the organ (eg, the intestine, respiratory tract, or kidney tubules—topologically continuous with the external world). The **basolateral compartment** lies between epithelial cells and is continuous with the underlying connective tissue and its blood vessels. Tight junctions restrict diffusion of solutes between the apical and basolateral compartments of the extracellular space. The extent of this seal varies from very tight to leaky. Tight junctions also separate the plasma membrane into apical and basolateral domains, restricting the movement of integral membrane proteins between these domains.

Glucose Transport in the Intestine, Kidney, Fat, and Muscle

A chemiosmotic cycle transports glucose from food uphill from the lumen of the intestine to the blood (Fig. 17.2). Tight junctions restrict movement of glucose between the epithelial cells, so all the glucose must move through the epithelial cells using the following components:

- Na^+K^+-adenosine triphosphatase (ATPase), located in the basolateral plasma membrane
- SGLT1 (sodium glucose cotransporter 1) Na^+/glucose symporter, restricted to the apical plasma membrane
- GLUT5 (glucose transporter 5) glucose uniporter, restricted to the basolateral plasma membrane

Na^+K^+-ATPases use ATP hydrolysis to produce Na^+ and K^+ gradients across the plasma membrane by continuously pumping Na^+ out of and K^+ into the cell. SGLT Na^+/glucose symporters use Na^+ moving inward down its electrochemical gradient across the apical plasma membrane to accumulate high internal concentrations of glucose from the gut lumen. In this step, energy is expended (dissipation of the Na^+ gradient) to move glucose uphill. GLUT (glucose transporter) uniporters in the basolateral membrane simply facilitate movement of cytoplasmic glucose down its concentration gradient out of the cell. The kidney uses a similar strategy to recapture glucose filtered from blood, transporting it across the renal proximal tubule cell and back into the blood.

Fat and muscle cells use the GLUT4 D-glucose uniporter to take up glucose from the blood when it is plentiful following a meal. These cells store GLUT4 internally in membrane vesicles. After a meal, high blood glucose stimulates secretion of insulin into blood. Signal transduction mechanisms (see Fig. 27.7) lead to fusion of the GLUT4 vesicles with the plasma membrane. This increases the rate of glucose transport into fat and muscle by fivefold to 20-fold, lowering the blood glucose concentration and providing the cells with glucose to convert into glycogen and triglycerides for storage.

Salt and Water Transport in the Kidney

In a section of the kidney tubule called the loop of Henle, the epithelium uses Na^+K^+-ATPase pumps and $Na^+/K^+/2Cl^-$ symporters to reabsorb NaCl that is filtered from blood into the excretory pathway (Fig. 17.3). Without this mechanism, salt would be lost in urine. Tight junctions seal this epithelium, so that salt must pass through the cells to return to the blood. $Na^+/K^+/2Cl^-$ symporters in the apical plasma membrane allow NaCl from the urine to enter the cell down its concentration gradient. Abundant Na^+K^+-ATPases in the basolateral plasma membrane (5000/μm²) create a Na^+ gradient to drive the symporter and to clear the cytoplasm of Na^+ accumulated from the tubule lumen. KCl that enters with Na^+ through the $Na^+/K^+/2Cl^-$ symporter leaves the cell through channels: K^+ channels in apical and basolateral membranes and Cl^- channels in basolateral membranes.

Furosemide, a drug used to treat congestive heart failure, inhibits the $Na^+/K^+/2Cl^-$ symporter in the loop of Henle. A weak heart leads to accumulation of fluid in the lungs (causing shortness of breath) and other tissues (causing swelling of the ankles). Inhibiting the $Na^+/K^+/2Cl^-$ symporter reduces NaCl reabsorption, so the kidney produces larger quantities of urine, clearing excess fluid from the body and relieving symptoms.

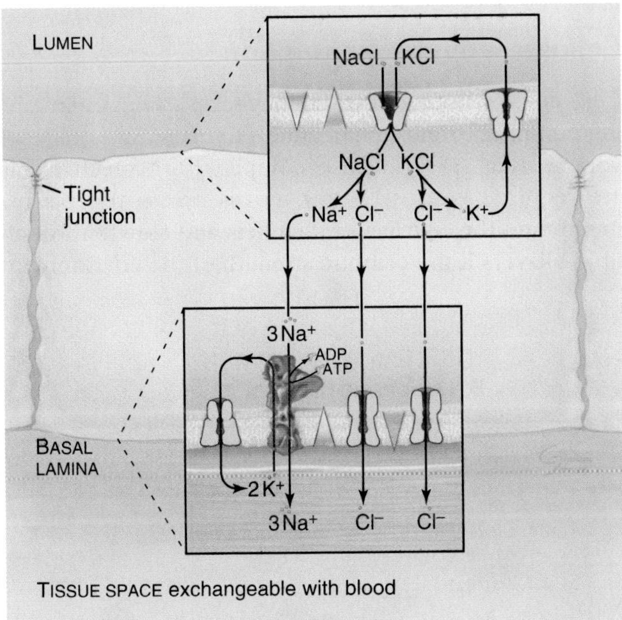

FIGURE 17.3 SODIUM CHLORIDE TRANSPORT BY THE EPITHELIUM OF THE KIDNEY TUBULE. Tight junctions seal the space between these polarized epithelial cells of the thick ascending limb of the loop of Henle. Na^+K^+-ATPase pumps (space-filling model) in the basolateral plasma membrane drive $Na^+/K^+/2Cl^-$ symporters in the apical plasma membrane. K^+ channels in the apical plasma membrane and K^+ channels and Cl^- channels in the basolateral plasma membrane provide paths for K^+ to circulate and for Cl^- to follow Na^+ across the cell from the lumen of the tubule to the blood compartment.

Cystic Fibrosis as a Transporter Disease

Cystic fibrosis results from loss of function mutations of the cystic fibrosis transmembrane regulator (CFTR), an unorthodox ABC transporter that functions as a Cl^- channel (see Chapter 9). Patients suffer from lung infections and impaired secretion of digestive enzymes by the pancreas. Understanding the pathology requires knowledge of the mechanisms that produce the fluid layer containing NaCl on the apical surface of the epithelial cells lining the airways of the lung (Fig. 17.4). Water and Cl^- move through the cell, while Na^+ moves between the cells.

The complicated process depends on familiar pumps and carriers. Cl^- moves into the base of the cell along with Na^+ and K^+ through $Na^+/K^+/2Cl^-$ symporters powered by the electrochemical gradient of Na^+ created by Na^+K^+-ATPases in the basolateral membrane. This brings excess potassium chloride into the cell. K^+ then

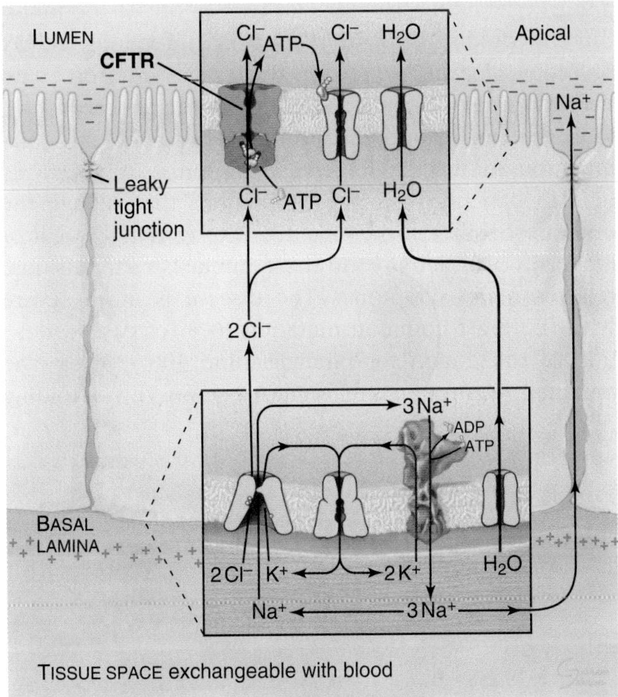

FIGURE 17.4 SALT AND WATER TRANSPORT ACROSS THE EPITHELIUM LINING THE RESPIRATORY TRACT. Leaky tight junctions partially seal the space between these polarized epithelial cells. Na^+K^+-ATPase pumps in the basolateral plasma membrane drive $Na^+/K^+/2Cl^-$ symporters in the basolateral plasma membrane. CFTR (cystic fibrosis transmembrane regulator) Cl^- channels in the apical plasma membrane allow Cl^- to move into the lumen, creating a negative electrical potential that pulls Na^+ between the cells into the lumen. CFTR also releases ATP, which activates additional Cl^- channels. Water follows sodium chloride into the lumen through water channels and between the cells. Basolateral K^+ channels allow K^+ to circulate.

recycles out of the cell by way of channels in the basolateral plasma membrane, leaving behind excess Cl^- inside the cell. When activated by phosphorylation of its regulatory domain and ATP binding, CFTR in the apical plasma membrane opens a channel for Cl^- and bicarbonate. Cl^- moves down its electrochemical gradient out of the cell, carrying charge to the outside. The whole epithelium becomes polarized, with the lumen electrically negative relative to the basolateral fluid compartment. This electrical driving force draws Na^+ between cells through leaky tight junctions from the extracellular fluid compartment to the surface of the epithelium. Sodium chloride on the apical surface creates an osmotic force that draws water down its concentration gradient across the cells to the lumen through water channels (see Fig. 11.15). CFTR also appears to inhibit transport mechanisms that reabsorb fluid from the lumen of the epithelium. A balance between this fluid secretion and fluid reabsorption normally keeps a layer of water on the surface of the epithelium allowing the cilia to clear secretions and bacteria from the lung.

Loss of CFTR function leaves the lungs of cystic fibrosis patients too dry. This situation is life threatening because cilia in the respiratory tract cannot move sticky, dry, mucus containing bacteria and viruses out of the lungs, thereby predisposing to respiratory infections. Sticky secretions in the pancreatic ducts also interfere with the secretion of digestive enzymes by the pancreas.

The severity of the disease depends on the particular mutation in the CFTR gene. Symptoms are relatively mild in patients with point mutations that reduce the open probability of the channel. A drug called ivacaftor increases the activity of these mutant channels and relieves many symptoms. The disease is more severe with the most common mutation (67% of cases), deletion of the codon for phenylalanine 508 (F508). The mutant Δ508 protein is temperature-sensitive, not folding properly at 37°C, so it fails to negotiate the secretory pathway to the plasma membrane. Patients with two copies of the Δ508 mutation have classic cystic fibrosis. Heterozygotes with the Δ508 mutation and a normal gene (approximately 5% of humans) have no symptoms. Combining the ΔF508 mutation with one of more than a thousand different mutations in the other copy of the gene causes the typical lung disease and a range of severity in the pancreatic problems. Treating patients with the Δ508 mutation is challenging, but new therapies are being explored, including activation of other plasma membrane Cl^- channels.

Transport Mechanisms to Improve Food Production

Plants depend on transport mechanisms to take up nutrients, CO_2, and water, and to export toxic materials, but their capacities vary widely. Thus some plants are more tolerant of salty conditions or concentrate higher concentrations of micronutrients such as iron. Carriers participate in most of these reactions, so providing food crops with optimal carriers is a promising strategy to improve production. For example, providing plants with carriers to export compounds that neutralize toxic ions such as Al^{3+} allows them to grow in acidic soils while supplying them with a channel-like Na^+ transporter improves salt tolerance. Similarly, providing variants of carriers with a high affinity for iron results in higher concentrations of this micronutrient in rice.

Cellular Volume Regulation

Cells employ both short- and long-term strategies involving pumps, carriers, and channels to maintain a constant volume (Fig. 17.5). These compensatory mechanisms are required because water moves across the plasma membrane through water channels and slowly through lipid bilayers if the osmotic strength of the environment

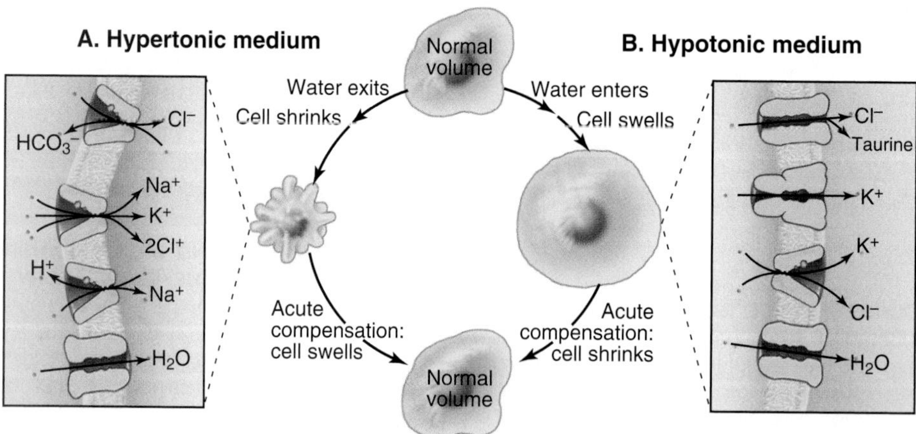

FIGURE 17.5 ACUTE CELLULAR VOLUME CONTROL. A, Cell is placed in hypertonic medium. **B,** Cell is placed in hypotonic medium. Cells compensate for volume changes by activating channels and carriers to move inorganic ions into or out of the cell. Swelling-sensitive LRRC8 channels also release taurine from the cell. Water follows passively through channels and across the lipid bilayer.

differs even slightly from that inside the cell. Water moves to maintain an osmotic equilibrium, as is illustrated for red blood cells in Fig. 8.8. In a hypotonic medium, water moves into the cell to dilute the cytoplasm. In a hypertonic medium, water moves out to concentrate the cytoplasm. The mechanisms employed to compensate for these volume changes are well defined, but the mechanisms that sense volume changes and trigger these responses are still being investigated.

Animal cells respond acutely to loss of water in a hypertonic environment by activating Na^+/H^+ antiporters, Cl^-/HCO_3^- antiporters, and/or $Na^+/K^+/2Cl^-$ symporters that bring KCl and NaCl into the cell. Water follows, returning the cell to its original volume in minutes.

Swelling in a hypotonic environment quickly activates plasma membrane K^+ channels, Cl^- channels, and/or a K^+/Cl^- symporter, releasing potassium chloride, osmolytes, and water from the cell. Osmolytes are small organic molecules, including amino acids and metabolically inactive polyalcohols (sorbitol and inositol) and methylamines. Bestrophins are the Cl^- channels activated by swelling in flies (see Fig. 11.13), but vertebrates use LRRC8 channels related to gap junction connexins (see Fig. 31.7) that release the osmolyte taurine along with Cl^-.

Compensation by moving water along with ions and osmolytes works in the short run, but changes in the internal concentrations of K^+, Na^+, and Cl^- affect the membrane potential and other physiological processes. In the long term, cells maintain their volume by adjusting the osmotic strength of cytoplasm with osmolytes. Adjusting osmolyte concentrations takes longer than ion concentrations, as it requires synthesis or degradation of osmolytes and their transport proteins, such as Na^+/osmolyte symporters.

Excitable Membranes

Regulation of membrane potential is particularly important in higher organisms, which use electrical signals generated by membrane channels for communication in their nervous and muscular systems. For example, reading and understanding this page depend on rapid creation and processing of electrical and chemical signals by cells in the visual system and brain. Ion channels produce the key event, a transient change in electrical potential of the plasma membrane, called an **action potential**. These energy-efficient electrical signals are the fastest means of communication in the body, spreading over the plasma membrane at tens of meters per second. Similarly, action potentials trigger skeletal muscle contraction, control the timing of the heartbeat, and coordinate the peristaltic motions of the gut and contractions of the uterus.

Electrical excitability is not limited to nerves and muscles. Eggs use a form of action potential as an early step in blocking fertilization by more than one sperm. Chemotaxis by macrophages and secretion of insulin and other hormones both depend on electrical excitability. The reader should be familiar with the appendixes in Chapter 11 to appreciate the following material.

Description of an Action Potential

If a microelectrode (see Fig. 11.16B) drives a small positive or negative current into a cell, a second microelectrode a short distance away detects a small voltage response. These electrotonic potentials decline rapidly with distance if the cell is not excitable.

The plasma membrane of an excitable cell, such as **neuron** or **muscle,** reacts much differently to depolarization. Rather than responding with a small local current, an excitable cell generates a large, reproducible change in membrane potential, called an action potential when depolarized beyond a certain level (called a **threshold**) (Fig. 17.6). Voltage-gated ion channels (see Fig. 11.7) generate this powerful electrical signal that spreads rapidly (10 m/sec) over the entire plasma membrane. During an action potential, the membrane potential can reach a peak of +40 to 50 mV before repolarizing to the resting potential. Because action potentials are self-triggering, they travel without dissipation over long

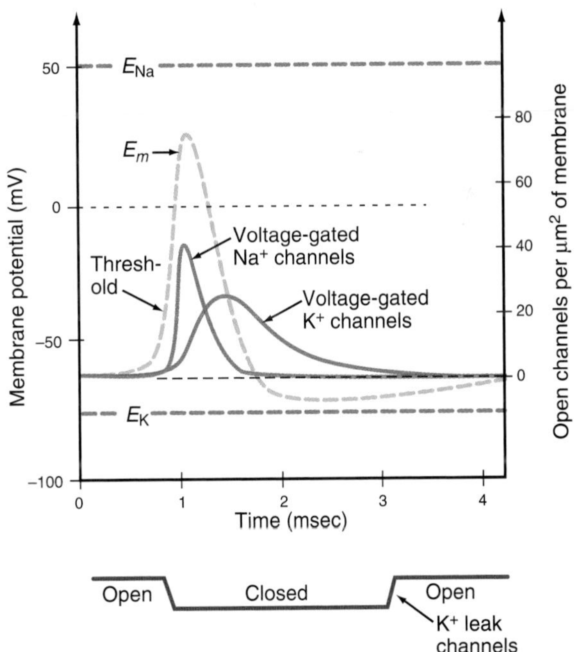

FIGURE 17.6 THE TIME COURSE OF AN ACTION POTENTIAL PASSING A MEASURING ELECTRODE INSERTED THROUGH THE PLASMA MEMBRANE OF A SQUID GIANT AXON. Spread of an action potential from an adjacent area of the membrane brings the membrane potential E_m to threshold, triggering the action potential at this point on the membrane. The other curves show the conductance of the membrane at this point for Na^+ and K^+ expressed as the concentration of open channels. The lower trace shows the times during which K^+ leak channels open and close at this point. E_{Na} is the Na^+ equilibrium potential, and E_K is the K^+ equilibrium potential.

distances. This high-speed transmission is very efficient, requiring movement of very few ions across the membrane.

The molecular events during an action potential were first characterized around 1950 in squid giant axons using microelectrodes coupled to an electronic feedback circuit. This clever "voltage clamp" holds the membrane potential constant by providing the cell with electrical current to compensate for changes in ion currents. Investigators discovered that changes in permeability to Na^+ and K^+ ions produced action potentials. Changing one variable at a time, they determined the time and voltage dependence of ion-specific conductance. They also determined the relationship between conductance and voltage. From these relationships, measured under controlled conditions, they could calculate the membrane response to virtually any experimental condition. To explain these changes in permeability, they postulated the existence of ion channels. The voltage clamp provided a direct measure of this channel activity.

Three Channels Generating Action Potentials

Voltage-gated Na^+ and K^+ channels open and close in sequence to produce action potentials. Depending on the type of open channel, the membrane potential varies in time between the K^+ equilibrium potential (E_K) and the Na^+ equilibrium potential (E_{Na}) (see Fig. 11.18). Because membrane depolarization activates these ion channels, and because the response spreads this depolarization, triggering an action potential initiates a cascade of reactions that moves over the membrane, first to depolarize and then to repolarize the membrane. In nerves, just three types of voltage-gated channels are required to generate action potentials:

- K^+-selective leak channels of the Kir family and the TWIK family (see Fig. 11.2) are open at resting potentials. Cytoplasmic Mg^{2+} blocks Kir channels when the membrane depolarizes.
- Voltage-gated Na^+ channels are closed at the resting potential but open if the membrane depolarizes to approximately −40 mV. They open only transiently, because a first-order inactivation reaction closes the pore, even if the membrane potential is at or above zero (see Fig. 16.5). These channels return to the closed state without passing through the open state.
- Delayed-rectifier voltage-gated K^+ channels have a low probability of being open at the resting potential. They respond to membrane depolarization by opening, but more slowly than Na^+ channels do. They stay open long enough to allow the repolarizing membrane potential to approach E_K.

The properties of these channels explain the time course of an action potential as follows:

Stage 1: At rest, the membrane is slightly permeable to K^+ but not to other ions, so the resting potential is near E_K, approximately −70 mV. K^+-selective leak channels and a few open voltage-gated K^+ channels contribute to this basal K^+ permeability.

Stage 2: If the membrane is depolarized by an oncoming action potential and reaches the threshold potential, K^+-selective leak channels *close* and voltage-gated Na^+ channels *open*. Because the membrane is permeable only to Na^+ and because many Na^+ channels open, Na^+ moves into the cell and the membrane potential rapidly approaches E_{Na}, approximately +45 mV.

Stage 3: After 1 to 2 msec, Na^+ channels spontaneously *inactivate* and slowly responding delayed-rectifier K^+ channels *open*. Now the membrane is strongly and selectively permeable to K^+, so K^+ moves out of the cell, and the membrane potential reverses all the way to E_K, approximately −80 mV. K^+ channels are less synchronized than Na^+ channels, so the membrane potential falls slower than it rises.

Stage 4: Delayed-rectifier K^+ channels close progressively as the membrane repolarizes, and K^+-selective leak channels open, returning the membrane potential to the resting voltage, just above E_K.

During an action potential, the membrane voltage changes by 100 to 150 mV in 1 to 2 msec. The membrane bilayer is approximately 7 nm thick, so this voltage corresponds to a field variation on the order of 150,000 volts/cm in 1 to 2 msec. Such strong forces elicit conformational changes in membrane proteins, such as voltage-gated ion channels.

Membrane Depolarization: The Stimulus for Action Potentials

The initial depolarization of the plasma membrane that triggers an action potential can arise from activation by a neurotransmitter (see "Synaptic Transmission" below) or spreading of an action potential from an adjacent region of the membrane or from an adjacent cell through a gap junction. Membrane depolarization must exceed a certain threshold to trigger an action potential. The threshold arises directly from the properties of the ion channels. Depolarization less than threshold activates a few Na^+ channels, producing a small inward Na^+ current, but it also activates some delayed-rectifier K^+ channels, resulting in K^+ efflux. If the Na^+ conductance is small in relation to the K^+ conductance, outward currents predominate and the membrane repolarizes. Depolarization greater than threshold activates additional Na^+ channels, yielding inward Na^+ currents greater than outward K^+ currents, at least briefly. This positive feedback loop further depolarizes the membrane, amplifying activation of Na^+ channels and producing the cascade of channel activation that makes action potentials an all-or-nothing event.

Myelin Sheaths Speed Action Potentials

Action potentials naturally spread rapidly over muscle cells and along extensions of neurons called axons, but some axons in the central and peripheral nervous system have insulation that speeds their propagation up to 10-fold. Supporting cells make the insulation by wrapping layers of plasma membrane around the axon to form a **myelin sheath** (Fig. 17.7). Gaps between these insulated sections expose the plasma membrane of the axon with voltage-gated ion channels. Action potentials jump at high speed from one gap to the next. Activity of the neuron can stimulate supporting cells to increase the thickness and extent of the sheath, which raises the speed of action potentials and contributes to learning certain motor skills.

Synaptic Transmission

Most neurons use chemical messengers called **neurotransmitters** (Fig. 17.8) to communicate rapidly with each other and with effector cells, such as skeletal muscle and glands. This chemical communication occurs at sites called **synapses** (Figs. 17.9 and 17.10), where the sending cell is specialized to secrete a particular neurotransmitter and the receiving cell is specialized to respond to that neurotransmitter. The sending side of a synapse is referred to as **presynaptic,** whereas the receiving side is designated **postsynaptic.** Small vesicles containing neurotransmitter pack the presynaptic nerve terminal. **Neurotransmitter receptors** concentrate in the postsynaptic plasma membrane. Modest changes in either the presynaptic release of neurotransmitter or postsynaptic receptor activation can profoundly influence how a neuron processes this information. Analysis of synaptic transmission has revealed much about the mechanisms of secretion (see Chapter 22), signal transduction, and psychoactive drugs that affect behavior. Not all synapses use chemical transmitters. In special cases, gap junctions (see Fig. 31.6) connect neurons at "electrical synapses," where current moves directly between the two cells.

All seven neurotransmitters shown in Fig. 17.8 are small organic molecules with an amino group. Secretory

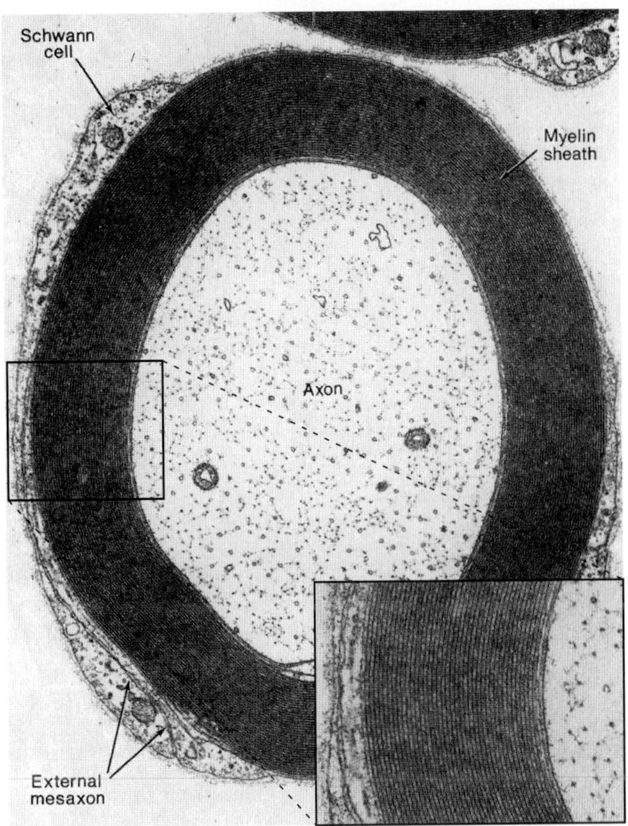

FIGURE 17.7 MYELIN SHEATH ON COCHLEAR NERVE THAT TRANSMITS IMPULSES FROM THE EAR TO THE BRAIN. Electron micrograph of a thin section showing a Schwann cell and the myelin sheath that is wrapped around the axon. The inset shows a portion of the sheath at a higher magnification. (Courtesy Enrico Mugnaini. From Fawcett DW. *The Cell,* 2nd ed. Philadelphia: WB Saunders; 1981.)

Transmitter		Acetylcholine	Dopamine	γ-Aminobutyric acid (GABA)	Glutamate	Glycine	Norepinephrine	Serotonin
Structure		O CH_3 C O CH_2 CH_2 H_3C–$^+$N–CH_3 CH_3	OH HO CH_2 ^+H_3N–CH_2	O O^- C CH_2 CH_2 ^+H_3N–CH_2	COO^- CH_2 CH_2 ^+H_3N–C–C(=O)O^- H	H ^+H_3N–C–C(=O)O^- H	OH HO HO–CH ^+H_3N–CH_2	NH HO CH_2 ^+H_3N–CH_2
Receptors	Channels	Excitatory (nicotinic) Na^+/K^+ channel	—	Inhibitory Cl^- channel	Excitatory Na^+/K^+ channel or $Na^+/K^+/Ca^{2+}$ channel	Inhibitory Cl^- channel	—	Excitatory Na^+/K^+ channel
Receptors	Seven-helix	Muscarinic receptor	Dopamine receptor	β-type GABA receptor	Metabotropic glutamate receptor	—	Adrenergic receptor	Serotonin receptor

FIGURE 17.8 NEUROTRANSMITTERS AND THEIR LIGAND-GATED ION CHANNELS AND SEVEN-HELIX RECEPTORS.

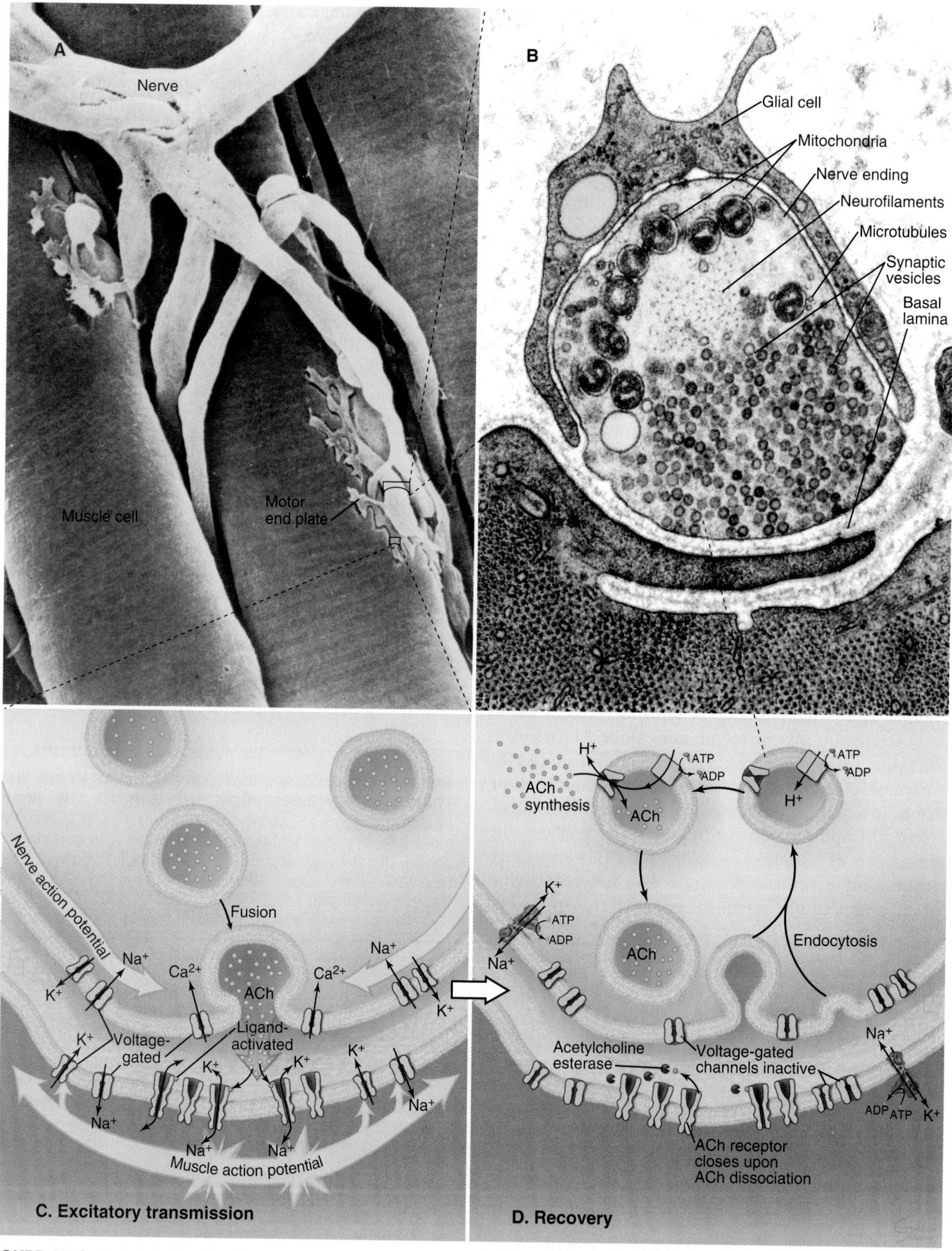

FIGURE 17.9 NEUROMUSCULAR JUNCTION. A, A scanning electron micrograph of a motor nerve and the skeletal muscle cells that it innervates. **B,** An electron micrograph of a thin section of a frog neuromuscular junction. **C,** Excitatory synaptic transmission. The nerve action potential opens voltage-gated calcium channels. Entry of Ca^{2+} triggers fusion of a synaptic vesicle containing acetylcholine (ACh) with the plasma membrane. ACh binds and opens postsynaptic channels on the muscle cell, which trigger an action potential. **D,** Recovery includes ACh hydrolysis, recycling of synaptic vesicle membranes, and loading of synaptic vesicles with new ACh. (**A,** Courtesy Don Fawcett, Harvard Medical School, Boston, MA. **B,** Courtesy J.E. Heuser, Washington University, St. Louis, MO.)

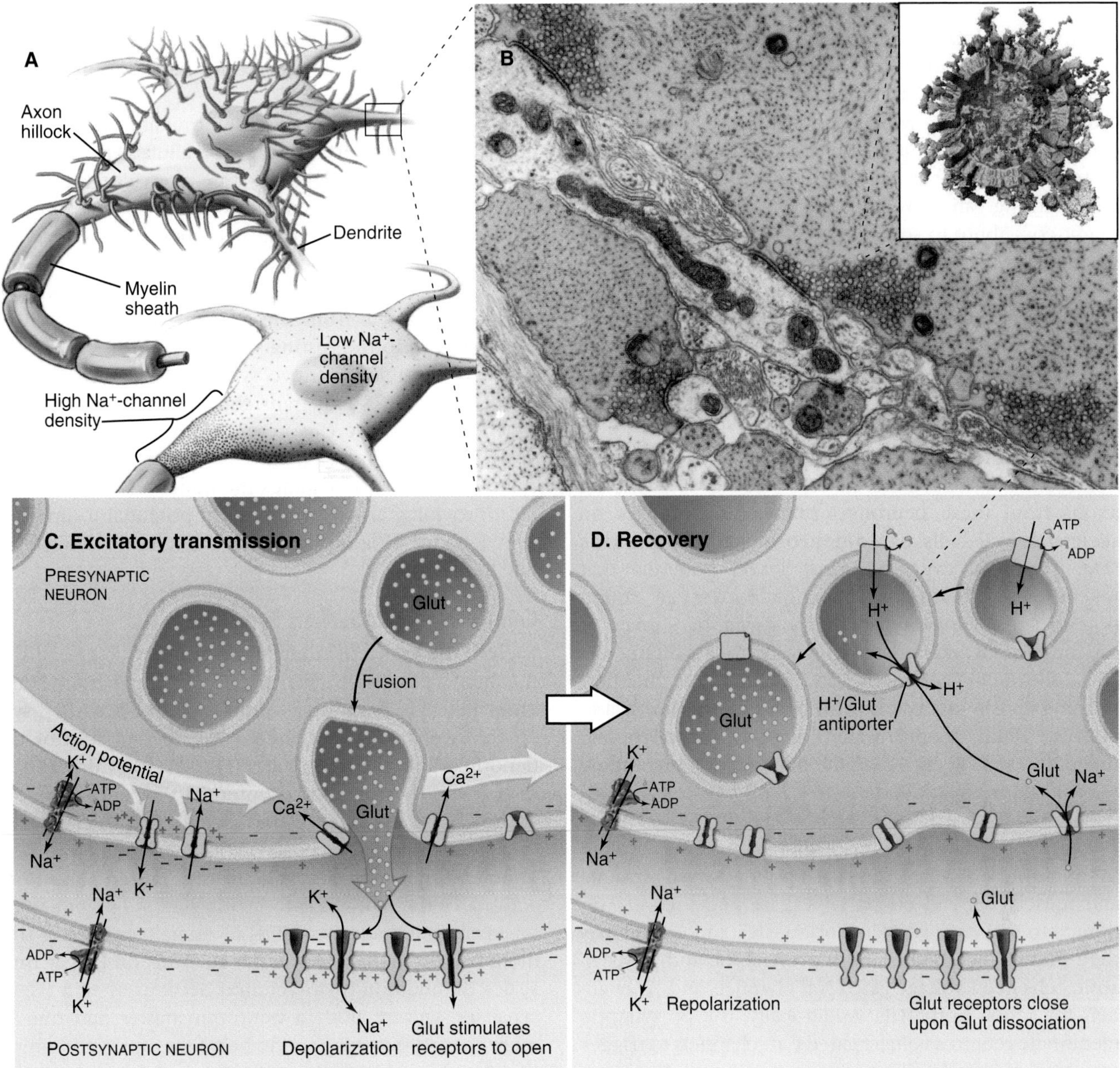

FIGURE 17.10 CENTRAL NERVOUS SYSTEM SYNAPSES. A, A neuron with its cell body and dendrites covered with a mixture of excitatory and inhibitory synapses. A high density of voltage-gated Na^+ channels in the proximal part of the axon, called the axon hillock, favors the generation of an action potential when the sum of postsynaptic potentials brings the axon hillock to threshold. **B,** An electron micrograph of a thin section of brain showing synapses with vesicles *(green)* clustered in the presynaptic axon. The *inset* shows an anatomically correct molecular model of a synaptic vesicle. **C,** Synaptic transmission at a central nervous system (CNS) excitatory synapse. A presynaptic action potential opens voltage-gated Ca^{2+} channels. Entry of Ca^{2+} stimulates fusion of synaptic vesicles filled with glutamate (Glut) with the plasma membrane. Glutamate binds and opens postsynaptic α-amino-3-hydroxy-5-methylisoxazole-4-propionate (AMPA) receptors that generate a local postsynaptic potential change. **D,** Recovery from excitatory stimulation includes retrieval of glutamate by a presynaptic Na^+/glutamate symporter and concentration of glutamate in synaptic vesicles by a H^+/glutamate antiporter. (**B,** Courtesy Don Fawcett, Harvard Medical School, Boston, MA. *Inset,* Modified from Takamori S, Holt M, Stenius K, et al. Molecular anatomy of a trafficking organelle. *Cell.* 2006;127:831–846.)

mechanisms are similar at all synapses, but each neurotransmitter requires its own biochemical machinery for synthesis, packaging in synaptic vesicles, and reception by postsynaptic cells. Such distinctive features of synapses using a particular transmitter make it possible to modify synaptic transmission selectively in the clinic, such as in treatment with psychoactive drugs.

In addition to activating ligand-gated ion channels, most neurotransmitters also stimulate particular seven-helix receptors (Fig. 17.8; see also Fig. 24.3). For example, acetylcholine stimulates the seven-helix **muscarinic acetylcholine receptor,** which uses a trimeric G-protein intermediary to activate Kir3.1 K^+ channels (Fig. 39.21). Glutamate stimulates seven-helix **"metabotropic" receptors,** which also act through trimeric G proteins. Disruption of the gene for metabotropic glutamate receptors leaves mice with defects in coordination and learning, and overstimulation of these receptors might contribute to some forms of intellectual disability in humans.

The following sections compare synapses at the neuromuscular junction and in central nervous system. Both illustrate how pumps, carriers, and channels work together at synapses.

Neuromuscular Junction

Motor neurons in the spinal cord and brainstem control contraction of skeletal muscle cells (see Fig. 39.14). Long axons from these neurons terminate in synapses on skeletal muscle cells, called **neuromuscular junctions** (Fig. 17.9A–B). Every neuronal action potential that reaches a neuromuscular junction evokes an action potential that spreads over the postsynaptic surface of the muscle cell and initiates contraction. This highly reliable, one-to-one communication depends on chemical transmission by **acetylcholine** between the nerve and muscle. Highly concentrated nicotinic acetylcholine receptors (see Fig. 11.12) in the postsynaptic membrane (~$20,000/\mu m^2$) transduce the arrival of extracellular acetylcholine into membrane depolarization.

Figure 17.9C illustrates the membrane proteins required for neuromuscular transmission. Both the nerve terminal and muscle depend on Na^+K^+-ATPase and Ca^{2+}-ATPase pumps to maintain gradients of Na^+, K^+, and Ca^{2+} across their plasma membranes. Both presynaptic and postsynaptic cells need voltage-gated Na^+ channels and K^+ channels for action potentials. Additionally, the presynaptic membrane requires voltage-gated Ca^{2+} channels to trigger secretion of acetylcholine.

A neuronal action potential initiates synaptic transmission by admitting Ca^{2+} into the presynaptic terminal through voltage-gated Ca^{2+} channels. In less than 1 msec, Ca^{2+} triggers fusion of **synaptic vesicles** containing acetylcholine with the plasma membrane. Within microseconds, acetylcholine released into the synaptic cleft between the cells reaches millimolar concentrations and binds postsynaptic acetylcholine receptors.

Weak binding of acetylcholine to two subunits of the **acetylcholine receptor** (see Fig. 11.12) opens a nonselective cation channel. The open pore is about equally permeable to K^+ and Na^+ and less permeable to Ca^{2+}, so the membrane potential collapses toward a reversal potential (see Fig. 16.19) of approximately 0 mV. This is above the threshold for triggering a self-propagating action potential in the muscle plasma membrane, which occurs with nearly 100% efficiency. The action potential traveling over the muscle plasma membrane activates voltage-gated Ca^{2+} channels that trigger Ca^{2+} release from smooth endoplasmic reticulum, resulting in muscle contraction (see Fig. 39.15).

Two different mechanisms terminate activation of acetylcholine receptors. An extracellular enzyme, **acetylcholinesterase,** degrades free acetylcholine in the synaptic cleft in a few milliseconds. In parallel, acetylcholine receptors automatically undergo a conformational change that increases the affinity for bound acetylcholine and *closes* the channel. Acetylcholine then dissociates slowly from these **desensitized** receptors, which return to the resting state.

Nerve terminals retrieve synaptic vesicle membrane by pinching off some transiently opened vesicles and by endocytosis (see Chapter 22). Cytoplasmic enzymes synthesize new acetylcholine. A V-type ATPase proton pump (see Fig. 9.6) acidifies the lumen of synaptic vesicles, providing an electrochemical potential to drive an acetylcholine/H^+ antiporter that concentrates acetylcholine in vesicles.

Central Nervous System Synapses

Each of the approximately 100 billion (10^{11}) neurons in the human brain receives synaptic inputs from many other neurons, forming a grand total of about 10^{15} synapses. Inputs at synapses covering the surfaces of the dendrites and the cell body (Fig. 17.10A) drive local changes in the membrane potential that are integrated at the base of the axon to start an action potential. Some synapses excite the postsynaptic cell, while others inhibit. Furthermore, most neurotransmitters (Fig. 17.8) activate both channels, producing point-to-point signals on a fast time scale, as well as seven-helix receptors that modulate the behavior of other neurons on longer time scales. In addition, many central nervous system (CNS) synapses secrete both a neurotransmitter and one of more than 100 neuropeptide hormones for communication through seven-helix receptors on neighboring cells. Finally, some neurons can secrete two neurotransmitters or switch their neurotransmitters during development. In all these ways synaptic transmission between neurons in the CNS (Fig. 17.10) differs fundamentally from the stable, efficient, one-to-one, excitatory coupling at neuromuscular junctions.

Transmission at chemical synapses in the CNS depends on cooperation of pumps, carriers, and channels (Fig. 17.10C–D) similar to the neuromuscular junction. Incoming information takes the form of action potentials that arrive at synapses and open voltage-gated Ca^{2+} channels in the presynaptic membrane. The transient rise in cytoplasmic Ca^{2+} can trigger the fusion of synaptic vesicles with the presynaptic plasma membrane, releasing transmitter, but the probability of successful fusion is lower than at neuromuscular junctions. Neurotransmitters secreted by the presynaptic cell activate ligand-gated channels that control the postsynaptic membrane potential. Carriers in the presynaptic membrane and adjacent supporting cells terminate transmission by removing neurotransmitter from the synaptic cleft.

Excitatory synapses using the neurotransmitters **glutamate, acetylcholine,** or **serotonin** activate receptor channels (see Figs. 16.11 and 16.12) permeable to **cations** that depolarize the membrane. Opening these channels causes a local, short-lived change in membrane potential, called a **postsynaptic potential** (PSP). Individual PSPs do not fire action potentials for two reasons. First, individual PSPs raise the membrane potential only a few millivolts, so they do not bring the postsynaptic membrane to threshold. Second, the plasma membrane of dendrites and the cell body contains few voltage-gated Na^+ channels. Furthermore, **inhibitory synapses** on the same cell counteract excitatory synapses by secreting **glycine** or **γ-aminobutyric acid** (GABA) to activate **Cl^- channels** that hyperpolarize the membrane, taking it farther from threshold (see Fig. 11.18).

Neurons spatially average excitatory and inhibitory PSPs as they spread passively over the postsynaptic membrane and generate an action potential only when their sum at a particular time brings the membrane potential to threshold in the proximal part of the axon, called the **axon hillock** (Fig. 17.10A). This part of the plasma membrane is particularly sensitive to voltage, owing to a high concentration of voltage-gated Na^+ channels. At threshold, they open, depolarizing the membrane (Fig. 17.6). Delayed-rectifier K^+ channels then repolarize the membrane, resetting it in preparation for subsequent action potentials. The frequency of postsynaptic action potentials is proportional to the intensity of the presynaptic input above a threshold. Each action potential is identical and propagates down the axon. Both the pattern and frequency of action potentials carry information in the brain.

Removal of neurotransmitters from the synaptic cleft terminates activation of postsynaptic receptors (Fig. 17.9D). The Na^+ gradient across the plasma membrane drives symporters that return neurotransmitters to their presynaptic cells. Within the presynaptic cell, a V-type, proton-translocating ATPase acidifies the lumen of the synaptic vesicle and establishes a proton electrochemical gradient across the vesicle membrane to drive the antiporter that concentrates transmitter inside the vesicle.

Modification of Central Nervous System Synapses by Drugs and Disease

The duration of synaptic stimulation depends on the rate of clearance of neurotransmitters from the synaptic cleft, so inhibiting these transport processes with drugs prolongs stimulation at particular classes of CNS synapses, with profound effects on brain function and behavior. **Cocaine** inhibits a plasma membrane dopamine transporter as well as transporters for serotonin and norepinephrine. Tricyclic **antidepressants** inhibit norepinephrine uptake, and other drugs inhibit serotonin uptake. These drugs have dramatic effects on the symptoms of depression as well as other psychiatric disorders.

Excess stimulation of *N*-methyl-D-aspartate (NMDA) receptors rapidly kills postsynaptic neurons, most likely owing to the deleterious effects of excess cytoplasmic Ca^{2+}. This occurs when glutamate is released from ischemic brain tissue during a stroke caused by compromising the blood supply to a region of the brain. Such damage might also contribute to neuron death in degenerative diseases of the nervous system, such as amyotrophic lateral sclerosis and Alzheimer disease.

Nicotinic acetylcholine receptors in the CNS are found on both the postsynaptic and *presynaptic* membranes along with glial cells, so the effects of acetylcholine secreted by neurons and **nicotine** from tobacco are widespread. In some cases they carry out fast, excitatory synaptic transmission as at the neuromuscular junction. Nicotinic acetylcholine receptors in the *presynaptic* plasma membrane are highly permeable to Ca^{2+}, so their stimulation admits Ca^{2+} into the presynaptic terminal. This enhances both the spontaneous release of neurotransmitter and release in response to action potentials. The isoform composition of CNS acetylcholine receptors differs from that of muscles (see Fig. 11.12). Some are homopentamers of α-subunits. Others are heteropentamers of α- and β-subunits. Activation of these ligand-gated channels in different regions of the brain may account for the enhancing effects of nicotine on learning and memory but also for tobacco addiction. Loss of CNS neurons that secrete acetylcholine might contribute to dementia in Alzheimer disease.

Modification of Central Nervous System Synapses by Use

Memories are thought to depend on structural changes that modify the strength or numbers of synapses between neurons in the brain, processes called **synaptic plasticity**. Particular patterns of stimulation can produce long-term changes that enhance or reduce the efficiency of transmission at various glutamate-mediated synapses (Fig. 17.11). The **hippocampus,** a region of the vertebrate cerebral cortex known to participate in some forms of learning and memory, is often used for observing a simple form of cellular learning. Intense stimulation of excitatory glutamate synapses (20 pulses over a period of 200 msec) can increase synaptic strength for days or weeks. This is called **long-term potentiation (LTP).** Conversely, slow, prolonged stimulation of glutamate synapses reduces the response for hours. This is called **long-term depression (LTD).** The mechanisms of LTP and LTD have been investigated thoroughly, because

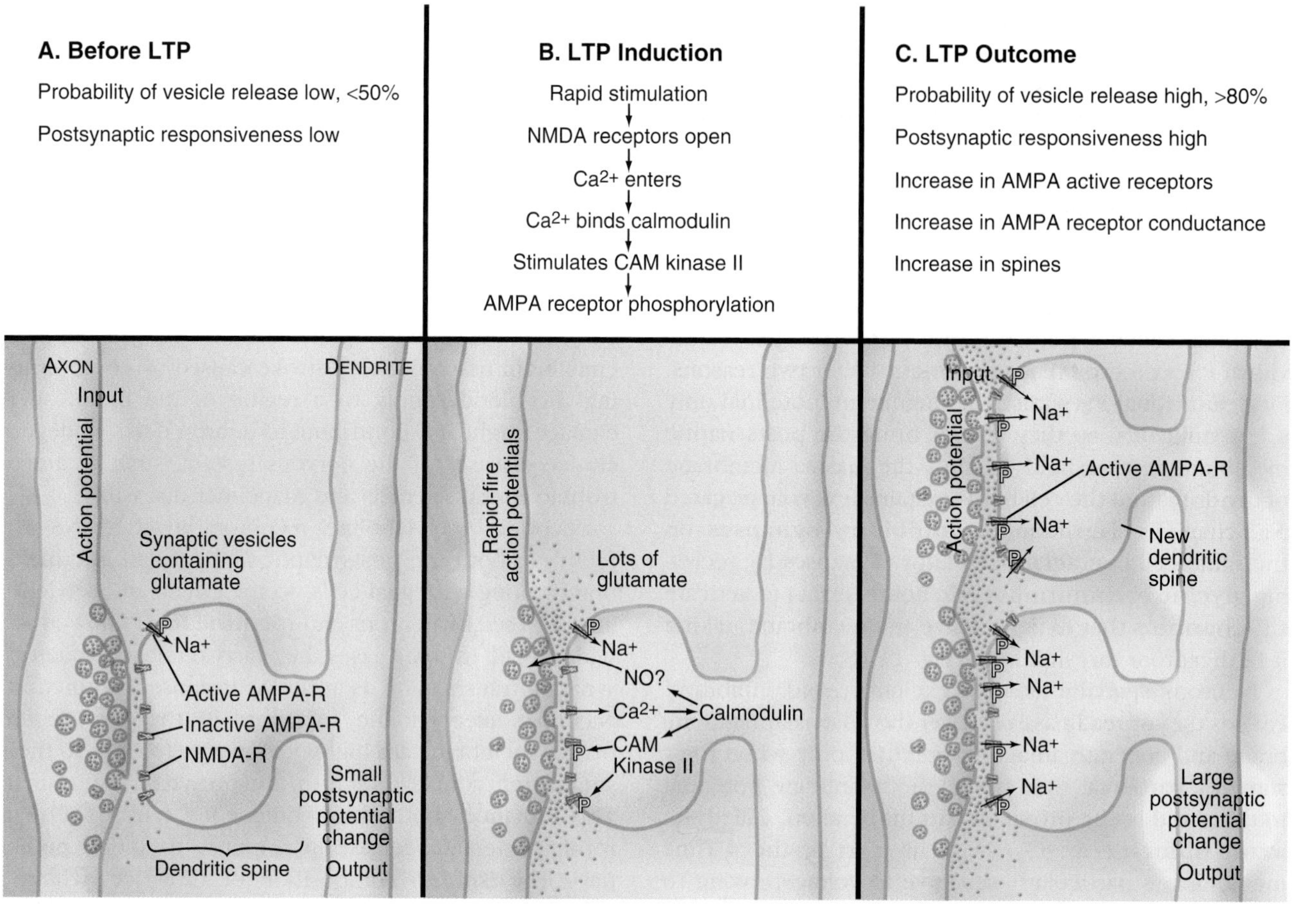

FIGURE 17.11 LONG-TERM POTENTIATION OF SYNAPTIC TRANSMISSION AT EXCITATORY SYNAPSES IN THE HIPPOCAMPUS. **A,** Prior to long-term potential (LTP), postsynaptic responses to presynaptic action potentials are unreliable and small. **B,** Some acute responses to vigorous stimulation. **C,** After induction of LTP, postsynaptic responses are more reliable and larger. AMPA-R (α-amino-3-hydroxy-5-methylisoxazole-4-propionate receptor) and NMDA-R (*N*-methyl-D-aspartate receptor) are two classes of glutamate receptors; NO is nitric oxide, a candidate for the retrograde signaling molecule; CAM kinase II is calcium-calmodulin kinase II.

they occur on appropriate time scales to modify synapses during development and high-order brain functions, such as learning and memory.

Induction of LTP involves two types of glutamate receptor channels (see Fig. 11.11) located in postsynaptic specializations called dendritic spines (Fig. 17.11). **AMPA (α-amino-3-hydroxy-5-methylisoxazole-4-propionate) receptors** open and close rapidly in response to glutamate and depolarize the membrane by admitting Na^+. **NMDA receptors** respond slowly to glutamate and also depolarize the membrane by admitting Ca^{2+}. The response of NMDA receptors to glutamate depends on the membrane potential, as partial depolarization is required to displace an extracellular Mg^{2+} ion blocking the channel. This dual dependence on glutamate and membrane potential makes NMDA receptors coincidence detectors, responsive to rapid stimulation or stimulation at nearby excitatory synapses.

In principle, LTP and LTD might alter the efficiency of synaptic transmission by changing glutamate release from the presynaptic cell or the responsiveness of the postsynaptic cell to glutamate. Investigators debate the relative importance of presynaptic and postsynaptic processes, but both likely contribute. Inconsistencies in observations likely reflect the fact that LTP involves multiple processes (Fig. 17.11), and the protocols used to study LTP can emphasize one feature over others.

One presynaptic factor is the probability that an action potential will stimulate the fusion of a glutamate-containing synaptic vesicle with the plasma membrane. In the resting state, glutamate release by these vesicles is unreliable. LTP increases the probability of exocytosis from less than 0.5 to greater than 0.8. The mechanism of enhanced transmitter release is still being investigated. Candidates include NMDA receptors in the presynaptic membrane, diffusive messengers such as nitric oxide (see Fig. 26.17) providing feedback from the postsynaptic side, and strengthened adhesion between the pre- and postsynaptic membranes.

At least two factors on the postsynaptic side influence the response to glutamate: the number of active AMPA receptors at a synapse; and the number and sizes of synapses with the stimulating axon. Inactive synapses with only NMDA receptors do not respond to weak stimulation with glutamate, but intense presynaptic release of glutamate during LTP arouses such silent synapses, especially if coordinated with membrane depolarization from neighboring synapses. Ca^{2+} enters the postsynaptic dendritic spine through active NMDA receptors, binds calmodulin (see Fig. 3.12C) and activates the processes that initiate and maintain LTP. Within seconds, calcium-calmodulin activates calcium-calmodulin (CAM)-kinase II (see Fig. 25.4A), which phosphorylates AMPA receptors. Phosphorylated AMPA receptors are more responsive to glutamate. In addition LTP shifts AMPA receptors from an intracellular pool of recycling endosomes to the postsynaptic membrane where they are anchored by proteins in the postsynaptic density on the inside of the plasma membrane.

Within minutes, induction of LTP triggers signaling cascades that maintain the increased efficacy, leading to structural changes and increased protein synthesis. These changes may induce dendrites to stabilize existing spines or sprout new filopodia and spines that increase the number of synapses within an hour or so. Extension of these processes and remodeling of the shape of dendritic spines depend on actin filament assembly (see Fig. 38.7). Direct imaging in the brain of live mice has documented the formation of new spines associated with learning a motor skill, a change enhanced by sleep. Over the longer term, the postsynaptic cell initiates gene transcription and protein synthesis, bringing about further changes that stabilize enhanced synaptic transmission.

LTD appears, in many ways, to be the reverse of LTP with multiple factors contributing. Small molecules released from the dendritic spine activate seven-helix cannabinoid receptors on the presynaptic membrane and reduce the reliability of exocytosis. In addition, postsynaptic AMPA receptors are less responsive owing to sequestration in internal compartments.

Armed with the growing body of knowledge about synaptic plasticity at the cellular level, many neuroscientists are investigating the short- and long-term changes in the brain that account for learning and memory.

ACKNOWLEDGMENTS

We thank Pietro De Camilli for his suggestions on revisions to this chapter.

SELECTED READINGS

Birren SJ, Marder E. Plasticity in the neurotransmitter repertoire. *Science*. 2013;340:436-437.

Dineley KT, Pandya AA, Yakel JL. Nicotinic ACh receptors as therapeutic targets in CNS disorders. *Trends Pharmacol Sci*. 2015;36:96-108.

Dorwart M, Thibodeau P, Thomas P. Cystic fibrosis: Recent structural insights. *J Cyst Fibros*. 2004;3:91-94.

Feldman DE. Synaptic mechanisms for plasticity in the neocortex. *Annu Rev Neurosci*. 2009;32:33-55.

Hoffmann EK, Lambert IH, Pedersen SF. Physiology of cell volume regulation in vertebrates. *Physiol Rev*. 2009;89:193-277.

Hogg RC, Bertrand D. What genes tell us about nicotine addiction. *Science*. 2004;306:983-984.

Keating MT, Sanguinetti MC. Molecular and cellular mechanisms of cardiac arrhythmias. *Cell*. 2001;104:569-580.

Mall MA, Galietta LJ. Targeting ion channels in cystic fibrosis. *J Cyst Fibros*. 2015;14:561-570.

Park P, Volianskis A, Sanderson TM, et al. NMDA receptor-dependent long-term potentiation comprises a family of temporally overlapping forms of synaptic plasticity that are induced by different patterns of stimulation. *Philos Trans R Soc Lond B Biol Sci*. 2013;369:20130131.

Qiu Z, Dubin AE, Mathur J, et al. SWELL1, a plasma membrane protein, is an essential component of volume-regulated anion channel. *Cell*. 2014;157:447-458.

Riordan JR. CFTR function and prospects for therapy. *Annu Rev Biochem*. 2008;77:701-726.

Strange K. Cellular volume homeostasis. *Adv Physiol Educ*. 2004;28:155-159.

Vincent GM. The long QT syndrome: Bedside to bench to bedside. *N Engl J Med*. 2003;348:1837-1838.

Voss FK, Ullrich F, Münch J, et al. Identification of LRRC8 heteromers as an essential component of the volume-regulated anion channel VRAC. *Science*. 2014;344:634-638.

Yang G, Lai CS, Cichon J, et al. Sleep promotes branch-specific formation of dendritic spines after learning. *Science*. 2014;344:1173-1178.

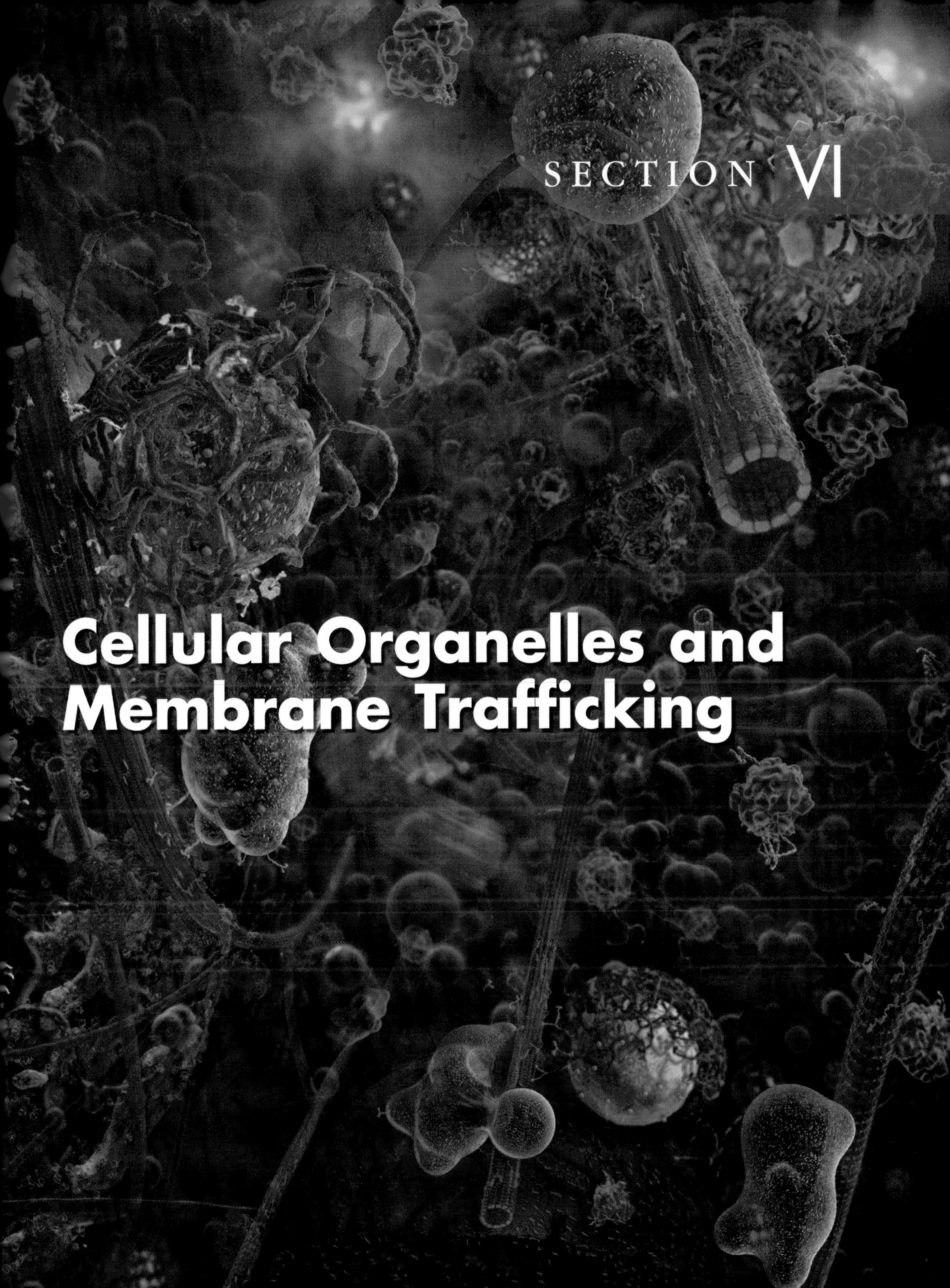

SECTION VI

Cellular Organelles and Membrane Trafficking

SECTION VI OVERVIEW

Eukaryotic cells evolved membrane-bounded compartments specialized to provide energy; to synthesize lipids, carbohydrates, proteins, and nucleic acids; and to degrade cellular constituents. These subcellular compartments, called **organelles,** have distinctive chemical compositions. Organelles vary in abundance and size in different cell types, including multicellular organisms, where each tissue and organ has specialized functions. An organelle often holds a monopoly on performing a given task; for example, **endoplasmic reticulum** (ER) synthesizes membrane proteins and certain membrane lipids, lysosomes contain enzymes to degrade many macromolecules, and mitochondria convert energy derived from the covalent bonds of nutrients into adenosine triphosphate (ATP) to provide energy for diverse cellular functions.

A semipermeable membrane surrounds each organelle and establishes an internal microenvironment with concentrated enzymes, cofactors, and substrates to favor particular macromolecular interactions. Pumps (Chapter 14), carriers (Chapter 15), and channels (Chapter 16) in each organelle membrane establish an internal chemical environment (pH, divalent cation concentration, reduction–oxidation [redox] potential) that is appropriate for particular biochemical functions. **Mitochondria** and **chloroplasts** use many enzymes embedded in their membranes to catalyze reactions that depend on the separation of reactants across the membrane or involve hydrophobic substrates and products soluble in the lipid bilayer (Chapter 19). Compartments also protect the rest of the cell from potentially dangerous activities, such as degradative enzymes in lysosomes and oxidative enzymes in **peroxisomes**.

This division of labor among organelles has many advantages but also presents cells with challenges in terms of coordination of cellular activities, organelle biosynthesis, and cell division. Organelles are not autonomous, so their activities must be integrated to benefit the whole cell. Therefore, mechanisms are required to transport material between compartments and across the membranes that surround them. Many functional pathways require macromolecules and lipids to move

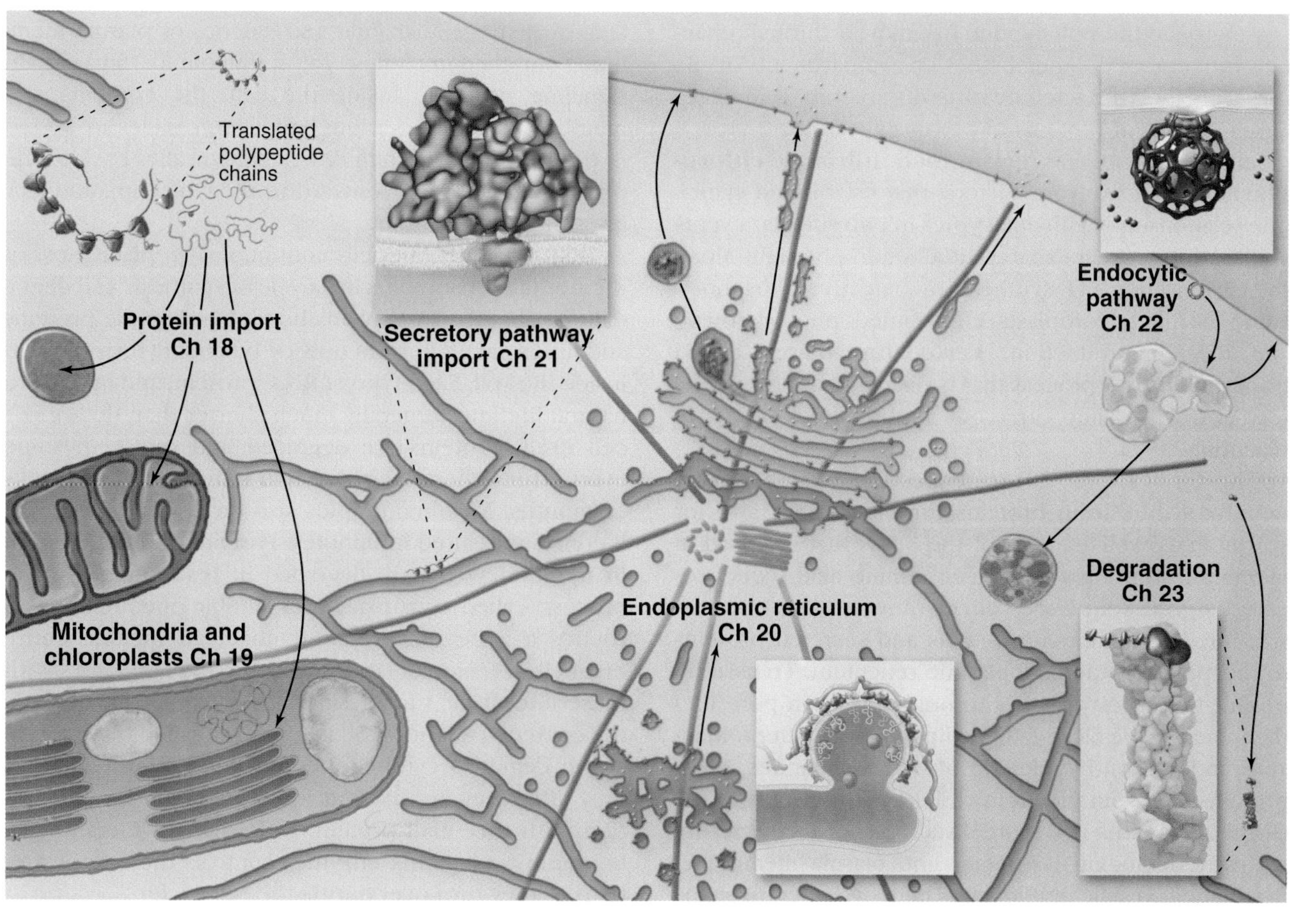

from one organelle to another in a vectorial manner. This transport between organelles generally involves budding of vesicles from one membrane-bounded compartment followed by fusion with another, in a process collectively termed **vesicular trafficking.** In addition, contact sites between some membranes allow for exchange of lipids.

This section focuses on two important processes as they pertain to the biogenesis and functions of the various organelles: the targeting of proteins, either during or after translation to their home organelle, and the bidirectional movement of vesicular traffic between organelles and the plasma membrane. The exocytic or **secretory pathway** from the endoplasmic reticulum to the plasma membrane and lysosomes coordinates organelle biosynthesis and secretion. The **endocytic pathway** takes in molecules and microscopic particles from outside the cell along with plasma membrane components. Operating together, the two pathways coordinate the distribution and turnover of membrane proteins and lipids.

Proteins that are synthesized in the cytoplasm either remain there or move to their final destinations in the nucleus (Chapter 9), mitochondria, chloroplasts, and peroxisomes (Chapter 18). Hundreds of proteins destined for mitochondria and chloroplasts are synthesized in the cytoplasm and directed to these organelles by zip codes built into their polypeptide sequences. Usually, these guide sequences are removed by proteolytic processing once the polypeptide has moved through channels into one of the membranes or compartments inside these organelles. Different sorts of targeting sequences target dozens of proteins to peroxisomes.

Chapter 19 explains how **mitochondria** and **chloroplasts** descended from bacteria that established symbiotic relationships with eukaryotes in two singular events about a billion years apart. Mitochondria brought along the capacity for ATP synthesis by oxidative phosphorylation, while chloroplasts contributed photosynthesis and oxygen production. **Peroxisomes** are derived from the ER by a process that is distinct from the secretory pathway. They carry out a number of oxidative reactions.

The **ER** (Chapter 20) generates the secretory pathway by synthesizing proteins for membranes and for secretion as well as many of the lipids that are used in membranes throughout the cell. Amino acid sequences called **signal sequences** direct ribosomes that synthesize integral membrane proteins and secreted proteins to receptors on the endoplasmic reticulum. Translation pushes these polypeptides through a protein pore into the lumen of the ER or into the lipid bilayer. After folding and modification by addition of oligosaccharides, these proteins exit from the ER in vesicles for transport to the **Golgi apparatus** and more distal parts of the secretory pathway, including lysosomes and plasma membrane. Retrograde vesicle traffic mediated by other proteins retrieves membranes and proteins from the Golgi apparatus back to the ER. Despite this heavy bidirectional traffic between organelles, accurate sorting mechanisms allow each organelle to maintain its identity.

Chapter 21 explains the mechanisms used for membrane trafficking. Cells use three different types of coat proteins with common evolutionary origins for budding membrane vesicles. Under the direction of membrane-associated guanosine triphosphatases (GTPases), these proteins form a coat on a donor membrane that distorts the lipid bilayer into a vesicle that buds from the surface. The vesicle carries membrane proteins and lipids and any material in the lumen. Sorting signals direct some proteins into these transport vesicles. After the vesicle moves by diffusion or by active transport along the cytoskeleton to a target membrane, different GTPases and peripheral proteins facilitate fusion of the vesicle with a target membrane.

Cells employ at least five distinct mechanisms to internalize plasma membrane, along with a wide range of extracellular materials (Chapter 22). Ingestion of small particles, including bacteria, takes place by **phagocytosis,** in which a veil of plasma membrane surrounds the particle and takes it into a vacuole inside the cell. Fusion of vesicles containing lysosomal enzymes initiates the degradation of the contents. A second endocytic pathway takes receptors and their ligands into cells in small vesicles coated with **clathrin.** Other forms of endocytosis take up extracellular fluid and patches of plasma membrane enriched in cholesterol, sphingolipids, and certain signaling proteins. Inside the cell, the contents and membranes of these various endocytic vesicles are sorted in **endosomes** and then directed in vesicles back to the plasma membrane or onward to the Golgi apparatus or lysosomes for recycling.

DNA is stable, but cells continuously replace most of their other constituents in a cycle of synthesis and degradation. Chapter 23 explains how cells degrade proteins and lipids, taken in from outside by endocytosis or from inside the cell. Each type of RNA, protein, and lipid has a natural lifetime, generally much shorter than that of the cell itself. Proteins are degraded and replaced, some every hour, others every day, and some every few weeks or months. Membrane lipids also turn over; some with lifetimes measured in minutes. Proteins and lipids taken in by endocytosis are degraded in **lysosomes.** In the process called **autophagy,** a double membrane surrounds a zone of cytoplasm, that can include entire organelles. Fusion of late endosomes and lysosomes with these autophagic vacuoles delivers enzymes that degrade the contents. A large protein complex called the **proteasome** degrades cytoplasmic and nuclear proteins, but only after they are marked for degradation by conjugation with the small protein, **ubiquitin.** A hierarchy of ubiquitin-conjugating enzymes controls the fate of proteins as they turn over during the cell cycle.

CHAPTER 18

Posttranslational Targeting of Proteins

Protein synthesis is largely carried out by cytoplasmic ribosomes that provide all the proteins for the nucleus, cytoplasm, peroxisomes, and secretory pathway. Mitochondria and chloroplasts import most of their proteins from the cytoplasm, even though they originated as bacterial endosymbionts and have retained the capacity to synthesize a few of their proteins. Most of the original bacterial genes moved to the nucleus of the eukaryotic host.

Given a common site of synthesis, accurate addressing is essential to direct proteins to their sites of action and to maintain the unique character of each cellular compartment. This is achieved by "zip codes" built into the structure of each protein (Fig. 18.1). Residues in the sequence of each protein—often, but not necessarily, contiguous amino acids—form a signal for targeting.

Targeting signals are both necessary and sufficient to guide proteins to their final destinations. Transplantation of a targeting signal, such as a presequence from a mitochondrial protein, to a cytoplasmic protein reroutes the hybrid protein into the organelle specified by the targeting sequence, mitochondria in this example. Some targeting signals are transient parts of the protein. For example, most mitochondrial proteins are synthesized with N-terminal extensions that guide them to mitochondria and then are removed. Alternatively, signals may be a permanent part of the mature protein, in some cases serving repeatedly to target a mobile protein between different destinations. Permanent nuclear targeting

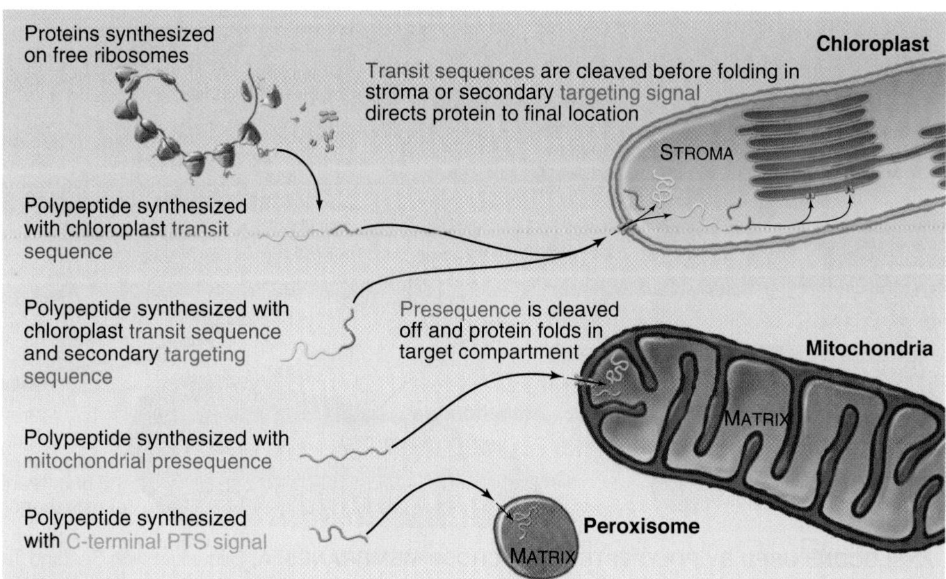

FIGURE 18.1 TARGETING SIGNALS DIRECT POLYPEPTIDES SYNTHESIZED ON CYTOPLASMIC RIBOSOMES TO CHLOROPLASTS, MITOCHONDRIA, AND PEROXISOMES. Signal peptidases remove some signals after the polypeptide enters the organelle.

signals can be located at the N-terminus, the C-terminus, or even the middle of a protein (see Chapter 9). Some proteins have more than one targeting signal: a primary code that directs the protein to the target organelle or pathway, and a second signal that steers the protein to its specific site of residence within the organelle or pathway.

Targeting signals direct proteins to their destination by binding to organelle-specific receptors or using soluble escort factors as intermediaries. When necessary, proteins cross membranes via channels called **translocons** formed by integral membrane proteins (Fig. 18.2). Like ion channels (see Chapter 16), these protein-translocating channels are gated to prevent indiscriminate transport of cellular constituents when not occupied by a polypeptide. Polypeptides fit so tightly in these channels during translocation that ions do not leak through. Ions traverse ion channels in a microsecond, whereas polypeptides take tens of seconds to move through translocons. Protein synthesis, adenosine triphosphate (ATP) hydrolysis, or the membrane potential provides the energy to power protein translocation across membranes.

Three families of protein translocation channels are found in all three domains of life. Sec translocons direct proteins into the endoplasmic reticulum in eukaryotes and out of prokaryotes. The Tat family of pores translocates folded proteins into chloroplast thylakoids and out of prokaryotes. Membrane proteins related to Oxa1p help insert proteins synthesized in the mitochondrial matrix and prokaryotic cytoplasm into membranes. Mitochondria (Fig. 18.4), chloroplasts (Fig. 18.6), and prokaryotes (Fig. 18.10) have additional families of protein translocation channels.

Primary targeting can occur either cotranslationally, coincident with protein synthesis, or posttranslationally, after polypeptide synthesis. Chapter 20 covers protein targeting to the endoplasmic reticulum (ER) where, with a few exceptions, targeting is cotranslational. This chapter covers **posttranslational targeting** mechanisms that move proteins across membrane bilayers into mitochondria, chloroplasts, and peroxisomes, and out of Bacteria. Eukaryotes also secrete a few proteins directly across the plasma membrane. Chapter 9 covers posttranslational movements of proteins into and out of the nucleus through a large aqueous channel in the nuclear pore.

Transport of Proteins Into Mitochondria

Mitochondrial outer and inner membranes define two spaces: one between the **outer** and **inner membranes** (termed the **intermembrane space**) and an interior space termed the **matrix** (Fig. 18.3). Each membrane and space has distinct functions and protein compositions, which are discussed in Chapter 19. Targeting signals and specific translocation machinery guide more than 1000 imported proteins selectively to these compartments.

Genetic and biochemical experiments on fungi defined the molecular machinery for proteins to enter mitochondria. The **TOM40** complex (translocase of the outer mitochondrial membrane) is the main portal into the mitochondrion. Thereafter proteins take one of four routes to different compartments: the outer membrane

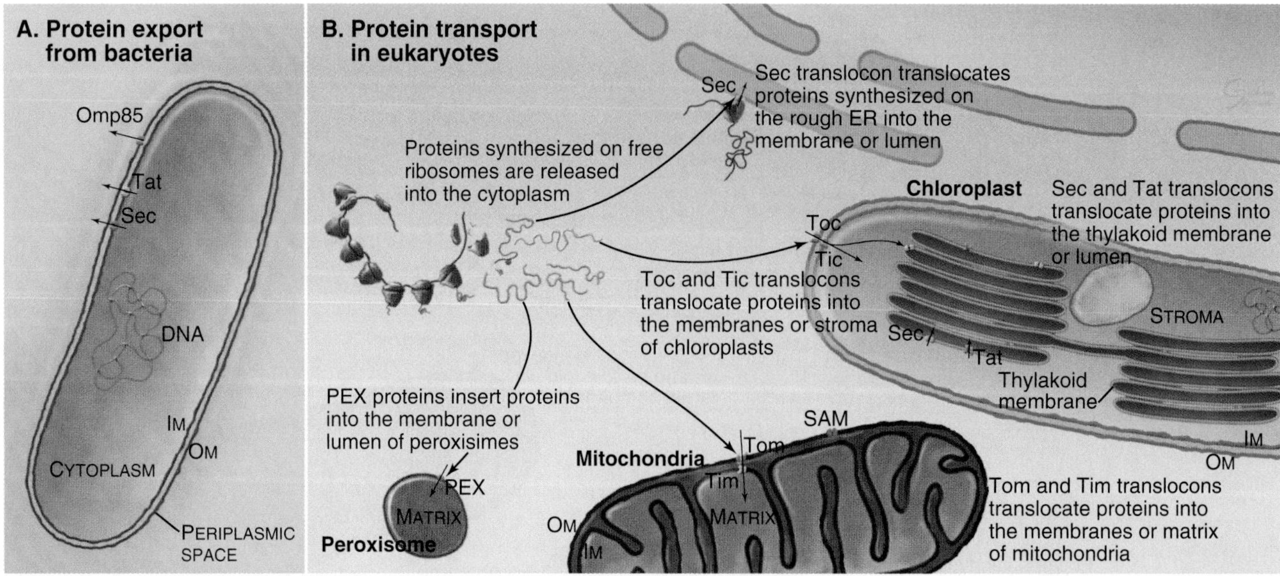

FIGURE 18.2 TRANSLOCONS USED BY POLYPEPTIDES TO CROSS MEMBRANES. A, Bacterium with Sec and Tat translocons in the inner membrane and Omp85 in the outer membrane. **B,** Eukaryote translocons including Sec in the endoplasmic reticulum and thylakoid membrane of chloroplasts, Toc in the outer membrane of chloroplasts, Tic in the inner membrane of chloroplasts, Tat in the thylakoid membrane of chloroplasts, Tom and SAM in the outer membrane of mitochondria, Tim in the inner membrane of mitochondria, and PEX in peroxisomes.

using the **SAM** complex (sorting and assembly machinery of the outer membrane); the intermembrane space; the inner membrane via the **TIM22** complex (translocase of the inner mitochondrial membrane); and the matrix using the **TIM23** complex. Targeting signals direct proteins to the TOM40 complex and then on to other locations. Translocation requires energy and assistance from protein chaperones both outside and inside mitochondria.

Delivery of Protein to Mitochondria

After synthesis by cytoplasmic ribosomes, most proteins destined for mitochondria bind cytosolic chaperones of the **Hsp70** (heat shock protein 70) family (see Fig. 12.12). This interaction maintains proteins in unfolded configurations competent for import (Fig. 18.3). Some imported proteins require additional factors for targeting to the translocation machinery.

Targeting motifs for matrix proteins are called **presequences,** because they are usually removed by proteolytic cleavage in the mitochondrial matrix. Presequences are generally located at the N-termini of precursor polypeptides as contiguous sequences of 10 to 70 amino acids. They are rich in basic, hydroxylated, and hydrophobic amino acids, but share no defined sequences in common. The targeting sequences of many mitochondrial membrane proteins are in the middle of the polypeptide and are not cleaved after import. Cytochrome c, a component of the electron transport chain in the intermembrane space (see Fig. 19.5), also has an internal signal for import into mitochondria.

A succession of weak interactions with outer membrane receptors Tom20, Tom22, and Tom70 guides presequences and other target signals to the outer membrane translocon. The presequence initially contacts Tom20. Eight residues of the presequence fold into an amphipathic (hydrophobic on one side, hydrophilic on the other) α-helix that binds in a shallow hydrophobic groove on Tom20 with favorable electrostatic interactions (Fig. 18.4D). Other parts of the presequence are thought to interact with Tom40, the translocon itself. Although these associations are weak, collectively, they distinguish mitochondrial presequences from other proteins in the cytoplasm with high fidelity.

Translocation Across the Outer Membrane

Tom40, a β-barrel protein similar to a porin (see Fig. 13.9C), forms the translocon of the outer membrane. Six proteins with single transmembrane helices associate around the periphery of Tom40: the three receptor subunits and three small subunits. These Tom40 complexes form dimers or trimers in the plane of the membrane (Fig. 18.3E) with pores approximately 2 nm in diameter.

The outer membrane receptors transfer presequences and other targeting sequences to the translocon channel. Chemical crosslinking showed that both hydrophobic

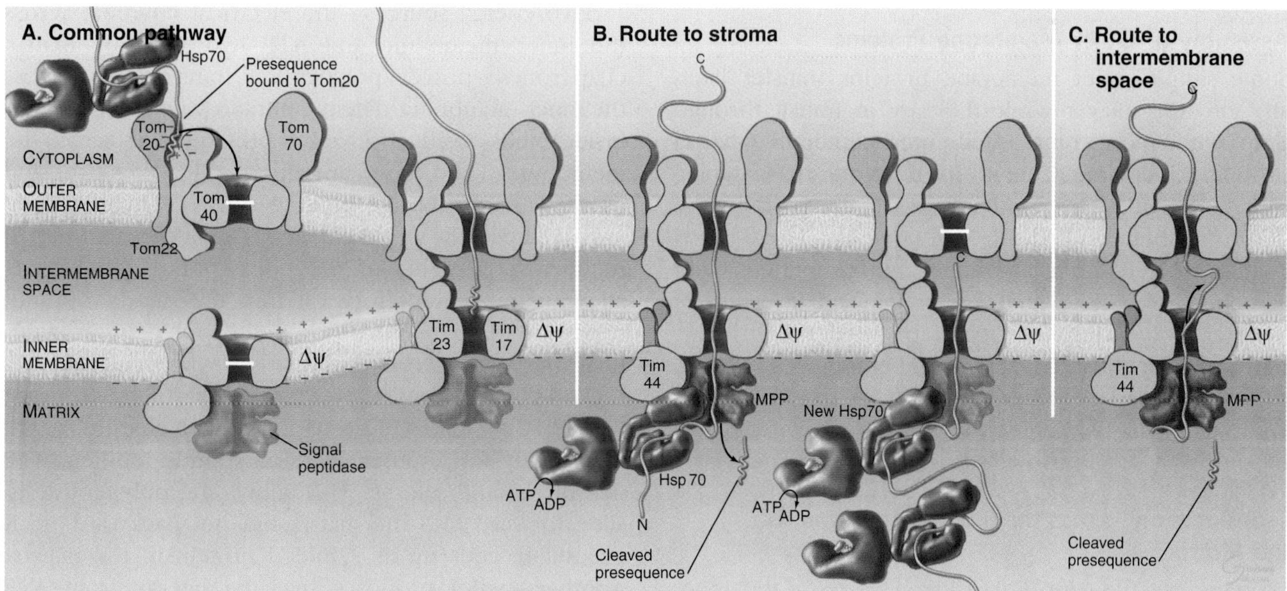

FIGURE 18.3 IMPORT OF MATRIX PROTEINS INTO MITOCHONDRIA. Models of the Tom and Tim complexes. **A,** Common pathway across the outer membrane. Hsp70 (heat shock protein 70) escorts polypeptides synthesized on cytoplasmic ribosomes to mitochondria where the presequence associates with Tom20/22. The basic presequence leads the polypeptide through Tom40, the translocase of the outer membrane, to the intermembrane space. **B,** Route to the stroma. The presequence enters the translocase of the inner membrane (Tim). The potential across the inner membrane (Δψ) pulls the presequence through Tim into the matrix, where it is cleaved by MPP (matrix processing protease). Cycles of Hsp70 binding to the peptide followed by adenosine triphosphate (ATP) hydrolysis and dissociation of Hsp70 from Tim44 ratchet the translocating peptide into the matrix, where it folds. **C,** Route to the intermembrane space. After cleavage of the presequence, the polypeptide backs up into the intermembrane space.

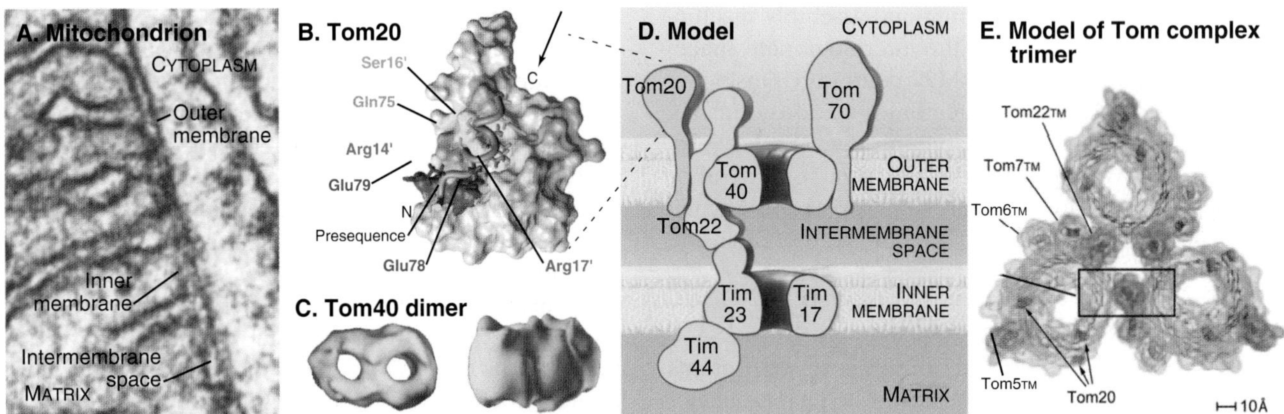

FIGURE 18.4 MITOCHONDRIAL IMPORT COMPONENTS. A, Electron micrograph of a thin section of a mitochondrion. **B,** Structure determined by nuclear magnetic resonance spectroscopy of a presequence peptide bound to a hydrophobic patch on Tom20, a receptor from the mitochondrial outer membrane. Space-filling model of a cytoplasmic domain of Tom20. The presequence forms two turns of α-helix with two arginines exposed on the surface. N is the N-terminus and C is the C-terminus of the peptide. *Yellow* is a hydrophobic patch; *orange* is Gln-rich; *red* is Glu-rich. **C,** Three-dimensional reconstruction from electron micrographs of a dimer of Tom40, the translocase of the outer mitochondrial membrane. **D,** Schematic of the mitochondrial import apparatus, including Tom complex in the outer membrane and Tim complex in the inner membrane. **E,** Model of the active trimer of Tom complex. This face view shows cross sections of the transmembrane helices of four accessory subunits. (**A,** Courtesy Don W. Fawcett, Harvard Medical School, Boston, MA. **B,** Courtesy D. Kohda, Kyushu University. From Abe Y, Shodai T, Muto T, et al. Structural basis of presequence recognition by the mitochondrial protein import receptor Tom20. *Cell.* 2000;100:551–560 and Protein Data Bank [PDB; www.rcsb.org] file 1OM2. **C,** From Ahting U, Thun C, Hegerl R, et al. The Tom core complex: the general protein import pore of the outer membrane of mitochondria. *J Cell Biol.* 1999;147:959–968, copyright The Rockefeller University Press. **E,** From Shiota T, Imai K, Qiu J, et al. Molecular architecture of the active mitochondrial protein gate. *Science.* 2015;349:1544–1548.)

and charged polypeptides move through the central pore of the β-barrel. Proteins must be largely unfolded to fit through the 2-nm pore. Like other translocons Tom channels are likely to be gated, and they close when not occupied by a translocating polypeptide.

Assembly of Outer Membrane Proteins

Some simple outer membrane proteins transfer laterally into the bilayer while they are in transit through Tom, while more complicated outer membrane β-barrel proteins, including Tom40 itself, require assistance. This is provided by another protein complex in the outer membrane called the SAM complex. The β-barrel protein SAM50 forms a channel and cooperates with a second subunit to mediate folding and insertion of β-barrel proteins into the membrane, similar to its bacterial counterpart. The TOM and SAM complexes interact, so the unfolded polypeptide can move from the TOM complex to the SAM complex with assistance from TIM chaperones associated with the inner membrane.

Translocation Across the Inner Membrane to the Matrix

Proteins use the Tim23 translocon to cross the inner membrane into the matrix. The integral membrane proteins Tim23 and Tim17 form the channel across the inner membrane and associate with three other subunits (Fig. 18.3). These proteins consist of four transmembrane helices rather than β-strands as in Tom40. Interactions of the N-terminal presequences of matrix proteins with Tim50 and Tim23 guide the presequence into the translocation channel. Physical interactions of Tom and Tim complexes may facilitate the transfer of matrix proteins across both membranes. The MPP peptidase (matrix processing protease) cleaves off the presequences once they enter the matrix.

Two energy sources—the electrical potential across the inner membrane and ATP hydrolysis by matrix chaperones—power polypeptide translocation across the inner membrane. The membrane potential (negative inside) pulls positively charged presequences across the membrane. Then the chaperone Hsp70 takes over and uses cycles of peptide binding and ATP hydrolysis to move the peptide into the matrix. One idea is that Hsp70 rectifies movements of the polypeptide in the pore, allowing movement forward into the matrix but not backward. Hsp70 binds when the polypeptide slides forward. After ATP hydrolysis, Hsp70 dissociates from the polypeptide and the exchange factor mGrp1 (see Fig. 12.12 for a related protein) rapidly recharges it with ATP, ready for another cycle of peptide binding, ATP hydrolysis, and release. This allows the polypeptide to slide forward into the matrix but not backward, so it eventually ends up as a folded protein in the matrix. Another model proposes that the energy from ATP hydrolysis is used to actively pull the polypeptide across the inner membrane.

Proteins destined for the intermembrane space take the same route as those going to the matrix. However, after their presequences are cleaved by the MPP peptidase, they reverse into the space between the two membranes rather than continuing into the matrix.

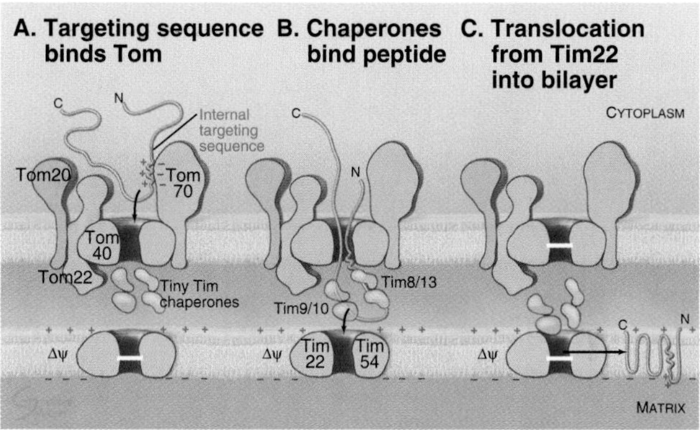

FIGURE 18.5 IMPORT OF THE ADP/ATP ANTIPORTER ACC AND ITS INSERTION INTO THE INNER MEMBRANE BILAYER. A *white bar* across a translocon pore indicates that it is closed. **A,** An internal targeting sequence binds the ACC polypeptide to Tom70, which directs it into the Tom channel. **B,** In the intermembrane space, Tim9/10 and Tim8/13 capture the polypeptide and direct it to the Tim22/54 translocon that is used for import of matrix proteins. **C,** Tim22/54, in conjunction with the inner membrane potential ($\Delta\psi$), promotes insertion of the six transmembrane helices into the inner membrane bilayer.

Translocation Into the Inner Membrane Bilayer

The integral proteins of the inner membrane lack cleavable targeting signals, depending instead on targeting information contained in the intact protein to reach their destination. One example is the most abundant protein of the inner membrane, the adenosine diphosphate (ADP)/ATP antiporter that spans the inner membrane six times (see Fig. 15.4). Its signal sequence is located in the middle of the polypeptide.

A family of small "tiny Tim" chaperone proteins (Tim8, Tim9, Tim10, Tim12, and Tim13) guide inner membrane proteins from the TOM complex across the intermembrane space to the Tim22 translocon in the inner membrane (Fig. 18.5). Complexes of Tim9/10 or Tim8/13 bind to hydrophobic segments of polypeptides during transit to the inner membrane.

Many inner membrane proteins use the Tim22 translocon to access the lipid bilayer. Tim22 has three or four transmembrane helices and forms the channel, but little is known about its structure. It associates with Tim54, Tim12, and Tim18 in a 300-kD complex. Insertion of transmembrane segments into the bilayer depends on membrane potential.

Export From the Matrix

Insertion of proteins synthesized in the matrix into the inner membrane depends on an inner membrane protein called Oxa1p, which forms a translocon similar to bacterial YidC and chloroplast Alb3 (see later sections). Mitochondrial ribosomes are anchored to Oxa1p, allowing hydrophobic transmembrane segments to insert directly into the bilayer. At least one other protein complex participates in export of proteins from the matrix.

Transport of Proteins Into Chloroplasts

Eukaryotes acquired chloroplasts through symbiosis with a photosynthetic cyanobacterium (see Figs. 2.4 and 19.7). Over time, most of the bacterial genes moved to the nucleus, so more than 3000 chloroplast proteins are synthesized on cytoplasmic ribosomes and imported into one of three chloroplast membranes or the three compartments that they surround (Fig. 18.2). The innermost thylakoid membranes contain the photosynthetic apparatus inherited from cyanobacteria. The outer membrane likely came from the eukaryotic host, whereas the inner envelope membrane has both bacterial and eukaryotic features. Some organisms acquired their photosynthetic plastids by secondary or even tertiary rounds of endosymbiosis (see Fig. 2.7). These secondary or tertiary plastids are bounded by one or more additional membranes and have more complicated mechanisms to import the proteins expressed from nuclear genes.

Although the protein import systems of chloroplasts and mitochondria both use zip codes on the imported proteins and translocons in the membranes (Fig. 18.6), the two systems share no common proteins. Another striking difference is that chloroplasts use specialized versions of some import components to acquire different proteins. This flexibility is important, as these organelles (collectively called plastids), have tissue-specific and age-specific functions, ranging from photosynthesis to starch storage (see Chapter 19). Mitochondria are more homogeneous, so a restricted set of import proteins suffices.

In plants, N-terminal signal sequences called **transit sequences** target chloroplast proteins to the import machinery in the outer envelope. When added experimentally to the N-terminus of a test protein, transit sequences suffice to guide the test protein into the stroma of chloroplasts. These N-terminal targeting sequences vary in length from 13 to 146 residues, and their amino acid sequences have little in common beyond a net positive charge. The current understanding is that transit sequences are heterogeneous, because they contain many different types of zip codes that bind selectively to the specialized import receptors that plants express to allow different plastids to import the proteins appropriate for chloroplasts, starch storage organelles, and other specialized functions.

All imported proteins use the same "general import pathway" to cross the outer and inner envelope membranes. The machinery consists of different protein complexes in each membrane called **Toc** (translocon at the outer envelope membrane of chloroplasts) and **Tic** (translocon at the inner envelope membrane of chloroplasts) (Fig. 18.6). These complexes were identified by chemical crosslinking of imported proteins to translocon proteins. Both Toc and a "super complex" of Toc with

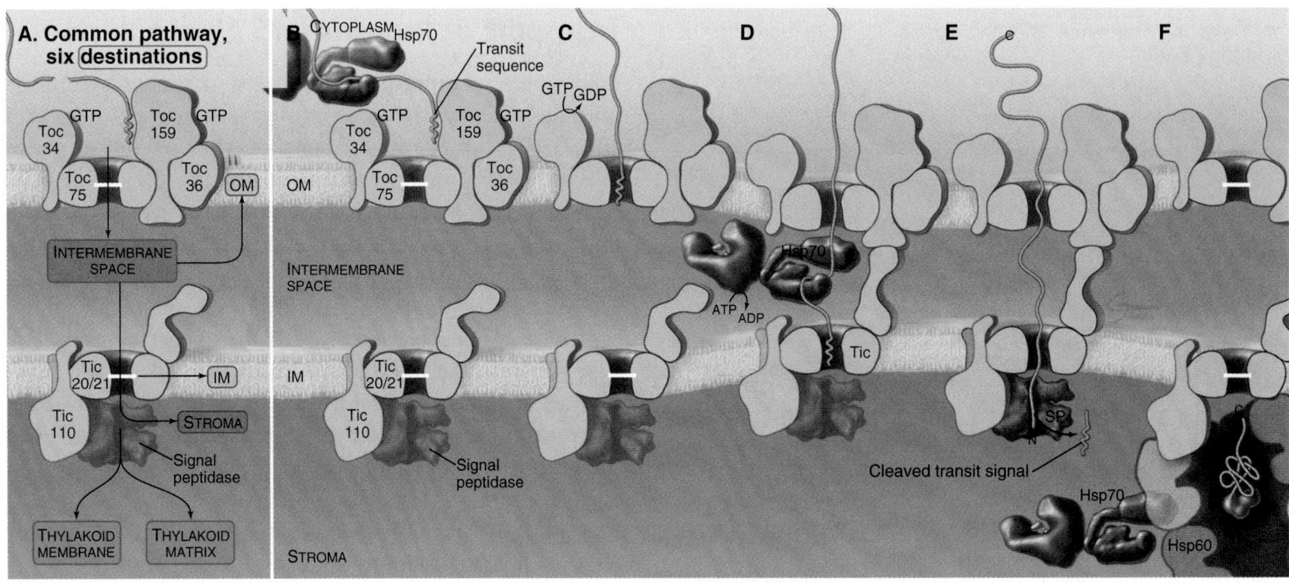

FIGURE 18.6 CHLOROPLAST PROTEIN IMPORT VIA TOC AND TIC COMPLEXES. A, Common pathway across the outer membrane leading to six different destinations. **B–F,** Five stages in the movement of proteins into the stroma. **B,** Energy-independent binding of the transit sequence to outer membrane lipids and proteins, especially Toc159. **C,** Then guanosine triphosphate (GTP) hydrolysis by Toc34 and perhaps Toc159 promote insertion of the transit sequence through the outer membrane pore composed of Toc75. **D,** ATP hydrolysis by Hsp70 facilitates delivery to the inner membrane translocon Tic20/21. **E,** A stromal protease SP removes the N-terminal transit sequence. **F,** Hsp60 (heat shock protein 60) and Hsp70 promote folding of stromal proteins, while other proteins are rerouted to other compartments, including thylakoids. Tic, translocon at the inner envelope membrane of chloroplasts; Toc, translocon at the outer envelope membrane of chloroplasts. (Modified from Chen X, Schnell D. Protein import into chloroplasts. *Trends Cell Biol.* 1999;9:222–227; and May T, Soll J. Chloroplast precursor protein translocon. *FEBS Lett.* 2000;452:52–56.)

Tic can be isolated for analysis of their composition. Mutations that compromise chloroplast import have also contributed to understanding the process.

The journey of a protein from its site of synthesis in cytoplasm into the stroma is understood in broad outline. The Toc159 and Toc34 proteins are receptors for transit sequences on the surface of chloroplasts. Plant cells express different isoforms of Toc159 and Toc34 to accommodate the import of different proteins appropriate to their differentiated state, although the details of these interactions are not yet clear. Both Toc159 and Toc34 have cytoplasmic guanosine triphosphatase (GTPase) domains similar to other small GTPases (see Fig. 25.7). Bound guanosine triphosphate (GTP) favors binding of transit sequences, and a bound transit sequence stimulates GTP hydrolysis and transfer of the transit sequence to the translocon.

The β-barrel protein Toc75 forms the translocon across the outer membrane for all imported proteins. A homologous protein Omp85 translocates proteins in the opposite direction across the outer membrane of gram-negative bacteria. These proteins have a variable number of N-terminal POTRA (polypeptide transport associated) domains that may interact with many different transit sequences. Polypeptides are thought to be unfolded during transit through the narrow pore. Proteins destined for the outer membrane insert after passing through Toc75, similar to mitochondria.

The pore across the inner membrane consists of a complex of at least seven Tic proteins, but many details regarding their structures and functions are not yet settled. The abundant protein Tic110 not only forms some or all of a pore but also binds Hsp70 chaperones on the stromal side of the membrane. Smaller proteins called Tic20 and Tic21 appear to be distantly related to mitochondrial Tim23/17, so they may form a second type of channel composed of transmembrane helices in the inner membrane.

As proteins emerge into the stroma, a signal peptidase cleaves off the transit peptide before the proteins fold or redistribute to their final locations. Hydrolysis of hundreds of ATP molecules in the stroma is required to complete the translocation and folding of imported proteins. The contributing enzymes include an AAA adenosine triphosphatase (ATPase), heat shock protein 93 (Hsp93), chaperone Hsp70, and a chaperonin heat shock protein 60 (Hsp60) (see Chapter 12). Some proteins remain in the stroma. Other proteins move on to thylakoid membranes or the thylakoid lumen using at least four different pathways.

Some photosynthesis proteins insert directly into thylakoid membranes from the stroma. Others require help from proteins homologous to parts of the signal recognition particle (SRP) system used for export from bacteria (Fig. 18.10) and into the ER of eukaryotes (see Fig. 20.6). Although chloroplasts lack SRP RNA, GTPases similar to an SRP protein and the SRP receptor cooperate with a protein that is homologous to translocon Oxa1p to mediate insertion into the thylakoid membrane.

Hydrophilic proteins destined for the thylakoid lumen retain a secondary N-terminal signal sequence after the transit sequence is cleaved in the stroma. Some move across the thylakoid membrane into the thylakoid lumen through a translocon homologous to bacterial SecYE, powered by ATP hydrolysis by a homolog of SecA (Fig. 18.9). Other proteins with tightly bound redox factors cross the thylakoid membrane while compactly folded using translocon factors similar to the bacterial Tat system (Fig. 18.2). Secondary signal sequences with two arginine residues direct these proteins to a Tat translocon and the proton gradient drives the polypeptide across the membrane. After translocation, a peptidase in the thylakoid lumen removes both types of secondary signal sequences.

Transport of Proteins Into Peroxisomes

Peroxisomes are simple organelles with a single membrane limiting a lumen containing many **oxidative enzymes** (see Fig. 19.10). Nuclear genes encode all proteins found in the membrane and lumen of peroxisomes. Their messenger RNAs are translated on cytoplasmic ribosomes, and the proteins are incorporated posttranslationally into peroxisomes (Fig. 18.1).

Two types of targeting signals direct proteins from the cytoplasm to the peroxisome lumen (called matrix). The **type 1 peroxisomal targeting signal (PTS1)** is found at the extreme C-terminus of most peroxisomal enzymes (Fig. 18.7). PTS1 is just three amino acids long with consensus sequence of serine-lysine-leucine-COOH, or a conservative variant. For example, alanine or cysteine can substitute at the −3 position, arginine or histidine can function at the penultimate position, and methionine can substitute for the C-terminal leucine. Amidation of the C-terminal carboxylate inactivates the signal. The type 2 peroxisomal targeting signal **(PTS2)** also targets a few proteins (only four are known in humans, one in yeast) to the peroxisome matrix. PTS2 sequences are located at or near the N-terminus and have a loose consensus sequence of RLXXXXXH/QL (where X is any amino acid).

Proteins called **peroxins** deliver proteins from the cytoplasm to the peroxisomal membrane or lumen (Fig. 18.8 and Table 18.1). Loss-of-function mutations in humans and yeast revealed genes for more than 20 peroxins that participate in the biogenesis and proliferation of peroxisomes. Mutations of human *PEX* genes cause devastating diseases (Table 18.1).

The PEX5 import receptor carries proteins with a PTS1 signal to the peroxisomal lumen by a highly unusual mechanism. The soluble protein PEX5 is proposed to bind the PTS1 motif and insert into the lipid bilayer surrounding the peroxisome, where it forms a transient pore to deliver the cargo protein into the lumen. Membrane proteins including a receptor called PEX14 participate in delivery of proteins into the lumen. Then PEX5 returns to the cytoplasm for further rounds of import. Recycling PEX5 depends on conjugation with ubiquitin and ATP hydrolysis by an AAA ATPase, a process with some similarities to ER-associated degradation (ERAD; see Fig. 20.12). A similar mode of action is proposed for PEX7, the import receptor for PTS2 proteins.

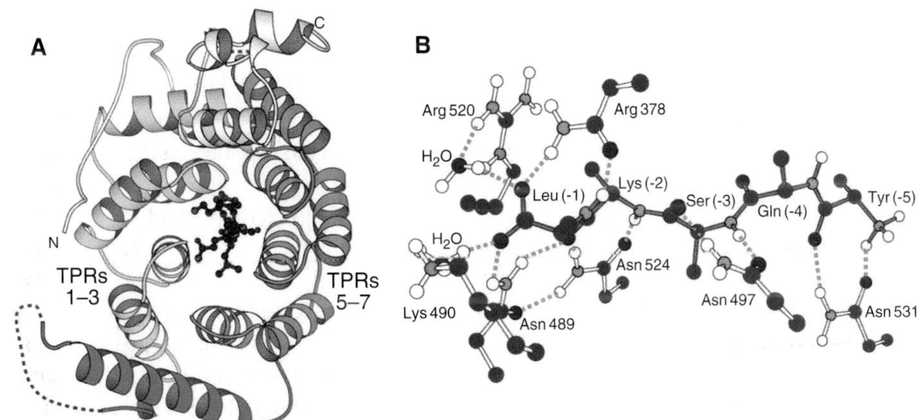

FIGURE 18.7 BINDING OF A PEROXISOMAL TARGETING SIGNAL TYPE 1 (PTS1) TARGETING SIGNAL TO PEX5. A, PTS1 binds to the C-terminal tetratricopeptide repeat (TPR) domain of PEX5. The C-terminal, 40-kD TPR domain of PEX5, shown as a ribbon diagram, surrounds the PTS1 peptide, shown as a stick figure. Note TPRs 1 to 3 *(yellow ribbons)* and TPRs 5 to 7 *(blue ribbons).* An α-helical span *(green ribbon)* links the two triplet TPRs at the bottom of this structure; the C-terminal extension *(white ribbons)* also connects the two triplet TPRs. **B,** Detailed view of PEX5–PTS1 interactions between the PTS1 backbone *(brown bonds)* and PEX5 side chains *(white bonds);* the putative hydrogen bonds are shown as *dashed green lines.* This structure revealed the chemical basis of PEX5-PTS1 binding, as well as the sequence constraints of PTS1. (**A,** From PDB file 1FCH and Gatto GJ Jr, Geisbrecht BV, Gould SJ, Berg JM. Peroxisomal targeting signal-1 recognition by the TPR domains of human PEX5. *Nat Struct Biol.* 2000;7:1091–1095.)

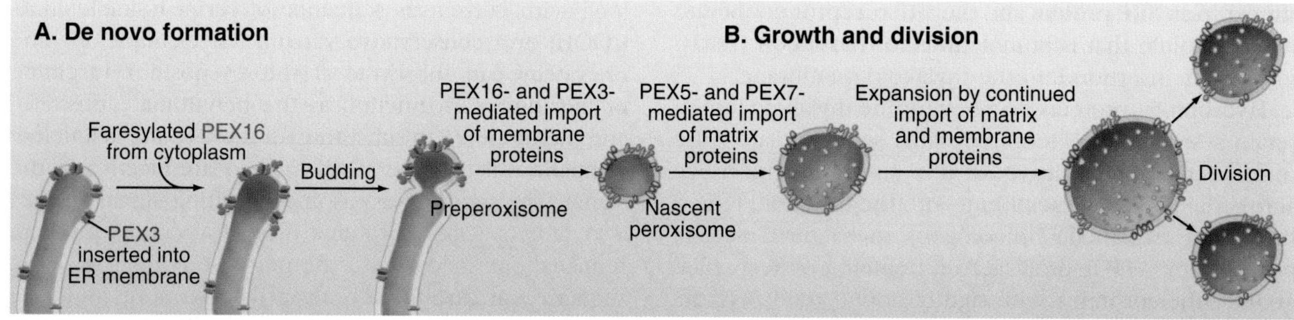

FIGURE 18.8 PEROXISOME BIOGENESIS. A, De novo formation by budding of a vesicle containing PEX3 and PEX16 from endoplasmic reticulum to form a preperoxisome. **B,** Growth and division of peroxisomes. PEX3 and PEX16 mediate the import of membrane proteins. The PEX5–PTS1 receptor, PEX7, and other peroxins mediate the import of proteins with PTS1 and PTS2 into peroxisomes.

TABLE 18.1 Peroxins: Proteins for the Assembly of Peroxisomes

Groups	Peroxins	Functions	Diseases
Peroxisome Matrix Protein Import			
	PEX5	PTS1 receptor and channel	NALD, ZS
	PEX13, PEX14, PEX17, PEX33	Membrane docking complex for PEX5	NALD, ZS
	PEX8 (PEX23)	Membrane partners of PEX5 & PEX14	
	PEX7	PTS2 receptor	RCDP
	PEX18, PEX20, PEX21	PTS2 coreceptors with PEX7	
Recycling Matrix Import Machinery			
	PEX1, PEX6	AAA ATPases for recycling PEX5	IRD, NALD, ZS
	PEX15, PEX26	Membrane anchors for PEX1 & PEX6	
	PEX2, PEX4, PEX10, PEX12	Ubiquitin-conjugating enzyme for cycling PEX5	IRD, NALD, ZS
	PEX22	Membrane anchor for PEX4	
Peroxisome Membrane Protein Import			
	PEX19	Cytoplasmic PMP receptor & chaperone	ZS
	PEX3 (PEX16)	Membrane receptor for imported protein	IRD, ZS
Peroxisome Biogenesis From Endoplasmic Reticulum			
	PEX16	Recruits PMPs from ER membrane	ZS
	PEX23, PEX30	Regulate de novo peroxisome formation	
	PEX25	Recruits Rho1	
Peroxisome Fusion and Fission			
	PEX11, PEX25, PEX27, PEX34	Regulate membrane elongation & fission	Mild ZS
	PEX1, PEX6	AAA ATPases membrane fission & fusion	IRD, NALD, ZS

ATPase, adenosine triphosphatase; ER, endoplasmic reticulum; IRD, infantile Refsum disease; NALD, neonatal adrenoleukodystrophy; PBD, peroxisome biogenesis disorder; PMP, peroxisome membrane protein; PTS1/PTS2, peroxisomal targeting signal type 1/type 2; RCDP, rhizomelic chondrodysplasia punctate; ZS, Zellweger syndrome.

For reference, see Smith JJ, Aitchison JD. Peroxisomes take shape. *Nat Rev Mol Cell Biol.* 2013;14:803–817.

Peroxisomal membranes form from lipids synthesized in the ER and proteins imported from the cytoplasm. The lipids are delivered in vesicles. Peroxisomal membrane proteins use one or more **membrane peroxisomal targeting sequences (mPTSs)** for delivery to peroxisomes. The sequences of mPTS motifs vary widely but include basic amino acids along with a transmembrane domain. PEX19 is the cytoplasmic receptor for mPTS motifs. It stabilizes these membrane proteins in the cytoplasm and delivers them to PEX3 and PEX16 on the peroxisome for insertion into the membrane (Fig. 18.8). Cells deficient in any of these three peroxins lack peroxisomal membranes and the peroxisomal membrane proteins are degraded or mislocalized to other cellular membranes, particularly mitochondria.

Peroxisomes may arise by either of two pathways (Fig. 18.8). Some peroxisomes form de novo by budding from the ER. PEX3 is inserted into the ER, where it recruits PEX16 and other peroxins. This specialized domain of ER then pinches off to form a nascent peroxisome de novo. By originating from the ER in this manner, peroxisomes can arise in cells that lack them without a preexisting peroxisome as template. Preexisting peroxisomes can grow by fusing with nascent peroxisomes and importing proteins and lipids. They also divide by a fission process that depends the GTPase dynamin (see Fig. 22.9).

Translocation of Eukaryotic Proteins Across the Plasma Membrane by ABC Transporters

Most proteins that are secreted by eukaryotic cells travel to the cell surface through the classical secretory pathway, including the ER and Golgi apparatus (see Chapters 20 and 21). But budding yeast use an **ABC transporter** (see Fig. 8.9) to transport their a-type mating factor directly from the cytoplasm across the plasma membrane. The a-factor is synthesized in the cytoplasm as part of a precursor, excised from the precursor by proteolytic cleavage, and then prenylated on its C-terminus before transport across the plasma membrane. This mechanism has been invoked to explain the secretion of a few mammalian proteins that lack the "signal sequences" that direct proteins to the classic ER secretory pathway. These include some cytokines, fibroblast growth factor, and some blood-clotting factors. This is a well-characterized route for secretion of some bacterial proteins.

Targeting to the Surfaces of the Plasma Membrane

Many proteins synthesized in the cytoplasm are targeted to the cytoplasmic side (known as the cytoplasmic leaflet) of organelle and plasma membranes (see Fig. 13.9). These include peripheral membrane proteins that bind to cytoplasmic domains of integral membrane proteins or bind directly to the lipid bilayer.

Other proteins are tethered to membrane bilayers by a covalently attached lipid added as a posttranslational modification following synthesis on cytoplasmic ribosomes. Lipid modifications on tethered proteins include long-chain, saturated fatty acids and isoprenoids. The saturated fatty acids are either myristate (14 carbons), which is added through amide linkage to aminoterminal glycine residues, or palmitate (16 carbons), which is usually added through a thioether linkage to cysteine residues found toward the C-terminus. The isoprenoids farnesyl (15 carbons) and geranylgeranyl (20 carbons) are added through thioether linkages to cysteine residues located at or near the C-terminus in specific structural motifs. Attachment of a lipid helps stabilize membrane association, but does not guarantee permanent anchoring to the membrane. Some proteins, such as the catalytic subunit of cyclic adenosine monophosphate–dependent protein kinase (PKA), are fatty acylated but remain mostly soluble in cytoplasm.

Proteins attached to the external surface of plasma membranes by glycosylphosphatidylinositol anchors arrive by a different route. These proteins are synthesized on ribosomes associated with the ER and then translocated into the ER lumen anchored by a C-terminal transmembrane segment. Inside the ER the protein is cleaved from its membrane anchor and transferred enzymatically to glycosylphosphatidylinositol before transport to the cell surface (see Fig. 20.8C).

Prokaryotic Protein Export

Bacteria and Archaea employ at least 10 distinct strategies to transport proteins from the cytoplasm across the inner membrane and beyond. Seven of these pathways use a common pore across the inner membrane called the Sec translocon. These pathways are important because some contribute to human disease. In addition, they serve as important model systems, since eukaryotes use a homologous translocon to move proteins into the bilayer or lumen of the ER (see Fig. 20.7). This section begins with a discussion of six branches of the Sec secretory pathway and finishes with three distinct pathways.

Pathways Dependent on the SecYE Translocon

Organisms in all three domains of life use Sec translocons to move proteins synthesized in the cytoplasm across membranes. Translocons in the plasma membranes of Bacteria and Archaea consist of two transmembrane proteins called SecY and SecE in Bacteria (Fig. 18.9). The translocons of the ER of eukaryotes consist of homologous protein subunits called Sec61α and γ (see Fig. 20.7). The narrow pore for translocating the secreted polypeptide is located in the middle of a bundle of α helices. Loss-of-function mutations of SecY or SecE compromise the secretion of most proteins by Bacteria or Archaea. Several accessory subunits assist in translocation, but they are not essential in Bacteria and are not present in eukaryotes.

Posttranslational Protein Translocation

Bacteria use **Sec-signal sequences** to direct many proteins to the SecYE translocon for transport across the plasma membrane or for insertion into the plasma membrane. Gram-positive bacteria such as *Bacillus subtilis* lack an outer membrane, so the proteins leave the cell after crossing the plasma membrane. In gram-negative bacteria, translocated proteins enter the periplasm, insert into the outer membrane, or leave the cell.

Proteins targeted to the Sec translocon are synthesized in the cytoplasm with an N-terminal Sec-signal sequence. These targeting sequences consist of approximately 25 residues beginning with methionine, followed by a few basic residues, 10 to 15 hydrophobic residues, and a site for cleavage by a proteolytic enzyme called signal peptidase after translocation across the inner membrane. Chaperones such as **SecB** bind newly synthesized proteins to prevent folding and maintain a state that is

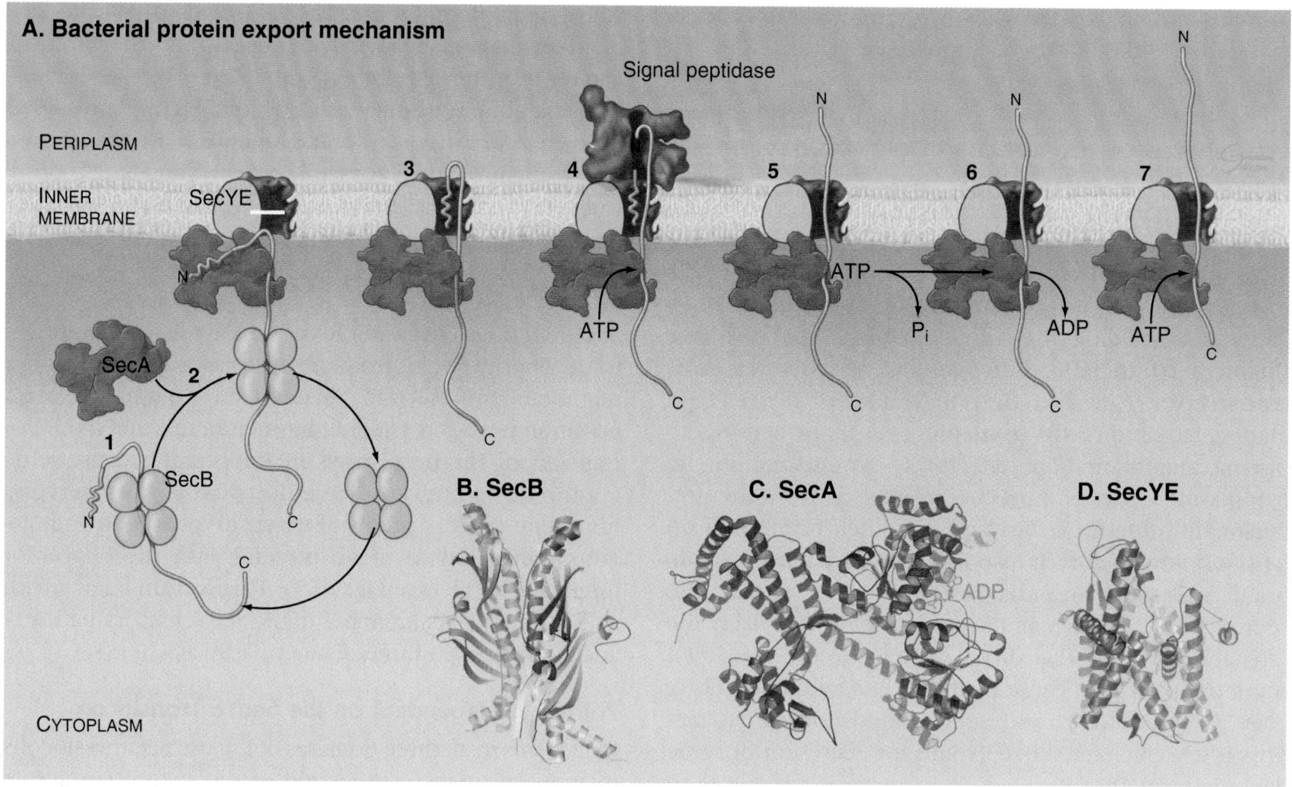

FIGURE 18.9 SECRETION OF PROTEIN FROM BACTERIA THROUGH THE SecYE TRANSLOCON. A, Pathway of secretion. *1,* After synthesis by a cytoplasmic ribosome, the polypeptide associates with the SecB chaperone. *2,* SecA binds the presequence (*blue*) and docks on the SecYE translocon. *3,* The presequence inserts into the translocon. *4,* ATP-binding to SecA promotes insertion of the associated polypeptide into the translocon, followed by cleavage of the signal sequence. *5–7,* The membrane potential and cycles of ATP hydrolysis by SecA drive the polypeptide across the inner membrane. **B,** Ribbon diagram of *Haemophilus influenzae* SecB. **C,** Ribbon diagram of *Bacillus subtilis* SecA. **D,** Ribbon diagram of *Methanococcus jannaschii* SecY complex translocon. (**A,** Modified from Danese PN, Silhavy TJ. Targeting and assembly of periplasmic and outer-membrane proteins in *E. coli*. *Annu Rev Genet*. 1999;32:59–94. For reference, see PDB file IOZB and Zhou J, Xu Z. Structural determinants of SecB recognition by SecA in bacterial protein translocation. *Nat Struct Biol*. 2003;10:942–948 [**B**]; PDB file 1TF2 and Osborne AR, Clemons WM, Rapoport TA. A large conformational change of the translocation ATPase SecA. *Proc Natl Acad Sci U S A*. 2004;101:10937–10942 [**C**]; and PDB file 1RHZ and van de Berg B, Clemons WM, Collinson I, et al. X-ray structure of a protein-conducting channel. *Nature*. 2004;427:36–44 [**D**].)

competent for translocation (Fig. 18.9). Unlike most other chaperones (see Figs. 12.13 and 12.14), SecB does not require ATP hydrolysis for cycles of interaction with substrates. Hsp70 homologs (DnaK) have a secondary role in chaperoning precursors for translocation.

Translocation of many bacterial membrane and secreted proteins with cleavable signal sequences depends on the ATPase **SecA.** A system reconstituted from purified SecA, SecY, and SecE can translocate precursor proteins across lipid membranes in the presence of ATP. Remarkably, Archaea lack SecA, although they depend on translocon components that are homologous to SecYE. Eukaryotes use SecA only for translocation into chloroplast thylakoids (Fig. 18.2).

SecA binds proteins associated with SecB in the cytoplasm and targets the signal sequence to the SecY translocon. SecY "proofreads" the signal sequence associated with SecA, releasing those with imperfect matches to the consensus prior to translocation. Binding to SecY changes the conformation of SecA, bringing together two domains that form a clamp around the substrate peptide. Inside the clamp the peptide substrate binds weakly through hydrogen bonds to the edge of a small β-sheet. Cycles of ATP hydrolysis by SecA cause a rocking motion of a lever arm that pushes the unfolded peptide through the channel in small steps. The protein then folds after translocation.

Signal peptidases located on the outer surface of the plasma membrane cleave signal peptides from translocated proteins soon after they cross the plasma membrane. Some bacterial signal peptidases are similar to eukaryotic homologs. Other bacterial signal peptidases are specialized to cleave lipoproteins just before an invariant cysteine. This cysteine is then conjugated to diacylglycerol, which anchors the lipoprotein to the outer surface of the plasma membrane or to the outer membrane of gram-negative bacteria. Signal peptidases also degrade cleaved signal peptides.

Translocation Dependent on the Signal Recognition Particle

In eukaryotes, the **signal recognition particle** (SRP) is the adapter between signal sequences and the translocon of ER (see Fig. 20.5), but in bacteria, only a minority of integral membrane proteins and secreted proteins depend on SRP for targeting to the Sec translocon. Eukaryotic and Archaeal SRPs consist of a 7S RNA and several proteins, whereas *Escherichia coli* SRP consists of a smaller 4.5S RNA and a single protein called Ffh (for "fifty-four homolog," after its eukaryotic counterpart) (see Fig. 20.5). SRP binds Sec-signal sequences and signal-anchor sequences as they emerge from the ribosome. This interaction slows translation until SRP docks on the cytoplasmic surface of the inner membrane with its receptor FtsY and the Sec translocon. Resumption of translation drives the polypeptide through the translocon. See Chapter 20 for more details on SRP and eukaryotic cotranslational translocation.

Insertion of Inner Membrane Proteins

Proteins inserted into the inner membrane depend on YidC, a protein with six transmembrane helices related to Oxa1p and Alb3, that direct proteins into the inner membrane of mitochondria and thylakoid membranes of chloroplasts. Work is being done to determine if YidC inserts membrane proteins on its own or accepts them from the Sec translocon.

Insertion of Proteins in the Outer Membrane of Gram-Negative Bacteria

Outer membrane proteins are synthesized in the cytoplasm and directed to the Sec translocon by signal sequences. The signal sequence is cleaved from the unfolded protein after crossing the inner membrane into the periplasm. Several periplasmic chaperones and assembly factors participate in protein folding, including enzymes that catalyze the isomerization of proline peptide bonds and oxidation/reduction of cysteine thiol groups. Insertion into the outer membrane depends on β-barrel proteins, but little is known about the targeting signals.

Outer Membrane Autotransporter Pathway

Some proteins, including secreted proteolytic enzymes and toxins as well as membrane-anchored adhesins and invasins, hitch a ride to the cell surface on their own outer membrane transporters (Fig. 18.10A). These proteins have an N-terminal secreted domain and a C-terminal domain that forms a transmembrane β-barrel like a porin (see Fig. 13.9). The protein uses the Sec pathway to cross the inner membrane and the β-barrel inserts into the outer membrane. The N-terminal functional domain then translocates across the outer membrane through its β-domain pore. An outer membrane protease releases toxins and proteases, whereas adhesins that follow this route remain on the surface attached to the β-domain.

Outer Membrane Single Accessory Pathway

Some hemolysins and hemagglutinins move to the periplasm through the Sec pathway and then use a single accessory protein to translocate across the outer membrane. The accessory protein forms a β-barrel in the outer membrane with a pore like autotransporters and the porins that transport peptides across the outer membranes of chloroplasts.

Chaperone/Usher Pathway

Gram-negative bacteria use a novel mechanism, downstream of the Sec pathway, to transport and assemble pili on their outer surface. Pili are appendages involved with bacterial pathogenesis, including urinary tract infections. A periplasmic chaperone binds the pillus peptide and promotes folding. The pilus subunit is folded similar to an immunoglobulin domain (see Fig. 3.13), but lacks the seventh β-strand. This exposes core hydrophobic residues. The chaperone consists of two immunoglobulin-like domains, one of which donates a strand to complete the immunoglobulin domain of the pilus subunit. The chaperone delivers a pilus subunit to an outer membrane translocon called usher (Fig. 18.10A). There it transfers its bound subunit to the end of a growing chain of pilus subunits, all bound together, head to tail, by strands that complete the seven-strand β-sheet of the adjacent subunit. On the outer surface, the pilus subunits rearrange into a helical pilus. The assembly reaction is thought to provide the energy for translocation. The chaperone prevents premature assembly of the pilus.

Type II Secretion

Bacteria use an alternate route downstream of the Sec pathway to secrete other toxins and enzymes with cleaved signal sequences (Fig. 18.10A). At least a dozen protein subunits participate in this complicated pathway. The pore in the outer membrane is composed of a secretin, a protein with relatives that also participate in type III secretion, phage biogenesis, and formation of one type of pilus. The secretin pore is a ring of 12 to 14 subunits around a large gated channel that is 5 to 10 nm in diameter.

Type IV Secretion

Bacteria secrete a few proteins using an apparatus similar to that used for DNA transfer between two bacteria during conjugation and for DNA injection into plant cells by *Agrobacterium*. DNA is transferred directly from the cytoplasm of one bacterium to the cytoplasm of another bacterium or plant cell. Proteins that are secreted by this pathway include pertussis toxin by *Bordetella pertussis* and another toxin by *Helicobacter pylori*. This pathway starts with synthesis in the cytoplasm and translocation across the plasma membrane by the Sec translocon. If present, the signal sequence is cleaved

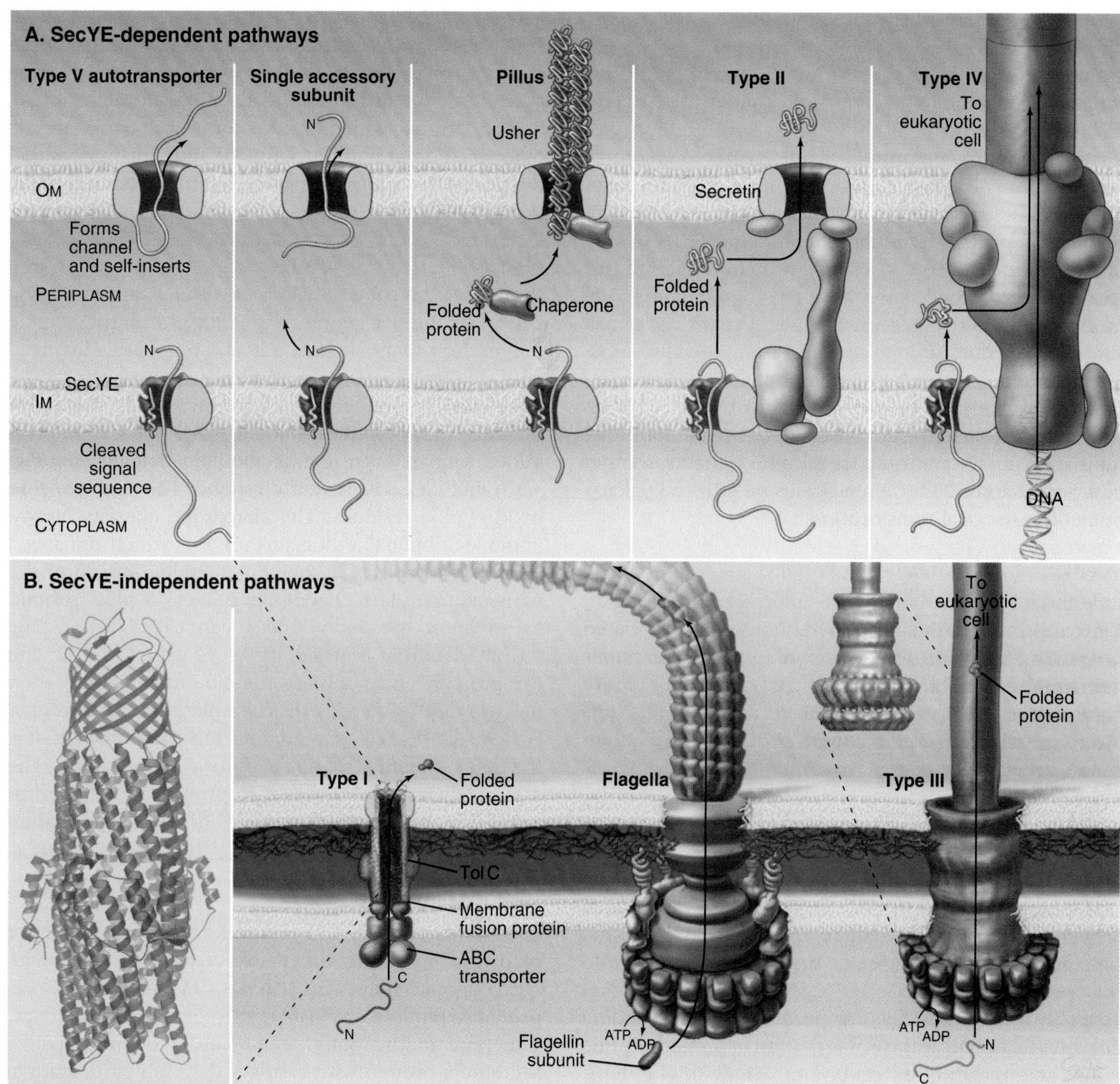

FIGURE 18.10 SECRETION ACROSS THE OUTER MEMBRANE OF GRAM-NEGATIVE BACTERIA. A, Pathways dependent on SecYE. The cleaved signal sequence is shown in *blue.* The β-domain of autotransporters forms a pore for the translocation of part of its own chain, which may remain attached, as shown, or be cleaved for escape from the cell. Single accessory proteins form a pore for secretion of separate proteins. Usher forms a pore for the translocation and assembly of pili. Type II secretion uses a secretin pore for translocation. Type IV secretion employs a large translocon similar to that used by *Agrobacterium* for secretion of DNA. **B,** Pathways independent of SecYE. Type I secretion uses an ABC transporter to cross the inner membrane and additional subunits to cross the periplasm and outer membrane. *Left panel,* Ribbon model of TolC, one type of translocon that spans the periplasm and outer membrane. *Right panel,* Each TolC subunit contributes four β-strands to a porin-like structure that spans the outer membrane. α-Helical continuations of these β-strands form a tube having an internal diameter of 3.5 nm for transport of proteins across the periplasm. Bacterial flagella transport flagellin subunits across both membranes and then through the central channel of the flagellar filament for incorporation at the growing tip. Type III secretion uses components similar to the basal body of flagella. *Gray* illustration *(far right)* shows a three-dimensional reconstruction of the type III secretion apparatus from *Salmonella typhimurium.* IM, inner membrane; OM, outer membrane. (**A–B,** Modified from Thanassi DG, Hultgren SJ. Multiple pathways allow protein secretion across the bacterial outer membrane. *Curr Opin Cell Biol.* 2000;12:420–430. **B,** TolC ribbon diagram based on PDB file 1EK9. For reference, see Koronakis V, Sharff A, Koronakis E, et al. Crystal structure of the bacterial membrane protein TolC central to multidrug efflux and protein export. *Nature.* 2000;405:914–919. Reconstruction of the type III secretion complex from *S. typhimurium* based on Marlovits TC, Kubori T, Sukhan A, et al. Structural insights into the assembly of the type II secretion needle complex. *Science.* 2004;306:1040–1042.)

before translocation across the outer membrane by the type IV secretion system (Fig. 18.10A).

Pathways Independent of the Sec Translocon

Type I ABC Transporters

Bacteria use ABC transporters (see Fig. 8.9) to secrete a small number of toxins (eg, *E. coli* hemolysin), proteases, and lipases. C-terminal signal sequences of 30 to 60 residues target these proteins to the ABC transporter, the only component required for secretion by gram-positive bacteria. Gram-negative bacteria require not only a transporter in the inner membrane but also two proteins that form a continuous channel across the periplasm and outer membrane (Fig. 18.10B). ATP hydrolysis by the ABC transporter provides energy for translocation. Protein conduits across the periplasm and outer membrane engage ABC transporters presenting substrates for export and then disengage when translocation is complete. Genes for secreted proteins are generally in the same operon as the export machinery.

Flagellar and Type III Secretion Systems

The basal bodies of bacterial flagella transport flagellin subunits through a central pore that crosses both membranes (Fig. 18.10B) and extends the length of the flagellar shaft to the tip, where subunits add to the distal end (see Fig. 5.8). This flagellar pathway transports a few other proteins, including a phospholipase that contributes to the virulence of *Yersinia,* the cause of the black plague.

Pathogenic gram-negative bacteria, such as *Yersinia,* use the syringe-like type III apparatus, similar to a bacterial flagellum, to transport toxins from the cytoplasm into the medium or directly into target cells. In the target cell, these toxins disrupt cellular physiology, in part by forming pores in target cell membranes. The type III secretion complex consists of approximately 20 different protein subunits including some with homology to rotary ATPases (see Fig. 8.4). A complex base consisting of several protein rings spans the periplasm and both membranes. A polymer of a single type of protein forms a hollow needle up to 40 nm long for injection of toxins directly into target animal or plant cells.

Several signals direct proteins to this pathway. One such signal is a noncleavable signal sequence in the secreted protein that binds a chaperone dedicated to targeting toxins to the type III pathway. A cytoplasmic ATPase separates secreted proteins from chaperones, and the transmembrane electrochemical gradient of protons provides most of energy for transport.

Double Arginine Pathway

Many but not all Bacteria and Archaea use proteins homologous to chloroplast Tat proteins to translocate folded proteins across the plasma membrane. In both prokaryotes and chloroplasts, some of these cargo proteins participate in redox reactions and have bound cofactors such as flavins or FeS clusters. These cofactors are incorporated as the proteins fold in the cytoplasm or chloroplast stroma. In contrast to the Sec translocon, the Tat translocon accommodates folded proteins. The N-terminal signal sequences for this pathway have a pair of arginines (RR) in a conserved sequence (Ser/Thr-Arg-Arg-X-Phe-Leu-Lys, where X is any amino acid) adjacent to a stretch of at least 13 uncharged residues. Translocation of these proteins in *E. coli* requires three Tat proteins. One forms the transmembrane pore, and the others appear to participate in targeting. Virtually all Archaeal proteins that move through Tat remain anchored to the cell surface.

SELECTED READINGS

Berks BC. The twin-arginine protein translocation pathway. *Annu Rev Biochem*. 2015;84:843-864.

Chacinska A, Koehler CM, Milenkovic D, et al. Importing mitochondrial proteins: machineries and mechanisms. *Cell*. 2009;138:628-644.

Chatzi KE, Sardis MF, Karamanou S, Economou A. Breaking on through to the other side: protein export through the bacterial Sec system. *Biochem J*. 2013;449:25-37.

Costa TR, Felisberto-Rodrigues C, Meir A, et al. Secretion systems in gram-negative bacteria: structural and mechanistic insights. *Nat Rev Microbiol*. 2015;13:343-359.

Dautin N, Bernstein HD. Protein secretion in Gram-negative bacteria via the autotransporter pathway. *Annu Rev Microbiol*. 2007;61:89-112.

Demarsy E, Lakshmanan AM, Kessler F. Border control: selectivity of chloroplast protein import and regulation at the TOC-complex. *Front Plant Sci*. 2013;5:483.

Diepold A, Armitage JP. Type III secretion systems: the bacterial flagellum and the injectisome. *Philos Trans R Soc Lond B Biol Sci*. 2015;370:20150020.

Gutensohn M, Fan E, Frielingsdorf S, et al. Toc, Tic, Tat et al.: Structure and function of protein transport machines in chloroplasts. *J Plant Physiol*. 2006;163:333-347.

Kedrov A, Kusters I, Driessen AJ. Single-molecule studies of bacterial protein translocation. *Biochemistry*. 2013;52:6740-6754.

Li HM, Chiu CC. Protein transport into chloroplasts. *Annu Rev Plant Biol*. 2010;61:157-180.

Li HM, Teng YS. Transit peptide design and plastid import regulation. *Trends Plant Sci*. 2013;18:360-366.

Neupert W, Herrmann JM. Translocation of proteins into mitochondria. *Annu Rev Biochem*. 2007;76:723-749.

Shiota T, Imai K, Qiu J, et al. Molecular architecture of the active mitochondrial protein gate. *Science*. 2015;349:1544-1548.

Smith JJ, Aitchison JD. Peroxisomes take shape. *Nat Rev Mol Cell Biol*. 2013;14:803-817.

Szabo Z, Pohlschroder M. Diversity and subcellular distribution of archaeal secreted proteins. *Front Microbiol*. 2012;3:207.

Thanassi DG, Hultgren SJ. Multiple pathways allow protein secretion across the bacterial outer membrane. *Curr Opin Cell Biol*. 2000;12:420-430.

CHAPTER 19

Mitochondria, Chloroplasts, Peroxisomes

This chapter considers three organelles formed by posttranslational import of proteins synthesized in the cytoplasm. Mitochondria and chloroplasts both arose from endosymbiotic bacteria, two singular events that occurred about 1 billion years apart (see Fig. 2.4B). Both mitochondria and chloroplasts retain remnants of those prokaryotic genomes but depend largely on genes that were transferred to the nucleus of the host eukaryote. Both organelles brought biochemical mechanisms that allow their eukaryotic hosts to acquire and use energy more efficiently. In **oxidative phosphorylation** by mitochondria and **photosynthesis** by chloroplasts, energy from the breakdown of nutrients or from absorption of photons is used to energize electrons. As these electrons tunnel through transmembrane proteins, energy is extracted to create proton gradients. These proton gradients drive the rotary adenosine triphosphate (ATP) synthase (see Fig. 14.5) to make ATP, which is used as energy currency to power the cell. Peroxisomes contain no genes and depend entirely on nuclear genes to encode their proteins. Their evolutionary origins are obscure. Peroxisomes contain enzymes that catalyze a wide range of oxidation reactions that are essential for cellular homestasis. Patients who lack peroxisomes have severe neural defects.

Mitochondria

Evolution of Mitochondria

Mitochondria (Fig. 19.1) arose about 2 billion years ago when a bacterium fused with an archaeal cell (see Fig. 2.4B and associated text). The bacterial origins of mitochondria are apparent in their many common features (Fig. 19.2). The closest extant relatives of the bacterium that gave rise to mitochondria are *Rickettsia,* aerobic α-proteobacteria with a genome of 1.1 megabase (Mb) pairs. These intracellular pathogens cause typhus and Rocky Mountain spotted fever. It now appears likely that the actual progenitor bacterium had the genes required for both aerobic and anaerobic metabolism.

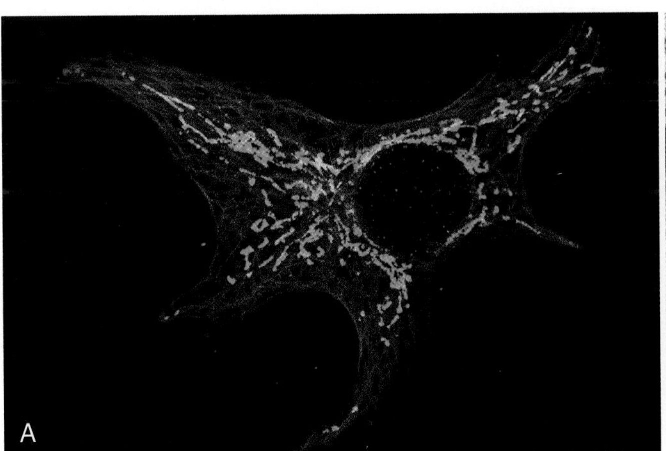

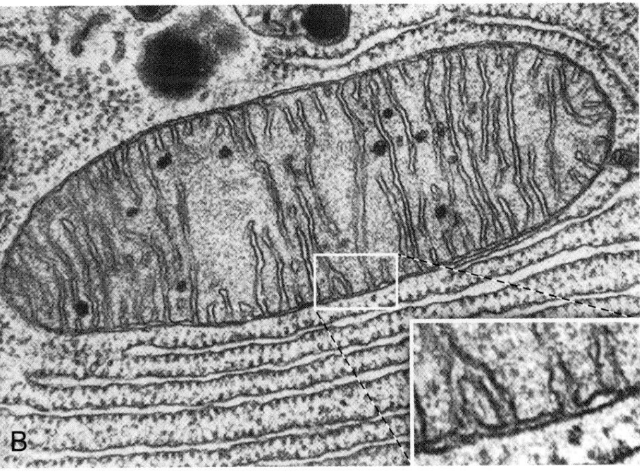

FIGURE 19.1 CELLULAR DISTRIBUTION AND STRUCTURE OF MITOCHONDRIA. A, Fluorescence light micrograph of a Cos-7 tissue culture cell with mitochondria labeled with *green* fluorescent antibody to the β-subunit of the F1-ATPase (adenosine triphosphatase) and microtubules labeled *red* with an antibody. **B,** Electron micrograph of a thin section of a mitochondrion. (**A,** Courtesy Michael Yaffee, University of California–San Diego. **B,** Courtesy Don Fawcett, Harvard Medical School, Boston, MA.)

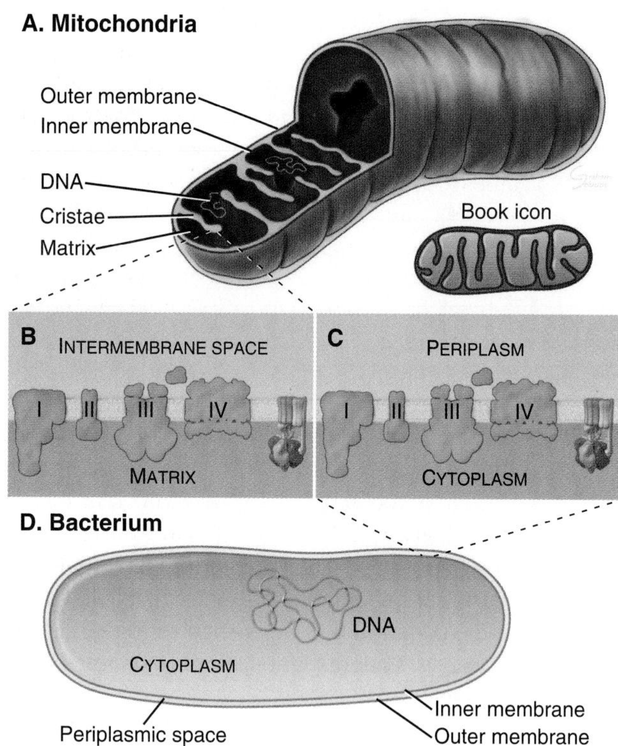

FIGURE 19.2 The compartments of a mitochondrion **(A–B)** compared with a bacterium **(C–D).** Respiratory chain complexes I to IV are labeled with roman numerals.

By the time of the last eukaryotic common ancestor (LECA), most of the bacterial genes were lost or moved to the nucleus, but all known mitochondria retain some bacterial genes. A few eukaryotes that branched from the last eukaryotic common ancestor, such as *Entamoeba,* subsequently lost the organelle, leaving behind a few mitochondrial genes in the nucleus.

Chromosomes of contemporary mitochondria vary in size from 366,924 base pairs (bp) in the plant *Arabidopsis* to only 5966 bp in *Plasmodium.* These small, usually circular genomes encode RNAs and proteins that are essential for mitochondrial function, including some subunits of proteins responsible for ATP synthesis. The highly pared-down human mitochondrial genome with 16,569 bp encodes only 13 mitochondrial membrane proteins, two ribosomal RNAs, and just enough transfer RNAs (tRNAs) (22) to translate these genes. The number of proteins encoded by other mitochondrial genomes ranges from just three in *Plasmodium* to 97 in a protozoan. Nuclear genes encode more than 1000 other mitochondria proteins, including those required to assemble ribosomes and synthesize proteins in the matrix. All mitochondrial proteins encoded by nuclear genes are synthesized in the cytoplasm and imported into mitochondria (see Figs. 18.3 and 18.4).

Structure of Mitochondria

Mitochondria consist of two membrane-bounded compartments, one inside the other (Fig. 19.2). The **outer membrane** surrounds the **intermembrane space.** The **inner membrane** surrounds the **matrix.** Each membrane and compartment has a distinct protein composition and functions.

Porins in the outer membrane provide nonspecific channels for passage of molecules of less than 5000 Da, including most metabolites required for ATP synthesis. Some proteins in the intermembrane space participate in ATP synthesis but, when released into the cytoplasm, trigger programmed cell death (see Fig. 46.16). The matrix is the site of fatty acid oxidation, the citric acid cycle, and mitochondrial protein synthesis.

The highly impermeable inner membrane has two domains. The boundary domain next to the outer membrane is specialized for protein import at contacts with the outer membranes (see Fig. 18.4). The rest of the inner membrane forms folds called *cristae* that are specialized for converting energy provided by breakdown of nutrients in the matrix into ATP. Cristae may be tubular or flattened sacs and vary in number and shape, depending on the species, tissue, and metabolic state. Four complexes (I to IV) of integral membrane proteins use the transport of electrons to create a gradient of protons across the inner membrane (Fig. 19.2B). The F-type rotary ATP synthase (see Fig. 14.5) uses this proton gradient to synthesize ATP. A complex of five transmembrane proteins and two soluble proteins stabilizes the junction of cristae with the inner membrane. The complex is called MICOS for mitochondrial contact site and cristae organizing system, because loss of these proteins results in disorganized cristae. MICOS links the inner and outer membranes and separates the transmembrane proteins of the inner membrane into two domains.

Biogenesis of Mitochondria

Mitochondria grow by importing most of their proteins from the cytoplasm and by internal synthesis of some proteins and replication of their own genome (Fig. 19.3). Targeting and sorting signals built into the mitochondrial proteins that are synthesized in the cytoplasm direct them to their destinations (see Fig. 18.4).

Similar to cells, mitochondria divide, but unlike most cells, they also fuse with other mitochondria. These fusion and division reactions were first observed nearly 100 years ago. Now it is appreciated that a balance between ongoing fusion and division determines the number of mitochondria within a cell. Both fusion and division depend on proteins with guanosine triphosphatase (GTPase) domains related to dynamin (see Fig. 22.8). In fact, eukaryotes might have acquired their dynamin genes from the bacterium that became the mitochondrion.

Mitochondrial division starts when a tubule of endoplasmic reticulum (ER) membrane encircles a mitochondrion to mark the site for division. Then a dynamin-related

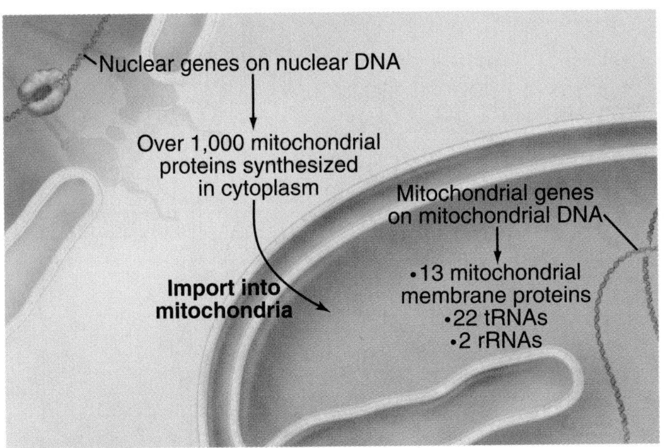

FIGURE 19.3 BIOGENESIS OF HUMAN MITOCHONDRIA. The relative contributions of nuclear and mitochondrial genes to the protein composition. rRNA, ribosomal RNA; tRNA, transfer RNA.

GTPase forms a spiral around the mitochondrion and cooperates with actin filaments to pinch mitochondria in two. During apoptosis (see Chapter 46), this GTPase also participates in the fragmentation of mitochondria.

Mitochondrial fusion involves two GTPases, one anchored in the outer membrane and the other in the inner membrane, both linked by an adapter protein in the intermembrane space. Fusion of the outer membranes requires a proton gradient across the inner membrane, whereas fusion of the inner membranes depends on the electrical potential across the inner membrane. Loss-of-function mutations in fusion proteins lead to cells with numerous small mitochondria, some lacking a mitochondrial DNA (mtDNA) molecule. Human mutations in the genes for fusion proteins result in defects in the myelin sheath that insulates axons (one form of Charcot-Marie-Tooth disease) and the atrophy of the optic nerve. Mitochondrial fusion proteins are also required for apoptosis.

Synthesis of ATP by Oxidative Phosphorylation

Mitochondria use energy extracted from the chemical bonds of nutrients to generate a proton gradient across the inner membrane. This proton gradient drives the F-type rotary ATP synthase to produce ATP from adenosine diphosphate (ADP) and inorganic phosphate. Enzymes in the inner membrane and matrix cooperate with pumps, carriers, and electron transport proteins in the inner membrane to move electrons, protons, and other energetic intermediates across the impermeable inner membrane. This is a classic chemiosmotic process (see Fig. 17.1).

Mitochondria receive energy-yielding chemical intermediates from two ancient metabolic pathways, **glycolysis** and **fatty acid oxidation** (Fig. 19.4), that both evolved in the common ancestor of living things. Both pathways feed into the equally ancient **citric acid cycle** of energy-yielding reactions in the mitochondrial matrix:

- The glycolytic pathway in the cytoplasm converts the six-carbon sugar glucose into pyruvate, a three-carbon substrate for pyruvate dehydrogenase, a large, soluble, enzyme complex in the mitochondrial matrix. The products of pyruvate dehydrogenase (carbon dioxide, the reduced form of nicotinamide adenine dinucleotide **[NADH]**, and acetyl coenzyme A [CoA]) are released into the matrix. NADH is a high-energy electron carrier. **Acetyl-CoA** is a two-carbon metabolic intermediate that supplies the citric acid cycle with energy-rich bonds.
- Breakdown of lipids yields fatty acids linked to acetyl-CoA by a thioester bond. These intermediates are transported across the inner membrane of mitochondria, using carnitine in a shuttle system. In the matrix, acyl-carnitine is reconverted to acyl-CoA. Enzymes in the matrix degrade fatty acids two carbons at a time in a series of oxidative reactions that yield NADH, the reduced form of flavin adenine dinucleotide **($FADH_2$**, another energy-rich electron carrier associated with an integral membrane enzyme complex), and acetyl-CoA for the citric acid cycle.

Breakdown of acetyl-CoA during one turn of the citric acid cycle produces three molecules of NADH, one molecule of $FADH_2$, and two molecules of carbon dioxide. Energetic electrons donated by NADH and $FADH_2$ drive an **electron transport pathway** in the inner mitochondrial membrane that powers a chemiosmotic cycle to produce ATP (Fig. 19.5). Electrons use two routes to pass through three protein complexes in the inner mitochondrial membrane. Starting with NADH, electrons pass through complex I to complex III to complex IV. Association of these three complexes in a "super complex" may facilitate electron transfer. Electrons from $FADH_2$ pass through complex II to complex III to complex IV. Along both routes, energy is used to transfer multiple protons (corresponding to at least 10 electrons per NADH oxidized) across the inner mitochondrial membrane from the matrix to the inner membrane space. The resulting electrochemical gradient of protons drives ATP synthesis (see Fig. 14.5).

This process is called **oxidative phosphorylation,** because molecular oxygen is the sink for energy-bearing electrons at the end of the pathway, and because the reactions add phosphate to ADP. Eukaryotes that live in environments with little or no oxygen use other acceptors for these electrons and produce nitrite, nitric oxide, or other reduced products rather than water. Oxidative phosphorylation is understood in remarkable detail, thanks to atomic structures of F-type ATP synthase and each electron transfer complex. Nuclear genes encode most of the protein subunits of these complexes, but mitochondrial genes encode a few key subunits.

Bacteria and mitochondria share homologous proteins for the key steps in oxidative phosphorylation (Fig. 19.2), although the machinery in mitochondria is usually

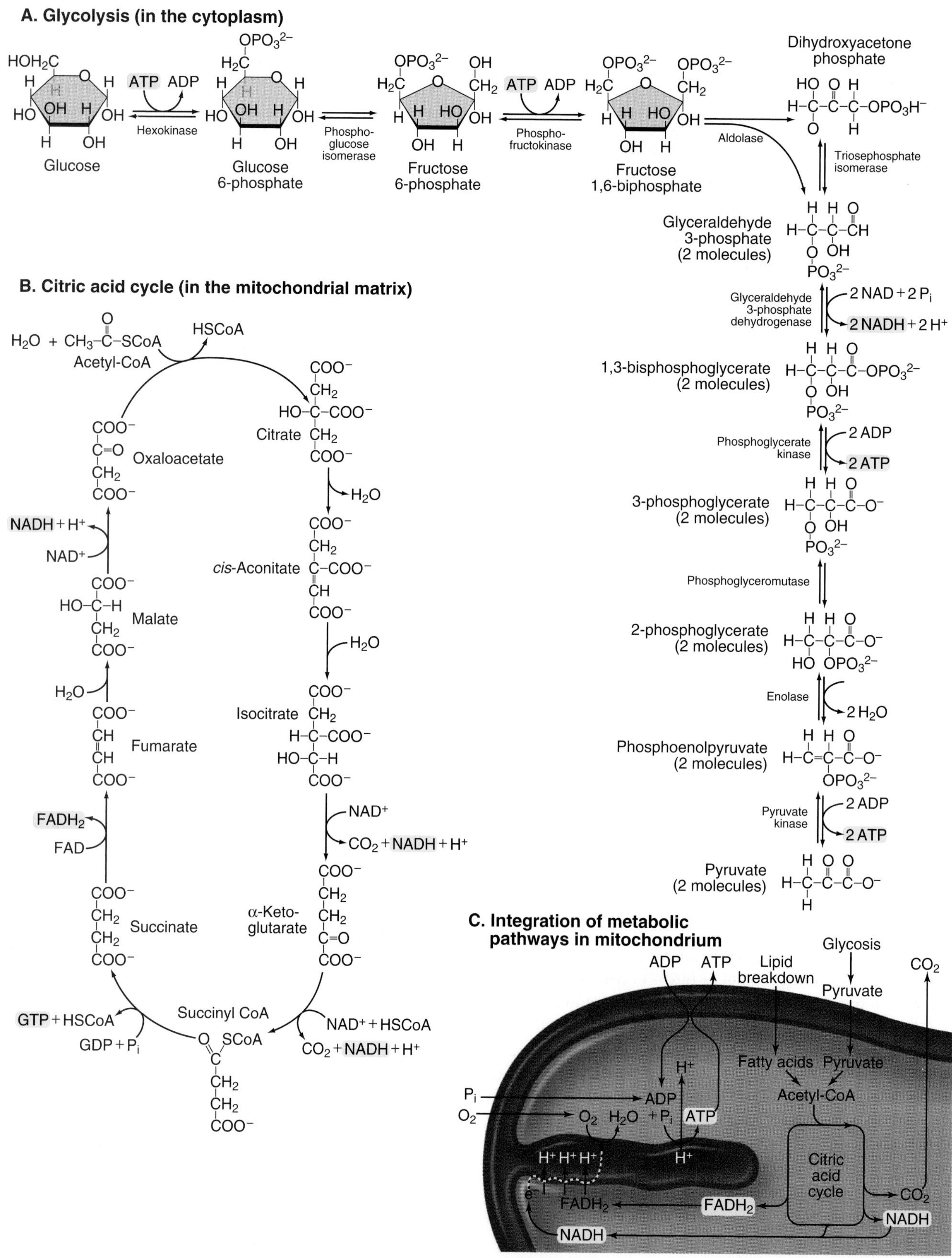

FIGURE 19.4 METABOLIC PATHWAYS SUPPLYING ENERGY FOR OXIDATIVE PHOSPHORYLATION. A, Glycolysis. ADP, adenosine diphosphate; ATP, adenosine triphosphate; NAD, nicotinamide adenine dinucleotide; NADH, reduced form of nicotinamide adenine dinucleotide. **B,** Citric acid cycle. Production of acetyl-coenzyme A (CoA) by the glycolytic pathway in cytoplasm and fatty acid oxidation in the mitochondrial matrix drive the citric acid cycle in the mitochondrial matrix. This energy-yielding cycle is also called the Krebs cycle after the biochemist H. Krebs. NADH and FADH$_2$ (reduced form of flavin adenine dinucleotide) produced by these pathways supply high-energy electrons to the electron transport chain. GTP, guanosine triphosphate; HSCoA, reduced coenzyme A. **C,** Overview of metabolic pathways. Note energy-rich metabolites *(yellow)*.

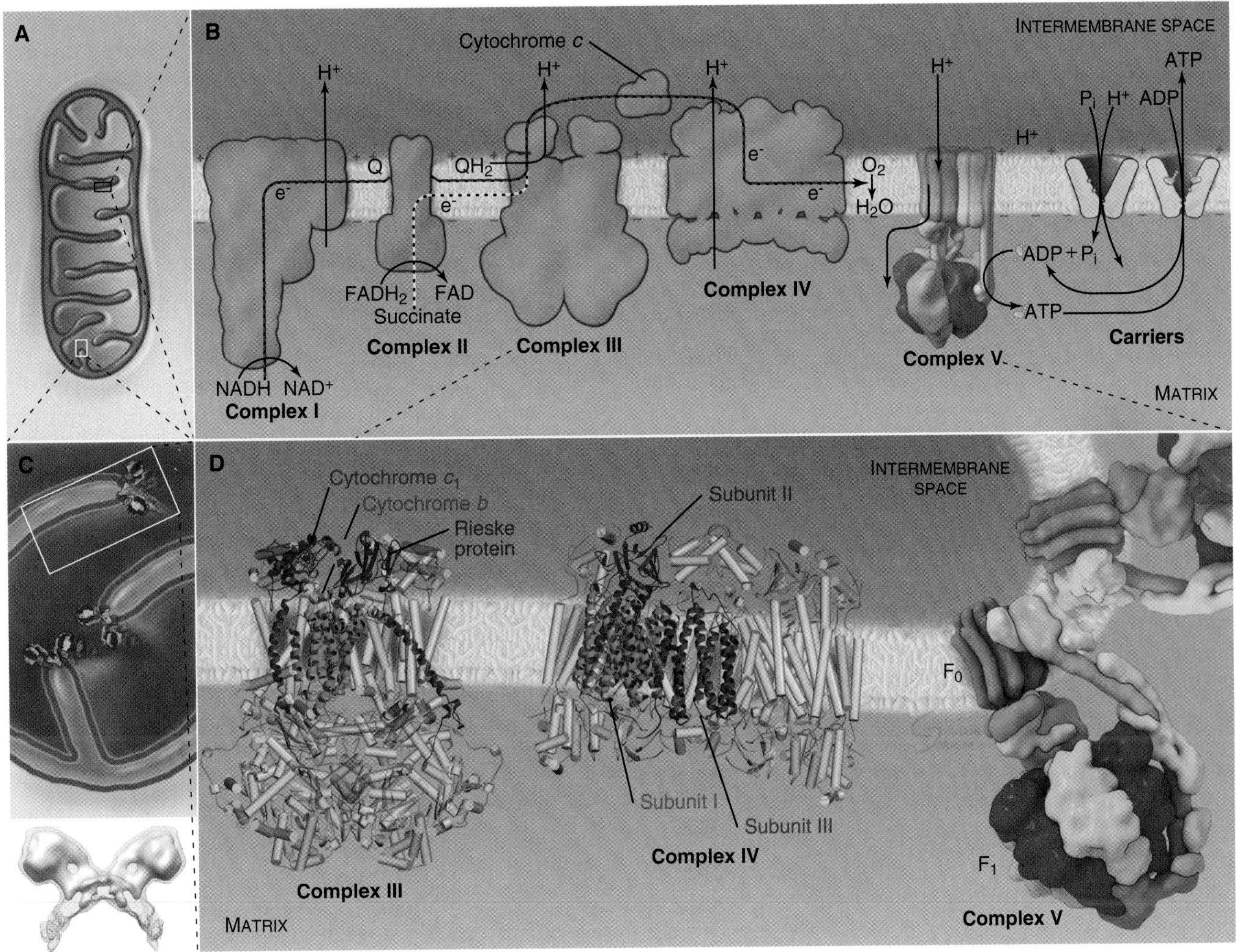

FIGURE 19.5 CHEMIOSMOTIC CYCLE OF THE RESPIRATORY ELECTRON TRANSPORT CHAIN AND ADENOSINE TRIPHOSPHATE (ATP) SYNTHASE. A, A mitochondrion for orientation. **B,** The electron transport system of the inner mitochondrial membrane. Note the pathway of electrons through the four complexes (*red* and *yellow arrows*) and the sites of proton translocation between the matrix to the intermembrane space *(black arrows)*. The stoichiometry is not specified, but at the last step, four electrons are required to reduce oxygen to water. The F-type ATP synthase uses the electrochemical proton gradient produced by the electron transport reactions to drive ATP synthesis. FAD, flavin adenine dinucleotide. **C,** Arrangement of F-type rotary ATP synthases and the electron transport chain in the inner membrane. **D,** Some atomic structures of the electron transport chain components. In the cytochrome bc_1 complex III, the 3 of 11 mitochondrial subunits used by bacteria are shown as ribbon models. The supporting subunits found in mitochondria are shown as cylinders. The four subunits of complex IV encoded by the mitochondrial genome are shown as ribbon models. They form the functional core of the complex, which is supported by additional subunits shown as cylinders. See Fig. 14.4 for further details of ATP synthase (complex V). (**D,** Images of complex III and complex IV courtesy of M. Saraste, European Molecular Biology Laboratory, Heidelberg, Germany. For reference, see Zhang Z, Huang L, Schulmeister VM, et al. Electron transfer by domain movement in cytochrome bc_1. *Nature*. 1998;392:677–684; and Yoshikawa S, Shinzawa-Itoh K, Nakashima R, et al. Redox-coupled crystal structural changes in bovine heart cytochrome *c* oxidase. *Science*. 1998;280:1723–1729. Also see Protein Data Bank [PDB; www.rcsb.org] file 2OCC.)

more complex. Plasma membranes of bacteria and inner membranes of mitochondria have equivalent components, and the bacterial cytoplasm corresponds to the mitochondrial matrix (Fig. 19.2). Thus, bacteria are useful model systems with which to study the common mechanisms.

Energy enters this pathway in the form of electrons that are produced when NADH is oxidized to the oxidized form of nicotinamide adenine dinucleotide (NAD^+), releasing one H^+ and two electrons (Fig. 19.5B). If the proton and electrons were to combine immediately with oxygen, their energy would be lost as heat. Instead, these high-energy electrons are separated from the protons and then passed along the electron transport pathway before finally rejoining molecular oxygen to form water. Along the pathway, electrons associate transiently with a series of oxidation-reduction acceptors, generally metal ions associated with organic cofactors, such as hemes in cytochromes and iron-sulfur centers (2Fe2S) and copper centers in complex IV. Electrons move along the transport pathway at rates of up to $1000\ s^{-1}$. To travel at this rate through a transmembrane protein complex spanning a 35-nm lipid bilayer, at least three reduction–oxidation (redox) cofactors are required in each complex, because the efficiency of quantum mechanical tunneling of electrons between redox cofactors falls off rapidly with distance. Two cofactors, even with optimal orientation, would be too slow.

Electrons give up energy as they move step by step along the transport pathway. In three complexes along the pathway, this energy is used to pump protons from the matrix to the inner membrane space. This establishes an electrochemical proton gradient across the inner mitochondrial membrane that is used by the F-type rotary **ATP synthase** to drive ATP production. Direction is provided to the movements of electrons by progressive increases in the electron affinity of the acceptors. The final acceptor, oxygen (at the end of the pathway), has the highest affinity.

The first component of the electron transport pathway, **complex I** (or NADH:ubiquinone oxidoreductase), handles electrons obtained from NADH. Vertebrate mitochondrial complex I with 46 different protein subunits is much more complex than bacterial complex I with 14 subunits. NADH donates two electrons to flavin mononucleotide associated with protein subunits located on the matrix side of the inner membrane. A crystal structure of the cytoplasmic domain of the bacterial complex shows the path for the electrons from flavin mononucleotide through seven iron sulfur clusters to quinone in the lipid bilayer. For each molecule of NADH oxidized, the transmembrane domains of complex I transfer four protons from the matrix into the inner membrane space.

The second component of the electron transport pathway is **complex II** or succinate:ubiquinone reductase, a transmembrane enzyme that makes up part of the citric acid cycle. Complex II couples oxidation of succinate (a four-carbon intermediate in the citric acid cycle) to fumarate with reduction of flavin adenine dinucleotide (FAD) to $FADH_2$. Complex II does not pump protons but transfers electrons from $FADH_2$ to ubiquinone. Reduced ubiquinone carries these electrons to complex III.

The third component of the electron transport pathway is **complex III,** also called **cytochrome bc_1.** This well-characterized, transmembrane protein complex consists of 11 different subunits. The homologous bacterial complex has only three of these subunits, the ones that participate in energy transduction in mitochondria. Eight other subunits surround this core. Complex III couples the oxidation and reduction of ubiquinone to the transfer of protons from the matrix across the inner mitochondrial membrane. Energy is supplied by electrons from both complex I and complex II that move through the cytochrome b subunit to a subunit with a 2Fe2S redox center. This subunit then rotates into position to transfer the electron to cytochrome c_1, another subunit of the complex. Cytochrome c_1 then transfers the electron to the water-soluble protein cytochrome c in the intermembrane space (or periplasm of bacteria).

Cytochrome oxidase, complex IV, takes electrons from four cytochrome c molecules to reduce molecular oxygen to two waters and to pump four protons out of the matrix. Mitochondrial genes encode the three subunits that form the core of this enzyme, carry out electron transfer, and translocate protons. Nuclear genes encode the surrounding 10 subunits.

The electrochemical proton gradient produced by the electron transport chain provides energy to synthesize ATP. Chapter 14 explained how the F-type rotary ATP synthase (complex V) can either use ATP hydrolysis to pump protons or use the transit of protons down an electrochemical gradient to synthesize ATP (see Figs. 14.4 and 14.5). The proton gradient across the inner mitochondrial membrane drives rotation of the γ-subunit. The rotating γ-subunit physically changes the conformations of the α- and β-subunits, bringing together ADP and inorganic phosphate to make ATP. An antiporter in the inner membrane exchanges cytoplasmic ADP for ATP synthesized in the matrix (see Fig. 15.4A).

Cryoelectron tomography revealed that dimers of F-type rotary ATP synthases are located in rows along the crests of the cristae, where they are responsible for the sharp bend in the inner membrane (Fig. 19.5C). The other components of the electron transfer machinery occupy the flat sides of the cristae. This arrangement of proteins and the small volume inside cristae facilitates the movements of protons and cytochrome c between the components.

Mitochondria and Disease

As expected from the central role of mitochondria in energy metabolism, mitochondrial dysfunction contributes to a remarkable diversity of human diseases (Fig. 19.6), including seizures, strokes, optic atrophy, neuropathy, myopathy, cardiomyopathy, hearing loss, and type 2 diabetes mellitus. These disorders arise from mutations in genes for mitochondrial proteins encoded by both mtDNA and nuclear DNA. Many of the known disease-causing mutations are in genes for mitochondrial tRNAs.

The existence of approximately 1000 copies of mtDNA per vertebrate cell influences the impact of deleterious mutations. A mutation in one copy would be of no consequence, but segregation of mtDNAs may lead to cells in which mutant mtDNAs predominate, yielding defective proteins. For example, a recurring point mutation in a subunit of complex I causes some patients to develop sudden onset of blindness in middle age owing to the death of neurons in the optic nerve. Patients with the same mutation in a larger fraction of mtDNA

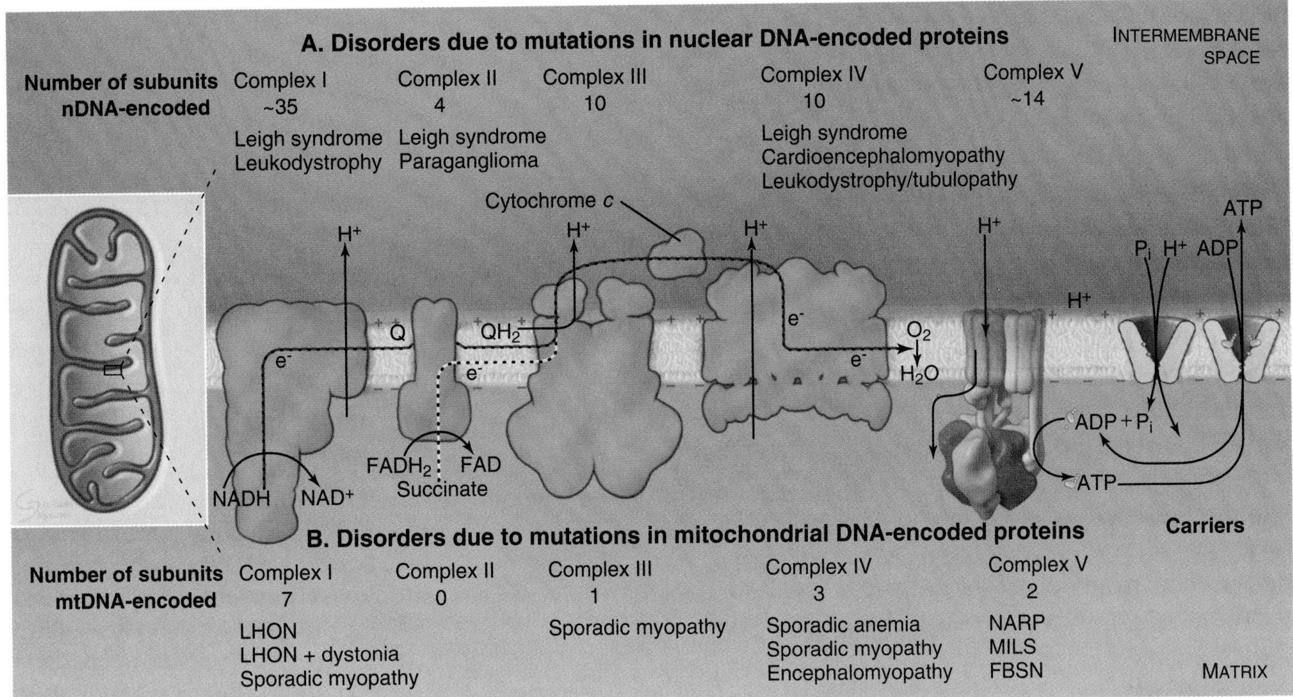

FIGURE 19.6 Mutations in both mitochondrial and nuclear genes for mitochondrial proteins cause a variety of diseases by compromising the function of particular mitochondrial subsystems. FBSN, familial bilateral striatal necrosis; LHON, Leber hereditary optic neuropathy; MILS, maternally inherited Leigh syndrome; mtDNA, mitochondrial DNA; NARP, neurogenic muscle weakness, ataxia, retinitis pigmentosa; nDNA, nuclear DNA; P_i, inorganic phosphate. (Modified from Schon EA. Mitochondrial genetics and disease. *Trends Biochem Sci*. 2000;25:555–560.)

molecules suffer from muscle weakness and intellectual disability as children. Mutations in the genes for subunits of ATP synthase cause muscle weakness and degeneration of the retina. Slow accumulation of mutations in mtDNA may contribute to some symptoms of aging.

Mutations in mitochondrial DNA are passed from a mother to her children, as sperm mitochondria do not contribute to the embryo. Methods are being developed to combine genetically normal nuclei from affected mothers with enucleated cytoplasm from healthy donors to eliminate these mutations in infants conceived by in vitro fertilization.

Mutations in nuclear genes for mitochondrial proteins cause similar diseases (Fig. 19.6A). A mutation in one subunit of the protein import machinery (see Fig. 18.5), Tim8, causes a type of deafness.

Chloroplasts

Structure and Evolution of Photosynthesis Systems

Photosynthetic bacteria and chloroplasts of algae and plants (Fig. 19.7) use **chlorophyll** to capture the remarkable amount of energy carried by single photons to boost electrons to an excited state. These high-energy electrons drive a chemiosmotic cycle to make nicotinamide adenine dinucleotide phosphate (NADPH) and ATP. Photosynthetic organisms use ATP and the reducing power of NADPH to synthesize three-carbon sugar phosphates from carbon dioxide. Glycolytic reactions (Fig. 19.4) running backward use this three-carbon sugar phosphate to make six-carbon sugars and more complex carbohydrates for use as metabolic energy sources and structural components. Some bacteria and Archaea, such as *Halobacterium halobium,* use a completely different light-driven pump lacking chlorophyll to generate a proton gradient to synthesize ATP (see Fig. 14.3). In that case, retinol associated with bacteriorhodopsin absorbs light to drive proton transport.

Photosynthesis originated approximately 3.5 billion years ago in a bacterium, most likely a gram-negative purple bacterium (see Fig. 2.4). These bacteria evolved components to assemble a transmembrane complex of proteins, pigments, and oxidation/reduction cofactors called a **reaction center** (Fig. 19.8). Reaction centers absorb light and initiate an electron transport pathway that pumps protons out of the cell. Such photosystems turn sunlight into electrical and chemical energy with 40% efficiency, better than any human-made photovoltaic cell. Given their alarming complexity and physical perfection, it is remarkable that photosystems emerged only a few hundred million years after the origin of life itself.

Broadly speaking, photosynthetic reaction centers of contemporary organisms can be divided into two different groups (Fig. 19.8). The reaction centers of purple bacteria and green filamentous bacteria use the pigment pheophytin and a quinone as the electron acceptor, similar to **photosystem II** of cyanobacteria

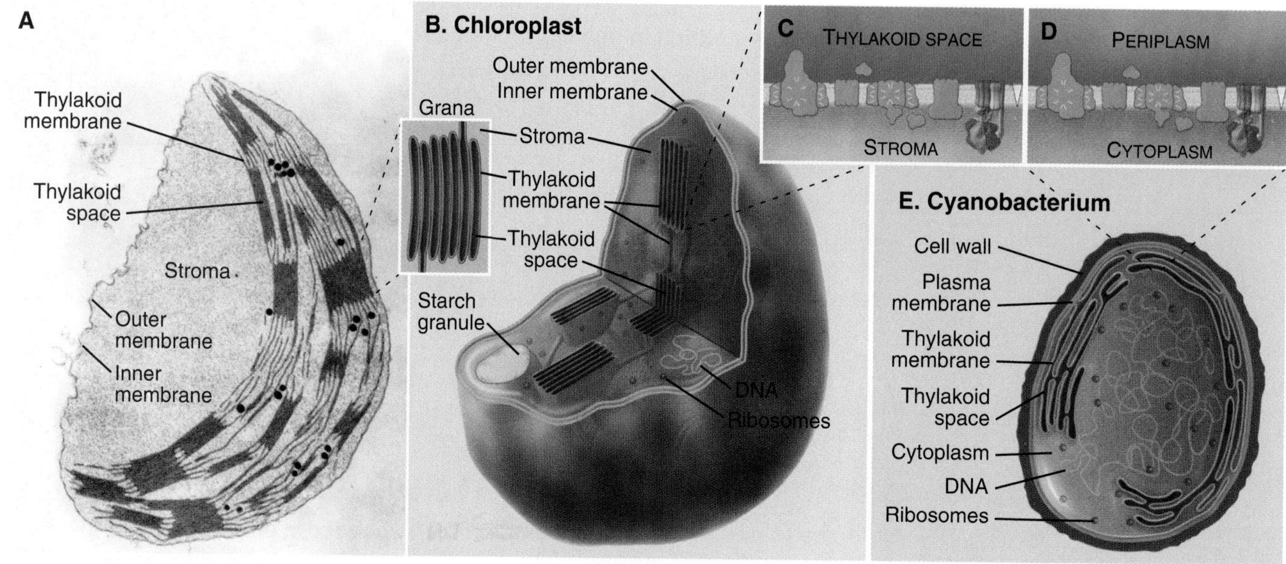

FIGURE 19.7 MORPHOLOGY OF CHLOROPLASTS AND CYANOBACTERIA. A, Electron micrograph of a thin section of a spinach chloroplast. **B,** Chloroplast. **C–D,** Comparison of the machinery in the photosynthetic membranes of chloroplasts and cyanobacteria. **E,** Drawing of a cyanobacterium illustrating the internal folds of the plasma membrane to form photosynthetic thylakoids. (**A,** Courtesy K. Miller, Brown University, Providence, RI.)

and chloroplasts. The reaction centers of green sulfur bacteria and heliobacteria have iron-sulfur centers as electron acceptors, similar to **photosystem I** of cyanobacteria and chloroplasts.

Cyanobacteria are unique among bacteria in that they have both types of photosystems as well as a manganese-containing enzyme that splits water, releasing from two water molecules four electrons, four protons, and oxygen (Fig. 19.7E). Coupling this enzyme to photosynthesis was a pivotal event in the history of the earth, as this reaction is the source of most of the oxygen in the earth's atmosphere.

Chloroplasts of eukaryotic cells arose from a symbiotic cyanobacterium (see Fig. 2.7). Much evidence indicates that this event occurred just once, giving all chloroplasts a common origin. Thereafter chloroplasts moved by lateral transfer to various organisms that diverged prior to the acquisition of chloroplasts, for example, from a green alga to *Euglena.*

Chloroplasts have retained up to 250 original bacterial genes on circular genomes. As in the case of mitochondria, many bacterial genes were lost or moved to the nucleus of host eukaryotes. Chloroplast genomes encode subunits of many proteins responsible for photosynthesis and chloroplast division, ribosomal RNAs and proteins, and a complete set of tRNAs. More than 2000 chloroplast proteins encoded by nuclear genes are synthesized in the cytoplasm and transported posttranslationally into chloroplasts (see Fig. 18.6).

The organization of cyanobacterial membranes explains the architecture of chloroplasts (Fig. 19.7C–E). In cyanobacteria, light-absorbing pigments, as well as protein complexes involved with electron transport and ATP synthesis, are concentrated in invaginations of the plasma membrane. The F1 domain of the F-type rotary ATP synthase faces the cytoplasm, and the lumen of this membrane system is periplasmic. This internal membrane system remains in chloroplasts but is separated from the inner membrane (the former plasma membrane). These **thylakoid membranes** contain photosynthetic hardware and enclose the thylakoid membrane space. Like the bacterial plasma membrane, the chloroplast **"inner membrane"** is a permeability barrier, containing carriers for metabolites. The inner membrane surrounds the **stroma,** the cytoplasm of the original symbiotic bacterium, a protein-rich compartment devoted to synthesis of three-carbon sugar phosphates, chloroplast proteins, and all plant fatty acids. The stroma also houses the genomes and stores starch. The **outer membrane,** like the comparable bacterial and mitochondrial membranes, has large pore channels that allow free passage of metabolites.

Plastids

Chloroplasts are just one manifestation of a class of organelles called plastids. Depending on the developmental stage and tissue type plastids have a range of compositions and functions made possible by selective synthesis and import of proteins. Chloroplasts are specialized for photosynthesis in green plant tissues but differ considerably in composition and physiology in developing and senescent plant tissues. Some plastids, such as the starch-storing amyloplasts in potatoes, lack the photosynthetic machinery.

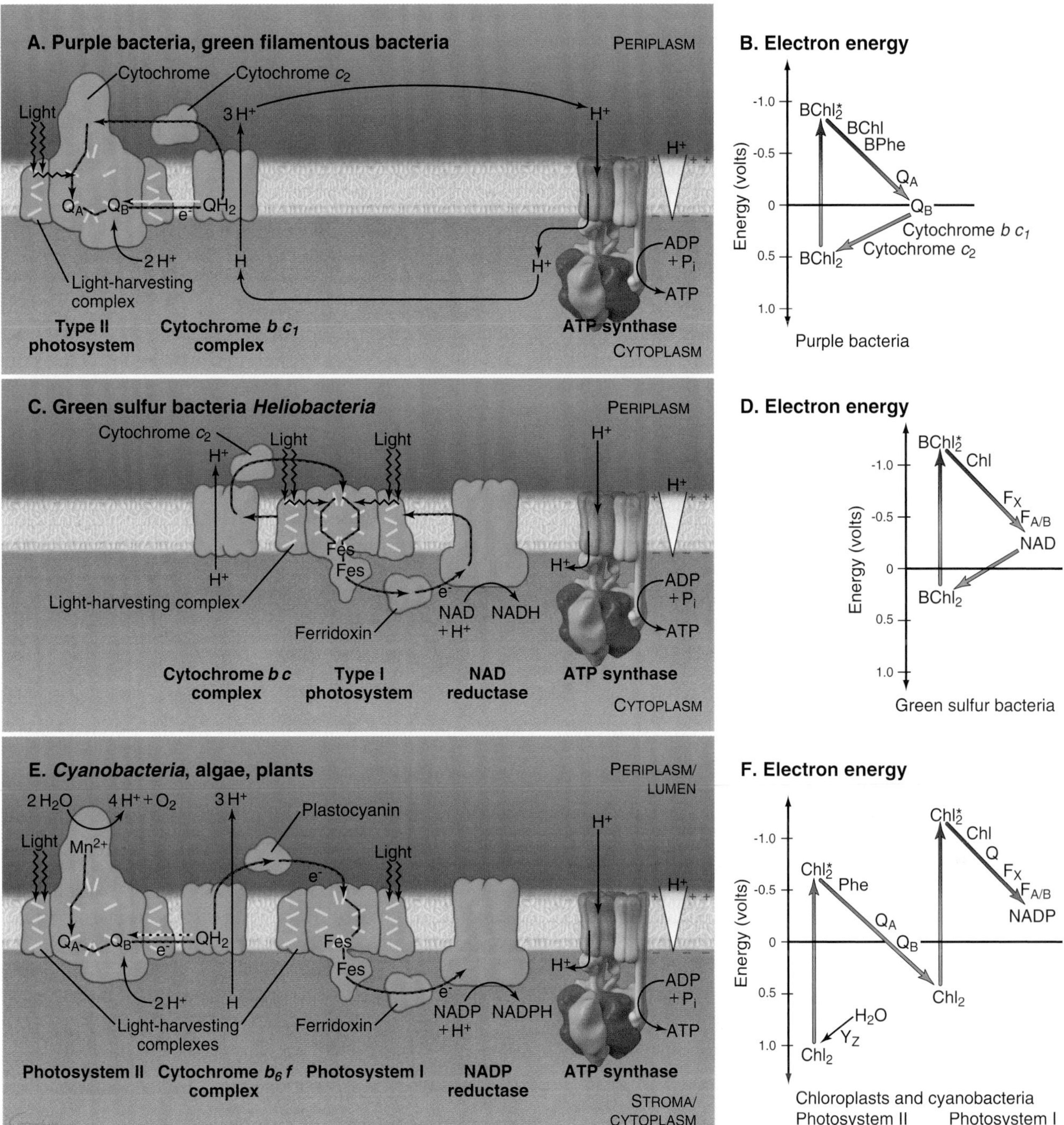

FIGURE 19.8 COMPARISON OF PHOTOSYNTHETIC COMPONENTS, ELECTRON TRANSPORT PATHWAYS, AND CHEMIOSMOTIC CYCLES TO MAKE ADENOSINE TRIPHOSPHATE. A–B, Type II photosystem only. **C–D,** Type I photosystem only. **E–F,** Both photosystem II and photosystem I. *Right diagrams,* The energy levels of electrons in the three types of photosynthetic organisms, showing excitation of an electron by an absorbed photon *(vertical arrows),* electron transfer pathways through each reaction center *(arrows sloping right),* and electron transfer steps outside the reaction centers *(arrows sloping left).* (**A, C,** and **E,** For reference, see Kramer DM, Schoepp B, Liebl U, et al. Cyclic electron transfer in *Heliobacillus mobilis. Biochemistry*. 1997;36:4203–4211. **B, D,** and **F,** For reference, see Allen JP, Williams JC. Photosynthetic reaction centers. *FEBS Lett*. 1998;438:5–9.)

Light and Dark Reactions

Photosynthetic mechanisms capture energy from photons to drive two types of reactions:

- **Light reactions** depend on continuous absorption of photons. These reactions occur in or on the surface of thylakoid membranes. They include the generation of high-energy electrons, electron transport to make NADPH, creation of a proton gradient across the thylakoid membrane for the chemiosmotic synthesis of ATP, and generation of oxygen.

- **Dark reactions** convert carbon dioxide into three-carbon sugar phosphates. These reactions continue for some time in the dark. However, they depend on ATP and NADPH produced by light reactions, so they eventually stop when ATP and NADPH are exhausted in the dark. These reactions account for most of the carbon dioxide converted to carbohydrates on earth. (Alternatively specialized prokaryotes drive carbon fixation by oxidation of hydrogen sulfide and other inorganic compounds.)

All photosynthetic systems use similar mechanisms to capture energy from photons (Fig. 19.8). Pigments associated with transmembrane proteins in photosynthetic reaction centers absorb photons and use the energy to boost electrons to a high-energy **excited state.** Subsequent electron transfer reactions partition this energy in several steps to generate a proton gradient across the membrane. Generation of this proton electrochemical gradient and chemiosmotic production of ATP are similar to oxidative phosphorylation (Fig. 19.5).

Specific photosynthetic systems differ in the complexity of the hardware, the source of electrons, and the products (Fig. 19.8). Most photosynthetic bacteria use either a type I photosystem or a type II photosystem to create a proton gradient to synthesize ATP. Cyanobacteria and green plants use both types of reaction centers in series to raise electrons to an energy sufficient to make NADPH in addition to ATP. These advanced systems also use water as the electron donor and produce molecular oxygen as a by-product.

Energy Capture and Transduction by Photosystem II

The reaction center from the purple bacterium *Rhodopseudomonas viridis* (Fig. 19.9A) serves as a model for the more complex photosystem II of cyanobacteria and chloroplasts. This bacterial reaction center consists of just four subunits. A cytochrome subunit on the periplasmic side of the membrane donates electrons. Two core subunits form a rigid transmembrane framework to bind 10 cofactors in orientations that favor transfer of high-energy electrons from two "special" **bacteriochlorophylls** through **chlorophyll b** and **bacteriopheophytin b.**

Photosynthesis begins with absorption of a photon by the special pair bacteriochlorophylls. Photons in the visible part of the spectrum are quite energetic, 40 to 80 kcal mol^{-1}, enough to make several ATPs. The purple bacterium reaction center absorbs relatively low-energy, 870-nm red light. The energy elevates an electron in the special pair bacteriochlorophylls to an excited state (Fig. 19.8B). This excited state can decay rapidly (10^9 s^{-1}), causing the energy to dissipate as heat or emission of a less-energetic photon by fluorescence or phosphorescence. However, reaction centers are optimized to transfer excited-state electrons rapidly and efficiently from the special pair bacteriochlorophylls to bacteriopheophytin (3×10^{-12} s) and then to tightly bound quinone A (200×10^{-12} s). Transfer is by **quantum mechanical tunneling** right through the protein molecule. Because the tunneling rate falls off sharply with distance, four redox centers must be spaced close together to allow an energetic electron to transfer across the lipid bilayer faster than spontaneous decay of the excited state.

On the cytoplasmic side of the membrane, two electrons transfer from quinone A to quinone B (100×10^{-9} s), where they combine with two protons to make a high-energy **reduced quinone,** QH_2 (Fig. 19.8A). In purple bacteria, these cytoplasmic protons are taken up through water-filled channels in the reaction center, contributing to the proton gradient.

QH_2 has a low affinity for the reaction center and diffuses in the hydrophobic core of the bilayer to the next component in the pathway, the chloroplast equivalent of the mitochondrial cytochrome bc_1 complex III (Fig. 19.8A). As in mitochondria, passage of energetic electrons through this complex releases protons from QH_2 on the periplasmic side of the membrane, adding to the electrochemical gradient. The electron circuit is completed by transfer of low-energy electrons from complex bc_1 to a soluble periplasmic protein, cytochrome c_2. Electrons then move to the cytochrome subunit of the reaction center, which supplies special pair chlorophylls with electrons for the photosynthetic reaction cycle.

The net result of this cycle is the conversion of the energy of two photons into transport of three protons to the periplasm. A diagram of the energy levels of the various intermediates in the cycle (Fig. 19.8B) shows how energy is partitioned after an electron is excited by a photon and then moves, step by step, through protein-associated redox centers back to the ground state.

The proton electrochemical gradient established by photosynthetic electron transfer reactions is used to drive an F-type rotary ATP synthase (see Fig. 14.5) similar to those of nonphotosynthetic prokaryotes and mitochondria.

Light Harvesting

Reaction center chlorophylls absorb light, but both chloroplasts and bacteria increase the efficiency of light collection with proteins that absorb light and transfer the energy to a reaction center. Most of these **light-harvesting complexes** are small, transmembrane proteins that cluster around a reaction center, although some bacteria and algae also have soluble light-harvesting proteins. Transmembrane, light-harvesting proteins consist of a few α-helices associated with multiple chlorophyll and carotenoid pigments (Figs. 19.8A and C and 19.9B). Using multiple pigments broadens the range of wavelengths absorbed and increases the efficiency of photon capture. Leaves are green because chlorophylls

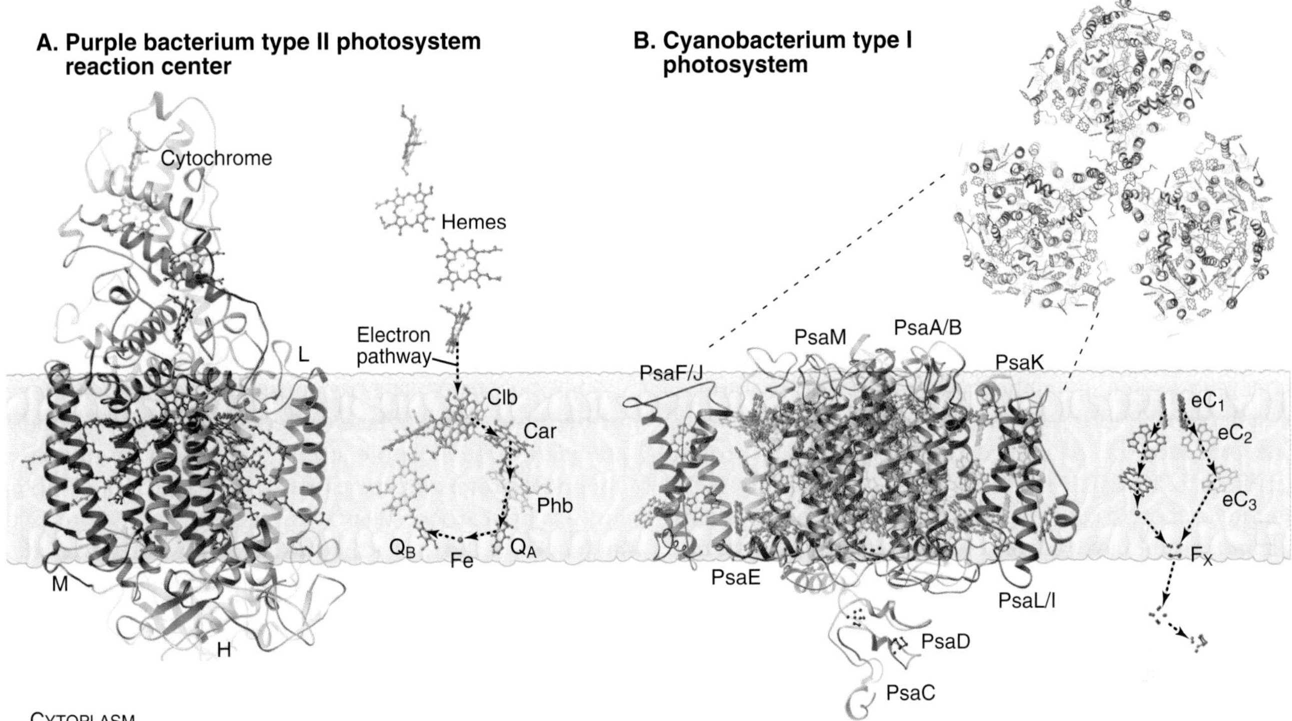

FIGURE 19.9 STRUCTURES OF PHOTOSYSTEM HARDWARE. A, Ribbon diagram of type II photosystem from the purple bacterium *Rhodopseudomonas viridis,* with ball-and-stick models of bacteriochlorophyll and other cofactors to the right in their natural orientations. Similar core subunits L and M each consists of five transmembrane helices. This pair of subunits binds four molecules of chlorophyll b (Clb), two molecules of bacteriopheophytin b (Phb), one nonheme iron (Fe), two quinones (Q_A, Q_B), and one carotenoid (Car) in a rigid framework. A cytochrome with four heme groups binds to the periplasmic side of the core subunits. Subunit H associates with the core subunits via one transmembrane helix and with their cytoplasmic surfaces. The atomic structure of this photosynthetic reaction center was the Nobel Prize work of J. Diesenhofer, R. Huber, and H. Michel. **B,** Ribbon diagram of photosystem I of *Synechococcus elongatus,* with ball-and-stick models of chlorophyll and other cofactors to the *right* in their natural orientations. This trimeric complex consists of three identical units, each composed of 11 polypeptide chains. Within each of these units, this 4-Å resolution structure includes 43 α-helices, 89 chlorophylls, 1 quinone, and 3 iron-sulfur centers, but other details (eg, amino acid side chains) are not resolved. The photosynthetic reaction center consists of the C-terminal halves of the two central subunits *(PsaA/PsaB, red-brown)* associated with six chlorophylls, one or two quinones, and a shared iron-sulfur cluster. Plastocyanin or cytochrome c_6 on the lumen side donates electrons to reduce the P700 special pair chlorophylls (eC_1) of the reaction center. Light energizes an electron, which passes successively through two other chlorophylls, a quinone, and the shared iron-sulfur cluster *(red),* F_X. The electron then transfers to the iron-sulfur clusters of the accessory subunit PsaC on the stromal side of the membrane. The surrounding eight subunits *(red, gray),* associated with approximately 80 chlorophylls, compose the core antenna system, forming a nearly continuous ring of α-helices around the reaction center. Absorption of light by additional light-harvesting complexes and these antenna subunits puts chloroplast electrons into an excited state. This energy passes from one pigment to the next until it eventually reaches the reaction center. (**A,** Copyright Diesenhofer & Michel, Nobel Foundation, 1988. For reference, see PDB file 1PRC and Diesenhofer J, Michel H. The photosynthetic reaction center from the purple bacterium *Rhodopseudomonas viridis. Science*. 1989;245:1463–1473. A 3.5-Å crystal structure of the photosystem II complex from the cyanobacterium *Thermosynechococcus elongatus,* including 19 subunits, is now available. Also see Ferreira KN, Iverson TM, Maghlaoui K, et al. Architecture of the photosynthetic oxygen-evolving center. *Science*. 2004;303:1831–1838. **B,** For reference, see PDB file 2PPS and Schubert W-D, Klukas O, Krauss N, et al. Photosystem I of *Synechococcus elongatus* at 4 Å resolution: comprehensive structure analysis. *J Mol Biol*. 1997;272:741–769.)

and carotenoids absorb purple and blue wavelengths (<530 nm) as well as red wavelengths (>620 nm), reflecting only yellow-green wavelengths in between.

Light absorbed by light-harvesting proteins boosts pigment electrons to an excited state. This energy (but not the electrons) moves without dissipation by **fluorescence resonance energy transfer** from one closely spaced pigment molecule to another and eventually to the special pair chlorophylls of a reaction center. This rapid (10^{-12} s), efficient process transfers energy captured over a wide area to a reaction center to initiate a cycle of electron transfer and energy transduction.

Energy Capture and Transduction by Photosystem I

The reaction centers of green sulfur bacteria and heliobacteria are similar to photosystem I of cyanobacteria and chloroplasts. Generation of a proton gradient by photosystem I has many parallels with photosystem II. Direct absorption of light or resonance energy transfer from surrounding light-harvesting complexes excites special-pair chlorophylls in photosystem I (Fig. 19.8C–D). Excited-state electrons move rapidly within the reaction center from these chlorophylls through two accessory chlorophylls to an iron-sulfur center. The pathway includes a quinone in cyanobacteria and chloroplasts. Electrons then move to the iron-sulfur center of a subunit on the cytoplasmic side of the membrane. The subsequent events in green sulfur bacteria and heliobacteria include electron transfer by the soluble protein ferredoxin to an NAD reductase, followed by transfer by a lipid intermediate to cytochrome bc complex, and then back to the reaction center via a cytochrome c.

Oxygen-Producing Synthesis of NADPH and ATP by Dual Photosystems

Chloroplasts and cyanobacteria combine photosystem II and photosystem I in the same membrane to form a system capable of accepting low-energy electrons from the oxidation of water and producing both a proton gradient to drive ATP synthesis and reducing equivalents in the form of NADPH (Fig. 19.8E–F). Both photosystems are more elaborate in dual systems than in single systems. Although plant photosystem II, with more than 25 protein subunits, is much more complicated than is the homologous reaction center of purple bacteria, the arrangement of transmembrane helices and chlorophyll cofactors in the core of the plant reaction center is similar to the simple reaction center of purple bacteria.

Photosynthesis involves a tortuous electron transfer pathway powered at two waystations by absorption of photons. This process begins when the special pair chlorophylls of photosystem II are excited by direct absorption of light or by resonance energy transfer from surrounding light-harvesting complexes (Fig. 19.8E–F). Electrons come from splitting two waters into molecular oxygen and four protons. Excited-state electrons tunnel through the redox cofactors and combine with protons from the stroma (or cytoplasm in bacteria) to reduce quinone QB to QH_2, a high-energy electron donor. QH_2 diffuses to complex b_{6f}, the chloroplast equivalent of the mitochondrial bc_1 complex. Passage of electrons through complex b_{6f} releases protons from QH_2 into the thylakoid lumen (or bacterial periplasm), contributing to the proton gradient across the membrane.

Complex b_{6f} donates electrons from QH_2 to photosystem I. Direct absorption of 680-nm light or resonance energy transfer from surrounding light-harvesting complexes boosts special pair chlorophyll electrons to a very high-energy, excited state (Fig. 19.8F). Excited-state electrons pass through chlorophyll and iron-sulfur centers of photosystem I to the iron-sulfur center of the redox protein, ferredoxin, on the cytoplasmic/stromal surface of the membrane. The enzyme nicotinamide adenine dinucleotide phosphate (NADP) reductase combines electrons from ferredoxin with a proton to form NADPH, the final product of this complex electron transfer pathway powered at two waystations by absorption of photons. Uptake of stromal protons during NADPH formation contributes to the transmembrane proton gradient for the synthesis of ATP. Antiporters in the inner membrane exchange ATP for ADP, as in mitochondria.

Synthesis of Carbohydrates

ATP and NADPH produced by light reactions drive the unfavorable conversion of carbon dioxide into sugars. This is the first step in the earth's annual production of approximately 10^{10} tons of carbohydrates by photosynthetic organisms. This process is very expensive, consuming three ATPs and two NADPHs for each carbon dioxide added to the five-carbon sugar ribulose 1,5-bisphosphate. The responsible enzyme, **ribulose phosphate carboxylase** (called RUBISCO), is the most abundant protein in the stroma and might be the most abundant protein on the earth. The products of combining the five-carbon sugar with carbon dioxide are two molecules of the three-carbon sugar 3-phosphoglycerate.

An antiporter in the inner chloroplast membrane exchanges 3-phosphoglycerate for inorganic phosphate, so 3-phosphoglycerate can join the glycolytic pathway in the cytoplasm (Fig. 19.4). Driven by this abundant supply of 3-phosphoglycerate, the glycolytic pathway runs backward to make six-carbon sugars, which are used to make disaccharides such as sucrose to nourish nonphotosynthetic parts of the plant, the glucose polymer **starch** to store carbohydrate, and **cellulose** for the extracellular matrix (see Figs. 3.25A and 32.13).

Peroxisomes

Peroxisomes are organelles bounded by a single membrane (Fig. 19.10), named for their content of enzymes that produce and degrade hydrogen peroxide, H_2O_2. Oxidases produce H_2O_2 and peroxidases such as catalase break it down. Peroxisomes also contain diverse enzymes for the metabolism of lipids and other metabolites, including the β-oxidation of fatty acids and oxidation of bile acids and cholesterol. All peroxisomal proteins are encoded by nuclear genes, translated on cytoplasmic ribosomes, and then subsequently incorporated into peroxisomes (see Fig. 18.8).

Peroxisomes form in two different ways: de novo synthesis by budding from the ER and growth and division of preexisting peroxisomes (see Fig. 18.8). Cells that lack preexisting peroxisomes can form peroxisomes without a template by differentiation and budding of ER membranes. Two key proteins known as peroxins, PEX3 and PEX16, are targeted to the ER, where they recruit other peroxins to form a specialized domain that pinches off to form a nascent peroxisome.

Defects in peroxisomal biogenesis cause a spectrum of lethal human diseases known as the **peroxisomal biogenesis disorders** (see Table 18.1). These diseases include Zellweger syndrome, neonatal adrenoleukodystrophy, infantile Refsum disease, and rhizomelic

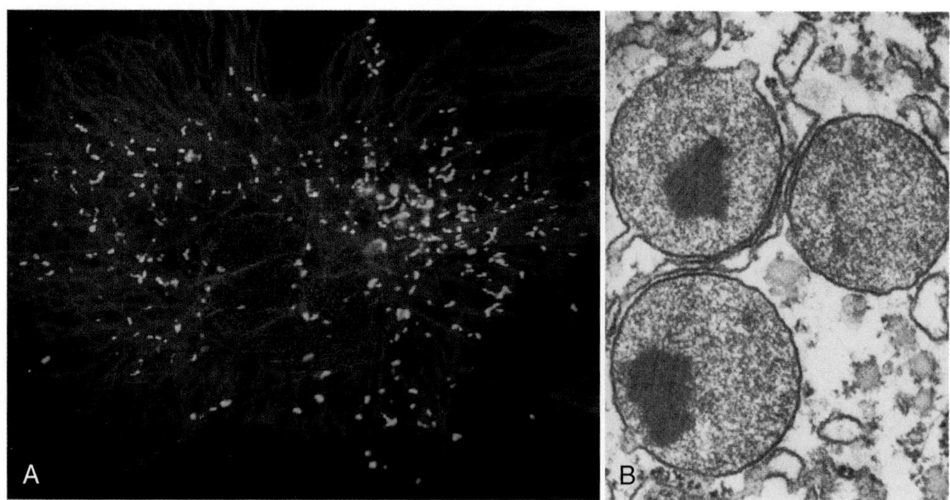

FIGURE 19.10 PEROXISOMES. A, Fluorescence micrographs of a CV1 cell expressing *green* fluorescent protein fused to PTS1, which labels peroxisomes *green.* Microtubules are stained *red* with labeled antibodies, and nuclear DNA is stained *blue* with propidium iodide. **B,** Electron micrograph of a thin section of a tissue culture cell showing three peroxisomes. Peroxisomes have a single bilayer membrane and a dense matrix, including a crystal (in some species) of the enzyme urate oxidase. (**A,** Courtesy S. Subramani, University of California–San Diego. For reference, see Wiemer EAC, Wenzel T, Deernick TJ, et al. Visualization of the peroxisomal compartment in living mammalian cells. *J Cell Biol.* 1997;136:71–80. **B,** Courtesy Don W. Fawcett, Harvard Medical School, Boston, MA.)

chondrodysplasia punctata. These diseases are moderately rare, occurring in approximately 1 in 50,000 live births. Most patients with peroxisomal biogenesis disorders display no defect in peroxisome membrane synthesis or import of peroxisomal membrane proteins, but they do have mild-to-severe defects in matrix protein import. However, in rare cases, patients lack peroxisome membranes altogether. Studies of both yeast *pex* mutants and cells from patients with peroxisomal biogenesis disorders have provided clues regarding peroxisome biogenesis (see Table 18.1).

SELECTED READINGS

Chacinska A, Koehler CM, Milenkovic D, et al. Importing mitochondrial proteins: machineries and mechanisms. *Cell.* 2009;138:628-644.

Demarsy E, Lakshmanan AM, Kessler F. Border control: selectivity of chloroplast protein import and regulation at the TOC-complex. *Front Plant Sci.* 2013;5:483.

Emma F, Montini G, Parikh SM, et al. Mitochondrial dysfunction in inherited renal disease and acute kidney injury. *Nat Rev Nephrol.* 2016;doi: 10.1038/nrneph.2015.214; [Epub ahead of print].

Hohmann-Marriott MF, Blankenship RE. Evolution of photosynthesis. *Annu Rev Plant Biol.* 2011;62:515-548.

Hosler JP, Ferguson-Miller S, Mills DA. Energy transduction: Proton transfer through the respiratory complexes. *Annu Rev Biochem.* 2006;75:165-187.

Jarvis P, López-Juez E. Biogenesis and homeostasis of chloroplasts and other plastids. *Nat Rev Mol Cell Biol.* 2013;14:787-802.

Keeling PJ. The number, speed, and impact of plastid endosymbiosis in eukaryotic evolution. *Annu Rev Plant Biol.* 2013;64:583-607.

Kühlbrandt W. Structure and function of mitochondrial membrane protein complexes. *BMC Biol.* 2015;13:89.

Kühlbrandt W, Davies KM. Rotary ATPases: A new twist to an ancient machine. *Trends Biochem Sci.* 2016;41:106-116.

Labbé K, Murley A, Nunnari J. Determinants and functions of mitochondrial behavior. *Annu Rev Cell Dev Biol.* 2014;30:357-391.

Nelson N, Junge W. Structure and energy transfer in photosystems of oxygenic photosynthesis. *Annu Rev Biochem.* 2015;84:659-683.

Pfanner N, van der Laan M, Amati P, et al. Uniform nomenclature for the mitochondrial contact site and cristae organizing system. *J Cell Biol.* 2014;204:1083-1086.

Poole AM, Gribaldo S. Eukaryotic origins: How and when was the mitochondrion acquired? *Cold Spring Harb Perspect Biol.* 2014;6:a015990.

Schon EA, DiMauro S, Hirano M. Human mitochondrial DNA: roles of inherited and somatic mutations. *Nat Rev Genet.* 2012;13:878-890.

Smith JJ, Aitchison JD. Peroxisomes take shape. *Nat Rev Mol Cell Biol.* 2013;14:803-817.

CHAPTER 20

Endoplasmic Reticulum

The **endoplasmic reticulum (ER)** is the largest membrane-delineated intracellular compartment within eukaryotic cells, having a surface area up to 30 times that of the plasma membrane (Fig. 20.1). The ER performs many essential cellular functions, including protein synthesis and processing, lipid synthesis, compartmentalization of the nucleus, calcium (Ca^{2+}) storage and release, detoxification of compounds, and lipid transfer and signaling to other organelles (Table 20.1). It also has roles in the biogenesis of the Golgi apparatus, peroxisomes and lipid droplets, and helps mitochondria to divide. Approximately one-third of all cellular proteins are imported into the lumen of the ER or integrated into ER membranes. Import occurs at rates of 2 to 13 million new proteins synthesized per minute. The ER retains some of these imported proteins for its own functions, some are degraded, and others are exported into the secretory pathway (see Chapter 21) for targeting to other compartments within the cell.

The ER is organized as an extensive array of tubules and flat saccules called cisternae (*cisterna* means "reservoir") that form an interconnected and contiguous three-dimensional network (a reticulum) stretching from the nuclear envelope to the cell surface. This system has several structural domains. The ER that flattens around the cell nucleus to form a double membrane bilayer barrier is called the **nuclear envelope** (see Fig. 9.5). The peripheral ER that extends from the nuclear envelope is comprised of both a polygonal network of tubules and flat, stacked membrane cisternae close to the nucleus. The stacked cisternae are covered with ribosomes for the synthesis, import, and folding of membrane, luminal, and secreted proteins. The tubule network extends throughout the cytoplasm (Fig. 20.1C); its functions include lipid synthesis, Ca^{2+} storage and release, and making contacts with the membranes of other organelles and the plasma membrane.

This chapter describes (a) the overall functions and organization of the ER, (b) insertion of proteins into and

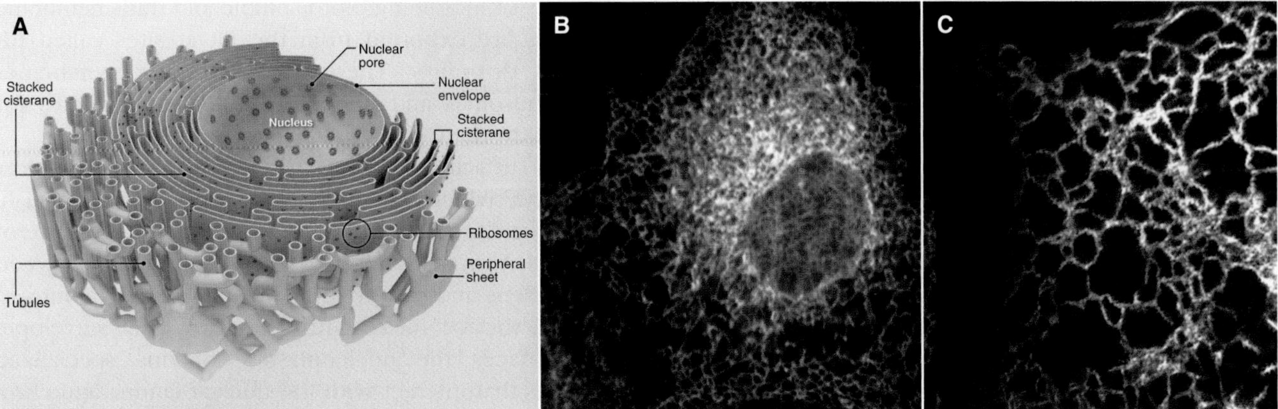

FIGURE 20.1 OVERVIEW OF THE ENDOPLASMIC RETICULUM/NUCLEAR ENVELOPE. A, Diagram of the endoplasmic reticulum (ER) and nuclear envelope. **B–C,** Fluorescence micrographs of cells expressing an ER marker tagged with *green* fluorescent protein (*white* in this image). **B,** The expansive character of ER is emphasized. **C,** Peripheral ER tubules. (**A,** From Goyal U, Blackstone C. Untangling the web: mechanisms underlying ER network formation. *Biochim Biophys Acta.* 2013;1833:2492–2498. **B–C,** Courtesy Drs. Chris Obara and Aubrey Weigel, Janelia Research Campus, Ashburn, VA.)

TABLE 20.1 Subdomains of the Endoplasmic Reticulum

ER Domain	Function	Associated Proteins
Rough ER	Protein translocation Protein folding and oligomerization Carbohydrate addition ER degradation	Sec61 complex, TRAP, TRAM, BiP PDI, Calnexin, Calreticulin, BiP Oligosaccharide transferase EDEM, Derlin1
Smooth ER	Detoxification Lipid metabolism Heme metabolism Calcium release	Cytochrome P450 enzymes HMG-CoA reductase Cytochrome b_5 IP_3 receptors
Nuclear envelope	Nuclear pores Chromatin anchoring	POM121, GP210 Lamin B receptor
ER export sites	Export of proteins and lipids into secretory pathway	Sar1p, Sec12p, Sec16p
ER contact zones	Transport of lipids	LTPs

BiP, biding immunoglobulin protein; EDEM, ER degradation-enhancing α-mannosidase-like protein; ER, endoplasmic reticulum; HMG-CoA, β-hydroxy-β-methylglutaryl-coenzyme A; IP_3, inositol triphosphate; LTP, lipid-transfer protein; PDI, protein disulfide isomerase; TRAM, translocating chain-associating membrane protein; TRAP, translocon-associated protein.

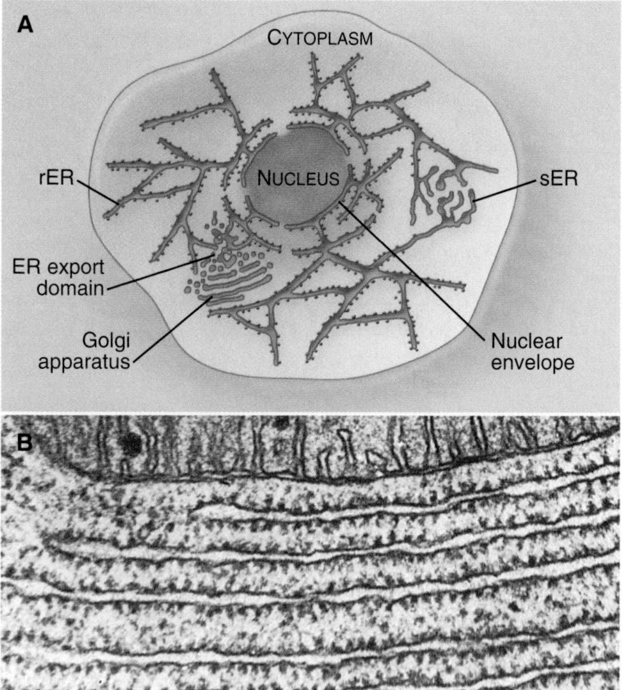

FIGURE 20.2 ENDOPLASMIC RETICULUM SUBDOMAINS. **A,** Drawing of a cell with specialized regions of the endoplasmic reticulum (ER). Rough ER (rER) with bound ribosomes extends from the nuclear envelope to the cell periphery. Smooth ER (sER) without ribosomes is specialized for drug metabolism and steroid synthesis and includes tubulovesicular elements composing ER exit sites. The nuclear envelope consists of ER membrane wrapped around the chromosomes and other nuclear elements. **B,** Electron micrograph of a thin section of rough ER and neighboring mitochondrion from the pancreas. The lumen is colored blue. (Micrograph by Keith R. Porter; courtesy Don W. Fawcett, Harvard Medical School, Boston, MA.)

across the ER membrane, (c) the mechanisms of folding, assembly, and degradation of proteins in the ER, and (d) the synthesis and metabolism of lipids by the ER. Chapter 21 covers the secretory pathway, which begins at the ER. Chapter 26 explains how the ER stores and releases Ca^{2+}.

Overview of Endoplasmic Reticulum Functions and Organization

The foremost function of the ER is producing most proteins that are secreted from the cell as well as lipids that make up the membranes of the other organelles, including the Golgi apparatus, endosomes, lysosomes, and plasma membrane. The ER also supplies much of the lipid for the membranes of mitochondria and peroxisomes.

The area of ER specialized for protein synthesis, folding, and degradation is called the **rough ER,** because its cytoplasmic surface is studded with ribosomes (Fig. 20.2B). The rough ER contains specialized receptors and channels that transfer proteins synthesized by ribosomes in the cytoplasm across ER membranes. Inside the ER lumen, newly synthesized proteins are exposed to a dense meshwork of chaperones and other modifying enzymes (estimated to be 200 mg/mL) that catalyze their folding and assembly. Proteins that are incorrectly folded or misfolded can be exported back into the cytoplasm, where they are degraded. Misfolded proteins, when accumulated in the ER at high levels, can trigger an unfolded protein response. This activates specific genes in the nucleus whose products help modify or destroy the misfolded proteins and compensate for the decreased capacity of ER folding. Both soluble and transmembrane proteins are exported from the ER at sites called **ER export domains**. These tubulovesicular membranes lack ribosomes and bud off vesicle intermediates for delivery to the Golgi apparatus (see Chapter 21).

The surface of the ER forming the outer nuclear envelope, which faces the cytoplasm, is indistinguishable from the rest of the ER except for the presence of nuclear pores that allow passage of molecules between the nucleus and cytoplasm (see Fig. 9.18). By contrast, the ER surface forming the inner nuclear envelope, which faces the nucleoplasm, contains specialized proteins that interact with the nuclear lamina and chromatin (see Fig. 9.8). In mitosis, the ER maintains its morphology as an interconnected network, whereas the nuclear envelope either disassembles (in most animal cells; see Fig. 44.6) or remains intact (in yeasts and most other fungi). In cells where the nuclear envelope disassembles during mitosis, nuclear pores disassemble and integral membrane proteins of the nuclear envelope diffuse into surrounding ER membranes.

Peroxisomes, lipid droplets, and the Golgi apparatus all depend on the ER for their biogenesis and maintenance. During peroxisome biogenesis, the ER provides the initial scaffold for recruiting core components (ie, Pex16p and Pex3p) involved in peroxisomal protein import (see Fig. 18.8). Biogenesis of the Golgi apparatus depends on the ER to synthesize resident enzymes and on constitutive cycling of membrane between the ER and Golgi apparatus. Consequently, perturbing the export of proteins from the ER impacts the structure and function of the Golgi apparatus. Lipid droplets emerge directly from the ER by the accumulation of neutral lipids between the leaflets of the ER bilayer. This forms an oil droplet that eventually buds toward the cytoplasm. Subsequent growth of lipid droplets may take place by neutral lipid synthesis either on their surface or at the ER.

The ER lumen is the major Ca^{2+} storage site in cells, owing to calcium pumps in the membrane (see Figs. 14.7 and 26.12) and many Ca^{2+}-binding proteins in the lumen. The second messenger IP_3 (inositol triphosphate) releases Ca^{2+} from the lumen by activating calcium release channels (see Fig. 26.12). Carefully regulated release and uptake of Ca^{2+} by the ER control muscle contraction (see Fig. 39.15) and many other cellular processes. The ER can also release Ca^{2+} next to neighboring organelles at ER–organelle contact sites.

The **smooth ER,** composed of tubular elements lacking ribosomes, is dedicated to enzyme pathways involved in drug metabolism (hepatocytes), steroid synthesis (endocrine cells), or calcium uptake and release (see Fig. 26.12). The **cytochrome P450** family of heme-containing membrane proteins resides in the smooth ER. These enzymes use an electron transfer process to detoxify endogenous steroids, carcinogenic compounds, lipid-soluble drugs, and environmental xenobiotics. The lumen of the ER is an oxidizing environment that favors disulfide bond formation, which helps stabilize proteins that are exported from the ER to the outside of the cell.

The abundance of a particular ER region varies in specialized cells. Cells dedicated to the production, storage, and regulated secretion of proteins (such as exocrine cells and activated B cells) are rich in rough ER. By contrast, smooth ER is abundant in endocrine cells that synthesize steroid hormones and in muscle cells owing to their requirement to store and release Ca^{2+} to control contraction.

Endoplasmic Reticulum Shape Generation

Several classes of proteins determine the unique morphology of the ER as it undergoes continual rearrangements. These proteins ensure that the ER maintains itself as a single polygonal network of tubules and stacked cisternae extending from the nuclear envelope to the cell periphery (Fig. 20.3). Proteins that mediate ER tubule formation include the reticulons and DP1/Yop1 proteins

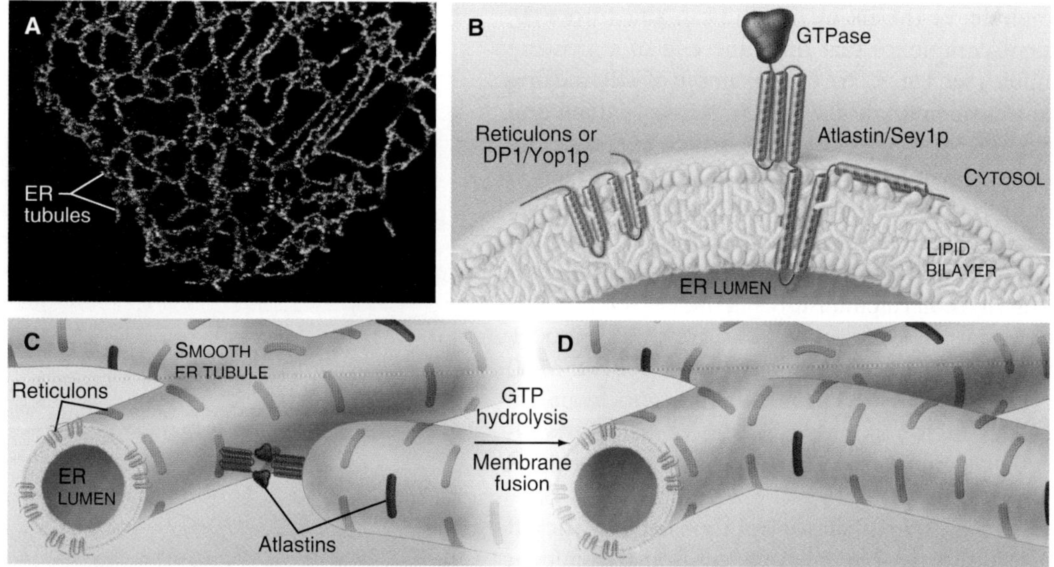

FIGURE 20.3 ENDOPLASMIC RETICULUM SHAPING AND FUSION PROTEINS. A, Superresolution fluorescence micrograph of the peripheral network of tubular ER membranes stained with an antibody to reticulon. See Fig. 6.5 for details. **B,** Insertion of reticulons and atlastin into the ER membrane bilayer. **C,** Drawings of tubular ER membranes showing reticulon proteins in the cytoplasmic leaflet of the bilayer and atlastin proteins connecting two tubules. **D,** Proposal for the fusion of ER membranes by atlastin with energy coming from guanosine triphosphate (GTP) hydrolysis. (**B,** Modified from Chen S, Novick P, Ferro-Novick S. ER structure and function. *Curr Opin Cell Biol.* 2013;25:428–433. **C–D,** Modified from Goyal U, Blackstone C. Untangling the web: mechanisms underlying ER network formation. *Biochim Biophys Acta.* 2013;1833:2492–2498. **D,** Modified from Lin S, Sun S, Hu J. Molecular basis for sculpting the endoplasmic reticulum membrane. *Int J Biochem Cell Biol.* 2012;44:1436–1443.)

(Fig. 20.3B–C). Although not related by sequence, these proteins share hydrophobic segments predicted to form α-helical hairpins that partially span the lipid bilayer (Fig. 20.3B). Insertion of these hydrophobic segments into the cytoplasmic leaflet of the ER membrane bilayer is thought to help generate the highly curved, tubular ER morphology, along with directly scaffolding reticulons on the surface of the ER. Reticulons and DP1/YOP1 proteins also participate in nuclear pore formation, most likely by stabilizing curved membranes.

Members of the atlastin/RHD3/Sey1p family of dynamin-related guanosine triphosphatases (GTPases) localize to highly curved ER membranes and mediate the formation of three-way junctions responsible for the polygonal structure of the tubular ER network (Fig. 20.3B–C). Atlastins consist of an N-terminal cytoplasmic GTPase domain and a three-helix bundle followed by two very closely spaced transmembrane segments and a C-terminal amphipathic helix. Guanosine triphosphate (GTP) binding stimulates interactions between atlastin oligomers in two adjacent membranes, forming a tethered complex. Fusion between ER tubules depends on a conformational change in the cytosolic domain linked to GTP hydrolysis that pulls the membranes together, leading to oligomerization of the transmembrane segments, and interaction of the C-terminal tail with the membranes undergoing fusion. After membrane fusion, atlastin releases GDP before another round of membrane fusion.

Interactions with microtubules can remodel the ER in two ways: motor proteins can pull ER along the side of a microtubule; or the ER membrane can attach to +TIP attachment complexes that track the end of a growing microtubule (see Fig. 37.6). One example of TIP tracking uses the transmembrane ER protein stromal interaction molecule 1 (STIM1) (see Fig. 26.12), which binds directly to the +TIP protein EB1 (see Fig. 34.4C). This interaction allows ER tubules containing STIM1 to reach the plasma membrane and activate Orai (see Fig. 26.13A), the store-operated Ca^{2+}-channel in the plasma membrane that admits Ca^{2+} to refill calcium stores in the ER. Interactions with the actin cytoskeleton can also drive ER remodeling. This is particularly important in plants (see Fig. 37.9) and yeast cells (see Fig. 37.11), but also occurs in animal cells such as neurons, where myosin-Va transports ER along actin filaments (see Fig. 36.8) into the dendritic spines of neurons.

Defects in proteins that control the shape of the ER often lead to disease. For example, autosomal dominant mutations in atlastin or reticulons result in length-dependent degeneration of the distal parts of the axons of corticospinal upper motor neurons in hereditary spastic paraplegias, and contribute to the pathogenesis of amyotrophic lateral sclerosis, which involves degeneration of both upper and lower motor neurons. The large size and highly polarized geometry of neurons makes shaping and distributing the ER network especially important in these cells.

Endoplasmic Reticulum–Organelle Contacts

The ER makes many types of contacts with other membranes, which are mediated by specific proteins. Chapter 26 describes contacts between Ca^{2+}-sensing proteins in the ER membrane with plasma membrane Ca^{2+} channels that cells use to resupply the ER with Ca^{2+} (see Fig. 26.12). Other **membrane contact sites** (Fig. 20.4 and Table 20.2) depend on a conserved receptor on the cytoplasmic surface of the ER called **VAP** (for VAMP [vesicle-associated membrane protein]-associated protein; see Fig. 20.17). The cytoplasmic domain of VAP binds a wide range of proteins with an **FFAT motif** containing two phenylalanines (FF) in an acidic tract. FFAT proteins link the ER to mitochondria, endosomes, Golgi, peroxisomes or the plasma membrane. Within the gap of 10 to 30 nm between the organelles, tethered proteins with lipid binding domains transfer lipid molecules between the lipid bilayers (see Fig. 20.17). Mutations of one of the two genes for VAP cause some cases of neurodegenerative diseases (amyotrophic lateral sclerosis, Parkinson disease). Membrane contact sites are also important for Ca^{2+} signaling. Because Ca^{2+} diffuses only about 100 nm from its release site (see Chapter 26), Ca^{2+} signals from the ER to other organelles are more efficient at tight interfaces. ER-organelle contacts also control organelle division and many aspects of cytoplasmic and plasma membrane organization.

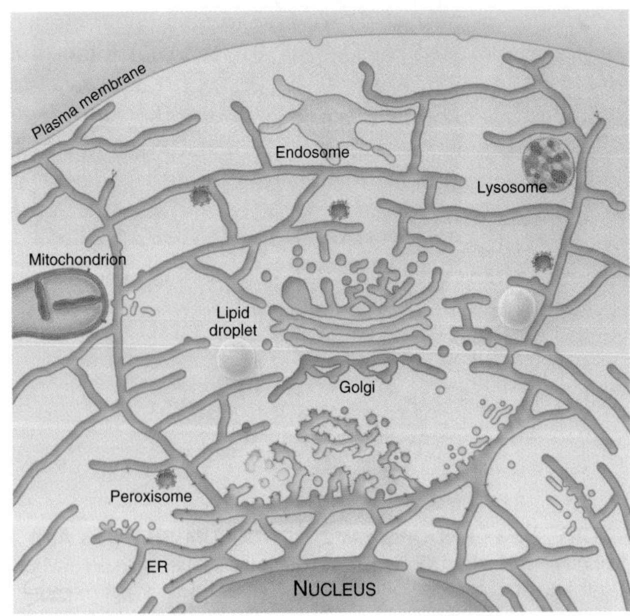

FIGURE 20.4 ENDOPLASMIC RETICULUM–ORGANELLE CONTACT SITES. Drawing of a cell showing contacts of the endoplasmic reticulum (ER) with endosomes, Golgi apparatus, lipid droplets, mitochondria, peroxisomes, and plasma membrane.

TABLE 20.2 Location and Proposed Functions of Proteins at Endoplasmic Reticulum–Organelle Contacts

ER-Mitochondria Contacts	
MFN1-MFN2	Calcium transfer
VAPs-PTPIP51	Lipid transfer
ERMES complex	Tethers and possibly transfers lipids
ER-Endosome	
VAP-A-ORP1L	Senses sterol levels and regulates endosome positioning
VAP-A-STARD3	Transfers sterols
VAP-A-Protrudin	Transfers kinesin-1 from the ER to late endosomes to facilitate their to the cell periphery
ORP5-NPC1	ORD domain of ORP5 transfers cholesterol
ER-Golgi Contacts	
VAP-OSBP	Transfers PtdIns(4)P from the Golgi apparatus to ER and sterols from ER to Golgi apparatus
VAP-CERT	Transfers ceramide
VAP-NIR2	Maintains diacylglycerol levels in the Golgi
ER-Lipid Droplet Contacts	
FATP1-DGAT2	Coordinates lipid droplet expansion at lipid droplet-ER MCSs

CERT, ceramide transport protein; ER, endoplasmic reticulum; MCS, membrane contact site; OSBP, oxysterol-binding protein; VAP, VAMP (vesicle-associated membrane protein)-associated protein.

Overview of Protein Translocation Into the Endoplasmic Reticulum

A major function of the ER is to import and process newly synthesized proteins from the cytoplasm. This is necessary for growth of other organelles, including the Golgi apparatus, nucleus, endosomes, lysosomes, and plasma membrane, as well as for production of nearly all proteins secreted from the cell. All proteins are synthesized in the cytoplasm and must be targeted specifically to the ER, where they are either fully translocated across the ER membrane and released into the ER lumen (soluble proteins) or only partly translocated across the ER membrane and transferred into the lipid bilayer of the ER membrane (transmembrane proteins). The orientation of a protein in the lipid bilayer or its localization to the lumen is established during initial protein translocation and maintained as vesicles transfer the protein between membranes of the secretory pathway (see Fig. 21.2). Thus, domains of transmembrane proteins to be exposed on the cell surface must be in the ER lumen when their transmembrane domains are inserted into the ER membrane. Similarly, secreted soluble proteins must be fully translocated into the lumen of the ER.

Because the ER lumen is topologically equivalent to the extracellular space, transport of proteins into the ER is analogous to transport into or across a prokaryotic plasma membrane (see Fig. 18.9). In both cases transported substrates must be recognized, targeted from the cytoplasm to the membrane, and translocated across the membrane through a protein channel without other molecules leaking across the membrane. Overcoming these obstacles involves specialized and regulated factors in the cytosol and target membrane.

Insertion of proteins in the ER can occur either as the protein is being made by membrane-bound ribosomes (**cotranslational translocation**) or after synthesis is complete (**posttranslational translocation**), each by distinct mechanisms. In the posttranslational pathway, the protein is first fully synthesized in the cytoplasm and then translocated independently of the ribosome.

Cotranslational Translocation

Signal Sequences

All soluble and membrane proteins destined for ER translocation using the cotranslational pathway contain a hydrophobic "leader" sequence that serves as a recognition signal for direction to the ER membrane. N-terminal leader sequences (termed **signal sequences**) are typically 15 to 35 amino acids long and contain a hydrophobic core of at least 6 residues. For many membrane proteins, the first transmembrane segment (a hydrophobic stretch of 16 to 25 residues) serves as a signal sequence. Aside from hydrophobicity, these signal sequences have no other features in common. Nevertheless, when attached to proteins that are not normally targeted to the ER, these signal sequences direct the protein to the ER and not to other organelles such as to mitochondria or peroxisomes, which use different targeting signals (see Fig. 18.1).

Signal Recognition Particle and Signal Recognition Particle–Receptor

The cotranslational pathway begins once the first hydrophobic residues of either a signal sequence or a transmembrane segment emerges from the ribosome and is recognized by the **signal recognition particle (SRP),** a large ribonucleoprotein complex (Fig. 20.5). SRP binds these hydrophobic peptides, slows translation by that ribosome and delivers the ribosome to a protein-conducting channel in the ER membrane called

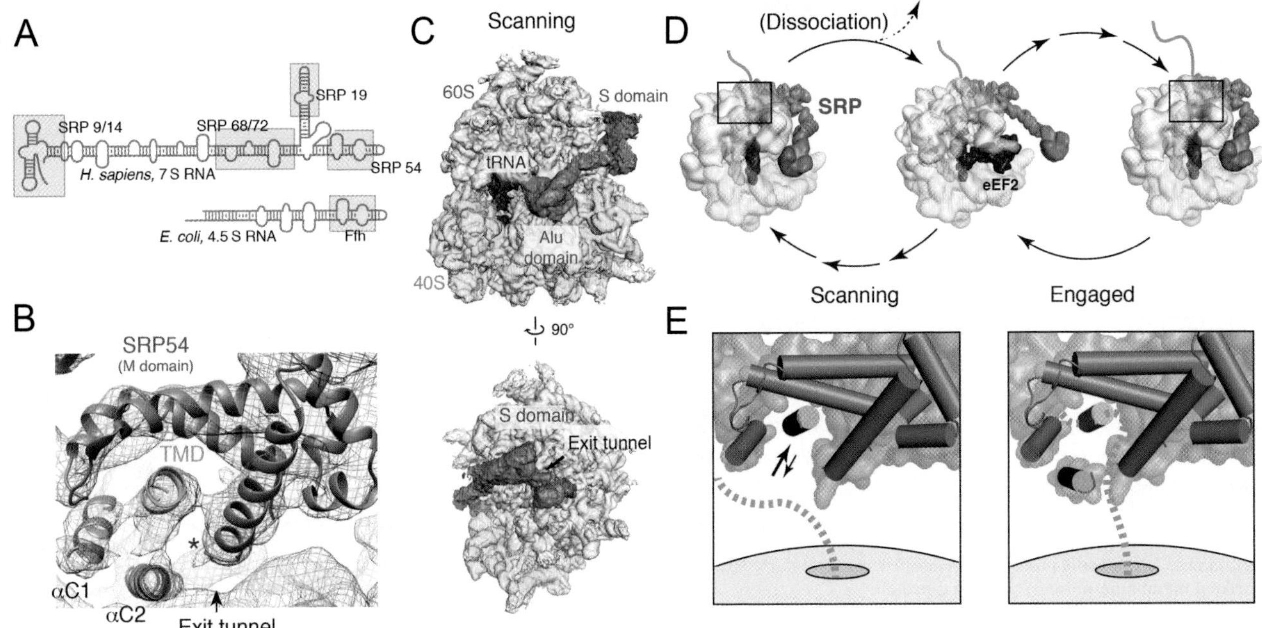

FIGURE 20.5 STRUCTURE AND MECHANISM OF THE SIGNAL RECOGNITION PARTICLE. A, Secondary structure of human and *Escherichia coli* signal recognition particle (SRP) RNAs with sites of protein interactions in blue. **B–E,** Cryoelectron microscopic structures of SRP bound to mammalian ribosomes. SRP *(green)* wraps around the ribosome from the exit tunnel of the large subunit to the guanosine triphosphatase (GTPase) center near transfer RNA (tRNA) in the P-site. **B,** Structure with the SRP54 subunit *(green ribbon diagram)* engaging the transmembrane domain (TMD) of a nascent polypeptide that has just emerged from the exit tunnel of the ribosome. **C,** Two views of the structure of the SRP–ribosome complex in the scanning mode. **D,** Steps in the interaction of SRP with the ribosome, nascent polypeptide, and eukaryotic elongation factor 2 (eEF2). **E,** In the engaged mode the transmembrane domain of the nascent chain displaces one of the SRP54 helices. (**B–E,** From Voorhees RM, Hegde RS. Structures of the scanning and engaged states of the mammalian SRP-ribosome complex. *eLife*. 2015;4:e07975.)

a **translocon**. Then translation resumes, driving the polypeptide through the translocon into the lumen of the ER or inserting it into the membrane bilayer.

Human SRP is composed of six proteins (named by their apparent molecular weights) assembled on a 300-nucleotide RNA that spans the distance from the exit tunnel on the large ribosome subunit to the GTPase center where translation factor GTPases (eEF1, eEF2, eEF3) bind. The SRP54 protein subunit binds signal sequences in a deep, hydrophobic groove lined by the flexible side chains of several methionines. Like bristles of a brush, the methionines accommodate the various hydrophobic side chains of different signal sequences. Phosphates of the SRP RNA near one end of the hydrophobic groove interact with basic residues that are often (but not always) adjacent to the hydrophobic core of signal sequences and transmembrane signal segments.

Once SRP54 has engaged a signal sequence, the opposite end of SRP associates with the other side of the large ribosomal subunit at the GTPase center, where it can compete with binding of elongation factors eEF1 and eEF2 thereby slowing translation (Fig. 20.5E). This modest slowing of translation provides time to target the ribosome to the translocation channel in the ER before excessive polypeptide synthesis precludes cotranslational transport.

The complex consisting of ribosome, nascent chain, and SRP targets selectively to the ER membrane by binding the **SRP receptor,** a heterodimer consisting of one subunit that binds SRP and another that spans the ER membrane (Fig. 20.6). Both the SRP54 subunit and the SRP receptor have GTPase domains (Fig. 20.6B) similar to Ras (see Fig. 4.6). Neither SRP nor SRP receptor hydrolyzes GTP on its own. Instead, association of the two GTPase domains that accompanies successful targeting completes both of their active sites, resulting in hydrolysis of both bound GTPs. Dissociation of the γ-phosphates reduces the affinity of SRP for its receptor. The net result is that they dissociate from each other as well as from the ribosome and nascent chain. Dissociation of SRP54 from its binding site near the ribosome exit tunnel now allows the protein-conducting Sec61 channel to bind at the same site. The ribosome is now successfully targeted to and docked at the translocon, providing an opportunity for subsequent translocation or membrane insertion of the nascent polypeptide. SRP is released into the cytoplasm for another round of targeting, while SRP receptor diffuses away in the membrane to capture the next SRP–ribosome complex.

The targeting cycle thus delivers the ribosome-nascent chain complex to the translocon and recycles the SRP and SRP receptor. GTP binding and hydrolysis

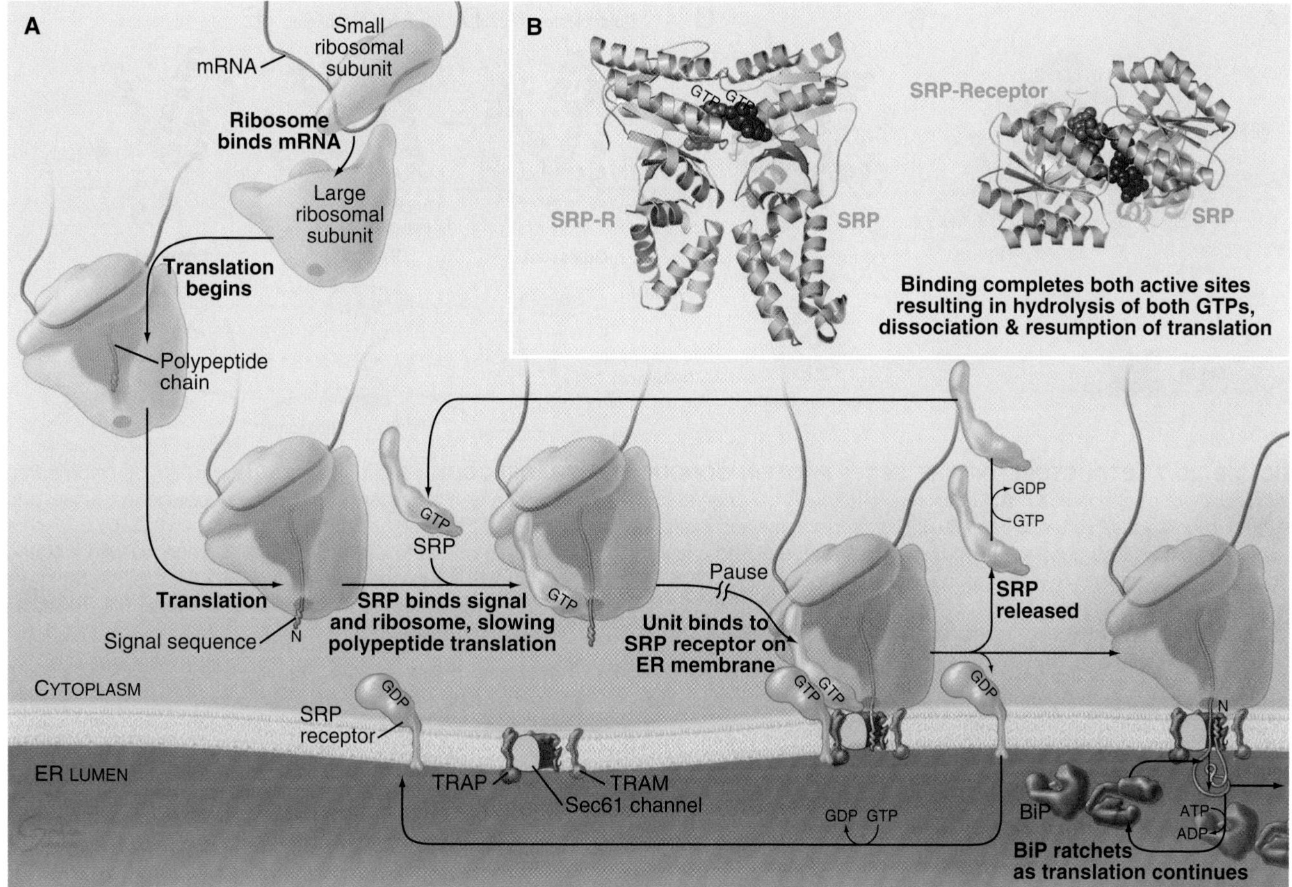

FIGURE 20.6 COTRANSLATIONAL PATHWAY FROM RIBOSOME TO THE ENDOPLASMIC RETICULUM LUMEN. A, Signal recognition particle (SRP) and SRP-receptor use a cycle of recruitment and GTP hydrolysis to control delivery of a ribosome with an messenger RNA (mRNA) and nascent chain with a signal sequence to the Sec61 translocon in the ER membrane. SRP binds a signal sequence emerging from a ribosome and slows polypeptide translation. SRP also directs the ribosome to the SRP-receptor on the ER membrane, where the ribosome docks on the translocon and continues translation. **B,** Ribbon diagrams of two views of the complex between SRP and SRP-receptor showing the close association between the two GTPase domains, which activates GTPase hydrolysis. ADP, adenosine diphosphate; ATP, adenosine triphosphate; BiP, binding immunoglobulin protein; GDP, guanosine diphosphate; TRAM, translocating chain-associating membrane protein; TRAP, translocon-associated protein.

by SRP and SRP receptor provide directionality and order to the sequence of reactions that bring the nascent chain to the translocation channel.

Sec61 Complex: The Protein-Conducting Channel

The nascent chain emerging from the ribosome engages and opens the translocon for transport across the ER membrane. The translocon is an evolutionarily conserved complex of three transmembrane proteins associated with proteins inside the ER. The eukaryotic **Sec61 complex** consists of an α subunit with 10 transmembrane helices and smaller β and γ subunits with single helices crossing the membrane (Fig. 20.7). The subunits of the homologous bacterial and archaeal SecY complex are called SecY, SecE, and SecG (see Fig. 18.9D).

The Sec61 complex provides a high-affinity docking site for the ribosome–nascent chain complex. Additional factors thought to facilitate binding of the signal sequence to the Sec61 complex include the protein TRAM (translocating chain-associating membrane protein) and the protein complex TRAP (translocon-associated protein). Bacterial SecY interacts with fully translated peptides associated with the chaperone SecA (see Fig. 18.9A).

Prokaryotic and eukaryotic translocons have nearly identical structures. The 10 transmembrane helices of Sec61α are arranged around a narrow, hourglass-shaped pore that is most constricted near the center of the lipid bilayer (Fig. 20.7B). A seam between two pairs of helices, termed the lateral gate, can potentially separate to open the pore toward the membrane like a clamshell. The small accessory subunits form supporting transmembrane helices.

Interactions with the ribosome and signal sequence control opening of the Sec61 translocon across or toward the membrane. Without these ligands translocons are quiescent with closed pores and lateral gates. A ring of hydrophobic side chains fills the pore and a short helix

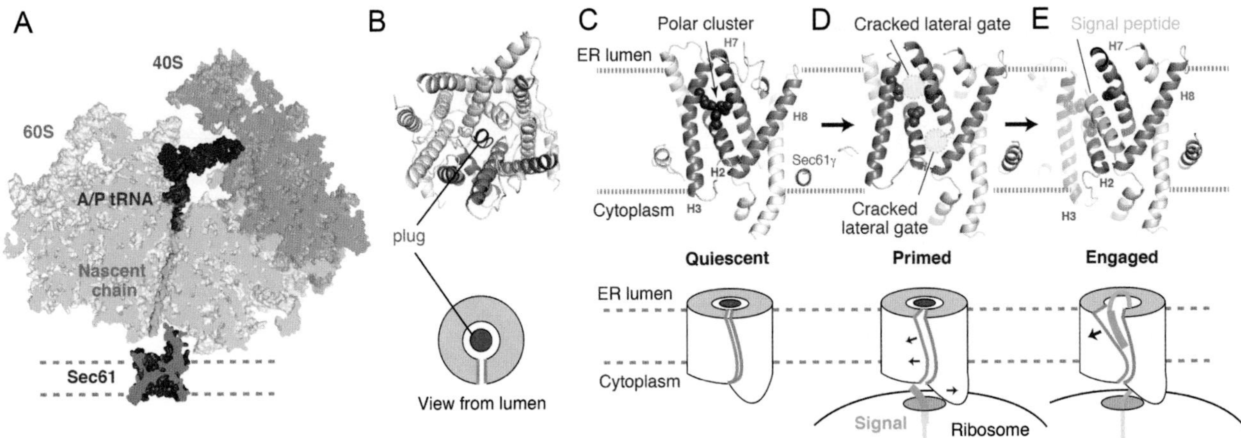

FIGURE 20.7 STRUCTURE OF THE SEC61 PROTEIN-CONDUCTING TRANSLOCON. Structures were determined by cryoelectron microscopy of single particles. **A,** Cross section through the large subunit of a ribosome with a nascent chain in the ribosome exit tunnel and Sec61 associated with the exit pore. **B–E,** Ribbon diagrams and interpretative drawings of the three functional states of Sec61: **B–C,** quiescent without a translocating polypeptide and with a plug in the closed channel; **D,** primed while bound to a ribosome; and **E,** engaged with a bound signal sequence helix (in *blue*), an open channel and open lateral gate. (**A,** From Voorhees RM, Fernandez IS, Scheres SH, Hedge RS. Structure of the mammalian ribosome-Sec61 complex to 3.4 Å resolution. *Cell.* 2014;157;1632–1643. **B–E,** From Voorhees RM, Hegde RS. Toward a structural understanding of co-translational protein translocation. *Curr Opin Cell Biol.* 2016;41:91–99.)

called a plug further blocks the channel. When the cytoplasmic loops of Sec61α interact with the ribosome, the translocon undergoes a subtle conformational change to a "primed" state that contains a partially opened lateral gate.

A signal sequence emerging from the ribosome exploits this cracked lateral gate to fully open the channel. The signal sequence entering the pore of the translocon forms an α-helix that binds to a hydrophobic patch in the lateral gate. The bound signal sequence is oriented with its N-terminus toward the cytoplasm, so it forms a loop in the pore of the translocon extending back to the exit site on the ribosome. This interaction leads to parting of the lateral gate, which is held open by the bound signal peptide. A parted lateral gate widens the translocon pore, leading to displacement of the plug in preparation for protein translocation.

The ability to recognize signal sequences allows Sec61 to discriminate substrates for translocation from other proteins. Traditionally, this was thought to be a constitutive process predetermined by the sequences on the substrate. However, various cell types differ in the efficiency with which they recognize particular signal sequences. This can be explained if additional proteins at the translocation site influence signal sequence recognition. For example, proteins Sec62, Sec63, p180, p34, TRAM, and TRAP complex may stimulate or inhibit the translocation of selected substrates by recognizing diversity within the signal sequence. Selective changes in expression or modifications of these accessory components in different cell types could then affect the outcome of translocation for different substrates.

Polypeptides can take one of two possible paths through the engaged translocon as translation proceeds. Secretory proteins and soluble parts of membrane proteins move through the translocon pore into the ER lumen. Transmembrane domains of integral membrane proteins leave the translocon pore through the lateral gate to enter the hydrophobic environment of the lipid bilayer. Elongation of the polypeptide chain by the ribosome provides the energy for the nascent chain to pass through the channel across the membrane or into the bilayer. Thus, the energy used for protein synthesis is harnessed to drive translocation of the polypeptide to its destination.

Translocation of Soluble Proteins Into the Lumen of the Endoplasmic Reticulum

Soluble proteins destined for secretion or retention in the ER lumen are translocated through the central pore of the translocon (Fig. 20.8A). The small, flexible side chains lining the pore fit snugly around a translocating peptide, preventing passage of ions or other small molecules. Thus the nascent chain has a continuous path from the peptidyl transferase center in the ribosome, through the translocation channel into the ER. Inside the ER lumen binding and release of chaperones such as binding immunoglobulin protein (BiP) may help translocate the polypeptide across the membrane, although this has not been firmly established.

Once the translocating polypeptide has grown to approximately 150 residues, a **signal peptidase** associated with the translocon in the ER lumen cleaves off the signal sequence. After signal peptide cleavage, the new

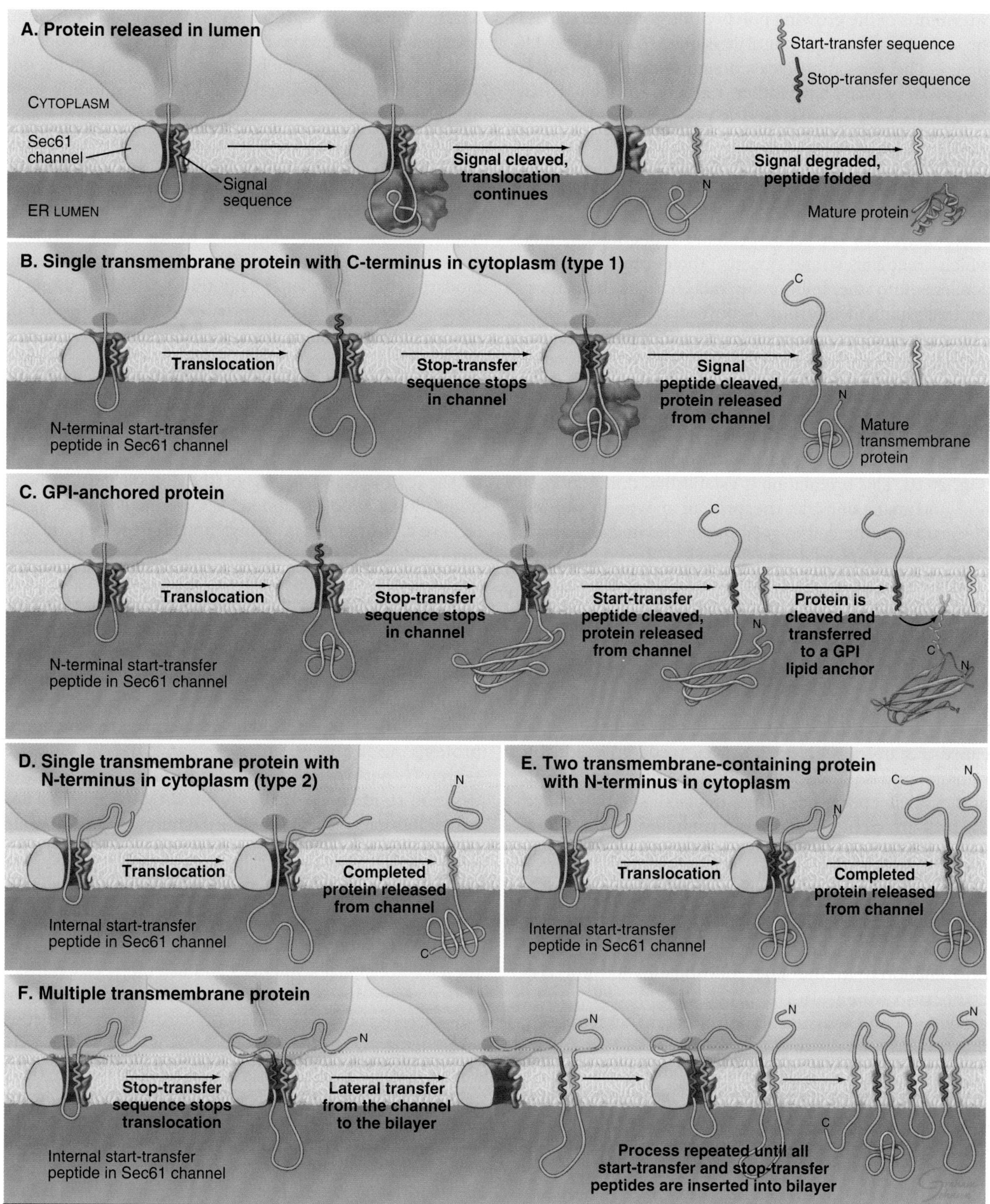

FIGURE 20.8 COMPARISON OF PATHWAYS USING SEC61 TO TARGET PROTEINS TO THE ENDOPLASMIC RETICULUM LUMEN OR MEMBRANE. The drawings show how signal sequences, start-transfer sequences and stop-transfer sequences target protein to Sec61 channel and then to the lumen and membrane of the endoplasmic reticulum (ER). GPI, glycosylphosphatidylinositol.

N-terminus of the growing polypeptide continues to pass through the translocon until it is released into the ER lumen. The remaining cleaved signal peptide either is degraded or may have other functions elsewhere in the cell.

Insertion of Membrane Proteins Into the Endoplasmic Reticulum Bilayer

Most proteins destined for insertion into the ER membrane bilayer use the Sec61 protein-conducting channel, which recognizes and inserts their transmembrane domains into the membrane bilayer. The machinery for targeting and insertion is physically coupled to the ribosome near the polypeptide exit tunnel, so the transmembrane domains are shielded from the aqueous cytoplasm.

Transmembrane proteins are categorized as type 1, type 2, or polytopic depending on the orientation of their transmembrane domains across the lipid bilayer (Fig. 20.8). This orientation is established during translation and maintained as the protein moves to its final destination in the cell by membrane budding and fusion events (see Chapters 21 and 22).

The N-terminal signal sequence of **type 1 transmembrane proteins** initiates translocation, similar to soluble proteins by engaging two helices adjacent to the lateral gate (Fig. 20.8B). As translation proceeds, a stop-transfer signal (usually a transmembrane domain) stops the transfer process before the polypeptide chain is completely translocated. After the signal sequence (also called a start-transfer signal) is cleaved off by the ER signal peptidase, the transmembrane segment slides out of the translocon through the lateral exit site. This leaves the protein oriented with its N-terminus in the ER lumen, its transmembrane segment spanning the ER membrane and its C-terminus in the cytoplasm.

Some type 1 proteins, called **glycosylphosphatidylinositol (GPI)-anchored proteins,** exchange their C-terminal transmembrane segment for an oligosaccharide anchored to the lipid phosphatidylinositol (Fig. 20.8C; also see Fig. 13.10). An enzyme in the ER lumen cleaves off the transmembrane segment and transfers the new C-terminus to a preassembled GPI membrane anchor. Many cell-surface proteins are attached to the plasma membrane in this manner. This allows them to be readily released from the cell when specific phospholipases in the plasma membrane are activated. For example, during sperm capacitation, many GPI-anchored proteins are cleaved and released from the sperm plasma membrane. This reorganization of the sperm's cell surface is essential for a sperm to fertilize an egg.

Type 2 transmembrane proteins use a transmembrane domain in the middle of the polypeptide as an internal signal sequence (Fig. 20.8D). Once such an internal signal sequence emerges from a ribosome, it is recognized by SRP and brought to the ER membrane, where it serves as a start-transfer signal to initiate protein translocation. When the protein is fully synthesized, the start-transfer signal, which is not cleaved off, slides out of the translocation channel into the surrounding lipid bilayer, where it serves as a transmembrane anchor.

Polytopic proteins that span the membrane multiple times (such as ion channels and carriers) use multiple stop-transfer signals, none of which are cleaved by signal peptidases (Fig. 20.8E–F). SRP is probably required to target the first signal sequence to the ER membrane. Thereafter, the dynamics of the channel must accommodate sequences that specify translocation of loops in the cytoplasm or lumen alternating with the transfer of transmembrane segments to the lipid bilayer.

Association of Lipid-Anchored Proteins With the Cytoplasmic Surface of the Endoplasmic Reticulum

Many classes of lipid-anchored proteins, including N-Ras and H-Ras GTPases, are targeted from the cytoplasm to the cytoplasmic leaflet of the ER bilayer by posttranslational modification of a C-terminal CAAX motif (where C is cysteine, A is any aliphatic residue, and X is any residue). First, a soluble enzyme attaches a farnesyl or geranylgeranyl lipid to the cysteine residue in CAAX with a thioether bond, anchoring the protein to the cytoplasmic surface of the ER membrane. Then, a prenyl-CAAX protease in the ER bilayer cleaves off the AAX residues, leaving the prenylcysteine as the new C terminus. Finally, another enzyme in the ER bilayer methylates the carboxyl group of the modified cysteine. Lipid-anchored proteins reach the plasma membrane by following the secretory pathway out of the ER. One or two other cysteine residues upstream of the CAAX motif of N-Ras and H-Ras can be tagged with palmitic acid via a labile thioester bond.

Posttranslational Translocation by Sec61 Translocons

Proteins targeted for posttranslational translocation into the ER have signal sequences and use the Sec61 translocon for entering the ER, but use proteins other than SRP or SRP receptor to guide them to the translocons after being released from the ribosome into the cytoplasm (Fig. 20.9). Posttranslational translocation is most common in yeast and bacteria. Mammalian cells use it primarily for translocating small proteins (<200 amino acids), which are inefficiently recognized by SRP, because their signal sequence is exposed only briefly (or not at all) before translation is complete. Specific chaperones including the cytoplasmic adenosine triphosphatase (ATPase) TRC-40 and calmodulin hold these polypeptides in a largely unfolded state in the cytoplasm until they are delivered to the tetrameric Sec62/63 protein complex in the ER membrane. There the signal sequence engages and opens the Sec61 channel in a fashion similar to cotranslational translocation.

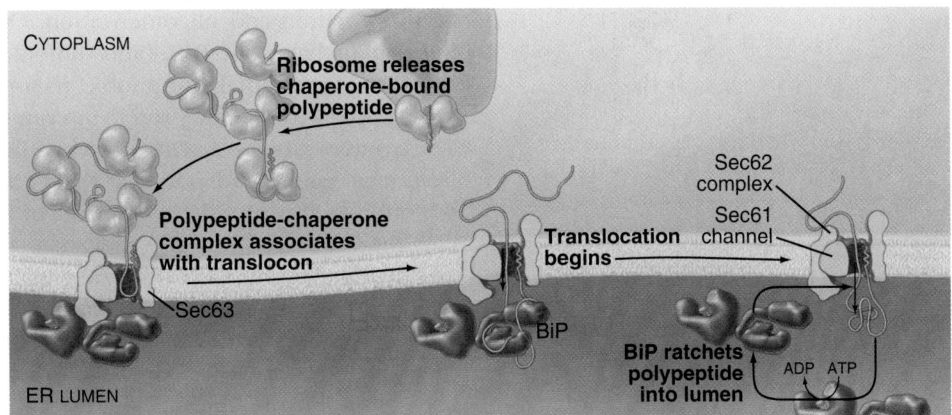

FIGURE 20.9 POSTTRANSLATIONAL PATHWAY FROM RIBOSOME TO ENDOPLASMIC RETICULUM LUMEN. As the polypeptide *(purple thread)* emerges from the ribosome in the cytoplasm, it is bound by chaperones *(green)* that prevent it from aggregating and guide it to the Sec62/63 protein complex at the ER membrane. After signal sequence *(blue)* engages with the Sec61 protein-conducting channel, cycles of BiP binding to and release in the lumen pull the polypeptide across the membrane.

Because protein synthesis is already complete, another energy source is exploited to move these polypeptides through the translocon into the ER. **BiP,** a luminal ER chaperone of the heat shock protein 70 (Hsp70) family, binds the substrate in the ER lumen, thereby preventing it from sliding back into the cytoplasm. Repeated rounds of substrate binding and release, driven by ATP hydrolysis, allow BiP to act as a molecular ratchet to bias the diffusion of the substrate into the lumen. The transmembrane protein Sec63 helps recruit BiP to the translocon and regulates BiP ATPase activity. Thus, the peptide is "pulled" across the membrane from the luminal side instead of being "pushed" from the cytoplasmic side, as during cotranslational translocation. Other Hsp70 family members function similarly during import of proteins into mitochondria and chloroplasts (see Figs. 18.3 and 18.6).

Tail-Anchored Proteins

Tail-anchored proteins represent approximately 3% to 5% of all membrane proteins. They have diverse roles in membrane biogenesis and traffic, as well as in cell metabolism (Table 20.3). Tail-anchored proteins include SNARE (soluble *N*-ethylmaleimide-sensitive factor attachment protein receptor) proteins, such as syntaxins and synaptobrevins, which are responsible for membrane fusion events within cells (see Chapter 21); Sec61γ and Sec61β of the translocation channel; cytochrome b_5, which participates in lipid metabolism; and Bcl and Bax, which are found on the mitochondrial outer membrane and regulate apoptosis.

Tail-anchored proteins are inserted into membranes by a posttranslational translocation pathway that avoids the Sec61 translocation channel. A single stretch of hydrophobic amino acids close to the C-terminus anchors these proteins in the bilayer with only two to three hydrophilic residues in the ER lumen and N-terminal functional domains in the cytoplasm (Fig. 20.10). Tail-anchored proteins lack an N-terminal signal sequence and their membrane-interacting region is so close to the C-terminus that it emerges from the ribosome only on termination of translation. Consequently, tail-anchored proteins do not bind to SRP, which recognizes only signal peptides or transmembrane domains as part of a nascent polypeptide chain (ie, still attached to the ribosome). Once in the cytoplasm tail-anchored proteins are not directed posttranslationally to the Sec61 channel. Instead, a cytoplasmic ATPase and membrane proteins insert the hydrophobic C-terminal anchor into the membrane (Fig. 20.10).

TABLE 20.3 Examples of Tail-Anchored Proteins

Protein	Function
ER-Inserted	
Target SNAREs (syntaxin)	Target membrane for vesicle insertion
Vesicle SNAREs (synaptobrevin)	Target membrane for vesicle insertion
Giantin	Golgi tethering protein
Sec61γ, Sec61β	ER protein translocation
Cytochrome b_5	Heme metabolism
Heme oxygenases I and II	Heme metabolism
UBC 6	ER degradation
Mitochondrial-Inserted	
Bcl-2	Apoptosis
Bax	Apoptosis
Tom5, Tom6	Mitochondrial protein translocation

ER, endoplasmic reticulum; SNARE, soluble *N*-ethylmaleimide-sensitive factor attachment protein receptor.

First the hydrophobic transmembrane domain of tail-anchored proteins binds to Get3, a homodimeric, cytoplasmic ATPase with an α helical domain. With bound

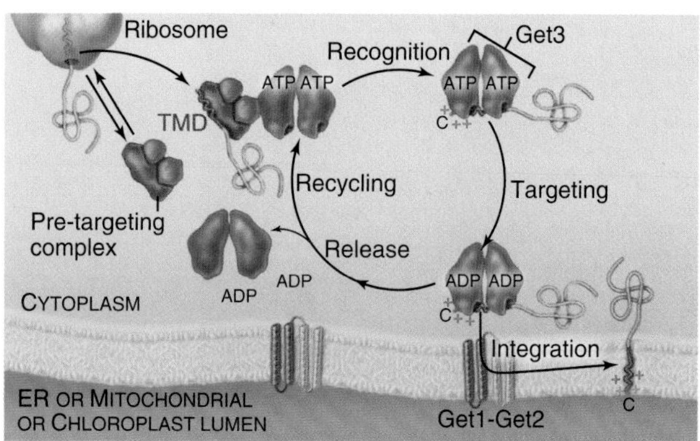

FIGURE 20.10 TARGETING OF TAIL-ANCHORED PROTEINS TO THE ER OR MITOCHONDRIAL MEMBRANE. Tail-anchored proteins have a short, hydrophobic transmembrane domain flanked on one or both sides by positively charged residues. After the protein emerges from the ribosome, the polypeptide binds to a pretargeting complex that direct it to a dimer of the Get3 adenosine triphosphatase (ATPase). Hydrolysis of ATP closes the transmembrane helix in a hydrophobic groove for delivery to the target membrane bilayer. ATP binding releases the tail-anchored protein for insertion into the target membrane bilayer. After insertion, the majority of the protein remains in the cytoplasm with only the C-terminal charged residues in the lumen. ADP, adenosine diphosphate; TMD, transmembrane domain. (Modified from Hegde RS, Keenan RJ. Tail-anchored membrane protein insertion into the endoplasmic reticulum. *Nat Rev Mol Cell Biol.* 2011;12:787–798. See Protein Data Bank [www.rcsb.org] files 4XTR, 4XVU, and 4XWO.)

ATP the Get3 subunits adopt a conformation, exposing a large hydrophobic groove for binding hydrophobic substrates. While the target protein is bound to Get3 in the cytoplasm, additional cofactors help maintain the solubility of its hydrophobic tail and prevent binding to other chaperones.

Next the complex of Get3 with bound tail-anchored protein is recognized by the Get1-Get2 receptor on the ER surface (Fig. 20.10). This interaction promotes hydrolysis of ATP bound to Get3 and release of the tail-anchored protein into the membrane bilayer. This two-step mechanism is reminiscent of the SRP-mediated cotranslational insertion pathway, which also involves nucleotide binding (albeit GTP) followed by receptor interaction. Finally, Get3 dissociates from Get1 and is recycled back to the cytoplasm for handling the next substrate. Insertion of a tail-anchored protein in a bilayer translocates many fewer C-terminal hydrophilic residues across the membrane than the translocation of most other transmembrane and soluble proteins.

Protein Folding and Oligomerization in the Endoplasmic Reticulum

Once nascent polypeptides translocate across the ER bilayer through the Sec61 translocon, they emerge into the ER lumen and interact with a wealth of proteins. Those proteins remove the signal sequence, add oligosaccharides, and direct folding by catalyzing disulfide bond formation and oligomerization. One such factor, the Hsp70 chaperone **BiP,** binds unfolded polypeptides by interacting with hydrophobic regions that are normally sequestered in the protein interior (see Fig. 12.12). These interactions prevent newly synthesized proteins from aggregating and promote their folding. Cycles of BiP binding also bias the movement of the polypeptide into the ER lumen but not back out. Another enzyme called **oligosaccharyl transferase** adds core sugars to the growing chain when an asparagine in an appropriate sequence context is detected. A fourth enzyme, **protein disulfide isomerase (PDI),** catalyzes disulfide exchange between sulfhydryl groups on cysteines allowing the formation of disulfide (S-S) bonds. The oxidizing equivalents to form disulfide bonds flow from flavin adenine dinucleotide (FAD) through two pairs of cysteines of an ER membrane protein that oxidizes a pair of cysteines in the active site of PDI. PDI then mediates correct formation of disulfide bonds by forming and breaking mixed disulfides with polypeptide substrates until the correct disulfides are formed.

Retention of these folding factors in the ER depends on the sequence lysine-aspartic acid-glutamic acid-leucine (KDEL) at the C-termini of these enzymes. If this sequence is deleted, the mutated protein is transported to the Golgi apparatus and secreted from the cell. Remarkably, addition of KDEL to a normally secreted protein results in its accumulation in the ER.

Folding and assembly factors interact with proteins throughout their lifetimes in the ER. The following sections describe the machinery that controls protein folding and assembly in the ER, mechanisms for sensing correctly folded or misfolded proteins, and pathways for disposing of misfolded proteins that accumulate in the ER.

N-Linked Glycosylation

Most proteins synthesized in association with the ER are glycoproteins with covalently attached carbohydrates. One class of oligosaccharides is added to asparagine residues during translocation of the protein into the ER (Fig. 20.11). These asparagine or **N-linked oligosaccharides** form flexible hydrated branches that can extend 3 nm from the polypeptide. They frequently make up a sizable portion of the mass of a glycoprotein and cover a large fraction of its surface. These polar oligosaccharides make proteins more hydrophilic and less likely to aggregate. By avoiding aggregation, the protein has a higher probability of folding correctly. Once correctly folded, the protein can leave the ER and move through the secretory pathway, where the sugars can be modified further. The great diversity of oligosaccharides found on secreted proteins is also crucial for their functions outside the cell.

Preformed oligosaccharides are added to asparagine residues. These precursors are composed of 14 sugars

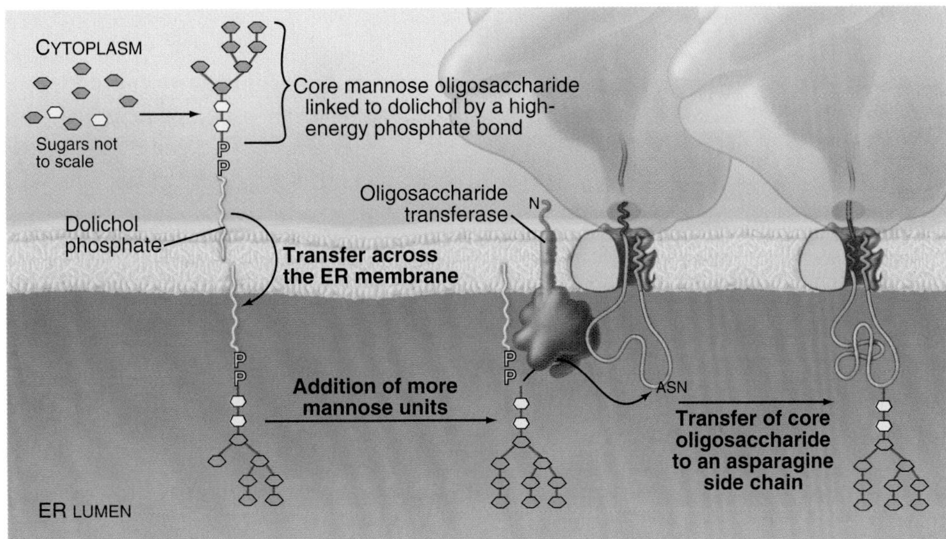

FIGURE 20.11 DOLICHOL PATHWAY. A core oligosaccharide consisting of mannose and *N*-acetylglucosamine is synthesized in the cytoplasm, attached through high-energy pyrophosphate bonds to dolichol in the endoplasmic reticulum (ER) membrane. Following transfer across the ER bilayer, the addition of sugars imported into the ER completes the core structure. The oligosaccharide-transferase complex transfers the completed oligosaccharide to the consensus Asn-X-Ser/Thr motif of a nascent chain as it enters the lumen of the ER.

(three glucoses, nine mannoses, and two *N*-acetyl glucosamines) ($Glc_3Man_9GlcNAc_2$). They are synthesized in a stepwise fashion while attached to the ER membrane by **dolichol phosphate** (a long-chained, unsaturated isoprenoid alcohol with pyrophosphate at one end; Fig. 20.11). Assembly of the oligosaccharide precursor involves 14 separate transfer reactions: seven on the cytoplasmic face of the ER and seven in the lumen. Midway in this process, the glycolipid flips from facing the cytoplasm to facing the lumen of the ER. Once facing the ER lumen, the enzyme oligosaccharyltransferase transfers the complete oligosaccharide from dolichol to asparagine side chains on the nascent polypeptide contained in the sequences Asn-X-Ser/Thr, where X is any amino acid other than proline. This sequence is recognized after it has emerged a distance of 12 to 14 residues out of the translocon into the ER lumen.

Calnexin/Calreticulin Cycle

Once the core oligosaccharide has been added, the glycoprotein begins a cycle of modifications that help it achieve its fully folded state. This **calnexin cycle** (Fig. 20.12) starts when glucosidases I and II remove the first two glucose residues of the core glycan. The resulting monoglucosylated transmembrane or soluble proteins bind to **calnexin,** a type I transmembrane protein in the ER lumen (Fig. 20.12). Monoglucosylated soluble proteins also bind to **calreticulin,** a soluble protein in the ER lumen similar to calnexin. Both are monomeric, calcium-binding proteins related to sugar-binding lectin proteins from legumes. Binding to calnexin or calreticulin prevents glycoproteins from aggregating and exposes them to glycoprotein to ERp57, a thiol-disulfide oxidoreductase, closely related to PDI. ERp57 bound to calnexin and calreticulin catalyzes intramolecular disulfide bond interchange during protein folding.

Once glucosidase II removes the remaining glucose residue on the core glycan, the glycoprotein is released from calnexin/ERp57. The glycoprotein is now free to leave the ER unless it is recognized by a soluble enzyme, uridine diphosphate (UDP)-Glc:glycoprotein glucosyltransferase (GT). GT reglucosylates incompletely folded glycoproteins, so it serves as a folding sensor in the cycle. When reglucosylated by GT, the glycoprotein goes through another cycle of binding to calnexin or calreticulin. The glycoprotein stays in the cycle until it is properly folded and oligomerized, at which point it enters the secretory pathway. If the protein fails to fold or oligomerize properly, it is removed from the cycle by being translocated out of the ER into the cytoplasm, where it is degraded. If participation in the cycle is inhibited, for example, by blocking the action of the glucosidases, the folding efficiency decreases. In this case, the glycoprotein may associate with BiP, which cooperates with the calnexin cycle in helping the protein to fold correctly.

Many polypeptides transferred to the ER are subunits of multiprotein complexes that assemble before leaving the ER. Because each polypeptide is synthesized on a separate ribosome and because the synthesis of the chains composing a complex may be unbalanced, BiP and other chaperones must protect hydrophobic surfaces and promote subunit interactions before export. For example, chaperones play a critical role in loading antigenic peptides onto major histocompatibility complex type I proteins in the ER (see Fig. 27.8D). These

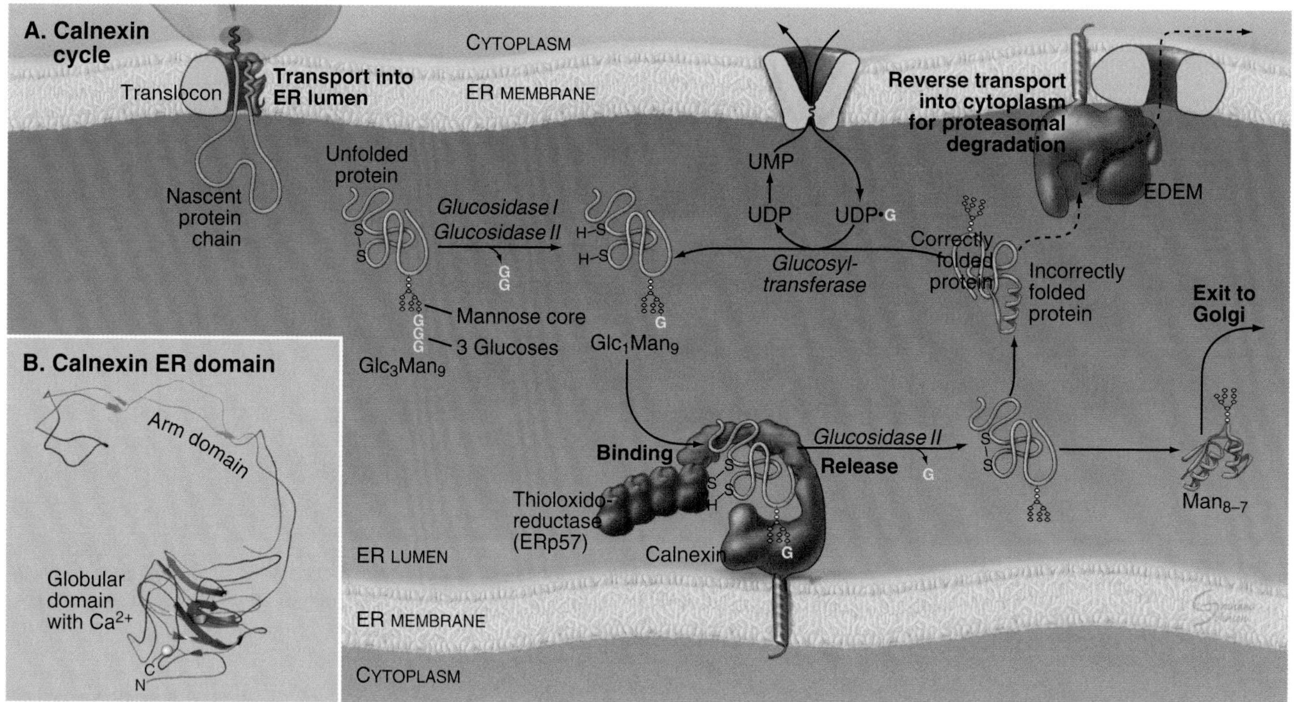

FIGURE 20.12 CALNEXIN CYCLE OF PROTEIN FOLDING IN THE LUMEN OF THE ER AND PROTEIN DEGRADATION FROM THE ER. A, Glucosidases I and II rapidly remove two of three glucoses from newly synthesized, unfolded glycoprotein. The calnexin-thioloxidoreductase complex binds the monoglucosylated protein. Glucosidase II removes the remaining glucose, releasing the protein. If the released protein is folded, it can exit the ER. If unfolded, it is recognized by glucosyltransferase and reglucosylated so that it reenters the folding cycle until folding is complete, or it enters the degradation pathway by interacting with EDEM (ER degradation-enhancing α-mannosidase-like protein) and a retrograde translocation channel, which deliver the unfolded polypeptide to the proteasome for degradation. Thioloxidoreductases catalyze rearrangements of disulfide bonds during folding. (Glc, glucose; Man, mannose; UDP, uridine diphosphate; UMP, uridine monophosphate.) **B,** Three-dimensional structure of calnexin showing a globular, lectin-binding domain and an extended arm composed of four repeat modules that folds around a sugar residue after it binds to the globular domain.

chaperones also protect cell surface receptors and secreted proteins from binding potential ligands that are also imported into the ER and that could activate them prematurely.

Protein Degradation in the Endoplasmic Reticulum and the Unfolded Protein Response

Protein Degradation in Endoplasmic Reticulum

Improperly folded polypeptides, excess subunits of oligomeric assemblies, or incorrectly assembled oligomers are degraded rather than being exported from the ER (Fig. 20.12). The degradation process, termed **ER-associated degradation (ERAD),** prevents accumulation of unsalvageable, misfolded proteins in the ER. Misfolded glycoproteins are recognized, retrotranslocated across the ER membrane to the cytoplasm, ubiquitylated (see Fig. 23.3), and then degraded by the proteasome in the cytoplasm.

The ERAD mechanism distinguishes misfolded or unassembled glycoprotein subunits from bona fide folding intermediates by linking degradation to the trimming of mannoses (Fig. 20.12). The concentration of mannosidases in the ER is low, so the terminal mannoses of the core oligosaccharide are unlikely to be trimmed from newly synthesized proteins if they fold promptly. On the other hand, if folding is prolonged, the terminal glucose and mannose residues are lost from the core oligosaccharide making them a substrate for a membrane-bound ER protein called **EDEM** (for "ER degradation-enhancing α-mannosidase-like protein"). EDEM directs the glycoprotein to the retrotranslocation channel for extraction to the cytoplasm. Sec61 is one candidate for the retrotranslocon, but the nature and mechanism of this pathway are still being investigated. In the cytoplasm the Bag6 complex (comprised of Bag6, TRC35α, and ubl4A) prevents aggregation of these proteins until they are ubiquitinated and degraded by proteasomes.

Sometimes proteins with signal sequences fail to translocate into the ER and are mislocalized to the cytoplasm. This can happen if there are mutations in the signal sequence, certain stresses, or intrinsic inefficiencies in their translocation. The Bag6 complex and proteasomes dispose of these proteins.

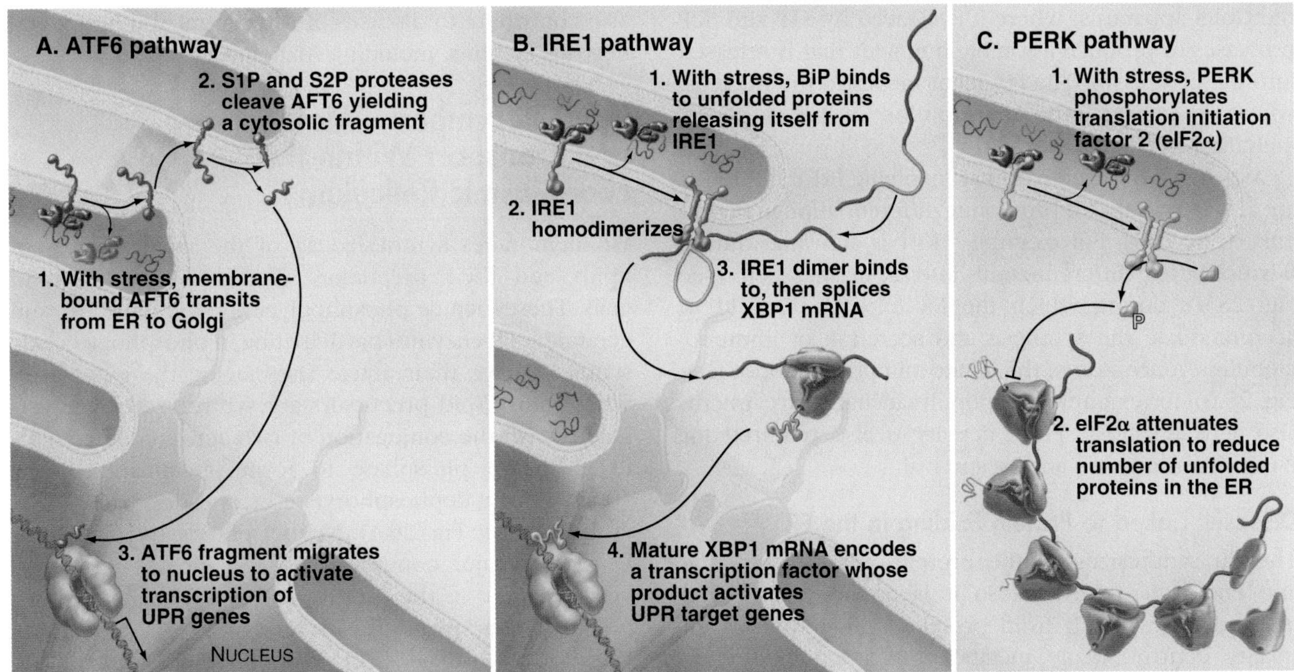

FIGURE 20.13 UNFOLDED PROTEIN RESPONSE (UPR) PATHWAYS TO STRESS IN THE LUMEN OF THE ENDOPLASMIC RETICULUM (ER). A, ATF6 (activating transcription factor 6) pathway. **B,** IRE1 (inositol requiring 1) pathway. **C,** PERK (PKR-like endoplasmic reticulum kinase) pathway. mRNA, messenger RNA.

Stress Responses in the Endoplasmic Reticulum

Conditions that flood the ER with excess protein or result in accumulation of misfolded proteins trigger the **unfolded protein response (UPR;** Fig. 20.13). Essentially, any condition in which protein import exceeds the protein-folding capacity of the ER triggers the UPR: misfolding of mutant proteins, inhibition of ER glycosylation (eg, by the drug tunicamycin), inhibition of disulfide formation (by reducing agents), or even overproduction of normal proteins. To compensate for these events, this stress-induced signaling pathway upregulates genes that are required to synthesize the entire ER, including its folding machinery. In yeast, the UPR activates more than 300 genes involved with all aspects of ER function, including lipid synthesis, protein translocation, protein folding, glycosylation, and degradation, as well as export to and retrieval from the Golgi apparatus. Developmental programs in metazoans may work through the same genetic controls to determine the abundance of ER in differentiated cells, producing, for example, extensive ER in secretory cells such as plasma, liver, and pancreatic acinar cells.

The UPR depends on three ER transmembrane proteins that sense and respond to stress in the ER: **ATF6** (activating transcription factor 6); **IRE1** (inositol requiring 1); and **PERK/PEK** (PKR-like endoplasmic reticulum kinase/pancreatic eIF2a kinase) (Fig. 20.13). Each of these proteins has a different mechanism but all initiate signaling pathways that regulate the production of proteins required for ER function. This allows cells to adjust the capacity of the ER to promote protein folding depending on the demand. Metazoan cells have all three pathways, but yeast have only IRE1.

Under normal conditions the concentration of BiP in the ER lumen exceeds the concentration of unfolded proteins, so free BiP is available to bind to the luminal domains of three ER transmembrane proteins, ATF6, IRE1 and PERK (PKR-like endoplasmic reticulum kinase). These interactions with BiP retain ATF6 in the ER and prevent the dimerization of IRE1 and PERK, keeping them inactive.

If unfolded proteins in the lumen of the ER bind to all of the BiP, IRE1 is free to dimerize. This activates the endoribonuclease activity of the cytoplasmic domain of IRE1, allowing it to remove a small intron from the messenger RNA (mRNA) for XBP1, a **bZIP (basic leucine zipper) domain–containing transcription factor** (see Fig. 10.14). Removing this intron alters the translational reading frame of XBP1 allowing the mRNA to encode a potent transcriptional activator of genes for ER proteins.

Similarly, without free BiP in the ER lumen, PERK dimerizes and phosphorylates **eukaryotic translation initiation factor 2 (eIF2).** This reduces the frequency of AUG codon recognition and slows the rate of translation initiation on many mRNAs. However, mRNAs for many proteins involved in cell survival and ER functions are preferentially translated under these conditions.

Accumulated unfolded proteins release BiP from ATF6, causing ER transmembrane protein to be transported to the Golgi apparatus, where it is cleaved by S1P and S2P proteases to produce a soluble fragment that is released into the cytoplasm. The fragment moves to the nucleus, where it activates the transcription of target genes, including XBP1.

Aspects of the UPR pathway involving IRE1 and PERK are also important for promoting differentiation in higher eukaryotic cells. For example, IRE1 is activated during B-lymphocyte differentiation into a plasma cell (see Fig. 28.9), during which the ER expands fivefold to accommodate the synthesis and secretion of immunoglobulins. Activation of the innate immune response (see Fig. 28.6), for example by lipopolysaccharide treatment, also activates IRE1. PERK activity is also required for B-cell differentiation and/or survival.

Diseases Linked to Protein Folding in the ER

The ER synthesizes all the proteins for the exocytic and endocytic pathways, so it is not surprising that many diseases result from proteins failing to pass ER quality control. Many metabolic disorders, including some lysosomal storage diseases (see Appendix 23.1), are a direct consequence of key enzymes failing to be exported from the ER. The most common form of **cystic fibrosis** results from the inability of cells to export a mutant form of the cystic fibrosis transmembrane regulator to the cell surface, where it normally functions as a chloride channel in the respiratory system and pancreas (see Fig. 17.4). Similarly, the inability of the liver to secrete mutated forms of α_1-antitrypsin into the blood predisposes affected individuals to the lung disease emphysema. Normally, α_1-antitrypsin protects tissues by inhibiting extracellular proteases such as elastase, which is produced by neutrophils. Mutations causing disease prevent α_1-antitrypsin folding in the ER, resulting in its degradation. The resulting deficiency in α_1-antitrypsin circulating in the blood allows elastase to destroy lung tissue, leading to emphysema. In severe cases, mutant forms of the protein not only fail to be exported from the ER but also elude degradation pathways, accumulating as insoluble aggregates that induce stress responses and liver failure.

In some conditions, the UPR helps compensate, in part, for mutations in cargo proteins. In congenital **hypothyroidism,** mutant thyroglobulin (the precursor of thyroid hormone) is not exported efficiently from the ER. Excess protein accumulates as insoluble aggregates in the ER. Feedback pathways trigger massive proliferation of ER in an attempt to produce normal levels of circulating hormone. Similarly, in mild forms of **osteogenesis imperfecta** (see Chapter 32), osteoblasts assemble and secrete defective procollagen chains for bone synthesis. As a result, the resulting bone tissue is weak. The alternative, complete loss of procollagen by retention and degradation of the mutant procollagen in the ER, would be lethal. Faulty ER quality control may also contribute to diseases of the central and peripheral nervous systems, including Alzheimer disease.

Lipid Biosynthesis, Metabolism, and Transport Within the Endoplasmic Reticulum

ER membranes synthesize all of the major classes of lipids and their precursors that are formed within cells. These include phosphoglycerides, cholesterol, and ceramide. ER enzymes participating in phosphoglyceride synthesis have their active sites facing the cytoplasm, where most lipid precursors are synthesized. Synthesis begins with the conjugation of two activated fatty acids to glycerol-3 phosphate to form phosphatidic acid, which can be dephosphorylated to produce diacylglycerol (DAG; see Fig. 26.4). Neither phosphatidic acid nor DAG is a major component of membranes; however, both are used in the synthesis of the four major phosphoglycerides: **phosphatidylcholine** (PC), **phosphatidylethanolamine** (PE), **phosphatidylserine** (PS), and **phosphatidylinositol** (PI) (see Fig. 13.2). The most abundant phospholipids, PC and PE, are produced with the activated head groups, cytidine diphosphate (CDP)-choline and CDP-ethanolamine. PI synthesis is by a distinct route using inositol and CDP-DAG produced from phosphatidic acid (see Fig. 26.7). PS synthesis (in mammalian cells) is an energy-independent exchange of polar head groups of PE. Head group exchange of phospholipids occurs primarily in the ER but may also occur in other organelles.

Biosynthesis of lipids in the cytoplasmic leaflet of the ER creates a topologic problem by restricting membrane growth to that side of the membrane. Individual lipids move to the leaflet facing the ER lumen by flipping across the bilayer (Fig. 20.14). This transfer across the ER bilayer is much faster than in vesicles of pure lipids, owing to protein translocators called **flippases**. The flippases in the ER membrane catalyze movements of most phospholipids in both directions without consuming metabolic energy (Fig. 20.14A). Thus they distribute lipids symmetrically across the ER bilayer. These key enzymes have not yet been identified.

Pumps using ATP hydrolysis as their energy source redistribute lipids across the membrane bilayers of the Golgi apparatus and plasma membrane (Fig. 20.14B). **P4-type ATPases** (see Fig. 14.7; also called aminophospholipid translocases) mediate fast exchange of PS and PE from the extracellular leaflet to cytoplasmic leaflet of the plasma membrane. Keeping PS in the extracellular leaflet low is important because PS on the cell surface is a signal for macrophages to engulf apoptotic cells (see Fig. 46.7) and activates blood platelets (see Fig. 30.14). Keeping PE levels high in the *cytoplasmic* leaflet facilitates endocytic budding events at the plasma membrane because of PE's cone-like shape, which

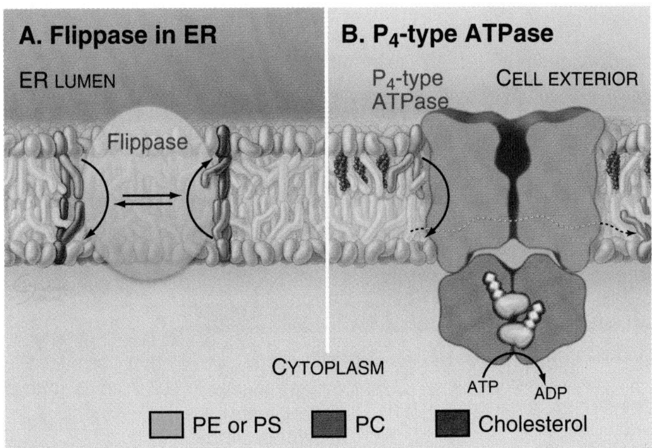

FIGURE 20.14 MOVEMENTS OF LIPIDS BETWEEN LEAFLETS OF THE MEMBRANE BILAYER. A, Uncharacterized flippases in the endoplasmic reticulum (ER) catalyze the exchange of lipids between leaflets promoting lipid symmetry across the ER bilayer. **B,** ABC transporter ATPases in the plasma membrane use ATP hydrolysis to transfer phosphatidylserine (PS) and phosphatidylethanolamine (PE) from the extracellular leaflet to the cytoplasmic leaflet of the plasma membrane. PC, phosphatidylcholine.

expands the cytoplasmic leaflet relative to the extracellular leaflet. Certain **ABC transporters** (see Fig. 14.10; also called **floppases**) facilitate the translocation of PC, glycosphingolipids, and cholesterol from the extracellular leaflet to the cytoplasmic leaflet. In liver cells, this process translocates lipids to the extracellular leaflet of the apical plasma membrane where they are released into the bile.

Cholesterol Synthesis and Metabolism

Cholesterol is maintained in animal cells by a combination of de novo synthesis (Fig. 20.15) in the ER and receptor-mediated endocytosis of lipoprotein particles containing esterified cholesterol (see Chapter 22). Coordinated regulation of these two processes precisely maintains the physiological level of cholesterol in cellular membranes. Peroxisomes also may participate in aspects of cholesterol synthesis and metabolism.

Twenty-two sequential steps are required to synthesize cholesterol from acetate (Fig. 20.15). Cytoplasmic enzymes catalyze the initial steps, using water-soluble molecules to produce **farnesyl-pyrophosphate** from acetyl coenzyme A (acetyl-CoA). An important exception is the step going from 3-hydroxy-3methylglutaryl CoA (HMG-CoA) to mevalonate. A carefully regulated, integral membrane protein of the ER (Fig. 20.15), **HMG-CoA reductase,** catalyzes this key step. Millions of people take drugs called **statins** that inhibit HMG-CoA reductase and lower cholesterol levels in the blood, an effective treatment to reduce cardiovascular disease owing to arteriosclerosis. Enzymes in the ER bilayer catalyze the subsequent condensation of farnesyl-pyrophosphate to make squalene, and cyclization to cholesterol, progressively less-polar molecules. Although the final steps

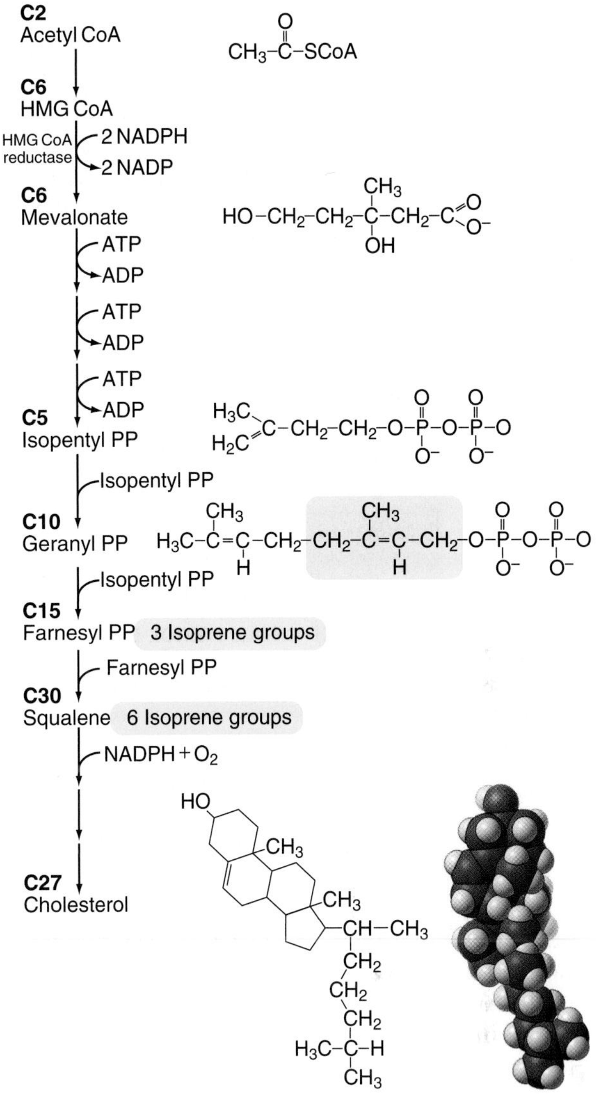

C2
Acetyl CoA — $CH_3-C(=O)-SCoA$
C6
HMG CoA
HMG CoA reductase: 2 NADPH → 2 NADP
C6
Mevalonate — $HO-CH_2-CH_2-C(CH_3)(OH)-CH_2-COO^-$
ATP → ADP
ATP → ADP
ATP → ADP
C5
Isopentyl PP — $H_3C(H_2C=)C-CH_2-CH_2-O-P(=O)(O^-)-O-P(=O)(O^-)-O$
Isopentyl PP
C10
Geranyl PP — $H_3C-C(CH_3)=CH-CH_2-CH_2-C(CH_3)=CH-CH_2-O-P(=O)(O^-)-O-P(=O)(O^-)-O$
Isopentyl PP
C15
Farnesyl PP 3 Isoprene groups
Farnesyl PP
C30
Squalene 6 Isoprene groups
$NADPH + O_2$
C27
Cholesterol

FIGURE 20.15 BIOSYNTHESIS OF CHOLESTEROL. Cholesterol is synthesized from acetyl coenzyme A (CoA). Synthesis of mevalonate from 3-hydroxy-3-methylglutaryl CoA (HMG-CoA) is closely regulated by controlling the concentration of the enzyme, HMG-CoA reductase, as shown in Figure 20.16. Five-carbon isoprene groups are the building blocks *(yellow)* for making a 10-carbon geranyl-pyrophosphate (PP), a 15-carbon farnesyl-pyrophosphate, and 30-carbon squalene intermediates. Several reactions add a hydroxyl group and join squalene into four rings to make cholesterol. NADP/NADPH, nicotinamide adenine dinucleotide phosphate.

leading to cholesterol take place in the ER, cholesterol is not a resident ER lipid and is rapidly exported to post-ER membranes, including the plasma membrane, where it constitutes up to 50% of the bilayer. Chapter 21 discusses the distribution of cholesterol in different organelles and current ideas regarding its transport.

Feedback loops sense the cholesterol content of the ER membrane and regulate both the synthesis and degradation of enzymes that synthesize cholesterol (Fig. 20.16). High cholesterol levels inhibit the synthesis and stimulate the destruction of key synthetic enzymes. A

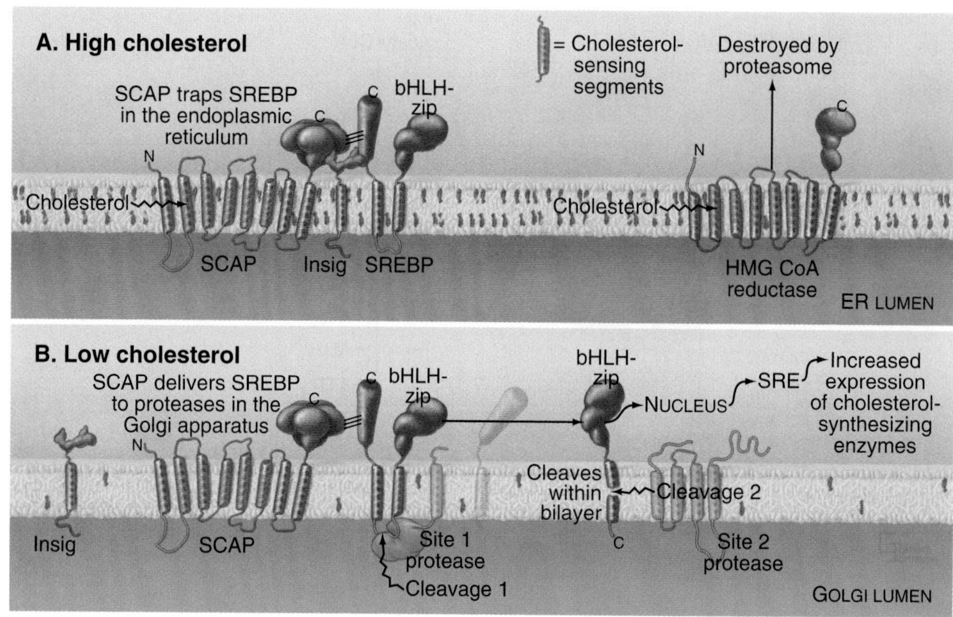

FIGURE 20.16 CONTROL OF CHOLESTEROL BIOSYNTHESIS BY PROTEOLYSIS. A, High-cholesterol conditions. Cholesterol-sensing transmembrane segments *(pink)* of SCAP (SREBP cleavage-activation protein) retain intact SREBP (sterol regulatory element-binding protein) in the endoplasmic reticulum (ER) through an interaction with Insig. Similar cholesterol-sensing transmembrane segments of β-hydroxy-β-methylglutaryl-coenzyme A (HMG-CoA) reductase stimulate its destruction by proteolysis. **B,** Low-cholesterol conditions. Insig dissociates from SCAP and SREBP, allowing these molecules to move to the Golgi apparatus where the membrane-anchored site 1 serine protease cleaves the loop of SREBP in the lumen. Subsequently, a transmembrane zinc-protease cleaves SREBP at a second site within the bilayer, releasing the basic helix-loop-helix-zip (bHLH-zip) transcription factor. This factor enters the nucleus, where it activates genes for cholesterol biosynthetic enzymes by steroid-response elements (SREs).

transcription factor precursor, steroid regulator element-binding protein **(SREBP),** controls the expression of these genes (see Fig. 23.10). When cholesterol is abundant, SREBP cleavage-activating protein **(SCAP),** a partner protein with cholesterol-sensing transmembrane segments (see Fig. 23.11), retains SREBP in the ER owing to interactions with the protein Insig. When the membrane cholesterol level is low, SCAP and Insig do not interact. The SREBP–SCAP complex is therefore free to move to the Golgi apparatus, where two successive proteolytic cleavages release the N-terminal domain of SREBP, a basic helix-loop-helix leucine zipper transcription factor, into the cytoplasm. (The Golgi proteases that are responsible for cleaving SREBP are SP1 and SP2, the same ones used to process ATF6 during the UPR [Fig. 20.13].) In the nucleus, the transcription factor binds steroid regulatory elements, enhancers for a wide range of genes encoding enzymes that synthesize cholesterol and other lipids, as well as low-density lipoprotein receptors that take up cholesterol from outside the cell (see Fig. 23.10). Cholesterol also regulates degradation of HMG-CoA reductase, which has cholesterol-sensing transmembrane domains similar to SCAP. Abundant cholesterol targets HMG-CoA reductase for degradation by the proteolytic pathway that disposes of unfolded proteins through the proteasome (Fig. 20.12; also see Chapter 23). The enzyme **acyl-CoA-cholesterol transferase** helps lower cholesterol levels in the ER bilayer by catalyzing the formation of cholesterol esters, a storage form of cholesterol.

Ceramide Synthesis

Ceramide, the backbone of all sphingolipids (see Fig. 13.3), also begins its synthesis on the cytoplasmic face of ER membranes. It is made through sequential condensation of the amino acid serine with two fatty acids. Ceramide is transported to the Golgi apparatus by a nonvesicular pathway (see the next section), where enzymes on the luminal leaflet either add oligosaccharide chains to form glycosphingolipids or add a choline head group to form sphingomyelin (see Fig. 13.3 and Chapter 21).

Lipid Movements Between Organelles

Lipids diffuse rapidly in the plane of the bilayer and slowly flip-flop across the bilayer (see Chapter 13), but their hydrophobic nature makes free diffusion between adjacent bilayers through the aqueous cytoplasm thermodynamically unfavorable and very slow. Budding and fusion of vesicles is a major pathway for moving lipids between organelles (see Chapter 21). However, **lipid-transfer proteins (LTPs)** can mediate direct exchange of individual lipid molecules between bilayers at membrane contact sites (Fig. 20.17A–B). Plant LTPs are tiny

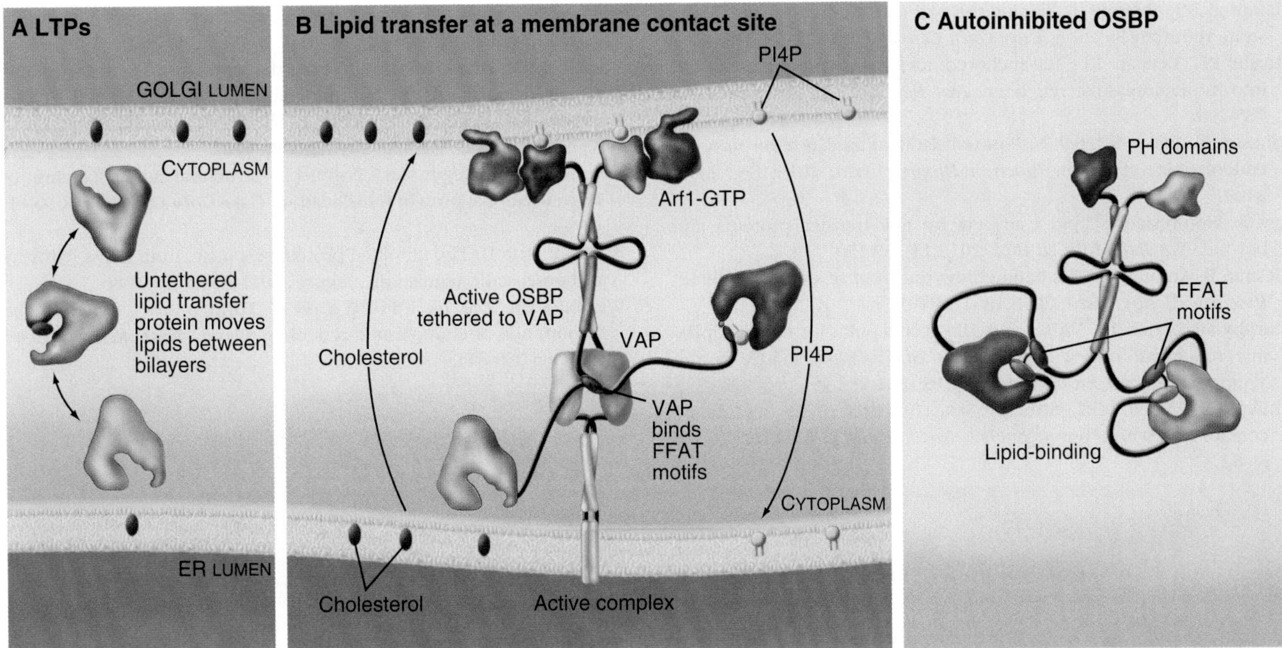

FIGURE 20.17 MECHANISMS OF LIPID MOVEMENTS BETWEEN MEMBRANE BILAYERS. A, Cytoplasmic lipid transfer proteins bind membrane lipids and transfer them to other membranes. **B,** Model for lipid transfer at contact sites between the endoplasmic reticulum (ER) and Golgi apparatus membranes. The VAP protein anchored to the ER binds the FFAT motif of OSBP, while the PH domain of OSBP associates with phosphatidylinositol (PI)-4P in the Golgi apparatus membrane. The lipid binding domains of OSBP transfer cholesterol from the ER to the Golgi apparatus and PI4P in the opposite direction while swinging on tethers between the two membranes. **C,** Domain organization of autoinhibited OSBP (oxysterol-binding protein). (Modified from Mesmin B, Antonny B. The counterflow transport of sterols and PI4P. *Biochim Biophys Acta.* 2016;1861:940–951.)

(9-kD) proteins with a hydrophobic pocket that binds a range of lipids and facilitates their movements between membranes.

Animals have several families of LTPs with different structures but related strategies for binding particular lipid molecules. Often a hinged "cover" protects the lipid-binding pocket, allowing the lipid-binding domain to embed partially into the cytoplasmic leaflet of a membrane bilayer.

Some LTPs consist of a single lipid-binding domain, so they extract a lipid from a membrane and then diffuse through the cytoplasm to deliver the lipid to another membrane like plant LTPs. Other LTPs have additional domains (Fig. 20.17B). Many have a ligand-binding domain such as a PH domain (see Fig. 25.10) that binds phosphoinositides and an FFAT motif that binds **VAP** on the cytoplasmic surface of the ER. These interactions bring two membranes close together (Fig. 20.4), setting up conditions that favor lipid transfer. For example, VAP in the ER anchors one end of the LTP and the PH domain binds phosphoinositides in the Golgi apparatus or plasma membrane. The lipid-binding domain exchanges lipids between the closely apposed membranes as it moves back and forth on a mobile tether (see Fig. 12.10B). The result is lipid exchange but not net lipid transfer.

One example of an LTP is **ceramide transport protein (CERT)**, which extracts newly synthesized ceramide from the ER and carries it to the Golgi apparatus. Another example is the oxysterol-binding protein (OSBP) that binds to VAP-A in the ER and Arf1-GTPase in the Golgi apparatus. The lipid-binding domain promotes the counter-flow of cholesterol from the ER to the Golgi apparatus and PI4P in the opposite direction. This helps the cell tune its cholesterol levels according to the PI4P level in the Golgi. VAP also recruits other proteins with FFAT-like motifs to ER membrane contact sites.

ACKNOWLEDGMENTS

We thank Craig Blackstone, Ramanujan Hegde, and Rebecca Voorhees for their suggestions on revisions to this chapter.

SELECTED READINGS

Borgese N, Fasana E. Targeting pathways of C-tail-anchored proteins. *Biochim Biophys Acta.* 2011;1808:937-946.

Brambilla Pisoni G, Molinari M. Five questions (with their answers) on ER-associated degradation. *Traffic.* 2016;17:341-350.

Cherepanova N, Shrimal S, Gilmore R. N-linked glycosylation and homeostasis of the endoplasmic reticulum. *Curr Opin Cell Biol.* 2016;41:57-65.

Chung J, Torta F, Masai K, et al. PI4P/phosphatidylserine countertransport at ORP5- and ORP8-mediated ER-plasma membrane contacts. *Science.* 2015;349:428-432.

Goyal U, Blackstone C. Untangling the web: Mechanisms underlying ER network formation. *Biochim Biophys Acta.* 2013;1833:2492-2498.

Hampton RY, Sommer T. Finding the will and the way of ERAD substrate retrotranslocation. *Curr Opin Cell Biol.* 2012;24:460-466.

Hegde RS, Keenan RJ. Tail-anchored membrane protein insertion into the endoplasmic reticulum. *Nat Rev Mol Cell Biol.* 2011;12:787-798.

Johnson N, Powis K, High S. Post-translational translocation into the endoplasmic reticulum. *Biochim Biophys Acta.* 2013;1833:2403-2409.

Lev S. Non-vesicular lipid transport by lipid-transfer proteins and beyond. *Nat Rev Mol Cell Biol.* 2010;11:739-750.

Mesmin B, Antonny B. The counterflow transport of sterols and PI4P. *Biochim Biophys Acta.* 2016;1861:940-951.

Murphy SE, Levine TP. VAP, a versatile access point for the Endoplasmic reticulum: review and analysis of FFAT-like motifs in the VAPome. *Biochim Biophys Acta.* 2016;1861:952-961.

Phillips MJ, Voeltz GK. Structure and function of ER membrane contact sites with other organelles. *Nat Rev Mol Cell Biol.* 2016;17:69-82.

Shao S, Hegde RS. Membrane protein insertion at the endoplasmic reticulum. *Annu Rev Cell Dev Biol.* 2011;27:25-56.

Voorhees RM, Hegde RS. Structures of the scanning and engaged states of the mammalian SRP-ribosome complex. *eLife.* 2015;4:e07975.

Voorhees RM, Hegde RS. Structure of the Sec61 channel opened by a signal sequence. *Science.* 2016;351:88-91.

Voorhees RM, Hegde RS. Toward a structural understanding of co-translational protein translocation. *Curr Opin Cell Biol.* 2016;41:91-99.

Walter P, Ron D. The unfolded protein response: from stress pathway to homeostatic regulation. *Science.* 2011;334:1081-1086.

Westrate LM, Lee JE, Prinz WA, et al. Form follows function: the importance of endoplasmic reticulum shape. *Annu Rev Biochem.* 2015;84:791-811.

CHAPTER 21

Secretory Membrane System and Golgi Apparatus

Eukaryotic cells transport newly synthesized proteins destined for the extracellular space, the plasma membrane, or the endocytic/lysosomal system through a series of functionally distinct, membrane-bound compartments, including the **endoplasmic reticulum (ER), Golgi apparatus,** and vesicular transport intermediates. This secretory membrane system (Fig. 21.1) allows eukaryotic cells to perform three major functions: (1) distribute proteins and lipids synthesized in the ER to the cell surface and other cellular sites; (2) modify and/or store protein and lipid molecules after their export from the ER; and (3) generate and maintain the unique identities and functions of the ER, Golgi apparatus, and plasma membrane.

The eukaryotic secretory membrane system offers advantages over the simpler secretory process in prokaryotic cells that inserts newly synthesized proteins directly into or across the plasma membrane (see Fig. 18.9). First, distinct compartments of the secretory pathway provide protective environments to synthesize, fold, assemble, and modify membrane and secretory proteins before they are exposed on the cell surface.

Second, eukaryotic cells have the capacity to store secretory proteins in membrane compartments. This is an advantage because it allows the cell to regulate the release of proteins from the cell surface in response to internal or external signals.

Third, membranes with distinct lipid compositions allow for specialized functions in the various compartments. For example, the lipid composition of the ER is favorable for folding of transmembrane proteins. On the other hand, ordered, flexible arrays of cholesterol and sphingolipids in the plasma membrane are mechanically stable and impermeable to water-soluble molecules. These properties are advantageous for the cell surface as well as for membrane protrusion for phagocytosis (see Chapter 22) and for crawling (see Chapter 38).

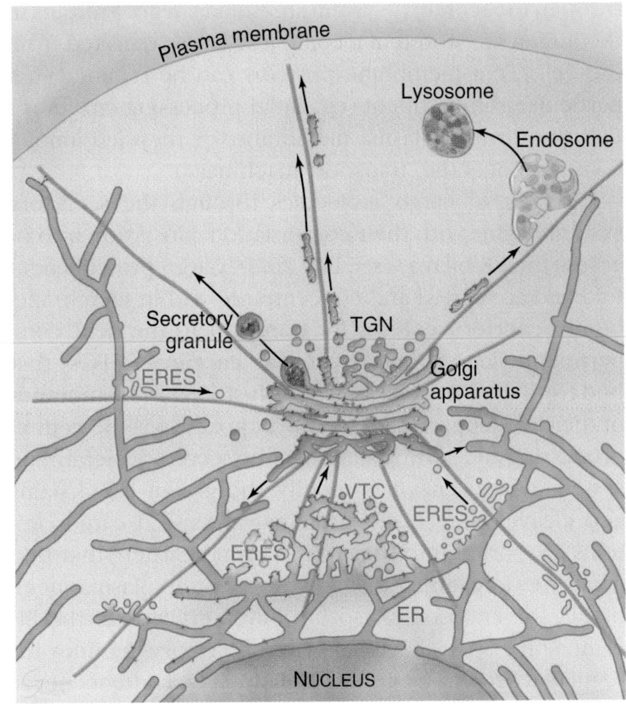

FIGURE 21.1 OVERVIEW OF THE SECRETORY MEMBRANE SYSTEM. Transport carriers move lipids and proteins between the three principal organelles of the secretory pathway—the endoplasmic reticulum (ER), Golgi apparatus, and plasma membrane. The carriers are either small vesicles or larger vesicle–tubule elements that move along microtubules *(brown lines)* between membranes. The anterograde pathway takes newly synthesized proteins, called cargo, from the ER to the Golgi apparatus, and from the Golgi apparatus to the plasma membrane or to the endosome/lysosomal system. Carriers form at ER export sites (ERES), producing vesicular tubular carriers (VTCs) that move to the Golgi apparatus. Retrograde transport carriers bud from the VTC or Golgi apparatus to retrieve proteins and lipids back to the ER. A different set of retrograde carriers retrieve proteins and lipids from the endosomal/lysosomal system back to the *trans*-Golgi network. The spaces inside carriers and organelles of the secretory membrane system are all topologically equivalent to the outside of the cell.

Prokaryotic cells must keep their plasma membrane pliable enough with loosely packed phosphoglycerides for newly synthesized transmembrane proteins to fold in a hydrophobic environment.

This chapter describes how the secretory membrane system is organized and operates to fulfill these functions. It also provides a detailed description of the Golgi apparatus whose conserved features are central for the operation of the secretory membrane system.

Overview of the Secretory Membrane System

The secretory membrane system uses membrane-enclosed transport carriers to move thousands of diverse macromolecules—including proteins, proteoglycans, and glycoproteins—efficiently and precisely among different membrane-bound compartments (ie, the ER, Golgi apparatus, and plasma membrane). Proteins transported through the secretory system are called **cargo.** Proteins in the lumen are stored in a compartment or secreted from the cell. Transmembrane proteins can be retained in a particular compartment (eg, Golgi processing enzymes), delivered to the plasma membrane, or recycled among compartments (eg, transport machinery).

Transfer of cargo molecules through the secretory system begins with their cotranslational insertion into or across the ER bilayer (see Fig. 20.8). Cargo proteins next fold and are sorted and concentrated within membrane-bound **carrier vesicles** for transport to the next compartment. The **vesicular tubular carriers (VTCs)** that bud from the ER are destined for fusion with membranes of the Golgi apparatus. The Golgi apparatus is the central processing and sorting station in the secretory membrane system where enzymes modify the glycan side chains and cleave some cargo proteins. Processed cargo proteins are sorted into membrane-bound carriers that bud from the Golgi apparatus and move to the plasma membrane, the endosome/lysosomal system, back to the ER or, in some specialized cells, into secretory granules for regulated secretion later on. Remarkably, contents never leak during these steps in cargo transport.

Membrane-enclosed **carriers** that transport cargo through the secretory system are shaped as tubules, vesicles, or larger structures (Fig. 21.2). Carriers continuously shuttle between ER, Golgi apparatus, endosomes, and plasma membrane, distributing cargo to its appropriate target organelle. Carriers are too large to diffuse in the crowded cytoplasm but are transported along microtubules or actin filaments by molecular motor proteins.

The flow of cargo and lipid forward through the secretory system toward the plasma membrane **(anterograde traffic)** is balanced by selective **retrograde traffic** of cargo proteins and lipids back toward the ER (Fig. 21.1) for repeated use. Retrograde traffic also returns proteins

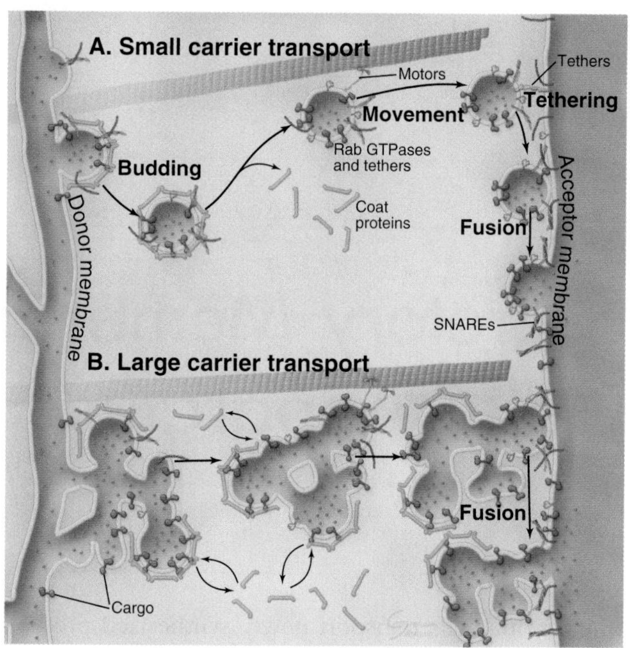

FIGURE 21.2 PROTEIN MACHINERY FOR SECRETORY TRANSPORT. Coat proteins help sort soluble cargo and transmembrane proteins into a coated bud that pinches off a donor membrane as **(A)** a coated vesicle or **(B)** larger vesicular tubular carriers. Motor proteins move carriers along either microtubules (shown here) or actin filaments. Long coiled-coil tethers or multimeric tethering complexes attach carriers to an acceptor membrane. SNARE (soluble *N*-ethylmaleimide-sensitive factor attachment protein receptor) proteins on the carrier and acceptor membrane then form a complex that drives membrane fusion and delivery of the carrier's membrane and content to the acceptor membrane. During this process, the relative topology of the lipids and transmembrane proteins is maintained.

that are inadvertently carried forward through the secretory system for redirection to their proper location. Both anterograde and retrograde flows of membrane within the secretory system are necessary for the ER, Golgi apparatus, and plasma membrane to generate and maintain their distinct functional and morphologic identities. Despite this continuous two-way traffic, all the organelles in the secretory system maintain their distinct characters, including different lipid and protein compositions.

The orientation (topology) of lipids and proteins in the membrane bilayer is established during synthesis in the ER and maintained during transport by a carrier (Fig. 21.2). Hence, the same side of the membrane always faces the cytoplasm. The other side initially faces the lumen of the ER. This side remains inside each membrane compartment along the secretory pathway or is exposed on the cell surface if the carrier fuses with the plasma membrane.

Cargo freely enters, concentrates, or is excluded from each transport carrier. This chapter describes the various mechanisms that facilitate such sorting of cargo, including the role of lipid gradients and protein-based sorting machinery. It next explains proteins that ensure carriers fuse only with the correct target compartment. Finally, it describes the Golgi apparatus and its diverse membrane trafficking pathways.

Role of Lipid Gradients in Membrane Trafficking

A conserved feature of the secretory membrane system is the differential distribution of **glycerophospholipids** (phosphoglycerides), **sphingolipids** (eg, sphingomyelin and glycosphingolipids), and **cholesterol** (see Figs. 13.2, 13.3, and 13.4) along the pathway (Fig. 21.3A). The concentrations of cholesterol and sphingolipids range from low in the ER membranes to intermediate in the Golgi apparatus and high in the plasma membrane (Fig. 21.3). Membranes containing these lipids differ in thickness, one property used to sort transmembrane proteins, which tend to partition into membranes with a thickness that matches the lengths of their transmembrane segments.

Generation and Maintenance of the Lipid Gradient

Cells establish and actively maintain a lipid gradient across the secretory pathway with a thin ER bilayer, progressively thicker bilayers in intermediate compartments and the thickest bilayer of the plasma membrane. The gradient results from the self-organizing capacity of sphingolipids, cholesterol, and phosphoglycerides; synthesis of these lipids at different sites; and sorting of lipids by vesicular and nonvesicular mechanisms. These three factors produce lipid circulation that generates the lipid gradient (Fig. 21.3A). Physical interactions among the various classes of lipids can concentrate or exclude specific membrane lipids and proteins in *microdomains*. The different lipid compositions within the secretory system facilitate sorting of membrane proteins.

Mixtures of purified lipids in artificial membrane lipid bilayers can self-organize into distinct **lipid domains** with unique properties. For example if sphingolipid is added to an artificial bilayer containing glycerophospholipids and cholesterol, the cholesterol and sphingolipid partition into domains separate from the glycerophospholipids (Fig. 21.4A–B). Cholesterol and sphingolipids associate in the plane of the membrane because of van der Waals interactions between the sphingolipid's long, saturated hydrocarbon chain and cholesterol's rigid, flat-cylindrical steroid backbone. Glycerophospholipids are largely excluded from cholesterol and sphingolipid domains, because unsaturated, kinked hydrocarbon chains do not interact as closely with cholesterol. The domains enriched in cholesterol and sphingolipid are thicker than the surrounding membrane composed of shorter, unsaturated, kinked glycerophospholipids (Fig. 21.4C). Tension on the bilayer (ie, from binding of proteins that bend or curve the membrane or from pulling the membrane into a narrow tubule) enhances the tendency of lipids with different physical properties to separate into distinct phases (see Fig. 13.6).

Cholesterol can also affect a bilayer composed of glycerophospholipids alone (Fig. 21.4D; also see Fig. 13.6) by filling in the spaces between the floppy hydrocarbon chains of glycerophospholipids. This forces the glycerophospholipids into a tighter alignment and increases the distance between their head groups. As a result, the bilayer is thicker and ordered. This is a characteristic of bilayers enriched in sphingomyelin alone or sphingomyelin plus cholesterol. Indeed, disordered and ordered lipid phases separate in a giant lipid vesicle and the disordered lipid phase spontaneously partitions into a narrow tubule that can be pulled from the vesicle (see Fig. 13.6). This effect of shape could promote the partitioning of lipids along the secretory pathway.

Synthesis of the major membrane lipids at different locations contributes to the lipid gradient. The ER produces cholesterol and glycerophospholipids while the Golgi apparatus synthesizes glycosphingolipids and sphingomyelin. Newly synthesized cholesterol is

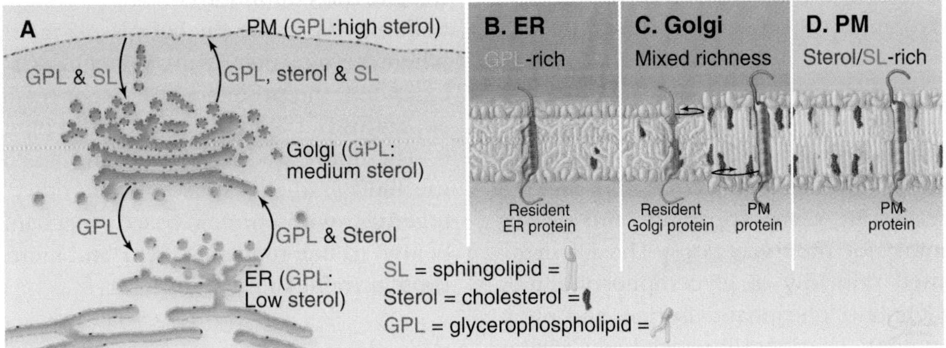

FIGURE 21.3 GRADIENT OF LIPIDS ACROSS THE SECRETORY PATHWAY. The gradient arises from the self-organizing properties of the lipids and their synthesis at different sites. This gradient helps sort transmembrane proteins during membrane traffic. **A,** Lipid circulation and sorting within the secretory membrane system. Glycerophospholipids (GPLs) and cholesterol (sterol) are synthesized in the ER, whereas sphingolipids (SLs), including sphingomyelin and glycosphingolipids, are synthesized in the Golgi apparatus. Cholesterol moves to the Golgi apparatus where it associates with SL and is carried to the plasma membrane. **B–D,** Sorting of transmembrane proteins based on the length of their transmembrane helices: short helices in the thin bilayer of the ER composed of GPLs; medium helices in mixed bilayers of the Golgi apparatus; and long helices in the thick bilayer of the plasma membrane enriched in SL and sterols.

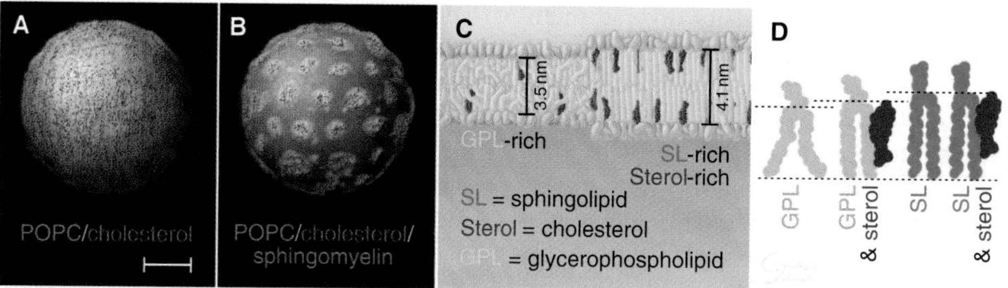

FIGURE 21.4 PHYSICAL PROPERTIES OF MEMBRANE LIPIDS. A–B, Illustrations of vesicles with artificial bilayers. **A,** A 2:1 mixture of POPC/cholesterol. **B,** A 2:1:1 mixture of POPC/cholesterol/sphingomyelin. The *blue circular spots* in part **B** are domains enriched in cholesterol and sphingomyelin that have segregated from POPC. **C,** A bilayer enriched in cholesterol and sphingomyelin is thicker than a bilayer composed mainly of glycerophospholipids, owing to the long saturated hydrocarbon chains of glycosphingolipids. **D,** Heights of membrane lipids: short, glycerophospholipids alone; medium, glycerophospholipids with high concentrations of cholesterol; and tall, sphingolipids and sterols. POPC, 1-palmitoyl-2-oleoyl-*sn*-glycero-3-phosphocholine.

continually removed from the ER by both vesicular and nonvesicular pathways (see Fig. 20.17) and redistributed to the Golgi apparatus, where interactions with sphingolipids prevent it from returning to the ER. Association of cholesterol with sphingolipids in the Golgi apparatus triggers the lateral differentiation of domains enriched in these lipids.

Movement of lipids by protein-based sorting and trafficking (see the next section) is a third factor contributing to the lipid gradient. Anterograde movements of vesicles containing glycerophospholipids, sphingolipids, and cholesterol from the Golgi apparatus to the plasma membrane is balanced by selective retrograde flow. Vesicular transport recycles glycerophospholipids transferred from the ER to the Golgi apparatus back to the ER. Similarly, sphingolipids delivered to the plasma membrane from the Golgi apparatus are returned to the Golgi apparatus. Cholesterol, in contrast, is not returned through these retrograde pathways to either the ER or Golgi apparatus; instead, it circulates within the endocytic pathway leading to lysosomes. From lysosomes, cholesterol can be passed back to the ER by lipid transfer proteins to inform sterol sensors in the ER of the cellular cholesterol balance (see Fig. 20.16). Excess cholesterol sensed in the ER can be esterified and stored in lipid droplets.

Functions of the Lipid Gradient

The lipid gradient along the secretory pathway serves two important functions. First, it generates different lipid environments in the ER, Golgi apparatus, and plasma membrane favorable for their functions. The ER membrane is composed primarily of glycerophospholipids (ie, phosphatidylcholine, phosphatidylserine, and phosphatidylethanolamine). The loosely packed acyl chains of phosphatidylcholine, phosphatidylcholine, and phosphatidylethanolamine are readily deformable, permitting newly synthesized membrane proteins to insert into and fold in the ER bilayer. By contrast, abundant sterols and sphingolipids make the plasma membrane bilayer thicker and less permeable to small molecules, suitable as the surface barrier. The intermediate concentration of sterols and sphingolipids in the Golgi apparatus may have advantages for a membrane-sorting station.

The lipid gradient also directs transmembrane proteins to specific destinations where the lengths of transmembrane helices match the thickness of the target membranes (Fig. 21.3B–D). This matching avoids the energy cost of exposing hydrophobic residues to water or hydrophilic residues to the lipid bilayer. Hence, resident membrane proteins in the ER and Golgi apparatus typically have shorter transmembrane segments (approximately 15 residues) than resident plasma membrane proteins (approximately 20 to 25 residues). Retention or transport of these proteins occurs because the lipid bilayers of carriers budding out from either the ER (toward the Golgi apparatus) or the Golgi apparatus (toward the plasma membrane) are thicker than the bilayers of the donor organelles. Only transmembrane proteins with transmembrane segments long enough to span this thickness enter such carriers.

If the transmembrane segment of a plasma membrane protein is shortened experimentally by recombinant DNA techniques, the new protein is retained in the thinner bilayers of the ER and/or Golgi apparatus rather than moving to the thicker plasma membrane. Similarly, when the transmembrane segment of a Golgi protein is extended, the protein is no longer retained in the Golgi apparatus but is transported to the plasma membrane.

The lipid-based protein sorting mechanism based on the lipid gradient across the secretory pathway works together with protein-based mechanisms, described below, to fine tune specificity and increase efficiency in protein trafficking within cells.

Proteins Involved in Membrane Trafficking

Sorting and transporting proteins within the secretory membrane system depend on a variety of proteins: specialized coat proteins that help generate transport carrier vesicles containing specific cargo proteins; "tethering factors" that attach carriers to their target membranes;

proteins that mediate fusion of the carrier vesicle with an acceptor membrane; and membrane-bending proteins that detect and help drive membrane curvature. Activated GTPases or specific lipids such as phosphoinositides recruit these sorting proteins to the cytoplasmic surfaces of appropriate membranes. Motor proteins move the carrier vesicles along cytoskeletal tracks to the appropriate cellular location.

Guanosine Triphosphatases That Regulate Membrane Traffic

Two major families of small guanosine triphosphatases (GTPases)—Arf (adenosine diphosphate [ADP]-ribosylation factor) and Rab (Ras-related in brain)—regulate membrane traffic; all are related to Ras (Figs. 4.6 and 4.7). Like Ras, their intrinsic GTPase cycles are slow, being limited by the rate of GTP hydrolysis, and even more by the rate of guanosine diphosphate (GDP) dissociation. Each GTPase has guanine nucleotide exchange factors (GEFs) that not only catalyze the exchange of GDP for GTP. This both activates these proteins and targets them to particular membranes. Once located on the appropriate membrane and activated with bound GTP, these GTPases bind effector proteins that form membrane buds or proteins that target carrier vesicles to their destinations. Like Ras, GTPase activating proteins (GAPs) turn off these membrane traffic GTPases by stimulating GTP hydrolysis.

ADP-Ribosylation Factor and Sar1 GTPases

Arf family **GTPases,** including **Arf1** to **Arf6** and **Sar1** (secretory and Ras-related) GTPase, recruit specific protein effectors to the membrane surface. The effectors function in vesicle formation and tethering, organelle biogenesis and dynamics, nonvesicular lipid transport and cytoskeletal regulation. Multiple GEFs on specific membranes and GAPs coordinate the spatial and temporal regulation of these small GTPases. Whereas Arf1-6 proteins function at various sites throughout the endomembrane system, Sar1 functions exclusively at ER exit sites.

A key feature of Arfs is a myristoylated N-terminal amphipathic helix that facilitates binding to membranes (Fig. 21.5). Upon GTP binding, this helix is released from a hydrophobic pocket on Arf and inserts into the lipid bilayer, resulting in strong membrane association.

Sequence comparisons separate Arf proteins into three classes. All eukaryotes have class I (Arfs 1, 2, and 3), animals alone have class II (Arf4 and 5), and both animals and plants have class III (Arf6). Arfs 1, 3, 4, and 5 function primarily in the secretory pathway, whereas Arf6 functions in the endocytic pathway.

Arf GEFs recruit specific Arf proteins to particular membrane surfaces and then catalyze the exchange of GDP for GTP. All six subfamilies of Arf GEFs contain a conserved Sec7-like domain, which catalyzes GTP exchange for GDP. Some Arf GEFs function in the secretory pathway (GBF/Gea GEFs), others work at the *trans*-Golgi and *trans*-Golgi network (TGN) (BIG/Sec7 GEFs), and still others function in the endocytic system (cytohesin/Arno GEFs). The toxic fungal metabolite **brefeldin A** inhibits Arf1 activation in the secretory pathway by trapping a Sec7 domain–Arf–GDP complex (Fig. 21.5D). This prevents Arf1 activation by GTP, blocks the secretory pathway and causes the Golgi apparatus to disassemble.

Hydrophobic residues of effector proteins then bind Arf-GTP in an area centered on a triad of aromatic residues (Fig. 21.5). Effectors are usually dimers or pseudodimers, so they bind two Arf-GTP molecules symmetrically. This may orientate effectors on the membrane to create multivalent membrane-binding platforms.

The highly diverse Arf effectors include coat complexes, lipid transfer proteins, membrane tethers, scaffold proteins, and actin regulators. Arf1's effectors include the coat protein I (COPI) coat in the early secretory pathway; the adaptor protein 1 (AP1), and AP3 coats in post-Golgi trafficking; long coiled-coil proteins associated with Golgi membranes; and lipid transfer proteins at contact sites between ER and Golgi. Arf-GTP both recruits and mediates conformational changes of effector proteins.

While associated with a membrane, active Arfs bind their effectors until an Arf GAP induces hydrolysis of the bound GTP, reversing membrane association and effector binding. The locations of 11 subfamilies of GAPs determine where each type of Arf is inactivated. Arf GAPs use a conserved domain with a zinc finger to catalyze the hydrolysis of GTP (Fig. 21.5C). Some Arf GAPs contain a BAR domain, which both senses and induces membrane curvature. Other Arf GAPs have amphipathic lipid packing sensor (ALPS) motifs that mediate binding to highly curved membranes. Still other Arf GAPs contain domains such as Rho GAP domains, ankryin repeats or Src homology-3 (SH3) domains for interactions with the actin cytoskeleton and signaling networks.

Rab GTPases

Rab family GTPases control protein–protein interactions between transport carriers and docking complexes on target membranes (Fig. 21.6). Mammals express approximately 70 different Rab proteins to specify numerous transport steps in the secretory membrane system. The Rab proteins localize to different membrane compartments and trafficking pathways, functioning as membrane organizers by regulating a plethora of soluble effector molecules (Fig. 21.6). Researchers use Rab proteins as markers for specific organelles (eg, Rab5 for early endosomes) or, when expressed in mutant forms, to perturb the system (see Fig. 22.12B).

The bound nucleotide controls the cycling of Rab proteins between membranes and the cytoplasm

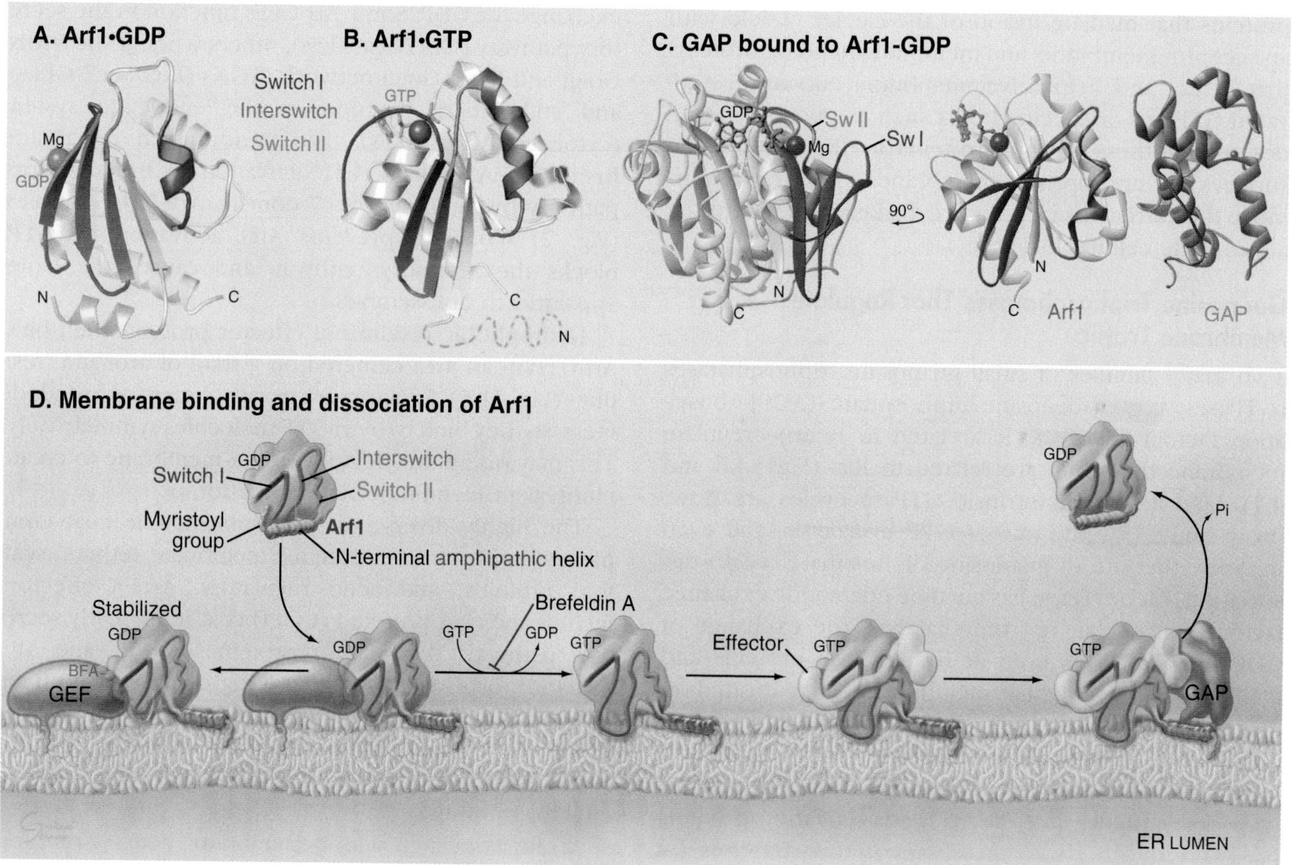

FIGURE 21.5 ARF-GTPASE CYCLE. A–C, Ribbon diagrams. **A,** Arf1-GDP; **B,** Arf1-GTP; and **C,** Arf1 bound to its GAP. **D,** Membrane binding and dissociation of Arf1. In the cytoplasm, Arf1 exists in its GDP-bound form with its N-terminal amphipathic helix tucked into a hydrophobic pocket. An N-terminal myristoyl group allows Arf1 to bind membranes for activation by a GEF. The exchange of GDP for GTP changes the conformation of switch 1 and switch 2, as well as the interswitch loop and displaces the N-terminal helix *(striped blue and pink)* from its pocket so it can anchor Arf1-GTP to the membrane. Arf1-GTP then recruits effectors. Association of a GAP with the Arf1–GTP–effector complex stimulates GTP hydrolysis. Arf1-GDP returns to the cytoplasm, and GAP and effector proteins dissociate from the membrane. The drug BFA stabilizes the association of Arf1-GDP with its GEF. This interferes with exchange of GDP for GTP, so Arf1 cannot recruit effectors to the membrane and traffic between the ER and Golgi apparatus is disrupted. Arf, adenosine diphosphate–ribosylation factor; BFA, brefeldin A; ER, endoplasmic reticulum; GAP, GTPase-activating protein; GDP, guanosine diphosphate; GEF, guanine nucleotide exchange factor; GTP, guanosine triphosphate; GTPase, guanosine triphosphatase.

(Fig. 21.6). Rab-specific GEFs associated with membranes recruit Rabs and catalyze the exchange of GDP for GTP. Geranylgeranyl lipids coupled to two conserved, C-terminal cysteine residues facilitate the association of Rab-GTP with the membrane bilayer and are essential for their functions. The cysteines are included in a variable segment of 30 amino acids that targets each Rab to its correct subcellular location.

Like Ras, switch I and switch II regions of Rabs differ in conformation when GDP or GTP is bound. In the GDP-bound state, the switch regions are unfolded, whereas when GTP is bound they adopt well-defined conformations, allowing effector binding.

Membrane-associated Rab-GTP recruits the targeting and docking proteins used to recognize the target membrane and initiate bilayer fusion or other membrane trafficking steps. Effectors that interact specifically with Rab-GTP include fusion regulators, molecular tethers, motors, sorting adaptors, kinases, phosphatases, Rab regulators, and components of membrane contact sites. Recruitment of such effectors occurs in a spatiotemporally controlled manner in part due to the ability of Rab proteins to bind to overlapping or nonoverlapping sites on the same effector. In the early endosome, for example, Rab4, Rab5, and Rab33 all interact with rabaptin-5.

Following membrane fusion, a Rab-specific GAP stimulates GTP hydrolysis, recycling Rab-GDP back to **GDI (guanine nucleotide dissociation inhibitor)** in the cytoplasm. Rabs intrinsically hydrolyze GTP relatively rapidly, so Rab GAPs are less crucial than GEFs in modulating Rabs on membranes.

In the cytoplasm, the carrier protein GDI prevents exchange of GDP bound to Rab for GTP and sequesters the hydrophobic geranylgeranyl groups. Proteins called displacement factors recruit Rab-GDP to membranes by displacing GDI.

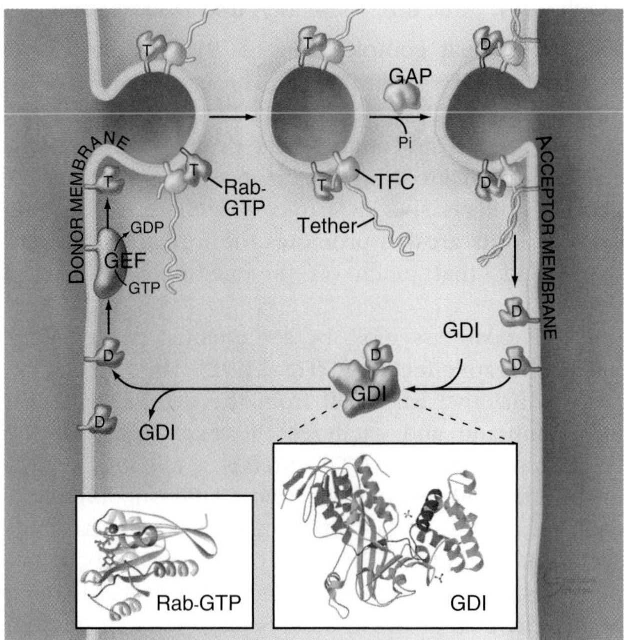

FIGURE 21.6 RAB GTPASE CYCLE. Rab-GDP is bound to GDI in the cytoplasm. A GTPase dissociation factor (not shown) on the membrane separates Rab-GDP from GDI so it can be activated by a membrane-associated, Rab-specific GEF. Rab-GTP recruits effectors, such as coiled-coil tethers and TFCs, which aid in targeting and docking the vesicle. A GAP stimulates GTP hydrolysis and returns Rab-GDP to GDI in the cytoplasm. *Insets* show ribbon diagrams of Rab-GTP and GDI. GDI, guanine nucleotide dissociation inhibitor; GDP, guanosine diphosphate; GTP, guanosine triphosphate; GTPase, guanosine triphosphatase; P_i, inorganic phosphate; TFC, tethering factor complex.

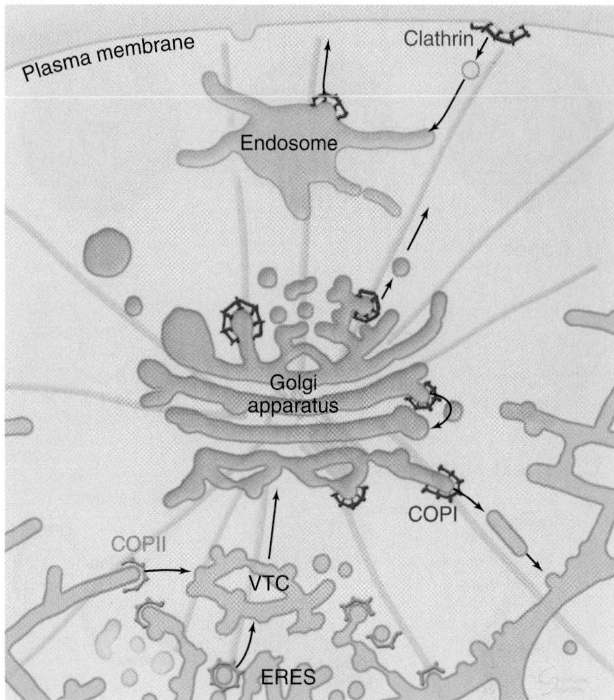

FIGURE 21.7 DISTRIBUTIONS OF COAT PROTEINS ALONG THE SECRETORY PATHWAY. Clathrin, coat protein I (COPI), and coat protein II (COPII) coats are color coded.

This GTPase cycle allows Rab proteins to regulate the timing of the assembly and disassembly of diverse multiprotein complexes involved in the trafficking of transport containers. Some Rab effectors have Rab GAP or GEF activity. They facilitate membrane traffic by promoting the rapid accumulation or removal of Rab GTPases on membranes.

Mutations in the genes for Rab GTPases cause diseases. Rab27a normally controls exocytosis of lysosome-related organelles, such as melanosomes and lytic granules. Rab27a mutations cause Griscelli syndrome type 2 characterized by albinism and immunodeficiency. Other Rab mutations cause neuropathies, mental retardation and ciliopathies.

Rab GTPases play a crucial role in phagocytosis and phagosome maturation, so many intracellular pathogens target Rab GTPases to interfere with the ability of the host phagocytes to ingest and degrade pathogens. The bacterial pathogenesis factors can either act as Rab GEFs or GAPs that activate or inactivate specific Rab GTPases, or they can interfere with Rab functions through enzymatic modifications, such as proteolytic cleavage, lipidation, or AMPylation. These activities allow the pathogen to evade destruction in the phagolysosome and establish a replicative niche within the host.

Membrane Budding Coat Complexes

Coat complexes form a basketlike lattice or cage covering the cytoplasmic surface of many transport intermediates. The cage deforms the membrane into a curved shape, helping to drive the formation of a transport carrier. Soon after the carrier is formed and loaded with cargo, the coat lattice disassembles, allowing the carrier to fuse with a target membrane.

Three different types of coats, coat protein II (COPII), COPI, and clathrin, are used for different transport steps (Fig. 21.7). COPII and COPI coats function in the early secretory pathway, with COPII coats budding membrane with cargo from the ER, and COPI coats budding cargo from pre-Golgi and Golgi compartments. Clathrin operates later in the secretory pathway and in the endosomal system, budding cargo from the *trans*-Golgi network and from the plasma membrane to endosomes.

X-ray crystallography and cryoelectron microscopy revealed that the three coats have underlying similarities as well as obvious differences (Fig. 21.8A). In all three cases an inner "adaptor" layer interacts with cargo and small GTPases (or phosphatidylinositol[4,5]bis-phosphate [$PI(4,5)P_2$]). An outer "cage" shapes the carrier, enabling it to bud from the bilayer. Clathrin and COPII coats form these layers sequentially with adaptor protein complexes binding the membrane before recruiting cage components. Both layers of the COPI coat are recruited together as a complex called "coatomer."

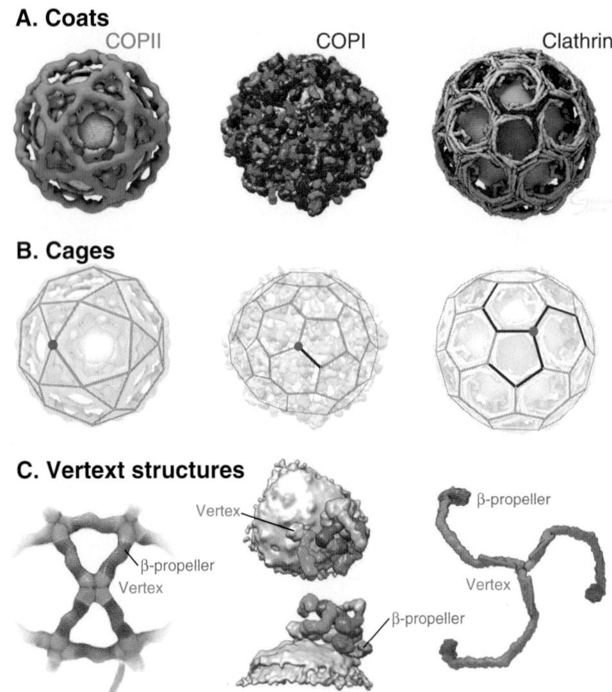

FIGURE 21.8 THREE HOMOLOGOUS VESICLE COATS. The panels compare the organization of coat protein I (COPI), coat protein II (COPII) and clathrin coats, each with an inner and outer layer. The homologous outer cage components are α-COP and β′-COP of COPI, Sec31 of COPII, and clathrin. They impart geometric order to the coat and each has a β-propeller domain next to the membrane for binding the sorting motifs of cargo proteins at the end of a rod consisting of an α-solenoid. Both the clathrin and COPI cages are assembled from three elongated chains joined at a central hub. Leg segments intertwine to form each edge of the lattice. The COPII cage is assembled from four rods joined at a central hub. COPII lattices have triangular, pentagonal or hexagonal faces, whereas COPI and clathrin lattices have hexagonal or pentagonal faces. The homologous inner components are the γ–ζ–β–δ-COP subcomplex of COPI, Sec23p/Sec24p of COPII, and the AP1 and AP2 clathrin adaptors.

Outer-cage components impart geometric order to each coat complex, and each of the outer-cage proteins has a domain that binds transmembrane cargo proteins (Fig. 21.8B–C). Clathrin and COPI cages assemble from three legged units joined at a central hub (Fig. 21.8B–C). Leg segments from adjacent trimers intertwine to form each edge of the icosahedral lattice of hexagons and pentagons (see Fig. 5.4). By contrast, the COPII cage assembles from units with four-legged chains joined at a central hub (Fig. 21.8B–C). Each leg forms an edge of a COPII cage composed of triangular, pentagonal or hexagonal faces (Fig. 21.5C).

A specific GEF and GTPase usually set the time and place of coat-mediated vesicle formation. The exception is clathrin-mediated endocytosis, where the adaptor protein 2 (AP2) complex is recruited to membranes by $PI(4,5)P_2$. The following sections explain the coats in the order they act in the secretory pathway.

COPII Coat: Structure, Assembly, and Disassembly

The COPII coat complex (Fig. 21.9) assembles cages that bud vesicles from the ER. The machinery consists of the Sar1p GTPase, Sec23p•Sec24p subcomplex, and Sec13p•Sec31p subcomplex. The proteins self-assemble a polymeric scaffold on the cytoplasmic surface of the ER that collects specific cargo molecules. The coat is curved, so its growth promotes the formation of membrane buds that pinch off the membrane as coated vesicles.

COPII coats assemble by a sequential process with built in negative feedback (Fig. 21.9E). The GEF Sec12p recruits inactive Sar1-GDP from the cytoplasm to the ER membrane and catalyzes the exchange GDP for GTP. Activated Sar1p-GTP extends a tail into the ER membrane and recruits the two COPII subcomplexes from the cytoplasm. Sec23p•Sec24p subcomplexes bind Sar1p and recruit Sec13p•Sec31p subcomplexes, which polymerize into a mesh-like scaffold that coats the membrane. The coat grows as more Sec23p•Sec24p subcomplexes anchored by Sar1p diffuse in from surrounding membrane and become crosslinked by Sec13p•Sec31p subcomplexes recruited from the cytoplasm. This lattice serves as the outer layer of the COPII coat. Growth of the lattice bends the patch of membrane into a coated bud that pinches off as a coated carrier (containing concentrated proteins) or remains as a "metastable" coated structure that participates in the differentiation of membrane domains at ER exit sites.

Hydrolysis of GTP bound to Sar1p initiates COPII coat disassembly. The coat subunit Sec23p is the GAP for Sar1p. Together, the coat protein Sec13p and membrane curvature stimulate the GAP activity of Sec23p, so assembly of the coat automatically leads to its disassembly as inactive Sar1p-GDP releases from the COPII lattice.

Localization of the Sec12p GEF to the ER membrane and the Sec23p GAP for Sar1p in the coat provides a mechanism for the continuous, self-regulated assembly and disassembly of the COPII coat. Sar1p-GTP guides addition of coat subcomplexes to the rim of the lattice, thereby deforming the membrane into a coated carrier. At the same time Sar1p-GDP and associated subcomplexes are released from the lattice interior. Sec12p is left behind in the ER membrane when the coated vesicle detaches. Then the two COPII complexes dissociate from the carrier vesicle.

Transmembrane cargo proteins are concentrated in the bud in the ER membrane by interacting with the Sec24p component of COPII coats (Fig. 21.9E). Sec24p recognizes several types of sorting signals in the cytoplasmic domains of the cargo proteins including diacidic motifs with the consensus sequence asp-glu-x-asp-glu (DExDE) and short hydrophobic motifs such as phe-phe (FF), phe-tyr (FY), leu-leu (LL), or ile-leu (IL). Transmembrane cargo proteins lacking these sorting signals collect in COPII-coated buds at ER exit sites either by associating

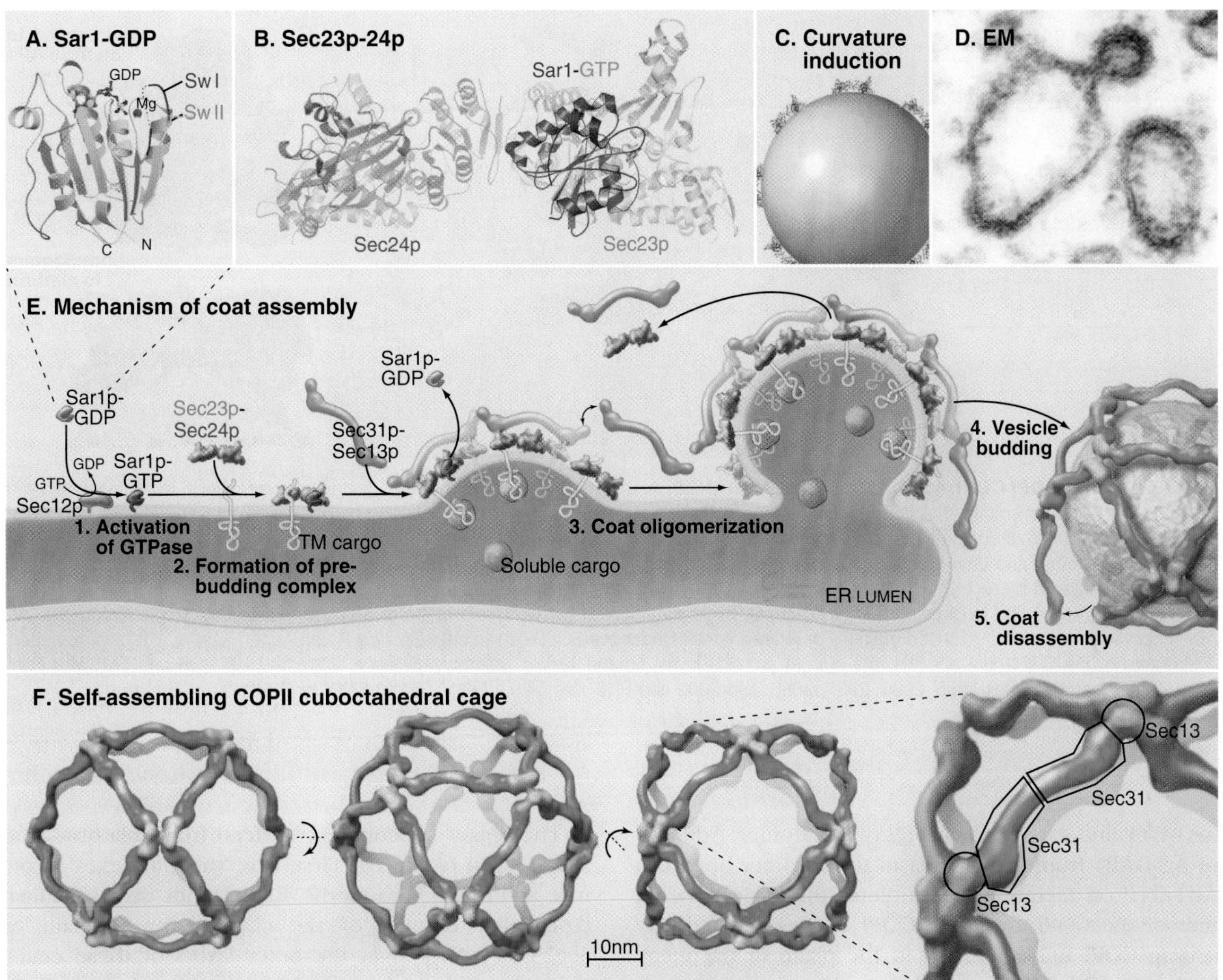

FIGURE 21.9 COPII COAT ASSEMBLY ON THE ENDOPLASMIC RETICULUM. A, Ribbon diagram of Sar1-GDP. **B,** Ribbon diagram of the Sec23p-Sec24p complex bound to Sar1-GTP. **C,** Sec23p-Sec24p has a concave, positively charged surface that promotes curvature when it binds a lipid bilayer. **D,** Electron micrograph of a thin section of a COPII vesicle formed in vitro when ER membranes were incubated with cytosol and ATP. **E,** The membrane-anchored Sec12p GEF activates Sar1 by bringing Sar1 to the membrane and promoting exchange of GDP for GTP. Sar1p-GTP then recruits the Sec23p•Sec24p subcomplex, which assembles the Sec31/Sec13p cage. Transmembrane cargo with suitable sorting motifs binds Sec24p. Coat complexes dissociate from the lattice after Sar1-GTP converts to Sar1-GDP and releases into the cytosol. The lattice grows into a coated bud that can pinch off the membrane as a coated vesicle, leaving Sec12 behind on the ER membrane. **F,** Three-dimensional reconstruction of COPII cage at 30-Å resolution from cryoelectron microscopy of single particles. ATP, adenosine triphosphate; COPII, coat protein II; ER, endoplasmic reticulum; Sar1, secretory and Ras-related 1. (**A,** For reference, see Protein Data Bank [PDB; www.rcsb.org] files 1F6B, 1M2B and 1M2O. **C,** Modified from Bickford LC, Mossessova E, Goldberg J. A structural view of the COPII vesicle coat. *Curr Opin Struct Biol.* 2004;14:147–153. **D,** Courtesy W. Balch, Scripps Research Institute, La Jolla, CA. **F,** Modified from Stagg SM, Gurkan C, Fowler DM, et al. Structure of the Sec13/31 COPII cage. *Nature.* 2006;439:234–238.)

with other cargo proteins with sorting signals or by partitioning into the cholesterol-rich bilayers of ER exit sites as a result of their long transmembrane segments. Large, soluble cargo proteins like protocollagen (300 nm long; see Fig. 29.4) are too large to fit into conventional COPII transport carrier vesicles of 60 to 90 nm in diameter. Instead, these molecules leave the ER in enlarged COPII structures containing special packing machinery.

COPI Coat: Structure, Assembly, and Disassembly

The COPI coat complex guides protein sorting and retrograde transport from the Golgi apparatus and pre-Golgi membranes (also called VTC) back to the ER (Fig. 21.1). This transfer of membrane and proteins is crucial for Golgi structures to differentiate functionally and morphologically from the ER. Like the COPII coat complex, the COPI coat complex assembles into a lattice (ie, COPI coat) on a patch of membrane, recruits specific proteins and deforms the membrane into a coated bud that either pinches off as a coated carrier or remains as a "metastable" coated structure.

The COPI **coatomer complex** consists of seven subunits (α, β, β[1], ε, γ, ζ), ranging in mass from 25 kD to more than 100 kD (Fig. 21.10). Membrane-bound

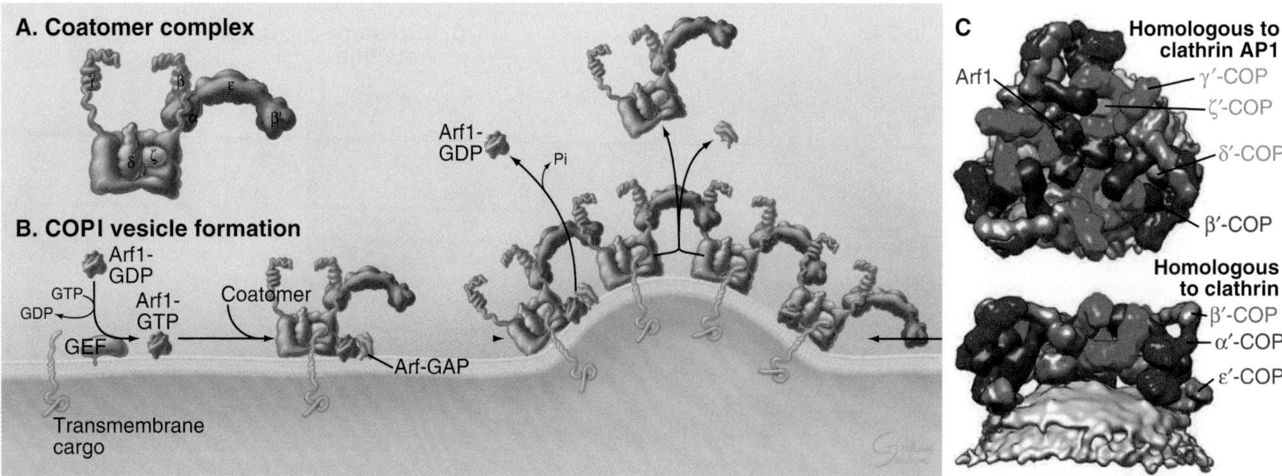

FIGURE 21.10 COPI COAT ASSEMBLY ON MEMBRANES. A, Schematic model of the COPI coatomer complex with the subunits labeled. **B,** Pathway of coat assembly. An Arf1-specific GEFs recruits whole coatomer complexes which bind cargo, vesicle tethering factors, and fusion factors. Polymerization of coatomers bends a patch of membrane and activates the GAP that stimulates Arf1 to hydrolyze GTP. After Arf1-GDP is released, coatomer and GAP are destabilized and dissociate from the coat. The continuous cycle of coatomer binding, polymerization, and dissociation mediated by Arf1 GTPase activity can form a coated bud that pinches off the membrane or remain as a meta-stable coated bud that imparts curvature and tension to the membrane. **C,** Reconstruction of the two layers of the COPI coat from cryoelectron micrographs. Arf, ADP-ribosylation factor; COPI, coat protein I. (For reference, see Dodonova SO, Diestelkoetter-Bachert P, von Appen A, et al. Vesicular transport. A structure of the COPI coat and the role of coat proteins in membrane vesicle assembly. *Science*. 2015;349:195–198 and EM Data Bank [www.emdatabank.org] files 2985, 2986, 2987, 2988, and 2989 and PDB files 5A1U, 5A1V, 5A1W, 5A1X, and 5A1Y.)

Arf1-GTP and cargo proteins recruit coatomer. Addition of Arf-GAP1 from the cytoplasm to coatomer bound to Arf1-GTP on membranes completes the basic building unit of the COPI coat. The COPI coat polymerizes by adding COPI units diffusing in the plane of the membrane. Growth of the coat bends the membrane as it forms a basket-like lattice.

Although recruited *en bloc* onto membranes, coatomer consists of a cage-like subcomplex and an adaptor subcomplex. The adaptor complex composed of the ε, γ, and ζ subunits is close to the membrane and binds two copies of Arf1-GTP. This adaptor subcomplex makes multiple contacts with the cage subcomplex. The cage subcomplex composed of the α, β, and β^1 subunits forms an outer layer of the coat, analogous to clathrin or the outer layer of COPII (Sec13/Sec31). Multiple interactions of both Arf1-GTP and cargo in the membrane drive simultaneous recruitment of cage and adaptor subcomplexes.

Transmembrane cargo proteins recycling from the Golgi apparatus to the ER are concentrated in COPI-coated buds by interacting with the cage subcomplex. Most of these proteins have a **dilysine sorting motif** generally found at their C-terminus (Lys-Lys-x-x-COOH or KKxx, where x is any amino acid). Two arginine residues substitute for the lysines in some proteins. These dilysine motifs bind on the top surface of the N-terminal β-propeller domains of α- and β^1-COPI, which have highly charged basic and acidic patches facing the bilayer.

The cage subcomplex differs from clathrin and Sec13/31 in two ways. First, cage subcomplexes recognize and bind cargo motifs. Cargo binds the N-terminal β-propeller domain of the clathrin heavy chain or Sec23/24 rather than the outer layers of these coats. Second, clathrin and Sec13/31 polymerize to form regular lattices with global point group symmetry, while cage subcomplexes show significant conformational heterogeneity and flexibility.

Arf-GAP1 is inactive during its diffusion with individual coatomer units, but curved membranes favor its interaction with Arf1-GTP and hydrolysis of the GTP (Fig. 21.10B). Thus, assembly of the COPI coat on a membrane automatically inactivates Arf1. Arf1-GDP is released from the membrane, destabilizing the lattice of coatomer and Arf-GAP1 and leading to disassembly of the coat.

As in the case of COPII these dynamic events result in the assembly of COPI around the periphery of the lattice and their release from the interior (Fig. 21.11A). The flux of coat units through the lattice continues whether or not a coated vesicle detaches from the membrane and allows for several outcomes (Fig. 21.11A-C). The lattice can grow, disassemble (after budding off the membrane as a coated vesicle), or persist as a coated bud (creating tension in the bilayer) with coat units associating and dissociating at equal rates. These behaviors of the coat lattice orchestrate protein sorting and morphologic events at ER export domains that allow the VTC to form (Fig. 21.11D).

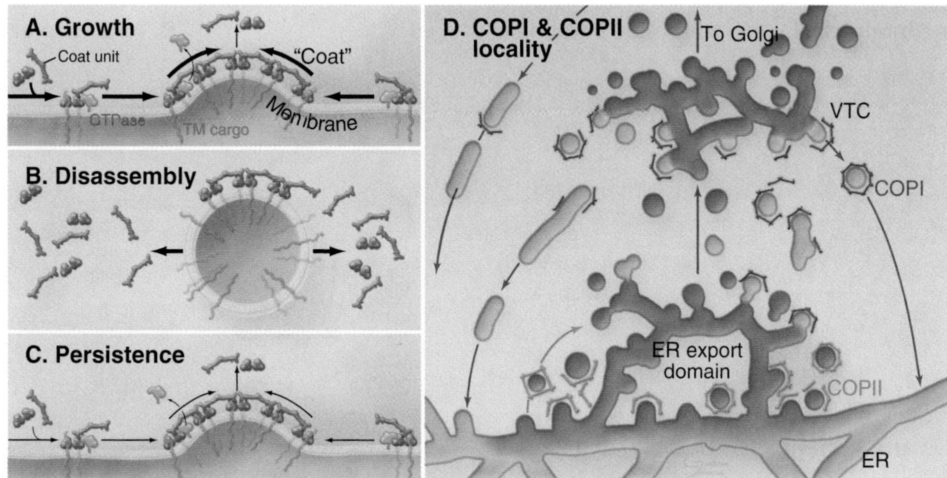

FIGURE 21.11 FATES OF COPI COAT COMPLEXES ON MEMBRANES. A, When binding of coat units is faster than release, the coat grows and forms a coated vesicle. **B,** After a coated vesicle forms, coat binding becomes slower than release (owing to GEF not being incorporated into the coated bud), and the coat disassembles. **C,** When coat units bind at the same rate as they release, then the coat is metastable (it neither shrinks nor grows but imparts curvature to the membrane). By increasing curvature in the membrane, metastable coats increase membrane tension, which can cause lipid partitioning. **D,** Cartoon diagram of the distribution of COPII and COPI coats on ER export domains and VTCs. COPII coats are restricted to ER membranes, where they recruit cargo into the ER export domain. COPI coats are present on the vesicular/tubular elements of the VTC and Golgi apparatus, where they orchestrate retrieval of proteins back to the ER. COPI/II, coat protein I/II; ER, endoplasmic reticulum; GEF, guanine nucleotide exchange factor; VTC, vesicular tubular carrier.

Clathrin Coats: Structure, Assembly, and Disassembly

Clathrin forms a three-legged structure termed a triskelion (Fig. 21.12A). The three 190-kD heavy chains form the legs, which consist of α-helical zigzags associated with one of two extended light chains of approximately 30 kD (LCα or LCβ). Under special conditions this hexameric unit can self-assemble an empty cage built like a soccer ball with clathrin forming the ribs between adjacent faces. Each rib of the cage incorporates portions of four different triskelions, which are, in turn, arranged in pentagons and hexagons. (Fig. 5.4 explains how pentagons and hexagons form closed shells.) Together with clathrin, these molecules help drive curvature of the underlying membrane and promote vesicle formation.

The assembly of clathrin cages depends on **assembly proteins (APs)** that link clathrin to membrane lipids and proteins (Fig. 21.12B-E). Two classes of structurally and functionally distinct APs exist: the monomeric AP AP180/CALM and **heterotetrameric adapter protein complexes (AP1-4).** AP2 is the only heterotetrameric AP involved in clathrin-coated vesicle formation at the plasma membrane and binds to membranes without the need of a small GTPase because it binds $PI(4,5)P_2$. Other heterotetrameric APs that assemble clathrin-coated vesicles on other membranes require Arf GTPases.

The four subunits of the AP2 complex have distinct functions. The large α adaptin subunit binds $PI(4,5)P_2$ in the plasma membrane and cooperates with the small σ_2 adaptin subunit to bind **dileucine-based sorting motifs** (D/ExxxLL where D is aspartic acid, E is glutamic acid and L is leucine) that serve as internalization signals in the cytoplasmic tails of transmembrane receptors. The large β_2 subunit binds directly to clathrin and promotes its assembly. The medium sized μ_2 subunit interacts with **tyrosine-based sorting motifs** (YxxΦ where Y is tyrosine and Φ is any bulky hydrophobic amino acid), the internalization signals for other transmembrane receptors. All these ligands, as well as AP180/CALM and synaptotagmin, recruit AP2 to the plasma membrane.

Additional **adaptor proteins** help coordinate clathrin coat formation by recruiting cargo and other proteins. Adaptor proteins, including epsin, amphiphysin, Hrs/Vps27p, and β-arrestin, consist of one or more folded domains connected by long, flexible linkers with multiple weak-binding sites (Fig. 21.12C–E). This "string and knots" design allows the flexible linkers to sweep through cytoplasm hunting for multiple ligands that bind reversibly to the string-like regions. Cooperative networks of weak interactions between multivalent binding partners create a positive amplification cascade that, once initiated, drives clathrin-coat formation. The large GTPase **dynamin** (see Fig. 22.8) forms a collar at the neck of deeply invaginated coated pits and helps control the fission and internalization of clathrin-coated vesicles.

After a coated vesicle pinches from the membrane, multiple mechanisms disassemble coat components, which are recycled, leaving the vesicle free to fuse with other membranes to form, for example, endosomes. Endophilin recruits the lipid phosphatase synaptojanin, which dephosphorylates $PI(4,5)P_2$ and weakens the

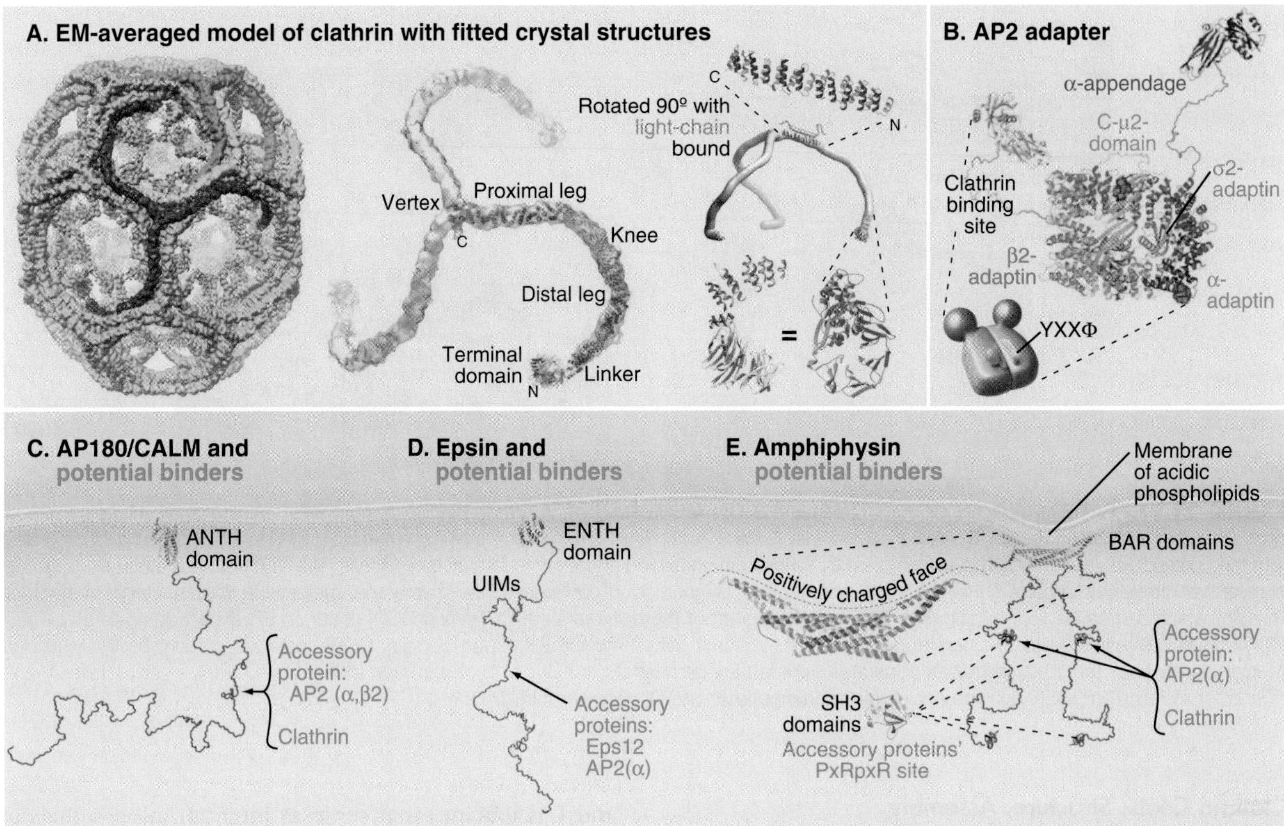

FIGURE 21.12 STRUCTURE OF THE CLATHRIN COAT. A, Structure of clathrin triskelions assembled into a clathrin cage at 2-nm resolution. The legs of clathrin triskelions are formed from repeating units of five helix hairpins. The N-terminal domain is a seven-bladed β-propeller similar to a trimeric G-protein β-subunit. **B,** Model of the AP2 complex from high-resolution structures of its individual subunits. The large subunits are helical solenoids of approximately 30 repeats of six- to eight-turn helices connected by short loops, whereas σ_2 and μ_2 are each five-stranded β-sheets flanked by α-helices. **C,** Model of AP180/CALM from its high-resolution structure. The α-helical solenoid domain at the N-terminus (called the ANTH domain) binds PI(4,5)P_2 and has a similar structure to the epsin ENTH domain. The long C-terminal tail has no predicted secondary structure but contains binding motifs for Eps15, clathrin, and Dx[FW]. **D,** Model of epsin from a high-resolution structure. The ENTH domain binds PI(4,5)P_2, attaching epsin to the membrane. The long flexible arm contains a UIM and can bind AP2 and clathrin. When ubiquitin is bound to this motif, it serves as a signal for directing the membrane through the endocytic pathway, ending in incorporation into internal vesicles of multivesicular bodies that ultimately are degraded by hydrolytic enzymes stored in lysosomes. **E,** High-resolution structural model of amphiphysin. The molecule contains an N-terminal BAR domain and a C-terminal SH3 domain. The positively charged, concave surface of the dimer binds to curved membrane bilayers. The central extended region binds clathrin and AP2α, and C-terminal SH3 domain bind polyproline motifs. AP2, adaptor protein 2; PI(4,5)P_2, phosphatidylinositol(4,5)bis-phosphate; SH3, Src homology region-3; UIM, ubiquitin-interacting motif. (**A,** Courtesy Corinne Smith, Medical Research Council Laboratory of Molecular Biology, Cambridge, United Kingdom. **B–C,** Courtesy Frances Brodsky, University of California, San Francisco; Tomas Kirchhausen, Harvard Medical School, Boston, MA; and David Owen, Medical Research Council Laboratory of Molecular Biology, Cambridge, United Kingdom.)

attachment of coat proteins such as dynamin, APs, and clathrin to the membrane. The heat shock chaperone protein HSC70 uses adenosine triphosphate (ATP) hydrolysis and cooperates with the protein auxilin to disassemble clathrin cages. In parallel the network of adapters dissociates in a self-propagating fashion.

Tethering Factors

Tethering factors target carrier vesicles to specific organelles (Fig. 21.13) prior to fusion directed by SNARE (soluble *N*-ethylmaleimide-sensitive factor attachment protein receptor) proteins (Fig. 21.14). Tethering proteins also play structural roles as components of a Golgi matrix or scaffold for the assembly of other factors important for fusion and/or cargo sorting. Tethering factors can be divided into two general classes.

1. Coiled-coiled tethering factors
   - These rod-shaped proteins are homodimeric, coiled-coils with globular heads at both ends that extend more than 150 nm from membranes into the cytoplasm (Fig. 21.13). Active Rabs anchor these tethers to the carrier vesicle. They promote formation of SNARE complexes on vesicle and target membranes. An internal hinge-like region in the tail collapses once the tether brings the carrier vesicle close to the acceptor membrane.
   - The membranes of the Golgi apparatus are decorated with several large coiled-coil tethering

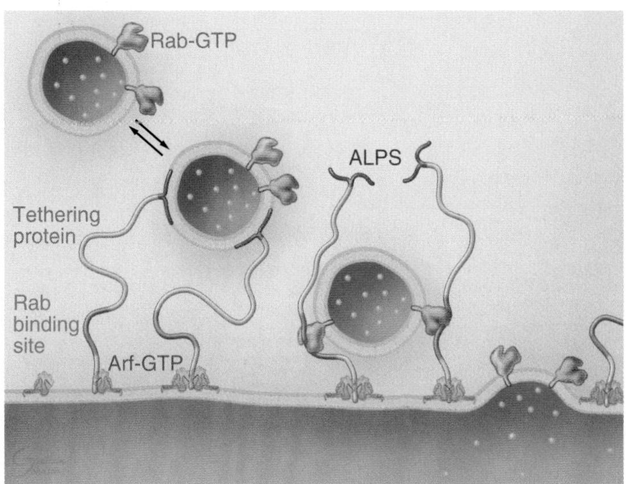

FIGURE 21.13 TETHERING PROTEINS. Golgins, such as GMAP-210, emanate from the Golgi like tentacles, capturing vesicles through their ALPs motif. The captured vesicle undergoes fast on and off reactions to penetrate into the Golgin matrix, interacting with Rab proteins with specific binding sites along the coiled-coil region of golgins. This brings the vesicle close to the surface so that SNARE (soluble *N*-ethylmaleimide-sensitive factor attachment protein receptor) proteins, operating at a closer distance (10–15 nm), can engage it and drive fusion. In addition to their role in membrane trafficking, golgins provide binding sites for regulators of the cytoskeleton.

proteins called "golgins" (including GM130, GMAP-210 and p115). Golgins have multiple binding sites for Rab GTPases along their lengths of up to 300 nm. Golgins may emanate like tentacles from the Golgi membranes, capturing vesicles, so SNARE proteins operating at a closer distance (10–15 nm) can engage the vesicles and drive fusion (Fig. 21.14).

2. Multisubunit tethering factors
   - These tethers are more compact than the coiled-coil tethers but also direct carrier vesicles to acceptor membranes. TRAPP I (transport protein particle) is a flat complex of seven subunits that participates in fusion of vesicles from ER with the Golgi apparatus. The eight subunits of COG (conserved oligomeric Golgi) extend only approximately 30 nm from the membrane. The exocyst consists of eight subunits that tether secretory vesicles to the plasma membrane. All three of these tethering complexes bind inactive Rabs and function as GEFs to exchange GDP for GTP. Cells defective in COG function have pleiotropic defects in glycosylation reactions in the Golgi, consistent with a role in vesicular trafficking across the Golgi system. The HOPS (homotypic fusion and vacuolar protein sorting) complex has two subunits for interaction with Rab7 and shares four subunits with another tether. HOPS participates in the fusion of endosomal and lysosomal vesicles by bridging Rab GTPases on two membranes.

Membrane Fusion Machinery: SNAP Receptor Components

Fusing two membranes requires several steps (Fig. 21.14). The membranes must approach each other and bend to get destabilized. Then, the proximal bilayer leaflets fuse to form a "stalk intermediate," followed by fusion of the distal leaflets to form and subsequently expand a fusion pore. Theoretical modeling suggests the total energy to drive fusion ranges from 40 to 100 K_BT.

The **SNARE** family of proteins drives the fusion of carrier vesicles with acceptor membranes using energy from assembling an alpha-helical bundle to pull the membranes together (Fig. 21.14). Two mechanisms assure that each type of vesicle (or tubule) carrier fuses with the right target membrane. First, Rab GTPases recruit specific tethering factors that link carrier vesicles to the correct target membranes. Then particular SNAREs help ensure the specificity of fusion.

Most SNAREs are transmembrane proteins with C-termini anchored in the bilayer and N-terminal domains in the cytoplasm. Each contains a sequence of 60 to 70 residues with a heptad repeat (ie, "SNARE motif") that can form a coiled-coil (see Fig. 3.10) but is largely disordered on its own. Fusion requires SNARE proteins on both transport vesicles (**v-SNAREs**) and target membranes (**t-SNAREs**). When cognate v- or t-SNAREs interact on opposing membranes, they form a *trans*-SNARE complex or **SNAREpin**, comprised of four SNARE motifs assembled in a four-helix bundle. Assembly of the SNAREpin complex forces the opposing membranes together.

Synaptic vesicle fusion (Figs. 17.9 and 17.10) is a well-studied example of SNARE activity. The synaptic vesicle carries one v-SNARE (synaptobrevin, also called VAMP) and the target membrane has a binary complex of t-SNAREs syntaxin-1 and SNAP-25 (soluble N-ethylmaleimide sensitive factor attachment protein). Synaptobrevin and SNAP-25 are partially disordered on their own and form α-helices during fusion, while syntaxin-1 folds back on itself in an antiparallel coiled-coil before contributing a single helix to the bundle. Six proteins regulate the assembly of SNAREs, which initially form a primed complex that zips together into the complete four-helix bundle when triggered by Ca^{2+}.

The process starts with syntaxin-1 folded in a closed, inhibited conformation because of bound Munc18-1. Then Munc13s opens syntaxin-1, allowing the N-terminal domains of the three SNAREs to form a primed but inactive complex with Munc18-1 still bound. The regulatory proteins synaptotagmin-1 and complexin prevent primed, partially assembled SNARE complexes from fusing the membrane.

Entry of Ca^{2+} into the cell through voltage-gated Ca^{2+} channels causes membrane fusion in less than 1 millisecond. Ca^{2+} binds synaptotagmin-1 and allows completion of the four-helix bundle, likely by relieving the inhibition by complexin. This assembly process brings the

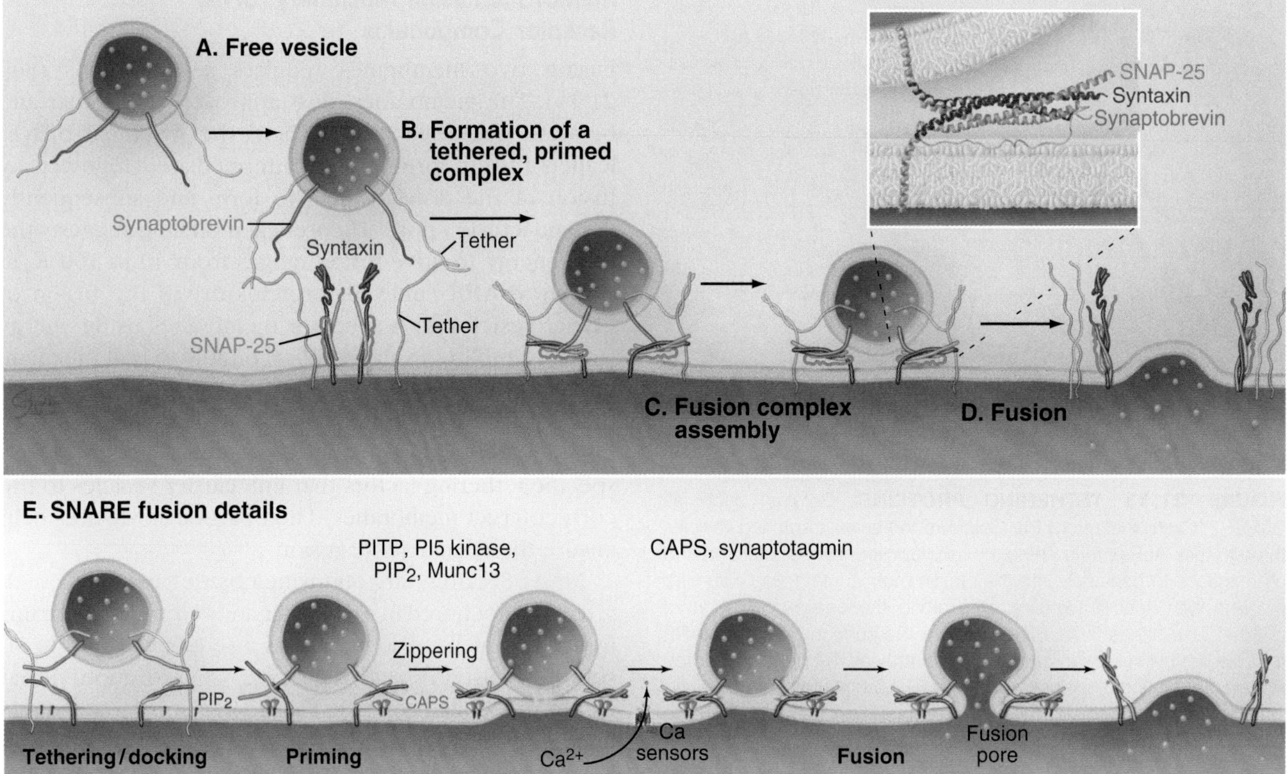

FIGURE 21.14 MEMBRANE FUSION BY SNARE PROTEINS. The example is neurotransmitter release. **A,** A synaptic vesicle with tethers and v-SNARE synaptobrevin *(blue)* approaches the plasma membrane, which has tethers and complexes of the t-SNAREs syntaxin *(red)* and SNAP-25 *(green)*. **B,** The v- and t-SNAREs dock to form a primed but inactive complex. **C,** When stimulated by an action potential (see Fig. 17.10), the SNAREs zip together to form a trans-SNARE complex that pulls the membranes together. The inset in the upper right shows a ribbon diagram of the fully assembled SNAREs anchored to the membranes. **D,** The vesicle fuses with the plasma membrane, leaving cis-SNARE complexes on the plasma membrane. **E,** Overview of SNARE-mediated fusion. SNARE, soluble *N*-ethylmaleimide-sensitive factor attachment protein receptor. (*Top right inset,* Modified from Ossig R, Schmitt HD, de Groot B, et al. Exocytosis requires asymmetry in the central layer of the SNARE complex. *EMBO J*. 2000;19:6000–6010.)

membranes together to overcome the energy barrier to fusion. The *trans*-SNARE complex between two membranes is transformed into a *cis*-SNARE complex on the cytoplasmic face of the fused membrane (Fig. 21.14D).

Most SNAREs are tail-anchored proteins (see Fig. 20.10) with a transmembrane segment inserted into ER membranes after translation. SNAREs without a transmembrane domain are attached to the membrane by lipid modifications such as palmitoylation. SNAREs reach their destinations in cells either by traversing the secretory pathway to reach specific organelles, or by directly inserting into the organelle membrane (ie, plasma membrane or mitochondria) via their tail anchors.

After the lipid bilayers fuse, SNAP proteins recruit a ubiquitous AAA adenosine triphosphatase (ATPase) (see Box 36.1) called **NSF** (for ***N*-ethylmaleimide [NEM]–sensitive factor**) to disassemble the *cis*-SNARE complex. NSF uses the energy from ATP hydrolysis to dissociate the bundle of SNARE helices and recycles the SNAREs for another round of membrane fusion. The sulfhydryl alkylating reagent NEM inactivates NSF and can prevent all carrier transport in the cell.

SNAREs often cycle between compartments for repeated use. For example, SNAREs involved in ER to Golgi transport are packaged into COPII coats at ER export sites for delivery to the Golgi apparatus, where they mediate homotypic fusion (ie, fusion of two like transport containers that have identical *cis*-SNARE pairs) among incoming carriers as well as heterotypic fusion (ie, fusion of two distinct membrane structures that have different *cis*-SNARE pairs) of these carriers with the Golgi membrane. The SNAREs are then packaged into COPI coats for retrieval to the ER. This allows them to function repeatedly in ER-to-Golgi apparatus transport.

Membrane Bending Proteins

Proteins with BAR domains influence membrane trafficking by inducing membrane curvature, stabilizing curvature generated by other forces, and recruiting cytoplasmic proteins to membranes of a particular size or shape

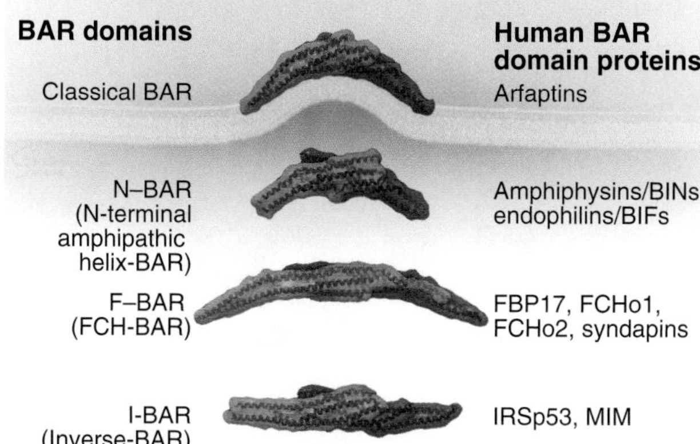

FIGURE 21.15 MEMBRANE-BENDING PROTEINS. Structures of subclasses of proteins with BAR domains. Proteins with BAR domains play important roles in membrane trafficking by inducing membrane curvature, stabilizing curvature generated by other forces, or detecting curvature in order to recruit cytosolic factors to membranes of a particular size or shape. (For reference, see PDB files 1I49 [classical bar], 1ZWW [N-BAR], 2EFK [F-BAR], and 1WDZ [I-BAR].)

(Fig. 21.15). Most BAR domains have a coiled-coil core of three α helices approximately 20 nm long. Classical BAR domains are shaped like a banana and are adapted to bind membranes with highly positive curvatures (ie, small vesicles). F-BAR and I-BAR domains fit on membranes with less-positive curvature or even on concave, negatively curved membranes. Most BAR domain proteins have at least one additional domain, such as an SH3 domain (see Fig. 25.10) that binds polyproline sequences in other proteins.

BAR domains use two mechanisms to bend membranes. In one, the intrinsic curvature of BAR domain dimers simply imposes its shape on the membrane substrate. In the second, amphipathic wedges present in the BAR domain insert into the bilayer, promoting membrane curvature by concerted displacement of lipids in the leaflet proximal to the site of insertion. For example, N-BARs have an N-terminal amphipathic helix within the BAR domain that can wedge into one leaflet of the bilayer to push the lipids apart. These two mechanisms are not mutually exclusive. The local deformations caused by insertion of the BAR domain can facilitate binding of additional BAR domains from other proteins and thereby generate a positive-feedback cycle for curvature propagation.

BAR domain proteins are important at membrane trafficking hubs where the generation of membrane curvature is coupled tightly to reorganization of the actin cytoskeleton and signaling through small GTPases. At the TGN, a BAR protein regulates biogenesis of transport carriers for regulated secretion. Other BAR proteins participate in clathrin-mediated endocytosis where they coordinate bud neck constriction, actin filament assembly and recruitment of proteins for fission and uncoating.

Secretory Transport From the Endoplasmic Reticulum to the Golgi Apparatus

Transport of newly synthesized proteins out of the ER takes place in specialized areas called **ER export sites.** In fluorescence images these structures appear as dispersed, punctate structures approximately 1 to 2 μm in diameter scattered over the surface of the ER (Fig. 21.16A). Individual ER export sites are organized into two zones: a region of smooth ER membrane studded with COPII-coated buds and uncoated tubules; and a central cluster of vesicles and tubules (Fig. 21.16B–C). The ER membrane is continuous between these two zones until the vesicle-tubule cluster and its associated cargo detach from the ER and move to the Golgi apparatus as transport intermediates, called **vesicular tubular carriers (VTCs)** (Fig. 21.11D).

Cargo molecules use three distinct modes of transport out of the ER: bulk flow, signal-mediated sorting, and partitioning within the lipid bilayer. In bulk flow, cargo passively distributes between the ER and the transport vesicles it generates. Soluble proteins that do not interact with proteins having export signals use this pathway to exit the ER.

In signal-mediated transport, sorting motifs in the cytoplasmic domains of transmembrane proteins interact with COPII machinery to concentrate the cargo (and any associated molecules) in transport carriers. These export signals are found on proteins that constitutively cycle between the ER and Golgi (such as the p24 family of proteins) as well as on a portion of secretory membrane proteins.

Many membrane proteins partition into the bilayer of transport carriers based on their collective physicochemical properties without the help of specific sorting signals or receptors (Figs. 21.3 and 21.4). The p24 family of proteins facilitates this sorting by interacting with COPII coats and concentrating at ER exit sites, where they increase the thickness of the bilayer by sequestering cholesterol. Plasma membrane proteins with long transmembrane helices partition into the thicker membranes of transport vesicles, while ER-resident proteins with shorter transmembrane domains are retained in the thinner bilayer of the bulk ER. This process allows thousands of diverse types of membrane cargo to efficiently move out of the ER into vesicular carriers destined for the Golgi apparatus.

The sequential actions of the Sar1, Rab1, and Arf1 GTPases and their effectors orchestrate the differentiation of the ER export sites into mobile VTCs (Fig. 21.17). Sar1 GTPase initiates the formation of the ER export site by assembling the COPII coat that forms coated buds and concentrates specific integral membrane proteins (including the p24 proteins and v-SNAREs) in the bud. The thicker membrane bilayer promotes partitioning of proteins with long transmembrane domains into the ER

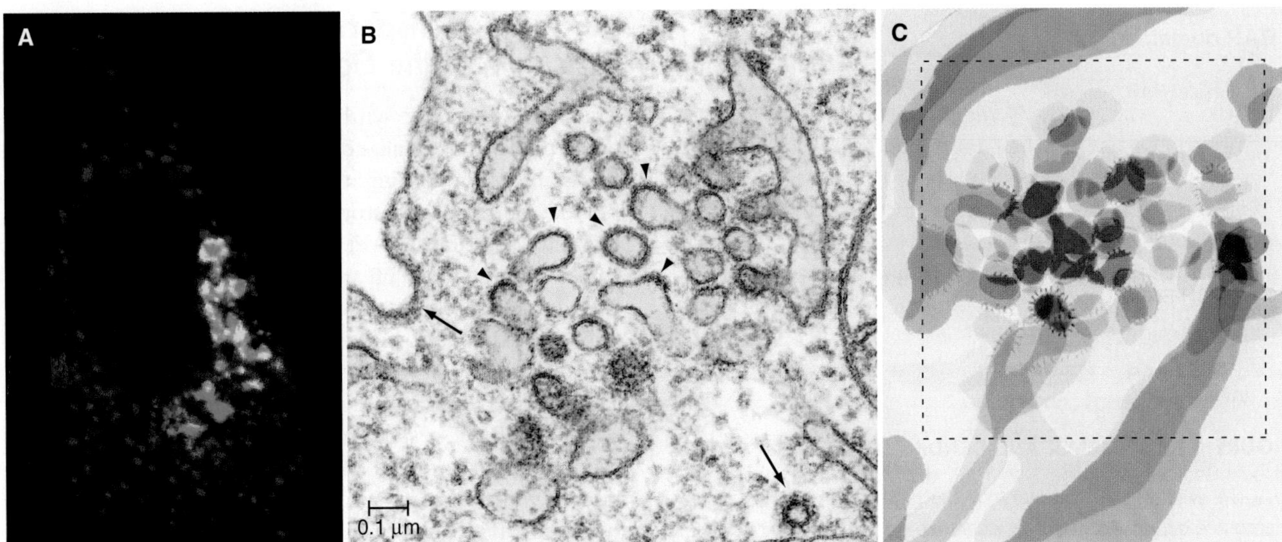

FIGURE 21.16 ENDOPLASMIC RETICULUM EXPORT DOMAINS AND VESICLE TUBULE CARRIERS. A, Fluorescence micrograph showing the distribution of ER export domains and Golgi apparatus in a fibroblast expressing an ER export domain marker, Sec31-YFP *(red)* and a Golgi marker, galactosyltransferase-CFP *(green)*. The ER exit sites are distributed throughout the cytoplasm as punctate structures, whereas the Golgi apparatus is localized next to the nucleus. **B,** Electron micrograph of a thin section of a typical ER export domain containing a central VTC that can detach and traffic to the Golgi apparatus. ER is *green,* ER-associated-coated buds are *blue,* the VTC is *red; arrowheads* mark COPI coats, and *arrows* mark clathrin-coated vesicles from the plasma membrane. **C,** Reconstruction from four consecutive serial-thin sections illustrating the three-dimensional organization of an ER export domain demarcated by the box. CFP, cyan fluorescent protein; COPI, coat protein I; ER, endoplasmic reticulum; VTC, vesicular tubular carrier. (**A,** Modified from Altan-Bonnet N, Sougrat R, Liu W, et al. Golgi inheritance in mammalian cells is mediated through endoplasmic reticulum export activities. *Mol Biol Cell*. 2006;17:990–1005. **B–C,** From Bannykh SI, Nishimura N, Balch WE. Getting into the Golgi. *Trends Cell Biol*. 1998;8:21–25.)

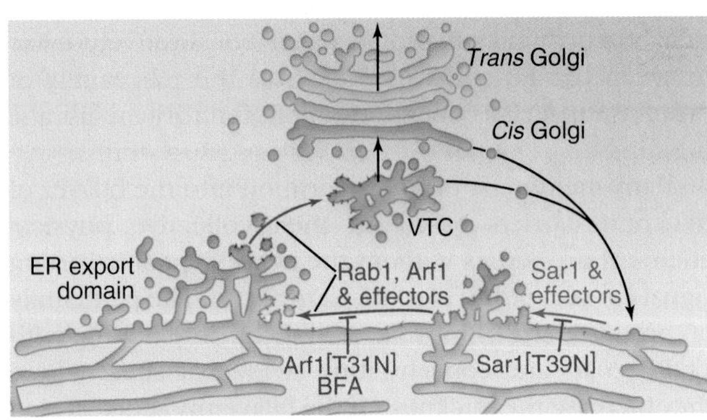

FIGURE 21.17 TRANSPORT FROM THE ENDOPLASMIC RETICULUM TO THE GOLGI APPARATUS. ER to Golgi transport is orchestrated by the combined activities of many molecules. Sar1 and its effectors initiate COPII-coated bud formation and clustering of cargo at regions called ER export sites. This induces p115 and Rab1 to bind to these regions, which in turn recruits GBF1, the GEF for Arf1. Subsequent recruitment of Arf1 and its effectors further differentiates the ER export domain into a VTC. The VTC detaches from the ER and targets the Golgi apparatus, where it fuses with the *cis* face of the Golgi. The cargo in the VTC is then released into the Golgi and moves to the *trans*-Golgi (where it will exit from the TGN). Expression of a constitutively inactive Sar1 mutant, Sar1[T39N], blocks COPII recruitment, and no ER exit sites form. Expression of an inactive Arf1 mutant, Arf1[T31N], or BFA treatment blocks recruitment of Arf1 effectors, which prevents ER exit sites from differentiating into VTCs. This causes the shrinkage and disappearance of the Golgi apparatus because new membrane from the ER cannot be delivered to the Golgi. BFA, brefeldin A; COPII, coat protein II; ER, endoplasmic reticulum; GEF, guanine nucleotide exchange factor; Sar1, secretory and Ras-related 1; TGN, *trans*-Golgi network; VTC, vesicular tubular carrier.

export site, including those without COPII recognition motifs. The SNARE proteins allow vesicles and membrane tubules that bud from the ER export site to fuse with themselves to form a tubule cluster. The tethering factors GM130 and giantin anchor these membranes to the cytoskeleton. The GEF GBF1 activates Arf1, which recruits effectors that initiate the retrieval of proteins in vesicles back to the ER. Activated Arf1 also recruits ankyrin and spectrin (see Fig. 13.11), which form a scaffold for other cytoskeletal proteins, including dynactin and dynein.

A mutation that locks Sar1 in the inactive GDP state (causing COPII to be released from membranes) prevents ER export domain formation. Likewise, trapping Arf1 in the GDP state with a mutation or by brefeldin A (BFA) treatment blocks VTC formation (Fig. 21.17A). Both the mutations and BFA also cause the Golgi apparatus to disappear, because the Golgi's existence depends on continuous membrane input from the ER.

The dynein-dynactin motor complex (see Fig. 37.2) helps detach VTCs from ER export domains and transports VTCs toward the minus ends of microtubules located at the centrosome near the Golgi apparatus (Fig. 21.17). The VTC matures as activated Arf1 recruits dozens of cytoplasmic proteins to its membranes. Lipid-modifying enzymes such as phosphatidylinositol kinases and phosphatases change the lipid composition, permitting tethering factors and matrix proteins to bind the membrane of the motile VTC. The COPI coat binds to and clusters specific proteins, enabling them to be retrieved back to the ER.

Upon arriving at the centrosome VTCs fuse with membranes on the ***cis* or entry face** of the Golgi apparatus. This region is also called the *cis*-Golgi network and is characterized by an elaborate tubular appearance. Membrane fusion releases cargo proteins and lipids carried by the VTC into the Golgi system for processing by enzymes that modify the cargo's oligosaccharide side chains.

Golgi Apparatus

The Golgi apparatus (Fig. 21.18) performs three primary functions within the secretory membrane system. First, it is a factory to synthesize the carbohydrate chains of glycoproteins, proteoglycans, and polysaccharides secreted by plants (eg, inulin) in preparation for their biological functions at the cell surface. Second, the Golgi apparatus is a protein-sorting station for the delivery to many cellular destinations. This includes transport to the plasma membrane, secretion to the cell exterior, sorting to the endosome/lysosomal system, or retrieval back to the ER. Third, the Golgi apparatus synthesizes sphingomyelin and glycosphingolipids. These lipids associate with cholesterol and influence protein sorting in the Golgi apparatus and plasma membrane.

Golgi Morphology and Dynamics

The Golgi apparatus in many animal cells appears as a ribbon-like structure adjacent to the nucleus and close to the centrosome, the main microtubule-organizing center of the cell (Fig. 21.18A). Electron micrographs of thin sections show that the Golgi apparatus consists of stacked, flattened, membrane-enclosed cisternae that resemble a stack of pancakes (Fig. 21.18B). Crosslinking of cisternae by Golgi-associated tethering factors results in their tight, parallel alignment within the stack. Tubules and vesicles at the rims of the stacks interconnect many stacks into a single ribbon-like structure by a process dependent on microtubules. If microtubules are experimentally depolymerized, the ribbon-like Golgi structure reorganizes into single stacks found at ER exit sites (Fig. 21.19). This distribution resembles the distribution of Golgi stacks in plant cells, where, hundreds of single stacks are located adjacent to ER exit sites rather than being joined together as a single ribbon.

Stacks of Golgi cisternae in animal and plant cells all exhibit a *cis*-to-*trans* polarity reflecting the passage of cargo through the organelle. Proteins and lipids from the ER enter the *cis* face (entry face) of the stack. After passing through the stack of cisternae, cargo leaves from the *trans* face at the opposite side of the stack. Membrane sorting and transport activities of the Golgi are thought to be especially high at the *cis* and *trans* faces and within the tubular-vesicular elements (noncompact zone) that interconnect the stacks (Fig. 21.18B).

Three proposed mechanisms explain transport of secretory cargo proteins through the Golgi apparatus (Fig. 21.20). In one model, the cisternae comprising the Golgi stack are relatively stable structures and secretory cargo transits from cisterna to cisterna across the stack in tubules or vesicles that bud from one cisternae and

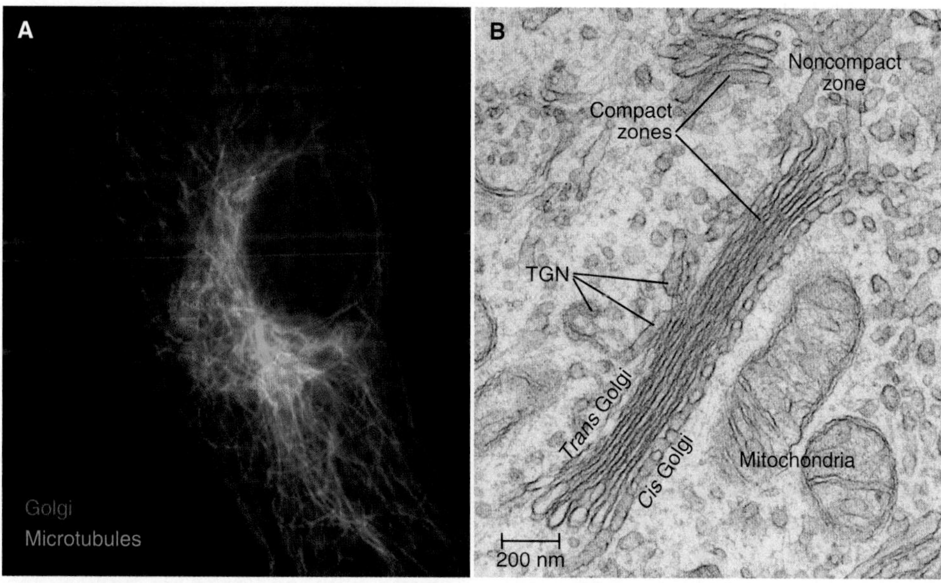

FIGURE 21.18 LOCALIZATION AND MORPHOLOGY OF THE GOLGI APPARATUS IN ANIMAL CELLS. A, Fluorescence micrograph of a rat fibroblast stained with antibodies to galactosyltransferase (a Golgi enzyme) *(red)* and antibodies to tubulin *(green)*. The Golgi apparatus typically extends as a ribbon-like structure around the centrosome located on one side of the nucleus. **B,** Electron micrograph of a rat epithelial cell showing a single stack of cisternae cut transversely. The *cis* and *trans* faces of the Golgi are at opposite sides of the stack, with the *trans*-Golgi network (TGN) extending from the *trans* face. (Courtesy J. Lippincott-Schwartz and Rachid Sougrat, National Institutes of Health, Bethesda, MD.)

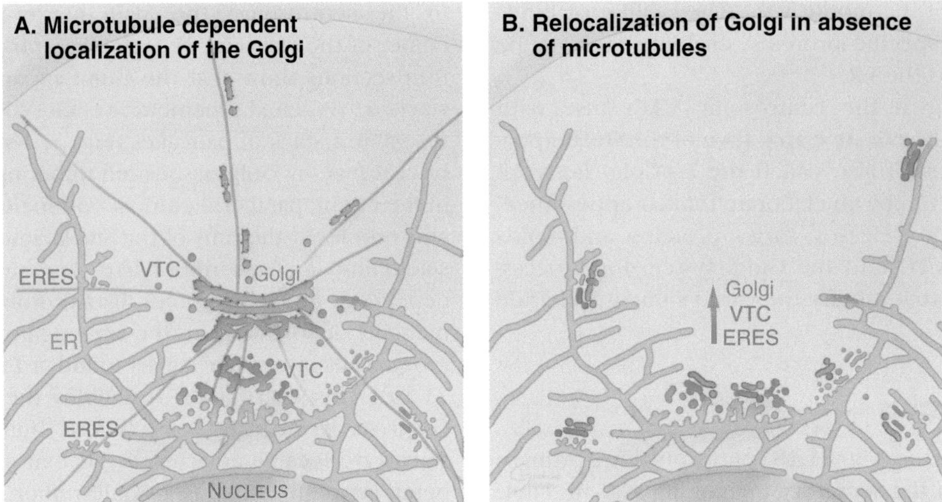

FIGURE 21.19 EFFECT OF MICROTUBULE DEPOLYMERIZATION ON THE DISTRIBUTION OF THE GOLGI APPARATUS. A, Microtubules *(red)* radiating from the centrosome *(red lines)* with their plus ends at the cell periphery help localize the Golgi apparatus in many animal cells by serving as tracks for the inward movement of membrane-bound carriers (vesicular tubular carrier [VTC]) derived from the endoplasmic reticulum (ER). The carriers deliver secretory cargo, as well as Golgi enzymes, to the Golgi apparatus. Retrograde transport of Golgi enzymes back to the ER does not depend on microtubules (since the ER is widely distributed throughout the cytoplasm). Because of this, when microtubules are disassembled **(B),** the Golgi apparatus reforms as separate stacks at sites adjacent to ER export sites, owing to the accumulation of cycling Golgi enzymes at these sites.

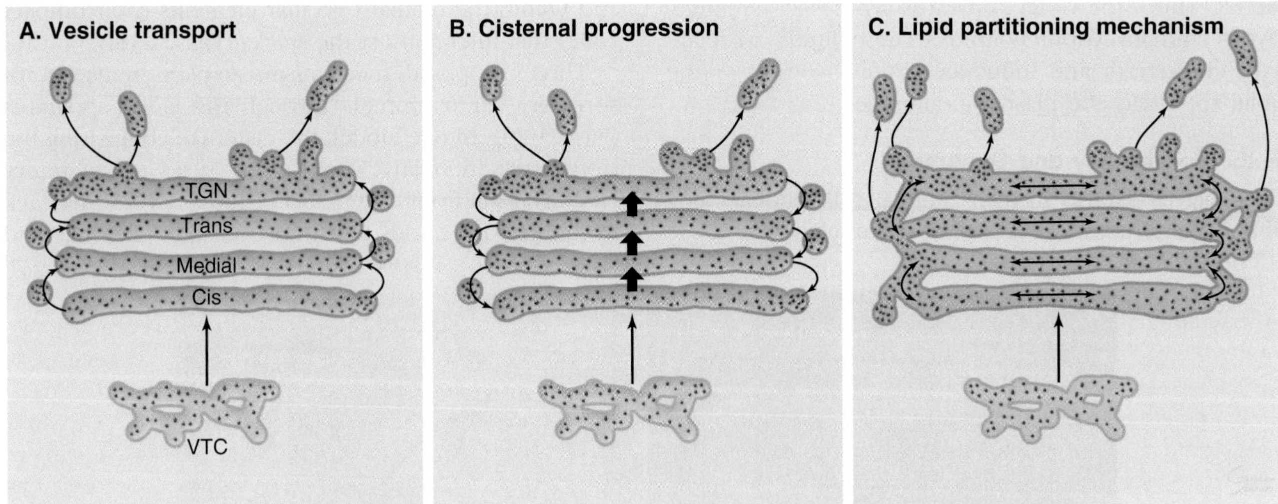

FIGURE 21.20 MODELS FOR CARGO MOVEMENT THROUGH THE GOLGI APPARATUS. A, Vesicle budding and fusion as the cargo moves forward. **B,** Cisternal progression as cisternae with cargo form on the *cis* side and mature as the Golgi enzymes move back through the stack. **C,** Lipid partitioning mechanism.

fuse with the next. Directional flow is achieved by the cargo proteins having preferential affinity for the membranes comprising the tubular/vesicular transport intermediates budding out from the Golgi toward the plasma membrane. In a second mechanism, called cisternal progression, secretory cargo is transported across the stack in continuously progressing cisternae. New cisternae form at the *cis* face of the stack by coalescence of VTCs and then progress across the stack to the *trans* side. Secretory cargo molecules are confined within a given cisterna until it passes from the *cis* face to the *trans* face and exits from the Golgi apparatus in transport carriers. Support for cisternal progression derives from studies in yeast showing markers in individual Golgi cisterna mature from early to late forms over time. Live cell kinetic measurements in mammalian cells show that cargo exits from the Golgi over an exponential time course with no lag. This finding, together with the observation that resident enzymes and cargo partition into distinct domains within the Golgi apparatus, in addition to having overlapping distributions, have led to a third model of Golgi trafficking. In this model, partitioning of cargo proteins into lipid domains depleted of Golgi enzymes provides a mechanism for their export out of the Golgi (Fig. 21.20).

The size, appearance, and even existence of the Golgi apparatus depend on the amount and rate of cargo movement through the secretory pathway. The yeast *Saccharomyces cerevisiae,* for example, has a poorly developed Golgi apparatus because secretory transport is normally too fast for elaborate Golgi structures to accumulate. However, conditions that slow cargo transport out of the Golgi apparatus in yeast cells lead to the Golgi apparatus enlarging and rearranging into compact stacks similar to those seen in most animal and plant cells.

The Golgi apparatus is a dynamic rather than permanent cellular structure, because both its proteins and lipids move continuously along various pathways. No class of Golgi protein is stably associated within this organelle. Integral membrane proteins, including processing enzymes and SNAREs, continuously exit and reenter the Golgi apparatus by membrane-trafficking pathways leading to and from the ER. Peripheral membrane proteins associated with the Golgi apparatus (including Arf1, coatomer, Rab proteins, matrix proteins, tethering factors, and GEFs) exchange constantly between Golgi membranes and cytoplasmic pools.

The transient and dynamic association of molecules with the Golgi apparatus makes this organelle sensitive to the functions of many cellular systems. For example, in the absence of microtubules the Golgi apparatus in mammalian cells relocates adjacent to ER export sites (Fig. 21.19). This arises because Golgi enzymes that are continuously recycling back to the ER cannot return to a centrosomal location without microtubules. Instead, they accumulate together with Golgi scaffolding, tethering, and structural coat proteins at ER exit sites distributed across the ER, forming Golgi ministacks.

BFA disperses the Golgi apparatus by a different mechanism. The drug prevents Arf1 from exchanging GDP for GTP (Fig. 21.5), thereby preventing the membrane from recruiting Arf1 effectors from the cytoplasm. Within minutes, resident transmembrane proteins of the Golgi are recycled to the ER where they are retained, and the Golgi apparatus vanishes. If BFA is removed, the Golgi apparatus reforms by outgrowth of membrane from the ER.

The Golgi apparatus disassembles during mitosis in many eukaryotic cells and then reassembles in interphase (Fig. 21.21). This process superficially resembles the effects of BFA application and washout, since many Golgi enzymes return to the ER or to ER exit sites during mitosis and reemerge from the ER at the end of mitosis. This is triggered both by Arf1 being inactivated and by tethering factors/matrix proteins of the Golgi apparatus being phosphorylated by mitotic kinases (see Chapter 40) during mitosis.

Although the Golgi apparatus is highly dynamic and continually exchanges its protein and lipid components with other cellular compartments, it maintains a unique biochemical and morphologic identity. This allows the Golgi apparatus to participate in several major biosynthetic and processing pathways in the cell, as is discussed next.

Glycoprotein and Glycolipid Processing

A primary function of the Golgi apparatus is the **glycosylation** (ie, sugar modification) of proteins and lipids called **glycoproteins** and **glycolipids**. Most cell-surface proteins and lipids and secreted proteins are glycosylated. Their glycans participate in numerous biological functions, including cell–cell and cell–matrix interactions, intracellular and intercellular trafficking, and signaling.

The most widely recognized glycosylation event occurring within the Golgi involves the modification of *N*-linked oligosaccharides on glycoproteins (Fig. 21.22). These *N*-linked sugar chains are added as preformed complexes of 14 sugar residues to asparagine side chains of proteins in the ER (see Figs. 3.26 and 20.11). Following delivery to the Golgi, the *N*-linked sugar chains of the glycoprotein undergo extensive modifications in an ordered sequence. First, some of the mannose residues

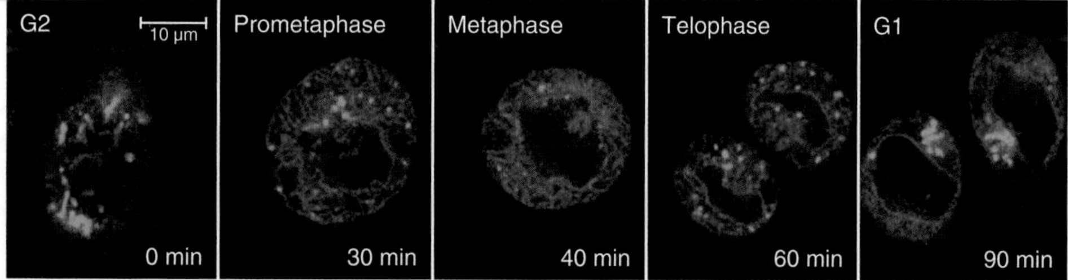

FIGURE 21.21 FLUORESCENCE MICROGRAPHS OF A CELL EXPRESSING A TAGGED GOLGI ENZYME, GALACTOSYLTRANSFERASE–GREEN FLUORESCENCE PROTEIN (GFP), DURING MITOSIS. Time lapse images. As the cell in the *left* of the image passes through prophase and metaphase, its Golgi membranes fragment and then disperse into the ER. During cytokinesis, the Golgi membranes reappear as fragments. These fragments then coalesce into a juxtanuclear Golgi ribbon at the end of mitosis. (From Sengupta P, Sateite-Lrosjnan P, Seo AY, et al. ER trapping reveals Golgi enzymes continually visit the ER through a recycling pathway that controls Golgi organization. *Proc Natl Acad Sci USA.* 2015;112:6752–6761.)

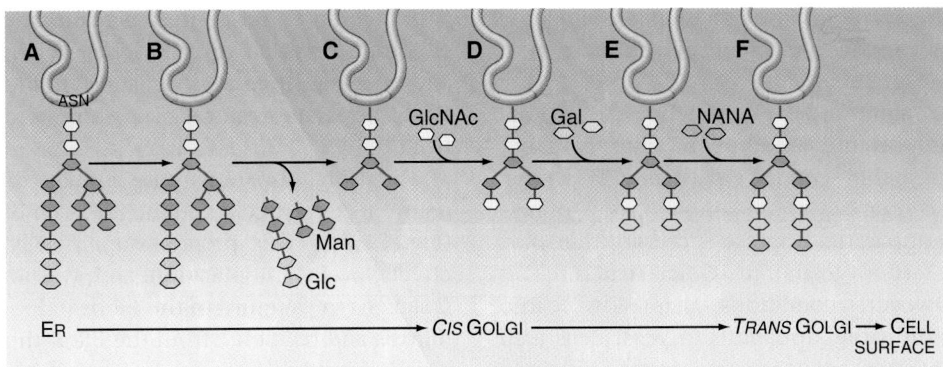

FIGURE 21.22 PROCESSING OF *N*-LINKED CORE OLIGOSACCHARIDES IN THE GOLGI APPARATUS. A–F, Sequential steps trim the mannose (Man)/glucose (Glc) core and then add *N*-acetylglucosamine (GlcNAc), galactose (Gal), and sialic acid (NANA) to form a variety of complex oligosaccharides, one of which is shown here. ASN, asparagine.

are removed. This is followed by the sequential addition of *N*-acetylglucosamine, removal of more mannoses, addition of fucose and more *N*-acetylglucosamine, and finally addition of galactose and sialic acid residues. Cell biologists have used the *N*-linked glycan-processing steps that take place in the mammalian Golgi apparatus as experimental signposts for the passage of glycoproteins through the secretory pathway.

After growing by simple addition of monosaccharide units, many oligosaccharides are modified by enzymes that add phosphate, sulfate, acetate, or methyl groups or isomerize specific carbons. These modifications, as well as differential processing of *N*-linked oligosaccharide structures (producing high-mannose type, complex type, and hybrid structures), contribute to the diversity of sugar residues on the cell surface and can impart specific functions to the sugar chains.

More than 200 Golgi enzymes participate in the biosynthesis of glycoproteins and glycolipids. **Glycosyltransferases** add specific sugar residues to glycans, while **glycosidases** remove specific sugar residues. All these enzymes are type II transmembrane proteins with a short cytoplasmic amino terminal domain followed by a transmembrane segment and catalytic luminal domain within the Golgi cisternae.

Carrier proteins transfer sugar-nucleotides made in the cytoplasm into the lumen of the Golgi apparatus for elongation of glycan chains (see Chapter 15). These carriers are antiporters (see Fig. 15.2) that exchange nucleotide sugars (such as uridine diphosphate [UDP]-*N*-acetylglucosamine, UDP-galactose, and cytidine monophosphate [CMP]-*N*-acetylneuramic acid) for nucleoside monophosphates formed during glycosyl transfer. Glycosyltransferases then use the high-energy sugar-nucleotides as substrates to add new sugars to an oligosaccharide chain. Most glycosyltransferases are specific for sugar-nucleotide donors and particular oligosaccharide acceptors, but the oligosaccharides are synthesized without a template, so their structures vary more than polypeptides and polynucleotides, which are synthesized on templates. Glycosidases trim sugars from the branched core oligosaccharides prior to addition of other sugars. They include mannosidases I and II, which clip outer-branch mannose residues on *N*-linked oligosaccharides prior to the addition of *N*-acetylglucosamine.

The Golgi enzymes also add oligosaccharides to the hydroxyl groups of serine and threonine residues of selected proteins, such as **proteoglycans,** heavily glycosylated proteins in secretory granules, and the extracellular matrix (see Figs. 29.12 and 29.13). This process, called **O-linked glycosylation,** begins with the addition of one of three short oligosaccharides to selected serine and threonine residues of a proteoglycan core protein (see Fig. 3.26). Glycosyltransferases in the Golgi apparatus then add many copies of the same disaccharide unit to the growing polysaccharide. Other enzymes then add sulfates to a few of the sugar residues before the molecule exits the Golgi system.

Enzymes in the Golgi apparatus also mark specific proteins for transport to lysosomes by phosphorylating the 6-hydroxyl of mannose. This modification is the sorting signal that directs lysosomal enzymes to mannose 6-phosphate receptors in the *trans*-Golgi apparatus for targeting to lysosomes. The *N*-linked oligosaccharides on these lysosomal enzymes are initially processed in the ER by trimming of glucose and mannose residues. However, in the Golgi apparatus, they are the unique substrates for two enzymes that act sequentially to generate terminal mannose 6-phosphates. Human patients with the lethal disease mucolipidosis II (called I-cell disease) fail to phosphorylate the mannose residues required for targeting to lysosomes (see Chapter 23). As a result, their lysosomal enzymes are secreted from the cell, and lysosomes fail to degrade waste materials. Lysosomes become engorged with undigested substrates, leading to fatal cell and tissue abnormalities.

Enzymes in the Golgi stacks also load noncovalently associated cholesterol and phospholipids onto high-density and **low-density lipoproteins** for secretion by liver cells into the blood. In plant cells, Golgi enzymes also synthesize complex polysaccharides that are important constituents of the plant cell wall (see Fig. 32.13).

Proteolytic Processing of Protein Precursors

A number of proteins, including peptide hormones, are cleaved into active fragments in the Golgi apparatus and its secretory vesicles. These proteins are synthesized as large precursors with one or more small hormones embedded in long polypeptides. One example is a yeast mating pheromone. Another is proopiomelanocortin, the precursor to no fewer than six small peptide hormones including endogenous opioids in vertebrates. Proteolytic enzymes called **prohormone convertases** cleave the precursor proteins into active hormones in the TGN and post-TGN transport intermediates. The mixture of products depends on the prohormone convertases expressed in particular cells. Proteolysis in the Golgi also affects the final folding state and activity of many other proteins. Inherited defects in these processing pathways lead to a number of diseases, including hormone insufficiency and a hereditary amyloid disease.

Lipid Biosynthesis and Metabolism

The Golgi apparatus also synthesizes sphingolipids, including **sphingomyelin** and the **glycosphingolipids** glucosylceramide and galactosylceramide (see Fig. 13.3). These sphingolipids are key components of the lipid gradient across the secretory pathway (Fig. 21.4) and contribute to the function of the Golgi apparatus as a sorting station. Sphingolipids spontaneously form microdomains or **lipid rafts** in the luminal leaflet of Golgi membranes. Glycosylphosphatidylinositol (GPI)-anchored proteins (see Fig. 13.10), doubly acylated proteins, and transmembrane proteins physically partition into these thicker microdomains and are thereby enriched in vesicles destined for the plasma membrane.

Ceramide, the backbone of all sphingolipids (see Chapter 20 and Figs. 13.2 and 26.4), is synthesized in the ER and then transported to the Golgi complex, where it is modified to form glucosylceramide and sphingomyelin. Glucosylceramide synthesis is catalyzed on the cytoplasmic surface of Golgi membranes by the enzyme UDP-glucose:ceramide glucosyltransferase. Glucosylceramide can then be transported to the plasma membrane or translocated to the luminal leaflet of Golgi membranes, where galactosylation of the head group results in the formation of lactosylceramide. Sequential glycosylation of lactosylceramide by glycosyltransferases of the Golgi lumen generates complex glycolipids and gangliosides for the plasma membrane. Ceramide and cholesterol can be transferred between the ER and Golgi apparatus by a **nonvesicular transport pathway** involving the ER-localized ceramide transfer protein (CERT) and Golgi-localized oxysterol-binding protein (OSBP) (see Fig. 20.17).

Sphingomyelin synthase, an enzyme on the luminal leaflet of Golgi membranes, catalyzes the synthesis of sphingomyelin. The enzyme transfers phosphorylcholine from phosphatidylcholine to ceramide, releasing the signaling lipid diacylglycerol (DAG) in the process. This mechanism therefore couples consumption of the signaling lipid ceramide (see Fig. 26.11) with the production of the signaling lipid DAG (see Fig. 26.8). If DAG accumulates in the Golgi apparatus, phosphorylcholine can be transferred from sphingomyelin back to DAG, forming phosphatidylcholine and ceramide. Alternatively, DAG can be digested by lipases.

Sorting From the *Trans*-Golgi Network

After transport through the Golgi system, cargo leaves the *trans* or exit face of the Golgi apparatus (Fig. 21.23). The exit region is called the **TGN** (trans-Golgi network) because of its tubular network organization. This organization is characteristic of other sorting compartments, such as the VTC, the *cis*-Golgi, and sorting endosomes (see Fig. 22.12). Cargo that arrives in the TGN can be distributed, via distinct transport carriers, to three main intracellular locations: the plasma membrane and cell exterior; the endosome/lysosomal system, polarized cell surfaces, or specialized secretory organelles or granules.

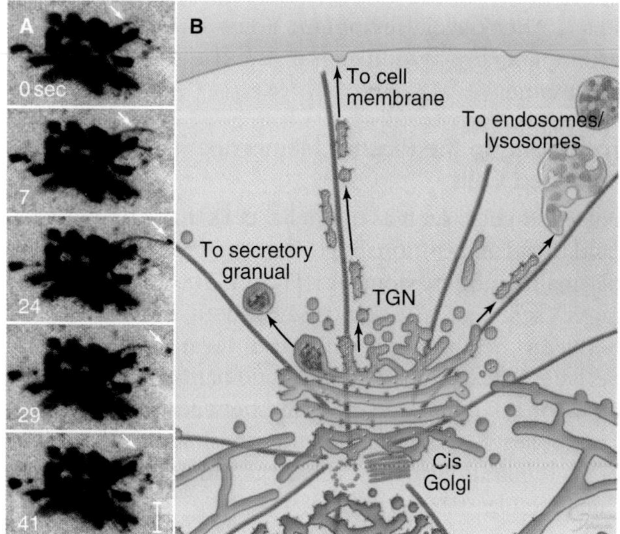

FIGURE 21.23 PATHWAYS FROM THE *TRANS*-GOLGI NETWORK. A, Time series of fluorescence micrographs of a tissue culture cell expressing green fluorescence protein (GFP)-tagged vesicular stomatitis G-protein (VSVG-GFP). The images show long tubules enriched in the labeled protein *(arrows)* emanating from the Golgi apparatus. The tubules later detach from the Golgi and traffic to the plasma membrane. Scale bar is 5 μm. **B,** Cargo is sorted and packaged into distinct transport carriers for targeting to the plasma membrane, endosomes/lysosomes, and secretory granules. (**A,** From Hirschberg K, Miller CM, Ellenberg J, et al. Kinetic analysis of secretory protein traffic and characterization of Golgi to plasma membrane transport intermediates in living cells. *J Cell Biol*. 1998;143:1485–1503, copyright by the Rockefeller University Press.)

Specialized cells emphasize one or more of these pathways. The intracellular route taken by each protein depends on many factors, including whether the protein aggregates, prefers specialized lipid environments, or has sorting properties encoded in the polypeptide chain.

Constitutive Transport to the Plasma Membrane

A steady stream of proteins and lipids from the TGN to the cell surface occurs constitutively through tubular transport carriers that bud out from the TGN (Fig. 21.23B). No known coat proteins form these structures. Instead, cargo proteins conveyed to the plasma membrane by these structures have transmembrane segments that partition into lipid domains containing sphingolipids and cholesterol. Activation of specific lipid-modifying enzymes such as phosphatidylinositol 4-kinase in the sorting domain of the TGN enriched in sphingolipids and cholesterol results in the domains forming tubules that pinch off the TGN. In addition to membrane components, bulk soluble markers are also carried to the plasma membrane by these structures. Motors moving on microtubules and/or actin filaments facilitate extension of these tubules, while dynamin-2 mediates tubule severing. In mammalian cells, motor proteins such as kinesins move the constitutive membrane carriers outward from the Golgi apparatus along microtubules. Fusion of the carriers with the plasma membrane releases cargo carried in the lumen of the carrier vesicle into the extracellular space. After fusion, membrane lipids and proteins redistribute laterally by diffusion in the plane of the plasma membrane.

Trafficking to the Plasma Membrane in Polarized Cells

Polarized cells, such as epithelial cells, have functionally (and thus compositionally) distinct apical and basolateral plasma membrane domains (Fig. 21.24A–B). Tight junctions (see Figs. 31.2 and 31.3) both seal the space between neighboring cells and form the boundary between these two membrane domains by preventing diffusion of lipids and proteins between the domains. This polarity is essential for the physiological functions of epithelia in virtually every organ including the intestine (see Fig. 17.2) and kidney (see Fig. 17.3).

Most of our knowledge of membrane sorting in polarized cells has come from studying epithelial cells. As expected, the trafficking complexity increases as destination options increase, and cells use at least three distinct mechanisms for the polarized sorting of plasma membrane proteins (Fig. 21.24C–D). Most epithelial cells use combinations of these three mechanisms to generate and maintain cell polarity.

The first mechanism (Fig. 21.24C) involves selective packaging of proteins destined for the apical or basolateral membranes into distinct carrier vesicles at the TGN for delivery to the appropriate surface. Direct targeting to the basolateral membrane uses sorting signals in the cytoplasmic domains of membrane proteins for sorting during secretory transport, often canonical motifs such as YXXΦ (tyrosine–any two amino acids–hydrophobic) and [DE]XXXL[LI] (aspartic acid or glutamic acid–any three amino acids–leucine–leucine or isoleucine). These or other sorting signals direct receptors for low-density lipoprotein, transferrin, mannose-6-phosphate receptors (MPRs), and polymeric immunoglobulins from the TGN into basolaterally directed carriers. The μ-1B subunit of AP1 has a higher affinity for these targeting motifs than μ-1A and packages cargo proteins into clathrin-coated vesicles for delivery to basolateral membranes. The μ-1A subunit of AP1 is involved in nonpolarized trafficking. AP4 can also mediate packaging of some proteins for basolateral membranes.

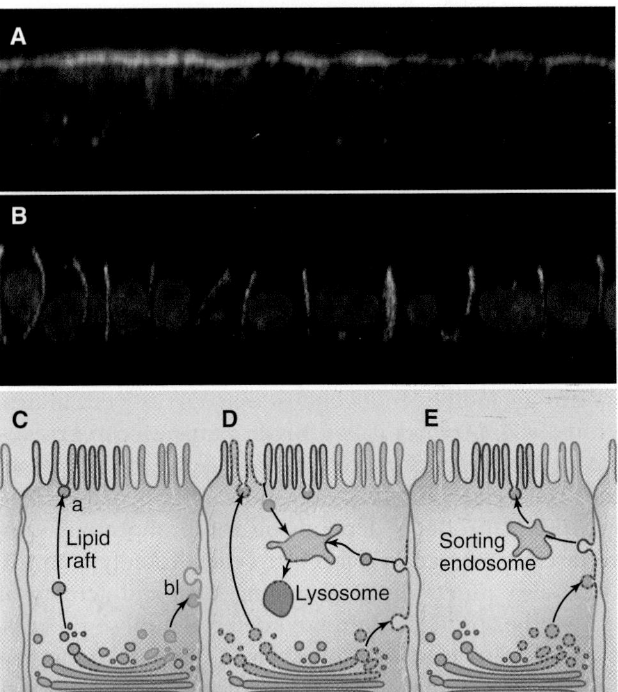

FIGURE 21.24 PATHWAYS THAT ESTABLISH DISTINCT PLASMA MEMBRANE DOMAINS IN POLARIZED EPITHELIAL CELLS. A–B, Fluorescence micrographs of polarized epithelial cells showing tight junctions (marked with *red* fluorescence in both **A** and **B**) that seal the boundary between apical and basolateral domains. **A,** E-cadherin *(green)* is restricted to the apical plasma membrane. **B,** Syntaxin-3 *(green)* is restricted to the basolateral surface. Nuclei are stained *red*. **C–D,** Three pathways for the distribution of integral membrane proteins destined for either the apical *(a [red])* or basolateral *(bl [blue])* membranes of polarized epithelial cells. **C,** Direct sorting from the TGN to either the apical or basolateral surface. Apical transport involves inclusion into lipid rafts, whereas proteins destined for direct transport to the basolateral surface carry a cytoplasmic sorting motif for inclusion into specific transport vesicles. **D,** Indirect pathway. Newly synthesized proteins are randomly targeted to both surfaces followed by selective retention and/or selective degradation from one surface or the other, resulting in a polarized distribution. **E,** Indirect pathway. Newly synthesized proteins are transported to the basolateral surface, followed by retention of basolateral proteins and selective transcytosis of apical proteins to the apical surface. (**A–B,** Courtesy T. Weimbs and S.H. Low, Cleveland Clinic Foundation, Cleveland, OH.)

Direct targeting of proteins from the TGN to the apical domain of polarized epithelial cells is less-well defined, but may depend on partitioning into lipid rafts enriched in sphingomyelin and cholesterol (see Fig. 13.7) formed in the TGN. Proteins concentrate in lipid rafts in the TGN followed by sorting into carrier vesicles directed to the apical cell surface. Examples include proteins with a GPI anchor, *N*- or *O*-glycans that bind to galectins (a multivalent carbohydrate-binding protein) and influenza virus hemagglutinin and neuraminidase. Other GPI-anchored proteins depend on MAL (myelin and lymphocyte protein, VIP17), or annexin2, to target to apical membranes.

The second mechanism for generating cell polarity (Fig. 21.24D) delivers newly synthesized proteins randomly to the apical and basolateral surfaces, followed by their selective retention or depletion at those sites. At steady state the proteins become concentrated at the domain where they are more stable. This mechanism is used for establishing polarity during cellular differentiation. In this case, uniformly distributed proteins that preexist on a nonpolarized cell will redistribute in a polarized fashion in response to cell–cell contacts that initiate polarization. Often, this involves the selective retention of a specific protein in the appropriate domain through intracellular (cytoskeletal) or extracellular (cell–cell or cell–matrix) interactions, or both. Proteins that are not actively retained in the other plasma membrane domain are internalized and degraded in lysosomes. Examples of proteins that are polarized in this way include Na^+K^+-ATPase and the immunoglobulin cell adhesion molecule (IgCAM) adhesion molecule uvomorulin.

A third mechanism for generating cell polarity (Fig. 21.24E) delivers all newly synthesized proteins to the basolateral surface, followed by selective internalization of apical proteins, which are sorted in the endosomal compartment and delivered to the apical surface in a process termed transcytosis. Proteins with motifs such as YXXΦ and [DE]XXXL[LI] are retained at the basolateral surface through local recycling, while proteins that associate with lipid rafts, including GPI-anchored proteins, use transcytosis to reach the apical domain.

Sorting to the Endosome/Lysosomal System

Coats capture specialized cargo for the endolysosome system in carrier vesicles budding off the TGN. Approximately 50 different acid hydrolases follow this pathway, including glycosidases, proteases, lipases, nucleases, and sulfatases. These enzymes are marked with mannose 6-phosphate in the *cis*-Golgi apparatus to divert them from the constitutive secretory pathway. Marked proteins bind to **MPRs** (mannose-6-phosphate receptors) (Fig. 21.25A) in the lumen of the *trans*-Golgi apparatus.

Two sorting motifs in the cytoplasmic tails of MPRs target the receptor and bound lysosomal enzymes to clathrin-coated vesicles in the TGN for delivery to endosomes and lysosomes. The YXXΦ sorting motifs bind heterotetrameric AP1 (Fig. 21.12B). "Acidic cluster dileucine signals" bind the coat adaptor protein GGA.

AP1 and GGA cooperate to package lysosomal enzymes into clathrin-coated vesicles that bud from TGN membranes. AP1 interacts with Arf1-GTP, phosphatidylinositol 4-phosphate (PI4P) and clathrin. GGA recruits clathrin. Its VHS domain recognizes the acidic-cluster-dileucine motif of MPRs. GGA proteins have several

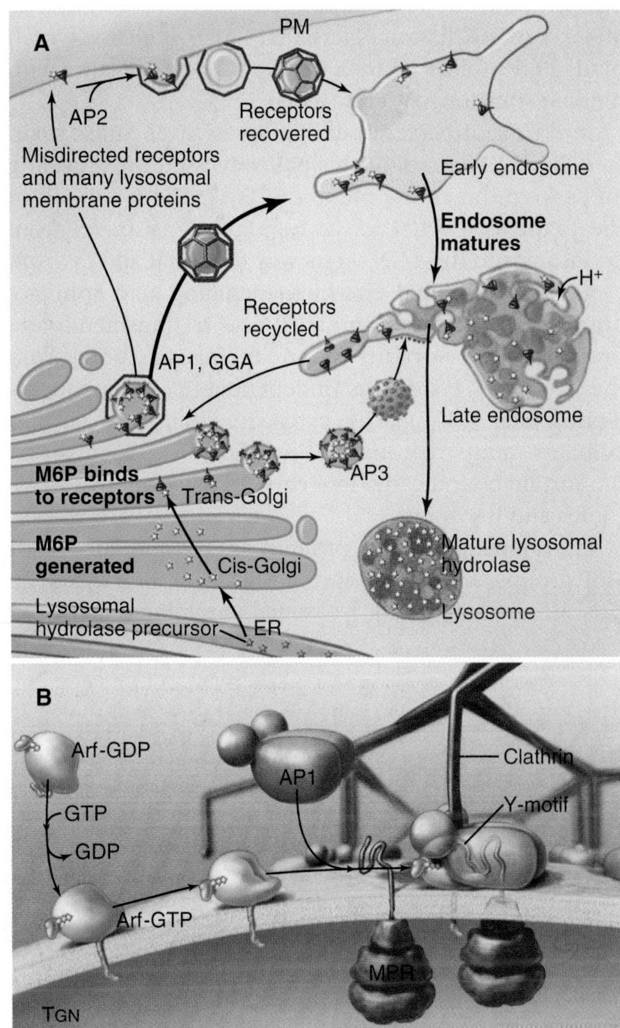

FIGURE 21.25 SORTING PATHWAYS USED BY MANNOSE-6-PHOSPHATE RECEPTORS AND COAT ASSEMBLY AT THE *TRANS*-GOLGI NETWORK. A, Mannose-6-phosphate receptors (MPRs) carry newly synthesized lysosomal hydrolases with mannose-6-phosphate (M6P) from the *trans*-Golgi network (TGN), via endosomes, to lysosomes, after which the MPRs return to the TGN. Receptors misdirected to the cell surface are recovered by endocytosis and returned to the pathway in endosomes. GGA (Golgi-localizing, γ-adaptin ear domain homology, Arf-binding protein), clathrin adapter proteins. **B,** Coordination of coat assembly and cargo recruitment at the TGN. An exchange factor activates the small GTPase Arf to bind GTP, which triggers recruitment of adaptor protein 1 (AP1) coat constituents to the TGN membrane. The MPR is concentrated in the emerging coated vesicle through interactions between a tyrosine-based sorting motif in its cytoplasmic domain and the μ-subunit of AP1.

other conserved domains: the GAT domain interacts with Arf-GTP and PI4P on *trans*-Golgi membranes; and the GAE domain binds accessory proteins that contribute to vesicle formation, movement and fusion with post-TGN membranes.

Carrier vesicles with MPRs move from the TGN to discharge their cargo in the acidic environment in endosomes. Then carrier vesicles transfer unoccupied MPRs back to the TGN (Fig. 21.25A). This process is controlled both by retromer and EpsinR complexes. **Retromer** is a heteropentameric complex that binds cargo such as MPRs. **EpsinR** has an N-terminal ENTH domain (which binds PI4P) followed by a long, unfolded domain (with binding sites for AP1 and GGAs).

Internalized bacterial exotoxins such as Shiga toxin (see Fig. 22.10) exploit this pathway used for recycling MPRs from the endosome to the TGN. The B subunit of the toxin uses retromer and/or EpsinR to travel from endosomes to the TGN on its way to the ER and cytosol.

Certain lysosomal enzymes, including acid sphingomyelinase and cathepsin D/H, use a transmembrane protein called **sortilin** as an alternative route from the TGN to lysosomes independently of mannose-6-phosphate (M6P). The cytosolic domain of sortilin contains motifs that bind GGAs and APs, which direct the sortilin-bound cargo into carriers that target to endosomes and lysosomes.

Lysosomal membrane proteins are not modified with M6P groups, so they use sorting signals in their cytosolic tails to mediate both lysosomal targeting and rapid endocytosis at the cell surface. These lysosomal-targeting signals belong to YXXΦ or acidic-cluster-dileucine types, but have subtle features (ie, placement close to the transmembrane segment of the protein) that direct the protein to lysosomes. These sorting signals interact with AP1 at the TGN, AP2 at the plasma membrane, and AP3 and AP4 in endosomes. Unlike the other AP proteins, AP4 does not recruit clathrin and forms a non-clathrin coat.

Lysosomal membrane proteins (including lysosomal-associated membrane protein [LAMP]-1, LAMP-2, and CD63) follow either a direct or indirect pathway from the TGN to lysosomes. The direct pathway includes early or late endosomes before lysosomes. The indirect pathway involves transport to the plasma membrane followed by endocytosis and passage through endosomes before eventual delivery to lysosomes. Because depletion of plasma membrane-localized AP2 and its partner clathrin inhibits transport of LAMPs to lysosomes more profoundly than depletion of AP1, the indirect pathway may be more heavily used than the direct pathway for targeting LAMPs to lysosomes.

Secretory Granule Formation, Transport, and Fusion

Endocrine, exocrine, and neuronal cells use a third sorting pathway from the TGN to concentrate and package selected proteins in storage granules before they are discharged from the cell in response to hormonal or neural stimulation. This **regulated secretory pathway** (Fig. 21.26) stores and discharges on command many proteins needed intermittently such as polypeptide

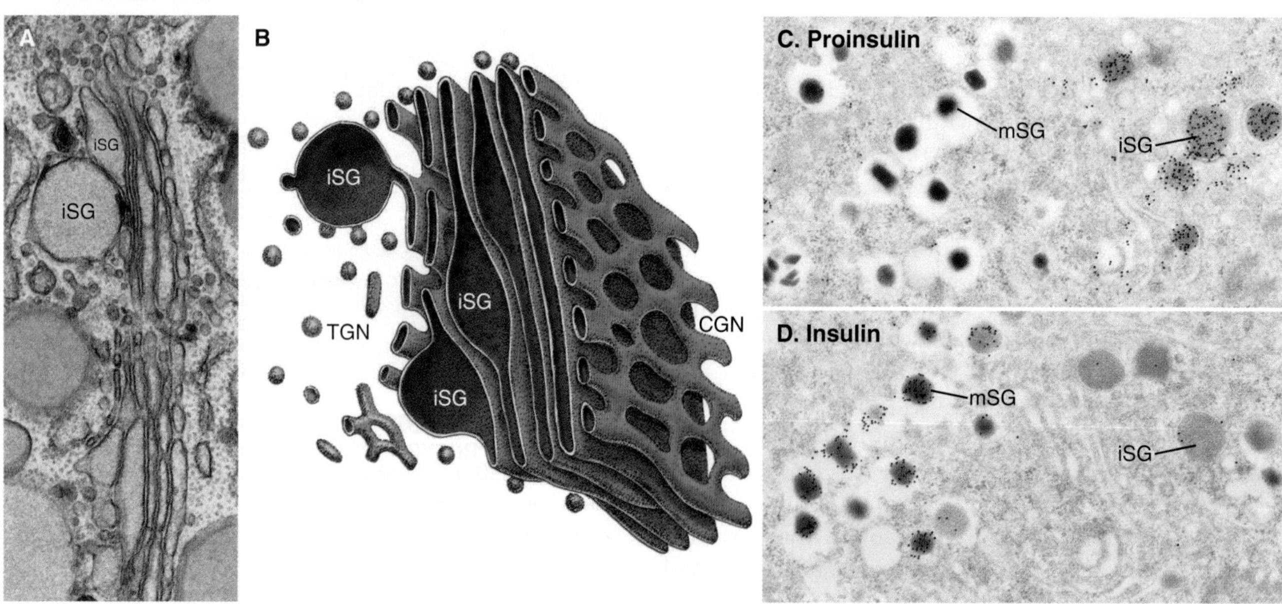

FIGURE 21.26 FORMATION OF SECRETORY GRANULES. Transmission electron micrograph of a thin section **(A)** and a diagram **(B)** show immature secretory granules (iSG) as they emerge from the *trans*-Golgi network (TGN). Much of the TGN surface is consumed by forming immature secretory granules. **C–D,** Cryoelectron micrographs of frozen sections reacted with gold-labeled antibodies to proinsulin **(C)** or insulin **(D).** Proinsulin is concentrated in immature secretory granules. After processing, insulin is concentrated in mature, dense-core secretory granules (mSG). (**A,** Courtesy Y. Clermont, McGill University, Montreal, Quebec, Canada, with permission of Wiley-Liss, Inc. **B,** Modified from Clermont Y, Rambourg A, Hermo L. *Trans*-Golgi network (TGN) of different cell types: three-dimensional structural characteristics and variability. *Anat Rec*. 1995;242:289–301. **C–D,** Courtesy L. Orci, University of Geneva, Switzerland.)

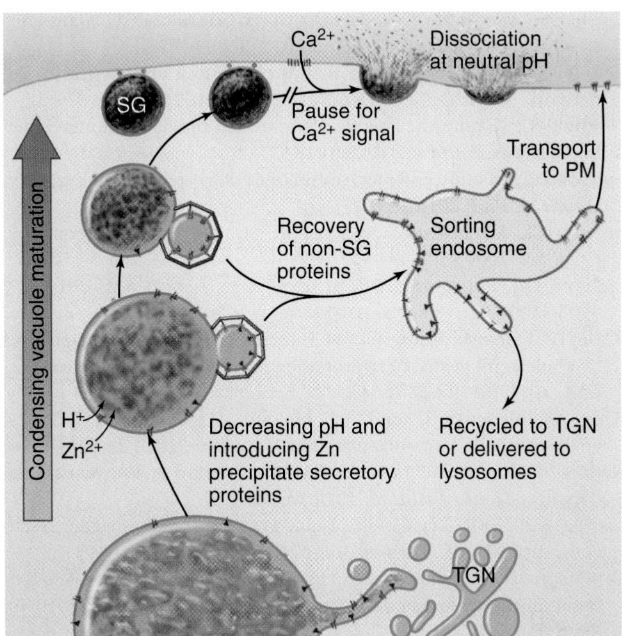

FIGURE 21.27 MATURATION OF NASCENT SECRETORY GRANULES/CONDENSING VACUOLES. The vacuolar H^+-ATPase in the secretory granule (SG) membrane lowers the internal pH. This drives condensation and concentration of the contents. Dense-core, mature secretory granules are stored in the cytoplasm until a Ca^{2+}-mediated signaling event triggers fusion and release of their contents. Proteins inadvertently included in large, immature secretory granules emerging from the *trans*-Golgi network (TGN) are captured by clathrin-coated vesicles and recycled to endosomes and the TGN. PM, plasma membrane.

hormones and digestive enzymes. These secreted proteins apparently lack a universal sorting signal. Instead, secretory granule formation appears to involve physical sorting, selective retention, and condensation of the secretory proteins through charge neutralization, protein aggregation and active extrusion of ions (Fig. 21.27).

Aggregation can be selective during formation of secretory granules. For example, the polypeptide hormone **insulin** is cleaved from a precursor called proinsulin. Proteolytic enzymes cleave proinsulin at two sites in immature granules, generating insulin and C-peptide. Zinc ions selectively condense insulin in the granule core surrounded by C-peptide. Subsequently, budding of vesicles from immature granules removes more C-peptide than insulin. Carefully regulated insulin secretion controls the glucose concentration in the blood plasma by a process that adjusts the rate of glucose taken into muscle and fat cells (see Fig. 27.7). Either the production of insulin or the regulation of glucose uptake is compromised in **diabetes mellitus.**

The proteolytic enzyme carboxypeptidase E (CPE) may act as a sorting signal for sorting of hormones and neuropeptides at the TGN into secretory granules. A short α-helical domain on CPE interacts with the granule membrane, allowing cargo proteins that bind to CPE to be codelivered with CPE into a secretory granule.

A P-type calcium ATPase pump called SPCA1 (secretory pathway Ca^{2+}/Mn^{2+}-ATPase pump type 1) helps aggregate proteins for secretory granules in the TGN. Binding of actin filaments on the cytosolic face of the TGN is thought to concentrate SPCA1 into a sorting domain. Ca^{2+} pumped into the lumen of the TGN by SPCA1 activates a Ca^{2+} binding protein called Cab45 to bind soluble cargo proteins and helps segregate them into secretory vesicles.

Fusion of secretory granules with the plasma membrane and release of their contents is carefully regulated to ensure that they fuse only on demand. Regulated fusion has been studied most extensively in the context of synaptic vesicle release (see Figs. 17.9 and 17.10), in endocrine cells (Fig. 21.26), and in mast cells (see Fig. 28.8).

Regulated secretion can be divided into three steps: docking, priming, and fusion (Fig. 21.14). Docking is the slowest step and is believed to involve interactions of v-SNARE and t-SNAREs regulated by Rab GTPases. In vitro reconstitution studies suggest that additional proteins are required for the priming step in neuroendocrine cells. They include a phosphatidylinositol transfer protein and a phosphatidylinositol 5-kinase and its product phosphatidylinositol 4,5-bisphosphate (PIP_2).

In most cases a transient local increase in cytoplasmic Ca^{2+} triggers fusion of the secretory granule membrane with the plasma membrane, a process called *calcium-secretion coupling.* Diverse signals produce the calcium influx that triggers fusion, including activation of seven-helix receptors on neuroendocrine cells, activation of immunoglobulin E receptors in mast cells (see Fig. 28.8), and membrane depolarization in neurons (see Figs. 17.10 and 17.11). Calcium-binding proteins called synaptotagmins are believed to act as clamps on the fusion machinery, inhibiting fusion until calcium triggers their release (Fig. 21.14). In endocrine cells and neurons a cytoplasmic protein, CAPS (calcium activator protein for secretion), is recruited to the secretory vesicle by interacting with PIP_2 and is required for calcium-triggered fusion of dense core secretory granules.

The fine-tuned control of regulated secretion is illustrated by exocytosis of the glucose carrier GLUT4, which mediates insulin-stimulated glucose uptake in fat and muscle (see Fig. 27.7). GLUT4 delivery to the cell surface requires: mobilization of GLUT4-containing storage vesicles from intracellular sites; docking at the plasma membrane; and, finally, fusion of the two membranes. Fig. 27.7 illustrates how insulin controls this pathway by activating phosphatidylinositide 3′-kinase (PI3K) and Akt/PKB, which phosphorylates the Rab-GAP protein AS160. Unphosphorylated AS160 retains GLUT4 vesicles within the cell by inactivating Rab10. AS160 loses its GAP activity when phosphorylated in response to insulin. This turns on the Rab proteins, which interact with myosin-Va to translocate the GLUT4 vesicle to the cell surface. Once docked at fusion sites, the SNARE machinery drives fusion of the vesicle with the plasma membrane, delivering GLUT4 to the cell surface.

ACKNOWLEDGMENTS

We thank Juan Bonifacino, Catherine Jackson, and Yongli Zhang for their suggestions on revisions to this chapter.

SELECTED READINGS

Altan-Bonnet N, Sougrat R, Lippincott-Schwartz J. Molecular basis for Golgi maintenance and biogenesis. *Curr Opin Cell Biol*. 2004;16(4):364-372.

Antonny B. Membrane deformation by protein coats. *Curr Opin Cell Biol*. 2006;18:1-9.

Bigay J, Antonny B. Curvature, lipid packing, and electrostatics of membrane organelles: defining cellular territories in determining specificity. *Dev Cell*. 2012;23:886-895.

Bonifacino JS. Adaptor proteins involved in polarized sorting. *J Cell Biol*. 2014;204:7-17.

Bonifacino JS, Hurley JH. Retromer. *Curr Opin Cell Biol*. 2008;20:427-436.

Borgese N. Getting membrane proteins on and off the shuttle bus between the endoplasmic reticulum and the Golgi complex. *J Cell Sci*. 2016;129:1537-1545.

Cherfils J. Arf GTPases and their effectors: assembling multivalent membrane-binding platforms. *Curr Opin Struct Biol*. 2014;29:67-76.

Cherfils J, Zeghouf M. Regulation of small GTPases by GEFs, GAPs and GDIs. *Physiol Rev*. 2013;93:269-309.

Dancourt J, Barlowe C. Protein sorting receptors in the early secretory pathway. *Annu Rev Biochem*. 2010;79:777-802.

Debuke ML, Munson M. The secret life of tethers: the role of tethering factors in SNARE complex regulation. *Front Cell Dev Biol*. 2016;4:1-8.

De Matteis MA, Wilson C, De Angelo G. Phosphatidylinositol-4-phosphate: the Golgi and beyond. *Bioessays*. 2013;35:612-622.

Faini M, Prinz S, Beck R, et al. The structures of COPI-coated vesicles reveal alternate coatomer conformations and interactions. *Science*. 2012;336:1451-1454.

Faini M, Beck R, Wieland FT, et al. Vesicle coats: structure, function, and general principles of assembly. *Trends Cell Biol*. 2013;23:279-288.

Gillingham AK, Munro S. Finding the Golgi: Golgin coiled-coil proteins show the way. *Trends Cell Biol*. 2016;26:399-408.

Guo Y, Sirkis DW, Schekman R. Protein sorting at the *trans*-Golgi network. *Annu Rev Cell Dev Biol*. 2014;30:169-206.

Holthuis JC, Menon AK. Lipid landscapes and pipelines in membrane homeostasis. *Nature*. 2014;510:48-57.

Jackson LP. Structure and mechanism of COPI vesicle biogenesis. *Curr Opin Cell Biol*. 2014;29:67-73.

Jackson CL. Mechanisms for transport through the Golgi complex. *J Cell Sci*. 2009;122:443-452.

Jackson CL, Bouvet S. Arfs at a glance. *J Cell Sci*. 2014;127:4103-4109.

Kaiser HJ, Orlowski A, Rog T, et al. Lateral sorting in model membranes by cholesterol-mediated hydrophobic matching. *Proc Natl Acad Sci USA*. 2011;108:16628-16633.

Khan AR, Menetrey J. Structural biology of Arf and Rab GTPases' effector recruitment and specificity. *Structure*. 2013;21:1284-1296.

Kienzle C, von Blume J. Secretory cargo sorting at the *trans*-Golgi network. *Trends Cell Biol*. 2014;24:584-593.

Leto D, Saltiel AR. Regulation of glucose transport by insulin: traffic control of GLUT4. *Nat Rev Mol Cell Biol*. 2012;13:383-395.

Lippincott-Schwartz J, Phair RD. Lipids and cholesterol as regulators of traffic in the endomembrane system. *Annu Rev Biophys*. 2010;39:559-578.

Miller EA, Schekman R. COPII—a flexible vesicle formation system. *Curr Opin Cell Biol*. 2013;25:420-427.

Patterson GH, Hirschberg K, Polishchuk RS, et al. Transport through the Golgi apparatus by rapid partitioning within a two-phase membrane system. *Cell*. 2008;133:1055-1067.

Rizo J, Xu J. The synaptic vesicle release machinery. *Annu Rev Biophys*. 2015;44:339-367.

Santos AJ, Raote I, Scarpa M, et al. TANGO1 recruits ERGIC membranes to the endoplasmic reticulum for procollagen export. *Elife*. 2015;4:e10982.

Sengupta P, Satpute-Krishnan P, Seo AY, et al. ER trapping reveals Golgi enzymes continually revisit the ER through a recycling pathway that controls Golgi organization. *Proc Natl Acad Sci USA*. 2015;112:6752-6761.

Wu B, Guo W. The exocyst at a glance. *J Cell Sci*. 2015;128:2957-2964.

Zanetti G, Prinz S, Daum S, et al. The structure of the COPII transport-vesicle coat assembled on membranes. *eLife*. 2013;2:1-15.

Zhen Y, Stenmark H. Cellular functions of Rab GTPases at a glance. *J Cell Sci*. 2015;128:3171-3176.

CHAPTER 22

Endocytosis and the Endosomal Membrane System

Regulated entry of molecules into eukaryotic cells occurs at the plasma membrane, the interface between the intracellular and extracellular environments. Small molecules such as amino acids, sugars, and ions traverse the plasma membrane through the action of integral membrane protein pumps (see Chapter 14), carriers (see Chapter 15), or channels (see Chapter 16). On the other hand, macromolecules can enter cells only by being captured and enclosed within membrane-bound carriers that invaginate and pinch off the plasma membrane in a process known as **endocytosis.** Cells use endocytosis to feed themselves, to defend themselves, and to maintain homeostasis. Some toxins, viruses, pathogenic bacteria, and protozoa "hijack" this process to enter cells.

Endocytosis was discovered more than a century ago in white blood cells (macrophages and neutrophils), the body's "professional phagocytes" (see Fig. 28.6). Endocytosis by these cells is very active, and they internalize the equivalent of their entire plasma membrane surface every hour. When macrophages internalize particles of blue litmus paper, the particle color changes, revealing that endocytic vacuoles are acidic. Investigators still use molecules tagged with fluorescent dyes, green fluorescent protein, or electron-dense markers to follow endocytosis in living or fixed cells by light or electron microscopy. Subcellular fractionation, sometimes aided by loading cells with tracers that alter the density of the endocytic compartments or with ferromagnetic tags, has allowed the isolation and biochemical characterization of distinct classes of endocytic structures. In vitro reconstitution systems have also helped to decipher the mechanisms governing membrane trafficking along the endocytic pathway.

Cells use many different mechanisms for endocytosis (Fig. 22.1). These differ in mode of uptake and in the type and intracellular fate of internalized cargo. The mechanisms include **phagocytosis, macropinocytosis, clathrin-mediated endocytosis, caveolae-dependent uptake,** and **nonclathrin/noncaveolae endocytosis.** The protrusions or invaginations of the plasma membrane that are formed during these diverse endocytic processes all require coordinated interactions between a variety of protein and lipid molecules that dynamically link the plasma membrane and cortical actin cytoskeleton inside the cell.

In phagocytosis and clathrin-mediated endocytosis, cell surface receptors selectively bind macromolecules (ligands) to be internalized. Ligands can be proteins, glycoproteins, or carbohydrates. In phagocytosis, the ligands are usually membrane constituents of other cells, bacteria, or viruses, whereas in clathrin-mediated endocytosis the ligands are often growth factors or other soluble components. After ligand-receptor complexes are concentrated into patches in the membrane, the membrane is then either pinched off to form small vesicles (in clathrin-mediated endocytosis) or zipped up around the particle to form a large vacuole inside the cell (in phagocytosis).

Other forms of endocytosis are less selective. In some cases, cells take up bulk fluid through small, pinocytic vesicles or through macropinocytosis, in which the cell extends its membrane and engulfs extracellular fluid indiscriminately. Alternatively, ligands and molecules associated with lipid rafts are taken up at the plasma membrane through caveolae-mediated or nonclathrin/noncaveolar endocytosis.

Endocytic carriers produced by the various endocytic mechanisms are transported into the cytoplasm away from the plasma membrane, where they fuse with each other and with other membrane compartments comprising the **endosomal membrane system.** Among the various compartments of the endocytic membrane system are **early/recycling endosomes, multivesicular bodies, late endosomes,** and **lysosomes.** Each has a distinct role in the sorting, processing, and degradation of internalized cargo, and they communicate with each

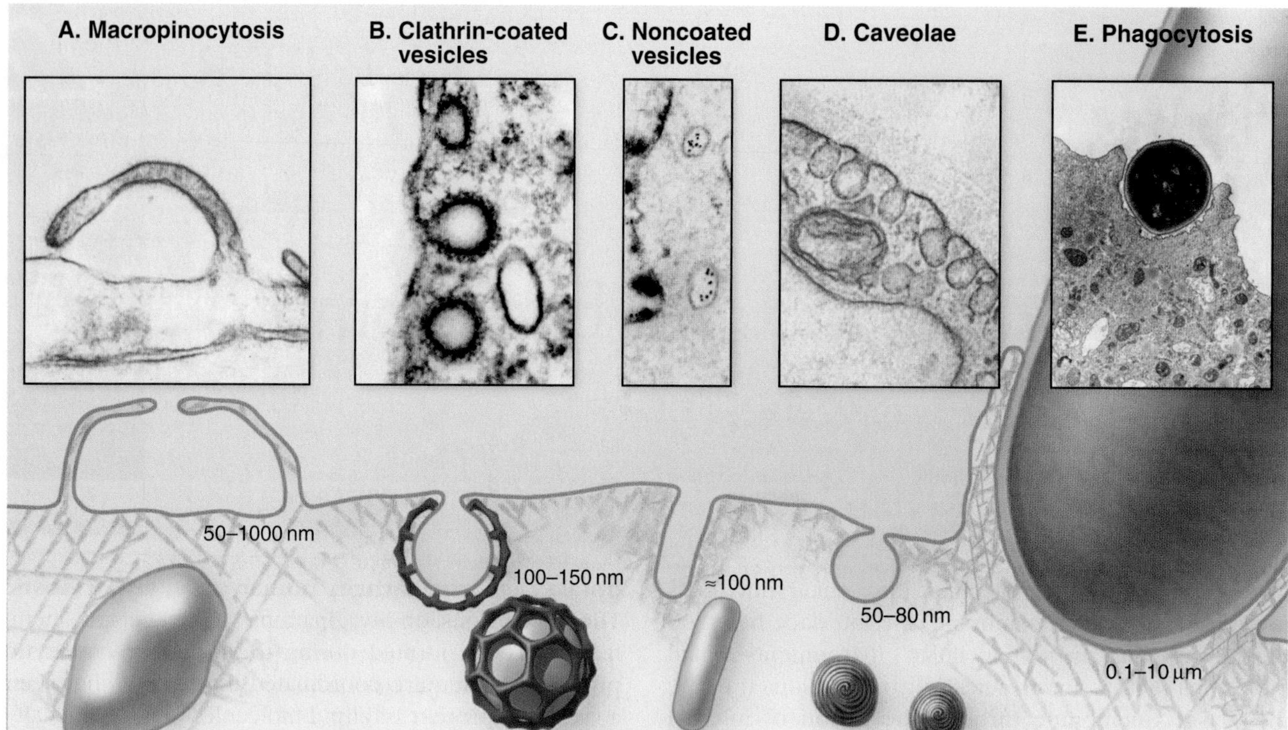

FIGURE 22.1 **A–E,** Electron micrographs and diagrams illustrating five structurally and mechanistically distinct pathways for entry into the cell. The endocytic vesicles that are generated differ in size and structure, as shown. (**A** and **D,** Courtesy D. Fawcett, Harvard Medical School, Boston, MA. **B,** Courtesy C-M Chang and S. Schmid, Scripps Research Institute, La Jolla, CA. **C,** Courtesy S. Hansen and B. van Deurs, University of Copenhagen, Denmark. **E,** Courtesy Blair Bowers, National Institutes of Health, Bethesda, MD.)

other and/or the plasma membrane by mechanistically diverse and highly regulated pathways.

This chapter describes the molecular mechanisms of the major types of endocytosis and the functions of the endosomal system.

Phagocytosis

Phagocytosis is the ingestion of large particles such as bacteria, foreign bodies, and remnants of dead cells (Fig. 22.2). Cells use the actin cytoskeleton to push a protrusion of the plasma membrane that surrounds these particles (Fig. 22.3).

Some cells, including macrophages, dendritic cells, and neutrophils (see Chapter 28), are specialized for phagocytosis. The presence of bacteria or protozoa in tissues attracts phagocytes from the blood (see Fig. 30.13). They then ingest the microorganisms and initiate inflammatory and immune responses. Other cell types use phagocytosis to remove dead neighboring cells, while amoeba use phagocytosis for feeding.

Phagocytosis proceeds through four steps: attachment, engulfment, fusion with lysosomes, and degradation (Fig. 22.3). These steps are highly regulated by cell surface receptors, polyphosphatidylinositides, and signaling cascades mediated by Rho-family guanosine triphosphatases (GTPases).

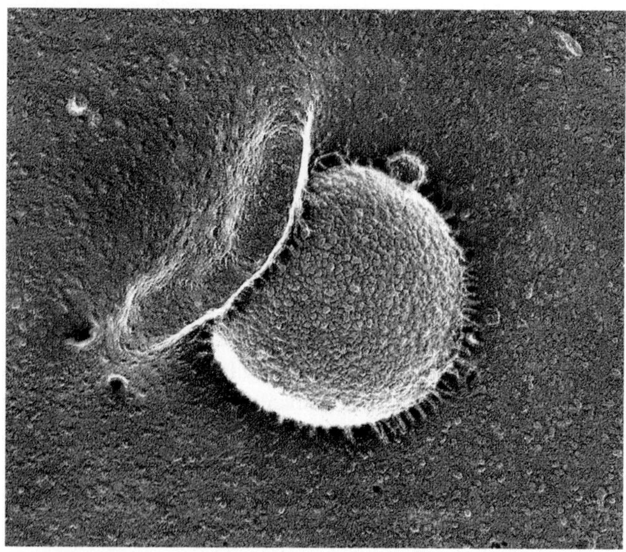

FIGURE 22.2 ELECTRON MICROGRAPH OF AN AMOEBA INGESTING A LATEX BEAD BY PHAGOCYTOSIS. Note the numerous sites of attachment between the amoeba cell surface and the bead. (Courtesy John Heuser, Washington University, St. Louis, MO.)

Attachment and Engulfment

Attachment depends on the ability of the phagocytic cell to recognize the particle to be ingested. Such specific interactions trigger ingestion of the particle. Vertebrates use proteins, collectively called "opsonins," to mark

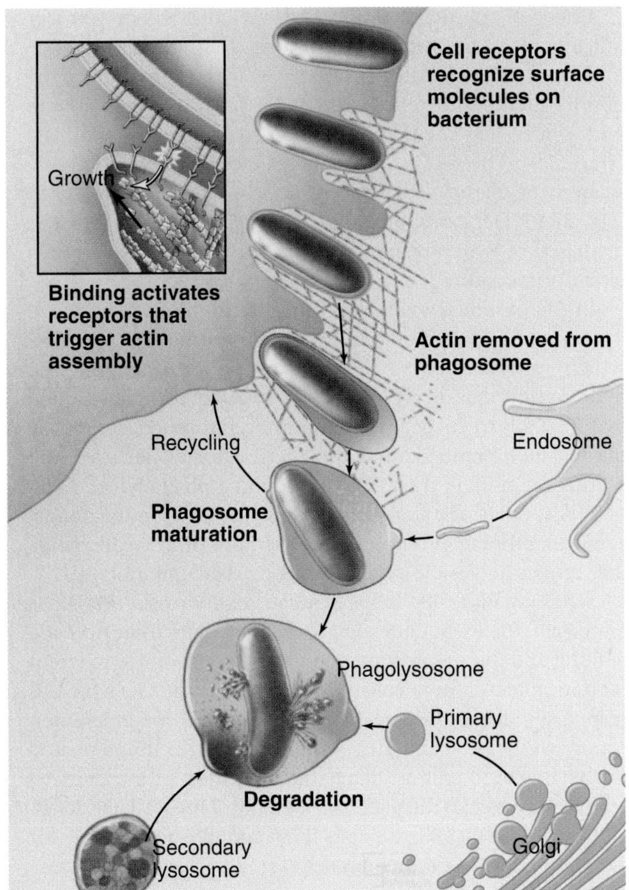

FIGURE 22.3 MOLECULAR MECHANISM FOR PHAGOCYTOSIS OF A BACTERIUM BY A MACROPHAGE. Macrophage surface receptors are activated by contact with a bacterium; this triggers actin filament assembly that leads to protrusion of the plasma membrane to engulf the bacterium. The actin filaments encasing the newly formed phagosome depolymerize, and membrane traffic to and from the phagosome leads to its maturation. Hydrolytic enzymes are delivered to the phagosome by fusion with primary lysosomes or more mature secondary lysosomes, and the bacterium is degraded.

bacteria and other foreign particles for phagocytosis. Opsonins include antibodies, which bind to foreign antigens on bacteria, and complement proteins, which tag infected or dying cells. Phagocytes such as macrophages use plasma membrane receptors to bind particles coated with opsonins. For example, **immunoglobulin Fc receptors** bind to the constant regions of immunoglobulin G molecules (see H3 and H4 domains in Fig. 3.13B) coating pathogenic bacteria and viruses.

Engulfment begins by binding of receptors such as the Fc receptor to a foreign particle with the appropriate target on its surface to generate localized signals on the cytoplasmic side of the plasma membrane. These signals trigger the assembly of actin filaments immediately adjacent to the particle to be ingested. Growth of these actin filaments supports the plasma membrane as it zips tightly around the particle to form a cup-like protrusion, called the **phagocytic cup.** These events depend on polyphosphoinositides and phosphatidylinositol (PI) kinases (Box 22.1 and Figs. 22.4 and 26.7). In the phagocytic cup, PI(3) kinase generates **PI(3,4,5)P_3 (PIP_3),** a ligand for PH-domains of guanine nucleotide exchange factors for the small GTPases Rac1, Arf6, and Cdc42. Once activated, these GTPases stimulate actin assembly, leading to phagocytic cup growth.

Closure of the phagocytic cup occurs when the membrane zips up around the particle and then fuses together. Phagosome closure coincides with local depletion of PIP_3 by PI phosphatases and phospholipase Cγ. The phosphatase produces PI(4,5)P_2 in the phagosome membrane. This promotes assembly of actin filaments that drive the vesicle away from the plasma membrane into the cytoplasm.

Both the plasma membrane and internal membranes contribute to make a phagocytic cup. Internal membranes from recycling endosomes, late endosomes, and possibly endoplasmic reticulum (ER) contribute to the phagocytic cup by fusing with the plasma membrane in a process called **focal exocytosis.** When secretory lysosomes fuse at the forming phagocytic cup, they release cytokines that contribute to inflammation. This couples phagocytosis to the immune response. Focal exocytosis relies on the same steps that are involved in other membrane fusion events, including transport of internal membranes along cytoskeletal tracks and their fusion by compartment-specific SNARES (soluble N-ethylmaleimide-sensitive factor [NSF] attachment protein receptors) under the control of Rab GTPases (see Fig. 21.6).

Fusion With Lysosomes and Alternative Fates

After closure and movement of the phagosome away from the plasma membrane, the actin filaments surrounding it disassemble, and motors direct the phagosome along microtubules deep into the cell during a process termed **directed maturation.** A series of fusion and fission reactions remove plasma membrane components and replace them with endosome-specific components including proteins (eg, SNAREs) required for selective fusion with acidic lysosomes containing active hydrolytic enzymes. Fusion with lysosomes creates a hybrid vacuole called a **phagolysosome** (Fig. 22.3).

A combination of factors kills ingested microorganisms in phagolysosomes. The reduced form of nicotinamide-adenine dinucleotide phosphate oxidase in the phagolysosomal membrane produces a lethal barrage of toxic oxidants. Small peptides called **defensins** bind and disrupt microbial membranes. Proteases and acid hydrolases in the lumen of the phagolysosome degrade the ingested organism to its constituent amino acids, monosaccharides and disaccharides, nucleotides, and lipids. These small products of digestion are transported across the phagolysosomal membrane into the cytoplasm, for reuse to synthesize new macromolecules.

Phosphatidylinositol (PI) is a glycerophospholipid with a cyclohexanol head group (Fig. 22.4D) that can be phosphorylated on carbons 3, 4, and 5 either singly or in combination to produce **polyphosphoinositides** (see Fig. 26.7). Polyphosphoinositides are minor lipids in the cytoplasmic leaflet of the plasma membrane (~1% of total lipids) and endocytic membranes, but lipid kinases and phosphatases can change their levels rapidly at local sites in membranes (Fig. 22.4E). This local synthesis of particular polyphosphoinositides regulates membrane remodeling during exocytosis, endocytosis, and vesicular trafficking by recruiting and/or activating proteins that sense the curvature of the lipid bilayer, form scaffolds on the membrane (eg, clathrin and dynamin), or regulate actin assembly.

The most important polyphosphoinositide for endocytosis is phosphatidylinositol(4,5)bis-phosphate **($PI(4,5)P_2$)** with phosphates on carbons 4 and 5 of the head group. Two lipid kinases synthesize $PI(4,5)P_2$ by adding phosphate first to the hydroxyl on carbon 4 and then to the hydroxyl on carbon 5 (Fig. 22.4E; also see Fig. 26.7). The second enzyme, **phosphatidylinositol-4-P-5 kinase,** is activated by another glycerophospholipid, phosphatidic acid (PA; see Fig. 13.2). Because $PI(4,5)P_2$ activates the phospholipase D (see Fig. 26.4) that makes PA, the two enzymes form a positive feedback loop that enriches $PI(4,5)P_2$ locally in the membrane. Interactions of $PI(4,5)P_2$ with proteins from the cytoplasm retards its mobility in the plane of the membrane, raising its local concentration until it is depleted by removal of the head group or by dephosphorylation (see Fig. 26.7).

$PI(4,5)P_2$ participates in clathrin-mediated endocytosis, phagocytosis, and macropinocytosis (Fig. 22.4A–C). The formation of the clathrin lattice and its tethering to the plasma membrane relies on several proteins that interact with $PI(4,5)P_2$, including AP180/CALM, epsin, and AP2 (Fig. 22.9). The GTPase dynamin, which is essential for the scission of clathrin-coated vesicles, also binds to $PI(4,5)P_2$ (Fig. 22.9). Dephosphorylation of $PI(4,5)P_2$ to form PI(4)P is mediated by synaptojanin, which plays an important role in clathrin uncoating.

In phagosome biogenesis, high-affinity binding between ligands and plasma membrane receptors attracts PI(3)kinase, which produces PIP_3 (PI(3)P, phosphorylated on the hydroxyl on carbon 3). Activation of Rac, Arf6, and Cdc42 by PIP_3 leads to cortical actin assembly and protrusion of the plasma membrane around the phagocytosed particle. The plasma membrane then zips up around a phagocytosed particle. After plasma membrane closure, $PI(4,5)P_2$ in the phagosome membrane promotes the assembly of actin filaments that move the vesicle away from the plasma membrane.

Whereas $PI(4,5)P_2$ helps regulate endocytosis, PI(3)P is important for early endosome dynamics. It is found on the limiting and intraluminal membranes of endosomes, where it recruits effector molecules. These include EEA1, which is responsible for endosome-endosome fusion through its interaction with Rab5, and Hrs, which recognizes ubiquitinated endocytic cargo and facilitates the formation of intraluminal endosomal vesicles through the assembly of ESCRT-I, ESCRT-II, and ESCRT-III. PI(3)kinase class II or class III is responsible for generating PI(3)P on membranes (Fig. 22.4E).

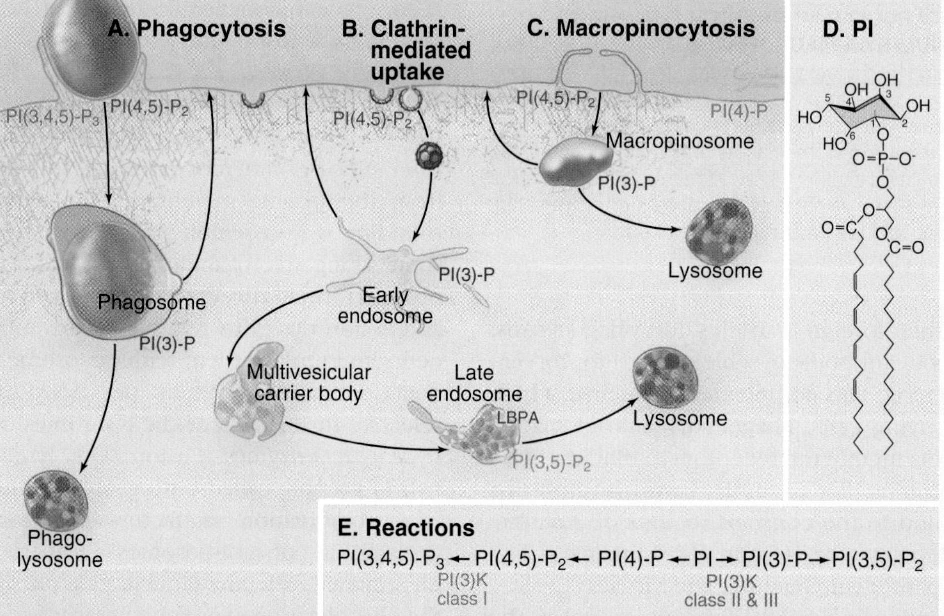

FIGURE 22.4 DISTRIBUTION OF PHOSPHOINOSITIDES AMONG ENDOCYTIC COMPARTMENTS. A–C, Polyphosphoinositol territories within the endocytic system. Localized $PI(4,5)P_2$ in the plasma membrane plays a role in phagocytosis **(A),** clathrin-mediated endocytosis **(B),** and macropinocytosis **(C).** $PI(3,4,5)P_3$ in the plasma membrane participates in phagocytosis. PI(3)P is enriched in endosomes, whereas $PI(3,5)P_2$ and lysobisphosphatidic acid (LBPA) are enriched in late endosomes. Proteins, including clathrin adapters, bind to the specific polyphosphoinositides depicted here, providing a mechanism for their targeting. **D,** Phosphatidylinositol can be phosphorylated on the 3, 4, or 5 position of its inositol ring, with all seven combinations possible. The polyphosphatidylinositides that are so generated embed in the cytoplasmic leaflets of membranes. **E,** Biochemical pathways that generate different polyphosphatidylinositides. Three classes of PI(3)-kinases participate: Class I PI(3)-kinase uses $PI(4,5)P_2$ as substrate yielding $PI(3,4,5)P_3$ (involved in phagocytosis); Class II and III PI(3)-kinases use PI yielding PI(3)P (involved in endosome maturation). PI(3)-kinase inhibitors such as wortmannin and 3-methyladenine have helped to characterize the function of these PI(3)-kinases. The inhibitors compete for adenosine triphosphate (ATP) binding in the active site of the kinase domain.

BOX 22.2 Survival Strategies for Intracellular Pathogens

"Escape"

- Secretion of toxins that disrupt phagosomal membrane (*Shigella flexneri, Listeria monocytogenes, Rickettsia rickettsii*)

"Dodge"

- Entrance through alternative, pathogen-specific pathway (*Salmonella typhimurium, Legionella pneumophila, Chlamydia trachomatis*)
- Inhibition of phagosome-lysosome fusion (*S. typhimurium, Mycobacterium tuberculosis*)
- Inhibition of phagolysosome maturation and acidification (*Mycobacterium* species)

"Stand and Fight"

- Low pH-dependent replication (*Coxiella burnetii, S. typhimurium*)
- Enhancement of DNA repair to survive oxidative stress (*S. typhimurium*)
- Protective pathogen-specific virulence factors (*C. burnetii, S. typhimurium*)
- Prevention of the processing and presentation of bacterial antigens (*S. typhimurium*)

Any undegraded material remains in the lysosome, which is then called a **residual body.**

Antigen-presenting phagocytic cells, such as dendritic cells, cleave proteins of ingested microorganisms into small peptides for loading onto membrane receptors called major histocompatibility complex (MHC) class II molecules. This transfer occurs in phagolysosomes called the **antigen-presenting compartment** in these cells. MHC Class II molecules loaded with peptides recycle back to the surface of phagocytic cells, where they activate CD4+ T-lymphocytes (see Fig. 27.8).

Some pathogens have counterstrategies to avoid destruction by phagolysosomes. These include mechanisms either to inhibit fusion of phagosomes with lysosomes, to resist the low pH environment of the lysosome, or to escape to the cytoplasm by lysing the phagolysosome membrane (Box 22.2). For example, in the lungs of patients with tuberculosis, macrophages ingest the bacterium ***Mycobacterium tuberculosis*** but the bacterium evades destruction by secreting a phosphatase that dephosphorylates PIP_3 and thus halts phagosome maturation.

Macropinocytosis

Many cells ingest extracellular fluid in large endocytic structures called **macropinosomes.** Growth factors or other signals stimulate actin-driven protrusions of the plasma membrane in the form of ruffles (Fig. 22.5).

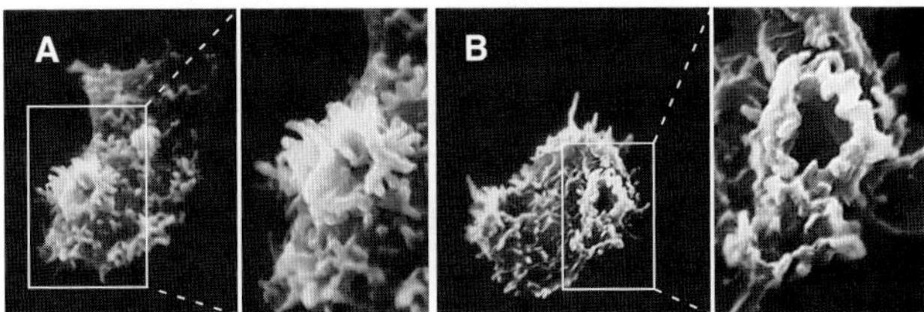

FIGURE 22.5 MACROPINOCYTOSIS. A–B, Scanning electron micrographs of *Acanthamoeba castellanii* showing membrane ruffling and macropinocytosis, the major pathway for nutrient uptake in this organism. (Courtesy of Steve Doberstein, Johns Hopkins Medical School, Baltimore, MD.)

These protrusions close around extracellular fluid, forming a macropinosome, which is then transported along microtubules toward the center of the cell. This process allows cells to internalize unconcentrated extracellular fluid, which is useful for bulk nutrient uptake.

Macropinosomes persist inside cells for 5 to 20 minutes, during which their membrane components either recycle back to the plasma membrane, potentially bypassing other organelles within the cell, or are delivered to lysosomes (Fig. 22.4C). Although the composition of macropinosome membranes resembles the plasma membrane ruffles from which they were derived, the ruffles themselves are believed to be enriched in both specific polyphosphoinositides and lipid raft markers. Internalization of this unique part of the membrane during macropinocytosis, therefore, alters the overall composition of the plasma membrane. This might influence cellular motility and responses to external stimuli.

Formation of macropinosomes depends on many of the same proteins used for phagocytosis. PI kinases and GTPases recruit and activate proteins that assemble the actin filaments supporting membrane ruffles. For example, the GTPase Arf6 activates phosphatidyl-4-phosphate-kinase, leading to production of $PI(4,5)P_2$ at plasma membrane sites of macropinocytosis (Fig. 22.4C). $PI(4,5)P_2$ then activates factors that nucleate or promote actin assembly (see Fig. 33.13). Overexpressing a constitutively active form of Arf6 increases ruffling and accumulation of macropinosomes enriched in $PI(4,5)P_2$.

Macropinocytosis serves diverse cellular functions. In some cases, it is induced by activation of plasma membrane receptors. Removal of these same receptors from the cell surface by macropinocytosis downregulates their signaling activity. Constitutive macropinocytosis allows cells to take up molecules from the medium. Examples include uptake of nutrients by amoeba, thyroglobulin by thyroid cells, and bulk extracellular fluid by dendritic cells for immune surveillance. Some migrating cells use macropinocytosis for motility to coordinate insertion and uptake of plasma membrane with their direction of motion. Certain pathogenic bacteria (eg, *Salmonella typhimurium*) inject toxins into cells to trigger macropinocytosis and provide a pathway for the bacterium to enter the cell. Once in a macropinosome, the bacteria can replicate and avoid being destroyed by other cells engaged in phagocytosis.

Endocytosis Mediated by Caveolae

Caveolae are small (~50 nm wide) patches of plasma membrane ("little caves") enriched in cholesterol and coated on the cytoplasmic surface with the proteins caveolin and cavin. Cavins are only found in vertebrates, while caveolins are more widely distributed in metazoa. Caveolae also contain diverse signaling molecules (eg, H-Ras), glycosphingolipids and membrane transporters, including calcium pumps (Fig. 22.6). Caveolae vary in shape from flat to flask-shape invaginations of the membrane.

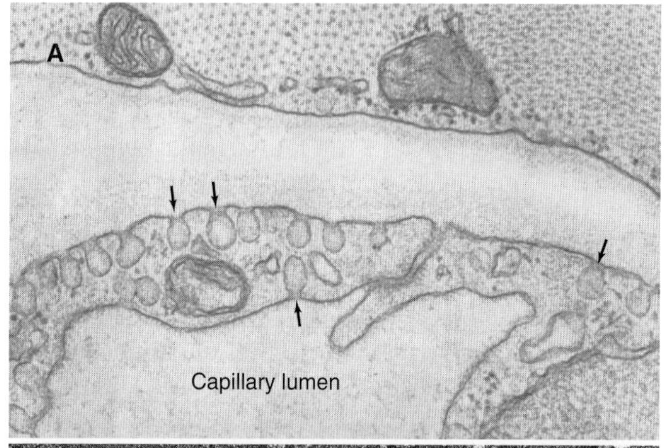

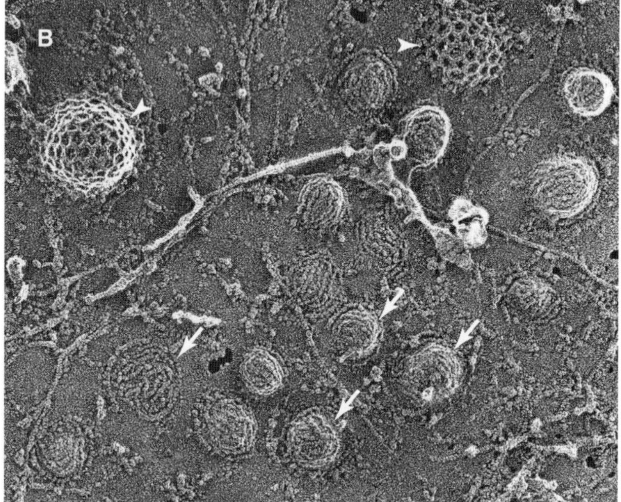

FIGURE 22.6 CAVEOLAE. A, Electron micrograph of a thin section of a muscle capillary showing caveolae ("little caves"). These structures are abundant in endothelial cells that mediate transcytosis. *Arrows* show "cave" openings. **B,** Electron micrograph of the inside surface of a fibroblast prepared by quick-freezing, deep-etching, and rotary shadowing. The protein cavin forms the rope-like coats on the caveolae *(white arrows)*. Caveolae are typically smaller than clathrin-coated pits *(arrowheads)*. (**A,** Courtesy D. Fawcett, Harvard Medical School, Boston, MA. **B,** Courtesy John Heuser, Washington University, St. Louis, MO.)

The coat proteins, **caveolin** and **cavin,** stabilize the unique caveolae micro-domains (Fig. 22.7). Cells devoid of either protein lack morphologically evident caveolae. Caveolin inserts as a loop into the inner leaflet of the plasma membrane, with both its C- and N-terminus facing the cytoplasm (Fig. 22.7B). Single caveolae contain approximately 150 molecules of caveolin. Oligomers of 14 to 16 caveolin molecules are stabilized by palmitoylation, binding cholesterol in a 1 : 1 ratio, and also by interactions with complexes of peripheral membrane proteins called cavin that confine caveolin to caveolae. Cavins form the rope-like polymers that coat the cytoplasmic surface of caveolae (Fig. 22.7D). They are trimers of subunits with two coiled-coil domains. Some cell types express multiple cavin isoforms that assemble heterotrimers. Interactions with phosphatidylserine and PIP_2 direct cavins to caveolae, as they do not seem to bind directly to caveolin. The rope-like polymers of cavins contrast with the geometrical assemblies of the coat proteins on clathrin-coated pits and COP-coated buds (see Fig. 21.8).

The density of caveolae on the cell surface can vary widely. For example, the flow of blood over endothelial cells causes caveolae to flatten. Other conditions that flatten caveolae include phosphorylation of caveolin (ie, in cells expressing viral-sarcoma tyrosine kinase, v-src) or depletion of membrane cholesterol. When caveolae collapse, cavins are degraded and caveolin and associated lipids redistribute across the plasma membrane (Fig. 22.7C). Dispersal of caveolin and lipid from caveolae in this manner during different stresses increases the surface area and elasticity of the plasma membrane. This property allows cells with abundant caveolae, such as endothelial and muscle cells, to respond to physiological stretch and fluid flow.

Caveolae are most abundant on endothelial cells (>10% of the plasma membrane area) where they mediate transcellular shuttling of serum proteins and nutrients from the bloodstream into tissues. Transmembrane cargo proteins in the plasma membrane associate with caveolae through interactions with caveolin and/or with components of the cholesterol-enriched membrane. Internalization of caveolae requires rearrangements of the actin cytoskeleton as well as the action of the GTPase **dynamin** (Fig. 22.8). The vesicles that form during this process are small (~60 nm in diameter) and interact transiently with endosomes or fuse with each other while retaining their cytoplasmic coat of caveolin and cavin. Caveolae also concentrate at the basolateral surface of other epithelial cells and at the rear of migrating cells.

Caveolae contribute less to endocytosis in cells other than endothelial cells but serve additional roles including storing microdomain-associated lipids, and regulating both noncaveolar endocytic pathways and a variety of signaling pathways.

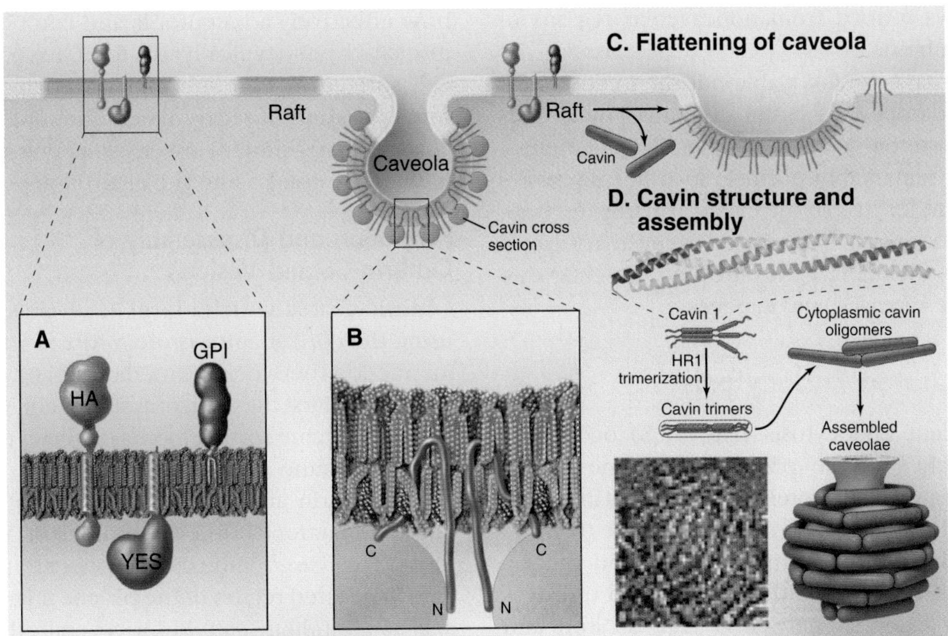

FIGURE 22.7 COMPARISON OF LIPID RAFTS AND CAVEOLAE. A, Lipid rafts enriched in cholesterol and sphingolipids can exist independently of caveolin. Lipid rafts are enriched in proteins anchored to the outer leaflet by glycosylphosphatidylinositol (GPI) tails and some integral membrane proteins (HA, influenza virus hemagglutinin, and YES, a Src-family tyrosine kinase), depending on the composition of their transmembrane domains. **B–D,** Caveolae are stabilized by caveolin *(blue schematic),* which inserts into the inner leaflet and binds cholesterol *(red)* and by rope-like polymers of cavin *(purple).* **C,** Dissociation of cavin results in flattening of the membrane bud. **D,** Cavin structure. Ribbon diagram of the first cavin helical segment, which forms a triple coiled-coil along the with second helical domain. Cavin trimers assemble oligomers the associate with membrane lipids to coat caveolae as shown by the detail from the electron micrograph in Fig. 22.6. (For reference, see Protein Data Bank [www.rcsb.org] file 4QKV and Kovtun O, Tillu VA, Ariotti N, et al. Cavin family proteins and the assembly of caveolae. *J Cell Sci.* 2015;128: 1269–1278.)

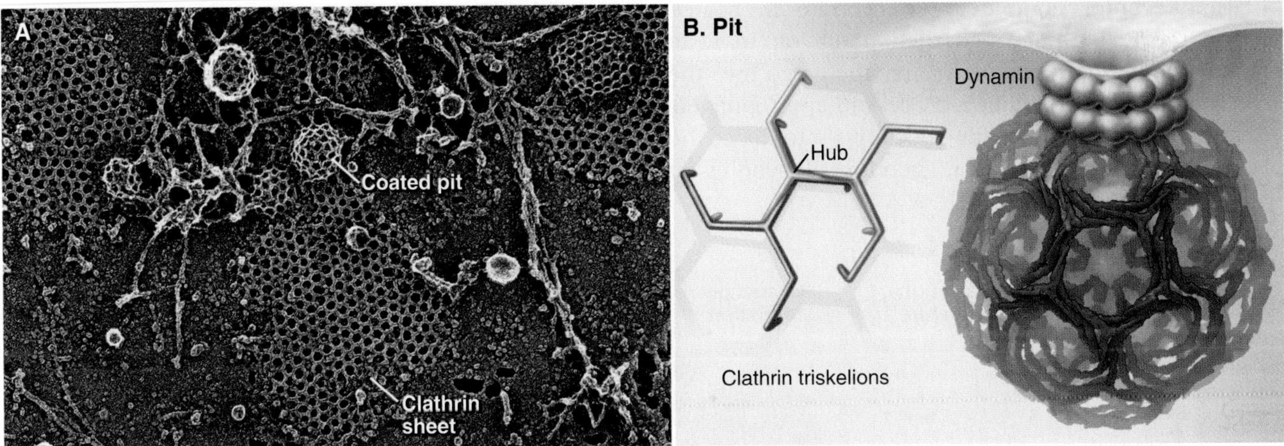

FIGURE 22.8 CLATHRIN-COATED PITS. A, Electron micrograph of the cytoplasmic surface of the plasma membrane with sheets of clathrin and clathrin-coated pits. The cytoplasm was washed away before the membrane was prepared by rapid-freezing, deep-etching and rotary shadowing with platinum. **B,** Drawing of a clathrin-coated pit and the arrangement of clathrin triskelions in the coat. (Micrograph courtesy John Heuser, Washington University, St. Louis, MO.)

Proteins in caveolae regulate cell growth and division, mitogen-activated protein (MAP) kinase signaling and cell-cell contact inhibition. The N-terminus of caveolin interacts with a variety of signaling molecules, including src family kinases (see Fig. 24.3), H-Ras and MAP kinase pathways (see Fig. 27.6) and endothelial nitric oxide synthase (see Fig. 26.17). Interactions of caveolin with the kinase domain of epithelial growth factor receptor (see Fig. 24.5) and the Gα subunit of heterotrimeric G proteins (see Fig. 25.9) inhibit their signaling. However, the relation of the effects on signaling and endocytosis is unclear. Indeed, caveolin sometimes marks other uptake pathways formed from noncaveolar regions of the plasma membrane.

Mice lacking caveolin are viable but their endothelial cells cannot bind or take up serum albumin (which binds fatty acids and steroids) from the blood. Nonetheless, these mice are remarkably normal, so other pathways must compensate for loss of the caveolar pathway. Loss-of-function mutations of caveolin or cavin genes cause human diseases, including lipodystrophy, muscular dystrophies, cardiac diseases, and cancer.

Clathrin-Mediated Endocytosis

Clathrin-dependent endocytosis (Fig. 22.8) occurs on specialized patches of the plasma membrane, called **coated pits,** formed by a protein lattice of **clathrin** and adapter molecules on their cytoplasmic surface (see Fig. 21.12 for details about clathrin structure and mechanism). Eukaryotic cells use **clathrin-mediated endocytosis** to obtain essential nutrients, such as iron and cholesterol, and to remove activated receptors from the cell surface. The process also controls the activation of signaling pathways and participates in the turnover of membrane components. Clathrin-coated vesicles also retrieve synaptic vesicle membrane at synapses following neurotransmitter release (see Fig. 17.9). In addition to its role in endocytosis at the plasma membrane, clathrin also participates in cargo sorting and membrane budding at other sites in cells, including endosomes and the *trans*-Golgi network (TGN).

Ligand-receptor complexes concentrate in coated pits on the cell surface and then pinch off to form **clathrin-coated vesicles** that carry the cargo into the cell, completing the budding process in approximately 1 minute. Coated pits typically occupy 1% to 2% of the plasma membrane surface area. Therefore, depending on how effectively a receptor-ligand complex concentrates in coated pits (typically 10- to 20-fold), 20% to 40% of cell surface receptors can be internalized per minute. Internalization of receptors that are not concentrated in coated pits is much slower, reflecting the rate of bulk membrane uptake into the clathrin-dependent pathway.

Formation and Disassembly of Clathrin-Coated Vesicles

Clathrin-coated vesicles form in several steps (Fig. 22.9) using the core adaptor protein AP2 as the major hub for interactions between the membrane and the clathrin coat (Fig. 22.8). First the α-adaptin subunit of AP2 binds PI(4,5)P_2 and dileucine sorting motifs on the cytoplasmic tails of transmembrane receptors. Then the β_2 subunit of AP2 binds clathrin and initiates assembly of the polyhedral coat. Interactions of the μ_2 subunits of AP2 with sorting motifs on cargo molecules concentrate them in the clathrin-coated region of the plasma membrane. AP2 also acts as a binding scaffold for several other components that assist with or regulate coated vesicle invagination including Eps15, amphiphysin, and intersectin (Fig. 22.9).

Coated pits concentrate a wide range of cargoes, because diverse accessory adaptor proteins link different membrane proteins to AP2 (see Fig. 21.12). Most cargo adaptors also help bend the vesicle membrane.

Once the coated pit is deeply invaginated, the neck narrows and pinches off a clathrin-coated vesicle from the plasma membrane. The large (100-kD) GTPase **dynamin** binds to PI(4,5)P_2 in the membrane tubule and assembles a spiral "collar" around the necks of deeply invaginated coated pits. GTP hydrolysis during interactions between turns of the dynamin spiral constricts the collar and promotes scission of the membrane. In cells without dynamin, vesicle formation arrests at the stage of clathrin coat formation or before vesicle scission.

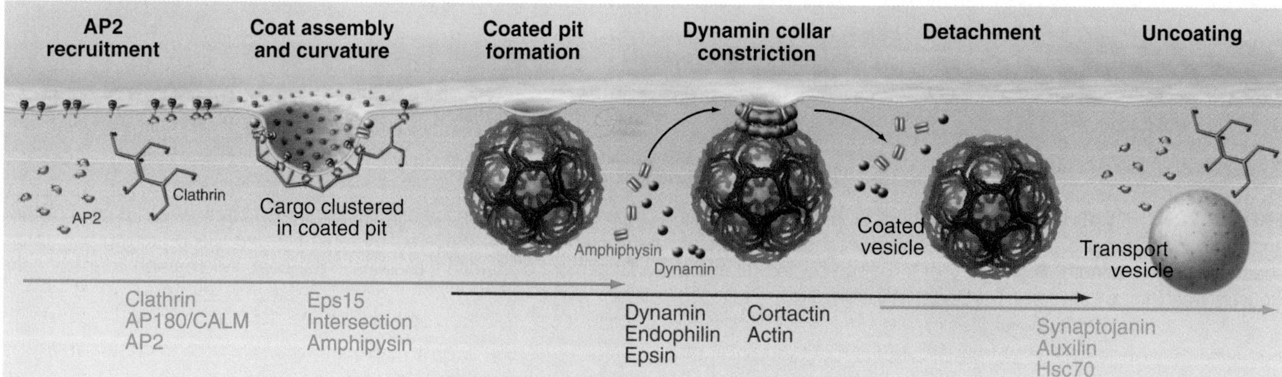

FIGURE 22.9 TIMELINE OF RECEPTOR-MEDIATED ENDOCYTOSIS DRIVEN BY THE CLATHRIN-COATED VESICLE. AP2 complexes are targeted to the plasma membrane and interact with tyrosine-based sorting motifs in the cytoplasmic domains of transmembrane proteins. AP2 also initiates the assembly of a polygonal lattice of clathrin, which concentrates cargo molecules in coated pits. AP2 and clathrin interact with the BAR-domain protein amphiphysin, which attracts the guanosine triphosphatase (GTPase) dynamin. After assembly of a coated pit and fission of the coated vesicle from the plasma membrane, auxilin and the adenosine triphosphatase (ATPase) Hsc70 dissociate clathrin and release the vesicle carrying cargo into the cell for fusion with endosomes.

The proline-rich domain of dynamin binds a number of proteins with SH3 domains, including endophilin, cortactin, and amphiphysin. These proteins, together with dynamin, help orchestrate coated pit invagination and budding. For example, crescent-shaped dimers of the BAR proteins endophilin and amphiphysin (see Fig. 21.12) induce membrane curvature during coated pit constriction and coated vesicle release by binding to highly curved, negatively charged membranes. Epsin's lipid-binding ENTH domain and extended tail help retain clathrin and AP2 on membranes. Intersectin promotes actin assembly by recruiting the GTPase Cdc42 and the nucleation-promoting factor N-WASP (neural Wiskott-Aldrich syndrome protein). Together with cortactin they stimulate assembly of actin filaments that help drive internalization.

Soon after a clathrin-coated vesicle pinches off the plasma membrane, the chaperone adenosine triphosphatase (ATPase) Hsc70, the J domain protein auxilin (see Fig. 12.12B) and cyclic GMP–dependent kinase begin to disassemble the clathrin coat (Fig. 22.9). Endophilin recruits the lipid phosphatase synaptojanin, which contributes to uncoating by dephosphorylating $PI(4,5)P_2$. The loss of $PI(4,5)P_2$ weakens the attachment of dynamin, AP2, and clathrin to the membrane. Uncoating recycles coat components and frees the vesicle to fuse with similar vesicles or with early endosomes. Thereafter, the content and membrane components of the coated vesicle are routed to the various destinations (Fig. 22.4).

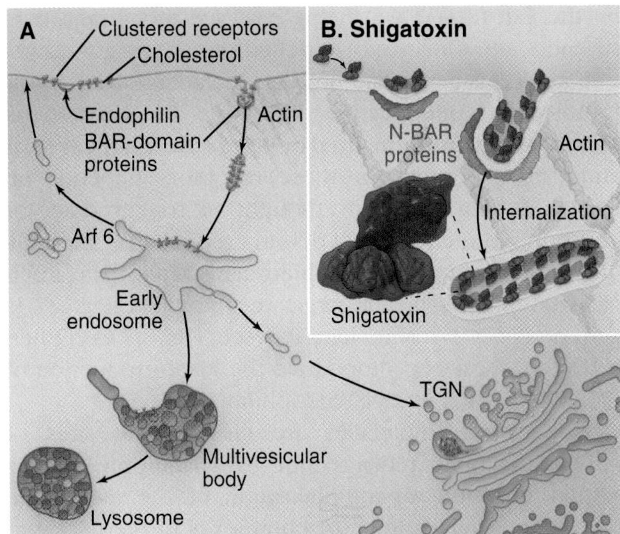

FIGURE 22.10 NONCLATHRIN/NONCAVEOLAR ENDOCYTOSIS PATHWAY. A, Clustering of membrane receptor proteins initiates membrane deformation in an inward bud. Recruitment of the BAR domain protein endophilin helps stabilize the curved membrane. Actin polymerization drives inward protrusion of the tubule. After separation from the plasma membrane tubular carrier vesicles fuse with early endosomes and the internalized receptors sort into different pathways toward lysosomes, *trans*-Golgi network (TGN) or back to the cell surface. Arf6 regulates receptor recycling back to the plasma membrane. **B,** Shigatoxin uptake. Shigatoxin binds the ganglioside Gb3. Clustering of toxin–Gb3 complexes results in invagination of a membrane tubule, in which other lipids sort passively. BAR-domain proteins like endophilin A2 bind to the tubule and recruit motor proteins to extend the tubule along microtubules followed by scission of a vesicle.

Nonclathrin/Noncaveolar Endocytosis

Surprisingly, cells continue to take up certain proteins and lipids even when clathrin coat assembly is disrupted by overexpression of domains from Eps15. Eps15 normally links several clathrin assembly proteins to AP2, so individual Eps15 domains can interfere with coat assembly. Since this discovery, this process of "nonclathrin/noncaveolar endocytosis" has been shown to take up extracellular fluid, membrane lipids, and many other types of cargoes. It also plays important roles in generating cell polarity, in maintaining lipid homeostasis, and as an entry portal for pathogens (Fig. 22.10).

Nonclathrin/noncaveolar endocytosis takes up highly diverse cargo molecules using several mechanisms to concentrate them on the membrane. The best understood of these depends on clustering the extracellular domains of glycosylated membrane lipids and proteins. For example, extracellular lectins (eg, galectins) can recognize and crosslink glycolipids and glycoproteins into tightly packed nano-environments. Shiga toxin, from *Shigella* bacterium that causes diarrhea, uses this strategy to enter cells. The toxin binds to glycosphingolipid globotriaosylceramide (Gb3) on the cell surface and creates lipid clusters that reorganize the lipid bilayer in a way that favors membrane bending without internal coat proteins (Fig. 22.10B). Cholera and Shiga toxins are taken up by other endocytic pathways, but only exert their toxic effects after entering the cell through the nonclathrin/noncaveolar pathway. Shiga toxin's A subunit is transferred from the endocytic vesicle to the Golgi apparatus and the ER. After moving across the ER membrane into the cytoplasm, the A subunit inhibits protein synthesis by removing an adenine from the 28S RNA of the large ribosomal subunit.

Glycosylphosphatidylinositol (GPI)-anchored proteins serving as hydrolytic enzymes, adhesion molecules, complement regulators, receptors, and prion proteins can enter cells by nonclathrin/noncaveolar endocytosis. Their lipid anchors reside in the outer leaflet of the membrane bilayer (see Fig. 13.10), so no part of these molecules can interact with the cytoplasmic proteins used to recognize cytoplasmic sorting signals of transmembrane proteins during clathrin-mediated and caveolar endocytosis. Some GPI-linked proteins can partition into membrane domains for nonclathrin/noncaveolar endocytosis but others depend on proteins such as the lipid raft-associated protein flotillin to cluster efficiently. During this process, actin polymerization (usually driven by the Cdc42 GTPase) helps coalesce membrane constituents into cholesterol-enriched microdomains.

Some transmembrane proteins including MHC1 Class I molecules, interleukin-2 receptors, and seven-helix receptors can be internalized by nonclathrin/noncaveolar endocytosis. Uptake of these receptors depends on ligand binding, which is thought to trigger receptor clustering into cholesterol-rich microdomains. The multifunctional adaptor, endophilin, assists with the internalization of the β_1-adrenergic receptor (see Fig. 27.3) and receptor tyrosine kinases (see Fig. 27.6). When activated, these receptors drive the local production of $PI(4,5)P_2$, which attracts endophilin.

Once cargo molecules are clustered at sites of nonclathrin/noncaveolar endocytosis, multiple mechanisms contribute to internalization of the membrane bud. Rather than using coat complexes to recruit cargo and to bud from the membrane, this pathway is believed to exploit heterogeneity in the lipid and protein composition of the plasma membrane to form lipid microdomains with cargo that bud into the cell (Fig. 22.10). Like lipid rafts, the microdomains are enriched in cholesterol and glycosphingolipids and cholesterol depletion disrupts the uptake of all cargo molecules. In some cases the membrane microdomains recruit N-BAR proteins or helix-inserting epsins that bend the membrane and serve as scaffolding modules (see Fig. 21.15). For example, the endophilin BAR domain induces membrane curvature and supports scission of the vesicle, while its SH3 domain binds cytoplasmic proteins. Differential recruitment of lipid kinases and phosphatases along the tube and its narrow neck initiates a kinetic segregation of lipids in the neck region, triggering scission of the tubule carrier from the plasma membrane. This can occur either with or without dynamin. As the membrane invaginates, actin polymerization and traction by myosin motors are thought to elongate it into a tube.

These components cooperate to drive the formation of tubular endocytic structures that separate from the plasma membrane (Fig. 22.10). Once inside the cell, tubular carriers sort to early endosomes or the Golgi apparatus by various mechanisms requiring Rab and Arf GTPases and scaffolds for connecting to microtubule motors. Routing is also affected by ubiquitination tags placed on the cargo.

In addition to being a portal for entry of GPI-anchored proteins, toxins, receptors, and other proteins, the nonclathrin/noncaveolar pathway serves as a reservoir of membrane required for cell spreading, delivering key adhesion proteins (ie, integrins) and nutrient pumps to the plasma membrane. It also helps differentiate the plasma membrane of epithelial cells into polarized apical and basolateral domains (see Fig. 21.25). Indeed, enrichment of GPI-anchored proteins and lipid rafts (see Fig. 13.7) in apical domains requires the nonclathrin/noncaveolar endocytic pathway, socholesterol depletion disrupts cell polarity. By contrast, most proteins in basolateral domains, such as transferrin receptors and low-density lipoprotein (LDL) receptors, are excluded from lipid rafts and do not enter the nonclathrin/noncaveolar endocytic pathway. Instead, they are taken up by clathrin-mediated endocytosis and recycle back to basolateral domains after releasing their ligand.

Endocytic Compartments Associated With Clathrin-Dependent Endocytosis

Endocytic transport intermediates formed by clathrin-dependent or clathrin-independent mechanisms fuse with and deliver their cargo to endosomes. Like the TGN in the biosynthetic pathway, endosomes are the major sorting compartments along the endocytic pathway toward lysosomes and other destinations. Consistent with their sorting function, endosomes are a pleomorphic collection of vesicles, vacuoles, tubules, and multivesicular bodies. The following sections describe these compartments in relation to clathrin-mediated endocytosis, the most extensively studied endocytic pathway.

In clathrin-mediated endocytosis, four classes of endosomes are distinguished based on the kinetics with which they accumulate endocytic tracers, their morphology, localization within the cell, and the presence of specific marker proteins (Fig. 22.11). Newly internalized proteins are first delivered to so-called **early endosomes,** an anastomosing network of tubules and vacuoles near the plasma membrane. They serve primarily as a sorting compartment for two pathways, one recycling receptors and other components in tubules and vesicles back to the plasma membrane and the other leading to destruction in lysosomes (Fig. 22.12A). Receptors returning to the cell surface first move from early endosomes to **recycling endosomes,** tubular portions of early endosomes located in the perinuclear Golgi region of the cell. Alternatively, early endosomes transform into **multivesicular bodies**, transport intermediates that mature into **late endosomes**. Late endosomes ultimately fuse with **lysosomes** (discussed in Chapter 23), whose acid hydrolases degrade internalized cargo.

These four endosomal compartments are not distinct, stable organelles, but rather exist as stages of a continuum of compartments engaged in sorting of endocytic cargo. Each compartment uses specific sorting mechanisms to separate cargo, receptors, and lipids for trafficking into different routes. These sorting mechanisms are linked to membrane differentiation events that allow particular compartments to fuse together, move apart, extend tubules, form invaginated intraluminal vesicles, or remain as vacuolar structures. The endosomal compartments are constantly remodeled according to the quantity and type of receptors and cargoes that traffic through them.

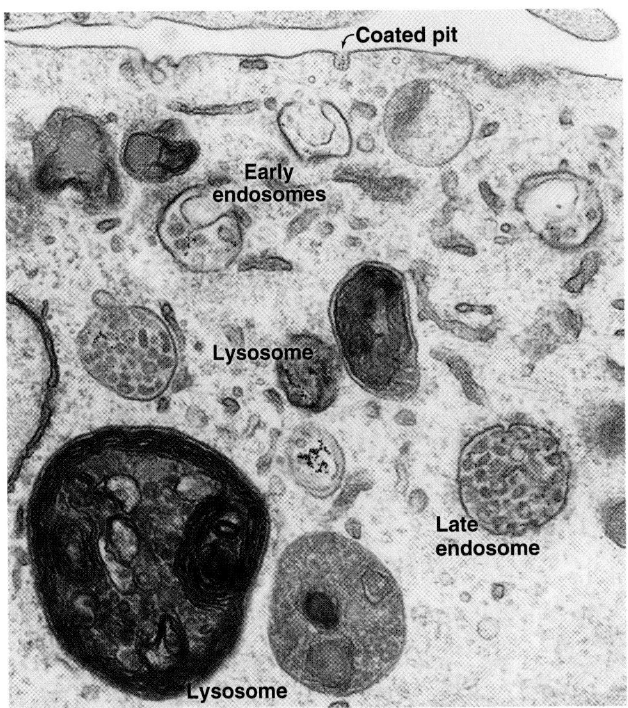

FIGURE 22.11 ENDOSOMAL COMPARTMENTS. Electron micrograph showing internalized gold-conjugated protein being transported through the structurally diverse organelles of the endosomal compartment. A gradient of colors shows the maturation of early endosomes to late endosomes and lysosomes. Gold particles *(tiny black dots)* are first delivered to early or sorting endosomes *(yellow)* that have both tubular and vacuolar regions and few intraluminal membranes. Tubular portions are recycled to the plasma membrane, whereas vacuolar portions undergo maturation. Late endosomes *(light brown)* are vacuolar and contain increasing amounts of intraluminal membranes. Lysosomes *(dark brown)* are very dense organelles, packed with internal vesicles and membrane whorls. (Courtesy Mark Marsh, University College, London, United Kingdom.)

Early Endosomal Compartment

Early endosomes receive membrane proteins and lipids delivered into the cell by clathrin-dependent endocytosis. The amount of protein and lipid entering the early endosomal compartment is enormous, given that bulk plasma membrane is internalized at rates as high as 2% per minute and nutrient receptors such as transferrin receptor (which binds iron-bound ferritin) are internalized at rates exceeding 20% per minute. Remarkably, the majority of this internalized material (approximately 90% of internalized protein and lipid and 60% to 70% of all internalized fluid) is rapidly recycled to the cell surface from early endosomes. The rest of the membrane and contents are routed to lysosomes for degradation. This sorting of proteins and lipids in endosomes depends on the following features.

First, early endosomal membranes extend long membrane tubules that can move apart and fuse with other compartments. This allows the detached membrane tubules to fuse with the plasma membrane or recycling endosomes.

Second, the geometry of the early endosome assists with sorting proteins and lipids. Soluble ligands accumulate in the volume-rich vacuolar portions of the early endosome, whereas receptors accumulate in tubular portions with a high ratio of surface area to volume (Fig. 22.12D). This physically separates membrane components for recycling back to the plasma membrane either directly or after passing through recycling endosomes or the TGN, depending on when they detach from the endosomal membrane. On the other hand, vacuolar portions of early endosomes mature into multivesicular bodies and late endosomes before fusing with lysosomes where their contents are degraded.

Third, **V-type rotary ATPases** pump protons (see Fig. 14.6) into endosomes, progressively lowering the luminal pH along the endocytic pathway from approximately pH 6.5 in early endosomes to approximately pH 5.0 in late endosomes (Fig. 22.12D). The interaction of many ligands with their transmembrane sorting receptors is sensitive to pH (Fig. 22.12C), so the pH gradient controls where each cargo molecule dissociates from its membrane receptor along the endosomal pathway (Fig. 22.12D). When receptor–ligand complexes reach their threshold pH for dissociation, the ligands are released into the lumen of the endosome and are carried in the vacuolar portion of the endosome toward lysosomes, while the transmembrane receptors are returned to the plasma membrane in tubules.

Fourth, oligomerization and aggregation of transmembrane proteins influence sorting in endosomes. For example, monomeric Fc receptors recycle from endosomes to the plasma membrane, but Fc receptors cross-linked by binding antigen-antibody complexes on the cell surface are targeted from endosomes to lysosomes for degradation. Processing of internalized epidermal growth factor (EGF) receptors (see Fig. 27.6) is another example. Binding of EGF causes receptors to dimerize, followed by addition of a single ubiquitin (see Fig. 23.3). After internalization, EGF does not dissociate until the pH is less than 5.0. As a result, the EGF–EGF receptor complex is targeted to lysosomes for degradation. This process, termed *downregulation* (see Fig. 23.5), is a negative feedback loop that allows cells to adapt to continuous stimulation (see Fig. 27.6).

In addition to these features, the membranes of early endosomes exhibit a "mosaic" of specialized domains with unique lipids, protein-lipid, and protein-protein complexes (Fig. 22.12B). These domains orchestrate membrane tubulation, invagination, and fusion. The differentiated domains are maintained dynamically by localized production of PI(3)P by PI(3)-kinase in early endosomal membranes. PI(3)P recruits the protein **early endosome antigen 1 (EEA1)**, which has a **FYVE-domain** to bind PI(3)P. Homodimers of EEA1 have

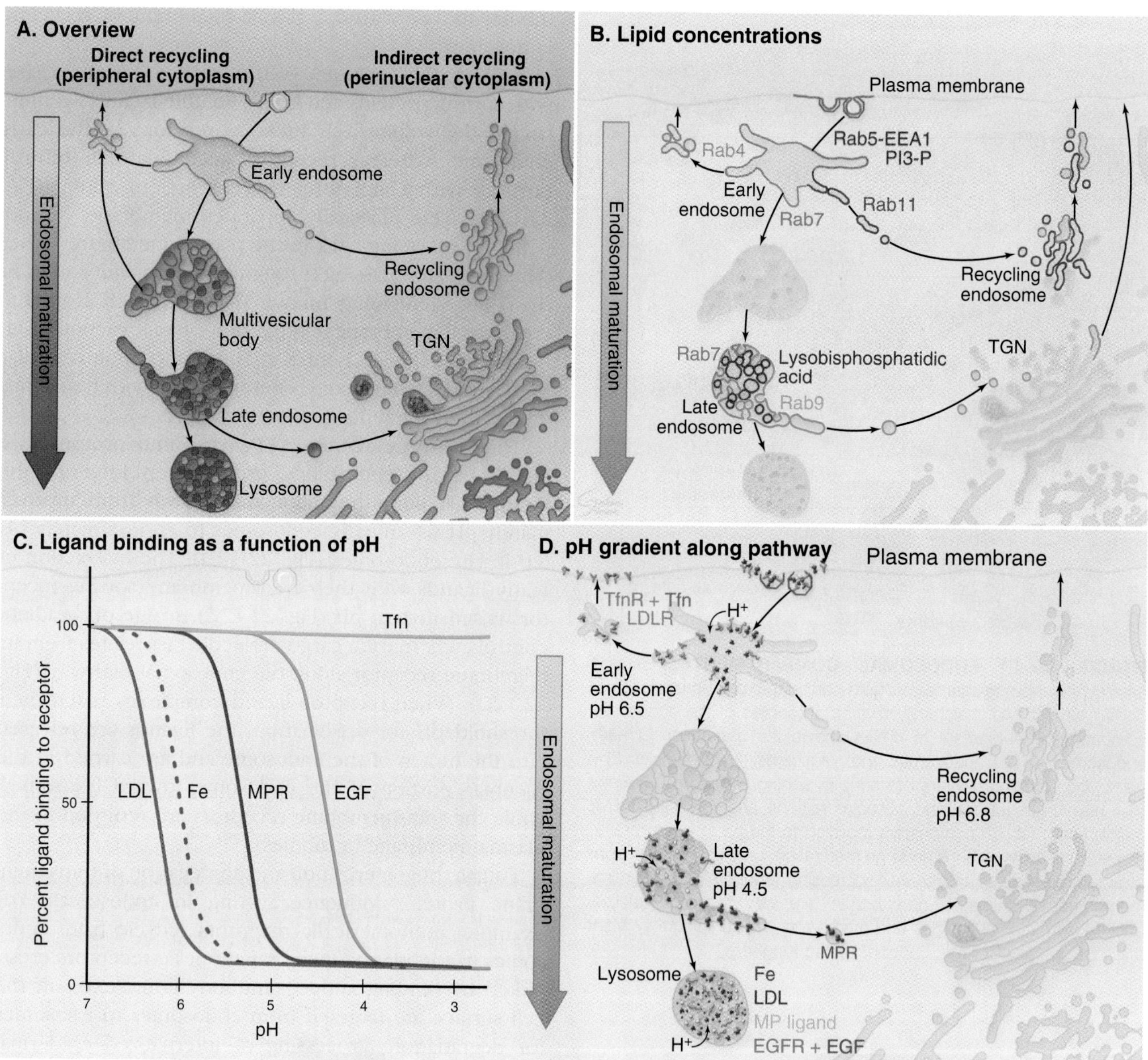

FIGURE 22.12 MEMBRANE TRAFFIC AND CONTENT SORTING ALONG THE ENDOCYTIC PATHWAY. A, Overview. Cargo and membrane taken up by clathrin-mediated endocytosis are delivered to tubulovesicular early endosomes, which mature into multivesicular bodies and late endosome before fusion with lysosomes. Each compartment sorts membrane containing receptors and other proteins into tubules and vesicles that are recycled to the plasma membrane either directly or indirectly through perinuclear recycling endosomes or the *trans*-Golgi network (TGN). Endosomes mature by accumulating internal membranes, delivery of lysosomal hydrolases from the TGN, and acquisition of targeting and fusion machinery. **B,** Domain organization of the endocytic pathway. Local production of PIs in subregions of endosomal membranes recruit specific Rabs and Rab-binding proteins such as EEA1 (depicted in different colors). **C,** Interactions of many cargo molecules with their trans-membrane receptors depend on pH. **D,** V-type rotary ATPase proton pumps progressively acidify the lumens of endosomal compartments. This gradient of pH facilitates protein sorting by regulating where along the pathway that each ligand dissociates from its transmembrane receptor. Dissociated ligands in the luminal space are targeted to lysosomes, whereas receptors in the membrane can be recycled to the cell surface in tubular endosomes. Iron dissociates from transferrin (Tfn) in early endosomes and apotransferrin without bound iron recycles with its receptor. Mannose-6-phosphate proteins dissociate from their receptors (MPRs) in late endosomes. Some ligands, such as epidermal growth factor (EGF), remain bound and are delivered with their receptor (epidermal growth factor receptor [EGFR]) to lysosomes.

binding sites for the **Rab5 GTPase** at both ends of a long coiled-coil, so they can link two membranes marked with Rab5, such as two early endosomes or an endocytic vesicle and an early endosome. Rab5 recruits SNARE machinery (eg, syntaxin 13 and *N*-ethylmaleimide-sensitive factor [NSF]; see Fig. 21.15) to membranes forming dynamic oligomeric complexes that promote homotypic and heterotypic fusions of early endosomes. Inhibiting PI(3)-kinase with the drug wortmannin prevents the formation of such fusion assemblies, as PI(3)P is required to recruit EEA1. Early endosomes without PI(3)P cannot fuse.

The binding of Rab5 effectors to early endosomes and the presence of guanine nucleotide exchange factors for Rab5 in the effector complexes establishes a positive-feedback loop that amplifies the recruitment and the activation of Rab5 in specific regions of early endosomes. This feedback generates and maintains domains that contain Rab5-EEA1-PI(3)P in early endosomes. Other Rabs are thought to operate in a similar fashion in the endosomal system, with Rab4 and Rab11 in early/recycling endosomes and Rab7 and Rab9 in late endosomes (Fig. 22.12B). Like Rab5, these other Rab proteins are thought to organize membrane domains with distinct functions within a single endosome compartment.

Multivesicular Bodies

Sorting endosomes transform into **multivesicular bodies**, which continue the sorting process. Most residual plasma membrane proteins and recycling receptors are returned to the plasma membrane, while proteins destined for degradation are sorted into intraluminal vesicles and tubules. Multivesicular bodies gain lysosomal hydrolases from vesicles delivered from the TGN along the biosynthetic pathway.

Multivesicular bodies begin to form from the vacuolar portions of early endosomal membranes by invaginating membrane with receptors destined for late endosomes or lysosomes (Fig. 22.13; see Fig. 23.5). Many downregulated receptors are tagged with ubiquitin (Fig. 23.2) for sorting into multivesicular bodies. Two types of proteins contribute to sorting ubiquitinated receptors: **Hrs (hepatocyte-growth-factor-regulated tyrosine kinase substrate)** and **ESCRT-I, ESCRT-II, and ESCRT-III (endosomal sorting complexes required for transport-I, transport-II, and transport-III).**

Hrs binds ubiquitinated receptors when they arrive in early endosomes. Association with PI(3)P retains Hrs on the cytoplasmic surface of endosomes, so the PI 3-kinase inhibitor wortmannin inhibits the formation of intraluminal vesicles in MVBs. Through interactions with clathrin, Hrs sorts ubiquitinated receptors into clathrin-coated domains on new multivesicular bodies. These domains do not form clathrin-coated vesicles but instead sort ubiquitinated membrane proteins for transfer to ESCRT-1. Sequential transfer to ESCRT-II and -III on the cytoplasmic surface of the multivesicular body drives both receptor incorporation into the membrane invaginations, and invagination of the membrane itself. As the intraluminal membrane system matures, the compartment is called a **late endosome**. Late endosomes subsequently fuse with preexisting lysosomes.

ESCRT-I is also involved in retrovirus budding at the plasma membrane (eg, **Ebola and HIV**), a process that is topologically equivalent to membrane invagination occurring in multivesicular bodies. The HIV Gag protein highjacks the ESCRT-I complex by mimicking the binding activity of HRS.

Under certain conditions multivesicular bodies fuse with the plasma membrane, releasing their intraluminal vesicles, termed **exosomes,** outside the cell. Exosomes have regulatory functions in the immune system. For example, antigen-presenting cells such as B-lymphocytes and dendritic cells secrete exosomes during exocytic fusion of MHC class II compartments with the plasma membrane. The released exosomes can stimulate

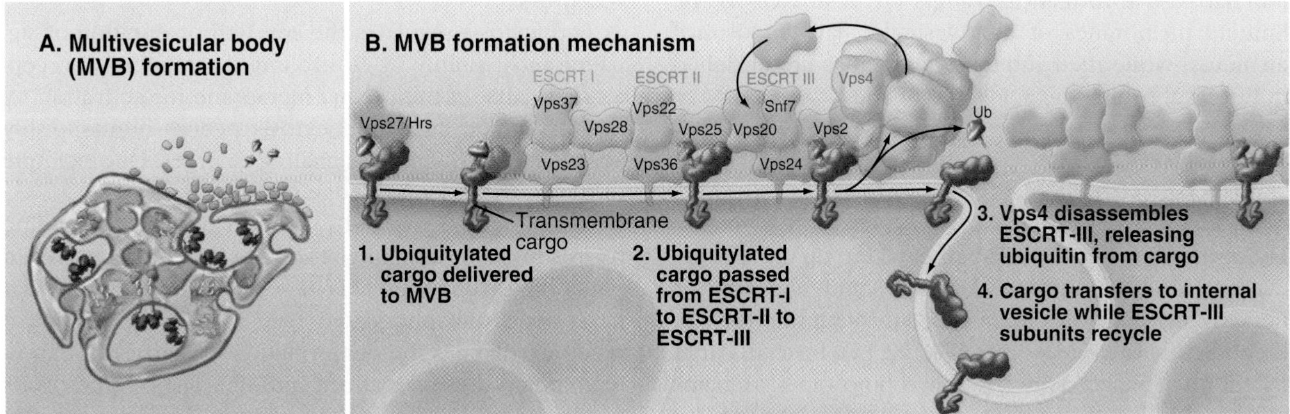

FIGURE 22.13 PROTEIN SORTING IN MULTIVESICULAR BODIES. A, MVBs accumulate small vesicles and tubules targeted for lysosomal degradation by invagination of the limiting membrane. MVB formation and sorting require interactions among several proteins. Interactions with Vps/Hrs sort ubiquitinated receptors that are sorted to MVBs. The ubiquitinated receptor is then delivered to ESCRT-I by an interaction between Hrs and the Vps23 subunit of ESCRT-I. The receptor is then relayed to ESCRT-II and then to ESCRT-III. Invagination of an intraluminal vesicle containing the receptor is mediated through polymerization of ESCRT-III complexes, which are small, highly charged coiled-coil proteins. An AAA ATPase, Vps4, disassembles the multimeric ESCRT subunits, allowing them to be reutilized.

proliferation of T-lymphocytes. Exosomes might also represent a novel method of intercellular communication. For example, HIV particles found in exosomes could represent a type of "Trojan horse" capable of transmitting the HIV virus when taken up by other cells.

Late Endosomes

Late endosomes (Fig. 22.11) serve as a final sorting station for determining which membrane constituents in the endocytic pathway will be degraded and which will be recycled (Fig. 22.12A). For example, mannose 6-phosphate receptors for acid hydrolases and transmembrane enzymes involved in the proteolytic processing of precursor proteins (eg, the dibasic endopeptidase furin and carboxypeptidase D) are recycled from late endosomes to the TGN and back to the plasma membrane. Transport from late endosomes to TGN is thought to involve budding of membrane-enclosed carriers from late endosomes followed by fusion with the TGN.

Genetic and biochemical analyses of this process revealed an important role of **retromer,** a complex composed of five proteins that includes members of the **sorting nexin** family of proteins. Mutations in any of the retromer proteins prevent acid hydrolase receptors from being retrieved to the TGN and result in secretion of acid hydrolases at the plasma membrane.

Late endosomes are structurally distinct from multivesicular bodies by being pleomorphic, with cisternal, tubular, and multivesicular regions. They also contain large amounts of lysosomal glycoproteins (in particular, **lysosome-associated membrane proteins [Lamp] 1 and Lamp2**) and the lipid **lysobisphosphatidic acid** (Fig. 22.12B). This lipid prefers to be in a hexagonal phase that can promote positive or negative bilayer curvatures that favor membrane invagination and intraluminal vesicle formation. Lamps are restricted to the limiting membranes of multivesicular, late endosomal elements, while their internal membranes are enriched in lysobisphosphatidic acid.

Other Endocytic Compartments and Pathways

Endocytic cargo and membrane taken up by phagocytosis, macropinocytosis, caveolae, and nonclathrin/noncaveolar pathways do not pass through multivesicular bodies or late endosomes (Fig. 22.14). Instead, cargo taken into phagosomes or macropinosomes remains within these structures as they mature into and/or fuse with lysosomes. Endocytic structures formed by caveolae serve as a conduit for movements of molecules to other regions of the plasma membrane. Nonclathrin/noncaveolar endocytic structures deliver selected components to the TGN (resupplying it with glycosphingolipids) and lysosome, as well as recycling other molecules back to the plasma membrane.

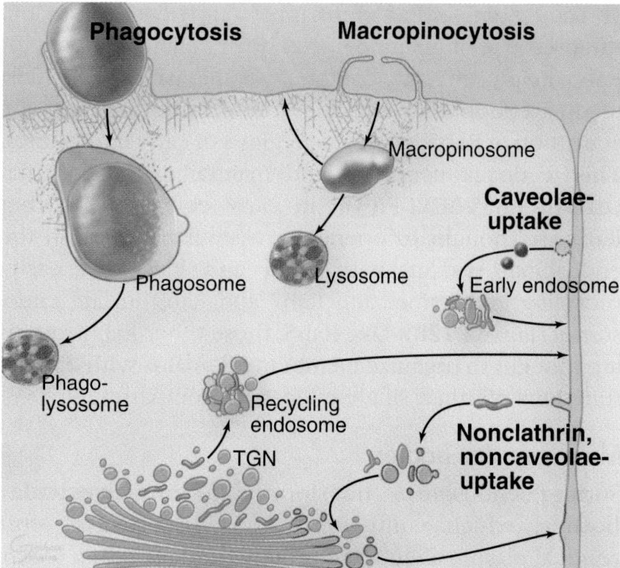

FIGURE 22.14 PATHWAYS FOLLOWED BY MOLECULES TAKEN UP BY PHAGOCYTOSIS, MACROPINOCYTOSIS, CAVEOLAE, AND NONCLATHRIN/NONCAVEOLAE UPTAKE. *Arrows* depict the routes followed by membrane-bound and soluble cargo molecules after uptake by each endocytic mechanism. TGN, *trans*-Golgi network.

Endocytosis and Signaling

Endocytosis regulates signaling by cell-surface receptors in all phases of animal development, from cell proliferation to cell migration and survival. Understanding this regulation is important for disease prevention; for example, many cancer cells are sensitive to therapeutics that target the endocytic pathways of specific signaling receptors.

Endocytosis regulates the strength or duration of signaling most simply by controlling the number of receptors available in the plasma membrane for activation by ligands. When such endocytosis occurs nonuniformly across the cell, localized signaling emerges. One example of this is at the leading edge of migrating cells.

Receptors are marked for internalization by binding β-arrestin or by ubiquitylation. For example, β-arrestin binds the cytoplasmic tails of seven-helix receptors that have been phosphorylated (see Fig. 24.3). Binding a receptor changes the conformation of β-arrestin, exposing binding sites for clathrin and AP2. These interactions trigger internalization of the receptor. β-arrestin can also coordinate components of the MAP kinase pathway on the internalized membranes to facilitate downstream signaling. Ubiquitylation of ligand bound receptors promotes their recruitment into clathrin-coated pits, presumably through interactions with ubiquitin-binding domains of the clathrin-coat adaptors Epsin and EPS15. After internalization, the HRS complex targets active, ubiquitinylated receptors to lysosomes for degradation (Fig. 22.13).

Some receptors continue to signal after endocytosis, prolonging the duration and intensity of the signal when ligand is limiting. For example, activated EGF receptors continue activating the MAP kinase cascade (see Fig. 27.6). Another example is transforming growth factor-β–receptor signaling (see Fig. 27.10) from early endosomes. This signaling causes the transcription factor SMAD2 (Sma- and Mad-related protein 2) to dissociate from SARA (SMAD anchor for receptor activation) on endosomes. SMAD2 then translocates into the nucleus to regulate gene transcription. Another way to sustain signaling after endocytosis is to return internalized receptors back to the plasma membrane, rather than to lysosomes (Fig. 22.14).

Viruses and Protein Toxins as "Opportunistic Endocytic Ligands"

The threat of infectious diseases throughout the world has made research on the survival tactics of intracellular pathogens and cellular defenses against them particularly important. Many enveloped viruses (ie, those with a membrane bilayer) enter cells by catching a ride on membrane proteins internalized by endocytosis. Once inside an endosome, the low pH causes conformational changes in viral membrane proteins that promote their insertion into and fusion with the organelle membrane. This transfers the viral nucleocapsid to the cytoplasm, where it has access to the cell's synthetic machinery to replicate itself (Fig. 22.15).

Protein toxins produced by bacteria and plants enter cells by binding to cell surface integral proteins or glycolipids and hitchhiking into cells when these receptors are taken in by endocytosis (Fig. 22.15). Various toxins enter the cytoplasm from different intracellular compartments, because their requirements for translocation differ. When pH is the trigger, toxins can be translocated directly across the endosomal membrane. Other toxins travel back to the ER and use the cell's translocation machinery in reverse to enter the cytoplasm. Once inside these toxins can kill animal cells by inhibiting cytoplasmic functions, such as protein translation.

Both clinicians and basic researchers benefit by studying these highly evolved "hitchhikers." From them, much can be learned about which properties, sequences, and motifs to look for in endogenous, fusogenic proteins. Understanding something as esoteric as the action of a plant toxin can also have medical benefits, as in the treatment of cancer through coupling of the catalytic (A) subunits of toxins such as ricin to antibodies and targeting of the toxic subunit to malignant cells. These chimeric proteins are called **immunotoxins.** Similarly, the discovery of efficient viral mechanisms for delivering their genomes into host cells has made different viruses powerful candidates for therapeutic delivery of genes.

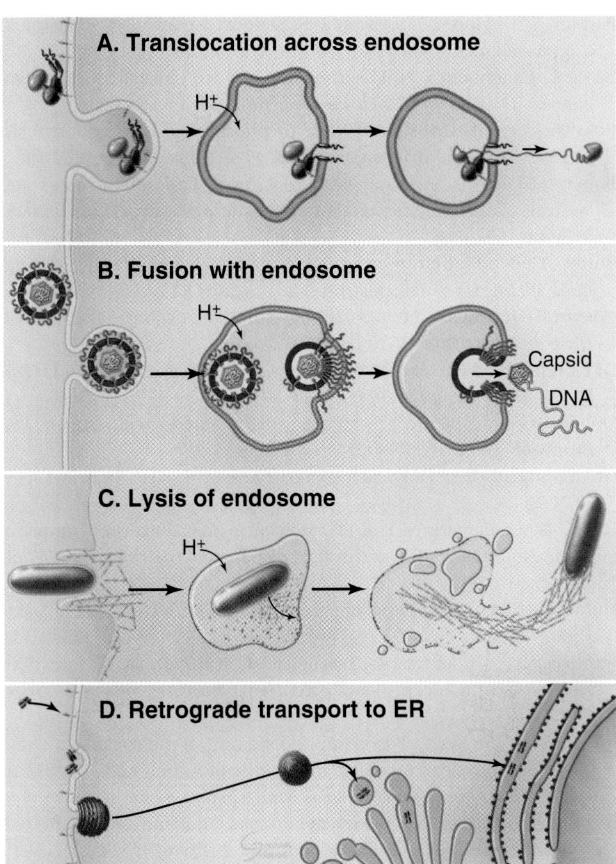

FIGURE 22.15 ENTRY OF VIRUSES AND TOXINS INTO THE CELLS. Many viruses and toxins bind to cell surface receptors that are efficiently internalized. **A,** Once in endosomes, pH-dependent conformational changes can trigger the translocation of toxin subunits across the endosomal membrane into the cytoplasm. **B,** pH-dependent conformational changes can activate fusogenic viral coat proteins to mediate fusion of the viral envelope with the endosomal membrane, releasing the nucleocapsid into the cytoplasm. **C,** Some bacteria secrete toxins after entering the endosome/phagosome; these intercalate into the membrane, creating large pores that disrupt endosomal compartments. Once in the cytoplasm, the bacterium can usurp the cell's actin *(yellow filaments)* assembly machinery for propulsion. (See Fig. 37.12.) **D,** Some toxins enter through caveolae and are transported to the endoplasmic reticulum (ER), where they can use the cell's translocation machinery—in reverse—to enter the cytoplasm.

ACKNOWLEDGMENTS

We thank Juan Bonifacino, Satyajit Mayor, and Julie Donaldson for their suggestions on revisions to this chapter.

SELECTED READINGS

Asrat S, de Jesus DA, Hempstead AD, et al. Bacterial pathogen manipulation of host membrane trafficking. *Annu Rev Cell Dev Biol.* 2014;30:79-109.

Barbieri E, Di Fiore PP, Sigismund S. Endocytic control of signaling at the plasma membrane. *Curr Opin Cell Biol.* 2016;39:21-27.

Bissig C, Gruenberg J. Lipid sorting and multivesicular endosome biogenesis. *Cold Spring Harb Perspect Biol.* 2013;5:a016816.

Bohdanowicz M, Grinstein S. Role of phospholipids in endocytosis, phagocytosis, and macropinocytosis. *Physio Rev.* 2013;93:69-106.

Boucrot E, Ferreira A, Almeida-Souze L, et al. Endophilin marks and controls a clathrin-independent endocytic pathway. *Nature.* 2015; 517:460-465.

Burd C, Cullen PJ. Retromer: a master conductor of endosome sorting. *Cold Spring Harb Perspect Biol.* 2014;6:a016774.

Cossart P, Helenius A. Endocytosis of viruses and bacteria. *Cold Spring Harb Perspect Biol.* 2014;6:a016972.

Di Fiore PP, von Zastrow M. Endocytosis, signaling, and beyond. *Cold Spring Harb Perspect Biol.* 2014;6:a016865.

Doherty GJ, McMahon HT. Mechanisms of endocytosis. *Annu Rev Biochem.* 2009;78:857-902.

Goldenring JR. Recycling endosomes. *Curr Opin Cell Biol.* 2015;35: 117-122.

Henne WM, Senmark H, Emr SD. Molecular mechanisms of the membrane sculptin ESCRT pathway. *Cold Spring Harb Perspect Biol.* 2013;5:a016766.

Huotari J, Helenius A. Endosome maturation. *EMBO J.* 2011;30:3481-3500.

Irannejad R, Tsvetanova NG, Lobingier BT, von Zastrow M. Effects of endocytosis on receptor-mediated signaling. *Curr Opin Cell Biol.* 2015;35:137-143.

Johannes L, Wunder C, Bassereau P. Bending "on the rocks"—a cocktail of biophysical modules to build endocytic pathways. *Cold Spring Harb Perspect Biol.* 2014;6:a016741.

Johannes L, Parton RG, Bassereau P, et al. Building endocytic pits without clathrin. *Nat Rev Mol Cell Biol.* 2015;16(5):311-321.

Kirchhausen T, Owen D, Harrision SC. Molecular structure, function, and dynamics of clathrin-mediated membrane traffic. *Cold Spring Harb Perspect Biol.* 2014;6:a016725.

Klumperman J, Raposo G. The complex ultrastructure of the endolysosomal system. *Cold Spring Harb Perspect Biol.* 2014;6:a016857.

Kovtun O, Tillu VA, Ariotti N, et al. Cavin family proteins and the assembly of caveolae. *J Cell Sci.* 2015;128:1269-1278.

Maxfield FR. Role of endosomes and lysosomes in human disease. *Cold Spring Harb Perspect Biol.* 2014;6:a016931.

Mayor S, Parton RG, Donaldson JG. Clathrin-independent pathways of endocytosis. *Cold Spring Harb Perspect Biol.* 2014;6:a016758.

McMahon HT, Boucrot E. Molecular mechanism and physiological functions of clathrin-mediated endocytosis. *Nat Rev Mol Cell Biol.* 2011;12:517-533.

Merrifield CJ, Kaksonen M. Endocytic accessory factors and regulation of clathrin-mediated endocytosis. *Cold Spring Harb Perspect Biol.* 2014;6:a016733.

Parton RG, del Pozo MA. Caveolae as plasma membrane sensors, protectors and organizers. *Nat Rev Mol Cell Biol.* 2013;14:98-112.

Raiborg C, Rusten TR, Stenmark H. Protein sorting into multivesicular endosomes. *Curr Opin Cell Biol.* 2003;15:446-455.

Renard H-F, Simunovic M, Lemiere J, et al. Endophilin-A2 functions in membrane scission in clathrin-independent endocytosis. *Nature.* 2015;517:493-496.

Robinson MS. Forty years of clathrin-coated vesicles. *Traffic.* 2015; 16:1210-1238.

Schink KO, Tan K-W, Stenmark H. Phosphoinositides in control of membrane dynamics. *Annu Rev Cell Dev Biol.* 2016;32:24.1-24.29.

Wandinger-Ness A, Zerial M. Rab proteins and the compartmentalization of the endosomal system. *Cold Spring Harb Perspect Biol.* 2014;6:a022616.

CHAPTER 23

Processing and Degradation of Cellular Components

Cells can synthesize nearly all the building blocks required to make proteins, nucleic acids and other macromolecules, but they are also masters at recycling. Thus, even though an individual cell might live for weeks, months, years, or even the entire lifetime of the organism, its proteins, lipids, and RNA turn over continuously. Catabolic processes ensure the regular replacement of misfolded, mislocalized, or otherwise damaged molecules with newly synthesized ones. These processes are also triggered under starvation conditions, when cells perceive a shortage of raw materials, such as amino acids. Lastly, targeted degradation of specific molecules in response to physiological signals is critical for signal transduction, cell-cycle regulation, and cell and tissue remodeling during development. This chapter focuses primarily on mechanisms that govern protein catabolism (degradation) and turnover, as these are the best studied; lipid turnover is also discussed. Chapter 11 considers RNA turnover.

Characteristics of Constitutive Protein Turnover

The levels of cellular constituents at steady state are dictated by a careful balance of the rates of synthesis and degradation. Once translated on the ribosome, each protein turns over at a characteristic rate. Turnover can be massive. For example, approximately 40% of total cellular protein in rat liver is degraded every day. The time course of degradation of the population of any specific protein follows a single exponential—suggesting that chance determines which copies of the protein are degraded (see Fig. 4.1 for first-order reactions). The rate is usually expressed as a half-life, the time for half of the molecules to be degraded. Measured, protein half-lives range from longer than 100 hours to less than 10 minutes. For human HeLa (Henrietta Lacks) cells the average half-life of a typical protein is 24 hours—the generation time of the cells. This is an indication that most proteins are relatively stable.

The intrinsic rate of degradation of a given protein is determined by many factors, including its size, overall charge, thermal instability, flexibility, hydrophobicity, folding, posttranslational modifications, and assembly with other protein subunits (if it is multimeric). Multimeric proteins are more stable when associated with their partners in complexes than on their own. A protein may have specific sequences or structural motifs (**degrons**) that are recognized by the proteolytic machinery. Phosphorylation marks some proteins for destruction. The rate at which a protein is degraded can be altered either by increasing the activity of its degradative pathway or by exposure or creation of degrons to initiate destruction.

Tor Kinase—A Master Regulator of Cell Physiology

Tor kinase—the target of rapamycin—is, more than any other protein, the master regulator of cell physiology. (Rapamycin is an important immunosuppressant drug used in transplantation and cancer therapy that was isolated from soil of Rapa Nui, Easter Island). This serine-threonine kinase, which is known as mTor (mechanistic target of rapamycin) in humans, is the subject of thousands of research studies because it sits at several nodes that control cellular pathways activated by a wide range of surface receptors. These include both heterotrimeric G protein signaling pathways (mTor is activated by phosphatidylinositol 3,4,5-trisphosphate [PIP_3]; see Fig. 26.7) and intrinsic pathways initiated by the cell to coordinate its growth with its physiological state (Fig. 23.1). Tor kinase regulates protein synthesis and cell growth, and mutants in this kinase pathway are characterized by having small cells. Mutants in several regulators of mTor activity are strongly associated with cancer.

mTor kinase functions in complexes that have characteristic components and cellular roles. For example, in

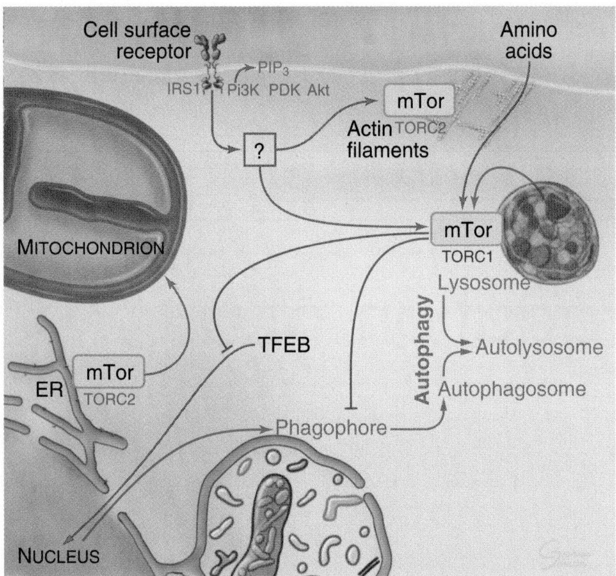

FIGURE 23.1 TOR KINASE IS A MASTER REGULATOR OF CELL PHYSIOLOGY. mTor (mechanistic target of rapamycin) kinase is present in two complexes, TORC1, which regulates catabolism in response to amino acid levels in lysosomes, and TORC2 which is responsible for cytoskeletal-based signaling. Both complexes have many other functions, with many other inputs and regulators not indicated here. The black box downstream of the cell surface receptor indicates other intermediate steps that have been omitted here for the sake of simplicity. IRS, insulin receptor substrate; PDK, phosphoinositide-dependent protein kinase; Pi3K, phosphatidylinositol 3′-kinase; TFEB, transcription factor EB; TORC, target of rapamycin complex.

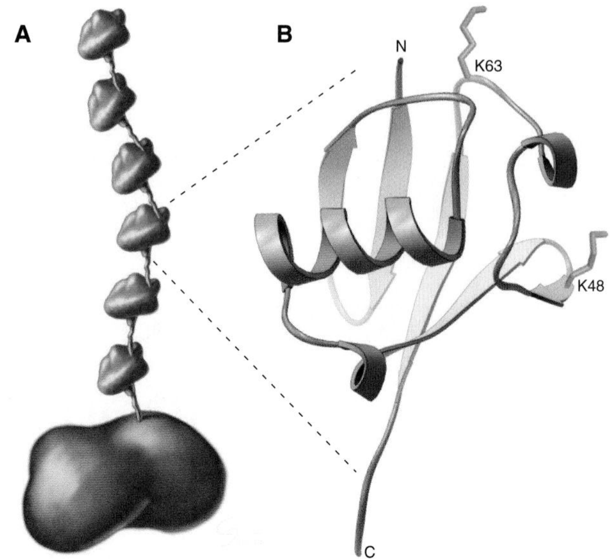

FIGURE 23.2 A polyubiquitin chain **(A)** is generated on a protein by sequential conjugation of ubiquitin **(B),** shown in a ribbon diagram. This typically targets the protein for degradation by the proteasome. Lysines 48 and 63 (K48, K63), two of the most common sites of linkage.

association with Raptor (regulatory-associated protein of mTor) and other proteins, mTor forms the TORC1 complex (target of rapamycin complex-1), which is a master regulator of catabolic pathways. TORC1 associates with the surface of lysosomes, where its regulatory subunits read the content of free amino acids in the lysosomal lumen in a mechanism facilitated by the lysosomal v-ATPase (vacuolar adenosine triphosphatase, the same enzyme that establishes the acidic environment in the lysosome interior). Active mTor inhibits cellular catabolism by phosphorylating a transcription factor called TFEB (transcription factor EB), thereby causing it to be trapped on the surface of the lysosomes. TFEB is the master regulator of lysosome assembly and autophagy (see Cell SnapShot 2). A drop in lysosomal amino acid levels results in mTor inactivation. This allows TFEB dephosphorylation, releasing the protein to enter the nucleus, where it activates the transcription of more than 400 genes.

The TORC2 complex of mTor with Rictor (rapamycin-insensitive companion of mTor) is associated with the plasma membrane and endoplasmic reticulum and activated by interacting with ribosomes. TORC2 is a key regulator of the cytoskeleton and signaling pathways that receive their input from it. It is also an important regulator of cell death pathways.

Modification of Proteins by Ubiquitylation

Ubiquitin is a very abundant and highly conserved 76-residue protein that marks proteins for degradation by proteasomes and lysosomes in addition to other functions (Fig. 23.2). Enzymes attach the C-terminus of ubiquitin covalently to lysine side chains of target proteins. In addition to its roles in proteolysis, reversible ubiquitylation contributes to DNA repair, chromosome structure, and the assembly of ribosomes, proteasomes, and other multimeric complexes. Proteins with ubiquitin-binding domains direct ubiquitylated proteins to their various fates.

Ubiquitylation of protein substrates proceeds through a tightly regulated multistep pathway that was elucidated through biochemical purification of mammalian components and in vitro reconstitution of partial reactions. The overall scheme can be subdivided into three stages (Fig. 23.3):

- *Activation of ubiquitin:* A **ubiquitin-activating enzyme** E1 catalyzes the formation of a covalent thioester bond between the side chain of one of its own cysteine residues and the carboxyl group of the C-terminal glycine of ubiquitin. Humans have only two E1 enzymes for ubiquitin.
- *Transfer of ubiquitin to an E2 enzyme:* Activated ubiquitin is transferred to a cysteine residue of an E2 or **ubiquitin-conjugating** (or carrier) **enzyme.** Humans have approximately 35 E2 enzymes.
- *Ubiquitylation of target proteins:* E3 **ubiquitin ligases** facilitate the transfer of ubiquitin from an E2-conjugate to the protein substrate, either directly or in two steps through an E3-ubiquitin intermediate. The ubiquitin C-terminus is usually attached to the

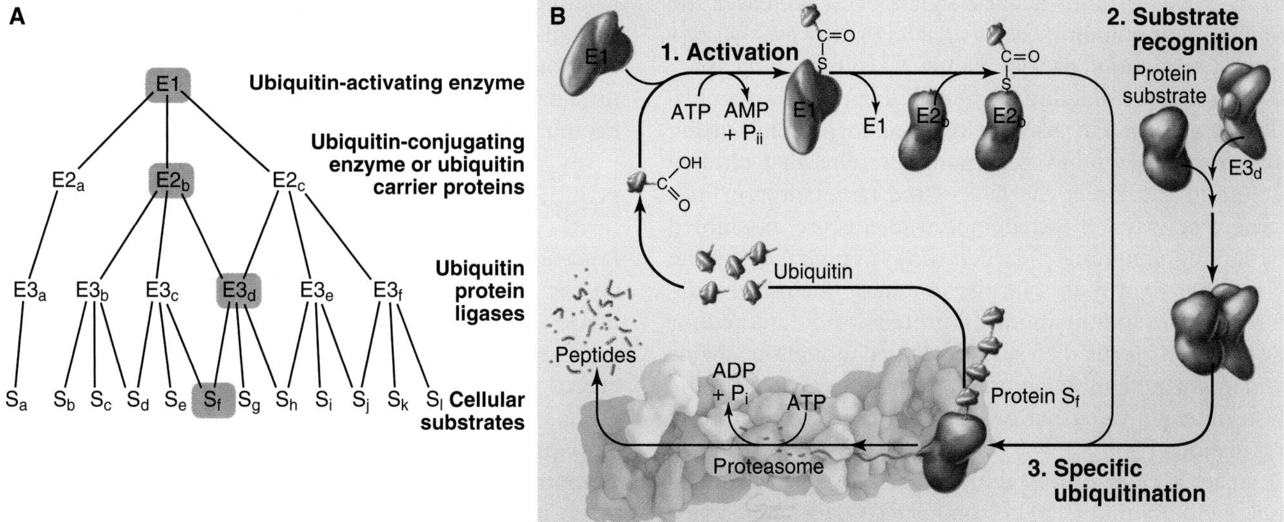

FIGURE 23.3 UBIQUITIN CONJUGATION MECHANISM. A, A hierarchy of ubiquitin-conjugating enzymes and ubiquitin protein ligases work together to recognize and ubiquitylate specific cellular substrates in a highly regulated manner. **B,** The three stages of ubiquitylation for one representative set of enzymes. A single ubiquitin-activating enzyme (E1) serves all downstream pathways. One of more than approximately 35 (in human) E2 enzymes serves as an intermediate to transfer activated ubiquitin and works with one of more than 600 (in human) E3 enzymes to recognize the appropriate target protein (S_f) and to transfer the first ubiquitin molecule. Subsequent polyubiquitylation targets the protein for degradation in the proteasome. ADP, adenosine diphosphate; AMP, adenosine monophosphate; P_{ii}, pyrophosphate.

TABLE 23.1 Ubiquitin/Proteasome Targeting Systems

Recognition Determinant (Degron)	Cellular Substrates	Ubiquitin Ligases (E3)
Phosphorylated signals DS*GXXS* S*P, T*P sequences phosphorylated by CDKs	Transcription factors (eg, IκB, β-catenin) Cell cycle regulators (eg, CDK inhibitors)	SCF complexes
Destruction box R(A/T)(A)L(G)X(I/V)(G/T)(N)	Mitotic cyclins Cell-cycle regulators	APC
KEN Box … KEN …	Cell-cycle regulators	APC
The N-end rule … N-terminal aa F,L,W,Y,R,K,H	General applicability unknown Cleaved cohesion subunit Scc1	E3α (Ubr1p)
Auxin-inducible degron	Any protein with the degron attached	SCF with plant F-box protein Tir1
Amphipathic or hydrophobic peptides	Misfolded or damaged endoplasmic reticulum proteins	Hrd1, gp78

APC, anaphase-promoting complex; CDK, cyclin-dependent kinase.
*Phosphorylation sites on serine residues.

target protein by an amide bond to the ε-amino group of a lysine residue or to the N-terminal amino group.

Humans have more than 600 E3 enzymes that confer specificity to the ubiquitylation reaction. Many of these contain a protein structural motif of 40 to 60 amino acids called a RING (really interesting new gene) finger. This is a specialized type of Zn^{2+} finger (see Fig. 10.14) that recognizes E2-ubiquitin conjugates. E3 ligases recognize the substrate and position the E2-ubiquitin conjugate so that ubiquitin transfer to the target lysine is efficient. Additional ubiquitin molecules may subsequently be conjugated by other E2 enzymes to create a **polyubiquitin chain** (Figs. 23.2 and 23.3). When acting as a signal for protein destruction, the C-terminus of the incoming ubiquitin usually links to lysine 48 of the previous one, but links to all seven lysines of ubiquitin (or the NH_2-terminus) have been observed.

The primary responsibility for substrate selectivity lies with the E3 family of enzymes, which can bind either directly to protein substrates or indirectly through adapter molecules. Although the E2 enzymes can interact directly with substrate, in general, they recognize an E3 substrate complex. One E2 enzyme can cooperate with several different E3 enzymes in the ubiquitylation reaction. The E2-ubiquitin/E3 pair usually recognizes a specific degron sequence on the target protein (Table 23.1).

Humans also have approximately 80 deubiquitylating enzymes (**DUBs**) that remove ubiquitin from target proteins, allowing the target protein fate to be changed in response to other cellular signals. This increases the flexibility of ubiquitin-based signaling pathways. Disruption of the ubiquitylation machinery is lethal in yeast.

In addition to acting as a signal for protein degradation, ubiquitin can also be involved in protein sorting and protein–protein interactions. Proteins that bind to ubiquitin can affect the fate of the target protein in a variety of ways. If polyubiquitin chains are assembled by linking ubiquitins between the C-terminal and lysine 63, rather than lysine 48, the ubiquitin chain has a different shape and the modified protein is not targeted for destruction. Instead, lysine 63–linked ubiquitin chains are implicated in membrane protein targeting to the lysosome, signaling pathways, DNA repair, and mitotic regulation.

At least eight other highly conserved small proteins related to ubiquitin can be conjugated to ε-amino groups of lysine on target proteins by E1, E2, and E3 enzymes that are distinct from those involved in conjugating ubiquitin. These modifications are generally not involved in protein degradation and instead serve a variety of signaling functions. The best known of these are the three SUMO (small ubiquitin-like modifier) proteins. SUMO can be conjugated to the same residues on target proteins as ubiquitin, but typically regulates protein–protein interactions rather than promoting degradation. In some specialized instances, SUMO can recruit ubiquitin E3 ligases and trigger degradation of the target protein.

Proteolysis: A Compartmentalized Process

Unregulated proteolysis within a cell would be lethal. Therefore, cells compartmentalize intracellular proteolytic activity in two distinct ways so that only appropriate substrates are degraded. **Lysosomes** are membrane-bound compartments that sequester various hydrolases, including **proteases,** and provide a low pH environment in which the enzymes are optimally active. **Proteasomes** are proteolytic machines assembled from multiple protein subunits with the proteolytically active sites corralled inside on the walls of a narrow cylindrical chamber. The constricted internal diameter of the cylinder and regulatory complexes that guard the openings allow access only to unfolded polypeptide chains. Energy in the form of adenosine triphosphate (ATP) is required for proteasomes to unfold and degrade proteins, even though hydrolysis of a peptide bond actually releases energy.

Intracellular proteolysis typically depends on specific recognition of protein substrates and their translocation into a proteolytic compartment. Generally speaking, long-lived cytosolic proteins and integral membrane proteins circulating within the secretory and endosomal systems are degraded by lysosomes, whereas short-lived cytoplasmic proteins and endoplasmic reticulum (ER) membrane proteins are degraded by the proteasome. Ubiquitin targets most (though not all) molecules for degradation by proteasomes and can also target proteins to lysosomes for degradation. Ubiquitin or a polyubiquitin chain is recognized by specific receptors on the proteolytic machinery.

Degradation in Lysosomes

Lysosomes, the major digestive organelles, contain at least 60 distinct hydrolytic enzymes, including proteases, peptidases, phosphatases, lipases, phospholipases, glycosidases, sulfatases, and nucleases. Lysosomal hydrolases are tagged in the Golgi apparatus with mannose-6-phosphate groups on their N-linked oligosaccharides. **Mannose-6-phosphate receptors** in the *trans*-Golgi network then divert the tagged hydrolases to **endosomes** and lysosomes (see Chapter 21). Most lysosomal hydrolases are synthesized as inactive precursors and are activated by proteolysis on arrival in lysosomes. The low pH of lysosomes, maintained by the v-ATPase proton pump (see Fig. 14.6), is essential for efficient degradation. Most lysosomal enzymes have maximal hydrolytic activity at pH 4 to 5 rather than at the cytoplasmic pH of 6.5 to 7.0. Under these conditions, some of the hydrolases carry a positive charge and adhere to the lipid bilayers of lysosomal membranes. This renders them more resistant than most non-lysosomal macromolecules to the harsh environment. Lysosomal hydrolases degrade proteins, lipids, and nucleic acids to amino acids, lipids, sugars, and nucleotides that are transported across the lysosomal membrane to the cytoplasm, where they are reused to synthesize new macromolecules.

Lysosomes degrade substrates that originate both outside and inside the cell. Extracellular substrates taken into the cell by endocytosis are delivered to lysosomes via the endocytic pathway (see Chapter 22). Lysosomes also degrade cellular constituents, accounting for 50% to 70% of cellular protein turnover. Cytoplasmic substrates are delivered to lysosomes by dedicated autophagosomes and by direct capture from the cytoplasm in a process termed **autophagy**. In certain hematopoietic cell lineages, lysosomes can also act as secretory organelles.

The essential role of lysosomes as the primary site for constitutive degradation is revealed by the more than 50 distinct human lysosomal storage diseases (Appendix 23.1). Patients with these diseases lack the function of one or more lysosomal hydrolases, membrane proteins, or other factors involved in lysosomal function. Consequently, undigested material accumulates in lysosomes (Fig. 23.4). This disrupts homeostasis and can kill the cell. The manifestations of lysosomal storage diseases are exceedingly complex, and although approximately two-thirds of patients suffer from a range of neurological symptoms, any organ system can be affected by these diseases.

Delivery to Lysosomes Via the Endocytic Pathway

Lysosomal degradation of plasma membrane proteins internalized by endocytosis plays an important role in remodeling the plasma membrane in response to various stimuli. For example, the half-life of the receptor for the epidermal growth factor (EGF) is normally approximately 10 hours. However, when circulating EGF binds, the activated receptor is more efficiently internalized and degraded in lysosomes with a half-time of less than 1 hour. This downregulates the biological response (Fig. 23.5; also see Fig. 27.6).

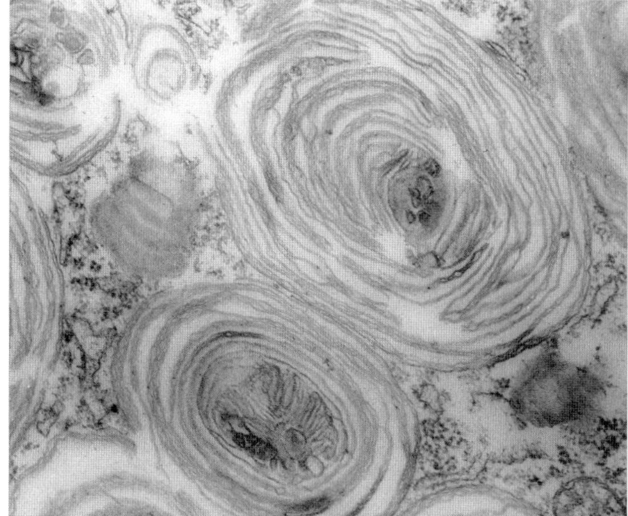

FIGURE 23.4 ELECTRON MICROGRAPH OF ABNORMAL LYSOSOMES IN THE NEURONS OF A PATIENT WITH GM_1 GANGLIOSIDOSIS. Similar lysosomes, called *membranous cytoplasmic bodies*, accumulate in the neurons of patients with GM_2 gangliosidosis (Tay-Sachs disease). (Courtesy Kinuko Suzuki, University of North Carolina, Chapel Hill.)

The degradation of lipids and membrane proteins in lysosomes poses a topologic problem, as fusion of lysosomes with other membrane compartments would simply merge the two membranes. This problem is solved by segregating the components to be degraded into regions of membrane that bud into endosomes, forming the intraluminal vesicles or tubules of **multivesicular** bodies (see Fig. 22.13). Fusion of a multivesicular body with a lysosome delivers the intraluminal vesicles to the lumen of the lysosome for digestion (Fig. 23.5).

Multivesicular bodies form via a sorting process in endosomes (see Chapter 22). Proteins destined for degradation are tagged with single ubiquitin molecules, a process distinct from the polyubiquitylation reaction required for targeting to proteasomes. These monoubiquitinated proteins are gathered together in endosomes by ubiquitin-binding proteins, which are localized there by interaction with phosphatidylinositol 3-phosphate (PI3-P) formed by PI3-kinase (see Fig. 26.7). A sequence of multiprotein ESCRT complexes (endosomal complex

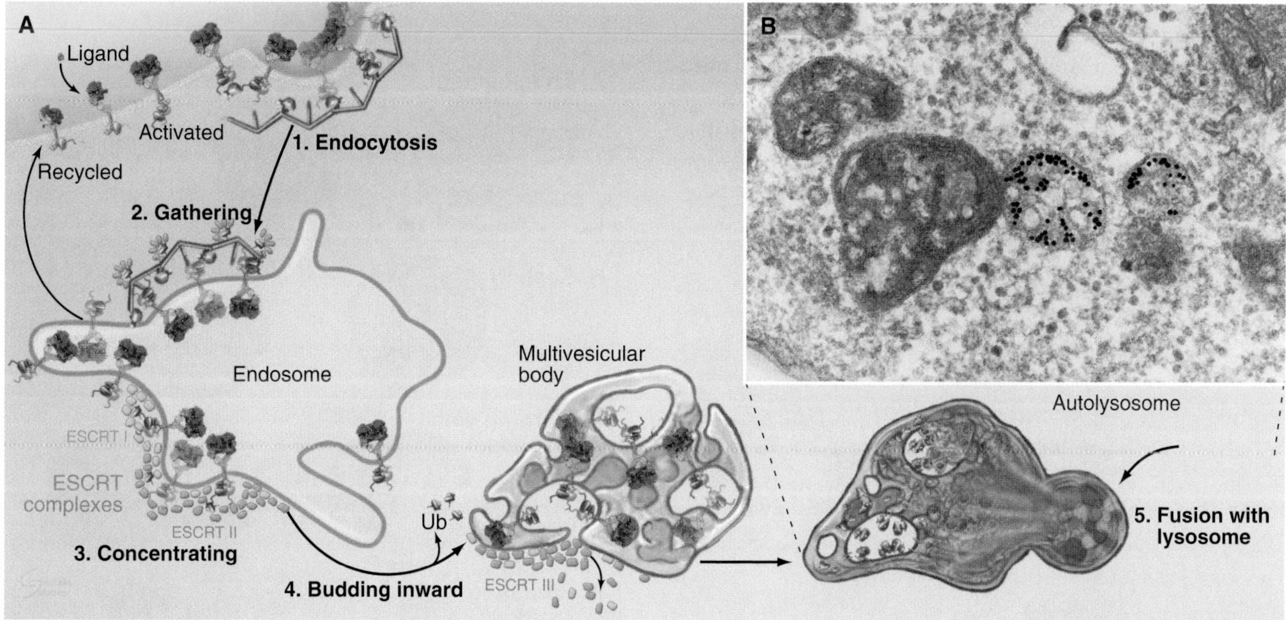

FIGURE 23.5 RECEPTOR DOWNREGULATION. A, To limit the time course of epidermal growth factor (EGF) stimulation, activated EGF receptors are internalized and delivered to endosomes and sequestered with other membrane proteins to be destroyed within intraluminal vesicles that accumulate to form multivesicular bodies. Proteins destined for internalization and destruction are marked with single ubiquitin molecules. Sorting of the ubiquitin-tagged proteins and invagination of the late endosomal membrane are driven by the three ESCRT (endosomal sorting complexes required for transport) complexes. The ubiquitins are recycled prior to budding into the multivesicular body interior After fusion with lysosomes, lysosomal proteases and lipases ensure that both the extracellular and cytoplasmic domains of the EGF receptor are degraded along with the intraluminal vesicles. Cytosolic proteins are also incorporated into the internal vesicles of multivesicular late endosomes and are degraded in a constitutive process termed microautophagy. **B,** Electron micrograph of EGF receptor marked with gold-labeled antibodies in multivesicular endosomes fusing with a lysosome. (**B,** Courtesy Colin Hopkins, MRC Laboratory, University College, London, United Kingdom.)

required for transport [Fig. 22.13]) further concentrates the monoubiquitinated proteins to membrane regions that bud into the vesicle interior. ESCRT complexes initially bind monoubiquitin but later recruit enzymes that remove and recycle the ubiquitin. Constriction of the bud necks driven by the assembly of helical filaments of the ESCRT-III complex releases the intraluminal vesicles into the interior of the endosome, creating a multivesicular body. Interestingly, budding of a number of viruses, including Ebola and HIV-1, the virus that causes AIDS, resembles the process of multivesicular body formation. Indeed, HIV hijacks two of the ESCRT complexes for this purpose.

Autophagy

Macroautophagy, microautophagy, and chaperone-mediated autophagy are processes leading to the breakdown of cytoplasmic constituents within lysosomes. Autophagy first evolved as a defense of single-celled organisms against starvation, but now has key functions in cellular homeostasis.

Macroautophagy (usually simply referred to as *autophagy*) involves the engulfment of large regions of cytoplasm. An initiating signal leads to the formation of a specialized membrane in association with specific cargo. This membrane expands, curls and seals itself off, enclosing the cargo and forming a double-layered **autophagosome,** which then fuses with a lysosome, leading to the degradation and ultimate recycling of all internal contents (Fig. 23.6). Cargo may include glycogen granules, ribosomes, and even organelles such as mitochondria and peroxisomes.

Genetic screens in budding yeast have identified 36 so-called *Atg* genes required for autophagosome formation and development. Although the details are still being determined, it appears that the process may initiate at more than one site in the cytoplasm–including the ER, ER exit sites, mitochondria or *trans*-Golgi. There, numerous small vesicles containing the membrane protein Atg9, interact with the multi-protein Atg1/ULK scaffold complex to fuse, forming a flattened membrane sheet. This membrane curls to form a cup-like structure, the **phagophore,** which grows and ultimately is sealed off to form a double-membrane autophagosome (Fig. 23.7). Two ubiquitin-like systems recruit components to the growing phagophore. One conjugates Atg8 to phosphatidylethanolamine, anchoring it to the membrane surface, where it recruits other membrane vesicles to the phagophore. The other forms a conjugate of two Atg proteins that recruits other pathway components to the phagophore. Both conjugation reactions use a chain of enzymes analogous to those that attach ubiquitin to its target proteins targeted (see later).

Fusion of a nascent autophagic vacuole with late endosomes and lysosomes involves a specialized SNARE (soluble *N*-ethylmaleimide-sensitive factor [NSF] attachment protein receptor) protein (see Fig. 21.15) and forms an **autolysosome** with acid hydrolases in the lumen. These degrade the contents, including the internalized membrane, releasing amino acids, lipids, sugars, and nucleotides back to the cytoplasm (Fig. 23.7). The end stage of an autolysosome is a **residual body** with a

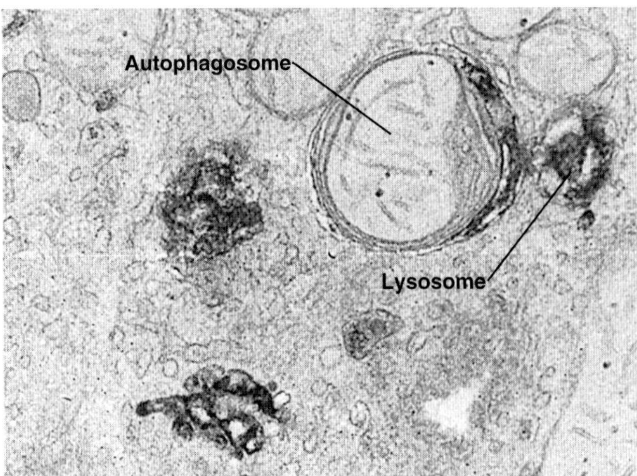

FIGURE 23.6 ELECTRON MICROGRAPH OF AN AUTOPHAGOSOME. From starved rat liver shows an autophagosome containing a mitochondrion in the process of fusing directly with a secondary lysosome. (From Dunn WA Jr. Studies on the mechanisms of autophagy: Maturation of the autophagic vacuole. *J Cell Biol*. 1990;110:1935–1945, copyright The Rockefeller University Press.)

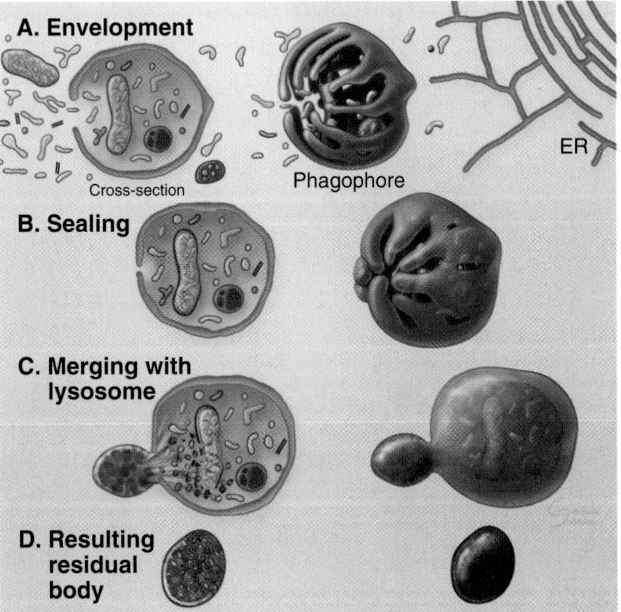

FIGURE 23.7 THE FOUR STAGES OF AUTOPHAGY. A, A phagophore membrane forms, enveloping a region of cytoplasm containing target proteins and organelles. **B,** Membrane fusion results in formation of a nascent autophagosome. **C,** The nascent autophagosome fuses with a primary or secondary lysosome, which delivers hydrolytic enzymes that degrade the autophagosome contents. **D,** Undigested material remains in residual bodies.

dense core of undegraded material. The process of formation and degradation of autophagic vacuoles in the liver requires less than 15 minutes.

Macroautophagy is regulated by mTor, which phosphorylates both the master transcriptional regulator, TFEB, and key Atg components. Intracellular signals that trigger macroautophagy are tied to intracellular levels of particular amino acids. Amino acids and some circulating peptide hormones are potent inhibitors of autophagy, promoting both cytoplasmic sequestration of TFEB and phosphorylation (and inactivation) of Atg1. Starvation increases circulating levels of the hormone glucagon, which releases TFEB and also leads to Atg1 dephosphorylation and stimulates autophagy in liver cells. Feeding produces the opposite reaction due to the action of insulin (see Fig. 27.7), and ultimately reduces autophagy.

Budding yeast use autophagy as a cell survival pathway during starvation, but autophagy can kill some cells. For example, autophagy can lead to death when a cell receives a lethal insult under conditions in which apoptotic pathways (see Chapter 46) are not functional. Autophagy has also been implicated in the developmental programs of a number of higher organisms and in the destruction of intracellular protein aggregates. Protein aggregates and damaged organelles, such as mitochondria are tagged with ubiquitin and recognized by one of two autophagy receptors that bind both ubiquitin and Atg8-containing vesicles. This results in their envelopment by growing phagophore membranes, and delivery to autolysosomes via autophagy.

Microautophagy is a byproduct of multivesicular body formation (Fig. 23.5; also see Fig. 22.13). Small volumes of cytoplasm are captured in the intraluminal vesicles and tubules that invaginate within endosomal or lysosomal membranes. The cytoplasmic components are degraded as the vesicles surrounding them are consumed. Not all proteins in intraluminal vesicles are chosen randomly—some cytosolic proteins are selected, packaged and delivered to lysosomes by this route. When starved for glucose, *Saccharomyces cerevisiae* expresses several cytoplasmic enzymes and membrane transporters that are required to process more complex sugars. When glucose becomes available, the yeast switches metabolic pathways and degrades the enzymes it no longer needs. Transporters in the plasma membrane are internalized and delivered to the vacuole (the yeast lysosome equivalent) through the formation of multivesicular bodies and microautophagy. Unneeded cytoplasmic enzymes are selectively packaged into small vesicles that deliver their contents to the vacuole by fusion.

Chaperone-mediated autophagy (CMA) is a quality control mechanism to eliminate soluble cytoplasmic proteins that are incorrectly folded or assembled. The process selects for degradation the 20% to 30% of cytoplasmic proteins with the linear target sequence KFERQ. The molecular chaperone Hsc70 recognizes KFERQ when the protein is unfolded or—if the protein is part of a complex—not associated with its normal partners. Hsc70 binds and escorts the protein to lysosomes where an integral membrane protein transfers it across the membrane aided by a second chaperone functioning inside the lysosome. Prolonged starvation also activates CMA, apparently to supply cells with amino acids for the synthesis of essential proteins. In both Parkinson and Alzheimer diseases neurons accumulate proteins with KFERQ sequences, so defects in CMA may contribute to these neurodegenerative diseases. CMA is unable to combat those diseases once they start, as the process functions only on soluble proteins and not proteins in complexes or cytoplasmic aggregates.

Degradation by Proteasomes

The proteasome is a mobile self-contained compartment for regulated proteolysis. Proteasomes are highly abundant structures about half the size of a ribosome that are located in both the cytoplasm and nucleoplasm (Fig. 23.8). Proteasomes contain an array of proteolytic active sites arrayed on the interior wall of a cylindrical chamber. They degrade target proteins down to small peptides (Fig. 23.9). Proteins targeted to the proteasome can be selected because they are abnormal or misfolded but may also be selected in response to signaling cascades or at key transitions of the cell cycle. One class of proteasomes processes intracellular antigens for presentation by the immune system.

The proteasome has two major structural components: the core and the regulatory particle, the latter comprising base and lid subassemblies. The cylindrical core, referred to as the 20S proteasome (named according to its sedimentation coefficient; see Chapter 6) is structurally conserved from bacteria to mammals, although the subunit composition varies. Mammals have three distinct regulatory particles that feed different types of substrates to the core.

In mammals, the core consists of four stacked seven-member rings. Each ring consists of seven distinct polypeptides: two rings of β-type subunits form a central chamber lined by the proteolytic active sites; and rings of α-type subunits form antechambers on either end of the central chamber. The noncatalytic α-subunits gate the access of the substrate into the proteolytic chamber. The narrow lumen of the antechamber only allows access to unfolded polypeptide chains.

Three of the β-type subunits of eukaryotic proteasomes have hydrolytic activities (Fig. 23.9). The caspase-like activity of the β_1-subunit cleaves after acidic residues, the trypsin-like activity of the β_2-subunit cleaves after basic residues, and the chymotrypsin-like activity of the β_5-subunit cleaves after hydrophobic residues (see Fig. 46.11 for a description of caspases). These three

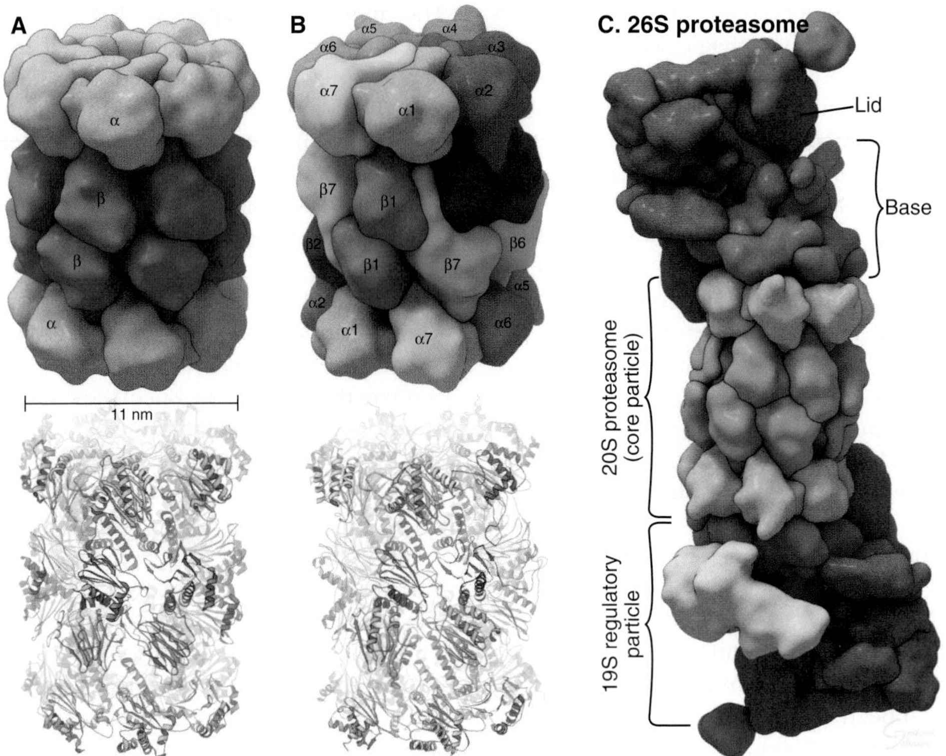

FIGURE 23.8 STRUCTURE OF THE PROTEASOME. **A** and **B,** Crystal structure of 20S proteasomes from *Thermoplasma acidophilum* **(A)** and from *Saccharomyces cerevisiae* **(B).** Note the conservation of structure from *Archaea* to yeast. **C,** Structure of the 26S proteasome from *Schizosaccharomyces pombe*. (For reference, see Protein Data Bank [PDB; www.rcsb.org] file 1PMA from Lowe J, Stock D, Jap B, et al. Crystal structure of the 20S proteasome from the archaeon *T. acidophilum* at 3.4 Å resolution. *Science*. 1995;268:533–539 [**A**], PDB file 1RYP from Groll M, Ditzel L, Lowe J, et al. Structure of 20S proteasome from yeast at 2.4 Å resolution. *Nature*. 1997;386:463–471 [**B**], and PDB file 4CR2 from Lasker K, Förster F, Bohn S, et al. Molecular architecture of the 26S proteasome holocomplex determined by an integrative approach. *Proc Natl Acad Sci U S A*. 2012;109:1380–1387 [**C**].)

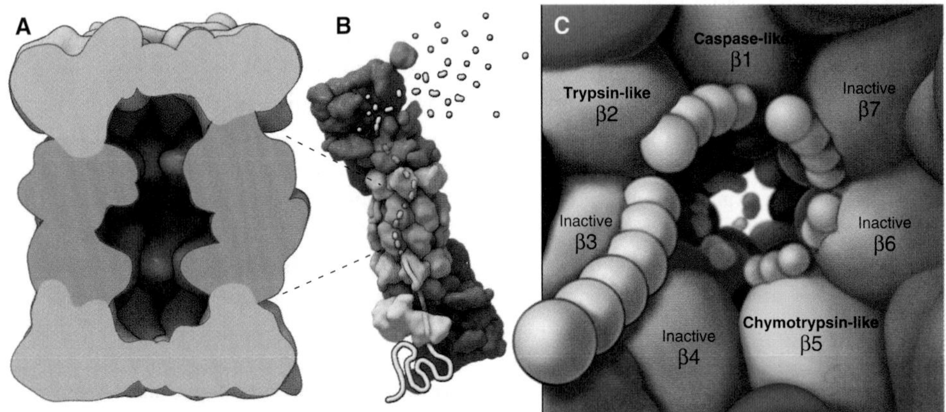

FIGURE 23.9 A POLYPEPTIDE MOVING THROUGH THE CENTRAL CHANNEL OF THE 20S PARTICLE. Note that only three of the seven subunits are catalytically active, and each has a distinct catalytic activity. View **C** is down the barrel in **A**.

activities combine to give the proteasome broad specificity, allowing it to cleave diverse substrates into short peptides of seven to nine residues. An N-terminal threonine residue of the β-subunits is exposed by autocatalytic proteolysis and serves as the key active site residue for proteolysis. The antibiotic **lactacystin** reacts covalently and selectively with these threonine residues to inactivate the proteasome. The remaining four β-type subunits in eukaryotic proteasomes are not posttranslationally processed to mature, catalytically active enzymes.

The cylindrical cores of eukaryotic and archaeal proteasomes are capped on one or both ends by the base and lid complexes of the regulatory particle, forming the 26S proteasome. The type of regulatory particle varies depending on the function of the proteasome. The 19S regulatory particle associated with proteasomes that degrade most proteins has two functions. First, the 10 lid proteins include receptors that bind ubiquitin and deubiquitinating enzymes, **DUBs**, that clip off ubiquitin chains. Removal of the ubiquitin enables its recycling for later reuse and is critical for proteolysis, because the substrate cannot fit into the channel with ubiquitin attached. Second, the 10 subunit base includes an AAA-ATPase (see Box 36.1) composed of six distinct polypeptides. This enzyme uses energy from ATP hydrolysis to unfold protein substrates and thread them into the central channel for proteolysis. ATP hydrolysis is not required for the proteolytic cleavages in the central chamber.

Cells of higher vertebrates have a distinct regulatory particle, the 11S cap, associated with a subpopulation of 20S proteasome cores. The cytokine **α-interferon** induces the synthesis of this specialized **"immunoproteasome,"** which cleaves intracellular antigens, such as those derived from an infecting virus, into peptides suitable in size for presentation on the surface of antigen-presenting cells (see Fig. 27.8). The 11S cap does not recognize ubiquitylated protein substrates and may be a docking site for specific molecular chaperones. Specialized catalytic β-subunits in the immunoproteasome 20S core generate somewhat longer peptides that are better suited for antigen presentation. The immunoproteasome is physically and functionally coupled to an **ABC transporter** (see Fig. 14.10) called the **TAP** (transporter associated with antigen presentation) that translocates the peptides it generates into the ER. Another integral membrane protein of the ER directly loads translocated peptides onto **class I major histocompatibility antigen (MHC) molecules** for transport to the cell surface, where the MHC-peptide complex stimulates T-cells (see Fig. 27.8). To avoid detection by the immune system, some viruses block this pathway and force the translocation of MHC molecules backward through TAP, out of the ER, and into the waiting maw of the proteasome.

Motifs That Specify Ubiquitylation

Ubiquitylation directs the selective degradation of many different proteins. These include abnormally folded proteins, regulatory proteins (including some that control cell-cycle progression), components of signal transduction systems, and regulators of transcription.

Polyubiquitin chains are most commonly linked through lysine 48, but all other linkages, except lysine 63, also appear to be involved in proteasomal targeting. It was thought that chains of four or more ubiquitins are required for targeting to the proteasome, but it now seems that the number of ubiquitins bound to the target protein (possibly at multiple sites) rather than the length of individual chains may be the critical determinant.

Regulated proteolysis is critical in controlling cell-cycle progression and transcription activation. Here, targeting signals for degradation are often generated by specific phosphorylation events (Table 23.1). For example, phosphorylation of a conserved sequence near the N-terminus of several transcription factors or their regulatory subunits generates a phospho-degron recognized by the SCF complex (a family of modular E3 enzymes named for their three core components: *s*kp1, *c*dc53/cullin, and an *F*-box-containing protein; see Fig. 40.16). SCF then ubiquitylates the target, marking it for destruction. This phosphorylation is often performed by *c*yclin-*d*ependent *k*inases (Cdks [see Fig. 40.14]) and is thereby tightly linked to the cell cycle.

A second multisubunit class of E3 enzymes, designated the APC/C (anaphase-promoting complex/cyclosome), recognizes a degenerate, nine-residue "destruction box" sequence near the N-terminus of several cell-cycle regulatory proteins that targets these proteins for degradation (see Chapter 40). An even shorter amino acid sequence lysine, glutamic acid, asparagine (KEN), the "KEN box," can also serve as a destruction signal for the APC/C. Deletion of the destruction box or the KEN box stabilizes the target protein, whereas transferring these sequences to a normally stable protein may result in cell-cycle–dependent ubiquitylation and rapid degradation. Destruction or KEN boxes are not themselves regulated, but APC/C activity and specificity are regulated by Cdk phosphorylation (see Chapter 40).

The simplest natural degron is described by the **"N-end rule."** Certain destabilizing amino acids at the N-terminus of a protein can be recognized for ubiquitylation by a specific E3, leading to subsequent destruction of the protein. Although several apparent N-end rule substrates have been identified, it is unlikely that this simple rule applies generally in vivo as proteins with "destabilizing" NH_2-terminal residues (after processing to remove methionine) do not tend to have shorter half-lives than most proteins.

The plant hormone auxin uses targeted protein destruction to induce expression of the genes that it regulates. Auxin binds to a specific F-box protein which changes its conformation and can, as a result, recognize a degron motif on a repressor protein that normally holds auxin-responsive genes in an inactive state. SCF-binding results in ubiquitylation and destruction of the repressor leading to activation of the auxin-responsive genes. When attached to animal proteins, the auxin-induced degron can be used experimentally to trigger the rapid destruction of the target proteins in cells induced to express the plant F-box protein.

Amphipathic or hydrophobic stretches of amino acids also function as general recognition determinants for ubiquitylation. Because hydrophobic surfaces are often buried in a folded protein or at the interface between subunits, exposure of this class of determinant is thought to assist in targeting misfolded proteins or excess subunits of oligomeric proteins for degradation. This pathway is especially prevalent in controlling the degradation of proteins that fail to fold in the ER (see later).

Because proteolysis is key for cell-cycle progression, interference with the proteasome has been adopted as a strategy for treatment of cancer. One proteasome inhibitor, bortezomib, is now used in the clinic to treat advanced multiple myeloma, a leukemia of B-lymphocytes.

Elimination of Misfolded Proteins From the Endoplasmic Reticulum

Integral membrane proteins and secretory proteins fold and assemble in the lipid bilayer or lumen of the ER (see Fig. 20.8). Proteins that fail to fold or assemble are retrieved from the ER and degraded by the proteasome in a pathway known as **ERAD** (ER-associated degradation). The ERAD pathway also regulates levels of a number of ER resident proteins. ERAD target proteins are detected either by a chaperone in the ER lumen, or directly by a large multi-protein complex inserted in the ER membrane. Either way, the substrate is retro-translocated by that complex back to the cytoplasmic surface of the ER where it either has its trans-membrane domains cleaved in the plane of the membrane by specific proteases or is captured, forcibly extracted from the membrane by an AAA-ATPase and ubiquitylated by one of two dedicated E3 ligases prior to degradation by proteasomes.

Medical interest in the ERAD pathway arises because defects in ubiquitylation of particular proteins are associated with the pathology of Parkinson disease. Furthermore, the most common form of cystic fibrosis results from ERAD-mediated degradation of a slow-folding (but catalytically competent) variant of the CFTR (cystic fibrosis transmembrane regulator) ABC (adenosine triphosphate binding cassette) transporter (see Fig. 17.4) before it can be exported to the cell surface.

Other Regulated Intracellular Proteolysis

Another form of regulated intracellular proteolysis is activation of inactive proenzymes or transcription factors by proteolytic cleavage (see Fig. 10.21 for NFκB [nuclear factor κB]). An important example of activation by proteolytic cleavage is provided by **caspases.** Extracellular or intracellular signals trigger the cleavage of procaspases, turning on their proteolytic activity and initiating a cascade that leads to an inflammatory response or apoptosis (see Figs. 46.17 and 46.18). In all cases, intracellular proteolysis is tightly regulated through a combination of triggered activation of the protease, specific substrate recognition, and compartmentalization.

Lipid Turnover and Degradation

Lipids and membrane proteins destined for degradation enter the lysosome either via the endocytic pathway or via autophagy. Distinct pathways exist for the turnover of the three classes of cellular lipids: phosphoglycerides, glycolipids, and cholesterol. Glycolipids, which are restricted to the extracellular leaflet of lipid bilayers, are degraded primarily in lysosomes, and they accumulate in lysosomal storage diseases (Appendix 23.1). Sphingomyelin and gangliosides are delivered to lysosomes via vesicular transport and degraded to the level of ceramide, sugars, and fatty acids by a series of lysosomal hydrolases. Their degradation requires association with an activator protein to extract them from membranes and render them accessible to the catabolic enzymes. The lysosomal membrane has a specialized lipid composition including **lysobisphosphatidic acid** (see Fig. 22.14), which is also enriched in intraluminal vesicles of multivesicular bodies and lysosomes. Lysobisphosphatidic acid may play a role in activating sphingomyelinases and restricting their hydrolytic activity to the intraluminal side of the membranes. Other sphingomyelinases also exist on the cell surface, and their activation triggers production of the lipid second messenger ceramide (see Fig. 26.11).

The turnover of phosphoglycerides is much more varied in mechanism and location. Phosphoglycerides from the outer leaflet of the plasma membrane, are degraded in lysosomes to their fatty acids, head group, and glycerol constituents. Often, phosphoglyceride degradation is only partial, and the degradative products (eg, fatty acids, lysophospholipids, and diacylglycerol) are salvaged and reutilized in "short-circuit pathways." In this way, "old" phospholipids are "remodeled," forming new ones with the same or altered properties. These phospholipid-remodeling reactions are catalyzed by a variety of **phospholipases** that cleave the phospholipid to generate distinct products (see Fig. 26.4). Localized lipid remodeling can generate specialized lipid subdomains required for vesicle fusion or fission or the selective recruitment of proteins to the membrane. In addition, molecules released from partial degradation of phosphoglycerides, fatty acids, diacylglycerol, and some head groups function as second messengers in signaling cascades (see Fig. 26.5).

Cholesterol Homeostasis

Cholesterol metabolism in mammals involves multiple organs (see Fig. 20.15 for the synthetic pathway). Approximately 90% of the free cholesterol in animal cells is in the plasma membrane. Cholesterol is the precursor for steroid hormones, which are synthesized in specialized cells but used throughout the body for myriad essential functions. Cholesterol is also the precursor for bile acids, which are synthesized by the liver and transported to the gut, where they aid in the digestion of dietary fat. Unlike the case with virtually all other cellular molecules, individual cells cannot degrade cholesterol. Instead, cellular levels of cholesterol are regulated by a complex balance of endogenous synthesis, uptake of extracellular cholesterol, and efflux of intracellular

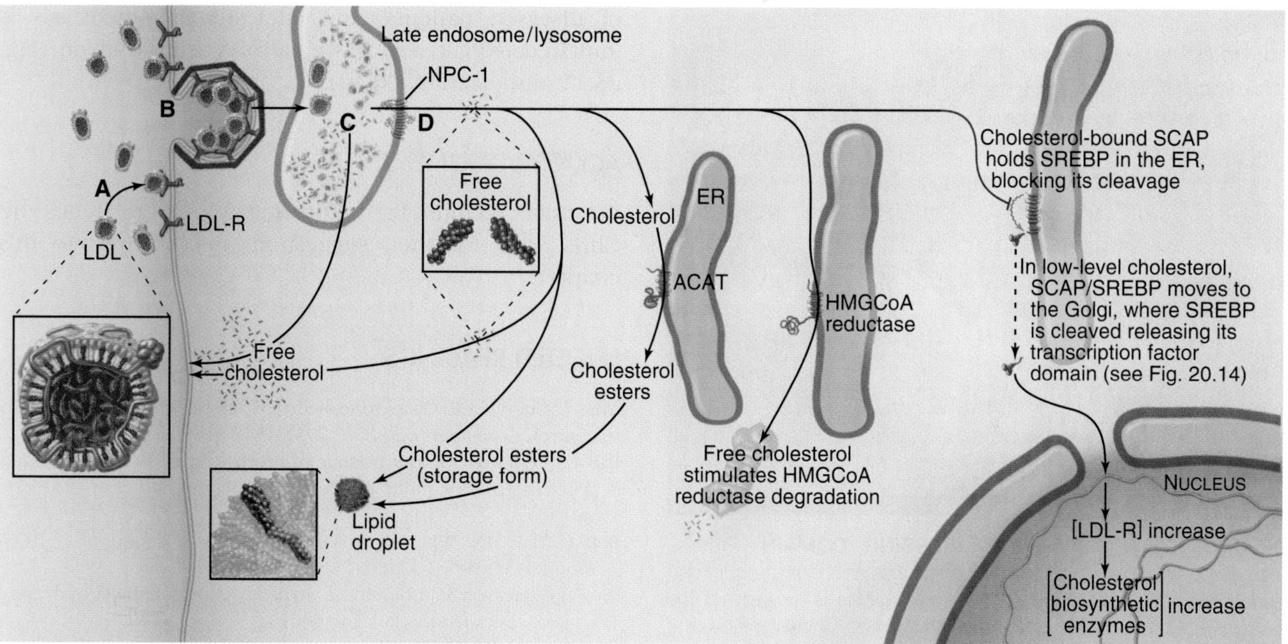

FIGURE 23.10 THE INTRACELLULAR PROCESSING AND REGULATION OF CHOLESTEROL BIOSYNTHESIS. A, Dietary cholesterol is delivered to cells in low-density lipoprotein (LDL) particles. **B,** LDL particles are taken up by clathrin-mediated endocytosis. **C,** Free cholesterol is released in late endosomes/lysosomes and transported to the cell surface or internal membranes, depending, in part, on the activity of the Niemann-Pick disease type C1 (NPC-1) integral membrane protein **(D)**. Excess cholesterol can be acylated by acyl coenzyme A cholesterol acyltransferase (ACAT) activity and stored in cytoplasmic lipid droplets as cholesterol esters. ACAT activity is increased by high intracellular cholesterol levels. At the same time, high cholesterol in the membrane decreases new cholesterol synthesis by triggering the proteasome-dependent degradation of the enzyme β-hydroxy-β-methylglutaryl-coenzyme A (HMG-CoA) reductase. Finally, high cellular cholesterol decreases the uptake of LDL particles and dietary cholesterol by blocking proteolytic processing of the transcription factor sterol regulatory element-binding protein (SREBP), which is required for LDL-receptor (LDL-R) expression (see Fig. 20.16). Genetic defects that perturb steps **A** to **D,** which are required to maintain the delicate balance of cholesterol homeostasis, cause several human diseases. Familial hypercholesterolemia is caused either by a lack of LDL-R **(A)** or by LDL-R that is defective in endocytic activity **(B).** Wolman disease is a lysosomal storage disease that is caused by defective lysosomal cholesterol esterase activity; Niemann-Pick disease type C, another lysosomal storage disease, results in defective trafficking of cholesterol out of late endosomes and lysosomes caused by mutations in NPC-1 **(D).** ER, endoplasmic reticulum; SCAP, SREBP cleavage activating protein.

cholesterol to vascular fluids (Fig. 23.10). When present in excess, cholesterol accumulates as plaques in the walls of major arteries, contributing to atherosclerosis.

Cholesterol is transported through the body as cholesterol esters packaged with other lipids and proteins. Cholesterol esters are even more hydrophobic than cholesterol, because they have a fatty acid esterified to the hydroxyl group. The intestine assembles dietary cholesterol into particles called chylomicrons, which are transported through the blood and eventually taken up by the liver, which is also the major site of cholesterol synthesis in mammals. The liver packages dietary and de novo–synthesized cholesterol into **low-density lipoproteins** (LDLs), which are secreted into the blood for transport to other tissues. Other cells take up LDL particles via receptor-mediated endocytosis and deliver them along the endocytic pathway to lysosomes (Fig. 23.10). Within the lysosome, cholesterol esters are hydrolyzed, and the bulk of free LDL-derived cholesterol is transported by a cytoplasmic carrier protein back to the plasma membrane. Importantly, a small portion of cholesterol is also transported to the ER, where the cholesterol level controls the activity of transcription factors that regulate genes involved with cholesterol metabolism.

Two key enzymes in the ER have sterol-sensing domains that allow them to respond to the cholesterol content of the membrane and control intracellular free cholesterol levels (Fig. 23.11). Accumulation of LDL-derived cholesterol in the ER activates acyl coenzyme A cholesterol acyltransferase (ACAT), the enzyme that converts free cholesterol to cholesterol esters for storage. Substantial increases in the levels of free cholesterol (or an oxygenated metabolite of it) also triggers the destruction of the enzyme that catalyzes the first step in cholesterol biosynthesis, β-hydroxy-β-methylglutaryl-coenzyme A (HMG-CoA) reductase (Fig. 23.11). Cholesterol triggers the degradation of HMG-CoA reductase through a pathway that depends on ubiquitin and proteasomes. In addition, the membrane-bound transcription activator sterol regulatory element-binding protein (SREBP) is trapped in the ER by SREBP cleavage activating protein (SCAP), a third protein with a sterol-sensing domain. This limits, the expression of the genes for both HMG-CoA reductase and the LDL receptor. When ER cholesterol is low, SCAP/SREBP escapes to the Golgi apparatus, where

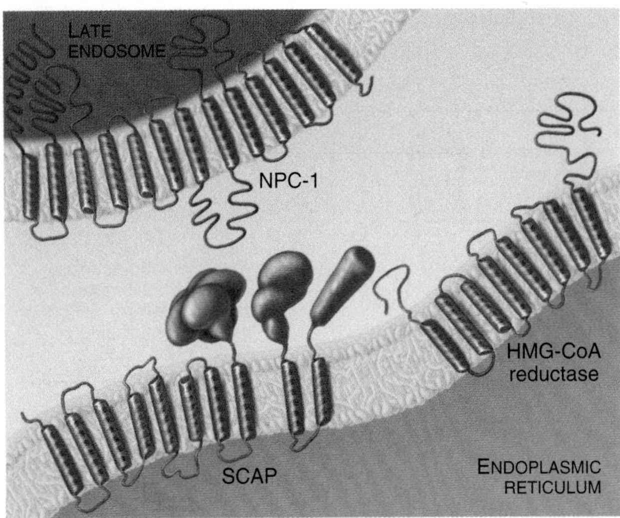

FIGURE 23.11 THE STEROL-SENSING DOMAIN. Proteins involved in cholesterol trafficking and homeostasis share a common sequence motif—the sterol-sensing domain. This is a region of the protein *(red)* that spans the membrane five times. Cholesterol binding to this domain in the protein Niemann-Pick disease type C1 (NPC-1) is required for normal cholesterol trafficking, whereas binding to the analogous domain inhibits the function of β-hydroxy-β-methylglutaryl-coenzyme A (HMG-CoA) reductase (which is ubiquitylated and destroyed) and of SCAP (which is retained in the ER as a complex with sterol regulatory element-binding protein [SREBP]), thereby downregulating sterol production and uptake.

proteolytic cleavage liberates the SREBP activation domain (see Fig. 20.16). This then travels to the nucleus to drive the expression of LDL receptor, HMG-CoA reductase, and other proteins involved with cholesterol metabolism. This negative feedback mechanism reduces cholesterol input both from de novo synthesis and from extracellular sources (Fig. 23.10).

Cholesterol homeostasis is critical to human health, and a number of genetic diseases result from defects in cholesterol metabolism. Defects in the LDL receptor reduce or eliminate LDL uptake, and LDL builds up in the blood, leading to cholesterol deposition in the walls of arteries and atherosclerosis. Rare defects in the enzyme that hydrolyzes cholesterol esters in lysosomes lead to Wolman disease, which causes death within the first year of life. Niemann-Pick type C disease is a devastating neurodegenerative disorder that results from mutations in a multitransmembrane domain NPC-1 protein that is required for transport of LDL-derived cholesterol from late endosomes to both the plasma membrane and ER. Cholesterol accumulates in lysosomes of diseased patients, and cholesterol homeostasis is impaired. NPC-1 also has a sterol-sensing domain (Fig. 23.11; also see Fig. 20.10).

ACKNOWLEDGMENTS

We thank Klaudia Brix, Ron Hay, Margarete Heck, and Chris Scott for their suggestions on revisions to this chapter.

SELECTED READINGS

Arias E, Cuervo AM. Chaperone-mediated autophagy in protein quality control. *Curr Opin Cell Biol.* 2011;23:184-189.

Babst M, Odorizzi G. The balance of protein expression and degradation: an ESCRTs point of view. *Curr Opin Cell Biol.* 2013;25:489-494.

Betz C, Hall MN. Where is mTOR and what is it doing there? *J Cell Biol.* 2013;203:563-574.

Brown JS, Jackson SP. Ubiquitylation, neddylation and the DNA damage response. *Open Biol.* 2015;5:150018.

Christianson JC, Ye Y. Cleaning up in the endoplasmic reticulum: ubiquitin in charge. *Nat Struct Mol Biol.* 2014;21:325-335.

Deshaies RJ, Joazeiro CA. RING domain E3 ubiquitin ligases. *Annu Rev Biochem.* 2009;78:399-434.

Earnshaw WC, Martins LM, Kaufmann SH. Mammalian caspases: Structure, activation, substrates, and functions during apoptosis. *Annu Rev Biochem.* 1999;68:383-442.

Fernández ÁF, López-Otín CJ. The functional and pathologic relevance of autophagy proteases. *J Clin Invest.* 2015;125:33-41.

Goldstein JL, Brown MS. A century of cholesterol and coronaries: from plaques to genes to statins. *Cell.* 2015;161:161-172.

Huber LA, Teis D. Lysosomal signaling in control of degradation pathways. *Curr Opin Cell Biol.* 2016;39:8-14.

Hurley JH. ESCRTs are everywhere. *EMBO J.* 2015;34:2398-2407.

Inobe T, Matouschek A. Paradigms of protein degradation by the proteasome. *Curr Opin Struct Biol.* 2014;24:156-164.

Kimura H, Caturegli P, Takahashi M, Suzuki K. New insights into the function of the immunoproteasome in immune and nonimmune cells. *J Immunol Res.* 2015;2015:541984.

Kraft C, Martens S. Mechanisms and regulation of autophagosome formation. *Curr Opin Cell Biol.* 2012;24:496-501.

Matyskiela ME, Martin A. Design principles of a universal protein degradation machine. *J Mol Biol.* 2013;425:199-213.

Nakatsukasa K, Kamura T, Brodsky JL. Recent technical developments in the study of ER-associated degradation. *Curr Opin Cell Biol.* 2014;29:82-91.

Parenti G, Andria G, Ballabio A. Lysosomal storage diseases: from pathophysiology to therapy. *Annu Rev Med.* 2015;66:471-486.

Settembre C, Ballabio A. Lysosome: regulator of lipid degradation pathways. *Trends Cell Biol.* 2014;24:743-750.

Simons K, Ikonen E. How cells handle cholesterol. *Science.* 2000;290:1721-1726.

Sontag EM, Vonk WI, Frydman J. Sorting out the trash: the spatial nature of eukaryotic protein quality control. *Curr Opin Cell Biol.* 2014;26:139-146.

APPENDIX 23.1

Lysosomal Storage Diseases

Disease(s)	Enzyme Defect	Accumulated Material
Sphingolipidosis GM_1 ganglioside	β-Galactosidase	GM_1 gangliosidosis glycoproteins
Tay-Sachs GM_2 gangliosidosis	Hexosaminidase A	GM_2 gangliosides
Sandhoff GM_2 gangliosidosis	Hexosaminidase A and B	GM_2 gangliosides
Krabbe (galactosylceramide lipidosis)	Galactosyl ceramide β-Galactosidase	Galactocerebrosides
Niemann-Pick A and B (sphingomyelin lipidosis)	Sphingomyelinase	Sphingomyelin Cholesterol
Gaucher Glucosylceramide lipidosis	β-Glucocerebrosidase	Glucosylceramide
Fabry	α-Galactosidase A	Trihexosylceramide
Glycoprotein storage diseases	α-Fucosidase α-Mannosidase α-Aspartylglycosamine	Glycopeptides Glycolipids Oligosaccharides
Mucopolysaccharidosis Several types	α-Iduronidase Iduronosulfate sulfatase N-acetyl-α-glucosaminidase Heparan sulfatase β-Glucuronidase	Heparan sulfate
Sialidosis	Neuraminidase	Sialyloligosaccharides
Mucolipidosis II I-cell disease	UDP-*N*-acetylglucosamine (GlcNAc): glycoprotein GlcNAc-1-phosphotransferase	Glycoproteins Glycolipids

SECTION VII

Signaling Mechanisms

SECTION VII OVERVIEW

Cells depend on signaling systems to adapt to changing environmental conditions. Free-living organisms, such as yeast and bacteria, respond to changes in temperature, osmotic stress, and nutrients by synthesizing the proteins required to optimize their survival. Motile cells respond to chemicals by migrating toward attractants and away from repellants. In vertebrate animals, the hormone adrenaline stimulates cellular energy metabolism, and growth factors stimulate cells to duplicate their genomes and divide. Developmentally regulated genetic programs equip each cell with the molecular hardware that is required to adapt to remarkably diverse stimuli.

The first three chapters in this section introduce the main molecular components of signaling pathways: receptors, protein messengers, and second messengers. With this background, the reader can appreciate the nine well-characterized signal transduction pathways presented in Chapter 27 without being distracted by descriptions of molecular components.

Cells use molecular **receptors** (Chapter 24) to detect chemical and physical stimuli. Physical interaction of the stimulus with the receptor provides energy to modify the structure of the receptor and initiate a signaling pathway. With the exception of RNA "riboswitches," all receptors are proteins. A few stimuli, including light, steroid hormones, and gases, penetrate the plasma membrane and react with receptors inside the cell. Most stimuli from outside the cell, including proteins, peptides, and charged organic molecules, cannot penetrate the plasma membrane. These extracellular ligands bind transmembrane receptors on the cell surface that transfer the signal across the lipid bilayer.

Most stimuli act through one of approximately 25 families of receptor proteins, each coupled to a distinct signal transduction pathway (see Fig. 24.1). Multiple isoforms within each family provide thousands of different receptors, each with specificity for particular stimuli. For example, of 20,447 genes in the nematode genome, nearly 800 encode a large family of receptors with seven transmembrane helices. Presumably, all members of each receptor family arose from a common ancestor and acquired new specificities by multiple rounds of gene duplication and divergent evolution.

Active receptors generate a chemical signal inside the cell by interacting with one or more cytoplasmic proteins (Chapter 25). This **transduction** step converts one type of signal (the stimulus) into another signal (the messenger) and often amplifies the signal. Some receptors have a cytoplasmic domain with **protein kinase** activity or associate with a separate protein kinase. These enzymes transfer phosphate from adenosine triphosphate (ATP) to specific amino acids on target proteins. The cytoplasmic domains of active seven-helix receptors catalyze the exchange of guanosine diphosphate (GDP) for guanosine triphosphate (GTP) on signal-transducing guanosine triphosphatases (GTPases), called G-proteins. GTP binding activates these **G-proteins,** allowing them to bind and regulate target proteins. Adapter proteins may link active receptors to downstream effector proteins, including kinases and GTPases. Cytoplasmic signaling proteins often act in cascades, passing a signal from one to another. Amplification along these pathways allows small stimuli to rapidly generate large biochemical responses inside the cell.

Many signaling pathways regulate the concentrations of small molecules, called **second messengers** (Chapter 26). The most widely used second messengers are **Ca^{2+}, cyclic nucleotides,** and **lipids.** They modify cellular behavior by binding to and activating a wide range of effector proteins, regulating membrane physiology, cellular metabolism, motility, and gene expression.

Signaling pathways regulate most cellular processes (Chapter 27). Effector systems include transcription factors that control gene expression, proteins that regulate secretion, metabolic enzymes, structural elements of the cytoskeleton and associated motors, cell-surface receptors, regulators of the cell cycle, and membrane ion channels. Multiple signaling pathways converge on each of these effector systems. Integration of these diverse signals determines the behavior of the cell, whether it secretes, moves, grows, divides, or differentiates.

Understanding signaling pathways can be challenging. First, cells employ hundreds of distinct signaling pathways, involving hundreds to thousands of different proteins. Second, few signal transduction mechanisms involve simple linear pathways from a stimulus to a change in behavior. Rather, most pathways branch and converge multiple times. Thus information from several inputs can influence each effector system. This provides for integration of regulatory mechanisms but makes it difficult to predict how information flows through a system. Third, most pathways have positive or negative feedback loops that can either augment or inhibit responses. These **feedback loops** can either prolong or foreshorten the signal or even make it oscillate. Fourth, the response of some pathways depends on both the strength and the temporal pattern of the stimulus. Ultimately, signaling pathways must be understood as integrated systems, like complex electrical circuits.

Biochemistry/pharmacology and genetic analysis have revealed much about signaling mechanisms. The biochemical approach generally starts with identification of a naturally occurring or synthetic chemical, such as a hormone, that modifies the activity of an organism, organ, or cell. These compounds are called **agonists.** Characterization of the biological effects of agonists is often aided by the discovery of chemicals that antagonize their action. In many cases, such **antagonists** prove to be useful as drugs, even before their mechanisms are understood. Aspirin is just one historic example. To define the mechanism, it is necessary to purify and characterize the receptor that binds the chemical and then trace the biochemical steps from receptor to effectors. Pioneering research established these pathways for each class of receptor. Subsequently the primary structure of each new receptor has revealed (by homology with known receptors) the type of transduction mechanism that lies between the receptor and the effector systems in the cell.

The genetic approach involves characterization of mutations that affect the flow of information through a signaling pathway. By collecting enough mutants and testing for a hierarchy of effects, investigators can often define the flow of information through a pathway. Cloning and sequencing the mutated genes then reveals the proteins involved. Because one need make no assumptions about the nature of the biochemical components or how they are connected, completely novel molecules emerge from genetic screens just as easily as familiar ones. One particularly fruitful genetic approach has been to analyze genes that predispose individuals to cancer or cause naturally occurring heritable diseases in humans, mice, or other species. Many proteins responsible for regulating cell growth and proliferation cause cancer when constitutively activated by mutations. Inactivating mutations in other signaling proteins cause cancer, developmental defects or endocrine diseases.

To understand the dynamics of a signaling system, one must learn enough about all the pathways and the rates of the reactions to formulate mathematical models that can explain how the system responds to the intensity and pattern of the stimuli. This is best understood for the system that controls bacterial chemotaxis.

CHAPTER 24

Plasma Membrane Receptors

Cells use approximately 25 different families of receptor proteins (Fig. 24.1) to detect and respond to a myriad of chemical and physical stimuli (Appendix 24.1). Most receptors are plasma membrane proteins that interact with chemical **ligands** or are stimulated by physical events such as light absorption. A few chemical stimuli, including steroid hormones and the gas nitric oxide, cross the plasma membrane and bind receptors inside the cell. Some ligands arise inside cells. These include the cyclic nucleotides that stimulate cyclic nucleotide-gated channels (see Fig. 16.10) and metabolites that bind riboswitches (see Fig. 3.22).

Gene duplication and divergent evolution within each family produced genes for multiple **receptor isoforms** that interact with different ligands. Members of each family share one or more structurally homologous domains. Isoforms in some families share both ligand-binding and signal-transducing strategies (eg, seven-helix receptors and cytokine receptors). Members of other families share either a similar ligand-binding structure (tumor necrosis factor [TNF] receptor family) or a common signal-transducing method (receptor tyrosine kinases) but differ in other respects. Amino acid substitutions in common structural scaffolds allow isoforms to recognize their specific ligands.

In multicellular organisms, selective expression of certain receptors and the associated transduction molecules allows differentiated cells to respond specifically to particular ligands but not others. Fortunately, the mechanisms of the best-characterized receptors usually apply to the rest of their family. Thus, learning about a few examples provides a working knowledge of many related receptors.

One cannot predict the type of receptor, signal transduction mechanism, or nature of the response from the chemical nature of a stimulus (Appendix 24.1). Although proteins and peptides are the only known ligands for receptor kinases and kinase-linked receptors, proteins

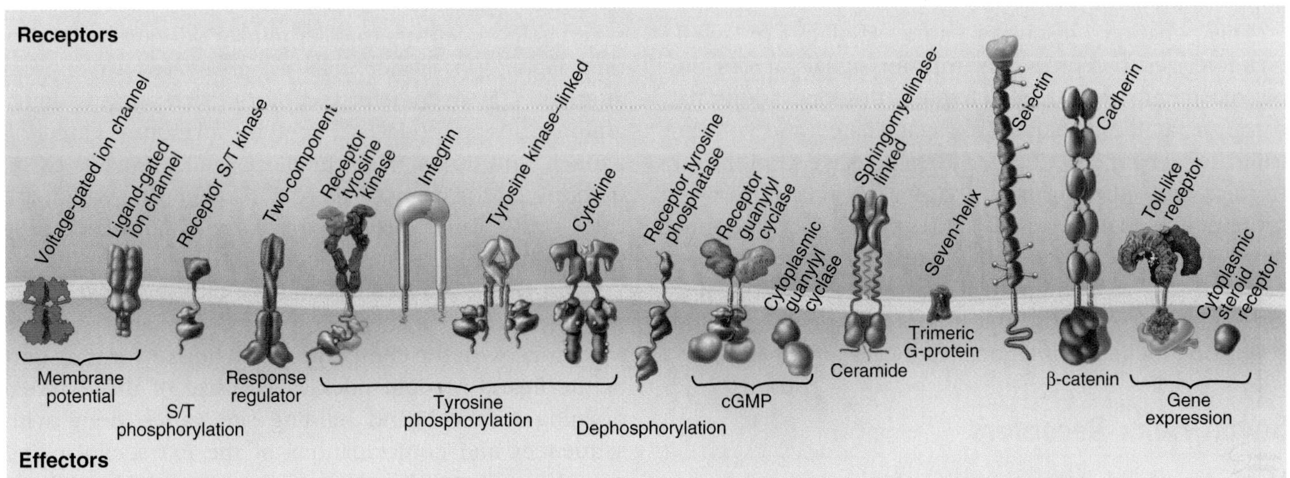

FIGURE 24.1 SIXTEEN CLASSES OF RECEPTORS AND THEIR SIGNAL TRANSDUCTION MECHANISMS. S/T, serine/threonine.

and peptides also stimulate some seven-helix receptors and guanylyl cyclase receptors. A particularly wide range of stimuli activates seven-helix receptors. This includes photons, amino acids, nucleotides, biogenic amines, lipids, peptides, proteins, and hundreds of different organic molecules. Some ligands bind distinct receptors on different cells. For example, acetylcholine activates skeletal muscle contraction by opening a ligand-gated ion channel (see Fig. 16.12). It also binds seven-helix receptors on other cells, activating signaling pathways mediated by guanosine triphosphate (GTP)–binding proteins (see Fig. 39.21). Some ligands with similar names bind to different types of receptors. For example, several interleukins (IL-2 through IL-6) bind to cytokine receptors, but IL-1 activates a TNF family receptor, and IL-8 binds a seven-helix, G-protein–coupled receptor.

Energy from ligand binding is used to change the conformations of receptors and transfer the signal across the plasma membrane to activate cytoplasmic signals. In the simplest case ligand binding on the cell surface changes the conformation of seven-helix receptors, including parts of the receptor exposed in the cytoplasm. More commonly ligand binding to extracellular domains of a receptor aligns cytoplasmic domains in a manner that stimulates enzyme activity or favors binding of proteins that propagate the signal.

Most signal-transducing pathways include one or more enzymes that amplify signals. In some receptor families, an enzyme is part of the receptor protein itself (receptor tyrosine kinases), but in others, the receptor interacts with a separate cytoplasmic enzyme (trimeric G-proteins, cytoplasmic protein kinases).

If extracellular stimulation is sustained, most signaling systems downregulate their response. The literature variously calls this **adaptation,** attenuation, desensitization, tachyphylaxis, or tolerance. For example, rhodopsin and odorant receptors turn off within a second of continuous stimulation (see Figs. 27.1 and 27.2). This allows one to distinguish rapidly changing visual information and concentrations of odors.

This chapter discusses nine families of well-characterized receptors that transfer signals across the plasma membrane. Other chapters describe additional receptor families: Chapter 16, ligand-gated and voltage-gated ion channels; Chapter 10, nuclear receptors for steroids and other ligands; Chapter 25, receptors with protein-phosphatase activity; Chapter 26, cytoplasmic nitric oxide receptors with guanylyl cyclase activity; Chapter 27, two-component receptors and tyrosine kinase-linked receptors; and Chapter 30, cell adhesion receptors, including integrins, cadherins, and selectins.

Seven-Helix Receptors

Members of the largest family of plasma membrane receptors are built from a serpentine arrangement of seven transmembrane α-helices (Fig. 24.2). When activated, these diverse receptors are guanine nucleotide exchange proteins (GEFs; see Fig. 4.6) for cytoplasmic **trimeric GTP-binding proteins** (see Fig. 25.9). For this reason they are also called G-protein–coupled receptors. GTP binding activates the trimeric G-proteins, which relay signals to effector proteins inside cells. Eight Nobel Prizes have been awarded for work on this pathway.

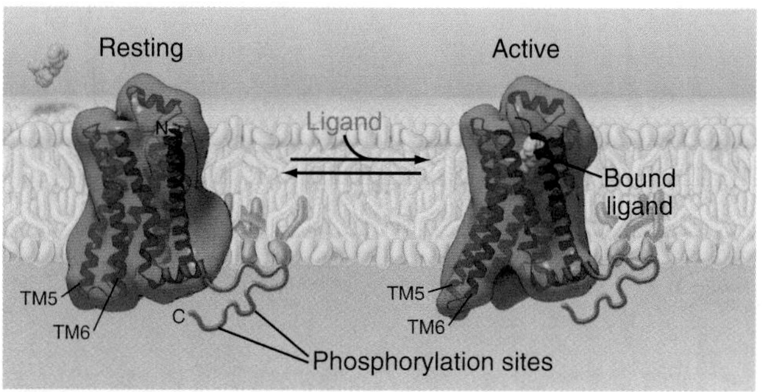

FIGURE 24.2 STRUCTURE OF SEVEN-HELIX RECEPTORS. Ribbon diagrams from crystal structures of resting and active human β_2-adrenergic receptors illustrating the differences in the conformation brought about by binding the ligand, adrenaline. The structure of active receptor bound to a trimeric G-protein is nearly identical. The N-terminal segment outside and the C-terminal segment inside the cell are modeled. (For reference, see Protein Data Bank [PDB; www.rcsb.org] files 2RH1 and Cherezov V, Rosenbaum DM, Hanson MA, et al. High-resolution crystal structure of an engineered human beta2-adrenergic G protein-coupled receptor. *Science*. 2007;318:1258–1265; and PDB file 4LDO and Ring AM, Manglik A, Kruse AC, et al. Adrenaline-activated structure of β2-adrenoceptor stabilized by an engineered nanobody. *Nature*. 2013;502:575–579.)

Seven-helix receptors have the same topology and ligand-binding site as bacteriorhodopsin (see Fig. 13.8), so the genes may be very ancient. Slime molds have seven-helix receptors linked to trimeric G-proteins, so the eukaryotic genes for these proteins are at least 1 billion years old. Four percent (790) of the genes of the nematode *Caenorhabditis elegans* encode seven-helix receptors, the largest family of proteins in the worm. In mammals, olfactory cells alone use 500 to 1000 different seven-helix receptors to discriminate odorant molecules (see Fig. 27.1). Other cells are estimated to express another 375 seven-helix receptors to respond to light, amino acids, peptide and protein hormones, catecholamines, and lipids. The chemical ligand remains to be determined for many of these 375 receptors, which are therefore termed *orphan receptors.* Many medically useful drugs bind seven-helix receptors.

More than 75 crystal structures established that seven-transmembrane helices are packed similarly in these receptors, with the centrally located helix 3 forming part of the ligand binding pocket. The size of the external opening to the ligand binding site varies along with sequences and conformations of the extracellular and cytoplasmic loops. The N-termini are outside the cell and vary from seven to 6000 residues. Some large N-terminal domains participate in ligand binding. The cytoplasmic loops between helices 1–2, 3–4, and 5–6 interact with trimeric G-proteins. The C-terminal segment of the polypeptide extends into the cytoplasm but is anchored to the bilayer by two covalently attached fatty acids. It is probably intrinsically disordered and varies in length from 12 to more than 350 residues. The figures in this book show seven-helix receptors as monomers, but many seven-helix receptors function as dimers or larger oligomers, allowing for crosstalk between the subunits.

Soluble chemical ligands activate most seven-helix receptors by binding in a central pocket among the extracellular ends of the helices approximately one-third of the way across the membrane. Residues lining this pocket are highly variable between isoforms, providing specificity for each receptor to bind a particular ligand. Drugs also bind between the helices. The light-absorbing pigment **11-*cis* retinal** is covalently bound in a similar location to the photoreceptor protein **rhodopsin** (Fig. 24.2B) and is activated by absorbing a photon that changes its conformation (see Fig. 27.2). Peptide hormones bind deep in the helical pocket but probably also interact with residues that are more exposed on the cell surface. Receptors for some large ligands (pituitary glycoprotein hormones, such as luteinizing hormone, follicle-stimulating hormone, and thyroid-stimulating hormone) and some small ligands (glutamate, γ-aminobutyric acid, calcium) bind with high affinity to extracellular N-terminal domains of their seven-helix receptor. The N-terminal domain with bound ligand then stimulates the transmembrane domain of the receptor. The blood-clotting enzyme thrombin activates its receptor on platelets by proteolysis of the receptor rather than by direct binding (see Fig. 30.14). The N-terminal peptide cleaved from the receptor dissociates and activates other receptors; what is left of the newly truncated N-terminus folds back and activates its own receptor.

Seven-helix receptors exist in an equilibrium between two conformations (Fig. 24.2), a **resting state** and an **activated state** with the ability to catalyze the exchange of nucleotide bound to trimeric G-proteins (Fig. 24.3). Without bound ligand, the resting state is strongly favored. Ligand binding to the receptor (or the isomerization of retinal after absorbing light) shifts the equilibrium to the active state and initiates signal transduction. Activation involves reorganizing of contacts between the helices in the core of the protein and large movements of transmembrane helices 5 and 6 (Fig. 24.2B). These changes create a binding site for the α-subunit of the target G-protein composed of the cytoplasmic ends of transmembrane helices 3, 5, and 6 (Fig. 24.3).

Active receptors catalyze the dissociation of guanosine diphosphate (GDP) bound to an inactive G_α subunit of the trimeric G-protein. Cytoplasmic GTP then binds and activates G_α (see Fig. 25.9). A single active seven-helix receptor can amplify the signal by activating up to 100 G-proteins. After dissociating from the receptor and each other, both G_α-GTP and $G_{\beta\gamma}$ stimulate downstream

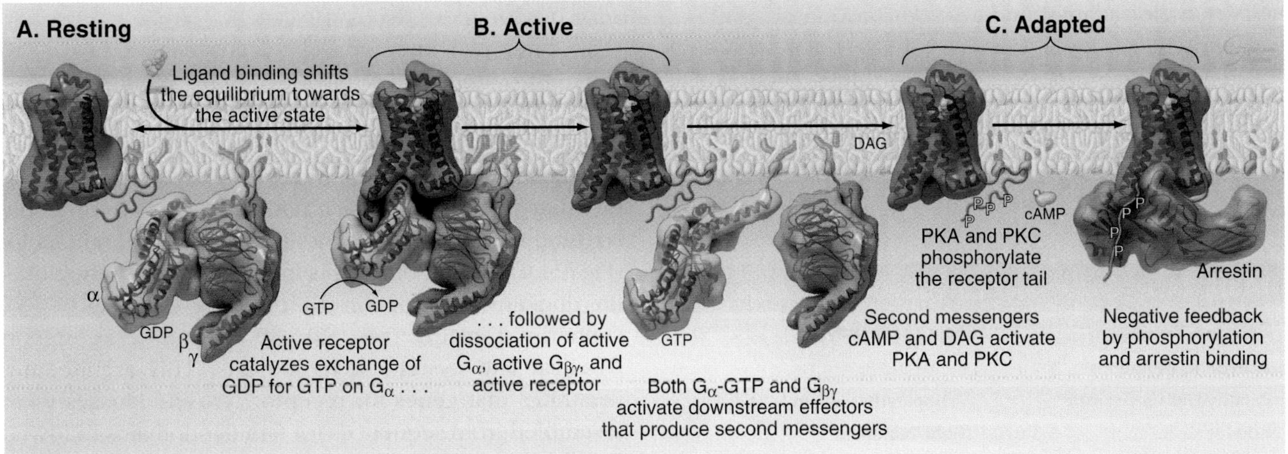

FIGURE 24.3 ACTIVATION AND ADAPTATION OF A SEVEN-HELIX RECEPTOR. A, Ligand binding shifts the equilibrium from the resting conformation toward the active conformation. **B,** The active receptor binds trimeric G-protein, and promotes dissociation of guanosine diphosphate (GDP), allowing GTP to bind. This dissociates G_α from $G_{\beta\gamma}$, allowing both to activate downstream effectors that produce, for example, the second messengers cyclic adenosine monophosphate (cAMP), which activates protein kinase A (PKA), and diacylglycerol (DAG), which activates protein kinase C (PKC). **C,** Both kinases phosphorylate active receptors on their C-terminus, which attracts arrestin, putting the receptor into the inactive adapted state. (For reference, see PDB file 2RH1 [inactive BAR with bound carazolol] and Cherezov V, Rosenbaum DM, Hanson MA, et al. High-resolution crystal structure of an engineered human beta2-adrenergic G protein-coupled receptor. *Science*. 2007;318:1258–1265; PDB file 4LDO [active BAR with bound adrenaline] and Ring AM, Manglik A, Kruse AC, et al. Adrenaline-activated structure of β2-adrenoceptor stabilized by an engineered nanobody. *Nature*. 2013;502:575–579; PDB file 3SN6 [active BAR bound to G_s] and Rasmussen SG, DeVree BT, Zou Y, et al. Crystal structure of the β2 adrenergic receptor–Gs protein complex. *Nature*. 2011;477:549–555; and PDB file 4JQI [β-arrestin bound to a phosphorylated C-terminal peptide] and Shukla AK, Manglik A, Kruse AC, et al. Structure of active β-arrestin-1 bound to a G-protein-coupled receptor phosphopeptide. *Nature*. 2013;497:137–141.)

effector proteins, further amplifying the signal (see Fig. 27.3 for an example of amplification).

Most seven-helix receptors adapt to sustained stimulation by **negative feedback** from the signaling pathway. Receptor activation produces **second messengers** that stimulate multiple kinases. These kinases phosphorylate the C-terminal tail of the receptor, inhibiting interactions of G-proteins with receptors still bound to their ligands (Fig. 24.3C). The kinases include cyclic adenosine monophosphate (cAMP)–activated protein kinase A and protein kinase C (see Fig. 25.4). Furthermore, $G_{\beta\gamma}$ subunits released in response to receptor stimulation activate **G-protein–coupled receptor kinases** that modify the receptors themselves. These pathways allow for cross-talk between receptors, as activation of one class of receptors can inactivate other receptors.

Phosphorylation of the receptor tail creates a binding site for **arrestin,** a protein with multiple functions (Fig. 24.3). First, arrestin blocks interactions of the receptor with G-proteins, terminating signaling through the main pathway downstream of most seven-helix receptors. In some cases, arrestin initiates a new signal through the mitogen-activated protein (MAP) kinase pathway (see Figs. 27.6 and 27.7 for two other pathways). Arrestin also promotes the removal of seven-helix receptors from the plasma membrane by endocytosis in clathrin-coated vesicles. Some internalized receptors recycle to the plasma membrane, some continue to activate trimeric G-proteins from endosomes, and others are modified by ubiquitin and directed to lysosomes for destruction. Chapter 27 discusses three dramatic examples of seven-helix receptor adaptation.

Hundreds of mutations have been documented in seven-helix receptors. Some are harmless, such as the recessive mutations in the gene for melanocortin 1 receptor that confer red hair and fair skin. Others have been linked to human diseases (Table 24.1). Mutations can inactivate seven-helix receptors by every conceivable means from failure to synthesize the full-length protein to reduced affinity for ligands to failure to activate G-proteins. For example, loss-of-function mutations in rhodopsin cause retinitis pigmentosa, a degeneration of photoreceptor cells. Mutations of the melanocortin 4 receptor cause human obesity. More than a hundred different mutations produce seven-helix receptors that are constitutively active without ligand. Particular mutations of rhodopsin cause night blindness, and mutations in a calcium receptor cause dysfunction of the parathyroid gland. The physiology of these activating mutations is complicated, because cells use feedback mechanisms to compensate for the continually active receptors. These inherited loss-of-function mutations are generally recessive.

TABLE 24.1 Seven-Helix Receptors and Disease

Defective Receptor	Disease Phenotype
Activating Mutations	
Parathyroid Ca^{2+} sensor	Hypoparathyroidism
Rhodopsin	Night blindness
Thyroid hormone receptor	Hyperthyroidism, thyroid cancer
Loss-of-Function Mutations	
Cone cell opsin	Color blindness
Parathyroid Ca^{2+} sensor	Hyperparathyroidism, failure to respond to serum Ca^{2+}
Rhodopsin	Retinitis pigmentosa, retinal degeneration
Thyroid hormone receptor	Hypothyroidism
Vasopressin receptor	Nephrogenic diabetes insipidus; kidneys fail to resorb water

Receptor Tyrosine Kinases

Polypeptide growth factors control cellular proliferation and differentiation by binding plasma membrane receptors with cytoplasmic protein tyrosine kinase activity (Fig. 24.4). For example, **epidermal growth factor (EGF)** stimulates proliferation and differentiation of epithelial cells. **Platelet-derived growth factor (PDGF)** stimulates growth of smooth muscle cells, glial cells, and fibroblasts (see Fig. 32.11). Growth factors and their tyrosine kinase receptors were discovered by three strategies: biochemical purification of proteins that stimulate cellular growth or differentiation; genetic analysis of development of flies and nematodes; and searches for genes that cause cancer. Cytokine receptors (Fig. 24.6) and immune cell receptors (see Fig. 27.8) signal through separate tyrosine kinase subunits.

Genome sequencing established that humans have 58 genes for 20 families of receptor tyrosine kinases, each with distinct structural features. Most have an extracellular ligand-binding domain connected to a cytoplasmic tyrosine kinase domain by a single transmembrane helix (Fig. 24.4). Ligand binding is mediated by immunoglobulin domains, fibronectin III domains (see Fig. 3.13), cadherin domains (see Fig. 30.5), and less-common β-helical and cysteine-rich domains. This architecture illustrates that genes for receptor tyrosine kinases were assembled from sequences for familiar domains followed by divergence to allow for interactions with diverse ligands.

Ligand binding activates receptor tyrosine kinases by bringing together a pair of kinase domains on the cytoplasmic face of the membrane. The juxtaposition of kinase domains allows the partners to activate each other by direct interaction or by phosphorylating each other on tyrosine residues. In most cases, phosphorylation of the **activation loop** of the catalytic domain converts the kinase from an inactive to an active conformation (see Fig. 25.3E–F). Phosphorylation of tyrosines between

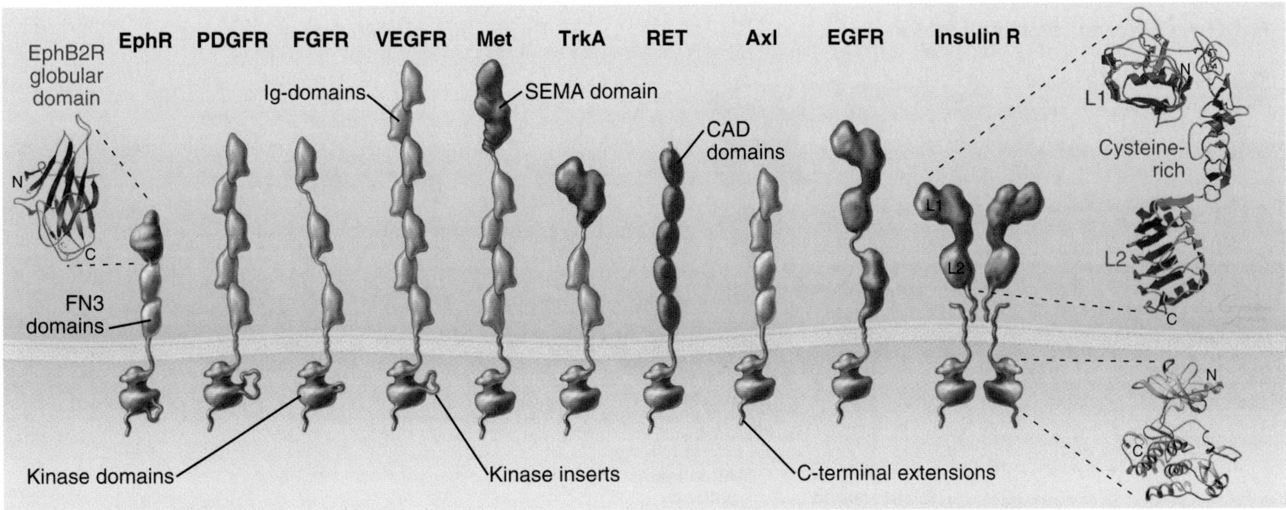

FIGURE 24.4 RECEPTOR TYROSINE KINASES. Domain architecture of nine of the 20 families of receptor (R) tyrosine kinases, with ribbon models of several domains. The globular domain of the EphB2 receptor is a β sandwich with a ligand-binding site that includes the exposed loop on the front of this model (see PDB file 1IGY). The extracellular part of the insulin-like growth factor consists of two similar β-helical domains connected by cysteine-rich domains (see PDB file 1IGR). The cytoplasmic kinase domain from the insulin receptor is similar to most typical kinases (see PDB file 1IRK). Kinase inserts and C-terminal extensions contain tyrosine phosphorylation sites. Receptor names: Axl, receptor for the growth factor Gas6; EGFR, epidermal growth factor receptor; EphR, receptor for ephrin membrane-bound ligands in the nervous system, the largest class of receptor tyrosine kinases; FGFR, fibroblast growth factor receptor; Met, receptor for hepatocyte growth factor; PDGFR, platelet-derived growth factor receptor; RET, a cadherin adhesion receptor; TrkA, receptor for nerve growth factor; VEGFR, vascular endothelial growth factor. Domain names: CAD, cadherin; F3, fibronectin-III; Ig, immunoglobulin. (For reference, see Lemmon MA, Schlessinger J. Cell signaling by receptor tyrosine kinases. *Cell*. 2010;141:1117–1134.)

the membrane and the kinase domain contribute to activating or inhibiting some receptors.

Ligands juxtapose tyrosine kinase domains in three ways. Dimeric ligands such as PDGF and stem cell factor recruit a pair of receptors from the pool of subunits diffusing in the plane of the membrane and link them physically (Fig. 24.5A). This **induced dimerization** juxtaposes two kinase domains in the cytoplasm.

The extracellular domains of EGF receptors fold in an autoinhibited conformation that precludes dimerization (Fig. 24.5B). EGF binding stabilizes a rearrangement of extracellular domains that favors their dimerization without EGF directly crosslinking the subunits. Receptor dimerization brings together the cytoplasmic kinase domains. One kinase domain interacts physically with the other to turn on its activity without requiring phosphorylation of the kinase activation loop.

Insulin induces a conformational change in a preformed receptor dimer (see Fig. 24.4; also see Fig. 27.7). The conformational change brings together the kinase domains, which activate each other by transphosphorylation of activation loops (see Fig. 25.3E–F).

Receptor tyrosine kinases activate effector proteins in two different ways. All activated kinases phosphorylate tyrosines on inserts and C-terminal extensions of their own kinase domains, creating phosphotyrosine-binding sites for downstream effector and adapter proteins with **Src homology 2 (SH2)** and **phosphotyrosine-binding (PTB)** domains (Fig. 24.5; also see Figs. 27.6 and 27.7). Each SH2 and PTB domain binds preferentially to a certain phosphotyrosine site by virtue of a pocket that recognizes both the phosphotyrosine and several adjacent residues (see Fig. 25.10). For example, each of five phosphotyrosines of the PDGF receptor binds a different effector or adapter protein. Binding an effector protein to a receptor phosphotyrosine favors its phosphorylation by the receptor kinase. In the case of **phospholipase Cγ,** tyrosine phosphorylation both activates its catalytic activity and dissociates the enzyme from its PTB site, allowing it to move to its site of action on the membrane. Alternatively, binding to the receptor may promote activity by bringing an effector protein near its substrate. This applies to both **phosphoinositide 3-kinase,** which acts on lipid substrates in the membrane bilayer (see Fig. 27.7), and the nucleotide exchange protein that activates the **Ras** guanosine triphosphatase (GTPase), which is anchored to the membrane bilayer (see Fig. 27.6).

Multiple mechanisms silence receptor tyrosine kinases. In the short term, some intracellular transducers activated by the receptor provide negative feedback to turn off the receptor. For example, lipid second messengers produced by phospholipase Cγ activate protein kinase C, which inhibits the receptor tyrosine kinase by phosphorylation. In the longer term, endocytosis in clathrin-coated vesicles removes active dimeric receptors from the cell surface. Receptors may continue to signal from endosomes. Some internalized receptor tyrosine kinases recycle to the plasma membrane. Others

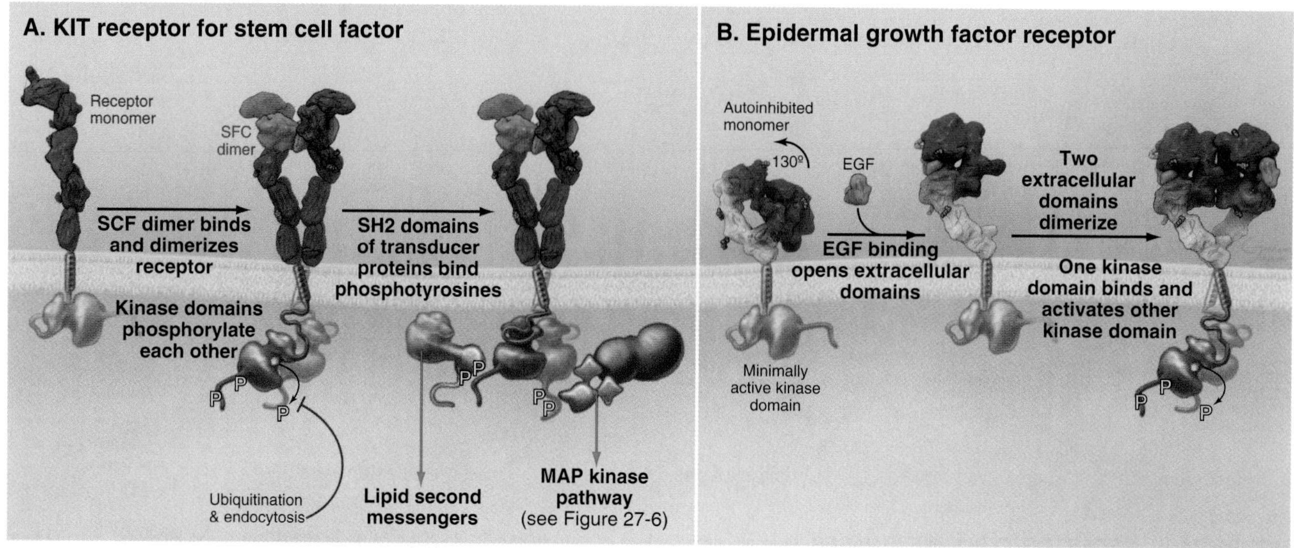

FIGURE 24.5 SUBUNIT DIMERIZATION MECHANISMS FOR ACTIVATING RECEPTOR TYROSINE KINASES. A, KIT receptor for stem cell factor (SCF). (Note that this is not the SCF Skp1-Cullin-F box ubiquitin E3 ligase that is important for cell cycle regulation.) In the absence of SCF, receptor monomers diffuse in the plasma membrane. Binding of an SCF dimer brings together two receptor molecules, bringing into proximity their cytoplasmic kinase domains. This allows for transphosphorylation of their activation loops and creation of phosphotyrosine binding sites for the Src homology 2 (SH2) and phosphotyrosine-binding (PTB) domains of downstream signal transduction proteins. **B,** Epidermal growth factor (EGF) receptor. In the absence of EGF, intramolecular interactions preclude dimerization. EGF binding changes the conformation of the extracellular domains allowing dimerization of two receptors, bringing together two cytoplasmic kinase domains. One kinase domain binds and activates the other, which creates phosphotyrosine binding sites for SH2 and PTB domains of downstream signal transduction proteins. (For reference, see PDB files 2EC8 [KIT monomer], 2E9W [KIT dimer SCF complex], 2AHX and 2A91 [EGF receptor] and Yuzawa S, Opatowsky Y, Zhang Z, et al. Structural basis for activation of the receptor tyrosine kinase KIT by stem cell factor. *Cell*. 2007;130:323–334; and Kovacs E, Zorn JA, Huang Y, et al. A structural perspective on the regulation of the epidermal growth factor receptor. *Annu Rev Biochem*. 2015;84:739–764.)

are modified by the addition of ubiquitins to one or more lysine side chains and degraded (see Fig. 23.3).

Mutations of genes for receptor tyrosine kinases cause human disease. Many cancers have activating mutations or overexpression of EGF family receptors. Activating mutations are found in all parts of these receptors and work by most conceivable mechanisms including dimerizing receptors without bound ligands and reversing autoinhibition. Antibodies against extracellular domains or drugs that inhibit kinase activity are useful therapeutically. Activating mutations in a fibroblast growth factor receptor lead to a variety of congenital abnormalities of the skeleton, including a form of dwarfism and premature fusion of the sutures between the bones of the skull. Some of these mutations activate by promoting receptor dimerization through disulfide bonds or association of transmembrane helices. Others change ligand specificity.

Cytokine Receptors

Cytokines are a diverse family of polypeptide hormones and growth factors that bind transmembrane receptors associated with cytoplasmic tyrosine kinases. These kinases activate cytoplasmic transcription factors called STATs (signal transducer and activator of transcriptions) that regulate many cellular processes. Although they differ in detail, all cytokines are four-helix bundles. Pituitary **growth hormone** controls body growth and development of mammals, so loss of function receptor mutations cause one type of dwarfism. **Erythropoietin** regulates the proliferation and differentiation of red blood cell precursors (see Fig. 28.4). **Interleukins** modulate cells of the immune system, so loss-of-function receptor mutations result in deficient immune cells.

The 30 human cytokine receptors are homodimers or heterodimers with extracellular fibronectin III domains that bind the ligand and a single transmembrane helix (Fig. 24.6). The large (40-kD), intrinsically disordered cytoplasmic domains have a binding site near the membrane for one of several protein tyrosine kinases called **JAKs** ("just another kinase"). JAKs have both a kinase domain and an inactive pseudokinase domain. When associated with the cytoplasmic domains of inactive receptor dimers the pseudokinase domains inhibit the kinase domains of the partner JAK.

The best-characterized cytokine receptor binds growth hormone. Binding of this monomeric ligand rotates the extracellular and transmembrane domains of the receptor relative to each other and activates the two JAK kinases bound to the cytoplasmic domains by separating the inhibitory pseudokinase domains from the kinase domains (Fig. 24.6). This allows the JAKs to activate each other by transphosphorylation. Active JAKs

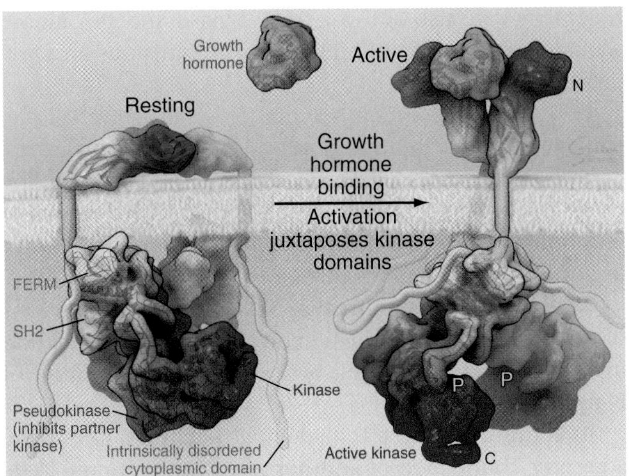

FIGURE 24.6 STRUCTURE AND ACTIVATION OF THE GROWTH HORMONE RECEPTOR. Resting dimeric receptors have a JAK2 kinase bound to each of the intrinsically disordered cytoplasmic tails near the inside of the plasma membrane. The pseudokinase domains interact with and inhibit the kinase domain of the partner JAK2. Growth hormone binding rearranges the two receptor molecules relative to each other and separates the JAK2 kinases so the kinase domains transactivate each other to initiate the intracellular signaling cascade. (For reference, see Brown RJ, Adams JJ, Pelekanos RA, et al. Model for growth hormone receptor activation based on subunit rotation within a receptor dimer. *Nat Struct Mol Biol*. 2005;12:814–821; and Brooks AJ, Dai W, O'Mara ML, et al. Mechanism of activation of protein kinase JAK2 by the growth hormone receptor. *Science*. 2014;344:1249783.)

also phosphorylate tyrosines in the receptor cytoplasmic tails as well as transcription factors called **STATs,** which migrate to the nucleus to regulate gene expression (see Fig. 27.9 for more on the pathways).

Receptor Serine/Threonine Kinases

A third class of growth factor receptors uses cytoplasmic serine/threonine kinase domains to transduce signals (Figs. 24.7 and 24.8). Dimeric protein ligands bring together two different types of receptor subunits to turn on kinase activity. Active receptors phosphorylate transcription factors called **SMADs** (Sma- and Mad-related proteins), stimulating their movement from cytoplasm into the nucleus, where they regulate genes controlling cellular proliferation and differentiation (see Fig. 27.10).

Forty different proteins bind human receptor serine/threonine kinases. These dimeric growth factors are particularly important during embryonic development. **Transforming growth factor-β (TGF-β)** both inhibits proliferation of most adult cells and also stimulates production of extracellular matrix, including collagen, proteoglycans, and adhesive glycoproteins (see Chapter 29). **Activin** was discovered as a releasing factor for pituitary follicle-stimulating hormone but also has a strong influence on the differentiation of early embryonic cells into primitive germ layers. A family of **bone morphogenetic proteins (BMPs)** influence the differentiation of osteoblasts, which lay down bone matrix, as well as many other cells.

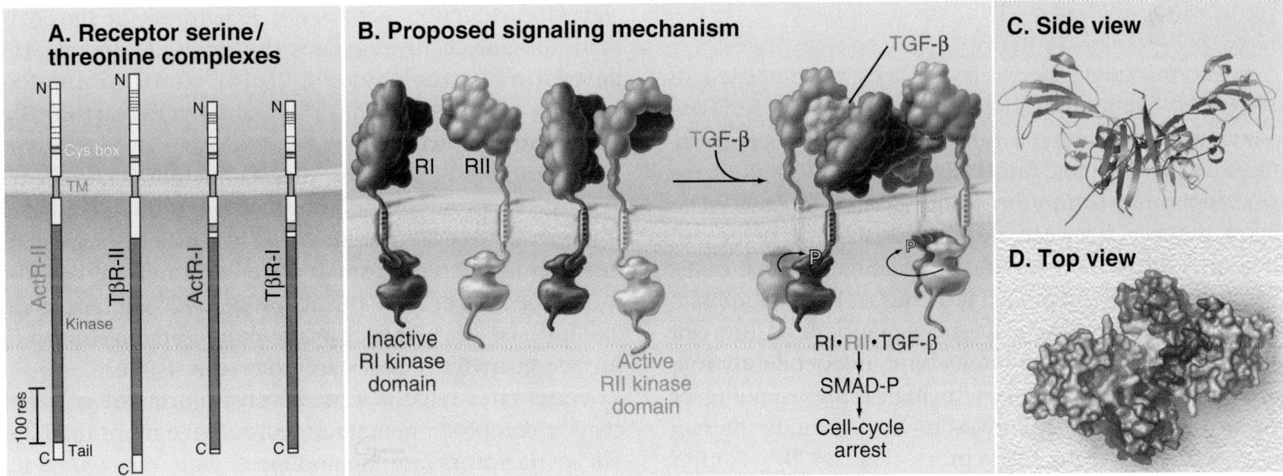

FIGURE 24.7 RECEPTOR SERINE/THREONINE KINASES. A, Drawing of domain architecture of activin (Act) and transforming growth factor-β (TGF-β) receptors. TM, transmembrane domain. **B,** Mechanism of activation of receptor serine/threonine kinases. Ligand binds two type II receptors (RII) and two type I receptors (RI). Within this hexameric complex, the type II receptor phosphorylates and activates the type I receptors, which, in turn, phosphorylate cytoplasmic transcription factors called SMADs (Sma- and Mad-related proteins). Phosphorylated SMADs move to the nucleus to activate particular genes. See Fig. 27.10 for details. **C–D,** Ribbon and space-filling models of bone morphogenetic protein 7 (BMP7) bound to type I and type II receptors. This model is based on crystal structures of dimeric BMP7 bound to two extracellular domains of the type II activin receptor and of BMP2 bound to two extracellular domains of BMP type I receptor. (For reference, see PDB files 1LX5 and 1S4Y and Greenwald J, Groppe J, Gray P, et al. The BMP7/ActRII extracellular domain complex provides new insights into the cooperative nature of receptor assembly. *Mol Cell*. 2003;11:605–617; and Shi Y, Massague J. Mechanisms of TGFβ signaling from cell membranes to the nucleus. *Cell*. 2003;113:685–700.)

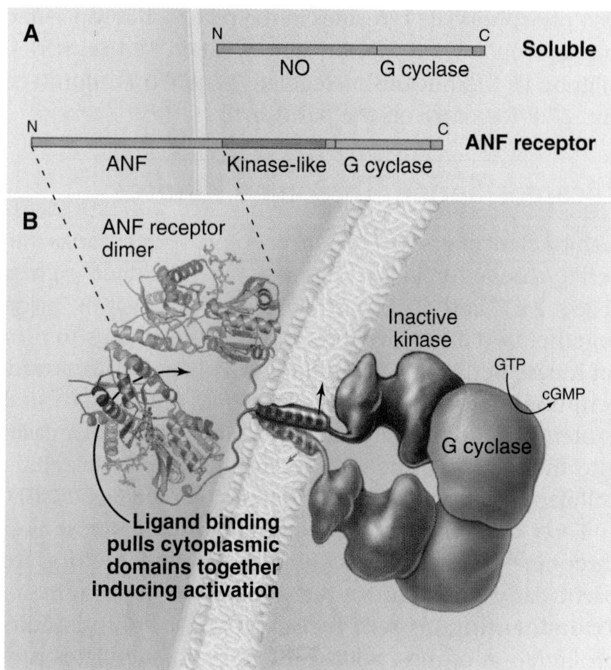

FIGURE 24.8 GUANYLYL CYCLASE RECEPTORS. A, Comparison of the domain architecture of the transmembrane atrial natriuretic factor (ANF) receptor and the cytoplasmic nitric oxide receptor. **B,** Ribbon model of the extracellular domains of a dimer ANF receptor with a model of the juxtaposed cytoplasmic domains. ANF binding causes tilting of the transmembrane domains, which aligns the active sites of the guanylyl cyclase domains and turns on enzyme activity. (For reference, see PDB file 1JDN and Ogawa H, Qiu Y, Ogata CM, et al. Crystal structure of hormone-bound atrial natriuretic peptide receptor extracellular domain: rotation mechanism for transmembrane signal transduction. *J Biol Chem*. 2004;279:28625–28631.)

Serine/threonine kinase receptors are composed of two types of subunits. Humans have genes for seven type I receptors and five type II receptors. Single transmembrane sequences link the small ligand-binding domains to cytoplasmic serine/threonine kinase domains. The linker residues between the transmembrane helix and the type I kinase domain inhibit its activity. Signal transduction requires the kinase activities of both subunits.

Receptors are present in small numbers on the cell surface. Four receptor subunits bind independently to a dimeric ligand, beginning with high-affinity binding of two type II receptors followed by lower-affinity interactions with two type I receptors (Fig. 24.7B). Within this complex, the constitutively active type II receptor kinases activate the type I receptors by phosphorylating serine and threonine residues in the inhibitory linker region.

The signal is propagated by type I receptor kinases phosphorylating cytoplasmic transcription factors called SMADs. Phosphorylated SMADs move to the nucleus, where they cooperate with other transcription factors to regulate gene expression (see Fig. 27.10). These receptors can also activate MAP kinase and PI3 kinase signaling pathways and phosphorylate tyrosines on a few proteins.

Loss of TGF-β receptors makes some tumors unresponsive to growth inhibition by TGF-β and contributes, in part, to their ability to replicate autonomously. On the other hand, inappropriate regulation of TGF-β signaling may contribute to overproduction of extracellular matrix in chronic inflammatory diseases and pathological fibrosis that scars diseased organs. Mice with null mutations of one of their three TGF-β genes die with inflammation in multiple organs caused by excessive proliferation of lymphocytes.

In addition to these transducing receptors, cells have a greater abundance of another plasma membrane TGF-β-binding protein (type III receptors) that lacks signal transduction activity. A single transmembrane sequence links the large extracellular proteoglycan domain to a small cytoplasmic domain. This receptor may concentrate TGF-β on the cell surface. Even in the absence of TGF-β, type II receptors phosphorylate type III receptors, creating a binding site for β-arrestin and promoting endocytosis of both type II and type III receptors.

Guanylyl Cyclase Receptors

Animals have a family of cell surface receptors (Fig. 24.8) with intracellular domains that catalyze the formation of 3′-5′-cyclic guanosine monophosphate (cGMP) from GTP. A second type of guanylyl cyclase is found in the cytoplasm and activated by the gases **nitric oxide** and carbon monoxide (see Fig. 26.1 for more details of these signal transduction pathways). Regardless of its enzymatic origin, cGMP regulates the same targets: **cGMP-gated ion channels** (see Fig. 16.10), **cGMP-stimulated protein kinases** (see Fig. 25.4), and **cyclic nucleotide phosphodiesterases** (see Fig. 26.1).

All known ligands for plasma membrane guanylyl cyclase receptors are peptides, although the ligands for some receptors are not known. Most insights regarding function have come from knowledge about the ligands, the tissue distribution of receptors, and receptor gene disruptions, as highly specific inhibitors of the cell-surface guanylyl cyclases are not yet available.

Vertebrates have at least seven isoforms of guanylyl cyclase receptors; nematode worms have more than 25. These receptors are homodimers with characteristic, ligand-binding extracellular domains, single transmembrane helices, and two cytoplasmic domains—an enzymatically inactive kinase domain and a guanylyl cyclase domain (Fig. 24.8). The cyclase domain is closely related to adenylyl cyclases with the active sites shared between two domains (see Fig. 26.2).

In the absence of ligand the cytoplasmic guanylyl cyclase domains are inactive. Ligand binding in the cleft between the extracellular domains causes a scissor-like motion that changes the relationship of cytoplasmic domains in a way that stimulates guanylyl cyclase activity.

Guanylyl cyclase receptor A (GC-A) binds **atrial natriuretic factor,** a polypeptide hormone that is secreted mainly by the heart to control blood pressure. It stimulates excretion of salt and water by the kidney and dilates blood vessels. Mice with null mutations for GC-A have high blood pressure and enlarged hearts and fail to respond when overloaded with fluid and salt administered intravenously. Intestinal guanylyl cyclase receptor C (GC-C) binds bacterial **enterotoxin,** the mediator of fluid secretion in bacterial dysentery. The bacterial toxin mimics endogenous intestinal peptides that regulate fluid secretion motility of the gut. Mice with null mutations for GC-C are completely resistant to enterotoxin without other apparent physiological defects. Guanylyl cyclase receptors E and F (GC-E, GC-F) are restricted to the eye; a null mutation for GC-E results in loss of cone visual receptor cells. Guanylyl cyclase receptor D (GC-D) is restricted to olfactory neuroepithelium. Sea urchin spermatozoa use a guanylyl cyclase receptor to respond to peptides secreted by eggs.

Tumor Necrosis Factor Receptor Family

TNF and its receptor are the prototypes for a diverse group of cell-signaling partners (Fig. 24.9) that regulate the expression of genes for a wide range of developmental and inflammatory processes as well as cell death (see Fig. 46.18). Lymphocytes produce three isoforms of TNF (also called lymphotoxin and cachectin), a trimeric lymphokine. TNF plays many roles in shock and inflammation, protection from bacterial infections, killing tumor cells, and wasting in chronic disease. In humans, the family includes 19 ligands and 29 related receptors.

TNF is synthesized as a transmembrane protein and released from the cell surface by proteolysis. Mice with a genetic deletion of the lymphotoxin-α (TNF) gene have no lymph nodes, so TNF participates in the development of the immune system. Other ligands for the TNF class of receptors are also trimers of subunits composed of β strands, such as the membrane-bound **Fas-ligand** on killer T cells (see Fig. 46.18). **Nerve growth factor** binds with low affinity to a member of this family in addition to the tyrosine receptor kinase A (TrkA; see Fig. 24.4).

Human cells express two types of TNF receptors that bind the same ligands but generate different responses. The two receptors have similar ligand-binding domains coupled by single transmembrane segments to different cytoplasmic domains. The extracellular part of these receptors consists of four similar repeats of approximately 40 amino acids, each stabilized by three disulfide bridges (Fig. 24.9B). In the absence of ligand, individual receptor subunits are presumed to diffuse independently in the plane of the membrane.

Three finger-like receptors grasp one trimeric TNF molecule by binding along the interfaces between TNF subunits. The tapered shape of TNF brings together the transmembrane segments and cytoplasmic domains of three receptors. Clustering three cytoplasmic domains allows an active TNF receptor to assemble a complex signaling platform of adapter proteins that

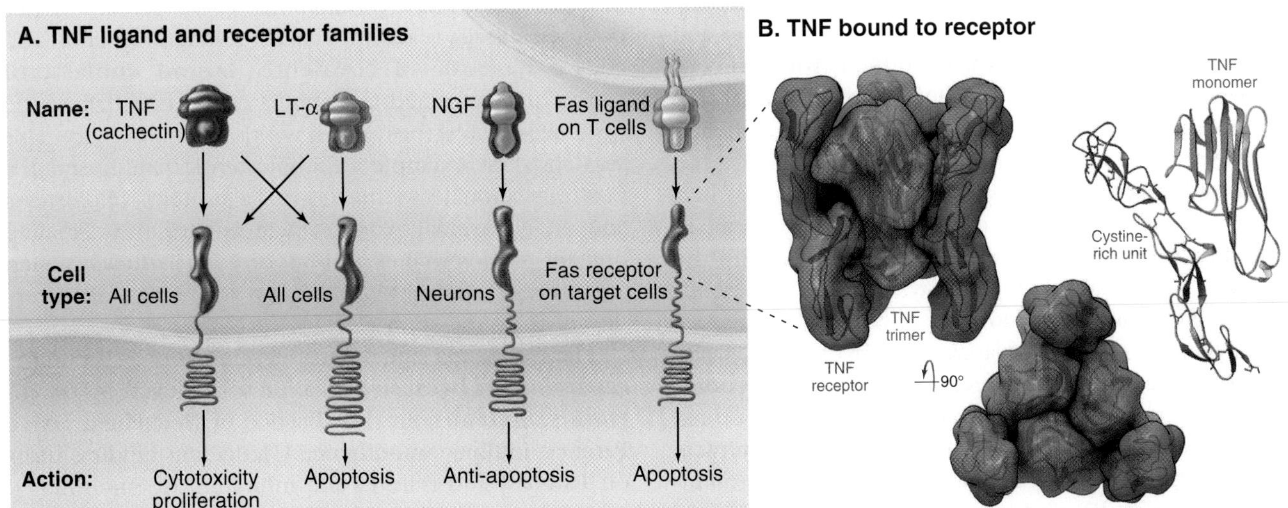

FIGURE 24.9 TUMOR NECROSIS FACTOR (TNF) RECEPTOR FAMILY. A, Domain architecture of a sample of members from the TNF receptor family. LT-α, lymphotoxin-α; NGF, nerve growth factor. **B,** Atomic structure of TNF bound to its receptor. TNF is a trimer of three identical β sandwich subunits arranged in a pear-like structure. The four extracellular cysteine-rich domains of the receptor grasp TNF like prongs. This clusters three cytoplasmic domains and allows the assembly of adapter proteins that transduce the signal. (**A,** Modified from Beutler B, van Huffel C. Unraveling function in the TNF ligand and receptor family. *Science*. 1994;264:667–668, copyright American Association for the Advancement of Science. **B,** For reference, see PDB file 1TNR and Banner DW, D'Arcy A, Janes W, et al. Crystal structure of the soluble human 55 kD TNF receptor-human TNF beta complex: implications for TNF receptor activation. *Cell*. 1993;73:431–445.)

become modified with polyubiquitin chains and ultimately lead to activation of the transcription factor nuclear factor κB (NF-κB; see Fig. 10.21). Active TNF receptors also turn on a plasma membrane phospholipase that hydrolyzes sphingomyelin, producing the second messenger **ceramide** (see Fig. 26.11).

TNF participates in the inflammation associated with autoimmune diseases such as rheumatoid arthritis. Intercepting TNF before it reaches its receptor is remarkably successful in blunting inflammation in these diseases. This is accomplished by injections of monoclonal antibodies to TNF or constructs containing just the extracellular domains of the TNF receptor.

Related receptors also bind to multimeric ligands leading to assembly of cytoplasmic signaling complexes. For example, Fas ligand triggers cell death by clustering Fas and assembling a signaling platform that activates a cascade of intracellular proteolysis (see Fig. 46.18).

Toll-Like Receptors

Metazoan organisms and plants use a family of receptors named Toll-like receptors (TLRs) to sense and respond to infection by a wide variety of microorganisms including viruses, bacteria, fungi, and protozoa. They mediate responses by regulating the expression of genes for inflammatory factors. Fig. 28.6 provides details on their structures and functions.

Notch Receptors

Components of the Delta/Notch signaling pathway were identified by analysis of mutations affecting early development in flies and nematodes. Ligands are transmembrane proteins called **Delta** in flies and vertebrates and LAG-2 in worms. These ligands and their **Notch receptors** regulate cellular fates during early embryonic development. Typically, cells expressing Delta interact with Notch receptors on adjacent cells to force the neighboring cells to choose a fate different from their own. The actual outcome depends on the context; in each tissue, Delta/Notch signals are integrated with the actions of other signaling pathways. As a general point, Delta/Notch signals tend to reinforce differences between cells in a particular tissue. For example, Delta on the earliest neurons directs adjacent cells to other fates. Defects in Delta or Notch result in excess neurons.

Delta is active as a cell surface protein that interacts locally with adjacent cells, but a matrix metalloproteinase (see Fig. 29.19) cleaves some Delta from the membrane, allowing it to act at a distance from its cell of origin. The receptors, called Notch (flies, vertebrates) or Lin-12 (worms), consist of 36 extracellular EGF-like domains and leucine-rich repeats, a single transmembrane span, and an intracellular region of ankyrin repeats that lacks enzyme activity. Notch is synthesized as a single polypeptide chain and is cleaved once near the base of the extracellular domain in the Golgi apparatus before transport to the plasma membrane. The two polypeptides remain associated through noncovalent interactions.

Cells carrying Delta activate Notch receptors on adjacent cells. This leads to proteolytic cleavage that frees the intracellular domains from the membrane. These cytoplasmic domains move into the nucleus and join a complex of proteins, including CSL, that activate transcription of certain genes.

Hedgehog Receptors

Genetic studies of *Drosophila* revealed a novel class of protein ligands, called **Hedgehog,** and two membrane proteins, called **Patched** and **Smoothened,** that are required for signal transduction during development. The Hedgehog receptor Patched consists of 12 transmembrane segments related to proton-driven bacterial antiporters, but the transported substrate, if any, is not known. Smoothened is an unusual seven-helix receptor that is constitutively active. Substoichiometric quantities of Patched inhibit the activity of Smoothened, perhaps by transporting out of the cell a ligand that binds and inhibits Smoothened. In flies, the Hedgehog pathway cooperates with the **Wnt** system (see Fig. 30.7) to establish boundaries between segments of the embryo and maintain a pool of stem cells.

Every aspect of this novel signaling pathway established new principles. A signal sequence guides Hedgehog protein into the secretory pathway, but before it reaches the cell surface, the protein cleaves itself in two pieces. The C-terminal half of the protein carries out the cleavage reaction. This autocatalytic reaction also adds a molecule of **covalently bound cholesterol** to the newly formed C-terminus of the first half of the protein, which is the domain with signaling activity. This was the first example of cholesterol being used for posttranslational modification of a protein. Cholesterol and an N-terminal palmitic acid anchor the signaling domain to membranes and lipoprotein particles, which are secreted and act on cells up to 30 cell diameters distant from the source.

The Hedgehog signal transduction pathway is complicated, in part because activation is achieved by *inactivating inhibitors.* In the absence of Hedgehog, active Patched inhibits Smoothened. Hedgehog binding turns off Patched and relieves the inhibition of Smoothened (the first inactivation of an inhibitor in this pathway). Active Smoothened assembles a complex of several proteins that inhibits the proteolytic inactivation of a transcription factor called Ci (the second inactivation of an inhibitor in this pathway). Active Ci controls the expression of several genes required for cell fate specification and differentiation including Patched itself.

Vertebrate orthologs of the proteins that form the insect Hedgehog pathway have similar functions, regulating cellular differentiation in many tissues, including formation of the neural tube. Mutations in one of three mammalian Hedgehog genes (sonic hedgehog) cause widespread developmental defects that range from mild to grotesque, including a single eye in the middle of the face. Mutations in the gene for Patched cause basal cell carcinoma of the skin, the most common cancer in fair-skinned people. Human Smoothened is a protooncogene; activating mutations prevent its inhibition by Patched and cause skin tumors. The Ci ortholog Gli1 is an oncogene that was originally discovered in brain tumors.

ACKNOWLEDGMENTS

We thank Dan Leahy for suggestions on revisions of this chapter.

SELECTED READINGS

Chillakuri CR, Sheppard D, Lea SM, Handford PA. Notch receptor-ligand binding and activation: insights from molecular studies. *Semin Cell Dev Biol*. 2012;23:421-428.

Derbyshire ER, Marletta MA. Structure and regulation of soluble guanylate cyclase. *Annu Rev Biochem*. 2012;81:533-559.

Ingham PW, Nakano Y, Seger C. Mechanisms and functions of Hedgehog signalling across the metazoa. *Nat Rev Genet*. 2011;12: 393-406.

Kang JS, Liu C, Derynck R. New regulatory mechanisms of TGF-beta receptor function. *Trends Cell Biol*. 2009;19:385-394.

Kovacs E, Zorn JA, Huang Y, et al. A structural perspective on the regulation of the epidermal growth factor receptor. *Annu Rev Biochem*. 2015;84:739-764.

Krzysztof P. G protein-coupled receptor rhodopsin. *Annu Rev Biochem*. 2006;75:743-767.

Lefkowitz RJ, Whalen EJ. Beta-arrestins: Traffic cops of cell signaling. *Curr Opin Cell Biol*. 2004;16:162-168.

Lemmon MA, Schlessinger J. Cell signaling by receptor tyrosine kinases. *Cell*. 2010;141:1117-1134.

Massagué J. TGFβ signalling in context. *Nat Rev Mol Cell Biol*. 2012; 13:616-630.

Misono KS, Philo JS, Arakawa T, et al. Structure, signaling mechanism and regulation of the natriuretic peptide receptor guanylate cyclase. *FEBS J*. 2011;278:1818-1829.

Park PS, Filipek S, Wells JW, et al. Oligomerization of G protein-coupled receptors: past, present, and future. *Biochemistry*. 2004;43:15643-15656.

Schoneberg T, Schulz A, Biebermann H, et al. Mutant G-protein-coupled receptors as a cause of human diseases. *Pharmacol Ther*. 2004; 104:173-206.

Thompson MD, Hendy GN, Percy ME, et al. G protein-coupled receptor mutations and human genetic disease. *Methods Mol Biol*. 2014; 1175:153-187.

Venkatakrishnan AJ, Deupi X, Lebon G, et al. Molecular signatures of G-protein-coupled receptors. *Nature*. 2013;494:185-194.

Waters MJ. The growth hormone receptor. *Growth Horm IGF Res*. 2016;28:6-10.

Zhang G. Tumor necrosis factor family ligand-receptor binding. *Curr Opin Struct Biol*. 2004;14:154-160.

Zhang X, Gureasko J, Shen K, et al. An allosteric mechanism for activation of the kinase domain of epidermal growth factor receptor. *Cell*. 2006;125:1137-1149.

APPENDIX 24.1

Receptors and Ligands

Classes of Receptors	Nature of Activation	Examples of Biological Function
Activators		
Voltage-Gated Ion Channels → Membrane Depolarization or Repolarization		
Voltage-gated potassium channel	Electrical	Membrane repolarization
Voltage gated sodium channel	Electrical	Action potential
Ligand-Gated Ion Channels → Changes in Membrane Permeability		
Acetylcholine (nicotinic)	Biogenic amine	Action potential
Adenosine triphosphate	Nucleotide	Change in membrane potential
Glutamate (*N*-methyl-D-aspartate)	Amino acid	Change in membrane potential
Glutamate (non-*N*-methyl-D-aspartate)	Amino acid	Change in membrane potential
Glycine	Amino acid	Change in membrane potential
Serotonin	Biogenic amine	Change in membrane potential
Seven-Helix Receptors → Trimeric G Proteins → Diverse Responses		
Acetylcholine (muscarinic)	Biogenic amine	Slows heart; stimulates intestinal secretion
Adrenocorticotropic hormone	Peptide	Stimulates adrenal cortisol production
Adenosine	Nucleoside	Dilates blood vessels
Angiotensin II	Peptide	Stimulates aldosterone secretion; contracts smooth muscle
Bradykinin	Protein	Stimulates intestinal secretion
Calcitonin	Protein	Inhibits calcium resorption from bone
Cholecystokinin	Peptide	Stimulates intestinal secretion

Continued

APPENDIX 24.1

Receptors and Ligands—cont'd

Classes of Receptors	Nature of Activation	Examples of Biological Function
Complement (C5a, C3a)	Protein	Leukocyte chemoattractant
Dopamine	Biogenic amine	Neurotransmitter; inhibits prolactin secretion
Eicosanoids (prostaglandins)	Lipid	Promote or inhibit platelet aggregation, many other actions
Endothelins	Protein	Vasoconstriction
Epinephrine	Biogenic amine	Glycogenolysis; increases cardiac contractility
F-met-leu-phe	Peptide	Leukocyte chemotaxis
Follicle-stimulating hormone	Protein	Growth of ovarian follicle
γ-Aminobutyric acid	Amino acid	Inhibitory neurotransmitter; stimulates intestinal secretion
Glucagon	Peptide	Glycogenolysis; stimulates intestinal secretion
Glutamate	Amino acid	Modulates synaptic transmission
Growth hormone–releasing factor	Peptide	Stimulates secretion of growth hormone
Histamine	Amino acid	Allergic responses; vasodilation; stimulates secretion
IL-8	Protein	Chemotaxis of leukocytes
Luteinizing hormone	Protein	Steroid production by ovarian granulosa cells
Light absorption by rhodopsin	Photon	Vision
Lysophosphatidic acid	Lipid	Fibroblast proliferation, neurite retraction
Melanocyte-stimulating hormone	Protein	Melanin synthesis
Neurokinins (substance P)	Peptide	Stimulates gastrointestinal and pancreatic secretion; neurotransmitter
Norepinephrine	Biogenic amine	Smooth muscle relaxation
Odorants	Organics	Olfaction
Opioids	Alkaloids	Alter mood, deaden pain, inhibit intestinal motility
Oxytocin	Peptide	Contraction of uterus
Parathyroid hormone	Protein	Bone calcium resorption
Peptide-releasing factors	Protein	Secretion of pituitary hormones
Platelet-activating factor	Lipid	Platelet activation
Serotonin	Biogenic amine	Stimulates intestinal secretion
Somatostatin	Peptide	Inhibits secretion of growth hormone, insulin, and glucagon
Thrombin	Protein	Activates platelets
Thyroid-stimulating hormone	Protein	Thyroid hormone secretion
Vasoactive intestinal peptide	Peptide	Stimulates intestinal secretion
Vasopressin	Peptide	Regulates the permeability of the renal tubule to water
Wingless (Wnt)	Protein	Modulates gene expression
Two-Component Systems: Receptor/Histidine Kinase → Response Regulator → Diverse Responses		
Aspartate	Amino acid	Controls flagellar motor and chemotaxis
Osmotic pressure	Physical	Regulates gene expression
Receptor Tyrosine Kinase → Ras, MAP Kinase, PLC, PI3 Kinase → Alter Gene Expression		
Epidermal growth factor	Protein	Epithelial cell proliferation and differentiation
Fibroblast growth factor-α	Protein	Mesoderm differentiation; fibroblast mitogen
Fibroblast growth factor-β	Protein	Fibroblast mitogen
Hepatocyte growth factor (scatter factor)	Protein	Epithelial cell mitogenesis, motility
Insulin	Protein	Glucose uptake; cell growth
Insulin-like growth factor I	Protein	General body growth
Macrophage colony-stimulating factor	Protein	Growth and differentiation of monocytes
Neurotrophins (nerve growth)	Protein	Neural growth; neuron survival
Platelet-derived growth factor	Protein	Smooth muscle, fibroblast, glial growth and differentiation
Steel ligand	Protein	Development of melanocytes, germ cells
Transforming growth factor-α	Protein	Differentiation of connective tissue
Vascular endothelial cell growth factor	Protein	Endothelial cell growth

APPENDIX 24.1

Receptors and Ligands—cont'd

Classes of Receptors	Nature of Activation	Examples of Biological Function
Cytokine Receptors → JAK Kinase → STAT Transcription Factors → Gene Expression		
Ciliary neurotrophic factor	Protein	Survival/differentiation of neurons and glial cells
Erythropoietin	Protein	Growth and differentiation of red cell precursors
Granulocyte colony-stimulating factor	Protein	Growth and differentiation of granulocyte precursors
Granulocyte-monocyte colony-stimulating factor	Protein	Growth and differentiation of leukocyte precursors
Growth hormone	Protein	Cell growth and differentiation of somatic cells
IL-2	Protein	Growth factor for lymphocytes
IL-3	Protein	Growth factor for hematopoietic stem cells
IL-4	Protein	Regulates gene expression
IL-5	Protein	Regulates gene expression
IL-6	Protein	Regulates gene expression
Interferon α/β	Protein	Regulates gene expression
Interferon γ	Protein	Macrophage and lymphocyte gene expression
Prolactin	Protein	Stimulates milk synthesis
Tyrosine Kinase–Linked Receptors → Cytoplasmic Tyrosine Kinase → Gene Expression		
MHC-peptide complex→T-cell receptor	Protein	Growth and differentiation of T lymphocytes
Antigens→B-cell receptor	Various	Growth and differentiation of B lymphocytes
Receptor Serine/Threonine Kinase → SMAD Transcription Factors → Control of Gene Expression		
Activin	Peptide	Mesoderm development
Bone morphogenetic protein	Protein	Mesoderm development
Inhibins	Protein	Inhibition of gonadal stromal mitogenesis
Transforming growth factor-β	Protein	Growth arrest, mesoderm development
Membrane Guanylyl Cyclase Receptors → cGMP → Regulation of Kinases and Channels		
Atrial natriuretic peptide	Peptide	Vasodilation; sodium excretion; intestinal secretion
Heat-stable endotoxin, guanylin		Unknown
Sea urchin egg peptides	Peptide	Fertilization
Sphingomyelinase-Linked Receptors → Ceramide-Activated Kinases → Gene Expression		
IL-1	Protein	Inflammation, wound healing
Tumor necrosis factor	Protein	Inflammation, tumor cell death
Integrins → Nonreceptor Tyrosine Kinases → Diverse Responses		
Fibronectin, other matrix proteins	Protein	Cell motility, gene expression
Selectins		
Mucins	Glycoproteins	Cell adhesion
Cadherins		
Like cadherins on another cell	Protein	Contact inhibition
Notch		
Delta	Cell-surface protein	Cell fate determination
Cytoplasmic Guanylyl Cyclase Receptors → cGMP → Kinases, cGMP-Gated Channels		
Nitric oxide	Gas	Smooth muscle relaxation
Cytoplasmic Steroid Receptors → Active Transcription Factor → Gene Expression		
Retinoic acid	Organic	Cell growth and differentiation
Steroid hormones	Steroids	Cell growth and differentiation
Thyroid hormone	Amino acid	Cell growth and differentiation

cGMP, cyclic guanosine monophosphate; IL, interleukin; JAK, "just another kinase," later renamed Janus kinase; MAP, mitogen-activated protein; MHC, major histocompatibility complex; PI3, phosphatidylinositol 3; PLC, phospholipase C; SMAD, Sma- and Mad-related protein; STAT, signal transducer and activator of transcription.

CHAPTER 25

Protein Hardware for Signaling

This chapter introduces proteins that transduce signals in the cytoplasm: protein kinases, protein phosphatases, guanosine triphosphatases (GTPases), and adapter proteins. Remarkably, kinases and GTPases use the same strategy to operate molecular switches that carry information through signaling pathways: the simple covalent addition and removal of inorganic phosphate. Protein kinases add phosphate groups to specific protein targets, and phosphatases remove them. GTPases bind guanosine triphosphate (GTP) and hydrolyze it to guanosine diphosphate (GDP) and inorganic phosphate (P_i), which dissociates. In both cases, the presence or absence of a single phosphate group switches a protein between active and inactive conformations.

Addition of phosphate is reversible, so both types of switches can be used as molecular timers that cycle on and off at tempos determined by the intrinsic properties of the switch and its environment. GTPases are active with bound GTP and switch off when they hydrolyze GTP to GDP. Similarly, phosphorylation activates many proteins but can inhibit others.

These molecular switches are often linked in series to form a **signaling cascade** that can both transmit and refine signals. Enzymes along signaling pathways (including kinases) often act as amplifiers. Turning on the binary switch of one enzyme molecule can produce many product molecules, each of which, in turn, may continue to propagate and amplify the original signal by activating downstream molecules. Many pathways include negative feedback loops. Few signaling pathways are linear; instead, most branch and intersect, allowing cells to integrate information from multiple receptors and to control multiple effector systems simultaneously. Chapter 27 illustrates the functions of molecular switches in several signaling pathways.

Protein Phosphorylation

Phosphorylation is an extremely common posttranslational modification of proteins; it regulates the activity of one or more proteins along most signaling pathways. Among other things, phosphorylation and dephosphorylation control metabolic enzymes, cell motility, membrane channels, assembly of the nucleus, and cell-cycle progression. Phosphorylation turns some processes on and others off. In either case, both the addition of a phosphate by a **protein kinase** and its removal by a **protein phosphatase** are required to achieve regulation.

For historical and practical reasons, it has been easier to study protein kinases than protein phosphatases, so most research and accounts of regulation by phosphorylation emphasize kinases (witness the 522,950 PubMed hits for "protein kinase" compared with 72,899 for "protein phosphatase" in January 2016). Readers should understand that both directions are important on this two-way street.

In eukaryotes, more than 99% of protein phosphorylation occurs on **serine** and **threonine** residues, but phosphorylation of **tyrosine** residues regulates many processes in animals (Fig. 25.1). Bacteria and Archaea use **histidine** and **aspartate** phosphorylation for signaling (see Fig. 27.11), but these modifications are little known in eukaryotes. Phosphohistidine and phosphoaspartate are more difficult to assay than are other phosphorylated residues, so pathways using these phosphoamino acids might have escaped detection. Less appreciated is that many serines and threonines (including some modified by phosphorylation) are reversibly modified by *O*-linked β-*N*-acetylglucosamine, which is much larger than phosphate. The numerous effects of this modification are slowly emerging.

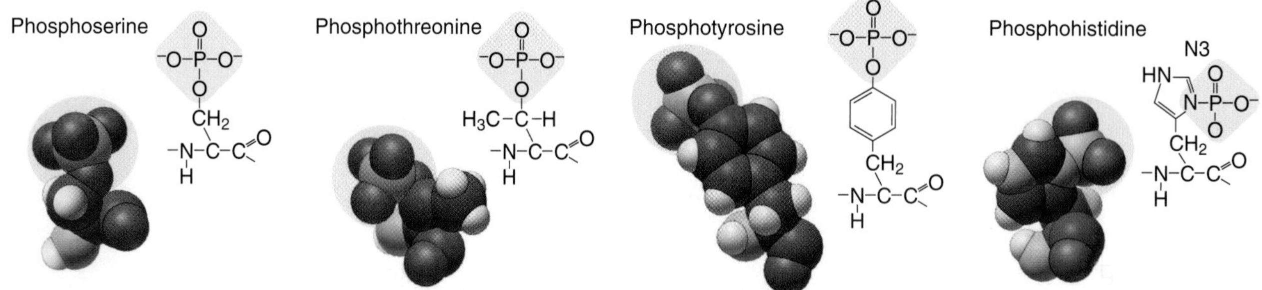

FIGURE 25.1 STRUCTURES OF PHOSPHOAMINO ACIDS. In addition to the N1 nitrogen illustrated, histidine may be phosphorylated on N3.

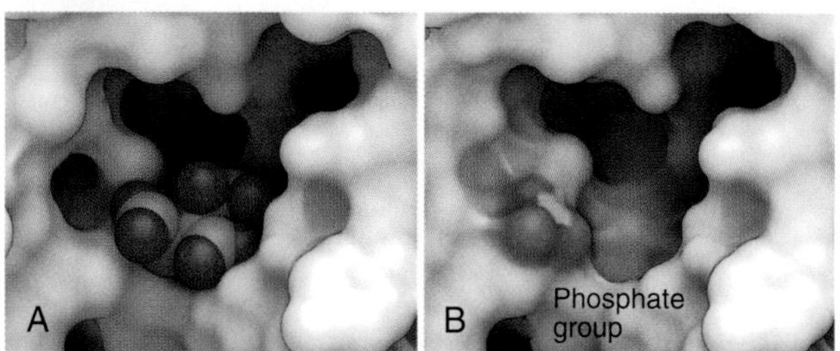

FIGURE 25.2 PHOSPHORYLATION BLOCKS SUBSTRATE BINDING TO ISOCITRATE DEHYDROGENASE. A, Surface representation with isocitrate *(red/gray)* bound to the active site. **B,** Phosphorylation of serine 113 *(yellow)* blocks isocitrate binding. (For more information, see Protein Data Bank [PDB; www.rcsb.org] files 3ICD and 4ICD.)

Effects of Phosphorylation on Protein Structure and Function

Despite its small size, phosphate is well suited to regulate the activity of proteins. The addition of a phosphate group with two negative charges to a single amino acid can alter the activity of a protein in several ways:

- *Direct interference.* Addition of a phosphate group to the active site of the metabolic enzyme isocitrate dehydrogenase blocks substrate binding (Fig. 25.2) by both direct steric hindrance and electrostatic repulsion of negatively charged substrates. Phosphorylation also can directly block protein interactions, such as the polymerization of intermediate filaments (see Fig. 44.6) and binding of cofilin to actin monomers and filaments (see Fig. 33.16). Electrostatic repulsion caused by phosphorylation regulates interactions between kinetochores and microtubules in mitosis.
- *Conformational change.* A phosphate group can participate in hydrogen bonds and electrostatic interactions distinct from those of the hydroxyl group that it replaces on an amino acid side chain. These interactions of phosphorylated residues may change the conformation of the protein. For example, phosphorylation activates the insulin receptor tyrosine kinases by inducing a dramatic change in a polypeptide loop in the active site (Fig. 25.3E). Phosphorylation of a single serine activates the metabolic enzyme glycogen phosphorylase by stabilizing a compact, active conformation.
- *Creation of binding sites.* Phosphorylated residues participate in many protein interactions (see Fig. 25.10A). Phosphorylated tyrosines act as MARKS recognized by READERS, such as SH2 (Src homology 2) domains and phosphotyrosine-binding (PTB) domains to regulate protein-protein interactions. A phosphoserine is required on protein ligands for binding 14-3-3 domains, some WW domains, and some FHA domains. Phosphotyrosine and SH2 domains were the first MARK/READER pair to be identified (see Chapter 8).

Protein Kinases

Protein kinases catalyze the transfer of the γ-phosphate from adenosine triphosphate (ATP) to amino acid side chains of proteins. Eukaryotes have numerous kinase genes: 116 in budding yeast (second to transcription factor genes), 409 in nematode worms (second only to seven-helix receptor genes), and 544 in humans. Most protein kinases in eukaryotes are either **serine/threonine kinases** or **tyrosine kinases** (Appendix 25.1), but a few kinases phosphorylate all three amino acids. Some of these kinases (55 in humans) lack residues normally required for kinase activity. Some pseudokinases have residual enzyme activity, but usually they regulate active kinases.

Most serine/threonine and tyrosine kinases had a common evolutionary origin and share similar structures and catalytic mechanisms, despite extensive sequence divergence and differences in substrate specificity. Tyrosine kinases emerged in animals. Nevertheless, fungi have phosphotyrosine owing to two families of serine/threonine kinases that also phosphorylate tyrosine. Three protein tyrosine phosphatases reverse these reactions. A family of 40 "atypical" protein kinases had a separate origin from the major family. **Lipid kinases** have a catalytic domain related to typical protein kinases. They phosphorylate inositol phospholipids (see Fig. 26.7) or a few proteins.

The catalytic domain of eukaryotic protein kinases consists of approximately 260 residues in two lobes

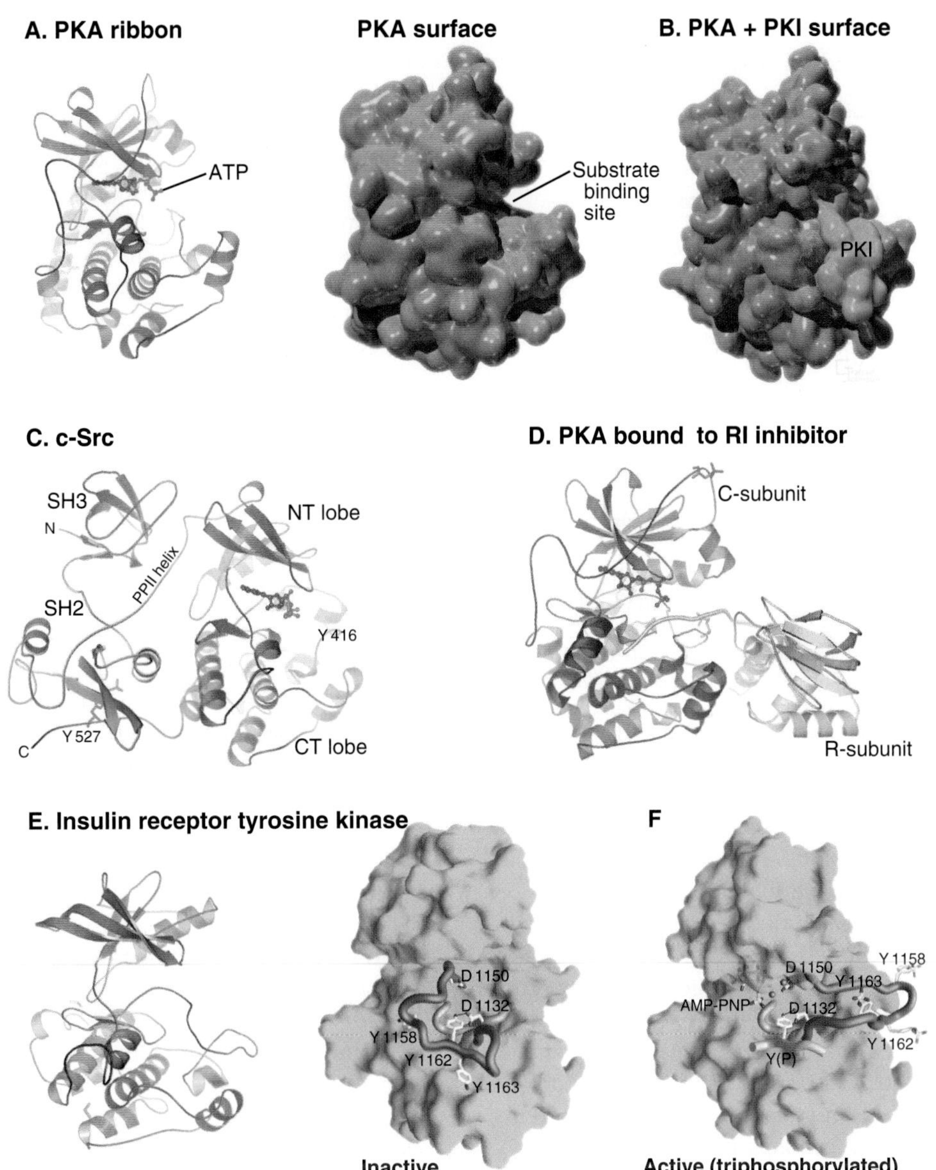

FIGURE 25.3 PROTEIN KINASE STRUCTURES. A, Ribbon diagram and space-filling model of cyclic adenosine monophosphate (cAMP)-dependent protein kinase with a nonhydrolyzable adenosine triphosphate (ATP) analog *(red)* bound to the active site. The adenine base of the ATP fits into a hydrophobic cleft formed by β-sheets lining the interface of the two lobes. The phosphates bind to conserved residues in loops connecting the β-strands. (See PDB file 2CPK.) **B,** Space-filling model of protein kinase A (PKA) with bound protein kinase inhibitor (PKI). The location of this inhibitory peptide revealed the binding site for protein substrates. (See PDB file 1FMO.) **C,** Ribbon diagram of c-Src. When tyrosine-527 is phosphorylated, the SH2 (Src homology 2) domain binds intramolecularly to the C-terminus, locking the kinase in an inactive conformation. The N-terminal SH3 (Src homology 3) domain binds intramolecularly to a proline-rich sequence (PPII helix) connecting the SH2 and kinase domains. NT and CT are the N- and C-terminal lobes of the kinase domain. **D,** Ribbon diagram of PKA bound to the RIα regulatory subunit. The pseudo substrate peptide *(yellow)* sits in the active site. Binding of cAMP to two sites on the RIα subunit causes a conformational change that dissociates RIα from the catalytic subunit. (See PDB file 3FHI.) **E,** Insulin receptor tyrosine kinase. Ribbon diagram and space-filling model with the catalytic loop in *orange* and the activation loop in *green.* (See PDB file 1IRK.) **F,** Space-filling model of insulin receptor tyrosine kinase triphosphorylated on the activation loop. This rearranges the activation loop, allowing substrates *(pink with a white tyrosine side chain)* access to the active site. AMP-PNP is a nonhydrolyzable analog of ATP with nitrogen bridging the β- and γ-phosphates. (See PDB file 1IR3.) (**E–F,** Space-filling models courtesy Steven Hubbard, New York University, New York.)

surrounding the ATP-binding pocket (Fig. 25.3). Each kinase phosphorylates a restricted range of protein substrates that bind to a groove on the surface of the kinase with the side chain of the residue to be phosphorylated positioned in the active site (Fig. 25.3B). Typically, all substrates that bind a particular kinase have similar residues surrounding the target serine, threonine, or tyrosine (a **consensus target sequence**). For example, the consensus sequence for protein kinase A (PKA) is Arg-Arg-Gly-**Ser/Thr**-Ile. The arginines and isoleucine

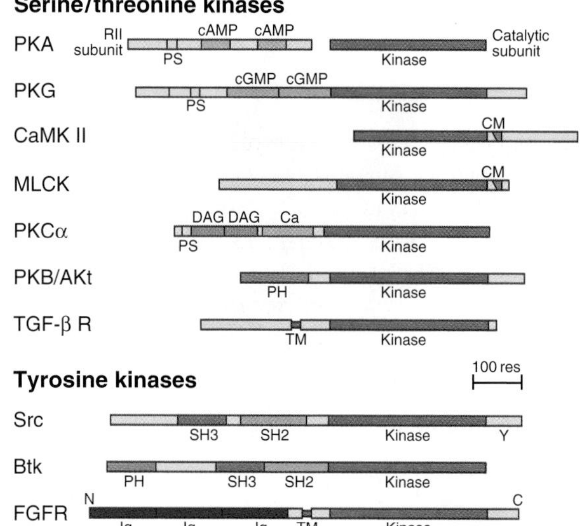

FIGURE 25.4 PROTEIN KINASE DOMAIN ARCHITECTURE. See Appendix 25.1 for definitions of the kinase names. Btk, Bruton tyrosine kinase; Ca, calcium-binding site; cAMP and cGMP, cyclic nucleotide–binding sites; CM, overlapping pseudosubstrate/calmodulin-binding site; DAG, diacylglycerol-binding site; FGFR, fibroblast growth factor receptor; Ig, immunoglobulin domains; PH, pleckstrin homology domain; PS, pseudosubstrate sequences; SH2 and SH3, Src homology domains; TM, transmembrane domain.

flanking the target serine or threonine residue specify binding to PKA. Interactions outside the catalytic site may also contribute to specific binding.

In addition to the catalytic domain, most protein kinases have other domains for regulation or localization (Fig. 25.4). Single transmembrane helices anchor receptor kinases to membranes. Many receptor tyrosine kinases have tyrosine residues inserted within the kinase domain and at the C-terminus. Phosphorylation of these tyrosines creates binding sites for effector proteins with SH2 (Src homology 2) domains (see Figs. 27.6, 27.7, and 27.8).

Prokaryotes generally lack serine/threonine/tyrosine kinases but use a large family of **histidine kinases** for signal transduction (see Fig. 27.11). These prokaryotic kinases differ in structure, mechanism and evolutionary origin from eukaryotic kinases. A few bacteria have acquired eukaryotic kinases by lateral transfer of genes.

Regulation of Protein Kinases

Each kinase has its own regulatory mechanism, but most involve multiple strategies including phosphorylation and interactions with intrinsic peptides or extrinsic proteins.

Phosphorylation

Phosphorylation by the same type of kinase or another kinase can either activate or inhibit protein kinases. When linked in series, different types of kinases form signaling cascades that can amplify and sharpen the response to a stimulus (Fig. 27.5):

- *Activation by phosphorylation.* This is the most common way to regulate kinases. For example, phosphorylation of three tyrosines on an **activation loop** activates the insulin receptor kinase. Phosphorylation refolds the activation loop, allowing substrates access to the active site and bringing together the residues required for catalysis (Fig. 25.3D). Activation loop phosphorylation also turns on other receptor tyrosine kinases, Src-family tyrosine kinases (Fig. 25.3C; see also Box 27.5), mitogen-activated protein (MAP) kinases (Fig. 27.5), cyclin-dependent kinases (see Fig. 40.14), and calcium-calmodulin–dependent kinases (Fig. 25.4).
- *Inhibition by phosphorylation.* Phosphorylation of myosin light-chain kinase by PKA reduces its affinity for its protein substrate, and phosphorylation of platelet-derived growth factor receptor tyrosine kinase by protein kinase C (PKC) inhibits its activity. Phosphorylation of a C-terminal tyrosine inhibits Src-family tyrosine kinases by creating an intramolecular binding site for an SH2 domain at the N-terminus (Fig. 25.3C). This interaction traps the kinase in an inactive conformation. Phosphorylation also inhibits cyclin-dependent cell-cycle kinases by blocking ATP binding (see Fig. 40.14).

Regulation of Substrate Binding

Peptides that are intrinsic to the kinase or part of a separate protein can inhibit kinases by competing with protein substrates for binding to the enzyme (Figs. 25.3B and D and 25.4):

- *Extrinsic regulation by inhibitory subunits.* Separate **regulatory (R) subunits** inhibit PKA by blocking the protein substrate site with a **pseudosubstrate** (Figs. 25.3D and 25.4). Pseudosubstrates have consensus target sequences lacking the phosphorylated residue. For example, RI pseudosubstrate has the sequence Arg-Arg-Gly-**Ala**-Ile, which binds in the substrate groove but is not phosphorylated. The RII pseudosubstrate has a serine, which is phosphorylated by PKA but then does not dissociate from the catalytic subunit like phosphorylated substrates. Cyclic adenosine monophosphate **(cAMP)** regulates the affinity of these regulatory subunits for the PKA catalytic subunit. In resting cells, the regulatory subunit is free of cAMP and binds the catalytic subunit with high affinity. With a rise in cAMP concentration (see Fig. 26.3), cAMP binds the regulatory subunit, dissociates it from the catalytic subunit, and allows substrates access to the active site. Regulatory proteins inhibit cyclin-dependent kinases by blocking access of ATP or cyclin (see Fig. 40.14).
- *Autoinhibition.* Many kinases have an intrinsic **pseudosubstrate** sequence (Fig. 25.4) that binds intramolecularly to the active site, autoinhibiting the enzyme (Fig. 25.3B). **Ca^{2+}-calmodulin** activates myosin

light-chain kinase and calmodulin-activated kinase (CaMK) by displacing the pseudosubstrate from the active site. Cyclic guanosine monophosphate (cGMP) binding to protein kinase G (PKG) displaces the autoinhibitory peptide from the catalytic domain, activating the enzyme.

- *Extrinsic regulation by activating proteins.* Regulatory proteins activate some protein kinases. For example, **cyclins** activate cyclin-dependent cell-cycle kinases **(Cdks)** (see Fig. 40.14).

Targeting

Several mechanisms target kinases to specific cellular locations, bringing them close to their substrates. Targeting helps explain how kinases with broad specificity can have specific effects in particular cells:

- The intracellular location of PKA is determined by both its RI and RII subunits and a family of **A kinase–anchoring proteins (AKAPs).** When the cAMP concentration is low, regulatory subunits bind and inhibit PKA. RII subunits also bind to AKAPs, which target the inhibited PKA catalytic subunit to cellular locations, including centrosomes, actin filaments, microtubules, endoplasmic reticulum, peroxisomes, mitochondria, or plasma membrane. An increase in cytoplasmic cAMP releases active PKA in close proximity to particular substrates. Once freed from RI or RII subunits by cAMP, the active PKA catalytic subunit can migrate into the nucleus, where it encounters a different array of substrates and regulates gene transcription (see Fig. 26.3F). The inhibitory protein PKI (protein kinase inhibitor) (Fig. 25.3B) is capable of capturing PKA in the nucleus and targeting it for transport back to the cytoplasm. Some AKAPs bind other protein kinases, such as PKC and phosphatases (PP2B). Similar to AKAPs, a three-protein module, including INCENP, survivin, and borealin, binds and targets aurora-B kinase to centromeres early in mitosis and the central spindle and cleavage furrow late in mitosis (see Fig. 44.10).
- **Pleckstrin homology (PH)** domains (Fig. 25.10A) and lipid tags target some kinases to lipid bilayers. A PH domain directs **PKB/Akt** to membrane polyphosphoinositides. This interaction with lipids opens up sites on the catalytic domain for phosphorylation and activation by phosphoinositide-dependent protein kinase 1 (PDK1), another kinase with a PH domain. An N-terminal myristic acid anchors Src tyrosine kinase to the plasma membrane (see Fig. 7.10).
- Phosphorylation induces the MAP kinase extracellular signal-regulated kinase 2 (ERK2) to form homodimers, triggering movement of the dimer into the nucleus, where it regulates gene expression (see Fig. 27.6).
- A scaffolding protein called STE5, first identified in yeast, brings together three protein kinases that form part of the cascade of kinases that activate MAP kinases (see Fig. 27.5).

Kinases and Disease

Unregulated kinases—for example, the receptor tyrosine kinase RET in endocrine cancers and cyclin-dependent kinase 4 (Cdk4) in melanoma—predispose individuals to cancer. Most patients with chronic myelogenous leukemia have a gene rearrangement that produces a fusion of the protein bcr and c-abl, a nonreceptor tyrosine kinase. The constitutively active fusion protein promotes the transformation of white blood cell precursors into cancer cells. The most effective treatment is with a drug called imatinib mesylate (Gleevec), which inhibits bcr-abl kinase activity. Additional drugs are being developed to inhibit other hyperactive protein kinases that drive the proliferation of other cancer cells. Their therapeutic potential is great, but cancer cells acquire mutations in the target kinase that readily prevent binding of the drugs and result in relapse of the tumor.

Protein Phosphatases

Eukaryotes have several families of protein phosphatases that remove phosphate from amino acid side chains (Table 25.1 and Fig. 25.5). Like protein kinases, most protein phosphatases are active toward either phosphoserine-threonine or phosphotyrosine, although several dual-specificity phosphatases can dephosphorylate all three residues. Each of the 90 human tyrosine phosphatases is thought to act on a limited number of substrates. The 20 human serine-threonine phosphatases achieve specificity by associating with an array of accessory subunits, which regulate enzyme activity and target catalytic subunits to particular substrates. Domains flanking the catalytic domains may also regulate enzyme activity (Fig. 25.6). For example, phosphorylation regulates binding of some phosphatases to their targeting subunits.

PPP Family of Serine/Threonine Phosphates

Members of the PPP family of **serine/threonine phosphatases** are found in Bacteria, Archaea, and all tissues of eukaryotes. All three PPP subfamilies (Table 25.1) share the same catalytic fold with a two-metal ion cluster (Fe^{2+} and Zn^{2+} in vivo) in the active site (Fig. 25.5A). **PP1** and **PP2A** are two of the most evolutionarily conserved enzymes.

Diverse regulatory subunits determine the substrates for PP1 and PP2A by targeting catalytic subunits to specific sites in the cell (Table 25.1). For example, more than 50 proteins target a 38-kD catalytic subunit of PP1 to specific substrates. M subunits not only target PP1 to smooth muscle myosin-II but also create an active site specific for the regulatory light chains relative to other substrates. Dephosphorylation of light chains relaxes smooth muscle (see Fig. 39.24). G subunits regulate

TABLE 25.1 Protein Phosphatases

Catalytic Subunit	Regulatory Elements	Inhibitors	Regulated Functions
Serine/Threonine Phosphatases			
PPP Family			
PP1C subfamily	>50 regulatory subunits target and regulate catalytic subunit	Okadaic acid Microcystin	Glycogen metabolism, muscle contraction, cell cycle, mRNA splicing
PP2A subfamily = catalytic, A and B subunits	B subunits target and regulate core enzyme	Okadaic acid Microcystin	MAP kinase pathway, metabolism, cell cycle
PP2B (calcineurin) = catalytic A subunit + B subunit	Calcium-calmodulin activates by binding autoinhibitory peptide	Cyclosporine-cyclophilin FK506 (tacrolimus)-FKBP	T-lymphocyte activation, brain NMDA receptor signaling
PPM Family			
PP2C	Integral N- or C-terminal peptides	Unsaturated fatty acids	Antagonism of stress-activated kinases
Protein Tyrosine Phosphatases			
PTP Family			
Cytosolic PTPs (PTP1B, SHP1, SHP2)	SH2 and other domains target to substrates	Vanadate	Various signaling pathways
Transmembrane PTPs (CD45, RPTPμ, RPTPα)	Homodimerization may inhibit activity	Vanadate	Lymphocyte activation
Dual-Specificity Family (MAP kinase phosphatases, etc.)			MAP kinase pathway
Cdc25 Family	Polo kinase, CDKs, phosphatases	Sulfuretin, coscinosulfate	Cell cycle
Low Molecular Weight Family (acid phosphatases)	Located in lysosomes	—	Unknown

CDK, cyclin-dependent kinase; FKBP, FK-binding protein; MAP, mitogen-activated protein; mRNA, messenger RNA; NMDA, *N*-methyl-D-aspartate; SH2, Src homology 2.

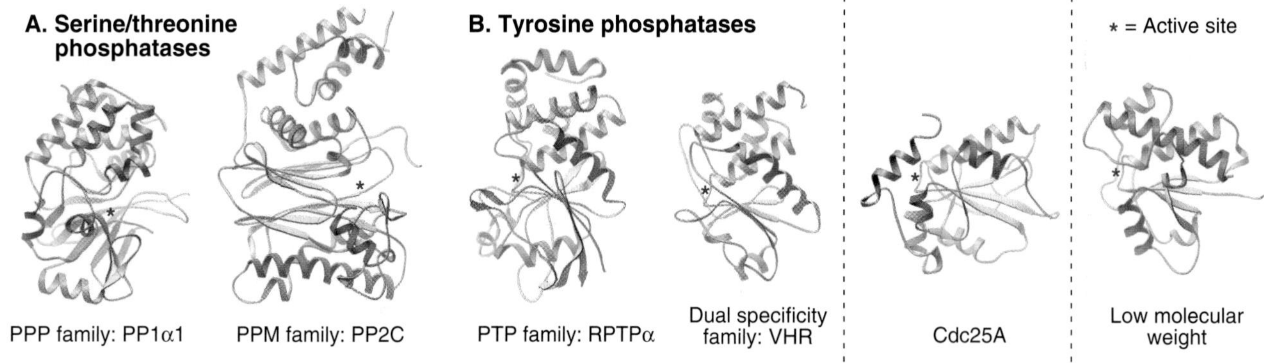

FIGURE 25.5 PROTEIN PHOSPHATASE STRUCTURES (*RED ASTERISKS* MARK THE ACTIVE SITES). A, Serine-threonine phosphatases: PP1α1 (PDB file 1FJM) and PP2C (PDB file 1LR4). **B,** Four families of protein tyrosine phosphatases: receptor tyrosine phosphatase RPTPα (PDB file 1YFO), dual-specificity phosphatase VHR (PDB file 1VHR), Cdc25A (PDB file 1C25), and low-molecular-weight phosphatase (PDB file 1PNT).

glucose metabolism by targeting PP1 to glycogen particles, where it dephosphorylates two enzymes that control glycogen metabolism. Fig. 27.3 explains how the hormone adrenaline activates PKA, which phosphorylates the G subunit, allowing enzymes to break down glycogen into glucose-6-phosphate. A 65-kD scaffold subunit and several B subunits regulate PP2A, which dephosphorylates many substrates, including kinases in the MAP kinase cascade (see Fig. 27.5). The inhibitors okadaic acid (a polyketide from dinoflagellates) and microcystin (a cyclic peptide from cyanobacteria) block access of substrates to the active sites of PP1 and PP2A.

PP2B, also known as **calcineurin,** is the only cytoplasmic phosphatase regulated by Ca^{2+}. The A subunit has the phosphatase active site. The A subunit has an autoinhibitory segment that blocks its own active site. An increase in cytoplasmic Ca^{2+} activates PP2B by first binding to calmodulin. Calcium-calmodulin then binds the autoinhibitory segment and displaces it from the active site. The transcription factor **NF-AT** (nuclear factor-activated T cells) is the best known substrate.

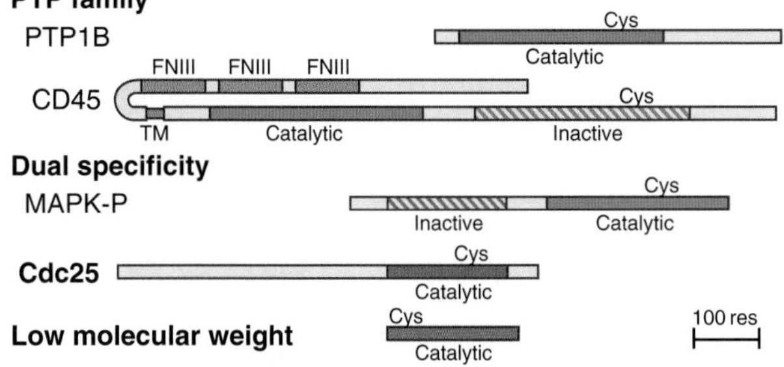

FIGURE 25.6 PROTEIN PHOSPHATASE DOMAIN ARCHITECTURE: SCALE LINEAR MODELS. A, Serine-threonine phosphatases. **B,** Protein tyrosine phosphatases. The five different catalytic domain folds are coded with different colors. CB, calcium binding; CM, calmodulin-binding; FNIII, fibronectin III; MAPK, mitogen-activated protein kinase; TM, transmembrane segment.

Fig. 27.8 explains how activated T-cell receptors turn on PP2B, which dephosphorylates NF-AT, allowing it to enter the nucleus and turn on expression of several lymphocyte growth factors. PP2B is the indirect target of two potent drugs that inhibit the immune rejection of transplanted organs. These drugs—**cyclosporine** and **FK506**—bind two different small proteins: **cyclophilin** and **FK-binding protein.** Both drug-protein complexes inhibit PP2B by blocking the active site. This prevents expression of genes regulated by NF-AT. Cyclosporine revolutionized human organ transplantation by suppressing the immune response.

PPM Family of Serine/Threonine Phosphates

Members of the large PPM family of serine/threonine phosphatases are found in bacteria, plants, fungi, and animals. The catalytic domain is incorporated into a variety of polypeptides with additional domains that confer specificity toward substrates such as stress-activated kinases and mitochondrial dehydrogenases. The structures of PPMs and PPPs are unrelated, but both have two metal ions in the active site (Mg^{2+} for PPM), and both catalyze the same phosphomonoester hydrolytic reaction. This is thought to be an example of **convergent evolution** toward similar active sites.

Protein Tyrosine Phosphatases

The 107 human protein tyrosine phosphatases (PTPs [Table 25.1]) regulate lymphocyte activation, the cell cycle, and many other processes by reversing the actions of protein tyrosine kinases. Some PTPs are tumor suppressors regulating cell proliferation, so somatic mutations that inactivate these enzymes are common in cancer cells. On the other hand, dephosphorylation of tyrosine activates other proteins, such as Src tyrosine kinase (Fig. 25.3C) and cyclin-dependent protein kinases (see Fig. 40.14). Eleven human PTP genes encode proteins that are missing key catalytic residues, so they must have other functions.

The four families of PTPs represent a remarkable evolutionary tale. PTPs and dual-specificity phosphatases derived from a common ancestor and have similar three-dimensional structures, while Cdc25 and low-molecular-weight tyrosine phosphatases have different folds, so they arose from different ancestors (Fig. 25.5B). Nevertheless, all four families converged to have similar active sites that bind phosphotyrosine in a deep, narrow pocket, transfer the phosphate to the sulfur atom of a cysteine in the sequence Cys-x-x-x-x-x-Arg, and release the phosphate in the rate-limiting step, when water attacks the phosphocysteine intermediate.

Some PTPs have multiple substrates including phosphoinositides, while others are limited to a few protein substrates. Localization contributes additional specificity. For example, the transmembrane segment that anchors CD45 to the plasma membrane (Fig. 25.6) enhances its access to some substrates and restricts its access to other substrates.

PTP Subfamily

The 38 human PTPs have catalytic domains with about 230 residues that favor phosphotyrosine as a substrate by a factor of 10^5 over phosphoserine or phosphothreonine. Transient oxidation of the catalytic cysteine by hydrogen peroxide accompanies activation of some signaling pathways that depend on phosphorylation of tyrosines, such as the insulin pathway (see Fig. 27.7).

Many PTPs are located in the cytoplasm where they may be bound to cellular partners. Adapter domains, such as SH2 domains, bind the phosphatases SHP-1 and SHP-2 to phosphotyrosines. C-terminal hydrophobic residues bind the phosphatase PTP1B to the endoplasmic reticulum. Eight families of human PTPs are transmembrane proteins. A single transmembrane segment links a variety of extracellular domains to one or two PTP domains in the cytoplasm (Fig. 25.6). The membrane proximal PTP domain has phosphatase activity. In most cases, the second PTP domain is inactive. Extracellular domains of some transmembrane PTPs bind extracellular matrix glycoproteins or the same protein on other cells. Some of these extracellular ligands can inhibit phosphatase activity, but much remains to be learned about the regulation of these PTPs.

CD45 (also called RPTPc), the best-characterized transmembrane PTP, constitutes a remarkable 10% of the plasma membrane protein of human lymphocytes. CD45 is required for antigens to activate B and T lymphocytes (see Fig. 27.8) where it dephosphorylates inhibitory phosphotyrosine residues of Src-family tyrosine kinases (Fig. 25.3C). Lymphocytes that lack CD45 fail to release intracellular Ca^{2+}, secrete lymphokines, or proliferate in response to antigen stimulation.

Dual-Specificity Subfamily

The dual-specificity phosphatases prefer phosphotyrosine as a substrate, but can also dephosphorylate serine and threonine at approximately 1% of that rate. The most studied members of this group, the MAP kinase phosphatases (MKPs), inactivate **MAP kinases** by dephosphorylating both phosphotyrosine and phosphothreonine residues (see Fig. 25.5B). Divergent members of this family, such as the **PTEN (phosphatase and tensin)** phosphatase, remove phosphate from the D-3 position of polyphosphoinositides (see Fig. 26.7).

Cdc25 Subfamily

Cdc25 removes inhibitory phosphates from threonine and tyrosine residues on the master cell-cycle kinases Cdk1 and Cdk2, releasing these enzymes to promote cell-cycle progression (see Fig. 43.7). This is an example of a phosphatase activating a biological process.

Cooperation Between Kinases and Phosphatases

Some protein phosphatases are stably associated with their substrate proteins. One example is the dual-specificity MAP kinase phosphatase-3 (MKP-3), which is bound to MAP kinase (see Fig. 27.6). Following activation by upstream kinases, this MAP kinase is active only transiently, owing to dephosphorylation by the associated phosphatase. These have been called *self-correcting signal complexes,* but more broadly speaking, this is an example of a biological timer.

Pharmacological Agents for Studying Protein Phosphatases

Inhibitors of protein phosphatases (Table 25.1) are used to explore the biological functions of these enzymes, but few of these inhibitors are entirely specific for one phosphatase. Development of specific inhibitors of protein phosphatases is challenging because these enzymes have similar active sites and achieve their specificity primarily via targeting subunits rather than substrate recognition.

Guanosine Triphosphate–Binding Proteins

Cells use GTP-binding proteins (called **GTPases** or **G-proteins**) to regulate a host of functions, including protein synthesis, signal transduction from plasma membrane receptors, cytoskeleton assembly, membrane traffic, and nuclear transport (Appendix 25.2). Genes for GTPases are ancient, as all forms of life use GTPases to regulate protein synthesis. The other three classes of

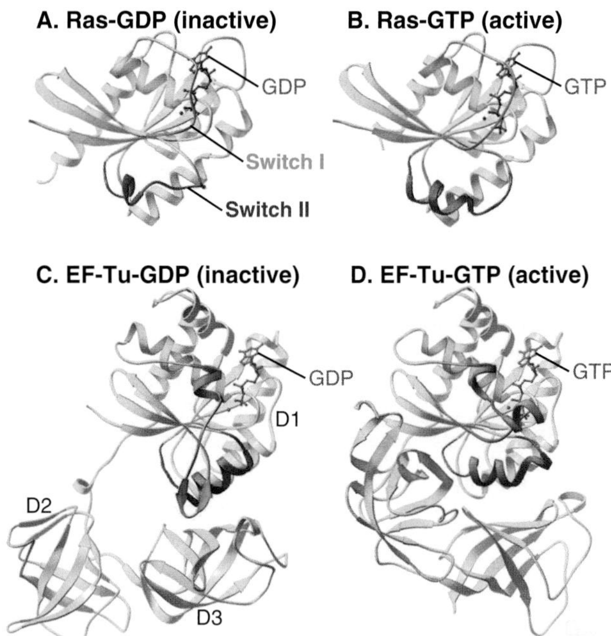

FIGURE 25.7 GTPase ATOMIC STRUCTURES. Ribbon models with ball-and-stick models of bound nucleotides. Switch I is *green*, and switch II is *red*. **A,** Ras-GDP (PDB file 1Q21). **B,** Ras-GTP (PDB file 121P). **C,** EF-Tu-GDP (PDB file 1TUI). **D,** EF-Tu-GTP. GTP (PDB file 1EFT) hydrolysis and phosphate dissociation cause major changes in the conformations of the switch loops of both proteins and of the orientations of the D2 and D3 domains of EF-Tu.

GTPases are small GTPases related to Ras, trimeric G-proteins, and dynamin-related GTPases. The small GTPases appeared during evolution in *Lokiarchaeota*, the archaeal progenitor of eukaryotes (see Fig. 2.4B). Gene duplication and divergence created multiple isoforms within these classes to provide more specificity.

All these GTPases share a core domain that binds a guanine nucleotide (Fig. 25.7) and use a common enzymatic cycle of GTP binding, hydrolysis, and product dissociation to switch the protein on and off (Fig. 25.8). The core GTP-binding domain consists of approximately 200 residues folded into a β-sheet of six strands sandwiched between five α-helices (Fig. 25.7). GTP binds in a shallow groove formed largely by loops at the ends of elements of secondary structure. A network of hydrogen bonds between the protein and guanine base, ribose, triphosphate, and Mg^{2+} anchor the nucleotide. The architecture of the GTPase domain was maintained during evolution despite differences between the major classes in approximately 80% of the residues.

All GTPases use the same enzyme cycle. The conformation of a GTPase depends on whether GTP or GDP is bound. The active GTP-bound conformation interacts with effector proteins. The GDP conformation is inactive, because it does not bind effectors (Fig. 4.7). Chapter 4 explains how the mechanism was determined.

The GTPase cycle (Fig. 25.8) consists of four steps: (1) Rapid binding of GTP is coupled to changes in the

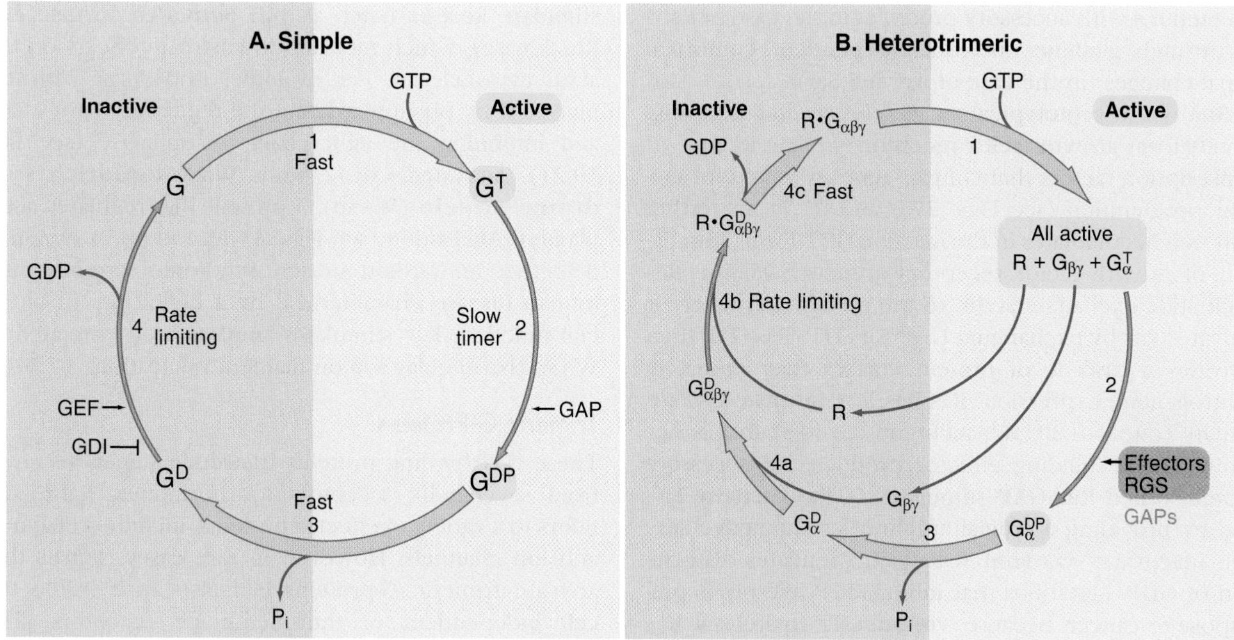

FIGURE 25.8 COMPARISON OF THE GTPASE CYCLES OF RAS AND A TRIMERIC G-PROTEIN. The size of the arrows indicates the relative rates of the reactions. **A,** GTPase cycle of Ras. **B,** GTPase cycle and subunit cycle of a trimeric G-protein. R is a seven-helix receptor. Regulators of G-protein signaling (RGS) and some effector proteins stimulate GTP hydrolysis. GAP, GTPase-activating protein; G^D, GTPase with bound GDP; GDI, guanine nucleotide dissociation inhibitor; G^{DP}, GTPase with bound GDP and inorganic phosphate; GEF, guanine nucleotide exchange factor; G^T, GTPase with bound GTP.

conformations of three segments of the polypeptide called switch-I, switch-II, and switch-III. In the active GTP state, these switch loops form a binding site for downstream target proteins. (2) GTP hydrolysis is slow and irreversible. (3) Dissociation of the γ-phosphate is fast and coupled to the return of the switch loops to the inactive conformation. (4) GTPases tend to accumulate in the inactive GDP-state, because GDP dissociates slowly and GTP cannot bind until GDP dissociates.

Diverse intrinsic or extrinsic protein modules regulate the GTPase cycle. Most GTPases depend on other proteins, called **guanine nucleotide exchange factors (GEFs),** to accelerate dissociation of GDP (Appendix 25.2). Although diverse in structure, these **GEFs** have similar mechanisms. They distort the P loop, the part of the nucleotide binding site that interacts with the β phosphate, allowing GDP to escape, and then bind tightly to the nucleotide-free GTPase. Most small GTPases require extrinsic **GTPase-activating proteins (GAPs)** to stimulate the GTP hydrolysis that turns them off.

Elongation Factors

These GTPases act as timers to ensure the fidelity of protein synthesis (see Fig. 12.9). **Elongation factor Tu (EF-Tu)** has two accessory domains hinged to the GTPase core (Fig. 25.7C–D). GTP EF-Tu binds and delivers aminoacyl-tRNAs (transfer RNAs) to the A site of a ribosome. If the tRNA anticodon matches the messenger RNA (mRNA) codon at the A site, it remains bound long enough for GTP to be hydrolyzed. This releases EF-Tu from the ribosome and allows the correct amino acid to be added to the growing polypeptide chain. The accessory factor EF-Ts is the GDP exchange factor for this system.

Small Guanosine Triphosphatases

All six families of small GTPases consist of a single domain similar to Ras (Fig. 25.7A–B). The Arf, Rab, and Sar GTPases act as switches for intracellular traffic of membrane vesicles (see Chapter 21). The Ran family regulates nuclear transport (see Fig. 9.18) and assembly of the mitotic spindle (see Fig. 44.8). This chapter introduces the Ras and Rho families of small GTPases that transduce signals from cell surface receptors.

Covalently attached lipids anchor many small GTPases to membrane bilayers. These hydrophobic chains are required for activity. Ras and some Rho and Rab proteins are modified with a C-15 or C-20 prenyl chain on a C-terminal cysteine. Arfs are myristoylated, but other small GTPases, such as Ran, are not modified and are soluble in the cytoplasm. Membrane-associated GTPases may cycle into the cytoplasm when their lipid tails turn over or when they bind cytoplasmic regulatory proteins such as **Rho-GDI** (Rho-guanine nucleotide dissociation inhibitor [see Fig. 21.13]). Rab, Arf, and Sar GTPases also recycle through the cytoplasm as they direct vesicular traffic from one compartment to another. Membrane association can be regulated either by interactions with accessory proteins (in the case of Rab) or through guanine nucleotide-dependent conformational changes (in the case of Arf and Sar).

Ras is the prototypical small GTPase. Ras transmits signals from growth factor receptor tyrosine kinases to transcription factors that control genes required for cellular proliferation (see Figs. 27.6 and 27.7). In resting cells, Ras accumulates in the inactive GDP form. Stimulation of growth factor receptors attracts SOS, the Ras nucleotide exchange factor, to the membrane, where it activates Ras by exchanging GDP for GTP. Ras-GTP then activates a cascade of protein kinases that ultimately controls gene expression. Ras has low intrinsic GTPase activity (rate = 0.005 s^{-1}; half-time = 140 s) that is not stimulated by binding effector proteins. An accessory protein called **Ras-GAP** stimulates GTPase activity 10^5-fold by providing a crucial arginine for the active site. This inactivates Ras until SOS again stimulates dissociation of GDP. Mutations that inhibit GTP hydrolysis predispose to cancer, because without GTP hydrolysis, Ras is continually active and stimulates cellular growth. (Fig. 41.12). These mutations are extremely common, occurring in 20% to 30% of all human cancers.

The 16 members of the human **Rho** family include Rho itself, Rac, and Cdc42. They regulate the actin and microtubule cytoskeletons, cellular growth, and cellular polarity. Stimulation of certain seven-helix receptors and receptor tyrosine kinases activates Rho-family GTPases by turning on exchange factors that catalyze the exchange of GDP for GTP. Activated Rho-family proteins stimulate kinases (such as **p21-activated kinase** and Rho-kinase), which mediate downstream effects on the actin cytoskeleton. For example, Rho-kinase activates myosin-II by phosphorylating the regulatory light chain and inhibiting the light chain phosphatase (see Fig. 39.21). Activated Cdc42 binds **Wiskott-Aldrich syndrome protein** (WASp), a protein that regulates actin filament nucleation (see Figs. 33.13 and 38.7). WASp is defective in Wiskott-Aldrich syndrome, an inherited human disease characterized by a deficiency in blood cell function. Rac stimulates another protein related to WASp that regulates actin filament nucleation.

Trimeric G-Proteins

These GTP-binding proteins transduce signals received from seven-helix receptors for hormones, light, and odors to a variety of effector proteins, including enzymes and ion channels. However, in rare cases, signals that activate trimeric G-proteins can also arise inside the cell independent of transmembrane receptors. The best-characterized examples of intracellular activation are G-proteins that regulate asymmetrical cell division.

Trimeric G-proteins consist of three subunits Gα, Gβ, and Gγ (Fig. 25.9). The Gα subunits have a GTP-binding domain similar to small GTPases linked by two strands to a domain of α-helices. The helical domain protects the nucleotide-binding site when GDP is bound to the inactive protein. Gβ and Gγ subunits bind tightly to each other and reversibly to Gα. Gβ is a torus-shaped molecule composed of seven modules, each folded into an

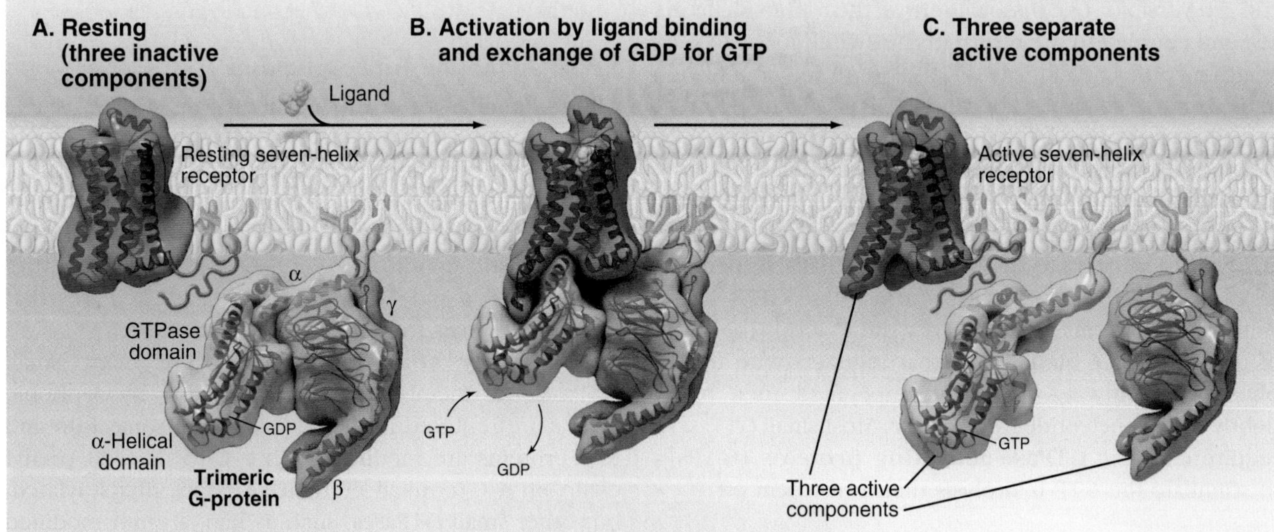

FIGURE 25.9 ACTIVATION OF A TRIMERIC G-PROTEIN BY A SEVEN-HELIX RECEPTOR. These examples are β_2-adrenergic receptor and the G_s-protein complexes. **A,** Receptor resting state without a ligand and a separate trimeric G-protein in its inactive GDP-Gαβγ state. Both Gα and Gγ are anchored to the lipid bilayer. **B,** Interaction of active β_2-adrenergic receptor with a trimeric G-protein. This interaction catalyzes the exchange of GDP for GTP on Gα. **C,** Active Gα and Gβγ dissociate from each other and the receptor and are available to interact with effector proteins. GTPase, guanosine triphosphatase. (For reference, see PDB file 2RH1 and Cherezov V, Rosenbaum DM, Hanson MA, et al. High-resolution crystal structure of an engineered human beta2-adrenergic G protein-coupled receptor. *Science*. 2007;318:1258–1265 [**A**]; PDB file 3SN6 and Rasmussen SG, DeVree BT, Zou Y, et al. Crystal structure of the β2 adrenergic receptor–Gs protein complex. *Nature*. 2011;477:549–555 [**B**]; and PDB file 4LDO and Ring AM, Manglik A, Kruse AC, et al. Adrenaline-activated structure of β2-adrenoceptor stabilized by an engineered nanobody. *Nature*. 2013;502:575–579 [**C**].)

antiparallel β-sheet (Fig. 25.9). These so-called **WD repeats** are found in many proteins. Loops on one face of the torus interact with Gα-GDP, blocking effector interaction sites on both partners. Loops on the other face of Gβ bind tightly to Gγ.

A variety of lipid anchors attach trimeric G-proteins to the inside of the plasma membrane. Most Gα subunits are anchored by an N-terminal fatty acid, usually myristate, and some have an additional palmitic acid anchor on a reactive cysteine. Most Gγ subunits have a C-20 prenyl group on a C-terminal cysteine.

Subunit Diversity

Yeasts have genes for just two Gα proteins, one of which participates in responses to sex pheromones, but mammals have genes for 20 Gα, 5 Gβ, and 12 Gγ subunits. Alternative splicing of Gα mRNAs creates additional diversity. If these subunits were combined in all possible ways, mammals could make more than 1000 different trimeric G-proteins, but only a subset of these combinations have been detected.

Given more than 1000 genes for seven-helix receptors, many receptors use the same G-proteins to transduce signals. For example, hundreds of different odorant receptors in the nose activate $G_{olf}α$ (see Fig. 27.1). Trimeric G-proteins act on a limited variety of downstream effectors, including ion channels, kinases, and enzymes that produce second messengers (Table 25.2).

Guanosine Triphosphatase Cycle

Seven-helix receptors activate trimeric G-proteins by catalyzing the exchange of GDP for GTP on the Gα subunit. GTP binding from the cytoplasm changes the conformation of Gα and releases Gβγ (Fig. 25.9). This generates two signals, as both Gα and Gβγ can engage downstream effector proteins. It is convenient to think about this process as two cycles: a GTPase cycle coupled to a subunit cycle (Fig. 25.8B).

The intrinsic rate of GDP dissociation from trimeric G-proteins is near zero (half-life approximately 10 to 60 minutes), but interaction of an active receptor with Gα triggers conformational changes that allow GDP to dissociate many orders of magnitude faster. Conformational changes in Gα include rotation of the α-helical domain away from the nucleotide-binding site. Then GTP in the cytoplasm binds nucleotide-free Gα on a millisecond-to-second time scale, fast enough for vision (see Fig. 27.2). Gβγ does not interact with the receptor, but must be bound to Gα-GDP before active membrane receptors can trigger dissociation of GDP.

GTP draws the three switch loops together around the γ-phosphate in a conformation that favors dissociation of Gβγ and association with effector proteins. This conformational change is the physical manifestation of the transfer of a signal from an activated receptor to Gα.

Like all GTPases, the duration of the signal carried by trimeric G-proteins depends on the rate of GTP hydrolysis. Thanks to a strategically placed arginine (on a linker to the helical domain), trimeric G-proteins hydrolyze GTP faster than small GTPases. The half-time of the active state is approximately 10 to 20 seconds, adequate for activation of a target protein. Note that the helical domain acts both as an intrinsic inhibitor of GDP dissociation and as an activator of GTP hydrolysis, similar to two separate regulators of small GTPases (guanine nucleotide dissociation inhibitor [GDI] and GAP).

Two different types of proteins promote GTP hydrolysis and limit the lifetime of the active state: effector proteins and extrinsic GTPase activators, called **RGS proteins** (regulators of G-protein signaling). For example, $G_qα$-GTP binds and stimulates phospholipase Cβ, which produces two second messengers: inositol 1,4,5-triphosphate (IP_3) and diacylglycerol (see Fig. 26.12). At the same time, phospholipase Cβ accelerates GTP hydrolysis by $G_qα$, providing negative feedback to turn off the signal.

The RGS family includes more than 20 proteins that can stimulate the GTPase activity of G-proteins approximately 100-fold, yielding half-times of less than 1 second. These RGS proteins work by stabilizing the transition

TABLE 25.2 Receptors and Effectors for G-Protein Isoforms

Family (No. of Human Members)	Receptors	Effectors
$G_iα$ (7)	α-Adrenergic amines, acetylcholine, chemokines, various neurotransmitters, tastants	Inhibit adenylyl cyclase, open potassium channels, close calcium channels
$G_qα$ (5)	α-Adrenergic amines, acetylcholine, various neurotransmitters	Activate phospholipase Cβ to produce IP_3, which releases Ca^{2+}
$G_sα$ (3)	β-Adrenergic amines, hormones (corticotropin, glucagon, parathyroid, thyrotropin, others)	Stimulate adenylyl cyclase to produce cAMP, receptor kinase
$G_tα$ (2)	Rhodopsin, which absorbs light	Activate cGMP phosphodiesterase to break down cGMP, receptor kinase
$G_{13}α$ (2)	Thrombin and others	Rho and others
$G_{olf}α$	Odorant receptors	Activate adenylyl cyclase, receptor kinase

cAMP, Cyclic adenosine monophosphate; cGMP, cyclic guanosine monophosphate; IP_3, inositol triphosphate.

state between GTP and GDP-P_i. Such rapid adaption is important in dynamic physiological systems such as the heart and eye. Some RGS proteins also have a Rho–GEF domain, so they may connect seven-helix receptors and trimeric G-proteins to the Rho family of small GTPases.

Subunit Cycle

The GTPase cycle is linked a subunit cycle (Fig. 25.8B). Gα cycles on and off both Gβγ and the receptor as it traverses its GTPase cycle. The conformational changes in Gα induced by GTP binding reduce the affinity of Gα for both its receptor and associated Gβγ subunits, so the three molecular partners separate on the cytoplasmic face of the membrane (Fig. 25.9). Once dissociated, the receptor, Gα, and Gβγ are each free to interact with other partners to independently propagate the signal (Table 25.2). The conformation of Gβγ is the same whether bound to Gα or free, but dissociation from Gα unmasks binding sites on Gβγ for activating membrane targets such as ion channels. Gβγ also provides a membrane anchor to enhance the interaction of cytoplasmic effectors (such as kinases that phosphorylate seven-helix receptors) with membrane targets.

The linked GTPase and subunit cycles allow activation of a single receptor to generate a massive signal. Although most seven-helix receptors are active only briefly (owing to rapid ligand dissociation and rapid inactivation [see Figs. 27.1 and 27.2]), they can turn on multiple G-proteins and activate several downstream effector proteins. These effectors often include enzymes or channels that rapidly amplify the signal. Reassociation of Gβγ with Gα-GDP terminates the signal at the end of the cycle.

Mechanisms of Effector Activation

Activated Gα and Gβγ subunits may act individually, but they may also act synergistically and even antagonistically. Two examples, elaborated upon in Chapter 27, illustrate how G-proteins activate effector proteins.

In the eye, phototransduction takes place on a millisecond time scale. When the seven-helix photoreceptor rhodopsin absorbs a photon, the G-protein $G_t\alpha$ (also called transducin) relays and amplifies a signal (see Fig. 27.2). Each activated rhodopsin generates approximately 500 $G_t\alpha$-GTPs, which bind and inactivate an inhibitory subunit of the enzyme cGMP phosphodiesterase. This stimulates the activity of the phosphodiesterase (see Fig. 26.1), which lowers the cytoplasmic concentration of cGMP and closes an ion channel. The signal is transient, because both an RGS protein and the inhibitory subunit promote hydrolysis of GTP bound to $G_t\alpha$. Inactive $G_t\alpha$-GDP dissociates from the inhibitory subunit, terminating the signal that flowed through $G_t\alpha$ to the enzyme.

In the heart, the β-adrenergic receptor activates $G_s\alpha$, releasing both $G_s\alpha$-GTP and Gβγ to bind effector proteins (see Fig. 27.3). During its 10-second lifetime, $G_s\alpha$-GTP binds and stimulates the enzyme adenylyl cyclase to produce the second messenger cAMP. Gβγ assists in receptor inactivation by binding β-ARK, a kinase that phosphorylates and turns off the β-adrenergic receptor, terminating the signal.

Trimeric G-Proteins in Disease

Either abnormal activation or inactivation of trimeric G-proteins can cause disease (Table 25.3). Mutations that interfere with GTP hydrolysis cause Gα to accumulate in the GTP state and persistently activate downstream effectors. For example, mutations in arginine or glutamine residues of Gα that are crucial for GTP hydrolysis can cause tumors by prolonging the activation of pathways responsible for cell proliferation. Common variants in the sequence of other G-proteins are associated with high blood pressure and other common diseases.

Some bacterial toxins mediate their effects by acting on G-proteins. The cholera bacterium causes diarrhea by secreting **cholera toxin**, an enzyme that enters the cytoplasm after a complicated journey from the cell surface.

TABLE 25.3 Guanosine Triphosphatases and Disease

Disease	GTPase	Mechanism
Excess Signal		
Cholera	$G_s\alpha$	Cholera toxin ADP-ribosylation of R201 inhibits GTP hydrolysis in intestinal epithelium.
Pituitary and thyroid adenomas	$G_s\alpha$	Somatic point mutations of R201 or Q227 inhibit GTP hydrolysis; constitutive activity mimics signal from hormones that stimulate proliferation and secretion by these glands.
Various cancers	Ras	Point mutations inhibit GTP hydrolysis, generating persistent signals for cell proliferation.
Deficient Signal		
Whooping cough	$G_i\alpha$	Pertussis toxin ADP-ribosylation of $G_i\alpha$ C347 in the bronchial epithelium blocks receptor activation; connection to coughing not established.
Night blindness	$G_t\alpha$	Germline point mutation in G38.
Pseudohypoparathyroidism type Ia	$G_s\alpha$	Point mutations result in loss of $G_s\alpha$ or may block its activation by receptors.

ADP, adenosine diphosphate; C, cysteine; G, glycine; GTP, guanosine triphosphate; GTPase, guanosine triphosphatase; Q, glutamine; R, arginine. Modified from Farfel Z, Bourne HR, Iiri T. The expanding spectrum of G protein diseases. *N Engl J Med.* 1999;340:1012–1020.

The enzyme catalyzes the addition of an adenosine diphosphate (ADP)-ribose to the arginine required for $G_s\alpha$ to hydrolyze GTP. Activated $G_s\alpha$-GTP accumulates, prolonging the production of high levels of cAMP, which causes life-threatening diarrhea by stimulating salt and water secretion into the intestine. **Pertussis toxin** from the whooping cough bacterium secretes an enzyme that adds ADP-ribose to a cysteine residue of $G_i\alpha$ or other $G\alpha$ subunits. This inhibits the interaction of the trimeric G-protein with activated receptors, so the G-protein accumulates in the inactive GDP state. One consequence is airway irritability. Similarly, *Clostridium botulinum* C3 toxin ADP-ribosylates and inhibits Rho-GTPases, whereas a *Clostridium difficile* toxin uses uridine diphosphate (UDP)-glucose to glucosylate and turn off the whole class of Rho proteins.

Dynamin-Related Guanosine Triphosphatases

These large GTPases have an N-terminal GTP-binding domain and a C-terminal GTPase-activating domain that allows self-assembly into spiral polymers around the membrane tubules that form at sites of endocytosis (see Fig. 22.8). The GTPase cycle drives interaction of the subunits along the spiral to constrict the tubule. Other dynamin family members regulate vacuolar trafficking in yeast and the division of mitochondria.

Experimental Tools

Mutations, especially those that constitutively activate GTPases (by inhibiting GTP hydrolysis) or inactivate GTPases, have been used to investigate their functions in cells. For biochemical experiments, slowly hydrolyzed analogs of GTP, such as GTPγS (with a sulfur substituted for one of the γ-phosphate oxygens), are used to prolong the active state of GTPases. Similarly, aluminum fluoride and beryllium fluoride bind very tightly in place of the hydrolyzed γ-phosphate, keeping $G\alpha$ in an active GDP-P_i state similar to GTP. The fungal metabolite brefeldin A blocks nucleotide exchange on some Arfs, disrupting membrane traffic between the Golgi complex and the endoplasmic reticulum (see Fig. 21.11).

Molecular Recognition by Adapter Domains

During the characterization of signaling pathways, related sequences of amino acid were discovered in different proteins, such as Src (Figs. 25.3C and 25.4B). These turned out to be compactly folded domains (Fig. 25.10) that are incorporated into a variety of proteins (Fig. 25.11), including many with no role in signaling. These domains mediate interactions between proteins and with membrane lipids (Table 25.4).

The names of adapter domains generally came from the proteins where they were discovered. For example, **Src homology (SH)** domains were first recognized in the Src tyrosine kinase (Fig. 25.3C). SH1 is the tyrosine kinase domain; the SH2 domain binds phosphotyrosine peptides; and the SH3 binds polyproline type II helices. Chapter 27 provides detailed examples of how adapter domains function in signalling pathways. This section provides an overview of their structure and ligand-binding properties. The following points apply to adapter proteins in general.

Adapter domains mediate interactions that are required to assemble proteins into multimolecular functional units that typically carry out a series of reactions. To facilitate these interactions, many signaling proteins have more than one adapter domain or bind more than one ligand. In signal transduction, these physical associations make transmission from receptors to effectors more reliable, like an integrated machine rather than one that relies solely on diffusion and random associations. Mutations in experimental organisms and roles in various human diseases have verified the importance of adapter domains for many pathways. Interactions mediated by

TABLE 25.4 Adapter Domains

Domain Name	Size (Residues)	Consensus Ligands	Example of Proteins With Domain
EH (Eps15 homology)	95	S/T-N-P-F′-Φ	Clathrin adapter proteins, synaptojanin I
EVH1 (Ena-VASP homology)	110	D/E-Φ-P-P-P-P	WASp, VASP, Ena
PH (Pleckstrin homology)	100	PIP_2, PIP_3	Kinases, scaffolds, GEFs, GAPs, PLCδ, dynamin
PDZ	100	-x-x-S/T-x-V-COOH -x-x-Φ-x-Φ-COOH	Scaffolds for channels and transduction enzymes
PTB (phosphotyrosine binding)	125	-Φ-x-N-P-x-pY-	IRS1, Shc scaffold proteins
SH2 (Src homology 2)	100	-pY-x-x-Φ-	Transduction enzymes and scaffold proteins
SH3 (Src homology 3)	60	(+) -R/K-x-x-P-x-x-P- (−) -x-P-x-x-P-x-R/K-	Tyrosine kinases, phosphatases, Grb2, PLCγ, spectrin, myosin I
WW	38–40	-P-P-x-Y-	Peptidyl prolyl isomerase, ubiquitin ligase
14-3-3	250	-R-S-X-pS-x-P-	14-3-3 isoforms

Φ, hydrophobic residue; COOH, C-terminus; Ena, enabled gene; GAP, GTPase-activating protein; GEF, guanine nucleotide exchange factor; Grb2, adapter protein; IRS1, insulin receptor substrate 1; PLC, phospholipase C; pS, phosphoserine; pY, phosphotyrosine; Shc, receptor tyrosine kinase substrate; VASP, vasodilator-stimulated phosphoprotein, WASp, Wiskott-Aldrich syndrome protein; (−), minus orientation; (+), plus orientation.

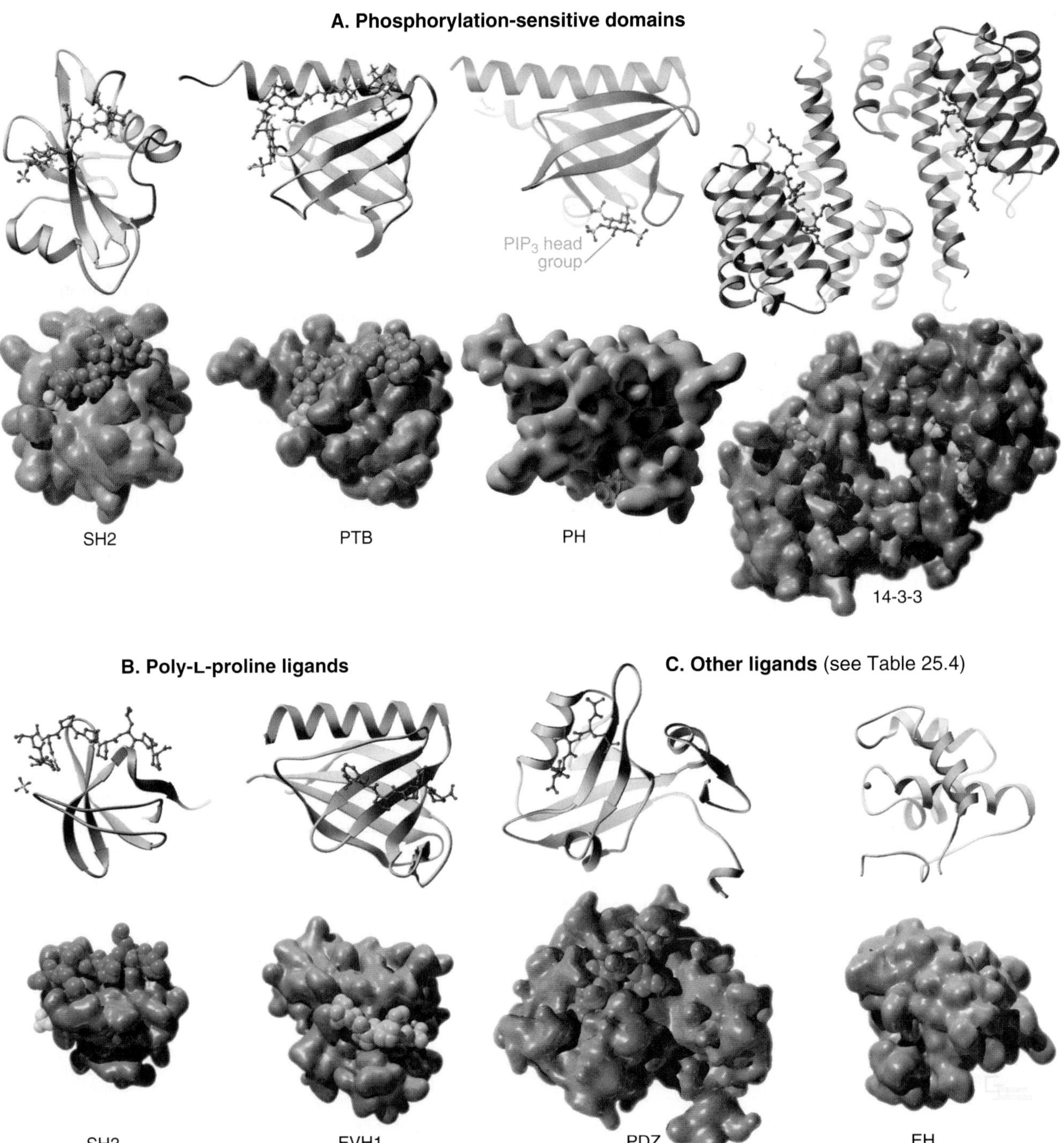

FIGURE 25.10 ATOMIC MODELS OF ADAPTER PROTEIN DOMAINS. Ribbon diagrams show their architecture, and the surface renderings show how ligands bind. **A,** Domains with phosphorylation-sensitive interactions. **B,** Domains with poly-l-proline ligands. **C,** Domains with other ligands. (For reference, see PDB files 1HCS [SH2], 1IRS [PTB], 1DYN [PH], 1A38 [14-3-3], 1ABO [SH3], 1EVH [EVH1], 1BEQ [PDZ], and 1EH2 [EH].) EH, Eps15 homology; EVH1, Ena-Vasp homology 1; PDZ, postsynaptic density protein 95, disc large tumor suppressor (Dlg1), and zona occludens 1; PH, pleckstrin homology; PTB, phosphotyrosine-binding; SH2/3, Src homology 2/3.

adapter domains complement the organizing activities of anchoring proteins, such as STE5 and AKAPs (see "Targeting" under "Regulation of Protein Kinases" above).

All members of each family of adapters have similar structures (and common evolutionary origins) but differ in their affinities for a range of similar ligands. For example, all SH2 domains bind peptides with a sequence phosphotyrosine-X-X-hydrophobic residue. All require phosphotyrosine but differ in their affinity for peptides depending on the hydrophobic residue and the intervening residues. This lock-and-key strategy creates specificity with many particular combinations, as in macroscopic locks and keys. Although lacking in sequence similarity, three of these domains—PH, PTB, and EVH1 (Ena-VASP homology 1)—have similar folds (Fig. 25.10) and likely had a common ancestor. Nevertheless, their ligands are

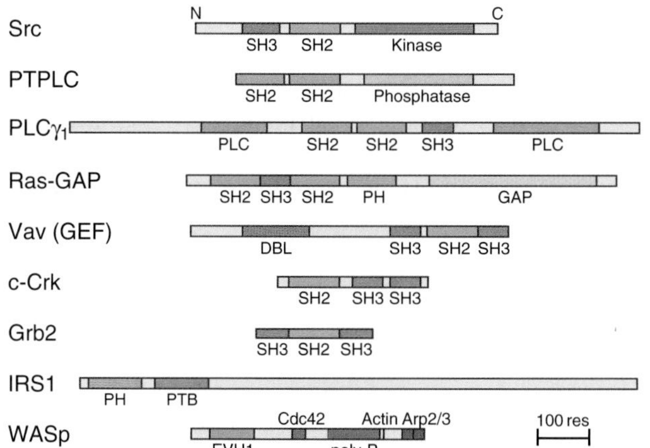

FIGURE 25.11 SCALE DRAWINGS OF PROTEINS WITH ADAPTER DOMAINS. Actin and Arp2/3 indicate binding sites for actin monomers and Arp2/3 complex; Cdc42, a Rho-family GTPase; c-Crk and Grb2, multidomain adapter proteins; DBL, a guanine nucleotide exchange factor domain; EVH1, Ena-VASP homology 1 domain; GAP, GTPase activating; IRS1, insulin receptor substrate; PLC, phospholipase C; poly-P, polyproline; PTPLC, protein tyrosine phosphatase; Ras-GAP, Ras GTPase activating protein; SH, Src homology; Vav, a guanine nucleotide exchange protein; WASp, Wiskott-Aldrich syndrome protein.

quite distinct from each other and bind to different sites on the commonly folded domains.

Some interactions depend on reversible covalent modifications of ligands: tyrosine phosphorylation for SH2 and some PTB domains, serine phosphorylation for some 14-3-3 and WW domains, and 3-phosphorylation of inositol for some PH domains. This allows these adapters to form and dissipate in response to signals that modulate phosphorylation of their ligands.

Many interactions of adapter domains with their ligands are tenuous, so associations are reversible on a time scale of seconds, allowing rapid rearrangements in response to signals. Frequent dissociation is also required for covalent modifications of the ligands, such as access of phosphatases to their substrates.

Phosphorylation-Sensitive Adapters

SH2 Domains

SH2 domains bind short peptide sequences beginning with a phosphotyrosine. Like a two-pronged plug, these peptides insert a phosphotyrosine and a hydrophobic side chain into two cavities in the SH2 socket. A phosphate on the tyrosine increases the affinity of a peptide for its partner SH2 domain by orders of magnitude by virtue of an extensive network of hydrogen bonds between the phosphate and its deep binding pocket. This allows reversible phosphorylation to control interactions between SH2 domains and their ligands. This switching mechanism is used in growth factor signaling and lymphocyte activation (see Figs. 27.6 to 27.8). The hydrophobic side chain inserts into a cavity on the surface of the SH2 domain. The size of this side chain is a major determinant of binding specificity. Two residues between these plug residues straddle the β-sheet of the SH2 with their side chains exposed to solvent.

SH2 interactions with target peptides and proteins are relatively weak, with K_ds in the range of 0.1 to 1 μM. This allows for rapid exchange of partners and dephosphorylation of the phosphotyrosine. Nevertheless, these interactions are specific, as the 115 human proteins with SH2 domains each engage a limited number of target phosphoproteins. SH2 domains target several signal transduction enzymes to phosphotyrosines on receptor tyrosine kinases (see Fig. 27.6) as the first step in propagating signals to effectors, including second messengers and transcription factors.

Phosphotyrosine-Binding Domains

Most PTB domains require a phosphotyrosine at the C-terminal end of the peptide ligand. Adjacent hydrophobic residues contribute to binding specificity. The bound peptide is hydrogen-bonded onto the edge of a β-sheet, and phosphotyrosine interacts with basic residues. PTB domains target adapter proteins to phosphotyrosines on receptor tyrosine kinases, such as insulin receptor (see Fig. 27.7). A few PTB domains bind unphosphorylated peptide ligands.

14-3-3 Proteins

Vertebrates have at least seven genes for 14-3-3 subunits that assemble into homodimers or heterodimers. Protein ligands with an appropriate sequence and a central phosphoserine bind each subunit with submicromolar affinity. Peptides with two appropriate sequences bind much more tightly. The 14-3-3 proteins regulate protein kinases, including the Ras-activated kinase Raf (see Fig. 27.6) and cellular death pathways (see Chapter 46). During interphase or after DNA damage, a 14-3-3 protein inhibits the cell cycle regulatory phosphatase Cdc25, when it is phosphorylated on a serine (see Fig. 43.4).

WW Domains

These tiny adapter domains are found in more than 100 proteins. Most bind certain phosphoserine or phosphothreonine peptides, but some bind proline-rich ligands. These phosphorylation-dependent interactions regulate Cdc25 and ubiquitin-mediated protein destruction (see Chapter 23).

PH Domains

These compact domains of approximately 100 residues are named for the PH domain in **pleckstrin,** the major substrate for PKC in platelets. PH domains bind **polyphosphoinositides** (see Fig. 26.7). PH domains of dynamin and phospholipase Cδ prefer phosphatidylinositol (PI) (4,5) P_2 (PIP_2), whereas the PH domain of the Bruton tyrosine kinase (Btk) favors PI(3,4,5)P_3 (PIP_3). These interactions target proteins with PH domains to membrane bilayers rich in PIP_2 and PIP_3 and make these membrane interactions responsive to the activities of phosphoinositide kinases (see Figs. 26.7 and 38.10). PH-domain proteins include kinases (PKB/Akt, PDK1; see Fig. 27.7), signaling scaffolding proteins (insulin receptor substrate 1 [IRS1]; see Fig. 27.7), enzymes (phospholipase Cγ1 [PLCγ1]; see Figs. 26.4 and 26.12), and GEFs. Mutations in the PH domain of the Btk tyrosine kinase reduce affinity for PI(3,4,5)P_3 and cause a failure of B-lymphocyte development (see Fig. 28.9), resulting in immunodeficiency from a lack of antibodies.

Adapters With Proline-Rich Ligands

SH3 Domains

SH3 domains found in more than 250 different human proteins bind to proline-rich sequences of numerous target proteins that help assembly signalling pathways. For example, SH3 domains of the adapter protein Grb2 (Fig. 25.11) link activated growth factor receptors to the nucleotide exchange protein for Ras (see Fig. 27.6). SH3 domains are also found in tyrosine kinases and cytoskeletal proteins, including myosin-I, spectrin, and cortactin.

The ligands for SH3 domains form left-handed, **type II polyproline helices** (see Fig. 29.1) that make hydrophobic interactions with aromatic residues in a shallow groove on the SH3 domain, as well as hydrogen bonds contributed by ligand peptide carbonyl oxygens. Residues flanking the central proline helix contribute to binding specificity. Even optimal peptide ligands bind relatively weakly (K_ds in the micromolar range), so they exchange rapidly. When incorporated into proteins, type II poly-L-proline helices with appropriate sequences bind somewhat more tightly, owing to secondary interactions.

EVH1 Domains

EVH1 domains (Fig. 25.10) are found in WASp (see Fig. 33.13) and other signaling proteins that regulate actin polymerization. EVH1 domains are folded like PH and PTB domains, but have a groove that binds type II proline-rich helices of target proteins. This site also differs from that for phosphotyrosine peptides on PTB domains. Thus, a common scaffold has diverged to form three completely different binding sites.

Other Adapter Domains

PDZ Domains

PDZ domains are found in one to seven copies in scaffolding proteins that cluster together ion channels and signal transduction proteins at synapses, in photoreceptors, and in polarized epithelial cells. PDZ domains bind specific sequence motifs, usually found at the very C-terminus of proteins or, more rarely, at the end of β hairpin structures. PDZ domains bind their ligands, in a manner reminiscent of PTB domains, by incorporating them through hydrogen bonds as an extra strand in a β-sheet.

EH Domains

EH domains are small bundles of four α-helices that bind peptides with the sequence asparagine-proline-phenylalanine. Flanking residues contribute to specificity. The best-characterized EH-mediated interactions are involved with endocytosis (see Chapter 22).

SELECTED READINGS

Kinases

Endicott JA, Noble MEM, Johnson LN. The Structural Basis for Control of Eukaryotic Protein Kinases. *Annu Rev Biochem*. 2012;81:587-613.

Langeberg LK, Scott JD. Signalling scaffolds and local organization of cellular behaviour. *Nat Rev Mol Cell Biol*. 2015;16:232-244.

Wang Z, Gucek M, Hart GW. Cross-talk between GlcNAcylation and phosphorylation: site-specific phosphorylation dynamics in response to globally elevated O-GlcNAc. *Proc Natl Acad Sci USA*. 2008;105:13793-13798.

Zhang J, Yang PL, Gray NS. Targeting cancer with small molecule kinase inhibitors. *Nat Rev Cancer*. 2009;9:28-39.

Phosphatases

Alonso A, Pulido R. The extended human PTPome: A growing tyrosine phosphatase family. *FEBS J*. Nov 17, 2015;[Epub ahead of print].

Aramburu J, Heitman J, Crabtree GR. Calcineurin: A central controller of signalling in eukaryotes. *EMBO Rep*. 2004;5:343-348.

Barford D, Das AK, Egloff M-P. Structure and mechanism of protein phosphatases: Insights into catalysis and regulation. *Annu Rev Biophys Biomol Struct*. 1998;27:133-164.

Gallego M, Virshup DM. Protein serine/threonine phosphatases: Life, death, and sleeping. *Curr Opin Cell Biol*. 2005;17:197-202.

Mohebiany AN, Nikolaienko RM, Bouyain S, Harroch S. Receptor-type tyrosine phosphatase ligands: looking for the needle in the haystack. *FEBS J*. 2013;280:388-400.

Guanosine Triphosphatases

Bockoch GM. Biology of the p21-activated kinases. *Annu Rev Biochem*. 2003;72:743-781.

Garcia-Mata R, Boulter E, Burridge K. The 'invisible hand': regulation of RHO GTPases by RHOGDIs. *Nat Rev Mol Cell Biol*. 2011;12:493-504.

Kach J, Sethakorn N, Dulin NO. A finer tuning of G-protein signaling through regulated control of RGS proteins. *Am J Physiol Heart Circ Physiol*. 2012;303:H19-H35.

Rasmussen SG, DeVree BT, Zou Y, et al. Crystal structure of the β_2 adrenergic receptor-Gs protein complex. *Nature*. 2011;477:549-555.

Rauen KA. The RASopathies. *Annu Rev Genomics Hum Genet*. 2013;14:355-369.

Schmid SL, Frolov VA. Dynamin: Functional design of a membrane fission catalyst. *Annu Rev Cell Dev Biol*. 2011;27:79-105.

Vigil D, Cherfils J, Rossman KL, Der CJ. Ras superfamily GEFs and GAPs: validated and tractable targets for cancer therapy? *Nat Rev Cancer*. 2010;10:842-857.

Adapters

Nourry C, Grant SG, Borg JP. PDZ domain proteins: Plug and play! *Sci STKE*. 2003;2003(179):RE7.

Pawson T, Nash P. Assembly of cell regulatory systems through protein interaction domains. *Science*. 2003;300:445-452.

Scott JD, Dessauer CW, Taskén K. Creating order from chaos: Cellular regulation by kinase anchoring. *Annu Rev Pharmacol Toxicol*. 2013;53:187-210.

APPENDIX 25.1

Families of Protein Kinases

Groups	Substrate	Bacterial Genes	Yeast Genes	Human Genes	Examples	Regulation	Targets or Regulated Function
AGC	S, T	0	17	63	PKA	cAMP	Metabolic enzymes, TFs, channels
					PKB	PI3K, PDKs	GSK3/metabolism, survival
					PKC	Ca^{2+}, lipids	Receptor tyrosine kinases, channels, TFs
					PKG	cGMP	IP_3R, CFTR, VASP
					RSK	MAPK, PDKs	Ribosome/synthesis of translation machinery
					GRK	G proteins	Seven-helix receptor downregulation
CaMK	S, T	0	16	74	CaMK	Ca^{2+}, calmodulin	Synaptic transmission, cytoskeleton, TFs
					AMP-PK	AMP	Fatty acid, cholesterol synthesis
					MLCK	Ca^{2+}, calmodulin	Myosin-II/contraction
CK1	S, T, (Y)	0	4	10	CK-I, CK-II		Circadian clocks, Wnt
CMGC	S, T, (Y)	0	21	33	Cdks	Cyclins, phosphorylation	Many/cell cycle
					MAPK	Phosphorylation	TFs/proliferation
					GSK3	PKB	Glycogen metabolism, survival
RGC		0	0	5			
STE	S, T, Y	0	18	47	MAPKK	Phosphorylation	MAPK/proliferation
					PAK	Small GTPases	LIM kinase/cytoskeleton
Tyrosine kinase	Y	0	0	90	Receptor tyrosine kinases	Growth factors	PLCγ, Ras, MAPK pathway/cell proliferation
					Src family	Phosphorylation	Many/proliferation, lymphocyte activation, cytoskeleton, adhesion
Tyrosine kinase-like	S	0	0	43	Raf	Ras	MAPKK/proliferation
					TGF-βR	TGF-β	Smads/differentiation
Other	S, T, Y		38	83	Wee1p	Phosphorylation	Cdks/cell cycle
					Polo-like	Phosphorylation	Several/mitosis, cytokinesis
Atypical	S	0	15	48	ATM, ATR	DNA damage	p53, Chk1/cell-cycle arrest
Histidine kinases	H	0 to >30*	2	?	Tar	Aspartate	Bacterial chemotaxis, gene expression

AGC, protein kinase A, G, and C family; AMP-PK, adenosine monophosphate (AMP)-activated protein kinase; ATM, ataxia telangiectasia mutated; ATR, ataxia telangiectasia and Rad3-related protein; CaMK, calmodulin-activated protein kinase; cAMP, cyclic adenosine monophosphate; Cdk, cyclin-dependent kinase; CFTR, cystic fibrosis transmembrane regulator; cGMP, cyclic guanosine monophosphate; CK, casein kinase; CMGC, CDK, MAPK, GSK3, and CLK family; GF, growth factor; GRK, G protein–coupled receptor kinase; GSK, glycogen synthase kinase; H, histidine; IP_3R, inositol trisphosphate receptor, a Ca^{2+} release channel; LIM, Lin11, Isl-1, and Mec-3; MAPK, mitogen-activated protein kinase (also called ERK, for extracellular signal–regulated kinase); MAPKK, MAP kinase kinase; MLCK, myosin light-chain kinase; PAK, p21 (small GTPase)-activated kinase; PDK, 3-phosphoinositide-dependent protein kinase; PI3K, phosphatidylinositide 3′-kinase; PKA, cyclic AMP–dependent protein kinase; PKB, protein kinase B (also called Akt); PKC, calcium-dependent protein kinase; PKG, cyclic guanosine monophosphate (GMP)–dependent protein kinase; polo, a *Drosophila* gene; PLCγ, phospholipase C-gamma; Raf, cellular homologue of retroviral oncogene; RGC, Receptor guanylate cyclase family; RSK, ribosomal subunit 6 kinase; S, serine; Smads, Sma- and Mad-related proteins; STE, homologs of yeast STE7; T, threonine; Tar, aspartic acid receptor; TFs, transcription factors; TGF, transforming growth factor; VASP, vasodilator-stimulated phosphoprotein; Wee1p, fission yeast kinase; (Y), tyrosine.

*Number of histidine kinases in prokaryotes. **Bacteria:** *Bacillus subtilis* 37; *Escherichia coli* 7; *Borrelia burgdorferi*—Lyme disease spirochaeta 2; *Myobacterium tuberculosis* 14—also has 11 eukaryotic serine/threonine kinase genes, likely derived by lateral gene transfer from eukaryotic hosts. **Archaea:** *Methanococcus jannaschii* 0; *Aquifex aeolicus* 0; *Archaeoglobus fulgidus* 3.

APPENDIX 25.2

Parallels Among Guanosine Triphosphate-Binding Proteins

Family	Bacterial Genes	Yeast Genes	Worm Genes	Functions	GDP Dissociation Inhibitors	Receptors	GTP Exchange Factors	GTPase-Activating Factors	Direct Effectors
Small Gtpases									
Arf	0	6	11	Vesicule formation		Arf-GEFs	Sec-7/ARNO	Arf-GAP	COPI coat proteins
Rab	0	10	24	Vesicle targeting and fusion	Rab-GDI	Rab-GEFs	Rab-GEFs	Rab-GAP	Docking and fusion factors
Ran	0	2	2	Nuclear transport, mitotic spindle			Ran-GDF1, RCC1	RanBP1, RanGAP1	Importin β
Ras	0	4	8	Transduction of growth factor signals		Receptor tyrosine kinases	SOS	Ras-GAP	Raf
Rho	0	7	10	Regulation of actin cytoskeleton	Rho-GDI	Receptor tyrosine kinases, 7-helix receptors	Dbl/PH-GEFs	Rho-GAP	p65 PAK, Rho kinase, WASp
Sar	0	1	3	Vesicule formation		Sec12 GEF	Sec12	Sec23	COPII coat proteins
Trimeric G Proteins									
	0	2	20	Transduction of a wide variety of signals	Gβγ	Seven-helix receptors	7-Helix receptors	Effector proteins, RGS proteins	Many enzymes, channels
Elongation Factors									
EF-Tu/EF1α	1–2	4	5	Protein synthesis		Ribosome	EF-Ts/EF1β	Ribosome	
EF-G/EF2	1–2	5	4	Protein synthesis		Ribosome	—	Ribosome	
RF1,2/eRF	1–2	1	1–12	Protein synthesis	Ribosome				
Dynamin									
	0	2	1–3	Endocytosis	?		Not required	Dimerization	Membrane fission factors
Translocation GTPases									
Ffh/SRP54		2	?	Polypeptide translocation into ER			Nascent polypeptide chains	SRP receptor	Sec 61 translocon

EF-Tu/Ts, elongation factor Tu/Ts; ER, endoplasmic reticulum; GAP, GTPase-activating protein; GDI, guanine nucleotide dissociation inhibitor; GEFs, guanine nucleotide exchange factors; RGS, regulators of G-protein signaling; SRP, signal recognition particle; WASp, Wiskott-Aldrich syndrome protein.

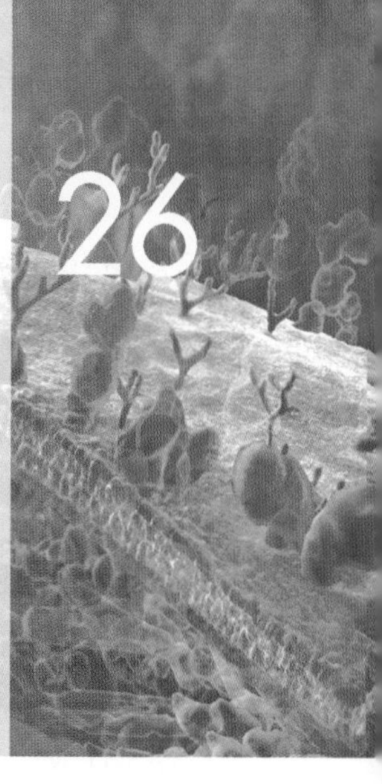

CHAPTER 26

Second Messengers

This chapter considers the remarkable variety of small molecules that carry signals inside living cells. These second messengers are chemically diverse, ranging from hydrophobic lipids confined to membrane bilayers, to an inorganic ion (Ca^{2+}), to nucleotides (cyclic adenosine monophosphate [cAMP] and cyclic guanosine monophosphate [cGMP]), to a gas (nitric oxide). The messages that these molecules carry are encoded by their concentrations. In the simplest case, a rise or fall in the concentration of the second messenger conveys a signal from its source to its target. In other cases, the signal depends on the rate or frequency of the fluctuations in the concentration of the second messenger. The local concentration of a second messenger depends on the rate of production, the rate of diffusion from the site of production, and the rate of removal. Most second messengers are produced by enzymes that switch on and off rapidly, allowing modulation of the concentration of second messengers on a millisecond time scale. In the case of Ca^{2+}, the cytoplasmic concentration is determined by channels that release the ion from membrane-delimited stores and by pumps that remove it from cytoplasm.

The physical state of second messengers has important consequences. Lipid-derived second messengers reach different targets in the cell depending on whether they are more soluble in lipid bilayers or in water. Similarly, Ca^{2+} acts only locally in cytoplasm, where a high concentration of binding sites limits its free diffusion. Cyclic nucleotides diffuse rapidly through cytoplasm, but their concentrations may rise and fall locally, owing to restricted sites of synthesis combined with rapid hydrolysis at particular sites in the cell.

The complexity of signaling pathways is determined by the number of sources and targets of each second messenger. Generally, multiple signal sources and multiple second messenger targets generate a remarkable complexity. Chapter 27 considers a few model systems in which it is possible to understand how signals are integrated and transduced.

This chapter presents second messengers in four sections: Cyclic Nucleotides, Lipid-Derived Second Messengers, Calcium, and Nitric Oxide. All these topics are interrelated, as multiple second messengers participate in many signaling systems. For example, nitric oxide controls the production of cGMP, and inositol triphosphate derived from a membrane lipid controls the release of Ca^{2+} into cytoplasm.

Cyclic Nucleotides

Two cyclic nucleoside monophosphates—adenosine 3′,5′-cyclic monophosphate **(cAMP)** and guanosine 3′5′-cyclic monophosphate **(cGMP)**—are employed as second messengers (Fig. 26.1). Both act by binding reversibly to specific proteins. Enzymes that produce and degrade cyclic nucleotides determine the concentrations of these messengers available to bind targets. These enzymes turn over their substrates rapidly, so they can amplify signals massively on a millisecond time scale, under the control of diverse signaling pathways (see three examples in Chapter 27). Cyclases make cyclic nucleotides in a single step from the corresponding nucleoside triphosphate, either adenosine triphosphate (ATP) or guanosine triphosphate (GTP).

Cyclic nucleotides diffuse in the cytoplasm at about the same rate as in free solution, activating a selected repertoire of downstream targets, including **protein kinases** (see Figs. 25.3 and 25.4), **cyclic nucleotide–gated ion channels** (see Fig. 16.10), and, in the case of cAMP, one class of guanine nucleotide exchange factors (Epac) for small guanosine triphosphatases (GTPases) (Rap1 and Rap2). The components of this system are quite ancient. The protein domains of the effector proteins that bind cAMP or cGMP are homologous to the cAMP-binding domain of CAP (cAMP binding protein), a bacterial transcription factor. Enzymes called cyclic nucleotide **phosphodiesterases** hydrolyze cAMP and cGMP to inactive nucleoside 5′-monophosphates.

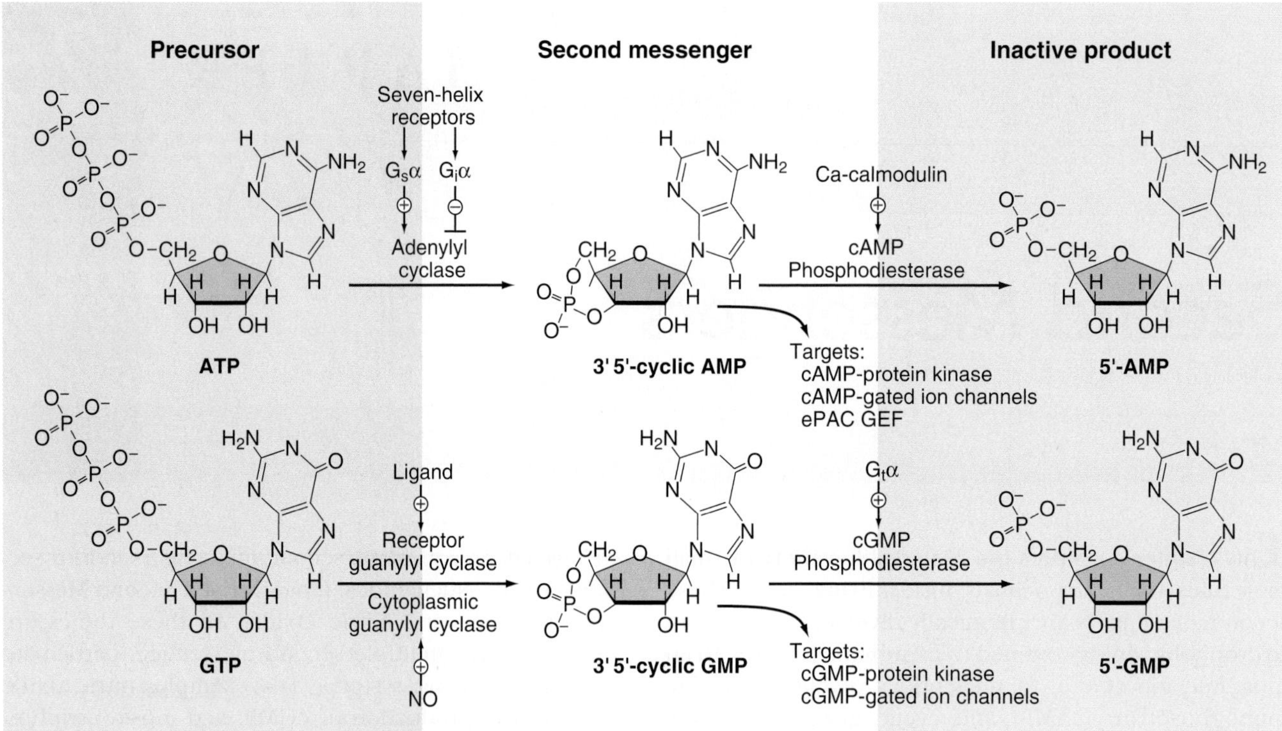

FIGURE 26.1 CYCLIC NUCLEOTIDE METABOLISM. Synthesis and degradation of cyclic adenosine monophosphate (cAMP) and cyclic guanosine monophosphate (cGMP), including regulatory inputs and targets. $G_s\alpha$, $G_i\alpha$, and $G_t\alpha$ are trimeric guanosine triphosphatase (GTPase) α subunits (see Fig. 25.9). AMP, adenosine monophosphate; ATP, adenosine triphosphate; Ca, calcium; GMP, guanosine monophosphate; GTP, guanosine triphosphate; NO, nitric oxide.

Eleven genes encode more than 40 different phosphodiesterases, which vary in their specificities for the two cyclic nucleotides, expression in various tissues, and localization to cellular compartments.

A family (10 human genes) of enzymes called **adenylyl cyclases** synthesize cAMP from ATP. One adenylyl cyclase is a soluble enzyme that can concentrate in the nucleus. However, most of these enzymes are anchored to the plasma membrane by multiple transmembrane segments. Two homologous catalytic domains reside in the cytoplasm (C_1 and C_2 in Fig. 26.2). These cytoplasmic domains can be produced experimentally as soluble proteins separately from the transmembrane domains and can then be recombined to make a fully active enzyme. Both domains are necessary because the active site lies at the interface between them. Generally, the concentration of adenylyl cyclases is very low relative to the trimeric G-proteins that regulate their activity.

Multiple regulatory mechanisms act synergistically to regulate adenylyl cyclases. **GTP-$G_s\alpha$,** the GTPase subunit of a trimeric G-protein, activates many membrane-associated adenylyl cyclases by binding far from the active site (Fig. 26.2) and inducing a conformational change. Ca^{2+}-calmodulin or protein kinase C (PKC) activate other adenylyl cyclases. **GTP-$G_i\alpha$,** the GTPase subunit of another trimeric G-protein, and protein kinase A (PKA) each inhibit some cyclases. Gβγ subunits of trimeric G-proteins activate some adenylyl cyclases but inhibit others. These diverse regulatory mechanisms allow adenylyl cyclases to integrate a variety of input signals. The diterpene **forskolin** from a *Coleus* plant binds to and activates adenylyl cyclases. Forskolin is not only useful to manipulate the cAMP concentration in cells experimentally, but it is also used to purify adenylyl cyclases by affinity chromatography. Regulation of the soluble form of adenylyl cyclase differs from that of all other isoforms. It is activated by bicarbonate, an essential step in sperm maturation.

The concentration of cAMP in resting cells is too low, approximately 10^{-8} M, to bind its targets. Stimulation of appropriate receptors (such as the seven-helix β-adrenergic receptor, see Fig. 27.3) increases the cytoplasmic cAMP concentration more than 100-fold, enough to saturate the regulatory subunits of protein kinase A (PKA) (Fig. 26.3). Dissociation of the regulatory (R) subunits frees the active PKA catalytic subunits (see Fig. 25.3D) to phosphorylate cytoplasmic and membrane substrates and to move into the nucleus to activate the transcription factor CREB (cyclic nucleotide regulatory element–binding protein) (see Fig. 10.21).

Outside the animal kingdom, cAMP has many functions. In bacteria, cAMP controls gene expression in response to nutritional conditions. The cellular slime mold *Dictyostelium* uses cAMP as an extracellular signal,

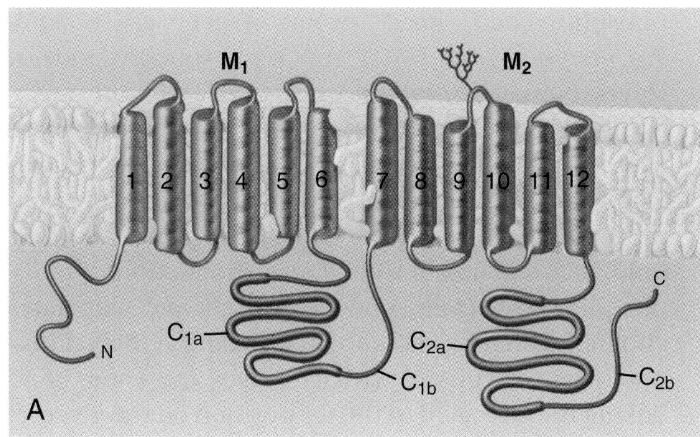

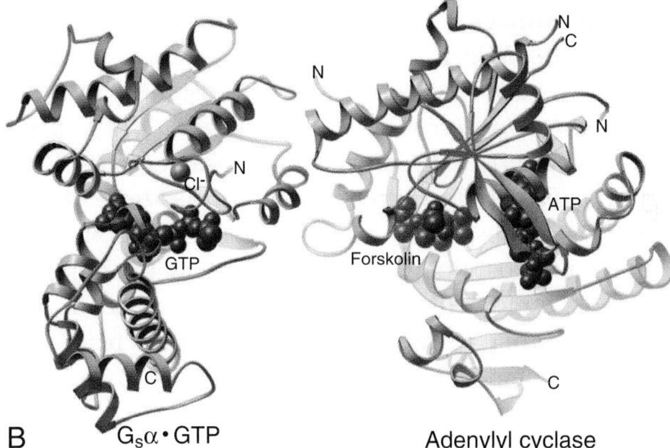

FIGURE 26.2 ADENYLYL CYCLASE. A, Topology of the polypeptide. The C_{1a} and C_{2a} regions fold together to form the active enzyme. **B,** Atomic structure of the catalytic domains of adenylyl cyclase associated with $G_s\alpha$. ATP is bound to the active site. Breaks in the chain are caused by disordered regions. (For reference, see Protein Data Bank [PDB; www.rcsb.org] file 1CJK. From Tesmer JJ, Sunahara RK, Gilman AG, et al. Crystal structure of the catalytic domains of adenylyl cyclase in a complex with $G_{s\alpha}$ GTP$_\gamma$S. *Science.* 1997;278:1907–1916. Copyright 1997 American Association for the Advancement of Science.)

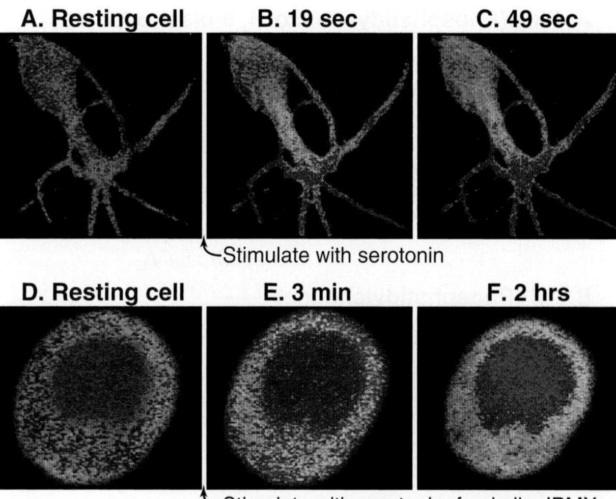

FIGURE 26.3 IMAGES OF cAMP TRANSIENTS IN CULTURED APLYSIA SENSORY NEURONS. Neurons were injected with protein kinase A (PKA) labeled on the catalytic subunit with fluorescein and on the regulatory subunit with rhodamine. Fluorescence energy transfer between the dyes on the two subunits provides an assay for cAMP, which dissociates the subunits and reduces energy transfer. Fluorescent dyes also allow detection of the subunits inside the neuron. **A,** Free cAMP in the resting cell is <50 nM *(blue).* **B–C,** Stimulation with serotonin activates adenylyl cyclase and increases cytoplasmic cAMP to the micromolar range *(red),* especially within fine processes. Images 120 μm wide were taken just inside the cell near the coverslip. **D,** Another resting neuron imaged at the level of the nucleus. At the low resting level of cAMP, <50 nM *(blue)* labeled PKA is excluded from the nucleus. **E,** Stimulation with serotonin plus forskolin (to stimulate adenylyl cyclase) and isobutylmethylxanthine (IBMX) to inhibit phosphodiesterases that catalyze cAMP breakdown, raises the concentration of cAMP around the nucleus *(yellow).* **F,** Two hours later, the free catalytic subunit of PKA *(pink)* accumulates in the nucleus. (Courtesy R.Y. Tsien, University of California, San Diego. **F,** From Bacskai BJ, Hochner B, Mahaut-Smith M, et al. Spatially resolved dynamics of cAMP and protein kinase A in *Aplysia* neurons. *Science.* 1993;260:222–226. Copyright 1993 American Association for the Advancement of Science.)

acting through a seven-helix receptor, for its social interactions (see Fig. 38.10).

Guanylyl cyclases are dimeric enzymes similar to adenylyl cyclases. In fact, mutation of just two amino acid residues can convert a guanylyl cyclase to an adenylyl cyclase. Vertebrates express two types of guanylyl cyclases (see Fig. 24.8). Members of a family of transmembrane receptors with cytoplasmic cyclase domains respond to ligand binding by aligning the two cytoplasmic domains of the enzyme to produce cGMP. The gases nitric oxide and carbon monoxide activate cytoplasmic guanylyl cyclases when they bind a heme group in a regulatory domain (see "Nitric Oxide" below).

Lipid-Derived Second Messengers

Phosphoglycerides (see Fig. 13.2) and sphingolipids (see Fig. 13.3) not only form cellular membranes, but also participate in a wide range of signaling mechanisms. The long list of intracellular and extracellular second messengers derived from lipids is undoubtedly still incomplete. Three membrane lipids are the primary sources of these signaling molecules (Fig. 26.4):

1. **Phosphatidylinositol** and its various phosphorylated derivatives, discussed later, are minor lipids of the cytoplasmic leaflet of the plasma membrane and organelle membranes.
2. **Phosphatidylcholine** is a major membrane phosphoglyceride found in both leaflets of the plasma membrane and organelle membranes (see Fig. 13.2).
3. **Sphingomyelin,** the major membrane sphingolipid, concentrates in the outer leaflet of the plasma membrane (see Fig. 13.3).

Enzyme Reactions That Produce Lipid Second Messengers

Three kinds of enzymes—phospholipases, lipid kinases, and lipid phosphatases—produce most lipid-derived second messengers. Remarkably, most conceivable

A. PIP_2 (Phosphatidylinositol 4, 5-bisphosphate)

Phosphatidic acid
Diacylglycerol
IP_3
Arachidonic acid
Phospholipase D
Phospholipase C
Phospholipase A_2

B. PC (Phosphatidylcholine)

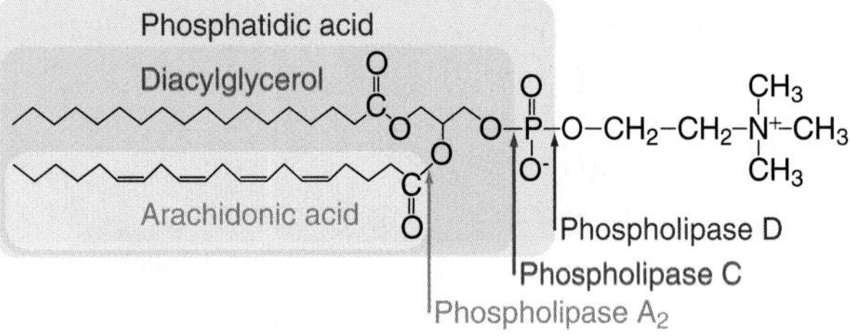

C. Sphingomyelin

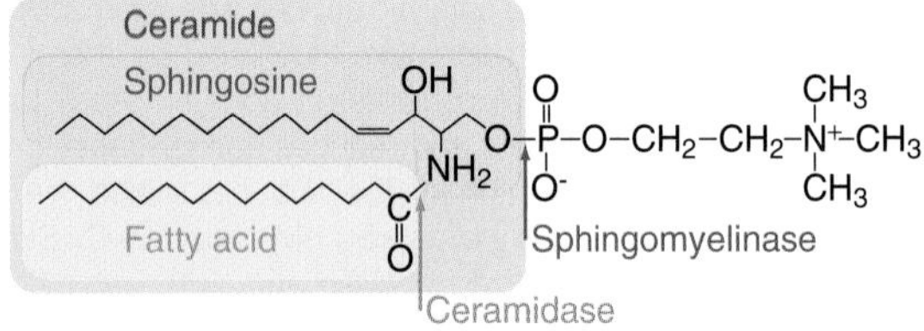

FIGURE 26.4 PRODUCTION OF LIPID SECOND MESSENGERS BY ENZYMATIC ATTACKS ON THREE PARENT LIPIDS. A, PIP_2 (phosphatidylinositol 4,5-bisphosphate). IP_3, inositol triphosphate. **B,** PC (phosphatidylcholine). **C,** Sphingomyelin. Second messengers are named and are surrounded by *colored boxes.*

products that could be produced by these enzymes from the three parent lipids do indeed participate in signaling reactions, either directly or indirectly. The second messengers produced by these reactions partition between the aqueous phase of the cytoplasm (products designated by *italics* in this section) and the hydrophobic phase of the membrane bilayer (products designated by ***bold italics*** in this section). In the following paragraphs, the details of the various enzymatic reactions and the structures of the lipid derivatives are less important than is the broader principle that cells use the full range of chemical diversity in their membrane lipids to create chemical signals to regulate cellular activities.

Three types of enzyme reactions generate lipid second messengers:

- Three types of **phospholipases** cleave three of the four ester bonds of phosphoglycerides (Fig. 26.4). Corresponding enzymes attack the two ester bonds and single amide bond of sphingomyelin. **Phospholipase A_2** (PLA_2) removes the C2 fatty acid, yielding a *free fatty acid,* which partitions into the cytoplasm with the aid of fatty acid–binding proteins, and a ***lysophosphoglyceride.*** The corresponding ceramidase removes the fatty acid from sphingomyelin. When **phospholipase C** (PLC) cleaves the phosphorylated head group (such as *inositol 1,4,5-triphosphate* [IP_3]) from a phosphoglyceride, it leaves behind ***diacylglycerol (DAG)*** in the membrane bilayer. The corresponding sphingomyelinase leaves behind ***ceramide*** in the membrane bilayer. **Phospholipase D** (PLD) cleaves the polar head group from phosphoglycerides, producing ***phosphatidic acid*** that remains in the bilayer. Production of phosphatidic acid from phosphatidylcholine generates choline, which has no known signaling activity. Phospholipase A_1 (PLA_1), which cleaves the ester bond linking the fatty acid to the C1 position on the glycerol is also not yet known to participate in signaling.
- **Lipid kinases** add phosphate groups to DAG to make ***phosphatidic acid*** and to phosphatidylinositol to make a variety of polyphosphoinositides, including ***phosphatidylinositol 4-phosphate (PIP), phosphatidylinositol 4,5-bisphosphate (PIP_2),*** and ***phosphatidylinositol 3,4,5-trisphosphate (PIP_3).*** PIP_2 is a substrate for a family of phosphoinositide-specific phospholipase Cs that produce the important signaling molecules ***DAG*** and *IP_3* (Fig. 26.4A).
- **Lipid phosphatases** remove phosphate from phosphatidic acid (another way to make ***DAG***) and phosphates from inositol head groups.

Cells also use transferases that add or exchange head groups or fatty acids on membrane phosphoglycerides (see Fig. 13.2). These enzymes are essential for lipid biosynthesis but have not been implicated in signaling. For example, transferases add choline (as the nucleotide conjugate cytidine diphosphate [CDP]-choline) to DAG to make phosphatidylcholine). In the case of phosphatidylinositol, the head group is provided by dephosphorylation of IP_3 to inositol, which is recombined enzymatically with a CDP conjugate of DAG to make phosphatidylinositol.

Sequential action of two or more enzymes produces some lipid second messengers. For example, cells can make ***DAG*** in two steps from phosphatidylcholine: PLD first makes phosphatidic acid, which is then dephosphorylated by phosphatidic acid phosphatase. PLD followed by PLA2 makes *lysophosphatidic acid* from phosphatidylcholine. PLA2 also initiates the production of a huge family of fatty acid derivatives called *eicosanoids* by cleaving the fatty acid arachidonic acid from phosphatidylcholine. Cyclooxygenases and lipoxygenases then modify arachidonic acid to make *prostaglandins, thromboxanes,* and *leukotrienes* (Figs. 26.9 and 26.10), signaling molecules that leave the cell to interact with surface receptors on other cells.

Agonists and Receptors

Most signaling pathways have the potential to produce lipid second messengers, depending on the expression of the appropriate enzymes. Some pathways activate enzymes directly. For example, receptor tyrosine kinases bind and activate certain **phosphatidylinositol phospholipase Cs** (PI-PLCs). Seven-helix receptors signal through trimeric G-proteins to activate another PI-PLC. Other pathways are less direct; one leads from G-protein–coupled receptors or receptor tyrosine kinases to DAG and PKC, activating an isoform of PLD.

Targets of Lipid Second Messengers

Downstream targets of the lipid second messengers are diverse but can be generalized to some extent into three categories (Fig. 26.5):

1. Most lipid second messengers derived from phosphoglycerides and retained in the membrane bilayer exert their physiological effects on PKC isozymes (Figs. 26.5 and 26.6). Ceramide is also retained in the bilayer, where it activates another protein kinase and a protein phosphatase (Fig. 26.11). Phosphatidic acid activates a lipid kinase, PI5 kinase, which phosphorylates phosphatidylinositol (Fig. 26.7).
2. IP_3 and sphingosine-1-phosphate each use different mechanisms to release Ca^{2+} from vesicular stores in the cytoplasm (Fig. 26.12).
3. All water-soluble lipid second messengers containing or derived from fatty acids (platelet-activating factor [PAF], lysophosphatidic acid [LPA], eicosanoids) cross the plasma membrane and leave the cell. Then they bind and activate seven-helix receptors on the surface of target cells.

With so much potential for information transfer (multiple agonists, multiple membrane transduction

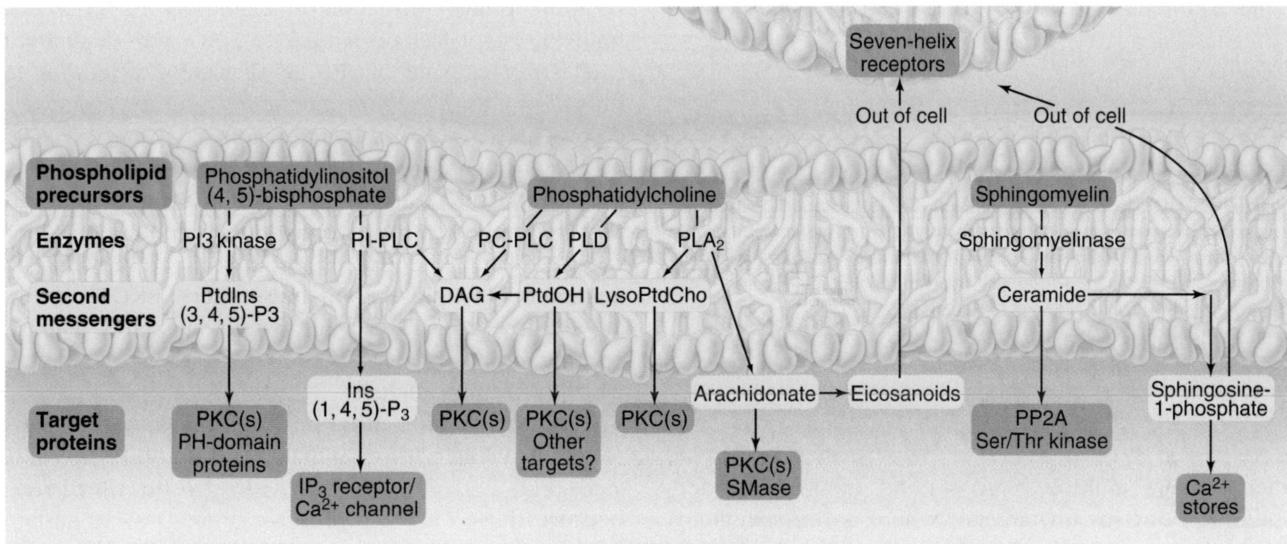

FIGURE 26.5 GENERATION OF LIPID SECOND MESSENGERS AND THEIR CELLULAR TARGETS. LysoPtdCho, lysophosphatidyl choline; PKC, protein kinase C; PLA_2, phospholipase A_2; PP2A, protein phosphatase 2A; PtdIns (3,4,5)-P_3, phosphatidylinositol 3,4,5-trisphosphate (PIP_3); PtdOH, phosphatidic acid. (Modified from Liscovitch M, Cantley LC. Lipid second messengers. *Cell.* 1994;77:329–334.)

A. Structure

Regulatory domain — DAG/Phorbol esters (C1); Calcium/acidic lipids (C2); PS; N … C — Kinase domain; C2-like (δ)

A. Structure	B. Tissue distribution	C. Activators
α	Ubiquitous	Ca^{2+}, DAG, PS, FFA, LysoPC
β	Many tissues	Ca^{2+}, DAG, PS, FFA, LysoPC
γ	Brain only	Ca^{2+}, DAG, PS, FFA, LysoPC
δ	Ubiquitous	DAG, PS
ε	Brain, others	DAG, PS
η	Lung, skin, heart	?
ζ	Ubiquitous	PIP_3

FIGURE 26.6 PROTEIN KINASE C (PKC) FAMILY. A, Domain organization of PKC isozymes with ligand-binding sites. PS *(yellow box)* indicates the pseudosubstrate sequence that binds to and inhibits the kinase catalytic site. **B,** Tissue distribution. **C,** Activators. DAG, diacylglycerol; FFA, free fatty acid; LysoPC, lysophosphatidylcholine; PIP_3, phosphatidylinositol 3,4,5-trisphosphate; PS, phosphatidylserine.

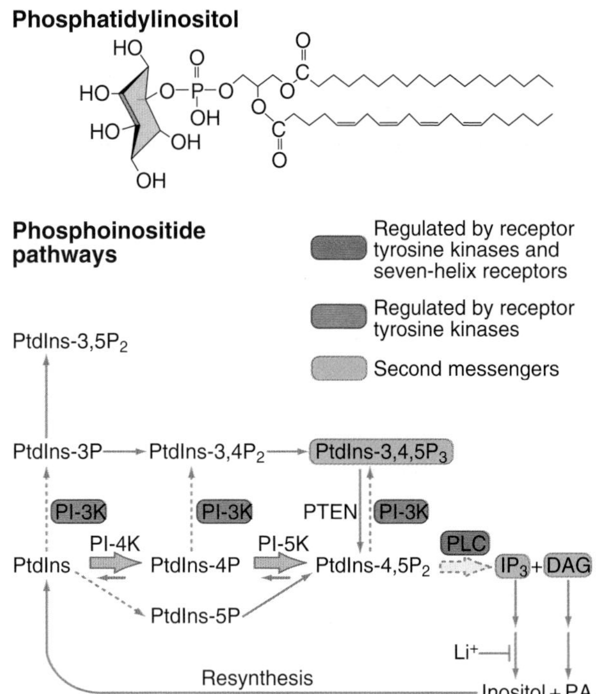

FIGURE 26.7 SYNTHESIS AND TURNOVER OF PHOSPHOINOSITIDES. Enzymes regulated by both receptor tyrosine kinases and seven-helix receptors are *purple*. Enzymes regulated only by receptor tyrosine kinases are *blue*. Second messengers are *green*. PA, phosphatidic acid; PI-3k, phosphatidylinositol 3′-kinase; PI-4k, phosphatidylinositol 4′-kinase; PI-5k, phosphatidylinositol 5′-kinase; PLC, phospholipase C; PLD, phospholipase D; PtdIns, phosphatidylinositol; PTEN, phosphatase and tensin homolog deleted on chromosome 10 (a tumor suppressor).

mechanisms, multiple lipid second messengers, and multiple downstream targets), where is the specificity in these systems? Do all cells respond in all possible ways? The answer is no because the protein hardware required for these reactions is selectively expressed in differentiated cells and carefully localized at particular sites in cells, such as the plasma membrane, nucleus, or cytoskeleton. In addition, targeting of PKC isozymes to particular cellular compartments ensures that only selected substrates are phosphorylated in response to lipid second messenger production. Thus, each cell uses a limited number of items from the lipid second messenger menu to produce a customized response to each agonist.

Other parts of this book present examples of lipid second messengers participating in many processes. For example, regulated secretion (see Chapter 21) in response to an agonist binding a seven-helix or tyrosine kinase receptor requires a transient Ca^{2+} signal in the cytoplasm. These receptors activate a PI-PLC to produce IP_3 from PIP_2. IP_3 releases Ca^{2+} from the endoplasmic reticulum (ER). IP_3-mediated Ca^{2+} release from the ER also controls smooth muscle contraction (see Fig. 39.24).

Protein Kinase C

Many lipid second messengers, including DAG, PIP_3, arachidonic acid, phosphatidic acid, and lysophosphatidylcholine activate one or more of the 10 PKC isozymes expressed by vertebrate cells (Fig. 26.6). These diverse PKC isozymes provide a selective response to various lipid second messengers. Some, but not all, PKC isozymes also require Ca^{2+} for activation. Sphingosine may inhibit some PKC isozymes.

Lipid second messengers activate PKC by dissociating an intramolecular **pseudosubstrate sequence** from the active site. Pseudosubstrates have alanine at the phosphorylation site instead of the serine found in substrates. DAG and other lipid second messengers bind C1 regions adjacent to the pseudosubstrate. DAG binding depends on phosphoglycerides, such as phosphatidylserine. **Phorbol esters,** pharmacological activators of PKC that promote tumor formation in laboratory experiments, bind PKC in a fashion similar to DAG. Ca^{2+}-dependent PKC isozymes have C2 regions that mediate binding to phospholipids in the presence of Ca^{2+}. During apoptosis, caspases cleave off this regulatory domain (see Fig. 46.12) producing constitutively active PKC isoforms.

Activated PKCs have many potential targets in cells and are implicated in the regulation of cellular activities ranging from gene expression to cell motility to the generation of lipid second messengers. PKC isozymes are selective toward certain protein substrates. The C2 regions target PKC isozymes to the plasma membrane, cytoskeleton, or nucleus.

PKCs can provide either positive or negative feedback to the signaling pathways that turn them on. PKC activates PLD and PLA_2 and provides positive feedback, because those enzymes produce more DAG to sustain the activation of PKC. On the other hand, PKC provides negative feedback when it phosphorylates and inhibits both growth factor receptors and PI-PLCγ_1. PKC also phosphorylates and inhibits PI-PLCβ, generating negative feedback after activation of seven-helix receptors. Negative feedback makes both of these signaling events transient.

Phosphoinositide Signaling Pathways

Although minor in terms of mass in biological membranes, phosphoinositides are major players in signaling (Fig. 26.7). The parent compound, PI, is a phosphoglyceride with a cyclohexanol head group called inositol. Specific lipid kinases phosphorylate the 4 and 5 hydroxyl groups of phosphatidylinositol to form PI(4-)P and PI(4,5-)P_2, usually referred to simply as PIP and PIP_2. A specific phosphatase can remove the D5 phosphate. Inactivation of the single copy of this gene on the human X-chromosome causes Lowe syndrome with cataracts, renal failure, and intellectual disability.

PIP_2 is a substrate for a family of receptor-controlled PI-PLCs that cleave off the phosphorylated head group, producing two potent second messengers: IP_3 and DAG. Water-soluble IP_3 activates Ca^{2+} release channels in the ER (Fig. 26.12), and lipid-soluble DAG activates PKC (Fig. 26.5). In contrast to Ca^{2+}, which diffuses slowly and acts locally, IP_3 diffuses rapidly through the cytoplasm, triggering Ca^{2+} release. DAG is confined to membranes but diffuses laterally to bind and activate PKC. Enzymes inactivate both IP_3 and DAG. DAG is inactivated by phosphorylation to make phosphatidic acid, a second messenger in its own right but also an intermediate in the resynthesis of phosphoinositides. IP_3 is dephosphorylated to inositol, which is inactive as a second messenger. Lithium chloride (LiCl) inhibits the final step, dephosphorylation of inositol-1-phosphate. Li^+ is clinically useful as a treatment for bipolar disorder, but it is not certain that it has its effects through phosphoinositide signaling. Some tissues inactivate IP_3 by phosphorylation to inositol 1,3,4,5-tetrakisphosphate (IP_4).

Vertebrates use 10 classes of PI-PLCs to provide tissue-specific coupling of various receptors to the production of IP_3 and DAG. An Src homology 2 (SH2) domain targets PI-PLCγ_1 to a phosphotyrosine on a receptor tyrosine kinase, which then activates the newly bound PLC by tyrosine phosphorylation (Fig. 26.12). Trimeric G-protein $G_q\alpha$ activates another isozyme, PI-PLCβ.

When a PI-PLC hydrolyzes PIP_2, phosphatidylinositol-4 kinase and phosphatidylinositol-5 kinase respond to replace the pool of PIP_2 at the expense of membrane phosphatidylinositol. This generates a transient flux of lipid molecules from phosphatidylinositol to PIP to PIP_2 to DAG. On a longer time scale, phosphatidylinositol is replaced by synthesis from phosphatidic acid and inositol.

Receptor tyrosine kinases activate another family of lipid kinases, **PI-3 kinases,** which phosphorylate the 3-hydroxyl group of PI, PIP, and PIP_2. The products are PI(3-)P, PI(3,4-)P_2, and PI(3,4,5-)P_3. PI(3,4,5-)P_3 is not a substrate for the PI-PLCs that produce IP_3 and DAG, but it activates some PKC isozymes and binds specifically to certain pleckstrin homology domains (see Fig. 25.10), bringing enzymes such as protein kinase B **(PKB/Akt)** to the plasma membrane. Association with the membrane leads to phosphorylation and activation of PKB as part of insulin signaling (see Fig. 27.7). A phosphatase that removes the D_3 phosphate from PIP_3 is a tumor suppressor called **PTEN** (phosphatase and tensin homolog deleted on chromosome 10). Loss of PTEN function in tumors allows PIP_3 to build up, activating PKB and growth-promoting downstream pathways involving Tor kinase. A steroid-like molecule called **wortmannin** inhibits PI-3 kinase relatively specifically and is used to investigate the physiological roles of PIP_3.

Phosphatidylcholine Signaling Pathways

Phosphatidylcholine is not only a major structural lipid of the plasma membrane but also an important source of DAG and a large family of other signaling molecules (Fig. 26.5). In response to agonist stimulation, cells produce two waves of DAG (Fig. 26.8). Within seconds, PI-PLCs activated by seven-helix or tyrosine kinase receptors produce the first wave of DAG from PIP_2. Then, over a period of minutes, a second wave of DAG is derived from phosphatidylcholine, either directly by a phosphatidylcholine (PC)-PLC or in two steps by PLD (to remove choline) and a phosphatidic acid phosphatase (to remove the phosphate from phosphatidic acid). The first wave of DAG may contribute to the second wave as PKC activates one PLD isoform, and Ca^{2+} produced in the first wave may also activate PLD.

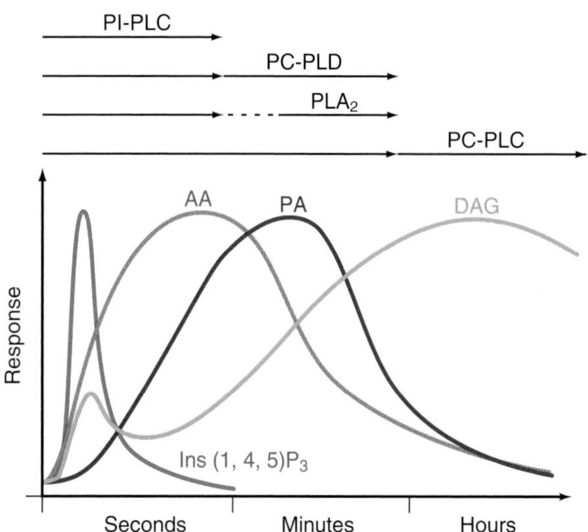

FIGURE 26.8 TIME COURSE OF LIPID SECOND MESSENGERS PRODUCED BY VARIOUS PHOSPHOLIPASES FOLLOWING ACTIVATION OF RECEPTOR TYROSINE KINASES OR SEVEN-HELIX RECEPTORS. AA, arachidonic acid; DAG, diacylglycerol; PA, phosphatidic acid; PLA_2, phospholipase A_2.

Phosphatidylcholine is also the main source of fatty acid–derived second messengers. PLA_2 releases **arachidonic acid,** a C20 unsaturated fatty acid that is found predominantly at the C2 position of phosphoglycerides. Arachidonic acid activates some PKC isozymes and, together with DAG, provides positive feedback to PLA_2 and PLD to sustain the production of arachidonic acid and DAG.

Lipid-Derived Second Messengers for Intercellular Communication

Cells produce an amazing array of bioactive compounds from phosphatidylcholine and arachidonic acid, forming second messengers that escape from the cell and mediate their effects by binding to receptors either on the cell of origin or neighboring cells. Thus these compounds are locally active hormones. This sets them apart from classical second messengers, which (with the exception of nitric oxide) act inside the cell of origin. All the compounds presented here activate target cells by binding

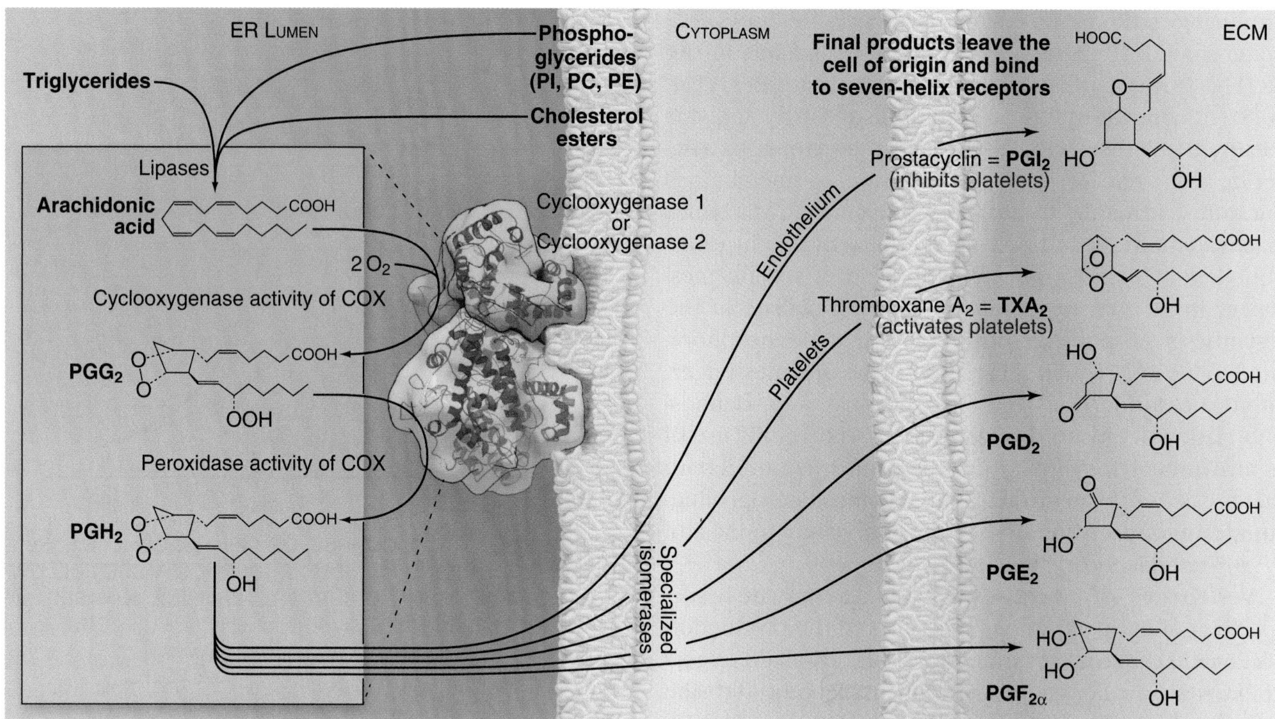

FIGURE 26.9 PATHWAY OF PROSTAGLANDIN SYNTHESIS. Cyclooxygenase-1 (ribbon diagram with space-filling surface) is a homodimer that is bound to the inner surface of the endoplasmic reticulum (ER) membrane by hydrophobic membrane-binding helices. This enzyme has two active sites that convert arachidonic acid to prostaglandin H_2. A hydrophobic channel 2.5 nm long leads from the bilayer to the active sites. Nonsteroidal antiinflammatory drugs compete with arachidonic acid for binding to the cyclooxygenase active site, and aspirin covalently modifies serine 530 in the active site. Specific prostaglandin synthases on the cytoplasmic surface of the ER convert prostaglandin H_2 into various products, which then diffuse out of the cell. COX, cyclooxygenase; PC, phosphatidylcholine; PE, phosphatidylethanolamine; PI, phosphatidylinositol; TXA_2, thromboxane A_2. (For reference, see PDB file 1CQE.)

seven-helix receptors. Vertebrates make a particularly rich variety of these lipid-derived signaling molecules, but slime molds and algae use some of the same compounds for communication.

Eicosanoids (*eicosa* is Greek for "20") are a diverse family of metabolites derived from the 20-carbon fatty acid arachidonic acid. They include **prostaglandins, thromboxanes, leukotrienes,** and **lipoxins** (Figs. 26.9 and 26.10). Depending on the particular receptor and G-protein, eicosanoids selectively activate or inhibit the synthesis of cAMP, release Ca^{2+}, activate PKC, or regulate ion channels. Biological consequences are diverse, depending on the specific eicosanoid and target cell. Eicosanoids are important medically as mediators of inflammation.

Prostaglandins and thromboxanes are synthesized from arachidonic acid by pairs of enzymes, the first being generic and the second being specific for a particular product (Fig. 26.9). Most differentiated cells express predominantly one second-step enzyme and thus produce just one of these local hormones.

The first enzyme, **prostaglandin H synthetase** (also called cyclooxygenase), has two active sites that catalyze successive reactions that convert arachidonic acid into prostaglandin G_2 and then into prostaglandin H_2. Most cells express cyclooxygenase-1 (COX-1) constitutively as a housekeeping enzyme. Inflammatory stimuli induce expression of the closely related enzyme **cyclooxygenase-2** (COX-2). Cyclooxygenases are located inside the ER and intermembrane space of the nuclear envelope. Second-tier enzymes are specific **prostaglandin isomerases** that convert prostaglandin H into various prostaglandin and thromboxane products on the cytoplasmic surface of the ER.

Physiological responses to each eicosanoid depend on the selective expression of specific seven-helix receptors. For example, receptors for prostaglandin F_{2a} prepare the uterus, but not other organs of pregnant mammals, for delivery. Counteracting eicosanoids and their receptors control the response of platelets to blood vessel damage (see Fig. 30.14). Normally, endothelial cells lining blood vessels produce prostaglandin I_2 (prostacyclin), which inhibits interaction of platelets with the vessel wall and promotes blood flow by relaxing smooth muscle cells in the wall of the vessel. If blood vessels are damaged, the blood-clotting protein thrombin activates platelets to aggregate and plug gaps in the endothelium (see Fig. 30.14). Activated platelets produce thromboxane A_2, which diffuses locally to promote irreversible platelet aggregation and blood vessel constriction by triggering smooth muscle cells to produce IP_3, release Ca^{2+}, and contract (see Fig. 39.24). This secondary response

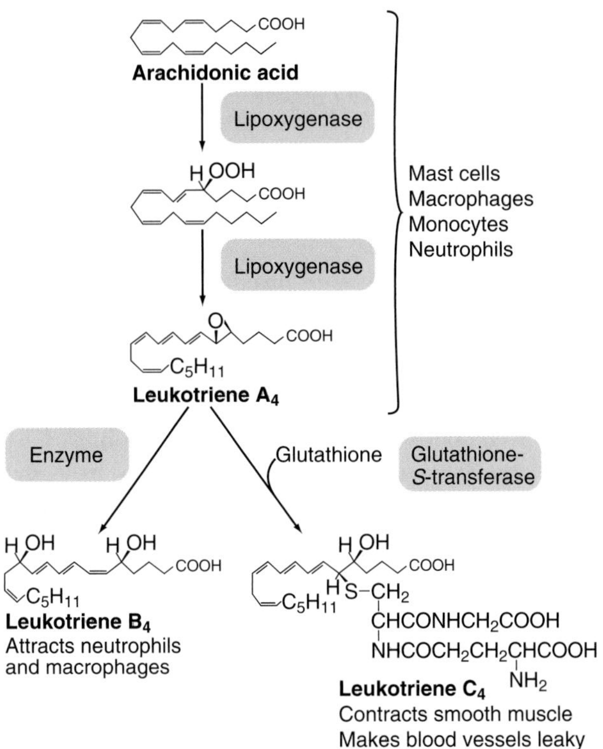

FIGURE 26.10 PATHWAYS OF LEUKOTRIENE SYNTHESIS.

mediated by positive feedback through thromboxane A_2 minimizes blood loss but can contribute to pathological formation of clots in vital blood vessels, resulting in heart attacks and strokes. Tissue injury also provokes synthesis of prostaglandins E_2 and I_2, which mediate inflammation locally by dilating blood vessels, sensitizing pain receptors, and causing fever.

Nonsteroidal antiinflammatory drugs, including aspirin and ibuprofen, target both cyclooxygenase isozymes. Most of these drugs competitively inhibit arachidonic acid binding to both enzymes, but aspirin covalently and irreversibly acetylates a serine residue in their active sites. Either way, these drugs inhibit the synthesis of all prostaglandins and thromboxanes. Aspirin reduces the incidence of heart attacks and strokes by inhibiting the synthesis of thromboxane A_2 by platelets, which reduces platelet aggregation and pathological clotting in blood vessels. Low doses of aspirin are selective and effective for platelets, as platelets have only COX-1 and lack the capacity for protein synthesis to replace inactivated enzyme. Low-dose aspirin also protects against colon cancer. High doses of aspirin and other nonsteroidal antiinflammatory drugs reduce inflammation, pain, and fever by inhibiting the synthesis of other eicosanoids, but this is not without potential side effects, including gastrointestinal bleeding. Drugs that selectively inhibit COX-2 relieve inflammation without the gastrointestinal side effects caused by inhibition of COX-1. However, they inhibit the synthesis of prostaglandin I_2 by endothelium. The lack of prostaglandin I_2 to inhibit platelet aggregation increases the risk of heart attacks and strokes, so most of these drugs have been withdrawn from clinical use.

Macrophages and white blood cells produce an enzyme called **5-lipoxygenase** that synthesizes leukotrienes and lipoxins by addition of oxygen to specific double bonds of arachidonic acid (Fig. 26.10). The first biologically active product, leukotriene A_4, can be modified by addition and subsequent trimming of glutathione to yield a variety of active leukotrienes. Leukotrienes mediate inflammatory reactions by constricting blood vessels, allowing plasma to leak from small vessels and attracting white blood cells into connective tissue. These effects, together with constriction of the respiratory tract, contribute to the symptoms of asthma. Drugs that inhibit 5-lipoxygenase and leukotriene receptors can relieve symptoms of asthma, but other agents are more effective. Lipoxins are another family of mediators that act on blood vessels. They are formed by the addition of oxygen to both the C-5 and C-15 positions of arachidonic acid.

Phosphatidylcholine is also the starting point for production of two other water-soluble, intercellular, second messengers: lysophosphatidic acid (**LPA**) and platelet activating factor (**PAF**). Activated platelets and injured fibroblasts produce LPA from phosphatidylcholine by the action of the phospholipases PLD and PLA_2 (Fig. 26.4). A variety of cells synthesize PAF in two steps, using PLA_2 to remove the C2 fatty acid from phosphatidylcholine molecules with an ether-bonded fatty alcohol (rather than an ester-bonded fatty acid) on the C1 position and a second enzyme to acetylate the C2 hydroxyl group. LPA and PAF escape cells and stimulate target cells by binding to seven-helix receptors. Depending on their signaling hardware, cells respond to LPA in different ways: Activation of the PLC/IP_3 pathway releases intracellular Ca^{2+} in some cells; activation of a mitogen-activated protein (MAP) kinase pathway (see Fig. 27.6) stimulates some cells to divide; and activation of Rho-family small GTPases stimulates formation of actin bundles in cultured cells (see Fig. 33.19). As is expected from its name, PAF activates platelets, but it also modifies the behavior of other blood cells, inhibits heart contractions, and stimulates contraction of the uterus.

Endocannabinoids are a family of fatty acid amides that activate the seven-helix receptors that also respond to Δ^9-tetrahydrocannabinol, the active ingredient in marijuana. Many tissues, including the brain, synthesize the classic endocannabinoid, *N*-arachidonyl-ethanolamine, from ethanolamine and arachidonic acid. Brain cannabinoid receptors are found exclusively on the axonal side of synapses. Stimulation of the postsynaptic neuron results in Ca^{2+} entry, activating the enzymes that synthesize endocannabinoids. When these lipid second messengers diffuse from the postsynaptic cell back to the presynaptic terminal, they bind class 1 cannabinoid receptors, which suppress transmitter release by blocking Ca^{2+} entry. Some of these presynaptic terminals secrete the inhibitory neurotransmitter γ-aminobutyric acid (GABA), so the endocannabinoid increases the activity of the postsynaptic neuron, including those that suppress the sensation of pain. The functions of cannabinoid receptors in other tissues are less-well understood. Another lipid amide, oleamide (*cis*-9,10-octadecenoamide), induces sleep in mammals, most likely through GABA receptors (see Fig. 11.8). An enzyme called *fatty acid amide hydrolase* degrades all these signaling molecules.

Sphingomyelin/Ceramide Signaling Pathways

Activation of plasma membrane receptors for **TNF** and **IL-2** stimulates formation of the lipid-soluble second messenger ***ceramide*** from sphingomyelin (Fig. 26.11) in parallel with the activation of the transcription factor nuclear factor kappa B (NF-κB) (see Fig. 10.21). Sphingomyelin is concentrated in the external leaflet of the plasma membrane (see Fig. 26.4). Ligand binding to TNF or IL-2 receptors activates a plasma membrane **sphingomyelinase** (a specialized PLC) that removes phosphorylcholine from sphingomyelin, producing ceramide. Ceramide flips across the lipid bilayer to the cytoplasmic surface and activates a proline-directed, serine/threonine protein kinase associated with the plasma membrane. Target proteins have the sequence X-Ser/Thr-Pro-X and include epidermal growth factor (EGF) receptor and Raf kinase. Downstream events include activation of MAP kinase, which activates transcription factors and other effectors (see Fig. 27.6). Ceramide also activates a protein phosphatases 1 and 2A.

Sphingosine and ***sphingosine-1-phosphate*** are produced from sphingomyelin by a succession of enzymatic reactions (Fig. 26.4). Ceramidase removes the fatty acid, and sphingomyelinase removes the phosphorylcholine head group to form sphingosine. Then sphingosine kinase adds phosphate to C1 to make sphingosine-1-phosphate. Sphingosine-1-phosphate escapes from cells and activates seven-helix receptors on other cells. Several parallel signaling pathways influence motility of lymphocytes and growth of blood vessels, as well as smooth muscle contraction. Antagonists of sphingosine-1-phosphate are being tested as antiinflammatory drugs.

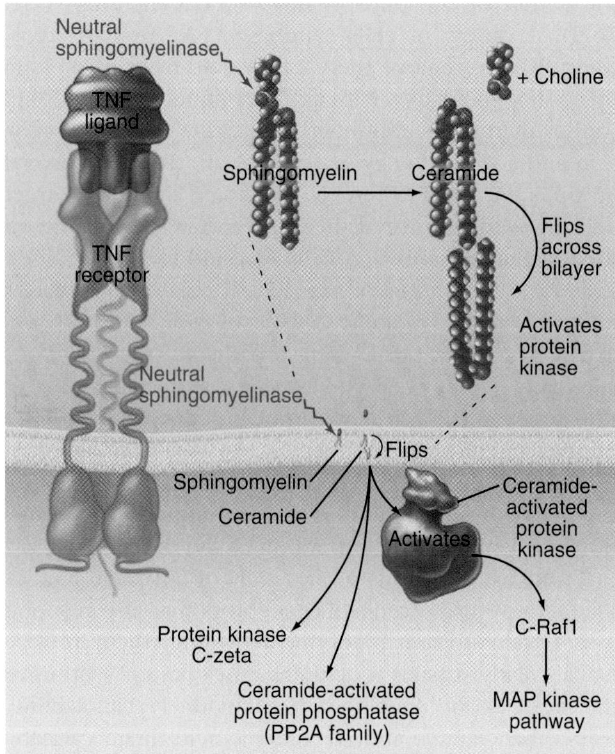

FIGURE 26.11 SPHINGOMYELIN/CERAMIDE SIGNALING PATHWAY. Stimulation of the tumor necrosis factor (TNF) receptor activates a neutral sphingomyelinase, which cleaves choline from sphingomyelin. Ceramide flips across the bilayer and activates a cytoplasmic kinase as well as protein kinase C (PKC)-ζ and a protein phosphatase. MAP, mitogen-activated protein.

Cross Talk

Interaction (cross talk) among lipid second messenger pathways is thought to integrate signals from different agonists. Consequently, it is difficult to define distinct linear pathways from an agonist to an individual effector through lipid second messengers. The following are some examples of cross talk using enzymes defined in Fig. 26.4:

1. Positive feedback. PI-PLCs produce DAG, which activates PKC, which phosphorylates and activates PLA_2 and PLD. These activated enzymes produce additional lipid second messengers (arachidonic acid and phosphatidic acid) to amplify and diversify the initial response.
2. Convergence of two pathways on the same target. Arachidonic acid (produced by activated PLA_2) and DAG (produced by either a PI-PLC or PLD and phosphatidic acid phosphatase) synergistically activate some PKC isozymes.
3. Conversion of a messenger into another messenger. DAG and phosphatidic acid are readily interconverted by the appropriate kinase and phosphatase.

Calcium

Overview of Calcium Signaling

Calcium ion, Ca^{2+}, is a versatile second messenger that regulates many processes, including synaptic transmission, fertilization, secretion, muscle contraction, and cytokinesis. All eukaryotes (but not prokaryotes) use Ca^{2+} signals. Nature probably chose Ca^{2+} for signaling by default. Given that cells depend on phosphate for energy metabolism (ATP), nucleic acid structure, and many other functions, early cells evolved mechanisms to extrude Ca^{2+} from cytoplasm to avoid precipitation of calcium phosphate. Cells took advantage of the resulting Ca^{2+} gradient between cytoplasm and ocean water (or extracellular space in animals) to drive tiny pulses of Ca^{2+} into cytoplasm for signaling. This movement of Ca^{2+} between compartments via ion channels is very fast.

Rather than being synthesized and metabolized like all other second messengers, Ca^{2+} is released into and removed from the cytoplasm (Fig. 26.12). ATP-driven **Ca^{2+} pumps** in the plasma membrane and ER keep cytoplasmic Ca^{2+} levels low. A variety of stimuli, operating through different receptors (Table 26.1), open **Ca^{2+} channels** in the plasma membrane or ER, allowing a concentrated burst of Ca^{2+} to enter the cytoplasm. Cellular responses to Ca^{2+} signals depend on the repertoire of Ca^{2+}-sensitive proteins and effector systems (Appendix 26.1). Some proteins respond directly by binding Ca^{2+}, whereas others respond to Ca^{2+} bound to a small protein called calmodulin.

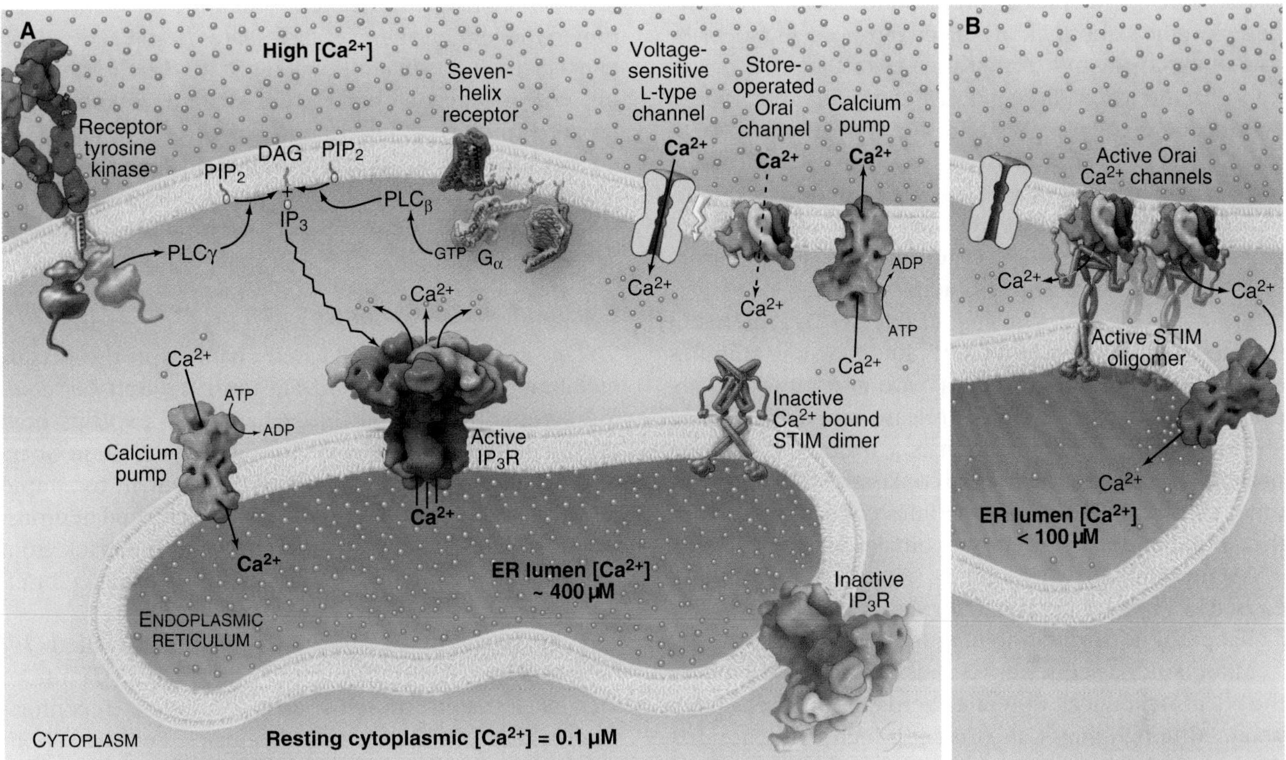

FIGURE 26.12 PATHWAYS OF Ca^{2+} RELEASE AND UPTAKE. A, Ca^{2+} pumps in the plasma membrane and endoplasmic reticulum (ER) membrane use ATP hydrolysis to pump Ca^{2+} out of the cytoplasm. A variety of receptors activate phospholipase C (PLC) isozymes to produce IP_3 from phosphatidylinositol 4,5-bisphosphate (PIP_2). IP_3 diffuses to the ER, where it opens IP_3 receptors (IP_3Rs), releasing Ca^{2+} from the lumen into cytoplasm. Voltage-sensitive L-type Ca^{2+} channels respond to membrane depolarization by admitting extracellular Ca^{2+}. ADP, adenosine diphosphate; STIM, stromal interaction molecule. **B,** Store-operated Ca^{2+} entry. If signaling activities deplete Ca^{2+} from the lumen of the ER, Ca^{2+} dissociates from binding sites on STIM located in the lumen. This results in aggregation of STIM and interactions of its cytoplasmic domains with plasma membrane Orai Ca^{2+} channels, This opens a channel that allows extracellular Ca^{2+} to move slowly into the cytoplasm, so the calcium pumps in the ER can replenish the Ca^{2+} stores.

TABLE 26.1 Stimuli for Ca^{2+} Signals

Stimulus	Receptor Class	Second Messenger	Distribution
Action potentials	Voltage-sensitive Ca^{2+} channels	None	Nerve, muscle
Adenosine triphosphate (ATP)	P2X plasma membrane channels	None	Many cells
	P2Y seven-helix G-protein-coupled receptors	Inositol triphosphate (IP_3)	Many cells
Peptide growth factors	Receptor tyrosine kinases	IP_3	Many cells
Antigens	T-cell receptor, tyrosine kinases	IP_3	Lymphocytes
Peptide hormones	Seven-helix G-protein-coupled receptors	IP_3	Endocrine cells, other epithelial cells
Neurotransmitters	Ligand-gated cation channels	None	Neurons
	Seven-helix G-protein-coupled receptors	IP_3	Neurons
Light	Rhodopsin, G-proteins	IP_3	Photoreceptors

Removal of Ca^{2+} From Cytoplasm

P-type ATP-driven Ca^{2+} pumps (see Fig. 14.7) in the plasma membrane and ER keep cytoplasmic Ca^{2+} levels low and generate a 5000-fold concentration gradient across these membranes (Fig. 26.12). Elevated cytoplasmic Ca^{2+} concentrations activate Ca^{2+} pumps until the cytoplasmic Ca^{2+} concentration falls to about 0.1 μM, the resting level. Three different genes and alternative splicing produce at least five different Ca^{2+}-ATPase pumps.

A remarkably high concentration of SERCA (sarcoplasmic-endoplasmic reticulum calcium ATPase) pumps in the smooth ER of striated muscle cells (specially named **sarcoplasmic reticulum**) gives them the capacity to handle the millisecond Ca^{2+} transients that control contraction (see Fig. 39.15). In the heart, the activity of the Ca^{2+}-ATPase is modulated by **phospholamban,** a 6-kD integral membrane protein of the sarcoplasmic reticulum. Phosphorylation by PKA and calmodulin-activated kinase (CaM kinase) dissociates phospholamban from the Ca^{2+} pump, stimulating its activity.

ER stores Ca^{2+} for signaling (Table 26.2). Proteins in the lumen bind much of the Ca^{2+}, but their low affinities ensure that bound and free Ca^{2+} are in a rapid equilibrium, providing free Ca^{2+} for release when membrane channels open. **Calsequestrin** is a major Ca^{2+}-binding protein of striated muscle sarcoplasmic reticulum. In other cells, the ER lumen sometimes contains calsequestrin, but more commonly, it contains **calreticulin,** a 47-kD protein with a low affinity (K_d = 250 μM) but high capacity (25 moles per mole) for Ca^{2+}. Calreticulin is also a chaperone for protein folding (see Fig. 20.12).

Mitochondria sequester Ca^{2+}, using carriers driven by the electrochemical potential across the inner membrane. Although their Ca^{2+} content is high, mitochondria do not participate in signal transduction by regulated release of Ca^{2+} into cytoplasm.

Refilling Endoplasmic Reticulum by Store-Operated Ca^{2+} Entry

Repeated stimulation can deplete Ca^{2+} from intracellular stores, because plasma membrane pumps remove from the cell part of the Ca^{2+} released into the cytoplasm from the ER. Experimentally, ER stores can also be depleted by **thapsigargin,** a lactone isolated from plants that inhibits most known ER Ca^{2+} pumps.

A process called **store-operated Ca^{2+} entry** replenishes Ca^{2+} stores in the ER by admitting extracellular Ca^{2+} through plasma membrane channels called *Orai* (Fig. 26.12B). These low-conductance Ca^{2+}-selective channels are formed by a hexamer of identical subunits with four transmembrane domains unrelated to other channels described in Chapter 16. An integral protein of the ER called STIM senses the Ca^{2+} concentration in the lumen and controls Orai. Stromal interaction molecule (STIM) has Ca^{2+}-binding EF hands (Ca^{2+} binding sites in calmodulin consisting of α-helices E and F) (see Fig. 3.12C) in the lumen, one transmembrane helix and a cytoplasmic domain with coiled-coils that interacts with Orai. STIM dimers are dispersed in the membrane when the Ca^{2+} level is high inside the lumen. When the Ca^{2+} concentration in the lumen falls, Ca^{2+} dissociates from STIM, which clusters into units that interact directly with the cytoplasmic end of Orai and open the Ca^{2+}-channel. Pumps in the ER take up the Ca^{2+} entering the cytoplasm, refilling the Ca^{2+} stores and dissociating STIM clusters from Orai. Mutations in Orai cause immune deficiency, because store-operated Ca^{2+} entry is essential for lymphocyte activation (see Fig. 27.8).

Calcium-Release Channels

Voltage-gated and agonist-gated channels in the plasma membrane (Table 26.3 and Fig. 26.12) admit Ca^{2+} into the cytoplasm from outside. Chapter 16 explains how the membrane potential or agonists open these channels. Voltage-gated channels are essential for rapid responses in excitable cells such as muscles and neurons. Owing to rapid inactivation by negative feedback from the released Ca^{2+}, most of these channels produce brief, self-limited Ca^{2+} pulses.

Two types of agonist-gated channels—called *IP_3 receptors* and *ryanodine receptors*—release Ca^{2+} from the ER. Striated muscles uses ryanodine receptors, whereas smooth muscle and nonmuscle cells have both types of release channels (Table 26.2). In excitable muscle cells, plasma membrane Ca^{2+} channels trigger ryanodine-receptor channels to release Ca^{2+} from the ER. In nonexcitable cells, stimulation of either seven-helix receptors or receptor tyrosine kinases produces IP_3, which triggers IP_3 receptors to release Ca^{2+} from the ER.

Inositol 1,4,5-Trisphosphate Receptor Ca^{2+} Channels

Numerous signal transduction pathways generate IP_3 (Fig. 26.7), which activates IP_3 receptors to release Ca^{2+} from the ER in animal cells. Plants and fungi appear to lack IP_3 receptors.

IP_3 receptors are tetramers of giant 313-kD polypeptides with multiple domains mostly in the cytoplasm (Fig. 26.13B). Six transmembrane segments near the

TABLE 26.2 Molecular Components of the Calcium-Sequestering Compartments

Cell Type	Ca^{2+} Pump	Sequestering Proteins	Release Channel
Striated muscles	SR Ca-ATPase	Calsequestrin	Ryanodine receptor
Smooth muscle	Ca-ATPase	Calreticulin > calsequestrin	IP_3 receptor and ryanodine receptor
Nonmuscle cells	One of five Ca-ATPases	Calreticulin > calsequestrin	IP_3 receptor and/or ryanodine receptor

TABLE 26.3 Ca^{2+}-Release Channels

Type	Distribution	Control	Features
Plasma Membrane Ca^{2+} Channels			
ATP-activated channel	Smooth muscle	Extracellular ATP	
Cyclic adenosine monophosphate (cAMP)-activated channel	Sperm	Cytoplasmic cAMP	
L-type Ca^{2+}-channel	Skeletal and cardiac muscle, brain, other nonmuscle cells	Voltage	Excitation-contraction coupling, defective in muscular dysgenesis. High threshold, dihydropyridine (DHP)-sensitive, regulated by PKA
N-type Ca^{2+}-channel	Neurons, endocrine cells	Voltage	Neurotransmitter release, modulated by G-proteins. High threshold, conotoxin sensitive
P-type Ca^{2+}-channel	Purkinje neurons	Voltage	Insensitive to DHP and conotoxin
T-type Ca^{2+}-channel		Voltage	Low threshold
Endoplasmic Reticulum Ca^{2+} Channels			
IP_3 receptors	Most cells including brain and smooth muscle	IP_3, Ca^{2+}	Heparin-sensitive
Type I ryanodine receptor	Skeletal muscle	DHP-receptor, Ca^{2+}	Ca^{2+} release stimulates contraction
Type II ryanodine receptor	Cardiac muscle, other cells	Ca^{2+}, cyclic adenosine diphosphate (cADP)-ribose	Ca^{2+} release stimulates contraction
Type III ryanodine receptor	Smooth muscle, other cells	Ca^{2+}, cADP-ribose	Ca^{2+} release stimulates contraction

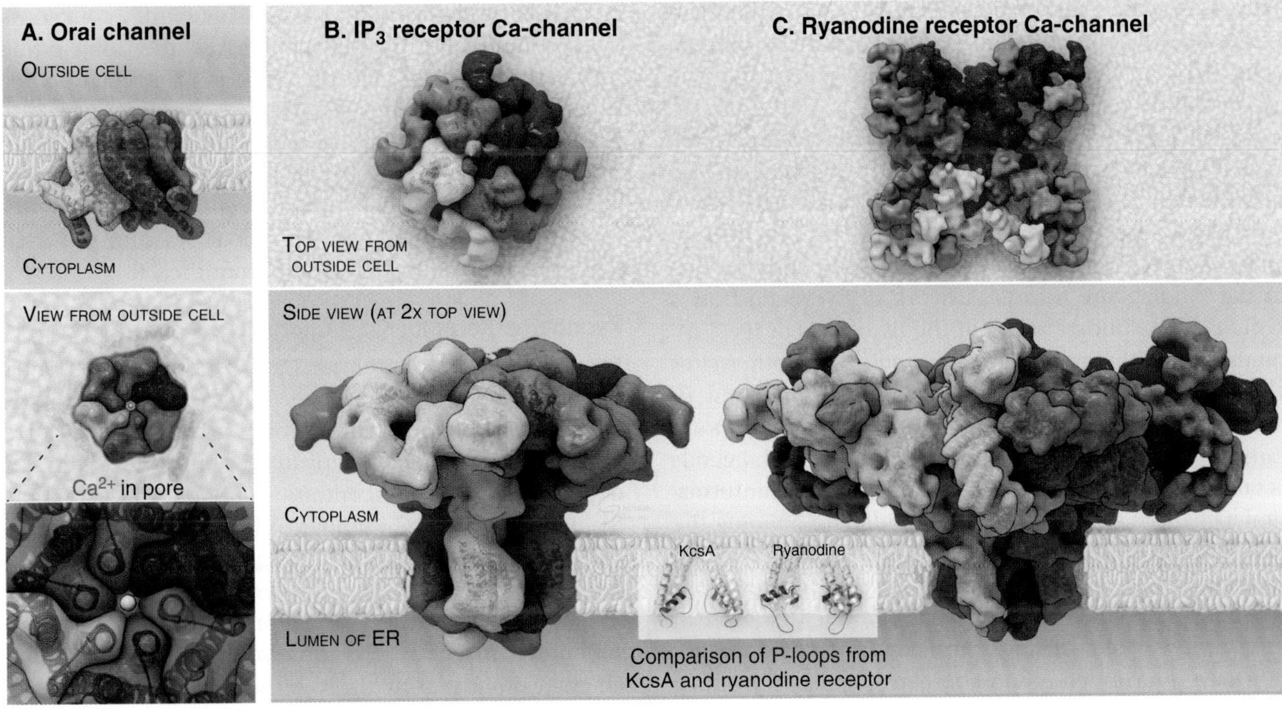

FIGURE 26.13 STRUCTURES OF CALCIUM CHANNELS. Face views and side views of ribbon diagrams with space-filling surfaces based on crystal and cryoelectron microscopy structures. **A,** Crystal structure of the Orai channel with a detail of Ca^{2+} in the selectivity filter. **B,** Cryo-electron microscopy structure of the inositol triphosphate (IP_3) receptor–channel consisting of four identical subunits of 2750 residues. Six trans-membrane segments and a P-loop facing the lumen of the endoplasmic reticulum (ER) are similar to other channels. **C,** Cryoelectron microscopy structure of the ryanodine receptor–channel consisting of four identical subunits of 5037 residues. The six helices and P-loop forming the channel are located in the middle of the transmembrane domain. TRPV1, transient receptor potential vanilloid-1. (For reference, see PDB file 4HKR and Hou X, Pedi L, Diver MM, et al. Crystal structure of the calcium release-activated calcium channel Orai. *Science*. 2012;338:1308–1313 [**A**]; EMData Bank file 6369, PDB file 3JAV, and Fan G, Baker ML, Wang Z, et al. Gating machinery of InsP3R channels revealed by electron cryomicroscopy. *Nature.* 2015;527:336–341 [**B**]; and Yan Z, Bai XC, Yan C, et al. Structure of the rabbit ryanodine receptor RyR1 at near-atomic resolution. *Nature.* 2015;517:50–55 [**C**].)

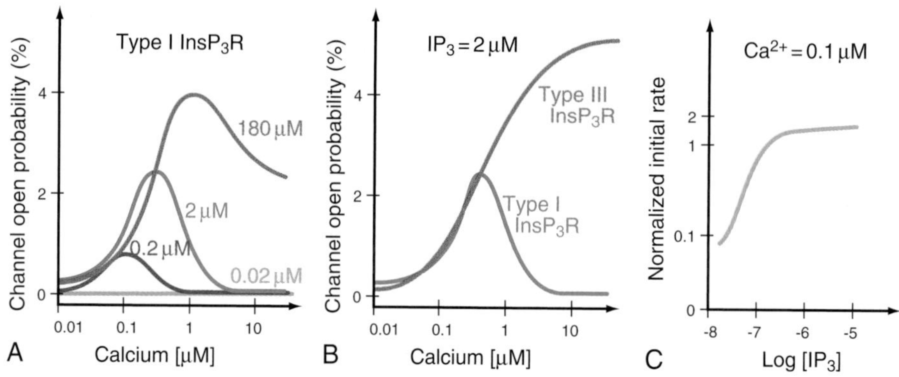

FIGURE 26.14 GATING OF INOSITOL TRIPHOSPHATE (IP_3) RECEPTOR Ca^{2+} RELEASE CHANNELS. A, Dependence of channel open probability of type I receptors on the concentrations of IP_3 (next to each curve) and Ca^{2+}. **B,** Comparison of the dependence of open probability of type I and type III receptors on the concentration of Ca^{2+} at a fixed concentration of 2 μM IP_3. High concentrations of Ca^{2+} inhibit type I receptors but not type III receptors. **C,** Dependence of open probability of type I receptors on the concentration of IP_3 at a fixed concentration of 0.1 μM Ca^{2+}. $InsP_3R$, inositol 1,4,5-trisphosphate receptor. (**A,** Data from Kaftan EJ, Ehrlich BE, Watras J. $InsP_3$ and Ca^{2+} interact to increase the dynamic range of $InsP_3$ receptor-dependent Ca^{2+} signaling. *J Gen Physiol.* 1997;110:529–538. **B,** Data from Hagar RE, Burgstahler AD, Nathanson MH, et al. Type III $InsP_3$ receptor channel stays open in the presence of increased calcium. *Nature.* 1998;396:81–84. **C,** Data from Hirota J, Michikawa T, Miyawaki A, et al. Kinetics of calcium release by IP_3 receptor in reconstituted lipid vesicles. *J Biol Chem.* 1995;270: 19046–19051.)

C-terminus form a tetrameric Ca^{2+} channel similar to other cation channels, including a P-loop between segments 5 and 6 facing the lumen of the ER (see Fig. 10.3). The pore is large and nonselective, but Ca^{2+} is the main ion crossing the open channel, owing to the large concentration gradient between the ER lumen and the cytoplasm.

Cytoplasmic IP_3 and Ca^{2+} cooperate to open and close these channels, with IP_3 setting the sensitivity of the channel to Ca^{2+} (Fig. 26.14). IP_3 binds with submicromolar affinity between domains near the N-terminus of each subunit far from the membrane pore. Basic amino acids in the binding site form a network of hydrogen bonds with all three phosphates and the hydroxyls of IP_3. Ca^{2+} binds to the IP_3-binding domains and several other sites along the polypeptide. IP_3 must occupy at all four of these sites to open the channel. Channels respond rapidly, because binding and dissociation of both ligands is fast ($k_+ = 33\ \mu M^{-1}\ s^{-1}$, $k_- = 6\ s^{-1}$ for IP_3). The conformational changes that couple ligand binding to opening the gate in the pore are still being investigated. High concentrations of Ca^{2+} in the ER lumen sensitize receptors to IP_3. Phosphorylation by PKA, PKC, and CaM kinase can raise or lower the sensitivity to IP_3.

Three human genes and alternative splicing produce a variety of IP_3 receptors with different physiological properties for various cell types. Type I and type II IP_3 receptors open in response to Ca^{2+} with a bell-shaped concentration dependence. A channel is most likely to be open when the cytoplasmic Ca^{2+} concentration is about 0.3 μM. Below 0.1 μM and above 100 μM Ca^{2+}, the channel is generally closed. When IP_3 activates a channel, Ca^{2+} release provides rapid *positive feedback* as its local cytoplasmic concentration rises into the micromolar range, stimulating channel opening and then slow *negative feedback* as the local Ca^{2+} concentration climbs higher. The result is a short, self-limited pulse of Ca^{2+} release in response to a modest change in IP_3 concentration. Calmodulin probably mediates the long-lasting inhibitory effects of high Ca^{2+}. Type III IP_3 receptors are different; Ca^{2+} activates them, but high Ca^{2+} concentrations do not compete with IP_3 for binding the channel or inhibiting Ca^{2+} release. This lack of negative feedback (or very slow feedback) allows cells with type III IP_3 receptors to produce a large, global pulse of Ca^{2+} that can ultimately drain Ca^{2+} stores from the ER.

Ryanodine Receptor Ca^{2+} Channels

Ryanodine receptors release Ca^{2+} from the ER to trigger contraction of striated muscles. The name came from the high affinity of the channel for a plant alkaloid called ryanodine, which can activate or block Ca^{2+} release, depending on its concentration and the target tissue. Ryanodine has no physiological function, but the name has stuck, because binding of radioactive ryanodine was the key assay for isolating the protein.

Ryanodine receptors are homotetramers of 565-kD subunits with a massive cytoplasmic domain and a cation channel domain near the C-terminus (Fig. 26.13C), an architecture similar to that of IP_3 receptors. Three ryanodine receptor genes encode proteins that are about 60% identical and expressed in different cells (see Table 26.3). Ryanodine receptors are the primary release channels in striated muscles where they are activated by physical contact with voltage-gated Ca^{2+} channels (see Fig. 39.15). In cardiac muscle the ryanodine receptors are activated by calcium induced calcium release where the initiating calcium comes through plasma membrane

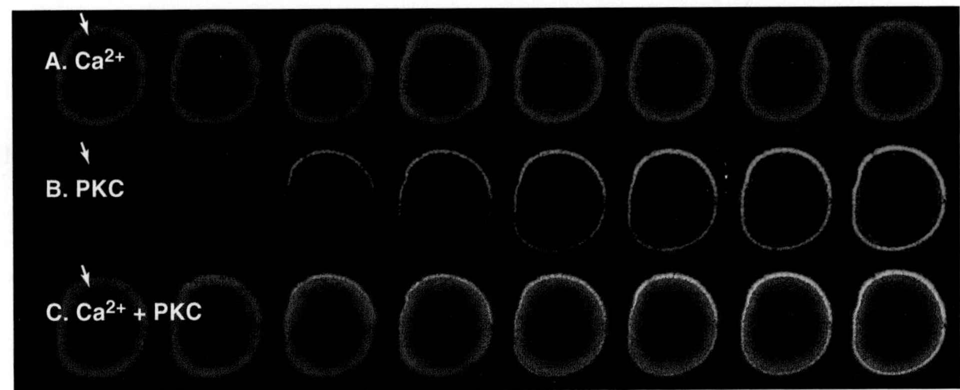

FIGURE 26.15 WAVE OF Ca^{2+} RELEASE AND PROTEIN KINASE C (PKC) ACTIVATION SPREADING FROM THE SITE OF ARTIFICIAL ACTIVATION OF A *XENOPUS* EGG. A, Ca^{2+} signal. **B,** PKC activation. **C,** Superimposition of the two signals. The egg was injected with calcium *red* (a fluorescent dye sensitive to the concentration of Ca^{2+}) and a fusion protein, consisting of *green* fluorescent protein and PKC, which produces *green* fluorescence when PKC is activated. The egg was activated by a needle prick *(arrows)* and imaged at intervals of 20 seconds. A wave of Ca^{2+} (more intense *red*) precedes a wave of active PKC *(green)* from the site of activation. (Courtesy Carolyn Larabell, Lawrence Berkeley Laboratory, Berkeley, CA.)

voltage-gated Ca^{2+} channels. In smooth muscle and nonmuscle cells, ryanodine receptors augment IP_3 receptor Ca^{2+}-release channels.

As in the case of type I and type II IP_3 receptors, Ca^{2+} activates ryanodine receptors with a bell-shaped concentration dependence. This Ca^{2+}-induced Ca^{2+} release allows a local wave of transient activation to spread from one ryanodine receptor to the next.

Cyclic adenosine diphosphate–ribose (cADP-ribose), sets the Ca^{2+} sensitivity of ryanodine receptors, similar to IP_3 for its receptor. At low concentrations of cADP-ribose, high levels of cytoplasmic Ca^{2+} are required to open the channel, whereas at high concentrations of cADP-ribose, even resting Ca^{2+} concentrations open the channel. A single enzymatic step produces cADP-ribose from the metabolite nicotinamide adenine dinucleotide, NAD^+. cGMP regulates **ADP-ribosyl cyclase,** presumably through a cGMP-dependent protein kinase. cADP-ribose is implicated in the Ca^{2+} transient that triggers secretion of insulin from pancreatic β cells in response to glucose. In fertilization of echinoderm eggs, cADP-ribose releases Ca^{2+} through ER ryanodine receptors in parallel with IP_3-mediated Ca^{2+} release. Vertebrate eggs depend entirely on the IP_3 release mechanism (Fig. 26.15). Nicotinic acid adenine dinucleotide phosphate, a metabolic product of β-NADP (nicotinamide-adenine dinucleotide phosphate), also releases Ca^{2+} from internal stores. Its mechanism is not yet certain, but may involve a protein that modulates the activity of Ca^{2+} release channels.

Physiological and pharmacological agents can stimulate or inhibit ryanodine receptor activity. Phosphorylation by PKA increases ryanodine receptor channel activity and may contribute to the effects of β-adrenergic receptor stimulation on the heart (see Fig. 39.21). Caffeine activates Ca^{2+} release by ryanodine receptors and is used to stimulate sperm in fertility tests. Numerous agents suppress the spontaneous release of Ca^{2+} by ryanodine receptors in heart and skeletal muscle. These include FKBP (FK-binding protein, the protein that binds the immunosuppressant drug FK506), micromolar ryanodine, the local anesthetic procaine, and calmodulin.

Point mutations in the RyR1 ryanodine receptor gene (expressed in skeletal muscle) cause **malignant hyperthermia** in humans and pigs. The most common human mutation is autosomal dominant with an incidence of approximately 1 in 50,000. Mutant ryanodine receptors are unusually sensitive to activation by general anesthetics, which trigger Ca^{2+} release, sustained skeletal muscle contraction, and pathological heat generation. If not treated promptly, the fever can be lethal. The pig mutation is autosomal recessive, and stress can trigger lethal attacks.

Calcium Dynamics in Cells

Only about 1 in 100 cytoplasmic Ca^{2+} ions is free to diffuse; the other 99 are bound to Ca^{2+}-binding proteins, estimated to be approximately 300 μM. At a concentration of 0.1 μM in the cytoplasm of a resting cell, the half-time for a free Ca^{2+} ion is approximately 30 μs, and its range of diffusion is only about 0.1 μm. When bound to a protein messenger such as **calmodulin,** Ca^{2+} has a wider range in the cytoplasm of approximately 5 μm.

When cells are stimulated, Ca^{2+} pours into the cytoplasm through release channels at approximately 10^6 ions per second reaching peak cytoplasmic Ca^{2+} concentrations in the micromolar range. The temporal and spatial patterns depend on the stimulus, type of release channel, rate of Ca^{2+} pumping out of cytoplasm, and rates of binding and dissociation on target proteins. These Ca^{2+} signals last for milliseconds to minutes. They can rise and fall locally, flood the whole cytoplasm, or travel in waves across a cell.

Methods to visualize Ca^{2+} concentrations inside living cells revealed this amazing temporal and spatial complexity of Ca^{2+} signals. The original experimental Ca^{2+} sensor was a jellyfish protein called **aequorin,** which emits light when it binds Ca^{2+} (see Fig. 39.15). An evolving series of **Ca^{2+}-sensitive fluorescent dyes** (Fura-2, calcium green, calcium red, Fluo-4) largely replaced aequorin and made possible the observations described in the following paragraphs. Genetically encoded Ca^{2+} biosensors, constructed by fusing calmodulin to fluorescent proteins, offer advantages, including targeting to particular cellular compartments and the ability to adjust the range of Ca^{2+} concentrations that generate a response by mutating the Ca^{2+}-binding site.

Voltage-dependent Ca^{2+} channels in excitable cells such as neurons and striated muscles respond rapidly (<1 ms) to an action potential to admit extracellular Ca^{2+}. At chemical synapses, this produces a brief, locally confined Ca^{2+} pulse that triggers the release of synaptic vesicles (see Fig. 11.9). In striated muscles, each action potential causes a transient, global increase in cytoplasmic Ca^{2+} lasting tens of milliseconds (see Fig. 39.15). The details differ in skeletal and cardiac muscle, but the elementary events are similar. Voltage-sensitive Ca^{2+} channels in T tubules (invaginations of the plasma membrane; see Fig. 39.10) activate one or a few ryanodine receptor channels in the adjacent ER through direct physical interaction in skeletal muscle and through Ca^{2+} release in the heart. Ca^{2+} released by these ryanodine receptors triggers Ca^{2+} release from nearby ryanodine receptors, generating a local pulse of Ca^{2+} called a **Ca^{2+} spark.** Because T tubules penetrate throughout the muscle cytoplasm, thousands of these sparks are produced simultaneously, yielding a transient global rise in Ca^{2+}. The brief duration of the excitatory signal and the robust Ca^{2+} pumping activity of the ER limit the duration of the signal.

Nonexcitable cells generally rely on slower, receptor-mediated production of IP_3 to produce transient changes in cytoplasmic Ca^{2+}. For example, activation of a frog egg at the point of sperm entry or artificial activation by pricking with a needle (Fig. 26.15) results in a self-propagating wave of cytoplasmic Ca^{2+} that spreads slowly around the cell. This produces a wave of secretion and activation of downstream effectors, such as PKC.

Many cells with type I and II IP_3 receptors respond to constant agonist stimulation with transient bursts of cytoplasmic Ca^{2+} (Fig. 26.16). The agonist concentration sets the level of cytoplasmic IP_3 (and, possibly, cADP-ribose), which determines the sensitivity of release channels to cytoplasmic Ca^{2+}. A region of the cell with a high density of the most sensitive channels then initiates the release of Ca^{2+} at a focus. The resulting Ca^{2+} transient may spread through the cytoplasm as a planar or spiral wave that is driven locally by Ca^{2+}-induced Ca^{2+} release from either ryanodine receptors or IP_3 receptors. Such transients are self-limited, because cytoplasmic Ca^{2+} concentrations exceeding 0.5 to 1.0 μM inhibit both type I and type II IP_3 receptors. This negative feedback closes release channels and allows Ca^{2+} pumps to clear the cytoplasm of Ca^{2+}. Release channels recover slowly from this negative feedback, creating a pause between Ca^{2+} transients. In this way, the cell decodes the concentration of agonist as a function of the frequency of Ca^{2+} pulses. Colliding waves annihilate each other, owing to negative feedback by high concentrations of Ca^{2+} or local depletion of Ca^{2+} stores.

Cells with type III IP_3 receptors respond differently. Agonists produce a flood of Ca^{2+} that fills the entire cytoplasm for seconds, owing to less negative feedback by high Ca^{2+}.

Participation of multiple second messengers and channel types can produce complex responses to stimulation. For example, acetylcholine and cholecystokinin stimulate polarized epithelial cells of the pancreas to secrete digestive enzymes. Stimulation produces a wave of cytoplasmic Ca^{2+} originating at the apical surface and spreading throughout the cell. The frequency of these waves depends on agonist concentration and depends on IP_3 and cADP-ribose and their receptors in the ER, which set the sensitivity of the release mechanisms to cytoplasmic Ca^{2+}.

Ca^{2+} Targets

Ca^{2+} signals have widespread effects owing to the diversity of target proteins (Appendix 26.1). The response depends on the available targets, as well as modulating effects of parallel signaling pathways. Most target proteins have a signature fold called an EF hand that binds Ca^{2+} and/or other divalent cations (see Fig. 3.12C). Some of these proteins bind Ca^{2+} directly, including Ca^{2+}-activated plasma membrane channels for K^+ and Cl^-, troponin-C in striated muscles, synaptotagmin (a Ca^{2+}-sensing synaptic vesicle protein), and calpain (a Ca^{2+}-activated protease). The proteins parvalbumin and calbindin D28K buffer the cytoplasmic Ca^{2+} concentration.

Ca^{2+} regulates many other proteins indirectly by first binding and activating calmodulin (see Fig. 3.12C), which binds target proteins by wrapping around an α-helix (see Fig. 3.12C). In many cases Ca-calmodulin stimulates the activity of the target protein, such as CaM kinase II, which modifies multiple substrate proteins, greatly amplifying the effect of Ca^{2+}. On the other hand, Ca-calmodulin inhibits the activity of other target proteins such as cyclic nucleotide-gated ion channels (see Fig. 16.10). PKA phosphorylation activates a small protein called "regulator of calmodulin signaling," which inhibits calmodulin. As expected for a highly conserved protein with diverse functions, few viable mutations of calmodulin have been linked to human disease, with the exception of rare cases of cardiac arrhythmias.

Owing to the oscillatory, transient nature of Ca^{2+} signals, some cellular responses depend on the frequency

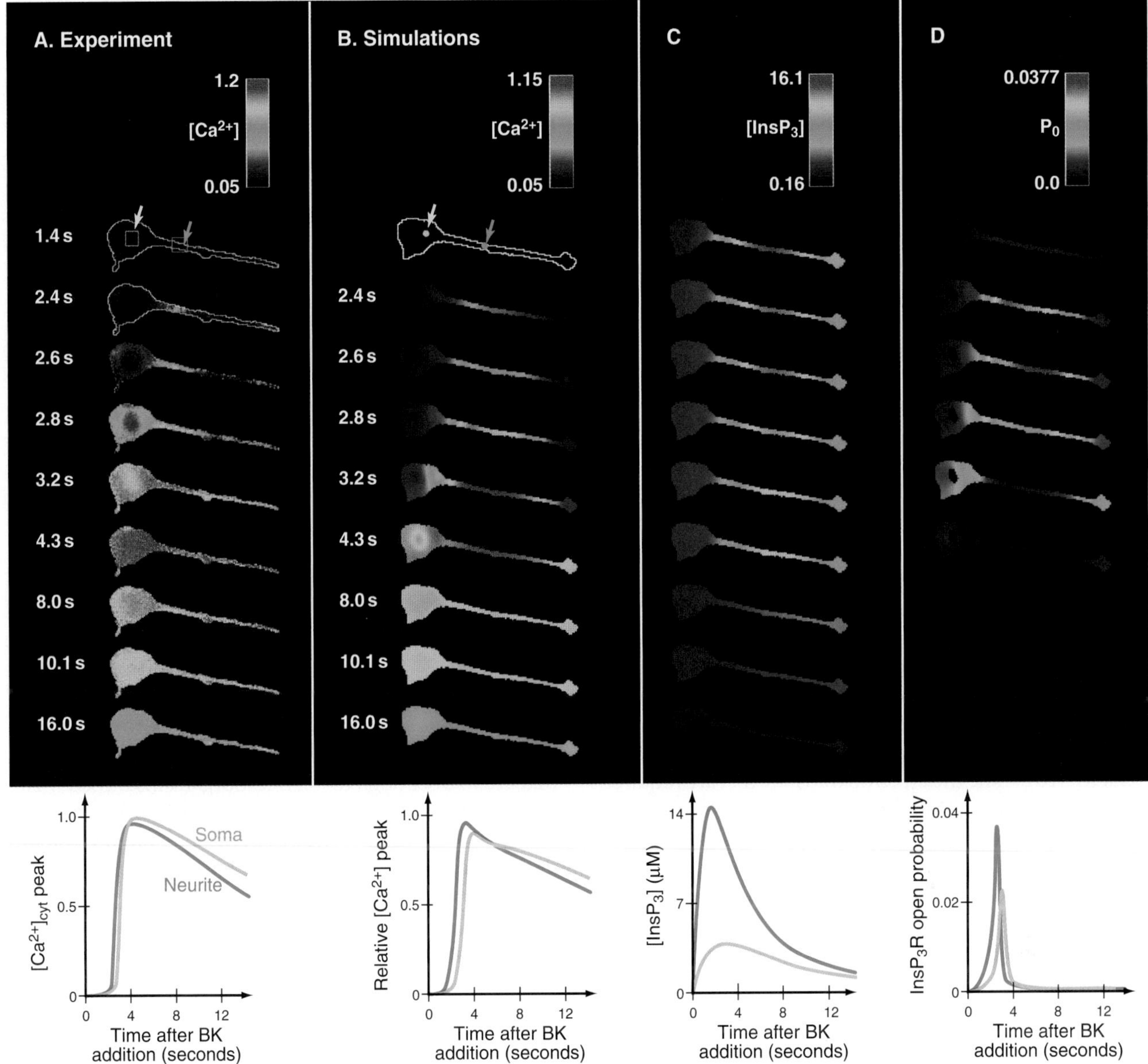

FIGURE 26.16 WAVE OF Ca^{2+} RELEASED IN THE CYTOPLASM OF A CULTURED NEUROBLASTOMA CELL STIMULATED AT TIME ZERO WITH BRADYKININ (BK) AND FOLLOWED AT INTERVALS FOR 16 SECONDS. False colors indicate Ca^{2+} concentration or modeled inositol triphosphate (IP_3) concentration or modeled open probability of IP_3 receptor channels. The model was made with Virtual Cell software, taking into account measured local concentrations of IP_3 receptors and their biochemical properties. Graphs show the time course of various parameters in the neurite and cell body. **A,** Experimental data with Ca^{2+} (micromolar) measured with the intracellular indicator calcium *green*. **B,** Model of the local cytoplasmic Ca^{2+} concentration (micromolar). **C,** Model of the local cytoplasmic IP_3 concentration (micromolar). **D,** Model of the open probability of IP_3 receptor channels. InsP3, IP_3. (Courtesy L. Loew, University of Connecticut Health Center. From Fink CC, Slepchenko B, Moraru II, et al. Morphological control of IP_3-dependent signals. *J Cell Biol.* 1999;147:929–936. For reference, see Fink CC, Slepchenko B, Moraru II, et al. An image-based model of calcium waves in differentiated neuroblastoma cells. *Biophys J.* 2000;79:163–183.)

of Ca^{2+} transients. At least one target protein, CaM kinase II, decodes frequency information into a prolonged adjustment in its level of activity.

Nitric Oxide

The free radical gas nitric oxide (NO˙ [the dot represents an unpaired electron on the nitrogen]) provides cells with a unique way to transmit signals. It diffuses rapidly through membranes, allowing the signal to spread from cell to cell like some lipid second messengers rather than being confined to the cell of origin like most other second messengers. Nitric oxide has long been known as a mildly toxic air pollutant, so it was easy to accept the finding that macrophages produce nitric oxide to the kill microorganisms and tumor cells. On the other hand, it was surprising to learn in the late 1980s that nitric oxide is a diffusible messenger: Nitric oxide made by endothelial cells lining blood vessels relaxes smooth muscle in the walls of arteries and serves as an unconventional neurotransmitter for some neurons in the central and peripheral nervous systems.

Nitric oxide is not a single chemical species. $NO^{\bullet}$ is only one of several readily interconvertible reduction–oxidation (redox) states of nitrogen monoxide, including nitrosonium (NO^+) and nitroxyl (NO^-) ions. Although nitric oxide is stable in water, it reacts readily with oxygen and has a half-life of only a few seconds in the body. Consequently, nitric oxide must be produced continuously to provide a sustained effect. Much nitric oxide is inactivated by binding to the heme of hemoglobin. Nitric oxide is eventually metabolized to nitrate and nitrite and excreted from the body. Note that the stable anesthetic gas nitrous oxide, N_2O, also known as laughing gas, is not part of this family of nitrogen monoxides.

Enzymes called **nitric oxide synthases** produce nitric oxide by converting L-arginine and molecular oxygen into citrulline and nitric oxide (Fig. 26.17). Nicotinamide adenine dinucleotide phosphate (NADPH) provides reducing equivalents for the reaction. Nitric oxide synthase (NOS) is really two enzymes in one. The N-terminal oxidase domain has a heme group that participates directly in oxidation of arginine. The C-terminal reductase domain supplies electrons to the oxidase domain. NOS depends on an extraordinary number of cofactors to handle the electron transfers that are required to produce nitric oxide: heme and tetrahydrobiopterin bound to the oxidase domain, and flavin adenine dinucleotide, flavin mononucleotide, and NADPH in the reductase domain. Some gram-positive bacteria have an NOS with only the N-terminal oxidase domain that is presumably the evolutionary ancestor of eukaryotic NOS. These bacterial enzymes catalyze nitration of substrates rather than production of nitric oxide.

Vertebrates express three NOS isozymes selectively in various tissues. NOS1 and NOS3 are synthesized constitutively. Inducible NOS (iNOS or NOS2) is found in macrophages, liver, and fibroblasts. Macrophages produce NOS2 only when stimulated by endotoxin, interferon-γ, or other factors. Endothelial NOS (eNOS or NOS3) is also made by some neurons. NOS3 is targeted to membranes by N-terminal myristoylation and palmitoylation. Neuronal NOS (nNOS or NOS1) is made by approximately 1% of neurons in the cerebral cortex as well as skeletal muscle and epithelial cells. NOS1 associates with the plasma membrane dystrophin complex in skeletal muscle (see Fig. 39.17) and is lost from the membrane in patients with muscular dystrophy.

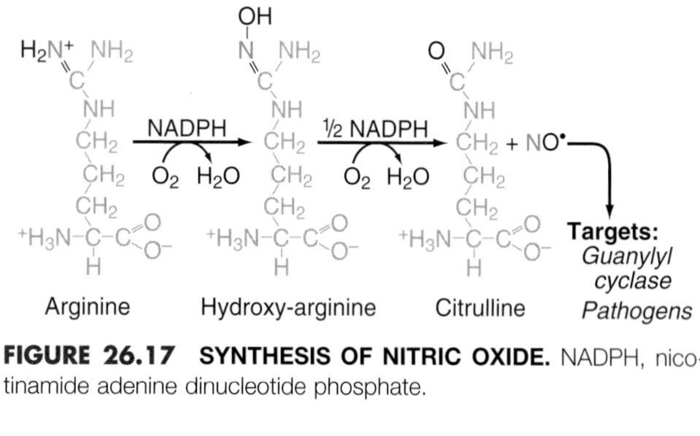

Ca^{2+}-calmodulin and phosphorylation regulate NOS activity independently. Ca^{2+}-calmodulin activates NOS by binding a short regulatory sequence between the two enzyme domains. Ca^{2+} signals activate NOS in most tissues, although macrophage NOS2 binds Ca^{2+}-calmodulin so tightly that it is permanently activated. PKB/Akt, the kinase activated by PIP_3, phosphorylates and activates NOS3.

The main target for the low concentrations of nitric oxide used for signaling is soluble guanylyl cyclase, the cytoplasmic enzyme that makes cGMP (see Fig. 24.8). Nitric oxide binds reversibly to iron in the heme group of guanylyl cyclase, causing a conformational change that activates the enzyme. Nitric oxide also reacts with cysteine side chains of proteins (*S*-nitrosylation). This covalent posttranslational modification can modify the activity of sensitive proteins such as NSF (*N*-ethylmaleimide-sensitive factor), a protein that is required for membrane trafficking (see Fig. 21.12).

Macrophages produce concentrations of nitric oxide that are high enough to kill microorganisms directly. The nitric oxide combines with superoxide anion (O_2^-) to form peroxynitrite ($OONO^-$), which rapidly breaks down into $OH^{\bullet}$ and $NO_2^{\bullet}$, both very toxic oxidants that kill ingested microorganisms. Plants also produce nitric oxide as part of their defense mechanism against pathogens.

Nitric oxide from three different sources regulates blood flow and blood pressure in response to local physiological demands. During exercise, the repeated release of Ca^{2+} that stimulates skeletal muscle contraction (see Fig. 39.15) also binds calmodulin and activates NOS to produce nitric oxide. Nitric oxide diffuses out of the skeletal muscle and relaxes smooth muscle cells surrounding nearby blood vessels by activating cGMP production and thereby protein kinase G (PKG) (see Fig. 25.4). PKG reduces membrane excitability by opening Ca-activated K-channels and inhibits contraction by lowering the internal Ca^{2+} concentration by inhibiting production of IP_3. This increases local blood flow to supply oxygen and nutrients. Endothelial cells lining the inside of blood vessels also use nitric oxide to regulate vascular smooth muscle. Mechanical shear stress from blood flow continuously stimulates endothelial cell phosphatidylinositol-3 kinase to produce PIP_3, which stimulates PKB/Akt to phosphorylate and activate NOS3. This provides a sustained, long-term signal to relax vascular smooth muscle. Acutely, hormones such as acetylcholine and bradykinin can stimulate endothelial cells to

FIGURE 26.17 SYNTHESIS OF NITRIC OXIDE. NADPH, nicotinamide adenine dinucleotide phosphate.

produce nitric oxide by causing Ca^{2+} release from ER. Mice that lack NOS3 have high blood pressure.

Mono- and di-N^G-methylated arginines strongly inhibit NOS and therefore have been used experimentally to reveal many biological functions of nitric oxide. Inhibition of NOS compromises killing of bacteria by macrophages. Similarly, inhibition of endothelial nitric oxide production causes vascular smooth muscle cells to contract, which raises blood pressure and demonstrates a constitutive role for nitric oxide in relaxation of these cells. Nitric oxide produced by autonomic nerves causes erection of the penis by stimulating the production of cGMP and relaxing vascular smooth muscle cells. The drug sildenafil (Viagra) is used to treat erectile dysfunction, as it inhibits the phosphodiesterase that breaks down cGMP. Sildenafil and related drugs bind in the active site, excluding cGMP. Other autonomic nerves use nitric oxide to control the smooth muscle cells in the walls of the intestines. Nitric oxide is the active metabolite of nitroglycerin, a drug that is widely used to relieve pain (angina pectoris) associated with compromised blood flow in the heart. It dilates coronary arteries and improves circulation. In addition to regulating cerebral blood flow, nitric oxide produced by nerve cells in the brain may contribute to certain types of learning by reinforcing the release of neurotransmitters (see Fig. 17.11).

Carbon monoxide (CO) is another gaseous intercellular second messenger under some circumstances. It too regulates soluble guanylate cyclase and influences the activity of some enzymes.

ACKNOWLEDGMENTS

We thank Barbara Ehrlich and Richard Lewis for their suggestions on revisions to this chapter.

SELECTED READINGS

Amcheslavsky A, Wood ML, Yeromin AV, et al. Molecular biophysics of Orai store-operated Ca^{2+} channels. *Biophys J*. 2015;108:237-246.

Baillie GS, Scott JD, Houslay MD. Compartmentalisation of phosphodiesterases and protein kinase A: Opposites attract. *FEBS Lett*. 2005;579:3264-3270.

Berridge MJ, Lipp P, Bootman MD. The versatility and universality of calcium signaling. *Nat Rev Mol Cell Biol*. 2000;1:11-21.

Bos JL. EPAC: A new cAMP target and new avenues in cAMP research. *Nat Rev Mol Cell Biol*. 2003;4:733-738.

Chandrasekharan NV, Simmons DL. The cyclooxygenases. *Genome Biol*. 2004;5:241.

De Caterina R, Zampolli A. From asthma to atherosclerosis: 5-lipoxygenase, leukotrienes and inflammation. *N Engl J Med*. 2004;350:4-7.

Fan G, Baker ML, Wang Z, et al. Gating machinery of $InsP_3R$ channels revealed by electron cryomicroscopy. *Nature*. 2015;527:336-341.

Futerman AH, Hannun YA. The complex life of simple sphingolipids. *EMBO Rep*. 2004;5:777-782.

Garavito RM, Mulichak AM. The structure of mammalian cyclooxygenases. *Annu Rev Biophys Biomol Struct*. 2003;32:183-206.

Geiger M, Wrulich OA, Jenny M, et al. Defining the human targets of phorbol ester and diacylglycerol. *Curr Opin Mol Ther*. 2003;5:631-641.

Guse AH. Calcium mobilizing second messengers derived from NAD. *Biochim Biophys Acta*. 2015;1854:1132-1137.

Hill BG, Dranka BP, Bailey SM, et al. What part of NO don't you understand? Some answers to the cardinal questions in nitric oxide biology. *J Biol Chem*. 2010;285:19699-19704.

Hou X, Pedi L, Diver MM, Long SB. Crystal structure of the calcium release-activated calcium channel Orai. *Science*. 2012;338:1308-1313.

Mak DO, Foskett JK. Inositol 1,4,5-trisphosphate receptors in the endoplasmic reticulum: A single-channel point of view. *Cell Calcium*. 2015;58:67-78.

Marshall CB, Nishikawa T, Osawa M, et al. Calmodulin and STIM proteins: Two major calcium sensors in the cytoplasm and endoplasmic reticulum. *Biochem Biophys Res Commun*. 2015;460:5-21.

McKinney MK, Cravatt BF. Structure and function of fatty acid amide hydrolase. *Annu Rev Biochem*. 2005;74:411-432.

McLaughlin S, Wang J, Gambhir A, Murray D. PIP_2 and proteins: Interactions, organization, and information flow. *Annu Rev Biophys Biomol Struct*. 2002;31:151-175.

Newton J, Lima S, Maceyka M, Spiegel S. Revisiting the sphingolipid rheostat: Evolving concepts in cancer therapy. *Exp Cell Res*. 2015;333:195-200.

Olson KR, Donald JA, Dombkowski RA, Perry SF. Evolutionary and comparative aspects of nitric oxide, carbon monoxide and hydrogen sulfide. *Respir Physiol Neurobiol*. 2012;184:117-129.

Patterson RL, Boehning D, Snyder SH. Inositol 1,4,5-trisphosphate receptors as signal integrators. *Annu Rev Biochem*. 2004;73:437-465.

Radmark O, Samuelsson B. Regulation of 5-lipoxygenase enzyme activity. *Biochem Biophys Res Commun*. 2005;338:102-110.

Rhee SG. Regulation of phosphoinositide-specific phospholipase C. *Annu Rev Biochem*. 2001;70:281-312.

Soboloff J, Rothberg BS, Madesh M, et al. STIM proteins: dynamic calcium signal transducers. *Nat Rev Mol Cell Biol*. 2012;13:549-565.

Toyoshima C, Inesi G. Structural basis of ion pumping by Ca^{2+}-ATPase of the sarcoplasmic reticulum. *Annu Rev Biochem*. 2004;73:269-292.

Usman MW, Luo F, Cheng H, et al. Chemopreventive effects of aspirin at a glance. *Biochim Biophys Acta*. 2015;1855:254-263.

Vanhaesebroeck B, Leevers SJ, Ahmadi K, et al. Synthesis and function of 3-phosphorylated inositol lipids. *Annu Rev Biochem*. 2001;70:535-602.

Van Petegem F. Ryanodine receptors: allosteric ion channel giants. *J Mol Biol*. 2015;427:31-53.

Wilson RI, Nicoll RA. Endocannabinoid signaling in the brain. *Science*. 2002;296:678-682.

Worby CA, Dixon JE. PTEN. *Annu Rev Biochem*. 2014;83:641-669.

Zaccolo M, Magalhães P, Pozzan T. Compartmentalisation of cAMP and Ca^{2+} signals. *Curr Opin Cell Biol*. 2002;14:160-166.

APPENDIX 26.1

Examples of Ca^{2+} Regulated Proteins

Protein	Binding Site	Function
First-Order Proteins That Bind Ca^{2+} Directly		
Membrane Proteins		
Annexins	Novel	Promote membrane interactions
Bestrophin Ca^{2+}-activated Cl^- channels	Novel	Participate in secretion
Ca^{2+}-activated K^+ channels	Novel	Control membrane excitability
IP_3 receptor	Novel	Ca^{2+}-release channel; activated and inhibited by Ca^{2+}
Ryanodine receptor	Novel	Ca^{2+}-release channel, activated and inhibited by Ca^{2+}
Synaptotagmin	Novel	A synaptic vesicle Ca^{2+} sensor
TMEM16A Ca^{2+}-activated Cl^- channels	Novel	Intestinal motility
Enzymes		
Calmodulin-domain protein kinases	EF hand*	Plant protein kinases
Calpain	EF hand	Ca^{2+}-dependent protease
Protein kinase C, some isozymes	C2 domain	Multifunctional protein kinases activated by Ca^{2+}
Cytoskeletal Proteins		
α-Actinin (some isoforms)	EF hand	Actin filament crosslinking protein
Centrin/caltractin	EF hand	Ca^{2+}-sensitive contractile fibers
Gelsolin, villin	Novel	Actin filament severing and capping proteins
Molluscan myosin light chains	EF hand	Regulate muscle contraction; activated by Ca^{2+}
Troponin C	EF hand	Ca^{2+}-activated regulator striated muscle contraction
Calcium-Binding Proteins		
Calmodulin	EF hand	Ca^{2+}-activated regulator of many proteins
Calbindin-D28K	EF hand	Cytoplasmic Ca^{2+} buffer
Calretinin	EF hand	Activates guanylyl cyclase
Parvalbumin	EF hand	Cytoplasmic Ca^{2+} buffer
S100 proteins (18 human isoforms)	EF hand	Diverse regulatory functions; some isoforms may be secreted
S100 calbindin-D9K	EF hand	Cytoplasmic Ca^{2+} buffer
Recoverin	EF hand	Regulates visual phototransduction
Second-Order Proteins Activated by Ca^{2+}-Calmodulin		
Membrane Proteins		
Adenylyl cyclase (some isoforms)		Produces cAMP
Ca^{2+}-dependent Na^+ channels		Na^+ currents
cGMP-gated cation channels		Phototransduction
Plasma membrane Ca^{2+}-ATPase pumps		Clears cytoplasm of Ca^{2+}
Enzymes		
Calcineurin		Protein phosphatase 2B
CaM kinase (several isozymes)		Multifunctional protein kinase
cAMP phosphodiesterase		Degrades cAMP
IP_3 kinase		Phosphorylates IP_3
Myosin light-chain kinase		Activates smooth muscle and cytoplasmic myosin
NAD kinase		Phosphorylates NAD
Nitric oxide synthetase		Makes nitric oxide
Phosphorylase kinase		Phosphorylates phosphorylase
Cytoskeletal Proteins		
MARCKS		Actin filament crosslinking protein

CaM, calmodulin; cAMP, cyclic adenosine monophosphate; cGMP, cyclic guanosine monophosphate; IP_3, inositol triphosphate; NAD, nicotinamide adenine dinucleotide.

*EF hand is the abbreviation for the Ca^{2+} binding site in calmodulin consisting of α-helices E and F.

CHAPTER 27

Integration of Signals

This chapter summarizes how a variety of well-characterized signal transduction pathways work at the cellular and molecular levels. These examples illustrate diverse mechanisms, but common strategies, for carrying information about changing environmental conditions into cells and eliciting adaptive responses. Chapters 24 to 26 describe the molecular hardware used in these pathways. Here, the focus is on the flow of information, including examples of branching and converging pathways. For each pathway, the key events are reception of the stimulus, transfer of the stimulus into the cell, amplification of a cytoplasmic signal, modulation of effector systems over time, and adaptation through negative feedback loops. Few signaling pathways operate in isolation; physiological responses usually depend on the integration of several pathways.

Although each pathway illustrated here is the best characterized of its kind, gaps usually exist in our knowledge about some aspects of the dynamical behavior. These gaps are expected, as most pathways are complicated and few pathways are amenable to quantitative analysis in live cells. Ongoing investigations will continue to refine the schemes presented in the following sections, particularly with respect to how each operates as an integrated system.

Signal Transduction by G-Protein–Coupled, Seven-Helix Transmembrane Receptors

The first three signaling pathways use seven-helix receptors coupled to trimeric G-proteins. Two highly specialized sensory systems—olfactory and visual reception—are particularly well characterized, because their outputs are electrophysiological events that can be monitored at the level of single cells and single membrane ion channels on a rapid time scale. On a millisecond time scale, these two sensory transducers amplify minute stimuli to produce changes in the membrane potential that initiate a signal to the central nervous system. The response of cells to the hormone epinephrine through the β-adrenergic receptor provides an interesting contrast to these rapidly responding sensory systems. Although many of the protein components are similar, the response is slower and much more global, affecting cellular metabolism and responsiveness as a whole.

Detection of Odors by the Olfactory System

Metazoan sensory systems detect external stimuli with extremely high sensitivity and specificity. The chemosensory processes of smell and taste have evolved into highly specialized biochemical and electrophysiological pathways that connect individuals with their environment. Of these two sensory modalities, olfaction is more sensitive, allowing mammals to detect **odorants** at concentrations of a few parts per trillion in air and to distinguish among more than 10^{10} different combinations of odorants. Most volatile chemicals with molecular weights of less than 1000 are perceived to have some odor. The olfactory system shares features with many systems that eukaryotes use for communication, such as external pheromone signals of yeast and insects as well as internal hormonal signals, such as adrenaline, that circulate in the blood between organs of metazoans.

Volatile odorant compounds first dissolve in the mucus that bathes the sensory tissue in the nasal cavity. Animals use a family of **odorant-binding proteins** to solubilize odorants in mucus. Odorant-binding proteins typically bind a number of related compounds with low affinities, so odorants exchange rapidly on and off. When odorants dissociate, they can interact with receptor proteins located on specialized cilia of the olfactory neurons.

Sensory Neurons

Olfactory sensory neurons located in the nasal epithelium of vertebrate organisms detect specific odorants and respond by sending action potentials to the brain (Fig. 27.1). These neurons have three specialized zones. An apical dendrite extends to the surface of the epithelium and sprouts approximately 12 **sensory cilia** specialized for responding to particular extracellular odorants. The response depends on high concentrations of four proteins in the ciliary membrane: a single type of odorant receptor per cell, the trimeric G-protein G_{olf}, type III adenylyl cyclase, and cyclic nucleotide-gated ion channels. The cell body contains the nucleus, protein-synthesizing machinery, and plasma membrane pumps and channels that set the resting electrical potential of the plasma membrane. An **axon** projects from the base of each neuron to secondary neurons in the **olfactory bulb** at the front of the brain.

Olfactory sensory neurons are unique among adult neurons in their ability to replace themselves from precursor cells in the epithelium within 30 days after they are destroyed. If protected from viruses and environmental toxins, they turn over much more slowly, so the ability of self-renewal appears to be an adaptation to the hazards associated with exposure to the environment. Loss of olfactory function with age results, in part, from a decreased ability to maintain this neuronal replacement process. The presence of neurons in an epithelium might seem odd, but recall that the entire central nervous system derives from the embryonic ectoderm.

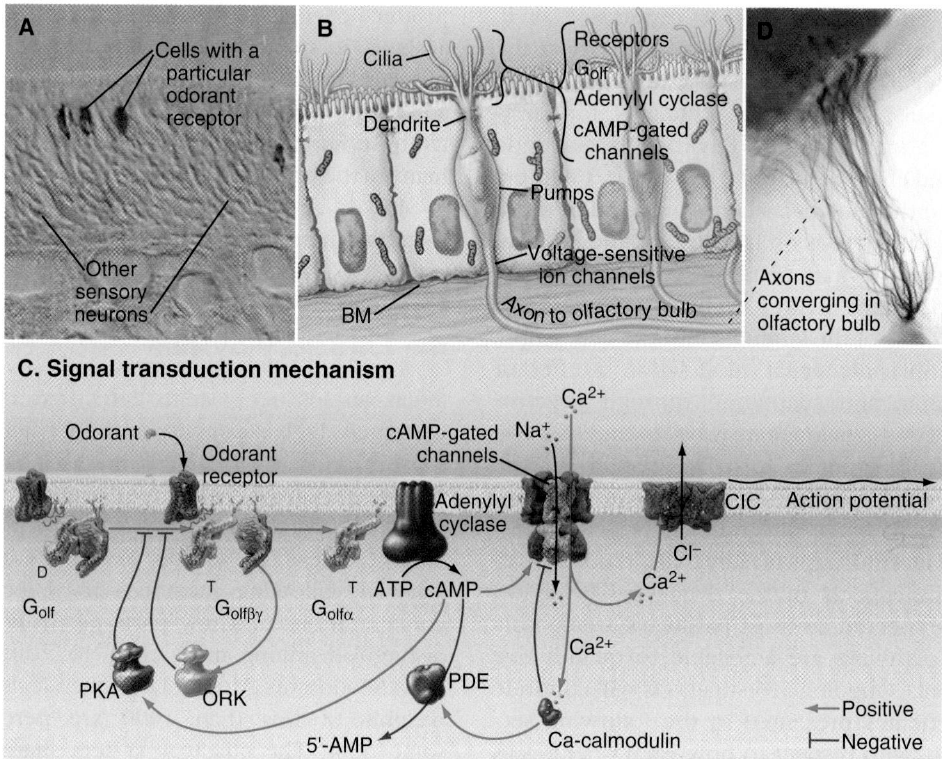

FIGURE 27.1 OLFACTION. A, Light micrograph of a section of sensory epithelium from the nasal passage of a mouse after staining *(purple)* for one olfactory receptor messenger RNA. Note that only a few cells express this gene. **B,** Drawing of the sensory epithelium, highlighting the locations of signal-transducing proteins in three parts of one olfactory sensory neuron. BM, basement membrane. **C,** Signal transduction mechanism. An odorant molecule binds and activates a specific seven-helix plasma membrane receptor. The activated receptor catalyzes the exchange of guanosine diphosphate (GDP) for guanosine triphosphate (GTP) on multiple trimeric G-proteins, causing dissociation of $G_{olf\alpha}$ from $G\beta\gamma$. $G_{olf\alpha}$-GTP activates adenylyl cyclase to produce multiple cyclic adenosine monophosphates (cAMPs). cAMP binds to and opens cyclic nucleotide–gated ion channels, depolarizing the plasma membrane. Ca^{2+} admitted by the cAMP-gated channel opens chloride channels (ClCs), which augment membrane depolarization. Membrane depolarization triggers an action potential at the base of the axon that travels along the axon to secondary neurons in the brain. Multiple negative feedback loops *(red)* terminate stimulation. cAMP activates protein kinase A (PKA), and $G\beta\gamma$ activates odorant receptor kinase (ORK *[green]*), both of which phosphorylate and inhibit the receptor. Ca^{2+} binds calmodulin, which inhibits the cAMP-gated channel and activates phosphodiesterase (PDE) to break down cAMP. **D,** Light micrograph of the olfactory bulb of a mouse expressing a marker enzyme (β-galactosidase) driven by the promoter for one odorant receptor. A histochemical reaction marks the axons of these cells *blue.* Axons from many olfactory sensory neurons expressing the same receptor converge on the same glomerulus in the olfactory bulb. (**A,** From Ressler KJ, Sullivan SL, Buck LB. A zonal organization of odorant receptor gene expression in the olfactory epithelium. *Cell.* 1993;73:597–609. **D,** Courtesy Charles Greer, Yale University, New Haven, CT. For reference, see Zou D-J, Feinstein P, Rivers AL, et al. Postnatal refinement of peripheral olfactory projections. *Science.* 2004;304:1976–1979.)

Overview of the Pathway

The process of reception and transduction of olfactory stimuli takes place in five steps (see Fig. 27.1):

1. Extracellular odorants bind and activate seven-helix receptors on the cilia of olfactory neurons.
2. Activated receptors amplify the signal by catalyzing the exchange of guanosine diphosphate (GDP) for guanosine triphosphate (GTP) on multiple molecules of G_{olf}.
3. Each GTP-$G_{olf\alpha}$ dissociates from $G\beta\gamma$ and amplifies the signal by activating an adenylyl cyclase molecule to produce many molecules of cyclic adenosine monophosphate (cAMP).
4. cAMP binds to and opens cyclic nucleotide–gated ion channels, depolarizing the plasma membrane.
5. Membrane depolarization opens voltage-gated ion channels, triggering an action potential (a self-propagating wave of membrane depolarization; see Fig. 17.6) that starts at the base of the axon of the sensory neuron and moves to secondary neurons in the olfactory bulb in the brain.

As one can appreciate from everyday experience, olfactory signal transduction is very sensitive but adapts quickly to the continued presence of an odorant as explained in the following account of each step.

Odorant Receptors

The olfactory system uses a large family of seven-helix receptors to detect a wide range of ligands present at low concentrations in nasal mucus. These receptors were identified by cloning their complementary DNAs (see Fig. 6.11 for cDNAs) from the olfactory epithelium. Genome sequencing established that mice have approximately 1000 functional odorant receptor genes (approximately 4% of total genes!), whereas humans have approximately 350 functional genes and fish have 100. Sequence comparisons and mutagenesis established that odorants bind among transmembrane helices 3, 5, 6, and 7, the most variable part of these proteins. Each sensory neuron typically expresses a single type of odorant receptor (Fig. 27.1A), using negative feedback from the receptor itself to suppress the expression of other types of odorant receptors. The 1000 cells that express each receptor are scattered in zones throughout the olfactory epithelium.

G-Protein Relay

Odorant binding changes the conformation of the receptor, allowing it to catalyze the exchange of nucleotide on G_{olf} on the cytoplasmic face of the plasma membrane. In fewer than 100 msec, an activated receptor can produce 10 to 100 activated GTP-$G_{olf\alpha}$ molecules, which then dissociate from their $G\beta\gamma$ subunits.

Production of Cyclic Adenosine Monophosphate

In the third step, GTP-$G_{olf\alpha}$ binds to and activates a specialized olfactory isozyme of **adenylyl cyclase.** While active, each enzyme generates nearly 100 cAMP molecules. By 75 milliseconds after stimulation, the concentration of cAMP inside the cilium peaks at greater than 10 μM, returning to baseline within 500 milliseconds. This dramatic fluctuation in cAMP concentration is made possible by the high surface-to-volume ratio of cilia, which ensures that membrane-associated signaling components interact rapidly and confines the cAMP within a small volume. The cAMP transient is short-lived, owing to hydrolysis of cAMP by a **phosphodiesterase** with a high turnover rate.

Cyclic Nucleotide–Gated Channels Trigger an Action Potential

Experiments on isolated olfactory neurons established that the fast cAMP transient depolarizes the plasma membrane by activating **cyclic nucleotide**-gated cation channels. The density of these channels in ciliary membranes (>2000/μm^2) is much higher than that in the cell body (6/μm^2). Binding of at least two cAMP molecules increases the probability that the channel is open from near 0 to about 0.65. Because the probability is not 1.0, individual activated channels flicker open and closed on a millisecond time scale. The ensemble of many activated channels admits enough Na^+ and Ca^{2+} to depolarize the membrane. Ca^{2+}-activated chloride channels carry additional current. The lag of 200 to 500 milliseconds between binding of the odorant and peak membrane depolarization is attributable to the relatively slow binding of cAMP to the channel. Triggering an action potential differs from the role of cAMP in most other tissues, where the main target is protein kinase A (PKA [see Fig. 25.3]).

Depolarization of the ciliary membrane initiates an action potential (see Fig. 17.6) by activating voltage-gated sodium channels (see Fig. 16.2) in the cell body. The action potential propagates along the axon to a chemical synapse with the second neuron in the pathway located in the olfactory bulb of the brain. The two stages of amplification downstream of the receptor allow a few active receptors to produce an action potential.

Adaptation

Desensitization—the waning of perceived odorant intensity despite its continued presence—results from a combination of processes in both olfactory neurons and the brain. Even with constant exposure to odorant the system adapts at each step in the signaling pathway owing to the transient G-protein activation, the self-limited increase in cAMP, and the brief membrane depolarization (Fig. 27.1C). The sequential nature of many of these feedback circuits implies that they have intrinsic delays and therefore serve not only to alter the magnitude of the response but also to shape its time course.

G-protein-coupled receptors are desensitized by protein kinases that phosphorylate the receptor and by proteins called **arrestins** that bind phosphorylated receptors (see Fig. 24.3). These modifications inhibit the interaction of activated receptors with G-proteins and provide negative feedback at the first stage of signal amplification. Negative feedback is coupled to receptor stimulation, because the **olfactory receptor kinase** is brought to the plasma membrane by binding the Gβγ subunits released by receptor-induced G-protein dissociation.

Ca^{2+} entering the cell through cyclic nucleotide-gated channels binds calmodulin, which provides two types of negative feedback. Calcium-calmodulin activates the cAMP phosphodiesterase, which rapidly converts cAMP to inactive 5′-adenosine monophosphate (AMP). It also binds to the cyclic nucleotide-activated channels, reducing their affinity for cAMP by 10-fold and also reducing the probability of opening at less than saturating cyclic nucleotide concentrations. These two effects of Ca^{2+} alter the responsiveness of the neuron to initial odorant exposure, limit the time course of the response, extend the dynamic range over which the cell can respond, and make a cell transiently refractory to additional stimulation.

Processing in the Brain

Mammals discriminate vastly more odorants than the number of available receptors. This is achieved by combining information from multiple types of receptors in their central nervous systems. The molecular specificity established in the olfactory epithelium between an odorant and its receptor is preserved at the first step in the brain, because all the axons from the sensory neurons that express a particular odorant receptor converge on only two to three target areas in the olfactory bulb called glomeruli (Fig. 27.1D). There the axons from like sensory neurons form synapses with dendrites of about 50 secondary neurons in the pathway. Given approximately 1000 odorant receptors in the mouse, each mouse olfactory bulb has approximately 2000 glomeruli. Most of the secondary neurons send their axons to higher levels, where they terminate in a combinatorial manner on cortical neurons. Of special interest, the odorant receptor itself is an important determinant of axon targeting to the glomeruli. Expression of different odorant receptors results in the axons selecting new glomerular targets.

The discrimination of a particular odorant is achieved in two stages: At the first stage, each odorant activates several different receptors, and each receptor can bind a group of related odorants. Therefore, each odorant activates a particular pattern of olfactory sensory neurons and their coupled glomeruli. At the next level, neurons in the cerebral cortex receive information from a combination of glomeruli, leading to eventual discrimination of many different smells at higher levels of the brain. Box 27.1 has information on our second olfactory system.

> **BOX 27.1 Sex and the Second Olfactory System**
>
> Animals use olfaction to find their mates, identify their offspring, and mark their territories. Some of the odorants used for social interactions are volatile chemicals that stimulate the main olfactory system. A second accessory olfactory system detects other social odorants. Some are volatile chemicals found in urine; others are not volatile, including major histocompatibility complex (MHC) class II peptide complexes that are shed from the surfaces of cells into the urine and other secretions. Accessory sensory neurons are located in a special part of the epithelium lining nasal cavity called the vomeronasal organ. Each of these neurons expresses one of approximately 300 seven-helix receptors from a different family than the main odorant receptors. Odorant binding activates a signal transduction pathway distinct from main olfactory neurons, dependent on a Trp channel (see Fig. 16.9) rather than a cyclic nucleotide-gated channel. The axons project to the accessory olfactory bulb in the brain.

Photon Detection by the Vertebrate Retina

Overview of Visual Signal Processing

Photons are energetic but unconventional agonists. They are tiny, move very fast, and penetrate most biochemical materials. These properties create a formidable challenge for detecting photons and transducing their intensities and wavelengths into a signal that can be transmitted to the brain. Nevertheless, vertebrate photoreceptor cells capture single photons and convert this energy into a highly amplified electrical response (Fig. 27.2). Phototransduction is the best-understood eukaryotic sensory process, because the system is amenable to sophisticated biophysical, biochemical, and physiological analysis. Single-cell organisms use similar mechanisms to respond to light (see Fig. 38.16).

Vertebrate **photoreceptor cells** are neurons located in a two-dimensional array in the retina, an epithelium inside the eye. The cornea and lens of the eye form an inverted real image of the outside world on the retina, so the intensity of the light across the field of view is encoded by the array of geographically separate photoreceptor cells. Photoreceptor cells lie at the base of a complex neural processing system. Having detected the rate of photon stimulation at a particular place in the visual field, photoreceptor neurons communicate this information to higher levels of the visual system. Initial processing of the information takes place in the retina, where secondary and tertiary neurons take input from multiple photoreceptors to derive local information regarding image contrast, as well as color and intensity. Neuroscience texts present more detailed information

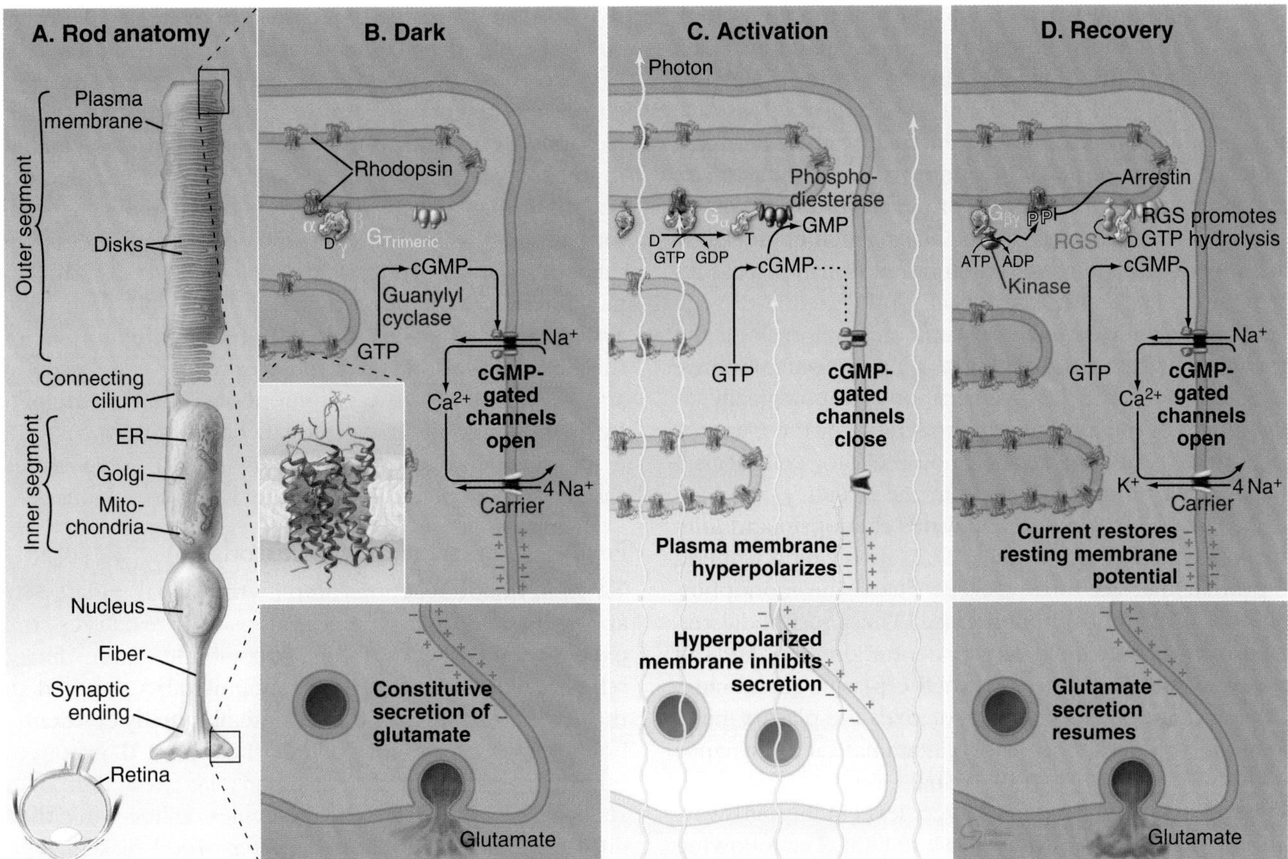

FIGURE 27.2 VERTEBRATE VISUAL TRANSDUCTION. A, Drawing of a rod cell. Disks in the outer segment are rich in rhodopsin. ER, endoplasmic reticulum. **B–D,** Drawings of small portions of an outer segment *(upper panels)* and the synaptic terminal of a rod cell *(lower panels)* in three physiological states. Active components are highlighted by bright colors. **B,** Resting cell in the dark. Constitutive production of cyclic guanosine monophosphate (cGMP) keeps a subset of the plasma membrane cGMP-gated channels open most of the time, allowing an influx of Na^+ and Ca^{2+}. At this membrane potential, the synaptic terminal constitutively secretes the neurotransmitter glutamate. Ca^{2+} leaves the outer segment via a sodium/calcium exchange carrier in the outer segment, whereas Na^+ leaves the cell via a sodium pump in the plasma membrane of the inner segment. **C,** Absorption of a photon activates one rhodopsin, allowing it to catalyze the exchange of GTP for GDP bound on many molecules of transducin (G_T). This dissociates $G_T\alpha$ from $G\beta\gamma$. Each $G_T\alpha$-GTP binds and activates one molecule of phosphodiesterase (attached to the disk membrane by N-terminal isoprenyl groups), which rapidly converts cGMP to guanosine monophosphate (GMP). As the concentration of free cGMP declines, the cGMP-gated channels close, leading to hyperpolarization of the plasma membrane and inhibition of glutamate secretion at the synaptic body. **D,** Recovery is initiated when rhodopsin kinase phosphorylates activated rhodopsin. Binding of arrestin to phosphorylated rhodopsin prevents further activation of G_T. Phosphodiesterase and an RGS protein cooperate to stimulate hydrolysis of GTP bound to G_T, returning G_T to the inactive $G_T\alpha$-GDP state. Synthesis of cGMP by guanylyl cyclase returns the cytoplasmic concentration of cGMP to resting levels and opens the cGMP-gated channels. Constitutive secretion of glutamate resumes. ADP, adenosine diphosphate; ATP, adenosine triphosphate.

on higher levels of visual processing in the retina and brain.

The response of photoreceptor cells depends on the intensity of the light, that is, the flux of photons. Vertebrate retinas detect light with intensities that range over 10 orders of magnitude. **Rod photoreceptors** (Fig. 27.2A) detect low levels of light from approximately 0.01 photon/μm^2/s (dim stars) to 10 photons/μm^2/s but do not discriminate light of different colors. **Cone photoreceptors** (cones) respond to more intense light, up to about 10^9 photons/μm^2/s (full sunlight). Three classes of cones with chromophores sensitive to different wavelengths of light allow humans to encode wavelength. This is the basis for color vision (Box 27.2).

BOX 27.2 Color Vision

Three types of cones allow us to discriminate colors. Cones are organized much like rods but express one of three different seven-helix photoreceptors, each with a distinct visual pigment. The absorption spectra of the photoreceptors overlap, so the central nervous system can perceive colors from deep purple to deep red by comparing the relative activation of the three types of cones at each point in the visual field. As in rods, signal transduction in cones depends on degradation of cyclic guanosine monophosphate (cGMP) and closure of cGMP-gated channels. Activation of cones requires much more intense light, but cones respond faster than rods.

Rods and cones have three specialized regions with different molecular components and functions. The nucleus and the organelles in the **inner segment** maintain the cell's structure and metabolism. A vestigial cilium connects the inner segment to the **outer segment,** which consists of a stack of internal membrane **disks** containing the photoreceptor protein, **rhodopsin** in the case of rods, surrounded by the plasma membrane. Disks form by invagination and pinching off of flattened sacks of plasma membrane. Rhodopsin synthesized in the cell body is transported to the plasma membrane along the secretory pathway and segregated into disk membranes. The lumen of the disks corresponds topologically to the lumen of the endoplasmic reticulum or the extracellular space. The base of the photoreceptor cell forms a synapse with the next neuron in the circuit.

Absorption of a photon activates rhodopsin and initiates a signaling cascade (Fig. 27.2) involving a trimeric G-protein that activates a cyclic guanosine monophosphate (cGMP) phosphodiesterase. The phosphodiesterase lowers the cytoplasmic concentration of cGMP and closes cGMP-gated channels in the plasma membrane. Closing these channels hyperpolarizes the plasma membrane and reduces the release of glutamate at the synapse with the next neuron in the visual circuit.

Feedback loops operate at every level in this pathway, turning off the response to a flash of light. The following sections explain how these reactions achieve their spectacular sensitivity in rods.

Rhodopsin

Rhodopsin is a seven-helix, G-protein-coupled receptor with a light-absorbing chromophore, **11-*cis* retinal,** covalently attached to lysine 296 through a protonated Schiff base (Fig. 27.2B). Although 11-*cis* retinal is bound to a site in the bundle of transmembrane helices similar to sites where ligands bind other seven-helix receptors, this conformation of rhodopsin is inactive with respect to catalyzing nucleotide exchange on its trimeric G-protein. Thus, rhodopsin is a seven-helix receptor with a covalently attached, but inactive, ligand.

The ability of rods to detect single photons depends on two favorable properties. First, the noise level is very low, owing to the stability of the 11-*cis* retinal. In vertebrate rods, fewer than one molecule in 4×10^{10} isomerizes spontaneously every second, so the background level of activated rhodopsin is very low, even with more than 10^8 molecules of rhodopsin per cell. Thus, one does not perceive spots of light in the dark. Second, rods absorb photons very efficiently by virtue of the high density of rhodopsin in disks (~25,000 rhodopsins/μm^2) and stacking thousands of disks on top of each other in the direction of incoming photons. Rhodopsin constitutes 90% of the disk membrane protein and 45% of the disk membrane mass. About half of the photons that traverse the outer segment are absorbed, and about two-thirds of absorbed photons produce an electrical change in the plasma membrane.

Absorption of light initiates the signal transduction pathway. The energy from the absorbed photon isomerizes the 11-*cis* retinal chromophore to **all-*trans* retinal** in picoseconds, creating a cascade of intramolecular reactions that activate rhodopsin by changing its conformation. The active conformation called **metarhodopsin II** catalyzes nucleotide exchange on **transducin,** its trimeric G-protein partner (see Fig. 25.9). Following activation, rhodopsin is inactivated by hydrolysis of the Schiff base linking all-*trans* retinal to the protein and dissociation of the chromophore. Attachment of a fresh molecule of 11-*cis* retinal, derived from vitamin A or regenerated from all-*trans* retinal, regenerates rhodopsin.

Positive Arm of the Signal Cascade

Two enzymatic reactions amplify the signal initiated by absorption of light. Metarhodopsin II catalyzes the exchange of GDP for GTP on the α subunit of transducin, which then dissociates its βγ subunits also attached to the cytoplasmic face of the disk membrane by covalently bound lipid groups. Each metarhodopsin II produces hundreds of activated transducins in a fraction of a second, nearly as fast as the molecules collide while they diffuse in the plane of the very crowded disk bilayer. Nevertheless, nucleotide exchange on transducin is rate-limiting in the whole transduction cascade, even with a 10^7 acceleration by metarhodopsin II.

Transducin α-GTP activates **phosphodiesterase** associated with the disk membrane by binding the enzyme's two inhibitory γ subunits. This frees the catalytic α and β subunits of phosphodiesterase to break down cGMP to guanosine monophosphate (GMP) at a high rate. The cytoplasmic concentration of cGMP depends largely on its rate of destruction by light-activated phosphodiesterase, as it is made continuously by **guanylyl cyclase.**

As the concentration of cGMP falls, **cGMP-gated cation channels** (see Fig. 16.10) in the plasma membrane close resulting in hyperpolarization of the membrane. This change in membrane potential inhibits glutamate release at the synapse. The light-induced decline in glutamate release has opposite effects on the two types of "bipolar neurons" connected to rods: It stimulates "on-type" bipolar neurons to fire action potentials and hyperpolarizes the "off-type" bipolar neurons. This combination of responses is the first step in the discrimination of contrast in our visual world.

Amplification in this pathway is spectacular. Within 1 second after absorption of a single photon, rhodopsin activates 1000 transducins and a similar number of phosphodiesterases, which break down 50,000 cGMPs. This change in concentration closes hundreds of cGMP-gated channels, each of which blocks the entry of more

BOX 27.3 Electrical Circuits in the Photoreceptor

Absorption of light changes currents flowing through electrical circuits in photoreceptor cells. In the dark, the resting cyclic guanosine monophosphate (cGMP) concentration in the outer segment keeps open 1% of the cGMP-gated channels. These open channels produce an inward "dark current" of Na^+ and Ca^{2+}, which is balanced by an outward current of K^+ through channels in the inner segment. Sodium-potassium adenosine triphosphatase (ATPase) pumps in the inner segment compensate for the accumulation of Na^+ and the depletion of K^+. Ca^{2+} entering the outer segment is exported by a carrier in the plasma membrane of the outer segment that exchanges Ca^{2+} and K^+ for Na^+.

Following absorption of a photon, both the cGMP concentration and the probability of the cGMP-gated channels being open declines on a millisecond time scale. In parallel, the cation current into the outer segment falls, hyperpolarizing the plasma membrane. The cytoplasmic Ca^{2+} concentration also declines from about 300 nM to 50 nM. The magnitudes of these responses depend on the number of photons absorbed and the size of the amplified signal.

Two useful properties emerge from the fact that the extracellular concentration of Ca^{2+} largely blocks open photoreceptor channels, similar to the cyclic nucleotide-gated channel of olfactory neurons. First, this reduces the burden on the pumps that maintain the ionic gradients in the cell. Second, using multiple channels with low ionic conductance improves the signal-to-noise ratio. For example, if only two channels carried the dark current, the statistical opening or closing of one channel would create large fluctuations in the current. If 100 partially blocked channels carried the same current, then opening or closing single channels has a modest effect on total current.

than 10,000 cations. Box 27.3 provides more details about the electrical circuit in the rod cell.

Recovery and Adaptation

After a dim flash of light, the dip in cytoplasmic cGMP concentration and hyperpolarization of the plasma membrane are short-lived, on the order of 2 seconds in rods, and even less in cones. Rods reset the signaling pathway by inhibiting metarhodopsin II, inactivating transducin α-GTP, and stimulating the synthesis of cGMP.

Phosphorylation turns off active metarhodopsin II. Transducin βγ subunits activate **rhodopsin kinase** (called GRK1 for G-protein–coupled receptor kinase 1), which phosphorylates several residues near the C-terminus of the receptor. Phosphorylation not only reduces the ability of rhodopsin to activate transducin, but it also creates a binding site for **arrestin**, which prevents further production of transducin α-GTP (see Fig. 24.3).

BOX 27.4 Second Visual System to Set Circadian Clocks

Many organisms, including humans, use the regular variation in light during the day and night to entrain a network of transcription factors that control a 24-hour circadian cycle of metabolic activities throughout the body. For example, mice that are kept in complete darkness continue to run (searching for food) during the hours corresponding to night and sleep during the hours corresponding to day. Without light input, they gradually drift from a precise 24-hour cycle. Neither rods or cones are required to receive the light that synchronizes the internal cycle with the 24-hour day. Instead, a subset of retinal ganglion cells absorbs the light and sends signals to the hypothalamic region of the brain. At least two different photoproteins absorb the light: melanopsin, an opsin family member, and cryptochromes, proteins with a flavin chromophore. So the eye is two photodetectors in one.

The second level of negative feedback comes from the hydrolysis of GTP bound to transducin α in less than 1 second. Both phosphodiesterase (just activated by transducin α) and an **RGS protein** (regulator of G-protein signaling; see Fig. 25.8) stimulate transducin α to hydrolyze the bound GTP. Humans with mutations that disable the retinal RGS protein cannot adapt to rapid changes in light, so they are blinded for several seconds when they step out of a dark room into full sunlight. Dissociation of transducin α-GDP from phosphodiesterase inhibitory subunits terminates cGMP breakdown.

The reduction in cytoplasmic Ca^{2+} that accompanies closure of cGMP-gated cation channels stimulates the guanylyl cyclase that rapidly restores the cGMP concentration. This change opens the cation channels and returns the membrane potential to the resting level.

Box 27.4 has information on a second visual system that mammals use to set circadian clocks.

Regulation of Metabolism Through the β-Adrenergic Receptor

Epinephrine, a catecholamine that is also called *adrenaline* (Fig. 27.3), is secreted by the neuroendocrine cells of the adrenal gland and other tissues when an animal is startled, is stressed, or otherwise needs to respond vigorously. **Norepinephrine,** a closely related catecholamine, is secreted by sympathetic neurons, including those that regulate the contractility of the heart. These hormones flow through the blood and stimulate cells of many types throughout the body to heighten their metabolic activity.

The particular physiological response in each tissue depends on selective expression of a family of nine **adrenergic receptors** and their associated signaling hardware in differentiated cells (Table 27.1). Epinephrine binding

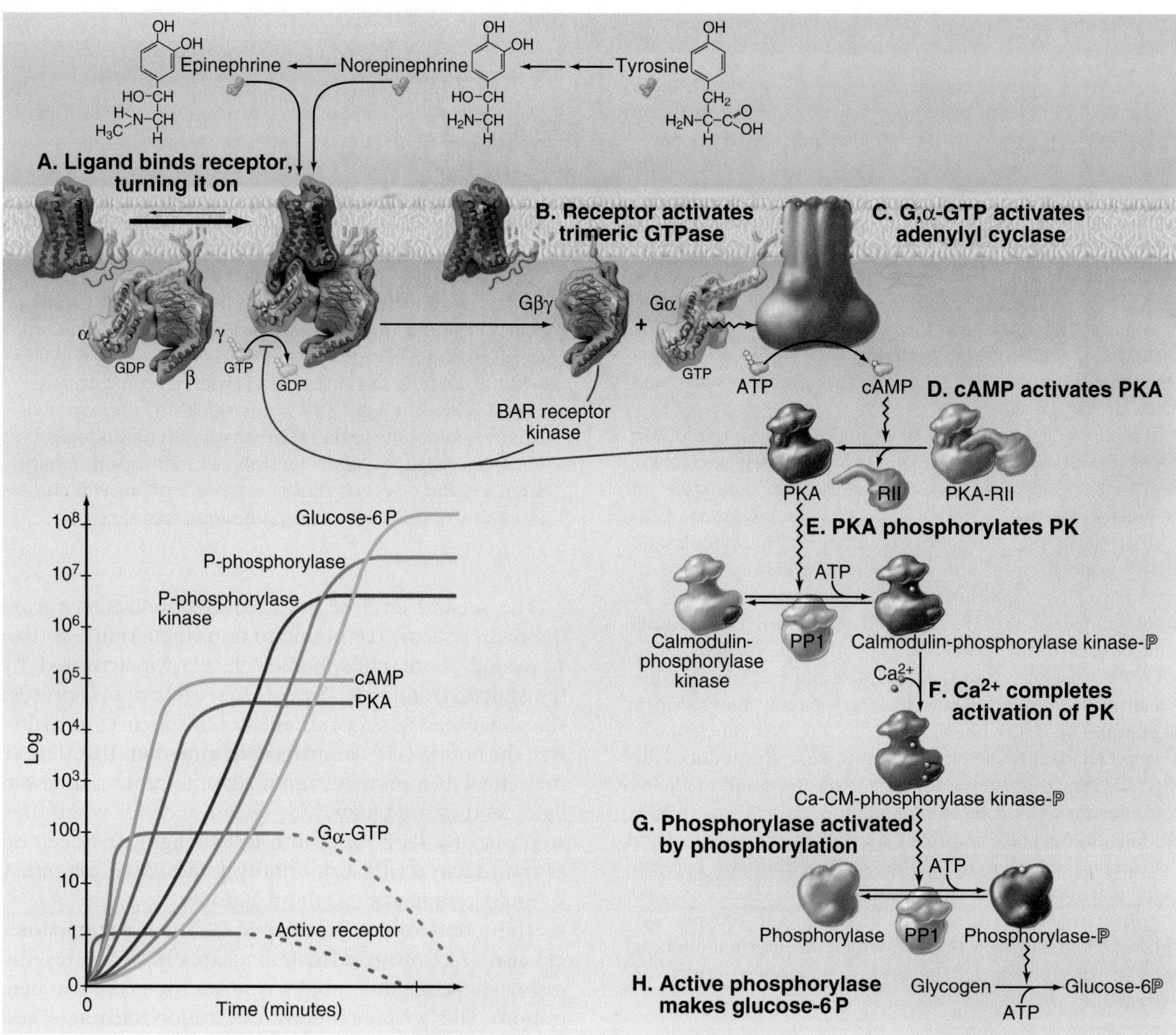

FIGURE 27.3 β-ADRENERGIC SIGNALING MECHANISM. Active components are shown in bright colors. *Upper right,* Pathway of epinephrine synthesis. *Lower left,* Time course of the amplification of the signal by the catalytic cascade of signal-transducing enzymes. **A,** Epinephrine binds the seven-helix β-adrenergic receptor, shifting its equilibrium to the active conformation. **B,** Active receptor catalyzes the exchange of GDP for GTP on $G_S\alpha$, dissociating $G_S\alpha$-GTP from Gβγ. **C,** $G_S\alpha$-GTP activates adenylyl cyclase, which produces multiple cAMPs. **D,** cAMP activates protein kinase A (PKA) by dissociating the regulatory subunit, RII. **E,** PKA phosphorylates and partially activates multiple molecules of phosphorylase kinase (PK). **F,** Ca^{2+} binds to calmodulin (CM) associated with PK, completing activation. **G,** PK phosphorylates and activates phosphorylase. **H,** Phosphorylase catalyzes the conversion of glycogen to glucose-6 phosphate. Negative feedback loops *(red)* terminate stimulation. cAMP activates PKA and Gβγ activates β-adrenergic receptor kinase, both of which phosphorylate and inhibit the receptor from catalyzing nucleotide exchange. PP1, protein phosphatase 1.

to these β-adrenergic receptors is the classic example of a pathway using a seven-helix receptor (see Fig. 24.2), a trimeric G-protein (see Fig. 25.9), and adenylyl cyclase (see Fig. 26.2) to produce cAMP. This second messenger mediates a wide variety of cellular responses by activating PKA (see Fig. 25.3), which phosphorylates many different cellular proteins, thereby changing their activities.

Differentiated cells vary in their responses to epinephrine and norepinephrine, because they express different targets for PKA. In the heart, PKA phosphorylates voltage-gated Ca channels, increasing cytoplasmic Ca^{2+}, and phospholamban, a small membrane protein that stimulates Ca^{2+} pumps in the smooth endoplasmic reticulum to clear Ca^{2+} from the cytoplasm (see Fig. 39.14). These changes stimulate the heart to contract more frequently and with greater force (see Fig. 39.21). On the other hand, PKA in smooth muscle cells of arteries phosphorylates and inhibits myosin light chain kinase, which inhibits contraction and increases blood flow (see

TABLE 27.1 Four Examples of Adrenergic Receptors and Physiological Responses

Receptor	Tissue	Signaling Pathway	Responses
α_1	Smooth muscle, blood vessels	G_q, PLC-β, IP_3, Ca^{2+}, MLCK	Contraction
	Smooth muscle, GI tract	G_q	Relaxation
	Liver	G_q	Glycogenolysis
α_2	Smooth muscle, blood vessels	G_i, Ca^{2+}	Contraction
	Pancreatic islets	G_i, inhibit A-cyclase, K^+ channel open	Secretion inhibition
β_1	Heart	G_s, A-cyclase, cAMP, PKA, phospholamban	Increased contraction
β_2	Liver	G_s, A-cyclase, cAMP, PKA, phosphorylase	Glycogenolysis
	Skeletal muscle	G_s, A-cyclase, cAMP, PKA, phosphorylase	Glycogenolysis

cAMP, cyclic adenosine monophosphate; GI, gastrointestinal; IP_3, inositol triphosphate; MLCK, myosin light-chain kinase; PKA, protein kinase A; PLC-β, phospholipase Cβ.

Fig. 39.24). Skeletal muscle and liver cells respond to epinephrine and PKA by breaking down glycogen to release glucose into the circulation (Fig. 27.3), while PKA in brown fat cells dissipates energy as heat (see Fig. 28.3).

This section explains how β-adrenergic receptors regulate the production of glucose-6 phosphate from glycogen (Fig. 27.3), a process termed glycogenolysis. As with vision and olfaction, the response to epinephrine is sensitive, highly amplified (see the graph in Fig. 27.3), and subject to negative feedback control. Five stages of amplification along the seven-step pathway allow binding of a single molecule of epinephrine to a receptor to activate millions of enzyme molecules that produce many millions of molecules of glucose-6 phosphate:

1. Epinephrine binds to a β-adrenergic receptor, trapping it in its active conformation. The resting system is on the verge of activation, because the ligand-free receptor is in a rapid equilibrium between activated and unactivated states. Even without agonist, a small fraction of receptors is active at any given time. Thus, experimentally increasing the total concentration of receptors (and therefore the concentration of spontaneously active receptors) can maximally activate downstream pathways, even in the absence of agonists.
2. Each activated receptor catalyzes the exchange of GDP for GTP on many molecules of the trimeric protein $\mathbf{G_s}$, causing the dissociation of GTP-$G_s\alpha$ from the Gβγ subunits and amplifying the signal up to 100-fold. The G-protein subunits remain attached to the membrane by their lipid anchors but separate to activate different targets. This is the first branch point in the pathway.
3. GTP-$G_s\alpha$ binds and activates **adenylyl cyclase,** an integral membrane protein (see Fig. 26.2) that produces many molecules of cAMP. This is the second stage of amplification.
4. cAMP activates **PKA** by binding and dissociating its inhibitory RII subunit (see Fig. 25.3). Activation of PKA mediates most effects of cAMP, but cAMP also activates cyclic nucleotide-gated ion channels in some cells.
5. Each activated PKA amplifies the signal by phosphorylating many substrate molecules, including **phosphorylase kinase.** This enzyme requires Ca^{2+} for activity, but phosphorylation by PKA reduces the Ca^{2+} requirement, so the kinase is active even at resting (0.1 μM) Ca^{2+} concentrations. On the other hand, high cytoplasmic Ca^{2+} concentrations alone, as during muscle contraction, can activate phosphorylase kinase without phosphorylation.
6. Each activated phosphorylase kinase further amplifies the signal by phosphorylating and activating many molecules of the enzyme **phosphorylase b.** PKA enhances the phosphorylation of both phosphorylase kinase and phosphorylase b by inhibiting the regulatory G subunit of protein phosphatase 1 (see Fig. 25.5), which dephosphorylates both enzymes.
7. Each activated phosphorylase molecule (called phosphorylase a) removes many glucose subunits from glycogen, one at a time. This fifth stage of amplification produces glucose-6-phosphate, which can enter the energy-releasing glycolytic pathway of the cell (see Fig. 19.4) or, in the case of liver, can be dephosphorylated and released into the bloodstream to provide an energy source for other cells in the body.

If the concentration of epinephrine in the blood declines, epinephrine dissociates rapidly from β-receptors and their equilibrium shifts promptly toward the inactive state. Even in the continued presence of epinephrine the cell adapts owing to reactions that counterbalance the positive arm of the pathway as follows.

First, each activating reaction is reversible, either spontaneously or catalyzed by specific enzymes. Activated GTP-$G_s\alpha$ hydrolyzes its bound nucleotide slowly, at a rate of approximately 0.05 s^{-1}, then rebinds Gβγ, returning the complex to its inactive state. A Ca^{2+}-calmodulin-activated phosphodiesterase degrades cAMP to inactive 5′-adenosine monophosphate (AMP). This convergence allows signaling pathways that release Ca^{2+} (see Fig. 26.12) to modulate the β-adrenergic pathway.

Second, the system has several negative feedback loops that operate on a range of time scales. As in olfaction and vision, active β-adrenergic receptors are inhibited by phosphorylation. In seconds to minutes inhibitory sites in the C-terminal cytoplasmic tail are phosphorylated by PKA activated by cAMP and **β-adrenergic receptor kinase** (now called GRK2) recruited by Gβγ subunits released from $G_s\alpha$-GTP. **β-arrestin** binding to phosphorylated receptors rapidly blocks interactions of the active receptor with G-proteins and attracts cAMP phosphodiesterase to the membrane. On a time scale of many minutes, interactions of β-arrestin with clathrin and adapter proteins (see Fig. 22.9) mediate removal of receptors from the cell surface by endocytosis. Prolonged stimulation by epinephrine results in receptor ubiquitination (see Fig. 23.2), endocytosis, and degradation.

In addition to these effects on glucose metabolism, active β-adrenergic receptors produce at least two other signals. Gβγ subunits activate calcium channels in some cells. This Ca^{2+} can augment glycogen breakdown at the phosphorylase kinase step. In addition to its negative effects, β-arrestin activates the mitogen-activated protein (MAP) kinase pathway (Fig. 27.6) by binding to phosphorylated receptors and also serving as a membrane-anchoring site for the cytoplasmic tyrosine kinase, c-Src.

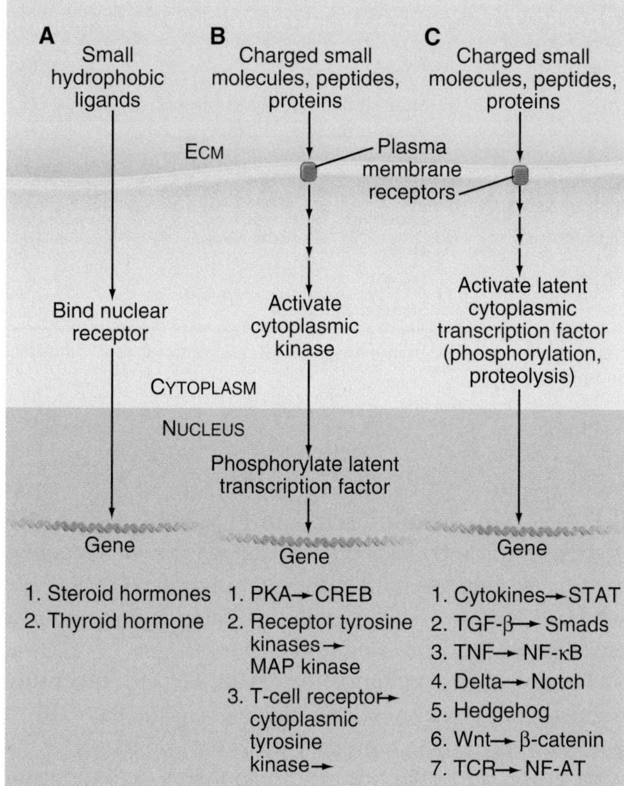

FIGURE 27.4 THREE SIGNALING PATHWAYS BY WHICH EXTRACELLULAR LIGANDS INFLUENCE GENE EXPRESSION. **A,** Nuclear receptor pathway for small hydrophobic ligands that penetrate the plasma membrane (see Fig. 10.21A for an example). **B,** Pathways employing a plasma membrane receptor and a cytoplasmic protein kinase that enters the nucleus to activate a latent transcription factor. (See Fig. 10.21B for the PKA pathway, Fig. 27.6 for a receptor tyrosine kinase pathway, and Fig. 27.8 for a cytoplasmic tyrosine kinase pathway.) **C,** Pathways employing a plasma membrane receptor and activating a latent transcription factor in the cytoplasm. The list includes six known pathways of this type. (See Fig. 27.8 for nuclear factor–activated T cells [NF-AT]; Fig. 27.9 for a signal transducer and activator of transcription [STAT] pathway; Fig. 27.10 for a SMAD [Sma- and Mad-related protein] pathway; Fig. 10.21C for the nuclear factor κB [NF-κB] pathway; Chapter 24 for the Notch and Hedgehog pathways; and Fig. 30.7 for the β-catenin pathway.) CREB, cAMP response element–binding protein; ECM, extracellular matrix; TCR, T-cell receptor; TNF, tumor necrosis factor.

Signaling Pathways Influencing Gene Expression

Many extracellular ligands influence gene expression, via just three kinds of generic pathways (Fig. 27.4). The ligands for the first generic pathway are small and hydrophobic, such as steroids, vitamin A, and thyroid hormone. These ligands penetrate the plasma membrane and bind nuclear receptors in the cytoplasm. The ligands for the other two generic pathways include small charged molecules, peptides, and proteins that cannot penetrate the plasma membrane. Therefore, they must bind receptors on the cell surface to initiate pathways that activate transcription factors. In all cases, activated transcription factors cooperate with other nuclear proteins to regulate the expression of specific genes:

1. *Nuclear receptor pathways:* Ligands such as steroid hormones cross the plasma membrane into the cytoplasm, where they bind latent transcription factors called **nuclear receptors.** Ligand-bound receptors move from the cytoplasm into the nucleus and, in combination with other proteins, activate transcription of specific genes (see Fig. 10.21A).
2. *Pathways activating mobile kinases in the cytoplasm:* Some plasma membrane receptors turn on pathways to activate cytoplasmic protein kinases, which enter the nucleus, where they phosphorylate latent transcription factors. These **mobile kinases** include PKA (see Fig. 10.21B) and several MAP kinases (Figs. 27.5 through 27.8).
3. *Pathways activating latent transcription factors in the cytoplasm:* Other plasma membrane receptors activate latent transcription factors in the cytoplasm, generally by phosphorylation or proteolysis. These activated transcription factors then enter the nucleus. These **mobile transcription factors** include nuclear factor–activated T cells (NF-AT) (Fig. 27.8); signal transducer and activator of transcriptions (STATs) (Fig. 27.9), SMADs [Sma- and Mad-related proteins] (Fig. 27.10); nuclear factor κB (NF-κB) (see Fig.

10.21C); Notch (see Chapter 24); Hedgehog (see Chapter 24); and β-catenin (see Fig. 30.7).

Mitogen-Activated Protein Kinase Pathways to the Nucleus

Cascades of three protein kinases terminating in a **MAP kinase** relay signals from diverse stimuli and receptors to the nucleus (Fig. 27.5). The first kinase activates the second kinase by phosphorylating serine residues. The second kinase activates MAP kinase by phosphorylating both a tyrosine and a serine residue in the activation loop (see Fig. 25.3F). Active MAP kinase enters the nucleus and phosphorylates transcription factors, which regulate gene expression. Key targets include genes that advance or restrain the cell cycle, depending on the system. MAP kinases also regulate the synthesis of nucleotides required for making RNA and DNA.

A variety of cell surface receptors initiate pathways that activate MAP kinase cascades. Many of these pathways pass through the small guanosine triphosphatase (GTPase) **Ras,** allowing cells to integrate diverse growth-promoting signals to control the cell cycle (see Fig. 41.9). Receptor tyrosine kinases for growth factors (Fig. 27.6) and insulin (Fig. 27.7) send signals through Ras. Other receptors use nonreceptor tyrosine kinases coupled to Ras and MAP kinase, such as T-lymphocyte receptors via zeta-associated protein kinase (ZAP-kinase) (Fig. 27.8). Seven-helix receptors can also activate MAP kinase pathways. For example, β-arrestin not only inactivates β-adrenergic receptors, it also couples them to a MAP kinase pathway (see Fig. 24.3). Budding yeast activates MAP kinase pathways in two ways. Mating pheromones bind seven-helix receptors that release Gβγ subunits of trimeric G-proteins, which activate the first kinase in the cascade. Alternatively, osmotic shock activates a two-component receptor (Fig. 27.11) upstream of another MAP kinase pathway that regulates the synthesis of glycerol, which is used to adjust cytoplasmic osmolarity.

Animal cells have multiple MAP kinase cascades with particular isoforms of the three kinases linked in series and leading to different effectors (Fig. 27.5). The kinases that make up these pathways are expressed selectively in various cells and tissues. Deletion of single MAP kinases in mice is generally not lethal, so crosstalk between pathways is likely to be extensive.

A cascade of kinases provides opportunities to integrate inputs from converging pathways and to amplify signals. Amplification can be so strong that a MAP kinase cascade can act like an all-or-nothing switch. For example, frog oocytes that are arrested in the G_2 stage of the cell cycle react to the hormone progesterone by either remaining arrested or entering the cell cycle at full speed. Progesterone activates a MAP kinase cascade consisting of Mos, MEK1, and the p42 MAP kinase. In individual cells, the MAP kinase is either unphosphorylated and inactive or doubly phosphorylated and fully active. This bistable switch-like response depends in part on the fact that both MEK1 and MAP kinase require two independent phosphorylation events for activation. In addition, active MAP kinase provides two types of positive feedback (Fig. 27.5B). MAP kinase both activates Mos by phosphorylation and also drives Mos expression. Consequently, a marginal stimulus turns on some cells strongly and others not at all rather than producing a partial response in all the cells.

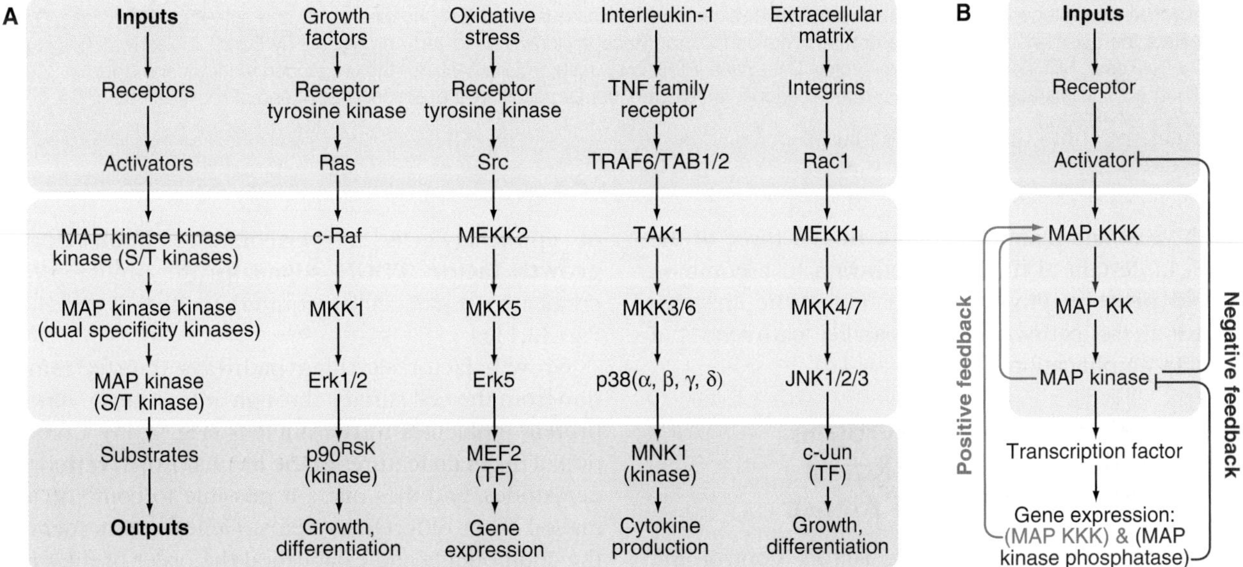

FIGURE 27.5 A, Four mitogen-activated protein (MAP) kinase pathways. The pathways are arranged vertically. The levels of the pathways are arranged horizontally with the nuclear compartment at the bottom. Dual-specificity protein kinases phosphorylate S/T and Y residues. **B,** Positive and negative feedback loops along a generic MAP kinase pathway. K, kinase; TF, transcription factor.

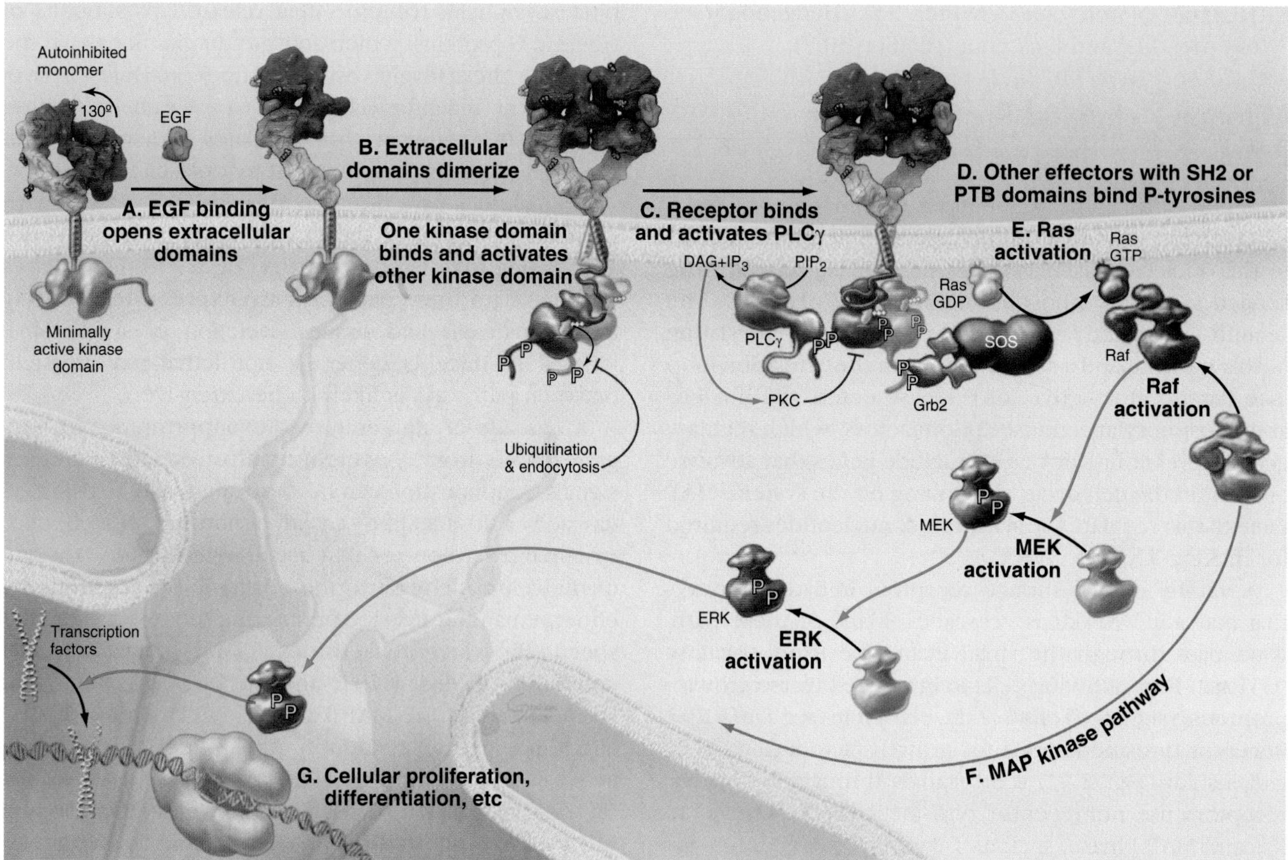

FIGURE 27.6 EPIDERMAL GROWTH FACTOR (EGF) RECEPTOR TYROSINE KINASE SIGNALING PATHWAY THROUGH MITOGEN-ACTIVATED PROTEIN (MAP) KINASE. A, Ligand binding changes the conformation of the extracellular domains of the receptor. **B,** Extracellular domains dimerize, bringing together the tyrosine kinase domains of two receptor subunits in the cytoplasm. Direct interactions and transphosphorylation activate the kinases and create specific docking sites for effector proteins with SH2 (Src homology 2) domains. **C,** Phospholipase Cγ (PLCγ) binds one phosphotyrosine and is activated by phosphorylation to break down PIP_2 into diacylglycerol and inositol triphosphate (IP_3). **D,** A complex of the adapter protein Grb2 and the nucleotide exchange factor SOS binds another phosphotyrosine. (The gene for SOS protein got its name—"son of sevenless"—as a downstream component of the *sevenless* growth factor receptor gene required for the development of photoreceptor cell number seven in the fly eye.) **E,** SOS catalyzes the exchange of GDP for GTP on the membrane-associated small GTPase Ras. Ras-GTP attracts the cytoplasmic serine/threonine kinase Raf to the plasma membrane. **F,** Raf phosphorylates and activates the dual-function kinase mitogen activated protein/extracellular signal-related kinase (MEK). MEK phosphorylates and activates MAP kinase. **G,** MAP kinase enters the nucleus and activates latent transcription factors. (Receptor drawings based on originals by Daniel J. Leahy, Johns Hopkins University, Baltimore, MD. For reference, see Protein Data Bank [PDB; www.rcsb.org] file 2AHX for the unliganded receptor and Bouyain S, Longo PA, Li S, et al. The extracellular region of ErbB4 adopts a tethered conformation in the absence of ligand. *Proc Natl Acad Sci U S A.* 2005;102:15024–15029.)

Both yeast and mammals anchor two or three of the kinases in certain MAP kinase pathways to a common **scaffold protein.** Physical association of the enzymes insulates these pathways from parallel pathways but precludes amplification.

Growth Factor Receptor Tyrosine Kinase Pathway Through Ras to Mitogen-Activated Protein Kinase

Protein and polypeptide growth factors control the expression of genes required for growth and development. For example, the protein **epidermal growth factor** (EGF) controls proliferation and differentiation of epithelial cells in vertebrates. **Platelet-derived growth factor** (PDGF) stimulates the proliferation of connective tissue cells required to heal wounds (see Fig. 32.11).

Growth factor signaling pathways transfer information from the cell surface through at least seven different protein molecules to the nucleus (Fig. 27.6). Conservation of the main features of the mechanism in vertebrates, nematodes, and flies made it possible to combine information from different systems. Genetic tests identified the components and established the order of their interactions. Many components were identified independently as **oncogenes** and by biochemical isolation and reconstitution of individual steps.

Information flows from growth factors to the nucleus as follows:

1. Growth factors bind to the extracellular domain of their receptors, bringing two receptors together either by linking them (see Fig. 24.5A) or inducing a conformational change (see Fig. 24.5B).
2. Dimerization of receptors activates their cytoplasmic tyrosine kinase domains either by **transphosphorylating** tyrosine residues on the **activation loop** of their partner (see Fig. 24.5A) or by allosteric interactions (see Fig. 24.5B). The active kinases phosphorylate other tyrosines on the cytoplasmic domain of the receptor.
3. Each newly created **phosphotyrosine** is flanked by unique amino acids that form specific binding sites for SH2 domains of downstream effectors, including **phospholipase Cγ1, phosphatidylinositol 3-kinase** (PI3K) and a preformed complex of the adapter protein **Grb2** with **SOS.** Grb2 consists of three Src homology domains: SH3/SH2/SH3 (see Fig. 25.10). The SH2 domain binds tyrosine-phosphorylated growth factor receptors. The SH3 domains anchor proline-rich sequences (PPPVPPRR) of SOS, a guanine nucleotide exchange factor for the small GTPase **Ras** (see Fig. 4.6).
4. Association of Grb2-SOS with the receptor raises its local concentration near Ras, which is anchored to the bilayer by farnesyl and palmitoyl groups. Proximity alone suffices for SOS to activate Ras, by exchanging GDP for GTP, as forced experimental targeting of SOS to the plasma membrane by other means also activates Ras. Ras-GTP is active for some time, as it hydrolyzes GTP very slowly (0.005 s^{-1}). Two mechanisms inactivate Ras-GTP: a GTPase-activating protein **(Ras-GAP)** binds to the receptor and stimulates GTP hydrolysis; and removal of the palmitate releases Ras from the plasma membrane.
5. Ras-GTP binds and activates **Raf-1,** a serine/threonine kinase (a MAP kinase kinase kinase).
6. Active Raf-1 phosphorylates and activates the dual-function protein kinase **MEK** (also called MAP kinase kinase or MKK1).
7. MEK activates MAP kinase by phosphorylating both serine and tyrosine residues on the activation loop.
8. Active MAP kinase has some cytoplasmic substrates, and also enters the nucleus to phosphorylate and activate transcription factors already bound to DNA (Fig. 27.5). These transcription factors control the expression of genes for proteins that drive the cell cycle (see Fig. 41.9), as well as phosphatases that generate negative feedback by inactivating the kinases along these pathways (Fig. 27.5B).

The routes from the cell surface through Ras and MAP kinase to nuclear transcription factors are not simple linear pathways. The signal is amplified at some steps and influenced by both positive and negative feedback loops at multiple levels (Fig. 27.5). For example, negative feedback comes from pathways through phospholipase Cγ1 and PI3K that produce Ca^{2+} and lipid second messengers that activate protein kinase C (PKC) isoforms (see Fig. 26.6). PKC inhibits growth factor receptors by phosphorylation. Active receptors are also modified by addition of a single ubiquitin, a signal for inactivation by endocytosis (see Figs. 22.13 and 23.2).

Growth factor pathways are essential for normal growth and development, but malfunctions can cause disease by inappropriate cellular proliferation. One example is the release of PDGF at the sites of blood vessel injury. Normally, PDGF stimulates wound repair (see Fig. 32.11), but excess stimulation of the proliferation of smooth muscle cells in the walls of injured blood vessels is an early event in the development of **arteriosclerosis.**

Many components of growth factor signaling pathways were discovered during the search for genes that cause cancer. As Jean Marx put it, "growth pathways are liberally paved with oncogene products."* Several of the genes were identified in cancer-causing viruses as oncogenes capable of transforming cells in tissue culture. Oncogenes include *sis,* a retroviral homolog of PDGF; *erbB,* a homolog of the EGF receptor; and *raf* kinase. Subsequently, the normal homologs of these genes were found to have mutations in human cancers. Cancer-causing mutations typically make the protein constitutively active, producing a positive signal for growth in the absence of external stimuli (see Fig. 41.12).

Two types of mutations increase the concentration of active Ras-GTP and transmit positive signals for growth in the absence of external stimuli, predisposing individuals to malignant disease (see Fig. 41.12). Point mutations (such as substitution of valine for glycine-12) can constitutively activate Ras by reducing its GTPase activity. Alternatively, mutations that inactivate GTPase-activating proteins (GAPs), such as NF1 (the gene causing neurofibromatosis, the so-called elephant man disease; see Fig. 4.7), reduce the rate that Ras hydrolyzes GTP.

Insulin Pathways to GLUT4 and Mitogen-Activated Protein Kinase

The **insulin receptor tyrosine kinase** triggers an acute response that allows muscle and adipose cells to take up blood glucose following a meal (Fig. 27.7). High blood glucose levels stimulate β cells in the islets of Langerhans of the pancreas to secrete insulin, a small protein hormone. Insulin receptors are found on many cells. Insulin is also a growth factor acting through the MAP kinase pathway.

The insulin receptor is a stable, dimeric tyrosine kinase composed of two identical subunits, each consisting of two polypeptides covalently linked by a disulfide bond. In the absence of insulin the extracellular domains hold

*Marx J: Forging a path to the nucleus. *Science.* 1993;260:1588–1590.

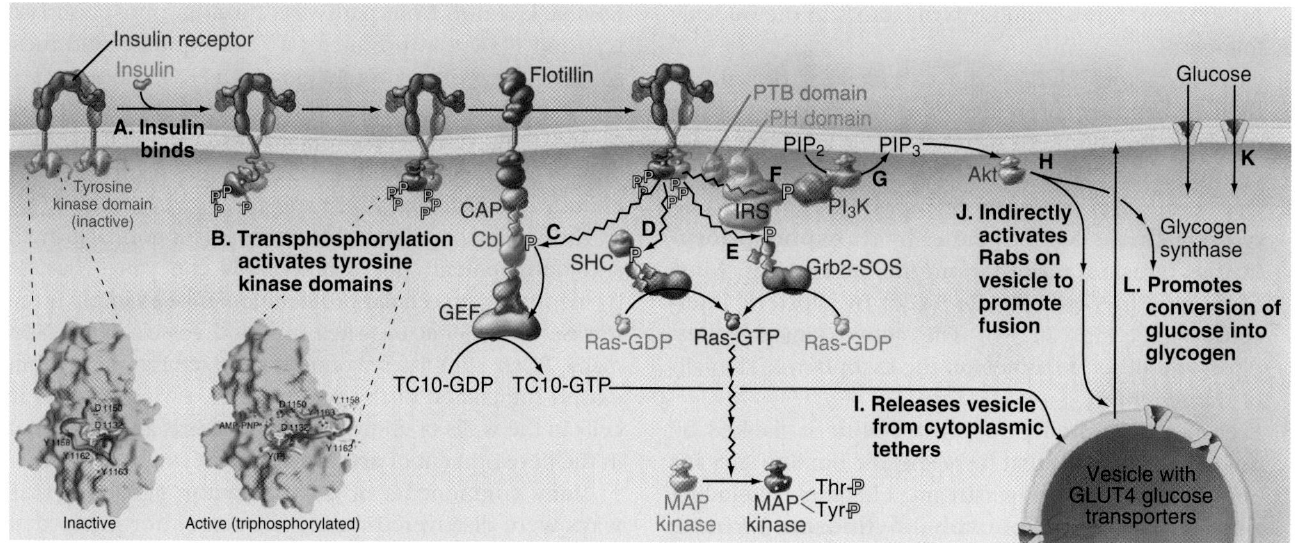

FIGURE 27.7 INSULIN SIGNALING PATHWAYS IN AN ADIPOSE CELL. A, Insulin binds the preformed dimeric receptor, allowing the association of the transmembrane domains and bringing together the tyrosine kinase domains in the cytoplasm. **B,** The tyrosine kinase domains activate each other by transphosphorylation of activation loops (Fig. 27.3F). Receptor kinases then phosphorylate a variety of downstream targets: **(C)** the adapter protein Cbl, which activates a guanine nucleotide exchange protein (GEF) that in turn activates the small GTPase TC10; **(D)** the adapter protein SHC, which binds Grb2-SOS and slowly initiates the MAP kinase pathway; **(E)** the adapter protein IRS, which binds Grb2-SOS and rapidly initiates the MAP kinase pathway; and **(F)** another IRS phosphotyrosine, which binds phosphatidylinositol 3-kinase (PI3K). **G,** PI3K phosphorylates PIP_2 to make phosphatidylinositol 3,4,5-trisphosphate (PIP_3). **H,** PIP_3 binds and activates several protein kinases: Akt (protein kinase B [PKB]), protein kinase Cλ (PKCλ), and PKCζ. **I,** Activated TC10 initiates the release of glucose transporter 4 (GLUT4) storage vesicles from cytoplasmic tethers. **J,** Akt phosphorylates and inactivates Rab GTPase-activating proteins (GAPs), activating Rab GTPases on GLUT4 storage vesicles, stimulating their fusion with the plasma membrane. **K,** GLUT4 transports glucose into the cell. **L,** Akt indirectly activates glycogen synthase. CAP binds Cbl to the plasma membrane protein flotillin. The phosphotyrosine-binding (PTB) domain of IRS binds phosphotyrosine and the PH domain binds PIP_3. (For reference, see Kavran JM, McCabe JM, Byrne PO, et al. How IGF-1 activates its receptor. *Elife*. 2014;25:3.)

the transmembrane helices and cytoplasmic tyrosine kinase domains apart. Insulin binding to the extracellular domains allows the transmembrane domains to come together. This brings the tyrosine kinase domains into proximity so they can transphosphorylate each other (see Fig. 25.3F), turning on their kinase activities (Fig. 27.7D). The kinases propagate the signal by phosphorylating adapter proteins including **IRS (insulin receptor substrate,** isoforms 1 to 4), **SHC** (for SH2 and collagen-like), and **Cbl.** Each plays a distinct role in the ensuing response. This strategy differs from growth factor receptors, which phosphorylate tyrosines on the receptor itself to create binding sites for effector enzymes with SH2-domains.

The short-term effects of insulin are to stimulate glucose uptake from blood (particularly into skeletal muscle and white fat) and the synthesis of glycogen, protein, and lipid. Glucose uptake in these cells depends on the glucose carrier, **GLUT4** (see Fig. 15.1 for closely related carriers). In the absence of insulin glucose uptake is limited, because these cells have few GLUT4 uniporters in their plasma membranes. Most GLUT4 molecules are stored in the membranes of vesicles trapped near the Golgi apparatus inside the cell. Insulin stimulates fusion of these GLUT4 storage vesicles with the plasma membrane, increasing the capacity to transport glucose into the cell.

Delivery of GLUT4 to the plasma membrane requires two separate signals from the insulin receptor, one to release the storage vesicles from their tethers and the other to target the vesicles for fusion with the plasma membrane. The first signal begins with phosphorylation of the adapter protein Cbl, which activates a **guanine nucleotide exchange factor** (GEF). The GEF activates the small GTPase **TC10**, leading to the release of GLUT4 vesicles trapped near the Golgi apparatus. The second signal begins with PI3K binding to a phosphotyrosine on IRS. PI3K synthesizes phosphatidylinositol 3,4,5-trisphosphate (PIP_3), which activates protein kinase PKB/Akt and other kinases. Akt indirectly activates Rab GTPases associated with the GLUT4 vesicles by phosphorylating and inactivating a Rab GAP. Active Rab-GTP promotes fusion of GLUT4 storage vesicles with the plasma membrane. Akt also stimulates the conversion of glucose entering the cell through GLUT4 into its storage form, glycogen, by releasing **glycogen synthase** (the enzyme that makes glycogen) from inhibition by glycogen synthase kinase (GSK) 3. Inactivating mutations of Akt and the downstream Rab-GAP cause rare forms of diabetes, showing their importance in the response to insulin.

Insulin is also a growth factor for some cells, acting through the Ras/MAP kinase pathway to nuclear

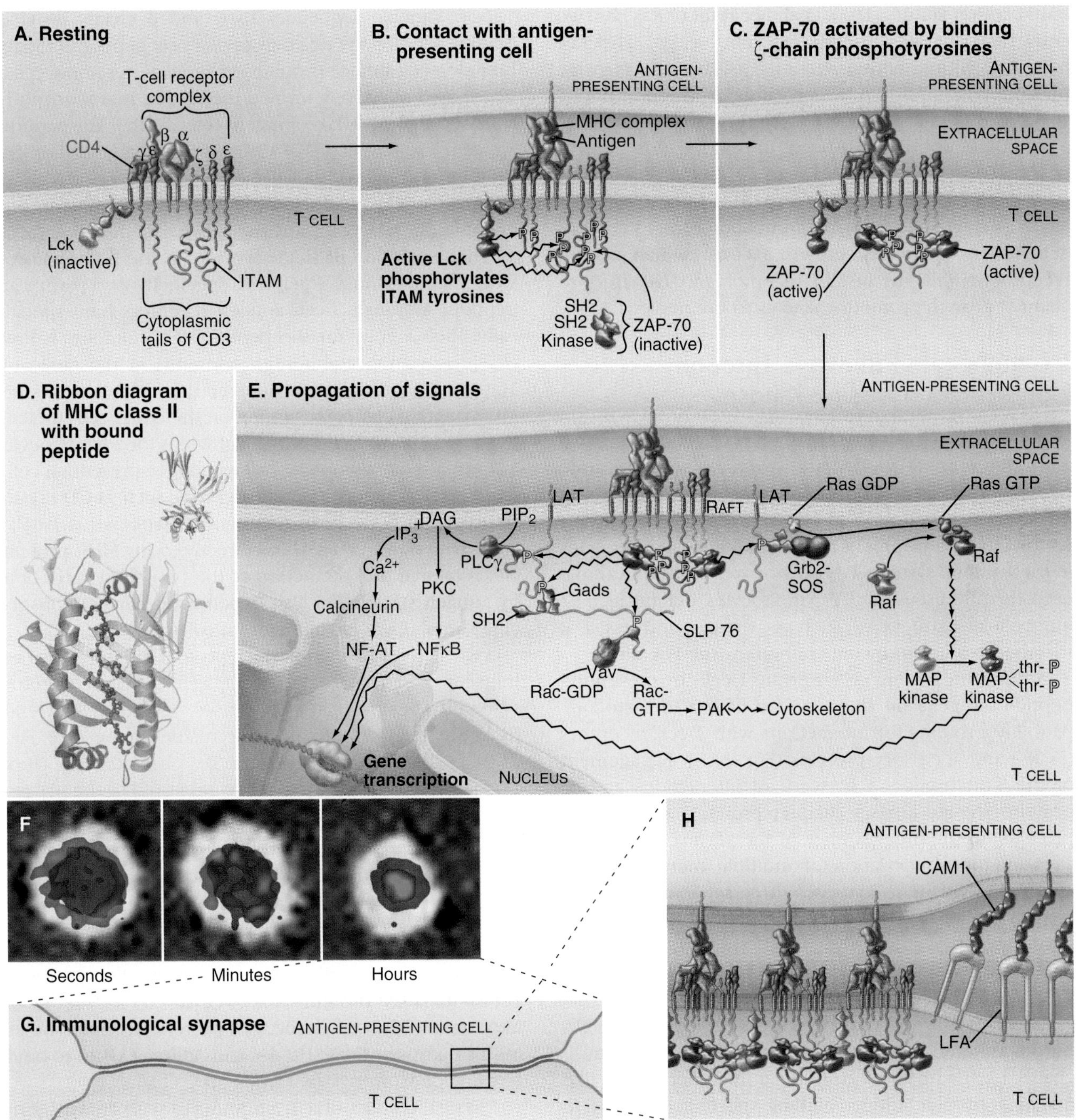

FIGURE 27.8 T-LYMPHOCYTE ACTIVATION. A, Drawing of the T-cell receptor (TCR) complex on a resting T lymphocyte including inactive nonreceptor tyrosine kinase Lck and unphosphorylated phosphorylation sites (ITAMs [immunoreceptor tyrosine-based activation motifs]) on the cytoplasmic tails of CD3 subunits. **B,** An encounter with an antigen-presenting cell with a major histocompatibility complex (MHC)-antigenic peptide complex complementary to the particular TCR initiates signaling. Active Lck phosphorylates various ITAMs. **C,** The nonreceptor tyrosine kinase ZAP-70 (zeta-associated protein of 70 kD) is activated by binding via its two SH2 (Src homology 2) domains to phosphorylated ITAMs on the zeta chains. **D,** Ribbon diagrams of MHC II *(green)* with bound peptide from moth cytochrome c *(orange).* The main model is reduced in size and tilted 90 degrees forward in the view in the *upper right corner,* the same orientation as in the panels **B, C, E,** and **H. E,** Active ZAP-70 phosphorylates various targets, including the transmembrane protein LAT and the adapter protein SLP76, which then propagate the signal. Phospholipase Cγ binds a LAT phosphotyrosine and produces inositol triphosphate (IP_3) and diacylglycerol (DAG). IP_3 releases Ca^{2+} from vesicular stores. Ca^{2+} activates calcineurin (protein phosphatase 2B), which activates the latent transcription factor nuclear factor–activated T cells (NF-AT). DAG and Ca^{2+} activate protein kinase C (PKC), which activates latent transcription factor nuclear factor κB (NFκB). Vav, the nucleotide exchange factor of the small GTPase Rac, is activated by binding to SLP76. Grb2-SOS binds another phosphorylated ITAM and initiates the MAP kinase cascade. **F,** Micrographs of the time course of the interaction of a T cell with an artificial membrane mimicking a specific antigen-presenting cell. Each image comprises a superimposition interference reflection micrograph, showing the closeness of contact as shades of *gray* (with *white* being closest apposition), and a fluorescence micrograph, showing TCRs *(green)* and intercellular adhesion molecule 1 (ICAM1) *(red).* The stable arrangement of ICAM1 around concentrated TCRs is called an immunologic synapse. **G–H,** Immunologic synapse with a central zone of TCRs bound to MHC complexes and peripheral ICAM1 bound to the integrin LFA. Gads is an adapter protein; RAFT is a lipid raft. (**D,** For reference, see PDB file 1KT2 and Fremont DH, Dai S, Chiang H, et al. Structural basis of cytochrome c presentation by IE^k. *J Exp Med.* 2002;195:1043–1052. **F,** Courtesy M. Dustin, New York University, New York.)

transcription factors. The signaling circuit to Ras has two arms that operate on different time scales. The fast pathway, acting within seconds, is through tyrosine phosphorylation of IRS, which binds Grb2-SOS and initiates the MAP kinase pathway. The slow arm, acting over a period of minutes, is through phosphorylation of SHC, which binds larger quantities of Grb2-SOS and slowly initiates a sustained response of the MAP kinase pathway. Normal growth and tissue differentiation of many animals depend on insulin-like growth factors, which act on receptors similar to insulin receptor and use IRS1 to channel growth-promoting signals to the nucleus.

T-Lymphocyte Pathways Through Nonreceptor Tyrosine Kinases

Some signaling pathways that control cellular growth and differentiation operate through **cytoplasmic protein tyrosine kinases** separate from the plasma membrane receptors. The best-characterized pathways control the development and activation of lymphocytes in the immune system. **T lymphocytes** are the example used in this discussion. T lymphocytes defend against intracellular pathogens, such as viruses, and assist B lymphocytes in producing antibodies (see Fig. 28.9).

Antigen-presenting cells activate T cells by presenting peptide antigens on their surface bound to histocompatibility proteins for interactions with T-cell receptors (TCRs) and accessory proteins (Fig. 27.8). Engagement of the TCR triggers a network of interactions among protein tyrosine kinases, adapter proteins, and effector proteins on the inner surface of the plasma membrane. Tyrosine phosphorylation of multiple membrane and cytoplasmic proteins activates three separate pathways to the nucleus. Two pathways activate cytoplasmic transcription factors; the third uses the Ras/MAP kinase pathway to activate transcription factors in the nucleus.

The **T-cell antigen receptor** is a complex of eight transmembrane polypeptides (Fig. 27.8A). The α and β chains, each with two extracellular immunoglobulin-like domains, provide antigen-binding specificity. Similar to antibodies (see Fig. 28.10), one of these immunoglobulin domains is constant and one is variable in sequence. The genes for TCRs are assembled from separate parts, similar to the rearrangement of antibody genes (see Fig. 28.10). Genomic sequences for variable domains are spliced together randomly in developing lymphocytes from a panel of sequences, each encoding a small part of the protein. Assembly of TCRs in the endoplasmic reticulum requires six additional transmembrane polypeptides, each with one or more short sequence motifs, called **immunoreceptor tyrosine activation motifs (ITAMs),** in their cytoplasmic domains.

This combinatorial strategy creates a diversity of T-cell antigen receptors, with one type expressed on any given T cell. Variable sequences of α and β chains provide binding sites for a wide range of different peptide antigens bound to cell surface proteins on antigen-presenting cells. These are collectively termed the **major histocompatibility complex (MHC)** antigens (Fig. 27.8D). The peptide antigens are fragments of viral proteins or other foreign proteins that are degraded inside the cell, inserted into compatible MHC molecules during their assembly in the endoplasmic reticulum, and transported to the cell surface.

The expression of single types of α and β chains provides each individual T cell with specificity for a particular peptide. Although T-cell antigen receptors bind specifically, their affinity for the peptide-MHC complex is low (K_d [dissociation equilibrium constant] in the range of 10 μM). Given the small number (hundreds) of unique MHC-peptide complexes found on the target cell surface, this low affinity would not be sufficient for a lymphocyte to form a stable complex with an antigen-presenting cell. Accessory proteins called coreceptors, such as **CD4** (also the receptor for human immunodeficiency virus [HIV]) and **CD8** (see Fig. 30.3), bind directly to any MHC protein and reinforce the interaction of the two cells. Activation by antigen stimulation also depends on parallel nonspecific stimuli from inflammatory mediators.

Two classes of cytoplasmic protein tyrosine kinases transmit a signal from the engaged TCR to effector systems. The first class of kinases, including **Lck** and **Fyn,** are relatives of the Src tyrosine kinase (see Fig. 25.3C), the first oncogene to be characterized (Box 27.5). These tyrosine kinases are anchored to the plasma membrane by myristoylated N-terminal glycines and inhibited by a phosphotyrosine near the C-terminus (see Fig. 25.3C). This tyrosine is phosphorylated by a kinase, Csk, and dephosphorylated by the transmembrane protein tyrosine phosphatase, **CD45** (see Fig. 25.6B). Apparently, CD45 keeps Lck partially dephosphorylated and therefore partially active in resting lymphocytes. Zeta-associated protein–70 kD **(ZAP-70)** is the most important of the second class of protein tyrosine kinases in this pathway. Two SH2 domains allow ZAP-70 to bind tyrosine-phosphorylated ITAMs on ζ chains.

Physical contact of a T lymphocyte with an **antigen-presenting** cell carrying an MHC-peptide specific for its TCR generates multiple signals as follows:

1. Engagement of TCRs leads to activation of Lck by dephosphorylation of the inhibitory C-terminal tyrosine and phosphorylation of its activation loop.
2. Active Lck phosphorylates ITAMs on the various TCR accessory chains.
3. ZAP-70 is activated by binding phosphorylated ITAMs and phosphorylation of its activation loop by Lck.
4. Active ZAP-70 phosphorylates targets that include the key transmembrane protein LAT and the adapter protein SLP76, which then propagate the signal.
5. Signals reach the nucleus by three pathways. First, phospholipase Cγ1 is activated by binding a

BOX 27.5 Src Family of Protein Tyrosine Kinases

The founding member of the Src family of protein tyrosine kinases has a prominent place in modern biology. During the 1920s, Peyton Rous discovered the first cancer-causing virus in a mesodermal cancer of chickens called a sarcoma. Later, the Rous sarcoma virus was found to be a virus with an RNA genome (a retrovirus). Comparisons with similar viruses that did not cause cancer revealed that one gene, named *src*, is responsible for transforming cells into cancer cells. Many years later, a gene very similar to *src* was found in normal chicken cells. The cellular protein product, c-Src, is a carefully regulated protein tyrosine kinase that controls of cellular proliferation and differentiation. Mutations in the gene for viral *src*, *v-src*, activate its protein product constitutively, driving cells to proliferate and contributing to the development of cancer.

The family of Src-like proteins shares a common structure (see Fig. 25.3C). Five functionally distinct segments are recognized in the sequences. An N-terminal myristic acid anchors the protein to the plasma membrane. Without this modification, the protein is inactive. The next domains are the founding examples of Src homology domains SH3, which bind proline-rich peptides, and SH2, which bind peptides containing a phosphorylated tyrosine (see Fig. 25.10). The kinase domain is followed by a tyrosine near the C-terminus. Phosphorylation of this tyrosine and its intramolecular binding to the SH2 domain lock the kinase in an inactive conformation. Dephosphorylation of the C-terminal tyrosine and phosphorylation of the activation loop activate the kinase.

Expression of c-Src is highest in brain and platelets, but a null mutation in mice produces relatively few defects, except in bones, where a failure of osteoclasts to remodel bone leads to overgrowth, a condition called osteopetrosis (see Fig. 32.6).

phosphotyrosine on LAT and by its own tyrosine phosphorylation. Active phospholipase Cγ1 produces inositol triphosphate (IP_3) and diacylglycerol. Release of Ca^{2+} from vesicular stores by IP_3 activates **calcineurin** (protein phosphatase 2B [see Fig. 25.6A]). This activates the latent transcription factor **nuclear factor–activated T cells (NF-AT).** Second, Grb2-SOS binds another phosphotyrosine on LAT and initiates the MAP kinase cascade by activating Ras. Third, Vav, a nucleotide exchange factor for small GTPases, is anchored indirectly to LAT. This initiates a pathway that degrades IκB, freeing nuclear factor κB (NF-κB) (see Fig. 10.21C) to enter the nucleus.

6. The signal ultimately reaches the cytoskeleton via Vav, which catalyzes the exchange of GTP for GDP on the Rho-family GTPase Rac, which leads to actin filament assembly (see Fig. 38.7).

When a T cell recognizes an antigen-presenting cell with an appropriate peptide bound to MHC on its surface, the TCRs and adhesion proteins in the interface between the cells rearrange to form an **"immunologic synapse"** (Fig. 27.8F). TCRs initially gather around a region of contact between integrins (lymphocyte function-associated antigen) on the T cell and immunoglobulin-cellular adhesion molecules (ICAMs) on the antigen-presenting cell. Over time, actin polymerization around the periphery of this zone drives TCR complexes into a central region where the plasma membranes are close together surrounded by a ring of relatively tall adhesion molecules (Fig. 27.8G–H). Active TCR clusters in this central region generate a signal to the nucleus and are then internalized and degraded.

The best **immunosuppressive drugs** used in human organ transplantation block lymphocyte proliferation by inhibiting calcineurin, the phosphatase that activates NF-AT. These drugs completely block T-cell activation providing that they are given within 1 hour of the stimulus; that is, before gene expression has been initiated. **Cyclosporine** and FK506 bind to separate cytoplasmic proteins, cyclophilin, and FK-binding protein. Both of these drug-protein complexes bind calcineurin and inhibit its phosphatase activity. Considering that many cells express calcineurin, the effects of these drugs on lymphocytes is amazingly specific, with relatively few side effects. Specificity arises from the low concentration of calcineurin in lymphocytes: only 10,000 molecules in T cells compared with 300,000 in other cells. Hence, low concentrations of inhibitor can selectively block calcineurin in T lymphocytes. Cyclosporine made human heart and liver transplantation feasible.

The response to TCR activation depends on the particular state of differentiation of the T cell that encounters its partner antigenic peptide. Stimulation causes some T cells to secrete toxic peptides that kill the antigen-presenting cell, others synthesize and secrete lymphokines (immune system hormones), others proliferate and differentiate, and yet others commit to apoptosis (see Fig. 46.9).

Cytokine Receptor, JAK/STAT Pathways

Many polypeptide hormones and growth factors (collectively called **cytokines**) regulate gene expression through a three-protein relay without a second messenger—the most direct signal transduction pathway from extracellular ligands to the nucleus (Fig. 27.9). Growth hormone uses this mechanism to drive overall growth of the body, erythropoietin directs the proliferation and maturation of red blood cell precursors, and several interferons and interleukins mediate antiviral and immune responses. Slime molds and animals use these pathways, but these proteins are not present in fungi or plants.

The three components in these pathways are a dimeric plasma membrane receptor that lacks intrinsic enzymatic activity, a tyrosine kinase **(JAK)** constitutively

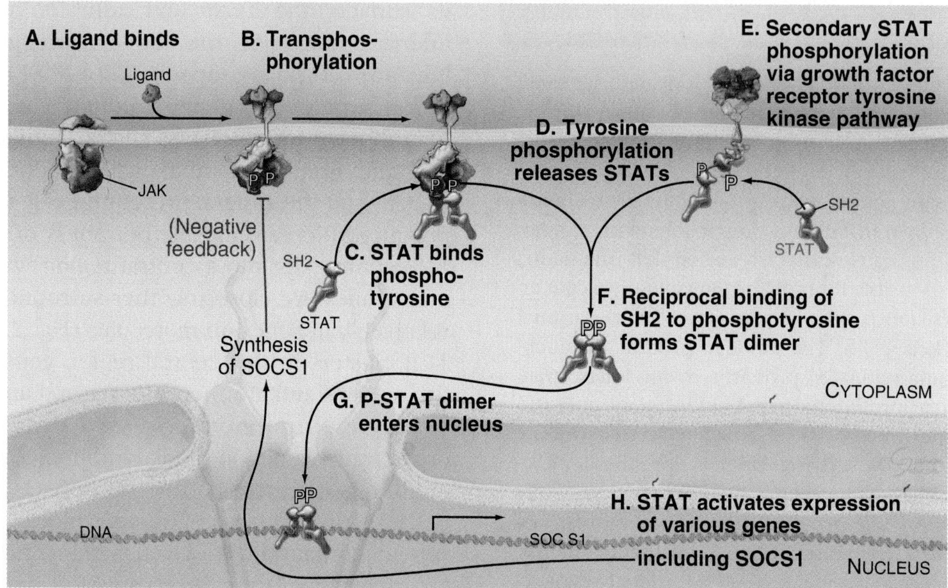

FIGURE 27.9 CYTOKINE JAK KINASE/SIGNAL TRANSDUCER AND ACTIVATOR OF TRANSCRIPTION (STAT) SIGNALING PATHWAY. **A,** Cytokine binds a preformed receptor dimer (see Fig. 24.7), changing arrangement of the JAK tyrosine kinases bound to the cytoplasmic domains of the receptor. **B,** Active JAKs phosphorylate each other and tyrosines on the receptor. **C,** The SH2 (Src homology 2) domain of the latent transcription factor STAT binds a receptor phosphotyrosine. **D,** JAK phosphorylates the STATs, which then dissociate from the receptor. **E,** Growth factor receptor tyrosine kinases can also activate STATs. **F,** STATs form active dimers by reciprocal SH2-phosphotyrosine interactions. **G,** The STAT dimer enters the nucleus. **H,** The STAT dimer activates the expression of various genes. One of these genes encodes SOCS1, which creates negative feedback by inhibiting further STAT activation. JAK, just another kinase.

associated with each receptor subunit, and a latent, cytoplasmic transcription factor called a **STAT** (signal transducer and activator of transcription). JAKs, originally given the lighthearted name "just another kinase," are often called Janus kinases (for the Greek god who opens doors). The N-terminal halves of JAKs mediate their association with receptors. In the absence of a ligand the kinases are inactive owing to interactions with the partner kinase (Fig. 24.6). Some cytokine receptors bind and activate a single type of JAK; others are promiscuous.

The signal from ligand binding moves from the cytokine receptor to JAK to STAT to the nucleus (Fig. 27.9) as follows:

1. Ligand binding changes the conformation of the receptor and relieves the mutual inhibition of the associated JAKs (Fig. 24.6).
2. The JAKs activate each other by transphosphorylation and phosphorylate tyrosines on the cytoplasmic tails of the receptors, creating docking sites for STATs.
3. SH2 domains target preformed STAT dimers to phosphotyrosine binding sites on the receptor.
4. The JAK phosphorylates the STAT, changing the orientation of the subunits. This dissociates the STAT from the receptor.
5. Active STAT dimers enter the nucleus and activate expression of various genes.

Three different mechanisms turn down the response to cytokine activation. Phosphatases inactivate the receptor, kinase, and intranuclear STATs. Endocytosis also turns off active receptors. A slowly acting, negative feedback loop limits the duration of the response. One of the genes expressed in response to STAT encodes SOCS1. Once synthesized, the SOCS1 protein inhibits further STAT activation by interacting with the cytokine receptor.

Selective expression of specific cytokine receptors, four JAKs, and six STATs prepares differentiated mammalian cells to respond specifically to various cytokines. Active STATs are either homodimers or heterodimers of two different STATs. A variety of STATs, with some unique and some common subunits, binds regulatory sites of genes required to activate the target cells. The products of genes controlled by STATs not only contribute to differentiated cellular functions, but some also drive proliferation. Accordingly, loss of JAK function causes certain immune deficiencies, while patients with mutations in STAT5b are resistant to growth hormone and fail to grow. The opposite effect follows from a mutation that renders JAK2 constitutively active: red blood cells proliferate out of control, independent of stimulation by erythropoietin.

The three-protein pathway from a cytokine receptor to JAK to activated STAT is appealing in its simplicity, but in reality, these pathways do not operate in isolation. On one hand, converging signals from EGF- and PDGF-receptors can phosphorylate and activate STATs, a second input to STAT-responsive genes (Fig. 27.9). On the other hand, some cytokine receptors can regulate gene expression by acting through Shc and Grb2-SOS to Ras and MAP kinases and other pathways.

Serine/Threonine Kinase Receptor Pathways Through SMAD

All metazoans use a family of dimeric polypeptide growth factors related to **transforming growth factor-β (TGF-β** [see Fig. 24.7]) to specify developmental fates during embryogenesis and to control cellular differentiation in adults. In humans more than 40 genes in this family of ligands are divided into two classes: (a) those closely related to TGF-β and **activins** and (b) a large family of **bone morphogenetic proteins.** All activate a short pathway consisting of **receptor serine/threonine kinases** and a family of eight mobile transcription factors called **SMADs (Sma- and Mad-related proteins).** The receptors consist of two types of subunits called RI (seven isoforms in human) and RII (five isoforms). Various combinations of these receptors bind about 30 different ligands, some of which antagonize each other.

Ligand binding brings together two RI and two RII receptors, allowing the RII receptors to activate the RI receptors by transphosphorylation (see Fig. 24.7). Active RI receptors phosphorylate "regulated SMADs" (R-SMADs), such as SMAD2 and SMAD3 (Fig. 27.10). Phosphorylated R-SMADs dissociate from RI and associate with SMAD4, called a co-SMAD, because it is not subjected to phosphorylation itself. A trimer consisting of two R-SMADs and one co-SMAD enters the nucleus and associates with other DNA binding proteins to activate or inhibit transcription of specific genes as well as influencing chromatin structure. The number of targeted genes varies from a handful to hundreds depending on the other transcription factors produced by the cell and epigenetic modifications of the target gene chromatin. Other SMADs regulate these pathways by inhibiting phosphorylation of R-SMADs.

The SMAD pathway activated by TGF-β regulates cellular proliferation and differentiation of many cell types, including epithelial and hematopoietic cells. Although its name implies that it should drive transformation, TGF-β actually stops the cell cycle of normal cells in G_1 by promoting expression of negative regulators of cyclin-dependent kinases (see Fig. 41.3). On the other hand, TGF-β can stimulate the growth of some cancers.

In accord with the ability of TGF-β pathways to inhibit cellular growth, many human tumors have loss-of-function mutations in the genes for TGF-β receptors or SMADs. Mutations in an accessory receptor for TGF-β cause malformed blood vessels in the human disease hereditary hemorrhagic telangiectasia. Mice with homozygous

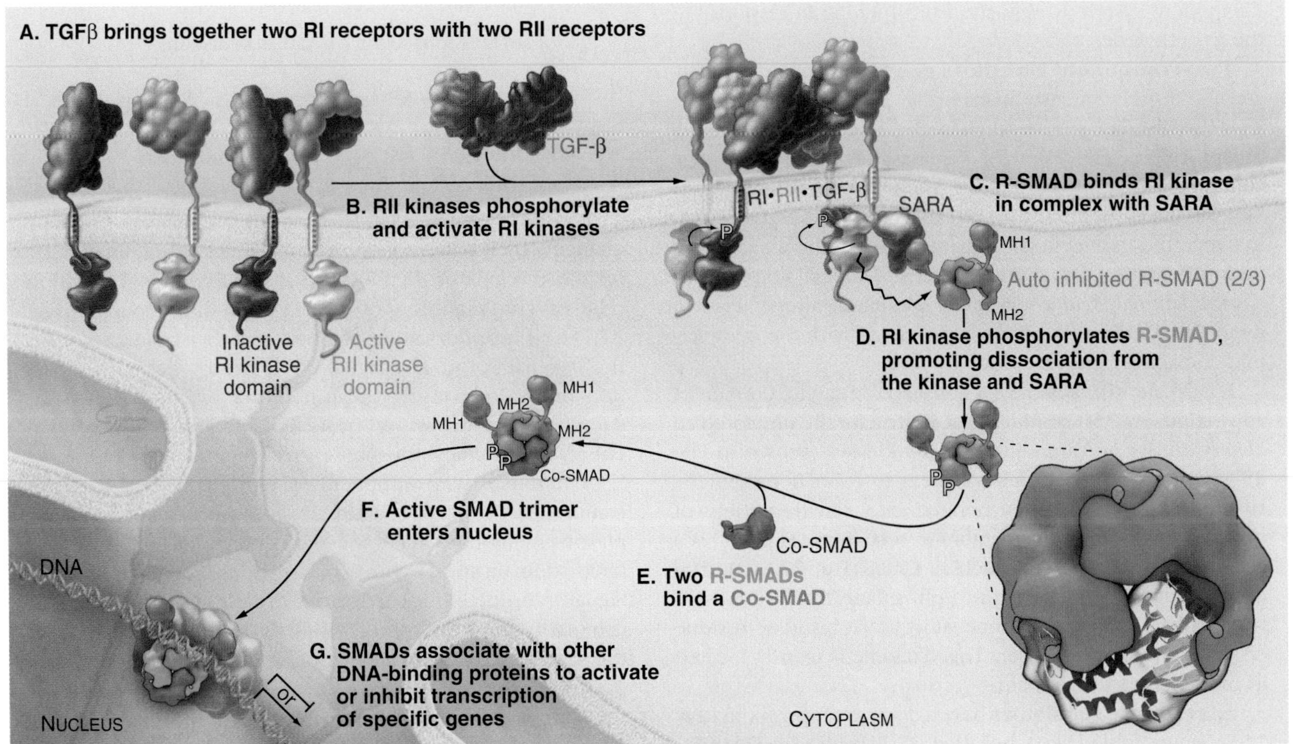

FIGURE 27.10 TRANSFORMING GROWTH FACTOR (TGF)-β/SMAD SIGNALING PATHWAY. A, Binding of a TGF-β dimer assembles a complex consisting of two RII receptors and two RI receptors. **B,** RII phosphorylates and activates RI. **C,** An autoinhibited R-SMAD binds RI in a complex with the adapter SARA. **D,** RI kinase phosphorylates R-SMAD, which promotes its dissociation from RI. **E,** A co-SMAD binds an R-SMAD dimer to form an active trimer. **F,** The SMAD trimer enters the nucleus. **G,** The SMAD trimer associates with other DNA-binding proteins to activate or inhibit transcription of specific genes.

loss-of-function mutations in genes for the components of the TGF pathway die during embryonic development.

Like other signaling pathways, these receptors and SMADs do not operate in isolation. MAP kinases and cyclin dependent kinases can phosphorylate SMADs, and active RI receptors can activate MAP kinase pathways and other signaling pathways.

Bacterial Chemotaxis by a Two-Component Phosphotransfer System

The two-component system (Box 27.6) regulating bacterial chemotaxis (Fig. 27.12) is the best-understood signaling pathway of any kind. *E. coli* cells use five types of plasma membrane receptors to sense a variety of

BOX 27.6 "Two-Component" Signaling

Prokaryotes, fungi, and plants transduce stimuli ranging from nutrients to osmotic pressure using signaling systems consisting of as few as two proteins, a receptor-linked histidine kinase, and a "response regulator" activated by phosphorylation of an aspartic acid (Fig. 27.11). Extensive collections of mutants in these pathways and sensitive single-cell assays for responses, such as flagellar rotation, provide tools for rigorous tests of concepts and mathematical models derived from biochemical experiments on isolated components.

Two-component systems are abundant in bacteria with 32 response regulators and 30 histidine kinases in *Escherichia coli.* Archaea have genes for up to 24 response regulators. Some eukaryotes have a few two-component systems, but these genes were lost in metazoans. The slime mold *Dictyostelium* has more than 10 histidine kinases, whereas fungi have just one or two of these systems. Plants use a two-component system to regulate fruit ripening in response to the gas ethylene.

Two-component receptors either may include a cytoplasmic histidine kinase domain (Fig. 27.11C) or may bind a separate histidine kinase, such as the aspartate chemotactic receptor **Tar** (Fig. 27.11B). Tar consists of two identical subunits. Three Tar dimers are anchored together at their bases in the cytoplasm (Fig. 27.11B). Binding of aspartic acid between the extracellular domains of two subunits changes their orientation by a few degrees. Transmission of this conformational change across the membrane alters the activity of **CheA,** a histidine kinase associated with the cytoplasmic domains of the receptor.

Histidine kinases have a conserved catalytic domain of approximately 350 residues that is structurally unrelated to eukaryotic serine/threonine/tyrosine kinases (shown in Fig. 25.3). Another domain allows them to form homodimers. Histidine kinases are incorporated into a wide variety of proteins, including transmembrane receptors (Fig. 27.11C) and cytoplasmic proteins such as CheA (Fig. 27.11B). The catalytic domain transfers the γ-phosphate from adenosine triphosphate (ATP) to just one substrate, a histidine residue of its homodimeric partner. This histidine is usually located in the dimerization domain.

All **response regulators** have a domain of approximately 120 residues folded like **CheY** (Fig. 27.11B; also see Fig. 3.7). Transfer of phosphate from the phosphohistidine of a kinase to an invariant aspartic acid changes the conformation of the response regulator. Most response regulators such as CheB (Fig. 27.11B) and OmpR (Fig. 27.11C) are larger than CheY, having C-terminal effector domains. Many effector domains, including OmpR, bind DNA and regulate transcription of specific genes when the response regulator is activated by aspartate phosphorylation. Other response regulators are included as a domain of the histidine kinase itself.

Reversible phosphorylation transfers information through two-component systems. The mechanism differs fundamentally from eukaryotic kinase cascades, which transfer phosphate from ATP to serine, threonine, or tyrosine, forming phosphoesters at every step. By contrast, two-component systems first transfer a phosphate from ATP to a nitrogen of a histidine of the kinase, the first of the two protein components. The high-energy his~P phosphoramidite bond is unstable, so the phosphate is readily transferred to the side chain of an aspartic acid of the response regulator (RR):

$$\text{ATP} + \text{kinase-his} \leftrightarrow \text{ADP} + \text{kinase-his}{\sim}\text{P}$$

$$\text{kinase-his}{\sim}\text{P} + \text{RR-asp} \leftrightarrow \text{kinase-his} + \text{RR}^*\text{-asp}{\sim}\text{P}$$

$$\text{RR}^*\text{-asp}{\sim}\text{P} + H_2O \leftrightarrow \text{RR-asp} + \text{phosphate}$$

Phosphorylation activates response regulators (RR*) by changing their conformation. Details differ depending on the response regulator. In the case of OmpR, phosphorylation relieves autoinhibition of the DNA-binding domain (Fig. 27.11C). Phosphorylation of CheY reveals a binding site for the flagellar rotor. The signal dissipates by dephosphorylation of the response regulator, either by autocatalysis or stimulated by accessory proteins. Lifetimes of the high-energy aspartic acylphosphate vary from seconds to hours.

A minimal two-component system, such as a bacterial osmoregulatory pathway (Fig. 27.11C), consists of a dimeric plasma membrane receptor with a cytoplasmic histidine kinase domain and a cytoplasmic response regulator protein. Signal transduction is carried out in four steps. A change in osmolarity alters the conformation of the receptor, activating the kinase activity of its cytoplasmic domain. The kinase phosphorylates a histidine residue on the other subunit of the dimeric receptor. This phosphate is transferred from the receptor to an aspartic acid side chain of the response regulator protein OmpR. Phosphorylation changes the conformation of the response regulator domain of OmpR, allowing its DNA-binding domain to activate the expression of target genes.

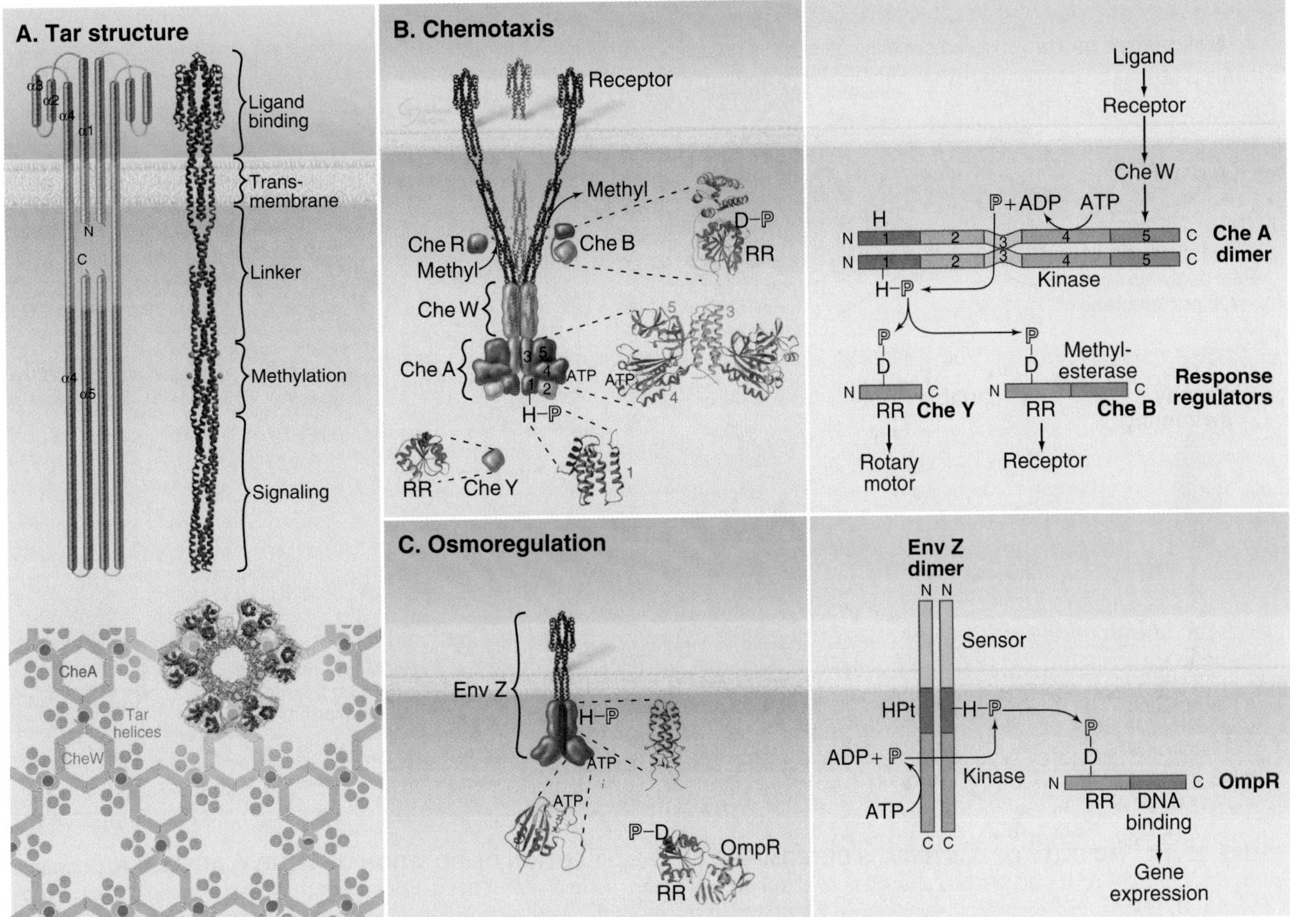

FIGURE 27.11 TWO-COMPONENT BACTERIAL SIGNALING SYSTEMS. A, Atomic model of the aspartate receptor Tar. The atomic structures of the extracellular and cytoplasmic domains were determined by X-ray crystallography. The transmembrane α-helices are models based on the primary structure. The two polypeptides are shown in *red* and *blue.* Each polypeptide starts in the cytoplasm and passes twice through the lipid bilayer. **B,** Bacterial chemotaxis signaling proteins. Scale models of the molecular components and pathway of information transfer. The ribbon diagrams of domains shown on the *right* are color coded in the molecular models on the *left*. An accessory protein, CheW, facilitates binding of the histidine kinase CheA to the aspartate receptor Tar. CheY and CheB are response regulators. CheR is a methyl transferase. **C,** Bacterial osmoregulation. The histidine kinase forms the cytoplasmic domain of the receptor. OmpR is the response regulator with a DNA-binding domain. Scale models of the molecules *(left)* and pathway of information transfer *(right)*. (**A,** Modified from material courtesy S.H. Kim, University of California, Berkeley. For reference, see PDB file 1QU7 and Kim KK, Yokota H, Kim SH. Four helical-bundle structure of the cytoplasmic domain of a serine chemotaxis receptor. *Nature*. 1999;400:787–792. **B–C,** Based on material courtesy A.M. Stock, Robert Wood Johnson Medical School. For reference, see Stock AM, Robinson WL, Goudreau PN. Two-component signal transduction. *Annu Rev Biochem*. 2000;69:183–215.)

different chemicals, which range from nutrients to toxins. These receptors are also called methyl-accepting chemotaxis proteins, as they are regulated by methylation. The most abundant receptor, with thousands of copies per cell, is Tar (Fig. 27.11A–B), the receptor for the nutrients aspartic acid (Tar-D) and maltose, protons (as part of pH sensing), temperature, and the repellent nickel. The receptors form clusters of various sizes over the cell membrane, with the highest concentration at one end of the cell (Fig. 27.12A). This polarized distribution facilitates interactions between receptor molecules but has nothing to do with sensing the direction of chemical gradients.

The chemotactic signaling system guides swimming bacteria toward attractive chemicals and away from repellents in a biased random walk. Environmental chemicals influence the behavior of the cell by biasing the rotation direction of the flagellar motor (see Figs. 38.24 and 38.25 for details of the motor itself). In the absence of a response regulator, the motor turns counterclockwise, and the bacterium swims smoothly in a more or less linear path. When the flagella turn clockwise, the bacterium tumbles about in one place. A **tumble** allows a bacterium to reorient its direction randomly, so when it resumes **smooth swimming,** it usually heads in a new direction. In the absence of

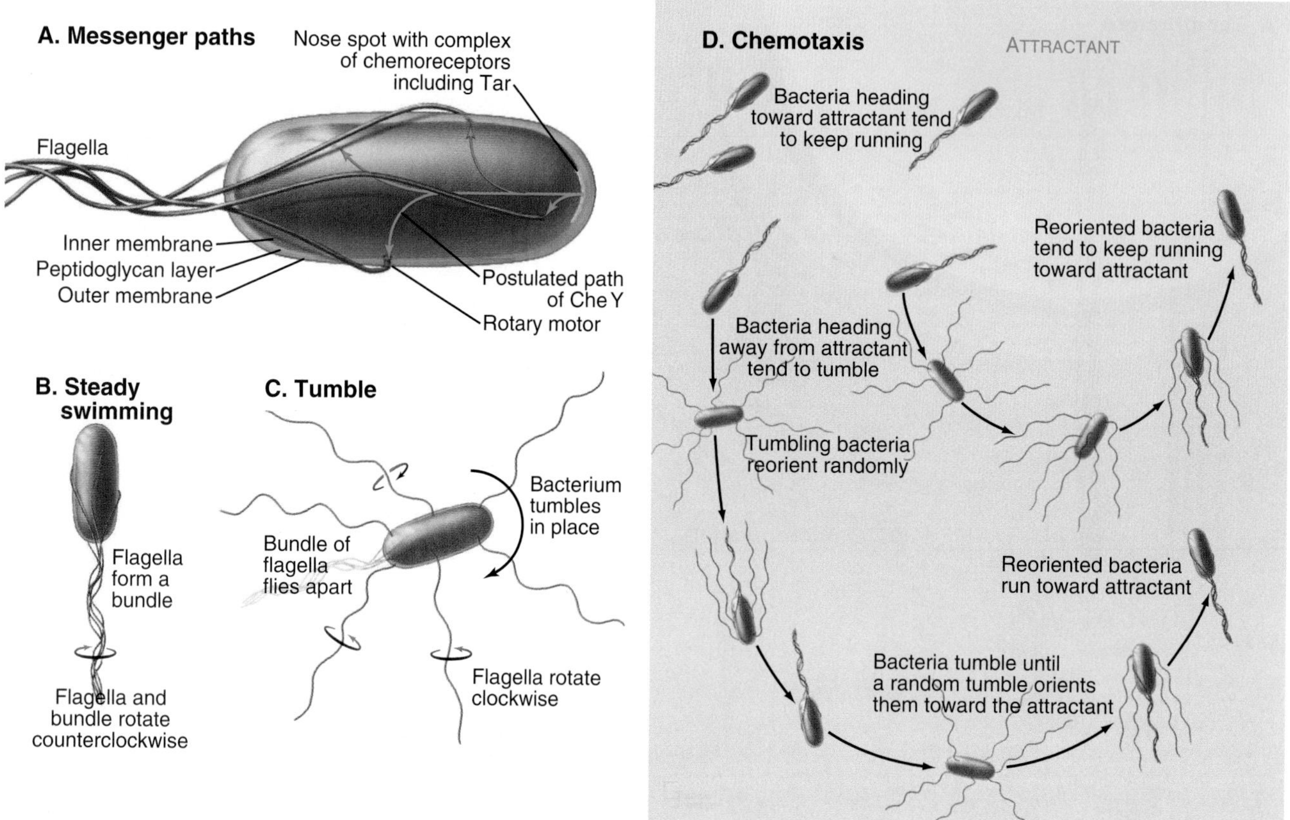

FIGURE 27.12 MOTILITY OF BACTERIA IS DETERMINED BY THE DIRECTION OF ROTATION OF THEIR FLAGELLA. A, Drawing of *Escherichia coli* showing the flagella and arrangement of signal transduction components with the highest concentration of receptors at one pole of the cell. **B,** Flagella that rotate counterclockwise (viewed from the tip of the flagella) form a bundle that pushes the cell smoothly forward. **C,** When flagella rotate clockwise, the bundle flies apart, and the cell tumbles in place. **D,** An attractive chemical biases movement toward its source by modulating the frequency of runs and tumbles.

chemoattractants, bacteria randomly switch between periods of swimming that last about 0.9 second and brief tumbles lasting about 0.1 second, allowing for random reorientations every second.

A gradient of chemical attractant promotes the length of runs up the gradient by suppressing tumbling if the concentration of attractant increases over time (Fig. 27.12D). A two-component signaling pathway senses the concentration of attractant and controls the frequency of tumbling through phosphorylation of the response regulator **CheY**, which acts on the flagellar motor. Most components of the system were discovered by mutagenesis and named "Che" for chemotaxis gene with a lowercase "p" to indicate phosphorylation.

Ligand-free Tar receptors stimulate the phosphorylation of the associated histidine kinase **CheA**, which is bound to the receptor by a "scaffold" protein CheW. CheAp activates the response regulator, CheY, by transferring phosphate from histidine to aspartic acid 57 (D57) of CheY. CheYp has a higher affinity for the flagellar motor than CheY, so ligand-free receptors maintain a steady state with the rotors partially saturated with CheYp. With several bound CheYps, the motor switches from its free-running, counterclockwise state to a brief clockwise tumble approximately once per second.

Information about aspartate in the environment flows rapidly through the pathway as changes in the *concentrations* of the phosphorylated species CheAp and CheYp. A key point is that Tar with bound aspartate, Tar-D, ceases to activate histidine phosphorylation of CheA. Hence aspartate binding to Tar *reduces* the saturation of the flagellar motors with CheYp and the frequency of tumbles.

For the cell to respond to aspartate on a subsecond time scale, an accessory protein, **CheZ,** is required to increase the rate of CheYp dephosphorylation more than 100-fold from its slow spontaneous rate of 0.03 s^{-1}. Given this fast dissipation of CheYp, phosphate must flow constantly from adenosine triphosphate (ATP) to CheAp to CheYp to maintain a tumbling frequency of about 1 s^{-1} in the absence of an attractant (Fig. 27.13A). (Most other two-component pathways respond in minutes rather than milliseconds, because

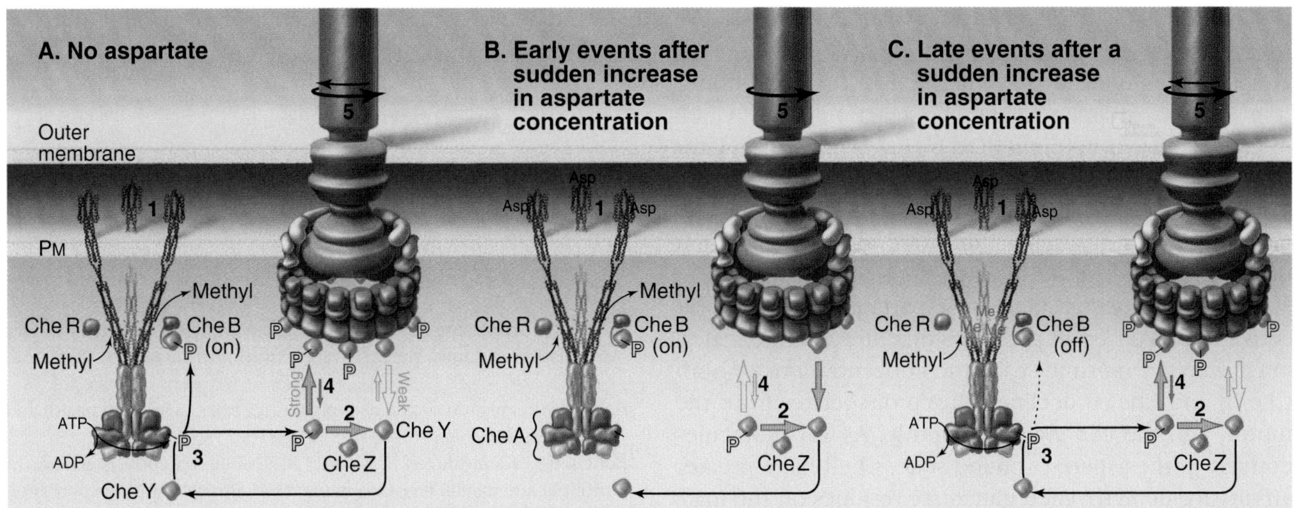

FIGURE 27.13 SIGNALING DURING BACTERIAL CHEMOTAXIS. Drawing of the signaling system under three conditions. **A,** Absence of aspartate. Ligand-free Tar *(1)* allows CheA to phosphorylate CheY and CheB *(3)*. Constant dephosphorylation of CheYp drives a cycle of phosphorylation *(2)*. The steady-state concentration of CheYp keeps the motor partially saturated *(4)*. The partially saturated motor turns counterclockwise 90% of the time (runs *[thick arrow]*) and clockwise 10% of the time (tumbles *[thin arrow]*) *(5)*. **B,** Rapid response to the presence of aspartate. Aspartate binding turns off Tar *(1)*. Constant dephosphorylation depletes CheYp on a time scale of tens of milliseconds *(2)*. CheY dissociates from the motor *(4)*. The motor, without CheYp, rotates counterclockwise, so the bacterium runs continuously *(5)*. **C,** Slower, adaptive response to the presence of aspartate. Inactive CheA stops phosphorylating CheB, allowing dephosphorylation of CheBp on a time scale of seconds; this inactivation of CheB and the higher affinity of CheR for inactive Tar result in higher methylation of Tar *(1)*. Even with bound aspartate, methylated Tar is partially active, allowing phosphorylation of CheA *(2)*. CheY is phosphorylated *(3)*. CheYp rebinds the motor *(4)*. The flagella turn clockwise part of the time, returning to the steady-state frequency of runs and tumbles *(5)*.

dephosphorylation of the response regulator is much slower. The following sections examine chemotaxis on the system level.)

Temporal Sensing of Gradients

Unlike eukaryotic cells, most bacteria are too small and move too fast to detect a spatial gradient along the length of the cell body. Instead, they sense the gradient as a perceived change in concentration of attractant or repellent as a function of time. When a bacterium swims up a gradient of chemoattractant, the concentration of attractant increases with time, and the signaling mechanism suppresses tumbling. Fewer tumbles result in a biasing of smooth swimming toward the attractant. When a cell swims down the gradient, tumbling is more frequent, allowing for reorientation.

A sudden increase in the concentration of aspartate yields a smooth swimming response within 200 milliseconds as a result of rapid reequilibration of the concentrations of all the cytoplasmic signaling components (Fig. 27.13B). Aspartate binds Tar and inhibits autophosphorylation of CheA. CheYp has a half-life of less than 100 milliseconds, so the concentrations of CheAp and CheYp decrease rapidly. CheYp dissociates from the flagellar motor and the tendency of the motor to stay in the counterclockwise, smooth swimming direction increases. If the concentration change with time is due to a persistent gradient of aspartate, the bacterium tends to move on average up the gradient toward the source, although it will make several runs up and down the gradient during this time.

The opposite sequence of events takes place if a bacterium swims down a gradient of aspartate. The fraction of Tar with bound aspartate declines, CheAp and CheYp concentrations rise, and tumbling is more frequent, providing opportunities to reorient and swim back up the gradient. Such is the case with avoiding a repellant gradient, as well.

Adaptation

If the aspartate concentration suddenly increases everywhere, bacteria respond quickly with smooth swimming, but within tens of seconds to minutes, they return to their normal frequency of intermittent tumbling. Thus, the steady-state tumbling frequency depends on changes in the concentration of aspartate relative to background levels rather than the absolute concentration. This remarkable capacity to adapt is accomplished by a feedback control mechanism provided by reversible **methylation** of the receptor (Fig. 27.13C).

Two relatively slow enzymes determine the level of Tar methylation (Fig. 27.11B). **CheR** adds methyl groups to four glutamic acid residues on each receptor polypeptide, whereas the response regulator **CheB** removes them. Methylated Tar has a somewhat lower affinity for aspartate than unmethylated Tar, but Me-Tar with bound aspartate is more effective at stimulating CheA phosphorylation than Tar with bound aspartate.

CheR is constitutively active but sensitive to the overall metabolic state of the cell, as it depends on the concentration of *S*-adenosyl methionine, a methyl donor that is used in many metabolic reactions. The CheB methylesterase is autoinhibited by its response regulator domain but activated by phosphorylation by CheAp. Thus, CheB methylesterase activity depends on the concentration of CheAp, and rate of demethylation determines the level of receptor methylation.

Adaptation occurs because aspartate binding to Tar activates two different pathways on different time scales. On a *millisecond* time scale, the concentrations of both CheAp and CheYp decline, CheYp dissociates from the motor, and the cell swims smoothly. As Tar molecules convert to the aspartate-bound state, CheR has a greater affinity for demethylated glutamate residues on the inactive Tar-D and begins to methylate them, which occurs on a time scale of *seconds*. As Tar molecules accumulate methyl groups, the reduced activity associated with ligand binding is slowly reversed. In this way, Tar molecules with bound aspartate have their activity restored to the baseline level that they had before the aspartate concentration increased. This **robust adaptation** mechanism is an integral feedback system, just like a thermostat on a heater. The system works in part, because CheR preferentially methylates inactive receptors and CheBp preferentially demethylates active receptors.

Extended Range of Response

An amazing feature of this system is its ability to respond with fast changes in flagellar rotation and slow adaptation to changes of *just a few percentage points* in aspartate concentration over a range of five orders of magnitude. Clearly, a simple bimolecular reaction of aspartate with Tar cannot change the fractional saturation of Tar over such an extended range of concentrations. This extended range of sensitivity is valuable for the survival of the bacterium and depends on amplification at the level of the receptor whereby aspartate binding to one Tar activates many surrounding Tars in the receptor clusters at the end of the cell. This physical communication is achieved by a lattice of CheA and CheW between the cytoplasmic tips of the receptors.

Bacterial chemotaxis illustrates some of the classical features of signaling pathways, including high sensitivity due to amplification at the level of CheA phosphorylation, feedback control through methylation of Tar, and branching networks that respond on different time scales to the same stimulus. The mechanism has been tested thoroughly by mutating all the signaling components and observing the consequences. Furthermore, random variations in the numbers of these proteins allow individuals in populations of bacterial cells to take advantage of a wide range of environments and improve the fitness of the community.

ACKNOWLEDGMENTS

We thank Jonathan Bogan and Nicholas Frankel for suggestions on revisions of this chapter.

SELECTED READINGS

Babon JJ, Lucet IS, Murphy JM, Nicola NA, Varghese LN. The molecular regulation of Janus kinase (JAK) activation. *Biochem J*. 2014;462:1-13.

Bogan JS. Regulation of glucose transporter translocation in health and diabetes. *Annu Rev Biochem*. 2012;81:507-532.

Boucher J, Kleinridders A, Kahn CR. Insulin receptor signaling in normal and insulin-resistant states. *Cold Spring Harb Perspect Biol*. 2014;6:a009191.

Bray D. The propagation of allosteric states in large multiprotein complexes. *J Mol Biol*. 2013;425:1410-1414.

Call ME, Wucherpfennig KW. The T cell receptor: Critical role of the membrane environment in receptor assembly and function. *Annu Rev Immunol*. 2005;23:101-125.

Capra EJ, Laub MT. Evolution of two-component signal transduction systems. *Annu Rev Microbiol*. 2012;66:325-347.

Dustin ML, Groves JT. Receptor signaling clusters in the immune synapse. *Annu Rev Biophys*. 2012;41:543-556.

Ferrell JE Jr. Self-perpetuating states in signal transduction: Positive feedback, double-negative feedback and bistability. *Curr Opin Cell Biol*. 2002;14:140-148.

Frankel NW, Pontius W, Dufour YS, et al. Adaptability of non-genetic diversity in bacterial chemotaxis. *Elife*. 2014;3:doi:10.7554/eLife.03526.

Goh LK, Sorkin A. Endocytosis of receptor tyrosine kinases. *Cold Spring Harb Perspect Biol*. 2013;5:a017459.

Hubbard SR. The insulin receptor: both a prototypical and atypical receptor tyrosine kinase. *Cold Spring Harb Perspect Biol*. 2013;5:a008946.

Johnson GL, Lapadat R. Mitogen-activated protein kinase pathways mediated by ERK, JNK and p38 protein kinases. *Science*. 2002;298:1911-1912.

Jones CW, Armitage JP. Positioning of bacterial chemoreceptors. *Trends Microbiol*. 2015;23:247-256.

Krzysztof P. G protein-coupled receptor rhodopsin. *Annu Rev Biochem*. 2006;75:743-767.

Massagué J. TGFβ signalling in context. *Nat Rev Mol Cell Biol*. 2012;13:616-630.

Mombaerts P. Genes and ligands for odorant, vomeronasal and taste receptors. *Nat Rev Neurosci*. 2004;5:263-278.

Ridge KD, Abdulaev NG, Sousa M, et al. Phototransduction: Crystal clear. *Trends Biochem Sci*. 2003;28:479-487.

Rieke F, Baylor DA. Single photon detection by rod cells of the retina. *Rev Mod Phys*. 1998;70:1027-1036.

Ritter SL, Hall RA. Fine-tuning of GPCR activity by receptor-interacting proteins. *Nat Rev Mol Cell Biol*. 2009;10:819-830.

Shi Y, Massague J. Mechanisms of TGF-β signaling from cell membranes to the nucleus. *Cell*. 2003;113:685-700.

Tu Y. Quantitative modeling of bacterial chemotaxis: signal amplification and accurate adaptation. *Annu Rev Biophys*. 2013;42:337-359.

van der Merwe PA, Dushek O. Mechanisms for T cell receptor triggering. *Nat Rev Immunol*. 2011;11:47-55.

Venkatakrishnan AJ, Deupi X, Lebon G, et al. Molecular signatures of G-protein-coupled receptors. *Nature*. 2013;494:185-194.

SECTION VIII

Cellular Adhesion and the Extracellular Matrix

SECTION VIII OVERVIEW

This section covers the variety of extracellular materials that provide mechanical support for the tissues of multicellular organisms. After their divergence about 1 billion years ago, animals and plants evolved completely different macromolecules to construct their extracellular matrices. The main biopolymer in animals is the protein collagen, whereas plants use the polysaccharide cellulose. Both can make impressively strong structures, including cartilage and bone in animals and wood that supports giant trees. This section also explains the mechanisms that cells of all sorts use to adhere to each other and the objects in their environments, including the extracellular matrix. Cell surface adhesion proteins enable cells to establish intimate relationships with each other and macromolecules in the extracellular matrix. These interactions are essential for tissue integrity and intercellular communication in complex tissues, including the brain, heart, and other organs.

Chapter 28 describes the cells that are found in the **extracellular matrix** of vertebrate animals. **Fibroblasts** synthesize and secrete the macromolecules that form the extracellular matrix. Fat cells store high-energy lipid molecules. Specialized **phagocytic cells** and immune system cells patrol the extracellular matrix of connective tissues, seeking out and destroying foreign cells and molecules throughout the body.

Chapter 29 describes the biosynthesis of the macromolecules that form the extracellular matrices of vertebrates. In connective tissue, fibroblasts secrete the protein subunits of **collagen fibrils** and **elastic fibers** as well as adhesion proteins and complex polysaccharides that reinforce the protein fibers in the extracellular matrix. The proportions of these macromolecules vary considerably. Tendons, ligaments, and some layers of the intestinal wall are composed largely of massive collagen fibrils with relatively few cells. The vitreous body of the eye is composed mostly of gelatinous polysaccharides with few fibers. The simplest type of extracellular matrix is the **basal lamina**, a thin layer of matrix that is secreted as rug beneath epithelial cells and a sheath around muscle cells and neurons.

Remarkably, just four families of plasma membrane adhesion proteins, **immunoglobulin cell adhesion molecules (IgCAMs)**, **cadherins**, **integrins**, and **selectins**, account for much of cellular adhesion. Chapter 30 introduces these adhesion proteins and explains some of their diverse structures and functions. IgCAMs and cadherins make specific interactions with

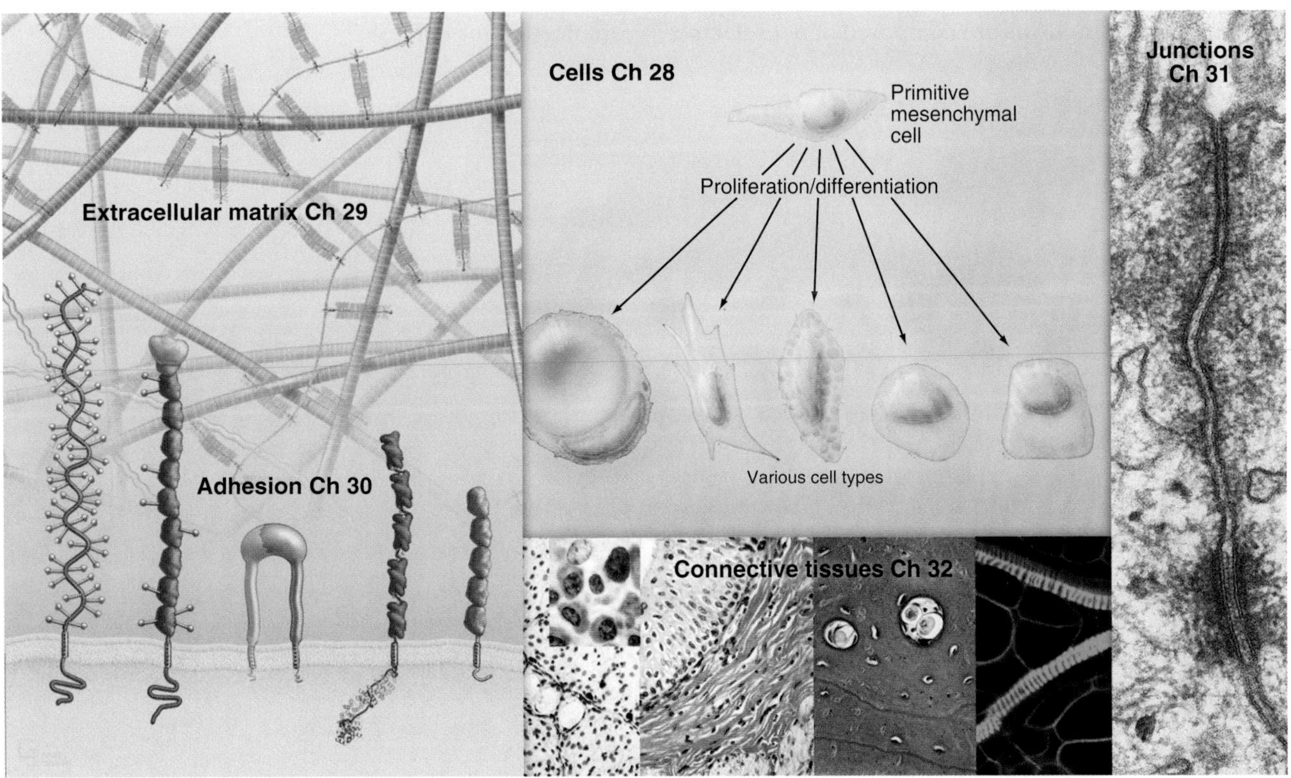

complementary proteins on the surface of partner cells. Most integrins bind to extracellular matrix molecules, but some engage adhesion proteins on other cells. Selectins interact with glycoproteins called mucins on the surfaces of other cells.

Expression of a limited repertoire of adhesion protein isoforms allows cells of multicellular organisms to establish specific interactions with appropriate partner cells while avoiding inappropriate interactions. The specificity provided by these adhesion molecules is required to form epithelia during embryonic development, assemble specialized connective tissues (Chapter 32), heal wounds, and transmit the force of muscle contraction to the extracellular matrix. Chapter 30 features two examples of dynamic, selective adhesion: adhesion of platelets to each other during the repair of damage to small blood vessels and blood clotting, and adhesion of white blood cells to the endothelial cells lining blood vessels of inflamed tissues.

Even unicellular organisms require molecular mechanisms to adhere to other cells and objects that they encounter in their environments. For example, unicellular algae and yeast adhere to each other during mating, and slime mold amoebas adhere to each other as they develop into fruiting bodies. Bacteria also form complex biofilms to help them survive in hostile environments.

Intercellular junctions are specialized sites of adhesion between cells in some tissues (Chapter 31). **Tight junctions** allow a sheet of epithelial cells to create semipermeable barriers between tissue compartments such as the lumen and the wall of the intestine. Such barriers allow epithelia to concentrate materials on one side or the other. **Gap junctions** are composed of nonselective channels that connect the cytoplasms of two cells. These channels enable action potentials to spread directly from one cell to the next, as in the heart. Gap junction channels also allow solutes that are less than 1000 Da in molecular weight to move between the coupled cells. Cadherins connect adjacent cells to the cytoskeleton at two types of adhesive junctions. The cytoplasmic domains of cadherins are anchored to actin filaments at **adherens junctions** and to intermediate filaments at **desmosomes**.

The abundance, organization, and proportions of macromolecular components determine the mechanical properties of the extracellular matrix (Chapter 32). Plant cells secrete cellulose, a polymer of glucose units, plus a mixture of other polysaccharides, glycoproteins, and organic molecules to make a wall around each cell. These cell walls form materials ranging from soft cotton fibers to the wood that supports giant Sequoia trees.

Connective tissues of animals also exhibit striking variety, owing to their particular mixtures of matrix molecules. Skin and blood vessels are resilient, because of numerous elastic fibers. Tendons have great tensile strength, owing to the high density of collagen fibrils. Bone is incompressible and rigid, because of its calcified collagen matrix. On the level of gross anatomy, cells and fibers form fascia, tendons, cartilage, and bones that support the organs of the body. Connective tissue also provides avenues for communication and supply within the body. Both the circulatory system and the peripheral nervous system run through connective tissue compartments of each organ. The vascular system transports phagocytic and immune system cells to sites where they are needed for defense.

CHAPTER 28

Cells of the Extracellular Matrix and Immune System

A remarkable variety of specialized cells populate the connective tissues of animals. These cells manufacture extracellular matrix, defend against infection, and maintain energy stores in the form of lipid (Fig. 28.1). Some of these cells arise in connective tissue and remain there. These **indigenous cells** are specialized: Fibroblasts make the collagen, elastic fibers, and proteoglycans of the extracellular matrix; fat cells store lipids; chondrocytes secrete the matrix for cartilage; and osteoblasts manufacture the calcified matrix of bone. The remaining cells arise elsewhere, travel through blood and lymph, and enter connective tissue as needed, so they are known as **immigrant cells**. These visitors are part of the immune system, which defends against pathogens. This chapter introduces all these cells.

Indigenous Connective Tissue Cells

Mesenchymal Stem Cells

During embryogenesis the indigenous cells of connective tissues (fibroblasts, fat cells, chondrocytes, and osteoblasts) derive from multipotential mesenchymal stem cells (see Box 41.2 and Fig. 28.1). In adults, small numbers of these inconspicuous precursor cells associate with small blood vessels but are difficult to identify by light microscopy. By electron microscopy (Fig. 28.2), mesenchymal cells resemble fibroblasts but with fewer organelles of the secretory pathway. The nature of the mesenchymal stem cells in adult tissues is still being investigated.

Fibroblasts

Fibroblasts are the connective tissue workhorses, synthesizing and secreting most macromolecules of the extracellular matrix (Fig. 28.2). Chapter 29 considers the synthesis of these matrix molecules in detail. As appropriate for a secretory cell, mature fibroblasts have abundant rough endoplasmic reticulum and a large Golgi apparatus. They are generally spindle-shaped, with a flattened, oval nucleus, but can assume many other shapes depending on the mechanical forces in the surrounding matrix. The migratory patterns of the fibroblasts determine the patterns of collagen fibrils in tissues.

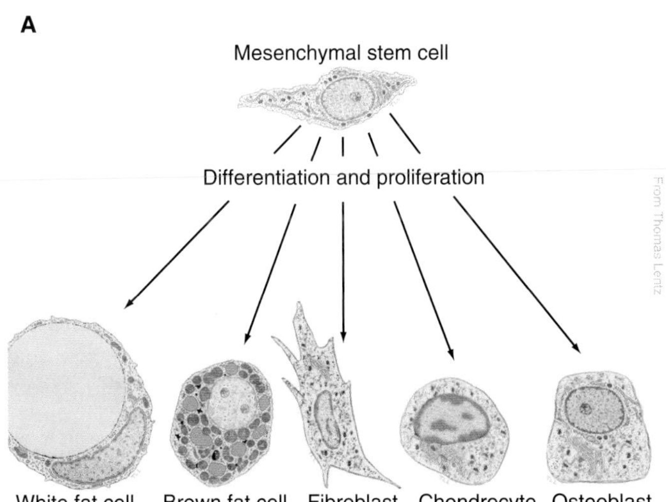

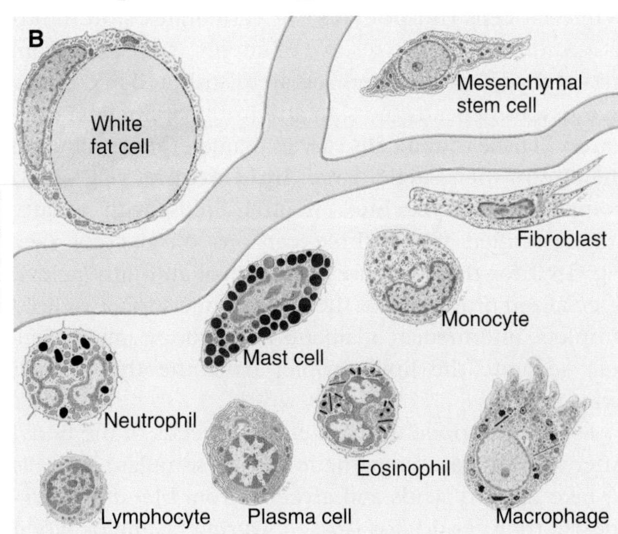

FIGURE 28.1 CONNECTIVE TISSUE CELLS. A, Indigenous connective tissue cells all originate from a stem cell called a primitive mesenchymal cell. **B,** Connective tissue near a small blood vessel showing indigenous cells in *pink* and immigrant cells in *green*.

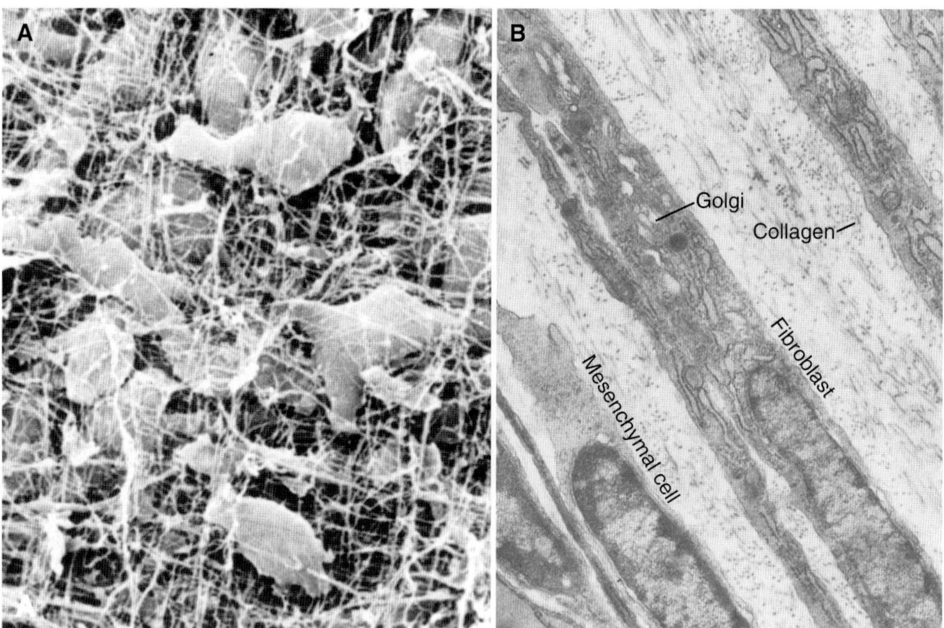

FIGURE 28.2 FIBROBLASTS. A, Scanning electron micrograph of fibroblasts migrating through collagen fibrils. **B,** Transmission electron micrograph of a thin section of a fibroblast illustrating the abundant organelles of the secretory pathway (endoplasmic reticulum and Golgi apparatus) and extracellular collagen fibrils. A primitive mesenchymal cell is shown in the *lower left.* (**A,** Courtesy E.D. Hay, Harvard Medical School, Boston, MA. **B,** Courtesy D.W. Fawcett, Harvard Medical School, Boston, MA.)

In response to tissue damage, fibroblasts proliferate and migrate into the wound, where they synthesize new matrix to restore the integrity of the tissue (see Fig. 32.11). This process can get out of control if inflammatory cells secrete excessive transforming growth factor (TGF)-β (see Fig. 27.10) and other factors that stimulate fibroblasts to produce matrix molecules. Fibrosis, excess accumulation of extracellular matrix, can compromise the functions of the heart, liver, lung, or skin.

White Fat Cells

White fat cells **(adipocytes)** of vertebrates store lipids as a readily accessible reserve of energy. They arise from mesenchymal progenitors and are distributed in connective tissues beneath the skin and in the abdominal mesentery. These round cells vary in diameter depending on the size of their single, large, **lipid droplet** (Fig. 28.3) containing **triglycerides,** neutral lipids with a fatty acid esterified to all three carbons of glycerol (see Fig. 13.2 for the structures of glycerol and fatty acids). Specialized proteins coat the cytoplasmic surface of lipid droplets. Intermediate filaments and endoplasmic reticulum separate the lipid droplet from the thin rim of cytoplasm.

Fat cells respond to the metabolic needs of the body. After a meal, parasympathetic nerves stimulate fat cells to take up fatty acids and glycerol from blood and synthesize triglycerides for storage. During fasting or when the body requires energy, sympathetic nerves acting through β-adrenergic receptors (see Fig. 27.3), stimulate adipocytes to hydrolyze fatty acids from triglycerides for release into the blood for use by other organs. If a mammal ingests excess calories, white fat cells enlarge their lipid stores and increase in number. White fat cells are long lived, with a half-life approximately 9 years, so the excess cells persist in obese individuals.

White fat tissue is also an endocrine organ that secretes several cytokines and polypeptide hormones including **leptin**. Many factors including nutritional status, mass of fat cells, exercise and sleep influence the rate of leptin secretion. Leptin suppresses appetite and hunger by stimulating cytokine receptors on neurons in the brain. These neurons respond by secreting other polypeptide hormones that regulate appetite. Massive obesity results from congenital absence of leptin or defects in its receptor.

A variety of mutations cause inherited lipodystrophies, conditions with loss of fat in one or more tissues. Examples include loss of function mutations of an enzyme required to synthesize triglycerides or a nuclear receptor that stimulates differentiation of fat cells. It is an ongoing mystery why mutations in the gene for nuclear lamins A and C (see Fig. 9.7) or a protease that processes lamin A should cause selective loss of fat from the trunk and limbs.

Brown and Beige Fat Cells

Brown and beige fat cells of placental mammals derive their color from cytochromes in numerous mitochondria, which they use to generate heat in response to cold

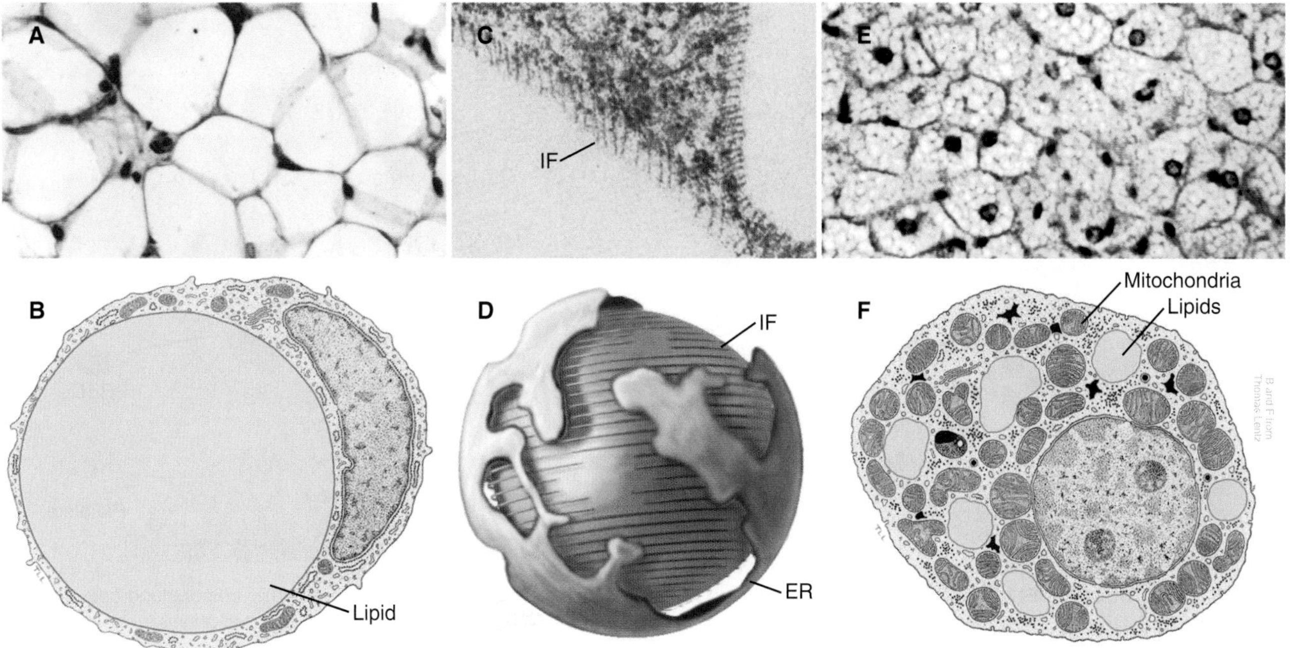

FIGURE 28.3 FAT CELLS. A, Light micrograph of a section of white adipose (fat) cells stained with hematoxylin and eosin. **B,** Drawing of a white adipose cell. **C,** Transmission electron micrograph of a thin section of the edge of a lipid droplet showing the circumferential sheath of vimentin intermediate filaments (IF [see Chapter 35]). **D,** Interpretive drawing of a lipid droplet with its associated filaments and endoplasmic reticulum (ER). **E,** Light micrograph of a section of brown fat. **F,** Drawing of a brown fat cell. (**A, C,** and **E,** Courtesy D.W. Fawcett, Harvard Medical School, Boston, MA. **B** and **F,** Modified from T. Lentz, Yale Medical School, New Haven, CT. **D,** Modified from Werner Franke, University of Heidelberg, Germany.)

or (in lean rodents) excess food intake. Fat is stored in multiple, small droplets (Fig. 28.3F). Brown fat is less abundant than white fat, being concentrated in connective tissue between the scapulae in mammals. Newborn humans have more brown fat than do adults in order to generate heat during the adjustment to a new environment after birth. Hibernating animals use brown fat to raise their temperatures when emerging from hibernation. Beige fat cells are found among white fat cells and increase in numbers when a mouse is exposed to cold. A fourth type of fat cell is found in bone marrow.

Both brown and beige fat cells generate heat by short-circuiting the proton gradient that is usually used to generate adenosine triphosphate (ATP) in mitochondria (see Fig. 19.5). Sympathetic nerves acting through β-adrenergic receptors (see Fig. 27.3) and protein kinase A stimulate brown fat cells to break down lipids to provide fatty acids for oxidation by mitochondria. β-Adrenergic receptors also drive expression of "uncoupling protein," a carrier in the inner mitochondrial membrane similar in structure to the mitochondrial ATP/adenosine diphosphate (ADP) antiporter (see Fig. 15.4A). Uncoupling protein-1 dissipates the proton electrochemical gradient across the inner mitochondrial membrane (perhaps acting as a fatty acid/proton symporter), so energy is lost as heat rather than being used to synthesize ATP. Beige fat cells also run a futile cycle of creatine phosphorylation and dephosphorylation to produce heat. **Thermogenesis** may be an "energy buffer" that, when defective in animals, can contribute to obesity. Therefore, stimulating thermogenesis is one goal in the treatment of obesity.

Brown fat cells express thermogenic proteins constitutively, whereas β-adrenergic stimulation drives expression of these proteins in beige fat cells. The thermogenic fat cells have different precursors. Brown fat cells arise from the same mesenchymal stem cells as skeletal muscle, while beige fat cells are more closely related to white fat cells.

Origin and Development of Blood Cells

The blood of vertebrates contains a variety of cells, each with a specialized function (Fig. 28.4 and Table 28.1). Red blood cells transport oxygen, platelets repair damage to blood vessels, and various types of white blood cells defend against infections. All blood cells derive ultimately from **pluripotent stem cells** (Fig. 28.4B; also see Box 41.2). These hematopoietic stem cells can restore the production of all blood cells in mice that have been irradiated to destroy their own blood cell precursors or after transplantation of human bone marrow. Destruction of stem cells (eg, by drugs such as

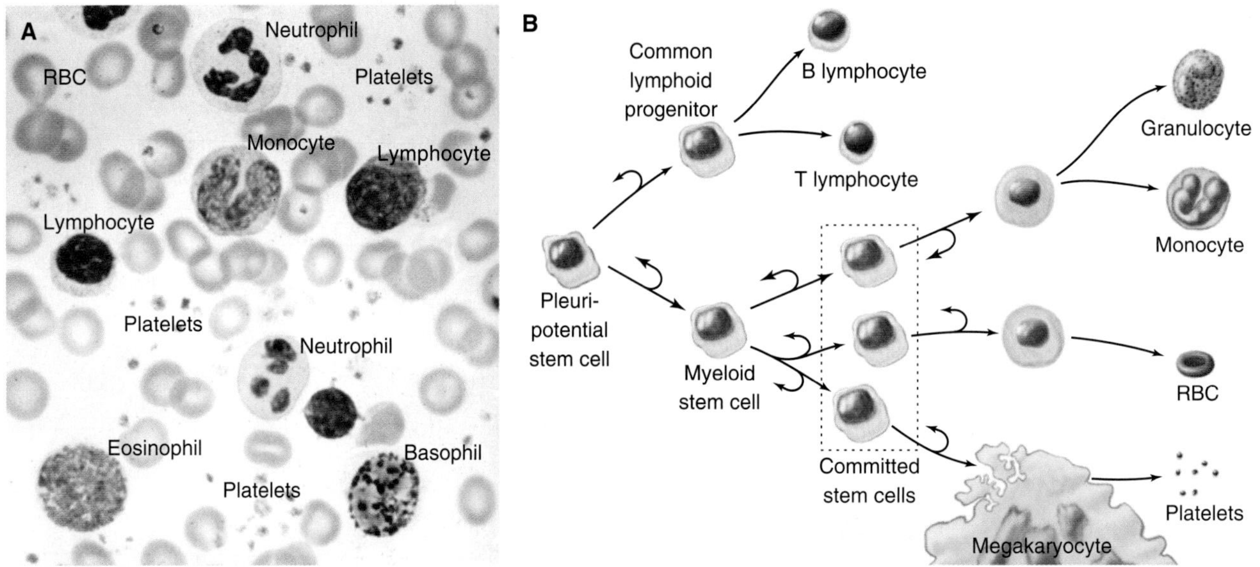

FIGURE 28.4 BLOOD CELLS. A, Light micrograph of a dried blood smear prepared with Wright stain. **B,** Family tree of blood cells showing the developmental relationships of the various lineages. *Looping-back arrows* indicate renewal of the cell type. *Forward-oriented arrows* indicate differentiation and proliferation. RBC, red blood cell. (**A,** Courtesy J.-P. Revel, California Institute of Technology, Pasadena.)

TABLE 28.1 Blood Cells (as Seen on a Stained Smear of Blood)

Type	Concentration	Features
Platelets	300,000/μL	Anucleate; 2-3 μm wide; purple granules
Erythrocytes	~5 × 10^6/μL	7 μm wide biconcave disks; no nucleus; pink cytoplasm
Neutrophils	~60% of total WBCs	10-12 μm wide; multilobed nucleus; many unstained granules; few azurophilic (blue) granules
Eosinophils	~2% of total WBCs	Bilobed nucleus; numerous, large, refractile, pink-stained granules; ~12 μm wide
Basophils	~0.5% of total WBCs	Lobed nucleus; large, blue-stained granules; ~10 μm wide
Lymphocytes	~30% of total WBCs	Small, round, intensely stained nucleus; some small azurophilic granules; variable amount of clear blue cytoplasm, so may be classified as either small (~7-8 μm wide), medium, or large
Monocytes	~5% of total WBCs	Up to 17 μm wide; large, indented nucleus and gray-blue cytoplasm with a few azurophilic granules

WBCs, white blood cells.

chloramphenicol) leads to **aplastic anemia,** a condition in which few blood cells are produced, owing to a lack of precursors.

Hematopoietic stem cells do not grow well in tissue culture, because they require a special environment provided in the bone marrow. In these niches located next to small blood vessels, endothelial cells, mesenchymal cells, and sympathetic nerves provide contacts and growth factors. Sympathetic nerves also drive the release of small numbers of hematopoietic stem cells into the blood every day following a circadian rhythm.

Pluripotent hematopoietic stem cells give rise to much larger numbers of more mature stem cells that proliferate and produce blood cells. At several stages along the differentiation pathway, precursors undergo irreversible differentiation that commits them to a particular lineage. The first branch in the pathway of differentiation separates the precursors of lymphocytes from the precursors of the other blood cells called myeloid stem cells. Next, myeloid stem cells differentiate into committed stem cells that reside in different locations in bone marrow. One branch gives rise to red blood cells, megakaryocytes, and platelets. In the other branch a common committed stem cell differentiates into monocytes and the three types of granulocytes (neutrophils, eosinophils, and basophils). Mast cells also arise from myeloid stem cells, although the lineage is uncertain. Through differentiation, each mature cell type acquires unique functions. Platelets, red cells, granulocytes, and monocytes develop in bone marrow. Lymphocytes develop in both bone marrow and lymphoid tissues (thymus, spleen, and lymph nodes).

Minute quantities of specific glycoprotein **growth factors** control the balance between self-renewal and proliferation at each stage of blood cell development, starting with pluripotential stem cells. Feedback mechanisms control production of these growth factors. For example, the oxygen level in the kidney controls the synthesis of **erythropoietin,** the growth factor for the red blood cell series. (See Fig. 27.9 for the cytokine signaling pathway.) A dimeric transcription factor called hypoxia inducible factor (HIF)-1α/HIF-1β regulates the expression of erythropoietin. When oxygen is abundant, HIF-1α is hydroxylated on a proline residue, marking it for ubiquitination and destruction (see Fig. 23.3), turning down the expression of erythropoietin. When kidney cells lack oxygen (owing to low levels of red blood cells, poor blood circulation in the kidney, or high altitude) HIF-1α/HIF-1β accumulates, and erythropoietin is expressed and secreted to stimulate red blood cell production by bone marrow. The reciprocal relationship between oxygen and erythropoietin that is achieved by this feedback mechanism sets red blood cell production at a level required to deliver oxygen to the tissues. Many other cells use the HIF-1α/HIF-1β system to adjust gene expression to local oxygen levels.

Mutations altering the growth control (see Fig. 41.12) of a stem cell can give rise to proliferative disorders, such as **leukemia.** In chronic myelogenous leukemia, a chromosomal rearrangement in a single white blood cell precursor creates a fusion between the genes for BCR (breakpoint cluster region) and ABL (a Src family cytoplasmic tyrosine kinase; see Fig. 25.3 and Box 27.5). The BCR-ABL protein is constitutively active and drives proliferation of a clone of immature white blood cells that crowd out and inhibit the production of other blood cells, leading to anemia and platelet deficiency. Affected individuals are prone to infection, because the immature white blood cells are ineffective phagocytes. Fortunately, a small-molecule inhibitor of the kinase activity of BCR-ABL (imatinib mesylate [GLEEVEC]) suppresses this clone in many patients. Mutations rendering the JAK2 (just another kinase 2) kinase (see Fig. 24.6) constitutively active drive proliferation in other leukemias. Uncontrolled proliferation of a clone of red blood cell precursors causes a similar condition, characterized by excess red cells, called **polycythemia vera.**

Cells Confined to the Blood

Erythrocytes (Red Blood Cells)

Red blood cells (RBCs) (Fig. 28.4; also see Fig. 13.8) contain more than 300 mg/mL of hemoglobin to carry oxygen from the lungs to tissues and carbon dioxide from tissues to the lungs. As they proliferate and differentiate in bone marrow, RBC precursors accumulate hemoglobin and shed all their organelles. A resilient, spectrin-actin membrane cytoskeleton (see Fig. 13.11) maintains the biconcave shape even after the cell is heavily distorted each time it squeezes through a small capillary. The elasticity of the membrane skeleton allows it to regain its shape.

Human bone marrow produces about 100 billion RBCs each day. After circulating in blood for 120 days, erythrocytes abruptly become senescent, and phagocytes in the spleen, liver, and bone marrow remove them from the blood. The biochemical basis of this precise cellular aging and clearance process is still being investigated.

Mutations in many different genes cause RBC diseases. In **hereditary spherocytosis** (and other hemolytic anemias), the membrane cytoskeleton loses its resiliency as a result of mutations, causing deficiencies or molecular defects of spectrin or other component proteins. These defective cells are easily damaged and eventually become smaller and rounder than normal. Many different mutations of the globin genes may compromise the synthesis of a particular globin gene or decrease the stability or oxygen-carrying capacity of hemoglobin. In **sickle cell disease,** hemoglobin S is prone to assemble into tubular polymers that distort the cell and clog up the circulation.

Platelets

Platelets are small cellular fragments without a nucleus that contribute to blood clotting and repair of minor defects in the sheet of endothelial cells that lines blood vessels. A long, coiled microtubule presses out against the plasma membrane, like a spring, to maintain the platelet's disk shape (Fig. 28.5). The most prominent organelles are two types of membrane-bound granules. Dense granules contain ADP and serotonin. Alpha granules contain stores of adhesive glycoproteins including fibrinogen, fibronectin (see Fig. 29.14), and thrombospondin as well as the potent protein hormone called **platelet-derived growth factor.** Platelet-derived growth factor has a role in wound healing (see Fig. 32.11), but can contribute to atherosclerosis by stimulating the abnormal proliferation of smooth muscle cells in the walls of damaged arteries. Another secreted cytokine kills malaria parasites inside RBCs.

Platelets containing a full complement of organelles bud from the tips of protrusions on the surface precursor cells—giant polyploid **megakaryocytes** in the bone marrow. **Thrombopoietin,** a protein hormone related to erythropoietin, controls platelet production by stimulating the thrombopoietin cytokine receptor. Liver and kidney cells secrete thrombopoietin at a constant rate. Receptors on circulating platelets bind part of the thrombopoietin. Consequently, the blood concentration of thrombopoietin available to stimulate megakaryocyte maturation and platelet formation is inversely related to the total number of platelets. This feedback loop stimulates platelet production if the platelet supply is low.

Like RBCs, platelets are confined to the blood where they circulate for about seven days. Two pools of platelets freely exchange with each other: About two-thirds of the total platelets circulate, whereas one-third of the platelets are stored in the blood vessels of the spleen. The stored pool may increase when the spleen is enlarged, decreasing the platelet count in the blood.

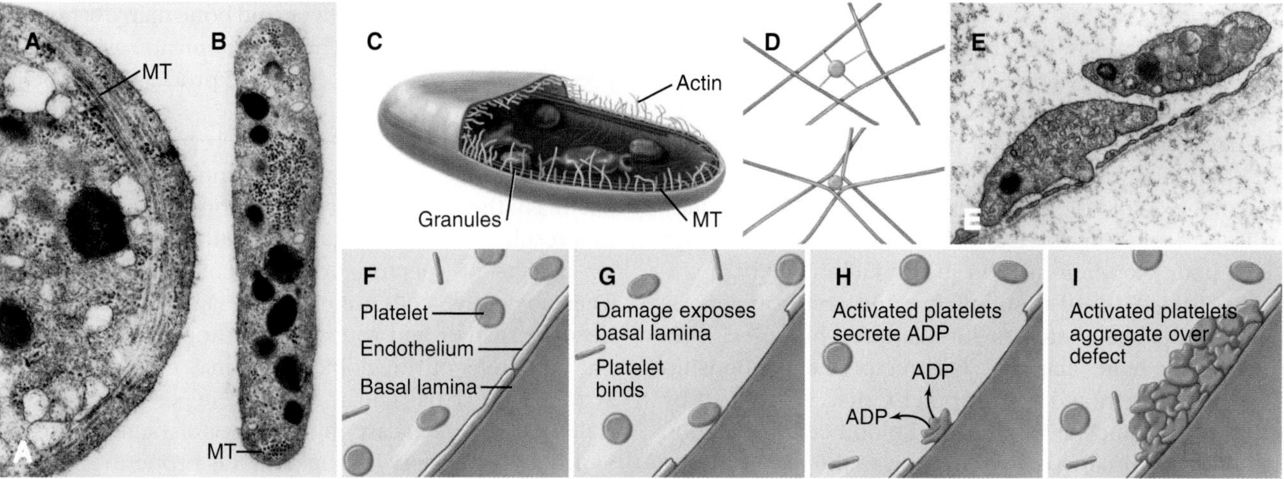

FIGURE 28.5 PLATELETS AND THEIR ROLE IN HEMOSTASIS. A–B, Transmission electron micrographs of thin sections of platelets. **C,** Interpretive drawing showing the circumferential band of microtubules (MT), actin filaments in the cortex, and granules in the cytoplasm. **D,** Role of platelets in blood clot retraction. Filopodia grasp strands of fibrin in a blood clot *(upper panel)* and pull them together *(lower panel)*. **E,** Electron micrograph of a thin section showing platelets adhering to the basal lamina through a small defect in the endothelium and to a second platelet in the lumen. **F–I,** Stages in the repair of a defect in the endothelial lining of a blood vessel. **F,** Circulating platelets do not bind to normal endothelial cells. **G,** Damage to the endothelium exposes the basal lamina, and a platelet binds to the collagen. **H,** Collagen activates the platelet to secrete adenosine diphosphate (ADP), which activates passing platelets. **I,** Activated platelets bind together, covering the defect in the endothelium. (**A–B,** Courtesy O. Behnke, University of Copenhagen, Denmark.)

Platelets control bleeding in three ways. First, they adhere and change shape to cover damaged vascular surfaces. Second, platelets stimulate **blood clotting.** A surface protein on activated platelets stimulates a cascade of proteolytic reactions culminating in the cleavage of plasma fibrinogen to form fibrin, which polymerizes to clot blood. The fibrin gel stops the flow of blood from damaged blood vessels. Third, platelets help close holes in damaged blood vessels by pulling on fibrin strands and contracting the clot.

When the mild trauma of daily existence produces defects in the endothelial cells lining blood vessels, platelets repair the damage. Platelets bind to von Willebrand factor and collagen in the basal lamina when it is exposed by damage to the endothelium. This triggers one of the best-understood examples of regulated cellular adhesion (for more detail, see Fig. 30.14). Activated platelets aggregate, extend actin-containing filopodia, and secrete the contents of their granules. The secreted ADP sets up a positive feedback loop, activating more platelets that form a cluster to fill the defect in the endothelium. Other platelets are consumed by phagocytic cells in the liver and spleen.

Patients with defective platelets or reduced circulating platelets (a complication of bone marrow disease and cancer chemotherapy) bruise easily, owing to unrepaired damage in small blood vessels, and may even bleed spontaneously. Conversely, hyperactive platelets may initiate pathological clots in the blood vessels of the heart, causing heart attacks or thrombosis in the veins of the legs.

Cellular Basis of Innate Immunity

All multicellular animals use two forms of **innate immunity** to defend themselves against infection by microorganisms. Phagocytic cells track down, ingest, and kill bacteria and fungi (see Fig. 22.3). A second line of defense is secretion of cytokines and small antimicrobial proteins by white blood cells and epithelial cells of the skin and intestine.

The following sections describe the main mammalian phagocytes, macrophages and neutrophils (Fig. 28.4). They originate in bone marrow from a common committed stem cell (see Fig. 28.4B) and acquire unique functions as they differentiate.

Innate immune cells detect the presence of microorganisms using **Toll-like receptors (TLRs)** (Fig. 28.6). The Toll gene was discovered in *Drosophila* encoding a receptor that was first linked to dorsal–ventral polarity in early development and later shown to be required for resistance to fungal infections.

TLRs are called "pattern recognition receptors," because they bind generic, repeating structures of polymeric macromolecules rather than recognizing fine molecular details like antibodies. These "pathogen-associated molecular patterns" or PAMPs are found on components essential for the functions of the microorganisms such as viral double-stranded RNA, bacterial flagellin, lipopolysaccharide from the outer membrane of gram-negative bacteria, and zymosan from the cell walls of fungi.

Mammals have about a dozen TLRs in three classes. All TLRs have cytoplasmic TIR (Toll interleukin-1 receptor)

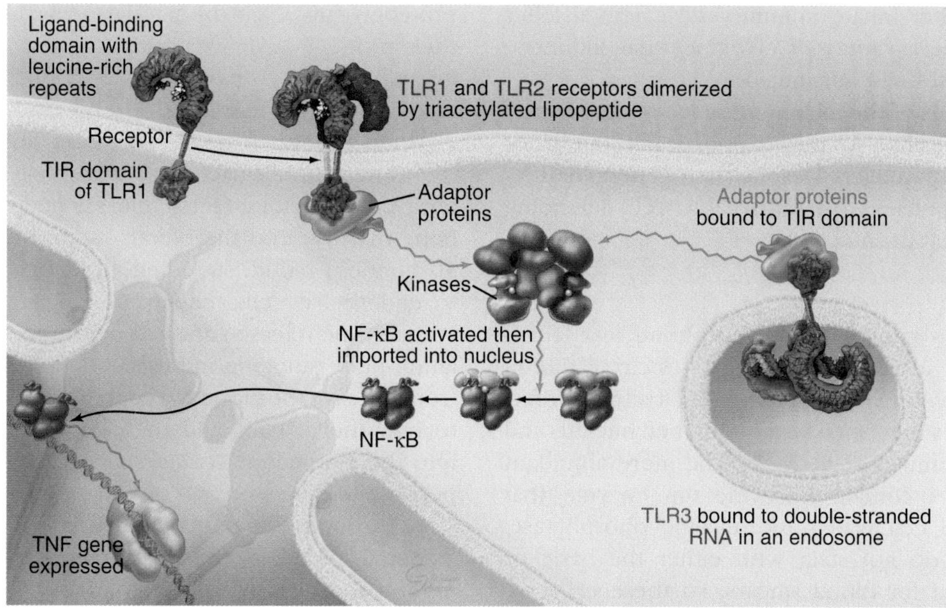

FIGURE 28.6 TOLL-LIKE RECEPTORS SIGNALING FROM THE PLASMA MEMBRANE AND ENDOSOMES. Binding of a triacetylated bacterial lipopeptide stabilizes a heterodimer of Toll-like receptor (TLR)-1 and TLR2 in the plasma membrane shown as ribbon diagrams with space-filling surfaces. Dimers of TLR3 in the membranes of endosomes bind double-stranded RNAs released from viruses. Ligand binding to receptor dimers aligns their cytoplasmic Toll-interleukin receptor (TIR) domains and allows binding of adapter proteins that initiate a signal transmitted to kinases, which activate cytoplasmic transcription factors including nuclear factor κB (NF-κB). NF-κB moves to the nucleus and stimulates expression of tumor necrosis factor (TNF) and other inflammatory mediators. (Molecular structures are based on Protein Data Bank [www.rcsb.org] file 2Z2X. For reference, see Jin MS, Kim SE, Heo JY, et al. Crystal structure of the TLR1-TLR2 heterodimer induced by binding of a triacylated lipopeptide. *Cell.* 2007;130:1071–1082; and Liu L, Botos I, Wang Y, et al. Structural basis of toll-like receptor 3 signaling with double-stranded RNA. *Science.* 2008;320:379–381.)

domains, but differ in other respects. Classic TLRs have ligand binding domains consisting of leucine rich repeats (Fig. 28.6). Interleukin (IL)-1 receptors have extracellular immunoglobulin (Ig) domains.

PAMP binding to a dimeric TLR on the plasma membrane of white blood cells or antigen-processing dendritic cells stimulates the secretion of inflammatory mediators such as tumor necrosis factor (TNF) and IL-1 and IL-6. TNF and ILs then alert distant cells to respond to the infection. The signaling pathway from TLRs to TNF (Fig. 28.6) involves cytoplasmic adapter proteins and kinases that activate transcription factors including nuclear factor κB (NF-κB) (see Fig. 10.22C) and expression of some long noncoding RNAs. Plasma membrane TLRs also activate mitogen-activated protein (MAP) kinase pathways (see Fig. 27.5).

The response also includes secretion of approximately 40 small proteins called **chemokines** that attract motile phagocytic cells. Chemokines with many unrelated names (IL-8, RANTES [regulated upon activation, normal T-cell expressed and secreted], eotaxin, monocyte chemotactic protein [MCP]-1, etc.) have similar structures and bind to a family of 14 different **chemokine receptors** expressed selectively by lymphocytes, monocytes, and granulocytes. These seven-helix receptors are coupled to trimeric G-proteins (see Fig. 24.3) that mediate chemotaxis (see Fig. 38.10) toward the source of the chemokine. Secretion of antimicrobial peptides such as defensins and cathelicidins not only kill pathogens by interacting with their membranes but also attract and activate cells of both the innate and the adaptive immune systems.

Cytoplasmic receptors with TIR domains bind nucleic acids. TLRs 3, 7, 8, and 9 are located in endosomes, where they bind double-stranded RNA released from viruses and single stranded RNAs from viruses and bacteria. Signaling pathways using some of the same components as the plasma membrane TLRs stimulate the expression and secretion of TNF and interferon (IFN)-α. The cytoplasmic system uses a helicase called RIG-I and other proteins to recognize foreign RNAs and to respond via NF-κB to produce IFN-β.

Other pattern recognition receptors reside in the cytoplasm of innate immune cells and other cells. For example, nucleotide-binding oligomerization domain (NOD)-like receptor proteins activate a kinase that stimulates both the transcription factor NF-κB and the MAP kinase pathways. Other NOD-like receptor proteins activate the proteolytic enzyme caspase-1 (see Fig. 46.11), which processes inflammatory cytokine precursors for secretion.

In addition to phagocytes, mammals have special lymphocytes called **natural killer cells** that express a variety of receptors to detect and kill cells infected with a virus. Like other innate immune cells, natural killer cells also secrete a variety of cytokines that influence immune responses and inflammation.

These elements of the innate immune system are programmed genetically, so they respond without prior exposure to the pathogen. Despite their generic nature, these innate responses work remarkably well, defending all metazoans against infection.

Neutrophils

Neutrophils, also known as polymorphonuclear leukocytes or "polys," are the main phagocytes circulating in blood, ready to enter connective tissues at sites of infection. They are distinguished by a multilobed nucleus and two types of granules (Fig. 28.7). The more abundant specific granules contain lysozyme (an enzyme that digests bacterial cell walls) and alkaline phosphatase. These granules do not stain with either the basic or acidic dyes used for blood smears, so these cells are called neutrophils. Azurophilic granules are true lysosomes containing hydrolytic enzymes bound to acidic proteoglycans. Neutrophils have few mitochondria, so they produce ATP in poorly oxygenated wounds by breaking down stores of glycogen by glycolysis. They are among the most motile cells in the body.

Human bone marrow produces about 100 billion (80 g) neutrophils each day. In response to infection or injury, a circulating factor releases neutrophils from the bone marrow into the blood. Neutrophils spend about 10 hours in blood, spending part of the time adherent to endothelial cells, chiefly in the lung. Exercise and epinephrine release adherent neutrophils into the circulating pool; smoking increases the adherent pool. Neutrophils leave the blood by receptor-mediated attachment to endothelial cells and then crawling between them into the connective tissue (see Fig. 30.13), where they perish after a day or two of phagocytosis.

Neutrophils are humans' first line of defense against bacterial infection, and they are highly specialized for finding and destroying bacteria. Provided that the concentration of neutrophils is high enough (approximately 10 million cells per milliliter), these motile phagocytes

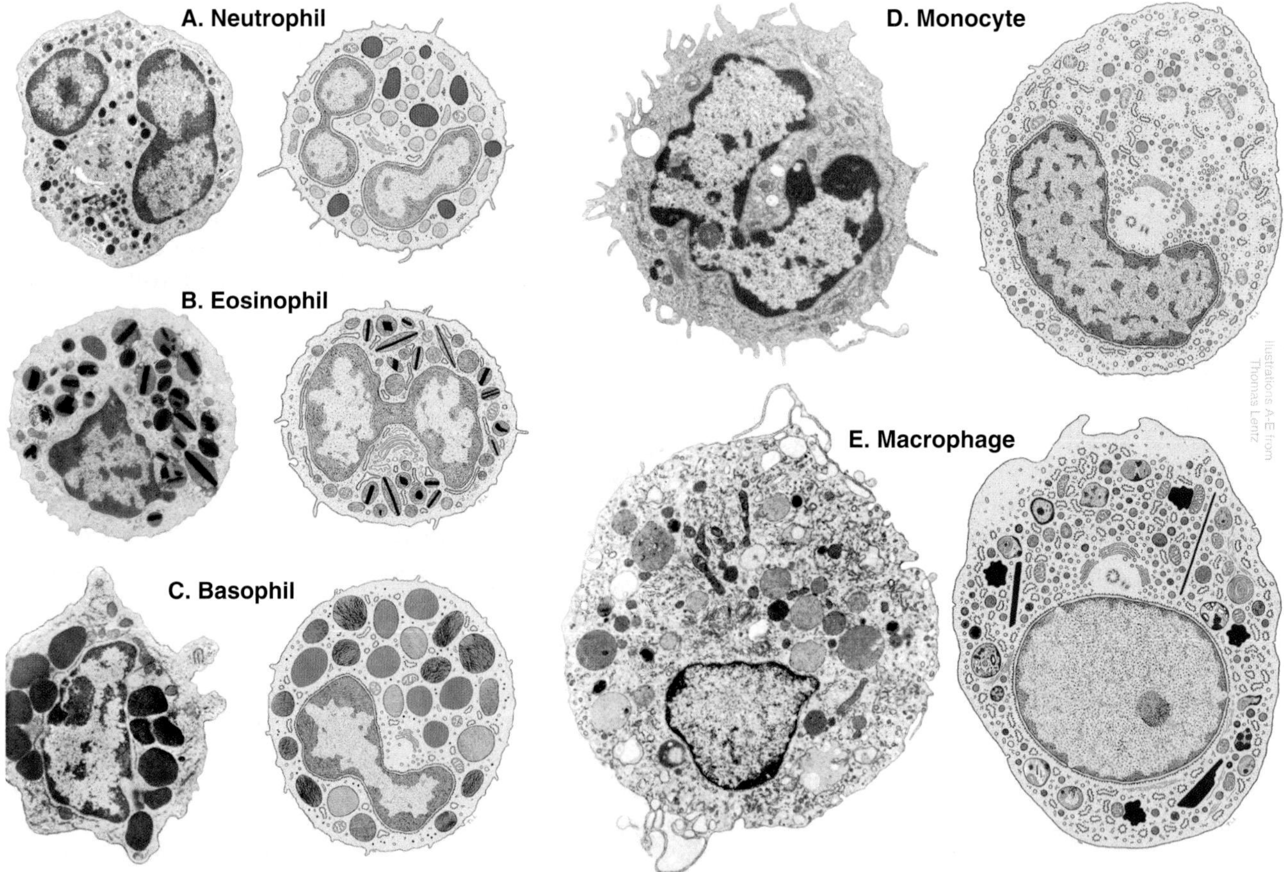

FIGURE 28.7 WHITE BLOOD CELLS. Transmission electron micrographs of thin sections of each cell and interpretive drawings with lysosomes shown in *brown*. **A,** A neutrophil showing the multilobed nucleus (the connections between lobes are in other sections) and the two classes of granules. **B,** An eosinophil showing the bilobed nucleus and the large, specific granules containing a darkly stained crystalloid. **C,** Basophil with large specific granules colored *blue*. **D,** Blood monocyte. **E,** Macrophage grown in tissue culture. (Micrographs courtesy D.W. Fawcett, Harvard Medical School, Boston, MA. Drawings modified from T. Lentz, Yale Medical School, New Haven, CT.)

can find and destroy bacteria faster than the invaders can reproduce. Bacterial products, especially *N*-formylated peptides, attract neutrophils by binding plasma membrane receptors and stimulating locomotion, similar to chemotaxis by other cells (see Fig. 38.10). Neutrophils bind and ingest bacteria by phagocytosis (see Fig. 22.3). Both types of granules fuse with phagosomes, delivering antibacterial proteins and proteolytic enzymes that kill the ingested bacteria. Some granules fuse with the plasma membrane, releasing antibacterial proteins outside the cell. Phagosome membranes produce millimolar concentrations of superoxide (O_2^-) radicals and other reactive oxygen species that help disperse the granule enzymes and contribute to killing bacteria. These toxic oxygen species may also cause collateral damage to the neutrophil. Genetic defects in the enzymes that produce superoxide cause chronic granulomatous disease, a serious human disease, because neutrophils cannot kill ingested bacteria and fungi.

Eosinophils

Eosinophils are members of the granulocyte lineage, present in low numbers in the blood. They are identified in blood smears as cells with a bilobed nucleus and large specific granules that stain brightly red with eosin (Fig. 28.7B). Specific granules contain a cationic protein crystalloid, a ribonuclease and peroxidase, in addition to a crystalloid of a basic protein.

Like neutrophils, eosinophils transit the blood for hours on their way to connective tissues, especially in the gastrointestinal tract, where they survive for a few days. Chemotactic factors generated by the complement system, basophils, some tumors, parasites, and bacteria all attract eosinophils to tissues. Many of the same factors attract other leukocytes, but particular chemokines are specialized for eosinophils. Eosinophils accumulate in blood and in tissues infected with parasites, but experts do not agree on whether eosinophils kill bacteria or parasites. Activated eosinophils contribute to inflammation in some allergic disorders such as asthma but they also secrete factors that promote immune responses by lymphocytes.

Macrophages

Macrophages are a diverse group of professional phagocytes with many common features but two different origins. All have a receptor tyrosine kinase (colony-stimulating factor receptor) that drives their differentiation into phagocytes. Macrophages in brain, liver, lung and some other tissues arise from cells in the embryonic yolk sac, while adult tissue macrophages develop from monocytes that develop in bone marrow and circulate in the blood. Monocytes are the largest blood cells with an indented nucleus and a small number of azurophilic granules (Figs. 28.4A and 28.7D). After about 3 days in the blood, monocytes enter tissues and differentiate into macrophages under the influence of local growth factors, including lymphokines secreted by lymphocytes (Fig. 28.9). Macrophages enlarge and amplify their machinery for locomotion, phagocytosis, and killing microorganisms and tumor cells. Tissue macrophages may divide and survive for months. Local growth factors in bone stimulate monocytes to fuse and differentiate into multinucleated **osteoclasts** that degrade bone matrix during bone remodeling (see Fig. 32.6).

Macrophages generally follow neutrophils to wounds or infections to clean up debris and foreign material. Plasma membrane receptors for antibodies allow macrophages to recognize foreign matter marked with antibodies and to facilitate its ingestion. Primary lysosomes fuse with phagosomes to degrade the contents. Eventually, the cytoplasm fills with residual bodies containing the remains of ingested material (see Fig. 23.4). When confronted with large foreign bodies, macrophages can fuse together to form **giant cells.** Giant multinucleated microphages will even try to ingest a Petri dish if it is coated with antibody.

Macrophages and their cousins, called *dendritic cells* (see below), participate in the immune response by degrading ingested protein antigens and presenting antigen fragments on their surface bound to major histocompatibility complex (MHC) class II proteins (Fig. 28.9). This complex activates helper T lymphocytes carrying the appropriate T-cell receptors (see Fig. 27.8). Activated T cells proliferate and secrete growth factors that stimulate B lymphocytes to produce antibodies.

Engagement of TLRs stimulates macrophages, which secrete dozens of factors involved with host defense, inflammation, and normal development. Among these, IL-1, TGF-α, TGF-β, and platelet-derived growth factor stimulate the proliferation and differentiation of the cells required to heal wounds (see Fig. 32.11). Chemokines attract cells of the immune system to sites of inflammation.

Mast Cells and Basophils

Basophils (Fig. 28.6C) and mast cells (Fig. 28.8) both have histamine-containing granules that are secreted when antigens bind cell-surface IgE molecules. The large, abundant **granules** contain, by mass, 30% heparin-basic protein complex, 10% **histamine,** and 35% basic proteins, including proteases. Plasma membrane receptors bind a random selection of **IgE** antibodies made by the immune system in response to exposure to antigens. Binding of the corresponding antigen to IgE on the surface of a basophil or mast cell activates a signaling pathway with a Src family tyrosine kinase similar to those in T lymphocytes (see Fig. 27.8). A cytoplasmic Ca^{2+} pulse triggers fusion of granules with the plasma membrane (see Fig. 21.16) and other pathways stimulate production of cytokines and lipid second messengers. Mechanical trauma, radiant energy (heat, X-rays),

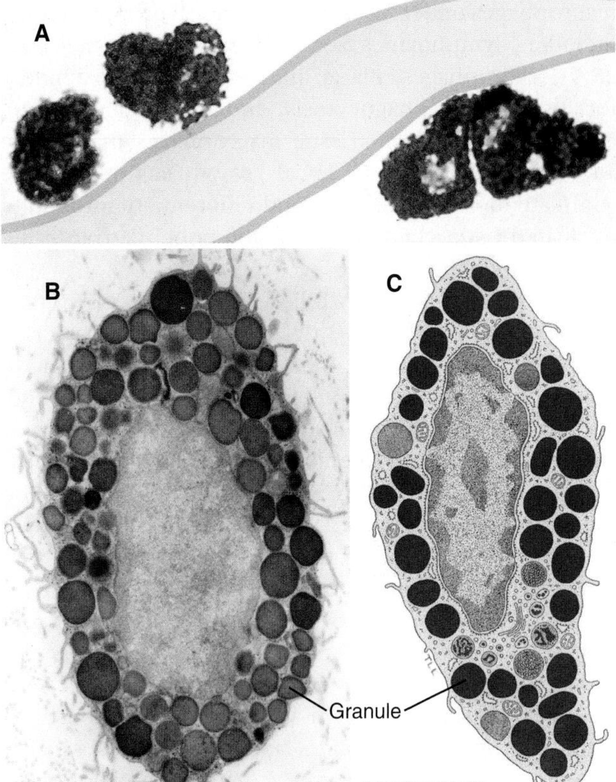

FIGURE 28.8 MAST CELLS. A, Light micrograph of loose connective tissue, stained with toluidine blue, illustrating mast cells scattered along a blood vessel (drawn in to enhance the contrast). Large mast cell granules stain intensely *purple* with basic thiazine dyes such as toluidine blue. **B,** Transmission electron micrograph of a thin section of a mast cell. **C,** Drawing of a mast cell. (**A–B,** Courtesy D.W. Fawcett, Harvard Medical School, Boston, MA. **C,** Modified from T. Lentz, Yale Medical School, New Haven, CT.)

bacterial toxins, and venoms are less-specific stimuli, but can also trigger secretion. Outside the cell, the carrier proteins release heparin and histamine.

On the positive side, secretion of histamine and other granule contents rapidly attracts other cells to fight infections as part of "immediate hypersensitivity" reactions. On the negative side, histamine binds to cellular receptors, causing plasma to leak from blood vessels, contraction of smooth muscle, and itching sensations. This results in congestion and constriction of the respiratory tract in allergic reactions and swelling of the skin after an insect bite. Secreted fibrinolysin and heparin inhibit blood clotting. Stimulated mast cells also secrete TNF-α and eicosanoids, contributing to the activation of other inflammatory cells in chronic conditions including asthma and arthritis.

Basophils and mast cells both arise from bone marrow myeloid stem cells (Fig. 28.4B), with mast cells differentiating on a different, and still uncertain, pathway from that of basophils and other granulocytes. After arising in the bone marrow immature mast cells move quickly to other tissues where they mature and distribute along connective tissue blood vessels (Fig. 28.8) and beneath epithelial cells lining airways and the gastrointestinal tract. Basophils circulate in small numbers in the blood. They look much like neutrophils, but have a bilobed nucleus and large, basophilic, specific granules containing serotonin and all the histamine in the blood (Fig. 28.7C). Basophils are weak phagocytes.

Humans have both circulating basophils and tissue mast cells, but this is not universal. Mice, for example, have mast cells but no basophils, and turtles have basophils but no mast cells.

Cellular Basis of Adaptive Immunity

Starting with cartilaginous fish, vertebrates developed a sophisticated **adaptive immune system.** The response is slower than innate immunity, because it depends on the selection and multiplication of lymphocytes that produce soluble antibodies or cell surface receptors precisely targeted to foreign molecules. This response depends on rearrangement and mutation of genes to produce highly selective antibodies and receptor proteins. Although this adaptive response takes about a week to mobilize, it produces specialized lymphocytes that survive for years, providing the host with a faster adaptive response when exposed to the pathogen a second time.

In response to infection, lymphocytes of the immune systems of vertebrates produce two kinds of adaptive responses: humoral (in the body fluids) and cellular. **B lymphocytes** produce the humoral response by secreting **antibodies (immunoglobulins),** soluble proteins that diffuse in the blood and tissue fluids. Many types of T lymphocytes mediate the cellular arm of the adaptive immune response. Of these, **cytotoxic T lymphocytes (killer T cells)** destroy cells infected with viruses, whereas **helper T cells** regulate other lymphocytes. These responses protect against infection but fail in AIDS when the HIV kills helper T cells. A blood smear reveals lymphocytes of various sizes and shapes (Fig. 28.4), but not their remarkable heterogeneity at the molecular level (Fig. 28.9).

Antibodies produced by B cells provide a chemical defense against viruses, bacteria, fungi, and toxins. Antibodies, or immunoglobulins, are an incredibly diverse family of proteins, each with a binding site that accommodates one of millions of different ligands termed **antigens.** Antigens include proteins, polysaccharides, nucleic acids, lipids, and small organic molecules produced biologically or chemically. Antibody binding can mark an antigen for phagocytosis or neutralize its toxicity.

The huge repertoire of antigen-binding sites present in the collection of antibodies that circulate in a single individual arises through **rearrangement** and **somatic mutations** of Ig genes (Fig. 28.10). This remarkable

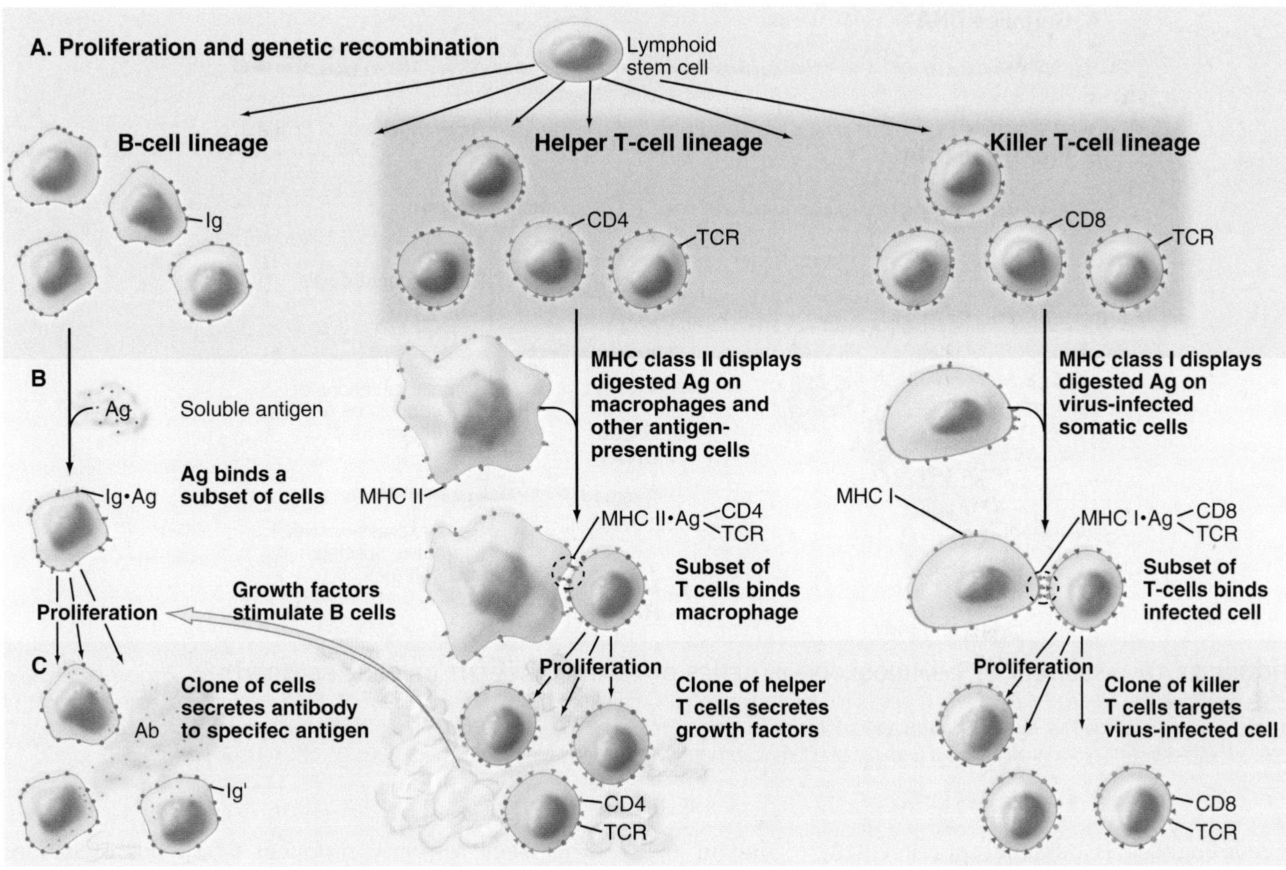

FIGURE 28.9 THE IMMUNE RESPONSE BY THREE CLASSES OF LYMPHOCYTES THROUGH THREE PARALLEL STEPS. A, Genetic recombination produces populations of cells with a wide variety of antigen specificities provided by cell-surface immunoglobulins (Ig) or T-cell receptors (TCR). **B,** The binding of specific antigens (Ag) to surface immunoglobulins or TCRs selects a subset of the cells for proliferation. MHC, major histocompatibility complex. **C,** Proliferation of clones of selected cells yields many cells specialized to produce antibody (Ab) to soluble antigens, secretion of growth factors by helper T cells in response to ingested and degraded antigens, or killing of virus-infected cells identifiable by the viral peptides on their surface. The helper and killer T cells use a common set of TCRs and are guided to the appropriate target cells by the CD4 and CD8 accessory molecules. See Figs. 27.8 and 46.9 for details on T-lymphocyte activation and selection, respectively.

process was exploited during evolution specifically for the use of the immune system. Immunoglobulins of most mammals are composed of four polypeptide chains—two identical heavy chains and two identical light chains—each encoded by different genes (Fig. 28.10). Light chains and heavy chains both contribute to the antigen-binding site. Camels and llamas are an exception; their antibodies consist of a single polypeptide.

Immunoglobulin genes exist in segments aligned along a vertebrate chromosome. Several of these gene segments must be combined in the proper order to make a functional antibody gene. Some gene segments encode the framework of the antibody protein, which is essentially identical within each antibody class. Other gene segments, present in many variations, encode the part of the polypeptide chain that forms the antigen-binding site.

During maturation of a particular B cell, **recombination enzymes** (RAG1 and RAG2) assemble immunoglobulin gene segments into one unique full-length gene for a heavy chain and one for a light chain. As a result of random gene arrangements, each B cell assembles and expresses novel immunoglobulin genes. The process is precise in that the right number of segments is always chosen to make a heavy chain or a light chain, but it is also random in that any one of the variable segments may be chosen. The resulting antibody contains two identical but unique antigen-binding sites. The gene segments can be assembled in many different combinations, and most heavy chains can assemble with most light chains. The diversity arising from the combinatorial process is expanded further in two ways. First, the recombination process inserts a variable number of nucleotides between the gene segments. Second, pre–B cells use enzymes to mutate codons for amino acids in the antigen-binding site, creating unique variations in the antigen-binding specificity in different cells. In principle, approximately 3000 different light chains and 60,000 heavy chains can combine to produce approximately 100 million different antibodies—even without taking mutations into account.

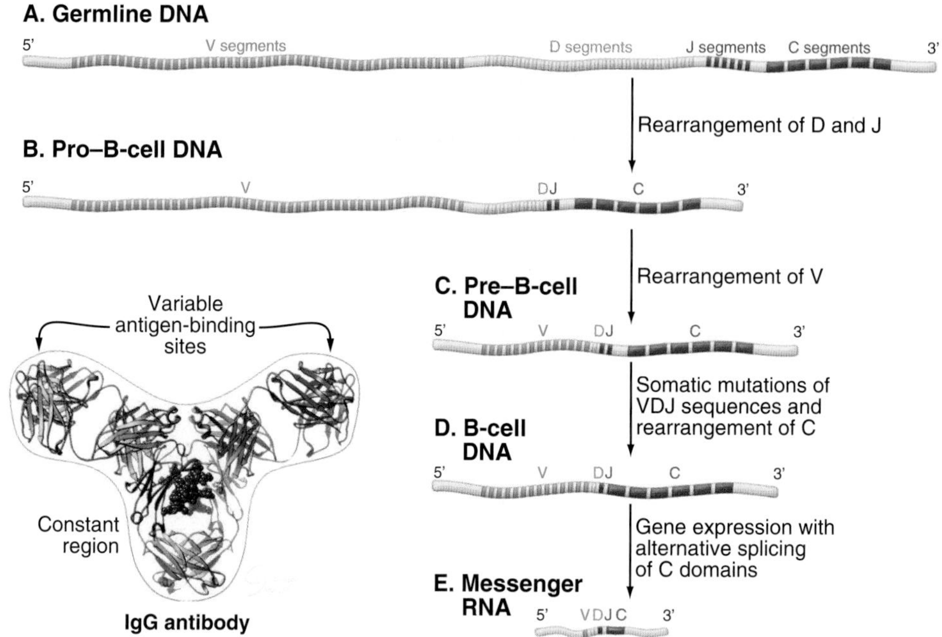

FIGURE 28.10 ASSEMBLY OF IMMUNOGLOBULIN GENES BY REARRANGEMENT OF GENE SEGMENTS. A, Region of germline DNA with multiple, tandem V, D, J, and C segments for assembling an immunoglobulin (Ig) heavy chain. **B,** Same region after rearrangement of D and J segments in a Pro–B cell. **C,** After rearrangement of V in a Pre–B cell. **D,** B cell. **E,** Immunoglobulin messenger RNA produced by a B cell. (Modified from Chiorazzi N, Rai KR, Ferrarini M. Chronic lymphocytic leukemia. *N Engl J Med.* 2005;352:804–815.)

Accordingly, it is possible experimentally to find in a mouse an antibody that is specific for almost any naturally occurring or synthetic chemical.

Infection by a pathogen results in the selection of antibodies that bind to the pathogen, but not to any of the individual's own molecules. This response comes from activation and proliferation of preexisting B cells making antibodies that bind to molecules of the pathogen. Activation requires a chance encounter of particular B cells with T cells presenting antigenic molecules from the pathogen (see later) and stimulates the cell to mature into a factory for secreting antibodies. Alternate splicing of messenger RNA (mRNA [see Fig. 11.6]) selects domains required to direct the same antibody to either the plasma membrane or the secretory pathway.

Antigen binding to the surface immunoglobin displayed on a particular B cell stimulates that cell to go through multiple rounds of cell division, giving rise to a clone of cells, all making the same antibody. In an animal, this process of **clonal expansion** amplifies multiple clones of B cells. All the clones produce antibodies that react with the antigen, but most clones make novel antibodies to that antigen. Therefore, an animal makes many different antibodies to the same antigen. By isolating single B cells and expanding their numbers in the laboratory, one may obtain enough identical cells to make useful amounts of the same exact antibody, called a **monoclonal antibody**, because it arose from the proliferation of a single cell into a clone of cells.

The B-cell response produces two types of mature cells. **Plasma cells** are each highly specialized to secrete large amounts of one specific antibody. Long-lived **memory cells** display a specific antibody on their surface, ready to mount an amplified response on subsequent exposure to the same antigen. This immunologic memory explains why exposure to a particular pathogen or vaccination against a pathogen results in protection, in the form of antibodies, for many years.

Specialized B lymphocytes and plasma cells secrete different antibody isoforms or isotypes. Formation of immunoglobulins with the various isotypes requires further recombination events to join the variable region containing the antigen-binding site to the isotype constant domain. **IgG** isotypes, produced in lymph nodes and spleen, circulate in blood and tissue fluids. **IgA** isotypes, secreted by B cells in lymphoid nodules of the respiratory and gastrointestinal tracts and by mammary glands, are taken up by epithelial cells and resecreted into the lumens of these organs (transcytosis [see Fig. 22.6]). **IgE** isotypes bind to receptors on the surface of mast cells and basophils (see earlier discussion).

T lymphocytes provide cellular responses to pathogens. Cytotoxic T cells kill tumor cells and virus-infected cells. Helper T cells stimulate antibody production by B cells. The specificity of these responses is provided by variable cell surface receptors called **T-cell receptors** (see Fig. 27.8). A set of segmented genes analogous to immunoglobulin genes encodes T-cell receptors. In contrast to antibodies, T-cell receptors do not bind free antigens but rather recognize peptide antigens bound to proteins called **MHC** antigens on the surface of a target cell (see Fig. 27.8). These highly variable MHC proteins are responsible for the rejection of tissue grafts from nonidentical individuals.

The two types of MHC proteins—class I and class II—acquire their antigenic peptides differently. All somatic cells produce class I MHC proteins. In cells that have been infected by a virus, cytoplasmic proteasomes degrade some viral proteins to peptides (see Chapter 23), which ABC transporters (TAP1, TAP2) move from the cytoplasm (see Fig. 14.9) into the endoplasmic reticulum (ER). In the lumen of the ER, peptides insert into the binding site of compatible class I molecules and the complex moves to the plasma membrane. In contrast, macrophages and other **antigen-presenting cells,** such as **dendritic cells,** ingest foreign matter and degrade it in endosomes and lysosomes. Such peptide fragments bind to class II proteins in endosomes and thence move to the cell surface of these antigen-presenting cells.

T lymphocytes patrol the body, inspecting the surfaces of other cells. A chance encounter with a cell displaying a peptide-MHC complex complementary to its T-cell receptor stimulates the T cell (see Fig. 27.8). The response is proliferation and expansion of a clone of identical T cells. Accessory membrane proteins CD4 and CD8 on the T-cell surface cooperate with T-cell receptors to direct the two types of T cells to target cells with the appropriate MHC proteins. T-cell receptors provide antigen specificity. Immature T-cells express both CD4 and CD8 but lose one of them as they mature into cytotoxic ($CD8^+$) or helper ($CD4^+$) T cells.

CD8-positive cytotoxic T cells are specialized to kill cells infected with viruses. The presence of virus inside is revealed by MHC class I proteins displaying vital peptides on the surface of the infected cell. CD8 binds to a constant region of MHC class I proteins carrying viral peptides. Cytotoxic T cells use three weapons to kill the target cell: First, T cells carry a ligand for the Fas receptor on the target, which stimulates apoptosis of the target cell (see Fig. 46.18). Second, activated T cells bind to the target cell, forming an immunological synapse (see Fig. 27.8) into which the T cell secretes **perforin,** a protein that inserts into the plasma membrane of the target cell, forming large (10 nm) pores that leak cytoplasmic contents and ultimately lyse the cell. Third, T cells secrete toxic enzymes into the synapse that enter target cells through the plasma membrane pores.

CD4 on helper T cells binds a constant part of the MHC class II protein and targets helper T cells to cells presenting ingested antigens. The progeny of stimulated helper T cells secrete growth factors (lymphokines or interleukins) in the vicinity of B cells with the foreign antigen bound to immunoglobulins on their surface. Helper T cells are required for B cells to make antibodies against most antigens. This explains how **HIV** causes AIDS. The virus uses CD4 as a receptor to infect and eventually kill helper T cells. Loss of helper T cells severely limits the capacity of B cells and cytotoxic T cells (which also require T-cell help) to mount antibody and cellular responses to microorganisms. Infections that the immune system normally dispatches with ease then become life-threatening.

Genetic defects cause a wide variety of **immunodeficiency diseases.** For example, defects in Bruton tyrosine kinase result in failure to produce B cells. Remarkably, humans who lack function of the enzyme adenosine deaminase have no B cells or T cells, but are otherwise normal. Deficiencies of many specialized lymphocyte proteins (cytokine receptors, interleukin receptors, Lck tyrosine kinase, ZAP-70 [zeta-associated protein of 70 kD] tyrosine kinase, RAG1 or RAG2, TAP1 or TAP2) also lead to immunodeficiencies.

ACKNOWLEDGMENTS

We thank Matthew Rodeheffer for suggestions on revisions to this chapter.

SELECTED READINGS

Berry R, Rodeheffer MS, Rosen CJ, et al. Adipose tissue residing progenitors. *Curr Mol Biol Rep*. 2015;1:101-109.

Beutler B. Inferences, questions and possibilities in Toll-like receptor signalling. *Nature*. 2004;430:257-263.

Bianco P. "Mesenchymal" stem cells. *Annu Rev Cell Dev Biol*. 2014;30: 677-704.

Boes M, Ploegh HL. Translating cell biology in vitro to immunity in vivo. *Nature*. 2004;430:264-271.

Busiello RA, Savarese S, Lombardi A. Mitochondrial uncoupling proteins and energy metabolism. *Front Physiol*. 2015;6:36.

Call ME, Wucherpfennig KW. The T cell receptor: Critical role of the membrane environment in receptor assembly and function. *Annu Rev Immunol*. 2005;23:101-125.

Chakraborty AK, Weiss A. Insights into the initiation of TCR signaling. *Nat Immunol*. 2014;15:798-807.

Eaves CJ. Hematopoietic stem cells: concepts, definitions, and the new reality. *Blood*. 2015;125:2605-2613.

Gay NJ, Symmons MF, Gangloff M, Bryant CE. Assembly and localization of Toll-like receptor signalling complexes. *Nat Rev Immunol*. 2014;14:546-558.

Grinnell F. Fibroblast biology in three-dimensional collagen matrices. *Trends Cell Biol*. 2003;13:264-269.

Hargreaves DC, Medzhitov R. Innate sensors of microbial infection. *J Clin Immunol*. 2005;25:503-510.

Hartwig J, Italiano J Jr. The birth of the platelet. *J Thromb Haemost*. 2003;1:1580-1586.

Ho MS, Medcalf RL, Livesey SA, et al. The dynamics of adult haematopoiesis in the bone and bone marrow environment. *Br J Haematol*. 2015;170:472-486.

Howell WM. HLA and disease: guilt by association. *Int J Immunogenet*. 2014;41:1-12.

Hubbi ME, Semenza GL. Regulation of cell proliferation by hypoxia-inducible factors. *Am J Physiol Cell Physiol*. 2015;309:C775-C782.

Lacy P, Rosenberg HF, Walsh GM. Eosinophil overview: structure, biological properties, and key functions. *Methods Mol Biol.* 2014; 1178:1-12.

Mitchell WB, Bussel JB. Thrombopoietin receptor agonists: a critical review. *Semin Hematol.* 2015;52:46-52.

Morrison SJ, Scadden DT. The bone marrow niche for haematopoietic stem cells. *Nature.* 2014;505:327-334.

Murray PJ, Wynn TA. Protective and pathogenic functions of macrophage subsets. *Nat Rev Immunol.* 2011;11:723-737.

Perez-Lopez A, Behnsen J, Nuccio SP, et al. Mucosal immunity to pathogenic intestinal bacteria. *Nat Rev Immunol.* 2016;16:135-148.

Raje N, Dinakar C. Overview of Immunodeficiency Disorders. *Immunol Allergy Clin North Am.* 2015;35:599-623.

Rockey DC, Bell PD, Hill JA. Fibrosis—a common pathway to organ injury and failure. *N Engl J Med.* 2015;372:1138-1149.

Rosen ED, Spiegelman BM. What we talk about when we talk about fat. *Cell.* 2014;156:20-44.

Rot A, von Adrian UH. Chemokines in innate and adaptive host defense: Basic chemokinese grammar for immune cells. *Annu Rev Immunol.* 2004;22:891-928.

Schmetzer O, Valentin P, Church MK, Maurer M, Siebenhaar F. Murine and human mast cell progenitors. *Eur J Pharmacol.* 2016;778: 2-10.

Schroder K, Tschopp J. The inflammasomes. *Cell.* 2010;140:821-832.

Segal AW. How neutrophils kill microbes. *Annu Rev Immunol.* 2005; 23:197-223.

Trombetta ES, Mellman I. Cell biology of antigen processing in vitro and in vivo. *Annu Rev Immunol.* 2005;23:975-1028.

Varol C, Mildner A, Jung S. Macrophages: development and tissue specialization. *Annu Rev Immunol.* 2015;33:643-675.

Waggoner SN, Reighard SD, Gyurova IE, et al. Roles of natural killer cells in antiviral immunity. *Curr Opin Virol.* 2015;16:15-23.

Wynn TA, Chawla A, Pollard JW. Macrophage biology in development, homeostasis and disease. *Nature.* 2013;496:445-455.

Yang D, Biragyn A, Hoover DM, et al. Multiple roles of antimicrobial defensins, cathelicidins, and eosinophil-derived neurotoxin in host defense. *Annu Rev Immunol.* 2004;22:181-215.

CHAPTER 29

Extracellular Matrix Molecules

This chapter introduces the macromolecules of the extracellular matrix. Although the extracellular matrix is composed of only five classes of macromolecules—collagens, elastin, proteoglycans, hyaluronan, and adhesive glycoproteins—it can take on a rich variety of different forms with vastly different mechanical properties. This is possible for two reasons. First, each of these classes of macromolecule comes in a number of variants (encoded by different genes or produced by alternative splicing), each with distinctive properties. Second, the cells that constitute the extracellular matrix secrete different proportions of these isoforms in various geometrical arrangements. As a result, the extracellular matrix in different tissues is adapted to particular functional requirements, which vary widely in tendons, blood vessel walls, cartilage, bone, the vitreous body of the eye, and subcutaneous fat. Beyond providing mechanical support, the extracellular matrix also strongly influences embryonic development, provides pathways for cellular migration, provides essential survival signals, and sequesters important growth factors.

Collagen

The collagen family is the most abundant and versatile classes of proteins in the human body. Collagens form a wide range of different structures with remarkable mechanical properties. Weight for weight, fibrous collagens are as strong as steel. Their name, which comes from the Greek words for "glue" and "producing," reflects the long-known adhesive properties of denatured collagen extracted from animal tissues.

The defining feature of collagens is a rod-shaped domain composed of a **triple helix** of polypeptides (Fig. 29.1). Each polypeptide folds into a left-handed polyproline II helix that repeats every third residue with the side chains on the outside. Three of these helices associate to form a triple helix that may be up to 420 nm long. The triple helical domains have a repeating amino acid sequence: glycine-X-Y, where X is most often proline and Y is most often hydroxyproline. The small glycine residues allow tight contact between the polypeptides in the core of the triple helix. Larger residues, even alanine, interfere with packing. Poly-L-proline has a strong tendency to form a left-handed helix like individual collagen chains but does not form a triple helix, owing to steric interference. The triple helix is most stable if all X residues are proline and all Y residues are hydroxyproline,

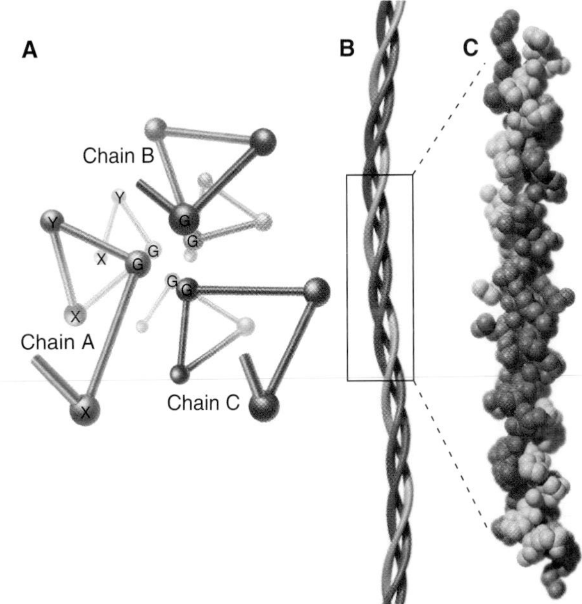

FIGURE 29.1 COLLAGEN TRIPLE HELIX. A, End-on view of three left-handed polyproline type II helices with glycines (G) in the core. **B,** Longitudinal view of the strands of a triple helix. **C,** Space-filling model of the structure of a short collagen triple helix. (**A,** Modified from van der Rest M, Garrone A. Collagen family of proteins. *FASEB J.* 1991;5:2814–2823. **C,** See Protein Data Bank [PDB; www.rcsb.org] file 1BKV.)

but other residues at some of these positions are essential for collagen to assemble higher-order structures.

Humans have approximately 100 genes with collagen triple repeats, and more than 20 specialized collagen proteins have been characterized (Fig. 29.2 and Appendix 29.1). Most are components of the extracellular matrix, but a few are transmembrane proteins (see Fig. 31.8C). Collagen proteins were named with Roman numerals in the order of their discovery. The polypeptides are called **α-chains** and numbered separately. Appendix 29.1 groups collagens according to function.

The size and shape of collagens vary according to function. Some collagens are homotrimers of three identical α-chains. Others are heterotrimers of two or three different α-chains. Some chains (eg, [α_1(II)]) are used in more than one type of collagen.

Other proteins, including the extracellular enzyme acetylcholine esterase (see Fig. 17.9) and some cell surface receptors, have similar triple helical domains but are not classified as collagens. To be a collagen, a protein must also form fibrils or other assemblies in the extracellular matrix. Nematodes, which lack connective tissue, seem to have lost the genes for fibrillar collagens but have elaborated a family of 160 genes for collagens that form their cuticle.

Collagen biochemistry is challenging, because many tissue collagens are insoluble, owing to covalent cross-linking between proteins. Historic purification protocols began with proteolytic digestion to liberate

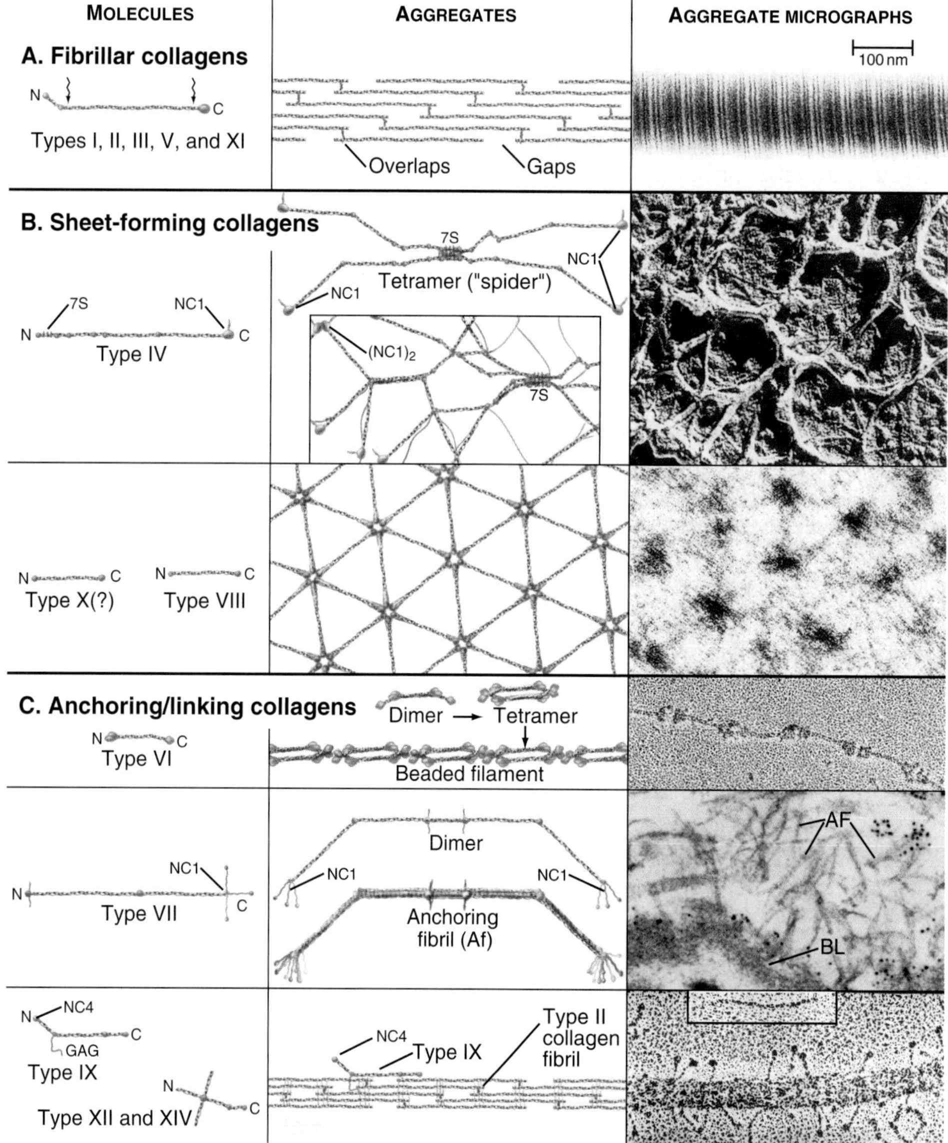

FIGURE 29.2 COMPARISON OF MAJOR COLLAGEN FAMILIES. Scale drawings and micrographs of collagen molecules and their assembly into higher-order structures. AF, anchoring fibrils; BL, basal lamina; NC1, noncollagenous domain 1; NC4, noncollagenous domain 4; 7S, a domain of type IV collagen. (Modified from van der Rest M, Garrone R. Collagen family of proteins. *FASEB J.* 1991;5:2814–2823.)

protease-resistant triple helical fragments. Now intact collagens can be isolated after secretion by cells in tissue culture.

Fibrillar Collagens

Triple helical rod-shaped collagen molecules about 300 nm long self-associate to form strong but flexible banded fibrils (Fig. 29.2) that reinforce all the tissues of the body. Collagen fibrils form a variety of higher-order structures. Loose connective tissue (see Fig. 32.1A) has an open network of individual fibrils or small bundles of fibrils that support the cells. In many tissues, the fibrils of type I and associated collagens aggregate to form the so-called collagen *fibers* that are visible by light microscopy (Fig. 29.3A). In extreme cases, such as in tendons, the extracellular matrix consists almost exclusively of tightly packed, parallel bundles of collagen fibers (see Fig. 32.1B). In bone, type I collagen fibrils form regular layers reinforced by calcium phosphate crystals (see Fig. 32.4). Layers of orthogonal collagen fibers make the transparent cornea through which one sees (Fig. 29.3C). In cartilage and the vitreous body of the eye, type II collagen fibrils trap glycosaminoglycans and proteoglycans, which retain enough water for the matrix to resist compression (see Fig. 32.3) and, in the case of the eye, to provide an optically clear path for light.

Fibrillar collagens are widespread in nature and have been highly conserved during evolution, so the homologs from sponges to vertebrates are similar. Each fibrillar collagen can form homopolymers in vitro; but in vivo, most form heteropolymers with at least one other type of fibrillar collagen (Appendix 29.1). This mix of the fibrillar collagen subunits is one factor that regulates the size of collagen fibers. Proteoglycans also participate in regulating collagen assembly (Appendix 29.2).

Biosynthesis and Assembly of Fibrillar Collagens

The biosynthesis of collagen is noteworthy for the extensive number of processing steps required to prepare the protein for assembly in the extracellular matrix (Fig. 29.4). Fibroblasts synthesize type I collagen. Collagen follows the exocytic pathway used by other secreted proteins (see Chapter 21), but along the way it undergoes several rounds of precise proteolytic cleavage, glycosylation, catalyzed folding, and chemical crosslinking. The final product is a smooth fibril with staggered molecules crosslinked to their neighbors. Other fibrillar collagens are likely to be produced by similar mechanisms.

Large genes with 42 exons encode the α-chains of type I collagen. All the exons for the triple helical domain were derived during evolution by duplication and divergence from a primordial exon of 54 base pairs (bp) coding for 18 amino acids or six turns of polyproline helix. Approximately half of the exons consist of 54 bp; a few with 45 bp have lost one Gly-X-Y; and the rest are 108 (2 × 54) or 162 (3 × 54) bp. Distinctive exons encode the N- and C-terminal globular domains.

The initial transcript, referred to as **preprocollagen,** translocates into the lumen of the rough endoplasmic reticulum, where intracellular processing begins (Fig. 29.4). First, removal of the N-terminal signal sequence yields **procollagen** with unfolded α-chains with N- and C-terminal nonhelical propeptides. Second, enzymes hydroxylate most prolines and some lysines in the Y-position. Third, enzymes add sugars (gal-glu or gal) to the delta-carbon of some *lysines,* by a mechanism distinct from the typical glycosylation of asparagine or serine (Fig. 3.26).

A novel mechanism initiates the folding of collagen in the endoplasmic reticulum: the **C-terminal propeptides** of three α-chains form a globular structure stabilized by cysteines linked with disulfide bonds. The enzyme **protein disulfide isomerase** catalyzes the formation of these disulfides. Formation of this globular domain has three important consequences. First, it ensures the correct selection of α-chains (two α_1-chains and one α_2-chain in the case of type I collagen). Second, it aligns the three polypeptides with their C-terminal

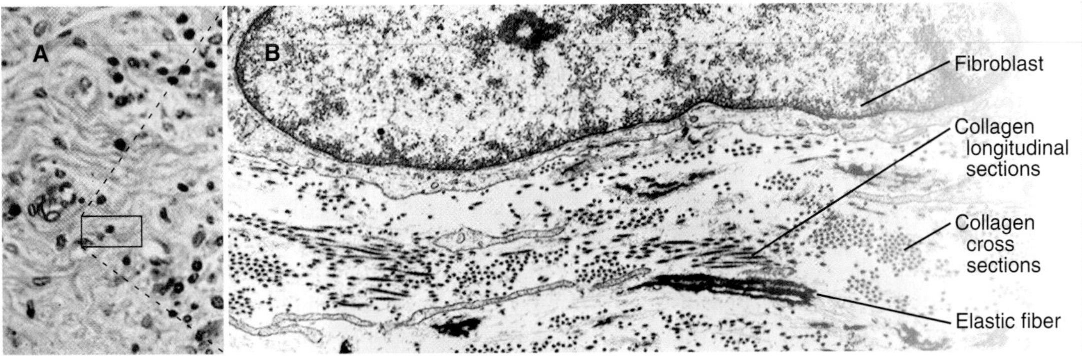

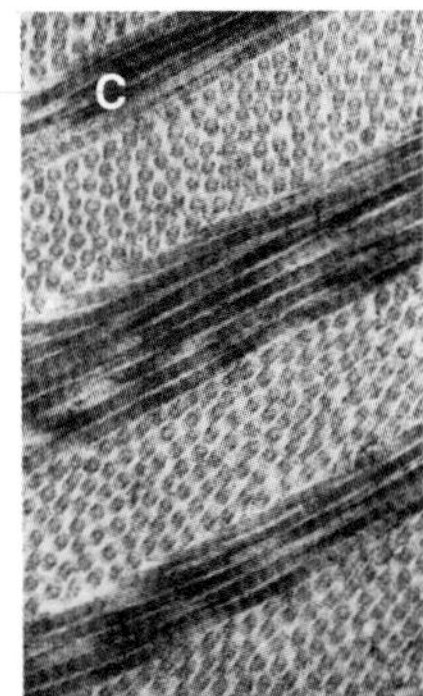

FIGURE 29.3 MICROGRAPHS OF COLLAGEN FIBRILS IN CONNECTIVE TISSUES. A, Collagen fibrils *(pink)* in the dense connective tissue of the dermis. **B,** Electron micrograph of a thin section of a fibroblast, collagen fibrils, and elastic fibers. **C,** Orthogonal layers of collagen fibrils in the cornea of the eye. (**A,** Courtesy D.W. Fawcett, Harvard Medical School, Boston, MA. **B,** Courtesy J. Rosenbloom, University of Pennsylvania, Philadelphia. **C,** Courtesy E.D. Hay, Harvard Medical School, Boston, MA.)

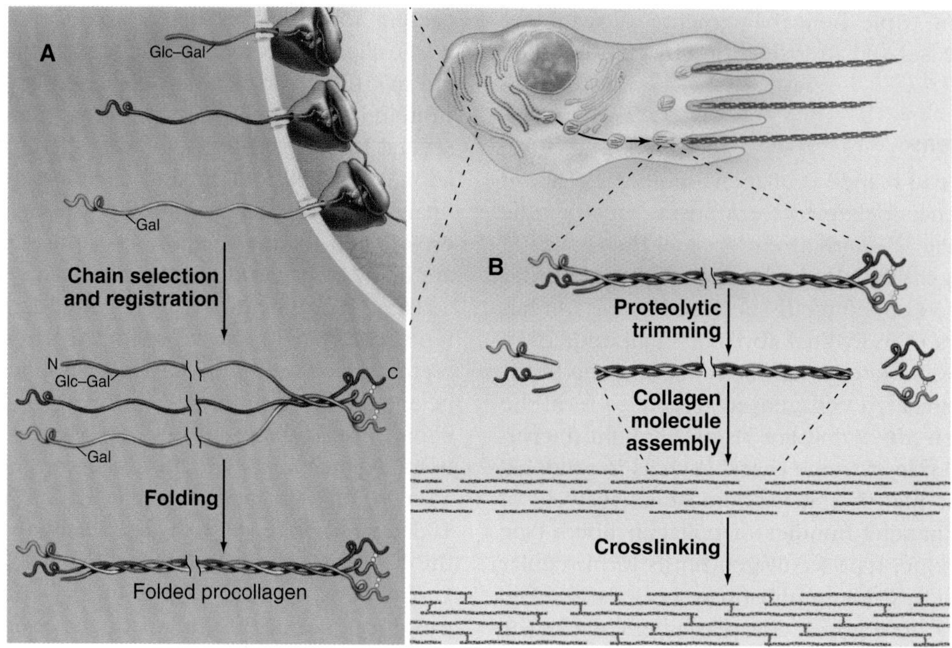

FIGURE 29.4 BIOSYNTHESIS AND ASSEMBLY OF FIBRILLAR COLLAGEN ILLUSTRATING DETAILS COVERED IN THE TEXT. **A,** Translation of α-chains, chain registration, and folding. **B,** Secretion, assembly, and crosslinking. (Modified from Prokop DJ. Mutations in collagen genes as a cause of connective tissue diseases. *N Engl J Med.* 1992;326:540–546. Copyright 1992 Massachusetts Medical Society. All rights reserved.)

Gly-X-Y repeats in register, ensuring that the triple helix forms with all three chains in phase. Third, the globular propeptides prevent assembly of procollagen into insoluble fibrils during transit through the secretory pathway. Given their repeating Gly-X-Y structure, separated collagen chains without propeptides associate indiscriminately and out of register with other chains. For example, gelatin is simply a mixture of collagen chains without propeptides. Boiling dissociates the chains from each other. When cooled, the chains randomly associate *out of register* at random positions along their lengths, forming a branching network that solidifies into the gel that is used in food preparation.

Following selection and registration of the three α-chains, the helical rod domains zip together, beginning at the C-terminus. Correct folding of the triple helix requires all-*trans* peptide bonds. Because proline forms *cis* and *trans* peptide bonds randomly, the slow isomerization of *cis* prolyl-peptide bonds to *trans* limits the rate of triple helix folding in vitro. The enzyme **prolyl-peptide isomerase** catalyzes the interconversion of these prolyl-peptide bonds and speeds up folding of the triple helix in vivo. The resulting rod-shaped, triple-helix glycoprotein is called *procollagen*.

Procollagen is too large (>300 nm long) to fit into conventional COPII coated vesicles that bud from the endoplasmic reticulum (ER) with cargo destined for the Golgi apparatus (see Fig. 21.3), so accessory proteins are required. These transmembrane proteins interact with both procollagen inside the ER and the forming COPII coat on the cytoplasmic side of the membrane. They allow the COPII vesicle to grow large enough to accommodate protocollagen. The COPII vesicles deliver protocollagen to the Golgi apparatus. Humans with mutations in the gene for the COPII protein Sec23A have defects in collagen secretion and bone formation.

Less is known about the path of procollagen through the Golgi apparatus and vesicles that transport protocollagen to the cell surface, where it is secreted. Some cells have specialized collagen assembly sites (Fig. 29.4). Like a spider trailing its silk web behind, fibroblasts help determine the arrangement of collagen fibrils as they move through tissues (Fig. 29.3C). Outside the cell, matrix metalloproteinases (Fig. 29.19) cleave the propeptides from the triple helical domain, forming the mature collagen molecule (formerly called tropocollagen).

Relieved of its inhibitory propeptides, collagen self-assembles into fibrils by a classical entropy-driven process (Fig. 29.5). Weak, noncovalent bonds between collagen molecules specify the self-assembly of fibrils but provide little tensile strength. Adjacent collagen molecules are staggered by 67 nm, so a 35-nm gap is required between the ends of the collagen molecules (5 staggers at 67 nm = 335 nm = 1 molecular length of 300 nm + a 35-nm gap). The size of the fibrils is influenced by incorporation of minor fibrillar collagens (eg, collagen V) and interactions of FACIT (fibril-associated collagens with interrupted triple helices) collagens and other matrix molecules with their surfaces.

The great tensile strength of mature collagen fibrils comes from **covalent crosslinking** between the inextensible triple helices. For most fibrillar collagens, the enzyme **lysyl oxidase** catalyzes the formation of covalent bonds between the ends of collagen molecules (Figs. 29.4 and 29.6). The enzyme oxidizes the ε amino groups of selected lysines and hydroxylysines to aldehydes. These aldehydes react spontaneously with nearby lysine and hydroxylysine side chains to form a variety of covalent crosslinks between two or three polypeptides. Disulfide bonds, rather than modified lysine side chains, crosslink type III collagen fibrils.

Point mutations or deletions in collagen genes or lack of function of one of the enzymes that processes collagen (lysyl hydroxylase, lysyl oxidase, or procollagen proteases) can each cause defective collagen fibrils (Appendix 29.1). These defects cause a number of deforming and even lethal human diseases: brittle bones (osteogenesis imperfecta), fragile cartilage (several forms of dwarfism), and weak connective tissue (Ehlers-Danlos syndrome). Chapter 32 discusses these diseases in more detail.

Sheet-Forming Collagens

Collagens in this second group polymerize into sheets rather than fibrils (Fig. 29.2). These sheets surround organs, epithelia, or even whole animals. Six different human genes for type IV collagen encode proteins that form net-like polymers that assemble into the **basal lamina** beneath epithelia (Fig. 29.7) and around muscle and nerve cells. The concluding section of this chapter provides details about basal lamina structure, function, and diseases. Hexagonal nets of type VIII collagen form a special basement membrane (Descemet membrane) under the endothelium of the cornea. Related collagens form the cuticle of earthworms and the organic skeleton of sponges.

Linking Collagens

Specialized connecting and anchoring collagens (also called FACIT) link fibrillar and sheet-forming collagens

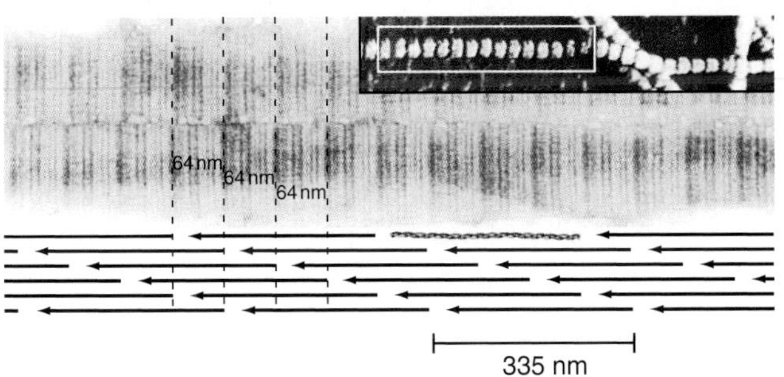

FIGURE 29.5 STRUCTURE OF COLLAGEN FIBRILS. Electron micrographs and drawing of molecular packing. (Micrographs courtesy Alan Hodges, Marine Biological Laboratory, Woods Hole, MA.)

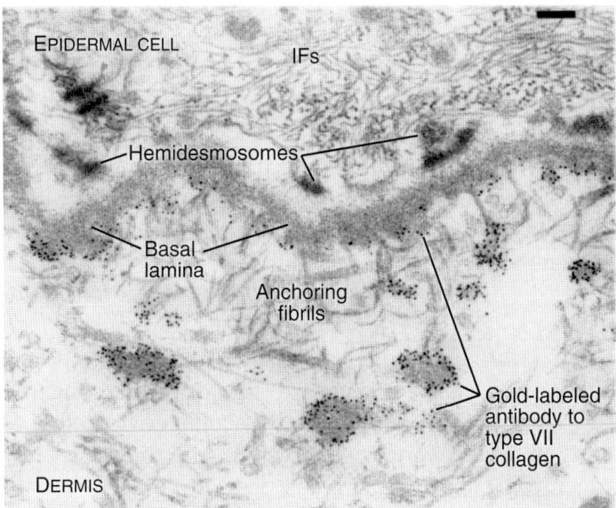

FIGURE 29.7 ANCHORING FIBRILS OF TYPE VII COLLAGEN. Electron micrograph of a thin section of human skin reacted with a gold-labeled antibody to the C-terminal domain of type VII collagen. *Top to bottom,* Basal epithelial cell with keratin intermediate filaments (IFs) attached to hemidesmosomes, which link to the basal lamina. Short fibrils of type VII collagen link the basal lamina to plaques in the dermis. Both ends of these bipolar fibrils (Fig. 29.2) are labeled with gold. Bar is 0.1 μm. (Courtesy D.R. Keene, Portland Shriners Hospital, OR.)

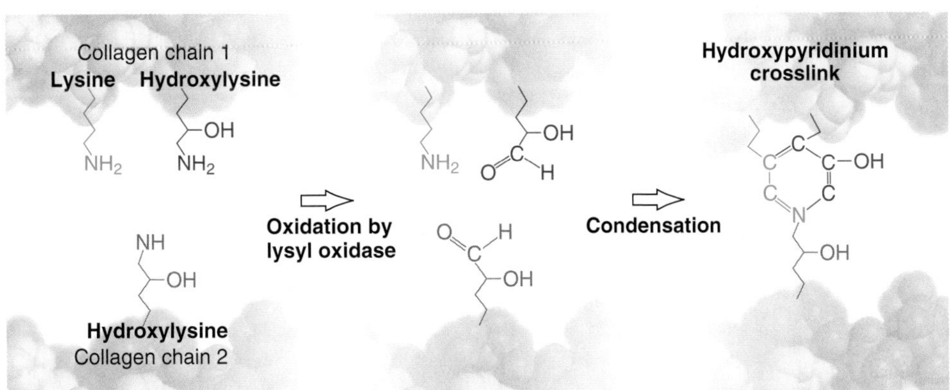

FIGURE 29.6 COVALENT CROSSLINKING OF COLLAGEN MOLECULES. After lysyl oxidase oxidizes hydroxylysine side chains, the aldehydes condense with each other and a lysine to form two- and three-membered (shown) crosslinks between adjacent collagen molecules.

to other structures (Fig. 29.2). For example, type VII collagen forms **anchoring fibrils** that link of the type IV collagen in the basal lamina of stratified epithelia to plaques in the underlying connective tissue (Fig. 29.7). The type VII collagen homotrimer has an exceptionally long triple-helix domain with nonhelical domains at the N-terminus of each chain. The tails of type VII molecules overlap to form antiparallel dimers that associate laterally to form anchoring fibrils. Mutations in type VII collagen cause both dominant and recessive forms of a severe blistering disease, dystrophic **epidermolysis bullosa.** In heterozygotes, mutated chains interfere with the assembly of anchoring fibrils by normal type VII collagen chains. Without anchoring fibrils, the basal lamina adheres weakly to the connective tissue matrix. Even mild physical trauma to the skin causes the epithelium to pull away from the connective tissue, forming a blister. Mutations in intermediate filaments cause similar defects (see Fig. 35.6).

Type IX collagen heterotrimers do not polymerize. Instead, they associate laterally with type II collagen fibrils with an N-terminal helical segment and a glycosaminoglycan on a serine projecting from the surface (Fig. 29.2). In the vitreous body of the eye, these polysaccharides fill most of the extracellular space.

Elastic Fibers

Rubber-like elastic fibers are found throughout the body and are prominent in the connective tissue of skin, the walls of arteries (Fig. 29.8), and the lung. They are entropic springs that recoil passively after tissues are stretched. For example, each time the heart beats, pressurized blood flows into and stretches the large arteries. Energy stored in elastic fibers pushes blood through the circulation between heartbeats.

Elastic fibers are composite materials; a network of **fibrillin microfibrils** is embedded in an amorphous core of crosslinked **elastin,** which makes up 90% of the organic mass (Fig. 29.9). Fibroblasts produce both components. Loose bundles of microfibrils initiate assembly. A third protein, called fibulin, is required for elastin subunits to assemble between the microfibrils.

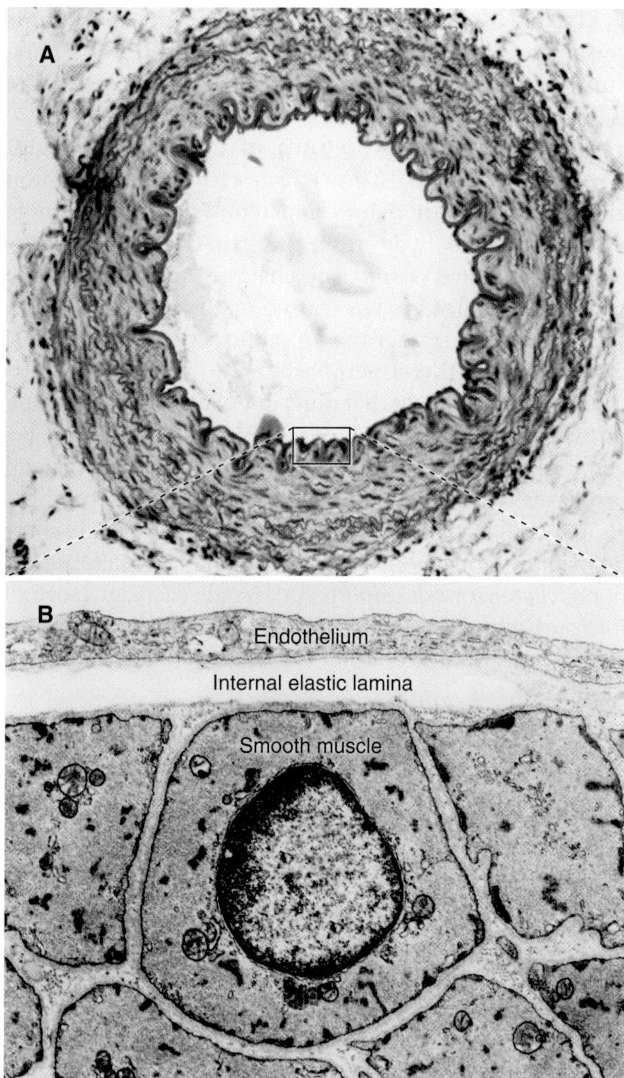

FIGURE 29.8 ELASTIC FIBERS IN THE WALL OF A SMALL ARTERY. A, Light micrograph of a cross section stained to bring out the internal elastic lamina (box) and wavy elastic fibers among the muscle cells. The boxed area includes the internal elastic lamina between the endothelial cells lining the lumen and the underlying smooth muscle cells. **B,** Electron micrograph of a longitudinal thin section illustrating the internal elastic lamina. In such standard preparations, elastic fibers stain poorly and appear amorphous except for occasional 10-nm microfibrils on the surface. (Courtesy Don W. Fawcett, Harvard Medical School, Boston, MA.)

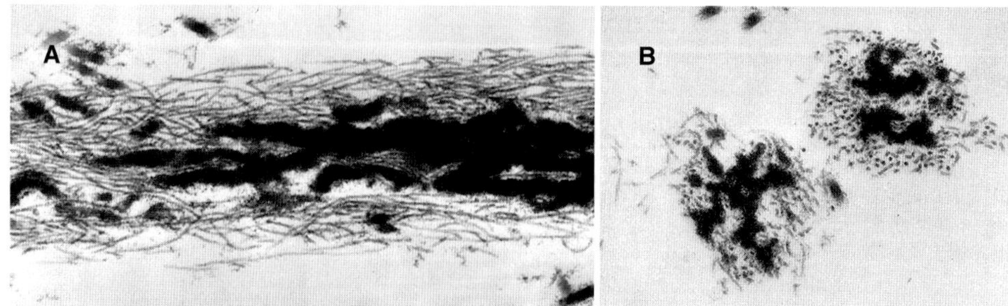

FIGURE 29.9 ELECTRON MICROGRAPHS OF DEVELOPING ELASTIC FIBERS FROM A FETAL CALF. A, Longitudinal section. **B,** Cross section. Fibrillin microfibrils form a scaffolding for elastin, which stains darkly in this preparation. (Courtesy J. Rosenbloom, University of Pennsylvania, Philadelphia.)

Fibrillin is the primordial component of elastic fibers, having arisen in Cnidarians (see Fig. 2.8). It is a long, floppy protein consisting of a tandem array of domains some of which are glycosylated (Fig. 29.10). Humans have three fibrillin genes. Fibrillin-1 is the main component of 10-nm microfibrils, along with several glycoproteins. Microfibrils are composed of parallel fibrillin molecules that interact head to tail, reinforced by disulfide bonds made by the first hybrid domains. Parts of neighboring subunits overlap in globular beads connected by flexible arrays of domains. Microfibrils are about 100 times stiffer than elastin, and they stretch by rearrangement of molecules and domains rather than unfolding. Fibrillins and related proteins called latent-TGFβ (transforming growth factor-β)-binding proteins, act as repositories for TGFβ family proteins in connective tissues.

Elastin subunits are a family of closely related 60-kD proteins called **tropoelastins,** the products of alternative splicing from a single elastin gene. Long sequences rich in hydrophobic residues are interrupted by short sequences with pairs of lysines separated by two or three small amino acids (Fig. 29.11). Lysine-rich sequences are thought to form α-helices with pairs of lysines adjacent on the surface.

As tropoelastin assembles on the surface of elastic fibers, lysyl oxidase oxidizes paired lysines of tropoelastin to aldehydes. Oxidized lysines condense into a **desmosine** ring that covalently crosslinks tropoelastin molecules to each other (Fig. 29.11). The four-way crosslinks, involving pairs of lysines from two tropoelastin molecules, are unique to elastin. The same enzyme catalyzes the crosslinking of collagen, but it forms only two- and three-way crosslinks.

Elastic fibers are similar to rubber except that elastic fibers require water as a lubricant. Hydrophobic segments between the crosslinks are thought to form extensible random coils that extend and become aligned when an elastic fiber is stretched (Fig. 29.11C). A difference in entropy of the polypeptide in the contracted and stretched states is thought to be the physical basis for the elasticity (see the Gibbs-Helmhotz equation in Chapter 4). Stretched fibers store energy, owing to ordering (low entropy) of the polypeptide chains. Fibers shorten when the resistance is reduced, because the

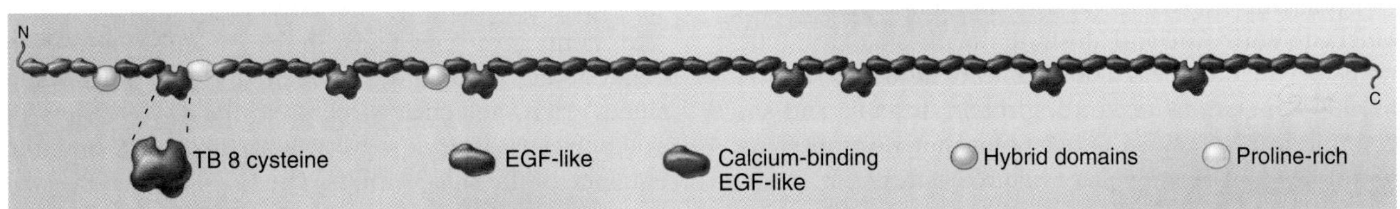

FIGURE 29.10 DOMAIN ORGANIZATION OF HUMAN FIBRILLIN-1. A tandem array of independently folded domains, including 47 epidermal growth factor–like (EGF-like) domains, forms a linear molecule. (Modified from Rosenbloom J, Abrams WR, Mecham R. Extracellular matrix 4: the elastic fiber. *FASEB J.* 1993;7:1208–1218.)

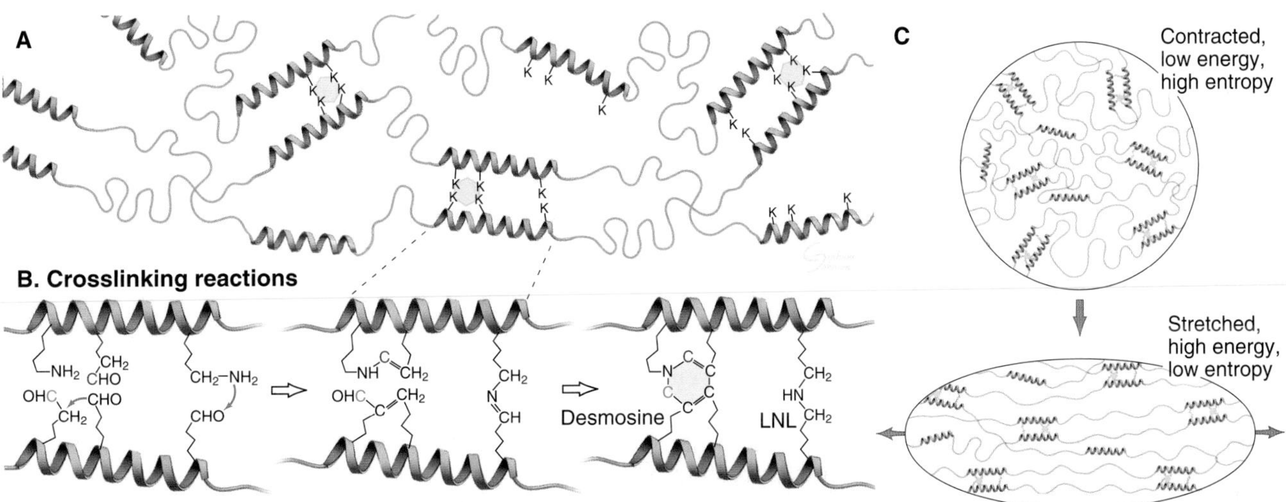

FIGURE 29.11 ELASTIN POLYPEPTIDES AND CROSSLINKING REACTIONS. A, Lysine-rich helical domains separate random chains rich in hydrophobic residues. **B–C,** Lysyl oxidase converts lysine amino groups to aldehydes, which react with other lysines to form simple linear crosslinks or six-membered rings linking two polypeptides. If the peptide bonds are hydrolyzed experimentally (not shown here), the linear crosslink is released as leucyl-norleucine (LNL) and the six-membered crosslink is released as the amino acid desmosine. **C,** Comparison of the contracted state with low-energy disordered chains having high entropy with the stretched state with high-energy ordered chains having low entropy. Elastin polypeptides form a continuous, covalently bonded network. Application of force stretches the chains between the crosslinks. This is a low-entropy, high-energy state. Reduced force allows the chains to contract into a more disordered, higher-entropy state with lower energy. (Modified from Rosenbloom J, Abrams WR, Mecham R. Extracellular matrix 4: the elastic fiber. *FASEB J.* 1993;7:1208–1218.)

polypeptide chains return to their disordered, lower-energy, higher-entropy state.

Only embryonic and juvenile fibroblasts synthesize elastic fibers, which turn over slowly, if at all, in adults. Consequently, adults must make do with the elastic fibers that are formed during adolescence. Fortunately, these fibers are amazingly resilient. Arterial elastic fibers withstand more than 2.5 billion cycles of stretching and recoil during a human life. Many tissues become less elastic with age, particularly the skin, which is subjected to damage from ultraviolet irradiation. Compare, for example, how readily the skin of a baby recoils from stretching compared with that of an aged person. The loss of elastic fibers in skin is responsible for wrinkles. Dominant mutations in the elastin gene cause a human disease called cutis laxa. The skin and other tissues of patients with this disease lack resilience.

Collagens are found across the phylogenetic tree, but only vertebrates are known to produce elastin. Invertebrates evolved two completely different elastic proteins. Mollusks have elastic fibers composed of the protein abductin. Insects use another protein, called resilin, to make elastic fibers.

Dominant mutations in the fibrillin-1 gene cause **Marfan syndrome** and illustrate the physiological functions of elastic fibers. Most of the thousand known fibrillin-1 mutations make the protein unstable and susceptible to proteolysis. Other point mutations interfere with folding. All patients are heterozygotes.

Elastic fibers of patients with Marfan syndrome are poorly formed, accounting for most of the pathological changes. Most dangerously, weakness of elastic fibers in the aorta leads to an enlargement of the vessel, called an aneurysm, which is prone to rupture, with fatal consequences. Prophylactic replacement of the aorta with a synthetic graft and medical treatment with drugs that block β-adrenergic receptors (see Fig. 27.3) allow patients a nearly normal life span. In some patients, a floppy mitral valve in the heart causes reflux of blood from the left ventricle back into the left atrium. Weak elastic fibers that suspend the lens of the eye result in dislocation of the lens and impaired vision. Weak elastic fibers result in lax joints and curvature of the spine. Most affected patients are tall, with long limbs and fingers, but the connection of these features to fibrillin is not known. The manifestations of the disease are quite variable, even within one family, for reasons that are not understood. Mutations in the fibrillin-2 gene cause congenital contractural arachnodactyly, a disease characterized by joint stiffness.

Glycosaminoglycans and Proteoglycans

Glycosaminoglycans (GAGs, formerly called mucopolysaccharides) are long polysaccharides made up of repeating disaccharide units, usually a hexuronic acid and a hexosamine (Fig. 29.12). With one important exception—hyaluronan—GAGs are synthesized as covalent, posttranslational modifications of a large family of proteins called proteoglycans. These proteins vary in structure and function, but their associated GAGs confer some common features.

All vertebrate cells synthesize proteoglycans. Most are secreted into the extracellular matrix, where they are major constituents of cartilage, loose connective tissue, and basement membranes. Mast cells package the proteoglycan serglycin, along with other molecules in secretory granules. A few proteoglycans, including syndecan and CD44, are transmembrane proteins with their GAGs exposed on the cell surface.

Of the known GAGs, **hyaluronan** (formerly called hyaluronic acid) is exceptional in two regards. First, enzymes on the cell surface synthesize the alternating polymer of [D-glucuronic acid β (1 → 3) D-*N*-acetyl glucosamine β (1 → 4)]$_n$ (Fig. 29.12). Other GAGs are synthesized as posttranslational modifications of a core protein. Second, hyaluronan is not modified postsynthetically, as are all other GAGs. The linear polymer, often exceeding 20,000 disaccharide repeats (a length >20 μm) is released into the extracellular space.

In contrast to proteins, nucleic acids, and even *N*-linked oligosaccharides, which are precisely determined macromolecular structures, the GAG chains of proteoglycans appear to vary both in length and the sequence of the sugar groups. The four-step synthesis of GAGs (Fig. 29.12) explains this variability:

1. Ribosomes associated with ER synthesize the **core protein,** which enters the secretory pathway.
2. In compartments between the ER and the *trans*-Golgi apparatus, **glycosyltransferases** initiate GAG synthesis by adding one of three different, short, *link oligosaccharides* to serine or asparagine residues of the core proteins (Fig. 29.12A–B). The structural clues identifying these sites are not understood, as they do not have a common amino acid sequence motif. A tetrasaccharide attached to serine anchors dermatan sulfate, chondroitin sulfate, and heparan sulfate. Branched oligosaccharides anchor keratan sulfate to serine or asparagine.
3. In the *trans*-Golgi network, other glycosyltransferases elongate the polysaccharide by adding, sequentially, *two alternating sugars* to the growing chain (Fig. 29.12D–F). The primary products are homogeneous, linear polymers, each with one pair of alternating sugars.
4. Enzymes modify some but not all the residues along these alternating sugar polymers by adding sulfate to hydroxyl or amino groups, or by isomerizing certain carbons to convert D-glucuronic acid to its epimer L-iduronic acid (Fig. 29.12D–F). The result is a heterogeneous polymer. The mechanisms that select particular sites for modification are not understood.

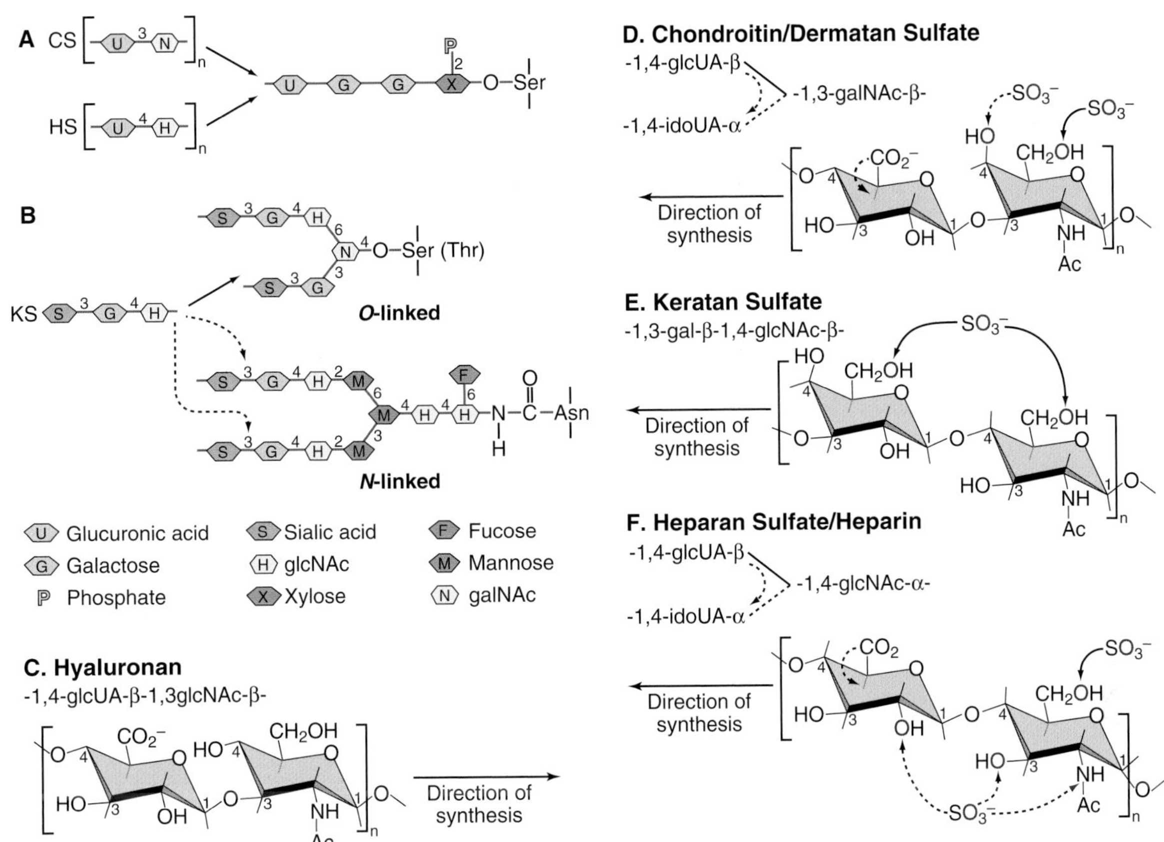

FIGURE 29.12 SYNTHESIS OF GLYCOSAMINOGLYCANS (GAGS). A–B, Three short oligosaccharides link GAGs *(left)* to proteoglycan core proteins *(right)*. **A,** A tetrasaccharide anchors chondroitin sulfate (CS), dermatan sulfate, and heparan sulfate (HS) to serine residues. **B,** Two different, branched oligosaccharides link keratan sulfate (KS) to either serine or asparagine. **C–F,** Four parent polymers and postsynthetic modifications. **C,** Hyaluronan [D-glucuronic acid β (1 → 3) D-*N*-acetylglucosamine β (1 → 4)]$_n$ (*n* ≥25,000) is not modified postsynthetically. **D,** Chondroitin sulfate and dermatan sulfate are synthesized as [D-glucuronic acid β (1 → 3) D-*N*-acetylgalactosamine β (1 → 4)]$_n$ (*n* usually <250) and then modified. Some *N*-acetylgalactosamines are sulfated. In dermatan sulfate, D-glucuronic acids are epimerized to L-iduronic acid. **E,** Keratan sulfate is synthesized as [D-galactose β (1 → 4) D-*N*-acetylglucosamine β (1 → 3)]$_n$ (*n* usually = 20–40) and then modified by sulfation. **F,** Heparan sulfate/heparin is synthesized as [D-glucuronic acid β (1 → 4) D-*N*-acetylglucosamine α (1 → 4)]$_n$ (*n* usually <100) and then modified by sulfation and by epimerization of D-glucuronic acid to L-iduronic acid. galNAc, *N*-acetylgalactosamine; glcNAc, *N*-acetylglucosamine. (Modified from Wright TN, Heinegard DK, Hascall VC. Proteoglycans, structure and function. In Hay ED, ed. *Cell Biology of the Extracellular Matrix,* 2nd ed. New York: Plenum Press; 1991:45–78.)

The present nomenclature for proteoglycans is based on the core protein. The historic nomenclature based on the identity of the GAGs was imprecise, as more than one type of proteoglycan can carry the same GAG. The weakness of the new system is that the protein name reveals nothing about the associated GAGs. This information is important, because various cells add different GAGs to the same core protein or can modify the same GAG in different ways.

Cells secrete many proteoglycans into the extracellular matrix, but they retain some types on the plasma membrane through transmembrane polypeptides or a glycosylphosphatidylinositol anchor (Appendix 29.2 and Fig. 29.13). The core proteins vary in size from 100 to 4000 amino acids. Many are modular, consisting of familiar structural domains found in epidermal growth factor (EGF), complement regulatory protein, leucine-rich repeats, or lectin. Three collagens carry GAG side chains: Types IX and XII have chondroitin sulfate chains, and type XVII has heparin sulfate chains.

The number of GAGs attached to the core protein varies from one (decorin) to more than 200 (aggrecan) (Fig. 29.13). A particular core protein can have identical (fibroglycan, glypican, versican) or different (aggrecan, serglycin, syndecan) types of GAGs. Some cell types add different GAGs to the same core protein or secrete a core protein without GAGs.

Given their physical properties and distribution among the fibrous elements of the extracellular matrix, proteoglycans and hyaluronan are thought to be elastic water-trapping space fillers. Each hydrophilic disaccharide unit bears a carboxyl or sulfate group or both, so GAGs are highly charged polyanions that extend themselves by electrostatic repulsion in solution and attract up to 50 g of water per gram of proteoglycan. Hyaluronan, the largest GAG, occupies a vast volume. A single

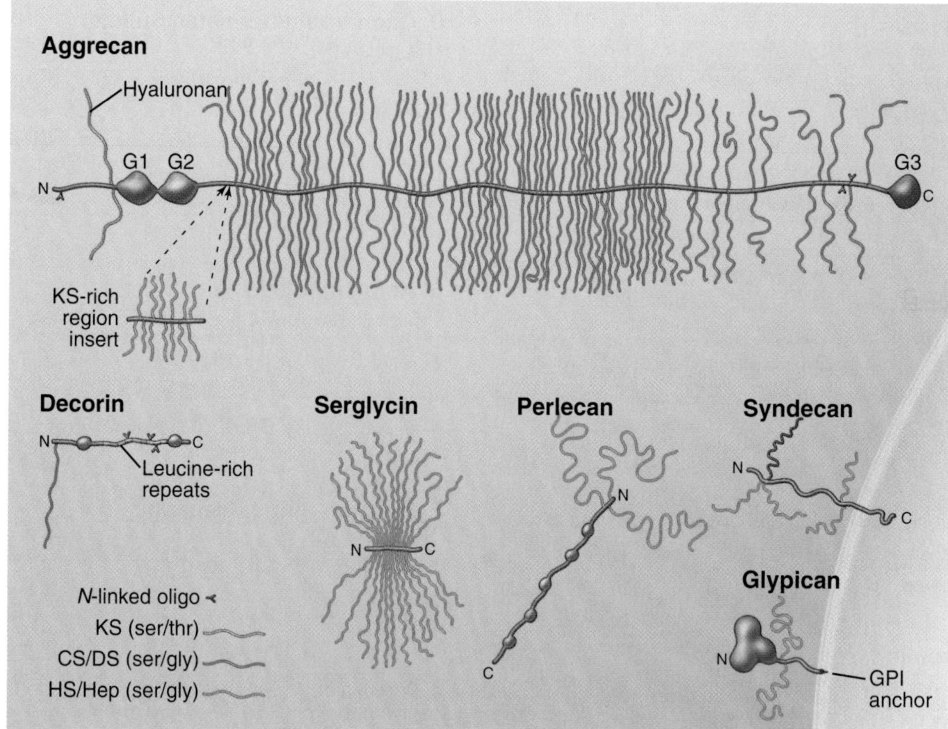

FIGURE 29.13 SCALE DRAWINGS OF A VARIETY OF PROTEOGLYCANS. Core proteins are *purple* except for the *red* leucine-rich repeats of decorin and the glycosaminoglycans are color-coded. Proteins were named according to the following: aggrecan aggregates along hyaluronan; decorin decorates collagen fibrils; perlecan resembles a string of pearls; serglycin has 24 Ser-Gly repeats; syndecan (syndein = link) links cells to the matrix; glypican has a glycosylphosphatidylinositol (GPI) membrane anchor. CS, chondroitin sulfate; DS, dermatan sulfate; Hep, heparin; HS, heparan sulfate; KS, keratan sulfate. (Modified from Wright TN, Heinegard DK, Hascall VC. Proteoglycans, structure and function. In Hay ED, ed. *Cell Biology of the Extracellular Matrix,* 2nd ed. New York: Plenum Press; 1991:45–78.)

hydrated molecule of 25,000 kD has a diameter of 200 nm, larger than a synaptic vesicle. Retention of water by hyaluronan and **aggrecan**-keratan sulfate/chondroitin sulfate proteoglycan is essential for the mechanical properties of cartilage (see Fig. 32.3). In the extracellular matrix of other tissues, networks of densely charged hyaluronan restrict water flow, limit diffusion of solutes (especially macromolecules), and impede the passage of microorganisms. Hyaluronan and proteoglycans also act as lubricants in joint cavities and as an optically transparent, space-filling medium in the vitreous body of the eye. Exceptionally high concentrations of hyaluronan in the tissues of subterranean naked mole rats seem to protect them from cancer and give them life spans much longer than any other rodent. The mechanism is not known.

Beyond these mechanical functions, proteoglycans influence cellular behavior such as adhesion or motility. Transmembrane proteoglycans can link cells to fibronectin and connective tissue collagens. **Syndecan** provides a particularly clear example. Lymphocytes express syndecan twice: early in their maturation, when they adhere to matrix fibers in the bone marrow, and later, when, as mature plasma cells, they adhere to the matrix of lymph nodes. In between, syndecan expression is lower while the lymphocytes circulate in the blood.

Carefully regulated expression allows proteoglycans to influence embryonic development and wound healing in at least three different ways. First, both decorin and fibromodulin regulate assembly of collagen fibrils. Second, many polypeptide growth factors (including platelet-derived growth factor and TGFβ) bind to proteoglycans in the extracellular matrix. This allows the matrix to concentrate circulating growth factors at specific locations and to release them locally over time. Third, membrane-bound proteoglycans, including syndecan and glypican, act as coreceptors for growth factors.

The well-known anticoagulant effects of heparin and heparan sulfate are attributable to their ability to bind both thrombin (the proteolytic enzyme that converts fibrinogen to fibrin) and a thrombin inhibitory protein. This promotes interaction of the inhibitor with thrombin and inactivates the clotting cascade. A short sequence of five modified sugars has the anticoagulant activity.

Adhesive Glycoproteins

In principle, the macromolecules of the extracellular matrix and the constituent cells might interact relatively nonspecifically, but the evidence suggests that specific molecular interactions mediate most of the interactions that organize the matrix and the associated cells. Most interactions are between proteins, but some are between proteins and sugars. Although some of these interactions are direct (with some cell surface receptors binding collagen directly), adapters called adhesive glycoproteins mediate many of the interactions (Appendix 29.3).

Adhesive glycoproteins were discovered by using biochemical assays for factors that favor particular interactions, such as adherence of cells to a matrix component. Further work revealed that adhesive glycoproteins are more than molecular glue; they also provide cells with signals required for the development and repair of tissues. Cells receive these signals when they bind to the matrix components. Chapter 30 focuses on their receptors.

Adhesive glycoproteins provide specific molecular interactions in the matrix by binding to cells, matrix macromolecules, or both. Adhesive proteins with multiple binding sites for cell surface receptors link cells together. For example, fibrinogen aggregates platelets during blood clotting (see Fig. 30.14). Other adhesive proteins link cells to the extracellular matrix. Thus, fibronectin mediates the attachment of cells to fibrin and collagen (Fig. 29.14). A third group of adhesive proteins links matrix macromolecules together. For instance, nidogen attaches laminin to collagen and link protein attaches aggrecan-proteoglycan to hyaluronan.

The repertoire of adhesive proteins extends far beyond the number of named proteins (Appendix 29.3). Multiple genes or, more commonly, alternative splicing of the product of a single gene (see Fig. 11.6), generate multiple, functionally distinct isoforms of most of the proteins. Particular isoforms are often expressed in specific tissues at predictable times during development.

Most adhesive glycoproteins are constructed of a series of compact modules (see Fig. 3.13 and Appendix 29.3). During evolution, duplication and recombination of the coding sequences for these domains produced the genes encoding these large proteins. In addition to the familiar domains, each of these proteins also contains a significant fraction of unique sequences.

Heterodimeric transmembrane receptors called **integrins** bind most adhesive glycoproteins that interact with cells (see Fig. 30.9). Remarkably, the integrin-binding sites of many adhesive proteins include the simple tripeptide arginine-glycine-aspartic acid (RGD; Fig. 29.14), which acts as a universal "zip code."

Establishing the biological functions of adhesive glycoproteins is challenging because of their overlapping functions and large sizes. Initial hypotheses were based on the identification of binding partners and the time and place of expression of each protein. Later, antibodies or peptides were used to disrupt specific molecular interactions in live organisms. Disruption of the gene for each protein or its receptors provides definitive data, and the consequences can be surprising. Some phenotypes proved to be milder than expected from earlier studies. These results argue that the adhesive glycoproteins function as a complementary system with partially overlapping functions. Two examples illustrate what we know about adhesive glycoproteins.

Fibronectin

Fibronectins are large proteins composed of two polypeptides of approximately 235 kD linked by disulfide bonds near their C-termini (Fig. 29.14). In electron micrographs, fibronectin appears as a V-shaped pair of long, flexible rods connected at one end. In solution, the molecule is probably more compact. Each polypeptide is a linear array of three types of domains called FN-I, FN-II, and FN-III (for fibronectin-I, -II and -III). All three types of fibronectin domains consist of antiparallel β strands with conserved residues in their hydrophobic cores. Two disulfide bonds stabilize FN-I and FN-II domains, whereas FN-III domains have no disulfide

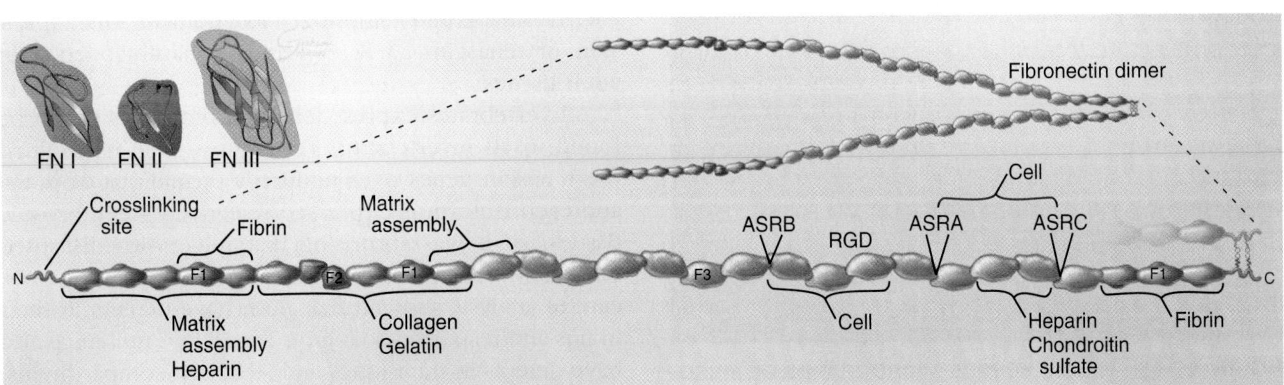

FIGURE 29.14 DOMAIN ORGANIZATION OF FIBRONECTIN. A linear array of FN-I (45 residues), FN-II (45 residues), and FN-III (90 residues) domains forms a rod-shaped subunit. Disulfide bonds near the C-termini covalently link two identical subunits in the dimeric molecule. Ligand-binding sites are indicated. The FN-III domain 10 contains the RGD (arginine-glycine-aspartic acid) sequence that binds cell surface integrins. Binding sites for fibrin, collagen, and glycosaminoglycans are indicated. Alternative splicing at sites ASRB, ASRA, and ASRC creates different isoforms. (For reference, see PDB files 1PDC and 1FNA and Potts JR, Campbell ID. Fibronectin structure and assembly. *Curr Opin Cell Biol.* 1994;6:648–655.)

bonds. FN-I and FN-II domains consist of approximately 45 residues; FN-III domains are twice as large. FN-I and FN-II domains are present in a few other proteins, whereas the human genome contains about 170 genes with FN-III domains, including proteins in the extracellular matrix (Appendix 29.3), on the cell surface (human growth hormone receptor; see Fig. 24.6), and inside cells (titin; see Fig. 39.7).

Fibronectin binds a variety of ligands, including cell surface receptors, collagen, proteoglycans, and fibrin (another adhesive protein). Thus, it contributes to the adhesion of cells to the extracellular matrix, guides the assembly of collagens and fibrillins and may also cross-link matrix molecules. The RGD sequence that contributes to binding integrins is on an exposed loop of FN-III domain 10, but some binding sites are exposed only when the protein is stretched. The variably spliced V domain included in plasma fibronectin has a second integrin-binding site. Chapter 30 provides additional details on integrins.

Two pools of fibronectin have different distributions and solubility properties. *Tissue fibronectin* forms insoluble fibrils in connective tissues throughout the body, especially in embryos and healing wounds. Fibroblasts use an integrin-dependent process to assemble fibronectin dimers into fibrillar aggregates large enough to visualize by light microscopy (Fig. 29.15). Denaturing agents and disulfide reduction are required to solubilize these fibrils. Fibronectin fibrils seem to bind cells more efficiently than soluble fibronectin, and may have additional activities important for biological functions.

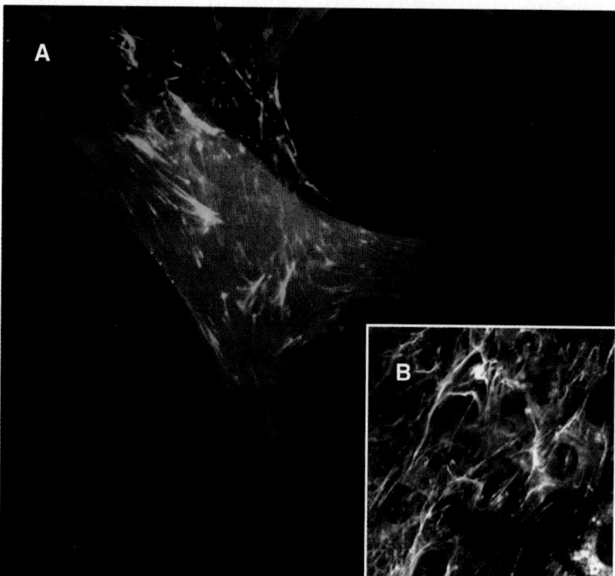

FIGURE 29.15 FLUORESCENCE MICROGRAPHS OF FIBRONECTIN NETWORKS IN TISSUE CULTURE. A, This fibroblast expressed fibronectin-YFP (fibronectin fused to yellow fluorescent protein, appearing *yellow-green*) and moesin-CFP (moesin fused to cyan fluorescent protein, appearing *red*). The fibronectin assembled an extracellular network. Moesin is associated with actin filaments in stress fibers. **B,** Lower magnification of a fibronectin network. (Courtesy T. Ohashi and H.P. Erickson, Duke University, Durham, NC.)

Soluble *plasma fibronectin* dimers circulate in the body fluids. The protein differs from tissue fibronectin as a result of alternate splicing of the messenger RNA (mRNA). In blood clots, the enzyme transglutaminase covalently couples plasma fibronectin to fibrin, forming a provisional matrix for wound repair (see Fig. 32.11).

Early embryos form an initial extracellular matrix from fibronectin that is replaced by collagen as the embryo matures. Embryonic cells, such as neural crest cells (precursors of pigment cells, sympathetic neurons, and adrenal medullary cells), migrate along tracks in the extracellular, fibronectin-rich matrix. Antibodies or fibronectin fragments that interfere with the adhesion of cells to fibronectin inhibit neural crest cell migration, gastrulation, and the formation of many embryonic structures derived from mesenchymal cells.

Surprisingly, mouse embryos without fibronectin can develop almost normally up to about day 8 (when the basic body plan is already determined). Thereafter homozygous null mutant mice die from defects in mesodermal structures, including the notochord, muscles, heart, and blood vessels. Mice with null mutations in the main fibronectin receptor, integrin α_5, have similar but slightly milder defects. Other adhesive glycoproteins and receptors must compensate for fibronectin during the first few days of development.

Tenascin

Tenascins are a family of four giant proteins with six arms (Fig. 29.16), found in the extracellular matrix of many embryonic tissues, wounds, and tumors. The N-terminal ends of three subunits self-associate in a triple helical coiled-coil. Disulfides covalently link two of these three-chain units to make the hexameric molecule. The arms of the four isoforms consist of different numbers of EGF and FN-III domains, terminated by three similar FN-III domains and a fibrinogen-like domain. The expression of tenascins-C, -R, -W, and -X hardly overlap in adult tissues.

All vertebrates express tenascin, but it has yet to be found in an invertebrate. Vertebrates have maintained the tenascin genes over hundreds of millions of years, and each isoform is expressed selectively in embryonic tissues, so it was surprising that mice with disrupted tenascin-C or tenascin-R genes are viable. However, careful analysis showed that both have defects in their brains and responses to injury. Tenascin-C null mice also have defects in their lungs and stem cell compartments. Genetic deficiency of tenascin-X causes Ehlers-Danlos syndrome, a human condition with hyperextensible skin and lax joints that is most often caused by mutations in collagen type V gene.

Tenascins bind to fibronectin, integrins, proteoglycans, and immunoglobulin-superfamily receptors on the

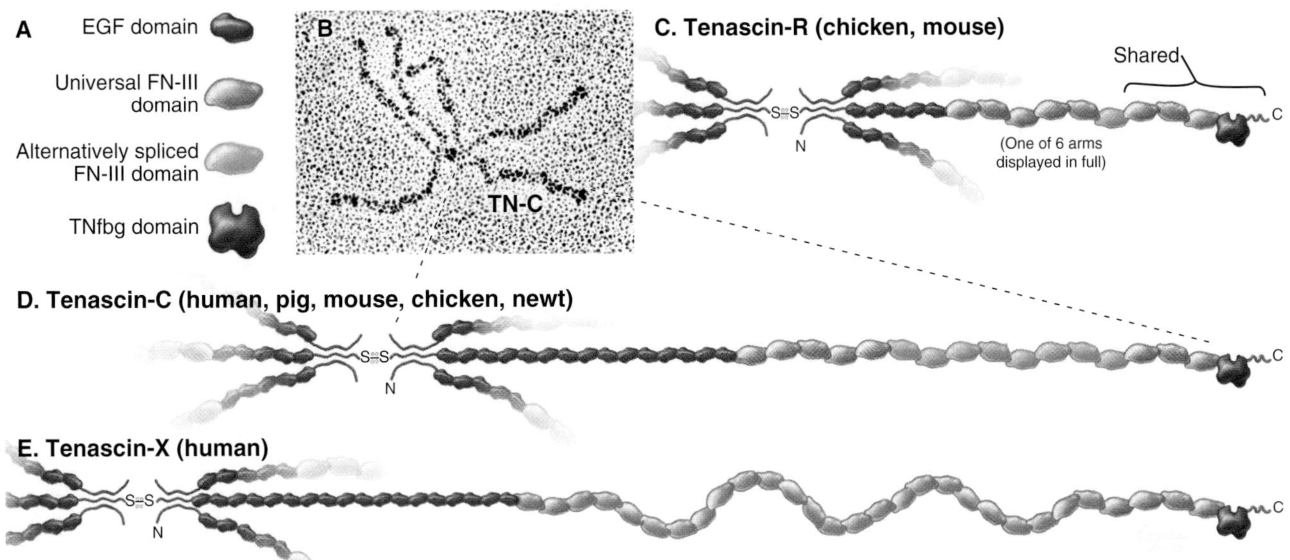

FIGURE 29.16 DOMAIN ORGANIZATION OF THE THREE ISOFORMS OF TENASCIN. A, Domains. TNfbg, fibrinogen-like domain. **B,** Electron micrograph of tenascin-C. **C,** Tenascin-R. **D,** Tenascin-C. **E,** Tenascin-X. Each of these tenascin molecules has six identical chains. One is shown in its entirety. Five chains are represented only by a few of their N-terminal EGF domains. (Courtesy H.P. Erickson, Duke University, Durham, NC.)

cell surface. Depending on the cell and the experimental situation, tenascin can promote or inhibit adhesion of cells to various substrates. This may contribute to spreading of cancer cells.

Basal Lamina

The basal lamina, a thin, planar assembly of extracellular matrix proteins, supports all epithelia, muscle cells, and nerve cells outside the central nervous system (Fig. 29.17). This two-dimensional network of protein polymers forms a continuous rug under epithelia and a sleeve around muscle and nerve cells. In addition, basal laminae can act as semipermeable filters for macromolecules, a particularly important role that they play in the conversion of blood plasma into urine in the kidney. The genes for basal lamina components are very ancient, having arisen in early metazoans.

In electron micrographs of thin sections of tissues prepared by chemical fixation, the basal lamina is a homogenous, finely fibrillar material close to the plasma membrane (Fig. 29.17D). Collagens type VI, VII, XV, and XVIII connect the lamina to the underlying connective tissue. The basal lamina and associated collagen fibrils form the "basement membrane" that is observed in histologic preparations of epithelia. A basal lamina alone cannot be seen by light microscopy without special labels, such as those used in Fig. 29.17A and C.

Although many proteins contribute to the stability of the basal lamina (Fig. 29.18), only the adhesive glycoprotein **laminin** is essential for the initial assembly of basal laminae during embryogenesis. The C-terminal end of the cross-shaped laminin molecule binds to cell surface receptors (integrins, see Fig. 30.9; **dystroglycan;** see Fig. 39.17). Laminins self-assemble into continuous, two-dimensional networks through noncovalent interactions of their short arms. Mouse embryos that lack dystroglycan or laminin die early in development, owing to failure to make basal laminae.

The subsequent addition of other proteins reinforces the laminin network. A two-dimensional network of **collagen IV** self-assembles through head-to-head interactions of the N-termini of four molecules and tail-to-tail interactions of the C-terminal NC1 domains of two molecules (Fig. 29.2). The networks of both laminin and collagen IV lie relatively parallel to the cell surface. Mouse embryos that lack collagen IV make nascent basal laminae composed of laminin but eventually die from defects in basal lamina functions.

Other proteins reinforce the collagen IV and laminin in basal laminae. The rod-shaped protein **nidogen** crosslinks laminin to type IV collagen. **Perlecan,** a heparan sulfate proteoglycan, provides additional crosslinks. It binds to itself in addition to laminin, nidogen, and collagen IV. These crosslinks help determine the porosity of the basal lamina and thus the size of molecules that can filter through it. Fibrillin and an associated protein, fibulin, are also present.

Epithelial and muscle cells secrete laminin and the other components of basal lamina. Two different cells can cooperate to produce a basal lamina between two tissues. For example, epithelial cells make laminin, and mesenchymal cells contribute nidogen to the same basal lamina.

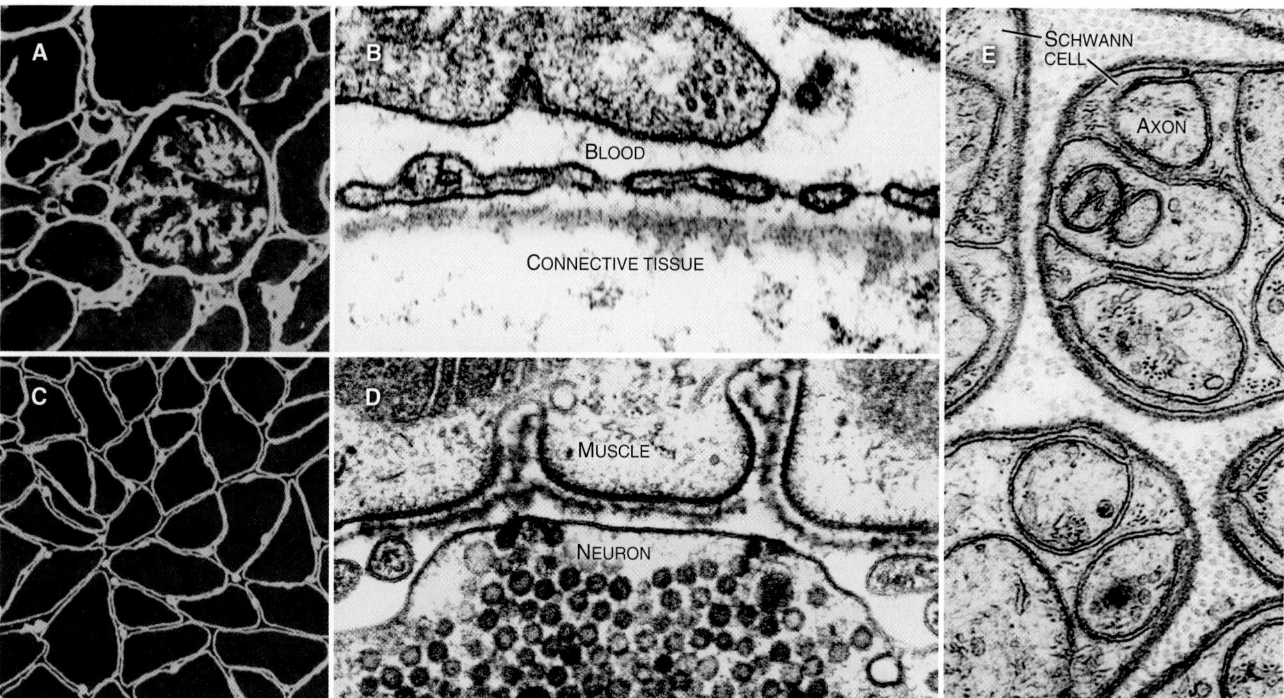

FIGURE 29.17 MICROGRAPHS OF THE BASAL LAMINA. A and **C,** Fluorescence micrographs of tissue sections stained with fluorescent antibodies to type IV collagen, a major component of basal laminae. **A,** Kidney with basal laminae around the tubules and blood vessels, including those of the glomerulus in the center. **C,** Skeletal muscle with basal laminae around the muscle cells. **B, D,** and **E,** Electron micrographs of thin sections showing basal laminae (colored *pink*). **B,** Endothelial cell lining a blood vessel with a platelet in the lumen. **D,** Neuromuscular junction. **E,** Unmyelinated nerve with numerous axons *(yellow)* surrounded by invaginations of Schwann cells *(blue)*. (**A** and **C**, From Odermatt BF, Lang AB, Ruttner JR, et al. Monoclonal antibody to human type IV collagen. *Proc Natl Acad Sci U S A.* 1984;81:7343–7347. **B** and **E,** Courtesy Don W. Fawcett, Harvard Medical School, Boston, MA. **D,** Courtesy J. Heuser, Washington University, St. Louis, MO.)

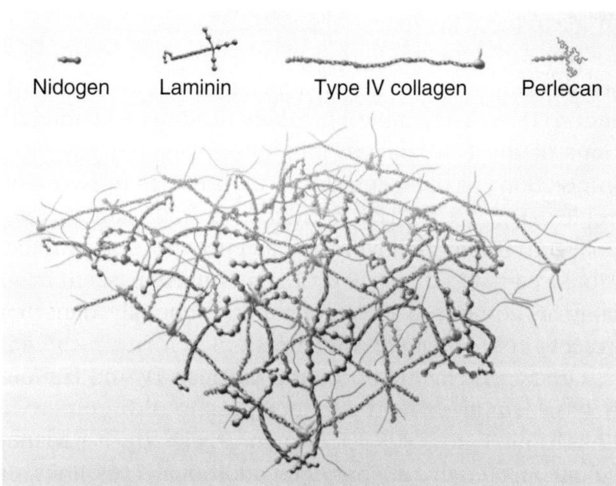

FIGURE 29.18 MOLECULAR MODEL OF THE BASAL LAMINA. The drawing shows the sizes and shapes of the component molecules and their postulated three-dimensional arrangement in the basal lamina. (From Yurchenco P, Cheng YS, Colognato H. Laminin forms an independent network in basement membranes. *J Cell Biol.* 1992;117:1119–1133.)

The interwoven network of protein fibers provides the physical basis for the two main functions of the basal lamina: mechanical support and selective permeability. The basal lamina is a scaffold that anchors epithelial, muscle, and nerve cells. In epithelia, all the basal cells attach to the underlying basal lamina, which is, in turn, attached to the underlying connective tissue. Thus, force applied to an exposed epithelial surface, such as skin, is transmitted through the basal lamina to the connective tissue. Similarly, all epithelial cells in tubular structures, such as blood vessels and glands, adhere to a cylindrical basal lamina that contributes to the integrity of the tube. In muscle, the basal lamina around each cell transmits the contractile forces between cells and to tendons.

The fibrous network in the basal lamina also acts as a filter for macromolecules and a permeability barrier for cellular migration. In kidney, a basal lamina sandwiched between two sheets of epithelial cells filters the blood plasma to initiate the formation of urine. The molecular weight threshold for the filter is approximately 60 kD, so most serum proteins are retained in the blood, whereas salt and water pass into the excretory tubules. The high charge of basal lamina proteoglycans contributes to filtering by electrostatic repulsion. Basal laminae also confine epithelial cells to their natural compartment. If neoplastic transformation occurs in an epithelium, the basal lamina prevents the spread of the tumor until matrix metalloproteinases (see "Matrix Metalloproteinases" below) break down the basal lamina.

The major basement membrane type IV collagen consists of two α_1(IV) chains and one α_2(IV) chain. No

TABLE 29.1 Inherited Diseases or Mutant Phenotypes of Basal Lamina Components

Protein Subunit	Distribution	Disease or Mutant Phenotype
Collagen α_3IV	Many tissues	Human autoantibodies cause Goodpasture syndrome of renal failure.
Collagen α_5IV	Kidney, muscle	Human mutation causes Alport syndrome of renal failure.
Laminin α_1	Many tissues	Fly null mutation is lethal during embryogenesis.
Laminin α_2	Muscle, heart	Mouse dy mutation causes muscular dystrophy.
Laminin γ_2	Epidermis	Human mutation causes Herlitz junctional epidermolysis bullosa.
Perlecan	Many tissues	Worm unc-52 mutation disrupts myofilament attachment to membrane.

human mutations in the two major type IV collagen genes have been observed, presumably because they are lethal, as observed in *Drosophila.*

Restricted human tissues express four additional type IV collagens. Remarkably, each has been implicated in human disease (Table 29.1). More than 200 different point mutations and deletions in the α_5(IV) collagen gene cause Alport X-linked familial nephritis. These mutations interfere with folding of the collagen molecule and disrupt the basement membranes that form the blood filtration barrier in the glomerulus of the kidney, causing progressive kidney failure. They also cause defects in the eye and ear, other places where the α_5(IV) collagen gene is expressed. Patients with autosomally inherited Alport syndrome have mutations in their α_3(IV) or α_4(IV) collagen genes. In Goodpasture syndrome, the immune system produces autoantibodies to the C-terminal NC1 domain of α_3(IV) collagen. The protein sequences that elicit autoantibody production are buried in the NC1 domain, so they may be exposed by bacterial infections or organic solvents, predisposing events that trigger the syndrome. Antibodies bound to basement membranes in the kidney and lung cause inflammation that leads to kidney failure and bleeding in the lungs.

Matrix Metalloproteinases

Many physiological processes depend on the controlled degradation of the extracellular matrix. Examples include tissue remodeling during embryogenesis (eg, resorption of a tadpole tail), wound healing, involution (massive shrinkage secondary to loss of cells and extracellular matrix) of the uterus after childbirth, shedding of the uterine endometrium during menstruation, and invasion of the uterine wall by the embryonic trophoblast during implantation. Conversely, uncontrolled destruction of extracellular matrix contributes to degenerative diseases, such as emphysema and arthritis. In addition to their roles in remodeling, many of these enzymes cleave and release biologically active fragments from matrix or membrane proteins. Three classes of Zn-dependent proteases account for both the physiological and pathological degradation of diverse extracellular matrix and cell surface proteins.

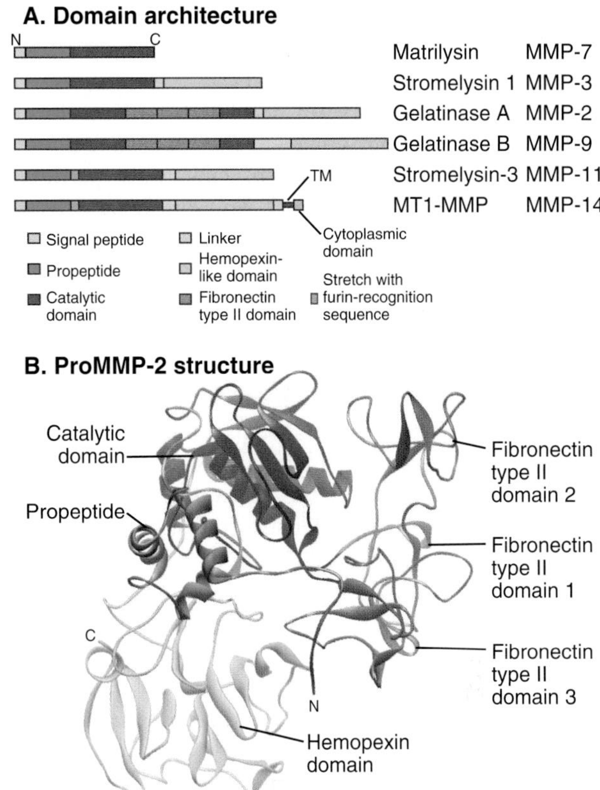

FIGURE 29.19 MATRIX METALLOPROTEINASE (MMP) STRUCTURES. A, Domain organization of MMP isoforms. All have an N-terminal cleaved signal sequence, a propeptide that binds to the active site and inhibits the protease activity, and a catalytic domain. Gelatinases have fibronectin (FN)-II domains inserted in the catalytic domain. Matrilysin lacks the C-terminal domain. MMP-14 has a transmembrane segment near its C-terminus. **B,** Atomic structure of MMP-2 showing the arrangement of the domains and the propeptide occupying the active site. (For reference, see PDB file 1CK7.)

The first class is called **matrix metalloproteinases (MMPs).** These 24 homologous enzymes share a zinc-protease domain (Fig. 29.19) similar to bacterial thermolysin. All have an N-terminal signal sequence and are processed through the secretory pathway. Between the signal sequence and catalytic domain, all MMPs have an **autoinhibitory propeptide,** including a conserved cysteine that binds to the zinc ion in the catalytic site. Most MMPs have a C-terminal regulatory domain that influences the substrate specificity of the catalytic domain. Some MMPs have three FN-II domains inserted into the sequence of the catalytic domain (Fig. 29.10B).

Most inactive pro-MMPs are secreted and then bind directly or indirectly to cell surface receptors. However, several MMPs are anchored to the plasma membrane by a C-terminal transmembrane domain or a glycosylphosphatidylinositol tail.

MMP activity is carefully regulated at three levels, normally restricting proteolysis to sites of tissue remodeling or physiological breakdown. First, only particular connective tissue, inflammatory, and epithelial cells are genetically programmed to express MMP genes and to respond to growth factors and cytokines to increase production under appropriate circumstances. Second, autoinhibited MMPs on the cell surface require propeptide cleavage for activation. Proteolytic cleavage and dissociation of the propeptide activate the enzyme. Cellular movements then deliver the active protease to specific substrates. For example, membrane-anchored MMP-14 directly activates MMP-2, which then degrades basement membrane collagen and other substrates. Third, secreted proteins called **tissue inhibitors of metalloproteinases (TIMPs)** and the plasma protein α_2-macroglobulin bind to the active site of MMPs, keeping their activity in check.

Each MMP is selective for targets in the extracellular matrix. In some cases, proteolysis disrupts the mechanical integrity of the matrix. In others, cleavage of a collagen isoform or other matrix protein releases a fragment that favors or inhibits the formation of blood vessels. For example, the N-terminal domain cleaved from collagen XVIII is an inhibitor of angiogenesis that is called endostatin. Mice survive null mutations in any one of several MMPs tested, but the loss of an MMP may alter the susceptibility to disease dramatically. Mice without MMP-12 (macrophage elastase) are resistant to cigarette smoke, which causes emphysema in normal mice. Without MMP-12, smoke fails to stimulate elastin degeneration, which weakens lung tissue and mediates inflammation. MMPs contribute to the spread of tumors in mice, but disappointingly small-molecule inhibitors of MMPs have not proven useful for treatment of advanced tumors in humans.

The second class of Zn-dependent proteases consists of approximately 20 proteases called **ADAMs** (a disintegrin and metalloproteinase). These enzymes are anchored to the plasma membrane by a single transmembrane domain. Like other metalloproteinases, they are inhibited by TIMPs. ADAMs cleave and release extracellular domains of cell surface proteins, some of which are important informational molecules (eg, tumor necrosis factor [TNF]-α; transforming growth factor [TGF]-α). ADAM-17 null mutations are lethal during embryogenesis, owing to a lack of TGF-α or other ligands for EGF receptors. A polymorphism in the ADAM-33 gene is associated with some types of human asthma, although the mechanism is not understood.

A third class of Zn-dependent proteases is called **ADAMTS,** ADAMs with a thrombospondin domain. These secreted proteases cleave specific matrix substrates, such as the cartilage proteoglycan aggrecan. Experiments with mice show that inactivation of the protease domain of ADAMTS5 reduces the development of the common joint disease osteoarthritis.

SELECTED READINGS

Buehler MJ. Nature designs tough collagen: Explaining the nanostructure of collagen fibrils. *Proc Natl Acad Sci USA*. 2006;103:12285-12290.

Capila I, Linhardt RJ. Heparin-protein interactions. *Angew Chem Int Ed Engl*. 2002;41:390-412.

Chiquet-Ehrismann R, Tucker RP. Tenascins and the importance of adhesion modulation. *Cold Spring Harb Perspect Biol*. 2011;3: a004960.

Hacker U, Nybakken K, Perrimon N. Heparan sulphate proteoglycans: The sweet side of development. *Nat Rev Mol Cell Biol*. 2005;6: 530-541.

Halper J, Kjaer M. Basic components of connective tissues and extracellular matrix: elastin, fibrillin, fibulins, fibrinogen, fibronectin, laminin, tenascins and thrombospondins. *Adv Exp Med Biol*. 2014; 802:31-47.

Hudson BG, Tryggvason K, Sundaramoorthy M, Neilson EG. Alport's syndrome, Goodpasture's syndrome and type IV collagen. *N Engl J Med*. 2003;348:2543-2556.

Iozzo RV. Basement membrane proteoglycans: From cellar to ceiling. *Nat Rev Mol Cell Biol*. 2005;6:646-656.

Kim EB, Fang X, Fushan AA, et al. Genome sequencing reveals insights into physiology and longevity of the naked mole rat. *Nature*. 2011; 479:223-227.

Lu P, Takai K, Weaver VM, Werb Z. Extracellular matrix degradation and remodeling in development and disease. *Cold Spring Harb Perspect Biol*. 2011;3:a005058.

Malhotra V, Erlmann P. The pathway of collagen secretion. *Annu Rev Cell Dev Biol*. 2015;31:109-124.

Mienaltowski MJ, Birk DE. Structure, physiology, and biochemistry of collagens. *Adv Exp Med Biol*. 2014;802:5-29.

Mithieux SM, Weiss AS. Elastin. *Adv Protein Chem*. 2005;70:437-461.

Ricard-Blum S. The collagens family. *Cold Spring Harb Perspect Biol*. 2011;3:a004978.

Saito K, Katada T. Mechanisms for exporting large-sized cargoes from the endoplasmic reticulum. *Cell Mol Life Sci*. 2015;72: 3709-3720.

Schwarzbauer JE, DeSimone DW. Fibronectins, their fibrillogenesis, and in vivo functions. *Cold Spring Harb Perspect Biol*. 2011;3: a005041.

Sugahara K, Mikami T, Uyama T, et al. Recent advances in the structural biology of chondroitin sulfate and dermatan sulfate. *Curr Opin Struct Biol*. 2003;13:612-620.

Taylor KR, Gallo RL. Glycosaminoglycans and their proteoglycans: Host-associated molecular patterns for initiation and modulation of inflammation. *FASEB J*. 2005;20:9-22.

Timpl R, Sasaki T, Kostka G, Chu ML. Fibulins: A versatile family of extracellular matrix proteins. *Nat Rev Mol Cell Biol*. 2003;4: 479-489.

Yurchenco PD. Basement membranes: cell scaffoldings and signaling platforms. *Cold Spring Harb Perspect Biol*. 2011;3:a004911.

APPENDIX 29.1

Collagen Families

Type	Chains	Assembly	Interactions	Distribution	Disease Mutations
Fibrillar Collagens					
I	$\alpha_1.\alpha_1.\alpha_2$(I)	Fibrils	Self; types III and V collagen; types XII and XIV collagen	Bone, tendons, ligaments, skin, dentin	Osteogenesis imperfecta; Ehlers-Danlos syndrome type VII; Kniest dysplasia; Stickler syndrome
II	[α_1(II)]3	Fibrils	Self; type IX and XI collagen	Hyaline cartilage, vitreous body	Spondyloepiphyseal dysplasia; hypochondrogenesis; achondrogenesis; Kniest dysplasia; Stickler syndrome
III	[α_1(III)]3	Fibrils	Self; type I collagen	Skin, blood vessels	Ehlers-Danlos syndrome type IV
V	$\alpha_1.\alpha_1.\alpha_2$(V) $\alpha_1.\alpha_2.\alpha_3$(V)	Fibrils	Self; type I collagen	Fetal membranes, skin, bone, placenta, synovial membranes	Ehlers-Danlos syndrome
XI	α_1(XI)α_2(XI)α_1(II)	Fibrils	Self; type II collagen	Hyaline cartilage	Stickler's syndrome
Sheet-Forming Collagens					
IV	$\alpha_1.\alpha_1.\alpha_2$(IV) $\alpha_3.\alpha_4.\alpha_5$(IV) $\alpha_5.\alpha_5.\alpha_6$(IV)	Nets	Self; perlecan laminin, nidogen, integrins	Basement membranes	Autoantigen in Goodpasture syndrome; α_3(IV), α_4 (IV), & α_5(IV) mutated in Alport nephritis & porencephaly
VIII	[α_1(VIII)]? [α2(VIII)]?	Hexagonal net	Self	Descemet membrane (cornea)	Posterior polymorphous corneal dystrophy
X	[α_1(X)]3	?	?	Hypertrophic cartilage	Schmid metaphyseal chondrodysplasia
Connecting and Anchoring Collagens (Fibril-Associated Collagens With Interrupted Triple Helices [FACIT])					
VI	$\alpha_1.\alpha_2.\alpha_3$(VI)	Beaded fibrils	Self; type IV collagen	Vessels, skin, intervertebral disk	Bethlem myopathy, atopic dermatitis
VII	[α_1(VII)]3	Anchoring fibril	Self; type IV collagen	Epidermal-dermal junction	Dystrophic epidermolysis bullosa
IX	$\alpha_1.\alpha_2.\alpha_3$(IX)	Linker	Covalent GAG; type II collagen	Hyaline cartilage, vitreous body	Multiple epiphyseal dysplasia
XII	[α_1(XII)]3	? Linker	GAG; type I collagen	Embryonic tendon, skin	Not known
XIV	[α1(XIV)]3	? Linker	GAG; ? type I collagen	Fetal tendon, skin	Not known
XVIII	[α_1(XVIII)]3	? Linker	GAG	Basal lamina	Not known
Transmembrane					
XIII	[α_1(XIII)]3	Transmembrane	Integrin $\alpha_1\beta_1$, fibronectin		Not known
XVII	[α_1(XVII)]3	Transmembrane	Basal lamina	Hemidesmosomes, epidermal-dermal junction	Blistering conditions; antigen in bullous pemphigoid

APPENDIX 29.2

Proteoglycans

Name	Core Protein	Glycosaminoglycans	Expression	Functions
Secreted Proteoglycans				
Aggrecan	One-gene, 250-kD protein with link protein, EGF, lectin & complement regulatory domains	100–150 keratin sulfate chains >150 chondroitin sulfate chains	Cartilage	Binds hyaluronan and link protein; hydrates and fills the ECM; no known diseases
Biglycan	One-gene, 38-kD protein	2 Chondroitin sulfate or dermatan sulfate chains	Developing muscle, bone, cartilage, and epithelia	Associated with cell surfaces; no known ligands or functions
Decorin	One-gene, 38-kD protein	1 Chondroitin sulfate or dermatan sulfate chain	Connective tissue fibroblasts	Binds collagen fibrils and modifies their assembly
Fibromodulin	One-gene, 43-kD protein	4 Asparagine-linked keratin sulfate chains	Cartilage, skin, tendon	Binds collagen I and II; limits size of collagen fibrils
Perlecan	One-gene, 400-kD protein	3 Heparan sulfate chains of 30–60 kD	All cells making basement membranes (epithelia, muscle, peripheral nerve)	Self-associates; binds laminin in basal lamina; binds basic fibroblast growth factor
Serglycin	One-gene, 12-kD protein; 24 serine-glycine repeats	Heparin or chondroitin sulfate	White blood cells, mast cells	Binds histamine in secretory granules
Versican	One-gene, 260-kD protein with link-protein, GAG attachment, 2 EGF, lectin & complement regulatory domains	12–15 chondroitin sulfate chains *N*- and *O*-linked oligosaccharides	Fibroblasts; ? other cells	May bind hyaluronan; functions unknown
Membrane-Associated Proteoglycans				
Fibroglycan	One-gene, integral membrane protein of 20 kD	Heparan sulfate chains	Fibroblasts	Binds collagen I and fibronectin; cell adhesion to ECM
Glypican	62-kD protein with glycosylphosphatidylinositol anchor to membrane on C-terminus	4 Heparan sulfate chains	Lung, skin, epithelia, endothelium, smooth muscle	Binds fibronectin, collagen I, and antithrombin III; cell adhesion to ECM
Syndecan	One-gene, integral membrane protein of 33 kD	Variable number of heparan sulfate and chondroitin sulfate chains	Embryonic epithelia and mesenchyme, developmentally regulated in adult lymphocytes	Binds fibronectin; collagens I, III, and V; thrombospondin; basic fibroblast growth factor; cell adhesion to ECM

ECM, extracellular matrix; EGF, epidermal growth factor; GAG, glycosaminoglycan.

APPENDIX 29.3

Adhesive Glycoproteins

Name(s)	Composition	Expression	Ligands	Functions and Diseases
Agrin	One-gene, 205-kD protein with cysteine-rich, EGF, and Kazal protease inhibitor domains	Motor neurons secrete into basal lamina of neuromuscular junction	? Acetylcholine receptor	Aggregates acetylcholine receptors
Fibrinogen	2 × 67 kD Aα chains, 2 × 56 kD Bβ chains, 2 ×-47 kD γ chains, joined by disulfide bonds; N-linked CHO on Bβ & γ chains	Hepatocytes secrete into blood	Platelet integrin GPIIb/GPIIIa	Thrombin cleavage releases fibrin, which polymerizes into fibrils stabilized by covalent crosslinking by transglutaminase; deficiency or defects cause bleeding
Fibronectin	One gene; RNA splicing isoforms of 235–270 kD; dimers disulfide-bonded; 12 FN-I, 2 FN-II and 15–17 FN-III domains; *N*- and *O*-linked CHO	Many tissues; increased with wounding	Fibrin, heparin, cells via integrins, collagen	Assembles fibrils in ECM; promotes cellular adhesion to ECM and migration
Fibulin	One gene; RNA splicing generates monomeric isoforms of 566, 601 and 683 residues with 9 EGF and 3 complement-like domains; *N*- and *O*-linked CHO	Fibroblasts; present in plasma and some basement membranes	Ca^{2+}, fibronectin, fibrinogen	Required for elastin assembly; mutated in some patients with macular degeneration
HB-GAM (heparin-binding, growth-associated molecule)	136 residues, 5 internal disulfide bonds	Brain, uterus, intestine, kidney, muscle, lung, skin	Heparin	? Neuronal differentiation
Laminin	1 × 200–400 kD A chain, 1 × 200 kD B_1 chain, 1 × 200 kD B_2 chain, several isoforms of each; poly-*N*-acetyl galactosamine	Epithelium, endothelium, smooth & striated muscle, peripheral nerve, myotendinous junction	Nidogen, 6 integrins perlecan, collagen IV, α-dystroglycan	Self-associates into network in basement membrane linked to collagen IV network by nidogen; promotes cell adhesion and migration
Laminin-binding protein (Mac-2)	Soluble S-type monomeric lectin of 29–35 kD	Macrophages, epithelia, fibroblast, trophoblast, cancer cells	Poly-*N*-acetyl galactosamine on laminin	?
Link protein	One gene; alternative splicing generates isoforms of 41, 46, & 51 kD with two 4-cysteine & one immunoglobulin domains; CHO content variable	Cartilage	Aggrecan and hyaluronan	Links aggrecan to hyaluronan
Mucins	Heterogeneous secreted & transmembrane glycoproteins	GI tract, salivary glands	Selectins	Lubricates mucous membranes
Nidogen (entactin)	One-gene, 148-kD monomeric protein with 8 EGF & 2 EF hand domains; *N*- & *O*-linked CHO	Basement membranes of epithelia, muscles, and nerves	Laminin, collagen IV, Ca^{2+}	Links collagen IV to laminin in basement membrane
Osteopontin (secreted phosphoprotein)	Monomer of ~300 residues, phosphorylated, *N*- & *O*-linked CHO, including sialic acid	Bone, milk, kidney, uterus, ovary	? Vitronectin receptor; ? hydroxyapatite	Promotes cell adhesion to ECM, including osteoclasts to bone

Continued

APPENDIX 29.3

Adhesive Glycoproteins—cont'd

Name(s)	Composition	Expression	Ligands	Functions and Diseases
Restrictin	3 × 180 kD chains linked by disulfide bonds; each has 1 cysteine-rich, EGF, 9 FN-III and 1 fibrinogen-like domains	A limited number of neurons in embryonic nervous systems	Cells, receptor unknown	? Cell adhesion
SPARC (secreted protein rich in cysteine); osteonectin	One-gene; 32-kD monomer	Bone, skin, connective tissue	Collagens III and V, Ca^{2+}, hydroxyapatite, cells	? Wound healing, development
Tenascin (cytotactin)	Four genes (C, R, X, Y) and alternate splicing; EGF, FN-III, and fibrinogen-like domains; chains linked by disulfides	Embryonic mesenchyme; adult perichondrium, periosteum, tendon, ligament, myotendinous junction, wounds	Integrins, Ig-CAMs, proteoglycans	Tenascin-X mutated in some patients with Ehlers-Danlos syndrome; mice with tenascin-C null mutation develop normally
Thrombospondin	420-kD protein with procollagen, EGF, and complement-like domains	Platelets, fibroblasts, embryonic heart, muscle, bone, brain	Integrin $\alpha_v\beta_3$, CD36 cell surface receptor, syndecan, Ca^{2+}	Platelet aggregation; stimulates proliferation of smooth muscle; inhibits proliferation of endothelium
Vitronectin	One-gene, 75-kD protein with *N*-linked CHO, phosphorylated, sulfated Secreted by liver into blood	Integrin $\alpha_V\beta_3$, heparin, glass, plastic	Promotes cell adhesion, inactivates heparin, stabilizes plasminogen activator inhibitor	
von Willebrand factor	One-gene, 2050-residue protein that forms head to head and tail to tail disulfide-linked oligomers, *N*- & *O*-linked CHO	Endothelium, platelets	Factor VIII, heparin, collagen, platelet integrin GPIIb/GPIIIa	Promotes adhesion of platelets to collagen and each other; required to control bleeding; deficiency causes most common congenital blood clotting abnormality

CHO, oligosaccharide chains; ECM, extracellular matrix; EF hand, calcium binding motif of calmodulin family; EGF, epidermal growth factor; GPI, glycosylphosphatidylinositol; FN, fibronectin; Ig-CAM, immunoglobulin cell adhesion molecule.

CHAPTER 30

Cellular Adhesion

All cells interact with molecules in their environment, in many cases relying on cell-surface adhesion proteins to bind these molecules. Multicellular organisms are particularly dependent on adhesion of cells to each other and the extracellular matrix (ECM). During development, carefully regulated genetic programs specify cell-cell and cell-matrix interactions that determine the architecture of each tissue and organ. Some adhesive interactions are stable. Muscle cells must adhere firmly to each other and to the connective tissue of tendons to transmit force to the skeleton (see Chapter 39). Skin cells must also bind tightly to each other and the underlying connective tissue to resist abrasion (see Fig. 35.6). On the other hand, many cellular interactions are transient and delicate. At sites of inflammation, leukocytes roll along endothelial cells lining small blood vessels and then interact weakly with the ECM as they migrate through connective tissue (look ahead to Figs. 30.13 and 30.14).

Cells use a relatively small repertoire of adhesion mechanisms to interact with matrix molecules and each other. A conceptual breakthrough came when comparisons of amino acid sequences showed that most adhesion proteins fall into five large families (Fig. 30.1). Within each of these distinctive families, ancestral genes duplicated and diverged during evolution, giving rise to adhesion proteins with the many different specificities required for embryonic development, maintenance of organ structure, and migrations of cells of our defense systems. General principles emerged from characterizing a few examples. Several important adhesion proteins fall outside the five major families, and additional families may emerge from continued research.

Many adhesion proteins were named before they were classified into families. Tables 30.1 through 30.5 are designed to help the reader with the challenging nomenclature. Many adhesion proteins are named

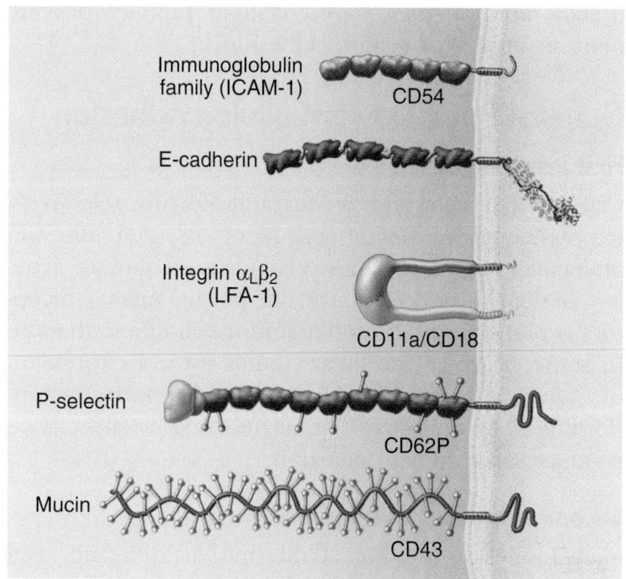

FIGURE 30.1 CLASSES OF ADHESION PROTEINS. Immunoglobulin cell adhesion molecules (IgCAMs) have one or more extracellular domains folded like an immunoglobulin domain *(orange)*. Cadherins have five or more extracellular CAD (cadherin) domains *(maroon)* that typically bind the same class of cadherin on a neighboring cell. Ca^{2+} stabilizes the interaction of neighboring CAD domains. Their cytoplasmic tails bind β-catenin *(blue)* and other adapter proteins. Integrins are heterodimers of α and β subunits that bind a wide range of matrix molecules by combinations of 1 of 16 α-chains and 1 of 8 β-chains. Selectins have a Ca^{2+}-dependent lectin (carbohydrate-binding) domain *(green)*, an epidermal growth factor (EGF)-like domain *(blue)*, and a variable number of complement regulatory domains *(red)*. Mucins use multiple carbohydrates for interactions with other cells. CD numbers refer to the names of representative adhesion molecules using the "clusters of differentiation" nomenclature (see the text). Single transmembrane helices anchor all these receptors to the plasma membrane. Adapter proteins link the cytoplasmic tails of most adhesion proteins to the actin cytoskeleton or, in the case of specialized cadherins and integrins, to intermediate filaments. ICAM, intercellular adhesion molecule; LFA, lymphocyte function–associated antigen. (Modified from van der Merwe PA, Barclay AN. Transient intercellular adhesion: the importance of weak protein-protein interactions. *Trends Cell Biol.* 1994;19:354–358.)

"CD" followed by a number. This stands for "clusters of differentiation," a term used to classify cell-surface antigens recognized by monoclonal antibodies. These names were given without knowledge of the structure or function of the antigen. Hence, members of the four major families of adhesion proteins have CD numbers.

This chapter first highlights some general features of adhesion proteins and then introduces the five major families: immunoglobulin–cell adhesion molecules **(IgCAMs), cadherins, integrins, selectins, and mucins**. While learning about each family, the reader should bear in mind that these receptors rarely act alone. Rather, they usually function as parts of multicomponent systems. Two examples at the end of the chapter illustrate the cooperation that is required for leukocytes to respond to inflammation and for platelets to repair damaged blood vessels. Chapter 31 on intercellular junctions, Chapter 32 on specialized connective tissues, and Chapter 38 on cellular motility provide more examples of cellular adhesion.

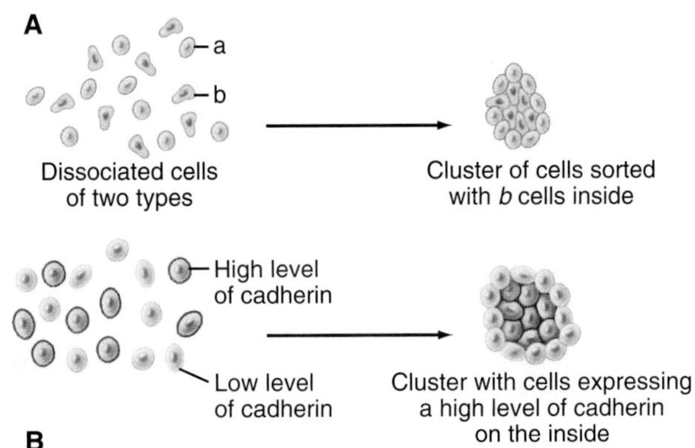

FIGURE 30.2 SORTING OF EMBRYONIC CELLS. A, When cells from different tissues are dissociated and mixed together, they spontaneously sort themselves into two layers, with the more adherent cells inside the less-adherent cells. **B,** Cells with high concentrations of cadherin sort inside less-adherent cells. (Based on the work of M. Steinberg, Princeton University, Princeton, NJ.)

General Principles of Cellular Adhesion

First Principle of Adhesion

Cells define their capacity for adhesion by selectively expressing plasma membrane receptors (cell adhesion molecules [CAMs]) with selective ligand-binding activity. Generally, expression of the proper mix of receptors is part of a genetic program for cell differentiation. In some cases, extracellular stimuli control expression of adhesion receptors. For example, inflammatory hormones or endotoxin stimulate endothelial cells to produce E-selectin on demand.

Second Principle of Adhesion

Many adhesion proteins bind one main ligand, and many ligands bind a single type of receptor (refer to Tables 30.1 through 30.5). If this one-to-one pairing were the rule, adhesion would be simple indeed. However, many exceptions exist, particularly in the integrin family of receptors (Table 30.3), which generally bind more than one ligand, and some ligands, such as fibronectin, bind more than one integrin. One can generalize about the ligands for the several families of CAMs:

- Most cadherins prefer to bind themselves, so they promote the adhesion of like cells. These **homophilic interactions** (association of like receptors on two cells) require Ca^{2+}.
- Most IgCAMs bind other cell-surface adhesion proteins. These **heterophilic interactions** (association of unlike receptors) may occur between the same or different cell types.
- Selectins bind anionic polysaccharides like those on mucins. Generally, such interactions bind together two different types of cells.
- Integrins stand apart because they bind a variety of ligands: matrix macromolecules, such as fibronectin (see Fig. 29.14) and laminin (see Fig. 39.9); soluble proteins, such as fibrinogen in blood; and adhesion proteins on other cells, including IgCAMs and one cadherin.

Third Principle of Adhesion

Cells modulate adhesion by controlling the surface density, state of aggregation, and state of activation of their adhesion receptors. Surface density reflects not only the level of synthesis but also the partitioning of adhesion molecules between the plasma membrane and intracellular storage compartments. For example, endothelial cells express P-selectin constitutively but store it internally in membranes of cytoplasmic vesicles. When inflammatory cytokines activate endothelial cells, these vesicles fuse with the plasma membrane, exposing P-selectin on the cell surface, where it binds white blood cells (Fig. 30.13). The importance of surface density is illustrated by an experiment with mixtures of cells that express different levels of the same cadherin. Over time, they sort from each other, with the more adherent cells forming a cluster surrounded by the less-adherent cells (Fig. 30.2). Such differential expression of cadherin determines the position of the oocyte in *Drosophila* egg follicles. A variety of extracellular stimuli activate intracellular signaling pathways in lymphocytes, platelets, and other cells. These pathways enhance or inhibit the ligand-binding activity of integrins already located on the cell surface.

Fourth Principle of Adhesion

The rates of ligand binding and dissociation are important determinants of cellular adhesion. Many cell-surface adhesion proteins (including IgCAMs, cadherins, integrins, and selectins) bind their ligands weakly compared with other specific macromolecular interactions, such as antigens and antibodies, hormones and receptors, or transcription factors and DNA. The measured dissociation equilibrium constants for adhesion receptors are in the range of 1 to 100 μM, reflecting high rate constants ($>1\ s^{-1}$) for dissociation of ligand. This actually makes good biological sense. Rapidly reversible interactions allow white blood cells to roll along the endothelium of blood vessels (Fig. 30.13). Transient adhesion also allows fibroblasts to migrate through connective tissue. In contrast, the interactions of cells in epithelia and muscle appear to be more stable, perhaps owing to multiple weak interactions between clustered adhesion proteins cooperating to stabilize adherens junctions and desmosomes (see Fig. 31.8). The combined strength of these bonds is said to increase the "avidity" of the interaction.

Fifth Principle of Adhesion

Many adhesion receptors interact with the cytoskeleton inside the cell. Adapter proteins link cadherins and integrins to actin filaments or intermediate filaments. These interactions provide mechanical continuity from cell to cell in muscles and epithelia, allowing them to transmit forces and resist mechanical disruption.

Sixth Principle of Adhesion

Association of ligands with adhesion receptors activates intracellular signal transduction pathways, leading to changes in gene expression, cellular differentiation, secretion, motility, receptor activation, and cell division. Signaling through adhesion receptors allows cells to adjust their behavior based on physical interactions with the surrounding matrix or cells.

Identification and Characterization of Adhesion Receptors

The ability of mixed populations of cells to sort into homogeneous aggregates revealed that mechanisms exist to bind like cells together. Similar experiments showed that cells also bind matrix macromolecules, such as fibronectin, laminin, collagen, and proteoglycans. Biochemical isolation of the responsible adhesion proteins was challenging, but progressed rapidly once monoclonal antibodies (see Fig. 28.10) that inhibit adhesion were available. These antibodies provided assays for purification of adhesion proteins and cloning of their complementary DNAs (cDNAs). With representatives from each family in hand, cloning cDNAs for related proteins was straightforward. Once the first crystal structures were determined, the structures of other family members could be modeled using the structures of shared functional domains.

Insights about the functions of adhesion receptors came in several steps. Localization of a protein on specific cells frequently provided the first clues. Typically, the expression of each protein is restricted to a subset of cells or to a specific time during embryonic development or both. Next, investigators used specific antibodies to test for the participation of the adhesion protein in cellular interactions in vitro or in tissues. Both human genetic diseases and mutations in mice and other organisms produce defects caused by the absence of adhesion proteins. Blistering skin diseases called pemphigus illustrate the serious consequences when pathological autoantibodies disrupt adhesion between skin cells expressing the antigen (see Fig. 31.8). In leukocyte adhesion deficiency, white blood cells lack the β_2-integrin that is required to bind the endothelial cells that line blood vessels. These defective white blood cells fail to bind to blood vessel walls or to migrate into connective tissue at sites of infection. Patients with a bleeding disorder called Bernard-Soulier syndrome lack one of the adhesion receptors for von Willebrand factor, a protein that promotes platelet aggregation. Loss of cadherins contributes to the spread of some cancer cells.

Immunoglobulin Family of Cell Adhesion Molecules

The IgCAM family includes hundreds of adhesion proteins that bind ligands on the surfaces of other cells. Some interactions are homophilic binding to the same IgCAM on another cell; others are heterophilic with different IgCAMs, integrins, other proteins or proteins with sialic acid. These interactions help specify interactions between different cell types in developing and mature animals.

IgCAMs have one to seven extracellular immunoglobulin domains anchored to the plasma membrane by a single transmembrane helix (Fig. 30.3 and Table 30.1). The compact immunoglobulin (Ig) domains consist of 90 to 115 residues folded into seven to nine β-strands in two sheets, usually stabilized by an intramolecular disulfide bond. The N- and C-termini are at opposite ends of the Ig domains, so they can form linear arrays. Some nervous system IgCAMs have three or four fibronectin III (FN-III) domains (see Fig. 13.13) between the Ig domains and the membrane anchor. Most IgCAMs consist of a single polypeptide, but others are multimeric, with two (CD8) or four subunits (see Fig. 27.8 for the T-cell receptor).

The C-terminal cytoplasmic tails of these receptors vary in sequence and binding sites for intracellular ligands. The cytoplasmic domains of the lymphocyte accessory receptors CD4 and CD8 bind protein tyrosine kinases required for cellular activation (see Fig. 27.8 for CD4). The cytoplasmic domains of neuronal IgCAMs bind PDZ domain proteins or the membrane skeleton (see Fig. 13.11). An adapter protein links IgCAMs in the nectin family to cytoplasmic actin filaments.

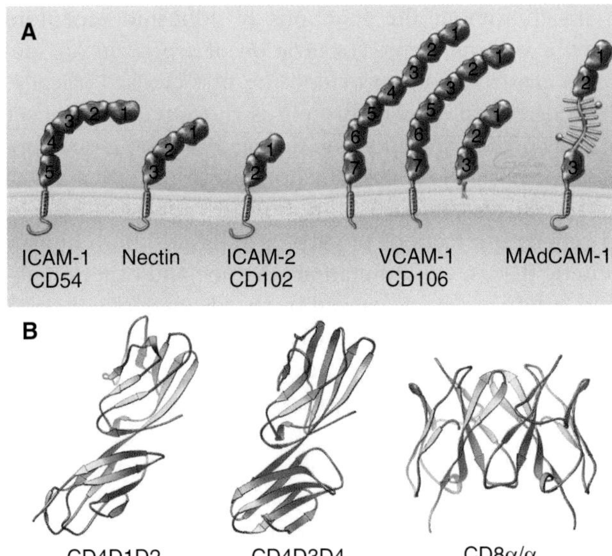

FIGURE 30.3 STRUCTURES OF REPRESENTATIVE IMMUNOGLOBULIN CELL ADHESION MOLECULES. A, Domain maps of examples with their common names and CD numbers. **B,** Ribbon diagrams of the lymphocyte coreceptors CD4 (domains 1 and 2 on the *left* and domains 3 and 4 in the *middle*) and CD8. ICAM, intercellular adhesion molecule; MAdCAM, mucosal addressin cell adhesion molecule; VCAM, vascular cell adhesion molecule. (For reference, see Protein Data Bank [PDB; www.rcsb.org] files 3CD4, 1CID, and 1CD8.)

Differentiated metazoan cells express IgCAMs selectively, especially during embryonic development, when they contribute to the specificity of cellular interactions required to form the organs. Neurons and glial cells express specific IgCAMs that guide the growth of neurites, mediate synapse formation, and promote the formation of myelin sheaths. In adults, interaction of endothelial cell intercellular adhesion molecule (ICAM)-1 with a white blood cell integrin is essential for adhesion and movement of the leukocytes into the connective tissue at sites of inflammation (Fig. 30.13).

Like other cell adhesion proteins, IgCAMs participate in signaling processes. Best understood are interactions of lymphocytes with antigen-presenting cells during immune responses. IgCAMs reinforce the interaction of T-cell receptors with major histocompatibility complex molecules carrying appropriate antigens on other cells (see Fig. 27.8). Although individual interactions are weak, the combination of specific (T-cell receptor) and nonspecific (CD2 and CD4) interactions with the target cell suffices to initiate signaling.

Cadherin Family of Adhesion Receptors

The complex architecture of organs in vertebrates relies on Ca^{2+}-dependent associations between the cells mediated by more than 80 cadherins (Table 30.2). Their name derives from "calcium-dependent adhesion" protein.

TABLE 30.1 Immunoglobulin Family of Cell Adhesion Molecules*

Examples	Structure	Extracellular Ligands	Intracellular Ligands	Expression	Functions
CD2[†]	2Ig-1TM	LFA-3 (CD58)		T cells	T-cell activation
CD4[†]	4Ig-1TM	Class II MHC	Lck	T cells, macrophages	T-cell coreceptor
CD8[†]	Dimer: 1Ig-1TM	Class I MHC	Lck	Cytotoxic; other T cells	T-cell coreceptor
C-CAM	4Ig-1TM	Self		Liver, intestine, WBCs	Cell adhesion
F11 (contactin)	6Ig-4FN-II-1TM			Neurons	Neurite fasciculation
ICAM-1[†]	5Ig-1TM	LFA-1, MAC-1		Epithelia, WBCs	WBC adhesion
ICAM-2	2Ig-1TM			Endothelium, WBCs	
L1 (Ng-CAM) [mouse]	6Ig-3FN-III-1TM	Self	Ankyrin	Neurons, Schwann cells	Adhesion
LFA-3 (CD58)	2Ig-1TM or GPI anchor	CD2		WBCs, epithelia, fibroblasts	Adhesion
MAG	5Ig-1TM	Neurons		Glial cells	Myelin formation
NCAM	5Ig-3FN-III-1TM	Self		Neurons, other cells	Adhesion
Neurofascin [chick]	6Ig-4FN-III-1TM	? Self	Ankyrin	Neurites	Bundling neurites
PECAM-1 (CD31)	6Ig-1TM	Self		Platelets, endothelium, myeloid cells	Adhesion
TAG-1	6Ig-4FN-III-GPI anchor	? Self		Neurons	Neuron migration
VCAM-1	7Ig-1TM	WBC α_4 integrin		Endothelium (regulated)	WBC/endothelium adhesion

CAM, cell adhesion molecule; CD, cellular differentiation antigen; FN-III, fibronectin-III domain; GPI, glycosylphosphatidylinositol; ICAM, intercellular adhesion molecule; Ig, immunoglobulin domain; Lck, nonreceptor tyrosine kinase; LFA, lymphocyte function–associated antigen; MAG, myelin-associated glycoprotein; MHC, major histocompatibility complex; NCAM, neural cell adhesion molecule; PECAM, platelet endothelial cell adhesion molecule; TAG, transient axonal glycoprotein; TM, transmembrane domain; VCAM, vascular cell adhesion molecule; WBC, white blood cells.

*Hundreds are known.

†Partial atomic structure.

TABLE 30.2 Cadherin Family of Adhesion Molecules*

Type (Examples)	Extracellular Ligands	Intracellular Ligands	Expression	Functions
Classic Cadherins				
E-cadherin	Self	Catenins (actin)	Epithelia, others	Adherens junctions
N-cadherin	Self	Catenins (actin)	Neurons, muscle, endothelium	Adhesion
R-cadherin	Self	Catenins (actin)	Retina, neurons	Adhesion
Desmosomal Cadherins				
Desmocollins	Self, desmogleins	Plakoglobin (desmoplakin, plackophilin, IF)	Epithelia	Desmosomes
Desmogleins	Self, desmocollins	Plakoglobin (desmoplakin, plackophilin, IF)	Epithelia, heart	Desmosomes
Atypical Cadherins				
T-cadherin	Self	None (GPI anchor)	Early embryos, neurons	Intercellular adhesion
Protocadherins				
α-, β-, and γ- Protocadherins	Self	Fyn tyrosine kinase (some)	Vertebrate neurons, other cells	Self-avoidance
Signaling Cadherins				
RET protooncogene	Self	None	Endocrine glands, neurons	Intercellular adhesion

GPI, glycosylphosphatidylinositol; IF, intermediate filament.
*More than 80 are known. Plakoglobin is also known as γ-catenin.

Genes for cadherin domains appeared in unicellular precursors of sponges, representing an early step toward the evolution of metazoan organisms.

Cadherins generally interact with like cadherins on the surfaces of other cells in a calcium-dependent fashion, but some cadherins form heterophilic interactions. Homophilic interactions of cadherins link epithelial and muscle cells to their neighbors, especially at specialized adhesive junctions called **adherens junctions** and **desmosomes** (Fig. 30.4; also see Fig. 31.8).

The structural hallmark of the cadherin family is the **CAD domain** (Figs. 30.5 and 30.6), which consists of approximately 110 residues folded into a sandwich of seven β-strands. This fold resembles Ig and FN-III domains, but appears to be a case of convergent evolution. N- and C-termini are on opposite ends of CAD domains. Ca^{2+} bound to three sites between adjacent CAD domains links them together into rigid rods. Without Ca^{2+}, the domains rotate freely around their linker peptides.

Classic cadherins interact head to head through their N-terminal CAD1 domains (Fig. 30.6D) forming strong "*trans*-interactions" with a partner on another cell. Cadherins on the same cell can interact laterally in "*cis*-interactions." Three-dimensional reconstructions of electron micrographs of desmosomes show both *trans*- and *cis*-interactions (Fig. 30.6B). Cadherins are synthesized with a small domain before the N-terminal interaction strand, which must be removed by proteolysis to allow binding to another cadherin.

A single α-helix links classic cadherins and desmosomal cadherins to the plasma membrane, but T-cadherin has a glycosylphosphatidylinositol (GPI) anchor (see Fig. 13.10). Cytoplasmic domains vary in size, sequence, and binding sites for associated proteins. The protooncogene RET is a cadherin with a cytoplasmic tyrosine kinase domain.

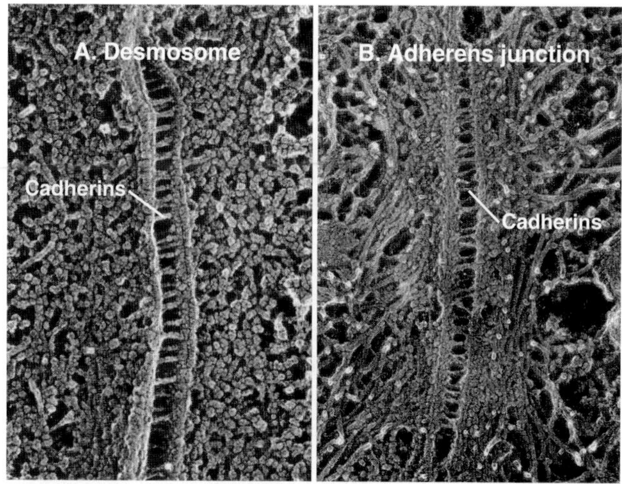

FIGURE 30.4 ELECTRON MICROGRAPHS OF ROD-LIKE CADHERINS CONNECTING THE PLASMA MEMBRANES OF ADJACENT CELLS. Intestinal epithelial cells were prepared by rapid freezing, freeze-fracture, deep etching, and rotary shadowing. **A,** Desmosome with associated intermediate filaments in the cytoplasm. **B,** Adherens junction with associated actin filaments. (Courtesy N. Hirokawa, University of Tokyo, Japan. Modified from Hirokawa N, Heuser J. Quick-freeze, deep-etch visualization of the cytoskeleton beneath surface differentiations of intestinal epithelial cells. *J Cell Biol.* 1981;91:399–409, copyright The Rockefeller University Press.)

Adapter proteins link the cytoplasmic domains of cadherins to actin filaments or intermediate filaments

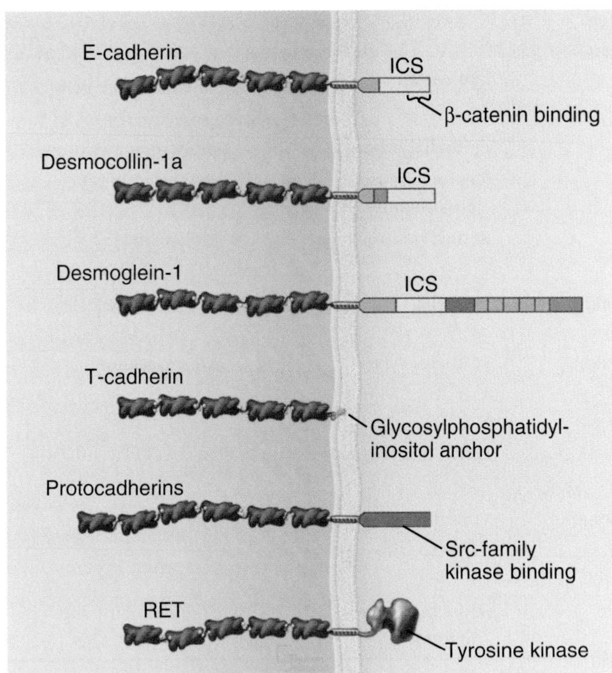

FIGURE 30.5 DOMAIN MAPS OF CADHERINS. All have extracellular CAD domains. A single transmembrane segment anchors five examples. A glycosylphosphatidylinositol (GPI) tail anchors T-cadherin. Intracellular cadherin segment (ICS) domains interact with catenin adapters to link E-cadherin to actin filaments and desmocollin and desmoglein to intermediate filaments.

to reinforce adhesion and maintain the physical integrity of tissues (see Fig. 31.8). The cytoplasmic tails of classic cadherins bind along the entire length of the adapter protein **β-catenin** (catenin is "link" in Greek), a long, twisted coil of 36 short α-helices (see Fig. 13.10F). Monomers of **α-catenin** link β-catenin to actin filaments, an interaction strengthened by tension. α-Catenin dimers also stabilize actin filaments. The more complicated cytoplasmic domains of desmosomal cadherins (desmocollins and desmogleins) interact with **γ-catenin** (a relative of β-catenin also called plakoglobin) and desmoplakin. Desmoplakin links these cadherins to keratin intermediate filaments (see Fig. 31.8B). The tails of some cadherins interact with formins, proteins that nucleate and elongate actin filaments (see Fig. 33.14).

Signaling by Cadherins and Catenins

In addition to helping with the mechanical sorting of embryonic cells, cadherins produce signals that influence cellular proliferation, migration, and differentiation. For example, interactions between cadherins on epithelial cells result in **"contact inhibition"** of both growth and motility (see Figs. 41.3 and 41.11). Rho-family guanosine triphosphatases (GTPases) (see Fig. 33.19) mediate contact inhibition of movements. The GTPase Rho stimulates contraction at the contact site, while the GTPase Rac drives protrusion of the side of the cell facing away from the contact site. Both cadherins and Eph receptor tyrosine kinases participate in this reaction to cell-cell contact. Engagement of cadherins at sites of contact also stops cellular proliferation through the Hippo signaling pathway of protein kinases. This pathway inhibits expression of genes required for cell cycle progression such as cyclin-dependent kinases (see Fig. 41.3). The mechanism involves phosphorylation of a transcription factor, excluding it from the nucleus.

Contact inhibition of growth and motility suppresses the spread of cancer cells, so mutations disabling E-cadherin can contribute to the transition from benign to invasive malignant tumors. For example, genetic defects in E-cadherin predispose people to stomach cancer. The oncogenic tyrosine kinase Src (see Box 27.5) phosphorylates both E-cadherin and β-catenin. This phosphorylation is associated with loss of adhesion of epithelial cells, suggesting one way in which transformation might alter cellular adhesion.

In addition to linking cadherins to actin filaments, β-catenin is an active component in the Wnt signal transduction pathway that regulates gene expression during differentiation of embryonic cells (Fig. 30.7). Extracellular signaling proteins called **Wnts** regulate the concentration of β-catenin available to regulate gene expression. The pathway was discovered in *Drosophila*, where it helps to determine the polarity of segments in early embryos. Wnts were named from the original *Drosophila* gene *Wingless* and the mouse protooncogene *Int-1*. The human genome encodes 29 Wnts.

Most β-catenin in cells is bound to cadherins, but a second pool exchanges between the cytoplasm and the nucleus, where it recruits transcription factors to regulate the expression of genes for cellular proliferation and tissue differentiation. Cells synthesize β-catenin continuously, but it turns over rapidly in resting cells, so little accumulates in the nucleus. A cytoplasmic complex controls degradation of β-catenin. The complex consists of two kinases—glycogen synthase kinase (GSK) and casein kinase-1α—and the product of the **APC** gene (defective in patients with familial adenomatous polyposis coli, giving rise to multiple precancerous polyps in the large intestine). Phosphorylated β-catenin is ubiquitylated and degraded by proteasomes.

Wnts suppress the degradation of β-catenin by binding to seven-helix receptors and another class of receptors in the plasma membrane. Several steps downstream in an incompletely characterized pathway, the Wnt signal inhibits GSK and casein kinase-1α. Inhibition of the kinases stops proteolysis of β-catenin, so ongoing synthesis of β-catenin raises the concentration that is free to accumulate in the nucleus. Stem cell proliferation is one of many developmental events influenced by Wnt signaling and adhesion by cadherins (see Box 41.2). Wnt binding its receptors also activates a parallel "noncanonical pathway" involving Rho-family GTPases that control the cytoskeleton during cell migration.

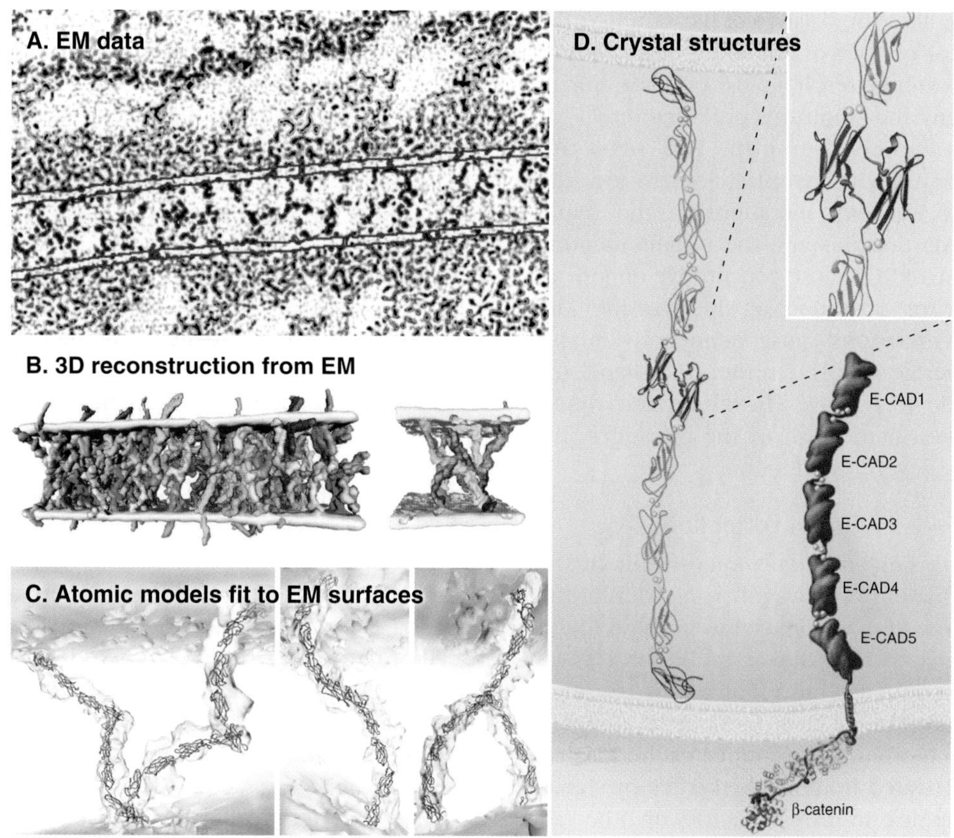

FIGURE 30.6 ADHESION BY CADHERINS. A, Electron micrograph of a thin section of a desmosome, colorized to emphasize the plasma membranes *(red)* and cadherin extracellular domains *(blue)*. **B,** Three-dimensional reconstructions of the plasma membrane and cadherin extracellular domains. **C,** Crystal structure of the C-cadherin extracellular domains fit into electron microscopic reconstructions of intercellular links between the cells. **D,** Ribbon diagrams of the crystal structure of a dimer of C-cadherin extracellular domains compared with the book icon for cadherins. The *inset* highlights the antiparallel intermolecular interaction of the two CAD1 domains mediated by flexible N-terminal peptides. A conserved tryptophan fits into a hydrophobic pocket of the partner CAD1 domain, forming the reciprocal interactions. Calcium ions *(blue)* stabilize interactions between CAD domains. (**A–C,** From He W, Cowin P, Stokes DL. Untangling desmosomal knots with electron tomography. *Science*. 2003;302:109–113, copyright the American Association for the Advancement of Science. **D,** For reference, see PDB file 1L3W and Boggen TJ, Murray J, Chappuis-Flament S, et al. C-Cadherin ectodomain structure and implications for cell adhesion mechanisms. *Science*. 2002;296:1308–1313.)

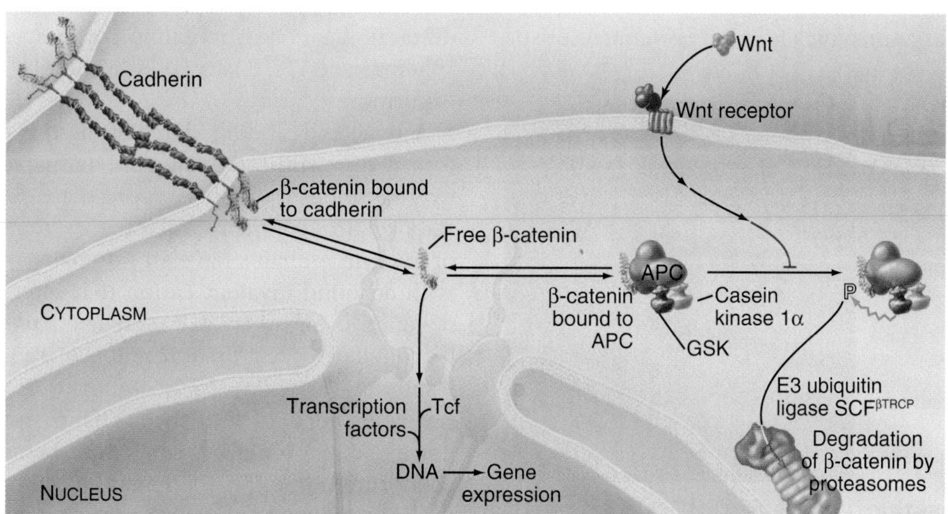

FIGURE 30.7 PARTICIPATION OF β-CATENIN IN GENE EXPRESSION. Free β-catenin is in equilibrium with binding sites on cadherins and APC (adenomatous polyposis coli) and may also enter the nucleus, where it combines with Tcf/LEF-1 transcription factors. When β-catenin concentrations are low in the nucleus, Tcf/LEF-1 represses gene expression, but the complex of β-catenin with Tcf/LEF-1 activates the expression of genes for cellular growth and differentiation. Synthesis of β-catenin is constant, so degradation determines its concentration in cytoplasm: glycogen synthetase kinase (GSK) and casein kinase-1α phosphorylate β-catenin bound to APC, triggering its ubiquitinylation and degradation. Extracellular Wnt acts through a seven-helix receptor and another receptor to promote gene expression by inhibiting the phosphorylation and degradation of β-catenin.

Mutations can alter the balance of β-catenin synthesis and turnover. Loss of APCs or mutations of phosphorylation sites on β-catenin result in excess β-catenin that enters the nucleus and stimulates proliferation.

The cadherin expressed from the **RET protooncogene** signals through its cytoplasmic tyrosine kinase domain (Fig. 30.5). Point mutations in the segment between the CAD domains and the plasma membrane or tyrosine kinase of RET cause constitutive dimerization of the receptor or activation of the tyrosine kinase or both. These mutations cause dominantly inherited cancers of endocrine glands. On the other hand, mutations that disable RET cause **Hirschsprung disease.** Autonomic nerves in the wall of the intestines fail to develop, causing severe dysfunction.

Roles of Cadherins in Organ Formation

Differential expression and regulation of cadherins help guide organ formation during embryonic development (Fig. 30.8). Cells with matching cadherins bind together and exclude cells without those cadherins (or other appropriate adhesion receptors), although the mechanism is more complicated than differential affinities of cadherins for each other. For example, cadherins can be activated or inactivated from inside the cell by signaling pathways in response to growth factors or other adhesion proteins. In other situations, IgCAMs facilitate the assembly of cadherins in adhesive junctions.

All cells of early embryos express several different cadherins, but as soon as the embryo forms three germ layers, the ectoderm on the outside surface expresses E-cadherin. In its absence, embryos die. Subsequently, when ectoderm folds inward to form the neural tube, those cells switch to expressing N-cadherin. Neurons use protocadherins to avoid making synapses with themselves (see below). Later in development, cells in specialized organs typically express characteristic cadherins, such as those in osteoblasts (OB-cadherin), kidney (K-cadherin), and muscle (M-cadherin). A giant-sized cadherin links sensory stereocilia on the hair cells in the inner ear. These tip links pull open ion channels when the stereocilia move in response to sound waves.

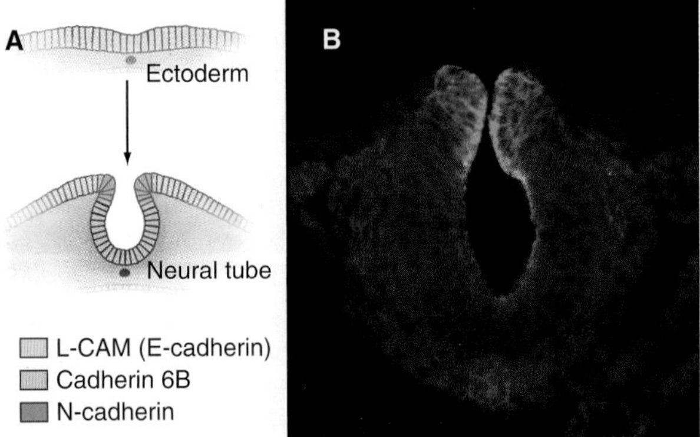

FIGURE 30.8 RESTRICTED EXPRESSION OF CADHERINS DURING FORMATION OF THE NEURAL TUBE. A, Distribution of three cadherins before and after the neural tube forms. **B,** Fluorescent antibody staining reveals the selective expression of cadherin 6B *(green)* and N-cadherin *(red)* in the neural tube of a developing chick embryo. (Courtesy M. Takeichi, Kyoto University, Japan.)

Integrin Family of Adhesion Receptors

Integrins are the main cellular receptors for ECM molecules (Table 30.3), but some integrins bind adhesion molecules on other cells. Their genes arose early in metazoan evolution, so they are present in sponges and corals that branched early in the evolution of animals (see Fig. 2.8). Fibroblasts and white blood cells use integrins to adhere to fibronectin and collagen as they move through the ECM. Integrins bind epithelial and muscle cells to laminin in the basal lamina, providing the physical attachments necessary to transmit internal forces to the matrix and to resist external forces (see Fig. 31.8). These interactions generate signals that control cell growth and structure. When defects in small blood vessels need repair, integrins allow platelets to adhere to basement membrane collagen and to each other via plasma fibrinogen (see Fig. 30.14). Together, these interactions are essential for tissue development and integrity in multicellular organisms. Genetic losses of integrin function result in several human diseases.

Structure of Integrins

Integrins are heterodimers of two transmembrane polypeptides called α- and β-chains, which both contribute to ligand-binding specificity (Fig. 30.9). Vertebrate cells use a combinatorial strategy to establish their integrin repertoire by selectively expressing a subset of 18 different α-chains and eight β-chains. These chains combine to form at least 24 different kinds of dimers that bind different ligands. Alternative messenger RNA (mRNA) splicing (see Fig. 11.6) adds to the diversity of integrin isoforms.

The ligand-binding domains of the α- and β-chains form a globular head connected to the plasma membrane by 16-nm legs (Fig. 30.9). More than 25 disulfide bonds stabilize these domains. All integrin β-chains and a subset of integrin α-chains have an I domain (inserted domain) with a bound divalent cation that interacts with acidic residues of ligands. All α-chains have an N-terminal β-propeller domain similar to a trimeric G protein β-subunit (see Fig. 25.9). Interaction of the α-chain propeller domain with the β-chain I domain holds the integrin dimer together. Single transmembrane segments anchor both integrin chains to the cell membrane. Short ($\alpha \leq 77$ residues; $\beta = 40$ to 60 residues, except $\beta_4 = 1000$ residues) C-terminal cytoplasmic tails contribute to efficient heterodimer assembly.

The I domains and the β-propeller bind at least two sites on ligands. For instance, integrin $\alpha_5\beta_1$ binds two

TABLE 30.3 Integrin Family of Cell Adhesion Molecules*

Examples	Structure	Extracellular Ligands	Me^{2+}	Intracellular Ligands	Expression	Function
Fibronectin receptors	$\alpha_5\beta_1$, others	Fibronectin	Ca	Talin, paxillin	Fibroblasts, other cells	Cell-matrix adhesion
GPIIb/GPIIIa	$\alpha IIb\beta_3$	Fibrinogen, von Willebrand factor	Ca	Talin, paxillin	Platelets	Platelet aggregation
Laminin receptor	$\alpha_6\beta_1$, $\alpha_7\beta_1$	Laminin	Yes	Talin, paxillin	Epithelia, muscle	Cell-matrix adhesion
LFA-1 (CD11/CD18)	$\alpha_L\beta_2$	Ig-CAM-1, -2, -3	Mg	Talin, paxillin	All WBCs	WBC/endothelium adhesion
MAC-1	$\alpha_M\beta_2$	Ig-CAM-1, fibrinogen	Yes	Talin, paxillin	WBCs except lymphocytes	WBC/endothelium adhesion
Vitronectin receptor	$\alpha_V\beta_3$	Vitronectin, fibronectin	Ca	Talin, paxillin	Endothelium, smooth muscle, others	
VLA-4	$\alpha_2\beta_1$	Collagen, laminin	Mg	Talin, paxillin	WBCs, epithelium, endothelium	WBC/matrix adhesion

CD, cellular differentiation antigen; GP, glycoprotein; ICAM, intercellular adhesion molecule; LFA, lymphocyte function-associated antigen; Me^{2+}, divalent cation dependence; VLA, very late antigen; WBC, white blood cell.
*Twenty-four are known.

sites on fibronectin: an arginine-glycine-aspartic acid (RGD) sequence on a surface loop of FN-III domain 10 and a secondary site on the adjacent FN-III domain 9 (see Fig. 29.14). Both sites are required for binding, so simple RGD peptides can compete fibronectin from the integrin. Integrin binding sites of some ligands are on separate polypeptide chains. The RDD binding site for integrin $\alpha_1\beta_1$ is on three different polypeptide chains of the type IV collagen triple helix.

Binding of extracellular ligands (outside-in signaling) and intracellular ligands (inside-out signaling) influences the conformation and activity of integrins (Fig. 30.10). Without bound ligands integrins bend over on closely spaced legs held together by interactions of the transmembrane helices. This closed state has a low affinity for extracellular ligands owing to an occluded binding site. Ligand binding inside or outside the cell favors an open state with widely spaced, extended legs holding the head above the membrane. In this conformation the exposed ligand-binding site has the highest affinity for extracellular ligands.

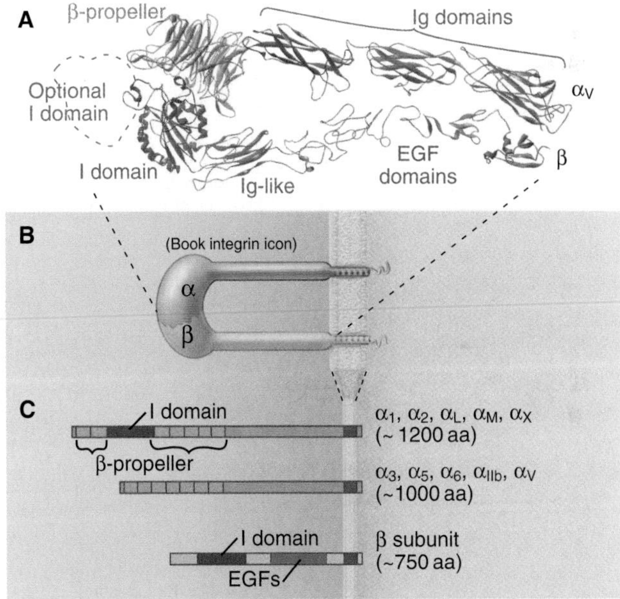

FIGURE 30.9 INTEGRIN ARCHITECTURE. A, Ribbon diagram of integrin $\alpha_V\beta_3$ based on a crystal structure of the extracellular domain. The I domain is inserted into the sequence of an immunoglobulin-like domain. **B,** Integrin icon used throughout this book. **C,** Domain models of integrin polypeptides. Both α-chains and β-chains have single transmembrane segments and cytoplasmic tails that vary in length. All β-chains and some α-chains have an I domain *(red)* that binds a divalent cation and participates in ligand binding. The seven blades of the α-chain β-propeller domains are shown in *orange*. The α-chain I domain, if present, is inserted between the second and third of the seven blades of its propeller domain. (**A,** Based on an atomic model. For reference, see PDB file 1JV2 and Xiong JP, Stehle T, Diefenbach B, et al. Crystal structure of the extracellular segment of integrin $\alpha_V\beta_3$. *Science*. 2001;294:339–345. **C,** Modified from Kuhn K, Eble J. The structural basis of integrin-ligand interactions. *Trends Cell Biol*. 1994;4:256–261.)

Extracellular Ligands

Integrins are more promiscuous than most adhesion receptors, as some bind to several protein ligands, and many matrix molecules bind to more than one integrin. For example, fibronectin binds to at least nine different integrins, and both laminin and von Willebrand factor bind at least five different integrins. This promiscuity reflects common motifs in the ligands. Approximately one-third of matrix ligands for integrins involve the sequence motif RGD or other simple sequences in otherwise unrelated proteins.

Even in the open state (legs apart), integrins generally have a low affinity for extracellular ligands. For example, the micromolar K_d for integrin $\alpha_5\beta_1$ binding fibronectin

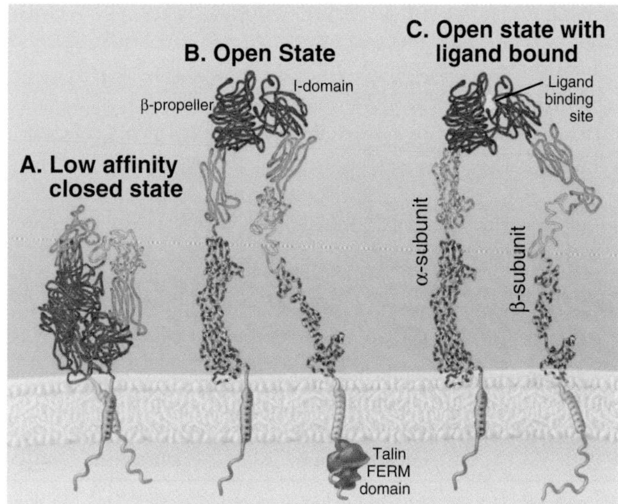

FIGURE 30.10 CONFORMATIONAL STATES OF INTEGRINS. Drawings based on atomic models derived from crystal structures and electron microscopy. The bent closed conformation observed in all crystal structures is inactive. Binding of either an extracellular ligand to the head or an activated signal transduction protein such as talin to the cytoplasmic domains can favor the open active state. (Modified from Xiao T, Takagi J, Coller BS, et al. Structural basis for allostery in integrins and binding to fibrinogen-mimetic therapeutics. *Nature*. 2004;432:59–67.)

results in rapid association and dissociation, allowing cells to adjust their grip on fibronectin in the matrix as they move through connective tissue. Nonadhesive RGD proteins, such as tenascin (see Fig. 29.16), may modulate these interactions by competing with fibronectin and other ligands for binding integrins.

Intracellular Ligands

Cytoplasmic tails of integrins interact directly or indirectly with a remarkable variety of signaling and structural proteins (Fig. 30.11). These interactions are best understood at **focal contacts,** specialized sites where integrins cluster together to transduce transmembrane signals and link actin filaments across the plasma membrane to the ECM. The adapter proteins **talin** and **vinculin** link the cytoplasmic domains of β-integrins to actin filaments at the ends of stress fibers. **Paxillin** links integrins to signaling proteins, forming a scaffold for Src family tyrosine kinases (see Fig. 25.3) and **focal adhesion kinase** (an essential tyrosine kinase).

Several types of integrins associate laterally, in the plane of the bilayer, with other transmembrane proteins. The best characterized is CD47 (also called integrin-associated protein), an IgCAM with five transmembrane

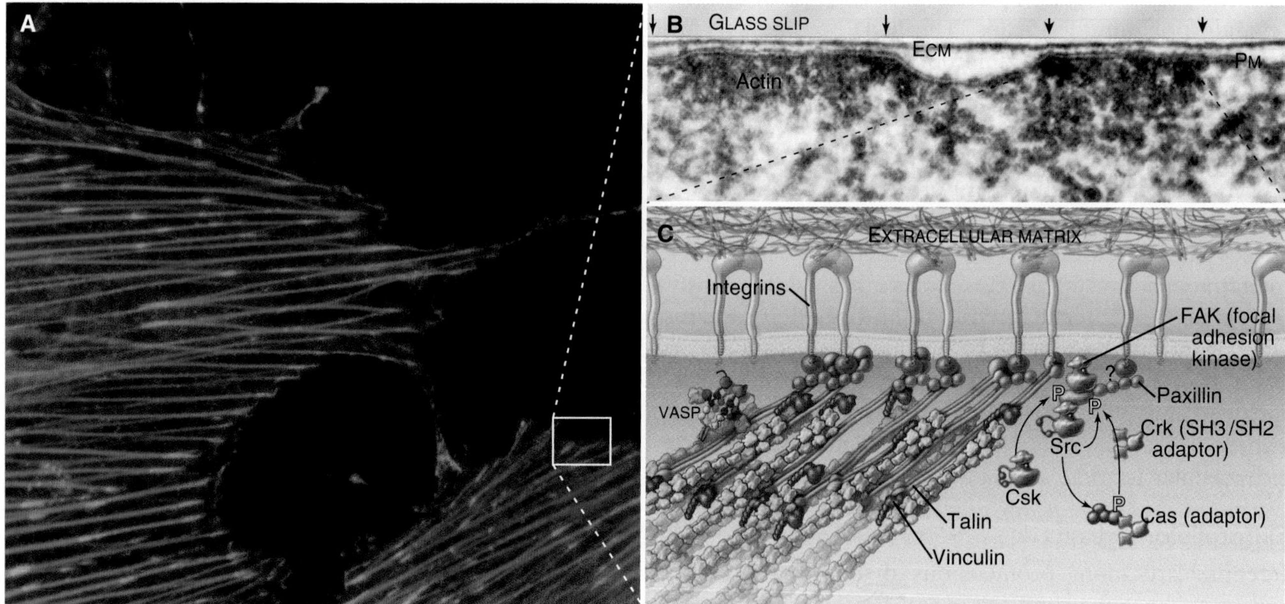

FIGURE 30.11 FOCAL CONTACTS OF EPITHELIAL CELLS WITH THE EXTRACELLULAR MATRIX (ECM). A, Fluorescence micrograph of parts of two vertebrate tissue culture cells with focal contacts labeled with a fluorescent antibody to phosphotyrosine *(orange)*. Actin filament stress fibers are stained *green* with phalloidin. **B,** Electron micrograph of a thin section of two focal contacts showing fine connections to the ECM deposited on the surface of the glass coverslip and cross-sections of actin filaments in the cytoplasm. This HeLa (Henrietta Lacks) cell was grown on a glass coverslip, fixed, and cut perpendicular to the substrate. **C,** Drawing of the interactions of some of the proteins concentrated on the cytoplasmic face of the membrane at focal contacts. For clarity, the actin filament interactions *(left)* are shown separately from some signaling proteins *(right)*. The rod-shaped dimeric protein talin interacts with the cytoplasmic domains of β-integrins and actin filaments. Vinculin interacts with membrane phospholipids, actin filaments, and talin. An unidentified protein (the question mark) links the adapter protein paxillin to integrins. Paxillin anchors tyrosine kinases (FAK and Src) and, after phosphorylation, the adapter proteins Crk and Cas. (**A,** Courtesy K. Burridge, University of North Carolina, Chapel Hill. **B,** Courtesy Pamela Maupin, Johns Hopkins University, Baltimore, MD. From Maupin P, Pollard TD. Improved preservation and staining of HeLa cell actin filaments. *J Cell Biol*. 1983;96:51–62, copyright the Rockefeller University Press. **C,** For reference, see Turner C. Paxillin and focal adhesion signaling. *Nat Cell Biol*. 2000;2:E231–E236; and Critchley DR. Focal adhesions—the cytoskeletal connection. *Curr Opin Cell Biol*. 2000;12:133–139.)

segments. Binding of the adhesive glycoprotein, thrombospondin, to the extracellular Ig-like domain of CD47 generates a transmembrane signal through trimeric G-proteins that contributes to neutrophil and platelet activation.

Outside-in Signaling From Integrins

Integrin binding to matrix ligands initiates signals that modify cellular adhesion, locomotion, and gene expression, with the responses depending on the particular integrin and cell. Extracellular ligands stabilize the open state with wide spacing of the cytoplasmic domains. Presumably this physical change influences the activities of signal transduction proteins associated with the cytoplasmic domains, but the details are not known. Within seconds, the cytoplasmic tyrosine kinases shown in Fig. 30.11 phosphorylate several focal adhesion proteins, including paxillin, tensin, and focal adhesion kinase, which has a central role in transducing these signals. Within a minute, some cells raise their cytoplasmic Ca^{2+} concentration high enough to initiate many calcium-dependent processes (see Chapter 26).

Over a period of minutes, ligand binding to integrins also activates Rho-family GTPases that stimulate actin assembly (see Fig. 33.19) and spreading of the cell on ligand-coated surfaces. Other Rho-family GTPases drive contraction of the trailing edge of moving cells (see Fig. 38.6). Integrins cluster together in small "focal complexes" at the leading edge and grow into mature focal contacts (Fig. 30.11A), also called *focal adhesions,* which anchor actin filament stress fibers to the cell membrane. Contraction of stress fibers applies tension to the focal contacts, which remain stationary as the cell advances past them. Rapid rearrangements of the linker proteins between the integrins and actin serve as a molecular clutch to transmit forces, even as actin assemble and disassembles. A Ca^{2+}-mediated signal inactivates obsolete attachments at the rear of the cell.

The adhesiveness of a cell for its substrate (a function of integrin density on the cell, ligand density on the substratum, and their affinity) determines the rate of movement. The maximum rate occurs at intermediate adhesiveness. Rapid association and dissociation of integrins on matrix ligands allow cells to rearrange their hold on the matrix as they move.

After several hours of integrin engagement, activation of the Ras/mitogen-activated protein kinase pathway (see Fig. 27.6) turns on expression of selected genes. These changes in gene expression contribute to cellular differentiation during development. Integrins allow cells to include the ECM as a signaling input along with other stimuli operating through different receptors.

Inside-Out Signaling to Integrins

Cells fine-tune their interactions with matrix molecules by regulating the activity of cell-surface integrins. For example, integrins on white blood cells (Fig. 30.13) and platelets (Fig. 30.14) require "inside-out" activation before they can bind their extracellular ligands. Integrin activation also regulates cellular interactions during development. Cytoplasmic proteins, talin and kindlins, activate integrins by binding the cytoplasmic tail of the β-integrin and separating the two transmembrane domains. One pathway downstream from seven-helix receptors uses a membrane-bound GTPase and an adapter protein to bring together talin and the β-integrin.

Some cells can mobilize a reserve pool of integrins stored in cytoplasmic vesicles within minutes. For example, chemoattractants stimulate white blood cells to fuse storage vesicles containing integrins with the plasma membrane (Fig. 30.13). Both intracellular and extracellular ligands can cluster integrins in focal complexes and focal adhesions and increase their activities.

Biological Functions of Integrins

With the exception of red blood cells, integrins are present in the plasma membranes of most animal cells. Experiments with neutralizing antibodies, genetic diseases, and experimental gene disruptions revealed the functions of integrin isoforms. For example, many vertebrate cells express β_1 and β_3 integrins for adhesion to the ECM, so integrin antibodies inhibit cell migration and embryonic development by competing with fibronectin. Like null mutations in the fibronectin gene (see Fig. 29.14), homozygous disruption of the integrin α_4 or α_5 genes is lethal during development. Only white blood cells express β_2-integrins, which they use to bind endothelial cells lining the walls of blood vessels.

Other integrins bind adhesion molecules on other cells. For example, mouse sperm bind integrins on the egg membrane during fertilization. Integrins cooperate with adhesion receptors of the IgCAM, mucin, and selectin families to facilitate the adhesion of white blood cells to endothelial cells at sites of inflammation (Fig. 30.13). Other cells supplement the functions of integrins with structurally distinct matrix adhesion proteins, such as muscle dystroglycans and platelet GPIb-IX-V.

Integrins also participate in the decision of cells to undergo apoptosis, programmed cell death (see Chapter 46). Normal epithelial cells require anchorage to the basal lamina by β_4-integrins to grow and divide. When forced to live in suspension or when dissociated from the matrix by RGD peptides, these cells arrest in the G_1 phase of the cell cycle (see Chapter 41) and eventually undergo apoptosis. Loss of contact with the basal lamina may contribute to the terminal differentiation and death of cells in the upper levels of stratified epithelia, such as skin (see Figs. 35.6 and 40.1). Epithelial cancers typically lose this integrin-mediated, anchorage dependence for growth, which is one of the normal limitations on uncontrolled proliferation in inappropriate locations.

Snake venoms contain small, monomeric RGD proteins that inhibit blood clotting by competing with fibrinogen for binding the integrins that activated platelets use for aggregation. These "disintegrins" are potential inhibitors of the pathological thrombosis that contributes to heart attacks and strokes. Both small-molecule and antibody antagonists for integrins are now used as clinical treatments for heart attacks and stroke.

Selectin Family of Adhesion Receptors

White blood cells and platelets use three selectin proteins to interact with vascular endothelial cells and each other. In lymph nodes or at sites of inflammation, selectins snare circulating white blood cells, allowing them to roll over the surface of endothelial cells and eventually to exit the blood (Fig. 30.13). Selectins (Table 30.4) also contribute to adhesion in other systems, including the initial binding of early mammalian embryos to the wall of the mother's uterus.

The defining feature of selectins is a calcium-dependent lectin domain (Fig. 30.12) that binds *O*-linked sulfated oligosaccharides containing sialic acid and fucose. The lectin domain sits at the end of a rod-shaped projection composed of complement regulatory domains that is anchored to the plasma membrane by a single transmembrane sequence.

Natural ligands for selectins are mucin-like glycoproteins expressed on endothelial and white blood cells. Specific binding to mucins requires selectins to interact with both the oligosaccharide and mucin protein. The affinity is low (millimolar K_ds), and not highly selective among oligosaccharides. Interaction with the mucin protein is less well understood, but one or more sulfated tyrosine residues on the leukocyte mucin called P-selectin glycoprotein ligand (PSGL)-1 participate in binding P-selectin.

Bonds between selectins and their mucin ligands have high tensile strength (withstanding forces greater than 100 piconewtons [pN]) but form and dissociate rapidly,

TABLE 30.4 Selectin Family (Lec-CAM) of Cell Adhesion Molecules

Examples	Structure	Extracellular Ligands	Me^{2+}	Expression	Functions
E-selectin (CD62E, ELAM-1)	Lectin-EGF-6CR-1TM	L-selectin	Ca	Endothelium (regulated)	WBC-endothelium adhesion
L-selectin (CD62L, gp90M)	Lectin-EGF-2CR-1TM	E-selectin, mucins	Ca	Lymphocytes, other WBCs	WBC-endothelium adhesion
P-selectin (CD62P, GMP-140)	Lectin-EGF-9CR-1TM	Mucins	Ca	Endothelium, platelets	WBC-endothelium adhesion

CD, cellular differentiation antigen; CR, complement regulatory domain; EGF, epidermal growth factor; ELAM-1, endothelial-leukocyte adhesion molecule 1; GMP-140, granule membrane protein 140; gp, glycoprotein; Me^{2+}, divalent cation dependence; TM, transmembrane domain; WBC, white blood cell.

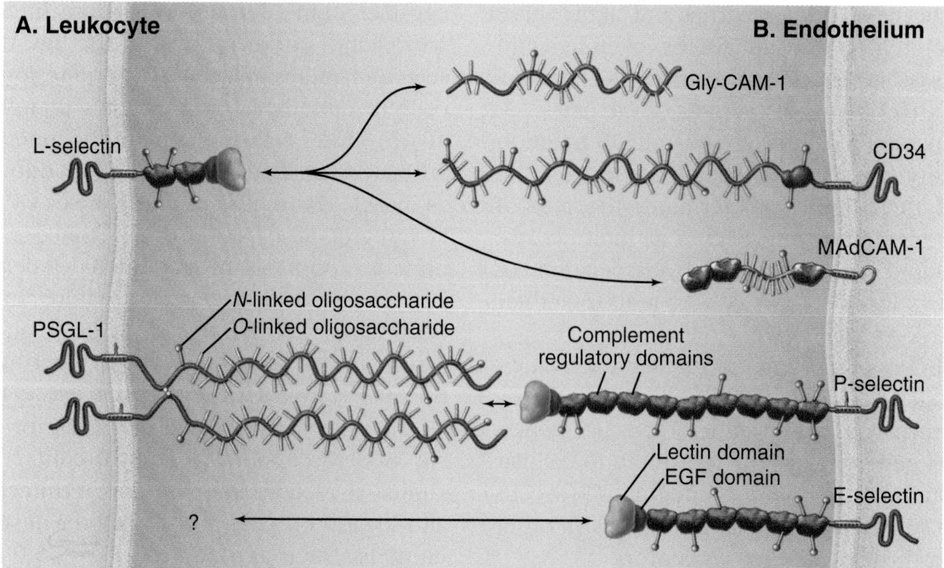

FIGURE 30.12 STRUCTURE OF SELECTINS AND THEIR MUCIN LIGANDS. Domain architecture of selectins and mucins exposed on the surfaces of leukocytes **(A)** and endothelial cells **(B)**. Complement regulatory domains of the selectins are shown in *red*. EGF, epidermal growth factor; MAdCAM, mucosal addressin cell adhesion molecule; PSGL, P-selectin glycoprotein ligand. (Modified from Rosen SD, Bertozzi CR. The selectins and their ligands. *Curr Opin Cell Biol*. 1994;6:663–673.)

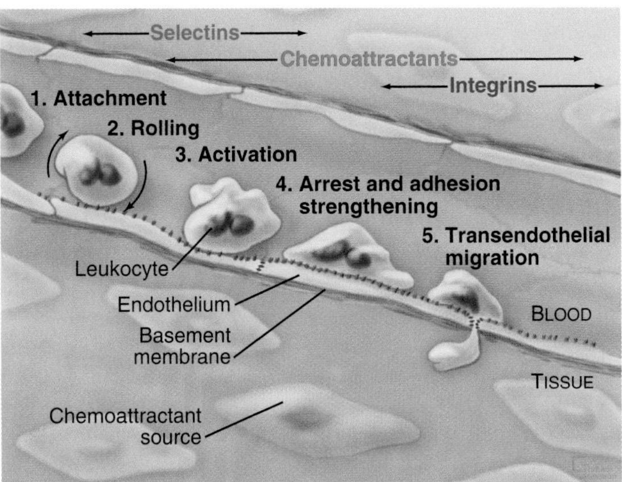

FIGURE 30.13 MIGRATION OF A NEUTROPHIL FROM THE BLOOD TO THE CONNECTIVE TISSUE. Endothelial cells exposed to inflammatory agents like histamine move selectins to their surfaces and snare mucins on neutrophils flowing in the bloodstream *(1)*. As a neutrophil rolls along the surface *(2)*, chemotactic factors and engagement of mucins activate their integrins *(3)*, causing the neutrophil to bind tightly to immunoglobulin cell adhesion molecules (IgCAMs) on the endothelium *(4)*. The neutrophil then migrates between the endothelial cells into the connective tissue *(5)*. (For reference, see Springer T. Traffic signals for lymphocyte and leukocyte emigration: the multi-step paradigm. *Cell*. 1994;76:301–314.)

on a second time scale. Low forces on these bonds prolong their lifetimes modestly by altering the conformation of the selectin, whereas high forces promote dissociation. Consequently, few selectin-mucin bonds are required to tether white blood cells to the endothelium, whereas the brief lifetime of the bonds allows blood flow to propel the cells with a rolling motion over the surface of the endothelium (Fig. 30.13).

Engagement of selectins with mucins stimulates signals in both cells that activate integrins and promote adhesion. The process resembles T-cell activation and includes Src-family kinases and adapter proteins with tyrosine phosphorylation sites. However, the links from the cytoplasmic domains of selectins and mucins to the signaling proteins is not established.

Inflammatory mediators regulate selectins in multiple ways. Activation of endothelial cells with histamine or platelets with thrombin causes vesicles storing P-selectin to fuse with the plasma membrane, exposing the selectins on the cell surface. Various inflammatory agents stimulate endothelial cells to synthesize E-selectin and P-selectin. Activation of white blood cells increases the affinity of L-selectin for mucins and later leads to its proteolytic release from the cell surface.

Mucins

The extracellular segments of mucins are rich in serine and threonine, which are heavily modified with acidic oligosaccharide chains (Fig. 30.12). Because of their strong negative charge, these proteins extend like rods up to 50 nm from the cell surface. Mucins on endothelial cells or white blood cells interact with complementary selectins on the other cell type. Endothelial mucin CD34 interacts with white blood cell L-selectin, whereas endothelial P-selectin interacts with white blood cell PSGL-1 mucin. These interactions depend on anchoring of the cytoplasmic domain of PSGL-1 to the actin cytoskeleton. Other mucins are displayed on the surface of or secreted by epithelia lining the respiratory and gastrointestinal tracks.

Other Adhesion Receptors

Table 30.5 lists a variety of adhesion receptors that fall outside the five main families. See Fig. 25.6 for the CD45 phosphatase, Fig. 31.3 for claudins and Fig. 31.7 for connexins.

Galactosyltransferase

The enzyme galactosyltransferase is also an adhesion receptor. This enzyme is a resident protein in the Golgi apparatus where it glycosylates proteins (see Chapter 21). However, the mRNA for galactosyltransferase has two alternative initiation sites, one of which adds 13 amino acids to the cytoplasmic, N-terminus of this transmembrane protein. The longer enzyme moves to the cell surface rather than being retained in the Golgi apparatus. On the cell surface, the enzyme can bind oligosaccharides that terminate in *N*-acetylglucosamine, which is found on both cell surface and matrix proteins. The complex of transferase and ligand oligosaccharide is stable, because the galactose-nucleotide substrate added to the oligosaccharide in the Golgi apparatus is not available outside the cell to complete the reaction.

During fertilization, a surface galactosyltransferase mediates the initial contact of mouse sperm with the matrix surrounding the egg (called the zona pellucida). This association induces secretion of the contents of the sperm acrosomal vesicle, including an enzyme that destroys the transferase binding site on the matrix so that the sperm can proceed through the zona to fuse with the egg. The enzyme is present on the surface of many cells that migrate during embryogenesis and may contribute to their interactions with the matrix.

Adhesion Receptors With Leucine-Rich Repeats (GPIb-IX-V)

The platelet receptor for the adhesive glycoprotein called von Willebrand factor (Fig. 30.14) is a disulfide-bonded complex of four transmembrane polypeptides: GPIbα, GPIbβ, GPIX, and GPV. Leucine-rich repeats at the end of a long stalk bind von Willebrand factor (see Fig. 28.6 for other receptors with leucine-rich repeats).

TABLE 30.5 Other Cell Adhesion Molecules

Examples	Structure	Extracellular Ligands	Me^{2+}	Intracellular Ligands	Expression	Functions
CD44	Link protein-1TM	Hyaluronan		Ankyrin	Lymphocytes	Adhesion to endothelium
CD44E	Link protein-HS/CS-1TM	Fibronectin, hyaluronan	No		Many epithelial cells	Adhesion to matrix
Claudins	Four TM	Claudins	No	ZO-1, ZO-2, ZO-3, cingulin	Epithelia	Tight junctions
Connexins	Multispan, hexamer	Self	No	ZO-1	Epithelia, muscle, nerve	Gap junctions
Dystroglycans	Multisubunit, TM	Laminin, agrin	Ca	Dystrophin	Muscle	Adhesion, synapse formation
Galactosyl-transferase	Galactose transferase-1TM	*N*-acetylglucosamine	No	? Actin filaments	Many cells, including sperm	Adhesion to cells and matrix
Glypican	4HS-GPI anchor	Fibronectin,	No	None	Endothelium, smooth muscle, epithelium	Adhesion to matrix
GPIB-IX	7 leucine-rich-1TM	von Willebrand factor		Filamin, actin	Platelets, endothelium	Adhesion
LCA (CD45)	50 kD-1TM-tyrosine phosphatase				WBCs	Tyrosine phosphatase
Mucins (CD34, CD43)	Sialylated oligosaccharide-1TM	Selectins	No		Epithelia, leukocytes	Intercellular adhesion

CD, cellular differentiation antigen; CS, chondroitin sulfate; GPI, glycosylphosphatidylinositol; HS, heparan sulfate; LCA, leukocyte common antigen; Me^{2+}, divalent cation dependence; TM, transmembrane domain; WBC, white blood cell.

Platelets bind to von Willebrand factor to initiate the repair of damaged blood vessels. This interaction also generates an intracellular signal that enhances affinity of integrin $\alpha_{IIb}\beta_3$ for fibrinogen and reorganizes the cytoskeleton.

Dystroglycan/Sarcoglycan Complex

In muscles, a complex of transmembrane glycoproteins links a network of dystrophin and actin filaments on the inside of the plasma membrane to two proteins of the extracellular basal lamina, α_2 laminin and agrin (see Fig. 39.17). These protein associations stabilize the muscle plasma membrane from inside and outside, similar to the actin-spectrin network of red blood cells (see Fig. 13.11). Genetic defects or deficiencies in dystrophin, transmembrane linker proteins of the dystroglycan/sarcoglycan complex, or α_2-laminin cause muscular dystrophy in humans, most likely owing to the mechanical instability of the membrane, leading to cellular damage and eventual atrophy of the muscle. Chapter 39 provides details on their role in muscle function and disease.

Nonmuscle cells in other tissues express many of these proteins (or their homologs), where they may contribute to adhesion to the ECM. Some pathogens use the dystroglycan complex to bind their cellular targets. Arenavirus, the cause of Lassa fever, binds directly to α-dystroglycan, and the leprosy bacterium binds laminin-2.

Examples of Dynamic Adhesion

Adhesion of Leukocytes to Endothelial Cells

Movement of white blood cells from blood into connective tissue illustrates how cells integrate the activities of selectins, mucins, integrins, IgCAMs, and chemoattractant receptors. Infection or inflammation in connective tissue attracts lymphocytes as well as neutrophils and monocytes, the main phagocytes circulating in blood (see Fig. 28.7). Blood cell precursors use a similar mechanism to enter lymphoid organs and bone marrow.

In the absence of inflammation, neutrophils flow over endothelial cells without binding, because the appropriate pairs of adhesion molecules are not exposed or activated. Infection or other inflammation in nearby tissues causes neutrophils to bind to the vascular endothelium and to move out of the blood into the tissue. Neutrophils adhere to the endothelium in three sequential but overlapping steps (Fig. 30.13):

1. Locally generated inflammatory molecules, including histamine (secreted by mast cells), bind to seven-helix receptors on endothelial cells and stimulate fusion of cytoplasmic vesicles (called Weibel-Palade bodies) with the plasma membrane. This exposes P-selectin, formerly stored in the vesicle membranes, on the cell surface facing the blood. Selectins bind mucins that are constitutively exposed on the surface of neutrophils, tethering them to the surface. The bonds form and break rapidly, allowing the neutrophil

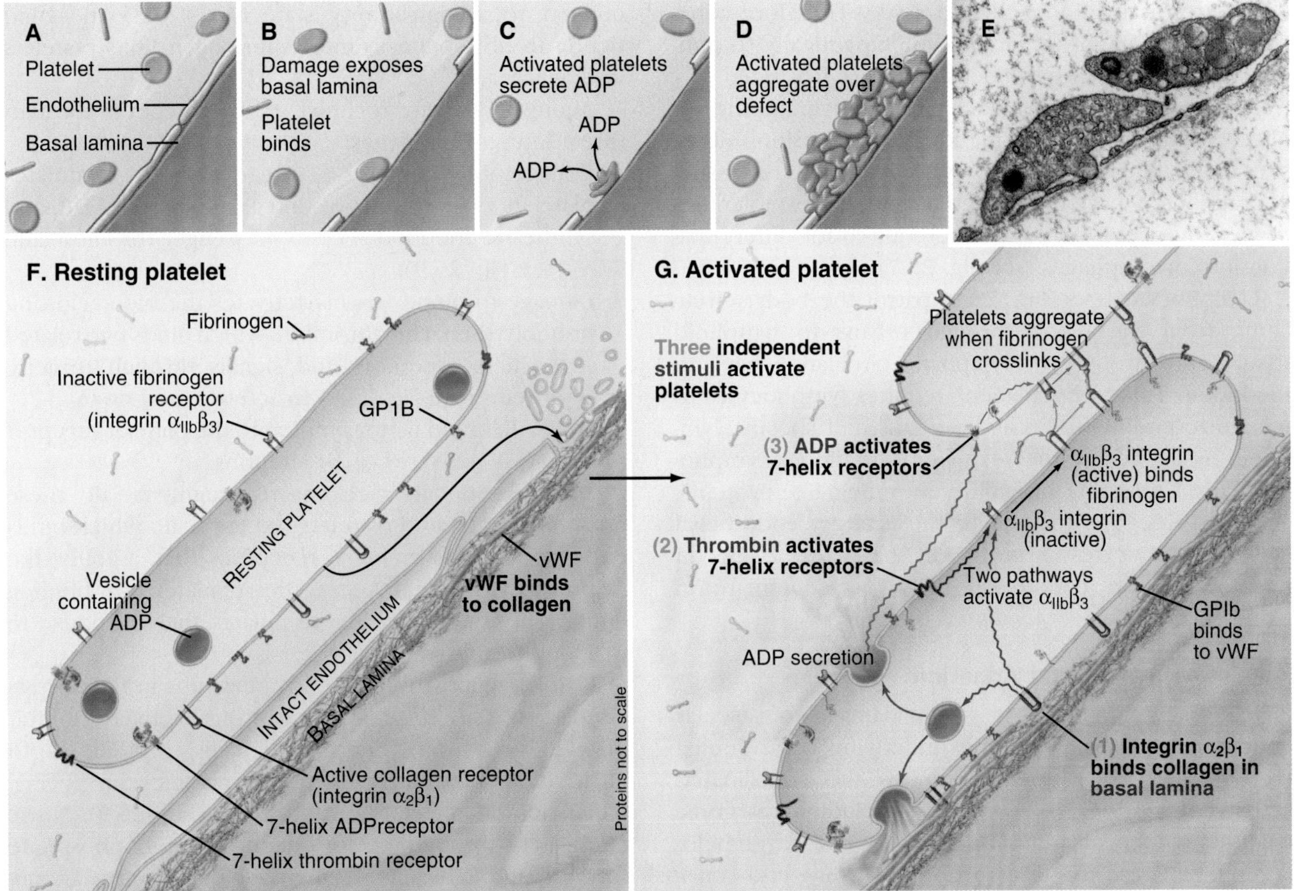

FIGURE 30.14 PLATELET ACTIVATION AND AGGREGATION AT A DEFECT IN THE ENDOTHELIUM. A–D, Steps in platelet activation and aggregation. **E,** Electron micrograph of a thin section of a platelet adhering to the basal lamina through a tiny defect in the endothelium. **F,** Resting platelets circulate in the blood without interacting with the intact endothelium lining the vessel. **G,** Platelets are activated in three ways. (1) Binding of $\alpha_2\beta_1$ integrins to collagen results in firm adhesion. Where the basal lamina is exposed, von Willebrand factor (vWF) binds the collagen. (2) Platelet GPIb-IX binds weakly to von Willebrand factor, allowing platelets to adhere to the exposed matrix. (3) Thrombin activates seven-helix receptors. These interactions stimulate secretion of adenosine diphosphate (ADP), which binds seven-helix receptors and activates the $\alpha_{IIb}\beta_3$ integrins; then $\alpha_{IIb}\beta_3$ integrins bind dimeric fibrinogen and aggregate platelets together. Platelet proteins are not to scale.

to roll along the surface of the endothelium at rates greater than 10 μm/s as the blood flow pushes them along.

2. Two signaling pathways (seven-helix receptors for chemotactic factors and P-selectin glycoprotein ligand [PSGL]) activate leukocyte integrins from inside the cell (Fig. 30.10). Activation of approximately 10% of the neutrophil integrins increases their affinity for their ligand by 200-fold, making the third step possible.
3. Activated integrins bind tightly to IgCAMs on the surface of endothelial cells, immobilizing the leukocyte despite the force of the blood flow. Within 2 minutes, the leukocyte opens the tight junctions between endothelial cells (see Chapter 31) and squeezes into connective tissue toward the source of the chemoattractant. The leukocyte and endothelial cells interact closely during this passage, because they share a self-associating IgCAM called platelet endothelial cell adhesion molecule (PECAM).

Defects in either the weak or strong interactions compromise the movement of leukocytes into connective tissue, increasing the risk of acute and chronic infections. One type of human leukocyte adhesion deficiency is caused by a genetic defect in fucose metabolism that interferes with the synthesis of a carbohydrate ligand on leukocytes that binds endothelial selectins. Cells cannot roll, so they fail to initiate the emigration process. A genetic deficiency of β_2-integrins causes a second type of leukocyte adhesion deficiency. White blood cells that lack β_2-integrins roll on the endothelium through the selectin mechanism but do not bind tightly enough to migrate out of the circulation. Consequently, these individuals are susceptible to bacterial infections.

On the other hand, neutrophils also generate reactive oxygen species that can damage tissues at sites of inflammation or at sites that are temporarily deprived of oxygen. Thus, movement of white blood cells into tissues contributes to damage that occurs when blood flow is restored to an ischemic tissue. Therefore adhesion proteins might be targeted therapeutically to mitigate damage after heart attacks or severe frostbite.

A similar mechanism and a partially overlapping set of receptors attract blood monocytes and eosinophils to sites of inflammation. Once they are in connective tissue, interactions of monocyte integrins with matrix molecules trigger the expression of genes required for differentiation into macrophages (see Fig. 28.7).

Lymphocytes (see Fig. 28.9) patrol the body, circulating from the blood through organs to lymphoid tissues and through the lymphatic circulation back to the blood. This "recirculation" requires lymphocytes to recognize endothelial cells in organs and specific lymphoid tissues where they exit from the blood. Lymphocytes use L-selectin, three different mucin-like proteins, and $\alpha_4\beta_2$ integrins to bind to these target endothelial cells. Lymphocytes from mice that lack L-selectin do not roll on endothelial cells or accumulate in lymph nodes.

Platelet Activation and Adhesion

Platelets aggregate at sites where damage to vascular endothelial cells exposes the underlying basal lamina (Fig. 30.14). This process requires the coordinated activity of a variety of receptors, including integrins, leucine-rich repeat adhesion proteins, and seven-helix receptors. These reactions prevent bleeding and bruising, but inappropriate activation of platelets produces clots in blood vessels, causing heart attacks and strokes. To understand the good effects and avert the bad, investigators have studied platelet activation and adhesion in great detail.

Resting platelets have a low tendency to aggregate, even though they circulate in a sea of ligands, including fibrinogen and the adhesive glycoprotein von Willebrand factor. Multiple mechanisms limit the reactivity of resting platelets. First, the major integrin, $\alpha_{IIb}\beta_3$, has a low affinity ($K_d \gg \mu M$) for its plasma ligand, fibrinogen. Similarly, the GPIb-IX-V complex has a low affinity for soluble von Willebrand factor. Third, the endothelium masks potential ligands, collagen, and von Willebrand factor in the basal lamina. The concentrations of soluble activators, such as adenosine diphosphate (ADP) and thrombin, are low under physiological conditions.

Damage to the endothelium usually initiates platelet activation by exposing platelets to von Willebrand factor and collagen in the basal lamina. Under conditions of high shear, GPIb-IX-V interacts strongly with von Willebrand factor bound to basal lamina collagen. This interaction transiently tethers platelets to the basal lamina and favors binding of integrin $\alpha_2\beta_1$ to collagen. Exposure to soluble agonists such as ADP or thrombin also activates platelets and promotes their aggregation. Within seconds of activation, platelet $\alpha_{IIb}\beta_3$ integrins convert to a high-affinity state ($K_d < \mu M$) and bind tightly to fibrinogen. Dimeric fibrinogen links platelets into aggregates.

Agonists activate platelet $\alpha_{IIb}\beta_3$ integrins through three different pathways:

1. Collagen binding to $\alpha_2\beta_1$ integrin directly stimulates platelets to activate $\alpha_{IIb}\beta_3$ integrins, secrete ADP, and synthesize the lipid second messenger thromboxane A_2 (see Fig. 26.9).
2. Damage to blood vessels activates the blood-clotting proteolytic enzyme thrombin, which binds two related seven-helix receptors and signals through trimeric G-proteins (see Fig. 25.9) to activate integrin $\alpha_{IIb}\beta_3$.
3. von Willebrand factor binding to the platelet receptor GPIb-IX-V activates $\alpha_{IIb}\beta_3$ integrins.

Two additional mechanisms augment all these responses. Activated platelets secrete ADP, which binds two types of seven-helix receptors that amplify the response to thrombin. Aggregation of platelets by binding dimeric fibrinogen further stimulates their response to ADP and thrombin.

Platelet aggregation is disadvantageous in the normal circulation, so several mechanisms actively inhibit platelet activation. Endothelial cells produce both nitric oxide and an eicosanoid, prostacyclin (PGI_2), which inhibit platelet activation (see Fig. 26.9). Nitric oxide acts through cyclic guanosine monophosphate (cGMP), and PGI_2 acts through cyclic adenosine monophosphate (cAMP; see Fig. 26.1). Antibodies and small molecule drugs that inhibit $\alpha_{IIb}\beta_3$ are used to treat heart attacks.

The most common human bleeding disorder is von Willebrand disease, caused by mutations in von Willebrand factor or its receptor, the GPIbα subunit of GPIb-IX-V. Some mutations reduce the concentration of the factor in blood or reduce the affinity of the factor for its receptor. Remarkably, mutations in either the factor or receptor that increase their affinity for each other also cause bleeding. These high-affinity interactions cause platelets to aggregate and be removed from the blood. Loss-of-function mutations in GPIbα cause the human bleeding disorder called Bernard-Soulier syndrome. Individuals with Glanzmann thrombasthenia bleed abnormally because $\alpha_{IIb}\beta_3$ integrin is absent or defective, and their platelets do not aggregate.

Self-Avoidance in the Nervous System

Surprisingly, neurons use adhesion proteins to avoid forming synapses with themselves by repelling axons and dendrites from the same cell (Fig. 30.15). Insects use a family of DSCAM1 IgCAMs for this self-avoidance. Alternative splicing of the pre-mRNA in each cell generates a subset of the many thousands of different DSCAM1 isoforms. These large IgCAMs have 10 Ig-domains and six FN-III domains that make homophilic interactions between three Ig-domains of each protein.

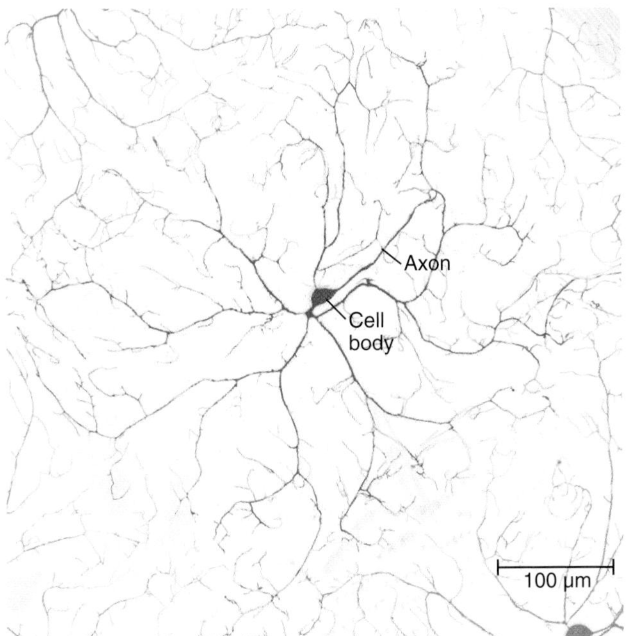

FIGURE 30.15 SELF-AVOIDANCE OF NEURITES OF CLASS IV SENSORY NEURONS OF A *DROSOPHILA* LARVA. Stack of fluorescence micrographs ~20 μm thick of neurons expressing a mouse transmembrane protein called CD8 tagged with GFP. False coloring is used to distinguish the cells. The central neuron is red. (Courtesy Sujoy Ganguly and Jonathan Howard, Yale University, New Haven, CT.)

Contacts between neurites with the same isoforms cause repulsion.

Self-avoidance of vertebrate neurons depends on a family of 48 cadherins called protocadherins. Alternative splicing of transcripts from three gene clusters produces a unique mixture of protocadherins on each neuron. These protocadherins form random dimers in the membrane and the extracellular domains bind homophilically to like protocadherins. When a process from a neuron contacts another part of itself, extensive interactions between like protocadherins somehow create a signal that repels the neurite. Weaker interactions between mixtures of protocadherins on different cells do not generate a repulsive signal and allow N-cadherins and other adhesion proteins to specify synaptic connections. A point mutation in one protocadherin gene is a common cause of human deafness and blindness.

SELECTED READINGS

Campbell ID, Humphries MJ. Integrin structure, activation, and interactions. *Cold Spring Harb Perspect Biol*. 2011;3:a004994.

Case LB, Waterman CM. Integration of actin dynamics and cell adhesion by a three-dimensional, mechanosensitive molecular clutch. *Nat Cell Biol*. 2015;17:955-963.

Chen WV, Maniatis T. Clustered protocadherins. *Development*. 2013;140:3297-3302.

Constantin B. Dystrophin complex functions as a scaffold for signalling proteins. *Biochim Biophys Acta*. 2014;1838:635-642.

Gérard C, Goldbeter A. Dynamics of the mammalian cell cycle in physiological and pathological conditions. *Wiley Interdiscip Rev Syst Biol Med*. 2016;8:140-156.

Gumbiner BM. Regulation of cadherin-mediated adhesion in morphogenesis. *Nat Rev Mol Cell Biol*. 2005;6:622-634.

Harwood A, Coates JC. A prehistory of cell adhesion. *Curr Opin Cell Biol*. 2004;16:470-476.

Hernández AR, Klein AM, Kirschner MW. Kinetic responses of β-catenin specify the sites of Wnt control. *Science*. 2012;338:1337-1340.

Livne A, Geiger B. The inner workings of stress fibers—from contractile machinery to focal adhesions and back. *J Cell Sci*. 2016;129:1293-1304.

Logan CY, Nusse R. The Wnt signaling pathway in development and disease. *Annu Rev Cell Dev Biol*. 2004;20:781-810.

McEver RP. Selectins: initiators of leucocyte adhesion and signalling at the vascular wall. *Cardiovasc Res*. 2015;107:331-339.

McEver RP, Zhu C. Rolling cell adhesion. *Annu Rev Cell Dev Biol*. 2010;26:363-396.

Roycroft A, Mayor R. Molecular basis of contact inhibition of locomotion. *Cell Mol Life Sci*. 2016;73:1119-1130.

Rosen SD. Ligands for L-selectin: Homing, inflammation, and beyond. *Annu Rev Immunol*. 2004;22:129-156.

Samanta D, Almo SC. Nectin family of cell-adhesion molecules: structural and molecular aspects of function and specificity. *Cell Mol Life Sci*. 2015;72:645-658.

Shattil SJ, Kim C, Ginsberg MH. The final steps of integrin activation: the end game. *Nat Rev Mol Cell Biol*. 2010;11:288-300.

Yu FX, Zhao B, Guan KL. Hippo Pathway in Organ Size Control, Tissue Homeostasis, and Cancer. *Cell*. 2015;163:811-828.

Zipursky SL, Grueber WB. The molecular basis of self-avoidance. *Annu Rev Neurosci*. 2013;36:547-568.

CHAPTER 31

Intercellular Junctions

The mechanical integrity of animal tissues such as epithelia, nerves, and muscles depends on the ability of the cells to interact with each other and the extracellular matrix. Plasma membrane specializations, called cellular junctions, mediate these interactions. Physical connections from the extracellular matrix or adjacent cells through these junctions and the associated cytoskeletal filaments inside cells impart mechanical strength to tissues.

Investigation of junctions began when microscopists and physiologists recognized that epithelial and muscle cells adhere to each other and the underlying extracellular matrix. They also discovered that some epithelia form a tight barrier between the luminal surface and the underlying tissue spaces. The physical basis of these interactions became clear during the 1960s, when electron micrographs of thin sections of vertebrate tissues revealed four types of intercellular junctions that connect the plasma membranes of adjacent cells (Table 31.1 and Fig. 31.1) and two types of junctions to bind to the extracellular matrix. Subsequent research established the molecular architecture of these junctions, each based on a different transmembrane protein:

Adherens junctions: Transmembrane proteins called cadherins (see Fig. 30.5) link neighboring cells and connect to actin filaments in the cytoplasm.

Desmosomes: Another type of cadherin links cells together and connects to cytoplasmic intermediate filaments.

Tight junctions: Transmembrane proteins called claudins join the plasma membranes of two cells to create a barrier that limits diffusion of ions and solutes between the cells and molecules between apical and basolateral domains of the plasma membrane.

Gap junctions: Transmembrane proteins called connexins form channels for small molecules to move between the cytoplasms of neighboring cells.

Hemidesmosomes: Integrins (see Fig. 30.9) connect cytoplasmic intermediate filaments to the basal lamina across the plasma membrane.

Focal adhesions: Integrins associated with cytoplasmic actin filaments adhere to the extracellular matrix.

Each tissue uses a selection of junctions suited to its physiological functions. Columnar epithelial cells in the intestine interact with their neighbors using all four types of intercellular junctions (Fig. 31.1B–D). Belt-like tight junctions and adherens junctions encircle the apex of the cell. Desmosomes and gap junctions form patch-like lateral connections between the cells. Hemidesmosomes anchor the cells to the basal lamina. Stratified epithelial cells in the skin (Fig. 31.1A) use desmosomes and intermediate filaments (Fig. 31.1B) to resist mechanical forces but also interact via claudins and adherens junctions. Desmosomes and adherens junctions link muscle cells to the surrounding basal lamina (see Fig. 29.17C). Gap junctions connect heart and smooth muscle cells, but not skeletal muscle cells. Most nerve cells communicate chemically, but some use gap junctions for electrical communication.

Invertebrate animals assemble junctions from homologous proteins but with different organization than the junctional complex of vertebrate epithelia. Insect epithelia have apical adherens junctions and more basal "septate junctions" built from claudins and cytoplasmic proteins with sequence homology to the tight junction proteins ZO-1 and ZO-2. Nematode epithelia have one type of junction with adherens functions and claudins.

Tight Junctions

Tight junctions form a belt-like adhesive seal that selectively limits the diffusion of water, ions, and larger solutes between epithelial cells (Fig. 31.2). This allows epithelia to separate the interior of the body from the

TABLE 31.1 Molecular Components of Cell-Cell and Cell-Matrix Junctions

Junction	Target Molecule	Adhesive Protein	Cytoplasmic Proteins	Cytoskeletal Filaments
Sealing of the Extracellular Space				
Tight junction	Claudin	Claudin	ZO-1, ZO-2, cingulin, spectrin	Actin
Communication Between Cells				
Gap junction	Connexin	Connexin	ZO-1, drebrin	Actin
Adhesion to Other Cells				
Zonula adherens	Cadherin	Cadherin	Catenins, plakoglobin	Actin
Desmosome	Desmoglein Desmocollin	Desmoglein Desmocollin	Plakoglobin, desmoplakin	Intermediate
Adhesion to the Extracellular Matrix				
Hemidesmosome	Laminin	Integrin	Plectin, BP 180	Intermediate
Focal contact	Fibronectin	Integrin	Talin, vinculin, α-actinin	Actin

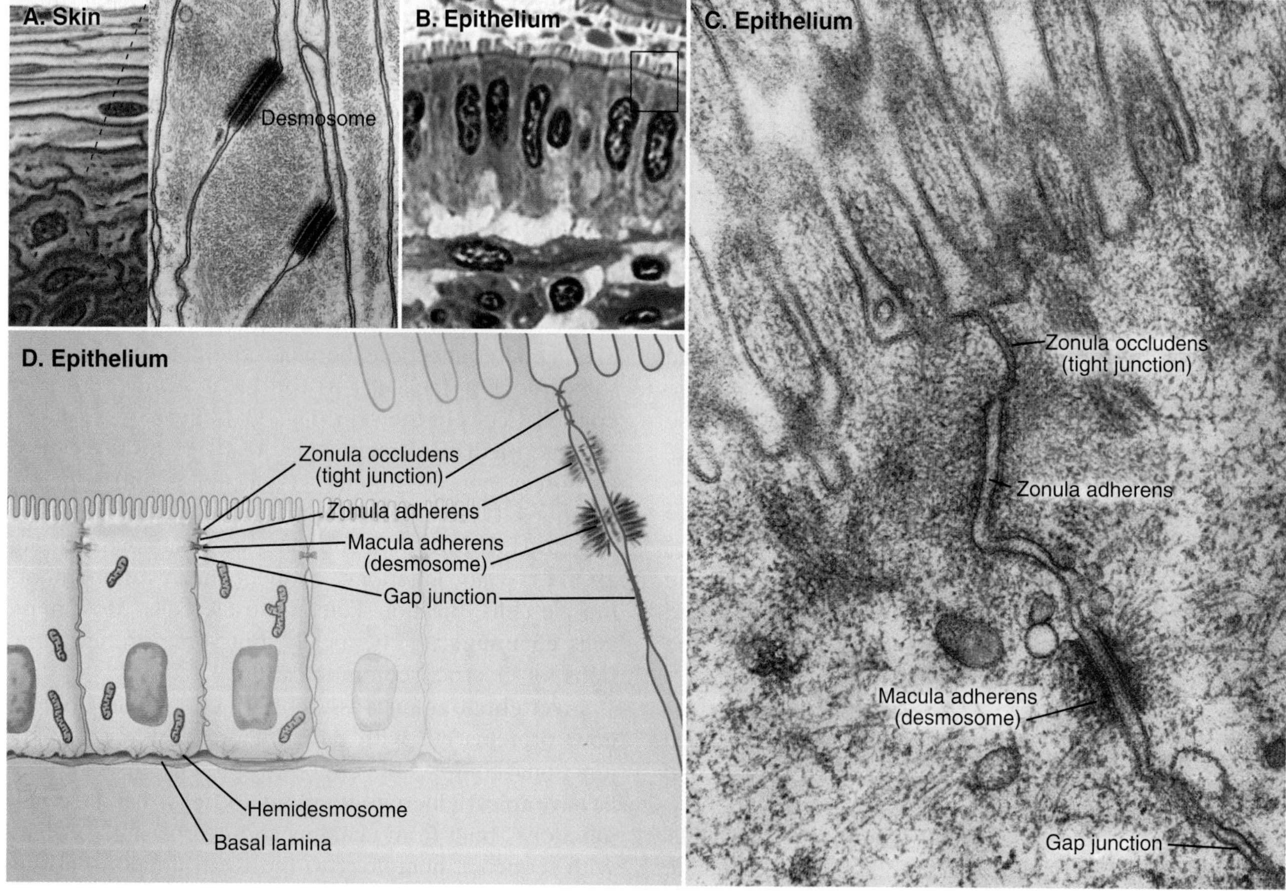

FIGURE 31.1 LIGHT AND ELECTRON MICROGRAPHS OF JUNCTIONS. A, Desmosomes. *Left,* Light micrograph of a section of skin showing numerous desmosomes as *pink dots* between the cells. *Right,* Electron micrograph of a thin section of skin showing desmosomes. **B,** Light micrograph of a section of intestinal epithelium stained with hematoxylin and eosin, showing the junctional complex (also called "terminal bars") as *bright pink dots* between the cells near their apex, just below the microvilli of the brush border. **C,** Electron micrograph of a thin section of intestinal epithelial cells, showing the junctional complex consisting of a belt-like tight junction (also called the zonula occludens), a belt-like adherens junction (also called the zonula adherens), and desmosomes (also called the macula adherens), all in their characteristic relation to each other. The circumferential tight junction seals the extracellular space. The zonula adherens is anchored to the actin cytoskeleton. Desmosomes are attached to cytoplasmic intermediate filaments. **D,** Drawing showing the position of the junctional complex in the cell and the locations of gap junctions, basal lamina, and hemidesmosomes. (**A,** Courtesy Don W. Fawcett, Harvard Medical School, Boston, MA. **C,** Courtesy Marilyn Farquhar, University of California, San Diego.)

external world. Tight junctions also define the boundary between the biochemically distinct **apical** and **basolateral domains** of the plasma membrane of polarized epithelial cells.

Tight junctions were first recognized in electron micrographs of thin sections as places where the plasma membranes of adjacent cells appear to fuse together in one or more contacts (Fig. 31.2). Freeze-fracture images revealed that these contacts correspond to continuous strands of intramembranous particles that form a branching network in the plane of the lipid bilayer.

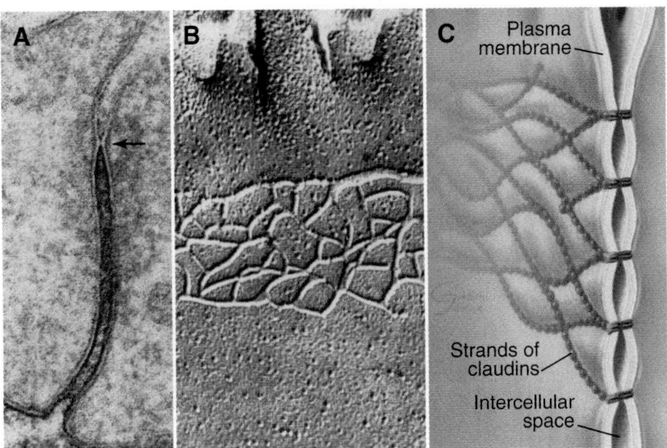

FIGURE 31.2 EPITHELIAL TIGHT JUNCTIONS. A, Electron micrograph of a thin section of endothelial cells, showing a point of contact between the plasma membranes at a tight junction *(arrow)*. **B,** Electron micrograph of a replica of a freeze-fractured cell. This method exposes proteins within the lipid bilayer and reveals strands aligned along the points of contact between the plasma membranes. **C,** Drawing showing the strands of transmembrane proteins at points of contact. (**A,** Courtesy George Palade, University of California, San Diego. **B,** Courtesy Don W. Fawcett, Harvard Medical School, Boston, MA.)

Transmembrane proteins forming the strands observed by freeze-fracture were difficult to identify until investigators found a monoclonal antibody that bound to the cytoplasmic side of the plasma membrane at tight junctions. They used this antibody to isolate an integral membrane protein and named it occludin. The amino acid sequence of occludin suggested four transmembrane strands and two hydrophobic extracellular loops. However, mice lacking their single occludin gene survive with normal tight junctions. Subsequently, these scientists discovered **claudins**, the main structural proteins of tight junction strands (Fig. 31.3A). Humans have a family of 27 homologous claudin genes that are expressed selectively in various tissues.

Claudins consist of four highly conserved transmembrane helices with two extracellular loops that fold into a small β-sheet and single helix. Close associations between claudins form barriers to diffusion in the plane of the membrane and *between the cells*.

- *The barrier in the membrane bilayer:* Intimate lateral interactions of claudins within the lipid bilayer (Fig. 31.3B) block diffusion of lipids and proteins in the plane of the membrane. This barrier separates different pumps, carriers, receptors and lipids in the apical and basolateral domains of the plasma membrane.
- *The barrier between cells:* The small extracellular domains of claudins interact with their neighbors in intramembrane strands and with claudins in similar strands on adjacent cells to make barriers interrupted by rows of pores. These pores block diffusion of solutes larger than approximately 1 nm in diameter, but selectively allow the passage of small cations or anions. Most tight junctions are more permeable to cations than to anions. Permeability in the two directions across the junction is identical.

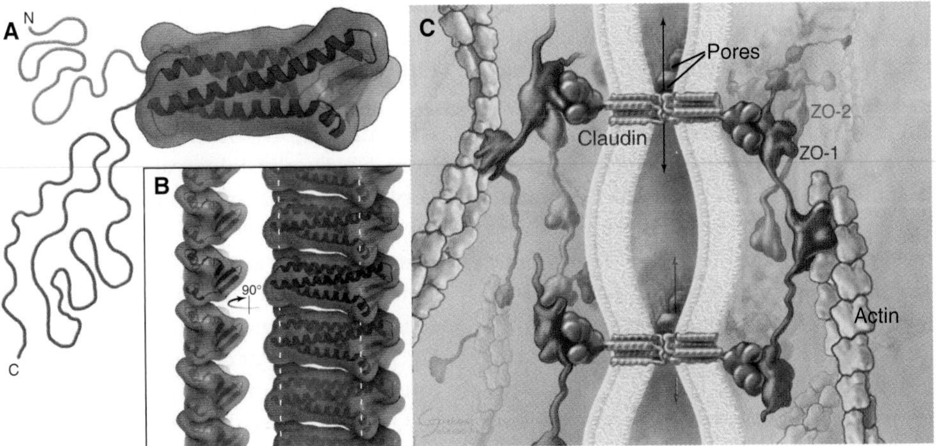

FIGURE 31.3 STRUCTURE OF TIGHT JUNCTIONS. A, Ribbon diagram of the transmembrane domains of claudin-15. **B,** Interactions between the molecules in crystals of claudin-15 that are thought to represent contacts in tight junctions. **C,** Model of tight junction structure with claudins linking the two membranes together and peripheral protein ZO-1 linking the cytoplasmic tail of claudins to actin filaments. (For reference, see Protein Data Bank [PDB; www.rcsb.org] file 4P70 and Suzuki H, Nishizawa T, Tani K, et al. Crystal structure of a claudin provides insight into the architecture of tight junctions. *Science*. 2014;344:304–307.)

Adapter proteins with PDZ protein-interaction domains (see Fig. 25.10) link tight junctions to the cytoskeleton. These adapters are called **ZO-1, ZO-2,** and **ZO-3**. They interact with the long C-terminal cytoplasmic tail of claudin and JAM (junctional adhesion molecule) in the membrane. In the cytoplasm they interact with actin filaments, a small guanosine triphosphatase (GTPase) and other proteins that regulate actin polymerization and the adapter protein cingulin. ZO-2 and cingulin are specific for tight junctions, but ZO-1 also associates with cadherins in adherens junctions and connexins in gap junctions.

These two barrier functions of tight junctions set up the two conditions required for many physiological processes. The actions of pumps, carriers, and channels located selectively in the apical or basolateral domains of the plasma membrane allow polarized cells to create different extracellular environments on the two sides of the epithelium. Maintaining these environments depends on the selective permeability across the epithelium through the extracellular pores of tight junctions. For example, tight junctions are essential for intestinal epithelial cells to take up nutrients from the lumen of the intestine and transport them into the extracellular space beneath the cells (see Fig. 17.2) and for other physiological processes (see Figs. 17.3 and 17-4). Restricting the diffusion of macromolecules across epithelia can regulate some types of signaling. For instance, airway epithelial cells release the growth factor heregulin from the apical surface, but restrict its receptor tyrosine kinase erbB2 (see Fig. 24.5B for a related receptor) to the basolateral surface. Thus, the receptor is activated only if the epithelium is damaged or the tight junctions compromised.

Sheets of epithelial cells vary by several orders of magnitude in the quality of the seal created by circumferential tight junctions between adjacent cells. The tightness of the barrier to diffusion of ions in the extracellular space determines the electrical resistance across the epithelium and depends on the claudin isoform and the number and continuity of the strands. Epithelia also appear to differ in the ability of water to flow through tight junctions. Extremely tight barriers with many strands and distinct claudin forms are found where epithelia must maintain high ion gradients, such as in the distal tubules of the kidney, where urine is concentrated. Leaky tight junctions with fewer strands and different claudins are found where ion gradients across epithelia are small but a barrier is required for large solutes, proteins, and leukocytes (eg, in most blood vessels).

Intracellular and extracellular factors regulate the transepithelial barrier established by tight junctions. These regulators include hormones such as vasopressin and aldosterone and cytokines such as tumor necrosis factor (see Fig. 24.9) that act through second messengers (eg, Ca^{2+} and cyclic adenosine monophosphate [cAMP]; see Chapter 26), and effectors (eg, protein kinases A and C; see Chapter 25). The mechanisms are not yet well understood, but posttranslational modifications of tight junctions might modulate their assembly. Tension on associated actin filaments may physically open passages through tight junctions. The metabolic state of the cell also influences tight junctions; depletion of adenosine triphosphate (ATP) causes tight junctions to leak without destroying the barrier between the apical and basolateral domains of the plasma membrane. White blood cells migrating across epithelia from the blood to the connective tissue, open tight junctions locally without disrupting the tight seal across the epithelium (see Fig. 30.13). A localized increase in cytoplasmic Ca^{2+} in the epithelial cells is required to open the tight junctions.

Several bacterial toxins affect the tight junction barrier. The ZO-toxin of *Vibrio cholerae* induces diarrhea by loosening tight junctions, independent of the classic cholera toxin, which induces secretion of salt and water. *Helicobacter pylori* injects a protein toxin into the cells lining the stomach. This toxin disrupts tight junctions, breaking the barrier that protects the underlying tissues and predisposing to ulcers.

Mutations of human claudin genes cause highly selective defects in epithelial barriers. One example is reduced ability of the kidney to reabsorb potassium (claudin-16). Another is deafness due to loss of ion gradients in the inner ear (claudin-14).

Gap Junctions

The idea that channels might couple cells arose relatively late, because electrophysiological experiments on nerves and skeletal muscles reinforced the widespread belief that cells were autonomous. However, nerve and skeletal muscle cells later turned out to be exceptions to the general principle that cells in animal tissues communicate with each other by gap junctions. Cells in plant tissues also communicate with each other, but they use direct cytoplasmic connections, called **plasmodesmata,** rather than gap junctions (Box 31.1 and Fig. 31.4).

The first convincing evidence for direct electrical communication between cells came around 1960 from electrophysiological experiments on the synapses between giant axons and the motor neurons that drive the flipper muscles of crayfish. These **electrical synapses** transmit action potentials (see Fig. 17.6) directly from one cell to the next without the delay required for secretion and reception of a chemical transmitter (see Fig. 17.10). Similar electrical junctions were subsequently found to connect heart muscle cells (see Fig. 39.18).

Over the next decade, physiologists used microelectrodes to establish that plasma membrane depolarization of one cell can be transmitted with little resistance to adjacent epithelial cells (Fig. 31.5B), although the amplitude of the response declined with distance. Similarly, fluorescent molecules, radioactive tracers, and essential

BOX 31.1 Plasmodesmata

Most cells in plant tissues maintain cytoplasmic continuity with their neighbors through plasmodesmata, channels across the cell wall lined by plasma membrane across the cell wall (Fig. 31.4). These connections form by incomplete cytokinesis, but secondary plasmodesmata can form independently. Plasmodesmata are essential for plant viability.

A narrow tubule of modified endoplasmic reticulum (ER) fills most of the pore and is linked to the surrounding plasma membrane by a number of proteins. The protein spokes connecting the ER and plasma membrane are not yet well characterized, but candidates include proteins that participate in membrane contact sites in animals and fungi such as synaptotagmins (see Chapter 21), synaptobrevin (see Chapter 21), lipid exchange proteins (see Fig. 20.17), junctophilins, and stromal interaction molecules (STIMs) (see Fig. 26.12).

Molecules smaller than about 1 kD diffuse freely through the narrow cylinder of cytoplasm in plasmodesmata, but larger molecules, even nucleic acids pass selectively through these channels. Constitutive diffusion of small molecules allows exchange of metabolites and hormones between cells. Regulated passage of larger molecules, including double-stranded RNAs and proteins such as transcription factors, allows developmental signals to move between cells and tissues. Specialized viral "movement proteins" allow some viruses to move their whole genomes between cells.

Permeability varies among tissues and with physiological states and developmental stages. For example, all cells in plant embryos are connected, whereas cells in some adult tissues are isolated. Reversible deposition of callose and other molecules in the cell wall surrounding plasmodesmata regulates their diameter and permeability. Actin filaments contribute to regulation of the pore size, but the mechanism and signals controlling permeability are not known.

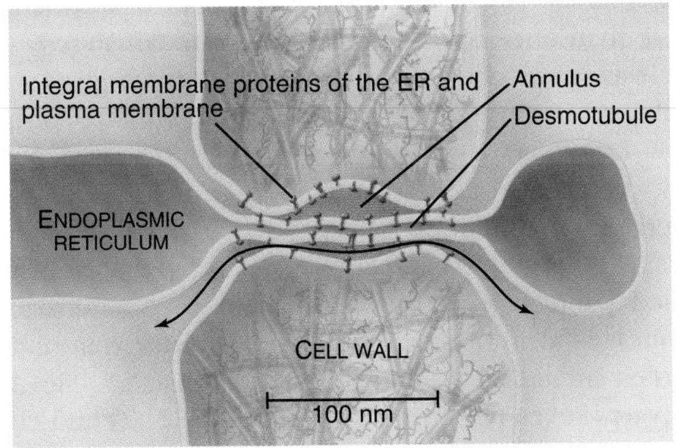

FIGURE 31.4 A PLASMODESMATA CONNECTING TWO PLANT CELLS. The plasma membrane is continuous between the two cells. The space between the tubule of endoplasmic reticulum and the surrounding plasma membrane allows molecules in the cytoplasm to move between the two cells. ER, endoplasmic reticulum.

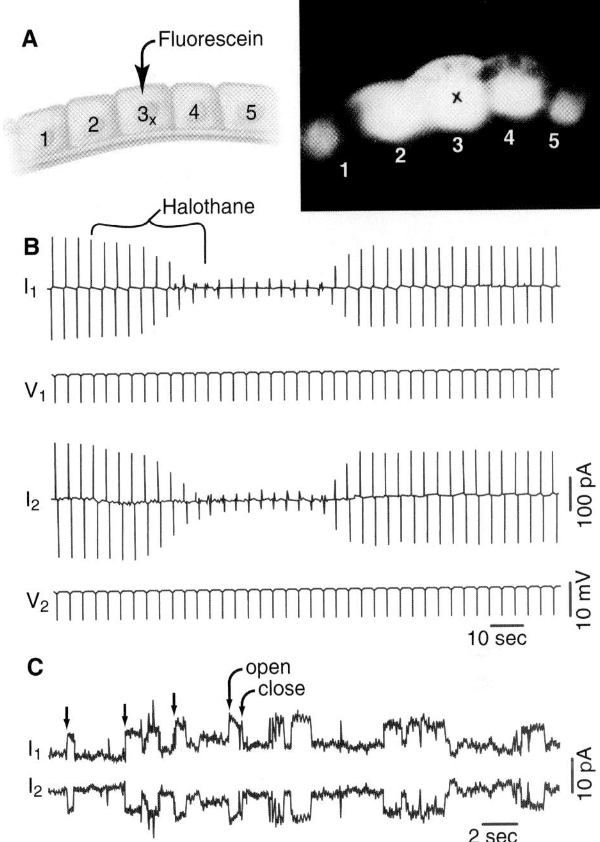

FIGURE 31.5 GAP JUNCTION PHYSIOLOGY. A, Drawing and fluorescence micrograph, showing the movement of a tracer dye between epithelial cells from the salivary gland of *Chironomus.* Cell 3 was injected with fluorescein (molecular weight: 330), which spread to adjacent cells via gap junctions. **B–C,** Electrical recordings from pairs of cells coupled by gap junctions. **B,** Two cells (1 and 2) were voltage-clamped (see the text that describes Fig. 17.6) and subjected alternately to small depolarizing voltage changes (V_1, V_2). Being electrically coupled, they responded with opposite currents (I_1, I_2). Transient exposure to the anesthetic halothane *(horizontal bar)* closes most of the channels, reducing the current in response to depolarization. **C,** When the cells are held at a constant depolarizing voltage in the presence of halothane, current records reveal the opening and closing of individual gap junction channels as opposite step changes in current. (**A,** From Lowenstein W. Junctional intercellular communication: the cell-to-cell membrane channel. *Physiol Rev*. 1981;61:829. **B** and **C,** From Eghbali B, Kessler JA, Spray DC. Expression of gap junction channels in communication-incompetent cells. *Proc Natl Acad Sci U S A*. 1990;87:1328–1331.)

nutrients can pass from the cytoplasm of one cell to the cytoplasm of neighboring cells.

Electron microscopy revealed that low-resistance communication between cells is associated with the presence of plasma membrane specializations that were called gap junctions owing to the regular 2- to 4-nm separation of the adjacent cell membranes (Fig. 31.6). Gap junctions are plaques composed of large *intercellular* channels that connect the cytoplasms of a pair of cells. These plaques exclude other transmembrane proteins and contain a few to thousands of channels. Half

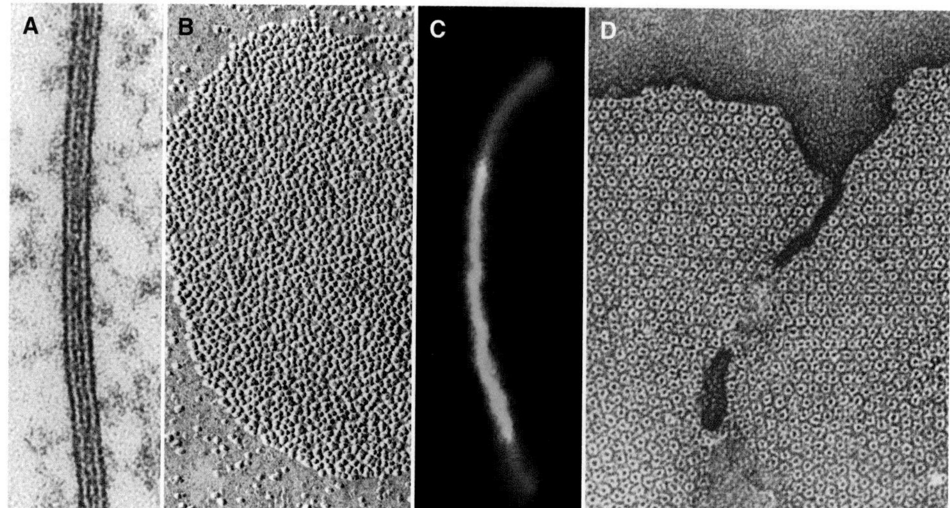

FIGURE 31.6 LIGHT AND ELECTRON MICROGRAPHS OF GAP JUNCTIONS. A, Thin section of embedded cells, showing the closely apposed membranes of adjacent cells separated by a gap of 2 nm. **B,** Replica of a freeze-fractured cell, showing an irregular array of particles exposed in the plane of the lipid bilayer. **C,** Fluorescence micrograph of a gap junction plaque *(red and green)* between cultured HeLa (Henrietta Lacks) cells expressing connexin-43 with a tetracysteine peptide tag. The cells were first exposed to a *green* fluorescent dye that binds tightly to the tetracysteine tag and then, after 4 hours of growth without the *green* dye, the same cells were incubated with a second *red* fluorescent dye that binds to the tetracysteine tag on newly synthesized connexin-43. The older central part of this plaque is *green*. The newer peripheral regions of the plaque are *red*. **D,** Negative staining of an isolated gap junction reveals the intercellular connexon channels packed together in a regular, two-dimensional array. Each connexon has a central channel filled with stain. (**A–B** and **D,** Courtesy Don W. Fawcett, Harvard Medical School, Boston, MA; from the work of N.B. Gilula, Scripps Research Institute, La Jolla, CA. **C,** Courtesy Mark Ellisman, University of California, San Diego and from Gaietta G, Deernick TJ, Adams SR, et al: Multicolor and electron microscopic imaging of connexin trafficking. *Science*. 2002;296:503–507.)

channels in each membrane called **connexons** are each formed from six protein subunits, named **connexins** (Fig. 31.7).

Structure of Gap Junction Channels

A hexagonal ring of connexins forms a central aqueous channel across the lipid bilayer and pairs with a connexon in an adjacent cell to connect their cytoplasms. Four transmembrane α-helices of each connexin subunit span the lipid bilayer and two loops between the helices form small extracellular domains (Fig. 31.7). Connexons on adjacent cells dock to form a continuous pore between the pair of cells. Tight interactions between the subunits seal the pore and preclude leakage of ions out of either cell.

The transmembrane pore begins with a funnel from the cytoplasm that narrows to a diameter of 1.4 nm, larger than the pores of tetrameric S5-P-S6 ion channels (see Fig. 16.5) or pentameric ligand-gated ion channels (see Fig. 16.12). This pore passes hydrophilic molecules up to approximately 1 kD in size, including ions (to establish electrochemical continuity between the cells), second messengers (to establish a common network of information), small peptides, and metabolites (to allow sharing of resources). Connexon hemichannels (the ring of six connexins in one plasma membrane) also open infrequently for the nonspecific release of ions and solutes as large as ATP from the cell.

Connexin Gene Families and Evolution

Humans have genes for 21 connexin isoforms, ranging in size from 26 to 60 kD. Connexins are named by molecular weight; for instance, connexin-43 (Cx-43) is the name for the 43-kD isoform. All connexins have the conserved features required to form the connexon channel but variable N- and C-terminal cytoplasmic sequences. The various connexin isoforms make channels that differ in their permeability and charge selectivity.

Remarkably, gap junction genes seem to have arisen more than once during evolution. Connexins are found exclusively in chordates, while invertebrate gap junctions are composed of innexins (invertebrate connexins). Innexins have four transmembrane helices but lack any sequence similarity to connexins and form intercellular junctions from two octameric hemichannels. Vertebrates have a few genes related to innexins called pannexins. Rather than forming gap junctions, pannexins form plasma membrane channels that release solutes including ATP from the cytoplasm. The ATP activates seven-helix receptors (see Fig. 24.2) and channels activated by adenine nucleotides (see Fig. 16.2) as part of local signaling pathways in the immune and nervous systems. Many animals have another channel, calcium homeostasis modulator 1, that shares structural features with connexins, pannexins, and innexins.

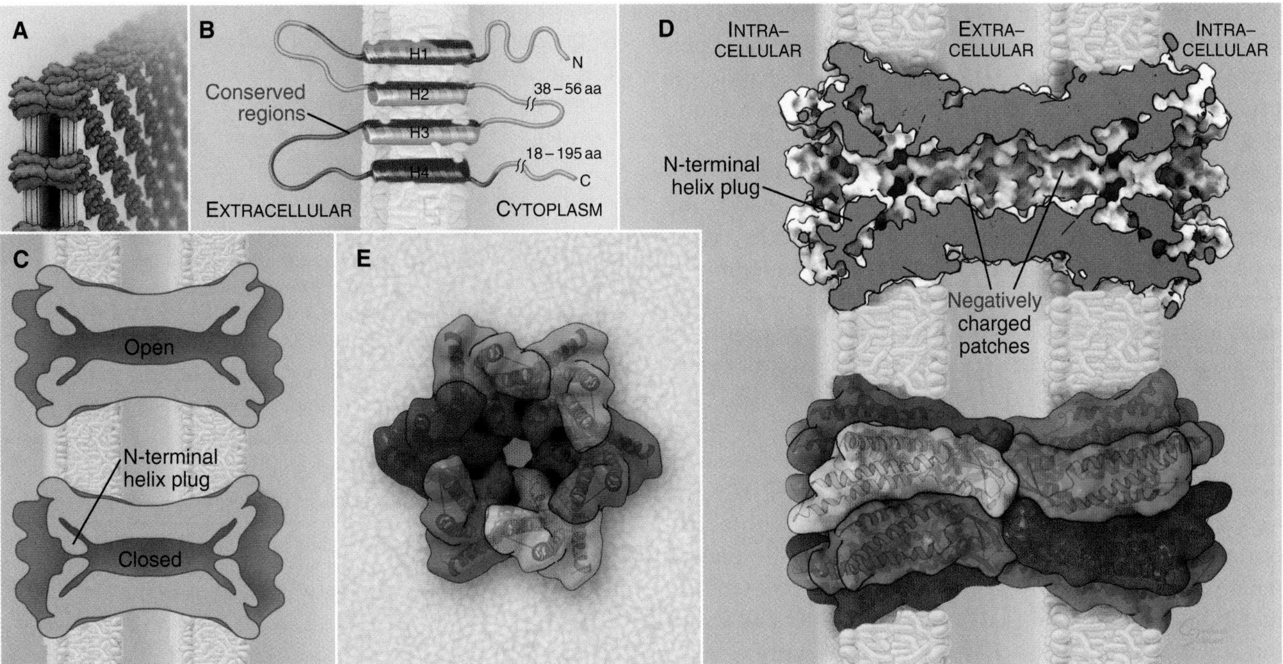

FIGURE 31.7 STRUCTURE OF THE GAP JUNCTION CONNEXON CHANNEL. A, Drawing of gap junction connexons forming channels between the cytoplasms of adjacent cells. **B,** Transmembrane topology of connexins. Four α-helices cross the lipid bilayer. Conserved residues *(maroon)* form the transmembrane and extracellular loops are required for channel assembly. Cytoplasmic loops between helix 2 and helix 3 and the C-terminal tails vary in length among connexin isoforms. Removal of the C-terminal tail from connexin-43 alters its gating properties. **C,** Diagram showing how the N-terminal α-helices of Cx26 may form a plug that blocks the pore of the closed channel. **D–E,** Crystal structure of the connexin 26 the gap junction channel. **C–D,** Top and side views of a ribbon diagram with each of the six subunits a different color. The two extracellular loops of each subunit associate with the other half channel to span the 4-nm gap between the membranes. The upper panel of **D** shows as a space-filling model cut through the middle of the transmembrane pore. (For reference, see PDB file 2ZW3 and Maeda S, Nakagawa S, Suga M, et al. Structure of the connexin 26 gap junction channel at 3.5 A resolution. *Nature*. 2009;458:597–602.)

However, the lack of sequence homology suggests that this gene may have arisen separately.

Assembly of Gap Junctions

Connexin proteins turn over on a time scale of several hours and are replaced by new connexons that assemble in vesicles along the secretory pathway. New connexons add around the periphery of gap junction plaques in the plasma membrane and old connexons are removed from the middle of plaques (Fig. 31.6C).

Many gap junctions are composed of one connexin isoform and pass molecules equally well in both directions. However, some connexons assemble from mixtures of subunits creating hybrid gap junctions with novel properties. Furthermore, some connexons can pair with a different type of connexon on the neighboring cell. Such hybrid channels may pass fluorescent tracers more readily in one direction than the other or react more sensitively to the transjunctional potential of one polarity than the other. This might explain the asymmetrical coupling that is sometimes observed between both excitable and nonexcitable cells, such as neuronal gap junctions, which pass action potentials in one direction but not the other.

Regulation of Gap Junction Permeability

Electrophysiological measurements showed that connexons alternate between open and closed states (Fig. 31.5C). The structural basis for this gating is not established definitively, but a plug-gating mechanism is plausible. In this model N-terminal helices can either form a plug that blocks the pore or pull back against the wall of the channel to create the 1.4-nm pore (Fig. 31.7E). This would allow the two connexins to gate the pore independently; both must be open to connect the two cytoplasms. Cytoplasmic loops, C-terminal tails or the extracellular loops may also contribute to gating.

The pores of gap junction channels are large enough to pass all common inorganic ions as well as nucleotides, amino acids, and even small peptides and RNAs. The conductance of the open state depends on the connexin isoform and varies from about 30 psec to 300 psec. Given the permeability of gap junctions to relatively large solutes, it is surprising that their conductance is in the same range as narrower ligand- and voltage-gated ion channels. Both the greater length and the arrangement of charged residues lining the channel may contribute to the low conductance of connexons.

Gap junctional communication depends on both the number of channels and the fraction of those channels that is open or closed. The fraction of open channels is usually less than 1.0; it is approximately 0.2 in heart and as low as 0.01 in one nerve cell that was tested.

Many factors regulate the opening and closing of connexon channels. The transjunctional potential (ie, the potential difference between the coupled cells) gates most connexons, regardless of the plasma membrane potentials of these cells. Like other voltage-gated channels, individual transitions are fast, but the response to potential changes, on the scale of seconds, is very slow in comparison with other channels (see Fig. 16.7). Unphysiological concentrations of cytoplasmic Ca^{2+} (100 to 500 μM) and cytoplasmic acidification also close connexons. These effects of membrane potential, H^+, and Ca^{2+} allow cells to terminate communication with neighboring cells that are damaged (depolarizing the plasma membrane and admitting high concentrations of Ca^{2+}) or metabolically compromised (allowing Ca^{2+} to leak out of intracellular stores and acidifying the cytoplasm).

Chemicals also modulate gap junctions. Oleamide, a fatty acid amide produced by the brain, blocks gap junctional communication and induces sleep in animals. Organic alcohols (heptanol and octanol) and general anesthetics (halothane) can also close gap junction channels reversibly (Fig. 31.5), but these agents are not specific for gap junctions.

Signaling pathways control gap junction activity through phosphorylation of numerous sites by several kinases. For example, on a time scale of seconds, cAMP activates protein kinase A, which phosphorylates the C-terminal tail of some connexins, increasing or decreasing the fraction of open channels (depending on the connexin isoform and the cell type). On a time scale of hours cAMP also promotes the assembly of gap junctions.

Physiological Functions of Gap Junctions

Cells in most metazoans communicate by gap junctions. Coupled cells in vertebrates include epithelial cells of the skin, endocrine glands, exocrine glands, gastrointestinal tract, and renal-urinary tract as well as smooth muscle, cardiac muscle, bone, some neurons, and glial cells. Epithelial cells can coordinate their activities with their neighbors. This is used to synchronize the beats of cilia (see Fig. 38.12C). Fragments of viral proteins can spread from infected cells to neighboring cells, which then become targets for cytotoxic T lymphocytes (see Fig. 28.9). Gap junctions allow osteocytes buried deep in bone to maintain a cellular supply line to acquire nutrients from distant blood vessels (see Fig. 32.4). Passage of action potentials between cardiac and smooth muscle cells sets off waves of contraction (see Fig. 39.18). Electrical synapses between neurons can transmit action potentials at very high frequencies (>1000 per second). In some parts of the brain, gap junctions also coordinate action potentials in groups of neurons. Even white blood cells may form transient gap junctions with endothelial cells.

TABLE 31.2 Phenotypes of Humans With Mutations in Gap Junction Subunits

Connexin	Phenotype
Cx-26β_2	Dominant and recessive mutations with deafness; skin disease
Cx-30β_6	Recessive deafness; skin disease
Cx-31β_3	Recessive deafness; skin disease
Cx-32β_1	Point mutations, defective myelin, peripheral nerve degeneration in X-linked Charcot-Marie-Tooth disease; deafness
Cx-37α_4	Female infertility, defect in communication of granulosa cells with oocyte
Cx-40α_5	Partial block of impulse conduction in heart
Cx-43α_1	Deafness; many mutations may be lethal as in mice
Cx-46α_3	Cataracts in lens of the eye
Cx-50α_8	Cataracts in lens of the eye

Note: Mutations are homozygous loss of function mutations unless noted otherwise. The nomenclature used here combines the Cx-"molecular mass in kD" and molecular phylogeny αβ-number systems.

Gap Junctions in Disease

Point mutations in connexin genes cause remarkably specific defects in humans (Table 31.2), considering that most connexins are expressed in several tissues. Recessive mutations in the connexin-26 gene are the most common causes of inherited human **deafness.** As many as 1 in 30 people are carriers, and their mutations may contribute to hearing loss late in life. Connexin-26 participates in the transport of K^+ in the epithelia supporting the sensory hair cells in the ear. Patients with one of more than 100 different mutations in the connexin-32 gene can suffer from degeneration of the myelin sheath around axons, an X-linked variant of **Charcot-Marie-Tooth disease.** Many human tissues express connexin-32, but the pathology is confined to myelin. The stability of myelin might depend on intracellular gap junctions between layers of the myelin sheath that provide a pathway between the metabolically active cell body and the deep layers of the sheath near the axon. Defects in myelin membrane proteins cause other forms of Charcot-Marie-Tooth disease.

Adherens Junctions

Adherens junctions use homophilic (like-to-like) interactions of **E-cadherins** (see Fig. 30.5) to bind epithelial cells to their neighbors. Adherens junctions are essential for viability from the earliest stages of animal embryonic development. In mature epithelia, a belt-like adherens junction, called the **zonula adherens,** encircles the cells near their apical surface (Fig. 31.1D) and maintains the physical integrity of the epithelium. Adherens junctions also anchor muscle cells to the extracellular matrix.

Adherens junctions can transmit mechanical forces between cells and reinforce tissues, because the cytoplasmic domains of the E-cadherins are linked to the actin cytoskeleton. Adapter proteins connect cadherins to actin filaments and signaling proteins including guanine nucleotide exchange proteins for the Rho-family GTPases that promote actin assembly and force generation by myosin (see Fig. 33.19F). The adapter proteins include β-catenin, a related protein called plakoglobin, p120-catenin, and the actin-binding protein α-catenin. Moderate physical forces stabilize this link from the cadherin tail through β-catenin and α-catenin to actin filaments.

Adherens junctions are the first connections established within developing sheets of epithelial cells. Contact begins when cadherins on the tips of filopodia engage partner cadherins of the same type on another cell. The contact spreads laterally as more cadherins are recruited along with associated actin filaments, as illustrated by dorsal closure of the ectoderm by *Drosophila* embryos (see Fig. 38.5). These pioneering adherens junctions eventually allow like cells to associate in epithelial sheets (see Fig. 30.8) and to influence the maturation of the epithelium. Adherens junctions are a prerequisite for the assembly of tight junctions that allow epithelial cells to establish polarity with different proteins and lipids in the apical and basal plasma membranes. The shapes of cells in epithelial sheets depend on Rho family GTPases and protein kinases associated with the adherens junction, which regulate the assembly and contraction of the associated actin cytoskeleton.

The junctions and polarity of the cells determine the orientation of the mitotic spindle and the plane of division. This allows for asymmetrical division of stem cells, such as those at the base of stratified epithelia (see Figs. 35.6 and 41.4).

Desmosomes

Desmosomes (*desmos* = "bound," *soma* = "body") use cadherins to provide strong adhesions reinforced by intermediate filaments between epithelial and muscle cells. In epithelia, these junctions are small, disk-shaped, "spot welds" between adjacent cells. Desmosomes in the heart are more complicated because they are mixed with adherens junctions (see Fig. 39.18).

Two families of desmosomal cadherins, named **desmogleins** and **desmocollins**, mediate cellular adhesion at desmosomes (see Fig. 30.5 and Table 30.2). The most distal of five extracellular CAD domains interact head to head with CAD1 domains from the partner cells and laterally with other cadherins in a dense tangle midway between the two plasma membranes (see Fig. 30.6).

Desmosomal cadherins connect to cytoplasmic intermediate filaments via adapter proteins analogous to those that connect adherens junction cadherins to actin filaments. Two proteins related to β-catenin, **plakoglobin** and **plakophilin**, bind to cytoplasmic domain of desmosomal cadherins and form a physical link to **desmoplakin,** a dimeric protein related to plectin (see Fig. 35.7). The C-terminus of desmoplakin binds directly to the N-terminal, nonhelical domains of **epidermal keratin intermediate filaments.** Mutations in this part of epidermal keratins cause blistering skin diseases by compromising the integrity of desmosomes (see Fig. 35.6).

Although all desmosomes share a common plan, selective expression of isoforms of their component proteins give desmosomes unique molecular compositions in various cells. For example, in epidermis, desmoglein-1 and desmocollin-1 are found only in the upper layers, whereas desmoglein-3 is in the basal layers. This explains the pathology in autoimmune blistering diseases. Patients with **pemphigus foliaceus** make antibodies that react with desmoglein-1 and disrupt desmosomes in the upper layers of the epidermis, whereas patients with **pemphigus vulgaris** produce autoantibodies to desmoglein-3 that disrupt the basal layers. These antibodies are directly responsible for the disease; transfusion of human autoantibodies into a mouse reproduces the disease. Other organs are spared, owing to the restricted expression of these two isoforms. Mutations in the corresponding desmoglein genes in mice compromise desmosomes and cause skin blisters similar to pemphigus.

The development of animal tissues depends on desmosomes. Loss-of-function mutations can lead to mechanical failures. For example, mutations in the plakoglobin gene can be lethal in mice and humans during embryogenesis, owing to disruption of the heart. Similarly, mutations in the desmoplakin gene cause skin and cardiac defects that can be fatal. Desmosomal proteins also participate in signal transduction. For example, plakoglobin and desmoglein suppress proliferation and promote differentiation by competing with β-catenin for binding DNA (see Fig. 30.7) and inhibiting the mitogen-activated protein (MAP) kinase pathway. Loss of desmosomes is associated with the spread of epithelial cancer cells.

Adhesion to the Extracellular Matrix: Hemidesmosomes and Focal Contacts

Adhesion to the extracellular matrix is fundamentally different from intercellular adhesion because integrins, rather than homophilic interactions of cadherins, provide the transmembrane link between the cytoskeleton and ligands in the extracellular matrix (see Fig. 30.9). At focal contacts and related assemblies, transmembrane integrins link cytoplasmic actin filaments to the extracellular matrix (see Fig. 30.11).

Hemidesmosomes are another type of integrin-based adhesive junction that links cytoplasmic intermediate filaments to the basal lamina. The morphologic resemblance of hemidesmosomes to half of a conventional desmosome belies the fact that they are fundamentally different at the molecular level (Fig. 31.8C). Like desmosomes, hemidesmosomes have a dense plaque on the cytoplasmic surface of the plasma membrane that anchors loops of intermediate filaments. The similarity ends there.

The hemidesmosomes of simple epithelia use **$\alpha_6\beta_4$ integrin** to adhere to **laminin-5** in the basal lamina. Plectin (see Fig. 35.7) links the large cytoplasmic domain of β_4-integrin to keratin intermediate filaments.

More complex hemidesmosomes of stratified epithelial cells have, in addition to $\alpha_6\beta_4$-integrin, a second transmembrane adhesion protein, **type XVII collagen**. The type XVII collagen trimer forms an extracellular collagen triple helix (see Fig. 29.1) that anchors the

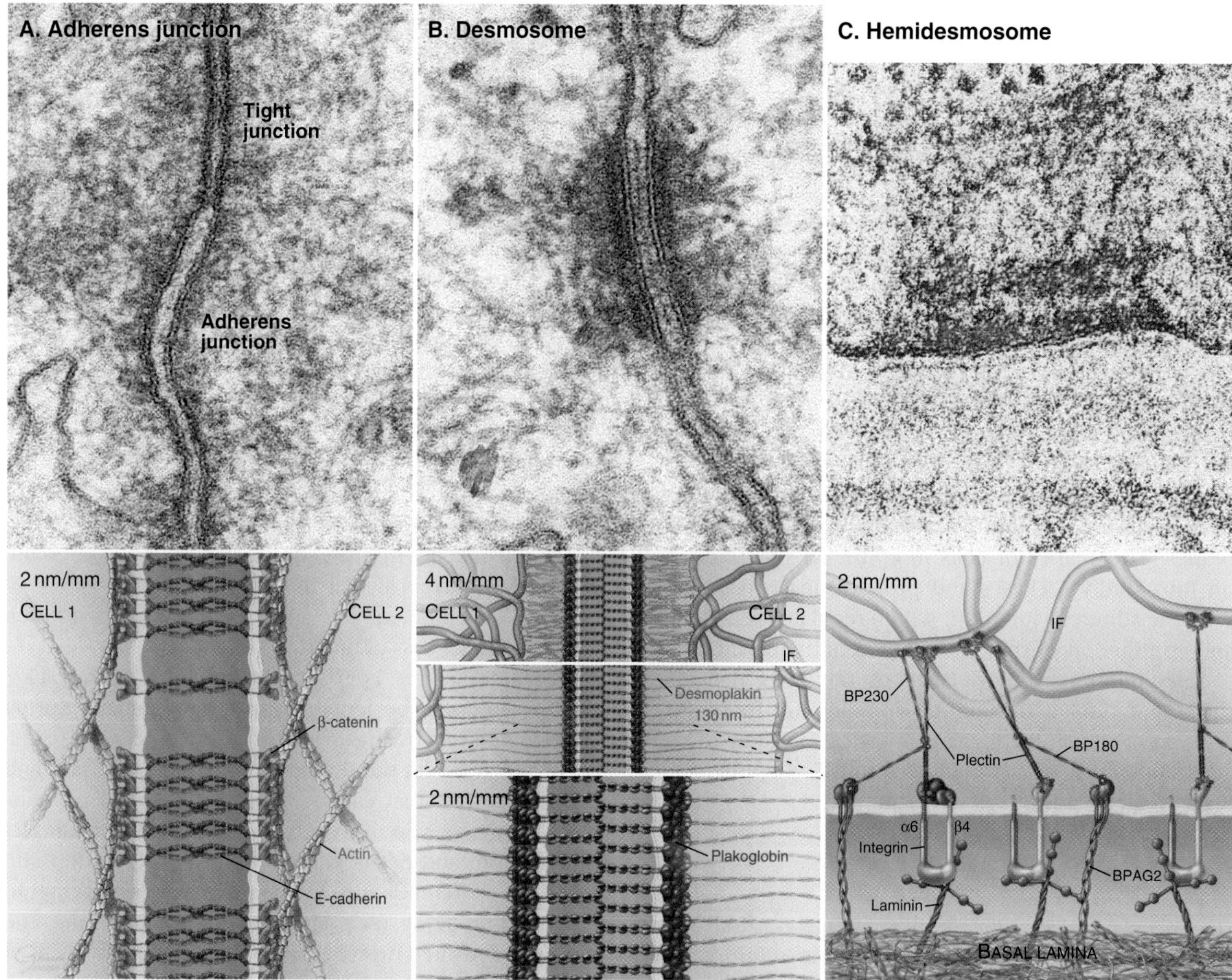

FIGURE 31.8 COMPARISON OF ADHERENS JUNCTION, DESMOSOME, AND HEMIDESMOSOME. *Top,* Electron micrographs of thin sections. *Bottom,* Molecular models. **A,** Adherens junction. Electron micrograph from the intestinal epithelium. E-cadherins link two cells together. β-Catenin and α-catenin link the cytoplasmic domain of E-cadherin to actin filaments. **B,** Desmosome. Two types of cadherins—desmoglein and desmocollin—link adjacent cells together. The central dense stratum seen in the micrograph presumably corresponds to the interaction sites of the cadherins, although accessory proteins may participate. Desmoplakin and other accessory proteins link the cadherins and associated plakoglobin (related to catenin) to keratin intermediate filaments. Desmoplakin molecules are shown extended to their full length in the *middle drawing,* whereas in desmosomes, they must be kinked or folded (as shown in the *upper drawing*) because the thickness of the desmoplakin layer is half that expected from extended molecules. **C,** Hemidesmosome. Integrin $\alpha_6\beta_4$ and type XVII collagen (also called BPAG2 [bullous pemphigoid antigen-2]) attach to the basal lamina. Plectin, BP230, and BPAG1 (bullous pemphigoid antigen-1) link the membrane proteins to keratin intermediate filaments. (**A–B,** Micrographs courtesy Hilda Pasolli and Elaine Fuchs, Rockefeller University, New York and from Perez-Moreno M, Jamora C, Fuchs E. Sticky business: orchestrating cellular signals at adherens junctions. *Cell.* 2003;112:535–548. **C,** Micrograph courtesy Jonathan Jones, Northwestern University, Chicago, IL.)

membrane to the basal lamina. In a blistering skin disease called **bullous pemphigoid,** autoantibodies attack type XVII collagen, so the protein is also called bullous pemphigoid antigen-2, or BPAG2. This clinical observation and genetic deletions established that both $\alpha_6\beta_4$-integrin and type XVII collagen are required for assembly of stable hemidesmosomes in skin. BPAG1 (bullous pemphigoid antigen-1; also called BP230) is related to plectin and helps connect the integrin to intermediate filaments.

Mutations in the genes for any of the hemidesmosome proteins cause blistering skin diseases known as epidermolysis bullosa. Pathology can also occur in other tissues that depend on hemidesmosomes, including the cornea, gastrointestinal tract, and muscles. Mutations in keratin genes also cause epidermolysis bullosa (see Fig. 35.6).

SELECTED READINGS

Broussard JA, Getsios S, Green KJ. Desmosome regulation and signaling in disease. *Cell Tissue Res*. 2015;360:501-512.

Buckley CD, Tan J, Anderson KL, et al. Cell adhesion. The minimal cadherin-catenin complex binds to actin filaments under force. *Science*. 2014;346:1254211.

Evans WH. Cell communication across gap junctions: a historical perspective and current developments. *Biochem Soc Trans*. 2015; 43:450-459.

Gumbiner BM. Regulation of cadherin-mediated adhesion in morphogenesis. *Nat Rev Mol Cell Biol*. 2005;6:622-634.

Johnson JL, Najor NA, Green KJ. Desmosomes: regulators of cellular signaling and adhesion in epidermal health and disease. *Cold Spring Harb Perspect Med*. 2014;4:a015297.

Lee JY. Plasmodesmata: a signaling hub at the cellular boundary. *Curr Opin Plant Biol*. 2015;27:133-140.

Nielsen MS, Axelsen LN, Sorgen PL, et al. Gap junctions. *Compr Physiol*. 2012;2:1981-2035.

Niessen CM, Leckband D, Yap AS. Tissue organization by cadherin adhesion molecules: dynamic molecular and cellular mechanisms of morphogenetic regulation. *Physiol Rev*. 2011;91:691-731.

Oshima A. Structure and closure of connexin gap junction channels. *FEBS Lett*. 2014;588:1230-1237.

Oshima A, Matsuzawa T, Murata K, et al. Hexadecameric structure of an invertebrate gap junction channel. *J Mol Biol*. 2016;428: 1227-1236.

Padmanabhan A, Rao MV, Wu Y, et al. Jack of all trades: functional modularity in the adherens junction. *Curr Opin Cell Biol*. 2015;36:32-40.

Powell AM, Sakuma-Oyama Y, Oyama N, et al. Collagen XVII/BP180: A collagenous transmembrane protein component of the dermoepidermal anchoring complex. *Clin Exp Dermatol*. 2005;30:682-687.

Scemes E. Nature of plasmalemmal functional "hemichannels." *Biochim Biophys Acta*. 2012;1818:1880-1883.

Siebert AP, Ma Z, Grevet JD, et al. Structural and functional similarities of calcium homeostasis modulator 1 (CALHM1) ion channel with connexins, pannexins, and innexins. *J Biol Chem*. 2013;288: 6140-6153.

Stahley SN, Kowalczyk AP. Desmosomes in acquired disease. *Cell Tissue Res*. 2015;360:439-456.

Suzuki H, Nishizawa T, Tani K, et al. Crystal structure of a claudin provides insight into the architecture of tight junctions. *Science*. 2014;344:304-307.

Tilsner J, Nicolas W, Rosado A, et al. Staying tight: plasmodesmal membrane contact sites and the control of cell-to-cell connectivity in plants. *Annu Rev Plant Biol*. 2016;67:337-364.

Van Itallie CM, Anderson JM. Architecture of tight junctions and principles of molecular composition. *Semin Cell Dev Biol*. 2014;36:157-165.

Walko G, Castañón MJ, Wiche G. Molecular architecture and function of the hemidesmosome. *Cell Tissue Res*. 2015;360:529-544.

Yap AS, Gomez GA, Parton RG. Adherens junctions revisualized: organizing cadherins as nanoassemblies. *Dev Cell*. 2015;35:12-20.

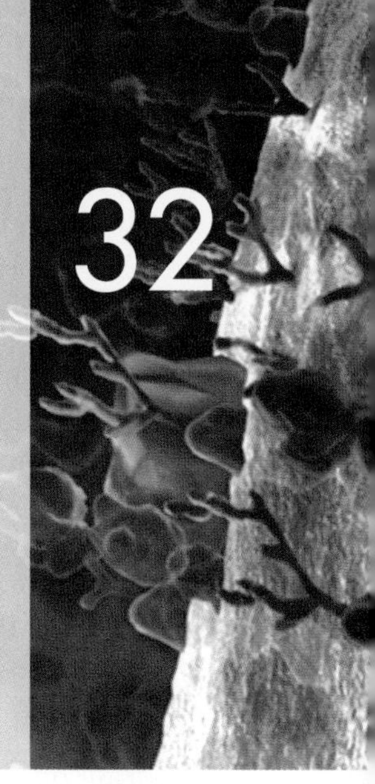

CHAPTER 32

Connective Tissues

Animals use different proportions of matrix macromolecules to construct connective tissues with a range of mechanical properties to support their organs. Bone is a stiff, hard solid; blood vessel walls are flexible and elastic; and the vitreous body of the eye is a watery gel. Plant and fungal cell walls are functionally similar to the animal extracellular matrix but are composed of completely different molecules. This chapter begins with a discussion of simple connective tissues then concentrates on cartilage, bone, development of the skeleton, and the mechanisms that repair wounds, finishing with a discussion of the plant cell wall.

Loose Connective Tissue

Loose connective tissue consists of a sparse extracellular matrix of **hyaluronan** and **proteoglycans** supported by a few **collagen fibrils** and **elastic fibrils.** In addition to fibroblasts, the cell population is heterogeneous, including both indigenous and emigrant connective tissue cells (see Fig. 28.1). The loose connective tissue underlying the epithelium in the gastrointestinal tract is a good example of this heterogeneity (Fig. 32.1A), with lymphocytes, plasma cells, macrophages, eosinophils, neutrophils, and mast cells, as well as fibroblasts and occasional fat cells (see Chapter 28 for details on these cells). This variety of defensive cells is appropriate for a location near the lumen of the intestine, which contains microorganisms and potentially toxic materials from the outside world. Loose connective tissue is also found in and around other organs. In the optically transparent vitreous body of the eye, fibroblasts produce a highly hydrated gel of hyaluronan and proteoglycans, supported by a loose network of type II collagen. Few defensive cells are required, as the interior of the eye is sterile.

Dense Connective Tissue

Collagen fibers, with or without elastic fibers, make up the bulk of dense connective tissue (Fig. 32.1B). Sparse fibroblasts are present to manufacture extracellular matrix. Other connective tissue cells are even rarer, as these tissues are not usually exposed to microorganisms. Collagen fibers can be arranged precisely, as in tendons or cornea (see Fig. 29.3), or less so, as in the wall of the intestine or the skin. Tendons consist nearly exclusively of type I collagen fibers, all aligned along the length of the tendon to provide the tensile strength that is required to transmit forces from muscle to bone. The cornea that forms the transparent front surface of the eye is also well organized into orthogonal layers of collagen fibrils.

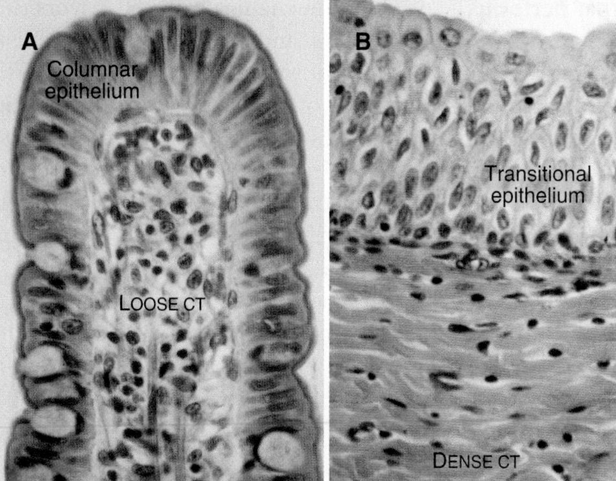

FIGURE 32.1 CONNECTIVE TISSUES. A, Loose connective tissue (CT) underlying the columnar epithelium of the small intestine. Light micrograph of a section stained with Masson trichrome stain. **B,** Dense connective tissue (CT) underlying transitional epithelium in the wall of the ureter. Light micrograph of a section stained with hematoxylin-eosin. (Courtesy D.W. Fawcett, Harvard Medical School, Boston, MA.)

Dense connective tissues can also be elastic. For example, the walls of arteries (see Fig. 29.8) and the dermal layer of skin consist of both collagen and elastic fibers. Energy from each heartbeat stretches the elastic fibers in the walls of arteries. Recoil of these elastic fibers propels blood between heartbeats and affects the blood pressure.

Approximately one in 5000 humans inherits a mutation in a gene for fibrillar collagens type I, type III, or type V, which causes a range of connective tissue defects called **Ehlers-Danlos syndrome.** Most affected individuals have thin skin and lax joints. Severe mutations lead to rupture of arteries, bowel, or uterus, often with fatal consequences. Ehlers-Danlos syndrome illustrates the importance of these collagens with regard to the integrity of the affected tissues. Inheritance is dominant, as these collagens consist of trimers of three identical subunits. Given one mutant gene, only one in eight (½ × ½ × ½) procollagen molecules is normal.

Cartilage

Cartilage (Fig. 32.2) is tough, resilient connective tissue that performs a variety of mechanical roles. It covers the articular surfaces of joints and supports the trachea, other large airways, the nose, and ears. Cartilage also forms the entire skeleton of sharks and the embryonic precursors of many bones in higher vertebrates. The mechanical properties of cartilage are attributable to abundant extracellular matrix consisting of fine collagen fibrils and high concentrations of glycosaminoglycans and proteoglycans (Fig. 32.3).

Chondrocytes synthesize and secrete macromolecules for the cartilage matrix, which eventually surrounds them completely. Chondrocytes replenish the matrix as the macromolecules turn over slowly, but their ability to remodel and repair the matrix is limited. No blood vessels penetrate cartilage, owing to production of several inhibitors of endothelial cell growth by chondrocytes. Thus, all nutrients must diffuse into cartilage from the nearest blood vessel in the **perichondrium,** a dense capsule of fibrous connective tissue that covers the surface of cartilage. This capsule contains mesenchymal stem cells (see Box 41.2 and Fig. 28.1) that are capable of differentiating into chondrocytes.

A meshwork of **type II collagen fibrils,** accounting for approximately 25% of the dry mass, fills the extracellular matrix. These slender collagen fibrils are hard to see even in electron micrographs but are quite stable, with lifetimes estimated to be many years. Fibrils tend to line up parallel to surfaces but otherwise are arranged randomly. Minor collagen type IX crosslinks type II collagen fibrils and collagen type XI binds to the surface of type II fibrils. Expression of type X collagen is restricted to cartilage that is undergoing conversion to bone. The matrix contains several minor adhesive proteins, and other proteins inhibit invasion of blood vessels.

Glycosaminoglycans, including hyaluronan, constitute the second major class of matrix macromolecules. Molecules of the proteoglycan **aggrecan** attach to a hyaluronan backbone like the bristles of a test tube brush, forming so-called **megacomplexes** (see Fig. 29.13). Aggrecan also binds type II collagen. Highly charged

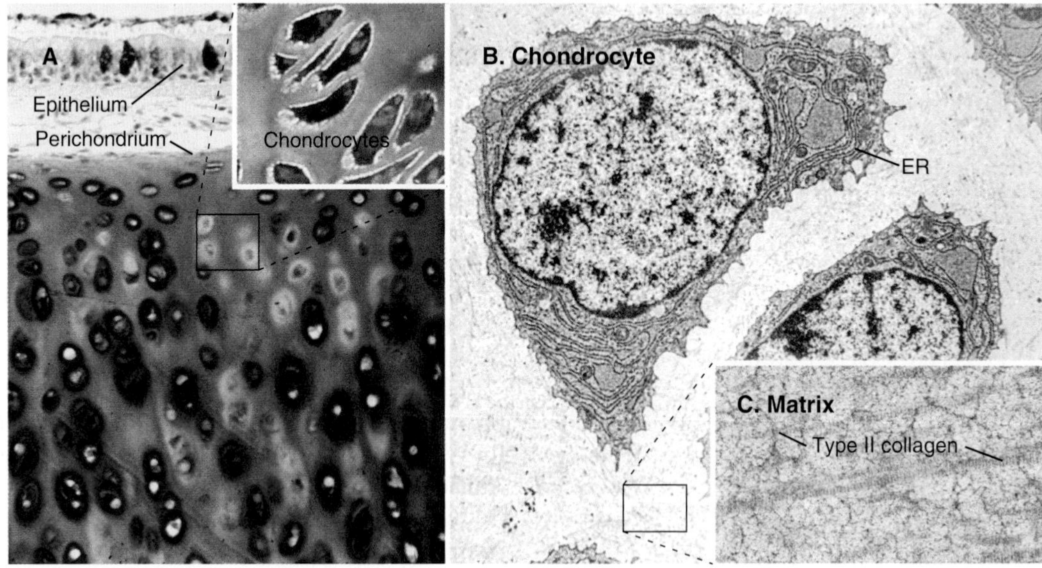

FIGURE 32.2 CARTILAGE AND CHONDROCYTES. A, Light micrograph of a section of hyaline cartilage in the wall of the respiratory tree stained with periodic acid–Schiff stain and Alcian *blue.* The cartilage capsule of dense connective tissue (perichondrium) and the columnar epithelium lining the respiratory passage are at the top. **Inset,** Light micrograph of hyaline cartilage stained with toluidine *blue.* The proteoglycans in the matrix stain *pink.* The rough endoplasmic reticulum stains *blue.* Shrinkage during fixation and embedding creates the artifactual cavity or lacuna around each cell. **B,** Electron micrograph of a thin section of hyaline cartilage showing chondrocytes embedded in dense extracellular matrix. **C,** Electron micrograph of cartilage matrix at high magnification. This specimen was rapidly frozen and prepared by freeze-substitution to avoid collapse of the proteoglycans during dehydration and embedding. ER, endoplasmic reticulum. (**A,** Courtesy D.W. Fawcett and E.D. Hay, Harvard Medical School, Boston, MA. **B,** Courtesy of E.D. Hay, Harvard Medical School, Boston, MA. **C,** Courtesy E.B. Hunziker, M. Müller Institute, University of Bern, Switzerland.)

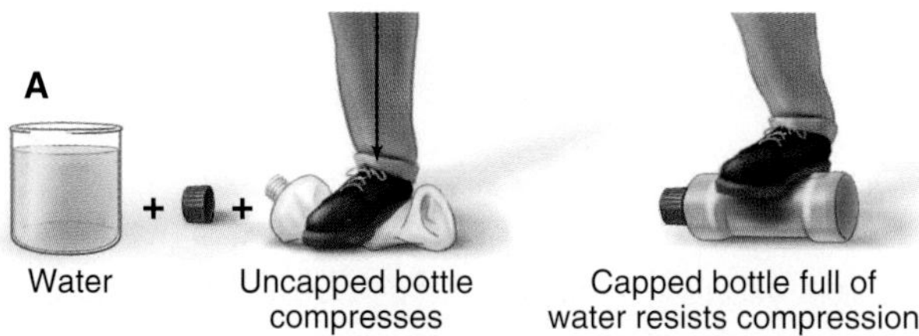

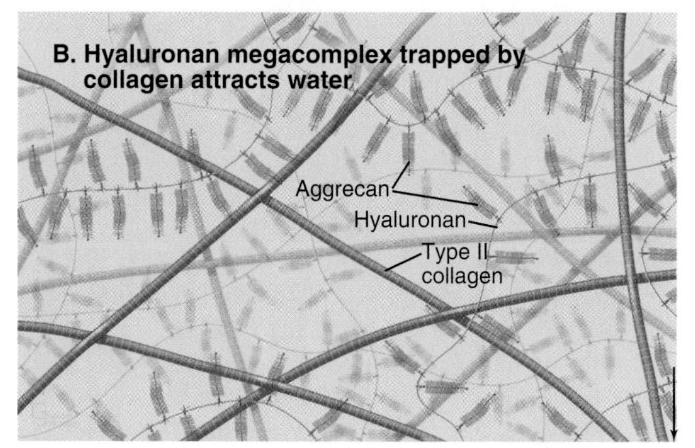

FIGURE 32.3 MACROMOLECULAR STRUCTURE AND MECHANICAL PROPERTIES OF HYALINE CARTILAGE MATRIX. **A,** Hydrostatic model of the mechanical properties of cartilage. Water trapped in the extracellular matrix resists compression. Neither water alone *(in beaker)* nor a pliable container *(uncapped plastic bottle)* resists compression. However, if water fills a capped bottle, it resists compression. **B,** In the cartilage matrix, flexible strands of type II collagen trap proteoglycans, which attract large amounts of water. Trapped water resists compression because its "container," the network of collagen fibrils, does not stretch.

glycosaminoglycans fill the extracellular space and attract water, the most abundant component of the matrix.

A hydrostatic mechanism allows cartilage to resist deformation (Fig. 32.3). Collagen fibrils provide tensile strength (ie, resistance to stretching) but do not resist compression or bending. Glycosaminoglycans strongly attract water, resulting in an internal swelling pressure that pushes outward against collagen fibrils aligned parallel to the surface of the cartilage. The force of internal hydrostatic swelling pressure balances the force produced by tension on the collagen fibrils. Remarkably, this internally stressed material can resist strong external forces such as those on the articular surfaces of joints. A macroscopic analog is a thin-walled plastic bottle filled with water. One can stand on the bottle provided that it is sealed, whereas neither the empty bottle nor the water could separately support any weight.

Specialized Forms of Cartilage

Hyaline cartilage provides mechanical support for the respiratory tree, nose, articular surfaces, and developing bones. Elastic cartilage has abundant elastic fibers in addition to collagen, making the matrix much more elastic than hyaline cartilage. Elastic cartilage supports structures subjected to frequent deformation, including the larynx, epiglottis, and external ear. Fibrocartilage has features of both dense connective tissue (an abundance of thick collagen fibers) and cartilage (a prominent glycosaminoglycan matrix). It is tough and deformable, appropriate for its role in intervertebral disks and insertions of tendons.

Differentiation and Growth of Cartilage

Cartilage grows by expansion of the extracellular matrix either from within or on the surface. For surface growth, mesenchymal cells in the perichondrium differentiate into chondrocytes that synthesize and secrete matrix materials. For internal growth, chondrocytes trapped in the matrix divide and manufacture additional matrix, which is sufficiently deformable to allow for internal expansion. Cartilage has a limited capacity to repair damage, but stem cell transplants can help some patients.

Many growth factors and their receptors cooperate to influence the differentiation of precursor cells into chondrocytes, the proliferation of chondrocytes, and the production of cartilage matrix molecules. These include Indian hedgehog (Ihh), members of transforming growth factor-β family (TGF-β and bone morphogenetic proteins [BMPs]), multiple fibroblast growth factors (FGFs), parathyroid hormone–related protein (PTHrP), and insulin-like growth factors (IGF-1 and IGF-2). Chondrocytes produce some of these growth factors (TGF-β, FGFs, and IGFs). During development, adjacent tissues can induce cartilage formation by secreting TGF-β and FGF. SOX9 is the key transcription factor mediating expression of cartilage-specific genes.

Diseases of Cartilage

Cartilage fails in common human diseases, including arthritis and ruptured intervertebral disks. Osteoarthritis, degeneration of cartilage on joint surfaces, is very common in older people and has a complex genetic component attributable to variations in many genes. Rarely, human diseases are caused by mutations in single genes for cartilage proteins, growth factors or growth factor receptors (Appendix 32.1). For example, more than 25 different mutations of the human gene for type II collagen cause disorders of cartilage, ranging in severity from death in utero to dwarfism or osteoarthritis. Mutations in genes for minor cartilage-associated collagens cause a variety of symptoms, including degenerative joint disease. A premature stop codon in chicken aggrecan causes lethal skeletal malformations.

Bone

For most vertebrates, bones provide mechanical support and serve as a storage site for calcium. The great strength and light weight of bones are attributable both to the mechanical properties of the extracellular matrix and to efficient overall design, including tubular form and

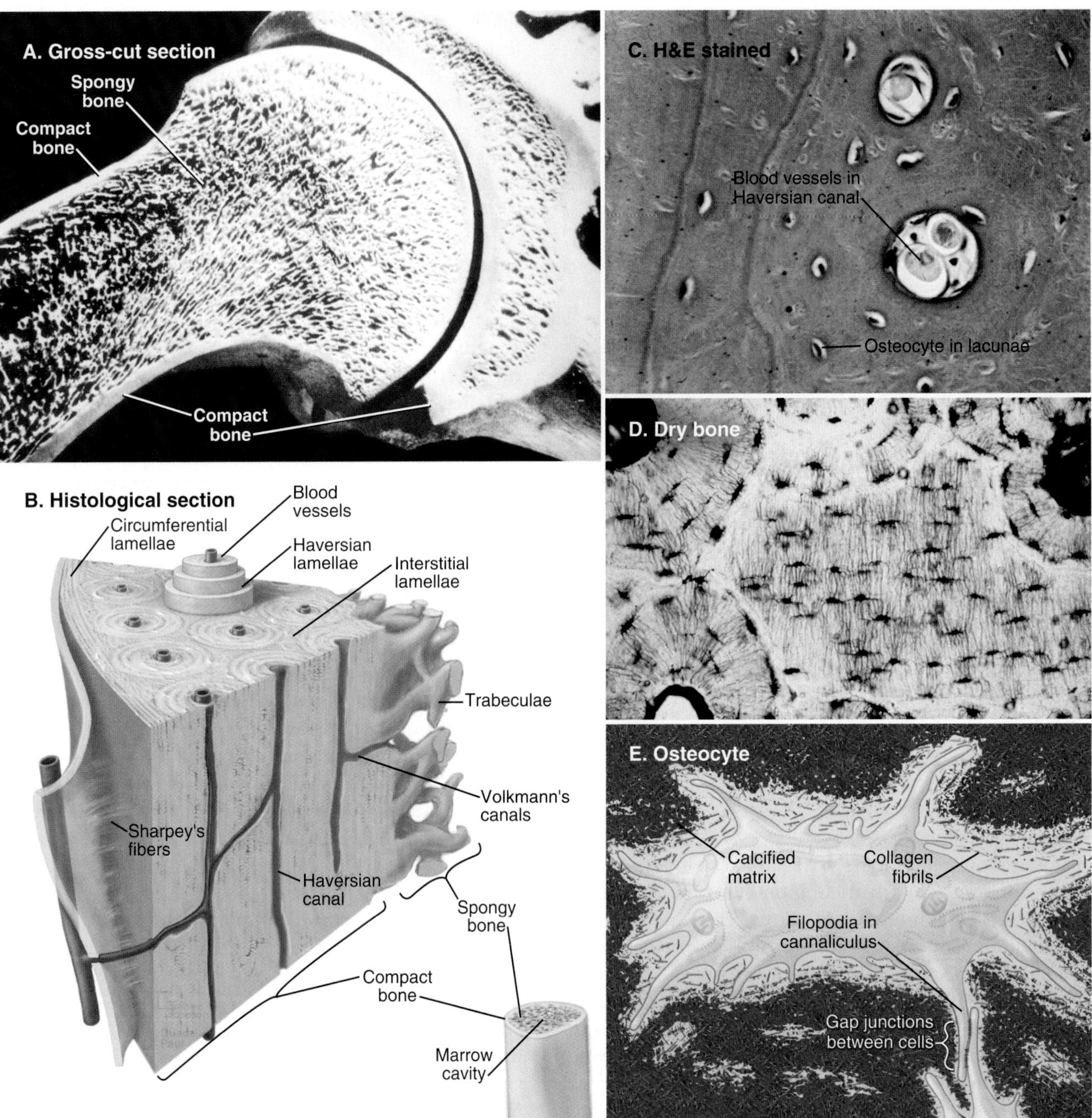

FIGURE 32.4 ORGANIZATION OF LONG BONES. A, Longitudinal section of a shoulder joint of a dried bone specimen. Struts of trabecular spongy bone reinforce compact bone in the cortex. **B,** A wedge of long bone. Circumferential lamellae form the outer layer just beneath the periosteum *(blue)* covering the surface. Osteons (Haversian systems) consist of concentric lamellae of calcified matrix and osteocytes arranged around a channel containing one or two capillaries or venules. Interstitial lamellae are fragments of osteons that remain after remodeling (Fig. 32.10). Radial vascular channels connect longitudinal vascular channels to the medullary cavity or periosteum. **C,** Light micrograph of a cross section stained with hematoxylin and eosin (H&E) showing circumferential lamellae on the *left* and two Haversian canals. **D,** Light micrograph of a cross section of dried bone showing a central interstitial lamella surrounded by three osteons. Narrow canaliculi connect the lacunae housing osteocytes. **E,** An osteocyte surrounded by calcified matrix and extending filopodia into canaliculi. (Micrographs courtesy D.W. Fawcett, Harvard Medical School, Boston, MA.)

lamination (Fig. 32.4). A superficial layer of compact bone surrounds a central cavity that is filled with marrow, fat, or both and is supported by struts of bone arranged precisely along lines of mechanical stress. External surfaces of bones are covered either by dense connective tissue, called **periosteum,** or by cartilage at joint surfaces. Two cell types make bone matrix: osteoblasts covering the internal surfaces and osteocytes embedded in the bone. A third cell type, called the osteoclast, degrades bone, recycling matrix components. Blood vessels penetrate compact bone through a network of channels to supply the central cavity. Although bone is durable and strong, continuous remodeling makes bone much more dynamic than it appears.

Extracellular Matrix of Bone

Bone is a composite material consisting of type I collagen fibrils (providing tensile strength) embedded in a matrix of calcium phosphate crystals (providing rigidity) (Fig. 32.4E). The calcium-phosphate crystals are similar to **hydroxyapatite** [$Ca_{10}(PO_4)_6(OH)_2$] and make up about two-thirds of the dry weight of bone. Macroscopic analogs of the bone matrix are concrete reinforced by steel rods and fiberglass consisting of a brittle plastic reinforced by glass fibers. Each of these composites is stronger than its separate components. Simple extraction experiments illustrate the contributions of the two components of bone. After removal of calcium phosphate with a calcium chelator, bone is so rubbery that it bends easily. After destruction of collagen by heating, bone is hard but brittle.

Fibrils of type I collagen, the dominant organic component of the matrix (Table 32.1), are arranged in sheets or a meshwork. Covalent crosslinks between the collagen molecules in fibrils make them inextensible. The matrix also contains more than 100 minor proteins, including growth factors, proteins that promote hydroxyapatite deposition and adhesive glycoproteins, but few proteoglycans.

Cells that make bone lay down type I collagen as the substrate for crystallization of calcium phosphate. Some calcium phosphate crystallizes directly in the collagen matrix. Other crystals form in small "matrix vesicles" that bud from the plasma membranes of osteoblasts and use pumps and carriers to concentrate calcium and phosphate. After being released from these vesicles, tiny crystals associate with collagen fibrils. The crystals grow and eventually fill spaces between the collagen molecules within the fibrils.

Bone Cells

Overview

A balance among the activities of osteoblasts, osteocytes, and osteoclasts forms, grows, and maintains bones. Osteoblasts and osteocytes produce extracellular matrix and establish conditions for its calcification. Osteoclasts resorb and remodel bone. An imbalance of these opposing cellular activities causes human diseases.

Properties of Osteoblasts

A monolayer of **osteoblasts** on the surface of growing bone tissue uses a well-developed secretory pathway to synthesize and secrete the organic components of the matrix (Fig. 32.5). Osteoblasts also act as endocrine cells, secreting growth factors that control the differentiation of osteoclasts (Fig. 32.6), as well as cells in other organs. They also help to form the niche in the bone marrow for hematopoietic stem cells (see Fig. 41.4).

Regulation of Osteoblast Development

Osteoblasts arise from the same mesenchymal stem cells that give rise to fibroblasts and chondrocytes (see Fig. 28.1). Growth factors control the differentiation from mesenchymal cells. They include Ihh, a subset of BMPs

TABLE 32.1 Bone Proteins

Name	Content	Functions
Bone morphogenic proteins	Minor	Transforming growth factor (TGF)-β homologs; cartilage stimulation and bone development and repair
Collagen type I	90%	Forms fibrils in the bone matrix
Osteocalcin	1%–2%	Network of aspartic acid and γ-carboxylated glutamic acid side chains bind hydroxyapatite; promotes calcification; attracts osteoclasts and osteoblasts
Osteonectin	2%	Synthesized in developing and regenerating bone; binds collagen and hydroxyapatite; may nucleate hydroxyapatite crystallization in bone matrix
Osteopontin	Minor	Arginine-glycine-aspartic acid (RGD) sequence; binds osteoclast integrins to bone surface
Proteoglycans	Minor	Decorin, biglycan, osteoadherin; may bind TGF-β
Sialoproteins	2%	RGD sequence; binds osteoclast integrins to bone surface

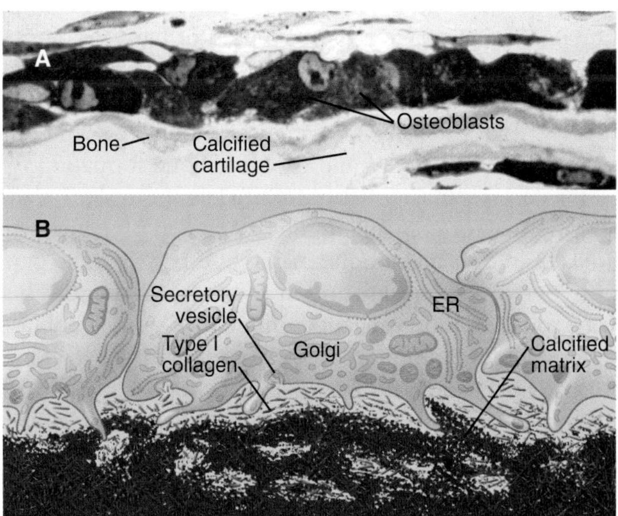

FIGURE 32.5 OSTEOBLASTS. A, Light micrograph of a section of forming bone stained with toluidine *blue.* Osteoblasts with abundant, *blue-stained,* rough endoplasmic reticulum lay down bone matrix *(light green)* on the surface of calcified cartilage *(light pink)*. **B,** Drawing of osteoblasts. ER, endoplasmic reticulum. (**A,** Courtesy R. Dintzis and from the work of D. Walker, Johns Hopkins Medical School, Baltimore, MD.)

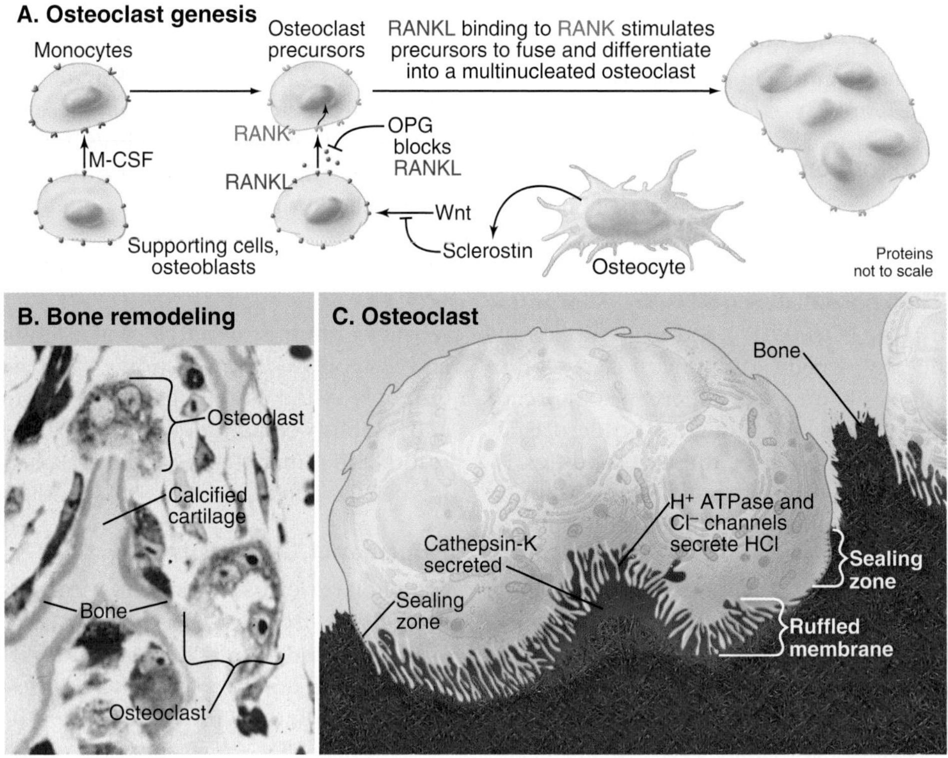

FIGURE 32.6 OSTEOCLASTS. A, Formation of a multinucleated osteoclast by fusion of monocytes stimulated by the receptor activator of nuclear factor κB ligand (RANKL), macrophage colony-stimulating factor (M-CSF), and other factors. **B,** Light micrograph of a section of forming bone stained with toluidine *blue* showing two osteoclasts degrading bone and calcified cartilage. **C,** An osteoclast attached to the bone matrix by a sealing zone, forming a resorption cavity *(pink).* The cell pumps H^+ and secretes lysosomal enzymes into this cavity to resorb the surface of the matrix. ATPase, adenosine triphosphatase; OPG, osteoprotegerin (OPG); RANK, receptor activator of nuclear factor κB. (**B,** Courtesy R. Dintzis and from the work of D. Walker, Johns Hopkins Medical School, Baltimore, MD.)

(see Fig. 24.7), and some Wnts (see Fig. 30.7). Humans with loss-of-function mutations in a Wnt coreceptor have few osteoblasts and low bone density, whereas loss-of-function mutations in a BMP competitor have the opposite effect. On the other hand, osteocytes secrete a protein called **sclerostin** that blocks the Wnt coreceptor (Fig. 32.6), so loss-of-function mutations of sclerostin strengthen Wnt signals and promote bone formation.

Inside osteoblasts, **Runx2/Cbfa1** is the master transcription factor controlling the expression of genes that are required to make bone matrix. Runx2/Cbfa1 is part of a network of transcription factors and microRNAs with positive and negative influences on osteoblast differentiation and function. Mouse embryos lacking Runx2/Cbfa1 have no osteoblasts or osteoclasts. They make a cartilage skeleton that never transforms to bone. Humans and mice with just one active Runx2/Cbfa1 gene lack collarbones and experience a delay in the fusion of joints between skull bones. This syndrome is the most common human skeletal defect.

Osteocyte Properties

Once an osteoblast has enclosed itself within bone matrix, it is called an **osteocyte.** Long-lived, metabolically active osteocytes are connected to each other by many long, slender filopodia that run through narrow channels in the matrix (Fig. 32.4D–E). Gap junctions between the processes of osteocytes provide a continuous network of intercellular communication that stretches from cells adjacent to blood vessels to the most deeply embedded osteocyte.

Osteocytes can either lay down or resorb matrix in their immediate vicinity. Circulating hormones influence the activity of osteoblasts and osteocytes. In response to the calcium concentration in blood, parathyroid glands secrete **parathyroid hormone,** which stimulates osteocytes to mobilize calcium from the surrounding matrix. This feedback loop maintains a constant concentration of calcium in the blood.

Osteoclast Properties

Osteoclasts form by fusion of blood monocytes and resorb bone, as required for growth and remodeling. **Osteoclasts** are multinucleated giant cells specialized for bone resorption (Fig. 32.6). They attach like a suction cup to the surface of bone. Interactions of a plasma membrane integrin ($\alpha_V\beta_3$) with bone matrix proteins (osteopontin and sialoprotein) help to create a leakproof compartment on the bone surface. Osteoclasts amplify the plasma membrane lining this closed space, forming a "ruffled border" composed of microvilli enriched with H^+-transporting **V-type** rotary adenosine triphosphatase (ATPase) pumps (see Fig. 14.6) and chloride channels (see Fig. 16.13). The combined activities of the H^+ pump and chloride channels allow the cell to secrete hydrochloric acid into the sealed extracellular compartment on the bone surface. This closed space acts like an extracellular lysosome: Acid dissolves calcium phosphate crystals, and secreted proteolytic enzymes, including **cathepsin K,** digest collagen and other organic components. Degradation products are taken up by endocytosis and transported across the cell in vesicles for secretion on the free surface. Amino acids are reused, but collagen crosslinking groups are not, so they are excreted in the urine, where their concentration is a measure of bone turnover.

Osteoclast Formation

Bone marrow supporting cells, osteoblasts, and activated T lymphocytes produce two proteins that stimulate blood monocytes to fuse and differentiate into multinucleated osteoclasts (Fig. 32.6). These key factors are **macrophage colony-stimulating factor (M-CSF)** and **RANKL** (receptor activator of nuclear factor kappa B [NF-κB] ligand, also called osteoprotegerin ligand [OPGL] or tumor necrosis factor–related activation-induced cytokine [TRANCE]). Both factors are produced locally in bone marrow as transmembrane proteins with the growth factor domain on the cell surface. These proteins control differentiation through binding to their receptors on monocytes by either direct cell-to-cell contact or release of the active domain by proteolytic cleavage. First, M-CSF activates a cytokine receptor (see Fig. 24.6) on monocytes, stimulating a JAK (just another kinase) kinase–signal transducer and activator of transcription (JAK-STAT) pathway (see Fig. 27.9) and turning on expression of genes required for the monocyte to differentiate into a preosteoclast. An important change is the expression of a receptor called RANK (receptor for activation of NF-κB, a member of the tumor necrosis factor [TNF] receptor family; see Fig. 24.9). Once this receptor is expressed, RANKL can activate preosteoclasts through the transcription factor NF-κB (see Fig. 10.21C) to express the proteins required for cell fusion and further differentiation into an osteoclast. Mice that lack RANKL form no osteoclasts, so bone resorption fails.

Other growth factors, including TNF itself, contribute to this process by acting directly on osteoclasts, but many stimulators of osteoclast differentiation (eg, parathyroid hormone, vitamin D, leptin) act indirectly by stimulating supporting cells to make RANKL. For example, **leptin,** a satiety hormone secreted by fat cells, acts on neurons of the hypothalamus in the brain that regulate not only appetite but also bone metabolism indirectly via the sympathetic nervous system. Norepinephrine released by sympathetic nerves activates osteoblasts to secrete RANKL. This explains why animals and people that lack leptin or its receptor not only are obese but also have dense bones. Osteoclast growth factors RANKL, TNF, and interleukin-1 mediate excess bone resorption at sites of chronic inflammation in rheumatoid arthritis and gum diseases.

Differentiation of osteoclasts is subject to negative regulation by a soluble **decoy receptor** for RANKL called OPG (osteoprotegerin), which binds RANKL and competes for activation of RANK (Fig. 32.6). Estrogens inhibit osteoclast differentiation by stimulating osteoblasts to produce OPG, so circulating OPG declines in parallel with estrogen levels after menopause. The resulting increase in osteoclasts contributes to bone loss (osteoporosis) in older women.

Formation and Growth of the Skeleton

Both genetic and environmental information direct formation of the skeleton. Genetic information predominates in the master plan and initial development of skeletal tissues, as the size and shape of bones are characteristic for each species. Subsequently, environmental information is important in remodeling of the skeleton in response to use. Mutations in genes for structural and informational molecules have provided valuable clues about the genetic blueprint for the skeleton (Appendix 32.1).

Genetic information is read out on at least two levels. First, master genetic regulators—including transcription factors encoded by **HOX (homeobox)** and **PAX (paired box)** genes—specify the developmental fate of each embryonic segment. Homeoboxes are DNA sequences that encode a family of 60-residue protein domains that bind DNA (see Fig. 15.14). The human genome contains 39 HOX genes arrayed in four linear arrays, similar to those in other animals. HOX genes were discovered in flies as a result of mutations that cause "homeotic conversion," whereby the fate of one segment is converted into another, such as the substitution of a leg for an antenna. The same thing happens in vertebrates: Mouse embryos express Hoxd-4 in the second cervical (neck) vertebra and more posterior segments. Mutation of Hoxd-4 results in the second cervical vertebra taking on some of the features of the first cervical vertebra. Mutations in other HOX genes cause congenital malformations in humans. The pathways from HOX genes to determinants of three-dimensional architecture are still incompletely understood.

Both systematically circulating and locally secreted growth factors control the proliferation and differentiation of the cells of cartilage and bone. Mutations in these factors and their receptors also cause surprisingly specific human skeletal malformations (Appendix 32.1). Circulating **growth hormone** produced by the pituitary gland is a major determinant of skeletal size. Individuals deficient in growth hormone are short in stature. Locally produced growth factors, including BMPs and FGFs and their receptors, control the development and growth of cartilage and bone during embryogenesis, in addition to stimulating repair after fractures. FGF receptors are tyrosine kinases (see Fig. 24.4). BMPs are related in structure and mechanism to TGF-β and are expressed in tissues other than bone and cartilage (see Fig. 24.7). BMPs are part of a system of positive and negative factors that regulates formation of cartilage, bone, and joints. For example, a BMP called GDF-5 specifies the position of joints, but joints form only if noggin protein, an inhibitor of other BMPs, is present.

Embryonic Bone Formation

Bone always forms by replacement of preexisting connective tissue. During embryonic development, flat bones, such as the skull and shoulder blades, form from **neural crest cell** precursors in loose connective tissue (Fig. 32.7). Somehow, information in the genome is read out as the three-dimensional pattern of a skull. Growth factors, vitamins (eg, retinoic acid), and local matrix molecules, such as glycosaminoglycans, all influence the differentiation of these cells into osteoblasts at specific locations in connective tissue. Osteoblasts lay down struts of bone matrix in the loose connective tissue. As new bone is laid down on the surface of these bone spicules, some osteoblasts are trapped and become osteocytes.

During embryonic and postnatal development, genetic information precisely controls changes in the size and proportions of flat bones. For example, for the skull to increase in size both externally and internally, osteoclasts on the outer surface lay down new bone at the same rate as osteoclasts resorb old bone inside (Fig. 32.8A). These cellular activities are carefully coordinated to change the proportions of the skull as the individual matures.

Long bones, such as the humerus, begin as cartilage models that are replaced by bone (Fig. 32.9). Multiple, genetically programmed factors induce clusters of mesenchymal cells at specific locations to differentiate into chondrocytes that secrete type II collagen and glycosaminoglycans. This produces a miniature cartilaginous version of the adult bone.

Bone replaces this cartilage precursor in a series of steps that are coordinated locally by production of growth factors. Perichondrial cells and proliferating chondrocytes secrete PTHrP, which promotes chondrocyte division and growth. In supporting roles, BMPs promote and FGFs inhibit the growth and differentiation of chondrocytes by acting upon populations of cells that express particular receptors for these molecules (Appendix 32.1). Active FGF receptors stimulate STAT transcription factors (see Fig. 27.9) and/or mitogen-activated protein (MAP) kinase pathways (see Fig. 27.6). More mature chondrocytes produce Ihh, which directs the terminal differentiation of neighboring chondrocytes.

For a long bone to maintain its shape as it grows in size, deposition and removal of bone tissue must be highly selective. For the shaft to grow in diameter, new bone is laid down on the outer surface by osteoblasts at the same time as old bone is removed inside by osteoclasts (Figs. 32.8B and 32.9).

Bones grow longer as a result of interstitial growth of cartilage in the **epiphyseal plate** and its continual replacement by bone. Chondrocytes contribute to the elongation of bones in two ways: chondrocytes continuously proliferate in one zone and then rapidly increase in mass and swell in the adjacent zone next to forming bone (Fig. 32.9B). The **hypertrophic chondrocytes** secrete type X collagen and use matrix metalloproteinases to resorb some of their surrounding matrix. They also direct the calcification of the cartilage matrix before undergoing apoptosis or differentiating into osteoblasts. Osteoblasts lay down bone matrix on the surface of the cavities in the calcified cartilage. Hypertrophic cartilage

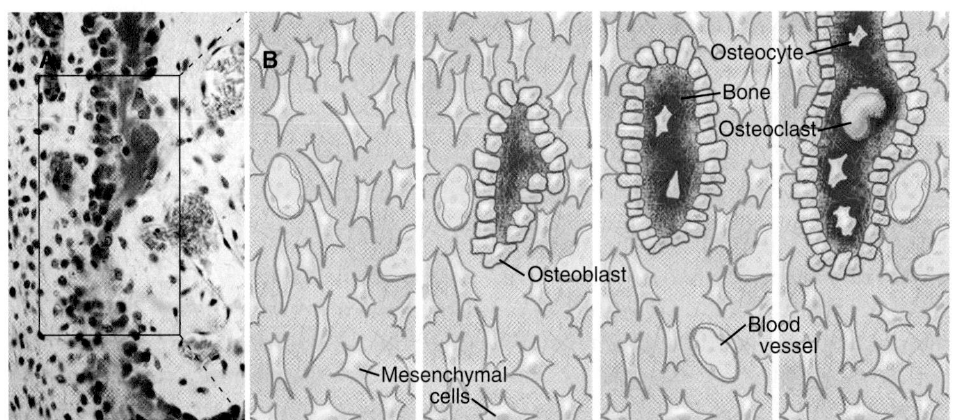

FIGURE 32.7 BONE FORMATION BY INTRAMEMBRANOUS OSSIFICATION. A, Light micrograph of a section of forming bone stained with hematoxylin and eosin. Calcified bone matrix is *maroon.* **B,** Interpretive drawings. Connective tissue mesenchymal cells differentiate into osteoblasts, which lay down bone matrix *(blue).* Osteoblasts become trapped as the matrix grows. (**A,** Courtesy D.W. Fawcett, Harvard Medical School, Boston, MA.)

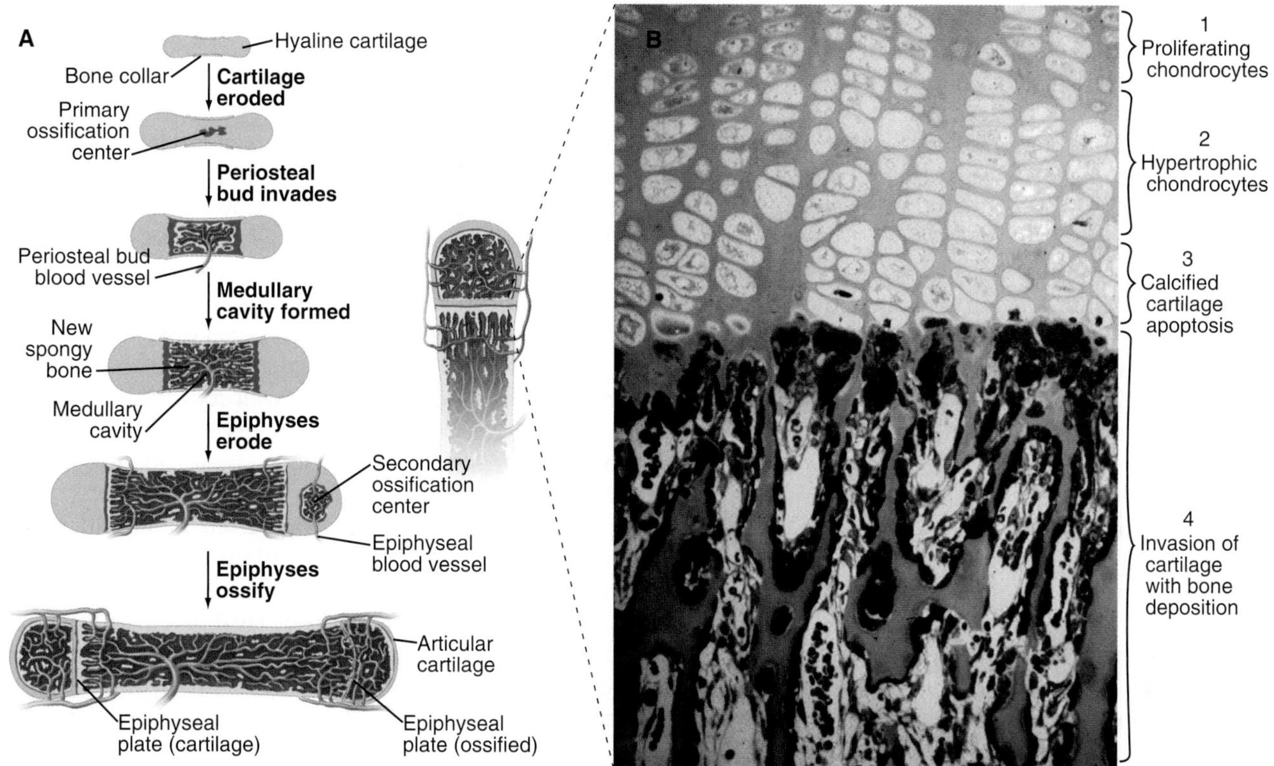

FIGURE 32.8 FORMATION OF A LONG BONE BY REPLACEMENT OF CARTILAGE. A, The shaft grows in diameter as osteoblasts lay down bone *(tan)* on the outer surface of the primary collar of bone and osteoclasts remove bone from the inner surface to form and maintain the marrow cavity. The bone grows in length by interstitial expansion of the cartilage in the epiphyseal plate and its replacement by bone. **B,** Light micrograph of a section of an epiphyseal plate stained with toluidine *blue.* Cartilage growth, differentiation, and replacement by bone occur in several zones. Proliferation of chondrocytes and their production of matrix *(pink)* containing type II collagen are solely responsible for the longitudinal growth of the bone *(1).* Hypertrophic chondrocytes enlarge and make type X collagen, as well as matrix metalloproteinases that resorb some of the surrounding matrix *(2).* Chondrocytes die by apoptosis (see Chapter 46), and the matrix calcifies *(3).* Blood vessels and osteoblasts move into spaces vacated by chondrocytes and lay down bone *(blue)* on the surface of calcified cartilage *(4).* (Micrograph courtesy R. Dintzis and from the work of D. Walker, Johns Hopkins Medical School, Baltimore, MD.)

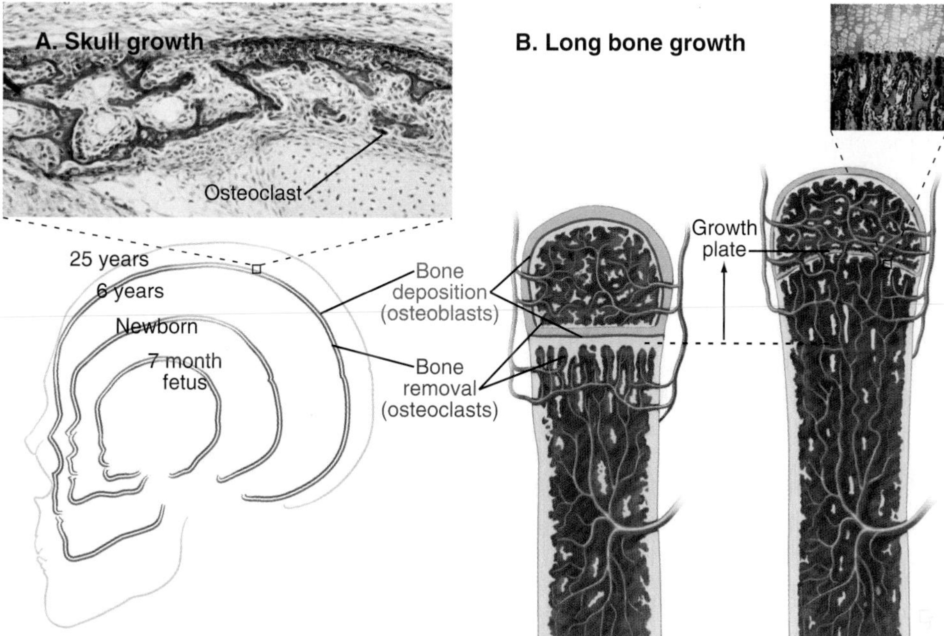

FIGURE 32.9 BONE GROWTH. A, Light micrograph of a section of skull stained with Mallory's trichrome stain and an interpretive drawing. The skull expands during fetal development and growth to adulthood as osteoblasts lay down new bone on the outer surface *(blue)* and osteoclasts resorb bone *(pink)* on the inner surface. **B,** Long bones grow entirely by expansion of cartilage in the epiphyseal plate and its replacement by bone *(tan),* followed by resorption *(pink).* (**A,** Courtesy D.W. Fawcett, Harvard Medical School, Boston, MA.)

ceases to make the factors that inhibit endothelial cell growth, allowing FGF-2, TGF-β, and vascular endothelium growth factor to attract capillaries as part of the transformation of cartilage to bone.

Growth of long bones stops at puberty, when high concentrations of estrogen and testosterone stop proliferation of epiphyseal chondrocytes so that bone replaces this cartilage. This closure of the epiphyses throughout the body occurs over several years in a predictable order, so one can judge the maturity of a child by examining epiphyses by radiographic studies. Genetic variations in this process of maturation give rise to differences in stature. Metabolic and endocrine disorders can also affect the timing of epiphyseal closure.

Bone Remodeling

Bone is amazingly dynamic and is remodeled continuously in response to stresses. Bone cells and matrix turn over every few years. Reorganization of bone requires two carefully coordinated steps: breakdown of preexisting bone by osteoclasts and replacement with new bone by osteoblasts. More than 100 years ago, Wolff realized that the strength of a bone depends on use. For example, bones of the racquet arm of tennis players are more robust than the bones of their other arm. Thus, mechanical forces on the bones must generate modulatory signals that control remodeling. Sensory nerves are involved in some way, but most research has focused on how cells detect fluid flow through canaliculi.

Primary cilia have been implicated in both the differentiation of bone cells and their responses to mechanical forces. Part of their function must be in hedgehog signaling (see Chapter 38), but they probably also sense fluid flowing through canaliculi as a result of mechanical force on the bone. Accordingly, some mutations in genes for components of the intraflagellar transport machinery (see Fig. 38.18) cause severe skeletal defects.

Formation of the cylindrical units of long bones called **osteons** is a good example of well-coordinated remodeling. The process involves two steps (Fig. 32.10). First, osteoclasts resorb preexisting bone to form long, cylindrical, resorption channels in the same way that a plumber's snake clears debris from drain pipes. The second step is slower, as osteoblasts take weeks to fill in these channels by depositing concentric layers of lamellar bone against the walls. They lay down matrix at a rate of about 1 μm of thickness per day until bone completely surrounds the blood vessels trapped in the middle of the newly formed osteon. Because resorption channels cut randomly through the bone, fragments of older osteons are left behind during the remodeling of mature bone. These fragments are called **interstitial lamellae.**

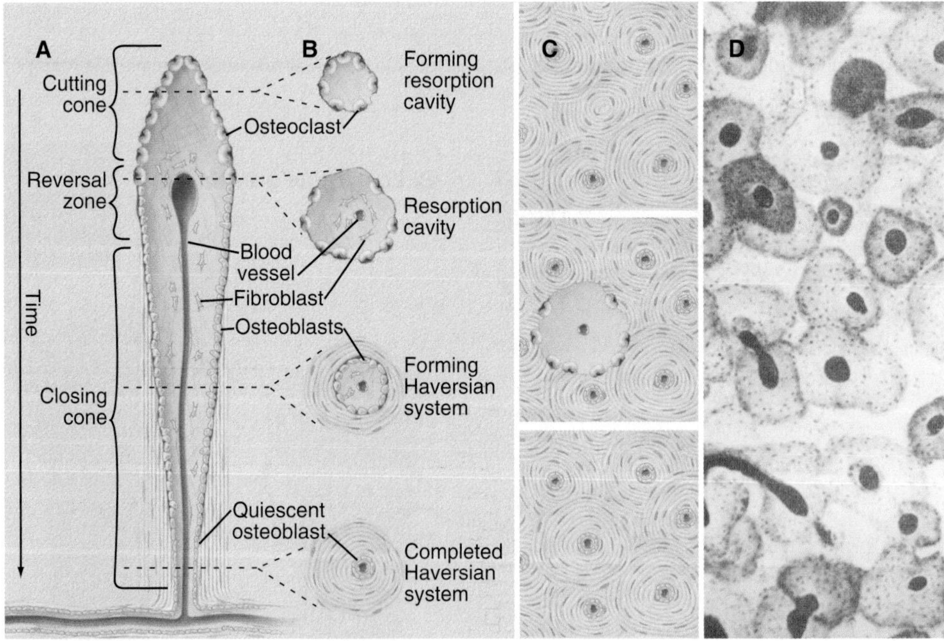

FIGURE 32.10 BONE REMODELING. A–B, Longitudinal and cross sections of a time line illustrating the formation of an osteon. Osteoclasts cut a cylindrical channel through bone. Osteoblasts follow, laying down bone on the surface of the channel until the matrix surrounds the central blood vessel of the newly formed osteon. **C,** Steps in the formation of a new osteon. Parts of older osteons are left behind as interstitial lamellae. **D,** Microradiograph of a cross section of a long bone, illustrating the range of ages of the structures. A section of bone is placed on x-ray film, exposed to x-rays, developed, and examined by light microscopy. Older parts of the bone, such as the interstitial lamellae, are more heavily calcified and therefore absorb more of the x-rays, appearing lighter. Newly formed osteons appear the darkest, as they are the least calcified. Vascular spaces are empty and fully exposed by the x-rays. (**A,** Modified from Parfitt AM. The action of parathyroid hormone on bone. *Metabolism.* 1976;25:809–844. **D,** Courtesy D.W. Fawcett, Harvard Medical School, Boston, MA.)

Resorption may release growth and differentiation factors from the mineralized matrix that provide a local stimulus for the next round of bone formation by new osteoblasts.

Bone Diseases

Osteoporosis, a thinning of bones, is common in elderly people as a result of an imbalance of bone resorption over renewal. In the United States, osteoporosis results in 1.5 million painful fractures, costing nearly $20 billion annually. Almost half of women suffer from such a fracture at some time in their lives, typically as estrogen levels decline after menopause. Osteoporosis also occurs at reduced gravitational forces during space flight. The pathogenesis is not understood, but both behavioral (eg, inactivity, poor nutrition, smoking) and multiple genetic factors have modest effects. Among many genetic factors, one might be naturally occurring variants of the nuclear receptor for vitamin D. This receptor is a transcription factor required for vitamin D to stimulate intestinal calcium uptake and calcification of bone. Variations in the genes for type I collagen or bone growth factors may also contribute.

Two strategies are used to treat osteoporosis. The first is to reduce bone resorption using with bisphosphonates (pyrophosphate mimics) or injection of either OPG or antibodies to RANKL, which interfere with osteoclast formation. New inhibitors of cathepsin K are also being tested. The other approach is to promote bone formation with vitamin D, estrogen, calcium, or strontium, but these measures are only partially effective. More promising is injection of an analog of parathyroid hormone or antibodies to sclerostin, which promote osteoblast activity.

Osteopetrosis is failure of bone resorption, leading to an imbalance of renewal over resorption. This rare disease of osteoclasts is fatal in humans, owing to bone marrow failure. Recessive mutations in the gene for the V-type proton-ATPase pump (60%), two genes for a chloride channel (~15%), and two genes for proteins involved with the secretory pathway account for most human cases. Naturally occurring or engineered mutations in the genes for essentially any protein required for osteoclast differentiation or function cause osteopetrosis in mice. Osteoclasts are present in bone but fail to function properly. The disease can be cured in humans and mice by transplantation of bone marrow stem cells to replace defective osteoclast precursors, an early example of stem cell therapy. If the mutation is in the gene encoding RANKL, replacement of this growth factor cures the disease.

Osteogenesis imperfecta is the name of a variety of congenital fragile bone syndromes. Severely affected fetuses die in utero from multiple broken bones. Mildly affected individuals are born but suffer multiple fractures resulting in skeletal deformities. All of the patients have mutations in the gene for type I collagen. Some are deletions or insertions, which may be mild. Most patients with severe disease have point mutations leading to replacement of a glycine by a larger amino acid. This prevents the zipper-like folding of the collagen triple helix (see Fig. 29.1), even if only one chain is defective per molecule. This poisons assembly and accounts for the dominant phenotype. No one knows why these mutations in type I collagen do not affect other tissues, such as skin, which are rich in type I collagen.

Repair of Wounds and Fractures

Healing of minor skin wounds is a familiar occurrence that illustrates the mechanisms controlling the assembly of connective tissue. Repair of connective tissue in the dermis underlying the epithelium proceeds in three stages: formation of a blood clot, assembly of provisional connective tissue, and remodeling of the connective tissue (Fig. 32.11).

Tissue damage ruptures blood vessels, releasing blood that clots to stem the hemorrhage and fill the damaged area. **Thrombin**, a proteolytic enzyme in blood plasma, drives the clotting reaction by cleaving the plasma protein **fibrinogen** to form **fibrin.** Fibrin spontaneously polymerizes and is crosslinked to itself and to **plasma fibronectin.** This provisional extracellular matrix of fibrin and fibronectin provides physical integrity for the clot and an environment for wound repair. Thrombin also activates seven-helix receptors on platelets (see Fig. 30.14), stimulating them to secrete matrix molecules (thrombospondin, fibrinogen, fibronectin, and von Willebrand factor) and growth factors (platelet-derived growth factor [PDGF], TGF-β, and TGF-α) that initiate the cellular events required to complete wound repair.

Chemotactic factors attract phagocytes from the blood into the wound. These factors include PDGF, chemokines, peptides cleaved from fibrinogen by thrombin, and peptides from any contaminating bacteria. Neutrophils arrive first from the nearby blood vessels, having attached to activated endothelial cells (see Fig. 30.13) and migrated into the connective tissue and clot. They ingest any bacteria. Then monocytes (using a similar mechanism) migrate into the clot and clear foreign material and any dead neutrophils. The environment in a wound promotes transformation of monocytes into macrophages (see Fig. 28.6), which synthesize and secrete cytokines and growth factors that mediate the cellular events that complete the repair process. In this way, platelets, monocytes, and fibroblasts form a relay, passing information from one cell to the next.

During the next phase of repair, macrophages, fibroblasts, and capillary endothelial cells migrate into the fibrin clot and reestablish the connective tissue. Endothelial cells form capillary loops that allow blood to flow and

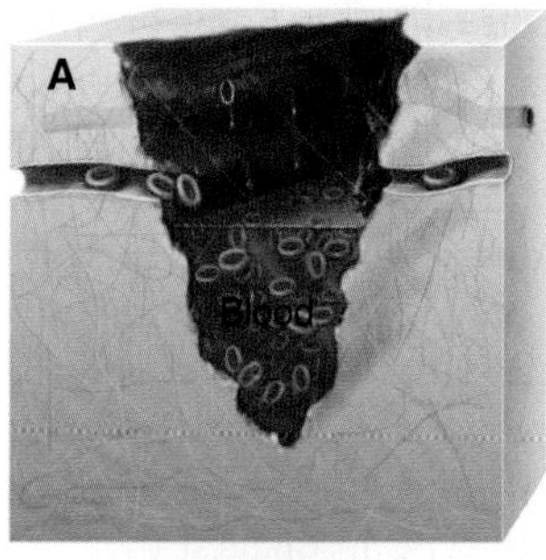

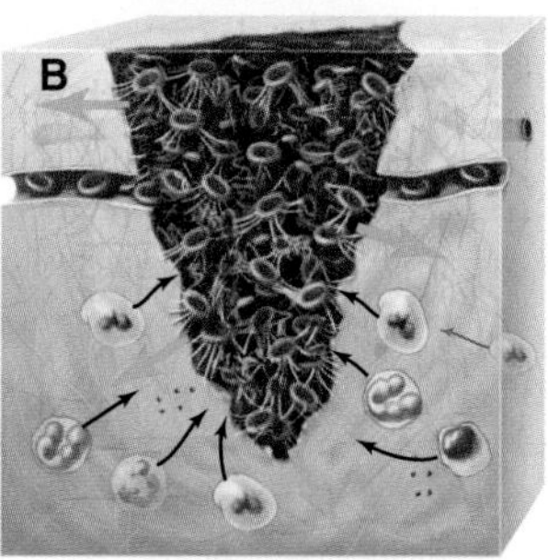

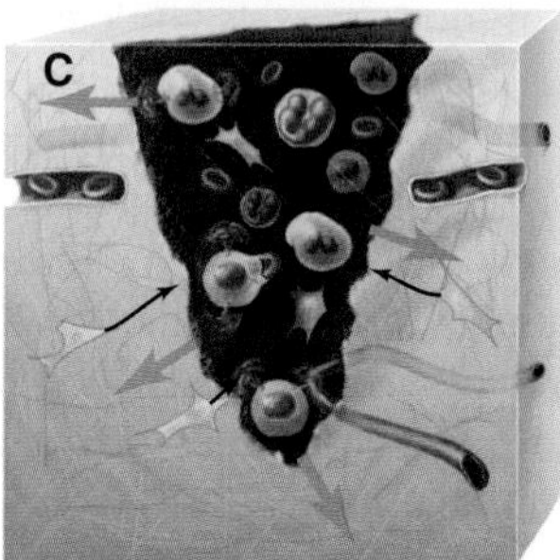

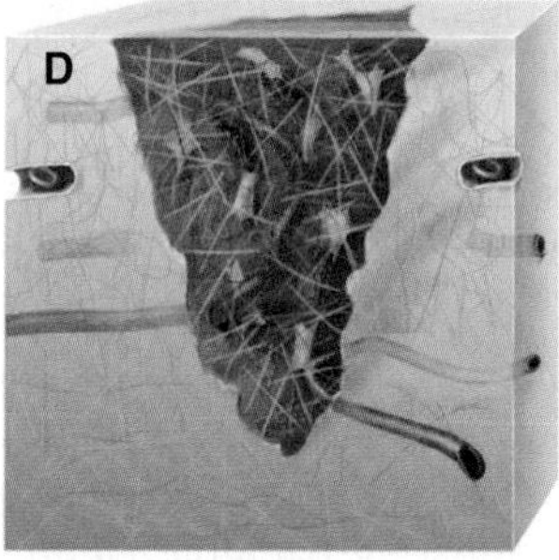

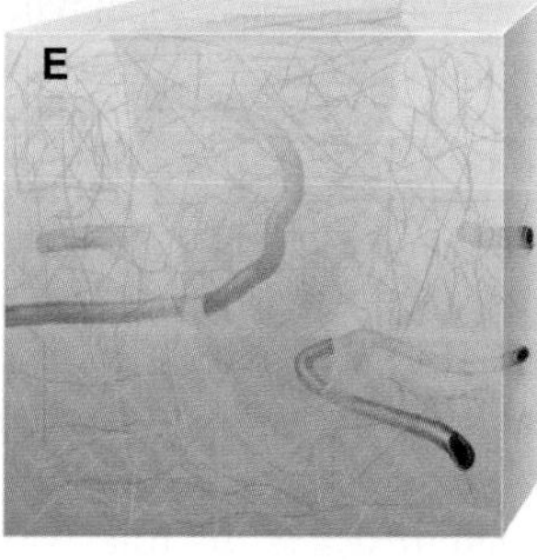

FIGURE 32.11 REPAIR OF A WOUND IN CONNECTIVE TISSUE. A, Wounding removes some tissue and damages blood vessels, releasing blood into the defect. **B,** Blood forms a clot of fibrin and fibronectin, releasing fibrin peptides, and platelets secrete platelet-derived growth factor (PDGF) and transforming growth factor (TGF)-β, all of which attract neutrophils and monocytes. **C,** Neutrophils ingest any bacteria. Monocytes clean up debris and differentiate into macrophages, which secrete cytokines, attracting fibroblasts and blood vessels. **D,** Fibroblasts secrete type III collagen and hyaluronan, which, in turn, replace the fibrin clot. **E,** Fibroblasts remodel the provisional connective tissue with type I collagen, and blood vessels grow back into the new tissue.

to provide oxygen. Initially, the endothelial cells are attracted by growth factors released by platelets, but macrophages and dissolution of fibrin provide a more sustained supply of chemoattractants and growth factors. Integrin receptors for fibronectin allow fibroblasts to migrate into the clot. They secrete more fibronectin as they move. Within the clot, PDGF and TGF-β from macrophages stimulate fibroblasts to secrete type III collagen, hyaluronan, SPARC (secreted protein acidic and rich in cysteine), and tenascin. Initially, this loose connective tissue is disorganized and weak. Hyaluronan predominates transiently, but after about five days, it is gradually replaced by proteoglycans and type I collagen.

Two events complete the repair of the matrix. First, fibroblasts differentiate into (smooth muscle-like) **myofibroblasts,** which contract the collagen matrix, closing the edges of the wound. This step is particularly important for large wounds. Second, fibroblasts remodel the provisional connective tissue to restore its original architecture with nearly normal physical strength. This requires resorption of provisional collagen fibrils by metalloproteinases (see Fig. 29.19) and assembly of more robust type I collagen fibrils.

While fibroblasts repair the connective tissue, the epithelium bordering the wound spreads by cell division and migration to cover the defect. This process of migration is initiated within hours of wounding. Both the loss of contacts with neighboring cells at the edge of the wound and the release of growth factors in the wound are thought to transform the static epithelial cells into migrating cells. Keratin filaments that predominate in the cytoskeleton of skin epithelial cells are replaced with actin filaments. Hemidesmosomes that anchor the skin epithelial cells to the basal lamina are lost, and the cells migrate over the surface of the underlying matrix, which consists initially of fibrin and fibronectin and later of collagen. As they go, epithelial cells lay down a new basal lamina. Depending on the size of the defect, proliferation of epithelial cells might be required to complete coverage of the surface. When it is covered, the cells begin to differentiate into stratified epithelium.

Many parallels exist between repair of a fractured bone and repair of a skin wound. Blood escapes from damaged blood vessels and clots at the fracture site. PDGF released by platelets stimulates mesenchymal cells to proliferate in the surrounding tissue. These cells migrate into the clot along with blood vessels and macrophages. Stimulated by growth factors released initially by platelets and in a more sustained fashion by macrophages, mesenchymal cells differentiate into chondrocytes and osteoblasts that recapitulate the development of new bone to fill in the defect. Although the bone that is initially produced to join the fractured ends is poorly organized, fractures are mechanically stable within approximately 6 weeks. The fibrin clot is converted directly into bone if the broken bone is immobilized. A cartilage intermediate may form first if the fracture is allowed to move. Over a period of months, remodeling reestablishes the normal pattern of the bone. With time, remodeling can even straighten out bones that are mildly bent at fracture sites.

In all of these examples, wound healing is coordinated by a variety of growth factors and cytokines and is supported by the environment provided by the extracellular matrix. For example, PDGF from platelets stimulates the proliferation of fibroblasts and attracts them to the fibrin clot at the site of a wound. TGF-β inhibits fibroblast proliferation but stimulates fibroblasts to make matrix molecules. The actions of cytokines and growth factors depend on the local environment in the matrix. In a fibrin clot, TGF-β binds to its receptor on cells rather than the matrix. In the normal connective tissue matrix, TGF-β binds to proteoglycans in preference to its cell surface receptors, limiting its effects. In a fibrin/fibronectin clot, cellular fibronectin receptors bind the matrix, stimulating production of matrix metalloproteinases that are appropriate for remodeling the matrix. In normal connective tissue with less fibronectin, cells produce less metalloproteinase.

The mechanisms that mediate physiological wound repair can also contribute to disease. For example, PDGF that is released from activated platelets in clots at the sites of wounds initiates the cellular events that are required for repair. On the other hand, when the endothelium lining of large arteries is damaged, binding to the exposed basal lamina activates platelets. This stimulates them to release PDGF, which promotes proliferation of fibroblasts and smooth muscle cells in the artery wall, an early step in the development of arteriosclerosis.

Plant Cell Wall

The cell walls of land plants are composite materials consisting of cellulose, other polysaccharides, and glycoproteins (Figs. 32.12 and 32.13). Wood and cotton are two familiar examples of cell wall material that is left behind after plant cells have died. Like the extracellular matrix of animals, plant cell walls not only provide mechanical support but also may influence development. Because of these robust cell walls, plant cells are

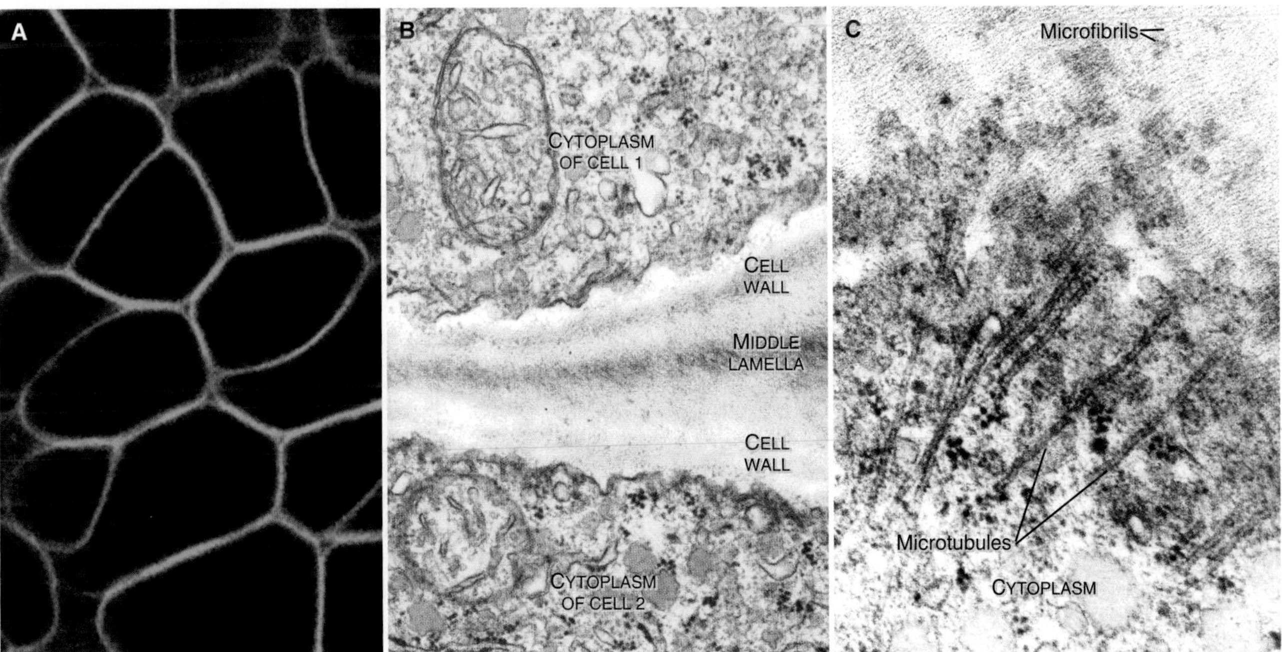

FIGURE 32.12 PLANT CELL WALL. A, Confocal fluorescence micrograph of an *Arabidopsis* leaf with cell walls stained by the periodic acid–Schiff reaction using Acriflavine as the Schiff reagent. **B–C,** Electron micrographs of thin sections of cell walls in the root-like appendages of the parasitic weed dodder. **B,** Two cells are separated by an electron-translucent cell wall consisting of cellulose, xyloglycan, and pectins. The darker area between the two cell walls is the middle lamella, which contains a high concentration of pectins. **C,** At high magnification, an oblique section through the plasma membrane and cell wall shows cellulose microfibrils aligned roughly parallel to cortical microtubules inside the plasma membrane. (**A,** Courtesy Steven E. Ruzin, University of California, Berkeley. **B–C,** Courtesy K.C. Vaughn, U.S. Department of Agriculture, Stoneville, MD.)

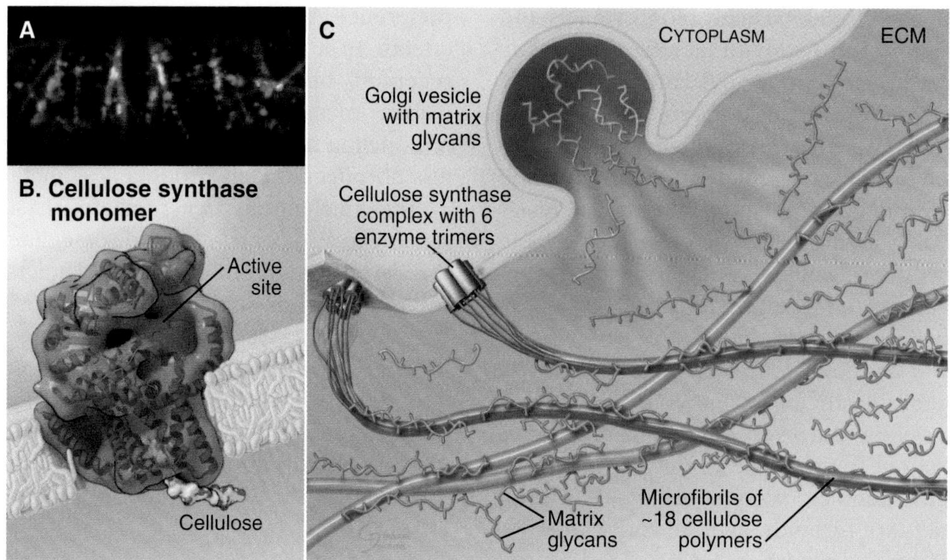

FIGURE 32.13 CELL WALL SYNTHESIS. A, Confocal fluorescence micrograph of an *Arabidopsis* hypocotyl epidermal cell expressing tubulin tagged with cyan fluorescent protein (CFP, shown in *magenta*) and cellulose synthase CESA6 tagged with yellow fluorescent protein (YFP, shown in *green*). This is a superimposition of five successive images taken at 10-second intervals to show green particles of CESA aligned with the magenta microtubules. **B,** Ribbon diagram and space-filling module of the crystal structure of a bacterial cellulose synthase showing the eight transmembrane helices, the glycosyltransferase domain in the cytoplasm between helices 4 and 5, and the growing cellulose polymer *(white and red)* threading across the membrane. **C,** Schematic showing the biosynthesis of the cell wall. ECM, extracellular matrix. (**A,** Courtesy R. Gutierrez, J. Lindeboom, and D. Erhardt, Stanford University. For reference, see Paredez AR, Somerville CR, Ehrhardt DW. Visualization of cellulose synthase demonstrates functional association with microtubules. *Science.* 2006;312:1491–1495. **B,** For reference, see Protein Data Bank [www.rcsb.org] file 4HG6 and Morgan JL, Strumillo J, Zimmer J. Crystallographic snapshot of cellulose synthesis and membrane translocation. *Nature.* 2013;493:181–186. **C,** Modified from Cosgrove DJ. Loosening of plant cell walls by expansins. *Nature.* 2000;407:321–326.)

not motile. Therefore the morphology of plants is established by the orientation of the cell divisions that occur during their development. Two types of forces act on cell walls. Internally, the vacuole of the plant cell applies a high turgor pressure on the order of one atmosphere. Cell walls also resist a variety of external mechanical forces that tend to deform the cell.

The main constituent of cell walls is **cellulose,** the most abundant biopolymer on earth. It is a long, unbranched polymer of glucose (see Fig. 3.25). Cellulose polymers associate laterally into 5- to 7-nm bundles called **microfibrils** (Fig. 32.13C). Two types of branched polysaccharides—**hemicelluloses** and **pectins**—associate with cellulose microfibrils along with many proteins. Plasma membrane enzymes synthesize cellulose, while the other components come from the secretory pathway and associate with cellulose outside the cell. Products of more than 1000 genes are thought to participate in cell wall synthesis.

Plants inherited their genes for **cellulose synthases** from bacteria. *Arabidopsis* has genes for approximately 10 different cellulose synthases. These enzymes consist of eight transmembrane helices with a cytoplasmic β-glycosyltransferase domain similar to hyaluronan synthase and chitin synthase (Fig. 32.13B). The active site is exposed to the cytoplasm to provide access to uridine diphosphate (UDP)-glucose that supplies the glucose added to the polymer. The transmembrane helices form a channel for the cellulose polymer across the membrane. These transmembrane enzymes form a rosette of six particles that are visible by electron microscopy, each particle likely to consist of three enzyme subunits.

Outside the cell the cellulose polymers assemble into linear crystals called microfibrils. The number of polymers per microfibril was long thought to be 36, but 18 is now the accepted number. Hydrogen bonds constrain the glucose units to face in alternate directions in planar ribbons (Fig. 3.25A). These ribbons self-assemble laterally into planar crystalline sheets, which stack vertically into paracrystalline bundles that are held together by C-H•••O hydrogen bonds.

Cellulose synthesis moves the rosettes of cellulose synthase in the plane of the plasma membrane along paths defined by cytoplasmic microtubules (Fig. 32.13A). Typically, cytoplasmic microtubules, the tracks of cellulose synthases, and the cellulose microfibrils are all aligned like barrel hoops perpendicular to the axis of cellular growth to allow for directed (or anisotropic) expansion of the cell wall. Cellulose synthesis continues without microtubules but it is not so well organized. Newly synthesized microfibrils are deposited between the cell surface and older cell wall components.

A large number of glycosyltransferases and other enzymes in the Golgi apparatus synthesize pectins and hemicelluloses, which are transported in vesicles to the surface for secretion. Pectins are acidic polysaccharides that form a gel between microfibrils and play important roles in cell wall expansion during growth and development. Hemicelluloses are branched polysaccharides that coat microfibrils but are less important than pectins. Primary cell walls, laid down at the time of cellular growth and expansion, mature with the addition of glycoproteins and organic molecules, such as **lignins** (polymers of phenylpropanoid alcohols and acids), which contribute to the integrity of the "secondary" cell wall. Covalent and noncovalent bonds are thought to link cellulose and these other matrix molecules. The great strength and flexibility of tree branches illustrate the remarkable mechanical properties of mature cell walls.

Cellulose microfibrils are flexible and have a tensile strength greater than that of steel, so they do not stretch. For a plant tissue to expand, microfibrils and bonds among wall components must rearrange, processes facilitated in plants and bacteria by matrix proteins called **expansins**. The mechanism is still being investigated, but expansins may break noncovalent links between the polymers transiently, allowing turgor pressure to expand the volume of the cell. Genetic defects in expansins inhibit the growth of plant tissues and the ripening of some fruits, such as tomatoes. Expansins in grass pollen are one allergen responsible for hay fever. Localized chemical modifications of pectins can also loosen the cell wall and contribute to expansion of plant cells.

Little is known about the molecular basis of plant cells adhering to their cell walls. By virtue of their physical connection with their product, cellulose synthases offer one means of attachment. Other plasma membrane proteins, including a family of serine/threonine kinases and some proteins with glycosylphosphatidylinositol anchors, may contribute to adhesion by binding cell walls. Integrins are conspicuously missing from plant cells.

ACKNOWLEDGMENT

We thank David Ehrhardt for his suggestions on revisions of this chapter.

SELECTED READINGS

Capulli M, Paone R, Rucci N. Osteoblast and osteocyte: games without frontiers. *Arch Biochem Biophys*. 2014;561:3-12.

Charles JF, Aliprantis AO. Osteoclasts: more than "bone eaters." *Trends Mol Med*. 2014;20:449-459.

Cosgrove DJ. Loosening of plant cell walls by expansins. *Nature*. 2000; 407:321-326.

Georgelis N, Nikolaidis N, Cosgrove DJ. Bacterial expansins and related proteins from the world of microbes. *Appl Microbiol Biotechnol*. 2015;99:3807-3823.

Goldring MB, Berenbaum F. Emerging targets in osteoarthritis therapy. *Curr Opin Pharmacol*. 2015;22:51-63.

Golub EE. Role of matrix vesicles in biomineralization. *Biochim Biophys Acta*. 2009;1790:1592-1598.

Martin TJ. Bone biology and anabolic therapies for bone: current status and future prospects. *J Bone Metab*. 2014;21:8-20.

McFarlane HE, Döring A, Persson S. The cell biology of cellulose synthesis. *Annu Rev Plant Biol*. 2014;65:69-94.

Newman RH, Hill SJ, Harris PJ. Wide-angle x-ray scattering and solid-state nuclear magnetic resonance data combined to test models for cellulose microfibrils in mung bean cell walls. *Plant Physiol*. 2013; 163:1558-1567.

Ornitz DM, Marie PJ. Fibroblast growth factor signaling in skeletal development and disease. *Genes Dev*. 2015;29:1463-1486.

Peaucelle A, Wightman R, Höfte H. The control of growth symmetry breaking in the *Arabidopsis* hypocotyl. *Curr Biol*. 2015;25: 1746-1752.

Sobacchi C, Schulz A, Coxon FP, Villa A, Helfrich MH. Osteopetrosis: genetics, treatment and new insights into osteoclast function. *Nat Rev Endocrinol*. 2013;9:522-536.

Tao J, Battle KC, Pan H, et al. Energetic basis for the molecular-scale organization of bone. *Proc Natl Acad Sci USA*. 2015;112:326-331.

Yuan X, Serra RA, Yang S. Function and regulation of primary cilia and intraflagellar transport proteins in the skeleton. *Ann N Y Acad Sci*. 2015;1335:78-99.

Zhong R, Ye ZH. Secondary cell walls: biosynthesis, patterned deposition and transcriptional regulation. *Plant Cell Physiol*. 2015;56: 195-214.

APPENDIX 32.1

Genetic Defects of Cartilage and Bone

Protein	Species	Mutation	Phenotype
Growth Factors			
BMP-4	Human	Overexpression	Fibrodysplasia progressiva; ectopic bone formation
BMP-5	Mouse	Null	Defective ears and sternum *(short ear mutation)*
CSF-1	Mouse	Null	Osteopetrosis; reduced osteoclasts *(osteopetrotic mutation)*
GDF-5 (TGF-β family)	Mouse	Null	Reduced size of long bones; no joints *(brachypodism mutation)*
Growth hormone	Human	Null	Reduced size of bones
OPG (osteoprotegerin)	Human	Null	Recessive juvenile Paget disease with excess bone remodeling
PTHrP	Human	Null	Reduced chondrocyte growth; epiphyseal plates fused at birth
RANKL	Mouse	Null	Osteopetrosis; no osteoclasts
Sclerostin	Human	Loss of function	Dense bones, van Buchem disease
Wnt1	Human	Loss of function	Osteoporosis
Signal Transduction Components			
c-Src	Mouse	Null	Osteopetrosis; osteoclasts fail to attach to or degrade bone
Connexin 43	Human	Point mutations	Dominant oculodentodigital dysplasia
FGF receptor 1	Human	Point mutation	Pfeiffer syndrome; cranial synostosis; long bone defects
FGF receptor 2	Human	Point mutation	Jackson-Weiss syndrome; cranial synostosis; long bone defects
FGF receptor 2	Human	Point mutation	Crouzon disease; cranial synostosis
FGF receptor 3	Human	Point mutation	Gain-of-function mutation; achondroplasia; short, wide bones
LRP5 Wnt coreceptor	Human	Loss of function	Osteoporosis-pseudoglioma syndrome
Transcription Factors			
c-fos	Mouse	Null	Osteopetrosis; no osteoclasts
hoxa-2	Mouse	Null	Deletion of the second branchial arch; duplication of first branchial arch
hoxd-13	Mouse	Null	Deletion fourth sacral derivatives; duplication third sacral derivatives
msx-1	Mouse	Null	Cleft palate
msx-2	Mouse	Null	Craniosynostosis (fusion of skull bones)
Runx2/Cbfa-1	Human Mouse	+/- Null	Dominant skeletal defects (cleidocranial dysplasia) No osteoblasts or bone
SOX9	Human	Point mutations	Dominant cartilage and skeletal defects (campomelic dysplasia)
Collagen and Other Structural Components of Cartilage and Bone			
Aggrecan	Mouse	Missense	Recessive cartilage deficiency; dwarfism; cleft palate
Cathepsin-K	Mouse	Deletion	Osteopetrosis
CLC7	Human	Point mutations	Osteopetrosis
COL1	Human	Missense, deletions	Dominant osteogenesis imperfecta; fragile bones
COL2	Human	Nonsense	Dominant Stickler syndrome; chondrodysplasia, eye defects
	Human	Point mutations	Dominant chondrodysplasia and osteoarthritis of variable severity
COL9A2	Human	Splicing mutation	Defective cartilage with degeneration of knee joint
COL10A1	Human	Point mutations	Dominant Schmid metaphyseal chondrodysplasia with short bones
COL11A2	Human	Exon skipping	Dominant Stickler syndrome; chondrodysplasia, eye defects
	Human	Point mutation	Recessive severe chondrodysplasia; deafness; cleft palate
Lysyl hydroxylase	Human	Point mutation	Bruck disease; fragile bones
Perlecan	Mouse	Deletion	Recessive defects in cartilage and bone formation
Proton ATPase	Human	Point mutations	Osteopetrosis
Sulfate transporter	Human	DTDST gene	Recessive cartilage defects; short limbs; joint deformation

ATPase, adenosine triphosphatase; BMP, bone morphogenetic protein; CSF, colony stimulating factor; FGF, fibroblast growth factor; PTHrP, parathyroid hormone-related protein; RANKL, receptor activator of nuclear factor κB ligand; TGF, transforming growth factor.

SECTION IX

Cytoskeleton and Cellular Motility

SECTION IX OVERVIEW

The seven chapters in this section cover the cytoskeleton and cellular motility. These topics are intimately related, because two of the protein polymers constituting the cytoskeleton—the internal scaffolding of the cell—are also tracks for motor proteins that power many cellular movements. Assembly and disassembly of the cytoskeletal polymers also produce some types of cell movements.

Most organisms depend on motility to sustain life itself. Without a motile sperm the egg would not be fertilized. Without cellular motility a fertilized egg would not progress past the single-cell stage. Without active changes in cell shape and cellular migrations complex embryos would not form. Without cellular motility white blood cells would neither accumulate at sites of inflammation nor ingest invading microorganisms. Without active and rapid movements of organelles in axons and large plant cells the peripheral parts of these cells would not be nourished. Without muscle contractions we would be paralyzed and unable to move. Even a yeast, prevented from locomotion by its rigid cell wall, depends on internal movements for cell division and endocytosis. Many prokaryotes use rotary flagella for locomotion. Consequently, an understanding of the basis of cellular motility is central to our understanding of the functioning of all cells and organisms.

This section starts with Chapters 33 to 35, which introduce the three cytoskeletal polymers, and Chapter 36, which explains the mechanisms of motor proteins. Three concluding chapters show how cells use

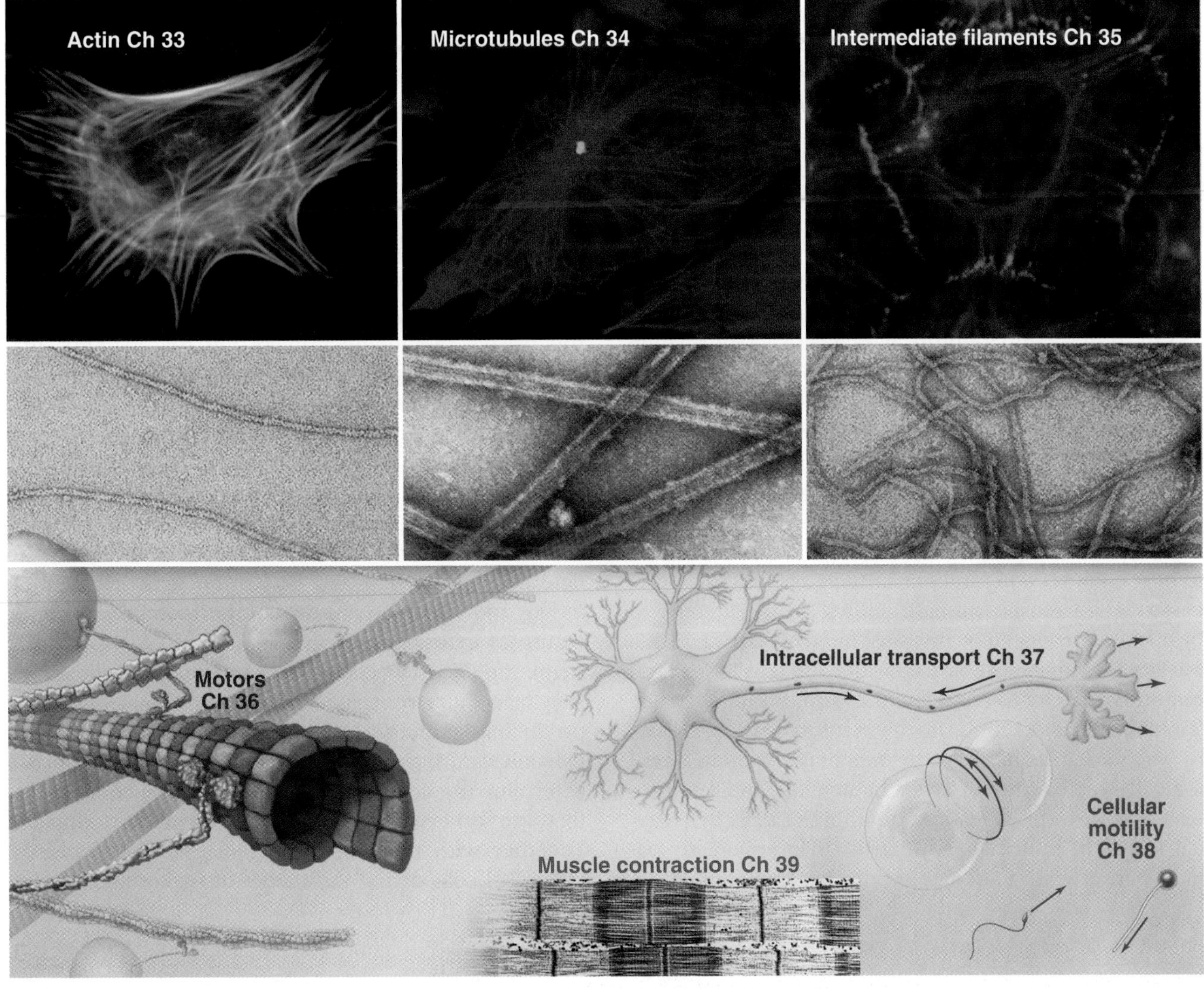

cytoskeletal polymers and motors to produce a vast variety of movements: intracellular movements (Chapter 37); cell shape changes, cellular locomotion, and swimming (Chapter 38); and muscle contraction (Chapter 39). Mitosis and cytokinesis appear in our discussion of the cell cycle (Chapter 44).

Actin filaments (Chapter 33) and **microtubules** (Chapter 34) have much in common, including their evolutionary origins in prokaryotes. Both assemble spontaneously into polymers that are used as tracks by molecular motors. The protein subunits of both polymers bind a nucleoside triphosphate: adenosine triphosphate (ATP) in the case of actin and guanosine triphosphate (GTP) for tubulin. After polymerization, hydrolysis of these bound nucleotides and dissociation of the γ-phosphate destabilize the polymer, much more so in the case of microtubules than in the case of actin filaments. Both polymers can turn over rapidly in cells or remain as stable components. Cells use many proteins to regulate the assembly of these polymers: Some proteins bind to the cytoplasmic pools of the subunit proteins; others initiate the assembly; some stabilize the polymers, others sever or depolymerize them; still others link the polymers together or to other cellular constituents. During divergent evolution from the common ancestor, contemporary organisms came to use actin filaments and microtubules for some of the same functions. For example, microtubules separate chromosomes in eukaryotes, while homologs of actin separate plasmids in bacteria. Actin filaments are tracks for long distance transport in plants, whereas microtubules serve the same purpose in animal nerve cells.

Actin filaments and microtubules cooperate with a third polymer called **intermediate filaments** (Chapter 35) to form a **cytoskeleton,** which resists deformation and transmits mechanical forces. Microtubules are rigid, hollow reinforcing rods that sustain both compression and tension. These mechanical properties make microtubules useful for supporting asymmetrical cellular processes and for bidirectional traffic generated by the motor proteins kinesin and dynein. Actin filaments are more flexible, so they must be crosslinked into bundles to bear compression forces or support asymmetrical processes. High tensile strength allows actin filaments to bear forces produced by myosins. Intermediate filaments are flexible cables that have considerable tensile strength but little capacity to resist compression. Both intermediate filaments and actin filaments reinforce whole tissues by anchoring cadherins, transmembrane proteins used for cell-to-cell adhesion (see Chapter 30). Intermediate filaments prevent excessive stretching of cells by external forces in multicellular animals. If intermediate filaments are defective, tissues are mechanically fragile.

Most movements of eukaryotic cells depend on actin filaments and microtubules. Assembly and disassembly of actin filaments and microtubules produce force for several types of cellular movements (Chapter 37). Actin polymerization drives extension of pseudopods at the leading edge of motile cells. Hydrolysis of ATP bound to actin regulates recycling of subunits rather than being used directly to produce force. Growth of microtubules supports the extension of some asymmetrical cellular processes, including nerve cell processes.

Many other cellular movements result from the physical movements of protein motors (Chapter 36) along actin filaments and microtubules in cytoplasm. Different motors move along these two polymers: **myosins** move on actin filaments, and **dyneins** and **kinesins** move along microtubules. These motors use energy released from the hydrolysis of ATP to take nanometer steps along their protein polymer tracks. These small steps apply force and move cargo attached to the motor. The cargo includes membrane-bound organelles, macromolecular complexes, and cytoskeletal polymers. Microtubule motors power most organelle movements in animal cells (Chapter 37), chromosomal movements during mitosis (Chapter 44), and beating of cilia and flagella (Chapter 38). The actin–myosin system is responsible for cytokinesis (see Chapter 44), some organelle movements (especially in plants and fungi [Chapter 37]), and muscle contraction (Chapter 39).

Several motility systems do not depend on actin filaments or microtubules (Chapter 38). Nematode sperm use the reversible assembly of another protein to make pseudopods for their movements. Calcium-sensitive contractile fibers cause rapid contractions of some protozoa. A proton or sodium ion gradient across the plasma membrane powers the rotary motor that turns bacterial flagella. Although not usually considered to be molecular motors, nucleic acid polymerases and helicases use ATP hydrolysis to move along polymers of DNA or RNA.

The ability of actin filaments and microtubules to resist mechanical deformation and to transmit forces from motors allows the cytoskeletal-motility system to determine cell shape and hence the structure of both tissues and whole organisms. Furthermore, the dynamic nature of cytoskeletal polymers allows cells to change shape rapidly, in a time frame of seconds. Active extension of cellular processes and active changes in shape produce asymmetrical cell shapes. Movements of chromosomes during mitosis and organelles in cytoplasm determine the cellular distribution of these components that are otherwise too large to move by diffusion. Together with the extracellular matrix, the shapes of individual cells define the shapes of tissues and organs.

CHAPTER 33

Actin and Actin-Binding Proteins

Actin filaments form cytoskeletal and motility systems in all eukaryotes (Fig. 33.1). Crosslinked actin filaments resist deformation, transmit forces, and restrict diffusion of organelles. A network of cortical actin filaments excludes organelles (Fig. 33.2C), reinforces the plasma membrane, and restricts the lateral motion of some integral membrane proteins. The **cortex** varies in thickness from a monolayer of actin filaments in red blood cells (see Fig. 13.11) to more than 1 μm in amoeboid cells (Fig. 33.2C). Like fingers in a glove, bundles of actin filaments support slender protrusions of plasma membrane called **microvilli** or **filopodia** (Fig. 33.2B). Microvilli expand the cell surface for transport of nutrients and participate in sensory processes, including hearing. The actin cytoskeleton complements and interacts physically with cytoskeletal structures composed of microtubules (see Chapter 34) and intermediate filaments (see Chapter 35).

Actin contributes to cell movements in two ways. First, polymerization and depolymerization of the network of actin filaments just inside the plasma membrane contribute to the extension of pseudopods, cell locomotion (Fig. 33.2D–E), and phagocytosis (see Fig. 22.3). Second, actin filaments are tracks for movements of the myosin family of motor proteins (see Fig. 36.7). Actin filaments and myosin filaments form the highly ordered, stable contractile apparatus of muscles (Fig. 33.3B; also see Fig. 39.3), as well as the transient **contractile ring** that helps separate the two daughter cells at the end of mitosis (Fig. 33.3A; also see Fig. 44.24). Myosins also power movements of membranes and other cargo along actin filaments, complementing

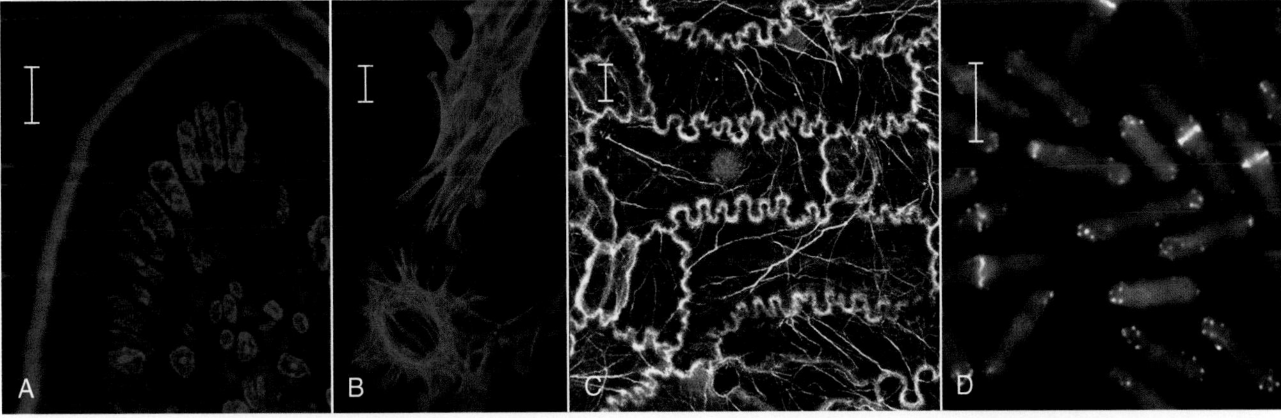

FIGURE 33.1 FLUORESCENCE MICROGRAPHS ILLUSTRATING THE DISTRIBUTION OF ACTIN FILAMENTS IN CELLS. A, Intestinal epithelial cells stained *red* with rhodamine-labeled phalloidin, a cyclic peptide that binds actin filaments. Actin filaments are concentrated in a band of microvilli bordering the intestinal lumen. Nuclei are stained *blue* with DAPI (4,6-diamidino-2-phenylindole). **B,** Cultured vascular smooth muscle cells. Actin filaments, stained *red* with a fluorescent antibody, are concentrated in stress fibers and in the cortex around the edges of these cells. **C,** Maize epidermis with actin filaments stained with rhodamine-labeled phalloidin in the cortex and in cytoplasmic bundles. **D,** Fission yeast *Schizosaccharomyces pombe*, stained with rhodamine-labeled phalloidin. Actin filaments are found in patches at the tips of growing cells and in the cleavage furrow of dividing cells. Scale bars are 10 μm. (**A,** Courtesy C. Rahner, Yale University, New Haven, CT. **B,** Courtesy I. Herman, Tufts Medical School, Boston, MA. **C,** Courtesy M. Frank, University of California, San Diego. **D,** Courtesy W.-L. Lee, Salk Institute, La Jolla, CA.)

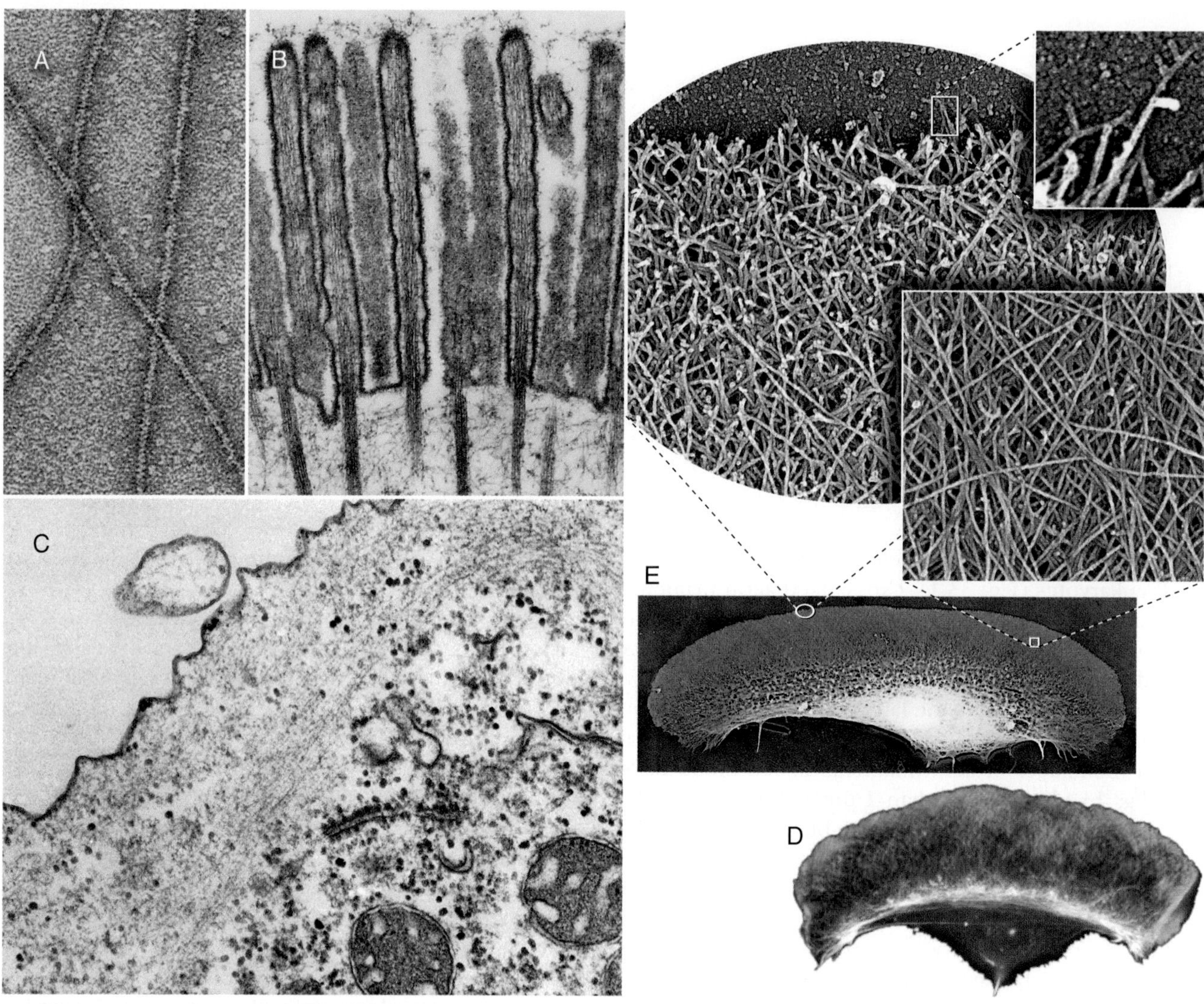

FIGURE 33.2 ELECTRON MICROGRAPHS OF ACTIN FILAMENTS. A, Filaments of purified actin prepared by negative staining. **B,** A thin section of an intestinal epithelial cell illustrating finger-like microvilli with tightly packed bundles of actin filaments linked to the surrounding plasma membrane by myosin-I. The barbed ends of these filaments (see Fig. 33.8) are located at the tips of the microvilli. **C,** A thin section of *Acanthamoeba* showing the actin filament meshwork in the cortex (the *gray* area here) beneath the plasma membrane. **D,** Fluorescence micrograph of a cultured fish scale keratocyte fixed while actively migrating toward the *top* of the figure. Actin filaments are stained blue with phalloidin and myosin II stained red with antibodies. **E,** Electron micrographs of actively migrating fish scale keratocytes fixed and extracted to remove the plasma membrane and soluble components prior to coating the cytoskeleton with platinum. A meshwork of branched filaments concentrates near the leading edge with longer, unbranched filaments deeper in the cytoplasm. Most filaments are oriented with their barbed ends forward. (**A,** Courtesy U. Aebi, University of Basel, Switzerland. **B,** Courtesy M. Mooseker, Yale University, New Haven, CT. **D–E,** Courtesy T. Svitkina and G. Borisy, University of Wisconsin, Madison.)

organelle movements along microtubules powered by other motors (see Fig. 37.1). Actin, myosin, and accessory proteins form intracellular bundles called **stress fibers** (Fig. 33.1B) that apply tension between adhesive junctions on the plasma membrane (see Fig. 30.11), where cells attach to each other or to the extracellular matrix. Stress fibers are prominent in tissue culture cells grown on glass or plastic and in endothelial cells lining major arteries.

Actin is often the most abundant protein in eukaryotic cells, composing up to 15% of total protein, and the many types of actin-binding proteins may account for another 10% of cellular protein. In muscle, actin and myosin constitute more than 60% of the total protein. Given this abundance, it is curious that actin was discovered in muscle only in the 1940s and in nonmuscle cells in the late 1960s. Since the 1970s, scientists have discovered new actin-binding proteins every year, but the inventory is probably still incomplete. Genetic defects in components of the actin cytoskeletal and motility system cause many human diseases, including muscular dystrophy (see Table 39.3), hereditary fragility of red blood cells (ie, hemolytic anemias, see Fig. 7.11), and hereditary heart diseases called cardiomyopathies (see Table 39.1).

Actin Molecule

Actin is folded into two domains that are stabilized by an adenine nucleotide lying in between (Fig. 33.4). The polypeptide of 375 residues crosses twice between the two domains, with the N- and C-termini located near each other. The two domains are folded similarly, suggesting that the actin gene arose by duplication of an ancestral gene.

Actin binds adenosine triphosphate (ATP) or adenosine diphosphate (ADP) and a divalent cation, Mg^{2+}, in cells, with nanomolar affinity. Actin binds ATP with higher affinity than ADP, so given the higher concentration of ATP in cells, unpolymerized actin is saturated with ATP. The bound nucleotide exchanges relatively slowly with nucleotide in the medium (Fig. 33.11). Different actin monomer-binding proteins can inhibit or promote nucleotide exchange. Bound nucleotide stabilizes the molecule but is not required for polymerization in vitro. However, ATP-actin and ADP-actin polymerize at different rates.

Posttranslational modifications of actins include acetylation of the N-terminus and (in most cases) methylation of histidine-68. In some insect flight muscles, the small protein ubiquitin (see Fig. 23.2) is attached covalently to approximately one in six actin molecules, yielding a 55-kD polypeptide that is incorporated with unmodified actin into filaments. Some invertebrate actins are phosphorylated on tyrosine-211. The functional significance of these modifications is still being investigated.

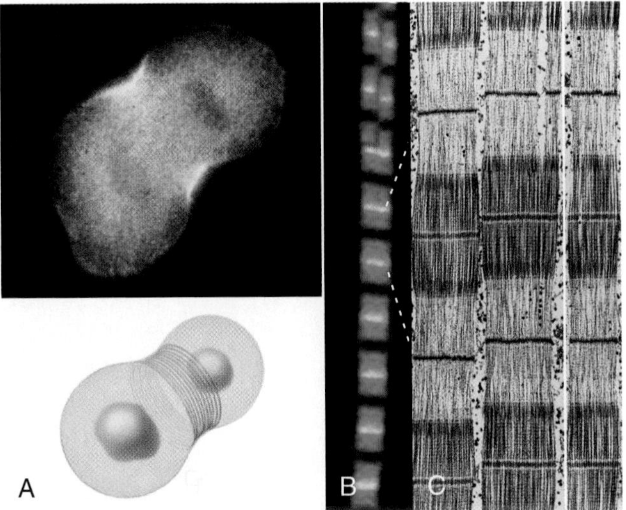

FIGURE 33.3 MICROGRAPHS OF CONTRACTILE BUNDLES OF ACTIN FILAMENTS. A, Fluorescence micrograph of a dividing normal rat kidney cell stained with fluorescein-phalloidin to mark actin filaments in the contractile ring of the cleavage furrow. The drawing illustrates the filaments in the contractile ring. **B,** Fluorescence micrograph of a myofibril isolated from skeletal muscle and stained with fluorescein-phalloidin to label actin filaments *(green)* and rhodamine-antibody to α-actinin to label Z disks *(yellow)*. **C,** Electron micrograph of a thin section of skeletal muscle. (**A,** Micrograph courtesy Y.-L. Wang, University of Massachusetts, Worcester. **B,** Courtesy V. Fowler, Scripps Research Institute, La Jolla, CA. **C,** Courtesy H.E. Huxley, Brandeis University, Waltham, MA.)

Evolution of the Actin Family

The actin gene is ancient, with roots before the universal common ancestor. The primordial gene apparently encoded a nucleotide-binding protein and gave rise to actin, the glycolytic enzyme hexokinase (see Fig. 3.12) and the heat shock protein Hsc70. The three proteins have similar folds with a central ATP binding site, but different functions. The actin genes diversified extensively in prokaryotes giving rise to families of genes for actins with distinct functions: MreB involved with cell shape,

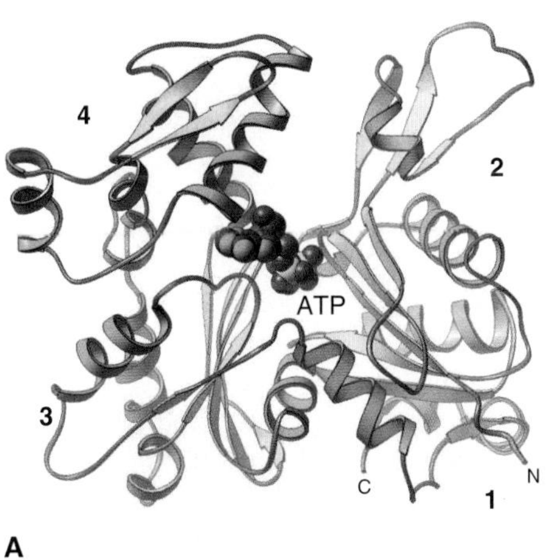

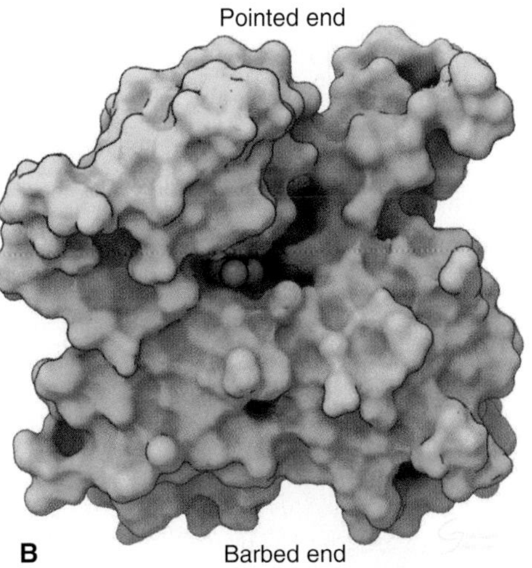

FIGURE 33.4 STRUCTURE OF ACTIN. A, Ribbon model showing the polypeptide fold and the location of Mg-ATP (magnesium–adenosine triphosphate), shown as space-filling. Numbers 1 to 4 indicate the four subdomains. **B,** Surface rendering. ATP is almost completely buried in the cleft between the two lobes of the protein. The barbed end of the molecule (this is a way to describe the polarity of the actin filament; see Fig. 33.8) is at the *bottom* in this orientation. (For reference, see Protein Data Bank [PDB; www.rcsb.org] file 1ATN and Kabsch W, Mannherz HG, Suck D, et al. Atomic structure of the actin-DNase I complex. *Nature.* 1990;347:37–44.)

FtsA that participates in cytokinesis, ParM that separates plasmids, and more than 30 other actin-like proteins expressed from plasmids or by bacteriophages.

The original eukaryotic actin gene came from an archaeal cell related to Lokiarchaeota (see Fig. 2.4B). Eukaryotic actin genes are highly conserved, likely because of constraints imposed by the interactions required to form polymers and numerous regulatory proteins. However, through divergent evolution, the genes encode subtly different proteins, some with novel functions. Most organisms have multiple actin genes, and all known actin isoform diversity arises from multiple genes rather than from alternative splicing of messenger RNAs (mRNAs). Humans have six actin genes; *Dictyostelium* has more than 10; but budding yeast has only one. Muscle actin genes diverged from cytoplasmic actins in primitive chordates (see Fig. 2.8). To fulfill special developmental functions, plant actin genes diverged among themselves more than animal actin genes.

The biochemical similarities of eukaryotic **actin isoforms** are more impressive than their differences. The sequences of pairs of actins are generally more than 90% identical, even between highly divergent eukaryotes. Humans express β and γ isoforms in nonmuscle cells and four different α and β isoforms in various muscle cells. Many nonmuscle cells express both the β and γ isoforms, but red blood cells use only β-actin.

In every case examined, actin isoforms copolymerize in the test tube, so it is remarkable and still unexplained that cells can sort actin isoforms into different structures. For example, β-actin is concentrated near the plasma membrane of cultured cells, whereas γ-actin is concentrated in stress fibers (Fig. 33.5). In muscle, α-actin forms the thin filaments of the contractile apparatus, whereas γ-actin localizes around mitochondria.

Actin-Related Proteins

Genes for **actin-related proteins (Arps)** diverged from actin genes before the last eukaryote common ancestor (Fig. 33.6). Arps share with actin the fold of the polypeptide chain and residues forming the nucleotide-binding site, but fewer than 60% of the overall residues are

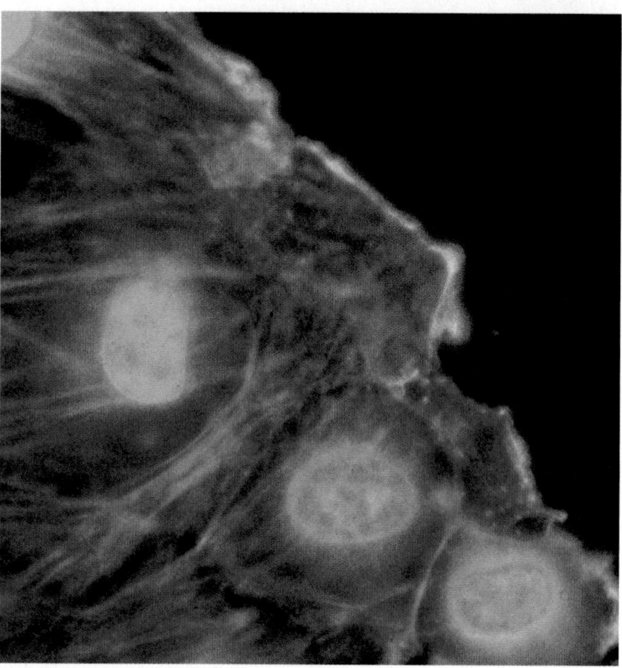

FIGURE 33.5 SORTING OF ACTIN ISOFORMS IN CELLS. Fluorescence micrograph of cultured cells doubly stained with fluorescent antibodies specific for β-actin concentrated at the leading edge *(orange)* and γ-actin concentrated in stress fibers *(green)*. Nuclei are stained *blue* with DAPI (4,6-diamidino-2-phenylindole). (Courtesy I. Herman, Tufts Medical School, Boston, MA.)

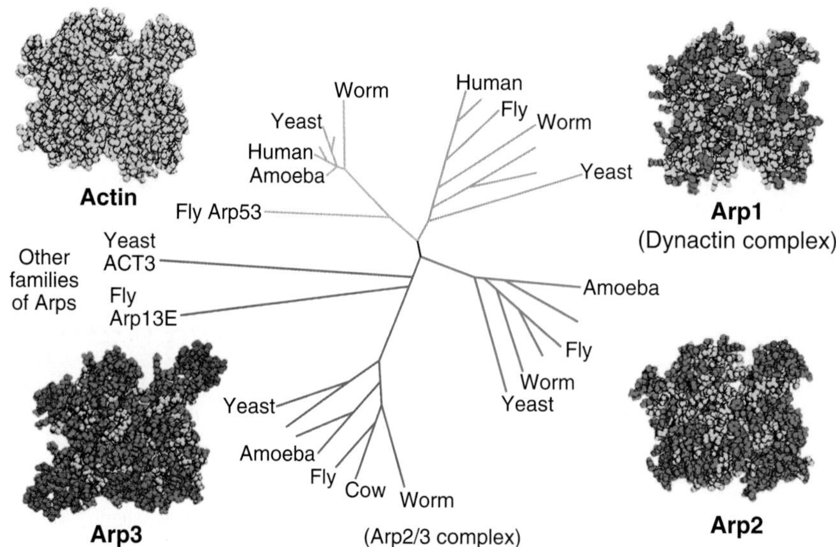

FIGURE 33.6 COMPARISON OF ACTIN AND ACTIN-RELATED PROTEINS. Space-filling models of actin and Arps showing residues identical to actin *(yellow)*, conservative substitutions compared with actin *(green)*, nonconservative substitutions *(blue)*, and insertions *(red)*. The phylogenetic tree, based on sequence comparisons, shows divergence from a common ancestor. *Arps,* Actin-related proteins. (Modified from Mullins RD, Kelleher JF, Pollard TD. Actin' like actin. *Trends Cell Biol.* 1996;6:208–212.)

identical to actin. Divergent surface residues allow Arps to participate in molecular interactions different from actin. Arp1 forms a short filament as part of the dynactin complex that promotes cargo movement by the microtubule motor dynein (see Fig. 37.2). Arp2 and Arp3 are two of seven subunits in a protein complex that nucleates branched actin filaments in the cell cortex (Fig. 33.12). Eight additional types of Arps are widespread in eukaryotes. Several participate in complexes that regulate chromatin structure.

Actin Polymerization

Actin filaments are polarized, owing to the uniform orientation of the asymmetrical subunits along the polymer (Fig. 33.7). The subunits, all pointed in the same direction, form a double helix with the subunits in the two strands staggered by half. One end is called the barbed end, the other is called the pointed end. This nomenclature arises from the asymmetrical arrowhead pattern seen when myosin heads bind along the length of actin filaments (Fig. 33.8).

Actin self-assembles into filaments by means of a series of bimolecular reactions (Fig. 33.9; see also Fig. 5.6). Actin is isolated from cells as a monomer at low salt concentrations. Physiological concentrations of monovalent and divalent cations bind to low-affinity sites on actin and promote polymerization. In vitro, actin trimers appear to be the nucleus that initiates polymer growth. The reactions required to form trimers are very unfavorable in comparison with reactions for elongation of polymers larger than trimers. To initiate new filaments, cells use regulatory proteins to overcome these unfavorable nucleation reactions.

Actin filaments grow and shrink by the addition and loss of subunits at the two ends of the polymer. The reactions at the two ends have different rate constants (Fig. 33.8). Subunit association is a diffusion-limited reaction (see Chapter 4) at the rapidly growing barbed end and somewhat slower at the pointed end. Subunit dissociation is relatively slow at both ends, between 0.3 and 8 subunits per second. The rates of these reactions depend on the nucleotide bound to the monomer associating with or dissociating from a filament.

In the presence of ATP, purified actin assembles almost completely, leaving as monomers the **critical concentration** of ~0.1 μM ATP-actin. The critical concentration is the monomer concentration at which equal rates of association and dissociation occur, 1.4 s^{-1} at the barbed end (see Fig. 5.6). The critical concentration

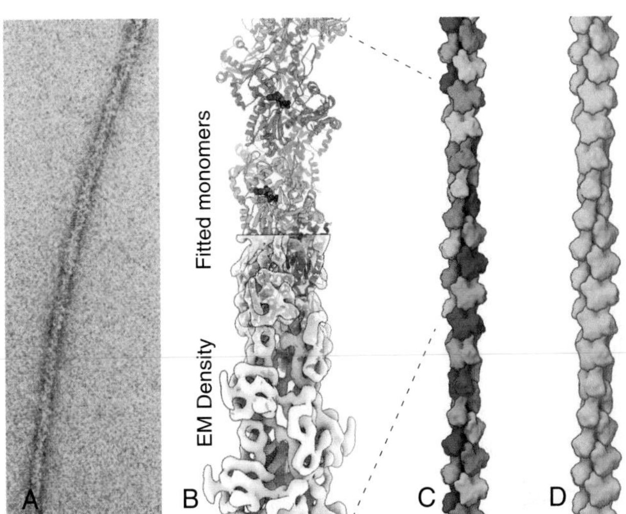

FIGURE 33.7 STRUCTURE OF THE ACTIN FILAMENT. A, Electron micrograph of a negatively stained actin filament. **B,** Reconstruction of the actin filament from electron cryomicrographs at 0.66 nm resolution. **C,** Surface rendering of the molecular model. Subunits in the two long-pitch helices are shown as *yellow-orange* and *blue-green* (see Fig. 5.5 for nomenclature). The short pitch helix, including every subunit, follows a *yellow-green-orange-blue* pattern. **D,** Scale drawing used throughout this text. (**B,** Data from Fujii T, Iwane AH, Yanagida T, Namba K. Direct visualization of secondary structures of F-actin by electron cryomicroscopy. *Nature*. 2010;467:724–728.)

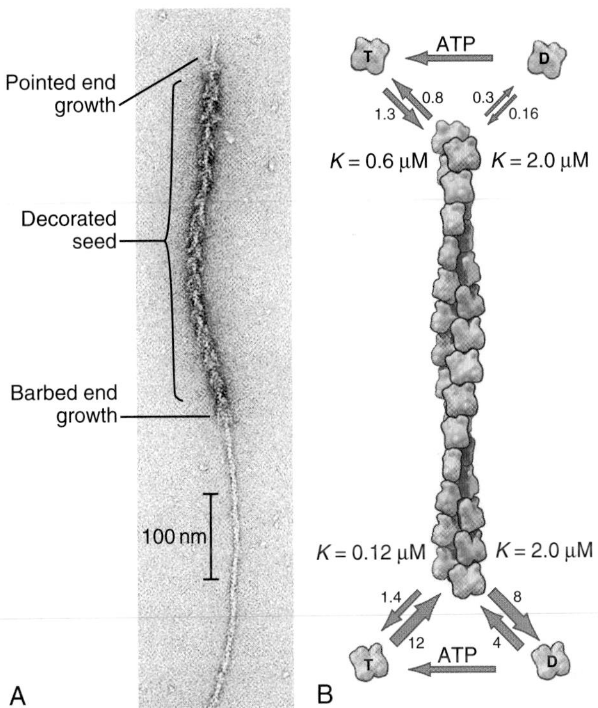

FIGURE 33.8 ACTIN FILAMENT ELONGATION. A, Electron micrograph of growth from an actin filament "seed" decorated with myosin heads to reveal the polarity. Growth is faster at the barbed end than at the pointed end. **B,** Rate constants for association (units: $\mu M^{-1}\ s^{-1}$) and dissociation (units: s^{-1}) for Mg-ATP-actin (T) and Mg-ADP-actin (D) were determined by measuring the rate of elongation at the two ends as a function of monomer concentration. Ratios of the rate constants yield critical concentrations (*K*, units: μM) for each reaction. Critical concentrations at the two ends are the same for ADP-actin but differ for ATP-actin. ADP, adenosine diphosphate. (**A,** Courtesy M. Runge, Johns Hopkins Medical School, Baltimore, MD. **B,** Data from Pollard TD. Rate constants for the reactions of ATP- and ADP-actin with the ends of actin filaments. *J Cell Biol*. 1986;103:2747–2754.)

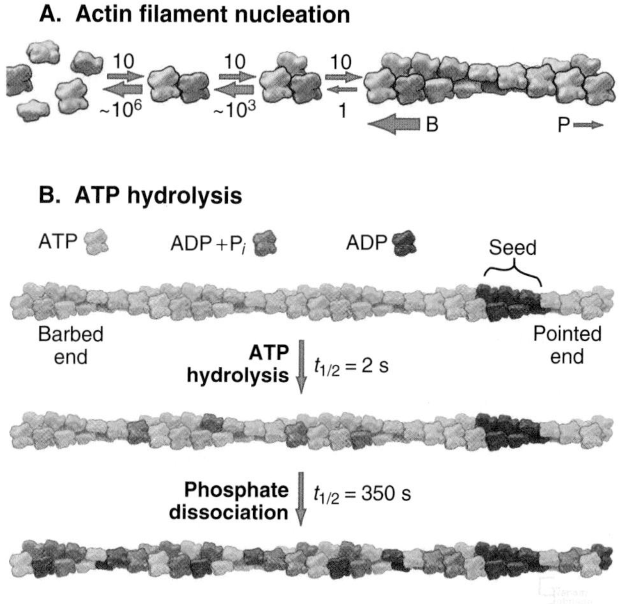

A. Actin filament nucleation

B. ATP hydrolysis

FIGURE 33.9 ACTIN FILAMENT NUCLEATION, GROWTH, AND NUCLEOTIDE HYDROLYSIS. A, Nucleation. Formation of dimers and trimers is very unfavorable, owing to rapid dissociation of subunits. Actin trimers are called nuclei, because they initiate the highly favorable elongation reactions. Rate constants estimated by kinetic modeling have units of μM^{-1} s^{-1} for association reactions and s^{-1} for dissociation reactions. **B,** ATP hydrolysis by a polymer of ATP-actin *(yellow subunits)* is random and irreversible at a rate of 0.3 s^{-1}, yielding subunits with bound ADP and inorganic phosphate *(orange)*. Phosphate (P*i*) dissociates slowly at a rate of 0.002 s^{-1}, converting half of the newly polymerized subunits to ADP-actin *(pink)* in 6 minutes. ADP bound to polymerized subunits does not exchange with nucleotide in the medium. Phosphate binding is reversible, but the affinity is low, so most subunits dissociate phosphate. (For reference, see Pollard TD, Blanchoin L, Mullins RD. Biophysics of actin filament dynamics in nonmuscle cells. *Annu Rev Biophys Biomol Struct*. 2000;29: 545–576.)

for ADP-actin is approximately 20 times higher than for ATP-actin.

Hydrolysis of bound ATP and dissociation of the γ-phosphate after assembly modify the behavior of actin filaments, including their affinity for regulatory proteins. Following incorporation of an ATP-actin subunit into a filament, bound ATP is hydrolyzed irreversibly to ADP and phosphate with a half-time of 2 s (Fig. 33.9). These ADP-P_i subunits behave much like ATP subunits. Phosphate dissociates slowly and reversibly over several minutes. This yields filaments with a core of subunits with tightly bound ADP. At the millimolar concentrations of phosphate in cytoplasm, phosphate is bound to some ADP-actin subunits.

The critical concentrations for ATP-actin differ at the two ends of the filament. This results from differences in the probability of nucleotide hydrolysis and phosphate release at the two ends, which is more likely to expose ADP-subunits at the pointed end. At steady state in the presence of ATP, the actin monomer concentration falls between the critical concentrations at the two ends. Though the polymer and monomer concentrations remain constant, net addition of subunits at the barbed end and net loss of subunits at the pointed end result in the slow migration of subunits through the polymer from the barbed end to the pointed end. This slow process is called *treadmilling*. Neither end exhibits rapid fluctuations in length like those of microtubules (see Fig. 34.6).

Actin-Binding Proteins

In contrast to the slow turnover of filaments of purified actin at steady state, actin filaments in live cells can polymerize and depolymerize rapidly. This requires the control of more than 60 families of actin-binding proteins (Fig. 33.10 and Appendix 33.1). Broadly, these proteins fall into families that bind monomers, sever filaments, cap filament ends, nucleate filaments, promote polymerization, crosslink filaments, stabilize filaments, or move along filaments. Like actin, genes for many of these actin-binding proteins arose in early eukaryotes and are found in protozoa, yeast, plants, and vertebrates.

This section introduces examples of each class of actin-binding protein. However, no actin-binding protein functions in isolation, so the chapter concludes by explaining how ensembles of these proteins work together to regulate actin filament dynamics in cells.

Actin Monomer–Binding Proteins

Proteins that bind actin monomers cooperate with capping proteins to maintain a pool of unpolymerized actin in cells and regulate the nucleotide bound to actin.

Profilins are abundant proteins, found in all branches of the eukaryotic tree. Cells require three different profilin activities for viability: binding actin monomers, catalyzing the exchange of nucleotides bound to actin (Fig. 33.11) and binding to polyproline sequences on other protein such as formins (Fig. 33.14). Profilins also bind membrane polyphosphoinositides. The nucleotide bound to actin monomers determines the affinity for profilin: highest for nucleotide-free actin monomers, followed by ATP-actin and ADP-actin. Profilins bind to the barbed end of actin monomers. This sterically blocks nucleation and pointed end elongation but not association of the profilin-actin complex with the barbed end of filaments. Profilin dissociates rapidly after the profilin-actin complex binds to a barbed end.

Found only in vertebrates, **β-thymosins** are extended peptides of just 43 residues. α-Helices at both ends block the barbed and pointed ends of the actin monomer. They bind ATP-actin monomers with higher affinity than ADP-actin monomers and inhibit both actin polymerization and nucleotide exchange. Thymosin-β_4 is the most abundant actin-binding protein in some animal cells, where it sequesters most of the unpolymerized actin.

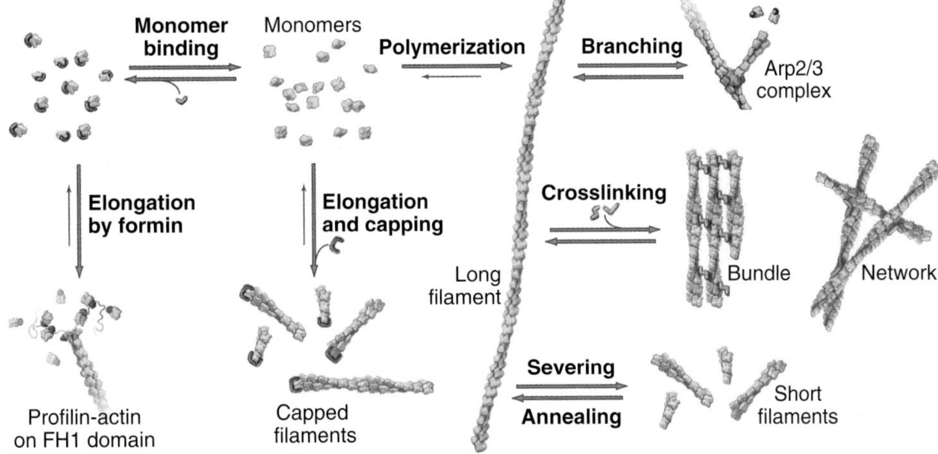

FIGURE 33.10 FAMILIES OF ACTIN-BINDING PROTEINS. Summary of the reactions of regulatory proteins with actin monomers and filaments.

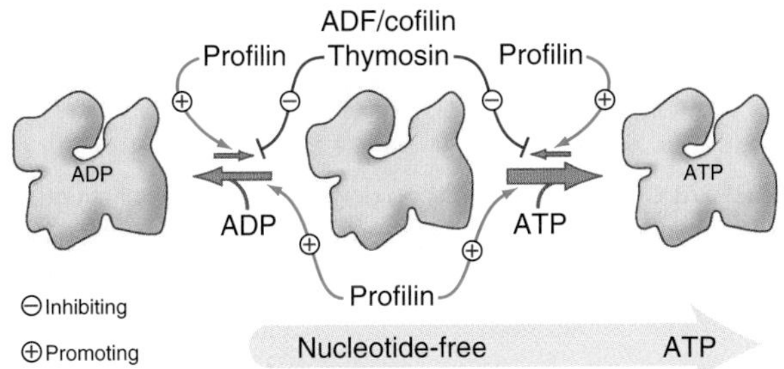

FIGURE 33.11 REGULATION OF ACTIN NUCLEOTIDE EXCHANGE BY ACTIN-BINDING PROTEINS. The rate-limiting step is dissociation of the bound nucleotide from the nucleotide cleft. ADF/cofilins and β-thymosins inhibit dissociation of both ATP and ADP. Profilin competes with both inhibitors for binding actin monomers and increases the rates of both nucleotide dissociation and binding. The ability of profilin to promote nucleotide exchange, the higher affinity of actin for ATP than for ADP, and the higher concentration of ATP than ADP in cytoplasm drive the reactions to the right, so essentially all unpolymerized actin in cells has bound ATP.

Members of the **ADF/cofilin** family bind ADP-actin monomers with higher affinity than ATP-actin and inhibit nucleotide exchange but not polymerization. As explained below, their main function is to sever ADP-actin filaments (Fig. 33.16). These proteins are essential for the viability of many eukaryotes.

Actin Filament Nucleating Proteins

Spontaneous nucleation of filaments from actin monomers is intrinsically unfavorable and inhibited by profilin, so cells use proteins to specify when and where new filaments form. The best characterized are Arp2/3 complex, formins and proteins with multiple WH2 (WASp homology 2) domains such as spire. Each of these has a different evolutionary origin, mechanism of action, and physiological functions.

Arp2/3 complex consists of Arp2 and Arp3 tightly bound to five novel proteins (Fig. 33.12). Arp2/3 complex binds to the side of actin filaments and forms branches with the Arps as the first two subunits at the pointed end of the new filament. Growth of the free barbed ends of these branches produces force that pushes the plasma membrane forward at the leading edge of motile cells (Fig. 33.2E).

Arp2/3 complex is intrinsically inactive, but is stimulated to initiate a branch by interactions with actin filaments and nucleation-promoting factors such as **WASp (Wiskott-Aldrich syndrome protein)**. The name comes from Wiskott-Aldrich syndrome, a genetic disorder with defects in blood cells causing immunodeficiency and bleeding. Other tissues use a closely related nucleation-promoting factor called N-WASP. The C-terminal VCA (verprolin homology, connecting, and acidic) motifs of WASp activate Arp2/3 complex. The V motif (also called WH2 for WASp homology 2) binds an actin monomer and the CA motif binds to two sites on Arp2/3 complex (Fig. 33.13). These interactions promote the binding of Arp2/3 complex to the side of a filament, bringing about a large conformational change that positions Arp3 and Arp2 to initiate a branch. Signaling molecules overcome intramolecular interactions that autoinhibit WASp and N-WASP (Fig. 33.13); Rho-family GTPases (guanosine triphosphatases) and membrane polyphosphoinositides bind near the middle of the protein, and proteins with SH3 domains bind proline-rich sequences.

Additional nucleation promoting factors with VCA motifs regulate Arp2/3 complex in specific cellular locations. For example, Scar/WAVE is active at the leading edge of motile cells, WASH participates in membrane traffic from endosomes and WHAMM is associated with the Golgi apparatus. The small GTPase Rac regulates Scar/WAVE and WASH by freeing them of inhibition by homologous regulatory complexes of four proteins.

Spire was discovered in *Drosophila* as a gene required for the development of eggs and embryos and later found in other metazoans but not fungi or protozoa.

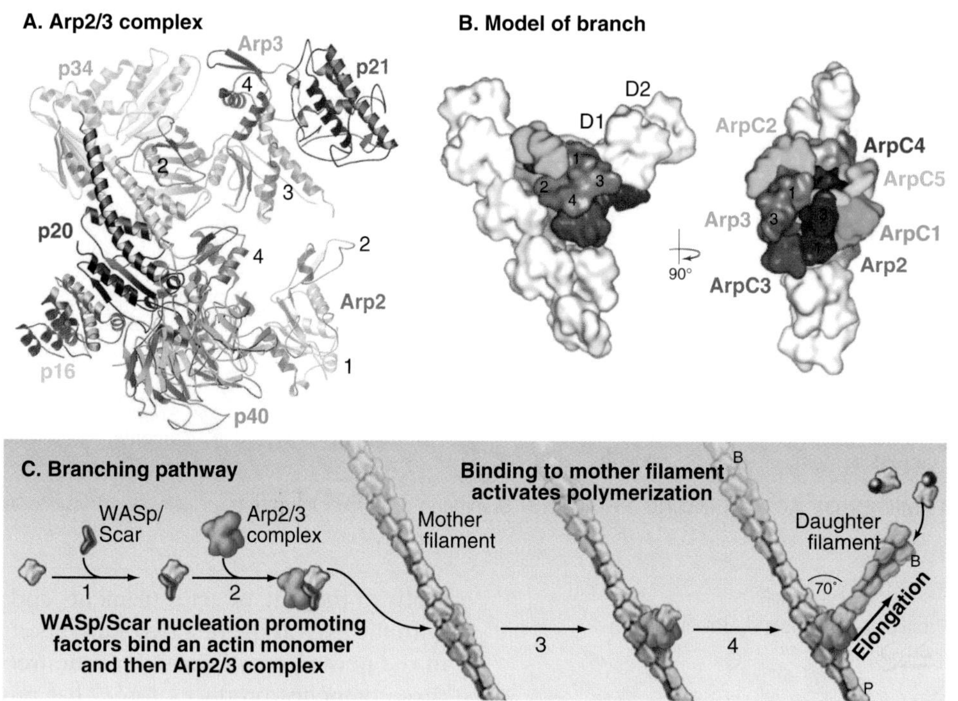

FIGURE 33.12 NUCLEATION OF BRANCHED ACTIN FILAMENTS BY ARP2/3 COMPLEX. A, Ribbon diagram of the crystal structure of Arp2/3 complex. The seven subunits are color coded and labeled. Numbers label the subdomains of Arp2 and Arp3. **B,** Model of a branch based on a three-dimensional reconstruction from electron micrographs. **C,** Steps in branch formation: *(1)* A nucleation-promoting factor binds an actin monomer. *(2)* These binary complexes bind two sites on Arp2/3 complex, bringing together two actin subunits with Arp2 and Arp3. *(3)* The ternary complex binds to the side of an actin filament, completing the activation process. *(4)* A new daughter filament grows at its barbed end from the side of the older mother filament. *B* is the barbed end. *P* is the pointed end. WASH, Wiskott-Aldrich syndrome protein (WASP) and Scar homolog; WHAMM, WASP homologue associated with actin, membranes, and microtubules. (**A,** For reference, see PDB file 1K8K and Robinson R, Turbedsky K, Kaiser DA, et al. Crystal structure of Arp2/3 complex. *Science*. 2001;294:1679–1684. **B,** From Rouiller I, Xu X-P, Amann KJ et al. The structural basis of actin filament branching by the Arp2/3 complex. *J Cell Biol*. 2008;180:887–895.)

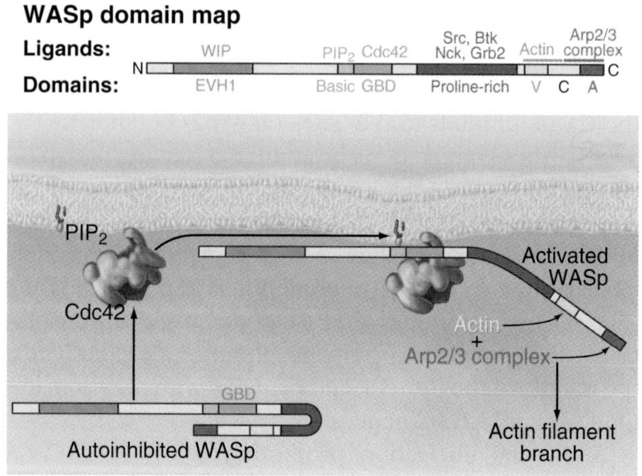

FIGURE 33.13 WASP (WISKOTT-ALDRICH SYNDROME PROTEIN) NUCLEATION PROMOTING FACTOR. The V (verprolin homology) motif binds an actin monomer, and the CA (connecting and acidic) motifs bind Arp2/3 complex. Intramolecular association of C with the GBD (GTPase-binding domain) autoinhibits the protein. Membrane-bound Rho-family guanosine triphosphatases (GTPases) and polyphosphoinositides compete C from GBD, releasing VCA to interact with actin and Arp2/3 complex. SH3 domain proteins such as Nck also activate WASp by binding the proline-rich domain. The N-terminal EVH1 (Ena-VASP [vasodilator-stimulated protein] homology) domain binds WIP (WASp interacting protein).

Spire proteins have multiple WH2 domains (corresponding to the V domains in Fig. 33.13) that bind formins, actin monomers, and nucleate unbranched filaments. One function is to build a meshwork of actin filaments in the oocyte cytoplasm. Other proteins with WH2 domains also nucleate filaments.

Actin Filament Polymerases

Formins not only initiate unbranched actin filaments but remain associated with the elongating barbed end and promote its growth. Filaments produced by formins are incorporated into the contractile ring and bundles in yeast (see Fig. 37.11) and stress fibers in animal cells (Fig. 33.1B). The three formins in fission yeast each assemble actin filaments for specific structures, cytokinetic contractile ring, interphase actin cables, or mating structures. The specific functions of the 15 different mammalian formins are still being investigated.

The characteristic feature of formins is a formin homology-2 (FH2) domain (Fig. 33.14) that forms a doughnut-shaped homodimer around the barbed end of an actin filament. Interactions of the FH2 dimer with two actin monomers nucleate a filament. Then the FH2 domain tracks the growing barbed end by faithfully

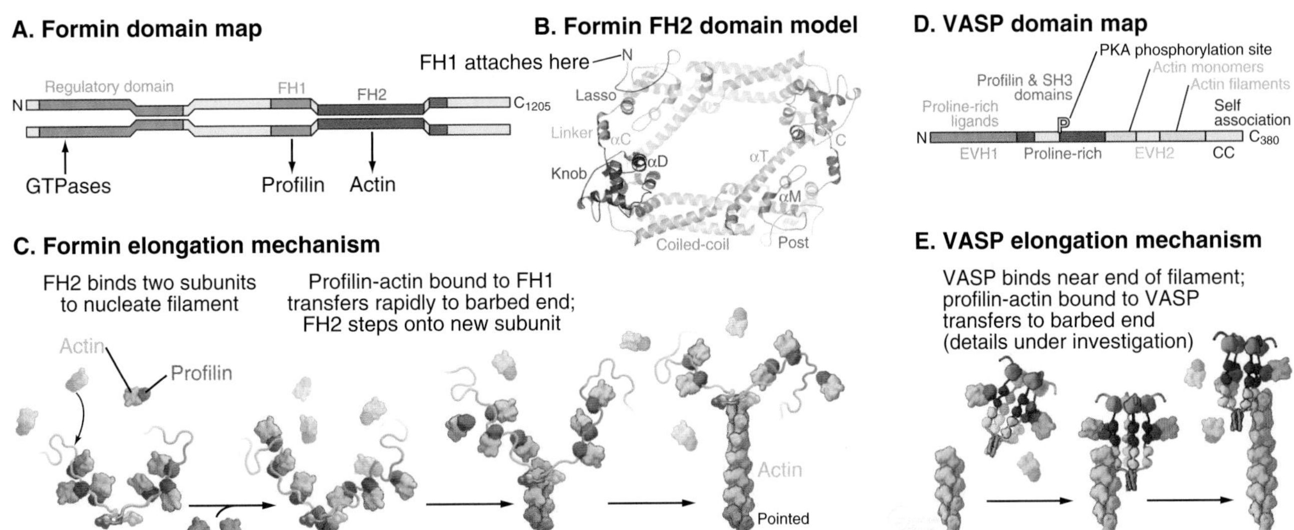

FIGURE 33.14 ACTIN FILAMENT POLYMERASES. A–C, Formins. **A,** Domain structure of a generic, dimeric formin similar to mouse mDia1. Binding of Rho-family guanosine triphosphatases (GTPases) activates formins by disrupting intramolecular associations that autoinhibit the formin homology-2 (FH2) domains. **B,** Ribbon diagram of the structure of the homodimer of FH2 domains of budding yeast Bni1p. The linker segments extend when the dimer fits around an actin filament. **C,** Pathway of actin filament nucleation and elongation by formin FH1 and FH2 domains. Multiple polyproline sequences in FH1 bind profilin-actin for rapid transfer to the barbed end of the filament. **D,** VASP (vasodilator-stimulated phosphoprotein). The EVH1 domain (see Fig. 25.11B) binds polyproline ligands, including vinculin and zyxin, in focal contacts; the proline-rich domain binds profilin-actin complexes; the EVH2 domain binds both actin monomers and filaments; and the C-terminal coiled-coil (CC) mediates the formation of VASP tetramers. **E,** The diagram shows a VASP tetramer delivering profilin-action onto the barbed end of the filament. (**B,** See PDB file 1UX5.)

stepping onto each new subunit as it is added. The FH2 domain also protects the barbed end from capping proteins. Polyproline sequences in the flexible FH1 domain next to FH2 bind multiple profilin-actin complexes that transfer rapidly to the barbed end, enhancing the rate of elongation. Some formins are autoinhibited by interactions between parts of the polypeptide flanking the FH1-FH2 domains. Rho family GTPases activate formins by overcoming this autoinhibition.

An unrelated protein called **VASP** (vasodilator-stimulated phosphoprotein) also stimulates the growth of actin filament barbed ends and protects them from capping in filopodia and at the leading edge of motile cells (Fig. 33.14D–E). VASP tetramers bind near the barbed end of actin filaments, allowing their proline-rich domains to bind and deliver profilin-actin complexes to the growing barbed end, similar to formins. EVH1 domains allow VASP and related proteins to bind proline-rich ligands in focal adhesions and on the surface of the bacterium *Listeria* where it stimulates actin filament elongation.

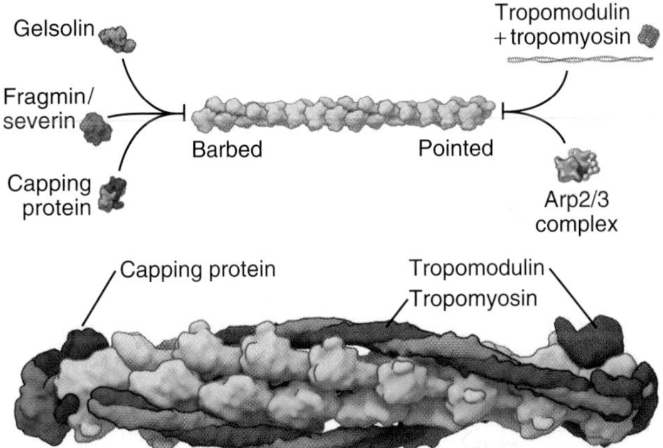

FIGURE 33.15 ACTIN FILAMENT–CAPPING PROTEINS. Interactions of capping proteins with the ends of actin filaments. Most of these proteins bind with high affinity to a filament end. Tropomodulin requires tropomyosin for high-affinity binding. (Roberto Dominguez of the University of Pennsylvania created the PDB file used for the space-filling model of the capped filament with tropomyosin.)

Actin Filament–Capping Proteins

Capping proteins bind to and stabilize either the barbed or pointed end of actin filaments (Fig. 33.15). Some capping proteins also stimulate the formation of new filaments and/or sever actin filaments.

Gelsolins consist of six domains with similar folds but different sequences and activities. They bind tightly to the sides and barbed ends of actin filaments, blocking both the association and dissociation of actin subunits. Gelsolin also binds actin dimers, forming a nucleus that grows at the pointed end. Fragmin and severin, are similar in structure and function to the first three domains of gelsolin. Genes for these capping proteins, found widely among eukaryotes, are likely to have duplicated during evolution to give rise to gelsolin genes.

Heterodimeric capping proteins consist of two homologous subunits of ~30 kD. They cap barbed ends with high affinity and promote nucleation of new pointed ends by stabilizing small actin oligomers. Heterodimeric capping proteins are found in most eukaryotic cells. In striated muscle, they cap the barbed end of actin filaments in the Z disk (see Fig. 39.5). Several different proteins regulate capping protein by interfering directly or indirectly with its binding to barbed ends.

Tropomodulin caps the pointed end of stable actin filaments in muscle, red blood cells, and other animal cells. High-affinity binding to pointed ends requires tropomyosin, an α-helical protein that binds along the length of actin filaments (Fig. 33.15).

Actin Filament–Severing Proteins

Four classes of proteins just introduced—ADF/cofilin, some formins, fragmin/severin, and gelsolin—also sever actin filaments into short fragments (Fig. 33.16). ADF/cofilins bind ADP-actin subunits in filaments and promote severing and depolymerization. A few formin isoforms can sever actin filaments. Ca^{2+} triggers gelsolin, fragmin, and severin to sever actin filaments. Domain 2 of gelsolin binds to the side of an actin filament, positioning domain 1 to bind between subunits and disrupt the filament. One gelsolin isoform is found inside cells; another is secreted into blood plasma, where it may sever actin filaments released from damaged cells.

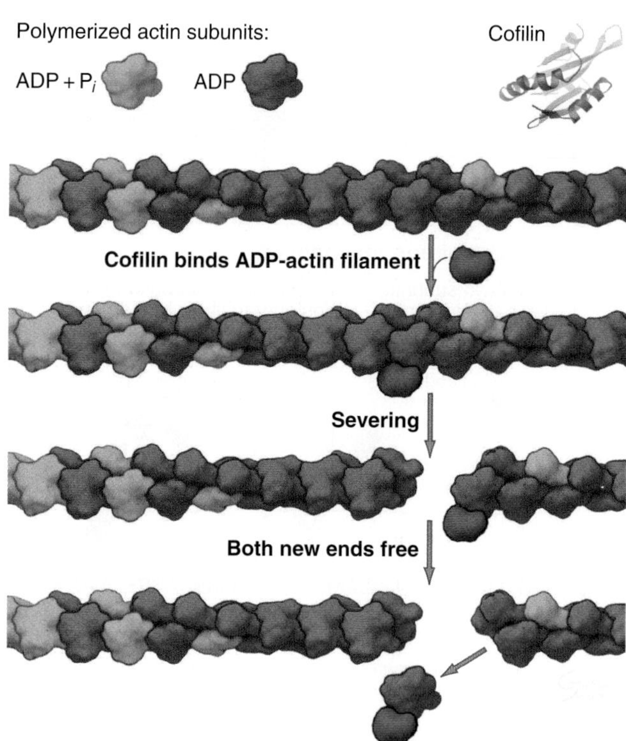

FIGURE 33.16 ACTIN FILAMENT SEVERING BY COFILIN. ADF/cofilins bind to ADP-actin filaments and stabilize a rare, but naturally occurring, tighter helical twist. This destabilizes and severs the filament. The products are two uncapped ends that are available for subunit association and dissociation. *Pi*, phosphate. (For reference, see Elam WA, Kang H, De la Cruz EM. Biophysics of actin filament severing by cofilin. *FEBS Lett.* 2013;587:1215–1219.)

Proteins That Bind the Side of Actin Filaments

Tropomyosin, nebulin, and **caldesmon** are extended proteins that bind along the sides of actin filaments. Tropomyosin increases the tensile strength of actin filaments in striated muscles, it is also an essential component of the Ca^{2+}-sensitive regulatory machinery that controls the interaction of myosin and actin (see Fig. 39.4). Nebulin is one factor that determines the length of the actin filaments in skeletal muscle (see Chapter 39). Caldesmon, together with tropomyosin and Ca^{2+}-calmodulin, regulates interaction of actin and myosin in smooth muscle (see Fig. 39.24) and nonmuscle cells.

Actin Filament Crosslinking Proteins

Possession of two actin filament-binding sites enables crosslinking proteins (Fig. 33.17) to bridge filaments and to stabilize higher-order assemblies of actin filaments. Some have a greater tendency to crosslink filaments in regular bundles, like those in microvilli (Fig. 33.2B), but depending on protein concentrations and filament lengths, most of these proteins can form both random networks of crosslinked filaments and regular bundles of filaments.

Many of these proteins have actin binding domains consisting of two calponin-homology domains. **α-Actinin** is found in the cortical actin network, at intervals along stress fibers, on the cytoplasmic side of cell adhesion plaques (see Fig. 30.11), and in the Z-disk of striated muscles (see Fig. 39.5). Ca^{2+} binding to EF hands (see Chapter 26) of some α-actinins inhibits binding to actin. Fimbrin and villin (a relative of gelsolin with an extra actin-binding site) stabilize the regular actin filament bundles in microvilli. Filamin crosslinks filaments in the cortex of many cells and also anchors these filaments to an integrin, a plasma membrane receptor for adhesive glycoproteins (see Fig. 30.11). Actin filament crosslinking proteins of the plasma membrane skeleton, such as spectrin (see Fig. 7.11) and dystrophin (see Fig. 39.9), are anchored to integral membrane proteins.

Functional Redundancy of Actin-Binding Proteins

The diversity and the apparent redundancy of actin-binding proteins are striking. Why should most organisms retain genes for 60 or more actin-binding proteins if they have such a limited repertoire of functions: monomer binding, nucleation, capping, severing, crosslinking, stabilizing, and motility? Null mutations show that some proteins are essential for normal physiology. These include myosin-II, Arp2/3 complex, profilin, and cofilin in yeast. On the other hand, organisms can survive genetic deletion of some actin-binding proteins, suggesting that parts of the system are redundant. For example, in a laboratory environment, *Dictyostelium* tolerates the loss of crosslinking proteins (α-actinin or

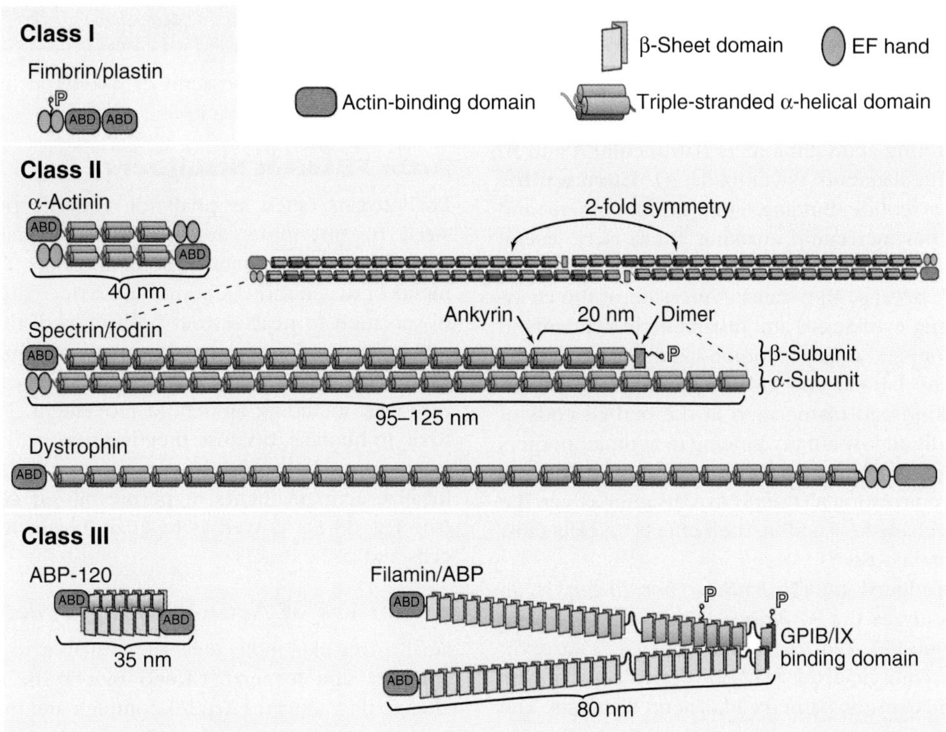

FIGURE 33.17 Actin filament crosslinking proteins sharing homologous actin-binding domains *(red)*. Crosslinking requires two actin-binding sites, which can be part of one polypeptide (fimbrin) or on different subunits of dimeric proteins (α-actinin, filamin). Dystrophin has a second actin-binding site in the middle of the tail of triple helical repeats. (Modified from Matsudaira P. Modular organization of actin cross-linking proteins. *Trends Biochem Sci*. 1991;16:87–92.)

ABP-120), severin, or one of two profilin genes with only minor defects in behavior and growth. Humans who lack dystrophin develop and grow normally for a few years but later succumb to muscle wasting (see Table 39.3). Mice without their single gelsolin gene reproduce normally, with only mild defects in platelets and other cells.

These observations suggest that each actin-binding protein has a distinct function, conferring a small selective advantage. Multiple proteins sharing overlapping functions make the actin system relatively failsafe so that it is difficult to detect the phenotypic consequences of the loss of particular proteins in mutant animals. Alternatively, these proteins may be retained owing to unknown functions distinct from actin binding.

Actin Dynamics in Live Cells

Cellular actin filaments vary widely in stability. Filament lifetimes are seconds in motile cells such as amoebae and white blood cells, tens of minutes in stress fibers and microvilli, and days in red blood cells and striated muscles where capping protein on the barbed end and tropomodulin on the pointed end limit the exchange of subunits and tropomyosin provides mechanical stability (see Figs. 7.11, 33.15, and 39.4).

Biochemical and genetic experiments identified the proteins that orchestrate actin filament assembly and disassembly. Essential features are a pool of unpolymerized actin, mechanisms to initiate and terminate new filaments and control disassembly. Reconstitution of the actin filament comet tail of the intracellular bacterium *Listeria* (see Fig. 37.12) showed that the minimal requirements for rapid assembly and turnover are actin, ADF/cofilin, Arp2/3 complex, a capping protein, and profilin, in addition to a nucleation promoting factor on the surface of the bacterium.

Methods to Document Actin Filament Turnover

The original approach to observe actin turnover was treatment with drugs that bind actin monomers such as **cytochalasin** or **latrunculin** (Box 33.1). Neither disassembles actin filaments directly, but both interfere with assembly from monomers, so their effects reveal if filaments are turning over naturally. Muscle actin filaments are relatively resistant to these drugs, but they disrupt the cortical networks of actin filaments in other cells in seconds (Fig. 33.18). When the drug is removed, the cortical actin network reforms rapidly from the leading edge. This response indicates that many cellular filaments turn over rapidly.

Much more has been learned about actin filament turnover by light microscopy of live cells. Although individual actin filaments are not resolved when crowded together in cytoplasm, networks or bundles of actin filaments can be imaged by differential interference

BOX 33.1 Tools to Study the Actin System

Actin Filament Destabilizers

Sponges synthesize toxins that destabilize actin filaments in cells by sequestering actin monomers (latrunculin A and B) or severing actin filaments (swinholide A). **Latrunculins** bind in the nucleotide binding cleft of monomers and prevent their polymerization, making them very useful experimentally. **Cytochalasins** (meaning "cell relaxing") were so named, because they cause regression of the cleavage furrow during cytokinesis and disrupt actin filaments in cells. These complex organic compounds synthesized by fungi bind in the barbed end groove of actin and inhibit subunit association and dissociation at the barbed ends of filaments. In addition low-affinity binding to actin monomers promotes their dimerization and the hydrolysis of ATP bound to one subunit, converting ATP-actin to ADP-actin. Given the complicated mechanism of action, their effects on cells must be interpreted cautiously.

C2 toxin produced by *Clostridium botulinum* is an enzyme that catalyzes the ADP-ribosylation of cytoplasmic actins on arginine-177. *Clostridium perfringens* iota toxin does the same to muscle actin. ADP-ribosylated actin polymerizes poorly and caps the barbed end of actin filaments. The ability of these protein toxins to penetrate live cells, cap actin filaments, and alter actin polymerization accounts for their disruption of the actin cytoskeleton in cells and may contribute to their toxicity.

Actin Filament Stabilizers

Phallotoxins (such as phalloidin), cyclic peptides synthesized by poisonous mushrooms, and the sponge toxin jasplakinolide both stabilize actin filaments. They bind to filaments between three subunits and reduce the rate of subunit dissociation to near zero at both ends of the polymer. Jasplakinolide can enter cells but phalloidin must be microinjected. They inhibit processes that depend on actin filament turnover, including amoeboid movement. Phallotoxins are toxic to humans, because they interfere with bile secretion. Fluorescent derivatives of phallotoxins are widely used to localize actin filaments in permeabilized cells and tissues (see Fig. 33.1), as well as to quantify polymerized actin in cells and cell extracts.

Inhibitors of Actin-Binding Proteins

Small, drug-like molecules are available to inhibit Arp2/3 complex and formins. CK666 blocks the conformational change that activates Arp2/3 complex and inhibits actin filament branch formation in cells. A molecule called SMIFH2 inhibits nucleation and elongation by many different formins.

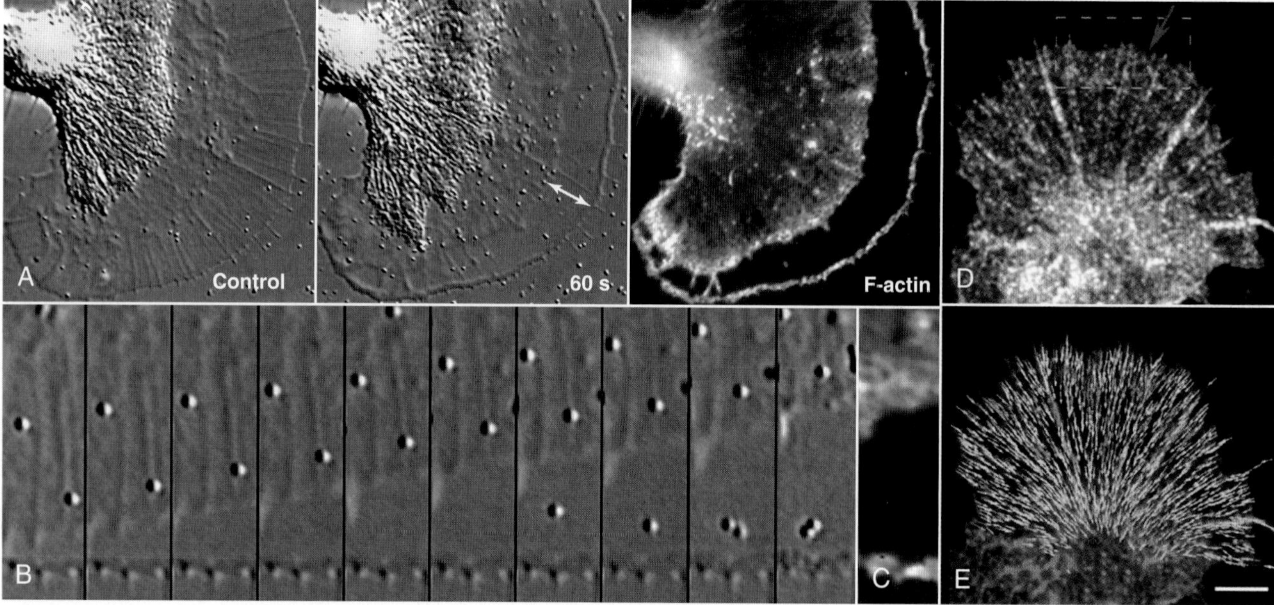

FIGURE 33.18 ACTIN FILAMENT DYNAMICS AT THE LEADING EDGE OF A GIANT GROWTH CONE OF A NEURON ISOLATED FROM THE MOLLUSK *APLYSIA*. A network of actin filaments forms continuously at the leading edge of the growth cone, moves inward by retrograde flow, and disassembles near the central organelle-rich zone. **A–C,** Effect of the drug cytochalasin D on the growth cone. **A,** *(left and middle)* Differential interference contrast (DIC) micrographs before and 60 s after applying the drug, which disrupts the network of actin filaments at the leading edge. The *double-headed arrow* marks the zone cleared of actin filaments, stained with rhodamine phalloidin in the *right panel.* A narrow rim of filaments survives at the leading edge. **B,** Time series of DIC micrographs at 6-s intervals, showing that retrograde flow of existing filaments (and small beads on the surface) continued toward the cell body after cytochalasin blocked the formation of new filaments at the leading edge. **C,** Detail of this growth cone after fixing after 60 s and staining with rhodamine-phalloidin. If cytochalasin is removed from a live cell, the actin filament network recovers, beginning near the leading edge. **D–E,** Fluorescence speckle microscopy of a growth cone injected with a low concentration of Alexa 488 actin monomers. **D,** Distribution of actin in lamellar regions between radial bundles. **E,** Vectors showing the velocities of actin speckles. False color indicates velocities ranging from *(dark blue)* zero to *(red)* 7 μm per second. (Courtesy Paul Forscher, Yale University, New Haven, CT, and modified from Forscher P, Smith SJ. Actions of cytochalasins on the organization of actin filaments and microtubules in a neuronal growth cone. *J Cell Biol.* 1988;107:1505–1516, and Yang Q, Zhang XF, Pollard TD, Forscher P. Arp2/3 complex-dependent actin networks constrain myosin II function in driving retrograde actin flow. *J Cell Biol.* 2012;197:939–956.)

contrast or phase contrast microscopy. For example, networks of actin filaments in nerve growth cones constantly assemble and move away from the leading edge (Fig. 33.18).

Fluorescence microscopy is even more informative if actin is labeled with a fluorescent probe. Purified actin can be labeled with a fluorescent dye and microinjected into live cells, where it is incorporated into the actin-containing structures (Fig. 33.18D). Expression of actin tagged with fluorescent protein is more convenient, although the fluorescent protein tag interferes with polymerization by formins. Alternatively, the fluorescent protein can be attached to an actin binding domain to label cellular actin filaments indirectly.

Fluorescent actin incorporates quickly into most of the filaments in nonmuscle cells. If low levels of fluorescent actin are used, random incorporation into filaments can result in fluorescent "speckles" that can be used to follow the movements and turnover of these subunits (Fig. 33.18D). Fig. 38.9 illustrates how bleaching or activating fluorescent actin in a live cell can reveal where filaments assemble and disassemble at the leading edge of motile cells.

Pool of Unpolymerized Actin

Cells can respond rapidly to stimuli such as chemoattractants by assembling actin filaments where needed, because they have a large pool of unpolymerized actin to grow the new filaments. Roughly half of the actin in the cytoplasm of cells other than muscle is unpolymerized, corresponding to 50 to 100 μM monomers, 500 to 1000 times higher than the critical concentration.

The combination of monomer binding to profilin and capping barbed ends allows cells to maintain a large pool of actin subunits ready to elongate any barbed ends created by uncapping, severing, or nucleation. In vertebrate cells, thymosin-β_4 augments the effects of profilin by sequestering a fraction of the actin monomers. The concentrations of profilin and thymosin-β_4 exceed the concentration of unpolymerized actin, and these proteins bind tightly enough to reduce the free monomer concentration to the micromolar level. Actin monomers bound to profilin or thymosin-β_4 do not nucleate new filaments. However, for profilin and thymosin-β_4 to maintain a monomer pool, most actin filament barbed ends must be capped as rapid addition of actin-profilin complexes to free barbed ends would quickly deplete the pool of unpolymerized actin. Cells contain enough heterodimeric capping protein (augmented by gelsolin in some cells) to cap the barbed ends of most filaments.

Initiation and Termination of Actin Filaments

A variety of external agonists and internal signals stimulate the assembly of actin filaments. Examples include the ability of chemoattractants to direct pseudopod formation in amoebas (see Fig. 38.10) and white blood cells (see Fig. 30.13), and the influence of the mitotic spindle on the assembly of the cytokinetic contractile ring (see Fig. 44.24). Polymerization depends on creation of barbed ends, which grow rapidly at rates estimated to be 50 to 500 subunits per second, depending on the concentration of actin-profilin.

Three mechanisms are thought to create free barbed ends: uncapping, severing, and de novo formation of new barbed ends. In many cases, new barbed ends appear to form de novo. At the leading edge of motile cells, Rho-family GTPases associated with the plasma membrane and polyphosphoinositides activate nucleation promoting factors, which stimulate Arp2/3 complex to form branches with free barbed ends (Figs. 33.2D and 33.12). Formins nucleate filaments for the cleavage furrow and initiate or sustain the growth of actin filaments in filopodia (see Fig. 38.3A). When thrombin activates platelets (see Fig. 30.14), plasma membrane polyphosphoinositides uncap barbed ends by dissociating gelsolin. Dephosphorylation activates ADF/cofilin proteins, which can sever and nucleate filaments, creating free barbed ends.

The duration of the growth of a cellular actin filament depends on the nucleation mechanism and the local environment. At the leading edge, new branches nucleated by Arp2/3 complex grow rapidly but transiently, as the concentration of free capping protein is high enough to terminate growth by capping barbed ends in a few seconds. On the other hand, barbed ends growing in association with a formin are protected from capping and grow persistently, as is observed at the tips of filopodia and the barbed ends of actin cables located in the buds of yeast cells.

Actin Filament Turnover and Subunit Recycling

Actin filaments are long lived if protected by tropomyosin and capping, as in muscle and stress fibers, but many actin filaments, such as those at the leading edge of motile cells, turn over quickly (see Fig. 38.7). Turnover starts with hydrolysis of ATP bound to the actin subunits and dissociation of the γ-phosphate, reactions that provide a timer to mark older filaments for depolymerization. ADF/cofilin proteins bind ADP-actin subunits in filaments and sever these older filaments (Fig. 33.16). This creates more ends available for dissociation of ADP-actin. Profilin stimulates the exchange of ADP for ATP on actin monomers and restores the pool of ATP-actin monomers bound to profilin available for polymerization. In cells with a high concentration of thymosin-β_4, much of the ATP-actin is stored bound to thymosin. Rapid exchange allows actin monomers to move from thymosin-β_4 to profilin, ready for polymerization.

How Do Cells Organize Actin Assemblies?

Cells organize actin filaments in a variety of structures, including cortical networks, microvilli or filopodia, and contractile bundles (Figs. 33.1 to 33.3). Each cell in a population is unique, but all cells of a particular type achieve a similar pattern of organization. The mechanisms appear to depend on expression of an appropriate mixture of actin-binding proteins to assemble particular structures. For example, actin forms bundles similar to microvilli and filopodia when polymerized in the presence of fimbrin and villin, the major crosslinking proteins in microvilli. Overexpression of villin induces cells to extend existing filopodia and form new ones. Thus, the pool of villin and fimbrin and other components may set the number of microvilli.

Many actin filament barbed ends are oriented toward the plasma membrane allowing their growth to push on the inside of the membrane (Fig. 33.2). Other barbed ends are anchored in structures including adhesion sites (see Fig. 30.11), the cleavage furrow (Fig. 33.3), and Z-disks of muscles (see Fig. 39.5). These anchors transmit forces on these filaments produced by myosin to the plasma membrane or the rest of the contractile apparatus in muscles. This makes mechanical sense, as actin filaments sustain tension better than compression, and because (with one interesting exception) all known myosins pull filaments in a direction away from the barbed end.

The **Rho-family GTPases Cdc42, Rac**, and **Rho** regulate the assembly of many actin filaments (Fig. 33.19). A complex of proteins, including Cdc42, in the bud of yeast cells anchors formins, which mediate the continuous assembly of a cable of actin filaments (see Fig. 37.11). The formins anchor the growing barbed ends as the pointed ends extend into the mother cell. These cables are tracks for myosin-V to move cargo into the bud. In motile cells, signals downstream of chemotactic receptors activate Cdc42 and Rac (see Fig. 38.7), which activate nucleation-promoting factors. They, in turn, stimulate Arp2/3 complex to generate the branched filament network that pushes the membrane forward. *Listeria* use a surface protein, ActA, to activate Arp2/3 complex, which generates a "comet tail" of actin filaments to push the bacterium through the cytoplasm (see Fig. 37.12). Proteins that mediate endocytosis activate homologs of WASp and other proteins to assemble actin patches in budding yeast (see Fig. 37.11) and fission yeast (Fig. 33.1D).

Physical forces also help organize actin filaments. Tension generated by **myosin II** contributes to the alignment of actin filaments in stress fibers (Fig. 33.1) and the contractile ring during cytokinesis (Fig. 33.3; also see Fig. 44.24). Rho activates myosin II by stimulating two kinases that phosphorylate its regulatory light chain and inhibit a phosphatase that reverses the light chain phosphorylation (see Fig. 39.24). Crosslinking proteins, such as α-actinin, help maintain the integrity of these bundles under mechanical stress.

Mechanical Properties of Cytoplasm

Actin filaments account for many of the mechanical properties of cytoplasm, a complicated, viscoelastic material. *Viscoelastic* means that cytoplasm can both resist flow, like a viscous liquid (eg, molasses), and store mechanical energy when stretched or compressed, like a spring.

The physical properties of actin filaments depend on their lengths and their interactions. At physiological concentrations, purified actin filaments are viscoelastic. At high concentrations, actin filaments also align spontaneously into large parallel arrays called liquid crystals. Crosslinking actin filaments increases both their viscosity and stiffness. Severing actin filaments decreases their viscoelasticity. On the other hand, shorter filaments have a higher tendency to form bundles in the presence of crosslinking proteins, so severing can actually promote the formation of rigid actin filament bundles.

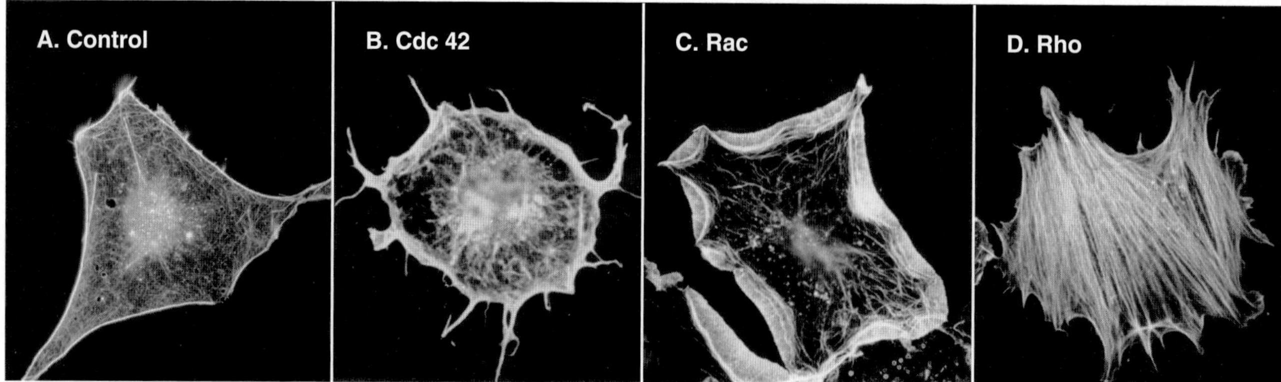

FIGURE 33.19 RHO-FAMILY GTPASES PROMOTE THE ASSEMBLY OF ACTIN-BASED STRUCTURES. Fluorescence micrographs of Swiss 3T3 fibroblasts stained with rhodamine-phalloidin to reveal actin filaments. **A,** Resting cells. **B,** Cells microinjected with activated Cdc42 form many filopodia. **C,** Cells microinjected with activated Rac have a thick cortical network of actin filaments around the periphery. **D,** Stress fibers anchored at their ends by focal contacts are abundant in cells microinjected with an activated form of Rho. (Courtesy Alan Hall, University of London, United Kingdom.)

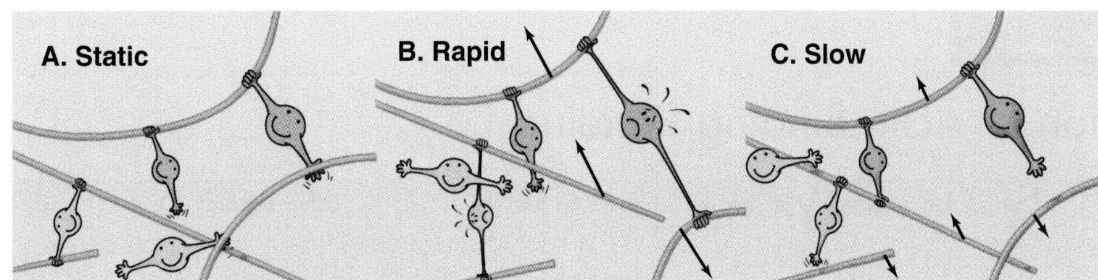

FIGURE 33.20 DYNAMIC CROSSLINKING OF ACTIN FILAMENTS. Rapid binding and dissociation of crosslinking proteins allow networks of actin filaments to resist rapid deformations but to change shape passively when force is applied for a prolonged time. **A,** Crosslinked network in a static region. **B,** Crosslinking proteins resist deformation if force is applied rapidly. **C,** Crosslinking proteins provide little resistance to deformation if force is applied slowly because the crosslinks rearrange faster than the filaments are displaced. (Modified from Pollard TD, Satterwhite L, Cisek L, et al. Actin and myosin biochemistry in relation to cytokinesis. *Ann N Y Acad Sci.* 1990;582:120–130.)

Many crosslinking proteins, including α-actinin, have low affinities for actin filaments with K_ds in the micromolar range. At steady state in vitro, these crosslinking proteins bind to and dissociate from actin filaments on a second or subsecond time scale. Consequently, gels of actin filaments and α-actinin are much more rigid when deformed rapidly than when the deformations are slower (Fig. 33.20), because crosslinks resisting the displacement of the filaments can rearrange if given sufficient time. Dynamic crosslinks between filaments allow actin networks to remodel passively as cells move. Cells also remodel the actin cytoskeleton actively by nucleating, severing or depolymerizing filaments.

SELECTED READINGS

Allingham JS, Klenchin VA, Rayment I. Actin-targeting natural products: structure, properties and mechanisms of action. *Cell Mol Life Sci.* 2006;63:2119-2134.

Bravo-Cordero JJ, Magalhaes MA, Eddy RJ, Hodgson L, Condeelis J. Functions of cofilin in cell locomotion and invasion. *Nat Rev Mol Cell Biol.* 2013;14:405-415.

Campellone KG, Welch MD. A nucleator arms race: cellular control of actin assembly. *Nat Rev Mol Cell Biol.* 2010;11:237-251.

Chen Z, Borek D, Padrick SB, et al. Structure and control of the actin regulatory WAVE complex. *Nature.* 2010;468:533-538.

Derman AI, Becker EC, Truong BD, et al. Phylogenetic analysis identifies many uncharacterized actin-like proteins (Alps) in bacteria: regulated polymerization, dynamic instability and treadmilling in Alp7A. *Mol Microbiol.* 2009;73:534-552.

Dominguez R, Holmes KC. Actin structure and function. *Annu Rev Biophys.* 2011;40:169-186.

Edwards M, Zwolak A, Schafer DA, et al. Capping protein regulators fine-tune actin assembly dynamics. *Nat Rev Mol Cell Biol.* 2014;15:677-689.

Elam WA, Kang H, De la Cruz EM. Biophysics of actin filament severing by cofilin. *FEBS Lett.* 2013;587:1215-1219.

Gunning PW, Ghoshdastider U, Whitaker S, Popp D, Robinson RC. The evolution of compositionally and functionally distinct actin filaments. *J Cell Sci.* 2015;128:2009-2019.

Higgs HN, Peterson KJ. Phylogenetic analysis of the formin homology 2 domain. *Mol Biol Cell.* 2005;16:1-13.

Janmey PA, Weitz DA. Dealing with mechanics: mechanisms of force transduction in cells. *Trends Biochem Sci.* 2004;29:364-370.

Löwe J, van den Ent F, Amos LA. Molecules of the bacterial cytoskeleton. *Annu Rev Biophys Biomol Struct.* 2004;33:177-198.

McCullagh M, Saunders MG, Voth GA. Unraveling the mystery of ATP hydrolysis in actin filaments. *J Am Chem Soc.* 2014;136:13053-13058.

Nag S, Larsson M, Robinson RC, Burtnick LD. Gelsolin: the tail of a molecular gymnast. *Cytoskeleton (Hoboken).* 2013;70:360-384.

Paul A, Pollard TD. Review of the mechanism of processive actin filament elongation by formins. *Cell Motil Cytoskeleton.* 2009;66:606-617.

Pollard TD. Actin and actin binding proteins. *Cold Spring Harb Perspect Biol.* 2016;8:8.

Quinlan M, Heuser JE, Kerkhoff E, Mullins RD. *Drosophila* Spire is an actin filament nucleation factor. *Nature.* 2005;433:382-388.

Rotty JD, Wu C, Bear JE. New insights into the regulation and cellular functions of the ARP2/3 complex. *Nat Rev Mol Cell Biol.* 2013;14:7-12.

Stossel TP, Condeelis J, Cooley L, et al. Filamins as integrators of cell mechanics and signalling. *Nat Rev Mol Cell Biol.* 2001;2:138-145.

Svitkina T. Actin cytoskeleton and actin-based motility. In: *The Cytoskeleton.* Cold Spring Harbor, NY: Cold Spring Harbor Laboratory Press; 2016;281-302.

APPENDIX 33.1

Classification of Actin-Binding Proteins

Protein	Distribution	Subunits (N × kD)	K_d Actin Binding	Other Ligands	Diseases
Monomer Binding					
β Thymosins	An	1 × 5	0.7 μM monomer	—	—
DNase I	An	1 × 29	0.1 nM monomer and pointed end	Ca^{2+}, DNA	—
Profilin	Eu	1 × 13-15	0.1 μM monomer	PIP_2, VASP, formins	Amyotrophic lateral sclerosis
Vitamin D-binding protein (Gc globulin)	An	1 × 58	1 nM monomer	Vitamin D, C5A complement	Obstructive lung diseases
Small Severing					
ADF/cofilin (actophorin, depactin, destrin)	Eu	1 × 15-19	0.1 μM ADP monomer, 0.5 μM ADP filament	PIP_2	Cancer, neurodegenerative diseases
Nucleation and Elongation					
Arp2/3 complex	Eu	1 each of 49, 44, 40, 35, 21, 20, 16	10 nM pointed end 0.5 μM filament side	Profilin, Scar, WASp, cortactin	Psychiatric diseases (?)
Formins	Eu	Dimers range of sizes	Barbed end	Profilin, spire	Glomerular sclerosis
Spire	An	1 × 84	Side	Formins	—
Capping					
Capping protein (CapZ)	Eu	1 × 32-36(α) + 1 × 28-32(β)	<1 nM barbed end	CARMIL, V-1	—
Fragmin (severin, gCAP39)	Am, An	1 × 40	1 nM barbed end	—	—
Gelsolin (scinderin)	Am, An	1 × 80 or 83	50 nM barbed end, μM dimers, sides	Ca^{2+}, PIP_2	Finnish amyloidosis
Tropomodulin	An	1 × 40	Pointed end, 1 nM with TM	TM	—
Villin (Cap 100)	Am, An	1 × 93	7 μM filament, 0.3 μM filament for head domain	Ca^{2+}, PIP_2	—
Filament Side Binding					
Abp1p	An, Fu	1 × 67		—	Yst
Adducin	An	1 each 100 and 105	0.3 μM filament	Spectrin	Hypertension
Caldesmon	An	1 × 90 or 1 × 61	1 μM filament with TM	CM, TM, myosin	—
Calponin	An	1 × 34	0.2 μM filament with TM	CM, TM	—
Coronin	Am, An, Fu	1 × 51	5-40 nM filament	—	—
Drebrin	An	1 × 95	0.1 μM filament	—	—
Nebulin	An	1 × 750	Filament		Nemaline myopathy
Tropomyosin	Am, An, Fu	Homodimers of 28-32	Filament, cooperative	Self, troponin, caldesmon, calponin	Cardiomyopathy
Troponin	An	1 each 18 (TNC), 21 (TNI), 31 (TNT)	Filament	TM, Ca^{2+}	Cardiomyopathy
Cross Linking					
α-Actinin (actinogelin)	Am, An, Fu	2 × 100	1-5 μM filament	Vinculin, zyxin, integrin, NMDAR, selectin	Muscular dystrophy
Anillin	An, Fu	1 × 132	Filament	Myosin, IQGAP	—
Dematin (Band 4.9)	An	3 × 43-45	Filament	GLUT1	—
Espin	An	1 × 95 or 29	0.1-0.2 μM filaments	—	Deafness
Fascin	An	1 × 56	Filament	β-catenin	—

APPENDIX 33.1

Classification of Actin-Binding Proteins—cont'd

Protein	Distribution	Subunits (N × kD)	K_d Actin Binding	Other Ligands	Diseases
Filamin (ABP-280)	An	2 × 240–280	0.5 μM filament	GP1B^h^1X, several receptors	Skeletal & lung dysphasias, periventricular heterotopia
Fimbrin (plastin)	Eu	1 × 68	Filament	Ca^{2+}	Yst
Scruin	An	1 × 102	Filament	CM	—
Membrane Associated					
Annexin-II	An, Pl	2 × 38 + 2 × 10	0.2 μM filament	Ca^{2+}, acidic, phospholipids	—
Dystrophin/ utrophin	An	1 × 427/1 × 395	Filament, head 44 μM, tail 0.5 μM	β-Dystroglycan	Muscular dystrophies
Ezrin/moesin/ radixin	An	1 × 68 + oligomers	Filament	Self	—
Protein 4.1	An	1 × 80	Filament	Spectrin, band 3, glycophorin	Hereditary elliptocytosis
Spectrin (fodrin, calspectin)	Am, An	2 × 280 (α) + 2 × 246 (β)	1-25 μM filament	Ca^{2+}, ankyrin, CM, band 4.1, adducin Hs	Hereditary spherocytosis
Talin	Am, An	1 × 272	Filament, 0.25 μM monomer	Vinculin, integrins, p125FAK	—
Microtubule Binding					
MAP-2	An	1 × 210	Filament sides	Microtubules, PKA, IF	—
Tau	An	1 × 43-86	Filament sides	Microtubules	Alzheimer disease
Intermediate Filament Binding					
BPAG1	An	? × 280 or ? × 230	0.2 μM filaments	IF	Bullous pemphigoid
Motors					
Myosins I-XII	Eu	Various	1-100 μM with ATP, 4nM without ATP	Various (self, membranes)	Cardiomyopathy, deafness, retinitis

ADF, actin depolymerizing factor; ADP, adenosine diphosphate; Am, amoebas; An, animals; Arp, actin-related protein; CARMIL, capping protein, Arp2/3, and myosin I linker; CM, calmodulin; DNase, deoxyribonuclease; Eu, all eukaryotes; Fu, fungi; GLUT1, glucose transporter 1; Hs, Homo sapiens; IF, intermediate filaments; MAP-2, microtubule-associated protein 2; NMDAR, *N*-methyl-D-aspartate receptor; PIP_2, phosphatidylinositol 4,5-bisphosphate; PKA, protein kinase A; Pl, plants; TM, tropomyosin; TNC, troponin C; TNI, troponin I; TNT, troponin T; VASP, vasodilator-stimulated phosphoprotein; WASp, Wiskott-Aldrich syndrome protein; Yst, yeast.

CHAPTER 34

Microtubules and Centrosomes

Microtubules are stiff, cylindrical polymers of **α-tubulin** and **β-tubulin** (Fig. 34.1) that provide support for a variety of cellular structures and tracks for movements powered by motor proteins called kinesins and dyneins. Microtubules help organize the cytoplasm during interphase and form the **mitotic spindle**, which separates duplicated chromosomes.

Microtubules are 25 nm in diameter and can grow longer than 20 μm in cells and much longer in vitro. The head-to-tail arrangement of dimers of α-tubulin and β-tubulin in the walls of microtubules give the polymer a molecular polarity. The **"plus" (β-tubulin) end** grows faster than the **"minus" (α-tubulin) end.**

A simplifying principle is that **microtubule organizing centers (MTOCs)** containing γ-tubulin nucleate and anchor the minus ends of many microtubules. In the simplest case, the **centrosome** is the main MTOC in many animal cells, so the microtubules are arranged radially with plus ends in the periphery (Fig. 34.2A). Some of these radially arranged microtubules are associated with the nearby Golgi apparatus. However, microtubules are not arranged radially in neurons, muscles, and many epithelial cells. Microtubules in polarized epithelial cells originate from a broad zone with γ-tubulin near the apical surface and extend the length of the cell (Fig. 34.2B). Microtubules in dendrites of nerve cells have mixed orientations. Most plant cells lack centrioles and centrosomes. During interphase, γ-tubulin and microtubules are found throughout the cortex, and they form a bipolar **mitotic spindle** during cell division (Fig. 34.2C).

The centrosome (covered in detail later in "Centrosomes and Other Microtubule Organizing Centers") consists of a pair of centrioles and a surrounding matrix containing the active component in microtubule nucleation—a complex of proteins including the specialized tubulin isoform **γ-tubulin.** The centrosome duplicates during interphase, and as cells enter mitosis the two centrosomes separate to form the poles of the mitotic spindle. Animal cells and protozoa with cilia and flagella use centrioles as **basal bodies** (Fig. 34.3B) to nucleate the assembly of microtubules for their motile structure, called an axoneme (see Fig. 38.14). Microtubules in fungi grow from a **spindle pole body** (Fig. 34.2D), an organizing center containing γ-tubulin associated with the nuclear envelope (see Fig. 34.20).

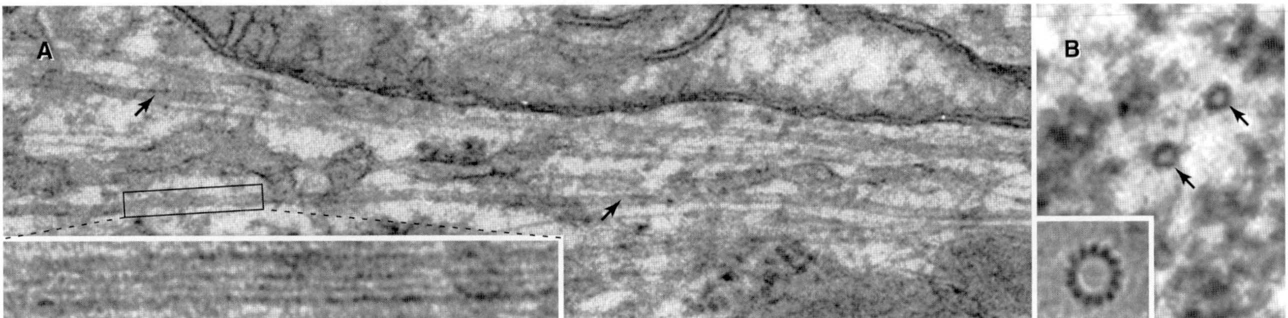

FIGURE 34.1 MICROTUBULES *(ARROWS)* VISUALIZED IN ELECTRON MICROGRAPHS OF THIN SECTIONS OF A CULTURED MAMMALIAN CELL. A, Longitudinal section. **B,** Cross section. *Insets,* Electron micrographs of microtubules in frozen hydrated rat hepatoma cells. (*Main panels,* Courtesy R. Goldman, Northwestern University, Evanston, IL. *Insets,* Courtesy Cédric Bouchet-Marquis and Jacques Dubochet, University of Lausanne, Switzerland.)

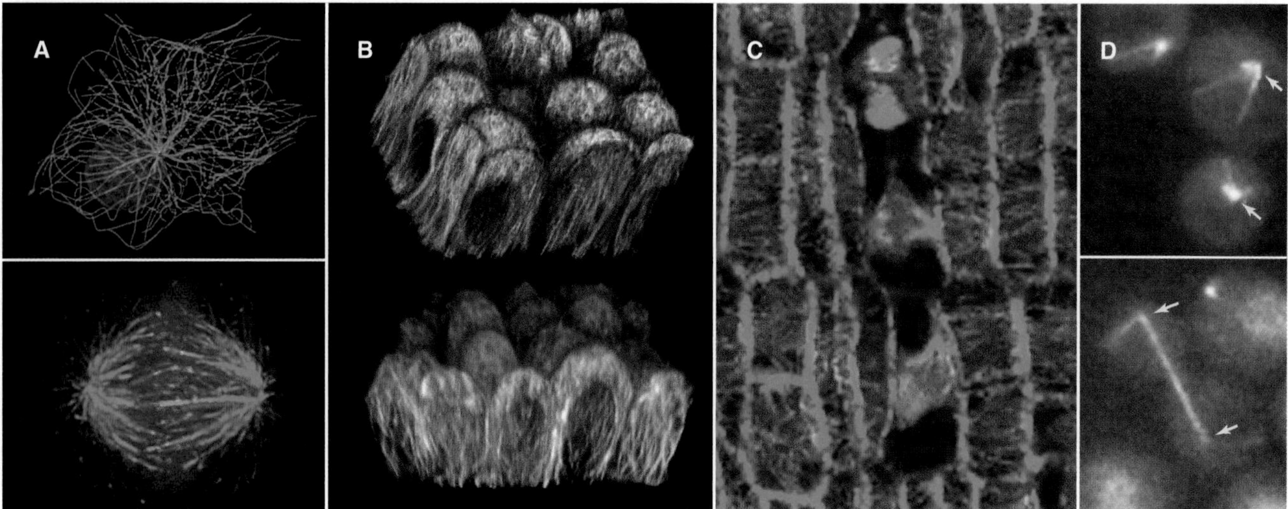

FIGURE 34.2 ARRANGEMENTS OF MICROTUBULES IN VARIOUS CELLS. A–C, Fluorescence micrographs of microtubules stained with antibodies to tubulin. **A,** Vertebrate tissue culture cells. *Green* microtubules radiate from the *red* centrosome near the *blue* nucleus of an interphase cell *(upper panel).* A HeLa (Henrietta Lacks) cell in mitosis with *green* microtubules radiating from the two poles toward the *blue* chromosomes and *red* centromeres (stained with anticentromere antibody) *(lower panel).* **B,** Columnar epithelial cells in tissue culture. Three-dimensional reconstructions show microtubules oriented along the long axis of the cell. **C,** Plant cells. Maize epidermal cells with *green* microtubules in the cortex and in the mitotic spindles of three dividing cells near the middle of the field. Chromosomes are stained *orange.* **D,** Live budding yeast. Two interphase cells with microtubules radiating from the spindle pole bodies *(arrows)* associated with the nuclear envelope; these microtubules are marked with dynein fused to *green* fluorescent protein *(upper panel).* A cell late in mitosis with a bundle of microtubules extending from one spindle pole body to the other inside the nucleus; these microtubules are marked with tubulin fused to *green* fluorescent protein *(lower panel).* (**A,** *Top,* Courtesy A. Khodjakov, Wadsworth Center, Albany, NY; *bottom,* courtesy D.W. Cleveland, University of California, San Diego. **B,** Courtesy R. Bacallao, University of Indiana Medical School, Indianapolis. **C,** Courtesy L. Smith, University of California, San Diego. **D,** Courtesy P. Maddox, University of North Carolina, Chapel Hill.)

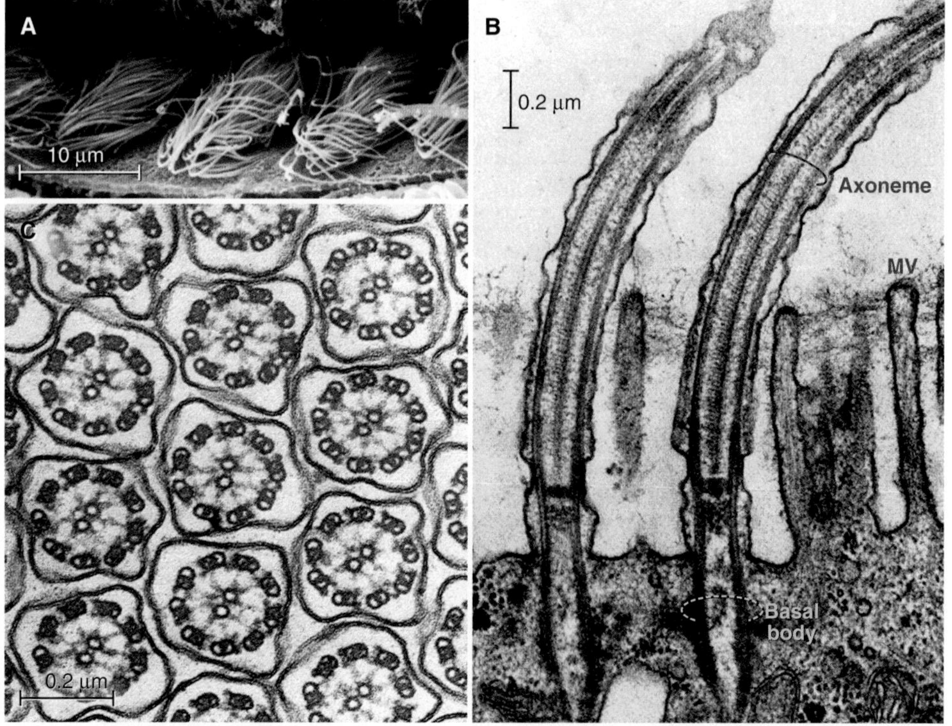

FIGURE 34.3 CILIA ON MUSSEL GILL EPITHELIAL CELLS. A, A scanning electron micrograph reveals how the bending movements of the thread-like cilia are coordinated in waves. **B–C,** Transmission electron micrographs of cilia. **B,** Longitudinal section showing basal bodies and proximal parts of two cilia along with adjacent microvilli (MV). **C,** Cross section of many cilia showing the nine outer doublets and the central pair of microtubules. (Courtesy P. Satir, Albert Einstein College of Medicine, Bronx, NY. For reference, see Satir P. How cilia move. *Sci Am.* 1974;231:44–52.)

BOX 34.1 Pharmacologic Tools for Studying Tubulin and Microtubules

Tubulin binds several therapeutically active plant alkaloids and synthetic chemicals, including two of the most successful drugs used to treat cancer. *Vinblastine* (from periwinkle) interferes with microtubule dynamics by binding between tubulin dimers at the ends of microtubules. This inhibits mitosis. *Taxol* (from the bark of the Western yew) binds β-tubulin and stabilizes microtubules. At substoichiometric concentrations, vinblastine and Taxol are effective in cancer chemotherapy because they interfere with the dynamic instability of mitotic spindle microtubules and block cell division. *Colchicine* (from the autumn crocus) and *nocodazole* (a synthetic chemical) inhibit microtubule assembly by binding dissociated tubulin dimers. Colchicine binds dimers between the two subunits, stabilizing a bent conformation that cannot fit into the microtubule lattice. Colchicine is used empirically to treat gout, a painful condition that results from the accumulation of uric acid crystals in joints and other tissues, but no one knows precisely how it works or why it is not more toxic.

Microtubules vary considerably in stability. Microtubules in **axonemes** of eukaryotic cilia and flagella are stable for days to weeks. Cytoplasmic microtubules turn over much more rapidly, within minutes during interphase and tens of seconds in the mitotic spindle. These dynamic microtubules randomly undergo periods of rapid depolymerization and then regrow over a period of seconds to minutes. This **"dynamic instability"** constantly remodels the network of microtubules in cytoplasm and contributes to the assembly of the mitotic spindle (see Fig. 44.12). Drugs that depolymerize or stabilize microtubules (Box 34.1) are useful for exploring their functions in cells and in cancer therapy.

The same tubulin dimers can form dynamic single microtubules in the cytoplasm and stable doublet microtubules in axonemes, a difference explained by posttranslational modifications and accessory proteins. Various families of **microtubule-associated proteins (MAPs)** bind tubulin dimers, stabilize polymers, associate with microtubule ends, or sever cytoplasmic microtubules. In the axonemes of cilia and flagella, more than 100 accessory proteins organize and stabilize the array of nine outer doublet microtubules and two central single microtubules (see Fig. 38.14).

Microtubule motor proteins (see Figs. 36.13 and 36.14) power movements ranging from the slow movements of chromosomes on the mitotic spindle (see Fig. 44.14) to the rapid beating of cilia and flagella (see Fig. 38.14). Different motors move toward the plus and minus ends of microtubules. The main minus-end-directed motor, dynein, drives the beating of cilia and flagella. Dynein cooperates with the kinesin family of plus-end motors to move membrane-bound organelles, RNA particles, viruses, and other cargo along microtubules (see Fig. 37.7). These active movements determine, to a great extent, the distribution of cellular organelles and the shape of cells (see Chapter 37).

Structure of Tubulin and Microtubules

The tubulin molecule is a heterodimer of α and β subunits that share a common fold and 40% identical residues (Fig. 34.4B). Dimers of α-tubulin and β-tubulin are stable and rarely dissociate at the 10- to 20-μM concentrations of tubulin found in cells.

Each tubulin subunit binds a guanine nucleotide, either guanosine triphosphate (GTP) or guanosine diphosphate (GDP). Neither the overall fold nor the GTP-binding site of tubulin resembles those of the large family of guanosine triphosphatases (GTPases) related to Ras (see Figs. 4.6 and 25.7). GTP on α-tubulin is buried in the dimer, so it does not exchange with GTP in solution. Its binding site is called the nonexchangeable site. GTP on β-tubulin is exposed in the dimer and exchanges slowly (K_a = 50 nM), so this is known as the exchangeable site. When incorporated into a microtubule, contacts with the adjacent α subunit bury the GTP on the β subunit. This promotes its hydrolysis. Subsequent dissociation of the γ-phosphate profoundly affects microtubule stability. Neither bound guanine nucleotide can exchange in the wall of a microtubule.

Tubulin Evolution and Diversity

The tubulin gene arose in the common ancestor of life on earth. Today, most *Archaea* and *Bacteria* have a gene for the protein **FtsZ** with the same fold as tubulin but with highly divergent sequences. FtsZ polymers are required for cytokinesis of prokaryotes (see Fig. 44.21). All eukaryotes have genes for α-tubulin and β-tubulin. One group of *Bacteria* lost their FtsZ but acquired genes for both α-tubulin and β-tubulin, most likely by lateral transfer from a eukaryote. Genes for δ-tubulin, ε-tubulin, ζ-tubulin, and η-tubulin are found in protozoa, algae, and animals where they are components of centrioles and basal bodies. Their absence from fungi and most plants may explain their lack of centrioles.

The sequences of eukaryotic tubulins are remarkably conserved, with more than 75% of the residues of animal α- or β-tubulins identical to their plant homologs. On the other hand, species differ considerably in the number of tubulin genes. Vertebrates have six to eight genes for both α-tubulin and β-tubulin, whereas unicellular ciliates such as *Tetrahymena* assemble a greater variety of microtubule-based structures than humans with only one α-tubulin and one β-tubulin polypeptide. Most vertebrate cells express several tubulin isoforms, but exceptional cases, such as bird red blood cells, express a single α-tubulin and β-tubulin.

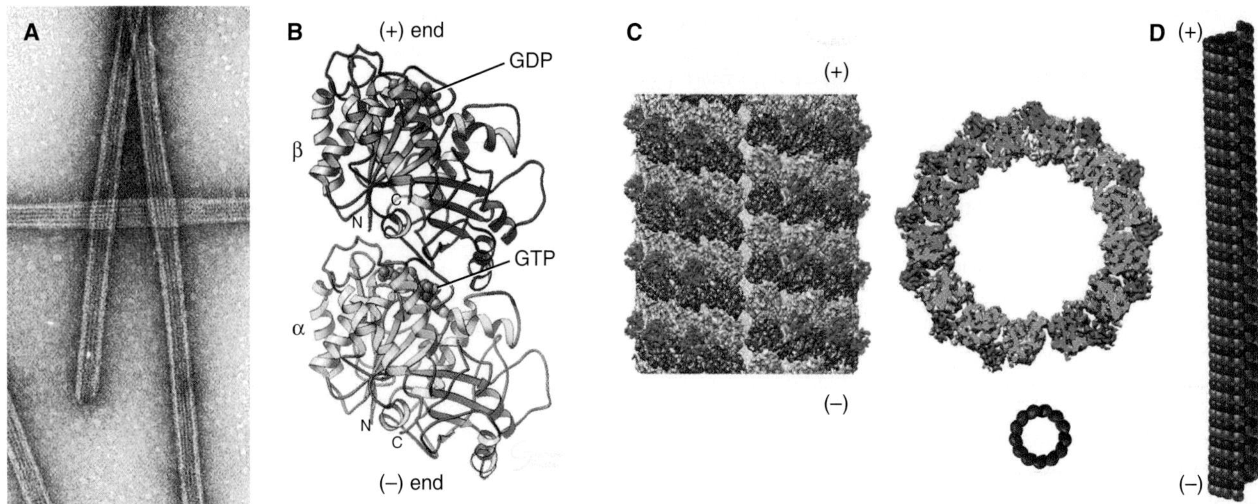

FIGURE 34.4 STRUCTURE OF THE α-TUBULIN/β-TUBULIN DIMER AND THE MICROTUBULE. A, Electron micrograph of negatively stained microtubules. **B,** Ribbon diagram with space-filling guanosine triphosphate (GTP) on α-tubulin and guanosine diphosphate (GDP) on β-tubulin. This historic structure was the first high-resolution structure of a nonmembrane protein to be determined by electron crystallography, using sheets of tubulin protofilaments as the specimen. Each subunit consists of approximately 450 residues arranged in two domains. Each domain is a β-sheet flanked by α-helices. The nucleotides bind in pockets similar to the binding site for nicotinamide adenine dinucleotide (NAD) on the enzyme glyceraldehyde-3-phosphate dehydrogenase. **C,** Reconstructions from electron micrographs of frozen GTPγS (a slowly hydrolyzed analog of GTP) microtubules decorated with the small protein EB3 *(blue)* showing a side view and cross section. **D,** Drawing of the microtubule used throughout this book. Both **C** and **D** show the longitudinal seam between two protofilaments, which breaks the helical repeat of tubulin dimers. (**A,** Courtesy D.B. Murphy, Johns Hopkins Medical School, Baltimore, MD. **B,** For reference, see Protein Data Bank [PDB; www.rcsb.org] file 1TUB and Nogales E, Downing K. Structure of the α/β tubulin dimer by electron crystallography. *Nature*. 1998;391:199–203. **C,** For reference, see Electron Microscopy Data Bank [www.emdatabank.org] file 6347 and Zhang R, Alushin GM, Brown A, Nogales E. Mechanistic origin of microtubule dynamic instability and its modulation by EB proteins. *Cell*. 2015;162:849–859.)

Two views rationalize the significance of multiple α-tubulin and β-tubulin isoforms. On the one hand, isoforms have different assembly properties and affinities for MAPs that may confer specific properties to the microtubules. This is illustrated by the failure of paralogous tubulin genes to substitute for each other in flies. On the other hand, the proteins themselves may be largely interchangeable (and all isoforms appear to copolymerize), but different genes may be required to ensure precise control of biosynthesis in particular cells at appropriate times during development. For example, the two α-tubulin proteins of the filamentous fungus *Aspergillus* can substitute for each other, but two genes are required to control the expression of tubulin at specific times in the life cycle.

The most variable regions of both α-tubulin and β-tubulin are the negatively charged C-terminal tails where most posttranslational modifications (some unique to tubulins) are made. Over time, a carboxypeptidase removes the C-terminal tyrosine from α-tubulin, leaving a glutamic acid exposed on many, but not all, stable microtubules. Another enzyme, tyrosine-tubulin ligase, can replace the tyrosine. The presence or absence of this tyrosine is a code that can be read by the motor protein CENP-E (see Chapter 45). Both α-tubulin and β-tubulin can also be modified by the addition of a polymer of up to six glutamic acid residues to the γ-carboxyl groups of glutamic acid residues in the C-terminal tails. Addition of one or more glycines to the γ-carboxyl of other glutamate residues stabilizes the central pair of microtubules in axonemes (see Fig. 38.16). Lysine-40 in the microtubule lumen is acetylated on α-tubulin in axonemes, basal bodies, and a few cytoplasmic microtubules, but the impact of this modification on function is not yet clear. Stable microtubules accumulate several of these modifications, but much work will be required to establish cause and effect and learn how each modification influences interactions with motors and MAPs.

Structure and Physical Properties of Microtubules

Microtubules are cylinders constructed of longitudinally oriented **protofilaments** with a 4-nm longitudinal repeat arising from the tubulin subunits (Fig. 34.5). Most cytoplasmic microtubules have 13 protofilaments (~1600 tubulin dimers per μm), but microtubules in some cells have 11, 15, or 16 protofilaments. Microtubules assembled in vitro can have 11 to 15 protofilaments, but 13 is the favored number. Microtubules with 13 protofilaments have a longitudinal seam or discontinuity between two of the protofilaments with α-tubulin next to β-tubulin, disrupting the helical lattice of the subunits

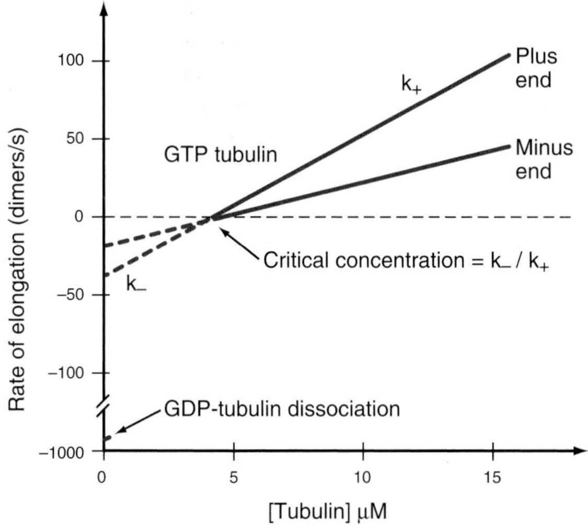

FIGURE 34.5 ELONGATION OF PURE TUBULIN MICROTUBULES. The rate of elongation at plus and minus ends depends on the concentration of GTP-tubulin dimers. Slopes give apparent association rate constants, *x*-intercepts give the critical concentrations, and the *y*-intercepts (after linear extrapolation *[dashed lines]* from positive growth rates to the *y*-axis) give the apparent dissociation rate constants for GTP tubulin. The dissociation rate of GDP tubulin is shown as a single point at 733 s^{-1} on the *y*-axis. (Data from Walker RA, O'Brien ET, Pryer NK, et al. Dynamic instability of individual microtubules. *J Cell Biol.* 1988;107:1437–1448.)

(Fig. 34.5D). Throughout the microtubule, including the seam, the contacts along the length of the protofilaments are much more robust than the lateral contacts between protofilaments.

Microtubules are polar, because all the dimers have the same longitudinal orientation. β-Tubulin is oriented toward the plus end; α-tubulin is oriented toward the minus end. The plus end grows faster and is less stable than the minus end. For the purpose of determining microtubule polarity in electron micrographs, microtubules in extracted cells can be decorated with either exogenous dynein or with excess tubulin, which can form curved "hooks." These hooks provided the original evidence that the minus ends are usually associated with MTOCs.

As expected from their cylindrical structure, microtubules are much stiffer than are either actin filaments or intermediate filaments. If enlarged 1 million-fold to a diameter of 25 mm, microtubules would have mechanical properties similar to those of a plastic pipe: quite stiff locally but flexible over distances of several meters (micrometers in the cell). On the same scale, actin filaments would be like an 8-mm plastic rod, and intermediate filaments would be akin to 10-mm braided plastic rope.

The stiffness, length, and polarity of microtubules make them valuable both for cytoskeletal support and as tracks for microtubule-based motors. Because they resist compression, microtubules were adapted more frequently than actin filaments or intermediate filaments to support asymmetrical cellular structures, including axonemes, the mitotic spindle, and elaborate surface processes of some protozoa (see Fig. 38.4). Plants provide a spectacular example of the influence of microtubules on morphology: Point mutations in tubulin influence whether climbing plants wrap in a left-handed or right-handed helix around their supports. Interactions of microtubules with both actin filaments and intermediate filaments reinforce the cytoskeleton (see Fig. 35.7).

Microtubule Assembly and Dynamic Stability

Pioneering light microscopic observations of live cells established that mitotic spindle fibers (later shown to be microtubules) are sensitive to both cold and high hydrostatic pressure. This cold sensitivity makes it possible to purify microtubules and tightly associated proteins by cycles of depolymerization in the cold and repolymerization when rewarmed to higher temperatures. Pelleting in a centrifuge separates microtubules and any associated proteins from soluble contaminants. This cycle of cooling → pelleting; rewarming/reassembly → pelleting → recooling, and so on is repeated until the desired degree of purity is achieved. Adaptations of tubulins or accessory proteins allow cold-water organisms to assemble microtubules at temperatures near freezing.

Pure GTP-tubulin subunits assemble microtubules much like adenosine triphosphate (ATP)-actin polymerizes into filaments (see Figs. 5.6 and 33.8). However, the spontaneous nucleation of microtubules from tubulin dimers is so unfavorable that cells must use various templates to initiate most microtubules.

Superficially, microtubule elongation is a simple bimolecular reaction of GTP-tubulin dimers with the ends of the polymer. Growth is faster at the plus end than at the minus end (Fig. 34.5), and assembly of GTP-tubulin is vastly more favorable than that of GDP-tubulin. Above the **critical concentration** at each end the rate of elongation depends on the concentration of GTP-tubulin dimers.

Elongation of microtubules differs from actin in two ways. First, GDP-tubulin depolymerizes very fast, on the order of 1000 subunits per second. Second, the rate constant for GTP-tubulin dissociation from a growing end depends on the concentration of GTP-tubulin dimers. This surprising feature may arise from an effect of the elongation rate on the structure of the growing end. At high growth rates individual protofilaments may extend further beyond the tip than at slow growth rates. If so, GTP-tubulin dimers may dissociate faster owing to the lack of support from adjacent protofilaments.

Steady-State Dynamics of Microtubules in Vitro

If the GTP-tubulin reactions were all that contributed to assembly, microtubules would grow until the concentration of free tubulin dimers decreased to the critical concentration, after which polymers would be relatively stable, as the critical concentrations are similar at the two ends. This appeared to be what happens in experiments looking at bulk properties of pure GTP-tubulin, as the overall microtubule polymer and monomer concentrations are stable over time. However, it was observed that the microtubule number declines as some microtubules disappear, and the survivors grow longer. Direct observation by light microscopy (Fig. 34.6A) showed that this redistribution of tubulin is attributable to random fluctuations in the lengths of microtubules. Amazingly, growing and shrinking microtubules coexist at steady state; some shrinking microtubules disappear while others grow. This behavior is called **dynamic instability.**

At steady-state in excess GTP, individual microtubules grow slowly until they undergo a random transition to a phase of rapid shortening. This transition is called a **catastrophe.** As the microtubule shortens, tubulin is lost from the end at a rate of nearly 1000 dimers per second, so the polymer shrinks more than 0.5 μm per second. Electron micrographs of rapidly shortening microtubules show curved segments of protofilaments peeling out from the end (Fig. 34.6B). Dimers dissociate from these curved protofilaments and sheets before or after they break free from an end.

Rapid shortening can be terminated by another stochastic event called a **rescue,** after which the microtubule grows again at a steady rate. Occasionally, a microtubule disappears completely during a shortening phase if its length reaches zero before a rescue event occurs.

Hydrolysis of GTP bound to the (exchangeable) E-site on β-tubulin drives dynamic instability. Dimeric tubulin

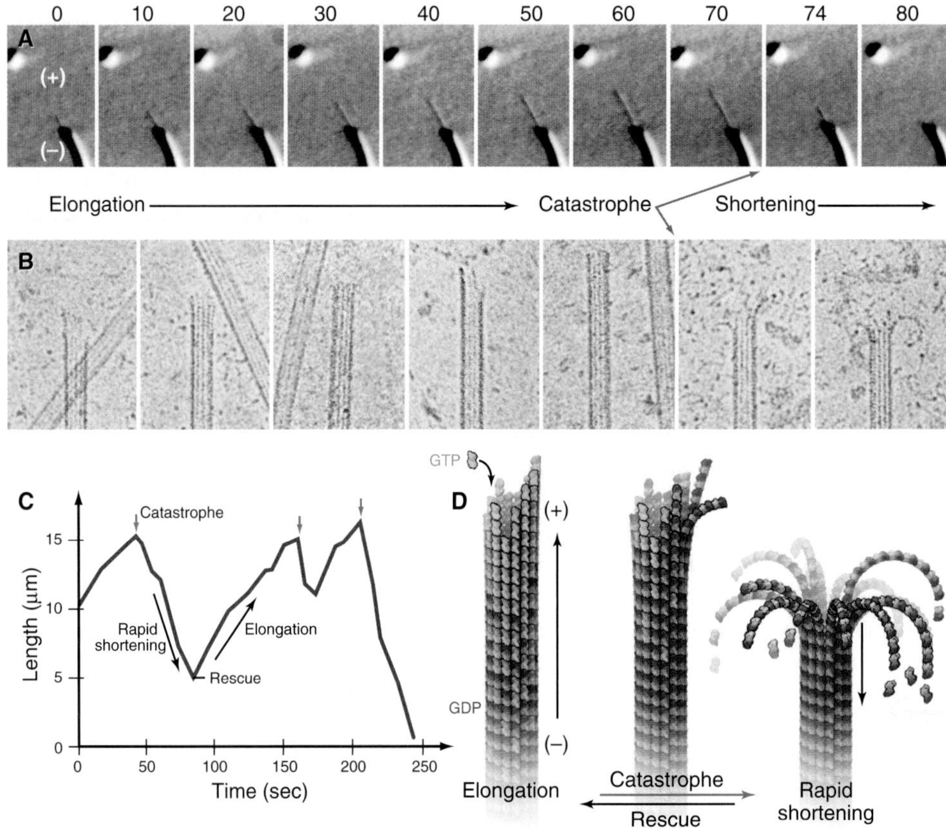

FIGURE 34.6 DYNAMIC INSTABILITY OF MICROTUBULES IN VITRO. A, Time series of differential interference contrast light micrographs of a microtubule growing from the broken end of an axoneme in the presence of 11 μM GTP tubulin. At 70 seconds, a catastrophe occurs, and the microtubule shortens rapidly. **B,** Electron micrographs of frozen samples. Growing microtubules have flat or oblique ends interpreted as sheets of polymerized tubulin that have not yet closed up into a tube. Rapidly shortening microtubules have protofilaments and sheets peeling from the end. **C,** Time course of the fluctuations in the length of a microtubule from an experiment similar to **A. D,** Model for the transitions during steady-state dynamic instability of microtubules in vitro. GTP tubulin is *blue-green;* GDP tubulin is *red-orange.* (**A,** From Walker RA, O'Brien ET, Pryer NK, et al. Dynamic instability of individual microtubules. *J Cell Biol*. 1988;107:1437–1448, copyright The Rockefeller University Press. **B,** From Mandelkow E-M, Mandelkow E, Milligan RA. Microtubule dynamics and microtubule caps: a time-resolved cryo-electron microscopy study. *J Cell Biol*. 1991;114:977–991, copyright The Rockefeller University Press. **C,** For reference, see Erickson HP, O'Brien E. Microtubule dynamic instability and GTP hydrolysis. *Annu Rev Biophys*. 1992;21:145–166.)

hydrolyzes GTP very slowly, because the active site on the β subunit depends on residues from a neighboring α subunit. However, each new dimer added to the plus end of a protofilament stimulates GTP hydrolysis 250-fold ($k = 0.3\ s^{-1}$, corresponding to a half time of ~2 seconds) on the adjacent β subunit already incorporated into the polymer. At the minus end, the same mechanism presumably stimulates hydrolysis of GTP bound to the incoming subunit. The γ-phosphate then dissociates ($k = 0.02\ s^{-1}$, corresponding to a half time of 35 seconds). Neither dimeric nor polymeric tubulin hydrolyzes GTP bound to the α subunit, because the β subunit does not provide the residues required to complete the active site. Thus, "GDP-tubulin" has GDP on the β subunit and GTP on the α subunit. Microtubules composed of GDP-tubulin are extremely unstable in comparison to microtubules assembled from dimers with slowly hydrolyzed GTP analogs bound to the β subunit.

Converting GTP-tubulin to GDP-tubulin after assembly creates a polymer that is poised for explosive disassembly. Tubulin dimers with bound GDP pack more compactly than GTP-tubulin in the wall of the microtubule, and protofilaments of GDP-tubulin tend to be curved rather than straight. These structural changes following GTP hydrolysis and γ-phosphate dissociation create microtubules with an unstable, strained GDP-tubulin core capped at both ends with less-dynamic GTP-tubulin.

When exposed at the end of a microtubule during a catastrophe, these strained GDP-protofilaments curve and peel from the cylinder, promoting disassembly. The energy derived from GTP hydrolysis is released during disintegration of the end of the microtubule, so hydrolysis prepares the polymer for rapid depolymerization. Michael Caplow of the University of North Carolina provides this apt description: "A catastrophe of an elongating microtubule is like removing the cork from a shaken bottle of champagne."

Because GDP-tubulin is constrained in a straight lattice but would prefer to be in curved protofilaments, the lattice actually stores the energy released by GTP hydrolysis. During disassembly, the curving protofilaments release enough energy to do significant work, such as moving huge chromosomes through the cytoplasm of dividing cells.

Loss of the GTP cap causes a catastrophe, but the number of terminal subunits with GDP required for a catastrophe is still being studied. The size of the GTP cap is difficult to measure, as few subunits are involved compared with the large number of GDP subunits in the core of a microtubule. GTP caps are maintained at steady state by exchange of GTP-tubulin dimers at the ends and by direct exchange of GTP onto the β-tubulin subunits exposed at the plus end. If the GTP cap is lost, the end shortens rapidly, because of the large dissociation rate constant of GDP tubulin. Catastrophes liberate GDP-tubulin subunits that exchange their bound GDP for

TABLE 34.1 Rate Constants for the Assembly of Microtubules in Vitro and in Cells

Reaction	Plus End	Minus End
In Vitro Elongation of Purified Tubulin		
Association of guanosine triphosphate (GTP) tubulin	$9\ \mu M^{-1}\ s^{-1}$	$4\ \mu M^{-1}\ s^{-1}$
Dissociation of GTP tubulin	$44\ s^{-1}$	$23\ s^{-1}$
Association of guanosine diphosphate (GDP) tubulin	Unknown	Unknown
Dissociation of GDP tubulin	$733\ s^{-1}$	$915\ s^{-1}$
Steady-State Dynamic Instability		
Frequency of catastrophe in vitro at 7 μM tubulin	$0.0045\ s^{-1}$	$0.003\ s^{-1}$
Frequency of rescue in vitro at 7 μM tubulin	$0.02\ s^{-1}$	$0.06\ s^{-1}$
Microtubule Dynamic Instability in Live Cells		
Frequency of catastrophe in vivo (interphase)	$0.014\ s^{-1}$	
Frequency of catastrophe in vivo (mitosis)	$0.017\ s^{-1}$	
Frequency of rescue in vivo (interphase)	$0.046\ s^{-1}$	
Frequency of rescue in vivo (mitosis)	0	

Data from Walker RA, O'Brien ET, Pryer NK, et al. Dynamic instability of individual microtubules. *J Cell Biol*. 1988;107:1437–1448.

GTP. This maintains a pool of GTP-tubulin dimers to support elongation of the surviving microtubules. Proteins discussed in the next section ("Microtubule Dynamics in Cells") can promote catastrophes, while others stabilize microtubules.

If rapid dissociation of GDP-tubulin drives shortening, it is logical to assume that recapping with GTP-tubulin might rescue shortening microtubules. This hypothesis is likely to be true, but the actual mechanism is complicated and not well understood. For example, the frequency of rescue depends only weakly on the concentration of GTP tubulin.

Dynamic instability has been studied more thoroughly at the plus end than the minus end, where catastrophes are less frequent and rescues are more likely than at the plus end (Table 34.1).

Microtubule Dynamics in Cells

Microtubule dynamics in cells (Fig. 34.7A) are remarkably similar to those in simple buffers in vitro (Table 34.1). At any moment in time, the plus ends of many microtubules are likely to be growing, as revealed by the presence of tip-binding proteins (Fig. 34.14). Despite catastrophes approximately once each minute ($k = 0.01\ s^{-1}$), during which they shorten rapidly ($-0.28\ \mu m\ s^{-1}$), many interphase microtubules are long, both because the chance of rescue is high ($k = 0.05\ s^{-1}$) and because steady growth during the elongation phase (at a rate of 0.11 μm s^{-1}) restores their length before the next catastrophe. As a result of dynamic instability, the bulk of interphase microtubules have a half-life of approximately 10 minutes. Many microtubule minus ends are stabilized by γ-tubulin ring complexes (Fig. 34.16) in MTOCs or by other proteins (Fig. 34.8), but some minus ends are free to elongate and shorten.

Dynamic instability allows the plus ends of microtubules to explore the entire cytoplasm as they search for targets such as kinetochores on chromosomes during mitosis (Fig. 34.2A). It also concentrates tip-binding proteins, including signaling proteins, at the cell cortex. Fluctuations in microtubule length can use the energy stored in the lattice to do mechanical work such as moving cargo away from the center of the cell (see Fig. 37.6), positioning the nucleus in the center of cells as in fission yeast, or locating the mitotic spindle off center in cells that divide asymmetrically. Reciprocally, the local environment can influence the behavior of the microtubules as they explore the cytoplasm. For example, signaling pathways in the cortex involving Rho-family GTPases and kinases can stabilize microtubules locally and polarize the microtubule array.

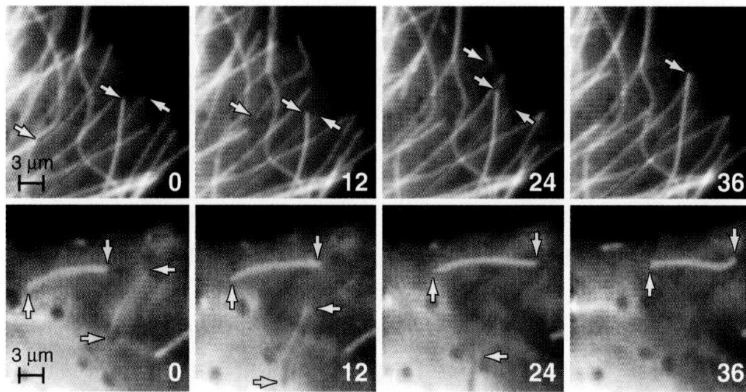

FIGURE 34.7 DYNAMIC INSTABILITY OF MICROTUBULES IN VIVO. Time series of fluorescence micrographs of a cultured vertebrate cell microinjected with rhodamine-labeled tubulin. *Upper row,* The plus ends of microtubules anchored at the centrosome grow and shrink randomly in the same cell *(arrows). Lower row,* Free microtubules treadmill, growing at their plus end at about the same rate as shortening proceeds at their minus end. Marking a spot on such microtubules shows that this is treadmilling rather than transport of a stable microtubule through cytoplasm. (Courtesy G. Borisy, University of Wisconsin, Madison.)

Microtubules can also treadmill in the cytoplasm, growing at one end and shrinking at the other. This behavior is common in plants in which dynamic instability at the plus end is biased toward net growth and minus ends depolymerize steadily. Treadmilling is seen in some animal cells when a microtubule detaches from the centrosome (Fig. 34.7B), but the minus ends of many microtubules are stable.

Regulation by Microtubule-Associated Proteins

The properties of pure tubulin cannot explain all microtubule behavior in cells. For example, microtubules in ciliary axonemes are stable for days, even under conditions that would depolymerize cytoplasmic microtubules, even though tubulin purified from axonemes forms microtubules that are just as dynamic as cytoplasmic microtubules. The explanation is that **MAPs** stabilize microtubules in axonemes. Other MAPs regulate microtubule initiation, elongation, shortening, catastrophes, and rescues (Appendix 34.1 and Fig. 34.8). Cells can modulate the activity of MAPs to change the dynamics of microtubules, as when dynamic instability becomes more pronounced during mitosis (Fig. 34.2A). Developmentally programmed gene expression establishes the mix of MAPs in each cell type.

MAPs were discovered as proteins that copurified along with microtubules from vertebrate brains during cycles of microtubule assembly and disassembly. Copurification selects a subset of MAPs that bind tightly to microtubules, but it misses important proteins that bind weakly. Functional assays yielded additional MAPs. For example, a light microscopic assay for microtubule fragmentation led to the discovery of the severing protein katanin. Genetic screens have also uncovered proteins that regulate microtubules, such as specialized kinesins

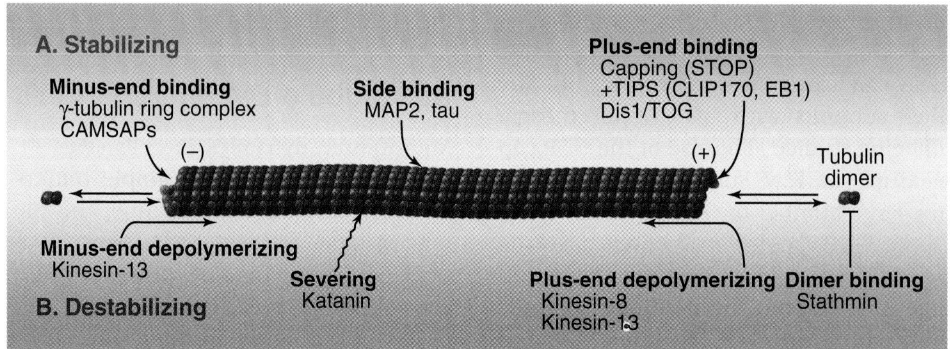

FIGURE 34.8 FAMILIES OF MICROTUBULE-BINDING PROTEINS. A, Proteins that stabilize microtubules. **B,** Proteins that destabilize microtubules. These microtubule-associated proteins (MAPs) can bind to the end or the side of the polymer or to tubulin dimers. CAMSAP, calmodulin-regulated spectrin-associated protein.

and several plus-end-binding proteins. The following sections group MAPs according to their activities.

Microtubule-Stabilizing Proteins

At least a dozen distinct MAPs stabilize microtubules (Appendix 34.1). Most bind along the length of microtubules, but some interact only at or near ends. Some are expressed widely, but others are restricted to specialized cells. Members of the **tau** family, including **MAP2** and **MAP4,** differ in size and pattern of expression but share many common features (Fig. 34.9). These MAPs are abundant in brain, the historical tissue of choice for isolating microtubules.

- They share similar tubulin-binding motifs consisting of three or four, short, imperfect, tandem repeats that each binds independently to tubulin subunits.
- N-terminal domains of tau family members project from the surface of microtubules (Fig. 34.10). The long side arm of MAP2 excludes other structures, including microtubules (see Fig. 35.9), accounting for the wider spacing of microtubules in dendrites compared with axons, where the smaller tau predominates. Neither tau nor MAP2 crosslinks microtubules, but MAP2 links microtubules to actin filaments.
- Tau family MAPs stabilize microtubules. For example, in the presence of tau, microtubules grow three times faster, shorten slower, and have catastrophes 50 times less frequently than pure tubulin microtubules. The rapid equilibrium of individual tubulin-binding repeats with the microtubule surface might allow tau to dampen microtubule dynamics without stopping tubulin association and dissociation altogether.
- Phosphorylation of the microtubule-binding motifs of these MAPs inhibits microtubule binding and destabilizes microtubules. The negatively charged phosphate groups are repelled from the negatively charged surface of the microtubule.

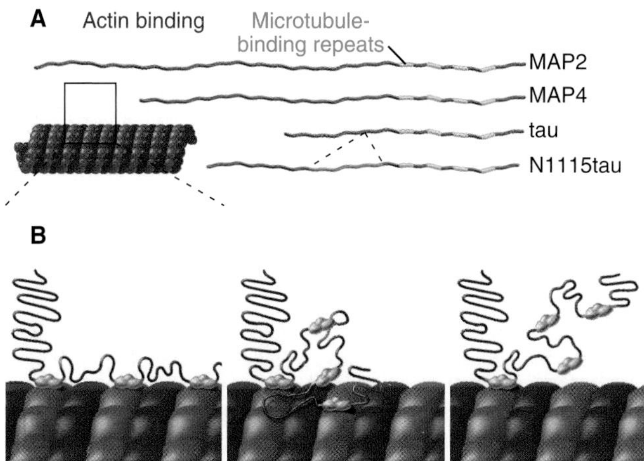

FIGURE 34.9 TAU, MICROTUBULE-ASSOCIATED PROTEIN (MAP)-2, AND MAP4. A, Comparison of the domain organization, showing the homologous tubulin-binding motifs. **B,** The tubulin-binding motifs bind to either α-tubulin or β-tubulin and are thought to exchange rapidly among tubulin subunits along a protofilament. (Modified from Butner KA, Kirschner M. Tau protein binds to microtubules through a flexible array of distributed weak sites. *J Cell Biol*. 1991;115:717–730. For reference, see Al-Bassam J, Ozer RS, Safer D, et al. MAP2 and tau bind longitudinally along the outer ridges of microtubule protofilaments. *J Cell Biol*. 2002;157:1187–1196.)

Tau, named for tubulin-associated protein, is the major MAP in the axons of neurons in vertebrate brains (Fig. 34.11B). It is also present in some neuronal cell bodies and glial cells. Alternate splicing produces seven tau isoforms from a single tau gene. Most of these isoforms have molecular weights of 40 to 50 kD, but a higher-molecular-weight tau is found in peripheral nerves. As is expected from its ability to stabilize microtubules in vitro, reduction in the tau concentration in cultured nerve cells by depletion of the messenger RNA

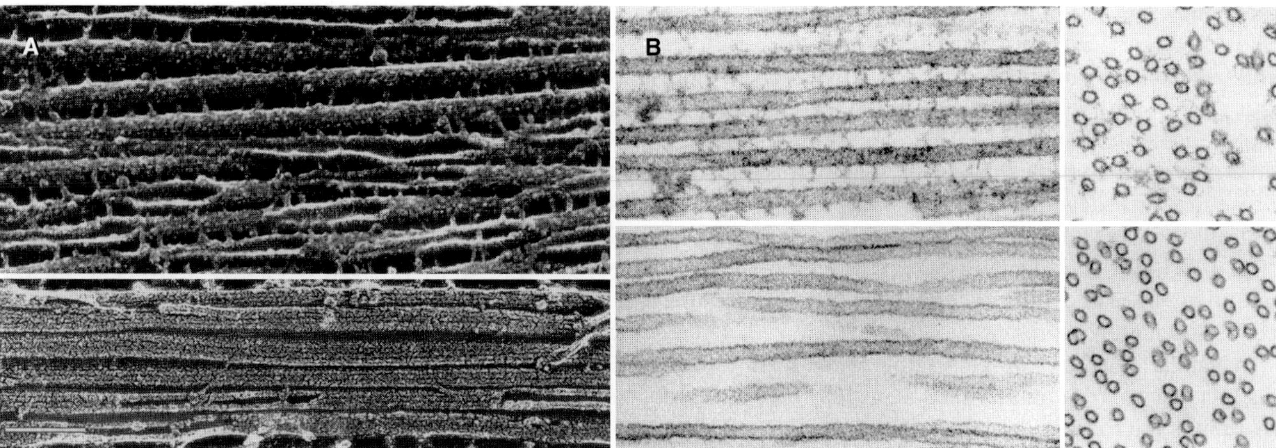

FIGURE 34.10 ELECTRON MICROGRAPHS OF TAU AND MAP-2 BOUND TO MICROTUBULES. A, Frozen, deep-etched, and shadowed specimens of microtubules with tau *(top)* and pure tubulin microtubules *(bottom)*. **B,** Thin sections of microtubules with MAP2 showing 40- to 100-nm projections *(top)* and bare pure tubulin microtubules *(bottom)*. (**A,** Courtesy N. Hirokawa, Tokyo University, Japan. For reference, see Hirokawa N, Shiomura Y, Okabe S. Tau proteins: the molecular structure and mode of binding on microtubules. *J Cell Biol*. 1988;107:1449–1459. **B,** Courtesy D.B. Murphy, Johns Hopkins Medical School, Baltimore, MD.)

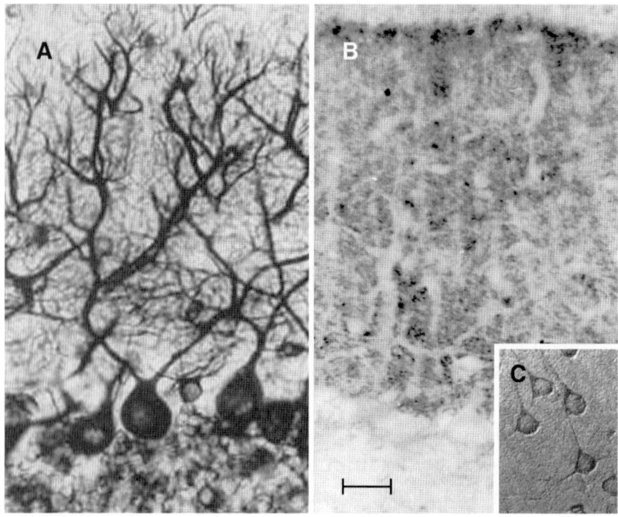

FIGURE 34.11 DISTRIBUTIONS OF TAU AND MAP-2 IN SECTIONS OF THE CEREBELLUM FROM RAT BRAIN LABELED WITH ANTIBODIES AND SUBJECTED TO HISTOCHEMICAL STAINING. A, MAP2 *(black)* is concentrated in the cell bodies and dendrites of Purkinje cells. **B,** Tau is concentrated in the axons of granule cells, which appear here as *small dark dots.* **C,** Tau staining in the cell body and neurites of a pyramidal cell from another part of the brain. Scale bar is 20 μm. (Courtesy L. Binder, Northwestern University Medical School, Evanston, IL.)

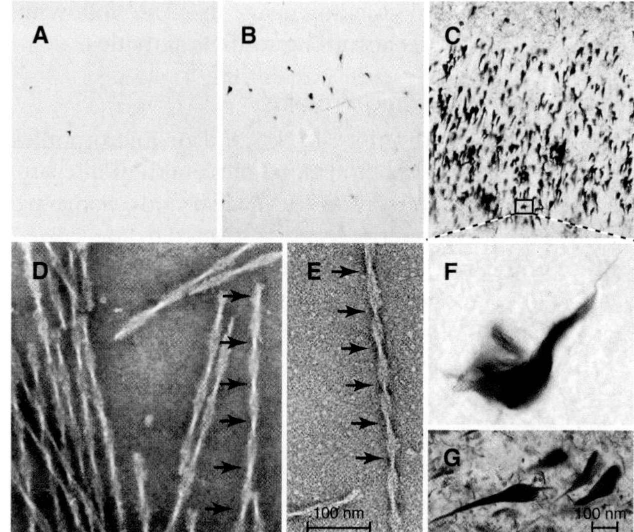

FIGURE 34.12 NEUROFIBRILLARY TANGLES OF PAIRED HELICAL FILAMENTS IN THE BRAINS OF PATIENTS WITH ALZHEIMER DISEASE. A–C, Light micrographs of sections of the hippocampus of human brains stained with silver for neurofibrillary tangles. **A,** Stage I with few tangles. **B,** Stage III with moderate numbers of tangles. **C,** Stage V, advanced Alzheimer disease, showing abundant tangles. **D,** Electron micrograph of paired helical filaments isolated from Alzheimer neurofibrillary tangles and prepared by negative staining. **E,** Electron micrograph of a negatively stained, paired, helical filament reassembled in vitro from recombinant tau protein. **F–G,** High-powered micrographs of neurofibrillary tangles from the brain of an Alzheimer patient stained *brown* with an antibody to tau. (**A–C** and **G,** Courtesy E. and H. Braak, University of Frankfurt, Germany. **D–E,** Courtesy E.-M. Mandelkow, Max Planck Institute, Hamburg, Germany. **F,** Courtesy L. Binder, Northwestern University Medical School, Evanston, IL.)

reduces the numbers of microtubules. Nevertheless, mice survive the loss of their single tau gene with only minor alterations of their neurons. Compensation by other MAPs is postulated to compensate for the chronic loss of tau.

Fragments of tau form intracellular **paired helical filaments** (Fig. 34.12) that aggregate in "neurofibrillary tangles" in neurodegenerative diseases including **Alzheimer disease,** the most common dementia of older persons. The number of tangles judged by light microscopy correlates with the severity of the dementia. Dozens of different tau mutations cause rare inherited dementias, but for individuals with normal tau, excess phosphorylation can dissociate tau from microtubules and lead to its fragmentation by proteolysis. Over time, short, phosphorylated tau fragments assemble into highly insoluble paired helical filaments and tangles. It is not known which molecular species along this pathway cause neuronal degeneration.

Unlike tau, MAP2 concentrates in dendrites of neurons (Fig. 34.11A). The mechanism giving preferential distribution of MAP2 in dendrites and tau in axons of the same cell is still largely a mystery.

A variety of other proteins stabilize microtubules in distinct ways. Vertebrates have a protein called STOP that binds microtubules, making them resistant to depolymerization by cold, dilution, or drugs. Bird red blood cells express syncolin, which stabilizes the marginal band of microtubules. Tektins are fibrous proteins that stabilize microtubules in axonemes (see Fig. 38.14) and centrioles.

Microtubule-Destabilizing Proteins

Cells destabilize microtubules in three different ways. First, the small protein **stathmin** destabilizes microtubules by sequestering tubulin dimers. Stathmin has a long α-helix that binds laterally to a pair of tubulin dimers, blocking their polymerization. This sequestration of dimers may promote catastrophes. Thus, overexpression of stathmin in cell lines reduces tubulin available for polymerization. Mitotic kinases phosphorylate tubulin-interacting sites on stathmin. This releases tubulin and promotes assembly of the mitotic spindle. Many malignant cells express unusually high levels of stathmin (hence its synonym Op18, for oncoprotein 18).

In a second mechanism, three classes of microtubule motor proteins, **kinesin-8, kinesin-13,** and **kinesin-14**, promote microtubule disassembly by removing subunits from the polymer ends. Independent discovery of these kinesins in several systems led to a variety of historic names. Kinesin-8 walks to the plus end of the microtubule and uses energy from ATP hydrolysis to remove tubulin. Rather than using ATP hydrolysis to walk along a microtubule (see Fig. 36.13), kinesin-13 accumulates at both ends, probably by diffusion. There it uses energy from ATP hydrolysis to remove tubulin subunits. This promotes catastrophes and regulates the lengths of the microtubules in the mitotic spindle (see Chapter 44).

In the third mechanism, AAA adenosine triphosphatases (ATPases; see Box 36.1) called **katanin** and **spastin,** use energy from ATP hydrolysis to disrupt the noncovalent bonds between the subunits in the microtubule wall. This destabilizes and severs the microtubule. Tubulin dimers dissociate from the cut ends and return to the cytoplasmic pool for reassembly. Activation of severing at the onset of mitosis might contribute to remodeling the interphase microtubule network. Autosomal dominant mutations in the human spastin gene cause hereditary spastic paraplegia.

Proteins Associated With Microtubule Ends

A number of proteins bind near microtubule ends. They use different mechanisms and impact assembly in different ways.

Proteins called **+TIPs** concentrate at sites of microtubule growth, which are largely plus ends in cells where the minus ends are anchored by MTOCs (Fig. 34.13). Two families of +TIPs concentrate at growing microtubule ends independent of other proteins, while other "hitchhiker" proteins locate at ends by binding to one of these core proteins.

The "end binding protein" **(EB)-1** (and homologs EB2 and EB3) are dimers held together by a coiled-coil. N-terminal calponin homology domains (similar to the actin binding domain of α-actinin; see Fig. 33.1) bind the ends of a growing microtubule at a site between protofilaments composed of newly incorporated GTP-tubulin (Fig. 34.6D). This interaction can stabilize plus ends by reducing the frequency of catastrophes, but it may also stimulate GTP hydrolysis under some circumstances. After tubulin hydrolyzes its bound GTP, EBs dissociate from the microtubule owing to lower affinity for the more compact lattice of older parts of microtubules with bound GDP. The C-terminal domain of EBs interacts with hitchhiker +TIPs in two ways. Several have a microtubule tip localization sequence (S/T*x*IP [serine or threonine–any amino acid–isoleucine–proline]) that binds to receptor sites at the C-terminus of EB proteins. One example is stromal interaction molecule (STIM)-1, a transmembrane protein of the endoplasmic reticulum that anchors the membrane to EB1 for transport on the growing microtubule tip (see Fig. 37.6). Alternatively, an EEY/F (glutamic acid-glutamic acid-tyrosine or phenylalanine) sequence motif of EB binds to domains of proteins such as **cytoplasmic linker protein (CLIP)**-170. This interaction allows the fission yeast EB protein to anchor CLIP-170 and an associated fission yeast protein, Tea1p, to the plus ends microtubules (see Fig. 6.3D). A kinesin delivers this complex to the ends of the cell where Tea1p directs polar growth. EBs or CLIPs target cytoplasmic linker-associated proteins (CLASPs) to plus ends, where they promote microtubule elongation near the cell cortex in interphase and at kinetochores during mitosis.

The **Dis1/TOG** family of proteins (called XMAP215 in frogs) associate with microtubule plus ends independently of EBs. The TOG domains have higher affinity for curved tubulin dimers than straight tubulin. These proteins have multiple TOG domains allowing them to associate with both curved protofilaments at growing microtubule plus ends and multiple tubulin dimers. Transfer of tethered tubulin dimers to the growing end promotes elongation. These proteins help to organize the mitotic spindle poles in animals and the cortical array of microtubules in plants. Mixtures of tubulin, XMAP215, and a depolymerizing kinesin-13 or an EB can recapitulate microtubule dynamic instability in test tubes very similar to that seen in extracts from mitotic cells. Members of the Dis1/TOG family may also stabilize microtubule plus ends by reducing the frequency of catastrophes.

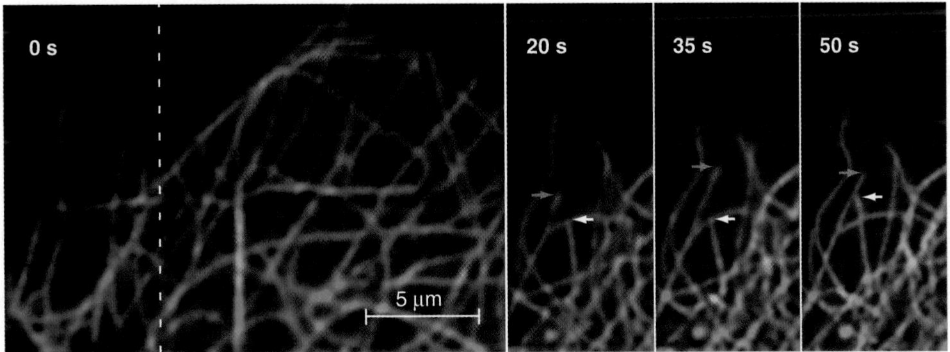

FIGURE 34.13 Time series of fluorescence micrographs of a Chinese hamster ovary (CHO) cell expressing the microtubule tip-binding protein CLIP-170 tagged with green fluorescence protein (GFP) *(red)* and microinjected with Cy3-tubulin to mark microtubules *(green)*. CLIP-170 concentrates at the plus ends of growing microtubules but dissociates from shrinking microtubules. CLIP, cytoplasmic linker protein. (Courtesy Yulia Komarova, Northwestern University, Evanston, IL. Modified from Komarova Y, Vorobjev IA, Borisy GG. Life cycle of microtubules: persistent growth in the cell interior, asymmetric transition frequencies and effects of the cell boundary. *J Cell Sci.* 2002;115:3527–3539.)

Several other proteins that function at or near the plus ends of microtubules are not classical +TIPs. These include the chromosomal passenger protein INCENP (inner centromere protein; see Fig. 44.10), which targets the Aurora-B kinase to the central spindle and regulates its kinase activity toward spindle components.

The γ-tubulin ring complex (Fig. 34.15) nucleates microtubules and caps their minus ends, but if minus ends arise from severing or other means, proteins called CAMSAPs (calmodulin-regulated spectrin-associated proteins) bind along microtubule walls near the minus ends. This association depends on growth of the minus end, but the mechanism is still being investigated. Minus ends associated with CAMSAPs are stabilized from depolymerization.

Microtubule Linker Proteins

Protein connectors link microtubules to many other cellular structures. For example, MAP2 and spectraplakins link microtubules to actin filaments, and plectin links microtubules to intermediate filaments. Gephyrin binds microtubules and is required for clustering glycine receptors (see Fig. 16.12) in the plasma membrane of neurons.

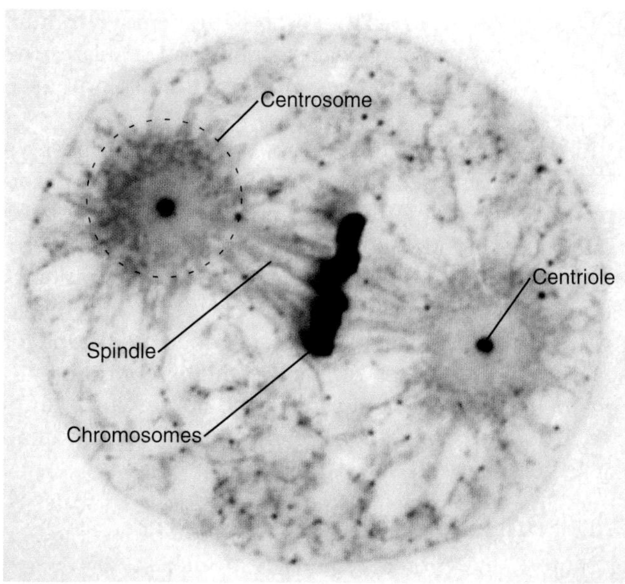

FIGURE 34.14 METAPHASE SPINDLE IN AN EGG OF *PARASCARIS EQUORUM*. In this micrograph taken from a classic slide prepared by Boveri in the early 20th century, the centrosomes are particularly prominent, and the centrioles are clearly visible within them. (Courtesy Joseph Gall, Carnegie Institution, Baltimore, MD.)

Centrosomes and Other Microtubule Organizing Centers

Most microtubules arise from assemblies of γ-tubulin located in **MTOCs**. The main organizing centers in animal cells are **centrosomes** (Figs. 34.2A, 34.14, and 34.15). A typical centrosome in animal cell consists of a "centriole" that organizes the **pericentriolar material** where most γ-tubulin resides. In addition to organizing the pericentriolar material, **centrioles** also directly initiate microtubules for cilia and flagella (Fig. 34.3 and see Fig. 38.17). Some cells, such as mature eggs in animals and all cells in higher plants, lack centrosomes. The MTOCs of fungi ("spindle pole bodies") lack centrioles. Microtubules are also nucleated by γ-tubulin recruited to the Golgi apparatus, nuclear envelope, or in the cortex of epithelial cells (Fig. 34.2B). Additional microtubules also arise within the mitotic spindle. This process depends on a complex of proteins called augmin that interacts with both microtubules and γ-tubulin, but the nucleation mechanism, in which the new microtubules branch from the sides of existing spindle microtubules, is still being investigated.

Overview of Centrosomes

When Flemming and Van Beneden discovered the centrosome in 1875, it was regarded as one of the three main cellular components, together with the nucleus and the cell body. In fact, it was correctly identified as the organelle that regulates mitosis and was called the "dynamic centre" of the cell by Boveri, who coined the term *centrosome.* Large centrosomes of certain eggs appeared to contain one or two central granules, which Boveri called "centrioles" (Fig. 34.14). Electron microscopy later showed that centrioles are cylinders of nine microtubule triplets or doublets approximately 0.5 μm long (Fig. 34.15). Centrioles help organize centrosomes, although centrosomes in some cells lack centrioles. Among the major cytoplasmic organelles, centrosomes are unique in lacking a bounding membrane and in duplicating semiconservatively in a process that takes more than one complete cell cycle.

The **centrosome** is the dominant MTOC during interphase and mitosis in most animal somatic cells. Centrosome-associated microtubules emanate from **γ-tubulin ring complexes** (usually abbreviated **γTuRC**) located in **pericentriolar material** surrounding two centrioles. Centrosomes also have roles in cytokinesis, cell polarity, and organization of primary cilia. Many cancer cells have structural and numerical centrosome aberrations.

The last common ancestor of eukaryotes had centrioles to serve as basal bodies for its flagella (see Fig. 2.4B). Many eukaryotes inherited the genes for centrioles, but multiple lineages lost these genes. The male gametes of lower plants such as ferns have basal bodies, but most plants and fungi lost centrioles, so other structures serve as MTOCs. Metazoa have centrioles, which organize centrosomes in most cells. Some multinucleated animal cells such as megakaryocytes (see Fig. 28.4) and osteoclasts (see Fig. 32.6) have multiple centrosomes, but mature oocytes and multinucleated vertebrate muscle cells have none.

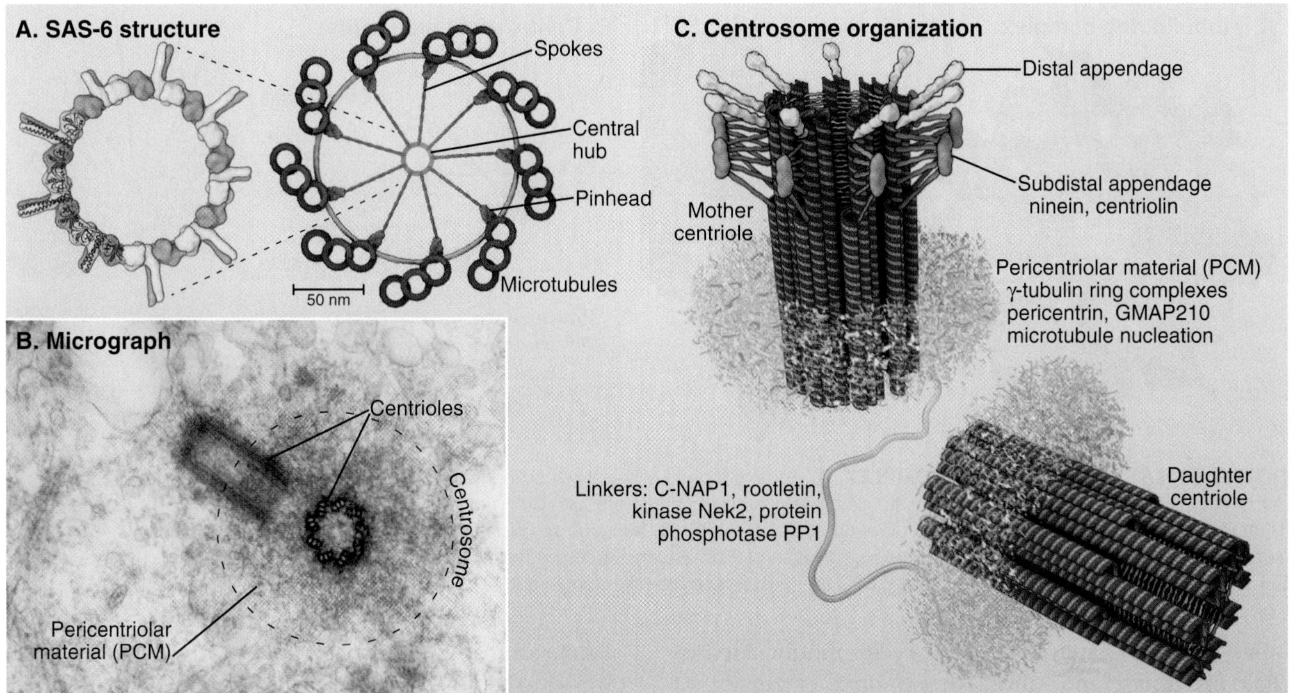

FIGURE 34.15 STRUCTURE OF CENTRIOLES. A, Model for how the crystal structure of the protein SAS-6 can form a unit with ninefold symmetry and a drawing showing how this unit may fit into the center of the centriole. **B,** Electron micrograph of a thin section of paired centrioles surrounded by pericentriolar material at telophase in a PtK1 (rat kangaroo) cell. **C,** Diagram of the structure of the centrioles with the locations of some of the constituent proteins. (**A,** Based on the work of Kitagawa D, Vakonakis I, Olieric N, et al. Structural basis of the 9-fold symmetry of centrioles. *Cell*. 2015;144:364–475, for reference, see PDB file 4CKP. **B,** Courtesy Conly Rieder, Wadsworth Center, Albany, NY.)

Protein Composition of Centrioles

Centrioles are cylindrical structures typically composed of nine triplet microtubules (Fig. 34.15), but some are simpler: doublet microtubules in flies and single microtubules in nematodes. The older **mother centriole** in each pair has appendages lacking on the new **daughter centriole**. Electron microscopy revealed the overall structure and crystal structures of component proteins are filling in details. The most abundant protein components of centrioles are α-tubulin and β-tubulin. Centriolar microtubules are more stable than most cytoplasmic microtubules, exchanging only approximately 10% of their tubulin per cell cycle. Like other stable microtubules, the α- and β-tubulins of centrioles are highly modified by polyglutamylation. Microinjection of cells with antibodies to glutamylated tubulin disassembles both centrioles and the pericentriolar material.

Less-abundant proteins form other centriolar structures (Appendix 34.2). Oligomers of the protein SAS-6 form a ring or spiral with ninefold symmetry that templates the cartwheel pattern of triplet microtubules during centriole formation (Fig. 34.15A). This framework is usually absent in mature mother centrioles. Distal appendages anchor the mother centriole to the plasma membrane when it serves as the basal body for cilia (see Fig. 38.17). Other minor proteins form subdistal appendages on mother centrioles (Fig. 34.15C). Ninein helps anchor the minus ends of microtubules to the centrosome and centriolin appears to anchor proteins that control the cell cycle and regulate microtubule stability. Two large coiled-coil proteins, C-NAP1 and rootletin, form fibers that connect the pericentriolar material around the two centrioles (Fig. 34.15C).

Centrioles also contain multiple isoforms of centrin (see Fig. 38.21), an EF-hand, Ca^{2+} binding protein similar to calmodulin (see Fig. 3.12). Centrin filaments link centrioles to the nucleus in *Chlamydomonas*, but most centrins are in the cytoplasm, where their functions are not established.

Protein Composition of the Pericentriolar Material

Large coiled-coil proteins surrounding the mother centriole organize the pericentriolar material. This scaffold accommodates a network of γTuRC, kinases, phosphatases, and more than 100 other proteins. Electron micrographs of thin sections show that the pericentriolar material excludes organelles but do not reveal any internal regular structure (Fig. 34.15A). Superresolution fluorescence microscopy and crystal structures of individual components are uncovering the organization of the proteins.

The size of the pericentriolar material changes across the cell cycle. During interphase, a single layer of the protein pericentrin forms the scaffold. One end of pericentrin binds to centriolar microtubules and the other end extends radially to interact with the γTuRC and

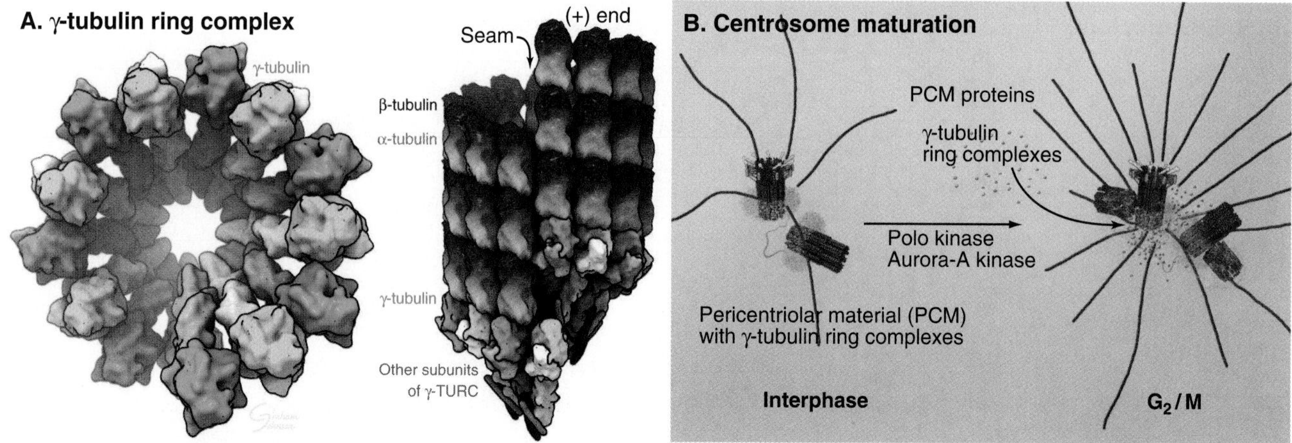

FIGURE 34.16 γ-TUBULIN RING COMPLEX. A, Reconstruction of yeast γ-tubulin small complex from electron micrographs and *(right)* a model of active γ-tubulin ring complex (γTuRC) on the minus end of a microtubule. Color code: *yellow,* γ-tubulin; *light gray*, α-tubulin; *dark gray*, β-tubulin; *dark blue, light blue*, and *green* other subunits of γTuRC. **B,** Microtubule nucleation by γTuRC in the pericentriolar material. As cells enter mitosis γTuRC are recruited in response to activation of Polo-like and Aurora-A kinases. (**A,** Based on the work of Kollman JM, Greenberg CH, Li S, et al. Ring closure activates yeast γTuRC for species-specific microtubule nucleation. *Nat Struct Mol Biol.* 2015;22:132–137.)

other proteins. Later in the cell cycle, mitotic kinases (Polo-like kinase and Aurora kinase) drive expansion of the pericentriolar material, which accumulates more copies of γTuRC and other pericentriolar proteins. The larger mitotic centrosome nucleates new microtubules at higher rates than during interphase.

The γTuRC is an assembly of 14 γ-tubulin molecules supported by accessory proteins that form a dislocated ring similar to a left-handed lock washer (Fig. 34.16). The conserved building block of γTuRC is a small complex of two γ-tubulins with two of the supporting proteins. This γTuRC template initiates assembly of α/β-tubulin dimers into a microtubule with 13 protofilaments and a seam (Fig. 34.16A). The γTuRC caps the minus end of the new microtubule as it grows at its free plus end.

Centrosome Duplication During the Cell Cycle

The cycle of semiconservative centrosome duplication and division is closely linked to the cell cycle (Fig. 34.18; see Chapter 40 for an introduction to the cell cycle). Centrosome replication begins during S phase, when the nuclear DNA is replicated (see Chapter 42). The process depends on cyclin dependent kinase Cdk2 (activated by cyclin E and/or cyclin A). These kinases turn off ubiquitin E3 ligase APC/C^{CDH1} (see Fig. 40.15), allowing accumulation of proteins to replicate the centriole. Equally important a polo-like kinase (PLK4 in humans) turns on at this time.

Each centriole initiates formation of a new centriole oriented at right angles to its proximal end. Thus the previous daughter centriole becomes into a new mother centriole, M^{new}. An adapter protein located on one side of each mother centriole recruits a polo-like kinase, which activates proteins that allow SAS-6 to assemble a large cartwheel with ninefold radial symmetry (Fig. 34.15A). Depending on the species, SAS-6 forms a stack of rings or a spiral that cooperates with another protein (SAS-4) to template the assembly of a ring of nine single microtubules, the procentriole (Fig. 34.18). The ring is converted into an array of nine triplet microtubules in a process that depends on δ-tubulin and ε-tubulin in most organisms. The absence of δ-tubulin and ε-tubulin in *Drosophila* and *Caenorhabditis elegans* likely contributes to the fact that most of their centrioles are built of doublet or singlet microtubules. Daughter centrioles elongate gradually during the remainder of the cell cycle, reaching their mature length just before mitosis. Then M^{new} acquires distal and subdistal appendages and becomes morphologically and functionally equivalent to M^{old}.

Thus cells about to enter mitosis have three types of centrioles that differ in age, structure, and activity (Fig. 34.17). Two distinct mother centrioles, M^{old} and M^{new}, are each paired to one of two identical daughter centrioles. M^{old} was assembled at least two cell cycles previously. M^{new} was assembled in the previous cell cycle, and the daughter centrioles were assembled during the S phase of the present cell cycle. The SAS-6 cartwheel disappears from the daughter centriole at this stage in human cells.

The relationships among the four centrioles change as the cell enters mitosis. Around the time of the transition from G2 to M the kinase Nek2 phosphorylates the fiber protein C-NAP1 and fibers linking the two pairs of replicated centrioles break down, allowing them to separate. During prophase kinesin-5 and other mechanisms drive the two pairs of centrioles apart over the surface of the nuclear envelope to set up the two poles of the mitotic spindle. During anaphase a cascade of mitotic kinases (Cdk1 → Aurora A → polo-like kinase 1) releases the new daughter centriole from the side of their mother centriole, but the two remain attached by flexible linkers between their proximal ends (Fig. 34.15C).

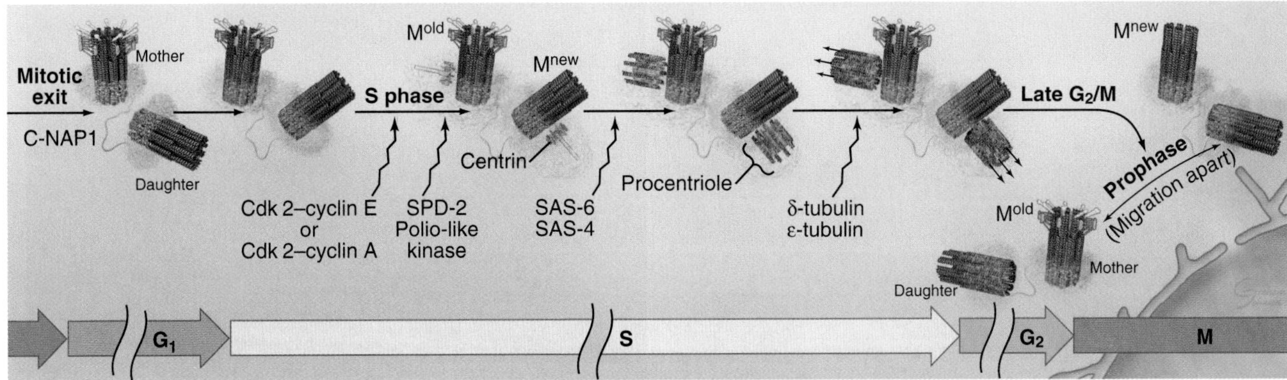

FIGURE 34.17 THE PATHWAY OF CENTRIOLE DUPLICATION IS LINKED TO THE CELL CYCLE.

If centrosomes are surgically removed from normal cells, the cell enters mitosis and assembles a bipolar mitotic spindle using motor proteins to bundle and remodel the microtubules. These cells slowly complete mitosis and about 40% divide by cytokinesis. These cells assemble new centrioles de novo during S-phase of the next cell cycle and recover. Continued suppression of centrosome formation with a drug interferes indirectly with proliferation, because of errors in chromosome segregation.

Functions of Centrosomes

The main functions of centrosomes are to nucleate and anchor microtubules that organize the cytoplasm during both interphase and mitosis. Other mechanisms, including motor proteins, must be active because plants and some animal cells can organize mitotic spindles and other microtubule structures without centrosomes. Similarly, much of the development of a fruit fly can take place without centrosomes, but the lack of cilia is eventually lethal.

The γTuRC in the pericentriolar material around the mother centriole nucleates microtubules with capped (stabilized) minus ends and plus ends probing the cytoplasm as they undergo dynamic instability. The distal and subdistal appendages of centriole M^{old} help anchor the minus ends of these microtubules. This creates a radial array of microtubules with centriole M^{old} at the center (Fig. 34.2A). The pericentriolar material expands as cells enter mitosis, increasing the capacity to nucleate microtubules for the mitotic spindle (Fig. 34.16B).

After experimental disruption of centrosomes, cytoplasmic microtubules emanating from the centrosome disappear, but other cytoplasmic microtubules persist (Fig. 34.18). In many differentiated cells, microtubules initially nucleated at centrosomes subsequently detach and become anchored back to the centrosome or elsewhere in the cytoplasm; for example, at the apical membrane in epithelial cells (Fig. 34.2B). In neurons, microtubules detach from the centrosome and are transported along axons and dendrites (see Fig. 37.5).

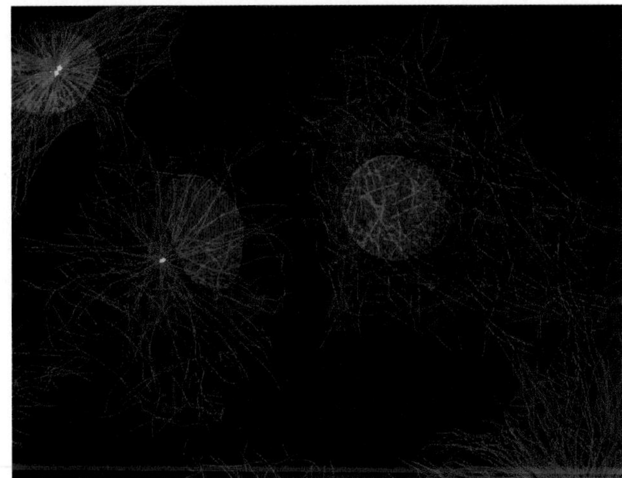

FIGURE 34.18 Destruction of the centrosome with a laser does not prevent the assembly of cytoplasmic microtubules, but they are disorganized. Centrioles, *green;* nuclei, *blue;* microtubules, *red.*

The array of polarized, dynamic microtubules radiating from the centrosome strongly influences the distributions of membrane bound organelles in the cytoplasm. The endoplasmic reticulum is typically dispersed throughout the peripheral cytoplasm by transport away from the centrosome by kinesins and growing microtubule plus ends (see Fig. 37.6). Thus the position of the centrosomes and their associated microtubules can polarize the secretory pathway. For example, cytotoxic T lymphocytes move the centrosome to the cell cortex adjacent to their target cells so that secretion of pore-forming proteins can lyse target cells (see Fig. 27.8). On the other hand, the Golgi apparatus and endosomes accumulate near the centrosome by transport toward the minus ends of the microtubules by cytoplasmic dynein (see Fig. 21.20). Close association of centrosomes with the nucleus allows the microtubule network to anchor the nucleus to the cell cortex during cell migration. Anchoring to the nucleus is important for nuclear positioning and migration in fungi, as well as during brain development in vertebrates.

Centrosomes and their radiating microtubules also help specify the position of the contractile ring during

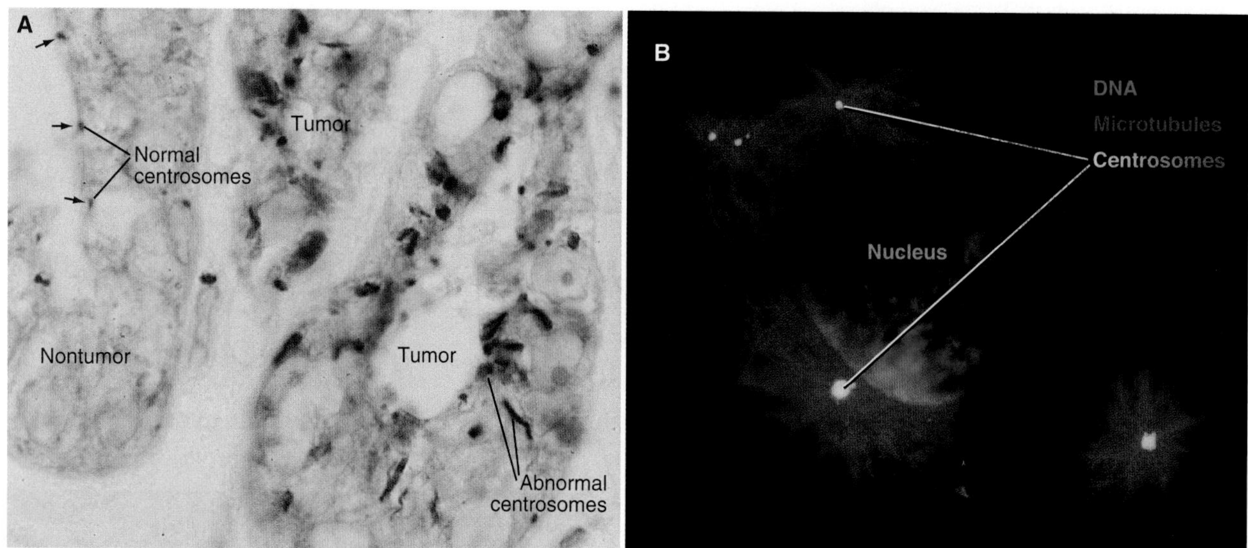

FIGURE 34.19 CENTROSOMES AND CANCER. A, Centrosomes are often abnormal in human tumors. Prostate tissue labeled by antibodies coupled to an enzyme (peroxidase) that produces a *brown* stain shows single uniform centrosomes in normal tissue *(left, arrows)* and abnormal centrosomes in tumor cells *(right).* Abnormal centrosomes are greater in number, elongated, and much larger than those in normal cells. Centrosome defects lead to spindle abnormalities, mistakes in the segregation of chromosomes, and abnormal numbers of chromosomes, a hallmark of tumors. **B,** Tumor cells have abnormal centrosomes. Fluorescence micrograph of a single prostate tumor cell: DNA is stained *blue* with DAPI (4,6-diamidino-2-phenylindole), microtubules are stained *red* with a fluorescent antibody, and pericentrin in the centrosomes is stained *green* (showing as *yellow* where it overlaps *red*) with a second fluorescent antibody. Cells were treated with nocodazole to depolymerize microtubules and were then released briefly to allow microtubule regrowth before processing. Tumor cells have abnormal numbers of centrosomes, which are heterogeneous in size, but they remain competent to nucleate microtubules. Normal cells typically have a single centrosome with a single focus of microtubules (not shown). (Courtesy S. Doxsey, G. Pihan, and A. Purohit, University of Massachusetts, Worcester.)

cytokinesis (see Fig. 44.21). As the cleavage furrow forms during late anaphase, mother and daughter centrioles detach from each other. In some cells, the mother centriole moves transiently into the intercellular bridge between the two daughter cells just before they separate. These movements of the mother centriole might somehow influence the cleavage of the bridge.

Centrioles serve as basal bodies that template and anchor axonemes of cilia (Fig. 34.3B). Centrosomal centrioles in most vertebrate tissues have the capacity to grow a nonmotile **primary cilium,** which serves as a sensory organelle (see Fig. 38.19). When the cell exits from the cell cycle, the mother centriole docks on the inside of the plasma membrane and nucleates an axoneme consisting of nine doublet microtubules. Defects in primary cilia cause polycystic kidney disease and a spectrum of other diseases (see Chapter 38).

Centrioles also form basal bodies for motile cilia of many protists, most animal sperm and ciliated epithelia (Fig. 34.3; also see Fig. 38.17). These cells assemble a basal body for each cilium de novo, rather than only templating on the side of a preexisting mother centriole only once per cycle.

Centrosomes and Cancer

Human cancer cells rarely have mutations in genes for centrosomal proteins, but often have abnormal centrosomes. One study found that 96% of high-grade tumors had abnormal centrosomes, including larger sizes (5 to 10 times normal), unusual shapes, and increased numbers (Fig. 34.19). Researchers are investigating how these defects in centrosomes might contribute to tumor formation or if they accumulate as a secondary consequence of other mutations that actually drive malignancy.

The most likely mechanism whereby defective centrosomes might promote cancer is mistakes in chromosome segregation during mitosis. The most drastic alternation of chromosome numbers results from failure to complete cytokinesis, producing progeny with twice the normal content of DNA, number of chromosomes, and number of centrosomes. The loss or gain of chromosomes can contribute to uncontrolled cell proliferation, if the balance between growth-promoting oncogenes and growth-regulating tumor-suppressor genes is upset.

Most vertebrate cells with abnormal numbers of chromosomes arrest the cell cycle or die by apoptosis (see Chapter 46). However, many human cancer cells lack this control, owing to loss-of-function mutations of the tumor-suppressor p53 (see Figs. 41.15 and 43.11). Cells lacking this transcription factor survive with abnormal centrosomes and/or chromosomes. In these examples the centrosome defects appear to be a consequence of the underlying mutation rather than a cause of the cancer.

Several other mechanisms may contribute to defects in centrosomes and contribute to cancer. For example, overexpression of enzymes that regulate the centriole cycle might produce extra centrosomes. Indeed, some cancer cells amplify the Aurora-A genetic locus. The resulting overexpression of the Aurora-A protein kinase can cause centrosomal abnormalities and interfere with normal growth regulation. Such cells form tumors when injected into mice. Centrosomal defects may also compromise regulation of proliferation by primary cilia, promote the migration of metastatic cancer cells or alter the behavior of cancer stem cells.

Spindle Pole Body, the "Centrosome" of Yeasts

For many fungi, including budding and fission yeasts, the **spindle pole body** (SBP) plays the role of the centrosome by initiating microtubules, particularly during mitosis. Rather than being in the cytoplasm like the centrosome, SPBs are plaque-like structures embedded in the nuclear envelope for the entire cell cycle in budding yeast (Fig. 34.20) and most of the cell cycle in fission yeast. SPBs contain centrin and γ-tubulin

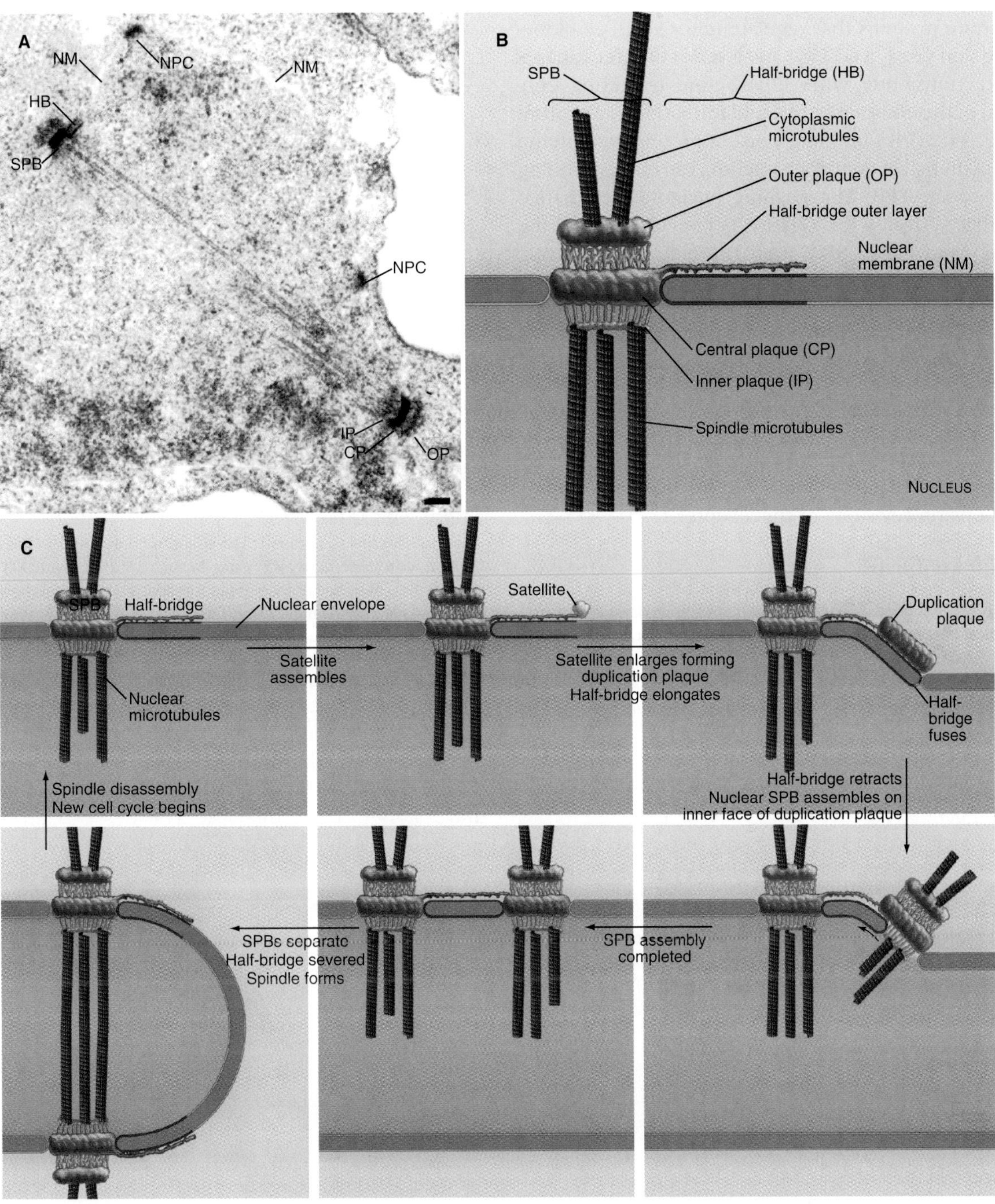

FIGURE 34.20 STRUCTURE AND DUPLICATION OF THE BUDDING YEAST SPINDLE POLE BODY. A, Electron micrograph of a thin section of the mitotic spindle of *Saccharomyces cerevisiae,* with both poles ending in an spindle pole body (SPB). **B,** Diagram of the parts of the SPB. **C,** Pathway of duplication of the budding yeast SPB. (**A,** Courtesy John Kilmartin. For reference, see Adams IR, Kilmartin JV. Spindle pole body duplication: a model for centrosome duplication? *Trends Cell Biol*. 2000;10:329–335.)

complexes, but few of the other 45 proteins identified in SPBs have been found in vertebrate centrosomes. SPB duplication is tightly linked to cell cycle progression. Similar to centrioles, a new SPB forms adjacent to and attached to the original SPB during the S phase of the cell cycle. The slime mold *Dictyostelium* has a centrosome with features of both yeast SPBs (integrated into the nuclear envelope during mitosis) and animal centrosomes (free in the cytoplasm).

In another similarity to centrosomes, fungal SPBs concentrate proteins that regulate mitosis and cytokinesis. In fission yeast, a GTPase and a series of three kinases associate transiently with SPBs before triggering constriction of the contractile ring and formation of a septum (see Fig. 44.24). In budding yeast, a protein resembling part of mammalian centriolin anchors the corresponding GTPase to the SPB. The guanine nucleotide exchange factor that activates the GTPase is concentrated in the bud, far from either SPB, until the elongating mitotic spindle relocates one SPB to the bud during anaphase. Only then can the guanine nucleotide exchange factor activate the GTPase and trigger a signaling cascade that ultimately drives the cell out of mitosis.

ACKNOWLEDGMENTS

We thank Anna Akhmanova, Holly Goodson, Alexey Khodjakov, Laurence Pelletier, and Jordan Raff for their suggestions on revisions of this chapter.

SELECTED READINGS

Akhmanova A, Steinmetz MO. Control of microtubule organization and dynamics: two ends in the limelight. *Nat Rev Mol Cell Biol*. 2015; 16:711-726.

Al-Bassam J, Chang F. Regulation of microtubule dynamics by TOG-domain proteins XMAP215/Dis1 and CLASP. *Trends Cell Biol*. 2011; 21:604-614.

Bowne-Anderson H, Zanic M, Kauer M, Howard J. Microtubule dynamic instability: a new model with coupled GTP hydrolysis and multistep catastrophe. *Bioessays*. 2013;35:452-461.

Conduit PT, Wainman A, Raff JW. Centrosome function and assembly in animal cells. *Nat Rev Mol Cell Biol*. 2015;16:611-624.

Coombes CE, Yamamoto A, Kenzie MR, Odde DJ, Gardner MK. Evolving tip structures can explain age-dependent microtubule catastrophe. *Curr Biol*. 2013;23:1342-1348.

Dong G. Building a ninefold symmetrical barrel: structural dissections of centriole assembly. *Open Biol*. 2015;5:pii: 150082.

Fu J, Hagan IM, Glover DM. The centrosome and its duplication cycle. *Cold Spring Harb Perspect Biol*. 2015;7:a015800.

Gardner MK, Charlebois BD, Janosi IM, et al. Rapid microtubule self-assembly kinetics. *Cell*. 2011;146:582-592.

Gönczy P. Centrosomes and cancer: revisiting a long-standing relationship. *Nat Rev Cancer*. 2015;15:639-652.

Gupta KK, Li C, Duan A, et al. Mechanism for the catastrophe-promoting activity of the microtubule destabilizer Op18/stathmin. *Proc Natl Acad Sci USA*. 2013;110:20449-20454.

Honnappa S, Gouveia SM, Weisbrich A, et al. An EB1-binding motif acts as a microtubule tip localization signal. *Cell*. 2009;138:366-376.

Kapitein LC, Hoogenraad CC. Building the neuronal microtubule cytoskeleton. *Neuron*. 2015;87:492-506.

Kollman JM, Merdes A, Mourey L, et al. Microtubule nucleation by γ-tubulin complexes. *Nat Rev Mol Cell Biol*. 2011;12:709-721.

Lin TC, Neuner A, Schiebel E. Targeting of γ-tubulin complexes to microtubule organizing centers: conservation and divergence. *Trends Cell Biol*. 2015;25:296-307.

Mandelkow EN, Mandelkow E. Biochemistry and cell biology of tau protein in neurofibrillary degeneration. *Cold Spring Harb Perspect Med*. 2012;2:a006247.

Mennella V, Agard DA, Huang B, et al. Amorphous no more: subdiffraction view of the pericentriolar material architecture. *Trends Cell Biol*. 2014;24:188-197.

Sanchez-Huertas C, Luders J. The Augmin connection in the geometry of microtubule networks. *Curr Biol*. 2015;25:R294-R299.

Suozzi KC, Wu X, Fuchs E. Spectraplakins: master orchestrators of cytoskeletal dynamics. *J Cell Biol*. 2012;197:465-475.

Yu I, Garnham CP, Roll-Mecak A. Writing and reading the tubulin code. *J Biol Chem*. 2015;290:17163-17172.

Zhang R, Alushin GM, Brown A, et al. Mechanistic origin of microtubule dynamic instability and its modulation by EB proteins. *Cell*. 2015;162:849-859.

APPENDIX 34.1

Some Microtubule-Associated Proteins

Name (Synonyms)	Distribution	Composition	Properties	Functions (Diseases)
Destabilizers				
Stathmin/Op18	Vertebrate cells	1 × 18 kD	Binds tubulin dimers	Sequesters tubulin; enhances dynamic instability; null mice viable without defects
Kinesin-8	Eukaryotes	2 × 82 kD	Plus end motor	Removes tubulin from plus ends
Kinesin-13 (MCAK)	Eukaryotes	2 × 82 kD	Lacks motor activity	Removes tubulin from both ends
Kinesin-14 (NCD)	Eukaryotes	2 × 82 kD	Minus end motor	Removes tubulin from minus ends
Severing				
Katanin	Metazoans, plants	6 × 84 kD, 6 × 60 kD	AAA ATPase	ATP-dependent MT severing
Spastin	Metazoans, plants	6 × 86 kD	AAA ATPase	ATP-dependent MT severing (hereditary spastic paraplegia)
Stabilizers				
MAP2a, b, c	Dendrites of vertebrate neurons	One gene, 4 isoforms 42–200 kD	Rod with three or four 18-residue MT-binding repeats	Promotes MT assembly; binds regulatory subunit of PKA; binds actin
MAP4 (MAP3, MAPU)	Vertebrate brain glia, many other cell types	1 × 135 kD	Rod with three or four 18-residue MT-binding repeats	Promotes MT assembly and stability
STOP	Vertebrate cells	100 kD	Inhibited by Ca-calmodulin or phosphorylation	Stabilizes MT against cold depolymerization
Tau	Axons of vertebrate neurons	One gene, 6 isoforms 1 × 40–80 kD	Three or four 18-residue MT-binding repeats	Promotes MT assembly and stability (Alzheimer disease)
Tektin	Metazoan axonemes and cytoplasmic MTs	2 × 47–53 kD	Coiled-coil protein	Stabilizes MTs in axonemes and centrioles
Linkers				
Gephyrin	Vertebrate neurons	? × 93 kD		Anchors glycine receptors to MTs
Spectraplakin family	Metazoans	>500 kD (multiple genes & splice isoforms)	Calponin homology domain, various other domains	Crosslinks MT to actin and intermediate filaments
+Tips (Plus End Binding Proteins)				
APC	Vertebrates, insects	1 × 300 kD	Binds EB1 and β-catenin	Regulates β-catenin; tumor suppressor (colon cancer)
CLASP (Mast/Orbit)	Eukaryotes	? × 165 kD	Binds CLIP-170 and EB1	Regulates MT dynamics in the cell cortex and at kinetochores
CLIP-170	Eukaryotes	170 kD	Binds EB; phosphorylation inhibits MT binding	Binds endosomes to plus ends of MT
Dis1/TOG family (XMAP215, others)	Eukaryotes	215 kD, other variants		Regulates MT dynamics and spindle pole
EB-1, -2, -3 (Bim1p, Mal3p)	Eukaryotes	30 kD	Binds MT plus ends and APC	Promotes MT assembly
KMN network	Eukaryotes	Many subunits including KNL-1, Mis12 complex, NCD80 complex		Kinetochore MT binding complex
–Tips (Minus End Binding Proteins)				
Augmin		435,000	Eight different subunits	Binds γ-tubulin ring complex
CAMSAP-1, -2, -3 (Patronin)	Metazoa	>200 kD	Calponin homology domain, coiled-coils	Binds near plus ends; stabilizes end
γ-Tubulin ring complex	Eukaryotes	14 × 50 kD γ-tubulin + 8 other proteins	Polymeric lockwasher	Nucleates MT assembly from minus end

ATP, adenosine triphosphate; ATPase, adenosine triphosphatase; CAMSAP, calmodulin-regulated spectrin-associated protein; CLIP, cytoplasmic linker protein; EB, end-binding protein; MAP, microtubule-associated protein; MT, microtubule; PKA, protein kinase A.

APPENDIX 34.2

Centrosomal Structural Proteins

Name	Distribution	Composition (Subunit Size)	Properties	Functions (Diseases)
C-NAP1	Metazoa	2 × 250 kD	Coiled-coil protein	Fibers connecting PCM of two centrioles, regulated by the protein kinase Nek2
Centriolin	Metazoa	2 × 240 kD	Coiled-coil protein	Located in in subdistal appendages (stem cell myeloproliferative disorder)
γ-Tubulin ring complex	Eukaryotes	14 × 50 kD γ-tubulins + eight other proteins	Polymeric lockwasher	Nucleates MT assembly from minus end
Ninein	Metazoa	2 × 236 kD	Coiled-coil protein	Located in in subdistal appendages; MT anchoring (Seckel syndrome; prenatal dwarfism)
Pericentrin (Kendrin)	Animals Plants	2 × 380 kD	Coiled-coil protein	PCM scaffold; binds calmodulin, dynein, γ-tubulin ring complex, kinases, and phosphatases (human autoantigen; microcephalic osteodysplastic primordial dwarfism)
Rootletin	Many eukaryotes	2 × 230 kD	Coiled-coil protein	Fibers connecting PCM of two centrioles and anchoring basal bodies
SAS-4	Many eukaryotes	2 × 92 kD	Coiled-coil protein	Part of centriole scaffold with SAS-6
SAS-6	Many eukaryotes	2 × 74 kD	Coiled-coil protein with globular domains	Forms 9-fold scaffold for centrioles (microcephaly 14)

MT, microtubule; PCM, pericentriolar material.

CHAPTER 35

Intermediate Filaments

Intermediate filaments (Fig. 35.1) are strong but flexible polymers that provide mechanical support for metazoan cells. These filaments are composed of many different but homologous proteins. The filaments were named *intermediate,* because their 10-nm diameters are intermediate between those of the thick and thin filaments in striated muscles, where they were first recognized (see Figs. 39.3 and 39.8). They are not found in plants, fungi, or prokaryotes, although one bacterial species has a coiled-coil protein with some properties of intermediate filaments. Cytoplasmic intermediate filaments, in particular keratin filaments, tend to cluster into wavy bundles that vary in compactness, forming a branching network between the plasma membrane and the nucleus. Intercellular junctions called desmosomes anchor intermediate filaments to the plasma membrane (see Fig. 31.8B) and thereby transmit mechanical forces between adjacent cells. Hemidesmosomes connect intermediate filaments across the plasma membrane to the extracellular matrix (see Fig. 31.8C).

The continuum of intermediate filaments and junctions prevents excessive stretching of cells and gives tissues such as epithelia and heart muscle their mechanical integrity. Skin appendages built from crosslinked

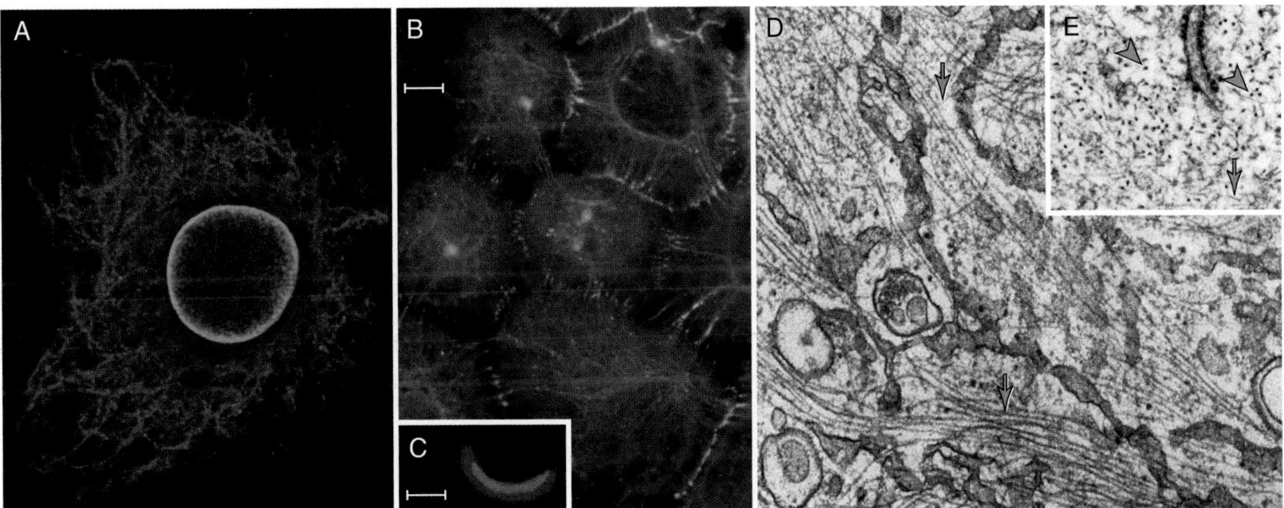

FIGURE 35.1 LIGHT AND ELECTRON MICROGRAPHS OF INTERMEDIATE FILAMENTS. A, Fluorescence light micrograph of a cultured fibroblast stained with antibodies to vimentin *(red)* and nuclear lamins *(green)* and with DAPI (4,6-diamidino-2-phenylindole) for DNA *(blue)*. **B,** Fluorescence micrograph of cultured epithelial cells stained with antibodies to keratin intermediate filaments *(orange).* Desmosomes are stained *green.* Scale bar is ~10 μm. **C,** Fluorescence micrograph of crescentin labeled with a *red* fluorescent dye in the bacterium *Caulobacter crescentus.* Scale bar is 2 μm. **D–E,** Electron micrographs of thin sections of a cultured baby hamster kidney cell showing longitudinal *(arrows)* and cross sections *(arrowheads)* of vimentin intermediate filaments. (**A,** Courtesy U. Aebi, Biozentrum, University of Basel, Switzerland, and H. Herrmann, German Cancer Research Center, Heidelberg, Germany. **B,** Courtesy E. Smith and E. Fuchs, University of Chicago, IL. **C,** Courtesy M. Cabeen and C. Jacobs-Wagner, Yale University, New Haven, CT. **D–E,** Courtesy R.D. Goldman, Northwestern University, Chicago, IL.)

keratin intermediate filaments such as hair and whale baleen, illustrate their flexibility and high tensile strength. Molecular defects in cytoplasmic intermediate filaments or junctions associated with them result in rupture of skin cells and blistering diseases. Defects in lamins associated with the nuclear envelope cause a bewildering array of diseases (see Fig. 9.10).

Structure of Intermediate Filament Subunits

Spectroscopic data and X-ray fiber diffraction of materials composed of intermediate filaments, like wool, established the α-helical coiled-coil as their basic structure. However, the proteins forming intermediate filaments can vary greatly in size, in contrast to the uniform sizes of actins and tubulins. Eventually amino acid sequences of several intermediate filament proteins in the 1980s established that they all have a central α-helical segment that forms a **rod domain** flanked by variable N- and C-terminal end domains (Fig. 35.2). Gene sequences allow grouping of the proteins into five amino acid sequence homology classes (Table 35.1).

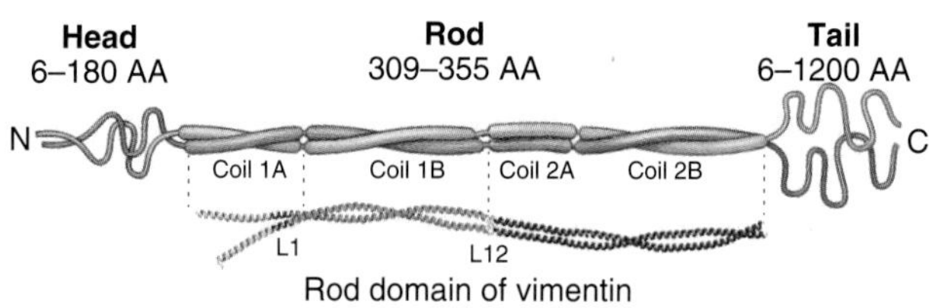

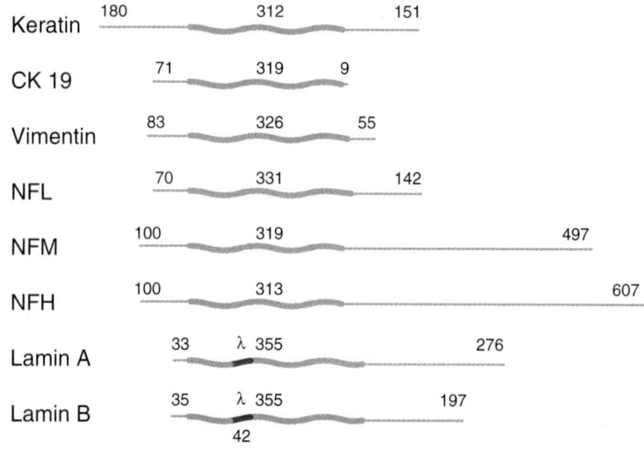

FIGURE 35.2 INTERMEDIATE FILAMENT (IF) PROTEINS HAVE A CENTRAL ROD DOMAIN FLANKED BY HEAD AND TAIL DOMAINS OF VARIABLE LENGTHS. The ribbon diagram shows a crystal structure of the vimentin rod domain (see Protein Data Bank [www.rcsb.org] file 1GK7). Rod domains consist of an α-helical coiled-coil of 310 residues and are 46 nm long. Lamins have an additional 42 residues in the rod domain (λ). The residues that are most important for assembly are at the beginning and end of the rod. End domains differ in sequence and size from 6 to 1200 residues. (Model of vimentin courtesy H. Herrmann, German Cancer Research Center, Heidelberg, Germany.)

The characteristic feature of all intermediate filament proteins is a dimeric, parallel, α-helical coiled-coil rod domain that forms the backbone of the filaments. The rod domains of cytoplasmic intermediate filament proteins are approximately 46 nm long. Those of nuclear lamins are 6 nm longer (see Fig. 35.2 and Fig. 35.3A). Like other coiled-coils (see Fig. 3.10), intermediate filament rod domains have a heptad repeat pattern of amino acids with the first and fourth residues providing a continuous row of hydrophobic interactions along the interface of the two α-helices (see Fig. 3.10). The rod domains have two highly conserved sites with interruptions in the coiled-coil termed L1 and L12. Zones of positive and negative charge alternate along the rod. When staggered appropriately, these zones provide complementary electrostatic bonds for assembly into filaments. Approximately 20 highly conserved residues at each end of the rod are essential for filament elongation through head-to-tail interactions of dimeric molecules. Studies with mutant proteins show that assembly of filaments depends on both head-to-tail overlaps and lateral associations between rod domains.

The N- and C-terminal end domains flanking the rod are largely unstructured and vary considerably in size (see Fig. 35.2). The N-terminal end ("head") domains are essential for assembly whereas the C-terminal end ("tail") domains protrude from the filament surface (see Fig. 35.3C) to control filament diameter and/or interact with other cellular components.

Each class of intermediate filament molecule forms in a characteristic manner. **Keratins** are obligate heterodimers of one acidic (class I) and one basic (class II) keratin polypeptide. Vimentin and desmin (class III) and nuclear **lamins** (class V) form parallel dimers of identical polypeptides (see Fig. 35.3A). The intermediate filament proteins in the nervous system (class IV) form complex mixtures of filaments, and it is not yet clear if they are mainly homodimers or if some are heterodimers.

Two molecules of cytoplasmic intermediate filament proteins associate in an antiparallel, half-staggered manner to form stable apolar dimers, sometimes called "tetramers" because they consist of four polypeptides (see Fig. 35.3D). These tetramers are, at least in vitro, the principal intermediates in filament assembly as they further associate laterally and longitudinally.

Evolution of Genes for Intermediate Filament Proteins

Well after the genes for intermediate filament proteins were discovered in the higher branches of the animal lineage, whole-genome sequencing established their presence in a wide range of other organisms that descended from the last common eukaryotic ancestor (see Fig. 2.4B). Genes for animal intermediate filament proteins arose in early metazoan cells from genes

TABLE 35.1 Classification of Intermediate Filament Proteins Based on Rod Domain Sequences

Class	Type	Number of Human Genes	Molecule	Distribution	Diseases
I	Acidic keratin	28	40–65 kD, obligate heterodimer with class II	Epithelial cells and their appendages	Blistering skin, corneal dystrophy, brittle hair and nails
II	Basic keratin	26	51–68 kD, obligate heterodimer with class I	Epithelial cells and their appendages	Similar to class I
III	Desmin	1	53 kD, homopolymers	Muscle cells	Cardiac and skeletal myopathies
	GFAP	1	50 kD, homopolymers	Glial cells	Alexander disease; mouse null viable
	Peripherin	1	57 kD	Peripheral > CNS neurons	
	Synemin	1	190 kD, interacts with other class III IFs	Muscle cells	
	Vimentin	1	54 kD, homopolymers and heteropolymers	Mesenchymal cells	Mouse null viable
IV	Neurofilament				
	NFL	1	Obligate heteropolymers with NFM, NFH	Neurons	Mouse null viable; human neuropathies
	NFM	1	Obligate heteropolymers with NFL, NFH	Neurons	
	NFH	1	Obligate heteropolymers with NFL, NFM	Neurons	Mutations a risk factor in amyotrophic lateral sclerosis
	Nestin	1	230 kD, homopolymers	Embryonic neurons, muscle, other cells	
	α-Internexin	1	55 kD, homopolymers	Embryonic neurons	
V	Lamins	3	7 Isoforms, 62–72 kD, homodimers	Metazoan nuclei, some protozoa	Cardiomyopathy, lipodystrophy, one form of Emery-Dreifuss muscular dystrophy, two forms of progeria, plus many others

CNS, central nervous system; GFAP; glial fibrillary acidic protein; IF, intermediate filament; NFH, neurofilament heavy; NFL, neurofilament light; NFM, neurofilament medium.
For reference, see Omary MB, Coulombe PA, McLean WH: Intermediate filament proteins and their associated diseases. *N Engl J Med.* 2004;351:2087–2100. For current information see: http://www.interfil.org/.

encoding nuclear lamins (see Fig. 14.7). Most metazoans, including chordates, mollusks, insects, and nematodes (see Fig. 2.8), retain genes for lamins.

The gene for cytoplasmic intermediate filaments arose from a duplicated lamin gene in an invertebrate organism in the lineage leading to chordates. One copy of the duplicated gene was modified by deletion of the nuclear localization sequence and the CAAX box (a C-terminal prenylation site; see Fig. 13.10). After deletion of the codons for 42 residues (6 heptads) in coil 1B and the immunoglobulin domain in the "tail" in early chordates, further gene duplications and divergence produced the four families of genes for cytoplasmic intermediate filaments of vertebrates (see Table 35.1). The unique functional requirements for each class of intermediate filament protein have conferred strong selective pressure on their genes, so that orthologs are much more similar than the paralogs. For example, human desmin is much more similar to frog desmin than it is to human keratin.

The bacterium *Caulobacter crescentus* has a gene for a coiled-coil protein, crescentin, with some features of intermediate filament proteins (see Fig. 35.1C). However, it lacks some of the highly conserved residues in the rod domain of animal intermediate filament proteins, including those vital for filament elongation. Crescentin is required for the asymmetrical shape of *Caulobacter* cells and when expressed in *Escherichia coli* makes the cells spiral shaped. The origin of this gene is unknown, but lateral transfer from a eukaryote followed by divergence is possible.

Filament Structure and Assembly

Intermediate filaments are approximately 10 nm in diameter with wavy profiles in electron micrographs of thin sections of cells (see Fig. 35.1B) or after negative staining of isolated filaments (see Fig. 35.3B). In some cases, such as neurofilaments, parts of the head domains and most of the tail domains project radially from the filament core, forming a type of bottlebrush (see Fig. 35.3C). The most carefully studied intermediate filaments are built from octameric complexes (ie, two laterally associated molecular dimers) that associate end to end to form protofibrils like the strands of a rope (see Fig. 35.3D). In cross section, a standard intermediate filament has up to 16 coiled-coils, but their exact internal arrangement is not known. Because the molecular dimers lack polarity, intermediate filaments are considered to be apolar (ie, both ends of the filament are equivalent; see Fig. 35.3D). This is a striking difference from actin filaments (see Fig. 33.8) and microtubules (see Fig. 34.4), which depend

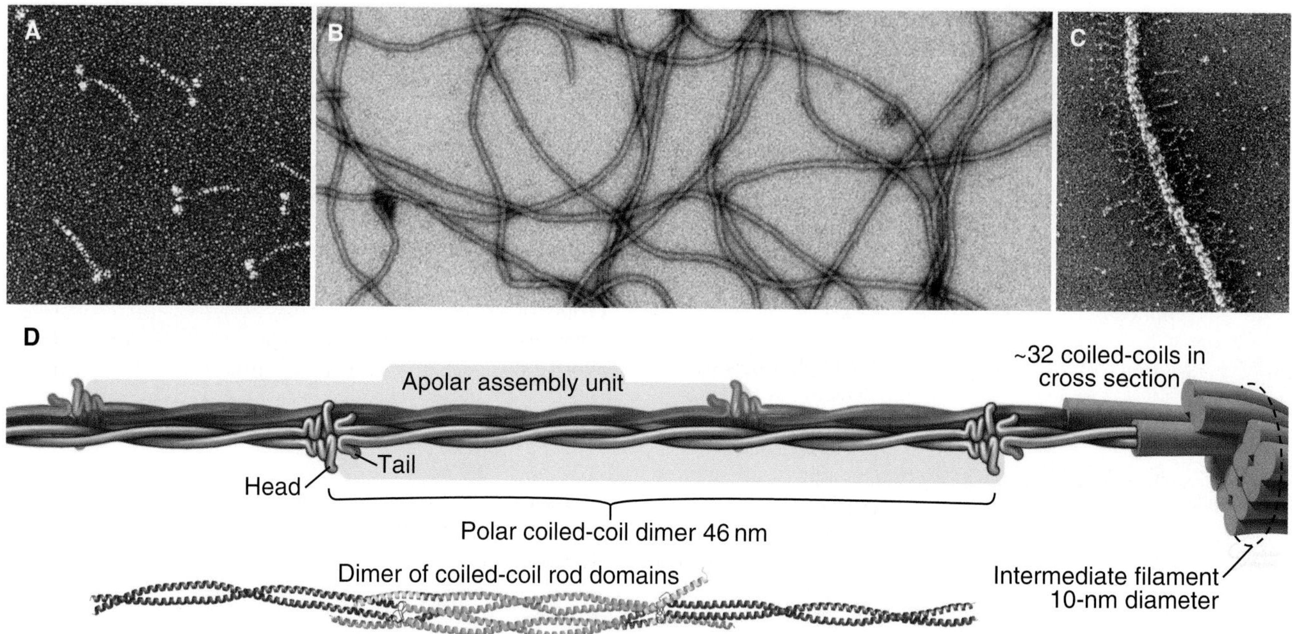

FIGURE 35.3 INTERMEDIATE FILAMENTS ARE CONSTRUCTED LIKE A MULTISTRAND ROPE. A–C, Electron micrographs. **A,** Metal-shadowed lamin molecules consisting of two polypeptides joined by a long α-helical coiled-coil with globular tail domains at the C-terminus. **B,** Negatively stained vimentin intermediate filaments. **C,** Rotary shadowed intermediate filament showing radial projections. **D,** A model for intermediate filament structure. The building blocks are antiparallel complexes of two coiled-coil molecular dimers. The ribbon diagram is a model of the dimer of vimentin rod domains. Assembly occurs via the formation of unit-length filaments, the products of the lateral association of eight antiparallel dimers, which then longitudinally anneal into intermediate filaments. This model is consistent with x-ray fiber diffraction patterns, chemical crosslinking, and other data, but details of the subunit packing remain to be determined. (**A** and **C,** Courtesy U. Aebi, University of Basel, Switzerland. **B,** Courtesy H. Herrmann, German Cancer Research Center, Heidelberg, Germany. **D,** *Top,* Data from Steinert P, Marekov LN, Parry DA. Conservation of the structure of keratin intermediate filaments. *Biochemistry*. 1993;32:10046–10056. **D,** *Bottom,* Model courtesy H. Herrmann, German Cancer Research center, Heidelberg, Germany. For reference, see Chernyatina AA, Nicolet S, Aebi U, Herrmann H, Strelkov SV. Atomic structure of the vimentin central α-helical domain and its implications for intermediate filament assembly. *Proc Natl Acad Sci U S A*. 2012;109:13620–13625.)

on their polarity for many functions, including the unidirectional motion of motor proteins. Furthermore, the number of protofilaments can vary along a single filament, making them much more heterogeneous than actin filaments or microtubules.

Intermediate filaments are insoluble under physiological conditions, but can be dissociated in buffers of low ionic strength and high pH. Under physiological conditions isolated subunits spontaneously repolymerize in a few minutes. The first assembly product observed is a "unit-length filament" consisting of eight laterally associated molecular dimers with the length (60 nm) of a molecular dimer. Intermediate filaments grow by longitudinal annealing of unit-length filaments at both ends, in contrast to growth of actin filaments and microtubules by addition of single subunits at their ends. The nucleation mechanism that initiates polymerization and the elongation reactions are still being investigated, but it is clear that no nucleotides or other cofactors are needed for assembly. Most of the head domain is required to assemble intermediate filaments in vitro and in vivo. The tail domain is dispensable for assembly, although more molecular dimers can pack laterally into a filament in its absence.

Intermediate filaments are among the most chemically stable cellular structures, resisting solubilization by extremes of temperature as well as high concentrations of salt and detergents (Fig. 35.4). Nevertheless, intermediate filaments in some cells exchange their subunits within minutes to hours during interphase. For example, if **vimentin** is labeled with a fluorescent dye and injected into live cells or GFP-vimentin is expressed, fluorescent vimentin incorporates into cytoplasmic filaments. After a spot of fluorescent filaments is photobleached with a laser, the fluorescence recovers over a period of several minutes, indicating that subunits along the length of the filaments exchange with a pool of unpolymerized molecules. (Fig. 38.8 shows a similar experiment with actin.)

Although no known motors move on the apolar intermediate filaments, motor proteins move intermediate filaments along microtubules. A spectacular example is found in nerve cells (see Fig. 37.5C).

Posttranslational Modifications

Phosphorylation dramatically affects polymer assembly and dynamics of many types of intermediate filaments.

The process is complex and incompletely understood, as several different kinases phosphorylate many different sites and these phosphates tend to turn over rapidly. The impact of phosphorylation depends critically on the particular residue modified.

The best example of phosphorylation destabilizing an intermediate filament is the breakdown of the nuclear lamina during mitosis (see Fig. 44.6). The mitotic kinase Cdkl-cyclin B phosphorylates two sites immediately flanking the rod domains of lamins, disrupting the head-to-tail overlap required for the interactions of molecules that mediate filament elongation and lateral association of subunits. Cytoplasmic vimentin filaments also disassemble in some cell types during mitosis (Fig. 35.5B), but the process is more complex. Vimentin lacks the Cdk1-recognition sites immediately flanking the α-helical rod domain and coassembly with another intermediate filament protein, nestin, appears to be a prerequisite for phosphorylation to mediate disassembly. In contrast the organization of keratins changes only subtly during mitosis (Fig. 35.5C). The role of phosphorylation of intermediate filaments during interphase is less clear, but it might influence the structure of the cytoskeleton in response to various signals.

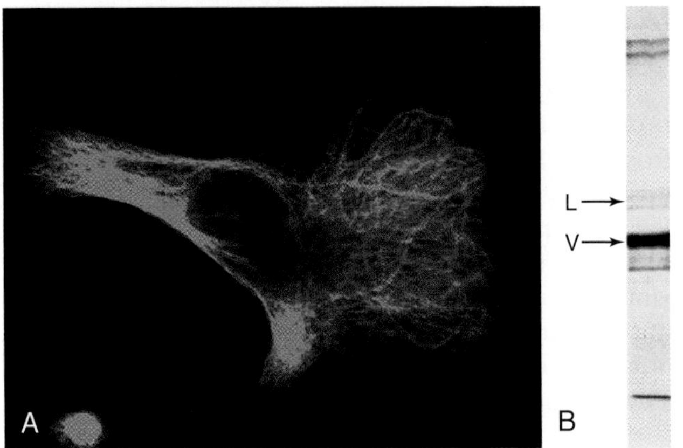

FIGURE 35.4 INTERMEDIATE FILAMENTS RESIST SOLUBILIZATION WHEN CELLS ARE EXTRACTED. A, A fluorescence micrograph shows the network of vimentin intermediate filaments remaining after extraction of a Chinese hamster ovary (CHO) cell with the detergent Triton X-100, DNase, and a high concentration of salt to remove lipids, DNA, and soluble proteins. **B,** Gel electrophoresis of these extracted cells reveals that lamins (L) and vimentin (V) are among the few proteins remaining in the detergent-resistant cytoskeletal fraction. (Courtesy R. Goldman, Northwestern University, Chicago, IL.)

Neurofilaments, abundant intermediate filaments in nerve axons and dendrites (see Fig. 35.9), are an exception to the rule that phosphorylation destabilizes intermediate filaments. The most stable neurofilaments are heavily phosphorylated in their large C-terminal tail end domain (see Fig. 35.2), whereas the pool of unpolymerized of NFM (neurofilament medium) and NFH (neurofilament heavy) molecules is not phosphorylated. The NFM and NFH end domains are not essential for assembly, so phosphorylation might influence other functions of neurofilaments.

Keratin intermediate filaments in hair are chemically crosslinked to each other and associated with matrix proteins by disulfide bonds and amide bonds between lysines and acidic residues, creating a tough composite material built on the same principles as fiberglass. Beauticians take advantage of these crosslinks to modify the

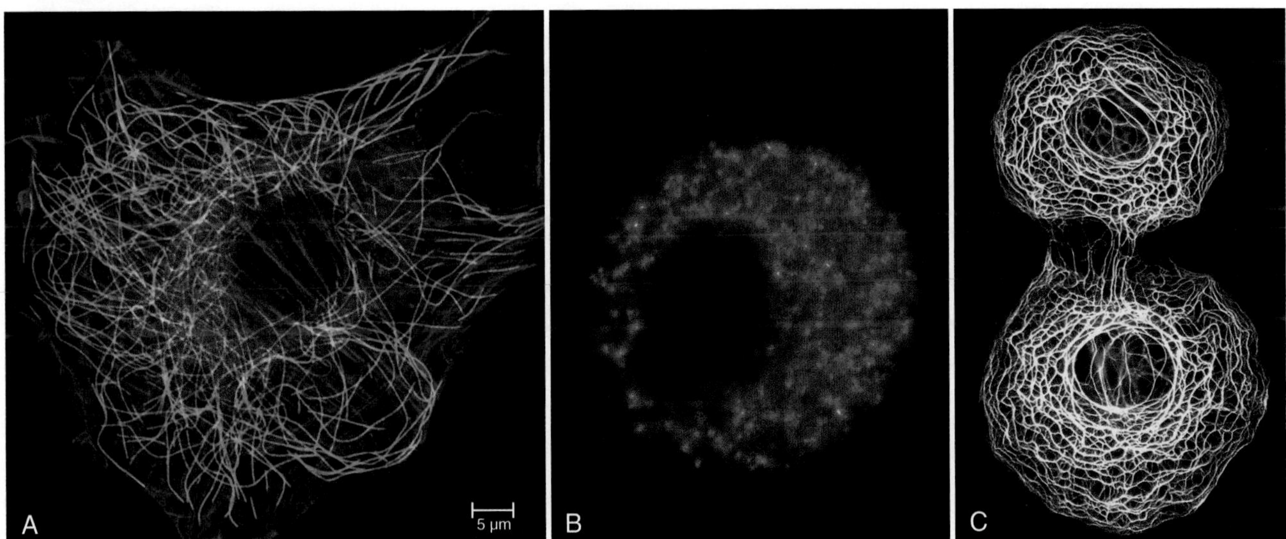

FIGURE 35.5 FLUORESCENCE MICROGRAPHS OF INTERMEDIATE FILAMENTS. A, A cultured fibroblast stained with antibodies to vimentin intermediate filaments *(red)* and microtubules *(green)* and fluorescent phalloidin for actin filaments *(blue)*. **B,** Vimentin intermediate filaments dispersed in mitosis. **C,** Dividing epithelial cells stained with an antibody to keratin, which remains polymerized during mitosis. (**A,** Courtesy U. Aebi, Biozentrum, University of Basel, Switzerland, and H. Herrmann, German Cancer Research Center, Heidelberg, Germany. **B,** Courtesy R.D. Goldman, Northwestern University, Chicago, IL. **C,** Courtesy H. Herrmann, German Cancer Research Center, Heidelberg, Germany.)

shape of hairs during "permanents." They first reduce disulfide bonds and then reform them after molding the hair into a new shape.

Expression of Intermediate Filaments in Specialized Cells

Animal cells express at least one of the three major nuclear lamins, whereas the repertoire of cytoplasmic intermediate filament proteins varies greatly in different cell types (see Table 35.1). Most cells express predominantly one class—or at the most two classes—of cytoplasmic intermediate filament proteins, presumably making use of their unique properties. For example, epithelial cells express class 1 and class 2 keratins, whereas muscle cells express desmin and mesenchymal cells express vimentin. A few cells, such as the basal myoepithelial cells of the mammary gland, express two types of intermediate filament proteins that sort into separate filaments with different distributions in the cytoplasm. Similarly, microinjection or expression of foreign intermediate filament subunits usually (but not invariably) results in correct sorting to the homologous class of filaments.

In tissues such as skin and brain, cells express a succession of intermediate filament isoforms as they mature and differentiate. For example, dividing cells at the base of the epidermis of skin express mainly keratins 5 and 14, whereas terminally differentiating cells express keratins 1 and 10 (Fig. 35.6). The switch in keratin expression is associated with a marked increase in filament bundling, a feature that might contribute to the resistance of the surface layers of the skin to chemical dissociation and mechanical rupture. In the nervous system, supporting glial cells express a class III intermediate filament protein, whereas embryonic neurons first express the class IV α-internexin and later express the three other class IV neurofilament isoforms (see Table 35.1). Although the smallest neurofilament isoform (NFL [neurofilament light]) can assemble on its own in vitro, the formation of intermediate filaments in neurons requires NFL and one of the larger isoforms NFM of NFH, which are encoded by distinct genes.

Tumors often express the intermediate filament protein that is characteristic of the differentiated cells from which they arose. This is helpful to pathologists in diagnosing poorly differentiated or metastatic cancers. For example, tumors of muscle cells express desmin rather than keratin (expressed in epithelial cells) or vimentin (expressed in mesenchymal cells). This rule is not absolute, as some tumors arising in epithelia turn down the expression of keratin and turn up the expression of vimentin before invading surrounding tissues.

Proteins Associated With Intermediate Filaments

A number of proteins bind intermediate filaments and link them to membranes and other cytoskeletal polymers

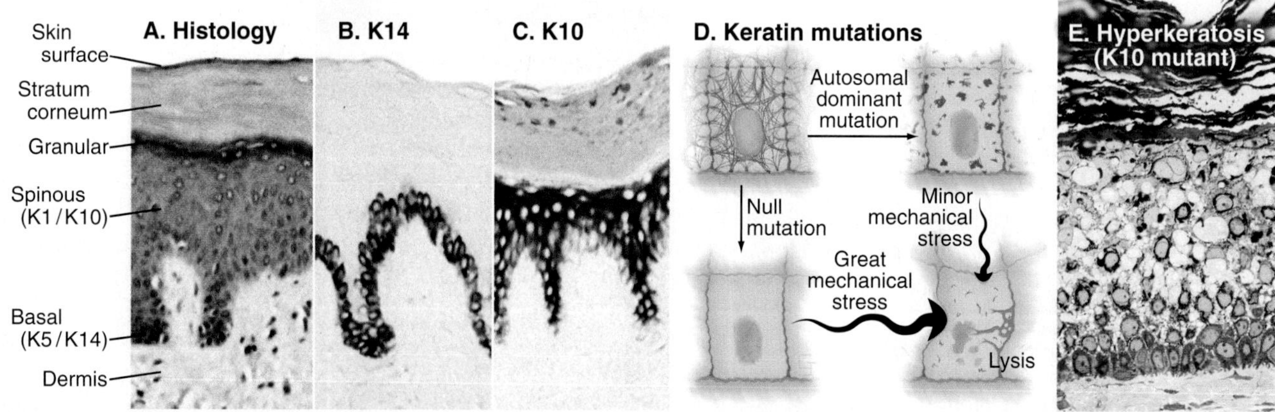

FIGURE 35.6 EXPRESSION OF KERATINS AND EFFECTS OF KERATIN MUTATIONS ON THE STRATIFIED SQUAMOUS EPITHELIUM OF SKIN. A, Light micrograph of a section of mouse skin stained with hematoxylin and eosin (H&E). **B,** Localization of keratin 14 in a section of skin using antibodies and a histochemical procedure that leaves a *brown* deposit. Proliferating cells in the basal layer express keratin 5 and keratin 14. **C,** Localization of keratin 10 to differentiating cells in intermediate layers of the epithelium. These cells eventually lose their nuclei and form the surface layers of cornified cells. **D,** Drawings illustrating the effects of keratin mutations on the structure of the epithelium. Dominant negative keratin mutations affect the assembly of keratin filaments wherever they are expressed. Human patients with epidermolysis bullosa simplex have point mutations in keratin 5 or keratin 14 that disrupt the filaments in the basal cells of the stratified epithelium, causing mechanical fragility and cellular rupture with mild trauma, resulting in blisters. Mutations in keratin 1 or keratin 10 cause cell rupture in the middle layers of the epithelium where they are expressed. Null mutations in keratin genes disrupt the epithelium to a lesser extent than dominant negative point mutations. **E,** Light micrograph of a histologic section of skin illustrating how a mutation in keratin 10 disrupts cells in the spinous layer of the epithelium and causes hyperkeratosis (excess scaling of surface layers). (**A–C** and **E,** Courtesy P. Coulombe, Johns Hopkins University, Baltimore, MD. **D,** Modified from Fuchs E, Cleveland DW. A structural scaffolding of intermediate filaments in health and disease. *Science*. 1998;279:514–519. Copyright 1998 American Association for the Advancement of Science.)

TABLE 35.2 Proteins Associated With Intermediate Filaments

Name	Molecule	Distribution	Partners	Diseases
Plakins				
BPAG-1	Multiple splice isoforms (a, b, e, n)	a: Hemidesmosomes b: Muscle, cartilage e: Epithelial hemidesmosomes n: Neurons	IFs, MTs, actin	Autoimmune bullous pemphigoid
Desmoplakin	Two splice isoforms	Desmosomes	IFs; cadherin and other desmosome proteins	Autoimmune pemphigus; genetic striate palmoplantar keratoderma
Plectin	Multiple splice isoforms	Most tissues except neurons	IFs, actin, MTs, spectrin, β_4-integrin	Autoimmune pemphigus; genetic epidermolysis bullosa with muscular dystrophy
Epidermal				
Filaggrin	Ten 37-kD filaggrins cut by proteolysis from profilaggrin	Cornified epithelia	Aggregates keratin	?
Lamin Associated				
LAP1	57–70-kD isoforms	Integral nuclear membrane proteins	Binds lamins to nuclear envelope	
LAP2	50 kD	Integral nuclear membrane protein	Binds lamins to nuclear envelope	
LBR	73 kD	Integral nuclear membrane protein		Pelger-Huët anomaly; Greenberg skeletal dysplasia
Emerin	34 kD	? Peripheral protein of the inner nuclear membrane	Binds actin filaments to the nuclear envelope	Emery-Dreifuss muscular dystrophy

ABD, actin binding domain; IFs, intermediate filaments; MTs, microtubules.

(Table 35.2). Integral membrane proteins, called nuclear envelope transmembrane proteins, anchor nuclear lamins to the nuclear membrane (see Fig. 9.8). Filaggrin mediates the aggregation of keratin filaments in the upper layers of skin.

Plakins are giant proteins that link cytoskeletal polymers to each other and to membranes by virtue of binding sites for cytoskeletal polymers and proteins of adhesive junctions. Like several other plakins, **plectin** has globular domains on both ends of a 200-nm coiled-coil. Binding sites in both globular domains enable plectin to crosslink intermediate filaments to each other, to actin filaments, and microtubules (Fig. 35.7). Recessive mutations in human plectin cause a rare form of muscular dystrophy associated with skin blisters. The null mutation in mice is lethal. Plectin and two other plakins link keratin filaments to three different plasma membrane adhesion proteins at desmosomes and hemidesmosomes (see Fig. 31.8). Similarly, **desmoplakin** anchors keratin filaments to cadherins at desmosomes (see Fig. 31.8B). At hemidesmosomes plectin 1a links keratin to β_4-integrins, while the plakin **BPAG1e** (bullous pemphigoid antigen 1-e) links keratin filaments to the transmembrane protein, BPAG2 (bullous pemphigoid antigen 2). BPAG1e is one of the many splice forms of the dystonin gene, which is mutated in some patients with neuropathies and one form of epidermolysis bullosa.

Functions of Intermediate Filaments in Cells

Intermediate filaments function primarily as flexible intracellular tendons (analogous to nylon rope) that prevent excessive stretching of cells that are subjected to external or internal physical forces. This function is complemented by their interactions with microtubules, actin filaments, and membranes. For example, if a relaxed smooth muscle is stretched, the intracellular network of desmin filaments between cytoplasmic dense bodies and the plasma membrane (see Fig. 39.23) is transformed from a polygonal three-dimensional network into a continuous strap that runs the length of the cell (Fig. 35.8). Up to the point at which this network is taut, the cell offers little resistance to stretching. Beyond this point, the cell strongly resists further stretching. Actin filaments anchored to dense bodies interact with myosin (see Fig. 39.23) to apply contractile force to the network of intermediate filaments.

Although the geometry of the network of intermediate filaments is different in striated muscles, the concept is remarkably similar to smooth muscle. Desmin filaments surround the Z disks in addition to forming a looser, longitudinal basket around the myofibrils (see Fig. 39.8). The ends of both skeletal and cardiac muscle cells must be anchored to transmit their contractile

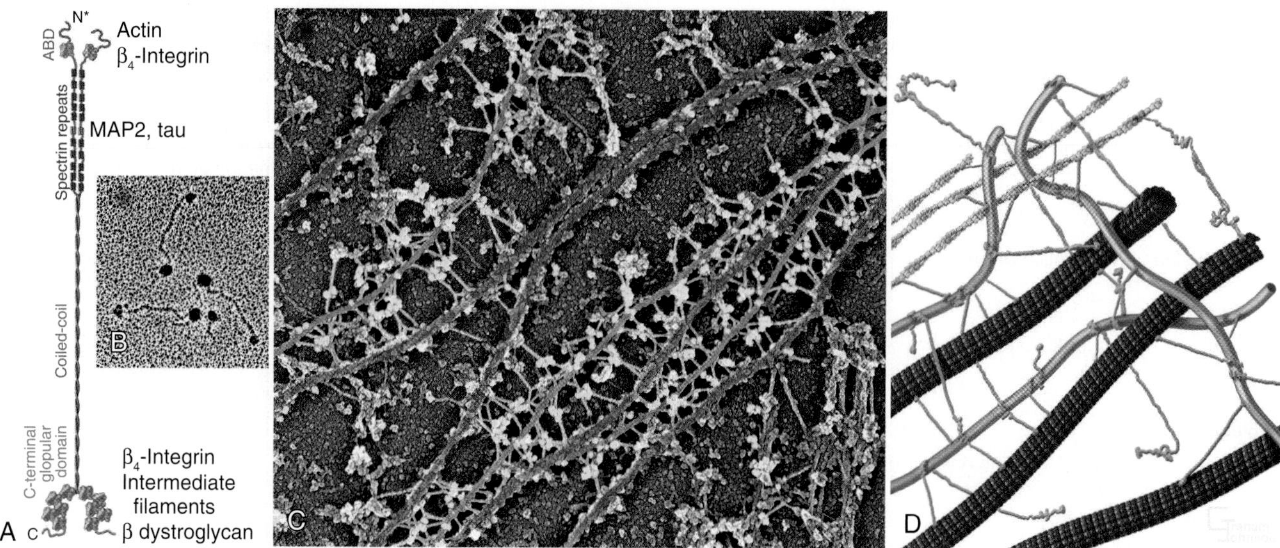

FIGURE 35.7 PLECTIN STRUCTURE AND ACTIVITIES. A, Domain structure of plectin with the some of its known ligands listed to the right: the N-terminal actin-binding domain and the spectrin repeats are similar to those of α-actinin (see Fig. 33.17); the 170-nm long, α-helical coiled-coil forms dimers; six C-terminal repeats form a large globular domain. **B,** Electron micrograph of plectin molecules. **C,** Electron micrograph of an extracted fibroblast reacted with gold-labeled antibodies to plectin. Gold particles *(yellow)* identify plectin molecules *(blue)* as linkers between intermediate filaments *(orange)* and microtubules *(red).* The specimen was prepared by rotary shadowing. The molecules are pseudocolored for clarity. To visualize these interactions numerous actin filaments were removed by incubation with a gelsolin fragment. **D,** Drawing of plectin *(blue)* connecting cytoskeletal polymers to each other. (**B,** Courtesy G. Wiche, University of Vienna, Austria. **C,** Courtesy G. Borisy, University of Wisconsin, Madison.)

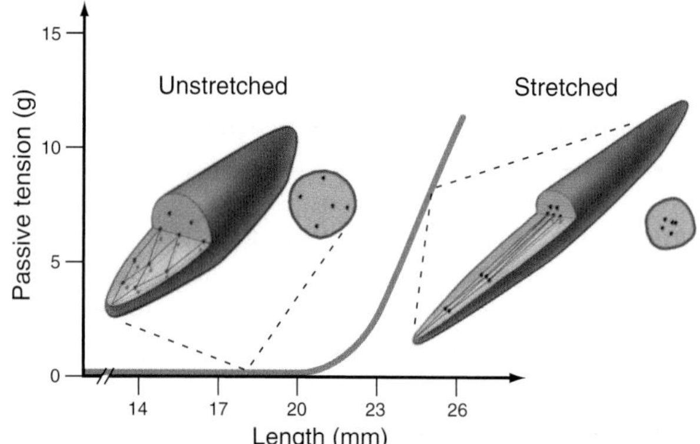

FIGURE 35.8 SMOOTH MUSCLE CELL INTERMEDIATE FILAMENTS FORM AN INTRACELLULAR TENDON THAT RESISTS EXCESSIVE STRETCHING. The graph shows that a relaxed smooth muscle resists stretching very little up to a length of 21 mm. Resistance (passive tension) increases dramatically with further stretching. At short lengths, the three-dimensional network of intermediate filaments and dense bodies is open, offering little resistance to stretching. At the inflection point of the resistance curve, the filaments are extended linearly from one end of the cell to the other and so resist further stretching. (Data from Cooke P, Fay R. Correlation between fiber length, ultrastructure, and the length tension relationship of mammalian smooth muscle. *J Cell Biol.* 1972;52:105–116.)

forces. This is accomplished by intercellular junctions that combine features of desmosomes or hemidesmosomes (anchoring intermediate filaments) and adherens junctions (anchoring actin filaments). In heart muscle cells, these hybrid junctions are called intercalated disks (see Fig. 39.18).

Keratin intermediate filaments are the major proteins in the epithelial cells of the skin, where they form a dense network connected to numerous desmosomes and hemidesmosomes (see Figs. 35.1 and 35.6). These junctions anchor a physically continuous network of intermediate filaments, imparting mechanical stability to the epithelium. If either the junctions or keratin filaments fail, cells pull apart or rupture, and the skin blisters.

Mutations that compromise keratin intermediate filament assembly or the junctions to which they are anchored illustrate the importance of this network. Point mutations near the ends of the keratin rod cause especially severe forms of skin diseases (such as **epidermolysis bullosa simplex**) characterized by blistering and sensitivity to mechanical stress. Similar mutations engineered in transgenic mice faithfully reproduce the human disease. The expression pattern of the defective keratin determines which epithelial cells are affected. For example, a mutation in the rod domain of keratin 14 or keratin 5 leads to disruption of the basal cells in the epidermis where these keratins are expressed. Similarly, mutations in keratin 10 or keratin 1 cause cellular rupture at higher cell layers in the epidermis where those keratins are found (see Fig. 35.6E). Similarly, mutations in keratin 12 or keratin 3 cause sores on the cornea of the eye, where they are expressed.

A mutant keratin can cause disease in heterozygotes that express one normal keratin gene. This is called a **dominant negative mutation** (or autosomal dominant mutation). Defective keratin subunits assemble

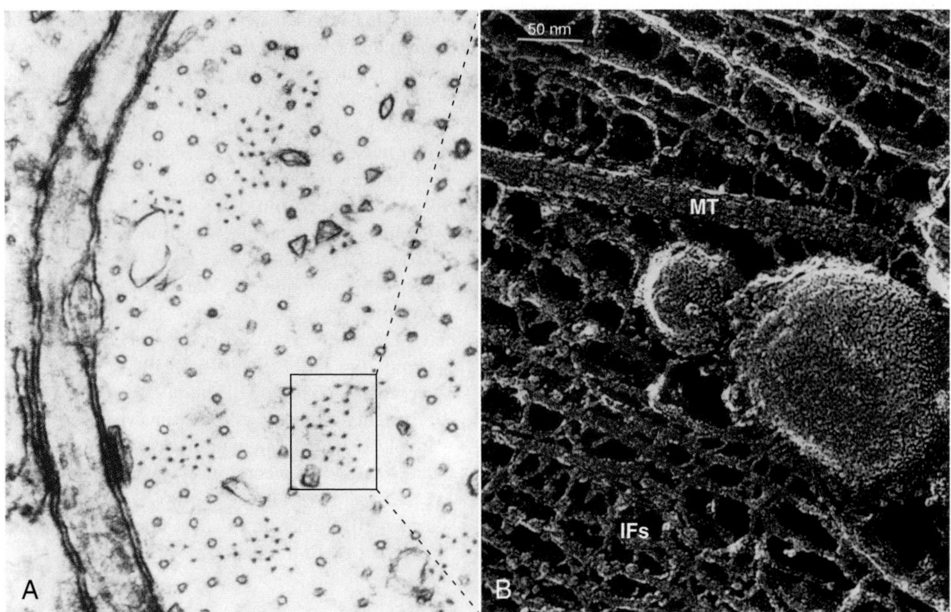

FIGURE 35.9 ELECTRON MICROGRAPHS OF INTERMEDIATE FILAMENTS (CALLED NEUROFILAMENTS) IN AXONS OF NERVE CELLS. A, A thin cross section shows clusters of intermediate filaments and microtubules. **B,** A longitudinal freeze-fracture preparation shows a microtubule (MT *[red]*) with associated vesicles and many intermediate filaments (IF *[orange]*). (**A,** Courtesy P. Eagle, Kings College, London, United Kingdom. **B,** Courtesy N. Hirokawa, University of Tokyo, Japan.)

imperfectly with normal keratin subunits and thereby compromise the physical integrity and strength of the filaments. The affected epithelial cells can grow, divide, and even form desmosomes with neighboring cells, but they tear apart physically when subjected to the shearing forces that affect the skin during normal activities. Young children are severely affected, but some patients improve with age. They learn to avoid physical trauma to their skin and may also adapt biochemically in some way.

In contrast to these dominant negative keratin mutations, complete loss of a keratin subunit by a null mutation can be less severe (see Fig. 35.6D). Mice and humans that lack keratin 14 suffer from milder blistering than do patients with dominant negative point mutations of keratin 14. Mice without functional keratin 8 or keratin 18 genes may die during embryonic development, but a few survive with only modest defects in their colon and liver. Remarkably, mice also survive deletion of both copies of the gene for desmin have only mildly disorganized muscle architecture, although vigorous exercise is fatal. In contrast humans heterozygous for many different desmin mutations may suffer severely from generalized muscle failure. Other desmin mutations cause severe dilated cardiomyopathy requiring heart transplantation.

In addition to providing mechanical stability neurofilaments have a second function of equal importance. Once a nerve cell forms synapses (see Figs. 17.9 and 17.10), it produces neurofilaments, apparently to expand the diameter of the axon (Fig. 35.9). This enhances electrical communication in the nervous system, because the velocity of action potentials (see Fig. 17.6) depends on the diameter of the axon. Japanese quail with a truncation mutation of NFL gene are viable, but the diameters of their axons are smaller than normal and their coordination is defective.

Lamins were originally thought to be a simple support network for the nuclear envelope, but they have other important functions. For example, mutations that create toxic fragments of lamins may perturb lamin assembly and thereby interfere with DNA replication. This effect may reflect a role for the lamina in organizing the chromosomal architecture in the interphase nucleus. Most remarkably, more than 400 human mutations in the lamin A/C gene *LMNA* cause diverse human diseases. These include premature aging (progeria) (see Fig. 9.10), the Emery-Dreifuss form of muscular dystrophy, and multiple disorders of fat tissue and nerves. These tissue-specific deficiencies are remarkable given the ubiquitous expression of lamins A and C in all tissues.

ACKNOWLEDGMENTS

We thank Ueli Aebi and Harald Herrmann for their detailed suggestions on the revision of this chapter and their contributions of new illustrations.

SELECTED READINGS

Bouameur JE, Favre B, Borradori L. Plakins, a versatile family of cytolinkers: roles in skin integrity and in human diseases. *J Invest Dermatol.* 2014;134:885-894.

Chernyatina AA, Guzenko D, Strelkov SV. Intermediate filament structure: the bottom-up approach. *Curr Opin Cell Biol.* 2015;32: 65-72.

Chernyatina AA, Nicolet S, Aebi U, Herrmann H, Strelkov SV. Atomic structure of the vimentin central α-helical domain and its implications for intermediate filament assembly. *Proc Natl Acad Sci USA*. 2012;109:13620-13625.

Clemen CS, Herrmann H, Strelkov SV, Schröder R. Desminopathies: pathology and mechanisms. *Acta Neuropathol*. 2013;125:47-75.

Erber A, Riemer D, Bovenschulte M, Weber K. Molecular phylogeny of metazoan intermediate filament proteins. *J Mol Evol*. 1998;47: 751-762.

Helfand BT, Chang L, Goldman RD. The dynamic and motile properties of intermediate filaments. *Annu Rev Cell Dev Biol*. 2003;19: 445-467.

Herrmann H, Aebi U. Intermediate filaments: molecular structure, assembly mechanism, and integration into functionally distinct intracellular scaffolds. *Annu Rev Biochem*. 2004;73:749-789.

Jefferson JJ, Leung CL, Liem RK. Plakins: goliaths that link cell junctions and the cytoskeleton. *Nat Rev Mol Cell Biol*. 2004;5:542-553.

Kirmse R, Portet S, Mücke N, et al. A quantitative kinetic model for the in vitro assembly of intermediate filaments from tetrameric vimentin. *J Biol Chem*. 2007;282:18563-18572.

Köster S, Weitz DA, Goldman RD, Aebi U, Herrmann H. Intermediate filament mechanics in vitro and in the cell: from coiled coils to filaments, fibers and networks. *Curr Opin Cell Biol*. 2015;32:82-91.

Leung CL, Green KJ, Liem RKH. Plakins: a family of versatile cytolinker proteins. *Trends Cell Biol*. 2002;12:37-45.

Lowery J, Kuczmarski ER, Herrmann H, Goldman RD. Intermediate filaments play a pivotal role in regulating cell architecture and function. *J Biol Chem*. 2015;290:17145-17153.

Moller-Jensen J, Löwe J. Increasing complexity of the bacterial cytoskeleton. *Curr Opin Cell Biol*. 2005;17:75-81.

Nöding B, Herrmann H, Köster S. Direct observation of subunit exchange along mature vimentin intermediate filaments. *Biophys J*. 2014;107:2923-2931.

Omary MB, Coulombe PA, McLean WH. Intermediate filament proteins and their associated diseases. *N Engl J Med*. 2004;351:2087-2100.

Peter A, Stick R. Evolutionary aspects in intermediate filament proteins. *Curr Opin Cell Biol*. 2015;32:48-55.

Szeverenyi I, Cassidy AJ, Chung CW, et al. The Human Intermediate Filament Database: comprehensive information on a gene family involved in many human diseases. *Hum Mutat*. 2008;29:351-360.

Wiche G. Role of plectin in cytoskeleton organization and dynamics. *J Cell Sci*. 1998;111:2477-2486.

Worman HJ, Courvalin J-C. The nuclear lamina and inherited disease. *Trends Cell Biol*. 2002;12:591-598.

CHAPTER 36

Motor Proteins

Molecular motors use adenosine triphosphate (ATP) hydrolysis to power movements of subcellular components, such as organelles and chromosomes, along the two polarized cytoskeletal fibers: actin filaments and microtubules. No motors are known to move on intermediate filaments. Motor proteins also produce force locally within the network of cytoskeletal polymers, which transmits these forces to determine the shape of each cell and, ultimately, the architecture of tissues and whole organisms. Chapters 37 to 39 and 44 illustrate how motors move cells and their internal parts.

Just three families of motor proteins—**myosin, kinesin,** and **dynein**—power most eukaryotic cellular movements (Fig. 36.1 and Table 36.1). During evolution, myosin, kinesin, and Ras family guanosine triphosphatases (GTPases) appear to have shared a common ancestor (Fig. 36.1), whereas dynein is a member of the **AAA adenosine triphosphatase (ATPase)** family (Box 36.1). Although the ancestral genes appeared in prokaryotes, and prokaryotes have homologs of both actin and tubulin, none of these motor proteins has been found in prokaryotes. Over time, gene duplication and divergence in eukaryotes gave rise to multiple genes for myosin, dynein, and kinesin, each encoding proteins with specialized functions. Even the slimmed down genome of budding yeast includes genes for five myosins, six kinesins, and one dynein. Table 36.1 lists other protein machines that produce molecular movements during protein and nucleic acid synthesis, proton pumping, and bacterial motility.

Motor proteins have two parts: a **motor domain** that uses ATP hydrolysis to produce movements and a **tail** that allows the motors to self-associate and/or to bind particular cargo. Within the three families, the tails are more diverse than the motor domains, allowing for specialized functions of each motor isoform.

All motor proteins are enzymes that convert chemical energy stored in ATP into molecular motion to produce force upon an associated cytoskeletal polymer (Fig. 36.2). If the *motor* is anchored, the polymer may move. If the *polymer* is anchored, the motor and any attached cargo may move. If *both* are anchored, the force stretches elastic elements in the molecules transiently, but nothing moves, and the energy is lost as heat. Cells use all these options (see Chapters 37 to 39).

Biochemists originally discovered and purified these motors using enzyme (eg, ATP hydrolysis) or in vitro motility assays (Fig. 36.11). With the prototype motors identified, investigators found further examples and

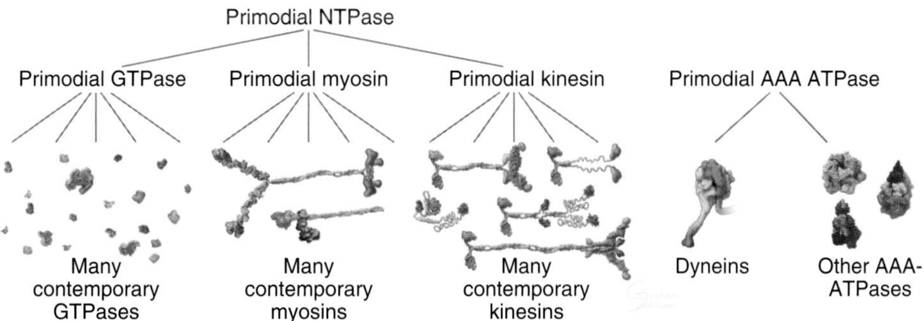

FIGURE 36.1 Evolution of myosin, kinesin, and dynein adenosine triphosphatase (ATPase) motors from genes that encoded two primordial proteins that bound and hydrolyzed nucleoside triphosphates. Gene duplication and divergence created genes for many contemporary motors.

TABLE 36.1 Mechanochemical Enzymes and Other Proteins That Produce Movements

Families	Track	Direction	Cargo	Energy
ATPases				
Myosins				
Muscle myosin	Actin	Barbed end	Myosin filament	ATP
Myosin II	Actin	Barbed end	Myosin, actin	ATP
Myosin I	Actin	Barbed end	Membranes	ATP
Myosin V	Actin	Barbed end	Organelles	ATP
Myosin VI	Actin	Pointed end	Endocytic vesicles	ATP
Dyneins				
Axonemal	Microtubule	Minus end	Microtubules	ATP
Cytoplasmic	Microtubule	Minus end	Membranes, chromosomes	ATP
Kinesins				
Kinesin-1	Microtubule	Plus end	Membranes, intermediate filaments	ATP
Kinesin-14	Microtubule	Minus end	? Microtubules	ATP
Other Mechanochemical Systems				
Polymerases and Helicases				
Ribosome	mRNA	5′ to 3′	None	GTP
DNA polymerase	DNA	5′ to 3′	None	ATP
RNA polymerase	DNA	5′ to 3′	None	ATP
CMG DNA helicase	DNA	—	DNA	ATP
RNA helicases	RNA	—	RNA	ATP
Conformational System				
Spasmin/centrin	None	None	Cell, basal body	Ca^{2+}
Polymerizing Systems				
Actin filaments	None	Barbed end	Membranes	ATP
Microtubules	None	Plus end	Chromosomes	GTP
Worm sperm MSP	None	Not polar	Cytoskeleton	
Rotary Motors				
Bacterial flagella	None	Bidirectional	Cell	H^+ or Na^+ gradient
F-type ATPase	None	Bidirectional	None	H^+ or ATP
V-type ATPase pump	None		None	ATP

ATP, adenosine triphosphate; ATPase, adenosine triphosphatase; GTP, guanosine triphosphate; mRNA, messenger RNA; MSP, major sperm protein. The terms "barbed" and "pointed" end refer to the appearance of actin filaments decorated with a myosin fragment (Fig. 33.8).

variant isoforms of each by purifying proteins, cloning complementary DNAs (cDNAs), sequencing genomes, or genetic screening.

Myosins

Myosins are the only motors that are known to use actin filaments as tracks. Members of the diverse myosin superfamily arose from a common ancestor and share a motor unit called a **myosin "head"** that produces force on actin filaments (Fig. 36.3). One or two heads are attached to various types of tails that are adapted for diverse purposes, including polymerization into filaments, binding membranes, and interacting with various cargos.

Myosin heads consist of two parts. A catalytic domain at the N-terminus of the myosin **heavy chain** binds and hydrolyzes ATP and interacts with actin filaments. Light chain domains consist of an α-helical extension of the heavy chain from the catalytic domain associated with one to seven **light chains.** Light chains are related to calmodulin (see Fig. 3.12), which also serves as a light chain for many myosins.

Myosin Mechanochemistry

Myosin was discovered in skeletal muscle and used to establish general principles that apply, with interesting variations, to energy transduction by all myosins. Muscle myosin is responsible for the forceful contraction of skeletal muscle (see Chapter 39). Like other types of myosin-II, it has two heads on a long tail formed from an α-helical coiled-coil. These tails polymerize into bipolar filaments (see Figs. 5.7 and 39.6).

The head of muscle myosin was originally isolated as a proteolytic fragment called subfragment-1 (Fig. 36.3). The N-terminal 710 residues of the heavy chain form the globular **catalytic domain.** The nucleotide binding site in the core of the catalytic domain is formed by a β-sheet

Generic motor with stretched spring

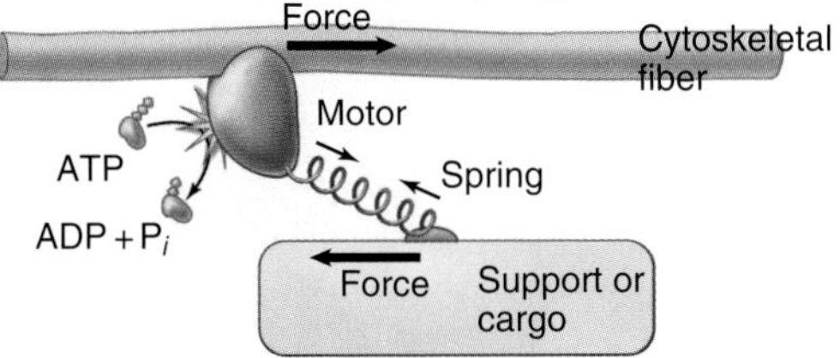

Resulting movement with anchored motor

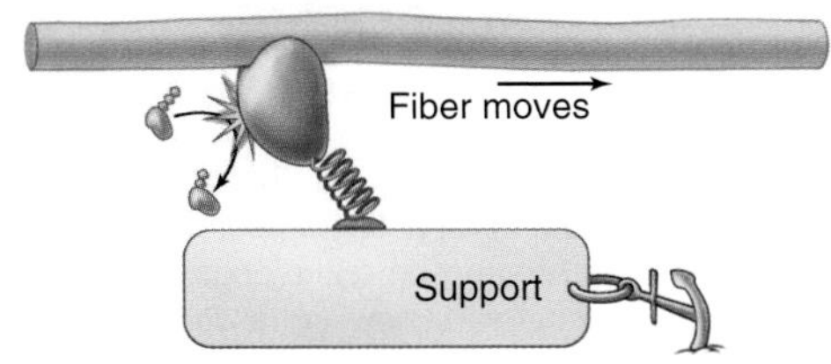

Resulting movement with anchored fiber

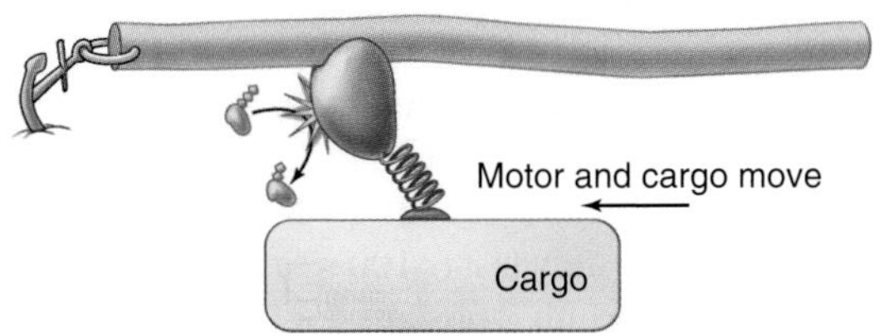

Result with anchored fiber and anchored motor

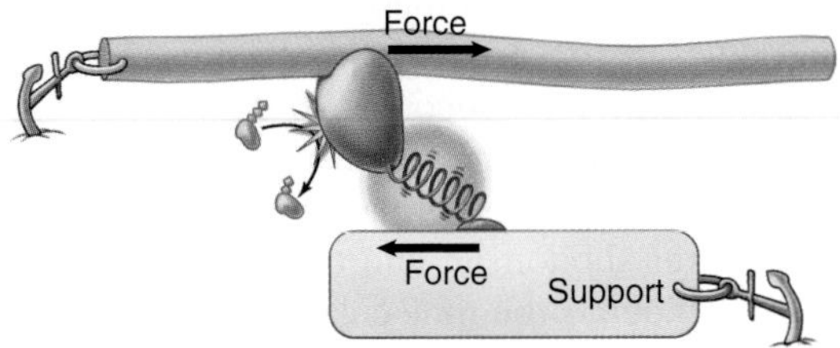

Spring stretched, force transmitted through fiber to anchoring sites, no movement, energy lost as heat

FIGURE 36.2 GENERAL FEATURES OF ATPase MOTORS. Motors bind stably to a support or cargo and transiently to a cytoskeletal fiber (actin filament or microtubule). Energy liberated by adenosine triphosphate (ATP) hydrolysis produces force to stretch an elastic element somewhere in the physical connection between the cargo and the cytoskeletal fiber. The resulting motion depends on whether the force in the spring exceeds the resistance of the fiber or the cargo.

flanked by α-helices with a topology similar to Ras GTPases (see Fig. 4.6) despite little sequence similarity. The γ-phosphate of ATP inserts deeply into the nucleotide-binding site with the adenine exposed on the surface. Actin binds more than 4 nm away from the nucleotide on the other side of the head. A region of the heavy chain called the converter subdomain is attached to the **light-chain domain** composed of an essential light chain and a regulatory light chain wrapped around and stabilizing a long α-helix formed by the heavy chain. The interaction of light chains with the heavy chain α-helix

BOX 36.1 AAA Adenosine Triphosphatases

The common ancestor of life on the earth had a gene for a versatile adenosine triphosphate (ATP)-binding domain. Through gene duplication and divergence this progenitor gave rise to the AAA family of adenosine triphosphatases (ATPases) in all branches of the phylogenic tree. Given the remarkable variety of functions of the contemporary proteins, the name "ATPases Associated with Diverse Activities" is apt. The family now includes regulatory subunits of proteasomes (see Fig. 23.8); proteases from prokaryotes, chloroplasts, and mitochondria; Hsp100 protein folding chaperones; dynein microtubule motors (Fig. 36.14); the microtubule severing protein katanin (see Fig. 34.8); activators of origins of replication (including ORC1, 4, and 5 and Mcm-7 [see Fig. 42.8]); clamp loader proteins for DNA polymerase processivity factors (see Fig. 42.12); two proteins required for peroxisome biogenesis (see Table 18.1); and proteins involved in vesicular traffic such as NSF (the N-ethylmaleimide-sensitive factor [see Fig. 21.15]).

AAA domains have a common fold with a catalytic site that binds and hydrolyzes ATP. A "Walker A" motif of conserved residues interacts with the β- and γ-phosphates of ATP, and "Walker B" motif residues participate in ATP hydrolysis. Many AAA ATPases form ring-shaped hexamers of identical subunits or up to six different AAA subunits although the dynein heavy chain has six AAA domains in one large polypeptide. Often, an arginine residue from the adjacent subunit in the hexamer inserts into the active site and facilitates conformational changes in response to ATP binding and release of the γ-phosphate.

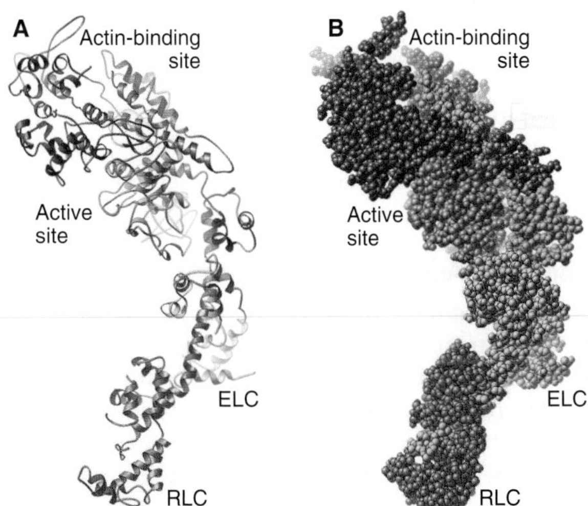

FIGURE 36.3 ATOMIC STRUCTURE OF THE HEAD OF MUSCLE MYOSIN. A, Ribbon drawing of the polypeptide backbones. **B,** Space-filling model. Heavy chain residues 4–204 *(green);* heavy chain residues 216–626 *(red);* heavy chain residues 647–843 *(purple):* essential light chain (ELC *[yellow]*); regulatory light chain (RLC *[orange]*). The myosin light chains consist of two globular domains connected by an α-helix, like calmodulin and troponin C. (For reference, see Protein Data Bank [PDB; www.rcsb.org] file 2MYS.)

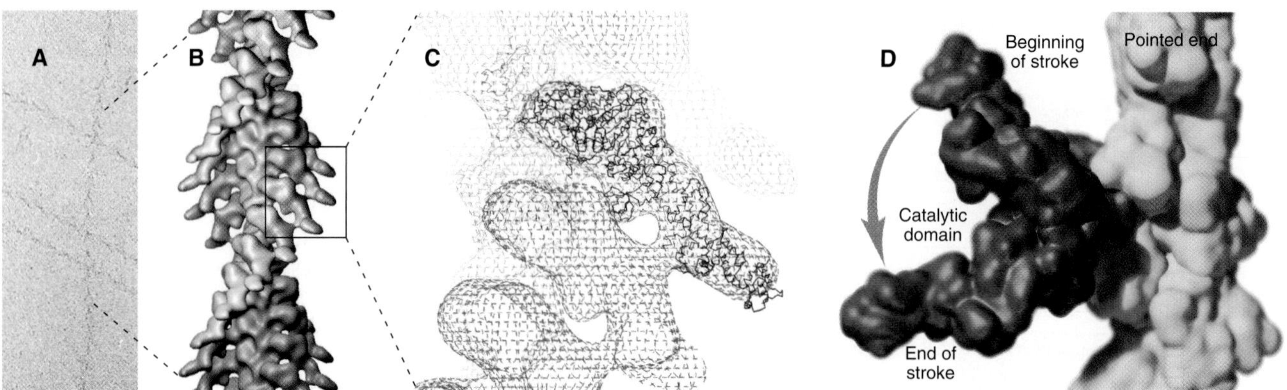

FIGURE 36.4 ACTIN FILAMENTS DECORATED WITH MYOSIN HEADS. A, Electron micrograph of frozen-hydrated actin filaments fully occupied with myosin heads. **B,** Three-dimensional reconstruction from electron micrographs of an actin filament saturated with myosin heads. **C,** Superimposition of atomic models of the actin filament and one myosin head on the reconstruction of the decorated filament (*blue* cage-like surface). **D,** Space-filling atomic model of an actin filament with one attached muscle myosin head showing the light-chain domain in two positions: (1) the end of the power stroke as observed in the absence of ATP *(blue),* and (2) the postulated beginning of the power stroke *(pink)* deduced from X-ray structures of isolated heads and spectroscopic studies. The catalytic domain *(red)* is fixed in one position on actin *(yellow).* (Courtesy R. Milligan, Scripps Research Institute, La Jolla, CA.)

resembles calmodulin binding to its target proteins (see Fig. 3.12).

Myosin heads bind tightly and rigidly to actin filaments in the absence of ATP. This is called a **rigor** complex, because it forms in muscle during rigor mortis when ATP is depleted after death. Myosin heads bound along an actin filament form a polarized structure, resembling a series of arrowheads when viewed from the side (Fig. 36.4). The heads bind at an angle and wrap around the filament. Their orientation defines the barbed and pointed ends of the actin filament (see Fig. 33.8). All known myosins, except myosin-VI, move toward the barbed end of the filament.

The atomic structures of the myosin head and actin filament fit nicely into the three-dimensional structure of the decorated filament determined by electron microscopy, providing the structural starting point for understanding the mechanics of force production (Fig. 36.4). Each myosin head contacts two adjacent actin subunits.

Actomyosin Adenosine Triphosphatase Cycle

Myosin uses energy from ATP hydrolysis to move actin filaments, so an appreciation of the mechanism requires an understanding of the biochemical steps along the reaction pathway. Fig. 36.5A looks intimidating, but working through it one step at a time reveals its logic and simplicity. Note that the mechanism consists of two parallel lines of chemical intermediates. First, consider the bottom line showing the reactions that explain why myosin alone turns over ATP remarkably slowly, at a rate of only approximately 0.02 s^{-1}:

Step 1. At physiological concentrations of ATP, myosin binds ATP in less than 1 millisecond. Energy from ATP binding allows a conformational change in the myosin that can be detected by a change in the intrinsic fluorescence of the protein itself.

Step 2. The enzyme catalyzes the hydrolysis of ATP. This reaction is moderately fast (>100 s^{-1}) and readily reversible. The equilibrium constant for hydrolysis on the enzyme is near 1, so each ATP is hydrolyzed to adenosine diphosphate (ADP) and inorganic phosphate and the triphosphate is resynthesized several times before the products dissociate from the enzyme. ATP splitting provides energy for a second conformational change, reflected in a further increase in the fluorescence of the myosin. These conformational changes reorient the converter subdomain and the light chain domain poised to undergo the molecular rearrangements that subsequently produce movement.

Step 3. Inorganic phosphate (P) slowly dissociates from the active site (at a rate of approximately 0.02 s^{-1}) by escaping through a narrow "back door" on the far side of the enzyme. This is the rate-limiting step in the pathway. The loss of phosphate is coupled to conformational changes that return myosin toward its basal state. The phosphate dissociation step has the largest negative free energy change, so it is presumed that energy derived from ATP binding and hydrolysis and stored in conformational changes in the myosin head is used to do work or dissipated as heat at this point in the reaction pathway.

Step 4. Once phosphate dissociates, ADP leaves rapidly from the "front door."

To summarize, in the absence of actin filaments, ATP binds rapidly to myosin and is rapidly but reversibly split, and the products slowly dissociate from the active site. The overall cycle of the enzyme is limited by the slow conformational change coupled to phosphate dissociation. Energy derived from ATP binding and hydrolysis is

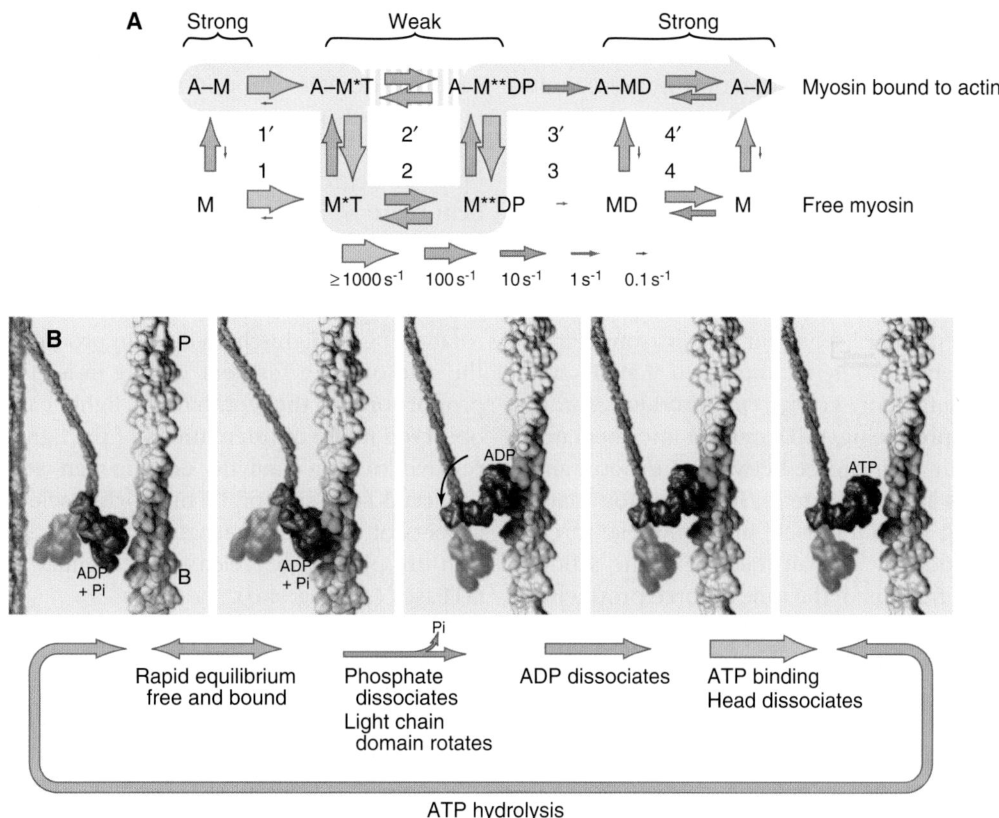

FIGURE 36.5 MYOSIN ATPase MECHANISMS. A, A diagram of the actomyosin ATPase cycle of striated muscle myosin-II showing the actin filament (A), myosin head (M), ATP (T), adenosine diphosphate (ADP) (D), and inorganic phosphate (P). Transient-state kinetics revealed the major chemical intermediates and the rate constants for their transitions. *Arrow sizes* are proportional to the rates of the reactions, with second-order reactions adjusted for physiological concentrations of reactants. *One* or *two asterisks* indicate conformational changes in the myosin head induced by ATP binding and hydrolysis. Myosin without nucleotide (M) and myosin with ADP (MD) bind much more tightly to actin filaments than do AMT and AMDP. The weakly bound AMT and AMDP intermediates are in a rapid equilibrium with free MT and MDP. The beige shading shows the main pathway through the reaction. **B,** The postulated force-producing structural changes in the orientation of the light-chain domain (*purple* and *blue*) coupled to the myosin ATPase cycle. (**B,** Data from R. Vale, University of California, San Francisco, and R. Milligan, Scripps Research Institute, La Jolla, CA.)

used for a conformational change in the myosin head that is dissipated when phosphate dissociates.

The upper line in Fig. 36.5A shows myosin associated with an actin filament. The chemical intermediates are the same, but some of the key rate constants differ for the actin-bound and free myosin. Steps 1 and 2 are similar to those of free myosin, but step 3—the dissociation of phosphate—is much faster when a head is bound to an actin filament. As a result, myosin bound to actin traverses the ATPase cycle approximately 200 times faster than myosin free in solution, and ATP hydrolysis becomes the rate-limiting step. This effect of actin is referred to as "actin activation of the myosin ATPase." A practical advantage of this mechanism is that the ATPase cycle is essentially turned off unless the head interacts with an actin filament.

Finally, consider the vertical arrows representing transitions between bound and free states of each myosin chemical intermediate. All myosin intermediates bind rapidly to actin filaments, but the dissociation rate constants vary over a wide range depending on the nucleotide that is bound to the active site of the myosin. Myosin with no nucleotide or with bound ADP alone dissociates very slowly and therefore binds tightly to actin filaments. Myosin with bound ATP or ADP+P_i dissociates rapidly from actin, so these states bind actin weakly.

One cycle of ATP hydrolysis takes about 50 milliseconds, but a single pathway cannot be drawn through the reaction mechanism of ATP, myosin, and actin owing to the rapid equilibrium of myosin intermediates (MT and MDP) hopping on and off actin filaments on a millisecond time scale. Starting with AM, ATP binds very rapidly and sets up a rapid, four-way equilibrium including AMT, MT, AMDP, and MDP—the major intermediates during steady-state ATP turnover in muscle (see Chapter 39). Because the products of ATP hydrolysis dissociate much more rapidly from AMDP than from MDP, the favored pathway out of this four-way equilibrium is through AMDP to AMD and back to AM. The overall ATPase rate depends on the actin concentration, which determines the fraction of myosin heads bound to actin in the

AMDP state. At the high actin concentrations in cells, a significant fraction of myosin heads is associated with actin (approximately 10% in contracting muscle), but each molecule continues to exchange on and off actin filaments.

Transduction of Chemical Energy Into Molecular Motion

Myosin heads produce force during the transition from the AMDP state to the AMD and AM states. Production of force at this step makes sense for two reasons: First, the large free-energy difference between AMDP and AMD provides sufficient energy to produce force; second, the force-producing AMD and AM intermediates bind tightly to actin, so any force between the motor and the actin track is not dissipated. However, for many myosins, including skeletal muscle myosin, these force-producing states occupy a small fraction of the whole ATPase cycle. The fraction of the time in force producing states is called the **duty cycle.** ADP dissociates rapidly from AMD, and ATP binds rapidly to AM, dissociating myosin from the actin filament and initiating another ATPase cycle.

Fifty years of research using a combination of mechanical measurements, static atomic structures of myosin heads with various bound nucleotides, and spectroscopic observations of contracting muscle revealed the structural basis for the conversion of free energy into force: a dramatic conformational change in the orientation of the light-chain domain associated with phosphate dissociation (Fig. 36.5B).

Elegant mechanical experiments measured the size of the mechanical step produced by a myosin during one cycle of ATP hydrolysis. These experiments on live muscles first suggested that each cycle of ATP hydrolysis moves an actin filament approximately 5 to 10 nm relative to myosin. Now one may observe myosin moving single actin filaments by fluorescence microscopy. An array of myosin heads attached to a microscope slide can use ATP hydrolysis to push actin filaments over the surface (Fig. 36.6A–C). Assays with single myosin molecules show that each cycle of ATP hydrolysis can move an actin filament up to 5 to 15 nm and develop a force of about 3 to 7 piconewtons (pN) (Fig. 36.6D). At low ATP concentrations, the interval between the force-producing step and the binding of the next ATP is relatively long, so single steps can be observed.

Further insights emerged from biophysical studies of muscle and purified proteins using x-ray diffraction (see Fig. 39.11), electron microscopy, electron spin resonance spectroscopy, and fluorescence spectroscopy. These experiments showed that the light-chain domain pivots around a fulcrum, the converter subdomain within the catalytic domain, which is stationary relative to the actin filament. For example, spectroscopic probes on light chains revealed a change in orientation when muscle is activated to contract, whereas probes on the catalytic domain do not rotate. Crystal structures of myosin heads with various bound nucleotides and nucleotide analogs show that the light-chain domain can pivot up to 90 degrees (Fig. 36.4D). The light-chain domain is bent more acutely in the AMT and AMDP intermediates and pivots to a more extended orientation when phosphate dissociates (Fig. 36.3). ADP dissociation extends this rotation of some classes of myosin. Consistent with rotation of the light-chain domain producing movement, the rate of actin filament gliding in an in vitro assay is proportional to the length of the light-chain domain. The observed range of orientations of the light-chain domain relative to the catalytic domain can account for the observed step size of 10 nm for muscle myosin. Some aspects of these conformational changes and their relation to phosphate release are similar to Ras family GTPases (see Fig. 4.6).

Rotation of the light-chain domain is believed to produce movement indirectly in the sense that force-producing intermediates stretch elastic elements in the system. This mechanism is represented by a spring in Fig. 36.2. The elastic elements in the myosin-actin complex are most likely to be mainly in the myosin head, with small contributions from the actin and myosin filaments. Movement of the light-chain domain tensions the spring transiently in the AMD and AM states. Dissociation of ADP and rebinding of ATP to the AM intermediate reverts the system to the rapid equilibrium of mostly dissociated weakly bound intermediates. Any force left in the spring is lost as soon as the head dissociates from the actin filament.

The actual motion produced depends on the mechanical resistance in the system (Fig. 36.2). If both myosin and actin are fixed, elastic elements are stretched for the life of the force-producing states (AMD and AM), and the energy is lost as heat when the head dissociates. This happens when one tries to lift an immovable object. If the resistance is less than the force in the stretched elastic elements, the actin filament moves relative to myosin, as in muscle contraction. The distance moved in each step depends on the resistance, as the spring stops shortening when the forces are balanced.

Myosin Superfamily

Eukaryotes have 35 classes of myosin and many other examples of unique myosins in single species (Fig. 36.7). All arose from a gene similar to myosin-I in the last eukaryote common ancestor more than a billion years ago. The primordial gene then gave rise to the gene for myosin-V, so this class is also widespread. Gene duplication, divergence, and acquisition of extra domains produced many other myosin genes encoding proteins specialized for particular biological functions made possible by variations of the mechanochemical ATPase cycle

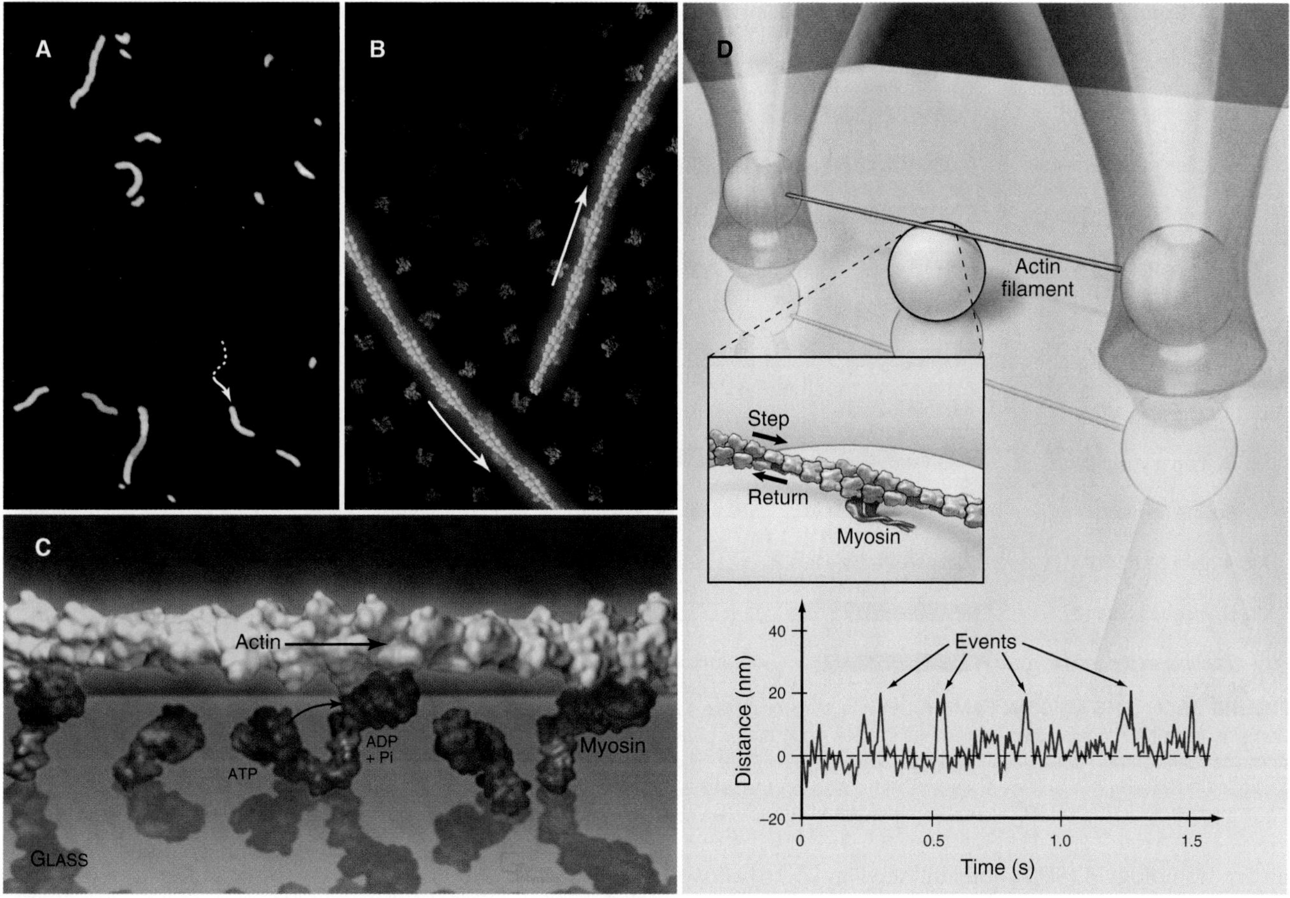

FIGURE 36.6 IN VITRO MOTILITY ASSAYS WITH PURIFIED MUSCLE MYOSIN AND ACTIN FILAMENTS. A–C, Actin filament gliding assays. **A,** Filaments are labeled with rhodamine-phalloidin to render them visible by light microscopy. ATP hydrolysis by myosin moves actin filaments over the surface with the pointed end leading as the myosins walk toward the barbed end of the filaments. **B–C,** Drawings of actin filaments moving over myosin heads immobilized on a glass coverslip. **D,** Measurement of the muscle myosin step size. An actin filament is attached between two plastic beads, which are suspended by laser optical traps. The optical traps move the filament near a myosin molecule on the surface of another bead attached to the microscope slide, allowing a myosin head to attach to the actin filament. When supplied with ATP, a single myosin head can move the actin filament a short distance corresponding to the step size. The graph shows the time course of displacements of the actin filament and attached beads. Brownian motion limits the precision of the measurement of the size of these steps to a range of 5 to 15 nm. The duration of the step depends on the ATP concentration, because ATP dissociates the force-producing AM state, allowing the force of the optical traps to return the beads and the actin filament to their original position. P_i, inorganic phosphate. (**A,** Courtesy A. Bresnick, Albert Einstein College of Medicine, New York. **D,** For reference, see Finer JT, Simmons RM, Spudich JA. Single myosin molecule mechanics: piconewton forces and nanometer steps. *Nature.* 1994;368:113–119.)

and acquisition of diverse tails to interact with cargo. Within a myosin class, the tails are similar to each other, but between classes, tails are diverse in terms of their ability to polymerize and interact with other cellular components including membranes and ribonucleoprotein particles.

No organism has genes for all 35 classes of myosin and a few species, including *Giardia lamblia,* have no myosin genes. Myosin-I is most widespread, but plants and related organisms lost this gene. A primitive myosin-V gene gave rise to plant myosin-VIII and myosin-XI, which move at very high speeds (see Fig. 37.9). Organisms on the branch including amoebas, yeast and animals have genes for myosins types I, II, and V, but myosin genes diversified in animals, so humans have 40 myosin genes from 13 classes. Gene duplications gave rise to multiple isoforms within most classes of myosin. For instance, the vertebrate smooth muscle myosin gene arose from duplication of a gene for a cytoplasmic myosin-II.

Establishing the biological functions of the various myosins has been challenging. Biochemical characterization of cargo and localization in cells provide some clues, but genetic or biochemical knockouts often have mild effects, probably owing to overlapping functions of the myosins and the capacity of some cells to adapt to their loss, at least under laboratory conditions.

Myosin-I was the first "unconventional myosin" discovered—unconventional in the sense that it differed from the type II myosin originally isolated from skeletal muscle. These myosins have one head and short tails with various types of domains, including a basic domain with affinity for acidic phospholipids. The presence of

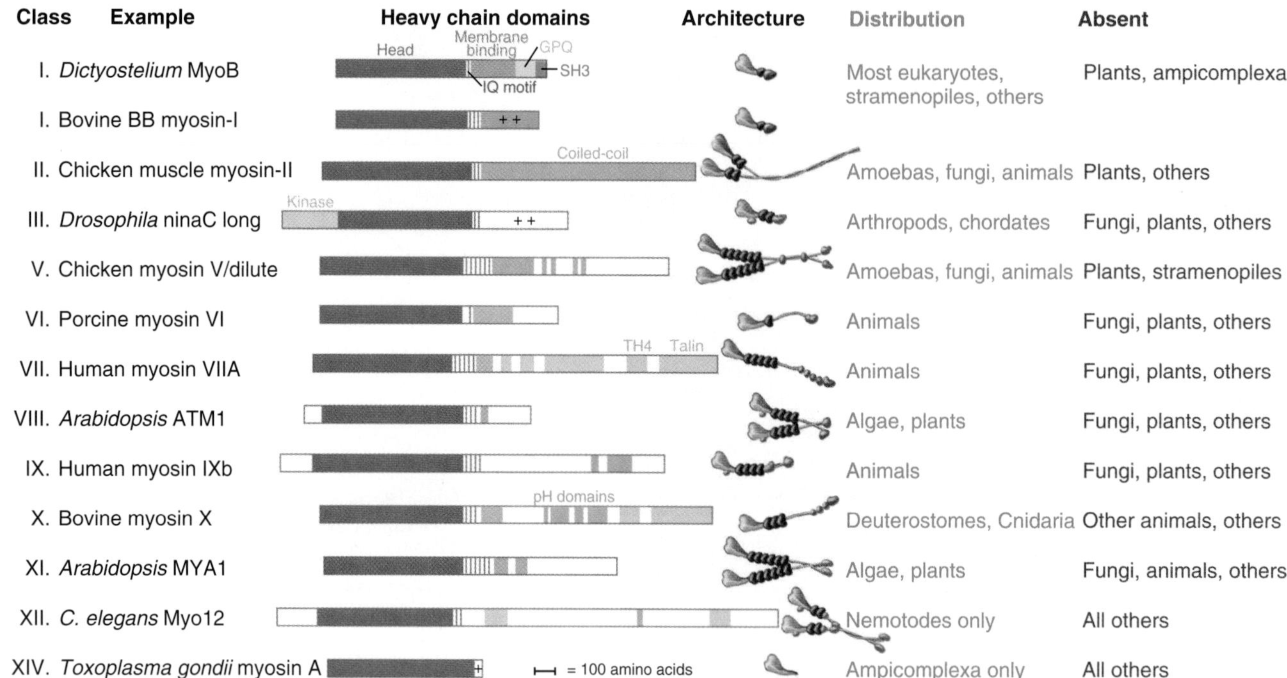

FIGURE 36.7 THE MYOSIN FAMILY. Drawing of myosin heavy chain domains and molecular models of myosin isoforms showing catalytic domains *(rose);* IQ motifs, light-chain–binding sites *(rose bars);* basic domains with affinity for membrane lipids *(violet);* SH3 (Src homology 3) domains *(dark green);* coiled-coil *(orange);* kinase domain *(light blue);* and pleckstrin homology domain *(blue).* (For reference, see Odronitz F, Kollmar M. Drawing the tree of eukaryotic life based on the analysis of 2,269 manually annotated myosins from 328 species. *Genome Biol.* 2007;8:R196. See also Myosin Home Page, available at http://www.mrc-lmb.cam.ac.uk/myosin/myosin.html.)

an Src homology 3 (SH3) domain (see Fig. 25.10) allows some type I myosins to bind proline-rich sequences in other proteins. Those with an actin filament–binding domain separate from the motor domain can crosslink actin filaments. With duty cycles of less than 10%, multiple myosin heads must work together in concert to move membranes. Mutations show that myosin-I participates in endocytosis, as expected from its concentration at sites of phagocytosis and macropinocytosis. In microvilli of intestinal epithelial cells, myosin-I links actin filaments laterally to the plasma membrane (see Fig. 33.2B). Heavy-chain phosphorylation activates myosin-I from lower eukaryotes, whereas calcium binding to calmodulin light chains regulates myosin-I from the intestinal brush border.

The **myosin-II** class includes various muscle and cytoplasmic myosins that also have two heads, two IQ motifs, and long coiled-coil tails. Assembly of tails into bipolar filaments (see Fig. 5.7) allows myosin-II to pull together oppositely polarized actin filaments during muscle contraction (see Chapter 39) and cytokinesis (see Fig. 44.24). As in smooth muscle (see Fig. 39.23), phosphorylation of the regulatory light chain activates myosin-II in animal nonmuscle cells. In addition, phosphorylation of the heavy chain regulates the enzyme activity and/or polymerization of some myosin-IIs.

Myosin-V moves pigment granules, ribonucleoprotein particles and other cellular components (see Fig. 37.11). A long light-chain domain with seven IQ motifs allows myosin-V to take long steps along the actin filament (Fig. 36.8). These steps are processive, because slow ADP dissociation from the AMD intermediate allows time for the other head to take a long step and bind an actin subunit 36 nm beyond the first head toward the barbed end of the filament. Mechanical strain after the step may modestly increase the rate of ADP dissociation from the trailing head. This cooperation between the heads initiates ATP binding and the next ATPase cycle, as the motor walks deliberately along the filament. These features make myosin-V a valuable model for the lever arm for movements of the whole myosin family.

Myosin-VI arose in metazoan cells and is the only myosin known to move toward the pointed end of actin filaments. Unique features of the converter domain result in the lever arm swinging opposite to the conventional direction. The lever arm is a long, single α-helix beginning with a single IQ-motif associated with calmodulin. Lacking coiled-coil, myosin-VI is a monomer unless adapter proteins bring together the C-terminal globular cargo-binding domains of two molecules. These dimers can take huge steps of approximately 30 nm, but myosin-VI can also act as a tether, because the force-producing AMD and AM states occupy a large fraction of the ATPase cycle, owing to slow ADP dissociation from AMD state and slow ATP binding to AM. These features allow myosin-VI to move endocytic vesicles from the plasma membrane into the cytoplasm and to contribute to the formation of autophagosomes. Myosin-VII and

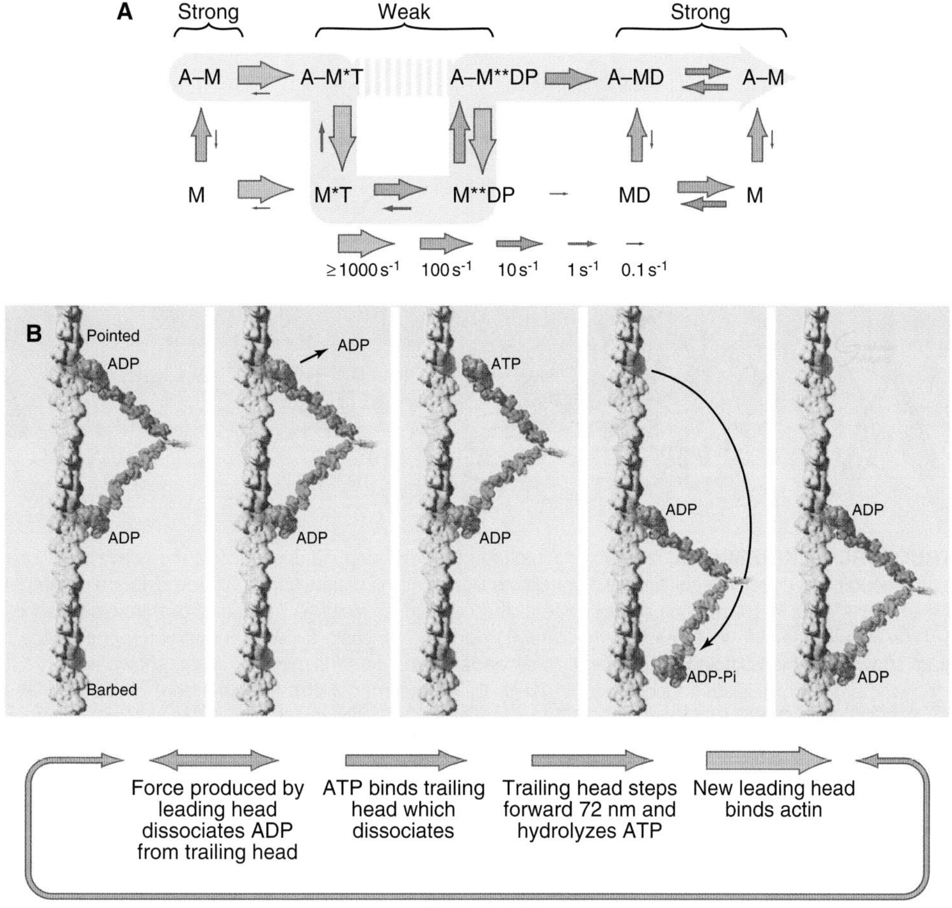

FIGURE 36.8 MYOSIN-V MECHANISM. A, ATPase cycle with ADP release as the rate-limiting step rather than phosphate dissociation as for muscle myosin (see Fig. 36.5A). **B,** Relationship of mechanical steps to the ATPase cycle. Shown are actin filament (A), myosin head (M), ATP (T), ADP (D), and inorganic phosphate (Pi). (For reference, see De La Cruz EM, Ostap EM. Relating biochemistry and function in the myosin superfamily. *Curr Opin Cell Biol.* 2004;16:61–67. For a movie of myosin-V stepping on actin filaments, see Kodera N, Yamamoto D, Ishikawa R, et al. Video imaging of walking myosin V by high-speed atomic force microscopy. *Nature.* 2010;468:72–76.)

myosin-X are also monomers with single chain α-helices as lever arms.

Myosin mutations cause human diseases. Loss-of-function mutations in the genes for myosins-IIA, -IIIA, -VI, -VIIA, and -XV cause deafness and vestibular dysfunction. Mutations in the genes for cardiac muscle myosin heavy and light chains are responsible for many cases of cardiomyopathies (see Table 39.1).

Microtubule Motors

The kinesin and dynein families of molecular motors use energy from ATP hydrolysis to move vesicles, membrane-bound organelles, chromosomes, and other cargo along microtubules (see Fig. 37.1). Dynein also powers bending motions of eukaryotic flagella and cilia (see Fig. 38.14). Dyneins move themselves and any cargo toward the minus end of microtubules. Most kinesins move in the opposite direction, toward the plus end, but some kinesin family members are minus-end-directed motors and others promote microtubule disassembly. Like myosins, microtubule motors have heads with ATPase activity and tails that interact with cargo.

Kinesins

Kinesin-1 is a processive motor that moves cargo, such as an organelle, continuously toward the plus end of a microtubule. The two heads are attached to an α-helical coiled-coil tail, much like myosin-II, except both the heads and coiled-coil are smaller (Fig. 36.9). Each head, consisting of approximately 340 residues, is a motor unit that binds microtubules and catalyzes ATP hydrolysis. Light chains associated with the C-terminal bifurcation of the tail bind cargo molecules (Fig. 36.9E).

Because the **kinesin head** is less than half the size of a myosin head and because the proteins lack appreciable sequence homology, the atomic structure of kinesin-1 (Fig. 36.9C) revealed a major surprise: The small kinesin head is folded like the core of the catalytic domain of myosin! In fact, this core, which consists of a central, mixed β-sheet flanked by helices, is similar to the

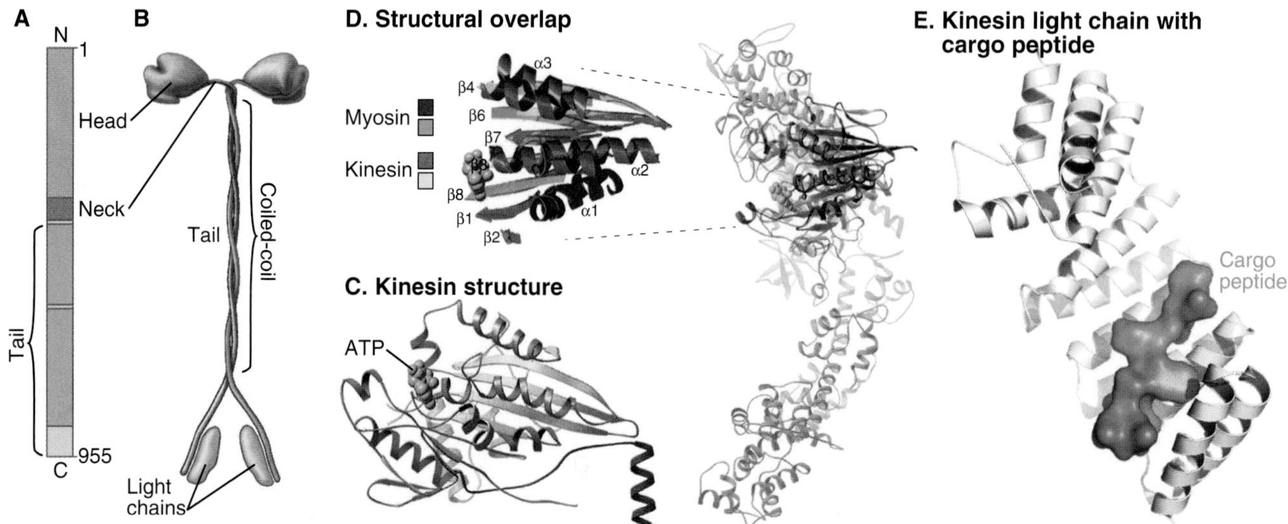

FIGURE 36.9 STRUCTURE OF KINESINS. A, Domain architecture of the polypeptide sequence of the heavy chain of kinesin-1. **B,** Sketch of kinesin-1 showing two heads and the coiled-coil tail with light chains bound at the distal end. **C,** Ribbon diagram of the polypeptide backbone of the kinesin head showing ATP as a space-filling model *(green),* the neck-linker residues *(red),* and the proximal part of the coiled-coil tail. **D,** Superimposition of the core of the kinesin-1 head on the catalytic domain of myosin showing the structural homology of the proteins. The detailed ribbon diagram shows only the homologous elements of secondary structure. The overview *(right)* shows kinesin-1 *(blue)* superimposed on the structure of the whole head of skeletal muscle myosin *(pink).* **E,** Ribbon model of a kinesin-1 light chain *(purple)* with a bound cargo peptide *(green)* with a DWED motif. (**C,** For reference, see PDB file 3KIN and Sack S, Muller J, Marx A, et al. X-ray structure of motor and neck domains from rat brain kinesin. *Biochemistry.* 1997;36:16155–16165. **D,** Superimposed ribbon diagrams courtesy of R. Vale, University of California, San Francisco. **E,** For reference, see PDB file 3ZFW and Pernigo S, Lamprecht A, Steiner RA, et al. Structural basis for kinesin-1: cargo recognition. *Science.* 2013;340:356–359.)

considerably smaller Ras family GTPases (see Fig. 4.6). This provided strong evidence that all three families of nucleoside triphosphatases evolved from a common ancestor (Fig. 36.1). ATP binds to a site on kinesin that is homologous to the guanosine triphosphate (GTP)-binding site of Ras, but the enzyme mechanisms differ in important ways. The microtubule-binding site is some distance from the ATP-binding site (Fig. 36.10).

Kinesin Mechanochemistry

Single kinesin-1 heads, produced experimentally from truncated cDNAs, traverse a microtubule-stimulated ATPase cycle much like myosin (Fig. 36.11A). Both bind and hydrolyze ATP rapidly followed by slower release of phosphate and ADP. However, in contrast to muscle myosin, kinesin heads may remain bound to the microtubule through multiple cycles of ATP hydrolysis rather than dissociating when bound to ATP or ADP-P_i.

In vitro motility assays (Fig. 36.12) revealed that a two-headed kinesin-1 can move along a single (or two parallel) microtubule protofilaments for long distances at 0.8 μm/s. The motor makes discrete steps of 8 nm, the spacing of successive tubulin dimers in a microtubule. Each step takes 10 milliseconds when kinesin is moving at full speed. This step is remarkably large for the small (<10 nm) kinesin heads linked together at the neck region. Kinesin is very powerful, stalling at a force of 6 pN. This allows single molecules to move large organelles through the cytoplasm.

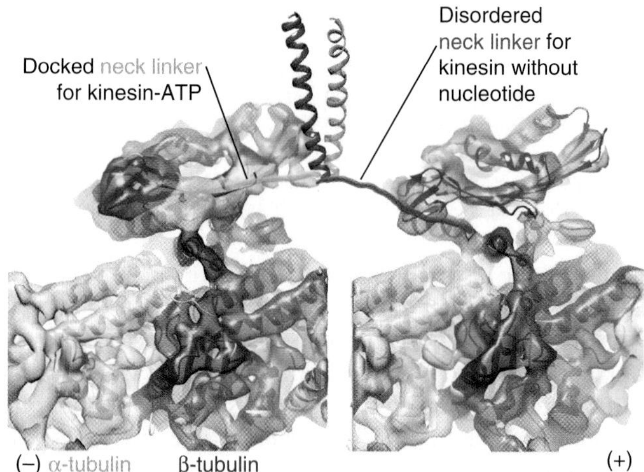

FIGURE 36.10 INTERACTION OF KINESIN-1 WITH MICROTUBULES. A, Three-dimensional reconstructions from electron micrographs of kinesin-1 heads *(blue)* bound to a microtubule *(yellow and red).* The kinesin head on the left has bound ATP and the neck linker *(green)* docked. The kinesin head on the right has no bound nucleotide and the neck linker *(red)* is disordered. Ribbon diagrams show how the neck linker of the leading head *(right)* must be unfolded to connect to the trailing head at the beginning of the coiled-coil tail. (From Charles Sindelar, Yale University, based on Shang Z, Zhou K, Xu C, et al. High-resolution structures of kinesin on microtubules provide a basis for nucleotide-gated force-generation. *Elife.* 2014;3:e04686.)

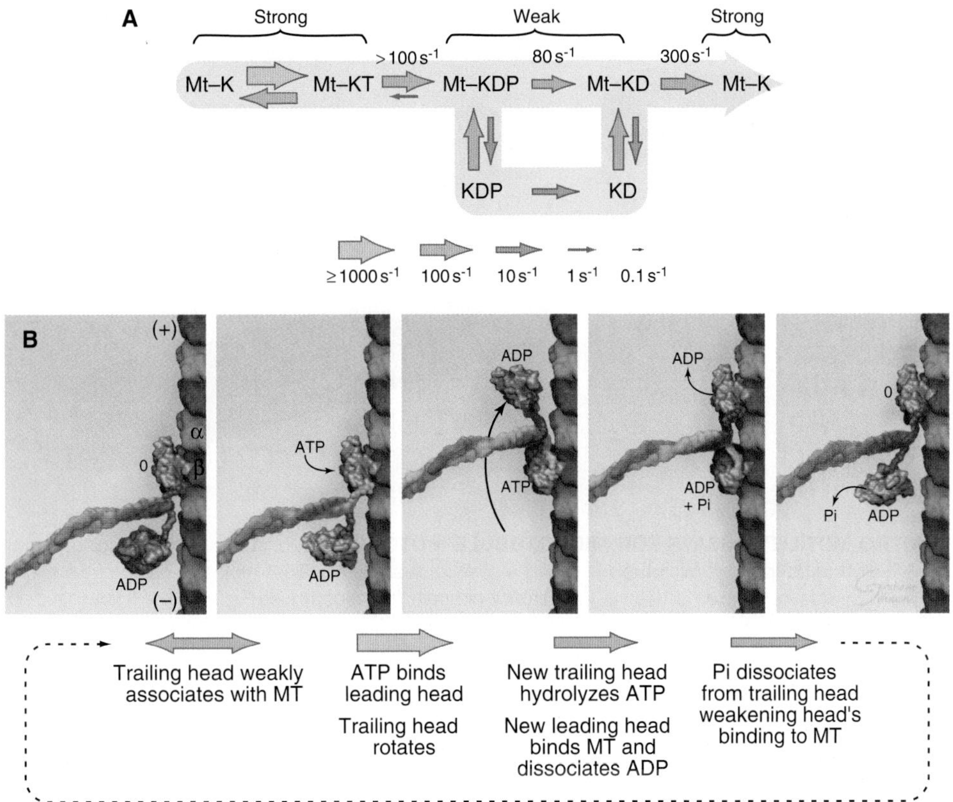

FIGURE 36.11 KINESIN-1 ATPase MECHANISM. A, A diagram of the kinesin-microtubule ATPase cycle for a single kinesin-1 head showing the kinesin (K), microtubule (Mt), ATP (T), ADP (D), and inorganic phosphate (P). *Arrow sizes* are proportional to the rates of the reactions, with second-order reactions adjusted for physiological concentrations of reactants. The *beige* shading shows two pathways through the reaction, one along the top line without dissociation, and the other with dissociation from the microtubule. **B,** Hand-over-hand, processive stepping of kinesin along a microtubule. The empty, microtubule-bound head binds and hydrolyzes ATP resulting in a conformational change favoring docking of its neck-linker *(green),* thereby moving forward the detached head with its undocked *(pink)* neck-linker. Binding of the new leading head to the microtubule causes a conformation change that dissociates ADP. Dissociation of the γ-phosphate from the trailing head results in its dissociation from the microtubule. This returns the heads to their original condition, but with the motor advanced 8 nm with the heads in the opposite chemical states. P_i, inorganic phosphate. (**B,** Data from R. Vale, University of California, San Francisco, and R. Milligan, Scripps Research Institute, La Jolla, CA. For reference, see Cao L, Wang W, Jiang Q, et al. The structure of apo-kinesin bound to tubulin links the nucleotide cycle to movement. *Nat Commun.* 2014;5:5364.)

Processive movement depends on the ability of kinesin to remain associated with the microtubule through more than a hundred cycles of ATP hydrolysis. This is made possible by cooperation between the two heads that ensures at least one head is bound to the microtubule throughout the ATPase cycle (Fig. 36.11B). Reciprocal affinities for nucleotide and microtubules allow the two heads to alternate between microtubule binding and dissociation. For example, if kinesin-1 with ADP bound to both heads is mixed with microtubules, one head binds a microtubule and rapidly dissociates its ADP, leaving the other head dissociated with bound ADP. Binding and hydrolysis of ATP on the open site of the head associated with the microtubule, drives a conformational change that repositions the trailing head forward so it can pass the bound head and bind the next tubulin dimer toward the plus end of the microtubule. Association of the new leading head with the microtubule promotes dissociation of its bound ADP. Dissociation of the γ-phosphate from the new trailing head weakens its affinity for the microtubule. Its detachment from the microtubule with bound ADP brings the cycle back to its starting point with kinesin having advanced 8 nm (Fig. 36.12). Experiments with single kinesin-1 molecules labeled with fluorescent dyes showed that they alternate steps on the right and left sides of the microtubule like a gymnast walking on a balance beam.

The mechanism of stepping is postulated to be the docking and undocking of a segment of the kinesin-1 heavy chain linking the motor domain to the coiled-coil neck and tail (Fig. 36.10). With ATP bound the motor domain has a conformation that favors association of the "neck-linker" peptide, as in the X-ray structure of dimeric kinesin (Fig. 36.9C). When either no nucleotide or ADP is bound, the conformation of the motor domain releases the neck-linker peptide.

Kinesin Superfamily

The last eukaryotic common ancestor had only a single myosin but already had at least 11 families of kinesins

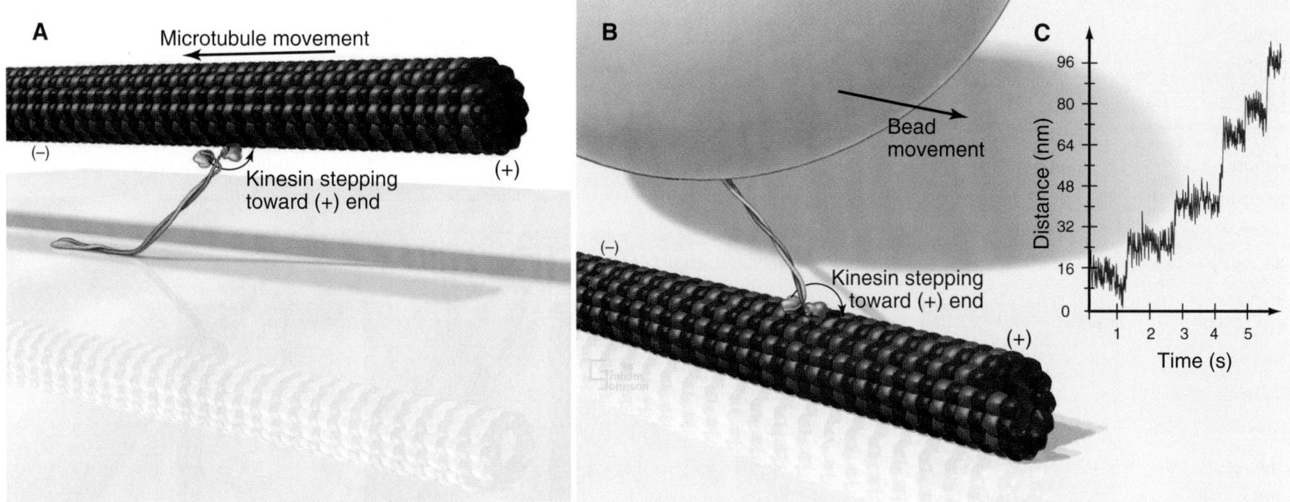

FIGURE 36.12 IN VITRO MOTILITY ASSAYS FOR MICROTUBULE MOTORS. A, Gliding assay. Kinesin or dynein that is attached to a microscope slide uses ATP hydrolysis to move microtubules over the surface. A single kinesin-1 molecule can move a microtubule in this assay. Microtubules can be imaged by video-enhanced differential interference contrast microscopy (see Fig. 34.6) or fluorescence microscopy. **B,** Bead assay. Kinesin or dynein that is attached to a plastic bead uses ATP hydrolysis to move the bead along a microtubule attached to the microscopic slide. **C,** Experimental measurement of the kinesin-1 step size using the bead assay. The bead is held in a laser optical trap so that 8-nm steps can be recorded, as a single, two-headed kinesin-1 moves a bead processively along a microtubule, as in **B.** The position of the bead is recorded with nanometer precision by interferometry. (For reference, see Svoboda K, Schmidt CF, Schnapp BJ, et al. Direct observation of kinesin stepping by optical trapping interferometry. *Nature.* 1993;365:721–727.)

Kinesin structures

N-terminal motor	**Example**	**Heavy chain domains**	**Architecture**
Kinesin-1	Hs Kif5B	Head Coiled-coil Tail	
Kinesin-2	Sp Krp85/95		85 95
Kinesin-3	Mm Kif1b		
Kinesin-4	Xl Klp1		
Kinesin-5	Dm Klp61f		
Kinesin-7	Hs CENP-E		
Internal motor			
Kinesin-13	Xl MCAK		
C-terminal motor			
Kinesin-14	Dm Ncd		

FIGURE 36.13 KINESIN FAMILY. A, Phylogenetic relationships of some of the kinesins based on the sequences of the motor domains. **B,** Drawing of kinesin heavy chain domains and molecular models of kinesin isoforms showing the catalytic domain *(red),* coiled-coil tail *(orange),* and tail piece *(blue).* (Data from R. Case and R. Vale, University of California, San Francisco. For reference, see Lawrence CJ, Dawe RK, Christie KR, et al. Standardized kinesin nomenclature. *J Cell Biol.* 2004;167:19–22; and Dagenbach EM, Endow SA. A new kinesin tree. *J Cell Sci.* 2004;117:3–7. See also the Kinesin Home Page at https://labs.cellbio.duke.edu/kinesin.)

with motor domains associated with a variety of coiled-coil stalks and tails (Fig. 36.13 and Table 36.2). Thus the microtubule system was much more developed than the actin system at this point in evolution. Contemporary eukaryotes have genes for 17 families of kinesins (often with multiple isoforms) plus a few more paralogs identified in isolated species. On the other hand, many eukaryotes have lost one or more kinesin genes; for example, amoebas lack kinesin-2 and alveolates lack kinesin-7.

TABLE 36.2 Kinesin Superfamily: Classification and Examples of Kinesin-Family Motor Proteins

Class	Examples	Subunits (kD)	Velocity (μm s^{-1})	Functions
N-Terminal Motor				
Kinesin-1	Human KHC	2 × 110, 2 × 70	+0.9	Organelle movement
Kinesin-2	Urchin KRP85/95	1 × 79, 1 × 84, 1 × 115	+0.4	Organelle movement
Kinesin-3	Mouse KIF1B	1 × 130	+0.7	Mitochondria movement
Kinesin-4	*Xenopus* Kp11	2 × 139	+0.2	Chromosome movement
Kinesin-5	Fly KLP61F	4 × 121	+0.04	Pole separation, mitosis
Kinesin-7	Human CENP-E	2 × 340	+0.1	Kinetochore-microtubule binding
Internal Motor				
Kinesin-13	MCAK	2 × 83	—	Microtubule disassembly
C-Terminal Motor				
Kinesin-14	Fly Ncd	2 × 78	−0.2	Mitotic/meiotic spindle

Modified from Vale RD, Fletterick RJ. The design plan of kinesin motors. *Annu Rev Cell Dev Biol.* 1997;13:745–777. More data on kinesins are available at the Kinesin Home Page, https://labs.cellbio.duke.edu/kinesin.

All members of the kinesin family have similar motor domains attached to a variety of tails that interact with cargo (Fig. 36.13). Motor domains are generally found at the N-terminus, but may be located in the middle (kinesin-13) or at the C-terminus (kinesin-14). Regardless of their location, motor domains have similar structures.

Most kinesins are dimeric, with two polypeptides joined in a coiled-coil. Most are homodimers, but kinesin-2 not only forms homodimers but also heterotrimers consisting of two different polypeptides with motor domains plus another large subunit. Tetrameric kinesin-V molecules bind to two microtubules with opposite polarities and move them apart.

Most kinesins move toward the plus end of the microtubule, but C-terminal kinesin-14 motors move toward the minus end. The reverse direction of kinesin-14 movement is not explained by either the architecture of the motor domain, the ATPase mechanism, which is similar to kinesin-1, or the attachment of the N-terminus of the motor domain to the stalk. Instead, the proximal part of the Ncd coiled-coil stalk rotates approximately 70 degrees toward the minus end of the microtubule when ATP binds to the active site.

Kinesins transport a variety of cargo, including chromosomes and organelles, along microtubules. Kinesin-1 moves membrane vesicles toward the plus end of microtubules away from the centrosome and along axons of neurons (see Fig. 37.1). The light chain links the end of the kinesin-1 tail to cargo proteins (Fig. 36.9E). Kinesin-2 is the motor for anterograde intraflagellar transport, movement of particles toward the tip of the axoneme in cilia and flagella (see Fig. 38.18). It also transports a variety of cargos over long distances in the cytoplasm. Kinesin-4 motors (also called **chromokinesins**) bind both DNA and microtubules. In neurons, they move cargo along axons. In dividing cells they participate in the formation of condensed mitotic chromosomes (see Fig. 44.7). Kinesin-7 (originally called **CENP-E**) concentrates at kinetochores where it helps move the chromosome toward the middle of the mitotic spindle prometaphase (see Fig. 44.5). Bipolar **kinesin-5** motors form an antiparallel tetramer of two dimeric kinesins that bridge a pair of oppositely polarized microtubules and push apart the poles of the mitotic spindle (see Fig. 44.7).

Both kinesin-8 and kinesin-13 use cycles of ATP hydrolysis to remove tubulin dimers from the ends of microtubules. Kinesin-8 motors to the plus end where it works, whereas kinesin-14 lacks motor activity but can diffuse on the microtubule surface to reach either end. This depolymerizing activity is important for mitosis.

Most kinesins appear to be constitutively active, but intramolecular interactions autoinhibit kinesin-1. The tail folds back and binds between the heads, shutting off the motor. Adapter proteins compete the tail from the head, freeing the motor to be active.

Dyneins

Dynein microtubule-based motors are AAA ATPases (Box 36.1), so their evolutionary origin differs from myosins and kinesins (Fig. 36.1). Most AAA ATPases consist of six separate ATPase domains, but in dynein these domains (AAA1–AAA6) are concatenated in a giant heavy chain of nearly 500 kD rather than separate polypeptides (Fig. 36.14A). Cytoplasmic dynein is a dimer of two heavy chains plus accessory polypeptides that bring the total molecular weight to approximately 1.4 MDa. The N-terminal quarter of the heavy chain forms a tail that interacts with intermediate chains, light intermediate chains, dimers of light chains, and cargo molecules (Fig. 36.14B). A linker domain connects the tail to AAA1. A segment of the dynein heavy chain within AAA4 forms an antiparallel coiled-coil stalk with a small microtubule-binding domain at the tip.

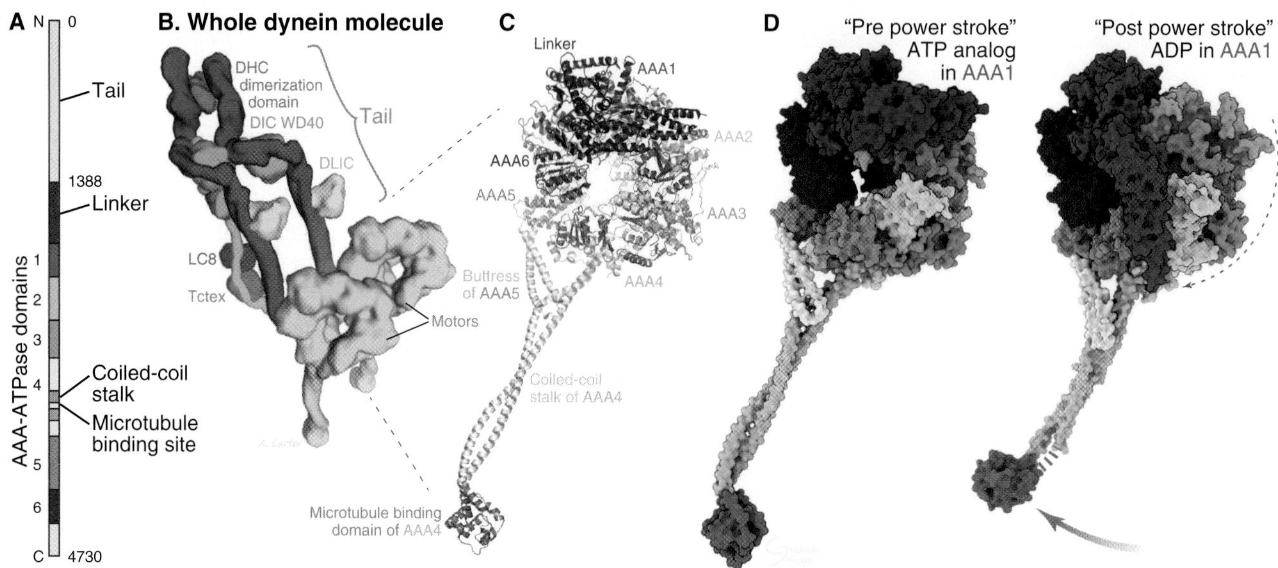

FIGURE 36.14 DYNEIN STRUCTURE. A, Domain organization of a dynein heavy chain showing the six AAA ATPase modules and two sequences that form an antiparallel coiled-coil stalk with an ATP-sensitive microtubule-binding site at the tip. The first AAA domain is the catalytic site. AAA domains 2, 3, and 4 bind ATP, but hydrolysis is not coupled to movement. AAA domains 5 and 6 do not bind ATP. **B,** Model for cytoplasmic dynein-1 based on crystal structures of motor domains and reconstructions of electron micrographs of the dynactin complex. **C,** Ribbon diagram of crystal structures of cytoplasmic dynein-1 motor domains (Apo) without nucleotide bound to AAA1. The linker domain is *purple*. AAA domains are color coded as in **A,** with the stalk and microtubule binding domain extending from AAA3. **D,** Space-filling models of dynein with (ADP-V_o) with ADP and the phosphate analog vanadate bound to AAA1 the "pre power stroke" state and ADP bound to AAA1 the "post power stroke state." The AAA hexamer is more compact with bound ADP-V_o. These two structures differ in the conformations of the AAA hexamer, linker domain and stalk. (For reference, see EMData Bank files 2861 and 2862 and Schmidt H, Zalyte R, Urnavicius L, et al. Structure of human cytoplasmic dynein-2 primed for its power stroke. *Nature.* 2015;518:435–438; and Urnavicius L, Zhang K, Diamant AG, et al. The structure of the dynactin complex and its interaction with dynein. *Science.* 2015;347:1441–1446.)

Dynein Mechanochemistry

Sufficient information is available from enzyme kinetics, crystal structures, and motility assays to construct a mechanochemical cycle for dynein interacting with a microtubule (Fig. 36.15A). The AAA1 domain binds and hydrolyzes ATP during force-producing interactions with microtubules. Full motor function requires ADP or ATP binding to AAA domains 2 to 4, but ATP hydrolysis by these domains is not coupled directly to motility. AAA domains 5 and 6 do not bind nucleotides. The dynein ATPase cycle of AAA1 resembles the actomyosin ATPase mechanism in broad outline.

When AAA1 is free of nucleotide, dynein binds tightly to a microtubule at a site between the α- and β-tubulin subunits with the stalk pointing toward the minus end of the polymer. ATP binding to AAA1 causes compaction of the whole AAA hexamer and produces two important conformational changes (Fig. 36.14). First, the "buttress" on AAA5 communicates the conformational change in the hexamer to the stalk by displacing the stalk helices relative to each other. This, in turn, changes the conformation of the microtubule-binding domain more than 20 nm distant from the ATP binding site and reduces its affinity for the microtubule. Thus, dynein-ATP dissociates from the microtubule. Second, the linker domain bends in the middle, moving to the "pre power stroke state." After ATP hydrolysis, dynein-ADP-P_i also binds weakly to microtubules, but phosphate dissociation during one of the transient interactions with a microtubule reverses both conformational changes produced by ATP. The linker domain straightens out, producing a power stroke that moves the tail and any associated cargo (including the other subunit of a dynein dimer) toward the minus end of the microtubule. In addition, the affinity for microtubules increases, allowing transmission of force from the microtubule to the cargo. Binding to a microtubule also stimulates the rate of ADP dissociation from dynein approximately 10-fold, from approximately 3 s^{-1} to approximately 33 s^{-1}, restarting the ATPase cycle.

Yeast dynein dimers can walk processively toward the minus end of a microtubule in in vitro motility assays. The mechanism likely involves steps by the two motor domains on adjacent protofilaments of the microtubule. The size of the mechanical step associated with each ATP hydrolysis in most often 8 nm, but cytoplasmic dynein can take larger steps up to 24 nm when the load is low. Yeast dynein produces a force of 5 to 7 pN, similar to kinesin.

Dynein Superfamily

Dynein genes are ancient, arising well before the last common eukaryotic ancestor (see Fig. 2.4B), but they were lost multiple times during evolution, so neither red

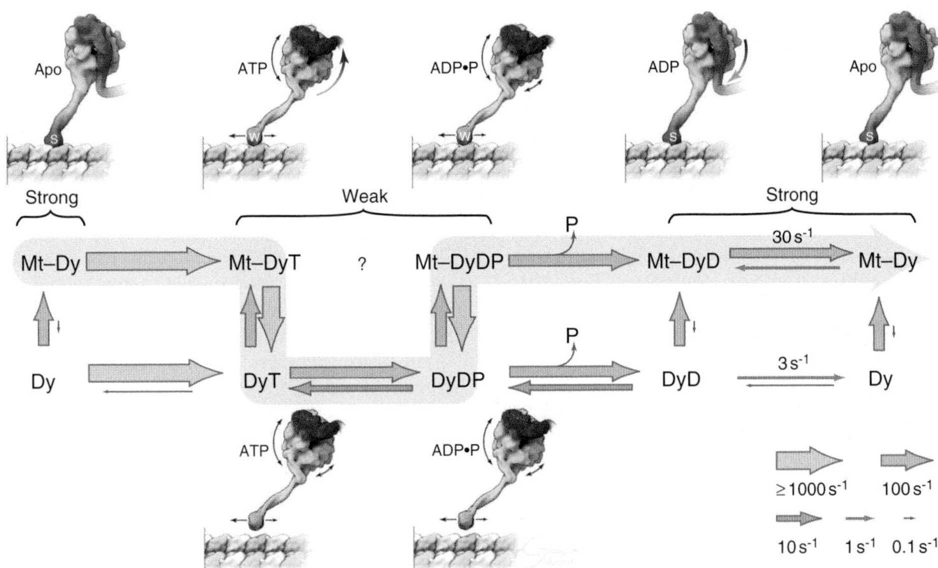

FIGURE 36.15 DYNEIN-MICROTUBULE ATPase MECHANISM. Chemical pathway and structures. *Arrow sizes* are proportional to the rates of the reactions, with second-order reactions adjusted for physiological concentrations of reactants. The *beige* shading shows the main pathway through the reaction. D, ADP; Dy, dynein; Mt, microtubule; P, inorganic phosphate; T, ATP. The drawings are interpretations of the intermediates in the cycle based on crystal structures. (Modified from Cianfrocco MA, DeSantis ME, Leschziner AE, et al. Mechanism and regulation of cytoplasmic dynein. *Annu Rev Cell Dev Biol.* 2015;31:83–108.)

algae nor flowering plants now have dynein genes. Animals have two genes for cytoplasmic dynein heavy chains and multiple isoforms of intermediate, light intermediate, and light chains. Alternative splicing, especially of intermediate chains, further increases the complexity.

Cytoplasmic dyneins are dimeric proteins (Fig. 36.14B). Cytoplasmic dynein-2 moves cargo for intraflagellar transport (see Fig. 38.18). Cytoplasmic dynein-1 has diverse functions around the entire cell cycle. During interphase, dynein-1 transports organelles, RNAs, and some viruses toward the minus ends of microtubules (see Fig. 37.2) generally moving cargo toward the cell center where the centrosome and Golgi apparatus are located. In neurons, dynein-1 moves cargo inside axons toward the cell body (see Fig. 37.3). During mitosis, dynein-1 in the cell cortex and bound to kinetochores of chromosomes applies forces to microtubules and helps to position the mitotic spindle (see Fig. 44.7).

Given these diverse activities, it is not surprising that dynein mutations cause serious phenotypes and contribute to disease. A null mutation in the gene for a mouse cytoplasmic dynein-1 heavy chain leaves the Golgi apparatus dispersed throughout the cytoplasm and is lethal during embryogenesis. A temperature-sensitive mutation in *Caenorhabditis elegans* dynein-1 causes defects in mitosis at the restrictive temperature. Mutations of dynein-1 and associated proteins contribute to human neurodegenerative diseases including some cases of Parkinson disease and spinal muscular atrophy.

Accessory proteins regulate the enzyme activity and movements of cytoplasmic dyneins. Transport by mammalian dynein depends on the **dynactin complex** (see Fig. 37.2), a huge complex of 23 subunits (11 different proteins) with a total molecular weight of approximately 1.2 MDa. Dynactin and an adapter protein not only link cytoplasmic dynein-1 to cargo, but also stimulate its motor activity, perhaps by positioning the two motors in a productive way.

A complex of proteins (**Lis-1** and Nudel/NudE) increases dynein's affinity for microtubules, slows the motor, and may increase force production. Nudel/NudE binds intermediate chains and tethers Lis-1 to dynein. Lis-1, a β-propeller protein, interacts directly with the AAA hexamer and sterically blocks a movement of the linker domain that is required for microtubule release. Genetic experiments suggest that Lis1 is a dynein activator, but how tight binding to microtubules activates dynein remains unclear. Loss-of-function mutations of the *Lis1* genes interfere with development of the cerebral cortex, which lacks gyres and is smooth in affected humans.

Humans have 14 genes for axonemal dyneins, which consist of one to three heavy chains. In axonemes of cilia and flagella, at least seven different dynein isoforms bind to unique sites on the outer doublets (see Fig. 38.14). Calcium and a cyclic adenosine monophosphate (cAMP)–dependent protein kinase (see Fig. 25.3D) regulate dynein in cilia and flagella.

ACKNOWLEDGMENTS

We thank Andrew Carter, Erika Holzbaur, Samara Reck-Peterson, and Lee Sweeney for their suggestions on revisions to this chapter.

SELECTED READINGS

Bhabha G, Johnson GT, Schroeder CM, et al. How dynein moves along microtubules. *Trends Biochem Sci*. 2016;41:94-105.

Bloemink MJ, Geeves MA. Shaking the myosin family tree: biochemical kinetics defines four types of myosin motor. *Semin Cell Dev Biol*. 2011;22:961-967.

Buss F, Spudich G, Kendrick-Jones J. Myosin VI: Cellular functions and motor properties. *Annu Rev Cell Dev Biol*. 2004;20:649-676.

Carter AP, Diamant AG, Urnavicius L. How dynein and dynactin transport cargos: a structural perspective. *Curr Opin Struct Biol*. 2016; 37:62-70.

Cianfrocco MA, DeSantis ME, Leschziner AE, et al. Mechanism and regulation of cytoplasmic dynein. *Annu Rev Cell Dev Biol*. 2015; 31:83-108.

Cross RA, McAinsh A. Prime movers: the mechanochemistry of mitotic kinesins. *Nat Rev Mol Cell Biol*. 2014;15:257-271.

De La Cruz EM, Ostap EM. Relating biochemistry and function in the myosin superfamily. *Curr Opin Cell Biol*. 2004;16:61-67.

Erzberger JP, Berger JM. Evolutionary relationships and structural mechanisms of AAA+ proteins. *Annu Rev Biophys Biomol Struct*. 2006;35:93-114.

Fu MM, Holzbaur EL. Integrated regulation of motor-driven organelle transport by scaffolding proteins. *Trends Cell Biol*. 2014;24: 564-574.

Geeves MA, Holmes KC. Structural mechanism of muscle contraction. *Annu Rev Biochem*. 1999;68:687-728.

Greenberg MJ, Ostap EM. Regulation and control of myosin-I by the motor and light chain-binding domains. *Trends Cell Biol*. 2013; 23:81-89.

Hammer JA 3rd, Sellers JR. Walking to work: roles for class V myosins as cargo transporters. *Nat Rev Mol Cell Biol*. 2011;13:13-26.

Hartman MA, Finan D, Sivaramakrishnan S, et al. Principles of unconventional myosin function and targeting. *Annu Rev Cell Dev Biol*. 2011;27:133-155.

Hartman MA, Spudich JA. The myosin superfamily at a glance. *J Cell Sci*. 2012;125:1627-1632.

King SM. Integrated control of axonemal dynein AAA(+) motors. *J Struct Biol*. 2012;179:222-228.

Kull FJ, Endow SA. Force generation by kinesin and myosin cytoskeletal motor proteins. *J Cell Sci*. 2013;126:9-19.

Milic B, Andreasson JO, Hancock WO, et al. Kinesin processivity is gated by phosphate release. *Proc Natl Acad Sci USA*. 2014;111: 14136-14140.

Odronitz F, Kollmar M. Drawing the tree of eukaryotic life based on the analysis of 2,269 manually annotated myosins from 328 species. *Genome Biol*. 2007;8:R196.

Preller M, Manstein DJ. Myosin structure, allostery, and mechanochemistry. *Structure*. 2013;21:1911-1922.

Roberts AJ, Kon T, Knight PJ, et al. Functions and mechanics of dynein motor proteins. *Nat Rev Mol Cell Biol*. 2013;14:713-726.

Schmidt H, Zalyte R, Urnavicius L, et al. Structure of human cytoplasmic dynein-2 primed for its power stroke. *Nature*. 2015;518:435-438. For video of conformational change see <http://www.nature.com/nature/journal/v518/n7539/fig_tab/nature14023_SV6.html>.

Scholey JM. Kinesin-2: a family of heterotrimeric and homodimeric motors with diverse intracellular transport functions. *Annu Rev Cell Dev Biol*. 2013;29:443-469.

Sun Y, Goldman YE. Lever-arm mechanics of processive myosins. *Biophys J*. 2011;101:1-11.

Tumbarello DA, Kendrick-Jones J, Buss F. Myosin VI and its cargo adaptors-linking endocytosis and autophagy. *J Cell Sci*. 2013;126: 2561-2570.

Walczak CE, Gayek S, Ohi R. Microtubule-depolymerizing kinesins. *Annu Rev Cell Dev Biol*. 2013;29:417-441.

Wickstead B, Gull K, Richards TA. Patterns of kinesin evolution reveal a complex ancestral eukaryote with a multifunctional cytoskeleton. *BMC Evol Biol*. 2010;10:110.

CHAPTER 37

Intracellular Motility

Virtually every component inside living cells moves to some extent (Fig. 37.1), but the magnitude and velocity of these movements vary by orders of magnitude (Table 37.1). At one extreme, the bulk cytoplasm of algae and giant amoebas streams tens of micrometers per second, but most cytoplasm is generally less dynamic. The cytoplasmic network of cytoskeletal polymers has a pore size of less than 50 nm (see Fig. 1.13). This allows small molecules and macromolecules to diffuse essentially unimpaired. Particles larger than the pores must be transported actively. For example, lysosomes, mitochondria, secretory vesicles, and endosomes all move actively in cytoplasm, frequently between the centrosome and the cell periphery. Similarly, messenger RNA (mRNA) moves from its site of synthesis in the nucleus through nuclear pores into the cytoplasm and then may be carried actively to specific parts of the cell. Intracellular pathogenic bacteria and viruses take advantage of the host cell's actin system to propel themselves through the cytoplasm. Virus particles can also move along microtubules.

Two ancient mechanisms (Fig. 37.1C) account for most intracellular movements in eukaryotes. Transport along microtubules by kinesin or dynein predominates in animal cells. Transport along actin filaments by myosin is more important in plants and fungi. Specialized isoforms of myosin, kinesin, and dynein are dedicated to particular movements. Experiments with drugs that depolymerize or stabilize actin filaments or microtubules (see Boxes 33.1 and 34.1) originally identified the cytoskeletal polymers that support various biological movements. Discovering the motors was more challenging, given multiple genes for myosin, kinesin, and dynein in most organisms and a limited choice of pharmacologic agents (Box 37.1). Even loss of function mutations and depletion by RNA interference (RNAi; see Fig. 11.12) have limitations, because some motors are essential for viability while others contribute redundantly to transport. For example, dynein contributes to mitosis, but *Drosophila* tissue culture cells depleted of dynein can complete mitosis, because other motors take over allowing mitosis to proceed, albeit more slowly.

The diversity of cargo, motors, and tracks poses a traffic control challenge, which cells manage by specifying the organization of microtubules and actin filaments, matching appropriate motors to specific cargo and regulating motor activity. Most cargos associate with more than one type of motor, so their activities must be coordinated. For example, some cargos with two motors reverse their direction either along the way or after reaching their destination. Local or global signaling pathways influence many of these choices, allowing cells to adjust transport to adapt to environmental conditions.

This chapter takes a broad view across biology, highlighting the wide range of mechanisms that produce intracellular transport. In most cases, transport involves single organelles on an actin filament or microtubule, but organelle transport can result in bulk movement of the cytoplasm. The chapter also considers mechanisms that produce special types of intracellular movements: cytoplasmic contractions generated by myosin and actin filaments that propel cytoplasmic streaming; and polymerization and depolymerization of microtubules and actin filaments. Chapters on membrane traffic (see Chapters 20 to 22) and mitosis (see Chapter 44) cover more examples of intracellular movements.

Rapid Intracellular Movements on Microtubules

Organelles in most eukaryotic cells can move along linear microtubule tracks at relatively high velocities, on the order of 1 $\mu m\ s^{-1}$ (Table 37.1) at least part of the time. The physical organization of microtubules determines the patterns of these movements (Fig. 37.1; also see Fig. 34.2). In cells with a radial arrangement of microtubules (Fig. 37.1) organelles and other cargos associated with kinesin-1 move toward microtubule plus ends located at the periphery of cells. Cargo associated with

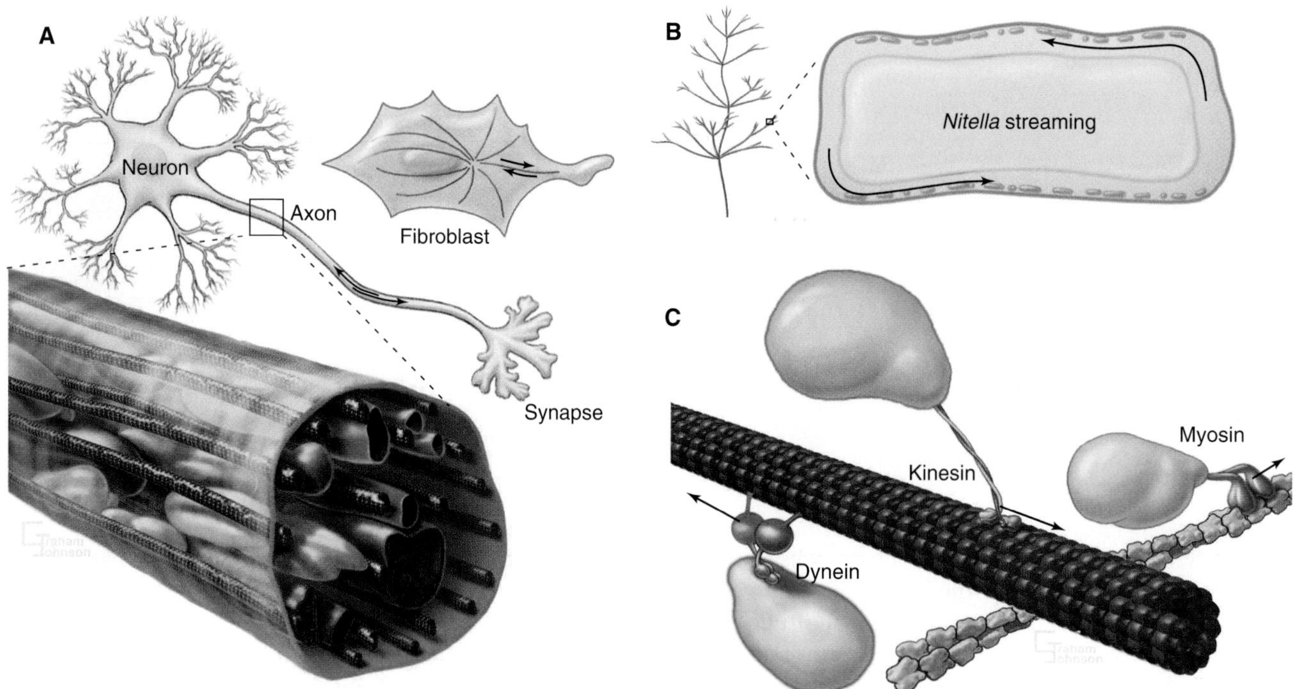

FIGURE 37.1 MECHANISMS OF INTRACELLULAR MOVEMENT. A, Fibroblasts and neurons move organelles bidirectionally along microtubules *(red)*. **B,** The green alga *Nitella* moves cytoplasmic organelles along bundles of actin filaments located in the cell cortex. **C,** Microtubule-based and actin filament–based motors.

TABLE 37.1 Velocities of Intracellular Movements

System	Velocity (μm s^{-1})	Mechanism
Microtubule Motors		
Anterograde fast axonal transport, squid	1	Individual kinesin motors
Retrograde fast axonal transport, squid	2	Individual dynein motors
Chromosome movement in anaphase of mitosis	0.003–0.2	Motors plus depolymerization
Endoplasmic reticulum sliding, newt cell	0.1	Individual kinesin motors
Slow axonal transport, rat nerves	0.002–0.1 net 1 (intermittent)	Motors on microtubules
Microtubule Polymerization		
Endoplasmic reticulum tip elongation, Newt cell	0.1	Microtubule polymerization
Actin-Myosin Motors		
Cytoplasmic streaming, *Nitella*	60	Myosin motors on tracks
Cytoplasmic streaming, *Physarum*	500	Actin–myosin contraction
Actin Polymerization		
Actin-propelled comet, *Listeria*	0.5	Actin polymerization

dynein moves in the opposite direction. These motors are processive (see Chapter 36), so they can work alone or in small numbers. Intraflagellar transport of proteins in cilia and flagella (see Fig. 38.18) shares many features with the movements of organelles.

The chemistry of individual microtubules subtly influences the delivery of cargo to specific locations, as some motors favor microtubules composed of particular tubulin isoforms or with posttranslational modifications. For example, kinesin-1 favors acetylated microtubules, and CENP-E favors tyrosinated microtubules in the center of the mitotic spindle. In addition, microtubule-associated proteins, such as tau, can influence the choice of tracks by reducing the run lengths of motors under some circumstances.

Matching Microtubule Motors With Cargo

Pairing motors with appropriate cargo and regulation of motor activity control the traffic along microtubules. Coupling can be as simple as kinesin binding to a

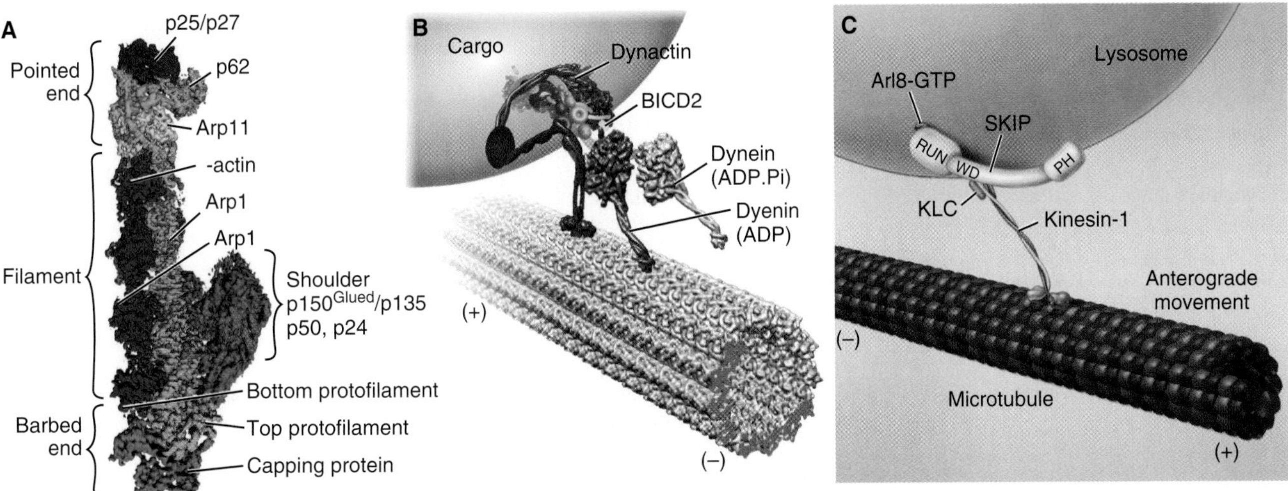

FIGURE 37.2 ATTACHMENT OF CYTOPLASMIC MICROTUBULE MOTORS TO MEMBRANES. A, Ribbon diagram of the structure of the core of the dynactin complex determined by electron cryomicroscopy. The short filament of Arp1 is capped by capping protein on the barbed end and by Arp11 on the pointed end. The shoulder is formed by part of the p150 molecule, most of which was disordered in these samples. **B,** Drawing showing dynein linked to a vesicle by dynactin complex and moving toward the minus end of a microtubule. **C,** Coupling of kinesin-1 to a lysosome. Guanosine triphosphatase (GTPase) Arl8 on the lysosome membrane engages the adapter protein SKIP (SifA-kinesin interacting protein), which binds to the kinesin light chain as shown in Fig. 36.9E. (**A,** From Urnavicius L, Zhang K, Diamant AG, et al. The structure of the dynactin complex and its interaction with dynein. *Science.* 2015;347:1441–1446. **B,** From Cianfrocco MA, DeSantis ME, Leschziner AE, Reck-Peterson SL. Mechanism and regulation of cytoplasmic dynein. *Annu Rev Cell Dev Biol.* 2015;31:83–108. **C,** Based on a drawing from Rosa-Ferreira C, Munro S. Arl8 and SKIP act together to link lysosomes to kinesin-1. *Dev Cell.* 2011;21:1171–1178.)

BOX 37.1 Tools for Studying Motor Proteins

A small compound named monastrol inhibits kinesin-5, resulting in monopolar mitotic spindles that fail to segregate the chromosomes. Higher-affinity inhibitors of kinesin-5 are being tested for cancer therapy. The small molecule blebbistatin inhibits cytoplasmic and skeletal muscle myosin-II (but not smooth muscle myosin-II) and blocks cytokinesis. Vanadate and ultraviolet light can inactivate dynein. Vanadate binds to the γ-phosphate site of dynein-adenosine diphosphate (ADP), but it binds similarly to other adenosine triphosphatases (ATPases), so it is not specific. However, ultraviolet light has a novel effect on the dynein-ADP-vanadate complex: It cleaves and inactivates the dynein heavy chain.

transmembrane protein on a cargo vesicle, but usually involves adapter proteins that recognize a molecule on the surface of the cargo and one or more motors. Dynein associates with the **dynactin complex** (Fig. 37.2B). About a dozen other known adapter proteins also interact with the dynactin complex, kinesins or both to direct traffic to specific locations (Table 37.2). Local conditions in the cell regulate both coupling of motors to cargo and the activity of the motors. Other proteins tether some organelles to immobilize them and prevent their transport.

The 1 mDa dynactin complex consists of a short filament of actin-related protein Arp1 capped on its barbed end by capping protein and on its pointed end by the most divergent actin-related protein, Arp11 (Fig. 37.2A). Other subunits at the pointed end serve as cargo adaptors. Association of the dimeric tail of dynein with the dynactin complex and adapter proteins activates its motor activity. The 150-kD dynactin subunit ($p150^{Glued}$) includes a long extension that links the Arp1 filament to both an intermediate chain of dynein and a microtubule. The two dynein motor domains take steps of variable size as they pull the cargo along the microtubule. Mutations of $p150^{Glued}$ cause developmental defects in the eye and brain of *Drosophila* plus some forms of Parkinson disease and motor neuron degeneration in humans.

Most proteins known to link organelles and ribonucleoprotein particles to microtubule motors interact with both dynein and kinesins, but some are specific for one or the other (Table 37.2). These interactions often activate or inhibit motor activity, coupling physical association with transport. For example, some cargo adapters overcome autoinhibitory interactions between the heads and tails of kinesins-1, -2, and -3 by binding to the kinesin tail or light chains.

Other than their linker and regulatory activities, adapter proteins have little in common, so they likely had independent evolutionary origins. The following sample illustrates the diversity of these adapters.

Motors and adapter proteins are targeted to cargo membranes by integral membrane proteins, peripheral proteins or lipids. Kinesin binds directly to a few transmembrane proteins, so no adapter is required. For example, calsyntenin and alcalpha on certain neuronal vesicles have short sequences with acidic residues flanking a key tryptophan that bind kinesin light chains

TABLE 37.2 Adapters for Microtubule Motors

Adapter Protein	Links to Motors	Links to Cargos	Cargo	Destination	Regulation	Disease Connection
Dynein Only						
Dynactin complex	Dynein IC, p150^{Glued}, other adapters listed here		Endosomes, mitochondria, Golgi membranes			Mutations cause neuron degeneration
RILP (Rab7-interacting lysosomal protein)	Dynactin	Rab7, spectrin	Late endosomes	*Trans*-Golgi network	Rab7 GTPase	
SNX6 (sorting nexin 6)	Dynactin	Other retromer proteins	Endosome vesicles with recycling receptors	*Trans*-Golgi network	PI4P on *trans*-Golgi membranes releases dynactin	
Kinesin Only						
DENN/MADD (Rab3 guanine nucleotide exchange factor)	Kinesin-3		Synaptic vesicles with Rab3	Synapse	Rab3 GTPase	
Mint1 (Munc18-interacting protein)	Kinesin-2 tail		Vesicles with NMDA glutamate receptors	Synapse	Phosphorylation by CaMKII	
SKIP (SifA-kinesin interacting protein)	Kinesin-1 LC	Arl8 GTP → lysosome	Lysosomes	Periphery	Arl8 GTPase	
Dynein and Kinesin						
Hook	Dynein, kinesin-3		Early endosomes	Bidirectional		
Huntingtin	Dynein IC, HAP1 → dynactin, kinesin-1 HC and LC	HAP40 → Rab5, Optineurin → myosin-VI	Endosomes, lysosomes, autophagosomes, vesicles with APP or GABA receptors, mRNA	Bidirectional	Phosphorylation of Huntingtin and motors	Traffic defects and massive neuronal degeneration in Huntington disease
JIP1 (JNK interacting protein)	Either dynactin or kinesin-1 stalk, tail & LC with aid of FEZ1	JNK, vesicles with amyloid precursor protein	Synaptic vesicles, autophagosomes, mitochondria, vesicles with APP	Bidirectional	Phosphorylation by JNK favors kinesin-1 binding and transport to plus end	
JIP3/JIP4 (JNK interacting protein)	Dynactin, kinesin-1 tail and LC, JIP1	JNK, Arf6	Synaptic vesicles to synapse, endosomes with nerve growth factor to cell body		Phosphorylation of DIC	
La	Dynein IC, kinesin-1 HC	RNPs	Ribonucleoprotein particles	Bidirectional	SUMOylation may reverse direction at periphery	
Milton/TRAK	Dynactin, kinesin-1 HC	Miro/RhoT2 transmembrane GTPase	Mitochondria	Bidirectional	Local concentrations of ATP and Ca^{2+}	Parkinson disease (?)

Data from Fu MM, Holzbaur EL: Integrated regulation of motor-driven organelle transport by scaffolding proteins. *Trends Cell Biol* 2014;24:564–574.
APP, amyloid precursor protein; ATP, adenosine triphosphate; CaMKII, calcium/calmodulin-dependent serine protein kinase; FEZ1, fasciculation and elongation protein zeta 1; GABA, γ-aminobutyric acid; GTP, guanosine triphosphate; GTPase, guanosine triphosphatase; HC, heavy chain; IC, intermediate chain; JNK, c-Jun N-terminal kinase; LC, light chain; mRNA, messenger RNA; NMDA, *N*-methyl-D-aspartate; RNP, ribonucleoprotein; PI4P, phosphatidylinositol 4-phosphate.

(see Fig. 36.9E). In other cases an adapter protein links a transmembrane protein to a motor. For instance, a transmembrane protein called Miro on the surface of mitochondria interacts with adapter protein Milton/TRAK1, which binds both the kinesin-1 heavy chain and dynactin for bidirectional transport in axons.

Guanosine triphosphatases (GTPases) on organelle membranes interact with other adapter proteins. For instance, different GTPases target kinesin and dynein to lysosomes. Lysosomes with Arl8-GTP (guanosine triphosphate) bind the adapter SKIP (SifA-kinesin interacting protein). An acidic/tryptophan sequence in SKIP interacts with kinesin light chains (Fig. 37.2C) during transport to the periphery. This interaction with SKIP also turns on the kinesin motor by overcoming the autoinhibitory interaction of the motor domains with the tail. Similarly, vesicles containing transmembrane receptors (mannose-6-phosphate receptor, for example) bud from late endosomes with the GTPase Rab7 on their surface. Rab7-GTP binds the adapter protein RILP (Rab7-interacting lysosomal protein), which anchors dynactin and dynein for vesicle transport to the *trans*-Golgi network (see Fig. 22.12). Phosphatidylinositol 4-phosphate (see Fig. 26.7) on the Golgi membranes binds sorting nexin 6 (part of the retromer complex; see Fig. 22.12) and releases dynactin and dynein.

Many cargo particles associate with both a kinesin and dynein. In this case, regulatory mechanisms responsive to local conditions must coordinate their activities to achieve net transport. One example is the adapter protein La, which links ribonucleoprotein particles to both dynein and kinesin. After being carried to the cell periphery by kinesin, covalent modification of La with a SUMO protein inactivates kinesin, allowing for reverse transport back to the cell center by dynein. Phosphorylation regulates the adapter protein JIP1 (JNK interacting protein-1), which binds either kinesin or dynein. Phosphorylated JIP1 binds and activates kinesin, favoring anterograde transport. Mitochondria move both directions in axons, but some stop moving in regions with limited adenosine triphosphate (ATP). A local high concentration of cytoplasmic calcium binds to EF-hands (Ca^{2+} binding site in calmodulin consisting of α-helices E and F) of the anchor protein Miro. Calcium binding turns off both kinesin and dynein, keeping mitochondria in place until ATP concentrations return to normal. Another protein anchors stationary mitochondria to microtubules.

Faulty axonal transport resulting in clumps of vesicles is observed in human degenerative diseases of the nervous system including Alzheimer and Parkinson diseases, but cause-and-effect relationships are still under investigation. One intriguing part of this story is that the transmembrane amyloid precursor protein not only binds directly to kinesin-1 light chain, but also is cleaved in Alzheimer disease to produce amyloid-β peptide—a toxic peptide that is implicated in neuronal death. Huntingtin, the giant (350 kD) adapter protein mutated in Huntington disease, normally participates in bidirectional axonal transport of multiple cargos (Table 37.2). Other membrane-associated scaffold proteins, including, JIP-1 and JIP-3, bind kinases from the mitogen-activated protein (MAP) kinase cascade (see Fig. 27.6) in addition to kinesin-1 light chains.

Fast Axonal Transport

Analysis of microtubule-based movements is particularly favorable in axons of neurons, because axons are long (up to 1 m) but narrow, the microtubules have a uniform polarity, and organelles move at steady rates in both directions. Furthermore, nerve cells contain high concentrations of microtubules and microtubule motors; indeed, cytoplasmic tubulin, cytoplasmic dynein, and kinesin were all originally isolated from brain.

High-contrast light microscopy of living axons reveals that most membrane-bound organelles move either toward (anterograde) or away from (retrograde) the end of the axon (Fig. 37.3) with some pauses and even

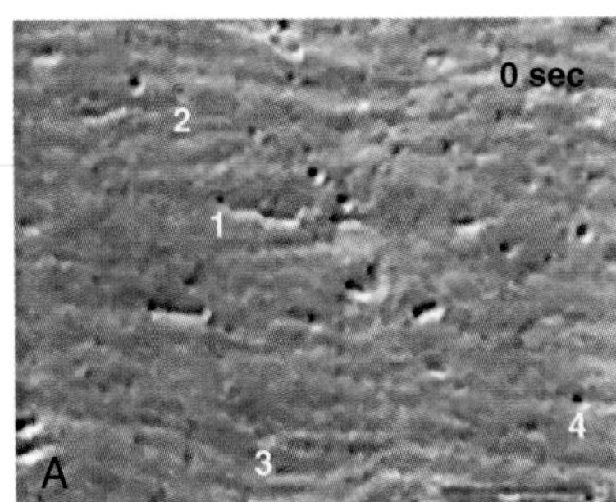

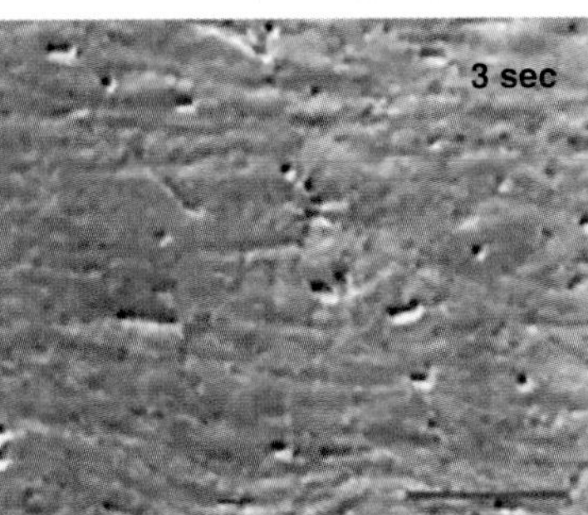

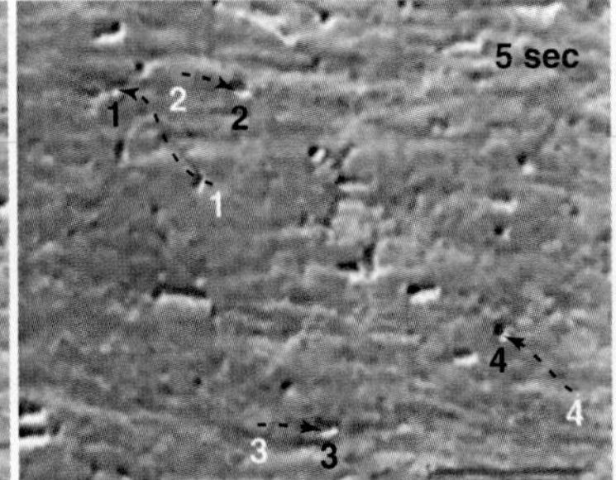

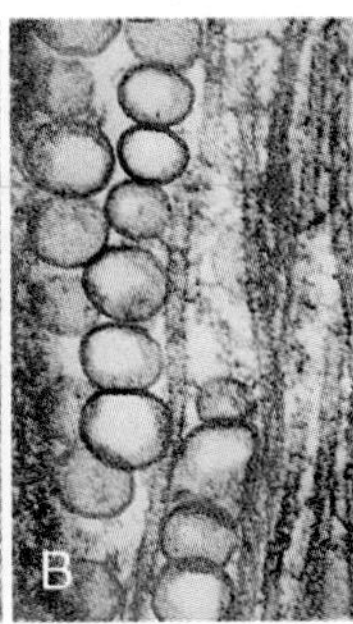

FIGURE 37.3 FAST TRANSPORT IN CYTOPLASM ISOLATED FROM SQUID GIANT AXONS. A, Three frames from a series of video-enhanced differential interference contrast micrographs show movement of organelles in both the anterograde (rightwards) and retrograde (leftwards) directions. Four large organelles are marked with numbers and colored *green* at zero time, *blue* at 3 seconds, and *red* at 5 seconds. Movement *(arrows in right panel)* is from the *white* to the *black number.* The original video record shows hundreds of smaller organelles moving steadily in either an anterograde or a retrograde direction at 1 to 2 μm/s. **B,** Electron micrograph of a thin section showing vesicles associated with microtubules in axoplasm. (**A,** Courtesy S. Brady, University of Texas Southwestern Medical School, Dallas. **B,** Courtesy R.H. Miller, Case Western Reserve Medical School, Cleveland, OH.)

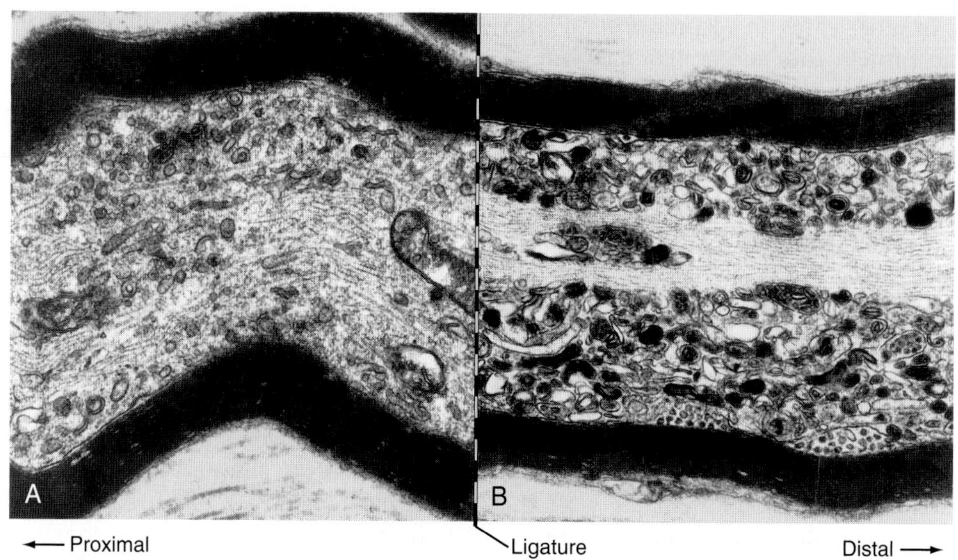

FIGURE 37.4 ELECTRON MICROGRAPHS SHOWING THE RESULT OF NERVE LIGATION. These are longitudinal sections of axons surrounded by a darkly stained myelin sheath. **A,** The cytoplasm proximal to the ligation demonstrates the accumulation of vesicles and mitochondria, which were being transported toward the nerve terminal to the right. **B,** The cytoplasm distal to the ligation shows the accumulation of lysosomes, multivesicular bodies, and mitochondria, which were being transported toward the cell body to the left. (**B,** From Hirokawa N, Sato-Yoshitake R, Yoshida T, et al. Brain dynein (MAP1C) localizes on both anterogradely and retrogradely transported membranous organelles in vivo. *J Cell Biol.* 1990;111:1027–1037.)

occasional changes of direction. **Retrograde movements** (2.5 $\mu m\ s^{-1}$ or 22 cm/day) are faster than **anterograde movements** (0.5 $\mu m\ s^{-1}$ or 4 cm/day). These rates are remarkable given the densely packed microtubules, intermediate filaments, and vesicles in the cytoplasm (Fig. 37.4; see also Fig. 35.9). At these rates, a round trip from a nerve cell body in the spinal cord of a human to the foot and back takes only 3 weeks. If a 0.1-μm vesicle were the size of a small car, it would move anterogradely at 50 miles per hour and retrogradely at 250 miles per hour. In the axons of vertebrate neurons, mitochondria move back and forth in both directions. Their net movement toward the nerve terminal or cell body depends on physiological conditions.

Biochemical reconstitution showed that the plus-end motors of the kinesin family are responsible for anterograde movements toward the nerve terminal and the minus-end motor dynein is responsible for movement in the retrograde direction. Accordingly, kinesin mutations in flies result in paralysis of the back half of larvae, because transport fails in the longest axons. Mutations in three different kinesin genes also cause human nerve degeneration.

Classic nerve ligation experiments revealed the cargo carried in each direction by fast transport (Fig. 37.4). Different organelles pile up on either side of a mechanical constriction that blocks transport. Small, round, and tubular vesicles, including components of synaptic vesicles (see Fig. 17.8), accumulate on the side near the cell body. Fast anterograde transport carries these cargoes from the cell body toward the end of the axon, where they enter the cycle of synaptic vesicle turnover (see Figs. 17.8 and 17.9). Endosomes and multivesicular bodies moving by fast retrograde transport pile up on the distal side of the constriction. Autophagic vesicles also move primarily in the retrograde direction. Retrograde transport can also move signals from nerve terminals to the cell body. For example, kinesin carries vesicles with the nerve growth factor TrkA receptor tyrosine kinase (see Fig. 24.4) to nerve terminals where it joins the plasma membrane. When activated by nerve growth factor, TrkA is taken up by endocytosis for transport by dynein back to the perinuclear region. Once the vesicle reaches the cell body TrkA activates the MAP kinase pathway to regulate cell growth (see Fig. 27.6). Herpes virus and rabies virus also use dynein for long-distance transport on microtubules from the terminals of sensory nerves to the cell body, where viral DNA enters the nucleus for replication.

Many questions remain regarding the control of microtubule motors during fast transport, including how kinesins remain active and dynein remains inactive during the long trip out to the nerve terminal, how cytoplasmic dynein on retrograde cargo is activated locally at the nerve terminal and kept active during movement to the cell body, and how defects in fast transport may contribute to neurodegenerative diseases.

Slow Transport of Cytoskeletal Polymers and Associated Proteins in Axons

Many neuronal proteins move slowly from their site of synthesis in the cell body toward the ends of axons and dendrites. This transport is essential, as most protein

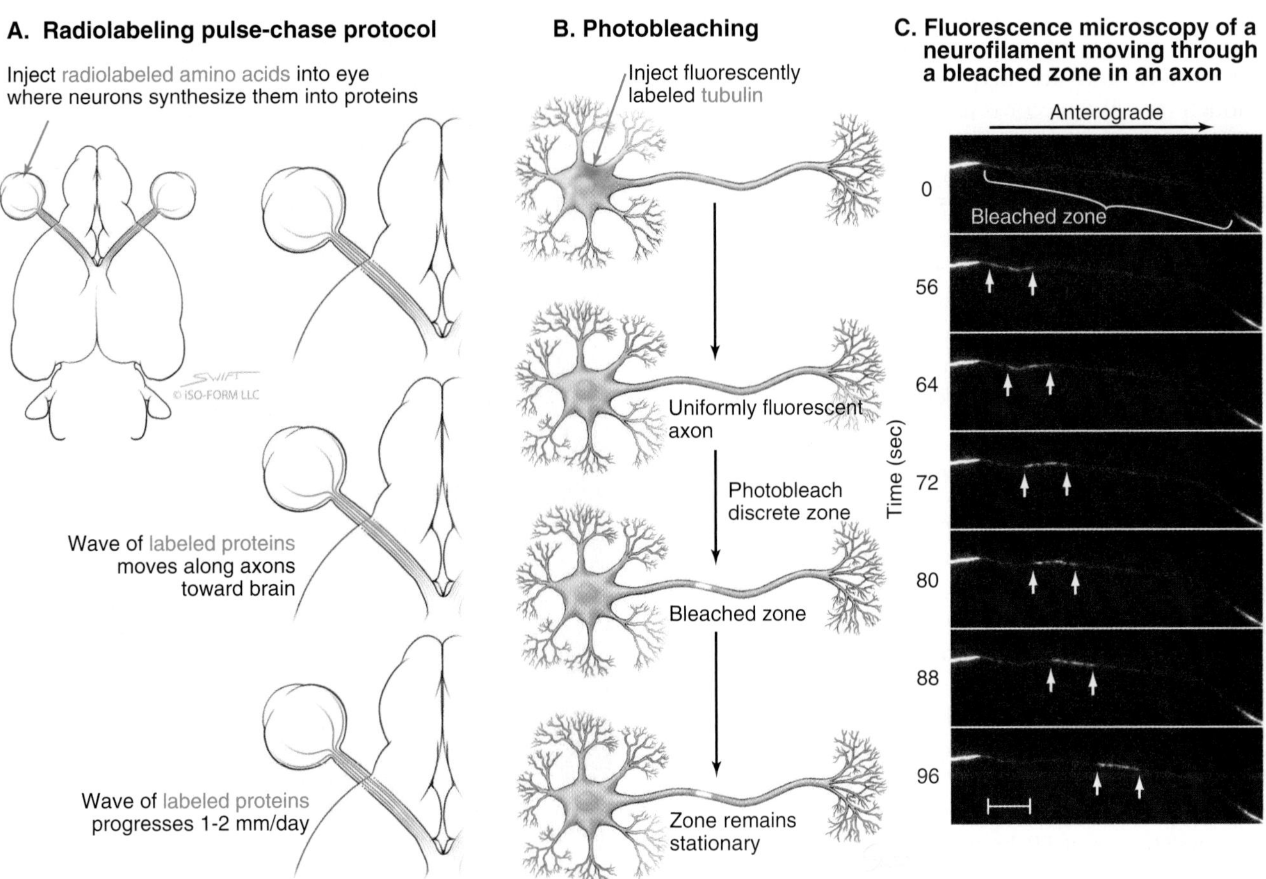

FIGURE 37.5 EXPERIMENTS ON SLOW AXONAL TRANSPORT. A, Pulse-chase experiment. Radioactive amino acids are injected into the eye of an experimental animal. In the nerve cell body, radioactive tracer is incorporated into proteins, which are transported along the axon. Some proteins are incorporated into stationary structures and are left along the way. **B,** Photobleaching experiment. A cultured nerve cell is injected with tubulin labeled with a fluorescent dye. Tubulin fills the cytoplasm and axon as it grows out. A section of the axon is then bleached with a strong pulse of light. This bleached zone is stationary over a period of minutes. **C,** Fluorescence micrographs of the axon of a cultured rat neuron showing rapid transport of a short neurofilament labeled with subunits fused to green fluorescence protein (GFP). Note the photobleached region *(bracket)* and the ends of the moving neurofilament *(arrows).* The neurofilament moves rapidly into the bleached region, but the bleached region does not move because most of the neurofilaments are stationary. Scale bar is 5 μm. (**A–B,** Modified from Cleveland DW, Hoffman PN. Slow axonal transport models come full circle. *Cell.* 1991;67:453–456. **C,** From Wang L, Brown A. Rapid intermittent movement of axonal neurofilaments observed by fluorescence photobleaching. *Mol Biol Cell.* 2001;12:3257–3267, 2001.)

synthesis occurs in the cell body, whereas more than 99% of cell volume can be in axons and dendrites. If nerve cells were smaller, shorter lived or less asymmetrical, we might not even notice such slow movements. Labeling proteins with radioactive amino acids during their synthesis in the cell body allows for tracking their movements along axons (Fig. 37.5A). Proteins that are moved by slow axonal transport are classified into two groups based on their velocities. Tubulin, intermediate filament proteins, and spectrin, which compose the "slow component-a," move exceedingly slowly, approximately 0.1 to 1.0 mm per day (or 1 to 10 nm s^{-1}). In a human, these molecules take more than 3 months to travel from their site of synthesis in the spinal cord to the foot. "Slow component-b" moves approximately 10 times faster and includes 10 times more protein than slow component-a. It is a heterogeneous mixture of many proteins, including clathrin, glycolytic enzymes, and actin.

Defining the mechanism of slow transport was challenging because various experimental approaches yielded apparently conflicting results. Radioactive labeling showed that the moving proteins are spread out and diluted as they move away from the cell body (Fig. 37.5A). Photobleaching of fluorescent tubulin and actin in axons of cultured neurons demonstrated that the bulk of those cytoskeletal polymers is stationary (Fig. 37.5B).

This puzzle was solved for slow component-a by imaging single fluorescent intermediate filaments in axons of live nerve cells. These filaments are stationary most of the time (up to 99%), but occasionally, they move rapidly (0.2 to 2 μm s^{-1}) for up to 20 μm. Most of these intermittent movements are away from the cell body, accounting for the net anterograde movement. These rapid but intermittent movements depend on microtubules and are presumably driven by kinesin, although how the motor is linked to the filaments is not known. A few intermediate filaments move in the retrograde direction explaining why waves spread in pulse-chase experiments (Fig. 37.5A). The movements of whole microtubules are similar to those of intermediate filaments. Thus, what appears to be slow bulk transport is caused by fast but intermittent movement of cargos. Slow component-b proteins seem move by alternatively hitchhiking transiently on some moving structures and diffusing locally between movements.

Other Microtubule-Dependent Movements

Other cells use the same molecular mechanisms as neurons to move organelles in the cytoplasm. Secretory vesicles use plus-end motors to move from the Golgi apparatus to the plasma membrane (see Fig. 21.1). Endosomes use dynein to move from the plasma membrane toward the centrosome.

The intracellular distributions of the endoplasmic reticulum (ER) depend on microtubules. Strands of the ER align with microtubules in cultured cells (Fig. 37.6). This codistribution is achieved in two ways: (a) Motors transport strands of ER bidirectionally on microtubules, and (b) other strands of the ER attach to the plus end of microtubules and ride the microtubule tip as it grows and shrinks during dynamic instability. The +TIP EB1 (see Fig. 34.8) interacts with transmembrane protein STIM1 to couple the ER to growing microtubules. STIM1 has another role in maintaining a supply of Ca^{2+} inside the ER (see Fig. 26.12). This is the best example of movement of an organelle driven by forces generated directly by microtubule assembly.

Microtubules and motors also distribute other organelles inside cells. The concentration of the Golgi apparatus near the centrosome depends on microtubules, because dynein motors transport Golgi vesicles toward the minus ends of the microtubules nucleated at centrosomes (see Fig. 21.20). Mitochondria move bidirectionally on microtubules in animal cells but depend on actin filaments in yeast. Thus, not only the dynamics of the organelles but also the overall organization of a cell depend on microtubules and associated motors. As a result, cellular architecture is determined actively, not passively.

The intracellular distribution of nucleoprotein complexes also depends on active movements. The most obvious example is the movement of chromosomes during mitosis, which depends on microtubule assembly and microtubule motors (see Fig. 44.7). Microtubules and motors can also produce asymmetrical distributions of mRNAs in cells. The adapter protein La links kinesins and dynein to specific ribonucleoprotein particles for

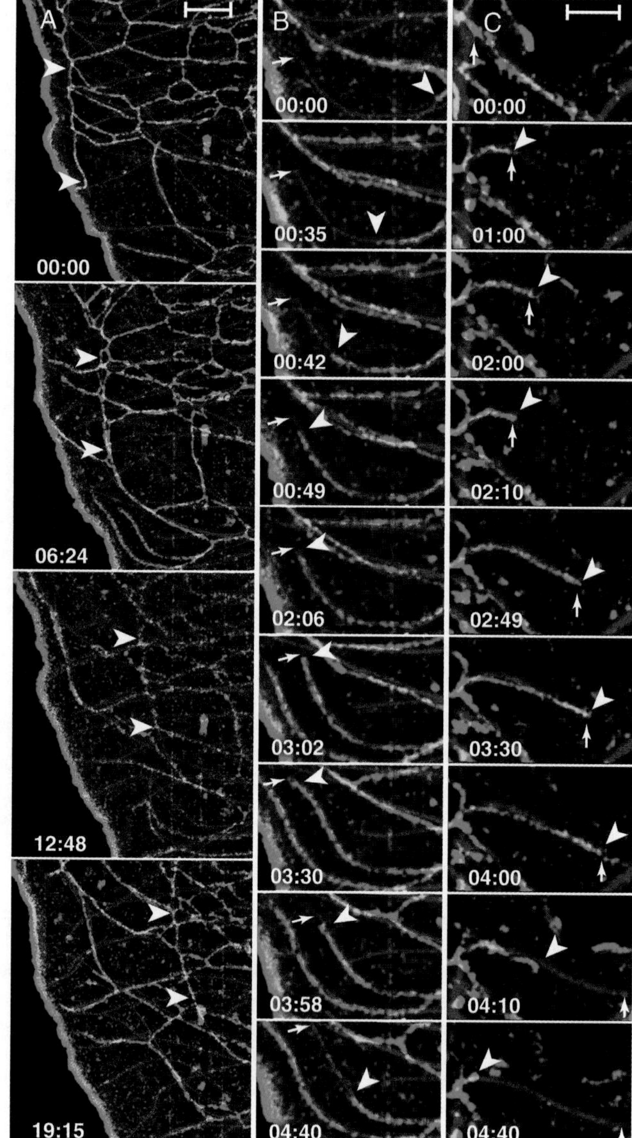

FIGURE 37.6 TWO MODES OF MICROTUBULE-DEPENDENT MOVEMENT OF THE ENDOPLASMIC RETICULUM IN A NEWT EPITHELIAL CELL. The cell was microinjected with rhodamine-labeled tubulin, which incorporates into microtubules, and a lipophilic green-fluorescent dye ($DiOC_6$ [3,3′-dihexyloxacarbocyanine iodide]), which labels endoplasmic reticulum (ER). Time series are indicated in minutes and seconds. Scale bar for all panels is 5 μM. **A,** This column of fluorescence micrographs illustrates the dynamics of microtubules *(red)* and ER *(green),* over a period of 19 minutes. Note the strand of ER moving away from the leading edge *(arrowheads).* **B,** Time course of the movement of a strand of ER toward the end of a microtubule, followed by retraction. This type of movement is thought to be driven by a kinesin motor attached to the tip of the elongating membrane *(arrowhead).* **C,** Time course of the movement of a strand of ER attached to the tip of a growing microtubule *(arrowhead),* followed by retraction of the membrane along the microtubule. (Courtesy C. Waterman-Storer and E.D. Salmon, University of North Carolina, Chapel Hill. For reference, see Waterman-Storer C, Salmon ED. Endoplasmic reticulum membrane tubules. *Curr Biol.* 1998;8:798–806.)

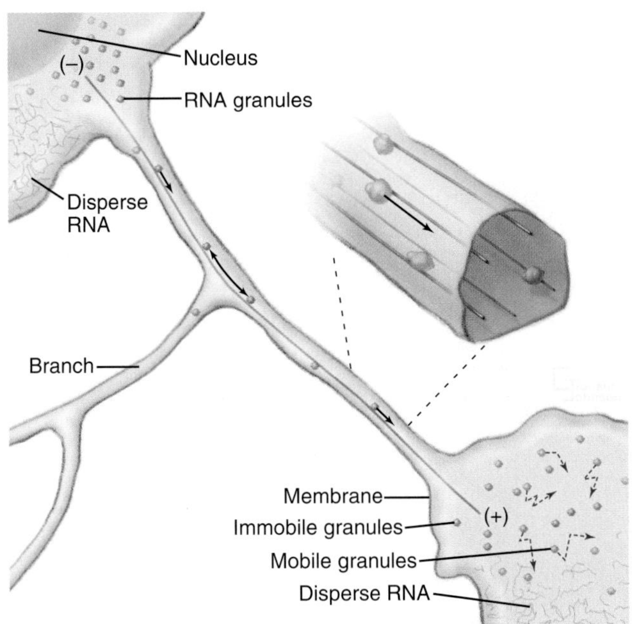

FIGURE 37.7 TRANSPORT OF MESSENGER RNA (mRNA) FOR MYELIN BASIC PROTEIN IN A CULTURED OLIGODENDROCYTE, A GLIAL CELL ISOLATED FROM BRAIN. mRNA synthesized in the cell body (or, in this case, labeled with a fluorescent dye and microinjected into the cell body) is packaged with proteins in a ribonucleoprotein particle, transported from the cell center along microtubules at a steady rate of 0.2 μm s^{-1}, and released at the periphery, where it moves randomly at 1 μm/s. (Modified from Ainger K, Avossa D, Morgan F, et al. Transport and localization of exogenous myelin basic protein mRNA microinjected into oligodendrocytes. *J Cell Biol.* 1993;123:431–441.)

long distance transport to sites of local protein synthesis (Fig. 37.7). For example, localization of certain mRNAs in fly oocytes helps establish the polarity of the embryos.

Intracellular Movements Driven by Microtubule Polymerization

Microtubule polymerization and depolymerization have long been known to play a central role in the assembly of the mitotic apparatus and the movement of chromosomes (see Fig. 44.7), as well as the establishment of cellular asymmetry (see Fig. 34.2). Polymerizing microtubules can exert substantial forces, but strong forces buckle microtubules longer than approximately 10 μm. Consequently, microtubule pushing mechanisms work best over short distances, such as positioning the ER (Fig. 37.6), the nucleus in fission yeast cells (see Fig. 6.3D) and the mitotic spindle in budding yeast cells (see Fig. 34.2D).

Remarkably, the end of a *depolymerizing* microtubule can also *pull* on attached cargo. An in vitro proof-of-principle experiment (Fig. 37.8) showed that chromosomes could ride on the end of a depolymerizing microtubule, even in the absence of ATP or GTP. Linker proteins associated with the kinetochore region of the chromosome make multiple weak bonds with the walls of the microtubule end (see Figs. 8.21 and 8.22). These interactions rearrange rapidly enough to maintain attachment, even as tubulin subunits dissociate from the end.

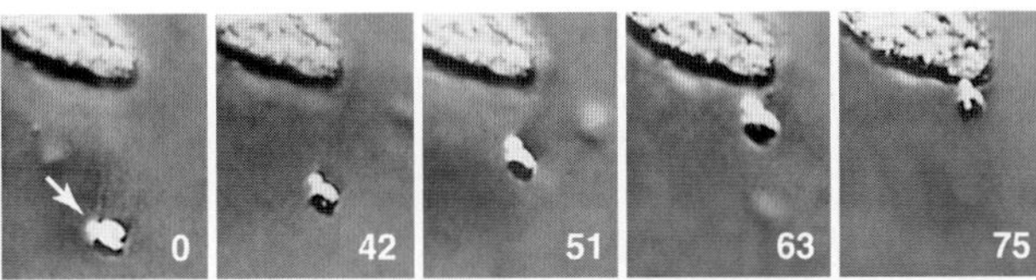

FIGURE 37.8 TRANSPORT OF AN ISOLATED CHROMOSOME ON A SHORTENING MICROTUBULE IN VITRO. A microtubule was grown from brain tubulin nucleated by a basal body in the extracted carcass of a ciliate, *Tetrahymena.* A chromosome *(arrow)* was added as a test cargo and captured by the end of the microtubule. When the concentration of tubulin was reduced, the microtubule shortened, carrying along the chromosome attached to its tip. This transport occurs in the absence of adenosine triphosphate (ATP) or guanosine triphosphate (GTP). (Courtesy J.R. McIntosh, University of Colorado, Boulder.)

Bulk Movement of Cytoplasm Driven by Actin and Myosin

Bulk streaming of cytoplasm is most spectacular in plant cells (Fig. 37.1B). Although confined within rigid walls, plant cell cytoplasm streams vigorously at very high velocities (up to 60 μm s^{-1}). At this rate, cytoplasm moves 5 m/day! Such **cytoplasmic streaming** is best understood in the giant cells of the green alga *Nitella.* Streaming occurs continuously in a thin layer of cytoplasm between the large central vacuole and chloroplasts immobilized in the cortex. On each side of the cell, a zone of stationary cytoplasm separates streams moving in opposite directions. The physiological function of this streaming is not clearly understood.

Bulk streaming in *Nitella* is brought about by movement of ER along tracks consisting of bundles of polarized actin filaments associated with chloroplasts (Fig. 37.9C). All the actin filaments in these bundles have the same polarity, and cytoplasm streams toward their barbed ends. In *Nitella* extracts, membrane vesicles move along actin filament bundles at the same high velocities as the cytoplasmic streaming. A type XI myosin (see Fig. 36.7) pulls the ER along these cortical actin tracks, dragging along other cytoplasmic components, including organelles and soluble molecules. This myosin moves nearly 10 times faster than the fastest muscle contraction, apparently by taking large steps rapidly and by the cooperation of several motors working rapidly on the same membrane.

A completely different actomyosin mechanism produces equally spectacular cytoplasmic streaming in the acellular slime mold *Physarum.* In these giant, multinucleated cells, cytoplasm flows back and forth rhythmically at high velocities through tubular channels (Fig. 37.10). Cycles of contraction and relaxation of

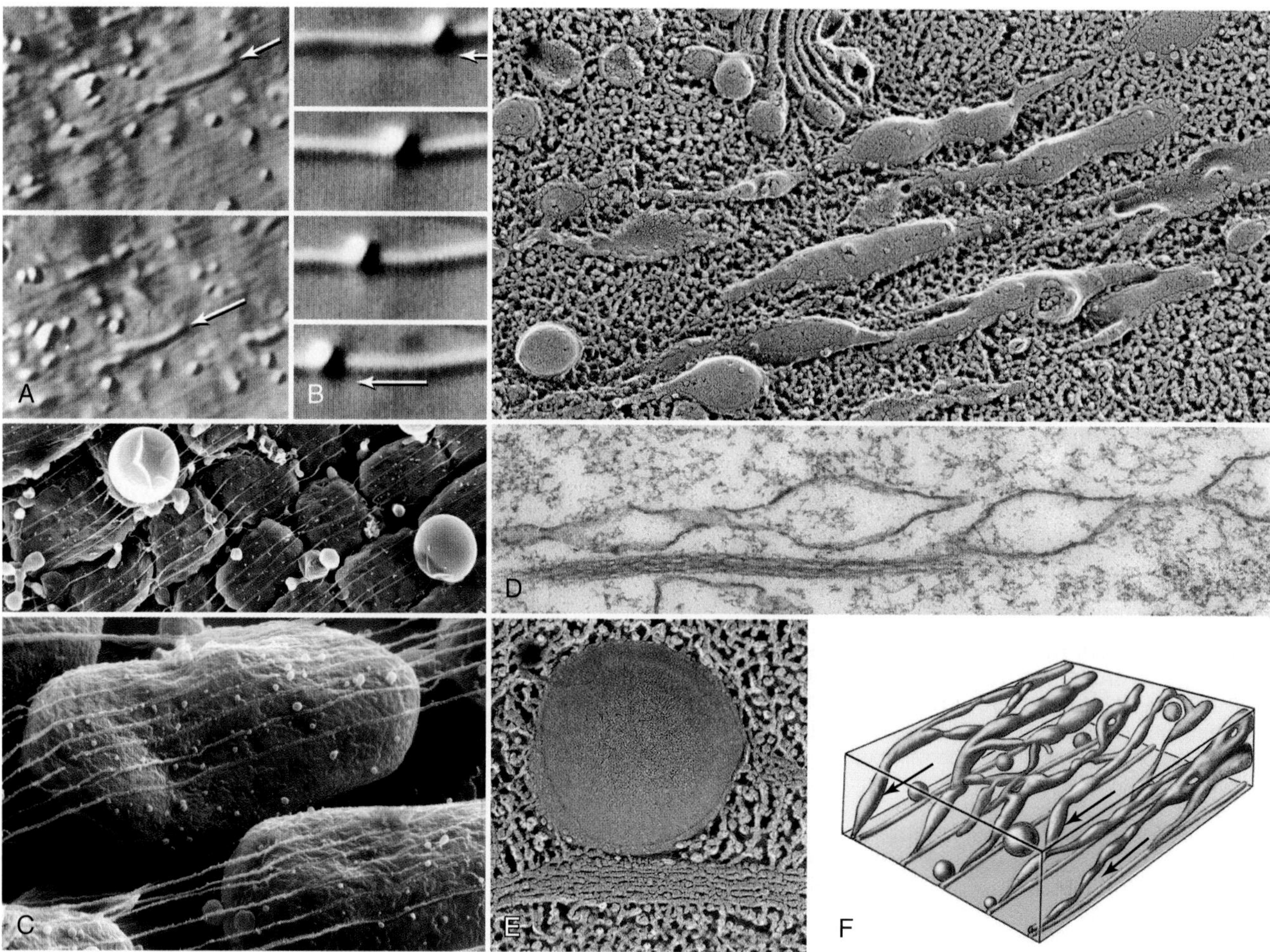

FIGURE 37.9 CYTOPLASMIC STREAMING IN THE GREEN ALGA *NITELLA*. A, A pair of differential interference contrast (DIC) light micrographs showing the movement of organelles in cytoplasm. Note the strand of endoplasmic reticulum (ER *[arrow]*). **B,** Time series of DIC light micrographs showing movement of a vesicle isolated from *Nitella* along a bundle of actin filaments isolated from *Nitella*. **C,** Scanning electron micrographs of the cortex isolated from *Nitella* showing the bundles of actin filaments associated with chloroplasts. **D,** Transmission electron micrographs of a freeze-fracture preparation *(upper)* and thin section *(lower)* showing ER associated with actin filament bundles. **E,** Freeze-fracture preparation of a vesicle associated with an actin filament bundle. **F,** Movement of ER along actin filament bundles dragging along bulk cytoplasm. (Courtesy B. Kachar, National Institutes of Health. For reference, see Kachar B, Reese T. The mechanism of cytoplasmic streaming in characean algal cells: sliding of endoplasmic reticulum along actin filaments. *J Cell Biol.* 1988;106:1545–1552.)

cortical actin filament networks push relatively fluid endoplasm back and forth in a manner akin to squeezing a toothpaste tube. Myosin-II is thought to generate the cortical contraction, as it is present in high concentration in this cell and can contract actin filament gels in vitro. (This was the first nonmuscle myosin to be purified in the late 1960s.) Cortical contractions similar to those of *Physarum* are used by giant amoebas for cell locomotion (see Fig. 38.1), and by other cells for cytokinesis (see Fig. 44.23), and movements of some embryonic tissues (see Fig. 38.5).

Actin-Based Movements of Organelles in Other Cells

Like *Nitella,* budding yeast cells transport vesicles along bundles of actin filaments from the mother to the bud (Fig. 37.11), although the movements of these solitary vesicles do not produce cytoplasmic streaming. Myosin-V is the motor (see Fig. 36.7), so vesicles fail to move from mother to bud in null mutants of myosin-V genes. Myosin-V also transports certain mRNAs along actin filament cables from the mother to the bud, where they determine cell fate. Similarly myosin-VII and myosin-X transport cytoplasmic and membrane proteins in filopodia of animal cells.

Animal cells generally use extended microtubules for long-distance movements and shorter actin filaments for local transport of vesicles and RNAs. Fish skin cells use both systems to change color: dynein aggregates and kinesin disperses pigment granules called melanophores along radial microtubule tracks. Myosin-V contributes by moving dispersed melanophores laterally between microtubules.

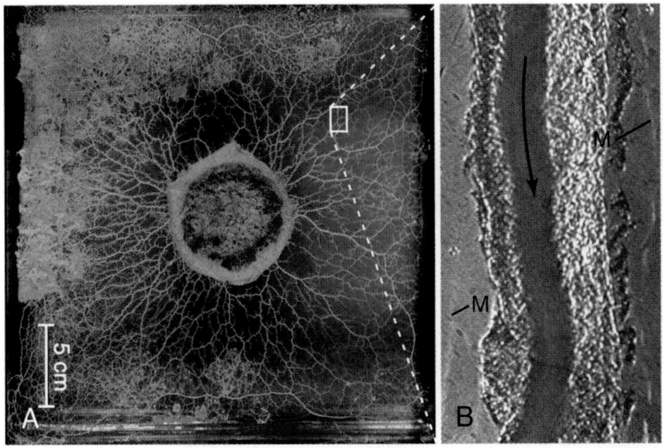

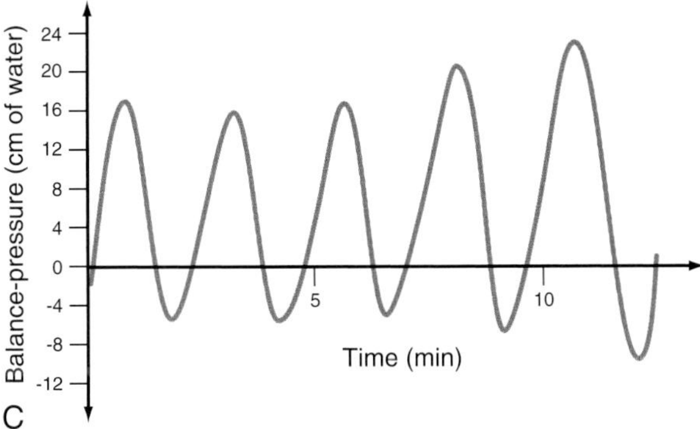

FIGURE 37.10 CYTOPLASMIC STREAMING IN THE ACELLULAR SLIME MOLD *PHYSARUM POLYCEPHALUM*. A, Photograph of *Physarum polycephalum,* a giant multinucleated single cell growing in a baking dish. **B,** Blur photomicrograph made with polarization optics by taking a time exposure showing the bulk streaming of the endoplasm in a cytoplasmic strand *(long arrow).* M, Mucus. **C,** Time course of pressure changes produced by shuttle streaming of cytoplasm through a strand. (**B,** From Nakajima H. The mechanochemical system behind streaming in *Physarum.* In Allen RD, Kamiya N, eds. *Primitive Motile Systems in Cell Biology.* New York: Academic Press; 1964:111–123. **C,** For reference, see Kamiya N. The mechanism of cytoplasmic movement in a myxomycete plasmodium. *Symp Soc Exp Biol.* 1968;22:199–214.)

The direction of movement depends on the motor and the organization of the actin filaments. Although many actin filaments in animal cells are not uniformly polarized, those in the cortex tend to have their barbed ends near the plasma membrane, allowing for local directional transport. For example, myosin-V moves recycling endosomes toward the plasma membrane. Similarly, myosin-V transports pigment granules called melanosomes within and between cells in the skin. In both cases a small GTPase and an adapter protein link melanosomes to the tail of myosin-V. Mutations of myosin-V, the GTPase, or the adapter protein in mice and humans cause not only pigmentation defects but also neurological problems owning to faulty mRNA transport in neurons. Other adapter proteins link myosin-VI to newly internalized endosomes with various types of receptors for transport away from the plasma membrane.

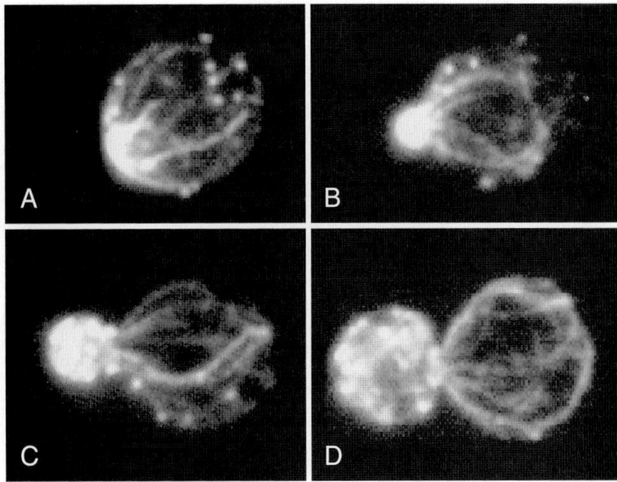

FIGURE 37.11 Fluorescence micrographs **(A–D)** showing actin filament bundles and patches at various stages in the cell cycle of the budding yeast *Saccharomyces cerevisiae*. Myosin-V uses these actin filament bundles to deliver vesicles (including the vacuole), certain messenger RNAs (mRNAs), and at least one enzyme (chitin synthase) from the mother to the bud. (Courtesy J.A. Cooper, Washington University, St. Louis, MO.)

Movements Driven by Actin Polymerization

Some intracellular pathogenic bacteria, including *Listeria* and *Shigella,* use actin polymerization to move through the cytoplasm of their animal cell hosts at about 0.5 μm s^{-1} (Fig. 37.12A). These bacteria hijack the machinery normally used to move the leading edge of motile cells to polymerize a comet tail of actin filaments that pushes the bacterium forward. One end of the bacterium has a concentration of proteins that directly *(Listeria)* or indirectly *(Shigella)* activate Arp2/3 complex to polymerize a network of branched actin filaments (see Fig. 33.12). Growth of this network at its trailing end pushes the bacterial cell forward. The comet tail of cross-linked actin filaments is stationary and depolymerizes distally at the same rate at which it grows next to the bacterium, so it remains a constant length.

Vaccinia viruses attached to the outer surface of animal cells move by a related mechanism. They use transmembrane proteins to activate the cytoplasmic actin assembly system to drive their movements at one stage in its life cycle (Fig. 37.12B). Placement of a plastic bead coated with adhesion proteins on the plasma membrane of some animal cells can induce similar propulsive actin comet tails in the cytoplasm.

Fungal and animal cells use Arp2/3 complex to assemble actin filaments at sites of clathrin-mediated endocytosis. Polymerization of these filaments assists with vesicle separation from the plasma membrane. Nucleation promoting factors associated with some endosomes stimulate Arp2/3 complex to assemble actin filament comets and move similar to *Listeria*.

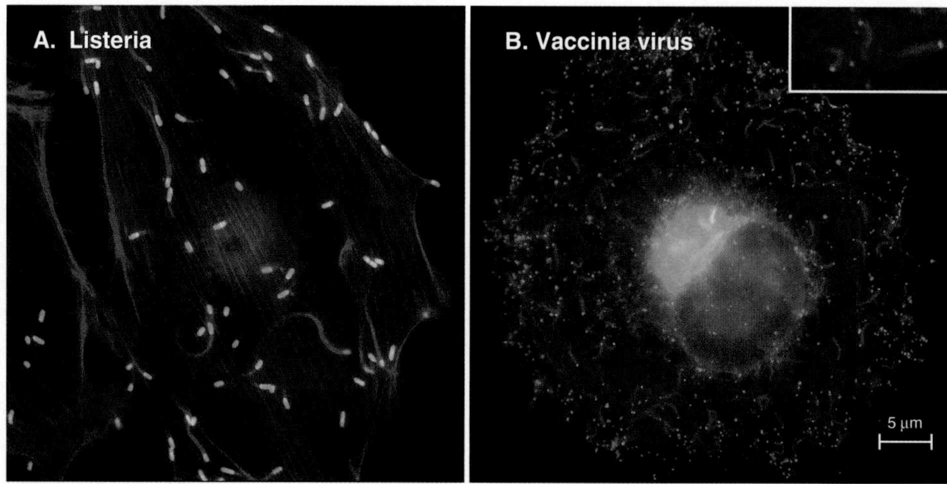

FIGURE 37.12 Fluorescence micrographs of actin filament comet tails in animal epithelial cells infected with the bacterium *Listeria monocytogenes* **(A)** or vaccinia virus **(B).** Both pathogens are stained *green* with fluorescent antibodies. They use host cell proteins to assemble a crosslinked network of actin filaments shaped like a comet tail. Actin filaments are stained *red* with rhodamine-phalloidin. **A,** The comet tail pushes *Listeria* in a PtK cell through the cytoplasm and into projections of the plasma membrane at the edge of the cell. **B,** When the replicated vaccinia viruses in this HeLa cell reach the cell surface 8 hours after infection, they activate Arp2/3 complex to assemble a cytoplasmic comet tail of actin filaments that are thought to enhance the spread of the virus from cell to cell. Actin-based motility of vaccinia virus depends on tyrosine phosphorylation of a viral transmembrane protein A36R that remains inserted in the plasma membrane. (**A,** Courtesy K. Skoble, D. Portnoy, and M. Welch, University of California, Berkeley. **B,** Courtesy T.P. Newsome and M. Way, Cancer Research UK, London. For reference, see Frischknecht F, Moreau V, Rottger S, et al. Actin-based motility of vaccinia virus mimics receptor tyrosine kinase signaling. *Nature.* 1999;401:926–929.)

ACKNOWLEDGMENTS

We thank Anthony Brown, Erika Holzbaur, and Margaret Titus for suggestions on revising this chapter.

SELECTED READINGS

Brown A. Slow axonal transport. Encyclopedia of Neuroscience, 2009. Update. October 28, 2014. Reference Module in Biomedical Sciences. Available at <http://dx.doi.org/10.1016/B978-0-12-801238-3.04765-6>, 2014.

Bullock SL. Messengers, motors and mysteries: sorting of eukaryotic mRNAs by cytoskeletal transport. *Biochem Soc Trans.* 2011;39:1161-1165.

Carter AP, Diamant AG, Urnavicius L. How dynein and dynactin transport cargos: a structural perspective. *Curr Opin Struct Biol.* 2016;37:62-70.

Cossart P, Pizarro-Cerdá J, Lecuit M. Invasion of mammalian cells by *Listeria monocytogenes:* Functional mimicry to subvert cellular functions. *Trends Cell Biol.* 2003;13:23-31.

Encalada SE, Goldstein LS. Biophysical challenges to axonal transport: motor-cargo deficiencies and neurodegeneration. *Annu Rev Biophys.* 2014;43:141-169.

Frank DJ, Noguchi T, Miller KG. Myosin VI: A structural role in actin organization important for protein and organelle localization and trafficking. *Curr Opin Cell Biol.* 2004;16:189-194.

Franker MA, Hoogenraad CC. Microtubule-based transport-basic mechanisms, traffic rules and role in neurological pathogenesis. *J Cell Sci.* 2013;126:2319-2329.

Fu MM, Holzbaur EL. Integrated regulation of motor-driven organelle transport by scaffolding proteins. *Trends Cell Biol.* 2014;24:564-574.

Hammer JA 3rd, Sellers JR. Walking to work: roles for class V myosins as cargo transporters. *Nat Rev Mol Cell Biol.* 2011;13:13-26.

Hancock WO. Bidirectional cargo transport: moving beyond tug of war. *Nat Rev Mol Cell Biol.* 2014;15:615-628.

Hartman MA, Finan D, Sivaramakrishnan S, et al. Principles of unconventional myosin function and targeting. *Annu Rev Cell Dev Biol.* 2011;27:133-155.

Hirokawa N, Niwa S, Tanaka Y. Molecular motors in neurons: transport mechanisms and roles in brain function, development, and disease. *Neuron.* 2010;68:610-638.

Lopez de Heredia M, Jansen R-P. mRNA localization and the cytoskeleton. *Curr Opin Cell Biol.* 2004;16:80-85.

Mooren OL, Galletta BJ, Cooper JA. Roles for actin assembly in endocytosis. *Annu Rev Biochem.* 2012;81:661-686.

Pruyne D, Legesse-Miller A, Gao L, et al. Mechanisms of polarized growth and organelle segregation in yeast. *Annu Rev Cell Dev Biol.* 2004;20:559-591.

Saxton WM, Hollenbeck PJ. The axonal transport of mitochondria. *J Cell Sci.* 2012;125:2095-2104.

Shimmen T, Yokota E. Cytoplasmic streaming in plants. *Curr Opin Cell Biol.* 2004;16:68-72.

Titus MA. Myosin-driven intracellular transport. In: Pollard TD, Goldman R, eds. *The Cytoskeleton.* Cold Spring Harbor Press; 2016;265-280.

Urnavicius L, Zhang K, Diamant AG, et al. The structure of the dynactin complex and its interaction with dynein. *Science.* 2015;347:1441-1446.

CHAPTER 38

Cellular Motility

Cells move at rates that range over four orders of magnitude (Fig. 38.1 and Table 38.1). At one extreme, ciliates, bacteria, and sperm swim rapidly through water, and giant amoebas crawl rapidly over solid substrates. At the other extreme, fungal, algal, and plant cells with rigid cell walls are immobile. However, even some plant cells move, such as pollen, which extends tubular pseudopods. Most cells, including white blood cells, nerve growth cones, and fibroblasts move at intermediate rates.

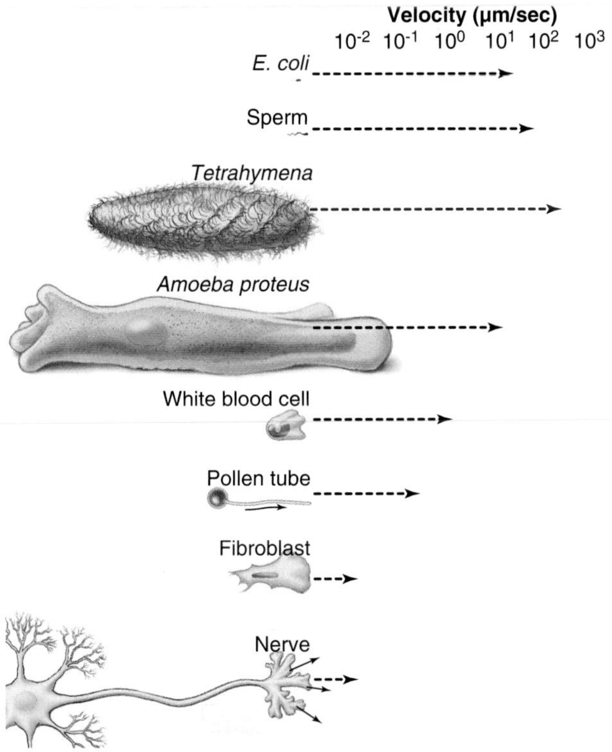

FIGURE 38.1 VELOCITIES OF MOVING CELLS SPAN MORE THAN FOUR ORDERS OF MAGNITUDE. Scale drawings of cells with a range of velocities. *E. coli*, *Escherichia coli*.

Cells produce forces for motility in many different ways, most commonly using the same four mechanisms that produce intracellular movements (see Chapter 37): contraction of actin–myosin networks, movement of motors on microtubules, reversible assembly of actin filaments, or reversible assembly of microtubules. These mechanisms often complement each other, even where movement depends mainly on one system. For example, microtubules contribute to actin-based pseudopod extension by helping to specify the polarity of the cell. This chapter compares these common mechanisms with a few more unusual, but informative mechanisms: contraction of calcium-sensitive fibers of ciliates, reversible assembly of novel cytoskeletal polymers of nematode sperm, and rotation of bacterial flagellar motors.

Most cells possess all proteins required for cellular motility, so the striking variation in their rates of movement arises from differences in the abundance, organization, and activities of this machinery. For example, both nonmotile yeasts and contractile muscle cells contain actin, myosin-II, heterodimeric capping protein, α-actinin, and tropomyosin. Yeasts use these proteins for cytokinesis (see Fig. 44.25), while muscle assembles high concentrations of similar proteins into sarcomeres (see Figs. 39.2 and 39.3) for powerful, fast contractions.

Cell Shape Changes Produced by Extension of Surface Processes

Alteration of cellular shape can be most simply brought about by assembly of new cytoskeletal polymers or by rearrangement of preexisting assemblies of actin filaments or microtubules.

TABLE 38.1 Velocities of Cellular Movements

System	Unitary Velocity (μm s⁻¹)	Summed Velocity (μm s⁻¹)	Motile Mechanism
Striated muscle contraction (biceps)	5–10	$4–8 \times 10^5$	Actin-myosin adenosine triphosphatase
Filopodium extension, *Thyone* sperm	10	10	Actin polymerization
Pseudopod extension, fibroblast	0.02	0.02	Actin polymerization
Pseudopod extension, human neutrophil	0.1	0.1	Actin polymerization
Pseudopod extension, *Amoeba proteus*	?	10	Actin-myosin ATPase
Pseudopod extension, nematode sperm	1	1	Assembly of major sperm protein
Retraction of axopodium, heliozoan	>100	>100	Disassembly of microtubules
Spasmoneme contraction, *Vorticella*	?	23,000	Calcium-induced conformational change
Swimming, *Escherichia coli*		25	Flagellum powered by rotary motor
Swimming, sea urchin sperm		15	Microtubule-dynein ATPase

Note: Unitary velocity refers to a single molecular unit. Summed velocity is the overall motion of the cell.

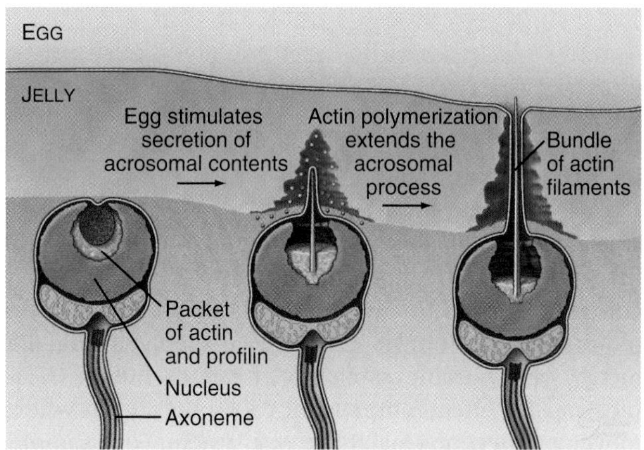

FIGURE 38.2 *THYONE* SPERM ACROSOMAL PROCESS. Actin polymerization drives the growth of the acrosomal process of the sperm of the sea slug, *Thyone*. The acrosome *(red)* is a membrane-bound secretory vesicle, which fuses with the plasma membrane and releases its hydrolytic enzymes prior to growth of the acrosomal process. When the acrosomal process reaches the egg, the plasma membranes of the two cells fuse. (Based on the work of L. Tilney, University of Pennsylvania, Philadelphia.)

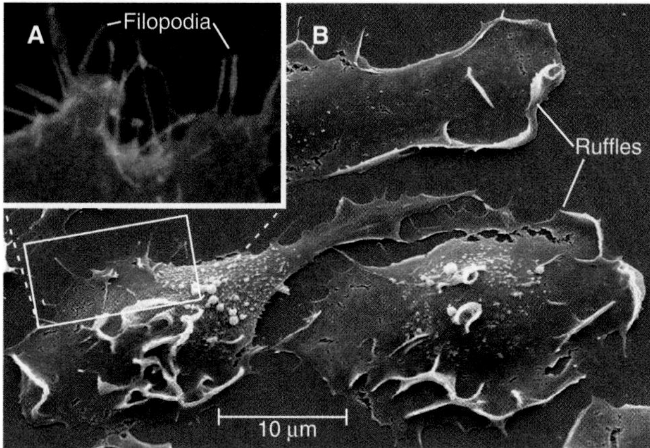

FIGURE 38.3 FILOPODIA. A, Fluorescence micrograph of the edge of a mouse NIH 3T3 cell expressing formin mDia2 and activated Rif, a Rho-family guanosine triphosphatase (GTPase) that activates mDia2. mDia2 concentrated at the tips of filopodia is stained *red* with fluorescent antibodies. **B,** Scanning electron micrograph of mouse macrophages spreading on a glass slide, illustrating the flat peripheral lamellae, wave-like "ruffles" on the upper surface, and finger-like filopodia. (**A,** From Pellegrin S, Mellor H. The Rho family GTPase Rif induces filopodia through mDia2. *Curr Biol.* 2005;15:129–133. **B,** From Chitu V, Pixley FJ, Macaluso F, et al. The PCH family member MAYP/PSTPIP2 directly regulates F-actin bundling and enhances filopodia formation and motility in macrophages. *Mol Biol Cell.* 2005;16: 2947–2959.)

Studies of echinoderm sperm revealed that actin polymerization drives the formation of cell surface projections called **filopodia**. To fertilize the egg the sperm extends a long filopodium to penetrate the protective jelly surrounding the egg (Fig. 38.2). Actin subunits for this acrosomal process are stored with profilin (see Figs. 1.4 and 33.11) in a novel concentrated packet near the nucleus. Contact with an egg stimulates actin filaments to polymerize, starting from a dense structure near the nucleus. Addition of subunits to the distal (barbed) end of growing filaments drives the elongation of the process and the surrounding membrane at a rate of 5 to 10 μm s^{-1} (an astounding maximum of 3700 subunits per second!). Actin subunits diffuse rapidly enough from their storage site to drive this rapid elongation. The number of filaments declines from approximately 150 near the base to less than 20 at the tip of the process, presumably from capping. The molecules (if any) guiding the growth of the filaments are not known.

Bundles of actin filaments support filopodia of a more modest size on many other cells. Filopodia on macrophages (Fig. 38.3), nerve growth cones (Fig. 38.6A), fibroblasts, and epithelial cells grow much more slowly and depend on formins and vasodilator-stimulated protein (VASP) at their tips (Fig. 38.3) to guide barbed-end assembly of actin filaments crosslinked by a protein called fascin. Microvilli of the brush border of epithelial cells (see Fig. 33.2) are short, stable filopodia. The

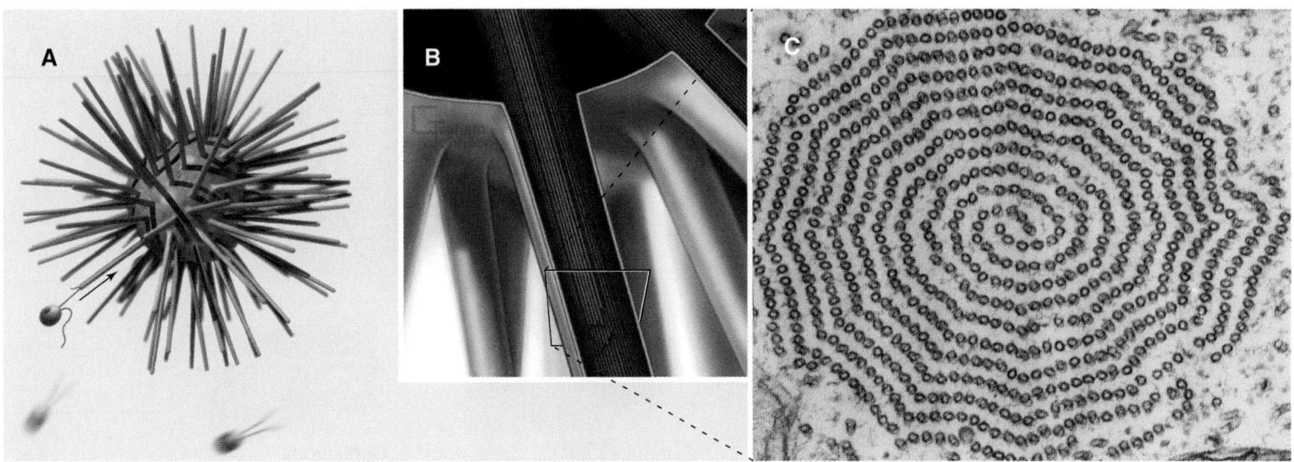

FIGURE 38.4 DYNAMIC CELL SURFACE PROJECTIONS SUPPORTED BY MICROTUBULES. A–B, Drawings of the radiolarian *Echinospherium* (a protozoan) showing projections called axopodia, which capture prey and draw them toward the cell body by transport on surface of the projection or collapse of the projection. **C,** Electron micrograph of a thin section across an axopodium, showing the double spiral array of microtubules. (Courtesy L. Tilney, University of Pennsylvania, Philadelphia.)

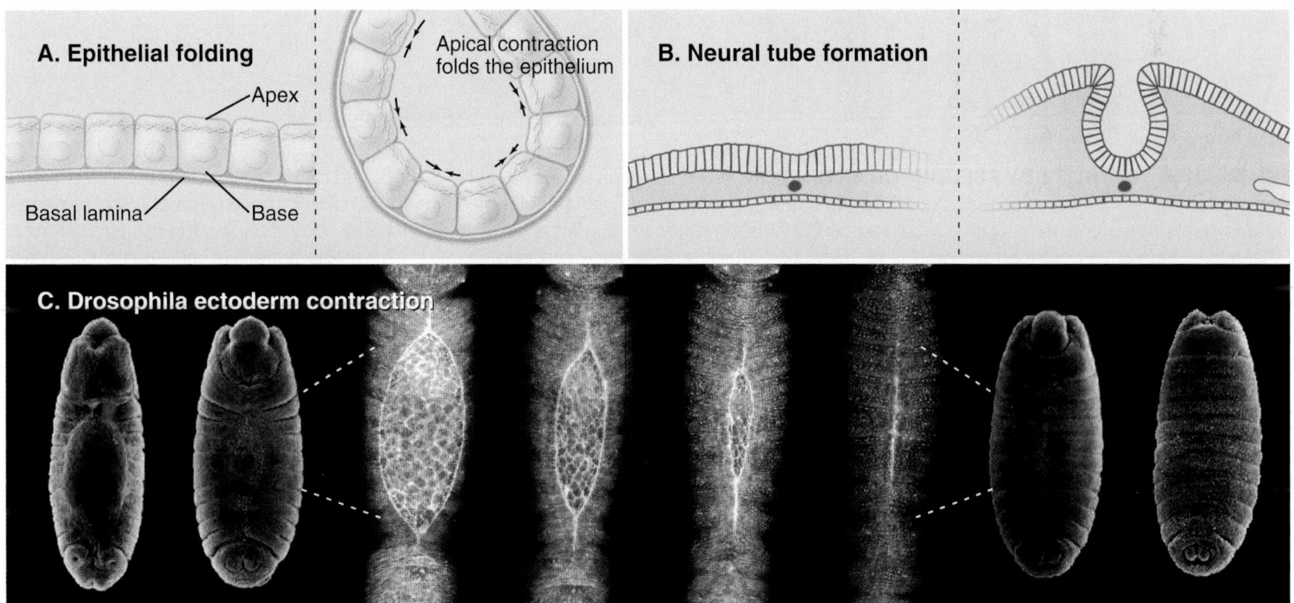

FIGURE 38.5 ACTOMYOSIN CONTRACTIONS MOLD THE SHAPE OF EPITHELIA DURING EMBRYONIC DEVELOPMENT. A, Folding of a planar epithelium into a tube. **B,** Formation of the neural tube by contraction of the apical pole of columnar epithelial cells resulting in shape change and invagination of the epithelium. **C,** Contraction around the margin of the ectoderm pulls this epithelium over the surface of a *Drosophila* embryo. The scanning electron micrographs (SEMs [*left* and *right*]) show the steps in the dorsal closure of the epithelium. The time series of fluorescence micrographs *(center)* shows live embryos expressing an actin-binding fragment of the protein moesin, fused to green fluorescent protein. (**C,** SEMs courtesy Thom Kaufman, Indiana University, Bloomington [see his movie "Fly Morph-o-genesis" at http://www.sdbonline.org/archive/dbcinema/kaufman/kaufman.html]; light micrographs courtesy D. Kiehart, Duke University, Durham, NC. For reference, see Kiehart DP, Galbraith CG, Edwards KA, et al. Multiple forces contribute to cell sheet morphogenesis for dorsal closure in *Drosophila. J Cell Biol.* 2000;149: 471–490.)

bundles of actin filaments supporting microvilli are crosslinked to each other by fimbrin and villin and to the plasma membrane by myosin-I. Synthesis of these accessory proteins triggers assembly of microvilli on epithelial cells as well as cells that normally have few microvilli.

Some embryonic cells grow long filopodia to contact cells micrometers distant. These filopodia create physical contacts that allow the cells to send or receive signals that influence developmental decisions. Long filopodia between some pairs of cells even make gap junctions (see Fig. 31.6).

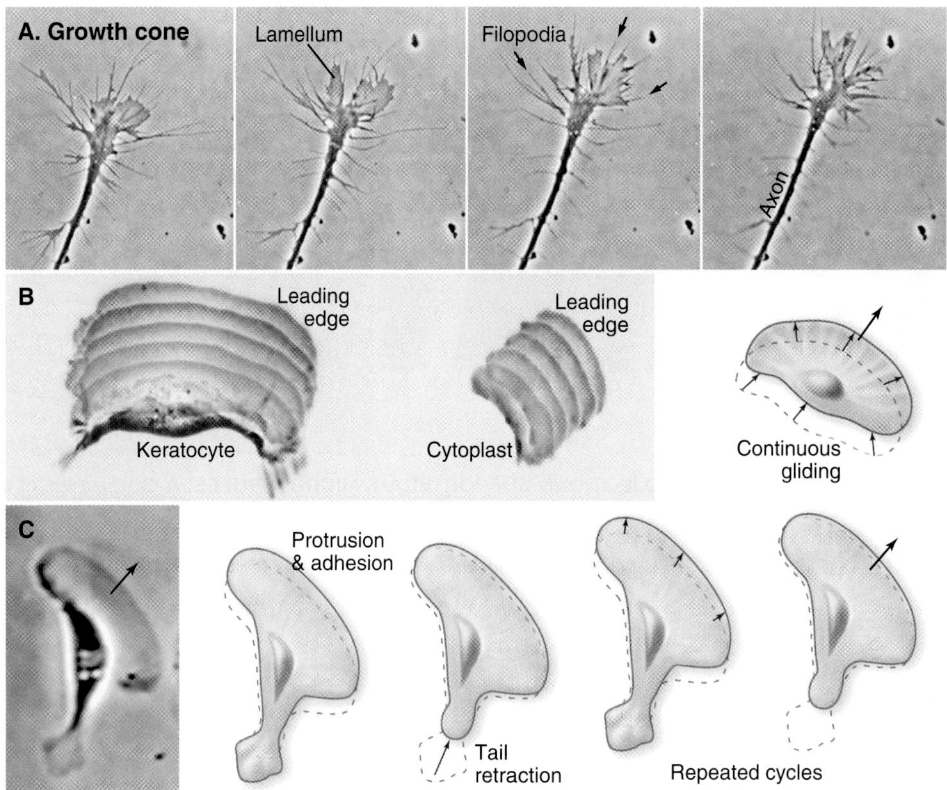

FIGURE 38.6 MOTILITY BY PSEUDOPOD EXTENSION. A, Phase-contrast micrographs of a cultured nerve cell's growth cone at 1-minute intervals. The growth cone extends filopodia and fills in the space between with an actin-filled lamella. **B,** Gliding movements of a fish epidermal keratocyte and a keratocyte cytoplast, a cell fragment consisting of the leading edge with most of the cell body including the nucleus removed. Differential interference contrast micrographs at 15-second intervals were superimposed. Drawing shows the pattern of movement. **C,** Phase-contrast micrograph of a keratocyte on glass. This cell moved toward the upper right using cycles of expansion of the broad leading lamella and retraction of the trailing edge from the surface as shown by the drawings. (**A,** Courtesy D. Bray, University of Cambridge, United Kingdom. **B,** From Pollard TD, Borisy GG. Cellular motility driven by assembly and disassembly of actin filaments. *Cell*. 2003;112:453–465. **C,** Courtesy J. Lee, University of Connecticut, Storrs.)

A group of protists called heliozoans, named for their similarity to a cartoon of the sun, are a rare example of using microtubules instead of actin filaments to extend, support, and retract long, thin processes bounded by the plasma membrane (Fig. 38.4). Microtubules in these **axopodia** are crosslinked into a precise geometrical array accounting for the rigidity of these long processes. Axopodia capture prey organisms and transport them toward the cell body for phagocytosis by two mechanisms. Some move along the surface of axopodia (similar to intraflagellar transport; Fig. 38.18). Other axopodia collapse owing to rapid depolymerization of the microtubules and drag the prey with them. Ca^{2+} influx appears to trigger depolymerization of the microtubules, but the details of the mechanism are not known.

Cell Shape Changes Produced by Contraction

Cells can change shape by localized or oriented cytoplasmic contractions. Muscle contraction (Chapter 39) and cytokinesis (Chapter 44) are the best examples, but contractions also remodel many embryonic tissues. Localized contractions at the base or apex of cells in a planar epithelium cause evaginations or invaginations (Fig. 38.5) such as those that form the neural tube (see Fig. 30.8) and glands that bud from the gastrointestinal tract and branches of the respiratory tract. Closure of the epidermis over a *Drosophila* embryo also requires contraction of a circumferential ring of cells (Fig. 38.5C). Tension generated by myosin-II and actin filaments deforms each cell and, collectively, the whole epithelium. Similarly, contraction of a ring of actin filaments associated with the zonula adherens of intestinal epithelial cells helps regulate the permeability of the tight junctions that seal sheets of epithelial cells (see Fig. 31.2).

Locomotion by Pseudopod Extension

The ability to crawl over solid substrates or through extracellular matrix is essential for many cells. Perhaps the most spectacular example is the slowly moving **growth cone** of a nerve axon (Fig. 38.6A). Although moving less than 50 nm s^{-1}, growth cones navigate

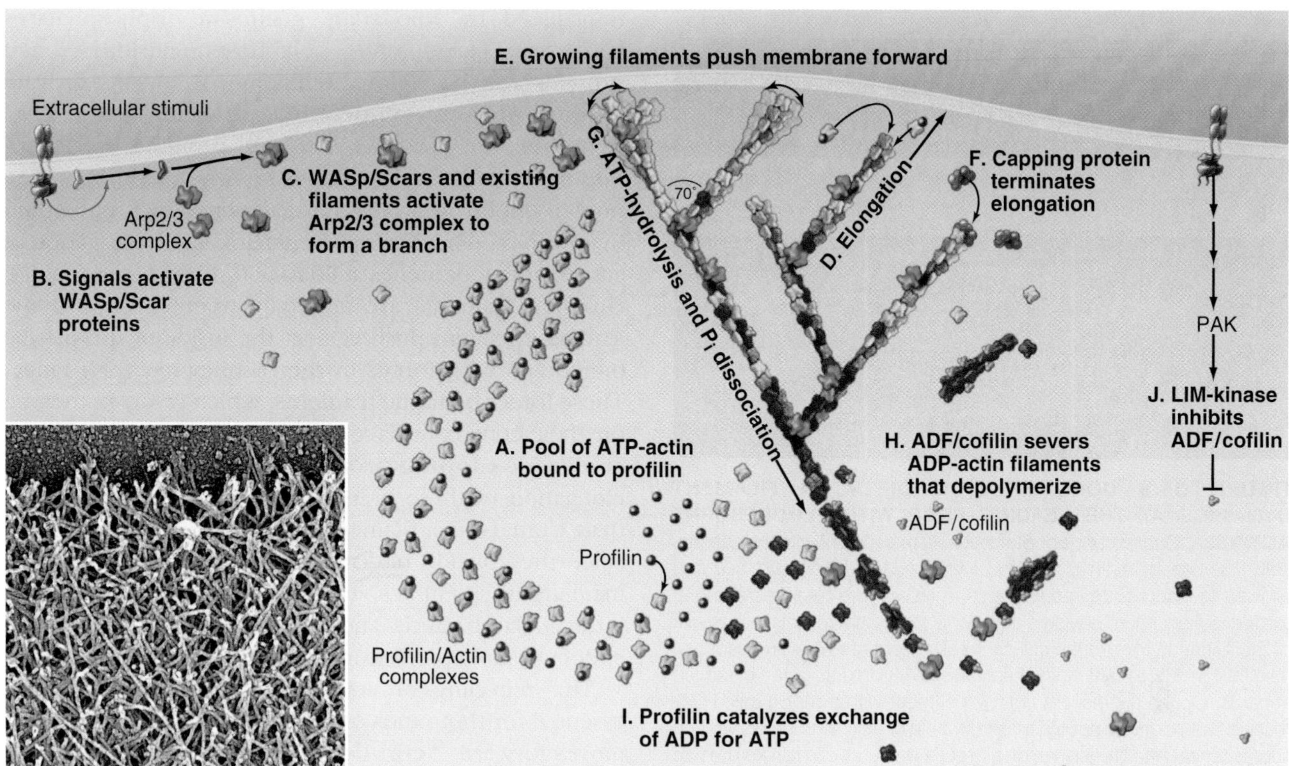

FIGURE 38.7 A MODEL FOR ACTIN FILAMENT ASSEMBLY AND DISASSEMBLY AT THE LEADING EDGE. The reactions are separated in space for clarity but actually occur together along the leading edge. **A,** Cells contain a large pool of unpolymerized actin bound to profilin. **B,** Stimulation of cell surface receptors produces activated Rho-family guanosine triphosphatases (GTPases) and other signals that activate Wiskott-Aldrich syndrome protein (WASp)/Scar proteins. **C,** These proteins, in turn, activate nucleation of new actin filaments by Arp2/3 complex on the side of existing filaments. **D,** The new filaments grow at their barbed ends until they are capped (see **F**). **E,** Growing filaments push the plasma membrane forward. **F,** Capping protein terminates elongation. **G,** Polymerized adenosine triphosphate (ATP)-actin *(yellow)* hydrolyzes the bound adenosine triphosphate (ATP) to adenosine diphosphate (ADP) and inorganic phosphate (P_i) *(orange),* followed by slow dissociation of phosphate yielding ADP-actin *(red).* **H,** ADF/cofilins bind and sever ADP-actin filaments and that disassemble, releasing ADP-actin. **I,** Profilin promotes the exchange of ADP for ATP, restoring the pool of unpolymerized ATP-actin bound to profilin. **J,** Some of the same stimuli that initiate polymerization can also stabilize filaments when LIM-kinase phosphorylates actin-depolymerizing factor (ADF)/cofilins, inhibiting their depolymerizing activity. **Inset,** Electron micrograph of the branched network of actin filaments at the leading edge. PAK, p21-activated kinase. (Modified from Pollard TD, Blanchoin L, Mullins RD. Biophysics of actin filament dynamics in nonmuscle cells. *Annu Rev Biophys Biomol Struct.* 2000;29:545–576. *Inset,* Courtesy T. Svitkina and G. Borisy, Northwestern University, Evanston, IL.)

precisely over distances ranging from micrometers to meters to establish all the connections in the human nervous system, which consists of billions of neurons and approximately 1.6 million kilometers of cellular processes. Some epithelial cells (Fig. 38.6B) and white blood cells move much faster, about 0.5 μm s^{-1}. These movements enable epithelial cells to cover wounds and allow leukocytes to move from the blood circulation to sites of inflammation (see Fig. 30.13) where they engulf microorganisms by phagocytosis (see Fig. 22.3). During vertebrate embryogenesis, neural crest cells migrate long distances through connective tissues before differentiating into pigment cells and sympathetic neurons. Fibroblasts lay down collagen fibrils as they move through the extracellular matrix (see Fig. 29.4).

The best-characterized motile system is animal and amoeboid cells moving on a flat surface such as a microscope slide that may be coated with extracellular matrix molecules. These cells use actin polymerization to extend the leading edge, then adhere to the underlying substrate, and (if the whole cell is to move) release and retract any attachments of its tail to the substrate. Animal cells move in other environments, such as three-dimensional extracellular matrix, using variations of this theme as described at the end of the section.

Lamellar Motility on Flat Surfaces by Pseudopod Extension

Pseudopods that lead the way in cell migration are filled with a dense network of branched actin filaments with their fast-growing barbed ends generally facing out towards the plasma membrane (Fig. 38.7). The leading lamella is autonomous for locomotion, and moves normally after amputation from the rest of the cell (Fig. 38.6B). Generally, the leading lamella is very flat, on the order of 0.25 μm thick, but some cells extend the lamellae up from the substrate into a wave-like fold called a ruffle (Fig. 38.3). Microtubules help maintain the polarized

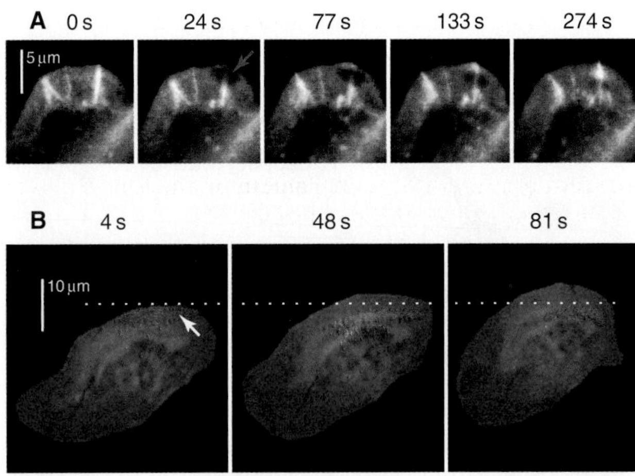

FIGURE 38.8 DOCUMENTATION OF ACTIN FILAMENT DYNAMICS AT THE LEADING EDGE WITH FLUORESCENT ACTINS. A, Fluorescence photobleaching experiment with a stationary cell. Fluorescent actin is injected into a cultured epithelial cell and allowed to incorporate into filaments. A laser pulse bleaches some of the fluorescent actin, leaving a dark spot *(arrow)* that reveals movement of the filaments toward the cell center. The spot recovers fluorescence as diffusing fluorescent actin assembles new filaments in the bleached zone. **B,** Caged fluorescent actin experiment with a motile cell. Fluorescent dye bound to actin is masked with a chemical group preventing fluorescence. After incorporation into actin filaments of a fish keratocyte (see Fig. 33.2E), dyes in one area of the cell are uncaged with a light pulse *(arrow),* and red fluorescence is followed with time. Fluorescent actin filaments are stationary with respect to the substrate as the cell moves forward (upward). The fluorescent spot of marked filaments fades with time, owing to depolymerization and dispersal of the fluorescent subunits. (**A,** From Wang Y-L. Exchange of actin subunits at the leading edge of living fibroblasts: possible role of treadmilling. *J Cell Biol*. 1985;101:597–602, copyright The Rockefeller University Press. **B,** From Theriot JA, Mitchison TJ. Actin microfilament dynamics in locomoting cells. *Nature*. 1991;352:126–131.)

shape that is required for persistent directional locomotion, but are not required for pseudopod extension.

Actin filaments assemble continuously near the **leading edge** of pseudopods and turn over rapidly deeper in the cytoplasm (Fig. 38.8). Thus, the inhibitor cytochalasin stops actin polymerization at the leading edge (see Fig. 33.18). Purified actin was labeled with a fluorescent dye and microinjected into live cells, where it incorporated into actin-containing structures, including the cortical network, pseudopods, stress fibers, and surface microspikes. Photobleaching (Fig. 38.8A) or photoactivating the fluorescent actin (Fig. 38.8B) showed that actin assembles at the leading edge and then turns over. Alternatively, if a low concentration of fluorescent actin is injected it will incorporate irregularly into filaments, producing tiny "speckles" of fluorescence that can be tracked over time to reveal sites of assembly and disassembly (see Fig. 33.18D–E).

Arp2/3 complex (see Fig. 33.12) and formins cooperate with VASP (see Fig. 33.14) to initiate and elongate actin filaments at the leading edge of motile cells. Chemotactic stimuli (Fig. 38.10) or intrinsic signals transduced by Rho-family guanosine triphosphatases (GTPases), membrane polyphosphoinositides, and proteins with SH3 (Src homology 3) domains activate **Wiskott-Aldrich syndrome protein (WASp)/Scar proteins** (see Fig. 33.13), which promote the formation of actin filament branches by Arp2/3 complex. The pool of unpolymerized actin maintained by profilin (and thymosin-β_4, where it is present) drives the elongation of actin filament branches at 50 to 500 subunits per second. Growing filaments are generally oriented toward the leading edge and push against the inside of the plasma membrane with forces in the piconewton (pN) range. These forces bend the filaments, which favors branching on their convex surface, facing the leading edge. Heterodimeric **capping protein** (see Fig. 33.15) terminates elongation of these branches before they grow longer than 1 μm. Longer filaments are less effective at pushing, since they buckle under piconewton forces. A similar mechanism assembles actin filaments in comet tails on intracellular bacteria and viruses (see Fig. 37.12) and endocytic actin patches in yeast (see Figs. 6.3 and 37.11).

The recycling of actin and accessory proteins is essential for thousands of rounds of assembly as the cell moves forward. **Actin-depolymerizing factor (ADF)/cofilins** (see Fig. 33.16) bind and sever aged adenosine diphosphate (ADP)-actin filaments located away from the leading edge. The fragments may be capped, in which case they slowly depolymerize at the pointed ends. This recycling mechanism is so efficient that the branched network is disassembled in seconds.

Formins assemble long unbranched filaments (see Fig. 33.2E) in the flat region of the cell just behind the branched network at the leading edge. Some of these filaments form bundles that protrude as filopodia beyond the leading edge (Fig. 38.3), where VASP (see Fig. 33.14) promotes their elongation. The tropomyosin binds along the sides of these long filaments, protecting them from severing by ADF/cofilins.

Actin filament crosslinking proteins stabilize pseudopods. Human melanoma cells that lack the crosslinking protein **filamin** (see Fig. 33.17) form unstable pseudopods all around their peripheries and locomote abnormally (Fig. 38.9). These tumor cells recover their normal behavior if provided with filamin. Similarly, *Dictyostelium* cells that lack a homolog of filamin form fewer pseudopods.

Influence of the Substrate on Lamellar Motility

The growing actin network at the leading edge will either push the membrane forward or slip backward depending on how well it is connected to the substrate across the plasma membrane. In highly motile cells such as epithelial cells from fish scales (Fig. 38.6B) transmembrane adhesion proteins anchor the actin filament network to the substrate, so the polymerization results in forward motion (Fig. 38.8B). In stationary cells (Fig. 38.8A;

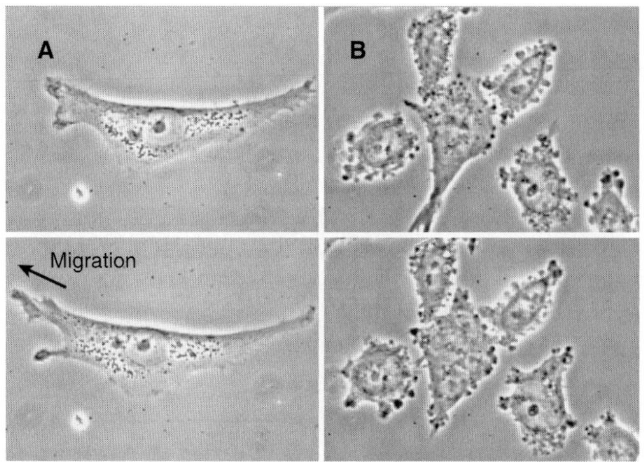

FIGURE 38.9 CONTRIBUTION OF THE ACTIN FILAMENT CROSSLINKING PROTEIN FILAMIN TO THE STABILITY OF THE LEADING EDGE OF HUMAN MELANOMA CELLS. Pairs of phase-contrast light micrographs, taken at different times, of living cells grown in serum-containing medium on a plastic surface. **A,** Melanoma cells expressing filamin have normal leading lamella. **B,** Melanoma cells lacking filamin form spherical blebs around their margins and migrate very little. (Courtesy C. Cunningham and T.P. Stossel, Harvard Medical School, Boston, MA. For reference, see Cunningham C, Gorlin JB, Kwiatkowski DJ, et al. Actin-binding protein requirement for cortical stability and efficient locomotion. *Science*. 1992;255:325–327.)

also see Fig. 33.18D–E), actin polymerizes at the edge of the cell causing the entire network to move en masse away from the membrane, a phenomenon called *retrograde flow*. Fibroblasts are an intermediate state, in which actin polymerization produces some forward movement but also considerable retrograde flow.

Movement driven by adhesion requires a compromise. Adhesion must be strong enough for the internal forces to propel the cell forward but not so strong that it prevents movement. Transmembrane adhesion proteins such as integrins (see Fig. 30.9) both anchor the cell and transmit the presence of their ligands and the stiffness of the environment to Rho-family GTPases that regulate actin assembly and myosin contractility. Rapidly reversible binding of integrins and other adhesion proteins to extracellular matrix molecules, such as fibronectin, allows adhesion without immobilization. Rapidly moving white blood cells attach weakly and transiently, whereas slowly moving fibroblasts form longer-lasting focal contacts (see Fig. 30.11). Cells tend to move up gradients of adhesiveness but stop if adhesion is too strong. This graded response to adhesion allows neural crest cells to migrate preferentially through regions of embryonic connective tissue marked by adhesive proteins.

Tail Retraction and Other Roles for Myosin in Motility

Growth cones of neurons draw out a long process from a stationary cell body, but most cells must break adhesions at their trailing edge to advance. Adherent, slowly moving cells such as fibroblasts exert significant tension on the underlying substrate, when myosin pulls on the bundles of actin filaments associated with focal contacts (see Fig. 30.11). When tension overcomes the attachments, the rear of the cell shortens elastically and then contracts further (Fig. 38.6C). Myosin also contributes to the retrograde flow of actin filaments in the zone between the leading edge and the cell body.

Other Modes of Motility

Cells that move in two-dimensions by pseudopod extension and tail retraction repurpose the same proteins to behave differently in other environments. In the extracellular matrix of connective tissues (see Chapter 29) cells must squeeze between three-dimensional networks of fibers using forces produced by myosin to generate hydrostatic pressure to push forward and to pull the relatively stiff nucleus forward. In other circumstances cells can expand by forming blebs of plasma membrane lacking networks of actin filaments. Nerve growth cones (Fig. 38.6A) extend filopodia and then fill the spaces in between them with a lamella containing new, branched actin filaments. Superfast giant amoebas (Fig. 38.1) use myosin to generate contractions in the cortex or the front of the pseudopod to drive the bulk streaming of cytoplasm into advancing pseudopods.

Chemotaxis of Motile Cells

Extracellular chemical clues direct locomotion by influencing the formation and persistence of pseudopods. Movement toward a positive signal is called **chemotaxis**. The best-characterized example is the attraction of *Dictyostelium* to cyclic adenosine monophosphate (cAMP) (Fig. 38.10), the extracellular chemical that these amoebas use to communicate as they form colonies before making spores. Remarkably, these cells can sense a gradient of cAMP corresponding to a concentration difference of less than 2% along their length. This small difference is amplified into strong internal signals that control motility. Binding of cAMP to seven-helix receptors in the plasma membrane activates trimeric G-proteins inside the cell (see Fig. 25.9). The G-proteins activate pathways that regulate the activity of enzymes that control the concentration of the lipid second messenger phosphatidylinositol 3,4,5-trisphosphate (PIP_3) in the plasma membrane: phosphatidylinositol 3-kinase (PI3K) synthesizes PIP_3 and PTEN (phosphatase and tensin homolog) degrades it. (See Fig. 26.7 for details on polyphosphoinositides.) A fast positive pathway that is sensitive to local receptor occupancy and a slower global negative signal that is proportional to total receptor occupancy produce a gradient of PI3K activity inside the cell that is highest near the external source of cAMP. These pathways concentrate PTEN on the plasma membrane away from the source of cAMP. This complementary regulation of the kinase and phosphatase creates an internal gradient of PIP_3 three to seven times steeper

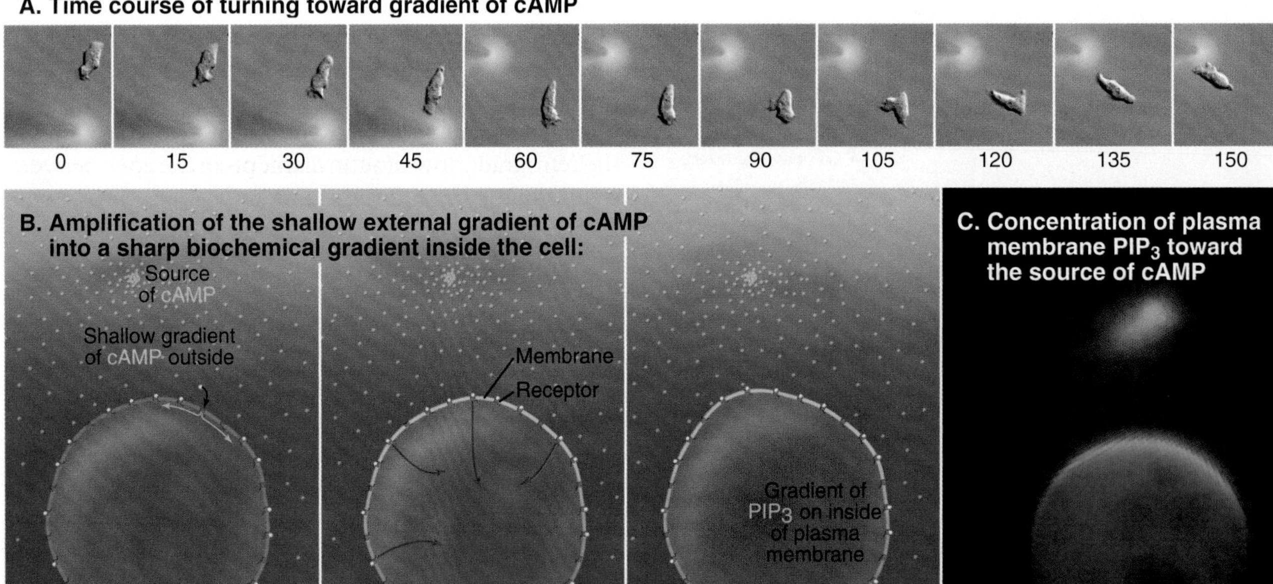

FIGURE 38.10 CHEMOTAXIS OF A *DICTYOSTELIUM* AMOEBA TOWARD CYCLIC ADENOSINE MONOPHOSPHATE. A, Live cell attracted to cyclic adenosine monophosphate (cAMP) *(gold)* released from a micropipette. A time series of differential interference micrographs shows the rapid formation of a new pseudopod and reorientation of the direction of movement when the position of the micropipette is moved at the 60-second time point. **B,** Cells have a uniform distribution of cAMP receptors (*yellow* and *red dots*) over their surface. A shallow gradient of cAMP activates these seven-helix receptors *(red),* which activate a trimeric G-protein and phosphatidylinositol 3-kinase, an enzyme that rapidly converts phosphatidylinositol 4-phosphate (PIP_2) to phosphatidylinositol 3,4,5-trisphosphate (PIP_3). On a slower time scale, the active G-protein activates phosphatase and tensin homolog (PTEN), a PIP_3 phosphatase, throughout the cell. The combination of these two signals creates a steep gradient of PIP_3 across the cell. **C,** Fluorescence micrograph of a cell exposed to a point source of cAMP *(yellow).* A green fluorescence protein (GFP)-pleckstrin homology (PH) domain fusion protein *(green)* inside the cell binds to PIP_3 on the inside of the plasma membrane, revealing the steep gradient of PIP_3. (**A,** Courtesy Susan Lee and Richard Firtel, University of California, San Diego. **B,** Modified from a sketch by Pablo Iglesias, Johns Hopkins University, Baltimore, MD. **C,** Courtesy Pablo Iglesias, Johns Hopkins University, Baltimore, MD. For reference, see Janetopoulos C, Ma L, Devreotes PN, Iglesias PA. Chemoattractant-induced phosphatidylinositol 3,4,5 trisphosphate accumulation is spatially amplified and adapts, independent of the actin cytoskeleton. *Proc Natl Acad Sci U S A.* 2004;101:8951–8956.)

than the external gradient of cAMP. Transduction of this internal gradient of PIP_3 into motility requires Rho-family GTPases and formation of new actin filaments. Local polymerization and crosslinking of these actin filaments expand the cortex facing the source of cAMP into a new pseudopod and move the cell toward the cAMP.

Leukocytes are attracted to chemokines and bacterial metabolites at sites of infection (see Fig. 30.13), especially small peptides derived from the N-termini of bacterial proteins, such as ***N*-formyl-methionine-leucine-phenylalanine** (referred to as FMLP in the scientific literature). Similar to *Dictyostelium,* activation of seven-helix receptors and trimeric G-proteins amplifies shallow external gradients of FMLP into steeper internal gradients of PIP_3 and other signals that control pseudopod formation.

Negative signals also influence pseudopod persistence and the direction of motility. A classic example is the negative effect of contact with another cell. Loss of **contact inhibition** of motility by tumor cells contributes to their tendency to migrate among other cells and spread throughout the body.

Growth Cone Guidance: A Model for Regulation of Motility

Growth cones of embryonic nerve cells use a combination of positive and negative cues to navigate with high reliability to precisely the right location to create a synapse (Fig. 38.11). This combinatorial strategy is much more complex than the simple chemoattraction of *Dictyostelium* to cAMP, as expected for the more complicated task of connecting billions of neurons to each other or specific muscle cells. Cues for growth cone guidance come from soluble factors and cell surface molecules, each with a specific receptor on the growth cone. As in other systems, extracellular matrix molecules provide the substrate for growth cone movements. Precisely positioned expression of cue molecules and their

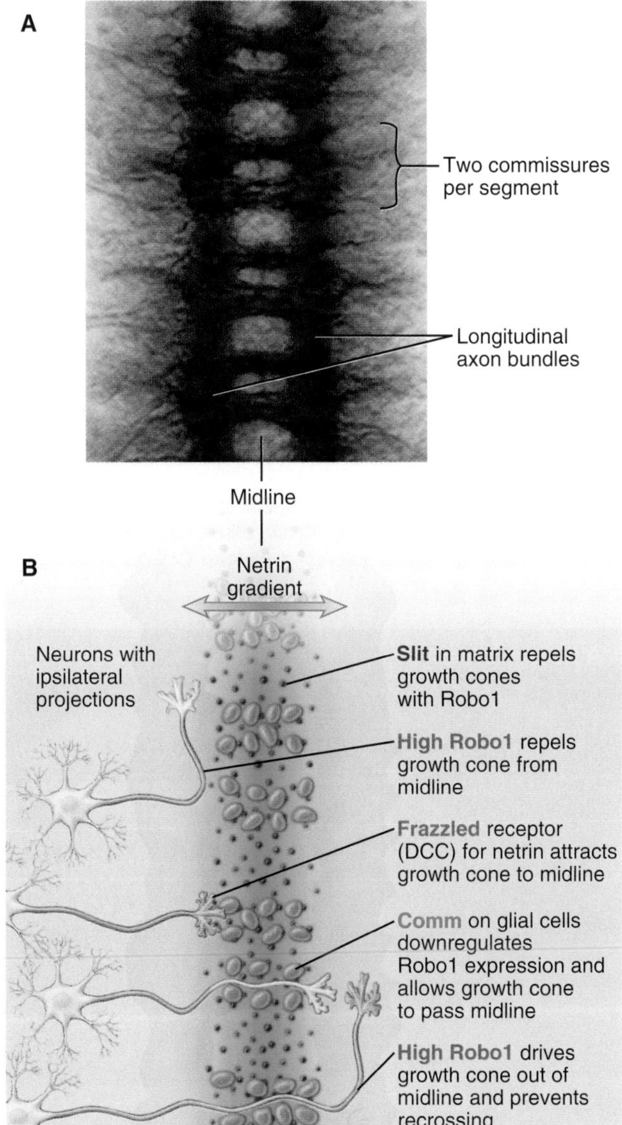

FIGURE 38.11 ***DROSOPHILA*** **GROWTH CONE GUIDANCE.** **A,** Light micrograph of a filleted embryo showing the nerve cord stained *brown* with an axon marker. The axons of approximately 90% of neurons cross the midline a single time in a transverse nerve bundle called a commissure before running longitudinally in fascicles on each side of the midline. **B,** Drawing showing the ligands and receptors that guide growth cones across the midline and prevent their return to the ipsilateral (original) side. Frazzled receptors for netrin attract the growth cone to the midline where Comm downregulates the activity of Robo1, a repulsive receptor for Slit, allowing axons to cross the midline. (**A,** Courtesy John Thomas, Salk Institute, La Jolla, CA.)

receptors guides growth cones along a staggering number of different pathways, as illustrated by the following well-characterized examples.

Localized cells in the nervous system, such as those in the floor plate of the developing spinal cord, secrete soluble guidance proteins such as **netrin.** Gradients of netrin provide long-range guidance for growth cones of cells that possess netrin receptors. Some netrin receptors, such as members of the DCC family, steer the growth cone toward a netrin source by activating actin polymerization by VASP. Other netrin receptors repel the growth cone from netrin. Growth cones without these receptors are insensitive to this cue.

Slit, a large extracellular matrix protein, repels growth cones with Slit receptors, which are immunoglobulin cell adhesion molecules (IgCAMs) called Robo1, Robo2, and Robo3. Mutations in the genes for these receptors cause growth cones to ignore Slit.

IgCAM cell surface adhesion proteins (see Fig. 30.3), such as **fasciculin II,** prompt growing axons to bundle together in bundles called fascicles by homophilic interactions. Growth cones can be attracted out of these bundles to particular targets, such as muscle cells, that secrete chemoattractants or proteins that antagonize fasciculin II adhesion.

The effects of mutations in genes for receptors and their ligands revealed how growth cones in *Drosophila* embryos navigate using multiple guidance cues (Fig. 38.11). Growth cones of neurons on one side of the nerve cord migrate across the midline to the opposite side and then navigate faithfully to their targets. Netrins secreted by cells at the midline attract growth cones expressing the DCC (Frazzled) netrin receptor. However, midline cells also secrete high levels of the matrix protein Slit, which repels growth cones. Growth cones cross the midline by downregulating the slit receptor. Once across the midline, the growth cones upregulate the slit receptor, so they never cross back to the side of origin. Local cues alert particular growth cones of motor neurons to branch off of fascicles to innervate individual muscle cells. Path finding by capillaries uses some of the same guidance mechanisms to grow blood vessels.

Eukaryotic Cilia and Flagella

Axonemes built from microtubules and powered by dynein produce the beating of cilia and flagella (Fig. 38.12). These exceedingly complex structures are remarkably ancient. More than 1 billion years ago the last common eukaryotic ancestor (see Fig. 2.4) had motile flagella with all the essential features of human cilia and flagella, as evidenced by motile axonemes in most branches of eukaryotes. However, most fungi and plants lost the genes for axonemal proteins.

Animal cells make three types of axonemes, all with nine outer doublet microtubules anchored by a basal body at the cortex of the cell and surrounded by plasma membrane (Fig. 38.12B). Motile axonemes have dynein motors, a central pair of single microtubules and radial spokes (Figs. 38.14 and 38.15). Rotating nodal cilia (see below for the biological context) have dynein arms but no central pair or radial spokes. Immobile primary cilia lack dynein, central pairs and radial spokes. This section begins with motile cilia and flagella.

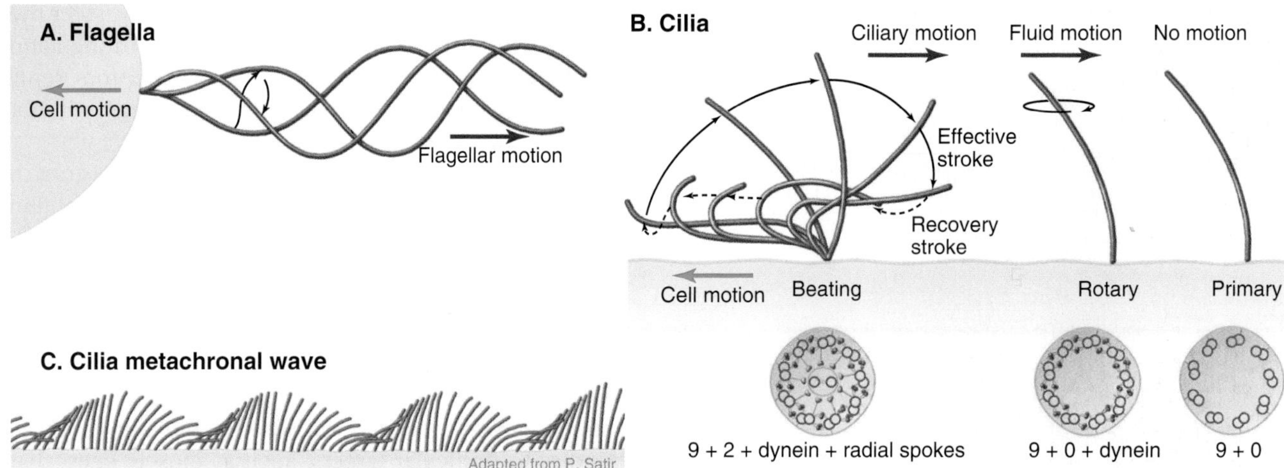

FIGURE 38.12 BEATING PATTERNS OF CILIA AND FLAGELLA. A, Waves of a sperm flagellum. **B,** Behavior of three types of cilia with cross sections of their axonemes below. *Left*, Ciliary power and recovery strokes of a beating cilium. *Middle*, Rotary cilium. *Right*, Primary cilium. **C,** Coordinated beating of cilia on the surface of an epithelium. (Modified from a drawing by P. Satir, Albert Einstein College of Medicine, New York, NY.)

Properties of Cilia and Flagella

Cilia and flagella are distinguished from each other by their beating patterns (Fig. 38.12), but are nearly identical in structure. In fact, the flagella of the green alga *Chlamydomonas* can alternate between propagating waves typical of flagella and the oar-like rowing motion of cilia. So one can use the terms cilia and flagella interchangeably. Subtle differences in the mechanism that converts the dynein-powered sliding of the axonemal microtubules into movements determine which beating pattern is produced.

Both cilia and flagella can propel cells as they cycle rapidly, beating up to 100 times per second. Propagation of bends along the length of individual flagella pushes a cell such as sperm forward (Figs. 38.1 and 38.15). Coordinated beating of many cilia can also move large cells (Fig. 38.13A). Reversal of the direction of the power stroke allows a unicellular ciliate to swim forward or backward. Alternatively, if the cell is immobilized, like the epithelial cells lining an animal respiratory tract or forming the embryonic skin (Fig. 38.13B), coordinated beating of cilia propels fluid and particles over their apical surface. Ctenophores fuse the membranes of many cilia together to make macrocilia that propel the organism in a manner similar to that of fins.

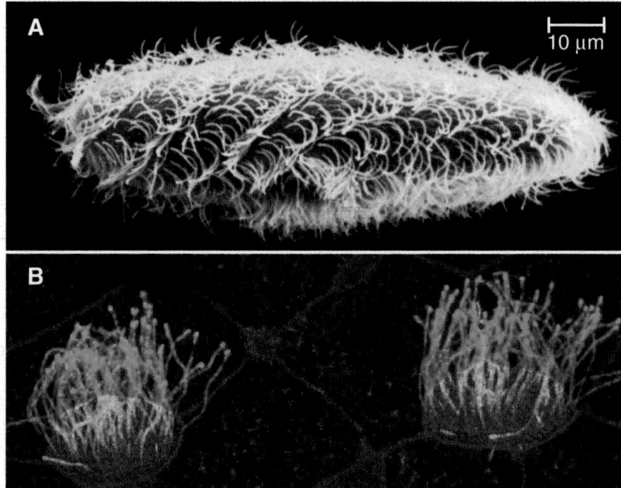

FIGURE 38.13 IMAGES OF CILIA. A, Scanning electron micrograph of Paramecium showing waves of effective strokes passing regularly over the cell surface from one end to the other to keep the cell moving steadily forward. **B,** Fluorescence micrograph of cilia (*green*, stained with fluorescent antibodies to tubulin) on epithelial cells of frog skin also stained with Alexa 647-phalloidin for actin filaments. (**A,** Courtesy T. Hamasaki, Albert Einstein College of Medicine, New York, NY. From Lieberman SJ, Hamasaki T, Satir P. Ultrastructure and motion analysis of permeabilized *Paramecium*. *Cell Motil Cytoskeleton*. 9:73–84, 1988. **B,** Courtesy Brian J Mitchell of Northwestern University. From Werner ME, Hwang P, Huisman F, et al. Actin and microtubules drive differential aspects of planar cell polarity in multiciliated cells. *J Cell Biol*. 2011;195:19–26.)

Structure of the Axoneme

The nine **outer doublet microtubules** of all axonemes consist of one complete A-microtubule with the usual 13 protofilaments, bearing an incomplete B-microtubule composed of 10 protofilaments attached to its side by distinct junctional structures on either side. The filamentous protein tektin associates with one protofilament in the wall of the A-microtubule. The **central pair** are typical 13-protofilament microtubules. The plus ends of all axonemal microtubules are at the distal tip.

More than 200 accessory proteins associate with the 9 + 2 microtubules (Fig. 38.14A), making axonemes stiff but elastic. Pioneering genetic analyses and more recent proteomic studies established the locations of many of these polypeptides, such as the 17 proteins that make up the radial spokes between the central sheath and the outer doublets and the 11 protein subunits of the nexin-dynein regulatory complex that links outer doublets to each other. A long coiled-coil protein acts as a molecular ruler to specify the longitudinal positions of

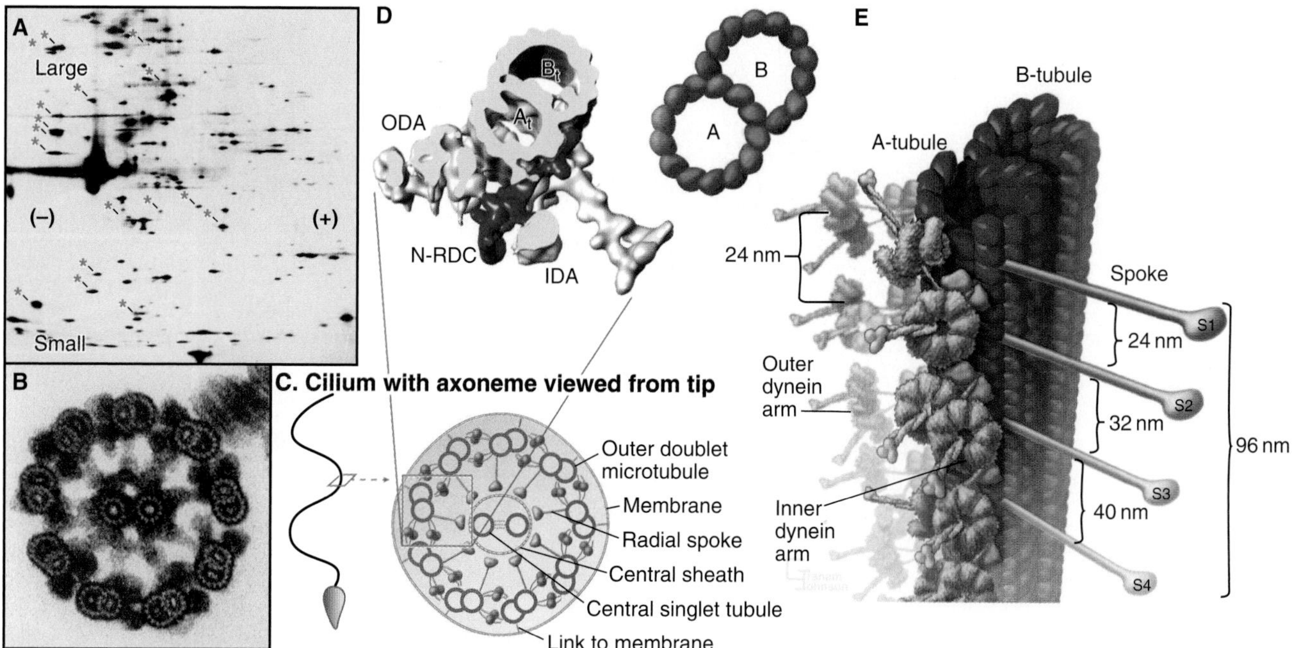

FIGURE 38.14 COMPOSITION AND STRUCTURE OF THE AXONEME. A, Two-dimensional gel electrophoresis separating more than 100 polypeptides of the axoneme of *Chlamydomonas.* Marked polypeptides *(blue asterisks)* are components of radial spokes. **B,** Electron micrograph of a thin cross section of a ciliary axoneme stained with tannic acid. **C,** Drawing of a cross section of a cilium. **D,** Three-dimensional reconstruction of one outer doublet and radial spoke based on electron microscopy tomography showing inner (IDA) and outer dynein arms (ODA) and a nexin-dynein regulatory complex (N-DRC; *blue*). **E,** A short section of an outer doublet. In this example, the outer arm dyneins have two heads. In some species, they have three heads. The dimensions indicate the longitudinal spacing between dynein arms and radial spokes. (**A,** Courtesy B. Huang, Scripps Research Institute, La Jolla, CA. **B,** Courtesy R. Linck, University of Minnesota, Minneapolis. **D,** From Song K, Awata J, Tritschler D, Bower R, et al. In situ localization of N and C termini of subunits of the flagellar nexin-dynein regulatory complex (N-DRC) using SNAP tag and cryo-electron tomography. *J Biol Chem.* 2015;290:5341–5353. **E,** From a Chlamydomonas axoneme modified from Amos LA, Amos WB. *Molecules of the Cytoskeleton.* New York: Guilford Press; 1991.)

many of these proteins with a 96 nm periodicity along the outer doublet microtubules. Central pair microtubules are connected by a bridge and decorated by elaborate projections. Mutations of these genes compromise axonemal function in experimental animals and human patients.

A family of axonemal **dyneins** bound to outer doublets generates force for movement. Each dynein consists of a large heavy chain forming a globular AAA ATPase (adenosine triphosphatase) domain and a tail anchored to an A-tubule by light and intermediate chains. A thin stalk projecting from the catalytic domain exerts force on the adjacent B-tubule during part of the ATPase cycle (see Figs. 36.14 and 36.15). Like other axonemal proteins the pattern of dynein arms repeats every 96 nm along the A-tubule of each outer doublet (Fig. 38.14D). The outer row of dynein arms consists of four copies of three-headed molecules in each repeat. The inner dynein arms are more complicated: seven different dynein heavy chains form six arms with one head plus one arm with two heads in each repeat.

Mechanism of Axoneme Bending

Dynein-powered sliding of outer doublets relative to each other bends axonemes. Sliding was first inferred from electron micrographs of the distal tips of microtubules in bent cilia. Later, sliding was observed directly by loosening connections between outer doublets with proteolytic enzymes and then adding adenosine triphosphate (ATP) to allow dynein to push the microtubules past each other (Fig. 38.15B). Sliding can be followed in axonemes stripped of their membrane by marking outer doublets with small gold beads (Fig. 38.15A). As outer doublets slide past each other, the relative positions of the beads change. Dynein attached to one doublet "walks" toward the base of the adjacent microtubule, pushing its neighbor toward the tip of the axoneme.

Biochemical extraction or genetic deletion of specific dynein isoforms alters the frequency and waveform of axonemal bending. Inner dynein arms are required for flagellar beating, and deletion of even a single type of inner-arm dynein can alter the waveform. Outer dynein arms are not essential but influence the beat frequency and add power to the inner arms. Humans with **Kartagener syndrome** lack visible dynein arms and have immotile sperm and cilia. As a result, affected males are infertile, and both men and women have serious respiratory infections, owing to poor clearance of bacteria and other foreign matter from the lungs.

The mechanism of beating is intrinsic to the axoneme. Thus, sperm tail axonemes swim normally when provided

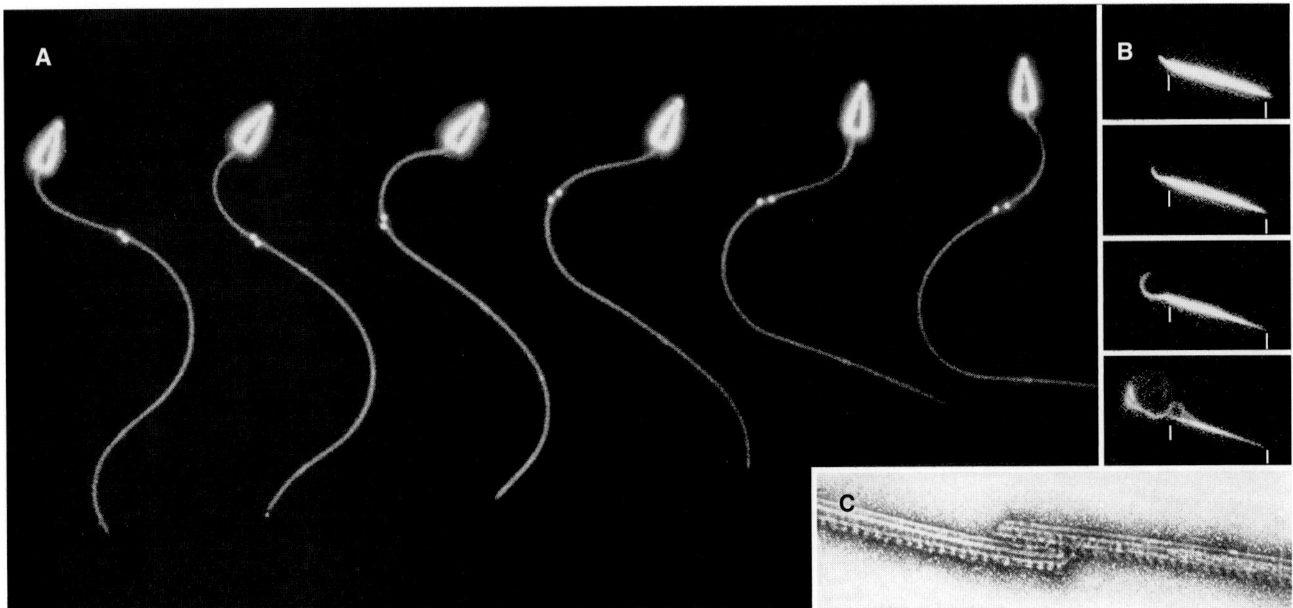

FIGURE 38.15 SLIDING MOVEMENTS OF OUTER DOUBLETS OF AXONEMES. A–B, Time series of darkfield light micrographs. **A,** Sea urchin sperm extracted with the detergent Triton X-100 and reactivated with ATP. Gold microbeads attached to two different outer doublets allow the visualization of their displacement as the tail bends. **B,** Fragment of a sea urchin flagellar axoneme treated with trypsin. The addition of ATP results in outer doublets sliding past each other out of the ends of the axonemal fragment. **C,** Electron micrograph of two outer doublets that have slid past each other in an experiment similar to that in panel B. (**A,** From Brokaw CJ. Microtubule sliding in swimming sperm flagella. *J Cell Biol*. 1991;114:1201–1215, copyright The Rockefeller University Press. **B,** Courtesy Ian Gibbons, University of California, Berkeley. For more information, see Summers KE, Gibbons I. ATP-induced sliding of tubules in trypsin-treated flagella of sea-urchin sperm. *Proc Natl Acad Sci U S A*. 1971;68:3092–3096. **C,** Courtesy P. Satir, Albert Einstein College of Medicine, New York, NY. For more information, see Sale WS, Satir P. Direction of active sliding of microtubules in *Tetrahymena* cilia. *Proc Natl Acad Sci U S A*. 1977;74:2045–2049.)

with ATP, even without the plasma membrane or soluble cytoplasmic components (Fig. 38.15A). Experiments with these demembranated sperm models revealed that the dynein ATPase activity is tightly coupled to movement. The beat frequency is proportional to ATPase activity, regardless of whether the frequency is limited by increasing the viscosity of the medium or the enzyme activity is limited by decreasing the ATP concentration.

The bending that produces the bending waves of flagella or the power and recovery strokes of cilia results from local variation in the rate of sliding of pairs of outer doublet microtubules along the length of an axoneme. Coordination of these events is still being investigated, but at least two factors are involved. Mutations show that the central pair and radial spokes help coordinate the activity of the dyneins around the circumference of the axoneme as it bends. Mechanical constraints are also required to convert microtubule sliding into local bending. Destruction of the links between outer doublets frees them to slide past each other rather than bending the axoneme.

Although axonemes function autonomously, signal transduction pathways regulate their activities. Phototaxis of *Chlamydomonas* is a particularly clear example of how fluctuations in intracellular Ca^{2+} can modify flagellar activity. The release of Ca^{2+} affects the two flagella of the organism differentially and allows a cell to steer toward or away from light (Fig. 38.16). Ciliates also have mechanosensitive channels that depolarize the plasma membrane when the organism collides with something. Depolarization opens voltage-sensitive plasma membrane Ca^{2+} channels, admitting Ca^{2+} into the cell. This reverses the direction of ciliary beat. Both calcium and cAMP-dependent phosphorylation of outer-arm dynein can change the beat frequency (all the way to zero) or alter the waveform.

Basal Bodies and Axoneme Formation

The mother centriole matures into a **basal body** that anchors and templates the axoneme (Fig. 38.17; see also Fig. 34.3B). After migrating to the cortex during interphase, its distal appendages dock on the plasma membrane and the nine outer doublet microtubules of the axoneme grow directly from the nine outer triplet microtubules of the basal body. This differs from microtubule nucleation in the pericentriolar material during interphase (see Fig. 34.15). Some protozoa use basal bodies as centrioles during mitosis. Multiciliated cells form a basal body for each axoneme de novo. Basal bodies are typically anchored by protein fibers called rootlets. Cilia seem to function normally in mice lacking the major rootlet protein, but are unstable over the long term. Mutations of genes for basal body proteins compromise axonemal function in experimental animals and cause human diseases.

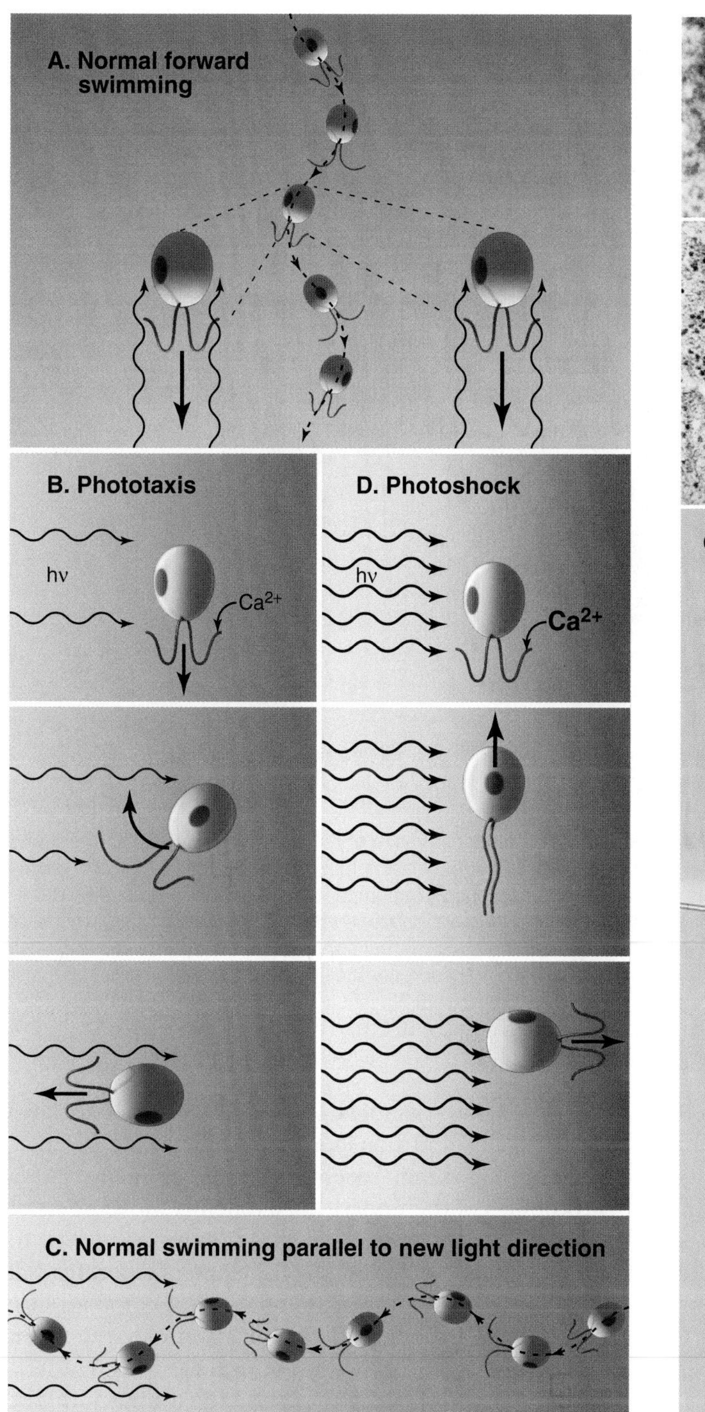

FIGURE 38.16 ***CHLAMYDOMONAS*** **PHOTOTAXIS. A,** Normal swimming toward the light using a cilia-like rowing motion of the flagella. Absorption of light by a sensory rhodopsin (related to sensory rhodopsins in *Archaea*) in the eyespot keeps the cell oriented. **B,** Moderate-intensity light from the side causes Ca^{2+} to enter the cytoplasm from outside the cell. The two flagella react differently, causing the cell to turn toward the light. **C,** Once the cell is reoriented, the flagella beat equally, and the cell swims toward the light. **D,** High-intensity light releases a high concentration of Ca^{2+} and causes transient wave-like motion of the flagella. This backward swimming allows the cell to reorient and to swim away from the light.

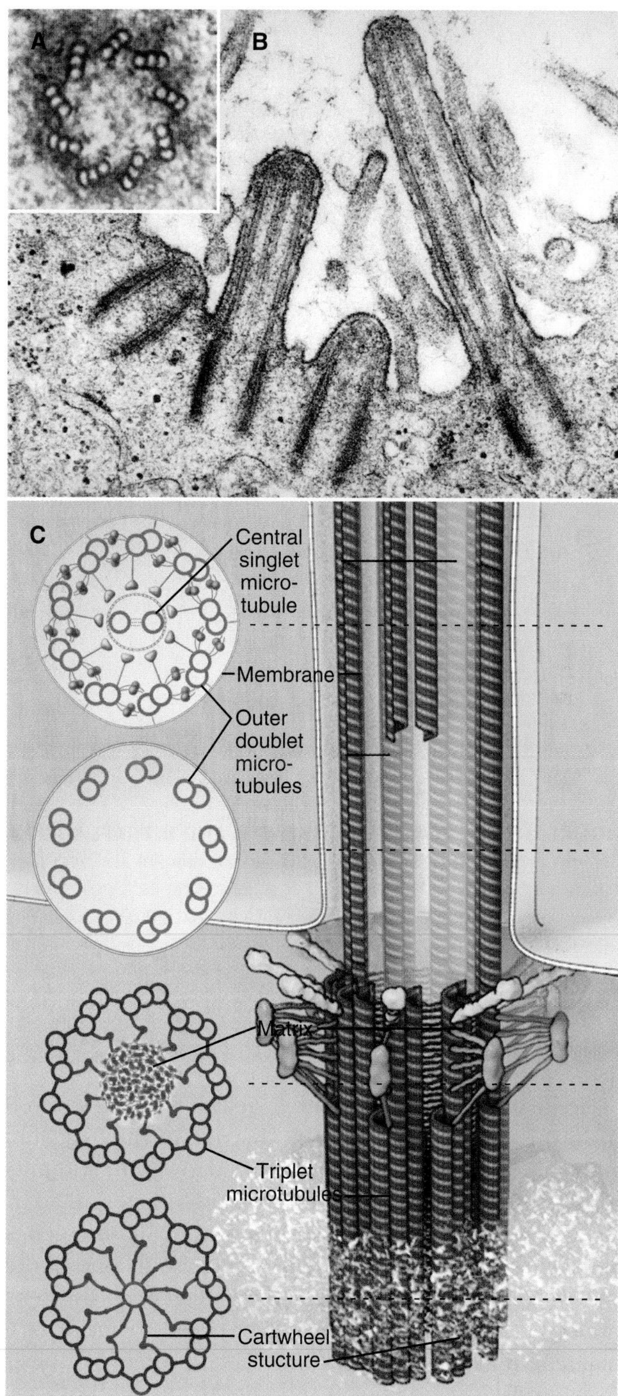

FIGURE 38.17 BASAL BODIES. A, Electron micrograph of a thin cross section of a basal body. **B,** Electron micrograph of thin longitudinal section of basal bodies and proximal axonemes of cilia. **C,** Drawings of cross sections and a three-dimensional (3D) drawing of the basal body and proximal flagella of *Chlamydomonas.* In the 3D drawing, the near side outer doublets are cut away to reveal the central pair microtubules. (**A–B,** Courtesy D.W. Fawcett, Harvard Medical School, Boston, MA. **C,** Modified from Amos LA, Amos WB. *Molecules of the Cytoskeleton*. New York: Guilford Press; 1991.)

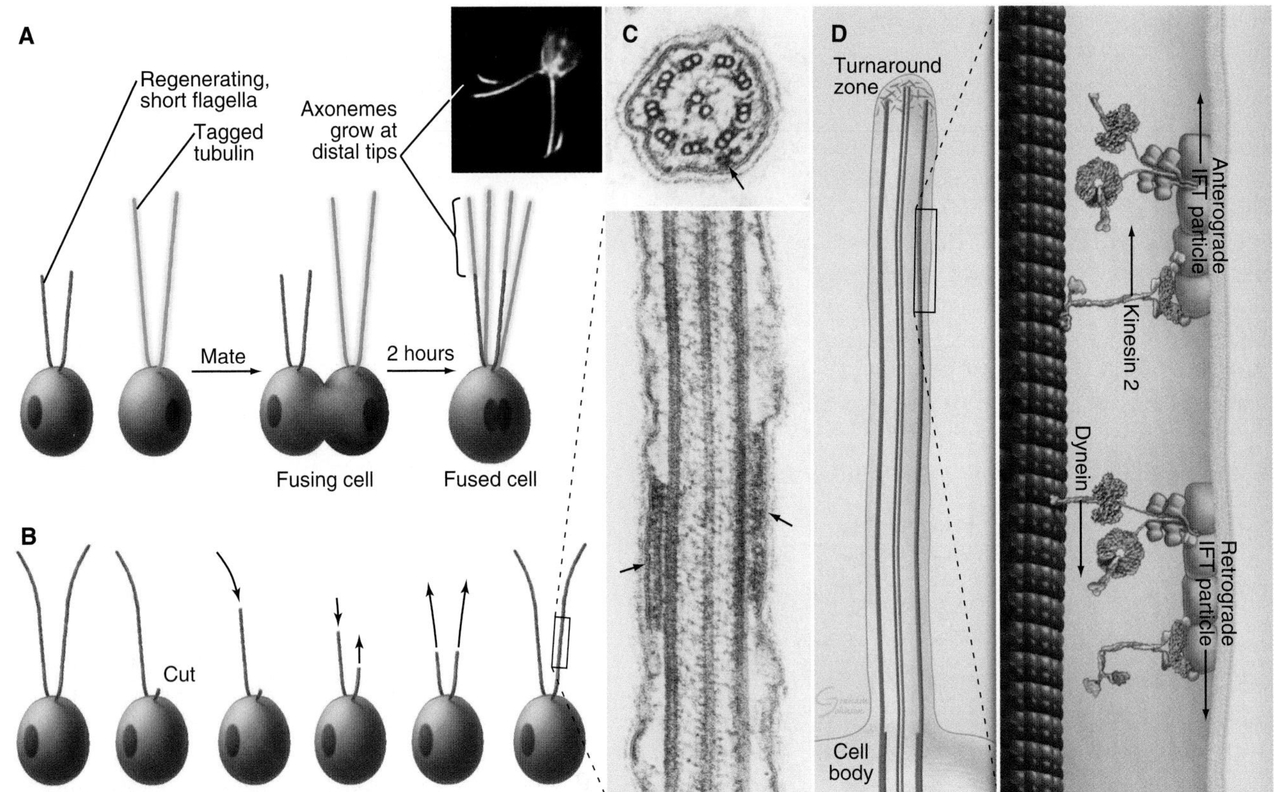

FIGURE 38.18 FLAGELLAR GROWTH AND INTRAFLAGELLAR TRANSPORT. A, Incorporation of protein subunits at the tip of growing *Chlamydomonas* flagella is revealed by an experiment involving the fusion of two cells, one expressing tubulin with an epitope tag that reacts with a specific antibody and the other regenerating its flagella. As is shown in the fluorescence micrograph, tagged tubulin is incorporated only at the distal tips of the growing flagella. Cells with paralyzed flagella made this experiment more convenient. **B,** Time course of regeneration of *Chlamydomonas* flagellum following amputation of one flagellum. The surviving flagellum shortens transiently before both grow out together. **C,** Electron micrographs of thin sections of *Chlamydomonas* flagella showing intraflagellar transport particles *(arrows)*. **D,** Model for intraflagellar transport (IFT). (**A,** Courtesy K. Johnson, Haverford College, Haverford, PA. *Inset,* From Johnson KA, Rosenbaum JL. Polarity of flagellar assembly in *Chlamydomonas*. *J Cell Biol*. 1992;119:1605–1611, copyright The Rockefeller University Press. **B,** Based on the work of J. Rosenbaum, Yale University, New Haven, CT. **C,** Courtesy Joel Rosenbaum, Yale University, New Haven, CT.)

Some organisms regenerate flagella if they are severed from the cell (Fig. 38.18A–B). Absence of the flagellum activates expression of genes required to supply subunits for regrowth of the axoneme. In approximately 1 hour, the cell regrows a replacement flagellum, and the genes are turned off. Even more remarkably, if only one of the two flagella is lost, the remaining flagellum shortens rapidly to provide components required to make two half-length flagella (Fig. 38.18B). Then protein synthesis slowly provides additional subunits to restore both flagella to full length.

Axonemes grow at their tips by incorporation of subunits synthesized in the cytoplasm (Fig. 38.18A). A process called **intraflagellar transport** (IFT) (Fig. 38.18C–D) carries individual proteins and subassemblies such as radial spokes to the growing tip. These cytoplasmic cargo proteins bind one of two IFT complexes, which associate 1 : 1 to form larger "trains" visible of electron microscopy (Fig. 38.18C). Kinesin-2 moves IFT trains toward the tip of the axoneme along the outer doublets just beneath the plasma membrane. Cytoplasmic dynein 1b transports particles back toward the cell body. A separate complex, called a BBSome, associates with IFT trains and transports transmembrane proteins (including signaling receptors) bidirectionally along microtubules of the underlying axoneme. Movements of transmembrane proteins allows *Chlamydomonas* to glide on surfaces including the flagellum of a partner cell during mating. Because this motion does not require beating of the axoneme, the mechanism may represent an early stage in the evolution of flagella.

IFT is remarkably similar to fast axonal transport (see Figs. 37.1 and 37.3) but on a smaller scale. Cargo proteins are loaded onto IFT complexes at the base of the cilium and then must pass through filters located just above the basal body. These filters have a cutoff of approximately 50 kD but pass much larger IFT complexes. Proteins, including the Ran GTPase, importins, and proteins of the nuclear pore complex (see Chapter 9), participate in this filtration system. After transport cargo proteins dissociate at the flagellar tip. Phosphorylation of kinesin at the tip of the axoneme may reverse transport for the return trip to the cell body. Trains are full of tubulin and other axonemal proteins in growing cilia; they keep moving bidirectionally but are largely empty when cilia are not growing.

Rotary Cilia

Single cilia on the epithelial cells of the "ventral node" of vertebrate embryos are required for the asymmetric location of some internal organs, such as the heart and liver, on opposite sides of the body. These nodal cilia lack the central pair microtubules and radial spokes. Rather than beating, the activity of the dynein arms causes the tip of the cilia to rotate clockwise. This rotary motion propels the extracellular fluid carrying certain growth factors toward the left side of the embryo. The absence of this flow explains why patients with Kartagener syndrome and mice missing a single dynein heavy chain have an equal chance of having their internal organs, such as heart and liver, positioned normally or on the opposite side, a condition called **situs inversus.** Rotary cilia may provide clues about an intermediate stage in the evolution of axonemes.

Primary Cilia

Except for blood cells, differentiated cells in vertebrate tissues produce a single **primary cilium** by growth of an axoneme from their older mother centriole (Fig. 38.19).

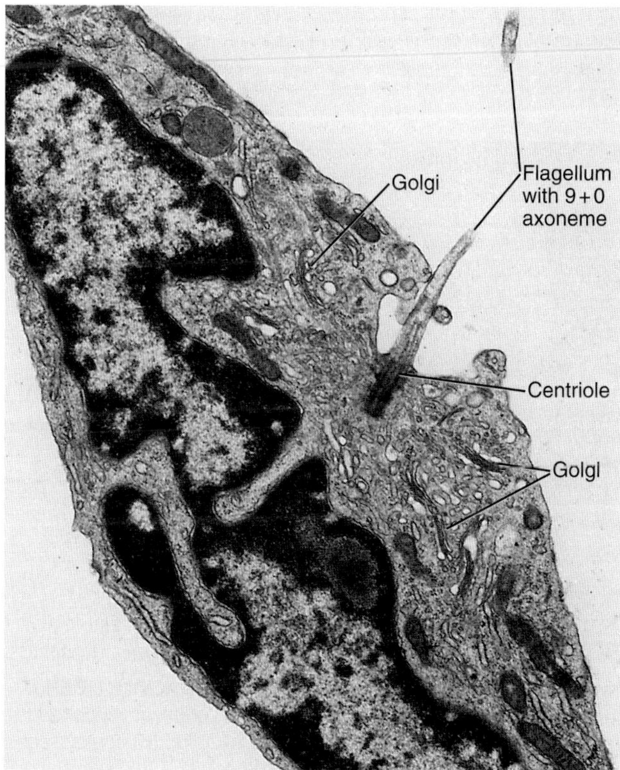

FIGURE 38.19 PRIMARY CILIUM. Electron micrograph of a thin section of a mesenchymal cell with a primary cilium assembled from one of the two centrioles, which serves as the basal body. (From Fawcett DW. *The Cell*. Philadelphia, PA: WB Saunders; 1981.)

The axonemes lack the central pair, and most lack dynein, so they are immotile (Fig. 38.12B).

Primary cilia are sensory organelles for both chemicals and mechanical forces. For example, odorant receptors of nematode olfactory neurons concentrate in the membranes of primary cilia. Rod and cone photoreceptors in the eye are modified cilia with a basal body and a vestigial axoneme (see Fig. 27.2). Primary cilia on the epithelial cells of kidney tubules act as flow sensors, admitting Ca^{2+} into the cilium through mechanosensitive channels in the plasma membrane when bent. Many receptors concentrate in primary cilia including those for developmental morphogens such as sonic hedgehog and Wnt, growth factors including platelet-derived growth factor, and hormones like somatostatin. Some ligands activate local Ca^{2+} release into the cilium.

Ciliopathies

Genetic deficiencies in the assembly of cilia or IFT result in a remarkably wide range of human disease syndromes, known collectively as **ciliopathies**. The underlying mutations are found in more than 20 genes encoding proteins for axonemes, basal bodies, and IFT. The defects can appear in virtually any organ, illustrating the diverse functions of primary and motile cilia. An early example was polycystic kidney disease, the most common cause of kidney failure. The most frequent mutations are in genes for a Trp-family calcium channel, but other patients have mutations in genes for IFT proteins. Kidney epithelial cells form abnormal cysts rather than tubules, perhaps as a result of abnormal cell division. Other ciliopathy mutations cause defects in the central and peripheral nervous systems, olfactory neurons, ear, liver, retina, skeleton, and reproductive organs. Patients with some ciliopathy syndromes are obese or have extra digits.

Box 38.1 discusses exotic eukaryotic motility systems.

Bacterial Flagella

Bacteria use a reversible, high-speed, rotary motor driven by H^+ or Na^+ gradients to power their flagella (Figs. 38.24 and 38.25). Bacterial flagella differ in every respect from eukaryotic cilia and flagella. The bacterial flagellum is an extracellular protein wire (see Fig. 5.8), not a cytoskeletal structure like an axoneme inside the plasma membrane. Bacteria with multiple flagella are more common than those with single flagellum. The principles derived from studies of *Escherichia coli* and a few other bacteria apply generally, although other species exhibit many variations on this theme.

A motor, embedded in the plasma membrane, turns the bacterial flagellum either clockwise or counterclockwise (viewed from the tip of the flagellum) like the propeller of a motorboat. In contrast to a motorboat, moving bacteria have no momentum, so they stop in a fraction of a nanometer if the motor stops. When

BOX 38.1 Exotic Eukaryotic Motility Systems

In contrast to conventional motility systems used by eukaryotic cells discussed in the main text, a few eukaryotes evolved exotic motility systems, four examples of which are described here.

A Preformed Actin Filament Spring

Sperm of the horseshoe crab, *Limulus*, use a novel acrosomal process to fertilize an egg (Fig. 38.20). They preassemble a coiled bundle of actin filaments crosslinked by a protein called scruin. This bundle is a tightly coiled spring. An encounter with an egg stimulates rearrangement of the crosslinks, causing the actin bundle to unwind. Uncoiling drives the bundle through a channel in the nucleus followed by extension of a process surrounded by plasma membrane that literally screws its way through the egg jelly to fuse with the egg plasma membrane.

Calcium-Sensitive Contractile Fibers

The ciliate *Vorticella* avoids predators by contracting a stalk that anchors the cell to leaves or other supports (Fig. 38.21). The contractile fibril, called a **spasmoneme,** contracts faster than any muscle. Ca^{2+} released from tubular membranes associated with the spasmoneme triggers contractions, when it binds to spasmin, a calmodulin-like protein that forms 3-nm filaments. Ca^{2+} binding changes the conformation of spasmin and results in rapid shortening, because many spasmin subunits are assembled in series. The spasmoneme relaxes when Ca^{2+} dissociates. Energy for contraction is supplied indirectly when ATP-driven pumps create a Ca^{2+} gradient between the lumen of the membrane system and cytoplasm. Movement of Ca^{2+} down this gradient drives contraction.

Proteins similar to spasmin are found in other ciliates, algae, fungi, and animals, where they are called **centrin** or **caltractin.** These calmodulin-like proteins form fibrils that anchor centrosomes and the basal bodies of cilia and flagella. Mutations that inactivate caltractin in algae or yeast compromise the functions of the microtubule organizers (centrosomes or spindle pole bodies; see Figs. 34.15 and 34.20) used for mitosis.

Major Sperm Protein, an Actin Substitute in Nematode Sperm

Nematode sperm use amoeboid movements to find an egg rather than swimming with flagella like other sperm (Fig. 38.22). The behavior of these sperm is so similar to a small amoeba cell that anyone would have guessed that it is based on the assembly of actin filaments. However, actin is a minor protein in nematode sperm. Instead, sperm pseudopods are filled with 10-nm wide, apolar filaments assembled from dimers of a 14-kD protein with an immunoglobulin-like fold called **major sperm protein.** Proteins in the cytoplasm and associated with the plasma membrane guide the assembly of the filaments, which function remarkably like actin, even though they have no bound nucleotide and no known associated motor protein. The 10-nm filaments assemble at the leading edge of the pseudopod and remain stationary with respect to the substrate as the expanding pseudopod

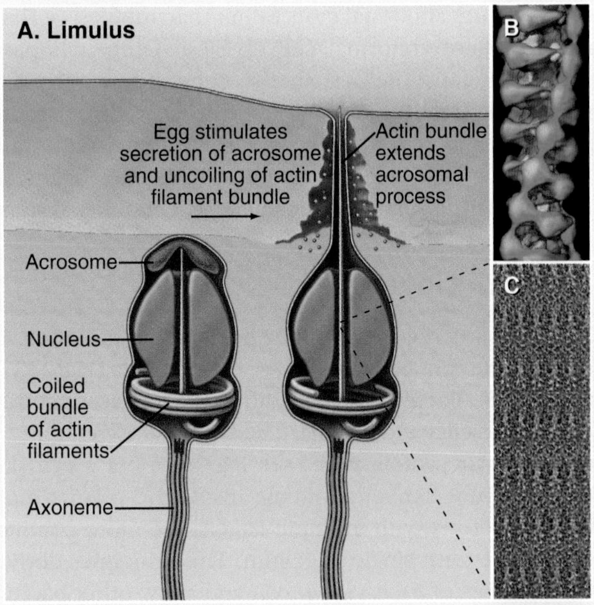

A. Limulus
B
C
Egg stimulates secretion of acrosome and uncoiling of actin filament bundle
Actin bundle extends acrosomal process
Acrosome
Nucleus
Coiled bundle of actin filaments
Axoneme

FIGURE 38.20 LIMULUS SPERM ACROSOMAL PROCESS. **A,** Uncoiling of a bundle of actin filaments extends the acrosomal process of the sperm of the horseshoe crab, *Limulus*. **B–C,** Electron micrograph of the actin filament bundle from the acrosomal process of *Limulus* and a three-dimensional reconstruction of one filament *(yellow)* decorated with crosslinking proteins *(green)*. (Based on the work of L. Tilney, University of Pennsylvania, Philadelphia. **B–C,** Courtesy W. Chiu, Baylor College of Medicine, Houston, TX.)

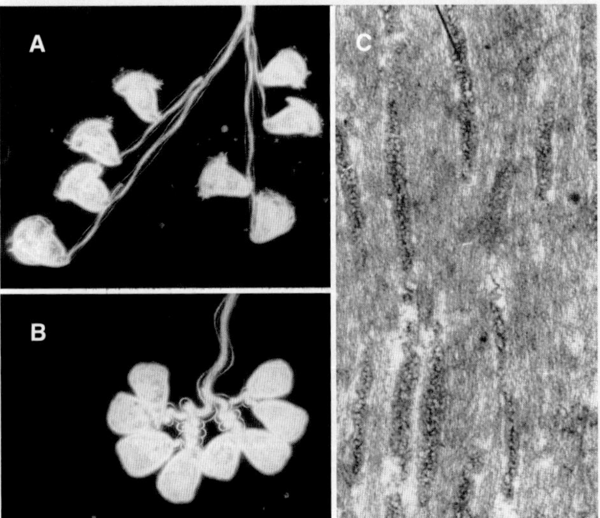

FIGURE 38.21 CALCIUM-SENSITIVE CONTRACTILE FIBERS. **A–B,** Light micrographs of a group of vorticellid protozoa suspended from the bottom of a leaf, taken before **(A)** and after **(B)** contraction of their spasmonemes. **C,** Electron micrograph of a thin section of contractile fibers and tubular membranes that store and release calcium. (Courtesy W.B. Amos, MRC Laboratory of Molecular Biology, Cambridge, United Kingdom.)

BOX 38.1 Exotic Eukaryotic Motility Systems—cont'd

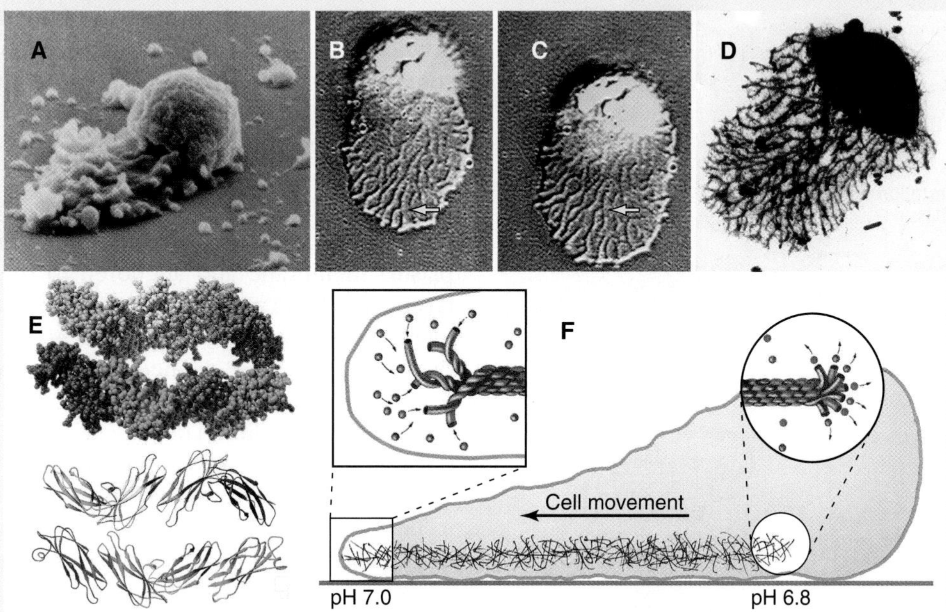

FIGURE 38.22 MOTILITY OF NEMATODE SPERM. A, Scanning electron micrograph of an amoeboid sperm showing the anterior pseudopod and trailing cell body. **B–C,** Time series of differential interference contrast light micrographs showing movement of a live sperm by assembly of a network of fibers at the leading edge. *Arrows* mark the same point in the network, which is stationary with respect to the substrate. **D,** Transmission electron micrograph of an extracted sperm showing the fibers. **E,** Atomic model of a short segment of the sperm filaments consisting of a polymer of major sperm protein (MSP). **F,** Cycle of MSP assembly at the leading edge and disassembly at the cell body. (Courtesy T. Roberts, Florida State University, Tallahassee, and M. Stewart, MRC Laboratory of Molecular Biology, Cambridge, United Kingdom.)

advances. Filament bundles depolymerize at the interface between the pseudopod and the spherical cell body. A pH gradient promotes assembly of major sperm protein at the front and disassembly at the rear of the pseudopod. This highly efficient motility system is still unknown in other parts of the phylogenetic tree.

Axostyles, Specialized Microtubular Organelles

Some protozoa use dynein to generate beating movements of large arrays of cytoplasmic microtubules called **axostyles** (Fig. 38.23). The mechanism seems to be similar to an axoneme, although the organization clearly differs. Crosslinking structures hold together sheets of singlet microtubules, which slide past each other as a result of the action of dynein motors on adjacent sheets. Coordinated beats of the axostyle distort the whole organism, allowing it to wiggle about.

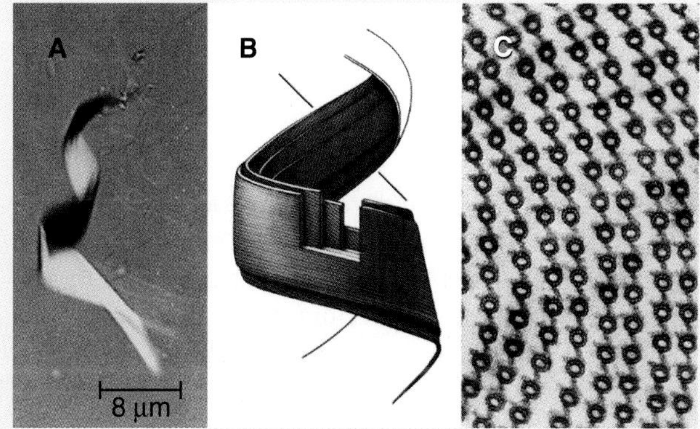

FIGURE 38.23 MOTILE AXOSTYLE OF *SACCINOBACULUS*, A PROTOZOAN PARASITE OF TERMITES. The twisting motions of this intracellular assembly of microtubules cause the whole parasite to twist and turn in the gut of termites. **A,** Polarization light micrograph of an isolated axostyle. **B,** Drawing of part of the axostyle showing the arrangement of sheets of crosslinked microtubules. **C,** Transmission electron micrograph of a cross section of the axostyle showing microtubules crosslinked into sheets with dynein arms between the sheets. (Courtesy R. Linck, University of Minnesota, Minneapolis. From Woodrum D, Linck R. Structural basis of motility in the microtubular axostyle. *J Cell Biol.* 1980;87:404–414, copyright The Rockefeller University Press.)

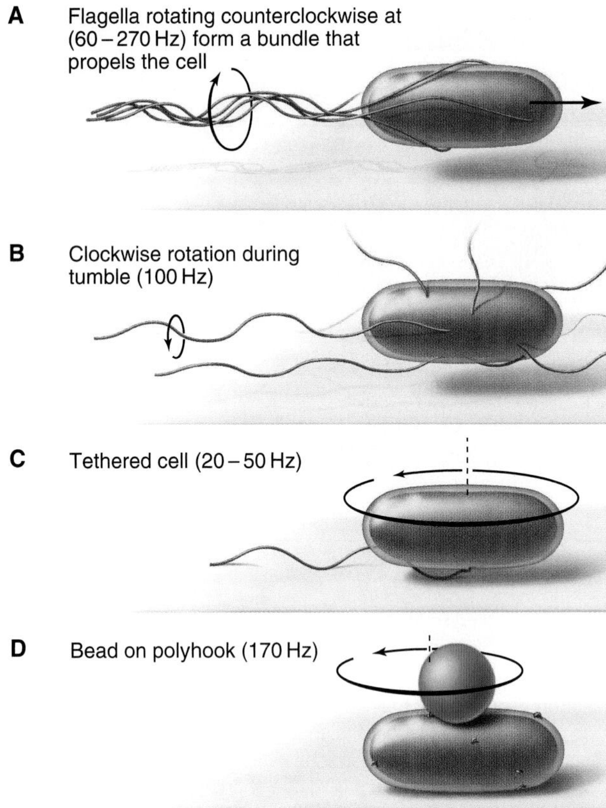

FIGURE 38.24 DIFFERENT MANIFESTATIONS OF THE ROTATION OF FLAGELLA. A, If the flagella rotate counterclockwise, they form a bundle that propels the cell forward. **B,** If one or more flagella rotate clockwise, the bundle falls apart and the cell tumbles in one place. **C,** If a flagellum is tethered to a surface, the bacterium rotates. **D,** If the flagellar filament is replaced by an elongated hook region with an attached bead, the bead rotates. (Modified from Schuster SD, Khan S. The bacterial flagellar motor. *Annu Rev Biophys Biomol Struct*. 1994;23:509–539.)

multiple flagella are present, counterclockwise rotation forms a bundle. Four flagella propel *E. coli* 30 μm s^{-1}, a velocity of 15 cell lengths per second, equivalent to 400 miles per hour if the bacterium were the size of an automobile. When one or more flagella reverse their direction and rotate clockwise, the bundle flies apart, and the cell tumbles in one place. Figs. 27.12 and 27.13 explain how chemotactic stimuli control the probability of clockwise rotation, favoring steady runs toward nutrients and allowing for more frequent tumbles to change direction to avoid harm.

Assays for rotation of single flagella provide insights about the mechanism of flagellar motion (Fig. 38.24). When a flagellum is attached to a glass slide by means of antibodies to the flagellar filament, the bacterium rotates, providing decisive evidence for rotation of flagella. Similarly, beads attached to short flagella are observed to rotate. The rotational speed depends on the resistance. The motor of a single immobilized flagellum can rotate a whole *E. coli* 10 to 50 times per second, whereas in some species unloaded motors rotate up to 1600 times per second (100,000 rpm)!

The rotary engine driving the flagellar filament is constructed from a rotor and stator. The cylindrical **basal body** on the end of the filament rotates inside the stator, a ring of stationary proteins embedded in the plasma membrane and anchored to the peptidoglycan layer (Fig. 38.25). Genetic screens for motility mutants identified all the protein components of the motor, and their functions were defined by analysis of the behavior of these mutants. Most of these proteins are present in isolated basal bodies. The functional units of the stator consist of four MotA subunits and two MotB subunits. MotA has four hydrophobic segments that are believed to be transmembrane helices and a substantial cytoplasmic domain. MotB has one transmembrane segment and a large periplasmic domain anchored to the peptidoglycan layer. Flagella and basal bodies assemble but are immotile in cells lacking either of these transmembrane proteins. If the missing protein is replaced by initiating its biosynthesis, the paralyzed flagellum begins to turn, increasing its speed of rotation in a stepwise fashion, as 10 to 12 independent, torque-producing $MotA_4MotB_2$ units are added one after another.

The energy to turn the motor comes from protons (or, in some bacteria, Na^+ ions) that move down an electrochemical gradient from outside the bacterium through the $MotA_4MotB_2$ units to the cytoplasm. Transfer of one proton across the membrane provides approximately the same energy as the hydrolysis of an ATP. Pumps driven by light, oxidation, or ATP hydrolysis (see Table 14.1) generate the **proton gradient.** MotA is part of the proton channel, because mutations in its gene inhibit both flagellar rotation and proton permeability. The MotB transmembrane helix has a conserved aspartic acid residue that interacts with the proton crossing the membrane.

The mechanism producing rotation is still under investigation, but it involves interaction of the cytoplasmic domain of MotA with FliG subunits on the top of the C-ring. The transfer of a proton across the plasma membrane results in a conformational change in MotA that moves the C-ring. Roughly 1000 protons cross the membrane for each rotation, corresponding to two protons for each tiny rotational step. Proton transfer is tightly coupled to rotation of the basal body, and the efficiency is near 100%.

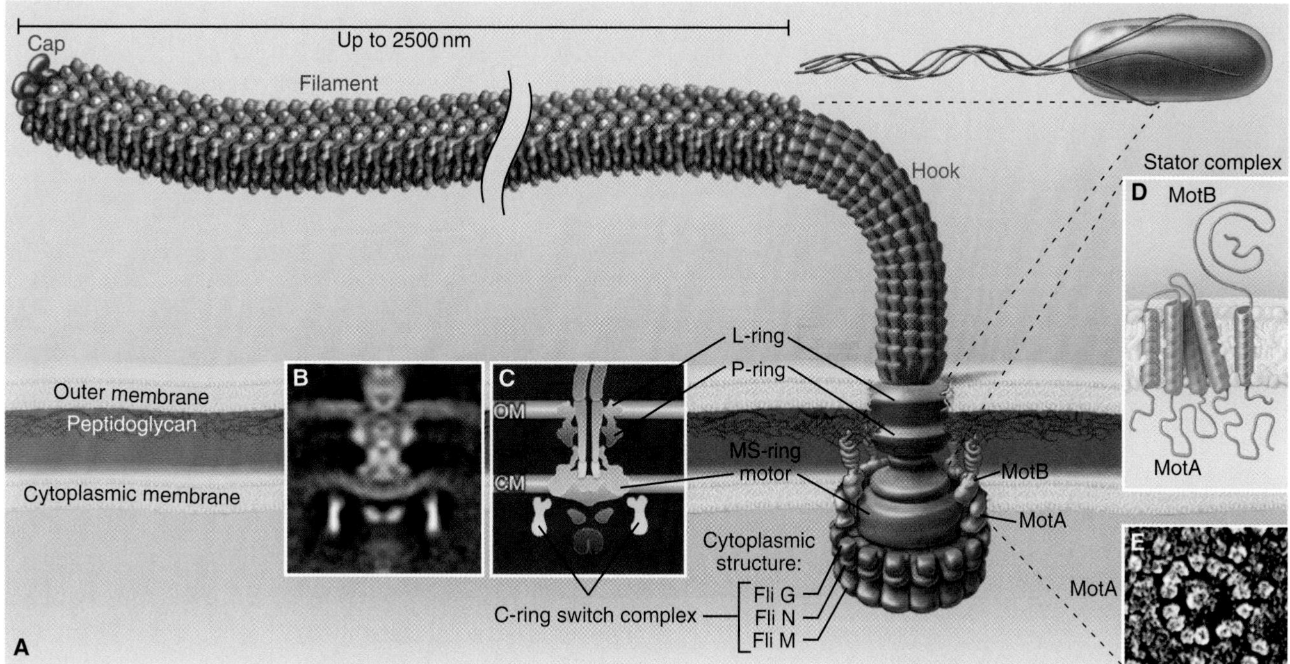

FIGURE 38.25 BACTERIAL ROTARY MOTOR. A, Drawing of the flagellar filament and rotary motor. **B–C,** Three-dimensional reconstruction and schematic cross section of basal body of the flagellar motor from *Escherichia coli*. **D,** Details of the stator, with schematic diagrams of the transmembrane and cytoplasmic parts of $MotA_4MotB_2$ hexamer. **E,** Electron micrograph of a freeze-fractured bacterium illustrating the ring of intramembranous particles comprising the stator of $MotA_4$ $MotB_2$ hexamers. (**A,** Modified from Schuster SD, Khan S. The bacterial flagellar motor. *Annu Rev Biophys Biomol Struct*. 1994;23:509–539. **B–C,** From Zhao X, Norris SJ, Liu J. Molecular architecture of the bacterial flagellar motor in cells. *Biochemistry*. 2014;53:4323–433. **D,** Stator unit with a ribbon diagram of the C-terminal domain of MotB. Schematic diagrams of MotA and MotB based on Kojima S, Imada K, Sakuma M, et al. Stator assembly and activation mechanism of the flagellar motor by the periplasmic region of MotB. *Mol Microbiol*. 2009;73:710–718. **E,** Courtesy S. Khan, Albert Einstein College of Medicine, New York, NY.)

SELECTED READINGS

Blanchoin L, Boujemaa-Paterski R, Sykes C, et al. Actin dynamics, architecture, and mechanics in cell motility. *Physiol Rev*. 2014;94: 235-263.

Carmeliet P, Tessier-Lavigne M. Common mechanisms of nerve and blood vessel wiring. *Nature*. 2005;436:193-2000.

Condeelis J, Singer RH, Segall JE. The great escape: When cancer cells hijack the genes for chemotaxis and motility. *Annu Rev Cell Dev Biol*. 2005;21:695-718.

Daniels DR. Effect of capping protein on a growing filopodium. *Biophys J*. 2010;98:1139-1148.

Devreotes P, Horwitz AR. Signaling networks that regulate cell migration. *Cold Spring Harb Perspect Biol*. 2015;3:a005959.

Gambardella L, Vermeren S. Molecular players in neutrophil chemotaxis—focus on PI3K and small GTPases. *J Leukoc Biol*. 2013;94:603-612.

Goetz SC, Anderson KV. The primary cilium: a signalling centre during vertebrate development. *Nat Rev Genet*. 2010;11:331-344.

Kim S, Dynlacht BD. Assembling a primary cilium. *Curr Opin Cell Biol*. 2013;25:506-511.

Kolodkin AL, Tessier-Lavigne M. Mechanisms and molecules of neuronal wiring: a primer. *Cold Spring Harb Perspect Biol*. 2011;3:a001727.

Lechtreck KF. IFT-cargo interactions and protein transport in cilia. *Trends Biochem Sci*. 2015;40:765-778.

Levin M. Left-right asymmetry in embryonic development: A comprehensive review. *Mech Dev*. 2005;122:3-25.

Lin J, Okada K, Raytchev M, et al. Structural mechanism of the dynein power stroke. *Nat Cell Biol*. 2014;16:479-485.

Martin AC, Goldstein B. Apical constriction: themes and variations on a cellular mechanism driving morphogenesis. *Development*. 2014; 141:1987-1998.

Mizuno N, Taschner M, Engel BD, et al. Structural studies of ciliary components. *J Mol Biol*. 2012;422:163-180.

Mogilner A, Rubinstein B. The physics of filopodial protrusion. *Biophys J*. 2005;89:782-795.

Moriyama Y, Okamoto H, Asai H. Rubber-like elasticity and volume changes in the isolated spasmoneme of giant *Zoothamnium* sp. under Ca^{2+}-induced contraction. *Biophys J*. 1999;76:993-1000.

Petrie RJ, Yamada KM. Fibroblasts lead the way: a unified view of 3D cell motility. *Trends Cell Biol*. 2015;25:666-674.

Pigino G, Ishikawa T. Axonemal radial spokes: 3D structure, function and assembly. *Bioarchitecture*. 2012;2:50-58.

Pollard TD, Borisy GG. Cellular motility driven by assembly and disassembly of actin filaments. *Cell*. 2003;112:453-465.

Reiter JF, Blacque OE, Leroux MR. The base of the cilium: roles for transition fibres and the transition zone in ciliary formation, maintenance and compartmentalization. *EMBO Rep*. 2012;13:608-618.

Ridge KD. Algal rhodopsins: Phototaxis receptors found at last. *Curr Biol*. 2002;12:R588-R590.

Ridley AJ, Schwartz MA, Burridge K, et al. Cell migration: Integrating signals from front to back. *Science*. 2003;302:1704-1709.

Rørth P. Reach out and touch someone. *Science*. 2014;343:848-849.

Smith HE. Nematode sperm motility. *WormBook*. 2014;4:1-15.

Witman G. Chlamydomonas phototaxis. *Trends Cell Biol*. 1993;3: 403-408.

Zhao X, Norris SJ, Liu J. Molecular architecture of the bacterial flagellar motor in cells. *Biochemistry*. 2014;53:4323-4433.

CHAPTER 39

Muscles

Muscles use actin and myosin to generate powerful, unidirectional movements (Fig. 39.1). The molecular strategies are specialized versions of those used by other cells to produce contractions, to adhere to each other and the extracellular matrix, and to control their activity.

Vertebrates have three types of specialized contractile cells: smooth muscle, skeletal muscle, and cardiac muscle. These muscles have much in common, but differ in their activation mechanisms, arrangement of contractile filaments, and energy supplies. The nervous system controls the timing, force, and speed of skeletal muscle contraction over a wide range. Cardiac muscle generates its own rhythmic contractions that spread through the heart in a highly reproducible fashion. Neurotransmitters, acting like hormones, regulate the force and frequency of heartbeats over a narrow range. Nerves, hormones, and intrinsic signals control the activity of smooth muscles, which contract slowly but maintain tension very efficiently. This chapter explains the molecular and cellular basis for these distinctive physiological properties of the three types of muscle.

Skeletal Muscle

Skeletal muscle cells (also called muscle fibers in the physiological literature) are among the largest cells of vertebrates. During development, mesenchymal stem cells give rise to progenitor cells with a single nucleus called myoblasts. A family of master transcription factors, including MyoD and myogenin, coordinates the expression of specialized muscle proteins. As they differentiate, myoblasts fuse and elongate to form muscle cells with multiple nuclei and lengths ranging from millimeters to tens of centimeters. The number of muscle cells is determined genetically and is relatively stable throughout life even as the size of the cells varies with the level of exercise and nutrition.

A basal lamina (see Fig. 29.17C) surrounds and supports each muscle cell. At the ends of each cell, actin thin filaments are anchored to the plasma membrane at myotendinous junctions, which are similar to adherens junctions (see Fig. 31.8). There, integrins spanning the membrane link actin filaments to the basal lamina and to collagen fibrils of tendons. These physical connections transmit contractile force to the skeleton.

Mature muscles harbor small numbers of long-lived stem cells (see Fig. 41.14) with the potential to repair damage. They are called satellite cells because they are located inside the basal lamina next to the muscle cells. Some of these cells are capable of both self-renewal and, when the muscle is injured, producing progeny that can differentiate into myoblasts that can fuse with each other or existing muscle cells to repopulate the muscle. These features have made satellite cells a focus of research to treat degenerative diseases of muscle.

Organization of the Actomyosin Apparatus

Skeletal muscle cells are optimized for rapid, forceful contractions. Accordingly, they have a massive concentration of highly ordered contractile units composed of actin, myosin, and associated proteins (Fig. 39.2). Actin and myosin filaments are organized into **sarcomeres,** aligned contractile units that give the cells a striped appearance in the microscope. For this reason, they are called striated muscles. Myosin uses adenosine triphosphate (ATP) hydrolysis to power contraction, which results from myosin-powered sliding of actin-based thin filaments past myosin-containing thick filaments. Speed of contraction is achieved by linking many sarcomeres in series. Power (force) is achieved by linking multiple sarcomeres in parallel. Nerve impulses stimulate a transient rise in cytoplasmic calcium that activates the contractile proteins.

Interdigitation of thick, bipolar, myosin filaments and thin actin filaments in the sarcomeres of living muscle

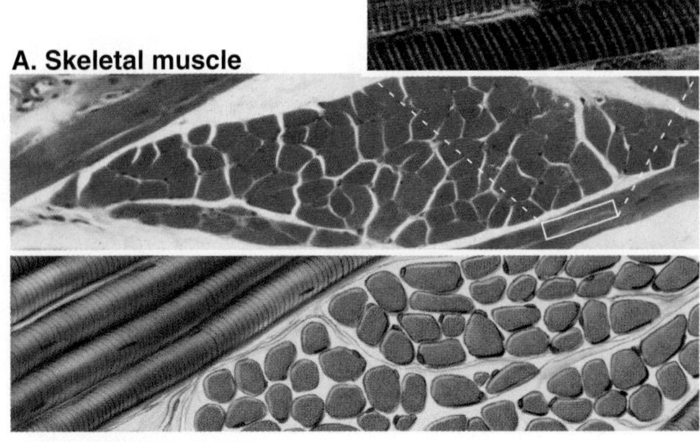

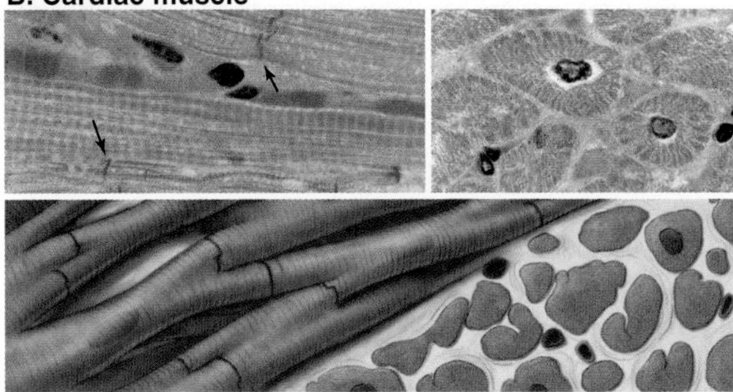

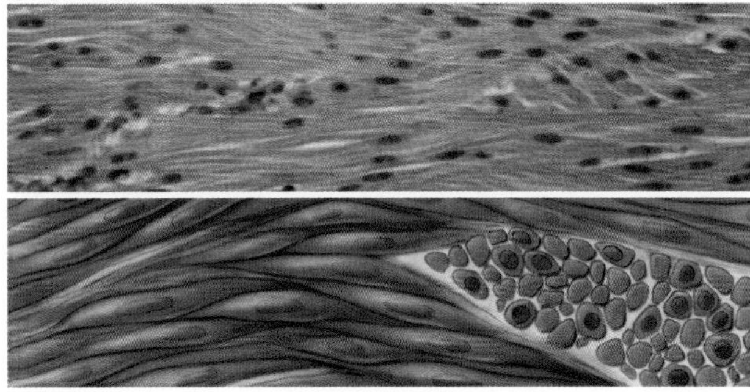

FIGURE 39.1 LIGHT MICROGRAPHS AND INTERPRETIVE DRAWINGS OF HISTOLOGIC SECTIONS OF SKELETAL, CARDIAC, AND SMOOTH MUSCLES. A, Skeletal muscle cells are shaped like cylinders and may be up to 50 cm long. Multiple nuclei are located at the periphery near the plasma membrane. Striations are seen in the inset, a longitudinal section at high magnification. **B,** Cardiac muscle cells are striated and have one or two nuclei. Adhesive junctions called intercalated disks (*bright pink vertical bars* in the longitudinal section, *top left arrows*) bind these short cells together end to end. **C,** Smooth muscle cells are spindle shaped with homogeneous cytoplasm and single nuclei.

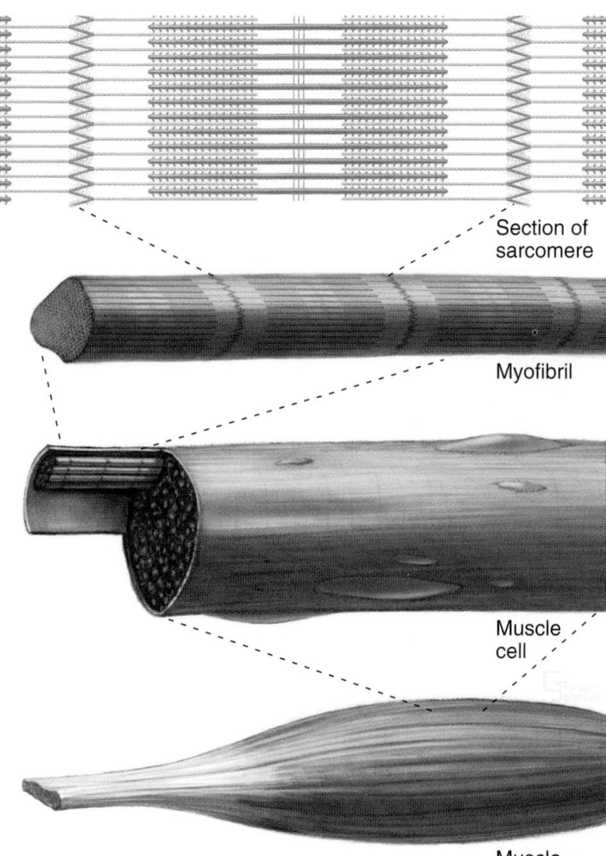

FIGURE 39.2 CONTRACTILE APPARATUS OF STRIATED MUSCLES. The contractile unit is the sarcomere, an interdigitating array of thick and thin filaments. Sarcomeres are arranged end to end into long, rod-shaped myofibrils that run the length of the cell. Mitochondria and smooth endoplasmic reticulum separate myofibrils, which can readily be isolated for functional and biochemical studies.

cells is so precise (Fig. 39.3) that it yields an X-ray diffraction pattern (see Fig. 39.11) revealing the spacing of the filaments and the helical repeats of their subunits to a resolution of about 3 nm. **Z disks** at both ends of the sarcomere anchor the barbed ends of the actin filaments, so their pointed ends are near the center of the sarcomere. Myosin heads project from the surface of **thick filaments,** whereas their tails form the filament backbone. Thick and **thin filaments** overlap, with the myosin heads only a few nanometers away from adjacent actin filaments. The alignment and interdigitation of the filaments facilitate the sliding interactions required to produce contraction.

An important, simplifying architectural feature is that sarcomeres are symmetrical about their middles (Fig. 39.3). Consequently, the polarity of myosin relative to the actin filaments is the same in both halves of the sarcomere, allowing the same force-generating mechanism to work at both ends of the bipolar myosin filaments. Sarcomeres are organized end to end into long, rod-shaped assemblies called **myofibrils** (Fig. 39.2) that retain their contractility even after isolation from muscle.

Thin Filaments

Thin filaments are a polymer of actin with tightly bound regulatory proteins **troponin** and **tropomyosin** (Fig. 39.4). When the cytoplasmic Ca^{2+} concentration is low, troponin and tropomyosin inhibit the actin-activated adenosine triphosphatase (ATPase) of myosin. Tropomyosin, a 40-nm long coiled-coil of two α-helical polypeptides (see Fig. 3.10), binds laterally to seven

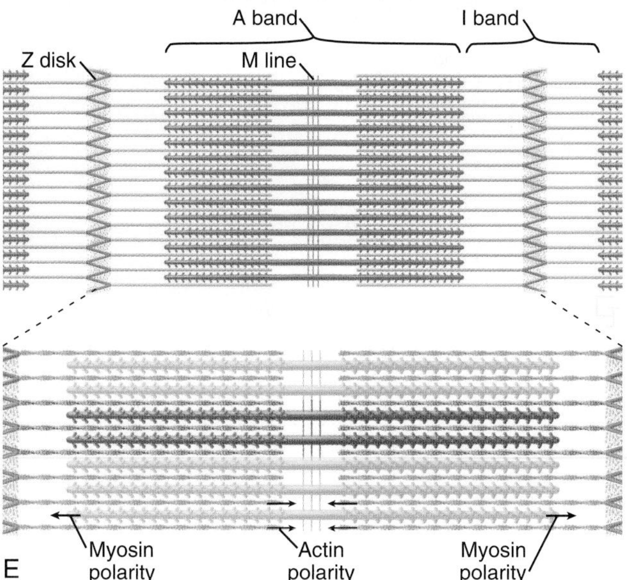

FIGURE 39.3 ELECTRON MICROGRAPHS AND DRAWINGS OF SARCOMERES. A, Longitudinal thin section showing the array of thin filaments anchored to Z disks and overlapping bipolar thick filaments crosslinked in the middle at the M line. **B,** Longitudinal freeze-fractured, etched, and shadowed sarcomere showing myosin cross-bridges attached to thin filaments near the bare zone in the center *(right)* of a sarcomere. **C–D,** Cross-sections of insect flight muscle and vertebrate skeletal muscle showing the double hexagonal arrays of thick and thin filaments. **E,** Drawings indicating the polarity of the thick and thin filaments. (**A** and **C,** Courtesy H.E. Huxley, Brandeis University, Waltham, MA. **B** and **D,** Courtesy J. Heuser, Washington University, St. Louis, MO.)

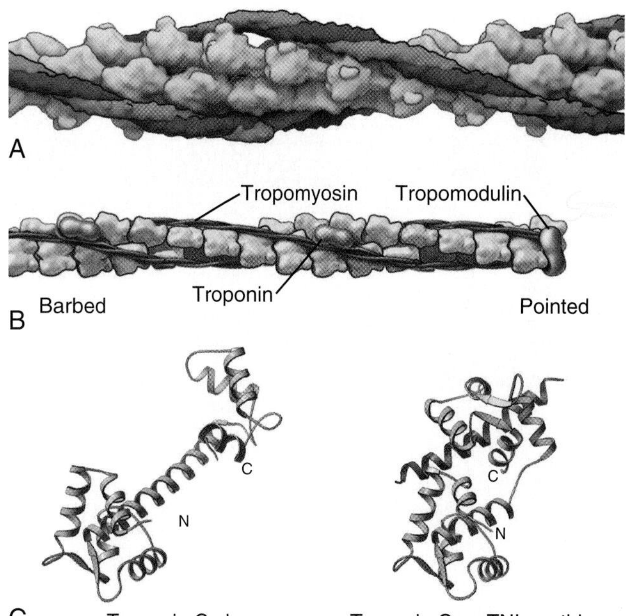

FIGURE 39.4 THIN FILAMENT STRUCTURE. A, Three-dimensional reconstruction from electron micrographs of a thin filament from vertebrate skeletal muscle showing actin and the position of tropomyosin in relaxed muscle. **B,** Drawing of a model of a thin filament from active muscle based on reconstructions of electron micrographs and crystal structures of troponin and tropomodulin. Each troponin–tropomyosin unit is associated with seven actin subunits. **C,** Ribbon diagrams of the atomic structures of troponin C, free and bound to a troponin I peptide. Two divalent cation-binding EF hands are found at each end, separated by a long α-helix. In cells, two high-affinity sites at the C-terminal end are permanently occupied with Mg^{2+}. Two low-affinity sites at the other end are unoccupied in relaxed muscle but bind Ca^{2+} when muscle is activated. (**A,** Based on Protein Data Bank [www.rcsb.org] file 1AX2, created by Roberto Dominguez, University of Pennsylvania.)

contiguous actin subunits as well as head to tail to neighboring tropomyosins, forming a continuous strand along the whole thin filament. Troponin (TN) consists of three different subunits called TNC, TNI, and TNT (see Table 39.1). TNT anchors one troponin complex to each tropomyosin coiled-coil. TNC is a dumbbell-shaped protein with four EF-hand motifs to bind divalent cations similar to calmodulin (see Fig. 3.12 and Chapter 26). In resting muscle, the C-terminal globular domain of TNC binds two Mg^{2+} ions and an α-helix of TNI, while the low-affinity sites in the N-terminal globular domain of TNC are empty. Ca^{2+} binding to the low-affinity sites (two in fast skeletal muscle; one in slow muscle) during muscle activation exposes a new binding site for TNI. The resulting conformational change in TNI allows tropomyosin to move away from the myosin-binding sites on the actin filament.

A protein meshwork in the Z disk anchors the barbed end of each thin filament (Fig. 39.5). Some crosslinks between actin filaments consist of α-actinin, a short rod with actin-binding sites on each end (see Fig. 33.17). At least a half dozen structural proteins stabilize the Z disk through interactions with α-actinin, actin, and titin. Some of these proteins also have signaling functions.

Proteins cap both ends of thin filaments. **Cap-Z,** the muscle isoform of capping protein (see Fig. 33.15), binds the barbed ends of thin filaments with high affinity, limiting actin subunit addition or loss. **Tropomodulin** associates with both tropomyosin and actin to cap and stabilize the pointed end of the thin filament (Fig. 39.4B).

Tropomyosin and a gigantic filamentous protein, **nebulin,** stabilize thin filaments laterally. Nebulin consists of 185 imperfect repeats of a 35-amino-acid motif

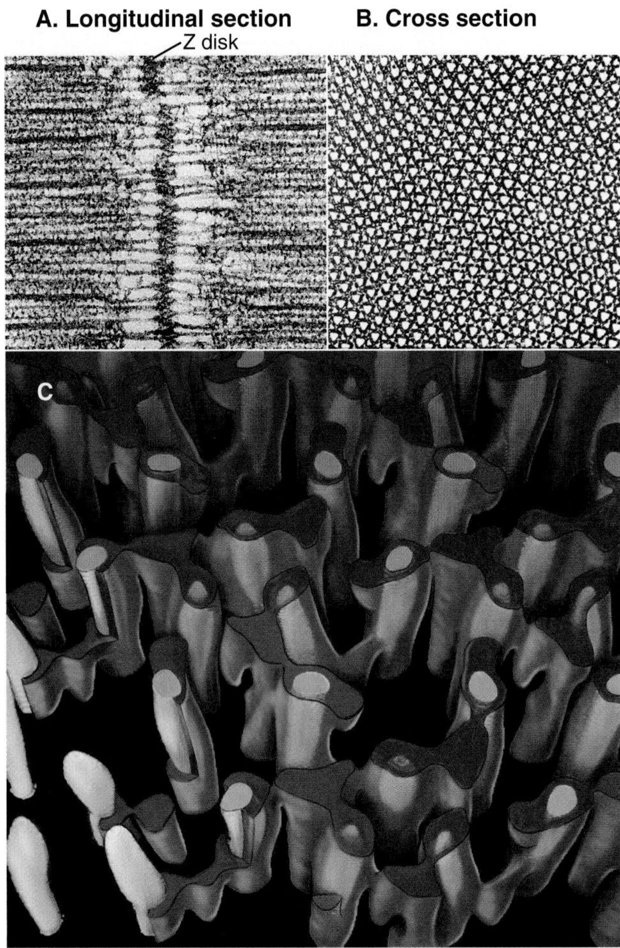

FIGURE 39.5 Z DISK STRUCTURE. A–B, Electron micrographs of thin sections perpendicular to and in the plane of the Z disk. **C,** Three-dimensional reconstruction, based on electron micrographs of the Z disk, showing the network of protein crosslinks that anchor the barbed ends of the *yellow* actin filaments. (Courtesy J. Deatherage, National Institutes of Health, Bethesda, MD and modified from Cheng NQ, Deatherage JF. Three dimensional reconstruction of the Z disk of sectioned bee flight muscle. *J Cell Biol.* 1989;108:1761–1774.)

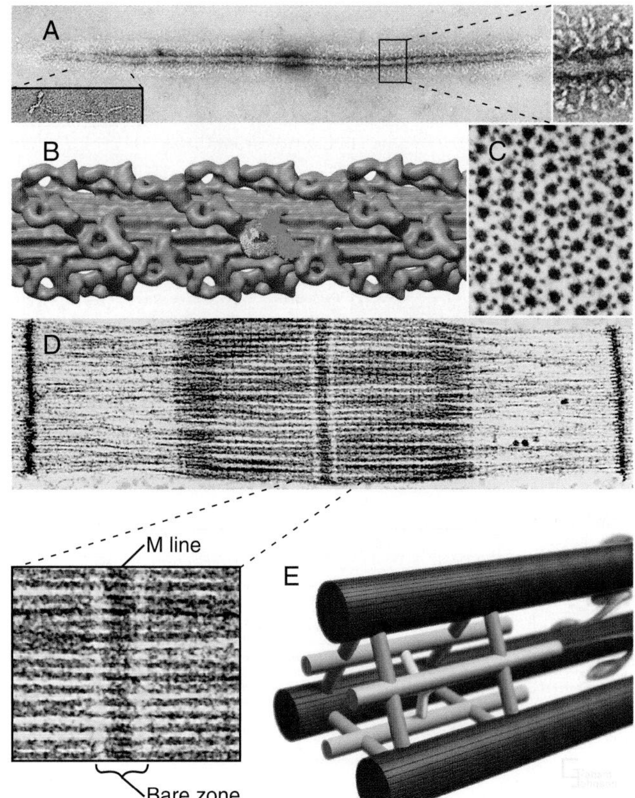

FIGURE 39.6 STRUCTURE OF BIPOLAR THICK FILAMENTS. A, Electron micrograph of a thick filament isolated directly from skeletal muscle and prepared by negative staining. A single myosin molecule is shown at the same magnification at the *lower left.* The myosin tails form the backbone of the thick filament and allow the two myosin heads to swing out from the side (see the enlarged *inset* on the *right*). **B,** Reconstruction from cryoelectron micrographs of part of a tarantula skeletal muscle thick filament with the bare zone (not shown) to the right. The tails form the backbone of the filament and the myosin heads are folded back toward the bare zone. Space-filling models of the two heads of one myosin molecule are superimposed on the reconstruction. The catalytic domains are green and blue; the light chains are pink, orange, yellow, and tan. **C,** Cross section of vertebrate skeletal muscle showing the double hexagonal arrays of thick and thin filaments. **D,** Electron micrograph of a highly stretched sarcomere with the M line in the middle. Note the nine faint stripes of myosin-binding protein C along both halves of the think filaments. **E,** Drawing of protein links between thick filaments in the M line. (**A,** Courtesy John Trinick, University of Bristol, United Kingdom. For reference, see Knight P, Trinick J. Structure of the myosin projections on native thick filaments from vertebrate skeletal muscle. *J Mol Biol.* 1984;177:461–482. **B,** Courtesy J. Woodhead and R. Craig, University of Massachusetts Medical School, Worcester, MA. For reference, see Woodhead JL, Zhao FQ, Craig R, et al. Atomic model of a myosin filament in the relaxed state. *Nature.* 2005;436:1195–1199. **C,** Courtesy J. Heuser, Washington University, St. Louis, MO. **D,** Courtesy H.E. Huxley, Brandeis University, Waltham, MA.)

that interact with each actin subunit, tropomyosin, and troponin along the length of thin filaments. Interactions with tropomodulin and Z disk proteins anchor nebulin at the two ends of the thin filament. These interactions influence the length of thin filaments, but nebulin does not act simply as a molecular ruler.

Thick Filaments

The self-assembly of myosin II (see Fig. 5.7) establishes the bipolar architecture of striated muscle thick filaments (Fig. 39.6). Some features of thick filaments are invariant, such as a superhelical backbone consisting of myosin tails, a surface array of myosin heads, the 14.3-nm stagger between rows of heads, and a central bare zone formed by antiparallel packing of tails. Phosphorylation of the myosin regulatory light chains favors release of the myosin heads from the surface of the filament in preparation for engagement with actin. Filaments may vary in length, diameter, and organization of the helical array of heads in various species. Invertebrate thick filaments have a core of paramyosin, a second coiled-coil protein, which is not found in vertebrates.

Accessory proteins stabilize thick filaments in striated muscles (Table 39.1). **Myosin-binding protein C**, with its fibronectin III and immunoglobulin domains, forms nine stripes along both halves of the thick filament (Fig. 39.6D) and also interacts with thin filaments. It fine-tunes the interactions of myosin heads with both filaments, maintaining the relaxed state but also favoring crossbridge formation when the muscle is activated. The **"M line"** in the center of the sarcomere of most types of muscle (Fig. 39.6D) is a three-dimensional array of protein crosslinks. These elastic crosslinks maintain the precise registration of thick filaments but also allow the filaments to move apart as the sarcomere shortens while maintaining a constant volume. At least three structural proteins and the enzyme MM-creatine phosphokinase (which transfers phosphate from creatine-phosphate to adenosine diphosphate [ADP]) are located in the M line.

Titin Filaments

Titin, the largest protein encoded by the human genome, forms a third array of filaments lying parallel to the thin and thick filaments and connecting the Z disk to the thick filaments and the M line (Fig. 39.7). Three titins flank each half thick filament. Although titin is the third most abundant protein in muscle, it is hard to preserve for electron microscopy, so these diaphanous filaments escaped notice for years. Each filament is a single polypeptide named after a mythological giant owing to its remarkable size: several splice isoforms consist of 27,000 to 33,000 amino acids folded into a linear array of up to 300 fibronectin III and immunoglobulin (Ig) domains measuring more than 1.2 μm long.

The elasticity of titin molecules provides passive resistance to stretching of relaxed muscle. Titin connections to the Z disk and thick filaments provide physical continuity from one sarcomere to the next and keep the thick filaments centered in the sarcomere during contraction. Differential splicing creates titin isoforms that differ in stiffness for various types of muscle. If titin molecules are broken experimentally, thick filaments slide out of register toward one Z disk during contraction. Two features provide the elasticity during short (~0.3 μm per titin), physiological stretching (Fig. 39.7). The irregular chain of Ig domains in the I band straightens out, and a segment of the polypeptide rich in proline, glutamic acid, valine, and lysine (the PEVK domain) partly unfolds. This decreases entropy and provides the energy for elastic recoil (an entropic spring; see also Fig. 29.11). Extreme stretching unfolds Ig domains one by one. Titin not only responds to phosphorylation but also binds signaling proteins that influence the performance of muscle in the short and long term.

Intermediate Filaments

Desmin intermediate filaments (see Chapter 35) help align the sarcomeres laterally (Fig. 39.8) by linking each Z disk to its neighbors and to specialized attachment sites on the plasma membrane called costameres. In

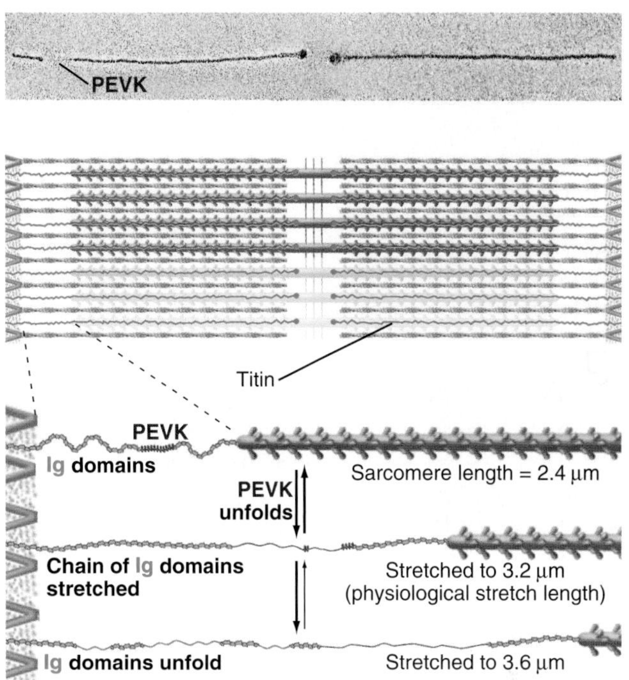

FIGURE 39.7 TITIN FILAMENTS. *Top panel,* Electron micrographs of single, isolated titin molecules prepared by heavy metal shadowing. Titin molecules are long enough to extend from the Z disk to the M line. *Middle panel,* Drawing of a sarcomere, to the same scale as the electron micrograph, with the thick filaments removed from the bottom half to illustrate how titin molecules anchor thick filaments to the Z disk and extend to the M line. *Bottom panel,* Drawing illustrating a model for the elasticity of titin. Modest stretches within the physiological range reversibly extend the chain of immunoglobulin (Ig) domains in the I-band and the PEVK domain. Extreme extension can unfold Ig domains. (Modified from Reif M, Gautel M, Oesterhelt F, et al. Reversible unfolding of individual titin immunoglobulin domains by AFM. *Science.* 1997;276:1090–1092. For reference, see Leake MC, Wilson D, Gautel M, Simmons RM. The elasticity of single titin molecules using a two-bead optical tweezers assay. *Biophys J.* 2004;87:1112–1135.)

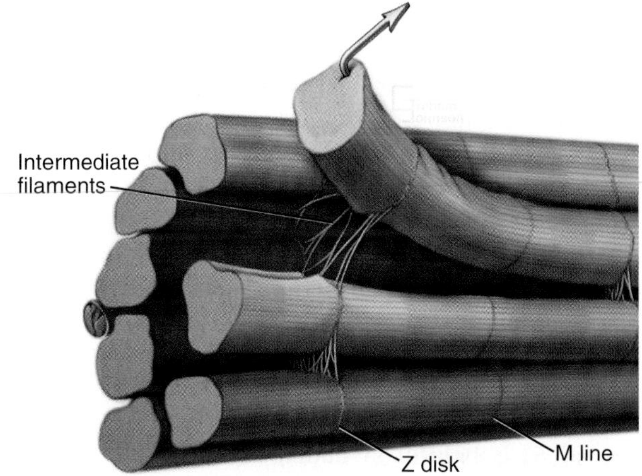

FIGURE 39.8 DESMIN INTERMEDIATE FILAMENTS IN SKELETAL MUSCLE. Desmin filaments connect Z disks laterally to each other and to the plasma membrane at specializations called costameres. (Modified from Lazarides E. Intermediate filaments as mechanical integrators of cellular space. *Nature.* 1980;283:249–256.)

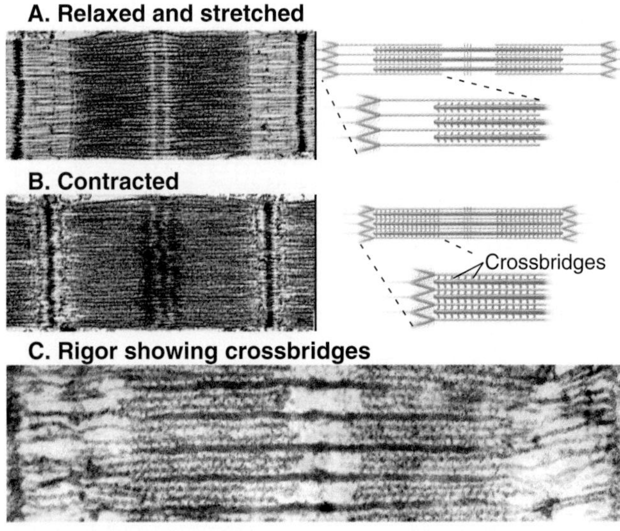

FIGURE 39.9 SLIDING FILAMENTS. Electron micrographs and interpretive drawings of longitudinal sections of a sarcomere from a relaxed muscle **(A)** and a contracted skeletal muscle **(B).** The lengths of the thin and thick filaments are constant as the sarcomere shortens, demonstrating that the filaments slide past each other during contraction. **C,** Crossbridges between thick and thin filaments from a muscle in rigor. (Micrographs courtesy H.E. Huxley, Brandeis University, Waltham, MA.)

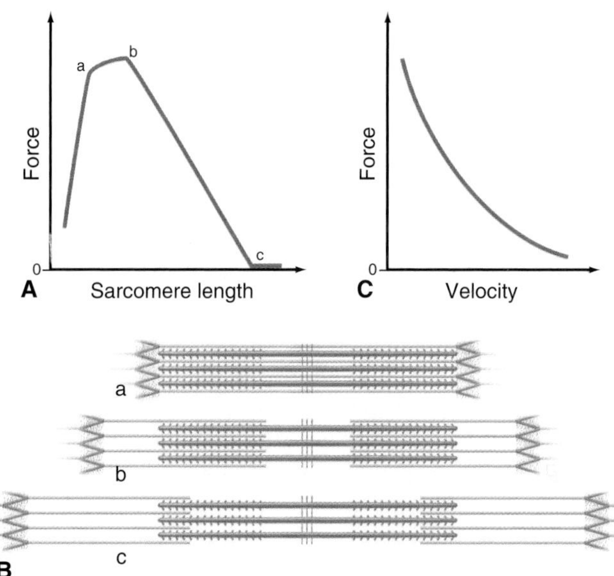

FIGURE 39.10 PHYSIOLOGICAL PROPERTIES OF SKELETAL MUSCLE. A, Dependence of maximum tension on the length of the sarcomeres. **B,** Interpretive drawings. Each relates to a point on **A. C,** Relationship of force and velocity during muscle contraction. (**A,** For reference, see Gordon AM, Huxley AF, Julian F. The variation in isometric tension with sarcomere length in vertebrate muscle fibres. *J Physiol*. 1964;171:28P–30P. **C,** From Ruch TC, Patton HD (eds). *Physiology and Biophysics*, 19th ed. Philadelphia: WB Saunders; 1965.)

addition to desmin, costameres contain clathrin plus several cytoskeletal proteins (vinculin, talin, spectrin, and ankyrin) found in focal contacts and adherens junctions of nonmuscle cells (see Figs. 30.11 and 31.8). Desmin mutations in humans cause disorganization of myofibrils, resulting in generalized muscle failure.

Molecular Basis of Skeletal Muscle Contraction

Sliding Filament Mechanism

The key to understanding muscle contraction was the discovery that thick and thin filaments maintain constant lengths and slide past each other as sarcomeres (and the muscle) shorten (Fig. 39.9). About the same time, it was discovered that crossbridges (now recognized to be myosin heads) can connect actin and myosin filaments, and that tension produced during contraction is proportional to the overlap of actin and myosin filaments (Fig. 39.10). Supported by biochemical and ultrastructural evidence for actin-myosin interaction, these pioneering observations led to the theory that crossbridges between the thick and thin filaments produce the force for contraction. Sixty years of research on crossbridges have yielded a detailed picture of the chemistry and molecular mechanics underlying the force-producing reactions. A review of the steps of the actomyosin-ATPase cycle (see Fig. 36.5) is helpful in understanding the contraction mechanism. Three different physiological states reveal information about crossbridge mechanisms.

Relaxed. One extreme is relaxed muscle. When the concentration of cytoplasmic Ca^{2+} is low, tropomyosin and troponin inhibit the interaction of myosin heads with actin filaments, resulting in few myosin heads being bound. Lacking long-lived physical connections between the filaments, muscle offers little resistance to passive stretching. X-ray diffraction (Fig. 39.11) shows that the myosin heads (with bound ATP or ADP and phosphate) are closely associated with the backbone of thick filaments and arranged in a helical array determined by the thick filament structure (Fig. 39.6B).

Rigor. The other extreme occurs after death. Depletion of ATP allows all myosin heads to bind tightly to actin filaments (Figs. 39.3B and 39.9C). By X-ray diffraction, the myosin heads bound to actin filaments contribute to the strength of the reflections from the actin filament helix. Strong physical connections between the filaments prevent stretching, making the muscle stiff (hence the term *rigor mortis*). This extreme condition illustrates what happens structurally and mechanically when all the crossbridges engage actin filaments.

Contracting. The most interesting, but most complicated, state is *actively contracting* muscle. Myosin heads "walk" along actin filaments toward their barbed ends, pulling Z disks toward the center of the sarcomere. Thousands of sarcomeres shorten in series, causing the whole muscle to shorten. ATP is consumed and force is produced. The thick filament

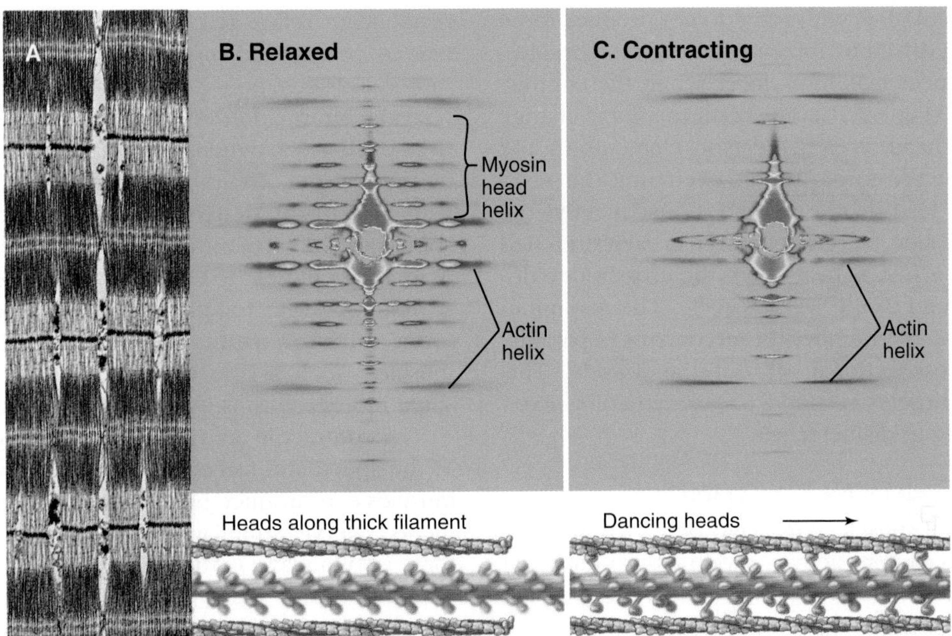

FIGURE 39.11 CROSSBRIDGE DYNAMICS REVEALED BY X-RAY DIFFRACTION PATTERNS OF WHOLE MUSCLE. A, Electron micrograph showing the orientation of the muscle in the x-ray beam in **B** and **C. B–C,** Fiber diffraction patterns from relaxed and contracting skeletal muscles with interpretive drawings of crossbridges in each state. Reflections from myosin heads arranged on the thick filament helix are strong in relaxed muscle. Reflections from the actin helix are stronger than the thick filament helix in contraction. The myosin and actin reflections are each labeled in only one of four equivalent quadrants. During contraction, a few myosin heads attach transiently to actin, increasing the strength of the actin helix reflections, but most are disordered. (Micrograph and x-ray patterns courtesy H.E. Huxley, Brandeis University, Waltham, MA.)

helical pattern is very weak by X-ray diffraction (Fig. 39.11C). Actin reflections are stronger than relaxed muscle but not as strong as rigor. Disordered myosin heads are distributed between the thick and thin filaments as each one dances asynchronously on and off of actin filaments.

Most myosin heads in contracting muscle have bound ATP or ADP-P_i (adenosine diphosphate–inorganic phosphate), allowing them to exchange rapidly among the four "weakly bound" states illustrated in Fig. 36.5. During some of the transient interactions of myosin–ADP-P_i with actin, phosphate dissociates from myosin, and the light-chain domain rapidly reorients (see Figs. 36.4 and 36.5). This stretches elastic elements in the myosin heads and both thick and thin filaments. Energy in these elastic elements can be used over a period of milliseconds to displace the actin filament relative to the crossbridge and contract the muscle. When ADP dissociates from the actin-myosin-ADP intermediate, ATP rapidly binds to the actin-myosin complex, dissociating the crossbridge and starting a new ATPase cycle.

Relationship of Crossbridge Behavior to the Mechanical Properties of Muscle

Normally, each sarcomere shortens less than 1 μm. However, the whole muscle shortens macroscopically, because it has thousands of sarcomeres in series. For example, a human biceps muscle 20 cm long has approximately 80,000 sarcomeres in series from end to end. When each contracts 0.25 μm, the muscle shortens 2 cm. Because the system maintains a constant volume, each sarcomere and the whole muscle increase in diameter as they shorten. Although the individual filaments slide past each other relatively slowly (about 2 $\mu m\ s^{-1}$ in both halves of each sarcomere), muscles contract rapidly because the motion of each sarcomere in the series is added together. In our example, without resistance, the biceps contracts approximately 3 cm in 100 ms during which most crossbridges pull just once.

Crossbridge behavior explains why the velocity of muscle contractions of an active muscle depends on the external load (Fig. 39.10). When opposed by no load the molecular motion stored in elastic elements of each crossbridge is largely converted into movement of actin filaments relative to myosin filaments and contraction velocity is maximal. Under these conditions, the filaments in muscle slide past each other at a rate of about 5 $\mu m\ s^{-1}$, the same speed observed for free actin filaments moving over myosin heads in vitro (see Fig. 36.6). For this rapid sliding to occur, myosin heads that do not produce force must not impede movement. If bound tightly to actin, they would interfere mechanically with rapid sliding. This is avoided by the rapid equilibrium of the myosin intermediates between being bound to actin and being free. Transient interactions of myosin heads bound to ATP or ADP-P_i with actin do not produce force or retard sliding driven by force-producing crossbridges.

Muscle produces maximum force when the contraction rate is zero (Fig. 39.10). The conformational change in the myosin head stretches elastic elements in the crossbridge, but the force cannot overcome the resistance from the load on the muscle. Consequently, the filaments do not slide, and energy stored in each stretched elastic element is lost as heat when the crossbridge dissociates at the end of the ATPase cycle. The maximum force depends on the numbers of sarcomeres in parallel, that is, the cross-sectional area of the muscle. This explains why muscles respond to strengthening exercises by growing in diameter.

Regulation of Skeletal Muscle Contraction

Although skeletal muscle *cells* have only two states—inactive (relaxed) or active (contracting)—skeletal *muscles* produce a wide range of contractions, varying from slow and delicate to rapid and forceful. These **graded contractions** are achieved by varying the *number* of muscle cells activated by voluntary or reflex signals from the nervous system (Fig. 39.12).

Control of Skeletal Muscle by Motor Neurons

Neural stimuli that activate skeletal muscles arise in two ways (Fig. 39.12). In organisms with well-developed central nervous systems, most neural signals that activate skeletal muscles result from conscious decisions, providing voluntary control over skeletal muscles. Other signals result from reflex responses to stimulation of sensory nerves. Specialized muscle cells innervated with both motor and sensory nerves function as stretch receptors, relaying information about length and tension back to the spinal cord, where reflexes coordinate the motor neuron output. Neural inputs from both sources converge on **motor neurons** located in the brainstem and spinal cord of vertebrates. Axons of these motor neurons branch in a muscle to contact one or more muscle cells. A motor neuron together with its target muscle cells forms a **motor unit.** In the most precisely controlled muscles, such as the extraocular muscles in the eye, some motor neurons innervate single muscle cells.

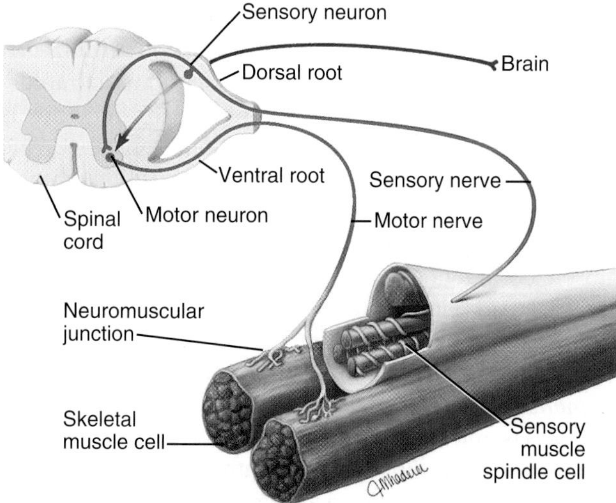

FIGURE 39.12 INNERVATION OF SKELETAL MUSCLE. Motor neurons in the spinal cord stimulate one or (usually) more skeletal muscle cells. Two neural pathways control motor neurons. Some stimuli come from neurons in higher centers of the brain. This pathway provides voluntary control over muscle contraction. Other stimuli come through local reflex circuits from sensory detectors, including muscle spindle cells. These signals help coordinate muscle contraction in response to changing forces on and within the muscle.

The contractile activity of a muscle is graded in terms of the speed and force of the contraction, so individual muscles can produce both delicate and powerful movements. Nerve stimulation determines the contractile force in two ways: (a) The number of active motor units determines how many muscle cells produce force, and (b) the rate of stimulation adjusts the force produced by active cells. Every time a muscle cell is stimulated, all the sarcomeres are activated, but the force that they produce increases as the rate of stimulation increases, up to a maximum of approximately 200 stimuli per second. The shortening velocity of an active muscle depends on the ratio between force produced and the resistance (Fig. 39.10C). If a large force or high velocity of contraction is required, many motor units are called into action. To sustain contraction, motor nerves fire repeatedly. By varying the number of active cells in a muscle and the rate of stimulation, the nervous system sets the force required for a particular movement.

Synaptic Transmission at Neuromuscular Junctions

The terminal branch of each motor neuron axon forms a large synapse called the **motor end plate** or **neuromuscular junction** on the muscle surface (see Fig. 17.9). These nerve endings are filled with synaptic vesicles containing the neurotransmitter **acetylcholine**. Arrival of an action potential at the nerve terminal stimulates fusion of synaptic vesicles with the nerve plasma membrane, releasing acetylcholine into the cleft between nerve and muscle. In less than a millisecond, acetylcholine diffuses across the extracellular space and binds to acetylcholine receptors concentrated in the adjacent muscle plasma membrane. Acetylcholine binding opens the receptor cation channel, initiating a new action potential that spreads over the muscle cell plasma membrane and intracellularly into the T tubules.

Coupling Action Potentials to Contraction

The plasma membrane of skeletal muscle cells, like that of nerve cells, is excitable (see Fig. 17.6) but, unlike that in nerves, it invaginates deeply to form **T tubules** that run across the entire cell (Fig. 39.13). Depending on the species and type of striated muscle (skeletal

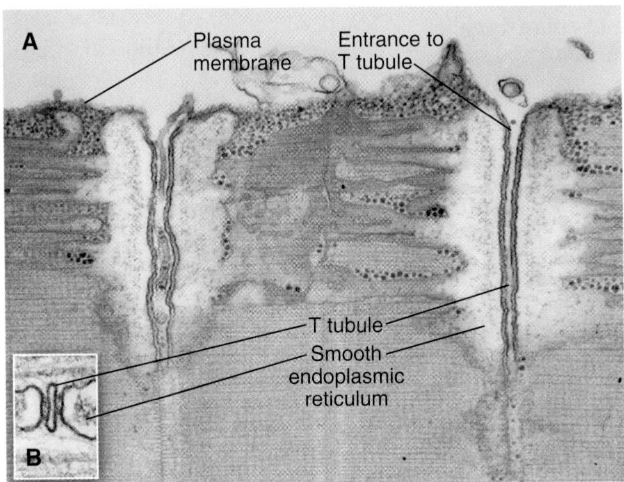

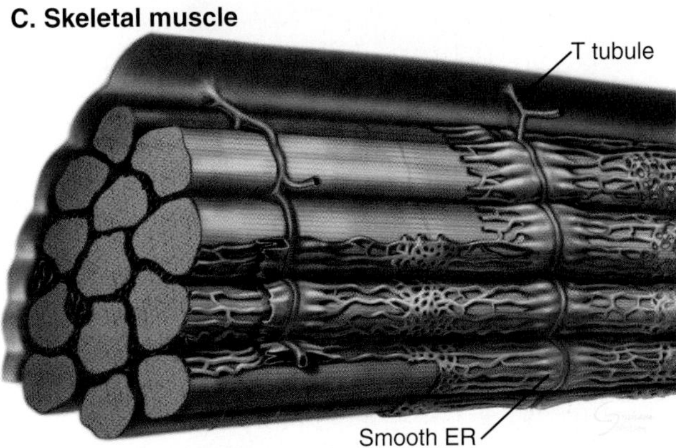

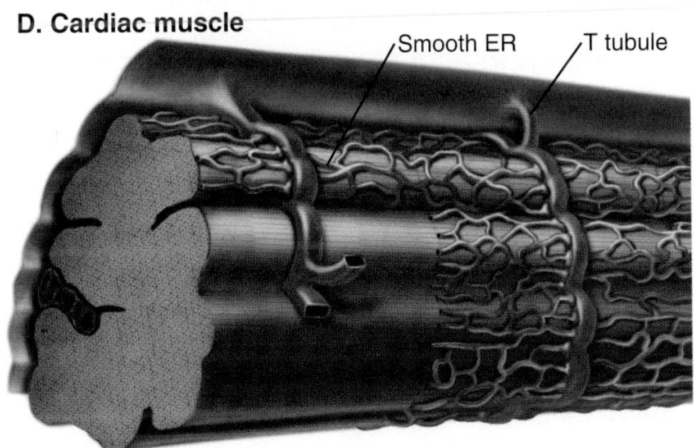

FIGURE 39.13 PLASMA MEMBRANE SPECIALIZATIONS OF STRIATED MUSCLES. A–B, Electron micrographs of thin sections of fish skeletal muscle showing plasma membrane invaginations called T tubules, which cross the whole muscle cell and associate closely with smooth endoplasmic reticulum (ER). The complex of a T tubule with smooth ER on both sides is called a triad. **A,** A longitudinal section of two T tubules. **B,** A cross section of a T tubule flanked on two sides by smooth ER. Foot processes, consisting of voltage-sensitive calcium channels in the T tubule paired with calcium release channels in the ER, connect the T tubule to the smooth ER (see Fig. 39.14 for molecular details). **C–D,** The three-dimensional arrangement of T tubules and smooth ER relative to the sarcomeres in skeletal and cardiac muscle. (**A–B,** Courtesy C. Franzini-Armstrong and K. Porter, University of Pennsylvania.)

versus cardiac), T tubules may be located either at the level of the Z disks or at the thick filament ends. Inside the muscle cell, T tubules interact extensively with the smooth endoplasmic reticulum (SER) surrounding each myofibril. Historically, this SER was called **sarcoplasmic reticulum.** Terminal cisternae of SER are closely associated with T tubules at foot processes that can be visualized by electron microscopy. Together, T tubules and SER constitute a signal-transducing apparatus that converts depolarizations of the plasma membrane into a spike of cytoplasmic Ca^{2+} to trigger contraction (Fig. 39.14).

An action potential moving through a T tubule triggers the release of Ca^{2+} from SER into the cytoplasm (Fig. 39.14). Ca^{2+} binding to troponin allows myosin to interact with the thin filament, initiating contraction. This signal transduction process is called **excitation–contraction coupling**. Three transmembrane proteins located in the T tubule and the terminal cisternae of the SER cooperate to generate the transient Ca^{2+} signal (Fig. 39.14).

1. A voltage-sensitive calcium channel (see Chapter 16) senses action potentials in the T tubule. These channels are called **dihydropyridine (DHP) receptors,** owing to their affinity for this class of drugs. The actual Ca^{2+} channel of DHP receptors is not essential for skeletal muscle contraction, as external Ca^{2+} is not required for contraction in the short term.
2. Ca^{2+} release channels (see Fig. 26.13), concentrated in the terminal cisternae of SER, release Ca^{2+} into the cytoplasm. A drug called ryanodine binds these channels and inhibits Ca^{2+} release. Every second **ryanodine receptor** is connected to cytoplasmic loops of four DHP receptors, forming bridges called foot processes between the T tubule and the endoplasmic reticulum (Fig. 39.13B).
3. The P-type **calcium-ATPase** (see Fig. 14.7) actively pumps Ca^{2+} from cytoplasm into the endoplasmic reticulum against a concentration gradient greater than 10^4. Inside the SER several low-affinity, high-capacity Ca^{2+}-binding proteins buffer the millimolar concentration of Ca^{2+}. For example, numerous carboxyl groups on the surface of calsequestrin bind Ca^{2+} with a millimolar K_d. This rapidly reversible reaction increases the Ca^{2+} storage capacity of endoplasmic reticulum without sacrificing the speed of Ca^{2+} release. Accessory subunits anchor calsequestrin to Ca^{2+} release channels, ensuring a local supply of Ca^{2+} for release into cytoplasm when muscle is activated.

An action potential in a T tubule results in a transient rise in cytoplasmic Ca^{2+}, from 0.1 μM to about 2 μM (Fig. 39.15), as follows. The action potential causes a short-lived conformational change in the DHP receptors that is transmitted mechanically to associated ryanodine receptor Ca^{2+} release channels. Many Ca^{2+} channels open transiently, allowing Ca^{2+} to diffuse down the steep

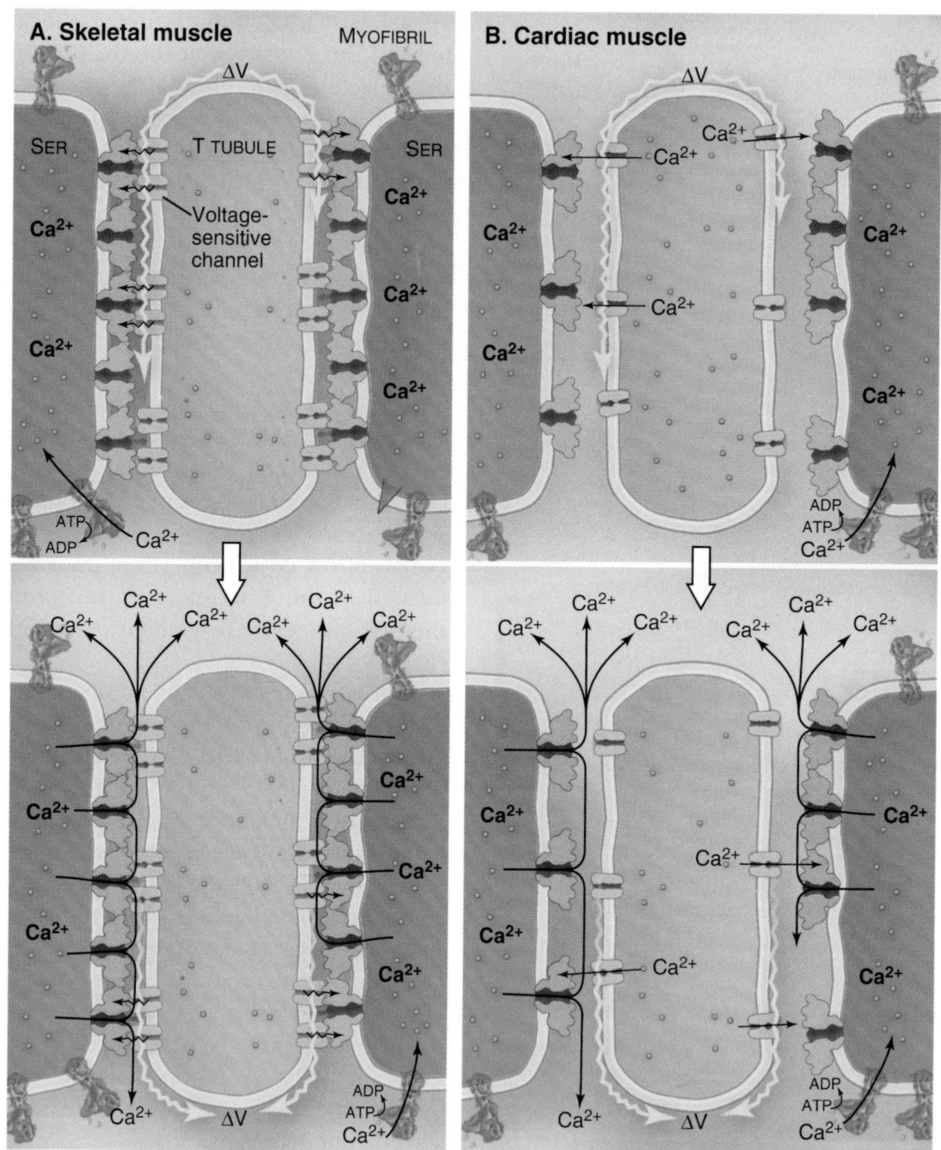

FIGURE 39.14 MECHANISM OF CALCIUM RELEASE IN SKELETAL AND CARDIAC MUSCLES. Both muscle types use voltage-sensitive calcium channels in the T tubule membranes and calcium release channels in the smooth endoplasmic reticulum (SER). **A,** Direct coupling in skeletal muscle. An action potential in the T tubule (ΔV) activates the voltage sensor (turning from *gray* to *blue*). This direct contact opens the calcium release channel (turning from *gray* to *pink*). Cytoplasmic Ca^{2+} levels rise only briefly because calcium–ATPase (adenosine triphosphatase) pumps Ca^{2+} back into the lumen of the SER. **B,** Calcium-induced Ca^{2+} release in cardiac muscle. An action potential opens the voltage-sensitive Ca^{2+} channel in the T tubule, releasing Ca^{2+} into the cytoplasm. This Ca^{2+} opens the calcium release channel in the SER. ADP, adenosine diphosphate; ATP, adenosine triphosphate.

concentration gradient from the SER lumen to the cytoplasm. Physical connections between ryanodine receptors may spread their activation laterally, ensuring synchronous activation of a patch of channels.

After a single action potential, the free Ca^{2+} in the cytoplasm rises for only a few milliseconds for three reasons. First, Ca^{2+} release channels close quickly. Second, cytoplasmic Ca^{2+} binds to troponin C and other proteins. Third, Ca^{2+} pumps efficiently transport cytoplasmic Ca^{2+} back into the lumen of the SER, even before the muscle develops maximum force. Ca^{2+} pumps are continuously active, keeping the cytoplasmic Ca^{2+} concentration low. This is why repeated action potentials are required to prolong the rise in cytoplasmic Ca^{2+} (Fig. 39.15B).

Transduction of the Calcium Spike Into Contraction

Troponin-tropomyosin on thin filaments cooperates with myosin to turn on contraction in response to a Ca^{2+} spike. At rest, two Ca^{2+}-binding sites of fast skeletal muscle troponin C are largely unoccupied (owing to their low affinity for Ca^{2+} and the low Ca^{2+} concentration). As a result, the troponin-tropomyosin complex partially blocks the binding site for myosin heads on actin (Fig. 39.16). This prevents most of the weak-binding

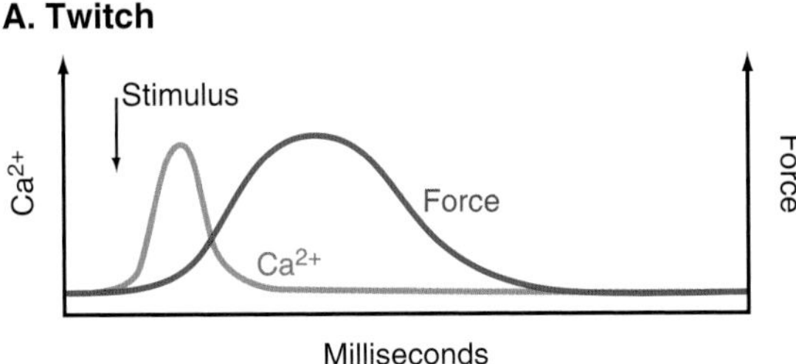

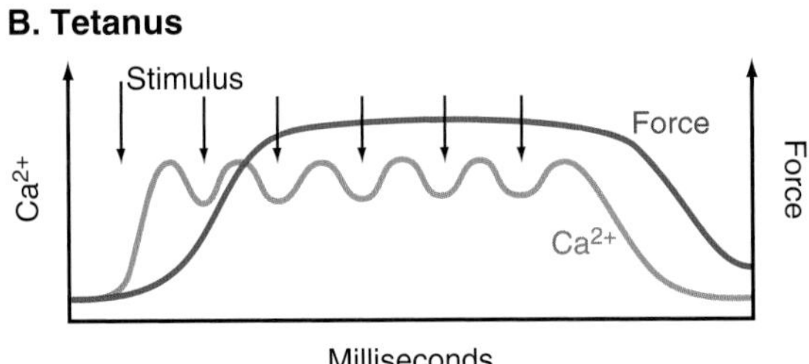

FIGURE 39.15 CA^{2+} TRIGGERS CONTRACTION OF SKELETAL MUSCLE. In these experiments, the Ca^{2+}-sensitive protein aequorin was injected into live muscle cells to provide a signal for the cytoplasmic Ca^{2+} concentration. **A,** Single stimulus. Cytoplasmic Ca^{2+} concentration increases transiently, followed by a short contraction. This brief contraction persists after cytoplasmic Ca^{2+} decreases to the resting level. **B,** Multiple stimuli. Each stimulus releases a new pulse of Ca^{2+}, prolonging the contraction in so-called tetanus. (For reference, see Ridgway EB, Ashley CC. Calcium transients in single muscle fibers. *Biochem Biophys Res Commun.* 1967;29:229–234.)

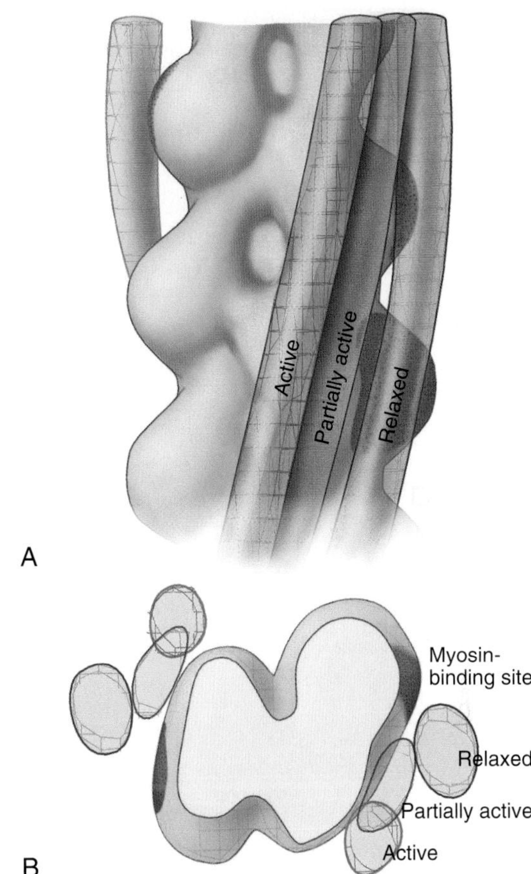

FIGURE 39.16 THIN FILAMENT ACTIVATION MECHANISM. Reconstructions from electron micrographs showing a short segment of thin filament **(A)** and a cross section of a thin filament **(B).** Ca^{2+} binding to troponin C partially activates the filament by moving tropomyosin away from its lateral position in relaxed muscle, where it overlaps the myosin-binding site on actin *(red).* Myosin binding to the partially activated filament shoves tropomyosin further out of the way into the active position. (Data from W. Lehman, Boston University, MA.)

myosin intermediates with ATP or ADP-P_i in the active site from binding the thin filament. When released into cytoplasm, Ca^{2+} binds troponin C, causing a conformational change that creates a binding site for a helical region of TNI. This interaction attracts the C-terminus of TNI away from actin and tropomyosin, allowing a small shift in the position of tropomyosin on the thin filament. This shift increases the probability that myosin-ADP-P_i heads will bind to the thin filament, dissociating their bound P_i and producing force. Activation is cooperative for three reasons: Ca^{2+} must occupy both binding sites on troponin C, the effects of Ca^{2+} binding and myosin binding are transmitted to neighboring tropomyosins through their end-to-end attachments, and every myosin that binds accentuates the response. This cooperativity makes the on–off switch respond very sharply to a relatively small, 10- to 20-fold change in the cytoplasmic Ca^{2+} concentration. The efficiency of this switch is underscored by the fact that the energy consumption of a muscle cell increases more than 1000-fold when it is activated. Activation of slow skeletal muscle (Table 39.3) and cardiac muscle is less cooperative, as their troponin C has only one Ca^{2+}-binding site.

Note that the muscle produces force well after the cytoplasmic Ca^{2+} concentration returns to resting levels (Fig. 39.15). The Ca^{2+}-sensitive switch is sharp but relatively slow owing to the slow response of thin filaments to Ca^{2+} binding. Ca^{2+} binds troponin C rapidly (milliseconds) but dissociates slowly (tens of milliseconds). Thus, after the Ca^{2+} spike saturates troponin C and the thin filament turns on, the muscle remains active even after free Ca^{2+} has returned to resting levels. Force declines slowly as Ca^{2+} dissociates from troponin C and returns to the SER without raising the cytoplasmic Ca^{2+} concentration.

A single action potential produces a short contractile "twitch" (Fig. 39.15). Maximum contractile force is produced by a series of closely spaced action potentials, leading to a sustained rise in cytoplasmic Ca^{2+} and prolonged activation of actomyosin. The extended contraction is called tetanus.

Regulation by Myosin Light Chains

The participation of skeletal muscle myosin light chains in the regulation of contraction varies among species. The skeletal muscles of mollusks are one extreme. Their

TABLE 39.1 Sarcomere Proteins of Vertebrate Striated Muscles

Name	Size (kD)	Domains	Functions	Disease Manifestations
Thick Filament				
Myosin	2 × 200	ATPase, coiled-coil	Motor, backbone of thick filament	HCM, HF, arrhythmias
Regulatory light chain	2 × 19	EF-hands	Stabilizes lever arm	HCM, arrhythmias
Essential light chain	2 × 18 or 25	EF-hands	Stabilizes lever arm	HCM, arrhythmias
Myosin-binding protein C	141	Ig, FNIII	Stabilizes thick filament	HCM, DCM, arrhythmias
Titin	3700	FNIII, IgC_2, kinase	Elastic connection from Z disk to M line	DCM, HF, muscular dystrophy
M Line				
MM-creatine phosphokinase	2 × 43		Glycolytic enzyme	
M protein	165	IgC_2, FNIII	M-line fast skeletal muscle	
Myomesin (skelemin)	185	IgC_2, FNIII	Link M-disk to desmin	None yet known
Thin Filament				
Actin	43		Thin filaments backbone	DCM, HF, myopathies
Tropomyosin	2 × 35	Coiled-coil	Blocks myosin binding to actin filament	HCM, DCM, arrhythmias, myopathies
Troponin C	18	EF-hands	Calcium-binding	DCM likely
Troponin I	21		Inhibitory component	HCM, arrhythmias, myopathies
Troponin T	31		Tropomyosin binding	HCM, DCM, arrhythmias, myopathies
Tropomodulin	43		Caps actin filament pointed end	None yet known
Nebulin	500-900	185 × 35 residues	Binds thin filament	Nemaline myopathy
Z Disk				
α-Actinin	2 × 100	Actin-binding, spectrin repeats	Crosslinks thin filaments in the Z disk	Focal segmental glomerulosclerosis
CapZ	31 + 32		Caps actin filament barbed end	None yet known
Desmin	2 × 53.5	Intermediate filament	Anchors Z disk	DCM, myopathy

DCM, dilated cardiomyopathy; EF, calcium-binding helices E and F of calmodulin; FN, fibronectin; HCM, hypertrophic cardiomyopathy; HF, heart failure; Ig, immunoglobulin; MLCK, myosin light-chain kinase.

myosin light chains bind Ca^{2+} and provide the main on-off switch for contraction. When the Ca^{2+} concentration is low in resting muscle, no Ca^{2+} binds to light chains, and the actin–myosin ATPase is off. Ca^{2+} that is released during activation binds to the light chains, turning on the ATPase and contraction. At the other extreme, the light chains of vertebrate skeletal muscle myosin do not bind Ca^{2+}, but their *phosphorylation* modulates contractile activity by increasing force production at suboptimal Ca^{2+} concentrations. Horseshoe crab skeletal muscle uses a dual system: Ca^{2+} binding to troponin–tropomyosin on thin filaments and Ca^{2+}-regulated phosphorylation of myosin light chains both stimulate contraction.

Specialized Skeletal Muscle Cells

All skeletal muscle cells are built on the same principles, but vertebrates actually have several different types of skeletal muscle cells, each with distinct isoforms of contractile protein and metabolic enzymes. The six myosin heavy chains and three actin isoforms are coded by different genes. In contrast, alternative splicing of one primary transcript (see Fig. 16.6) creates more than 50 isoforms of troponin T. Mutations in the genes for myosin, actin, desmin, nebulin, tropomyosin, troponin-I, and troponin-T can each cause defects in human skeletal muscles (Table 39.1).

TABLE 39.2 Muscle Cell Types

Physiological Type	Myosin Type	Mitochondria	Fatigue
Fast, white	Fast	Few	Rapid
Intermediate	Fast	Medium	Medium
Fast, red	Fast	Many	Slow
Slow, red	Slow	Many	Slow

Physiological properties, such as the speed of contraction and the rate of fatigue, provide criteria for classifying muscle cells (Table 39.2). The isoforms of myosin (and probably the other contractile proteins) determine the speed of contraction, whereas the content of mitochondria and the oxygen-carrying protein myoglobin determines the endurance and overall color of the muscle. White muscle cells depend largely on glycolysis to supply ATP, accounting for their rapid fatigue compared with red muscle cells, which are specialized for oxidative metabolism with abundant mitochondria and myoglobin.

Some muscles consist of only fast-twitch white muscle cells or slow-twitch red muscle cells, but most muscles are mixtures of two or more cell types. For example, in chickens, the leg muscles that are responsible for supporting the body, walking, and maintaining balance over long periods of time are rich in red muscle cells ("dark" meat). On the other hand, the chicken breast muscles, used for energetic flapping of the wings for short periods, are mainly white muscle cells ("light" meat).

Remarkably, the pattern of *nerve stimulation* determines the muscle cell type by controlling which genes are expressed (and, presumably, how the troponin T messenger RNA is processed). This was demonstrated by transplanting motor nerves between fast and slow muscles. Over a period of weeks, slow isoforms replaced fast isoforms and vice versa. Even more surprising, the same result is achieved by stimulating muscles electrically with fast or slow patterns of impulses. Chronic low-level stimulation biases gene expression toward the proteins that are found in slow muscle cells.

Calcium and calmodulin provide one prominent link between activity and gene expression. The concentration of active calmodulin tracks with the pattern of stimulation, because Ca^{2+} is released in the cytoplasm each time a muscle contracts. Among other things, calcium-calmodulin activates protein phosphatase PP2b (calcineurin; see Fig. 25.6), which dephosphorylates transcription factors (see Fig. 10.21). These activated transcription factors move into the nucleus and help to establish a transcription program that turns on expression of proteins found in slow muscles. These include contractile proteins and enzymes for oxidative metabolism.

The proportions of slow and fast muscle cells are determined genetically, so world-class sprinters (with a high proportion of fast, white fibers) and marathoners (with a high proportion of slow, red fibers) are born with advantages for their specialties. Training can lead to hypertrophy of specific muscle cell types and improved performance. Endurance training also leads to an increased proportion of slow cells. Without training, muscle strength declines with age; cell number remains constant, but each cell decreases in size.

Structural Proteins of the Plasma Membrane: Defects in Muscular Dystrophies

In addition to providing a permeability barrier, the plasma membrane of the muscle cell must maintain its integrity while being subjected to years of forceful contractions. Occasional breaches of the membrane are inevitable, so muscle cells also depend on a repair process that reseals holes. If membrane damage exceeds the repair capacity, muscle cells degenerate locally (segmental necrosis) or globally. Cell death beyond the ability of muscle stem cells to regenerate the tissue results in **muscular dystrophy.**

The proteins that stabilize muscle membranes were discovered in the late 1980s, when mutations in the **dystrophin** gene on the X-chromosome were linked to Duchenne muscular dystrophy, the most common human form of the disease. Dystrophin is an enormous member of the α-actinin superfamily of actin-binding proteins (see Fig. 33.17). The **dystroglycan–sarcoglycan complex** (Fig. 39.17 and Table 39.3) was found when it copurified with dystrophin after solubilizing the membrane with detergents.

More than 40 proteins are required to maintain the integrity of the plasma membrane as shown by mutations that cause muscular dystrophies (Table 39.3). Disease-causing mutations in genes for proteins of the dystroglycan–sarcoglycan complex typically lead to secondary loss of the other proteins in the complex. *O*-linked glycosylation by Golgi apparatus glycosyltransferases is

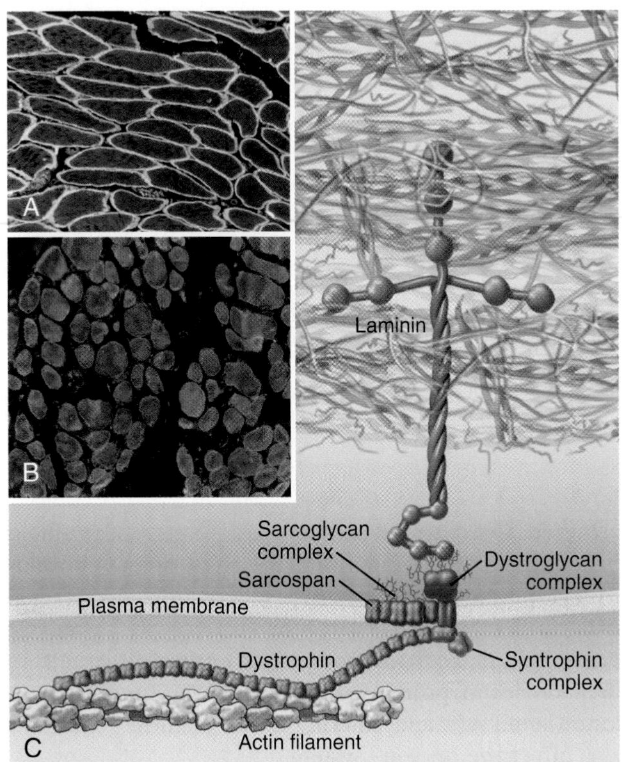

FIGURE 39.17 DYSTROPHIN AND ASSOCIATED PROTEINS STABILIZE THE PLASMA MEMBRANE OF SKELETAL MUSCLE. A–B, Fluorescent antibody staining of cross sections of human skeletal muscle showing the localization of dystrophin at the plasma membrane of a normal individual **(A)** and its absence in an individual with Duchenne muscular dystrophy **(B). C,** Model of the transmembrane complex of proteins that links dystrophin and actin filaments in cytoplasm to laminin in the basal lamina outside the cell. (**A–B,** Courtesy L. Kunkel, Harvard Medical School, Boston, MA. **C,** Based on a drawing by K. Amann and J. Ervasti, University of Wisconsin, Madison.)

TABLE 39.3 Proteins Required to Stabilize and Repair Muscle Plasma Membranes

Protein	Partners/Functions	Expression	Inheritance, Diseases
Membrane Skeleton			
Dystrophin	β-Dystroglycan, actin	Muscle, brain	X-linked, DMD, BMD
Utrophin	β-Dystroglycan, actin	Muscle, other tissues	
α-Syntrophins	Dystrophin	Muscle > other tissues	None detected in humans
β-Syntrophins	Dystrophin, utrophin	Muscle > other tissues	None detected in humans
Transmembrane Proteins			
Caveolin-3	Cholesterol	Muscle	AD, LGMD
α-Dystroglycan	Laminin, agrin	Many tissues	Embryonic lethal
β-Dystroglycan	Dystrophin, utrophin	Many tissues	Embryonic lethal
α-Sarcoglycan	Sarcoglycans, biglycan	Muscle	AR, LGMD, cardiomyopathy
β-Sarcoglycan	Sarcoglycans	Muscle	AR, LGMD
γ-Sarcoglycan	Sarcoglycans, biglycan	Muscle	AR, LGMD
Integrin α_7	Laminin	Many tissues	AR, CMD
Extracellular Matrix			
Collagen VI α_1, α_2, α_3	Biglycan	Muscle, other tissues	AD, Bethlem myopathy, Ulrich syndrome
α_2-Laminin	α-Dystroglycan	Muscle, other tissues	AR, CMD, dy/dy mouse
Agrin	α-Dystroglycan, AChR	Muscle	Null mouse perinatal lethal
Sarcomeric Proteins			
Titin	Myosin, Z-disk	Muscle	AR, LGMD, tibial MD
Myotilin	α-actinin, Z-disk	Muscle	AR, LGMD
Golgi Enzymes That Process Membrane and ECM Proteins			
Fukutin	Glycosyltransferase	Many tissues	AR, Fukuyama CMD
LARGE	Glycosyltransferase	Many tissues	AR, CMD
POMGnT1	Glycosyltransferase	Many tissues	AR, Muscle eye brain disease
POMTi	*O*-mannosyltransferase	Many tissues	AR, Walker-Warburg syndrome
Membrane Repair Machinery			
Dysferlin		Muscle	AR, LGMD, Miyoshi myopathy
Nuclear Envelope Proteins			
Emerin	Lamins, actin	All cells	XR, Emery-Dreifuss MD
Lamin A/C	Nuclear envelope	All cells	AD/AR, LGMD, Emery-Dreifuss MD

AD, autosomal dominant; AR, autosomal recessive; BMD, Becker muscular dystrophy; CMD, childhood muscular dystrophy; DMD, Duchenne muscular dystrophy; dy, dystrophia muscularis; ECM, extracellular matrix; LGMD, limb-girdle muscular dystrophy; MD, muscular dystrophy.

required for dystroglycan to bind to its extracellular ligands including α_2-laminin in the basal lamina. Proteins in cytoplasmic vesicles are used to repair damaged plasma membranes. The mechanical activity of muscle cells might make them more sensitive than other cells to deficiencies in proteins that support the nuclear envelope (lamin A/C and emerin). Some of these mutations also affect the nervous system.

Other than the X-linked dystrophin mutations, mutations causing muscular dystrophies are usually autosomal recessive. About one in several thousand humans develops some form of muscular dystrophy, because they inherit mutations in both copies of one of the sensitive genes. The mechanism of disease in muscular dystrophies is similar to that in hereditary spherocytosis, in which deficiencies of the membrane skeleton make red blood cells susceptible to mechanical damage (see Fig. 13.11). The age of onset and clinical features of inherited muscular dystrophies depend on the molecular defect. Patients with severe defects develop progressive muscle weakness as children. Ultimately, failure of respiratory muscles is fatal.

Dystroglycans and a dystrophin homolog, utrophin, participate in clustering acetylcholine receptors at the neuromuscular junction, the chemical synapse between motor neurons and skeletal muscle (see Fig. 17.9). When, during development, a motor neuron contacts the surface of its target muscle cell, the neuron secretes a proteoglycan called agrin, which is incorporated into the adjacent basal lamina. Agrin binds dystroglycan and a receptor tyrosine kinase in the muscle plasma membrane, which position associated acetylcholine receptors at the site where they receive acetylcholine secreted by the nerve in response to an action potential.

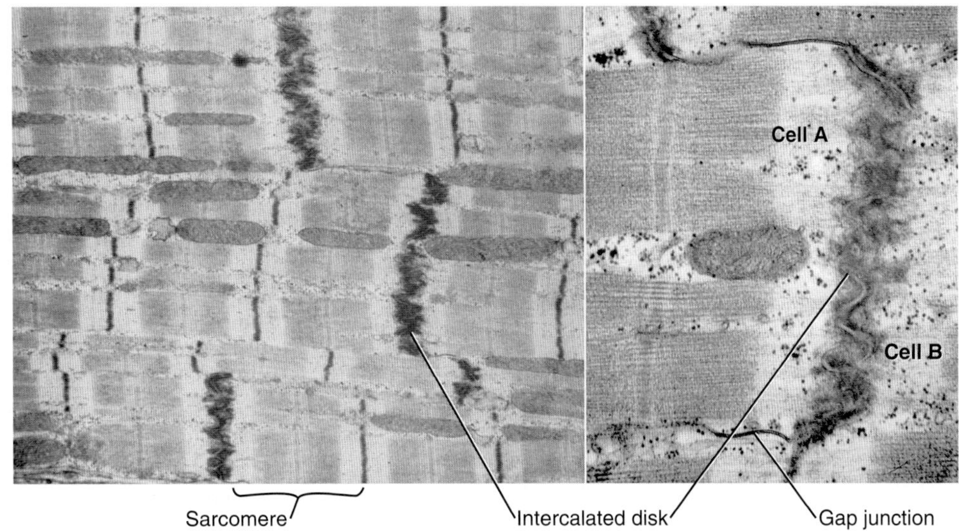

FIGURE 39.18 ELECTRON MICROGRAPHS OF A LONGITUDINAL SECTION OF TWO CARDIAC MUSCLE CELLS. Sarcomeres are similar to skeletal muscle. Intercalated disks anchor neighboring cells together, and gap junctions couple the cells electrically. (Courtesy D.W. Fawcett, Harvard Medical School, Boston, MA.)

Cardiac Muscle

To maintain the circulation of blood, heart muscle is specialized for repetitive (~100,000 times per day), fatigue-free contractions driven at regular intervals by action potentials from specialized pace-making cells within the heart. Gap junctions allow these action potentials to spread from one muscle cell to the next. Unlike skeletal muscle, the heart does not regenerate after injury, so efforts are being made to reprogram cardiac cells to recapitulate their normal development.

Contractile Apparatus of Cardiac Muscle

The architecture of the sarcomeres is very similar in cardiac and skeletal muscle, but cardiac muscle has more mitochondria, larger T tubules, and less SER (Fig. 39.18). The human atrium has one major myosin isoform. Two ventricular myosin isoforms (one shared with slow skeletal muscle) differ in ATPase activity and speed of contraction. Humans almost exclusively express only one of these isoforms. Myosin-binding protein C binds at intervals along the backbone of the thick filaments, interacts with actin, and modulates the myosin cross-bridges. The thin filaments are composed of a cardiac isoform of α-actin, tropomyosin, troponin, and a smaller version of nebulin called nebulette. When the cells are damaged by a heart attack or other disease, these proteins leak into the blood. Cardiac isoforms TNI and TNT in blood are a sensitive measure for cellular damage.

Short, modestly branched, cardiac muscle cells have centrally located nuclei and squared-off ends (Fig. 39.1) where cadherins (see Fig. 30.5) attach neighboring cells to each other at specialized adhesive junctions called **intercalated disks** (Fig. 39.18). These junctions have features of both adherens junctions (links to actin filaments) and desmosomes (links to intermediate filaments).

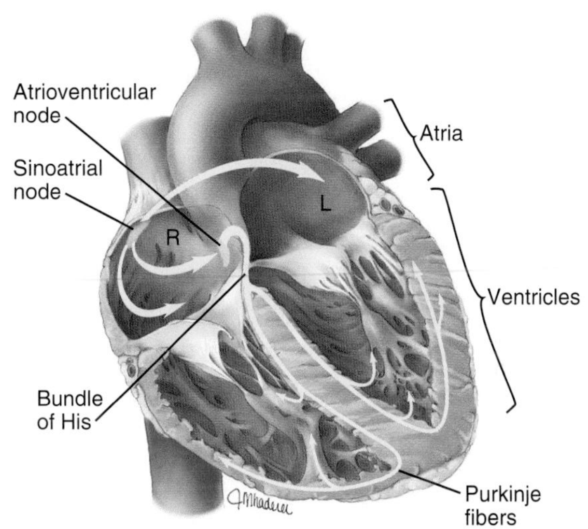

FIGURE 39.19 ACTIVATION OF CARDIAC CONTRACTION. An action potential (*yellow arrows*) starts at the sinoatrial node and travels through atrial muscle cells to the atrioventricular node. After a short delay at the atrioventricular node, the action potential spreads through the interventricular septum in modified cardiac muscle cells, called Purkinje fibers, and then through muscle cells to the whole ventricle. The action potential follows the same path each time, giving rise to electrical signals that can be detected on the body surface by electrocardiogram. Damage during myocardial infarctions changes the electrocardiographic pattern and may cause arrhythmias.

Pacemaker Cells

Intrinsically excitable **pacemaker cells** in the **sinoatrial node** drive rhythmic contractions of the heart (Fig. 39.19). The membrane potential of these cells drifts spontaneously toward threshold, setting off action potentials about once each second (Box 39.1). Each

BOX 39.1 Action Potentials in Cardiac Pacemaker Cells

Spontaneous action potentials of cardiac pacemaker cells in the sinoatrial node are more complicated than those of nerves (Fig. 39.20). Seven different plasma membrane channels determine their frequency:

1. *Voltage-gated Na^+ channels.* As in nerves, these channels rapidly activate at membrane potentials above threshold and then rapidly inactivate.

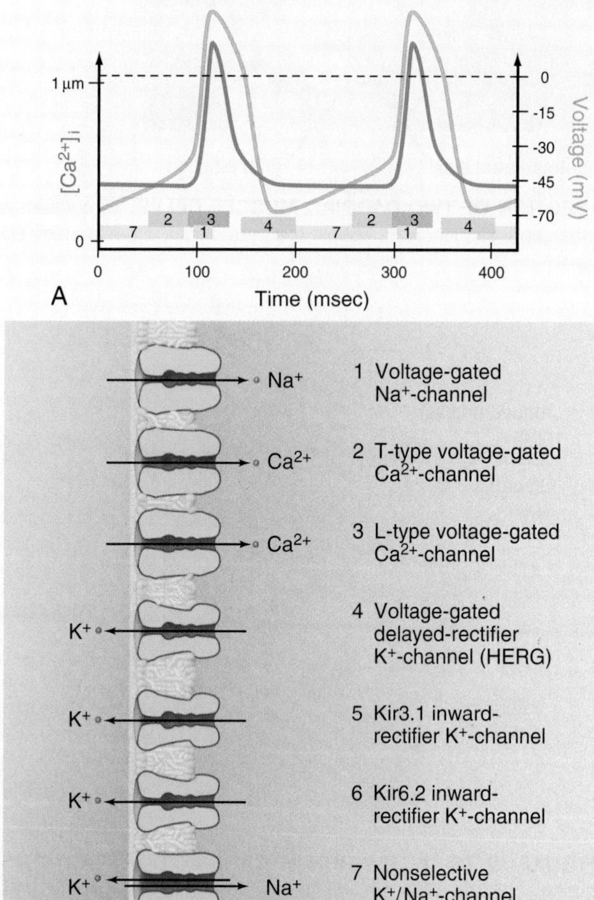

FIGURE 39.20 MECHANISM OF SPONTANEOUS CARDIAC PACEMAKER ACTION POTENTIALS. Sequential activation and inactivation of seven different plasma membrane channels account for the time course of action potentials and cytoplasmic Ca^{2+} transients in pacemaker cells of the sinoatrial node. **A,** Time courses of the fluctuations in membrane potential *(orange)* and cytoplasmic Ca^{2+} concentration *(blue)*. *Colored boxes* indicate when the various channels enumerated in part **B** are open. Kir3.1 and Kir6.2 are inactive under these conditions. **B,** Channels contributing to pacemaker activity.

2. *T-type, voltage-gated Ca^{2+} channels.* These low-conductance channels activate transiently at membrane potentials more negative than Na^+ channels, about −70 mV.
3. *L-type, voltage-gated Ca^{2+} channels.* These high-conductance channels slowly activate and inactivate when the membrane depolarizes to about −40 mV. Sympathetic nerve stimulation sensitizes these channels to membrane depolarization. Drugs called dihydropyridines block these channels.
4. *Delayed-rectifier, voltage-gated K^+ channels.* As in nerves, these HERG channels activate and inactivate slowly in response to membrane depolarization. Sympathetic nerves stimulate these channels.
5. *Kir3.1 inward-rectifier K^+ channels.* These channels conduct K^+ over a limited range of membrane potential, between about −30 and −80 mV. Parasympathetic nerve stimulation activates these channels.
6. *Kir6.2 inward-rectifier K^+ channels.* Normal levels of cytoplasmic ATP inhibit these channels. Depletion of cytoplasmic ATP activates these channels.
7. *Nonselective K^+/Na^+ channels.* Repolarization of the membrane activates these HCN (hyperpolarization-activated cyclic nucleotide-gated) channels, which slowly depolarize the membrane.

Acting together, these channels produce a spontaneous cycle of pacemaker action potentials. At the threshold potential (about −40 mV), voltage-gated Na^+ channels open synchronously and rapidly depolarize the membrane. As they inactivate, L-type Ca^{2+} channels open, prolonging the depolarization and admitting Ca^{2+}; this, in turn, triggers contraction by releasing more Ca^{2+} from internal stores (Fig. 39.14B). As these Ca^{2+} channels slowly inactivate, delayed-rectifier K^+ channels open and drive the membrane potential toward E_K, the K^+ equilibrium potential. As the membrane potential reaches a minimum, delayed-rectifier K^+ channels inactivate, but the two Kir channels open. In the absence of other channel activity, the membrane potential would remain near E_K, but the nonselective Na^+/K^+ channels open and the membrane slowly depolarizes, drifting toward threshold. T-type Ca^{2+} channels contribute to the slow, spontaneous depolarization. At threshold, the cycle repeats.

Channels similar to those in the sinoatrial node generates action potentials in cardiac muscle cells and stimulate contraction. Cardiac muscle cells can generate spontaneous action potentials, but they have fewer T-type Ca^{2+} channels and more Kir K^+ channels, so the rate of spontaneous action potentials is lower than that of pacemaker cells. Except in disease, pacemaker cells drive action potentials throughout the rest of the heart. Later in life, cells in the pulmonary veins can also generate action potentials. This is one cause of atrial fibrillation, a common arrhythmia in older humans.

action potential spreads from cell to cell through gap junctions (see Fig. 31.6), activating all cells in the atrium within a few hundred milliseconds. After a brief delay in the **atrioventricular node,** the action potential and contraction spreads from cell to cell through the ventricle. This highly reproducible pattern of electrical activity can be recorded on the surface of the body as an electrocardiogram.

Mutations in six different human ion channel genes including voltage-gated sodium channels (SCN5A) and potassium channels (HERG, KVLAT1, and minK) cause disorders of cardiac muscle electrophysiology. These

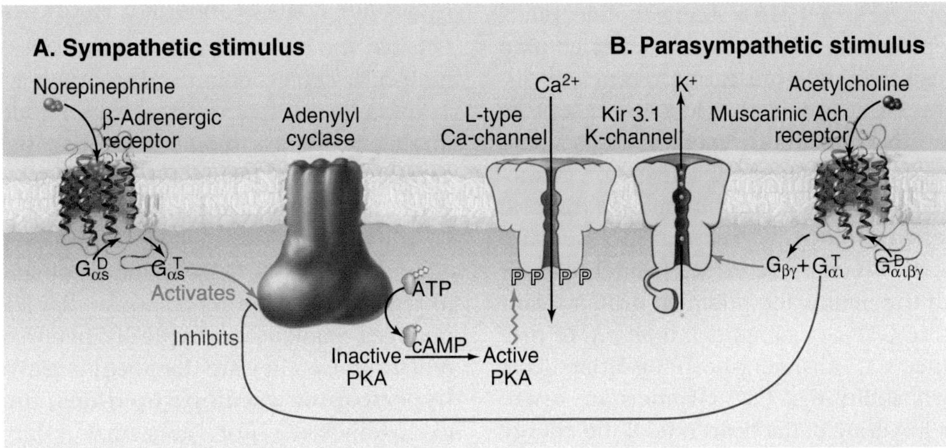

FIGURE 39.21 Regulation of the rate of cardiac pacemaker cells by sympathetic **(A)** and parasympathetic **(B)** nerves. GTP-Gα_s stimulates adenylyl cyclase. GTP-Gα_i inhibits adenylyl cyclase. D is GDP associated with G protein α-subunits; T is GTP. Ach, acetylcholine; cAMP, cyclic adenosine monophosphate; GDP, guanosine diphosphate; GTP, guanosine triphosphate; PKA, protein kinase A.

inherited diseases are called *long-QT syndrome* because the interval between the initial depolarization of the muscle cells and their relaxation is prolonged. This change predisposes the person to abnormal cardiac rhythms that are potentially fatal.

Excitation Contraction Coupling

As in skeletal muscle, plasma membrane action potentials stimulate cardiac muscle cells to contract by releasing cytoplasmic Ca^{2+} to activate troponin-tropomyosin (Fig. 39.14B). Calcium ATPase pumps (see Fig. 14.7) in SER and plasma membrane maintain the low cytoplasmic Ca^{2+} concentration with some help from plasma membrane Na^+/Ca^{2+} antiporters.

In both cardiac and skeletal muscle action potentials activate L-type voltage-sensitive Ca^{2+} channels (DHP receptors) in T tubules, but subsequent events differ as revealed by a requirement for extracellular Ca^{2+} in heart but not skeletal muscle (Fig. 39.14). Rather than interacting directly with ryanodine receptors as in skeletal muscle, the active cardiac L-type voltage-sensitive Ca^{2+} channels admit extracellular Ca^{2+}. This opens nearby ryanodine receptors in the SER, releasing a flood of Ca^{2+} to trigger contraction. This is called calcium-induced calcium release.

Excitation-contraction coupling can be defective when heart muscle cells grow larger in response to abnormal demands, such as high blood pressure. The defect may be explained by growth separating T tubules from SER, either physically or functionally, thereby decreasing the probability that Ca^{2+} entering through DHP receptors will trigger Ca^{2+} release from the endoplasmic reticulum.

Seven-Helix Receptors and Trimeric G-Proteins Regulate Heart Rate and Contractility

Motor nerves do not stimulate cardiac muscle directly. Instead, molecules secreted by autonomic nerves and the adrenal gland modulate the rate and force of contraction (Fig. 39.21). Acetylcholine from **parasympathetic nerves** slows the heartbeat, whereas norepinephrine from **sympathetic nerves** and **epinephrine** from the **adrenal gland** speed the rate and increase the strength of contraction (Fig. 39.21). These neurotransmitters target channels indirectly by activating two different seven-helix receptors and their associated trimeric G-proteins (see Fig. 25.9). The resting rate reflects a compromise in the competition between these two inputs.

Norepinephrine and epinephrine increase the heart rate and force of contraction by modulating L-type Ca^{2+} channels (Fig. 39.21A). Both bind β-adrenergic receptors that activate trimeric G proteins, which stimulate the production of cyclic adenosine monophosphate (cAMP) by **adenylyl cyclase** (see Fig. 26.2). cAMP activates protein kinase A (PKA) (see Fig. 25.3) that phosphorylates three types of channels. Phosphorylated L-type, voltage-gated Ca^{2+} channels are more likely to open in response to membrane depolarization and admit more Ca^{2+} to activate the contractile machinery more fully. Phosphorylated nonspecific HCN (hyperpolarization-activated cyclic nucleotide-gated) cation channels push the membrane potential toward threshold. Both increase the frequency of action potentials and the heart rate. Phosphorylated delayed-rectifier K^+ channels are more active in repolarizing the membrane, so they prevent activated Ca^{2+} channels from prolonging the action potential. Phosphorylation of phospholamban, the regulatory subunit of the calcium pump in the endoplasmic reticulum, increases its activity, which speeds relaxation These changes allows heart muscle cells to keep up with stimuli generated at a higher rate from pacemaker cells. PKA also targets sarcomeric proteins: phosphorylation of troponin-I and myosin-binding protein C each increases the rate of crossbridge cycling and the strength of contraction.

Acetylcholine released by parasympathetic nerves activates Kir3.1 inward-rectifier K^+ channels that slow the heartbeat (Fig. 39.21B). Acetylcholine binds seven-helix receptors, called muscarinic acetylcholine receptors (because they bind muscarine. These are distinct from pentameric nicotinic acetylcholine receptors (see Fig. 16.12). Active muscarinic receptors catalyze the exchange of guanosine diphosphate (GDP) for guanosine triphosphate (GTP) on the Gα_i subunit of a trimeric G protein, releasing the G$\beta\gamma$ subunits to activate Kir3.1/3.4 channels. When open, these channels reduce the rate at which the membrane potential drifts toward threshold. The free GTP-Gα_i subunits inhibit cAMP production and reduce Ca^{2+} channel phosphorylation. This decreases the probability that Ca^{2+} channels are open, contributing to a lowering of the heart rate. If the energy supply of the heart is compromised, ATP levels fall. This activates Kir6.2 channels, which reduce the rate of spontaneous depolarization and lower the heart rate until ATP levels are restored.

Therapeutic Effect of Digitalis in Congestive Heart Failure

In congestive heart failure, cardiac contraction fails to produce enough force to maintain adequate circulation of blood. Digitalis from the foxglove plant was the first drug to treat heart failure. Digitalis and related compounds strengthen cardiac contraction indirectly by inhibiting the α_2 isoform of Na^+K^+-ATPase in the plasma membrane (Fig. 39.22). Reduced sodium pump activity lowers the Na^+ gradient across the membrane, providing less driving force for Na^+/Ca^{2+} antiporters to exchange extracellular Na^+ for cytoplasmic Ca^{2+}. The slightly higher steady-state concentration of Ca^{2+} in cytoplasm strengthens contraction.

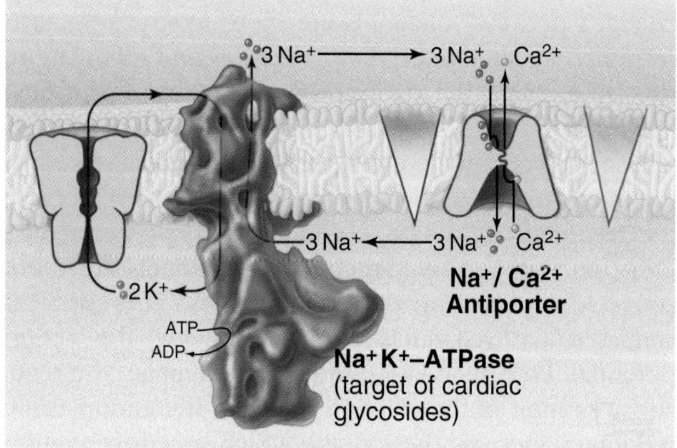

FIGURE 39.22 CHEMIOSMOTIC CYCLE THAT HELPS CLEAR CA^{2+} FROM THE CYTOPLASM OF CARDIAC MUSCLE CELLS. Na^+K^+-ATPase pumps create a Na^+ gradient *(purple triangle)* to drive the Na^+/Ca^{2+} antiporter to transport Ca^{2+} up its concentration gradient *(blue triangle)* out of the cell. Cardiac glycosides such as digitalis inhibit the cardiac isoform of Na^+K^+-ATPase, raising the concentration of cytoplasmic Ca^{2+} and strengthening cardiac contraction. K^+ channels *(left)* allow K^+ to circulate.

Molecular Basis of Inherited Heart Diseases

Because the heart is so vital to survival, relatively minor molecular defects command attention in humans. Nearly 1% of individuals carry an inherited or de novo mutation in a gene for a sarcomeric protein that compromises cardiac function. The most common mutations are in the genes for myosin heavy chain, myosin-binding protein C, and titin, but virtually every sarcomeric protein is affected (Table 39.1). Long, noncoding RNAs participate in regulating the stress response that leads to cardiomyopathies. Patients are typically heterozygous for these mutations (ie, they are dominant negative mutations). In **hypertrophic cardiomyopathies,** the heart attempts to compensate for abnormal contractility (either increased or decreased depending on the mutation) through hypertrophy, but the thickened heart wall compromises cardiac relaxation and refilling the chambers with blood. In **dilated cardiomyopathies,** the ventricles swell and their walls thin. Both types are associated with heart failure and abnormal cardiac rhythms that can be fatal. The rate of progress of these diseases depends on not only the particular mutation but also other factors that vary from person to person. Individuals with defects in myosin-binding protein C develop hypertrophy in their fifties but can live normal life spans. By contrast, those with defects in troponin T can be affected as teenagers and die of arrhythmias in their twenties. These severe mutations of cardiac contractile proteins account for about half of the deaths of apparently healthy young athletes.

Smooth Muscle

Contractile Apparatus

Smooth muscle cells are specialized for slow, powerful, efficient contractions under the control of a variety of involuntary mechanisms. Smooth muscle cells are generally confined to internal organs, such as blood vessels (where they regulate blood pressure), the gastrointestinal tract (where they move food through the intestines), and the respiratory system (where they control the diameters of the air passages; their excessive contraction contributes to asthma and other allergic reactions). The cytoplasm of spindle-shaped smooth muscle cells (Fig. 39.1) appears homogeneous by light microscopy, because the contractile proteins are not organized in regular arrays like sarcomeres of skeletal and cardiac muscle. A basal lamina and variable amounts of collagen and elastic fibers surround each cell. Smooth muscle cells rarely divide in adults, but they are capable of responding to local physiological conditions by remodeling their structure, as in atherosclerosis and hypertension.

In terms of organization and biochemistry, smooth muscle cells (Fig. 39.23) resemble nonmuscle cells more than skeletal or cardiac muscle. For example, the gene for smooth muscle myosin arose relatively recently from a cytoplasmic myosin II gene. These myosins also share

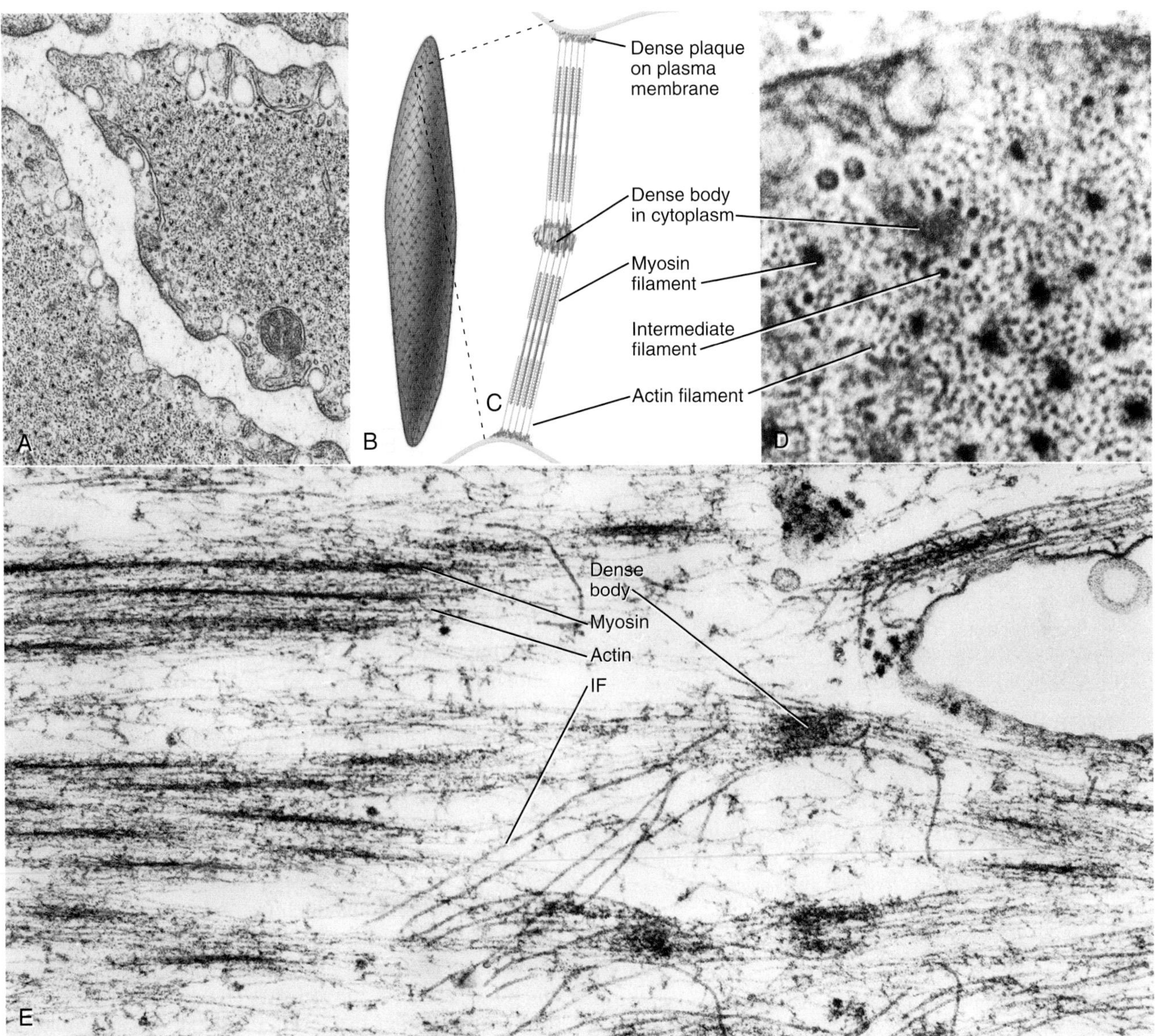

FIGURE 39.23 CONTRACTILE APPARATUS OF SMOOTH MUSCLE. A, Electron micrograph of a thin cross section of two smooth muscle cells. **B–C,** Organization of the contractile units, which stretch across the cell between plasma membrane attachment plaques. Contractile units consist of myosin filaments connecting thin filaments attached to a dense body or plasma membrane plaque. **D,** High-power electron micrograph showing a dense body and cross sections of three types of filaments. **E,** Electron micrograph of a longitudinal section of an extracted vascular smooth muscle cell illustrating associations of actin filaments and intermediate filaments (IF) with dense bodies, and myosin filaments interacting with actin filaments. (**A, D,** and **E,** Courtesy A.V. Somlyo and A.P. Somlyo, University of Virginia, Charlottesville. For reference, see Somlyo AP, Devine CE, Somlyo AV, Rice RV. Filament organization in vertebrate smooth muscle. *Philos Trans R Soc Lond B Biol Sci*. 1973;265:223–229; Bond M, Somlyo AV. Dense bodies and actin polarity in vertebrate smooth muscle. *J Cell Biol*. 1982;95:403–413.)

the same regulatory light chain. Long myosin thick filaments are interspersed among the thin filaments, but not in a regular way as in striated muscles. Thin filaments are composed of actin and tropomyosin, along with two regulatory proteins, caldesmon and calponin, rather than troponin. Thin filaments are arranged obliquely in the cell, some with their barbed ends attached to dense plaques on the plasma membrane, others to **dense bodies** in the cytoplasm. Like Z disks in striated muscles, dense bodies anchor desmin intermediate filaments, forming a continuous, inextensible, internal "tendon" running from end to end of the cell, preventing excess stretching (see Fig. 35.8).

Smooth muscle cells contract like a concertina (Fig. 39.24), because tension generated by myosin and actin is applied to discrete spots on the plasma membrane. This compression can be seen in light micrographs as irregular cells with "corkscrew" nuclei. Given that smooth muscle cells have less myosin than striated muscle cells do, it is remarkable that they develop the same force. This is explained by two factors. First, the force-generating unit, the myosin filament, is larger in

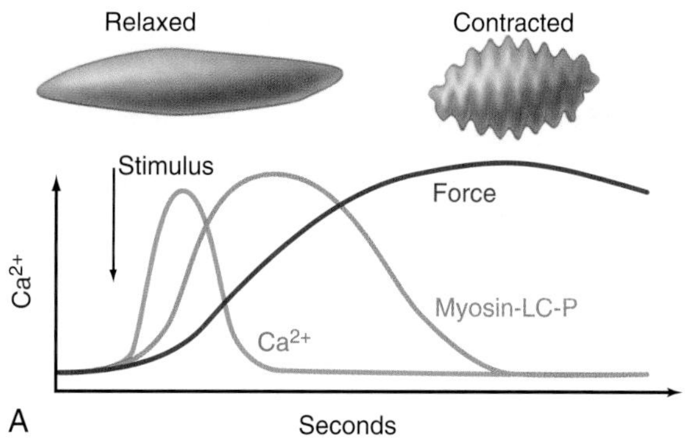

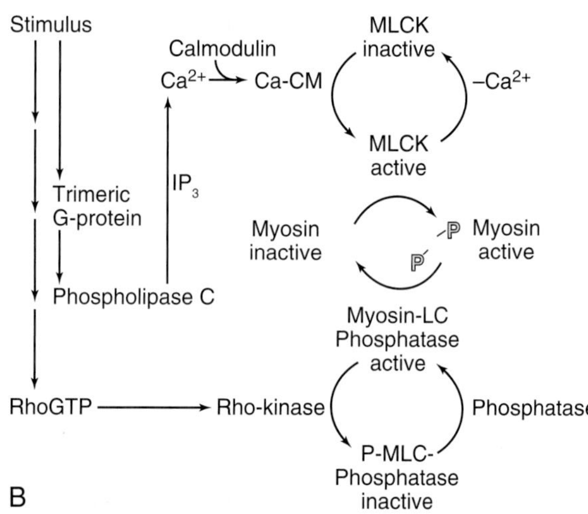

FIGURE 39.24 ACTIVATION OF SMOOTH MUSCLE CONTRACTION. A, The spindle-shaped smooth muscle cell develops accordion pleats as it contracts, owing to the attachment of the actin filaments at intervals along the plasma membrane. The graph shows the time course of activation, consisting of the release of Ca^{2+} into the cytoplasm, phosphorylation of myosin regulatory light chains, and then the slow development of force. Myosin light-chain phosphorylation (LC-P) is required to initiate, but not to sustain, the contraction of smooth muscle. **B,** Biochemical pathways controlling phosphorylation of myosin regulatory light chains. Receptor stimulation leads to production of IP_3 (inositol 1,4,5-triphosphate) by phospholipase C and release of Ca^{2+} into cytoplasm. Ca^{2+} binds calmodulin (CM), which activates myosin light-chain kinase (MLCK) by binding the kinase's autoinhibitory peptide and displacing it from the active site. Active MLCK phosphorylates activating sites on the regulatory light chain. Light-chain phosphatase reverses phosphorylation of myosin. Activation of the small GTPase (guanosine triphosphatase) Rho with GTP stimulates Rho-kinase, which phosphorylates and inactivates light-chain phosphatase. This makes the system more sensitive to Ca^{2+} levels, as light-chain phosphorylation is prolonged. LC, light chain; P-MLC-, phosphorylated myosin light chain. (**A,** Modified from the work of K. Kamm and J. Stull, University of Texas Southwestern Medical School, Dallas.)

smooth muscle than in skeletal muscle. Deploying a given amount of myosin in large, thick filaments in a long sarcomere produces more force than does the same myosin in smaller filaments arranged in a series of short sarcomeres. Second, individual smooth muscle myosin molecules produce a larger force than skeletal muscle myosin, at least in vitro assays.

Regulation of Smooth Muscle Contraction

A wide range of stimuli trigger smooth muscle contraction, but they all seem to act through seven-helix receptors coupled to trimeric G-proteins. Hormones stimulate contraction of the uterus, whereas motor nerves stimulate intrinsic eye muscles that close the pupil.

Depending on the particular smooth muscle, Ca^{2+} for contraction enters the cytoplasm through either voltage-dependent calcium channels in the plasma membrane or IP_3 (inositol 1,4,5-triphosphate) receptor Ca^{2+} release channels in the SER (Fig. 39.24). Drugs that block plasma membrane calcium channels can distinguish these two pathways experimentally. In intestines, parasympathetic nerves release acetylcholine to stimulate seven-helix muscarinic receptors (Fig. 39.21). Associated trimeric G-proteins activate cation channels that depolarize the plasma membrane and allow Ca^{2+} to enter through voltage-sensitive calcium channels. Consequently, calcium channel blockers strongly inhibit activation of gut smooth muscle. Gap junctions couple gut smooth muscle cells, allowing excitation to spread from cell to cell. At the other end of the spectrum, vascular smooth muscle depends on IP_3 to release Ca^{2+} from intracellular stores rather than depending on Ca^{2+} from outside the cell.

Following stimulation, intracellular Ca^{2+} increases rapidly but transiently, declining to a value above resting level as the receptors desensitize (see Fig. 24.3). Ca^{2+} pumps in both SER and plasma membrane clear the cytoplasm of Ca^{2+} so that Ca^{2+} levels decrease to resting levels and the muscle eventually relaxes when the activating stimulus declines. Relaxing agents, acting through cyclic guanosine monophosphate (cGMP) or cAMP (see Fig. 26.1), promote clearance of cytoplasmic Ca^{2+}. Epinephrine relaxes smooth muscles of the respiratory system by another mechanism. Stimulation of β-adrenergic receptors activates potassium channels that hyperpolarize the plasma membrane and reduce Ca^{2+} entry. This approach is widely used to treat asthma.

After a considerable delay (>200 ms) following the Ca^{2+} spike, contractile force develops slowly. The delay is attributable to the time required for a sequence of three biochemical reactions: Ca^{2+} binding to calmodulin, calcium-calmodulin activation of **myosin light-chain kinase** (see Fig. 25.4), and phosphorylation of myosin regulatory light chains, turning on the myosin-actin ATPase cycle (Fig. 39.24). Unphosphorylated myosin-II from smooth muscle and vertebrate nonmuscle cells is inactive.

Phosphorylation of myosin light chains is required to initiate but not maintain contraction, so slowly cycling, unphosphorylated myosins maintain peak force with little expenditure of energy. Regulation of unphosphorylated crossbridges is not well understood, but they appear to be activated cooperatively by a small population of phosphorylated myosin heads. Caldesmon, a calcium-calmodulin-binding protein associated with tropomyosin on actin filaments, may contribute to activation and/or allow myosin heads to cycle very slowly even in the presence of ATP.

The sensitivity of light-chain phosphorylation to Ca^{2+} depends on a parallel signaling pathway that partially inhibits myosin phosphatase, thus increasing the number of phosphorylated myosin cross-bridges and force at any given Ca^{2+} concentration (Fig. 39.24). Receptors coupled to trimeric G-proteins activate the small guanosine triphosphatase (GTPase) RhoA, which stimulates a protein kinase that inhibits myosin light-chain phosphatase.

ACKNOWLEDGMENTS

We thank Lee Sweeney and John Solaro for their suggestions on revisions to this chapter.

SELECTED READINGS

Agarkova I, Perriard J-C. The M-band: an elastic web that crosslinks thick filaments in the center of the sarcomere. *Trends Cell Biol.* 2005;15:477-485.

Alexander MR, Owens GK. Epigenetic control of smooth muscle cell differentiation and phenotypic switching in vascular development and disease. *Annu Rev Physiol.* 2012;74:13-40.

Bolton TB, Prestwich SA, Zholos AV, Gordienko DV. Excitation-contraction coupling in gastrointestinal and other smooth muscles. *Annu Rev Physiol.* 1999;61:85-115.

Butler T, Paul J, Europe-Finner N, Smith R, Chan EC. Role of serine-threonine phosphoprotein phosphatases in smooth muscle contractility. *Am J Physiol Cell Physiol.* 2013;304:C485-C504.

Cahill TJ, Ashrafian H, Watkins H. Genetic cardiomyopathies causing heart failure. *Circ Res.* 2013;113:660-675.

Clark KA, McElhinny AS, Beckerle MC, Gregorio CC. Striated muscle cytoarchitecture: an intricate web of form and function. *Annu Rev Cell Dev Biol.* 2002;18:637-706.

Doles JD, Olwin BB. Muscle stem cells on the edge. *Curr Opin Genet Dev.* 2015;34:24-28.

Franzini-Armstrong C, Protasi F, Ramesh V. Comparative ultrastructure of Ca^{2+} release units in skeletal and cardiac muscle. *Ann N Y Acad Sci.* 1998;853:20-30.

Gao N, Huang J, He W, et al. Signaling through myosin light chain kinase in smooth muscles. *J Biol Chem.* 2013;288:7596-7605.

Geeves MA, Holmes KC. Structural mechanism of muscle contraction. *Annu Rev Biochem.* 1999;68:687-728.

Gokhin DS, Fowler VM. A two-segment model for thin filament architecture in skeletal muscle. *Nat Rev Mol Cell Biol.* 2013;14:113-119.

Gordon AM, Homsher E, Regnier M. Regulation of contraction in striated muscle. *Physiol Rev.* 2000;80:853-924.

Guiraud S, Aartsma-Rus A, Vieira NM, et al. The pathogenesis and therapy of muscular dystrophies. *Annu Rev Genomics Hum Genet.* 2015;16:281-308.

Han R, Campbell KP. Dysferlin and muscle membrane repair. *Curr Opin Cell Biol.* 2007;19:409-416.

Hidalgo C, Granzier H. Tuning the molecular giant titin through phosphorylation: role in health and disease. *Trends Cardiovasc Med.* 2013;23:165-171.

Hinson JT, Chopra A, Nafissi N, et al. Titin mutations in iPS cells define sarcomere insufficiency as a cause of dilated cardiomyopathy. *Science.* 2015;349:982-986.

Hwang PM, Sykes BD. Targeting the sarcomere to correct muscle function. *Nat Rev Drug Discov.* 2015;14:313-328.

Hwang JH, Zorzato F, Clarke NF, Treves S. Mapping domains and mutations on the skeletal muscle ryanodine receptor channel. *Trends Mol Med.* 2012;18:644-657.

Lehrer SS, Geeves MA. The myosin-activated thin filament regulatory state, M^--open: a link to hypertrophic cardiomyopathy (HCM). *J Muscle Res Cell Motil.* 2014;35:153-160.

Marx SO, Marks AR. Dysfunctional ryanodine receptors in the heart: new insights into complex cardiovascular diseases. *J Mol Cell Cardiol.* 2013;58:225-231.

Moss RL, Fitzsimons DP, Ralphe JC. Cardiac MyBP-C regulates the rate and force of contraction in mammalian myocardium. *Circ Res.* 2015;116:183-192.

Murthy KS. Signaling for contraction and relaxation in smooth muscle of the gut. *Annu Rev Physiol.* 2006;68:345-374.

Myhre JL, Pilgrim D. A Titan but not necessarily a ruler: assessing the role of titin during thick filament patterning and assembly. *Anat Rec (Hoboken).* 2014;297:1604-1614.

Rao JN, Madasu Y, Dominguez R. Mechanism of actin filament pointed-end capping by tropomodulin. *Science.* 2014;345:463-467.

Somlyo AP, Somlyo AV. Ca^{2+}-sensitivity of smooth and non-muscle myosin II: modulation by G Proteins, kinases and myosin phosphatase. *Physiol Rev.* 2003;83:1325-1358.

Spudich JA. Hypertrophic and dilated cardiomyopathy: four decades of basic research on muscle lead to potential therapeutic approaches to these devastating genetic diseases. *Biophys J.* 2014;106:1236-1249.

Takamori M. Structure of the neuromuscular junction: function and cooperative mechanisms in the synapse. *Ann N Y Acad Sci.* 2012;1274:14-23.

Takeda S, Yamashita A, Maeda K, Maeda Y. Structure of the core domain of human cardiac troponin in the Ca^{2+}-saturated form. *Nature.* 2003;424:35-41.

Teekakirikul P, Padera RF, Seidman JG, Seidman CE. Hypertrophic cardiomyopathy: translating cellular cross talk into therapeutics. *J Cell Biol.* 2012;199:417-421.

Tskhovrebova L, Trinick J. Making muscle elastic: the structural basis of myomesin stretching. *PLoS Biol.* 2012;10:e1001264.

von der Ecken J, Müller M, Lehman W, et al. Structure of the F-actin-tropomyosin complex. *Nature.* 2014;519:114-117.

Wang YX, Dumont NA, Rudnicki MA. Muscle stem cells at a glance. *J Cell Sci.* 2014;127:4543-4548.

Wehrens XH, Lehnart SE, Marks AR. Intracellular calcium release and cardiac disease. *Annu Rev Physiol.* 2005;67:69-98.

SECTION X

Cell Cycle

SECTION X OVERVIEW

This last section of the book draws together principles from previous chapters to explain some of the rules that govern the lifestyles of cells. Cells exhibit a remarkable diversity in their patterns of growth, proliferation, and death. For example, some human cells (neurons) are born around the time of birth and live until the person dies—more than 100 years in a few cases. The fate of other cells is to live for only a day or two (eg, cells in the gut lining). Many differentiated cells form by elaborate pathways that employ a carefully choreographed series of cues from within the cell and from its neighbors. Other cells, such as many in the immune system, are spawned in excess, followed by selection of the few with correctly rearranged genes or with productive connections to partner cells. The unlucky majority of their siblings whose differentiation did not go so well ultimately commit suicide.

Very different strategies maintain populations of cells. Long-lived cells divide seldom, if at all. In contrast, the cells that are involved in producing the gut lining grow and divide at top speed. Most human cells differentiate to carry out specific functions and then no longer proliferate. How do cells decide whether to proliferate, to stop proliferating and differentiate, or to die? This section answers these and other questions.

Chapter 40 begins the section with an introduction to the language of the cell cycle. The cell cycle is driven by changing states of the cytoplasm created by shifting balances of protein phosphorylation, dephosphorylation, and degradation machinery. For the cell cycle, the key kinases are **cyclin-dependent kinases** (Cdks), which require an associated **cyclin** subunit for activity. Cdks are also regulated by phosphorylation and by additional protein cofactors that bind and inactivate them. Cdks are usually stable, but cyclin levels fluctuate, owing to targeted destruction at particular points in the cell cycle. In fact, targeted proteolytic destruction by the proteasome is a key aspect of cell-cycle control. Each cell-cycle phase is characterized by the activity of one or more E3 ubiquitin ligases. Each of these targets particular proteins for destruction by decorating them with chains of ubiquitin, a protein that was introduced

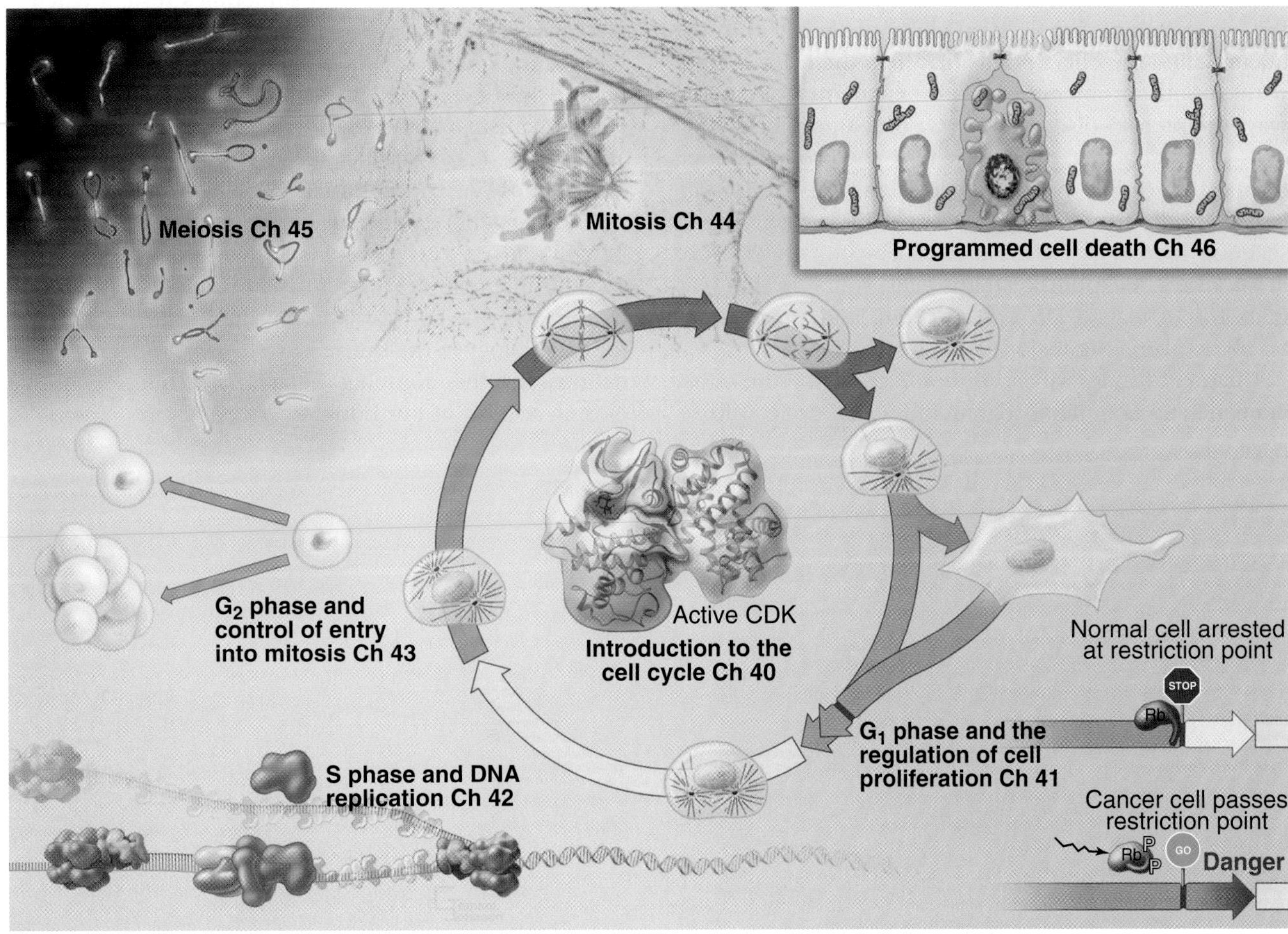

in Chapter 23. This sequential destruction of key factors gives cell-cycle transitions their irreversible character.

The chapters that follow explain how the cell-cycle machinery controls each step in the proliferation and differentiation of cells. Chapter 41 begins with newly born cells in the **G_1 phase** of the cell cycle. These cells must decide whether to commit themselves to a round of proliferation or to withdraw from the proliferation rat race and enter a nondividing differentiated state called **G_0**. Cells that will proliferate must first pass a control point known as the **restriction point**. This is a control circuit that determines whether internal and external conditions are suitable for proliferation. Malfunctions of the restriction point lead to one of the most terrifying perturbations of the cell cycle: cancer. If DNA damage is detected during G_1 phase, a **checkpoint** halts cell cycle progression until either the damage is repaired or the cell dies. The chapter includes a section on **stem cells** and concludes by considering the role of one of the most famous cell-cycle proteins, **p53,** in cell-cycle control.

Cells that decide to proliferate must replicate their DNA in a timely and accurate manner. Chapter 42 explains the mechanism of DNA replication during the **S phase,** including the selection of sites on DNA to initiate replication, the enzymes that copy the DNA, the regulation of replication by the cell-cycle machinery, the organization of replicating chromosomes within the nucleus, and the checkpoints that help the cells to cope with various problems that they encounter along the way. Chapter 43 discusses the **G_2 phase,** during which cells conduct a final "cockpit check" before embarking on the irreversible process of division. This is also the last point in the cell cycle at which the genome is scanned for damage so that it can be repaired before division. A checkpoint restrains cells from entering into mitosis if damaged DNA is detected, and the chapter briefly explains the major pathways of DNA repair.

Chapter 44 describes **mitosis,** certainly the most dramatic and complex program in the cell cycle. Mitosis has been studied since the 1800s, but technical advances have considerably advanced our understanding of how it is accomplished at the molecular level. Division requires wholesale reorganization of cellular structures, including chromosome condensation and the assembly of the mitotic spindle. In many cells, the nuclear envelope breaks down. Once the chromosomes are all attached to the microtubules of the mitotic spindle (yet another important checkpoint here), they are separated equally and form two daughter nuclei. Finally, **cytokinesis** separates the two daughter cells.

Chapter 45 considers **meiosis,** a specialized form of division that produces the gametes required for sexual reproduction. In this division, DNA recombination is key to segregation of the chromosomes. A number of arcane terms are used to describe the specialized structures and processes involved. The chapter then explains how problems with meiosis can lead to genetic diseases and how studies of chromosome segregation in yeast led to an understanding of why birth defects become more prevalent as human mothers age.

Chapter 46 closes the book with a discussion of what happens when cells commit suicide by **apoptosis, necroptosis,** and **autophagy.** This is not, strictly speaking, a cell-cycle event but instead represents several alternative pathways, each with its own machinery and signaling systems. Apoptosis sometimes results when it all "runs off the rails" and cells receive insults from which they cannot recover. But cell death is not always bad: apoptosis is an essential part of development of metazoan organisms, homeostasis of their organs and tissues, and can be a last-ditch defense against viral infection. Malfunctions of apoptotic pathways can lead to cancer.

The concepts that are discussed in this section of the book build on the ideas in earlier sections. Cells are wonderfully complex systems whose behavior is driven by the laws of chemistry and physics. A major challenge for cell biology in the future is to devise molecular explanations for the complex behaviors exhibited in this closing section of our book.

CHAPTER 40

Introduction to the Cell Cycle

The **cell cycle** is the series of events that leads to the duplication and division of a cell. Research on the molecular events of cell-cycle control revealed that variations of similar mechanisms operate the cell cycles of all eukaryotes from yeasts to humans. Furthermore, the components that regulate cell growth and division also play key roles in the cessation of cell division that is required for cells to differentiate. Control of the cell cycle is of major importance to human health because cancer is usually caused by perturbations of cell-cycle regulation. Based on 2010 to 2012 data, 40% of Americans will develop cancer during their lifetime.

Although animal cells have a wide variety of specialized cell cycles, the cells in the stratified epithelium that forms skin illustrate the most common types of cell cycles (Fig. 40.1). The basal layer of the epithelium is composed of **stem cells** that divide only occasionally (see Box 41.2). They can activate the cell cycle on demand and then return to a nondividing state. Stem cell populations can replenish themselves by symmetrical division, but when specific signals induce them to proliferate, usually one daughter cell remains a stem cell and the other enters a pool of rapidly dividing cells. The dividing cells populate the upper layers of the epithelium, stop dividing, and gradually differentiate into the specialized cells that cover the surface.

Like stem cells, fibroblasts of the connective tissue (see Fig. 28.2) typically are in a nondividing state, but they can be stimulated to enter the cell cycle following wounding or other stimuli (see Fig. 32.11). In the most extreme case, the nervous system contains a few stem cells and a few dividing glial cells, but most neurons, once differentiated, can live for more than 100 years without dividing again.

Principles of Cell-Cycle Regulation

The goal of the cell cycle in most cases is to produce two daughter cells that are accurate copies of the parent (Fig. 40.2). The cell cycle integrates a continuous **growth cycle** (an increase in cell mass) with a discontinuous **division** or **chromosome cycle** (the replication and partitioning of the genome into two daughter cells). The chromosome cycle is driven by a sequence of enzymatic cascades that produce a sequence of discrete biochemical "states" of the cytoplasm. Progress through the cell cycle is ratchet-like and irreversible because each new state arises not only by expression or activation of a new cohort of activities, but also by destruction or inactivation of key activities characteristic of the preceding state. Later sections of this chapter explain these mechanisms.

Phases of the Cell Cycle

In describing the cell cycle, it is convenient to divide the process into several phases. Recognition of these phases began in 1882, when Flemming named the process of nuclear division **mitosis** (from the Greek *mito,* or "thread") after he first observed the condensed chromosomes. Mitosis was a clear cell cycle landmark, and the rest of the cell cycle between mitoses was called **interphase** (Box 40.1).

Once DNA was recognized as the agent of heredity in the 1940s, it was deduced that it must be duplicated at some time during interphase so that daughter cells can each receive a full complement of genetic material. In 1953, a key experiment identified the relationship between the timing of DNA synthesis and the mitotic cycle (Fig. 40.3). This defined the four cell-cycle phases as they are known today (see Fig. 40.2).

Each cell is born at the completion of the **M phase,** which includes **mitosis,** the partitioning of the chromosomes and other cellular components, and **cytokinesis,** the division of the cytoplasm. The chromosomal DNA is replicated during **S phase** (synthetic phase). The remaining two phases are gaps between mitosis and the S phase. The **G_1 phase** (first gap phase) is the interval

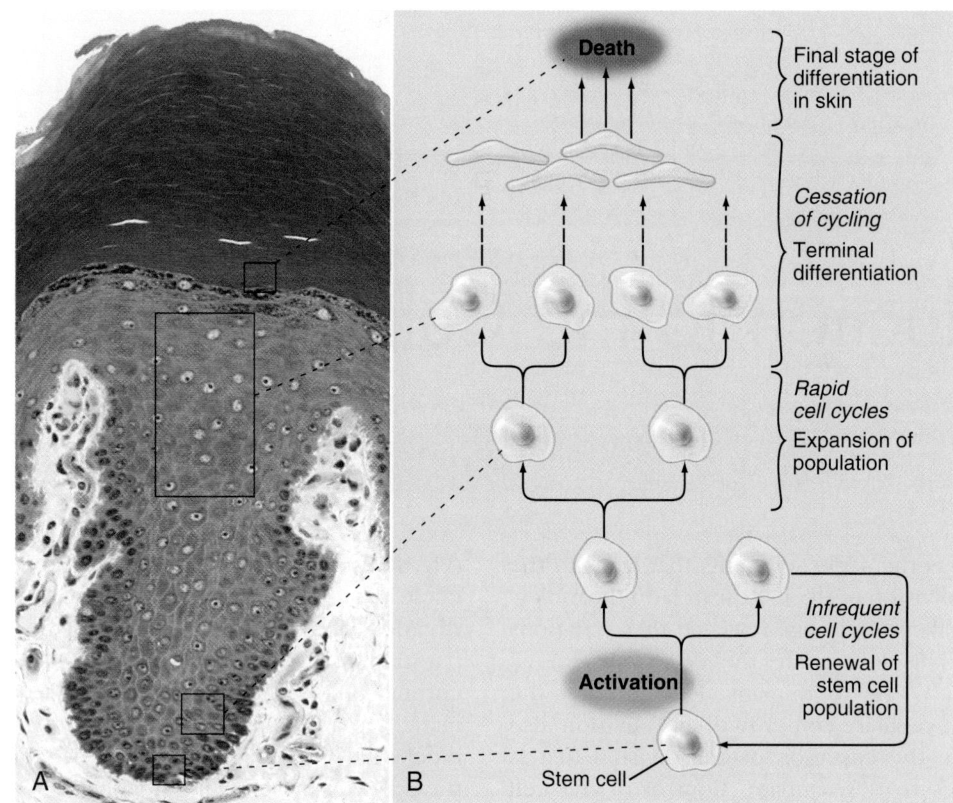

FIGURE 40.1 CELL CYCLES IN A STRATIFIED EPITHELIUM. A, Light micrograph of a section of skin, a stratified squamous epithelium, stained with hematoxylin and eosin (H&E). **B,** Diagram showing the different types of cell cycles at the various levels of this epithelium.

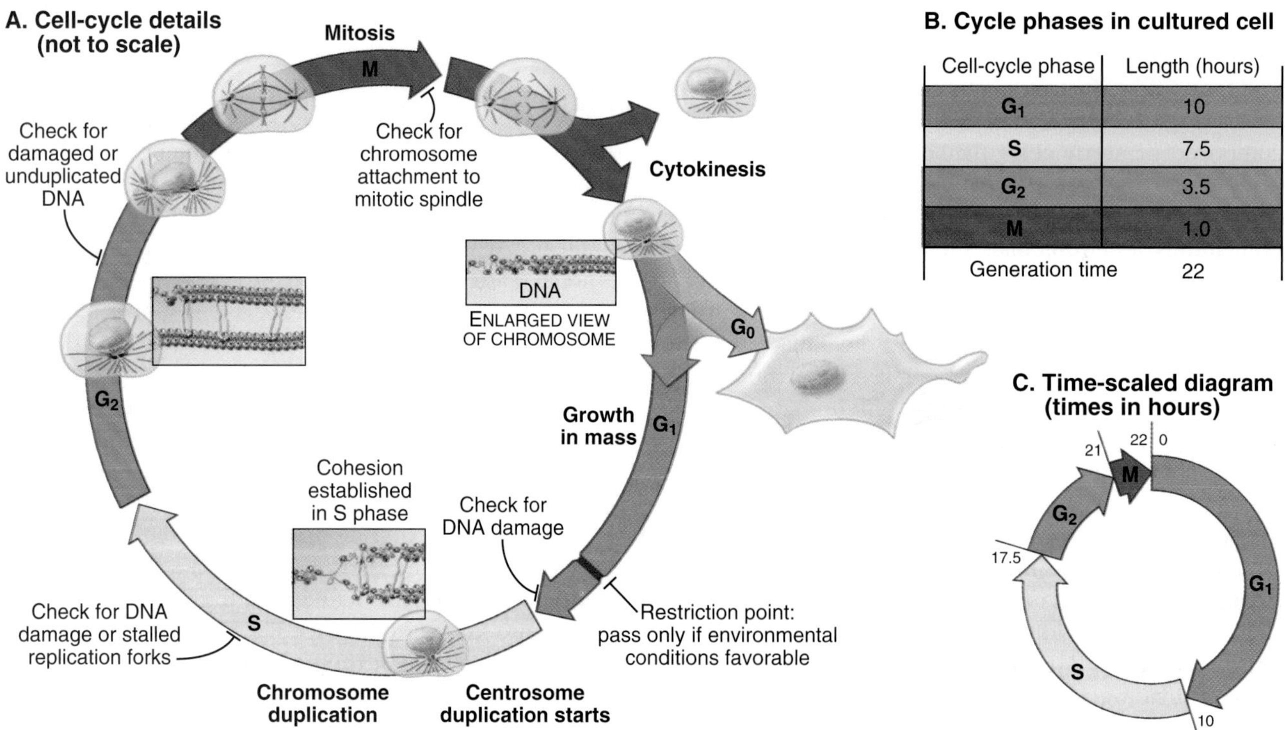

FIGURE 40.2 INTRODUCTION TO THE CELL-CYCLE PHASES. A, Diagrams of cellular morphology and chromosome structure across the cell cycle. **B,** Length of cell-cycle phases in cultured cells. **C,** Time scale of cell-cycle phases.

BOX 40.1 Selected Key Terms

M phase: Cell division, comprising *mitosis,* when a fully grown cell segregates the replicated chromosomes to opposite ends of a molecular scaffold, termed the spindle, and *cytokinesis,* when the cell cleaves between the separated chromosomes to produce two daughter cells. In general, each daughter cell receives a complement of genetic material and organelles identical to that of the parent cell.

Interphase: The portion of the cell cycle when cells grow and replicate their DNA. Interphase has three sections. The *G_1 (first gap) phase* is the interval between mitosis and the onset of DNA replication. The *S* (synthetic) *phase* is the time when DNA is replicated. The *G_2* (second gap) *phase* is the interval between the termination of DNA replication and the onset of mitosis. In multicellular organisms, many differentiated cells no longer actively divide. These nondividing cells (which may physiologically be extremely active) are in the *G_0 phase,* a branch of the G_1 phase.

Checkpoints: Biochemical circuits that regulate cell-cycle transitions in response to the physiological condition of the cell and signals from its environment. Checkpoints detect damage to the DNA due to external agents or problems that arise during DNA replication and trigger the DNA damage response. Other checkpoints detect problems that arise during attachment of chromosomes to the spindle.

between mitosis and the start of DNA replication. The **G_2 phase** (second gap phase) is the interval between the completion of DNA replication and mitosis. All cycling cells have an M phase and an S phase. The G_1 and G_2 phases vary in length and are very short in some early embryos. The following sections describe the stages of the cell cycle, starting just after the birth of the cell.

G_1 Phase

G_1 is typically the longest and most variable cell-cycle phase. When cells are "born" at cytokinesis, they are roughly half the size they were before mitosis, and during G_1, they grow back toward an optimal size. During this time, many activities involved in cell-cycle progression are repressed so that the cell cannot initiate a new round of proliferation. This repressive control system is called the **restriction point.** If the supply of nutrients is poor or if cells receive an antiproliferative stimulus such as a signal to embark on terminal differentiation, they delay their progress through the cell cycle in G_1 or exit the cycle to enter G_0 (see "G_0 and Growth Control" below). However, if appropriate positive stimuli are received, cells overcome the restriction point block and trigger a program of gene expression that commits them to a new cycle of DNA replication and cell division. Cancer cells often have defects in restriction point

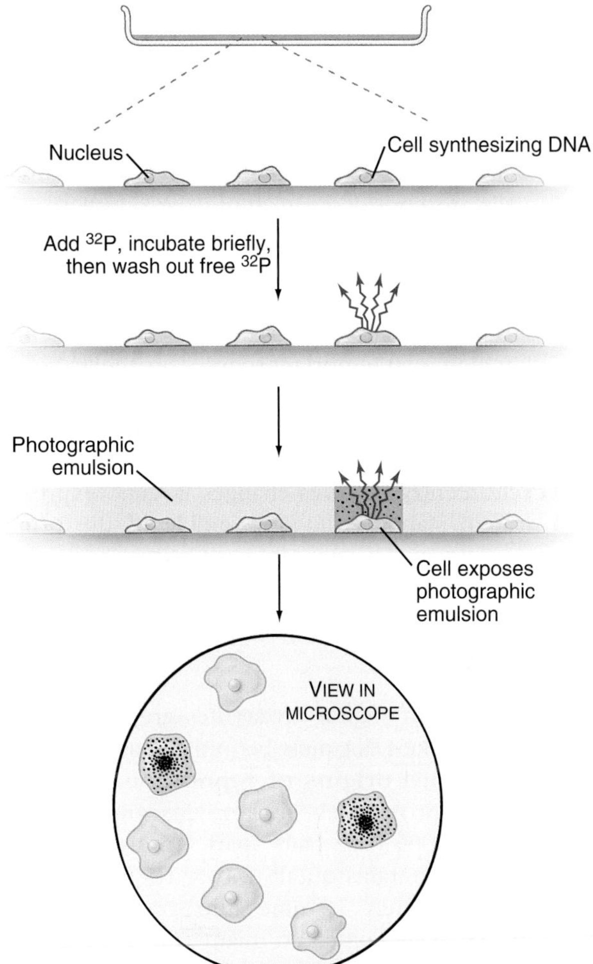

FIGURE 40.3 DISCOVERY OF CELL-CYCLE STAGES. To determine whether cells synthesize DNA during a defined portion of the cell cycle or constantly throughout the entire cycle (as is the case in bacteria, for example), Howard and Pelc fed a radioactive component of DNA (^{32}P) to onion root tip cells, spread the cells in a thin layer on a microscope slide, washed away the ^{32}P that had not become incorporated into DNA, and overlayered the slide with photographic emulsion. After incubation in the dark, the emulsion was developed like film and examined with a light microscope. The nuclei of cells active in replicating their DNA incorporated the radioactive ^{32}P into DNA and exposed the photographic emulsion above them. Two possible outcomes were predicted. If cells synthesized DNA constantly during interphase, then all cells would incorporate the radioactive label. Conversely, if each cell synthesized DNA only during a discrete portion of the cell cycle, then only cells engaged in active replication during the period of exposure to ^{32}P would expose the photographic emulsion. When the slides were examined, 20% of the interphase cell nuclei were labeled, proving that cells synthesize DNA only during a discrete portion of interphase. Mitotic cells were unlabeled. Assuming that the cells traverse the cycle at a more or less constant rate, it was possible to calculate the length of the synthetic phase. Overall, the time between successive divisions—the generation time—was approximately 30 hours in the root tip cells. If approximately 20% of the cells were labeled, then approximately 20% of the 30-hour generation time must be spent in DNA synthesis. Thus, 0.2 × 30, or 6 hours, was spent in replication. (Data from Pelc HA Sr. Synthesis of DNA in normal and irradiated cells and its relation to chromosome breakage. *Heredity Suppl.* 1953;6:261–273.)

control and continue to grow and attempt to divide even in the absence of appropriate environmental signals.

G_0 and Growth Control

Most cells of multicellular organisms differentiate to carry out specialized functions and no longer divide. Such cells are considered to be in the **G_0** phase. Cells often enter G_0 directly as they exit their last mitosis. G_0 cells are not dormant; indeed, they are often actively engaged in protein synthesis and secretion, and they may be highly motile. Many G_0 cells have a nonmotile primary cilium, which is an important sensory organelle (see Fig. 38.19). The G_0 phase is not necessarily permanent. In some specialized cases, G_0 cells may be recruited to reenter the cell cycle in response to specific stimuli. Cell-cycle reentry involves changes in gene expression and protein stability and disassembly of the primary cilium, if present. This process must be highly regulated, as the uncontrolled proliferation of cells in a multicellular organism can lead to cancer.

S Phase

Chromosomes of higher eukaryotes are so large that replication of the DNA must be initiated at many different sites, termed **origins of replication.** In budding yeast, the approximately 400 origins are spaced an average of 30,000 base pairs apart. An average human chromosome contains about 150×10^6 base pairs of DNA, approximately 10 times the size of the entire budding yeast genome, so many more origins are required. Each region of the chromosome that is replicated from a single origin is referred to as a **replicon**. Groups of neighboring replicons cluster in topologically associating domains (TADs) (see Chapter 8).

Proliferating diploid cells replicate their DNA once, and only once, each cell cycle. Each origin of replication is prepared for replication by the formation of a prereplication complex during G_1 (a process that is referred to as *licensing*). As each origin "fires" during S phase, the prereplication complex is dismantled and cannot be reassembled until the next G_1 phase. This ensures that each origin fires only once per cell cycle. The cyclic nature of origin licensing is driven at least in part by fluctuations in the activity of cyclin dependent kinases and protein destruction machinery (discussed later).

During replication, the duplicated DNA molecules, called **sister chromatids,** become linked to each other by a protein complex called cohesin (see Fig. 8.18). This pairing of sister chromatids is important for their orderly segregation later in mitosis (see Fig. 44.16).

G_2 Phase

In most cells of metazoans, G_2 is a relatively brief period during which key enzymatic activities that will trigger the entry into mitosis gradually accumulate and are converted to active forms. When these activities reach a critical threshold level, the cell enters mitosis. Along the way, the chromatin and cytoskeleton are prepared for the dramatic structural changes that will occur during mitosis. If damaged DNA is detected during G_2, a **checkpoint** activates the DNA damage response and delays entry of the cell into mitosis.

M Phase

During M phase (**mitosis** and the subsequent **cytokinesis**), chromosomes and cytoplasm are partitioned into two daughter cells. Mitosis is normally divided into five discrete phases.

Prophase is defined by the onset of chromosome condensation and is actually the final part of G_2 phase. TADs disassemble inside the intact nucleus and mitotic chromosomes begin to form their characteristic array of loops (see Chapter 8). In the cytoplasm, a dramatic change in the dynamic properties of the microtubules decreases their half-lives from approximately 10 minutes to approximately 30 seconds. The duplicated centrosomes (centrioles and associated pericentriolar material in animal cells; see Fig. 34.14) separate and form the two poles of the **mitotic spindle.**

Prometaphase begins when the nuclear envelope breaks down (in higher eukaryotes) and chromosomes begin to attach randomly to microtubules emanating from the two poles of the forming mitotic spindle. Other microtubules originate on chromosomes and within the mitotic spindle. As both kinetochores on a pair of sister chromatids attach to microtubules from opposite spindle poles, the pair of chromatids slowly moves to a point midway between the poles. When all chromosomes are properly attached, the cell is said to be in **metaphase.**

The exit from mitosis begins at **anaphase** with abrupt separation of the two **sister chromatids** from one another. Most cohesion molecules linking sister chromatids are removed without cleavage during prophase in a process initiated by mitosis-specific phosphorylation. Proteolytic cleavage of the remaining cohesin molecules triggers the metaphase-anaphase transition. During anaphase, the separated sister chromatids move to the two spindle poles **(anaphase A),** which themselves move apart **(anaphase B).** As the chromatids approach the spindle poles, the nuclear envelope reforms on the surface of the chromatin. At this point, the cell is said to be in **telophase.**

Finally, during telophase, a **contractile ring** of actin and myosin assembles as a circumferential belt at the cortex midway between spindle poles and constricts the equator of the cell. The separation of the two daughter cells from one another is called **cytokinesis.**

Control of Cell-Cycle Progression

Control networks and **checkpoints** regulate progression of the cell cycle. Checkpoints are biochemical circuits that detect external or internal stimuli and send appropriate signals to the cell-cycle system. The **restriction point** in G_1 phase is a control network that integrates the physiological state of the cell with its environment, including input from other cells and interactions with the surrounding extracellular matrix. Cells must receive appropriate growth stimuli from their environment to progress past this point in the G_1 phase; if not they may live on without dying or commit suicide by apoptosis (see Chapter 46).

DNA damage checkpoints operate throughout interphase. If damage is detected, the DNA damage response initiates a cascade of events that blocks cell-cycle progression and can also trigger cell death by apoptosis. Problems with DNA replication generally produce single-stranded DNA and activate the **DNA damage response**. This response stabilizes stalled replication forks so that they can be repaired. During mitosis, the **spindle assembly checkpoint** delays the onset of chromosome segregation until all chromosomes are attached properly to the mitotic spindle.

The DNA damage response regulates cell-cycle progression in a three-tier pathway (Fig. 40.4). First *sensors* detect DNA damage. These sensors activate *transducers*, which include both protein kinases and transcriptional activators. The transducers act on *effectors* that ultimately block cell-cycle progression and may also fulfill other functions. Two key protein kinases, ataxia-telangiectasia mutated (ATM) and ataxia-telangiectasia and Rad9 related (ATR), lie at the head of the pathway and may also act as sensors of DNA damage. They activate two transducer kinases, Chk1 and Chk2, as well as a transcription factor called p53 that induces the expression of a cohort of genes that halt cell-cycle progression by inhibiting cyclin-dependent kinases as well as genes

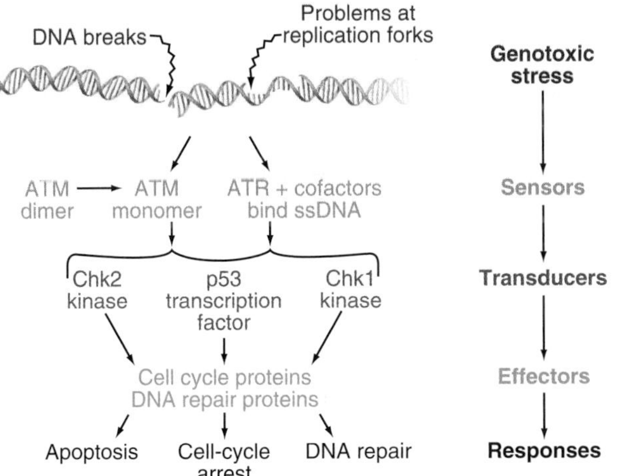

FIGURE 40.4 ELEMENTS OF THE DNA DAMAGE CHECKPOINT AND RESPONSE SYSTEM. ATM, ataxia-telangiectasia mutated; ATR, ataxia-telangiectasia and Rad9 related; ssDNA, single-stranded DNA.

that trigger cell death by apoptosis. Chapters 41 and 43 discuss these proteins in detail.

Biochemical Basis of Cell-Cycle Transitions

Transitions between cell-cycle phases are triggered by a network of protein kinases and phosphatases that is linked to the discontinuous events of the chromosome cycle by the periodic accumulation, modification, and destruction of several key components. This section provides a general introduction to the most important components of this network.

Cyclin-Dependent Kinases

Genetic analysis of the cell cycle in the fission yeast *Schizosaccharomyces pombe* identified a gene called cell division cycle-2$^+$ (*cdc2*$^+$) that is essential for cell-cycle progression during both the $G_1 \rightarrow S$ and $G_2 \rightarrow M$ transitions (Box 40.2). The product of this gene, a protein kinase of 34,000 Da originally called p34^{cdc2}, is the prototype for a family of protein kinases that is crucial for cell-cycle progression in all eukaryotes. This mechanism of cell-cycle control is so well conserved that a human homolog of p34^{cdc2} can replace the yeast protein, restoring a normal cell cycle to a *cdc2* mutant yeast. Boxes 40.3 and 40.4 present a number of the key experiments and experimental systems that led to the identification of the molecules that drive the cell cycle.

Humans have more than 10 distinct protein kinases related to p34^{cdc2}, although only a few are involved in cell-cycle control. To be active, each enzyme must associate with a regulatory subunit called a **cyclin**. Thus, they have been termed **cyclin-dependent kinases (Cdks)**. p34^{cdc2}, now termed Cdk1, seems to function primarily in the regulation of the $G_2 \rightarrow M$ transition in animal cells. Cdk2 (plus Cdk4 and Cdk6 in some cell types) is involved in passage of the restriction point during G_1. Cdk2 also contributes to the $G_2 \rightarrow M$ transition, although Cdk1 is the only Cdk absolutely essential for this step (Appendix 40.1). Cdk7 is important for activation of other Cdks, and also appears to participate in transcribing RNA and repairing damaged DNA. Other Cdks participate in diverse processes ranging from transcriptional regulation to neuronal differentiation. Surprisingly, fibroblasts from mice that lack Cdk2, Cdk4, or Cdk6 are viable; other Cdks can drive the cell cycle if necessary. The mice suffer developmental difficulties because those genes are needed for the differentiation of particular cell types.

Cyclins

Cdks require **cyclin** binding for catalytic activity (Fig. 40.13). Cyclins were discovered in rapidly dividing invertebrate embryos as proteins that accumulate gradually during interphase and are abruptly destroyed during mitosis (see Fig. 40.11). This process of cyclic accumulation and destruction inspired their name.

BOX 40.2 Use of Genetics to Study the Cell Cycle

Studies of the distantly related budding and fission yeasts *Saccharomyces cerevisiae* and *Schizosaccharomyces pombe* (see Fig. 2.8) were important for understanding the cell cycle for several reasons. First, the proteins that control the cell cycle are remarkably conserved between yeasts and mammals. Second, both yeast genomes are small, simplifying the discovery of important gene products. Third, genetic analysis is straightforward, as both yeasts can grow as haploids, and both efficiently incorporate cloned DNA into their chromosomes by homologous recombination.

These two yeasts evolved very different strategies for cell division. Budding yeasts divide by assembling a single bud on the surface of the cell every cell cycle. Fission yeasts divide by fission across the center of an elongated cell. A useful feature of using yeast to study the cell cycle is that the stage of the cell cycle is revealed by the cellular morphology in the light microscope. For budding yeast, unbudded cells are in G_1, cells with buds smaller than the mother cell are in S phase, and cells whose buds are similar in size to the mother cell are in G_2 or M. For fission yeast, cell length provides a yardstick for estimating cell-cycle position.

The cell cycles of both yeasts differ from those of animal cells. In budding yeast, much of the 90-minute cell cycle is spent in G_1. Thus, the networks controlling the $G_1 \rightarrow S$ transition are particularly amenable to study. In contrast, a fission yeast spends most of its 2-hour cell cycle in G_2. S phase follows separation of sister chromatids and occurs prior to cytokinesis. Thus, the control of the $G_2 \rightarrow M$ transition is readily studied in fission yeast. During mitosis, the nuclear envelopes of both yeasts remain intact, so chromosomes segregate on a spindle inside the nucleus.

Genetic studies revealed that the yeast cell cycle is a *dependent pathway* whereby events in the cycle occur normally only after earlier processes are completed. The cell cycle can be modeled as a line of dominoes, each domino corresponding to the action of a gene product that is essential for cell-cycle progression (Fig. 40.5) and the *n*th domino falling only when knocked down by the (*n*-1)th domino. According to the model, mutations in genes that are essential for cell-cycle progression cause an entire culture of yeast to accumulate at a single point in the cell cycle (the point at which the defective gene product *first becomes essential*). This is referred to as the **arrest point.** Fig. 40.5 shows this by including a "mutant" domino that does not fall over when struck by the upstream domino. Mutants that meet this criterion are called **cell division cycle mutants** or **CDC mutants.** Genetic screens for CDC mutants have identified many important genes involved in cell-cycle control.

Because CDC genes are essential for cell-cycle progression, it is impossible to propagate strains of yeast carrying CDC mutants unless the mutants have a **conditional lethal** phenotype. The most commonly used conditional lethal mutations are **temperature sensitive (ts).** Many yeast temperature-sensitive mutants are viable at 23°C (the **permissive temperature**), but cease dividing at 36°C (the **restrictive temperature**). Temperature-sensitive proteins often have an altered amino acid sequence, but occasionally, the lack of a gene product can cause a *ts* phenotype. More recently, the use of auxin-inducible degrons (see Chapters 6 and 23) has enabled experimenters to study the consequences of depleting an essential protein from yeast in a matter of minutes.

Fission yeasts with CDC mutants affecting the entry into mitosis have distinctive morphologies. Cells mutant in Wee1 (a kinase that keeps cyclin-dependent kinase-1 [Cdk1] inactive prior to mitosis) enter mitosis prematurely and are shorter than normal (Fig. 40.6B). In contrast, cells lacking Cdc25 (a phosphatase that counteracts Wee1 and activates Cdk1) are unable to undergo mitosis but continue their growth cycle, therefore becoming greatly elongated (Fig. 40.6C). This simple morphologic assay allowed straightforward classification of yeast CDC genes into those that stimulate progression through mitosis and those that retard entry into mitosis.

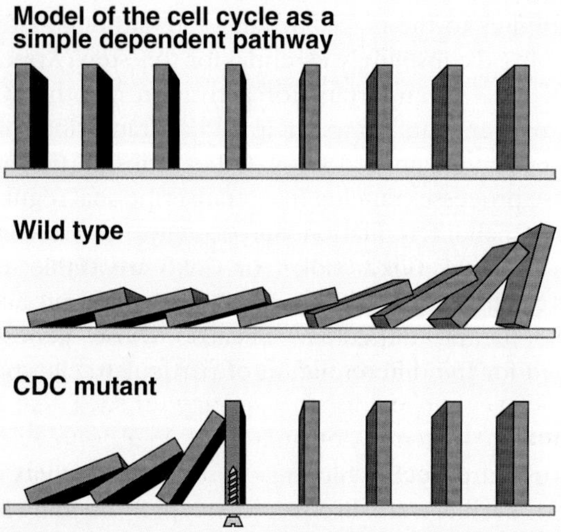

Model of the cell cycle as a simple dependent pathway

Wild type

CDC mutant

FIGURE 40.5 THE CELL CYCLE MAY BE MODELED AS A SIMPLE DEPENDENT PATHWAY. A cell division cycle (CDC) mutation can block further progression along the pathway at a characteristic point in the cell cycle.

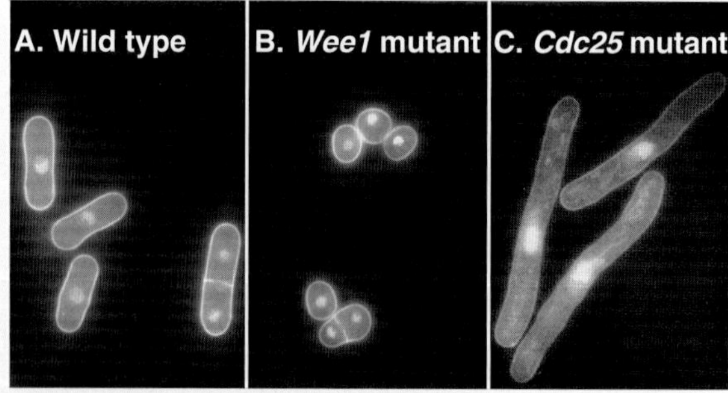

FIGURE 40.6 FLUORESCENCE MICROGRAPHS OF FISSION YEAST CELLS ILLUSTRATING PHENOTYPES OF CELL-CYCLE MUTATIONS. Cell walls and nuclei are stained. **A,** Wild-type cells. **B,** A *wee1* mutation that accelerates entry into mitosis at the restrictive temperature. **C,** A *cdc25* mutation that delays entry into mitosis at the restrictive temperature. (Courtesy H. Ohkura, Wellcome Trust Institute for Cell Biology, University of Edinburgh, United Kingdom.)

BOX 40.3 Studies of the Cell Cycle in Vitro

Amphibian oocytes and eggs are storehouses of most components needed for cell-cycle progression. Oocytes are arrested in G_2 until a surge of the hormone progesterone causes them to "mature" into eggs, which are then naturally arrested in metaphase of the second meiotic division (see Chapter 45). After fertilization, the embryo of the South African clawed frog *(Xenopus laevis)* undergoes a rapid burst of cell divisions. An initial cell cycle 90 minutes long is followed by a rapid succession of 11 cleavages spaced only 30 minutes apart to produce an embryo of 4096 cells (Fig. 40.7).

Thirty minutes per cycle is insufficient to transcribe and translate all the genes needed to make the daughter cells that are produced at each division. The frog solves this problem by making oocytes extremely large (~500,000 times the volume of a typical somatic cell) and storing within them vast stockpiles of the structural components needed to make cells. As a result, only DNA and a very few proteins need be synthesized during early embryonic divisions. In addition to structural components, many factors that regulate normal cell-cycle progression are also stockpiled in oocytes. These features make *Xenopus* oocytes an excellent source of material for biochemical analysis of the cell cycle.

Remarkably, it is possible to make cell-free extracts from *Xenopus* eggs that progress through the cell cycle in vitro (Fig. 40.8). Nuclei from G_1 cells or haploid sperm nuclei, when added to these extracts, efficiently replicate their DNA and proceed through the cell cycle into mitosis, complete with chromosome condensation, nuclear envelope breakdown, chromosome alignment on a spindle, and anaphase segregation of sister chromatids without any additions to the tube. Because these events occur in a cell-free milieu, they are readily accessible to biochemical manipulation. For example, antibodies and other proteins can be added to the extracts, and their effect on the cell cycle can readily be determined. Thus, the *Xenopus* extract system offers a powerful tool for testing the role of various proteins in the cell cycle in higher eukaryotes.

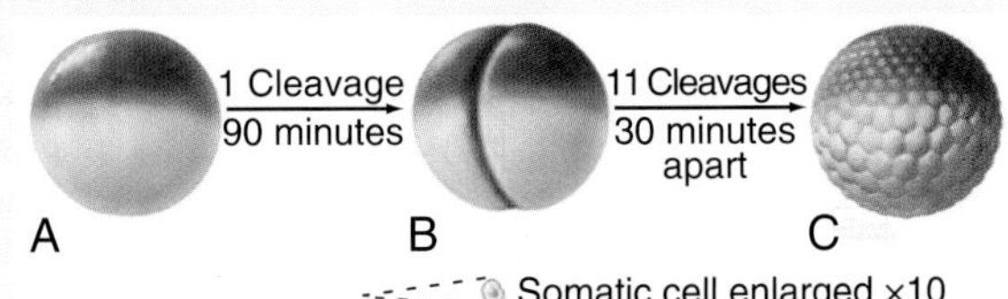

FIGURE 40.7 CLEAVAGES SUBDIVIDE THE EGG DURING *XENOPUS* EARLY DEVELOPMENT. A, Fertilized egg. **B,** Two-cell stage. **C,** Multicellular embryo. Compare size of somatic cell and egg.

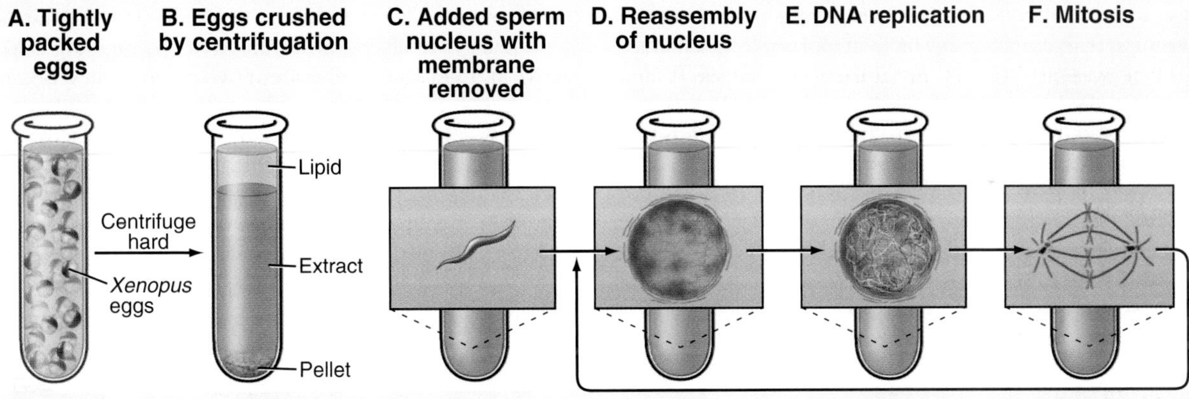

FIGURE 40.8 USE OF *XENOPUS* EGG EXTRACTS TO STUDY THE CELL CYCLE. A–B, Making a *Xenopus* egg extract that is competent to carry out cell-cycle oscillations in vitro. **C–F,** A cycling *Xenopus* extract undergoes alternating S and M phases. G_1 and G_2 phases are minimal (as they are during early development of the frog).

Humans have at least 16 different cyclin proteins that range in size from 35 to 130 kD. The highly conserved cyclin box domain, which docks with the Cdk partners, is the defining structural feature of these proteins. Only a handful of cyclin isoforms are involved in cell-cycle control. Of those that are, some function during G_1 phase, others during G_2 phase, and still others during M phase.

Positive Regulation of Cyclin-Dependent Kinase Structure and Function

Cdks monomers are intrinsically inactive, so they depend on activation by cyclins and are regulated by positive and negative controls. Like other eukaryotic protein kinases (see Fig. 25.3), Cdks have a bilobed structure with the active site in a deep cleft between a small N-terminal and larger C-terminal domain. Monomeric Cdks have a flexible T loop that blocks the mouth of the catalytic pocket. In addition, a short α-helix is oriented such that a glutamic acid required for adenosine triphosphate (ATP) hydrolysis points away from the catalytic cleft. As a result, ATP bound by the monomeric kinase cannot transfer its α-phosphate to protein substrates (see Fig. 40.13A).

Several different mechanisms regulate Cdk activity (Fig. 40.14). On one hand, cyclin binding and

BOX 40.4 Discovery of Factors Essential for Cell-Cycle Progression

The best early evidence for the existence of positive inducers of cell-cycle transitions in mammals was obtained in cell fusion experiments. When cultured cells in S phase were fused with cells in G_1, the G_1 nuclei initiated DNA replication shortly thereafter. In contrast, if S phase cells were fused with G_2 cells, the G_2 nuclei did not rereplicate their DNA until after passing through mitosis. The most dramatic results were obtained when mitotic cells were fused with interphase cells. This caused the interphase cells to enter into mitosis abruptly (as judged by nuclear envelope breakdown and chromosome condensation). The phenomenon was termed **premature chromosome condensation (PCC).** The mitotic inducer could work in any cell-cycle phase (Fig. 40.9). If mitotic cells were fused with cells in G_1 phase, interphase chromosomes condensed into long, single filaments. If the interphase cell was in G_2 phase, the duplicated chromosomes appeared as double filaments. If the interphase cell was in the S phase, the partially replicated chromosomes condensed into a complex pattern of single and double condensed regions separated by regions of decondensed chromatin corresponding to sites where DNA was actively replicating at the time of fusion.

Working independently, developmental biologists who were interested in the control of cell division during early development in frogs also discovered an activity that could cause interphase cells to enter the M phase. They used a micropipette to extract a tiny bit of cytoplasm from a mature egg that was arrested in metaphase of meiosis II and inject it into oocytes (which are in G_2 phase). The oocytes rapidly entered M phase, with concomitant chromosome condensation and nuclear envelope disassembly (Fig. 40.10). This stimulation to enter M phase is called *maturation,* and the unknown factor present in the egg cytoplasm that induced oocyte maturation was termed **MPF,** or **maturation-promoting factor** (now often referred to as **M phase**-promoting factor). It was realized early on that MPF might be related to the inducer of mitosis detected in the PCC experiments. In fact, extracts from mitotic tissue culture cells could induce meiotic maturation when injected into oocytes. Similar extracts from cells in other phases of the cell cycle did not cause the G_2/M phase transition in oocytes.

Other cell biologists studying protein synthesis in starfish and sea urchin embryos noticed a curious protein that seemed to accumulate across the cell cycle but was then destroyed during mitosis. They were well aware of the work on MPF, and immediately suspected that their protein, which they called cyclin, might be somehow involved in MPF activity (Fig. 40.11).

In a third line of investigation, geneticists working on yeasts realized that the cell cycle could be dissected through the isolation of cell division cycle (CDC) mutants (see Box 40.2). The analysis of the cell cycle with these mutants dominated cell-cycle research to such an extent that many human genes that are important in cell-cycle control bear the CDC name if they are related to well-characterized yeast genes. The best-known genes to emerge from this analysis were *Cdc2* (Cdk1) and *Cdc25,* both of which were determined genetically to encode proteins that actively promote the G_2/M transition. Other genes, such as *Wee1,* were found to encode activities that act as antagonists that inhibit the G_2/M transition.

When eventually purified from *Xenopus* eggs (Fig. 40.12), active MPF consisted primarily of two polypeptides of 32,000 and 45,000 Da. The smaller component of MPF is the *Xenopus* equivalent of the fission yeast Cdc2 gene product (now known as Cdk1). The larger component of *Xenopus* MPF is a B-type cyclin. Only 15 years later was it recognized that fully functional MPF also requires Greatwall kinase and its small substrates to inhibit protein phosphatase PP2A-B55δ and give Cdk activity a chance to take off and trigger mitotic entry.

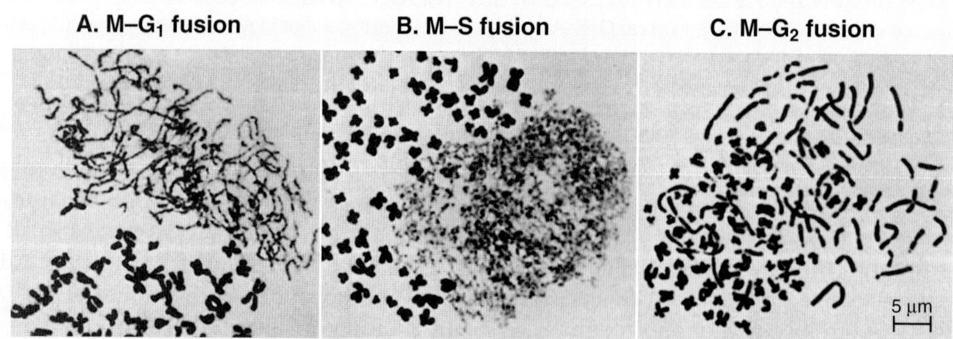

FIGURE 40.9 FUSION WITH MITOTIC CELLS CAUSES INTERPHASE CELLS TO ENTER MITOSIS PREMATURELY, NO MATTER WHERE THEY ARE IN THE CELL CYCLE. The resulting prematurely condensed chromosomes are single threads if the interphase cell was in G_1 phase **(A),** double threads if the cell was in G_2 phase **(C)** and a complex mixture of both interspersed with uncondensed regions if the cell was in S phase **(B).** M, mitosis. (From Hanks SK, Gollin SM, Rao PN, et al. Cell cycle-specific changes in the ultrastructural organization of prematurely condensed chromosomes. *Chromosoma.* 1983;88:333–342.)

BOX 40.4 Discovery of Factors Essential for Cell-Cycle Progression—cont'd

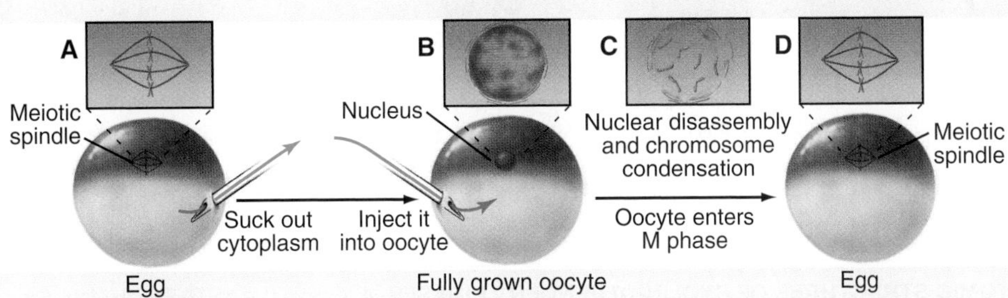

FIGURE 40.10 THE EXPERIMENT THAT IDENTIFIED MATURATION-PROMOTING FACTOR. A, The *box* shows the meiotic spindle in a *Xenopus* egg arrested in metaphase II of meiosis. **B,** The *box* shows the interphase nucleus in a mature oocyte. Following injection of MPF, the nucleus disassembles, mitotic chromosomes form **(C),** and the cell assembles a meiotic spindle **(D).** Disassembly of the oocyte nucleus and entry into M phase is called maturation, and the factor triggering this event was named maturation-promoting factor (MPF).

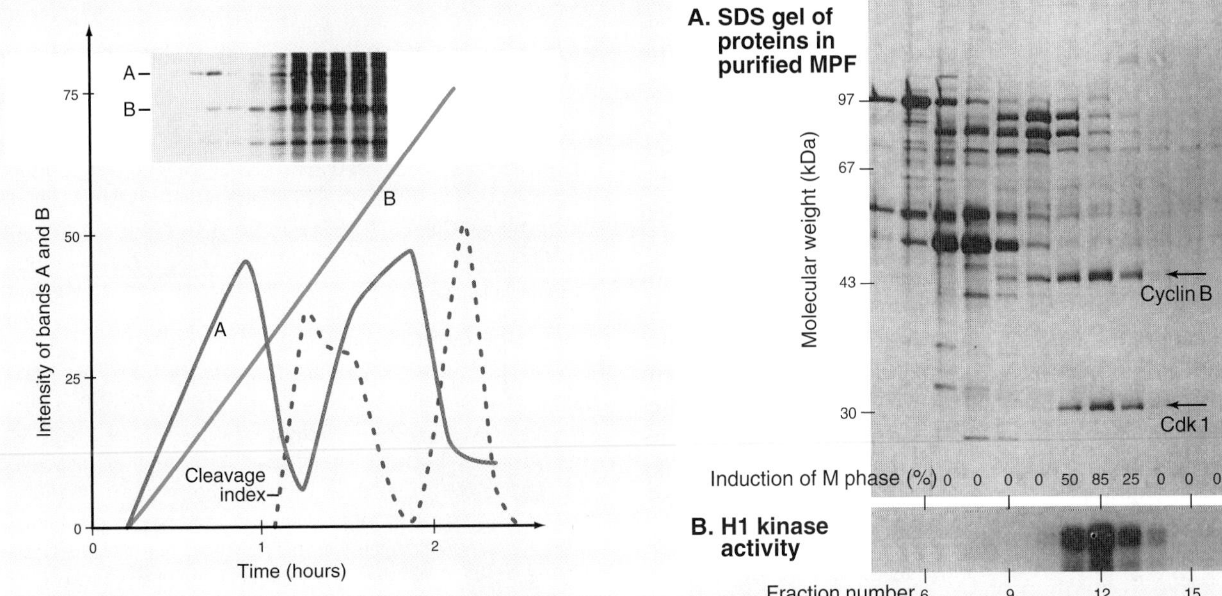

FIGURE 40.11 EXPERIMENTAL IDENTIFICATION OF A CYCLIN. Newly synthesized proteins (labeled with ^{35}S-methionine) in fertilized sea urchin eggs were separated by sodium dodecylsulfate (SDS) polyacrylamide gel electrophoresis. It was noted that the protein labeled A (which was named cyclin) first accumulated, was greatly reduced at the metaphase/anaphase transition, and then began to accumulate again. Protein B, which is not involved in cell-cycle regulation, accumulated progressively over this time. "Cleavage index" refers to the percentage of dividing cells observed in the microscope at varying times after fertilization. (From Evans T, Rosenthal ET, Youngblom J, et al. Cyclin: a protein specified by maternal mRNA in sea urchin eggs that is destroyed at each cleavage division. *Cell.* 1983;33:389–396.)

FIGURE 40.12 PURIFICATION OF MATURATION-PROMOTING FACTOR (MPF). A, Sodium dodecylsulfate (SDS) polyacrylamide gel electrophoresis of fractions from the final step of the purification. The numbers at the bottom show the percentage of oocytes that entered M phase when a portion of each column fraction was injected (the classical MPF assay). The roughly 32-kD band is Cdk1 ($p34^{cdc2}$). The roughly 45-kD band is cyclin B. **B,** Assay of the ability of the column fractions to phosphorylate histone HI. This is a standard assay for active Cdk enzymes. (Modified from Lohka M, Hayes MK, Maller JL. Purification of maturation-promoting factor, an intracellular regulator of early mitotic events. *Proc Natl Acad Sci U S A.* 1988;85:3009–3013.)

phosphorylation of the T loop stimulate enzyme activity. On the other, Cdks are inhibited by phosphorylation of residues adjacent to the ATP-binding site and binding of inhibitory proteins.

Cyclin binding causes the T loop to retract away from the mouth of the catalytic pocket (see Fig. 40.13B). In addition, the secondary structure of the N-terminal domain is altered, allowing the bound ATP to assume a conformation suitable for reaction with substrates.

Despite these changes, the Cdk–cyclin complex has low catalytic activity. Full activation of most Cdks requires a kinase called Cdk-activating kinase (CAK), which phosphorylates a threonine in the T loop of the Cdk (this threonine gives the loop its name). In

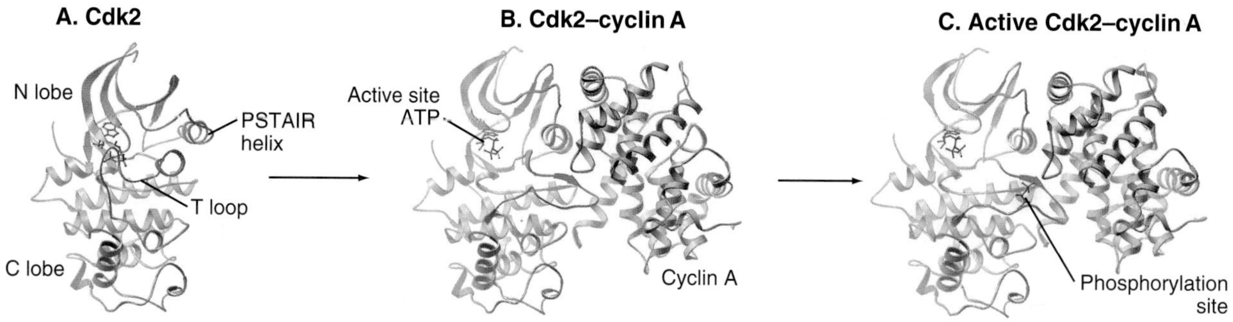

FIGURE 40.13 ATOMIC STRUCTURES OF CYCLIN-DEPENDENT KINASES. A, Cdk2. The PSTAIR helix, found in most Cdks, is named after a sequence of six amino acids that binds to cyclins (one letter code). (For reference, see Protein Data Bank [PDB; www.rcsb.org] file 1DM2.) **B,** Cdk2–cyclin A (kinase at basal activity level). (PDB file 1FIN.) **C,** Cdk2–cyclin (kinase fully active following phosphorylation of threonine[160]). (PDB file 1JST.) ATP, adenosine triphosphate.

vertebrates, CAK is composed of Cdk7-cyclin H. The phosphorylated threonine fits into a charged pocket on the surface of the enzyme, flattening the T loop back even farther from the mouth of the catalytic pocket (see Figs. 40.13C and 40.14A). This stimulates the catalytic activity up to 300-fold, in part because the flattened T loop forms part of the substrate-binding surface. Threonine phosphorylation also stabilizes the association of Cdk2 with cyclin A.

In addition to their cyclin partner, Cdk1 and Cdk2 bind an additional small *C*dc *k*inase *s*ubunit (Cks) protein to their C-terminal domain, away from the active site. Bound Cks enables the kinase to better hold onto its substrates and increases the efficiency with which Cdks can phosphorylate substrates at multiple sites (a hallmark of Cdk target phosphorylation).

Negative Regulation of Cyclin-Dependent Kinase Structure and Function

At least two mechanisms slow or stop the cell cycle by inactivating Cdks (see Fig. 40.14). During G_2 phase, the protein kinases Myt1 and Wee1 hold Cdk1 in check by phosphorylating threonine[14] and tyrosine[15] in the roof of the ATP-binding site. These phosphates interfere with ATP binding and hydrolysis. Threonine[14] and tyrosine[15] are accessible to the regulatory kinases only following cyclin binding, so this phosphorylation of Cdks depends, at least in part, on the availability of cyclins.

In mammals, three Cdc25 phosphatases (see Fig. 25.5) reverse these inhibitory phosphorylations. Cdc25A regulates both the $G_1 \rightarrow S$ and $G_2 \rightarrow M$ transitions and is essential for life of the cell. Ccd25B is dispensable for mitosis, but it is essential for the production of gametes in meiosis. Cdc25C is a target of the G_2 DNA damage checkpoint that prevents cells from undergoing mitosis with damaged DNA (see Fig. 43.11), but cells can survive without it.

A parallel mechanism for inactivating Cdks involves binding of subunits from two families of small inhibitory

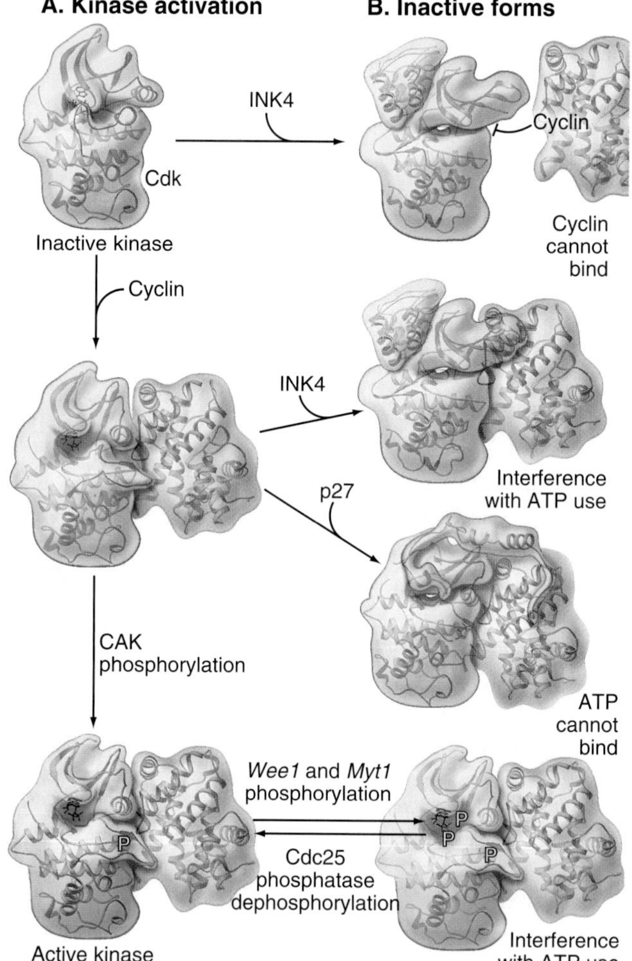

FIGURE 40.14 POSITIVE AND NEGATIVE REGULATION OF CYCLIN-DEPENDENT KINASES. A, Pathway of activation by cyclin binding and phosphorylation. **B,** Pathways of inactivation by inhibitor binding and phosphorylation. When INK4 binds, twisting of the Cdk upper lobe blocks cyclin binding or interferes with ATP hydrolysis. When p27 binds, a loop insinuates into the upper lobe of the Cdk and blocks ATP binding. (For reference, see PDB files 1B17 [Cdk2-INK4], 1FIN and 1BI7 [Cdk2-INK4-cyclin], and 1JSU [Cdk2-p27-cyclin A].) CAK, Cdk-activating kinase.

proteins called the cyclin-dependent kinase inhibitors (CKIs) and inhibitors of Cdk4 (INK4) (for their names, see Appendix 40.1). When activated in the DNA damage response pathway, p53 turns on transcription of the CKI p21, which inhibits Cdk-cyclin A. CKI $p27^{Kip1}$ inactivates complexes of Cdk2 and cyclin A by having a protein loop invade the N-terminal domain of the Cdk, disrupting its structure and competing with ATP for binding to the active site (see Fig. 40.14B).

Members of the INK4 family preferentially inactivate Cdk4 and Cdk6 in two ways (see Fig. 40.14B). First, binding to monomeric Cdk distorts the orientation of the N- and C-terminal lobes, so cyclin D does not bind. INK4 family inhibitors also inhibit preformed Cdk4/6–cyclin D complexes by binding the Cdk and distorting the ATP-binding site so that the kinase uses ATP much less efficiently.

Cdk inhibitors are important for growth regulation during the G_1 and G_0 phases of the cell cycle (see Chapter 41). They also play a critical role in the cell-cycle arrest that occurs in response to DNA damage and to antiproliferative signals. Mutations in the INK4 locus are strongly linked to cancer.

Role of Phosphatases in Counter-Balancing Cdk Activity

Two important phosphatases counter-balance Cdk activity in mitosis. Just as Cdks exhibit cyclic behavior and are activated in mitosis, these counteracting phosphatases must also be cyclic—but they are inhibited in mitosis. *P*rotein *p*hosphatase 1 (PP1), associates with numerous targets on chromosomes and the mitotic apparatus, and is highly active during mitotic exit. Many target proteins have a simple loop with the sequence RVS/TF that inserts into a groove on the enzyme. During mitosis, Cdk phosphorylation of adjacent sites often blocks this interaction with substrates. Cdk1-cyclin B also phosphorylates the PP1 catalytic subunit during mitosis, thereby inactivating the enzyme.

PP2A is more directly involved in Cdk regulation. It is a trimeric enzyme with catalytic, scaffolding, and regulatory subunits. The latter include B55α-δ and B56α-ε (see Appendix 40.1). PP2A-B55δ is largely responsible for removing phosphates added by Cdks. Consequently PP2A-B55δ must be inhibited to allow Cdks to drive the cell into mitosis. The Greatwall (Gwl) kinase regulates PP2A. Gwl is unusual, because 500 amino acids are inserted into its large lobe roughly adjacent to the T loop (see Fig. 40.13). Phosphorylation by Gwl allows two small proteins, Arpp19 (cyclic adenosine monophosphate [cAMP]-regulated phosphoprotein 19) and ENSA (α-endosulfine), to bind and inhibit PP2A-B55δ. Thus, Gwl confers the necessary cyclic behavior on PP2A. Understanding how Gwl is turned on and off is important for developing a full picture of mitotic regulation, and this is a subject of active study.

Role of Protein Destruction in Cell-Cycle Control

During mitosis active Cdk1-cyclin B-Cks phosphorylates key substrates leading to dramatic reorganization of the cell and, ultimately, to separation of sister chromatids on the mitotic spindle. Once chromatids are separated, the cell must return to a state with low levels of Cdk activity so that nuclear envelope reassembly, spindle disassembly, and cytokinesis can occur.

Exit from mitosis requires Cdk inactivation by the ubiquitin-directed proteolytic machinery. The destruction of A- and B-type cyclins inactivates Cdk1 and Cdk2. This allows PP1, PP2A-B55δ, and other phosphatases to reverse the action of Cdks and bring mitosis to a close. Ubiquitylation also results in proteolysis of a protein called securin, which regulates the onset of sister chromatid separation at anaphase.

Ubiquitin-mediated destruction of cyclins involves a cascade of three enzymes described in Chapter 23 (see Fig. 23.3). First, an E1 enzyme (**ubiquitin-activating enzyme**) activates the small protein **ubiquitin** by forming a thioester bond between the ubiquitin C-terminus and a cysteine on the enzyme. Activated ubiquitin is next transferred to another thioester bond on an E2 enzyme (**ubiquitin-conjugating enzyme**). The E2 often cooperates with an E3, which is important for imparting substrate specificity, to transfer ubiquitin to the ε-amino group of a lysine on a target protein. The resulting polyubiquitinated proteins are usually targets for destruction by the cylindrical 26S **proteasome** (see Fig. 23.8). This large multienzyme complex functions like a cytoplasmic garbage disposal, grinding target proteins down to short peptides and spitting out intact ubiquitin monomers for reuse in further rounds of protein degradation. Its role was originally thought to be the removal of damaged proteins from the cytoplasm; however, it is now recognized as a central factor in cell-cycle control.

The key E3 ligase regulating cyclin proteolysis is a large (15-subunit) complex called the **anaphase-promoting complex/cyclosome (APC/C)** (Fig. 40.15). The APC/C is inactive during the S and G_2 phases of the cell cycle. Phosphorylation by Cdk1-cyclin B-Cks1 and binding of the protein coactivator Cdc20 activate the APC/C in early mitosis. APC/C^{Cdc20} then triggers the metaphase-anaphase transition.

An important checkpoint, the **spindle assembly checkpoint**, regulates APC/C^{Cdc20} during mitosis, keeping it inactive until all kinetochores are productively attached to spindle microtubules. The checkpoint effector is **the mitotic checkpoint** complex whose formation is triggered by unattached kinetochores. This

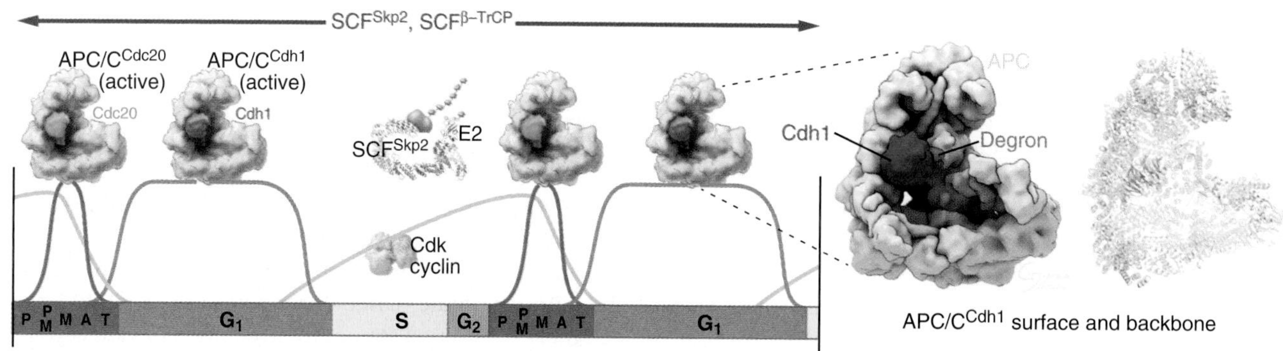

FIGURE 40.15 TWO FORMS OF THE ANAPHASE-PROMOTING COMPLEX/CYCLOSOME (APC/C) CONTROL THE CELL CYCLE. At the metaphase-anaphase transition, the APC/C with associated Cdc20 triggers the onset of anaphase by signaling the degradation of securin and cyclin B. During mitosis Cdh1 phosphorylated by Cdk1–cyclin B is unable to bind the APC/C, so APC/C^{Cdh1} activity is low. As Cdk1–cyclin B activity declines in anaphase, Cdh1 binds the APC/C and APC/C^{Cdh1} drives the exit from mitosis into G$_1$. APC/C^{Cdh1} remains active throughout G$_1$ but is inactivated following synthesis of the specific inhibitor Emi1. After the onset of S phase, SCF (shown here adding a ubiquitin chain to a docked substrate) directs the degradation of cell-cycle substrates such as p27^{Kip1}, following their phosphorylation by protein kinases.

complex inhibits APC/C^{Cdc20} by acting as a competitive substrate. As a result, cyclin B and securin are stable until the checkpoint is satisfied. A few substrates, including cyclin A, bind directly to the APC/C without requiring Cdc20, so they are marked for degradation even when the checkpoint is active.

Exiting from mitosis and allowing the G$_1$ cell to prepare chromatin for DNA replication (see Chapter 42) requires low Cdk activity and destruction of Cdc20. This is accomplished late in mitosis by Cdh1, a different co-activator of the APC/C. Phosphorylation of Cdh1 by Cdks blocks its binding to the APC/C early in mitosis. Thus, APC/C^{Cdh1} forms only after cyclin levels (and therefore Cdk activity) decline late in mitosis. As cells pass from G$_1$ into S phase, a newly synthesized inhibitory protein, Emi1, binds to and inactivates APC/C^{Cdh1}. This allows cyclins to accumulate during S and G$_2$. Remarkably, APC/C^{Cdh1} is also involved with regulating the activity of synapses in nondividing neurons.

After the G$_1$ $\rightarrow$ S transition and throughout the remainder of interphase members of a different family of E3 activities called SCF regulate the levels of proteins that control Cdk and other cell cycle factors. SCF is named after three of its four subunits: Skp1, cullin, and F-box protein (Fig. 40.16). SCF is a molecular toolbox built on a bow-shaped scaffold formed by the cullin subunit. The fourth subunit, Rbx1, binds near the C-terminus of cullin and uses a protein motif called a RING finger to dock to a ubiquitin-linked E2 enzyme. Skp1 binds to the other end of the cullin, where it provides a docking site for an F-box protein that recognizes and binds the substrate. (The F-box got its name because it was first discovered in cyclin F.) Humans have 78 F-box proteins, giving SCF enormous versatility. Two examples shown in Fig. 40.16 are Skp2, which targets the Cdk inhibitor p27 helping to drive the G$_1$ $\rightarrow$ S transition, and β-Trcp, which targets Cdc25A, Wee1, and Emi1.

A. Structure of SCF$^{Skp2\text{-}E2}$ complex

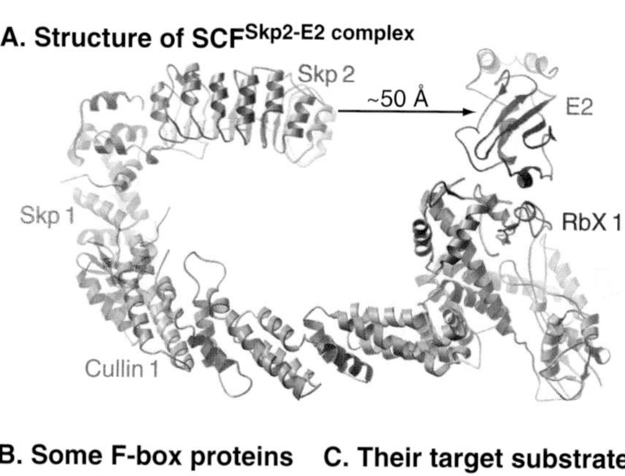

B. Some F-box proteins | **C. Their target substrates**

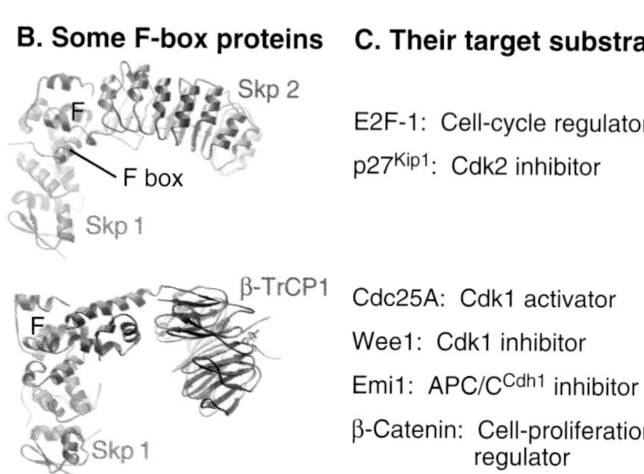

E2F-1: Cell-cycle regulator
p27^{Kip1}: Cdk2 inhibitor

Cdc25A: Cdk1 activator
Wee1: Cdk1 inhibitor
Emi1: APC/C^{Cdh1} inhibitor
β-Catenin: Cell-proliferation regulator

FIGURE 40.16 STRUCTURE AND FUNCTION OF SCF. **A,** Structure of SCF^{Skp2}. *Left,* SCF recognizes target proteins through its F-box subunit (Skp2 in this case). *Right,* Ubiquitin is then transferred from an E2 enzyme. The whole is assembled on a rigid bow-like scaffold composed of the cullin subunit. Structures of F-box proteins Skp2 and β-Trcp **(B)** and a list of several of their known target proteins **(C).**

SCF is fundamentally different from the APC/C, because it is constitutively active. However, it ubiquitylates substrates only after they have been phosphorylated, often by Cdks. This feature links SCF activity to the cell cycle.

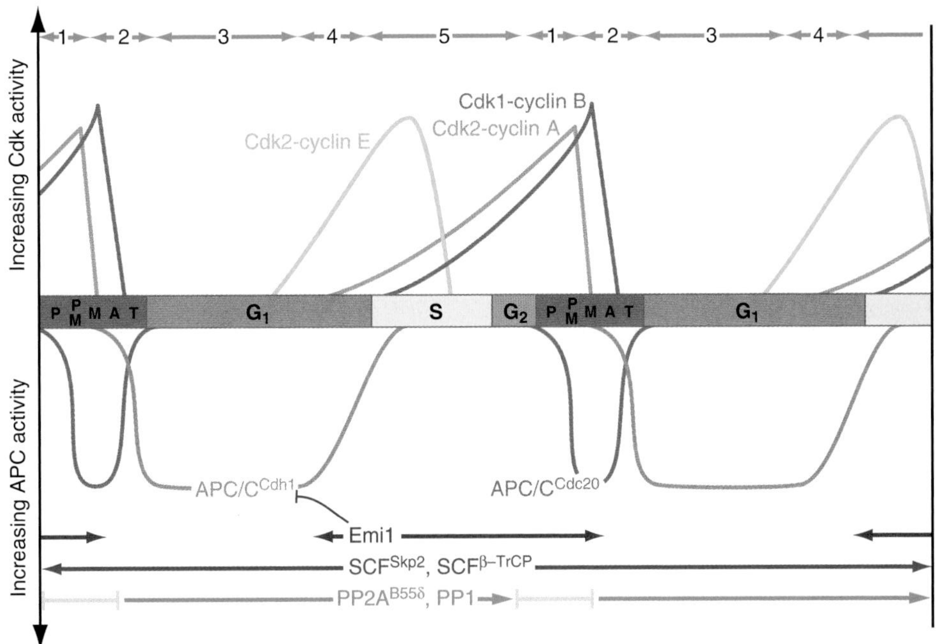

FIGURE 40.17 DIAGRAM SHOWING THE CHANGING STATES OF THE CYTOPLASM AS CELLS TRAVERSE THE CELL CYCLE. Between G_2 and G_1 are shown the various stages of mitosis: P, prophase; PM, prometaphase; M, metaphase; A, anaphase; T, telophase. The states of the cytoplasm discussed in the text are shown as *green arrows* across the top. Increased Cdk activity is shown as peaks upwards from the central bar. Increasing anaphase-promoting complex/cyclosome (APC/C) activity is shown as a mirror image, with peaks going down from the central bar. APC, anaphase-promoting complex; PP1, protein phosphatase 1; PP2A, protein phosphatase 2A.

Changing States of the Cytoplasm During the Cell Cycle

The cell cycle is characterized by five discrete physiological states of the cytoplasm (Fig. 40.17). Changing levels of Cdk activity drive transitions between these states, sometimes counteracted and sometimes reinforced by targeted proteolysis.

1. Early mitosis. Cdk2-cyclin A peaks in prophase, followed by Cdk1-cyclin B. When the nuclear envelope breaks down, the APC/C^{Cdc20} starts degrading cyclin A, but the spindle checkpoint inhibits destruction of other substrates. When the last chromosome has attached correctly to spindle microtubules (metaphase), the checkpoint is satisfied, and APC/C^{Cdc20} starts to degrade cyclin B and securin, an inhibitor of a key protease called separase. Their degradation continues throughout metaphase.
2. Anaphase and mitotic exit. When securin levels fall below a critical threshold, active separase cleaves a key component of the cohesin ring (see Fig. 8.18). This triggers sister chromatid separation. Cyclin destruction continues throughout anaphase and telophase, and falling Cdk1 activity allows the formation of APC/C^{Cdh1}, which marks Cdc20 for destruction along with the remaining B-type cyclins. $SCF^{\beta\text{-Trcp}}$ destruction of Emi1 allows APC/C^{Cdh1} to be active when it forms.
3. G_1 phase. APC/C^{Cdh1} and Cdk inhibitors of the CKI and Ink4 families cooperate to inhibit Cdk activity. Low Cdk activity is required for cytokinesis, spindle disassembly, chromosome decondensation, nuclear envelope reassembly, reactivation of transcription, reassembly of the Golgi apparatus, and assembly of prereplication complexes on the chromosomes.
4. G_1–S phase transition. Growth signals from the environment promote the transcription of Cyclin E. If the levels pass a critical threshold, a burst of Cyclin E-Cdk2 activity allows the cell to pass the restriction point, leading to synthesis of proteins required for DNA replication and cell-cycle progression. Cdk phosphorylation targets the CKI peptides for destruction by SCF allowing Cdk2 to become activated. In addition, APC/C^{Cdh1} is inactivated by newly synthesized Emi1.
5. S–G_2 phase. Cdk activity remains high throughout the remainder of the cell cycle, and SCF continues to degrade selected proteins tagged by Cdk phosphorylation. $SCF^{\beta\text{-Trcp}}$ destruction of Cdc25A keeps Cdk1 inactive, preventing a premature entry into mitosis. The APC/C remains inactive, allowing mitotic cyclins to accumulate. It is not known what ultimately triggers entry into mitosis, but an important factor may be a switch in the specificity of $SCF^{\beta\text{-Trcp}}$, which spares Cdc25A and instead degrades the Cdk-inhibitory kinase Wee1.

Although this sounds complicated, the underlying principles are actually quite straightforward. The following chapters discuss the cell-cycle transitions in greater detail and show how the process is modulated in response to a changing environment.

ACKNOWLEDGMENTS

We thank Tim Hunt, David Morgan, and Jonathon Pines for their suggestions on revisions to this chapter.

SELECTED READINGS

Bartek J, Lukas J. DNA damage checkpoints: from initiation to recovery or adaptation. *Curr Opin Cell Biol.* 2007;19:238-245.

Brown JS, Jackson SP. Ubiquitylation, neddylation and the DNA damage response. *Open Biol.* 2015;5:150018.

Craney A, Rape M. Dynamic regulation of ubiquitin-dependent cell cycle control. *Curr Opin Cell Biol.* 2013;25:704-710.

Hartwell LH, Weinert TA. Checkpoints: Controls that ensure the order of cell cycle events. *Science.* 1989;246:629-634.

Hunt T. On the regulation of protein phosphatase 2A and its role in controlling entry into and exit from mitosis. *Adv Biol Regul.* 2013; 53:173-178.

Lorca T, Castro A. The Greatwall kinase: a new pathway in the control of the cell cycle. *Oncogene.* 2013;32:537-543.

Morgan DO. *The Cell Cycle: Principles of Control.* London: New Science Press; 2007: 297p.

Nasmyth K. A prize for proliferation. *Cell.* 2001;107:689-701.

Nurse P. A long twentieth century of the cell cycle and beyond. *Cell.* 2000;100:71-78.

Primorac I, Musacchio A. Panta rhei: the APC/C at steady state. *J Cell Biol.* 2013;201:177-189.

Qian J, Winkler C, Bollen M. 4D-networking by mitotic phosphatases. *Curr Opin Cell Biol.* 2013;25:697-703.

Stukenberg PT, Burke DJ. Connecting the microtubule attachment status of each kinetochore to cell cycle arrest through the spindle assembly checkpoint. *Chromosoma.* 2015;124:463-480.

Wieser S, Pines J. The biochemistry of mitosis. *Cold Spring Harb Perspect Biol.* 2015;7:a015776.

APPENDIX 40.1

Inventory of the Enzymes of the Cell-Cycle Engine

Cyclin-Dependent Kinases and Their Cyclin Partners		
Kinase	**Cyclin (+ Other) Partner**	**Function**
Cdk1 ($p34^{cdc2}$)	A B_1, B_2 (*Xenopus* has 5 B-type cyclins) Cdk1-cyclin B binds Cks1 (Cdc kinase subunit)	Mammals: triggers $G_2 \rightarrow M$ transition. Yeasts: triggers $G_1 \rightarrow S$ and $G_2 \rightarrow M$ transitions. Cyclin A is synthesized in S and destroyed starting at prometaphase. Cyclins B are synthesized in S/G_2 and destroyed following the completion of chromosome attachment to the spindle. Cyclins A_1, B_3 function preferentially in meiosis.
Cdk2	A, E	Triggers $G_1 \rightarrow S$ transition. Can be replaced by other Cdks in mouse.
Cdk4, Cdk6	D_1–D_3	Phosphorylation of the retinoblastoma susceptibility protein (pRb) in G_1. Triggers passage of the restriction point and cyclin E synthesis in some cell types. Extracellular growth factors control synthesis of D cyclins. Can be replaced by other Cdks in mouse.
Cdk5	CDK5R1 or CDK5R2	Neuronal differentiation, sensory pathways.
Cdk7 (CAK)	H; also binds assembly factor MAT1	Cdk activation by phosphorylation of the T loop. Also in TFIIH, important for regulation of RNA polymerase II transcription and DNA repair.
Cdk8	C	Regulation of RNA polymerase II transcription.
Cdk9	T	Regulation of RNA polymerase II transcription.
Cyclin Inhibitors		
Inhibitor	**Cdk Substrates**	**Function**
CKI: $p21^{Cip1/Waf1}$ most Cdk-cyclin complexes	Most Cdk-cyclin complexes	Induced by p53 tumor suppresser. Cell-cycle arrest after DNA damage. Binds PCNA (proliferating cell nuclear antigen; see Chapter 42) and inhibits DNA synthesis. Promotes cell cycle arrest in senescence and terminal differentiation. At low levels, may help assemble active Cdk-cyclin complexes.
CKI: $p27^{Kip1}$	Most Cdk-cyclin complexes	Cell cycle arrest in response to growth suppressers like TGF-β and in contact inhibition and differentiation.
CKI: $p57^{Kip2}$	Most Cdk-cyclin complexes	Important in development of the palate.
INK4: $p15^{Ink4b}$	Cdk4, Cdk6	Cell-cycle arrest in response to transforming growth factor (TGF)-β. Altered in many cancers.
INK4: $p16^{Ink4a}$	Cdk4, Cdk6	Cooperates with the retinoblastoma susceptibility protein (pRb) in growth regulation. Cell-cycle arrest in senescence. Altered in a high percentage of human cancers. This gene overlaps the gene for $p19^{ARF}$, an important regulator of the p53 tumor-suppresser protein.
INK4: $p18^{Ink4c}$	Cdk4, Cdk6	Cell-cycle arrest in response to growth suppressers.
INK4: $p19^{Ink4d}$	Cdk4, Cdk6	Cell-cycle arrest in response to growth suppressers.

APPENDIX 40.1

Inventory of the Enzymes of the Cell-Cycle Engine—cont'd

Other Components

Enzyme	Substrates	Functions
Wee1 kinase	Cdk1 Y^{15}	Nuclear kinase. Inhibits Cdk1-cyclin B in G_2.
Myt1 kinase	Cdk1 $T^{14} + Y^{15}$	Cytoplasmic kinase. Inhibits Cdk1-cyclin B in G_2.
Greatwall (Gwl) kinase (MASTL in humans)	Arpp19, ENSA (α-endosulfine)	Phosphorylated Arpp19 and ENSA inhibit PP2a, allowing active Cdk1-cyclin B to accumulate and trigger mitotic entry
Cdc25A phosphatase	Cdk1 T^{14}, Y^{15}	Promotes $G_1 \rightarrow S$ transition and $G_2 \rightarrow M$ transition. Essential for life of the cell.
Cdc25B phosphatase	Cdk1 T^{14}, Y^{15}	Promotes $G_2 \rightarrow M$ transition. Essential in meiosis.
Cdc25C phosphatase	Cdk1 T^{14}, Y^{15}	Promotes $G_2 \rightarrow M$ transition. Dephosphorylates Cdk1 complexed to cyclins A, B at T^{14} and Y^{15}. Not essential for life.
PP2A phosphatase	Many proteins phosphorylated by Cdk1-cyclin B	Regulated by ENSA/Greatwall. With its targeting subunits B55α-δ and B56α-ε it regulates many activities during mitotic exit and cytokinesis.
PP1 phosphatase	Many targets	Associates with many "targeting subunits," which can be, for example, regulatory proteins, such as RepoMan or can be structural subunits of the kinetochore (see Chapter 8). It is inactivated by Cdk phosphorylation during mitosis, but has a key role in mitotic exit.
APC/C^{CDC20}	Cyclin B, securin many others	E3 ubiquitin ligase active during M. Requires high Cdk activity to function. Destruction of cyclins and other substrates essential for exit from mitosis. Contains 15 subunits plus the specificity factor Cdc20.
APC/C^{Cdh1}	Cyclins A, B, many others	E3 ubiquitin ligase active during G_1. Requires low Cdk activity to function. Keeps Cdk activity low in G_1 through cyclin proteolysis. Contains 15 subunits plus the specificity factor Cdh1.
SCF	Cyclin E, many others	Class of E3 ubiquitin ligases containing Skp1 + cullin + Rbx1 + an F-box protein. Humans have 78 F-box proteins (*Caenorhabditis elegans* has more than 300), acting as specificity factors for substrates phosphorylated at specific sites, including cyclin E and Cdk inhibitors.

CHAPTER 41

G_1 Phase and Regulation of Cell Proliferation

During the G_1 phase of the cell cycle, each cell makes a key decision: whether to continue through another cycle and divide or to remain in a nondividing state either temporarily or permanently. During development of metazoans, cells exit the cell cycle as the first step toward forming differentiated tissues. In adults, strict regulation of the timing and location of cell proliferation is critical to avoid cancer.

Cells enter G_1 phase at the end of a proliferation cycle, after completing mitosis. To be free to decide whether to proliferate or differentiate, the cell must inactivate the remnants of the proliferation machinery from the preceding cell cycle. This is initiated in late M-phase by inactivating cyclin-dependent kinases (Cdks [see Chapter 40]) via proteolytic destruction of their cyclin subunits. This continues in G_1 phase and is accompanied by synthesis and stabilization of Cdk-inhibitory proteins. The absence of Cdk activity activates a regulatory network that represses the transcription of many genes that promote cell-cycle progression. While this repressive network is active, the cell cannot proceed through the cell cycle. The repression can be switched off if the cell is stimulated by specific signals from the surrounding medium, extracellular matrix, and other cells (see Chapters 27 and 30). If these signals are diffusible substances, they are known as **mitogens**. Mitogens can trigger another round of DNA replication and mitosis, but first, the cell must pass a major decision point in G_1 called the **restriction point** (Fig. 41.1).

In metazoans, many cells cease cycling, either temporarily or permanently, exiting the cell cycle into a state known as G_0 (Fig. 41.1). This frequently accompanies their acquisition of specialized, differentiated characteristics. Occasionally, it is desirable in tissues for cells in G_0 to reenter the cell cycle to replace lost cells. Specialized cells called **stem cells** fulfill this role in tissue maintenance.

This chapter describes how cells decide whether to exit from the cell cycle into the G_0 phase, how they return to the cycle from G_0, and how they regulate their progress through the G_1 phase. It also considers some of the points at which defects in cell cycle control lead to cancer.

G_0 Phase and Growth Control

Cells stop cycling in three ways. First, they may receive external signals instructing them to withdraw from the cell cycle, enter into G_0, and differentiate, as discussed later. Second, cells may find themselves in an environment with insufficient mitogens to drive proliferation. Under these conditions many cell types enter a transient nondividing state known as **quiescence**, while they wait for conditions to improve. Third, cells that have suffered DNA damage because of a loss of cell-cycle control undergo **senescence,** or may, in some cases, commit suicide by apoptosis (see Chapter 46). Senescence is a permanent nondividing state that is physiologically distinct from G_0 and from which cells normally cannot exit. One physiological signal that can lead to senescence is a critical shortening of the telomere regions of the chromosomes in cells

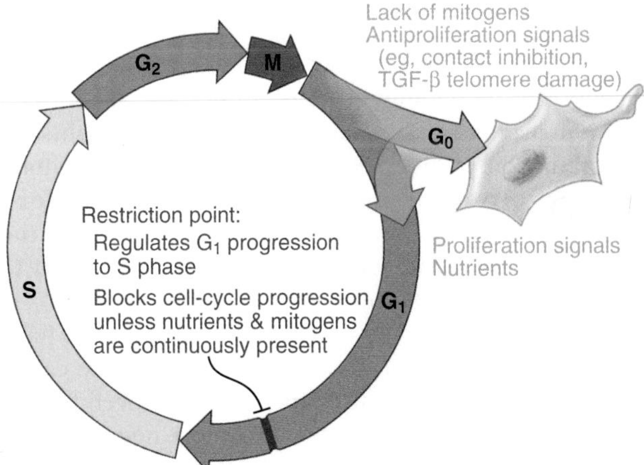

FIGURE 41.1 CELL CYCLE SHOWING MAJOR LANDMARKS IN THE G_1 PHASE. TGF-β, transforming growth factor–β.

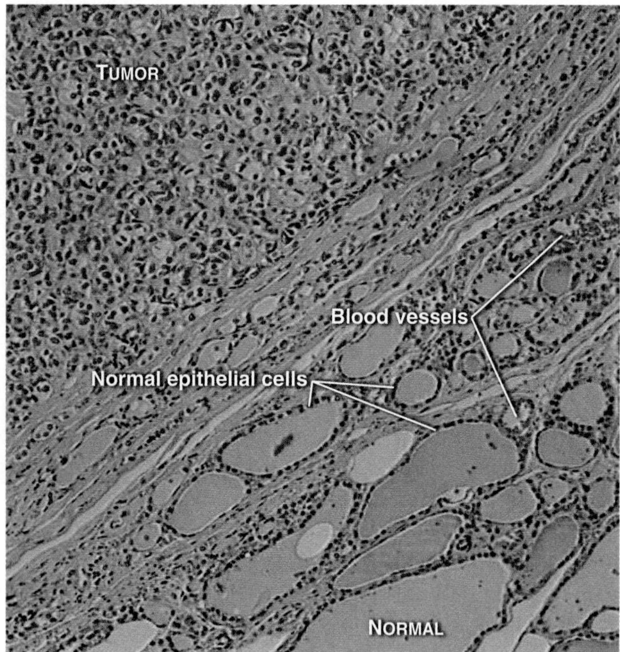

FIGURE 41.2 DISRUPTION OF NORMAL TISSUE ARCHITECTURE BY UNCONTROLLED PROLIFERATION OF CANCER CELLS. *Lower right,* Normal thyroid tissue. *Upper left,* A thyroid tumor with loss of the normal gland structure. (Courtesy Clara Sambade, IPATIMUP, Porto, Portugal.)

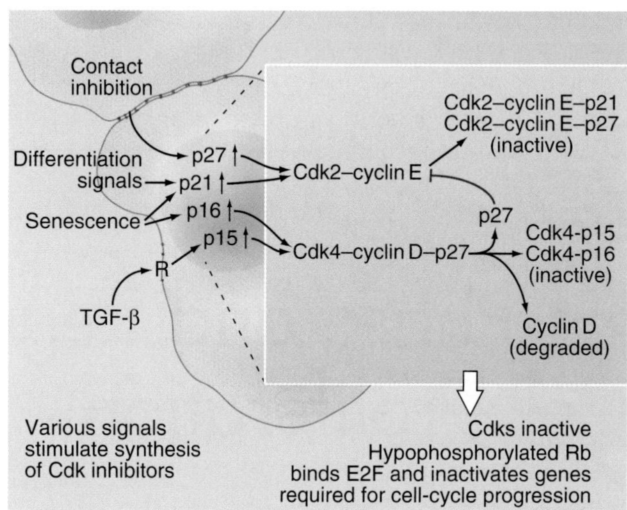

FIGURE 41.3 HOW EXTERNAL STIMULI ACT ON CDK INHIBITORS TO CAUSE CELLS TO ENTER A NONDIVIDING G_0 STATE. TGF-β, transforming growth factor–β.

that have divided more than a critical number of times. Overexpression of telomerase (see Fig. 7.14) can, when combined with suitable mitogenic stimuli, prevent cells from undergoing senescence in tissue culture.

Most cells in multicellular organisms are differentiated (adapted to carry out specialized functions) and no longer divide. They typically form specialized tissues, each with a distinctive structural organization that is important for its function. Unscheduled cell division can severely disrupt the organization of such tissues (Fig. 41.2). Accordingly, tissues strictly regulate both the location and the frequency of cell division. In most tissues, divisions normally occur at a low rate, producing new cells in numbers just sufficient to replace those that die. Under special circumstances, however, such as in response to wounding (see Fig. 32.11), the rate of cell division may increase dramatically. This highlights an important constraint on cell-cycle control in multicellular organisms: To make organized tissues, cells must exit from the cell cycle, but some cells must also retain the ability to reenter the active cell cycle when needed to repair injuries or replace worn-out cells.

Cells that stop cycling to differentiate are said to have left the cycle and entered a nonproliferating state called $\mathbf{G_0}$ (Fig. 41.1). G_0 may last hours or days, or even for the life of the organism, as it does for most neurons. It is important to note that *nondividing cells are not dormant:* G_0 cells can be biochemically very active and continue to expend large amounts of energy for many ongoing processes. Because of turnover, all cells must continuously synthesize housekeeping proteins. They must also expend energy to maintain intracellular pH and ionic composition and to power intracellular motility. In addition, many specialized G_0 cells consume large amounts of energy to synthesize and secrete protein products and generate action potentials. Energy metabolism is particularly dramatic in muscle cells that are responsible for all body movements. Thus, most G_0 cells should be regarded as active cells that just happen no longer to be engaged in cell division.

BOX 41.1 Cdk Inhibitor Scorecard

The regulation of G_1 progression requires the action of two families of Cdk inhibitory proteins (see Chapter 40). Cyclin-dependent kinase inhibitors (CKIs), which include $p21^{Cip1}$ and $p27^{Kip1}$, usually inhibit all Cdks and block cell-cycle progression, but under special circumstances, they can actually activate Cdk4/6-cyclin D and promote cell proliferation during G_1. Cell-cycle blocks imposed by CKIs tend to be temporary. INK inhibitors, which include $p15^{Ink4B}$ and $p16^{Ink4a}$, are specialized at inhibiting Cdk4/6-cyclin D. These can promote a profound and often permanent cell-cycle arrest by activating the Rb pathway of gene repression.

Transforming growth factor–β (TGF-β) is an example of an external signal that arrests progress through the cycle and regulates differentiation and tissue morphogenesis (Fig. 41.3). TGF-β stimulates a receptor serine/threonine kinase that activates SMAD transcription factors (see Fig. 27.10). SMADs suppress Cdk-4 synthesis and increase expression of CKI (cyclin-dependent kinase inhibitor) and Ink4 class Cdk inhibitors (Box 41.1). $p15^{Ink4B}$ preferentially inactivates Cdk4–cyclin D and Cdk6-cyclin D complexes. It also displaces CKI class inhibitors from the Cdk4–cyclin D, permitting them to transfer to Cdk2–cyclin E complexes in the nucleus. This further inhibits cell-cycle progression.

The CKI $p27^{Kip1}$ helps arrest the cell cycle of normal cells when they become crowded by neighboring cells (contact inhibition; see ahead, Fig. 41.11) or if their environment lacks mitogens. Genetic analysis in mice revealed that $p27^{Kip1}$ also regulates cell-cycle progression during development. Mice lacking $p27^{Kip1}$ are 30% larger than their normal littermates by several weeks of age. This is at least partly because cells in many organs undergo extra rounds of cell division.

An analogous mechanism limits proliferation during the differentiation of muscle. The transcription factor MyoD drives expression of the CKI inhibitor $p21^{Cip1}$, which helps arrest proliferation and start muscle differentiation (Fig. 41.3). $p21^{Cip1}$ stops cell-cycle progression in at least two ways. First, it binds Cdk–cyclin complexes and stops them from promoting cell-cycle progression. It also blocks DNA replication by inhibiting the DNA replication factor *p*roliferating *c*ell *n*uclear *a*ntigen (PCNA [see Chapter 42]) that is required for DNA polymerase δ activity.

Both $p21^{Cip1}$ and $p16^{Ink4a}$ contribute to the permanent cell-cycle arrest of senescent cells. $p16^{Ink4a}$ activates inhibitory proteins of the *r*etino*b*lastoma protein (Rb) and E2F families (see later) that bind to promoters and recruit histone methyltransferases, forming heterochromatin that permanently inactivates genes required for proliferation (see Fig. 8.7). Once cells exit the cycle, multiple redundant pathways block reentry by reinforcing the primary inhibition of Cdk activity. In addition, a specialized histone variant H1° replaces histone H1 in G_0 cells, resulting in more condensed chromatin. This represses transcription generally. However, not all gene expression is suppressed in differentiated cells, many of which synthesize large amounts of specific proteins (eg, digestive enzymes secreted by the pancreas).

Moving Into and Out of G_0: Stem Cells

Stem cells are professionals at moving back and forth between G_0 and proliferative cell cycles. One of their roles is to replace worn-out parts of tissues as differentiated cells age or die as a result of various misadventures. Box 41.2 provides a brief introduction to stem cells.

BOX 41.2 Stem Cells in Mammals

The defining feature of **stem cells** is their capacity to self-renew while producing daughter cells with the capacity to differentiate into more specialized cells under the control of intrinsic and environmental cues. Stem cells play a key role in the development of multicellular organisms in addition to providing cells for the renewal and regeneration of adult tissues.

Each multicellular organism begins as a single cell (a fertilized egg) with a genome encoding the information required to produce an adult. The divisions prior to implantation in the uterine wall produce a small group of **epiblast** cells that go on to form the embryo. The other cells that are produced at this stage are specialized to support the embryo. Epiblast cells are termed **pluripotent** because their progeny can form all the specialized cells of the adult. Although the pluripotent cells disappear as the embryo develops, epiblast cells can be propagated and maintained permanently in culture without losing their pluripotency if optimal conditions are provided. These cell lines are called **embryonic stem cells (ES cells)**.

Most adult tissues set aside a few **tissue stem cells** that have the capacity to renew themselves and to produce daughter cells that differentiate into a limited range of specialized cells (see Figs. 28.1, 28.4, and 40.1). Adult stem cells have diverse patterns of cell-cycle regulation. Some tissue stem cells continue the cell cycle throughout life. For example, epithelial stem cells give rise to mature cells that continuously replace the skin and the lining of the gastrointestinal tract. Hematopoietic stem cells in bone marrow give rise to both short-lived and long-lived differentiated blood cells. In other organs, such as liver and skeletal muscle, tissue stem cells are held in reserve unless the tissue is damaged, when they produce daughter cells to repair the damage. Stem cells are present even in organs that have a limited capacity for renewal and regeneration, such as the nervous system. The potential for regeneration from stem cells has stimulated research to find ways of using embryonic or tissue stem cells to repair damaged or diseased organs in human patients. Stem cells have also been useful for production of transgenic animals for scientific research (eg, knockout mice).

Discovery and Defining Features of Stem Cells

Pioneering work on blood cell development (see Fig. 28.4) established the existence of stem cells and defined many of the concepts that apply to all types of stem cells. The key experiment was to inject bone marrow cells from a normal mouse into a mouse that had been irradiated to kill all the cells that produce blood cells. Transplantation of bone marrow cells rescued the irradiated mice from death from anemia, bleeding, and infections. The transplanted bone marrow contained precursor cells that formed colonies of proliferating cells that regenerated the full range of blood cells. The blood-forming colonies in the spleen, each of which formed from a single stem cell, contained either one or, infrequently, several types of differentiating blood cells.

This experimental system first revealed the existence of several different types of progenitor cells in bone marrow with the dual capacity to proliferate and to give rise to more differentiated cells (see Fig. 28.4). These **committed progenitor cells** have a limited proliferation capacity and can

Continued

BOX 41.2 Stem Cells in Mammals—cont'd

give rise to only to specific subsets of blood cells, such as red blood cells, platelets, granulocytes, or lymphocytes. In contrast, there are a very few multipotent **hematopoietic stem cells** in bone marrow. They replenish the pool of committed progenitor cells, ultimately acting as a source of all types of blood cells, while maintaining themselves throughout the life of the individual. Antibodies for surface markers can now be used to distinguish and purify the various types of hematopoietic progenitors as well as stem cells from mice and humans. Once separated from the far more numerous mature and differentiating cells in bone marrow, stem cells can be used for transplantation into patients with bone marrow defects.

Most multipotent hematopoietic stem cells are in the G_0 phase of the cell cycle. A low level of metabolic activity is thought to contribute to their longevity, which can potentially exceed the life span of the individual. When stimulated by demand for more blood cells, growth factors drive multipotent stem cells into a cell cycle that culminates in an asymmetrical division. One daughter cell is another multipotent stem cell. The second daughter cell enters the proliferating pool of blood cell precursors as a committed progenitor cell. Committed progenitor cells and their progeny proliferate vigorously and differentiate into mature blood cells. An adult human produces more than one million blood cells every second.

Cytokines and other growth factors regulate proliferation and differentiation at every stage of blood cell production. The later stages are best understood. For example, the cytokine erythropoietin acts through a kinase-coupled receptor to activate a cytoplasmic transcription factor that stimulates the proliferation and differentiation of the red blood cell lineage. Other cytokines guide the differentiation of granulocytes and monocytes. Hematopoietic stem cells respond to the same families of growth factors that control other aspects of development, including Wnts (see Fig. 30.7), Notch (see Chapter 24), fibroblast growth factor (see Fig. 24.4), and insulin-like growth factor (see Fig. 24.4). However, too little is known about these regulatory mechanisms to grow hematopoietic stem cells in the laboratory.

Properties of Adult Stem Cells

Years of detailed analysis in the laboratory and clinic established hematopoietic stem cells as a model for stem cells in other tissues. General features include the capacity for self-renewal and the production of daughters that proliferate and differentiate. This dichotomy can be achieved by asymmetrical cell divisions guided by the same types of internal cues that control unequal divisions of cells in early embryos (Fig. 41.4). Symmetrical divisions yielding two daughter stem cells can also expand the numbers of stem cells during growth to maturity and during regeneration of damaged tissues.

Stem cells depend on local environmental cues to maintain their status as stem cells. These special environments, called **stem cell niches,** are created by tissue cells and the extracellular matrix. Niche cells anchor stem cells with adherens junctions and provide cell surface and secreted proteins that activate the signaling pathways that regulate the cell

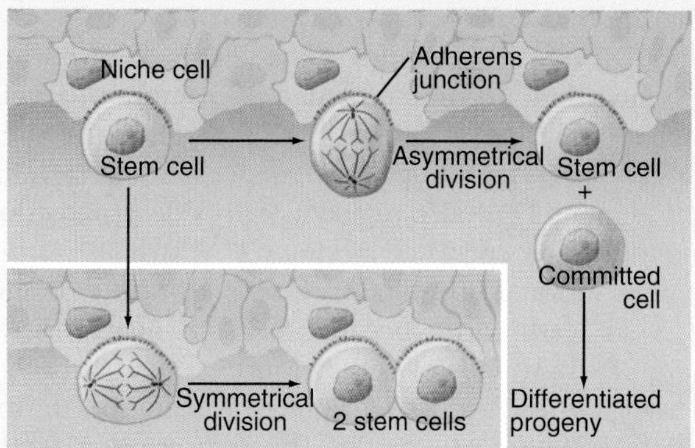

FIGURE 41.4 TWO PATTERNS OF STEM CELL DIVISION. Asymmetrical divisions create two daughter cells: a stem cell that remains associated with its niche cell to maintain the pool of stem cells and one that is committed to multiply and produce differentiated progeny. Symmetrical divisions produce two stem cells to expand the pool of stem cells.

cycle of the stem cell. Some of these factors stimulate division; others inhibit differentiation. The niches occupied by germ cells and neural stem cells from invertebrates are particularly well characterized. During asymmetrical divisions of these stem cells, the renewed stem cell stays behind in the niche, while the daughter that is destined to differentiate into an egg, sperm, or neuron is released. In bone marrow, osteoblasts (see Fig. 32.5) and endothelial cells (see Fig. 30.13) provide niches for hematopoietic stem cells.

Epidermal Stem Cells

Skin is an example of a continuously renewing organ with a considerable capacity for regeneration (see Fig. 40.1). Multipotential and committed stem cells contribute to both renewal and regeneration. Committed stem cells reside in the basal layer of the epidermis. Asymmetrical cell divisions oriented at right angles to the basal lamina produce two daughter cells. The daughter touching the basal lamina carries on as the stem cell. The apical daughter cell divides multiple times and differentiates into an ascending column of cells, forming the superficial layers of the epidermis (see Fig. 35.6). Multipotent stem cells associated with hair follicles give rise to all the cells of the hair follicle and also serve as a reserve for the committed epidermal stem cells in the event of injury (Fig. 41.5).

Skeletal Muscle Stem Cells

Small numbers of stem cells reside in a niche sandwiched between the basal lamina and the giant multinucleated muscle cells. If the muscle is damaged, these quiescent "satellite cells" multiply and produce muscle cells that regenerate the tissue. Positive signals for proliferation and differentiation come through receptor tyrosine kinases and the mitogen-activated protein (MAP) kinase pathway (see Fig. 27.6) and other pathways. Restraining signals are provided by myostatin, a member of the transforming growth factor (TGF)-β family (see Fig. 27.10). Inactivation of the myostatin pathway results in massive enlargement of muscles in mice and humans. Muscles are capable of regenerating multiple times,

BOX 41.2 Stem Cells in Mammals—cont'd

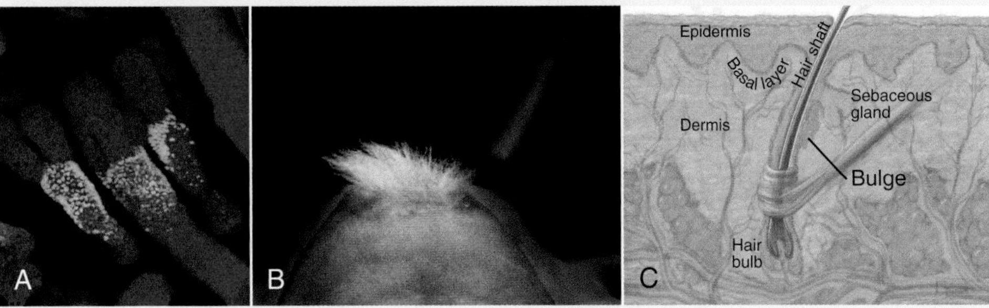

FIGURE 41.5 STEM CELLS FROM SKIN. Multipotent stem cells of the skin reside in the hair follicle bulge (*green* cells in **A**, diagram in **C**). They move up and repair the epidermis during wound healing, and they move down and generate new hair growth during the hair cycle. **B,** Depicts a *Nude* mouse grafted with the cultured cell progeny of a single "bulge" stem cell and displaying a large tuft of hair, all derived from that single stem cell. (**A** and **C,** From Fuchs E, Tumbar T, Guasch G. Socializing with the neighbors: stem cells and their niche. *Cell.* 2004;116:769–778. **B,** From Blanpain C, Lowry WE, Geoghegan A, et al. Self-renewal, multipotency, and the existence of two cell populations within an epithelial stem cell niche. *Cell.* 2004;118:635–648.)

so the stem cell population renews itself during regeneration or is augmented by stem cells that migrate through the blood from bone marrow or other tissues.

Cancer Stem Cells

Stem cells may play a role in cancer, acting as a source for proliferating cells that make up the bulk of the tumor. If true, this concept helps explain why it is relatively easy to reduce the size of tumors by targeting dividing cells but difficult to completely eliminate residual tumor stem cells, which may divide less frequently. It is thought that the spread of cancer from the primary tumor to other tissues (**metastasis**) requires circulating cancer cells to find locations (niches) where they can establish themselves as stem cells.

Meristematic Stem Cells in Plants

The growth of plants depends on carefully orchestrated proliferation and differentiation of cells derived from stem cells called meristems. Through asymmetrical divisions, these relatively inactive cells give rise to daughters that proliferate at the tips of shoots and roots. The proliferating cells differentiate into specialized tissues such as flowers, while the stem cells maintain a pool of slowly replicating cells in a special niche.

Use of Stem Cells to Make Transgenic Animals

Because embryonic stem cells grow in culture, they can be manipulated experimentally. They can be transfected with DNA, and if the proper sequences are present, this DNA can replace a region of the endogenous chromosome by homologous recombination. If the modified embryonic stem cells are subsequently injected into developing embryos at the blastocyst stage, they are, with low frequency, able to colonize the cell population that will produce germ cells. When such chimeric embryos grow to adulthood, a proportion of their gametes will carry a chromosome with the modification engineered in the embryonic stem cells. Furthermore, this chromosome will now be inherited by all progeny of that embryo, giving rise to a line of **transgenic** animals. This method is widely used in research to knock out genes by designing the original DNA construct so that when it enters the chromosome by homologous chromosome, a critical region of a target gene is deleted or disrupted. The use of knockout mice has revolutionized the study of developmental biology by allowing investigators to determine the function of specific genes in living animals.

Note the distinction between transgenic animals and reproductive cloning. "Cloned" animals are produced by introducing a somatic cell nucleus into an enucleated egg. Experiments first in frogs and later in mammals, such as Dolly the sheep, established that egg cytoplasm can reprogram gene expression of differentiated cell nuclei, and enable the development of a cloned animal. Transfer of nuclei from lymphocytes and olfactory neurons has been used to derive healthy adult mice. This approach involves the reversal of epigenetic changes in the nucleus that drove the differentiation of the adult cell. The molecular mechanisms underlying reprogramming are not well understood and the success rate at obtaining healthy animals through nuclear cloning is very low. This cloning does not involve the use of stem cells, but embryonic stem cells can be derived from the blastocyst stage of the cloned embryos.

Therapeutic Applications of Stem Cells

Where committed stem cells can be isolated from an adult organ, it is now possible to regenerate damaged tissues by transplanting these stem cells from patients themselves or from donors. The best example is transplantation of bone marrow stem cells to treat patients whose bone marrow has been damaged by cancer, chemotherapy, or other disease. Adverse immunologic reactions are a challenge for transplants from donors other than an identical twin. On one hand, the immune system of the recipient can reject the transplanted cells. On the other hand, lymphocytes contaminating the donor stem cells can mount an immunologic attack on the recipient. Using purified hematopoietic stem cells (ideally, the patient's own stem cells) rather than mixed bone marrow cells avoids this problem. Knowing how to expand hematopoietic stem cells in vitro would be helpful. This approach is already used for treating burns with epidermal stem cells. Normal skin is used as a source of committed

Continued

BOX 41.2 Stem Cells in Mammals—cont'd

skin stem cells, which are multiplied in culture and used to regenerate all the layers of the skin.

Stem cells might be used to regenerate other damaged tissues, including the insulin-producing cells that are lost in Type I diabetes, but appropriate stem cells are not available for many organs, including the pancreas, brain, and heart. Indeed, even with appropriate stem cells in hand, much remains to be learned about how to grow them and then direct them to differentiate into mature tissues.

Embryonic stem cells can potentially supply all cells necessary to replace any damaged tissue, but sources of human embryonic stem cells are limited, and acquiring them from early embryos discarded by fertility clinics is unacceptable to some people. Patient-specific (autologous) embryonic stem cells can be derived via somatic cell nuclear transfer (cloning) followed by expansion of epiblast cells from the embryo. However, the production of such "artificial" human embryos is also highly controversial.

Adult stem cells are an alternative to embryonic stem cells. This approach has the advantage that stem cells can be isolated from bone marrow, blood, and skin by using antibodies that recognize specific surface protein "markers"; however, these specialized stem cells normally do not produce differentiated cells for regeneration of other tissues.

An alternative method to generate pluripotent stem cells is based on the transient reintroduction into a differentiated cell of a group of specific transcription factors commonly expressed in stem cells. These include Sox2, Oct4, Klf4, and Myc. When these factors are ectopically expressed in differentiated cells, they can induce the expression of proteins required for pluripotency, as well as factors required to change the epigenetic landscape of the cell. This results in a stable perpetuation of pluripotency. These **induced pluripotent stem (iPS)** cells, offer the promise that they can be differentiated into any desired specific cell type for medical applications. This procedure potentially eliminates the problem of the availability of autologous stem cells as many cell types can be reprogrammed to iPS cells. This technology is rapidly developing, as researchers attempt to improve the generation of functional cell types from iPS cells

When injected into immunodeficient mice, iPS cells make tumors containing all cell types, called **teratocarcinomas**. This clearly shows that the iPS cells are pluripotent. They can also generate all cell types in tissue culture when appropriate growth factors necessary for self-renewal are removed from culture media. However, this differentiation is difficult to control. We cannot yet generate specific fully functional cell types necessary for therapies with high purity from pluripotent stem cells, except for a few cell types such as retinal pigment epithelial (RPE) cells. iPS cell-derived RPE cells were transplanted for the treatment of macular degeneration in 2014, and appeared to be successful in blocking the degeneration. However, as of 2016, this was the only clinical trial performed with iPS cell-derived cells. Investigation on how to control iPS cell differentiation is one of the main obstacles to be overcome for regenerative medicine.

Reentry Into the Cell Cycle

Cells in the G_0 phase may reenter the growth cycle in response to specific stimulation by mitogens, often induced by injury or normal cell turnover. Cultured fibroblasts are favored for laboratory studies of this process, as they readily enter a quiescent state mimicking G_0 phase when deprived of serum (ie, mitogens and growth factors) and rapidly reenter the cell cycle when serum is restored. This response reproduces that found in wounded tissues. When a living tissue is wounded (see Fig. 32.11), fibroblasts are exposed to serum that has leaked from damaged blood vessels. In response, they divide and colonize the wound, where they lay down new extracellular matrix to repair the damage.

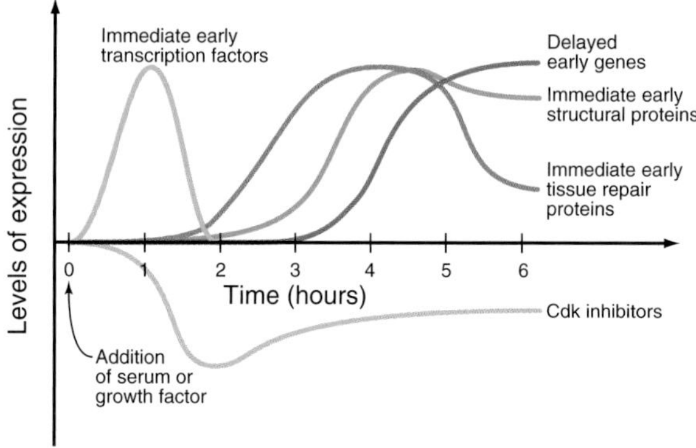

FIGURE 41.6 PATTERNS OF EXPRESSION OF IMMEDIATE AND DELAYED EARLY GENES DURING THE RETURN OF GROWTH-ARRESTED FIBROBLASTS FROM G_0 TO ACTIVE PROLIFERATION AND THE CELL CYCLE.

Serum stimulates three waves of gene expression in cultured quiescent fibroblasts (Fig. 41.6). The first includes more than 100 **"immediate early"** genes. These include transcription factors of the Jun, fos, myc, and zinc finger families (see Chapter 10) that activate numerous downstream genes required for cell growth and division. Other immediate early genes encode tissue remodeling factors, cytokines (growth factors), extracellular matrix components (fibronectin), plasma membrane adhesion proteins (integrins), and cytoskeletal proteins (actin, tropomyosin, vimentin), as well as activities involved in angiogenesis (blood vessel formation), inflammation, and coagulation. These proteins facilitate the movement of fibroblasts into wounds and initiate the repair of tissue damage.

Expression of a second wave of **"delayed early"** genes encoding a variety of proteins that are required for cell growth and proliferation, including cyclin D, precedes the onset of the S phase. Genes activated after the onset of the S phase are referred to as **"late"** genes. Both delayed early and late gene transcription require synthesis of the transcription factors encoded by immediate early genes.

These waves of transcription in response to mitogens enable the G_0 cells to pass through a "gate" and reenter the active cell cycle. This gate is analogous to the restriction point, a critical aspect of G_1 control that regulates the proliferation of all normal cells.

The Restriction Point: A Critical G_1 Decision Point

All eukaryotes have a mechanism that operates during the G_1 phase to ensure that cells duplicate their genome only when the environment is supportive and the chromosomes are undamaged. Healthy yeast cells do not embark on a round of DNA replication and division until they reach an appropriate minimum size (actually, they probably measure their ribosome content and ongoing rate of protein synthesis). This is important because after cell division, the daughter cell (bud) is smaller than the mother and needs more time to grow before it divides if the population is to maintain a constant cell size. Whether dividing mammalian cells also monitor their size is not yet settled.

The influence of cell size on the division cycle was first demonstrated in an elegant microsurgery experiment (Fig. 41.7). Two *Amoeba proteus* cells were grown under identical conditions in parallel cultures. Each day, a portion of the cytoplasm was amputated from one amoeba, and the other was left untouched as a control. Under those circumstances, the cell that suffered the amputations did not divide for 20 days. During this time, the control amoeba divided 11 times. When the amputations were stopped, the amoeba that had been operated on divided within 38 hours. The interpretation of this experiment was that the repeated amputations prevented the experimental amoeba from ever attaining a size sufficient to turn on the division program. Evidence suggests that some types of human cells have a similar size control while others do not.

An essential aspect of growth control during the G_1 phase involves monitoring the external environment for nutrient availability and for signals to proliferate (**mitogenic** signals) coming from other cells and from the extracellular matrix. In a classic experiment, when three flasks containing populations of cultured cells were starved by deprivation of amino acids, serum, or phosphate respectively, they all stopped cycling in G_1. When the missing ingredients were restored, cells in all three flasks resumed the cell cycle and entered the S phase at about the same time. This was surprising because amino acids are needed to make protein, serum provides growth factors and mitogens, and phosphate is needed for synthesis of DNA and phospholipids (needed to make membranes). This experiment was interpreted as evidence that all three types of starvation caused cells to arrest at a single point in the G_1 phase, termed the **restriction point.** The restriction point is now defined as the point after which the cell cycle will proceed even if mitogenic factors are withdrawn (Fig. 41.8). This supremely important aspect of cell-cycle control prevents cells from dividing at inappropriate times and in inappropriate places. Defects in restriction point control are among the most common causes of cancer.

Genetic analysis also revealed a point in the G_1 phase after which budding yeast cells appear to be committed to completion of the cycle. Cells that are starved for

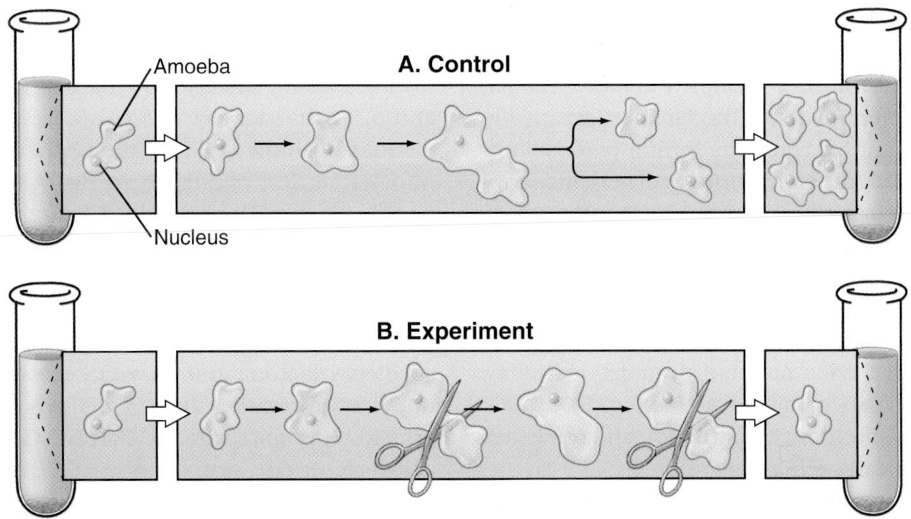

FIGURE 41.7 A MICROSURGERY EXPERIMENT DEMONSTRATES THAT AMOEBAE WILL NOT DIVIDE IF THEY ARE PREVENTED FROM ATTAINING A SUFFICIENT SIZE. A, Control cell continues to divide. **B,** Experimental cell does not divide. (For reference, see Prescott DM. Relation between cell growth and cell division. II: The effect of cell size on cell growth rate and generation time in *Amoeba proteus*. *Exp Cell Res*. 1956;11:86–98.)

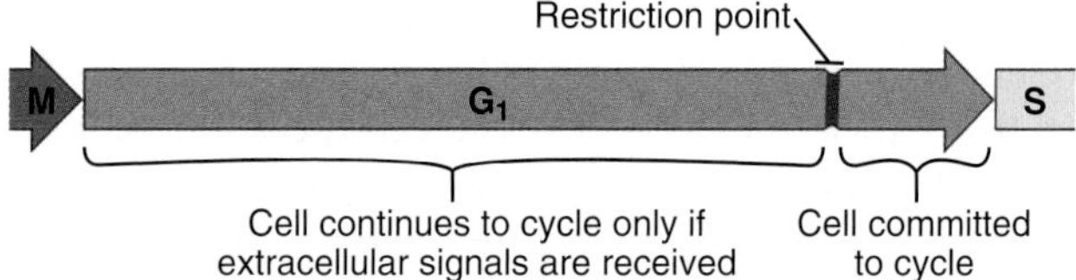

FIGURE 41.8 RESTRICTION POINT. During late G_1, cells assess external and internal stimuli and decide whether to commit to a further round of DNA replication and division.

nutrients arrest at, or just prior to, this point, termed **START.** The mammalian restriction point resembles yeast START in a number of aspects, but they are not exactly equivalent, owing to differences between animal and yeast cell cycles.

Regulation of Cell Proliferation by the Restriction Point

The restriction point is a molecular "gate" that regulates the expression of genes required for cell-cycle progression. The gate is based on proteins that are related to the **Rb** susceptibility protein and a family of essential transcription factors known as **E2F.**

In brief, E2F is a master transcriptional regulator that activates many of the genes whose products drive DNA replication and cell-cycle progression. Rb regulates the cell cycle by binding to E2F and converting it into a repressor of those same cell-cycle genes. When Rb is bound to E2F, the cell is said to be arrested at the restriction point. Escape from this arrest involves Cdk activation and subsequent Rb phosphorylation. This releases E2F, which then drives cell-cycle progression. An alternative way to pass the restriction point gate depends on a potent regulator called Myc, which is discussed separately later.

Mammals have three Rb-related proteins (pRb, p107, and p130) and approximately eight E2F family members. These together constitute a complex multifunctional network. This chapter refers to the families generically as Rb and E2F. Some E2F proteins form a heterodimer with one of three DP (differentiation-regulated transcription factor-1 protein) family members. This dimer associates with the promoter region of E2F target cell-cycle genes (Fig. 41.9A). Rb binding converts E2F/DP from a transcriptional activator to the Rb/E2F/DP repressor. Rb also recruits histone deacetylases, enzymes that remove acetyl groups from a wide range of proteins, including the aminoterminal tails of histones (see Fig. 8.7). This causes compaction of chromatin structure and represses genes required for cell-cycle progression.

Nonproliferating cells have low Cdk activity in G_1 before the restriction point. First, cyclin D messenger RNA (mRNA) levels are low, so little protein is made. Second, any cyclin D that is made is retained in the cytoplasm, where it is phosphorylated by glycogen synthase kinase (GSK)-β (see Fig. 30.7) and marked by SCF (Skp, Cullin, F-box containing complex) for degradation (see Fig. 40.16). Rb/E2F/DP represses the expression of the genes for cyclins E and A required for Cdk2 activation. Furthermore, high levels of the CKI class Cdk inhibitor $p27^{Kip1}$ inhibit any Cdk2–cyclin E or Cdk2–cyclin A that happens to be present (see Fig. 40.14).

Signals from mitogens and the extracellular matrix open the restriction point gate. Stimulation of receptor tyrosine kinases (see Chapters 25 and 27) or integrins (see Chapter 30) activates Ras and the mitogen-activated protein (MAP) kinase/extracellular signal-regulated kinase (ERK) cascade (see Fig. 27.6). The output of this cascade stimulates transcription of D-type cyclins (Figs. 41.9 and 41.10) and also inactivates GSK. This allows cyclin D to accumulate in nuclei. Nuclear cyclin D binds to and activates Cdk4 and Cdk6 (referred to hereafter as Cdk4/6–cyclin D), producing an initial pulse of Cdk activity that is later amplified by Cdk2–cyclin E and Cdk2–cyclin A. Cdk activity in early G_1 is regulated by adjusting the relative levels of the three D-type cyclins as well as the levels of Ink4 and CKI inhibitors. This regulation of cyclin D levels and Cdk inhibitors provides the crucial link between extracellular mitogens and the cell cycle.

Mitogens also stimulate transcription of the CKI class Cdk inhibitor $p27^{Kip1}$. This protein actually *activates* Cdk4/6–cyclin D complexes in two ways. First, the receptor associated tyrosine kinases Jak or Src phosphorylate p27, inducing a structural change within the Cdk4/6–cyclin D-$p27^{Kip1}$ complex that activates the kinase. This links mitogen signaling to Cdk-4/6 activation. It also promotes the nuclear import of Cdk4/6–cyclin D. This both leads to full activation of Cdks by Cdk-activating kinase, a nuclear enzyme (see Chapter 40), and increases the stability of cyclin D. All of this depends on the continuous presence of mitogenic signals; if these cease, then cyclin D stability rapidly declines again, since following dephosphorylation, p27 and p21 again act as Cdk4/6–cyclin D inhibitors.

Cdks push the cell past the restriction point by phosphorylating Rb, causing it to dissociate from E2F (Fig. 41.9B; see also Chapter 40 and Appendix 40.1). The E2F/DP heterodimer remains bound to promoter regions and now potently activates, rather than represses, the transcription of genes that stimulate cell proliferation. The proteins produced synthesize DNA (DNA polymerase α, accessory factors, and enzymes that synthesize nucleotide precursors; see Chapter 42), promote cell-cycle progression (cyclins E and A, Cdk1, and Cdc25), and regulate cell-cycle progression (pRb, p107, Emi1).

The chain of events as mitogens break the blockade on cell-cycle progression imposed by Rb involves a positive feedback loop as follows. Cdk4/6–cyclin D complexes begin to phosphorylate Rb. This releases some E2F and permits the initial expression of genes that

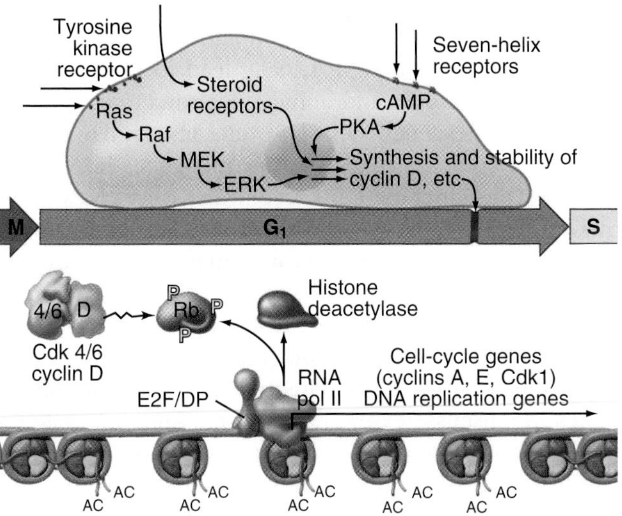

FIGURE 41.9 REGULATION OF CELL-CYCLE PROGRESSION BY THE E2F/DP/RB COMPLEX. A, The Rb/E2F/DP complex recruits histone deacetylases (see Chapter 8) and represses specific genes that are required for cell-cycle progression. This blocks cell-cycle progression at the restriction point. **B,** Phosphorylation of Rb by Cdks alleviates this block and permits passage of the restriction point. cAMP, cyclic adenosine monophosphate; MEK, mitogen-activated protein kinase kinase; PKA, protein kinase A.

encode cyclin E, cyclin A, and CDC25A. Active Cdk2-cyclin E can initiate $p27^{Kip1}$ degradation, permitting the rapid accumulation of active Cdks. The restriction point probably is passed here, and from this point on Cdks remain active until cyclin destruction in late mitosis.

Cdk2-cyclin E participates in a second wave of Rb phosphorylation on many sites, leading to the wholesale liberation of E2F/DP and a surge in transcription of genes that trigger the onset of DNA replication (S phase entry) and promote progression through the cell cycle. As the cell cycle proceeds, Rb phosphorylation is maintained first by Cdk2–cyclin A and then later by Cdk1–cyclin B until the exit from mitosis. Rb is dephosphorylated at the mitosis-G_0 or G_1 transition. This enables it once again to bind E2F and close the restriction point gate to exit from the next G_1.

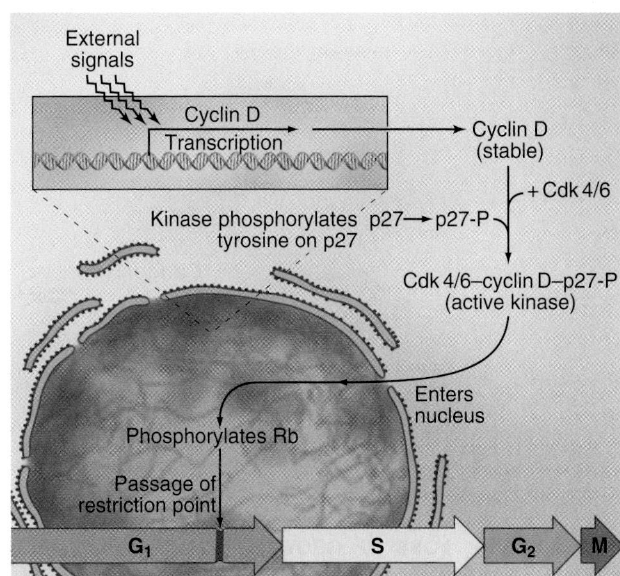

FIGURE 41.10 HOW GROWTH FACTORS REGULATE CDK4/6 ACTIVITY: THE ROLE OF D-TYPE CYCLINS AND P21. Mitogen-induced tyrosine phosphorylation of p27 determines if it assembles active or inactive CDK4/6–cyclin D complexes–linking assembly of active kinase to mitogens.

The transcriptional regulator **Myc** drives an alternative pathway for G_1 exit that is also stabilized by mitogenic signals. Association of myc with one partner activates the transcription of cyclins E and D2. Association with a different partner downregulates the transcription of Cdk inhibitors of both the CKI and INK classes. These activities partly explain why Myc can act as an **oncogene**—a protein that helps transform normal cells into cancer cells (explained further later).

Restriction Point and Cancer

Cancer is a complex class of diseases in which genetic changes within clones of cells lead to production of cell populations whose uncontrolled growth can disrupt tissue function and ultimately kill the individual. Two in five Americans will be affected by cancer during their lifetimes. This sounds very high, but considering the number of cell cycles that are required to produce a human composed of approximately 10^{14} cells, and considering the over 1 million cell divisions that occur per second in a healthy adult, the disease is actually remarkably rare on a per cell basis. This is at least partly because multiple genetic alterations are required to transform a normal cell into a cancer cell. Furthermore, the cell cycle is highly regulated by a web of negative feedback pathways that hold in check activities driving cellular

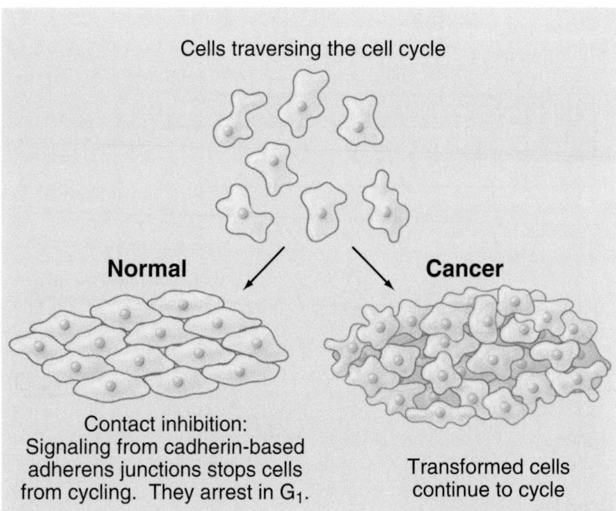

FIGURE 41.11 LOSS OF GROWTH CONTROL IN TRANSFORMED CELLS.

proliferation. In fact, many cancer-causing mutations that disturb growth control pathways are actually deleterious in normal cells and cause them to undergo senescence or commit suicide by apoptosis (see Chapter 46).

Dysregulation of cell proliferation in the G_1 phase causes most types of cancer. This is readily seen in the laboratory when cells are grown on plastic tissue culture dishes. Most normal cells proliferate until they cover the surface completely, forming a monolayer. When the monolayer is confluent (ie, when cells are touched by other cells on all sides), signaling initiated by cadherin proteins (see Fig. 30.7) causes cells to express $p27^{Kip1}$ and to arrest their cell-cycle progression in G_1. This is called **contact inhibition** of growth (see Chapter 30, in the section titled "Cadherin Family of Adhesion Receptors"). Cancer cells lack this control, so they keep proliferating and piling up on top of one another as long as nutrient and mitogen supplies last (Fig. 41.11). Cells that lose this aspect of growth regulation are said to be **transformed.**

Malfunction of the restriction point is an extremely common contributor to transformation. Indeed, one or more components of the p16/cyclin D/Cdk-4/Rb system are mutated in most human cancers. In addition, several cancer-causing viruses, such as simian virus 40 (SV40), papillomaviruses, and adenovirus, make proteins that facilitate the $G_1 \rightarrow S$ transition by binding Rb and liberating E2F.

Most cancer cells have abnormalities in the activities of two classes of genes. **Oncogenes** are genes whose inappropriate *activation* can cause oncogenic (cancerous) transformation of cells. The protein products of most oncogenes regulate cellular growth and proliferation. They are typically components of signal transduction pathways that are controlled by feedback mechanisms. **Tumor suppressors** are genes whose *inactivation* can lead to cancerous transformation. Their protein products typically inhibit products of oncogenes or negatively regulate cell proliferation. Several genes that are involved in restriction point control can act as either oncogenes or tumor suppressors.

More than 100 oncogenes have been identified thus far. Most normally function in signal transduction pathways that lie downstream of mitogens that stimulate cell-cycle progression. Their inappropriate activation can mimic the effects of persistent mitogenic stimulation, thereby uncoupling cells from normal environmental controls and leading to uncontrolled proliferation and cancer. For example, Ras activates the MAP/ERK kinase cascade and accumulation of cyclin D (Fig. 41.9). Ras genes are mutated in approximately 15% of human cancers. Inappropriate activation of Ras tricks the cell into thinking that it is receiving mitogenic signals, leading it to express cyclin D, phosphorylate Rb, and proliferate.

Fortunately, in most cells, mutations that result in uncontrolled proliferation usually lead to DNA damage (**oncogenic stress**) and activate a protective pathway leading to senescence. Other proteins involved in restriction point control can also act as oncogenes if hyperactivated. These include E2F1, cyclin D (overexpressed in 50% of breast cancers), and Cdk4. In each case, activation of the protein causes inappropriate transcription of genes promoting cell-cycle progression, bypassing the restriction point, and leading to uncontrolled cell cycles and cancerous transformation (Fig. 41.12).

Rb is one of the best-characterized tumor-suppressor genes. As discussed earlier, a primary function of Rb is to block cell-cycle progression until sustained mitogenic stimulation results in its inactivation. It is therefore not surprising that loss of Rb can lead to inappropriate cell-cycle progression and cancer. Rare individuals who inherit one defective Rb gene tend to develop retinoblastomas as children and osteosarcomas as adults. The cancer arises when the "good" allele is inactivated in a proliferating cell (this is called a somatic mutation). Such cancers are rare and occur only later in life in individuals who inherit two good Rb genes, as two independent somatic mutations (two "hits") are required in the same proliferating cell. Homozygous loss of Rb is lethal during embryogenesis. This is partly because under some circumstances, the unleashed E2F can act as a potent inducer of apoptotic cell death (see Chapter 46).

$P16^{Ink4a}$ is another important tumor suppressor involved in G_1 growth control. Normally, it suppresses Cdk4/6 activity in nondividing cells (see next section; also see Chapter 40), thereby reinforcing the ability of Rb to maintain the growth arrest of G_1 cells (Fig. 41.12). Mutations in the $p16^{Ink4a}$ gene are very common in cancer, but this is partly because this gene is fascinatingly complex (Fig. 41.16). Mutations in other INK4 Cdk inhibitors and the CKI protein $p27^{Kip1}$ are also found in cancer, although less frequently.

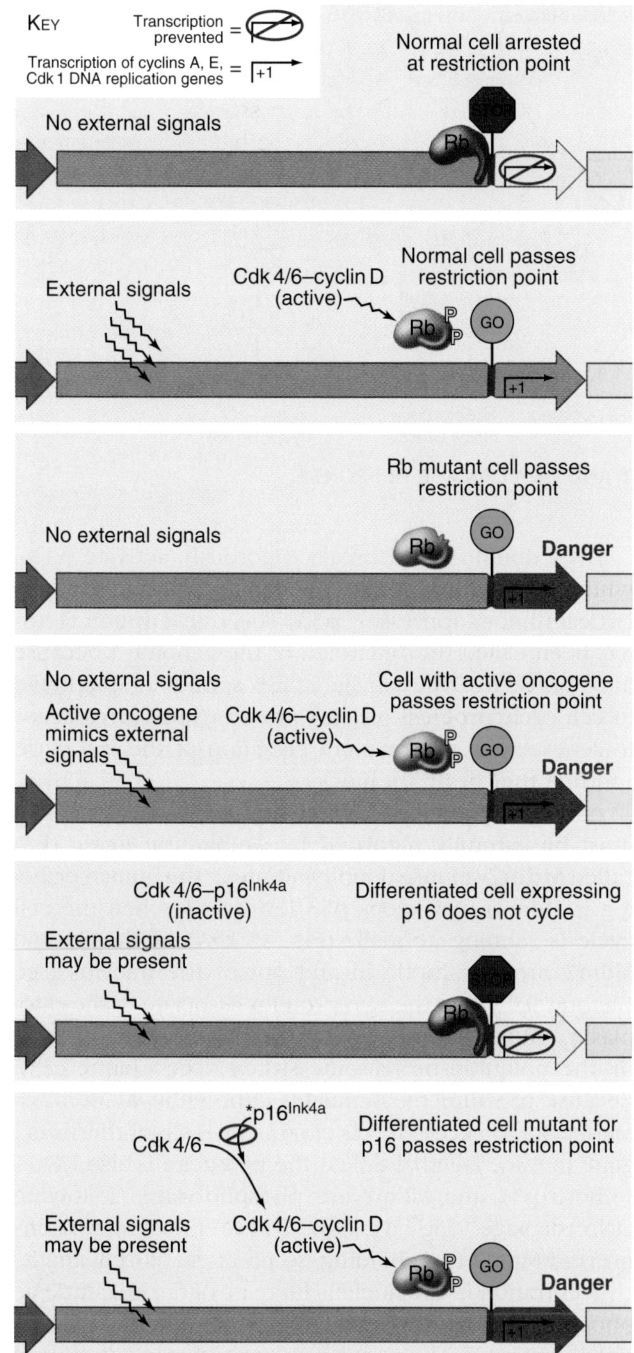

FIGURE 41.12 HOW ACTIVATED ONCOGENES OR MUTATIONS IN THE RB OR P16 TUMOR SUPPRESSOR PROTEINS CAN LEAD TO ABNORMAL PASSAGE OF THE RESTRICTION POINT AND CANCER.

Proteolysis and G_1 Cell-Cycle Progression

Just as controlled destruction of proteins is key to the transition of cells from mitosis to the G_1 phase (see Chapter 40), proteolysis also fulfills several key roles during progression from G_1 into the S phase (Fig. 41.13). For example, when Cdk2-cyclin E is activated following cyclin D synthesis, it phosphorylates its inhibitor $p27^{Kip1}$.

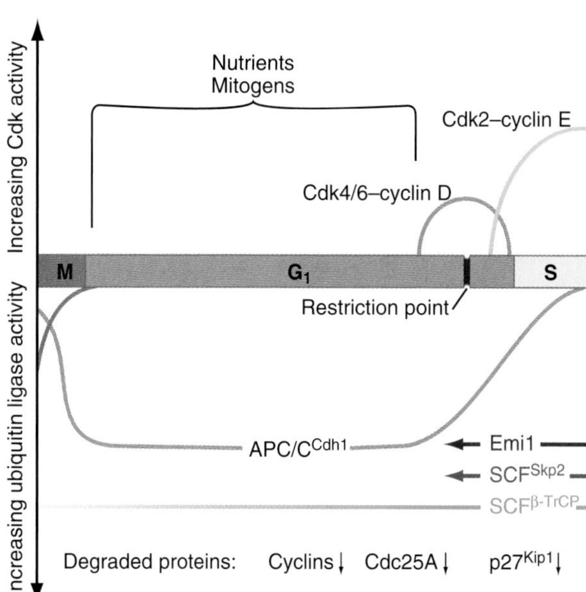

FIGURE 41.13 PROTEOLYTIC ACTIVITIES IN G_1 PHASE.

Phosphorylated $p27^{Kip1}$ is recognized by a specific E3 ubiquitin ligase called SCF^{Skp2} (see Figs. 40.15 and 40.16). The resulting destruction of $p27^{Kip1}$ permits a burst of Cdk2-cyclin E activation in a positive feedback loop that rapidly amplifies Cdk activity at the initiation of the S phase. Later in the S phase, phosphorylation of the DP subunit of E2F causes its dissociation from DNA, recognition by SCF, and destruction. This is essential to complete S phase. SCF also targets cyclins D_1 and E for destruction, the former when mitogens are limiting and the latter following autophosphorylation during progression through the S phase.

Progression throughout G_1 requires the activity of a second E3 ubiquitin ligase known as the *a*naphase-*p*romoting *c*omplex or *c*yclosome (APC/C). A specific cofactor, Cdc20, activates the APC/C to trigger the metaphase-to-anaphase transition (see Chapter 40). As the cell leaves mitosis, Cdh1 replaces Cdc20 and targets many cell-cycle regulatory proteins for degradation, including Cdc20, A- and B-type cyclins, and factors involved in DNA replication. Destruction of these target proteins during mitotic exit and G_1 (Fig. 41.13) likely contributes to the requirement for transcription and *de novo* synthesis of these proteins at the moment cells decide whether to enter or not a new cycle of genome replication.

Integrity of Cellular DNA Monitored by a G_1/S Checkpoint

The S phase is a point of no return in the history of any dividing cell. Given the semiconservative mechanism of DNA replication, whereby existing DNA strands serve as templates for newly synthesized strands, any DNA defect that passes unnoticed through the S phase becomes

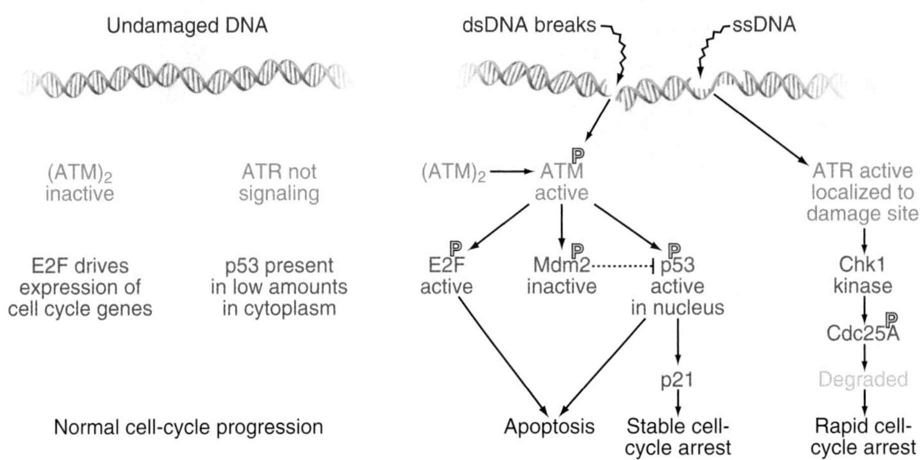

FIGURE 41.14 THE G_1/S CHECKPOINT AND DNA DAMAGE RESPONSE.

perpetuated as a mutation that is transmitted to all future progeny of the cell. Furthermore, any single-stranded nick in DNA may become a full-fledged chromosome break if present during replication. To avoid these problems, cells have a quality control mechanism to block entry into the S phase if damaged DNA is detected.

This quality control mechanism involves a **checkpoint** that operates throughout the G_1 and S phases (Fig. 41.14). Checkpoints are biochemical circuits superimposed on the normal cell cycle. When activated, the G_1/S checkpoint triggers a **DNA damage response** that blocks cell-cycle progression. The block may be temporary but, in some cases, checkpoint activation leads to senescence or cell death by apoptosis. Sensor proteins detect DNA damage and activate this checkpoint. In the subsequent DNA damage response, the sensor proteins activate protein kinases and a key transcriptional regulator that block cell-cycle progression (see Fig. 40.4).

The DNA damage response has fast and slow components: The former is analogous to applying the brakes in a car; the latter is analogous to removing the wheels and putting it up on blocks. Both components start with the protein kinases, ATM and ATR (see Fig. 40.4). ATM and ATR are related to the lipid kinase phosphatidylinositol 3-kinase (see Fig. 26.7), but their only known substrates are proteins (see Chapter 40). People lacking ATM have the disease ataxia-telangiectasia, which is characterized by immunodeficiency, photosensitivity, cerebellar degeneration, and an elevated incidence of leukemias and lymphomas. Loss of ATR is fatal.

DNA damage that disrupts ongoing DNA replication and produces single-stranded DNA activates ATR, which phosphorylates and activates a downstream kinase called Chk1. Chk1 targets the essential phosphatase CDC25A (Fig. 41.14), marking it for destruction. Because CDC25A is required to remove inhibitory phosphate groups from inactive Cdks, its destruction applies a rapid brake to cell-cycle progression.

DNA double-strand breaks vigorously activate ATM, which directly and indirectly stabilizes and activates a critical tumor suppressor, **p53.** This transcription factor has been called the "guardian of the genome," because in response to DNA damage it also applies a rapid brake to cell cycle progression. In some cases this arrest leads to senescence, a permanent cessation of the cell cycle (putting the car up on blocks).

p53 is very powerful medicine for the cell cycle and must be carefully regulated by a ubiquitin ligase (E3) called **Mdm2** (mouse double-minute 2; the human ortholog is Hdm2) that keeps p53 levels low when the cell cycle is running normally (Fig. 41.15A). Both p53 and Mdm2 protein shuttle in and out of the nucleus (see Chapter 9). When the two proteins associate in the cytoplasm, Mdm2 promotes rapid degradation of p53 by the ubiquitin/proteasome system (see Chapter 23). Because p53 directly stimulates expression of Mdm2, a feedback loop keeps levels of p53 low. Loss of the Mdm2 gene in mice is lethal unless the p53 gene is also lost.

Both p53 and Mdm2 are phosphorylated following DNA damage (Fig. 41.15B). These phosphorylations prevent Mdm2 from binding, so p53 is stabilized, and its concentration in the nucleus increases dramatically. The phosphorylations also make p53 a more potent transcriptional activator. The result is a burst of transcription of p53-regulated genes.

p53 is rapidly activated in response to DNA damage resulting from hyperproliferation of cells following loss of restriction point control (oncogenic stress). It also responds to DNA damage induced by the environment. If the damage is rapidly repaired, cells continue to cycle, but if the damage is too severe, p53 induces senescence or apoptotic cell death (see next paragraph).

In the case of oncogenic stress following loss of restriction point control (eg, by Ras mutations) E2F stimulates the expression of the tumor suppressor protein $p19^{Arf}$ (*a*lternate *r*eading *f*rame). This binds and sequesters Mdm2 in the nucleolus (Fig. 41.15C), allowing p53

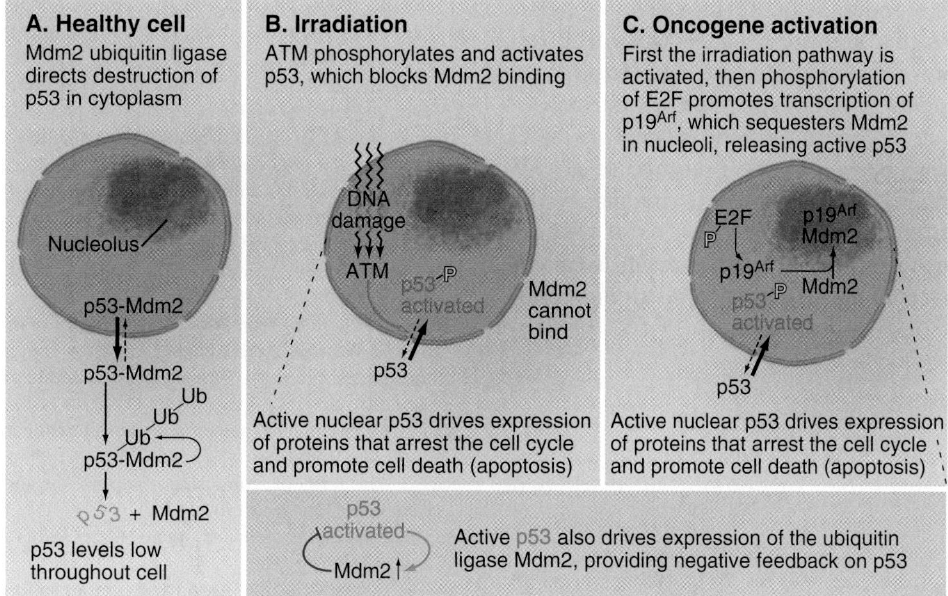

FIGURE 41.15 P53 REGULATION AND THE DNA DAMAGE CHECKPOINT IN G_1. A, Healthy cell. **B,** After irradiation, Mdm2 (mouse double-minute 2) can no longer bind p53, which accumulates in active form in the nucleus. **C,** After oncogene activation, Mdm2 is sequestered in the nucleolus, and active p53 accumulates in the nucleus. Activated p53 can induce either cell-cycle arrest or cell death.

to accumulate in the nucleoplasm. There, it activates transcription of the CKI p21, stopping the cell cycle. If this arrest is prolonged, $p16^{INK4A}$ is induced, activating Rb and leading to permanent cell cycle arrest (senescence). p53 activation can also induce apoptosis by activating transcription of genes for proapoptotic proteins, including *Bax*, BH3-domain proteins, *Puma*, CD95 (*Fas/Apo1*), and *Apaf-1* (Fig. 41.15; also discussed in Chapter 46). The decision whether to induce senescence or death is very complex and may be regulated by different posttranslational modifications of p53. One way or the other, p53 serves its function as guardian, as the outcome is that aberrantly proliferating cells are either permanently silenced or removed, and the body is protected.

p53 is mutated or deleted in about half of all human cancers. Families that carry a mutated p53 allele have Li-Fraumeni syndrome with an elevated risk of cancers. Mice lacking p53 are viable but a defective G_1 DNA damage checkpoint results in cancers while young. This illustrates a common theme that in many cases checkpoint components are not essential for life as long as nothing untoward occurs. Checkpoints exist primarily to deal with problems that arise during cell-cycle progression. However, the elevated cancer rates in Li-Fraumeni syndrome patients indicate that although p53 is not essential for the passage of every cell cycle, it is essential for long-term genetic stability and for maintaining a proper balance among cell proliferation, differentiation, and death during the lifetime of a mammal.

The $p19^{Arf}$ protein (in humans, the protein is smaller and so is called $p14^{Arf}$) is extremely unusual, as it is encoded in a common gene with the Cdk inhibitor

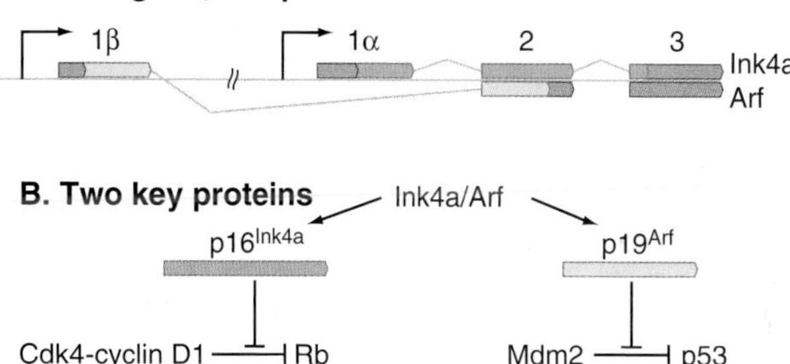

FIGURE 41.16 DUAL CONTROL OF G_1 PROGRESSION BY THE $P16^{INK4A}/P19^{ARF}$ GENE. This gene encodes two completely different proteins that are key to avoiding cancer. **A,** The intron–exon structure of the $p16^{Ink4A}/p19^{Arf}$ gene. **B,** $p16^{Ink4A}$ and $p19^{Arf}$ negatively regulate the restriction point via Rb and the DNA damage checkpoint via p53, respectively.

$p16^{Ink4a}$ (Fig. 41.16). Despite having different promoters (that respond to different stimuli), the two genes not only overlap but also share a common exon. Nevertheless, the two proteins have no common amino acid sequences because the shared exons are read in different frames in the mature mRNAs (mRNAs) for the two proteins. Thus, the $p16^{Ink4a}/p14^{Arf}$ locus encodes two vital protective factors with different jobs. It is not surprising that mutations in this key locus are found in between 25% and 70% of human cancers.

G_1 Regulation: A Matter of Life and Death

To commit to a new cycle of proliferation, cells must pass through the restriction point gate. The key to this gate is the phosphorylation of Rb by Cdks, so signals such as mitogens that activate Cdks set up a feedback loop that promotes passage of the gate. Of course, in the real world, accidents happen, and the G_1/S checkpoint and DNA damage response provide a way to block cell-cycle progression even in the presence of growth factors and mitogens. The complex G_1 regulatory networks have a potential impact on all of us. If they are disrupted by mutations or damage, the result is cancer. In fact, very few cancers have intact restriction point control networks.

ACKNOWLEDGMENTS

We thank Jiri Bartek, Ludger Hengst, Keisuke Kaji, and Marcos Malumbres for their suggestions on revisions to this chapter.

SELECTED READINGS

Blanpain C, Fuchs E. Epidermal stem cells of the skin. *Annu Rev Cell Dev Biol.* 2006;22:339-373.

Bryder D, Rossi DJ, Weissman IL. Hematopoietic stem cells: The paradigmatic tissue-specific stem cell. *Am J Pathol.* 2006;169:338-346.

Cardozo T, Pagano M. The SCF ubiquitin ligase: Insights into a molecular machine. *Nat Rev Mol Cell Biol.* 2004;5:739-751.

Chandler H, Peters G. Stressing the cell cycle in senescence and aging. *Curr Opin Cell Biol.* 2013;25:765-771.

Chen HZ, Tsai SY, Leone G. Emerging roles of E2Fs in cancer: an exit from cell cycle control. *Nat Rev Cancer.* 2009;9:785-797.

Cheung TH, Rando TA. Molecular regulation of stem cell quiescence. *Nat Rev Mol Cell Biol.* 2013;14:329-340.

Childs BG, Baker DJ, Kirkland JL, Campisi J, van Deursen JM. Senescence and apoptosis: dueling or complementary cell fates? *EMBO Rep.* 2014;15:1139-1153.

Clevers H. The intestinal crypt, a prototype stem cell compartment. *Cell.* 2013;154:274-284.

Dick FA, Rubin SM. Molecular mechanisms underlying RB protein function. *Nat Rev Mol Cell Biol.* 2013;14:297-306.

Fuchs E, Tumbar T, Gausch G. Socializing with the neighbors: Stem cells and their niche. *Cell.* 2004;116:769-778.

Goldstein M, Kastan MB. The DNA damage response: implications for tumor responses to radiation and chemotherapy. *Annu Rev Med.* 2015;66:129-143.

Jackson SP, Bartek J. The DNA-damage response in human biology and disease. *Nature.* 2009;461:1071-1078.

Johnson A, Skotheim JM. Start and the restriction point. *Curr Opin Cell Biol.* 2013;25:717-723.

Rando TA. Stem cells, ageing and the quest for immortality. *Nature.* 2006;441:1080-1086.

Sage J. The retinoblastoma tumor suppressor and stem cell biology. *Genes Dev.* 2012;26:1409-1420.

Scadden DT. The stem-cell niche as an entity of action. *Nature.* 2006;441:1075-1079.

Sherr CJ. The INK4a/ARF network in tumour suppression. *Nat Rev Mol Cell Biol.* 2001;2:731-737.

Shi X, Garry DJ. Muscle stem cells in development, regeneration and disease. *Genes Dev.* 2006;20:1692-1708.

Silverman JS, Skaar JR, Pagano M. SCF ubiquitin ligases in the maintenance of genome stability. *Trends Biochem Sci.* 2012;37:66-73.

Sperka T, Wang J, Rudolph KL. DNA damage checkpoints in stem cells, ageing and cancer. *Nat Rev Mol Cell Biol.* 2012;13:579-590.

Takahashi K, Yamanaka S. A decade of transcription factor-mediated reprogramming to pluripotency. *Nat Rev Mol Cell Biol.* 2016;17:183-193.

Veit B. Stem cell signalling networks in plants. *Plant Mol Biol.* 2006;60:793-810.

CHAPTER 42

S Phase and DNA Replication

Accurate replication of DNA, which is crucial for cellular propagation and survival, occurs during the S phase (DNA synthesis phase) of the cell cycle. This chapter begins with a brief primer on the events of replication and then discusses its regulation. Next, the chapter covers the proteins at origins of replication that ensure that each region of DNA is replicated once and only once per cell cycle. It closes by discussing how the structure of the nucleus influences replication.

DNA Replication: A Primer

One of the most exciting byproducts of the Watson-Crick model for the structure of DNA was a predicted mechanism for DNA replication. Because DNA strand pairing is determined by complementary base pairing, it was logical to propose the existence of "DNA polymerases," enzymes that would move along a single strand of DNA, recognize each base in turn, and insert the proper complementary base at the end of the growing chain. Thus, one *might* have surmised that only a single enzyme was required for DNA synthesis. In fact, DNA replication in cells involves a complex macromolecular machine.

In the chemical reaction of DNA replication, the 3′ hydroxyl at the end of the growing DNA strand makes a nucleophilic attack on the α-phosphate of the incoming nucleoside triphosphate to form a phosphodiester bond. This incorporates the nucleotide into the growing chain and releases pyrophosphate (Fig. 42.1). This reaction requires the presence of a template strand of DNA that specifies, via base pairing, which of the four nucleoside triphosphates is added to the growing molecule.

The exact site on the chromosomal DNA where replication begins is termed the **origin of bidirectional replication.** As the term *bidirectional* implies, two sets of DNA replication machinery head off in opposite directions from the origin. At the **replication fork**, one parental DNA molecule splits into two daughters (Fig. 42.2). The protein complex associated with the fork that is actively replicating the DNA is known as the **replisome**. Accumulating evidence suggests that the replisome is stationary at the fork as it "reels in" replicating DNA rather than moving along the DNA like a train on a track.

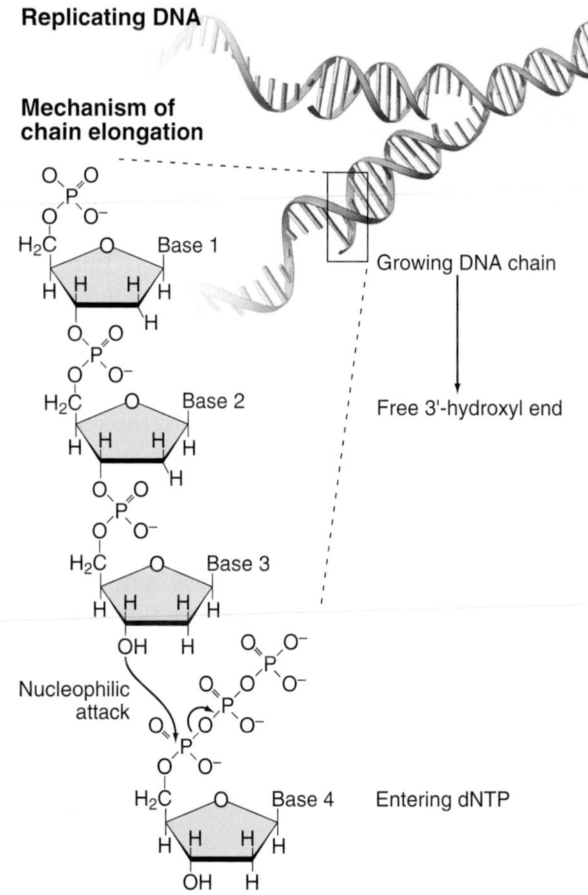

FIGURE 42.1 MECHANISM OF DNA POLYMERIZATION. A 3′ OH group at the end of a growing DNA chain makes a nucleophilic attack on the α-phosphate of a triphosphate precursor in the active site of polymerase (enzyme not shown here). dNTP, deoxynucleoside triphosphate.

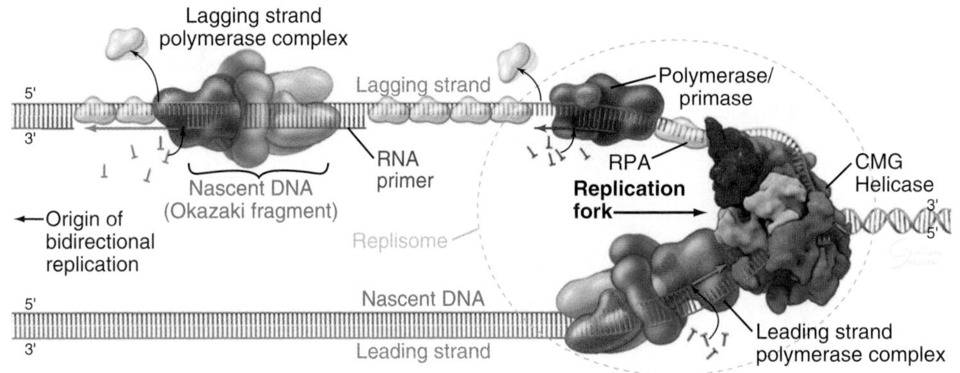

FIGURE 42.2 KEY COMPONENTS AND EVENTS AT THE REPLICATION FORK. All enzymes are closely associated in the replisome complex, but are shown separate here for clarity.

The bidirectional nature of DNA replication causes a fundamental problem, as the chemical reaction of DNA synthesis invariably proceeds in a 5′ to 3′ direction. Replication of the so-called **leading strand**, in which DNA polymerase ε moves in a 3′ to 5′ direction along the template (laying down nascent DNA in a 5′ to 3′ direction) poses no problems; the polymerase simply chases behind the replication fork (Fig. 42.2). However, the other template strand faces in the opposite direction, apparently requiring a DNA polymerase to synthesize DNA in the wrong direction as the replication fork progresses away from the origin (ie, adding nucleotides in a 3′ to 5′ direction). No DNA polymerase with this polarity has been found. Instead, this **lagging strand** replicates in a series of short segments. Every time the DNA strands have been peeled apart (unwound) by 200 nucleotides or so, probably corresponding to the unwinding of a single nucleosome, a polymerase/primase complex (Figs. 42.2 and 42.12) initiates DNA synthesis on the lagging strand, and DNA polymerase δ runs away from the fork, back toward the replication origin, again synthesizing nascent DNA in a 5′ to 3′ direction. Thus, lagging strand synthesis proceeds in bursts in a direction *opposite to the overall direction of fork movement.* Synthesis of each lagging strand fragment stops when DNA polymerase runs into the 5′ end of the previous fragment, displacing the RNA primer with which it was initiated (see later). Thus, the lagging strand is copied in a highly *discontinuous* fashion into short fragments known as **Okazaki fragments** (named after their discoverer [Fig. 42.2]). Fig. 42.12 describes the enzymes and events at the replication fork in greater detail.

Origins of Replication

Bacteria such as *Escherichia coli* replicate their circular chromosomes using two replication forks starting from a single **origin of replication** (Fig. 42.3A), but eukaryotes must use multiple origins of replication to duplicate

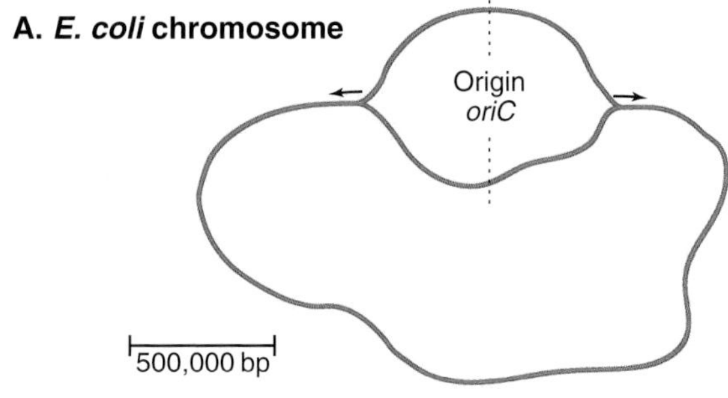

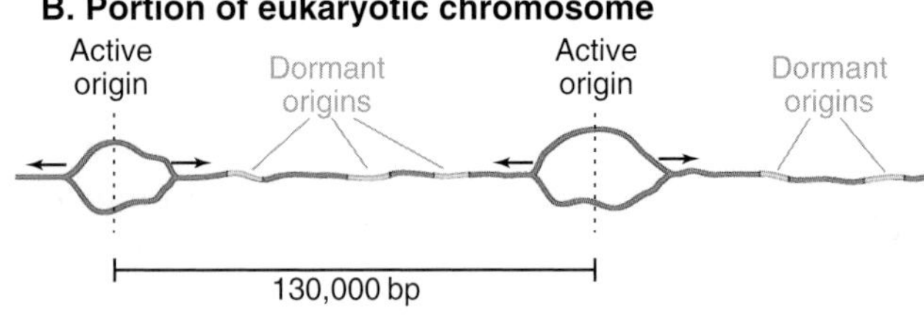

FIGURE 42.3 A, The *Escherichia coli* chromosome is a simple replicon with a single origin of replication. In cells, this chromosome has a complex, highly supercoiled structure. **B,** Eukaryotic chromosomes have multiple origins of replication, most of which remain dormant unless needed.

their large genomes during the S phase, which can be as short as a few minutes in some early embryos. These numerous origins are distributed along the chromosome: up to 600 to 700 in budding yeast and more than 100,000 in human cells. The origins are distributed so that all the DNA is replicated in the available time, and many more origins are prepared than are actually needed.

How is the "firing" of all these origins orchestrated so that each is used no more than once per S phase? Cells manage this problem by a mechanism termed *licensing,* which ensures that each segment of DNA is replicated just once per cell cycle. Replication of the origin removes the license, which cannot normally be renewed until the cell has completely traversed the cycle and has passed through mitosis.

The portion of chromosomal DNA replicated by the two bidirectional forks initiated at a single origin is termed a **replicon.** The classic replicon is the *E. coli* chromosome (which is 4×10^6 base pairs [bp] in size) with a single genetically defined **replicator** site called *oriC* (Fig. 42.3). An **initiator** protein (product of the *E. coli* DnaA gene [Fig. 42.13]) binds to this origin and either directly or indirectly promotes melting of the DNA duplex, giving the replication machinery access to two single strands of DNA. Other factors unwind the DNA, leading to the full assembly of the **replisome,** which powers a wave of DNA replication proceeding outward in both directions along the DNA (a replication "bubble") at approximately 750 to 1250 bases per second.

An average human chromosome contains approximately 150×10^6 bp of DNA. Because the replication machinery in mammals moves only approximately 20 to 40 bases per second (partly reflecting the fact that the DNA is packaged into chromatin and partly reflecting the slower speed of the eukaryotic replisome), it would take up to 2000 hours to replicate this length of DNA from a single origin. In most human cells the S phase takes approximately 10 hours. This means that at least 25 to 125 origins of replication are required to replicate an average chromosome in the allotted time. In fact, origins of replication are much more closely spaced than this. It is estimated that 30,000 to 50,000 origins of replication "fire" during each cell cycle to replicate the entire human genome, but many more dormant origins are licensed for use if necessary. These dormant origins are essential for resolving replication stress (see later).

To explain the events at origins of replication, the budding yeast *Saccharomyces cerevisiae* serves as a good example. Its DNA replication is better understood than that of any other eukaryote and although its origins of replication are specialized, the proteins that act on them are conserved across metazoa.

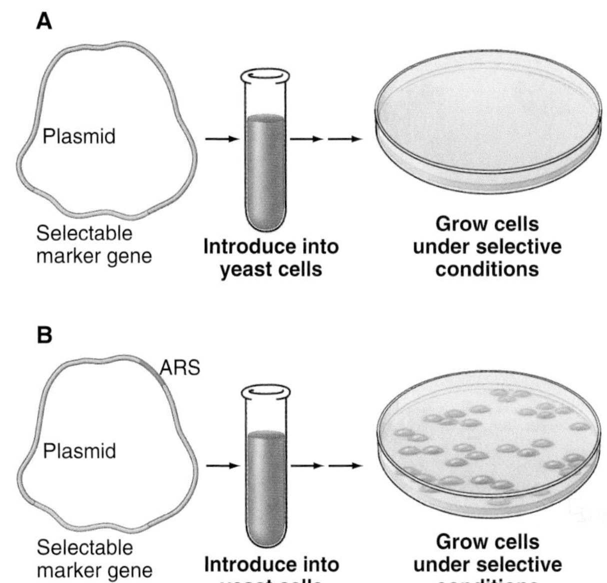

FIGURE 42.4 PLASMID ASSAY FOR IDENTIFICATION OF AN AUTONOMOUSLY REPLICATING SEQUENCE (ARS) ELEMENT (ORIGIN OF DNA REPLICATION) IN BUDDING YEAST. The plasmid has a selectable marker gene (eg, a gene required for the synthesis of an essential amino acid) plus (in panel **B**) an ARS element. This plasmid is transferred into growing yeast cells that are defective in the marker gene carried by the plasmid; these cells are then plated out on agar medium that lacks the essential amino acid. **A,** A plasmid lacking an ARS fails to replicate and is lost from the cells. These cells cannot grow into colonies on plates that lack the essential amino acid. **B,** If the plasmid contains an ARS element, it replicates along with the chromosomal DNA and is maintained in the population. These cells grow into colonies in the absence of the essential amino acid.

Replication Origins in *Saccharomyces cerevisiae*

Approximately 400 origins of replication participate in replicating the budding yeast genome. A major breakthrough in understanding DNA replication in *S. cerevisiae* was the identification of short (100 to 150 bp) segments of DNA that act as replication origins in vivo when cloned into a yeast plasmid (circular DNA molecule). These **autonomously replicating sequences** (or **ARS elements**) allow yeast plasmids to replicate in parallel with the cellular chromosomes (Fig. 42.4). ARS elements are often, although not always, bona fide replication origins in their native chromosomal context. Replication always initiates within ARS elements, but not all ARS elements act as origins of DNA replication in every cell cycle.

Yeast replication origins are spaced approximately every 30,000 bp, with a maximum separation of approximately 130,000 bp. Even this longest interval should replicate easily within the 30 minutes available during the S phase. Because the number of origins exceeds the number required to replicate the genome within the allotted time, some origins need not "fire" every cell cycle. The probability that any given origin will be used in a given cell cycle ranges from less than 0.2 to greater than 0.9. It is important to note that replication by a passing fork coming from an adjacent origin inactivates dormant origins. This prevents re-replication of genomic regions during the cell cycle.

The ARS element does two things to establish an origin of replication. First, its conserved sequences act as binding sites for a protein complex that marks it as a potential origin. Second, it has nearby sequences that are readily induced to unwind by separating the base-paired strands.

Budding yeast ARS elements share a common DNA sequence motif called the **ARS core consensus sequence:** 5′-(A/T)TTTAT(A/G)TTT(A/T)-3′ (Fig. 42.5). Single base mutations at several locations within this sequence completely inactivate ARS activity. Other, less well-conserved DNA sequences also contribute to the activity of the ARS as a replication origin. One of these, termed B1, together with the ARS core, forms the binding site for a complex of six proteins termed the **origin**

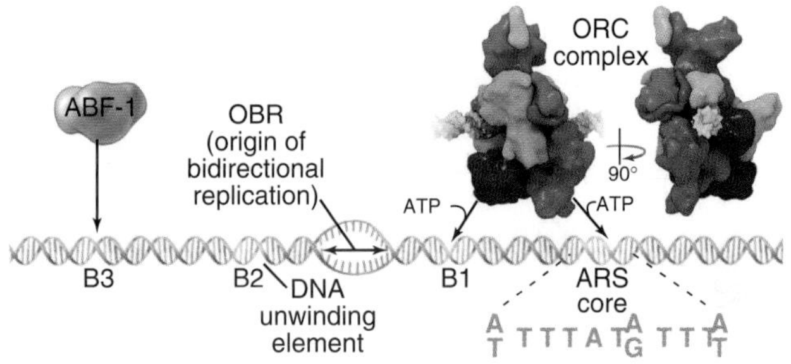

FIGURE 42.5 ORGANIZATION OF THE AUTONOMOUSLY REPLICATING SEQUENCE (ARS)-1 ELEMENT. The ORC (origin recognition complex) binds to the ARS core sequence plus element B1. B2 is a sequence that can readily be induced to unwind. The OBR (origin of bidirectional replication) is the site where DNA synthesis actually begins. B3 is a binding site for an auxiliary factor called ABF-1 that is both a transcriptional activator and an activator of the ARS element. ATP, adenosine triphosphate. (For reference, see Protein Data Bank [PDB; www.rcsb.org] file 4X6C.)

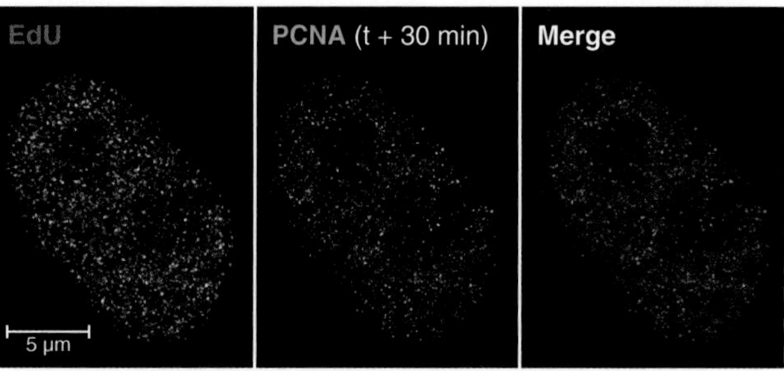

FIGURE 42.6 SUPERRESOLUTION VIEW OF ACTIVE REPLICONS IN A HELA (HENRIETTA LACKS) CELL. EdU *(red)* was used to label sites of active replication for 15 minutes followed by washing out for 15 minutes. Then antibody recognizing proliferating cell nuclear antigen (PCNA) was used to stain all active replicons in the cell. Thousands of replicons can be seen, and it is clear that in the 30 minutes between the labelling of the red and green channels, many new origins of replication have been activated. Scale bar = 5 μm. OMX superresolution. (Microscopy by Vadim Chagin and Cristina Cardoso, Technical University of Darmstadt, Germany.)

recognition complex (ORC [see later section]). The DNA unwinding element is another short sequence (B2) located a bit further along the DNA. DNA synthesis begins at an origin of bidirectional replication midway between the ORC binding site and the DNA unwinding element.

ORC was identified by its ability to bind the 11-bp ARS core sequence (Fig. 42.5). This binding has two noteworthy features. First, the subunits of the ORC complex are AAA adenosine triphosphatases (ATPases; see Box 36.1) and adenosine triphosphate (ATP) hydrolysis is required for ORC to bind ARS DNA. Second, in yeast, the ORC complex remains bound to the origins of replication across the entire cell cycle. Thus, something other than the presence of ORC must regulate the periodic activation of origins in the S phase (Fig. 42.14). In some metazoan cells, ORC behavior is more complex—for example, the largest subunit, Orc1, is degraded during part of the cell cycle.

ARS elements often contain binding sites for other sequence-specific DNA binding proteins, including transcription factors. For example, a transcription factor called ARS-binding factor 1 (ABF-1) binds to the B3 sequence within the ARS1 element (Fig. 42.5). Deletion of the ABF-1 binding site only slightly reduces the ability of ARS1 to act as a replication origin in vivo and other transcription factors can substitute for ABF-1.

In addition to their role in DNA replication, several ORC subunits also regulate heterochromatin formation and transcription (see Chapters 8 and 10). This crosstalk between the machinery used for transcription and DNA replication may explain why regions of chromosomes with actively transcribed genes typically replicate early in the S phase (see the next section below). In some metazoan cells, the Orc6 subunit also functions during cytokinesis, apparently via interactions with cytoskeletal filament proteins independent of its role at replication origins.

Replication Origins in Mammalian Cells

Less is known about the structure and function of mammalian origins of DNA replication than about ARS elements in budding yeast. Attempts to develop a mammalian equivalent to the yeast ARS assay had few successes. Over the years approximately 30 metazoan origins of replication were identified by painstaking methods, but more recently, high throughput methods, including the sequencing of short nascent strands of replicating DNA have identified thousands of mammalian replication origins. Superresolution microscopy can now resolve the thousands of replicons active at any one time in a human cell (Fig. 42.6).

Overall, the chromatin landscape influences mammalian origins. Important factors include DNA sequence, DNA modifications, chromatin structure, and nuclear organization. Although high throughput sequencing revealed no sequences as specific as the yeast ARS elements, many origins are associated with short regions of G-rich DNA. These G-rich regions can form specialized structures that have fewer nucleosomes and are therefore more accessible to the DNA replication machinery. Indeed, replication origins tend to be located near gene promoters, where the density of nucleosomes is low. Such actively transcribed regions of the genome are packaged into euchromatin with modified histones and tend to replicate earlier during S phase than regions of the genome that are not transcribed.

A typical mammalian replicon encompasses about 130 kb and contains approximately four or five licensed origins-only, one of which is typically used. The first replicon to be mapped lies just downstream of the hamster gene for dihydrofolate reductase, an enzyme that is essential for biosynthesis of thymidine. Investigators selected cells with this chromosomal region amplified as hundreds or even thousands of copies (Fig. 42.7) and looked for the first regions of the amplified DNA to

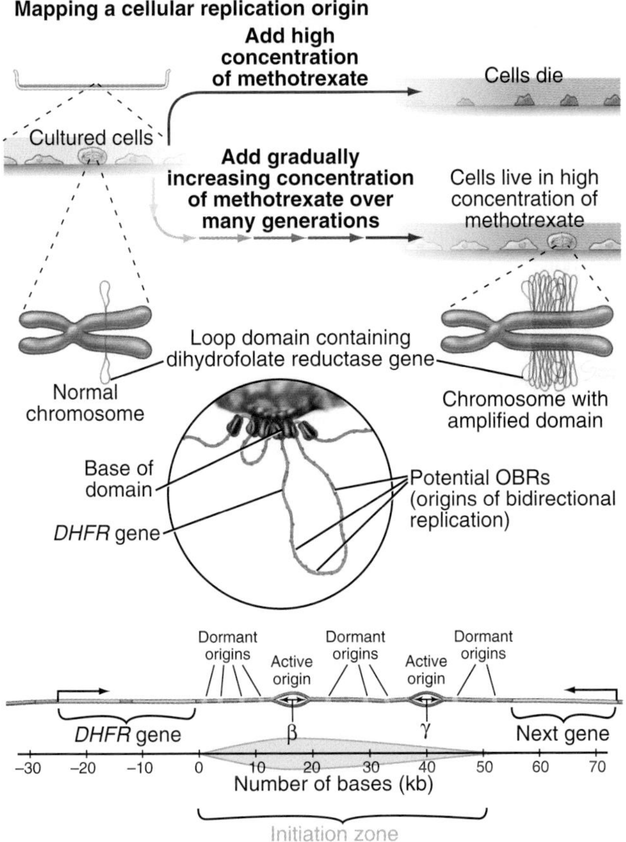

FIGURE 42.7 MAP OF DNA REPLICATION LANDSCAPE NEAR THE DIHYDROFOLATE REDUCTASE (DHFR) GENE. Normal cells are killed by exposure to methotrexate, but it is possible to select resistant cell lines by growing them in progressively increasing concentrations of the drug, selecting at each stage for cells that survive. Use of this procedure on hamster cells resulted in a cell line that contains approximately 1000 copies of a 230,000-bp chromosomal domain containing the *DHFR* gene. This region of DNA is replicated using origins found within a 55,000-bp region adjacent to the *DHFR* gene. Most initiation occurs at two specific origins, called β and γ, but other dormant origins are scattered throughout the entire 55,000-bp region.

replicate. They found that DNA replication can initiate with low efficiency at a number of sites distributed across a broad region of approximately 55,000 bp. Two of these sites, termed Ori-β and Ori-γ (Fig. 42.7) account for approximately 20% of all initiation in the region.

The emerging view is that the replication machinery is highly conserved between budding yeasts and vertebrates but that the location of replication origins is more flexible in vertebrates. This might reflect both the differing chromatin landscapes across metazoan genomes (see Chapter 8) and the diverse range of cell cycles required to make a complex metazoan.

Origin Licensing and Assembly of the Prereplication Complex

To preserve the integrity of the genome, each origin of replication must "fire" no more than once per cell cycle. We now have a reasonable understanding of the various solutions to this problem that have been reached by yeasts and vertebrates.

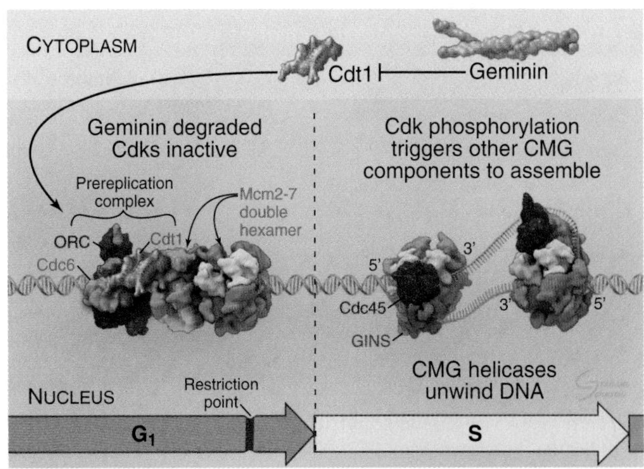

FIGURE 42.8 COMPONENTS OF THE PREREPLICATION COMPLEX BEFORE AND AFTER THE INITIATION OF DNA REPLICATION. CMG, Cdc45, Mcm2–7, and GINS (Go-Ichi-Ni-San). (For reference, see PDB files 3JA8 [for CMG helicase with Cdk45 bound], 3JC5 [for Mcm2–7], and 5F9R [for ORC].)

Recall that yeast ORC is stably bound to replication origins throughout the cell cycle. However, ORC is not the trigger for DNA replication. Rather, it acts as a "landing pad" for assembly of a **prereplication complex** of other proteins that initiates DNA replication.

Formation of the prereplication complex "licenses" each origin for a single initiation event as follows. During late anaphase or very early G_1 phase, several proteins, including **Cdc6** and **Cdt1,** bind to the ORC complex at origins of replication (Table 42.1). Before the onset of the S phase ORC-Cdc6-Cdt1 loads a double hexamer of **Mcm proteins** (*mini*chromosome *m*aintenance) on the origin DNA (Fig. 42.8). Mcm proteins were identified in a screen for genes of budding yeast that are required for the stability of small artificial chromosomes. Six of these *Mcm* genes encode a structurally related group of proteins, termed Mcm2–7, that is required for DNA replication. ORC-Cdc6–Cdt1 uses ATP hydrolysis to crack open two hexameric Mcm2-7 rings so they can wrap around DNA in a head-to-head orientation.

Replication starts when each Mcm2–7 hexamer associates with the GINS (Go-Ichi-Ni-San; 5-1-2-3 in Japanese) complex and the Cdc45 protein to form the replicative **CMG helicase** (*C*dc45-*M*cm-*G*INS). This helicase forms the core of the replisome and uses ATP hydrolysis to separate DNA strands (Fig. 42.8). As the two CMG helicases move away from each other, the origin is converted from a licensed to an unlicensed state. Because origin licensing by Mcm2–7 loading can only occur prior to entry into S phase, origins only fire once per cell cycle.

In mammals, licensing occurs in the early G_1 phase before passage of the restriction point (see Chapter 41). At the exit from mitosis, destruction of cyclins and synthesis of inhibitory proteins inactivates Cdks. This creates a window of time between anaphase and the restriction

TABLE 42.1 Biochemical Activities Required for Replication of DNA in Eukaryotes

Activity	Name of Protein
Origin recognition	ORC (origin *r*ecognition complex; five of six subunits are AAA ATPases)
Prereplication complex	Cdc6 (recruits Mcm2–7) Cdt1 (recruits Mcm2–7)
Origin activation	Cdc7-Dbf4 Cdk (in human, Cdk2-cyclin A) phosphorylation of Sld2/RecQ4 (yeast/human names) and Sld3/treslin recruits Dbp11/TopBP1 and initiates CMG assembly to trigger the actual start of replication
DNA unwinding (helicase)	CMG helicase is made up of: • Mcm2–7 (assemble double hexamer before other CMG components) • Cdc45 and the GINS complex (4 proteins)
Stabilization of single-stranded DNA	RPA (binds single-stranded DNA)
Replicative polymerases	DNA polymerase α/primase (no editing function) starts synthesis of the leading strand and each Okazaki fragment DNA polymerase δ replicates the lagging strand DNA polymerase ε replicates the leading strand (both δ and ε have 3′–5′ exonuclease editing capability)
Processivity factor	PCNA (ring-shaped clamp that slides along the DNA. Keeps polymerases δ and ε attached to the template strand so that they make longer chains; coordination of cell-cycle control and replication; role in repair)
PCNA loader	RF-C (Binds primer: template junction. AAA ATPase. Loading factor for PCNA, important for polymerase switch)
Closing Factors	
Removal of RNA primer	Fen1 5′–3′ exonuclease Dna2 helicase and endonuclease RNase H
Ligation of discontinuous DNA fragments	DNA ligase I
Releasing superhelical tension	DNA topoisomerase I
Disentangling daughter strands	DNA topoisomerase II

ATPase, adenosine triphosphatase; CMG, Cdc45, Mcm2–7, and GINS; GINS, Go-Ichi-Ni-San; PCNA, proliferating cell nuclear antigen; RPA, replication protein A.

point for licensing replication origins (Fig. 42.9). Mammalian cells pass the restriction point when the levels of Cdk2–cyclin E, and subsequently, Cdk2–cyclin A rise (see Fig. 40.17). This prevents the relicensing of replication origins until after the next mitosis. In yeasts, the single Cdk that is complexed with B-type cyclins inhibits origin licensing. Experimental inactivation of Cdk^{Cdc2} during the G_2 phase in the fission yeast *Schizosaccharomyces pombe* demonstrated the importance of Cdk activation: Cells lacking Cdk^{Cdc2} activity loaded Mcm2–7 onto already replicated DNA and then carried out further rounds of "illegal" DNA replication without division. Cdk regulation of origin licensing during G_1 phase actually provides one explanation for oncogenic stress (see Chapter 41), in which inappropriate activation of oncogenes leads to cell-cycle arrest followed by death or senescence. Inappropriate oncogene activation can lead to premature stabilization of cyclins D and cyclin E, resulting in premature activation of Cdks. This, in turn, can lead to an insufficient number of replication origins being licensed, thereby inhibiting the cell's ability to deal with replication stress (see later) and its associated DNA damage.

Different cell types use various combinations of several different pathways to downregulate the activity of the licensing system once cells enter S phase. In metazoans, the most important pathways reduce Cdt1 activity by degradation in S and G_2 phases. After the cell passes the restriction point, Cdks phosphorylate Cdt1. This promotes its ubiquitylation by the SCF^{Skp2} and degradation. In addition, the key replisome component proliferating cell nuclear antigen (PCNA) recruits another ubiquitin ligase that causes Cdt1 degradation during S phase. In some cell types, Cdks phosphorylate Cdc6 and Orc1, leading to their ubiquitylation and degradation.

Vertebrates use a protein called **geminin** as an alternative regulator of origin licensing by Cdt1 (Fig. 42.8). Geminin binds to Cdt1 and prevents it from loading Mcm proteins onto DNA. The anaphase-promoting complex/cyclosome (APC/C; see Fig. 40.15) triggers degradation of geminin, keeping its concentration very low from anaphase through the late G_1 phase, allowing prereplication complexes to assemble. Accumulation of geminin starting in the S phase inhibits the assembly of new prereplication complexes until after the next mitosis. Yeasts control origin licensing without geminin.

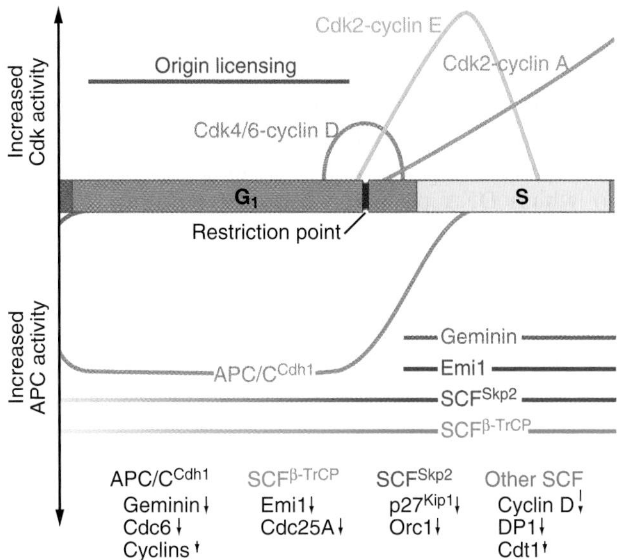

FIGURE 42.9 PROTEIN DEGRADATION REGULATES DNA REPLICATION. Degradation of geminin, Cdc6, Cdc25A, and cyclins during the G_1 phase keeps Cdk activity low and allows prereplication complex formation. Degradation of p27Kip1 and inactivation of the APC/C^{Cdh1} by Emi1 allows the activation of Cdks to levels sufficient for the initiation of the S phase. Once cells enter the S phase, the G1/S regulatory machinery (cyclins D and E and the E2F cofactor DP1) is degraded. Degradation of Cdt1 and accumulation of geminin block reassembly of prereplication complexes. APC, anaphase-promoting complex; SCF, Skp2 (S-phase kinase-associated protein), cullin, and F-box proteins.

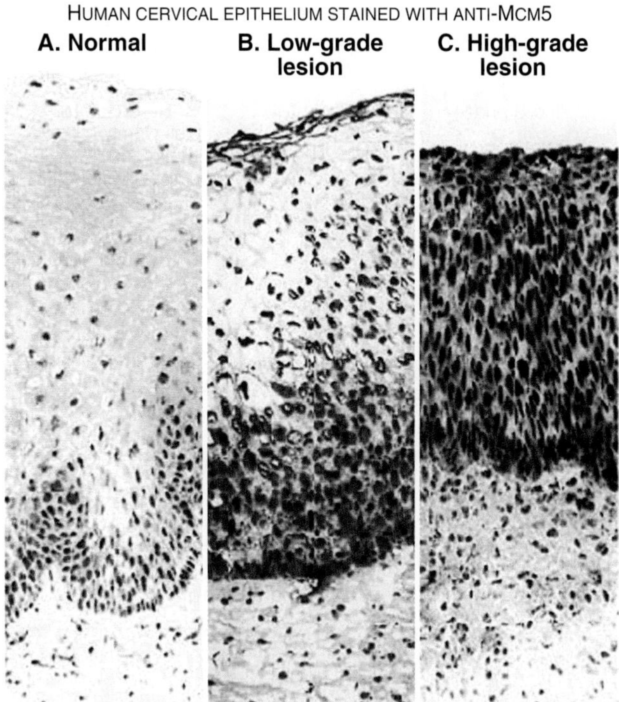

FIGURE 42.10 SECTIONS OF HUMAN CERVIX STAINED WITH ANTIBODIES TO MCM5. A, Normal G_0 cells in this stratified epithelium lack Mcm5 and other replication proteins. **B–C,** Cancer cells express Mcm5 at higher levels as they become more malignant. Bound antibodies were detected with peroxidase coupled to a secondary antibody (brown reaction product) and lightly counterstained with hematoxylin (blue). (Modified from Williams GH, Romanowski P, Morris L, et al. Improved cervical smear assessment using antibodies against proteins that regulate DNA replication. *Proc Natl Acad Sci U S A.* 1998;95:14932–14937.)

A third way to regulate origin licensing is to sequester molecules required to assemble the prereplication complex in the cytoplasm following the onset of S phase. This was first suggested by studies of DNA replication in *Xenopus* egg extracts (see Fig. 40.8) in which intact nuclei replicate their DNA only once but re-replicate if the nuclear membrane is disrupted. In different cell types, nuclear exclusion of ORC, Cdc6, Cdt1, or Mcm2-7 all contribute to limiting licensing during S and G_2 phases.

Components of the prereplication complex are absent from nondividing differentiated cells. In fact, detection of these proteins with antibodies in cells from cervical smears can be used as a sensitive method for the early detection of cancer cells (Fig. 42.10).

Signals That Start Replication

A classic experiment (Fig. 42.11) demonstrated that (a) a cytoplasmic inducer triggers the transition into the S phase and (b) this inducer triggers DNA replication in a G_1 nucleus but not in a G_2 nucleus. The inducer is a combination of protein kinases, including Cdk-cyclin pairs, as well as a specialized kinase, Cdc7-Dbf4. In mammals, Cdk2-cyclin E activity peaks at the G_1/S transition (Fig. 42.9) and phosphorylates pRb, thereby opening the restriction point "gate." This allows the E2F/DP dimer to function as a transcription factor and stimulate the transcription of genes involved in DNA replication (see Fig. 41.9). In addition to cyclin E itself, E2F drives expression of cyclin A, Cdc25A, enzymes required for synthesis of DNA precursors (dihydrofolate reductase, thymidine kinase, and thymidylate synthase), origin-binding proteins (Cdc6, Orc1, Cdt1 and its inhibitor geminin), and two components of the replication machinery (DNA polymerase α and PCNA; see Fig. 42.9).

In the S phase, the SCF^{Skp2} ubiquitin ligase complex targets the Cdk inhibitor p27^{Kip1} for destruction by proteasomes (see Fig. 40.16). SCF gets its name from three of its components: Skp2, cullin, and F-box proteins (see Fig. 40.16). Skp2, which is short for "S-phase kinase-associated protein," was identified in a complex with Cdk2-cyclin A. SCF recognizes and ubiquitylates many of its substrates only after they are phosphorylated by Cdk2-cyclin A. Cdk2-cyclin A also targets E2F/DP and cyclin E for degradation when cells enter the S phase (Fig. 42.9).

For DNA replication to start, the paired strands of the double helix must be separated so DNA polymerase can recognize the bases and begin synthesizing the daughter strand. The multi-subunit CMG DNA helicase uses ATP hydrolysis to peel apart the paired strands of the DNA double helix. Prior to S phase each origin of replication

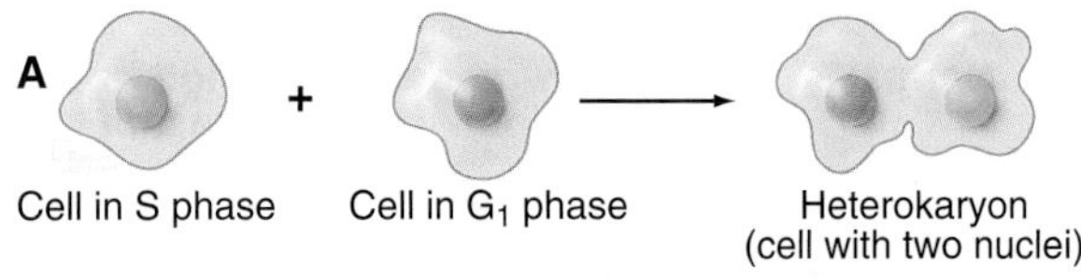

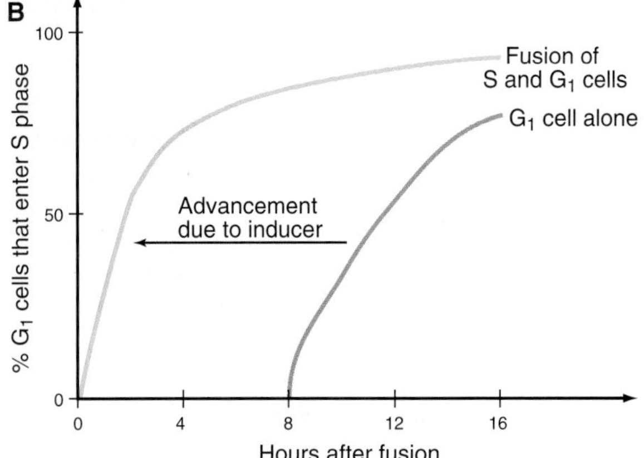

FIGURE 42.11 CELL FUSION EXPERIMENT REVEALS THE EXISTENCE OF A POSITIVE INDUCER OF THE S PHASE. **A,** Synchronized cells in different stages of the cycle were fused to yield two nuclei in a single cytoplasm. **B,** If the fusion involved G_1 and S cells, the G_1 nucleus was induced to enter the S phase sooner than expected. If the fusion involved S and G_2 cells, the G_2 nucleus failed to rereplicate its DNA (not shown). (Modified from Rao PN, Johnson RT: Mammalian cell fusion: studies on the regulation of DNA synthesis and mitosis. *Nature*. 1970;225:159–164.)

is licensed by assembly of a double hexamer of Mcm2-7 complexes. Cdc7 kinase with its associated regulatory subunit Dbf4 phosphorylates the Mcm complex. Next, Cdks phosphorylate other proteins that recruit GINS and Cdc45 to the phosphorylated Mcm2-7 hexamer, thereby producing the CMG helicase (Table 41.1 and Fig. 42.12A). Subsequent activation of the helicase initiates replication.

Mechanism of DNA Synthesis

Replication initiates when the activated CMG helicase starts to separate the DNA strands, which move outward in both directions as the replisome assembles at the origin of bidirectional replication. The newly unwound DNA binds the single-strand DNA-binding protein, **RPA** (replication protein A), ensuring that the separated strands do not base-pair with one another again (Fig. 42.12B). Table 42.1 describes several other proteins that also bind at this time. Box 42.1 provides an introduction to DNA replication in *E. coli*.

The separated DNA strands are ready for replication, but DNA synthesis always involves addition of an incoming nucleoside triphosphate to a free 3′ OH group at the terminus of a *preexisting* nascent polynucleotide (Fig. 42.1). In the absence of a nascent DNA chain, how does DNA polymerase get started? This problem is solved on the lagging strand by a DNA-dependent RNA polymerase called a primase, which, like other RNA polymerases, *can* initiate synthesis de novo without the need for a 3′OH group. In eukaryotes, all DNA chains are started by a complex of DNA polymerase α and a primase subunit, collectively known as **Pol α/Primase.** Primase synthesizes an RNA chain of approximately seven nucleotides to which DNA polymerase α adds another 20 to 25 nucleotides of so-called initiator DNA (iDNA) (Fig. 42.12D–E). These initiating reactions are potentially risky, because DNA polymerase α lacks proofreading ability. To avoid errors created by mismatching of an incoming base, the RNA primer and most or all the iDNA laid down by Pol α/Primase are subsequently replaced.

Once Pol α/Primase has done its job, two further essential factors act to complete replisome assembly. A pentameric protein complex called **replication factor C (RFC)** binds the 3′ end of the iDNA. RFC uses energy from ATP hydrolysis to load the trimeric protein **PCNA** onto the DNA (Fig. 42.12E–G). The PCNA trimer is doughnut-shaped, and when the DNA is inserted into its central hole, it is topologically locked onto the DNA. RFC binding and PCNA loading displace Pol α/Primase from the DNA, and PCNA then recruits **DNA polymerases δ** and ε to the DNA. With PCNA acting as a sliding platform, DNA is reeled through the replisome and polymerase ε synthesizes DNA continuously on the leading strand (Fig. 42.12G).

On the lagging strand, polymerase δ synthesizes DNA in bursts of approximately 250 bp, each initiated by Pol α/Primase and roughly correlating with the passage of one nucleosome by the replication fork. The replisome contains a chromatin remodeling activity that removes nucleosomes from the DNA ahead of the fork and replaces them after the fork passes.

Both polymerases δ and ε have associated exonuclease activities. This enables them to proofread the newly synthesized DNA and correct most mistakes that they have made. The combination of selecting the correct nucleotides and efficient correction of mistakes may explain the amazing fidelity of DNA replication, with typically only one error per 10^9 bp polymerized.

Locally, the final steps of DNA replication are removal of the RNA primer (and probably iDNA) and ligation of adjacent stretches of newly synthesized DNA. Removal of the primer can be accomplished in two ways (Fig. 42.12H). On one hand, an RNA exonuclease called **RNase H** can chew in from the 5′ end of the primer. However, this enzyme cannot remove the last ribonucleotide that is joined to iDNA. That requires the nucleases **Fen1 (Flap endonuclease)** or **Dna2**. Fen1 requires a helicase to peel the RNA (and possibly the iDNA) away from the template, creating a sort of flap. Fen1 then cleaves at the junction where the flap is anchored to the DNA template, removing the oligomer of unwanted nucleotides in one step. Alternatively, the Dna2 nuclease can do the whole job itself because it also has helicase

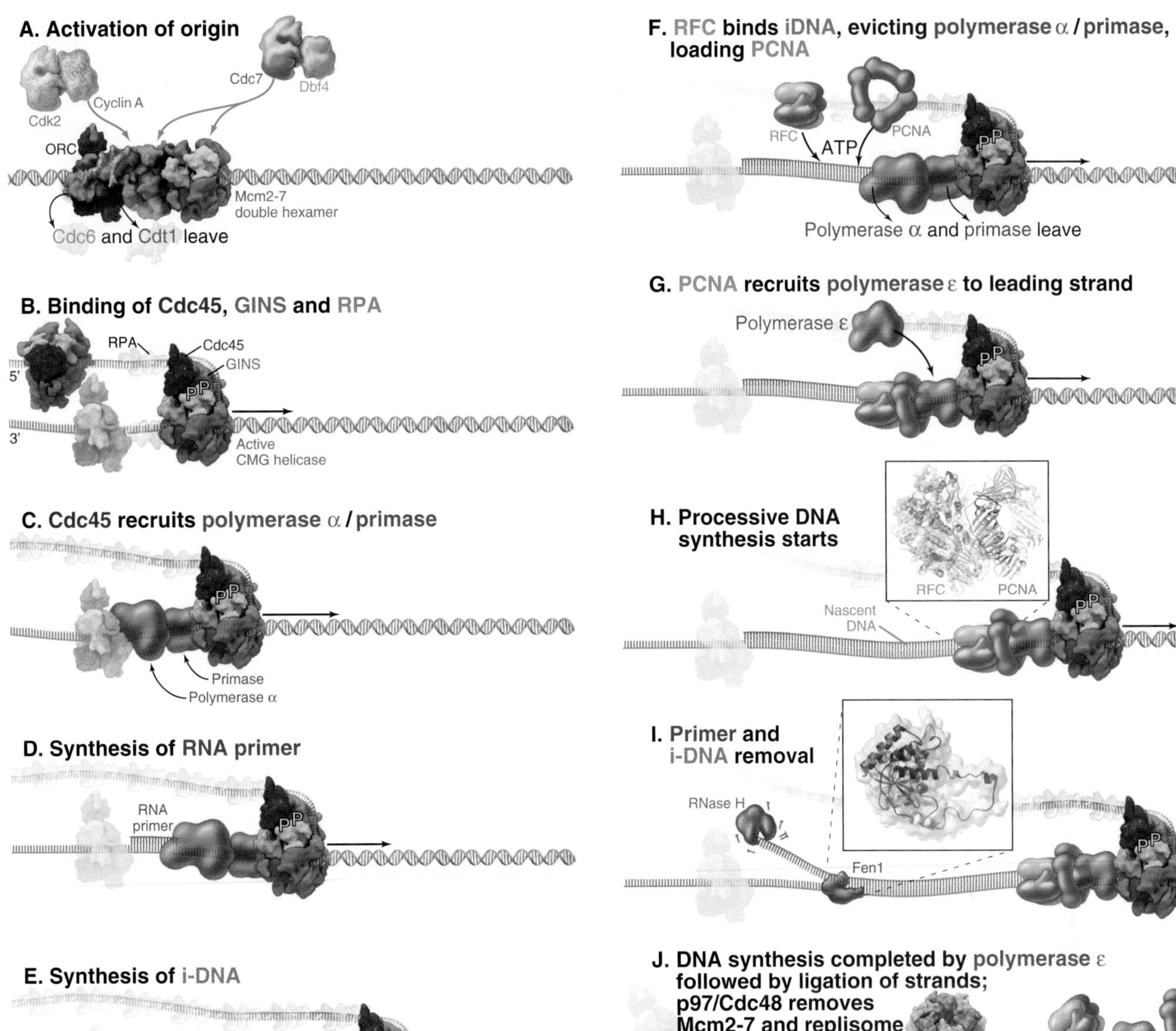

FIGURE 42.12 MAIN EVENTS OF DNA REPLICATION ON THE LEADING STRAND. For a more detailed description, including events on the lagging strand, see the text. (For reference, see PDB files 1A76 [for Fen1], 1SXJ [for RFC/PCNA], 2ZXX [for Cdt1], and 3CF2 [for p97/Cdc48].)

activity. Most of these maturation/processing enzymes are recruited by interacting with the PCNA trimer. The individual interactions are transient while PCNA remains as a stable loading platform.

Following removal of initiator RNA, the Pol δ/PCNA complex extends the upstream nascent chain until it runs into the downstream 5′ end created by Fen1. DNA ligase I then joins the two stretches of DNA together (Fig. 42.12I).

When the DNA is ligated, the replisome must be disassembled. This is an active process triggered by ubiquitylation of the Mcm7 subunit of the CMG helicase. The AAA-ATPase p97/Cdc48 (which also extracts proteins from the endoplasmic reticulum in the endoplasmic reticulum-associated degradation [ERAD] pathway; see Chapter 23) recognizes this ubiquitin and then actively separates the Mcm2-7 hexamer from Cdc45 and GINS, causing the replisome to fall apart.

Higher-Order Organization of DNA Replication in the Nucleus

The term *S phase* gives the impression that all DNA replicates more or less synchronously, but this is far from true. At any given time during the S phase, only 10% to 15% of the replicons actively synthesize DNA. Some replicate early, others late. This pattern of replication is not random; some segments of DNA consistently

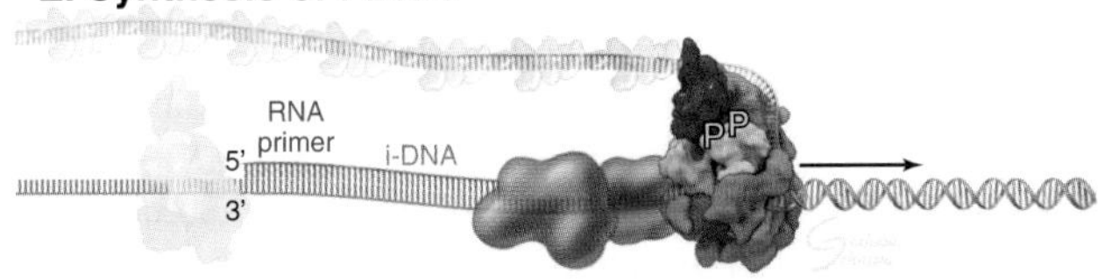

BOX 42.1 DNA Replication in *Escherichia coli*

The DNA replication system of *Escherichia coli* has been reconstituted entirely from purified components. Analysis of this system reveals many similarities with eukaryotic replication, indicating that this process is highly conserved. *E. coli* DNA replication can be subdivided into three phases: initiation, elongation, and termination. Thus far, at least 28 polypeptides are known to be involved.

Initiation

E. coli chromosomal DNA replication initiates within a 245-bp region, termed *oriC*. This region contains four 9-bp binding sites for the *E. coli* initiator protein, DnaA. Nearby are three repeats of a 13-bp A/T-rich sequence. *oriC* also contains specific binding sites for two small histone-like proteins called HU and IHF. Replication is initiated with the cooperative binding of 10 to 20 DnaA monomers to their specific binding sites (Fig. 42.13). To be active, these monomers must each have bound ATP. Binding of DnaA permits unwinding of the DNA at the 13-bp repeats, in a reaction that requires the histone-like proteins. Next, DnaC binds to DnaB and escorts it to the unwound DNA. DnaB is the key helicase that will drive DNA replication by unwinding the double helix, but it binds DNA poorly on its own in the absence of its DnaC escort. Once DnaB has docked onto the DNA, DnaC is released, and the helicase can then start to unwind the DNA, provided that ATP, SSB, and DNA gyrase are present. SSB is a single-stranded DNA binding protein that stabilizes the unwound DNA, and DNA gyrase is a topoisomerase (see Chapter 8) that removes the twist that is generated when the two strands of the double helix are separated.

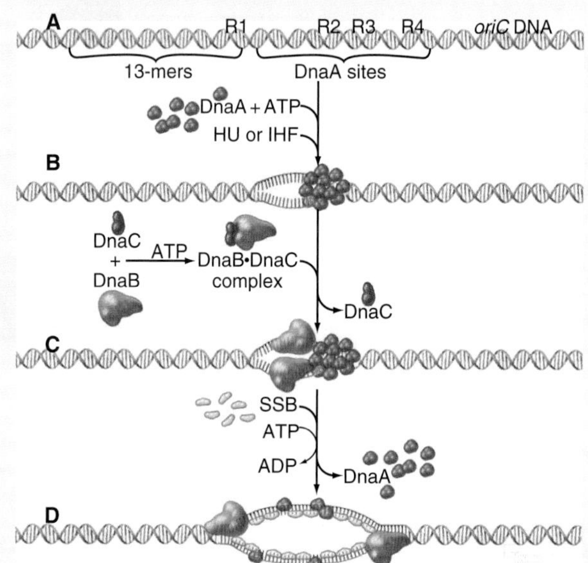

FIGURE 42.13 FACTORS INVOLVED IN THE INITIATION OF DNA REPLICATION IN *ESCHERICHIA COLI*. A, DNA sequences at OriC. **B,** Unwinding of the origin. **C,** Binding of helicase. **D,** The template, now ready for binding of DNA polymerase. ADP, adenosine diphosphate; ATP, adenosine triphosphate; SSB, single-stranded DNA binding protein. (Modified from Baker TA, Wickner SH. Genetics and enzymology of DNA replication in *Escherichia coli*. *Annu Rev Genet*. 1992;26:447–477.)

Elongation

As in eukaryotes, *E. coli* DNA replication involves a leading strand, with the daughter DNA synthesized as a single continuous molecule, as well as a lagging strand, with the DNA synthesized as discontinuous Okazaki fragments. All daughter strands are started by an RNA primase that deposits primers of 11 ± 1 nucleotides. The enzyme that actually synthesizes the DNA is the polymerase III holoenzyme, which has at least 10 subunits. This contains polymerase and proofreading subunits and is held to the DNA by a doughnut-like "sliding clamp" (β). The β is loaded onto the DNA by a pentameric complex in a process that requires ATP. The parallel with PCNA and RFC in eukaryotes is striking. Activities specific for the lagging strand include RNase H, which removes the RNA primers; DNA polymerase I, which fills in the gaps left behind by primer removal; and DNA ligase, which links the Okazaki fragments together. DNA replication in *E. coli* is significantly faster than it is in eukaryotes, with the fork moving at a rate of approximately 1000 bp per second. This higher speed is presumed to be at least partially attributable to the absence of nucleosomes on the bacterial chromosome, but the eukaryotic enzymes may also be intrinsically slower.

Termination

A specialized termination zone is found on the circular *E. coli* chromosome opposite *oriC*. This zone contains binding sites called *ter* sites, to which the *ter* binding protein binds. This protein appears to block the movement of DNA helicases, such as DnaB, thereby stalling the DNA replication fork. Following termination of replication, a specialized topoisomerase, the product of the *parC* and *parE* genes, is required to separate the daughter chromosomes from one another.

replicate early in the S phase, whereas others consistently replicate late in the S phase.

Up to 1000 sites of replication, called **replication foci** are active at any one time during the S phase in a mammalian cell nucleus (Figs. 42.6 and 42.14B–C and E–F). Given that each of these replication foci is active for only about 30 minutes out of the 8- to 10-hour S phase, a cell will replicate DNA at approximately 10,000 of these foci. There are roughly 60,000 origins in a mammalian cell, so each replication focus represents five or six replication origins that are activated coordinately. The replication foci correspond to structural domains of chromosomes that are now called TADs (*t*opologically *a*ssociating *d*omains) (see Fig. 8.13). The function of each TAD, measured by replication timing, chromatin composition and transcriptional output, can change

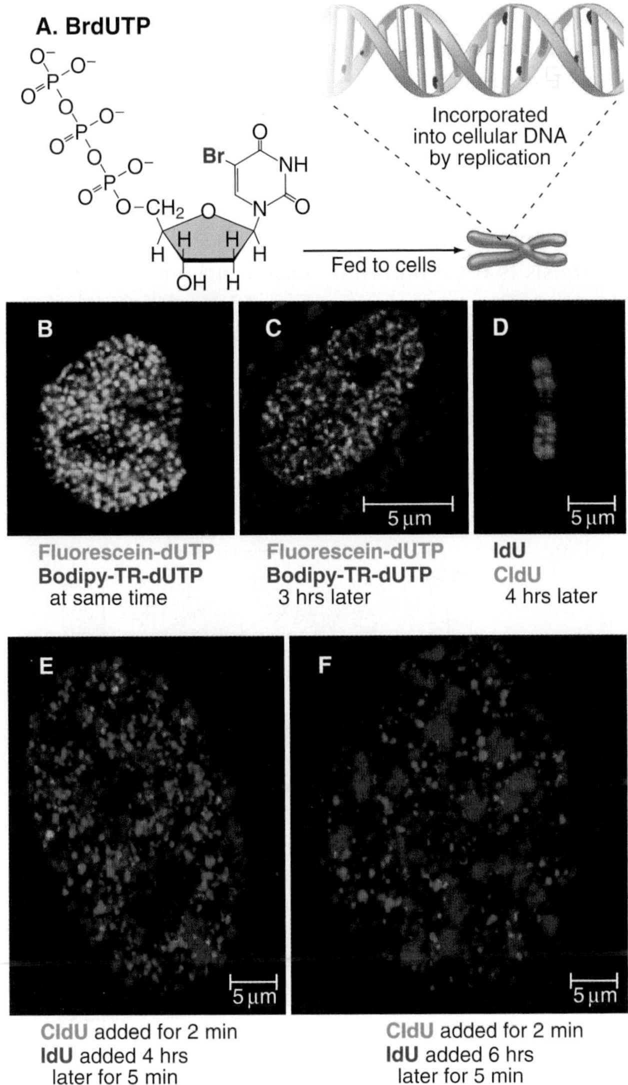

FIGURE 42.14 VISUALIZATION OF DNA REPLICATION WITHIN THE NUCLEUS. A, BrdUTP is introduced into DNA in place of dTTP to label newly replicated DNA. The incorporated BrdU (bromodeoxyuridine) molecules are detected with fluorescent antibodies. **B,** In a related approach, *green*-dUTP (deoxyuridine triphosphate) and *red*-dUTP, when added together, show the many sites of DNA replication in a cell nucleus. Because both UTP (uridine triphosphate) analogs are incorporated simultaneously into the DNA, the sites of replication appear *yellow.* **C,** *Green*-dUTP is followed by *red*-dUTP added 3 hours later. The later sites of DNA replication show very little overlap with the earlier sites. **D,** Mitotic chromosome from a cell that was labeled early in the S phase with IdU *(green),* and then 4 hours later with CldU *(red).* The late-replicating and early replicating regions of the chromosome are segregated into discrete bands. **E,** CldU *(green)* added early in the S phase and IdU *(red)* added 4 hours later show relatively little overlap. **F,** CldU *(green)* added early in the S phase and IdU *(red)* added 6 hours later show no overlap. The *large red blocks* of labeling seen with the IdU are characteristic of the pattern of replicating heterochromatin seen late in the S phase. Bodipy-TR-dUTP, a *red* fluorescent form of dUTP; BrdUTP, bromodeoxyuridine triphosphate; CldU, chlorine-dUTP; Fluorescein-dUTP, a *green* fluorescent form of dUTP; IdU, iodine-dUTP. All are used in place of dTTP (deoxythymidine triphosphate) in DNA synthesis. (**B–C,** Courtesy P.R. Cook, University of Oxford, United Kingdom. From Manders EMM, Kimura H, Cook PR. Direct imaging of DNA in living cells reveals the dynamics of chromosome formation. *J Cell Biol.* 1999;144:813–821, copyright The Rockefeller University Press. **D,** Courtesy A.I. Lamond, University of Dundee, United Kingdom. From Ferreira J, Paolella G, Ramos C, et al. Spatial organization of large-scale chromatin domains in the nucleus: a magnified view of single chromosome territories. *J Cell Biol.* 1997;139:1597–1610, copyright The Rockefeller University Press. **E–F,** From Ma H, Samarabandu J, Devdhar RS, et al. Spatial and temporal dynamics of DNA replication sites in mammalian cells. *J Cell Biol.* 1998;143:1415–1425. Copyright 1998 The Rockefeller University Press.)

coordinately during differentiation. Thus, TADs with similar function tend to end up close to each other forming larger compartments—for example, euchromatin and heterochromatin.

Evidence for the replication of clusters of replicons within the nucleus was first obtained by fiber autoradiography. Cells were fed radioactive precursors for DNA synthesis and then examined by electron microscopy. (For an explanation of this technique, see Fig. 40.3.) Clusters of replicating DNA regions were observed.

Several methods are available to observe the spatial distribution of DNA replication during the S phase. One can employ a pulse of nucleotide base analogs that are incorporated into DNA and later detected by fluorescence microscopy. For example, thymidine analogs such as BrdU (bromodeoxyuridine triphosphate), IdU (iododeoxyuridine), and CldU (chlorodeoxyuridine) can be detected in fixed cells by reaction with fluorescent antibodies (BrdU is called by its correct name BrdUTP in Fig. 42.14). Alternatively analogs labeled with a fluorescent dye allow the newly replicated DNA to be observed directly in living cells.

The time at which each replicon fires during the S phase can be seen clearly by synchronizing cells at the beginning of the S phase, releasing them from cell-cycle arrest, and then exposing them to BrdU at various times thereafter. This experiment reveals very distinctive patterns of DNA synthesis occurring at different times during the S phase (Fig. 42.14B–C and E–F). Early on, euchromatin replicates throughout the nucleus. Later, replicating regions are concentrated around nucleoli and other areas of constitutive and facultative heterochromatin (see Chapter 8). Toward the end of the S phase, replication is largely concentrated in blocks of heterochromatin. These observations show that DNA replication occurs throughout the nucleus, wherever DNA is located. DNA does not move to a small number of discrete sites to be replicated (as was once thought).

The timing of replication of particular replication origins was studied in detail in budding yeast using a modification of the BrdU labeling method. First, a procedure was developed whereby all cells in a population entered S phase synchronously. Next, the shift in the density of the DNA following BrdU incorporation was used to distinguish DNA that had replicated from DNA that had not (Fig. 42.15). Incorporation of BrdU into DNA makes the newly synthesized daughter DNA strand heavier, allowing its separation from the parental DNA by centrifugation on a cesium chloride density gradient (see Chapter 6). It was then relatively simple to take DNA probes from different regions of the chromosome and determine when each replicated (changed its density) during the S phase. This experiment demonstrated that each ARS element replicates at a characteristic time during the S phase. More recently, genome-wide maps of replication timing have been constructed by isolating newly replicated DNA at various times during S phase and identifying the chromosome regions involved by high-throughput DNA sequencing.

The most striking aspect of these patterns of DNA synthesis is their reproducibility from one cell cycle to the next. For example, regions of DNA labeled early in the S phase overlap little or not at all with DNA labeled

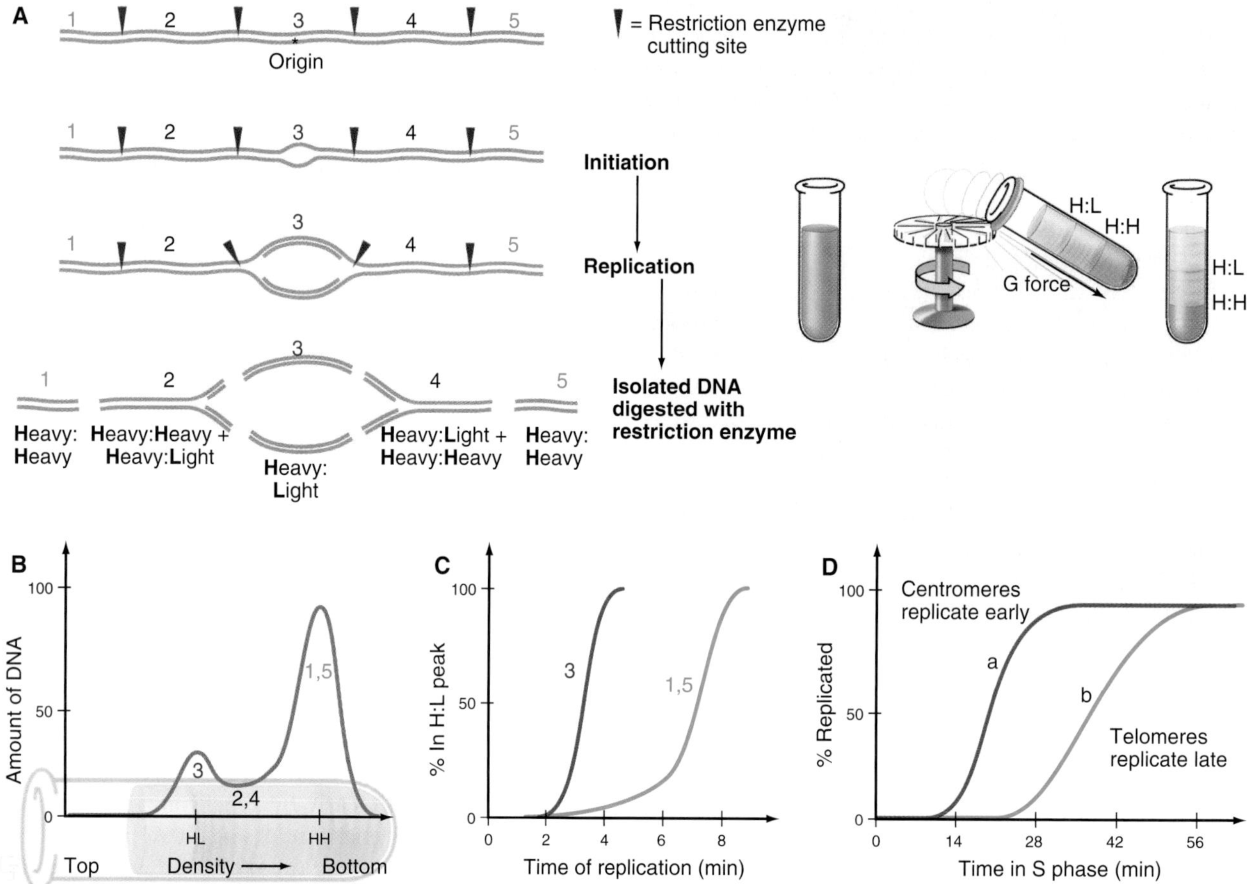

FIGURE 42.15 MEASUREMENT OF THE TIME OF REPLICATION OF PARTICULAR CHROMOSOMAL REGIONS IN *SACCHAROMYCES CEREVISIAE*. A–C, This protocol is based on a classic density shift experiment of Messelson and Stahl that proved that DNA replication is semiconservative. *S. cerevisiae* cells are grown for several generations in a medium containing ^{13}C and ^{15}N so that their DNA is fully substituted with heavy isotopes. At the beginning of the experiment, the cells are synchronized so that they enter the S phase in a single wave. At the same time, the heavy (H) isotope medium is removed and replaced with "light medium" (L) containing ^{12}C and ^{14}N. At various times after the initiation of the S phase, aliquots of cells are removed, and the DNA is isolated. The DNA is then cleaved with restriction enzymes so that the chromosomes are cut into many fragments. DNA from each time point is then subjected to CsCl density gradient centrifugation. When any local region of DNA is replicated, its density alters from heavy/heavy to heavy/light. After very short incubations with light isotopes, only DNA near the origin of replication will be heavy/light; all other DNA will be heavy/heavy. These two populations of molecules are separated from one another by density gradient centrifugation. To examine the timing of replication of a specific gene, a cloned segment of DNA corresponding to the region of interest is used to probe (by DNA hybridization) the heavy/heavy and heavy/light peaks from each gradient. The time of replication of each locus is the time at which the restriction fragment being detected by DNA hybridization moves from the heavy/heavy peak to the heavy/light peak. The numbers in panels **B** and **C** refer to the numbered regions of the chromosomes shown in **A. D,** Data from a replication timing experiment show that in budding yeast, centromeres replicate early in the S phase and telomeres replicate late. To generate *curve a,* fractions from a gradient like that shown in panel **B** were hybridized to a cloned centromere region. To generate *curve b,* fractions from the same gradient were hybridized to a cloned telomere region probe. Note that in mammalian cells, centromeres replicate late and telomeres replicate earlier. (Based on the work of the laboratory of B.J. Brewer and W.L. Fangman. For reference, see Meselson M, Stahl FW. The replication of DNA in *Escherichia coli. Proc Natl Acad Sci U S A.* 1958;44:671–682.)

3 hours later (Fig. 42.14C and E). However, DNA labeled at corresponding points of the S phase in two successive cell cycles superimposes almost entirely. This strongly suggests that particular TADs initiate DNA synthesis during preferred "windows" during S phase.

At least three mechanisms may explain the sequence of replication patterns for different chromosomal regions:

1. Local chromatin structures, established as a result of gene expression influence the time of replication, particularly in metazoans. Thus, transcriptionally active loci (where transcription factors are already bound) have a head start over other regions of the chromosomes, permitting them to initiate DNA replication first. In general, euchromatin (which has low DNA methylation and high histone acetylation) replicates first, followed by facultative heterochromatin (including the inactive X) and, finally, the constitutive pericentric heterochromatin (which has heavy DNA methylation and low histone acetylation). This mechanism can explain why a particular locus may replicate at different times in two cell types. For example, one origin in mammalian genomes is located just upstream of the DNA encoding the β-globin gene (encoding a protein subunit of hemoglobin). This region of more than 200-kb of DNA replicates early in erythroid cells that express the β-globin gene, but later in other cells where the gene is inactive.
2. The higher-order packing of chromatin in the nucleus may influence when origins replicate. It is likely that replication origins fire in a sequence that is imposed on them at the same time that the TAD organization of interphase chromosomes is reestablished during mitotic exit and early G_1 phase.
3. Limiting amounts of certain essential proteins may contribute to the sequential replication of different chromatin domains. These proteins accumulate on replisomes assembled on early-firing replicons and only become available to later-firing replicons following completion of DNA synthesis at the early replicons and disassembly of those replisomes.

Replication Stress and the Intra-S Checkpoint

The textbook-smooth DNA double-helix is actually littered with problems and obstacles, including DNA damage, ribonucleotides inserted by mistake into the DNA, and awkward secondary structures caused, for example, by base-pairing within single DNA strands as opposed to between strands. In addition, if cell-cycle regulation is perturbed, cells can simply run out of either the deoxynucleotide triphosphate (dNTP) precursors that they need to synthesize DNA or a number of key proteins that are also present in very low amounts. The replication fork may stall as result of these problems and often needs help to complete replication.

An intra-S checkpoint responds to stalled replication forks, probably as a result of detecting excessive single-stranded DNA (Fig. 42.16). Replication forks stall if they encounter a damaged DNA base, bases that the replisome cannot "read" or a DNA secondary structure that it cannot unfold. DNA damage can result from ultraviolet light, mutagens or chemicals such as aldehydes produced as a by-product of ethanol metabolism.

Replication blockage leads to a condition known as **replication stress** in which single-stranded DNA accumulates. Replication stress with stalled forks can also result from inappropriate activation of genes that promote cell cycle progression (oncogenic stress; see Chapter 41). Possible reasons include insufficient licensing of origins during the G_1 phase, depletion of nucleoside triphosphate pools, or insufficient pools of essential replication factors. Surprisingly, cells make a number of essential components—including the dNTPs—in amounts just sufficient for replication, so there is little margin for error.

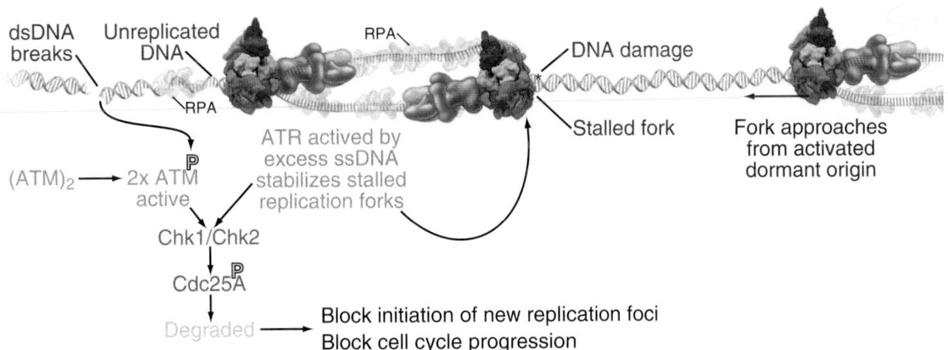

FIGURE 42.16 REPLICATION STRESS AND THE INTRA-S CHECKPOINT. If DNA breaks are detected, the ATM (ataxia-telangiectasia mutated) kinase activates downstream kinases Chk1 and Chk2, leading to phosphorylation of Cdc25A and its subsequent ubiquitin tagging and degradation. This blocks the initiation of new replication forks as well as cell-cycle progression more generally. If DNA persists in unreplicated form, or if replication forks stall, the ATR (ataxia-telangiectasia and Rad3–related) kinase activates a similar downstream response. ATR stabilizes stalled replication forks to give nearby dormant origins time to fire and replicate the DNA downstream of the block.

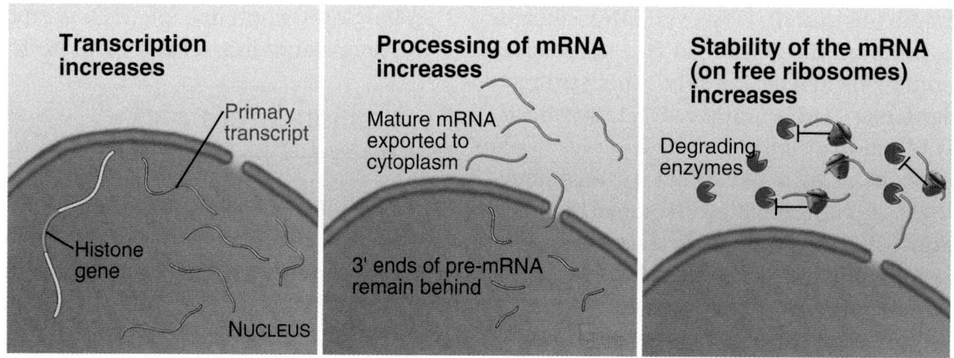

FIGURE 42.17 THREE MECHANISMS THAT ELEVATE HISTONE EXPRESSION DURING THE S PHASE.

In conditions of replication stress, the ATM (ataxia-telangiectasia mutated) and ATR (ataxia-telangiectasia and Rad3-related) kinases activate the DNA damage response (see Box 43.1). The activities of these kinases and their downstream effectors result in the degradation of Cdc25A, the phosphatase that triggers entry of the cell into mitosis. The resulting inactivation of Cdks during the S phase prevents replication initiation within other unreplicated domains.

A key aspect of this response unique to S phase is that ATR-activated by the presence of excessive levels of single-stranded DNA associated with RPA protects the stalled forks from disassembly, known as **replication fork collapse**. This response is critical, because replication forks with unwound and nicked DNA molecules can turn into breaks if the fork disassembles before replication is complete and all the DNA is ligated. Error-prone mechanisms tend to repair DNA damage that arises during replication stress. The resulting mutations are thought to contribute to oncogenic transformation.

Of course, stalled replication forks must not only be protected, but they must also somehow be rescued. At moderate levels of replication stress, ATR activation blocks the activation of new replication foci, but allows dormant origins to fire near stalled forks in replication domains that are already active. This results in the arrival of a converging fork on the downstream side of the damaged DNA, leaving only a very small gap that can be repaired by a specialized DNA polymerase in a process known as translesion synthesis. Thus, in the real world where replication forks routinely encounter problems that cause them to stall, a pool of dormant licensed replication origins is essential for integrity of the genome.

Synthesis of the Histone Proteins

Chromatin contains approximately equal masses of DNA and core histones. Human cells require about 62×10^6 copies of each core histone, assuming a genome size of 6.2×10^9 bp and 200 bp per nucleosome. Because approximately 90% of histone transcription occurs during the S phase, enormous amounts of these proteins are made during a relatively brief period. Histone synthesis apparently keeps pace, in part, because there are approximately 40 sets of histone genes.

Synthesis of histones during the S phase is tightly coupled to ongoing DNA replication and a controlled supply of histones is essential for normal replication. If replication is blocked either by addition of drugs or by temperature-sensitive mutants, histone synthesis declines abruptly shortly thereafter, possibly because nucleosome deposition appears to be linked to lagging strand synthesis. Indeed, the chaperone that assembles histones into nucleosomes is associated with the replisome. This link between histone synthesis and DNA replication appears to involve at least three processes (Fig. 42.17).

First, *transcription of the histone genes rises* threefold to fivefold as cells enter the S phase. Each histone gene has a cell-cycle-responsive element in its promoter to which a transcription factor binds specifically during the S phase.

Second, the *processing of histone messenger RNAs (mRNAs) increases* six- to 10-fold as cells enter the S phase. Histone mRNAs are not polyadenylated, and the primary transcripts are considerably longer than the mature forms. Processing of the 3′ end of histone pre-mRNAs involves the U7 small nuclear ribonucleoprotein (snRNP) (see Chapter 11), a portion of which recognizes histone mRNA and base-pairs with it during processing. Cell-cycle-dependent regulation of processing appears to involve changes in the accessibility of the necessary portion of U7 small nuclear RNA (snRNA). This region is inaccessible in G_0 cells but becomes accessible when cells that reenter the cycle and begin the S phase. The mechanism for this change in RNA conformation is not known.

Third, *changes in the stability of the mRNA* also regulate histone synthesis. Normally, the level of histone mRNA on free polysomes drops rapidly by approximately 35-fold as cells enter the G_2 phase. If DNA synthesis is interrupted during the S phase, a region at the 3′ end of the mature message targets the mRNA for degradation. If this region is removed from the 3′ terminus of the histone mRNA, the normal link between ongoing replication and mRNA stability is lost. Furthermore, this sequence, transposed onto the 3′ terminus of a globin mRNA, renders that mRNA sensitive to degradation if DNA synthesis is blocked. Degradation of histone mRNA requires ongoing protein synthesis, and it has been speculated that histones themselves participate in the control.

As discussed in Chapter 8, specialized variant forms of histones are synthesized and inserted into the chromatin outside of S phase. These histones are encoded by mRNAs with introns and normal poly(A) tails and are therefore not processed by the specific S phase–associated pathway (see Chapter 11). Their insertion into chromatin is typically correlated with RNA transcription rather than DNA replication and involves specific chromatin-remodeling factors.

Other Events of the S Phase

Although the bulk of attention on the S phase focuses on the duplication of the chromosomes, centrosomes also duplicate at this time. Duplication of the centrosomes is essential for stability of the genome, because they set up the poles of the mitotic spindle that is responsible for accurate partitioning of the replicated chromosomes. Interestingly, Orc1 functions independent of DNA replication with cyclin A to limit centriole duplication to once per cell cycle. (See Fig. 34.17 for a discussion of centrosome duplication.)

With the completion of DNA replication and duplication of the centrosomes, the cell is ready to divide. As the levels of Cdk activity rise toward the threshold that is sufficient to trigger mitotic entry and other factors necessary for mitosis accumulate, the cell continues to screen the integrity of the DNA to ensure that the genome has been replicated completely and that no harmful DNA damage is present. These checks, together with other ongoing preparations for mitosis, are the principal events of the G_2 phase (see Chapter 43).

ACKNOWLEDGMENTS

We thank Hiro Araki, Julian Blow, Cristina Cardoso, David Gilbert, and Julian Sale for suggestions on revisions to this chapter.

SELECTED READINGS

Boos D, Frigola J, Diffley JF. Activation of the replicative DNA helicase: breaking up is hard to do. *Curr Opin Cell Biol*. 2012;24:423-430.

Costa A, Hood IV, Berger JM. Mechanisms for initiating cellular DNA replication. *Annu Rev Biochem*. 2013;82:25-54.

Deegan TD, Diffley JF. MCM: one ring to rule them all. *Curr Opin Struct Biol*. 2016;37:145-151.

Hills SA, Diffley JFX. DNA replication and oncogene-induced replicative stress. *Curr Biol*. 2014;24:R435-R444.

Labib K. How do Cdc7 and cyclin-dependent kinases trigger the initiation of chromosome replication in eukaryotic cells? *Genes Dev*. 2010;24:1208-1219.

McIntosh D, Blow JJ. Dormant origins, the licensing checkpoint, and the response to replicative stresses. *Cold Spring Harb Perspect Biol*. 2012;4:a012955.

Méchali M, Yoshida K, Coulombe P, Pasero P. Genetic and epigenetic determinants of DNA replication origins, position and activation. *Curr Opin Genet Dev*. 2013;23:124-131.

O'Donnell M, Langston L, Stillman B. Principles and concepts of DNA replication in bacteria, archaea, and eukarya. *Cold Spring Harb Perspect Biol*. 2013;5:a010108.

O'Donnell M, Li H. The eukaryotic replisome goes under the microscope. *Curr Biol*. 2016;26:R247-R256.

Rhind N, Gilbert DM. DNA replication timing. *Cold Spring Harb Perspect Biol*. 2013;5:a010132.

Zeman MK, Cimprich KA. Causes and consequences of replication stress. *Nat Cell Biol*. 2014;16:2-9.

CHAPTER 43

G_2 Phase, Responses to DNA Damage, and Control of Entry Into Mitosis

The G_2 phase is the gap between the completion of DNA replication and the onset of mitosis. This chapter begins with the biochemical basis for the G_2/mitosis (M) transition and discusses how the G_2/M checkpoint delays this transition if DNA damage is detected. Finally, the chapter introduces the major pathways that cells use to repair damaged DNA.

Enzymology of the G_2/Mitosis Transition

The transition from the G_2 phase into mitosis is the most profound morphologic and physiological change that occurs during the life of a proliferating cell. Entry into mitosis is controlled by a network of stimulatory and inhibitory protein kinases and phosphatases, presided over by Cdk1-cyclin B1. (Chapter 40 introduced the components involved in the G_2/M transition.)

Cdk1, the driving force for entry into mitosis, is present at a constant level throughout the cell cycle. The multifaceted regulation of Cdk1 (see Fig. 40.14) includes binding of cyclin cofactors, inhibition and activation by phosphorylation, binding of inhibitory molecules, changes in subcellular localization and changes in the activities of competing protein phosphatases (Fig. 43.1). These various factors are finely balanced, until a positive feedback loop enables the cells to make an abrupt and decisive entry into mitosis.

Mammals have at least three **B-type cyclins:** B1, B2, and B3. Cyclin B_1 is essential for triggering the G_2/M transition. Cyclin B1, newly synthesized during the latter part of the cell cycle, binds Cdk1 and shuttles it in and out of the nucleus. Importin β carries the Cdk1-cyclin B1 complex into the nucleus, and then Crm1 rapidly exports it back to the cytoplasm (see Chapter 9). Cdk1-cyclin B2 associates with the Golgi apparatus during interphase and might function in Golgi disassembly during mitosis (see Fig. 44.4). Cyclin B3 may function only during meiosis in mammals.

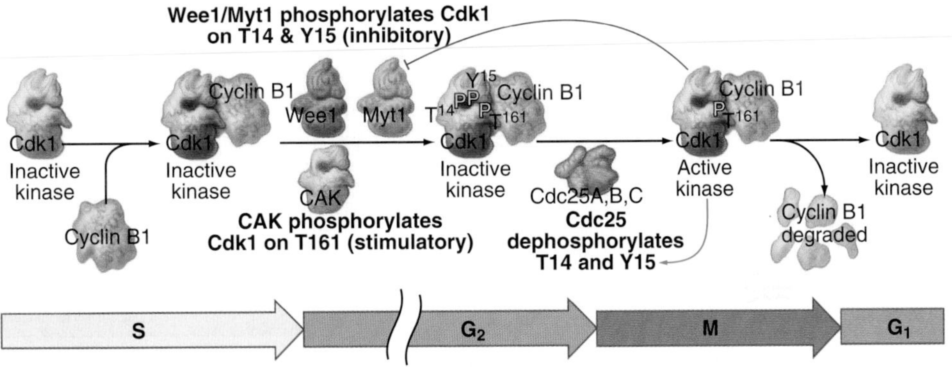

FIGURE 43.1 REGULATION OF CDK1 BY CYCLIN BINDING AND PROTEIN PHOSPHORYLATION FROM THE LATE S PHASE THROUGH MID-MITOSIS. (For reference, see Protein Data Bank [PDB; www.rcsb.org] file 1X8B and Squire CJ, Dickson JM, Ivanovic I, et al. Structure of human Wee1A kinase: kinase domain complexed with inhibitor PD0407824. *Structure.* 2005;13:541–550.)

As cells approach the G_2/M transition, phosphorylation simultaneously activates and inhibits Cdk1 bound to cyclin A or B. *Cdk-activating kinase* (CAK-actually Cdk7–cyclin H), located in the nucleus, phosphorylates Cdk1 on T^{161} (Fig. 43.1). This triggers a refolding of the active site cleft that allows the enzyme to bind substrates (see Fig. 40.13). This phosphorylation activates Cdk1-cyclin kinases in the nucleus (Fig. 43.1).

At the same time, Wee1 and Myt1 kinases phosphorylate T^{14} and Y^{15} to turn the enzyme off. Wee1 in the nucleus phosphorylates Cdk1 on Y^{15} adjacent to the adenosine triphosphate (ATP)-binding site. This stops the kinase from using ATP to phosphorylate substrates. While Cdk-cyclin complexes are in the cytoplasm, they are phosphorylated on T^{14} and Y^{15} by Myt1 kinase associated with the Golgi apparatus and endoplasmic reticulum. The role of Wee1 as a mitotic inhibitor is clearly demonstrated in *Schizosaccharomyces pombe:* overexpression of Wee1 delays or prevents the entry of cells into mitosis (Fig. 43.2). Together, Wee1 and Myt1 ensure that Cdk1 remains inactive as it shuttles into and out of the nucleus (Fig. 43.3).

This combination of stimulatory and inhibitory phosphorylations holds Cdk1–cyclin complexes poised for a burst of activation.

That burst occurs when one of three Cdc25 protein phosphatases removes the inhibitory phosphates from T^{14} and Y^{15} (Figs. 43.1 and 43.8). Cdc25s are "dual-specificity" protein phosphatases (see Fig. 25.5) that remove phosphates from serine (S), threonine (T), and tyrosine (Y) residues. Of the three Cdc25 phosphatases, only Cdc25A is indispensable for life. It functions at both the G_1/S and G_2/M transitions, whereas Cdc25B and Cdc25C have roles only in the G_2/M transition.

Because Cdc25 phosphatases trigger mitotic entry, their regulation is both important and elaborate. Cdc25A and Cdc25C are held inactive during interphase by mechanisms involving stimulatory and inhibitory

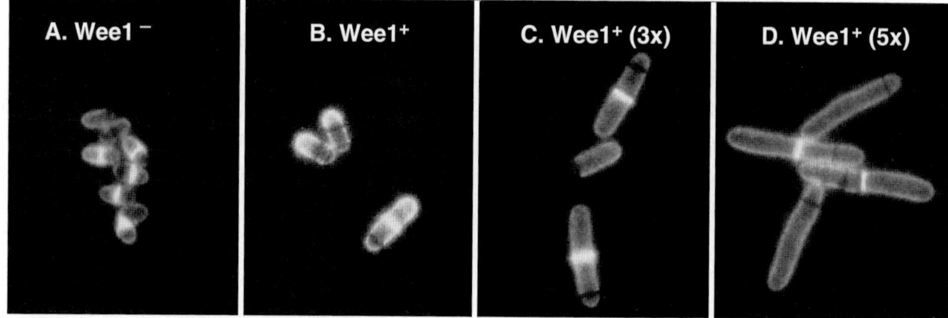

FIGURE 43.2 EFFECT OF CHANGING CELLULAR LEVELS OF WEE1 PROTEIN ON CELL-CYCLE PROGRESSION IN FISSION YEAST. A, Cells that lack functional Wee1 protein enter mitosis too soon in the cell cycle and are smaller than wild-type cells **(B). C–D,** Cells that express excess Wee1 protein are too effective at inactivating Cdk1 and are severely delayed in their ability to enter mitosis (hence their larger size). (From Russell P, Nurse P. Negative regulation of mitosis by Wee1⁺, a gene encoding a protein kinase homolog. *Cell.* 1987;49:559–567.)

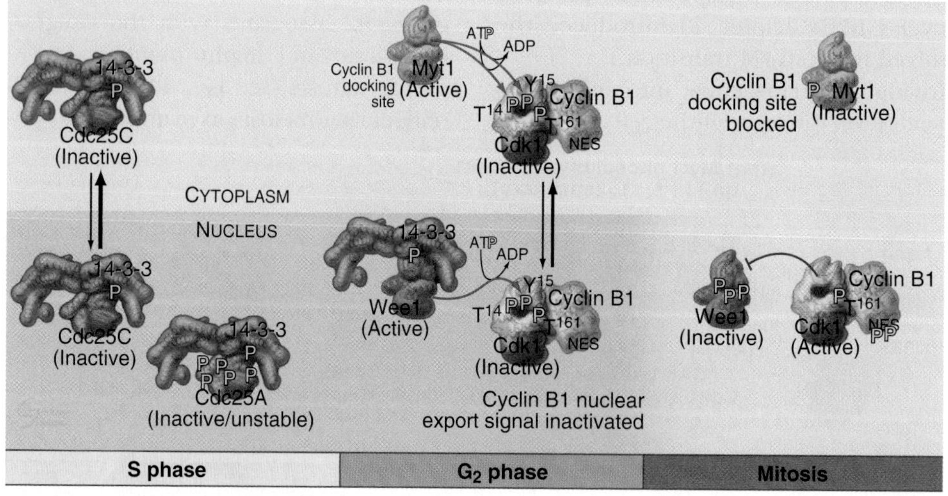

FIGURE 43.3 CDK REGULATION IN INTERPHASE AND MITOSIS. Inhibitory phosphorylation and shuttling of components between the nucleus and cytoplasm regulate Cdk1 activity in interphase and mitosis. Cdc25, which would remove the inhibitory phosphates, is held inactive until its activation by Polo kinase and Cdk1-cyclin B stimulates mitotic entry (not shown here). ADP, adenosine diphosphate; ATP, adenosine triphosphate; NES, nuclear export sequence. (Based on an original figure by Helen Piwnica-Worms. For reference, see PDB file 1YWT.)

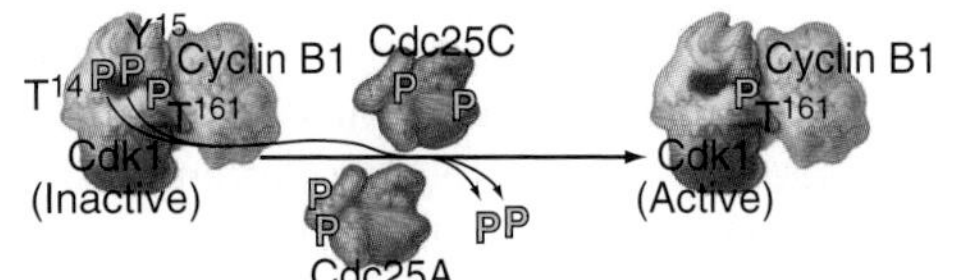

FIGURE 43.4 REGULATION OF CDC25A AND CDC25C ACTIVITY IN INTERPHASE AND MITOSIS. (Based on an original figure by Helen Piwnica-Worms.)

phosphorylation, binding to a 14-3-3 adapter protein, alterations in their subcellular localization, and ubiquitin-mediated proteolysis (Fig. 43.4).

Phosphorylation on a serine residue of Cdc25 creates a binding site for a 14-3-3 adapter protein that interacts with the phosphoserine and several flanking amino acids (see Fig. 25.10). Association with 14-3-3 interferes with Cdc25A binding to Cdk1-cyclin B1. Phosphorylation at other sites also targets Cdc25A for ubiquitination by $SCF^{\beta TrCP}$ and destruction by proteasomes. Cdc25C has a nuclear export sequence (see Fig. 9.17), so Cdc25C bound to 14-3-3 is primarily cytoplasmic (Fig. 43.3).

Following the theme of phosphorylation having both positive and negative effects, full activation of Cdc25s requires phosphorylation of other residues in the amino-terminal region. A protein kinase called Polo (see next paragraph) initiates this phosphorylation of Cdc25, followed by more extensive phosphorylation by Cdk1-cyclin B1, the substrate of Cdc25. This action of Cdk1-cyclin B1 on Cdc25 creates a powerful positive feedback amplification loop that provides a burst of Cdk activity and triggers entry into mitosis (Fig. 43.7).

A molecular trigger is required to start the amplification cycle in which Cdk1-cyclin B1 and Cdc25 activate each other. Members of the **Polo family of protein kinases** are candidates for this role, by activating Cdc25. These kinases possess amino acid motifs called "polo boxes" that bind to protein partners after they have been "primed" by phosphorylation by another kinase. Interestingly, the most common "priming kinases" for polo are Cdk-cyclin pairs. This creates another positive amplification loop that allows for additional levels of control and rapid activation of Cdk-cyclin complexes. In addition to activating Cdks by phosphorylating Cdc25, polo family kinases are involved in a variety of mitotic events, including formation of a bipolar spindle, cytokinesis, and passage through certain cell-cycle checkpoints.

For cells to enter and complete mitosis, protein phosphatase 2A (PP2A, a protein serine/threonine phosphatase; see Fig. 25.6) must be inactivated. PP2A removes the phosphates that Cdk1 puts on its target proteins. If PP2A is not inhibited, cells enter mitosis, but they then slip back into interphase. PP2A inhibition involves three steps. Newly active Cdk1 phosphorylates a protein kinase called Greatwall (a name from *Drosophila* genetics). Greatwall phosphorylates a small target protein that then binds to the substrate-binding pocket of PP2A, acting as a competitive inhibitor. Cdk1 activation at the onset of anaphase eventually leads to dephosphorylation of the small protein and activation of PP2A, thereby allowing cells to exit mitosis.

Changes in Subcellular Localization at the G_2/M Transition

Late in G_2, phosphorylation inactivates the nuclear export signal of cyclin B1 (see Chapter 9). As a result, Cdk1-cyclin B1 accumulates in the nucleus within 5 minutes (Figs. 43.3 and 43.5). Cdc25C also stops shuttling at the G_2/M transition, probably as a result of Polo kinase phosphorylation. The Cdk1-cyclin B that accumulates in the nucleus is thought to be already active, so the reason for this rapid nuclear localization is not entirely clear. The concentration of Cdk-cyclin B complexes together with Cdc25A and Cdc25C in the confined volume of the nucleus may contribute to the final burst of Cdk1-cyclin B1 activation.

Cdk1 Activity and the Initiation of Prophase

Cdk2-cyclin A plays a critical role during the S phase (see Chapter 42), but also helps trigger the G_2/M transition. Cdk-cyclin A activity peaks at G_2/M, before the peak of Cdk1-cyclin B1 activity, and inactivation of cyclin A in *Drosophila* or mammalian cultured cells arrests the cell cycle in G_2. Furthermore, G_2 cells enter mitosis prematurely if injected with active Cdk2-cyclin A complexes just completing S phase. Finally, microinjection of a selective inhibitor of Cdk-cyclin A causes prophase cells to return rapidly to interphase; chromosomes decondense, rounded prophase cells flatten, and the interphase microtubule network returns.

Cdk activity regulates several events during the G_2-to-prophase transition. In the cytoplasm, the half-life of microtubules drops dramatically from approximately 10 minutes to approximately 30 seconds late in G_2 (see Table 44.1). This, coupled with an enhanced ability of centrosomes to initiate microtubule polymerization, completely transforms the organization of the microtubule cytoskeleton. Centrosomes take on the appearance of spindle poles and migrate apart over the surface of the nucleus. While this is happening, chromatin begins to condense in the nucleus. A protein complex called condensin is required for this condensation to begin during prophase (see Fig. 8.18). Cdks activate condensin by phosphorylation of two of its subunits in late G_2.

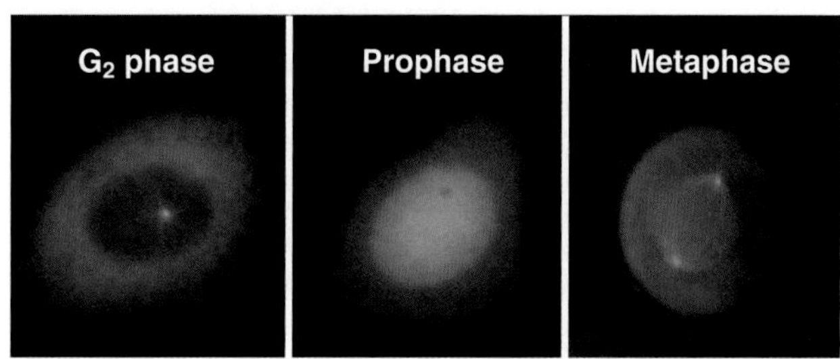

FIGURE 43.5 CYCLIN B1 LOCALIZATION DURING MITOSIS. Cyclin B1 moves rapidly from the cytoplasm into the nucleus at the onset of prophase and subsequently associates with the spindle during mitosis. (Courtesy Christina Karlsson and Jonathon Pines, Wellcome/CRC Institute, Cambridge, UK.)

These events occur while most Cdk1-cyclin B1 is in the cytoplasm. It appears most likely, therefore, that Cdk1-cyclin A triggers at least the nuclear events of prophase (Fig. 43.6). Commitment to mitosis appears to be irreversible only after Cdk1-cyclin B1 enters the nucleus.

Recap of the Main Events of the G_2/M Transition

Synthesis of cyclin B1 in the latter portion of the S and G_2 phases leads to assembly of Cdk1-cyclin B heterodimers that shuttle into and out of the nucleus, spending most of their time in the cytoplasm associated with microtubules. In late G_2 phase, Cdk1-cyclin A activity initiates mitotic prophase, beginning with changes in microtubule dynamics and chromosome condensation. Several events trigger entry into the active phase of mitosis. Cdc25A accumulates and no longer binds 14-3-3 proteins, allowing it to interact more effectively with Cdk1-cyclin B. The phosphoserine-binding site for 14-3-3 proteins on Cdc25C is dephosphorylated, allowing Cdc25C to accumulate in the nucleus. In addition, phosphorylation of cyclin B1 blocks its export from the nucleus and promotes its import, thus causing Cdk1-cyclin B1 to accumulate rapidly in the nucleus. The inhibitory kinase Wee1 is also phosphorylated, causing its activity to drop. Cdc25A and Cdc25C activate Cdk1-cyclin B1 by removing inhibitory phosphates on T^{14} and Y^{15}. This starts in the cytoplasm and then may be stimulated as the proteins concentrate in the nucleus. There, the action of Cdk1-cyclin B1 on the nuclear lamina triggers nuclear envelope breakdown and drives the cell into mitosis. Fig. 43.7 highlights the positive feedback loop between Cdc25 and Cdk1-cyclin B that drives the cell into mitosis.

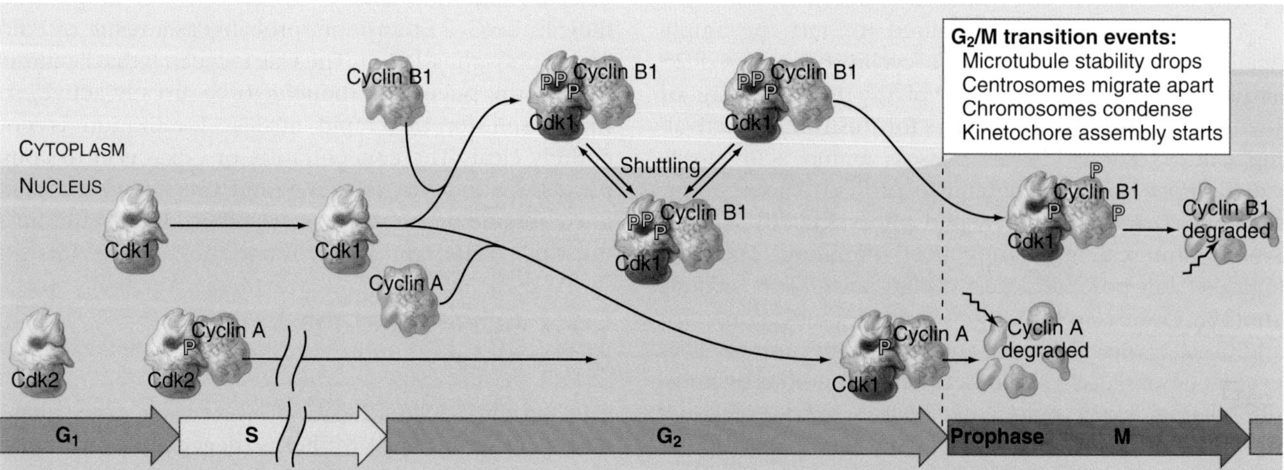

FIGURE 43.6 CDK1 REGULATION. Locations and patterns of activation of Cdk1 complexed to cyclin A versus cyclin B across the cell cycle.

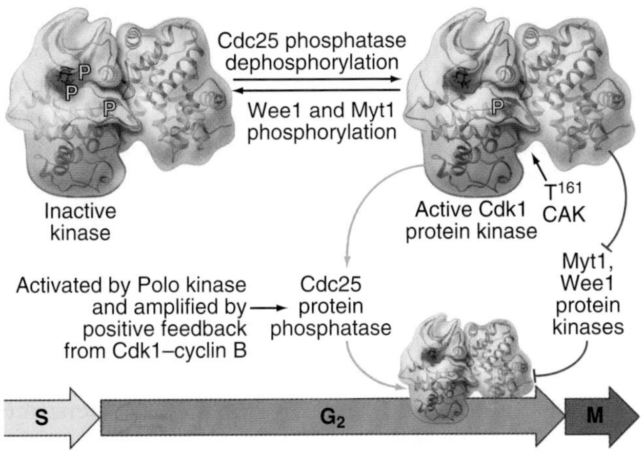

FIGURE 43.7 SUMMARY OF THE CDK1 FEEDBACK REGULATION MECHANISM AT THE G_2/M TRANSITION.

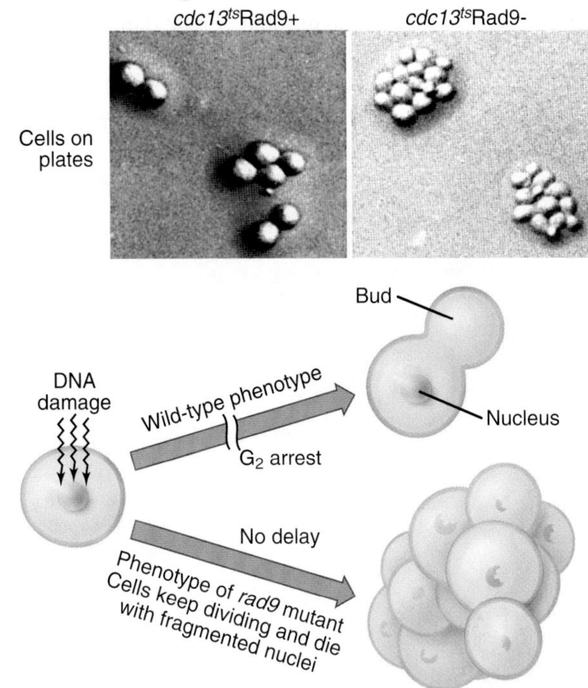

FIGURE 43.8 GENETIC IDENTIFICATION OF THE G_2/M CHECKPOINT. Cells defective in the G_2/M checkpoint (*Rad9* mutants of budding yeast) cannot delay their entry into mitosis in the presence of damaged DNA and therefore divide themselves to death. Budding yeast Rad9 is an ortholog of the ATM (ataxia-telangiectasia mutated) adapter protein 53BP1 (see text). Confusingly, budding yeast Rad9 is not related to fission yeast Rad9, which gives the 9-1-1 complex its name (see text). (Courtesy Ted Weinert, University of Arizona, Tucson.)

Why did such an elaborate system evolve to regulate the G_2/M transition? The answer appears to lie in the exquisite sensitivity provided by the interlocking network of stimulatory and inhibitory activities. On the one hand, this network ensures a rapid, almost explosive, final transition into mitosis. On the other, it provides a number of ways to delay the G_2/M transition if the cell detects damage to chromosomes. Attempting mitosis with chromosomal damage can lead to cell death or contribute to cancer.

G_2/M Checkpoint

Separation of sister chromatids during mitosis is a potential danger point for a cell. After DNA is replicated each chromosome consists of paired sister chromatids held together by cohesin. Therefore, if the DNA is damaged, the cell can use information present in the undamaged chromatid to guide the repair process. However, once sisters separate, this corrective mechanism can no longer operate. In addition, if a cell enters mitosis before completing replication of its chromosomes, attempts to separate sister chromatids damage the chromosomes. To minimize these hazards, a checkpoint operates in the G_2 phase to block mitotic entry if DNA is damaged or DNA replication is incomplete.

Just as DNA damage can arrest the cell cycle in G_1 phase, damaged or unreplicated DNA also halts the cell cycle temporarily in the G_2 phase. Interestingly, the G_1 checkpoint—which can be activated by a single DNA break in human cells—is more sensitive than the G_2/M checkpoint, which requires 10 to 20 breaks to block cell-cycle progression. The G_2/M checkpoint may be less sensitive then the G_1 checkpoint, because G_2 cells are already primed to enter mitosis. Consequently, human cells can enter mitosis with limited amounts of damaged or unreplicated DNA. These problem regions can be detected and repaired in the daughter cells after division (see later).

Studies of radiation-induced **G_2 delay** in budding yeast identified a major cell-cycle checkpoint that is sensitive to the status of the cellular DNA. Cells defective in this checkpoint are more sensitive than wild-type cells to radiation injury because they continue to divide, despite the presence of broken or otherwise damaged chromosomes (Fig. 43.8). The cells die, presumably from chromosomal defects or loss. In metazoans, the G_2/M checkpoint delays entry into mitosis until the damage is either fixed, triggers cell suicide by apoptosis, or causes cells to enter a nonproliferating (senescent) state. The checkpoint works by modulating the activities of the components that control the G_2/M transition.

DNA Damage Response

Considering that it is the blueprint for life, DNA is remarkably accident-prone, and organisms have an elaborate network of mechanisms to repair those accidents. This complex network is known as the **DDR**

(*D*NA *d*amage *r*esponse). This section discusses a few key components of the DDR, and Box 43.1 (see later) explains several mechanisms that repair damaged DNA.

The minimal machinery of a DNA damage checkpoint (see Fig. 40.4) involves **SENSORS** that detect DNA damage, **TRANSDUCERS** (usually protein kinases) that produce a biochemical signal as a result of the detected damage, and **EFFECTORS** (both protein kinases and transcriptional activators) that coordinate repair pathways to block cell-cycle progression and repair the damage.

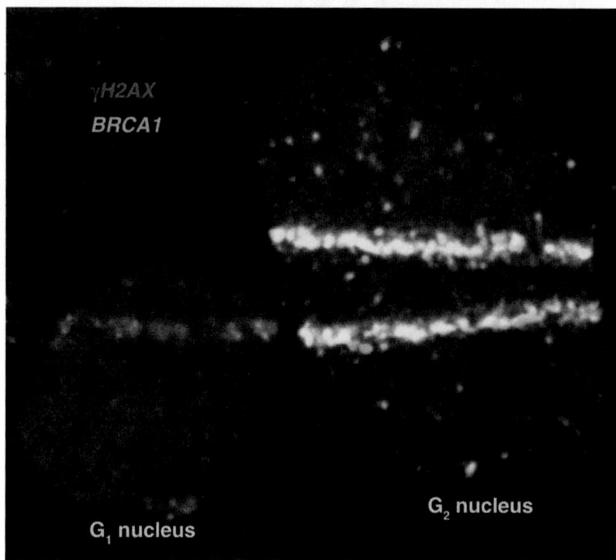

FIGURE 43.9 INDUCTION OF DNA DAMAGE FOCI USING A HIGH-ENERGY LASER. Laser-induced DNA damage results in activation of the DDR with production of γH2AX *(red)* at damage sites followed by binding of other factors, including BRCA1 *(green)*. DNA is *blue*. The left cell is not yellow because BRCA1 is not present in G_1-phase cells. Micrograph taken 30 minutes after induction of damage. (Micrograph courtesy Martin Mistrik and Jiri Bartek.)

When DNA is damaged, several proteins concentrate at the damage site within seconds to minutes, forming a focus that activates the G_2/M checkpoint and repairs the damage. The use of high-energy lasers to "draw" patterns of DNA damage onto cell nuclei revolutionized our understanding of the DDR by allowing a minute-by-minute mapping of the events during focus formation and leading to DNA repair (Fig. 43.9).

If a single strand of DNA is broken, the SENSOR enzyme poly(adenosine diphosphate [ADP]-ribose) polymerase, binds to the broken end and immediately starts polymerizing a chain of ADP-ribose (which it makes from nicotine adenine dinucleotide [NAD]). Several proteins bind to poly(ADP-ribose) chains and recruit the critical TRANSDUCER ATM kinase (*a*taxia-*t*elangiectasia *m*utated) to the break together with its partner NBS1, a subunit of the MRN complex (Fig. 43.10A). ATM is usually present in the nucleus as an inactive dimer. Interactions with MRN separate the dimer into ATM monomers that are activated by MRN docked to the broken end of a DNA molecule.

The MRN complex is a SENSOR that detects double-strand DNA breaks. In this case, the NBS1 subunit of MRN recruits ATM kinase to the break site. MRN is also a nuclease with a key role in repairing those breaks. Its job is to chew back (resect) one DNA strand in a 5′ → 3′ direction at the site of a DNA break, leaving a stretch

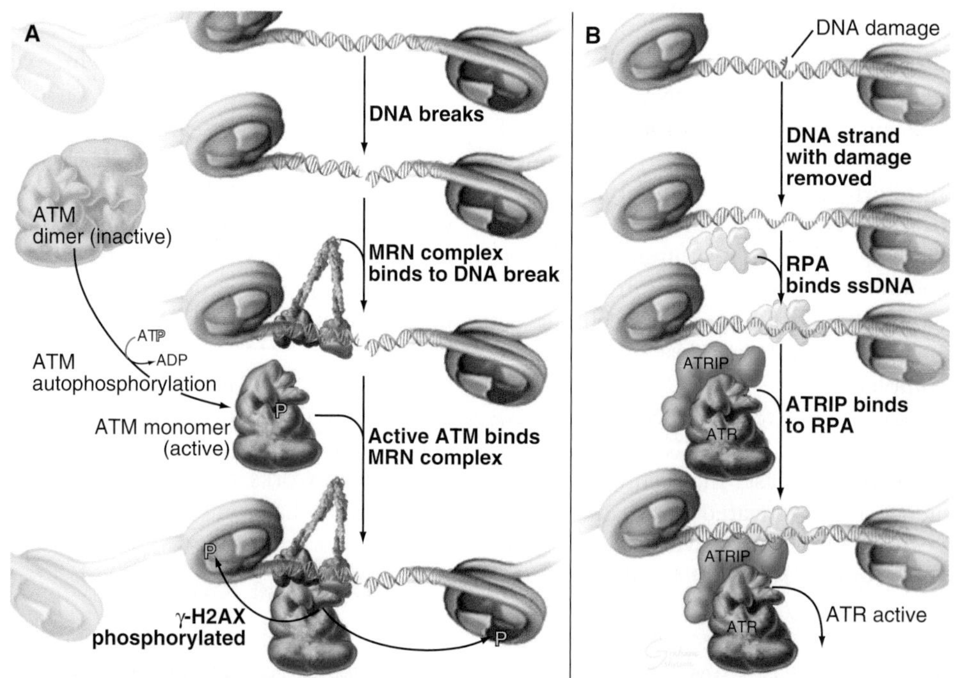

FIGURE 43.10 MECHANISMS FOR LOCALIZING ATM AND ATR TO SITES OF DNA DAMAGE. ATR, ataxia-telangiectasia and Rad3 related; ssDNA, single-stranded DNA.

of single-stranded DNA that becomes coated with RPA, the protein that also coats single-stranded DNA at replication forks (Fig. 42.12). RPA then attracts a protein called ATRIP plus its partner, the second key TRANSDUCER, ATR kinase (*a*taxia-*t*elangiectasia and *R*ad3 related).

ATM and ATR stay bound to the chromatin, which is remodeled by reactions involving protein phosphorylation, acetylation/deacetylation, methylation, and ubiquitylation. The response is very complex. For example, several hundred phosphorylation events have been detected during the DDR. These chromatin changes are likely part of a response to suppress transcription so that RNA polymerase does not collide with the damage and cause more problems. The histone modifications associated with the DDR generally tend to create "open" chromatin that is more accessible to the DNA repair machinery. Repair foci are often found just adjacent to heterochromatin, and indeed, DNA repair appears to be more efficient in euchromatin than heterochromatin.

ATR tends to stay on the single-stranded DNA close to the break, but ATM spreads along the chromatin through a cycle of reactions involving its most famous substrate, the specialized **histone isoform H2AX** (Figs. 43.9 and 43.11). H2AX phosphorylated by ATM is known as **γ-H2AX.** γ-H2AX forms very rapidly in a focus of modified chromatin immediately surrounding the damage. This amplifies and spreads the response to damage, because γ-H2AX binds a SENSOR protein, MDC1 that recruits more ATM and repair factors. The response spreads along the chromatin as ATM creates more γ-H2AX, so that after a few minutes the γ-H2AX focus covers about a megabase of DNA.

MDC1 has two BRCT domains that bind the phosphorylated site on γ-H2AX. BRCT domains were discovered in BRCA1, one of the first genes whose mutations were linked to familial breast cancer. BRCA1 is mutated in 90% of families where inherited predisposition to ovarian cancer coexists with breast cancer. BRCA1 is a very large, complex protein with numerous roles in DNA repair that are still being determined.

ATM (Table 43.1) is dispensable for life, though people lacking it have ataxia-telangiectasia—a disorder associated with degeneration of cerebellar neurons, dilation of blood vessels, a very high predisposition for cancer, and a number of other symptoms. The *NBN* gene is mutated in humans with Nijmegen breakage syndrome, a rare inherited disorder featuring chromosomal instability and a predisposition to cancer. ATR is essential for life, presumably because it has a key role in ensuring that replication forks are stabilized until all the single-stranded DNA created during S phase is replicated.

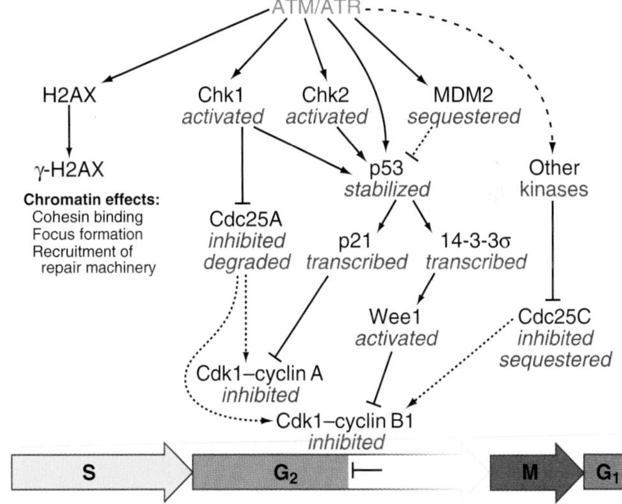

FIGURE 43.11 THE G_2/M CHECKPOINT BLOCKS THE G_2/M TRANSITION FOLLOWING ACTIVATION OF ATM AND/OR ATR BY DNA DAMAGE. *Dotted lines* show activities that are switched off by the checkpoint. The *dashed line* between ATM/ATR and the kinase that inhibits Cdc25C indicates that this pathway is not yet known. (Based on an original figure by Helen Piwnica-Worms.)

TABLE 43.1 Key DNA Repair Gene Defects Associated With Human Disease

Human Disease	Pathway	Defective Genes*
Ataxia-telangiectasia (AT)	Checkpoint	*ATM*
Seckel syndrome	Checkpoint	*ATR*
Xeroderma pigmentosum	NER	*XP-A, XP-B, XP-C, XP-D, DDB-2, XP-F, XP-G, POLH*
Cockayne syndrome	NER	*XP-B, XP-D, CSA, CSB*
Trichothiodystrophy	NER	*XP-D*
Hereditary nonpolyposis colon cancer	MMR	*MSH2, PMS2, MLH1*
AT-like disorder	DSB repair	*MRE11*
Nijmegen breakage syndrome	DSB repair	*NBN*
Breast cancer predisposition	HR	*BRCA1, BRCA2*
LIG4 syndrome	NHEJ	*LIGIV*
Severe combined immune deficiency	NHEJ	*ARTEMIS*

ATM, ataxia-telangiectasia mutated; ATR, ataxia-telangiectasia and Rad3 related; DSB, double-strand break; HR, homologous recombination; MMR, mismatch repair; NER, nucleotide excision repair; NHEJ, nonhomologous end joining.
*This list is an outline. For an updated list of the human DNA repair genes known to date, the reader is referred to http://sciencepark.mdanderson.org/labs/wood/.

Table 43.1 shows only a few of the diseases associated with mutations in proteins involved in repairing DNA breaks. (See Box 43.1 for a discussion of the major DNA repair pathways.) It is likely that new components of this pathway remain to be identified.

From the DNA Damage Response to the G_2/M Checkpoint

ATM and ATR at damage foci cooperate with adapters to phosphorylate and activate the two key EFFECTOR kinases

BOX 43.1 DNA Repair in Vertebrates*

Every human cell experiences approximately 10^5 DNA damage events each day. These are mostly repaired accurately, so only about 100 mutations are passed on in each new human generation. In cancer, the spontaneous mutation frequency can be at least 200-fold higher. Cell division must not occur with inaccurately replicated or damaged genomes, as this may cause cell death or heritable mutation. A number of systems have evolved to repair particular forms of DNA damage sustained by the cell (Fig. 43.12). These repair mechanisms act in concert with the apoptotic machinery to ensure that the cell will die if the DNA damage cannot be repaired (see Chapter 46). DNA damage checkpoints are a critical component of the cellular response to DNA damage (see Fig. 40.4), as they impose a delay in the cell cycle during which cells have a chance to repair their genomes. Promising new strategies for cancer treatment include use of agents that actually cause DNA damage, with the goal of overwhelming the already defective defenses in cancer cells, and causing them to activate cell death pathways.

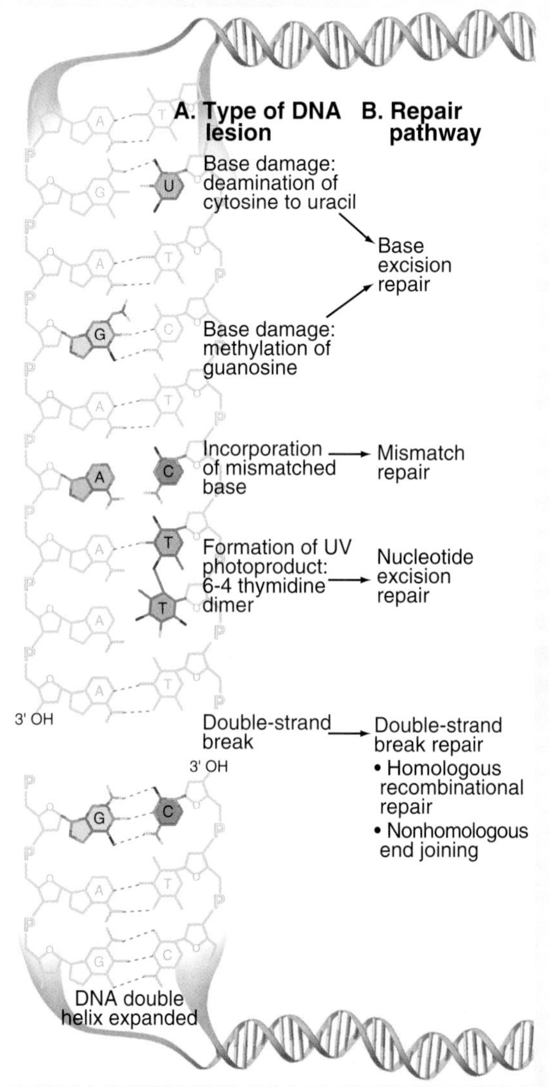

FIGURE 43.12 INTRODUCTION TO DNA REPAIR. Examples of DNA damage and the repair pathways that respond to different types of lesion.

Base Excision Repair

Bases in DNA can become oxidized, reduced, alkylated, or deaminated owing to endogenous activities or environmental stress. Damaged bases are cut away from the DNA sugar-phosphate backbone by a damage-recognizing **glycosylase,** leaving an abasic site (Fig. 43.13A). **Abasic sites,** which also can be generated directly by DNA damage, are then removed by cleavage of the sugar-phosphate backbone mediated by certain glycosylases and endonucleases. The missing sequence is then reconstructed from its complementary strand by DNA polymerase β, with DNA ligase III-Xrcc1 completing the repair by sealing the gaps in the backbone.

Nucleotide Excision Repair

Bulky DNA adducts caused by chemical agents or environmental stress (particularly UV radiation from sunlight) are excised by a complex, though well-understood, reaction (Fig. 43.13B). Defects in nucleotide excision repair genes cause the human genetic disease **xeroderma pigmentosum** (XP), which is characterized by hypersensitivity to sunlight and predisposition to skin cancer. Eight proteins encoded by genes mutated in xeroderma pigmentosum (Table 43.1) take part in nucleotide excision repair, providing one of the best examples in which human genetics has helped to unravel a complicated biological process. **Recognition** of the DNA lesion involves the heterotrimeric replication protein RPA, XPA, and XPC, and the nine-subunit transcription factor TFIIH, which contains XPB and XPD. ATP-dependent unwinding of the DNA by XPB and XPD forms a **preincision complex.** XPG, which replaces XPC in the complex, makes an **incision** six to nine bases 3′ of the damaged base, and XPF-Ercc1 cuts 20 to 25 bases 5′ of the damage site. This releases a short single-stranded DNA fragment containing the damaged DNA. After excision, DNA polymerases δ or ε fill in the gap by copying the undamaged strand. Prokaryotes have a similar system of adduct recognition, removal, and repair involving the UvrA, UvrB, and UvrC proteins; however, the enzymes that are involved are not conserved between kingdoms.

BOX 43.1 DNA Repair in Vertebrates*—cont'd

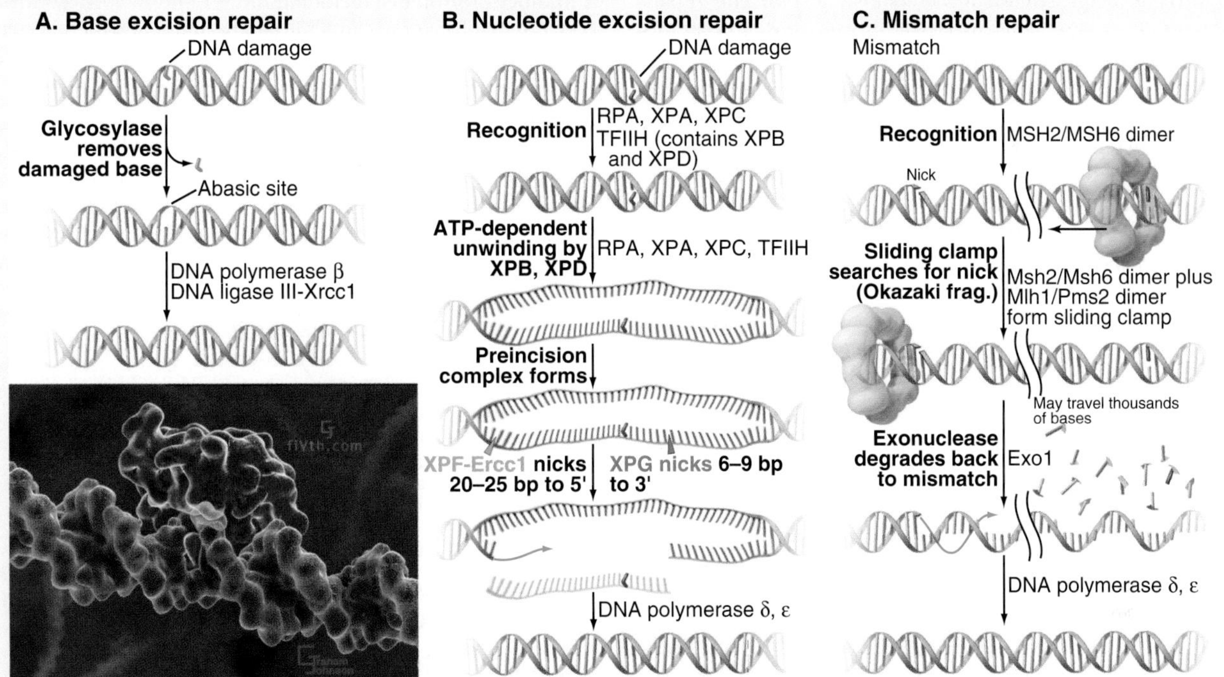

FIGURE 43.13 PATHWAYS FOR THE REPAIR OF BASE DAMAGE, BULKY ADDUCTS SUCH AS THYMIDINE DIMERS FORMED BY ULTRAVIOLET LIGHT, OR MISMATCHED BASES. For detailed descriptions, see the text. The *inset* in panel **A** shows human 3-methyladenine DNA glycosylase complexed to DNA. This enzyme scans the DNA for bases that are not strongly H-bonded, uses its "finger" to swing them up into the pocket for scanning, and, if they are damaged, catalyzes excision of that base. (Inset illustration by Graham Johnson [www.fivth.com] for the Howard Hughes Medical Institute, copyright 2004, all rights reserved. For reference, see PDB file 1BNK and Lau AY, Scharer OD, Samson L, et al. Crystal structure of a human alkylbase-DNA repair enzyme complexed to DNA: mechanisms for nucleotide flipping and base excision. *Cell.* 1998;95:249–258.)

Mismatch Repair

Errors in DNA replication missed by the proofreading activity of the DNA polymerase are recognized by a dimer consisting of the MSH2 and MSH6 proteins. When a mismatch is detected, this heterodimer undergoes an ATP-dependent transition to a **sliding clamp** and recruits a second heterodimer, consisting of MLH1 and PMS2 (Fig. 43.13C). To distinguish between the original ("correct") sequence and the newly synthesized DNA strand, this sliding clamp complex can then translocate along the DNA until a break is reached, such as that found between Okazaki fragments. The broken strand is therefore identified as the newly synthesized DNA strand. The mismatch repair complex then recruits the exonuclease EXO1 and degrades the newly synthesized DNA strand all the way back to the misincorporated base. The resultant long, single-stranded region is stabilized by binding of RPA and eventually filled in by the replicative DNA polymerases δ and/or ε. A second type of "mismatch" involves mistaken incorporation of ribonucleotides into DNA by either DNA polymerases δ and ε. Specialized ribonuclease H (RNAse H) enzymes detect and cleave DNA/RNA hybrids and remove the ribonucleotides.

Double-Strand Break Repair

DNA double-strand breaks can be caused by ionizing radiation or radiomimetic drugs or arise spontaneously after replication. They are particularly hazardous forms of damage, as they carry the risk of losing chromosomal material or, if misrepaired, causing chromosomal translocations. Two major pathways repair double-strand breaks. Homologous recombinational repair uses undamaged DNA as a template for the accurate repair of double-strand breaks, this sequence usually being derived from the sister chromatid after replication. Nonhomologous end joining (NHEJ) repairs double-strand breaks with no requirement for homology. NHEJ is the predominant activity that repairs double-strand breaks in the G_1 and early S phase, whereas homologous recombination becomes more important in the late S and G_2 phase. Homologous recombinational repair is normally extremely accurate, but NHEJ often introduces errors, as it can join both related and unrelated DNA ends together. Both pathways require the activity of the MRN protein complex (Mre11/RAD50/Nbs1), which localizes to DNA double-strand breaks and is also found at telomeres. The exonuclease activity of this complex resects (chews back) broken DNA ends

Continued

BOX 43.1 DNA Repair in Vertebrates*—cont'd

to provide single-stranded DNA substrates for the repair systems. These mechanisms repair also repair double-strand breaks during genome editing (see Fig. 6.16).

The key protein required for homologous recombinational repair in cells is **Rad51,** the eukaryotic homologue of *Escherichia coli* RecA (Fig. 43.14A). Rad51 forms an extended nucleoprotein filament on single-stranded DNA, replacing RPA, which acts like a placeholder when single-stranded DNA is produced. Rad51 catalyses the search for homologous sequences, strand pairing, and strand exchange. A number of proteins, including BRCA1 and BRCA2, control the activity of mammalian Rad51 in homologous recombinational repair. Inactivation of BRCA1 and BRCA2 predisposes to cancer. The helicase Rad54 facilitates strand invasion, when the single-stranded region forces its way into the complementary DNA duplex on the undamaged sister chromatid. Following invasion of the recombining DNA strands, polymerase activity extends the DNA beyond the site of the double-strand break, leading to the formation of a **Holliday junction** (Fig. 43.14B). Resolution of the Holliday junction

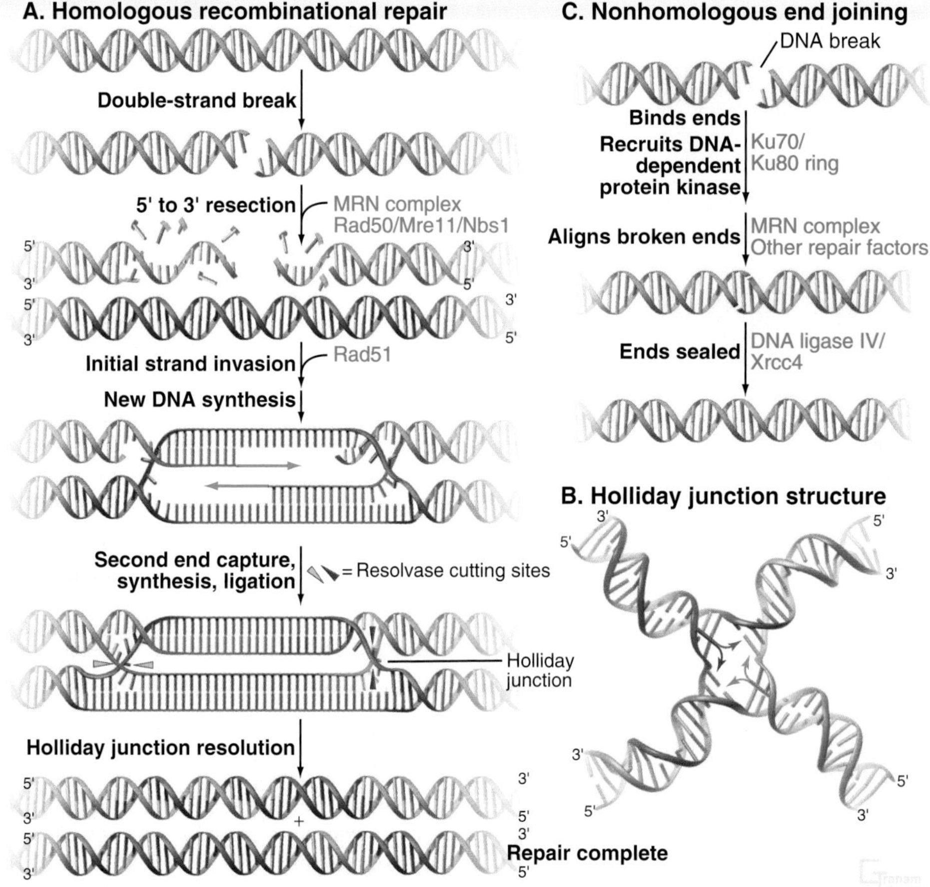

FIGURE 43.14 PATHWAYS FOR THE REPAIR OF DNA DOUBLE-STRAND BREAKS. A double-strand break is recognized by the MRN complex, which recruits ATM (ataxia-telangiectasia mutated). The break is then bound by repair factors involved in either the homologous recombination or nonhomologous end-joining pathway of DNA repair. **A,** Homologous recombination pathway of DNA repair. The MRN complex in conjunction with other nucleases chews back (resects) the DNA at a break, leaving a single-stranded overhang that is stabilized by RPA (not shown). This recruits ATR (ataxia-telangiectasia and Rad3 related) to the damage site. It is believed that the MRN complex also plays a role in keeping the broken ends close to one another. Next, RAD51 forms a nucleoprotein filament on the single-stranded DNA, displacing the RPA. The Rad51 nucleoprotein filament then initiates homology searching and repairs the DNA break by inserting the extended single-stranded DNA into homologous sequences (usually on the sister chromatid *[blue]*) and allowing homologous recombination and DNA repair/resynthesis to occur. Capture of the second single-stranded DNA end allows the formation of a joint molecule with a double Holliday junction. Resolution of this Holliday junction structure results in accurate, templated repair of the double-strand break. **B,** A Holliday junction formed by four complementary oligonucleotides complexed to the enzyme Cre (not shown). The Holliday junction is a dynamic structure *(arrows)* that can migrate along the DNA. **C,** Nonhomologous end-joining pathway of DNA repair. This pathway is initiated by break recognition by the Ku70/Ku80 heterodimer, which recruits DNA-PK and tethers the broken ends. The breaks are then processed in a reaction involving the MRN complex and other repair factors. DNA-PK's precise role is not yet entirely clear. Next, DNA ligase IV/XRCC4 is recruited to the processed double-strand break, which is ligated back together. (**B,** For reference, see PDB files 3CRX and 2CRX and Gopaul DN, Guo F, Van Duyne GD: Structure of the Holliday junction intermediate in Cre-loxP site-specific recombination. *EMBO J* 1998;17: 4175–4187.)

BOX 43.1 DNA Repair in Vertebrates*—cont'd

and filling-in of the repaired DNA sequences results in complete repair of the lesion. One of two known human Holliday junction resolvases is the Gen1 nuclease, a relative of the Fen1 flap endonuclease that functions during DNA replication (see Fig. 42.12). (The other more complex resolvase is the product of four genes.) Cells must have one or the other of these enzymes or they die during mitosis because of an inability to separate DNA strands that have undergone repair in the previous cell cycle. Note that Fig. 43.14 shows only a subset of the proteins involved in homologous recombinational repair. It is likely that new factors remain to be identified.

NHEJ is initiated at a DNA double-strand break by binding of the Ku70 and Ku80 heterodimer as a ring that binds the catalytic subunit of the **DNA-dependent protein kinase**, stimulating other repair factors and aligning the broken ends of the DNA (Fig. 43.14C). A complex of the protein XRCC4 with DNA ligase IV seals the ends of the broken DNA. NHEJ is also necessary for V(D)J recombination and therefore for the development of the immune system (see Fig. 28.10).

Given the importance of accurate transmission of the genetic material, deficiencies in DNA repair and checkpoint genes are associated with a number of diseases (Table 43.1). Note that several DNA repair activities are essential for life, so their inactivation has not been described in any human diseases.

*By Ciaran Morrison, National University of Ireland, Galway.

Chk1 and Chk2. Phosphorylation of the ATM adapter, 53BP1 (p*53*-*b*inding *p*rotein 1), recruits Chk2 for activation by ATM. The trimeric **9-1-1 complex** is required for Chk1 activation by ATR. This complex, which gets its name from its subunits Rad*9*, Hus*1*, and Rad*1*, resembles proliferating cell nuclear antigen (PCNA), the doughnut-shaped processivity factor that is indispensable during DNA replication (see Fig. 42.12). PCNA is loaded onto DNA by replication factor C (RFC, a pentameric AAA-ATPase) and anchors DNA polymerases and other factors to DNA. A similar ATPase composed of one special subunit, Rad17, plus the four small subunits of RFC loads the 9-1-1 complex onto DNA at or near sites of damage. Mutants in those four RFC subunits are defective in G_2/M checkpoint control in yeasts, *Drosophila* and *Caenorhabditis elegans.* RPA stimulates loading of the 9-1-1 complex at damage sites, making it specific for regions of single-stranded DNA.

Phosphorylation releases Chk1 and Chk2 from chromatin, so they can diffuse throughout the nucleus and cell to implement the DDR. They also trigger the G_2/M checkpoint response by altering the cell cycle machinery and inducing the transcription of key EFFECTORS. In some cases their actions trigger cell death by apoptosis.

Activation of Chk1 is important to establish the G_2/M checkpoint response because Chk1 phosphorylates the Cdc25A and Cdc25C protein phosphatases thereby blocking cell-cycle progression (Fig. 43.11). Phosphorylation produces binding sites for a 14-3-3 protein that blocks Cdc25A from activating Cdk1–cyclin B. Chk1 phosphorylation also targets Cdc25A for ubiquitin-mediated proteolysis ensuring that levels of Cdc25A remain low.

The transcription factor p53 (see Fig. 41.15), another EFFECTOR of the G_2/M checkpoint, is phosphorylated and activated following DNA damage (Fig. 43.11). Activated p53 stimulates transcription of the **Cdk inhibitor p21**. Although p21 is best known for promoting cell-cycle arrest in G_1 cells as part of the G_1 DNA damage checkpoint, it can also act in G_2. Expression of p21 is an effective way of blocking the initiation of prophase, because it inhibits Cdk1–cyclin A approximately 100-fold better than it inhibits Cdk1–cyclin B1.

Active p53 also drives the expression of 14-3-3σ, an adapter protein that interferes with shuttling of Cdk1–cyclin B1 between the nucleus and cytoplasm (Fig. 43.11). Binding of 14-3-3σ maintains the Wee1 inhibitory kinase in a *more* active state, ensuring that the Cdk1–cyclin B1 complex remains inactive. Disruption of the gene for 14-3-3σ is fatal for cells with DNA damage. Instead of activating their G_2/M checkpoint, they enter an aberrant state with characteristics of both mitosis and apoptosis, and then die.

Typically, G_2/M checkpoint activation has one of three outcomes. If DNA damage is so extensive that it cannot be repaired, the cell either enters a non-proliferating state known as senescence or commits suicide by apoptosis (see Chapter 46). Less-serious damage can be repaired by one of the systems described in Box 43.1.

Transition to Mitosis

The complex web of stimulatory and inhibitory activities in the G_2 phase poises Cdk1–cyclin B in a state ready for the explosive burst of activation that triggers the G_2/M transition. These complex regulatory pathways are the basis of the G_2/M checkpoint control that prevents cells from segregating their chromosomes if genomic DNA cannot meet stringent quality control standards. Eventually, however, if all goes well, Cdk1–cyclin A and Cdk1–cyclin B1 are activated, and the cell embarks on mitosis, probably the most dramatic event of its life.

ACKNOWLEDGMENTS

We thank Anton Gartner, Ciaran Morrison, and Ashok Venkitaraman for their suggestions on revisions to this chapter.

SELECTED READINGS

Ciccia A, Elledge SJ. The DNA damage response: making it safe to play with knives. *Mol Cell*. 2010;40:179-204.

Hartwell LH, Weinert TA. Checkpoints: Controls that ensure the order of cell cycle events. *Science*. 1989;246:629-634.

Jackson SP, Bartek J. The DNA-damage response in human biology and disease. *Nature*. 2009;461:1071-1078.

Mehta A, Haber JE. Sources of DNA double-strand breaks and models of recombinational DNA repair. *Cold Spring Harb Perspect Biol*. 2014;6:a016428.

Melo J, Toczyski D. A unified view of the DNA-damage checkpoint. *Curr Opin Cell Biol*. 2002;14:237-245.

Morgan DO. *The Cell Cycle: Principles of Control*. London: New Science Press; 2007.

Parrilla-Castellar ER, Arlander SJ, Karnitz L. Dial 9-1-1 for DNA damage: the Rad9-Hus1-Rad1 (9-1-1) clamp complex. *DNA Repair (Amst)*. 2004;3:1009-1014.

Paull TT. Mechanisms of ATM activation. *Annu Rev Biochem*. 2015;84: 711-738.

Pearl LH, Schierz AC, Ward SE, et al. Therapeutic opportunities within the DNA damage response. *Nat Rev Cancer*. 2015;15:166-180.

Polo SE, Jackson SP. Dynamics of DNA damage response proteins at DNA breaks: a focus on protein modifications. *Genes Dev*. 2011;25: 409-433.

Smits VA, Medema RH. Checking out the G(2)/M transition. *Biochim Biophys Acta*. 2001;1519:1-12.

Wieser S, Pines J. The biochemistry of mitosis. *Cold Spring Harb Perspect Biol*. 2015;7:a015776.

Wood RD, Mitchell M, Sgouros J, et al. Human DNA repair genes. *Science*. 2001;291:1284-1289.

CHAPTER 44

Mitosis and Cytokinesis

Mitosis is the division of a somatic cell (a vegetative cell in yeast) into two daughter cells. The daughters are usually identical copies of the parent cell, but the process can be asymmetrical. For example, division of some stem cells gives rise to one stem cell and another daughter cell that goes on to mature into a differentiated cell. See Box 41.2 for examples.

Traditionally, mitotic events are subdivided into six phases: **prophase, prometaphase, metaphase, anaphase, telophase,** and **cytokinesis** (Fig. 44.1). The dramatic reorganization of both the nucleus and cytoplasm during the mitotic phases is brought about by activation of a number of protein kinases, including Cdk1-cyclin B-cks (abbreviated here as "Cdk1 kinase"; see Chapter 40). After activation by Cdc25 phosphatase, Cdk1 kinase accumulates in the nucleus, where it joins Cdk1-cyclin A-cks, which was activated somewhat earlier (see Fig. 43.6). These two Cdk1 kinase complexes operate as both master controllers and workhorses that directly phosphorylate many proteins whose functional and structural status is altered during mitosis. Their progressive inactivation following the correct attachment of the chromosomes to spindle microtubules drives the orderly exit of cells from mitosis.

Mitosis is an ancient process, and a number of variations emerged during eukaryotic evolution. Many single-celled eukaryotes, including yeast and slime molds, undergo a **closed mitosis,** in which spindle formation and chromosome segregation occur within an intact nuclear envelope to which the spindle poles are anchored. This chapter focuses on **open mitosis,** as used by most plants and animals, in which the nuclear envelope disassembles before the chromosomes segregate. Fig. 44.2 summarizes some of the important events during the various mitotic phases.

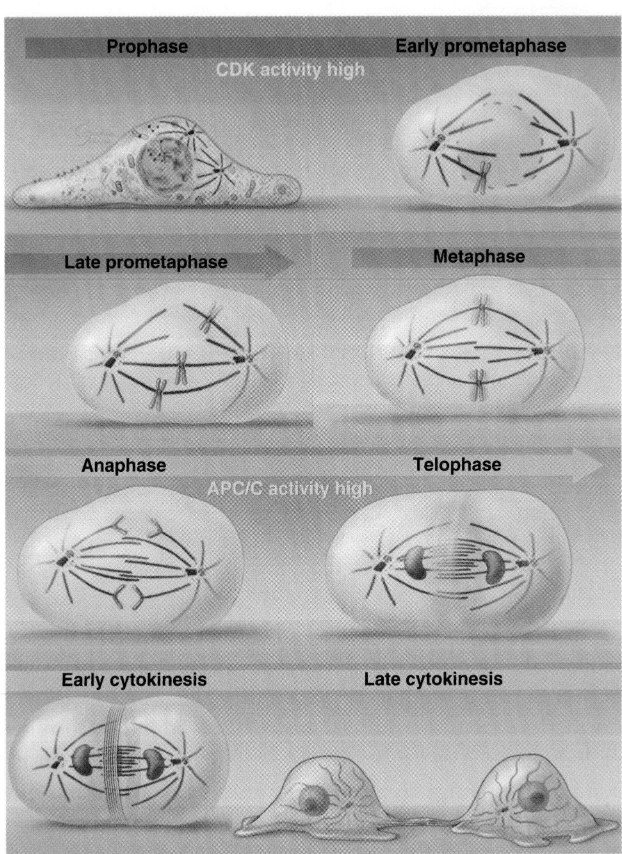

FIGURE 44.1 OVERVIEW OF THE PHASES OF MITOSIS. APC/C, anaphase-promoting complex/cyclosome.

Prophase

Prophase, the transition from G_2 into mitosis, begins with the first visible condensation of the chromosomes and disassembly of the nucleolus (Fig. 44.3). In the

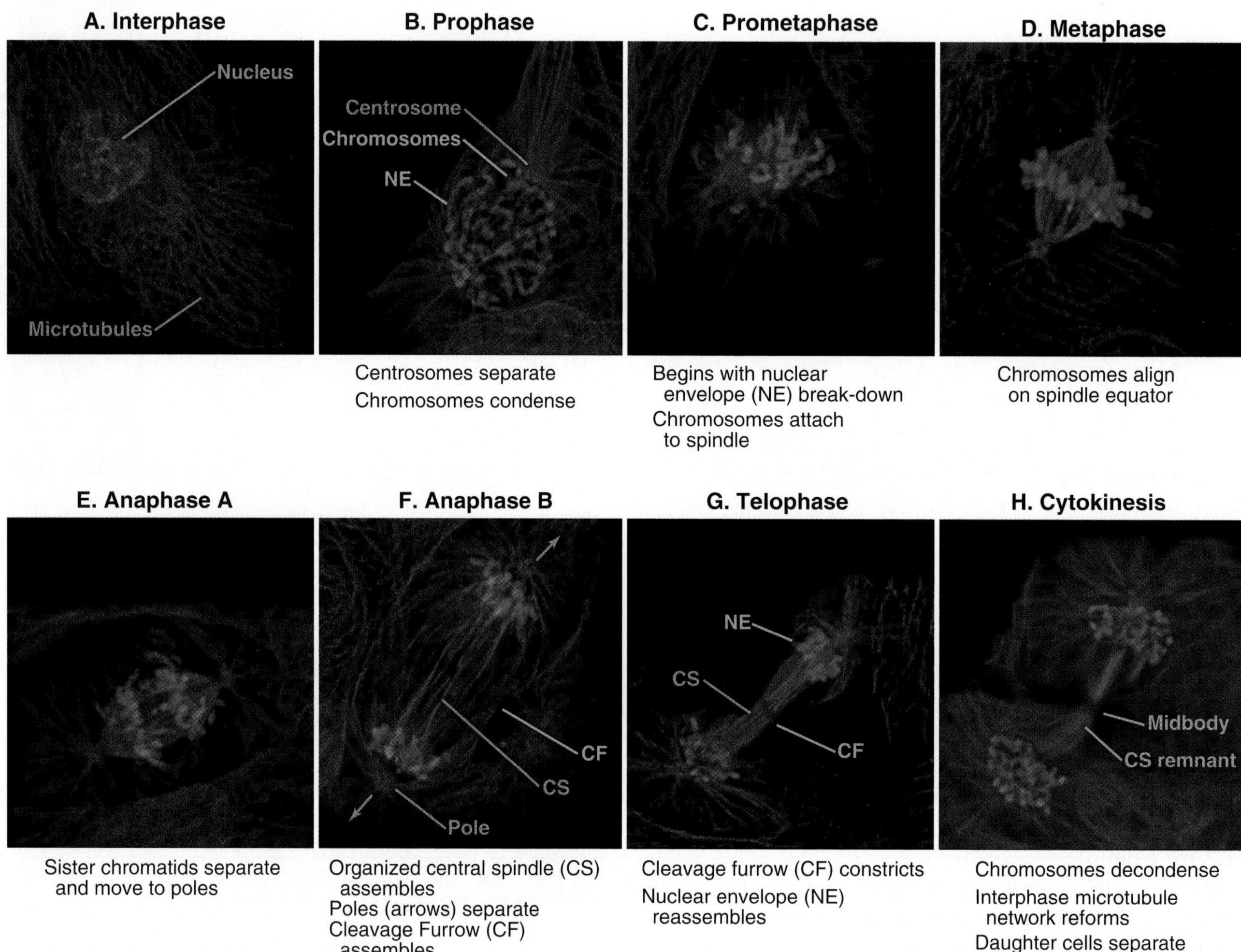

FIGURE 44.2 KEY EVENTS OF MITOSIS. A–C, *Prophase–prometaphase*: Cdk1 kinase triggers condensation of replicated sister chromatids, disassembly of the nuclear envelope and Golgi, and a dramatic reorganization of the cytoskeleton. As the barrier between the chromosomes and cytoplasm is abolished, microtubules contact the condensed chromosomes and attach at the kinetochores (see Fig. 8.21). Interaction of kinetochores with dynamic microtubules culminates with the chromosomes aligned at the midplane of the bipolar spindle. **D–F,** *Metaphase–anaphase*: Once all chromosomes achieve a bipolar attachment to the spindle, an inhibitory signal is silenced, leading to activation of a proteolytic network that destroys proteins responsible for holding sister chromatids together and also inactivates Cdk1 by destroying its cyclin B cofactor (see Fig. 40.15). Sister chromatids separate and move toward opposite spindle poles, which themselves move apart. **G–H,** *Telophase–cytokinesis*: Targeting of nuclear envelope components back to the surface of the chromatids leads to the reformation of two daughter nuclei. In most cells, the two daughter nuclei and the surrounding cytoplasm are partitioned by cytokinesis. (Micrographs courtesy William C. Earnshaw.)

cytoplasm, the interphase network of long microtubules centered on a single centrosome (see Fig. 34.18) is converted into two radial arrays of short microtubules called *asters.* Most types of intermediate filaments disassemble, the Golgi apparatus and endoplasmic reticulum fragment, and both endocytosis and exocytosis are curtailed.

Nuclear Changes in Prophase

Chromosome condensation, the landmark event at the onset of prophase, often begins in isolated patches of chromatin at the nuclear periphery. Later, chromosome condense into two threads termed **sister chromatids** that are closely paired along their entire lengths. Although chromosome condensation was first observed more than a century ago, the biochemical mechanism remains a mystery. Protein kinases trigger mitotic chromosome condensation and onset of condensation is correlated with phosphorylation of histones H1 by Cdk1 kinase and H3 by **Aurora-B** protein kinase. However, chromosomes still condense when both of these phosphorylation events are blocked. It is possible that a combination of histone modifications promotes mitotic chromatin condensation (see Fig. 8.3).

Two pentameric protein complexes, **condensin** I and condensin II are major regulators of mitotic chromosome architecture. These complexes share the SMC2 and SMC4 (*s*tructural *m*aintenance of *c*hromosomes) ABC adenosine triphosphatases (ATPases), but have two different sets of three auxiliary proteins (see Fig. 8.18).

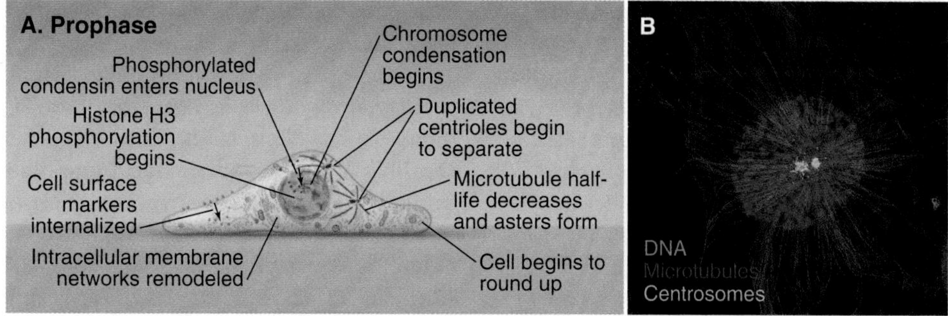

FIGURE 44.3 INTRODUCTION TO PROPHASE. A, Summary of the major events of prophase. **B,** Distribution of DNA *(blue)*, microtubules *(red)*, and γ-tubulin (centrosomes *[green]*) in a prophase human cell. (**B,** Images were recorded by Dr. Melpomeni Platani on the University of Dundee's School of Life Sciences Imaging Facility OMX 3DSIM Microscope and stored and processed in OMERO.)

Condensins are required to disassemble the topologically associating domains (TAD) of interphase chromatin (see Fig. 8.13) and promote the formation of the linear array of loops that characterizes mitotic chromosomes (see Fig. 8.14). The decisive experiment was to target SMC2 for rapid degradation, with the result that no well-organized mitotic chromosomes formed. The chromatin compacted but was less organized. Slower depletion of condensins by RNA interference (RNAi) or conditional gene disruption gave less-dramatic results.

Condensin complexes can encircle DNA and also promote its supercoiling in vitro, but how these activities help them to orchestrate the changes in chromatin architecture is not known. Condensin II is nuclear during interphase, and has an important role in prophase chromosome formation following its activation by phosphorylation. Condensin I acts both in prophase and prometaphase in further compacting the chromosomes. DNA topoisomerase II and other scaffold proteins also function during mitotic chromosome formation, but condensin appears to play the major role.

Cytoplasmic Changes in Prophase

Most of the cytoskeleton reorganizes during prophase. The microtubule array changes from an extensive network permeating the cytoplasm into two dense, radial arrays of short, dynamic microtubules around the duplicated centrosomes (see Fig. 34.17). Each of these **asters** eventually becomes one **pole** of the mitotic spindle. During prophase, the two asters usually migrate apart across the surface of the nuclear envelope, signaling the start of spindle assembly (Fig. 44.3).

Mitotic microtubules behave like interphase microtubules in many ways (see Chapter 34). They are mostly nucleated at their minus ends, they grow by addition of tubulin subunits at their plus ends, and they undergo random catastrophes during which they rapidly shorten. To a large extent, the prophase changes in microtubule organization can be explained by two simple biochemical changes: (a) increased microtubule-nucleating activity of centrosomes and (b) altered dynamic instability

TABLE 44.1 Comparison of Microtubule Dynamics in Interphase and Mitotic Newt Lung Cells

Parameter	Interphase	Mitosis
Elongation rate	7 μm/min	14 μm/min
Elongation time before catastrophe	71 s	60 s
Shortening rate	17 μm/min	17 μm/min
Probability of rescue from catastrophe*	0.046/s	0
Length	100 μm	14 μm

*Most cellular microtubules grow constantly by addition of subunits to their free ends but they occasionally stop growing and begin shrinking rapidly (a "catastrophe"). Unless shrinking is reversed (a "rescue"), the microtubule completely disappears.
Data from Gliksman NR, Skibbens RV, Salmon ED: How the transition frequencies of microtubule dynamic instability regulate microtubule dynamics in interphase and mitosis. *Mol Biol Cell.* 1993;4:1035–1050.

properties of the microtubules (Table 44.1; also see Chapter 34). Interphase microtubules have a high probability of recovering from catastrophes, so they grow quite long. Mitotic microtubules grow more rapidly but exist only transiently, because rescues are rare following a catastrophe. Thus, they usually shorten all the way back to the centrosome, with little chance of rescue. These differences in dynamic instability can be reproduced in vitro in mitotic and interphase cellular extracts. They appear to arise, at least in part, from counterbalancing interactions between microtubule-associated proteins that promote microtubule stability, and kinesin-13 (see Fig. 36.13), which promotes microtubule disassembly.

Other cytoskeletal elements that disassemble during prophase include many, but not all, classes of intermediate filaments (including the nuclear lamins) (see Fig. 35.1A) and specialized actin filament structures, such as stress fibers. However, the junctional complexes between adjoining cells are maintained in epithelial cells. As a result of the cytoskeletal reorganization, most cells round up during prophase. This is particularly evident for animal cells that are cultured on a flat substrate, but cells in tissues also change their shape dramatically during mitosis.

RNA transcription of the chromosomes stops during mitosis except for highly specialized transcription at centromeres. Phosphorylation of components of the transcriptional machinery by Cdk1 kinase appears to be responsible for this shutoff. Cdk1 kinase phosphorylation of ribosomal elongation factor 2a (EF2a) also stops most (but not all) ongoing protein synthesis and assembly of new ribosomes. Phosphorylation of several nucleolar proteins leads to disassembly of the nucleolus.

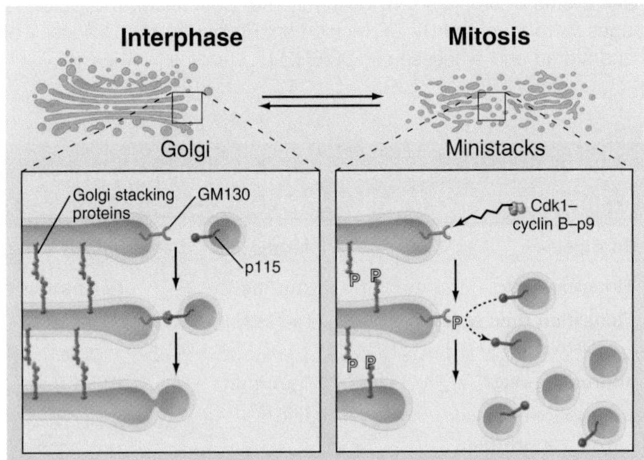

FIGURE 44.4 GOLGI APPARATUS DYNAMICS IN INTERPHASE AND MITOSIS. Disassembly in mitosis is driven by phosphorylation of components blocking fusion of Golgi membranes.

The Golgi apparatus and endoplasmic reticulum fragment and disperse during prophase (Fig. 44.4). Several kinases, including Cdk1 drives Golgi apparatus disassembly, the first step being fragmentation into smaller ministacks following phosphorylation of Golgi stacking proteins and tethers. Later steps are still being investigated. Many lines of evidence argue that Cdk1 phosphorylation of key components prevents the fusion of transport vesicles back into Golgi stacks (see Chapter 21), the net result being that the Golgi buds away into small vesicles that disperse throughout the mitotic cell cytoplasm. Other evidence suggests that an imbalance of vesicle flow between the Golgi and the endoplasmic reticulum results in the Golgi being absorbed into the endoplasmic reticulum during mitosis. Whatever the mechanism of its disassembly, Golgi reassembly begins again during late anaphase/early telophase, following inactivation of Cdk1 kinase.

Prometaphase

In cells that undergo an open mitosis, prometaphase begins abruptly with disassembly of the nuclear envelope (Fig. 44.5). Microtubules growing outward from the spindle poles penetrate holes in the nuclear envelope, make contact with the chromosomes, and attach to them

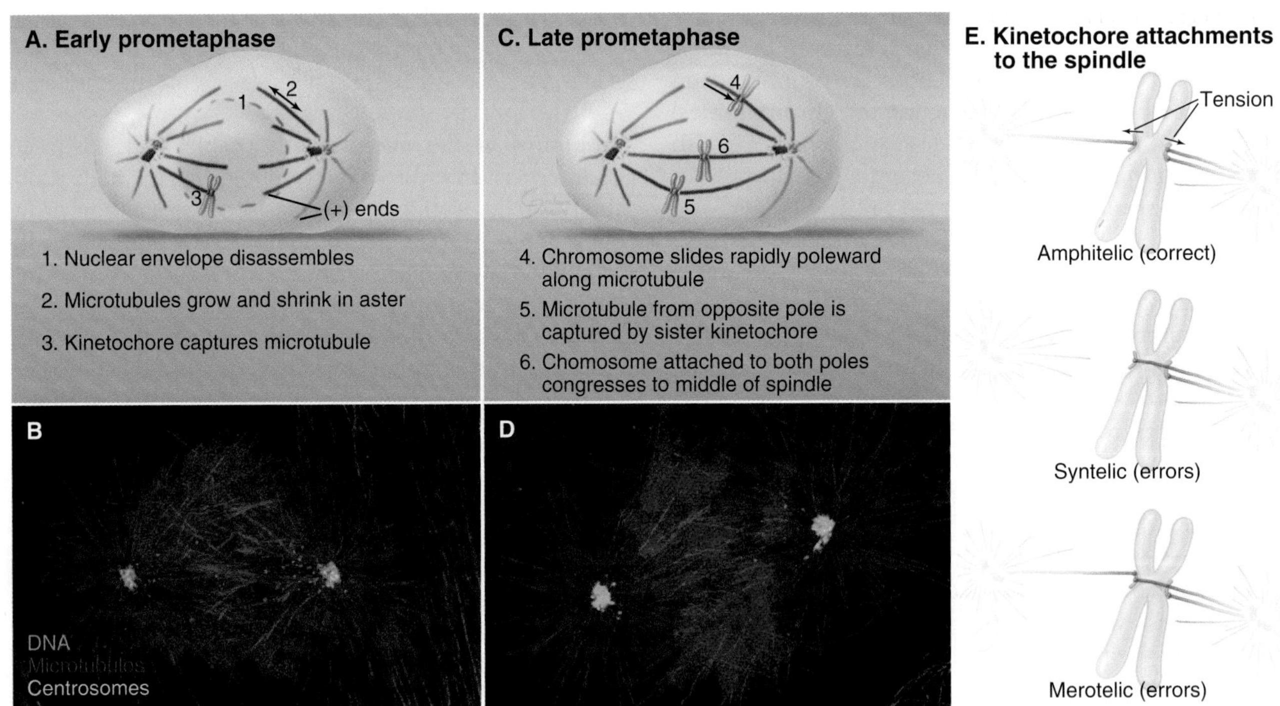

FIGURE 44.5 INTRODUCTION TO PROMETAPHASE. A, Summary of the key events of early prometaphase. **B,** Distribution of DNA *(blue),* microtubules *(red),* and γ-tubulin (centrosomes *[green]*) in early prometaphase human cells. **C,** Summary of the key events of late prometaphase. **D,** Distribution of DNA, actin, microtubules, and centrosomes in late prometaphase PtK1 cells. **E,** Terms used to describe the orientation of kinetochore attachments to the mitotic spindle. (**B** and **D,** Images were recorded by Dr. Melpomeni Platani on the University of Dundee's School of Life Sciences Imaging Facility OMX 3DSIM Microscope and stored and processed in OMERO.)

at specialized structures called **kinetochores** (see Fig. 8.19). Interactions of the two opposing kinetochores of paired sister chromatids with microtubules from opposite poles of the spindle ultimately result in alignment of the chromosomes in a group midway between the poles. An important cell-cycle checkpoint (see Chapter 40) known as the **spindle assembly checkpoint (SAC)** delays the onset of chromosome segregation until all kinetochores are attached to microtubules.

Nuclear Envelope Disassembly in Prometaphase

Nuclear envelope disassembly involves the removal of two membrane bilayers coupled with disassembly of the nuclear pores and the fibrous **nuclear lamina** meshwork that underlies the inner bilayer (Fig. 44.6). Phosphorylation causes the nucleoporin Nup98 to dissociate from nuclear pores. This removes the permeability barrier between nucleus and cytoplasm. Phosphorylation of other proteins causes the pore to disassemble to soluble subcomplexes. Phosphorylation of the nuclear lamins at two sites flanking the coiled-coil causes the lamina network to disassemble into subunits. Interaction between microtubules and dynein associated with the nuclear envelope can rip holes in the envelope, although this is not required for nuclear envelope disassembly.

Nuclear envelope membranes are dispersed in the cytoplasm from prometaphase until telophase (Fig. 44.6), but the mechanism is not settled. Some experiments suggest that the nuclear membranes break up into small vesicles that disperse in the cytoplasm. Other experiments suggest that the nuclear envelope is absorbed into the endoplasmic reticulum, which remains as an extensive tubular (or flattened cisternal network—another source of discussion) throughout mitosis. Further experiments are required to answer this question, and both mechanisms could contribute. Lamin B remains associated with the dispersed nuclear envelope, whereas lamins A and C and many proteins of the nuclear pore complexes disperse as soluble subunits.

During prophase, kinetochores transform from nondescript balls of condensed chromatin into structures on the surface of the chromosomes. By early prometaphase, the characteristic trilaminar disk structure (see Fig. 8.19) can be seen. Each sister chromatid has a kinetochore. **Sister kinetochores** are located on opposite faces of the mitotic chromosome.

Organization of the Mitotic Spindle

The mature metaphase spindle is a bilaterally symmetrical structure with centrally located chromosomes flanked by arrays of microtubules radiating from the poles (Fig. 44.7).

Three predominant classes of microtubules are present in the metaphase spindle (Fig. 44.12). **Kinetochore microtubules** have their plus ends embedded in the kinetochore and their minus ends at or near the spindle pole. They characteristically form bundles, called **kinetochore fibers,** which contain anywhere from 1 microtubule in the budding yeast to more than 200 microtubules in some higher plants. Each human kinetochore binds approximately 20 microtubules. Up to approximately 80% of the approximately 2200 spindle microtubules in humans may be present in kinetochore fibers, but not all microtubules in those fibers stretch all

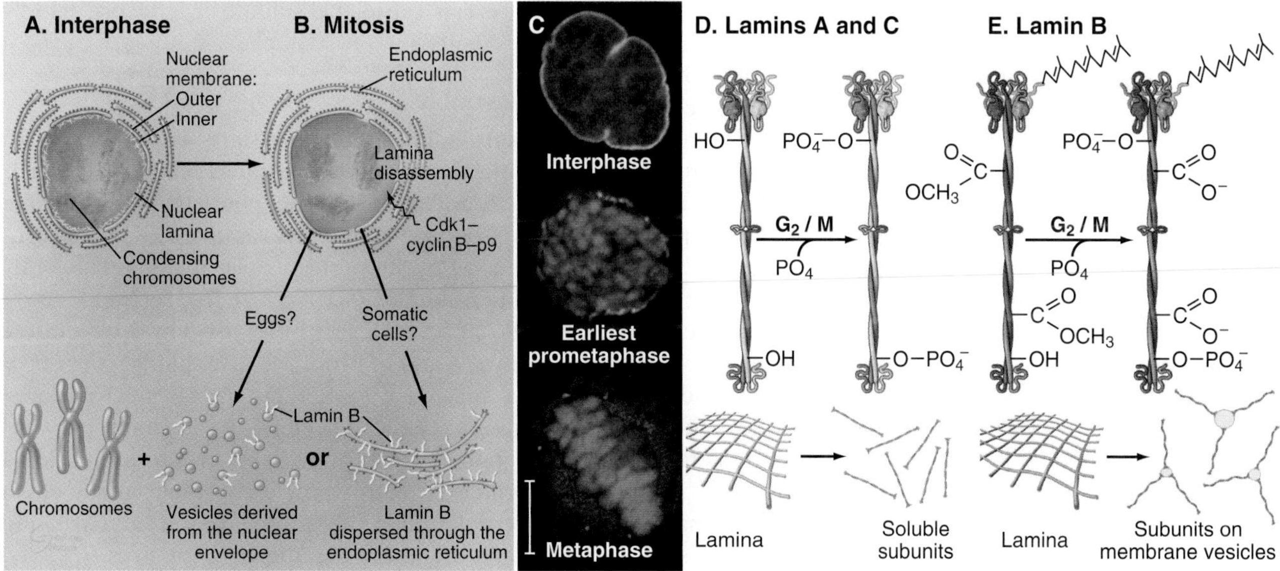

FIGURE 44.6 DISASSEMBLY OF THE NUCLEAR ENVELOPE DURING MITOSIS. A–B, Two contrasting models to explain the fate of the nuclear envelope during the transition from interphase to mitosis in a higher eukaryote. **C,** Micrographs showing solubilization of lamin A fused to green fluorescent protein (GFP) *(green)* during mitosis. DNA is *blue.* Scale bar is 10 μm. **D–E,** Reversible disassembly of lamins A, C, and B is driven by posttranslational modifications of the lamin polypeptides. (**C,** Courtesy William C. Earnshaw.)

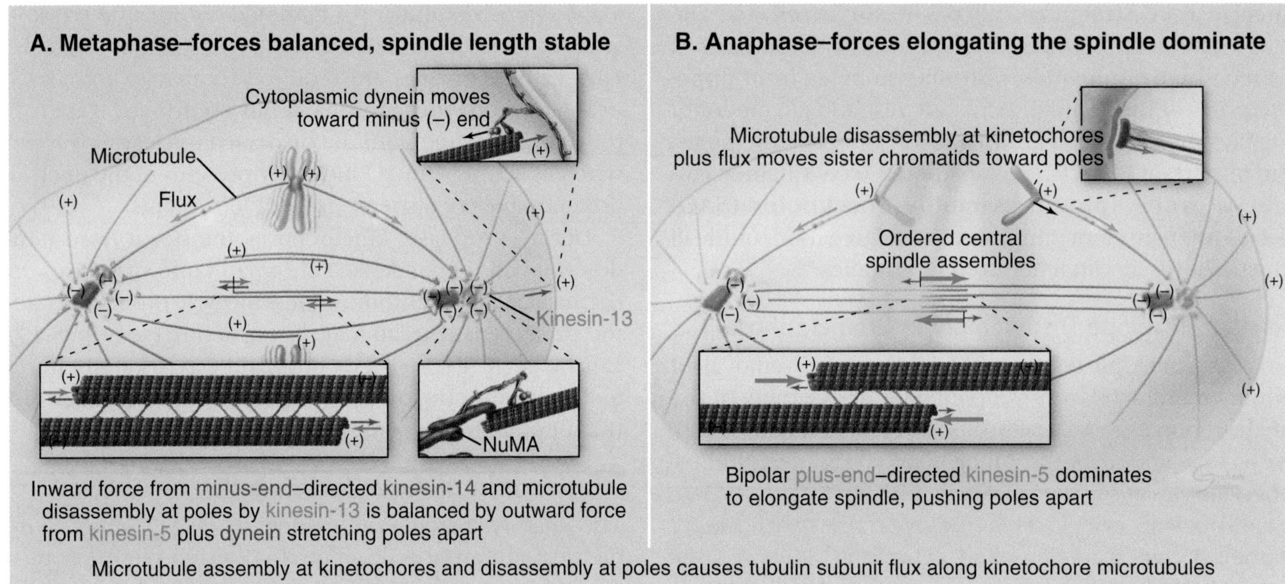

FIGURE 44.7 ROLE OF MOTOR PROTEINS IN SPINDLE DYNAMICS. Mitotic spindle structure depends on microtubule assembly/disassembly plus balanced forces that slide microtubules relative to one another and to pull the poles together or push them apart. **A,** In metaphase, the structure is at steady state. Forces that tend to elongate the spindle, including cytoplasmic dynein (which moves toward microtubule minus ends, pulling the poles out toward the cell cortex) and bipolar kinesin-5 (which moves toward microtubule plus ends, pushing the poles apart), are counterbalanced by cohesion between sister chromatids and kinesin-14, which moves toward microtubule minus ends (and pulls the poles together) and microtubule disassembly at the spindle poles. Dynein and its associated protein, NuMA (nuclear mitotic apparatus), also help to organize a focused spindle pole. **B,** In anaphase, sister chromatids separate, the balance of kinesin activity shifts, microtubule disassembly at the poles declines, and the spindle undergoes a dramatic elongation. During anaphase, bipolar kinesin-5, chromokinesin KIF4A, and protein regulated in cytokinesis 1 (PRC1) also have important roles in organizing the central spindle, which is essential for subsequent assembly and function of the cleavage furrow.

the way from the kinetochores to the spindle poles. **Interpolar microtubules** are distributed throughout the body of the spindle and do not attach to kinetochores. Many interpolar microtubules penetrate between and through the chromosomes and extend for some distance beyond them. Thus, the central spindle contains a large number of interdigitated antiparallel microtubules. Tracking these spindle microtubules by electron microscopy revealed a tendency for the interdigitated microtubules of opposite polarity to pack next to one another. During late anaphase, these antiparallel microtubules bundle to form a structure called the **central spindle** that plays important roles in signaling during cytokinesis. **Astral microtubules** project out from the poles and have a role in orienting and positioning the spindle in the cell through interactions with the cell cortex in somatic cells. All microtubules within each aster have the same polarity, with their minus ends proximal to the pole. Each unit of a spindle pole, with its associated kinetochore and interpolar and astral microtubules, is referred to as a **half-spindle.**

Spindle structure is largely determined by a combination of microtubule dynamics plus the action of at least seven different types of kinesins and cytoplasmic dynein (see Chapter 36). These motors often work in opposition to one another. Furthermore, forces exerted by motors can influence the dynamic assembly/disassembly of microtubules. As a result, the highly dynamic spindle changes shape as a delicate balance of forces shifts between the various motors. For example, inactivating one or more kinesins with drugs or switching a temperature-sensitive mutant to the nonpermissive temperature can cause the spindle to collapse rapidly on itself.

Spindle Assembly

In metazoans, spindle assembly starts in prophase with the separation of the asters. In most cells, each aster is organized around a centrosome, consisting of a centriole pair and associated pericentriolar material. **γ-Tubulin ring complexes** in the pericentriolar material efficiently nucleate microtubules (see Fig. 34.16), so each centrosome acts as a **microtubule organizing center** (MTOC). By the end of prophase, the spindle consists of two asters linked by a few interpolar microtubules. Cytoplasmic dynein at the cell cortex exerts an outward force separating the centrosomes, whereas kinesin-14 motors (which move toward microtubule minus ends) on the interpolar microtubules exert a counterbalancing force holding the asters together.

This balance of forces changes when the nuclear envelope breaks down. Bipolar kinesin-5 motors phosphorylated by Cdk1 kinase concentrate in the central spindle, where they crosslink adjacent antiparallel interpolar microtubules. Kinesin-5 moves toward the plus ends of microtubules, so when attached to two adjacent antiparallel microtubules, its action will cause them to slide and push the spindle poles apart (Fig. 44.7). However, the two half-spindles do not separate because they are physically linked via the chromosomes, with sister kinetochores attached to opposite spindle poles.

Also at this time, the asters mature into focused spindle poles. The pericentriolar material efficiently nucleates the assembly of new microtubules with their minus ends at the pole. Microtubule assembly at two other sites also contributes to spindle morphology. At kinetochores, a complex derived from interphase nuclear pores recruits γ-tubulin. This locally nucleates microtubules that grow by inserting subunits at the kinetochores, pushing the minus ends with γ-tubulin outwards toward the spindle poles. These preformed kinetochore fibers are then incorporated into the spindle. In the central spindle, microtubules are nucleated on the walls of other microtubules by γ-tubulin recruited by the multi-subunit **augmin** complex. The action of augmin creates branched microtubules throughout the central spindle, contributing to an even fir tree-like distribution of microtubules. The daughter microtubules may be released from their nucleation site, free at both ends, and move toward the spindle poles as part of the flux of tubulin from the center of the spindle toward the centrosomes.

The microtubule array focuses at the poles partly because centrosomes tether microtubules, and partly due to the concerted action of various motors and microtubule crosslinking proteins such as dynein and nuclear mitotic apparatus **(NuMA)** protein. NuMA is released from the nucleus when the nuclear envelope breaks down, and it accumulates near the poles at the minus ends of microtubules.

Large cells that lack centrosomes, such as eggs, form spindles by an alternative pathway that also functions in the background in cells with centrosomes (Fig. 44.8). This mechanism hijacks the nuclear trafficking system to enable chromosomes to control the activity of key spindle assembly factors. The nuclear import receptors importin α and β bind these factors, as though they were going to transport them into the nucleus. This blocks their spindle assembly activity. During mitosis, chromosomes bind high levels of RCC1, the guanine exchange factor for Ran (Ran-GEF in Fig. 9.18). This RCC1 creates a gradient of the active guanosine triphosphatase (GTPase) Ran-GTP (guanosine triphosphate) around the chromosomes. During interphase, Ran-GTP is confined to the nucleus, where it releases importins from their cargo. In mitosis the chromosome-associated cytoplasmic Ran-GTP gradient locally liberates the spindle assembly factors including motor proteins and NuMA from sequestration by importins. They then stabilize nearby microtubules and organize them into a bipolar spindle.

If centrosomes are removed or destroyed experimentally in cells about to enter mitosis, somatic cells can also use motor proteins to organize microtubules into bipolar spindles that lack asters but are otherwise remarkably normal. Most treated cells manage to complete mitosis successfully but normal mammalian cells then either arrest in the next cell cycle prior to replicating their DNA or commit suicide. Both outcomes depend on the presence of the important tumor suppressor protein p53 (see Fig. 43.11). Thus, centrosomes are not required to form spindles, but they contribute to cell-cycle progression in many cells. This dependence on centrosomes is not universal; *Drosophila,* for example, can live without centrosomes.

Chromosome Attachment to the Spindle

Dynamic microtubules of prometaphase asters scan the cytoplasm effectively "searching" for binding sites that will capture and stabilize their distal plus ends. Captured microtubules are approximately fivefold less likely to depolymerize catastrophically than free microtubules. When catastrophes do occur, the microtubules depolymerize back to the pole, recycling tubulin subunits for incorporation into other, growing microtubules.

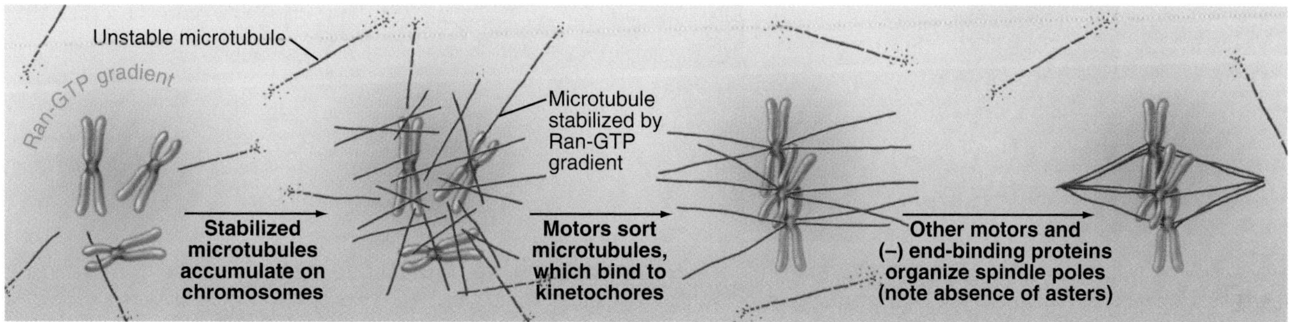

FIGURE 44.8 ASSEMBLY OF A BIPOLAR SPINDLE IN THE ABSENCE OF CENTROSOMES. A gradient of Ran-guanosine triphosphate (GTP) stabilizes microtubules around chromosomes, which contain high concentrations of bound Ran GEF RCC1. This releases spindle assembly factors from importin α and β. Microtubules that accumulate around the chromosomes are sorted, organized, and focused to make poles by motors and (–) end-binding proteins such as NuMA. These spindle poles lack prominent astral microtubules.

Breakdown of the nuclear envelope makes the condensed chromosomes accessible to the microtubules. Chance encounters allow kinetochores to capture microtubule plus ends. Capture probably involves the nine-component KMN network, which includes the rod-shaped Ndc80 complex (see Fig. 8.21) that binds along the sides of microtubules near their plus ends. Another member of the complex, the scaffolding protein Knl1 (its name in vertebrates—the "K" of KMN; see later), anchors Ncd80 in the kinetochore.

Historically, it was thought that forces generated by **bipolar attachment** of the kinetochores of sister chromatids center chromosomes midway between the two spindle poles. This hypothesis was based on the observation that when a kinetochore first attaches to a microtubule, the chromosome moves along the side of that microtubule toward the spindle pole (Fig. 44.9). Subsequent capture of a microtubule emanating from the opposite spindle pole by the sister kinetochore would provide a counterforce pulling the chromosome in the opposite direction. Chromokinesin family motor proteins distributed along the chromosome arms were also thought to contribute to the gradual movement of the chromosome toward the middle of the spindle. These movements are accompanied by coordinated shrinkage of the microtubules at the leading kinetochore and growth of microtubules at the trailing kinetochore.

More recent studies revealed that chromosomes attached to only one spindle pole can move away from that pole if the unattached kinetochore associates with the kinetochore fiber of a chromosome already aligned at the spindle equator. In this case, the kinetochore of the mono-oriented chromosome glides toward the equator, where it is more likely to capture microtubules emanating from the opposite pole. This motion of one chromosome along the kinetochore fiber of another chromosome requires the kinesin-7 motor centromere protein E (CENP-E) (see Fig. 36.13) associated with the kinetochore of the moving chromosome. Recognition of a tubulin posttranslational modification leads CENP-E to move the chromosome toward the spindle equator, rather than out into the aster.

The attachment of microtubules to kinetochores can be reconstituted in vitro from mixtures of chromosomes, isolated centrosomes, and tubulin subunits. The plus ends of microtubules grow out from centrosomes and

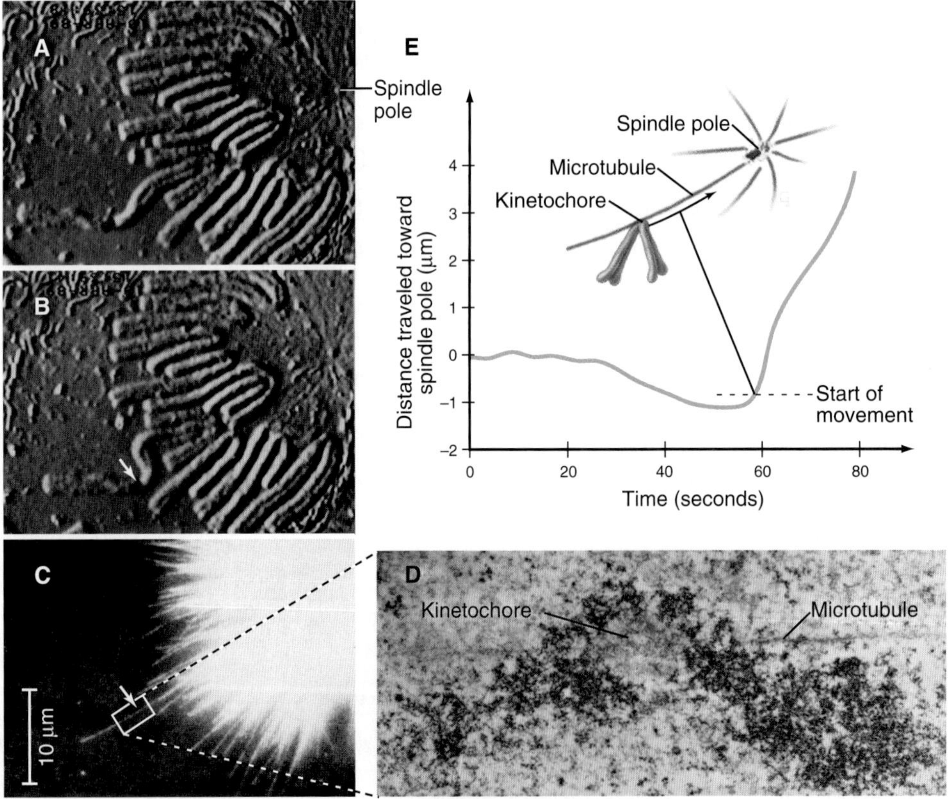

FIGURE 44.9 INITIAL CHROMOSOMAL MOVEMENTS DURING PROMETAPHASE. A–B, Capture of a microtubule by the kinetochore results first in movement along the side of the microtubule toward the pole from which that microtubule originated. These images come from a study in which living cells, observed by differential interference microscopy, were subjected to rapid chemical fixation just after a chromosome had attached to the spindle *(arrow).* **C,** Attachment of the chromosome to the spindle was confirmed by indirect immunofluorescence staining for tubulin and thin-section electron microscopy **(D). E,** The graph shows the movements of the chromosome before and after attachment. (From Rieder CL, Alexander SP, Rupp G. Kinetochores are transported poleward along a single astral microtubule during chromosome attachment to the spindle in newt lung cells. *J Cell Biol.* 1990;110:81–95, copyright The Rockefeller University Press.)

attach to the chromosomes. Surprisingly, chromosome-bound microtubules can either lengthen or shorten *at the attached end* without detaching from the chromosome. Similar experiments with kinetochores isolated from budding yeast cells showed that kinetochores can remain attached to a shortening microtubule plus end even against an applied force of 9 pN (piconewtons). Physiological levels of tension actually stabilize the attachments of kinetochores to microtubules in vitro, as in vivo. This tethering of kinetochores to disassembling microtubules is essential for chromosome movements during mitosis.

Correcting Errors in Chromosome Attachment to the Spindle

The goal of mitosis is to partition the replicated chromosomes accurately between two daughter cells. Therefore, all chromosomes must attach correctly to both spindle poles (known as **amphitelic** attachment; Fig. 44.5) before being segregated. Three other sorts of attachment are seen: (a) chromosomes with one kinetochore lacking attached microtubules (known as **monotelic** attachment; this is a normal intermediate), (b) chromosomes with both sister kinetochores attached to the same spindle pole (known as **syntelic** attachment), and (c) chromosomes with a single kinetochore attached simultaneously to both spindle poles (known as **merotelic** attachment). Correcting monotelic and syntelic errors takes time, and the **SAC** (see "Finding Time to Fix Chromosome Attachment Errors: The Spindle Assembly Checkpoint" below) delays mitotic progression to allow the correction process to occur.

When syntelic attachments occur, one or both kinetochores must detach for the chromosome to achieve a bipolar orientation. Chromosome attachment to opposite spindle poles is more stable than attachment to a single pole, because the *tension* generated by bipolar attachment (where forces pull a chromosome simultaneously toward opposite spindle poles) preferentially stabilizes microtubule connections to both kinetochores. Merotelic attachments are more dangerous, as the kinetochore is under tension and the attachments are therefore stable. Merotelic attachments are the most common cause of chromosome segregation errors in cultured mammalian cells.

Syntelic and merotelic chromosome attachments are corrected through the action of Aurora B protein kinase, which forms the **chromosomal passenger complex (CPC),** along with inner centromere protein (INCENP), survivin, and borealin (Fig. 44.10). The other subunits target Aurora B to its various sites of action during mitosis and regulate the kinase activity. The complex concentrates at inner centromeres (the heterochromatin beneath and between the two sister kinetochores) during prometaphase and metaphase. At anaphase onset, the CPC moves to the overlapping interpolar microtubules of the central spindle and to the cell cortex, where the cleavage furrow will form, ultimately winding up in the

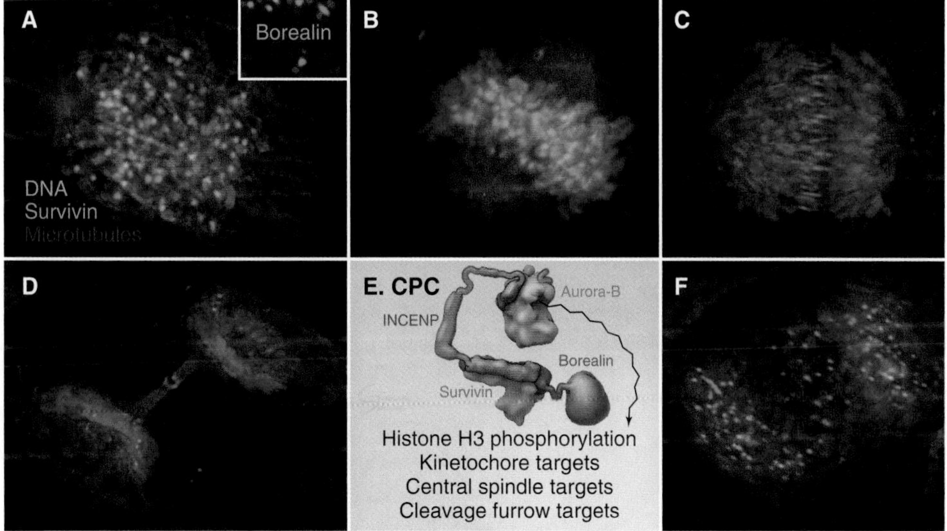

FIGURE 44.10 CHROMOSOMAL PASSENGER COMPLEX (CPC) REGULATES MITOTIC EVENTS. The CPC *(green)* is present at centromeres in prometaphase and metaphase **(B)**, but transfers to the spindle midzone at anaphase **(C)** and midbody at anaphase **(D)**. **E,** Diagram of Aurora B protein kinase complexed with INCENP (inner centromere protein), survivin, and borealin with some key targets of the CPC. **F,** If CPC function is inhibited (in this case by RNA interference [RNAi] depletion of borealin), chromosome attachment errors are common and many chromosomes fail to segregate properly in anaphase. Distribution of DNA *(blue),* microtubules *(red),* and survivin–green fluorescent protein (GFP) *(green)* in human mitotic cells. *Inset in A,* Distribution of kinetochores *(red),* and borealin *(green)* in a prometaphase cell. (**A–D,** Micrographs by Sally Wheatley and William C. Earnshaw. **F** and **Inset in A,** Micrographs by Ana Carvalho, Reto Gassmann, and William C. Earnshaw. **Insets in A** and **F,** From Gassmann R, Carvalho A, Henzing AJ, et al. Borealin: a novel chromosomal passenger required for stability of the bipolar mitotic spindle. *J Cell Biol*. 2004;166:179–191, copyright The Rockefeller University Press. **A–C,** From Wheatley SP, McNeish IA. Survivin: a protein with dual roles in mitosis and apoptosis. *Int Rev Cytol*. 2005;247:35–88.)

intercellular bridge during cytokinesis. The CPC regulates mitotic events from prophase through cytokinesis. Along the way it contributes to the correction of chromosome attachment errors and to the operation of the checkpoint that delays the cell cycle in response to those errors.

Aurora B corrects chromosome attachment errors by phosphorylating Ndc80 in the microtubule-binding KMN complex (see Fig. 8.21). Aurora B phosphorylation strongly inhibits Ndc80 binding to microtubules, causing the kinetochore to release attached microtubules. When a chromosome is correctly attached to both spindle poles, tension stretches the kinetochore away from the CPC buried in the chromatin beneath. This can stabilize chromosome-microtubule interactions by preventing the kinase from phosphorylating Ndc80.

Finding Time to Fix Chromosome Attachment Errors: The Spindle Assembly Checkpoint

Segregation of replicated chromosomes into daughter cells is extremely accurate. For example, budding yeasts lose a chromosome only once in 100,000 cell divisions. The frequency of chromosome loss may be 20-fold to 400-fold higher for human cells grown in culture. To achieve even this level of accuracy, most cells delay entry into anaphase until all chromosomes have achieved amphitelic attachment to the spindle. This delay is caused by the spindle assembly checkpoint (**SAC**), which senses the completion of microtubule binding to kinetochores at metaphase (Fig. 44.11). The spindle checkpoint differs from DNA damage checkpoints in that its default setting is "on" as cells enter mitosis. It is silenced only when every chromosome is properly attached to the spindle.

The SAC involves the products of the mitotic arrest-defective (MAD) genes, the budding-uninhibited-by-benzimidazole (BUB) genes, the monopolar spindle (Mps1) kinase and the CPC. The MAD and BUB genes were identified in yeast genetic screens for cells that continued to divide (and die) when the spindle was disassembled by drugs. These genes are conserved from yeast to humans. SAC proteins accumulate at

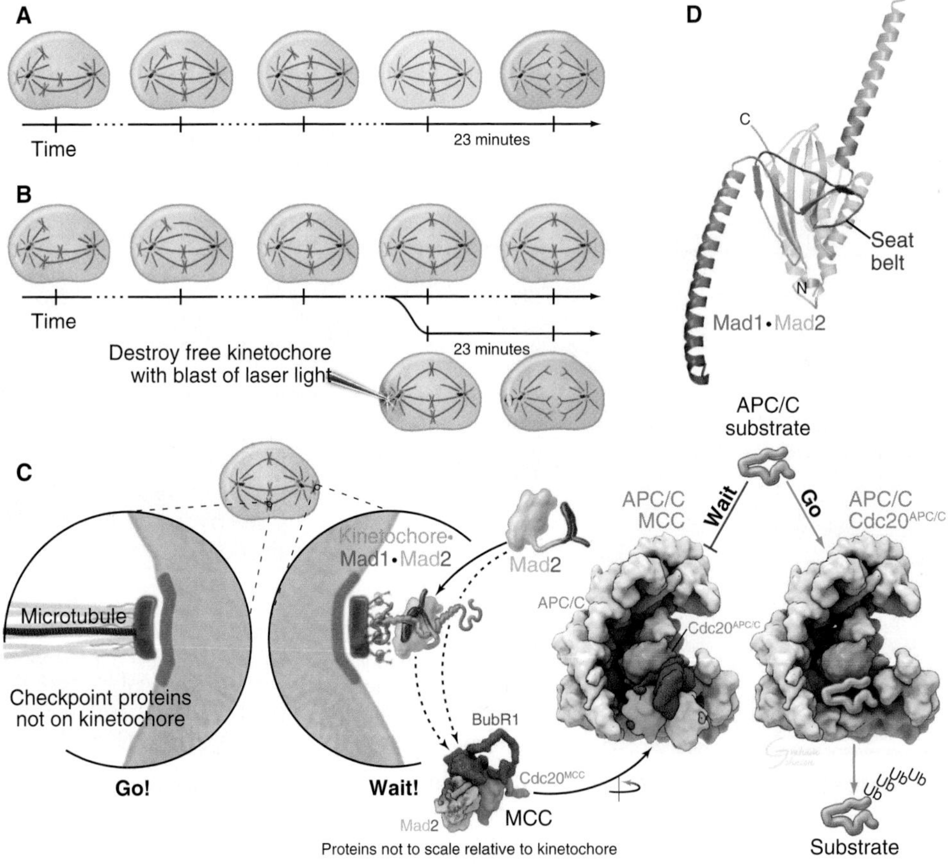

FIGURE 44.11 SPINDLE CHECKPOINT. Signaling by unattached kinetochores stops the cell from entering anaphase until all chromosomes have made a proper bipolar spindle attachment. **A,** As long as there is a chromosome that is not properly attached to the spindle *(beige cells)*, the cell does not enter anaphase. The cell enters anaphase approximately 20 minutes after chromosome attachment is complete *(green cells)*. **B,** In a cell with a persistently maloriented chromosome, anaphase entry is delayed *(beige cells)*. If the unattached kinetochore is destroyed with a high-powered laser, the cell enters anaphase about 20 minutes later. This proves that the unattached kinetochore sends an inhibitory signal. **C,** Overview of the spindle assembly checkpoint (see text for details). **D,** Structure of the Mad1/Mad2 complex.

kinetochores early in mitosis, when the checkpoint is "on" (ie, during prophase or prometaphase), and are gradually displaced as microtubules bind and the kinetochores come under tension.

The target of the SAC is the APC/C^{Cdc20}, the anaphase-promoting complex/cyclosome (APC/C) ubiquitin-protein ligase (an E3 enzyme; see Figs. 23.3 and 40.16) with its substrate recognition factor Cdc20 bound, APC/C^{Cdc20} ubiquitylates target proteins to mark them for destruction by proteasomes. Key APC/C^{Cdc20} substrates are proteins that must be degraded for the cell to move from metaphase to anaphase, including cyclin B and securin, an inhibitor of the enzyme that triggers separation of sister chromatids at anaphase (Fig. 44.16).

During mitosis, Cdk1 activates the APC/C by phosphorylating an auto-inhibitory loop, allowing Cdc20 to bind. The SAC is an additional regulatory circuit that inactivates APC/C^{Cdc20} until all kinetochores attach to spindle microtubules. Kinetochores without microtubules attract proteins that assemble the mitotic checkpoint complex (MCC), the inhibitor that inactivates APC/C^{Cdc20}.

Checkpoint activation starts when Aurora B in the CPC activates Mps1 kinase, allowing it to phosphorylate Knl1 in the kinetochore at several sites. Mps1 phosphorylation of Knl1 creates a binding site that results in Mad1 recruitment to the kinetochore. Mad1 then recruits Mad2 to form a stable complex (Fig. 44.11). A loop on Mad2 wraps around Mad1 like a safety belt making the complex particularly stable. This form of Mad 2 is known as "closed" Mad2. Mad1/Mad2 can transiently bind soluble Mad2 molecules (known as "open" Mad2), load them onto Cdc20 in the closed safety belt conformation (this loading probably occurs at kinetochores), then release them to form the soluble MCC of Mad2/Cdc20/BubR1/Bub3. The MCC associates with the APC/C^{Cdc20}, interfering with binding of cyclin B and other key substrates. As each chromosome becomes attached to both poles of the spindle it stops producing MCC. When the last chromosome has achieved a proper attachment, the last source of MCC is extinguished, and entry into anaphase can proceed.

Silencing of the checkpoint involves several pathways. Protein phosphatase 2A (PP2A) and protein phosphatase 1 (PP1) are recruited to the kinetochore in feedback loops involving the CPC and other checkpoint components. They dephosphorylate Knl1 so that it releases Mad1. When microtubules bind, cytoplasmic dynein motors actively strip checkpoint components from the kinetochore, dragging them away toward the centrosomes. In yeast, access of Mps1 to its target sites on Knl1 is physically blocked when microtubules bind. Exactly how this interaction is regulated in metazoans is still actively studied. In addition, the SAC appears to crosstalk with the DNA damage response (see Chapter 43), since DNA damage response components activate SAC components and vice versa. However, the network of interactions is very complex and details are still being worked out.

Experimental inactivation of the spindle checkpoint causes a catastrophic, premature entry into anaphase, regardless of the status of chromosome alignment. This leads to an unequal distribution of sister chromatids and genetic imbalance between daughter cells known as **aneuploidy.** Yeasts can live without the checkpoint genes, but their loss is lethal for mice, which die early during embryogenesis. Mice heterozygous for various checkpoint components show increased aneuploidy. Humans with mutations in BubR1 have mosaic variegated aneuploidy syndrome (extra copies or loss of various chromosomes in a variety of tissues), which is associated with microcephaly (decreased brain size) and an increased cancer risk.

Metaphase

When all the chromosomes have attained amphitelic orientations and moved to positions roughly midway between the two spindle poles, the cell is said to be in metaphase (Fig. 44.12). The compact grouping of chromosomes at the middle of the spindle is referred to as the **metaphase plate.** In many cells, even though chromosomes remain, on average, balanced at the middle of

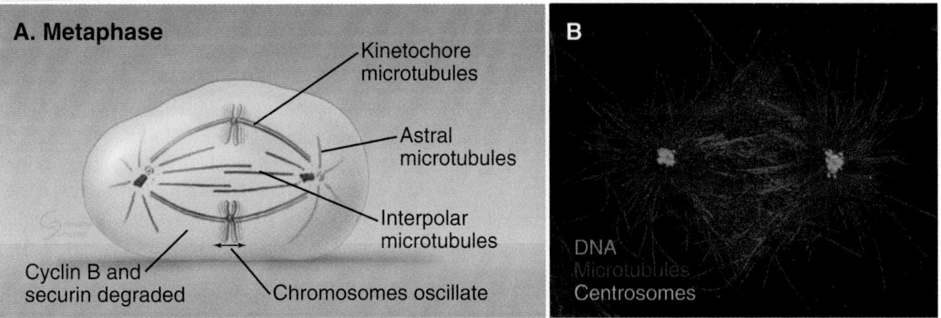

FIGURE 44.12 INTRODUCTION TO METAPHASE. A, Summary of the major events of metaphase. **B,** Distribution of DNA *(blue),* microtubules *(red),* and gamma tubulin (centrosomes *[green]*) in a metaphase human cell. (**B,** Images were recorded by Dr. Melpomeni Platani on the University of Dundee's School of Life Sciences Imaging Facility OMX 3DSIM Microscope and stored and processed in OMERO.)

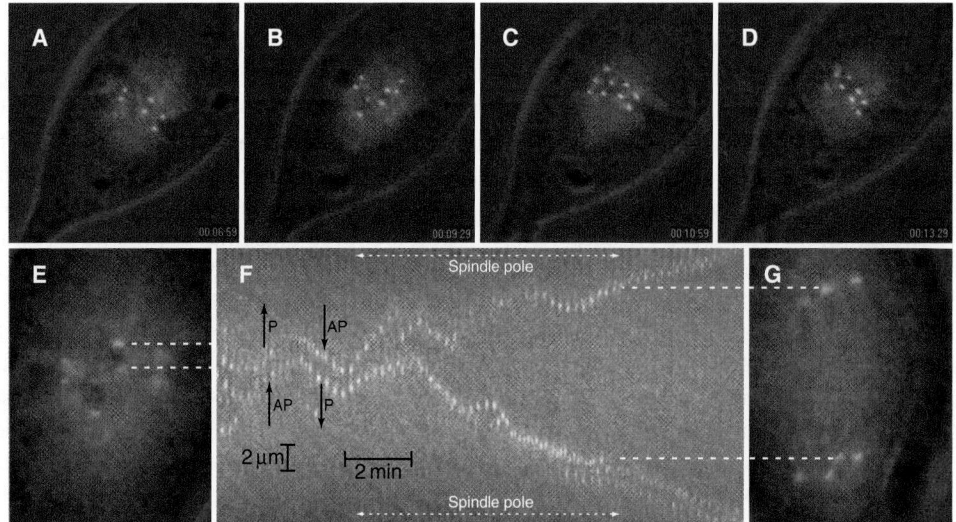

FIGURE 44.13 KINETOCHORE OSCILLATIONS BETWEEN P (POLEWARD) AND AP (AWAY FROM THE POLE) MOVEMENT DURING LATE PROMETAPHASE AND ANAPHASE IN PTK1 (RAT KANGAROO) CELLS. A–D, Images showing the movements of several pairs of sister kinetochores, labeled with green fluorescent protein (GFP)-Cdc20 *(green),* combined with phase-contrast images of the cell *(red).* **E** and **G,** Higher-magnification views of sister kinetochores (marked with *dashed lines*) in prometaphase and anaphase, respectively. **F,** Kymograph (collage of images of a vertical strip showing the same two kinetochores at various time points during the movie) showing the movements of these two kinetochores. Movements toward (P) and away from (AP) spindle poles are indicated. P movement involves microtubule shrinkage at the leading kinetochore and microtubule growth at the trailing kinetochore (which is undergoing AP movement away from its associated kinetochore). Spindle poles are near the top and bottom of panels **E** to **G.** (Micrographs courtesy E.D. Salmon, University of North Carolina, Chapel Hill.)

the spindle, they jostle one another and undergo numerous small excursions toward one pole or the other throughout metaphase (Fig. 44.13).

Metaphase can also be defined biochemically as the time of destruction of cyclin B and securin (Fig. 44.16), because this begins as soon as the last chromosome achieves amphitelic orientation. Degradation of cyclin A begins earlier, at the entry into prometaphase, and is largely complete before metaphase. Loss of securin initiates a process leading to the separation of sister chromatids and the onset of anaphase.

Microtubule Flux Within the Metaphase Spindle

Although the average length of the kinetochore microtubules is roughly constant during metaphase, the microtubules change continuously in three ways. First, there is constant net addition of new tubulin subunits (approximately 10 subunits per second) to the plus end of the microtubules, where they are attached to the kinetochore. Second, a comparable number of tubulin subunits is continuously lost from the minus end of the kinetochore tubules at the spindle poles. Therefore, tubulin subunits slowly migrate through kinetochore microtubules from the kinetochore to the pole (Fig. 44.14). This **subunit flux** or treadmilling in kinetochore microtubules is caused by microtubule depolymerization at the poles driven by kinesin-13 family members. In addition, tubulin moves toward the poles as by microtubules are transported towards the pole (many nucleated by the augmin complex) within the kinetochore fiber. All microtubules attached to each kinetochore

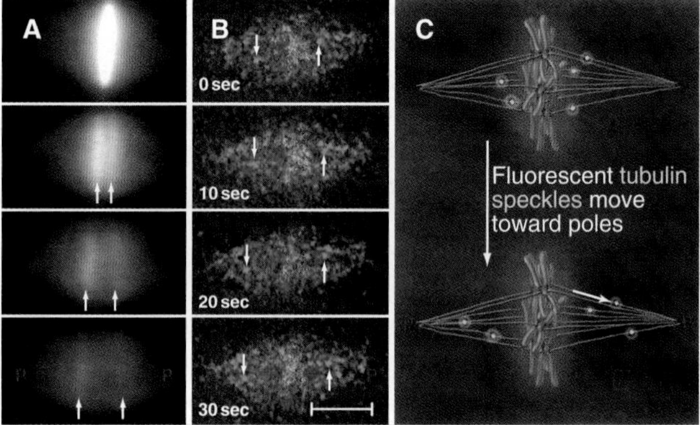

FIGURE 44.14 MICROTUBULE FLUX IN METAPHASE. A, Cells entering mitosis were injected with tubulin subunits modified chemically by attachment of a caged fluorescent dye. This dye becomes fluorescent after being irradiated with UV light. When cells entered metaphase, the spindle was illuminated with a narrow stripe of UV light, activating a narrow band of fluorescent tubulin subunits. With time, these subunits approach the spindle poles (P). Because the length of kinetochore microtubules is constant during this time, the labeled tubulin molecules must migrate along the microtubules toward the pole *(arrows).* This can occur if new subunits are added to the microtubule at the kinetochore and old subunits are removed at the pole. **B,** Microtubule flux at metaphase in a *Drosophila* embryo visualized by fluorescence speckle microscopy. Embryos were injected with very low levels of fluorescent tubulin, which appears as speckles distributed along the microtubules. If a very sensitive camera is used, these speckles can be seen to move toward the poles, reflecting the flux in the underlying microtubules. Scale bar is 5 µm. **C,** Movement of labeled tubulin speckles toward the spindle poles. (**A,** Courtesy Arshad Desai and the MBL Cell Division Group, Marine Biology Laboratory, Woods Hole, MA; from Mitchison TJ, Salmon ED. Mitosis: a history of division. *Nat Cell Biol.* 2001;3:E17–E21. **B,** Courtesy Paul Maddox and Arshad Desai, University of California, San Diego.)

change coordinately in length during chromosomal oscillations.

Anaphase

The separation of sister chromatids at anaphase is one of the most dramatic events of the entire cell cycle (Fig. 44.15). Sister chromatids move to opposite spindle poles **(anaphase A),** and the poles move apart **(anaphase B).** Anaphase is also the time when the mitotic spindle activates the cell cortex in preparation for cytokinesis.

Two forms of the APC/C (see Fig. 40.15) trigger the transition from metaphase to anaphase by degrading key proteins. APC/C^{Cdc20} targets cyclin B for degradation, causing Cdk activity to fall (see Fig. 40.17). This decline in Cdk activity allows for activation of APC/C^{Cdh1}, because Cdh1 phosphorylated by Cdk1 kinase cannot bind to the APC/C. APC/C^{Cdh1} targets polypeptides whose destruction by the proteasome is required for the cell to exit from mitosis and return to interphase. APC/C^{Cdh1} remains active during G_1 phase, where it is essential for the licensing of DNA replication origins (see Fig. 42.28).

Biochemical Mechanism of Sister Chromatid Separation

Separation of sister chromatids is regulated by the chromosomes themselves, not by the mitotic spindle. Under certain circumstances, sister chromatids can separate in the absence of microtubules, ruling out a requirement for forces from the spindle in the process.

Three factors regulate sister chromatid separation: a protein complex known as **cohesin,** a protease known as **separase,** and an inhibitor of separase known as **securin** (Fig. 44.16). This system is conserved from yeast to human. Chapter 8 discusses the functions of cohesin in interphase.

Cohesin is a complex of four proteins that resembles the condensin complex (see Fig. 8.18). Like condensin, cohesin has two large subunits from the SMC ATPase family. These proteins, SMC1 and SMC3, are complexed with proteins called Scc1 (which has other names omitted here for simplicity) and Scc3. Additional proteins are required to stabilize the loading of this complex onto DNA. Cells with mutations in cohesin components separate sister chromatids prematurely in mitosis, resulting in chaotic chromosome missegregation. This system is very ancient, as bacteria depend on an SMC-related protein for orderly chromosome segregation.

A variety of evidence suggests that cohesin forms a ring with a diameter of 35 nm, large enough to encircle two sister chromatids like a lasso. In yeast, the complex functions only if it binds chromosomes during DNA replication. Cohesin accumulates at preferred sites on the chromosomes, often near centromeres in budding yeast or in regions of heterochromatin in fission yeast. In vertebrates, most cohesin dissociates from the chromosome arms by late metaphase, owing to the action of the protein kinases Plk1 and Aurora B. Importantly, a critical fraction remains associated with heterochromatin flanking centromeres where it is protected from cleavage by shugoshin until the onset of anaphase (see following paragraphs and Chapter 45).

Sequential cleavage of two key proteins triggers sister chromatid separation at anaphase. This proteolysis makes anaphase onset an irreversible transition. The first target, securin, inhibits the separase protease. After the last chromosome forms an amphitelic attachment to the spindle, the spindle checkpoint is silenced. This allows APC/C^{Cdc20} to tag securin with ubiquitin, leading to its destruction by proteasomes throughout metaphase. When securin levels fall below a critical threshold, separase is unleashed to cleave the Scc1 subunit of cohesin. Cleavage of Scc1 breaks the cohesin ring, allowing the sister chromatids to separate triggering the onset of anaphase (Fig. 44.16B).

Efficient Scc1 cleavage requires that the protein be phosphorylated near its cleavage site. This allows a mode of regulation where shugoshin (Japanese for "guardian spirit") recruits PP2A to centromeres. PP2A keeps Scc1 dephosphorylated. This inhibits its cleavage and protects cohesin until shugoshin is released following amphitelic attachment of the chromosome. This

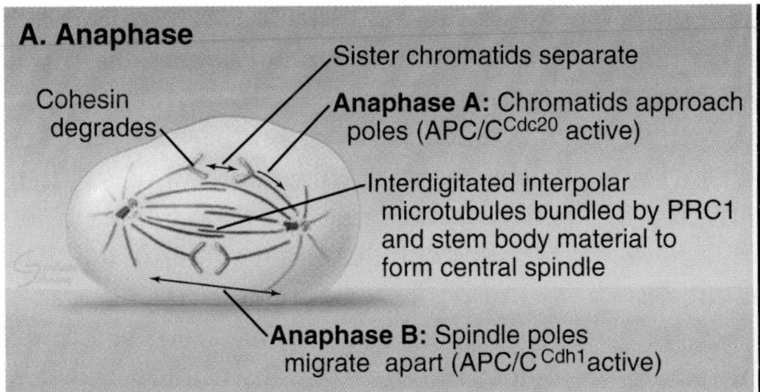

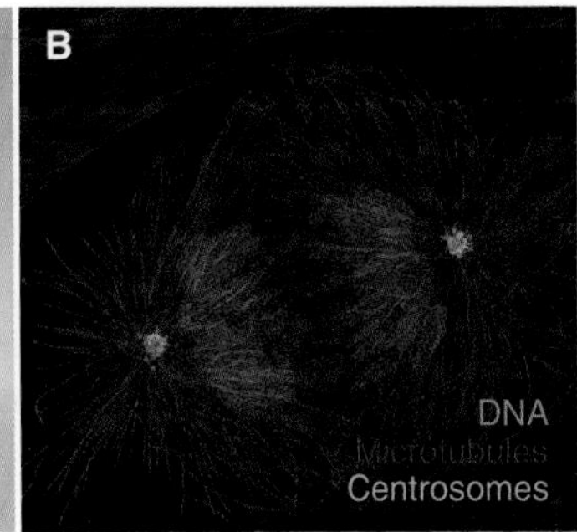

FIGURE 44.15 INTRODUCTION TO ANAPHASE. A, Summary of the major events of anaphase. **B,** Distribution of DNA *(blue),* microtubules *(red),* and gamma tubulin (centrosomes *[green]*) in a mid-anaphase human cell. APC/C, anaphase-promoting complex/cyclosome; PRC1, protein regulated in cytokinesis 1. (**B,** Images were recorded by Dr. Melpomeni Platani on the University of Dundee's School of Life Sciences Imaging Facility OMX 3DSIM Microscope and stored and processed in OMERO.)

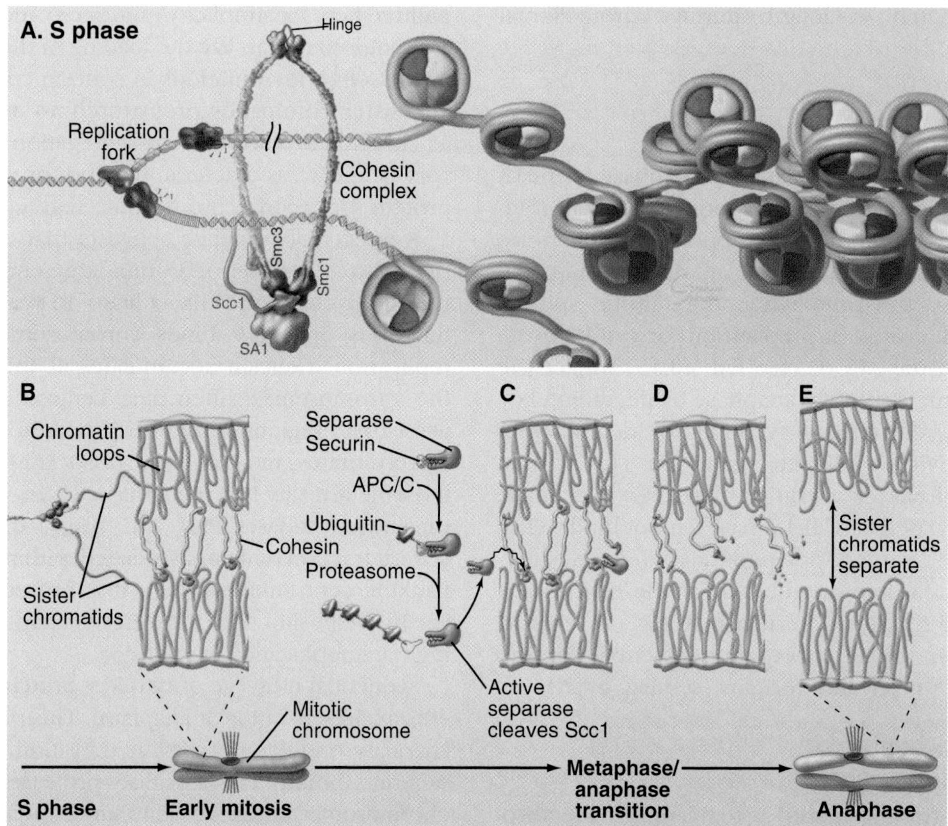

FIGURE 44.16 REGULATION OF SISTER CHROMATID PAIRING BY THE COHESIN COMPLEX. A–B, The cohesin complex forms a 35-nm diameter ring that links sister chromatids during DNA replication. At anaphase onset, degradation of its securin inhibitor liberates active separase enzyme. **C–E,** Separase then cleaves cohesin subunit Scc1, and the two sister chromatids are freed to separate from one another and move toward opposite spindle poles.

mechanism is absolutely essential during meiosis, as without it, it would not be able to segregate homologous chromosomes from each other (see Fig. 45.12).

Securin can act as an oncogene in cultured cells and is overexpressed in some human pituitary tumors. Overexpression of securin may disrupt the timing of chromosome segregation, leading to chromosome loss and, ultimately, contributing to cancer progression.

Mitotic Spindle Dynamics and Chromosome Movement During Anaphase

Anaphase is dominated by the orderly movement of sister chromatids to opposite spindle poles brought about by the combined action of motor proteins and changes in microtubule length. There are two components to anaphase chromosome movements (Fig. 44.15). **Anaphase A,** the movement of the sister chromatids to the spindle poles, requires a shortening of the kinetochore fibers. During **anaphase B,** the spindle elongates, pushing the spindle poles apart. The poles separate partially because of interactions between the antiparallel interpolar microtubules of the central spindle and partially because of intrinsic motility of the asters. Most cells use both components of anaphase, but one component may predominate in relation to the other.

Microtubule disassembly on its own can move chromosomes (see Fig. 37.8). Energy for this movement comes from hydrolysis of GTP bound to assembled tubulin, which is stored in the conformation of the lattice of tubulin subunits. Microtubule protofilaments are straight when growing, but after GTP hydrolysis protofilaments are curved, so they peel back from the ends of shrinking microtubules (see Fig. 34.6). Several kinesin "motors" influence the dynamic instability of the spindle microtubules. Members of the kinesin-13 class, which encircle microtubules near kinetochores and at spindle poles, use adenosine triphosphate (ATP) hydrolysis to remove tubulin dimers and promote microtubule disassembly rather than movement.

Kinetochores are remarkable in their ability to hold onto disassembling microtubules. In straight (growing) microtubules, the Ndc80 complex is mostly responsible for microtubule binding. It binds to the interface between α and β tubulin subunits. This interface bends in curved (shrinking) microtubules, so Ndc80 cannot bind. This could allow it to redistribute onto straight sections of the lattice and thereby move away from the curved protofilaments at the disassembling end. In metazoans the Ska complex in the outer kinetochore binds α and β tubulin subunits away from the interface, so it can bind to curved (disassembling) protofilaments. At yeast kinetochores the Dam1 ring (green in Fig. 8.21) couples the kinetochore to disassembling microtubules.

Anaphase A chromosome movement involves a combination of microtubule shortening and translocation of the microtubule lattice that result from flux of tubulin subunits (Fig. 44.14). The contributions of the two mechanisms vary among different cell types. When living vertebrate cells are injected with fluorescently labeled tubulin subunits, the spindle becomes fluorescent (Fig. 44.17). If a laser is used to bleach a narrow zone in the fluorescent tubulin across the spindle between the chromosomes and the pole early in anaphase, the chromosomes approach the bleached zone much faster than the bleached zone approaches the spindle pole. This shows that the chromosomes "eat" their way along the kinetochore microtubules toward the pole. In these cells, subunit flux accounts for only 20% to 30% of chromosome movement during anaphase A, and this flux is dispensable for chromosome movement. In *Drosophila* embryos, in which subunit flux accounts for approximately 90% of anaphase A chromosome movement, the chromosomes catch up with a marked region of the kinetochore fiber slowly, if at all.

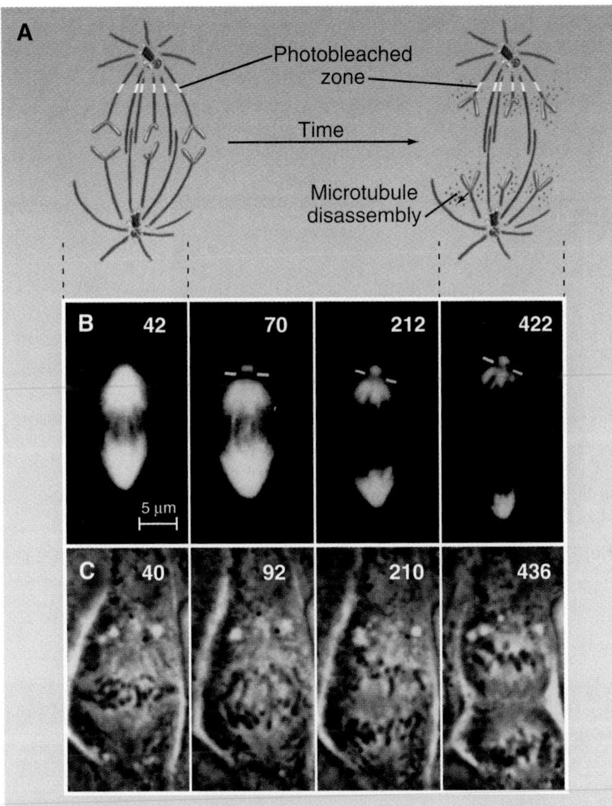

FIGURE 44.17 CHROMOSOMES MOVE ON SHRINKING MICROTUBULES DURING ANAPHASE. A, Mitotic cells were injected with a fluorescently labeled tubulin that was incorporated into the spindle. Just after anaphase onset, a laser was used to photobleach a stripe *(white)* across the spindle near the upper pole. The live cell was monitored over time by fluorescence **(B)** and phase-contrast **(C)** microscopy. In this mammalian cell, the chromosomes approach the bleached stripe much faster than the stripe approaches the spindle pole. In other organisms with higher rates of microtubule flux in their spindles, the bleached zone would also move appreciably toward the pole. The numbers are time in seconds. (**B–C,** From Gorbsky GJ, Sammak PJ, Borisy GG. Microtubule dynamics and chromosome motion visualized in living anaphase cells. *J Cell Biol*. 1988;106:1185–1192, copyright The Rockefeller University Press.)

Anaphase B appears to be triggered at least in part by the inactivation of the minus-end-directed kinesin-14 motors, so that all the net motor force favors spindle elongation. Four factors contribute to overall lengthening of the spindle: release of sister chromatid cohesion, sliding apart of the interdigitated half-spindles, microtubule growth, and intrinsic motility of the poles themselves (Fig. 44.7). During the latter stages of anaphase B, the spindle poles, with their attached kinetochore microtubules, appear to move away from the interpolar microtubules as the spindle lengthens. This movement of the poles involves interaction of the astral microtubules with cytoplasmic dynein molecules anchored at the cell cortex.

Anaphase B spindle elongation is accompanied by reorganization of the interpolar microtubules into a highly organized **central spindle** between the separating chromatids (Fig. 44.15). Within the central spindle, an amorphous dense material called **stem body matrix** stabilizes bundles of antiparallel microtubules and holds together the two interdigitated half-spindles. Proteins concentrated in the central spindle help regulate cytokinesis. One key factor, PRC1 (*p*rotein *r*egulated in *c*ytokinesis 1), is inactive when phosphorylated by Cdk kinase and functions only during anaphase when Cdk activity declines and phosphatases remove the phosphate groups placed on target proteins by Cdks and other mitotic kinases. PRC1 directs the binding of several kinesins to the central spindle. The kinesin KIF4A targets Aurora B kinase to a particular domain of the central spindle, where phosphorylation of key substrates then regulates spindle elongation and cytokinesis.

How can protein kinases such as Aurora B continue to function during anaphase while protein phosphatases are removing phosphate groups placed there by Cdks and, indeed, Aurora B during early mitosis? One answer is that the phosphatase activity is highly localized, controlled by specific targeting subunits. Cdk phosphorylation can inhibit targeting subunits such as the exotically named Repo-Man (*re*cruits *P*P1 *o*nto *m*itotic chromatin at *an*aphase) from binding protein phosphatase 1 or localizing to targets, such as chromatin in early mitosis. When Cdk activity drops, Repo-Man (and other similar targeting subunits) is dephosphorylated, and now targets PP1 to chromatin, where it removes phosphates placed there by Aurora B in the CPC. As long as phosphatases are not specifically targeted to the cleavage furrow, Aurora B can continue to control events there during mitotic exit by phosphorylating key target proteins required for cytokinesis.

Telophase

During telophase, the nuclear envelope reforms on the surface of the separated sister chromatids, which typically cluster in a dense mass near the spindle poles (Fig. 44.18). Some further anaphase B movement may still occur, but the most dramatic change in cellular structure at this time is the constriction of the cleavage furrow and subsequent cytokinesis.

Reassembly of the Nuclear Envelope

Nuclear envelope reassembly begins during anaphase and is completed during telophase (Fig. 44.19). As in spindle assembly, Ran-GTP promotes early steps of nuclear envelope assembly at the surface of the chromosomes by releasing key components sequestered by importin β. These include several nuclear pore components, and one of the earliest events in nuclear envelope reassembly involves binding of the nuclear pore scaffold protein ELYS to chromatin. ELYS can recognize DNA regions rich in A:T base pairs, so it is likely to bind directly to the DNA. ELYS then recruits other components of the nuclear pore scaffold and nuclear pore trans-membrane proteins. The pore subsequently matures as various peripheral components and elements of the permeability barrier are added.

The mechanism of nuclear membrane reassembly is debated. In cells where nuclear membranes fragments into vesicles during mitosis, a Ran-GTP–dependent pathway directs at least two discrete populations of vesicles to chromatin where they fuse to reform the nuclear envelope. In cells where the nuclear membrane is absorbed into the endoplasmic reticulum during mitosis, reassembly involves lateral movements of membrane components within the membrane network and their stabilization at preferred binding sites at the periphery of the chromosomes.

Lamin subunits disassembled in prophase are recycled to reassemble at the end of mitosis. Lamina reassembly is triggered by removal of mitosis-specific phosphate groups and methyl-esterification of several COOH side

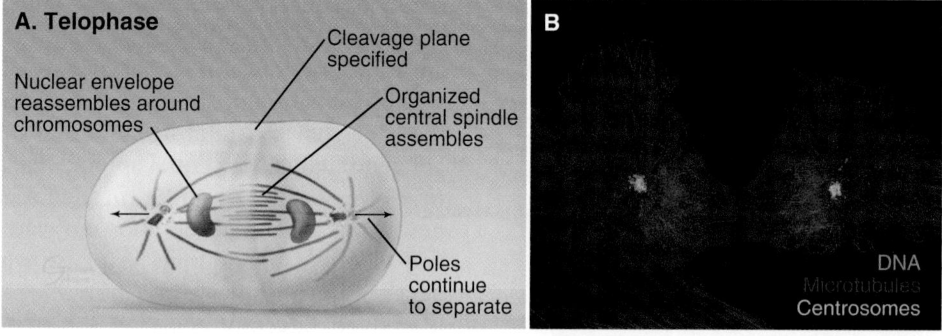

FIGURE 44.18 INTRODUCTION TO TELOPHASE. A, Summary of the major events of telophase. **B,** Distribution of DNA *(blue),* microtubules *(red),* and γ-tubulin (centrosomes *[green]*) in a telophase human cell. (**B,** Images were recorded by Dr. Melpomeni Platani on the University of Dundee's School of Life Sciences Imaging Facility OMX 3DSIM Microscope and stored and processed in OMERO.)

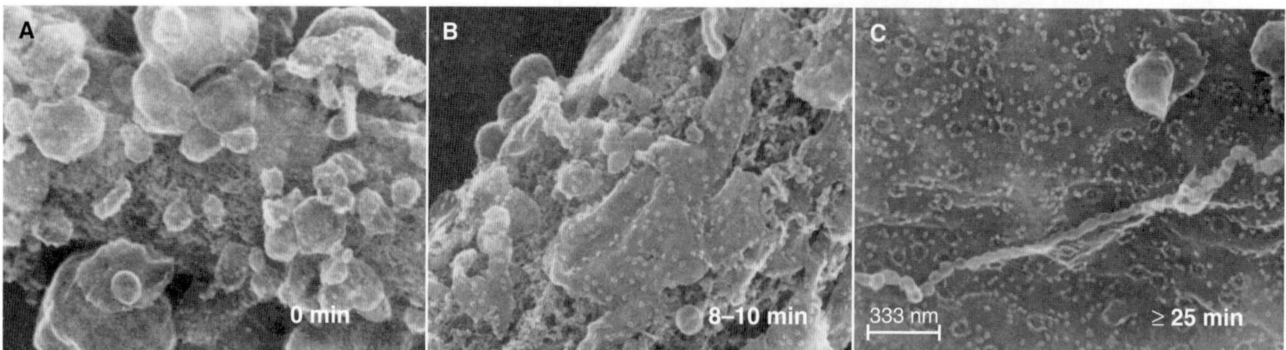

FIGURE 44.19 SCANNING ELECTRON MICROSCOPY OF THE STAGES OF ASSEMBLY OF MEMBRANE VESICLES ON THE SURFACE OF CHROMOSOMES IN A *XENOPUS* EGG CYTOSOLIC EXTRACT. A cell lysate containing membrane vesicles was added to isolated chromatin from *Xenopus* sperm, fixed, and then imaged by scanning electron microscopy. Each panel shows the time of incubation prior to fixation. (Micrographs courtesy of K.L. Wilson, Johns Hopkins Medical School, Baltimore, MD. **A** and **C,** From Wiese C, Goldberg MW, Allen TD, et al. Nuclear envelope assembly in *Xenopus* extracts visualized by scanning EM reveals a transport-dependent "envelope smoothing" event. *J Cell Sci*. 1997;110:1489–1502.)

chains on lamin B (Fig. 44.6). Together with ELYS, B-type lamins are among the earliest components of the nuclear envelope to target to the surface of the chromosomes during mid-anaphase. Either at this time or shortly thereafter, other proteins associated with the inner nuclear membrane, including BAF, LAP2, and lamin B receptor (see Fig. 9.10), join the forming envelope. Later during telophase when nuclear import is reestablished, lamin A enters the reforming nucleus and slowly assembles into the peripheral lamina over several hours in the G_1 phase. If lamin transport through nuclear pores is prevented, chromosomes remain highly condensed following cytokinesis, and the cells fail to reenter the next S phase.

Cytokinesis

Cytokinesis divides a mitotic cell into two daughter cells (Fig. 44.20). Cytokinesis depends on signals to specify the cleavage plane (Fig. 44.21), assembly and constriction of the contractile apparatus, specific alterations of the cell membrane, and the final separation (abscission) of the two daughter cells.

In animals, protozoa, and most fungi, a **contractile ring** of **actin** filaments and **myosin-II** guides the separation of daughter cells at the end of mitosis (Fig. 44.2). Myosin-II pulls on the ring of actin filaments, applying tension to the plasma membrane, much like contraction of smooth muscle (see Figs. 39.23 and 39.24). Because

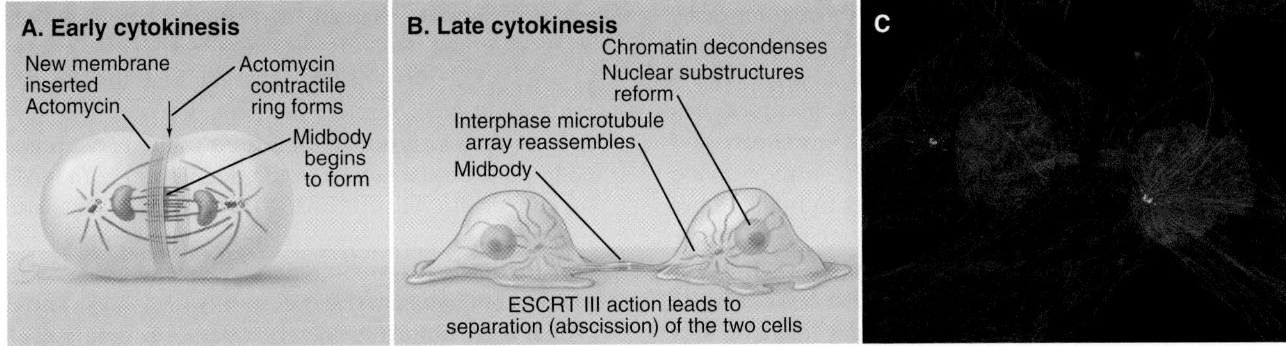

FIGURE 44.20 INTRODUCTION TO CYTOKINESIS. A–B, Summary of the major events of cytokinesis. **C,** Distribution of DNA *(blue),* microtubules *(red),* and γ-tubulin (centrosomes *[green]*) in a human cell undergoing cytokinesis. ESCRT, endosomal sorting complexes required for transport. (**C,** Images were recorded by Dr. Melpomeni Platani on the University of Dundee's School of Life Sciences Imaging Facility OMX 3DSIM Microscope and stored and processed in OMERO.)

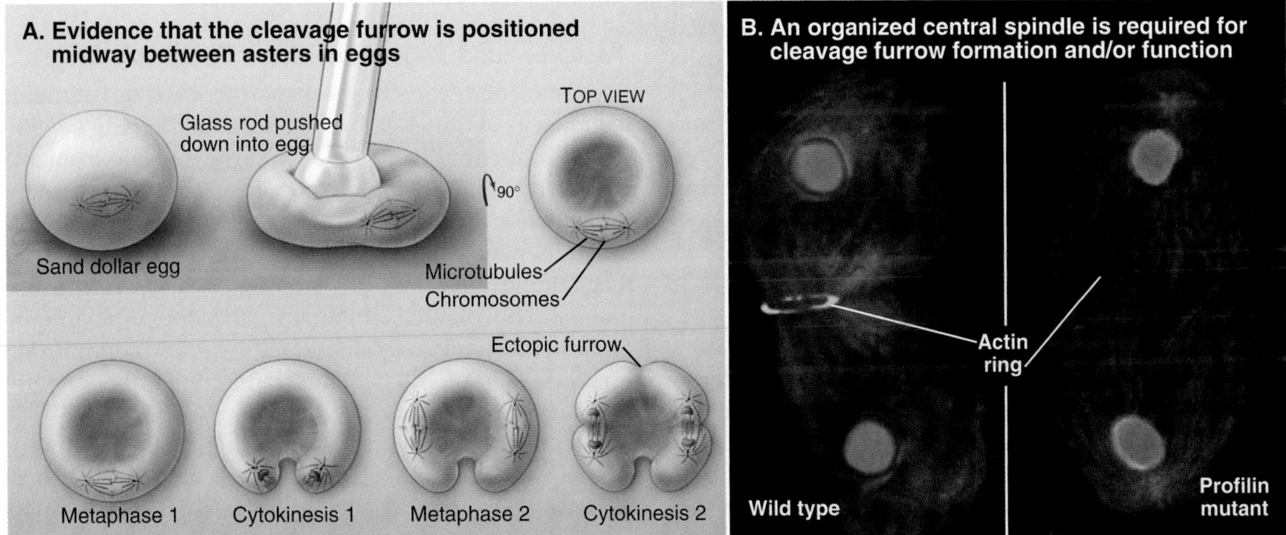

FIGURE 44.21 IN EGGS, A CLEAVAGE FURROW FORMS MIDWAY BETWEEN SPINDLE ASTERS. IN ANIMAL CELLS, THE CENTRAL SPINDLE IS IMPORTANT. A, A classic experiment in which a sand-dollar egg is caused to adopt a toroid shape. At cytokinesis 2, the egg cleaves into four cells, and a furrow forms between the back sides of the two spindles. (For a description of this and other classic experiments in cytokinesis, see the book by Rappaport in the "Selected Readings" list.) **B,** *Left,* A wild-type *Drosophila* spermatocyte undergoing cytokinesis, with the contractile ring stained in *yellow. Right,* In a profilin mutant, no central spindle forms, and the cell fails to form a contractile ring. (Micrographs courtesy Professor Maurizio Gatti, University of Rome, Italy. **B,** From Giansanti MG, Bonaccorsi S, Williams B, et al. Cooperative interactions between the central spindle and the contractile ring during *Drosophila* cytokinesis. *Genes Dev*. 1998;12:396–410.)

the contractile ring is confined to a narrow band of cortex around the equator, it forms a **cleavage furrow,** constricting the plasma membrane locally like a purse string (Fig. 44.20). Signals from the mitotic spindle and cell cycle machinery control the position of this ring (ie, the relative sizes of the two daughter cells) and the timing of its constriction.

Protozoa, animals, fungi, and plants use an evolutionarily conserved set of components to implement different strategies to separate daughter cells. For example, both fission yeast and metazoan cells use signals from polo kinase and a Rho-GTPase to direct the assembly of a contractile ring of actin, myosin-II, and other conserved components, even though the yeast has a closed mitosis and the metazoans have an open mitosis. In animal cells, contractile ring constriction provides the force that remodels the cortex to generate the two daughter cells. In contrast, in yeasts, which have a cell wall, contractile ring constriction is thought to guide the orderly centripetal growth of the cell wall septum, which contributes force to overcome turgor pressure and invaginate the plasma membrane. Plants lack myosin-II, so they divide by targeted fusion of membrane vesicles to build a new cell wall rather than constricting a cleavage furrow (Box 44.1 and Fig. 44.26). These differences reflect the fact that widely divergent eukaryotes use variations of similar themes for cytokinesis. Cytokinesis in prokaryotes is genuinely different, since completely different proteins are involved (Box 44.2 and Fig. 44.27).

Although cytokinesis has been studied for more than 100 years, it has posed a number of challenges due to its complexity at the molecular level. For example genetic analysis of fission yeast revealed more than 150 genes that contribute to cytokinesis. RNAi-based protein knockdown and molecular replacement analysis indicates that similar proteins participate in cytokinesis of *Caenorhabditis elegans, Drosophila,* and vertebrate tissue culture cells. Cytokinesis research typically employs living cells, although progress is being made toward reconstituting some aspects of the process in cell-free systems.

Signals Regulating the Position of the Cleavage Furrow

Elegant experimental data from classic studies on fertilized echinoderm eggs suggest that a **cleavage stimulus,** emitted by the mitotic spindle, specifies the position of the cleavage furrow midway between the poles and perpendicular to the long axis of the spindle, thereby ensuring that the cleavage process separates the daughter nuclei (Fig. 44.21). In fertilized eggs, the poles, with their large astral arrays of microtubules (see Fig. 6.4B), were regarded as the source of the cleavage stimulus, as furrows can be induced to form midway between two poles, even when no chromosomes are present. In addition, a signal emitted by the bundled microtubules of the central spindle appeared to modulate the behavior of the furrow signaled by the poles. We now know that the central spindle does emit a positive signal directing a cleavage furrow to form above it, while the poles contribute by focusing that furrow at a point on the cortex midway between them.

The molecular nature of the cleavage stimulus is now beginning to be understood in animals. The following is a simplified scenario:

1. During anaphase, overlapping microtubules between the separating chromatids establish an ordered array known as the central spindle. A key protein component of this array is a protein heterodimer known as **centralspindlin**. Centralspindlin is normally sequestered in the cytoplasm, but phosphorylation by the CPC enables it to target to the central spindle. *Drosophila* mutants that fail to form a central spindle cannot initiate cytokinesis (Fig. 44.21B). In contrast, *C. elegans* embryos that lack a central spindle can initiate but not complete the process.
2. One of the components of centralspindlin recruits a GEF (guanine exchange factor; see Figs. 4.6 and 4.7) for the small GTPase RhoA. This Rho-GEF, Ect2, also has a motif for targeting to the inside surface of the equatorial plasma membrane.
3. Membrane associated Ect2 locally activates RhoA, which then stimulates localized actin filament assembly and activation of myosin-II to begin assembly of the contractile ring.

Signals from the poles of the mitotic spindle contribute, particularly in large invertebrate embryos, by confining the zone of active RhoA to a narrow equatorial band between the separating sister chromatids.

Assembly and Regulation of the Contractile Ring

Exposure of the cell cortex to the cleavage stimulus culminates in the assembly of a **contractile ring** consisting of a very thin (0.1 to 0.2 μm) array of actin filaments attached to the plasma membrane at many sites around the equator (Fig. 44.22). Polymerization of the actin filaments depends on formins (see Fig. 33.14). Small, bipolar filaments of myosin-II are interdigitated with actin filaments. The plasma membrane adjacent to this actin-myosin ring undergoes alterations in its lipid composition that may help recruit proteins important for the function of the contractile ring.

Membrane furrowing requires actin and the motor activity of myosin-II (see Fig. 36.7). In animals, the small GTPase RhoA regulates actin polymerization by formins as well as constriction of the ring. Many other proteins are required for cytokinesis to go to completion. In their absence, furrowing begins, but the cleavage furrows ultimately regress, producing binucleated cells. These supporting proteins include anillin, actin filament cross-linking proteins, the CPC (Fig. 44.10) and the centralspindlin complex, among many others. Anillin helps

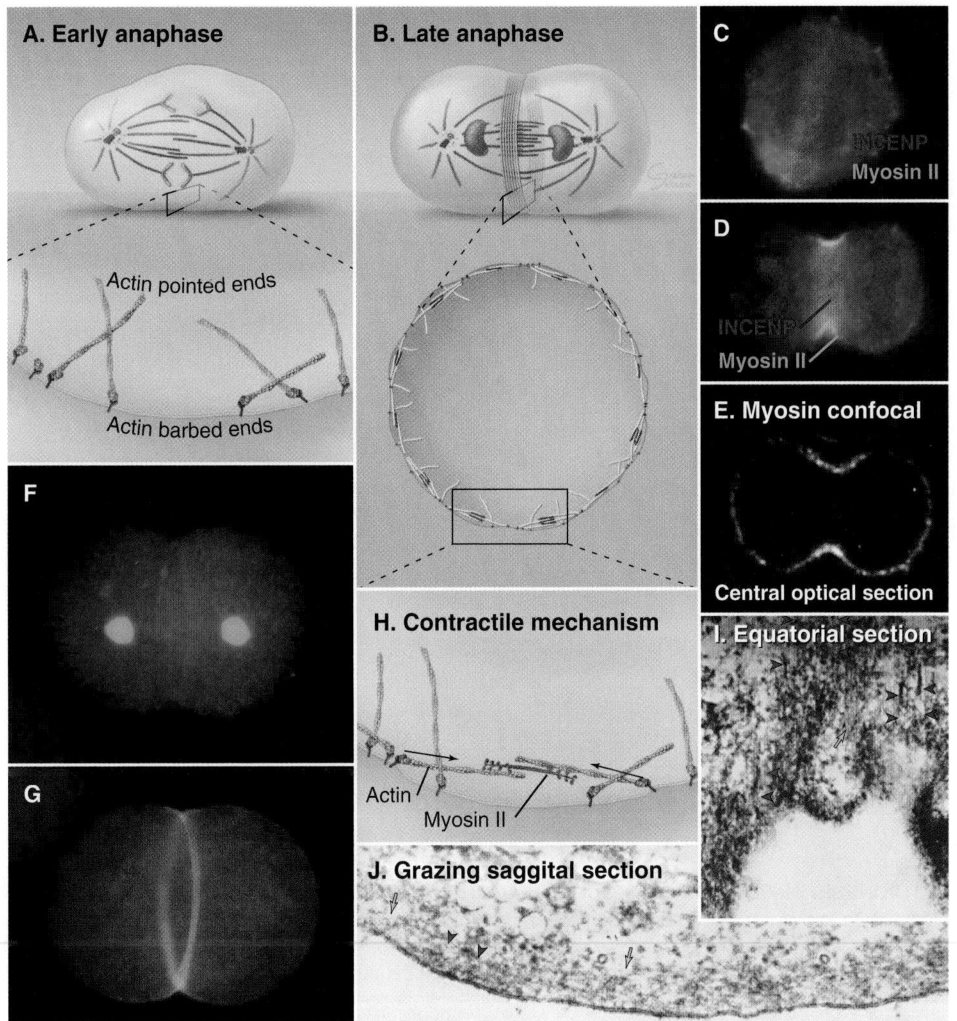

FIGURE 44.22 ORGANIZATION OF THE CONTRACTILE RING. A, Organization of actin at the cell cortex prior to cytokinesis. **B,** Distribution of actin and myosin at the start of ring contraction. **C,** INCENP (inner centromere protein) *(red)* concentrates at the site where the cleavage furrow will form just before myosin *(green)*. **D,** INCENP and myosin concentrate in the contractile ring during contraction. **E,** Confocal micrograph shows the distribution of myosin in an optical cross-section contracting contractile ring. **F–G,** Dividing invertebrate egg with DNA *(blue)* and actin *(red)* in the contractile ring. **H,** Organization of actin and myosin filaments during cytokinesis. **I–J,** Electron micrographs showing actin filaments in the contractile ring. Note the thick filaments that are thought to be myosin-II filaments *(red arrowheads)* and the thinner actin filaments *(yellow arrows)*. (**C–D,** Courtesy William C. Earnshaw. **E, I,** and **J,** Courtesy P. Maupin, Johns Hopkins Medical School, Baltimore, MD. **F–G,** Courtesy Professor Issei Mabuchi, University of Tokyo, Japan. For reference, see Maupin P, Pollard TD. Arrangement of actin filaments and myosin-like filaments in the contractile ring and actin-like filaments in the mitotic spindle of dividing HeLa cells. *J Ultrastruct Res.* 1986;94:92–103; Maupin P, Phillips CL, Adelstein RS, et al: Differential localization of myosin-II isozymes in human cultured cells and blood cells. *J Cell Sci.* 1994;107:3077–3090; and Eckley DM, Ainsztein AM, MacKay AM, et al. Chromosomal proteins and cytokinesis. *J Cell Biol.* 1997;136:1169–1183.)

keep active myosin-II focused into an organized contractile ring throughout cytokinesis.

The CPC and the centralspindlin complex are both required for animal cells to assemble the central spindle. Consequently, if either of the two complexes is eliminated in *C. elegans,* a contractile ring fails to form. Thus, they appear to contribute to the cleavage stimulus, and indeed, both require microtubules to localize to the site of cleavage furrow formation as originally shown for the cleavage stimulus. In addition, the CPC regulates the timely completion of cytokinesis by blocking premature activation of the ESCRT (endosomal sorting complexes required for transport) complex, which has a key role in the final separation of daughter cells (see later).

In fission yeast, with closed mitosis, the nucleus determines the position of cleavage. Fission yeast assemble a contractile ring along a well-defined pathway by recruiting proteins from cytoplasmic pools (Fig. 44.23). During interphase, assemblies of proteins called nodes form on the inside of the plasma membrane around the middle of cell. Prior to mitosis, an anillin-like protein leaves the nucleus and joins these nodes. During prophase, myosin-II, a formin and other contractile ring proteins join the nodes. When the formin polymerizes

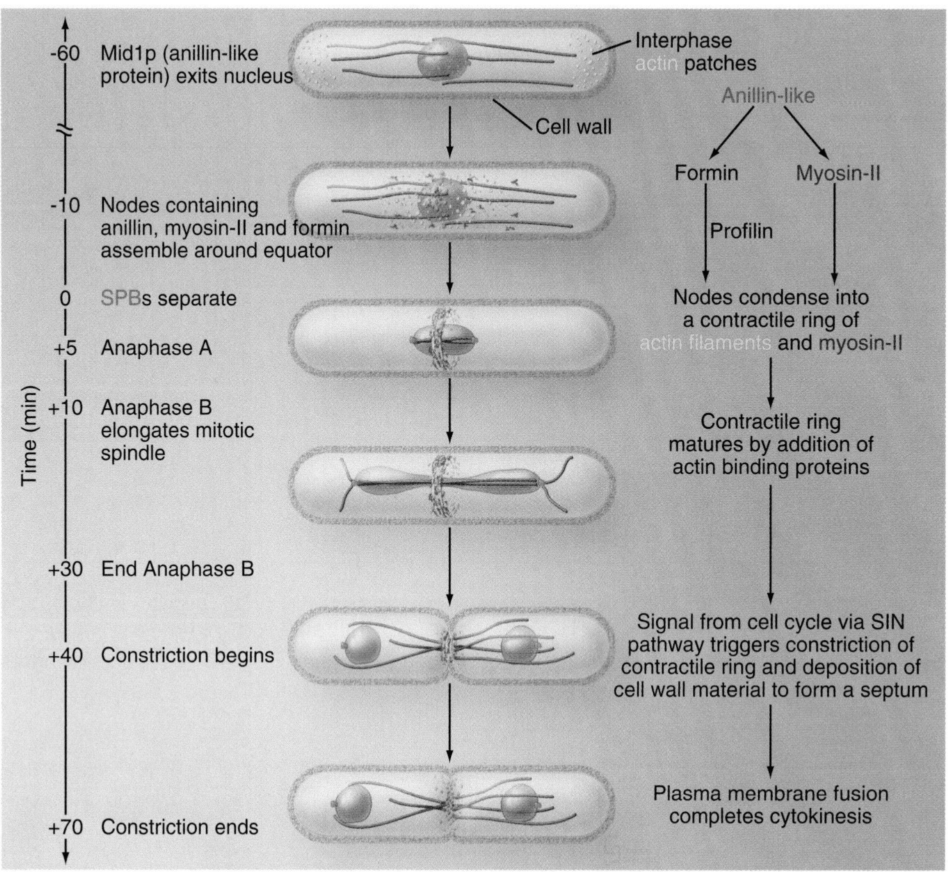

FIGURE 44.23 CYTOKINESIS IN FISSION YEAST *SCHIZOSACCHAROMYCES POMBE*. During interphase, microtubules *(red)* position the nucleus in the middle of the cell. Actin filaments concentrate in small patches *(yellow)* in the cortex at the two growing ends of the cell (see Fig. 33.1). The mitotic spindle is inside the nucleus, as the nuclear membrane does not break down during mitosis. As the cell enters mitosis, an anillin-like protein moves from the nucleus to the equatorial cortex, where it sets up nodes of proteins, including myosin-II and a formin. The formin grows actin filaments *(yellow),* and myosin-II pulls the nodes together into a continuous contractile ring. At the end of anaphase a signaling system consisting of a GTPase and three protein kinases (the septation initiation network [SIN]) triggers constriction of the contractile ring and associated synthesis of new cell wall to form a septum. The septum is a three-layered structure, with the primary septum flanked by two secondary septae. Digestion of the primary septum separates the daughter cells. (For reference, see Wu J-Q, Kuhn JR, Kovar DR, et al. Spatial and temporal pathway for assembly and constriction of the contractile ring in fission yeast cytokinesis. *Dev Cell*. 2004;5:723–734.)

actin filaments, myosin-II pulls the nodes together into a ring around the equator of the cell (Fig. 44.23).

Contractile ring assembly in animal cells shares many properties with fission yeast, but is less completely understood. The decline in the activity of cell cycle kinases at the onset of anaphase is part of the trigger, since they inhibit centralspindlin components through metaphase. INCENP and anillin move from the interphase nucleus to the cortex around the cell equator in early anaphase (Fig. 44.22). Formins and profilin polymerize some new actin filaments, but preexisting actin filaments are recruited into the contractile ring from adjacent areas of the cortex. Myosin-II is dispersed throughout the cytoplasm until anaphase, when it concentrates in the cortex, especially around the equator where the furrow forms. The myosin-II is derived from various interphase structures including stress fibers (see Fig. 33.1) that break down during prophase.

Constriction of the Cleavage Furrow

Contractile rings of echinoderm eggs produce enough force to invaginate the plasma membrane and form the cleavage furrow, although many details are still being studied. Constriction of the ring probably involves a sliding filament mechanism similar to muscle (see Figs. 39.9 and 39.23), but little is known about how the contractile ring is attached to the plasma membrane. During the early stages of furrowing, contractile rings maintain a constant volume, but then disassemble as they constrict further.

The role of myosin-II as the motor for cytokinesis was established by microinjection of inhibitory antibodies into echinoderm embryos and confirmed by genetic inactivation in the slime mold *Dictyostelium*. Slime mold amoebas lacking the myosin-II heavy chain round up during mitosis and complete nuclear division but cannot form a normal cleavage furrow. Mutant cells accumulate many nuclei, because the mitotic cycle continues. Mutant cells can divide on a substratum using pseudopods to pull themselves apart into smaller cells.

Constriction of the contractile ring is regulated so that it does not begin until after the onset of anaphase B, when sister chromatids are well separated. In fission yeast, a signaling pathway called the septation initiation network (SIN) initiates constriction. Much less is known in other cells.

Abscission

As the contractile ring pulls the cell membrane inward, the single cell that entered mitosis is gradually transformed into two daughters joined by a thin intercellular bridge (Fig. 44.20). This process requires a significant net increase in the surface area of the cell. New plasma membrane is inserted adjacent to the leading edge of the furrow. The source of the new membrane appears to be recycling endosomes (see Chapter 22), so addition of membrane to the cleavage furrow is a specialized form of exocytosis. Fusion of vesicles providing the new membrane depends on specific syntaxins, t-SNAREs (soluble *N*-ethylmaleimide-sensitive factor attachment protein receptors) (see Fig. 21.15) that promote vesicle fusion along the secretory pathway.

The plasma membrane in the cleavage furrow has a discrete composition. In budding yeast, this compartment is delineated by rings made from polymers of septins, a family of GTP-binding proteins recruited by anillin. Septins are essential for cytokinesis in *Saccharomyces cerevisiae* but not fission yeast.

In most animal cells, constriction of the cleavage furrow ultimately reduces the cytoplasm to a thin intercellular bridge between the two daughter cells. The intercellular bridge contains a highly ordered, antiparallel array of microtubules derived from the spindle with a dense knob, the **midbody,** at its center (Fig. 44.20). Isolated midbodies contain more than 160 proteins, with approximately one-third involved in various aspects of membrane trafficking.

The midbody is encircled by a dense ring of proteins that includes centralspindlin, anillin, and a centrosomal protein known as Cep55. Interactions between anillin and membrane-associated septin filaments tether the membrane to the ring. Cep55 recruits the ESCRT (endosomal sorting complexes required for transport) III complex, proteins with important roles in vesicle budding events such as formation of multivesicular bodies (see Chapters 22 and 23). In the intercellular bridge, ESCRT III forms a helical filament that spirals around the inner surface of the membrane, becoming more and more constricted as it grows away from the midbody (Fig. 44.24). ESCRT III also recruits factors that disassemble the bundled microtubules in the intercellular bridge, allowing the membrane to constrict further.

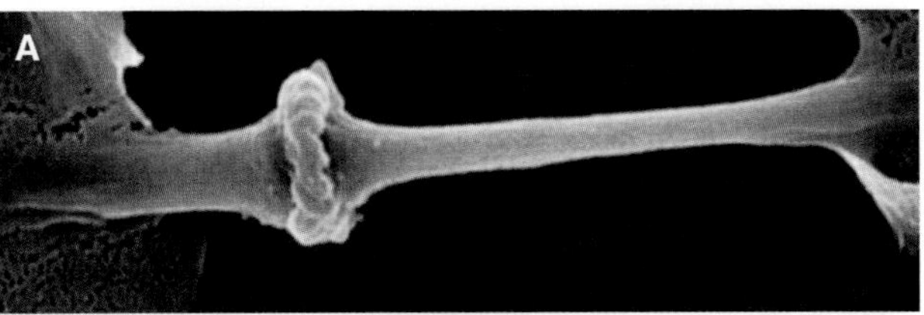

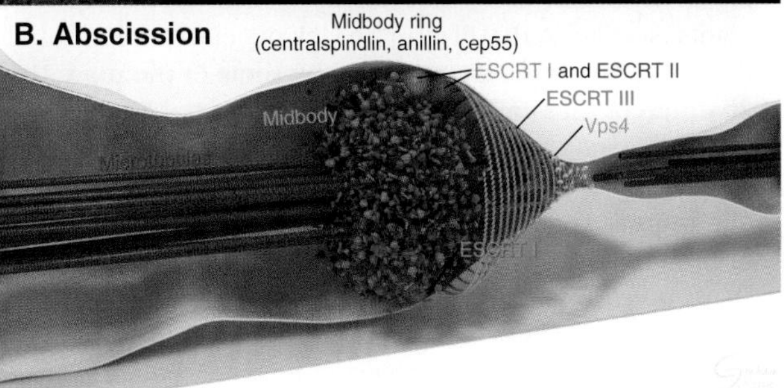

FIGURE 44.24 ABSCISSION IN ANIMAL CELLS. Proteins associated with overlapping bundles of microtubules form the midbody. The midbody ring links the midbody to the membrane, and nucleates the formation of ESCRT (endosomal sorting complexes required for transport) III filaments that constrict the membrane to divide the daughter cells.

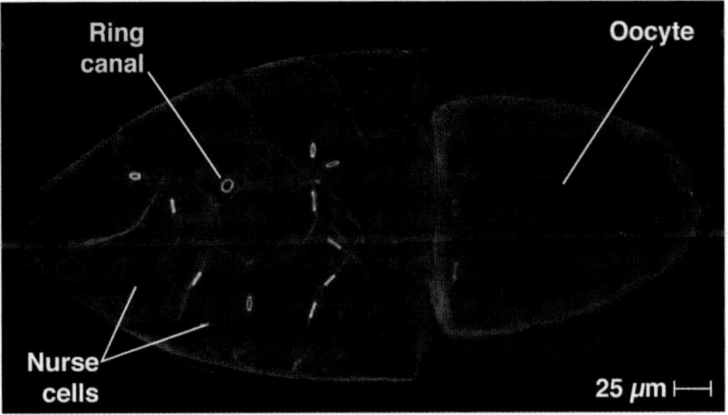

FIGURE 44.25 INCOMPLETE CYTOKINESIS IN A *DROSOPHILA* EGG CHAMBER LEAVES CELLS JOINED BY RING CANALS. Colocalization of actin *(red)* and the ring canal protein HtsRC *(green)* in the ring canals makes them appear *yellow.* In the *Drosophila* egg chamber, ring canals connect nurse cells to each other and to the oocyte. Late in oocyte development, nurse cell contraction forces their cytoplasmic contents through the ring canals and into the oocyte. This helps the oocyte gain the stockpile of components that is needed for early development of the fly embryo. (Courtesy Andrew Hudson and Lynn Cooley, Yale University, New Haven, CT.)

Ultimately, disassembly of the ESCRT filaments leads to separation of the two daughter cells—abscission.

In some tissues, intercellular bridges remain open as **ring canals.** After several rounds of nuclear division with incomplete cytokinesis, the network of cells maintains cytoplasmic continuity as each former contractile ring matures into a larger ring canal. During *Drosophila* oogenesis, four rounds of nuclear division with persistent ring canals creates 15 nurse cells, all in continuity with the oocytes (Fig. 44.25). The cytoplasmic continuity through ring canals allows nurse cells to transfer

BOX 44.1 Cytokinesis in Plants

Chromosome segregation is similar in plants and animals, but cytokinesis is very different because plants lack myosin-II and do not form a conventional contractile ring (Fig. 44.26). Myosin-II appeared during evolution in the common ancestor of amoebas, fungi and animals, after branching from plants (see Fig. 2.4B). Plants also lack dynein, so microtubule dynamics in mitosis are regulated by some of the more than 20 different plus-end- and minus-end-directed kinesins that are expressed in mitotic cells. In a further difference from animals, plants also lack centrosomes, and during interphase, microtubules radiate out from the surface of the cell nucleus in all directions. In mitosis, the spindle does not focus to sharp poles at metaphase; instead, it assumes a barrel shape with broad, flat poles. Early in mitosis, a band of microtubules and actin filaments forms around the equator of the cell adjacent to the nucleus. This so-called preprophase band disassembles as cells enter prometaphase. Because the entire cell cortex is covered by a meshwork of actin filaments, disassembly of the preprophase band actually leaves an actin-poor zone in a ring where cytokinesis will ultimately occur. This is called the cortical division site, and it is marked by the tethering of specific kinesin motors. In late anaphase, two nonoverlapping, antiparallel arrays of microtubules form over the central spindle. This structure, the phragmoplast, gradually expands laterally until it makes a mirror-symmetric double disk of short microtubules oriented parallel to the spindle axis with their plus ends abutting the plane of cell cleavage. Golgi vesicles, containing cell wall materials (see Fig. 32.13), move along phragmoplast microtubules to the equator, where they fuse due to the action of cytokinesis-specific soluble *N*-ethylmaleimide-sensitive factor (NSF) attachment protein receptor (SNARE) proteins (see Fig. 21.15), forming a membrane network that becomes the new plasma membrane and laying down the material that will become the new cell wall. Dynamin-related proteins also participate in shaping the newly forming plasma membrane. Thus, the membrane fusion machinery used for cytokinesis by eukaryotes likely came from the last eukaryotic common ancestor. Actin filaments polymerized by formins and myosin-VIII help position the phragmoplast in the cell. As the zone of newly deposited membrane expands radially, the ring of microtubules surrounding it similarly expands. Eventually, the new membrane reaches the lateral cell periphery, and fusion with the plasma membrane separates the two daughter cells. The cortical division site, not the spindle, determines the site of cleavage. This was shown by centrifuging mitotic cells to displace the spindle from the central location where it initially formed. Late in mitosis, the phragmoplast formed at the midzone of the displaced spindle, but this phragmoplast then migrated to the plane of the preprophase band, where cytokinesis occurred. Since plant cells have cell walls and do not move, the orientation of cleavage planes critically determines the morphology of the organism. The hormone auxin can influence cleavage, giving rise to asymmetric division of daughter cells, but the underlying mechanism is not yet known.

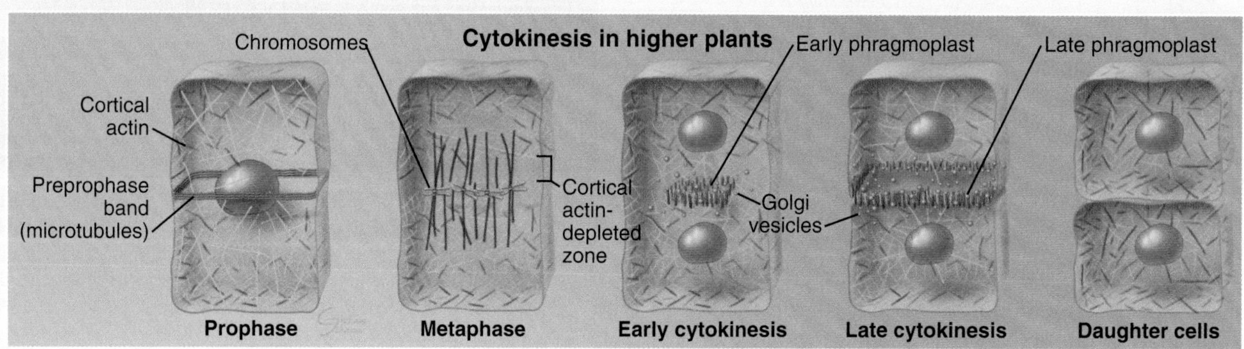

FIGURE 44.26 CYTOKINESIS IN HIGHER PLANTS. See the text for details.

their cytoplasm into the developing egg, thus greatly increasing its stockpile of proteins and messenger RNAs available for use in early development. In mammals, incomplete cytokinesis in the testis results in ring canals connecting several hundred developing sperm cells.

Exit From Mitosis

To exit from mitosis, cells must inactivate the Cdk1 kinase. This reverses the biochemical and structural changes that are characteristic of mitosis and prepares the cell for proliferation in the next cell cycle.

Yeast cells use a signaling pathway to terminate mitosis, promote contraction of the contractile ring, and initiate septation. These pathways, called the mitotic exit network (MEN) in budding yeast and the SIN in fission yeast, involve a small GTPase and protein kinases. Cdk kinase activity suppresses the pathway until anaphase, when Cdk activity drops sharply. The MEN GTPase is associated with one spindle pole body (the yeast version of the centrosome), while its key regulator, a GTP exchange factor, is located in the bud. Elongation of the mitotic spindle during anaphase B moves the GTPase into the bud, where it is activated.

BOX 44.2 Cytokinesis in Bacteria

The strategy for cytokinesis in bacteria is similar to that in animal cells (Fig. 44.27), but the molecules are completely different. Cleavage of most bacterial cells depends on a ring of the FtsZ protein (*filamentous temperature-sensitive;* mutants in *fts* genes cannot divide and make long filaments on cells). This is called the Z ring. FtsZ is the prokaryotic homolog of eukaryotic tubulins, but it assembles into filaments rather than tubules. As for tubulins (see Fig. 34.4), FtsZ polymerization requires bound GTP and hydrolysis of this GTP destabilizes the polymers. Although purified FtsZ forms rings that use energy from GTP hydrolysis to deform lipid vesicles, the main function of the Z ring seems to be to coordinate the assembly of a complex of proteins (divisome) including an actin homolog FtsA and number of transmembrane proteins. The transmembrane proteins synthesize cell wall materials to form the cleavage furrow.

The Z ring is positioned at the cell equator of *Escherichia coli* by the action of three gene products: MinC, MinD, and MinE (*minicell* mutants divide at inappropriate locations and give birth to tiny cells). MinD is an enzyme that recruits MinC to the cell cortex, where it inhibits Z-ring formation. MinE is an antagonist of MinC/MinD action. This system works in a truly remarkable way. MinE forms a ring at the cell equator that migrates along the inner surface of the cell membrane until it reaches the end of the cell, at which point it disassembles. The ring then reforms in the center of the cell and sweeps toward the other end of the cell. As it moves, MinE inactivates the MinC/MinD inhibitory complex on the cell cortex. The inhibitory complex rapidly reestablishes itself on the cell cortex behind the moving MinE ring. It takes approximately 2 minutes for each sweep of the MinE ring along half of the cell, and this cycle is repeated continuously until the FtsZ ring assembles at the cell center. *Bacillus subtilis* uses an alternative mechanism to position the Z ring for cytokinesis.

Chloroplasts use a homolog of FtsZ for their division, and FtsZ has been detected in mitochondria of certain primitive eukaryotes. Mitochondria of higher eukaryotes appear to use another GTPase, dynamin, to coordinate their fission (see "Biogenesis of Mitochondria" in Chapter 19).

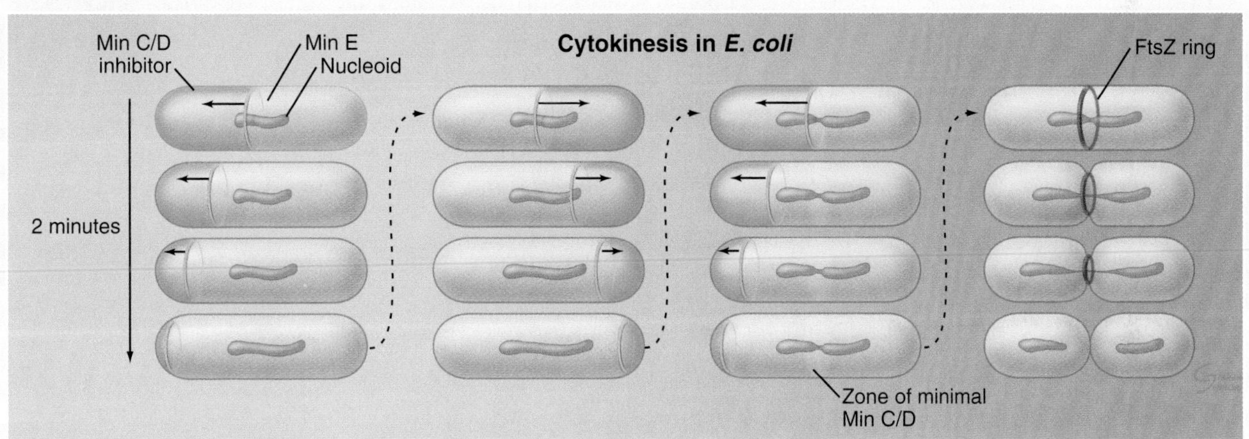

FIGURE 44.27 CYTOKINESIS IN THE BACTERIUM *ESCHERICHIA COLI*. See the text for details.

The MEN kinases downstream of the GTPase activate the phosphatase Cdc14p by releasing it from sequestration in the nucleolus. Cdc14p inhibits Cdk kinase activity in two ways: (a) it inhibits the degradation of a Cdk inhibitor protein, and (b) it dephosphorylates Cdh1, which binds the APC/C and triggers the degradation of B-type cyclins and other proteins. Cdc14p also triggers other events during anaphase, including the transfer of chromosomal passenger proteins to the central spindle.

In metazoans mitotic exit is triggered by the inactivation of Cdk1 and other mitotic kinases. This transition is irreversible, in part because cyclins and Aurora (and other kinases) are degraded. PP2A and its inhibitory kinase Greatwall (see Chapter 40) replace Cdc14 in mitotic regulation in metazoans. Greatwall activity requires Cdks, so when Cdk activity declines, PP2A is released from inhibition. When directed to targets throughout cells by their specificity, determining subunits PP2A and PP1 remove many of the phosphates placed on target proteins by the mitotic kinases. Targets include chromatin, where phosphorylation during mitosis had displaced factors involved in both gene activation and repression. Removal of those phosphates allows the interphase regulation of gene expression to resume. Dephosphorylation of other targets allows intermediate filaments to reform, nuclear envelope reassembly plus the resumption of RNA transcription, protein translation, and membrane trafficking.

ACKNOWLEDGMENTS

We thank David Burgess, Iain Cheeseman, Per Paolo D'Avino, Arshad Desai, Tatsuo Fukagawa, Gary Gorbsky, Karen Oegema, Jonathon Pines, and Graham Warren for their suggestions on revisions to this chapter. We thank the Dundee Imaging Facility for access to the OMX and help with microscopy.

SELECTED READINGS

Carmena M, Wheelock M, Funabiki H, et al. The chromosomal passenger complex (CPC): from easy rider to the godfather of mitosis. *Nat Rev Mol Cell Biol.* 2012;13:789-803.

Collas P, Courvalin J-C. Sorting nuclear membrane proteins at mitosis. *Trends Cell Biol.* 2000;10:5-8.

Glotzer M. Cytokinesis in metazoa and fungi. *Cold Spring Harb Perspect Biol.* 2016;(in press).

Green RA, Paluch E, Oegema K. Cytokinesis in animal cells. *Annu Rev Cell Dev Biol.* 2012;28:29-58.

Haeusser DP, Margolin W. Splitsville: structural and functional insights into the dynamic bacterial Z ring. *Nat Rev Microbiol.* 2016;14:305-319.

Jürgens G. Plant cytokinesis: Fission by fusion. *Trends Cell Biol.* 2005;15:277-283.

McIntosh JR. Mitosis. *Cold Spring Harb Perspect Biol.* 2016;(in press).

Müller S, Jürgens G. Plant cytokinesis—no ring, no constriction but centrifugal construction of the partitioning membrane. *Semin Cell Dev Biol.* 2016;53:10-18.

Nasmyth K, Haering CH. Cohesin: its roles and mechanisms. *Annu Rev Genet.* 2009;43:525-558.

Qian J, Winkler C, Bollen M. 4D-networking by mitotic phosphatases. *Curr Opin Cell Biol.* 2013;25:697-703.

Rappaport R. *Cytokinesis in Animal Cells: Developmental and Cell Biology Series.* Cambridge, England: Cambridge University Press; 1996.

Sánchez-Huertas C, Lüders J. The augmin connection in the geometry of microtubule networks. *Curr Biol.* 2015;25:R294-R299.

Sharp DJ, Rogers GC, Scholey JM. Microtubule motors in mitosis. *Nature.* 2000;407:41-47.

Stukenberg PT, Burke DJ. Connecting the microtubule attachment status of each kinetochore to cell cycle arrest through the spindle assembly checkpoint. *Chromosoma.* 2015;124:463-480.

CHAPTER 45

Meiosis

Meiosis (from the Greek, meaning "reduction") is a specialized program of two coupled cell divisions used by eukaryotes to maintain the proper chromosome number for the species during sexual reproduction. It also generates novel combinations of genes. Meiosis is an ancient process that occurs in virtually all eukaryotes, including the animal, fungal, and plant kingdoms, and is thought to have been present in the last eukaryotic common ancestor.

Sexually reproducing organisms are typically **diploid**, with pairs of **homologous chromosomes**, the two highly similar but nonidentical copies of each chromosome, one inherited from each parent. The number of chromosomes is halved during meiosis to form **haploid** gametes carrying just one set of chromosomes. The subsequent fusion of male and female gametes restores the diploid chromosome number. This pairing and subsequent separation of homologous chromosomes is made possible by genetic recombination, which occurs during the lengthy and complex prophase of the first meiotic division.

Each human somatic cell has 23 pairs of chromosomes (46 in all). Females have 23 homologous pairs, while males have 22 "autosomal" pairs and two different sex chromosomes that share a region of homology known as the **pseudoautosomal region**. One of each pair is contributed by each parent in the egg and sperm, respectively. The number of chromosome pairs, 23, is known as the **haploid** chromosome number. In animals, the only haploid cells are gametes (sperm and eggs). At fertilization, haploid gametes fuse to form a zygote, restoring the **diploid** chromosome number of 46. In plants, the haploid phase is represented by gametophytes, which produce ovules and pollen. In most fungi, such as yeasts, haploid and diploid forms are alternate phases of the life cycle, and both can propagate by mitosis.

Meiosis changes the genetic makeup of offspring relative to parents in two ways: the first round of meiotic segregation produces novel combinations of chromosomes from the two parents in each gamete, and recombination between parental chromosomes produces novel chromosomes. It works like this. During meiosis, one round of DNA replication and two rounds of chromosome segregation reduce the number of chromosomes from 2n to 1n. Each haploid gamete is endowed with a random set of the homologous chromosomes from the two parents. Prior to meiosis I the chromosomes duplicate (just like mitosis) (Fig. 45.1A), but during the first round of segregation (Fig. 45.1C) the duplicated chromatids remain paired and the homologous chromosomes separate randomly between the two daughter cells. Thus each daughter cells ends up with just one of each pair of homologous chromosomes. This differs from mitosis where the duplicated chromatids separate, so both daughter cells get the full set of homologous chromosomes from both parents. During meiosis II the duplicated chromatids separate and are partitioned equally between the two daughter cells. Equally important, homologous chromosomes exchange DNA sequences during meiotic prophase I, generating novel chromosomes.

The unique segregational events of meiosis usually occur in the first division, termed **meiosis I** (Figs. 45.1 and 45.2). Because it culminates in daughter cells carrying just one set of chromosomes instead of two, meiosis I is also known as the **reductional division.** The second division, **meiosis II,** is similar in most respects to mitosis: sister chromatids segregate, and the number of chromosomes remains the same (Box 45.1; see also Chapter 44). Meiosis II is called the **equational division.**

Meiosis: An Essential Process for Sexual Reproduction

Sexual reproduction is an important survival strategy that offers organisms an accelerated mechanism for altering the genetic makeup of offspring. Without meiosis, there would be no sex, because fusion of diploid gametes

Preparation for meiosis (This step duplicates each chromatid)

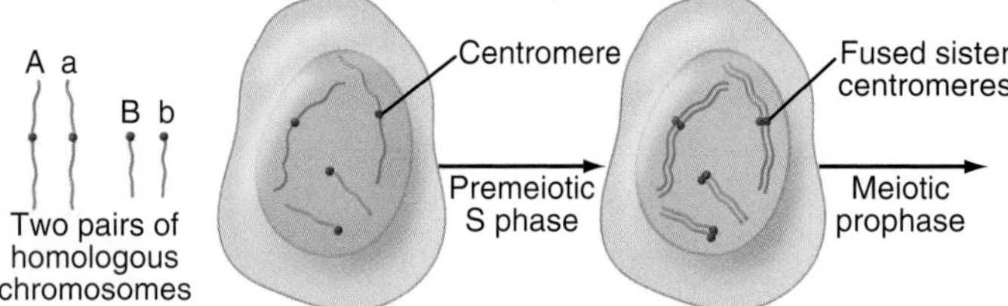

Meiotic prophase (Recombination drives pairing of homologous chromosomes and produces novel chromosomes)

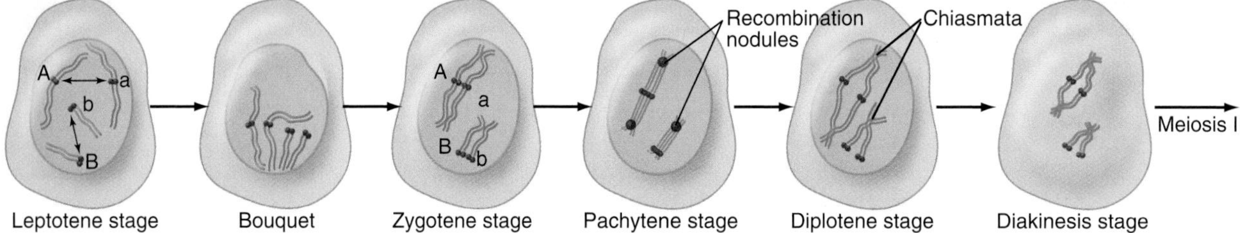

Meiosis I (Homologous chromosomes randomly separate from one another producing haploid progeny)

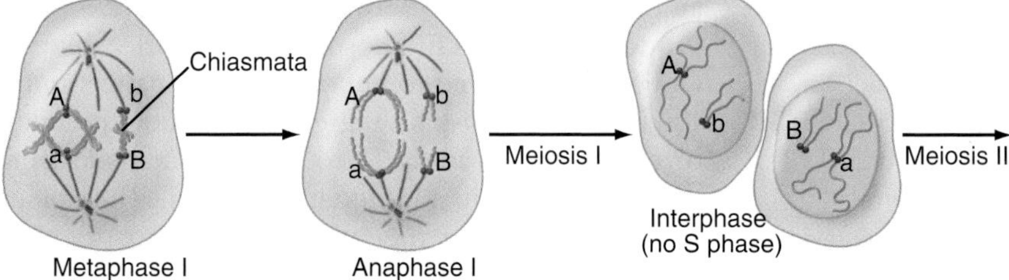

Meiosis II (Sister chromatids separate producing one or more gametes)

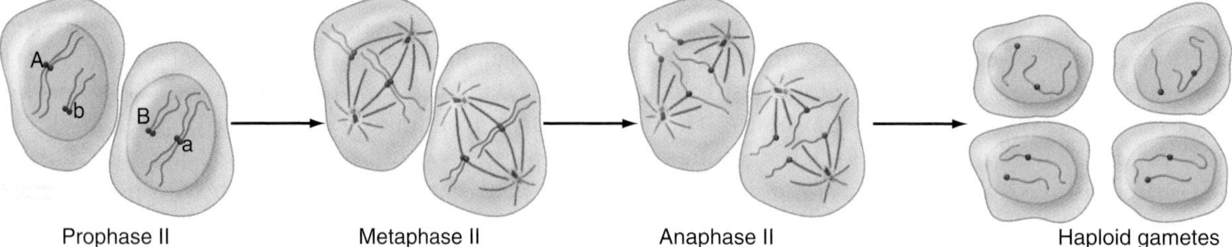

FIGURE 45.1 OVERVIEW OF THE PHASES OF MEIOSIS. Shown are important structures and the outcome of each stage for a homologous pair of metacentric chromosomes (A, a) and a homologous pair of telocentric chromosomes (B, b).

would double the number of chromosomes in the progeny at every generation.

Meiosis I produces random combinations of homologous maternal and paternal chromosomes. For each pair of homologs, orientation on the spindle is random during meiosis I (ie, each homolog has two equivalent options for the direction to migrate). Thus, for humans (with 23 pairs of homologous chromosomes), each gamete has one of 2^{23} (more than 8 million) possible complements of maternal and paternal chromosomes. This process does not create new versions of genes, but it guarantees the offspring will have novel *combinations* of subtly different (due to polymorphisms) chromosomes.

Meiosis I also produces novel *versions* of genes and chromosomes by recombinational exchange of DNA segments between homologs. This occurs because to segregate from one another, each pair of homologous chromosomes must first *find* each other. They do this by undergoing reciprocal recombination (crossover) events that then hold them together until anaphase of meiosis I. Chromosomes and the genes they carry vary hugely between individuals. In humans an average genome varies from the "reference genome" (see Chapter 7) at 4 to 5×10^6 sites. These include not only polymorphisms (differences of single base pairs), but also thousands of longer insertions, deletions, and rearrangements. Recombination events that result in a crossover and exchange chromosomal segments produce new chromosomes that are a patchwork of segments from the maternal and paternal homologs. The combined effects of recombination and random assortment of homologs in meiosis I yields a vast number of genetically different gametes. This genetic diversity increases the ability of eukaryotic populations to adapt to changing environmental conditions.

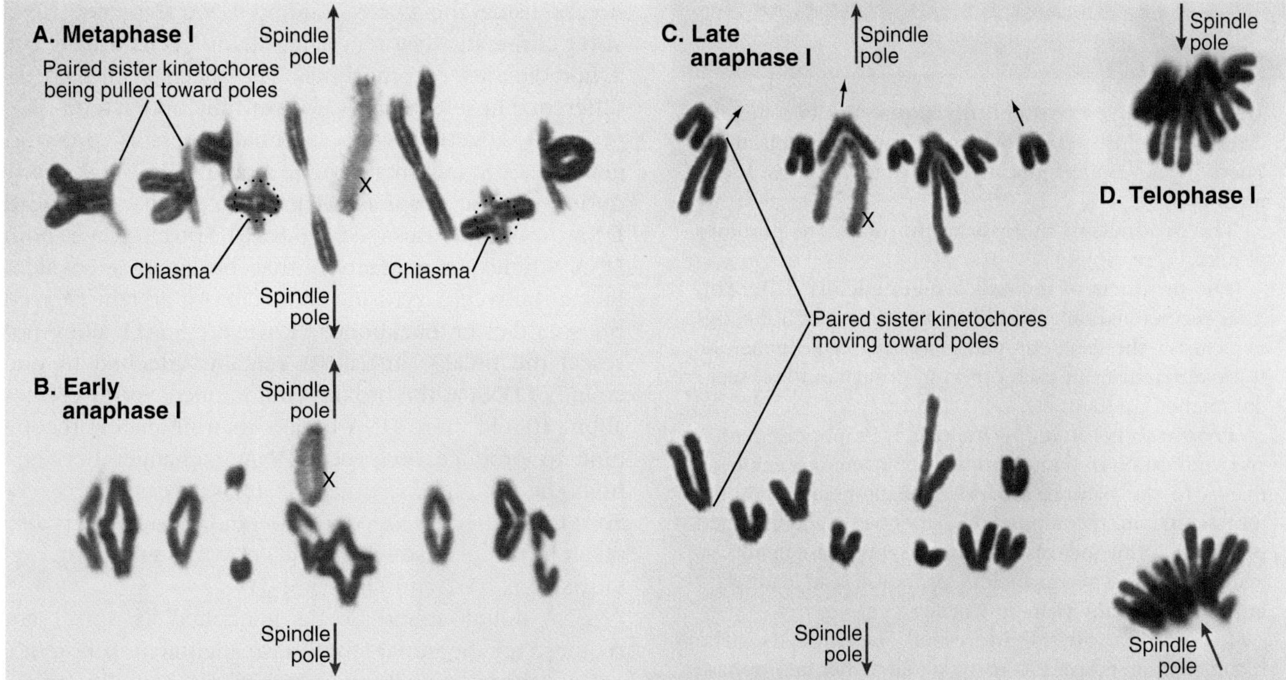

FIGURE 45.2 FIRST MEIOTIC DIVISION STAGES FROM THE GRASSHOPPER *PYRGOMORPHA CONICA* (2N IN MALES = 18 AUTOSOMES + 1 X CHROMOSOME). A, Metaphase I. **B,** Early anaphase I. **C,** Late anaphase I. **D,** Telophase I spermatocytes stained with lactopropionic orcein. All chromosomes are telocentric (see Fig. 7.2). Seven bivalents shown in the metaphase I spermatocyte have a single chiasma, whereas the two bivalents at the far right and far left have two chiasmata. The sex chromosome (X) remains unpaired and moves to a single spindle pole. (Courtesy José A. Suja and Julio S. Rufas, Universidad Autónoma de Madrid, Spain.)

The Language of Meiosis

Meiosis has a language of its own, characterized by a number of unusual terms, and is easiest to understand by focusing on the essential biological processes that are involved. This reduces the process to only three essential key terms: **pairing, homologous recombination,** and **segregation.** This chapter discusses each step in detail, so they are defined only briefly here.

Pairing is a two-step alignment of homologous chromosomes with one another in the nucleus. In *alignment,* corresponding DNA sequences on the homologous chromosome find each other among the billions of base pairs of DNA in the nucleus. In many organisms, early events of recombination drive the homologous pairing process. In the second stage, *synapsis,* the paired homologous chromosomes become intimately aligned along their entire lengths with one another separated by approximately 100 nm. A specialized scaffolding structure called the synaptonemal complex mediates this process.

Homologous recombination results in physical exchange of DNA between homologous chromosomes (a crossover event) and is a key determinant of chromosome behavior during meiotic prophase. Recombination drives the pairing process in many organisms and can occur without synapsis under certain circumstances. Crossover recombination sites are detected by microscopy as chromatin structures called **chiasmata** (singular: chiasma, from the Greek, meaning "X-shaped cross").

Segregation of homologous chromosomes in meiosis I differs from the segregation of sister chromatids during mitosis (Box 45.1), because the paternal and maternal homologous chromosomes segregate randomly to the two daughter cells. When homologs orient at the metaphase plate of the meiosis I spindle, centromeres belonging to the two sister chromatids are fused to form a single kinetochore that binds microtubules. Cohesion between chromosome arms distal to chiasmata (ie on the other side from the centromere; Figs. 45.2 and 45.10) keeps homologous chromosomes paired with one another until anaphase of meiosis I, counteracting the bipolar pulling force of the spindle on the homologs (Fig. 45.2). At anaphase I the distal cohesion is released from chromosomes allowing the chiasmata to separate, and the two sister chromatids (at least one of which has undergone a crossover exchange) move as a single unit toward the same spindle pole while the sister chromatids from other parent move to the other daughter cell. As a result, the two daughter cells produced in meiosis I have a haploid number of chromosomes derived randomly from the two parents, each with two sister chromatids. Each of the four daughter cells produced in meiosis II has one sister chromatid for each homologous chromosome (ie, half the number of chromatids as there are chromosomes in somatic cells).

BOX 45.1 Important Differences Between Meiosis and Mitosis

Meiosis involves two cell divisions. The two meiotic divisions are preceded by a round of DNA replication. There is no DNA replication between meiosis I and meiosis II.

The products of meiosis are haploid. The products of mitosis are diploid.

The products of meiosis are genetically different. After recombination and random assortment of homologs in meiosis I, the sister chromatids that segregate in meiosis II are different from each other. In normal mitosis, sister chromatids are identical.

Prophase is longer in meiosis I. Proper orientation and segregation of homologous chromosomes is achieved thanks to the pairing, synapsis (synaptonemal complex formation), and recombination that occur in a lengthened prophase during meiosis I. In humans, prophase in mitosis takes an hour, whereas meiotic prophase lasts many days in males and many years in females.

Recombination is increased in meiosis. The recombination rate is 100- to 1000-fold higher in prophase I of meiosis than in mitosis. The process has two main consequences: the formation of chiasmata and the introduction of genetic variation. Chiasmata are structures that physically link the homologous chromosomes after crossover and play an essential role in meiotic chromosome segregation.

Kinetochore behavior differs in meiosis. During meiosis I, kinetochores of sister chromatids attach to spindle microtubules emanating from the same pole. Homologous kinetochore pairs connect to opposite poles. In mitosis and meiosis II, sister kinetochores attach to spindle microtubules coming from opposite poles.

Chromatid cohesion differs in meiosis. Sister chromatid cohesion is essential for orientation of bivalents (paired homologous chromosomes) on the metaphase I spindle. During anaphase of meiosis I, cohesion is destroyed between sister chromatid arms, and chiasmata are released to allow segregation of homologs. Cohesion at sister centromeres persists until the onset of anaphase II, when it is lost to permit segregation of sisters. In prometaphase of meiosis II, sister chromatids are joined only by the centromeres, whereas at the beginning of mitotic prometaphase, sisters are joined all along the arms.

Recombination

Although meiotic recombination is similar to the process of homologous recombinational repair of double-strand DNA breaks in somatic cells (review Box 43.1 and Fig. 43.14 as a prelude to studying meiotic recombination), the two processes differ in two respects. First, meiotic cells use a specialized enzyme called Spo11 to create double-strand DNA breaks on purpose. Second, somatic cells with replicated chromosomes usually repair DNA breaks using the corresponding DNA sequence on a sister chromatid as a template. Meiotic cells usually use a homologous chromosome. The mechanism for this difference in selectivity is not yet fully understood.

Spo11, together with essential accessory proteins, generates programmed double-strand DNA breaks early during meiotic prophase (Fig. 45.3). Similar to type II DNA topoisomerases (see Fig. 8.16), Spo11 cleaves both DNA strands in a reaction that produces a covalent linkage between a tyrosine of the enzyme and the cleaved phosphodiester backbone. However, Spo11 does not reseal the breaks; instead it remains attached to one strand of DNA at the broken end. In mice, Spo11 creates about 10-fold more DNA breaks than ultimately recombine to produce reciprocal DNA exchanges between homologous chromosomes or **crossovers.** Repair of Spo11-mediated DNA double-strand breaks can also result in **noncrossover** events known as **gene conversions** (Box 45.2 and Fig. 45.3I–J).

DNA double-strand breaks generated by Spo11 are required for the initial lengthwise alignment of homologous chromosomes in many organisms, including mice, plants, and yeast. In mice lacking Spo11, recombination is not initiated, and synapsis, if it occurs at all, is aberrant, often involving nonhomologous chromosomes (Fig. 45.4). Gametes in these mutant mice die by apoptosis early in meiotic prophase. Spo11-induced double-strand breaks are not required for synapsis of homologous chromosomes in the nematode *Caenorhabditis elegans* and the fruit fly *Drosophila melanogaster*. How these organisms pair their homologs without recombination is still mysterious.

Once the DNA double-strand breaks are produced, the MRN (Mre11/Rad50/Nbs1) endonuclease nicks the single-stranded DNA, releasing Spo11. The DNA ends that lost Spo11 then undergo further processing, as Exo1 exonuclease, chews back the 5′ strands of the double helix (a process called **resection**) leaving single-stranded DNA tails with 3′ termini (Fig. 45.3C; see also Fig. 43.14). The MRN and Exo1 nucleases also function in somatic DNA repair.

Next, the Rad51 and Dmc1 proteins drive a search of the 3′ single-stranded DNA tails for complementary DNA sequences of the other chromosomes. Rad51 and Dmc1 are related to the *Escherichia coli* RecA protein used for homologous recombination in bacteria. Rad51 and Dmc1 coassemble along 3′ single-stranded DNA tails and use adenosine triphosphate (ATP) hydrolysis to catalyze a strand exchange reaction with an intact homologous DNA duplex. The Rad51-and-Dmc1-decorated nucleoprotein filament disrupts the targeted homologous double helix, displacing one of the two DNA strands. This allows formation of new Watson-Crick base pairs between the invading 3′ single-stranded DNA and the complementary strand of the target DNA. Following strand invasion and exchange, new DNA synthesis

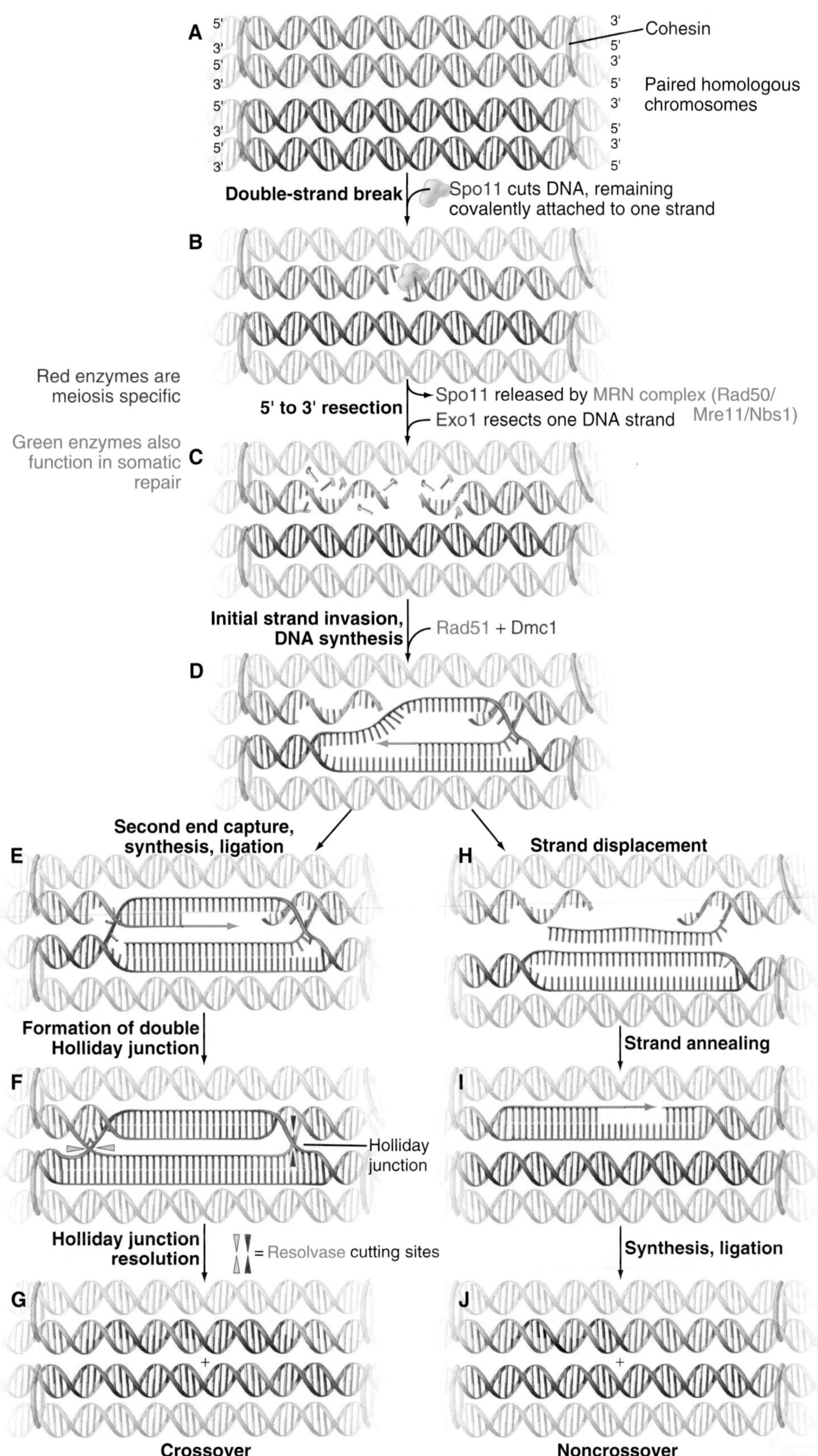

FIGURE 45.3 EVENTS OF RECOMBINATION. Recombination occurs between homologous chromosomes rather than between sister chromatids. **A,** Paired homologous chromosomes. Sister chromatids are held tightly together by cohesin, shown here schematically as hoops. **B,** Spo11 makes a double-strand break, remaining attached to the DNA. **C,** Removal of Spo11 and resection of the break. **D,** First strand invasion. At this point, the pathway splits in two, one outcome leading to a crossover and the other to a noncrossover. **Crossover pathway: E,** The second resected strand establishes base-pairing interactions with the displaced DNA strand of its homologous partner. New DNA synthesis fills the gaps. **F,** The resulting molecule contains a double Holliday junction in which the DNA is fully base-paired (see Fig. 43.14B). If the resolvase (nuclease) cuts the double Holliday junction asymmetrically as shown (ie, one vertical and one horizontal cut), the result is a crossover **(G).** If the cuts are symmetrical, a noncrossover molecule is produced. **Noncrossover pathway: H,** In most cases, the invading DNA, strand is ejected prior to stabilization and formation of a double Holliday junction. **I,** DNA gap-filling and ligation yield a noncrossover chromosome **(J).**

BOX 45.2 Brief Overview of Genetic Terminology

A comprehensive introduction to the field of genetics is beyond the scope of this text. However, here are a number of terms used by geneticists that will assist in the understanding of the discussion of genetic recombination and its role in meiosis (also see Box 6.1).

The **genotype** of an organism is the combination of genes present on the chromosomes of that organism. The **phenotype** is the physical manifestation of the action of these gene products (ie, the appearance and macromolecular composition of the organism). In discussing recombination, scientists typically refer to the presence or absence of specific genetic markers. Each **genetic marker** is a particular DNA sequence in or around a gene that can be monitored by examining the phenotypes of the cells that carry it. A genetic marker might be the presence of a functional gene, a mutation with altered activity, or simply a polymorphism of DNA sequence that has no known functional consequence.

A **haploid** organism has one copy of each chromosome. A **diploid** organism has two homologous copies of each chromosome. A diploid organism that is **homozygous** for a particular genetic marker has the same sequence of that particular region of the DNA on both the maternal and paternal homologous chromosomes. A **heterozygous** organism has different forms of the genetic marker on the two homologous chromosomes. Although the physical events of genetic recombination occur in both homozygotes and heterozygotes, they are most readily detected in the latter.

Two genetic markers located on different chromosomes will separate from one another in the anaphase of meiosis I 50% of the time as a result of the random distribution of chromosomes to the two spindle poles. If they are on the same chromosome, they will be *linked* to one another unless the chromosome undergoes a genetic recombination event between them. The greater the separation of two markers along one chromosome, the more likely it is for such an intervening recombination event to occur.

Two types of recombination events occur during meiosis (Fig. 45.3). The first of these—**noncrossover** events (frequently referred to as **gene conversion**)—may involve the *loss* of one or more genetic markers. Noncrossover events are the most common outcome of the programmed double-strand DNA breaks that occur during leptotene. They are thought to involve the invasion of a double helix by a region of single-stranded DNA with complementary sequence but then ejection of this sequence before assembly of a Holliday junction and completion of recombination.

The second type of recombination event—**crossing over**—involves the physical breakage and reunion of DNA strands on two different chromosomes, typically producing a balanced exchange of DNA sequences. This is what most people think of as recombination. In recombination by crossing over, the makeup of genetic markers remains constant; it is the linkage between different markers that changes.

The normal separation of chromosomes or chromatids is referred to as **disjunction** (disjoining). Mistakes in this separation are referred to as **nondisjunction.** Nondisjunction in meiosis I and II results in the production of gametes with either too many or too few chromosomes, a condition known as **aneuploidy.**

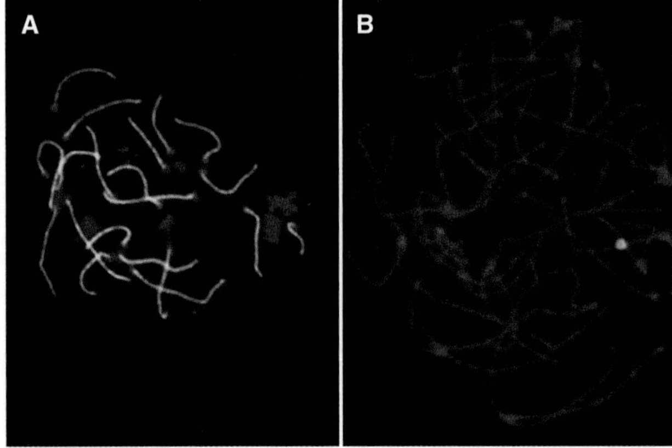

FIGURE 45.4 PAIRING OF HOMOLOGOUS CHROMOSOMES IS SEVERELY DISRUPTED IN THE *SPO11* MUTANT. Pachytene chromosomes from wild-type mice **(A)** and mice in which the *Spo11* gene has been disrupted **(B)**. SYC3 (axial elements) and centromeres are *red*. SYCP1 (in transverse filaments, which are seen only when synapsis has occurred) is *green*. (From Baudat F, Manova K, Yuen JP, et al. Chromosome synapsis defects and sexually dimorphic meiotic progression in mice lacking Spo11. *Mol Cell*. 2000;6:989–998.)

restores sequences that may have been lost or damaged at the position of the original DNA double-strand break.

Mutants lacking Dmc1 are defective in homologous chromosome pairing and interhomolog recombination. As a result, Dmc1 is thought to facilitate the search for homologous chromosomes as a DNA repair template, rather than sister chromatids as in somatic DNA repair. Rad51p and Dmc1p are found in structures called **early recombination nodules** that are distributed along the chromosome axes early in meiosis (Fig. 45.9). Dmc1 functions only in meiosis, but Rad51 has other essential functions.

It is now believed that noncrossover events arise primarily from recombination intermediates that involve a relatively transient single strand invasion of the homologous chromosome followed by restorative DNA synthesis and disassembly of the joint molecule intermediate. Crossovers, on the other hand, are thought to arise predominantly through a pathway that involves stable branched intermediates known as **Holliday junctions** (Fig. 45.3F–G; see also Fig. 43.14B), which are then cleaved by resolvases such as Gem1 to form (predominantly) crossover products.

Most sexually reproducing organisms depend on recombination during meiosis to produce haploid gametes, but fruit flies and yeast have other systems for segregating homologs in meiosis I. These mechanisms, collectively known as **achiasmate segregation,** allow the segregation of chromosomes that have *not* undergone crossover recombination. One model for the achiasmate segregation in flies proposes that nonrecombined chromosomes remain paired at the end of meiotic prophase owing to stickiness of heterochromatin and, as a result, segregate properly during anaphase I of meiosis. In a rare but notable example, the spermatocytes of *D. melanogaster* males do not recombine at all, yet still segregate their chromosomes happily during meiosis I. This might be regarded as a cruel joke of evolution by those students who find all the Greek terms of meiotic nomenclature to be daunting. However, meiosis without recombination is clearly the exception, and in most species meiosis depends on recombination in both males and females.

Tracking the Homologous Chromosomes Through the Stages of Meiotic Prophase I

Pairing and recombination of homologous chromosomes take place during **prophase** of meiosis I. Five stages of meiotic prophase are used to describe the process: leptotene, zygotene, pachytene, diplotene, and diakinesis (Fig. 45.1B).

Leptotene (from the Greek, meaning "thin ribbon") starts with the first visible condensation of the chromosomes. Paired sister chromatids become visible as linear arrays of loops flanking a single dense protein-containing axis (Fig. 45.5A–B). This axis consists of proteins that play a role in mitotic chromosome structure as well as proteins specialized for meiotic chromosomes. For example, the cohesin complex with several specialized meiosis-specific subunits is a prominent component of this axial structure (see Fig. 8.18). According to recent models, recombination begins during leptotene with the formation of DNA double-strand breaks. By the end of

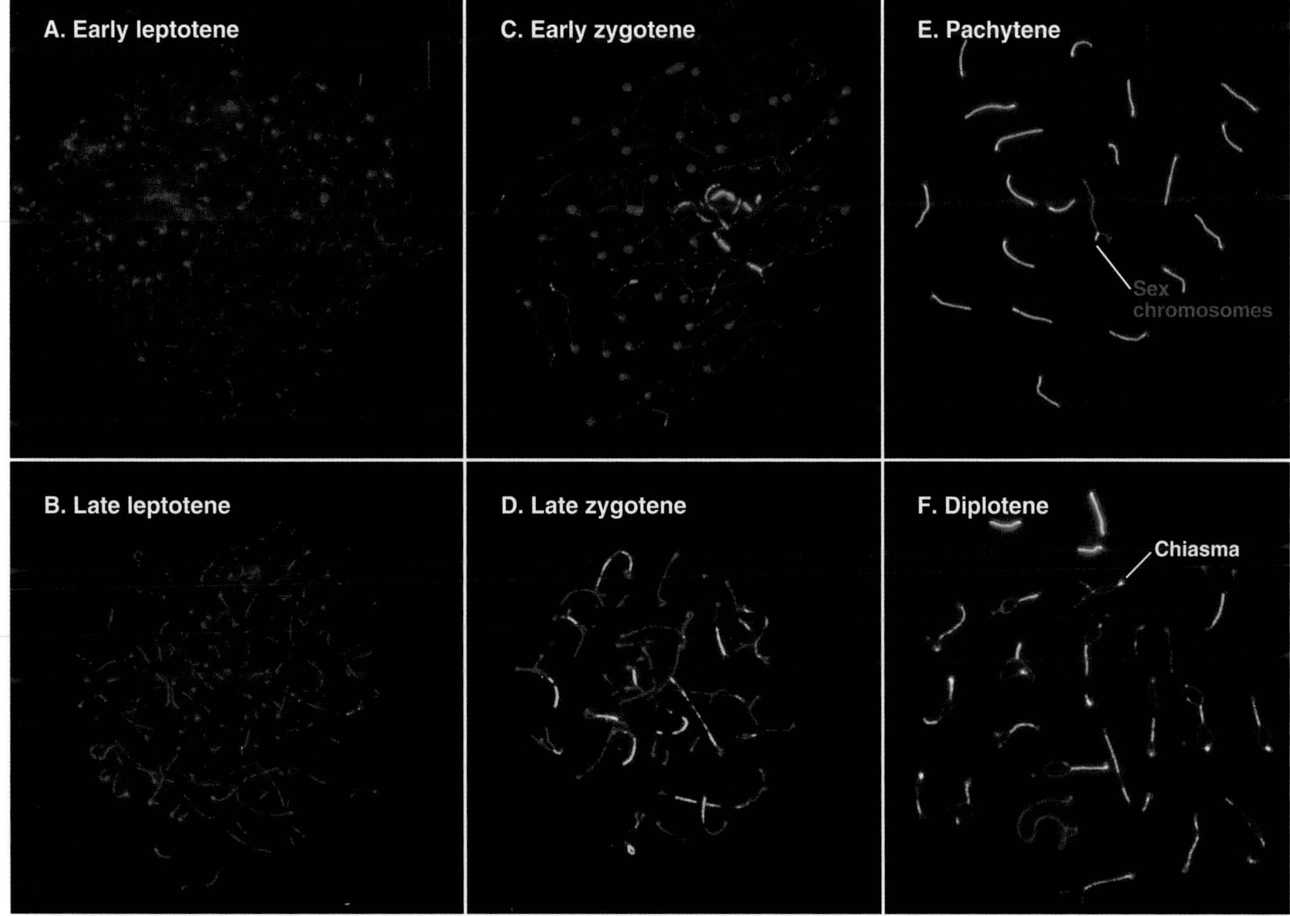

FIGURE 45.5 IMMUNOFLUORESCENCE IMAGES OF PROPHASE I SUBSTAGES IN MOUSE SPERMATOCYTES. These images demonstrate the pairing and synapsis of homologous chromosomes revealed by visualizing the synaptonemal complex proteins SYCP3 (a component of the axial elements *[red]*) and SYCP1 (a component of the transverse filaments that is present only when homologs are synapsed *[green]*). Centromeres are *blue.* (Courtesy Paula Cohen, Cornell University, Ithaca, NY.)

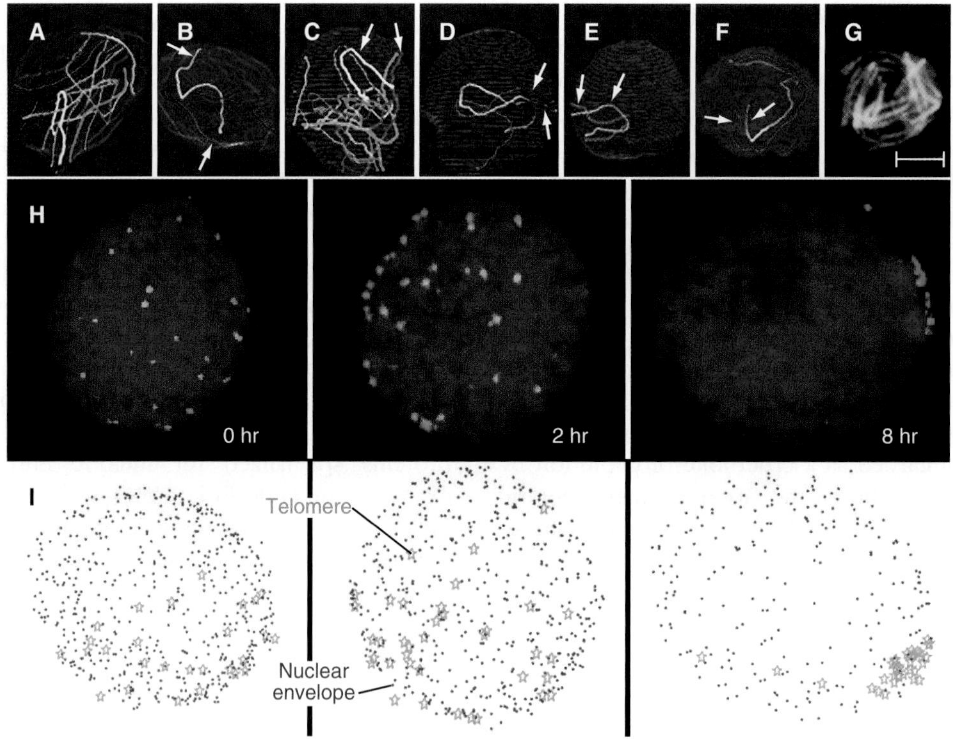

FIGURE 45.6 CHROMOSOMAL MOVEMENTS DURING EARLY MEIOTIC PROPHASE. A–G, Pairing of homologous chromosomes during leptotene in the ascomycete *Sordaria.* Scale bar is 1 µm in **A–F** and 5 µm in **G**. **A–B,** In early leptotene, homologous chromosomes (visualized in panels **A–F** by electron microscope reconstructions of serial-sectioned nuclei) are not yet aligned with one another. **C–E,** In mid-leptotene, regions of some homologs begin to align. (In panel **D,** only the telomeres have aligned. In panel **E,** the pair of homologs is fully aligned.) **F,** The alignment of homologs is complete by late leptotene. **G,** The alignment of homologs also can be seen by light microscopy using Spo76-GFP, a component of the chromosome axes. **H–I,** Stages of formation of the bouquet arrangement in rye. **H,** Telomeres *(green)* were detected in nuclei by in situ hybridization (see Fig. 8.10) after 0, 2, and 8 hours in culture. Chromatin is *red.* **I,** Three-dimensional models of the nuclei (nuclear periphery *[red dots],* telomere position [*green stars*]). (**A–G,** Modified from Tesse S, Storlazzi A, Kleckner N, et al. Localization and roles of Ski8p protein in *Sordaria* meiosis and delineation of three mechanistically distinct steps of meiotic homolog juxtaposition. *Proc Natl Acad Sci U S A.* 2003;100:12865–12870. Copyright 2003 National Academy of Sciences. **H–I,** Modified from Carlton PM, Cowan CR, Cande WZ. Directed motion of telomeres in the formation of the meiotic bouquet revealed by time course and simulation analysis. *Mol Biol Cell.* 2003;14:2832–2843.)

leptotene, homologous chromosomes are aligned loosely about 400 nm apart (Fig. 45.6D–G).

During leptotene, one or both telomeres attach to the inner surface of the nuclear envelope and move actively around the nuclear surface until they coalesce near the centrosome (spindle pole body in yeasts [Fig. 45.6]). These movements and clustering of telomeres depend on cytoplasmic microtubules. Telomeres are linked to microtubules through a pair of nuclear envelope proteins known as the LINC (*li*nker of *n*ucleoskeleton and *c*ytoskeleton) complex (see Fig. 9.8). Telomere clustering peaks at the leptotene-zygotene transition with the chromosomes radiating into the nuclear interior like a bouquet of flowers, hence the name "bouquet stage."

Bouquet formation is a nearly universal feature of this phase of meiosis and the movements of tethered telomeres help homologs find each other through physical alignment. Thus, telomere clustering per se may not be the goal of this movement. The details vary among different organisms. In fission yeast dynein motors and microtubule dynamics in the cytoplasm move the telomere cluster from one end of the cell to the other every 10 minutes or so. These "horsetail movements" stretch the chromosomes parallel to each other. In *C. elegans* special chromosome regions known as "pairing centers" mediate chromosome movement instead of telomeres; in budding yeast, the telomeres are linked to actin instead of microtubules, and *D. melanogaster* may have lost such a mechanism altogether.

During the transition from leptotene to the **zygotene** (Greek, "yoke ribbon") stage of prophase, clustering of chromosome ends at the nuclear envelope reaches its peak, with the "bouquet" arrangement of chromosomes. During this stage homologous chromosomes begin to achieve their maximal alignment as well, through the initiation of **synapsis** (Fig. 45.5C–D). Synapsis involves the assembly of the **axial element**. This protein scaffold forms part of the **synaptonemal complex** when pairing is complete.

In **pachytene** (from the Greek, meaning "thick ribbon"), synapsis is complete, with the homologous chromosome axes joined together along their lengths by synaptonemal complexes (Fig. 45.5E). During pachytene, crossover-designated recombination intermediates mature into Holliday junction-containing structures within the context of the full-length synaptonemal complex. The final resolution of these recombination intermediates into crossovers occurs close to the time of synaptonemal complex disassembly, dispersal of the bouquet of chromosomes and exit from pachytene. The crossovers then mature into structures called **chiasmata** that link homologous chromosomes through meiosis I metaphase.

Early in **diplotene** (from the Greek, meaning "double ribbon"), the synaptonemal complex disassembles, telomeres detach from the nuclear membrane, and chromosomes begin to condense in preparation for division (Fig. 45.5F). The duplicated sister chromatids remain closely associated, and chiasmata hold the homologous chromosomes together, although their axes tend to drift apart in the absence of synaptonemal complex. This part of meiotic prophase may last for days or years, depending on the sex and organism (up to 45 years or more in female humans).

Oocytes (immature eggs) actively transcribe their chromosomes during diplotene, as they store up materials for use during the first few divisions of embryonic development. Transcription can be so active that DNA loops are massively coated with nascent RNA transcripts whose associated proteins are visible by light microscopy in oocytes of most animals (except mammals). Chromosomes at this stage are known as **lampbrush chromosomes** (see Fig. 8.12).

Diakinesis (from the Greek, meaning "across movement") is the prometaphase of meiosis I. Following nuclear envelope breakdown, homologous chromosomes shorten and condense. At metaphase I, the bivalents (pairs of homologous chromosomes) are aligned at a metaphase plate (Figs. 45.1, 45.2, and 45.12). The two homologs (each a pair of tightly linked sister chromatids) are attached to opposite poles of the meiotic spindle, which applies force, attempting to pull them apart. Cohesion of the arms distal to chiasmata resists these pulling forces. The homologs separate and move to opposite spindle poles during anaphase I when the cohesion along the chromosome arms is released. The sister chromatids move together to one pole, because they remain linked by cohesion at their centromeres, where the cohesion complex is protected by a shugoshin protein (see later).

After telophase I, cells enter a brief **interkinesis** during which there is no DNA replication. The second meiotic division is mechanistically similar to mitosis except that the number of chromosomes is reduced by half. Additionally, in the eggs of most female vertebrates, meiosis is arrested at metaphase II until fertilization.

Pairing and Synapsis in More Detail

Pairing describes the side-by-side alignment of homologous chromosomes at a distance. Homologs are paired in nonmeiotic (somatic) cells in a few organisms, such as the fruit fly *D. melanogaster*, but not in vertebrates.

The earliest pairing events in meiosis involve a tendency of homologous chromosome territories to move together in the nucleus even before leptotene chromosome condensation. The mechanism is unknown. As double-strand breaks created by Spo11 initiate the recombination pathway during leptotene, the condensing homologous chromosomes align with one another at a distance of about 400 nm (Figs. 45.6 and 45.7). Genetic analysis in budding yeast revealed that mutants defective in the earliest stages of recombination are also defective in homolog pairing.

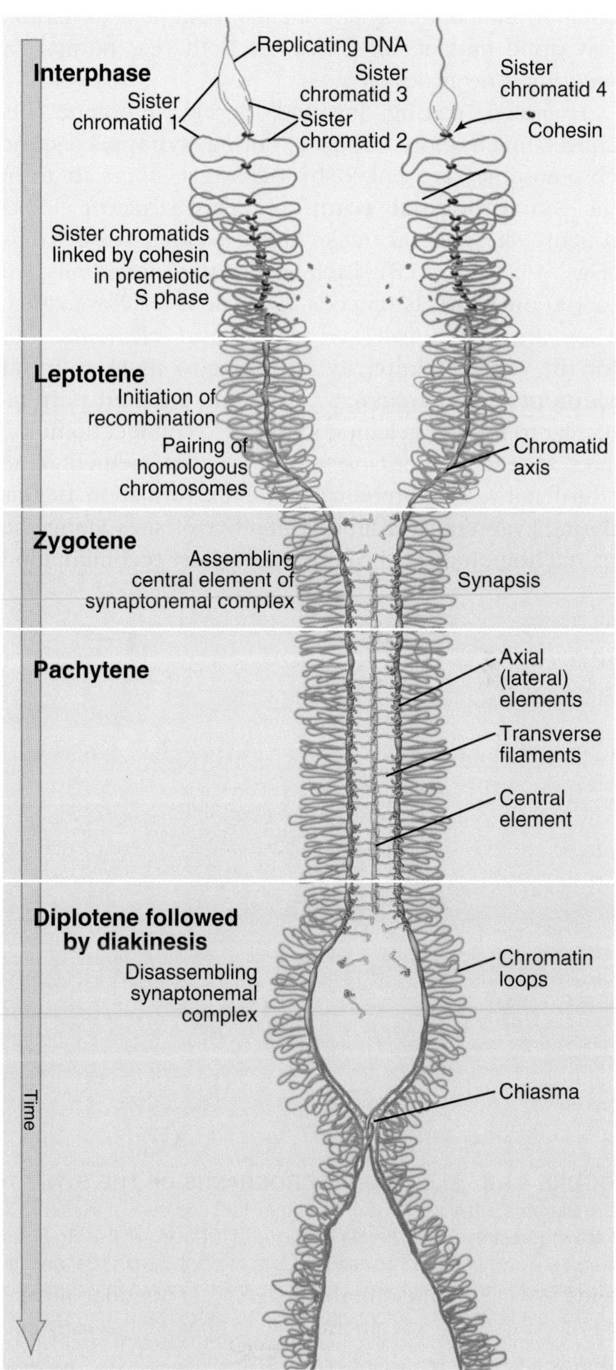

FIGURE 45.7 CHROMOSOMAL PAIRING IN MEIOTIC PROPHASE. Structural organization of homologous chromosomes and synaptonemal complex during various stages of meiotic prophase.

The process of homolog alignment almost certainly involves the invasion of neighboring DNA duplexes by single-stranded DNA complexed with Rad51 and Dmc1. Thus, recombination has important roles both in the exchange of genetic material and in the mechanics of chromosome behavior during meiotic prophase. Recombination is probably not the only factor driving homolog pairing, however. Pairing is reduced but not absent in yeast meiotic cells lacking both Rad51 and Dmc1, and homologous chromosomes still pair in some systems that lack recombination (eg, certain *D. melanogaster* recombination mutants), synaptonemal complex formation (asynaptic mutants in yeast), or both (eg, normal *D. melanogaster* males).

Homolog pairing initiated during leptotene becomes much more intimate during **synapsis** as the chromosomes are linked by transverse fibers to form the **synaptonemal complex.** This structure looks roughly like railroad tracks linked by transverse bands (Figs. 45.7 and 45.8). Each of the two outer rails, 90 to 100 nm apart, is the axis of a pair sister chromatids. They are traditionally termed *lateral elements,* but for the sake of simplicity, we refer to them as **axial elements**. Thin transverse filaments oriented perpendicular to the axial elements appear to connect homolog axes to each other and to the central element (the "third rail"). Synaptonemal complex formation begins during zygotene at a limited number of sites along the paired homologous chromosomes where recombination events will mature into crossovers. By pachytene, a continuous synaptonemal complex assembles along the full length of the aligned homologous chromosomes (Fig. 45.5C–E).

It was once thought that the synaptonemal complex aligns homologous chromosomes in preparation for recombination, but it is now clear that homolog pairing and (in many organisms) the initiation of recombination precedes synapsis. Thus, synapsis is a downstream consequence of early steps in recombination in some well-studied organisms including yeast and mammals. However, under certain artificial circumstances, even nonhomologous chromosomes can undergo synapsis.

Another longstanding model proposed that the synaptonemal complex promotes the resolution of crossover-designated recombination intermediates. However, analysis of budding yeast mutants missing certain synaptonemal complex proteins indicates that the structure per se is dispensable for the formation of crossovers and that the resulting chiasmata can hold homologous chromosomes paired until anaphase of meiosis I.

What then is the function of the synaptonemal complex? One possibility is that it may have a key role in crossover interference (see later), which ensures that crossovers are distributed broadly across the genome. Another interesting possibility is that the synaptonemal complex communicates information about meiotic chromosomes (such as homolog alignment and the formation of crossover-designated recombination

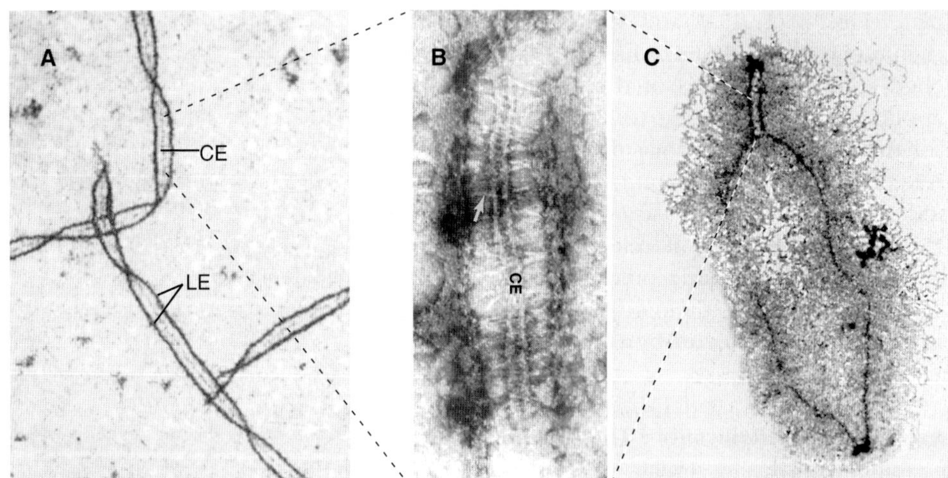

FIGURE 45.8 ELECTRON MICROGRAPHS OF THE SYNAPTONEMAL COMPLEX. A, Low-magnification view of maize synaptonemal complexes stained with silver. The lateral (LE) and central elements (CE) are clearly seen. **B,** A negatively stained cricket synaptonemal complex following treatment with deoxyribonuclease (DNase). The central element (CE) and transverse filaments *(arrow)* are visible. **C,** A whole mount of a silk moth zygotene chromosome. Cells in meiotic prophase were swollen and then lysed under gentle conditions with detergent. The chromosomes were then centrifuged onto thin carbon films so that they could be examined by electron microscopy. The axial elements are easily seen on this chromosome. Chromatin loops radiate outward from both the unpaired axial elements and the paired lateral elements (where synapsis has occurred). (**A,** Modified from Gillies CB. Electron microscopy of spread maize pachytene synaptonemal complexes. *Chromosoma.* 1981;83:575–591. **B,** Modified from Solari AJ, Moses MJ. The structure of the central region in the synaptonemal complexes of hamster and cricket spermatocytes. *J Cell Biol.* 1973;56:145–152, copyright the Rockefeller University Press. **C,** From Rattner JB, Goldsmith M, Hamkalo BA. Chromatin organization during meiotic prophase of *Bombyx mori*. *Chromosoma.* 1980;79:215–224.)

intermediates) to the cell-cycle pathways that control progression through the substages of meiosis.

Synaptonemal Complex Components

Both genetic and biochemical approaches have identified components of the synaptonemal complex. The budding yeast protein Zip1 (mammalian SYCP1) comprises the transverse filaments oriented perpendicular to chromosome axes in mature synaptonemal complex, between the axial elements (Fig. 45.9). Mammalian SYCP1 and Zip1 both consist of an extensive coiled-coil flanked by two globular domains but lack amino acid sequence similarity. Altering the length of the Zip1 coiled-coil changes the spacing between axial elements in the synaptonemal complex.

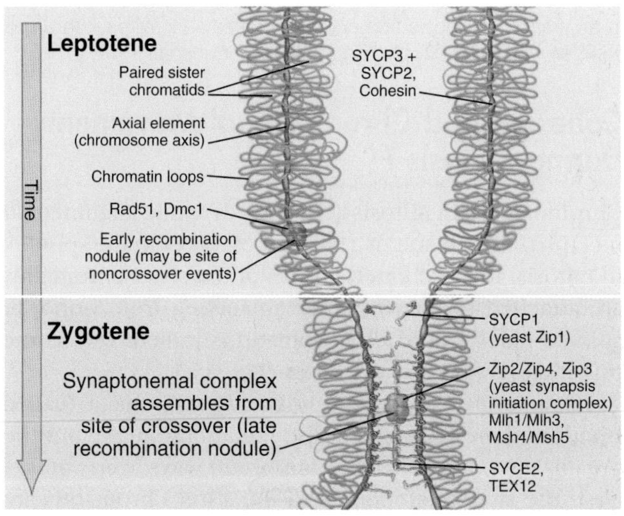

FIGURE 45.9 PROTEINS OF THE SYNAPTONEMAL COMPLEX. Homologous chromosomes and synaptonemal complex showing the locations of some protein constituents.

Several protein components of the axial elements (sister chromatid axes) have also been identified. One of these, SYCP3, interacts with both the cohesin complex (see Fig. 8.18) and Rad51p and Dmc1p. In SYCP3 knockout mice, the axial elements are much less prominent, and the axis of the condensed chromosome is about twofold longer. Other proteins of the synaptonemal complex, including SYCP1, do not assemble properly, and as a result chromosomes in male germ cells lack chiasmata, are unpaired, and cells die in pachytene/diplotene. Humans with mutations in genes for cohesin subunits lack chiasmata, fail to complete meiosis, and are infertile.

Chiasmata

Chiasmata are specialized chromatin structures that link homologous chromosomes together until anaphase I (Figs. 45.1 and 45.10). They form at sites where programmed DNA breaks generated by Spo11 undergo the full recombination pathway to generate crossovers.

It is not known how crossover events, which represent exchanges of DNA sequence information, are turned into chiasmata. The ultrastructure of chiasmata remains a mystery, but presumably each chiasma consists of two unperturbed sister chromatid arms intertwined with two recombinant arms in which the DNA molecules and their associated protein structures have been spliced. This DNA complex is held in place on the chromosome by cohesion of the distal sister chromatid arms between the chiasma and the telomeres. Chiasmata too close to telomeres can be unstable, presumably because the short length of sister chromatid arms between them and the telomeres is insufficient for stable cohesion. This can lead to failure of chromosome segregation in meiosis.

A single chiasma can link homologous chromosomes together during meiosis I. Humans have 39 such arms

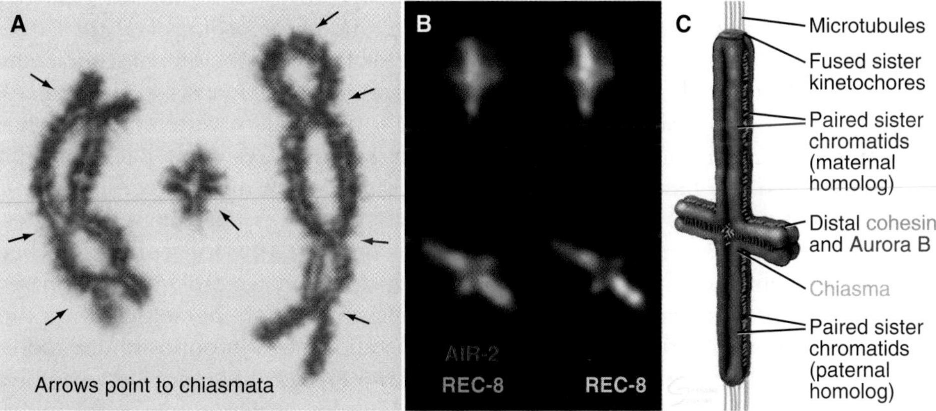

FIGURE 45.10 BIVALENTS (PAIRED HOMOLOGOUS CHROMOSOMES) ARE HELD TOGETHER BY CHIASMATA AFTER DISASSEMBLY OF THE SYNAPTONEMAL COMPLEX. A, Three diplotene bivalents from the grasshopper species *Chorthippus jucundus* are held together by three *(left),* one *(middle),* and four *(right)* chiasmata. The *middle* cross-shaped bivalent is telocentric; the other two longer bivalents are submetacentric. (For an explanation of the terminology, see Fig. 7.2.) Lactopropionic orcein staining. **B,** *Caenorhabditis elegans* chromosomes at metaphase I. Aurora B kinase AIR-2 *(red)* is located distal to chiasmata. Cohesin subunit REC-8 *(green)* is all along the chromosomes. **C,** Explanatory diagram. (**A,** Courtesy José A. Suja and Julio S. Rufas, Universidad Autónoma de Madrid, Spain. **B,** Courtesy Josef Loidl, Max Perutz Labs, Vienna Biocenter, Austria.)

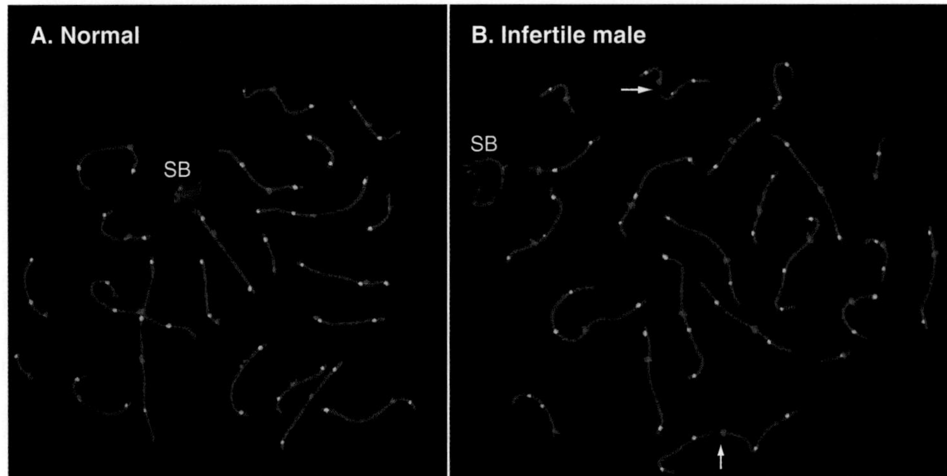

FIGURE 45.11 ABNORMAL PACHYTENE CHROMOSOMES IN AN INFERTILE MALE. A, Normal pachytene chromosome spread from a testis biopsy showing synaptonemal complexes *(red),* MLH1 foci (recombination sites *[green]*), and centromeres *(blue).* **B,** Abnormal pachytene spread from an infertile patient containing one synaptonemal complex with an area of asynapsis and one synaptonemal complex with a gap *(arrows).* SB, sex body (the paired X and Y chromosomes). (Courtesy Renée H. Martin, University of Calgary, Alberta, Canada.)

on the 23 pairs of homologous chromosomes, if one excludes the five acrocentric short arms, which do not normally undergo crossovers. Remarkably, there is typically only one chiasma produced for most arms; human males typically have 46 to 53 chiasmata (Fig. 45.11). Even more remarkably, that single chiasma can hold homologous stably paired for over 40 years in human females, yet still be released on schedule when the oocyte matures into an egg.

Only a small fraction of DNA breaks formed by Spo11 mature into full crossovers, because a mechanism called **crossover interference** decreases the likelihood that DNA breaks near a crossover-designated recombination event will also become crossovers. This interference tends to spread crossovers apart across the genome. If all breaks had an equal probability of forming crossovers, small chromosomes might be left without a crossover in a significant fraction of meiotic nuclei if large chromosomes used all of the structural components necessary to form crossovers and chiasmata.

Crossover interference has been defined genetically for almost 100 years, but its mechanism is not certain. Interference may be mediated by the synaptonemal complex. Organisms such as the fission yeast *Schizosaccharomyces pombe* and the mold *Aspergillus nidulans* that naturally lack synaptonemal complex also lack interference. Furthermore, the frequency of meiotic recombination is directly proportional to the length of the synapsed chromosome axis (ie, the length of the synaptonemal complex), rather than the actual length of DNA in the chromosome. For example, in human females, the synaptonemal complex is roughly 50% longer than in males, and females undergo recombination about twice as frequently as males. However, other observations suggest that interference may be established before the synaptonemal complex forms.

Cohesion and Chromosomal Movements During Meiosis I

Chromosomes in mitosis achieve a dynamic alignment at metaphase as a result of a balance of forces in the spindle. In mitosis, the two kinetochores of the sister chromatids are attached to microtubules emanating from opposite spindle poles, and each chromatid is pulled toward the pole that its kinetochore faces (Fig. 45.12A).

In meiosis I, homologs linked by chiasmata (called **bivalents**) are balanced at the metaphase plate, but the organization differs in three important ways from mitosis. First, the two kinetochores of the sister chromatids are fused and act as a single unit oriented toward one spindle pole. The structure of the meiosis I kinetochore is most easily explained if the two kinetochores are each rotated 90 degrees toward one another relative to their position on mitotic chromosomes and then fused (Fig. 45.12A). In yeast, this coorientation of sister kinetochores requires the presence of a meiosis-specific kinetochore protein—spo13 (meikin in vertebrates)—that associates with sister kinetochores from pachytene until anaphase of meiosis I. Spo13/meikin recruits polo kinase to kinetochores, but the critical kinase substrates are not known.

Second, the physical connection between the fused kinetochores is not broken in anaphase I. As a result, at anaphase the two homologs with their paired sister chromatids move in opposite directions.

A third major difference between bivalents in meiosis I and mitotic chromosomes is that cohesion of homologous chromosome arms distal to chiasmata (Figs. 45.2 and 45.10) rather than cohesion between sister chromatids at centromeres resists the poleward pulling of the kinetochores at the spindle midzone at metaphase I (Fig. 45.12B). This reflects specialized behavior of the meiotic **cohesin complex** in which the meiosis-specific

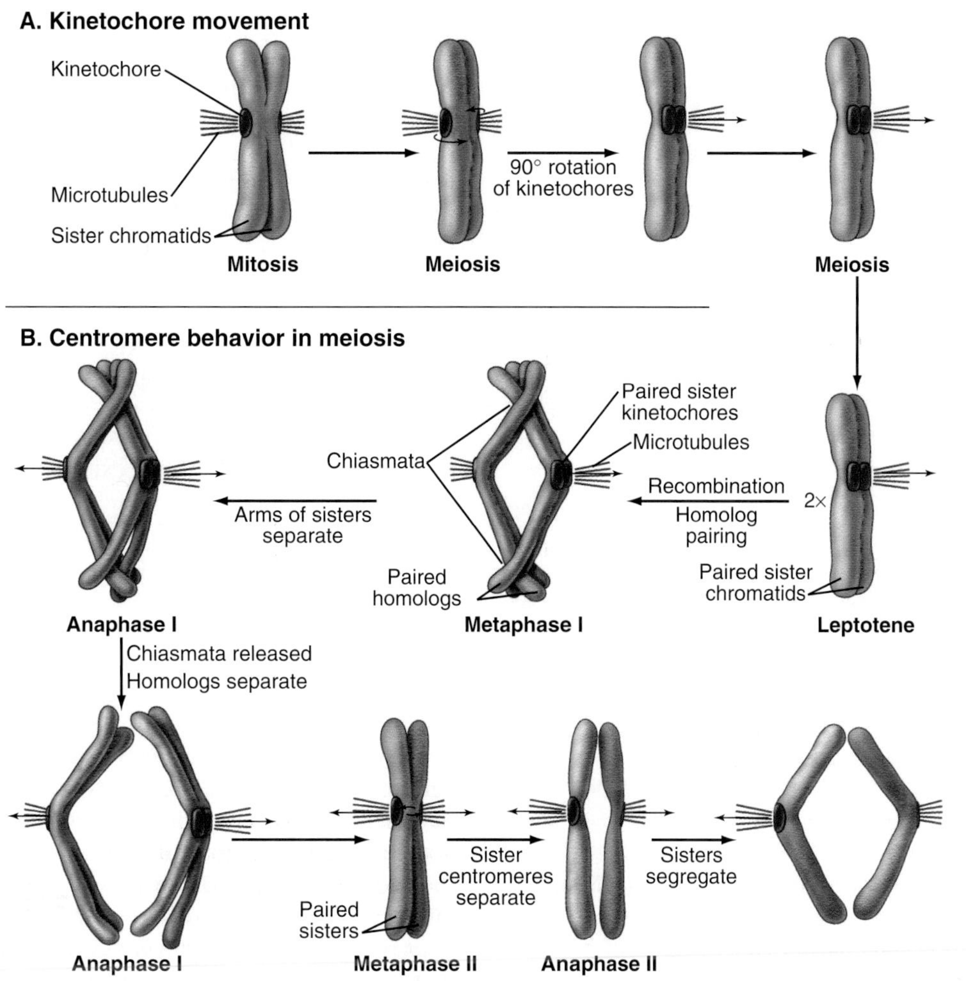

FIGURE 45.12 CHROMOSOMAL BEHAVIOR DURING MEIOSIS I AND II. During meiosis I, sister chromatids are tightly paired along their lengths, sister kinetochores are fused, and homologs are held together at the metaphase plate by chiasmata. During anaphase I, loss of cohesion between the arms of sister chromatids releases the chiasmata and allows homologous chromosomes to segregate to opposite spindle poles. During metaphase of meiosis II, sister chromatids are held together only at their centromeres. Release of centromeric cohesion at meiosis II allows the sister chromatids to segregate to opposite spindle poles.

Rec8 and Rad21L proteins replace Scc1 (see Figs. 8.18 and 44.16). After premeiotic DNA replication, the meiotic cohesion complex keeps sister chromatids together all along the arms. The cohesin complex plus synaptonemal complex proteins SYCP3 and SYCP2 are required for assembly of the dense axial elements that extend along the length of the chromosome during synapsis.

In mitosis, cohesion is released between sister chromatid arms during prometaphase, but in meiosis I it is retained distal to chiasmata until the onset of anaphase (Figs. 45.7 and 45.12), when Rec8 and Rad21L along the chromosome arms are cleaved by a protease called separase (see Fig. 44.16). Separation of sister chromatid arms allows the chiasmata to resolve (untangle), and the homologous chromosomes segregate to opposite spindle poles.

In the meantime, the Rec8 and Rad21L at centromeres are protected from cleavage and continue to hold the sister chromatid centromeres tightly paired until anaphase of meiosis II. This protection requires a class of proteins called Shugoshins (from the Japanese, meaning "guardian spirit"), which recruit the protein phosphatase 2A (PP2A). Rec8 must be phosphorylated for separase to cleave it efficiently, so the localized phosphatase can block the cleavage reaction. Cleavage of centromeric Rec8 releases sister chromatid cohesion at the onset of anaphase II similar to the release of cohesion during mitosis (see Fig. 44.16).

Behavior of the Sex Chromosomes in Meiosis

Of the 46 human chromosomes, two **sex chromosomes** carry genes that define the sex of the individual. The other 22 pairs of chromosomes are called **autosomes.**

If genetic recombination is required to stabilize homologous chromosomes at the metaphase plate in

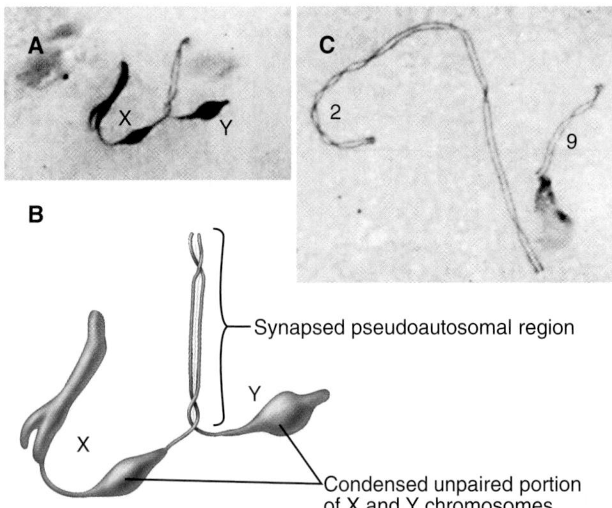

FIGURE 45.13 SEX CHROMOSOMES OF A CHINESE HAMSTER AT PACHYTENE. A–B, The X and Y chromosomes are synapsed at the pseudoautosomal region. Elsewhere, the unpaired chromatin adopts a highly condensed morphology. **C,** In the same preparation, autosomes are completely synapsed and show a lesser degree of condensation. (From Dresser ME, Moses MJ. Synaptonemal complex karyotyping in spermatocytes of the Chinese hamster *(Cricetulus griseus).* IV. Light and electron microscopy of synapsis and nucleolar development by silver staining. *Chromosoma*. 1980;76: 1–22.)

meiosis I, how is this accomplished for the X and Y chromosomes? The answer in most mammals is that the X and Y chromosomes have a short region of homologous sequence (approximately 2.6 million base pairs in humans) that does pair and undergo genetic recombination during meiosis. This **pseudoautosomal region** must undergo genetic recombination in every meiosis I cell for the X and Y chromosomes to be partitioned correctly. Thus, the X and Y chromosomes act like short homologous chromosomes with large regions of unrelated DNA attached (Fig. 45.13). Unpaired regions of the X and Y chromosomes acquire a distinct chromatin structure during late pachytene.

Cell-Cycle Regulation of Meiotic Events

Meiosis employs the full set of functions that regulate the division of somatic cells (see Chapters 40 to 43). However, the peculiarities of the meiotic cell cycle require additional regulation. One major difference from somatic cells is that meiotic chromosomes must undergo recombination and form chiasmata to segregate properly at the first meiotic division. Like somatic cells, meiotic cells have a DNA damage response that arrests cell-cycle progression in the presence of DNA breaks (see Fig. 43.11). In addition, they have a "crossover assurance" checkpoint that can detect the presence of stalled or abnormal recombination intermediates. Such intermediates accumulate if there are problems with the core recombination enzymes or if the assembly of the synaptonemal complex (required for the completion of recombination) is defective. When such problems are detected, cells arrest in meiotic prophase I. In yeast, this has been called the pachytene checkpoint, as cells arrest late in meiotic prophase with nuclear morphology reminiscent of the pachytene stage. Apoptosis eliminates mammalian germ cells that arrest owing to defects in recombination.

Suppression of DNA Replication Between Meiosis I and Meiosis II

Meiosis is unique in that it involves two M phases with no intervening S phase. On exit from meiosis I, Cdk1 kinase is reactivated immediately. This blocks assembly of prereplication complexes (see Fig. 42.8), thereby blocking DNA replication. At least two pathways contribute to reactivation of Cdk1.

The first involves downregulation of translation of Wee1 protein kinase in meiosis. Wee1 is a mitotic inhibitor (see Fig. 43.3) that inactivates Cdk1 by phosphorylation at Tyr^{15}. The absence of Wee1 in meiosis I was first observed in *Xenopus laevis* but this seems to be a universally conserved way of reactivating Cdk1 without an S phase. Ectopic expression of Wee1 in mature *X. laevis* oocytes prevents reactivation of Cdk1 immediately after the meiosis I division. As a result, the oocytes reenter interphase and replicate their DNA. Meiotic cells also express a specialized isoform of Cdc25, the phosphatase that counteracts Wee1 (see Fig. 43.1).

Metaphase II Arrest and the Mitogen-Activated Protein Kinase Pathway

Following their activation and release from the ovary (ovulation), oocytes of many vertebrates arrest in metaphase II of meiosis until they are fertilized. The activity that is responsible for this arrest was discovered in *X. laevis* eggs arrested in metaphase of meiosis II and is called **cytostatic factor** (CSF). Injection of cytoplasm containing CSF into one blastomere of a two-cell frog embryo blocks the next cell cycle at metaphase, just like the egg (Fig. 45.14). Therefore, CSF can even block somatic cells indefinitely in mitotic metaphase. CSF activity appears in meiosis II and disappears after fertilization.

One active component of CSF is the *X. laevis* homolog of a well-known viral oncogene, *v-mos,* the transforming gene of the Moloney murine sarcoma virus, which causes solid tumors in mice. The *v-mos* gene is a mutated form of the cellular *c-mos* gene. Vertebrates express *c-mos* (a mitogen-activated protein [MAP] kinase kinase kinase; see Fig. 27.5) exclusively in oocytes and eggs. Injection of either v-Mos or c-Mos proteins into dividing

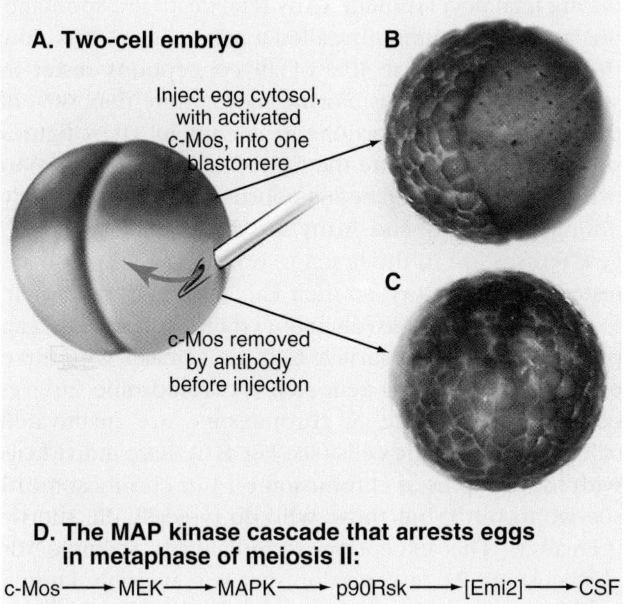

FIGURE 45.14 ROLE OF C-MOS IN THE CYTOSTATIC FACTOR PATHWAY. A, Diagram of the experiment that identified c-Mos as an essential component of cytostatic factor (CSF) required for arrest of eggs in meiotic metaphase. One blastomere of an *Xenopus laevis* embryo at the two-cell stage was injected with cytoplasm from a metaphase-arrested egg containing CSF activity. **B,** This blastomere (*right half* of the embryo) remained blocked in metaphase while the left blastomere divided many times. **C,** The same experiment was performed, but prior to injection, the c-Mos was removed from the egg cytoplasm by absorption with a specific antibody. Both the injected and uninjected blastomeres continued to divide normally. **D,** The mitogen-activated protein (MAP) kinase (MAPK) pathway leading to metaphase II arrest in vertebrate eggs. (**B–C,** Micrographs courtesy George Vande Woude, NCI, Frederick, MD. Modified from Sagata N, Watanabe N, Vande Woude GF, et al. The c-Mos proto-oncogene product is a cytostatic factor responsible for meiotic arrest in vertebrate eggs. *Nature*. 1989;342:512–518.)

blastomeres of early frog embryos arrests the cells at metaphase (Fig. 45.14). These experiments led to a proposal that c-Mos is CSF.

CSF arrest requires the MAP kinase (MAPK) signal transduction pathway (see Fig. 27.5). Mos activates the pathway by phosphorylating mitogen activated protein/extracellular signal-related kinase kinase (MEK), which then activates MAPK. MAPK then activates a downstream kinase called p90Rsk (Fig. 45.14D). Introduction of constitutively active c-Mos or p90Rsk into *X. laevis* eggs induces metaphase arrest like CSF. However, this is not the whole story, because metaphase arrest is maintained in extracts depleted of p90Rsk. Thus the pathway must include at least one unidentified step beyond p90Rsk.

The extra component of CSF is an anaphase-promoting complex/cyclosome (APC/C) inhibitor called Emi2. A burst of cytoplasmic Ca^{2+} released at fertilization (see Fig. 26.15) activates Ca^{2+}/calmodulin-dependent protein kinase II (CaMKII), which phosphorylates Emi2. This modification creates a binding site for polo kinase, which then also phosphorylates Emi2. Emi2 phosphorylated by polo kinase is recognized by $SCF^{\beta TrCP}$, which ubiquitylates it, marking it for destruction. This results in activation of the APC/C (see Fig. 40.15), termination of the CSF metaphase arrest, and completion of meiosis II.

Timing of Meiosis in Humans

The fate of cells undergoing meiosis, as well as the timing of meiotic events, differs significantly between human males and females.

Males produce approximately 100 million sperm a day in a process called **spermatogenesis.** This process continues throughout adult life. Spermatogenesis starts with the division of stem cells called **spermatogonia** and involves eight divisions prior to meiosis. These divisions are unusual in that cytokinesis is incomplete, and the cells remain connected by intercellular bridges. The process could produce up to 256 cells, but usually some cells die and others fail to divide, so a more typical number is around 200 cells arising from the initial stem cell division. When these cells pass through meiosis (they are then referred to as spermatocytes) the final result is approximately 800 postmeiotic **spermatids.** Spermatids then undergo a complex program of differentiation, resulting in the production of highly specialized **spermatozoa.** The entire process of spermatogenesis takes approximately 64 days, the bulk of which is spent in meiosis I. Approximately 16 days are spent in pachytene, the longest stage of the meiosis I prophase. In contrast, only about 8 hours are spent in meiosis II.

By the twentieth week of fetal life each ovary of a human female contains approximately 3×10^6 primordial follicles, each with an oocyte arrested in the diplotene stage of meiosis. This lengthy arrested stage is referred to as dictyate. Thereafter, arrested primordial follicles are recruited continuously to mature into growing primary follicles. However, successful follicular growth depends on follicle-stimulating hormone (FSH), so before puberty—when the hypothalamus-pituitary-ovarian axis matures—all activated oocytes undergo programmed cell death and degenerate in a process known as atresia. At birth, the ovary contains approximately 1,000,000 germ cells of which approximately 300,000 to 400,000 remain at puberty. Following puberty and in response to high levels of FSH each month, a cohort of follicles with their dictyate-arrested oocytes is activated to complete meiosis and grow. Only one of these activated oocytes matures fully and is shed from the ovary in response to a surge of luteinizing hormone. The other follicles degenerate by atresia. As the successful oocyte is shed from the ovary, it completes meiosis I and is arrested at metaphase of meiosis II by CSF. It remains arrested at this stage until fertilization occurs. By the time of menopause

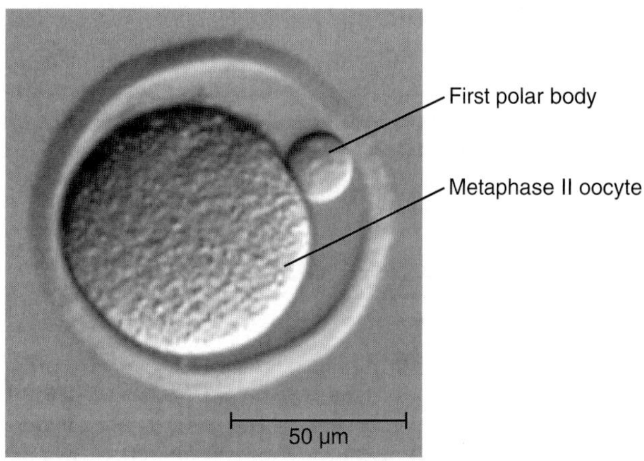

FIGURE 45.15 POLAR BODY FORMATION. Asymmetric cleavage during mouse oogenesis produces a large oocyte and small polar body. (Modified from Jiao ZX, Xu M, Woodruff TK. Age-associated alteration of oocyte-specific gene expression in polar bodies: potential markers of oocyte competence. *Fertil Steril*. 2012;98:480–486.)

the ovarian reserve is almost depleted, leaving approximately 1000 primordial follicles in the ovary.

In human females, meiosis produces only one mature egg. Both meiotic cell divisions are asymmetrical, producing one large and one very small and short-lived cell. The small cells are called **polar bodies** (Fig. 45.15).

Meiotic Defects and Human Disease

Abnormalities in meiosis are surprisingly common but are not widely observed in human populations because their consequences are extremely severe. In fact, meiotic abnormalities are a leading cause of fetal death, particularly during the first trimester of pregnancy in humans. The two major causes of problems are chromosome nondisjunction during the meiotic divisions and the generation of unbalanced chromosomal rearrangements via faulty recombination.

When chromosomes fail to segregate properly in one or both meiotic divisions (nondisjunction), the daughter cells lack the normal haploid complement of chromosomes. Embryos that have gained an entire set of chromosomes are referred to as **polyploid.** In human embryos, polyploidy is a common type of chromosomal abnormality. It is estimated that 1% to 3% of all conceptions are triploid (69 chromosomes; 23 from one parent and 46 from the other parent). Two-thirds of these arise from two sperm fertilizing one egg (nothing wrong with meiosis there). In other cases, they come from a diploid gamete, the result of a defective meiotic segregation. Very few triploid embryos survive to birth.

Most chromosomal abnormalities in human embryos result from the loss or gain of one or more chromosomes during meiosis. This condition is called **aneuploidy.** Most zygotes that arise from aneuploid gametes die during fetal development. (Any fetal death is a spontaneous abortion, commonly called a miscarriage.) It is now thought that at least 30% of all conceptions result in spontaneous abortions. Furthermore, more than 60% of those spontaneous abortions are aneuploid. These figures probably underestimate the frequency of meiotic abnormalities. Many spontaneous abortions occur very early during pregnancy and many are never detected at all. Few fetuses lost in the first 4 to 6 weeks of gestation are tested in a laboratory, so their karyotypes are unknown.

Meiotic errors involving certain autosomes can produce fetuses that survive to birth. Females with three or more copies of the gene-rich X chromosome survive, because all but one X chromosome are inactivated (silenced) in somatic cells (see Fig. 8.6). Rare individuals with three copies of chromosome 13 or chromosome18 survive to birth; but those who do typically die shortly thereafter. The exception is individuals trisomic for chromosome 21 (a condition that is commonly known as **Down syndrome**). These individuals have intellectual disability as well as other characteristic phenotypic features, including decreased life expectancy. Why do individuals with Down syndrome survive whereas others affected by aneuploidy do not? Perhaps the very small number (233) of coding genes on chromosome 21 includes none whose dosage is critical for survival.

The frequency of certain types of aneuploidy, such as trisomy for chromosome 21, increases with the ages of the mother and father. Only 0.04% of children of 20-year-old mothers old have trisomy 21. This number rises dramatically with maternal age; nearly 5% of the conceptions in mothers 45 years old have trisomy 21 (Fig. 45.16). This maternal age effect is a leading cause of human genetic disease. Some believe that during the many years of arrest of oocytes in meiosis I dictyate, chiasmata joining homologous chromosomes gradually dissociate. A mechanism to explain this might be the progressive loss of cohesion between sister chromatids as the mother ages. There appears to be no mechanism to replace cohesin complexes that are gradually lost in dictyate oocytes. Mice with a mutation in a key subunit of the cohesin complex (Fig. 45.7; see also Fig. 8.18) exhibit a pattern of chromosome nondisjunction with increasing maternal age that looks much like that seen in aging human mothers. Another potential source of errors lies in the mechanism of spindle assembly in oocytes, which does not involve centrosomes and appears to be more prone to errors than in somatic cells (see Chapter 44).

Not all cases of human aneuploidy originate from the mother. One of the most common aneuploidies, 45,X (see Table 45.1 for an explanation of nomenclature) involves the loss of the paternal X or Y chromosome 70% to 80% of the time. This aneuploidy accounts for nearly 10% of spontaneous abortions. In addition, about 7% of

TABLE 45.1 Aneuploidies Involving the Sex Chromosomes in Newborn Humans

Karyotype	Frequency	Sex	Comments
47,XXY*	1/1000	M	Klinefelter syndrome. Increased height, sterile, a proportion may have some learning difficulties.
47,XYY	1/1000	M	Increased height, generally fertile, typically with chromosomally normal offspring. A proportion may have some learning difficulties.
Other X or Y aneuploidy	1/1350		
			Total: 1 in 360 male births
47,XXX	1/900	F	Increased height, generally fertile, typically with chromosomally normal offspring. A proportion have serious learning difficulties.
45,X	1/4000	F	Turner syndrome. Reduced height, infertile, normal intelligence. Of 45,X embryos, 99% terminate as spontaneous abortions.
Other X or Y aneuploidy	1/2700		
			Total: 1 in 580 female births

*This number gives the total number of chromosomes, followed by the complement of sex chromosomes.
Modified from Nussbaum RL, McInnes RR, Willard HF. *Genetics in Medicine*. 6th ed. Philadelphia, PA: WB Saunders; 2001:150.

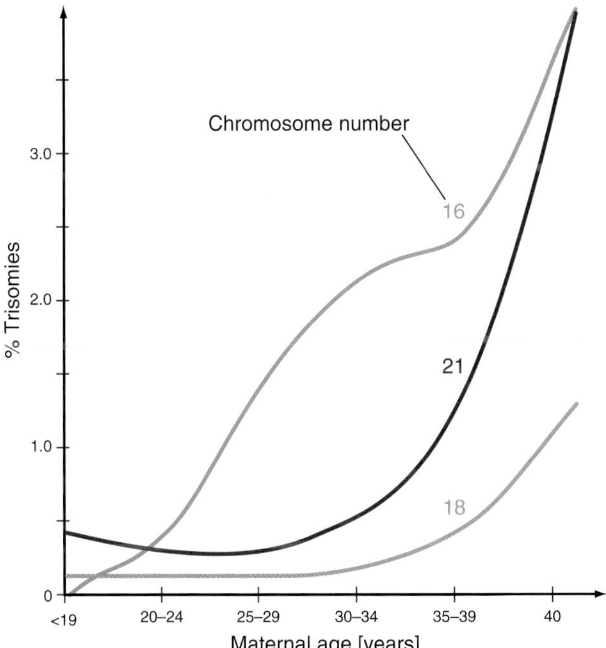

FIGURE 45.16 RELATIONSHIP BETWEEN MATERNAL AGE AND THE INCIDENCE OF HUMAN TRISOMIES. (Modified from Nagaoka SI, Hassold TJ, Hunt PA. Human aneuploidy: mechanisms and new insights into an age-old problem. *Nat Rev Genet*. 2012; 13:493–504.)

instances of trisomy 21 are of paternal origin. Clearly, more than one mechanism is responsible for the generation of aneuploid offspring.

These rather sobering statistics reveal two important facts about human reproduction. First, the production of gametes is error prone. This has been confirmed by direct studies, in which 20% of eggs and 3% to 4% of sperm were found to have chromosomal abnormalities. Second, the much lower rates of chromosomal abnormalities seen in live births (approximately 0.3% overall [Table 45.1]) reveal that spontaneous abortion is a highly efficient protective mechanism for the elimination of chromosomal imbalances that arise from errors in meiosis.

ACKNOWLEDGMENTS

We thank Abby Dernburg, Jim Haber, Scott Hawley, Amy MacQueen, Adele Marston, and Alberto Pendas for their suggestions on revisions to this chapter.

SELECTED READINGS

de Massy B. Initiation of meiotic recombination: how and where? Conservation and specificities among eukaryotes. *Annu Rev Genet*. 2013;47:563-599.

Duro E, Marston AL. From equator to pole: splitting chromosomes in mitosis and meiosis. *Genes Dev*. 2015;29:109-122.

Fraune J, Schramm S, Alsheimer M, Benavente R. The mammalian synaptonemal complex: protein components, assembly and role in meiotic recombination. *Exp Cell Res*. 2012;318:1340-1346.

Gutiérrez-Caballero C, Cebollero LR, Pendás AM. Shugoshins: from protectors of cohesion to versatile adaptors at the centromere. *Trends Genet*. 2012;28:351-360.

Hiraoka Y, Dernburg AF. The SUN rises on meiotic chromosome dynamics. *Dev Cell*. 2009;17:598-605.

Lam I, Keeney S. Mechanism and regulation of meiotic recombination initiation. *Cold Spring Harb Perspect Biol*. 2014;7:a016634.

Nagaoka SI, Hassold TJ, Hunt PA. Human aneuploidy: mechanisms and new insights into an age-old problem. *Nat Rev Genet*. 2012;13: 493-504.

Scherthan H. A bouquet makes ends meet. *Nat Rev Mol Cell Biol*. 2001;2:621-627.

Schmidt A, Rauh NR, Nigg EA, Mayer TU. Cytostatic factor: an activity that puts the cell cycle on hold. *J Cell Sci*. 2006;119:1213-1218.

Subramanian VV. Hochwagen A: The meiotic checkpoint network: step-by-step through meiotic prophase. *Cold Spring Harb Perspect Biol*. 2014;6:a016675.

Zickler D, Kleckner N. Recombination, Pairing, and Synapsis of Homologs during Meiosis. *Cold Spring Harb Perspect Biol*. 2015;7(6).

CHAPTER 46

Programmed Cell Death

Necessity for Cell Death in Multicellular Organisms

The ability to undergo **programmed cell death** (Box 46.1) is a built-in latent capacity in most cells of multicellular organisms. Cell death is important for embryonic development, maintenance of tissue homeostasis, establishment of immune self-tolerance, killing by immune effector cells, and regulation of cell viability by hormones and growth factors. Both extrinsic signals and internal imbalances can lead cells to kill themselves. Furthermore, many metazoan cells will die if they fail to receive survival signals from other cells. Abnormalities of the cell death program contribute to several important diseases, including cancer, Alzheimer disease, and AIDS. Cell death programs are ancient: much of the current network was present in the last eumetazoan common ancestor (see Fig. 2.8).

Programmed Cell Death Versus Accidental Cell Death: Apoptosis Versus Necrosis

Although cells die in many ways, it is useful to focus on two poles of this spectrum: apoptosis and necrosis. Fig. 46.1 summarizes the major pathways of programmed cell death. (Details are filled in as we progress through the chapter.) **Apoptosis**, the most widely studied pathway for *programmed* cell death, is cellular suicide resulting from activation of a dedicated intracellular program. Fig. 46.2 shows a detailed description of the events of apoptosis. At the other end of the spectrum **necrosis,** also called *accidental* cell death, occurs when cells sustain a structural or chemical insult that causes the cells to swell and undergo membrane lysis (Fig. 46.3). Examples of such insults include extremes of temperature and physical trauma. Cells can also initiate active programmed necrosis in response to certain stimuli, particularly when induction of apoptosis is inhibited. Programmed necrosis looks morphologically like accidental cell death. A third pathway leading to cell death involves **autophagy** (see Fig. 23.7). Although usually regarded as a protective response to starvation, autophagy has been implicated in certain examples of cell death, particularly during development.

Necrosis corresponds to what most of us naively imagine cell death would be like. Owing to lack of cellular homeostasis, water rushes into the dying cell, causing it to swell until the plasma and organelle membranes burst. As a result, the cell undergoes a generalized

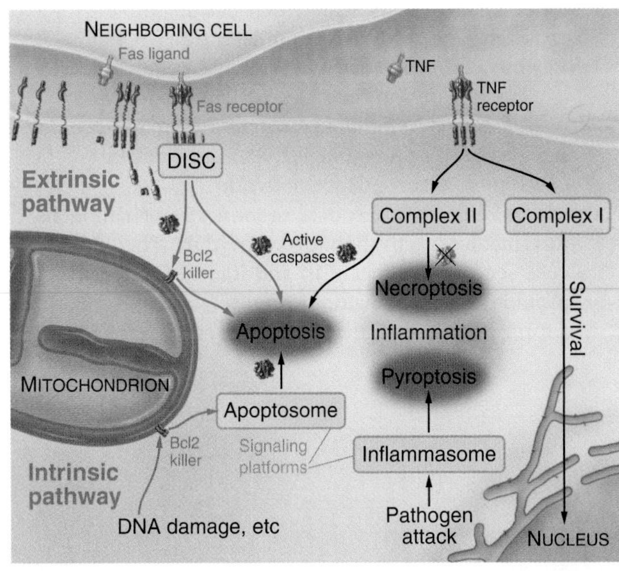

FIGURE 46.1 OVERVIEW OF PROGRAMMED CELL DEATH DISCUSSED IN THIS CHAPTER. All these types of death hinge on the assembly of a signaling platform (boxed) that is often involved in activating proteases called caspases. These pathways are all discussed in greater detail in the text.

BOX 46.1 Key Terms

Apoptosis: Type of programmed cell death that was identified due to a particular pattern of morphologic changes but now is defined by the action of molecular pathways involving cell surface receptors or mitochondria and resulting in the activation of specialized proteases. The name comes from the Greek, referring to shedding of the petals from flowers or leaves from trees. Apoptotic death occurs in two phases. During the *latent phase,* the cell looks morphologically normal but is committed to death. The *execution phase* is characterized by a series of dramatic structural and biochemical changes that culminate in fragmentation of the cell into membrane-enclosed *apoptotic bodies.* Activities that cause cells to undergo apoptosis are said to be *proapoptotic.* Activities that protect cells from apoptosis are said to be *antiapoptotic.*

Autophagic Cell Death: It is still debated how widely autophagy is used as a pathway for cell death, although it is accepted that the pathway (which is widely assumed to be primarily a survival pathway when cells are starved for nutrients) can promote cell death during development. Autophagy may also either promote or inhibit apoptosis under specialized circumstances.

Necroptosis: Programmed necrosis that occurs when tumor necrosis factor (TNF) and certain other cell-surface receptors are activated. Activation of these receptors normally leads to a proinflammatory response and cell survival, but can lead to apoptosis. If certain components of the apoptotic pathway are missing, cells instead undergo necroptosis, apparently as a backup pathway.

Necrosis (Accidental Cell Death): Death that results from irreversible injury to the cell. Cell membranes swell and become permeable. Lytic enzymes destroy the cellular contents, which then leak out into the intercellular space, leading to an inflammatory response.

Programmed Cell Death: Any active cellular process that culminates in cell death. This may occur in response to developmental or environmental cues or as a response to physiological damage detected by the cell's internal surveillance networks.

Pyroptosis: Often in response to intracellular pathogens, this involves activation of caspase 1. The infected cells secrete interleukin-1β and interleukin-18, which promote an inflammatory response, and also undergo a form of cell death that resembles necrosis.

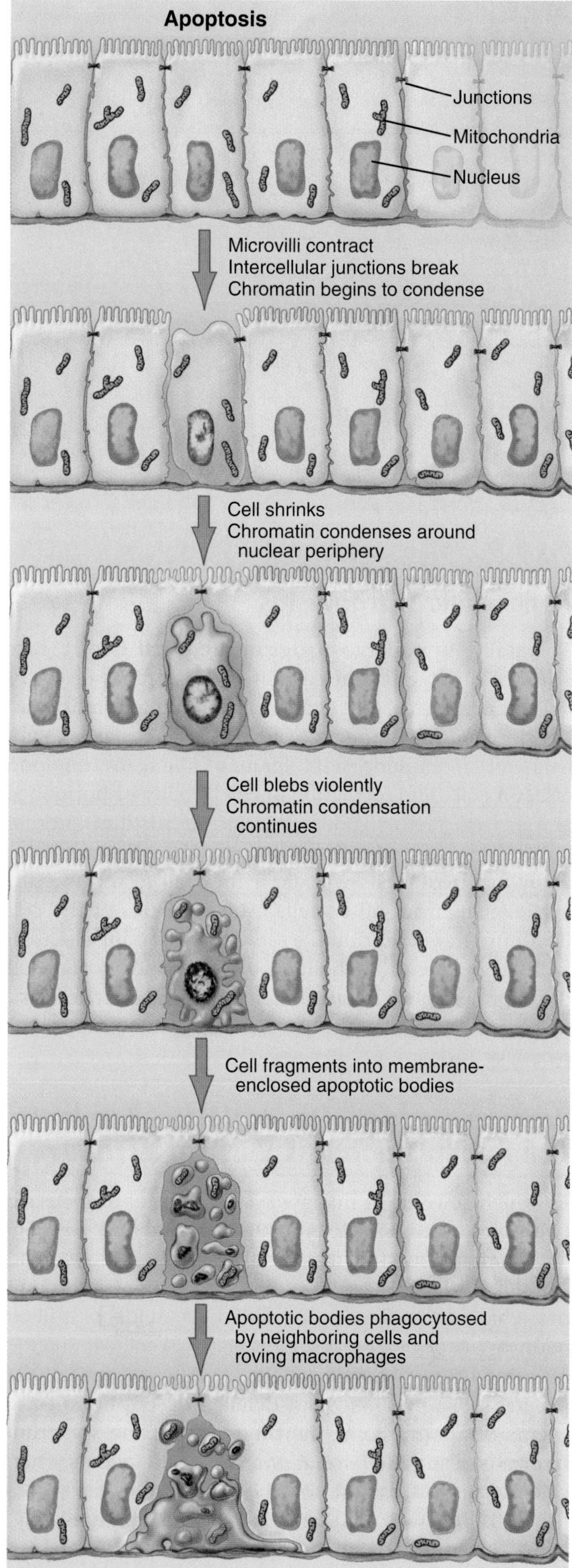

FIGURE 46.2 APOPTOSIS—ACTIVE SUICIDE—TYPICALLY AFFECTS SINGLE CELLS. Neighboring cells remain healthy. Apoptotic cell death usually does not lead to an inflammatory response.

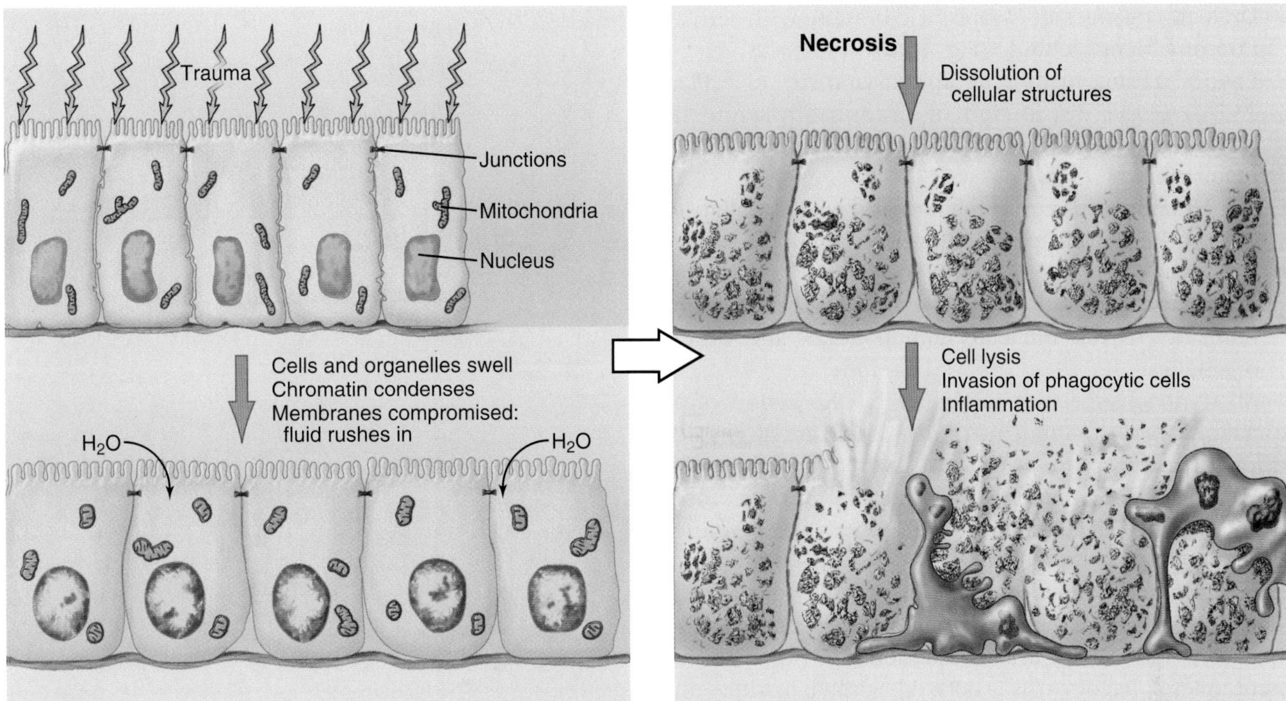

FIGURE 46.3 NECROSIS USUALLY RESULTS FROM IRREVERSIBLE INJURY TO CELLS. Typically, groups of cells are affected. In most cases, necrotic cell death leads to an inflammatory response (*red* "activated" macrophages).

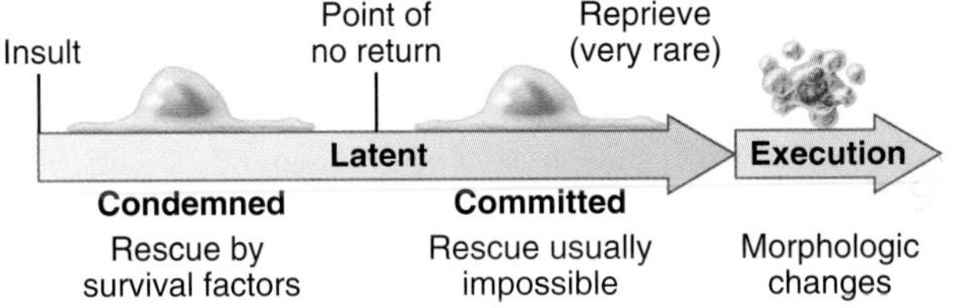

FIGURE 46.4 TWO PHASES OF APOPTOSIS. The latent phase can be subdivided into two stages: a *condemned* stage, during which the cell is proceeding on a pathway toward death but can still be rescued if it is exposed to antiapoptotic activities, and a *committed* stage, beyond which rescue is usually impossible.

process of autodigestion and dissolution, until it spills its cytoplasmic contents out into the surroundings (Fig. 46.3). This produces local inflammation as phagocytic cells are activated, flock to the site, and ingest the debris (see Figs. 22.3 and 30.13). Because agents that damage cells act over areas that are large in comparison to the size of a single cell, necrosis often involves large groups of neighboring cells.

In contrast to necrosis, apoptotic cells shrink rather than swelling and they typically do not lyse (Fig. 46.2). Apoptosis is a two-stage process. On receipt of the **proapoptotic** signal that triggers the pathway to death, cells enter a **latent phase** of apoptosis (Fig. 46.4). The duration of the latent phase of apoptosis, during which cells appear healthy, can be extremely variable, ranging from a few hours to several days.

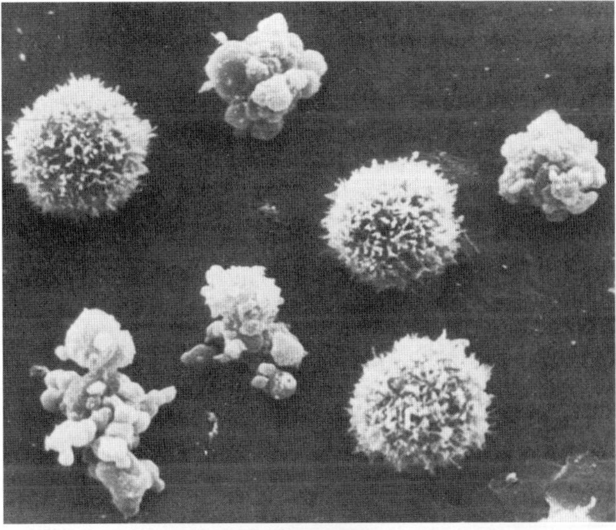

FIGURE 46.5 SCANNING ELECTRON MICROGRAPH OF INTACT AND APOPTOTIC MOUSE SARCOMA CELLS. Intact cells are covered with microvilli, whereas apoptotic cells have numerous smooth blebs. These cells were stimulated to undergo apoptosis as a result of interference with RNA metabolism. (From Wyllie AH, Kerr JFR, Currie AR. Cell death: the significance of apoptosis. *Int Rev Cytol.* 1980;68:251–306.)

Ultimately, the cells enter the **execution phase** of apoptosis, lasting about an hour, during which they undergo dramatic morphologic and physiological changes. These include:

- Loss of microvilli and intercellular junctions (Fig. 46.5)
- Shrinkage of the cytoplasm

- Dramatic changes in cytoplasmic motility with activation of violent blebbing (Fig. 46.6)
- Loss of plasma membrane lipid asymmetry (see Fig. 13.7), so the distribution of phosphatidylserine is randomized and it appears in the outer leaflet of the membrane
- Hypercondensation of the chromatin and its collapse against the nuclear periphery
- The "explosive" fragmentation of the cell into membrane-enclosed **apoptotic bodies** that contain remnants of the nucleus, mitochondria, and other organelles.

All these changes are instigated by the action of a specific set of death-inducing proteases and are discussed at length later.

In tissues, surrounding cells rapidly phagocytose apoptotic bodies in response to the phosphatidylserine and other markers on their surfaces (Fig. 46.7). Apoptosis can thus be considered to be the disassembly of the cell into "bite-sized" membrane-bound vesicles, so the cellular contents are not usually released into the environment unless phagocytosis is delayed. Surface markers on apoptotic bodies cause cells that ingest them to secrete antiinflammatory cytokines. As a result, apoptotic death usually does not lead to an inflammatory response and apoptotic cells seem to vanish from tissues without a trace.

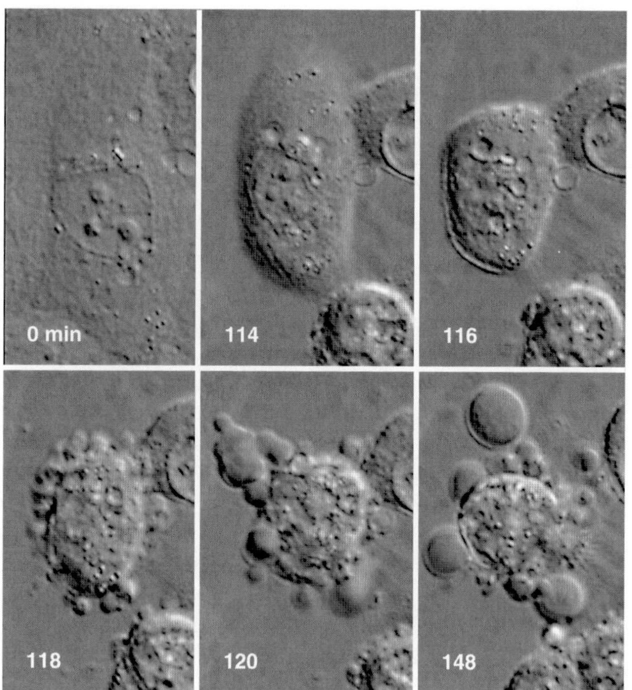

FIGURE 46.6 APOPTOSIS OF A TRANSFORMED PIG KIDNEY CELL FOLLOWING EXPOSURE TO ETOPOSIDE, A DRUG USED IN CANCER CHEMOTHERAPY. The dramatic cytoplasmic blebbing results in the disassembly of the cell into membrane-enclosed vesicles. (Courtesy L.M. Martins and K. Samejima, Wellcome Trust Institute for Cell Biology, University of Edinburgh, United Kingdom.)

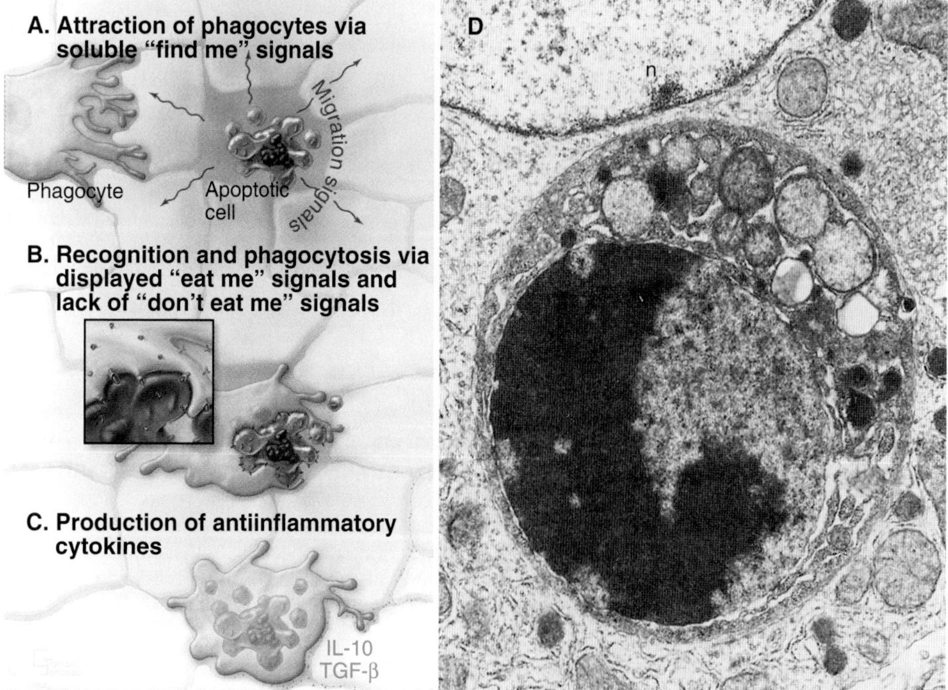

FIGURE 46.7 PHAGOCYTOSIS OF APOPTOTIC CELLS. A–C, Phagocytosis that occurs when cells express "eat me" signals results in the production of antiinflammatory cytokines. **D,** Electron micrograph of a phagocytosed apoptotic body containing a nuclear fragment. The nucleus *(n)* of the epithelial cell that engulfed this apoptotic body is shown at *top.* In this case, apoptosis occurred during allograft rejection in a pig. (**A,** Modified from Lauber K, Blumenthal SG, Waibel M, et al. Clearance of apoptotic cells: getting rid of the corpses. *Mol Cell.* 2004;12:277–287. **B,** From Wyllie AH, Kerr JFR, Currie AR. Cell death: the significance of apoptosis. *Int Rev Cytol.* 1980;68:251–306.)

Examples of Cells That Undergo Programmed Cell Death

The following are six common causes of programmed cell death (Fig. 46.8).

Developmentally Defective Cells

During molecular maturation of T-lymphocyte antigen receptors (see Figs. 27.8 and 28.9), immature T cells in the thymus (known as *thymocytes*) rearrange the genes encoding the receptor α and β chains. To function properly, the T-cell receptor must recognize major histocompatibility complex (MHC) glycoproteins on other thymic cells during antigen presentation (see Fig. 27.8). T lymphocytes whose T-cell receptors cannot interact with the spectrum of MHC glycoproteins expressed in a given individual are ineffective in the immune response. These cells die by apoptosis in a process known as **positive selection** (Fig. 46.9).

Many of the newly created receptors bind to foreign antigens, but others interact with self-antigens. The latter are potentially harmful and are eliminated through

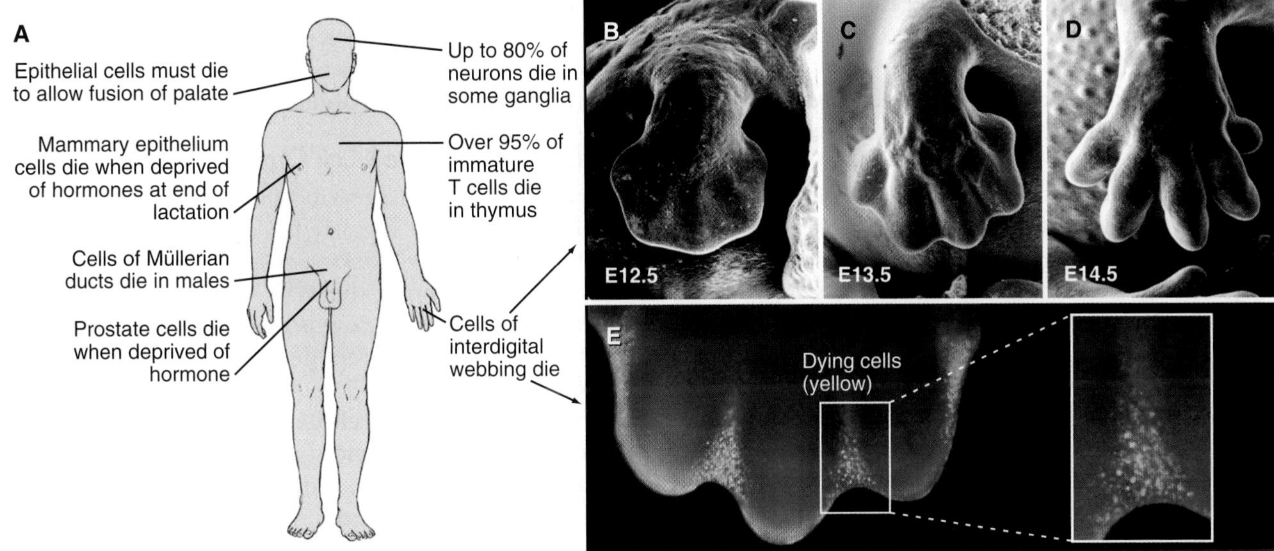

FIGURE 46.8 PROGRAMMED CELL DEATH DURING DEVELOPMENT. A, Examples of cells that undergo programmed cell death. **B–D,** Programmed cell death in the embryonic mouse paw. At day 12.5 of development, the digits are fully connected by webbing. By day 13.5, the webbing has started to die, and by day 14.5, all the webbing cells are gone. **E,** Cells undergoing programmed cell death take up acridine *orange,* whereas cells of the surrounding healthy tissue do not. (Micrographs courtesy William Wood and Paul Martin, Department of Anatomy and Developmental Biology, University College of London, United Kingdom.)

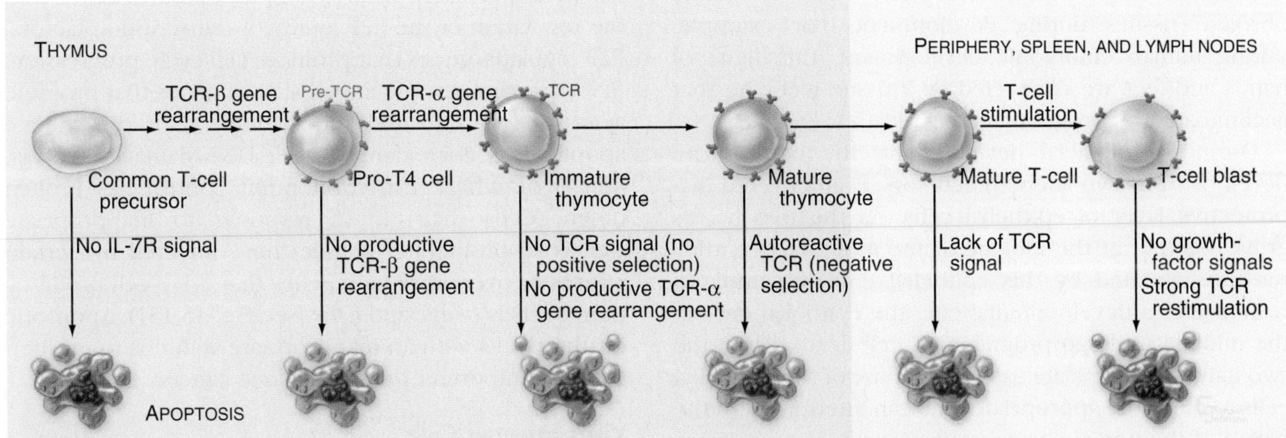

FIGURE 46.9 EXAMPLES OF SIGNALS THAT PROMOTE DIFFERENTIATION OR PROGRAMMED CELL DEATH OF IMMATURE THYMOCYTES IN THE THYMUS AND MATURE T CELLS IN THE PERIPHERY. Thymocytes that make functional T-cell receptors (TCRs) and do not recognize self-antigens mature, provided that they receive survival signals, such as interleukin-7. Pro-T cells are also known as double-negative thymocytes (referring to their lack of the two cell-surface markers CD4 and CD8). Immature thymocytes express CD4 and CD8 and are known as double-positive thymocytes. Thymocytes undergo apoptosis if they produce defective T-cell receptor, recognize self-antigens, suffer DNA damage, or receive a death stimulus (glucocorticoid hormone). More than 98% of immature thymocytes die without leaving the thymus. (Modified from Strasser A. The role of BH3-only proteins in the immune system. *Nat Rev Immunol.* 2005;5:189–200.)

apoptosis in a process known as **negative selection** (Fig. 46.9). The drug cyclosporine, which inhibits apoptosis in thymocytes, can cause autoimmune disease. Overall, defects in T-cell receptor assembly are extremely common, and up to 98% of immature T cells die by apoptosis without leaving the thymus.

Similar positive and negative selection steps occur during the maturation of B lymphocytes (see Fig. 28.10), which is accomplished by a combination of gene rearrangements and facilitated mutagenesis. B lymphocytes expressing antibodies directed against self-antigens or producing antibodies whose affinity for antigen is below a critical threshold are eliminated through apoptosis.

Excess Cells

Programmed cell death is also widely used for quality control during development. For example, in the brain, embryonic ganglia often have many more neurons than are required to enervate their target muscles. Production of excess cells is part of a Darwinian strategy to ensure that a sufficient number of axons reach their targets. Programmed cell death eliminates excess neurons that fail to make appropriate connections. Up to 80% of neurons in certain developing ganglia die in this way. Because of the importance of apoptosis during its development, the brain is often seriously affected in mice engineered to lack components of the apoptotic pathway.

Cells That Serve No Function

The elimination of obsolete cells whose function has been completed is most evident in organisms, such as insects and amphibians that undergo metamorphosis during development. For example, a burst of thyroid hormone initiates programmed cell death for resorption of the tadpole tail.

Mammals also use programmed cell death to eliminate obsolete tissues during development. For example, during human embryonic development, the digits of hands and feet are connected by a tissue webbing that is eliminated by programmed cell death (Fig. 46.8).

During craniofacial development, the hard palate develops from two lateral precursors, each covered in a protective layer of epithelial cells. As the two halves grow together at the midline of the nasopharynx, they remain separated by this epithelial covering until, in response to a developmental cue, the epithelial cells at the midline undergo programmed cell death. Then the two halves of the palate can fuse. Failure of the epithelial cells to die at the appropriate time can interfere with the fusion of the bone, causing cleft palate.

Populations of cells that are fully functional may become obsolete as a result of physiological changes in the status of an organism. For example, in male mammals, the prostate and other accessory glands of the reproductive system are regulated by the levels of circulating male hormone. If hormone levels fall below a critical threshold, these organs virtually disappear in a relatively brief time as their constituent cells undergo massive apoptotic death. Should levels of circulating androgens rise again, the remaining prostatic stem cells proliferate and reconstruct the gland. A similar cycle of growth and involution is seen in the mammary gland of female mammals, which exhibits substantial differences in size and cellular composition in the lactating and nonlactating states. Interference with survival signaling by sex hormones is one important strategy that is commonly used in the treatment of breast and prostate cancer.

Programmed cell death also eliminates certain populations of cells that never serve any function. The müllerian ducts develop into the female oviduct. Male embryos also develop progenitors of these ducts, which serve no purpose and are eliminated by apoptosis.

Cells With Perturbed Cell Cycles

Chapters 40 to 43 describe how biochemical circuits called checkpoints regulate the cell cycle. Cells respond to DNA damage by blocking cell-cycle progression while attempting to repair the DNA. The p53 transcription factor, an important downstream checkpoint EFFECTOR induces the expression of genes encoding proteins that arrest the cell cycle. p53 also turns on genes encoding proteins that induce cell death. It is generally thought that if the damage cannot be repaired quickly, the pro-death factors win out, and the outcome is apoptosis. Double-strand DNA breaks induced by ionizing radiation and DNA damage induced by chemotherapeutic agents frequently result in cell death rather than repair.

A second important cell-cycle checkpoint regulates the transition from the G_1 phase to S phase. Passing the restriction point (see Fig. 41.8) represents the commitment of the cell to undergo another cycle of DNA replication and division. Restriction point control centers on the regulation of the E2F family of transcription factors. E2F regulates genes that promote cell-cycle progression. E2F also regulates the expression of genes that promote apoptosis. If activated too strongly, E2F can initiate apoptosis as, for example, after DNA damage (see Fig. 41.14) or when restriction point control has broken down. Cells that die in response to inappropriate signals to proliferate include those infected by certain viruses or overexpressing genes that drive cell proliferation (such as *c-myc* and *c-fos* [see Fig. 46.15]). Apoptotic death of cells with an inappropriate stimulus to proliferate is an important defense against cancer.

Virus-Infected Cells

Cytotoxic T lymphocytes eliminate virus-infected cells by causing them to undergo programmed cell death either by apoptosis or by a second related pathway. For example, programmed cell death accounts for at least part of the loss of mature $CD4^+$ T-helper cells (see Fig. 28.9) in people infected with HIV-1. When exposed to agents that normally stimulate cell proliferation, these cells instead undergo apoptosis. Paradoxically, it appears that many of the dying cells are not themselves infected with HIV.

Chemotherapeutic Killing of Cells

Exposure of cancer cells to many of the agents used in chemotherapy does not kill the cells outright. Instead, they die because the drugs cause intracellular damage that acts as a signal for the induction of apoptotic cell death. Thus, the treated cancer cells kill themselves.

Genetic Analysis of Apoptosis

Several key components that are involved in the apoptotic execution of mammalian cells were discovered by genetic analysis of the nematode worm *Caenorhabditis elegans.* Because *C. elegans* is optically clear, it is possible to see every cell in a developing worm using differential interference contrast optics (see Fig. 6.2). This enabled investigators to trace the lineage of every cell in an adult worm back to the fertilized egg. These studies led to the surprising discovery that programmed cell death is one of the most common fates for newborn *C. elegans* cells. Of the 1090 somatic cells that are produced during embryogenesis of the *C. elegans* hermaphrodite, 131 undergo programmed cell death at reproducible locations and times.

Mutations in at least 14 *C. elegans* genes affect programmed cell death (Fig. 46.10). These are divided into three classes: (a) genes that mark cells for subsequent programmed death; (b) genes that are involved in cell killing and its regulation; and (c) genes that are involved in the phagocytosis and subsequent processing of the cell corpses. These mutants are collectively known as "cell death abnormal" (ced) **mutants**. Interestingly, this repertoire of nematode genes is vastly simplified from the apoptosis genes in the last eumetazoan common ancestor (see Fig. 2.8). By contrast, humans have approximately 300 genes involved in apoptosis.

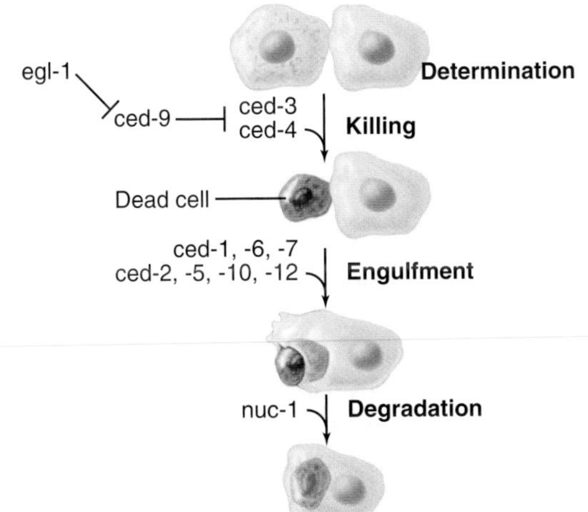

FIGURE 46.10 GENETIC DISSECTION OF PROGRAMMED CELL DEATH. The *ced* (cell death abnormal) mutants of the nematode worm *Caenorhabditis elegans* affect the killing, engulfment, and degradation stages of programmed cell death.

The three best-known worm cell death genes are ***ced-3, ced-4,*** and ***ced-9.*** If either *ced-3* or *ced-4* is inactivated, all cells throughout the organism that should die by apoptosis are reprieved. These cells remain alive and are apparently functional. Interestingly, these worms have normal life spans. This suggests that programmed cell death is not involved in the normal aging process, at least not in *C. elegans. Ced-9* negatively regulates *ced-3* and *ced-4.* In worms with *ced-9* loss-of-function mutations many cells die that normally stay alive. This is deleterious for the organism, so *ced-9* mutants die. The key gene triggering apoptosis is *egl-1.* Its regulation appears to govern apoptosis in the worm. All these genes have mammalian counterparts (discussed more fully later).

Ced-3 is a member of a specialized family of cell death proteases called **caspases**. *Ced-4* is a scaffolding/adapter protein that plays an essential role in the activation of *Ced-3* from its zymogen precursor. The mammalian counterpart of *Ced-4* is apoptotic protease-activating factor-1 **(Apaf-1).** *Ced-9* is a member of the **Bcl-2** family of cell death regulators (Fig. 46.14). Some *Bcl-2* family members protect against cell death, but others actively promote cell death. *Egl-1* is a BH3-only proapoptotic regulator.

Phagocytosis turned out to be surprisingly complex, with eight *C. elegans* genes encoding proteins that are involved in the engulfment and processing of cell corpses. Several are signaling proteins that reorganize the cytoskeleton to permit the cell to move toward and engulf its target. Another, the *nuc-1* gene, encodes one of several nucleases that digest the DNA of the dead cell in lysosomes of cells that ingest the corpse. In mammals, this digestion typically is initiated within the dying cell itself.

Signals and Pathways of Apoptosis

Two principal pathways lead to cell death by apoptosis. These are introduced only briefly here and expanded upon later. The **intrinsic pathway** (see Fig. 46.17) is activated by internal surveillance mechanisms or signals sent (or not sent) by other cells. Signals that induce this pathway include DNA damage, exposure to chemicals that interfere with a variety of cellular pathways and excessive activation of factors that promote cell-cycle progression. Withdrawal of nutrients or survival signals from the environment also activates the intrinsic pathway. Those survival signals include lymphokines, such as interleukin-2 and interleukin-3, which are essential for survival of thymocytes; nerve growth factor, which is required for survival of many neurons; and extracellular matrix, which is required for survival of epithelial cells. Signals that activate the intrinsic pathway converge on mitochondria, which release key factors that drive the apoptotic response.

Signals from other cells are the primary triggers of the **extrinsic pathway** (see Fig. 46.18). Direct contact of a killer cell with the target cell activates specific receptors that initiate this pathway, starting at the plasma membrane. Activation of the extrinsic pathway is one strategy that cytotoxic T lymphocytes use to kill cells that are recognized as foreign (or as harboring foreign pathogens). This pathway is also widely used to control cell populations in the immune system. In many cells, the extrinsic pathway activates the intrinsic pathway, which is then responsible for killing the cell.

Protein Regulators and Effectors of Apoptosis

Because the penalty for misregulation of apoptosis is death, it is not surprising that the process is carefully regulated. This is essential for cells but complicates matters for students. This section first lays out the overall strategy in generic terms and then fills in some important details.

A cascade of proteases called **caspases** drives apoptosis. Each caspase is harmless until activated (usually by dimerization for initiator caspases and proteolytic cleavage for effector caspases). Activation of a small number of initiator caspases triggers a cascade that amplifies the response and results in activation of numerous effector caspases. The ability of effector caspases to activate further initiators and effectors further amplifies the response.

This strategy of employing amplification and positive feedback has two powerful advantages. First, it can provide a very rapid change in the state of the cytoplasm, from pro-life to pro-death within seconds. Second, the relatively small number of initiator caspases that trigger the cascade constitutes a feasible target for negative regulators that can rapidly quell responses that are initiated under borderline conditions or by mistake. This is beneficial but also complicates the overall system. If initiator caspases start apoptosis and are then inactivated by suppressers, how does the response ever take hold? The answer is at least one more level of regulation: inhibitors of the inhibitors. This additional layer of regulation is thought to enable the rapid burst of caspase activation that triggers the onset of apoptotic execution.

The following sections discuss the workhorses of apoptosis—the caspases—followed by regulation of the response.

Caspases

Caspases (cysteine aspartases) are specialized proteases with a cysteine in their active site that cleave their targets on the C-terminal side of aspartate residues. Caspases generally inactivate cellular survival pathways and specifically activate other factors that promote cell death. In certain specialized instances caspases can participate in pathways leading to activation of nuclear factor κB (NF-κB; see Figs. 10.21 and 27.8) and the immune response and also in terminal differentiation. These nonlethal roles of caspases are not discussed further here.

C. elegans has three caspases, one of which *(Ced-3)* is essential for cell death. In contrast, mammals have as many as 17 caspase-related genes (Fig. 46.11A). Analysis based on sequence comparisons divides caspases into two major subfamilies. The caspase 1 subfamily encodes enzymes that process pro–interleukin-1β to yield mature interleukin-1β and pro–interleukin-18 to yield mature interleukin-18. Macrophages secrete these cytokines, which cause fever and inflammation. The second caspase 3 subfamily of enzymes participates almost exclusively in apoptotic cell death.

Like many proteases, caspases are synthesized as inactive zymogens, known as *procaspases*. All living vertebrate cells synthesize procaspases constitutively. Procaspases typically consist of three domains: an N-terminal prodomain followed by the large and small subunits of the mature enzyme (Fig. 46.11A–B). Aspartate residues between these three domains are targets for cleavage by caspases. Procaspases are activated either by multimerization or by cleavage and release of the prodomains. Following procaspase cleavage, the two large and two small subunits associate in a compact, block-like heterotetrameric molecule (Fig. 46.11B–D). Depending on the enzyme, multimerization or cleavage of the procaspase permits a major conformational change in the polypeptide, creating two stable active site pockets between the large and small subunits.

Two classes of caspases are involved in cell death. **Initiator caspases** have long prodomains (Fig. 46.11A). They exist as monomers in cells and become autoactivated when scaffolding cofactors promote their aggregation. Activation is induced by dimerization and does not require proteolytic cleavage (although cleavage typically follows activation). Sequences within the extended prodomains are involved in targeting the initiator procaspases to the appropriate cellular locations and in interactions with scaffolding factors.

Effector caspase zymogens are monomers with short prodomains. These inactive enzymes are typically incapable of autoactivation. Instead, they are activated through cleavage by initiator caspases or by other active effector caspases.

Scaffolding proteins and **adapters** play an essential role in activating initiator caspases by forming multiprotein activation platforms. For the *intrinsic pathway* of cell death, factors released from mitochondria (discussed later) activate the scaffold protein **Apaf-1**, which is related to AAA ATPases (adenosine triphosphatases) (see Box 36.1). Active Apaf-1 forms a ring-like structure with seven spokes called the **apoptosome** (Fig. 46.17).

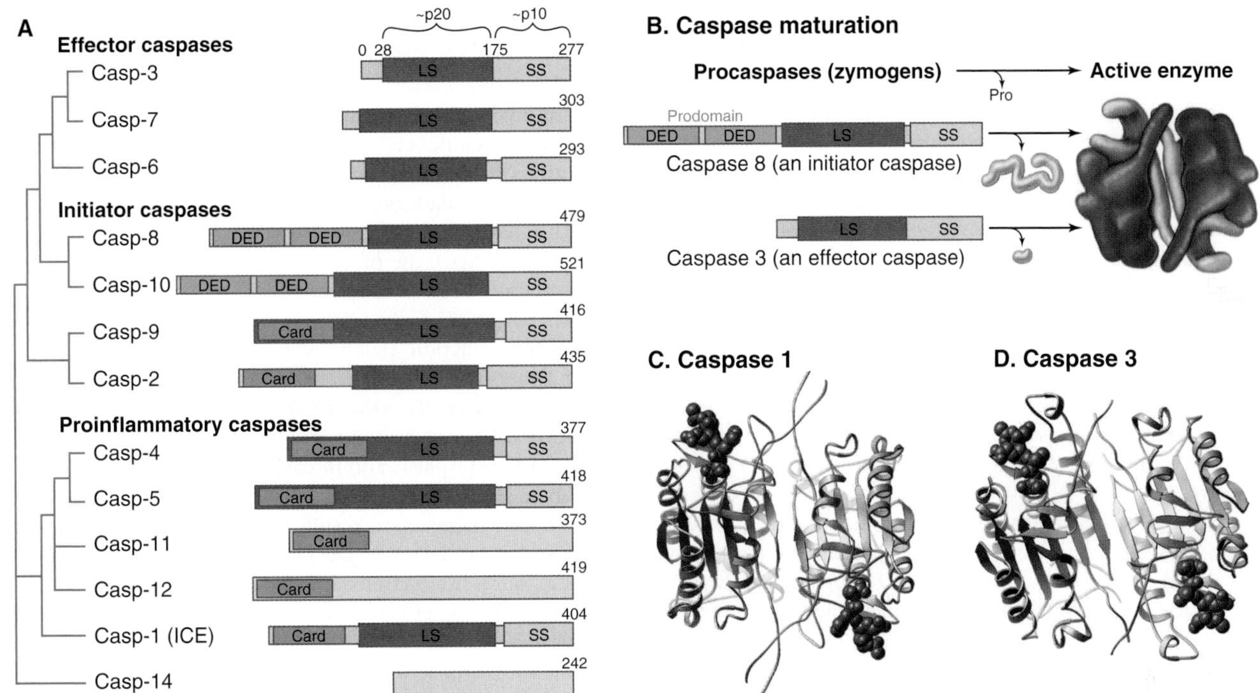

FIGURE 46.11 INTRODUCTION TO CASPASES. A, The 13 enzymatically active mammalian caspases fall into three groups. Where it has been determined, the portions of the zymogens that give rise to the large *(blue)* and small *(yellow)* subunits are shown. **B,** Initiator caspases have large prodomains that participate in subcellular targeting. Two initiator procaspases come together to form the active enzyme. **C–D,** Ribbon diagrams of crystal structures of caspases 1 and 3. The catalytic residues come primarily from the large subunit *(blue)*. The space-filling structure *(red)* represents a peptide inhibitor covalently bound in the active site of the enzyme.

Binding of procaspase 9 zymogen to this platform promotes dimerization and activation of the enzyme.

The scaffold proteins for the *extrinsic death pathway* are cytoplasmic domains of cell-surface receptors. When these receptors bind their ligands on the surface of other cells, they form stable trimeric complexes that recruit adapter proteins to form a signaling complex called the "**death-inducing signaling complex (DISC)**." Interactions involving multiple protein-protein interaction motifs link the procaspase 8 zymogen to the DISC (Box 46.2), resulting in dimerization, activation, self-cleavage, and release of active caspase 8.

Caspases cleave as many as 1000 cellular proteins (Fig. 46.12). Some targets are structural proteins, but many are involved in cellular signaling. For example, caspases cleave several protein kinases. Many kinases have pseudosubstrate motifs that autoinhibit the kinase and can be moved out of the way in response to physiological stimuli (see Fig. 25.4). Caspase cleavage often neatly clips off these regulatory domains, thereby producing constitutively active enzymes. These unregulated kinases may alter signaling pathways to promote cell death. Caspases also cleave and inactivate several proteins that normally function in the detection and repair of DNA damage.

Caspases can also create factors that directly promote cell death. The most obvious example is caspase activation of other caspases in the death cascade. Caspases also act indirectly to cause mitochondria to release factors that promote cell death through the intrinsic pathway. Caspase cleavage of an inhibitory chaperone is responsible for activation of the nuclease that ultimately destroys the chromosomal DNA of most cells undergoing apoptosis (see section on CAD nuclease).

Natural Caspase Inhibitors

Because most healthy cells express initiator procaspases with the potential to oligomerize and kill the cell, it is important to have a mechanism that dampens any potential "noise" in the pro-apoptotic pathway. The inhibitor of apoptosis protein (IAP) family is defined by the presence of a motif of approximately 80 amino acids known as a **baculovirus IAP repeat (BIR)** domain. This is a type of Zn^{2+} finger (see Fig. 10.14) that mediates protein–protein interactions. IAP proteins inhibit caspases in two ways. First, they bind the caspase and invade the active site, thereby blocking its access to substrates. Second, several IAPs are E3 ubiquitin ligases (see Fig. 23.3) that ubiquitylate caspases and tag them for destruction by proteasomes.

If IAP proteins inactivate caspases, then how is the apoptotic response ever initiated? Cells also express several molecules that can bind and inactivate IAPs. The best characterized of these, is the second mitochondrial activator of caspases (**Smac** or DIABLO). Smac is normally sequestered in mitochondria, but is released into

BOX 46.2 Matchmaking for Cell Death: the Key Is in the Domains

There are so many different proteins involved in apoptosis that even experts have a difficult time keeping them all straight. However, an understanding of the general principles is much simplified if we recognize that most proteins involved in apoptosis regulation are built from a relatively limited number of modules, which act as sites for protein–protein interactions. The following are the most important modules for apoptosis.

Bcl-2 family members are defined by the presence of short regions of conserved sequence, referred to as *BH domains* (Bcl-2 homology). One of these, the approximately 20-residue BH3 domain, found in all Bcl-2 family members, promotes complex formation between Bcl-2 family members by docking into a so-called BH3 binding groove.

Four domains regulate caspase targeting and activation. Although the amino acid sequences of these domains have diverged extensively, all are regions of approximately 80 to 90 residues that form a characteristic bundle of six α-helices. These domains all were present in the last eumetazoan common ancestor.

1. The *death domain* (DD) is found in many proteins that are involved in signaling pathways related to cell death. These include cell death receptors, such as Fas (and others), and adapter molecules, such as FADD (Fas-associated death domain).
2. The *death effector domain* (DED) is found in adapters such as FADD, the prodomains of caspases 8 and 10, and certain inhibitors of apoptosis.
3. The *Pyrin domain* (PYD) is found in proteins involved in inflammation (such as microbial sensors) and the activation of caspase 1.
4. The *caspase recruitment domain* (CARD) is found in several adapter proteins involved in cell death, including CED-4 and Apaf-1, and in caspases 1, 2, and 9.

The domains have similar folds, but different distributions of surface charge. As a result, they prefer to interact with themselves (ie, DD–DD, DED–DED, PYD–PYD, and CARD–CARD). Such interactions are said to be homophilic. When a new apoptosis effector protein is identified, it is possible to predict from an analysis of its amino acid sequence which of the known proteins it is most likely to interact with.

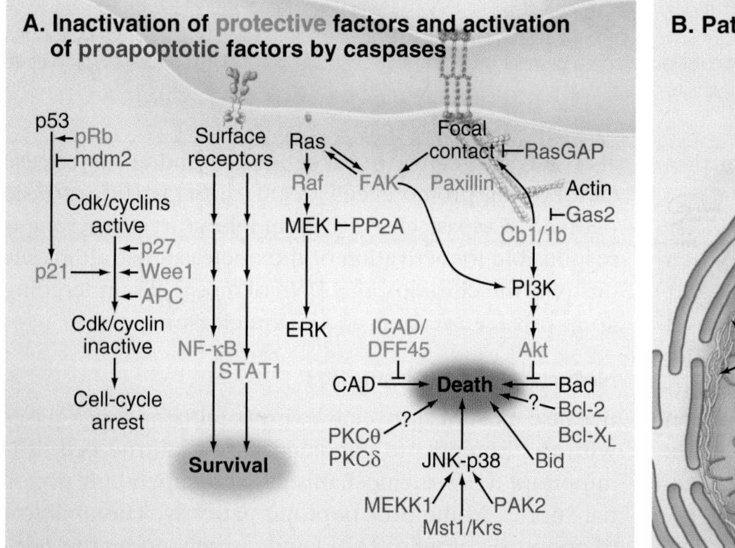

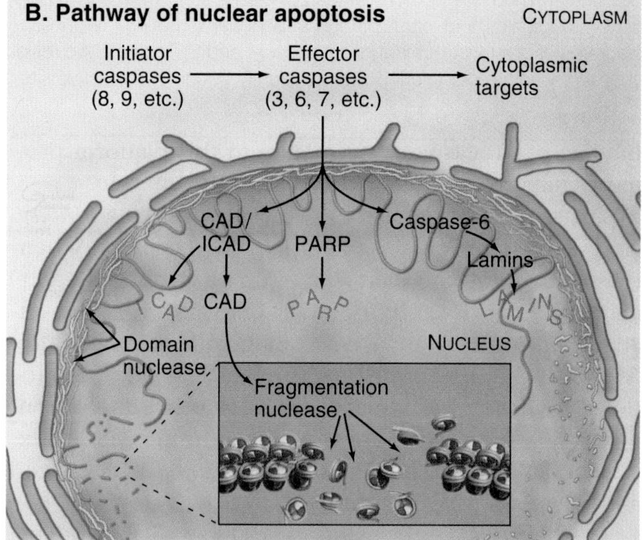

FIGURE 46.12 SOME WAYS THAT CASPASES PROMOTE CELL DEATH. A, A few of the many proteins cleaved by caspases in apoptotic cell death. Proteins shown in *green* normally have a role in keeping the cell alive and are inactivated by caspases. Proteins shown in *red* are turned into active death-promoting factors as a result of caspase cleavage. Proteins shown in *black* are not cleaved and are included to show the pathways that are affected by cleavage. Caspases inactivate a number of pathways that promote cell survival, thereby strongly reinforcing the decision of the cell to die. **B,** Some of the roles of caspases in disassembly of the nucleus. CAD, caspase-activated deoxyribonuclease; ERK, extracellular signal-regulated kinase; ICAD, inhibitor of caspase-activated deoxyribonuclease; JNK, c-Jun N-terminal kinase; MEK, mitogen activated protein/extracellular signal-related kinase kinase; NF-κB, nuclear factor κB; PI3K, phosphatidylinositol 3′-kinase; PKC, protein kinase C; STAT, signal transducer and activator of transcription.

the cytoplasm when the intrinsic pathway of apoptosis is initiated as a result of permeabilization of the mitochondrial outer membrane. It then binds to the BIR domain, inactivating the IAPs. Mathematical modeling shows that inactivation of IAPs by Smac is needed to make the proapoptotic signal act like a switch rather than a graded response.

IAPs were discovered in studies of the mechanisms viruses use to avoid being eliminated by cell death. When viruses infect cells and disassemble their capsids, they become vulnerable to suicide defense mechanisms: If cells can kill themselves before the virus has time to complete its life cycle, they will take the virus with them, and the organism will survive. To defend against this, viruses pilfer genes encoding cellular proteins and adapt them for their own means. For example, insect baculoviruses make two proteins that inhibit apoptosis, keeping the cell alive long enough for the virus to reproduce. One of these, IAP, was derived from a cellular gene. The origin of the second IAP inhibitor, p35, is less clear. p35 is a broad-spectrum caspase inhibitor that is thought to work by a serpin-like mechanism. Serpins are special protease substrates that, after cleavage, form a tight complex with the enzyme and inactivate it. Several mammalian pox viruses also make a serpin-like inhibitor of certain caspases called CrmA.

CAD Nuclease and Its Chaperone ICAD

The chromosomal DNA is destroyed during apoptosis. The nucleases involved in cleaving the cellular DNA during (and after) apoptotic cell death fall into two classes. **Cell autonomous nucleases** degrade the DNA within the dying cell (Fig. 46.13A). The best known is

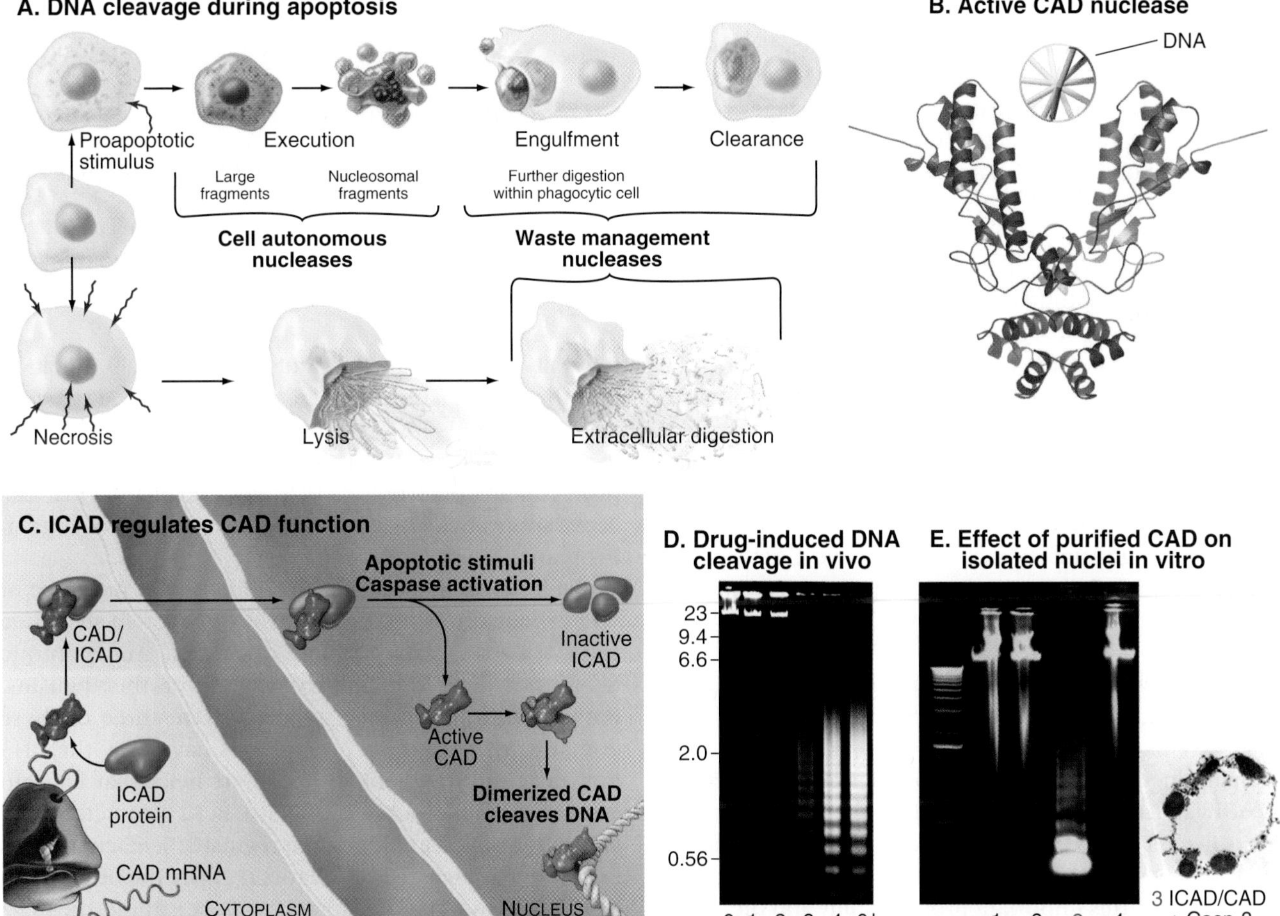

FIGURE 46.13 NUCLEASES DIGEST THE CELLULAR DNA DURING APOPTOSIS. A, In apoptosis, the DNA is digested first to large fragments and later to nucleosome-sized pieces (see Fig. 13.1) by cell autonomous nucleases expressed within the dying cell. Waste management nucleases made by other cells also have an essential role in cleaning up apoptotic and necrotic debris. **B,** The predominant cell autonomous nuclease (CAD) has a scissors-like structure. **C,** ICAD is an inhibitory chaperone for CAD, promoting its folding on the ribosome and continuing as an inhibitor when CAD is stored in the nucleus. ICAD cleavage leads to CAD activation. **D,** Cleavage of the chromosomal DNA by CAD during chemotherapy-induced apoptosis of a leukemia cell line. DNA separated according to size by electrophoresis on an agarose gel was stained with ethidium bromide. The ladder of bands reflects cleavage between adjacent nucleosomes. **E,** Activated CAD causes chromatin condensation and appearance of an apoptotic morphology in isolated cell nuclei. Cloned CAD and ICAD were expressed together in *Escherichia coli* and incubated with nuclei. ICAD cleavage with caspase 3 released active CAD, which degrades the nuclear DNA *(Lane 3).* Other lanes: DNA gel size markers *(left);* nuclei incubated with buffer or caspase 3 alone (*Lanes 1 and 2,* respectively); same experiment as in *Lane 3* but performed by using a mutant ICAD that could not be cleaved by caspase 3 *(Lane 4).* To the *right* is an electron micrograph of a thin section of one nucleus with condensed chromatin at the nuclear periphery. CAD, caspase-activated deoxyribonuclease; ICAD, inhibitor of caspase-activated deoxyribonuclease; mRNA, messenger RNA. (**A,** Modified from Samejima K, Earnshaw WC. Trashing the genome: The role of nucleases during apoptosis. *Nat Rev Mol Cell Biol*. 2005;6:677–688. **B,** For more information, see Protein Data Bank [PDB; www.rcsb.org] file 1V0D from Woo EJ, Kim YG, Kim MS, et al. Structural mechanism for inactivation and activation of CAD/DFF40 in the apoptotic pathway. *Mol Cell*. 2004;14:531–539. **D,** From Kaufmann SH. Induction of endonucleolytic DNA cleavage in human acute myelogenous leukemia cells by etoposide, camptothecin, and other cytotoxic anticancer drugs: a cautionary note. *Cancer Res*. 1989;49:5870–5878. **E,** Courtesy K. Samejima, Wellcome Trust Institute for Cell Biology, University of Edinburgh, United Kingdom.)

the caspase-activated DNase (CAD; see later discussion). A mitochondrial nuclease known as endonuclease G may also degrade DNA in some cell types. Cell autonomous nucleases are dispensable for cell death and for the life of the organism. They might have evolved to eliminate viral DNA as part of the suicide defense response described in the previous section.

Waste management nucleases clean up the debris after cells die. They either function within lysosomes of cells that have ingested apoptotic cell fragments or are secreted into the extracellular space. DNase II, one of the most important waste management nucleases, is essential for life. Mouse embryos that lack DNase II become overwhelmed with undegraded DNA and die.

Cell autonomous nucleases act in two stages. After an initial cleavage of the chromosomes into fragments of roughly 50,000 base pairs, DNA is usually cleaved between nucleosomes, producing a characteristic "ladder" of DNA fragments with a periodicity of approximately 200 base pairs. This ladder is seen when DNA isolated from apoptotic cells is subjected to gel electrophoresis. The responsible nuclease is **CAD.** CAD is normally present in a complex with **ICAD** (inhibitor of CAD [Fig. 46.13C]). The complex of CAD and ICAD is also known as DNA fragmentation factor (DFF). ICAD is a chaperone required to fold CAD into an active conformation during translation on the ribosome. However, ICAD also inhibits the nuclease activity of CAD. This dual function of ICAD guarantees that only inactive CAD can be synthesized in healthy cells. During apoptosis, caspase 3 cleaves ICAD and releases active CAD nuclease.

Bcl-2 Proteins and the Intrinsic Pathway of Apoptotic Cell Death

As mentioned previously, mitochondria are key players in a pathway to cell death that is triggered by a variety of toxic insults (Fig. 46.17). The Bcl-2 family of proteins regulates these mitochondrial events. The following sections describe this important protein family and the regulation of the intrinsic pathway of apoptosis.

Bcl-2 Proteins

Bcl-2 proteins can be grouped into three subfamilies (Fig. 46.14). Bcl-2 *protectors* protect cells against apoptosis. Bcl-2 *killers* (eg, Bax and Bak) are proapoptotic proteins that actively kill cells. Bcl-2 *regulators* (widely known as BH3-only proteins) promote cell killing by either interfering with the protectors or activating the killers. Bcl-2 proteins primarily regulate the release of death-promoting factors from mitochondria when cells receive signals that activate the intrinsic pathway.

C. elegans genetics identified a gene, *ced-9,* that protects cells against apoptosis and another gene, *egl-1,* that inactivates ced-9 protein and triggers apoptosis. In *ced-9* mutants, many cells die during development that

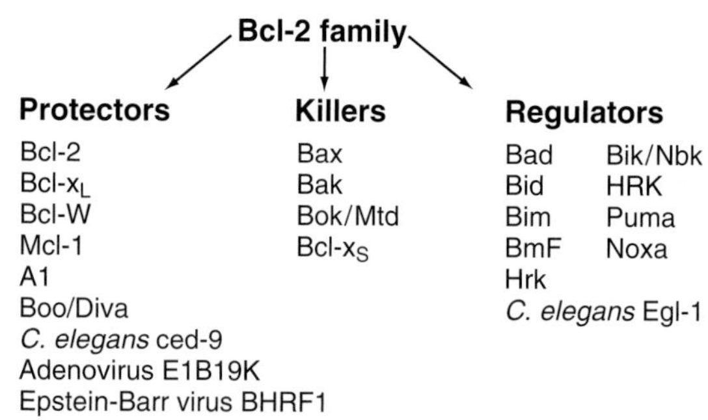

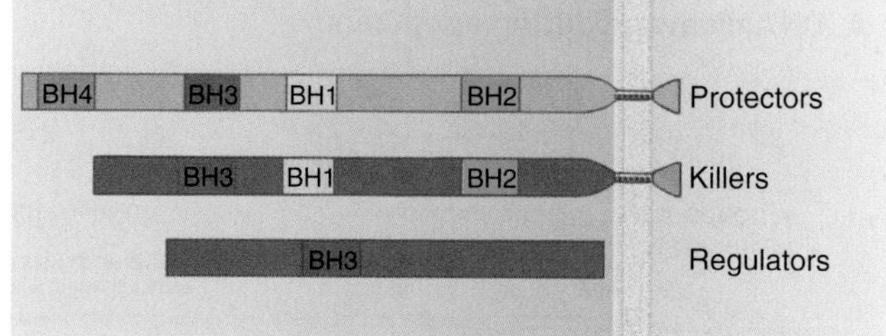

FIGURE 46.14 THE BCL-2 FAMILY OF APOPTOTIC REGULATORS. *C. elegans, Caenorhabditis elegans.*

normally survive into the adulthood. A Ced-9 mutation kills the worm. Human Bcl-2 is functionally and structurally homologous to *C. elegans* Ced-9 and can substitute for it in living worms. This ability of a human gene to protect nematodes reveals that the fundamental machinery of apoptotic cell death has been conserved over great evolutionary distances.

Bcl-2 family members are defined by the presence of one to four short blocks of conserved protein sequence called **BH** (Bcl-2 homology) **domains.** Antiapoptotic Bcl-2 protectors typically have four of the domains. Proapoptotic Bcl-2 killers typically have three of these domains, while the Bcl-2 regulators have only the BH3 domain. The BH3 domain is a short helix that fits into a groove on the surface of both Bcl-2 protectors and killers, forming complexes that regulate their activity. It is believed that the Bcl-2 protectors regulate the behavior of Bcl-2 killers by a similar interaction. For example, the Bcl-2 protein forms a complex with a proapoptotic Bcl-2 killer called Bax, thereby interfering with the ability of Bax to kill cells. Binding of BH3-only proteins to Bcl-2 protectors can inactivate their antiapoptotic functions. Egl-1 is a BH3-only protein and this is how it triggers apoptosis. A new generation of BH3 mimetic drugs induces apoptosis of cancer cells by mimicking this second mechanism.

Genetic experiments in mice revealed several different functions for Bcl-2 family members. Mice born without Bcl-2 have deficiencies of the immune system that are best understood if one role of this protein in vivo is to render lymphocytes resistant to proapoptotic signals during immune system maturation. Mice lacking another pro-life family member, Bcl-x_L, die during embryogenesis, apparently as a result of widespread death of neurons in the central and peripheral nervous systems and hematopoietic cells in the liver. In contrast, loss of the killers Bax plus Bak makes cells highly resistant to apoptosis by a wide variety of intrinsic pathway stimuli.

Bcl-2 Family Members and Cancer

A gene that prevents cells from dying poses a potential danger in multicellular organisms, in which rates of cell proliferation and death must be balanced carefully. In fact, the name *Bcl-2* comes from the discovery that this gene is the culprit responsible for certain types of B-cell lymphoma. These particular lymphomas arise when a chromosome translocation, involving chromosomes 14 and 18, moves the *Bcl-2* gene into the immunoglobulin heavy-chain gene cluster, a site of very active gene transcription in B lymphocytes. Elevated transcription of Bcl-2 is thought to be directly responsible for the cancerous phenotype in these patients, making Bcl-2 a cancer-promoting oncogene (see Chapter 41).

Unlike many other oncogenes, *Bcl-2* overexpression does not cause cell proliferation. Instead, it disrupts the balance of regulation between life and death. Cells that overexpress Bcl-2 protein actually grow somewhat slower than normal but are highly resistant to many stimuli that normally promote cell death. The net result is an accumulation of B cells: a lymphoma. Other anti-apoptotic Bcl-2 family members are amplified in a variety of solid tumors and loss of microRNAs directed at Bcl-1 is thought to play a major role in development of chronic lymphocytic leukemia. Fig. 46.15 shows an example of Bcl-2 conferring resistance to death when the cell cycle is perturbed by expression of an oncogene.

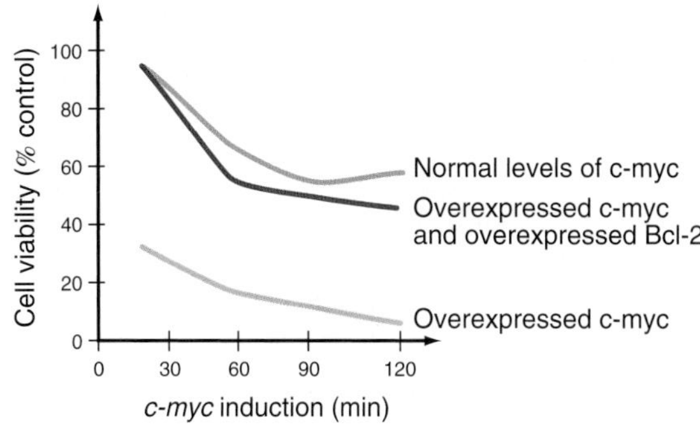

FIGURE 46.15 BCL-2 PROTECTS CELLS FROM ONCOGENE-INDUCED CELL DEATH. Chinese hamster ovary cells die when they are induced to express abnormally high levels of the c-myc protein, but simultaneous expression of the Bcl-2 protein rescues them from this effect. These cells contain many copies of the *c-myc* gene under control of a promoter that is activated when the cells are briefly exposed to high temperature (43°C for 90 minutes). The curves show the percentage of viable cells remaining at various times following the induction of *c-myc* expression. The *Bcl-2* gene was introduced into these cells on a plasmid under the control of a promoter that is always active. The *blue line* represents the parental cell line lacking either the cloned *c-myc* or *Bcl-2* genes. (Note that approximately 40% of these cells die following the heat treatment used to induce *c-myc* expression.) The *yellow line* shows that the cells that produce high levels of *c-myc* protein alone rapidly die (by apoptosis). The *red line* shows that cells expressing both the c-myc and Bcl-2 proteins survive the treatment almost as well as the parental cells. (From Bissonnette RP, Echeverri F, Mahboubi A, et al. Apoptotic cell death induced by c-myc is inhibited by *bcl-2*. *Nature*. 1992;359:552–554.)

Intrinsic Pathway of Apoptotic Death

In addition to their role in energy production, mitochondria have an essential role as sensors of the health of the cell. If cells sense insults from which they cannot recover, mitochondria trigger the **intrinsic pathway** of cell death (Fig. 46.17). Bcl-2 family members regulate this pathway. The regulation seems straightforward in *C. elegans,* in which the protector CED-9 (Bcl-2–like) binds to the CED-4 scaffolding protein (Apaf-1-like) and interferes with its activation of the CED-3 caspase. Apoptosis is induced when the regulator BH3-only protein EGL-1 binds to CED-9 and blocks it from inactivating CED-4.

In mammals, the situation is more complex, partly because Bcl-2 family members are more numerous and partly because they do not interact in such a straightforward fashion. In mammals, two of the killer proteins, Bax and Bak, are essential for activation of the intrinsic pathway. In healthy cells, Bak is loosely associated with the mitochondrial outer membrane, and Bax is in the cytoplasm (Fig. 46.17). On receipt of a proapoptotic stimulus, Bax and Bak insert deeply into the mitochondrial outer membrane and form oligomeric membrane pores that allow the release of pro-apoptotic factors from the mitochondrial intermembrane space. Binding of antiapoptotic Bcl-2 family members to Bax/Bak somehow prevents mitochondrial outer membrane permeabilization. BH3-only family members can either directly promote death by facilitating Bax/Bak oligomerization or indirectly promote it by binding and neutralizing anti-apoptotic Bcl-2 family members.

The proapoptotic factors released from the mitochondrial intermembrane space by Bax and Bak include the electron transport protein **cytochrome c** (see Fig. 19.5), Smac, and endonuclease G (Fig. 46.16). These mitochondrial proteins actively promote apoptotic cell death. In the cytoplasm, cytochrome c binds to the scaffolding protein **Apaf-1,** a mammalian homolog of *C. elegans* CED-4 protein, causing it to undergo a conformational change that exposes a hidden bound adenosine diphosphate (ADP) to solvent. Exchange of this ADP for deoxyadenosine triphosphate (dATP) allows Apaf-1 to form a wheel-like structure with seven spokes called the *apoptosome* (Fig. 46.17). Apaf-1 in the apoptosome binds caspase 9 through an N-terminal **caspase recruitment domain.**

Binding to the apoptosome elevates the catalytic activity of procaspase 9 approximately 2000-fold without the need for its cleavage. Thus, the active form of caspase 9 is an oligomeric complex of the procaspase with the apoptosome. Activated caspase 9 then cleaves

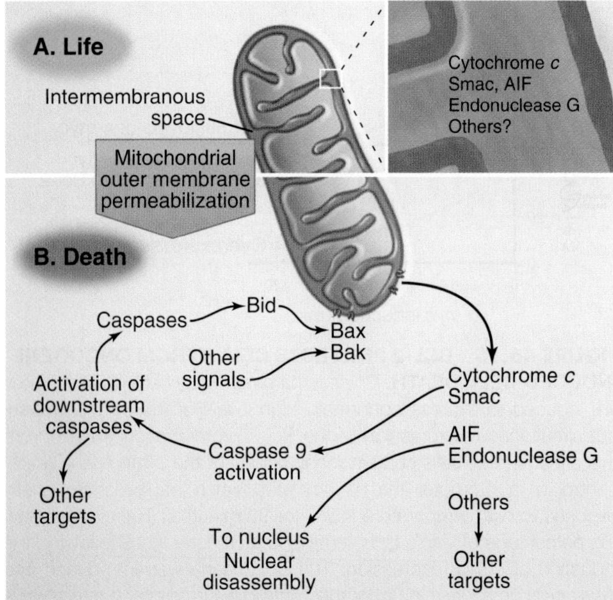

FIGURE 46.16 MITOCHONDRIA INTEGRATE A CELL'S LIFE AND DEATH DECISIONS. A, In healthy cells, a number of factors that promote apoptosis are stored in the intermembrane space of the mitochondria. **B,** In cells undergoing apoptosis, BH3-only proteins trigger the proapoptotic Bcl-2 family members to induce the release of these death-promoting factors; this initiates an amplifying cycle that ultimately leads to cell death. AIF, apoptosis-inducing factor; Smac, second mitochondrial activator of caspases.

multiple procaspase 3 zymogens, amplifying the cell death cascade. It also cleaves itself, triggering its release from the apoptosome with a resulting loss of activity. This autocleavage acts like a timer limiting apoptosome activity.

This cascade can be further amplified in at least two ways. First, caspase 3 cleaves other effector caspases, directly amplifying the cascade. In addition, active caspases cleave the BH3-only protein Bid, which then activates more Bax and Bak in a feedback loop, thereby promoting the release of more cytochrome c and Smac, and enhancing caspase 9 activation.

It was extremely surprising to find that an essential metabolic protein such as cytochrome c has a second function that is essential for death. Among the studies supporting the Jekyll-and-Hyde–like nature of this protein in life and death was the engineering of mice whose cytochrome c can function in electron transport but cannot bind Apaf-1. These mice die as a result of brain abnormalities caused by insufficient cell death.

Extrinsic Pathway of Apoptotic Death

Cells express at least six different cell surface molecules, collectively termed *death receptors,* that can trigger apoptotic death. Protein ligands on the surface of other

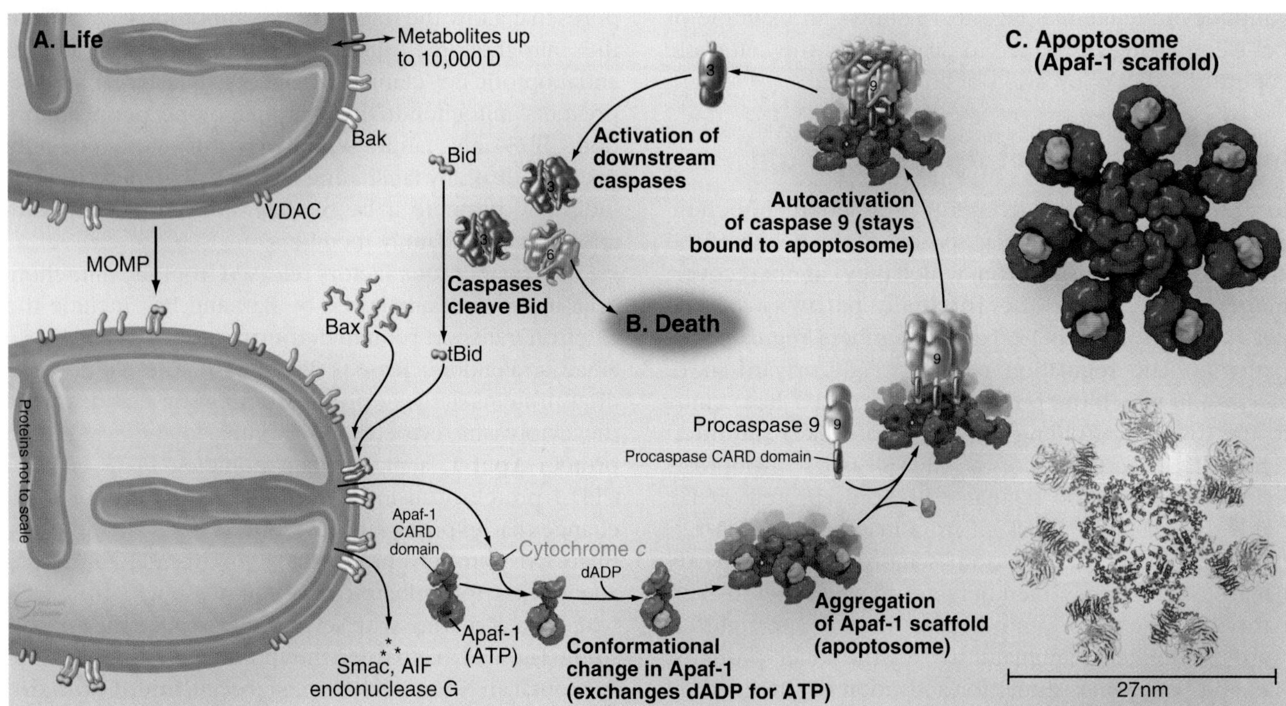

FIGURE 46.17 INTRINSIC CELL DEATH PATHWAY. A–B, Following activation by the BH3-only protein Bid, Bax and Bak form a pore that releases cytochrome c. Released cytochrome c binds to Apaf-1, inducing a conformational change leading to formation of the apoptosome, which binds and activates procaspase 9 via scaffold-induced oligomerization. Activated caspase 9 subsequently activates downstream effector caspases, leading to cell death. **C,** Molecular organization of the apoptosome. (**C,** Modified from PDB file 3JBT from Zhou M, Li Y, Hu Q, et al. Atomic structure of the apoptosome: mechanism of cytochrome c- and dATP-mediated activation of Apaf-1. *Genes Dev.* 2015;29:2349–2361.)

cells bind and activate these receptors, turning on pathways that can lead to apoptotic death.

One well-characterized death receptor called Fas (also known as Apo1 or CD95) is a member of the tumor necrosis factor receptor family (see Fig. 24.9). Fas is a type I membrane protein whose extracellular domain consists of three cysteine-rich domains (Fig. 24.9 shows the atomic structure of the related trimeric tumor necrosis factor receptor with bound ligand). The cytoplasmic domain of Fas contains a **death domain** of approximately 80 residues, which is shared by all the death receptors (see Box 46.2).

The **Fas ligand** is a trimeric 40-kD intrinsic membrane protein found on the surface of cells. Cytotoxic T lymphocytes use Fas ligand to rid the body of virally infected cells. When a cytotoxic T lymphocyte contacts a target cell, the Fas ligand on the lymphocyte surface binds to Fas on the target cell and initiates the extrinsic pathway of apoptotic death (Fig. 46.18). Ligand binding activates signaling from the intracellular death domain of Fas, possibly by stabilizing Fas trimers or by altering their conformation. Activated Fas binds an adapter protein called **FADD** (Fas-associated protein with a death domain), assembling the DISC. The DISC is an activation platform that binds procaspase 8 through interactions involving another type of motif called the **death effector domain,** which is present on both FADD and the prodomain of procaspase 8. Procaspase 8 monomers dimerize on the DISC and acquire catalytic activity. These dimers can cleave neighboring dimers, creating and releasing heterotetrameric active caspase 8, which can initiate the caspase cascade by activating downstream effector caspases. Frequently, caspase 8 cleaves the BH3-only protein Bid, thereby activating the intrinsic death pathway (Fig. 46.17).

The extrinsic pathway poses considerable risk for the cell. Fas is constitutively present in the cell membrane and can form at least transient trimers in the absence of binding by its ligand. How do cells avoid the accidental activation of apoptosis caused by chance binding of procaspase 8 to naturally occurring transient Fas trimers?

Cells express a caspase-related protein called cFLIP (FLICE-like inhibitory protein; FLICE is another name for caspase 8) that looks very much like a catalytically dead

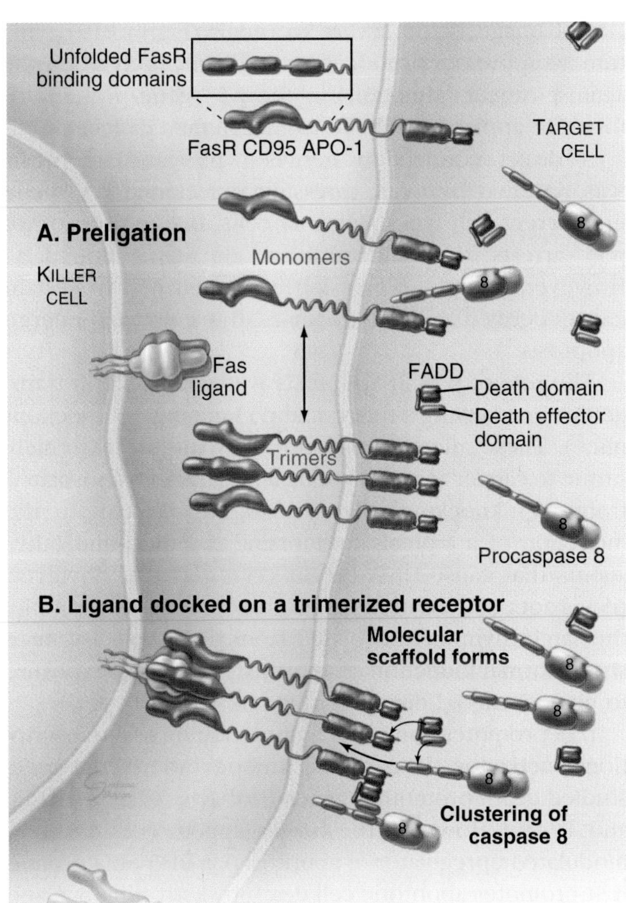

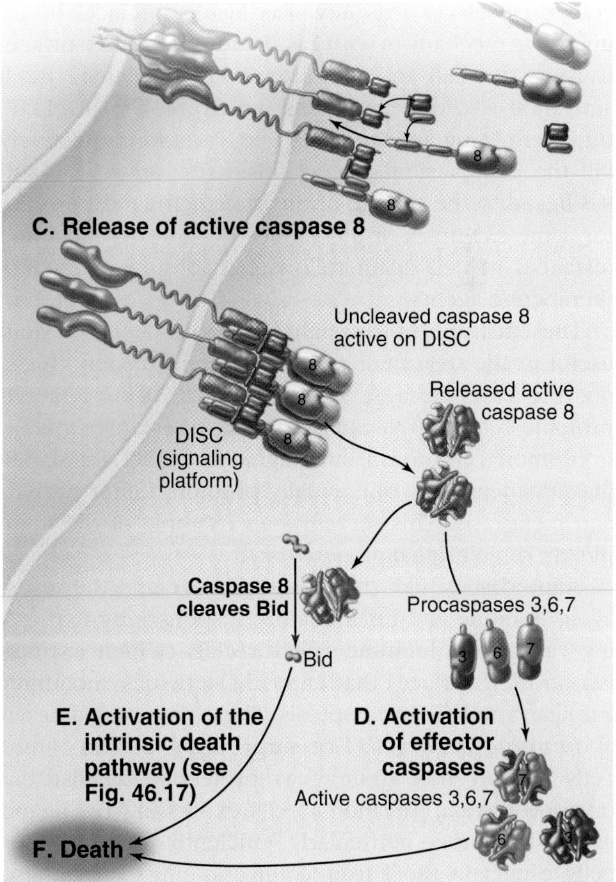

FIGURE 46.18 EXTRINSIC CELL DEATH PATHWAY. The pathways shown are downstream of the Fas cell death receptor. **A,** Preligation. **B,** Ligand docked on a trimerized receptor. **C,** Release of active caspase. **D,** Activation of effector caspases. **E,** Activation of the intrinsic death pathway (Fig. 46.17). **F.** Death (see the text for a detailed description). DISC, death-inducing signaling complex; FADD, Fas-associated death domain.

version of procaspase 8. When expressed at high levels, cFLIP coassembles with procaspase 8 monomers creating an enzyme with altered activity that does not trigger cell death. This role of cFLIP may be to dampen the Fas response locally to ensure that the cascade does not get activated by mistake. Another role of FLIP is to combine with caspase 8 to ensure that TNFα (tumor necrosis factor α) signaling, which occurs in many immune cells, promotes inflammation rather than killing cells by necroptosis (see later).

Role of the Fas Death Receptor in Normal and Diseased Cells

Fas is important for regulation of the immune system, but also has a very unexpected role in cancer cells. Mice with mutated *Fas* (the *lpr* mutation) or Fas ligand (the *gld* mutation) accumulate excessive lymphocytes. In the appropriate genetic background, these mice tend to develop autoimmune disorders.

Fas is important in regulating the life span of activated tissue T and B lymphocytes. Normally, T cells die within a few days of their activation during an immune response. Activation initiates the expression of Fas ligand on the T cells themselves. This new Fas ligand interacts by an unknown mechanism with Fas already on the cell surface, causing the cell to commit apoptotic suicide. T-cell activation also downregulates the expression of cFLIP, thus permitting activated caspase 8 to more effectively kill the cell. A similar mechanism (export of Fas and Fas ligand to the surface of the same cell) is responsible for some examples of p53-induced cell death and some instances of cell death following exposure to chemotherapeutic agents.

These features of Fas might seem to make this system useful in the treatment of cancer. Unfortunately this is not the case, because Fas does more than signal to promote cell death. It can also signal to several pathways to promote cell survival and migration. In fact, Fas signaling in cancer cells can actually promote tumor growth and metastasis. This serves as an example of the complexity of cell signaling networks.

Some tissues, like the lens of the eye and the testis, avoid immune and inflammatory responses by expressing Fas ligand. Immune effector cells (which express Fas on their surface) that enter these tissues encounter Fas ligand and die by apoptosis. These tissues are known as *immune-privileged.* Not surprisingly, certain tumor cells subvert this strategy as protection against the immune system. Melanoma cells expressing Fas ligand establish tumors particularly efficiently. Some tumor cells, especially those from colon and lung cancers, also defend themselves against immune surveillance with so-called **decoy receptors.** A secreted Fas decoy receptor blocks Fas ligand on cytotoxic T cells so that it cannot engage Fas on the surface of the tumor cells. Other decoy receptors remain membrane bound but do not signal cell death when they bind ligand because their intracellular domains lack functional death domains.

Linking Apoptosis to the Cell Cycle by p53

No obligate link exists between particular cell-cycle phases and apoptosis. Noncycling G_0 cells can undergo apoptosis, and cycling cells appear able to do so from any cell-cycle phase. However, one link between apoptosis and the cell-cycle machinery is firmly established. This involves the p53 tumor suppresser and DNA damage.

The p53 transcription factor is one of the downstream EFFECTORS of the DNA damage response pathway (see Fig. 43.11). When cells sense DNA damage induced by agents such as ionizing radiation, p53 levels rise dramatically. When stabilized and activated by phosphorylation (see Fig. 41.14) p53 upregulates the expression of a number of genes, including the Cdk inhibitor p21, which blocks the entry into the S and M phases. p53 also can trigger an apoptotic response in instances in which the DNA damage is too severe to repair (Fig. 46.19). This tumor-suppressor protein is critical in the body's defense against cancer. Mutations in the p53 gene/protein are found in approximately 50% of all human cancers.

A direct connection between p53 and apoptosis was revealed by overexpressing the cloned p53 gene in different cell types. In most cells, overexpression of p53 arrests the cell cycle at the G_1/S boundary. However, ectopic expression of cloned p53 in certain cancer-derived cell lines causes the cells to undergo apoptosis.

The role of p53 in apoptosis was confirmed in transgenic mice lacking a functional p53 gene (p53 knockout mice). These mice develop normally but are extremely prone to cancer at a very young age. Thymocytes isolated from p53 knockout mice are highly resistant to the induction of apoptosis by ionizing radiation and other agents that cause DNA breaks (Fig. 46.19B). However, p53 is not involved in all types of apoptosis. For example, the same thymocytes isolated from p53 knockout mice show normal induction of apoptosis following exposure to glucocorticoid hormone (Fig. 46.19).

p53 promotes apoptosis by functioning as a transcriptional activator. It controls, among others, the well-studied death-promoting genes *Bax*, *Fas* (*CD95*/*APO*-1), and *APAF-1*. However, the key target gene is *PUMA* (p53 modulated upregulator of apoptosis), a BH3-only protein that promotes apoptotic cell death by activating Bax and Bak. *PUMA* knockout mice show defects in cell death pathways that are essentially identical to those seen in p53 knockout mice and not seen in mice that lack *Bax*, *Fas*, or *Apaf-1*.

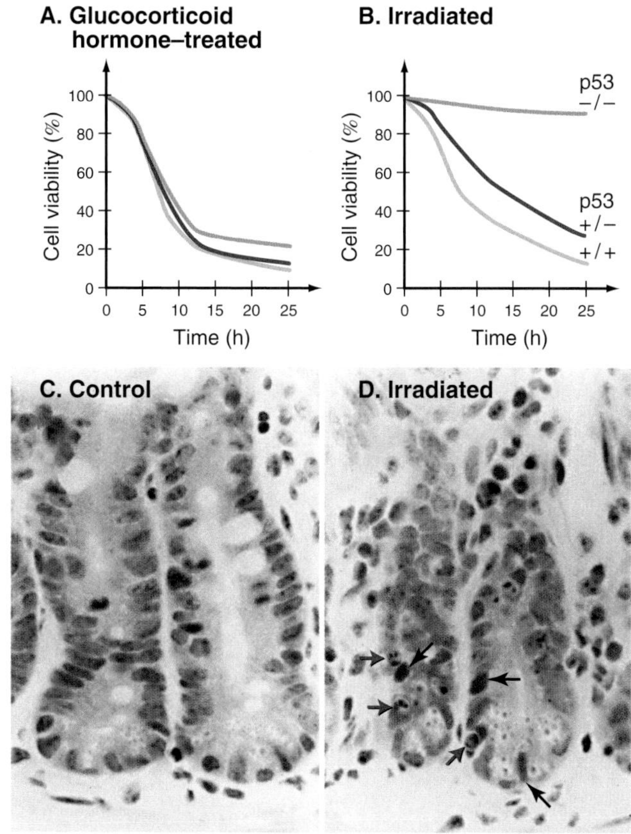

FIGURE 46.19 LINK BETWEEN p53 AND DNA DAMAGE-INDUCED APOPTOTIC DEATH. A–B, Survival of thymocytes from three strains of mice after exposure to glucocorticoids or irradiation. Cell death was from apoptosis. The strains were as follows: wild-type mice *(yellow),* heterozygous mice having one good copy of the *p53* gene and one defective copy *(red),* and mice lacking any functional copy of the *p53* gene *(blue).* Thymocytes that lack p53 are resistant to radiation-induced apoptosis but show normal induction of apoptosis following exposure to glucocorticoid hormone. **C–D,** Induction of p53 accumulation following radiation of the small intestine. *Black arrows* indicate cells with increased levels of p53. *Red arrows* indicate apoptotic cells. (**A–B,** From Lowe SW, Schmitt EM, Smith SW, et al. p53 is required for radiation-induced apoptosis in mouse thymocytes. *Nature.* 1993;362:847–849. **C–D,** Courtesy John Hickman, Molecular and Cellular Pharmacology Group, University of Manchester, United Kingdom.)

Other Types of Programmed Cell Death

The terms *apoptosis* and *programmed cell death* are sometimes viewed as synonymous. However, a number of specific examples of programmed cell death have been described that lack the features that classically define apoptosis.

Inflammatory Cell Death: Pyroptosis and Necroptosis

One of the defining features of apoptosis is that dying cells disappear without causing an inflammatory response. However, in certain instances, it is protective for the body to have cell death be accompanied by inflammation as this can trigger an immune response in the early phases of infection.

Pyroptosis (proinflammatory programmed cell death) occurs when signaling by a variety of cytoplasmic receptors in response to pathogen infection leads to the formation of the **inflammasome,** a cytosolic complex that contains numerous caspase 1 activation sites. Pyrin domains (PYDs) in these receptors are linked to caspase recruitment domains (CARDs), which in turn recruit caspases 1, 4, and 5 (1 and 11 in mice). Despite being the first caspase to be described, caspase 1 was the last to be shown to be involved in a programmed cell death response.

The function of caspase 1 is to produce interleukin-1β (IL-1β) and interleukin-18 (IL-18) by proteolytic processing of their precursors. These proteins are secreted together with caspase 1 by an unconventional pathway that is still being investigated and may involve cell lysis. IL-1β is a proinflammatory cytokine that triggers the fever response by acting on the hypothalamus. It also promotes the proliferation of immune cells, among other activities. IL-18 promotes increased immune cell recruitment and activation, as well as modifying local cells to facilitate adaptive immune responses. This proinflammatory signaling facilitates a ramping up of the immune response to the pathogen.

How pyroptosis leads to cell death is still being investigated, but the morphology of the dying cells resembles that seen in necrosis. As a result, the internal components of the cell are released into the extracellular space, provoking a strong inflammatory response (as in necrosis, above, and necroptosis, below).

Necroptosis (programmed necrosis) is a cell death pathway that may function as a backup when the extrinsic pathway of apoptosis is not functioning. Activation of the CD95 (Fas) and TNFR1 (tumor necrosis factor receptor 1) cell-surface receptors can have three possible outcomes. TNFR1 can signal via complex I (which contains protective IAP proteins) to activate the canonical NF-κB pathway to promote cell survival (Fig. 46.1). Alternatively TNFR1 can activate receptor interacting protein kinases 1 and 3 (RIPK1/RIPK3) and signal to form complex II (containing FADD, TRADD [tumor necrosis factor receptor-associated death domain], caspase 8, and cFLIP). When caspase 8 is associated with cFLIP, it inactivates RIPK3 and cells survive. Complex II activates necroptosis if either caspase 8 or cFLIP is missing or inactivated and if RIPK1/RIPK3 are active. RIPK1/RIPK3 activity leads to permeabilization of the cell membrane, cell lysis and death. As in other forms of necrosis, the released cellular contents cause a local inflammatory response.

Necroptosis is thought to be a backup pathway for apoptosis during viral infection. Mouse kidney cells infected with a virus that encodes a caspase 8 inhibitor proceed down the necroptosis death pathway. Removal of the gene from the virus restores apoptosis. Mice lacking RIPK3 show increased susceptibility to viral infection. Murine cytomegalovirus has proteins that help it to evade both apoptosis and necroptosis, supporting this argument.

Autophagic Death

Autophagy is a catabolic, energy producing process that allows cells to survive in adverse conditions by recycling amino acids liberated by degradation of cellular proteins and organelles (see Fig. 23.7). Autophagy can be activated by a wide range of stimuli, including nutrient and oxidative stress, accumulation of unfolded proteins, intracellular bacterial infection, and oncogenic stress. In addition to being a response to starvation, autophagy can also participate in antiviral responses by participating in antigen presentation by immune cells.

Connections between autophagy and cell death are complex and debated. In certain cases during development, autophagic pathways can lead directly to cell death. These dying cells lack chromatin condensation and exhibit extreme vacuolization of the cytoplasm. In some cells, autophagy can regulate apoptosis. For example, degradation of proapoptotic BH3-only proteins by autophagy provides a mechanism protecting cells against apoptosis. Alternatively, degradation of protective proteins can cause autophagy to actively promote apoptotic cell death.

Connections between autophagy and cancer are complex. It has been argued that by promoting cell survival under adverse conditions, autophagy might help promote the establishment of cancers particularly in avascular areas and by helping cancer cells to survive chemotherapy. However, mice heterozygous for Beclin1 (a scaffolding protein that functions early in assembly of the phagophore membrane; see Fig. 23.7) develop cancers, suggesting that Beclin-1 is a tumor suppressor and that autophagy may protect against tumorigenesis. Interestingly, Beclin-1 has a BH3 domain, and although it is not a canonical BH3-only proapoptotic protein, its function is negatively regulated by Bcl-2. Also, autophagy is activated in oncogene-induced senescence and inhibiting autophagy delays senescence. Thus autophagy might prevent cancer formation by helping damaged cells to become senescent.

Importance of Programmed Cell Death in Human Disease

Why have studies of programmed cell death so caught the scientific eye? One likely answer is that cell death is a point of intersection between cell signaling pathways, cell structure, the cell cycle, and, of course, human disease. This chapter has mentioned the roles that aberrations in programmed cell death play in the etiology of autoimmunity, AIDS, and cancer. Cell death is firmly established as a key factor in neurodegenerative diseases, such as Huntington disease and Alzheimer disease, as well as in myocardial infarction and stroke (Fig. 46.20). At a practical level, the realization that many successful chemotherapeutic agents act by inducing cancer cells to undergo apoptosis motivated searches for newer and better drugs that elicit this response. The generally disappointing outcome of those studies may largely be because other pathways (eg, necroptosis) kick in when apoptosis is inhibited. Alternatively, as more is learned about mechanisms of necroptosis and pyroptosis, these alternative types of programmed death become attractive therapeutic targets, as their induction may not only kill the target cells, but also provoke a localized immune response that might attack cancer cells or pathogens. With such important practical problems to be solved, programmed cell death will continue to occupy a prominent position in cell biology research over the coming years.

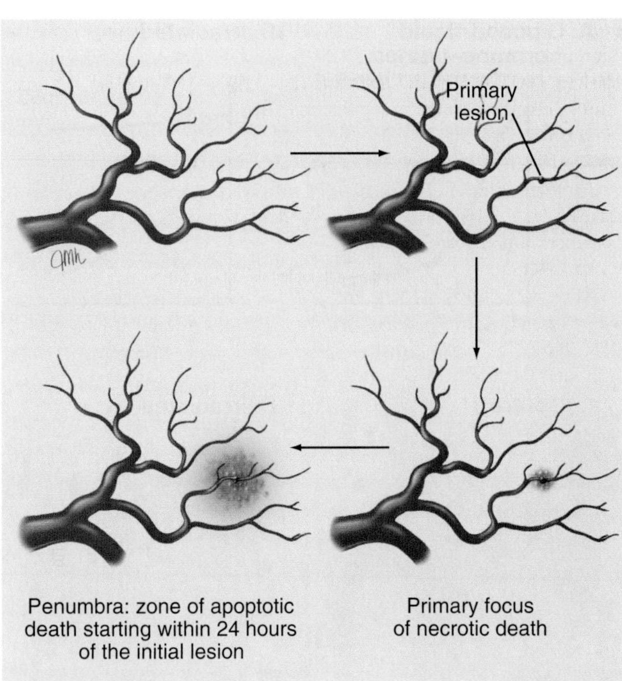

FIGURE 46.20 BRAIN DAMAGE IN STROKE. Secondary programmed cell death caused by oxygen deprivation in the penumbra greatly increases the size of the affected area of the brain in stroke.

ACKNOWLEDGMENTS

We thank Charles Earnshaw, Scott Kaufmann, Luis Miguel Martins, and Andrew Thorburn for their suggestions on revisions to this chapter.

SELECTED READINGS

Breckenridge DG, Xue D. Regulation of mitochondrial membrane permeabilization by BCL-2 family proteins and caspases. *Curr Opin Cell Biol.* 2004;16:647-652.

Czabotar PE, Lessene G, Strasser A, Adams JM. Control of apoptosis by the BCL-2 protein family. *Nat Rev Mol Cell Biol.* 2014;15:49.

Deretic V, Saitoh T, Akira S. Autophagy in infection, inflammation and immunity. *Nat Rev Immunol*. 2013;13:722-737.

Earnshaw WC, Martins LM, Kaufmann SH. Mammalian caspases: Structure, activation, substrates and functions during apoptosis. *Annu Rev Biochem*. 1999;68:383-424.

Fitzwalter BE, Thorburn A. Recent insights into cell death and autophagy. *FEBS J*. 2015;282:4279-4288.

Lamkanfi M, Festjens N, Declercq W, et al. Caspases in cell survival, proliferation and differentiation. *Cell Death Differ*. 2007;14:44-55.

Lauber K, Blumenthal SG, Waibel M, et al. Clearance of apoptotic cells: Getting rid of the corpses. *Mol Cell*. 2004;14:277-287.

Meier P, Finch A, Evan G. Apoptosis in development. *Nature*. 2000; 407:796-801.

Metzstein MM, Stanfield GM, Horvitz HR. Genetics of programmed cell death in *C. elegans:* Past, present and future. *Trends Genet*. 1998; 14:410-416.

Pasparakis M, Vandenabeele P. Necroptosis and its role in inflammation. *Nature*. 2015;517:311-320.

Raff MC. Social control on cell survival and cell death. *Nature*. 1992; 356:397-400.

Savill J, Fadok V. Corpse clearance defines the meaning of cell death. *Nature*. 2000;407:784-788.

Silke J, Meier P. Inhibitor of apoptosis (IAP) proteins–modulators of cell death and inflammation. *Cold Spring Harb Perspect Biol*. 2013; 5:a008730.

Taylor RC, Cullen SP, Martin SJ. Apoptosis: controlled demolition at the cellular level. *Nat Rev Mol Cell Biol*. 2008;9:231.

Wallach D, Kang TB, Dillon CP, et al. Programmed necrosis in inflammation: Toward identification of the effector molecules. *Science*. 2016;352:aaf2154.

Wyllie AH, Kerr JFR, Currie AR. Cell death: The significance of apoptosis. *Int Rev Cytol*. 1980;68:251-305.

Zmasek CM, Godzik A. Evolution of the animal apoptosis network. *Cold Spring Harb Perspect Biol*. 2013;5:a008649.

Cell SnapShots

SnapShot 1: Histone Modifications

See Chapter 8.

The histone proteins are decorated by a variety of protein posttranslational modifications (also called histone marks). Histone marks are critical to dynamic modulation of chromatin structure and function, contributing to the cellular gene expression program. In addition to the well-studied acetylation and methylation modifications, recent studies have revealed several new types of histone marks, including lysine propionylation, lysine butyrylation, lysine crotonylation, lysine 2-hydroxyisobutyrylation, lysine malonylation, and lysine succinylation. Preliminary studies on some of the new histone marks (eg, crotonylation and 2-hydroxyisobutyrylation) suggest that their effects on chromatin function are distinct from those of lysine acetylation. Given that the newly discovered lysine acylation reactions likely use the corresponding acyl-CoA (acyl-coenzyme A) molecules as cofactors, it is proposed that histone acylations provide a link between cellular metabolism and epigenetic mechanisms.

This SnapShot summarizes the reported human, mouse, and rat histone marks, including recently identified lysine acylation marks.

From Huang H, Sabari BR, Garcia BA, et al. SnapShot: histone modifications. *Cell*. 2014;159(2):458–458.e1.

SnapShot 2: Nuclear Transport

See Chapter 9.

Exchange of macromolecules between the nucleus and the cytoplasm occurs through nuclear pore complexes (NPCs), large proteinaceous structures embedded within the nuclear envelope. Small molecules (<30 kD) can passively diffuse through NPCs. Nuclear transport receptors are required to facilitate import and export of large molecules (>30 kD). This includes the karyopherin-β (Kapβ) receptor family wherein each associates directly with the NPC and has cargo binding controlled by the asymmetric distribution of Ran-GTP (guanosine triphosphate) and the Ran cycle. Nuclear localization of the Ran guanine-nucleotide exchange factor (Ran-GEF; Prp20 in *Saccharomyces cerevisiae*) via chromatin association maintains high concentrations of Ran-GTP (guanosine triphosphate) in the nucleus. Conversely, Ran-GDP (guanosine diphosphate) predominates in the cytoplasm through the combined action of cytoplasmically localized Ran-binding protein 1 (RanBP1; Yrb1 in *S. cerevisiae*) and Ran GTPase (guanosine triphosphatase)-activating protein (Ran-GAP; Rna1 in *S. cerevisiae*). During import, Kapβ receptors interact with cytoplasmic cargos and traverse the NPC. In the nucleus, Ran-GTP binds to the importing Kapβ and triggers cargo release. Conversely, during export, Ran-GTP stabilizes the association of Kapβ receptors with nuclear cargo. Hydrolysis of GTP to GDP at the cytoplasmic NPC face promotes disassembly of the export receptor-cargo complex. Transport receptors are categorized based on transport direction, the use of cargo adaptors, and transport of specialized cargo such as Ran-GDP, ribosomes, and/or messenger RNA (mRNA):

I. *Import receptor complexes with adaptors:* Kap60-Kap95 (importin-α-importin-β in metazoa) is the prototypical import receptor. Cargo with a classical nuclear localization signal (cNLS) is recognized by the adaptor Kap60, which binds the receptor Kap95. Whereas yeast has only one importin-α, metazoans have several importin-α homologs (not shown).

II. *Import receptors that bind cargo directly:* Many Kapβs, including Kap104, can bind directly to their cargo.

III. *Bidirectional transport receptors:* The receptor Msn5 has the striking capacity to function in both import and export.

IV. *Export receptor complexes:* The first identified export receptor, Crm1, recognizes leucine-rich nuclear export sequences (NES) in cargo proteins.

V. *Export of ribosomal subunits:* Large cargos, such as ribosomal subunits, require multiple receptors for transport. Export of the assembled 60S subunit requires Crm1 (with the adaptor Nmd3) and, in *S. cerevisiae*, the Mex67-Mtr2 heterodimer as well as Arx1. Export of the 40S subunit also requires Crm1 and possibly the adaptor Ltv1.

VI. *Export of mRNA:* Export of most mRNA is Ran independent and, instead, requires the nonkaryopherin Mex67-Mtr2 (TAP/NXF1 and p15/NXT1 in metazoans, respectively). Mex67-Mtr2 binds mRNA thorough association with mRNA-binding proteins, or some metazoan mRNAs can recruit TAP/NXF1 by directly binding a specific RNA element, the CTE (not shown). CTEs are also present in some viral RNAs. Directionality of mRNA export is determined by the adenosine triphosphatase (ATPase) Dbp5, whose activation is controlled by spatially restricted inositol hexakisphosphate (IP6)-bound Gle1.

Consensus Sequences

cNLS: PKKKRKV (monopartite) or KRX(10–12)KRRK (bipartite); PY-NLS: ϕ(G/S/A)ϕϕX(5–10)(R/H/K)X(2–5)PY (hPY-NLS) or basic-enriched(5–8)X(2–7) (R/H/K)X(2–5)PY

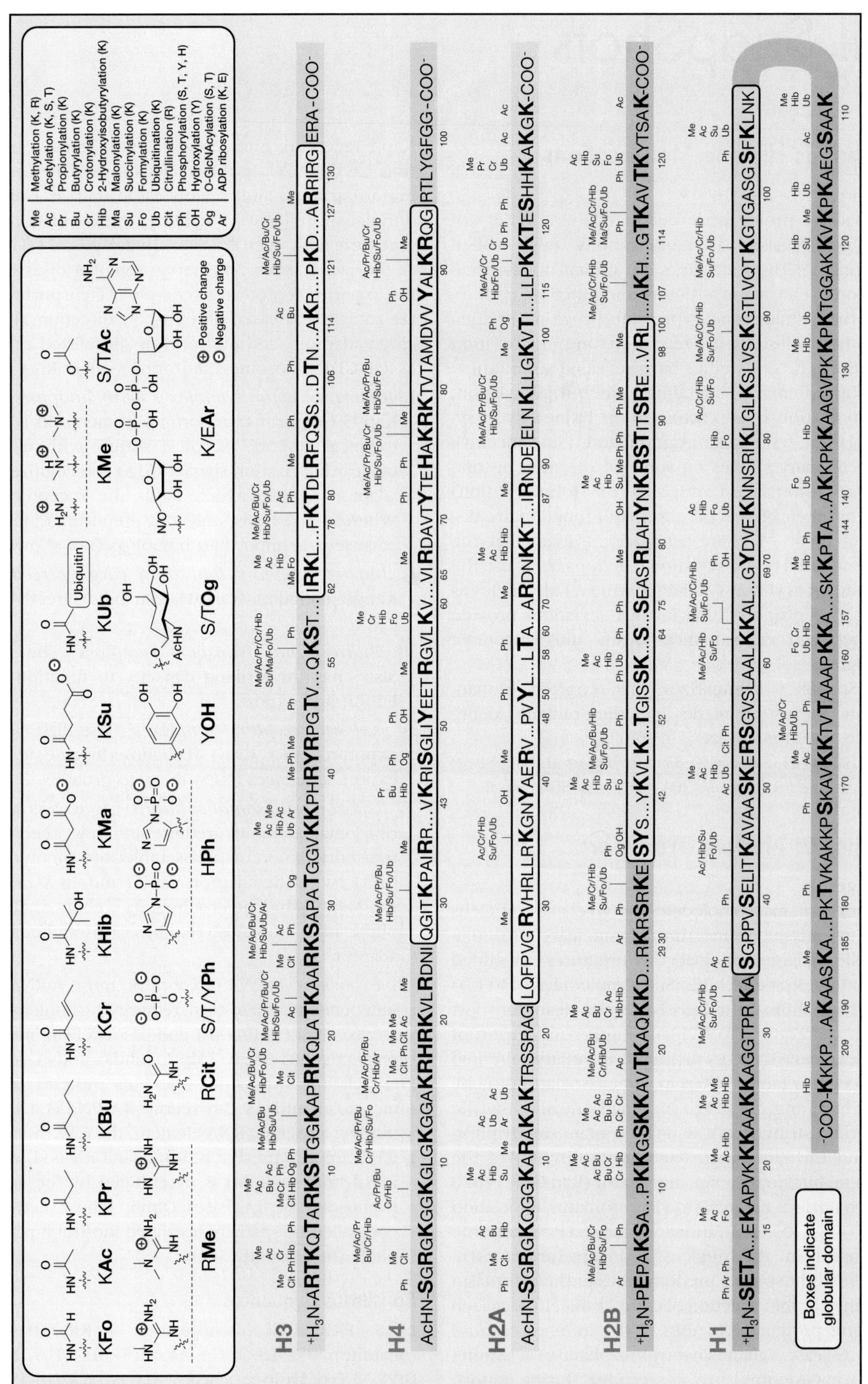

SNAPSHOT 1 HISTONE MODIFICATIONS.

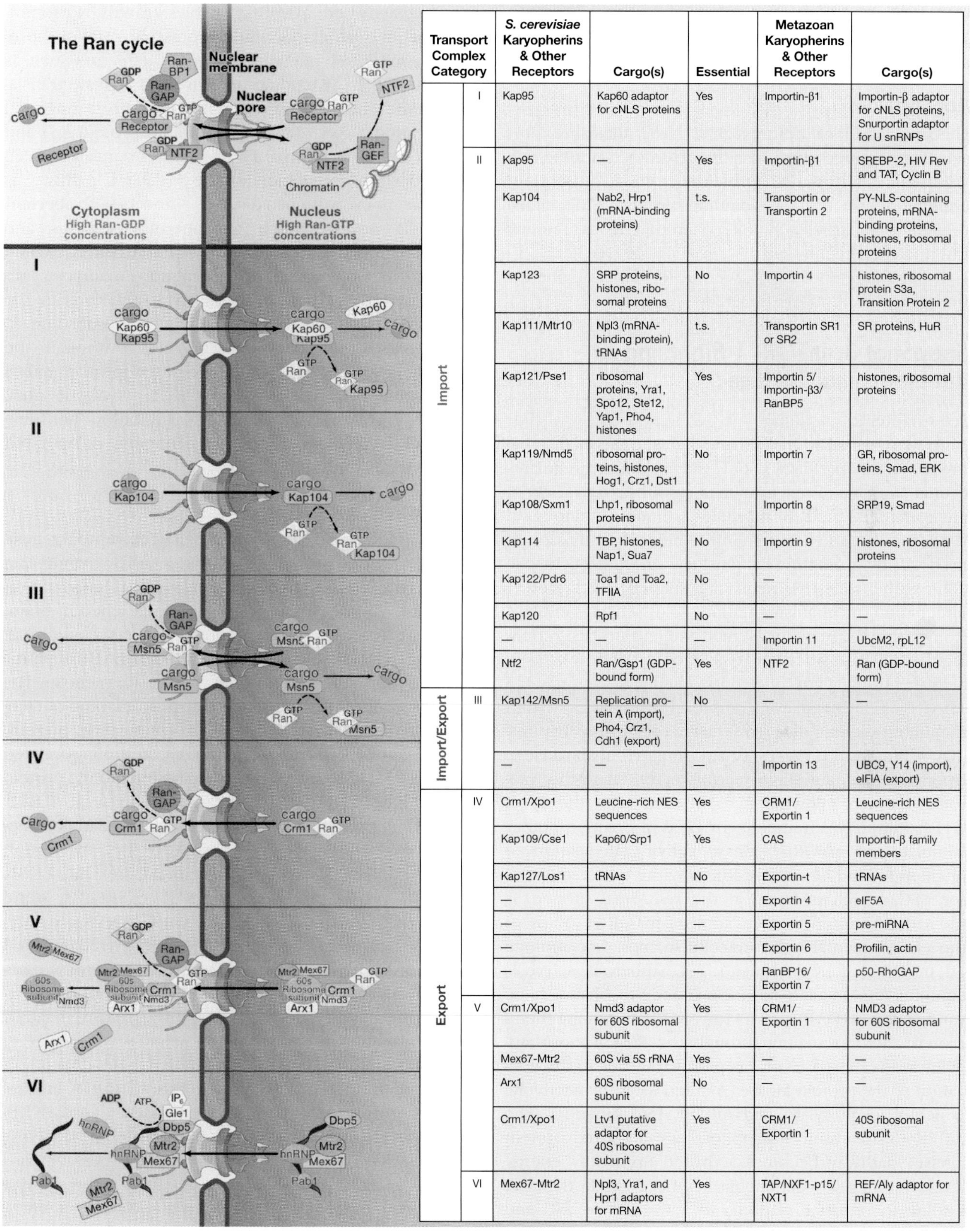

Transport Complex Category		*S. cerevisiae* Karyopherins & Other Receptors	Cargo(s)	Essential	Metazoan Karyopherins & Other Receptors	Cargo(s)
Import	I	Kap95	Kap60 adaptor for cNLS proteins	Yes	Importin-β1	Importin-β adaptor for cNLS proteins, Snurportin adaptor for U snRNPs
	II	Kap95	—	Yes	Importin-β1	SREBP-2, HIV Rev and TAT, Cyclin B
		Kap104	Nab2, Hrp1 (mRNA-binding proteins)	t.s.	Transportin or Transportin 2	PY-NLS-containing proteins, mRNA-binding proteins, histones, ribosomal proteins
		Kap123	SRP proteins, histones, ribosomal proteins	No	Importin 4	histones, ribosomal protein S3a, Transition Protein 2
		Kap111/Mtr10	Npl3 (mRNA-binding protein), tRNAs	t.s.	Transportin SR1 or SR2	SR proteins, HuR
		Kap121/Pse1	ribosomal proteins, Yra1, Spo12, Ste12, Yap1, Pho4, histones	Yes	Importin 5/ Importin-β3/ RanBP5	histones, ribosomal proteins
		Kap119/Nmd5	ribosomal proteins, histones, Hog1, Crz1, Dst1	No	Importin 7	GR, ribosomal proteins, Smad, ERK
		Kap108/Sxm1	Lhp1, ribosomal proteins	No	Importin 8	SRP19, Smad
		Kap114	TBP, histones, Nap1, Sua7	No	Importin 9	histones, ribosomal proteins
		Kap122/Pdr6	Toa1 and Toa2, TFIIA	No	—	—
		Kap120	Rpf1	No	—	—
		—	—	—	Importin 11	UbcM2, rpL12
		Ntf2	Ran/Gsp1 (GDP-bound form)	Yes	NTF2	Ran (GDP-bound form)
Import/Export	III	Kap142/Msn5	Replication protein A (import), Pho4, Crz1, Cdh1 (export)	No	—	—
		—	—	—	Importin 13	UBC9, Y14 (import), eIFIA (export)
Export	IV	Crm1/Xpo1	Leucine-rich NES sequences	Yes	CRM1/ Exportin 1	Leucine-rich NES sequences
		Kap109/Cse1	Kap60/Srp1	Yes	CAS	Importin-β family members
		Kap127/Los1	tRNAs	No	Exportin-t	tRNAs
		—	—	—	Exportin 4	eIF5A
		—	—	—	Exportin 5	pre-miRNA
		—	—	—	Exportin 6	Profilin, actin
		—	—	—	RanBP16/ Exportin 7	p50-RhoGAP
	V	Crm1/Xpo1	Nmd3 adaptor for 60S ribosomal subunit	Yes	CRM1/ Exportin 1	NMD3 adaptor for 60S ribosomal subunit
		Mex67-Mtr2	60S via 5S rRNA	Yes	—	—
		Arx1	60S ribosomal subunit	No	—	—
		Crm1/Xpo1	Ltv1 putative adaptor for 40S ribosomal subunit	Yes	CRM1/ Exportin 1	40S ribosomal subunit
	VI	Mex67-Mtr2	Npl3, Yra1, and Hpr1 adaptors for mRNA	Yes	TAP/NXF1-p15/ NXT1	REF/Aly adaptor for mRNA

SNAPSHOT 2 NUCLEAR TRANSPORT.

(bPY-NLS), ϕ denotes hydrophobic residue; NES: (L/V) X(2–3)(ψ/F)X(2–3)LXψ, ψ denotes large aliphatic residue (I, V, L, M).

Abbreviations

NLS, nuclear localization signal; NES, nuclear export signal; t.s., temperature sensitive; U snRNP, uridine-rich small nuclear ribonucleoprotein complex; SR, serine/arginine-rich; TBP, TATA-binding protein; SRP, signal recognition particle; hnRNP, heterogeneous nuclear ribonucleoprotein.

From Tran EJ, Bolger TA, Wente SR: SnapShot: nuclear transport. *Cell.* 2007;131(2):420.

SnapShot 3: mTORC1 Signaling at the Lysosomal Surface

See Chapter 23.

In mammals, the mTOR (mechanistic target of rapamycin) complex 1 (mTORC1) ser/thr kinase regulates cellular and organismal growth in response to a variety of environmental and intracellular stimuli. Amino acid levels mediate the first step in the bipartite activation of mTORC1 by promoting its translocation from a cytosolic compartment to the lysosomal surface. By a poorly understood mechanism, amino acid sensing initiates from within the lysosomal lumen and, in a process requiring the v-ATPase (vacuolar H+-adenosine triphosphatase), activates the guanine nucleotide exchange factor (GEF) activity of the Ragulator complex toward RagA within the heterodimeric Rag (ras-related guanosine triphosphate binding) GTPases (guanosine triphosphatases). Upon GTP binding, RagA recruits mTORC1 to the lysosomal surface, allowing it to interact with the small GTPase Rheb (ras homolog enriched in brain), a potent stimulator of mTORC1 kinase activity. Regulation of nucleotide binding state of Rheb by the tumor suppressor TSC, which is found at the lysosomal surface, is the second step in the activation of mTORC1. Many of the environmental and intracellular cues that impinge on mTORC1 funnel through TSC (tuberous sclerosis complex) and regulate its guanosine triphosphatase–activating protein (GAP) activity toward Rheb. Among them, growth factor signaling through the PI3K (phosphatidylinositide 3′-kinase) or Ras pathways leads to the activation of the protein kinases Akt and Rsk1, respectively, which phosphorylate and inhibit TSC function. The AMPK (5′-adenosine monophosphate–activated protein kinase) pathway becomes activated upon low energy levels and in a p53-dependent manner by DNA damage, leading to phosphorylation and activation of TSC and phosphorylation and inactivation of mTORC1. Reduction in oxygen levels induces Redd1 (protein regulated in development and DNA damage response 1) expression, which by an ill-defined process maintains TSC function.

Once activated, mTORC1 enables growth by promoting anabolic programs while repressing catabolic processes. mTORC1 phosphorylates key effectors such as z1 and 4EBP1 (4E-binding protein 1) to activate translation and inhibits autophagy by phosphorylating and inactivating ATG13 (autophagy-related protein 13) and ULK1 (unc-51-like kinase 1). As a master regulator of cell metabolism, deregulation of the mTORC1 pathway is common in many human diseases. Cancers with aberrant mTORC1 activity, such as tuberous sclerosis and advanced renal cell carcinoma, are increasingly treated with analogs of the mTORC1 inhibitor rapamycin. Furthermore, overactivation of this pathway leads to the downregulation of IRS1 (insulin receptor substrate 1) and progression of type 2 diabetes. Although the mTORC1 pathway is absolutely required for mammalian development, reduction of mTORC1 activity in mice models through pharmacological inhibition not only enhances adult stem cell numbers, function, or both, but also extends murine life span.

Abbreviations

mTOR, mechanistic target of rapamycin; raptor, regulatory associated protein of mTOR; mLST8, mammalian lethal with SEC13 protein 8; pras40, proline-rich Akt substrate 40 kDa; Rheb, ras homolog enriched in brain; TSC, tuberous sclerosis complex; Rag, ras-related GTP binding; MP1, MAPK scaffold protein 1; HBXIP, hepatitis B virus X-interacting protein; v-ATPase, vacuolar H+-adenosine triphosphatase ATPase; GEF, guanine nucleotide exchange factor; GAP, GTPase-activating protein; ULK1, unc-51-like kinase 1; ATG13, autophagy-related protein 13; FIP200, FAK family kinase-interacting protein of 200 kDA; S6K1, p70 ribosomal S6 kinase 1; 4EBP1, 4E-binding protein 1; Redd, protein regulated in development and DNA damage response 1; TFEB, transcription factor EB; HIF1a, hypoxia-inducible factor 1a; LKB1, serine/threonine-protein kinase STK11; SREBP1, sterol regulatory element binding protein-1; AMPK, 5′-AMP-activated protein kinase; PIP2, phosphatidylinositol 4,5-bisphosphate; PIP3, phosphatidylinositol 3,4,5-trisphosphate; PTEN, phosphatase and tensin homolog; PI3K, phosphatidylinositol 3-kinase; GRB2, growth factor receptor-bound protein 2; SOS, son-of-sevenless; NF1, neurofibromin 1; PDK1, phosphoinositide dependent kinase 1; IRS1, insulin receptor substrate 1; IGF, insulin-like growth factor; TNFa, tumor necrosis factor a; IKKB, inhibitor of nuclear factor k-B kinase subunit b; WNT, wingless; Dsh1, dishevelled 1; GSK3, glycogen synthase kinase 3; TK, tyrosine kinase; SLC1A5, solute carrier family 1 member 5; SLC7A5, solute carrier family 7 member 5; FKBP12, FK506-binding protein 12 KDa.

From Bar-Peled L, Sabatini DM: SnapShot: mTORC1 signaling at the lysosomal surface. *Cell.* 2012;151(6):1390–1390.e1.

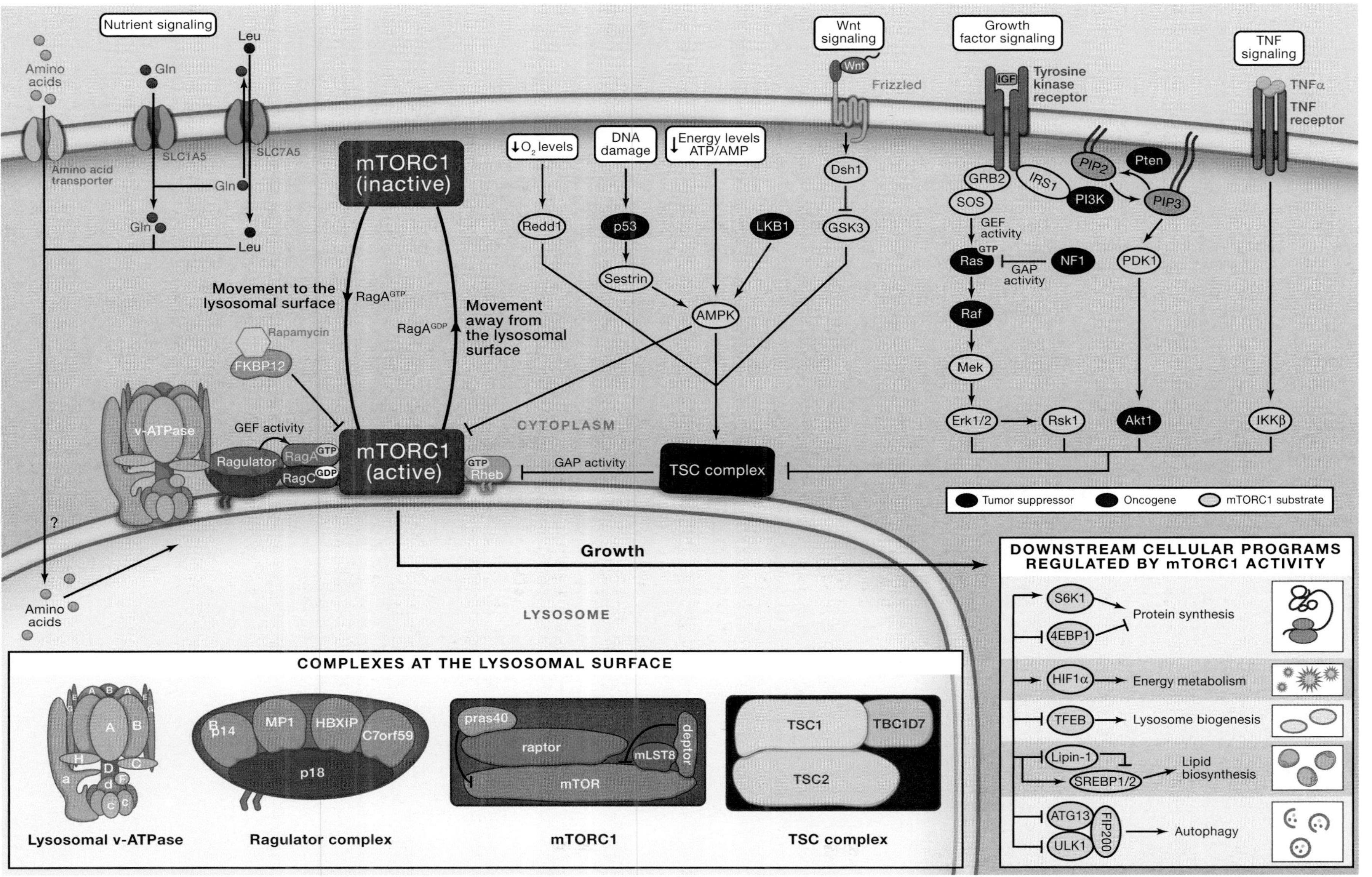

SNAPSHOT 3 mTORC1 SIGNALING AT THE LYSOSOMAL SURFACE.

Glossary

+TIPs. Family of proteins that associate with the plus ends of elongating microtubules

3′ untranslated region (3′ UTR). Region of an mRNA after the stop codon, often containing regulatory signals governing further RNA processing

5′ cap. Modified guanosine residue on the 5′ end of mRNAs that protects against degradation

5′ exonuclease. Enzyme that degrades RNA or DNA from the 5′ end

5-lipoxygenase. Enzyme that synthesizes leukotrienes and lipoxins from arachidonic acid

5S RNA. Smallest RNA component of the ribosome, transcribed by RNA polymerase III

9-1-1 complex. Trimeric doughnut-like complex that is responsible for loading enzymes onto sites of DNA damage

14-3-3 domain. Adapter domain that binds serine-phosphate ligands

30-nm fiber. Compacted filament of chromosomal DNA made up of closely packed nucleosomes

AAA ATPase. Family of multimeric enzymes that use ATP hydrolysis to do work in DNA replication, membrane trafficking, and microtubule-dependent motility

A-band. Region of striated muscle sarcomere with myosin thick filaments

ABC transporter. Family of enzymes that pumps diverse solutes and flips lipids across membranes

Abscission. Final separation of two daughter cells in cytokinesis

Acetylcholine. Neurotransmitter for the neuromuscular junction and other synapses

Acetylcholine esterase. Enzyme that degrades acetylcholine

Acrosomal process. Projection of the sperm plasma membrane supported by actin filaments

Actin. Subunit protein of cytoplasmic microfilaments and muscle thin filaments

Actin-related protein (Arp). Family of proteins sharing a common origin and fold with actin

Action potential. Self-propagating, transient change in the membrane potential

Activating transcription factor 6. A transmembrane protein of the ER that participates in the unfolded proteins response; it is cleaved to release a transcription factor that regulates expression of genes for ER proteins

Activation loop. Region of protein kinases that must be phosphorylated for the kinase to be fully activated

Active chromatin hub. Region of chromatin with a relatively high concentration of chromatin marks associated with active transcription

Acyl-CoA-cholesterol transferase (ACAT). Enzyme of ER membranes that catalyzes the formation of cholesterol esters

Adaptive immunity. Process that selects for proliferation antibody producing B-lymphocytes and T-lymphocytes with receptors for specific antigens

ADAR (adenosine deaminase acting on RNA). Enzyme converts adenine (which base-pairs with uracil) to inosine (which base-pairs with cytosine) by deamination, thereby potentially altering the protein encoded by the mRNA

Adenine. Purine base found in ATP, DNA, and RNA; H-bonds with thymine or uracil

Adenosine triphosphate (ATP). The common energy-carrying molecule for cellular metabolism; enzymes use the energy released from the hydrolysis of its γ-phosphate for many cellular processes

Adenylylcyclase. Enzyme that converts ATP into 3′ to 5′ cyclic AMP

ADF/cofilin. Protein that severs actin filaments and promotes depolymerization

Adherens junction. Intercellular junction that uses cadherins for adhesion and is anchored to actin filaments

Adipocyte. Fat cell

Adrenergic receptor. Seven helix receptor activated by epinephrine or norepinephrine

Affinity chromatography. Use of an immobilized ligand to purify interacting macromolecules

Aggrecan. Core protein for a proteoglycan that associates with hyaluronan in cartilage

Agonist. Ligand molecule that activates a receptor

Alpha-actinin. Actin filament cross-linking protein, found in striated muscle Z-disks

Alpha and beta tubulin. Isoforms of tubulin that form the heterodimeric building blocks of microtubules

Alpha-catenin. Adapter protein between cadherins, beta-catenin, and actin filaments

Alpha-helix. Common element of protein secondary structure, right-handed helix with 3.6 amino acid residues per turn

Alpha satellite. Repeated DNA sequences found in human centromeres, composed of monomers ~171 base pairs long, some with binding sites for centromeric protein CENP-B

ALT. Pathway using recombination to lengthen telomeres without telomerase

Alternative splicing. Production of more than one mRNA, and therefore more than one protein product, from a gene by alternative exclusion of exons or inclusion of introns from the initial transcript

Alzheimer disease. Most common dementia of older people, characterized by loss of neurons and formation of intracellular paired helical filaments of tau that aggregate in neurofibrillary tangles

Amide nitrogen. Nitrogen contributed by an amino acid to the peptide bonds of proteins

Amiloride-sensitive channels. Cation channels consisting trimers of subunits with two transmembrane helices

Amino acid. Building block of proteins, including an amino group, a central α-carbon with a side chain (or R group), and a carboxyl group

Aminoacyl-tRNA synthetases. Enzymes that catalyze covalent coupling of an amino acid to its cognate tRNA

Amino terminus. End of a polypeptide with a free amino group

Amphitelic attachment. Proper chromosome attachment to the spindle with sister kinetochores attached to opposite spindle poles

Anaphase A. The stage of mitosis when sister chromatids separate from each other by moving to the poles of the mitotic spindle, initiated by degradation of proteins that regulate sister chromatid cohesion

Anaphase B. The stage of mitosis when the poles of the mitotic spindle move apart

Anaphase-promoting complex/cyclosome (APC/C). A ubiquitin-conjugating (E3) enzyme complex that targets proteins, including cyclins and securin, for degradation during mitosis and G_1

Aneuploidy. Excess or missing chromosomes caused by errors in mitosis or meiosis

Anillin. Adapter protein for contractile ring organization and cytokinesis

Antagonist. Ligand molecule that inhibits receptor

Anterograde traffic. Movement of cargo and lipid forward through the secretory system toward the plasma membrane

Anterograde transport. Microtubule-based movements away from the cell center, generally powered by dynein

Antigen-presenting cell. Macrophages, dendritic cells and other cells that display antigenic peptides on cell surface major histocompatibility complexes

Antigen-presenting compartment. Phagolysosome specialized for loading MHC Class II molecules with peptides

Antiporter. Carrier proteins that catalyze movements of solutes across membranes up concentration gradients at the expense of transport of a second solute down its concentration gradient in the opposite direction

AP1 complex. Protein complex that directs clathrin coat assembly at the TGN

Apaf-1. *See* apoptotic protease activating factor 1

APC. Product of the adenomatous polyposis coli gene; mutations predispose to colon polyps and cancer

APC/C. *See* Anaphase-promoting complex/cyclosome

Apical plasma membrane. Region of plasma membrane on the free surface of epithelial cells separated from the basolateral membrane by a ring of tight junctions

Apoptosis. A type of programmed cell death triggered by internal signals or external stimuli and accompanied by characteristic morphologic and biochemical changes

Apoptosome. Seven-spoked ring-like structure containing Apaf-1 and procaspase 9 that starts the proteolytic cascade in the intrinsic (mitochondrial) pathway of apoptosis

Apoptotic bodies. Membrane-enclosed remnants of apoptotic cells

Apoptotic protease activating factor 1 (Apaf-1). Protein activated by cytochrome c that binds procaspase 9 to form the apoptosome, initiating the intrinsic pathway of apoptotic cell death

Aquaporins. Membrane channels selective for water

Arabidopsis thaliana. Mustard weed; popular genetic model organism favored by plant biologists

Arachidonic acid. A 20-carbon fatty acid with four double bonds; common constituent of membrane lipids and precursor of eicosanoids

Archaea. One of the three domains of life, along with Bacteria and Eukaryotes

ARE-mediated degradation. Pathway triggered by the presence of sequence motifs referred to as A+U-rich elements (AREs) that targets for rapid turnover of mRNAs that encode proteins for which limited and transient expression is important

Arf. Family of small GTPases, including Sar1 and Arf1-6, that mediate membrane traffic by associations of protein effectors with specific membranes; Arf1 recruits either COPI coat complexes or clathrin to Golgi-like membranes

Arp2/3 complex. Protein complex including Arp2 and Arp3 that nucleates branched actin filaments

Arrestin. Protein that binds and inhibits phosphorylated seven-helix receptors

ARS core consensus sequence. DNA sequence that binds the origin recognition complex (ORC) and defines the position of DNA replication origins in budding yeast (ARS, autonomously replicating sequence)

Assembly proteins (AP1, AP2). Clathrin coat constituents that regulate clathrin coat assembly

Aster. Radial array of dynamic microtubules emanating from the duplicated centrosomes at the poles of the mitotic spindle; has a role in orienting the spindle in the cell through interactions with the cell cortex

Astral microtubules. Microtubules that emanate from the poles of the mitotic spindle towards the cell cortex (ie, away from the central spindle)

Ataxia telangiectasia mutated (ATM). Large protein kinase that can initiate the response to DNA damage by activating the downstream kinases Chk1 and Chk2

ATF6 (atlastin). Dynamin-related GTPase that inserts into the ER membrane and promotes fusion of ER tubules

ATP. *See* adenosine triphosphate

ATP-gated channels. Family of cation channels gated by extracellular ATP; trimers of subunits with two transmembrane helices

ATP synthase. Reversible, rotary mitochondrial transmembrane protein that can either hydrolyze ATP to pump protons or use the passage of protons down a gradient across the membrane to synthesize ATP (*see* F-type ATPase)

ATR (ATM and Rad3 related). Protein kinase that responds to abnormal accumulation of single-stranded DNA by activating downstream kinases Chk1 and Chk2

Atresia. Programmed cell death and degeneration of activated oocytes that do not undergo full maturation to form eggs

A-type rotary ATPase. Archaeal rotary ATPase pumps

Augmin. Protein complex that promotes the nucleation of microtubules along the sides of other microtubules

Aurora A. Part of a network of mitotic kinases that regulate centriole division

Aurora B. Protein kinase in the chromosomal passenger complex that regulates chromosome attachment and cytokinesis during mitosis

Autolysosome. Compartment containing acid hydrolases for intracellular degradation formed by fusion of a nascent autophagic vacuole with a late endosome or lysosome

Autonomic nerve. Nerves of the sympathetic and parasympathetic peripheral nervous systems

Autonomously replicating sequences (ARS). Short (100 to 150 bp) DNA segments that act as replication origins in yeast

Autophagic vacuole. Compartment for intracellular degradation formed when a flattened membrane cisterna encloses a region of cytoplasm in a vesicle with two membranes

Autophagosome. Membrane compartment formed during autophagy in which cellular contents are enclosed and destined for degradation

Autophagy. Degradation of intracellular substrates in membrane-bounded compartments

Autosomes. Chromosomes that do not carry genes that define the sex of the individual

Auxin. Plant hormone that activates gene transcription by triggering the degradation of transcriptional inhibitors; used in a synthetic degron system for conditional destruction of proteins in animal cells

Axial element. Protein scaffold along the axis of paired sister chromatids in the synaptonemal complex (also known as lateral elements)

Axon. Process of a neuron capable of propagating an action potential and forming synapses with other neurons or muscles

Axoneme. Microtubule-based structural framework of eukaryotic cilia and flagella

Bacteriorhodopsin. Light-driven proton pump with seven transmembrane helices from the plasma membrane of a halophilic bacterium

Bag6 complex. Cytoplasmic protein complex involved with degradation of unfolded proteins transported out of the ER lumen

Barbed end. Fast-growing end of an actin filament

BAR domain protein. Dimeric, alpha-helical peripheral membrane proteins that induce or stabilize curved membranes

Barr body. Inactivated X chromosome forming a discrete patch of heterochromatin at the nuclear periphery

Basal body (axonemal). Cylindrical microtubule organizing center composed of nine triplet microtubules located at the base of cilia and flagella

Basal body (bacterial). A transmembrane complex of proteins forming the rotary motor for bacterial flagella

Basal lamina. A thin, planar specialization of extracellular matrix beneath epithelia and around muscle cells and peripheral nerve cells

Base excision repair. Process that replaces oxidized, reduced, alkylated, or deaminated DNA bases

Basolateral plasma membrane. Domain of the cell membrane of epithelial cells facing neighboring cells and the basal lamina, separated from the apical domain by a ring of tight junctions

Bax and Bak. Bcl-2 family proteins that activate the intrinsic pathway of apoptosis by inserting into mitochondrial outer membranes and releasing pro-apoptotic factors

Bcl-2 proteins. Three subfamilies of proteins with BH domains that regulate the release of death-promoting factors from mitochondria: Bcl-2 *protectors* inhibit apoptosis; Bcl-2 *killers* promote apoptosis; Bcl-2 *regulators* interfere with protectors or activate killers

Beige fat cells. Cells that store fat and dissipate energy as heat

Bestrophin. Family of pentameric chloride channels; mutations cause retinal degeneration

Beta-catenin. Adapter protein between the cytoplasmic domain of cadherins, α-catenin, and actin filaments; also a transcription factor regulated by Wnt signaling pathways

Beta-sheet. Common element of protein secondary structure consisting of parallel or antiparallel strands of polypeptides linked by backbone hydrogen bonds

BH (Bcl-2 homology) domains. Short blocks of conserved protein sequence characteristic of Bcl-2 family members

Bilayer. Planar assembly of lipids with hydrophobic fatty acid chains inside and hydrophilic head groups on both surfaces

BiP. HSP70 family chaperone protein that promotes folding of proteins in the ER lumen

Bipolar attachment. Attachment of the kinetochores of two sister chromatids to microtubules emanating from opposite poles of the mitotic spindle

BIR (baculovirus IAP repeat). Type of Zn^{2+} finger found in IAP proteins that promotes interactions with other proteins

Bivalents. Paired homologous chromosomes consisting of four sister chromatids held together by chiasmata during meiosis I

B-lymphocyte. Antibody producing cells

Bone morphogenetic proteins (BMP). Growth factors related to TGF-β

Branch point. Site of attachment of the 5′ end of an intron to an adenosine in the lariat intermediate during RNA splicing

BRdU. Uridine with a bromine linked to the 5-position of the pyrimidine ring is recognised by DNA polymerases as thymidine and can be detected in DNA using antibodies

Brefeldin A (BFA). Fungal metabolite used experimentally to disassemble the Golgi apparatus by preventing activation of Arf1 by GTP binding

Bright field. Light microscopic imaging system without optical elements to vary the phase or polarity of the light

Bromodomain. Protein motif that binds acetylated N-terminal histone tails

Brown fat. Fat cells with numerous mitochondria specialized for heat production

Budding uninhibited by benzimidazole (BUB). Conserved genes encoding proteins essential for the spindle checkpoint

Ca-ATPase pump. P-type membrane pump that uses ATP hydrolysis to transport Ca^{2+} into the endoplasmic reticulum or out of the cell

CAD domain. Extracellular domains of the cadherin family of adhesion proteins

Cadherin. Adhesion proteins that typically bind to like cadherins on other cells

Caenorhabditis elegans. Small nematode worm; genetic model organism favored by developmental biologists

Cajal bodies (coiled bodies). Nuclear structures that accumulate many factors involved in mRNA processing and within which specific nucleotides in snRNAs are modified

Calcineurin. *See* PP2B

Calcium-sensitive dye. Dye that changes its fluorescence on Ca^{2+} binding

Caldesmon. Protein associated with thin filaments in smooth muscle

Calmodulin. Small calcium ion binding protein that activates numerous effector proteins

Calmodulin-regulated spectrin associated proteins (CAMSAP). Proteins that associate with growing minus ends of microtubules

Calnexin. Sugar-binding, lectin-like protein in ER lumen

Calnexin cycle. Cycle of modifications that help glycoproteins fold in the ER

Calreticulin. Protein in the ER lumen that binds Ca^{2+} and acts as a chaperone for protein folding

Calsequestrin. Ca^{2+} binding protein in the ER lumen of striated muscle

cAMP. *See* Cyclic adenosine monophosphate

cAMP-gated channel. Family of cation ion channels activated by cAMP binding to a cytoplasmic domain

cAMP response element. Conserved DNA sequence that binds CREB, the mediator of the transcriptional response to cAMP

Cap recognition complex. Proteins that recognize 5′ caps on mRNAs, targeting them for export from the nucleus and to preinitiation complexes on ribosomes

Capping protein. Heterodimeric protein that caps the barbed ends of actin filaments

Cap Z. Isoform of heterodimeric capping protein that binds barbed ends of thin filaments in the Z-disk of striated muscles

Carbohydrate. Sugar molecules with the chemical composition $(CH_2O)_n$

Carbonyl oxygen. The oxygen atom on the carbon atom of peptide bonds

Carboxyl-terminal domain (CTD). Region of RNA polymerase II that participates in initiation and coordinates RNA splicing reactions

Carboxyl terminus. The end of a polypeptide with a free carboxyl group

Cardiac muscle. The striated muscle of the heart

Cardiomyopathy. Genetic diseases of heart muscle leading to heart failure or abnormal rhythms

Cargo selection. Mechanism for cargo sorting into membrane-bound transport carriers

Carrier vesicles. Tubules or larger membrane-enclosed structures that mediate transport among intracellular compartments

Cartilage. Specialized connective tissue consisting largely of collagen fibrils, proteoglycans, and water found in joints, respiratory tract, and developing bones

CAS. Nuclear export receptor that works with RanGTP to displace cargo from importin α

Caspase-activated DNase (CAD). Endonuclease activated during apoptotic execution to cleave the nuclear DNA into fragments of about 200 base pairs

Caspase recruitment domain (CARD). Protein interaction domain found in cell death adapter proteins and certain caspases

Caspases (cysteine aspartases). Proteases with an active site cysteine that cleave at aspartate residues and whose activation triggers apoptotic cell death

Catastrophe. Random change of state whereby an end of a microtubule stops growing and rapidly depolymerizes

Cathepsin K. Proteolytic enzyme secreted by osteoclasts to digest organic components of bone

Caveolae. Small (~50 nm) flask-shaped invaginations of plasma membrane enriched in caveolin, cavin, cholesterol, and signaling molecules

Caveolin. Major protein component of caveolae

Cavin. Alpha-helical protein that coats the cytoplasmic surface of caveolae

CD4. Ig-CAMs on helper T-lymphocytes that bind constant regions of class I major histocompatibility complexes; also an HIV receptor

CD8. Ig-CAMs on cytotoxic T-lymphocytes that bind constant regions of class II major histocompatibility complexes

CD45 (RPTPc phosphatase). Abundant transmembrane protein tyrosine phosphatase on while blood cells

Cdc20. Target of the spindle checkpoint; thought to be a substrate recognition factor for the APC/C

Cdc25. Three protein phosphatases that remove inhibitory phosphates from T^{14} and Y^{15} of Cdk-cyclin complexes, thereby triggering kinase activation

Cdc42. Small Rho family GTPase that regulates actin assembly

CDC mutants (cell division cycle mutants). Mutations in genes essential for cell cycle progression that cause yeast to accumulate at a single point in the cell cycle

Cdc45p. Protein recruited to active origins of replication to activate Mcm proteins, promote RPA binding, and recruit DNA polymerase

Cdk. *See* Cyclin-dependent kinase

Cdk-activating kinase (CAK). Complex of Cdk7 and cyclin H that phosphorylates Cdk1 on T^{161}, the final step in its activation

Cdk1–cyclin B. Cell cycle kinase with critical roles in the G_2/M transition and mitosis

Cdk2–cyclin A. Cell cycle kinase with critical roles in S phase and G_2/M transition

cDNA. A DNA copy of a messenger RNA

Ced (cell death abnormal) mutants. Mutations in *C. elegans* genes that affect programmed cell death

Cell cortex. Region of cytoplasm beneath the plasma membrane, typically rich in actin filaments

Cellulose. Long, unbranched polymer of glucose in plant cell walls; most abundant biopolymer on earth

Cellulose synthases. Plasma membrane enzymes that synthesize cellulose

Cell wall. Extracellular matrix of plants and fungi

CENP-A. Histone H3 variant, an integral component of kinetochores in species ranging from yeast to man

CENP-B. Protein that binds a 17-bp sequence in α-satellite DNA and establishes an epigenetic state favoring kinetochore assembly on α-satellite DNA arrays

CENP-C. Protein that appears to bridge between the inner and outer kinetochore

CEN sequences. Short DNA sequences in budding yeast chromosomes that specify protein-binding sites for assembly of kinetochores

Central pair. Two microtubules located in the middle of nine outer doublet microtubules in axonemes

Central spindle. Antiparallel bundle of microtubules that appears late in anaphase and has an important role during cytokinesis

Centralspindlin complex. Protein complex that recruits RhoGEF ect2 to help to establish the site where the cleavage furrow will form during telophase

Centrifuge. Machine that spins a sample holder to generate force to sediment particles in liquid samples

Centrin. Family of EF-hand, Ca^{2+} binding proteins similar to calmodulin that are essential for the biogenesis of centrioles (and spindle pole bodies of yeast)

Centrioles. Barrel-shaped structures composed of nine microtubule triplets that organize centrosomes (*see* Basal body)

Centromere. Chromosomal locus, defined by specific DNA sequences and associated proteins that regulates chromosomal movements during mitosis and meiosis

Centrosome. Pair of centrioles and surrounding matrix containing proteins including γ-tubulin that nucleate microtubules and serve as the microtubule-organizing center in most animal cells

Ceramide. Backbone of all sphingolipids that is converted to glucosylceramide and sphingomyelin in the Golgi apparatus

Ceramide transport protein (CERT). Removes newly synthesized ceramide from the ER for transport to the Golgi apparatus

cFLIP (FLICE-like inhibitory protein). Co-assembles with procaspase 8 monomers creating an enzyme with altered activity that does not trigger cell death; FLICE is another name for caspase 8

cGMP. *See* Cyclic guanosine monophosphate

Channelrhodopsin. Light-sensitive cation channel protein of green algae used in optogenetics

Chaperone. Protein that assists the folding of other proteins

Chaperone-mediated autophagy (CMA). Quality control mechanism to eliminate soluble cytoplasmic proteins that are incorrectly folded or assembled

Charcot-Marie-Tooth disease. Degeneration of peripheral motor and sensory nerves caused by a variety of mutations

Checkpoint. Biochemical circuit that, when active, blocks progression through the cell cycle, either temporarily or, in some cases, permanently

Checkpoint kinases 1/2 (Chk1/Chk2). Protein kinases activated by ATM and ATR that phosphorylate Cdc25A protein phosphatase and other substrates, thereby blocking cell cycle progression

Chemiosmotic cycle. A series of reactions across a lipid bilayer that couples the creation of an ion gradient by an energy-consuming transmembrane pump to energy-requiring transport or ATP synthesis by a second transmembrane protein

Chemotaxis. Process by which a cell moves up a concentration gradient of a chemical attractant

Chiasmata. Chromatin structures at sites where recombination has been completed; keep homologous chromosomes paired until anaphase of meiosis I

ChIP-seq. High throughput technique in which classes of chromatin are purified using antibodies recognizing histone MARKS or other chromatin proteins followed by sequencing the DNA

Chloride channel. Transmembrane ion channel selective for chloride ions

Chlorophyll. Organic molecule that absorbs photons and uses the energy to boost electrons to an excited state

Chloroplast. Eukaryotic organelle derived from a symbiotic cyanobacterium specialized for photosynthesis

Chloroplast stroma. Compartment inside the inner chloroplast membrane devoted to synthesis of three-carbon sugar phosphates, chloroplast proteins, and fatty acids

Cholesterol. Polycyclic lipid with one polar atom found in biological membranes; also the precursor for steroid hormones and bile acids

Chondrocyte. Cell that synthesizes and secretes the extracellular matrix of cartilage

Chromatid. A single chromosomal DNA molecule plus its attendant proteins

Chromatin. DNA plus the proteins that package it within the cell nucleus

Chromatography. Method to separate chemicals based on interactions with an immobile matrix such as a gel or paper; with the matrix in a tubular column or on a plate

Chromodomain (chromatin modification organizer). Motif of 50 amino acids that binds to histone H3 trimethylated on lysine

Chromokinesin. Kinesin that binds to chromatin and chromosomes; may have roles that do not involve movement along microtubules

Chromonema fiber. A chromatin fiber, 100 to 300 nm in diameter, thought to be an element of higher-order packing of chromatin within chromosomes

Chromosomal passenger complex. A complex of Aurora B kinase, inner centromere protein (INCENP), survivin, and borealin that is required to correct chromosome attachment errors, for the spindle checkpoint, and to complete cytokinesis

Chromosome. DNA molecule with its attendant proteins that behaves as an independent unit during mitosis and meiosis

Chromosome cycle. Replication and partitioning of chromosomes into two daughter cells

Chromosome scaffold. Biochemical fraction of non-histone proteins thought to play a role in organizing chromosome structure

Chromosome territories. Discrete regions of the interphase nucleus occupied by particular chromosomes

Ciliopathy. Diverse diseases caused by mutations in genes for ciliary proteins

Cilium. Cell surface organelle of eukaryotes based on a microtubule axoneme, usually capable of generating waves or other motions but sometimes immotile sensory structures

Citric acid cycle. Biochemical reactions in the mitochondrial matrix that derives energy by breaking down acetyl-CoA

Clathrin. Protein that forms a three-legged triskelion and a lattice on the cytoplasmic surface of membranes during the formation of buds

Clathrin-coated pit. Invaginated patch of membrane formed by a lattice of clathrin triskelions and adapter molecules on the cytoplasmic surface

Clathrin-mediated endocytosis. Selective uptake of ligands bound to receptors that concentrate in clathrin-coated pits

Claudins. Transmembrane proteins that link plasma membranes together at tight junctions

ClC chloride channel. Family of channels with 18 transmembrane helices selective for chloride

Cleavage furrow. Constriction of the plasma membrane that pinches a cell in two during cytokinesis as a result of action of a contractile ring of actin filaments and myosin-II

Cleavage stimulus. Signal emitted by the mitotic spindle that specifies the position of the cleavage furrow midway between the poles and perpendicular to the long axis of the spindle

CLIP-170. Protein that concentrates on plus ends of growing microtubules; involved in transport of membranes and behavior of microtubules at kinetochores

Clonal expansion. The process that produces a clone of identical lymphocytes arising from a single precursor cell

Closed mitosis. Form of mitosis in single-celled eukaryotes, including yeast and slime molds in which the mitotic spindle forms and chromosomes segregate within an intact nuclear envelope to which the spindle poles are anchored

CMG (Cdc45, MCM2-7, GINS) helicase. Helicase that separates DNA strands for replication

c-Mos. A MAP kinase kinase kinase and component in the cytostatic factor (CSF) pathway, which holds mature eggs in meiotic metaphase until they are fertilized by a sperm

Coactivator. Protein complex that facilitates loading of the transcriptional apparatus onto a gene, often by modifying N-terminal histone tails to "open" the chromatin

Coatomer complex. Proteins forming the COPI coat on vesicles budding from the ER

Codon. Three successive nucleic acid bases in mRNA that specify the position of a particular amino acid in a polypeptide during synthesis on a ribosome

Cohesin. Complex of four proteins that holds sister chromatids together from their replication during the S phase until their separation at the onset of anaphase

Coiled-coil. Left-handed helix of two α-helical polypeptides; either parallel or antiparallel

Colchicine. Drug isolated from the autumn crocus that inhibits microtubule assembly by binding dissociated tubulin dimers

Collagen. Chief fibrous protein of connective tissues, composed of three rod-shaped polypeptides, each folded in type II polyproline helix; many isoforms specialized for cartilage, basal lamina and other connective tissues

Committed progenitor cells. Stem cells with a limited proliferation capacity that can give rise to only specific subsets of differentiated cells

Complex I (NADH:ubiquinone oxidoreductase). Complex of proteins in mitochondrial inner membranes and bacterial plasma membranes, which takes electrons from NADH and transfers protons out of the mitochondrial matrix and bacterial cytoplasm

Complex II (succinate:ubiquinone reductase). A transmembrane enzyme complex from mitochondria and Bacteria that takes part in the citric acid cycle, by coupling oxidation of succinate to reduction of flavin adenine dinucleotide (FAD) to $FADH_2$

Complex III (cytochrome bc_1). Transmembrane protein complex from mitochondria and Bacteria that couples oxidation and reduction of ubiquinone to the transfer of protons out of the matrix

Complex IV (cytochrome oxidase). Transmembrane protein complex of mitochondria and Bacteria that takes electrons from four cytochrome c molecules to reduce molecular oxygen to two waters, as well as to pump four protons out of the mitochondrial matrix or bacterial cytoplasm

Condenser. Lenses in microscopes that focus the illuminating beam on the specimen

Condensin I and II. Two pentameric protein complexes with an essential role in chromosome architecture

Conditional mutation. A mutation that gives an abnormal phenotype only under certain conditions such as high temperature

Confocal microscope. Imaging system using pinholes to restrict the illumination to a thin plane in the specimen

Conformational change. Change in the shape of a macromolecule

Connexin. Protein subunit of gap junction channels

Connexon. Hexamer of connexin subunits making up gap junction channels connecting the cytoplasm of adjacent cells

Conserved oligomeric Golgi (COG). A multi-subunit tethering protein for Golgi apparatus trafficking

Constitutive heterochromatin. Inactive form of chromatin that remains condensed throughout the cell cycle owing to the presence of special proteins and modifications of the histone proteins

Constitutive secretion. Exocytosis without special stimuli being received by the cell

Contact inhibition. Arrest of cell movements and cell-cycle progression in G1 when cells growing in culture contact other cells, mediated by interactions of cadherins

Contractile ring. Band of actin filaments, myosin-II, and other proteins attached to the plasma membrane around the cortex midway between spindle poles that pinches daughter cells in two like a purse string during cytokinesis

Convergent evolution. The process that can produce genes encoding proteins with similar structures starting from unrelated ancestral genes

COPI coat complex. Assembly of Arf1 GTPase, coatomer, and a GAP on the cytoplasmic face of VTCs and Golgi membranes to mediate protein sorting, budding, and retrograde transport back to the ER

COPII coat complex. Assembly of Sar1p GTPase, Sec23p•Sec24p, and Sec13p•Sec31p on the cytoplasmic face of the ER to mediate sorting and trafficking of secretory cargo out of the ER

Core histones. Histones H2A, H2B, H3, and H4, which form the disk-like octameric core of nucleosomes

Core mannose oligosaccharide. Branched oligosaccharide rich in mannose that is transferred from dolichol phosphate to the side chain of an asparagine of a newly synthesized protein in the ER lumen

Core protein. Protein modified with glycosaminoglycans to make a proteoglycan

Cotranslational translocation. Movement of a protein across the ER membrane concurrent with its synthesis by a membrane-bound ribosome

Covariance method. Prediction of secondary structure in RNAs from comparisons of sequences in divergent organisms; bases that hydrogen bond usually vary together

CpG islands. Regions of DNA rich in CpG found in and around gene promoters

Crinophagy. Fusion of lysosomes directly with secretory vesicles resulting in the degradation of secretory proteins

CRISPR/Cas9 system. Method using bacterial proteins and a guide RNA to direct the Cas9 nuclease to make a double-strand break in a specific genomic DNA sequence

Cristae. Folds of the inner mitochondrial membrane

Critical concentration. The concentration of unpolymerized subunits giving equal rates assembly and disassembly at an end of a polymer

Crossbridge. Force-producing connection of a motor protein between its cytoskeletal track and its cargo

Crossover. Physical breakage and reunion of DNA strands on two different chromosomes, typically producing a balanced exchange of DNA sequences

Crossover interference. Uncharacterized mechanism that locally limits the number of DNA breaks that are processed to form crossovers and chiasmata during meiosis

c-Src. Nonreceptor tyrosine kinase important in signaling

Cyanobacteria. Photosynthetic Bacteria (formerly called blue-green algae) with both types of photosystems as well as a manganese enzyme that splits water

Cyclic adenosine monophosphate (cAMP). Nucleotide with an adenine base and a cyclic phosphodiester bond linking the 3′ and 5′ hydroxyls; an important signaling second messenger

Cyclic-ADP-ribose. Derivative of NAD that sets the sensitivity of ryanodine receptor calcium release channels to the cytoplasmic Ca^{2+} concentration

Cyclic guanosine monophosphate (cGMP). Nucleotide with a guanine base and a cyclic phosphodiester bond linking the 3′ and 5′ hydroxyls

Cyclic nucleotide–dependent protein kinases. Kinases regulated by binding of cyclic nucleotides to part of the enzyme or to a separate regulatory subunit

Cyclic nucleotide–gated channels. Cation channels regulated by binding of cyclic nucleotides to cytoplasmic domains

Cyclin. Class of subunits required for activity of cyclin-dependent kinases that undergo cyclic patterns of accumulation and destruction during the cell cycle

Cyclin-dependent kinase. Class of serine/threonine kinases regulated by cyclins. *See* Cdk

Cyclin-dependent kinase inhibitor (CKI). Class of proteins that negatively regulate Cdks to block cell cycle progression

Cyclooxygenase. Enzymes that convert arachidonic acid into prostaglandin H2

Cyclosporine. Drug based on a natural product that inhibits calcineurin and the cellular immune response

Cystic fibrosis transmembrane regulator (CFTR). An ABC transporter that acts as a chloride channel; mutated in cystic fibrosis

Cytochalasin. Fungal product used experimentally to depolymerize actin filaments in cells

Cytochrome c. Small heme protein, part of electron transport pathway of oxidative phosphorylation and, when released from mitochondria, a trigger for apoptosis

Cytochrome P450. Enzymes of the smooth ER that detoxify endogenous steroids, carcinogenic compounds, and lipid-soluble molecules from the environment

Cytokine. Diverse family of protein hormones and growth factors

Cytokine receptor. Transmembrane receptors for cytokines linked to JAK kinases in the cell

Cytokinesis. Division of the cytoplasm into two daughter cells at the end of mitosis

Cytoplasmic streaming. Bulk movement of organelles and cytoplasm

Cytosine. Pyrimidine base present in DNA and RNA; H-bonds with guanine

Cytoskeleton. The ensemble of protein polymers (including actin filaments, intermediate filaments, and microtubules) forming the mechanical scaffold for the cytoplasm

Cytotoxic T lymphocytes (killer T cells). Lymphocytes with receptors for specific antigens displayed by major histocompatibility complexes on target cells that they then kill

Cytostatic factor (CSF). A biochemical activity, including the APC/C inhibitor Emi2, that arrests vertebrate oocytes in metaphase II of meiosis until they are fertilized

Dam1 complex. Ring-like complex of proteins at the kinetochore of budding yeast that links the kinetochore to disassembling microtubules

Dark reactions. Biochemical reactions in chloroplasts that convert carbon dioxide into three-carbon sugar phosphates

Deadenylases. Enzymes that catalyze the stepwise removal of the poly(A) tail of mRNAs, signaling their degradation by removing binding sites for the poly(A)-binding protein (PABP), which when present inhibits cap removal at the other end of the RNA

Death domain (DD). Protein interaction domain in proteins from cell death signaling pathways, including Fas cell death receptors and FADD adapter molecules

Death effector domain (DED). Protein interaction domain in adapters such as FADD, the prodomains of caspases 8 and 10, and in certain inhibitors of apoptosis

Death-inducing signaling complex (DISC). Signaling complex formed on the intracellular domains of trimerised CD95 during the extrinsic cell death pathway to promote activation of caspase-8

Debranching enzyme. Linearizes the lariat mRNA splicing intermediate for degradation by exonucleases

Decapping complex. Removes the 5′ cap from mRNAs, triggering their rapid degradation

Deconvolution. Computational method to remove out of focus fluorescent light from a microscopic image

Decoy receptors. Receptors for surface proteins such as FAS/CD95 that lack transmembrane domains and are shed from tumor cells to protect them against killing by cytotoxic immune cells

Degron. Peptide sequence that targets the protein for ubiquitylation and degradation by proteasomes

Delta. Cell surface protein that stimulates Notch receptors on other cells; influence many developmental processes

Dendrite. Nerve cell process specialized for receiving synapses from other neurons

Dense body. Attachment sites for actin filaments and intermediate filaments in the cytoplasm of smooth muscle cells

Dense fibrillar component. Regions of nucleoli surrounding fibrillar centers

Dephosphorylation. Reaction that removes a phosphate from a protein side chain

Desmin. Intermediate filament isoform expressed by muscle cells

Desmocollin. Types of cadherins found in desmosomes

Desmoglein. Types of cadherins found in desmosomes

Desmoplakin. Protein link between desmosomal cadherins and intermediate filaments

Desmosome. Intercellular junction mediated by cadherins and anchored to cytoplasmic intermediate filaments

De-ubiquitylating enzyme (DUB). Enzymes that remove ubiquitin from target proteins

Diacylglycerol (DAG). Diglyceride with two fatty acids and no head group; activates PKC isoforms

Diakinesis. Prometaphase of meiosis I

Dicer. Double-strand-specific RNA endonuclease in the RNAi pathway that generates short RNA duplexes that are incorporated into the RISC complex

Dictyostelium discoideum. A cellular slime mold; model organism for studying chemotaxis, motility, and differentiation

Differential interference contrast (DIC). Light microscopy optics generating contrast from local differences in refractive index

Dihydropyridine (DHP) receptor. Voltage-gated calcium channels that couple plasma membrane depolarization to the release of calcium from internal stores in striated muscles

Dileucine-based sorting motifs. A sequence that directs proteins into clathrin-coated pits on the plasma membrane

Dilysine sorting motif. Sequence that directs proteins from the Golgi apparatus to the ER

Diploid chromosome number (2*n*). Total number of chromosomes in a diploid organism, comprising pairs of homologous chromosomes, one donated by the mother and the other by the father

Diplotene. Fourth stage of meiotic prophase with decondensed chromosomes held together by chiasmata (can last for decades in female humans)

Disjunction (disjoining). Normal separation of chromosomes or chromatids in meiosis or mitosis

Disks. Membrane compartments in photoreceptor cells rich in rhodopsin

Dis1/TOG family (called XMAP215 in frogs). Proteins that associate with microtubule plus ends to regulate microtubule assembly and dynamics; required for organization of mitotic spindle poles in animals and cortical arrays of microtubules in plants

Disulfide bond. S-S bond formed by oxidation between two cysteine residues

Dmc1. Together with Rad51, drives a search of 3′ single-stranded DNA tails for complementary DNA sequences of the other chromosomes during recombination

DNA. Polymer of phosphate-linked sugars (deoxyribose) linked to purine and pyrimidine bases that constitutes the genetic information for most organisms

DNA damage checkpoints. Biochemical pathways that detect damaged DNA and then either block cell cycle progression or trigger cell death by apoptosis

DNA damage response (DDR). Complex network that recognizes DNA damage and activates the cell and repair mechanisms

DNA-dependent protein kinase. Key factor in DNA repair by non-homologous end joining

DNA polymerases δ and ε. Enzymes that use PCNA to help them process along the DNA, synthesizing DNA continuously on the leading strand. On the lagging strand, they synthesize Okazaki fragments of about 250 bp

DNA replication. Synthesis of two complementary strands from a DNA double helix; duplication of the genome

DNA replication checkpoint. Biochemical mechanism that detects unreplicated DNA or stalled DNA replication forks, stabilizing the latter so that they can be repaired

DNA topoisomerase IIα. An enzyme found in the mitotic chromosome scaffold that alters DNA topology by passing one double-helix strand through another

Dolichol phosphate. Long-chained, unsaturated isoprenoid alcohol with pyrophosphate at one end that is the substrate for the synthesis of oligosaccharide precursors in the cytoplasm and subsequent transfer to asparagines of proteins in the ER lumen

Dominant negative mutation. Mutation giving rise to a deleterious phenotype even in the presence of a wild-type allele

Double-strand break repair. Processes that repair double-strand breaks in DNA either without a template (nonhomologous end-joining) or using undamaged DNA as a template (homologous recombinational repair) for accurate repair

Down syndrome. Common human aneuploidy with three copies of chromosome 21

DP1/YOP1. Protein that mediates formation of tubules by ER membranes

Drosha. Nuclear double-strand-specific endonuclease that releases individual pre-miRNAs from their precursors

Drosophila melanogaster. Fruit fly, genetic model organism popular for studying development

Dynactin complex. Protein complex linking dynein to membrane cargo

Dynamic instability. Behavior of microtubules with growing and shrinking microtubules coexisting at steady state

Dynamin. GTPase that coordinates the internalization of clathrin-coated vesicles

Dynein. Motor proteins that use ATP hydrolysis to move toward the minus ends of microtubules, members of AAA ATPase family

Dystroglycan/sarcoglycan complex. Transmembrane complex that stabilizes muscle plasma membranes through interactions with the cytoskeleton and the basal lamina

Dystrophin. Giant protein that links the dystroglycan/sarcoglycan complex to cytoplasmic actin filaments; mutations cause the most common form of muscular dystrophy

Early recombination nodules. Sites along chromosomes where DNA strand breaks have occurred and recombination is initiated early in meiosis

EB1. Protein that binds growing microtubule plus ends and associates with APC

E-cadherin. Isoform of cadherin adhesion protein expressed by epithelial cells

EDEM. *See* ER degradation-enhancing α-mannosidase-like protein

EDITOR. Enzymes that either place or remove a MARK on chromatin

EEA1. Tethering factor in the early endosomal membranes

E1 enzyme (ubiquitin-activating enzyme). Activates the small protein ubiquitin by forming a thioester bond between the C-terminus of ubiquitin and a cysteine on the enzyme

E2 enzyme (ubiquitin-conjugating enzyme). Either transfers ubiquitin directly to the ε amino group of a lysine of a target protein or combines with a third component (an E3 or ubiquitin-protein ligase) to do so

E3 enzyme (ubiquitin-protein ligase). Facilitates the transfer of ubiquitin from an E2 enzyme to the ε amino group of a lysine of a target protein

E2F. Family of 10 transcription factors in mammals that regulate genes promoting cell cycle progression at the restriction point and can trigger cell death by apoptosis

Effector caspases. "Downstream" caspases activated through cleavage by initiator caspases and responsible for most intracellular proteolysis during apoptosis

Ehlers-Danlos syndrome. Human genetic disease with thin skin and lax joints owing to mutations in the genes for fibrillar collagens type III or type IV

Eicosanoids. Diverse class of lipid second messengers derived from arachidonic acid

Elastin. Protein subunit of elastic fibers

Electrical potential. Voltage difference across a membrane

Electrical synapse. Site of rapid transmission of action potentials between neurons through gap junctions

Electron cryomicroscopy. Transmission electron microscopy of frozen specimens

Electron transport pathway. Sequence of reactions in the inner mitochondrial membrane and bacterial plasma membrane that uses energy from the passage of electrons to generate a proton gradient across the membrane to power a chemiosmotic cycle to synthesize ATP

Electrostatic interaction. Attraction of oppositely charged atoms

Elongation factors. Proteins that facilitate the synthesis of polypeptides on ribosomes

Elongation factor Tu (eEF-1 in animals). GTPase that delivers tRNAs charged with amino acids to ribosomes and is the timer for the proofreading reaction

ELYS protein. Nuclear pore protein whose binding to chromatin during telophase is one of the first steps in nuclear envelope reassembly in metazoans

Embryonic stem cells. Precursors of the entire embryo, produced by the first embryonic cell divisions

Emi2. An inhibitor of the APC/C and a key component of cytostatic factor

Endocannabinoids. Endogenous fatty acid amides that stimulate Δ^9-tetrahydrocannabinol receptors

Endocytosis. Process by which extracellular materials are captured and enclosed within membrane-bound carriers that invaginate and pinch off into the cytoplasm from the plasma membrane

Endoplasmic reticulum (ER). Large, membrane-delineated, intracellular compartment that collects proteins synthesized in the cytoplasm for modification and delivery into the secretory pathway

Endosome. A membrane-bounded compartment for processing materials taken in by endocytosis

Enhancer. Complex cluster of transcriptional regulator binding sites on DNA that increases the rate of initiation from a basal promoter; can function even if located up to 10 kb upstream or downstream from the promoter or in either orientation relative to it

Enthalpy. The internal energy of a system plus the product of its volume times its pressure; the pressure-volume term rarely applies to biological systems, so the internal energy corresponds to the heat contained in the chemical bonds

Entropy. Measure of the disorder in a system

Eosinophil. A type of white blood cell with large granules that stain with eosin; active against parasites

Epiblast. Embryonic cells produced by divisions prior to implantation in the uterine wall that form the embryo

Epidermal growth factor. Protein hormone that activates a receptor tyrosine kinase and promotes growth of epithelial cells

Epidermolysis bullosa. Genetic disease with mutations in keratins resulting in fragility and blistering of skin

Epifluorescence. Fluorescence microscopy method using the objective as both the condenser to focus the exciting light on the specimen and to image fluorescence emitted from the specimen

Epigenetic trait. Inheritable property of chromosomes carried by enzymatic modification of DNA or proteins associated with DNA rather than being encoded in the nucleotide sequence

Epiphyseal plate. Disk of cartilage whose expansion is responsible for the growth of long bones

Epithelial sodium channel. Channel protein formed by four subunits with two transmembrane segments

Equilibrium constant. Thermodynamic parameter describing the extent of a reaction at equilibrium; related to the change in free energy and to the ratio of the rate constants for the forward and reverse reactions

ERAD. *See* ER-associated protein degradation

ER-associated protein degradation (ERAD). Process to move misfolded proteins out of the ER into the cytoplasm for degradation by proteasomes

ER degradation-enhancing α-mannosidase-like protein (EDEM). Protein in the ER lumen that directs unfolded proteins into the cytoplasm for degradation

ER export domain. Site of secretory cargo export from ER

Erythropoietin. Cytokine driving red blood cell production, produced by the kidney

Escherichia coli. Gram-negative colon bacterium; popular genetic model organism

ESCRT III. Protein complex that pinches membranes apart in multivesicular body formation and in the final abscission stage of cytokinesis

ESCRT complex. Protein complex that sorts ubiquitinated receptors in early endosomes and drives them into the membrane invaginations of multivesicular bodies

E3 ubiquitin ligase. *See* E3 enzyme

Euchromatin. Transcriptionally active or potentially active chromatin, containing most of the genes

Eukaryote. Organism with the genome packaged in a nucleus

Excitable membrane. Plasma membrane with voltage-gated Na- and K-channels that propagates action potentials

Excited state. High-energy state of an electron achieved by absorption of a photon or a chemical reaction, used to generate proton gradients across membranes in photosynthesis and oxidative phosphorylation or to emit a photon during fluorescence

Execution phase. Final phase of apoptosis in which activated caspases stimulate the disassembly of the cell

Exocyst. A multi-subunit tethering protein on the plasma membrane for secretory vesicles

Exome. The portion of the genome that encodes proteins

Exonic splicing enhancers (ESEs). RNA sequences that bind SR-proteins and stimulate the use of the flanking 5′ and 3′ splice sites, promoting inclusion of the exon in the mRNA

Exon-junction complex (EJC). Protein complex deposited on mRNA during splicing in the nucleus and retained following export to the cytoplasm, where it marks mRNAs on which translation terminated prematurely

Exons. Regions of genes that appear in mature RNAs

Exosome (endocytosis). Vesicles of invaginated membrane inside multivesicular bodies that are released from cells when MVBs fuse with the plasma membrane

Exosome (RNA processing). Complex of multiple different 3′ to 5′ exonucleases that degrade nuclear RNAs and turnover mRNAs in the cytoplasm

Expansins. Plant proteins that break noncovalent links between cellulose polymers transiently, allowing turgor pressure to expand the volume of the cell

Expressed sequence tag (EST). DNA sequences collected by sequencing random cDNAs

Extracellular matrix. Fibrous proteins, polysaccharides, proteoglycans, and adhesive glycoproteins providing the mechanical support for tissues in animals and plants

Extrinsic pathway of cell death. Process initiated when cell surface ligands activate receptors on target cells, triggering the apoptotic pathway in the target cells

Facultative heterochromatin. DNA sequences located in heterochromatin in some cells and in euchromatin in others; X chromosome inactivation is a classic example of facultative heterochromatin in mammals.

FADD (Fas-associated protein with a death domain). Adapter protein that promotes association of activated Fas receptor with procaspase 8, triggering the extrinsic cell death pathway

Farnesyl. A15 carbon isoprenyl group used to anchor peripheral proteins to membrane bilayers; conjugated to a cysteine side chain near the C-terminus of the protein

Fas (also known as Apo1 or CD95). A death receptor of the tumor necrosis factor (TNF) receptor family with a death domain (DD) in the cytoplasmic tail

Fas ligand. Trimeric cell surface protein that initiates the extrinsic pathway of apoptotic death when it binds Fas on a target cell

Fast axonal transport. Bidirectional transport of membrane-bound vesicles and organelles on microtubules in nerve axons; anterograde movements powered by kinesin; retrograde movements powered by dynein

Fatty acid oxidation. Biochemical reactions that produce acetyl-CoA from the breakdown of fatty acids

Fen1 (flap nuclease). Nuclease that removes the RNA primer (and probably initiator DNA) during DNA replication

Fibrillar centers. Regions of nucleoli containing rRNA genes, RNA polymerase I, and associated transcription factors

Fibrillin microfibril. A polymer of the protein fibrillin, the scaffold for laying down elastin in elastic fibers

Fibrin. Blood protein that polymerizes to clot blood

Fibrinogen. Blood protein precursor of fibrin and also an adhesion protein for platelets

Fibroblast. Cell that synthesizes most of the extracellular matrix in connective tissues

Fibroblast growth factor receptor. A receptor tyrosine kinase for fibroblast growth factor

Fibronectin. Adhesive glycoprotein that connects fibers and cells in the extracellular matrix; multiple isoforms, one circulating in blood forms a provisional matrix in wounds

Filamin. Large actin filament crosslinking protein with Ig-domains linking actin-binding domains

Filopodium (also microspike). Finger-like extension of plasma membrane supported by a bundle of actin filaments

First-order reaction. A reaction with one reactant

Flagellin. Protein subunit of bacterial flagella

Flagellum (bacterial). Helical protein polymer forming the propeller for bacterial motility

Flagellum (eukaryotic). Motile cell surface organelle powered by an axoneme; plural = flagella

Flippase. P-type adenosine triphosphatase (ATPase) pumps that move lipids from one side of a bilayer to the other

Floppase. ABC transporter pumps that move lipids from one side of a bilayer to the other

Flotillin. Protein that clusters proteins in caveoli

Fluid-phase endocytosis. Ingestion of extracellular fluid in a vesicle formed by the plasma membrane

Fluorescence. Emission of light from a molecule after excitation by a shorter-wavelength, more energetic photon

Fluorescence correlation spectroscopy (FCS). A microscopy method to measure diffusion coefficients of molecules as they pass through a narrow beam of light

Fluorescence recovery after photobleaching (FRAP). A fluorescent probe in bleached with intense light in part of a living cell followed by observation of the redistribution of the remaining fluorescence molecules into the bleached area by diffusion or transport

Fluorescence resonance energy transfer (FRET). A method to measure distances on a nanometer scale by the transfer of energy from the excited state of a fluorescent donor molecule to an acceptor fluorescent molecule

F-met-leu-phe. A chemotactic tripeptide released by Bacteria that attracts white blood cells

Focal adhesion kinase. A nonreceptor tyrosine kinase active in focal contacts

Focal contact. Plasma membrane specialization with integrins to adhere to the extracellular matrix; associated with cytoplasmic signaling proteins and actin filaments

Formin. Dimeric protein that nucleates actin filaments and remains attached to the growing barbed end during assembly of contractile rings, filopodia, and other bundles of actin filaments

Free energy. Thermodynamic energy in a system available to do work

Freeze fracture. A method to prepare specimens for electron microscopy involving freezing, fracturing, etching (sublimation) of water from the fractured surface, and rotary shadowing with metal

FtsA. Prokaryotic actin homolog used during cytokinesis

FtsZ. Protein with the same fold as tubulin that participates in cytokinesis in Bacteria and Archaea

F-type ATPase. Reversible, rotary membrane pumps that use transport of protons down a concentration gradient to drive the synthesis of ATP or use ATP hydrolysis to pump protons

Fusion protein (membrane traffic). Protein that mediates fusion of a carrier with an acceptor membrane

Fusion protein (molecular biology). Fusion of coding sequences from two proteins to produce a hybrid protein

G_0 phase. Cell cycle state of nondividing (often terminally differentiated) cells

G_1 phase (first gap phase). Interval in cell cycle between mitosis and DNA replication

G_2 checkpoint. Cell cycle checkpoint operating in the G_2 phase to block mitotic entry if DNA is damaged or DNA replication is incomplete

G_2 delay. Temporary halt in cell cycle progression in cells with damaged DNA as a result of function of the G_2 checkpoint

G_2 phase (second gap phase). Interval between the completion of DNA replication and mitosis

G_α. Membrane-anchored GTP-binding subunit of trimeric G-proteins

G_β. Subunit of trimeric G-proteins

G_γ. Membrane-anchored subunit of trimeric G-proteins

Gamma-tubulin. Tubulin isoform crucial for microtubule nucleation by ring complexes in pericentriolar material

Gamma-tubulin ring complex (gTuRC). Complex of 10 to 13 γ-tubulin molecules and 8 associated polypeptides that nucleates microtubule assembly

GAP. *See* GTPase activating protein

Gap junction. Intercellular junction composed of connexons, channels that conduct molecules of <1 kD between cells

Gating. The process that controls the transitions between the open and closed states of channels

GEF. *See* Guanine nucleotide exchange factors

Gel electrophoresis. Use of an electric field to separate molecules according to their size and charge in a gel matrix

Gel filtration chromatography. Method to separate molecules based on their size (hydrodynamic radius)

Gelsolin. Calcium-sensitive actin filament severing and capping protein

Geminin. Cell cycle–regulated protein that regulates "licensing" of origins of DNA replication in metazoans by binding Cdt1 and preventing preinitiation complex assembly

Gene. Segment of DNA encoding a functional RNA or protein

General transcription factors (GTF). Conserved protein complex that recruits RNA polymerase to promoters

Genetic code. The correspondence between nucleotide triplets in mRNAs to amino acids in a polypeptide; one to six different triplet codons encode each amino acid

Genetic marker. Particular DNA sequence that can be monitored by examining the phenotypes of the cells that carry it

Genome. The entire DNA complement of an organism

Genome editing. Modification of genomic DNA using molecular biological methods

Genomics. The study of genomes by cloning and sequencing the DNA and by analyzing and comparing the sequences

Genotype. Combination of genes present on the chromosomes of an organism

GET1, 2, 3. Cytoplasmic proteins that direct tail-anchored proteins to organelle membranes

GLUT4. The glucose carrier (uniporter) that mediates the response of muscle and fat cells to insulin

Glutamate. Amino acid and neurotransmitter

Glutamate receptor. Cation channel activated by binding glutamate

Glycerol. Three-carbon molecule with a hydroxyl group on each carbon

Glycerolphospholipid. *See* Phosphoglyceride

Glycoconjugate. Protein modified with one or more sugars

Glycogen synthase kinase (GSK). Protein serine/threonine kinases with numerous substrates.

Glycolipid. Lipid modified with one or more sugars

Glycolysis. Biochemical reactions that derive energy from the breakdown of glucose to form ATP and other energy-carrying metabolites

Glycophorin. Glycoprotein of red blood cell membranes with a single transmembrane helix

Glycoprotein. Protein modified with one or more sugars

Glycosaminoglycan. Polysaccharide polymers, generally composed of a repeated pair of sugars

Glycosidases. Enzymes that remove sugars from glycoproteins

Glycosidic bond. Ether bond between sugar residues

Glycosphingolipid. Sphingolipid modified with one or more sugars

Glycosylase. Enzyme that cuts damaged bases from the DNA sugar-phosphate backbone leaving an abasic site

Glycosylation. Process that conjugates sugars with proteins and lipids

Glycosylphosphatidylinositol tail. Phosphoglyceride that is linked to the C-terminus of a protein by a short oligosaccharide and a phosphatidylinositol head group

Glycosyltransferases. Enzymes that add sugar residues to proteins

Golgi apparatus. Major compartment of the secretory membrane system for processing glycoproteins and sorting molecules in the lumen and lipid bilayer

G-protein. GTPase subunit of trimeric GTPases

G-protein-coupled receptor kinase. Serine/threonine kinases activated by G-protein beta-subunits, which phosphorylate and inactivate seven helix receptors

Granular component. Region of the nucleolus for ribosome subunit assembly

Grb2. Adapter protein consisting of two SH3 domains and one SH2 domain

Greatwall (Gwl) kinase. Regulates protein phosphatase 2A, the phosphatase that removes many phosphate groups placed on target proteins by Cdks

Green fluorescent protein. Jellyfish protein that absorbs blue light and emits green light; often fused to other proteins for observing their distribution in live cells

GroEL. Barrel-shaped chaperonins that use ATP hydrolysis to assist the folding of nascent polypeptides

Growth cone. Motile tip of a growing nerve cell process

Growth cycle. Increase in cellular mass

Growth factors. Proteins that promote growth of cell mass

Growth hormone. Pituitary hormone that controls body size

GST fusion protein. Hybrid protein consisting of a protein of interest fused to glutathione-S-transferase, which allows purification by binding to glutathione immobilized on beads

GTPase. Family of proteins activated by GTP binding (allowing interactions with effector proteins) and inactivated by GTP hydrolysis and γ-phosphate dissociation

GTPase activating protein (GAP). Proteins that stimulate GTP hydrolysis by small GTPases, reversing the association of the GTPase with effector proteins

GTP exchange factor. *See* Guanine nucleotide exchange factors (GEF)

Guanine. Purine base in DNA and RNA; H-bonds with cytosine

Guanine nucleotide dissociation inhibitor (GDI). Protein that prevents exchange of GDP for GTP on Rabs and other small GTPases

Guanine nucleotide exchange factors (GEF). Proteins that bind small GTPases and stimulate the dissociation of GDP, allowing GTP binding to activate the GTPase

Guanylylcyclase. Transmembrane and cytoplasmic enzymes that synthesize cGMP from GTP

Half-spindle. Portion of mitotic spindle consisting of a spindle pole with its associated kinetochore and interpolar and astral microtubules

Haploid chromosome number *(n)*. Number of chromosomes donated by the mother or the father in a diploid organism; only haploid cells in animals are sperm and eggs

Hedgehog. Signaling protein that helps to establish boundaries between cells of different fates during embryogenesis

Helicase. Enzyme that uses ATP hydrolysis to dissociate complementary strands of nucleic acids or remove secondary structure or bound proteins from nucleic acids

Helix-loop-helix proteins. Transcription factors with a basic region to recognize specific DNA sequences plus two helical dimerization domains separated by a loop region

Helix-turn-helix proteins. Transcription factors composed of two helices, one of which binds a recognition sequence of 6 bp in the major groove of DNA

Helper T cells. Lymphocytes selected by virtue of their receptors for specific antigens that produce growth factors for antibody-producing B-cells

Hematopoietic stem cells. Bone marrow cells that give rise to both short-lived and long-lived differentiated blood cells

Hemicellulose. Branched polysaccharide associated with cellulose in microfibrils in plant cell walls

Hemidesmosome. Plasma membrane specialization with integrins to bind the basal lamina and anchored to intermediate filaments in the cytoplasm

Hepatocyte growth factor–regulated tyrosine kinase substrate (HRS). Protein that sorts ubiquitinated proteins in multivesicular bodies

Hereditary spherocytosis. Condition with fragile red blood cells owing to defects in the membrane skeleton

Heterochromatin. Transcriptionally inert, condensed chromatin rich in histone H3 trimethylated on lysine 9

Heterochromatin protein 1 (HP1). Protein that binds nucleosomes containing histone H3 trimethylated on lysine 9 and recruits other components of heterochromatin

Heterogeneous nuclear RNA. Incompletely processed precursors of mRNA

Heterophilic interactions. Interactions between different molecules such as adhesion proteins on two cells

Heterozygous. In a diploid organism, the condition in which a particular genetic marker has different forms on the two homologous chromosomes

Hexose. A six-carbon sugar

HIF-1α. Hypoxia inducible (transcription) factor regulates the expression of erythropoietin and other genes

Histone acetyltransferase (HAT). An EDITOR that acetylates lysine residues of histones and other proteins

Histone code hypothesis. Proposal that posttranslational modifications of histones determine the level of functional activity of particular regions of chromatin

Histone isoform H2AX. Phosphorylation of this histone to create γ-H2AX to recruit repair factors to sites of DNA damage

HMG-CoA reductase. ER membrane enzyme that catalyzes a step in cholesterol biosynthesis; inhibited by statins

Holliday junctions. Intermediates in the recombination process consisting of branched DNA structures formed between two recombining DNA molecules

Holocentric chromosomes. Chromosomes with centromere activity distributed along the whole surface of the chromosome during mitosis

Homeodomain. DNA-binding domains of 60 amino acids found in transcription factors that specify body segments during development

Homogenize. Grind up or physically disrupt cells or tissues

Homologous chromosomes. Pairs of chromosomes (in diploid organisms) one donated by the mother and the other by the father

Homologous recombination. Physical exchange of regions of the genome between homologous chromosomes or between a plasmid and a chromosome

Homology-directed repair. Mechanism that repairs double strand breaks in DNA using a homologous DNA sequence as a guide

Homology model. Model of an atomic structure built by fitting an amino acid sequence into the known atomic structure of a related protein followed by refinement

Homophilic interactions. Interactions between like molecules such as adhesion proteins on two cells

Homotypic fusion and vacuolar protein sorting (HOPS). A multi-subunit tethering protein promoting fusion of endosomes and lysosomes

Homozygous. In a diploid organism, the condition in which a particular genetic marker has the same sequence on both the maternal and paternal homologous chromosomes

Hormone response elements. DNA sequences to which steroid hormone receptors bind to stimulate transcription

HOX (homeobox) gene. Genes for transcription factors that specify the development of embryonic segments

Hsp70. Protein chaperones that use ATP hydrolysis to drive a cycle of binding and release of short hydrophobic segments to promote polypeptide folding

Hsp90. Protein chaperones that maintain steroid receptors in an "open" state, ready to bind their ligands

hTERT. The reverse transcriptase subunit of telomerase

Hutchinson-Gilford progeria syndrome. Genetic disorder with premature ageing, caused by mutations in lamin genes

Hyaluronan. Very large glycosaminoglycans consisting of alternating D-glucuronic acid and D-N-acetylglucosamine

Hydrogen bond. Weak bonds between an H atom with a partial positive charge and an oxygen or nitrogen atom with a partial negative charge; contribute to stabilizing macromolecule secondary structure and interactions

Hydrophilic. Molecules or groups of atoms in molecules with favorable interactions with water

Hydrophobic. Molecules or groups of atoms in molecules with unfavorable interactions with water

Hydrophobic effect. Phenomenon whereby exclusion of water from complementary surfaces favors macromolecular associations; driven by increases in the entropy of water

Hydroxyapatite. Crystals of $[Ca_{10}(PO_4)_6(OH)_2]$ in bone matrix

I-band. Region of a striated muscle sarcomere with thin filaments but no thick filaments

ICAD. *See* Inhibitor of CAD

Icosahedron. Closed polyhedron with 12 fivefold vertices

Ig-CAM. Family of cell adhesion proteins with extracellular immunoglobulin-fold domains

IgG, IgA, IgE. Classes of immunoglobulin proteins produced by different B-lymphocytes

Image processing. Optical and computation methods to remove noise, average or make 3D reconstructions of micrographs

Immediate early genes. Genes expressed soon after stimulation of cultured fibroblasts in G_0 with serum

Immunologic synapse. Junction between T-lymphocytes and antigen presenting cells where major histocompatibility complexes engage and activate T-cell receptors

Immunoprecipitation. Method using antibodies to isolate specific proteins from cell extracts

Immunoproteasome. Specialized proteasomes that participate in ubiquitin-independent cleavage of intracellular antigens, such as viral proteins, into peptides of uniform length for presentation by MHC class I on the surface of antigen-presenting cells

Importin α. Adapter protein that recognizes small basic nuclear localization sequences and works with the transport receptor importin β in nuclear import

Importin β (karyopherin β). One class of receptor for proteins with nuclear localization sequences, regulated by Ran GTPase; also functions as a chaperone regulating the assembly of subcellular structures, including the mitotic spindle and nuclear envelope

Import receptor. Proteins that bind cargo proteins with nuclear localization sequences directly or in combination with adapter molecules and facilitate their transport into the nucleus (*see* Importin β)

Imprinting. Epigenetic mechanism that turns a gene off during formation of the egg or sperm and keeps the gene off in all cells that develop from that gamete

Inactivation. Processes operating in parallel with gating that stop the flux of ions through active channels

Inactivation peptide. A segment of polypeptide that blocks an open ion channel

Induced pluripotent (iPS) stem cells. Expression of a group of specific transcription factors in differentiated cells induces the expression of proteins required for pluripotency, creating cells that can be differentiated into many different specific cell types

Inhibitor of apoptosis protein (IAP). Family of proteins that inhibit caspases characterized by a motif of ~80 amino acids known as a baculovirus IAP repeat (BIR)

Inhibitor of CAD (ICAD). Chaperone required to fold caspase-activated DNase (CAD) that inhibits the nuclease until cleaved by a caspase, releasing active CAD

Initiation codon. Nucleotide triplet AUG in mRNA that specifies methionine, which begins polypeptide chains

Initiation factors. Proteins that coordinate assembly of mRNA and ribosomal subunits to begin the synthesis of a polypeptide

Initiator. DNA sequences in promoters near transcription start sites of many genes

Initiator caspases. Caspases that are autoactivated by association with scaffolding cofactors and propagate apoptosis by cleaving and activating effector caspase zymogens

Innate immunity. Defense against microorganisms by white blood cells using Toll-like receptors to recognize repeating structures of macromolecules

Inner nuclear membrane. The internal lipid bilayer surrounding the nucleus; associated with nuclear lamina

Inner segment. Part of photoreceptor cells between the cell body and light-absorbing outer segment

Inositol triphosphate (IP_3). Cyclohexanol with the 1, 4, and 5 hydroxyls phosphorylated

Insulators. DNA sequences that protect regions of a chromosome from the effects of neighboring regions

Insulin receptor. Receptor tyrosine kinase activated by insulin binding

Integral membrane protein. Protein embedded, at least in part, in a membrane lipid bilayer

Integrin. Family of heterodimeric plasma membrane adhesion proteins that generally bind extracellular matrix molecules and other cells

Intercalated disk. Adhesive junction linking the ends of cardiac muscle cells, anchored to cytoplasmic actin and intermediate filaments

Intercellular bridge. Thin connection between daughter cells at the end of cytokinesis containing an antiparallel array of microtubules derived from the mitotic spindle

Interchromatin granules. Intranuclear concentrations of factors for RNA processing

Interchromosomal domain. Region of nucleoplasm between adjacent chromosome territories

Interkinesis. Interphase period between the two meiotic divisions characterized by lack of an S phase

Interleukin. Family of proteins secreted by white blood cells that regulate the proliferation and differentiation of immune cells and inflammation

Intermediate filaments. Family of 10-nm filaments composed of α-helical subunits related to keratin

Interphase (or "resting stage"). Part of the cell cycle in which cells are not engaged in division

Interpolar microtubules. Microtubules distributed through the mitotic spindle, apparently free at both ends, that bundle to form the central spindle during anaphase and telophase

Intraflagellar transport. Bidirectional transport of proteins along the axonemes of cilia and flagella

Intra-S checkpoint. Stalling of the replication fork at a base that it cannot "read" results in a response that keeps the replication fork from disassembling until the situation is rescued by DNA replication moving upstream from an origin of replication further along the DNA

Intrinsically disorder protein. Segments of polypeptides or whole proteins with little or no secondary or tertiary structure

Intrinsic pathway of cell death. Apoptotic pathway triggered by release of pro-apoptotic factors from mitochondria and regulated by Bcl-2 family members

Introns. Regions of genes that are removed from immature RNA molecules by splicing

Inward rectifying channel. Channels with higher conductivity into than out of cells

Ion exchange chromatography. Separation of molecules based on their affinity for charged beads

Inositol requiring 1 (IRE1). A transmembrane protein of the ER that participates in the unfolded proteins response by splicing the mRNA for a transcription factor that regulates the expression of genes for ER proteins

Insulin receptor substrate (IRS). Cytoplasmic adaptor protein phosphorylated by the insulin receptor tyrosine kinase

IP_3 receptor. Channels activated by IP_3 to release calcium from the ER

Isoforms. Related proteins encoded by different genes or alternatively spliced mRNAs

Isoprenoid tail. Lipids consisting of three to six isoprenyl units (*see* Farnesyl)

JAK. Family of tyrosine kinases associated with cytokine receptors, playfully named "just another kinase"

Kartagener syndrome. Genetic disease with immobile cilia owing to a defect in dynein

Karyopherin. Another term for importins and exportins

KASH domain protein. Huge proteins of the outer nuclear envelope with KASH (Klarsicht, ANC-1, Syne Homology) domains that link SUN domain proteins to the cytoskeleton

Katanin. AAA ATPase that uses energy from ATP hydrolysis to sever microtubules

KcsA. Bacterial K-channel; first crystal structure of an ion channel

KDEL. C-terminal tetrapeptide sequence used to retain soluble proteins in the ER

KEN box. A degron recognized by the APC/C consisting of lysine-glutamic acid-asparagine

Keratin. Family of intermediate filament proteins expressed in epithelial cells

Kinesin. Family of motor proteins using ATP hydrolysis to walk along microtubules, generally toward their plus end

Kinesin isoforms. Family of motor proteins that share homologous catalytic domains coupled to a variety of tails.

Kinesin-13/kinesin-8. Kinesins that remove subunits from the ends of microtubules

Kinetochore. Structure at the surface of centromeric chromatin that binds microtubules and directs the movements of chromosomes in mitosis

Kinetochore fibers. Bundles of microtubules attached to kinetochores consisting of one microtubule in budding yeast to more than 200 microtubules in higher plants

Kinetochore microtubules. Class of mitotic spindle microtubules with their plus ends embedded in the kinetochore and minus ends at or near the spindle pole

K^+ leak channel. Potassium channel that helps to maintain the resting potential of excitable cell membranes

Kleisin. Protein family that links together the heads of SMC proteins in cohesin and condensin complexes

KMN network. Protein network composed of KNL1 and the Mis12 and NDC80 complexes that is responsible for microtubule binding by kinetochores

Lactacystin. Antibiotic that reacts covalently with threonine residues to inactivate proteasomes

Lagging strand. During DNA replication, the DNA strand along which replication occurs in a direction opposite to that of the replication fork, so that the newly synthesized DNA is laid down as a series of short discontinuous segments known as Okazaki fragments

Lamin A. Nuclear lamin encoded by a gene that is subject to many mutations that cause at least 16 human genetic diseases, including Emery-Dreifuss muscular dystrophy

Lamina associated domains (LAD). Regions of chromosomes localized near the nuclear lamina

Lamin B. Type V intermediate filament protein associated with the inner nuclear envelope

Lamin B receptor. Protein of the inner nuclear envelope that binds lamin B and heterochromatin protein HP1

Laminin. Adhesive glycoprotein of the basal lamina

Laminopathy. Inherited diseases associated with mutations in genes encoding lamin proteins or proteins that interact with lamins

Lampbrush chromosomes. Actively transcribed chromosomes with prominent loops visible during the diplotene stage of meiosis in animals

Lariat. Circular RNA molecule with a tail created early in mRNA splicing

Last eukaryotic common ancestor (LECA). Cell(s) that gave rise by divergent evolution to virtually all contemporary eukaryotic species

Late endosomes. Mature multivesicular bodies that have not yet fused with lysosomes

Latent phase. Phase of apoptotic death when a cell becomes committed to death but shows no phenotypic signs of this

Lateral gene transfer. Movement of entire genes between organisms

Latrunculin. Natural product the binds actin monomers and prevents their polymerization

Leading edge. Advancing pseudopod of motile cells driven by actin polymerization

Leading strand. During DNA replication, the DNA strand along which replication moves continuously in the same direction as the replication fork

Lectin. Protein that binds particular sugar molecules

Leptin. Satiety hormone secreted by fat cells and acting on neurons of the hypothalamus in the brain that regulate appetite and bone metabolism

Leptotene. First stage of meiotic prophase, defined by the first visible condensation of chromosomes, during which homologous chromosome pairing and alignment occur

Leucine zipper protein. Dimeric transcription factors with basic regions that recognize specific DNA sequences held together by a coiled-coil stabilized by leucine residues

Leukocyte. White blood cell

Leukotrienes. Lipid second messengers derived from arachidonic acid that participate in inflammatory responses

Ligand. Molecule that binds to a receptor

Light-harvesting complexes. Small, transmembrane proteins that absorb light and transfer energy to a photosynthetic reaction center

Light reactions. Steps in photosynthesis that depend on the continuous absorption of light, including production of high-energy electrons, electron transport to make NADPH, creation of a proton gradient for synthesis of ATP, and generation of oxygen

Light sheet microscopy. Illumination method for fluorescence microscopy that creates thin optical sections by focusing laser light into a sheet 2–8 μm thick and passing it through an illumination objective to focus onto the sample

Lignins. Polymers of phenylpropanoid alcohols and acids in "secondary" cell walls of plants

LINC complex. Nuclear envelope proteins that link telomeres in the nucleus to microtubules in the cytoplasm to enable chromosome movements that promote homolog pairing

Long interspersed nuclear elements (LINES). Common class of human retrotransposons with a consensus sequence of 6 to 8 kb that make up about 20% of the human genome

Linker DNA. DNA that links adjacent nucleosomes

Linker histone (H1). Binds to linker DNA, participates in formation of 30-nm fibers and chromatin compaction

Lipid gradient. Variation in the lipid composition of the membranes along the secretory pathway from the ER to the plasma membrane

Lipid raft. Microdomain in a lipid bilayer enriched in sphingolipids and cholesterol

Lipid-transfer protein (LTP). Protein catalyzes exchange but not net transfer of lipids between membranes

Lipoxins. Lipid second messengers related to leukotrienes and derived from arachidonic acid that participate in inflammatory responses

Listeria. Gram-negative intracellular pathogenic Bacterium that uses actin polymerization for motility

Localization microscopy (independently named FPALM, PALM and STORM). Super-resolution fluorescence microscopy method; a few widely separated, photoconvertable fluorescent molecules are activated and their positions are determined precisely until they are photobleached; an image is built up by thousands of cycles of photoactivation/conversion, imaging, and photobleaching

Locus control regions (LCRs). Short regions of DNA rich in binding sites for transcriptional regulators, that create "open" chromatin promoting the expression of nearby genes

Lokiarchaeota. Contemporary archaeon identified in an environmental DNA sample that is the closest known living relative of the ancient archaeon that became the eukaryote

Long noncoding RNA (lncRNA). RNA that does not encode any known protein

Long-QT syndrome. Prolonged action potentials in cardiac muscle resulting from mutations in sodium or potassium channels; predispose to arrhythmias

Long-term depression. Response of synapses to prolonged, slow stimulation that weakens neurotransmission for hours

Long-term potentiation. Response of synapses to strong stimulation that strengthens neurotransmission for days

Low-density lipoproteins (LDLs). Particles containing dietary and de novo–synthesized cholesterol, which are secreted into the blood for transport to other tissues

Lsm proteins ("like Sm"). Nuclear and cytoplasmic proteins that participate in decapping mRNA precursors that are destined for degradation

LUCA, last universal common ancestor. Cell(s) that gave rise to all forms of life on earth

Lymphocyte. White blood cell of the adaptive immune system

Lysobisphosphatidic acid. Lipid that promotes bilayer curvature during invagination and formation of intralumenal vesicles in multivesicular bodies

Lysosomal hydrolase. Enzymes concentrated in lysosomes that catalyze degradation of macromolecules by breaking covalent bonds by the addition of water

Lysosomal storage disease. Genetic diseases arising from absence of or defects in lysosomal hydrolases

Lysosome. Membrane-bound organelle containing acid hydrolases, including proteases; provides an acidic environment for digestion of contents

Lysyl oxidase. Extracellular enzyme that catalyzes the cross-linking of collagen and elastin

Macroautophagy. Engulfment into an autophagic vacuole of large volumes of cytoplasm that can include glycogen granules, ribosomes, and organelles

Macropinocytosis. Ingestion of extracellular fluid into a large endocytic structure

Major histocompatibility complex (MHC). Diverse family of cell surface proteins that carry antigenic peptides for activating immune cells

Major sperm protein. Nematode protein that assembles filaments for the movements of sperm

Mannose-6-phosphate receptors. Integral membrane proteins that bind lysosomal hydrolases modified with mannose-6-phosphate in the *trans*-Golgi network for delivery to endosomes and lysosomes

MAP-2. High-molecular-weight microtubule associated protein in the tau family

MAP kinase. Serine/threonine kinases activated by phosphorylation in the cytoplasm that move to the nucleus to activate transcription factors required for cell growth and division

MAP kinase kinase. Kinases that activate MAP kinases by phosphorylation of serine and tyrosine

MAP kinase kinase kinase. Serine kinases that activate MAP kinase kinases

MARK. A chemical modification (e.g. methylation, acetylation) placed on chromatin by an EDITOR

Mass spectrometry. Analytical method to measure the mass of molecules with high accuracy

Mast cell. Connective tissue cell activated by binding of antigens to cell surface immunoglobulins causing it to secrete granules containing histamine

Matrix metalloproteinase. Zinc proteases that digest the extracellular matrix of connective tissues

Matrix (mitochondrial). Innermost compartment of mitochondria with enzymes for the citric acid cycle and fatty acid oxidation

Maturation-promoting factor, M phase–promoting factor (MPF). Activity that causes interphase cells to enter mitosis; consists of an active Cdk with a cyclin partner

Mcm (minichromosome maintenance) proteins. Six AAA ATPases that form a hexameric complex thought to have helicase activity to separate DNA strands during replication

Mdm2. E3 ubiquitin ligase that regulates p53 stability (the human protein is Hdm2)

Mediator. Complex of 26 polypeptides that interacts with general transcription factors on promoters and recruits RNA polymerase II to initiate transcription

Megacomplex. Complex of many aggrecan proteoglycans with hyaluronan in cartilage

Megakaryocyte. Polyploid bone marrow cell that produces platelets by a budding process

Meiosis. Specialized program of two coupled cell divisions used by eukaryotes to maintain the proper chromosome number for the species during sexual reproduction

Meiosis I. First division of meiosis, in which homologous chromosomes separate, also known as the reductional division because the number of chromosomes is halved

Meiosis II. Second division of meiosis (also called the equational division); in this division, as in mitosis, sister chromatids segregate from each other and the number of chromosomes remains the same

Membrane carriers. Enzyme-like proteins that catalyze movements of solutes across membranes

Membrane channels. Protein pores for rapid movement of specific ions and solutes across membranes

Membrane contact sites. Regions of ER closely associated with the plasma membrane or organelle membranes with proteins that facilitate transfer of lipids between the membranes

Membrane peroxisomal targeting sequence (mPTS). Amino acid sequences that target proteins to peroxisomal membranes

Membrane potential. Voltage difference across a lipid bilayer

Membrane pumps. Transmembrane enzymes that use ATP hydrolysis or another energy source to move solutes across membranes up concentration gradients

Membrane skeleton. Network of actin filaments and accessory proteins associated with the cytoplasmic face of cellular membranes

Merotelic attachment. Defective chromosome attachment to the mitotic spindle with a single kinetochore attached to both spindle poles

Messenger RNA (mRNA). RNA molecules transcribed by RNA polymerase II and containing the sequence of bases that specify the sequences of amino acids in polypeptide chains

Metaphase. Third phase of mitosis with all chromosomes attached to both spindle poles and aligned near the equator of the mitotic spindle

Metaphase plate. Compact grouping of chromosomes at the middle of the mitotic spindle with all pairs of sister chromatids attached to both spindle poles

Metastasis. Exit of cancer cells from the primary tumor where the cancer started and growth elsewhere in the body

Microarray. Ordered pattern of spots of nucleic acids or proteins on a glass slide used for large-scale automated binding reactions such as measuring levels of mRNAs

Microelectrode. Glass capillary with a micrometer tip used to record the membrane potential of a single cell

Micro-RNA (miRNA). Small RNAs of about 22 nucleotides excised from larger precursors and associated with the RISC complex that represses translation of target mRNAs or directs their cleavage by the slicer endonuclease

Microtubule. Stiff cylindrical polymers of α- and β-tubulin that support a variety of cellular structures and serve as tracks for movements powered by motor proteins called kinesins and dyneins

Microtubule-associated proteins (MAPs). Proteins that regulate microtubule properties by binding tubulin dimers, by stabilizing or severing polymers, or by associating with microtubule ends

Microtubule-organizing center (MTOC). Structures containing γ-tubulin that nucleate microtubule assembly and usually anchor microtubule minus ends

Microvillus. Finger-like extension of plasma membrane supported by a bundle of actin filaments

Midbody. Dense knob surrounding antiparallel microtubules in the thin intercellular bridge between daughter cells following constriction of the cleavage furrow

Minus end. Slower-growing end of a microtubule terminating with α-tubulin

Mismatch repair. DNA repair process that removes errors that occur during DNA replication

Mitochondria. Eukaryotic organelle derived from a symbiotic proteobacterium specialized for oxidative phosphorylation to form ATP, fatty acid oxidation, and the citric acid cycle

Mitochondrial matrix. Innermost compartment of mitochondria with enzymes for citric acid cycle and fatty acid oxidation

Mitogen/mitogenic signals. Factors or signals coming from other cells and from the extracellular matrix that promote cell cycle progression

Mitosis (M phase). Cell cycle phase that partitions chromosomes and other cellular components to two daughter cells

Mitotic arrest-defective (MAD) genes. Encode proteins that execute the spindle checkpoint, delaying anaphase until all kinetochores are attached correctly to the spindle

Mitotic checkpoint complex (MCC). Complex of Mad2, Cdc20, BubR1 and Bub3 proteins that binds the APC/C and blocks mitotic progression if the spindle assembly checkpoint is activated

Mitotic exit network. Budding yeast signaling pathway that terminates mitosis, promotes constriction of the contractile ring, and initiates septation; called septation initiation network in fission yeast; related to Hippo pathway of animals.

Mitotic spindle. Framework of microtubules (between two centrosomes in animal cells or two spindle pole bodies in fungi) that segregates chromosomes during mitosis

M line. Network of proteins that crosslink the middles of thick filaments in striated muscles

Model organism. Species widely used in experimental biology, typically with fully a sequenced genome and facile methods to manipulate the genome including genetic crosses

Molecular dynamics simulation. Computational simulation of motions of atoms in macromolecules from thermal energy using Newton's laws of motion

Monoclonal antibody. Homogeneous antibodies produced by the progeny of a single B-lymphocyte

Monocyte. White blood cell, precursor of tissue macrophages and osteoclasts

Motor end plate. *See* Neuromuscular junction

Motor nerve. Axon of a neuron in the brain stem or spinal cord that controls muscle contraction

Motor proteins. Enzymes that used energy from ATP hydrolysis to produce force and motion on actin filaments or microtubules

Motor unit. All of the skeletal muscle cells controlled by one motor neuron

MreB. Family of prokaryotic actins

mRNA. *See* Messenger RNA

MRN complex. A SENSOR that detects double-strand DNA breaks and then chews back the DNA, leaving a single-stranded overhang that activates downstream repair pathways

mTor kinase (mechanistic target of rapamycin). Large protein kinase that is the lynchpin for key signaling pathways that control cell responses to nutrient levels

Mucin. Heavily glycosylated cell surface proteins, ligands for selectins

Multiple drug resistance proteins. ABC transporters that pump drugs and other hydrophobic molecules out of cells

Multivesicular body (MVB). Endocytic compartment derived from early endosomes by the inward invagination of vesicles for degradation of lipids and membrane proteins

Mus musculus. The mouse; commonly studied as a representative mammal with highly developed genetics

Mutation. Any alternation of genomic DNA sequences

Myc. Transcriptional factor that promotes expression of genes for cell cycle progression (cyclins E and D2) and represses expression of Cdk inhibitors (CKI and INK)

Myelin sheath. Insulating layer around some neuronal axons formed from the plasma membrane of glial cells

Myofibril. Contractile unit of striated muscle composed of many sarcomeres in series

Myofibroblast. Hypertrophied fibroblast with abundant contractile proteins to contract connective tissue damaged by a wound

Myosin. Family of motor proteins using ATP hydrolysis to apply tension to actin filaments

Myosin II. Principal myosin of muscles and contractile rings of dividing cells

Myosin-binding protein C. Stabilizes myosin thick filaments in striated muscles; mutations cause some cardiomyopathies

Myosin isoforms. Family of motor proteins that share homologous catalytic domains coupled to a variety of tails.

Myosin light chain kinase. Serine/threonine kinase activated by calcium-calmodulin that phosphorylates the regulatory light chain of myosin-II to trigger contraction of smooth muscle

Myosin light chains. Proteins related to calmodulin that reinforce the myosin lever arm

Myristoyl tail. Fourteen-carbon fatty acid added to the N-terminus of peripheral membrane proteins including c-Src

Myt1. Cytoplasmic kinase that phosphorylates at T^{14} and Y^{15} in the active site of Cdk1 inhibiting its activity until it is activated by dephosphorylation during M phase

NADH. Reduced form of nicotinamide adenine dinucleotide; an energy carrier in cells

Na^+K^+-ATPase. P-type pump using ATP hydrolysis to pump Na^+ out of and K^+ into animal cells

Natural killer cells. T-lymphocytes with receptors to recognize and attack cells infected with viruses

Nebulin. Giant protein that extents from end to end of striated muscle thin filaments

Necroptosis (programmed necrosis). Alternate backup pathway of cell death downstream of the CD95 (Fas) receptor that is triggered when key components of the CD95 pathway are missing or defective

Necrosis. Cell death resulting from irreversible injury involving leaking cell membranes, destruction of cellular contents lysosomal enzymes, and local inflammation

Negative selection. Apoptotic cell death of potentially harmful lymphocytes with T-cell receptors that recognize self-antigens

Negative staining. Method to prepare specimens for electron microscopy by drying in a puddle of heavy metal salts

N-end rule. Presence of certain amino acids at the N-terminus of a protein causes it to be ubiquitinated by a specific E3 enzyme and subsequently destroyed

Neocentromere. Functional centromeres that form rarely on noncentromeric DNA

Nernst equation. Relation between the concentrations of an ion on the two sides of a selectively permeable membrane and the equilibrium membrane potential

Nerve ending. *See* Synapse

***N*-ethyl maleimide (NEM)-sensitive factor (NSF).** AAA ATPase that uses the energy from ATP hydrolysis to dissociate cis-SNARE complexes and recycles the SNAREs for another round of membrane fusion

Netrin. Diffusible protein that signals attraction or repulsion of growth cones and capillaries

Neural crest cell. Embryonic cells that form sympathetic nervous system, pigment cells of skin, adrenal medullary cells, and many other cells

Neurofilament. Intermediate filaments of neurons

Neuromuscular junction. Synapse between a motor nerve and a skeletal muscle cell

Neurotransmitter. Small organic ions used for chemical communication between nerves and nerves and other cells

Neutrophil. White blood cell specialized for phagocytosis and destruction of bacteria

NF-κB. A family of transcription factors that controls diverse cellular processes including immune and inflammatory responses, development, cell growth, and apoptosis

N-formyl-methionine-leucine-phenylalanine (FMLP). Bacterial peptide that attracts white blood cells to sites of infection

Nicotinic acetylcholine receptor. Cation channel activated by binding acetylcholine

Nitric oxide. Gas that serves as a second messenger, by activating guanylylcyclase

Nitric oxide synthase. Enzyme that liberates nitric oxide from arginine

N-linked oligosaccharide. Sugar polymer conjugated to side chain of a protein asparagine residue

Nocodazole. Synthetic chemical that inhibits microtubule assembly by binding dissociated tubulin dimers

Nonclathrin/noncaveolae endocytosis. Endocytic events independent of clathrin and caveolin

Noncoding RNA (ncRNA). An RNA that does not encode any known protein

Noncrossover (gene conversion). Most common outcome of programmed double-strand DNA breaks during meiotic prophase; may involve loss of one or more genetic markers

Nondisjunction. Mistakes in the separation of chromosomes or chromatids in meiosis or mitosis resulting in aneuploidy, daughter cells with too many or too few chromosomes

Nonhistone proteins. Proteins of chromatin and chromosomes that are not histones

Nonhomologous end joining. Mechanism that repairs double strand breaks in DNA without using a homologous DNA sequence as a guide

Nonsense mediated decay (NMD). Process by which the presence of a premature translation termination signal (or nonsense codon) strongly destabilizes mRNA

Nonstop decay. Pathway for destruction of improperly processed mRNAs that lack a stop codon

Nonvesicular transport pathway. Transport of lipids between membranes at contact sites

Notch. Cell surface receptors for Delta ligand on other cells; influence many developmental processes

NSF. *See N*-ethyl maleimide (NEM)-sensitive factor

N-terminal tails. About 30 amino acids at the N-terminus of core histones that regulate chromatin compaction

Nuclear basket. Filaments that extend from the nuclear pore into the nuclear interior

Nuclear envelope. Double lipid bilayer that encloses the nucleus; outer membrane contiguous with the ER

Nuclear export sequence (NES). Short peptide sequence recognized by carrier proteins that direct a protein for transport out of the nucleus

Nuclear factor-activated T cell (NA-AT). Transcription factor activated by dephosphorylation by calcineurin

Nuclear lamina. Meshwork of intermediate filaments (nuclear lamins) that stabilizes the inner nuclear envelope

Nuclear lamins. Type V intermediate filament proteins that make up the nuclear lamina

Nuclear localization sequence (NLS). Short peptide sequence recognized by carrier proteins (transport receptors) that direct the protein for transport into the nucleus

Nuclear matrix or nucleoskeleton. Residual structures that remain when isolated nuclei are subjected to digestion with nucleases and extraction of the bulk of the histones

Nuclear mitotic apparatus protein (NuMA). Protein collaborates with dynein to focus microtubules at spindle poles

Nuclear pore complexes. Channels bridging both the inner and outer nuclear membranes for communication between the nucleus and cytoplasm during interphase

Nuclear receptor. Family of transcription factors activated by lipid soluble ligands, including steroid hormones; active receptors enter the nucleus to regulate gene expression

Nucleation. Initial steps in the assembly of polymeric macromolecular structures

Nucleic acid. Polymers of nucleotides linked by phosphodiester bonds, RNA, and DNA

Nucleolar organizer regions. Regions of chromosomes containing ribosomal RNA genes, whose transcription signals the formation of a nucleolus

Nucleolus. Nuclear subdomain for ribosome biogenesis

Nucleolus-organizing regions (NORs). Remnants of nucleolar fibrillar centers that remain associated with rRNA genes in condensed mitotic chromosomes

Nucleoplasm. The cellular region within the nuclear envelope

Nucleoporins. Family of about 30 structural proteins that regulate access through the nuclear pore

Nucleoside. Five-carbon sugar (ribose or deoxyribose) with a base on C1

Nucleosome. Complex of 165 base pairs of DNA wrapped twice around a protein core consisting of two copies each of the histones H2A, H2B, H3, and H4

Nucleosome core particle. 146 base pairs of DNA wrapped around a core consisting of a histone octamer

Nucleosome remodeling complex. Enzyme using energy from ATP hydrolysis to alter locations of nucleosomes on DNA

Nucleotide. Five-carbon sugar (ribose or deoxyribose) with a base on C1 and one to three phosphates on C5

Nucleotide excision repair. Process that replaces chemically modified bases in DNA

Nucleus. Membrane-bounded compartment in eukaryotes containing genomic DNA and machinery for RNA synthesis and processing

Objective. Microscope lens that collects light scattered by specimens

Occludin. Transmembrane protein subunit of tight junctions

Odorant. Vast array of volatile organic molecules that activate olfactory receptors to detect smells

Okazaki fragment. Short segments of newly synthesized DNA formed during replication of the lagging DNA strand

Olfactory sensory neuron. Cells in the nose that detect and signal the presence of odorant molecules by sending action potentials into the central nervous system

Oligosaccharyl transferase. ER enzyme associated with translocons that transfers core oligosaccharides from dolichol to an asparagine in an appropriate polypeptide sequence

***O*-linked oligosaccharide.** Glycosaminoglycans conjugated to a serine or threonine side chain

Oncogene. Gene that predisposes to oncogenic (cancerous) transformation of cells when the protein product is activated inappropriately; generally components of signal transduction pathways that regulate cellular growth and proliferation

Oncogenic stress. Excessive stimulation of cell cycle progression results in shortages of factors essential for DNA replication, thereby causing DNA damage

Open mitosis. Mitosis where the nuclear envelope disassembles before chromosomes segregate, as in most plants and animals

Open probability. Fraction of the time that an ion channel is open

Operon. Prokaryotic transcription units containing more than one gene, often encoding physiologically related proteins

Op18/stathmin. Protein that binds two tubulin dimers and inhibits their assembly

Optogenetics. Method to control the excitability of cells using light to open genetically-engineered and expressed channel rhodopsins

Orai channel. Plasma membrane calcium channels regulated by STIM proteins for store-operated Ca^{2+} entry

Origin of replication (defined genetically as a replicator element). Positions on the chromosome where DNA replication initiates

Origin recognition complex (ORC). AAA ATPase bound to origins of replication across the entire cell cycle

Orthologous genes. Homologous genes separated during evolution by a speciation event

Osteoblast. Cell secretes organic components of bone matrix

Osteoclast. Multinucleated cell formed by fusion of monocytes, specialized for bone resorption

Osteocyte. Cell surrounded by the bone matrix; can lay down or resorb matrix locally

Osteogenesis imperfecta. A variety of congenital fragile bone syndromes often caused by mutations in collagen

Osteon (Haversian system). Rod-shaped element of long bones formed by concentric layers of bone laid down on the inner surface of a resorption canal

Osteopetrosis. Disease with overgrown, dense bones owing to a failure of bone resorption due to lack of osteoclasts

Osteoporosis. Disease characterized by loss of bone tissue

Outer doublet. Pair of one complete and one incomplete microtubule in a ring of nine in axonemes

Outer nuclear membrane. Lipid bilayer continuous with the ER and sharing its functions

Outer segment. Part of photoreceptor cells with light-absorbing membranes; a modified cilium

Oxidative phosphorylation. Biochemical reactions in mitochondria and certain bacteria that utilize energy from the breakdown of nutrients to synthesize ATP from ADP

p21. Protein that blocks cell cycle progression by inhibiting Cdk1-cyclin A; its expression is turned on by p53 in response to DNA damage

p53. Transcription factor activated in response to DNA damage, which turns on expression of proteins that block cell-cycle progression or induce apoptosis

p62. The defining member of the FG-rich p62 complex of three nucleoporins

Pacemaker. Heart muscle cells that spontaneously generate action potentials and drive rhythmic contractions

Pachytene. Third stage of meiotic prophase, during which synapsis is complete and crossovers mature into chiasmata

Pairing. Side-by-side alignment of homologous chromosomes at a distance during meiotic prophase

Paralogous genes. Homologous genes created during evolution by duplication within a species

Parathyroid hormone. Stimulates osteocytes to mobilize calcium from bone matrix

ParM. Family of prokaryotic actins

Patch clamp. Glass micropipette that is applied to the surface of a cell or a piece of membrane to record electrical events in single ion channels

Patched. Cell surface receptor for Hedgehog proteins; participates in many developmental processes

Pathogen associated molecular patterns (PAMPs). Repeated features of pathogen macromolecules such as bacterial flagella

Pattern recognition receptors. See Toll-like receptors

PAX (paired box) genes. Class of genes that specify the development of embryonic segments

PDZ domain. Family of adapter domains that recognize C-terminal sequences of target proteins

Pectin. Branched polysaccharide associated with cellulose in microfibrils in plant cell walls

P element. Transposable element widely used in Drosophila genetic manipulations

Pentose. Five-carbon sugar

Peptide bond. Amide bond between the amino group of one amino acid and the carboxyl group of another amino acid

Peptidyl prolyl isomerase. ER enzyme catalyzes the interconversion of cis and trans peptide bonds involving proline

Pericentriolar material (PCM). Matrix surrounding pairs of centrioles that links them together and contains gamma-tubulin ring complexes, which nucleate microtubules

Perichondrium. Capsule of connective tissue that covers the surface of cartilage

Perichromatin fibrils. Nucleoplasmic structures on the surface of condensed chromatin that contain splicing factors and RNA packaging proteins

Perichromosomal layer. Ill-defined layer on the surface of mitotic chromosomes rich in nucleolar proteins and RNAs; may keep mitotic chromosomes from sticking to one another

Perinuclear space. Compartment continuous with the lumen of the ER that separates the inner and outer nuclear membranes

Periosteum. Connective tissue on the surface of bones

Peripheral membrane protein. Protein associated with either surface of a biological membrane by a covalently attached lipid, electrostatic interactions, or partial insertion into the bilayer; can be extracted by a basic solution

Peroxins. Proteins that deliver proteins to peroxisomes

Peroxisomal biogenesis disorders. Diseases arising from defects in peroxisome formation

Peroxisomal targeting signal type 1 (PTS1). Three C-terminal amino acids target enzymes to the lumen of peroxisomes

Peroxisomal targeting signal type 2 (PTS2). N-terminal sequences target proteins to the lumen of peroxisomes

Peroxisomes. Membrane-bounded organelles containing oxidative enzymes

Phagocytosis. Process by which cells ingest large particles such as bacteria, foreign bodies, and remnants of dead cells

Phagolysosome. Organelle formed on fusion of a phagosome with a lysosome

Phagophore. Cup-like membrane intermediate on the pathway of autophagosome formation

Phagosome. Membrane-bounded compartment containing ingested particles such as bacteria

Phalloidin. Cyclic peptide produced by poisonous mushrooms that has a high affinity for and stabilizes actin filaments; when conjugated with fluorescent dye, it is used to label actin filaments in cells

Phase contrast. Microscopic optical system generating contrast from differences in refractive index between a specimen and a reference beam

PH domain. Protein domains that bind polyphosphoinositides

Phenotype. Physical manifestation of the action of the genotype of an organism (refers to both the appearance and macromolecular composition of the organism)

Phorbol esters. Plant natural products stimulate protein kinase C

Phosphatidylcholine. Glycerolphospholipid with a choline head group

Phosphatidylethanolamine (PE). A phosphoglyceride with an ethanolamine head group

Phosphatidylinositol. Glycerolphospholipid with an inositol (cyclohexanol) head group that can be phosphorylated on carbons 3, 4, and 5 either singly or in combination to produce polyphosphoinositides

Phosphatidylinositol 4,5-bisphosphate (PIP_2). Phosphatidylinositol with phosphate esterified to the hydroxyls of inositol C4 and C5

Phosphatidylinositol 3-kinase. Lipid kinase that phosphorylates the C3 hydroxyl of phosphatidylinositol

Phosphatidylserine. Glycerolphospholipid with a serine head group

Phosphodiester bond. Phosphate esterified to two hydroxyls (can link either different molecules or within the same molecule in cyclic phosphodiesters)

Phosphodiesterase. Enzyme that catalyzes the conversion of 3′ 5′ cyclic nucleotides to the corresponding nucleoside monophosphate

Phosphoglyceride (glycerolphospholipid). Lipid with fatty acids esterified to the C1 and C2 hydroxyls of glycerol and phosphate on C3

Phospholamban. Regulatory subunit of P-type ATPase calcium pumps consisting of a single transmembrane domain, regulated itself by protein kinase A

Phospholipase A2. Enzyme that catalyzes cleavage of the ester bond between glycerol C2 and the fatty acid of a phosphoglyceride

Phospholipase C. Enzyme that catalyzes cleavage of the ester bond between glycerol C3 and the phosphate link to the head group of a phosphoglyceride

Phospholipase D. Enzyme that catalyzes cleavage of the ester bond between the phosphate and the head group of a phosphoglyceride

Phosphorylation. Formation of an ester bond between a phosphate and a hydroxyl of an amino acid, sugar, lipid, or other molecule

Photoreceptor cells. Sensory cells that express rhodopsin or other photoreceptor molecules and respond to absorption of photons

Photosynthesis. Biochemical reactions in chloroplasts and certain bacteria that utilize energy from absorption of photons to synthesize ATP

Photosynthetic reaction center. Transmembrane protein complex with light absorbing pigments that converts energy from absorbed photons into free electrons that reduce acceptor molecules during photosynthesis.

Photosystem I. Light-absorbing and electron carrier proteins in the membranes of purple bacteria, green filamentous bacteria, cyanobacteria, and chloroplasts that use energy from the absorption of photons to produce a proton gradient across the membrane to drive the synthesis of ATP

Photosystem II. Light-absorbing and electron carrier proteins in the membranes of green sulfur bacteria, heliobacteria, cyanobacteria, and chloroplasts that use energy from the absorption of photons to produce a proton gradient across the membrane to drive the synthesis of ATP

Phragmoplast. Arrays of microtubules formed by plant cells to deliver vesicles from the Golgi apparatus and ER to form new plasma membrane for cytokinesis

Physcomitrella patens. Experimentally tractable Moss used as a model organism among plants

Pinocytosis. Ingestion of extracellular fluids by vesicles formed from the plasma membrane

piRNAs. *See* Piwi-interacting RNAs

Piwi-interacting RNAs (piRNAs). Part of a silencing system in germ-line cells that blocks expression of transposons to protect the DNA against recombination and mutation

PKB/Akt. Protein kinase activated by binding PI3P; regulates metabolism

PKR-like endoplasmic reticulum kinase/pancreatic eIF2a kinase (PERK/PEK). A transmembrane protein of the ER that participates in the unfolded protein response by activating a translation initiation factor that stimulates the synthesis of ER proteins

Plakins. A family of giant proteins that link cytoskeletal polymers to each other and to membranes

Plakoglobin. Adapter protein linking cadherins and intermediate filaments in desmosomes

Plasma cell. Mature B-lymphocyte specialized to produce a particular antibody molecule

Plasmadesmata. Intercellular junction between plant cells providing continuity between their cytoplasms

Plasma membrane. Membrane forming the cellular boundary

Plasmid. Circular DNA molecule that can replicate and propagate through generations in cells

Plasmid cloning. Isolation of a piece of DNA by propagation in a plasmid from a single founder cell

Plastid. Organelles of plants and algae derived from cyanobacteria, including chloroplasts

Platelet. Small cellular fragments in blood responsible for patching defects in small blood vessels and promoting clotting

Platelet-derived growth factor. Protein growth factor released by activated platelets that stimulates proliferation of cells expressing the appropriate receptor tyrosine kinase

Pleckstrin homology (PH) domain. Adapter domains that bind phosphorylated inositides; found in multiple proteins

Plectin. Protein that links intermediate filaments to integrins, microtubules and actin filaments

P-loop. Polypeptide segments of cation ion channels that form selectivity pores

Pluripotent. Refers to stem cells that can develop into many different types of differentiated cells

Plus end. Faster-growing end of a microtubule terminating with β-tubulin

Point centromere. Budding yeast centromeres defined by sequence-specific binding of proteins

Pointed end. Slower-growing end of actin filaments

Polar bodies. Small cells produced by asymmetrical cell divisions during female meiosis

Polarized cell. Cell with functionally distinct apical and basolateral plasma membrane domains separated by tight junctions; internal contents are also polarized

Polo. Protein kinases regulating several aspects of cell cycle control; active on substrates that have been "primed" by prior phosphorylation by another kinase

Polyadenylation. Posttranscriptional addition of a chain of adenine residues to the 3' end of a mRNA - often important in regulation of RNA stability

Poly-A tail. Fifty to 200 adenine residues added posttranscriptionally to the 3′ end of most eukaryotic mRNAs

Polycomb repressive complex. One of several complexes that either EDIT or READ chromatin MARKS and promote a state repressive for transcription

Polymerase chain reaction (PCR). Method using heat-stable DNA polymerases to amplify DNA sequences by cycles of replication, strand dissociation, and further replication in the presence of excess primer sequences

Polymerase α/primase. Enzyme that initiates DNA replication by synthesizing an RNA chain of about 10 nucleotides to which DNA polymerase α adds another 20 to 30 nucleotides of "initiator DNA," all subsequently replaced

Polypeptide. Polymer of amino acids linked by peptide bonds

Polyploidy. Common chromosomal abnormality with an entire extra set of chromosomes

Polysomes. Complex of a mRNA with multiple ribosomes, each synthesizing the same polypeptide

Polytene chromosomes. Giant chromosomes in some tissues of insect larvae, consisting of more than 1000 identical DNA molecules packed side by side in register

Polytopic protein. Protein spans a membrane multiple times

Porins. Transmembrane proteins of the outer membrane of gram-negative bacteria and mitochondria forming channels for passage of molecules of less than 5000 D

Position effect. Repression of an actively transcribed gene when translocated into close proximity to constitutive heterochromatin, due to spreading of heterochromatin across the gene

Positive selection. Pathway of apoptotic death of lymphocytes that will be ineffective in immune responses because they express T-cell receptors that fail to interact with any of the MHC glycoproteins expressed by the individual

Postsynaptic. On the receiving side of a synapse

Posttranslational translocation. Translocation of a fully synthesized polypeptide from the cytoplasm into the ER, mitochondria, chloroplasts, or peroxisomes or out of prokaryotes

PP2B (calcineurin). Phosphatase activated by calcium-calmodulin; inhibition by cyclosporine blocks T-lymphocyte activation and allows organ transplantation

pRb. Family of transcriptional regulators (three in mammals) that control the activity of E2F family transcription factors and cell cycle progression at the restriction point

Preinitiation complex (transcription). Complexes of TATA box–binding protein and associated factors that promote the initiation of transcription

Preinitiation complex (translation). Assembly of mRNA and proteins on a small ribosomal subunit

Prenucleolar bodies. Particles containing nucleolar components that associate with NORs during nucleolar assembly

Prenylation. Lipid modification that anchors proteins on cytoplasmic surfaces of membranes (*see* Isoprenoid tail)

Preprocollagen. Precursor to collagen with a signal sequence and assembly domains at both ends

Prereplication complex. Protein complex of ORC, Cdc6p, CDT1, and Mcm proteins assembles at each replication origin once per cell cycle before the onset of S phase

Presequences. Amino acid sequences that target polypeptides to mitochondria

Presynaptic. On the sending side of a synapse

Primary cilium. Nonmotile cilium that serves as a sensory organelle; found on most animal cells

Primary constriction. Waist-like stricture at the centromere of mitotic chromosomes where the two sister chromatids are most intimately paired

Primitive mesenchymal cell. Stem cell for connective tissue cells

Proapoptotic. Signal or protein that triggers the apoptotic pathway of cell death

Processed pseudogenes. DNA sequences created by reverse transcription of mature mRNAs by a LINE reverse transcriptase and insertion back into the genome

Procollagen. Trimeric precursor of collagen held together by C-terminal assembly domains

Profilin. Protein that binds polyproline and actin monomers, catalyzes actin nucleotide exchange

Programmed cell death. Active cellular process that culminates in cell death in response to developmental signals, environmental cues, or physiological damage

Prohormone convertases. Proteolytic enzymes in the Golgi apparatus cleave small peptide hormones from large precursors

Proliferating cell nuclear antigen (PCNA). Doughnut-shaped trimer topologically locked onto the DNA by RFC that acts as a molecular "tool belt" to which numerous proteins involved with DNA replication and repair bind

Proliferation. Expansion of cell numbers by division

Prometaphase. Phase of mitosis beginning with nuclear envelope breakdown (in higher eukaryotes) and attachment of chromosomes to microtubules from the two poles of the forming mitotic spindle

Promoter. Assembly of DNA sequences required to form a preinitiation complex and initiate transcription

Promyelocytic leukemia (PML) bodies. Nuclear structures of unknown function containing an E3 ubiquitin ligase PML

Prophase. First phase of mitosis, defined by chromosome condensation inside an intact nuclear envelope accompained by changes in the dynamics of cytoplasmic microtubules

Prostaglandins. Family of lipid second messengers derived from arachidonic acid

Proteasome. Barrel-shaped multienzyme complex that degrades target proteins (typically tagged with chains of ubiquitin) into short peptides while recycling ubiquitin

Protein. One or more polypeptides folded into a functional three-dimensional structure

Protein coats. Polymeric structures on the cytoplasmic surface of membranes that promote formation of coated vesicles

Protein disulfide isomerase (PDI). Enzyme catalyzes exchange of protein of disulfide (S-S) bonds in the ER lumen

Protein domain. Independently folded part of a protein

Protein folding. Conversion a linear polypeptide into a particular three-dimensional structure

Protein kinase. Enzyme that catalyzes formation of phosphate esters on hydroxyl groups of proteins

Protein kinase A (PKA). Protein kinase regulated by cAMP binding to a regulatory subunit

Protein kinase C (PKC). Protein kinase regulated by binding of diacylglycerol or other lipids and calcium to its regulatory domains

Protein phosphatase. Enzyme that catalyzes the hydrolysis (removal) of phosphate from hydroxyl groups of proteins

Protein phosphatase 2A (PP2A). Family of protein phosphatases that antagonizes CDK activity and is inactivated by Greatwall Kinase to promote mitotic entry

Protein targeting. Mechanisms that deliver a protein to a particular location in a cell

Proteoglycans. Proteins modified with O-linked glycosaminoglycans, found in secretory granules and extracellular matrix

Protocadherin. A large class of cadherins that regulate cellular interactions between neurons

Protofilaments. Longitudinally oriented filaments of tubulin dimers; 13 make up the cylindrical wall of most microtubules

Pseudoautosomal region. Region of sequence homology between male and female sex chromosomes that must undergo genetic recombination in meiosis I for sex chromosomes to partition correctly

Pseudogene. Nonfunctional DNA sequences derived from gene transcripts that have been reverse transcribed and inserted back into the chromosome

Pseudopod. Cellular protrusion responsible for cellular locomotion

Pseudosubstrate. Region of polypeptide similar to a substrate in a kinase or a kinase regulatory subunit that inhibits access of substrates to the kinase by binding the active site

PTB domain. Adapter domain that binds particular peptides with a phosphotyrosine, found in multiple proteins

PTEN phosphatases. Family of phosphatases that remove phosphate from the D-3 position of polyphosphoinositides

P4-type ATPase (flippase). Variant P-type ATPase pump that flips lipids between the leaflets of bilayers

Pyrin domain (PYD). Protein interaction motif found in proteins involved in inflammation and caspase 1 activation

Pyroptosis (pro-inflammatory programmed cell death). Reaction to pathogen infection leading to formation of inflammasomes, cytosolic complexes containing caspase 1 that promote a necrosis-like death

Quiescence. A transient nondividing state from which cells exit given appropriate signals

Rab. Family of small GTPases that control protein-protein interactions between transport carriers and docking complexes on target membranes

Rac. Small GTPases related to Rho that regulate actin assembly

RAD51. Eukaryotic homolog of *E. coli* RecA, associates with single-stranded DNA and catalyzes the search for homologous sequences, strand pairing, and strand exchange during DNA recombination and repair

Radial spoke. Multiprotein connection between the central pair and outer doublets of axonemes

Raf. MAP kinase kinase kinase activated by Ras

Ran. Small GTPase that provides direction to nuclear transport; within the nucleus, Ran-GTP dissociates imported proteins from their carriers and binds proteins for export to their carriers; also functions in spindle assembly in eggs

Ran-GAP1. GTPase activating protein for Ran associated with cytoplasmic filaments nuclear pores; converts Ran-GTP from the nucleus into inactive Ran-GDP in the cytoplasm

RANKL (RANK ligand, also called osteoprotegerin ligand [OPGL] or TRANCE). "Receptor activator of nuclear factor κB ligand" a secreted protein that stimulates the differentiation of osteoclasts

Rapamycin. Fungal product from Rapa Nui (Easter Island) that inhibits mTOR kinase

Ras. Small GTPase couples activation of growth factor receptor tyrosine kinases to Raf at the start of the MAP kinase pathway

Ras-GEF. Nucleotide exchange factor called SOS (son of sevenless) that activates Ras

Rate constant. Proportionality constant between the concentration(s) of reactant(s) and the rate of a reaction

RCC1. *See* Regulator of chromosome condensation 1

Reaction center. Complex of proteins with light-absorbing chromophores and electron transfer cofactors that absorb light and initiate an electron transport pathway that pumps protons out of Bacteria and thylakoids of chloroplasts

READER. A protein or protein complex that binds to a MARK and then establishes a chromatin state, either directly or by recruiting other factors

Receptor. Macromolecule that selectively binds particular partner molecules (ligands), initiating a cellular response

Receptor-mediated endocytosis. Facilitated uptake of an extracellular ligand due to its binding to a receptor that undergoes endocytosis

Receptor serine/threonine kinase. Family of receptors that bind ligands related to transforming growth factor beta and initiate signaling through cytoplasmic serine/threonine kinase domains

Receptor tyrosine kinase. Family of receptors that bind growth factors and initiate signaling through cytoplasmic tyrosine kinase domains

Recombination. Physical exchange of DNA strands between homologous chromosomes, during meiosis 1 drives chromosomal pairing and formation of chiasmata that are critical for segregation of homologous chromosomes

Recycling endosome. Endosomes located in the perinuclear Golgi region where receptors returning to the cell surface accumulate

Reductional division. First division of meiosis when homologous chromosomes separate and the chromosome number halves

Reductionism. Experimental strategy relying on characterization and reconstitution of isolated molecular components of complex biological systems

Regional centromere. A centromere with the kinetochore histone CENP-A distributed across a broad region of DNA; defined by epigenetic marks rather than specific DNA sequences

Regulated secretory pathway. Route for concentrating and packaging proteins in storage granules for discharge from the cell in response to hormonal or neural stimulation

Regulator of chromosome condensation 1 (RCC1). GEF for the small GTPase Ran; associated with chromatin and insures a gradient of active Ran near chromatin

Repetitive DNA. Sequences present in many copies (thousands, in some cases) in eukaryotic genomes

Replication factor C (RFC). Protein complex that binds the 3′ end of initiator DNA and uses energy from ATP hydrolysis to load the trimeric protein PCNA onto the DNA

Replication foci. One thousand or more sites of replication in eukaryotic nuclei during S phase, each representing five or six coordinately activated replication origins

Replication fork. Site of DNA replication consisting of a parental DNA molecule unwound into two strands along with the replication machinery on both strands

Replication fork collapse. Disassembly of stalled replication forks

Replication stress. Situation in which stopping replication leads to an accumulation of single-stranded DNA, activating a pathway to try and prevent replication fork collapse

Replicator. Genetically defined site at which replication of a bacterial chromosome initiates

Replicon. Region of chromosomal DNA replicated from a single origin of replication

Replisome. Multiprotein complex containing helicases and polymerases that replicates the DNA

Rescue (in dynamic instability). Random transition of a microtubule from a phase of rapid shortening to regrowth

Resection. Process of trimming DNA back from the site of a Spo11 break to create 3' single-stranded DNA tails that can search for homologous sequences on other chromosomes

Residual body. Mature lysosome containing a large amount of undegraded material

Restriction point. Checkpoint in late G_1 phase that blocks cells from committing to a proliferation cycle unless nutrients and mitogens are present and the cell senses appropriate interactions with the surrounding extracellular matrix

Reticulon. Proteins that bend membranes by inserting into the lipid bilayer

Retina. Epithelium at the back of the vertebrate eye with an array of photoreceptor cells

Retinal. Vitamin A derivative that serves as the chromophore for rhodopsin, the photon receptor in the eye

Retrieval pathway. Recycling of proteins and lipids from the Golgi back to the ER

Retrograde traffic. Flow of cargo and lipids back toward the ER

Retrograde transport. Movement of membrane-bound particles toward the cell body of neurons

Retromer. Protein complex involved in transport from late endosomes to *trans*-Golgi network

Retrotranslocon. Channel proposed to export proteins from the ER lumen or membrane into the cytoplasm for degradation

Retrotransposons. Transposable elements of DNA that move via RNA intermediates

Reverse genetics. Study of gene function by engineering desired mutations into cloned coding regions of genes

Reverse transcriptase. Specialized DNA polymerase that copies RNA into DNA

RGD motif. Tripeptide (arginine-glycine-aspartic acid) used by several extracellular ligands to bind integrins

RGS proteins. Proteins that stimulate GTP hydrolysis by α-subunits of trimeric G-proteins

Rho. Family of small GTPases that regulate contraction mediated by myosin-II and other aspects of the actin cytoskeleton; essential for cytokinesis

Rhodopsin. Seven-helix receptor protein with covalently bound retinal that absorbs photons in the retina

Rho-GDI. Protein that binds Rho-GDP and blocks activation by GEFs

Ribose. Five-carbon sugar, component of RNA

Ribosomal RNA (rRNA). Three of the four RNA molecules forming the bulk of ribosomes including the catalytic site; precursor RNA cotranscribed by RNA polymerase and processed into 18S, 5.8S, and 25S/28S rRNAs

Ribosome. Complex of ribosomal RNAs with multiple proteins that catalyzes the synthesis of polypeptides

Riboswitch. RNA sequences that bind small molecules and regulate gene expression

Ribozymes. RNAs with catalytic activity independent of proteins, including group I and group II self-splicing introns

Ribulose phosphate carboxylase (RUBISCO). Most abundant protein on earth, catalyzes the combination of a five-carbon sugar with carbon dioxide to form two molecules of the three-carbon sugar 3-phosphoglycerate in chloroplasts

***Rickettsia*.** Alpha proteobacteria closely related to mitochondria; cause of typhus and other diseases

Ring canals. Intercellular bridges that remain open following incomplete cytokinesis to maintain cytoplasmic continuity between daughter cells in specialized tissues

RISC. *See* RNA-induced silencing complex

RNA. Polymer of phosphate-linked sugars (ribose) linked to purine and pyrimidine bases used to carry genetic information, catalyze reactions, bind ligands, or assemble with proteins in macromolecular complexes

RNA editing. Covalent modifications of individual nucleotides, which alter their base-pairing potential and thereby change the amino acid that is incorporated during protein synthesis; increasing the diversity of protein products that can be coded by the genome

RNA-induced silencing complex (RISC). Multienzyme complex that promotes the maturation of miRNAs and can cleave target RNAs, repress translation of mRNAs, or inhibit transcription of target genes via formation of heterochromatin

RNA interference (RNAi). Experimental use of double-stranded RNAs complementary to a target mRNA to trigger its cleavage by RISC complex

RNA polymerase. Enzyme that transcribes (synthesizes) RNA complementary to a DNA template

RNase H. Exonuclease that removes the RNA primer used to start DNA replication by chewing in from the 5′ end

RNase MRP. Eukaryotic enzyme that cleaves preribosomal RNA between the small and large subunit rRNAs

RNase P. RNA-protein enzyme that cleaves pre-tRNAs at the 5′ end of the mature tRNA sequence in all organisms

RNA world. Early stage of evolution of life on earth consisting of self-replicating RNA polymers sheltered inside lipid vesicles

Rod. Photoreceptor cell for sensitive detection of a broad range of wavelengths

Rotary ATPase. ATP-driven proton pumps with a rotor and stator; some use a proton gradient to synthesize ATP

Rough endoplasmic reticulum. Subdomain of ER with associated ribosomes synthesizing proteins for secretion and insertion into membranes and specialized for protein folding

RPA. Single-strand DNA-binding protein recruited by Cdc45 and Mcm proteins to stabilize separated strands of DNA during replication and repair

R (regulatory) subunits. Proteins that inhibit PKA but dissociate when they bind cAMP

Ryanodine receptor. Calcium release channel of the endoplasmic reticulum

S phase (synthetic phase). Portion of the cell cycle when DNA is replicated

Saccharomyces cerevisiae. Budding yeast; popular genetic model organism for studying basic cell biology

SAM complex (sorting and assembly machinery of the outer membrane). Protein complex for translocation of proteins into chloroplasts

Sar1. Small GTPase that recruits COPII coat complexes to the ER membrane

Sarcomere. Contractile unit of striated muscle consisting of a bipolar array of overlapping actin and myosin filaments

Sarcoplasmic reticulum. Smooth ER of striated muscles specialized for rapid release and reuptake of the calcium ions that regulate contraction

SAS-6. Forms a complex with 9-fold radial symmetry; templates the formation of centrioles

Satellite cells. Stem cells in striated muscles

Satellite DNAs. Repeated DNAs clustered in discrete areas of chromosomes, e.g., flanking centromeres

Scaffold/matrix attachment regions (SMARs). Regions of DNA that associate with the nuclear matrix and chromosome scaffold in biochemical fractionation experiments

Scanning electron microscope. Optical system to scan a fine electron beam over a metal-coated specimen and create an image from secondary electrons emitted from the surface

SCF. Class of E3 ubiquitin ligase containing Skp1/Skp2, a cullin, and an F-box protein; regulate the cell cycle by proteolysis

Schizosaccharomyces pombe. Fission yeast; popular genetic model organism for studying the cell cycle

Sec61 complex. *See* Translocon

SecA. Bacterial enzyme that uses ATP hydrolysis to promote the translocation of proteins through SecYE translocons

SecB. Bacterial chaperone for newly synthesized proteins to prevent folding before translocation

Second messengers. Calcium ions and small molecules including cyclic nucleotides and lipids that carry biochemical signals inside cells

Secondary constriction. Region of a chromosome associated with the nucleolus-organizing region

Secondary structure. Regular structures formed by polypeptides, especially α-helices and beta-sheets

Second-order reaction. Reaction with two reactants

Secretory cargo. Transmembrane and lumenal proteins transported through the secretory system

Secretory granule. Membrane-bounded packets of concentrated secretory proteins prepared for secretion

Secretory membrane system. Distributes proteins and lipids synthesized in the ER to other sites using vesicular intermediates for transport between the ER, Golgi apparatus, and plasma membrane

Securin. Inhibitor of the separase protease that cleaves proteins to trigger sister anaphase chromatid separation

Sedimentation coefficient. Ratio of the sedimentation velocity to the centrifugal force

Sedimentation equilibrium. Analytical ultracentrifuge method where the specimen is centrifuged until the molecules reach an equilibrium between diffusion and sedimentation

Sedimentation velocity. Analytical ultracentrifuge method measuring the rate that a molecule moves in a centrifugal force field

Segmental duplications. Regions of DNA ≥1000 base pairs with ≥90% sequence identity that are present in more than one copy but are not transposons

Selectin. Plasma membrane adhesion receptor for mucins on other cells

Self-assembly. Capacity of macromolecules to form large structures without guidance by templates

Self-cleaving ribozymes. RNAs with the capacity to cleave themselves in the absence of proteins

Senescence. Terminal G_0 state with viable but nondividing cells

Separase. Protease that cleaves a component of cohesin, triggering the onset of anaphase sister chromatid separation

Seven-helix receptor. A large class of receptor proteins composed of seven transmembrane α-helices, coupled to cytoplasmic trimeric G-proteins

Sex chromosomes. Chromosomes that carry genes that define the sex of an organism

SH2 domain. Adapter domains that bind peptides including a phosphorylated tyrosine, in many signaling proteins

SH3 domain. Family of adapter domains that bind proline-rich peptides, found in many signaling proteins

Shelterin. Protein complex that associates with and protects the ends of chromosomes in metazoans

Short interspersed nuclear elements (SINES). Retrotransposons of ~300 bp making up ~13% of the human genome

Shugoshin. From the Japanese "guardian spirit"; a protein complex involved in the protection of sister chromatid cohesion near centromeres

Side chain. Chemical group on the α-carbon of an amino acid

Signal peptidases. Enzymes that cleave signal peptides from proteins after translocation of proteins across membranes

Signal recognition particle (SRP). RNA-protein complex that recognizes signal sequences emerging from ribosomes and directs the ribosome to a translocon of endoplasmic reticulum

Signal sequence. N-terminal hydrophobic polypeptide signal that directs proteins to the ER

Signal transduction. Reactions that convert a stimulus into a change in the behavior of a cell

Sinoatrial node. Cluster of heart muscle cells that generate the action potentials that produce rhythmic contractions

siRNAs. *See* Small interfering RNAs

Sister chromatids. Products of DNA replication, two identical DNA molecules, each packaged by chromatin proteins

Situs inversus. Major organs are located on the opposite side of the body from normal

Skeletal muscle. Striated muscle cells controlled by motor neurons in the spinal cord and brain stem

Slicer (Ago2). Component of the RISC complex that cleaves target RNA sequences that are perfectly complementary to miRNAs associated with RISC

Slow axonal transport. Transport of structural proteins and cytoplasmic enzymes from their site of synthesis near the nucleus along nerve cells process; some of these components move by rare short bursts of fast transport along microtubules

Smac/DIABLO (second mitochondrial activator of apoptosis/ direct IAP binding protein with low pI). Inhibitor of IAP proteins released from mitochondria to promote apoptosis

Smad. Family of cytoplasmic transcription factors activated to enter the nucleus after phosphorylation by receptor serine/threonine kinases activated by binding ligands

Small heterochromatic RNAs (now more commonly known as siRNAs). RNAs generated by transcription of both strands of DNA that associate with a nuclear complex called RITS (RNA-induced transcriptional silencing), which is related to the cytoplasmic RISC complex and induce heterochromatin formation and silencing of the transcribed locus

Small interfering RNAs (siRNAs). 22-nucleotide double-stranded RNAs, either introduced into cells or expressed from a short hairpin RNA (shRNA) sequence to trigger the destruction of target complementary RNAs by the RISC (RNA-induced silencing) complex

Small nuclear RNAs (snRNAs). RNAs that function in complexes with proteins in small nuclear ribonucleoprotein (snRNP) particles to recognize signals in the pre-mRNA that identify introns and exons during splicing

Small nucleolar RNAs (snoRNAs). Small RNAs, mostly excised from introns of RNAs transcribed by RNA polymerase II, that are involved in selection of the sites of modification of RNA bases during the maturation of functional RNAs

Small ubiquitin-like modifier (SUMO). Ubiquitin-like protein conjugated to the same residues on target proteins as ubiquitin, but typically regulates protein-protein interactions rather than promoting degradation

SMC proteins (structural maintenance of chromosomes). ATPases that form part of the condensin and cohesin complexes that are essential for mitotic chromosome structure, regulation of sister chromatid pairing, DNA repair and replication, and regulation of gene expression

Smooth endoplasmic reticulum. Subdomain of ER lacking ribosomes and dedicated to drug metabolism, steroid synthesis, and calcium homeostasis

Smoothened. Unusual seven-helix receptor inhibited by the patched receptor for Hedgehog proteins

Smooth muscle. Muscle without highly organized sarcomeres found in the walls of blood vessels and internal organs

Sm proteins. Seven closely related proteins that assemble into a heptameric ring structure on snRNAs

SNAP receptor (SNARE). Family of proteins participates in fusion of carriers with appropriate acceptor compartments

Sodium dodecylsulfate (SDS). Ionic detergent used to solubilize proteins for separation by size by gel electrophoresis

Somatic mutation. Mutations in genomes of developing lymphocytes generate diversity in antibodies and T-cell receptors

Sortilin. Transmembrane protein directs proteins to lysosomes

Sorting nexin. Subunits of the retromer complex used for recycling protein from endosome to the Golgi apparatus

Spastin. AAA ATPase that severs microtubules; mutations cause autosomal dominant spastic paraplegia

Speckles. Clusters of nucleoplasmic interchromatin granules containing factors involved in RNA processing

Spectrin. Actin-binding protein of the membrane skeleton of the plasma membrane and some cytoplasmic organelles

Spermatids. Male germ cells that have completed meiosis but not yet differentiated into mature sperm

Spermatogenesis. Process that produces sperm

Spermatogonia. Stem cells that give rise to sperm

Spermatozoa. Mature sperm

Sphingomyelin. Sphingolipid with a choline or ethanolamine head group

Sphingomyelinase. A phospholipase C that removes the head group from sphingomyelin

Spindle assembly checkpoint (SAC). Signalling pathway that delays the onset of anaphase until all chromosomes have stable bipolar attachment to spindle microtubules

Spindle pole. One of the duplicated centrosomes that nucleates microtubules during mitosis

Spindle pole body. Plaque-like structure embedded in the nuclear envelope of fungi that contains γ-tubulin and acts as the microtubule organizing center for mitosis

Spire. Protein that promotes actin filament nucleation

Splice site. Sites where RNAs are cleaved during splicing

Splicing. RNA maturation reactions that cut out specific regions (introns) and rejoin the remaining RNA (exons)

Spo11. Enzyme that creates double-stranded DNA breaks to initiate recombination during meiosis

SREBP cleavage-activating protein (SCAP). Transmembrane ER protein that anchors SREBP when cholesterol is abundant in the membrane

sRNAs. *See* Small RNAs

SRP receptor. Transmembrane receptor in ER that binds the complex of ribosome, nascent polypeptide chain, and SRP prior to cotranslational translocation

SR proteins. Protein factors important for alternative splicing that contain domains rich in serine-arginine dipeptides

START. Point in the G_1 phase of budding yeast after which cells are committed to complete the cell cycle

STAT. Family of cytoplasmic transcription factors activated to enter nucleus after phosphorylation by JAK kinases

Stathmin. *See* Op18/stathmin

Statin. Family of drugs based on a natural product that inhibit HMG-CoA reductase and cholesterol biosynthesis

Stem body. Remnant of the central spindle composed of antiparallel microtubules in a dense matrix of proteins

Stem cell niche. Special environments created by tissue cells and the extracellular matrix that help stem cells maintain their status as stem cells

Stem cells. Cells with the capacity to produce, through intermittent asymmetrical cell division, both a self-renewing stem cell and a second cell with the capacity to differentiate into more specialized cells

Stem loop. RNA sequence that forms an antiparallel double helix with a loop at the end

Step size. Distance moved by a motor protein during one cycle of ATP hydrolysis

Steroid regulator element-binding proteins (SREBP). Family of transcription factors that regulate the expression of genes for proteins involved with steroid biosynthesis

STIM. Integral protein of the ER membrane that aggregates when Ca^{2+} is depeleted from the lumen of the ER and stimulates Orai channels to admit Ca^{2+} through the plasma membrane

Stimulated emission depletion (STED). Super resolution fluorescence microscopy method by scanning the specimen with two superimposed beams, one that suppresses emissions from all by a tiny spot in the middle of the second beam

Store-operated Ca^{2+} entry. Response of cells to depletion of Ca^{2+} from the lumen of the ER, acting through STIM and Orai channels

Stratified epithelium. Form of epithelium with multiple layers of cells on a basal lamina

Stress fiber. Bundle of actin filaments, myosin-II, and other proteins linking focal adhesions in nonmuscle cells

Striated muscle. Skeletal and cardiac muscles that have a striped appearance owing to alignment of the sarcomeres

Structured illumination microscopy (SIM). Super resolution fluorescence microscopy method using superimposition on the image of an intense, scanned and rotated bar pattern to improve the resolution

Subunit. Macromolecular building block of a larger structure

Subunit flux (treadmilling). Flow of actin or tubulin subunits through their polymers as a result of net addition of subunits to one end and loss of subunits from the other end

Sulfonylurea receptor. ABC transporters that regulate potassium channels that are sensitive to cytoplasmic ATP

SUN proteins. Proteins of the inner nuclear envelope with SUN (Sad1p, UNC-84) domains that link lamin A inside the nucleus to KASH domain proteins in the outer nuclear envelope and cytoplasm

Super resolution microscopy. Fluorescence microscopy methods with resolution better than the classical Abbe diffraction limit

Survival of motor neurons (SMN) protein. Subunit of a large protein complex that promotes the assembly of Sm-proteins on snRNA, gene mutated in spinal muscular atrophy

SWEET carrier. Plant carrier protein for glucose

Switch I/II. Regions of GTPases that change conformation depending on binding of GTP or GDP

Symporter. Carrier proteins that catalyze movements of solutes across membranes up concentration gradients at the expense of transport of a second solute down its concentration gradient in the same direction

Synapse. Specializations of nerve cells for rapid communication with other nerve and muscle cells in which the sending cell concentrates vesicles with a neurotransmitter prepared for secretion and the receiving cell concentrates receptors for that neurotransmitter

Synapsis. Intimate pairing of homologous chromosomes during zygotene of meiosis-I

Synaptic vesicle. Small vesicles filled with neurotransmitter concentrated in presynaptic endings

Synaptonemal complex. Protein scaffold assembled between homologous chromosomes during synapsis in meiotic prophase; looks like railroad tracks with a third rail running down the center

Synaptotagmin. Calcium-binding protein that triggers assembly of SNAREs and membrane fusion

Syntelic attachment. A form of defective chromosome attachment to the spindle with sister kinetochores both attached to a single spindle pole

Synthetic genetic interaction. Phenotype of double mutants more severe than expected from the phenotypes of the individual mutations

Tail-anchored protein. Protein inserted into membranes posttranslationally using a hydrophobic C-terminal anchor

Talin. Adapter protein between integrins and actin filaments in focal contacts

Targeting signals. Amino acid sequences that are necessary and sufficient to guide proteins to their destinations

TATA box. Promoter element for many genes transcribed by RNA polymerase II, with the consensus sequence TATAAAA, recognized by TATA box–binding protein (TBP)

TATA box–binding protein (TBP). Protein that binds the TATA box, bending the DNA and promoting the assembly of the RNA polymerase preinitiation complex

Tau. Family of microtubule-associated proteins, including tau, MAP2, and MAP4, characterized by conserved microtubule-binding motifs that stabilize microtubules

Taxol. Cancer chemotherapeutic drug from bark of the Western yew that binds β-tubulin and stabilizes microtubules

TBP-associated factors (TAFs). Proteins that associate with TATA binding protein (TBP) at promoters and activate transcription above basal levels

T-cell antigen receptor. Cell surface receptors with highly variable structures allowing interactions with major histocompatibility complexes carrying a specific peptide antigen

Telomerase. Specialized form of reverse transcriptase, containing both RNA and protein subunits that is responsible for maintaining DNA sequences at telomeres

Telomere. Structure at both ends of chromosomal DNA that protects ends and ensures their complete replication

Telophase. Fifth phase of mitosis, initiated by reformation of the nuclear envelope on the surface of the chromatin

Temperature-sensitive (ts) mutants. Conditional mutants, functional at a low *permissive* temperature but not at a higher *restrictive* temperature

Tenascin. Six-legged adhesive glycoproteins in the extracellular matrix

Teratocarcinoma. Tumor arising from germ cells and containing a wide variety of cell types

Termination. Reactions specified by a termination codon (UAA, UAG, or UGA) at the 3′ end of the coding sequence of an mRNA that complete the synthesis of a polypeptide and release the polypeptide and mRNA from a ribosome

Termination codons. The nucleotide triplets (UAA, UGA, UAG) that stop peptide synthesis

Terminator. Sequences in bacterial RNAs that trigger dissociation of a transcript and RNA polymerase when RNA polymerase reaches the end of a gene or operon

Tetanus. Maximal contraction of skeletal muscle achieved by repetitive stimulation by the motor neurons

Tethering factors. Proteins that tether membrane carriers to the cytoskeleton and target organelles prior to fusion

TGF-β. *See* Transforming growth factor–β

Thapsigargin. Plant lactone that inhibits membrane Ca^{2+} pumps

Thermogenesis. Dissipation of energy as heat by futile cycles of proton transfer across the mitochondrial inner membrane of brown and beige fat cells

Thick filament. Large bipolar filaments of myosin-II in striated muscles; interleaved with thin filaments

Thin filament. Actin-based filaments in muscle cells; interleaved with thick filaments

Thin section. Slice of plastic-embedded tissue for viewing by transmission electron microscopy

Threshold. Membrane potential required to activate voltage-gated Na-channels and initiate an action potential

Thrombin. Blood enzyme that cleaves the plasma protein fibrinogen into fibrin during clotting

Thromboxanes. Lipid second messengers derived from prostaglandin H that promote platelet aggregation

Thylakoid membranes. Chloroplast membranes containing proteins for photosynthesis

Thymine. Pyrimidine base found in DNA; H-bonds with adenine

Thymosin β-4. Small protein sequesters actin monomers

Tic (translocon at the inner membrane of chloroplasts). Integral membrane proteins specialized to transport proteins across the inner chloroplast membrane

Tight junction. Intercellular junction that occludes the extracellular space and regulates the passage of solutes between epithelial cells

Tim complexes (translocase of the inner mitochondrial membrane). Integral and peripheral proteins of the inner mitochondrial membrane that transport proteins into the matrix or inner membrane

Tissue inhibitors of metalloproteinases (TIMPs). Secreted proteins that bind and inhibit matrix metalloproteinases

Tissue stem cells. Stem cells with the capacity to renew themselves and to produce daughter cells that differentiate into a limited range of specialized cells

Titin. Giant striated muscle protein that extends between Z-disks and M-lines

T-lymphocyte. Immune cells that promote antibody production or kill cells infected with viruses

Toc (translocon at the outer membrane of chloroplasts). Integral membrane proteins that transport proteins across the outer chloroplast membrane

Toll-like receptors (TLRs). Plasma membrane receptors that bind certain repeated features of macromolecules

Tom complex (translocase of the outer mitochondrial membrane) Integral and peripheral membrane protein complex of the outer mitochondrial membrane that translocates polypeptides into mitochondria

Tomography. Computational reconstruction of 3D images from a series of images taken at a wide range of angles relative to optical axis of the microscope

Topologically associating domains (TADs). Chromosomal regions that localize near one another in the nucleus

TORC1 complex. Complex containing mTOR kinase that regulates catabolism in response to amino acid levels in lysosomes

Total internal reflection fluorescence. Illumination method for fluorescence microscopy; a beam of exciting light is reflected from the interface between the slide and the aqueous specimen, setting off an evanescent wave that penetrates the specimens only ~100 nm

Transcription. Synthesis of RNA complementary to a DNA template strand

Transcription elongation. Phase of transcription in which RNA polymerase synthesizes an RNA strand complimentary to the sequence of the template DNA

Transcription export (TREX) complex. Complex of mRNA and protein that docks on the inner surface of the nuclear pore, where the RNA they contain is subjected to quality control by the exosome prior to export to the cytoplasm

Transcription factor. Proteins that associate with promoter sequences and are necessary for specific transcription by purified RNA polymerases in vitro

Transcription initiation. First phase of transcription including formation of a preinitiation complex leading to an open complex with unwound DNA and formation of the first phosphodiester bond between the first two complementary ribonucleotides

Transcription termination. Final phase of transcription when RNA polymerase reaches a signal on DNA that causes an extended pause in elongation, release of the nascent transcript, and base pairing of the DNA

Transcription unit. Gene-coding and regulatory (*cis*-acting) DNA sequences that direct transcription initiation, elongation, and termination

Transcytosis. Vesicular transport of extracellular ligands through the cytoplasm across a cell

Transducin. Trimeric G-protein coupled to rhodopsin in photoreceptor cells

Transesterification. Reactions during RNA splicing coupling formation of a new phosphodiester bond with breaking old bonds, so, input of energy is not required

Transfer RNA (tRNA). Small RNA adaptors between amino acids and mRNA codons during protein carrying a particular amino acid on one end and with a sequence of bases (anticodon) on the other end complementary for the mRNA codons that specify that amino acid

Transformed cell. Cell lacking normal growth control that proliferates as long as nutrient and mitogen supplies last regardless of whether or not it is touching neighboring cells

Transforming growth factor-β (TGF-β). Activates a serine/threonine kinase receptor and pathways that promote the differentiation of mesenchymal cells

Transgenic. Organism with a genome containing foreign genes

***Trans*-Golgi network (TGN).** Exit face of the Golgi apparatus specialized for sorting cargo to various destinations

Transit sequences. Amino acid sequences that target polypeptides to chloroplasts

Translation. Protein synthesis catalyzed by ribosomes and guided by the sequence of nucleotides in mRNA that specifies the sequence of amino acids in a polypeptide

Translocating chain-associating membrane protein (TRAM). ER protein that facilitates docking of ribosomes with the Sec61 complex

Translocon. Protein-conducting channel composed of Sec61 complex in the ER and SecYE complex in Bacteria

Translocon-associated protein (TRAP). ER protein that facilitates docking of ribosomes with the Sec61 complex

Transmembrane segment. Part of a polypeptide that extends into or through a lipid bilayer

Transmission electron microscope. Optical system that uses a beam of electrons focused by electromagnetic lenses to produce an image

Transporter associated with antigen processing (TAP). ABC transporter that translocates peptides generated by the immunoproteasome into the ER for loading onto class I major histocompatibility complex I for presentation on the cell surface to immune cells

Transport protein particle (TRAPP I). A multi-subunit tethering protein on the Golgi apparatus for ER vesicles

Transposable elements (transposons). DNA segments dispersed throughout a genome, that either are now or were formerly capable of moving from place to place in the DNA

Transposons. *See* Transposable elements

Treadmilling. *See* Subunit flux

Trigger factor. Protein chaperone associated with bacterial ribosomes

Trimeric G-protein. Signal transduction complex consisting of a GTPase (alpha subunit), beta subunit and gamma subunit; GTP binding to the alpha subunit dissociates it from beta/gamma; both alpha and beta-gamma subunits can activate target molecules

Triskelion. A three-legged clathrin molecule

tRNA. *See* Transfer RNA

Tropomodulin. Protein that blocks the pointed end of actin filaments and binds tropomyosin

Tropomyosin. Alpha-helical coiled-coil protein that binds end to end along actin filaments

Troponin. Trimeric protein that binds calcium and cooperates with tropomyosin to regulate striated muscle contraction

TRP channels. Family of cation channels that serve as temperature sensors among other functions

t-SNARE. *See* SNAP receptor

T tubule. Plasma membrane in invaginations of striated muscles that communicate action potentials deep into the cytoplasm

Tumor necrosis factor. Inflammatory protein that activates trimeric receptors

Tumor suppressor. Gene that predisposes to cancer when inactivated; protein products are typically negative regulators of cell proliferation

Tunicamycin. Drug inhibits glycosylation of dolichol phosphate and therefore the formation of N-linked glycoproteins

Two-component signaling. Signaling pathways that consist of a minimum to proteins, a histidine kinase and its cytoplasmic substrate protein that regulates cellular processes including gene expression and flagellar rotation

Two-hybrid assay. Bioassay for protein interactions based on interacting proteins reconstituting a split transcription factor and activating a reporter gene

Type 1 transmembrane protein. Transmembrane protein with its N-terminus facing the ER lumen or cell exterior, and C-terminus in the cytoplasm

Type 2 transmembrane protein. Protein with its N-terminus in the cytoplasm, C-terminus facing the ER lumen or cell exterior and transmembrane segment acting as an internal signal sequence

Tyrosine-based sorting motif. A sequence that directs proteins into clathrin-coated pits on the plasma membrane

Ubiquitin. Small protein that when attached to the ε amino group of a lysine of a target protein, either signals the target protein for destruction or marks it for other interactions

Ubiquitin-activating enzyme. *See* E1 enzyme

Ubiquitin-conjugating enzyme. *See* E2 enzyme

Unfolded protein response (UPR). Response pathway triggered by excess misfolded proteins in the ER that leads to activation of genes controlling ER function

Uniporter. Carrier proteins that catalyze movements of solutes across membranes down concentration gradients

Unique-sequence DNA. DNA sequences typically present in a single copy per haploid genome, often coding regions of genes

Unprocessed pseudogenes. DNA sequences created by reverse transcription of unspliced precursor mRNAs or local duplications of the chromosome that generally occur as a result of recombination between transposable elements

Uracil. Pyrimidine base found in RNA; H-bonds with adenine

Vacuolar ATPase. V-type H^+ rotary ATPase pump that acidifies the compartments along the endocytic pathway

VAMP (vesicle associated membrane protein) associated protein (VAP). Transmembrane protein that anchors other proteins at membrane contact sites

van der Waals interaction. Distance-dependent attraction or repulsion of closely spaced atoms

VASP. An actin polymerase that promotes elongation of actin filament barbed ends

Vesicle-tubule carrier (VTC). Pleomorphic transport intermediates ferry secretory cargo from ER to Golgi apparatus

Vesicular transport. Delivery of cargo between donor and acceptor membrane-bound compartments involving small vesicles or tubular-vesicular carriers

Vimentin. Intermediate filaments in mesenchymal cells

Vinblastine. Drug from periwinkle that interferes with microtubule dynamics; useful in cancer treatment

Vinculin. Actin-binding adapter protein in focal contacts

Voltage-gated channel. Ion channels with a domain that senses the electrical potential across the membrane and opens the channel gate above a certain threshold

von Willebrand factor. Adhesive glycoprotein that participates in platelet aggregation and blood clotting

v-SNARE. *See* SNAP receptor

WASp. Protein that activates Arp2/3 complex to form branched actin filaments; product of the gene mutated in Wiskott-Aldrich syndrome, an X-linked immunodeficiency and bleeding disorder

Wee1. Nuclear protein kinase phosphorylates Y^{15} in the active site of Cdk1, inhibiting its function as part of a mechanism that holds Cdk1-cyclin B1 poised for a burst of activation

Wnt, wingless. Secreted growth factors that influence stem cell proliferation and many steps in tissue morphogenesis

Wortmannin. Inhibitor of phosphatidylinositol 3-kinase

WW domain. Adapter domains bind certain phosphoserine and phosphothreonine peptides; found in many proteins

Xeroderma pigmentosum (XP). Human genetic disease with hypersensitivity to sunlight and predisposition to skin cancer caused by defects in nucleotide excision repair genes

Xist. Large non-coding RNA expressed by the inactive X chromosome and playing a key role in its inactivation

Z-disk. Anchoring site for the barbed ends of striated muscle actin filaments

Zeta-associated protein–70 kD (ZAP-70). Cytoplasmic tyrosine kinases mediate signals from lymphocyte receptors

Zinc finger TF. Transcription factors composed of small domains stabilized by a single Zn^{2+} ion that each recognize three base pairs in DNA sequences

Zip1 (mammalian SYCP1). Forms transverse filaments oriented perpendicular to chromosome axes between the axial elements in mature synaptonemal complex

ZO-1, -2, and -3. Adapter proteins between claudins in tight junctions and the cytoskeleton

Zonula adherens. Ring-shaped adhesive junction around the apex of epithelial cells based on cadherins and anchored to cytoplasmic actin filaments

Zygotene. Second stage of meiotic prophase; pairing of homologous chromosomes and clustering of telomeres gives rise to a "bouquet" arrangement of chromosomes

Index

Page numbers followed by "*f*" indicate figures, "*t*" indicate tables, and "*b*" indicate boxes.

A

A kinase-anchoring proteins (AKAPs), 429
A site of tRNAs, 213
AAA ATPase family, 623, 623*f*, 625*b*, 636*f*
Aa-tRNA synthetases. *See* Aminoacyl-tRNA (aa-tRNA) synthetases.
A-band, 673*f*
Abasic sites, 750
ABC transporters, 244*t*, 250-252, 250*f*-251*f*, 346-347, 347*f*
  degradation by proteasomes and, 401
  protein translocation across plasma membrane by, 311, 314*f*
  type I, 314*f*, 315
ABP-280, 590*t*-591*t*
Abp1p, 590*t*-591*t*
Abscission, 771*f*, 775-776, 776*f*-777*f*
Accidental cell death, 798, 798*b*
  programmed cell death vs., 797-800, 797*f*-800*f*
Acetylated N-terminus, 34*f*
Acetylcholine (ACh), 291*f*, 295
  neuromuscular junctions and, 294
  skeletal muscle contraction and, 678
Acetylcholine (ACh) receptors, 294
  muscarinic, 293-294
  nicotinic, 273-275, 274*f*, 279
Acetylcholinesterase, 294
Acetyl-CoA, 319, 321*f*
ACh. *See* Acetylcholine (ACh).
ACh receptors. *See* Acetylcholine (ACh) receptors.
Achiasmate segregation, 785
Acrosomal process, 652*f*, 666*f*
Actin, 575-576, 576*f*
  bulk movement of cytoplasm driven by, 647-648, 648*f*-649*f*
  contractile ring of, 771-772
  critical concentration of, 579-580
  dynamics of, in live cells, 585-588, 586*b*, 586*f*, 588*f*
  evolution of, 577-578, 578*f*
  molecule of, 577, 577*f*
  movements of organelles based on, 648-649, 649*f*
  polymerization of, 579-580, 579*f*-580*f*, 640*t*
    cytoplasm movements driven by, 649, 650*f*
  of skeletal muscle, 682*t*
  substitute for nematode sperm, 666-667, 667*f*
  unpolymerized, pool of, 587
Actin filaments, 6*f*, 12-13, 13*f*, 574-575, 575*f*-577*f*, 579
  assembly of, 69-70, 69*f*
  exotic eukaryotic motility systems and, 666*b*-667*b*
Actin filaments *(Continued)*
  initiation of, 587
  mechanical properties of cytoplasm and, 588-589, 589*f*
  organization of, 587-588, 588*f*
  polymerization and, 579-580, 579*f*-580*f*, 582-583, 583*f*
  related proteins (Arp), 578-579, 578*f*, 581-582, 582*f*, 641, 641*f*
  stabilizers of, 586*b*
  subunit recycling and, 587
  termination of, 587
  turnover of, 585-587, 586*b*, 586*f*
Actin isoforms, 578, 578*f*
Actin-binding proteins, 580-585, 581*f*
  actin filament nucleation factors as, 581-582, 582*f*
  binding sides of filaments, 584
  capping, 583-584, 583*f*, 590*t*-591*t*
  classification of, 590*t*-591*t*
  crosslinking, 584, 584*f*, 590*t*-591*t*
  filament-severing, 583*f*, 584, 588
  functional redundancy of, 584-585
  inhibitors of, 586*b*
  intermediate filament binding, 590*t*-591*t*
  membrane associated, 590*t*-591*t*
  microtubule binding, 590*t*-591*t*
  monomer-binding, 580-581, 581*f*, 590*t*-591*t*
α-actinin, 584, 590*t*-591*t*, 682*t*
Actin-myosin motors, 640*t*
Actin-related proteins (Arps), 578-579, 578*f*, 581*f*
  actin filament nucleation factors and, 581-582, 582*f*
  adapter proteins and, 584-585
Action potentials, 261
  in cardiac pacemaker cells, 686*b*, 686*f*
  channels generating, 290
  description of, 289-290, 289*f*
  membrane depolarization and, 290
  myelin sheaths and, 291, 291*f*
  skeletal muscle contraction and, 678-680, 679*f*-681*f*
  triggering of, by cyclic nucleotide-gated channels, 465
Activated state of seven-helix receptors, 413, 413*f*
Activation loop, 428, 473*f*, 475
  of receptor tyrosine kinase catalytic domain, 414-415
Active chromatin hubs, 135
Active state, channels and, 265
Activins, 417
  signal transduction and, 481
Actomyosin apparatus, organization of, 671-676, 672*f*-673*f*
  intermediate filaments and, 675-676, 675*f*
Actomyosin apparatus, organization of *(Continued)*
  thick filaments and, 671-672, 674-675, 674*f*, 682*t*
  thin filaments and, 671-674, 673*f*-674*f*, 682*t*
  titin filaments and, 675, 675*f*
Actomyosin ATPase cycle, 626-628, 627*f*
Acute promyelocytic leukemia (APL), 145-146
ADAMa, 520
ADAMTS, 520
Adaptation, 412
  bacterial chemotaxis and, 483*f*, 485-486, 485*f*
  odor detection and, 464*f*, 465-466
  robust, 486
Adapter domains, molecular recognition by, 437-440, 437*t*, 438*f*-439*f*
  adapters with proline-rich ligands and, 439*f*, 440
  EH domains and, 440
  PDZ domains and, 440
  phosphorylation-sensitive adapters and, 439-440
Adapters, nuclear traffic and, 155-156, 157*f*, 804-805
Adaptive immune system, 500, 501*f*
Adaptive immunity, 500-503, 501*f*-502*f*
Adaptor proteins, 361, 362*f*
ADAR (adenosine deaminase acting on RNA), 195*f*, 204*f*
Adducin, 590*t*-591*t*
Adenine, 42, 44*f*
Adenosine deaminase acting on RNA (ARAD), 195*f*, 204*f*
Adenosine diphosphate (ADP), 42
Adenosine monophosphate (AMP), 42
Adenosine triphosphate (ATP), 3-4, 44*f*
  membrane pumps driven by, 243-248, 244*t*
    ABC transporters and, 244*t*, 250-252, 250*f*-251*f*
    A-type ATP synthases and, 245*f*, 247-248
    F_0F_1-ATPase family and, 244*t*, 246*f*
    F-type ATPases and, 245*f*-246*f*, 247
    P-type cation pumps and, 244*t*, 248-250, 248*f*-249*f*
    rotary ATPases in, 243-248, 244*t*, 245*f*-247*f*
    V-type ATPases and, 247, 247*f*
  synthesis of, 21. *See also* Photosynthesis.
    by dual photosystems, 328
    by oxidative phosphorylation, 317, 319-322, 320*f*-321*f*
Adenosine triphosphate-gated channels, 267

Adenylyl cyclases, 444, 445*f*, 687
metabolic regulation and, 471
odor detection and, 465
ADF/cofilins, 581, 584–585, 584*f*, 587, 590*t*–591*t*, 655*f*, 656
Adherens junctions, 490, 529, 529*f*, 529*t*, 543, 550–551, 552*f*
Adhesion
cellular. *See* Cellular adhesion.
pseudopods and, 654*f*, 656–657, 656*f*
Adhesion proteins, 9, 525, 525*f*
Adhesive glycoproteins, 523
in extracellular matrix, 514–517, 515*f*, 523
fibronectin as, 515–516, 515*f*–516*f*, 523
tenascin as, 516–517, 517*f*
Adipocytes, 491*f*, 492, 493*f*
ADP. *See* Adenosine diphosphate (ADP).
ADP-ribosyl cyclase, 457, 457*f*
ADP-ribosylation Factor GTPases, 355, 356*f*
Adrenal gland, 687
Adrenaline, metabolic regulation and, 469–472, 470*f*, 471*t*
β-adrenergic receptor kinase, 472
Adrenergic receptors, 469–470
metabolic regulation through, 469–472, 470*f*, 471*t*
Affinity chromatography, 91*b*, 91*f*, 96
Agarose gels, 89*b*, 108*f*
Aggrecan, 513–514, 514*f*, 522, 556–557
Aging, telomeres and, 121–122, 122*f*
Agonists, 410, 446–447
Agrin, 523, 684*t*
Agrobacterium, 313–315
AKAPs. *See* A kinase-anchoring proteins (AKAPs).
Alanine, structure of, 33*f*
Aldosterone, amiloride-sensitive channels and, 267
Algae
blue-green, 20
green, 22–23, 23*f*, 660
red, 22–23, 23*f*
Alleles
definition of, 86*b*
mutant, conditional, 87
All-*trans* retinal, 468
ALT (alternative lengthening of telomeres), 120
Alternative lengthening of telomeres (ALT). *See* ALT (alternative lengthening of telomeres).
Alternative splicing, 193–194, 195*f*
Alzheimer disease, 602, 602*f*
protein misfolding in, 219*b*
Amide nitrogen, 34*f*
Amide protons, 35–36
Amiloride-sensitive channels, 266–267
Amino acids, 29
properties of, 32–35, 33*f*–35*f*
sequence of, transmembrane sequence identification by, 234*b*
Amino groups, 34, 34*f*
Amino terminus, 36
Aminoacyl-tRNA (aa-tRNA) synthetases, 210–212, 211*f*
γ-aminobutyric acid (GABA), 291*f*
synaptic transmission and, 295
α-amino-3-hydroxy-5-methyl-4-isoxazole propionate (AMPA) receptors
glutamate and, 273
long-term potentiation and, 296–297, 296*f*
Ammonia channels, 276–277, 276*f*, 277*b*
Amoeba proteus, restriction point in, 719, 719*f*
AMP. *See* Adenosine monophosphate (AMP).
AMPA receptors. *See* α-amino-3-hydroxy-5-methyl-4-isoxazole propionate (AMPA) receptors.
Amphiphilic, fatty acids as, 229
Amphitelic attachment, 763, 763*f*
Amyloid fibrils, 219*b*
Anaphase, 700, 755, 755*f*–756*f*, 767–770, 767*f*
biochemical mechanism of sister chromatid separation in, 767–768, 768*f*
cytoplasm during, 709
mitotic spindle dynamics and chromosome movement during, 768–770, 769*f*
Anaphase A, 700, 767–769
Anaphase B, 700, 767–769
Anaphase-promoting complex/cyclosome (APC/C), 710*t*–711*t*
formation of, 708
geminin degradation by, 731*f*, 732
inactivation of, 707–708, 708*f*
spindle checkpoint and, 764–765, 764*f*
Anchoring fibrils, 506*f*, 509–510, 509*f*, 521
Anemia, aplastic, 493–494
Anesthetics, local, sodium channels and, 270
Aneuploidy, 765, 784*b*, 794–795, 795*t*
Anillin, 590*t*–591*t*
Animal cells, 4*f*
Golgi apparatus in, 367*f*
Animals, time line for divergence of, 16*f*
Annelid worms, 25
Annexin-II, 590*t*–591*t*
Anopheles gambiae, genome of, 109*t*
Antagonists, 410
Anterograde movements, 643–644
Anterograde traffic, 351*f*, 352
Anterograde transport, 640*t*, 643–646, 643*f*–645*f*
Antiapoptotic activities, 798*b*
Antibodies, 43*f*, 500, 502
Anticodon, 210
Antidiuretic hormone, 277
Antigen-presenting cells, 477*f*, 478–479, 503
Antigen-presenting compartment, 381
Antigens, 500
Rh, 276–277, 277*b*
Antiporters, 254, 254*f*–255*f*, 256–257, 257*t*
AP1 complex, 361
AP2 complex, 361
Apaf-1, 803–805, 803*f*, 809, 810*f*
APC gene, 530
APC/C. *See* Anaphase-promoting complex/cyclosome (APC/C).
Apical domain of tight junctions, 543–545
Apical plasma membrane, 551
tight junctions and, 543–546
trafficking to plasma membrane and, 372–373, 372*f*
APL. *See* Acute promyelocytic leukemia (APL).
Aplastic anemia, 493–494
Apoptosis, 12, 696, 798*b*, 798*f*
autophagy and, 399
execution phase of, 798*b*, 799–800, 799*f*
extrinsic pathway of, 804–805, 810–812, 811*f*
genetic analysis of, 803–807, 803*f*
p53 gene and, 812, 813*f*
protein regulators and effectors of, 804–807, 805*f*–806*f*, 806*b*
signals and pathways of apoptosis and, 803–804
human disease and, 814
intrinsic pathway of, 803–804, 803*f*, 808–810, 808*f*, 810*f*
latent phase of, 798*b*, 799, 799*f*
necrosis vs., 797–800, 797*f*–800*f*
Apoptosomes, 804–805
Apoptotic bodies, 798*b*, 799–800
APs. *See* Assembly proteins (APs).
Aquaglyceroporins, 277
Aquaporins, 276*f*, 277
Aqueous phase of cytoplasm, 51–52, 52*f*
Arabidopsis thaliana
cellulose synthase genes of, 568, 568*f*
genome of, 109*t*
mitochondria of, 318
Arachidonic acid, 449, 450*f*–451*f*
Archaea, 3, 3*f*–4*f*
evolution of, 15*f*, 19*f*
ARE-mediated decay, 190*f*
AREs. *See* A+U-rich elements (AREs).
Arf, 355, 356*f*, 366*f*, 373*f*
Arf1, 355, 356*f*, 365–367, 366*f*, 369
Arf 6, 355
Arf GTPases, 355, 356*f*, 360*f*, 361, 365–366, 442*t*
Arginine, structure of, 33*f*
Arp2/3 complex, 581, 582*f*, 590*t*–591*t*, 655*f*, 656
Arps. *See* Actin-related proteins (Arps).
Arrest point, 702*b*, 702*f*
Arrestins, 413*f*, 414, 466, 469
β-arrestins, 472
ARS core consensus sequence, 729–730, 730*f*
Arteriosclerosis, 475
Arthropods, 25
Asparagine, 33*f*, 342, 343*f*
Aspartate, phosphorylation and, 425
Aspartic acid, structure of, 33*f*
β-aspartyl phosphate intermediate, 248
Assembly proteins (APs), 361, 362*f*
Association rate, 54–55, 67
Asters in prophase of mitosis, 757, 757*f*
Astral microtubules, 759–760
AT-AC introns, 193, 193*f*
Ataxia-telangiectasia, 749*t*
ATF6, 345–346, 345*f*
Atlastins, 333*f*, 334
AT-like disorder, 749*t*

ATM, DNA damage response and, 701, 701*f*, 724, 724*f*–725*f*, 748*f*–749*f*, 750–753
ATP. *See* Adenosine triphosphate (ATP).
ATP synthases, 20, 321*f*, 322
  membrane pump driven by, 245*f*, 247–248
ATP-gated channels, 267
ATR, DNA damage response and, 724, 724*f*, 748*f*–749*f*, 750–753
Atrial natriuretic factor, 419
Atrial natriuretic peptide, 267
Atrioventricular node, 685–686, 685*f*
Attachment in phagocytosis, 378–379
A-type ATP synthases, membrane pumps and, 245*f*, 247–248
Atypical cadherins, 529*t*
A+U-rich elements (AREs), 190*f*, 197
Aurora-B protein kinase, 756
Autoinhibitory propeptides, 519
Autolysosomes, 398–399, 398*f*
Autonomously replicating sequences (ARS elements), 729–730, 729*f*–730*f*
Autophagic vacuoles, 398–399
Autophagosomes, 398, 398*f*
Autophagy, 302, 396, 696, 797, 798*b*, 814
  degradation by, 397*f*–398*f*, 398–399
Autosomal negative mutation, 620–621
Autosomes, 791
Axial elements, 788
Axon hillock, 295
Axonal transport
  fast, 640*t*, 643–644, 643*f*–644*f*
  slow, 640*t*, 644–646, 645*f*
Axonemes, 595, 659–665, 660*f*–664*f*
Axons, 464
  growth cone of, 654–655, 654*f*
Axopodia, 653*f*, 654
Axostyles, 667, 667*f*

B

B lymphocytes, 500, 501*f*
Bacillus subtilis, 47–48, 311
  cytokinesis in, 777*b*
  genome of, 109*t*
Bacteria, 3, 3*f*–4*f*. *See also specific Bacteria*.
  cytokinesis in, 774*f*, 777*b*
  evolution of, 15*f*, 19*f*
  flagella of
    assembly of, 70–71, 70*f*–71*f*
    motility by, 665–668, 668*f*–669*f*
  gram-negative, outer membrane of
    protein insertion in, 313
    secretion across, 314*f*
Bacterial chemotaxis, 483*f*–485*f*, 485–486
  adaptation and, 483*f*, 485–486, 485*f*
  extended range of response and, 486
  temporal sensing of gradients and, 485, 485*f*
Bacteriochlorophylls, 326
Bacteriophage T4, 68, 73–74, 73*f*
Bacteriopheophytin b, 326
Bacteriorhodopsin, light-driven proton pumping by, 235*f*, 242–243, 243*f*, 412
Baculovirus IAP repeat (BIR), 805
Barr bodies, 128*f*, 129–130
β-barrels, 39, 305–306
Basal bodies
  axonemal, 593, 594*f*, 595–596, 605, 608, 662–665, 663*f*–664*f*
  bacterial, 668, 669*f*
Basal lamina, 489, 509
  of extracellular matrix, 509*f*, 517–519, 518*f*, 519*t*
Base excision repair, 750, 751*f*
Bases of nucleotides, 42, 44*f*
Basic region zipper, 179*f*
Basolateral domains of plasma membrane, 543–545
Basolateral plasma membrane
  tight junctions and, 543–546
  trafficking to plasma membrane and, 372–373, 372*f*
Basophils, 494*t*, 497*f*, 499–500
Batrachotoxin, 279
Bax, 812
Bcl-2 proteins
  apoptosis and, 803, 803*f*, 808–809, 808*f*
  cancer and, 809, 809*f*
Beige fat cells, 491*f*, 492–493
Benzodiazepines, 275
Berns, Sam, 152*f*
Bestrophin channels, 275–276, 275*f*
B-form DNA, 43–44, 46*f*
BH domains, 806*b*, 808
Biglycan, 522
Bilayer, lipid. *See* Lipid bilayer.
Bim1p, 611
BiP, 341–342
Bipolar attachment, 762
Bipolar kinesin-5 motors, 635
BIR. *See* Baculovirus IAP repeat (BIR).
Bivalents, 790
Blood cells
  origin and development of, 493–495, 494*f*, 494*t*
  platelets, 493–496, 494*f*, 494*t*, 496*f*
  red, 494*f*, 494*t*, 495
  white, 497, 497*f*
Blood clotting, 496, 540
Blotting, 89*b*
Blue-green algae, 20
BMPs. *See* Bone morphogenetic proteins (BMPs).
Bonds. *See* Chemical bonds.
Bone morphogenetic proteins (BMPs), 417, 557
  signal transduction and, 481
Bones, 557–561, 558*f*
  bone cells and, 559–561, 559*f*–560*f*
  diseases of, 565, 570
  extracellular matrix of, 558*f*, 559, 559*t*
  fractures of, 566–567
  skeleton formation and growth and, 561–565
    embryonic, 562–564, 562*f*–563*f*
    remodeling and, 564–565, 564*f*
Bordetella pertussis, 313–315
Bouquet stage of meiosis, 780*f*, 785–786, 785*f*
BPAG1, 552–553, 552*f*, 590*t*–591*t*, 619*t*
BPAG2, 552–553, 552*f*
BPAG1e, 619
Brachiopods, 25
Brain, odor detection and, 464*f*, 466, 466*b*
Brain damage in stroke, 814, 814*f*
Branch point, 192
Breast cancer, predisposition to, 749*t*
Brefeldin A, 355, 356*f*
Bright field microscopy, 76, 76*t*, 77*f*
Bromodomains, 178, 178*t*
Brown fat cells, 491*f*, 492–493
B-type cyclins, 743, 743*f*
  initiation of prophase and, 745–746, 746*f*
  subcellular localization changes and, 744*f*, 745, 746*f*
BUB (budding uninhibited by benzimidazole), 764–765
Budding yeast. *See Saccharomyces cerevisiae*.
Bulky DNA adducts, 750, 751*f*
Bullous pemphigoid, 552–553, 552*f*
α-Bungarotoxin, 274, 279

C

Cl^- channels, 295
Ca^{2+} calmodulin, 428–429
Ca^{2+}-ATPase, 248–250, 248*f*–249*f*, 292*f*, 294
CAD domain, 529, 530*f*–531*f*
CAD nuclease, 807–808, 807*f*
Cadherins, 489–490, 526, 528–532, 529*f*–532*f*, 529*t*
CADPribose. *See* Cyclic adenosine diphosphate-ribose (cADPribose).
Caenorhabditis elegans
  apoptosis in, 803–804, 803*f*
  centromeres in, 140
  centrosomes of, 606
  cytokinesis in, 772
  dynein heavy chain of, 637
  genome of, 109–110, 109*t*
  meiosis in, 782
  seven-helix receptors of, 412
Cajal bodies, 145, 145*t*, 194, 202
CAK (Cdk-activating kinase), 705–706, 706*f*, 710*t*–711*t*, 744
Calcineurin, 429–430, 430*t*, 478–479
Calcium as second messenger, 409, 452–459
  calcium dynamics in cells and, 457–458, 459*f*
  calcium targets and, 458–459, 462
  calcium-release channels and, 453*f*, 454–457, 455*f*–457*f*, 455*t*
  overview of calcium regulation and, 452–453, 453*f*, 453*t*, 462
  removal from cytoplasm, 453*f*, 454, 454*t*
Calcium channels, 453, 453*f*
  agonist-gated, 453*f*, 454–457, 454*t*–455*t*
    inositol 1,4,5-triphosphate receptor and, 454–456, 455*f*–456*f*, 455*t*
    ryanodine receptor, 454, 455*f*, 455*t*, 456–457, 457*f*
  voltage-gated, 270–271, 271*t*, 292*f*, 294–295, 453*f*, 454, 455*f*, 455*t*, 456–457, 686*b*
Calcium spark, 458
Calcium spike, contraction and, 681, 690, 690*f*
Calcium-ATPase pump, 453, 453*f*, 679

Calcium-release channels, 271-272, 453*f*, 454-457, 455*f*-457*f*, 455*t*
Calcium-sensitive contractile fibers, 666, 666*f*
Calcium-sensitive fluorescent dyes, 458
Caldesmon, 584, 590*t*-591*t*
Calmodulin, 41, 42*f*, 272, 457, 466, 584, 683, 690-691, 690*f*
Calmodulin-regulated spectrin associated proteins. *See* CAMSAPs.
Calnexin, 343
Calnexin cycle, 343-344, 344*f*, 347*f*
Calponin, 590*t*-591*t*
Calreticulin, 343, 454
Calsequestrin, 454
Calspectin, 590*t*-591*t*
Caltractin, 666
Calveolae-mediated endocytosis, 377, 378*f*, 382-384, 382*f*-383*f*
Calveolin, 382
CAMP. *See* Cyclic adenosine 3′,5′-monophosphate (cAMP).
CAMP response element, 185-186
CAMP response element-binding (CREB) protein, 185-186
CAMP signaling. *See* Cyclic adenosine monophosphate (cAMP) signaling.
CAMP-gated channel, 272, 444*f*, 462
CAMSAPs, 604
Cancer
  Bcl-2 proteins and, 809, 809*f*
  breast, predisposition to, 749*t*
  centrosomes and, 608-609, 608*f*-609*f*
  G1 regulation and, 699-700, 725-726
  restriction point and, 699-700, 721-722, 722*f*-723*f*
  telomeres and, 121-122, 122*f*
Cancer stem cells, 717
Cap 100, 590*t*-591*t*
Cap recognition complexes, 214
Capacitance of membrane, 282
Capping proteins, 590*t*-591*t*, 655*f*, 656
  heterodimeric, 583-584
CapZ, 590*t*-591*t*, 673, 682*t*
Carbohydrates
  complex, 29, 48
  structure of, 48-51, 50*f*-51*f*
  synthesis of, 328
α-carbons, 34, 34*f*
Carbonyl oxygen, 34*f*, 35-36
Carboxyl groups, 34, 34*f*
Carboxyl terminus, 36
Carboxyl-terminal domain, 168-169
CARD. *See* Caspase recruitment domain (CARD).
Cardiac glycosides, 249-250
Cardiac muscle, 672*f*, 685-691
  congestive heart failure and, 688, 688*f*
  contractile apparatus of, 685, 685*f*
  excitation contraction coupling and, 679, 680*f*, 685*f*
  molecular basis of inherited heart disease and, 682*t*, 688
  pacemaker cells of, 685-687, 685*f*-686*f*, 686*b*
  smooth and, 672*f*, 688-691, 689*f*-690*f*
Cardiomyopathies, hypertrophic, 688
Cargo proteins, 352-353, 352*f*
  clathrin coats and, 361
  Golgi apparatus and, 367-369, 368*f*
  protein-based machinery and, in membrane trafficking, 352-353, 352*f*, 357-362, 358*f*-359*f*, 361*f*
  in secretory transport from endoplasmic reticulum, 365, 367
  sorting from *trans*-Golgi network and, 371-375, 371*f*, 373*f*
Carrier proteins, 226, 253-259, 370
  definition of, 241
  diversity of, 255*f*, 257-259
  physiology and mechanisms of, 253-257, 253*f*, 255*f*, 257*t*
    of antiporters, 254, 254*f*-255*f*, 256-257, 257*t*
    discovery of, 255*b*
    of symporters, 254-256, 254*f*-255*f*, 257*t*
    of uniporters, 253*f*-254*f*, 254, 256, 257*t*
  structures of, 255*f*
  tools for studying, 257*t*
Carrier vesicles, 352, 352*f*, 359, 366*f*
  multivesicular bodies and, 377-378, 386, 388*f*-389*f*, 389-390
  protein sorting and
    by lipid gradients, 354
    protein-based machinery for, 354-355, 357*f*, 358-359
Carriers, membrane-enclosed, 352, 352*f*
Cartilage, 556-557, 556*f*-557*f*
  diseases involving, 557, 570
  specialized forms of, 557, 557*f*
CAS, 158
Caspase recruitment domain (CARD), 806*b*, 809, 810*f*
Caspases
  apoptosis and, 804-805, 805*f*
  degradation by proteasomes and, 402
  effector, 804, 805*f*
  initiator, 804, 805*f*
  natural inhibitors of, 805-807, 806*b*, 806*f*
Catalytic cycle in transcription elongation, 173, 174*f*
Catalytic domain of myosin, 624-626
Catastrophe, 598
α-catenin, 529-530, 551, 552*f*
β-catenin, 529-530, 531*f*-532*f*, 532, 551, 552*f*
γ-catenin, 529-530
Cathepsin K, 560-561
Cation channels, synaptic transmission and, 295
Caulobacter crescentus, intermediate filaments of, 615
Caveolae, 377, 378*f*, 382-384, 382*f*-383*f*, 390, 390*f*-391*f*
Caveolae-dependent uptake, 377
Caveolin, 382
Caveolin-3, 684*t*
Caveosomes, 390*f*
Cavins, 382, 383*f*
Cbl, 475-476
C-CAM (C-cell adhesion molecule), 528*t*
CD45, 431-432, 478, 538*t*
CD2 cells, 528*t*
CD4 cells, 478, 503, 528*t*
CD8 cells, 478, 503, 528*t*
CD31 cells, 528*t*
CD34 cells, 538*t*
CD43 cells, 538*t*
CD44 cells, 538*t*
CD58 cells, 528*t*
Cdc2. *See* Cdk1.
Cdc6, 731, 732*t*
Cdc20, 707-709, 708*f*, 710*t*-711*t*, 764*f*, 765, 766*f*
Cdc25, 432, 704, 710*t*-711*t*, 744-745
Cdc42, 588, 588*f*
CDC mutants. *See* Cell division cycle (CDC) mutants.
Cdc25A, 744-745, 745*f*
Cdc25C, 744-745, 745*f*
  subcellular localization changes and, 744*f*, 745
CD11/CD18 cells, 533*t*
CD44E cells, 538*t*
CDE I, 114, 114*f*, 141
CDE II, 114, 114*f*, 141
CDE III, 114, 114*f*
Cdk1, 704, 710*t*-711*t*, 743, 743*f*-744*f*
  initiation of prophase and, 745-746, 746*f*
  subcellular localization changes and, 744*f*, 745
Cdk inhibitor p21, 714*b*, 753
Cdk inhibitors, 706-707, 710*t*-711*t*, 714*b*, 714*f*, 715
Cdks. *See* Cyclin-dependent kinases (Cdks).
CDNA. *See* Complementary DNA (cDNA).
Cdt1, 731, 732*t*
Ced (cell death abnormal) mutants, 803, 803*f*, 806*b*
Cell adhesion molecules
  immunoglobulin, 489-490, 526-528, 528*f*, 528*t*
  integrin, 532, 533*t*
  selectin, 536*t*
Cell adhesion proteins, 9, 525, 525*f*
  growth cones and, 659
Cell autonomous nucleases, 807-808, 807*f*
Cell cortex, 4*f*, 575, 640*f*, 647, 648*f*, 649
Cell culture, 86-87
Cell cycle, 7-8, 8*f*, 13-14, 63*f*, 697-710, 698*f*
  apoptosis and, 812, 813*f*
  biochemical basis of transitions during, 701-707
    cyclin-dependent kinases and, 701, 702*b*-704*b*, 702*f*-705*f*, 710*t*-711*t*
    cyclins and, 701-703, 705*f*-706*f*
    negative regulation of cyclin-dependent kinase structure and function and, 706-707, 706*f*, 710*t*-711*t*
    positive regulation of cyclin-dependent kinase structure and function and, 703-706, 706*f*
  centrosome duplication during, 8*f*, 605*f*-607*f*, 606-607
  changing states of cytoplasm during, 709-710, 709*f*
  factors essential for progression of, discovery of, 704*b*, 704*f*-705*f*

Cell cycle *(Continued)*
genetics for study of, 702*b*, 702*f*
perturbation of, programmed cell death and, 802
phases of, 697-700, 698*f*-699*f*, 699*b*
checkpoints and, 699*b*, 700-701, 701*f*
G_0 phase as, 696, 698*f*, 700, 713-715, 713*f*-714*f*, 714*b*, 718-719, 718*f*
G_1 phase as. *See* G_1 phase.
G_2 phase as. *See* G_2 phase.
M phase as, 697-700, 698*f*, 699*b*
S phase as. *See* S phase.
phosphatases in counter-balancing of Cdks and, 706*f*, 707, 710*t*-711*t*
progression, control of, 700-701, 701*f*
protein destruction in control of, 707-708, 708*f*
regulation of, principles of, 697, 698*f*
regulation of meiotic events, 792
in vitro studies of, 703*b*, 703*f*
Cell death
accidental, 798
programmed cell death vs., 797-800, 797*f*-800*f*
programmed. *See* Apoptosis; Programmed cell death.
Cell division cycle (CDC) mutants, 702*b*, 702*f*, 704
Cell functions, strategy for understanding, 59-60
Cell lines, 86-87
Cell shape, cellular motility in alteration of
contraction and, 653*f*, 654
extension of surface processes and, 651-654, 652*f*-653*f*
Cell walls of plants, 567-569, 567*f*-568*f*
Cells
eukaryotic. *See* Eukaryotic cells.
prokaryotic vs. eukaryotic, 4*f*, 8-9
universal principles of, 4-8, 5*f*-8*f*
Cellular adhesion, 525-541, 525*f*
adhesion receptors and
cadherin family of, 528-532, 529*f*-532*f*, 529*t*
dystroglycan/sarcoglycan complex as, 538, 538*t*
galactosyltransferase as, 537, 538*t*
identification and characterization of, 527
integrin family of, 532-536, 533*f*-534*f*, 533*t*, 537*f*, 539*f*
with leucine-rich repeats, 537-538, 538*t*, 539*f*
selectin family of, 536-537, 536*f*-537*f*, 536*t*
dynamic, 538-541, 541*f*
of leukocytes to endothelial cells, 537*f*, 538-540
platelet activation and, 539*f*, 540
self-avoidance in nervous system and, 540-541
general principles of, 525*f*-526*f*, 526-527
IgCam and, 527-528, 528*f*, 528*t*
Cellular motility, 651-669, 651*f*, 652*t*
by cell shape changes
produced by contraction, 653*f*, 654
Cellular motility *(Continued)*
produced by extension of surface processes, 651-654, 652*f*-653*f*
by cilia and flagella, 659-665, 660*f*-665*f*
bacterial flagella and, 665-668, 668*f*-669*f*
exotic eukaryotic motility systems and, 665, 666*b*-667*b*
primary cilia and, 660*f*, 665, 665*f*
by pseudopod extension, 654-659, 654*f*
chemotaxis of motile cells and, 657-658, 658*f*
growth cone guidance and, 658-659, 659*f*
lamellar motility on flat surfaces and, 654*f*-657*f*, 655-656
myosin and, 657
substrate and, 654*f*, 656-657, 656*f*
Cellulose, 49, 51*f*, 568
Cellulose synthases, 568, 568*f*
CEN sequences, 113-114, 114*f*
CENP-A, 115-117, 117*f*, 140, 140*f*
CENP-B, 111, 116, 116*f*, 140, 140*f*
CENP-C, 117*f*, 140-141, 140*f*
CENP-E, 634*f*, 635, 635*t*
CENPs (centromere proteins)
of budding yeast, 141, 141*f*
mammalian, 140-141, 140*f*
CENP-T, 140-141
Central nervous system, synapses of, 293*f*, 294-295
modification by drugs and disease, 295
modification by use, 295-297, 296*f*
Central pair, 660
Central spindle, 759-760
of microtubules, 769
Centralspindlin, 772
Centrifugation, 88-90
Centrin, 666
Centrioles, 604-605, 605*f*, 607*f*, 612
Centriolin, 612
Centromere anticentromere antibodies, 140*f*
Centromere proteins (CENPs). *See* CENPs (centromere proteins).
Centromeres, 10, 107, 108*f*, 113-117, 140, 140*f*
of budding yeasts, 114, 141, 141*f*
definition of, 108*b*
DNA of, 114-115, 114*f*, 141*f*
mammalian, 115-117, 116*f*-117*f*, 141
point centromere, 114
regional, 114
RNAi at, 141-142
variation in organization among species, 113-115, 114*f*
Centrosomes, 4*f*, 12-13, 63*f*, 593, 594*f*, 604-610, 604*f*-605*f*
cancer and, 608-609, 608*f*-609*f*
duplication of, 8*f*, 605*f*-607*f*, 606-607
functions of, 607-608, 607*f*-609*f*
proteins of, 605-606, 605*f*-606*f*, 612
of yeasts, 608*f*-609*f*, 609-610
Ceramide, 419-420, 446
synthesis of, in endoplasmic reticulum, 348
Ceramide signaling pathways, 452, 452*f*
Ceramide transport protein (CERT), 349
CERT. *See* Ceramide transport protein (CERT).
CFTR. *See* Cystic fibrosis transmembrane regulator (CFTR).
CGMP. *See* Cyclic guanosine 3′,5′-monophosphate (cGMP).
CGMP-gated ion channels, 418, 421-423, 468-469, 469*b*
CGMP-stimulated protein kinases, 418, 428*f*
CGN. *See Cis*-Golgi network (CGN).
α-chains in extracellular matrix, 506-508
Channelrhodopsins, 243
Channels, 226, 241-242. *See also* Calcium channels; Membrane channels.
Chaperone pathway, prokaryotic protein export via, 313, 314*f*
Chaperone-mediated autophagy (CMA), 399
Chaperones, 209
molecular, 68
prokaryotic protein export via, 313, 314*f*
protein folding assisted by, 218-221, 220*f*-222*f*
Chaperonins, 220-221, 220*f*, 222*f*
Charcot-Marie-Tooth disease, 550, 550*t*
Charge movement
net current through ion-selective channels and, 283
redistribution by electrical conduction, 284, 284*f*
for small cells, 282-283
through channels, rate of, 283
CheA, 482*b*, 483*f*, 484-485
CheB, 485-486, 485*f*
Checkpoints
cell cycle and, 14, 696, 699*b*, 700-701, 701*f*
G_1 phase and, 700-701, 723-725, 724*f*-725*f*
G_2 phase and, 700, 747-750, 747*f*-751*f*, 749*t*, 750*b*-753*b*
G_2/M, from DNA damage response to, 749*f*, 750-753
intra-S, 739-740, 739*f*
spindle, 764-765, 764*f*
Chemical bonds, 58-59, 58*f*
covalent, 58, 58*f*
in nucleic acids, 42-43, 44*f*-45*f*
electrostatic (ionic), 58*f*, 59
hydrogen, 58, 58*f*
peptide, 32, 34*f*
phosphodiester, 42, 44*f*
Chemical genetics, 99-100
Chemiluminescence, 89*b*
Chemiosmotic cycles, 256, 285-286, 285*f*, 321*f*, 325*f*
Chemoattractants, growth cones and, 658-659
Chemokine receptors, 497
Chemokines, 497
Chemotaxis
bacterial, 483*f*-485*f*, 485-486
adaptation and, 483*f*, 485-486, 485*f*
extended range of response and, 486
temporal sensing of gradients and, 485, 485*f*
of motile cells, 657-658, 658*f*
Chemotherapeutic killing of cells, 803

CheR, 485–486, 485*f*
CheY, 37*f*, 482*b*, 483*f*–485*f*, 484
CheZ, 484–485, 485*f*
Chiasmata, 785*f*, 786–787, 789–790, 789*f*–790*f*
ChIP-seq. *See* Chromatin immunoprecipitation coupled with high-throughput sequencing (ChIP-seq).
Chk1, 749*f*, 750–753
Chk2, 750–753
Chlamydomonas, 22–23
  axonemes of, 662, 663*f*–664*f*, 664
  centrosomes of, 605
  flagella of, 660
Chloride channels, 275–276, 275*f*–276*f*
Chlorophyll, 323
Chlorophyll b, 326
Chloroplast stroma, 324, 324*f*–325*f*, 327*f*, 328
Chloroplasts, 301–302, 323–328
  inner membrane of, 324
  origins and evolution of, 22–23, 23*f*
  outer membrane of, 324
  photosynthesis and
    carbohydrate synthesis and, 328
    energy capture and transduction by photosystem I, 327–328
    energy capture and transduction by type II photosystems and photosystem II, 326, 327*f*
    light and dark reactions and, 325–326, 325*f*
    light harvesting and, 326–327, 327*f*
    oxygen-producing synthesis of NADPH and ATP by dual photosystems, 328
    plastids and, 324
    structure and evolution of photosynthesis systems, 323–324, 324*f*–325*f*
  protein transport into, 303*f*, 307–309, 308*f*
  stroma of, 324, 324*f*–325*f*, 327*f*, 328
Cholera toxin, 436–437, 436*t*
Cholesterol, 229–230, 230*f*, 353–354, 353*f*–354*f*
  covalently bound, 420
  endoplasmic reticulum and, 347–348, 347*f*–348*f*
  homeostasis of, 402–404, 403*f*–404*f*
Chondrocytes, 491*f*
  in cartilage, 556, 556*f*
  hypertrophic, 562–564, 563*f*
Chordata, 25
Chromatids, 108*b*, 108*f*, 700
  separation of, in anaphase of mitosis, 767–768, 768*f*
  sister, 107, 108*b*, 700, 756
Chromatin, 4*f*, 123
  definition of, 108*b*
  functional compartmentation of, 128–130, 128*f*–129*f*, 131*f*, 134
  inner nuclear membrane proteins and, 150*f*
  macromolecular assembly and, 5–6
  30-nm fiber, 130–131, 131*f*, 135
  nuclear lamina and, 151*f*
Chromatin *(Continued)*
  nucleosomes and
    modification and regulation of function of, 123–125, 125*f*
    regulation of structure of, by histone N-terminal tails, 125–127, 125*f*–126*f*
  TADs and, 134–137, 134*f*
  transcription and, 177–184, 177*f*, 178*t*
    combinatorial control and, 183–184
    gene-specific eukaryotic GTFs and, 178–180, 179*f*
    transcription factor activity and, 180–184, 180*f*–184*f*
Chromatin immunoprecipitation coupled with high-throughput sequencing (ChIP-seq), 177, 177*f*
Chromatography, 90, 91*b*, 91*f*
Chromodomains, 129, 178, 178*t*
Chromokinesins, 635
Chromonema fibers, 130–131
Chromosomal passenger complex (CPC), 763–764, 763*f*
Chromosome banding, 137*f*
Chromosome conformation capture (3C), 133–134, 134*f*
Chromosome cycle, cell cycle and, 697
Chromosome number
  diploid, 86*b*, 779, 784*b*
  haploid, 86*b*, 779, 784*b*
Chromosome scaffold, 136–137, 138*f*
Chromosome territories, 143
Chromosomes
  acrocentric, 108*f*
  attachment to spindle, in prometaphase of mitosis, 761–764, 762*f*–763*f*
  centromeres and, 113–117, 116*f*–117*f*
    variation in organization among species, 113–115, 114*f*
  definition of, 108*b*
  dicentric, 117*f*
  DNA of, 107–108, 107*f*–108*f*
    packaging in interphase nuclei, 130–131, 131*f*
    telomeric replication of, 118–120, 118*f*–119*f*
  gene organization in, 108–110, 109*t*, 110*f*
  holocentric, 115
  homologous, 779
    in prophase I of meiosis, 785–787, 785*f*–786*f*
  human genome segmental duplications and, 112–113, 113*f*
  interphase, with clearly resolved loop structures, 132–133, 133*f*
  kinetochore and, 139–140, 139*f*
  lampbrush, 132, 133*f*, 787
  metacentric, 108*f*
  mitochondrial, 318
  mitotic, 135–136, 136*f*–137*f*, 147
  morphology of, 107, 108*f*
  movement of, in anaphase of mitosis, 640*t*, 768–770, 769*f*
  nomenclature for, 107
  nonhistone proteins and, 136–139, 136*f*, 138*f*–139*f*
  polytene, 132–133, 133*f*
Chromosomes *(Continued)*
  pseudogenes and, 112
  secondary constriction of, 147
  submetacentric, 108*f*
  telocentric, 108*f*
  telomeres and, 117–122
    aging and, 121–122, 122*f*
    cancer and, 121–122, 122*f*
    replication of ends of chromosomal DNA, 118–120, 118*f*–119*f*
    structural proteins of, 120–121, 120*f*–121*f*
    structure of telomeric DNA, 118
  transposons and, 110–112, 111*f*
ClC chloride channels, 275, 275*f*–276*f*
Cilia, 8, 12
  motility by, 659–665, 660*f*–661*f*, 663*f*, 665*f*
    exotic eukaryotic motility systems and, 665, 666*b*–667*b*
  primary, 608, 660*f*, 665, 665*f*
  rotary, 665
  sensory, 464
Ciliopathies, 665
Circadian cycle, visual system and, 469*b*
Cis-Golgi network (CGN), 367
Citric acid cycle, 319, 321*f*
CKIs (cyclin-dependent kinase inhibitors), 706–707, 710*t*–711*t*, 714–715
CLASPs, 603, 611
Clathrin coats, 302, 357–362, 357*f*–358*f*, 362*f*–365*f*
Clathrin-coated pits, 383*f*, 384–386
Clathrin-coated vesicles, 378*f*, 383*f*–384*f*, 384–385
Clathrin-mediated endocytosis, 377, 378*f*, 383*f*–384*f*, 384–385
  compartments associated with, 386–390, 387*f*–389*f*
Claudins, 538*t*, 544*t*, 545, 545*f*
Cleavage furrow
  constriction in cytokinesis, 774–775
  signals regulating position of, in cytokinesis, 771–772, 771*f*
Cleavage stimulus, 771*f*, 772
CLIP-170, 603, 611
Clonal expansion, 502
Cloning, 90–92, 93*f*
Closed complex, 169–170, 170*f*
Closed mitosis, 755
Clostridium botulinum, 436–437
Clostridium difficile, 436–437
CMA. *See* Chaperone-mediated autophagy (CMA).
CMG helicase, 731, 731*f*, 732*t*
C-NAP1, 612
Cnidarians, 25
Coactivators, 126, 126*f*, 182, 182*f*
Coat protein I. *See* COPI coat.
Coat protein II. *See* COPII coat.
Coated pits, 4*f*, 383*f*, 384
Coatomer complex, 359–360, 360*f*
α-Cobra toxin, 279
Cockayne syndrome, 749*t*
Codons, 209–210
Cohesin, 137–138, 139*f*, 767, 768*f*, 790–791
Cohesin complex, 790–791

Coiled-coils, 40, 40*f*, 70, 135, 672-674
Colchicine, 595*b*
Collagen fibrils, 489, 555
Collagens, 505-510, 505*f*-506*f*
  diseases involving, 519, 519*t*, 556
  fibrillar, 506*f*-507*f*, 507, 521-522
    biosynthesis and assembly of, 507-509, 508*f*-509*f*, 521
  linking, 506*f*, 509-510, 509*f*
  sheet-forming, 506*f*, 509, 509*f*, 521
  type I, 556, 559, 559*t*
  type II, 556, 556*f*
  type IV, 517-519, 518*f*, 519*t*
  type VI, 684*t*
  type XVII, 552-553
Color vision, 467*b*
Committed progenitor cells, 715-716
Committed stem cells, 494, 494*f*, 496, 715-716
Common ancestor, 3, 3*f*, 15, 15*f*
  last, divergent evolution from, 16*f*-17*f*, 17-18, 18*b*
Compartmentalization, eukaryotic, 21-22, 21*f*-22*f*
Compartments
  chromatin, 134
  endocytic, 386-390, 387*f*-390*f*
Complementary DNA (cDNA), 32, 88
  isolation of genes and, 90-92, 92*f*-93*f*
Complementary surfaces, specificity by multiple weak bonds on, 64-65
Complementation, 86*b*, 87
Complex carbohydrates, 29, 48
Complex I, 322
Complex II, 322
Complex III, 322
Concentration
  critical, 69-70
    of GTP-tubulin dimers, 597
  of reactants, 54*b*
Concentration gradients, 284
Condenser lens, 76, 77*f*
Condensin, 137, 138*f*-139*f*, 756-757
Conditional lethal phenotype, 702*b*
Conditional mutant alleles, 87
Conditional mutants, 99-100
Conditional mutations, 86*b*, 100
Conductance, 283
Cone photoreceptors, 467-468, 467*b*
Confocal microscopy, 79-81, 79*f*-81*f*
Conformational changes, 30, 41, 42*f*, 49*f*, 426, 427*f*
Congestive heart failure (CHF), 688, 688*f*
Connective tissues, 555-570
  bones as, 557-561, 558*f*
    cells of, 559-561, 559*f*-560*f*
    diseases of, 565, 570
    extracellular matrix of, 558*f*, 559, 559*t*
    fractures of, 566-567
    skeleton formation and growth of, 561-565, 562*f*-564*f*
  cartilage as, 556-557, 556*f*-557*f*, 570
  collagen fibrils in, 507*f*
  dense, 555-556, 555*f*
  loose, 555, 555*f*
  plant cell walls as, 567-569, 567*f*-568*f*
  wound repair in, 565-567, 566*f*
Connexins, 538*t*, 544*t*, 547-549
  mutations in, 550, 550*t*
Connexons, 547-548, 549*f*
ω-conotoxin, 271, 271*t*, 279
μ-Conotoxins, 279
Consensus target sequence, 426-428
Constitutive heterochromatin, 115, 129, 129*f*
Constitutive protein turnover, 393-396, 394*f*-395*f*, 395*t*
Constitutive secretion, 22*f*, 467*f*
Constriction
  primary, 113
  secondary, 147
Contact inhibition, 530, 658, 722
Contactin, 528*t*
Contractile ring, 575-576, 700, 771-772
  assembly and regulation of, in cytokinesis, 772-774, 775*f*
Contractions
  active, of muscles, 676-677, 677*f*
  of cardiac muscle, 685, 685*f*
  in cell shape changes, 653*f*, 654
  graded, 678, 678*f*
  of skeletal muscle
    molecular basis of, 676-678, 676*f*-677*f*
    regulation of, 678-682, 678*f*-681*f*
  of smooth muscle, 688-691, 689*f*-690*f*
COPI coat, 357-362, 357*f*-358*f*, 360*f*-361*f*
COPII coat, 357-362, 357*f*-359*f*
Corals, 25
Core enzyme, 167
Core histones, 123, 124*f*
Core mannose oligosaccharide, 343*f*-344*f*, 344
Core protein, 512
Corepressors, 126
Corona of kinetochore, fibrous, 139, 139*f*
Coronin, 590*t*-591*t*
Cortex, 4*f*, 575
Cotranslational translocation, 335-342
  insertion of membrane proteins into ER bilayer, 339*f*, 340
  lipid-anchored protein association with cytoplasmic surface of ER and, 340
  Sec61 complex and, 337-338, 338*f*
  signal recognition particle and signal recognition particle-receptor and, 335-337, 336*f*-337*f*
  signal sequences and, 335
  of soluble proteins into lumen of ER, 338-340, 339*f*
  tail-anchored protein association with ER membrane and, 335*t*, 341-342, 342*f*
Covalent bonds. *See* Chemical bonds.
Covalent crosslinking, 508*f*-509*f*, 509
Covalently attached lipids, 433-434
Covalently bound cholesterol, 420
Covariance method, 45-46
COX-2. *See* Cyclooxygenase-2 (COX-2).
CPC. *See* Chromosomal passenger complex (CPC).
CpG islands, 129-130
Crane, H. R., 64*b*
CREB protein. *See* CAMP response element-binding (CREB) protein.
CRISPR/Cas9 system, 94-95
Cristae, 318, 318*f*
Critical concentration, 69-70
  of actin, 579-580
  of GTP-tubulin dimers, 597
Crossbridges, skeletal muscle contraction and, 676*f*-677*f*, 677-678
Crossing over, 784*b*
Crosslinking of actin filaments, 584, 584*f*, 589, 589*f*, 590*t*-591*t*, 657*f*
Crossover interference, 790
Crossovers, 782, 783*f*, 784*b*
Cryomicroscopy, electron, 82
CSF. *See* Cytostatic factor (CSF).
C-Src, 427*f*, 472, 479*b*
CTCF, 130, 131*f*, 134-137
C-terminal domain (CTD), 190-191
C-terminal propeptides, 507-508
C-terminus, 36
Curare, 274
Cutis laxa, 512
Cyanobacteria, 19*f*, 20
  photosystems of, 324, 324*f*-325*f*
Cyclic adenosine diphosphate-ribose (cADPribose), 457-458, 457*f*
Cyclic adenosine 3′,5′-monophosphate (cAMP)
  odor detection and production of, 465
  as second messenger, 443-445, 444*f*-445*f*
Cyclic adenosine monophosphate (cAMP) signaling, 185-186
Cyclic guanosine 3′,5′-monophosphate (cGMP), 443-445, 444*f*
Cyclic nucleotide phosphodiesterases, 418
Cyclic nucleotide-gated ion channels, 272, 272*f*, 443-444, 465
Cyclic nucleotides, 409, 443-445, 444*f*-445*f*
Cyclin-dependent kinase inhibitors. *See* Cdk inhibitors.
Cyclin-dependent kinases (Cdks), 429, 695-696, 701, 702*b*-704*b*, 702*f*-705*f*, 710*t*-711*t*
  phosphatases' role in counter-balancing of, 706*f*, 707, 710*t*-711*t*
  regulation of structure and function of
    negative, 706-707, 706*f*, 710*t*-711*t*
    positive, 703-706, 706*f*
Cyclins, 429
  B-type, 743, 743*f*
    initiation of prophase and, 745-746, 746*f*
    subcellular localization changes and, 744*f*, 745, 746*f*
  cell-cycle transitions and, 701-703, 705*f*-706*f*, 710*t*-711*t*
Cyclooxygenase-2 (COX-2), 450
Cyclophilin, 429-430
Cyclosporine, 429-430, 479
Cysteine, structure of, 33*f*
Cysteine disulfide, 34*f*
Cystic fibrosis, 346, 402
Cystic fibrosis transmembrane regulator (CFTR), 250, 252, 287-288, 287*f*, 402
Cytochalasins, 585, 586*b*
Cytochrome b_{c1}, 322
Cytochrome C, 321*f*, 322, 323*f*, 325*f*, 326-328, 327*f*, 809, 810*f*
Cytochrome oxidase, complex IV, 322

Cytochrome P450, 333
Cytokine receptors, 416-417, 417*f*, 421-423
  signal transduction and, 479-481, 480*f*
Cytokines, 479, 566*f*
Cytokinesis, 8*f*, 14, 696-699, 698*f*, 755, 755*f*-756*f*, 771-776, 771*f*, 773*f*-774*f*
  in bacteria, 774*f*, 777*b*
  cleavage furrow constriction in, 774-775
  contractile ring assembly and regulation in, 772-774, 775*f*
  exit from mitosis and, 776-777
  membrane addition and abscission in, 771*f*, 775-776, 776*f*-777*f*
  in plants, 772, 773*f*, 776*b*
  signals regulating cleavage furrow position and, 771*f*, 772
Cytoplasm
  aqueous phase of, 51-52, 52*f*
  changes during prophase of mitosis, 757-758, 757*f*-758*f*, 757*t*
  changing states of, during cell cycle, 709-710, 709*f*
  mechanical properties of, 588-589, 589*f*
  movement of
    driven by actin and myosin, 647-648, 648*f*-649*f*
    driven by actin polymerization, 649, 650*f*
  traffic between nucleus and, 152-159, 154*f*-157*f*
    defective, disorders associated with, 159
    mRNA export and, 158
    nuclear import and export components and, 155-157, 157*f*
    regulation of transport across nuclear envelope, 158-159, 159*f*
    single import cycle and, 157-158, 157*f*
Cytoplasmic dyneins, 612, 624*t*, 637
Cytoplasmic linker protein (CLIP)-170. *See* CLIP-170.
Cytoplasmic polyadenylation, 196
Cytoplasmic protein tyrosine kinases
  JAKs, 416-417, 417*f*, 421-423, 479-481, 480*f*
  Src family, 427*f*-428*f*, 428-429, 431-432, 437, 479*b*
  T-lymphocyte pathways through, 477*f*, 478-479, 479*b*, 480*f*-481*f*
Cytoplasmic ring, nuclear pore complex, 151, 153*f*
Cytoplasmic streaming, 640*t*, 647-648, 648*f*-649*f*
Cytosine, 42, 44*f*
  deamination to uracil, 195, 195*f*
Cytoskeleton, 12-13, 12*f*-13*f*, 574-575, 576*f*, 586*b*, 619, 620*f*
Cytostatic factor (CSF), 792-793, 793*f*
Cytotactin, 523
Cytotoxic T lymphocytes, 500, 501*f*

D

DAG. *See* Diacylglycerol (DAG).
Dam1 complex, 141
Dark reactions, 326
Dark-field microscopy, 76-78, 76*t*
Daughter centrioles, 605, 605*f*
DDR. *See* DNA damage response (DDR).
Deadenylases, 197
Deafness, 546, 550
Death domain (DD), 806*b*, 811
Death effector domain (DED), 806*b*, 811
Death-inducing signaling complex (DISC), 805
Debranching enzyme, 192
Decapping complex, 197
Deconvolution, 79, 79*f*
Decorin, 514*f*, 522
Decoy receptors, 561, 812
DED. *See* Death effector domain (DED).
Defensins, 379-381
Degradation, 393-405
  by constitutive protein turnover, 393-396, 394*f*-395*f*, 395*t*
  lipid turnover and, 402-404, 403*f*-404*f*
  in lysosomes, 396-399, 397*f*
    autophagy and, 397*f*-398*f*, 398-399
    delivery to lysosomes via endocytic pathway in, 397-398, 397*f*
  of mRNAs, 190*f*, 196-197, 196*f*
  by proteasomes, 399-402, 400*f*
    elimination of misfolded proteins from endoplasmic reticulum and, 402
    intracellular proteolysis and, 402
    motifs specifying ubiquitination and, 395*t*, 401-402
    ubiquitination modification of proteins and, 394-396, 394*f*-395*f*, 395*t*
  of protein in endoplasmic reticulum, 344, 344*f*
Degrons, 99, 393
Delayed early" genes, 718-719, 718*f*
Delta proteins, 420
Dematin, 590*t*-591*t*
Dendritic cells, 499, 503
Dense bodies in smooth muscle, 688-689, 689*f*
Dense connective tissue, 555-556, 555*f*
Dense fibrillar component of nucleolus, 147
Deoxyribonucleic acid. *See* DNA (deoxyribonucleic acid).
Depactin, 590*t*-591*t*
Dephosphorylation, 429-432, 439
Depolarization, membrane, 290
Depression, long-term, 295-296
Desensitized state, 294
Desmin, 614, 615*t*, 675-676, 675*f*
Desmocollins, 529*t*, 544*t*, 551
Desmogleins, 529*t*, 544*t*, 551
Desmoplakin, 551, 619, 619*t*
Desmosine, 511
Desmosomes, 490, 529, 529*f*, 529*t*, 543, 544*f*, 544*t*, 551, 552*f*
Destrin, 590*t*-591*t*
Deubiquitylating enzymes (DUBs), 395-396, 399-400
Development, transcription factors in, 185*f*, 186
Developmentally defective cells, 801-802, 801*f*
DHP receptors. *See* Dihydropyridine (DHP) receptors.
Diabetes mellitus, 375
Diacylglycerol (DAG), 446, 452
Diakinesis, 787
DIC microscopy. *See* Differential interference contrast (DIC) microscopy.
Dicer, 202
Dictyostelium discoideum
  cAMP as extracellular signal in, 444-445
  chemotaxis in, 657-658, 658*f*
  histidine kinases in, 482
Differential interference contrast (DIC) microscopy, 76, 76*t*, 77*f*, 79*f*
Diffusion, facilitated, 254
Diffusion coefficients, 54-55
Diffusion potentials, 281, 281*f*
Digitalis in congestive heart failure, 688, 688*f*
Dihydropyridine (DHP) receptors, 679
Dihydropyridines, 279
Dilated cardiomyopathies, 688
Dileucine-based sorting motifs, 361
Dilysine sorting motif, 360
Dimerization, induced, 415, 416*f*
Diploid chromosome number, 86*b*, 779, 784*b*
Diplotene, 785*f*, 787
Directed maturation, 379
Directionality/recycling factors, 156-157
DISC. *See* Death-inducing signaling complex (DISC).
Diseases. *See also specific diseases.*.
  of bone, 565, 570
  of cartilage, 557, 570
  centrosomes and, 608-609, 608*f*-609*f*, 612
  of cilia, 665
  collagen and, 519, 519*t*, 556
  DNA repair defects associated with, 749*t*, 753
  elastic fibers and, 512
  endoplasmic reticulum and, 334, 346
  of endoplasmic reticulum folding, 346, 402
  gap junctions in, 550, 550*t*
  GTPases in, 436-437, 436*t*
  heart, molecular basis of, 682*t*, 688
  immunodeficiency, 503
  kinases and, 429, 430*f*, 430*t*
  meiotic defects and, 794-795, 795*f*, 795*t*
  mitochondria and, 322-323, 323*f*
  nuclear envelope defects leading to, 150-151, 152*f*
  peroxisomal biogenesis disorders as, 309, 328-329
  programmed cell death and, 814, 814*f*
  red blood cell, 495
  seven-helix receptors and, 414, 414*t*
  of skeletal muscle plasma membrane, 682*t*, 683-684, 683*f*, 684*t*
  tau and, 602, 602*f*
Disjunction, 784*b*
Disks of outer segment of photoreceptors, 468
Dissociation rate, 53-55, 67
Dis1/TOG proteins, 603, 611
Disulfide bonds, 34*f*
Disulfide isomerase, 342, 507-508
Divergent evolution, 41-42
DNA2. *See* Fen1 (Flap endonuclease).

DNA (deoxyribonucleic acid), 3-4, 5*f*
bases of, 42, 44, 44*f*
B-form, 43-44, 46*f*
chromosomal, 107-108, 107*f*-108*f*
packaging in interphase nuclei, 130-131, 131*f*
telomeric replication of, 118-120, 118*f*-119*f*
complementary, 32
damage to, checkpoints and. *See* Checkpoints.
linker, 127
of mammalian centromeres, 115-117, 116*f*-117*f*
protein binding to specific domains of, 179-180, 179*f*
repetitive sequences of, 110-111
replication. *See* DNA replication.
satellite, 111
secondary structure of, 43-45, 46*f*
sequences associated with chromosome scaffold and nuclear matrix, 136-137, 138*f*
sugar of, 42
synthesis of, mechanism of, 734-735, 735*f*, 736*b*
unique-sequence, 110
DNA damage checkpoints, 701, 701*f*
G_1 phase and, 723-725, 724*f*-725*f*
G_2 phase and, 700, 747-750, 748*f*-751*f*, 749*t*, 750*b*-753*b*
intra-S, 739-740, 739*f*
DNA damage response (DDR), 747-750, 748*f*-751*f*, 749*t*, 750*b*-753*b*
DNA "libraries", 90-92
DNA methyltransferases, 129*f*, 204
DNA polymerase δ, 728, 732*t*, 734, 735*f*
DNA polymerase ε, 728, 732*t*, 734
DNA polymerases, 90, 92*f*, 111, 118, 118*f*, 127, 624*t*, 727-728, 732*t*, 734-735, 735*f*, 736*b*, 747, 747*f*
DNA repair in vertebrates, 750*b*-753*b*, 750*f*
base excision repair, 750, 751*f*
of double-strand breaks, 749*t*, 750*f*, 751-753, 752*f*
mismatch, 751, 751*f*
nucleotide excision repair, 749*t*, 750, 751*f*
DNA replication, 727-741, 727*f*-728*f*
bidirectional, origin of, 727
in *Escherichia coli*, 728, 728*f*, 736*b*
higher-order organization of, in nucleus, 735-739, 737*f*-738*f*
histone protein synthesis and, 740-741, 740*f*
intra-S checkpoint and, 739-740, 739*f*
mechanism of DNA synthesis and, 734-735, 735*f*, 736*b*
between meiosis I and meiosis II, suppression of, 792
origins of, 728-731, 728*f*
in mammalian cells, 730-731, 730*f*-731*f*
in *Saccharomyces cerevisiae*, 729-730, 729*f*-730*f*
prereplication complex assembly and, 731-733, 731*f*, 732*t*, 733*f*
signals starting, 732*t*, 733-734, 733*f*-735*f*
DNA replication checkpoint, 701
DNA topoisomerase IIα, 138-139
DNA transcription, 163, 167
DNA-dependent protein kinase, 752*f*, 753
DNase I, 590*t*-591*t*
Dolichol phosphate, 342-343, 343*f*
Domains in proteins, modular, 41-42, 43*f*
Dominant mutations, 86*b*, 87
Dominant negative gene, 87
Dominant negative mutants, 100
Dominant negative mutation, 620-621
Dopamine, 291*f*
Double arginine pathway, prokaryotic protein export via, 315
Double helix, 43-44, 46*f*
Double-strand break repair, 749*t*, 750*f*, 751-753, 752*f*
Down syndrome, 794
Downregulation, 381-382
DP1/YOP1 proteins, 333-334, 333*f*
Drebrin, 590*t*-591*t*
Drosha, 202
Drosophila melanogaster
centromere organization of, 115
centrosomes of, 606
chromosomes of, 120
cytokinesis in, 772, 775-776, 777*f*
dynein in, 639, 641
genome of, 109-110, 109*t*, 110*f*
growth cone guidance, 659, 659*f*
Hedgehog receptors of, 420
meiosis in, 782, 787-788
DUBs. *See* Deubiquitylating enzymes (DUBs).
Duty cycle, myosins and, 628
Dynactin complex, 637, 640-641, 641*f*, 642*t*
Dynamic instability, 595, 598, 598*f*, 599*t*, 600*f*
Dynamin, 361, 384-385
Dynamin-related GTPases, 437, 442*t*
Dyneins, 574, 595, 624*t*, 631, 635-637, 636*f*, 639
axonemal, 661, 661*f*-662*f*
evolution of, 623, 623*f*
mechanochemistry of, 636, 636*f*-637*f*
superfamily of, 636-637, 636*f*
Dysferlin, 684*t*
Dystroglycans, 517, 538, 538*t*, 684, 684*t*
Dystroglycan/sarcoglycan complex, 538, 683, 683*f*, 684*t*
Dystrophin, 590*t*-591*t*, 683-684, 683*f*, 684*t*

E

E1 enzyme (ubiquitin-activating enzyme), 394, 395*f*, 396
E2 enzyme (ubiquitin-conjugating enzyme), 394-396, 395*f*
E3 enzyme (ubiquitin ligase), 394-396, 395*f*, 395*t*
E2F, 720-721, 721*f*, 802
E site of tRNAs, 213
Early endosomal compartment, 387-389, 388*f*
Early endosome antigen 1 (EEA1), 387-388, 388*f*
Early endosomes, 377-378, 386, 387*f*-388*f*
Early recombination nodules, 784, 789*f*
EB-1, 603, 611
Ebola, 389
E-cadherins, 529*t*, 550-551
Echinoderms, 25
ECM. *See* Extracellular matrix (ECM).
EDEM, 344
Editing
genome, 85*t*, 94
of mRNAs, 195-196, 195*f*
Edman degradation, 32
EEA1. *See* Early endosome antigen 1 (EEA1).
EEG. *See* Electroencephalogram (EEG).
Effector caspases, 804, 805*f*
Effectors
DNA damage checkpoints and, 701, 748, 753
for G-protein isoforms, 435*t*
EFs. *See* Elongation factors (EFs).
EF-Tu. *See* Elongation factor Tu (EF-Tu).
EGF. *See* Epidermal growth factor (EGF).
EH domains, 440
Ehlers-Danlos syndrome, 556
Eicosanoids, 446, 450, 450*f*
EJC. *See* Exon-junction complex (EJC).
Elastic cartilage, 557
Elastic fibers in extracellular matrix, 489, 510-512, 510*f*-511*f*, 555
Electrical circuits in photoreceptors, 469*b*
Electrical potential, 261
Electrical recordings, 279-280, 280*f*
Electrical synapses, 291, 546
Electroencephalogram (EEG), 280
Electromyography (EMG), 280
Electron cryomicroscopy, 82
Electron microscopy, 77*f*, 82-85, 83*f*-84*f*
Electron transport pathway, 319, 321*f*, 325*f*
Electrophoresis, gel, 88, 89*b*, 89*f*
pulsed-field, 108*f*
Electrostatic bonds, 58*f*, 59
Electrostatic interaction, 31-32, 32*f*, 58, 237
11-*cis* retinal, 412*f*, 413, 467*f*, 468
Elongation
actin filament, 579, 579*f*
in DNA replication, 736*b*
in transcription, 165-166, 165*f*, 171*f*, 172-173, 174*f*
in translation, 215-217, 216*f*-217*f*, 220*f*
Elongation factor Tu (EF-Tu), 432*f*, 433
Elongation factors (EFs), 215, 432*f*, 433, 442*t*
Embryonic stem (ES) cells, 715
Emerin, 619*t*, 684*t*
Emery-Dreifuss muscular dystrophy, nuclear envelope defects and, 150, 152*f*
EMG. *See* Electromyography (EMG).
Emi1, 708-709, 708*f*-709*f*
Endocannabinoids, 451-452
Endocytic pathway, 302, 397-398, 397*f*
Endocytosis, 377-392, 378*f*
calveolae-mediated, 377, 378*f*, 382-384, 382*f*-383*f*
clathrin-mediated, 377, 378*f*, 383*f*-384*f*, 384-385
compartments and pathways of, 386-390, 387*f*-390*f*
macropinocytosis and, 377, 378*f*, 380*f*-381*f*, 381-382

Endocytosis *(Continued)*
nonclathrin/noncaveolar, 377, 385-386, 385*f*
phagocytosis and, 377-381, 378*f*-379*f*
alternative fates of ingested particles and, 379-381, 381*b*
attachment in, 378-379
engulfment in, 378-379, 379*f*-380*f*, 380*b*
fusion with lysosomes and, 379-381, 379*f*
signaling and, 389*f*-390*f*, 390-391
viruses and protein toxins as "opportunistic endocytic ligands" in, 391, 391*f*
Endoplasmic reticulum (ER), 6, 7*f*, 8, 10-11, 11*f*, 301-302, 331-351, 331*f*
ceramide synthesis in, 348
cholesterol synthesis, metabolism, and, 347-348, 347*f*-348*f*
functions of, 331-333, 332*f*, 332*t*
lipids in
biosynthesis of, 346
transport of, 341*t*, 348-349, 349*f*
membrane contact sites of, 334, 334*f*, 341*t*
protein degradation in, 344, 344*f*
protein folding and oligomerization in, 342-344, 343*f*-344*f*
diseases and, 346, 402
protein translocation into, 335-342
insertion of membrane proteins into ER bilayer, 339*f*, 340
lipid-anchored protein association with cytoplasmic surface of ER and, 340
posttranslational, 335, 340-341, 341*f*
Sec61 complex and, 337-338, 338*f*
signal recognition particle and signal recognition particle-receptor and, 335-337, 336*f*-337*f*
signal sequences and, 335
of soluble proteins into lumen of ER, 338-340, 339*f*
tail-anchored protein association with ER membrane and, 335*t*, 341-342, 342*f*
refilling with calcium, 453*f*, 454
rough, 4*f*, 332, 332*f*
secretory transport to Golgi apparatus from, 352, 352*f*, 365-367, 366*f*
shaping and fusion proteins, 333-334, 333*f*
smooth, 333, 678-679, 679*f*-680*f*
stress responses of, 345-346, 345*f*
structure of, 331*f*-332*f*, 332-333, 332*t*, 336*f*
Endosomal membrane system, 377-378
Endosome/lysosomal system, sorting from *trans*-Golgi network and, 373-374, 373*f*
Endosomes, 11, 396, 498*f*
early, 377-378, 386, 387*f*-388*f*
early/recycling, 377-378, 386, 388*f*
late, 377-378, 386, 388*f*, 390
recycling, 377-378, 386, 388*f*
Endothelial cells, cellular adhesion of leukocytes to, 537*f*, 538-540
Engulfment in phagocytosis, 378-379, 379*f*-380*f*, 380*b*
Enhancer elements, 180-184, 181*f*-183*f*
Enhancer mutations, 96-97, 96*f*
Enhancers, 163
Entactin, 523
Entamoeba, mitochondria of, 318
Entamoeba histolytica, 21
Enterotoxin, 419
Enthalpy, 57
Entropy, 57
Entry face of Golgi apparatus, 367
Environmental conditions, regulation of assembly by, 68
Enzymes
debranching, 192
mechanism of, analysis of, 60-62, 60*f*-61*f*
oxidative, 309
reactions producing lipid second messengers, 445-446
recombination, 501-502
ubiquitin-activating, 394, 395*f*, 707
ubiquitin-conjugating, 394, 395*f*, 707
Eosinophils, 491*f*, 494*t*, 497*f*, 499
Epiblast cells, 715
Epidermal growth factor (EGF)
receptor tyrosine kinase activation by, 414
receptor tyrosine kinase pathway through Ras to MAP kinase, 473*f*-474*f*, 474-475, 483*f*
Epidermal keratins, 618, 618*f*, 619*t*
Epidermal stem cells, 716, 717*f*
Epidermolysis bullosa, 509-510, 620
Epifluorescence, 78-79, 79*f*
Epigenetic activation, 114-115
Epigenetic controls, 5
Epigenetic regulation, 117*f*, 125
Epinephrine, 469-472, 470*f*, 471*t*, 687
Epiphyseal plate, 562-564, 563*f*
Epithelial transport, 286-288, 286*f*
cystic fibrosis as transporter disease and, 287-288, 287*f*
food production and, 288
of glucose, in intestine. kidney, fat, and muscle, 286-287, 286*f*
of salt and water in kidney, 287
Epitope tags, 90
EpsinR, 374
Equational division, 779
Equilibrium, sedimentation, 88-90
Equilibrium constants, 29, 53, 55-56, 97-98
Equilibrium potential, 281
ER. *See* Endoplasmic reticulum (ER).
ER export domains, 332, 365, 366*f*
ERAD (ER-associated protein degradation), 344, 344*f*, 402
ER-associated degradation (ERAD), 344, 344*f*, 402
Erythrocytes, 494*f*, 494*t*, 495
Erythropoietin, 416, 494-495
ES cells. *See* Embryonic stem (ES) cells.
Escherichia coli
ABC transporters in, 250-251
bacteriophage T4 assembly and, 73-74, 73*f*
cytokinesis in, 774*f*, 777*b*
Escherichia coli (Continued)
DNA replication in, 728, 728*f*, 736*b*
flagella of, 665-668, 669*f*
genome of, 109*t*, 110
histidine kinases in, 482-483
Lac Y symporter of, 255*f*
poly(A) tails in, 198
SRP of, 313
ESCRT-I, -II, and -III (endosomal sorting complexes required for transport-I, -II, and -III), 389, 389*f*
E-selectin, 536*t*
ESEs. *See* Exonic splicing enhancers (ESEs).
Espin, 590*t*-591*t*
ESTs. *See* Expressed sequence tags (ESTs).
Eucarya, 3, 3*f*, 15*f*
Euchromatin, 128-130, 128*f*
Euglena, 23-24, 324
Eukaryotes
enhancer elements of, 180-184, 181*f*-183*f*
evolution of
compartmentalization in, 21-22, 21*f*-22*f*
divergence from last eukaryotic common ancestor, 23-24, 24*f*
in first billion years, 19*f*, 20-23, 21*f*-23*f*
multicellular, 3, 3*f*, 24-25, 24*f*
general transcription factors of, 170, 178-180, 179*f*
origin of, 3, 3*f*, 19*f*, 20, 21*f*
promoter proximal elements of, 176*f*, 180-184, 181*f*
RNA processing in. *See* MRNA (messenger RNA): synthesis of; RNA (ribonucleic acid)..
transcription factors of, gene-specific, 176-180, 176*f*, 179*f*
translation in, 214*f*, 216*f*
Eukaryotic cells
cell cycle of, 7-8, 8*f*, 13-14
cytoskeleton and motility apparatus of, 12-13, 12*f*-13*f*
endoplasmic reticulum of, 7*f*, 8, 10-11, 11*f*
exotic eukaryotic motility systems and, 665, 666*b*-667*b*
Golgi apparatus of, 8, 11, 11*f*
lysosomes of, 8, 11
mitochondria of, 3*f*, 8, 11-12, 11*f*
nucleus of, 10, 10*f*
organization and functions of, 9-14, 9*f*
peroxisomes of, 8, 12
plasma membrane of, 6*f*, 9, 9*f*
prokaryotic cells vs., 4*f*, 8-9
ribosomes and protein synthesis in, 7*f*, 10
EVH1 domains, 439*f*, 440
Evolution
of actin family, 577-578, 578*f*
connexin gene families and, 548-549
divergent, 41-42
of genes for intermediate filament proteins, 613*f*, 614-615, 615*t*
of membrane channels, 262, 263*f*
Evolution of life on earth, 15-26
of chloroplasts, 22-23, 23*f*

Evolution of life on earth *(Continued)*
divergent evolution from last common ancestor and, 16*f*-17*f*, 17-18, 18*b*
of eukaryotes, 19*f*, 20, 21*f*
divergence from last eukaryotic common ancestor, 23-24, 24*f*
in first billion years, 19*f*, 20-23, 21*f*-23*f*
multicellular, 24-25, 24*f*
of membrane-bounded organelles, 21-22, 21*f*-22*f*
of mitochondria, 20-21, 317-318, 317*f*-318*f*
prebiotic chemistry leading to RNA World and, 15-17, 16*f*
of prokaryotes, 18-20, 19*f*
Excess cells, programmed cell death and, 802
Excitable cells, 261
Excitable membranes, 289-291, 289*f*, 291*f*. *See also* Action potentials.
Excitation contraction coupling, 679, 680*f*, 685*f*
Excitatory neurotransmitter carriers, 258
Excitatory synapses, 295, 296*f*
Excited state, 326
Exocytosis, focal, 379
Exome, 110
Exon definition, 192
Exonic splicing enhancers (ESEs), 192
Exon-junction complex (EJC), 197
Exons, 109, 163, 192
5′ exonuclease, 190*f*
Exosome (RNA processing), 190*f*, 196-197, 196*f*
Exosomes, 389-390
Exotic eukaryotic motility systems, 665, 666*b*-667*b*
Expansins, 569
Expressed sequence tags (ESTs), 88
Extracellular electrical measurements, 280
Extracellular ligands, 538*t*
of integrins, 533-535, 533*t*
of selectins, 536*t*
Extracellular matrix (ECM), 9, 9*f*, 489, 505-523
adhesion to, 525, 551-553, 552*f*
adhesive glycoproteins in, 514-517, 515*f*, 523
fibronectin as, 515-516, 515*f*-516*f*, 523
tenascin as, 516-517, 517*f*
basal lamina of, 509*f*, 517-519, 518*f*, 519*t*
of bones, 558*f*, 559, 559*t*
collagen in, 505-510, 505*f*-506*f*, 521
fibrillar, 506*f*-509*f*, 507-509, 521
linking, 506*f*, 509-510, 509*f*
sheet-forming, 506*f*, 509, 509*f*, 521
elastic fibers in, 489, 510-512, 510*f*-511*f*
glycosaminoglycans in, 512-514, 513*f*-514*f*, 522
matrix metalloproteinases in, 519-520, 519*f*
proteoglycans in, 512-514, 513*f*-514*f*, 522
Extrinsic pathway of apoptosis, 804-805, 810-812, 811*f*
Ezrin, 590*t*-591*t*

F

Facilitated diffusion, 254
Facultative heterochromatin, 128-130, 128*f*
FADD (Fas-associated protein with a death domain), 842b, 811
$FADH_2$. *See* Flavin adenine dinucleotide, reduced form of ($FADH_2$).
Farnesyl, 149, 230
Farnesyl-pyrophosphate, 347, 347*f*
Fasciculin II, 659
Fascin, 590*t*-591*t*
Fas-ligand, 419, 419*f*, 811
in normal and diseased cells, 812
Fast axonal transport, 640*t*, 643-644, 643*f*-644*f*
Fat, glucose transport in, 286-287
Fat cells
brown and beige, 491*f*, 492-493
white, 491*f*, 492, 493*f*
Fatty acid oxidation, 319, 320*f*
Feedback loops, 409
Feedback mechanisms, 7-8, 8*f*
Fen1 (Flap endonuclease), 734-735, 750*f*, 752-753
FFAT motif, 334
Fibrillar centers, 147
Fibrillar collagens, 506*f*-507*f*, 507, 521-522
biosynthesis and assembly of, 507-509, 508*f*-509*f*, 521
Fibrillin microfibrils, 510, 510*f*
Fibrinogen, 523
Fibroblasts, 489, 491-492, 491*f*-492*f*, 566, 566*f*
Fibroglycan, 522
Fibromodulin, 522
Fibronectin receptors, 533*t*
Fibronectins
in extracellular matrix, 515-516, 515*f*-516*f*, 523
plasma, 516
tissue, 516, 516*f*
Fibrous corona of kinetochore, 139, 139*f*
Fibulin, 523
Filaggrin, 619*t*
Filamin, 590*t*-591*t*, 656, 657*f*
Filopodia, 575, 651-654, 652*f*-653*f*
First-order reactions, 53-54, 54*b*, 54*f*
FISH. *See* Fluorescence in situ hybridization (FISH).
Fission yeast. *See Schizosaccharomyces pombe*.
5′ cap, 189-190, 191*f*, 210, 210*f*
5′ splice site, 192-193, 193*f*
5S RNA, 167, 167*f*
FK506, 429-430
FK-binding protein, 429-430
Flagella, 12
bacterial
assembly of, 70-71, 70*f*-71*f*
motility by, 665-668, 668*f*-669*f*
eukaryotic, motility by, 659-660, 660*f*, 662, 662*f*-664*f*, 664
exotic motility systems and, 666*b*-667*b*
Flagellar pathway, prokaryotic protein export via, 314*f*, 315
Flap endonuclease. *See* Fen1 (Flap endonuclease).
Flavin adenine dinucleotide, reduced form of ($FADH_2$), 319
Flippases, 233
Fluorescence in situ hybridization (FISH), 78, 132*f*-133*f*
Fluorescence microscopy, 76*t*, 77*f*, 78, 79*f*-82*f*, 81*t*, 130-131, 131*f*
Fluorescence photoactivation localization microscopy (FPALM), 81-82, 81*t*
Fluorescence recovery after photobleaching (FRAP), 78, 146*f*, 237-238, 238*f*
Fluorescence resonance energy transfer (FRET), 78, 327
Fluorescent dyes
Ca^{2+}-sensitive, 458
for membrane potential measurement, 279-280
Fluorescent probes, 78, 79*f*
F-met-leu-phe, 421-423
FMLP. *See N*-formyl-methionine-leucine-phenylalanine (FMLP).
Focal adhesion kinase, 534, 534*f*
Focal adhesions, 534*f*, 535, 543
Focal contacts, 534, 534*f*, 544*t*, 551-552
Focal exocytosis, 379
Fodrin, 590*t*-591*t*
Food production, epithelial transport and, 288
Formins, 582-583, 583*f*, 656
N-formyl-methionine-leucine-phenylalanine (FMLP), 658
Forskolin, 444
FPALM, 81-82, 81*t*
Fractures, 566-567
Fragmin, 583-584, 583*f*, 590*t*-591*t*
FRAP. *See* Fluorescence recovery after photobleaching (FRAP).
Free energy, 56-57
Freeze-fracture method, 83*f*-84*f*, 84
FRET. *See* Fluorescence resonance energy transfer (FRET).
Fruit fly. *See Drosophila melanogaster*.
FtsZ, 595, 774*f*, 777*b*
F-type ATPases, membrane pumps and, 245*f*-246*f*, 247
Fukutin, 684*t*
Fungi, time line for divergence of, 16*f*
Fusion protein (molecular biology), 78
Fyn, 478
FYVE-domain, 387-388

G

G_0 phase, 696, 698*f*
exit from, 715
growth control and, 700, 713-715, 713*f*-714*f*, 714*b*, 718-719, 718*f*
G_1 phase, 696-700, 698*f*, 713-726, 713*f*
checkpoints and, 700, 723-725, 724*f*-725*f*
cytoplasm during, 709
proteolysis and, 723, 723*f*
regulation of, cancer and, 699-700, 725-726
restriction point and, 699-700, 713, 719-720, 719*f*-720*f*
cancer and, 699-700, 721-722, 722*f*-723*f*
regulation of cell proliferation by, 720-721, 721*f*
stem cells and, 713

G_2 checkpoint, 747-750, 747*f*-751*f*, 749*t*, 750*b*-753*b*
G_2 delay, 747, 747*f*
G_2 phase, 696-700, 698*f*
  cytoplasm during, 709
  G_2/mitosis transition and, 753
    Cdk1 activity and initiation of prophase in, 745-746, 746*f*
    enzymology of, 743-745, 743*f*-745*f*
    G_2 checkpoint and, 747, 747*f*, 749*f*, 750-753
    subcellular localization changes in, 744*f*, 745, 746*f*
    summary of main events of, 746-747, 747*f*
GAGs. *See* Glycosaminoglycans (GAGs).
Galactosyltransferase, 369*f*, 537, 538*t*
Gap junctions, 490, 543, 544*f*, 544*t*, 546-550, 547*b*, 547*f*-549*f*
  in disease, 550, 550*t*
GAPs (GTPase-activating proteins). *See* GTPase-activating proteins (GAPs).
Gating, 45, 262-265, 265*f*, 283
"Gating current", 268-269
GC globulin, 590*t*-591*t*
GCAP39, 590*t*-591*t*
GDI. *See* Guanine nucleotide dissociation inhibitor (GDI).
GEFs. *See* Guanine nucleotide exchange factors (GEFs).
Gel electrophoresis, 88, 89*b*, 89*f*
Gel filtration, 91*f*
Gel state, 231
Gelsolins, 583, 583*f*, 590*t*-591*t*
Geminin, 731*f*, 732
Gene conversions, 782, 783*f*, 784*b*
Gene deserts, 110
Gene duplication, 17-18, 41-42
Gene expression, 123, 126, 165-187
  β-catenin in, 530, 531*f*
  signaling pathways influencing, 472-473, 472*f*-474*f*, 476*f*-477*f*, 480*f*-481*f*
  transcription and. *See* Transcription; Transcription factors (TFs).
Gene silencing, 130, 131*f*
General transcription factors (GTFs), 170
Genes, 8, 86*b*. *See also specific genes.*
  divergence of, 17, 18*b*
  dominant negative, 87
  duplication of, divergence and, 17-18, 18*b*
  essential, 86*b*
  identification of, through mutations, 87-88
  for intermediate filament proteins, evolution of, 613*f*, 614-615, 615*t*
  isolation of cDNA and, 90-92, 92*f*-93*f*
  organization of, on chromosomes, 108-110, 109*t*, 110*f*
  tumor suppressor, 722, 723*f*
Gene-specific transcription, 175-177
  DNA-binding domains and, 179-180, 179*f*
  in eukaryotes, 176-180, 176*f*, 179*f*
  mapping transcription components on the genome, 177-178, 177*f*
  in prokaryotes, 175-176, 176*f*
Genetic code, 209-210, 210*f*
Genetic mapping, 87
Genetic markers, 784*b*
Genetically encoded control circuits, 5
Genetics
  chemical, 99-100
  key terms in, 86*b*
  reverse, 88
  in search for partners for genetic mutation, 96-97, 96*f*-97*f*
Genome engineering, 85*t*, 92-95, 94*f*
Genomes
  definition of, 86*b*
  DNA content of, 108-110, 109*t*, 110*f*
  human, 109*t*, 110, 113
    pseudogenes and, 112
    segmental duplications in, 112-113, 113*f*
    transposable elements of, 110-112, 111*f*
  mapping transcription components on, 177-178, 177*f*
  nuclear envelope in genome organization, 150, 151*f*
Genomics, 88
Genotype, 86*b*, 784*b*
Gephyrin, 611
Geranyl isoprenoids, 230
Geranylgeranyl isoprenoids, 230
GFAP, 615*t*
GFP. *See* Green fluorescent protein (GFP).
Giant cells, 499
Giardia lamblia, 21, 629
Glucocorticoid receptor, 179*f*
Glucose transport, in intestine, kidney, fat, and muscle, 286-287, 286*f*
GLUT4, 475-478, 476*f*
GLUT1 uniporter, 253*f*, 254-255
Glutamate, 291*f*
  synaptic transmission and, 295
Glutamate receptors, 273, 273*f*
Glutamic acid, structure of, 33*f*
Glutamine, structure of, 33*f*
Glutathione-*S*-transferase (GST), 90, 451*f*
Glycerophospholipids, 353-354, 353*f*-354*f*
Glycine
  structure of, 33*f*
  synaptic transmission and, 291*f*, 295
Glycoconjugates, 48-51, 51*f*
Glycogen, 49, 51*f*
Glycogen synthetase, 476
Glycolipids
  Golgi apparatus processing of, 369-371, 370*f*
  in lipid bilayer, 230
Glycolysis, 319, 320*f*
Glycophorin, transmembrane segment of, 235*f*
Glycoproteins, 50-51
  adhesive, 523
    in extracellular matrix, 514-517, 515*f*-517*f*, 523
  Golgi apparatus processing of, 369-371, 370*f*
  GPIIb/GPIIIa, 533*t*
Glycosaminoglycans (GAGs), 512-514, 513*f*-514*f*, 522
Glycosidases, 370
Glycosidic bonds, 48-50, 50*f*-51*f*
Glycosphingolipids, 229, 230*f*, 371
Glycosylase, 750, 751*f*
Glycosylation
  N-linked, 342-343, 343*f*
  O-linked, 370
Glycosylphosphatidylinositol (GPI) tails, 236-237
Glycosylphosphatidylinositols (GPIs), 339*f*, 340
Glycosyltransferases, 51, 370, 512, 513*f*
Glypican, 514*f*, 522, 538*t*
Golgi apparatus, 6, 8, 11, 11*f*, 302, 351
  functions of, 367, 367*f*
  Golgi-specific processing activities and, 369-371, 370*f*
  lipid biosynthesis and metabolism and, 371
  morphology and dynamics of, 367-369, 367*f*-369*f*
  proteolytic processing of protein precursors and, 371
  secretory transport from endoplasmic reticulum to, 365-367, 366*f*
Goodpasture syndrome, 519, 519*t*
GPI-anchored proteins, 233, 236-238, 236*f*
GPIb-IX-V, 537-538, 538*t*, 539*f*
GPIIb/GPIIIa, 533*t*
G-protein relay, 465
G-protein-coupled, seven-helix transmembrane receptors, signal transduction by, 463
G-protein-coupled receptor kinases, 414
Graded contractions, 678, 678*f*
Granular component of nucleolus, 147
Granules, secretory, 374-376, 374*f*-375*f*
Granules in mast cells, 499-500, 500*f*
Grb2, 431*f*, 439*f*, 440, 475, 476*f*-477*f*, 478-481
Green algae, 22-23, 23*f*, 660
Green fluorescent protein (GFP), 78, 79*f*, 87-88, 130-131
GRK1, 469
GroEL, 220-221, 222*f*
GroES, 220-221, 222*f*
Group I self-splicing introns, 205, 206*f*
Group II self-splicing introns, 205, 206*f*
Growth
  contact inhibition of, 722
  control of, G_1 phase and, 713-715, 713*f*-714*f*, 714*b*, 718-719, 718*f*
  in G_0 phase, 700
  loss of control of, cancer and, 722, 722*f*
  of membranes, 6, 7*f*
Growth cones
  of axons, 654-655, 654*f*
  guidance of, 658-659, 659*f*
Growth cycle, cell cycle and, 697
Growth factor receptor tyrosine kinase pathway through Ras to MAP kinase, 473*f*-474*f*, 474-475, 483*f*
Growth factors, blood cells and, 494-495
Growth hormone, 416-417, 417*f*
  skeletal growth and development and, 561-562
GST. *See* Glutathione-*S*-transferase (GST).
GTFs. *See* General transcription factors (GTFs).

GTPase-activating proteins (GAPs), 60*f*–61*f*, 62, 433, 433*f*, 435, 437*t*
GTPases, 409, 432–437, 432*f*–433*f*, 442*t*
analysis of enzyme mechanisms with, 60–62, 60*f*–61*f*
in disease, 436–437, 436*t*
dynamin-related, 437, 442*t*
elongation factors as, 432*f*, 433, 442*t*
ER shaping and, 333*f*, 334
experimental tools for, 437
G-protein relay and, 465
regulation of membrane trafficking, 355–357
Arf family, 355, 356*f*, 366*f*, 373*f*
Rab, 355–357, 357*f*
signal transduction by G-protein-coupled, seven-helix transmembrane receptors and, 463
small, 432*f*, 433–434, 442*t*
translocation, 442*t*
trimeric, 434–437, 434*f*, 442*t*
guanosine triphosphatase cycle and, 434*f*, 435–436
mechanisms of effector activation and, 436
subunit cycles and, 436
subunit diversity of, 435, 435*t*
GTP-$G_i\alpha$, 444
GTP-$G_s\alpha$, 444, 445*f*
Guanine, 44*f*, 47–48, 49*f*
Guanine nucleotide dissociation inhibitor (GDI), 356
Guanine nucleotide exchange factors (GEFs), 61–62, 157, 157*f*, 433, 476
Guanosine, 42
Guanosine triphosphatase cycle, 434*f*, 435–436
Guanosine triphosphatases (GTPases). *See* GTPases.
Guanylyl cyclase receptors, 418–419, 418*f*, 421–423
Guanylyl cyclases, 445, 468

H

H1 histone, 127, 128*f*
H19 locus, 131*f*
H2A:H2B heterodimers, 123
Half-spindles, 759–760
Halobacteria halobium, 242, 323
Haploid chromosome number, 86*b*, 779, 784*b*
Haploinsufficiency, 87
HB-GAM (heparin-binding, growth associated molecule), 523
Heart disease, inherited molecular basis of, 682*t*, 688
Heart rate and contractility, regulation of seven-helix receptors and trimeric G-proteins in, 687–688, 687*f*
Heavy chains of myosin, 624
Hedgehog receptors, 420–421
HeLa cells, 86–87, 131*f*, 730*f*
Helicases, 172, 214
α-helices, 36–40, 38*f*–40*f*
β-helices, 40
Helicobacter pylori, 546
Helix, 64*b*, 65*f*
Helix-loop-helix proteins, 179
Helix-turn-helix (HTH) proteins, 179–180
Helper T cells, 500, 501*f*
Hematopoietic stem cells, 494, 715–716
Hemicelluloses, 568
Hemidesmosomes, 543, 544*f*, 544*t*, 551–553, 552*f*, 566
Heptad repeats, 40
Hereditary nonpolyposis colorectal cancer, 749*t*
Hereditary spherocytosis, 495
Heterochromatin, 128–130, 128*f*–129*f*, 131*f*, 143*f*, 189
constitutive, 115, 129, 129*f*
facultative, 128–130, 128*f*
functional compartmentalization of chromatin and, 128–130, 128*f*–129*f*, 131*f*
Heterochromatin protein 1 (HP1), 129–130, 129*f*
Heterodimeric capping proteins, 583–584
Heterogeneous nuclear RNA (hnRNA), 167
Heterophilic interactions, 526
Heterotetrameric adapter protein complexes (AP1 to AP4), 361
Heterozygous organisms, 784*b*
Hexose, 48
H3:H4 heterodimers, 123
HI-C, 133–135, 134*f*
Hippocampus, synaptic transmission and, 295–296, 296*f*
Hirschsprung disease, 532
His-tag, 90
Histamine in mast cells, 499–500
Histidine
phosphorylation and, 425
structure of, 33*f*
Histidine kinases, 482*b*, 483*f*
phosphorylation and, 428, 441*t*
Histone code, 125, 125*f*
Histone deacetylase, 129*f*
Histone proteins, synthesis of, 740–741, 740*f*
Histones
3′ end formation on RNAs of, 191–192
core, 123, 124*f*
epigenetics and histone code and, 125
linker DNA and linker histone H1, 127, 128*f*
modifications, gene expression and, 178, 178*t*
nucleosome assembly and, 127
regulation of chromatin structure by histone N-terminal sites and, 125–127, 125*f*–126*f*
variants of, 127
HIV, 389–390, 503
HMG-CoA reductase, 347, 403–404, 404*f*
HnRNA. *See* Heterogeneous nuclear RNA (hnRNA).
Holliday junctions, 750*f*, 752–753, 752*f*, 783*f*, 784
Holocentric chromosomes, 115
Holoenzyme, 167, 169–170, 170*f*
Homeodomains, 179–180, 179*f*
Homo sapiens, genome of, 109*t*, 110, 113
pseudogenes and, 112
segmental duplications in, 112–113, 113*f*
transposable elements of, 110–112, 111*f*
Homogenizers, 88
Homologous chromosomes, 779, 782, 784*f*–786*f*, 785–787
Homologous proteins, 41–42
Homologous recombination, 85*t*, 94, 94*f*, 781
Homologous RNAs, 45–46
Homologs, 18*b*
Homology models, 37
Homology-directed repair, 92–94
Homophilic interactions, 526
Homozygous organisms, 784*b*
Hormone response elements, 184–185
HOX (homeobox) genes, 561
HP1. *See* Heterochromatin protein 1 (HP1).
HRS (hepatocyte-growth-factor-regulated tyrosine kinase substrate), 389, 389*f*
Hsp70, protein transport to mitochondria and, 305
Hsp70 chaperones, 220, 221*f*, 305
Hsp90 chaperones, 220, 221*f*
HTERC, 119–120
HTERT, 119–120, 122, 122*f*
HTH proteins. *See* Helix-turn-helix (HTH) proteins.
Human immunodeficiency virus (HIV), 389–390
Hutchinson-Gilford progeria syndrome, 150, 152*f*
Hyaline cartilage, 557, 557*f*
Hyaluronan, 512, 513*f*, 555
Hydrogen bonds, 58, 58*f*
Hydropathy, 234
Hydrophilic effect, 229
Hydrophobic effect, 31–32, 58–59, 58*f*, 229
Hydroxyapatite, 559
Hydroxyproline, 34*f*
Hyperthermia, malignant, 457
Hypertrophic cardiomyopathies, 688
Hypertrophic chondrocytes, 562–564, 563*f*
Hypothyroidism, 346

I

IAP (inhibitor of apoptosis protein), 805–807
I-band, 675*f*
ICAD, 807–808, 807*f*
ICAMs. *See* Intercellular adhesion molecules (ICAMs).
Icosahedrons, 66, 66*f*
IFT. *See* Intraflagellar transport (IFT).
IgA. *See* Immunoglobulin A (IgA).
IgCAMs (immunoglobulin cell adhesion molecules), 489–490, 526–528, 528*f*, 528*t*
IgE. *See* Immunoglobulin E (IgE).
IGF2 locus, 131*f*
IgG. *See* Immunoglobulin G (IgG).
Imaging, 75–76
electron microscopy in, 77*f*, 82–85, 83*f*–84*f*
light microscopy methods in. *See* Light microscopy methods.
"Immediate early" genes, 718, 718*f*
Immigrant cells, 491
Immunity
adaptive, 500–503, 501*f*–502*f*
innate, 496–500, 497*f*–498*f*, 500*f*

Immunoblot, 140*f*
Immunodeficiency diseases, 503
Immunofluorescence, 78, 79*f*
Immunoglobulin A (IgA), 502
Immunoglobulin cell adhesion molecules. *See* IgCAMs (immunoglobulin cell adhesion molecules).
Immunoglobulin E (IgE), 499-500, 502
Immunoglobulin Fc receptors, 379
Immunoglobulin G (IgG), 502
Immunoglobulins, rearrangement and somatic mutations of, 500-501, 502*f*
Immunological synapse, 477*f*, 479
Immunoprecipitation, 96
Immunoproteasomes, degradation by proteasomes, 401
Immunoreceptor tyrosine activation motifs (ITAMs), 478
Immunosuppressive drugs, 479
Immunotoxins, 391
Import receptors, 155
Importin β, 156, 159
Imprinting, 130, 131*f*
Inactivating inhibitors, 420
Inactivation of channels, 265-266, 265*f*
Inactivation peptide, 269, 269*f*
Incisions, 750
Indigenous cells, 491
  of connective tissue, 491-493, 491*f*-493*f*
Induced dimerization, 415, 416*f*
Induced pluripotent stem (iPS) cells, 718
Inflammation
  cellular adhesion between endothelial cells and leukocytes in response to, 537*f*, 538-540
  NF-KB and, 186
Ingested particles in endocytosis, alternative fates of, 379-381, 381*b*
Inhibitor of NF-KB, 185*f*, 186
Inhibitory synapses, 295
Initiation
  of DNA replication, 736*b*
  of prophase, Cdk1 and, 745-746, 746*f*
  of protein synthesis, 214-215, 216*f*
  of transcription, 165-166, 165*f*, 169-173, 170*f*, 170*t*
    in prokaryotes, 175-176, 176*f*
Initiation codons, 209-210
Initiator, 169
Initiator caspases, 804, 805*f*
Initiator proteins, 728-729
INK inhibitors, 714*b*
Innate immunity, cellular basis of, 496-500, 497*f*-498*f*
  eosinophils and, 499
  macrophages and, 499
  mast cells, basophils, and, 499-500, 500*f*
  neutrophils and, 494*t*, 497*f*, 498-499
Inner kinetochore, 139, 139*f*
Inner membrane
  of chloroplasts, 324
  mitochondrial, 304, 305*f*-306*f*, 306-307, 318, 318*f*
Inner nuclear membrane, 148
  proteins of, 149-150, 150*f*
Inner segment of photoreceptors, 468
Inositol 1,4,5-triphosphate (IP_3), 471*t*, 690, 690*f*
  signaling pathways and, 474*f*, 477*f*
  T lymphocyte activation and, 477*f*, 478-479
Inositol 1,4,5-triphosphate (IP_3) receptor calcium channels, 454-456, 455*f*-456*f*, 455*t*
Inositol 1,4,5-triphosphate (IP_3) receptors, 435, 435*t*, 454
Insulators, 130, 131*f*, 134-135
Insulin, synthesis of, 374*f*, 375
Insulin pathways to GLUT4 and MAP kinase, 475-478, 476*f*
Insulin receptor, 415*f*, 475-478, 476*f*
Insulin receptor substrates. *See* IRS (insulin receptor substrates).
Insulin receptor tyrosine kinase, 427*f*, 475, 476*f*
Integral membrane proteins, 227, 234-237, 234*b*, 235*f*, 237*f*
$\alpha_6\beta_4$-integrin, 552-553
Integrin α_7, 684*t*
Integrins, 489-490, 515, 515*f*, 526, 532-536, 533*f*-534*f*, 533*t*, 537*f*, 539*f*
Intercalated disks, 685, 685*f*
Intercellular adhesion molecules (ICAMs), 528*t*
Intercellular communication, lipid-derived second messengers for, 449-452, 450*f*-451*f*
Intercellular junctions, 543-553, 544*f*, 544*t*
  adherens, 490, 529, 529*f*, 529*t*, 543, 550-551, 552*f*
  desmosomes as, 490, 529, 529*f*, 529*t*, 543, 544*f*, 544*t*, 551, 552*f*
  focal contacts as, 534, 534*f*, 544*t*, 551-552
  gap, 490, 543, 544*f*, 544*t*, 546-550, 547*b*, 547*f*-549*f*, 550*t*
  hemidesmosomes as, 543, 544*f*, 544*t*, 551-553, 552*f*
  tight, 490, 543-546, 544*f*-545*f*, 544*t*
Interchromatin granules, 144, 144*f*, 194
Interchromosomal domain, 143
α-interferons, degradation by proteasomes and, 401
Interkinesis, 787
Interleukins, 416
Intermediate filaments, 574, 613-622, 613*f*, 675-676, 675*f*
  assembly of, 615-616, 617*f*
  expression in specialized cells, 615*t*, 618, 618*f*
  functions of, 619-621, 620*f*-621*f*
  posttranslational modifications of, 616-618, 617*f*
  proteins, evolution of genes for, 613*f*, 614-615, 615*t*
  proteins associated with, 618-619, 619*t*, 620*f*
  structure of, 613*f*, 615-616, 616*f*
    of subunits, 614, 614*f*, 615*t*, 616*f*
Intermembrane space, 304, 318
α-Internexin, 615*t*
Interphase, 63*f*, 697, 699*b*
  chromosomal DNA packaging in, 130-131, 131*f*
  chromosomes with clearly resolved loop structures, 132-133, 133*f*
Interpolar microtubules, 759-760
Interstitial lamellae, 564-565
Intestine, glucose transport in, 286-287, 286*f*
Intracellular ligands, 538*t*
  of integrins, 533*t*, 534-535, 534*f*
  of selectins, 536*t*
Intracellular motility, 639-650, 640*f*
  actin-based movements of organelles in other cells and, 647*f*, 648-649, 649*f*
  driven by microtubule polymerization, 640*t*, 646*f*-647*f*, 647
  in movement of cytoplasm
    driven by actin and myosin, 647-648, 648*f*-649*f*
    driven by actin polymerization, 649, 650*f*
  rapid movements on microtubules and, 639-643, 640*f*-641*f*, 640*t*, 642*t*
    fast axonal transport and, 640*t*, 643-644, 643*f*-644*f*
    slow transport of cytoskeletal polymers and associated proteins in axons and, 640*t*, 644-646, 645*f*
Intraflagellar transport (IFT), 651, 664-665, 664*f*
Intra-S checkpoint, 739-740, 739*f*
Intrinsic pathway of apoptosis, 803-804, 803*f*, 808-810, 808*f*, 810*f*
Intrinsic terminators, 174
Intrinsically disordered regions of proteins, 41
Intron branch point, 192
Introns, 109, 163, 192
  AT-AC, 193, 193*f*
Inward rectifiers, 268
Ion exchange chromatography, 91*b*
Ion flux, 264-265
Ionic bonds, 59
IP_3. *See* Inositol 1,4,5-triphosphate (IP_3).
IP_3 receptor calcium channels. *See* Inositol 1,4,5-triphosphate (IP_3) receptor calcium channels.
IP_3 receptors. *See* Inositol 1,4,5-triphosphate (IP_3) receptors.
IPS cells. *See* Induced pluripotent stem (iPS) cells.
IRE1, 345-346, 345*f*
IRS (insulin receptor substrates), 475-476
Isoelectric focusing, 89*b*
Isoforms, 194
  actin, 578, 578*f*
Isoleucine, structure of, 33*f*
Isomers, stereochemical, 48
Isopentyl, 230
Isoprenoid tails, 236
Isoprenoids, 230
ITAMs. *See* Immunoreceptor tyrosine activation motifs (ITAMs).

J

JAKs, 416-417, 417*f*, 421-423
JAK/STAT pathways, signal transduction and, 479-481, 480*f*
Jellyfish, 25, 78

K

Kainate, glutamate receptors and, 273
Kartagener syndrome, 661
Karyopherins, 156
Katanin, 603, 611
KcsA, 262-265, 264*f*-265*f*, 267-268
KDEL, 342
Kendrin, 612
Keratins, 614, 614*f*, 615*t*, 616*f*, 618, 618*f*, 619*t*
  epidermal, 551
Kidneys, glucose transport in, 286-287
Killer T cells, 500, 501*f*
Kinases
  cell-dependent, cell-cycle transitions and, 702*b*, 702*f*
  cooperation between phosphatases and, 432
  receptor, 417-418, 417*f*-418*f*, 421-423
Kinesin-1, 631-633, 632*f*, 634*f*, 635, 635*t*
Kinesin-5, 634*f*, 635, 635*t*
Kinesin-7, 634*f*, 635, 635*t*
Kinesin-8, 602-603
Kinesin-13, 602-603, 634*f*, 635*t*
Kinesin-14, 602-603, 634*f*, 635*t*
Kinesin head, 631-632, 632*f*
Kinesins, 13, 574, 611, 624*t*, 631-635
  evolution of, 623, 623*f*
  mechanochemistry of, 632-633, 632*f*-634*f*
  superfamily of, 633-635, 634*f*, 635*t*
Kinetics, 29
Kinetochore fibers, 759-760
Kinetochore microtubules, 759-760, 765*f*
Kinetochores, 113, 139-141, 139*f*-141*f*, 758-759
  definition of, 108*b*
  sister, 759
Kleisin, 137
KMN network, 611

L

L1, 528*t*
Lac operon, 175-176, 176*f*
Lactacystin, 399-400
LADs. *See* Lamina-associated domains (LADs).
Lagging strand, 728, 728*f*
Lamellar motility on flat surfaces by pseudopod extension, 654*f*-657*f*, 655-656
Lamin A, 148
Lamin AC, 684*t*
Lamin B, 148
Lamin C, 148
Lamina-associated domains (LADs), 150, 151*f*
Laminin receptor, 533*t*
Laminin-binding protein (Mac-2), 523
Laminins, 517, 518*f*, 519*t*, 523, 532-533, 533*t*, 538, 538*t*, 684*t*
Laminopathies, 150-151, 152*f*
Lamins, 614, 614*f*, 615*t*, 616*f*
Lamp1, 388*f*, 390
Lamp2, 388*f*, 390
Lampbrush chromosomes, 132, 133*f*, 787
LAP1, 619*t*
LAP2, 619*t*
LARGE, 684*t*
Large subunits of ribosomal RNA, 205, 212, 213*f*
Lariat (pre-mRNA splicing), 192, 193*f*
Last eukaryotic common ancestor (LECA), 19*f*, 20, 23-24
Last universal common ancestor (LUCA), 17-18, 17*f*, 18*b*
Late endosomes, 377-378, 386, 388*f*, 390
Lateral transfer, 18
Latrunculins, 585, 586*b*
Lax, Henrietta, 86-87
LBR, 619*t*
LCA (leukocyte common antigen), 538*t*
Lck, 468, 478
LCRs. *See* Locus control regions (LCRs).
LDLs. *See* Low-density lipoproteins (LDLs).
Leading edge, 655*f*-657*f*, 656
Leading strand, 728, 728*f*
LECA. *See* Last eukaryotic common ancestor (LECA).
Lectin, 536, 536*f*, 536*t*
Leptin, 492
  bone and, 561
Leptotene, 785-786, 785*f*-786*f*
Lethal mutations, synthetic, 96-97, 96*f*
Leucine, structure of, 33*f*
Leucine zipper proteins, 179-180, 179*f*
Leukemia, 495
Leukocytes, cellular adhesion between endothelial cells and, 537*f*, 538-540
Leukotrienes, 446, 450, 451*f*
LFA-1 (lymphocyte function associated antigen-1), 533*t*
LFA-3 cells (lymphocyte function associated antigen-3 cells), 528*t*
"Libraries" of DNA, 90-92
Licensing in DNA replication, 728
Liddle syndrome, 267
Lidocaine, 279
Life, evolution of. *See* Evolution of life on earth.
LIG4 syndrome, 749*t*
Ligands, 426, 437-440, 437*t*, 438*f*-439*f*
  for integrins, 533-535, 533*t*, 534*f*
  plasma membrane receptors and, 411-412, 411*f*, 421-423
    cytokine, 416-417, 417*f*, 421-423
    guanylyl cyclase, 418-419, 421-423
    Notch, 420-423
    receptor serine/threonine kinases and, 417-418, 417*f*, 421-423
    receptor tyrosine kinases and, 414-415, 415*f*, 421-423
    seven-helix, 412-414, 412*f*-413*f*, 421-423
    tumor necrosis factor family of, 419-420, 419*f*
  proline-rich, 439*f*, 440
  for selectins, 536-537, 536*f*, 536*t*
Light chains of myosin, 624, 681-682
Light microscopy methods, 76-78, 81*t*
  bright field, 76, 76*t*, 77*f*
  confocal, 79-81, 79*f*-81*f*
  dark-field, 76-78, 76*t*
  differential interference contrast, 76, 76*t*, 77*f*, 79*f*
Light microscopy methods *(Continued)*
  fluorescence, 76*t*, 77*f*, 78, 79*f*-82*f*, 81*t*
  phase-contrast, 76, 76*t*, 77*f*
  polarization, 76-78, 76*t*, 77*f*
Light reactions, 325
Light sheet microscopy, 80*f*-81*f*, 81
Light-chain domain of myosins, 624-626
Light-harvesting complexes, 326-327, 327*f*
LINES (long interspersed nuclear elements), 111-112, 111*f*
Link oligosaccharides, 512, 513*f*
Link protein, 523
Linked reactions, 57-58, 57*f*
Linker DNA, 127
Linker histone, 127, 128*f*
Linker proteins, 604, 611
Linking collagens, 506*f*, 509-510, 509*f*
Lipid asymmetry of organelle membranes, 232-233, 233*f*
Lipid bilayer, 225, 228-230
  electrostatic interaction with phospholipids in, 237
  glycolipids and, 230
  partial penetration of, 237
  peripheral membrane proteins with, 236, 236*f*
  phosphodiglycerides and, 228-229, 228*f*, 229*t*
  physical structure of, 230-234, 231*f*-233*f*
  sphingolipids and, 229, 230*f*
  sterols and, 229-230, 230*f*
  triglycerides and, 230
Lipid domains, 353-354, 353*f*
Lipid droplets, 492, 493*f*
Lipid kinases
  phosphorylation and, 426
  reactions producing lipid second messengers and, 446
Lipid phosphatases, 446
Lipid rafts, 371, 383*f*
Lipid turnover, degradation and, 402-404, 403*f*-404*f*
Lipid-anchored proteins, association with cytoplasmic surface of ER, 340
Lipid-derived second messengers, 445-452, 446*f*
  agonists and receptors and, 446-447
  cross talk and, 452
  enzyme reactions producing, 445-446
  for intercellular communication, 449-452, 450*f*-451*f*
  phosphatidylcholine signaling pathways and, 449, 449*f*
  phosphoinositide signaling pathways and, 448-449, 448*f*
  protein kinase C and, 447*f*, 448
  sphingomyelin/ceramide signaling pathways and, 452, 452*f*
  targets of, 447-448, 447*f*-448*f*
Lipids, 29, 228-230
  biosynthesis of
    in endoplasmic reticulum, 346
    in Golgi apparatus, 371
  covalently attached, 433-434
  gradient of, in membrane trafficking, 353-354, 353*f*-354*f*

Lipids *(Continued)*
movement between organelles, 341*t*, 348-349, 349*f*
as second messengers, 409
transport of, in endoplasmic reticulum, 341*t*, 348-349, 349*f*
Lipid-transfer proteins (LTPs), 341*t*, 348-349, 349*f*
Lipoproteins, low-density, 372, 403-404, 403*f*
Lipoxins, 450, 451*f*
5′-lipoxygenase, 451
Liquid disordered phase, 231, 231*f*
Lis-1, 637
Listeria, actin polymerization by, 640*t*, 649, 650*f*
LncRNAs. *See* Long noncoding RNAs (lncRNAs).
Local anesthetics, sodium channels and, 270
Localization
of molecules, 98
of pre-mRNA splicing, 194
Localization microscopy, 81-82, 81*t*, 82*f*
Locus control regions (LCRs), 135
Lokiarchaeota, 19*f*, 20, 432
Long bones, 558*f*, 562, 563*f*, 564
Long interspersed nuclear elements (LINES). *See* LINES (long interspersed nuclear elements).
Long noncoding RNAs (lncRNAs), 110
Long QT syndrome, 269-270, 686-687
Long-term depression (LTD), 295-296
Long-term potentiation (LTP), 295-297, 296*f*
Loose connective tissue, 555, 555*f*
Low-density lipoproteins (LDLs), 370-371, 403-404, 403*f*
LPA, 451
L-selectin, 536*t*
Lsm proteins, 202
LTD. *See* Long-term depression (LTD).
LTP. *See* Long-term potentiation (LTP).
LTPs. *See* Lipid-transfer proteins (LTPs).
LUCA. *See* Last universal common ancestor (LUCA).
Luminal ring, 151
Luminal space, 148
Lymphocytes, 491*f*, 494*t*
B, 500, 501*f*
cytotoxic T (killer T cells), 500, 501*f*
helper, 500, 501*f*
pathways through nonreceptor tyrosine kinases, 477*f*, 478-479, 479*b*, 480*f*-481*f*
Lysines, 33*f*, 507
Lysobisphosphatidic acid, 388*f*, 390, 402
Lysophosphatidic acid, 446
Lysophosphoglyceride, 446
Lysosomal hydrolase, 379-381, 386, 388*f*, 389, 396, 402
Lysosomal storage diseases, 405
Lysosome-associated membrane proteins (Lamp1 and Lamp2), 388*f*, 390
Lysosomes, 8, 11, 302, 386
degradation in, 396-399, 397*f*
autophagy and, 397*f*-398*f*, 398-399
delivery to lysosomes via endocytic pathway in, 397-398, 397*f*
Lysosomes *(Continued)*
fusion with, 379-381, 379*f*
proteolysis and, 396
Lysyl oxidase, 509, 509*f*

M

M2 channel, 266
M line, 674-675, 674*f*, 682*t*
M phase, 697-700, 698*f*, 699*b*
MAC-1, 533*t*
Macroautophagy, 398-399, 398*f*
Macromolecular assembly, 63-74
assembly pathways and reactions in, 67
regulation at multiple steps on sequential assembly pathways and, 67-74
by accessory proteins, 68-74, 69*f*, 71*f*-73*f*
by changes in environmental conditions, 68
by covalent modification of subunits, 68
of nucleation, 68-69, 69*f*
by subunit biosynthesis and degradation, 68
specificity by multiple weak bonds on complementary surfaces and, 64-65
subunits and, 63-64, 63*f*, 64*b*, 65*f*
symmetrical structures constructed from identical subunits with equivalent bonds and, 65-67, 66*f*
Macrophage colony-stimulating factor (M-CSF), 561
Macrophages, 491*f*, 497*f*, 499
Macropinocytosis, 377, 378*f*, 380*f*-381*f*, 381-382
Macropinosomes, 381-382, 381*f*
Mad1/Mad2 complex, 764*f*, 765
MAG (myelin associated glycoprotein), 528*t*
Major facilitator superfamily (MFS), 258-259
Major histocompatibility complex (MHC) antigens, 477*f*, 478, 502-503
class I, degradation by proteasomes and, 401, 501*f*, 503
class II, 503
Major sperm protein, 666-667, 667*f*
Malignant hyperthermia, 457
Mal3p, 611
Mammalian cells, replication origins in, 730-731, 730*f*-731*f*
Mannose-6-phosphate receptors (MPRs), 373, 373*f*
degradation in lysosomes and, 396
MAP2, 590*t*-591*t*, 601-602, 601*f*-602*f*, 611
MAP 3, 611
MAP 4, 601, 601*f*-602*f*, 611
MAP kinase kinase, 473*f*, 475
MAP kinase kinase kinase, 473*f*
MAP kinase pathways
metaphase II arrest and, 792-793, 793*f*
to nucleus, 473-474, 473*f*-474*f*, 476*f*-477*f*, 483*f*
MAP kinases, 432, 473, 474*f*
growth factor receptor tyrosine kinase pathway through Ras to, 473*f*-474*f*, 474-475, 483*f*
insulin pathways to, 475-478, 476*f*
MAPs. *See* Microtubule-associated proteins (MAPs).
MAPU, 611
Marfan syndrome, 512
Marx, Jean, 475
Mass spectrometry, 90
Mast cells, 491*f*, 499-500, 500*f*
Mathematical models of systems, 100
Matrix (mitochondrial), 304, 305*f*-306*f*, 306-307, 318, 318*f*, 320*f*-321*f*, 323*f*
Matrix metalloproteinases (MMPs), 519-520, 519*f*
Maturation-promoting factor (MPF), 704*b*, 705*f*
MCAK, 611
Mcm proteins (minichromosome maintenance), 731, 732*t*, 733-734, 733*f*
M-CSF. *See* Macrophage colony-stimulating factor (M-CSF).
Mdm2, 724-725, 724*f*-725*f*
MDR1, 251, 251*f*
MDR2, 251-252, 251*f*
Mechanosensitive channels, 266, 267*f*
Megacomplexes, 556-557
Megakaryocytes, 495
Meiosis, 696, 779-795, 780*f*-781*f*
bouquet stage of, 780*f*, 785-786, 785*f*
cell-cycle regulation of events of, 792
chiasmata and, 785*f*, 786-787, 789-790, 789*f*-790*f*
defects and human disease, 794-795, 795*f*, 795*t*
mitosis contrasted with, 782*b*
pairing in, 781, 786*f*-788*f*, 787-789
prophase I of, homologous chromosomes in, 785-787, 785*f*-786*f*
recombination and, 781-785, 783*f*-784*f*, 784*b*
sex chromosome behavior in, 791-792, 792*f*
sexual reproduction and, 779-780
synapsis in, 787-789, 787*f*-788*f*
synaptonemal complex and, 785*f*, 786-789, 787*f*-789*f*
terminology for, 781, 781*f*
timing of, in humans, 793-794, 794*f*
Meiosis I, 779, 780*f*-781*f*
cohesion and chromosomal movements during, 790-791, 791*f*
suppression DNA replication between meiosis II and, 792
Meiosis II, 779
metaphase II arrest and MAP kinase pathway and, 792-793, 793*f*
suppression DNA replication between meiosis I and, 792
MEK, 473, 475
Membrane capacitance, 282
Membrane carrier proteins, 226, 253-259
definition of, 241
diversity of, 255*f*, 257-259
physiology and mechanisms of, 253-257, 253*f*, 255*f*, 257*t*
of antiporters, 254, 254*f*-255*f*, 256-257, 257*t*
discovery of, 255*b*
of symporters, 254-256, 254*f*-255*f*, 257*t*
of uniporters, 253*f*-254*f*, 254, 256, 257*t*

Membrane carrier proteins *(Continued)*
structures of, 255*f*
tools for studying, 257*t*
Membrane channels
activity of, 265–266, 265*f*
ammonia, 276–277, 276*f*, 277*b*
blockers of, 279
charge movement and
net current through ion-selective channels and, 283
rate of, 283
for small cells, 282–283
charge redistribution by electrical conduction and, 284, 284*f*
chloride, 275–276, 275*f*–276*f*
definition of, 241–242
diversity of, 262
evolution of, 262, 263*f*
gated by extracellular ligands, 272–275
glutamate receptors and, 273, 273*f*
nicotine acetylcholine receptor and, 273–275, 274*f*
generating action potentials, 290
with one transmembrane segment, 266
opening simultaneously, consequences of multiple types of, 283–284, 284*f*
Piezo2 as, 278
porins as, 277–278
S5-S6 channel family, 267–272
calcium release channels, 271–272, 454–456, 455*f*–456*f*, 455*t*
channels gated by intracellular ligands, 272, 272*f*
channels with four transmembrane helices, 268
inward rectifiers, 268
transient receptor potential (TRP) channels, 271–272, 271*f*
voltage-gated cation channels. *See* Voltage-gated channels.
structure of, 262–265, 264*f*
TMBIM channels, 278
with two transmembrane segments, 266–267
amiloride-sensitive channels as, 266–267
ATP-gated channels as, 267
mechanosensitive channels as, 266, 267*f*
water, 276*f*, 277
Membrane contact sites, 334, 334*f*
Membrane depolarization, 290
Membrane peroxisomal targeting sequences (mPTSs), 310, 310*f*, 310*t*
Membrane potentials, 261, 284*f*
biophysical basis of, 280–282, 281*f*
diffusion potentials and, 281, 281*f*
Nernst potential and, 281–282, 282*f*
qualitative relationships and, 281–282
measurement of, 279–280, 280*f*
Membrane proteins, 234–239
dynamic behavior of, 237–239, 238*f*
electrostatic interaction with phospholipids, 237
insertion of, into ER bilayer, 339*f*, 340
integral, 227, 234–237, 234*b*, 235*f*, 237*f*
peripheral, 227, 236–237, 236*f*
Membrane pumps
ATP-driven, 243–248, 244*t*
ABC transporters and, 244*t*, 250–252, 250*f*–251*f*
A-type ATP synthases and, 245*f*, 247–248
F_0F_1-ATPase family and, 244*t*, 246*f*
F-type ATPases and, 245*f*–246*f*, 247
P-type cation pumps and, 244*t*, 248–250, 248*f*–249*f*
rotary ATPases and, 243–248, 244*t*, 245*f*–247*f*
V-type ATPases and, 247, 247*f*
diversity of, 242, 242*f*, 242*t*
light-driven pumping by bacteriorhodopsin and, 242–243, 243*f*
membrane permeability and, 241–242, 241*f*
Membrane skeleton, 237, 237*f*, 684*t*
Membrane structure, 227–239
development of ideas about, 227–228, 227*f*
lipid bilayer of, 228–230
amino acid sequences identifying transmembrane segments and, 234*b*
electrostatic interaction with phospholipids in, 237
glycolipids and, 230
partial penetration of, 237
peripheral membrane proteins with, 236, 236*f*
phosphodiglycerides and, 228–229, 228*f*, 229*t*
physical structure of, 230–234, 231*f*–233*f*
sphingolipids and, 229, 230*f*
sterols and, 229–230, 230*f*
triglycerides and, 230
proteins and, 234–239
dynamic behavior of, 237–239, 238*f*
electrostatic interaction with phospholipids, 237
integral, 227, 234–237, 234*b*, 235*f*, 237*f*
peripheral, 227, 236–237, 236*f*
Membranes
cellular volume regulation and, 288–289, 288*f*
chemiosmotic cycles and, 256, 285–286, 285*f*
endoplasmic reticulum and, 334, 334*f*, 341*t*
epithelial transport and, 286–288, 286*f*
cystic fibrosis as transporter disease and, 287–288, 287*f*
food production and, 288
of glucose, in intestine, kidney, fat, and muscle, 286–287, 286*f*
of salt and water in kidney, 287
excitable, 289–291, 289*f*, 291*f*
growth of, 6, 7*f*
secretory. *See* Secretory membrane system.
structure of. *See* Membrane structure.
synaptic transmission and, 291–297, 291*f*–293*f*
central nervous system synapses and, 293*f*, 294–297, 296*f*
neuromuscular junction and, 292*f*, 294
thylakoid, 324
Memory cells, 502
Menkes syndrome, 249–250
Meristematic stem cells, 717
Merotelic attachment, 763
Mesenchymal stem cells, 491, 491*f*–492*f*
Messenger RNA. *See* MRNA (messenger RNA).
Metabolism
endoplasmic reticulum and, 347–348, 347*f*–348*f*
Golgi apparatus and, 371
regulation of, through β-adrenergic receptors, 469–472, 470*f*, 471*t*
Metalloproteinases, matrix, 519–520, 519*f*
Metaphase, 133*f*, 700
of mitosis, 755, 755*f*–756*f*, 765–767, 765*f*–766*f*
Metaphase II arrest in meiosis, 792–793, 793*f*
Metaphase plate, 765–766
Metarhodopsin II, 468
Metastasis, 717
Metazoans, 24–25
Methionine, structure of, 33*f*
Methylation of receptors, 485, 485*f*
MFS. *See* Major facilitator superfamily (MFS).
MHC antigens. *See* Major histocompatibility complex (MHC) antigens.
Micro RNAs (miRNAs), 167, 202–204, 203*f*–204*f*
Microarrays, large-scale screening with, 97, 98*f*
Microautophagy, 399
Microdomains, 353
Microelectrodes, intracellular, 279–280, 280*f*
Microfibrils, 567*f*, 568–569
fibrillin, 510, 510*f*
Microscopy, 75–76
electron, 77*f*, 82–85, 83*f*–84*f*
light, 76–78, 81*t*
bright field, 76, 76*t*, 77*f*
confocal, 79–81, 79*f*–81*f*
dark-field, 76–78, 76*t*
differential interference contrast, 76, 76*t*, 77*f*, 79*f*
fluorescence, 76*t*, 77*f*, 78, 79*f*–82*f*, 81*t*, 130–131, 131*f*
phase-contrast, 76, 76*t*, 77*f*
polarization, 76–78, 76*t*, 77*f*
Microtubule motors, 631–633, 633*f*–634*f*
matching cargo with, 640–643, 641*f*, 642*t*
velocities of, 640*t*
Microtubule organizing centers (MTOCs), 593, 594*f*, 604, 760
Microtubule-associated proteins (MAPs), 595, 611
associated with growing microtubule plus ends, 600–601, 600*f*, 611
linker, 604, 611
MAP2, 590*t*–591*t*, 601–602, 601*f*–602*f*, 611
MAP3, 611
MAP4, 601, 601*f*–602*f*, 611
microtubule regulation by, 600–604, 600*f*, 611
associated with microtubule ends, 603–604, 603*f*

Microtubule-associated proteins (MAPs) *(Continued)*
linker, 604
microtubule-destabilizing, 602-603, 603*f*
microtubule-stabilizing, 601-602, 601*f*-602*f*, 611
Microtubules, 12-13, 13*f*, 63*f*, 139*f*, 140, 574, 593, 593*f*, 595
assembly from GTP tubulin, 597-599, 598*f*, 599*t*, 600*f*
astral, 759-760
axostyles as, 667, 667*f*
central spindle of, 769
dynamics of, 595
in cells, 599-600, 599*t*, 600*f*
steady-state, in vitro, 598-599, 598*f*, 599*t*, 600*f*
ER remodeling and, 334
flux within spindle, in metaphase of mitosis, 766-767, 766*f*
interpolar, 759-760
kinetochore, 759-760, 765*f*
outer doublet, 660
pharmacologic tools for studying, 595*b*
polymerization of, intracellular motility driven by, 640*t*, 646*f*-647*f*, 647
rapid movements on, 639-643, 640*f*-641*f*, 640*t*, 642*t*
fast axonal transport and, 640*t*, 643-644, 643*f*-644*f*
slow transport of cytoskeletal polymers and associated proteins in axons and, 640*t*, 644-646, 645*f*
regulation by microtubule-associated proteins, 600-604, 600*f*-603*f*, 611
structure of, 595-597, 596*f*-597*f*
Microvillii, 575
Midbody, 771*f*, 775
Minichromosome maintenance (Mcm) proteins. *See* Mcm proteins (minichromosome maintenance).
MinK molecules, 266
Minus End Binding Proteins, 611
Minus end of microtubules, 593, 599*t*
MiRNAs. *See* Micro RNAs (miRNAs).
Mismatch repair, 751, 751*f*
Mitochondria, 3*f*, 6, 8, 11-12, 11*f*, 301-302, 317-323
ATP synthesis by oxidative phosphorylation and, 317, 319-322, 320*f*-321*f*
biogenesis of, 318-319, 319*f*
disease and, 322-323, 323*f*
evolution of, 20-21, 317-318, 317*f*-318*f*
protein transport into, 303*f*, 304-307, 305*f*-307*f*
structure of, 318, 318*f*
Mitochondrial carrier proteins, 255*f*, 257-258
Mitochondrial inner membrane, 304, 305*f*-306*f*, 306-307, 318, 318*f*
Mitochondrial matrix, 304, 305*f*-306*f*, 306-307, 318, 318*f*
Mitochondrial outer membrane, 304, 306, 306*f*, 318, 318*f*
protein transport across, 305-306, 306*f*
Mitogenic signals, 719
Mitogens, 713
Mitosis, 7-8, 8*f*, 10, 63*f*, 696-697, 698*f*, 755-778, 755*f*-756*f*
anaphase of, 700, 755, 755*f*-756*f*, 767-770, 767*f*
biochemical mechanism of sister chromatid separation in, 767-768, 768*f*
mitotic spindle dynamics and chromosome movement during, 768-770, 769*f*
closed, 755
cytokinesis and. *See* Cytokinesis.
cytoplasm during, 709, 757-758, 757*f*-758*f*, 757*t*
exit from
in cytokinesis, 776-777
cytoplasm and, 709
meiosis contrasted with, 782*b*
metaphase of, 755, 755*f*-756*f*, 765-767, 765*f*-766*f*
nucleolus disassembly during, 146*f*, 147
open, 755
prometaphase of, 755, 755*f*-756*f*, 758-765, 758*f*
chromosome attachment to spindle in, 761-763, 762*f*
correction of errors in chromosome attachment to spindle in, 763-764, 763*f*
mitotic spindle organization in, 759-760, 760*f*, 765*f*
nuclear envelope disassembly in, 759, 759*f*
spindle assembly in, 760-761, 760*f*-761*f*
spindle checkpoint and, 764-765, 764*f*
prophase of, 755-758, 755*f*-757*f*
cytoplasmic changes in, 757-758, 757*f*-758*f*, 757*t*
nuclear changes in, 756-757
telophase of, 755, 755*f*-756*f*, 770-771, 770*f*
nuclear envelope reassembly in, 770-771, 770*f*
transition to, 753. *See also* G_2 phase: G_2/mitosis transition and..
Mitotic chromosomes, 135-136, 136*f*-137*f*, 147
Mitotic exit, 776-777
cytoplasm during, 709
Mitotic spindle, 13, 63*f*, 593, 594*f*, 595, 700
assembly of, in prometaphase of mitosis, 760-761, 760*f*-761*f*
chromosome attachment to
correction of errors in, 763-764, 763*f*
in prometaphase of mitosis, 761-763, 762*f*
dynamics of, chromosome movement and, in anaphase of mitosis, 768-770, 769*f*
microtubule flux within, in metaphase of mitosis, 766-767, 766*f*
organization of, in prometaphase of mitosis, 759-760, 760*f*, 765*f*
poles of, 757, 757*f*
spindle assembly checkpoint and, 701
spindle checkpoint, 764-765, 764*f*
MM-creatine phosphokinase, 682*t*
MMPs. *See* Matrix metalloproteinases (MMPs).
Mobile kinases, gene expression and, 472, 473*f*-474*f*, 476*f*-477*f*
Mobile transcription factors, gene expression and, 472-473, 477*f*, 480*f*-481*f*
Model organisms for biological research, 85-86, 86*b*
Modular domains in proteins, 41-42, 43*f*
Moesin, 590*t*-591*t*
Mold, 4*f*
Molecular chaperones, 68
Molecular clock, 18
Molecular movements, 6, 7*f*
Molecular structure, 95
Mollusks, 25
Monoclonal antibody, 502
Monocytes, 491*f*, 494*t*, 497*f*, 499
Mono-methylarginine, 34*f*
Monosaccharides, 29
Monotelic attachment, 763
Mother centrioles, 605, 605*f*
Motility. *See* Cellular motility; Intracellular motility.
Motor domain of motor proteins, 623
Motor end plate, synaptic transmission at, 678
Motor neurons, skeletal muscle contraction and, 678, 678*f*
Motor proteins, 6, 7*f*, 12, 13*f*, 595, 623-638, 623*f*, 624*t*, 625*f*. *See also* Intracellular motility.
dyneins as, 624*t*, 631, 635-637, 636*f*
evolution of, 623, 623*f*
mechanochemistry of, 636, 636*f*-637*f*
superfamily of, 636-637, 636*f*
kinesins as, 624*t*, 631-635
evolution of, 623, 623*f*
mechanochemistry of, 632-633, 632*f*-634*f*
superfamily of, 633-635, 634*f*, 635*t*
motor domain of, 623
myosins as, 624-631, 624*t*, 625*f*
actomyosin ATPase cycle and, 626-628, 627*f*
evolution of, 623, 623*f*
mechanochemistry of, 624-626, 625*f*-626*f*
superfamily of, 628-631, 630*f*-631*f*
transduction of chemical energy into molecular motion, 627*f*, 628, 629*f*
tail of, 623
tools for studying, 641*b*
MPF. *See* Maturation-promoting factor (MPF).
MPRs. *See* Mannose-6-phosphate receptors (MPRs).
MPTSs. *See* Membrane peroxisomal targeting sequences (mPTSs).
MRNA (messenger RNA), 5, 5*f*, 163-164, 167, 646-647, 647*f*
degradation of, 190*f*, 196-197, 196*f*
editing of, 195-196, 195*f*
export from nuclei, 158

MRNA (messenger RNA) *(Continued)*
polyadenylation and, 189-190, 191*f*, 196
pre-mRNA surveillance of, 190*f*
processing of
alternative splicing and, 195*f*
pre-mRNA splicing and, 194*f*-195*f*
role of snRNAs and, 194*f*
surveillance of, 197-198
synthesis of, 189-194, 190*f*
cytoplasmic polyadenylation and, 196
editing of mRNAs and, 195-196, 195*f*
links between RNA processing and transcription in, 190-192, 190*f*
mRNA capping and polyadenylation in, 189-190, 191*f*
mRNA degradation and surveillance and, 196-197, 196*f*
pre-mRNA splicing and, 192-194, 193*f*-195*f*
translation and, 209-210, 209*f*-210*f*
MRNA turnover, 190*f*
MTOCs. *See* Microtubule organizing centers (MTOCs).
Mucins, 523, 526, 536*f*, 536*t*, 537, 538*t*
Multiangle laser light scattering, 95
Multidrug carriers, 258
Multiple drug resistance proteins, 251-252, 251*f*
Multivesicular bodies (MVBs), 377-378, 386, 388*f*-389*f*, 389-390
Mus musculus, genome of, 109*t*
Muscarinic acetylcholine receptor, 293-294
Muscle myosin, 624*t*
Muscles, 671-691, 672*f*
action potentials and, 289
cardiac, 672*f*, 685-691
congestive heart failure and, 688, 688*f*
contractile apparatus of, 685, 685*f*
excitation contraction coupling and, 679, 680*f*, 685*f*
molecular basis of inherited heart disease and, 682*t*, 688
pacemaker cells of, 685-687, 685*f*-686*f*, 686*b*
glucose transport in, 286-287
skeletal, 671-684, 672*f*
actomyosin apparatus of, 671-676, 672*f*-675*f*, 682*t*
molecular basis of contraction of, 676-678, 676*f*-677*f*
regulation of contraction of, 678-682, 678*f*-681*f*
specialized cells of, 682-684, 682*t*, 683*f*, 684*t*
smooth, 672*f*, 688-691, 689*f*-690*f*
Muscular dystrophies, 682*t*, 683-684, 683*f*, 684*t*
Emery-Dreifuss, nuclear envelope defect and, 150, 152*f*
Mutagenesis, site-directed, 92, 93*f*
Mutant alleles, conditional, 87
Mutants
ced (cell death abnormal), 803, 803*f*
conditional, 99-100
definition of, 86*b*
dominant negative, 100
Mutations
basal lamina, 519, 519*t*
in collagen, 556
conditional, 86*b*, 100
definition of, 86*b*
dominant, 86*b*, 87, 620-621
enhancer, 96-97, 96*f*
in gap junction subunits, 550, 550*t*
gene identification through, 87-88
lethal, synthetic, 96-97, 96*f*
null, 92-94
recessive, 86*b*, 87
seven-helix receptors and, 414, 414*t*
suppressor, 96-97, 96*f*
temperature sensitive, 702*b*
MVBs. *See* Multivesicular bodies (MVBs).
Myasthenia gravis, 275
Myc, 721
Mycoplasma genitalium, 18
genome of, 108-109, 109*t*
Myelin sheaths, 291, 291*f*
Myosin filaments, assembly of, 70, 70*f*
Myosin head, 624, 625*f*, 626, 628
Myosin light chain kinase, 690, 690*f*
Myosin-binding protein C, 674-675, 682*t*
Myosin-I, 624*t*, 629-630, 630*f*
Myosin-II, 70, 70*f*, 588, 624*t*, 629-630, 630*f*, 771-772
Myosins, 12, 574, 590*t*-591*t*, 624-631, 624*t*, 625*f*
actomyosin ATPase cycle and, 626-628, 627*f*
bulk movement of cytoplasm driven by, 647-648, 648*f*-649*f*
catalytic domain of, 624-626
duty cycle and, 628
evolution of, 623, 623*f*
heavy chains of, 624
light chains of, 624
skeletal muscle contraction regulation by, 681-682
light-chain domain of, 624-626
mechanochemistry of, 624-626, 625*f*-626*f*
rigor complex of, 626
of skeletal muscle, 682*t*
superfamily of, 628-631, 630*f*-631*f*
tail retraction and, 657
transduction of chemical energy into molecular motion, 627*f*, 628, 629*f*
Myosin-V, 624*t*, 629-630, 630*f*-631*f*, 648-649, 649*f*
Myosin-VI, 624*t*, 630-631, 630*f*
Myotilin, 684*t*
Myristoyl tails, 236
Myt1, 706, 706*f*, 710*t*-711*t*, 743*f*-744*f*, 744, 747*f*

N

E-N-Acetyl lysine, 34*f*
NADH, 319, 320*f*, 321-322
NADPH. *See* Nicotinamide adenine dinucleotide phosphate (NADPH).
Na+K+-ATPase, 248
Natural killer cells, 498
N-cadherin, 529*t*
NCAM (neural cell adhesion molecule), 528*t*
NcRNAs. *See* Nonprotein coding RNAs (ncRNAs).
Nebulin, 584, 590*t*-591*t*, 673-674
Necroptosis, 696, 798*b*, 813-814
Necrosis, 798, 798*b*
apoptosis vs., 797-800, 797*f*-800*f*
Negative feedback, seven-helix receptors and, 413*f*, 414
Negative selection, 801-802, 801*f*
Nematode sperm, actin substitute in, 666-667, 667*f*
N-end rule, 401
Neocentromeres, 116-117, 117*f*
Nernst potential, 281-282, 282*f*
Nerve growth factor, 419
Nervous system, self-avoidance in, 540-541
NESs. *See* Nuclear export sequences (NESs).
Nestin, 615*t*
Netrin, 659
Neural crest cells, 561, 562*f*
Neurofascin, 528*t*
Neurofilaments, 615*t*, 617, 621, 621*f*
Neuromuscular junctions, 292*f*, 294
synaptic transmission at, 678
Neurons
action potentials and, 289, 464
growth cone, 657, 659, 659*f*
motor neurons, 678, 678*f*
olfactory, 464, 464*f*
photoreceptor, 466-468
sympathetic, 469, 687
Neurotransmitters, synaptic transmission and, 291-294, 291*f*
Neutrophils, 494*t*, 497*f*, 498-499, 537*f*
NF-AT. *See* Nuclear factor-activated T cells (NF-AT).
NFH, 615*t*
NF-KB. *See* Nuclear factor KB (NF-KB).
NFL, 615*t*
NFM, 615*t*
NhaA family of carriers, 258
NHEJ. *See* Nonhomologous end joining (NHEJ).
Nicotinamide adenine dinucleotide phosphate (NADPH), 328
Nicotinic acetylcholine receptor, 273-275, 274*f*, 279
Nidogen, 517, 518*f*, 523
Nijmegen breakage syndrome, 749*t*
Ninein, 612
Nitella, cytoplasmic streaming in, 640*t*, 647, 648*f*
Nitric oxide, 418
as second messenger, 459-461, 460*f*
Nitric oxide synthases, 460, 460*f*
N-linked oligosaccharides, 342, 343*f*, 370*f*, 513*f*
NLSs. *See* Nuclear localization sequences (NLSs).
NMD. *See* Nonsense-mediated decay (NMD).
NMDA receptors. *See* *N*-Methyl-D-aspartate (NMDA) receptors.
N-Methyl-D-aspartate (NMDA) receptors
glutamate and, 273
long-term potentiation and, 296-297, 296*f*
NMR spectroscopy. *See* Nuclear magnetic resonance (NMR) spectroscopy.

Nocodazole, 595*b*
Nonclathrin/noncaveolar endocytosis, 377, 385–386, 385*f*
Noncrossovers (gene conversions), 782, 783*f*, 784*b*
Nondisjunction, 784*b*
Nonhistone proteins, chromosome structure and, 136–139, 136*f*, 138*f*–139*f*
Nonhomologous end joining (NHEJ), 92–94, 121
Nonprotein coding RNAs (ncRNAs), 189
Nonreceptor tyrosine kinases. *See* Cytoplasmic protein tyrosine kinases.
Nonsense-mediated decay (NMD), 190*f*, 197
Nonsteroidal antiinflammatory drugs (NSAIDs), 451
Nonstop decay, 190*f*, 198
Nonstop mRNAs, 198
Nonvesicular transport pathways, 371
Norepinephrine, 291*f*, 469
NORs. *See* Nucleolus-organizing regions (NORs).
Notch receptors, 420–423
NSAIDs. *See* Nonsteroidal antiinflammatory drugs (NSAIDs).
NSF (*N*-ethylmaleimide [NEM]-sensitive factor), 364, 379
N-terminal tails, 123, 125–127, 125*f*–126*f*
N-terminus, 36
Nuclear actin-binding protein, 590*t*–591*t*
Nuclear basket, 151
Nuclear envelope, 8, 10, 10*f*, 143*f*, 331, 331*f*
  disassembly of, in prometaphase of mitosis, 759, 759*f*
  reassembly in telophase of mitosis, 770–771, 770*f*
  regulation of transport across, 158–159, 159*f*
  structure of, 147–152, 147*f*
    defects in, human diseases associated with, 150–151, 152*f*
    in genome organization, 150, 151*f*
    inner nuclear membrane proteins and, 149–150, 150*f*
    nuclear lamina and, 148–149, 148*f*–149*f*
    nuclear pore complexes and, 151–152, 153*f*
Nuclear export, 156–157, 156*f*
Nuclear export sequences (NESs), 155, 156*f*
Nuclear factor KB (NF-KB), 158–159, 159*f*, 472–473, 472*f*, 477*f*, 478–479
  inhibitor of, 185*f*, 186
Nuclear factor-activated T cells (NF-AT), 429–430, 478–479
Nuclear import, 156, 156*f*–157*f*
Nuclear lamina, 4*f*, 148, 151*f*, 759, 759*f*
  structure and assembly of, 148–149, 148*f*–149*f*
Nuclear lamins
  inner nuclear membrane proteins and, 149–150, 150*f*
  structure and assembly of, 148–149, 148*f*–149*f*
Nuclear localization sequences (NLSs), 154–155, 154*f*–155*f*
Nuclear magnetic resonance (NMR) spectroscopy, 35, 95
Nuclear matrix, 136–137
Nuclear membrane
  inner, 148
    proteins of, 149–150, 150*f*
  outer, 148
Nuclear mitotic apparatus (NuMA) protein, 761
Nuclear pore complexes, 147*f*–148*f*, 148, 151–152, 153*f*
Nuclear pores, 6, 10, 10*f*, 143*f*, 148*f*, 153*f*
Nuclear receptors, 472
Nuclear rings, 151, 153*f*
Nuclear speckles, 194
Nuclear transport receptors, 156, 157*f*
Nucleases, 807–808, 807*f*
Nucleation, 67, 69, 69*f*
  actin filament, 580*f*, 581–582, 582*f*
  regulation of assembly by, 68
Nucleic acids, 29
  building blocks of, 42, 44*f*
  structure of, 42–48
    covalent, 42–43, 44*f*–45*f*
    DNA secondary structure and, 43–45, 46*f*
    RNA secondary and tertiary structures and, 45–48, 47*f*–49*f*
Nucleolus, 10, 10*f*, 143*f*, 145*t*, 146–147, 146*f*
  disassembly during mitosis, 146*f*, 147
  ribosomal biogenesis in, 146*f*, 147
  ribosome synthesis in, 199, 200*f*
Nucleolus-organizing regions (NORs), 146*f*, 147
Nucleoporins, 151, 153*f*
Nucleoside, 42–43
Nucleoskeleton, 136–137
Nucleosome core particles, 123, 124*f*
Nucleosome remodeling complexes, 178
Nucleosomes, 123–127, 124*f*, 131*f*, 136*f*
  chromatin and
    modifications and regulation of function, 123–125, 125*f*
    regulation of structure by histone N-terminal tails, 125–127, 125*f*–126*f*
  core histones and, 123, 124*f*
  histone deposition during assembly of, 127
  histone variants and, 127
  linker DNA and linker histone H1 and, 127, 128*f*
Nucleotide excision repair, 749*t*, 750, 751*f*
Nucleotides, 29, 42, 44*f*
  cyclic, 443–445, 444*f*–445*f*
Nucleus, 6, 10, 10*f*, 143–160, 143*f*
  changes in, during prophase of mitosis, 746*f*, 756–757
  DNA replication in, higher-order organization of, 735–739, 737*f*–738*f*
  functional compartmentation of, 128–130, 128*f*–129*f*, 131*f*
  interphase, chromosomal DNA packaging in, 130–131, 131*f*
  MAP kinase pathways to, 473–474, 473*f*–474*f*, 476*f*–477*f*, 483*f*
Nucleus *(Continued)*
  nuclear envelope structure and, 147–152, 147*f*
    defects leading to human diseases and, 150–151, 152*f*
    in genome organization, 150, 151*f*
    inner nuclear membrane proteins and, 149–150, 150*f*
    nuclear laminar and, 148–149, 148*f*–149*f*
    nuclear pore complexes and, 151–152, 153*f*
  organization of, 143–147
    nucleolus and, 146–147, 146*f*
    specialized subdomains and, 143–146, 144*f*, 145*t*
  structural compartmentation of, 131–133, 132*f*–133*f*
  traffic between cytoplasm and, 152–159, 154*f*–157*f*
    defective, disorders associated with, 159
    mRNA export and, 158
    nuclear import and export components and, 155–157, 157*f*
    regulation of transport across nuclear envelope, 158–159, 159*f*
    single import cycle and, 157–158, 157*f*
Null mutation, 92–94
NuMA protein. *See* Nuclear mitotic apparatus (NuMA) protein.

O

Objective lens, 76–78, 77*f*
Occludin, 545
Odor detection, 463–466
  adaptation and, 464*f*, 465–466
  cAMP production and, 465
  cyclic nucleotide-gated channels and, 465
  odorant receptors and, 464*f*, 465
  processing in brain and, 464*f*, 466, 466*b*
  sensory neurons and, 464, 464*f*
Odorant-binding proteins, 463
Odorants, 463
Ohm's law, 283
Okazaki fragments, 728, 728*f*
Olfactory bulb, 464
Olfactory receptor kinase, 464*f*, 466
Olfactory sensory neurons, 464, 464*f*
Olfactory system
  odor detection by. *See* Odor detection.
  second, sex and, 466*b*
Oligosaccharides
  link, 512, 513*f*
  N-linked, 342, 343*f*, 370*f*, 513*f*
Oligosaccharyl transferase, 342
O-linked glycosylation, 370
O-linked oligosaccharide, 513*f*
Omega loops, 38*f*, 39
Oncogenes, 474, 721–722
Oncogenic stress, 722, 723*f*
Open complexes, 165–166, 169–170, 170*f*
Open mitosis, 755
Operons, 166, 166*f*
Optical trap, 237–238, 238*f*, 629*f*
Optogenetics, 243

ORC. *See* Origin recognition complex (ORC).
Ordered liquid phase, 232
Organelles, 6, 301-302
actin-based movements of, 648-649, 649*f*
lipid movement between, 341*t*, 348-349, 349*f*
membrane-bounded, evolution of, 21-22, 21*f*-22*f*
membranes of, lipid asymmetry of, 232-233, 233*f*
purification of, 88-90
Organisms for biological research, choice of, 85-87, 86*b*
Origin recognition complex (ORC), 729-731, 730*f*
Origins of bidirectional replication, 727
Origins of replication, 700, 728-731, 728*f*
in mammalian cells, 730-731, 730*f*-731*f*
in *Saccharomyces cerevisiae*, 729-730, 729*f*-730*f*
Orphan receptors, 412
Orthologs, 17, 18*b*
Oryza sativa japonica, genome of, 109*t*
Osteoblasts, 491*f*, 559-560, 559*f*-560*f*
Osteocalcin, 559*t*
Osteoclasts, 499, 560-561, 560*f*
Osteocytes, 558*f*, 560
Osteogenesis imperfecta, 346, 565
Osteonectin, 523, 559*t*
Osteons, 564-565
Osteopetrosis, 565
Osteopontin, 523, 559*t*
Osteoporosis, 565
Outer doublets, 660, 662*f*
Outer kinetochore, 139, 139*f*
Outer membrane
of chloroplasts, 324
of gram-negative bacteria
insertion of proteins in, 313
secretion across, 314*f*
of mitochondria, 304, 306, 306*f*, 318, 318*f*
protein transport across, 305-306, 306*f*
Outer membrane autotransporter pathway, 313, 314*f*
Outer membrane single accessory pathway, prokaryotic protein export via, 313
Outer nuclear membrane, 148
Outer plate, 139, 139*f*
Outer segment of photoreceptors, 468
Overexpression, 100
Oxidation, fatty acid, 319, 320*f*
Oxidative enzymes, 309
Oxidative phosphorylation, 317, 319-322, 320*f*-321*f*
Oxygen
atmospheric, 22
carbonyl, 34*f*, 35-36
oxidative phosphorylation, 317, 319-322, 320*f*-321*f*
photosynthesis, 328

P

P62, 152, 153*f*
P21-activated kinase, 434
P21 Cdk inhibitor, 714*b*
P element, 111
P loops, 262-264, 264*f*, 267-268
P site of tRNAs, 213
P53 tumor suppressor gene, 696
apoptosis and, 812, 813*f*
Pacemaker cells of cardiac muscle, 685-687, 685*f*-686*f*, 686*b*
Pachytene, 785*f*, 786-787
PAF, 451
Paired helical fragments, 602, 602*f*
Pairing in meiosis, 781, 786*f*-788*f*, 787-789
PALM. *See* Photoactivated localization microscopy (PALM).
Paralogs, 17-18, 18*b*
Parasympathetic nerves, heart rate and, 687, 687*f*
Parathyroid hormone, 560
Parkinson disease, protein misfolding in, 219*b*, 402
Passive transporters, 241-242
Patch clamps, 279, 280*f*
Patched protein, 420
PAX (paired box) genes, 561
Paxillin, 534, 534*f*
PCC. *See* Premature chromosome condensation (PCC).
PCNA, 715, 734, 735*f*
PCR. *See* Polymerase chain reaction (PCR).
PDGF. *See* Platelet-derived growth factor (PDGF).
PDI. *See* Protein disulfide isomerase (PDI).
PDZ domains, 440
PECAM-1 (platelet/endothelial cell adhesion molecule), 528*t*
Pectins, 568
Pedigree, definition of, 86*b*
Pemphigus foliaceus, 551
Pemphigus vulgaris, 551
Pentameric ligand-gated ion channels, 273-275, 274*f*
Pentose, 48
Peptide bonds, 32, 34*f*, 36
Peptides, inactivation, 269, 269*f*
Peptidyl prolyl isomerase, 437*t*
Perforin, 503
Pericentrin, 612
Pericentriolar material, 604-607, 605*f*-606*f*, 608*f*-609*f*
Perichondrium, 556
Perichromatin fibrils, 144
Perichromosomal layer, 147
Perinuclear space, 147*f*
Periosteum, 557-559
Peripheral membrane proteins, 227, 236-237, 236*f*
Peripherin, 615*t*
PERK/PEK, 345, 345*f*
Perlecan, 514*f*, 517, 518*f*, 519*t*, 522
Permeability, selective, 281
Permissive conditions, 87
Permissive temperature, 702*b*
Peroxins, 309, 310*f*, 310*t*
Peroxisomal biogenesis disorders, 309, 328-329
Peroxisomal targeting signal, type 1 (PTS1), 309, 309*f*, 310*t*
Peroxisomal targeting signal, type 2 (PTS2), 309
Peroxisomes, 8, 12, 301-302, 328-329, 329*f*
origins of, 22
protein transport into, 303*f*, 309-311, 309*f*-310*f*, 310*t*
Pertussis toxin, 436-437, 436*t*
PH domains. *See* Pleckstrin homology (PH) domains.
Phagocytic cells, 489
Phagocytic cup, 379
Phagocytosis, 302, 377-381, 378*f*-379*f*, 800*f*
alternative fates of ingested particles and, 379-381, 381*b*
engulfment in, 378-379, 379*f*-380*f*, 380*b*
fusion with lysosomes and, 379-381, 379*f*
Phagolysosomes, 379, 379*f*
Phagophores, 398, 398*f*
Phagosomes, 11
Phase-contrast microscopy, 76, 76*t*, 77*f*
Phenotype, 86*b*, 784*b*
Phenylalanine, structure of, 33*f*
Phorbol esters, 448
Phosphatases
cooperation between kinases and, 432
in counter-balancing Cdks, 706*f*, 707, 710*t*-711*t*
pharmacologic agents for studying, 432
protein tyrosine, 430*t*, 431-432, 431*f*-432*f*
Phosphatidic acid, 229, 446, 446*f*
Phosphatidylcholine, 229, 346, 347*f*, 445, 446*f*
Phosphatidylcholine signaling pathways, 449, 449*f*
Phosphatidylethanolamine, 229, 346, 347*f*
Phosphatidylglycerol, 229
Phosphatidylinositol (PI), 346, 380*b*, 380*f*, 445, 446*f*, 448*f*
Phosphatidylinositol 4,5-biphosphate (PIP_2), 380*b*, 380*f*, 446, 448-449, 448*f*
Phosphatidylinositol 3-kinase (PI3K), 475, 483*f*
Phosphatidylinositol 4-phosphate (PIP), 446
Phosphatidylinositol phospholipase Cs (PI-PLCs), 446-447, 452
Phosphatidylinositol 3,4,5-triphosphate (PIP_3), 379, 380*b*, 380*f*, 446
Phosphatidylinositol-4-P-5 kinase, 380*b*
Phosphatidylserine, 229, 346, 347*f*
Phosphodiester bonds, 42, 44*f*
Phosphodiesterases, 443-444, 444*f*-445*f*
cyclic nucleotide, 418
odor detection and, 465
visual signal processing and, 467*f*, 468, 469*b*
Phosphoglycerides, 228-229, 228*f*, 229*t*
Phosphohistidine, 34*f*, 426*f*
Phosphoinositide 3-kinase, 415
Phosphoinositide signaling pathways, 448-449, 448*f*
Phospholamban, 454
Phospholipase A_2 (PLA_2), 446, 446*f*
Phospholipase C (PLC), 446, 446*f*
Phospholipase Cβ, 435, 435*t*, 453*f*
Phospholipase Cγ, 415-416, 439-440, 439*f*, 453*f*, 475
Phospholipase D (PLD), 446, 446*f*

Phospholipases, 402, 446, 446*f*, 449*f*
Phospholipids in bilipid layer, electrostatic interaction with, 237
Phosphoprotein, secreted, 523
Phosphorylase b, metabolic regulation and, 471
Phosphorylase kinase, metabolic regulation and, 471
Phosphorylation
- 14-3-3 proteins and, 439
- actin-based motility of vaccinia virus and, 650*f*
- β-adrenergic receptor and, 469–472, 470*f*, 471*t*
- cell cycle and, 7–8
- cell cycle progression and, 723, 723*f*
- cell proliferation and, 720–721, 721*f*
- cell-cycle regulation of meiotic events and, 792
- changing states of cytoplasm during cell cycle and, 709–710, 709*f*
- condensin binding to chromosomes and, 137
- CTD, 168–169, 172, 190–191
- cyclin-dependent kinase structure and function and, 703–707, 706*f*
- cytoplasmic changes in prophase and, 757–758, 757*f*–758*f*, 757*t*
- dephosphorylation and, 380*b*
- detection of proteins introduced by, 89*b*
- of dynein, 662
- of glycogen phosphorylase, 41
- G_2/mitosis transition and, 743–747, 743*f*–747*f*
- inhibition of nuclear import by, 158
- integrity of cellular DNA monitored by a G_1 checkpoint and, 723–725, 724*f*–725*f*
- of light chains, 588
  - in vertebrate skeletal muscle, 681–682, 690–691, 690*f*
- of mannose 6-hydroxyl, 372
- by MAPs, 601–602, 611
- marking of proteins for destruction by, 393
- metaphase II arrest and MAP kinase pathway and, 792–793, 793*f*
- by mitotic kinases, intermediate filaments and, 616–617, 617*f*
- myosin-I activation by, 629–630
- nucleus disassembly and, 147
- oxidative, 317, 319–322, 320*f*–321*f*
- PH domains and, 439–440
- phosphotyrosine-binding domains and, 415, 415*f*, 439
- posttranslational histone modification and, 125–126, 125*f*
- protein destruction in cell-cycle control and, 707–708, 708*f*
- protein kinases and, 441*t*
- of proteins, 425–432, 426*f*
  - cooperation between kinases and phosphatases and, 432
  - pharmacologic agents for studying protein phosphatases and, 432
  - protein kinases and, 426–429, 427*f*–428*f*, 430*f*, 430*t*, 441*t*

Phosphorylation *(Continued)*
  - protein structure and function and, 426, 426*f*–427*f*
  - protein tyrosine phosphatases and, 430*t*, 431–432, 431*f*–432*f*
- P-type ATPases and, 248–250, 248*f*–249*f*
- receptor tyrosine kinases and, 414–416, 415*f*–416*f*
- reversible, 34–35
  - regulation of nuclear lamina assembly by, 68
- by rhodopsin kinase, 469
- seven-helix receptors and, 412–414, 412*f*–413*f*, 414*t*
- sex chromosomes in meiosis and, 791–792, 792*f*
- SH2 domains and, 439
- signaling cascades and, 399, 401
- signaling pathways and
  - cytokine receptor, JAK/STAT, 479–481, 480*f*
  - growth factor, 473*f*–474*f*, 474–475, 483*f*
  - influencing gene expression, 472–473, 472*f*–474*f*, 476*f*–477*f*, 480*f*–481*f*
  - serine/threonine kinase receptor, through Smads, 481–482, 481*f*
  - T-lymphocyte, through nonreceptor tyrosine kinase, 477*f*, 478–479, 479*b*, 480*f*–481*f*
- Src, of E-cadherin and β-catenin, 530
- suppression of DNA replication between meiosis I and meiosis II and, 792
- transcription factors and, 183–184, 184*f*–185*f*
- two-component phototransfer systems and, 482–486, 482*b*, 483*f*
  - bacterial chemotaxis and, 483*f*–485*f*, 485–486
- tyrosine, 382
- WW domains and, 439

Phosphoserine, 34*f*, 426*f*
Phosphothreonine, 34*f*, 426*f*
Phosphotransfer systems, two-component, 482–486, 482*b*, 483*f*
- bacterial chemotaxis and, 483*f*–485*f*, 485–486

Phosphotyrosine, 34*f*, 426*f*, 475
Phosphotyrosine-binding (PTB) domains, 415, 415*f*, 439
Photoactivated localization microscopy (PALM), 81–82, 81*t*
Photobleaching, 237–238, 238*f*, 656
Photon detection by vertebrate retina, 466–469
- overview of, 466–468, 467*b*, 467*f*
- positive arm of signal cascade and, 468–469, 469*b*
- recovery and adaptation and, 469, 469*b*
- rhodopsin and, 467*f*, 468

Photoreceptor cells, 466–467, 467*b*, 467*f*
- electrical circuits in, 469*b*

Photosynthesis, 317
- carbohydrate synthesis and, 328
- energy capture and transduction by photosystem I, 327–328

Photosynthesis *(Continued)*
- energy capture and transduction by type II photosystems and photosystem II, 326, 327*f*
- light and dark reactions and, 325–326, 325*f*
- light harvesting and, 326–327, 327*f*
- oxygen-producing synthesis of NADPH and ATP by dual photosystems, 328
- plastids and, 324
- structure and evolution of systems in, 323–324, 324*f*–325*f*

Photosynthetic reaction centers, 18–20, 323–324, 325*f*, 326, 327*f*
Photosystem I, 323–324
- energy capture and transduction by, 327–328
- oxygen-producing synthesis of NADPH and ATP by dual photosystems and, 328

Photosystem II, 323–324
- energy capture and transduction by, 326, 327*f*
- oxygen-producing synthesis of NADPH and ATP by dual photosystems and, 328

Phylogenetic tree, 3*f*–4*f*, 4, 15, 15*f*, 19*f*, 24, 24*f*
Physarum, cytoplasmic streaming in, 640*t*, 647–648, 649*f*
Physiological function tests
- anatomic, 98–99
- physiological, 99–100
- reconstitution as, 98

PI. *See* Phosphatidylinositol (PI).
PI-3 kinases, 449
Piezo2, 278
PI3K. *See* Phosphatidylinositol 3-kinase (PI3K).
PIKA. *See* Polymorphic interphase karyosomal association (PIKA).
PIP. *See* Phosphatidylinositol 4-phosphate (PIP).
PIP_2. *See* Phosphatidylinositol 4,5-biphosphate (PIP_2).
PIP_3. *See* Phosphatidylinositol 3,4,5-triphosphate (PIP_3).
PI-PLCs. *See* Phosphatidylinositol phospholipase Cs (PI-PLCs).
PiRNAs. *See* Piwi-interacting RNAs (piRNAs).
Piwi-interacting RNAs (piRNAs), 204
PKA. *See* Protein kinase A (PKA).
PKB/Akt, 429, 439–440, 449, 476, 476*f*
PKC. *See* Protein kinase C (PKC).
PKG. *See* Protein kinase G (PKG).
PKR, 112
PLA_2. *See* Phospholipase A_2 (PLA_2).
Plakins, 619, 619*t*, 620*f*
Plakoglobin, 529–530, 529*t*, 551
Plakophilin, 551
Plant cell walls, 567–569, 567*f*–568*f*
Plant cells, 4*f*, 547*f*
Plants
- cytokinesis in, 772, 773*f*, 776*b*
- time line for divergence of, 16*f*

Plasma cells, 491*f*, 502
Plasma fibronectin, 516

Plasma membrane, 6*f*, 9, 9*f*
  constitutive transport of cargo to, *trans*-Golgi network and, 371*f*
  depolarization by cyclic nucleotide-gated channels, 465
  posttranslational targeting of proteins to surfaces of, 311
  protein translocation across, by ABC transporters, 311, 314*f*
  structural proteins of, defects in muscular dystrophies and, 682*t*, 683-684, 683*f*, 684*t*
  trafficking to, in polarized cells, 372-373, 372*f*
Plasma membrane receptors, 411-423, 411*f*
  cytokine, 416-417, 417*f*, 421-423
  guanylyl cyclase, 418-419, 418*f*, 421-423
  Hedgehog, 420-421
  Notch, 420-423
  receptor serine/threonine kinases and, 417-418, 417*f*-418*f*, 421-423
  receptor tyrosine kinases and, 414-416, 415*f*-416*f*, 421-423
  seven-helix, 412-414, 412*f*-413*f*, 414*t*, 421-423
  toll-like, 420
  tumor necrosis factor family of, 419-420, 419*f*
Plasmid cloning, 93*f*
Plasmids, 86*b*, 87
Plasmodesmata, 546, 547*b*, 547*f*
Plasmodium, mitochondria of, 318
Plasticity, synaptic, 295-296, 296*f*
Plastids, 22, 324
Plastin, 590*t*-591*t*
Platelet-derived growth factor (PDGF), 495
  growth factor signaling pathways and, 474
  receptor tyrosine kinase activation by, 414
Platelets, 493-496, 494*f*, 494*t*, 496*f*
  activation and adhesion of, 539*f*, 540
  wound healing and, 565, 566*f*, 567
Platyhelminths, 28, 25
PLC. *See* Phospholipase C (PLC).
PLD. *See* Phospholipase D (PLD).
Pleckstrin, 439-440
Pleckstrin homology (PH) domains, 429, 439-440
Plectin, 619, 619*t*, 620*f*
Pluripotent stem cells, 493-494, 494*f*, 715
Plus End Binding Proteins, 600-601, 600*f*, 611
Plus end of microtubules, 593
PML (promyelocytic leukemia) bodies, 144*f*, 145, 145*t*
Point centromeres, 114
Pol α/Primase, 734, 735*f*
Polar bodies, 794, 794*f*
Polarity, 69
Polarization microscopy, 76-78, 76*t*, 77*f*
Polarized cells, 372-373, 372*f*, 543-546
Polo family of protein kinases, 745
Poly(A) tails, 198, 210
Polyacrylamide gels, 89*b*
Polyadenylation, 189-190, 191*f*, 196
Polycomb group bodies, 145*t*
Polycomb group proteins, 130
Polycythemia vera, 495
Polymerase chain reaction (PCR), 90, 92*f*
Polymerase I, 163
Polymerase II, 163, 176*f*
Polymerase III, 163
Polymorphic interphase karyosomal association (PIKA), 145*t*
Polypeptides, 5, 5*f*
  folding of, 36-40, 37*f*-40*f*
  posttranslational targeting and, 303*f*-304*f*, 304
  structure of, 32, 33*f*-34*f*, 34-36
Polyphosphoinositides, 229, 380*b*, 380*f*, 439-440
Polyploid, 794
Polyproline helices, type II, 440
Polysomes, 217
Polytene chromosomes, 132-133, 133*f*
Polytopic proteins, 339*f*, 340
Polyubiquitin chain, 394*f*, 395-396, 401
POMGnT1, 684*t*
POMTi, 684*t*
Pore helices, 262-264
Porifera, 24-25
Porins, 277-278, 318, 318*f*
Position effect, 129, 129*f*
Positive selection, 801-802, 801*f*
Postsynaptic, definition of, 291
Postsynaptic potential (PSP), 295
Posttranslational modifications, 32, 34-35, 34*f*, 125, 131*f*
  of intermediate filaments, 616-618, 617*f*
Posttranslational targeting of proteins, 303-315, 303*f*-304*f*
  prokaryotic protein export, 311-315
    pathways dependent on SecYE translocon in, 311-315, 312*f*, 314*f*
    pathways independent of SecYE translocon in, 314*f*, 315
  to surfaces of plasma membrane, 311
  translocation across plasma membrane by ABC transporters, 311, 314*f*
  transport into chloroplasts, 303*f*, 307-309, 308*f*
  transport into mitochondria and, 303*f*, 304-307, 305*f*-307*f*
  transport into peroxisomes, 303*f*, 309-311, 309*f*-310*f*, 310*t*
Posttranslational translocation, 335, 340-341, 341*f*
Potassium channels, voltage-gated, 269-270, 269*f*, 290, 295, 686*b*
PP1, 429-431, 430*f*, 430*t*, 707, 710*t*-711*t*
PP2A, 429-431, 430*f*, 430*t*, 707, 710*t*-711*t*, 745, 765
PP2B, 429-430, 430*t*
PPM family of serine/threonine phosphates, 430*t*, 431
PPP family of serine/threonine kinases phosphorylation and, 429-431, 430*f*, 430*t*
PRb, 720
Prebiotic chemistry, leading to RNA World, 15-17, 16*f*
Preincision complexes, 750
Preinitiation complexes, 165-166, 170-172, 171*f*-172*f*, 214
Premature chromosome condensation (PCC), 704, 704*f*
Pre-mRNA splicing, 192-194
  alternative splicing and, 193-194, 195*f*
  AT-AC introns and, 193, 193*f*
  localization of, 194
  signals for, 192, 193*f*
  splicing reaction and, 192-193, 193*f*-194*f*
Prenucleolar bodies, 147
Prenylation, 615
Preprocollagen, 507, 508*f*
Prereplication complex, 731-733, 731*f*, 732*t*, 733*f*
Presequences, 305
Presynaptic, definition of, 291
Primary active transporters, 241. *See also* Membrane pumps.
Primary cilia, 660*f*, 665, 665*f*
  of centrosomes, 608
Primary constriction, 113, 147
Prions, 219*b*
Proapoptotic activities, 798*b*, 799
Probability, 53-54
Probability of a channel being open (P_O), 266
Processed pseudogenes, 112
Procollagen, 507
Profilins, 580, 581*f*, 583*f*, 584-585, 587, 590*t*-591*t*, 655*f*
Programmed cell death, 797-815, 798*b*
  accidental cell death vs., 797-800, 797*f*-800*f*
  apoptotic. *See* Apoptosis.
  examples of cells undergoing, 801-803, 801*f*
    cells serving no function as, 801*f*, 802
    cells with perturbed cell cycles as, 802
    chemotherapeutic killing of cells as, 803
    developmentally defective cells as, 801-802, 801*f*
    excess cells as, 802
    virus-infected cells as, 802-803
  Fas death receptor and, in normal and diseased cells, 812
  importance of, in human disease, 814, 814*f*
  key terms in, 798*b*
  necessity for, 797
  nonapoptotic, 813-814
Prohormone convertases, 371
Prokaryotes. *See also* Bacteria.
  compartmentalization of, 21-22, 21*f*
  evolution of, 18-20, 19*f*
  transcription initiation in, regulation of, 175-176, 176*f*
Prokaryotic cells, eukaryotic cells vs., 4*f*, 8-9
Prokaryotic protein export, 311-315
  pathways dependent on SecYE translocon in, 311-315, 312*f*, 314*f*
  pathways independent of SecYE translocon and, 314*f*, 315
Proliferation, 720-721, 721*f*
Proline, structure of, 33*f*

Prolylpeptide isomerase, 508
Prometaphase of mitosis, 700, 755, 755*f*-756*f*, 758-765, 758*f*
  chromosome attachment to spindle in, 761-763, 762*f*
  correction of errors in chromosome attachment to spindle in, 763-764, 763*f*
  mitotic spindle organization in, 759-760, 760*f*, 765*f*
  nuclear envelope disassembly in, 759, 759*f*
  spindle assembly in, 760-761, 760*f*-761*f*
  spindle checkpoint and, 764-765, 764*f*
Promoter clearance, 173
Promoter proximal elements, 176*f*, 180-184, 181*f*
Promoters, 163, 165-166, 169, 169*f*
Promyelocytic leukemia (PML) bodies. *See* PML (promyelocytic leukemia) bodies.
Prophase, 700
  initiation of, Cdk1 activity and, 745-746, 746*f*
  of mitosis, 755-758, 755*f*-757*f*
    cytoplasmic changes in, 757-758, 757*f*-758*f*, 757*t*
    nuclear changes in, 756-757
Prophase I of meiosis, homologous chromosomes in, 785-787, 785*f*-786*f*
Prostaglandin H synthetase, 450-451, 450*f*
Prostaglandin isomerases, 450
Prostaglandins, 446, 450, 450*f*
Proteases, proteolysis and, 396
Proteasomes, 707
  degradation by, 399-402, 400*f*
    elimination of misfolded proteins from endoplasmic reticulum and, 402
    intracellular proteolysis and, 402
    motifs specifying ubiquitination and, 395*t*, 401-402
    ubiquitination modification of proteins and, 394-396, 394*f*-395*f*, 395*t*
  proteolysis and, 396
Protein 4.1, 590*t*-591*t*
Protein disulfide isomerase (PDI), 342, 507-508
Protein domains, 41-42, 43*f*
Protein folding, 36-40, 331-332, 332*t*
  chaperone-assisted, 218-221, 220*f*-222*f*
  in endoplasmic reticulum, 342-344, 343*f*-344*f*
    diseases and, 346
  misfolding in diseases and, 219*b*, 402
  spontaneous, 218, 219*b*
Protein kinase A (PKA), 185-186, 426-430, 427*f*-428*f*, 441*t*, 444, 445*f*, 457, 469-472, 470*f*
  consensus target sequence for, 426-428, 427*f*
Protein kinase C (PKC), 427*f*-428*f*, 428-429, 439-440, 441*t*, 444, 447*f*, 448-449, 452, 457*f*, 474*f*, 475, 476*f*-477*f*
Protein kinase G (PKG), 460-461
Protein kinases, 13, 34*f*, 409
  cGMP-stimulated, 418, 428*f*, 443-444, 444*f*

Protein kinases *(Continued)*
  DNA-dependent, 752*f*, 753
  phosphorylation and, 425-428, 427*f*-428*f*, 441*t*
    disease and, 429, 430*f*, 430*t*
    regulation of, 428-429, 430*f*
  PKR, 112
Protein phosphatases, phosphorylation and, 425, 429-431, 430*f*, 430*t*
  Cdc25 subfamily of, 432
  dual-specificity subfamily of, 432
  PPM family of serine/threonine phosphatases as, 430*t*, 431
  PPP family of serine/threonine phosphatases as, 429-431, 430*f*, 430*t*
  protein tyrosine phosphatases as, 431-432, 431*f*
  PTP subfamily of, 431-432, 431*f*
Protein targeting, 6, 7*f*, 354-356, 362-363, 372*f*, 373. *See also* Posttranslational targeting of proteins.
  protein translocation into endoplasmic reticulum and. *See* Protein translocation: into endoplasmic reticulum.
Protein toxins as "opportunistic endocytic ligands", 391, 391*f*
Protein translocation
  across plasma membrane by ABC transporters, 311, 314*f*
  into endoplasmic reticulum, 335-342
    insertion of membrane proteins into ER bilayer, 339*f*, 340
    Sec61 complex and, 337-338, 338*f*
    signal recognition particle and signal recognition particle-receptor and, 335-337, 336*f*-337*f*
    signal sequences and, 335
    of soluble proteins into lumen, 338-340, 339*f*
    tail-anchored protein association with ER membrane and, 335*t*, 341-342, 342*f*
  lipid-anchored protein association with cytoplasmic surface of ER and, 340
  into mitochondria, 305-306, 305*f*-306*f*
  posttranslational, 335, 340-341, 341*f*
Protein turnover, constitutive, 393-396, 394*f*-395*f*, 395*t*
Protein tyrosine kinases, Src family of, 479*b*
Protein tyrosine phosphatases, 430*t*, 431-432, 431*f*-432*f*
Proteins, 3-4, 29
  14-3-3, 439
  accessory, regulation of assembly by, 68-69
    of actin filaments, 69-70, 69*f*
    of bacterial flagella, 70-71, 70*f*-71*f*
    of bacteriophage T4, 73-74, 73*f*
    of tobacco mosaic virus, 68-69, 71-72, 71*f*
    of tomato bushy stunt virus, 72-73, 72*f*
  actin-binding. *See* Actin-binding proteins.
  adaptor, 361, 362*f*
  assembly, 361, 362*f*
  associated with intermediate filaments, 618-619, 619*t*, 620*f*

Proteins *(Continued)*
  Bcl-2
    apoptosis and, 803, 803*f*, 808-809, 808*f*
    cancer and, 809, 809*f*
  binding to specific DNA sequences, 179-180, 179*f*
  calcium-regulated, 462
  capping, 590*t*-591*t*, 655*f*, 656
  carrier, 253-259, 370
    definition of, 241
    discovery of, 255*b*
    diversity of, 255*f*, 257-259
    physiology and mechanisms of, 253-257, 253*f*, 255*f*, 257*t*
    structures of, 255*f*
    tools for studying, 257*t*
  cell adhesion, 9, 525, 525*f*, 659
  centromere
    of budding yeast, 141, 141*f*
    mammalian, 140-141, 140*f*
  of centrosomes, 605-606, 605*f*-606*f*, 612
  of chromosome scaffold, 136-137, 138*f*
  conformational changes of, 41
  CTCF, 130, 131*f*, 134-135
  degradation of
    in endoplasmic reticulum, 344, 344*f*
    selective, 99
  Delta, 420
  destruction of, cell cycle control and, 707-708, 708*f*
  Dis1/TOG, 603, 611
  dynamics of, 41, 42*f*
  ER shaping and fusion, 333-334, 333*f*
  folding and oligomerization in endoplasmic reticulum, 342-344, 343*f*-344*f*
    diseases and, 346
    unfolded protein response and, 345-346, 345*f*
  folding of, 36-40, 331-332, 332*t*
    chaperone-assisted, 218-221, 220*f*-222*f*
    misfolding in diseases and, 219*b*, 402
    spontaneous, 218, 219*b*
  functions of, phosphorylation and, 426, 426*f*-427*f*
  GPI-anchored, 233, 236-238, 236*f*
  green fluorescent, 78, 79*f*, 87-88
  GTP-binding. *See* GTPases.
  helix-loop-helix, 179
  helix-turn-helix, 179-180
  histone, synthesis of, 740-741, 740*f*
  of inner nuclear membrane, 149-150, 150*f*
  interactions with solvent, 40-41, 41*f*
  intermediate filaments, evolution of genes for, 613*f*, 614-615, 615*t*
  leucine zipper, 179-180, 179*f*
  lipid-anchored, 340
  lipid-transfer, 341*t*, 348-349, 349*f*
  major sperm, 666-667, 667*f*
  membrane, 234-239
    dynamic behavior of, 237-239, 238*f*
    electrostatic interaction with phospholipids, 237
    insertion of, into ER bilayer, 339*f*, 340
    integral, 227, 234-237, 234*b*, 235*f*, 237*f*
    peripheral, 227, 236-237, 236*f*

Proteins *(Continued)*
microtubule-associated. *See* Microtubule-associated proteins (MAPs).
misfolded
degradation of, 393, 399, 401-402
disease and, 219*b*
elimination from endoplasmic reticulum by proteasomes, 402
modular, transcription factors as, 180, 180*f*
motor. *See* Intracellular motility; Motor proteins.
multiple drug resistance, 251-252, 251*f*
NF-KB and, 186
nonhistone, chromosome structure and, 136-139, 136*f*, 138*f*-139*f*
odorant-binding, 463
Patched, 420
phosphorylation of. *See* Phosphorylation: of proteins.
polycomb group, 130
polytopic, 339*f*, 340
precursors of, proteolytic processing of, 371
RGS, 435, 469
scaffolding, 68, 474, 804-805
signal sequences in, 305, 307, 309-315, 309*f*, 312*f*, 314*f*
SMC, 137-138, 138*f*-139*f*
Smoothened, 420-421
SNARE, 363-364, 364*f*
soluble
cycling during protein synthesis, 213
purification of, 90, 91*b*, 91*f*
translocation of, into lumen of ER, 338-340, 339*f*
sorting of, across secretory membrane system
lipid gradients in, 353-354, 353*f*-354*f*
protein-based machinery in, 354-365, 356*f*
structural
of skeletal muscle plasma membrane, 682*t*, 683-684, 683*f*, 684*t*
of telomeres, 120-121, 120*f*-121*f*
structure of, 32-42, 33*f*
amino acid properties and, 32-35, 33*f*-35*f*
architecture of, 35-36, 36*f*
intrinsically disordered regions of, 41
modular domains in, 41-42, 43*f*
phosphorylation and, 426, 426*f*-427*f*
polypeptide folding in, 36-40, 37*f*-40*f*
protein dynamics and, 41, 42*f*
protein interactions with solvent and, 40-41, 41*f*
secondary, 36-39, 38*f*-39*f*
synthesis of, 209-213, 209*f*. *See also* Translation.
elongation phase of, 215-217, 217*f*, 220*f*
higher order features of, 217-218
initiation phase of, 214-215, 216*f*
mRNAs and, 209-210, 210*f*
ribosomes and, 7*f*, 10, 212-213, 212*f*-214*f*
soluble protein factors and, 213
Proteins *(Continued)*
termination phase of, 217, 217*f*
tRNAs and, 210-212, 211*f*
tail-anchored, association with ER membrane, 335*t*, 341-342, 342*f*
targeting and, 6, 7*f*, 354-356, 362-363, 372*f*, 373
TATA box-binding, 170
transmembrane, 684*t*
type 1, 339*f*, 340
type 2, 339*f*, 340
unfolded protein response and, 345-346, 345*f*
transport of, within secretory system, protein-based machinery for, 354-365, 356*f*
Arf family GTPases as, 355, 356*f*
clathrin coats as, 357-362, 357*f*-358*f*, 362*f*-365*f*
COPI coat as, 357-362, 357*f*-358*f*, 360*f*-361*f*
COPII coat as, 357-362, 357*f*-359*f*
GTPases as, 355-357
membrane budding coat complexes and, 357-362, 357*f*-358*f*
Rab GTPases as, 355-357, 357*f*
WASp/Scar, 655*f*, 656
zinc finger, 179-180, 179*f*
α-proteobacterium, 20, 21*f*
Proteoglycans, 370, 555, 559*t*
in extracellular matrix, 512-514, 513*f*-514*f*, 522
Proteolysis, 348*f*, 396, 402, 723, 723*f*
Protocadherins, 529*t*
Protofilaments, 596-597, 597*f*
Proton gradient, 242-244, 317-319, 318*f*, 321*f*, 322-323, 325-328, 668
Protons, amide, 35-36
P-selectin, 536*t*
Pseudoautosomal region, 791-792, 792*f*
Pseudogenes, 112
Pseudopod extension, locomotion by, 654-659, 654*f*
chemotaxis of motile cells and, 657-658, 658*f*
growth cone guidance and, 658-659, 659*f*
lamellar motility on flat surfaces and, 654*f*-657*f*, 655-656
myosin and, 657
substrate and, 654*f*, 656-657, 656*f*
Pseudosubstrate, 428-429, 428*f*
Pseudosubstrate sequence, 448
PSP. *See* Postsynaptic potential (PSP).
PTB domains. *See* Phosphotyrosine-binding (PTB) domains.
PTEN phosphatases, 432, 449
PTS1. *See* Type 1 peroxisomal targeting signal (PTS1).
PTS2. *See* Type 2 peroxisomal targeting signal (PTS2).
P4-type ATPase, 346-347, 347*f*
P-type cation pumps, 244*t*, 248-250, 248*f*-249*f*
Pulsed-field gel, 108*f*
Pumps, 225-226, 241. *See also* Membrane pumps.
Ca^{2+}, 453, 453*f*
Puromycin, 215-216, 220*f*
Pyrin domain (PYD), 806*b*
Pyroptosis, 798*b*, 813

Q

Quantum mechanical tunneling, 326
Quasi-equivalent packing, 72
Quiescence, 713-714
Quinone, reduced, 326

R

R groups, 34
R (regulatory) subunits. *See* Regulatory (R) subunits.
Rab5 GTPase, 387-388
Rab GTPases, 355-357, 357*f*
Rac, 588, 588*f*
RAD51, 752-753, 752*f*
Radial spoke, 659-662, 660*f*-662*f*, 664-665
Radixin, 590*t*-591*t*
Raf, 439, 441*t*-442*t*, 474*f*, 475, 477*f*
Rafts, 233, 233*f*
Ran, 155-159, 157*f*
Ran-BP1, 156-157
Ran-BP2, 156-157
Ran-GAP1, 156-157
Ran-GDP, 156-157, 157*f*, 159
Ran-GTP, 156-157, 157*f*, 159
RANKL, 561, 565
Ras, 415, 432-434, 432*f*-433*f*, 442*t*, 475
analysis of enzyme mechanism with, 60-62, 60*f*-61*f*
growth factor receptor tyrosine kinase pathway through, to MAP kinase, 473*f*-474*f*, 474-475, 483*f*
Ras-GAP, 434, 475
RAS-GTP, 474*f*, 475, 476*f*-477*f*
Rate constants, 53, 97-98
first-order, half-time related to, 54*b*
Rattus norvegicus, genome of, 109*t*
Rb genes, 722
RBCs. *See* Red blood cells (RBCs).
Rb-related proteins, 720-721, 721*f*
R-cadherin, 529*t*
RCC1. *See* Regulator of chromosome condensation 1 (RCC1).
Reaction centers, 323-324, 325*f*, 326, 327*f*
Reactions
first-order, 53-54, 54*b*, 54*f*
half-time of, first-order rate constant related to, 54*b*
light and dark, 325-326, 325*f*
linked, 57-58, 57*f*
reversible, 55-56
second-order, 54-55, 55*f*, 67
thermodynamics and, 56-57
Rearrangement of immunoglobulin genes, 500-501, 502*f*
Receptor isoforms, 411
Receptor kinases, G-protein-coupled, 414
Receptor serine/threonine kinases, 417-418, 417*f*-418*f*, 421-423, 481-482, 481*f*
Receptor tyrosine kinases, 414-416, 415*f*-416*f*, 421-423, 449*f*
pathway through Ras to MAP kinase, 473*f*-474*f*, 474-475, 483*f*

Receptor-mediated endocytosis, 384*f*, 403, 403*f*
Receptors, 6–7, 7*f*, 409, 446–447
decoy, 561, 812
glutamate, 273, 273*f*
for G-protein isoforms, 435*t*
list of, 421–423
nuclear transport, 156, 157*f*
odorant, 464*f*, 465
orphan, 412
plasma membrane. *See* Plasma membrane receptors.
signal recognition particle, 335–337, 337*f*
steroid, 179, 179*f*, 421–423
Recessive mutations, 86*b*, 87
Recognition of DNA lesions, 750
Recombination
homologous, 85*t*, 94, 94*f*
in meiosis, 781–785, 783*f*–784*f*, 784*b*
recessive, 86*b*
Recombination enzymes, 501–502
Recombination nodules, 784, 789*f*
Reconstitution of function from isolated components, 98
Rectification, 283
Recycling endosomes, 377–378, 386, 388*f*
Red algae, 22–23, 23*f*
Red blood cells (RBCs), 233–234, 233*f*, 235*f*, 237–238, 237*f*, 494*f*, 494*t*, 495
Reduced quinone, 326
Reductional division, 779, 780*f*–781*f*
suppression DNA replication between meiosis II and, 792
Reductionism, 30, 75
"Reference genome", 113
Regional centromere, 114
Regulated secretory pathway, 374–376, 374*f*–375*f*
Regulator of chromosome condensation 1 (RCC1), 157
Regulatory (R) subunits, 428–430, 430*t*, 444, 445*f*
Relaxed skeletal muscle, 676, 677*f*
Repetitive DNA, 110–111
Replication factor C (RFC), 734, 735*f*
Replication foci, 736–737, 737*f*
Replication fork, 727, 728*f*
Replication fork collapse, 740
Replication stress, 739–740, 739*f*
Replicators, 728–729, 728*f*
Replicons, 700, 728–729, 730*f*
Replisome, 728–729
Reporter gene, 181*f*
Rescue, 598–600, 598*f*, 599*t*
Research, 75–101
choice of organisms for, 85–87, 86*b*
gene and protein discovery and, 87–92
biochemical fractionation and, 88–90, 89*b*, 89*f*, 91*b*, 91*f*
genomics and reverse genetics and, 88
isolation of genes and cDNAs and, 90–92, 92*f*–93*f*
molecular structure and, 95
through mutations, 87–88
genome engineering and, 85*t*, 92–95, 94*f*
Research *(Continued)*
imaging in, 75–76
electron microscopy, 77*f*, 82–85, 83*f*–84*f*
light microscopy. *See* Light microscopy methods.
mathematical models of systems and, 100
partners and pathways and, 95–97
biochemical methods for, 96
genetics and, 96–97, 96*f*–97*f*
large-scale screening with microarrays and, 97, 98*f*
rates and affinities and, 97–98
tests of physiological function and
anatomic tests and, 98–99
physiological tests and, 99–100
reconstitution of function from isolated components and, 98
Residual bodies, 379–381, 398–399, 398*f*
Response regulators, 482, 482*b*, 483*f*
Resting state of seven-helix receptors, 413, 413*f*
Restrictin, 523
Restriction point, 699–700, 713, 713*f*, 719–720, 719*f*–720*f*
cancer and, 699–700, 721–722, 722*f*–723*f*
regulation of cell proliferation by, 720–721, 721*f*
Restrictive conditions, 87
Restrictive temperature, 702*b*
RET protooncogene, 529*t*, 530*f*, 532
Retina, 466–469
Retinal, 412*f*, 413
Retrograde movements, 643–644
Retrograde traffic, 351*f*, 352
Retrograde transport, 640*t*, 643–644
Retromer, 374, 390
Retrotransposons, 111, 111*f*
Reversal potential, 283
Reverse genetics, 88
Reverse transcriptase, 111
Reverse turns, 39
Reversible reactions, 55–56
RFC. *See* Replication factor C (RFC).
RGD sequence motif, 532–533, 535–536
RGS proteins, 435, 469
Rh antigens, 276–277, 277*b*
Rho, 588, 588*f*
Rho family GTPases, 434, 588, 588*f*
Rho-dependent terminators, 175*f*
Rhodopseudomonas viridis, reaction center of, 326, 327*f*
Rhodopsin, 412*f*, 413, 467*f*, 468
Rhodopsin kinase, 469
Rho-GDI, 433–434, 442*t*
Rho-independent terminators, 174, 175*f*
Rho-kinase, 434
Ribonucleic acid. *See* RNA (ribonucleic acid).
Ribose, 42, 44*f*
Ribosomal RNA. *See* RRNA (ribosomal RNA).
Ribosomes, 3–4, 46, 209, 209*f*
free, 4*f*
macromolecular assembly and, 5–6
protein synthesis and, 7*f*, 10, 212–213, 212*f*–214*f*
synthesis of, 199–201, 200*f*
Riboswitches, 42
Ribozymes, 16, 46, 48*f*, 205
large subunit rRNA as, 205
RNase P and RNase MRP as, 205
self-splicing introns as, 205, 206*f*
Ribulose phosphate carboxylase (RUBISCO), 328
Rickettsia prowazekii
genome of, 109, 109*t*
mitochondria and, 317
Rigor complex of myosin, 626
Rigor of skeletal muscle, 676, 676*f*
Ring canals, 775–776, 777*f*
RISC. *See* RNA-induced silencing complex (RISC).
RITS (RNA-induced transcriptional silencing), 204
RNA (ribonucleic acid), 3–4
bases of, 42, 44*f*
biogenesis of, 166–167, 167*f*
depletion of, 99
heterogeneous nuclear, 167
long noncoding, 110
messenger. *See* MRNA (messenger RNA).
micro, 167, 202–204, 203*f*–204*f*
Piwi-interacting, 204
ribosomal. *See* RRNA (ribosomal RNA).
secondary structure of, 45–48, 47*f*–48*f*
small nuclear, 166–167, 189
maturation of, 201–202, 201*f*
pre-mRNA splicing and, 192, 194*f*
sugar of, 42
tertiary structure of, 46–48, 48*f*–49*f*
transfer, 46, 47*f*, 164, 166–167
synthesis of, 198–199, 198*f*
translation and, 209–212, 209*f*, 211*f*
RNA helicases, 193
RNA interference (RNAi), 88, 99, 129, 141–142, 164, 189
RNA polymerase I, 172, 172*f*
RNA polymerase II, 170–172, 170*t*, 171*f*, 201*f*
RNA polymerase III, 172, 172*f*
RNA polymerase preinitiation complex, 165–166, 170–172, 171*f*–172*f*
transcriptional activation and, 182, 182*f*
RNA polymerases, 165–169, 165*f*, 168*f*, 624*t*
promoters of, 169, 169*f*
RNA processing, 144–145, 144*f*, 145*t*, 163–164
RNA World, 15–17, 16*f*
RNAi. *See* RNA interference (RNAi).
RNA-induced silencing complex (RISC), 202–204
RNase H, 734–735
RNase MRP, 205
RNase P, 205
Robust adaptation, 486
Rod domain, intermediate filament, 614, 614*f*, 615*t*
Rod photoreceptors, 467–468, 467*b*, 467*f*
Rotary ATPases and membrane pumps, 243–248, 244*t*, 245*f*–247*f*
Rotary cilia, 665
Rough endoplasmic reticulum, 4*f*, 332, 332*f*
RPA, 734, 735*f*
RPTPc. *See* CD45.

RRNA (ribosomal RNA), 146, 166, 212-213, 212*f*, 214*f*
5S rRNA, 166-167
RUBISCO. *See* Ribulose phosphate carboxylase (RUBISCO).
Runx2/Cbfa1, 560
Ryanodine receptor, 679
Ryanodine receptor calcium channels, 454, 455*f*, 455*t*, 456-457, 457*f*, 679

S

S phase, 696-700, 698*f*, 741
cytoplasm during, 709
DNA replication during. *See* DNA replication.
histone protein synthesis and, 740-741, 740*f*
intra-S checkpoint and, 739-740, 739*f*
SAC. *See* Spindle assembly checkpoint (SAC).
Saccharomyces cerevisiae
actin-based movement of organelles in, 649*f*
centromeres of, 116, 141, 141*f*
cytokinesis in, 775
genetics for study of, 702*b*
genome of, 109-110, 109*t*, 110*f*
Golgi apparatus of, 369
replication origins in, 729-730, 729*f*-730*f*
Salt transport in kidney, 287
SAM complex (sorting and assembly machinery of the outer membrane), 304-305
Sar1, 355
Sarcoglycans, 538, 684*t*
Sarcomeres, 671-672, 672*f*-673*f*
Sarcoplasmic reticulum, 454
SAS-6, 612
α-satellite DNA, 116-117, 116*f*
Satellite DNAs, 111, 116, 116*f*
Saxitoxin, 279
Scaffold proteins, 68, 474, 804-805
Scanning electron microscopes (SEMs), 85
SCAP. *See* SREBP cleavage-activating protein (SCAP).
ScaRNAs. *See* Small Cajal body RNAs (scaRNAs).
Scc1, 137-138
SCF, 708, 708*f*, 710*t*-711*t*, 723, 723*f*
Schizosaccharomyces pombe
cell cycle in, 701, 702*b*
centromere organization of, 113-115, 114*f*
cytokinesis in, 773-774, 775*f*
genome of, 109, 109*t*
RNAi at, 141-142
Scinderlin, 590*t*-591*t*
Scleroderma, 140*f*
Sclerostin, 559-560, 560*f*
Scorpion toxin, 279
Scruin, 590*t*-591*t*
SDS. *See* Sodium dodecylsulfate (SDS).
Sec61 complex, 337-338, 338*f*-339*f*
SecA, 312, 312*f*
SecB, 311-312, 312*f*
Seckel syndrome, 749*t*
Second messengers, 409, 443-462
calcium as, 409, 452-459
calcium dynamics in cells and, 457-458, 459*f*
calcium targets and, 458-459, 462
calcium-release channels and, 453*f*, 454-457, 455*f*-457*f*, 455*t*
overview of calcium regulation and, 452-453, 453*f*, 453*t*, 462
removal from cytoplasm, 453*f*, 454, 454*t*
cyclic nucleotides as, 443-445, 444*f*-445*f*
lipid-derived, 445-452, 446*f*
agonists and receptors and, 446-447
cross talk and, 452
enzyme reactions producing, 445-446
for intercellular communication, 449-452, 450*f*-451*f*
phosphatidylcholine signaling pathways and, 449, 449*f*
phosphoinositide signaling pathways and, 448-449, 448*f*
protein kinase C and, 447*f*, 448
sphingomyelin/ceramide signaling pathways and, 452, 452*f*
targets of, 447-448, 447*f*-448*f*
nitric oxide as, 459-461, 460*f*
seven-helix receptors and, 413*f*, 414
Secondary constriction of chromosomes, 147
Secondary structures, 36-39, 38*f*-39*f*
of DNA, 43-45, 46*f*
of RNA, 45-48, 47*f*-48*f*
Secondary transporters, 241
Second-order reactions, 54-55, 55*f*, 67
Secreted phosphoprotein, 523
Secretory granules, 374-376, 374*f*-375*f*
Secretory membrane system, 351-353, 351*f*-352*f*
lipid gradients in membrane trafficking of, 353-354, 353*f*-354*f*
protein-based machinery in membrane trafficking of, 354-365, 356*f*
Arf family GTPases as, 355, 356*f*
clathrin coats as, 357-362, 357*f*-358*f*, 362*f*-365*f*
COPI coat as, 357-362, 357*f*-358*f*, 360*f*-361*f*
COPII coat as, 357-362, 357*f*-359*f*
GTPases as, 355-357
membrane bending proteins as, 364-365, 365*f*
membrane budding coat complexes and, 357-362, 357*f*-358*f*
Rab GTPases as, 355-357, 357*f*
SNAP receptor components as, 363-364, 364*f*
tethering factors as, 362-363, 363*f*
Secretory pathway, 302, 374-376, 374*f*-375*f*
Sec-signal sequences, 311-312
Securin, 767, 768*f*
SecYE translocon, 311-315, 312*f*, 314*f*
Sedimentation coefficient, 95
Sedimentation equilibrium, 88-90
Sedimentation velocity, 88-90
Segmental duplications in human genome, 112-113, 113*f*
Segregation
achiasmate, 785
in meiosis, 781, 781*f*
Selectins, 489-490, 526, 536-537, 536*f*-537*f*, 536*t*
Selective permeability, 281
Selectivity filter, 262-264, 264*f*
Self-assembly, 5-6, 6*f*, 63. *See also* Macromolecular assembly.
Self-avoidance in nervous system, 540-541
Self-correcting signal complexes, 432
"Selfish DNA" hypothesis, 112
SemiSWEET sugar carriers, 258
SEMs. *See* Scanning electron microscopes (SEMs).
Senescence, 121, 713-714
Sensors, DNA damage checkpoints and, 701, 748
Sensory cilia, 464
Sensory neurons, olfactory, 464, 464*f*
Separase, 767, 768*f*
SER. *See* Smooth endoplasmic reticulum (SER).
Serglycan, 514*f*, 522
Serine
phosphorylation and, 425
structure of, 33*f*
Serine/threonine kinase receptor pathways through Smads, 481-482, 481*f*
Serine/threonine kinases phosphorylation and, 426, 427*f*-428*f*, 428
PPP family of, 429-431, 430*f*, 430*t*
Serine/threonine phosphates, PPM family of, 430*t*, 431
Serotonin, 291*f*, 295
Seven-helix receptors, 412-414, 412*f*-413*f*, 414*t*, 421-423, 434*f*, 449*f*, 463
cardiac muscle and, 687-688, 687*f*
Severe combined immunodeficiency, 749*t*
Severin, 583-585, 583*f*, 590*t*-591*t*
Sex, second olfactory system and, 466*b*
Sex chromosomes in meiosis, behavior of, 791-792, 792*f*
Sexual reproduction, meiosis and, 779-780
SH2. *See* Src homology 2 (SH2).
SH2 domains, 43*f*, 415, 437, 439, 474*f*, 475-476, 477*f*, 478, 479*b*, 480
SH3 domains, 43*f*, 439*f*, 440, 475, 479*b*
SHC, 475-476
Sheet-forming collagen, 506*f*, 509, 509*f*, 521
β-sheets, 36-37, 38*f*-39*f*, 39
Shelterin associates, 120, 120*f*
Shigella, actin polymerization and, 649
Short interspersed nuclear elements (SINES). *See* SINES (short interspersed nuclear elements).
ShRNAs. *See* Small heterochromatic RNAs (shRNAs).
Sialoproteins, 559*t*
Sickle cell disease, 495
Side chains, 34, 36
Signal peptidases, 312, 338-340
Signal recognition particle (SRP), 335-337, 336*f*-337*f*
translocation dependent on, 313
Signal recognition particle-receptor, 335-337, 337*f*

Signal sequences, 10-11
in proteins, 305, 307, 309-315, 309*f*, 312*f*, 314*f*, 335
Signal transduction, 6-7
cytokine receptors, JAK/STAT pathways and, 479-481, 480*f*
by G-protein-coupled, seven-helix transmembrane receptors, 463
growth factor receptor tyrosine kinase pathway through Ras to MAP kinase and, 473*f*-474*f*, 474-475, 483*f*
insulin pathways to GLUT4 and MAP kinase and, 475-478, 476*f*
MAP kinase pathways to nucleus and, 473-474, 473*f*-474*f*, 476*f*-477*f*, 483*f*
metabolic regulation through β-adrenergic receptor and, 469-472, 470*f*, 471*t*
odor detection by olfactory system and, 463-466
adaptation and, 464*f*, 465-466
cAMP production and, 465
cyclic nucleotide-gated channels and, 465
odorant receptors and, 464*f*, 465
processing in brain and, 464*f*, 466, 466*b*
sensory neurons and, 464, 464*f*
photon detection by vertebrate retina and, 466-469
overview of, 466-468, 467*b*, 467*f*
positive arm of signal cascade and, 468-469, 469*b*
recovery and adaptation and, 469, 469*b*
rhodopsin and, 467*f*, 468
serine/threonine kinase receptor pathways through Smads and, 481-482, 481*f*
signaling pathways influencing gene expression and, 472-473, 472*f*-474*f*, 476*f*-477*f*, 480*f*-481*f*
T-lymphocyte pathways through nonreceptor tyrosine kinases and, 477*f*, 478-479, 479*b*, 480*f*-481*f*
transcription factors and, 184-186
cAMP signaling and, 185-186
NF-KB signaling and, 185*f*, 186
steroid hormone receptors and, 184-185, 185*f*
transcription factors in development and, 185*f*, 186
two-component phosphotransfer systems and, 482-486, 482*b*, 483*f*
bacterial chemotaxis and, 483*f*-485*f*, 485-486
Signaling cadherins, 529*t*
Signaling cascade, 425
SIM. *See* Structured illumination microscopy (SIM).
SINES (short interspersed nuclear elements), 111-112, 111*f*
Sinoatrial node, 685-686, 685*f*
SiRNAs. *See* Small interfering RNAs (siRNAs).
Sister chromatids, 107, 108*b*, 700, 756
separation of, in anaphase of mitosis, 767-768, 768*f*
Sister kinetochores, 759
Site-directed mutagenesis, 92, 93*f*
Skelemin, 682*t*
Skeletal muscle stem cells, 716-717
Skeletal muscles, 671-684, 672*f*
actomyosin apparatus of, 671-676, 672*f*-673*f*
intermediate filaments and, 675-676, 675*f*
thick filaments and, 671-672, 674-675, 674*f*, 682*t*
thin filaments and, 671-674, 673*f*-674*f*, 682*t*
titin filaments and, 675, 675*f*
molecular basis of contraction of, 676-678
crossbridge behavior and, 676*f*-677*f*, 677-678
sliding filament mechanism and, 676-677, 676*f*-677*f*
plasma membrane system and
smooth endoplasmic reticulum and, 678-679, 679*f*-680*f*
structural proteins of, muscular dystrophy defects and, 682*t*, 683-684, 683*f*, 684*t*
regulation of contraction of, 678-682, 678*f*
action potential coupling to contraction and, 678-680, 679*f*-681*f*
calcium spike transduction into contraction and, 680-681, 681*f*
motor neuron control and, 678, 678*f*
by myosin light chains, 681-682
synaptic transmission at neuromuscular junctions and, 678
specialized cells of, 682-684, 682*t*, 683*f*, 684*t*
Skeleton, formation and growth of, 561-565, 562*f*-564*f*
Skin wounds, repair of, 565-567, 566*f*
Sliding clamp, 751
Sliding filament mechanism of skeletal muscle contraction, 676-677, 676*f*-677*f*
Slit, 659
Slow axonal transport, 640*t*, 644-646, 645*f*
Smads (Sma- and Mad-related proteins)
serine/threonine kinase receptor pathways through, 417, 421-423
signal transduction and, 481-482, 481*f*
Small Cajal body RNAs (scaRNAs), 202
Small cross-linking proteins, 590*t*-591*t*
Small heterochromatic RNAs (shRNAs), 204, 204*f*
Small interfering RNAs (siRNAs), 189, 203-204, 203*f*
Small nuclear ribonucleoprotein (snRNP) particles, 192
Small nuclear RNAs (snRNAs), 166-167, 189
maturation of, 201-202, 201*f*
pre-mRNA splicing and, 192, 194*f*
Small nucleolar RNAs (snoRNAs), 199, 200*f*-201*f*, 202
Small subunits of ribosomes, 212, 213*f*
SMC2, 137, 138*f*, 756-757
SMC proteins, 137-138, 138*f*-139*f*
SMN (survival of motor neurons) protein, 145, 145*t*, 201-202
Smooth endoplasmic reticulum (SER), 333, 678-679, 679*f*-680*f*
Smooth muscle, 672*f*, 688-691, 689*f*-690*f*
Smooth swimming, 483-484, 484*f*
Smoothened protein, 420-421
SNAP receptor, 363-364, 364*f*
SNARE pin, 363
SNARE proteins, 363-364, 364*f*
SnoRNAs. *See* Small nucleolar RNAs (snoRNAs).
SnRNAs. *See* Small nuclear RNAs (snRNAs).
SnRNP particles. *See* Small nuclear ribonucleoprotein (snRNP) particles.
Sodium channels, voltage-gated, 269*f*, 270, 290, 293*f*, 295, 686*b*
Sodium dodecylsulfate (SDS), 89*b*
Solvent, protein interaction with, 40-41, 41*f*
Somatic mutations of immunoglobulins, 500-501, 502*f*
Sorting nexins, 390
SOS, 475, 476*f*-477*f*
SPARC (secreted protein rich in cysteine), 523
Spasatin, 603
Spasmoneme, 666, 666*f*
SPBs. *See* Spindle pole bodies (SPBs).
Speckles (nuclear domains), 144-145, 144*f*, 145*t*
Spectrin, 237, 237*f*, 590*t*-591*t*
Sperm, nematode, actin substitute in, 666-667, 667*f*
Spermatids, 793
Spermatogenesis, 131*f*, 793
Spermatogonia, 793
Spermatozoa, 793
Spherocytosis, hereditary, 495
Sphingolipids, 353-354, 353*f*-354*f*
in lipid bilayer, 229, 230*f*
Sphingomyelin, 371
Sphingomyelinase, 452, 452*f*
Sphingomyelin/ceramide signaling pathways, 452, 452*f*
Sphingomyelins, 229, 230*f*, 445, 446*f*
Sphingosine, 229, 230*f*, 452
Sphingosine-1-phosphate, 452
Spindle assembly checkpoint (SAC), 701, 707-708, 758-759, 763-764, 764*f*
Spindle checkpoint, 764-765, 764*f*
Spindle pole, 757, 757*f*
in anaphase, 768-770, 769*f*
assembly of, 760-761, 760*f*-761*f*
chromosome attachment to, 761-764, 762*f*-763*f*
in metaphase, 766-767, 766*f*
organization of, 759-760, 760*f*, 765*f*
Spindle pole bodies (SPBs), 593, 593*f*, 608*f*-609*f*, 609-610
Spire, 581-582
Splice sites, 192-193, 193*f*
Spliceosomes, 42
Splicing, pre-mRNA, 192-194
alternative splicing and, 193-194, 195*f*
AT-AC introns and, 193, 193*f*
localization of, 194
signals for, 192, 193*f*
splicing reaction and, 192-193, 193*f*-194*f*

Spoke ring, 151, 153*f*
Sponges, 24-25
Spontaneous protein folding, 218, 219*b*
Src family protein tyrosine kinases, 479*b*
Src homology 2 (SH2), 415, 416*f*
Src homology (SH) domains, 437
SREBP. *See* Steroid regulator element-binding protein (SREBP).
SREBP cleavage-activating protein (SCAP), 347-348, 348*f*
SRNAs, 202
SRP. *See* Signal recognition particle (SRP).
SRP receptors, 336-337, 337*f*
SR-proteins, 192
Starch, 49, 328
START, 719-720
STAT (signal transducer and activator of transcription), 479-481, 480*f*
Stathmin, 602-603
STED microscopy. *See* Stimulated emission depletion (STED) microscopy.
Stem body matrix, 769
Stem cell factor. *See* SCF.
Stem cell niches, 716
Stem cells, 696-697, 713, 715, 715*b*-718*b*
  adult, properties of, 716
  cancer, 717
  committed, 715-716
  discovery and defining features of, 715-716
  embryonic, 715
  epidermal, 716, 717*f*
  hematopoietic, 494, 715-716
  meristematic, 717
  mesenchymal, 491, 491*f*-492*f*
  pluripotent, 493-494, 494*f*, 715
  skeletal muscle, 716-717
  therapeutic applications of, 717-718
  tissue, 715
  transgenic animals and, 717
Stem-loop structure, 46
Stereochemical isomers, 48
Steroid hormone receptors, gene expression regulation by, 184-185, 185*f*
Steroid receptors, 179, 179*f*, 421-423
Steroid regulator element-binding protein (SREBP), 347-348, 348*f*
Sterols, 229-230, 230*f*
Stimulated emission depletion (STED) microscopy, 81*t*, 82
Stochastic optical reconstruction microscopy (STORM), 81-82, 81*t*
Stokes radius, 91*b*
STOP, 611
Store-operated Ca^{2+} entry, 453*f*, 454
STORM. *See* Stochastic optical reconstruction microscopy (STORM).
Stratified epithelium, 543, 544*f*
Streptomyces lividans, as membrane channel model, 262-265, 264*f*
Stress fibers, 575-576, 576*f*
Stress responses of endoplasmic reticulum, 345-346, 345*f*
Striated muscle, 672*f*, 679*f*, 688-689
Stroke, brain damage in, 814, 814*f*
Stroma of chloroplasts, 324, 324*f*-325*f*, 327*f*, 328
Structured illumination microscopy (SIM), 80*f*-81*f*, 81*t*, 82
Strychnine, 275
Substrate, pseudopods and, 654*f*, 656-657, 656*f*
Subunit flux. *See* Treadmilling.
Subunit molecules, 63. *See also* Macromolecular assembly.
  assembly of macromolecular structures from, 63-64, 63*f*, 64*b*, 65*f*
    reactions in, 67
    of symmetrical structures, 65-67, 66*f*
  biosynthesis and degradation of, regulation of assembly by, 68
  covalent modification of, regulation of assembly by, 68
  in hexagonal arrays in plane sheets, 65, 66*f*
  polymerization of, 65-66
  proteins, composition of, 95
  regular polygons of, spherical assemblies formed by, 66*f*
Sugars, 29, 48-51, 50*f*-51*f*, 342-343, 343*f*-344*f*, 399, 512, 513*f*, 514-515
Sulfonylurea receptors (SURs), 268
Supercoils, 44-45, 46*f*
Superhelices, 44-45, 46*f*
Superresolution fluorescence microscopy, 81-82, 81*t*, 82*f*, 130-131, 131*f*
Suppressor mutations, 96-97, 96*f*
SURs. *See* Sulfonylurea receptors (SURs).
Surveillance complex, 197
Survival of motor neurons (SMN) protein. *See* SMN (survival of motor neurons) protein.
SWEET sugar carriers, 258
Switch I/II, 60-61, 60*f*, 432-433, 432*f*
Symmetry, laws of, 64*b*, 65*f*
Sympathetic nerves, heart rate and, 687, 687*f*
Symporters, 254-256, 254*f*-255*f*, 257*t*
Synapses, 291-297, 292*f*-293*f*
  of central nervous system, 293*f*, 294-295
    modification by drugs and disease, 295
    modification by use, 295-297, 296*f*
  excitatory, 295, 296*f*
  inhibitory, 295
  neuromuscular junction and, 292*f*, 294
Synapsis, 785*f*, 786-789, 787*f*-788*f*
Synaptic plasticity, 295-296, 296*f*
Synaptic transmission, 678
Synaptic vesicles
  , 292*f*-293*f*, 294-296, 296*f*
Synaptonemal complex, 785*f*, 786-789, 787*f*-789*f*
Syndecan, 514, 514*f*, 522
Synemin, 615*t*
Syntelic attachment, 763
Synthetic lethal mutations, 96-97, 96*f*

T

T cells. *See* Lymphocytes.
T tubules, 678-679, 679*f*-680*f*
Tacket, John, 152*f*
TADs, 134-138, 134*f*
TAFIIs. *See* TBP-associated factors (TAFIIs).
TAG-1 (transient axonal protein-1), 528*t*
Tail of motor proteins, 623
Tail retraction, myosin and, 657
Tail-anchored proteins, association with ER membrane, 335*t*, 341-342, 342*f*
Talin, 534, 534*f*, 590*t*-591*t*
TAP, 401
TAP1, 251
TAP2, 251
TAP (transporter associated with antigen presentation), 401
TAP tagging, 96
Tar receptors, 482*b*, 483*f*-485*f*, 485-486
Targeting signals, 303-304
TATA box, 169
TATA box-binding protein, 170
TATAAA, 170-171
Tau, 590*t*-591*t*, 601-602, 601*f*-602*f*, 611
Taxol, 595*b*
TBP-associated factors (TAFIIs), 170-171, 170*t*, 171*f*
TBSV. *See* Tomato bushy stunt virus (TBSV).
T-cadherin, 529*t*
T-cell antigen receptors, 477*f*, 478-479, 481*f*, 502-503
Tektin, 611
Telomerase, 112, 118-120, 119*f*-120*f*
Telomeres, 10, 10*f*, 107-108, 117-122
  aging and, 121-122, 122*f*
  cancer and, 121-122, 122*f*
  definition of, 108*b*
  DNA structure of, 118
  replication of ends of chromosomal DNA by, 118-120, 118*f*-119*f*
  structural proteins of, 120-121, 120*f*-121*f*
Telophase, 700, 755, 755*f*-756*f*, 770-771, 770*f*
  nuclear envelope reassembly in, 770-771, 770*f*
Temperature, 702*b*
Temperature sensitive (ts) mutations, 702*b*
Tenascin, 516-517, 517*f*
Teratocarcinomas, 718
Termination
  of DNA replication, 736*b*
  of transcription, 165-166, 165*f*, 173-175, 174*f*-175*f*
  of translation, 216*f*-217*f*, 217
Termination codons, 209-210
Terminators, 174, 175*f*
Tertiary structures of RNA, 46-48, 48*f*-49*f*
Tetanus, 681, 681*f*
Tethering factors, 362-363, 363*f*
Tetrahymena, linker histone genes in, 127
Tetramers, 614
Tetrodotoxin, 279
TFIIA, 170*t*, 171-172
TFIIB, 170*t*, 171*f*, 172
TFIID, 170-171, 170*t*
TFIIE, 170*t*, 172
TFIIF, 170*t*, 171*f*, 172
TFIIH, 170*t*, 172
TFs. *See* Transcription factors (TFs).
TGF-β. *See* Transforming growth factor-β (TGF-β).
TGN. *See* *Trans*-Golgi network (TGN).
Thapsigargin, 454
Thermodynamics, 29, 56-57

Thermogenesis, 493
Thick filaments of skeletal muscle, 671-672, 674-675, 674*f*, 682*t*
Thin filaments of skeletal muscle, 671-674, 673*f*-674*f*, 682*t*
Thin sections, 82, 83*f*
30-nm fibers, 123, 130-131, 131*f*, 135
3′ splice site, 192
Threonine
  phosphorylation and, 425
  structure of, 33*f*
Threshold, 289, 289*f*
Thrombopoietin, 495
Thrombospondin, 523
Thromboxanes, 446, 450, 450*f*
Thylakoid membranes, 324
Thymine, 42, 44*f*
β-thymosins, 580, 590*t*-591*t*
Thymosin-β_4, 580, 587, 590*t*-591*t*
Tic, 307-308, 308*f*
Tight junctions, 490, 543-546, 544*f*-545*f*, 544*t*
TIM complexes, 304-307, 305*f*-306*f*
TIMPs. *See* Tissue inhibitors of metalloproteinases (TIMPs).
+TIPs, 603-604, 603*f*, 611
-TIPs, 611
TIRF microscopy. *See* Total internal reflection fluorescence (TIRF) microscopy.
Tissue culture, 86-87
Tissue fibronectin in extracellular matrix, 516, 516*f*
Tissue inhibitors of metalloproteinases (TIMPs), 520
Tissue stem cells, 715
Titin filaments, 675, 675*f*, 682*t*
T-loop, 120*f*
TLRs. *See* Toll-like receptors (TLRs).
T-lymphocyte pathways through nonreceptor tyrosine kinases, 477*f*, 478-479, 479*b*, 480*f*-481*f*
TMBIM channels, 278
TMV. *See* Tobacco mosaic virus (TMV).
TNF. *See* Tumor necrosis factor (TNF).
Tobacco mosaic virus (TMV)
  assembly of, 68-69, 71-72, 71*f*
  conservation of genome by, 63-64
TOC, 307-308, 308*f*
Toll-like receptors (TLRs), 420, 496-497, 498*f*
TOM complexes, 304-305, 305*f*-306*f*, 307
Tomato bushy stunt virus (TBSV), 72-73, 72*f*
Tomography, 82-84, 83*f*
Topoisomerases, 45
Tor kinase, 393-394, 394*f*
Torpedo model (transcription termination), 191
Total internal reflection fluorescence (TIRF) microscopy, 79, 80*f*-81*f*
Transcription, 10, 165-169, 165*f*
  activation of, 182, 182*f*
  arrest of, 173-174, 174*f*
  catalytic cycle in, 173, 174*f*
  chromatin and, 177-184, 177*f*, 178*t*
    combinatorial control and, 183-184
Transcription *(Continued)*
    gene-specific eukaryotic GTFs and, 178-180, 179*f*
    transcription factor activity and, 180-184, 180*f*-184*f*
  closed complex, 169-170, 170*f*
  elongation in, 165-166, 165*f*, 171*f*, 172-173, 174*f*
  factor interactions and, 180, 180*f*
    transcriptional repressors and, 182, 182*f*
  gene-specific, 175-177
    DNA-binding domains and, 179-180, 179*f*
    in eukaryotes, 176-180, 176*f*, 179*f*
    mapping transcription components on the genome, 177-178, 177*f*
    in prokaryotes, 175-176, 176*f*
  initiation of. *See* Transcription initiation.
  mRNA processing and, 190-192, 190*f*
  open complex, 165-166, 169-170, 170*f*
  pausing of, 173-174, 174*f*
  preinitiation complex, 165-166, 170-172, 171*f*-172*f*
  repressors of, 182, 182*f*
  RNA biogenesis and, 166-167, 167*f*
  RNA polymerases and, 167-169, 168*f*-169*f*
  termination of, 165-166, 165*f*, 173-175, 174*f*-175*f*
  transcription unit and, 166, 166*f*
Transcription elongation, 165-166, 165*f*, 171*f*, 172-173, 174*f*
Transcription factors (TFs), 10, 125, 125*f*, 165
  binding to eukaryotic promoter proximal and enhancer elements, 180-184, 181*f*-183*f*
  in development, 185*f*, 186
  gene-specific, eukaryotic, 178-180
  as modular proteins, 180, 180*f*
  modulation of activity of, 184, 184*f*
  signal transduction and, 184-186
    cAMP signaling and, 185-186
    NF-KB signaling and, 185*f*, 186
    steroid hormone receptors and, 184-185, 185*f*
    transcription factors in development and, 185*f*, 186
Transcription initiation, 165-166, 165*f*, 169-173, 170*f*, 170*t*
  general eukaryotic transcription factors and, 170
  in prokaryotes, 175-176, 176*f*
  RNA polymerase I factors and, 172, 172*f*
  RNA polymerase II factors and, 170-172, 170*t*, 171*f*
  RNA polymerase III factors and, 172, 172*f*
  transcriptional activation and, 182, 182*f*
  transcriptional repressors and, 182, 182*f*
Transcription termination, 165-166, 165*f*, 173-175, 174*f*-175*f*
Transcription unit, 166, 166*f*
Transcytosis, 382*f*
Transducers, DNA damage checkpoints and, 701, 748
Transducin, 468
Transduction, 409. *See also* Signal transduction.
Transesterification, 193
Transfer RNAs. *See* TRNAs (transfer RNAs).
Transformed cells, 722, 722*f*
Transforming growth factor-β (TGF-β), 417-418, 481-482, 481*f*, 714-715
Transgenic animals, 717
Trans-Golgi network (TGN)
  to endosome/lysosomal system, 373-374, 373*f*
  secretory granules and, 374-376, 374*f*-375*f*
  sorting from, 371-376, 371*f*-375*f*
  trafficking to plasma membrane in polarized cells and, 372-373, 372*f*
  transport of cargo to the plasma membrane and, 371*f*
Transient receptor potential (TRP) channels, 271-272, 271*f*
Transit sequences, 307
Translation, 8-9, 166-167, 189-190, 190*f*, 209, 209*f*
  elongation phase of, 215-217, 217*f*, 220*f*
  higher order features of, 217-218
  initiation phase of, 214-215, 216*f*
  ribosomes and, 7*f*, 10, 212-213, 212*f*-214*f*
    mRNAs and, 209-210, 210*f*
  soluble protein factors and, 213
  termination phase of, 217, 217*f*
  tRNAs and, 210-212, 211*f*
Translocation. *See* Protein translocation.
Translocation GTPases, 442*t*
Translocons, 304, 304*f*, 307-308, 312*f*, 335-336
  SecYE, 311-315, 312*f*, 314*f*
Transmembrane proteins, 684*t*
  type 1, 339*f*, 340
  type 2, 339*f*, 340
  unfolded protein response and, 345-346, 345*f*
Transmembrane segments, 234-236, 235*f*
  amino acid sequence identification of, 234*b*
  channels with one transmembrane segment, 266
  channels with two transmembrane segments and, 266-267
    amiloride-sensitive, 266-267
    ATP-gated, 267
    mechanosensitive, 266, 267*f*
  channels with four transmembrane segments and, 268
Transmission electron microscopy, 82, 83*f*
Transphorylation, growth factor signaling pathways and, 473*f*, 475
Transposable elements, 111
Transposition, 41-42, 111*f*
Transposons, 110-112, 111*f*
Treadmilling, 580, 766-767, 766*f*
Trichothiodystrophy, 749*t*
Trigger factor, protein folding and, 218-220, 220*f*
Triglycerides, 492
  in lipid bilayer, 230

Trimeric G-proteins, 421-423, 434-437, 434*f*, 442*t*
  cardiac muscle and, 687-688, 687*f*
  in disease, 436-437, 436*t*
  guanosine triphosphatase cycle and, 434*f*, 435-436
  mechanisms of effector activation and, 436
  signal transduction by seven-helix receptors coupled to, 463
  subunit cycles and, 436
  subunit diversity of, 435, 435*t*
Trimeric GTP-binding proteins, 412
Tri-methyllysine, 34*f*
Triple helix of polypeptides in collagen, 505-506, 505*f*
Trisomy 21, 794-795, 795*f*
TRNAs (transfer RNAs), 164, 166-167
  crystal structures of, 46, 47*f*
  synthesis of, 198-199, 198*f*
  translation and, 209-212, 209*f*, 211*f*
Tropoelastins, 511
Tropomodulin, 584, 590*t*-591*t*, 673, 673*f*, 682*t*
Tropomyosin, 584, 590*t*-591*t*, 672-674, 673*f*, 682*t*
Troponins, 672-673, 673*f*, 682*t*
TRP channels. *See* Transient receptor potential (TRP) channels.
Trypanosoma brucei, 236-237
Tryptophan, structure of, 33*f*
T-SNAREs, 363, 375
α-tubulin, 593
  of centrioles, 605
  structure of, 595-596, 596*f*
β-tubulin, 593
  of centrioles, 605
  hydrolysis of GTP bound to, 598-599
  microtubule structure and, 596-597
  pharmacologic tools for studying, 595*b*
  structure of, 595-596, 596*f*
γ-tubulin, 593, 612, 760
  of centrioles, 604, 606
  microtubule-organizing centers and, 600*f*
γ-tubulin ring complexes (γTuRC), 604-607, 606*f*
Tubulins, 593, 593*f*
  evolution and diversity of, 595-596
  pharmacologic tools for studying, 595*b*
  structure of, 595-596, 596*f*
Tumble, 483-484, 484*f*
Tumor necrosis factor (TNF), 419-420, 419*f*
  sphingomyelin/ceramide signaling pathways and, 452, 452*f*
Tumor necrosis factor receptor family, 419-420, 419*f*
Tumor suppressor genes, 722, 723*f*
γTuRC. *See* γ-tubulin ring complexes (γTuRC).
Two-component phosphotransfer systems, 482-486, 482*b*, 483*f*
  bacterial chemotaxis and, 483*f*-485*f*, 485-486
Two-component receptors, 421-423, 482*b*
Two-hybrid assay, 97, 97*f*
Type 1 peroxisomal targeting signal (PTS1), 309, 309*f*, 310*t*
Type 2 peroxisomal targeting signal (PTS2), 309
Type 1 transmembrane proteins, 339*f*, 340
Type 2 transmembrane proteins, 339*f*, 340
Type I ABC transporters, 314*f*, 315
Type II collagen, 556, 556*f*
Type II photosystems, energy capture and transduction by, 326, 327*f*
Type II polyproline helices, 440
Type II secretion, prokaryotic protein export via, 313, 314*f*
Type III secretion, prokaryotic protein export via, 314*f*, 315
Type IV secretion, prokaryotic protein export via, 313, 314*f*
Type XVII collagen, 552-553
Tyrosine
  phosphorylation and, 425
  structure of, 33*f*
Tyrosine kinases, 426, 441*t*
  nonreceptor
    JAK, 416-417, 417*f*, 421-423, 479-481, 480*f*
    Src, 427*f*-428*f*, 428-429, 431-432, 437, 479*b*
    T-lymphocyte pathways through, 477*f*, 478-479, 479*b*, 480*f*-481*f*
  receptor, 414-416, 415*f*-416*f*, 421-423, 449*f*
Tyrosine-based sorting motifs, 361

U

U1 snRNA, 192-193
U2 snRNA, 192-193
U4 snRNA, 192-193
U5 snRNA, 192-193
U6 snRNA, 192-193
Ubiquitin, 302, 707
  in protein degradation, 394-396, 394*f*-395*f*, 395*t*
Ubiquitin ligases, 394-395, 395*f*, 395*t*
Ubiquitin-activating enzyme, 394, 395*f*, 707
Ubiquitin-conjugating enzyme, 394-395, 395*f*, 707
Ubiquitin-interacting motifs (UIMs), 388*f*
Ubiquitylation
  modification of proteins by, 394-396, 394*f*-395*f*, 395*t*
  motifs specifying, 395*t*, 401-402
UIMs. *See* Ubiquitin-interacting motifs (UIMs).
Unfolded protein response (UPR), 345-346, 345*f*
Uniporters, 253*f*-254*f*, 254, 256, 257*t*
Unique-sequence DNA, 110
Unprocessed pseudogenes, 112
3′ untranslated region (3′ UTR), 191-192
UPR. *See* Unfolded protein response (UPR).
Uracil, 42, 44*f*
  cytosine deamination to, 195, 195*f*
Usher pathway, prokaryotic protein export via, 313, 314*f*
Utrophin, 590*t*-591*t*, 684*t*

V

Vacuoles, autophagic, 398-399
Valine, structure of, 33*f*
Van der Waals interactions, 58-59
VAP, 334, 349
Vasodilator-stimulated protein. *See* VASP.
VASP, 583, 583*f*, 652-653, 656
VCAM-1 (vascular cell adhesion molecule-1), 528*t*
Versican, 522
Vesicles, clathrin-coated, 378*f*, 383*f*-384*f*, 384-385
Vesicular trafficking, 301-302
Vesicular tubular carriers (VTCs), 352, 352*f*, 359, 366*f*
Vibrio cholerae, ZO-toxin of, 546
Villin, 590*t*-591*t*
Vimentin, 614, 614*f*, 615*t*, 616-617, 617*f*
Vinblastine, 595*b*
Vinculin, 534, 534*f*
Viruses
  HIV, 389-390
  as "opportunistic endocytic ligands", 391, 391*f*
  programmed cell death and, 802-803
Viscoelasticity, 588
Visual signal processing
  overview of, 466-468, 467*b*, 467*f*
  positive arm of signal cascade and, 468, 469*b*
  recovery and adaptation and, 469, 469*b*
  rhodopsin and, 468
Vitamin D-binding protein, 590*t*-591*t*
Vitronectin, 523
Vitronectin receptor, 533*t*
VLA-4 (very late antigen-4), 533*t*
Voltage-gated channels, 268-269, 269*f*
  calcium, 270-271, 271*t*, 292*f*, 294-295, 453*f*, 454, 455*f*, 455*t*, 456-457, 686*b*
  potassium, 269-270, 269*f*, 290, 295, 686*b*
  sodium, 269*f*, 270, 290, 293*f*, 295, 686*b*
Voltage-sensing domains, 268-269, 269*f*
Volume regulation, 288-289, 288*f*
von Willebrand factor, 523
  platelet receptor for, 537-538, 539*f*
V-SNAREs, 363, 365-366, 375
VTCs. *See* Vesicular tubular carriers (VTCs).
V-type ATPases, 247, 560-561
V-type rotary ATPases, 387, 388*f*

W

WASp. *See* Wiskott-Aldrich syndrome protein (WASp).
WASp/Scar proteins, 655*f*, 656
Waste management nucleases, 808
Water
  channels, 276*f*, 277
  molecular structure of, 31-32
  transport of, 287
WD repeats, 434-435
Wee1 kinase, 702, 702*f*, 710*t*-711*t*, 744, 744*f*, 749*f*
  Cdk inactivation and, 706, 706*f*
  SCF and, 708, 708*f*
White blood cells, 497, 497*f*
White fat cells, 491*f*, 492, 493*f*
Wild type, 86*b*, 87
Williams-Beuren syndrome, 113, 113*f*
Wilson disease, 249-250

Wiskott-Aldrich syndrome protein (WASp), 434, 581, 582*f*
Wiskott-Aldrich syndrome protein (WASp)/ Scar proteins, 655*f*, 656
Wnts, 420-423, 530, 531*f*
Wortmannin, 449
Wound repair, 565-567, 566*f*
WW domains, 439

X

Xenopus laevis
  cell cycle in, 703*b*, 703*f*, 704, 705*f*
  DNA replication in, 733
  meiosis in, 792
Xeroderma pigmentosum (XP), 749*t*, 750, 751*f*
XP. *See* Xeroderma pigmentosum (XP).
X-ray crystallography, 35, 36*f*, 95, 357, 358*f*

Y

Yeasts, 4*f*
  budding. *See Saccharomyces cerevisiae*.
  fission. *See Schizosaccharomyces pombe*.
  spindle pole bodies and, 608*f*-609*f*, 609-610
Yersinia, 315

Z

Z disks, 671-672, 673*f*, 675-676, 682*t*
ZAP-70, 477*f*, 478, 503
Z-DNA, 44
Zinc finger proteins, 179-180, 179*f*
ZO-1, ZO-2, and ZO-3, 546
Zonula adherens, 544*f*, 544*t*, 550-551
Zonula occludens, 544*f*
Zygotene, 785*f*, 786